BIOCHEMISTRY

NINTH EDITION

Jeremy M. Berg

John L. Tymoczko

Gregory J. Gatto, Jr.

Lubert Stryer

macmillan
international
HIGHER EDUCATION

Vice President, STEM: Daryl Fox

Executive Program Director: Sandy Lindelof

Executive Editor, Life Sciences: Lauren Schultz

Marketing Manager: Maureen Rachford

Director of Development, STEM: Lisa Samols

Developmental Editors: Lisa Bess Kramer and Michael Zierler

Editorial Assistant: Justin Jones

Director of Content Management Enhancement: Tracey Kuehn

Senior Managing Editor: Lisa Kinne

Senior Content Project Manager: Peter Jacoby

Schedule Manager: Karen Misler

Senior Workflow Project Manager: Paul W. Rohloff

Production Supervisor: Jose Olivera

Director of Content: Jennifer Driscoll Hollis

Content Development Manager: Amber Jonker

Lead Content Developers: Angela Piotrowski and Cassandra Korsvik

Media Editor: Jennifer Compton

Media Project Manager: Daniel Comstock

Project Management: Lumina Datamatics, Inc.

Composition: Lumina Datamatics, Inc.

Illustrations: Jeremy Berg with Network Graphics,
    Gregory J. Gatto, Jr., Adam Steinberg

Illustration Coordinators: Janice Donnola and Emiko Paul

Photo Permissions Editor: Christine Buese

Photo Researcher: Jennifer Atkins

Director of Design, Content Management: Diana Blume

Design Services Manager: Natasha Wolfe

Interior Design: Marsha Cohen

Printing and Binding: LSC Communications

Gregory J. Gatto, Jr., is an employee of GlaxoSmithKline (GSK), which has not supported or funded this work in any way. Any views expressed herein do not necessarily represent the views of GSK.

ISBN-13: 978-1-319-11465-7
ISBN-10: 1-319-11465-2

2 3 4 5 6 7   26 25 24 23 22 21

W. H. Freeman and Company
One New York Plaza
Suite 4500
New York, NY 10004-1562
www.macmillanlearning.com

Q6 STR: BIO (C)

REFERENCE

To our teachers and our students

Courtesy Jeremy Berg

JEREMY M. BERG received his B.S. and M.S. degrees in Chemistry from Stanford University (where he did research with Keith Hodgson and Lubert Stryer) and his Ph.D. in Chemistry from Harvard with Richard Holm. He then completed a postdoctoral fellowship with Carl Pabo in Biophysics at Johns Hopkins University School of Medicine. He was an Assistant Professor in the Department of Chemistry at Johns Hopkins from 1986 to 1990. He then moved to Johns Hopkins University School of Medicine as Professor and Director of the Department of Biophysics and Biophysical Chemistry, where he remained until 2003. He then became Director of the National Institute of General Medical Sciences at the National Institutes of Health. In 2011, he moved to the University of Pittsburgh where he is now Professor of Computational and Systems Biology and Pittsburgh Foundation Chair and Director of the Institute for Personalized Medicine. He served as President of the American Society for Biochemistry and Molecular Biology from 2011–2013. He is a Fellow of the American Association for the Advancement of Science and a member of the Institute of Medicine of the National Academy of Sciences. He received the American Chemical Society Award in Pure Chemistry (1994) and the Eli Lilly Award for Fundamental Research in Biological Chemistry (1995), was named Maryland Outstanding Young Scientist of the Year (1995), received the Harrison Howe Award (1997), and received public service awards from the Biophysical Society, the American Society for Biochemistry and Molecular Biology, the American Chemical Society, and the American Society for Cell Biology. He also received numerous teaching awards, including the W. Barry Wood Teaching Award (selected by medical students), the Graduate Student Teaching Award, and the Professor's Teaching Award for the Preclinical Sciences. He is coauthor, with Stephen J. Lippard, of the textbook *Principles of Bioinorganic Chemistry*. Since 2016, he has been Editor-in-Chief for *Science* magazine and the *Science* family of journals.

Hai Ngo

JOHN L. TYMOCZKO is Towsley Professor of Biology Emeritus at Carleton College, where he has taught since 1976. He taught a variety of courses including Biochemistry, Biochemistry Laboratory, Oncogenes and the Molecular Biology of Cancer, and Exercise Biochemistry, and cotaught an introductory course, Energy Flow in Biological Systems. Professor Tymoczko received his B.A. from the University of Chicago in 1970 and his Ph.D. in Biochemistry from the University of Chicago with Shutsung Liao at the Ben May Institute for Cancer Research. He then had a postdoctoral position with Hewson Swift of the Department of Biology at the University of Chicago. The focus of his research has been on steroid receptors, ribonucleoprotein particles, and proteolytic processing enzymes.

Joe Lin, courtesy Greg Gatto

GREGORY J. GATTO, JR., received his A.B. degree in Chemistry from Princeton University, where he worked with Martin F. Semmelhack and was awarded the Everett S. Wallis Prize in Organic Chemistry. In 2003, he received his M.D. and Ph.D. degrees from the Johns Hopkins University School of Medicine, where he studied the structural biology of peroxisomal targeting signal recognition with Jeremy M. Berg and received the Michael A. Shanoff Young Investigator Research Award. He completed a postdoctoral fellowship in 2006 with Christopher T. Walsh at Harvard Medical School, where he studied the biosynthesis of the macrolide immunosuppressants. Dr. Gatto is currently a Scientific Leader and GSK Fellow in the Target Incubator Discovery Performance Unit at GlaxoSmithKline. While he enjoys losing at board games, attempting but not completing crossword puzzles, and watching baseball games at every available opportunity, he treasures most the time he spends with his wife Megan and sons Timothy and Mark.

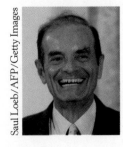

Saul Loeb/AFP/Getty Images

LUBERT STRYER is Winzer Professor of Cell Biology, Emeritus, in the School of Medicine and Professor of Neurobiology, Emeritus, at Stanford University, where he has been on the faculty since 1976. He received his M.D. from Harvard Medical School. Professor Stryer has received many awards for his research on the interplay of light and life, including the Eli Lilly Award for Fundamental Research in Biological Chemistry, the Distinguished Inventors Award of the Intellectual Property Owners' Association, and election to the National Academy of Sciences and the American Philosophical Society. He was awarded the National Medal of Science in 2006. The publication of his first edition of *Biochemistry* in 1975 transformed the teaching of biochemistry.

Forty-five years ago the first edition of *Biochemistry* hit the market and forever changed the way biochemistry is taught. In the ninth edition we remain true to the hallmarks of Lubert Stryer's vision while presenting molecular structure and function, metabolism, regulation, and laboratory techniques in new ways that engage students actively in the process of learning biochemistry.

THEMATIC THREADS

It has always been our goal to help students connect biochemistry to their own lives and the world around them. We use an evolutionary perspective to reveal the threads that bind organisms, clinical applications to show physiological relevance, and biotechnological applications to show how biochemistry is used to improve our lives.

- **Evolutionary perspective**. The French philosopher Pierre Teilhard de Chardin said, "Evolution is a light which illuminates all facts, a trajectory which all lines of thought must follow—this is what evolution is." We recognize that evolution shapes every pathway and molecular structure described in this text. Thus, we have integrated discussions of evolution throughout the narrative. To further emphasize the primacy of evolution in the context of the processes and molecules discussed, we use a phylogenetic tree as a marker to highlight important milestones in the evolution of life. (For a full list of these Molecular Evolution highlights, see p. xiii.)

- **Physiological relevance**. Pathways and processes are presented in a physiological context so students can see how biochemistry works in the body when it is at rest and during exercise, and in health, disease, and under stress. Many of these examples are highlighted with the caduceus symbol. (For a full list of these Clinical Applications, see p. xiv.)

- **New! Industrial Applications**. Biochemistry is an essential tool in numerous industries, such as pharmaceuticals, biotechnology, energy, and agriculture. Examples of industrial biochemistry are highlighted with the test tube icon. (For a full list of these highlighted Industry Applications, see p. xvi.)

Key Concepts in Biochemistry

Several chapters provide the foundation for students' understanding of key aspects of biochemistry.

- Chapter 7 (Hemoglobin) illustrates the intimate connection between function and structure.

- Chapters 8 through 10 (Enzymes) explain where enzymes derive the catalytic power, specificity, and control they have to drive chemical and energetic transformations.

- Chapter 15 is where students first learn the basic concepts and design of metabolic pathways before they dive into the chapters on specific pathways.

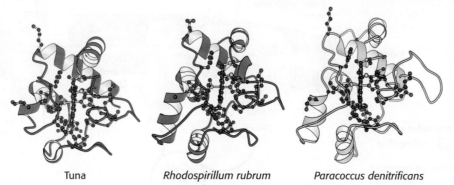

Tuna   *Rhodospirillum rubrum*   *Paracoccus denitrificans*

FIGURE 18.21 Conservation of the three-dimensional structure of cytochrome *c*. *Notice* the overall structural similarity of the different molecules from different sources. The side chains are shown for the 21 conserved amino acids as well as the centrally located planar heme. [Drawn from 3CYT.pdb, 3C2C.pdb, and 155C.pdb.]

• Chapter 27 presents an integration of metabolism focusing on healthy caloric homeostasis and its dysfunctional extremes: obesity and starvation. The chapter also presents the biochemical basis for the many benefits of regular exercise.

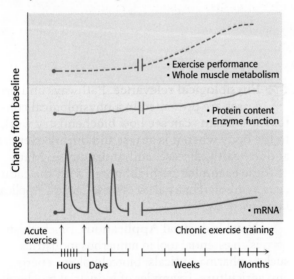

FIGURE 27.19 The molecular adaptation to exercise. Changes as a function of duration of exercise are shown for mRNA synthesis (bottom), protein synthesis (middle), and exercise performance (top). [Figure 1 from Egan, B., and Zierath, J. R. 2013. Exercise metabolism and the molecular regulation of skeletal muscle adaptation. *Cell Metab.* 17:162–184.]

Straightforward Art Program

• **Clear writing and straightforward illustrations**. The language of biochemistry is made accessible for students learning the subject for the first time. We present a logical flow of ideas, clearly written text, and informative figures to help students grasp and retain biochemical concepts and details.

• **One idea at a time**. Each figure illustrates a single concept, which helps students zoom in on the main points without the distraction of excess detail.

FOCUS OF THE NINTH EDITION

In developing a ninth edition of Berg/Tymoczko/Gatto/Stryer we stayed true to the integrity of the original Stryer text and focused on three specific areas to help biochemistry students manage the complexity of the course, engage with the material, and become more proficient problem solvers.

• Integrated text and media

• Promote effective problem solving

• Provide tools and resources for active learning

Integrated Text and Media

The ability to visualize concepts and how they are connected helps students gain a deeper conceptual understanding. Two innovations introduced in the first edition of *Biochemistry* were an open, uncluttered layout of text and figures that was easy on the eyes (and brain) and the presentation of one concept per figure. We continue these traditions in the ninth edition, while also presenting a revised art program that enhances the power of figures to help students learn.

Concise figures and ample margins are part of a layout that is well organized and easy to read. The art has been updated with greater shadowing and depth and a revised color palette.

In the ninth edition we are able to help students better visualize and understand biochemistry principles with the integrated SaplingPlus program.

• **Metabolic Map.** EXPLORE icons throughout the text denote topics where students are encouraged to take visual tours through the Metabolic Map within SaplingPlus to better see the connections between concepts and how pathways interrelate. With this

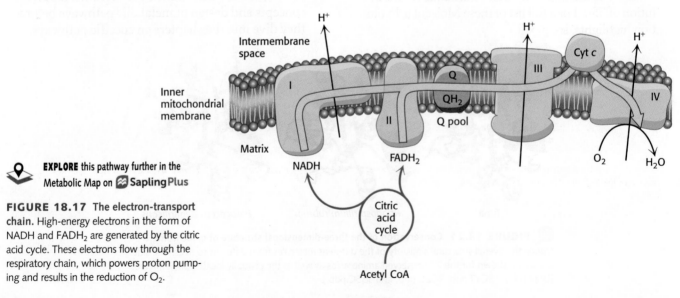

EXPLORE this pathway further in the Metabolic Map on **Sapling**Plus

FIGURE 18.17 The electron-transport chain. High-energy electrons in the form of NADH and $FADH_2$ are generated by the citric acid cycle. These electrons flow through the respiratory chain, which powers proton pumping and results in the reduction of O_2.

interactive tool, students can navigate and zoom between overviews and detailed views of the most commonly taught metabolic pathways. Embedded Tutorials take students through the pathways step by step. Such intimate interaction helps students visualize concepts in isolation as well as understand how concepts are interconnected, thereby helping students to grasp the ramifications of those connections. Assessment questions align with the interactive map to help students verify their understanding of the pathways and connections among the pathways. A screenshot of the Metabolic Map that is connected to Figure 18.17 (on the previous page) is shown below.

Metabolic Map Pathways include:

glycolysis, gluconeogenesis, the citric acid cycle, oxidative phosphorylation, the pentose phosphate pathway, glycogen metabolism, fatty acid synthesis, the urea cycle, and β oxidation.

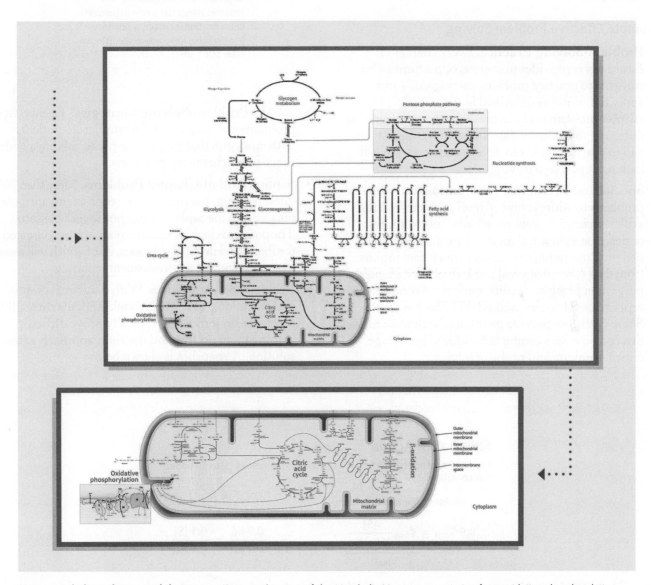

Major metabolic pathways and their connections. In this view of the Metabolic Map, we are entering from oxidative phosphorylation, seen in the inset in the lower left.

- **NEW! Tagged Learning Objectives** for resources included in SaplingPlus and an instructor guide make finding and assigning problems by learning objective easy and efficient.

- (☚) **Living Figures.** When the Living Figure icon is present, students can manipulate in Jmol an online, 3-D version of the molecule as it is rendered in the textbook.

- **Animated Technique Videos.** Whenever an EOC problem refers to a technique for which we have an animation, there is a link to those animations on SaplingPlus.

Promote Effective Problem Solving

- **Problem-Solving Practice.** Every chapter of *Biochemistry* provides numerous opportunities for students to practice problem-solving skills and apply the concepts described in the text. End-of-chapter problems vary in presentation and degree of difficulty to accommodate different learning styles and to coach students to develop different ways of looking at problem solving.

- **Specialized Problems.** There are four categories of problems to address higher-level problem-solving skills: *mechanism problems* ask students to suggest or describe a chemical mechanism; *data interpretation problems* require students to draw conclusions from data taken from real research papers; *chapter integration problems* require students to connect concepts across chapters; and **NEW!** *Think ▶ Pair ▶ Share Questions* provide instructors with tools to use this cooperative learning technique to encourage critical thinking and problem solving.

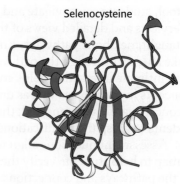

Selenocysteine

(☚) **FIGURE 24.24** Structure of glutathione peroxidase. This enzyme, which has a role in peroxide detoxification, contains a selenocysteine residue in its active site. [Drawn from 1GP1.pdb.]

- **NEW! Problem-Solving Strategies.** Twenty-five problems are "unpacked" to familiarize students with strategies that they can apply to solving a wide range of biochemistry problems.

- **Sapling End-of-Chapter Problems.** More than 200 questions from the end-of-chapter problems are incorporated into the Sapling online homework platform. The problems include unique module types, targeted feedback, and detailed solutions, all of which will assist in student success and assessment reporting.

- **Problem-Solving Videos.** With diagrams, graphs, and narration, these videos walk students through problems on topics that typically prove difficult, helping them understand the right approach to the solution. A snapshot is shown below.

Equation in Action → 0.9

The velocity of an enzyme is 90% V_{max}. What is the ratio of [S] to K_M under this condition?

Michaelis-Menton equation

$$\frac{0.9\,V_{max}}{V_{max}} = V_{max}\frac{\frac{[S]}{[S]+K_M}}{V_{max}}$$

$$([S]+K_M) \times 0.9 = \frac{[S]}{[S]+K_M} \times [S]+K_M$$

$$0.9[S] + 0.9K_M - 0.9[S] = [S] - 0.9[S]$$

$$\frac{0.9\,K_M}{0.1\,K_M} = \frac{0.1\,[S]}{0.1\,K_M}$$

$$9 = \frac{[S]}{K_M}$$

Tools and Resources for Active Learning

- **NEW! Think ▶ Pair ▶ Share Questions.** Found in every end-of-chapter problem set, these new questions are designed to facilitate group thinking and motivate peer-to-peer instruction and problem-solving discussion.

- **Clicker Questions** for Think ▶ Pair ▶ Share Questions with instructor notes and PowerPoint Questions and Biochemistry in Focus Questions.

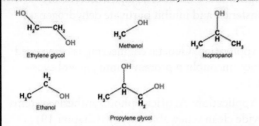

The ER staff places an intravenous catheter in the girl's arm and administers a dose of an alcohol dehydrogenase (ADH) inhibitor called fomepizole. Just as the girl is about to undergo gastric lavage (stomach pumping), her uncle arrives with the original antifreeze container. The ER doctor closely examines the bottle and then turns to the girl's father, "You're sure this is what she drank? How much again?"

"About a quarter of a cup, I think—just what's missing from the water bottle."

The doctor smiles as he turns towards the little girl, "Young miss, you're going to be just fine." To the nurse and father, "We'll cancel the lavage and dialysis. With the ADH inhibitor she's already on, we just need to give her more IV fluids and have her drink water to flush out her system. We'll monitor her for another couple of hours, but then you can both go home." The father, simultaneously relieved and confused, exclaims, "What? But, well, that's great! But Doc, how is it that this was so serious a minute ago and now we can just leave?"

The doctor points to the label on the bottle. "You see right here? It says 'pet-safe antifreeze.' The dose makes the poison, and this stuff isn't that bad in small doses."

Befuddled, but profoundly relieved, the father exclaims, "But it's still antifreeze! What's the difference!?"

In this case study, you will explore the toxicity of several alcohols commonly encountered in daily life. In particular, you will focus on the enzyme-catalyzed metabolic reactions in humans that largely determine the vast differences in alcohol toxicities. You have already been introduced to antifreeze and so-called "pet-safe" antifreeze. The main component of each is an alcohol, but the alcohol of regular antifreeze and the alcohol of "pet-safe" antifreeze have vastly different toxicities.

Note that the old adage "The dose makes the poison" is a reference to the biochemical fact that nothing is toxic at low enough concentrations, and everything is toxic at high enough concentrations, even water! What differentiates these compounds is the relative amounts of each that the human body can tolerate.

First, consider these five common alcohols encountered in daily life. How much do you know about these compounds?

Try to rank the the five alcohols in order of acute human toxicity, from most toxic to least toxic.

Laurence Mouton/AGE Fotostock

- **BLAST Database Activity.** An activity to introduce students to bioinformatic techniques that they would use in a research setting.

- **Case Studies.** Each case study introduces students to a biochemical mystery and allows them to determine what investigations will solve it. Accompanying assessments ensure that students have fully completed and understood the case study, showing that they have applied their knowledge of the concepts.

Topics include:

- A Likely Story: Enzyme Inhibition

- An Unexplained Death: Carbohydrate Metabolism

- A Day at the Beach: Lipid Metabolism

- Runner's Experiment: Integration of Metabolism

- **NEW!** pH Peril: pH and Amino Acid Structure

- Profound Relevance: Hemoglobin Regulation (Bohr Effect)

- Observation Over Intuition: Thermodynamics of Protein Folding

RECENT ADVANCES AND DISCOVERIES INCLUDED IN THE NEW EDITION

Biochemistry is an exciting and dynamic field of study. Two ways in which we highlight recent discoveries are:

- **Macromolecular Structures.** Beautifully rendered art depicting structures and structural models were prepared by Jeremy Berg and Gregory Gatto, based on the latest x-ray crystallographic, NMR, and cryo-electron microscopy data. For most molecular models, the **PDB number** at the end of the figure gives the student easy access to the file used in generating the structure (www.pdb.org).

- **NEW! Biochemistry in Focus.** Thirty-three vignettes focus on fascinating discoveries, current research methods, and curious phenomena illuminated by biochemistry. (For a full list of the Biochemistry in Focus features, see p. xvii.)

In this edition of *Biochemistry* we have updated examples and explanations to provide the latest and most interesting discoveries in biochemistry, along with material essential for understanding biochemical concepts. Some of these new topics are:

A review of functional groups that play major roles in biochemical reactions (Chapter 1)

Explanation of co-immunoprecipitation for isolating a protein and its binding partners (Chapter 3)

Explanation of cryo-electron microscopy for determining macromolecular structures (Chapter 3)

Extrachromosomal circular DNA molecules have been discovered in eukaryotic nuclei (Chapter 4)

An expanded discussion of the genome sequencing method, pyrosequencing (Chapter 5)

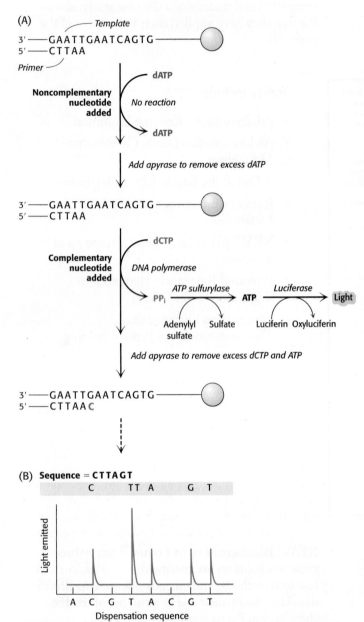

FIGURE 5.28 Pyrosequencing. (A) The template strand is attached to a solid support and an appropriate primer is added. A single deoxynucleotide is added: if this base is not complementary to the next base in the template strand, then no reaction occurs. However, if the added deoxynucleotide is complementary, pyrophosphate (PP_i) is released and coupled to the production of light by the enzymes ATP sulfurylase and luciferin. At the end of each step, apyrase is added to remove remaining nucleotide and ATP (if present), and then the steps are repeated with the next nucleotide. (B) Light is measured at each deoxynucleotide addition. A peak indicates the next base in the sequence and the height of the peak indicates how many of the same deoxynucleotide were incorporated in a row.

The CRISPR-Cas9 method and other naturally occurring gene editing methods (Chapter 5)

Updated discussion of SCID includes the use of Strimvelis gene therapy (Chapter 5)

Industry Application: Hydrolytic enzymes are being used to reduce the waste stream and generate methane for electricity (Chapter 8)

Clinical Application: Mutations in protein kinase A can cause Cushing Syndrome (Chapter 10)

Clinical Application: Exercise modifies the phosphorylation state of many proteins (Chapter 10)

Serpins can be degraded by an isozyme of trypsin (Chapter 10)

Clinical Application: Oligosaccharides in milk protect newborns against infection (Chapter 11)

Industrial Application: Chitin and chitosan have a thousand-and-one uses (Chapter 11)

Clinical Application: Myelination of neurons in health and disease (Chapter 12)

Updates on how the acetylcholine receptor opens and closes based on cryo-electron microscopic data (Chapter 13)

ATP has a role as a hydrotrope that prevents intracellular protein aggregation (Chapter 15)

Structural and biochemical data are revealing new examples of massive enzyme complexes in metabolic pathways (Chapters 15–18)

Clinical Application: Cancer cells hijack acetyl CoA acetyltransferase and inhibit pyruvate dehydrogenase (Chapter 17)

Industrial Application: Genetic engineering is being used to introduce uncoupling protein-1 into pig embryos (Chapter 18)

Industrial Application: Artificial photosynthesis systems may provide clean renewable energy (Chapter 19)

Updated discussion of the evolution and enzymology of C_4 plants (Chapter 20)

An isozyme of glycogen phosphorylase exists in the brain (Chapter 21)

Clinical Application: Liver's role in lipid metabolism is readily disrupted by ethanol (Chapter 22)

Ketogenic diets reduce epileptic seizures by altering the gut microbiome (Chapter 22)

Updated structural information on mammalian fatty acid synthase (Chapter 22)

Industrial Application: Enlisting microbes for the generation of triacylglycerols to be used to produce biodiesel (Chapter 22)

AMP-activated protein kinase is a key regulator of metabolism (Chapter 22)

Protein metabolism helps to power the flight of migratory birds (Chapter 23)

Amino acids are precursors for a number of neurotransmitters (Chapter 24)

Clinical Application: A Niemann-Pick disease is caused by failure to transport cholesterol from the lysosome (Chapter 26)

Five-carbon units are joined to form a wide variety of biomolecules (Chapter 26)

Industrial Application: Genetically engineered farnesene improves the profitability of its industrial uses (Chapter 26)

Updated model of the spliceosome based on cryo-electron microscopy (Chapter 30)

Cryo-electron microscopy reveals the existence of a bacterial expressome (Chapter 31)

Expanded discussion on toxins that block protein synthesis (Chapter 31)

Clinical Application: The problems with biofilms (Chapter 32)

Posttranslational control of ATP synthase production ensures proper stoichiometry (Chapter 32)

Cryo-electron microscopy is revealing the molecular details of mediator's interactions with the transcription complex (Chapter 33)

Updates on the histone code's control of transcriptional repression (Chapter 33)

RESOURCES IN SaplingPlus

Every Problem Counts

This comprehensive and robust online platform combines innovative high-quality teaching and learning features with Sapling Learning's acclaimed online homework.

Sapling Online Homework

Proven effective in raising students' comprehension and problem-solving skills and recently updated with hundreds of additional questions. A special effort has been made to increase the number of spectroscopy, synthesis, and mechanism problems. Sapling Learning's innovative online homework offers:

- **Immediate, Individualized Feedback.** Students get the guidance they need when they need it. Feedback is wrong-answer specific.

- **Efficient Course Management.** Automatic grading, tracking, and analytics helps instructors save time and tailor assignments to students needs.

- **Industry-Leading Peer-to-Peer Support and Service.** Each instructor is paired with a fully trained PhD or Master's-level colleague ready to help with everything from course setup, alignment of assignments to instructor's syllabus, and service during implementation.

SaplingPlus Resources for Berg/Tymoczko/Gatto/Stryer, *Biochemistry*, Ninth Edition

- **NEW! Interactive VitalSource e-book.** Mobile accessible with native apps on Mac, Windows, iOS, Android, Chrome, and Kindle. Allows students to read offline, have the book read aloud to them, highlight, take notes, and search for keywords. Instructors can also highlight and push notes for students to see.

- **LearningCurve.** Putting "testing to learn" into action, LearningCurve is the perfect tool to get students to engage before class and review after. With game-like quizzing, it creates individualized activities for each student, selecting questions—by difficulty and topic—according to their performance.

- **Interactive Metabolic Map.** Interactive tool designed to help students navigate and zoom between overview and detailed views of the most commonly taught metabolic pathways. Embedded Tutorials take students through the pathways step by step, helping students visualize concepts

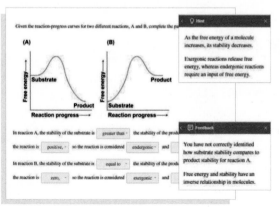

Hints attached to every problem encourage critical thinking by providing suggestions for completing the problem, without giving away the answer.

↓

Targeted Feedback: each question includes wrong-answer specific feedback targeted to students' misconceptions.

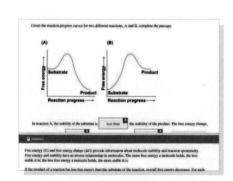

Detailed Solutions: fully-worked solutions reinforce concepts and provide an in-product study guide for every problem in the Sapling Learning system.

in isolation as well as understand how concepts are interconnected, and the ramifications of those connections.

- **Sapling Homework.** More than 250 selected end-of-chapter questions from *Biochemistry,* Ninth Edition and agnostic questions from Sapling's extensive biochemistry question library have been arranged to fit pedagogically with the text.

- **Readiness Assignment.** An optional prerequisite assignment to review math, general chemistry, and organic chemistry skills applicable for biochemistry students.

- **Problem-Solving Videos.** Videos designed to assist students in their understanding of data interpretation and in solving biochemical problems.

- **Case Studies.** Online biochemistry case studies that are assignable and assessable.

- **Clicker Questions.** Aligned with Think ▶ Pair ▶ Share; Biochemistry in Focus questions.

- **Animated Mechanisms.** Mechanisms of iconic enzymes in GIF format for easy use in lecture and study.

- **Animated Techniques.** Animations detailing lab techniques related to biochemistry. Designed to provide students with an overview of protocol, these animations give students a conceptual understanding of what is happening at the chemical level.

- **Living Figures.** Allow students to view textbook illustrations of protein structures online in interactive 3-D using Jmol. Students can zoom and rotate 54 "live"

structures to get a better understanding of their three-dimensional nature and can experiment with different display styles (space-filling, ball-and-stick, ribbon, backbone) by means of a user-friendly interface.

- **Instructor Resources:**

  - **Clicker Questions.**

  - **PPT Files.** These files offer suggested lectures, including key illustrations and summaries that instructors can adapt to their teaching styles.

  - **Updated textbook images and tables as high-resolution JPEG files.** The JPEGs are also offered in separate PowerPoint files.

  - **Test Bank.**

  - **Active Learning Tools.**

    - BLAST Database Activities

    - Think ▶ Pair ▶ Share Questions with instructor notes and PowerPoint Questions

    - Case Studies

Additional Resources

- Student Companion (solutions manual). For each chapter of the textbook, it includes: Chapter Learning Objectives and Summary, Self-Assessment Problems (including multiple-choice, short-answer, matching questions, and challenge problems) and their answers, and Expanded Solutions to text end-of-chapter problems. Instructors choose whether to make these files available to students.

MOLECULAR EVOLUTION

This icon signals the start of the many discussions that highlight protein commonalities or other molecular evolutionary insights.

Only L amino acids make up proteins (p. 31)

Why this set of 20 amino acids? (p. 37)

Codon bias (p. 134)

The genetic code is nearly universal (p. 135)

Sickle-cell trait and malaria (p. 221)

Additional human globin genes (p. 223)

Catalytic triads in hydrolytic enzymes (p. 281)

Major classes of peptide-cleaving enzymes (p. 283)

Common catalytic core in type II restriction enzymes (p. 298)

P-loop NTPase domains (p. 303)

Why do different human blood types exist? (p. 359)

Archaeal membranes (p. 379)

Ion pumps (p. 407)

P-type ATPases (p. 410)

ATP-binding cassettes (p. 411)

Sequence comparisons of Na^+ and Ca^{2+} channels (p. 420)

Small G proteins (p. 453)

Metabolism in the RNA world (p. 485)

Why is glucose a prominent fuel? (p. 493)

Evolution of glycolysis and gluconeogenesis (p. 530)

The α-ketoglutarate dehydrogenase complex (p. 552)

Evolution of the citric acid cycle (p. 563)

Mitochondrial evolution (p. 575)

Conserved structure of cytochrome c (p. 592)

Common features of ATP synthase and G proteins (p. 599)

Mitochondrial transporters have a tripartite structure (p. 602)

Pigs lack uncoupling protein 1 (UCP-1) and brown fat (p. 606)

Chloroplast evolution (p. 622)

Evolutionary origins of photosynthesis (p. 638)

Rubisco activase is a AAA ATPase (p. 649)

Evolution of the C4 pathway (p. 658)

The relationship of the Calvin cycle and the pentose phosphate pathway (p. 668)

Increasing sophistication of glycogen phosphorylase regulation (p. 692)

Glycogen synthase is homologous to glycogen phosphorylase (p. 694)

A recurring motif in the activation of carboxyl groups (p. 716)

Fatty acid synthase is a megasynthase (p. 738)

Prokaryotic counterparts of the ubiquitin pathway and the proteasome (p. 756)

A family of pyridoxal-dependent enzymes (p. 762)

Evolution of the urea cycle (p. 767)

The P-loop NTPase domain in nitrogenase (p. 790)

Conserved amino acids in transaminases determine amino acid chirality (p. 795)

Feedback inhibition (p. 806)

Recurring steps in purine ring synthesis (p. 827)

Ribonucleotide reductases (p. 833)

Increase in urate levels during primate evolution (p. 840)

Deinococcus radiodurans illustrates the power of DNA repair systems (p. 950)

DNA polymerases (p. 952)

Thymine and the fidelity of the genetic message (p. 972)

Sigma factors in bacterial transcription (p. 990)

Similarities in transcription between archaea and eukaryotes (p. 1001)

Evolution of spliceosome-catalyzed splicing (p. 1014)

Classes of aminoacyl-tRNA synthetases (p. 1030)

Composition of the primordial ribosome (p. 1032)

Homologous G proteins (p. 1036)

A family of proteins with common ligand-binding domains (p. 1063)

The independent evolution of DNA-binding sites of regulatory proteins (p. 1063)

Key principles of gene regulation are similar in bacteria and archaea (p. 1070)

CpG islands (p. 1084)

Iron-response elements (p. 1091)

miRNAs in gene evolution (p. 1093)

The odorant-receptor family (see p. 1099 on SaplingPlus)

Photoreceptor evolution (see p. 1110 on SaplingPlus)

The immunoglobulin fold (see p. 1127 on SaplingPlus)

Evolution of human MHC proteins (see p. 1141 on SaplingPlus)

Relationship of tubulin to prokaryotic proteins (see p. 1163 on SaplingPlus)

This icon signals the start of a clinical application in the text. Additional, briefer clinical correlations appear in the text as appropriate.

Osteogenesis imperfecta (p. 50)

Protein-misfolding diseases (p. 59)

Protein modification and scurvy (p. 60)

Antigen/antibody detection with ELISA (p. 86)

Synthetic peptides as drugs (p. 98)

Circular DNA is found in the nucleus (p. 124)

PCR in diagnostics and forensics (p.152)

A mouse model of ALS (p. 172)

RNAi can be used to treat disease (p. 177)

Gene therapy (p. 178)

Functional magnetic resonance imaging (p. 210)

2,3-BPG and fetal hemoglobin (p. 217)

Carbon monoxide poisoning (p. 217)

Sickle-cell anemia (p. 220)

Thalassemia (p. 221)

An antidote for carbon monoxide poisoning? (p. 227)

Aldehyde dehydrogenase deficiency (p. 247)

Action of penicillin (p. 258)

Protease inhibitors (p. 285)

Carbonic anhydrase and osteopetrosis (p. 287)

Isozymes as a sign of tissue damage (p. 318)

Protein kinase A mutations can cause Cushing's syndrome (p. 322)

Exercise modifies protein phosphorylation (p. 323)

Trypsin inhibitor helps prevent pancreatic damage (p. 327)

Emphysema (p. 327)

Blood clotting involves a cascade of zymogen activations (p. 328)

Vitamin K (p. 331)

Antithrombin and hemorrhage (p. 332)

Hemophilia (p. 333)

Monitoring changes in glycosylated hemoglobin (p. 347)

Human milk oligosaccharides protect newborns (p. 352)

Erythropoietin (p. 353)

Glycosylation functions in nutrient sensing (p. 354)

Hurler disease (p. 354)

Proteoglycans are important components of cartilage (p. 355)

Mucins (p. 356)

Blood groups (p. 359)

I-cell disease (p. 359)

Selectins (p. 362)

Influenza virus binding (p. 363)

Action potentials depend on myelin (p. 383)

Aspirin and ibuprofen (p. 387)

The Gram stain (p. 394)

Digitalis and congestive heart failure (p. 410)

Multidrug resistance (p. 411)

Long QT syndrome (p. 426)

Signal-transduction pathways and cancer (p. 455)

G-proteins, cholera and whooping cough (p. 457)

Vitamins (p. 479)

Triose phosphate isomerase deficiency (p. 497)

Excessive fructose consumption (p. 508)

Lactose intolerance (p. 510)

Galactosemia (p. 511)

Aerobic glycolysis and cancer (p. 517)

Substrate cycles affect health and disease (p. 528)

Isozymic forms of lactate dehydrogenase (p. 530)

Phosphatase deficiency (p. 558)

α-Ketoglutarate dehydrogenase deficiency is observed in Alzheimer's disease (p. 559)

Defects in the citric acid cycle and the development of cancer (p. 559)

Beriberi and mercury poisoning (p. 562)

Cancer cells alter acetyl CoA acetyltransferase function (p. 560)

Frataxin mutations cause Friedreich's ataxia (p. 582)

Reactive oxygen species (ROS) are implicated in a variety of diseases (p. 590)

ROS may be important in signal transduction (p. 591)

IF1 overexpression and cancer (p. 605)

Brown adipose tissue (p. 606)

Uncoupling proteins related to UCP-1 (p. 606)

Mild uncouplers sought as drugs (p.608)

The pentose phosphate pathway is required for rapid cell growth (p. 668)

Glucose 6-phosphate dehydrogenase deficiency causes drug-induced hemolytic anemia (p. 669)

Glucose 6-phosphate dehydrogenase deficiency protects against malaria (p. 670)

Brain glycogen phosphorylase (p. 690)

Developing drugs for type 2 diabetes (p. 699)

Glycogen-storage diseases (p. 700)

Chanarin-Dorfman syndrome (p. 714)

Hepatocytes are key in lipid metabolism (p. 714)

Carnitine deficiency (p. 717)

Zellweger syndrome (p. 724)

Some fatty acids may contribute to pathological conditions (p. 724)

Diabetic ketosis (p. 727)

Ketogenic diets to treat epilepsy (p. 727)

The use of fatty acid synthase inhibitors as drugs (p. 735)

Effects of aspirin on signaling pathways (p. 738)

Diseases resulting from defects in transporters of amino acids (p. 752)

Diseases resulting from defects in E3 proteins (p. 755)

Drugs target the ubiquitin-proteasome system (p. 757)

Using proteasome inhibitors to treat tuberculosis (p. 758)

Blood levels of aminotransferases indicate liver damage (p. 761)

Inherited defects of the urea cycle (hyperammonemia) (p. 767)

Alcaptonuria, maple syrup urine disease, and phenylketonuria (p. 776)

High homocysteine levels and vascular disease (p. 801)

Inherited disorders of porphyrin metabolism (p. 814)

Viral thymidine kinase binds acyclovir tightly (p. 826)

Anticancer drugs that block the synthesis of thymidylate (p. 835)

Ribonucleotide reductase is a target for cancer therapy (p. 838)

Adenosine deaminase and severe combined immuno-deficiency (p. 839)

Gout (p. 839)

Lesch–Nyhan syndrome (p. 840)

Folic acid and spina bifida (p. 841)

Enzyme activation in some cancers to generate phos-phocholine (p. 853)

Excess choline and heart disease (p. 853)

Gangliosides and cholera (p. 855)

Second messengers derived from sphingolipids and diabetes (p. 856)

Respiratory distress syndrome and Tay–Sachs disease (p. 856)

Ceramide metabolism stimulates tumor growth (p. 857)

Phosphatidic acid phosphatase and lipodystrophy (p. 858)

Hypercholesterolemia and atherosclerosis (p. 867)

Mutations in the LDL receptor (p. 868)

Niemann-Pick disease (p. 869)

LDL receptor cycling is regulated (p. 869)

The role of HDL in protecting against arteriosclerosis (p. 870)

Clinical management of cholesterol levels (p. 870)

Bile salts are derivatives of cholesterol (p. 871)

The cytochrome P450 system is protective (p. 874)

A new protease inhibitor also inhibits a cytochrome P450 enzyme (p. 875)

Aromatase inhibitors in the treatment of breast and ovarian cancer (p. 876)

Rickets and vitamin D (p. 877)

Caloric homeostasis is a means of regulating body weight (p. 890)

The brain plays a key role in caloric homeostasis (p. 892)

Diabetes is a common metabolic disease often resulting from obesity (p. 896)

Exercise beneficially alters the biochemistry of cells (p. 901)

Food intake and starvation induce metabolic changes (p. 905)

Ethanol alters energy metabolism in the liver (p. 909)

Antibiotics that target DNA gyrase (p. 960)

Blocking telomerase to treat cancer (p. 968)

Huntington disease (p. 973)

Defective repair of DNA and cancer (p. 973)

Translocations can result in diseases (p. 978)

Antibiotic inhibitors of transcription (p. 994)

Burkitt lymphoma and B-cell leukemia (p. 1001)

Diseases of defective RNA splicing (p. 1010)

Vanishing white matter disease (p. 1042)

Antibiotics that inhibit protein synthesis (p. 1043)

Diphtheria (p. 1044)

Ricin, a lethal protein-synthesis inhibitor (p. 1045)

Biofilms and human health (p. 1067)

Induced pluripotent stem cells (p. 1082)

Anabolic steroids (p. 1086)

Color blindness (see p. 1111 on SaplingPlus)

The use of capsaicin in pain management (see p. 1115 on SaplingPlus)

Immune-system suppressants (see p. 1133 on SaplingPlus)

MHC and transplantation rejection (see p. 1141 on SaplingPlus)

AIDS (see p. 1141 on SaplingPlus)

Autoimmune diseases (see p. 1143 on SaplingPlus)

Immune system and cancer (see p. 1144 on SaplingPlus)

Vaccines (see p. 1144 on SaplingPlus)

Charcot-Marie-Tooth disease (see p. 1162 on SaplingPlus)

Taxol (see p. 1164 on SaplingPlus)

This icon signals the start of a discussion which highlights an example from industry such as pharmaceuticals, biotechnology, and energy.

Genetically modified food crops (p. 178)

Improved biofuel production from genetically engineered algae (p. 180)

Aptamers are promising tools in medicine and biotechnology (p. 202)

Enzymes are important reagents in the clinic and laboratory (p. 261)

Chitin can be processed to a molecule with a variety of uses (p. 357)

Therapeutic applications for liposomes are being developed (p. 382)

Some drugs can inhibit the action of the K^+ channel hERG (p. 427)

Monoclonal antibodies can be used to inhibit signaling pathways in tumors (p. 456)

Protein kinase inhibitors can be effective anticancer drugs (p. 457)

Reintroduction of UCP-1 in pigs may be economically valuable (p. 607)

Artificial photosynthetic systems may provide clean, renewable energy (p. 639)

Triacylglycerols may become an important renewable energy source (p. 735)

Some isoprenoids have industrial applications (p. 880)

Effective vaccines must generate a sustained protective response (see p. 1144 on SaplingPlus)

Many of the Biochemistry in Focus features provide an opportunity for students to examine data and experimental details from classic and current research in biochemistry. This provides additional opportunities for students to practice analyzing data and thinking about the connections between experimental methods and the resulting data. The Biochemistry in Focus features also expose students to fascinating and fun aspects of biochemistry in their everyday lives.

Improved biofuel production from genetically engineered algae (Chapter 5)

Using sequence alignments to identify functionally important residues (Chapter 6)

A potential antidote for carbon monoxide poisoning? (Chapter 7)

The effect of temperature on enzyme-catalyzed reactions and the coloring of Siamese cats (Chapter 8)

Phosphoribosylpyrophosphate synthetase-induced gout (Chapter 10)

α-Glucosidase (maltase) inhibitors can help to maintain blood glucose homeostasis (Chapter 11)

The curious case of cardiolipin (Chapter 12)

Setting the pace is more than funny business (Chapter 13)

Gases get in on the signaling game (Chapter 14)

Triose phosphate isomerase deficiency (TPID) (Chapter 16)

Pyruvate carboxylase deficiency (PCD) (Chapter 16)

Diabetic neuropathy may result from inhibition of the pyruvate dehydrogenase complex (Chapter 17)

Leber hereditary optic neuropathy can result from defects in Complex I (Chapter 18)

Increasing the efficiency of photosynthesis will increase crop yields (Chapter 19)

Phosphoenolpyruvate carboxylase is a window into ancient ecosystems (Chapter 20)

Hummingbirds and the pentose phosphate pathway (Chapter 20)

McArdle disease results from a lack of skeletal muscle glycogen phosphorylase (Chapter 21)

Ethanol consumption results in triacylglycerol accumulation in the liver (Chapter 22)

Methylmalonic acidemia results from an inborn error of metabolism (Chapter 23)

Tyrosine is a precursor for human pigments (Chapter 24)

Uridine plays a role in caloric homeostasis (Chapter 25)

Excess ceramides may cause insulin insensitivity (Chapter 26)

Adipokines help to regulate fuel metabolism in the liver (Chapter 27)

Exercise alters muscle and whole-body metabolism (Chapter 27)

Monoclonal antibodies: Expanding the drug developer's toolbox (Chapter 28)

Identifying amino acids crucial for DNA replication fidelity (Chapter 29)

Discovering enzymes made of RNA (Chapter 30)

Selective control of gene expression by ribosomes (Chapter 31)

Regulating gene expression through proteolysis (Chapter 32)

A mechanism for consolidating epigenetic modifications (Chapter 33)

Binding many palatable tastants with a single receptor (Chapter 34, online on SaplingPlus)

The risks of class switching (Chapter 35, online on SaplingPlus)

Myosin motors appear to help us hear (Chapter 36, online on SaplingPlus)

ACKNOWLEDGMENTS

Writing a popular textbook is both a challenge and an honor. Our goal is to convey to our students our enthusiasm and understanding of a discipline to which we are devoted. Our students are our inspiration. Consequently, not a word was written nor an illustration constructed without the knowledge that bright, engaged students would immediately detect vagueness and ambiguity. We also thank our colleagues who supported, advised, instructed, and simply bore with us during this arduous task. We are grateful to our colleagues throughout the world who patiently answered our questions and shared their insights into recent developments.

We have had the privilege and pleasure of working with the publishing team at W. H. Freeman/Macmillan Learning for many years now, and our experiences have always been enjoyable and rewarding. This has held true once again with our work on the ninth edition of *Biochemistry*. Our Macmillan colleagues have a knack for coaxing without nagging, for critiquing while encouraging, for making our stressful and demanding undertaking both exhilarating and rewarding. We have many people to thank for this experience, some of whom are first timers to the *Biochemistry* project. We are delighted to work with Executive Editor, Lauren Schultz, for the second time. She was unfailing in her enthusiasm and generous with her support. Our two new developmental editors Lisa Bess Kramer and Michael Zierler joined the ranks of outstanding developmental editors we have worked with over the years. Lisa and Michael were thoughtful, insightful, and very efficient at identifying aspects of our writing and figures that were less than clear. Lisa Samols, a former developmental editor, now Director of Development, STEM served as our guide. With her extensive publishing knowledge and sense of diplomacy, Lisa provided consistent support and encouragement throughout the project from start to finish. It is almost impossible to put into words the amount of coordination and effort required to shepherd a textbook project from authors' first draft to publication. We have many to thank for directing us through this complex process. Debbie Hardin, Senior Developmental Editor, STEM, along with Project Managers Karen Misler, Valerie Brandenberg at Lumina Datamatics, and Peter Jacoby managed the flow of the entire project, from copyediting through bound book, with amazing efficiency. Irene Vartanoff, our manuscript copyeditor, enhanced the literary consistency and clarity of the text. Diana Blume, Director of Design, Content Management, and Natasha Wolfe, Design Services Manager, produced a design and layout that makes the book uniquely attractive while still emphasizing its ties to past editions. Photo Editor Christine Buese and Photo Researcher Jennifer Atkins found the photographs that we hope make the text not only more inviting, but also fun to look through. Janice Donnola, Illustration Coordinator, deftly directed the rendering of new illustrations. Paul Rohloff, Senior Workflow Project Manager, made sure that the significant difficulties of scheduling, composition, and manufacturing were smoothly overcome. Lumina Datamatics, Inc., provided composition. Jennifer Driscoll Hollis, Amber Jonker, Angela Piotrowski, Cassandra Korsvik, Jennifer Compton, and Daniel Comstock did a wonderful job in their management of the media program. Special thanks also to editorial assistant Justin Jones, who tirelessly juggled numerous tasks in support of our team. Maureen Rachford, Marketing Manager for the Physical Sciences, enthusiastically introduced this newest edition of *Biochemistry* to the academic world. We are deeply appreciative of Craig Bleyer and his sales staff for their support. Without their able and enthusiastic presentation of our text to the academic community, all of our efforts would be in vain.

Thanks also to our many colleagues at our own institutions as well as throughout the country who patiently answered our questions and encouraged us on our quest. Finally, we owe a debt of gratitude to our families—our wives, Wendie Berg, Alison Unger, and Megan Williams, and our children, especially Timothy and Mark Gatto.

Without their support, comfort, and understanding, this endeavor could never have been undertaken, let alone successfully completed.

We also especially thank those who served as reviewers for this ninth edition as well as those who reviewed the previous (eighth) edition. Their thoughtful comments, suggestions, and encouragement have been of immense help to us in maintaining the excellence of the preceding editions. These reviewers are:

Ninth Edition Textbook

Gerald Audette
York University

Donald Beitz
Iowa State University

Matthew Berezuk
Azusa Pacific University

Steven Berry
University of Minnesota Duluth

James Bouyer
Florida A&M University

David Brown
Florida Gulf Coast University

W. Malcolm Byrnes
Howard University

Chris Calderone
Carleton College

Naomi Campbell
Jackson State University

Weiguo Cao
Clemson University

Heather Coan
Western Carolina University

Kenyon Daniel
University of South Florida

Margaret Daugherty
Colorado College

Tomas T. Ding
North Carolina Central University

Nuran Ercal
Missouri University Science and Technology

Kirsten Fertuck
Northeastern University

Kathleen Foley Geiger
Michigan State University

Ronald Gary
University of Nevada, Las Vegas

Martina Gaspari
Dixie State University

Christina Goode
Western University of Health Sciences

Neena Grover
Colorado College

Donovan Haines
Sam Houston State University

Bonnie Hall
Grand View University

Robert Harris
Kansas University Medical Center

M. Nidanie Henderson-Stull
Augsburg College

Kirstin Hendrickson
Arizona State University

Newton Hilliard
Arkansas Tech University

Pat Huey
Georgia Gwinnett College

Jeba Inbarasu
Metropolitan Community College, South Omaha Campus

Lori Isom
University of Central Arkansas

Gerwald Jogl
Brown University

Jerry Johnson
University of Houston-Downtown

Anna Kashina
University of Pennsylvania

Bhuvana Katkere
Penn State University

Dmitry Kolpashchikov
University of Central Florida

Mark Larson
Augustana University

Sarah Lee
Abilene Christian University

Pan Li
University at Albany, State University of New York

Michael Lieberman
University of Cincinnati College of Medicine

Timothy Logan
Florida State University

Thomas Marsh
University of St. Thomas

Glover Martin
University of Massachusetts-Boston

Michael Massiah
George Washington University

Douglas McAbee
California State University, Long Beach

Karen McPherson
Delaware Valley University

John Means
University of Rio Grande

Nick Menhart
Illinois Institute of Technology

David Mitchell
College of St. Benedict and St. John's University

Fares Najar
University of Oklahoma

Li Niu
University at Albany, State University of New York

James Nolan
Georgia Gwinnett College

Suzanne O'Handley
Rochester Institute of Technology

Margaret Olney
Saint Martin's University

Pamela Osenkowski
Loyola University Chicago

Wendy Pogozelski
State University of New York at Geneseo

Tamiko Porter
Indiana University, Purdue University Indianapolis (IUPUI)

Ramin Radfar
Wofford College

Reza Razeghifard
Nova Southeastern University

Kevin Redding
Arizona State University

Tanea Reed
Eastern Kentucky University

Gillian Rudd
Georgia Gwinnett College

Matthew Saderholm
Berea College

Kavita Shah
Purdue University

Lisa Shamansky
California State University, San Bernardino

Rajnish Singh
Kennesaw State University

Jennifer Sniegowski
Arizona State University Downtown Phoenix Campus

Kathryn Tifft
Johns Hopkins University

Candace Timpte
Georgia Gwinnett College

Marc Tischler
University of Arizona

Marianna Torok
University of Massachusetts-Boston

Brian Trewyn
Colorado School of Mines

Vishwa Trivedi
Bethune Cookman University

Manuel Varela
Eastern New Mexico University

Grover Waldrop
Louisiana State University

Yuqi Wang
Saint Louis University

Kevin Williams
Western Kentucky University

Kevin Wilson
Oklahoma State University

Michael Wolyniak
Hampden-Sydney College

Chung Wong
University of Missouri-Saint Louis

Wu Xu
University of Louisiana at Lafayette

Brent Znosko
Saint Louis University

Ninth Edition Media

Heather Coan
Western Carolina University

Patricia Ellison
University of Nevada

Kirsten Fertuck
Northeastern University

Kathleen Foley Geiger
Michigan State University

Newton Hilliard
Arkansas Tech University

Mian Jiang
University of Houston

Michael Keck
Keuka College

Michael Massiah
George Washington University

Michael Mendenhall
University of Kentucky

Fares Najar
University of Oklahoma

Scott Napper
University of Saskatchewan

Margaret Olney
Saint Martin's University

Abdel Omri
Laurentian University

Sarah Robinson
University of Georgia

Gillian Rudd
Georgia Gwinnett College

Mark Saper
University of Michigan Medical School

Jennifer Sniegowski
Arizona State University Downtown Phoenix Campus

Candace Timpte
Georgia Gwinnett College

Xuemin Wang
University of Missouri

Yuqi Wang
Saint Louis University

Kevin Williams
Western Kentucky University

Chung Wong
University of Missouri-Saint Louis

Wu Xu
University of Louisiana at Lafayette

Eighth Edition Reviewers

Paul Adams
University of Arkansas, Fayetteville

Kevin Ahern
Oregon State University

Zulfiqar Ahmad
A.T. Still University of Health Sciences

Richard Amasino
University of Wisconsin

Young-Hoon An
Wayne State University

Kenneth Balazovich
University of Michigan

Donald Beitz
Iowa State University

Matthew Berezuk
Azusa Pacific University

Melanie Berkmen
Suffolk University

Steven Berry
University of Minnesota, Duluth

Loren Bertocci
Marian University

Mrinal Bhattacharjee
Long Island University

Elizabeth Blinstrup-Good
University of Illinois

Brian Bothner
Montana State University

Mark Braiman
Syracuse University

David Brown
Florida Gulf Coast University

Donald Burden
Middle Tennessee State University

Nicholas Burgis
Eastern Washington University

W. Malcolm Byrnes
Howard University

Graham Carpenter
Vanderbilt University School of Medicine

John Cogan
Ohio State University

Jeffrey Cohlberg
California State University, Long Beach

David Daleke
Indiana University

John DeBanzie
Northeastern State University

Cassidy Dobson
St. Cloud State University

Donald Doyle
Georgia Institute of Technology

Ludeman Eng
Virginia Tech

Caryn Evilia
Idaho State University

Kirsten Fertuck
Northeastern University

Brent Feske
Armstrong Atlantic University

Patricia Flatt
Western Oregon University

Wilson Francisco
Arizona State University

Gerald Frenkel
Rutgers University

Ronald Gary
University of Nevada, Las Vegas

Eric R. Gauthier
Laurentian University

Glenda Gillaspy
Virginia Tech

James Gober
UCLA

Christina Goode
California State University, Fullerton

Nina Goodey
Montclair State University

Eugene Gregory
Virginia Tech

Robert Grier
Atlanta Metropolitan State College

Neena Grover
Colorado College

Paul Hager
East Carolina University

Ann Hagerman
Miami University

Mary Hatcher-Skeers
Scripps College

Diane Hawley
University of Oregon

Blake Hill
Medical College of Wisconsin

Pui Ho
Colorado State University

Charles Hoogstraten
Michigan State University

Frans Huijing
University of Miami

Kathryn Huisinga
Malone University

Cristi Junnes
Rocky Mountain College

Lori Isom
University of Central Arkansas

Nitin Jain
University of Tennessee

Blythe Janowiak
Saint Louis University

Gerwald Jogl
Brown University

Kelly Johanson
Xavier University of Louisiana

Jerry Johnson
University of Houston-Downtown

Todd Johnson
Weber State University

David Josephy
University of Guelph

Michael Kalafatis
Cleveland State University

Marina Kazakevich
University of Massachusetts-Dartmouth

Jong Kim
Alabama A&M University

Sung-Kun Kim
Baylor University

Roger Koeppe
University of Arkansas, Fayetteville

Dmitry Kolpashchikov
University of Central Florida

Min-Hao Kuo
Michigan State University

Isabel Larraza
North Park University

Mark Larson
Augustana College

Charles Lawrence
Montana State University

Pan Li
University at Albany, State University of New York

Darlene Loprete
Rhodes College

Greg Marks
Carroll University

Michael Massiah
George Washington University

Keri McFarlane
Northern Kentucky University

Michael Mendenhall
University of Kentucky

Stephen Mills
University of San Diego

Smita Mohanty
Auburn University

Debra Moriarity
University of Alabama, Huntsville

Stephen Munroe
Marquette University

Jeffrey Newman
Lycoming College

William Newton
Virginia Tech

Alfred Nichols
Jacksonville State University

Brian Nichols
University of Illinois, Chicago

Allen Nicholson
Temple University

Brad Nolen
University of Oregon

Pamela Osenkowski
Loyola University Chicago

Xiaping Pan
East Carolina University

Stefan Paula
Northern Kentucky University

David Pendergrass
University of Kansas-Edwards

Wendy Pogozelski
State University of New York at Geneseo

Gary Powell
Clemson University

Geraldine Prody
Western Washington University

Joseph Provost
University of San Diego

Greg Raner
University of North Carolina, Greensboro

Tanea Reed
Eastern Kentucky University

Christopher Reid
Bryant University

Denis Revie
California Lutheran University

Douglas Root
University of North Texas

Johannes Rudolph
University of Colorado

Brian Sato
University of California, Irvine

Glen Sauer
Fairfield University

Joel Schildbach
Johns Hopkins University

Stylianos Scordilis
Smith College

Ashikh Seethy
Maulana Azad Medical College, New Delhi

Lisa Shamansky
California State University, San Bernardino

Bethel Sharma
Sewanee: University of the South

Nicholas Silvaggi
University of Wisconsin-Milwaukee

Kerry Smith
Clemson University

Narashima Sreerama
Colorado State University

Wesley Stites
University of Arkansas

Jon Stoltzfus
Michigan State University

Gerald Stubbs
Vanderbilt University

Takita Sumter
Winthrop University

Anna Tan-Wilson
State University of New York, Binghamton

Steven Theg
University of California, Davis

Marc Tischler
University of Arizona

Ken Traxler
Bemidji State University

Brian Trewyn
Colorado School of Mines

Vishwa Trivedi
Bethune Cookman University

Panayiotis Vacratsis
University of Windsor

Peter van der Geer
San Diego State University

Jeffrey Voigt
Albany College of Pharmacy and Health Sciences

Grover Waldrop
Louisiana State University

Xuemin Wang
University of Missouri

Yuqi Wang
Saint Louis University

Rodney Weilbaecher
Southern Illinois University

Kevin Williams
Western Kentucky University

Laura Zapanta
University of Pittsburgh

Brent Znosko
Saint Louis University

BRIEF CONTENTS

1 Biochemistry: An Evolving Science 1

2 Protein Composition and Structure 29

3 Exploring Proteins and Proteomes 69

4 DNA, RNA, and the Flow of Genetic Information 113

5 Exploring Genes and Genomes 145

6 Exploring Evolution and Bioinformatics 185

7 Hemoglobin: Portrait of a Protein in Action 207

8 Enzymes: Basic Concepts and Kinetics 233

9 Catalytic Strategies 273

10 Regulatory Strategies 309

11 Carbohydrates 341

12 Lipids and Cell Membranes 373

13 Membrane Channels and Pumps 403

14 Signal-Transduction Pathways 437

15 Metabolism: Basic Concepts and Design 463

16 Glycolysis and Gluconeogenesis 491

17 The Citric Acid Cycle 541

18 Oxidative Phosphorylation 573

19 The Light Reactions of Photosynthesis 619

20 The Calvin Cycle and the Pentose Phosphate Pathway 645

21 Glycogen Metabolism 679

22 Fatty Acid Metabolism 709

23 Protein Turnover and Amino Acid Catabolism 751

24 The Biosynthesis of Amino Acids 787

25 Nucleotide Biosynthesis 821

26 The Biosynthesis of Membrane Lipids and Steroids 849

27 The Integration of Metabolism 889

28 Drug Development 921

29 DNA Replication, Repair, and Recombination 949

30 RNA Synthesis and Processing 983

31 Protein Synthesis 1021

32 The Control of Gene Expression in Prokaryotes 1057

33 The Control of Gene Expression in Eukaryotes 1075

34 Sensory Systems 1097 ◉

35 The Immune System 1119 ◉

36 Molecular Motors 1151 ◉

◉ Chapters online only

CONTENTS

PREFACE   V

CHAPTER 1

Biochemistry: An Evolving Science   **1**

1.1 Biochemical Unity Underlies Biological Diversity   2

1.2 DNA Illustrates the Interplay Between Form and Function   4

DNA is constructed from four building blocks   4

Two single strands of DNA combine to form a double helix   5

DNA structure explains heredity and the storage of information   6

1.3 Concepts from Chemistry Explain the Properties of Biological Molecules   6

The formation of the DNA double helix as a key example   6

The double helix can form from its component strands   6

Covalent and noncovalent bonds are important for the structure and stability of biological molecules   7

The double helix is an expression of the rules of chemistry   10

The laws of thermodynamics govern the behavior of biochemical systems   11

Heat is released in the formation of the double helix   13

Acid–base reactions are central in many biochemical processes   14

Acid–base reactions can disrupt the double helix   15

Buffers regulate pH in organisms and in the laboratory   16

1.4 The Genomic Revolution Is Transforming Biochemistry, Medicine, and Other Fields   17

Genome sequencing has transformed biochemistry and other fields   18

Environmental factors influence human biochemistry   20

Genome sequences encode proteins and patterns of expression   21

APPENDIX

Visualizing Molecular Structures: Small Molecules   **23**

APPENDIX:

Functional Groups   **24**

CHAPTER 2

Protein Composition and Structure   **29**

2.1 Proteins Are Built from a Repertoire of 20 Amino Acids   31

2.2 Primary Structure: Amino Acids Are Linked by Peptide Bonds to Form Polypeptide Chains   37

Proteins have unique amino acid sequences specified by genes 39

Polypeptide chains are flexible yet conformationally restricted 40

2.3 Secondary Structure: Polypeptide Chains Can Fold into Regular Structures Such As the Alpha Helix, the Beta Sheet, and Turns and Loops 42

The alpha helix is a coiled structure stabilized by intrachain hydrogen bonds 42

Beta sheets are stabilized by hydrogen bonding between polypeptide strands 44

Polypeptide chains can change direction by making reverse turns and loops 46

2.4 Tertiary Structure: Proteins Can Fold into Globular or Fibrous Structures 46

Fibrous proteins provide structural support for cells and tissues 49

2.5 Quaternary Structure: Polypeptide Chains Can Assemble into Multisubunit Structures 51

2.6 The Amino Acid Sequence of a Protein Determines Its Three-Dimensional Structure 52

Amino acids have different propensities for forming α helices, β sheets, and turns 54

Protein folding is a highly cooperative process 55

Proteins fold by progressive stabilization of intermediates rather than by random search 56

Prediction of three-dimensional structure from sequence remains a great challenge 57

Some proteins are inherently unstructured and can exist in multiple conformations 58

Protein misfolding and aggregation are associated with some neurological diseases 59

Posttranslational modifications confer new capabilities to proteins 60

APPENDIX

Visualizing Molecular Structures: Proteins 64

CHAPTER 3

Exploring Proteins and Proteomes 69

The proteome is the functional representation of the genome 70

3.1 The Purification of Proteins Is an Essential First Step in Understanding Their Function 70

The assay: How do we recognize the protein we are looking for? 71

Proteins must be released from the cell to be purified 71

Proteins can be purified according to solubility, size, charge, and binding affinity 72

Proteins can be separated by gel electrophoresis and displayed 75

A protein purification scheme can be quantitatively evaluated 79

Ultracentrifugation is valuable for separating biomolecules and determining their masses 80

Protein purification can be made easier with the use of recombinant DNA technology 82

3.2 Immunology Provides Important Techniques with Which to Investigate Proteins 83

Antibodies to specific proteins can be generated 83

Monoclonal antibodies with virtually any desired specificity can be readily prepared 85

Proteins can be detected and quantified by using an enzyme-linked immunosorbent assay 86

Western blotting permits the detection of proteins separated by gel electrophoresis 87

Co-immunoprecipitation enables the identification of binding partners of a protein 88

Fluorescent markers make the visualization of proteins in the cell possible 89

3.3 Mass Spectrometry Is a Powerful Technique for the Identification of Peptides and Proteins 89

Peptides can be sequenced by mass spectrometry 92

Proteins can be specifically cleaved into small peptides to facilitate analysis 93

Genomic and proteomic methods are complementary 94

The amino acid sequence of a protein provides valuable information 95

Individual proteins can be identified by mass spectrometry 96

3.4 Peptides Can Be Synthesized by Automated Solid-Phase Methods 97

3.5 Three-Dimensional Protein Structure Can Be Determined by X-ray Crystallography, NMR Spectroscopy, and Cryo-Electron Microscopy 100

X-ray crystallography reveals three-dimensional structure in atomic detail 100

Nuclear magnetic resonance spectroscopy can reveal the structures of proteins in solution 102

Cryo-electron microscopy is an emerging method of protein structure determination 105

APPENDIX

Problem-Solving Strategies 108

CHAPTER 4

DNA, RNA, and the Flow of Genetic Information 113

4.1 A Nucleic Acid Consists of Four Kinds of Bases Linked to a Sugar–Phosphate Backbone 114

RNA and DNA differ in the sugar component and one of the bases 114

Nucleotides are the monomeric units of nucleic acids   115

DNA molecules are very long and have directionality   116

4.2 A Pair of Nucleic Acid Strands with Complementary Sequences Can Form a Double-Helical Structure   117

The double helix is stabilized by hydrogen bonds and van der Waals interactions   117

DNA can assume a variety of structural forms   119

Some DNA molecules are circular and supercoiled   120

Single-stranded nucleic acids can adopt elaborate structures   121

4.3 The Double Helix Facilitates the Accurate Transmission of Hereditary Information   122

Differences in DNA density established the validity of the semiconservative replication hypothesis   122

The double helix can be reversibly melted   123

Unusual circular DNA exists in the eukaryotic nucleus   124

4.4 DNA Is Replicated by Polymerases That Take Instructions from Templates   125

DNA polymerase catalyzes phosphodiester-bridge formation   125

The genes of some viruses are made of RNA   126

4.5 Gene Expression Is the Transformation of DNA Information into Functional Molecules   127

Several kinds of RNA play key roles in gene expression   127

All cellular RNA is synthesized by RNA polymerases   128

RNA polymerases take instructions from DNA templates   129

Transcription begins near promoter sites and ends at terminator sites   130

Transfer RNAs are the adaptor molecules in protein synthesis   131

4.6 Amino Acids Are Encoded by Groups of Three Bases Starting from a Fixed Point   132

Major features of the genetic code   133

Messenger RNA contains start and stop signals for protein synthesis   134

The genetic code is nearly universal   135

4.7 Most Eukaryotic Genes Are Mosaics of Introns and Exons   135

RNA processing generates mature RNA   136

Many exons encode protein domains   136

APPENDIX

Problem-Solving Strategies   140

CHAPTER 5

Exploring Genes and Genomes   145

5.1 The Exploration of Genes Relies on Key Tools   146

Restriction enzymes split DNA into specific fragments   147

Restriction fragments can be separated by gel electrophoresis and visualized   148

DNA can be sequenced by controlled termination of replication   149

DNA probes and genes can be synthesized by automated solid-phase methods   150

Selected DNA sequences can be greatly amplified by the polymerase chain reaction   151

PCR is a powerful technique in medical diagnostics, forensics, and studies of molecular evolution   152

The tools for recombinant DNA technology have been used to identify disease-causing mutations   153

5.2 Recombinant DNA Technology Has Revolutionized All Aspects of Biology   154

Restriction enzymes and DNA ligase are key tools in forming recombinant DNA molecules   154

Plasmids and λ phage are choice vectors for DNA cloning in bacteria   155

Bacterial and yeast artificial chromosomes   157

Specific genes can be cloned from digests of genomic DNA   158

Complementary DNA prepared from mRNA can be expressed in host cells   159

Proteins with new functions can be created through directed changes in DNA   160

Recombinant methods enable the exploration of the functional effects of disease-causing mutations   163

5.3 Complete Genomes Have Been Sequenced and Analyzed   163

The genomes of organisms ranging from bacteria to multicellular eukaryotes have been sequenced   164

The sequence of the human genome has been completed   165

Next-generation sequencing methods enable the rapid determination of a complete genome sequence   166

Comparative genomics has become a powerful research tool   168

5.4 Eukaryotic Genes Can Be Quantitated and Manipulated with Considerable Precision   169

Gene-expression levels can be comprehensively examined   169

New genes inserted into eukaryotic cells can be efficiently expressed   171

Transgenic animals harbor and express genes introduced into their germ lines   172

Gene disruption and genome editing provide clues to gene function and opportunities for new therapies   172

RNA interference provides an additional tool for disrupting gene expression   176

Tumor-inducing plasmids can be used to introduce new genes into plant cells   177

Human gene therapy holds great promise for medicine   178

APPENDIX

Biochemistry in Focus   180

CHAPTER 6

Exploring Evolution and Bioinformatics   185

6.1 Homologs Are Descended from a
Common Ancestor   186

6.2 Statistical Analysis of Sequence Alignments
Can Detect Homology   187

   The statistical significance of alignments can be
   estimated by shuffling   189

   Distant evolutionary relationships can be detected
   through the use of substitution matrices   190

   Databases can be searched to identify homologous
   sequences   193

6.3 Examination of Three-Dimensional
Structure Enhances Our Understanding of
Evolutionary Relationships   194

   Tertiary structure is more conserved than
   primary structure   194

   Knowledge of three-dimensional structures can aid
   in the evaluation of sequence alignments   195

   Repeated motifs can be detected by aligning sequences
   with themselves   196

   Convergent evolution illustrates common solutions
   to biochemical challenges   196

   Comparison of RNA sequences can be a source
   of insight into RNA secondary structures   197

6.4 Evolutionary Trees Can Be Constructed on the
Basis of Sequence Information   198

   Horizontal gene transfer events may explain unexpected
   branches of the evolutionary tree   199

6.5 Modern Techniques Make the Experimental
Exploration of Evolution Possible   200

   Ancient DNA can sometimes be amplified
   and sequenced   200

   Molecular evolution can be examined experimentally 200

APPENDIX

Biochemistry in Focus   203

APPENDIX:

Problem-Solving Strategies   204

CHAPTER 7

Hemoglobin: Portrait of a Protein in Action   207

7.1 Binding of Oxygen by Heme Iron   208

   Changes in heme electronic structure upon oxygen
   binding are the basis for functional imaging
   studies   210

   The structure of myoglobin prevents the release of
   reactive oxygen species   210

   Human hemoglobin is an assembly of four
   myoglobin-like subunits   211

7.2 Hemoglobin Binds Oxygen Cooperatively   212

   Oxygen binding markedly changes the quaternary
   structure of hemoglobin   213

   Hemoglobin cooperativity can be potentially
   explained by several models   214

   Structural changes at the heme groups are transmitted
   to the $\alpha_1\beta_1 - \alpha_2\beta_2$ interface 215

   2,3-Bisphosphoglycerate in red cells is crucial
   in determining the oxygen affinity of hemoglobin 216

   Carbon monoxide can disrupt oxygen transport
   by hemoglobin   217

7.3 Hydrogen Ions and Carbon Dioxide Promote the
Release of Oxygen: The Bohr Effect   217

7.4 Mutations in Genes Encoding Hemoglobin
Subunits Can Result in Disease   220

   Sickle-cell anemia results from the aggregation
   of mutated deoxyhemoglobin molecules   220

   Thalassemia is caused by an imbalanced production
   of hemoglobin chains   221

   The accumulation of free α-hemoglobin chains
   is prevented   222

   Additional globins are encoded in the human genome 223

APPENDIX

**Binding Models Can Be Formulated in Quantitative
Terms: The Hill Plot and the Concerted Model**   225

APPENDIX

Biochemistry in Focus   227

CHAPTER 8

Enzymes: Basic Concepts and Kinetics   233

8.1 Enzymes Are Powerful and Highly Specific
Catalysts   234

   Many enzymes require cofactors for activity   235

   Enzymes can transform energy from one form
   into another   236

8.2 Gibbs Free Energy Is a Useful Thermodynamic
Function for Understanding Enzymes   236

   The free-energy change provides information about
   the spontaneity but not the rate of a reaction   237

   The standard free-energy change of a reaction is related
   to the equilibrium constant   237

   Enzymes alter only the reaction rate and not the
   reaction equilibrium   239

8.3 Enzymes Accelerate Reactions by Facilitating the
Formation of the Transition State   240

   The formation of an enzyme–substrate complex is the
   first step in enzymatic catalysis   241

   The active sites of enzymes have some common features 242

   The binding energy between enzyme and substrate
   is important for catalysis   243

8.4 The Michaelis–Menten Model Accounts for the Kinetic Properties of Many Enzymes   244

Kinetics is the study of reaction rates   244

The steady-state assumption facilitates a description of enzyme kinetics   245

Variations in K_M can have physiological consequences   247

K_M and V_{max} values can be determined by several means   247

K_M and V_{max} values are important enzyme characteristics   248

k_{cat}/K_M is a measure of catalytic efficiency   249

Most biochemical reactions include multiple substrates   251

Allosteric enzymes do not obey Michaelis–Menten kinetics   252

8.5 Enzymes Can Be Inhibited by Specific Molecules   253

The different types of reversible inhibitors are kinetically distinguishable   254

Irreversible inhibitors can be used to map the active site   256

Penicillin irreversibly inactivates a key enzyme in bacterial cell-wall synthesis   258

Transition-state analogs are potent inhibitors of enzymes   260

Enzymes have impact outside the laboratory or clinic   261

8.6 Enzymes Can Be Studied One Molecule at a Time   261

APPENDIX

Enzymes Are Classified on the Basis of the Types of Reactions That They Catalyze   264

APPENDIX

Biochemistry in Focus   265

APPENDIX

Problem-Solving Strategies   266

CHAPTER 9

Catalytic Strategies   273

A few basic catalytic principles are used by many enzymes   274

9.1 Proteases Facilitate a Fundamentally Difficult Reaction   275

Chymotrypsin possesses a highly reactive serine residue   276

Chymotrypsin action proceeds in two steps linked by a covalently bound intermediate   277

Serine is part of a catalytic triad that also includes histidine and aspartate   278

Catalytic triads are found in other hydrolytic enzymes   281

The catalytic triad has been dissected by site-directed mutagenesis   282

Cysteine, aspartyl, and metalloproteases are other major classes of peptide-cleaving enzymes   283

Protease inhibitors are important drugs   285

9.2 Carbonic Anhydrases Make a Fast Reaction Faster   286

Carbonic anhydrase contains a bound zinc ion essential for catalytic activity   287

Catalysis entails zinc activation of a water molecule   288

A proton shuttle facilitates rapid regeneration of the active form of the enzyme   290

9.3 Restriction Enzymes Catalyze Highly Specific DNA-Cleavage Reactions   291

Cleavage is by in-line displacement of $3'$-oxygen from phosphorus by magnesium-activated water   292

Restriction enzymes require magnesium for catalytic activity   294

The complete catalytic apparatus is assembled only within complexes of cognate DNA molecules, ensuring specificity   295

Host-cell DNA is protected by the addition of methyl groups to specific bases   297

Type II restriction enzymes have a catalytic core in common and are probably related by horizontal gene transfer   298

9.4 Myosins Harness Changes in Enzyme Conformation to Couple ATP Hydrolysis to Mechanical Work   298

ATP hydrolysis proceeds by the attack of water on the gamma phosphoryl group   299

Formation of the transition state for ATP hydrolysis is associated with a substantial conformational change   300

The altered conformation of myosin persists for a substantial period of time   301

Scientists can watch single molecules of myosin move   303

Myosins are a family of enzymes containing P-loop structures   303

APPENDIX

Problem-Solving Strategies   306

CHAPTER 10

Regulatory Strategies   309

10.1 Aspartate Transcarbamoylase Is Allosterically Inhibited by the End Product of Its Pathway   310

Allosterically regulated enzymes do not follow Michaelis–Menten kinetics   311

ATCase consists of separable catalytic and regulatory subunits   312

Allosteric interactions in ATCase are mediated by large changes in quaternary structure   312

Allosteric regulators modulate the T-to-R equilibrium   316

10.2 Isozymes Provide a Means of Regulation Specific to Distinct Tissues and Developmental Stages   317

10.3 Covalent Modification Is a Means of Regulating Enzyme Activity   318

Kinases and phosphatases control the extent of protein phosphorylation   319

Phosphorylation is a highly effective means of regulating the activities of target proteins   321

Cyclic AMP activates protein kinase A by altering the quaternary structure   321

Mutations in protein kinase A can cause Cushing's syndrome   322

Exercise modifies the phosphorylation of many proteins   323

10.4 Many Enzymes Are Activated by Specific Proteolytic Cleavage   323

Chymotrypsinogen is activated by specific cleavage of a single peptide bond   324

Proteolytic activation of chymotrypsinogen leads to the formation of a substrate-binding site   325

The generation of trypsin from trypsinogen leads to the activation of other zymogens   325

Some proteolytic enzymes have specific inhibitors   326

Serpins can be degraded by a unique enzyme   328

Blood clotting is accomplished by a cascade of zymogen activations   328

Prothrombin must bind to Ca^{2+} to be converted to thrombin   329

Fibrinogen is converted by thrombin into a fibrin clot   329

Vitamin K is required for the formation of γ-carboxyglutamate   331

The clotting process must be precisely regulated   332

Hemophilia revealed an early step in clotting   333

APPENDIX

Biochemistry in Focus   335

APPENDIX

Problem-Solving Strategies   336

CHAPTER 11

Carbohydrates   341

11.1 Monosaccharides Are the Simplest Carbohydrates   342

Many common sugars exist in cyclic forms   344

Pyranose and furanose rings can assume different conformations   346

Glucose is a reducing sugar   347

Monosaccharides are joined to alcohols and amines through glycosidic bonds   348

Phosphorylated sugars are key intermediates in energy generation and biosyntheses   348

11.2 Monosaccharides Are Linked to Form Complex Carbohydrates   349

Sucrose, lactose, and maltose are the common disaccharides   349

Glycogen and starch are storage forms of glucose   350

Cellulose, a structural component of plants, is made of chains of glucose   350

Human milk oligosaccharides protect newborns from infection   352

11.3 Carbohydrates Can Be Linked to Proteins to Form Glycoproteins   352

Carbohydrates can be linked to proteins through asparagine (*N*-linked) or through serine or threonine (*O*-linked) residues   352

The glycoprotein erythropoietin is a vital hormone   353

Glycosylation functions in nutrient sensing   354

Proteoglycans, composed of polysaccharides and protein, have important structural roles   354

Proteoglycans are important components of cartilage   355

Mucins are glycoprotein components of mucus   356

Chitin can be processed to a molecule with a variety of uses   357

Protein glycosylation takes place in the lumen of the endoplasmic reticulum and in the Golgi complex   357

Specific enzymes are responsible for oligosaccharide assembly   358

Blood groups are based on protein glycosylation patterns   359

Errors in glycosylation can result in pathological conditions   359

Oligosaccharides can be "sequenced"   360

11.4 Lectins Are Specific Carbohydrate-Binding Proteins   361

Lectins promote interactions between cells and within cells   361

Lectins are organized into different classes   362

Influenza virus binds to sialic acid residues   363

APPENDIX

Biochemistry in Focus   365

APPENDIX

Problem-Solving Strategies   366

CHAPTER 12

Lipids and Cell Membranes   373

Many Common Features Underlie the Diversity of Biological Membranes   374

12.1 Fatty Acids Are Key Constituents of Lipids   374

Fatty acid names are based on their parent hydrocarbons   374

Fatty acids vary in chain length and degree of unsaturation   375

12.2 There Are Three Common Types of Membrane Lipids   376

Phospholipids are the major class of membrane lipids   376

Membrane lipids can include carbohydrate moieties 378

Cholesterol is a lipid based on a steroid nucleus 378

Archaeal membranes are built from ether lipids with branched chains 379

A membrane lipid is an amphipathic molecule containing a hydrophilic and a hydrophobic moiety 379

12.3 Phospholipids and Glycolipids Readily Form Bimolecular Sheets in Aqueous Media 380

Lipid vesicles can be formed from phospholipids 381

Lipid bilayers are highly impermeable to ions and most polar molecules 382

12.4 Proteins Carry Out Most Membrane Processes 383

Proteins associate with the lipid bilayer in a variety of ways 384

Proteins interact with membranes in a variety of ways 384

Some proteins associate with membranes through covalently attached hydrophobic groups 388

Transmembrane helices can be accurately predicted from amino acid sequences 388

12.5 Lipids and Many Membrane Proteins Diffuse Rapidly in the Plane of the Membrane 390

The fluid mosaic model allows lateral movement but not rotation through the membrane 391

Membrane fluidity is controlled by fatty acid composition and cholesterol content 391

Lipid rafts are highly dynamic complexes formed between cholesterol and specific lipids 392

All biological membranes are asymmetric 393

12.6 Eukaryotic Cells Contain Compartments Bounded by Internal Membranes 393

APPENDIX

Biochemistry in Focus 399

CHAPTER 13

Membrane Channels and Pumps 403

The expression of transporters largely defines the metabolic activities of a given cell type 404

13.1 The Transport of Molecules Across a Membrane May Be Active or Passive 404

Many molecules require protein transporters to cross membranes 405

Free energy stored in concentration gradients can be quantified 405

13.2 Two Families of Membrane Proteins Use ATP Hydrolysis to Pump Ions and Molecules Across Membranes 406

P-type ATPases couple phosphorylation and conformational changes to pump calcium ions across membranes 407

Digitalis specifically inhibits the Na^+-K^+ pump by blocking its dephosphorylation 410

P-type ATPases are evolutionarily conserved and play a wide range of roles 410

Multidrug resistance highlights a family of membrane pumps with ATP-binding cassette domains 411

13.3 Lactose Permease Is an Archetype of Secondary Transporters That Use One Concentration Gradient to Power the Formation of Another 413

13.4 Specific Channels Can Rapidly Transport Ions Across Membranes 415

Action potentials are mediated by transient changes in Na^+ and K^+ permeability 415

Patch-clamp conductance measurements reveal the activities of single channels 416

The structure of a potassium ion channel is an archetype for many ion-channel structures 417

The structure of the potassium ion channel reveals the basis of ion specificity 418

The structure of the potassium ion channel explains its rapid rate of transport 420

Voltage gating requires substantial conformational changes in specific ion-channel domains 421

A channel can be inactivated by occlusion of the pore: the ball-and-chain model 422

The acetylcholine receptor is an archetype for ligand-gated ion channels 423

Action potentials integrate the activities of several ion channels working in concert 424

Disruption of ion channels by mutations or chemicals can be potentially life-threatening 426

13.5 Gap Junctions Allow Ions and Small Molecules to Flow Between Communicating Cells 427

13.6 Specific Channels Increase the Permeability of Some Membranes to Water 429

APPENDIX

Biochemistry in Focus 431

APPENDIX

Problem-Solving Strategies 432

CHAPTER 14

Signal-Transduction Pathways 437

Signal transduction depends on molecular circuits 438

14.1 Epinephrine and Angiotensin II Signaling: Heterotrimeric G Proteins Transmit Signals and Reset Themselves 439

Ligand binding to 7TM receptors leads to the activation of heterotrimeric G proteins 440

Activated G proteins transmit signals by binding to other proteins 442

Cyclic AMP stimulates the phosphorylation of many target proteins by activating protein kinase A 443

G proteins spontaneously reset themselves through GTP hydrolysis 443

Some 7TM receptors activate the phosphoinositide cascade 444

Calcium ion is a widely used second messenger 445

Calcium ion often activates the regulatory protein calmodulin 447

14.2 Insulin Signaling: Phosphorylation Cascades Are Central to Many Signal-Transduction Processes 447

The insulin receptor is a dimer that closes around a bound insulin molecule 448

Insulin binding results in the cross-phosphorylation and activation of the insulin receptor 448

The activated insulin-receptor kinase initiates a kinase cascade 449

Insulin signaling is terminated by the action of phosphatases 451

14.3 EGF Signaling: Signal-Transduction Pathways Are Poised to Respond 451

EGF binding results in the dimerization of the EGF receptor 451

The EGF receptor undergoes phosphorylation of its carboxyl-terminal tail 452

EGF signaling leads to the activation of Ras, a small G protein 453

Activated Ras initiates a protein kinase cascade 453

EGF signaling is terminated by protein phosphatases and the intrinsic GTPase activity of Ras 454

14.4 Many Elements Recur with Variation in Different Signal-Transduction Pathways 454

14.5 Defects in Signal-Transduction Pathways Can Lead to Cancer and Other Diseases 455

Monoclonal antibodies can be used to inhibit signal-transduction pathways activated in tumors 456

Protein kinase inhibitors can be effective anticancer drugs 457

Cholera and whooping cough are the result of altered G-protein activity 457

APPENDIX

Biochemistry in Focus 459

CHAPTER 15

Metabolism: Basic Concepts and Design 463

15.1 Metabolism Is Composed of Many Coupled, Interconnecting Reactions 464

Metabolism consists of energy-yielding and energy-requiring reactions 464

A thermodynamically unfavorable reaction can be driven by a favorable reaction 465

15.2 ATP Is the Universal Currency of Free Energy in Biological Systems 466

ATP hydrolysis is exergonic 466

ATP hydrolysis drives metabolism by shifting the equilibrium of coupled reactions 468

The high phosphoryl potential of ATP results from structural differences between ATP and its hydrolysis products 469

Phosphoryl-transfer potential is an important form of cellular energy transformation 470

ATP may have roles other than in energy and signal transduction 472

15.3 The Oxidation of Carbon Fuels Is an Important Source of Cellular Energy 472

Compounds with high phosphoryl-transfer potential can couple carbon oxidation to ATP synthesis 473

Ion gradients across membranes provide an important form of cellular energy that can be coupled to ATP synthesis 474

Phosphates play a prominent role in biochemical processes 474

Energy from foodstuffs is extracted in three stages 474

15.4 Metabolic Pathways Contain Many Recurring Motifs 475

Activated carriers exemplify the modular design and economy of metabolism 475

Many activated carriers are derived from vitamins 478

Key reactions are reiterated throughout metabolism 480

Metabolic processes are regulated in three principal ways 483

Aspects of metabolism may have evolved from an RNA world 485

APPENDIX

Problem-Solving Strategies 487

CHAPTER 16

Glycolysis and Gluconeogenesis 491

Glucose is generated from dietary carbohydrates 492

Glucose is an important fuel for most organisms 493

16.1 Glycolysis Is an Energy-Conversion Pathway in Many Organisms 493

The enzymes of glycolysis are associated with one another 493

Glycolysis can be divided into two parts 493

Hexokinase traps glucose in the cell and begins glycolysis 495

Fructose 1,6-bisphosphate is generated from glucose 6-phosphate 496

The six-carbon sugar is cleaved into two three-carbon fragments 497

Mechanism: Triose phosphate isomerase salvages a three-carbon fragment 498

The oxidation of an aldehyde to an acid powers the formation of a compound with high phosphoryl-transfer potential 499

Mechanism: Phosphorylation is coupled to the oxidation of glyceraldehyde 3-phosphate by a thioester intermediate 500

ATP is formed by phosphoryl transfer from 1,3-bisphosphoglycerate 501

Additional ATP is generated with the formation of pyruvate 503

Two ATP molecules are formed in the conversion of glucose into pyruvate   504

NAD^+ is regenerated from the metabolism of pyruvate   505

Fermentations provide usable energy in the absence of oxygen   507

Fructose is converted into glycolytic intermediates by fructokinase   508

Excessive fructose consumption can lead to pathological conditions   508

Galactose is converted into glucose 6-phosphate   509

Many adults are intolerant of milk because they are deficient in lactase   510

Galactose is highly toxic if the transferase is missing   511

16.2 The Glycolytic Pathway Is Tightly Controlled   511

Glycolysis in muscle is regulated to meet the need for ATP   512

The regulation of glycolysis in the liver illustrates the biochemical versatility of the liver   513

A family of transporters enables glucose to enter and leave animal cells   516

Aerobic glycolysis is a property of rapidly growing cells   517

Cancer and endurance training affect glycolysis in a similar fashion   518

16.3 Glucose Can Be Synthesized from Noncarbohydrate Precursors   519

Gluconeogenesis is not a reversal of glycolysis   519

The conversion of pyruvate into phosphoenolpyruvate begins with the formation of oxaloacetate   521

Oxaloacetate is shuttled into the cytoplasm and converted into phosphoenolpyruvate   522

The conversion of fructose 1,6-bisphosphate into fructose 6-phosphate and orthophosphate is an irreversible step   523

The generation of free glucose is an important control point   523

Six high-transfer-potential phosphoryl groups are spent in synthesizing glucose from pyruvate   524

16.4 Gluconeogenesis and Glycolysis Are Reciprocally Regulated   525

Energy charge determines whether glycolysis or gluconeogenesis will be most active   525

The balance between glycolysis and gluconeogenesis in the liver is sensitive to blood-glucose concentration   526

Substrate cycles amplify metabolic signals and produce heat   528

Lactate and alanine formed by contracting muscle are used by other organs   528

Glycolysis and gluconeogenesis are evolutionarily intertwined   530

APPENDIX

BIOCHEMISTRY IN FOCUS   532

Biochemistry in Focus 1   532

Biochemistry in Focus 2   533

APPENDIX

Problem-Solving Strategies   533

CHAPTER 17

The Citric Acid Cycle   541

The citric acid cycle harvests high-energy electrons   542

17.1 The Pyruvate Dehydrogenase Complex Links Glycolysis to the Citric Acid Cycle   543

Mechanism: The synthesis of acetyl coenzyme A from pyruvate requires three enzymes and five coenzymes   544

Flexible linkages allow lipoamide to move between different active sites   546

17.2 The Citric Acid Cycle Oxidizes Two-Carbon Units   548

Citrate synthase forms citrate from oxaloacetate and acetyl coenzyme A   548

Mechanism: The mechanism of citrate synthase prevents undesirable reactions   549

Citrate is isomerized into isocitrate   550

Isocitrate is oxidized and decarboxylated to alpha-ketoglutarate   551

Succinyl coenzyme A is formed by the oxidative decarboxylation of alpha-ketoglutarate   552

A compound with high phosphoryl-transfer potential is generated from succinyl coenzyme A   552

Mechanism: Succinyl coenzyme A synthetase transforms types of biochemical energy   553

Oxaloacetate is regenerated by the oxidation of succinate   554

The citric acid cycle produces high-transfer-potential electrons, ATP, and CO_2 555

17.3 Entry to the Citric Acid Cycle and Metabolism Through It Are Controlled   557

The pyruvate dehydrogenase complex is regulated allosterically and by reversible phosphorylation   557

The citric acid cycle is controlled at several points   559

Defects in the citric acid cycle contribute to the development of cancer   559

An enzyme in lipid metabolism is hijacked to inhibit pyruvate dehydrogenase activity   560

17.4 The Citric Acid Cycle Is a Source of Biosynthetic Precursors   561

The citric acid cycle must be capable of being rapidly replenished   561

The disruption of pyruvate metabolism is the cause of beriberi and poisoning by mercury and arsenic 562

The citric acid cycle may have evolved from preexisting pathways 563

17.5 The Glyoxylate Cycle Enables Plants and Bacteria to Grow on Acetate 564

APPENDIX

BIOCHEMISTRY IN FOCUS 566

Biochemistry in Focus 1 566

Biochemistry in Focus 2 567

APPENDIX

Problem-Solving Strategies 568

CHAPTER 18

Oxidative Phosphorylation 573

18.1 Eukaryotic Oxidative Phosphorylation Takes Place in Mitochondria 574

Mitochondria are bounded by a double membrane 574

Mitochondria are the result of an endosymbiotic event 575

18.2 Oxidative Phosphorylation Depends on Electron Transfer 576

The electron-transfer potential of an electron is measured as redox potential 576

Electron flow from NADH to molecular oxygen powers the formation of a proton gradient 579

18.3 The Respiratory Chain Consists of Four Complexes: Three Proton Pumps and a Physical Link to the Citric Acid Cycle 579

Iron–sulfur clusters are common components of the electron-transport chain 581

The high-potential electrons of NADH enter the respiratory chain at NADH-Q oxidoreductase 582

Ubiquinol is the entry point for electrons from $FADH_2$ of flavoproteins 584

Electrons flow from ubiquinol to cytochrome c through Q-cytochrome c oxidoreductase 584

The Q cycle funnels electrons from a two-electron carrier to a one-electron carrier and pumps protons 584

Cytochrome c oxidase catalyzes the reduction of molecular oxygen to water 586

Most of the electron-transport chain is organized into a complex called the respirasome 589

Toxic derivatives of molecular oxygen such as superoxide radicals are scavenged by protective enzymes 590

Electrons can be transferred between groups that are not in contact 591

The conformation of cytochrome c has remained essentially constant for more than a billion years 592

18.4 A Proton Gradient Powers the Synthesis of ATP 592

ATP synthase is composed of a proton-conducting unit and a catalytic unit 594

Proton flow through ATP synthase leads to the release of tightly bound ATP: The binding-change mechanism 595

Rotational catalysis is the world's smallest molecular motor 597

Proton flow around the c ring powers ATP synthesis 597

ATP synthase and G proteins have several common features 599

18.5 Many Shuttles Allow Movement Across Mitochondrial Membranes 600

Electrons from cytoplasmic NADH enter mitochondria by shuttles 600

The entry of ADP into mitochondria is coupled to the exit of ATP by ATP-ADP translocase 601

Mitochondrial transporters for metabolites have a common tripartite structure 602

18.6 The Regulation of Cellular Respiration Is Governed Primarily by the Need for ATP 603

The complete oxidation of glucose yields about 30 molecules of ATP 603

The rate of oxidative phosphorylation is determined by the need for ATP 603

ATP synthase can be regulated 604

Regulated uncoupling leads to the generation of heat 605

Reintroduction of UCP-1 into pigs may be economically valuable 607

Oxidative phosphorylation can be inhibited at many stages 607

Mitochondrial diseases are being discovered 608

Mitochondria play a key role in apoptosis 609

Power transmission by proton gradients is a central motif of bioenergetics 609

APPENDIX

Biochemistry in Focus 611

APPENDIX

Problem-Solving Strategies 612

CHAPTER 19

The Light Reactions of Photosynthesis 619

Photosynthesis converts light energy into chemical energy 620

19.1 Photosynthesis Takes Place in Chloroplasts 621

The primary events of photosynthesis take place in thylakoid membranes 621

Chloroplasts arose from an endosymbiotic event 622

19.2 Light Absorption by Chlorophyll Induces Electron Transfer 622

A special pair of chlorophylls initiate charge separation 623

Cyclic electron flow reduces the cytochrome
of the reaction center   625

19.3 Two Photosystems Generate a Proton Gradient and NADPH in Oxygenic Photosynthesis   626

Photosystem II transfers electrons from water to plastoquinone and generates a proton gradient   626

Cytochrome *bf* links photosystem II to photosystem I   629

Photosystem I uses light energy to generate reduced ferredoxin, a powerful reductant   629

Ferredoxin–$NADP^+$ reductase converts $NADP^+$ into NADPH   630

19.4 A Proton Gradient across the Thylakoid Membrane Drives ATP Synthesis   631

The ATP synthase of chloroplasts closely resembles those of mitochondria and prokaryotes   632

The activity of chloroplast ATP synthase is regulated   633

Cyclic electron flow through photosystem I leads to the production of ATP instead of NADPH   634

The absorption of eight photons yields one O_2, two NADPH, and three ATP molecules   634

19.5 Accessory Pigments Funnel Energy into Reaction Centers   635

Resonance energy transfer allows energy to move from the site of initial absorbance to the reaction center   636

The components of photosynthesis are highly organized   637

Many herbicides inhibit the light reactions of photosynthesis   638

19.6 The Ability to Convert Light into Chemical Energy Is Ancient   638

Artificial photosynthetic systems may provide clean, renewable energy   639

APPENDIX

Biochemistry in Focus   641

APPENDIX

Problem-Solving Strategies   641

CHAPTER 20

The Calvin Cycle and the Pentose Phosphate Pathway   645

20.1 The Calvin Cycle Synthesizes Hexoses from Carbon Dioxide and Water   646

Carbon dioxide reacts with ribulose 1,5-bisphosphate to form two molecules of 3-phosphoglycerate   646

Rubisco activity depends on magnesium and carbamate   648

Rubisco activase is essential for rubisco activity   649

Rubisco also catalyzes a wasteful oxygenase reaction: Catalytic imperfection   649

Hexose phosphates are made from phosphoglycerate, and ribulose 1,5-bisphosphate is regenerated   651

Three ATP and two NADPH molecules are used to bring carbon dioxide to the level of a hexose   654

Starch and sucrose are the major carbohydrate stores in plants   654

20.2 The Activity of the Calvin Cycle Depends on Environmental Conditions   655

Rubisco is activated by light-driven changes in proton and magnesium ion concentrations   655

Thioredoxin plays a key role in regulating the Calvin cycle   655

The C_4 pathway of tropical plants accelerates photosynthesis by concentrating carbon dioxide   656

Crassulacean acid metabolism permits growth in arid ecosystems   659

20.3 The Pentose Phosphate Pathway Generates NADPH and Synthesizes Five-Carbon Sugars   659

Two molecules of NADPH are generated in the conversion of glucose 6-phosphate into ribulose 5-phosphate   659

The pentose phosphate pathway and glycolysis are linked by transketolase and transaldolase   660

Mechanism: Transketolase and transaldolase stabilize carbanionic intermediates by different mechanisms   662

20.4 The Metabolism of Glucose 6-Phosphate by the Pentose Phosphate Pathway Is Coordinated with Glycolysis   665

The rate of the oxidative phase of the pentose phosphate pathway is controlled by the level of $NADP^+$ 665

The flow of glucose 6-phosphate depends on the need for NADPH, ribose 5-phosphate, and ATP   666

The pentose phosphate pathway is required for rapid cell growth   668

Through the looking-glass: The Calvin cycle and the pentose phosphate pathway are mirror images   668

20.5 Glucose 6-Phosphate Dehydrogenase Plays a Key Role in Protection Against Reactive Oxygen Species   668

Glucose 6-phosphate dehydrogenase deficiency causes a drug-induced hemolytic anemia   669

A deficiency of glucose 6-phosphate dehydrogenase confers an evolutionary advantage in some circumstances   670

APPENDIX

BIOCHEMISTRY IN FOCUS   672

Biochemistry in Focus 1   672

Biochemistry in Focus 2   672

APPENDIX

Problem-Solving Strategies   673

CHAPTER 21

Glycogen Metabolism — 679

Glycogen metabolism is the regulated release and storage of glucose — 680

21.1 Glycogen Breakdown Requires the Interplay of Several Enzymes — 681

Phosphorylase catalyzes the phosphorolytic cleavage of glycogen to release glucose 1-phosphate — 681

Mechanism: Pyridoxal phosphate participates in the phosphorolytic cleavage of glycogen — 682

A debranching enzyme also is needed for the breakdown of glycogen — 684

Phosphoglucomutase converts glucose 1-phosphate into glucose 6-phosphate — 685

The liver contains glucose 6-phosphatase, a hydrolytic enzyme absent from muscle — 685

21.2 Phosphorylase Is Regulated by Allosteric Interactions and Reversible Phosphorylation — 686

Liver phosphorylase produces glucose for use by other tissues — 686

Muscle phosphorylase is regulated by the intracellular energy charge — 687

Biochemical characteristics of muscle fiber types differ — 688

Phosphorylation promotes the conversion of phosphorylase *b* to phosphorylase *a* — 688

Phosphorylase kinase is activated by phosphorylation and calcium ions — 689

An isozymic form of glycogen phosphorylase exists in the brain — 690

21.3 Epinephrine and Glucagon Signal the Need for Glycogen Breakdown — 690

G proteins transmit the signal for the initiation of glycogen breakdown — 690

Glycogen breakdown must be rapidly turned off when necessary — 692

The regulation of glycogen phosphorylase became more sophisticated as the enzyme evolved — 692

21.4 Glycogen Synthesis Requires Several Enzymes and Uridine Diphosphate Glucose — 693

UDP-glucose is an activated form of glucose — 693

Glycogen synthase catalyzes the transfer of glucose from UDP-glucose to a growing chain — 693

A branching enzyme forms α-1,6 linkages — 694

Glycogen synthase is the key regulatory enzyme in glycogen synthesis — 695

Glycogen is an efficient storage form of glucose — 695

21.5 Glycogen Breakdown and Synthesis Are Reciprocally Regulated — 696

Protein phosphatase 1 reverses the regulatory effects of kinases on glycogen metabolism — 696

Insulin stimulates glycogen synthesis by inactivating glycogen synthase kinase — 698

Glycogen metabolism in the liver regulates the blood-glucose concentration — 698

A biochemical understanding of glycogen-storage diseases is possible — 700

APPENDIX

Biochemistry in Focus — 702

APPENDIX

Problem-Solving Strategies — 703

CHAPTER 22

Fatty Acid Metabolism — 709

Fatty acid degradation and synthesis mirror each other in their chemical reactions — 710

22.1 Triacylglycerols Are Highly Concentrated Energy Stores — 710

Dietary lipids are digested by pancreatic lipases — 712

Dietary lipids are transported in chylomicrons — 713

22.2 The Use of Fatty Acids as Fuel Requires Three Stages of Processing — 713

Triacylglycerols are hydrolyzed by hormone-stimulated lipases — 713

Free fatty acids and glycerol are released into the blood — 714

Fatty acids are linked to coenzyme A before they are oxidized — 715

Carnitine carries long-chain activated fatty acids into the mitochondrial matrix — 716

Acetyl CoA, NADH, and $FADH_2$ are generated in each round of fatty acid oxidation — 717

The complete oxidation of palmitate yields 106 molecules of ATP — 718

22.3 Unsaturated and Odd-Chain Fatty Acids Require Additional Steps for Degradation — 719

An isomerase and a reductase are required for the oxidation of unsaturated fatty acids — 719

Odd-chain fatty acids yield propionyl CoA in the final thiolysis step — 720

Vitamin B_{12} contains a corrin ring and a cobalt atom — 721

Mechanism: Methylmalonyl CoA mutase catalyzes a rearrangement to form succinyl CoA — 721

Fatty acids are also oxidized in peroxisomes — 723

Some fatty acids may contribute to the development of pathological conditions — 724

22.4 Ketone Bodies Are a Fuel Source Derived from Fats — 724

Ketone bodies are a major fuel in some tissues — 725

Animals cannot convert fatty acids into glucose — 727

22.5 Fatty Acids Are Synthesized by Fatty Acid Synthase — 728

Fatty acids are synthesized and degraded by different pathways — 728

The formation of malonyl CoA is the committed step in fatty acid synthesis — 729

Intermediates in fatty acid synthesis are attached to an acyl carrier protein — 729

Fatty acid synthesis consists of a series of condensation, reduction, dehydration, and reduction reactions 729

Fatty acids are synthesized by a multifunctional enzyme complex in animals 731

The synthesis of palmitate requires 8 molecules of acetyl CoA, 14 molecules of NADPH, and 7 molecules of ATP 733

Citrate carries acetyl groups from mitochondria to the cytoplasm for fatty acid synthesis 733

Several sources supply NADPH for fatty acid synthesis 734

Fatty acid metabolism is altered in tumor cells 735

Triacylglycerols may become an important renewal energy source 735

22.6 The Elongation and Unsaturation of Fatty Acids Are Accomplished by Accessory Enzyme Systems 736

Membrane-bound enzymes generate unsaturated fatty acids 736

Eicosanoid hormones are derived from polyunsaturated fatty acids 737

Variations on a theme: Polyketide and nonribosomal peptide synthetases resemble fatty acid synthase 738

22.7 Acetyl CoA Carboxylase Plays a Key Role in Controlling Fatty Acid Metabolism 739

Acetyl CoA carboxylase is regulated by conditions in the cell 739

Acetyl CoA carboxylase is regulated by a variety of hormones 740

AMP-activated protein kinase is a key regulator of metabolism 740

APPENDIX

Biochemistry in Focus 743

APPENDIX

Problem-Solving Strategies 744

CHAPTER 23

Protein Turnover and Amino Acid Catabolism 751

23.1 Proteins Are Degraded to Amino Acids 752

The digestion of dietary proteins begins in the stomach and is completed in the intestine 752

Cellular proteins are degraded at different rates 753

23.2 Protein Turnover Is Tightly Regulated 753

Ubiquitin tags proteins for destruction 754

The proteasome digests the ubiquitin-tagged proteins 755

The ubiquitin pathway and the proteasome have prokaryotic counterparts 756

Protein degradation can be used to regulate biological function 757

23.3 The First Step in Amino Acid Degradation Is the Removal of Nitrogen 758

Alpha-amino groups are converted into ammonium ions by the oxidative deamination of glutamate 758

Mechanism: Pyridoxal phosphate forms Schiff-base intermediates in aminotransferases 759

Aspartate aminotransferase is an archetypal pyridoxal-dependent transaminase 761

Blood levels of aminotransferases serve a diagnostic function 761

Pyridoxal phosphate enzymes catalyze a wide array of reactions 762

Serine and threonine can be directly deaminated 763

Peripheral tissues transport nitrogen to the liver 763

23.4 Ammonium Ion Is Converted into Urea in Most Terrestrial Vertebrates 764

The urea cycle begins with the formation of carbamoyl phosphate 764

Carbamoyl phosphate synthetase is the key regulatory enzyme for urea synthesis 764

Carbamoyl phosphate reacts with ornithine to begin the urea cycle 765

The urea cycle is linked to gluconeogenesis 766

Urea-cycle enzymes are evolutionarily related to enzymes in other metabolic pathways 767

Inherited defects of the urea cycle cause hyperammonemia and can lead to brain damage 767

Urea is not the only means of disposing of excess nitrogen 768

23.5 Carbon Atoms of Degraded Amino Acids Emerge as Major Metabolic Intermediates 769

Pyruvate is an entry point into metabolism for a number of amino acids 770

Oxaloacetate is an entry point into metabolism for aspartate and asparagine 770

Alpha-ketoglutarate is an entry point into metabolism for five-carbon amino acids 770

Succinyl coenzyme A is a point of entry for several amino acids 771

Methionine degradation requires the formation of a key methyl donor, S-adenosylmethionine 772

Threonine deaminase initiates the degradation of threonine 772

The branched-chain amino acids yield acetyl CoA, acetoacetate, or propionyl CoA 772

Oxygenases are required for the degradation of aromatic amino acids 773

Protein metabolism helps to power the flight of migratory birds 775

23.6 Inborn Errors of Metabolism Can Disrupt Amino Acid Degradation 776

Phenylketonuria is one of the most common metabolic disorders 777

Determining the basis of the neurological symptoms of phenylketonuria is an active area of research 777

APPENDIX

Biochemistry in Focus 779

APPENDIX

Problem-Solving Strategies 780

CHAPTER 24

The Biosynthesis of Amino Acids — 787

Amino acid synthesis requires solutions to three key biochemical problems — 788

24.1 Nitrogen Fixation: Microorganisms Use ATP and a Powerful Reductant to Reduce Atmospheric Nitrogen to Ammonia — 788

The iron–molybdenum cofactor of nitrogenase binds and reduces atmospheric nitrogen — 790

Ammonium ion is assimilated into an amino acid through glutamate and glutamine — 791

24.2 Amino Acids Are Made from Intermediates of the Citric Acid Cycle and Other Major Pathways — 793

Human beings can synthesize some amino acids but must obtain others from their diet — 793

Aspartate, alanine, and glutamate are formed by the addition of an amino group to an alpha-ketoacid — 794

A common step determines the chirality of all amino acids — 795

The formation of asparagine from aspartate requires an adenylated intermediate — 796

Glutamate is the precursor of glutamine, proline, and arginine — 796

3-Phosphoglycerate is the precursor of serine, cysteine, and glycine — 797

Tetrahydrofolate carries activated one-carbon units at several oxidation levels — 798

S-Adenosylmethionine is the major donor of methyl groups — 799

Cysteine is synthesized from serine and homocysteine — 801

High homocysteine levels correlate with vascular disease — 801

Shikimate and chorismate are intermediates in the biosynthesis of aromatic amino acids — 801

Tryptophan synthase illustrates substrate channeling in enzymatic catalysis — 804

24.3 Feedback Inhibition Regulates Amino Acid Biosynthesis — 805

Branched pathways require sophisticated regulation — 805

The sensitivity of glutamine synthetase to allosteric regulation is altered by covalent modification — 807

24.4 Amino Acids Are Precursors of Many Biomolecules — 808

Glutathione, a gamma-glutamyl peptide, serves as a sulfhydryl buffer and an antioxidant — 809

Nitric oxide, a short-lived signal molecule, is formed from arginine — 809

Amino acids are precursors for a number of neurotransmitters — 810

Porphyrins are synthesized from glycine and succinyl coenzyme A — 810

Porphyrins accumulate in some inherited disorders of porphyrin metabolism — 814

APPENDIX

Biochemistry in Focus — 815

APPENDIX

Problem-Solving Strategies — 816

CHAPTER 25

Nucleotide Biosynthesis — 821

Nucleotides can be synthesized by de novo or salvage pathways — 822

25.1 The Pyrimidine Ring Is Assembled de Novo or Recovered by Salvage Pathways — 822

Bicarbonate and other oxygenated carbon compounds are activated by phosphorylation — 823

The side chain of glutamine can be hydrolyzed to generate ammonia — 823

Intermediates can move between active sites by channeling — 824

Orotate acquires a ribose ring from PRPP to form a pyrimidine nucleotide and is converted into uridylate — 824

Nucleotide mono-, di-, and triphosphates are interconvertible — 825

CTP is formed by amination of UTP — 825

Salvage pathways recycle pyrimidine bases — 826

25.2 Purine Bases Can Be Synthesized de Novo or Recycled by Salvage Pathways — 826

The purine ring system is assembled on ribose phosphate — 827

The purine ring is assembled by successive steps of activation by phosphorylation followed by displacement — 827

AMP and GMP are formed from IMP — 829

Enzymes of the purine synthesis pathway associate with one another in vivo — 830

Salvage pathways economize intracellular energy expenditure — 830

25.3 Deoxyribonucleotides Are Synthesized by the Reduction of Ribonucleotides Through a Radical Mechanism — 831

Mechanism: A tyrosyl radical is critical to the action of ribonucleotide reductase — 831

Stable radicals other than tyrosyl radical are employed by other ribonucleotide reductases — 833

Thymidylate is formed by the methylation of deoxyuridylate — 834

Dihydrofolate reductase catalyzes the regeneration of tetrahydrofolate, a one-carbon carrier — 835

Several valuable anticancer drugs block the synthesis of thymidylate — 835

25.4 Key Steps in Nucleotide Biosynthesis Are Regulated by Feedback Inhibition — 836

Pyrimidine biosynthesis is regulated by aspartate transcarbamoylase — 836

The synthesis of purine nucleotides is controlled by feedback inhibition at several sites — 837

The synthesis of deoxyribonucleotides is controlled by the regulation of ribonucleotide reductase — 837

25.5 Disruptions in Nucleotide Metabolism Can Cause Pathological Conditions — 838

The loss of adenosine deaminase activity results in severe combined immunodeficiency — 839

Gout is induced by high serum levels of urate — 839

Lesch–Nyhan syndrome is a dramatic consequence of mutations in a salvage-pathway enzyme — 840

Folic acid deficiency promotes birth defects such as spina bifida — 841

APPENDIX

Biochemistry in Focus — 842

APPENDIX

Problem-Solving Strategies — 843

CHAPTER 26

The Biosynthesis of Membrane Lipids and Steroids — 849

26.1 Phosphatidate Is a Common Intermediate in the Synthesis of Phospholipids and Triacylglycerols — 850

The synthesis of phospholipids requires an activated intermediate — 851

Some phospholipids are synthesized from an activated alcohol — 852

Phosphatidylcholine is an abundant phospholipid — 853

Excess choline is implicated in the development of heart disease — 853

Base-exchange reactions can generate phospholipids — 853

Sphingolipids are synthesized from ceramide — 854

Gangliosides are carbohydrate-rich sphingolipids that contain acidic sugars — 855

Sphingolipids confer diversity on lipid structure and function — 856

Respiratory distress syndrome and Tay–Sachs disease result from the disruption of lipid metabolism — 856

Ceramide metabolism stimulates tumor growth — 857

Phosphatidic acid phosphatase is a key regulatory enzyme in lipid metabolism — 857

26.2 Cholesterol Is Synthesized from Acetyl Coenzyme A in Three Stages — 858

The synthesis of mevalonate, which is activated as isopentenyl pyrophosphate, initiates the synthesis of cholesterol — 858

Squalene (C_{30}) is synthesized from six molecules of isopentenyl pyrophosphate (C_5) — 859

Squalene cyclizes to form cholesterol — 860

26.3 The Complex Regulation of Cholesterol Biosynthesis Takes Place at Several Levels — 861

Lipoproteins transport cholesterol and triacylglycerols throughout the organism — 863

Low-density lipoproteins play a central role in cholesterol metabolism — 866

The absence of the LDL receptor leads to hypercholesterolemia and atherosclerosis — 867

Mutations in the LDL receptor prevent LDL release and result in receptor destruction — 868

Inability to transport cholesterol from the lysosome causes Niemann-Pick disease — 869

Cycling of the LDL receptor is regulated — 869

HDL appears to protect against atherosclerosis — 870

The clinical management of cholesterol levels can be understood at a biochemical level — 870

26.4 Important Biochemicals Are Synthesized from Cholesterol and Isoprene — 871

Letters identify the steroid rings and numbers identify the carbon atoms — 872

Steroids are hydroxylated by cytochrome P450 monooxygenases that use NADPH and O_2 — 873

The cytochrome P450 system is widespread and performs a protective function — 874

Pregnenolone, a precursor of many other steroids, is formed from cholesterol by cleavage of its side chain — 875

Progesterone and corticosteroids are synthesized from pregnenolone — 875

Androgens and estrogens are synthesized from pregnenolone — 875

Vitamin D is derived from cholesterol by the ring-splitting activity of light — 877

Five-carbon units are joined to form a wide variety of biomolecules — 878

Some isoprenoids have industrial applications — 880

APPENDIX

Biochemistry in Focus — 882

APPENDIX

Problem-Solving Strategies — 882

CHAPTER 27

The Integration of Metabolism — 889

27.1 Caloric Homeostasis Is a Means of Regulating Body Weight — 890

27.2 The Brain Plays a Key Role in Caloric Homeostasis — 892

Signals from the gastrointestinal tract induce feelings of satiety — 892

Leptin and insulin regulate long-term control over caloric homeostasis — 893

Leptin is one of several hormones secreted by adipose tissue — 894

Leptin resistance may be a contributing factor
to obesity   895

Dieting is used to combat obesity   896

**27.3 Diabetes Is a Common Metabolic Disease
Often Resulting from Obesity   896**

Insulin initiates a complex signal-transduction
pathway in muscle   897

Metabolic syndrome often precedes type 2 diabetes   898

Excess fatty acids in muscle modify metabolism   899

Insulin resistance in muscle facilitates pancreatic
failure   899

Metabolic derangements in type 1 diabetes result
from insulin insufficiency and glucagon excess   901

**27.4 Exercise Beneficially Alters the Biochemistry
of Cells   901**

Mitochondrial biogenesis is stimulated by
muscular activity   902

Fuel choice during exercise is determined
by the intensity and duration of activity   902

**27.5 Food Intake and Starvation Induce Metabolic
Changes   905**

The starved–fed cycle is the physiological response
to a fast   905

Metabolic adaptations in prolonged starvation
minimize protein degradation   907

27.6 Ethanol Alters Energy Metabolism in the Liver   909

Ethanol metabolism leads to an excess of NADH   910

Excess ethanol consumption disrupts vitamin
metabolism   911

APPENDIX

BIOCHEMISTRY IN FOCUS   914

Biochemistry in Focus 1   914

Biochemistry in Focus 2   915

APPENDIX

Problem-Solving Strategies   916

CHAPTER 28

Drug Development   921

**28.1 Compounds Must Meet Stringent Criteria
to Be Developed into Drugs   923**

Drugs must be potent and selective   923

Drugs must have suitable properties to reach
their targets   924

Toxicity can limit drug effectiveness   929

**28.2 Drug Candidates Can Be Discovered by
Serendipity, Screening, or Design   930**

Serendipitous observations can drive
drug development   930

Natural products are a valuable source of drugs
and drug leads   932

Screening libraries of synthetic compounds expands
the opportunity for identification of drug leads   933

Drugs can be designed on the basis of three-dimensional
structural information about their targets   935

28.3 Genomic Analyses Can Aid Drug Discovery   937

Potential targets can be identified in the human
proteome   937

Animal models can be developed to test the validity
of potential drug targets   938

Potential targets can be identified in the genomes
of pathogens   939

Genetic differences influence individual responses
to drugs   939

**28.4 The Clinical Development of Drugs Proceeds
Through Several Phases   940**

Clinical trials are time-consuming and expensive   941

The evolution of drug resistance can limit the utility
of drugs for infectious agents and cancer   942

APPENDIX

Biochemistry in Focus   945

CHAPTER 29

DNA Replication, Repair, and Recombination   949

**29.1 DNA Replication Proceeds by the Polymerization
of Deoxyribonucleoside Triphosphates
Along a Template   951**

DNA polymerases require a template and a primer   951

All DNA polymerases have structural features
in common   951

Two bound metal ions participate in the polymerase
reaction   952

The specificity of replication is dictated
by complementarity of shape between bases   952

An RNA primer synthesized by primase enables
DNA synthesis to begin   953

One strand of DNA is made continuously, whereas
the other strand is synthesized in fragments   953

DNA ligase joins ends of DNA in duplex regions   954

The separation of DNA strands requires specific
helicases and ATP hydrolysis   955

**29.2 DNA Unwinding and Supercoiling Are
Controlled by Topoisomerases   956**

The linking number of DNA, a topological property,
determines the degree of supercoiling   956

Topoisomerases prepare the double helix for
unwinding   958

Type I topoisomerases relax supercoiled structures   959

Type II topoisomerases can introduce negative
supercoils through coupling to ATP hydrolysis   960

29.3 DNA Replication Is Highly Coordinated   961

DNA replication requires highly processive
polymerases   962

The leading and lagging strands are synthesized in a coordinated fashion 962

DNA replication in *Escherichia coli* begins at a unique site and proceeds through initiation, elongation, and termination 964

DNA synthesis in eukaryotes is initiated at multiple sites 965

Telomeres are unique structures at the ends of linear chromosomes 967

Telomeres are replicated by telomerase, a specialized polymerase that carries its own RNA template 968

29.4 Many Types of DNA Damage Can Be Repaired 968

Errors can arise in DNA replication 969

Bases can be damaged by oxidizing agents, alkylating agents, and light 969

DNA damage can be detected and repaired by a variety of systems 970

The presence of thymine instead of uracil in DNA permits the repair of deaminated cytosine 972

Some genetic diseases are caused by the expansion of repeats of three nucleotides 973

Many cancers are caused by the defective repair of DNA 973

Many potential carcinogens can be detected by their mutagenic action on bacteria 975

29.5 DNA Recombination Plays Important Roles in Replication, Repair, and Other Processes 975

RecA can initiate recombination by promoting strand invasion 976

Some recombination reactions proceed through Holliday-junction intermediates 976

APPENDIX

Biochemistry in Focus 980

CHAPTER 30

RNA Synthesis and Processing 983

RNA synthesis comprises three stages: Initiation, elongation, and termination 984

30.1 RNA Polymerases Catalyze Transcription 985

RNA chains are formed de novo and grow in the 5′-to-3′ direction 986

RNA polymerases backtrack and correct errors 988

RNA polymerase binds to promoter sites on the DNA template to initiate transcription 988

Sigma subunits of RNA polymerase recognize promoter sites 989

RNA polymerases must unwind the template double helix for transcription to take place 990

Elongation takes place at transcription bubbles that move along the DNA template 991

Sequences within the newly transcribed RNA signal termination 991

Some messenger RNAs directly sense metabolite concentrations 992

The *rho* protein helps to terminate the transcription of some genes 992

Some antibiotics inhibit transcription 994

Precursors of transfer and ribosomal RNA are cleaved and chemically modified after transcription in prokaryotes 995

30.2 Transcription in Eukaryotes Is Highly Regulated 996

Three types of RNA polymerase synthesize RNA in eukaryotic cells 997

Three common elements can be found in the RNA polymerase II promoter region 998

The TFIID protein complex initiates the assembly of the active transcription complex 999

Multiple transcription factors interact with eukaryotic promoters 1000

Enhancer sequences can stimulate transcription at start sites thousands of bases away 1001

30.3 The Transcription Products of Eukaryotic Polymerases Are Processed 1001

RNA polymerase I produces three ribosomal RNAs 1001

RNA polymerase III produces transfer RNA 1002

The product of RNA polymerase II, the pre-mRNA transcript, acquires a 5′ cap and a 3′ poly(A) tail 1002

Small regulatory RNAs are cleaved from larger precursors 1004

RNA editing changes the proteins encoded by mRNA 1004

Sequences at the ends of introns specify splice sites in mRNA precursors 1005

Splicing consists of two sequential transesterification reactions 1006

Small nuclear RNAs in spliceosomes catalyze the splicing of mRNA precursors 1007

Transcription and processing of mRNA are coupled 1010

Mutations that affect pre-mRNA splicing cause disease 1010

Most human pre-mRNAs can be spliced in alternative ways to yield different proteins 1011

30.4 The Discovery of Catalytic RNA Was Revealing in Regard to Both Mechanism and Evolution 1012

APPENDIX

Biochemistry in Focus 1017

CHAPTER 31

Protein Synthesis 1021

31.1 Protein Synthesis Requires the Translation of Nucleotide Sequences into Amino Acid Sequences 1022

The synthesis of long proteins requires a low error frequency 1022

Transfer RNA molecules have a common design 1023

Some transfer RNA molecules recognize more than one codon because of wobble in base-pairing 1025

31.2 Aminoacyl Transfer RNA Synthetases Read the Genetic Code 1026

Amino acids are first activated by adenylation 1027

Aminoacyl-tRNA synthetases have highly discriminating amino acid activation sites 1028

Proofreading by aminoacyl-tRNA synthetases increases the fidelity of protein synthesis 1029

Synthetases recognize various features of transfer RNA molecules 1029

Aminoacyl-tRNA synthetases can be divided into two classes 1030

31.3 The Ribosome Is the Site of Protein Synthesis 1031

Ribosomal RNAs (5S, 16S, and 23S rRNA) play a central role in protein synthesis 1031

Ribosomes have three tRNA-binding sites that bridge the 30S and 50S subunits 1032

The start signal is usually AUG preceded by several bases that pair with 16S rRNA 1034

Bacterial protein synthesis is initiated by formylmethionyl transfer RNA 1035

Formylmethionyl-tRNA$_f$ is placed in the P site of the ribosome in the formation of the 70S initiation complex 1035

Elongation factors deliver aminoacyl-tRNA to the ribosome 1036

Peptidyl transferase catalyzes peptide-bond synthesis 1037

The formation of a peptide bond is followed by the GTP-driven translocation of tRNAs and mRNA 1038

Protein synthesis is terminated by release factors that read stop codons 1040

31.4 Eukaryotic Protein Synthesis Differs from Bacterial Protein Synthesis Primarily in Translation Initiation 1041

Mutations in initiation factor 2 cause a curious pathological condition 1042

31.5 A Variety of Antibiotics and Toxins Can Inhibit Protein Synthesis 1043

Some antibiotics inhibit protein synthesis 1043

Diphtheria toxin blocks protein synthesis in eukaryotes by inhibiting translocation 1044

Some toxins modifiy 28S ribosomal RNA 1045

31.6 Ribosomes Bound to the Endoplasmic Reticulum Manufacture Secretory and Membrane Proteins 1045

Protein synthesis begins on ribosomes that are free in the cytoplasm 1046

Signal sequences mark proteins for translocation across the endoplasmic reticulum membrane 1046

Transport vesicles carry cargo proteins to their final destination 1048

APPENDIX

Biochemistry in Focus 1051

APPENDIX

Problem-Solving Strategies 1051

CHAPTER 32

The Control of Gene Expression in Prokaryotes 1057

32.1 Many DNA-Binding Proteins Recognize Specific DNA Sequences 1058

The helix-turn-helix motif is common to many prokaryotic DNA-binding proteins 1059

32.2 Prokaryotic DNA-Binding Proteins Bind Specifically to Regulatory Sites in Operons 1060

An operon consists of regulatory elements and protein-encoding genes 1061

The lac repressor protein in the absence of lactose binds to the operator and blocks transcription 1061

Ligand binding can induce structural changes in regulatory proteins 1062

The operon is a common regulatory unit in prokaryotes 1063

Transcription can be stimulated by proteins that contact RNA polymerase 1064

32.3 Regulatory Circuits Can Result in Switching Between Patterns of Gene Expression 1065

The λ repressor regulates its own expression 1065

A circuit based on the λ repressor and Cro forms a genetic switch 1066

Many prokaryotic cells release chemical signals that regulate gene expression in other cells 1066

Biofilms are complex communities of prokaryotes 1067

32.4 Gene Expression Can Be Controlled at Posttranscriptional Levels 1068

Attenuation is a prokaryotic mechanism for regulating transcription through the modulation of nascent RNA secondary structure 1068

APPENDIX

Biochemistry in Focus 1072

CHAPTER 33

The Control of Gene Expression in Eukaryotes 1075

33.1 Eukaryotic DNA Is Organized into Chromatin 1077

Nucleosomes are complexes of DNA and histones 1077

DNA wraps around histone octamers to form nucleosomes 1078

33.2 Transcription Factors Bind DNA and Regulate Transcription Initiation 1079

A range of DNA-binding structures are employed by eukaryotic DNA-binding proteins 1080

Activation domains interact with other proteins 1081

Multiple transcription factors interact with eukaryotic regulatory regions 1081

Enhancers can stimulate transcription in specific cell types   1082

Induced pluripotent stem cells can be generated by introducing four transcription factors into differentiated cells   1082

33.3 The Control of Gene Expression Can Require Chromatin Remodeling   **1083**

The methylation of DNA can alter patterns of gene expression   1084

Steroids and related hydrophobic molecules pass through membranes and bind to DNA-binding receptors   1084

Nuclear hormone receptors regulate transcription by recruiting coactivators to the transcription complex   1085

Steroid-hormone receptors are targets for drugs   1086

Chromatin structure is modulated through covalent modifications of histone tails   1087

Transcriptional repression can be achieved through histone deacetylation and other modifications   1089

33.4 Eukaryotic Gene Expression Can Be Controlled at Posttranscriptional Levels   **1090**

Genes associated with iron metabolism are translationally regulated in animals   1090

Small RNAs regulate the expression of many eukaryotic genes   1092

APPENDIX

Biochemistry in Focus   **1094**

◉ CHAPTER 34

Sensory Systems   **1097**

34.1 A Wide Variety of Organic Compounds Are Detected by Olfaction   **1098**

Olfaction is mediated by an enormous family of seven-transmembrane-helix receptors   1099

Odorants are decoded by a combinatorial mechanism   1101

34.2 Taste Is a Combination of Senses That Function by Different Mechanisms   **1102**

Sequencing of the human genome led to the discovery of a large family of 7TM bitter receptors   1103

A heterodimeric 7TM receptor responds to sweet compounds   1105

Umami, the taste of glutamate and aspartate, is mediated by a heterodimeric receptor related to the sweet receptor   1105

Salty tastes are detected primarily by the passage of sodium ions through channels   1105

Sour tastes arise from the effects of hydrogen ions (acids) on channels   1106

34.3 Photoreceptor Molecules in the Eye Detect Visible Light   **1106**

Rhodopsin, a specialized 7TM receptor, absorbs visible light   1107

Light absorption induces a specific isomerization of bound 11-*cis*-retinal   1108

Light-induced lowering of the calcium level coordinates recovery   1109

Color vision is mediated by three cone receptors that are homologs of rhodopsin   1110

Rearrangements in the genes for the green and red pigments lead to "color blindness"   1111

34.4 Hearing Depends on the Speedy Detection of Mechanical Stimuli   **1111**

Hair cells use a connected bundle of stereocilia to detect tiny motions   1112

Mechanosensory channels have been identified in *Drosophila* and vertebrates   1113

34.5 Touch Includes the Sensing of Pressure, Temperature, and Other Factors   **1113**

Studies of capsaicin reveal a receptor for sensing high temperatures and other painful stimuli   1114

APPENDIX

Biochemistry in Focus   **1116**

◉ CHAPTER 35

The Immune System   **1119**

Innate immunity is an evolutionarily ancient defense system   1120

The adaptive immune system responds by using the principles of evolution   1122

35.1 Antibodies Possess Distinct Antigen-Binding and Effector Units   **1124**

35.2 Antibodies Bind Specific Molecules Through Hypervariable Loops   **1126**

The immunoglobulin fold consists of a beta-sandwich framework with hypervariable loops   1127

X-ray analyses have revealed how antibodies bind antigens   1127

Large antigens bind antibodies with numerous interactions   1128

35.3 Diversity Is Generated by Gene Rearrangements   **1129**

J (joining) genes and D (diversity) genes increase antibody diversity   1130

More than 10^8 antibodies can be formed by combinatorial association and somatic mutation   1131

The oligomerization of antibodies expressed on the surfaces of immature B cells triggers antibody secretion   1132

Different classes of antibodies are formed by the hopping of V_H genes   1133

35.4 Major-Histocompatibility-Complex Proteins Present Peptide Antigens on Cell Surfaces for Recognition by T-Cell Receptors   **1134**

Peptides presented by MHC proteins occupy a deep groove flanked by alpha helices   1135

◉ Chapters online only

T-cell receptors are antibody-like proteins containing variable and constant regions 1136

CD8 on cytotoxic T cells acts in concert with T-cell receptors 1137

Helper T cells stimulate cells that display foreign peptides bound to class II MHC proteins 1139

Helper T cells rely on the T-cell receptor and CD4 to recognize foreign peptides on antigen-presenting cells 1139

MHC proteins are highly diverse 1141

Human immunodeficiency viruses subvert the immune system by destroying helper T cells 1141

35.5 The Immune System Contributes to the Prevention and the Development of Human Diseases 1142

T cells are subjected to positive and negative selection in the thymus 1142

Autoimmune diseases result from the generation of immune responses against self-antigens 1143

The immune system plays a role in cancer prevention 1144

Vaccines are a powerful means to prevent and eradicate disease 1144

APPENDIX

Biochemistry in Focus 1147

◉ CHAPTER 36

Molecular Motors 1151

36.1 Most Molecular-Motor Proteins Are Members of the P-Loop NTPase Superfamily 1152

Molecular motors are generally oligomeric proteins with an ATPase core and an extended structure 1153

ATP binding and hydrolysis induce changes in the conformation and binding affinity of motor proteins 1155

36.2 Myosins Move Along Actin Filaments 1156

Actin is a polar, self-assembling, dynamic polymer 1157

Myosin head domains bind to actin filaments 1158

Motions of single motor proteins can be directly observed 1159

Phosphate release triggers the myosin power stroke 1159

Muscle is a complex of myosin and actin 1160

The length of the lever arm determines motor velocity 1162

36.3 Kinesin and Dynein Move Along Microtubules 1162

Microtubules are hollow cylindrical polymers 1163

Kinesin motion is highly processive 1164

36.4 A Rotary Motor Drives Bacterial Motion 1166

Bacteria swim by rotating their flagella 1166

Proton flow drives bacterial flagellar rotation 1167

Bacterial chemotaxis depends on reversal of the direction of flagellar rotation 1168

APPENDIX

Biochemistry in Focus 1171

Answers to Problems A1

Selected Readings B1

Index C1

◉ Chapters online only

Biochemistry: An Evolving Science

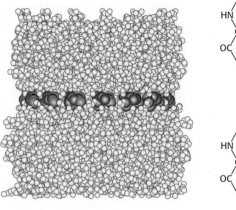
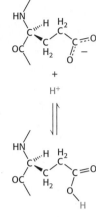

Chemistry in action. Human activities require energy. The interconversion of different forms of energy requires large biochemical machines comprising many thousands of atoms such as the complex shown above. Yet, the functions of these elaborate assemblies depend on simple chemical processes such as the protonation and deprotonation of the carboxylic acid groups shown on the right. The photograph is of Nobel Prize winners Peter Agre, M.D., and Carol Greider, Ph.D., who used, respectively, biochemical techniques to reveal key mechanisms of how water is transported into and out of cells and how chromosomes are replicated faithfully. [Keith Weller for Johns Hopkins Medicine.]

✓ LEARNING OBJECTIVES

By the end of this chapter, you should be able to:

1. Compare and contrast the unity of biology at the organismal and the biochemical levels.

2. Describe the double-helical structure of DNA including its formation from two component strands.

3. Discuss important interactions between atoms including covalent bonds, ionic interactions, hydrogen bonds, and van der Waals interactions.

4. Describe the structure of water and its contribution to interactions between molecules through the hydrophobic effect.

5. Discuss the conservation of energy and the first law of thermodynamics.

6. Discuss the concepts of entropy and enthalpy and the second law of thermodynamics.

7. Discuss acid-base reactions and perform calculations using the definitions of pH and pK_a as well as the Henderson–Hasselbalch equation.

8. Explain the roles of buffers in stabilizing the pH of solutions.

9. Briefly describe the role of gene sequences in biochemistry and the progress enabled by advances in genome sequencing technologies.

OUTLINE

1.1 Biochemical Unity Underlies Biological Diversity

1.2 DNA Illustrates the Interplay Between Form and Function

1.3 Concepts from Chemistry Explain the Properties of Biological Molecules

1.4 The Genomic Revolution Is Transforming Biochemistry, Medicine, and Other Fields

Biochemistry is the study of the chemistry of life processes. Scientists have explored the chemistry of life with great intensity since the 1828 discovery that biological molecules such as urea could be synthesized from nonliving components. Through these investigations, many of the most fundamental mysteries of how living things function at a biochemical level have now been solved. However, much remains to be investigated. As is often the case, each discovery raises at least as many new questions as it answers. Furthermore, we are now in an age of unprecedented opportunity for the application of our tremendous knowledge of biochemistry to problems in medicine, dentistry, agriculture, forensics, anthropology, environmental sciences, energy, and many other fields. We begin our journey into biochemistry with one of the most startling discoveries of the past century: namely, the great unity of all living things at the biochemical level.

1.1 Biochemical Unity Underlies Biological Diversity

The biological world is magnificently diverse. The animal kingdom is rich with species ranging from nearly microscopic insects to elephants and whales. The plant kingdom includes species as small and relatively simple as algae and as large and complex as giant sequoias. This diversity extends further when we descend into the microscopic world. Organisms such as protozoa, yeast, and bacteria are present with great diversity in water, in soil, and on or within larger organisms. Some organisms can survive and even thrive in seemingly hostile environments such as hot springs and glaciers.

The development of the microscope revealed a key unifying feature that underlies this diversity. Large organisms are built up of *cells,* resembling, to some extent, single-celled microscopic organisms. The construction of animals, plants, and microorganisms from cells suggested that these diverse organisms might have more in common than is apparent from their outward appearance. With the development of biochemistry, this suggestion has been tremendously supported and expanded. At the biochemical level, all organisms have many common features (Figure 1.1).

As mentioned earlier, biochemistry is the study of the chemistry of life processes. These processes entail the interplay of two different classes of molecules: large molecules such as proteins and nucleic acids, referred to as *biological macromolecules,* and low-molecular-weight molecules such as glucose and glycerol, referred to as *metabolites,* that are chemically transformed in biological processes.

Members of both these classes of molecules are common, with minor variations, to all living things. For example, *deoxyribonucleic acid* (DNA) stores genetic information in all cellular organisms. *Proteins,* the macromolecules that are key participants in most biological processes, are built from the same set of 20 building blocks in all organisms. Furthermore, proteins that play similar roles in different organisms often have very similar three-dimensional structures (Figure 1.1).

Key metabolic processes also are common to many organisms. For example, the set of chemical transformations that converts glucose and oxygen into carbon dioxide and water is essentially identical in simple bacteria such as *Escherichia coli (E. coli)* and human beings. Even processes that appear to be quite distinct often have common features at the biochemical level. Remarkably, the biochemical processes by which plants capture light energy and convert it into more-useful forms are strikingly similar to steps used in animals to capture energy released from the breakdown of glucose.

Glucose

Glycerol

Sulfolobus archaea   *Arabidopsis thaliana*   *Homo sapiens*

FIGURE 1.1 Biological diversity and similarity. The shape of a key molecule in gene regulation (the TATA-box-binding protein) is similar in three very different organisms that are separated from one another by billions of years of evolution. [(Left) Eye of Science/Science Source; (middle) Nigel Cattlin/Science Source; (right) Photo by J. R. Eyerman/The LIFE PictureCollection/Getty Images.]

These observations overwhelmingly suggest that all living things on Earth have a common ancestor and that modern organisms have evolved from this ancestor into their present forms. Geological and biochemical findings support a time line for this evolutionary path (Figure 1.2). On the basis of their biochemical characteristics, the diverse organisms of the modern world can be divided into three fundamental groups called *domains: Eukarya* (eukaryotes), *Bacteria,* and *Archaea.* Domain Eukarya comprises all multicellular organisms, including human beings as well as many microscopic unicellular organisms such as yeast. The defining characteristic of *eukaryotes* is the presence of a well-defined nucleus within each cell. Unicellular organisms such as bacteria, which lack a nucleus, are referred to as *prokaryotes.* The prokaryotes were reclassified as two separate domains in response to Carl Woese's discovery in 1977 that certain bacteria-like organisms are biochemically quite distinct from other previously characterized

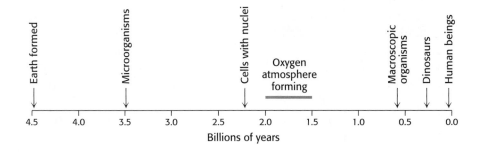

FIGURE 1.2 A possible time line for biochemical evolution. Selected key events are indicated. Note that life on Earth began approximately 3.5 billion years ago, whereas human beings emerged quite recently.

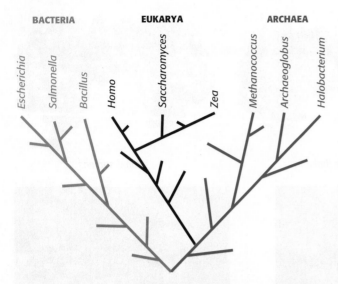

BACTERIA   **EUKARYA**   **ARCHAEA**

Escherichia
Salmonella
Bacillus
Homo
Saccharomyces
Zea
Methanococcus
Archaeoglobus
Halobacterium

FIGURE 1.3 Tree of life. A possible evolutionary path from a common ancestor approximately 3.5 billion years ago at the bottom of the tree to organisms found in the modern world at the top.

bacterial species. These organisms, now recognized as having diverged from bacteria early in evolution, are the *archaea*. Evolutionary paths from a common ancestor to modern organisms can be deduced on the basis of biochemical information. One possible such path is shown in Figure 1.3.

Much of this book will explore the chemical reactions and the associated biological macromolecules and metabolites that are found in biological processes common to all organisms. The unity of life at the biochemical level makes this approach possible. At the same time, different organisms have specific needs, depending on the particular biological niche in which they evolved and live. By comparing and contrasting details of particular biochemical pathways in different organisms, we can learn how biological challenges are solved at the biochemical level. In most cases, these challenges are addressed by the adaptation of existing macromolecules to new roles rather than by the evolution of entirely new ones.

Biochemistry has been greatly enriched by our ability to examine the three-dimensional structures of biological macromolecules in exquisite detail. Some of these structures are simple and elegant, whereas others are incredibly complicated. In any case, these structures provide an essential framework for understanding function. We begin our exploration of the interplay between structure and function with the genetic material, DNA.

1.2 DNA Illustrates the Interplay Between Form and Function

A fundamental biochemical feature common to all cellular organisms is the use of DNA for the storage of genetic information. The discovery that DNA plays this central role was first made in studies of bacteria in the 1940s. This discovery was followed by a compelling proposal for the three-dimensional structure of DNA in 1953, an event that set the stage for many of the advances in biochemistry and numerous other fields, extending to the present.

The structure of DNA powerfully illustrates a basic principle common to all biological macromolecules: the intimate relation between structure and function. The remarkable properties of this chemical substance allow it to function as a very efficient and robust vehicle for storing information. We start with an examination of the covalent structure of DNA and its extension into three dimensions.

DNA is constructed from four building blocks

DNA is a *linear polymer* made up of four different types of monomers. It has a fixed backbone from which protrude variable substituents, referred to as bases (Figure 1.4). The backbone is built of repeating sugar–phosphate

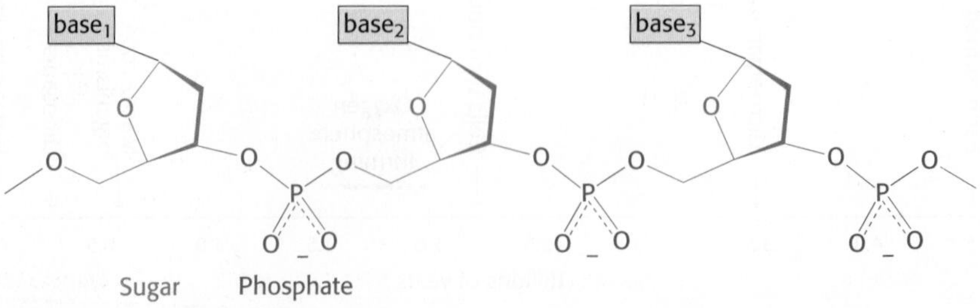

FIGURE 1.4 Covalent structure of DNA. Each unit of the polymeric structure is composed of a sugar (deoxyribose), a phosphate, and a variable base that protrudes from the sugar–phosphate backbone.

base$_1$ base$_2$ base$_3$

Sugar   Phosphate

units. The sugars are molecules of *deoxyribose* from which DNA receives its name. Each sugar is connected to two phosphate groups through different linkages. Moreover, each sugar is oriented in the same way, and so each DNA strand has directionality, with one end distinguishable from the other. Joined to each deoxyribose is one of four possible bases: adenine (A), cytosine (C), guanine (G), and thymine (T).

Adenine (A)   Cytosine (C)   Guanine (G)   Thymine (T)

These bases are connected to the sugar components in the DNA backbone through the bonds shown in black in Figure 1.4. All four bases are planar but differ significantly in other respects. Thus, each monomer of DNA consists of a sugar–phosphate unit and one of four bases attached to the sugar. These bases can be arranged in any order along a strand of DNA.

Two single strands of DNA combine to form a double helix

Most DNA molecules consist of not one but two strands (Figure 1.5). In 1953, James Watson and Francis Crick deduced the arrangement of these strands and proposed a three-dimensional structure for DNA molecules based, in part, on experimental data from Rosalind Franklin. This structure is a *double helix* composed of two intertwined strands arranged such that the sugar–phosphate backbone lies on the outside and the bases on the inside. The key to this structure is that the bases form *specific base pairs* (bp) held together by *hydrogen bonds* (Section 1.3): adenine pairs with thymine (A–T) and guanine pairs with cytosine (G–C), as shown in Figure 1.6. Hydrogen bonds are much weaker than *covalent bonds* such as the carbon–carbon or carbon–nitrogen bonds that define the structures of the bases themselves. Such weak bonds are crucial to biochemical systems; they are weak enough to be reversibly broken in biochemical processes, yet they are strong enough, particularly when many form simultaneously, to help stabilize specific structures such as the double helix.

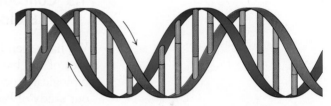

FIGURE 1.5 The double helix. The double-helical structure of DNA proposed by Watson and Crick. The sugar–phosphate backbones of the two chains are shown in red and blue, and the bases are shown in green, purple, orange, and yellow. The two strands are antiparallel, running in opposite directions with respect to the axis of the double helix, as indicated by the arrows.

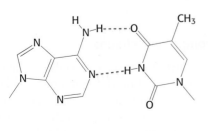

Adenine (A)   Thymine (T)

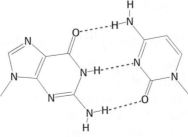

Guanine (G)   Cytosine (C)

FIGURE 1.6 Watson–Crick base pairs. Adenine pairs with thymine (A–T) and guanine with cytosine (G–C). The dashed green lines represent hydrogen bonds.

DNA structure explains heredity and the storage of information

The structure proposed by Watson and Crick has two properties of central importance to the role of DNA as the hereditary material. First, the structure is compatible with any sequence of bases. While the bases are distinct in structure, the base pairs have essentially the same shape (Figure 1.6) and thus fit equally well into the center of the double-helical structure of any sequence. Without any constraints, the sequence of bases along a DNA strand can act as an efficient means of storing information. Indeed, the sequence of bases along DNA strands is how genetic information is stored. The DNA sequence determines the sequences of the ribonucleic acid (RNA) and protein molecules that carry out most of the activities within cells.

Second, because of base pairing, the sequence of bases along one strand completely determines the sequence along the other strand. As Watson and Crick so coyly wrote: "It has not escaped our notice that the specific pairing we have postulated immediately suggests a possible copying mechanism for the genetic material." Thus, if the DNA double helix is separated into two single strands, each strand can act as a template for the generation of its partner strand through specific base-pair formation (Figure 1.7). The three-dimensional structure of DNA beautifully illustrates the close connection between molecular form and function.

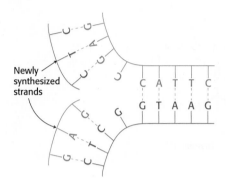

FIGURE 1.7 DNA replication. If a DNA molecule is separated into two strands, each strand can act as the template for the generation of its partner strand.

1.3 Concepts from Chemistry Explain the Properties of Biological Molecules

We have seen how a chemical insight into the hydrogen-bonding capabilities of the bases of DNA led to a deep understanding of a fundamental biological process. To lay the groundwork for the rest of the book, we begin our study of biochemistry by examining selected concepts from chemistry and showing how these concepts apply to biological systems. The concepts include the types of chemical bonds; the structure of water, the solvent in which most biochemical processes take place; the First and Second Laws of Thermodynamics; and the principles of acid–base chemistry.

The formation of the DNA double helix as a key example

We will use these concepts to examine an archetypical biochemical process—namely, the formation of a DNA double helix from its two component strands. The process is but one of many examples that could have been chosen to illustrate these topics. Keep in mind that, although the specific discussion is about DNA and double-helix formation, the concepts considered are quite general and will apply to many other classes of molecules and processes that will be discussed in the remainder of the book. In the course of these discussions, we will touch on the properties of water and the concepts of pK_a and buffers that are of great importance to many aspects of biochemistry.

The double helix can form from its component strands

The discovery that DNA from natural sources exists in a double-helical form with Watson–Crick base pairs suggested, but did not prove, that such double helices would form spontaneously outside biological systems. Suppose that two short strands of DNA were chemically synthesized to have complementary sequences so that they could, in principle, form a double helix with Watson–Crick base pairs. Two such sequences are CGATTAAT and ATTAATCG. The structures of these molecules in solution can be examined by a variety of techniques. In isolation, each sequence exists almost exclusively as a single-stranded molecule. However, when the two sequences are mixed, a double helix with Watson–Crick base pairs does form (Figure 1.8). This reaction proceeds nearly to completion.

What forces cause the two strands of DNA to bind to each other? To analyze this binding reaction, we must consider several factors: the types of interactions and bonds in biochemical systems and the energetic favorability of the reaction. We must also consider the influence of the solution conditions—in particular, the consequences of acid–base reactions.

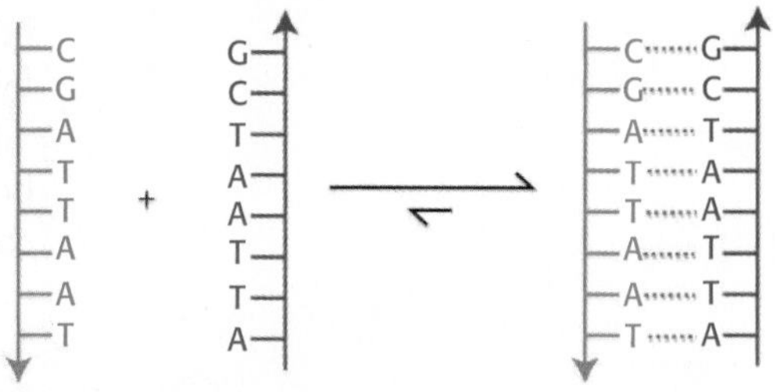

FIGURE 1.8 Formation of a double helix. When two DNA strands with appropriate, complementary sequences are mixed, they spontaneously assemble to form a double helix.

Covalent and noncovalent bonds are important for the structure and stability of biological molecules

Atoms interact with one another through chemical bonds. These bonds include the covalent bonds that define the structure of molecules as well as a variety of noncovalent bonds that are of great importance to biochemistry.

Covalent bonds. The strongest bonds are covalent bonds, such as the bonds that hold the atoms together within the individual bases shown on page 4. A covalent bond is formed by the sharing of a pair of electrons between adjacent atoms. A typical carbon–carbon ($C-C$) covalent bond has a bond length of 1.54 Å and bond energy of 355 kJ mol^{-1} (85 kcal mol^{-1}). Because covalent bonds are so strong, considerable energy must be expended to break them. More than one electron pair can be shared between two atoms to form a multiple covalent bond. For example, three of the bases in Figure 1.6 include carbon–oxygen ($C=O$) double bonds. These bonds are even stronger than $C-C$ single bonds, with energies near 730 kJ mol^{-1} (175 kcal mol^{-1}) and are somewhat shorter.

For some molecules, more than one pattern of covalent bonding can be written. For example, adenine can be written in two nearly equivalent ways called *resonance structures.*

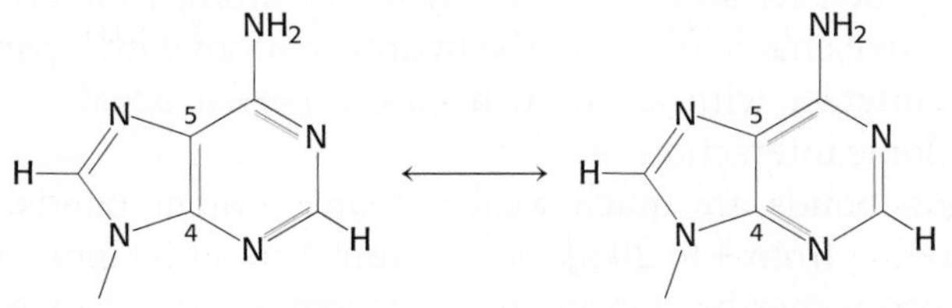

These adenine structures depict alternative arrangements of single and double bonds that are possible within the same structural framework. Resonance structures are shown connected by a double-headed arrow. Adenine's true structure is a composite of its two resonance structures. The composite structure is manifested in the bond lengths such as that for the bond joining carbon atoms C-4 and C-5. The observed bond length of 1.40 Å is between that expected for a $C-C$ single bond (1.54 Å) and a $C=C$ double bond (1.34 Å). A molecule that can be written as several resonance structures of approximately equal energies has greater stability than does a molecule without multiple resonance structures.

Noncovalent bonds. Noncovalent bonds are weaker than covalent bonds but are crucial for biochemical processes such as the formation of a double

Distance and energy units

Interatomic distances and bond lengths are usually measured in angstrom (Å) units:

$$1 \text{ Å} = 10^{-10} \text{ m} = 10^{-8} \text{ cm} = 0.1 \text{ nm}$$

Several energy units are in common use. One joule (J) is the amount of energy required to move 1 meter against a force of 1 newton. A kilojoule (kJ) is 1000 joules. One calorie is the amount of energy required to raise the temperature of 1 gram of water 1 degree Celsius. A kilocalorie (kcal) is 1000 calories. One joule is equal to 0.239 cal.

helix. Four fundamental noncovalent bond types are *ionic interactions, hydrogen bonds, van der Waals interactions,* and *hydrophobic interactions.* They differ in geometry, strength, and specificity. Furthermore, these bonds are affected in vastly different ways by the presence of water. Let's consider the characteristics of each type:

1. *Ionic Interactions.* A charged group on one molecule can attract an oppositely charged group on the same or another molecule. The energy of an ionic interaction (sometimes called an electrostatic interaction) is given by the *Coulomb energy*:

$$E = kq_1q_2/Dr$$

where E is the energy, q_1 and q_2 are the charges on the two atoms (in units of the electronic charge), r is the distance between the two atoms (in angstroms), D is the dielectric constant (which decreases the strength of the Coulomb depending on the intervening solvent or medium), and k is a proportionality constant ($k = 1389$, for energies in units of kilojoules per mole, or 332 for energies in kilocalories per mole).

By convention, an attractive interaction has a negative energy. The ionic interaction between two ions bearing single opposite charges separated by 3 Å in water (which has a dielectric constant of 80) has an energy of -5.8 kJ mol^{-1} (-1.4 kcal mol^{-1}). Note how important the dielectric constant of the medium is. For the same ions separated by 3 Å in a nonpolar solvent such as hexane (which has a dielectric constant of 2), the energy of this interaction is -232 kJ mol^{-1} (-55 kcal mol^{-1}).

2. *Hydrogen Bonds.* These interactions are largely ionic interactions, with partial charges on nearby atoms attracting one another. Hydrogen bonds are responsible for specific base-pair formation in the DNA double helix. The hydrogen atom in a hydrogen bond is partially shared by two electronegative atoms such as nitrogen or oxygen. The *hydrogen-bond donor* is the group that includes both the atom to which the hydrogen atom is more tightly linked and the hydrogen atom itself, whereas the *hydrogen-bond acceptor* is the atom less tightly linked to the hydrogen atom (Figure 1.9). The electronegative atom to which the hydrogen atom is covalently bonded pulls electron density away from the hydrogen atom, which thus develops a partial positive charge (δ^+). So, the hydrogen atom with a partial positive charge can interact with an atom having a partial negative charge (δ^-) through an ionic interaction.

Hydrogen bonds are much weaker than covalent bonds. They have energies ranging from 4 to 20 kJ mol^{-1} (from 1 to 5 kcal mol^{-1}). Hydrogen bonds are also somewhat longer than covalent bonds; their bond lengths (measured from the hydrogen atom) range from 1.5 Å to 2.6 Å; hence, a distance ranging from 2.4 Å to 3.5 Å separates the two nonhydrogen atoms in a hydrogen bond.

The strongest hydrogen bonds have a tendency to be straight, such that the hydrogen-bond donor, the hydrogen atom, and the hydrogen-bond acceptor lie along a straight line. This tendency toward linearity can be important for orienting interacting molecules with respect to one another. Hydrogen-bonding interactions are responsible for many of the properties of water that make it such a special solvent, as will be described shortly.

3. *van der Waals Interactions.* The basis of a van der Waals interaction is that the distribution of electronic charge around an atom fluctuates with time. At any instant, the charge distribution is not perfectly symmetric.

Hydrogen-   Hydrogen-
bond donor   bond acceptor

N—H- - - - - -N
δ^- δ^+ δ^-

N—H- - - - - -O

O—H- - - - - -N

O—H- - - - - -O

FIGURE 1.9 Hydrogen bonds. Hydrogen bonds are depicted by dashed green lines. The positions of the partial charges (δ^+ and δ^-) are shown.

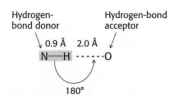

Hydrogen-   Hydrogen-bond
bond donor   acceptor

0.9 Å   2.0 Å

N—H - - - - -O

180°

This transient asymmetry in the electronic charge about an atom acts through ionic interactions to induce a complementary asymmetry in the electron distribution within its neighboring atoms. The atom and its neighbors then attract one another. This attraction increases as two atoms come closer to each other, until they are separated by the van der Waals *contact distance* (Figure 1.10). At distances shorter than the van der Waals contact distance, very strong repulsive forces become dominant because the outer electron clouds of the two atoms overlap.

Energies associated with van der Waals interactions are quite small; typical interactions contribute from 2 to 4 kJ mol^{-1} (from 0.5 to 1 kcal mol^{-1}) per atom pair. When the surfaces of two large molecules come together, however, a large number of atoms are in van der Waals contact, and the net effect, summed over many atom pairs, can be substantial.

We will cover the fourth noncovalent interaction, the hydrophobic interaction, after we examine the characteristics of water; these characteristics are essential to understanding the hydrophobic interaction.

Properties of water. Water is the solvent in which most biochemical reactions take place, and its properties are essential to the formation of macromolecular structures and the progress of chemical reactions. Two properties of water are especially relevant:

1. *Water is a polar molecule.* The water molecule is bent, not linear, and so the distribution of charge is asymmetric. The oxygen nucleus draws electrons away from the two hydrogen nuclei, which leaves the region around each hydrogen atom with a net positive charge. The water molecule is thus an electrically polar structure.

2. *Water is highly cohesive.* Water molecules interact strongly with one another through hydrogen bonds. These interactions are apparent in the structure of ice (Figure 1.11). Networks of hydrogen bonds hold the structure together; similar interactions link molecules in liquid water and account for many of the properties of water. In the liquid state, approximately one in four of the hydrogen bonds present in ice is broken. The polar nature of water is responsible for its high dielectric constant of 80. Molecules in aqueous solution interact with water molecules through the formation of hydrogen bonds and through ionic interactions. These interactions make water a versatile solvent, able to readily dissolve many species, especially polar and charged compounds that can participate in these interactions.

The hydrophobic effect. A final fundamental interaction called the *hydrophobic effect* is a manifestation of the properties of water. Some molecules (termed *nonpolar molecules*) cannot participate in hydrogen bonding or ionic interactions. The interactions of nonpolar molecules with water molecules are not as favorable as are interactions between the water molecules themselves. The water molecules in contact with these nonpolar molecules form "cages" around them, becoming more well-ordered than water molecules free in solution. However, when two such nonpolar molecules come together, some of the water molecules are released, allowing them to interact freely with bulk water (Figure 1.12). The release of water from such cages is favorable for reasons to be considered shortly. The result is that nonpolar molecules show an increased tendency to associate with one

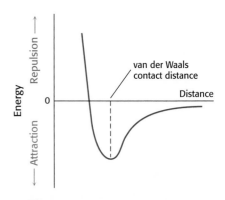

FIGURE 1.10 Energy of a van der Waals interaction as two atoms approach each other. The energy is most favorable at the van der Waals contact distance. Owing to electron–electron repulsion, the energy rises rapidly as the distance between the atoms becomes shorter than the contact distance.

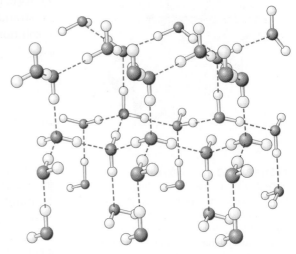

FIGURE 1.11 Structure of ice. Hydrogen bonds (shown as dashed green lines) are formed between water molecules to produce a highly ordered and open structure.

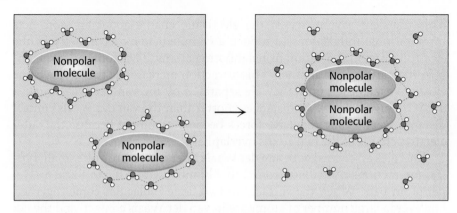

FIGURE 1.12 The hydrophobic effect. The aggregation of nonpolar groups in water leads to the release of water molecules, initially interacting with the nonpolar surface, into bulk water. The release of water molecules into solution makes the aggregation of nonpolar groups favorable.

another in water compared with other, less polar and less self-associating, solvents. This tendency is called the hydrophobic effect and the associated interactions are called *hydrophobic interactions*.

The double helix is an expression of the rules of chemistry

Let us now explore how these four noncovalent interactions work together in driving the association of two strands of DNA to form a double helix. First, each phosphate group in a DNA strand carries a negative charge. These negatively charged groups interact unfavorably with one another over distances. So, unfavorable ionic interactions take place when two strands of DNA come together. These phosphate groups are far apart in the double helix, with distances greater than 10 Å, but many such interactions take place (Figure 1.13). Thus, ionic interactions oppose the formation of the double helix. The strength of these repulsive ionic interactions is diminished by the high dielectric constant of water and the presence of ionic species such as Na^+ or Mg^{2+} ions in solution. These positively charged species interact with the phosphate groups and partly neutralize their negative charges.

Second, as already noted, hydrogen bonds are important in determining the formation of specific base pairs in the double helix. However, in single-stranded DNA, the hydrogen-bond donors and acceptors are exposed to solution and can form hydrogen bonds with water molecules.

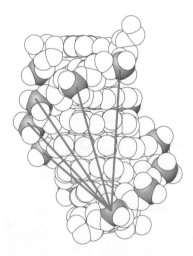

FIGURE 1.13 Ionic interactions in DNA. Each unit within the double helix includes a phosphate group (the phosphorus atom being shown in purple) that bears a negative charge. The unfavorable interactions of one phosphate with several others are shown by red lines. These repulsive interactions oppose the formation of a double helix.

When two single strands come together, these hydrogen bonds with water are broken and new hydrogen bonds between the bases are formed. Because the number of hydrogen bonds broken is the same as the number formed, these hydrogen bonds do not contribute substantially to driving the overall process of double-helix formation. However, they contribute greatly to the specificity of binding. Suppose two bases that cannot form Watson–Crick base pairs are brought together. Hydrogen bonds with water must be broken as the bases come into contact. Because the bases are not complementary in structure, not all of these bonds can be simultaneously replaced by hydrogen bonds between the bases. Thus, the formation of a double helix between noncomplementary sequences is disfavored.

Third, within a double helix, the base pairs are parallel and stacked nearly on top of one another. The typical separation between the planes of adjacent base pairs is 3.4 Å, and the distances between the most closely approaching atoms are approximately 3.6 Å. This separation distance corresponds nicely to the van der Waals contact distance (Figure 1.14). Bases tend to stack even in single-stranded DNA molecules. However, the base stacking and associated van der Waals interactions are nearly optimal in a double-helical structure.

Fourth, the hydrophobic effect also contributes to the favorability of base stacking. More-complete base stacking moves the nonpolar surfaces of the bases out of water into contact with each other.

The principles of double-helix formation between two strands of DNA apply to many other biochemical processes. Many weak interactions contribute to the overall energetics of the process, some favorably and some unfavorably. Furthermore, surface complementarity is a key feature: when complementary surfaces meet, hydrogen-bond donors align with hydrogen-bond acceptors and nonpolar surfaces come together to maximize van der Waals interactions and minimize nonpolar surface area exposed to the aqueous environment. The properties of water play a major role in determining the importance of these interactions.

The laws of thermodynamics govern the behavior of biochemical systems

We can look at the formation of the double helix from a different perspective by examining the laws of thermodynamics. These laws are general principles that apply to all physical (and biological) processes. They are of great importance because they determine the conditions under which specific processes can or cannot take place. We will consider these laws from a general perspective first and then apply the principles that we have developed to the formation of the double helix.

The laws of thermodynamics distinguish between a system and its surroundings. A *system* refers to the matter within a defined region of space. The matter in the rest of the universe is called the *surroundings*. *The First Law of Thermodynamics states that the total energy of a system and its surroundings is constant.* In other words, the energy content of the universe is constant; energy can be neither created nor destroyed. Energy can take different forms, however. Heat, for example, is one form of energy. Heat is a manifestation of the *kinetic energy* associated with the random motion of molecules. Alternatively, energy can be present as *potential energy*—energy that will be released on the occurrence of some process. Consider, for example, a ball held at the top of a tower. The ball has considerable potential energy because, when it is released, the ball will develop kinetic energy associated with its motion as it falls. Within chemical systems, potential energy is related to the likelihood that atoms can react with one another. For instance, a mixture of gasoline and oxygen has a large potential energy because these molecules may react to form carbon dioxide and water and release energy as heat. The First Law requires that any energy released in the formation of chemical bonds must be used to break other bonds, released as heat or light, or stored in some other form.

Another important thermodynamic concept is that of *entropy*, a measure of the degree of randomness or disorder in a system. *The Second Law of*

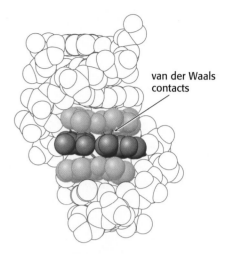

FIGURE 1.14 Base stacking. In the DNA double helix, adjacent base pairs are stacked nearly on top of one another, and so many atoms in each base pair are separated by their van der Waals contact distance. The central base pair is shown in dark blue and the two adjacent base pairs in light blue. Several van der Waals contacts are shown in red.

van der Waals contacts

Thermodynamics states that the total entropy of a system plus that of its surroundings always increases. For example, the release of water from nonpolar surfaces responsible for the hydrophobic effect is favorable because water molecules free in solution are more disordered than they are when they are associated with nonpolar surfaces. At first glance, the Second Law appears to contradict much common experience, particularly about biological systems. Many biological processes, such as the generation of a leaf from carbon dioxide gas and other nutrients, clearly increase the level of order and hence decrease entropy. Entropy may be decreased locally in the formation of such ordered structures only if the entropy of other parts of the universe is increased by an equal or greater amount. The local decrease in entropy is often accomplished by a release of heat, which increases the entropy of the surroundings.

We can analyze this process in quantitative terms. First, consider the system. The entropy (S) of the system may change in the course of a chemical reaction by an amount ΔS_{system}. If heat flows from the system to its surroundings, then the heat content, often referred to as the *enthalpy* (H), of the system will be reduced by an amount ΔH_{system}. To apply the Second Law, we must determine the change in entropy of the surroundings. If heat flows from the system to the surroundings, then the entropy of the surroundings will increase. The precise change in the entropy of the surroundings depends on the temperature; the change in entropy is greater when heat is added to relatively cold surroundings than when heat is added to surroundings at high temperatures that are already in a high degree of disorder. To be even more specific, the change in the entropy of the surroundings will be proportional to the amount of heat transferred from the system and inversely proportional to the temperature (T) of the surroundings. In biological systems, T [in kelvins (K), absolute temperature] is usually assumed to be constant. Thus, a change in the entropy of the surroundings is given by

$$\Delta S_{\text{surroundings}} = -\Delta H_{\text{system}}/T \tag{1}$$

The total entropy change is given by the expression

$$\Delta S_{\text{total}} = \Delta S_{\text{system}} + \Delta S_{\text{surroundings}} \tag{2}$$

Substituting equation 1 into equation 2 yields

$$\Delta S_{\text{total}} = \Delta S_{\text{system}} - \Delta H_{\text{system}}/T \tag{3}$$

Multiplying by $-T$ gives

$$-T\Delta S_{\text{total}} = \Delta H_{\text{system}} - T\Delta S_{\text{system}} \tag{4}$$

The function $-T\Delta S$ has units of energy and is referred to as *free energy* or *Gibbs free energy*, after Josiah Willard Gibbs, who developed this function in 1878:

$$\Delta G = \Delta H_{\text{system}} - T\Delta S_{\text{system}} \tag{5}$$

The free-energy change, ΔG, will be used throughout this book to describe the energetics of biochemical reactions. The Gibbs free energy is essentially an accounting tool that keeps track of both the entropy of the system (directly) and the entropy of the surroundings (in the form of heat released from the system).

Recall that the Second Law of Thermodynamics states that, for a process to take place, the entropy of the universe must increase. Examination of equation 3 shows that the total entropy will increase if and only if

$$\Delta S_{system} > \Delta H_{system}/T \qquad (6)$$

Rearranging gives $T\Delta S_{system} > \Delta H_{system}$ or, in other words, entropy will increase if and only if

$$\Delta G = \Delta H_{system} - T\Delta S_{system} < 0 \qquad (7)$$

Thus, the free-energy change must be negative for a process to take place spontaneously. *There is negative free-energy change when and only when the overall entropy of the universe is increased.* Again, the free energy represents a single term that takes into account both the entropy of the system and the entropy of the surroundings.

Heat is released in the formation of the double helix

Let's see how the principles of thermodynamics apply to the formation of the double helix (Figure 1.15). Suppose solutions containing each of the two single strands are mixed. Before the double helix forms, each of the single strands is free to translate and rotate in solution, whereas each matched pair of strands in the double helix must move together. Additionally, the free single strands exist in more conformations than possible when bound together in a double helix. Thus, the formation of a double helix from two single strands appears to result in an increase in order for the system, that is, a decrease in the entropy of the system.

On the basis of this analysis, we expect that the double helix cannot form without violating the Second Law of Thermodynamics unless heat is released to increase the entropy of the surroundings. Experimentally, we can measure the heat released by allowing the solutions containing the two single strands to come together within a water bath, which here corresponds to the surroundings. We then determine how much heat must be absorbed by the water bath or released from it to maintain it at a constant temperature. This experiment establishes that a substantial amount of heat is released— namely, approximately 250 kJ mol^{-1} (60 kcal mol^{-1}). This experimental result reveals that the change in enthalpy for the process is quite large, -250 kJ mol^{-1}, consistent with our expectation that significant heat would have to be released to the surroundings for the process not to violate the Second Law. We see in quantitative terms how order within a system can be increased by releasing sufficient heat to the surroundings to ensure that the entropy of the universe increases. We will encounter this general theme throughout this book.

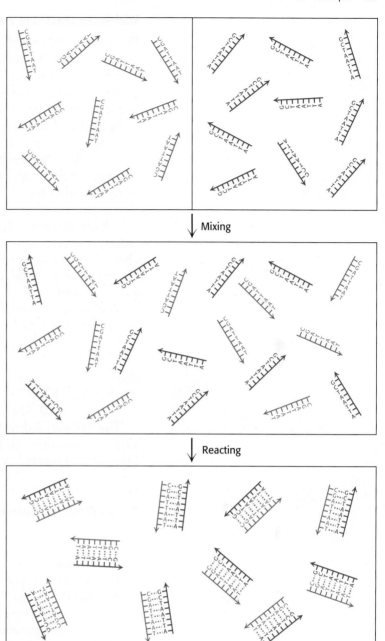

FIGURE 1.15 Double-helix formation and entropy. When solutions containing DNA strands with complementary sequences are mixed, the strands react to form double helices. This process results in a loss of entropy from the system, indicating that heat must be released to the surroundings to prevent a violation of the Second Law of Thermodynamics.

Acid–base reactions are central in many biochemical processes

Throughout our consideration of the formation of the double helix, we have dealt only with the noncovalent bonds that are formed or broken in this process. Many biochemical processes entail the formation and cleavage of covalent bonds. A particularly important class of reactions prominent in biochemistry is *acid–base reactions*.

In acid and base reactions, hydrogen ions are added to molecules or removed from them. Throughout the book, we will encounter many processes in which the addition or removal of hydrogen atoms is crucial, such as the metabolic processes by which carbohydrates are degraded to release energy for other uses. Thus, a thorough understanding of the basic principles of these reactions is essential.

A hydrogen ion, often written as H^+, corresponds to a proton. In fact, hydrogen ions exist in solution bound to water molecules, thus forming what are known as *hydronium ions*, H_3O^+. For simplicity, we will continue to write H^+, but we should keep in mind that H^+ is shorthand for the actual species present.

The concentration of hydrogen ions in solution is expressed as the pH. Specifically, the *pH* of a solution is defined as

$$pH = -\log[H^+]$$

where $[H^+]$ is in units of molarity. Thus, pH 7.0 refers to a solution for which $-\log[H^+] = 7.0$, and so $\log[H^+] = -7.0$ and $[H^+] = 10^{\log[H^+]} = 10^{-7.0} = 1.0 \times 10^{-7}\,M$.

The pH also indirectly expresses the concentration of hydroxide ions, $[OH^-]$, in solution. To see how, we must realize that water molecules can dissociate to form H^+ and OH^- ions in an equilibrium process.

$$H_2O \rightleftharpoons H^+ + OH^-$$

The equilibrium constant (K) for the dissociation of water is defined as

$$K = [H^+][OH^-]/[H_2O]$$

and has a value of $K = 1.8 \times 10^{-16}$. Note that an equilibrium constant does not formally have units. Nonetheless, the value of the equilibrium constant given assumes that particular units are used for concentration (sometimes referred to as standard states); in this case and in many others, units of molarity (M) are assumed.

The concentration of water, $[H_2O]$, in pure water is 55.5 M, and this concentration is constant under most conditions. Thus, we can define a new constant, K_W:

$$K_W = K[H_2O] = [H^+][OH^-]$$

$$K[H_2O] = 1.8 \times 10^{-16} \times 55.5$$

$$= 1.0 \times 10^{-14}$$

Because $K_W = [H^+][OH^-] = 1.0 \times 10^{-14}$, we can calculate

$$[OH^-] = 10^{-14}/[H^+] \text{ and } [H^+] = 10^{-14}/[OH^+]$$

With these relations in hand, we can easily calculate the concentration of hydroxide ions in an aqueous solution, given the pH. For example, at pH = 7.0, we know that $[H^+] = 10^{-7}\,M$ and so $[OH^-] = 10^{-14}/10^{-7} = 10^{-7}\,M$. In acidic solutions, the concentration of hydrogen ions is higher than 10^{-7} and, hence, the pH is below 7. For example, in 0.1 M HCl, $[H^+] = 10^{-1}\,M$ and so pH = 1.0 and $[OH^-] = 10^{-14}/10^{-1} = 10^{-13}\,M$.

Acid–base reactions can disrupt the double helix

The reaction that we have been considering between two strands of DNA to form a double helix takes place readily at pH 7.0. Suppose that we take the solution containing the double-helical DNA and treat it with a solution of concentrated base (i.e., with a high concentration of OH⁻). As the base is added, we monitor the pH and the fraction of DNA in double-helical form (Figure 1.16). When the first additions of base are made, the pH rises, but the concentration of the double-helical DNA does not change significantly. However, as the pH approaches 9, the DNA double helix begins to dissociate into its component single strands. As the pH continues to rise from 9 to 10, this dissociation becomes essentially complete. Why do the two strands dissociate? The hydroxide ions can react with bases in DNA base pairs to remove certain protons. The most susceptible proton is the one bound to the N-1 nitrogen atom in a guanine base.

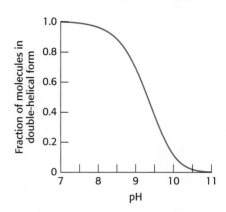

FIGURE 1.16 DNA denaturation by the addition of a base. The addition of a base to a solution of double-helical DNA initially at pH 7 causes the double helix to separate into single strands. The process is half complete at slightly above pH 9.

$pK_a = 9.7$

Guanine (G)

Proton dissociation for a substance HA (such as that bound to N-1 on guanine) has an equilibrium constant defined by the expression

$$K_a = [H^+][A^-]/[HA]$$

The susceptibility of a proton to removal by reaction with a base is often described by its pK_a *value*:

$$pK_a = -\log(K_a)$$

When the pH is equal to the pK_a, we have pH = pK_a

$$pH = pK_a$$

and so

$$-\log[H^+] = -\log([H^+][A^-]/[HA])$$

and

$$[H^+] = [H^+][A^-]/[HA]$$

Dividing by $[H^+]$ reveals that

$$1 = [A^-]/[HA]$$

and so

$$[A^-] = [HA]$$

So, when the pH equals the pK_a, the concentration of the deprotonated form of the group or molecule is equal to the concentration of the protonated form; the deprotonation process is halfway to completion.

The pK_a for the proton on N-1 of guanine is typically 9.7. When the pH approaches this value, the proton on N-1 is lost (Figure 1.16). Because this proton participates in an important hydrogen bond, its loss substantially destabilizes the DNA double helix. The DNA double helix is also destabilized by *low* pH. Below pH 5, some of the hydrogen bond *acceptors* that participate in base-pairing become protonated. In their protonated forms, these bases can no longer form hydrogen bonds and the double helix separates. So, acid–base reactions that remove or donate protons at specific positions on the DNA bases can disrupt the double helix.

Buffers regulate pH in organisms and in the laboratory

These observations about DNA reveal that a significant change in pH can disrupt molecular structure. The same is true for many other biological macromolecules; changes in pH can protonate or deprotonate key groups, potentially disrupting structures and initiating harmful reactions. Thus, systems have evolved to mitigate changes in pH in biological systems. Solutions that resist such changes are called *buffers*. Specifically, when acid is added to an unbuffered aqueous solution, the pH drops in proportion to the amount of acid added. In contrast, when acid is added to a buffered solution, the pH drops more gradually. Buffers also mitigate the pH increase caused by the addition of base and changes in pH caused by dilution.

Compare the result of adding a 1 M solution of the strong acid HCl drop by drop to pure water with adding it to a solution containing 100 mM of the buffer sodium acetate ($Na^+CH_3COO^-$; Figure 1.17). The process of gradually adding known amounts of reagent to a solution with which the reagent reacts while monitoring the results is called a *titration*. For pure water, the pH drops from 7 to close to 2 on the addition of the first few drops of acid. However, for the sodium acetate solution, the pH first falls rapidly from its initial value near 10, then changes more gradually until the pH reaches 3.5, and then falls more rapidly again. Why does the pH decrease so gradually in the middle of the titration? The answer is when hydrogen ions are added to this solution they react with acetate ions to form acetic acid. This reaction consumes some of the added hydrogen ions so that the pH does not drop. Hydrogen ions continue reacting with acetate ions until essentially all the acetate ion is converted into acetic acid. After this point, added protons remain free in solution and the pH begins to fall sharply again.

We can analyze the effect of the buffer in quantitative terms. The equilibrium constant for the deprotonation of an acid is

$$K_a = [H^+][A^-]/[HA]$$

Taking logarithms of both sides yields

$$\log(K_a) = \log([H^+]) + \log([A^-]/[HA])$$

Recalling the definitions of pK_a and pH and rearranging gives

$$pH = pK_a + \log([A^-]/[HA])$$

This expression is referred to as the *Henderson–Hasselbalch equation*.

We can apply the equation to our titration of sodium acetate. The pK_a of acetic acid is 4.75. We can calculate the ratio of the concentration of acetate ion to the concentration of acetic acid as a function of pH by using the Henderson–Hasselbalch equation, slightly rearranged.

$$[\text{Acetate ion}]/[\text{Acetic acid}] = [A^-]/[HA] = 10^{pH - pK_a}$$

At pH 9, this ratio is $10^{9 - 4.75} = 10^{4.25} = 17,800$; very little acetic acid has been formed. At pH 4.75 (when the pH equals the pK_a), the ratio is $10^{4.75 - 4.75} = 10^0 = 1$. At pH 3, the ratio is $10^{3 - 4.75} = 10^{-1.25} = 0.02$; almost all the acetate ion has been converted into acetic acid. We can follow the conversion of acetate ion into acetic acid over the entire titration (Figure 1.18). The graph shows that the region of relatively constant pH corresponds

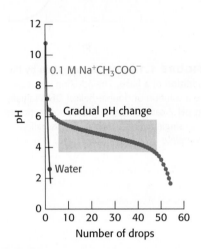

FIGURE 1.17 Buffer action. The addition of a strong acid, 1 M HCl, to pure water results in an immediate drop in pH to near 2. In contrast, the addition of the acid to a 0.1 M sodium acetate ($Na^+CH_3COO^-$) solution results in a much more gradual change in pH until the pH drops below 3.5.

precisely to the region in which acetate ion is being protonated to form acetic acid.

From this discussion, we see that a buffer functions best close to the pK_a value of its acid component. Physiological pH is typically about 7.4. An important buffer in biological systems is based on phosphoric acid (H_3PO_4). The acid can be deprotonated in three steps to form a phosphate ion.

$$H_3PO_4 \underset{pK_a = 2.12}{\overset{H^+}{\rightleftharpoons}} H_2PO_4^- \underset{pK_a = 7.21}{\overset{H^+}{\rightleftharpoons}} HPO_4^{2-} \underset{pK_a = 12.67}{\overset{H^+}{\rightleftharpoons}} PO_4^{3-}$$

At about pH 7.4, inorganic phosphate exists primarily as a nearly equal mixture of $H_2PO_4^-$ and HPO_4^{2-}. Thus, phosphate solutions function as effective buffers near pH 7.4. The concentration of inorganic phosphate in blood is typically approximately 1 mM, providing a useful buffer against processes that produce either acid or base. We can examine this utility in quantitative terms with the use of the Henderson–Hasselbalch equation. What concentration of acid must be added to change the pH of 1 mM phosphate buffer from 7.4 to 7.3? Without buffer, this change in $[H^+]$ corresponds to a change of $10^{-7.3} - 10^{-7.4}\,M = (5.0 \times 10^{-8} - 4.0 \times 10^{-8})\,M = 1.0 \times 10^{-8}\,M$. Let us now consider what happens to the buffer components. At pH 7.4,

$$[HPO_4^{2-}]/[H_2PO_4^-] = 10^{7.4-7.21} = 10^{0.19} = 1.55$$

The total concentration of phosphate, $[HPO_4^{2-}] + [H_2PO_4^-]$, is 1 mM. Thus,

$$[HPO_4^{2-}] = (1.55/2.55) \times 1\,mM = 0.608\,mM$$

and

$$[H_2PO_4^-] = (1/2.55) \times 1\,mM = 0.392\,mM$$

At pH 7.3,

$$[HPO_4^{2-}]/[H_2PO_4^-] = 10^{7.3-7.21} = 10^{0.09} = 1.23$$

and so

$$[HPO_4^{2-}] = (1.23/2.23) = 0.552\,mM$$

and

$$[H_2PO_4^-] = (1/2.23) = 0.448\,mM$$

Thus, $(0.608 - 0.552) = 0.056\,mM\,HPO_4^{2-}$ is converted into $H_2PO_4^-$, consuming $0.056\,mM = 5.6 \times 10^{-5}\,M\,[H^+]$. Thus, the buffer increases the amount of acid required to produce a drop in pH from 7.4 to 7.3 by a factor of $5.6 \times 10^{-5}/1.0 \times 10^{-8} = 5600$ compared with pure water.

1.4 The Genomic Revolution Is Transforming Biochemistry, Medicine, and Other Fields

Watson and Crick's discovery of the structure of DNA suggested the hypothesis that hereditary information is stored as a sequence of bases along long strands of DNA. This remarkable insight provided an entirely new way of thinking about biology. However, at the time that it was made, Watson and Crick's discovery, though full of potential, remained unconfirmed and many features needed to be elucidated. How is the sequence information

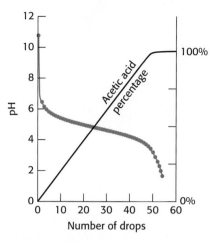

FIGURE 1.18 Buffer protonation. When acid is added to sodium acetate, the added hydrogen ions are used to convert acetate ion into acetic acid. Because the proton concentration does not increase significantly, the pH remains relatively constant until all the acetate has been converted into acetic acid.

read and translated into action? What are the sequences of naturally occurring DNA molecules and how can such sequences be experimentally determined? Through advances in biochemistry and related sciences, we now have essentially complete answers to these questions. Indeed, in the past decade or so, scientists have determined the complete genome sequences of thousands of different organisms, including simple microorganisms, plants, animals of varying degrees of complexity, and human beings. Comparisons of these genome sequences, with the use of methods introduced in Chapter 6, have been sources of many insights that have transformed biochemistry. In addition to its experimental and clinical aspects, biochemistry has now become an *information science*.

Genome sequencing has transformed biochemistry and other fields

The sequencing of a human genome was a daunting task, because it contains approximately 3 billion (3×10^9) base pairs. For example, the sequence

ACATTTGCTTCTGACACAACTGTGTTCACTAGCAACCTC
AAACAGACACCATGGTGCATCTGACTCCTG**A**GGAGAAGT
CTGCCGTTACTGCCCTGTGGGGCAAGGTGAACGTGGA . . .

is a part of one of the genes that encodes hemoglobin, the oxygen carrier in our blood. This gene is found on the end of chromosome 9 of our 24 distinct chromosomes. If we were to include the complete sequence of our entire genome, this chapter would run to more than 500,000 pages. The sequencing of our genome is truly a landmark in human history. This sequence contains a vast amount of information, some of which we can now extract and interpret, but much of which we are only beginning to understand. For example, some human diseases have been linked to particular variations in genomic sequence. Sickle-cell anemia, discussed in detail in Chapter 7, is caused by a single base change of an A (noted in boldface type in the preceding sequence) to a T. We will encounter many other examples of diseases that have been linked to specific DNA sequence changes.

Determining the first human genome sequences was a great challenge. It required the efforts of large teams of geneticists, molecular biologists, biochemists, and computer scientists, as well as billions of dollars, because there was no previous framework for aligning the sequences of various DNA fragments. One human genome sequence can serve as a reference for other sequences. The availability of such reference sequences enables much more rapid characterization of partial or complete genomes from other individuals. As we will discuss in Chapter 5, arrays containing millions of target single-stranded DNA molecules with sequences from the reference genome and known or potential variants are powerful tools. These arrays can be exposed to mixtures of DNA fragments for a particular individual and those single-stranded targets that bind to their complementary strands can be determined. This allows many positions within the genome of the individual to be simultaneously interrogated.

Methods for sequencing DNA have also been improving rapidly, driven by a deep understanding of the biochemistry of DNA replication and other processes. This has resulted in both dramatic increases in the DNA sequencing rate and decreases in costs (Figure 1.19). The availability of such powerful sequencing technology is transforming many fields, including medicine, dentistry, microbiology, pharmacology, and ecology, although a great deal remains to be done to improve the accuracy and precision of the interpretation of these large genomic and related data sets.

Each person has a unique sequence of DNA base pairs. How different are we from one another at the genomic level? Examination of genomic variation reveals that, on average, each pair of individuals has a different base in one position per 200 bases; that is, the difference is approximately 0.5%. This variation between individuals who are not closely related is quite substantial compared with differences in populations. The average difference between two people within one ethnic group is greater than the difference between the averages of two different ethnic groups.

The significance of much of this genetic variation is not understood. As noted earlier, variation in a single base within the genome can lead to a disease such as sickle-cell anemia. Scientists have now identified the genetic variations associated with thousands of diseases for which the cause can be traced to a single gene. For other diseases and traits, we know that variation in many different genes contributes in significant and often complex ways. Many of the most prevalent human ailments such as heart disease are linked to variations in many genes. Furthermore, in most cases, the presence of a particular variation or set of variations does not inevitably result in the onset of a disease but, instead, leads to a *predisposition* to the development of the disease.

Our own genes are not the only ones that can contribute to health and disease. Our bodies, including our skin, mouth, digestive tract, genitourinary tract, respiratory tract, and other areas, contain a large number of microorganisms. These complex communities have been characterized through powerful methods that allow DNA isolated from these biological samples to be sequenced without any previous knowledge of the organisms present. Many of these organisms had not previously been discovered because they can only grow as part of complex communities and thus cannot be isolated through standard microbiological techniques. Remarkably, it appears that we are outnumbered in our own bodies! Each of us contains approximately 10 times more microbial cells than human cells, and these microbial cells include many more genes than do our own genomes. These *microbiomes* differ from site to site, from one person to another, and over time in the same individual. They appear to play roles in health and in diseases such as obesity and dental caries (Figure 1.20).

In addition to the implications for understanding human health and disease, the genome sequence is a source of deep insight into other aspects of human biology and culture. For example, by comparing the sequences of different individuals and populations, we can learn a great deal about human history. On the basis of such analysis, a compelling case can be made that the human species originated in Africa, and the occurrence and even the timing of important migrations of groups of human beings can be demonstrated (Figure 1.21). Finally, comparisons of the human genome with the genomes of other organisms are confirming the tremendous unity that exists at the level of biochemistry and are revealing key steps in the course of evolution from relatively simple, single-celled organisms to complex, multicellular organisms such as human beings. For example, many genes that are key to the function of the human brain and nervous system

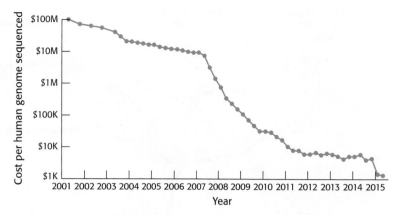

FIGURE 1.19 Decreasing costs of DNA sequencing. Through the Human Genome Project, the costs of DNA sequencing dropped steadily due to new methods and are now approaching $1000 for a complete human genome sequence. [National Human Genome Research Institute. www.genome.gov/sequencingcosts]

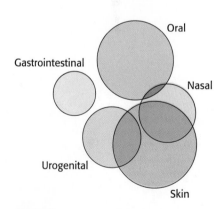

FIGURE 1.20 The human microbiome. Microorganisms cover the human body. Examination of the microbial communities using DNA sequencing methods revealed many previously uncharacterized species. The Venn diagrams represent populations of related species as determined by DNA sequence comparisons. The populations present on different body surfaces are largely distinct. [Adapted from www.nature.com/nature/journal/v486/n7402/fig_tab/nature11234_F1.html]

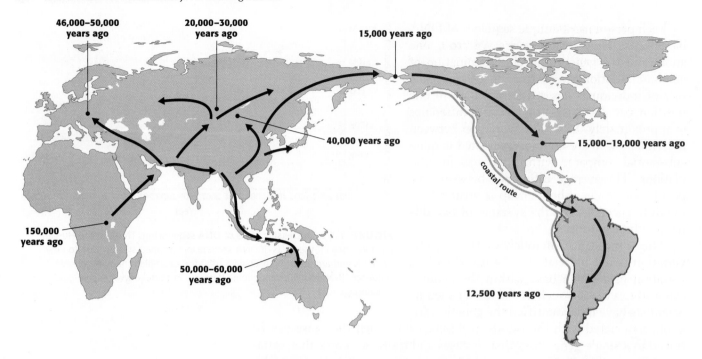

FIGURE 1.21 Human migrations supported by DNA sequence comparisons. Modern human beings originated in Africa, migrated first to Asia, and then to Europe, Australia, and North and South America. [Adapted from S. Oppenheimer, 2012 Out-of-Africa, the peopling of continents and islands: Tracing uniparental gene trees across the map. *Philos. Trans. R. Soc. Lond. B. Biol. Sci.* 367:770–784]

have evolutionary and functional relatives in the genomes of bacteria. Because many studies that are possible in model organisms are difficult or unethical to conduct in human beings, these discoveries have many practical implications. *Comparative genomics* has become a powerful science, linking evolution and biochemistry.

Environmental factors influence human biochemistry

Although our genetic makeup (and that of our microbiomes) is an important factor that contributes to disease susceptibility and to other traits, factors in a person's environment also are significant. What are these environmental factors? Perhaps the most obvious are chemicals that we eat or are exposed to in some other way. The adage "you are what you eat" has considerable validity; it applies both to substances that we ingest in significant quantities and to those that we ingest in only trace amounts. Throughout our study of biochemistry, we will encounter *vitamins* and *trace elements* and their derivatives that play crucial roles in some processes. In many cases, the roles of these chemicals were first revealed through investigation of *deficiency disorders* observed in people who do not take in a sufficient quantity of a particular vitamin or trace element. Despite the fact that the most important essential dietary factors have been known for some time, new roles for them continue to be discovered.

A healthful diet requires a balance of major food groups. In addition to providing vitamins and trace elements, food provides calories in the form of substances that can be broken down to release energy that drives other biochemical processes. Proteins, fats, and carbohydrates provide the building blocks used to construct the molecules of life (Figure 1.22). Finally, it is possible to get too much of a good thing. Human beings evolved under circumstances in which food, particularly rich foods such as meat, was scarce.

With the development of agriculture and modern economies, rich foods are now plentiful in parts of the world. Some of the most prevalent diseases in the so-called developed world, such as heart disease and diabetes, can be attributed to the large quantities of fats and carbohydrates present in modern diets. We are now developing a deeper understanding of the biochemical consequences of these diets and the interplay between diet and genetic factors.

Chemicals are only one important class of environmental factors. Our behaviors also have biochemical consequences. Through physical activity, we consume the calories that we take in, ensuring an appropriate balance between food intake and energy expenditure. Activities ranging from exercise to emotional responses such as fear and love may activate specific biochemical pathways, leading to changes in levels of gene expression, the release of hormones, and other consequences. Furthermore, the interplay between biochemistry and behavior is bidirectional. Our biochemistry is affected by our behavior, so, too, our behavior is affected, although certainly not completely determined, by our genetic makeup and other aspects of our biochemistry. Genetic factors associated with a range of behavioral characteristics have been at least tentatively identified.

Just as vitamin deficiencies and genetic diseases have revealed fundamental principles of biochemistry and biology, investigations of variations in behavior and their linkage to genetic and biochemical factors are potential sources of great insight into mechanisms within the brain. For example, studies of drug addiction have revealed neural circuits and biochemical pathways that greatly influence aspects of behavior. Unraveling the interplay between biology and behavior is one of the great challenges in modern science, and biochemistry is providing some of the most important concepts and tools for this endeavor.

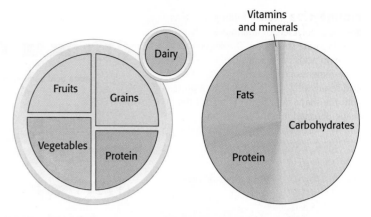

FIGURE 1.22 Nutrition. Proper health depends on an appropriate combination of food groups (fruits, vegetables, protein, grains, dairy) (left) to provide an optimal mix of biochemicals (carbohydrates, proteins, fats, vitamins, and minerals) (right). [Adapted from www.choosemyplate.gov]

Genome sequences encode proteins and patterns of expression

The structure of DNA revealed how information is stored in the base sequence along a DNA strand. But what information is stored and how is it expressed? The most fundamental role of DNA is to encode the sequences of proteins. Like DNA, proteins are linear polymers. However, proteins differ from DNA in two important ways. First, proteins are built from 20 building blocks, called *amino acids*, rather than just 4, as in DNA. The chemical complexity provided by this variety of building blocks enables proteins to perform a wide range of functions. Second, proteins spontaneously fold into elaborate three-dimensional structures, determined only by their amino acid sequences (Figure 1.23). We have explored in depth how solutions containing two appropriate strands of DNA come together to form a solution of double-helical molecules. A similar spontaneous folding process gives proteins their three-dimensional structure. A balance of hydrogen bonding, van der Waals interactions, and hydrophobic interactions overcomes the entropy lost in going from an unfolded ensemble of proteins to a homogenous set of well-folded molecules. Proteins and protein folding will be discussed extensively in Chapter 2.

The fundamental unit of hereditary information, the *gene*, is becoming increasingly difficult to precisely define as our knowledge of the complexities of genetics and genomics increases. The genes that are simplest to define encode the sequences of proteins. For these protein-encoding genes,

FIGURE 1.23 Protein folding. Proteins are linear polymers of amino acids that fold into elaborate structures. The sequence of amino acids determines the three-dimensional structure. Thus, amino acid sequence 1 gives rise only to a protein with the shape depicted in blue, *not* the shape depicted in red.

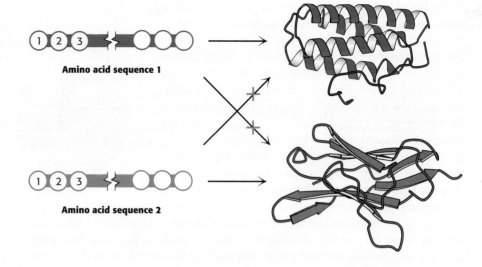

Amino acid sequence 1

Amino acid sequence 2

a block of DNA bases encodes the amino acid sequence of a specific protein molecule. A set of three bases along the DNA strand, called a *codon,* determines the identity of one amino acid within the protein sequence. The set of rules that links the DNA sequence to the encoded protein sequence is called the *genetic code.* One of the biggest surprises from the sequencing of the human genome is the small number of protein-encoding genes. Before the genome-sequencing project began, the consensus view was that the human genome would include approximately 100,000 protein-coding genes. The current analysis suggests that the actual number is between 19,000 and 23,000. We shall use an estimate of 20,000 throughout this book. However, additional mechanisms allow many genes to encode more than one protein. For example, the genetic information in some genes is translated in more than one way to produce a set of proteins that differ from one another in parts of their amino acid sequences. In other cases, proteins are modified after they have been synthesized through the addition of accessory chemical groups. Through these indirect mechanisms, much more complexity is encoded in our genomes than would be expected from the number of protein-encoding genes alone.

On the basis of current knowledge, the protein-encoding regions account for only about 3% of the human genome. What is the function of the rest of the DNA? Some of it contains information that regulates the expression of specific genes (i.e., the production of specific proteins) in particular cell types and physiological conditions. Essentially every human cell contains the same DNA genome, yet cell types differ considerably in the proteins that they produce. For example, hemoglobin is expressed only in precursors of red blood cells, even though the genes for hemoglobin are present in essentially every cell. Specific sets of genes are expressed in response to hormones, even though these genes are not expressed in the same cell in the absence of the hormones. The control regions that regulate such differences account for only a small amount of the remainder of our genomes. The truth is that we do not yet understand all the functions of much of the remainder of the DNA. Some of it is sometimes referred to as "junk"—stretches of DNA that were inserted at some stage of evolution and have remained. In some cases, this DNA may, in fact, serve important functions. In others, it may serve no function but, because it does not cause significant harm, it has remained.

Visualizing Molecular Structures: Small Molecules

The authors of a biochemistry textbook face the problem of trying to present three-dimensional molecules in the two dimensions available on the printed page. The interplay between the three-dimensional structures of biomolecules and their biological functions will be discussed extensively throughout this book. Toward this end, we will frequently use representations that, although of necessity rendered in two dimensions, emphasize the three-dimensional structures of molecules.

Stereochemical Renderings

Most of the chemical formulas in this book are drawn to depict the geometric arrangement of atoms, crucial to chemical bonding and reactivity, as accurately as possible. For example, the carbon atom of methane is tetrahedral, with H–C–H angles of 109.5 degrees, whereas the carbon atom in formaldehyde has bond angles of 120 degrees.

Methane   **Formaldehyde**

To illustrate the correct *stereochemistry* about tetrahedral carbon atoms, wedges will be used to depict the direction of a bond into or out of the plane of the page. A solid wedge with the broad end away from the carbon atom denotes a bond coming

toward the viewer out of the plane. A dashed wedge, with its broad end at the carbon atom, represents a bond going away from the viewer behind the plane of the page. The remaining two bonds are depicted as straight lines.

Fischer Projections

Although representative of the actual structure of a compound, stereochemical structures are often difficult to draw quickly. An alternative, less-representative method of depicting structures with tetrahedral carbon centers relies on the use of *Fischer projections*.

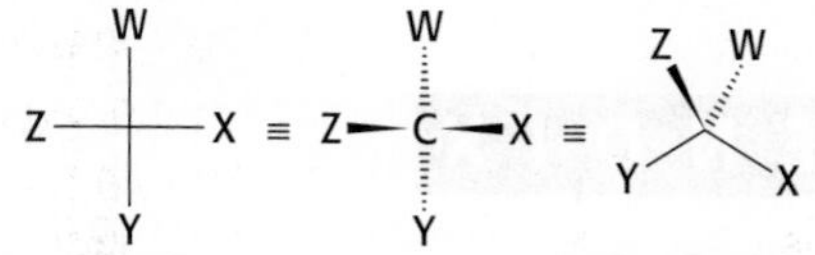

Fischer projection   **Stereochemical rendering**

In a Fischer projection, the bonds to the central carbon are represented by horizontal and vertical lines from the substituent atoms to the carbon atom, which is assumed to be at the center of the cross. By convention, the horizontal bonds are assumed to project out of the page toward the viewer, whereas the vertical bonds are assumed to project behind the page away from the viewer.

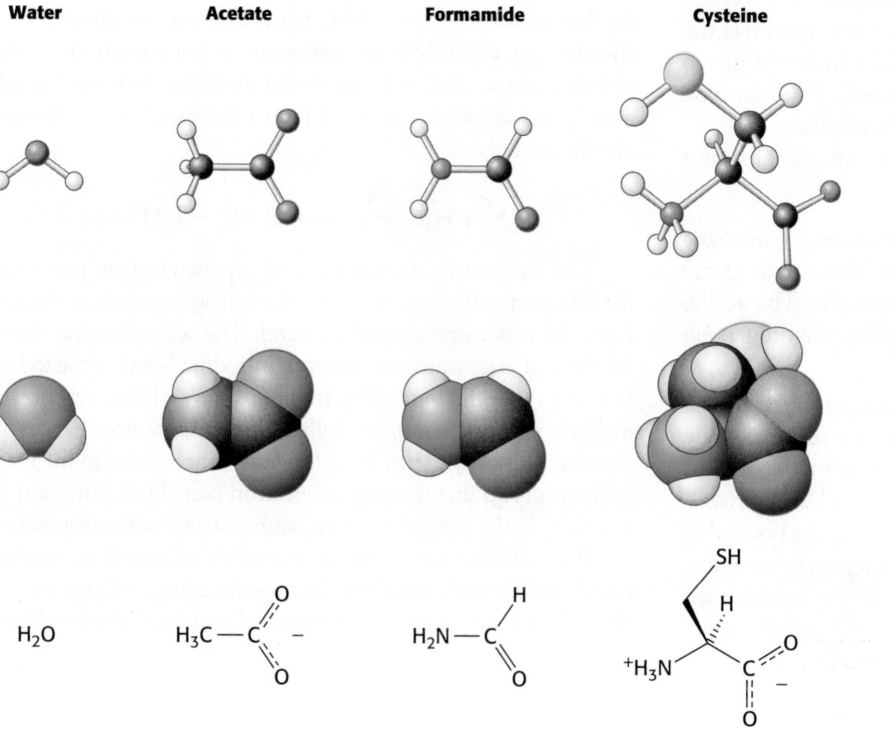

Water   **Acetate**   **Formamide**   **Cysteine**

H_2O $H_3C-C\big(\genfrac{}{}{0pt}{}{O}{O}\big)^-$ $H_2N-C\big(\genfrac{}{}{0pt}{}{H}{O}\big)$ Cysteine structure

FIGURE 1.24 Molecular representations. Structural formulas (bottom), ball-and-stick models (top), and space-filling representations (middle) of selected molecules are shown. Black = carbon, red = oxygen, white = hydrogen, yellow = sulfur, blue = nitrogen.

Molecular Models for Small Molecules

For depicting the molecular architecture of small molecules in more detail, two types of models will often be used: space filling and ball and stick. These models show structures at the atomic level.

1. *Space-Filling Models.* The space-filling models are the most realistic. The size and position of an atom in a space-filling model are determined by its bonding properties and van der Waals radius, or contact distance. A van der Waals radius describes how closely two atoms can approach each other when they are not linked by a covalent bond. The colors of the model are set by convention.

Carbon, black   Hydrogen, white   Nitrogen, blue
Oxygen, red   Sulfur, yellow   Phosphorus, purple

Space-filling models of several simple molecules are shown in Figure 1.24.

2. *Ball-and-Stick Models.* Ball-and-stick models are not as realistic as space-filling models, because the atoms are depicted as spheres of radii smaller than their van der Waals radii. However, the bonding arrangement is easier to see because the bonds are explicitly represented as sticks. In an illustration, the taper of a stick, representing parallax, tells which of a pair of bonded atoms is closer to the reader. A ball-and-stick model reveals a complex structure more clearly than a space-filling model does. Ball-and-stick models of several simple molecules are shown in Figure 1.24.

Molecular models for depicting large molecules will be discussed in the appendix to Chapter 2.

Functional Groups

Biochemistry depends on many concepts and principles from organic chemistry. The properties of individual atoms are substantially affected by the other atoms to which the individual atom is bonded. Specific combinations of a small number of atoms—which often have some properties that are relatively independent of other aspects of the molecules in which they are found—are referred to as *functional groups*. There are many functional groups in organic chemistry, but a relatively small number of these are central to biochemistry.

The first functional group we will discuss is the hydroxyl group, $-OH$. This functional group comprises an oxygen atom bonded to a hydrogen atom. The oxygen atom has the ability to bond to one additional atom. The hydroxyl group can both donate and accept hydrogen bonds. The hydroxyl group is both a very weak acid and a very weak base.

The remaining functional groups most important for biochemistry are summarized below:

▶ $-NH_2$ This is an amino group, comprising a nitrogen atom bonded to two hydrogen atoms. The amino group can both donate and accept hydrogen bonds. The amino group is a relatively strong base, accepting a hydrogen ion to form an ammonium group, $-NH_3^+$.

▶ $>C=O$ This is a carbonyl group, comprising a carbon atom double bonded to an oxygen atom. The carbonyl group can act as a hydrogen bond acceptor but not a donor. The carbonyl group is found in many different classes of compounds, depending on the other groups that are bonded to the carbon atom. The most important carbonyl-containing functional group in biochemistry is the amide, in which the carbonyl carbon is bonded to one carbon and one nitrogen.

Reaction mechanisms and "arrow pushing"

Many aspects of biochemistry depend on chemical reactions in which covalent bonds are broken and formed. These reactions involve the flow of electrons out of bonds and into spaces between atoms between which new bonds are formed. It is often useful to depict such reactions with arrows showing this electron flow. The process of analyzing reactions in this way is sometimes referred to informally as "arrow pushing" or "electron pushing."

Consider the reaction between ammonia, NH_3, and methyl iodide H_3C-I. NH_3 has a lone pair of electrons on the nitrogen atom that can participate in bond formation. The iodine atom in H_3C-I can accept electrons to form the relatively stable iodide ion, I^-. The reaction and its mechanism are shown below:

$$H_3\overset{..}{N}:+H_3C-I \longrightarrow H_3N^+-CH_3+I^-$$

The first arrow depicts the flow of the electron pair from the nitrogen to the space between the nitrogen and the carbon to form the new nitrogen-carbon bond. The second arrow shows the flow of electrons from the carbon-iodine bond to the iodine to form the iodide ion. The initial product of this reaction is methylammonium ion, methylamine in its protonated form. Another important term is *nucleophile,* which refers to the molecule or group that donates an electron pair during this sort of reaction. In the present reaction, ammonia is the nucleophile.

We will encounter many examples of reaction mechanisms, particularly when we discuss the actions of enzymes—the essential catalysts that facilitate so much of biochemistry.

KEY TERMS

biological macromolecule (p. 2)

metabolite (p. 2)

deoxyribonucleic acid (DNA) (p. 2)

protein (p. 2)

Eukarya (p. 3)

Bacteria (p. 3)

Archaea (p. 3)

eukaryote (p. 3)

prokaryote (p. 3)

double helix (p. 5)

covalent bond (p. 5)

resonance structure (p. 7)

ionic interaction (p. 8)

hydrogen bond (p. 8)

van der Waals interaction (p. 8)

hydrophobic effect (p. 9)

hydrophobic interaction (p. 8)

entropy (p. 11)

enthalpy (p. 12)

free energy (Gibbs free energy)
  (p. 12)

pH (p. 14)

pK_a value (p. 15)

buffer (p. 16)

predisposition (p. 19)

microbiome (p. 19)

amino acid (p. 21)

genetic code (p. 22)

PROBLEMS

1. *Donors and acceptors.* Identify the hydrogen-bond donors and acceptors in each of the four bases on page 5. ✓❸

2. *Resonance structures.* The structure of an amino acid, tyrosine, is shown here. Draw an alternative resonance structure.

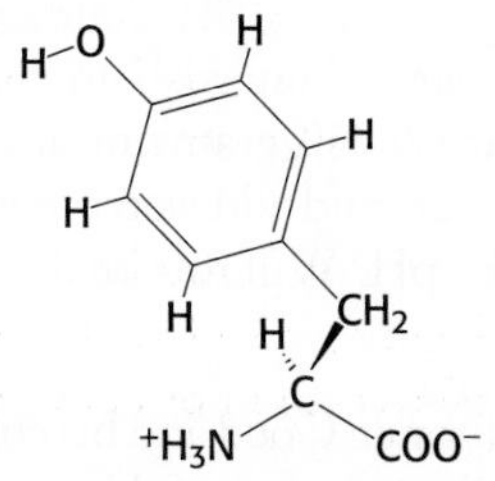

3. *It takes all types.* What types of noncovalent bonds hold together the following solids? ✓❸

(a) Table salt (NaCl), which contains Na^+ and Cl^- ions.

(b) Graphite (C), which consists of sheets of covalently bonded carbon atoms.

4. *Don't break the law.* Given the following values for the changes in enthalpy (ΔH) and entropy (ΔS), which of the following processes can take place at 298 K without violating the Second Law of Thermodynamics? ✓❻

(a) $\Delta H = -84 \text{ kJ mol}^{-1} (-20 \text{ kcal mol}^{-1})$,
  $\Delta S = +125 \text{ J mol}^{-1} \text{ K}^{-1} (+30 \text{ cal mol}^{-1} \text{ K}^{-1})$

(b) $\Delta H = -84 \text{ kJ mol}^{-1} (-20 \text{ kcal mol}^{-1})$,
  $\Delta S = -125 \text{ J mol}^{-1} \text{ K}^{-1} (-30 \text{ cal mol}^{-1} \text{ K}^{-1})$

(c) $\Delta H = +84 \text{ kJ mol}^{-1} (+20 \text{ kcal mol}^{-1})$,
  $\Delta S = +125 \text{ J mol}^{-1} \text{ K}^{-1} (+30 \text{ cal mol}^{-1} \text{ K}^{-1})$

(d) $\Delta H = +84 \text{ kJ mol}^{-1} (+20 \text{ kcal mol}^{-1})$,
  $\Delta S = -125 \text{ J mol}^{-1} \text{ K}^{-1} (-30 \text{ cal mol}^{-1} \text{ K}^{-1})$

5. *Double-helix-formation entropy.* For double-helix formation, ΔG can be measured to be $-54 \text{ kJ mol}^{-1} (-13 \text{ kcal mol}^{-1})$ at pH 7.0 in 1 M NaCl at 25°C (298 K). The heat released indicates an enthalpy change of $-251 \text{ kJ mol}^{-1} (-60 \text{ kcal mol}^{-1})$.

For this process, calculate the entropy change for the system and the entropy change for the surroundings. ✓❻

6. *Find the pH.* What are the pH values for the following solutions? ✓❼

(a) 0.1 M HCl

(b) 0.1 M NaOH

(c) 0.05 M HCl

(d) 0.05 M NaOH

7. *A weak acid.* What is the pH of a 0.1 M solution of acetic acid ($pK_a = 4.75$)? ✓❼

(Hint: Let x be the concentration of H^+ ions released from acetic acid when it dissociates. The solutions to a quadratic equation of the form $ax^2 + bx + c = 0$ are $x = (-b \pm \sqrt{b^2 - 4ac}/2a.)$

8. *Substituent effects.* What is the pH of a 0.1 M solution of chloroacetic acid ($ClCH_2 COOH$, $pK_a = 2.86$)? ✓❼

9. *Water in water.* Given a density of 1 g/ml and a molecular weight of 18 g/mol, calculate the concentration of water in water.

10. *Basic fact.* What is the pH of a 0.1 M solution of ethylamine, given that the pK_a of ethylammonium ion ($CH_3 CH_2 NH_3^+$) is 10.70? ✓❼

11. *Comparison.* A solution is prepared by adding 0.01 M acetic acid and 0.01 M ethylamine to water and adjusting the pH to 7.4. What is the ratio of acetate to acetic acid? What is the ratio of ethylamine to ethylammonium ion? ✓❼

12. *Concentrate.* Acetic acid is added to water until the pH value reaches 4.0. What is the total concentration of the added acetic acid? ✓❼

13. *Dilution.* 100 mL of a solution of hydrochloric acid with pH 5.0 is diluted to 1 L. What is the pH of the diluted solution? ✓❽

14. *Buffer dilution.* 100 mL of a 0.1 mM buffer solution made from acetic acid and sodium acetate with pH 5.0 is diluted to 1 L. What is the pH of the diluted solution? ✓**8**

15. *Find the pK_a.* For an acid HA, the concentrations of HA and A^- are 0.075 and 0.025, respectively, at pH 6.0. What is the pK_a value for HA? ✓**7**

16. *pH indicator.* A dye that is an acid and that appears as different colors in its protonated and deprotonated forms can be used as a pH indicator. Suppose that you have a 0.001 M solution of a dye with a pK_a of 7.2. From the color, the concentration of the protonated form is found to be 0.0002 M. Assume that the remainder of the dye is in the deprotonated form. What is the pH of the solution? ✓**7**

17. *What's the ratio?* An acid with a pK_a of 8.0 is present in a solution with a pH of 6.0. What is the ratio of the protonated to the deprotonated form of the acid? ✓**7**

18. *More ratios.* Through the use of nuclear magnetic resonance spectroscopy, it is possible to determine the ratio between the protonated and deprotonated forms of buffers. (a) Suppose the ratio of $[A^-]$ to $[HA]$ is determined to be 0.1 for a buffer with $pK_a = 6.0$. What is the pH? (b) For a different buffer, suppose the ratio of $[A^-]$ to $[HA]$ is determined to be 0.1 and the pH is 7.0. In this case, what is the pK_a of the buffer? (c) For another buffer with $pK_a = 7.5$ at pH 8.0, what is the expected ratio of $[A^-]$ to $[HA]$? ✓**7**

19. *Phosphate buffer.* What is the ratio of the concentrations of $H_2PO_4^-$ and HPO_4^{2-} at (a) pH 7.0; (b) pH 7.5; (c) pH 8.0? ✓**8**

20. *Neutralization of phosphate.* Given that phosphoric acid (H_3PO_4) can give up three protons with different pK_a values, sketch a plot of pH as a function of added drops of sodium hydroxide solution, starting with a solution of phosphoric acid at pH 1.0. ✓**8**

21. *Sulfate buffer?* Your laboratory is out of materials to make phosphate buffer and you are considering using sulfate to make a buffer instead. The pK_a values for the two hydrogens in H_2SO_4 are -10 and 2. (a) Will this approach work for making a buffer effective near pH 7? (b) Around what pH might a sulfate-based buffer be useful? ✓**8**

22. *Buffer capacity.* Two solutions of sodium acetate are prepared, one with a concentration of 0.1 M and the other with a concentration of 0.01 M. Calculate the pH values when the following concentrations of HCl have been added to each of these solutions: 0.0025 M, 0.005 M, 0.01 M, and 0.05 M. ✓**8**

23. *Buffer preparation.* You wish to prepare a buffer consisting of acetic acid and sodium acetate with a total acetic acid plus acetate concentration of 250 mM and a pH of 5.0. What concentrations of acetic acid and sodium acetate should you use? Assuming you wish to make 2 liters of this buffer, how many moles of acetic acid and sodium acetate will you need?

How many grams of each will you need (molecular weights: acetic acid 60.05 g mol^{-1}, sodium acetate, 82.03 g mol^{-1})? ✓**8**

24. *An alternative approach.* When you go to prepare the buffer described in problem 23, you discover that your laboratory is out of sodium acetate, but you do have sodium hydroxide. How much (in moles and grams) acetic acid and sodium hydroxide do you need to make the buffer? ✓**8**

25. *Another alternative.* Your friend from another laboratory who was out of acetic acid tries to prepare the buffer in problem 23 by dissolving 41.02 g of sodium acetate in water, carefully adding 180.0 ml of 1 M HCl, and adding more water to reach a total volume of 2 liters. What is the total concentration of acetate plus acetic acid in the solution? Will this solution have pH 5.0? Will it be identical with the desired buffer? If not, how will it differ? ✓**8**

26. *Blood substitute.* As noted in this chapter, blood contains a total concentration of phosphate of approximately 1 mM and typically has a pH of 7.4. You wish to make 100 liters of phosphate buffer with a pH of 7.4 from NaH_2PO_4 (molecular weight, 119.98 g mol^{-1}) and Na_2HPO_4 (molecular weight, 141.96 g mol^{-1}). How much of each (in grams) do you need? ✓**8**

27. *A potential problem.* You wish to make a buffer with pH 7.0. You combine 0.060 grams of acetic acid and 14.59 grams of sodium acetate and add water to yield a total volume of 1 liter. What is the pH? Will this be the useful pH 7.0 buffer you seek? ✓**8**

28. *Good buffers.* Norman Good and his colleagues developed a series of compounds that are useful buffers near neutral pH. Three of these buffers are abbreviated MES ($pK_a = 6.15$), MOPS ($pK_a = 7.15$), and HEPPS (8.00). (a) Which of these buffers would be most effective near pH 7.0? (b) Near 6.0? ✓**8**

29. *More good buffers.* One liter of a buffer is prepared consisting of 100 mM HEPPS at pH 8.0. What is the concentration of the protonated form of HEPPS in this solution? ✓**8**

30. *DNA in water.* Would you expect the double helix in a short segment of DNA to be more stable in 100 sodium phosphate buffer at pH 7.0 or in pure water? Why? ✓**2**

31. *Charge!* Suppose two phosphate groups in DNA (each with a charge of -1) are separated by 12 Å. What is the energy of the ionic interaction between these two phosphates, assuming a dielectric constant of 80? Repeat the calculation, assuming a dielectric constant of 2. ✓**2**

32. *Oil and water?* What force is primarily responsible for the fact that oil and water do not mix easily? ✓**4**

33. *Vive la différence.* On average, estimate how many base differences there are between two human beings. ✓**9**

34. *Epigenomics.* The human body contains many distinct cell types yet almost all human cells contain the same genome with 21,000 genes. The distinct cell types are primarily due to differences in gene expression. Assume that one set of 1000 genes is expressed in all cell types and that the remaining 20,000 genes can be divided into sets of 1000 genes that are either all expressed or all not expressed in a given cell type. How many different cell types are possible if each cell type expresses 10 sets of these genes? Note that the number of combinations of n objects into m sets is given $n!/(m!(n-m)!)$ where $n! = 1 \times 2 \times \cdots \times (n-1) \times n$. ✓ⓖ

35. *Predispositions in populations.* Assume that 10% of the members of a population will get a particular disease over the course of their lifetime. Genomic studies reveal that 5% of the population have sequences in their genomes such that their probability of getting the disease over the course of their lifetimes is 50%. What is the average lifetime risk of this disease for the remaining 95% of the population without these sequences? ✓ⓖ

Think ▸ Pair ▸ Share Problem

36. Many studies that have revealed important biochemical processes in human beings, including those that underpin the development of new drugs, have been conducted in other organisms. These organisms include mammals such as mice and rats but also extend to single-celled organisms such as yeast. Why is this approach often productive?

Protein Composition and Structure

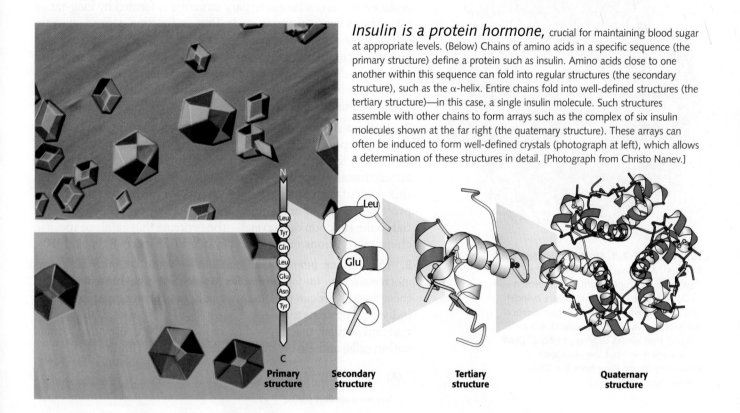

Insulin is a protein hormone, crucial for maintaining blood sugar at appropriate levels. (Below) Chains of amino acids in a specific sequence (the primary structure) define a protein such as insulin. Amino acids close to one another within this sequence can fold into regular structures (the secondary structure), such as the α-helix. Entire chains fold into well-defined structures (the tertiary structure)—in this case, a single insulin molecule. Such structures assemble with other chains to form arrays such as the complex of six insulin molecules shown at the far right (the quaternary structure). These arrays can often be induced to form well-defined crystals (photograph at left), which allows a determination of these structures in detail. [Photograph from Christo Nanev.]

Primary structure — **Secondary structure** — **Tertiary structure** — **Quaternary structure**

 LEARNING OBJECTIVES

By the end of this chapter, you should be able to:

1. Identify the 20 amino acids and their corresponding three-letter and one-letter abbreviations.

2. Distinguish between primary, secondary, tertiary, and quaternary structures.

3. Describe the properties of the principal types of secondary structure, including the α helix, the β sheet, and the reverse turn.

4. Explain how the hydrophobic effect serves as the primary driving force for folding of polypeptide chains into globular proteins.

OUTLINE

2.1 Proteins Are Built from a Repertoire of 20 Amino Acids

2.2 Primary Structure: Amino Acids Are Linked by Peptide Bonds to Form Polypeptide Chains

2.3 Secondary Structure: Polypeptide Chains Can Fold into Regular Structures Such As the Alpha Helix, the Beta Sheet, and Turns and Loops

2.4 Tertiary Structure: Proteins Can Fold into Globular or Fibrous Structures

2.5 Quaternary Structure: Polypeptide Chains Can Assemble into Multisubunit Structures

2.6 The Amino Acid Sequence of a Protein Determines Its Three-Dimensional Structure

Proteins are the most versatile macromolecules in living systems and serve crucial functions in essentially all biological processes. They function as catalysts, transport and store other molecules such as oxygen, provide mechanical support and immune protection, generate movement, transmit nerve impulses, and control growth and differentiation. Indeed, much of this book will focus on understanding what proteins do and how they perform these functions.

Several key properties enable proteins to participate in a wide range of functions.

1. *Proteins are linear polymers built of monomer units called amino acids,* which are linked end to end. The sequence of linked amino acids is called the primary structure. Remarkably, proteins spontaneously fold up into three-dimensional structures that are determined by the sequence of amino

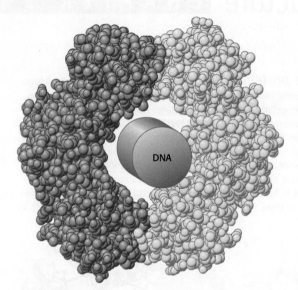

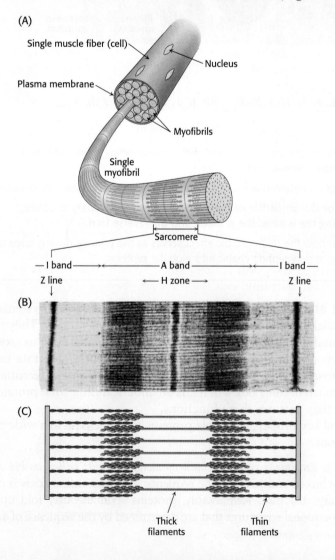 FIGURE 2.1 **Structure dictates function.** A protein component of the DNA replication machinery surrounds a section of DNA double helix depicted as a cylinder. The protein, which consists of two identical subunits (shown in red and yellow), acts as a clamp that allows large segments of DNA to be copied without the replication machinery dissociating from the DNA. [Drawn from 2POL.pdb.]

acids in the protein polymer. Three-dimensional structure formed by hydrogen bonds between amino acids near one another is called secondary structure, whereas tertiary structure is formed by long-range interactions between amino acids. Protein function depends directly on this three-dimensional structure (Figure 2.1). Thus, *proteins are the embodiment of the transition from the one-dimensional world of sequences to the three-dimensional world of molecules capable of diverse activities.* Many proteins also display quaternary structure, in which the functional protein is composed of several distinct polypeptide chains.

2. *Proteins contain a wide range of functional groups.* These functional groups include alcohols, thiols, thioethers, carboxylic acids, carboxamides, and a variety of basic groups. Most of these groups are chemically reactive. When combined in various sequences, this array of functional groups accounts for the broad spectrum of protein function. For instance, their reactive properties are essential to the function of *enzymes*—the proteins that catalyze specific chemical reactions in biological systems (Chapters 8 through 10).

3. *Proteins can interact with one another and with other biological macromolecules to form complex assemblies.* The proteins within these assemblies can act synergistically to generate capabilities that individual proteins may lack. Examples of these assemblies include macromolecular machines that replicate DNA, transmit signals within cells, and enable muscle cells to contract (Figure 2.2).

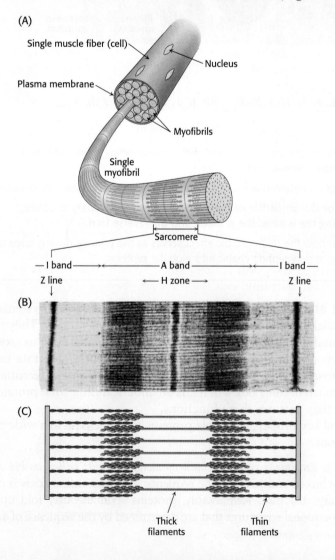

FIGURE 2.2 **A complex protein assembly.** (A) A single muscle cell contains multiple myofibrils, each of which is comprised of numerous repeats of a complex protein assembly known as the sarcomere. (B) The banding pattern of a sarcomere, evident by electron microscopy, is caused by (C) the interdigitation of filaments made up of many individual proteins. [(B) Courtesy of Dr. Hugh Huxley.]

4. *Some proteins are quite rigid, whereas others display considerable flexibility.* Rigid units can function as structural elements in the cytoskeleton (the internal scaffolding within cells) or in connective tissue. Proteins with some flexibility may act as hinges, springs, or levers. In addition, conformational changes within proteins enable the regulated assembly of larger protein complexes as well as the transmission of information within and between cells (Figure 2.3).

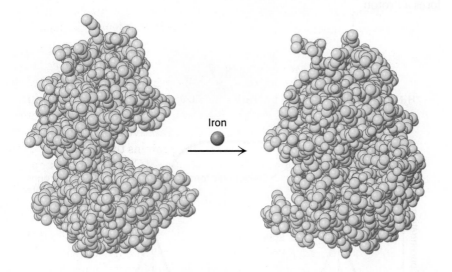

FIGURE 2.3 Flexibility and function. On binding iron, the protein lactoferrin undergoes a substantial change in conformation that allows other molecules to distinguish between the iron-free and the iron-bound forms. [Drawn from 1LFH.pdb and 1LFG.pdb.]

2.1 Proteins Are Built from a Repertoire of 20 Amino Acids

Amino acids are the building blocks of proteins. An *α-amino acid* consists of a central carbon atom, called the α *carbon,* linked to an amino group, a carboxylic acid group, a hydrogen atom, and a distinctive R group. The R group is often referred to as the *side chain.* With four different groups connected to the tetrahedral α-carbon atom, α-amino acids are *chiral:* they may exist in one or the other of two mirror-image forms, called the L isomer and the D isomer (Figure 2.4).

Only L *amino acids are constituents of proteins.* For almost all amino acids, the L isomer has *S* (rather than *R*) absolute configuration (Figure 2.5). What is the basis for the preference for L amino acids? The answer has been lost to evolutionary history. It is possible that the preference for L over D amino acids was a consequence of a chance selection. However, there is evidence that L amino acids are slightly more soluble than a racemic mixture of D and L amino acids, which tend to form crystals. This small solubility difference could have been amplified over time so that the L isomer became dominant in solution.

Amino acids in solution at neutral pH exist predominantly as *dipolar ions* (also called *zwitterions*). In the dipolar form, the amino group is protonated

Notation for distinguishing stereoisomers
The four different substituents of an asymmetric carbon atom are assigned a priority according to atomic number. The lowest-priority substituent, often hydrogen, is pointed away from the viewer. The configuration about the carbon atom is called *S* (from the Latin *sinister,* "left") if the progression from the highest to the lowest priority is counterclockwise. The configuration is called *R* (from the Latin *rectus,* "right") if the progression is clockwise.

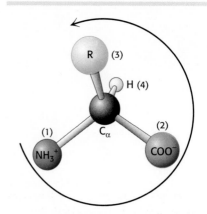

FIGURE 2.5 Only L amino acids are found in proteins. Almost all L amino acids have an *S* absolute configuration. The counterclockwise direction of the arrow from highest- to lowest-priority substituents indicates that the chiral center is of the *S* configuration.

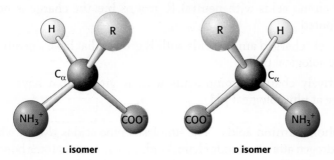

FIGURE 2.4 The L and D isomers of amino acids. The letter R refers to the side chain. The L and D isomers are mirror images of each other.

$(-NH_3^+)$ and the carboxyl group is deprotonated $(-COO^-)$. The ionization state of an amino acid varies with pH (Figure 2.6). In acid solution (e.g., pH 1), the amino group is protonated $(-NH_3^+)$ and the carboxyl group is not dissociated $(-COOH)$. As the pH is raised, the carboxylic acid is the first group to give up a proton, inasmuch as its pK_a is near 2. The dipolar form persists until the pH approaches 9.5, when the protonated amino group loses a proton.

FIGURE 2.6 Ionization state as a function of pH. The ionization state of amino acids is altered by a change in pH. At low pH, near the pK_a for the carboxylic acid, pK_1, the —COOH proton is lost from the fully protonated form. As the pH approaches physiological levels, the zwitterionic form predominates. At high pH, near the pK_a for the amino group, pK_2, one of the —NH_3^+ protons is lost to form the fully deprotonated species.

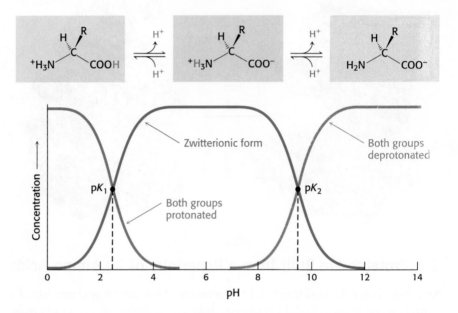

Twenty kinds of side chains varying in *size, shape, charge, hydrogen-bonding capacity, hydrophobic character,* and *chemical reactivity* are commonly found in proteins. Indeed, all proteins in all species—bacterial, archaeal, and eukaryotic—are constructed from the same set of 20 amino acids with only a few exceptions. This fundamental alphabet for the construction of proteins is several billion years old. The remarkable range of functions mediated by proteins results from the diversity and versatility of these 20 building blocks. Understanding how this alphabet is used to create the intricate three-dimensional structures that enable proteins to carry out so many biological processes is an exciting area of biochemistry and one that we will return to in Section 2.6.

Although there are many ways to classify amino acids, we will sort these molecules into four groups, on the basis of the general chemical characteristics of their R groups:

1. Hydrophobic amino acids with nonpolar R groups

2. Polar amino acids with neutral R groups but the charge is not evenly distributed

3. Positively charged amino acids with R groups that have a positive charge at physiological pH

4. Negatively charged amino acids with R groups that have a negative charge at physiological pH

Hydrophobic amino acids. The simplest amino acid is *glycine,* which has a single hydrogen atom as its side chain. With two hydrogen atoms bonded to the α-carbon atom, glycine is unique in being *achiral. Alanine,* the next simplest amino acid, has a methyl group (—CH_3) as its side chain (Figure 2.7).

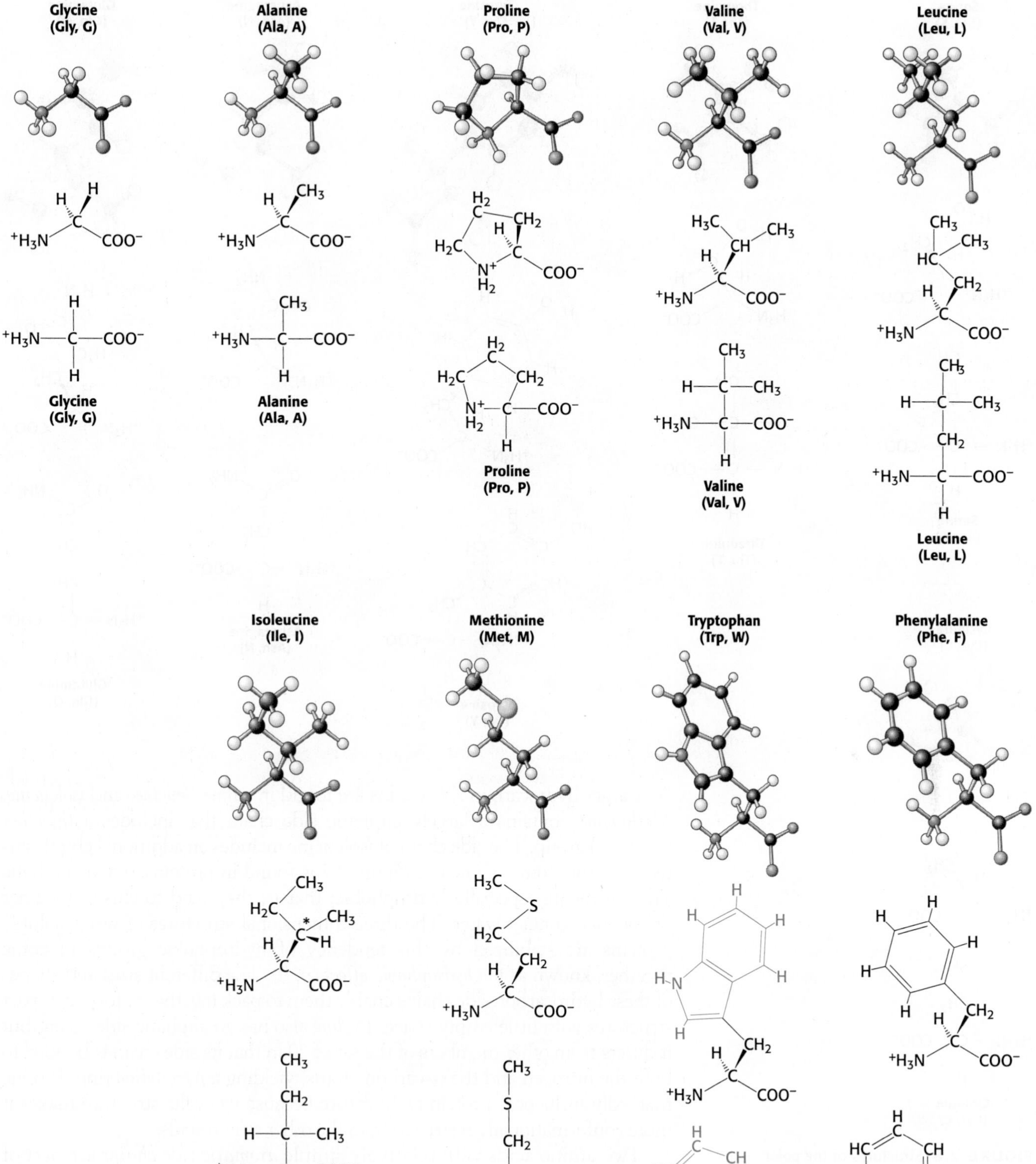

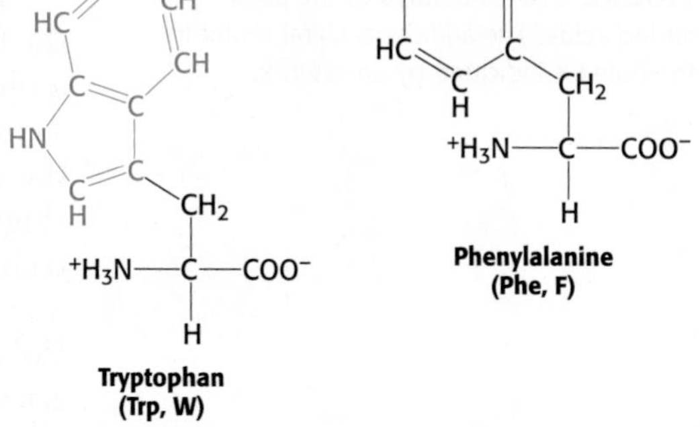

**Glycine
(Gly, G)**

**Alanine
(Ala, A)**

**Proline
(Pro, P)**

**Valine
(Val, V)**

**Leucine
(Leu, L)**

$$\begin{array}{c} H \quad\quad H \\ {}^+H_3N-\overset{|}{\underset{|}{C}}-COO^- \end{array}$$

Glycine
(Gly, G)

Alanine
(Ala, A)

Proline
(Pro, P)

Valine
(Val, V)

Leucine
(Leu, L)

**Isoleucine
(Ile, I)**

**Methionine
(Met, M)**

**Tryptophan
(Trp, W)**

**Phenylalanine
(Phe, F)**

Isoleucine
(Ile, I)

Methionine
(Met, M)

Tryptophan
(Trp, W)

Phenylalanine
(Phe, F)

FIGURE 2.7 Structures of hydrophobic amino acids. For each amino acid, a ball-and-stick model (top) shows the arrangement of atoms and bonds in space. A stereochemically realistic formula (middle) shows the geometric arrangement of bonds around atoms, and a Fischer projection (bottom) shows all bonds as being perpendicular for a simplified representation (see the Appendix to Chapter 1). The additional chiral center in isoleucine is indicated by an asterisk. The indole group in the tryptophan side chain is shown in red.

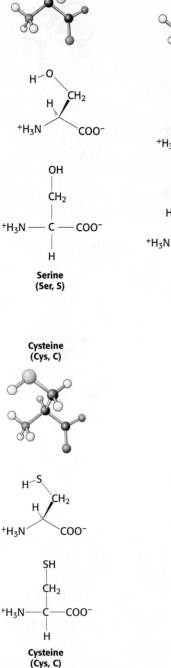

Serine
(Ser, S)

Threonine
(Thr, T)

Tyrosine
(Tyr, Y)

Asparagine
(Asn, N)

Glutamine
(Gln, Q)

Serine
(Ser, S)

Threonine
(Thr, T)

Tyrosine
(Tyr, Y)

Asparagine
(Asn, N)

Glutamine
(Gln, Q)

Cysteine
(Cys, C)

Cysteine
(Cys, C)

FIGURE 2.8 Structures of the polar amino acids. The additional chiral center in threonine is indicated by an asterisk.

Larger hydrocarbon side chains are found in *valine, leucine,* and *isoleucine. Methionine* contains a largely aliphatic side chain that includes a *thioether* ($-S-$) group. The side chain of isoleucine includes an additional chiral center; only the isomer shown in Figure 2.7 is found in proteins. These aliphatic side chains are especially hydrophobic; that is, they tend to cluster together rather than contact water. The three-dimensional structures of water-soluble proteins are stabilized by this tendency of hydrophobic groups to come together, known as *the hydrophobic effect* (p. 9). The different sizes and shapes of these hydrocarbon side chains enable them to pack together to form compact structures with little empty space. *Proline* also has an aliphatic side chain, but it differs from other members of the set of 20 in that its side chain is bonded to both the nitrogen and the α-carbon atoms, yielding a *pyrrolidine* ring. Proline markedly influences protein architecture because its cyclic structure makes it more conformationally restricted than the other amino acids.

Two amino acids with relatively simple *aromatic side chains* are part of the fundamental repertoire. *Phenylalanine,* as its name indicates, contains a phenyl ring attached in place of one of the hydrogen atoms of alanine. *Tryptophan* has an *indole* group joined to a methylene ($-CH_2-$) group; the indole group comprises two fused rings containing an NH group (Figure 2.7, shown in red). Phenylalanine is purely hydrophobic, whereas tryptophan is less so because of its side chain NH group.

Polar amino acids. Six amino acids are polar but uncharged. Three amino acids, *serine, threonine,* and *tyrosine,* contain *hydroxyl groups* ($-OH$) attached to a hydrophobic side chain (Figure 2.8). Serine can be thought of as a version of alanine with a hydroxyl group attached. Threonine resembles valine with a hydroxyl group in place of one of valine's methyl groups. Tyrosine is a version of phenylalanine with the hydroxyl group replacing a hydrogen atom on the aromatic ring. The hydroxyl group makes these amino acids much more *hydrophilic* (water loving) and *reactive* than their hydrophobic analogs. Threonine, like isoleucine, contains an additional asymmetric center; again, only one isomer is present in proteins.

In addition, the set includes *asparagine* and *glutamine*, two amino acids that contain a terminal *carboxamide*. The side chain of glutamine is one methylene group longer than that of asparagine.

Cysteine is structurally similar to serine but contains a *sulfhydryl*, or *thiol* ($-SH$), group in place of the hydroxyl ($-OH$) group. The sulfhydryl group is much more reactive. Pairs of sulfhydryl groups may come together to form disulfide bonds, which are particularly important in stabilizing some proteins, as will be discussed shortly.

Positively charged amino acids. We turn now to amino acids with complete positive charges that render them highly hydrophilic (Figure 2.9). *Lysine* and *arginine* have long side chains that terminate with groups that are *positively charged* at neutral pH. Lysine is capped by a primary amino group and arginine by a guanidinium group

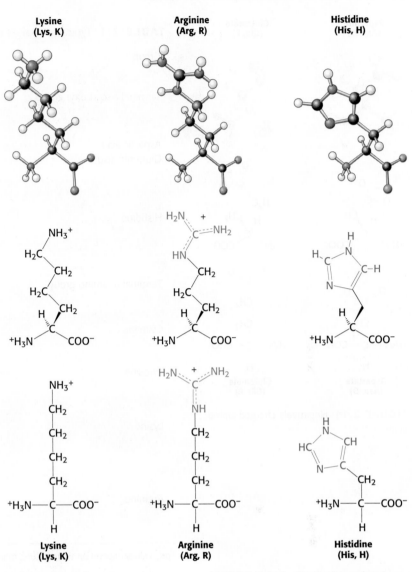

FIGURE 2.9 Positively charged amino acids lysine, arginine, and histidine. The guanidinium group in the arginine side chain is shown in red, while the imidazole group of histidine is shown in blue.

(Figure 2.9, shown in red). *Histidine* contains an imidazole group, an aromatic ring that also can be positively charged (Figure 2.9, shown in blue).

With a pK_a value near 6, the imidazole group can be uncharged or positively charged near neutral pH, depending on its local environment (Figure 2.10). Histidine is often found in the active sites of enzymes, where the imidazole ring can bind and release protons in the course of enzymatic reactions.

Negatively charged amino acids. This set of amino acids contains two with *acidic side chains: aspartic acid* and *glutamic acid* (Figure 2.11). These amino acids are charged derivatives of asparagine and glutamine (Figure 2.8), with a carboxylic acid in place of a carboxamide. Aspartic acid and glutamic acid are often called *aspartate* and *glutamate* to emphasize that, at physiological pH, their side chains usually lack a proton that is present in the acid form and hence are negatively charged. Nonetheless, these side chains can accept protons in some proteins, often with functionally important consequences.

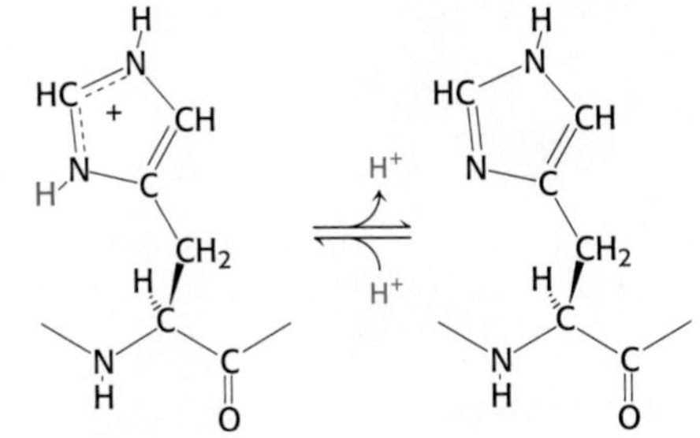

Seven of the 20 amino acids have readily ionizable side chains. These 7 amino acids are able to donate or accept protons to facilitate reactions as well as to form ionic bonds. Table 2.1 gives equilibria and typical pK_a values for ionization of the side chains of tyrosine, cysteine, arginine, lysine, histidine, and aspartic and glutamic acids in proteins. Two other groups in proteins— the terminal α-amino group and the terminal α-carboxyl group—can be ionized, and typical pK_a values for these groups also are included in Table 2.1.

FIGURE 2.10 Histidine ionization. Histidine can bind or release protons near physiological pH.

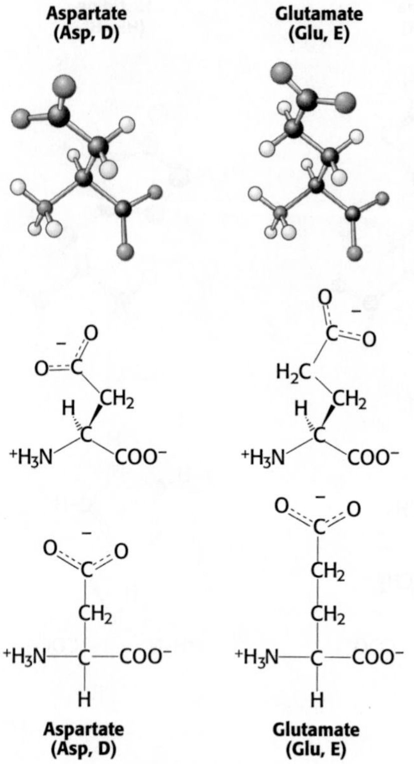

**Aspartate
(Asp, D)**

**Glutamate
(Glu, E)**

**Aspartate
(Asp, D)**

**Glutamate
(Glu, E)**

FIGURE 2.11 Negatively charged amino acids.

TABLE 2.1 Typical pK_a values of ionizable groups in proteins

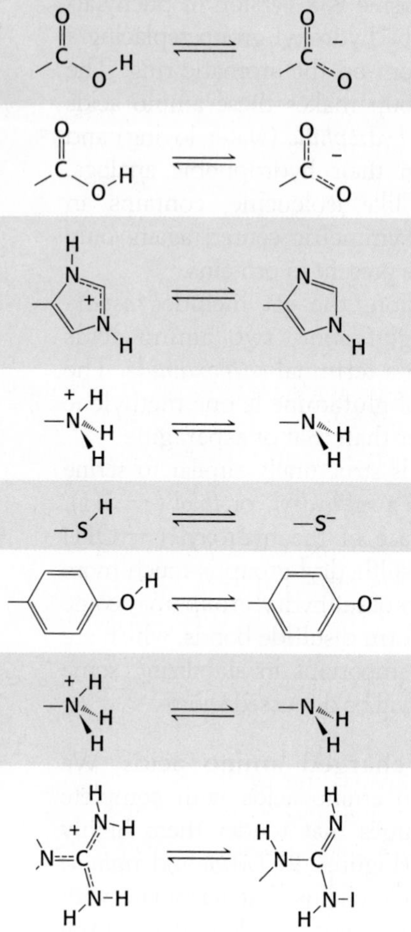

Group	Acid $\rightleftharpoons$ Base	Typical pK_a*
Terminal α-carboxyl group		3.1
Aspartic acid Glutamic acid		4.1
Histidine		6.0
Terminal α-amino group		8.0
Cysteine		8.3
Tyrosine		10.0
Lysine		10.4
Arginine		12.5

*pK_a values depend on temperature, ionic strength, and the microenvironment of the ionizable group

Amino acids are often designated by either a three-letter abbreviation or a one-letter symbol (Table 2.2). The abbreviations for amino acids are the first three letters of their names, except for asparagine (Asn), glutamine (Gln), isoleucine (Ile), and tryptophan (Trp). The symbols

TABLE 2.2 Abbreviations for amino acids

Amino acid	Three-letter abbreviation	One-letter abbreviation	Amino acid	Three-letter abbreviation	One-letter abbreviation
Alanine	Ala	A	Methionine	Met	M
Arginine	Arg	R	Phenylalanine	Phe	F
Asparagine	Asn	N	Proline	Pro	P
Aspartic acid	Asp	D	Serine	Ser	S
Cysteine	Cys	C	Threonine	Thr	T
Glutamine	Gln	Q	Tryptophan	Trp	W
Glutamic acid	Glu	E	Tyrosine	Tyr	Y
Glycine	Gly	G	Valine	Val	V
Histidine	His	H	Asparagine or aspartic acid	Asx	B
Isoleucine	Ile	I			
Leucine	Leu	L	Glutamine or glutamic acid	Glx	Z
Lysine	Lys	K			

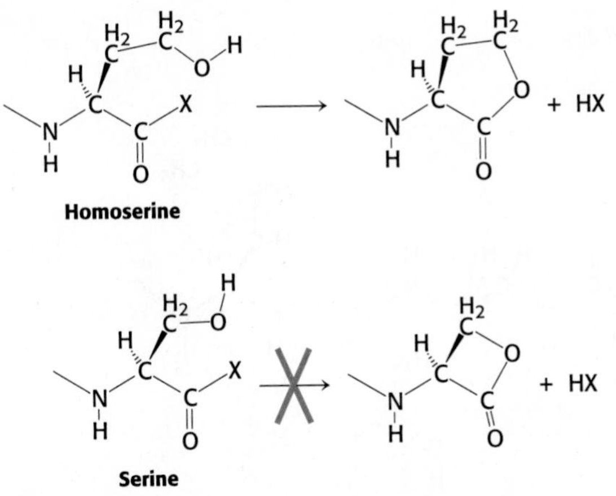

FIGURE 2.12 Undesirable reactivity in amino acids. Some amino acids are unsuitable for proteins because of undesirable cyclization. Homoserine can cyclize to form a stable, five-membered ring, potentially resulting in peptide-bond cleavage. The cyclization of serine would form a strained, four-membered ring and is thus disfavored. X can be an amino group from a neighboring amino acid or another potential leaving group.

for many amino acids are the first letters of their names (e.g., G for glycine and L for leucine); the other symbols have been agreed on by convention. These abbreviations and symbols are an integral part of the vocabulary of biochemists.

How did this particular set of amino acids become the building blocks of proteins? First, as a set, they are diverse: their structural and chemical properties span a wide range, endowing proteins with the versatility to assume many functional roles. Second, many of these amino acids were probably available from prebiotic reactions; that is, from reactions that took place before the origin of life. Finally, other possible amino acids may have simply been too reactive. For example, amino acids such as homoserine and homocysteine tend to form five-membered cyclic forms that limit their use in proteins; the alternative amino acids that are found in proteins—serine and cysteine—do not readily cyclize, because the rings in their cyclic forms are too small (Figure 2.12).

2.2 Primary Structure: Amino Acids Are Linked by Peptide Bonds to Form Polypeptide Chains

Proteins are *linear polymers* formed by linking the α-carboxyl group of one amino acid to the α-amino group of another amino acid. This type of linkage is called a *peptide bond* or an *amide bond*. The formation of a dipeptide from two amino acids is accompanied by the loss of a water molecule (Figure 2.13). Under most conditions, the equilibrium of this reaction lies on the side of hydrolysis rather than synthesis. Hence, the biosynthesis of peptide bonds requires an input of free energy. Nonetheless, peptide bonds are quite *stable kinetically* because the rate of hydrolysis is extremely slow; the lifetime of a peptide bond in aqueous solution in the absence of a catalyst approaches 1000 years.

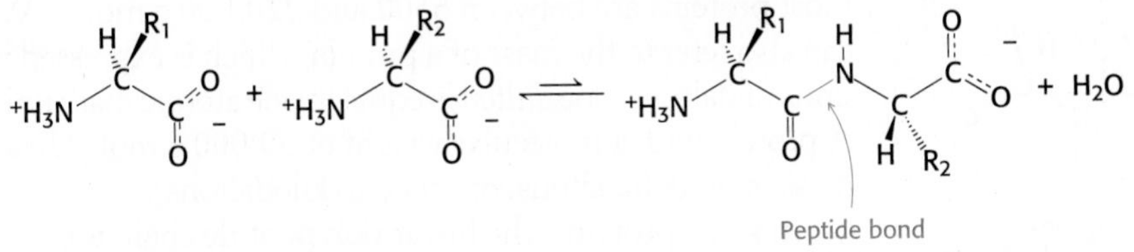

FIGURE 2.13 Peptide-bond formation. The linking of two amino acids is accompanied by the loss of a molecule of water.

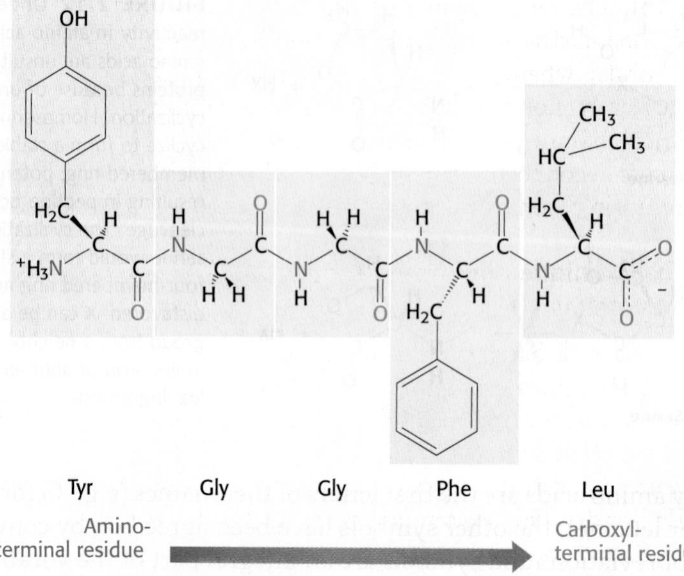

Tyr   Gly   Gly   Phe   Leu

Amino-terminal residue   Carboxyl-terminal residue

FIGURE 2.14 Amino acid sequences have direction. This illustration of the pentapeptide Tyr-Gly-Gly-Phe-Leu (YGGFL) shows the sequence from the amino terminus to the carboxyl terminus. This pentapeptide, Leu-enkephalin, is an opioid peptide that modulates the perception of pain. The reverse pentapeptide, Leu-Phe-Gly-Gly-Tyr (LFGGY), is a different molecule and has no such effects.

A series of amino acids joined by peptide bonds forms a *polypeptide chain*, and each amino acid unit in a polypeptide is called a *residue*. *A polypeptide chain has directionality because its ends are different*: an α-amino group is present at one end and an α-carboxyl group at the other. *The amino end is taken to be the beginning of a polypeptide chain;* by convention, the sequence of amino acids in a polypeptide chain is written starting with the amino-terminal residue. Thus, in the polypeptide Tyr-Gly-Gly-Phe-Leu (YGGFL), tyrosine is the amino-terminal (N-terminal) residue and leucine is the carboxyl-terminal (C-terminal) residue (Figure 2.14). Leu-Phe-Gly-Gly-Tyr (LFGGY) is a different polypeptide, with different chemical properties.

A polypeptide chain consists of a regularly repeating part, called the *main chain* or *backbone*, and a variable part, comprising the distinctive *side chains* (Figure 2.15). The polypeptide backbone is rich in hydrogen-bonding potential. Each residue contains a carbonyl group (C=O), which is a good hydrogen-bond acceptor, and, with the exception of proline, an NH group, which is a good hydrogen-bond donor. These groups interact with each other and with functional groups from side chains to stabilize particular structures (Section 2.3).

Most natural polypeptide chains contain between 50 and 2000 amino acid residues and are commonly referred to as *proteins*. The largest single polypeptide known is the muscle protein *titin*, which consists of more than 27,000 amino acids. Polypeptide chains made of small numbers of amino acids are called *oligopeptides* or simply *peptides*. The mean molecular weight of an amino acid residue is about 110 g mol^{-1}, and so the molecular weights of most proteins are between 5500 and 220,000 g mol^{-1}. We can also refer to the mass of a protein, which is expressed in units of daltons; one *dalton* is equal to one atomic mass unit. A protein with a molecular weight of 50,000 g mol^{-1} has a mass of 50,000 daltons, or 50 kDa (kilodaltons).

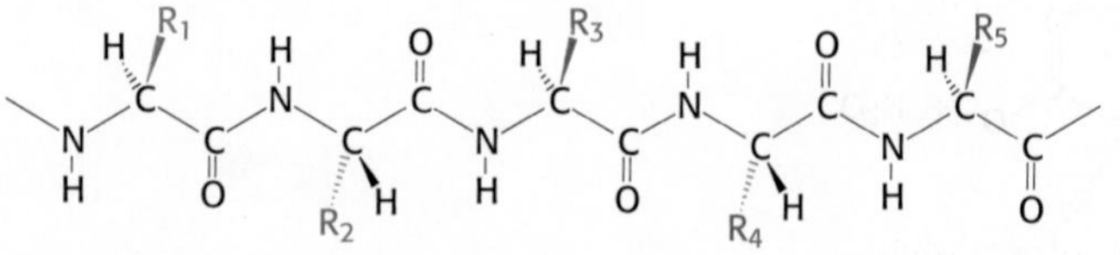

In some proteins, the linear polypeptide chain is cross-linked. The most common cross-links are *disulfide bonds*, formed by the oxidation of a pair of cysteine residues

Dalton

A unit of mass very nearly equal to that of a hydrogen atom. Named after John Dalton (1766–1844), who developed the atomic theory of matter.

Kilodalton (kDa)

A unit of mass equal to 1000 daltons.

FIGURE 2.15 Components of a polypeptide chain. A polypeptide chain consists of a constant backbone (shown in black) and variable side chains (shown in green).

(Figure 2.16). The resulting unit of two linked cysteines is called *cystine*. Extracellular proteins often have several disulfide bonds, whereas intracellular proteins usually lack them. Rarely, nondisulfide cross-links derived from other side chains are present in proteins. For example, collagen fibers in connective tissue are strengthened in this way, as are fibrin blood clots (Section 10.4).

Cysteine
Oxidation
Reduction
Cysteine
Cystine
+ 2 H⁺ + 2 e⁻

FIGURE 2.16 **Cross-links.** The formation of a disulfide bond from two cysteine residues is an oxidation reaction.

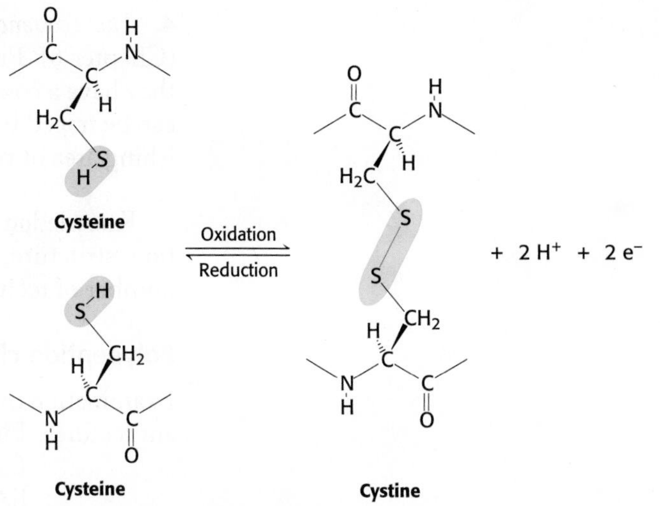

Proteins have unique amino acid sequences specified by genes

In 1953, Frederick Sanger determined the amino acid sequence of insulin, a protein hormone (Figure 2.17). *This work is a landmark in biochemistry because it showed for the first time that a protein has a precisely defined amino acid sequence* consisting only of L amino acids linked by peptide bonds. This accomplishment inspired other scientists to carry out sequence studies of a variety of proteins. Currently, the complete amino acid sequences of millions of proteins are known. *The striking fact is that each protein has a unique, precisely defined amino acid sequence.* The amino acid sequence of a protein is referred to as its *primary structure*.

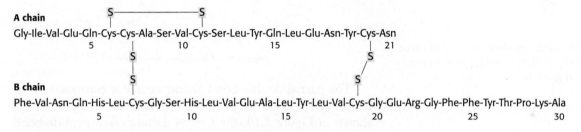

A chain
Gly-Ile-Val-Glu-Gln-Cys-Cys-Ala-Ser-Val-Cys-Ser-Leu-Tyr-Gln-Leu-Glu-Asn-Tyr-Cys-Asn
5 10 15 21
B chain
Phe-Val-Asn-Gln-His-Leu-Cys-Gly-Ser-His-Leu-Val-Glu-Ala-Leu-Tyr-Leu-Val-Cys-Gly-Glu-Arg-Gly-Phe-Phe-Tyr-Thr-Pro-Lys-Ala
5 10 15 20 25 30

FIGURE 2.17 Amino acid sequence of bovine insulin.

A series of incisive studies in the late 1950s and early 1960s revealed that the amino acid sequences of proteins are determined by the nucleotide sequences of genes. The sequence of nucleotides in DNA specifies a complementary sequence of nucleotides in RNA, which in turn specifies the amino acid sequence of a protein. In particular, each of the 20 amino acids of the repertoire is encoded by one or more specific sequences of three nucleotides (Section 4.6).

Knowing amino acid sequences is important for several reasons:

1. *Knowledge of the sequence of a protein is usually essential to elucidating its function (e.g., the catalytic mechanism of an enzyme).* In fact, proteins with novel properties can be generated by varying the sequence of known proteins.

2. *Amino acid sequences determine the three-dimensional structures of proteins.* The amino acid sequence is the link between the genetic message in DNA and the three-dimensional structure that performs a protein's biological function. Analyses of the relationships between amino acid sequences and three-dimensional structures of proteins are uncovering the rules that govern the folding of polypeptide chains.

3. *Alterations in amino acid sequence can lead to abnormal protein function and disease.* Severe and sometimes fatal diseases, such as sickle-cell anemia (Chapter 7) and cystic fibrosis, can result from a change in a single amino acid within a protein.

4. *The sequence of a protein reveals much about its evolutionary history* (Chapter 6). Proteins resemble one another in amino acid sequence only if they have a common ancestor. Consequently, molecular events in evolution can be traced from amino acid sequences; molecular paleontology is a flourishing area of research.

Knowledge of a protein's sequence is critical to understanding its function, structure, and evolutionary history. As we will discuss in Chapter 3, a number of techniques are available to determine this sequence.

Polypeptide chains are flexible yet conformationally restricted

Examination of the geometry of the protein backbone reveals several important features. First, *the peptide bond is essentially planar* (Figure 2.18). Thus, for a pair of amino acids linked by a peptide bond, six atoms lie in the same plane: the α-carbon atom and CO group of the first amino acid and the NH group and α-carbon atom of the second amino acid. The nature of the chemical bonding within a peptide accounts for the bond's planarity. The bond resonates between a single bond and a double bond. Because of this *partial double-bond character*, rotation about this bond is prevented, and conformation of the peptide backbone is constrained.

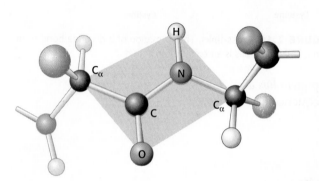

FIGURE 2.18 Peptide bonds are planar. In a pair of linked amino acids, six atoms (C_α, C, O, N, H, and C_α) lie in a plane. Side chains are shown as green balls.

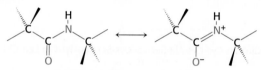

Peptide-bond resonance structures

The partial double-bond character is also expressed in the length of the bond between the CO and the NH groups. As shown in Figure 2.19, the C—N distance in a peptide bond is typically 1.32 Å, which is between the values expected for a C—N single bond (1.49 Å) and a C=N double bond (1.27 Å). Finally, the peptide bond is uncharged, allowing polymers of amino acids linked by peptide bonds to form tightly packed globular structures.

Two configurations are possible for a planar peptide bond. In the *trans* configuration, the two α-carbon atoms are on opposite sides of the peptide bond. In the *cis* configuration, these groups are on the same side of the peptide bond. *Almost all peptide bonds in proteins are trans.* This preference for trans over cis can be explained by the fact that steric clashes between groups attached to the α-carbon atoms hinder the formation of the cis configuration but do not arise in the trans configuration (Figure 2.20).

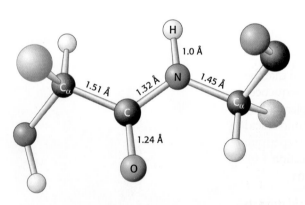

FIGURE 2.19 Typical bond lengths within a peptide unit. The peptide unit is shown in the trans configuration.

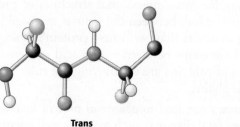

Trans

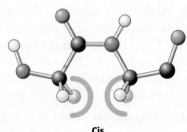

Cis

FIGURE 2.20 Trans and cis peptide bonds. The trans form is strongly favored because of steric clashes, indicated by the orange semicircles, that arise in the cis form.

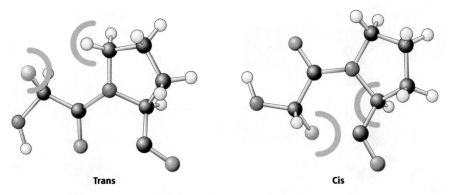

FIGURE 2.21 Trans and cis X–Pro bonds. The energies of these forms are similar to one another because steric clashes, indicated by the orange semicircles, arise in both forms.

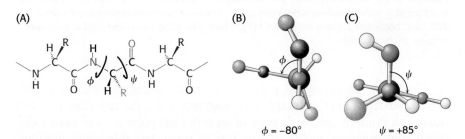

$\phi = -80°$ $\psi = +85°$

FIGURE 2.22 Rotation about bonds in a polypeptide. The structure of each amino acid in a polypeptide can be adjusted by rotation about two single bonds. (A) Phi (ϕ) is the angle of rotation about the bond between the nitrogen and the α-carbon atoms, whereas psi (ψ) is the angle of rotation about the bond between the α-carbon and the carbonyl carbon atoms. (B) A view down the bond between the nitrogen and the α-carbon atoms, showing how ϕ is measured. (C) A view down the bond between the α-carbon and the carbonyl carbon atoms, showing how ψ is measured.

By far the most common cis peptide bonds are X—Pro linkages. Such bonds show less preference for the trans configuration because the nitrogen of proline is bonded to two tetrahedral carbon atoms, limiting the steric differences between the trans and cis forms (Figure 2.21).

In contrast with the peptide bond, the bonds between the amino group and the α-carbon atom and between the α-carbon atom and the carbonyl group are pure single bonds. The two adjacent rigid peptide units can rotate about these bonds, taking on various orientations. *This freedom of rotation about two bonds of each amino acid allows proteins to fold in many different ways.* The rotations about these bonds can be specified by *torsion angles* (Figure 2.22). The angle of rotation about the bond between the nitrogen and the α-carbon atoms is called phi (ϕ). The angle of rotation about the bond between the α-carbon and the carbonyl carbon atoms is called psi (ψ). A clockwise rotation about either bond as viewed from the nitrogen atom toward the α-carbon atom or from the α-carbon atom toward the carbonyl group corresponds to a positive value. The ϕ and ψ angles determine the path of the polypeptide chain.

Are all combinations of ϕ and ψ possible? Gopalasamudram Ramachandran recognized that many combinations are forbidden because of steric collisions between atoms. The allowed values can be visualized on a two-dimensional plot called a *Ramachandran plot* (Figure 2.23). Three-quarters of the possible (ϕ, ψ) combinations are excluded simply by local steric clashes. *Steric exclusion, the fact that two atoms cannot be in the same place at the same time, can be a powerful organizing principle.*

Torsion angle

A measure of the rotation about a bond, usually taken to lie between -180 and $+180$ degrees. Torsion angles are sometimes called dihedral angles.

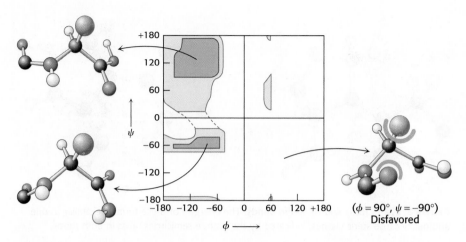

FIGURE 2.23 A Ramachandran plot showing the values of ϕ and ψ. Not all ϕ and values ψ are possible without collisions between atoms. The most favorable regions are shown in dark blue; borderline regions are shown in light blue. The structure on the right is disfavored because of steric clashes.

The ability of biological polymers such as proteins to fold into well-defined structures is remarkable thermodynamically. An unfolded polymer exists as a random coil: each copy of an unfolded polymer will have a different conformation, yielding a mixture of many possible conformations. The favorable entropy associated with a mixture of many conformations opposes folding and must be overcome by interactions favoring the folded form. Thus, highly flexible polymers with a large number of possible conformations do not fold into unique structures. *The rigidity of the peptide unit and the restricted set of allowed ϕ and ψ angles limits the number of structures accessible to the unfolded form sufficiently to allow protein folding to take place.*

2.3 Secondary Structure: Polypeptide Chains Can Fold into Regular Structures Such As the Alpha Helix, the Beta Sheet, and Turns and Loops

Ramachandran demonstrated that many conformations of a polypeptide backbone are disallowed on the basis of steric exclusion. Do the allowed conformations permit a polypeptide chain to fold into a regularly repeating structure? In 1951, Linus Pauling and Robert Corey proposed two periodic structures called the *α helix* (alpha helix) and the *β pleated sheet* (beta pleated sheet). Subsequently, other structures such as the *β turn* and *omega* (Ω) *loop* were identified. Although not periodic, these common turn or loop structures are well defined and contribute with α helices and β sheets to form the final protein structure. Alpha helices, β strands, and turns are formed by a regular pattern of hydrogen bonds between the peptide N—H and C=O groups of amino acids that are *near one another in the linear sequence.* Such folded segments are called *secondary structure.*

The alpha helix is a coiled structure stabilized by intrachain hydrogen bonds

In evaluating potential structures, Pauling and Corey considered which conformations of peptides were sterically allowed and which most fully exploited the hydrogen-bonding capacity of the backbone NH and CO groups. The first of their proposed structures, the *α helix,* is a rodlike structure (Figure 2.24). A tightly coiled backbone forms the inner part of the rod and the side chains

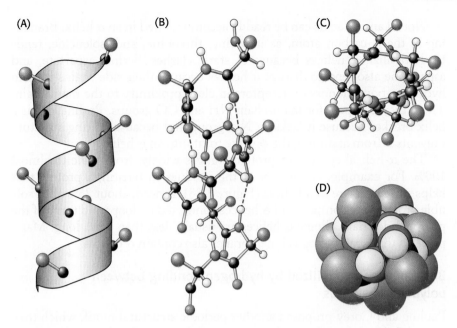

FIGURE 2.24 Structure of the α helix. (A) A ribbon depiction shows the α-carbon atoms and side chains (green). (B) A side view of a ball-and-stick version depicts the hydrogen bonds (dashed lines) between NH and CO groups. (C) An end view shows the coiled backbone as the inside of the helix and the side chains (green) projecting outward. (D) A space-filling view of part C shows the tightly packed interior core of the helix.

extend outward in a helical array. The α helix is stabilized by hydrogen bonds between the NH and CO groups of the main chain. In particular, the CO group of each amino acid forms a hydrogen bond with the NH group of the amino acid that is situated four residues ahead in the sequence (Figure 2.25). Thus, except for amino acids near the ends of an α helix, *all the main-chain CO and NH groups are hydrogen bonded.* Each residue is related to the next one by a *rise*, also called *translation*, of 1.5 Å along the helix axis and a rotation of 100 degrees; this gives 3.6 amino acid residues per turn of helix. Therefore, amino acids spaced three and four apart in the sequence are spatially quite close to one another in an α helix. In contrast, amino acids spaced two apart in the sequence are situated on opposite sides of the helix and so are unlikely to make contact. The *pitch* of the α helix is the length of one complete turn along the helix axis and is equal to the product of the rise (1.5 Å) and the number of residues per turn (3.6), or 5.4 Å. The *screw sense* of an α helix can be right-handed (clockwise) or left-handed (counterclockwise). The Ramachandran plot reveals that both the right-handed and the left-handed helices are among allowed conformations (Figure 2.26). However, right-handed helices are energetically more favorable because there is less steric clash between the side chains and the backbone. *Essentially all α helices found in proteins are right-handed.* In schematic representations of proteins, α helices are depicted as twisted ribbons or rods (Figure 2.27).

Screw sense
Describes the direction in which a helical structure rotates with respect to its axis. When viewed down the axis of a helix, if the chain turns in a clockwise direction, it has a right-handed screw sense. If the turning is counterclockwise, the screw sense is left-handed.

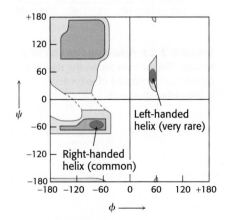

FIGURE 2.26 Ramachandran plot for helices. Both right- and left-handed helices lie in regions of allowed conformations in the Ramachandran plot. However, essentially all α helices in proteins are right-handed.

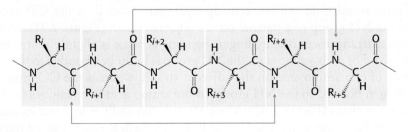

FIGURE 2.25 Hydrogen-bonding scheme for an α helix. In the α helix, the CO group of residue *i* forms a hydrogen bond with the NH group of residue *i* + 4.

(A)   (B)

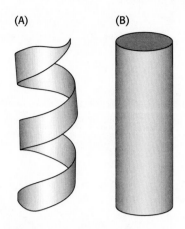

FIGURE 2.27 Schematic views of α helices. (A) A ribbon depiction. (B) A cylindrical depiction.

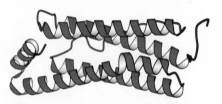

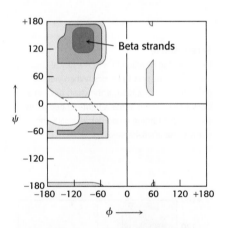

FIGURE 2.28 A largely α-helical protein. Ferritin, an iron-storage protein, is built from a bundle of α helices. [Drawn from 1AEW.pdb.]

FIGURE 2.29 Ramachandran plot for β strands. The red area shows the sterically allowed conformations of extended, β-strand-like structures.

Not all amino acids can be readily accommodated in an α helix. Branching at the β-carbon atom, as in valine, threonine, and isoleucine, tends to destabilize α helices because of steric clashes. Serine, aspartate, and asparagine also tend to disrupt α helices because their side chains contain hydrogen-bond donors or acceptors in close proximity to the main chain, where they compete for main-chain NH and CO groups. Proline also is a helix breaker because it lacks an NH group and because its ring structure prevents it from assuming the ϕ value to fit into an α helix.

The α-helical content of proteins ranges widely, from none to almost 100%. For example, about 75% of the residues in ferritin, a protein that helps store iron, are in α helices (Figure 2.28). Indeed, about 25% of all soluble proteins are composed of α helices connected by loops and turns of the polypeptide chain. Single α helices are usually less than 45 Å long. Many proteins that span biological membranes also contain α helices.

Beta sheets are stabilized by hydrogen bonding between polypeptide strands

Pauling and Corey proposed another periodic structural motif, which they named the *β pleated sheet* (β because it was the second structure they elucidated, the α helix having been the first). The β pleated sheet (or, more simply, the β sheet) differs markedly from the rodlike α helix. It is composed of two or more polypeptide chains called *β strands*. A β strand is almost fully extended rather than being tightly coiled as in the α helix. A range of extended structures are sterically allowed (Figure 2.29).

The distance between adjacent amino acids along a β strand is approximately 3.5 Å, in contrast with a distance of 1.5 Å along an α helix. The side chains of adjacent amino acids point in opposite directions (Figure 2.30).

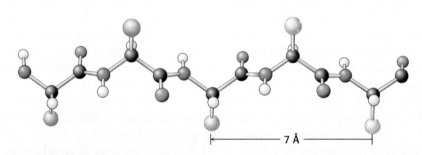

FIGURE 2.30 Structure of a β strand. The side chains (green) are alternately above and below the plane of the strand.

A β sheet is formed by linking two or more β strands lying next to one another through hydrogen bonds. Adjacent strands in a β sheet can run in opposite directions (antiparallel β sheet) or in the same direction (parallel β sheet). In the antiparallel arrangement, the NH group and the CO group of each amino acid are respectively hydrogen bonded to the CO group and the NH group of a partner on the adjacent chain (Figure 2.31). In the parallel arrangement, the hydrogen-bonding scheme is slightly more complicated. For each amino acid, the NH group is hydrogen bonded to the CO group of one amino acid on the adjacent strand, whereas the CO group is hydrogen bonded to the NH group on the amino acid two residues farther along the chain (Figure 2.32). Many strands, typically 4 or 5 but as many as 10 or more, can come together in β sheets. Such β sheets can be purely antiparallel, purely parallel, or mixed (Figure 2.33).

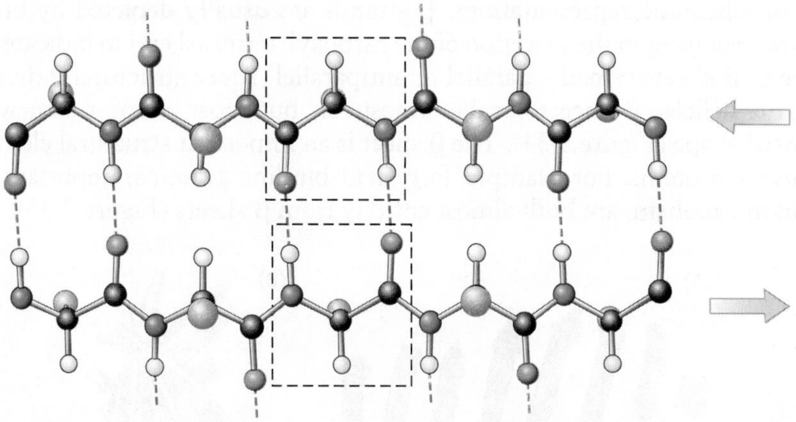

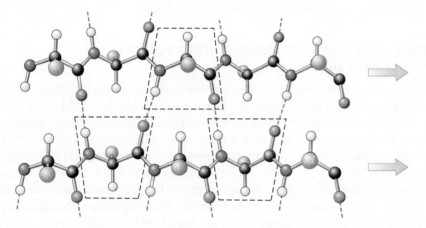

FIGURE 2.31 An antiparallel β sheet. Adjacent β strands run in opposite directions, as indicated by the arrows. Hydrogen bonds between NH and CO groups connect each amino acid to a single amino acid on an adjacent strand, stabilizing the structure.

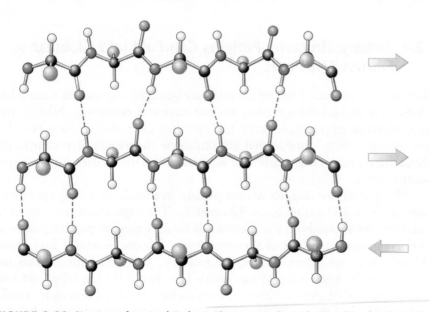

FIGURE 2.32 A parallel β sheet. Adjacent β strands run in the same direction, as indicated by the arrows. Hydrogen bonds connect each amino acid on one strand with two different amino acids on the adjacent strand.

FIGURE 2.33 Structure of a mixed β sheet. The arrows indicate directionality of each strand.

In schematic representations, β strands are usually depicted by broad arrows pointing in the direction of the carboxyl-terminal end to indicate the type of β sheet formed—parallel or antiparallel. More structurally diverse than α helices, β sheets can be almost flat but most adopt a somewhat twisted shape (Figure 2.34). The β sheet is an important structural element in many proteins. For example, fatty acid-binding proteins, important for lipid metabolism, are built almost entirely from β sheets (Figure 2.35).

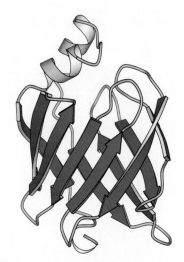

FIGURE 2.35 A protein rich in β sheets. The structure of a fatty acid-binding protein. [Drawn from 1FTP.pdb.]

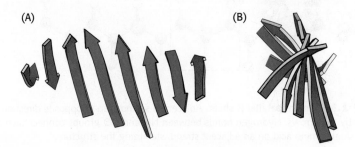

(A)　　　　　　　　(B)

FIGURE 2.34 A schematic twisted β sheet. (A) A schematic model. (B) The schematic view rotated by 90 degrees to illustrate the twist more clearly.

Polypeptide chains can change direction by making reverse turns and loops

As we will explore in Section 2.4, most proteins have compact, globular shapes owing to reversals in the direction of their polypeptide chains. Many of these reversals are accomplished by a common structural element called the *reverse turn* (also known as the *β turn* or *hairpin turn*), illustrated in Figure 2.36. In many reverse turns, the CO group of residue *i* of a polypeptide is hydrogen bonded to the NH group of residue *i* + 3. This interaction stabilizes abrupt changes in direction of the polypeptide chain. In other cases, more-elaborate structures are responsible for chain reversals. These structures are called *loops* or sometimes *Ω loops* (omega loops) to suggest their overall shape. Unlike α helices and β strands, loops do not have regular, periodic structures. Nonetheless, loop structures are often rigid and well-defined (Figure 2.37). Turns and loops invariably lie on the surfaces of proteins and often participate in interactions between proteins and other molecules.

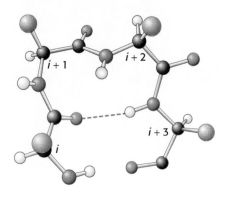

FIGURE 2.36 Structure of a reverse turn. The CO group of residue *i* of the polypeptide chain is hydrogen bonded to the NH group of residue *i* + 3 to stabilize the turn.

2.4 Tertiary Structure: Proteins Can Fold into Globular or Fibrous Structures

Let us now examine how amino acids are grouped together in a complete protein. X-ray crystallographic, nuclear magnetic resonance (NMR), and cryo-electron microscopic (cryo-EM) studies (Section 3.5) have revealed the detailed three-dimensional structures of thousands of proteins. We begin here with an examination of *myoglobin*, the first protein to be seen in atomic detail.

Myoglobin, the oxygen storage protein in muscle, is a single polypeptide chain of 153 amino acids (Chapter 7). The capacity of myoglobin to bind oxygen depends on the presence of *heme*, a nonpolypeptide *prosthetic (helper) group* consisting of protoporphyrin IX and a central iron atom. *Myoglobin is an extremely compact molecule.* Its overall dimensions are 45 × 35 × 25 Å, an order of magnitude less than if it were fully stretched out (Figure 2.38). About 70% of the main chain is folded into eight α helices, and much of the rest of the chain forms turns and loops between helices.

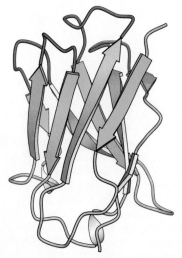

FIGURE 2.37 Loops on a protein surface. A part of an antibody molecule has surface loops (shown in red) that mediate interactions with other molecules. [Drawn from 7FAB.pdb.]

(A)

Heme group
Iron atom

(B)

Heme group

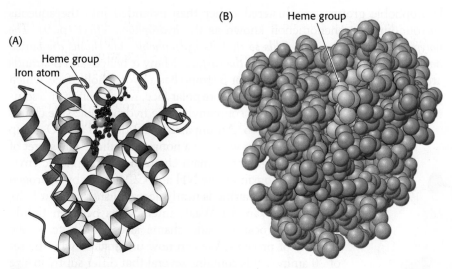

FIGURE 2.38 Three-dimensional structure of myoglobin. (A) A ribbon diagram shows that the protein consists largely of α helices. (B) A space-filling model in the same orientation shows how tightly packed the folded protein is. *Notice* that the heme group is nestled into a crevice in the compact protein with only an edge exposed. One helix is blue to allow comparison of the two structural depictions. [Drawn from 1A6N.pdb.]

The folding of the main chain of myoglobin, like that of most other proteins, is complex and devoid of symmetry. The overall course of the polypeptide chain of a protein is referred to as its *tertiary structure*. A unifying principle emerges from the distribution of side chains. Strikingly, *the interior consists almost entirely of nonpolar residues* such as leucine, valine, methionine, and phenylalanine (Figure 2.39). Charged residues such as aspartate, glutamate, lysine, and arginine are absent from the inside of myoglobin. The only polar residues inside are two histidine residues, which play critical roles in binding iron and oxygen. The outside of myoglobin, on the other hand, consists of both polar and nonpolar residues. The space-filling model shows that there is very little empty space inside. The tight packing of myoglobin into a highly compact structure, its lack of symmetry, and its solubility in water are characteristics of *globular proteins,* which perform a wide array of important functions, including regulatory, signaling, and enzymatic activities.

This contrasting distribution of polar and nonpolar residues reveals a key facet of protein architecture. In an aqueous environment, protein folding is driven by the strong tendency of hydrophobic residues to be excluded from water. Recall that a system is more thermodynamically stable when

(A)   (B)

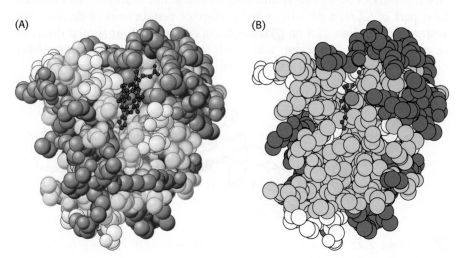

FIGURE 2.39 Distribution of amino acids in myoglobin. (A) A space-filling model of myoglobin with hydrophobic amino acids shown in yellow, charged amino acids shown in blue, and others shown in white. *Notice* that the surface of the molecule has many charged amino acids, as well as some hydrophobic amino acids. (B) In this cross-sectional view, *notice* that mostly hydrophobic amino acids are found on the inside of the structure, whereas the charged amino acids are found on the protein surface. [Drawn from 1MBD.pdb.]

hydrophobic groups are clustered rather than extended into the aqueous surroundings, a phenomenon known as the *hydrophobic effect* (p. 9). *The polypeptide chain therefore folds so that its hydrophobic side chains are buried and its polar, charged chains are on the surface.* Many α helices and β strands are amphipathic; that is, the α helix or β strand has a hydrophobic face, which points into the protein interior, and a more polar face, which points into solution. The fate of the main chain accompanying the hydrophobic side chains is important, too. An unpaired peptide NH or CO group markedly prefers water to a nonpolar milieu. The secret of burying a segment of main chain in a hydrophobic environment is to pair all the NH and CO groups by hydrogen bonding. This pairing is neatly accomplished in an α helix or β sheet. Van der Waals interactions between tightly packed hydrocarbon side chains also contribute to the stability of proteins. We can now understand why the set of 20 amino acids contains several that differ subtly in size and shape. They provide a palette from which to choose to fill the interior of a protein neatly and thereby maximize van der Waals interactions, which require intimate contact.

Some proteins that span biological membranes are "the exceptions that prove the rule" because they have the reverse distribution of hydrophobic and hydrophilic amino acids. For example, consider porins—proteins found in the outer membranes of many bacteria (Figure 2.40). Membranes are built largely of hydrophobic alkane chains (Section 12.2). Thus, porins are covered on the outside largely with hydrophobic residues that interact with the neighboring alkane chains. In contrast, the center of the protein contains many charged and polar amino acids that surround a water-filled channel running through the middle of the protein. Because porins function in hydrophobic environments, they are "inside out" relative to proteins that function in aqueous solution.

Certain combinations of secondary structure are present in many proteins and frequently exhibit similar functions. These combinations are called *motifs* or *supersecondary structures*. For example, an α helix separated from another α helix by a turn, called a *helix-turn-helix* unit, is found in many proteins that bind DNA (Figure 2.41).

Some polypeptide chains fold into two or more compact regions that may be connected by a flexible segment of polypeptide chain, rather like pearls on a string. These compact globular units, called *domains*, range in size from about 30 to 400 amino acid residues. For example, the extracellular part of CD4, a protein on the surface of certain cells of the immune system (See Section 35.4 on 🅂 Sapling Plus), comprises four similar domains of approximately 100 amino acids each (Figure 2.42). Proteins may have domains in common even if their overall tertiary structures are different.

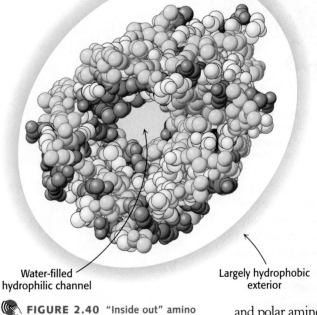

Water-filled hydrophilic channel

Largely hydrophobic exterior

FIGURE 2.40 "Inside out" amino acid distribution in porin. The outside of porin (which contacts hydrophobic groups in membranes) is covered largely with hydrophobic residues, whereas the center includes a water-filled channel lined with charged and polar amino acids. [Drawn from 1PRN.pdb.]

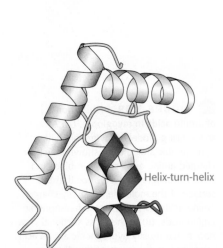

Helix-turn-helix

FIGURE 2.41 The helix-turn-helix motif, a supersecondary structural element. Helix-turn-helix motifs are found in many DNA-binding proteins. [Drawn from 1LMB.pdb.]

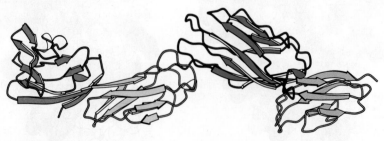

FIGURE 2.42 Protein domains. The cell-surface protein CD4 consists of four similar domains. [Drawn from 1WIO.pdb.]

(A)

(B)

FIGURE 2.43 An α-helical coiled coil. (A) Space-filling model. (B) Ribbon diagram. The two helices wind around one another to form a superhelix. Such structures are found in many proteins, including keratin in hair, quills, claws, and horns. [Drawn from 1C1G.pdb.]

Fibrous proteins provide structural support for cells and tissues

In contrast to globular proteins, *fibrous proteins* form long, extended structures that feature repeated sequences. These proteins, such as α-keratin and collagen, utilize special types of helices that facilitate the formation of long fibers that serve a structural role.

α-Keratin, which is an essential component of wool, hair, and skin, consists of two right-handed α helices intertwined to form a type of left-handed superhelix called an *α-helical coiled coil*. α-Keratin is a member of a superfamily of proteins referred to as *coiled-coil proteins* (Figure 2.43). In these proteins, two or more α helices can entwine to form a very stable structure, which can have a length of 1000 Å (100 nm, or 0.1 mm) or more. There are approximately 60 members of this family in humans, including intermediate filaments, proteins that contribute to the cell cytoskeleton (internal scaffolding in a cell), and the muscle proteins myosin and tropomyosin (See Section 35.2 on ✺Sapling Plus). Members of this family are characterized by a central region of 300 amino acids that contains imperfect repeats of a sequence of 7 amino acids called a *heptad repeat*.

The two helices in α-keratin associate with each other by weak interactions such as van der Waals forces and ionic interactions. The left-handed supercoil alters the two right-handed α helices such that there are 3.5 residues per turn instead of 3.6. As a result, the pattern of side-chain interactions can be repeated every seven residues, forming the heptad repeats. Two helices with such repeats are able to interact with one another if the repeats are complementary (Figure 2.44). For example, the repeating residues may be hydrophobic, allowing van der Waals interactions, or have opposite charge, allowing ionic interactions. In addition, the two helices may be linked by disulfide bonds formed by neighboring cysteine residues. The bonding of the helices accounts for the physical properties of wool, an example of an α-keratin. Wool is extensible and can be stretched to nearly twice its length because the α helices stretch, breaking the weak interactions between neighboring helices. However, the covalent disulfide bonds resist breakage and return the fiber to its original state once the stretching force is released. The number of disulfide bond cross-links further defines the fiber's properties. Hair and wool, having fewer cross-links, are flexible. Horns, claws, and hooves, having more cross-links, are much harder.

A different type of helix is present in collagen, the most abundant protein of mammals. Collagen is the main fibrous component of skin, bone, tendon, cartilage, and teeth. This extracellular protein is a rod-shaped molecule, about 3000 Å long and only 15 Å in diameter. It contains three helical polypeptide chains, each nearly 1000 residues long. Glycine appears at every third residue in the amino acid sequence, and the sequence

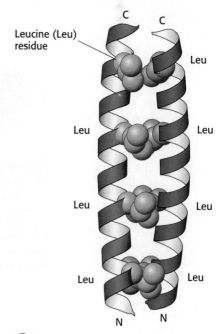

FIGURE 2.44 Heptad repeats in a coiled-coil protein. Every seventh residue in each helix is leucine. The two helices are held together by van der Waals interactions primarily between the leucine residues. [Drawn from 2ZTA.pdb.]

13
-Gly-Pro-**Met**-Gly-Pro-**Ser**-Gly-Pro-**Arg**-
22
-Gly-**Leu**-Hyp-Gly-Pro-Hyp-Gly-**Ala**-Hyp-
31
-Gly-Pro-**Gln**-Gly-**Phe**-**Gln**-Gly-Pro-Hyp-
40
-Gly-**Glu**-Hyp-Gly-**Glu**-Hyp-Gly-**Ala**-**Ser**-
49
-Gly-Pro-**Met**-Gly-Pro-**Arg**-Gly-Pro-Hyp-
58
-Gly-Pro-Hyp-Gly-**Lys**-**Asn**-Gly-**Asp**-**Asp**-

FIGURE 2.45 Amino acid sequence of a part of a collagen chain. Every third residue is a glycine. Proline and hydroxyproline (Hyp) also are abundant.

glycine-proline-hydroxyproline recurs frequently (Figure 2.45). Hydroxyproline is a derivative of proline that has a hydroxyl group in place of one of the hydrogen atoms on the pyrrolidine ring.

The collagen helix has properties different from those of the α helix. Hydrogen bonds within a strand are absent. Instead, *the helix is stabilized by steric repulsion of the pyrrolidine rings of the proline and hydroxyproline residues* (Figure 2.46). The pyrrolidine rings keep out of each other's way when the polypeptide chain assumes its helical form, which has about three residues per turn. Three strands wind around one another to form a *superhelical cable* that is stabilized by hydrogen bonds between strands. The hydrogen bonds form between the peptide NH groups of glycine residues and the CO groups of residues on the other chains. The hydroxyl groups of hydroxyproline residues also participate in hydrogen bonding.

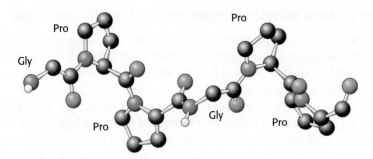

FIGURE 2.46 Conformation of a single strand of a collagen triple helix.

The inside of the triple-stranded helical cable is very crowded and accounts for the requirement that glycine be present at every third position on each strand (Figure 2.47A). *The only residue that can fit in an interior position is glycine.* The amino acid residue on either side of glycine is located on the outside of the cable, where there is room for the bulky rings of proline and hydroxyproline residues (Figure 2.47B).

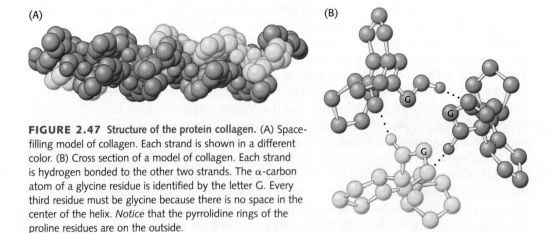

(A)

(B)

FIGURE 2.47 Structure of the protein collagen. (A) Space-filling model of collagen. Each strand is shown in a different color. (B) Cross section of a model of collagen. Each strand is hydrogen bonded to the other two strands. The α-carbon atom of a glycine residue is identified by the letter G. Every third residue must be glycine because there is no space in the center of the helix. *Notice* that the pyrrolidine rings of the proline residues are on the outside.

The importance of the positioning of glycine inside the triple helix is illustrated in the disorder osteogenesis imperfecta, also known as brittle bone disease. In this condition, which can vary from mild to very severe, other amino acids replace the internal glycine residue. This replacement leads to a delayed and improper folding of collagen. The most serious symptom is severe bone fragility. Defective collagen in the eyes causes the whites of the eyes to have a blue tint (blue sclera).

2.5 Quaternary Structure: Polypeptide Chains Can Assemble into Multisubunit Structures

Four levels of structure are frequently cited in discussions of protein architecture. So far, we have considered three. *Primary structure* is the amino acid sequence. *Secondary structure* refers to the spatial arrangement of amino acid residues that are nearby in the sequence. Some of these arrangements give rise to periodic structures. The α helix and β strand are elements of secondary structure. *Tertiary structure* refers to the spatial arrangement of amino acid residues that are far apart in the sequence and to the pattern of disulfide bonds. We now turn to proteins containing more than one polypeptide chain. Such proteins exhibit a fourth level of structural organization. Each polypeptide chain in such a protein is called a *subunit*. *Quaternary structure* refers to the spatial arrangement of subunits and the nature of their interactions. The simplest sort of quaternary structure is a *dimer,* consisting of two identical subunits. For example, this organization is present in the DNA-binding protein Cro, found in a bacterial virus called λ (Figure 2.48). More-complicated quaternary structures are also common. More than one type of subunit can be present, often in variable numbers. For example, human hemoglobin, the oxygen-carrying protein in blood, consists of two subunits of one type (designated α) and two subunits of another type (designated β), as illustrated in Figure 2.49. Thus, the hemoglobin molecule exists as an $\alpha_2\beta_2$ tetramer. Subtle changes in the arrangement of subunits within the hemoglobin molecule allow it to carry oxygen from the lungs to tissues with great efficiency (Chapter 7).

Viruses make the most of a limited amount of genetic information by forming coats that use the same kind of subunit repetitively in a symmetric array. The coat of rhinovirus, the virus that causes the common cold, includes 60 copies of each of four subunits (Figure 2.50). The subunits come together to form a nearly spherical shell that encloses the viral genome.

FIGURE 2.48 Quaternary structure. The Cro protein of bacteriophage λ is a dimer of identical subunits. [Drawn from 5CRO.pdb.]

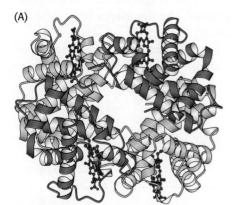

(A)

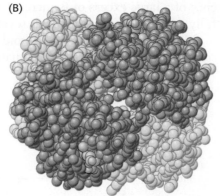

(B)

FIGURE 2.49 The $\alpha_2\beta_2$ tetramer of human hemoglobin. The structure of the two α subunits (red) is similar but not identical to that of the two β subunits (yellow). The molecule contains four heme groups (gray with the iron atom shown in purple). (A) The ribbon diagram highlights the similarity of the subunits and shows they are composed mainly of α helices. (B) The space-filling model illustrates how the heme groups occupy crevices in the protein. [Drawn from 1A3N.pdb.]

FIGURE 2.50 Complex quaternary structure. The coat of human rhinovirus, the cause of the common cold, comprises 60 copies of each of four subunits. The three most prominent subunits are shown as different colors.

2.6 The Amino Acid Sequence of a Protein Determines Its Three-Dimensional Structure

How is the elaborate three-dimensional structure of proteins attained? The classic work of Christian Anfinsen in the 1950s on the enzyme ribonuclease revealed the relationship between the amino acid sequence of a protein and its conformation. Ribonuclease is a single polypeptide chain consisting of 124 amino acid residues cross-linked by four disulfide bonds (Figure 2.51). Anfinsen's plan was to destroy the three-dimensional structure of the enzyme and to then determine what conditions were required to restore the structure.

(A)   (B)

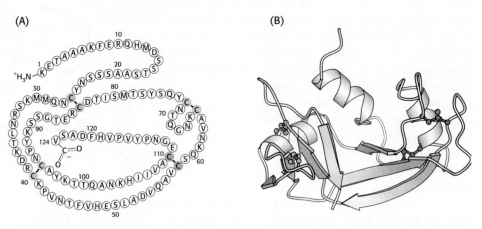

FIGURE 2.51 Amino acid sequence of bovine ribonuclease. (A) The four disulfide bonds are shown in color. (B) Three-dimensional structure of ribonuclease with the same four disulfide bonds highlighted. [(A) After C. H. W. Hirs, S. Moore, and W. H. Stein, *J. Biol. Chem.* 235:633–647, 1960; (B) Drawn from 1RBX.pdb.]

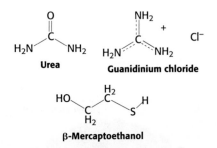

Agents such as *urea* or *guanidinium chloride* effectively disrupt a protein's noncovalent bonds. Although the mechanism of action of these agents is not fully understood, computer simulations suggest that they replace water as the molecule solvating the protein and are then able to disrupt the van der Waals interactions stabilizing the protein structure. The disulfide bonds in a protein can be cleaved reversibly by reducing them with a reagent such as *β-mercaptoethanol* (Figure 2.52). In the presence of a large excess of β-mercaptoethanol, the disulfides (cystines) are fully converted into sulfhydryls (cysteines).

Most polypeptide chains devoid of cross-links assume a *random-coil conformation* in 8 M urea or 6 M guanidinium chloride. When ribonuclease

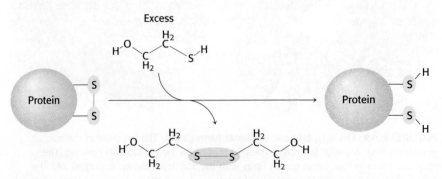

FIGURE 2.52 Role of β-mercaptoethanol in reducing disulfide bonds. *Notice* as the disulfides are reduced, the β-mercaptoethanol is oxidized and forms dimers.

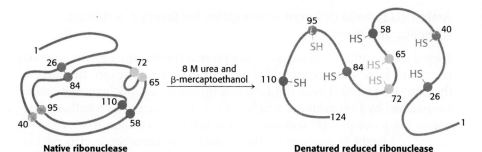

FIGURE 2.53 Reduction and denaturation of ribonuclease.

was treated with β-mercaptoethanol in 8 M urea, the product was a fully reduced, randomly coiled polypeptide chain *devoid of enzymatic activity*. When a protein is converted into a randomly coiled peptide without its normal activity, it is said to be *denatured* (Figure 2.53).

Anfinsen then made the critical observation that the denatured ribonuclease, freed of urea and β-mercaptoethanol by dialysis (Section 3.1), slowly regained enzymatic activity. He perceived the significance of this chance finding: the sulfhydryl groups of the denatured enzyme became oxidized by air, and the enzyme spontaneously refolded into a catalytically active form. Detailed studies then showed that nearly all the original enzymatic activity was regained if the sulfhydryl groups were oxidized under suitable conditions. All the measured physical and chemical properties of the refolded enzyme were virtually identical with those of the native enzyme. These experiments showed that *the information needed to specify the catalytically active structure of ribonuclease is contained in its amino acid sequence*. Subsequent studies have established the generality of this central principle of biochemistry: *sequence specifies conformation*. The dependence of conformation on sequence is especially significant because of the intimate connection between conformation and function.

A quite different result was obtained when reduced ribonuclease was reoxidized while it was still in 8 M urea and the preparation was then dialyzed to remove the urea. Ribonuclease reoxidized in this way had only 1% of the enzymatic activity of the native protein. Why were the outcomes so different when reduced ribonuclease was reoxidized in the presence and absence of urea? The reason is that the wrong disulfides formed pairs in urea. There are 105 different ways of pairing eight cysteine molecules to form four disulfides; only one of these combinations is enzymatically active. The 104 wrong pairings have been picturesquely termed "scrambled" ribonuclease. Anfinsen found that scrambled ribonuclease spontaneously converted into fully active, native ribonuclease when trace amounts of β-mercaptoethanol were added to an aqueous solution of the protein (Figure 2.54). The added β-mercaptoethanol catalyzed the rearrangement of disulfide pairings until the native structure was regained in about 10 hours. *This process was driven by the decrease in free energy as the scrambled conformations were converted into the stable, native conformation of the enzyme.* The native disulfide pairings of ribonuclease thus contribute to the stabilization of the thermodynamically preferred structure.

Similar refolding experiments have been performed on many other proteins. In many cases, the native structure can be generated under suitable conditions. For other proteins, however, refolding does not proceed efficiently. In these cases, the unfolded protein molecules usually become tangled up with one another to form aggregates. Inside cells, proteins called *chaperones* block such undesirable interactions. Additionally, it is now evident that some proteins do not assume a defined structure until they interact with molecular partners, as we will see shortly.

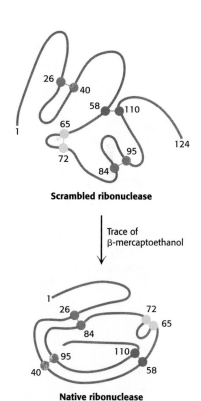

FIGURE 2.54 Reestablishing correct disulfide pairing. Native ribonuclease can be re-formed from scrambled ribonuclease in the presence of a trace of β-mercaptoethanol.

Amino acids have different propensities for forming α helices, β sheets, and turns

How does the amino acid sequence of a protein specify its three-dimensional structure? How does an unfolded polypeptide chain acquire the form of the native protein? These fundamental questions in biochemistry can be approached by first asking a simpler one: What determines whether a particular sequence in a protein forms an α helix, a β strand, or a turn? One source of insight is to examine the frequency of occurrence of particular amino acid residues in these secondary structures (Table 2.3). Residues such as alanine, glutamate, and leucine tend to be present in α helices, whereas valine and isoleucine tend to be present in β strands. Glycine, asparagine, and proline are more commonly observed in turns.

Studies of proteins and synthetic peptides have revealed some reasons for these preferences. Branching at the β-carbon atom, as in valine, threonine, and isoleucine, tends to destabilize α helices because of steric clashes. These residues are readily accommodated in β strands, where their side chains project out of the plane containing the main chain. Serine and asparagine tend to disrupt α helices because their side chains contain hydrogen-bond donors or acceptors in close proximity to the main chain, where they compete for main-chain NH and CO groups. Proline tends to disrupt both α helices and β strands because it lacks an NH group and because its ring structure restricts its ϕ value to near 60 degrees. Glycine readily fits into all structures, but its conformational flexibility renders it well-suited to reverse turns.

TABLE 2.3 Relative frequencies of amino acid residues in secondary structures

Amino acid	α helix	β sheet	Reverse turn
Glu	1.59	0.52	1.01
Ala	1.41	0.72	0.82
Leu	1.34	1.22	0.57
Met	1.30	1.14	0.52
Gln	1.27	0.98	0.84
Lys	1.23	0.69	1.07
Arg	1.21	0.84	0.90
His	1.05	0.80	0.81
Val	0.90	1.87	0.41
Ile	1.09	1.67	0.47
Tyr	0.74	1.45	0.76
Cys	0.66	1.40	0.54
Trp	1.02	1.35	0.65
Phe	1.16	1.33	0.59
Thr	0.76	1.17	0.96
Gly	0.43	0.58	1.77
Asn	0.76	0.48	1.34
Pro	0.34	0.31	1.32
Ser	0.57	0.96	1.22
Asp	0.99	0.39	1.24

Note: The amino acids are grouped according to their preference for α helices (top group), β sheets (middle group), or turns (bottom group).
Source: T. E. Creighton, *Proteins: Structures and Molecular Properties*, 2d ed. (W. H. Freeman and Company, 1992), p. 256.

Can we predict the secondary structure of a protein by using this knowledge of the conformational preferences of amino acid residues? Accurate predictions of secondary structure adopted by even a short stretch of residues have proved to be difficult. What stands in the way of more-accurate prediction? Note that the conformational preferences of amino acid residues are not tipped all the way to one structure (Table 2.3). For example, glutamate, one of the strongest helix formers, prefers α helix to β strand by only a factor of three. The preference ratios of most other residues are smaller. Indeed, some penta- and hexapeptide sequences have been found to adopt one structure in one protein and an entirely different structure in another (Figure 2.55). Hence, some amino acid sequences do not uniquely determine secondary structure. Tertiary interactions—interactions between residues that are far apart in the sequence—may be decisive in specifying the secondary structure of some segments. Context is often crucial: the conformation of a protein has evolved to work in a particular environment. Nevertheless, substantial improvements in secondary structure prediction have been achieved by using families of related sequences, each of which adopts the same structure.

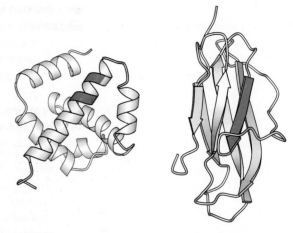

FIGURE 2.55 Alternative conformations of a peptide sequence. Many sequences can adopt alternative conformations in different proteins. Here the sequence VDLLKN shown in red assumes an α helix in one protein context (left) and a β strand in another (right). [Drawn from (left) 3WRP.pdb and (right) 2HLA.pdb.]

Protein folding is a highly cooperative process

Proteins can be denatured by any treatment that disrupts the weak bonds stabilizing tertiary structure, such as heating, or by chemical denaturants such as urea or guanidinium chloride. For many proteins, a comparison of the degree of unfolding as the concentration of denaturant increases reveals a sharp transition from the folded, or native, form to the unfolded, or denatured form, suggesting that only these two conformational states are present to any significant extent (Figure 2.56). A similar sharp transition is observed if denaturants are removed from unfolded proteins, allowing the proteins to fold.

The sharp transition seen in Figure 2.56 suggests that protein folding and unfolding is an *"all or none" process* that results from a *cooperative transition*. For example, suppose that a protein is placed in conditions under which some part of the protein structure is thermodynamically unstable. As this part of the folded structure is disrupted, the interactions between it and the remainder of the protein will be lost. The loss of these interactions, in turn, will destabilize the remainder of the structure. Thus, conditions that lead to the disruption of any part of a protein structure are likely to unravel the protein completely. The structural properties of proteins provide a clear rationale for the cooperative transition.

The consequences of cooperative folding can be illustrated by considering the contents of a protein solution under conditions corresponding to the middle of the transition between the folded and the unfolded forms. Under these conditions, the protein is "half folded." Yet the solution will appear to have no partly folded molecules but, instead, look as if it is a 50/50 mixture of fully folded and fully unfolded molecules (Figure 2.57). Although the protein may appear to behave as if it exists in only two states, this simple two-state existence is an impossibility at a molecular level. Even simple reactions go through

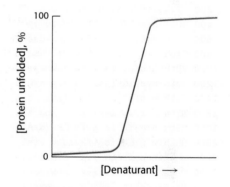

FIGURE 2.56 Transition from folded to unfolded state. Most proteins show a sharp transition from the folded to the unfolded form on treatment with increasing concentrations of denaturants.

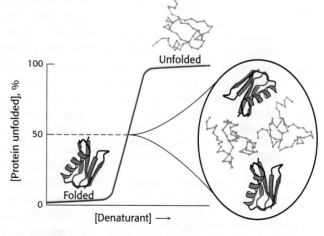

FIGURE 2.57 Components of a partly denatured protein solution. In a half-unfolded protein solution, half the molecules are fully folded and half are fully unfolded.

reaction intermediates, and so a complex molecule such as a protein cannot simply switch from a completely unfolded state to the native state in one step. Unstable, transient intermediate structures must exist between the native and denatured state. Determining the nature of these intermediate structures is an area of intense biochemical research.

Proteins fold by progressive stabilization of intermediates rather than by random search

How does a protein make the transition from an unfolded structure to a unique conformation in the native form? One possibility a priori would be that all possible conformations are sampled to find the energetically most favorable one. How long would such a random search take? Consider a small protein with 100 residues. Cyrus Levinthal calculated that, if each residue can assume three different conformations, the total number of structures would be 3^{100}, which is equal to 5×10^{47}. If it takes 10^{-13} s to convert one structure into another, the total search time would be $5 \times 10^{47} \times 10^{-13}$ s, which is equal to 5×10^{34} s, or 1.6×10^{27} years. In reality, small proteins can fold in less than a second. Clearly, it would take much too long for even a small protein to fold properly by randomly trying out all possible conformations. The enormous difference between calculated and actual folding times is called *Levinthal's paradox*. This paradox clearly reveals that proteins do not fold by trying every possible conformation; instead, they must follow at least a partly defined folding pathway consisting of intermediates between the fully denatured protein and its native structure.

The way out of this paradox is to recognize the power of *cumulative selection*. Richard Dawkins, in *The Blind Watchmaker*, asked how long it would take a monkey poking randomly at a typewriter to reproduce Hamlet's remark to Polonius, "Methinks it is like a weasel" (Figure 2.58). An astronomically large number of keystrokes, on the order of 10^{40}, would be required. However, suppose that we preserved each correct character and allowed the monkey to retype only the wrong ones. In this case, only a few thousand keystrokes, on average, would be needed. The crucial difference between these cases is that the first employs a completely random search, whereas, in the second, *partly correct intermediates are retained*.

The essence of protein folding is the tendency to retain partly correct intermediates. However, the protein-folding problem is much more difficult than the one presented to our simian Shakespeare. First, the criterion of correctness is not a residue-by-residue scrutiny of conformation by an omniscient observer but rather the total free energy of the transient species. Second, proteins are only marginally stable. The free-energy difference between the folded and the unfolded states of a typical 100-residue protein is 42 kJ mol^{-1} (10 kcal mol^{-1}), and thus each residue contributes on average only 0.42 kJ mol^{-1} ($0.1 \text{ kcal mol}^{-1}$) of energy to maintain the folded state. This amount is less than the amount of thermal energy, which is 2.5 kJ mol^{-1} ($0.6 \text{ kcal mol}^{-1}$) at room temperature. This meager stabilization energy means that correct intermediates, especially those formed early in folding, can be lost. The analogy is that the monkey would be somewhat free to undo its correct keystrokes. Nonetheless, the interactions that lead to cooperative folding can stabilize intermediates as structure builds up. Thus, local regions that have significant structural preference, though not necessarily stable on their own, will tend to adopt their favored structures and, as they form, can interact with one other, leading to increasing stabilization. This conceptual framework is often referred to as the *nucleation-condensation model*.

```
 200  ?T(\G{+s x[A.N5~,#ATxSGpn`e□@
 400  oDr'Jh7s DFR:W41'u+^v6zpJseOi
 600  e2ih'8zs n527x818d_ih=H1dseb.
 800  S#dh>}/s ]tZqC%1P%DK<|!^aseZ.
1000  V0th>nLs ut/is]l_kwojjwMasef.
1200  juth+nvs it is[lukh?SCw=ase5.
1400  Iithdn4s it isOl/ks/IxwLase~.
1600  M?thinrs it is 1Xk?T"_woasel.
1800  MSthinWs it is 1wkN7□Kw(asel.
2000  Mhthin`s it is likv,aww_asel.
2200  MMthinns it is lik+5avw1asel.
2400  MethinXs it is likydaqw)asel.
2600  Methin4s it is lik2dasweasel.
2800  MethinHs it is like□aTweasel.
2883  Methinks it is like a weasel.
```

```
 200  )z~hg)W4{{cu!kO{d6jS!N1EyUx}p
 400  "W hi\kR.<&CfA%4-Y1G!iT$6({|6
 600  .L=hinkm4(uMGP^1AWoE6k1wW=yiS
 800   AthinkaPa_vYH liR\Hb,Uo4\-"(
1000  OFthinksP)@fZO li8v] /+Eln26B
1200  6ithinksMVt -V likm+gl#K~)BFk
1400  vxthinksaEt □w like.S1Geutks.
1600  :Othinks<it MC likesN2[eaVe4.
1800  uxthinksqit 0r likeQh)weaoeW.
2000  Y/thinks it id like7a1wea)e&.
2200  Methinks it iW like a[weaWel.
2400  Methinks it is like a:weasel.
2431  Methinks it is like a weasel.
```

FIGURE 2.58 Typing-monkey analogy. A monkey randomly poking a typewriter could write a line from Shakespeare's *Hamlet*, provided that correct keystrokes were retained. In the two computer simulations shown, the cumulative number of keystrokes is given at the left of each line.

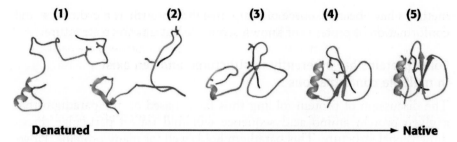

(1)   (2)   (3)   (4)   (5)

Denatured ————————————————————→ Native

FIGURE 2.59 Proposed folding pathway of chymotrypsin inhibitor. Local regions with sufficient structural preference tend to adopt their favored structures initially (1). These structures come together to form a nucleus with a nativelike, but still mobile, structure (4). This structure then fully condenses to form the native, more rigid structure (5). [From A. R. Fersht and V. Daggett. *Cell* 108:573–582, 2002; with permission from Elsevier.]

A simulation of the folding of a protein, based on the nucleation-condensation model, is shown in Figure 2.59. This model suggests that certain pathways may be preferred. Although Figure 2.59 suggests a discrete pathway, each of the intermediates shown represents an ensemble of similar structures, and thus a protein follows a general rather than a precise pathway in its transition from the unfolded to the native state. The energy surface for the overall process of protein folding can be visualized as a funnel (Figure 2.60). The wide rim of the funnel represents the wide range of structures accessible to the ensemble of denatured protein molecules. As the free energy of the population of protein molecules decreases, the proteins move down into narrower parts of the funnel and fewer conformations are accessible. At the bottom of the funnel is the folded state with its well-defined conformation. Many paths can lead to this same energy minimum.

Prediction of three-dimensional structure from sequence remains a great challenge

The prediction of three-dimensional structure from sequence has proved to be extremely difficult. The local sequence appears to determine only between 60 and 70% of the secondary structure; long-range interactions are required to stabilize the full secondary structure and the tertiary structure.

Investigators are exploring two fundamentally different approaches to predicting three-dimensional structure from amino acid sequence. The first is *ab initio* (Latin, "from the beginning") *prediction,* which attempts to predict the folding of an amino acid sequence without prior knowledge about similar sequences in known protein structures. Computer-based calculations are employed that attempt to minimize the free energy of a structure with a given amino acid sequence or to simulate the folding process. The utility of these methods is limited by the vast number of possible conformations, the marginal stability of proteins, and the subtle energetics of weak interactions in aqueous solution. The second approach takes advantage of our growing knowledge of the three-dimensional structures of many proteins. In these *knowledge-based methods,* an amino acid sequence of unknown structure is examined for compatibility with known protein structures or fragments therefrom. If a significant match is detected, the known structure can be used as an initial model. Knowledge-based

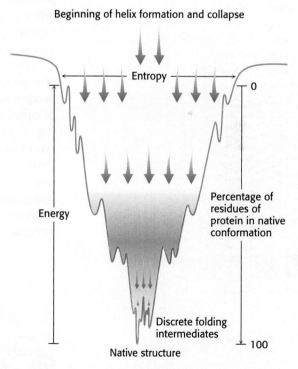

FIGURE 2.60 Folding funnel. The folding funnel depicts the thermodynamics of protein folding. The top of the funnel represents all possible denatured conformations—that is, maximal conformational entropy. Depressions on the sides of the funnel represent semistable intermediates that can facilitate or hinder the formation of the native structure, depending on their depth. Secondary structures, such as helices, form and collapse onto one another to initiate folding. [After D. L. Nelson and M. M. Cox, *Lehninger Principles of Biochemistry,* 5th ed. (W. H. Freeman and Company, 2008), p. 143.]

methods have been a source of many insights into the three-dimensional conformation of proteins of known sequence but unknown structure.

Some proteins are inherently unstructured and can exist in multiple conformations

The discussion of protein folding thus far is based on the paradigm that a given protein amino acid sequence will fold into a particular three-dimensional structure. This paradigm holds well for many proteins. However, it has been known for some time that some proteins can adopt two different structures, only one of which results in protein aggregation and pathological conditions. Such alternate structures originating from a unique amino acid sequence were thought to be rare—the exception to the paradigm. Recent work has called into question the universality of the idea that each amino acid sequence gives rise to one structure for certain proteins, even under normal cellular conditions.

Our first example is a class of proteins referred to as *intrinsically unstructured proteins* (IUPs). As the name suggests, these proteins, completely or in part, do not have a discrete three-dimensional structure under physiological conditions. Indeed, an estimated 50% of eukaryotic proteins have at least one unstructured region greater than 30 amino acids in length. Unstructured regions are rich in charged and polar amino acids with few hydrophobic residues. These proteins assume a defined structure on interaction with other proteins. This molecular versatility means that one protein can assume different structures and interact with the different partners, yielding different biochemical functions. IUPs appear to be especially important in signaling and regulatory pathways.

Another class of proteins that do not adhere to the paradigm is *metamorphic proteins*. These proteins appear to exist in an ensemble of structures of approximately equal energy that are in equilibrium. Small molecules or other proteins may bind to different members of the ensemble, resulting in various complexes, each having a different biochemical function. An especially clear example of a metamorphic protein is the chemokine lymphotactin. Chemokines are small signaling proteins in the immune system that bind to receptor proteins on the surface of immune-system cells, instigating an immunological response. Lymphotactin exists in two very different structures that are in equilibrium (Figure 2.61). One structure is a characteristic of chemokines, consisting of a three-stranded β sheet and a

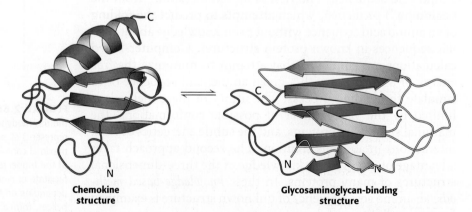

Chemokine structure   **Glycosaminoglycan-binding structure**

FIGURE 2.61 Lymphotactin exists in two conformations, which are in equilibrium. [R. L. Tuinstra, F. C. Peterson, S. Kutlesa, E. S. Elgin, M. A. Kron, and B. F. Volkman. *Proc. Natl. Acad. Sci. U.S.A.* 105:5057–5062, 2008, Fig. 2A.]

carboxyl-terminal helix. This structure binds to its receptor and activates it. The alternative structure is an identical dimer of all β sheets. When in this structure, lymphotactin binds to glycosaminoglycan, a complex carbohydrate (Chapter 11). The biochemical activities of each structure are mutually exclusive: the chemokine structure cannot bind the glycosaminoglycan, and the β-sheet structure cannot activate the receptor. Yet, remarkably, both activities are required for full biological activity of the chemokine.

Note that IUPs and metamorphic proteins effectively expand the protein-encoding capacity of the genome. In some cases, a gene can encode a single protein that has more than one structure and function. These examples also illustrate the dynamic nature of the study of biochemistry and its inherent excitement: even well-established ideas are often subject to modifications.

Protein misfolding and aggregation are associated with some neurological diseases

Understanding protein folding and misfolding is of more than academic interest. A host of diseases, including Alzheimer disease, Parkinson disease, Huntington disease, and transmissible spongiform encephalopathies (prion disease), are associated with improperly folded proteins. All of these diseases result in the deposition of protein aggregates, called *amyloid fibrils* or *plaques*. These diseases are consequently referred to as *amyloidoses*. A common feature of amyloidoses is that normally soluble proteins are converted into insoluble fibrils rich in β sheets. The correctly folded protein is only marginally more stable than the incorrect form. But the incorrect form aggregates, pulling more correct forms into the incorrect form. We will focus on the transmissible spongiform encephalopathies.

One of the great surprises in modern medicine was that certain infectious neurological diseases were found to be transmitted by agents that were similar in size to viruses but consisted only of protein. These diseases include *bovine spongiform encephalopathy* (commonly referred to as *mad cow disease*) and the analogous diseases in other organisms, including *Creutzfeld–Jakob disease* (CJD) in human beings, *scrapie* in sheep, and chronic wasting disease in deer and elk. The agents causing these diseases are termed *prions*. Prions are composed largely, if not exclusively, of a cellular protein called PrP, which is normally present in the brain; its function is still the focus of active research. The infectious prions are aggregated forms of the PrP protein termed PrP^{SC}.

How does the structure of the protein in the aggregated form differ from that of the protein in its normal state in the brain? The normal cellular protein PrP contains extensive regions of α helix and relatively little β strand. The structure of the form of PrP present in infected brains, termed PrP^{SC}, has not yet been determined because of challenges posed by its insoluble and heterogeneous nature. However, a variety of evidence indicates that some parts of the protein that had been in α-helical or turn conformations have been converted into β-strand conformations (Figure 2.62). The β strands of largely planar monomers stack on one another with their side chains tightly interwoven. A side view shows the extensive network of hydrogen bonds between the monomers. These fibrous protein aggregates are often referred to as *amyloid* forms.

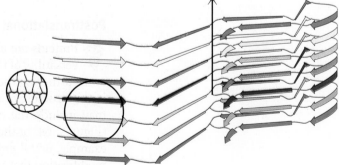

FIGURE 2.62 A model of the human prion protein amyloid. A detailed model of a human prion amyloid fibril deduced from spin labeling and electron paramagnetic resonance (EPR) spectroscopy studies shows that protein aggregation is due to the formation of large parallel β sheets. The black arrow indicates the long axis of the fibril. [N. J. Cobb, F. D. Sönnichsen, H. Mchaourab, and W. K. Surewicz. *Proc. Natl. Acad. Sci. U.S.A.* 104:18946–18951, 2007, Fig. 4E.]

FIGURE 2.63 The protein-only model for prion-disease transmission. A nucleus consisting of proteins in an abnormal conformation grows by the addition of proteins from the normal pool.

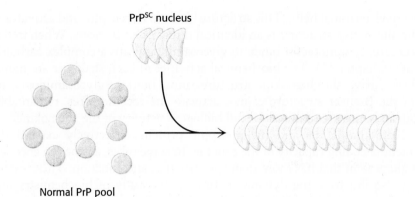

PrP^{SC} nucleus

Normal PrP pool

With the realization that the infectious agent in prion diseases is an aggregated form of a protein that is already present in the brain, a model for disease transmission emerges (Figure 2.63). Protein aggregates built of abnormal forms of PrP^{SC} act as sites of nucleation to which other PrP molecules attach. Prion diseases can thus be transferred from one individual organism to another through the transfer of an aggregated nucleus, as likely happened in the mad cow disease outbreak in the United Kingdom that emerged in the late 1980s. Cattle fed on animal feed containing material from diseased cows developed the disease in turn.

Amyloid fibers are also seen in the brains of patients with certain noninfectious neurodegenerative disorders such as Alzheimer and Parkinson diseases. For example, the brains of patients with Alzheimer disease contain protein aggregates called *amyloid plaques* that consist primarily of a single polypeptide termed Aβ. This polypeptide is derived from a cellular protein called *amyloid precursor protein* (APP) through the action of specific proteases. Polypeptide Aβ is prone to form insoluble aggregates. Despite the difficulties posed by the protein's insolubility, a detailed structural model for Aβ has been derived through the use of NMR techniques that can be applied to solids rather than to materials in solution. As expected, the structure is rich in β strands, which come together to form extended parallel β-sheet structures (Figure 2.62).

How do such aggregates lead to the death of the cells that harbor them? The answer is still controversial. One hypothesis is that the large aggregates themselves are not toxic but, instead, smaller aggregates of the same proteins may be the culprits, perhaps damaging cell membranes.

Posttranslational modifications confer new capabilities to proteins

Proteins are able to perform numerous functions that rely solely on the versatility of their 20 amino acids. This array of functions can be expanded even further by the incorporation of *posttranslational modifications*, alterations in the structure of a protein after its synthesis in the cell. These modifications may include the covalent attachment of groups to amino acid side chains (Figure 2.64) or the cleavage or modification of the polypeptide backbone. For example, *acetyl groups* are attached to the amino termini of many proteins, a modification that makes these proteins more resistant to degradation. As discussed earlier (p. 50), the addition of *hydroxyl groups* to many proline residues stabilizes fibers of newly synthesized collagen. The biological significance of this modification is evident in the disease scurvy: a deficiency of vitamin C results in insufficient hydroxylation of collagen, and the abnormal collagen fibers that result are unable to maintain normal tissue strength (Section 27.6). Another specialized amino acid is γ-*carboxyglutamate*. In vitamin K deficiency,

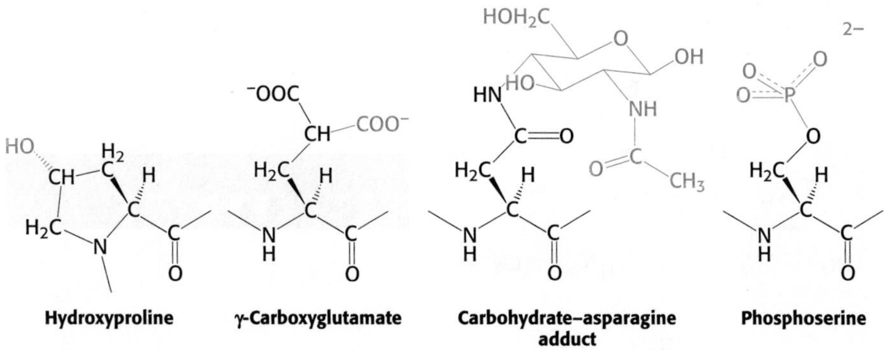

FIGURE 2.64 Finishing touches. Some common and important covalent modifications of amino acid side chains are shown.

insufficient carboxylation of glutamate in prothrombin, a clotting protein, can lead to hemorrhage (Section 10.4). Many proteins, especially those that are present on the surfaces of cells or are secreted, acquire *carbohydrate units* on specific asparagine, serine, or threonine residues (Chapter 11). The addition of sugars makes the proteins more hydrophilic and able to participate in interactions with other proteins. Conversely, the addition of a *fatty acid* to an α-amino group or a cysteine sulfhydryl group produces a more hydrophobic protein.

Many hormones, such as epinephrine (adrenaline), alter the activities of enzymes by stimulating the phosphorylation of the hydroxyl amino acids serine and threonine; *phosphoserine* and *phosphothreonine* are the most ubiquitous modified amino acids in proteins. Growth factors such as insulin act by triggering the phosphorylation of the hydroxyl group of tyrosine residues to form *phosphotyrosine*. The phosphoryl groups on these three modified amino acids are readily removed; thus, the modified amino acids are able to act as reversible switches in regulating cellular processes. The roles of phosphorylation in signal transduction will be discussed extensively in Chapter 14.

The preceding modifications consist of the addition of special groups to amino acids. Other special groups are generated by chemical rearrangements of side chains and, sometimes, the peptide backbone. For example, the jellyfish *Aequorea victoria* produces green fluorescent protein (GFP), which emits green light when stimulated with blue light. The source of the fluorescence is a group formed by the spontaneous rearrangement and oxidation of the sequence Ser-Tyr-Gly within the center of the protein (Figure 2.65A). Since the discovery of GFP, a number of mutants have been engineered which absorb and emit light across the entire visible spectrum (Figure 2.65B). These proteins are of great service to researchers as markers within cells (Figure 2.65C).

Finally, many proteins are cleaved and trimmed after synthesis. For example, digestive enzymes are synthesized as inactive precursors that can be stored safely in the pancreas. After release into the intestine, these precursors become activated by peptide-bond cleavage (Section 10.4). In blood clotting, peptide-bond cleavage converts soluble fibrinogen into insoluble fibrin. A number of polypeptide hormones, such as adrenocorticotropic hormone, arise from the splitting of a single large precursor protein. Likewise, many viral proteins are produced by the cleavage of large polyprotein precursors. We shall encounter many more examples of modification and cleavage as essential features of protein formation and function. Indeed, these finishing touches account for much of the versatility, precision, and elegance of protein action and regulation.

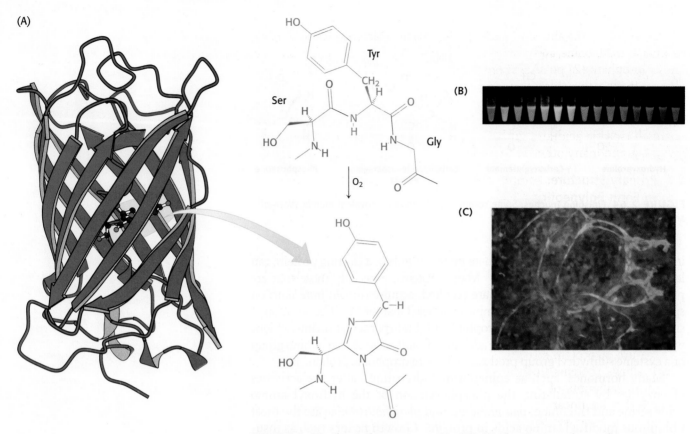

FIGURE 2.65 Chemical rearrangement in GFP. (A) The structure of green fluorescent protein (GFP). The rearrangement and oxidation of the sequence Ser-Tyr-Gly is the source of fluorescence. (B) Mutants of GFP emit light across the visible spectrum. (C) A melanoma cell line engineered to express one of these GFP mutants, red fluorescent protein (RFP), was then injected into a mouse whose blood vessels express GFP. In this fluorescence micrograph, the formation of new blood vessels (green) in the tumor (red) is readily apparent. [(A) Drawn from 1GFL.pdb; (B) Roger Y. Tsien, Review: Nobel Lecture: constructing and exploiting the fluorescent protein paintbox. *Integrative Biology*, 22 Feb. 2010. © The Nobel Foundation 2008. (C) Yang, M. et al. Dual-color fluorescence imaging distinguishes tumor cells from induced host angiogenic vessels and stromal cells. *Proc. Natl. Acad. Sci. U.S.A* 100, 14259–14262 (2003). Copyright 2003 National Academy of Sciences, U.S.A.]

SUMMARY

Protein structure can be described at four levels. The primary structure refers to the amino acid sequence. The secondary structure refers to the conformation adopted by local regions of the polypeptide chain. Tertiary structure describes the overall folding of the polypeptide chain. Finally, quaternary structure refers to the specific association of multiple polypeptide chains to form multisubunit complexes.

2.1 Proteins Are Built from a Repertoire of 20 Amino Acids

Proteins are linear polymers of amino acids. Each amino acid consists of a central tetrahedral carbon atom linked to an amino group, a carboxylic acid group, a distinctive side chain, and a hydrogen atom. These tetrahedral centers, with the exception of that of glycine, are chiral; only the L isomer exists in natural proteins. Nearly all natural proteins are constructed from the same set of 20 amino acids. The side chains of these 20 building blocks vary tremendously in size, shape, and the presence of functional groups. They can be grouped as follows: (1) hydrophobic side chains, including the aliphatic amino acids—glycine, alanine, valine, leucine, isoleucine, methionine, and proline—and aromatic side chains—phenylalanine, and tryptophan; (2) polar side chains, including hydroxyl-containing side chains—serine, threonine and tyrosine; the sulfhydryl-containing cysteine; and carboxamide-containing side chains—asparagine and glutamine; (3) basic side chains—lysine, arginine, and histidine; and (4) acidic side chains—aspartic acid and glutamic acid. These groupings are somewhat arbitrary, and many other sensible groupings are possible.

2.2 Primary Structure: Amino Acids Are Linked by Peptide Bonds to Form Polypeptide Chains

The amino acids in a polypeptide are linked by amide bonds formed between the carboxyl group of one amino acid and the amino group of the next. This linkage, called a peptide bond, has several important properties. First, it is resistant to hydrolysis, and so proteins are remarkably stable kinetically. Second, the peptide group is planar because the C — N bond has considerable double-bond character. Third, each peptide bond has both a hydrogen-bond donor (the NH group) and a hydrogen-bond acceptor (the CO group). Hydrogen bonding between these backbone groups is a distinctive feature of protein structure. Finally, the peptide bond is uncharged, which allows proteins to form tightly packed globular structures having significant amounts of the backbone buried within the protein interior. Because they are linear polymers, proteins can be described as sequences of amino acids. Such sequences are written from the amino to the carboxyl terminus.

2.3 Secondary Structure: Polypeptide Chains Can Fold into Regular Structures Such As the Alpha Helix, the Beta Sheet, and Turns and Loops

Two major elements of secondary structure are the α helix and the β strand. In the α helix, the polypeptide chain twists into a tightly packed rod. Within the helix, the CO group of each amino acid is hydrogen bonded to the NH group of the amino acid four residues farther along the polypeptide chain. In the β strand, the polypeptide chain is nearly fully extended. Two or more β strands connected by NH-to-CO hydrogen bonds come together to form β sheets. The strands in β sheets can be antiparallel, parallel, or mixed.

2.4 Tertiary Structure: Proteins Can Fold into Globular or Fibrous Structures

The compact, asymmetric structure that individual polypeptides attain is called tertiary structure. The tertiary structures of water-soluble globular proteins have features in common: (1) an interior formed of amino acids with hydrophobic side chains, and (2) a surface formed largely of hydrophilic amino acids that interact with the aqueous environment. The hydrophobic interactions between the interior residues are the driving force for the formation of the tertiary structure of water-soluble proteins. Some proteins that exist in a hydrophobic environment, such as in membranes, display the inverse distribution of hydrophobic and hydrophilic amino acids. In these proteins, the hydrophobic amino acids are on the surface to interact with the environment, whereas the hydrophilic groups are shielded from the environment in the interior of the protein. Fibrous proteins have extended, repetitive structures that provide strength and support to tissues such as hair, claws, and bone.

2.5 Quaternary Structure: Polypeptide Chains Can Assemble into Multisubunit Structures

Proteins consisting of more than one polypeptide chain display quaternary structure; each individual polypeptide chain is called a subunit. Quaternary structure can be as simple as two identical subunits or as complex as dozens of different subunits. In most cases, the subunits are held together by noncovalent bonds.

2.6 The Amino Acid Sequence of a Protein Determines Its Three-Dimensional Structure

The amino acid sequence determines the three-dimensional structure and, hence, all other properties of a protein. Some proteins can be unfolded completely yet refold efficiently when placed under conditions in which the folded form of the protein is stable. The amino acid sequence of a protein is determined by the sequences of bases in a DNA molecule. This one-dimensional sequence information is extended into the three-dimensional world by the ability of proteins to fold spontaneously. Protein folding is a highly cooperative process; structural intermediates between the unfolded and folded forms do not accumulate.

Some proteins, such as intrinsically unstructured proteins and metamorphic proteins, do not strictly adhere to the one-sequence–one-structure paradigm. Because of this versatility, these proteins expand the protein encoding capacity of the genome.

The versatility of proteins is further enhanced by posttranslational modifications. Such modifications can incorporate functional groups not present in the 20 amino acids. Other modifications are important to the regulation of protein activity. Through their structural stability, diversity, and chemical reactivity, proteins make possible most of the key processes associated with life.

APPENDIX

Visualizing Molecular Structures: Proteins

Scientists have developed powerful techniques for the determination of protein structures, as will be considered in Chapter 3. In most cases, these techniques allow the positions of the thousands of atoms within a protein structure to be determined. The final results from such an experiment include the x, y, and z coordinates for each atom in the structure. These coordinate files are compiled in the Protein Data Bank (http://www.pdb.org) from which they can be readily downloaded. These structures comprise thousands or even tens of thousands of atoms. The complexity of proteins with thousands of atoms presents a challenge for the depiction of their structure. Several different types of representations are used to portray proteins, each with its own strengths and weaknesses. The types that you will see most often in this book are space-filling models, ball-and-stick models, backbone models, and ribbon diagrams. Where appropriate, structural features of particular importance or relevance are noted in an illustration's legend.

Space-Filling Models

Space-filling models are the most realistic type of representation. Each atom is shown as a sphere with a size corresponding to the van der Waals radius of the atom (Section 1.3). Bonds are not shown explicitly but are represented by the intersection of the spheres shown when atoms are closer together than the sum of their van der Waals radii. All atoms are shown, including those that make up the backbone and those in the side chains. A space-filling model of lysozyme is depicted in Figure 2.66.

Space-filling models convey a sense of how little open space there is in a protein's structure, which always has many atoms in van der Waals contact with one another. These models are particularly useful in showing conformational changes in a protein from one set of circumstances to another. A disadvantage of space-filling models is that the secondary and tertiary structures of the protein are difficult to see. Thus,

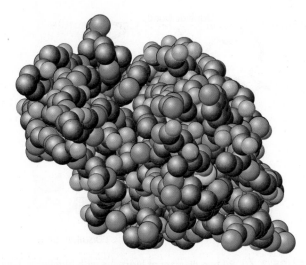

FIGURE 2.66 Space-filling model of lysozyme. *Notice* how tightly packed the atoms are, with little unfilled space. All atoms are shown with the exception of hydrogen atoms. Hydrogen atoms are often omitted because their positions are not readily determined by x-ray crystallographic methods and because their omission somewhat improves the clarity of the structure's depiction.

these models are not very effective in distinguishing one protein from another—many space-filling models of proteins look very much alike.

Ball-and-Stick Models

Ball-and-stick models are not as realistic as space-filling models. Realistically portrayed atoms occupy more space, determined by their van der Waals radii, than do the atoms depicted in ball-and-stick models. However, the bonding arrangement is easier to see because the bonds are explicitly represented as sticks (Figure 2.67). A ball-and-stick model reveals a complex structure more clearly than a space-filling model does. Yet, the depiction is so complicated that structural features such as α helices or potential binding sites are difficult to discern.

Because space-filling and ball-and-stick models depict protein structures at the atomic level, the large number of atoms in a complex structure makes it difficult to distinguish the relevant structural features. Thus, representations that are more schematic—such as backbone models and ribbon diagrams—have been developed for the depiction of macromolecular structures. In these representations, most or all atoms are not shown explicitly.

Backbone Models

Backbone models show only the backbone atoms of a polypeptide chain or even only the α-carbon atom of each amino acid. Atoms are linked by lines representing bonds; if only α-carbon atoms are depicted, lines connect α-carbon atoms of amino acids that are adjacent in the amino acid sequence (Figure 2.68). In this book, backbone models show only the lines connecting the α-carbon atoms; other carbon atoms are not depicted.

A backbone model shows the overall course of the polypeptide chain much better than a space-filling or ball-and-stick model does. However, secondary structural elements are still difficult to see.

Ribbon Diagrams

Ribbon diagrams are highly schematic and most commonly used to accent a few dramatic aspects of protein structure, such as the α helix (depicted as a coiled ribbon or a cylinder), the β strand (a broad arrow), and loops (thin tubes), to provide clear views of the folding patterns of proteins (Figure 2.69). The ribbon diagram allows the course of a polypeptide chain to be traced and readily shows the secondary structural elements. Thus, ribbon diagrams of proteins that are related to one another by evolutionary divergence appear similar (Figure 6.15), whereas unrelated proteins are clearly distinct.

In this book, coiled ribbons will be generally used to depict α helices. However, for membrane proteins, which are often quite complex, cylinders will be used rather than coiled

FIGURE 2.67 Ball-and-stick model of lysozyme. Again, hydrogen atoms are omitted.

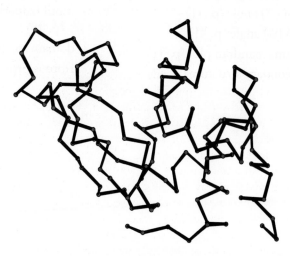

FIGURE 2.68 Backbone model of lysozyme.

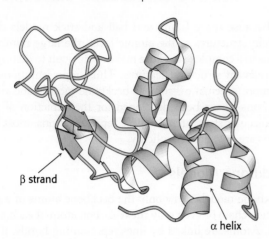

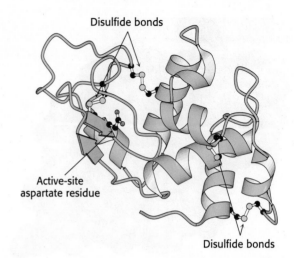

FIGURE 2.69 Ribbon diagram of lysozyme. The α helices are shown as coiled ribbons; β strands are depicted as arrows. More irregular structures are shown as thin tubes.

FIGURE 2.70 Ribbon diagram of lysozyme with highlights. Four disulfide bonds and a functionally important aspartate residue are shown in ball-and-stick form.

ribbons. This convention will also make membrane proteins with their membrane-spanning α helices easy to recognize (Figure 12.18).

Bear in mind that the open appearance of ribbon diagrams is deceptive. As noted earlier, protein structures are tightly packed and have little open space. The

openness of ribbon diagrams makes them particularly useful as frameworks in which to highlight additional aspects of protein structure. Active sites, substrates, bonds, and other structural fragments can be included in ball-and-stick or space-filling form within a ribbon diagram (Figure 2.70).

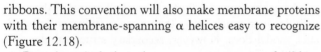

KEY TERMS

enzymes (p. 30)

side chain (R group) (p. 31)

L amino acid (p. 31)

dipolar ion (zwitterion) (p. 31)

peptide bond (amide bond) (p. 37)

disulfide bond (p. 38)

primary structure (p. 39)

torsion angle (p. 41)

phi (ϕ) angle (p. 41)

psi (ψ) angle (p. 41)

Ramachandran plot (p. 41)

secondary structure (p. 42)

α helix (p. 42)

rise (translation) (p. 43)

β pleated sheet (p. 44)

β strand (p. 44)

reverse turn (β turn; hairpin turn) (p. 46)

tertiary structure (p. 47)

globular protein (p. 47)

motif (supersecondary structure) (p. 48)

domain (p. 48)

fibrous protein (p. 49)

coiled coil (p. 49)

heptad repeat (p. 49)

subunit (p. 51)

quaternary structure (p. 51)

cooperative transition (p. 55)

intrinsically unstructured protein (IUP) (p. 58)

metamorphic protein (p. 58)

prion (p. 59)

posttranslational modification (p. 60)

PROBLEMS

1. *Identify.* Examine the following four amino acids (A–D):

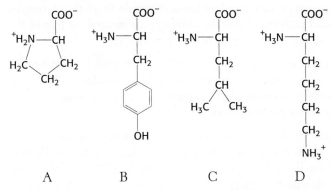

A   B   C   D

What are their names, three-letter abbreviations, and one-letter symbols? ✓❶

2. *Properties.* In reference to the amino acids shown in problem 1, which are associated with the following characteristics? ✓❶

(a) Hydrophobic side chain _____

(b) Basic side chain _____

(c) Three ionizable groups _____

(d) pK_a of approximately 10 in proteins _____

(e) Modified form of phenylalanine _____

3. *Match 'em.* Match each amino acid in the left-hand column with the appropriate side-chain type in the right-hand column. ✓❶

(a) Leu   (1) hydroxyl-containing

(b) Glu   (2) acidic

(c) Lys   (3) basic

(d) Ser   (4) sulfur-containing

(e) Cys   (5) nonpolar aromatic

(f) Trp   (6) nonpolar aliphatic

4. *Solubility.* In each of the following pairs of amino acids, identify which amino acid would be more soluble in water: (a) Ala, Leu; (b) Tyr, Phe; (c) Ser, Ala; (d) Trp, His.

5. *Bonding is good.* Which of the following amino acids have R groups that have hydrogen-bonding potential? Ala, Gly, Ser, Phe, Glu, Tyr, Ile, and Thr.

6. *Name those components.* Examine the segment of a protein shown here.

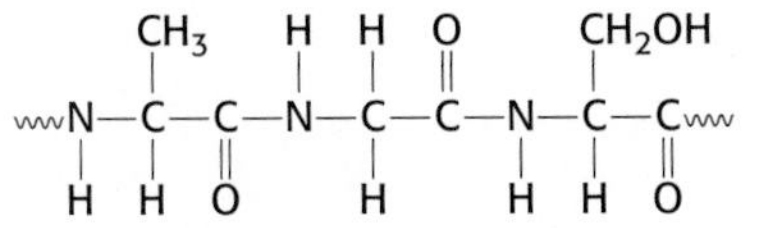

$$\text{\large wwwN--C--C--N--C--C--N--C--Cwww}$$

(a) What three amino acids are present?

(b) Of the three, which is the N-terminal amino acid?

(c) Identify the peptide bonds.

(d) Identify the α-carbon atoms.

7. *Who's charged?* Draw the structure of the dipeptide Gly-His. What is the charge on the peptide at pH 5.5? pH 7.5?

8. *Alphabet soup.* How many different polypeptides of 50 amino acids in length can be made from the 20 common amino acids?

9. *Sweet tooth, but calorie conscious.* Aspartame (NutraSweet), an artificial sweetener, is a dipeptide composed of Asp-Phe in which the carboxyl terminus is modified by the attachment of a methyl group. Draw the structure of Aspartame at pH 7.

10. *Isoelectricity.* On Figure 2.6, indicate the pH value at which a solution of this amino acid would have no net charge.

11. *One step further.* Figure 2.6 depicts the species profile for an amino acid, which carries a neutral side chain (e.g., alanine or glycine). What would this graph look like for histidine, assuming the pK_a of its carboxylic acid is 1.8, the pK_a of its amino group is 9.3, and the pK_a for its side chain is 6.0? Draw the structures of each of the individual species.

12. *Vertebrate proteins?* What is meant by the term *polypeptide backbone*?

13. *Not a sidecar.* Define the term *side chain* in the context of amino acid or protein structure.

14. *One from many.* Differentiate between *amino acid composition* and *amino acid sequence*. ✓❷

15. *Shape and dimension.* (a) Tropomyosin, a 70-kDa muscle protein, is a two-stranded α-helical coiled coil. Estimate the length of the molecule. (b) Suppose that a 40-residue segment of a protein folds into a two-stranded antiparallel β structure with a 4-residue hairpin turn. What is the longest dimension of this motif? ✓❸

16. *Contrasting isomers.* Poly-L-leucine in an organic solvent such as dioxane is α helical, whereas poly-L-isoleucine is not. Why do these amino acids with the same number and kinds of atoms have different helix-forming tendencies? ✓❸

17. *Exceptions to the rule.* Ramachandran plots for two amino acids differ significantly from that shown in Figure 2.23. Which two, and why?

18. *Active again.* A mutation that changes an alanine residue in the interior of a protein to valine is found to lead to a loss of activity. However, activity is regained when a second mutation at a different position changes an isoleucine residue to glycine. How might this second mutation lead to a restoration of activity?

19. *Exposure issues.* Many of the loops on proteins are composed of hydrophilic amino acids. Why might this be the case?

20. *Shuffle test.* An enzyme that catalyzes disulfide–sulfhydryl exchange reactions, called protein disulfide isomerase (PDI), has been isolated. PDI rapidly converts inactive scrambled ribonuclease into enzymatically active ribonuclease. In contrast, insulin is rapidly inactivated by PDI. What does this important observation imply about the relation between the amino acid sequence of insulin and its three-dimensional structure?

21. *Stretching a target.* A protease is an enzyme that catalyzes the hydrolysis of the peptide bonds of target proteins. How might a protease bind a target protein so that its main chain becomes fully extended in the vicinity of the vulnerable peptide bond?

22. *Often irreplaceable.* Glycine is a highly conserved amino acid residue in the evolution of proteins. Why?

23. *Potential partners.* Identify the groups in a protein that can form hydrogen bonds or electrostatic bonds with an arginine side chain at pH 7.

24. *Permanent waves.* The shape of hair is determined in part by the pattern of disulfide bonds in keratin, its major protein. How can curls be induced?

25. *Location is everything 1.* Most proteins have hydrophilic exteriors and hydrophobic interiors. Would you expect this structure to apply to proteins embedded in the hydrophobic interior of a membrane? Explain. ✓④

26. *Location is everything 2.* Proteins that span biological membranes often contain α helices. Given that the insides of membranes are highly hydrophobic (Section 12.2), predict what type of amino acids would be in such an α helix. Why is an α helix particularly suited to existence in the hydrophobic environment of the interior of a membrane? ✓④

27. *Neighborhood peer pressure?* Table 2.1 shows the typical pK_a values for ionizable groups in proteins. However, more than 500 pK_a values have been determined for individual groups in folded proteins. Account for this discrepancy.

28. *Greasy patches.* The α and β subunits of hemoglobin bear a remarkable structural similarity to myoglobin. However, in the subunits of hemoglobin, certain residues that are hydrophilic in myoglobin are hydrophobic. Why might this be the case?

29. *Maybe size does matter.* Osteogenesis imperfecta displays a wide range of symptoms, from mild to severe. On the basis of your knowledge of amino acid and collagen structure, propose a biochemical basis for the variety of symptoms.

30. *Issues of stability.* Proteins are quite stable. The lifetime of a peptide bond in aqueous solution is nearly 1000 years. However, the free energy of hydrolysis of proteins is negative and quite large. How can you account for the stability of the peptide bond in light of the fact that hydrolysis releases much energy?

31. *Minor species.* For an amino acid such as alanine, the major species in solution at pH 7 is the zwitterionic form. Assume a pK_a value of 8 for the amino group and a pK_a value of 3 for the carboxylic acid. Estimate the ratio of the concentration of the neutral amino acid species (with the carboxylic acid protonated and the amino group neutral) to that of the zwitterionic species at pH 7 (Section 1.3).

32. *A matter of convention.* All L amino acids have an *S* absolute configuration except L-cysteine, which has the *R* configuration. Explain why L-cysteine is designated as having the *R* absolute configuration.

33. *Hidden message.* Translate the following amino acid sequence into one-letter code: Glu-Leu-Val-Ile-Ser-Ile-Ser-Leu-Ile-Val-Ile-Asn-Gly-Ile-Asn-Leu-Ala-Ser-Val-Glu-Gly-Ala-Ser. ✓❶

34. *Who goes first?* Would you expect Pro—X peptide bonds to tend to have cis conformations like those of X—Pro bonds? Why or why not?

35. *Matching.* For each of the amino acid derivatives shown here (A–E), find the matching set of φ and ψ values (a–e).

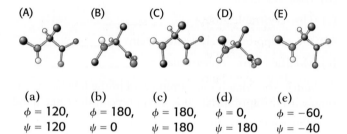

(a)
φ = 120,
ψ = 120

(b)
φ = 180,
ψ = 0

(c)
φ = 180,
ψ = 180

(d)
φ = 0,
ψ = 180

(e)
φ = −60,
ψ = −40

36. *Scrambled ribonuclease.* When performing his experiments on protein refolding, Christian Anfinsen obtained a quite different result when reduced ribonuclease was reoxidized while it was still in 8 M urea and the preparation was then dialyzed to remove the urea. Ribonuclease reoxidized in this way had only 1% of the enzymatic activity of the native protein. Why were the outcomes so different when reduced ribonuclease was reoxidized in the presence and absence of urea?

Think ▶ Pair ▶ Share Problem

37. A 2012 review article referred to posttranslational modifications as "Nature's escape from genetic imprisonment." Can you explain this rather lofty claim?

Exploring Proteins and Proteomes

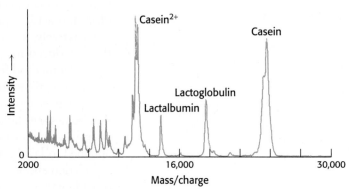

Milk, a source of nourishment for all mammals, is composed, in part, of a variety of proteins. The protein components of milk are revealed by the technique of MALDI–TOF mass spectrometry, which separates molecules on the basis of their mass-to-charge ratio. [bitt24/Shutterstock.]

✓ LEARNING OBJECTIVES

By the end of this chapter, you should be able to:

1. Explain the importance of protein purification and how it can be quantified.

2. Describe the different types of chromatography used to purify proteins.

3. Describe the different methods available for determining the mass of a protein.

4. Explain how antibodies can be used as tools for protein identification, purification, and quantitation.

5. Describe how proteins are sequenced, and explain why sequence determination is important.

6. Distinguish between the commonly used methods for protein structure determination, and describe the advantages and disadvantages of each.

OUTLINE

3.1 The Purification of Proteins Is an Essential First Step in Understanding Their Function

3.2 Immunology Provides Important Techniques with Which to Investigate Proteins

3.3 Mass Spectrometry Is a Powerful Technique for the Identification of Peptides and Proteins

3.4 Peptides Can Be Synthesized by Automated Solid-Phase Methods

3.5 Three-Dimensional Protein Structure Can Be Determined by X-ray Crystallography, NMR Spectroscopy, and Cryo-Electron Microscopy

Proteins play crucial roles in nearly all biological processes—in catalysis, signal transmission, and structural support. This remarkable range of functions arises from the existence of thousands of proteins, each folded into a distinctive three-dimensional structure that enables it to interact with one or more of a highly diverse array of molecules. A major goal of biochemistry is to determine how amino acid sequences specify the conformations, and hence functions, of proteins. Other goals are to learn how individual proteins bind specific substrates and other molecules, mediate catalysis, and transduce energy and information.

It is often preferable to study a protein of interest after it has been separated from other components within the cell so that the structure and function of this protein can be probed without any confounding effects from contaminants. Hence, the first step in these studies is the purification of the protein of interest. Proteins can be separated from one another on the basis of solubility, size, charge, and binding ability. After a protein has been purified, its amino acid sequence can be determined.

Many protein sequences, often deduced from genome sequences, are available in vast sequence databases. If the sequence of a purified protein has been archived in a publicly searchable database, the job of the investigator becomes much easier. The investigator need determine only a small stretch of amino acid sequence of the protein to find its match in the database. Alternatively, such a protein might be identified by matching its mass to those deduced for proteins in the database. Mass spectrometry provides a powerful method for determining the mass and sequence of a protein.

After a protein has been purified and its identity confirmed, the challenge remains to determine its function within a physiologically relevant context. Antibodies are choice probes for locating proteins in vivo and measuring their quantities. Monoclonal antibodies, able to recognize specific proteins, can be obtained in large amounts and used to detect and quantify the protein both in isolation and in cells. Peptides and proteins can be chemically synthesized, providing tools for research and, in some cases, highly pure material for use as drugs. Finally, x-ray crystallography, nuclear magnetic resonance (NMR) spectroscopy, and cryo-electron microscopy (cryo-EM) are the principal techniques for elucidating three-dimensional structure, the key determinant of function.

The exploration of proteins by this array of physical and chemical techniques has greatly enriched our understanding of the molecular basis of life. These techniques make it possible to tackle some of the most challenging questions of biology in molecular terms.

The proteome is the functional representation of the genome

As will be discussed in Chapter 5, the complete DNA base sequences, or *genomes*, of many organisms are now available. For example, the roundworm *Caenorhabditis elegans* has a genome of 97 million bases and about 19,000 protein-encoding genes, whereas that of the fruit fly *Drosophila melanogaster* contains 180 million bases and about 14,000 genes. The completely sequenced human genome contains 3 billion bases and about 23,000 genes. However, these genomes are simply inventories of the genes that *could* be expressed within a cell under specific conditions. Only a subset of the proteins encoded by these genes will actually be present in a given biological context. The *proteome*—derived from *prot*eins expressed by the gen*ome*—of an organism signifies a more complex level of information content, encompassing the types, functions, and interactions of proteins within its biological environment.

The proteome is not a fixed characteristic of the cell. Because it represents the functional expression of information, it varies with cell type, developmental stage, and environmental conditions. The proteome is much larger than the genome because almost all gene products are proteins that can be chemically modified in a variety of ways. Furthermore, these proteins do not exist in isolation; they often interact with one another to form complexes with specific functional properties. Whereas the genome is "hard wired," the proteome is highly dynamic. An understanding of the proteome is acquired by investigating, characterizing, and cataloging proteins. In some, but not all, cases, this process begins by separating a particular protein from all other biomolecules in the cell.

3.1 The Purification of Proteins Is an Essential First Step in Understanding Their Function

An adage of biochemistry is "never waste pure thoughts on an impure protein." Starting from pure proteins, we can determine amino acid sequences and investigate biochemical functions. From the amino acid sequences, we can map evolutionary relationships between proteins in diverse organisms

(Chapter 6). By using crystals grown from pure protein, we can obtain x-ray data that will provide us with a picture of the protein's tertiary structure—the shape that determines function.

The assay: How do we recognize the protein we are looking for?

Purification should yield a sample containing only one type of molecule—the protein in which the biochemist is interested. This protein sample may be only a fraction of 1% of the starting material, whether that starting material consists of one type of cell in culture or a particular organ from a plant or animal. How is the biochemist able to isolate a particular protein from a complex mixture of proteins?

A protein can be purified by subjecting the impure mixture of the starting material to a series of separations based on physical properties such as size and charge. To monitor the success of this purification, the biochemist needs a test, called an *assay*, for some unique identifying property of the protein. A positive result on the assay indicates that the protein is present. Although assay development can be a challenging task, the more specific the assay, the more effective the purification. For enzymes, which are protein catalysts (Chapter 8), the assay usually measures *enzyme activity*—that is, the ability of the enzyme to promote a particular chemical reaction. This activity is often measured indirectly. Consider the enzyme lactate dehydrogenase, which catalyzes the following reaction in the synthesis of glucose:

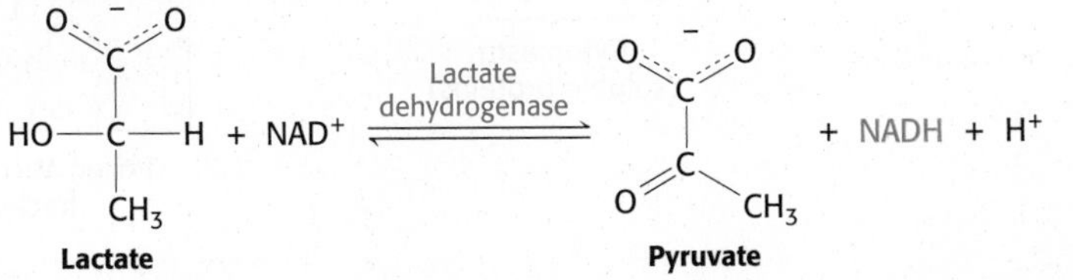

Reduced nicotinamide adenine dinucleotide (NADH, Figure 15.13) absorbs light at 340 nm, whereas oxidized nicotinamide adenine dinucleotide (NAD^+) does not. Consequently, we can follow the progress of the reaction by examining how much light-absorbing ability is developed by a sample in a given period of time—for instance, within 1 minute after the addition of the enzyme. Our assay for enzyme activity during the purification of lactate dehydrogenase is thus the increase in the absorbance of light at 340 nm observed in 1 minute.

To analyze how our purification scheme is working, we need one additional piece of information—the amount of protein present in the mixture being assayed. There are various rapid and reasonably accurate means of determining protein concentration. With these two experimentally determined numbers—enzyme activity and protein concentration—we then calculate the *specific activity*, the ratio of enzyme activity to the amount of protein in the mixture. Ideally, the specific activity will rise as the purification proceeds and the protein mixture will contain the protein of interest to a greater extent. In essence, the overall goal of the purification is to maximize the specific activity. For a pure enzyme, the specific activity will have a constant value.

Proteins must be released from the cell to be purified

Having found an assay and chosen a source of protein, we now fractionate the cell into components and determine which component is enriched in the protein of interest. In the first step, a *homogenate* is formed by disrupting the cell membrane, and the mixture is fractionated by centrifugation, yielding a dense pellet of heavy material at the bottom of the centrifuge tube and a lighter supernatant above (Figure 3.1). The supernatant is again centrifuged

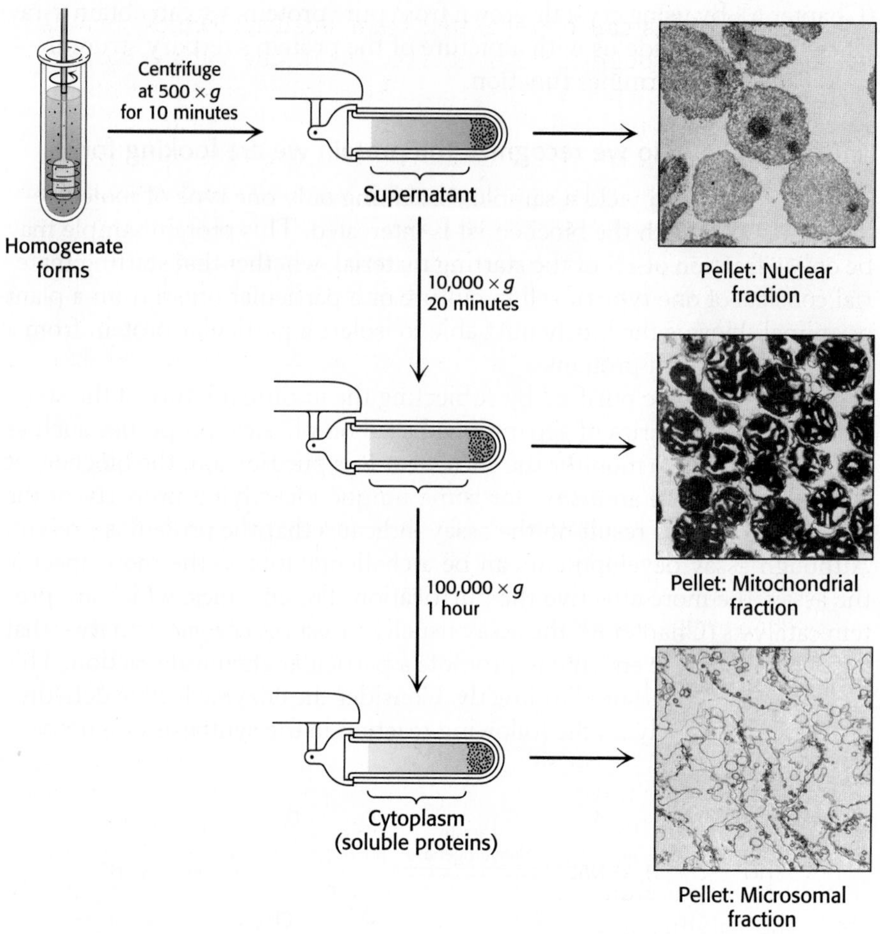

FIGURE 3.1 Differential centrifugation. Cells are disrupted in a homogenizer and the resulting mixture, called the homogenate, is centrifuged in a step-by-step fashion of increasing centrifugal force. The denser material will form a pellet at lower centrifugal force than will the less-dense material. The isolated fractions can be used for further purification. [Courtesy of S. Fleischer and B. Fleischer.]

at a greater force to yield yet another pellet and supernatant. This procedure, called *differential centrifugation*, yields several fractions of decreasing density, each still containing hundreds of different proteins. The fractions are each separately assayed for the desired activity. Usually, one fraction will be enriched for such activity, and it then serves as the source of material to which more-discriminating purification techniques are applied.

Proteins can be purified according to solubility, size, charge, and binding affinity

Several thousand proteins have been purified in active form on the basis of such characteristics as *solubility, size, charge,* and *binding affinity.* Usually, protein mixtures are subjected to a series of separations, each based on a different property. At each step in the purification, the preparation is assayed and its specific activity is determined. A variety of purification techniques are available.

Salting out. Most proteins are less soluble at high salt concentrations, an effect called *salting out.* The salt concentration at which a protein precipitates differs from one protein to another. Hence, salting out can be used to fractionate proteins. For example, 0.8 M ammonium sulfate precipitates fibrinogen, a blood-clotting protein, whereas a concentration of 2.4 M is needed to precipitate serum albumin. Salting out is also useful for concentrating dilute solutions of proteins, including active fractions obtained from other purification steps. Dialysis can be used to remove the salt if necessary.

Dialysis. Proteins can be separated from small molecules such as salt by *dialysis* through a semipermeable membrane, such as a cellulose membrane with pores (Figure 3.2). The protein mixture is placed inside the dialysis bag, which is then submerged in a buffer solution that is devoid of the small molecules to be separated away. Molecules having dimensions significantly greater than the pore diameter are retained inside the dialysis bag. Smaller molecules and ions capable of passing through the pores of the membrane diffuse down their concentration gradients and emerge in the solution outside the bag. This technique is useful for removing a salt or other small molecule from a cell fractionate, but it will not distinguish between proteins effectively.

Gel-filtration chromatography. More-discriminating separations on the basis of size can be achieved by the technique of *gel-filtration chromatography*, also known as molecular exclusion chromatography (Figure 3.3). The sample is applied to the top of a column consisting of porous beads made of an insoluble but highly hydrated polymer such as dextran or agarose (carbohydrates) or polyacrylamide. Sephadex, Sepharose, and Biogel are commonly used commercial preparations of these beads, which are typically 100 μm (0.1 mm) in diameter. Small molecules can enter these beads; large ones cannot. The result is that small molecules are distributed in the aqueous solution both inside the beads and between them, whereas large molecules are located only in the solution between the beads. *Large molecules flow more rapidly through this column and emerge first because a smaller volume is accessible to them.* Molecules of medium size occasionally enter the beads and will flow from the column at an intermediate position, while small molecules, which take a longer, tortuous path, will exit last.

Ion-exchange chromatography. To obtain a protein of high purity, one chromatography step is usually not sufficient, because other proteins in the crude mixture will likely co-elute with the desired material. Additional

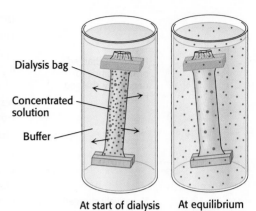

FIGURE 3.2 Dialysis. Protein molecules (red) are retained within the dialysis bag, whereas small molecules (blue) diffuse down their concentration gradient into the surrounding medium.

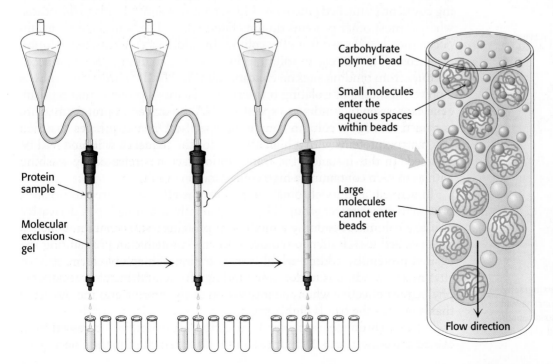

FIGURE 3.3 Gel-filtration chromatography. A mixture of proteins in a small volume is applied to a column filled with porous beads. Because large proteins cannot enter the internal volume of the beads, they emerge sooner than do small ones.

FIGURE 3.4 Ion-exchange chromatography. This technique separates proteins mainly according to their net charge.

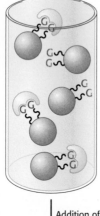

Glucose-binding protein attaches to glucose residues (G) on beads

Addition of glucose (G)

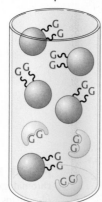

Glucose-binding proteins are released on addition of glucose

FIGURE 3.5 Affinity chromatography. Affinity chromatography of concanavalin A (shown in yellow) using a solid support containing covalently attached glucose residues (G).

purity can be achieved by performing sequential separations based on distinct molecular properties. For example, in addition to size, proteins can be separated on the basis of their net charge by *ion-exchange chromatography*. If a protein has a net positive charge at pH 7, it will usually bind to a column of beads containing carboxylate groups, whereas a negatively charged protein will not (Figure 3.4). The bound protein can then be eluted (released) by increasing the concentration of sodium chloride or another salt in the eluting buffer; sodium ions compete with positively charged groups on the protein for binding to the column. Proteins that have a low density of net positive charge will tend to emerge first, followed by those having a higher charge density. This procedure is also referred to as *cation exchange* to indicate that positively charged groups will bind to the anionic beads. Positively charged proteins (cationic proteins) can be separated by chromatography on negatively charged carboxymethylcellulose (CM-cellulose) columns. Conversely, negatively charged proteins (anionic proteins) can be separated by *anion exchange* on positively charged diethylaminoethylcellulose (DEAE-cellulose) columns.

Carboxymethyl (CM) group (ionized form)

Diethylaminoethyl (DEAE) group (protonated form)

Affinity chromatography. *Affinity chromatography* is another powerful means of purifying proteins that is highly selective for the protein of interest. This technique takes advantage of the high affinity of many proteins for specific chemical groups. For example, the plant protein concanavalin A is a carbohydrate-binding protein, or lectin (Section 11.4), that has affinity for glucose. When a crude extract is passed through a column of beads containing covalently attached glucose residues, concanavalin A binds to the beads, whereas most other proteins do not (Figure 3.5). The bound concanavalin A can then be released from the column by adding a concentrated solution of glucose. The glucose in solution displaces the column-attached glucose residues from binding sites on concanavalin A. Affinity chromatography is a powerful means of isolating transcription factors—proteins that regulate gene expression by binding to specific DNA sequences. A protein mixture is passed through a column containing specific DNA sequences attached to a matrix; proteins with a high affinity for the sequence will bind and be retained. In this instance, the transcription factor is released by washing with a solution containing a high concentration of salt.

In general, affinity chromatography can be effectively used to isolate a protein that recognizes group X by (1) covalently attaching X or a derivative of it to a column; (2) adding a mixture of proteins to this column, which is then washed with buffer to remove unbound proteins; and (3) eluting the desired protein by adding a high concentration of a soluble form of X or altering the conditions to decrease binding affinity. Affinity chromatography is most effective when the interaction of the protein and the molecule that is used as the bait is highly specific.

Affinity chromatography can be used to isolate proteins expressed from cloned genes (Section 5.2). Extra amino acids are encoded in the cloned gene that, when expressed, serve as an affinity tag that can be readily trapped. For example, repeats of the codon for histidine may be added such that the expressed protein has a string of histidine residues (called a *His tag*) on one end. The tagged proteins are then passed through a column of beads containing covalently attached, immobilized nickel(II) or other metal ions. The His tags bind tightly to the immobilized metal ions, binding the desired protein, while other proteins flow through the column. The protein can then be eluted from the column by the addition of imidazole or some other chemical that binds to the metal ions and displaces the protein.

High-performance liquid chromatography. A technique called *high-performance liquid chromatography* (HPLC) is an enhanced version of the column techniques already discussed. The column materials are much more finely divided and, as a consequence, possess more interaction sites and thus greater resolving power. Because the column is made of finer material, pressure must be applied to the column to obtain adequate flow rates. The net result is both high resolution and rapid separation. In a typical HPLC setup, a detector that monitors the absorbance of the eluate at a particular wavelength is placed immediately after the column. In the sample HPLC elution profile shown in Figure 3.6, proteins are detected by setting the detector to 220 nm (the characteristic absorbance wavelength of the peptide bond). In a short span of 10 minutes, a number of sharp peaks representing individual proteins can be readily identified.

Proteins can be separated by gel electrophoresis and displayed

How can we tell that a purification scheme is effective? One way is to ascertain that the specific activity rises with each purification step. Another is to determine that the number of different proteins in each sample declines at each step. The technique of electrophoresis makes the latter method possible.

Gel electrophoresis. A molecule with a net charge will move in an electric field. This phenomenon, termed *electrophoresis*, offers a powerful means of separating proteins and other macromolecules, such as DNA and RNA. The velocity of migration (v) of a protein (or any molecule) in an electric field depends on the electric field strength (E), the net charge on the protein (z), and the frictional coefficient (f).

$$v = Ez/f \qquad (1)$$

The electric force Ez driving the charged molecule toward the oppositely charged electrode is opposed by the viscous drag fv arising from friction between the moving molecule and the medium. The frictional coefficient f depends on both the mass and shape of the migrating molecule and the viscosity (η) of the medium. For a sphere of radius r,

$$f = 6\pi\eta r \qquad (2)$$

Electrophoretic separations are nearly always carried out in porous gels (or on solid supports such as paper) because the gel serves as a molecular sieve that enhances separation (Figure 3.7). Molecules that are small compared with the pores in the gel readily move through the gel, whereas molecules much larger than the pores are almost immobile. Intermediate-size molecules move through the gel with various degrees of facility. The electric field is applied such that proteins migrate from the negative to the positive electrodes, typically from top to bottom. *Gel electrophoresis* is performed in a thin, vertical slab of polyacrylamide gel. Polyacrylamide

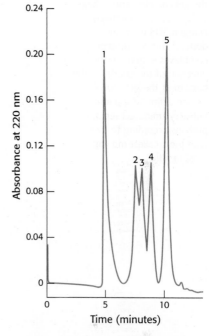

FIGURE 3.6 High-performance liquid chromatography (HPLC). Gel filtration by HPLC clearly defines the individual proteins because of its greater resolving power: (1) thyroglobulin (669 kDa), (2) catalase (232 kDa), (3) bovine serum albumin (67 kDa), (4) ovalbumin (43 kDa), and (5) ribonuclease (13.4 kDa). [Data from K. J. Wilson and T. D. Schlabach. In *Current Protocols in Molecular Biology*, vol. 2, suppl. 41, F. M. Ausubel, R. Brent, R. E. Kingston, D. D. Moore, J. G. Seidman, J. A. Smith, and K. Struhl, Eds. (Wiley, 1998), p. 10.14.1.]

FIGURE 3.7
Polyacrylamide gel
electrophoresis. (A) Gel-
electrophoresis apparatus.
Typically, several samples
undergo electrophoresis on
one flat polyacrylamide gel.
A microliter pipette is used
to place solutions of proteins
in the wells of the slab.
A cover is then placed over
the gel chamber and voltage
is applied. The negatively
charged SDS (sodium
dodecyl sulfate)–protein
complexes migrate in the
direction of the anode, at the
bottom of the gel. (B) The
sieving action of a porous
polyacrylamide gel separates
proteins according to size,
with the smallest moving
most rapidly.

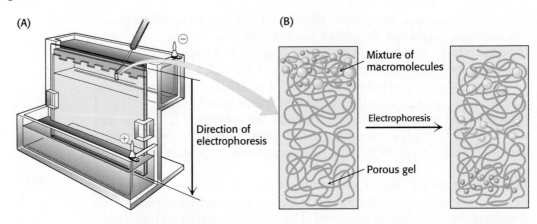

gels are choice supporting media for electrophoresis because they are chemically inert and readily formed by the polymerization of acrylamide with a small amount of the cross-linking agent methylenebisacrylamide to make a three-dimensional mesh (Figure 3.8). Electrophoresis is distinct from gel filtration in that, because of the electric field, all of the molecules, regardless of size, are forced to move through the same matrix.

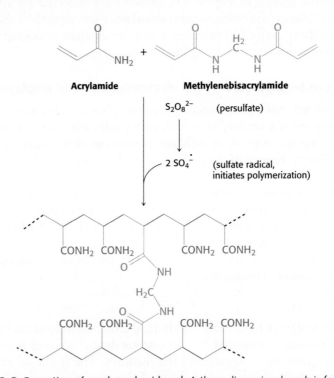

FIGURE 3.8 Formation of a polyacrylamide gel. A three-dimensional mesh is formed by copolymerizing activated monomer (blue) and cross-linker (red).

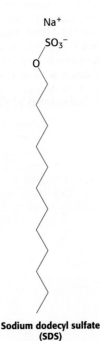

Proteins can be separated largely on the basis of mass by electrophoresis in a polyacrylamide gel under denaturing conditions. The mixture of proteins is first dissolved in a solution of sodium dodecyl sulfate (SDS), an anionic detergent that disrupts nearly all noncovalent interactions in native proteins. β-Mercaptoethanol (2-thioethanol) or dithiothreitol is added to reduce disulfide bonds. Anions of SDS bind to main chains at a ratio of about one SDS anion for every two amino acid residues. The negative charge acquired on binding SDS is usually much greater than the charge on the native protein; the contribution of the protein to the total charge of the SDS–protein complex is thus rendered insignificant. As a result, this complex of

SDS with a denatured protein has a large net negative charge roughly proportional to the mass of the protein. The SDS–protein complexes are then subjected to electrophoresis. When the electrophoresis is complete, the proteins in the gel can be visualized by staining them with silver nitrate or a dye such as Coomassie blue, which reveals a series of bands (Figure 3.9). Radioactive labels, if they have been incorporated into proteins, can be detected by placing a sheet of x-ray film over the gel, a procedure called *autoradiography*.

Small proteins move rapidly through the gel, whereas large proteins stay at the top, near the point of application of the mixture. The mobility of most polypeptide chains under these conditions is linearly proportional to the logarithm of their mass (Figure 3.10). Some carbohydrate-rich proteins and membrane proteins do not obey this empirical relation, however. SDS–polyacrylamide gel electrophoresis (often referred to as SDS-PAGE) is rapid, sensitive, and capable of a high degree of resolution. As little as 0.1 μg (~2 pmol) of a protein gives a distinct band when stained with Coomassie blue, and even less (~0.02 μg) can be detected with a silver stain. Proteins that differ in mass by about 2% (e.g., 50 and 51 kDa, arising from a difference of about 10 amino acids) can usually be distinguished with SDS-PAGE.

We can examine the efficacy of our purification scheme by analyzing a part of each fraction by SDS-PAGE. The initial fractions will display dozens to hundreds of proteins. As the purification progresses, the number of bands will diminish, and the prominence of one of the bands should increase. This band should correspond to the protein of interest.

Isoelectric focusing. Proteins can also be separated electrophoretically on the basis of their relative contents of acidic and basic residues. The *isoelectric point* (pI) of a protein is the pH at which its net charge is zero. At this pH, its electrophoretic mobility is zero because *z* in equation 1 is equal to zero. For example, the pI of cytochrome *c*, a highly basic electron-transport protein, is 10.6, whereas that of serum albumin, an acidic protein in blood, is 4.8. Suppose that a mixture of proteins undergoes electrophoresis in a pH gradient in a gel in the absence of SDS. Each protein will move until it reaches a position in the gel at which the pH is equal to the pI of the protein. This method of separating proteins according to their isoelectric point is called *isoelectric focusing*. The pH gradient in the gel is formed first by subjecting a mixture of *polyampholytes* (small multicharged polymers) having many different pI values to electrophoresis. Isoelectric focusing can readily resolve proteins that differ in pI by as little as 0.01, which means that proteins differing by one net charge can be separated (Figure 3.11).

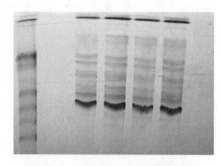

FIGURE 3.9 Staining of proteins after electrophoresis. Mixtures of proteins from cellular extracts subjected to electrophoresis on an SDS–polyacrylamide gel can be visualized by staining with Coomassie blue. The first lane contains a mixture of proteins of known molecular weights, which can be used to estimate the sizes of the bands in the samples. [© Dr. Robert Farrell.]

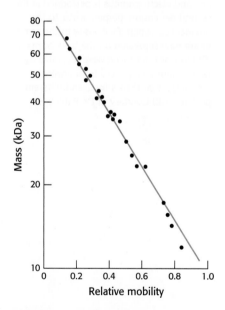

FIGURE 3.10 Electrophoresis can determine mass. The electrophoretic mobility of many proteins in SDS–polyacrylamide gels is inversely proportional to the logarithm of their mass. [Data from K. Weber and M. Osborn, *The Proteins*, vol. 1, 3d ed. (Academic Press, 1975), p. 179.]

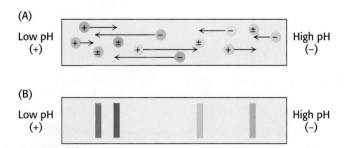

FIGURE 3.11 The principle of isoelectric focusing. A pH gradient is established in a gel before loading the sample. (A) Each protein, represented by the different colored circles, will possess a net positive charge in the regions of the gel where the pH is lower than its respective pI value and a net negative charge where the pH is greater than its pI. When voltage is applied to the gel, each protein will migrate to its pI, the location at which it has no net charge. (B) The proteins form bands that can be excised and used for further experimentation.

(A)

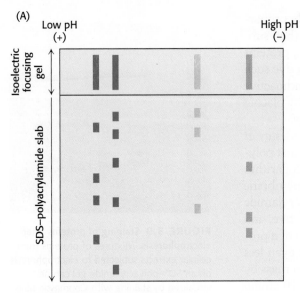

(B) Isoelectric focusing

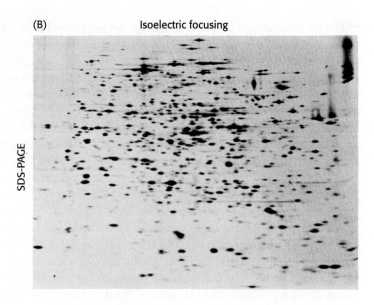

FIGURE 3.12 Two-dimensional gel electrophoresis. (A) A protein sample is initially fractionated in one dimension by isoelectric focusing as described in Figure 3.11. The isoelectric focusing gel is then attached to an SDS–polyacrylamide gel, and electrophoresis is performed in the second dimension, perpendicular to the original separation. Proteins with the same pI are now separated on the basis of mass. (B) Proteins from *E. coli* were separated by two-dimensional gel electrophoresis, resolving more than a thousand different proteins. [(B) Courtesy of Dr. Patrick H. O'Farrell.]

Two-dimensional electrophoresis. Isoelectric focusing can be combined with SDS-PAGE to obtain very high resolution separations by *two-dimensional electrophoresis*. A single sample is first subjected to isoelectric focusing. This single-lane gel is then placed horizontally on top of an SDS–polyacrylamide slab. The proteins are thus spread across the top of the polyacrylamide gel according to how far they migrated during isoelectric focusing. They then undergo electrophoresis again in a perpendicular direction (vertically) to yield a two-dimensional pattern of spots. In such a gel, proteins have been separated in the horizontal direction on the basis of isoelectric point and in the vertical direction on the basis of mass. Remarkably, more than a thousand different proteins in the bacterium *Escherichia coli* can be resolved in a single experiment by two-dimensional electrophoresis (Figure 3.12).

Proteins isolated from cells under different physiological conditions can be subjected to two-dimensional electrophoresis. The intensities of individual spots on the gels can then be compared, which indicates that the concentrations of specific proteins have changed in response to the physiological state (Figure 3.13). How can we discover the identity of a protein that is showing such responses? Although many proteins are displayed on a two-dimensional gel, they are not identified. It is now possible to identify proteins by coupling two-dimensional gel electrophoresis with mass spectrometric techniques. We will examine these powerful techniques shortly (Section 3.3).

FIGURE 3.13 Alterations in protein levels detected by two-dimensional gel electrophoresis. Samples of (A) normal colon mucosa and (B) colorectal tumor tissue from the same person were analyzed by two-dimensional gel electrophoresis. In the gel section shown, changes in the intensity of several spots are evident, including a dramatic increase in levels of the protein indicated by the arrow, corresponding to the enzyme glyceraldehyde-3-phosphate dehydrogenase. [Images courtesy Qingsong Lin © 2006 The American Society for Biochemistry and Molecular Biology.]

(A)

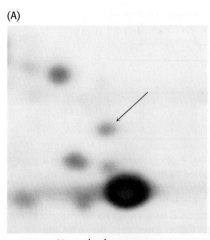

Normal colon mucosa

(B)

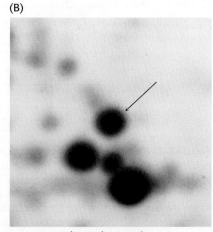

Colorectal tumor tissue

TABLE 3.1 Quantification of a purification protocol for a fictitious enzyme

Step	Total protein (mg)	Total activity (units)	Specific activity (units mg⁻¹)	Yield (%)	Purification level
Homogenization	15,000	150,000	10	100	1
Salt fractionation	4600	138,000	30	92	3
Ion-exchange chromatography	1278	115,500	90	77	9
Gel-filtration chromatography	68.8	75,000	1100	50	110
Affinity chromatography	1.75	52,500	30,000	35	3000

A protein purification scheme can be quantitatively evaluated

To determine the success of a protein purification scheme, we monitor each step of the procedure by determining the specific activity of the protein mixture and by subjecting it to SDS-PAGE analysis. Consider the results for the purification of a fictitious enzyme, summarized in Table 3.1 and Figure 3.14. At each step, the following parameters are measured:

Total Protein. The quantity of protein present in a fraction is obtained by determining the protein concentration of a part of each fraction and multiplying by the fraction's total volume.

Total Activity. The total activity for the fraction is obtained by measuring the enzyme activity in the volume of fraction used in the assay and multiplying by the fraction's total volume.

Specific Activity. This parameter is obtained by dividing total activity by total protein.

Yield. This parameter is a measure of the activity retained after each purification step as a percentage of the activity in the crude extract. The amount of activity in the initial extract is taken to be 100%.

Purification Level. This parameter is a measure of the increase in purity and is obtained by dividing the specific activity, calculated after each purification step, by the specific activity of the initial extract.

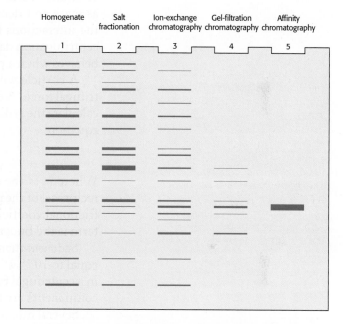

FIGURE 3.14 Electrophoretic analysis of a protein purification. The purification scheme in Table 3.1 was analyzed by SDS-PAGE. Each lane contained 50 μg of sample. The effectiveness of the purification can be seen as the band for the protein of interest becomes more prominent relative to other bands.

As we see in Table 3.1, the first purification step, salt fractionation, leads to an increase in purity of only 3-fold, but we recover nearly all the target protein in the original extract, given that the yield is 92%. After dialysis to lower the high concentration of salt remaining from the salt fractionation, the fraction is passed through an ion-exchange column. The purification now increases to 9-fold compared with the original extract, whereas the yield falls to 77%. Gel-filtration chromatography brings the level of purification to 110-fold, but the yield is now at 50%. The final step is affinity chromatography with the use of a ligand specific for the target enzyme. This step, the most powerful of these purification procedures, results in a purification level of 3000-fold but lowers the yield to 35%. The SDS-PAGE analysis in Figure 3.14 shows that, if we load a constant amount of protein onto each lane after each step, the number of bands decreases in proportion to the level of purification, and the amount of protein of interest increases as a proportion of the total protein present.

A good purification scheme takes into account both purification levels and yield. A high degree of purification and a poor yield leave little protein with which to experiment. A high yield with low purification leaves many contaminants (proteins other than the one of interest) in the fraction and complicates the interpretation of subsequent experiments.

Ultracentrifugation is valuable for separating biomolecules and determining their masses

We have already seen that centrifugation is a powerful and generally applicable method for separating a crude mixture of cell components. This technique is also valuable for the analysis of the physical properties of biomolecules. Using centrifugation, we can determine such parameters as mass and density, learn about the shape of a molecule, and investigate the interactions between molecules. To deduce these properties from the centrifugation data, we require a mathematical description of how a particle behaves when a centrifugal force is applied.

A particle will move through a liquid medium when subjected to a centrifugal force. A convenient means of quantifying the rate of movement is to calculate the sedimentation coefficient, s, of a particle by using the following equation:

$$s = m(1 - \bar{v}\rho)/f$$

Where m is the mass of the particle, $\bar{v}$ is the partial specific volume (the reciprocal of the particle density), ρ is the density of the medium, and f is the frictional coefficient (a measure of the shape of the particle). The $(1 - \bar{v}\rho)$ term is the buoyant force exerted by liquid medium.

Sedimentation coefficients are usually expressed in *Svedberg units* (S), equal to 10^{-13} s. The smaller the S value, the more slowly a molecule moves in a centrifugal field. The S values for a number of biomolecules and cellular components are listed in Table 3.2 and Figure 3.15.

Several important conclusions can be drawn from the preceding equation:

1. The sedimentation velocity of a particle depends in part on its mass. A more massive particle sediments more rapidly than does a less massive particle of the same shape and density.

TABLE 3.2 S values and molecular weights of sample proteins

Protein	S value (Svedberg units)	Molecular weight
Pancreatic trypsin inhibitor	1	6520
Cytochrome *c*	1.83	12,310
Ribonuclease A	1.78	13,690
Myoglobin	1.97	17,800
Trypsin	2.5	23,200
Carbonic anhydrase	3.23	28,800
Concanavalin A	3.8	51,260
Malate dehydrogenase	5.76	74,900
Lactate dehydrogenase	7.54	146,200

Source: T. Creighton, *Proteins,* 2nd ed. (W. H. Freeman and Company, 1993), Table 7.1.

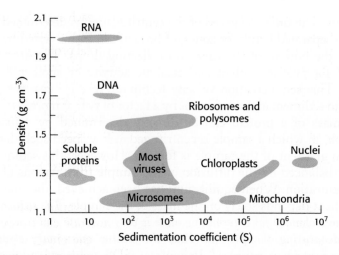

FIGURE 3.15 Density and sedimentation coefficients of cellular components. [Data from L. J. Kleinsmith and V. M. Kish, *Principles of Cell and Molecular Biology*, 2nd ed. (HarperCollins, 1995), p. 138.]

2. Shape, too, influences the sedimentation velocity because it affects the viscous drag. The frictional coefficient f of a compact particle is smaller than that of an extended particle of the same mass. Hence, elongated particles sediment more slowly than do spherical ones of the same mass.

3. A dense particle moves more rapidly than does a less dense one because the opposing buoyant force $(1 - \bar{\nu}\rho)$ is smaller for the denser particle.

4. The sedimentation velocity also depends on the density of the solution (ρ). Particles sink when $\bar{\nu}\rho < 1$, float when $\bar{\nu}\rho > 1$, and do not move when $\bar{\nu}\rho = 1$.

A technique called *zonal, band,* or most commonly *gradient* centrifugation can be used to separate proteins with different sedimentation coefficients. The first step is to form a density gradient in a centrifuge tube. Differing proportions of a low-density solution (such as 5% sucrose) and a high-density solution (such as 20% sucrose) are mixed to create a linear gradient of sucrose concentration ranging from 20% at the bottom of the tube to 5% at the top (Figure 3.16). The role of the gradient is to prevent convective flow. A small volume of a solution containing the mixture of proteins to be separated is placed on top of the density gradient. When the rotor is spun, proteins move through the gradient and separate according to their sedimentation

FIGURE 3.16 Zonal centrifugation. The steps are as follows: (A) form a density gradient, (B) layer the sample on top of the gradient, (C) place the tube in a swinging-bucket rotor and centrifuge it, and (D) collect the samples. [Information from D. Freifelder, *Physical Biochemistry*, 2nd ed. (W. H. Freeman and Company, 1982), p. 397.]

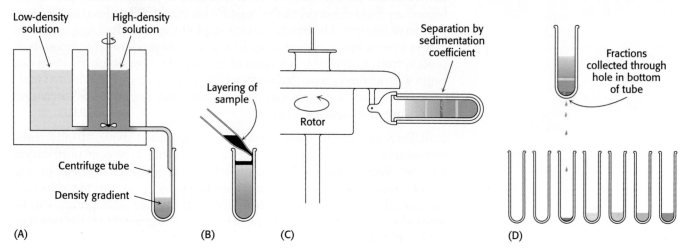

coefficients. The time and speed of the centrifugation is determined empirically. The separated bands, or zones, of protein can be harvested by making a hole in the bottom of the tube and collecting drops. The drops can be measured for protein content and catalytic activity or another functional property. This sedimentation-velocity technique readily separates proteins differing in sedimentation coefficient by a factor of two or more.

The mass of a protein can be directly determined by *sedimentation equilibrium,* in which a sample is centrifuged at low speed such that a concentration gradient of the sample is formed. However, this sedimentation is counterbalanced by the diffusion of the sample from regions of high to low concentration. When equilibrium has been achieved, the shape of the final gradient depends solely on the mass of the sample. *The sedimentation-equilibrium technique for determining mass is very accurate and can be applied without denaturing the protein. Thus the native quaternary structure of multimeric proteins is preserved.* In contrast, SDS–polyacrylamide gel electrophoresis provides an *estimate* of the mass of dissociated polypeptide chains under *denaturing* conditions. Note that, if we know the mass of the dissociated components of a multimeric protein as determined by SDS–polyacrylamide analysis and the mass of the intact multimer as determined by sedimentation-equilibrium analysis, we can determine the number of copies of each polypeptide chain present in the protein complex.

Protein purification can be made easier with the use of recombinant DNA technology

In Chapter 5, we shall consider the widespread effect of recombinant DNA technology on all areas of biochemistry and molecular biology. The application of recombinant methods to the overproduction of proteins has enabled dramatic advances in our understanding of their structure and function. Before the advent of this technology, proteins were isolated solely from their native sources, often requiring a large amount of tissue to obtain a sufficient amount of protein for analytical study. For example, the purification of bovine deoxyribonuclease in 1946 required nearly ten pounds of beef pancreas to yield one gram of protein. As a result, biochemical studies on purified material were often limited to abundant proteins.

Armed with the tools of recombinant technology, the biochemist is now able to enjoy a number of significant advantages:

1. *Proteins can be expressed in large quantities.* The homogenate serves as the starting point in a protein purification scheme. For recombinant systems, a host organism that is amenable to genetic manipulation, such as the bacterium *Escherichia coli* or the yeast *Pichia pastoris,* is utilized to express a protein of interest. The biochemist can exploit the short doubling times and ease of genetic manipulation of such organisms to produce large amounts of protein from manageable amounts of culture. As a result, purification can begin with a homogenate that is often highly enriched with the desired molecule. Moreover, a protein can be easily obtained regardless of its natural abundance or its species of origin.

2. *Affinity tags can be fused to proteins.* As described above, affinity chromatography can be a highly selective step within a protein purification scheme. Recombinant DNA technology enables the attachment of any one of a number of possible affinity tags to a protein (such as the "His tag" mentioned earlier). Hence, the benefits of affinity chromatography can be realized even for those proteins for which a binding partner is unknown or not easily determined.

3. *Proteins with modified primary structures can be readily generated.* A powerful aspect of recombinant DNA technology as applied to protein purification is the ability to manipulate genes to generate variants of a native protein sequence (Section 5.2). We learned in Section 2.4 that many proteins consist of compact domains connected by flexible linker regions. With the use of genetic-manipulation strategies, fragments of a protein that encompass single domains can be generated, an advantageous approach when expression of the entire protein is limited by its size or solubility. Additionally, as we will see in Section 9.1, amino acid substitutions can be introduced into the active site of an enzyme to precisely probe the roles of specific residues within its catalytic cycle.

3.2 Immunology Provides Important Techniques with Which to Investigate Proteins

The purification of a protein enables the biochemist to explore its function and structure within a precisely controlled environment. However, the isolation of a protein removes it from its native context within the cell, where its activity is most physiologically relevant. Advances in the field of immunology (See Chapter 35 on 🌀 SaplingPlus) have enabled the use of antibodies as critical reagents for exploring the functions of proteins within the cell. The exquisite specificity of antibodies for their target proteins provides a means to tag a specific protein so that it can be isolated, quantified, or visualized.

Antibodies to specific proteins can be generated

Immunological techniques begin with the generation of antibodies to a particular protein. An *antibody* (also called an *immunoglobulin,* Ig) is itself a protein (Figure 3.17); it is synthesized by vertebrates in response to the presence of a foreign substance, called an *antigen*. Antibodies have specific and high affinity for the antigens that elicited their synthesis. The binding of antibody to antigen is a step in the immune response that protects the animal from infection (See Chapter 35 on 🌀 SaplingPlus). Foreign proteins, polysaccharides, and nucleic acids can be antigens. Small foreign molecules, such as synthetic peptides, also can elicit antibodies, provided that the small molecule is attached to a macromolecular carrier. An antibody recognizes a specific group or cluster of amino acids on the target molecule called an *antigenic determinant* or *epitope*. The

FIGURE 3.17 Antibody structure. (A) Immunoglobulin G (IgG) consists of four chains—two heavy chains (blue) and two light chains (red)—linked by disulfide bonds. The heavy and light chains come together to form F_{ab} domains, which have the antigen-binding sites at the ends. The two heavy chains form the F_c domain. *Notice* that the F_{ab} domains are linked to the F_c domain by flexible linkers. (B) A more schematic representation of an IgG molecule. [Drawn from 1IGT.pdb.]

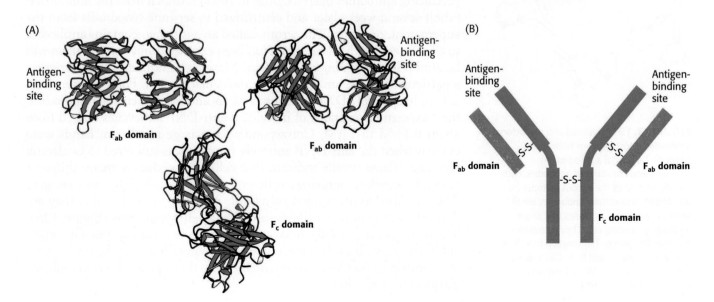

(A)

Antigen-binding site

Antigen-binding site

F_{ab} domain

F_{ab} domain

F_c domain

(B)

Antigen-binding site

Antigen-binding site

F_{ab} domain

F_{ab} domain

-S-S-

-S-S-

-S-S-

F_c domain

FIGURE 3.18 Antigen–antibody interactions. A protein antigen, in this case lysozyme, binds to the end of an F_{ab} domain of an antibody. *Notice* that the end of the antibody and the antigen have complementary shapes, allowing a large amount of surface to be buried on binding. [Drawn from 1YQV.pdb.]

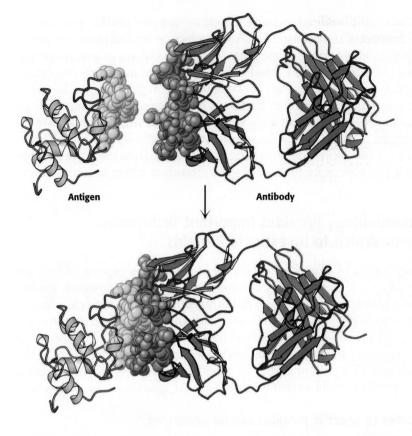

Antigen   Antibody

Polyclonal antibodies

Antigen

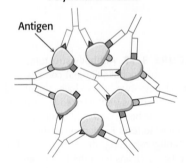

Monoclonal antibodies

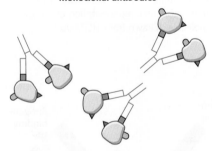

FIGURE 3.19 Polyclonal and monoclonal antibodies. Most antigens have several epitopes. Polyclonal antibodies are heterogeneous mixtures of antibodies, each specific for one of the various epitopes on an antigen. Monoclonal antibodies are all identical, produced by clones of a single antibody-producing cell. They recognize one specific epitope. [Information from R. A. Goldsby, T. J. Kindt, and B. A. Osborne, *Kuby Immunology*, 4th ed. (W. H. Freeman and Company, 2000), p. 154.]

specificity of the antibody–antigen interaction is a consequence of the shape complementarity between the two surfaces (Figure 3.18). Animals have a very large repertoire of antibody-producing cells, each producing an antibody that contains a unique surface for antigen recognition. When an antigen is introduced into an animal, it is recognized by a select few cells from this population, stimulating the proliferation of these cells. This process ensures that more antibodies of the appropriate specificity are produced.

Immunological techniques depend on the ability to generate antibodies to a specific antigen. To obtain antibodies that recognize a particular protein, a biochemist injects the protein into a rabbit twice, 3 weeks apart. The injected protein acts as an antigen, stimulating the reproduction of cells producing antibodies that recognize it. Blood is drawn from the immunized rabbit several weeks later and centrifuged to separate blood cells from the supernatant, or serum. The serum, called an *antiserum*, contains antibodies to all antigens to which the rabbit has been exposed. Only some of them will be antibodies to the injected protein. Moreover, antibodies that recognize a particular antigen are not a single molecular species. For instance, when 2,4-dinitrophenol (DNP) was used as an antigen to generate antibodies, the dissociation constants of individual anti-DNP antibodies ranged from about 0.1 nM to 1 μM. Correspondingly, a large number of bands were evident when the anti-DNP antibody mixture was subjected to isoelectric focusing. These results indicate that cells are producing many different antibodies, each recognizing a different surface feature of the same antigen. These antibodies are termed *polyclonal,* referring to the fact that they are derived from multiple antibody-producing cell populations (Figure 3.19). The heterogeneity of polyclonal antibodies can be advantageous for certain applications, such as the detection of a protein of low abundance, because each protein molecule can be bound by more than one antibody at multiple distinct antigenic sites.

Monoclonal antibodies with virtually any desired specificity can be readily prepared

The discovery of a means of producing *monoclonal antibodies* of virtually any desired specificity was a major breakthrough that intensified the power of immunological approaches. As with impure proteins, working with an impure mixture of antibodies makes it difficult to interpret data. Ideally, one would isolate a clone of cells producing a single, identical antibody. The problem is that antibody-producing cells isolated from an organism have short life spans.

Immortal cell lines that produce monoclonal antibodies do exist. These cell lines are derived from a type of cancer, *multiple myeloma,* which is a malignant disorder of antibody-producing cells. In this cancer, a single transformed plasma cell divides uncontrollably, generating a very large number of *cells of a single kind.* Such a group of cells is a *clone* because the cells are descended from the same cell and have identical properties. The identical cells of the myeloma secrete large amounts of *a single immunoglobulin* generation after generation. While these antibodies have proven useful for elucidating antibody structure, nothing is known about their specificity. Hence, they have little utility for the immunological methods described in the next pages.

César Milstein and Georges Köhler discovered that large amounts of antibodies of nearly any desired specificity can be obtained by fusing a short-lived antibody-producing cell with an immortal myeloma cell. An antigen is injected into a mouse, and its spleen is removed several weeks later (Figure 3.20;

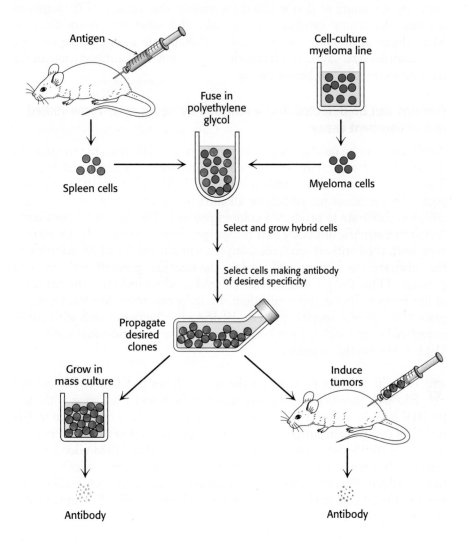

FIGURE 3.20 Preparation of monoclonal antibodies. Hybridoma cells are formed by the fusion of antibody-producing cells and myeloma cells. The hybrid cells are allowed to proliferate by growing them in selective medium. They are then screened to determine which ones produce antibody of the desired specificity. [Information from C. Milstein. Monoclonal antibodies. Copyright © 1980 by Scientific American, Inc. All rights reserved.]

FIGURE 3.21 Fluorescence micrograph of a developing *Drosophila* embryo. The embryo was stained with a fluorescence-labeled monoclonal antibody for the DNA-binding protein encoded by *engrailed,* an essential gene in specifying the body plan. [Courtesy of Dr. Nipam Patel and Dr. Corey Goodman.]

See animation in ⟨ SaplingPlus⟩). A mixture of plasma cells from this spleen is fused in vitro with myeloma cells. Each of the resulting hybrid cells, called *hybridoma* cells, indefinitely produces the identical antibody specified by the parent cell from the spleen. Hybridoma cells can then be screened by a specific assay for the antigen–antibody interaction to determine which ones produce antibodies of the preferred specificity. Collections of cells shown to produce the desired antibody are subdivided and reassayed. This process is repeated until a pure cell line, a clone producing a single antibody, is isolated. These positive cells can be grown in culture medium or injected into mice to induce myelomas. Alternatively, the cells can be frozen and stored for long periods.

The hybridoma method of producing monoclonal antibodies has had a dramatic impact on biology and medicine. Large amounts of identical antibodies with tailor-made specificities can be readily prepared. They are sources of insight into relations between antibody structure and specificity. Moreover, monoclonal antibodies can serve as precise analytical and preparative reagents. Proteins that guide development have been identified with the use of monoclonal antibodies as tags (Figure 3.21). Monoclonal antibodies attached to solid supports can be used as affinity columns to purify scarce proteins. This method has been used to purify interferon (an antiviral protein) 5000-fold from a crude mixture. Clinical laboratories use monoclonal antibodies in many assays. For example, the detection in blood of isozymes that are normally localized in the heart points to a myocardial infarction (heart attack). Blood transfusions have been made safer by antibody screening of donor blood for viruses that cause AIDS (acquired immune deficiency syndrome), hepatitis, and other infectious diseases. Monoclonal antibodies can be used as therapeutic agents (Chapter 28). For example, trastuzumab (Herceptin) is a monoclonal antibody useful for treating some forms of breast cancer.

Proteins can be detected and quantified by using an enzyme-linked immunosorbent assay

Antibodies can be used as exquisitely specific analytic reagents to quantify the amount of a protein or other antigen present in a biological sample. The *enzyme-linked immunosorbent assay* (ELISA) makes use of an enzyme, such as horseradish peroxidase or alkaline phosphatase, that reacts with a colorless substrate to produce a colored product. The enzyme is covalently linked to a specific antibody that recognizes a target antigen. If the antigen is present, the antibody–enzyme complex will bind to it and, on addition of the substrate, the enzyme will catalyze the reaction, generating the colored product. Thus, the presence of the colored product indicates the presence of the antigen. Rapid and convenient, ELISAs can detect less than a nanogram (10^{-9} g) of a specific protein. ELISAs can be performed with either polyclonal or monoclonal antibodies, but the use of monoclonal antibodies yields more-reliable results.

We will consider two among the several types of ELISA. *The indirect ELISA is used to detect the presence of antibody* and is the basis of the test for HIV infection. The HIV test detects the presence of antibodies that recognize viral core protein antigens. Viral core proteins are adsorbed to the bottom of a well. Antibodies from the person being tested are then added to the coated well. Only someone infected with HIV will have antibodies that bind to the antigen. Finally, enzyme-linked antibodies to human antibodies (e.g., enzyme-linked goat antibodies that recognize human antibodies) are allowed to react in the well, and unbound antibodies are removed by washing.

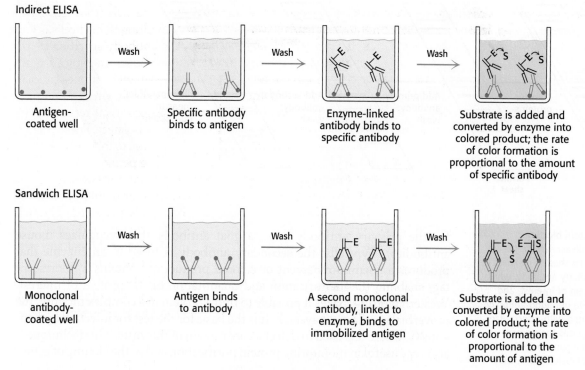

FIGURE 3.22 Indirect ELISA and sandwich ELISA. (A) In indirect ELISA, the production of color indicates the amount of an antibody to a specific antigen. (B) In sandwich ELISA, the production of color indicates the quantity of antigen. [Information from R. A. Goldsby, T. J. Kindt, and B. A. Osborne, *Kuby Immunology*, 4th ed. (W. H. Freeman and Company, 2000), p. 162.]

Substrate is then applied. An enzyme reaction yielding a colored product suggests that the enzyme-linked antibodies were bound to human antibodies, which in turn implies that the patient has antibodies to the viral antigen (Figure 3.22A). This assay is quantitative: the rate of the color-formation reaction is proportional to the amount of antibody originally present.

The sandwich ELISA is used to detect antigen rather than antibody. Antibody to a particular antigen is first adsorbed to the bottom of a well. Next, solution containing the antigen (such as blood or urine, in medical diagnostic tests) is added to the well and binds to the antibody. Finally, a second antibody that also recognizes the antigen is added. Importantly, this antibody binds to the antigen in a region distinct from, and thus does not displace, the first antibody. This antibody is enzyme-linked and is processed as described for indirect ELISA. In this case, the rate of color formation is directly proportional to the amount of antigen present. Consequently, it permits the measurement of small quantities of antigen (Figure 3.22B).

Western blotting permits the detection of proteins separated by gel electrophoresis

Very small quantities of a protein of interest in a cell or in body fluid can be detected by an immunoassay technique called *western blotting* (Figure 3.23). A sample is subjected to electrophoresis on an SDS–polyacrylamide gel. A polymer sheet is pressed against the gel, transferring the resolved proteins on the gel to the sheet, which makes the proteins more accessible for reaction. An antibody that is specific for the protein of interest, called the *primary antibody*, is added to the sheet and reacts with the antigen. The antibody–antigen complex on the sheet can then be detected by rinsing the sheet with a second antibody, called the *secondary antibody*, that is specific

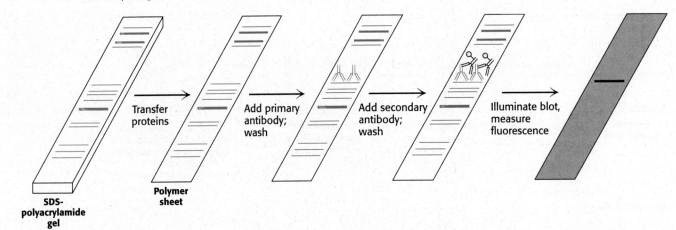

Transfer proteins

Add primary antibody; wash

Add secondary antibody; wash

Illuminate blot, measure fluorescence

SDS-polyacrylamide gel

Polymer sheet

FIGURE 3.23 Western blotting. Proteins on an SDS–polyacrylamide gel are transferred to a polymer sheet. The sheet is first treated with a primary antibody, which is specific for the protein of interest, and then washed to remove unbound antibody. Next, the sheet is treated with a secondary antibody, which recognizes the primary antibody, and washed again. Since the secondary antibody is labeled (here, with a fluorescent tag indicated by the yellow circle), the band containing the protein of interest can be identified.

for the primary antibody (e.g., a goat antibody that recognizes mouse antibodies). Typically, the secondary antibody is fused to an enzyme that produces a chemiluminescent or colored product or contains a fluorescent tag, enabling the identification and quantitation of the protein of interest. Western blotting makes it possible to find a protein in a complex mixture, the proverbial needle in a haystack. It is the basis for the test for infection by hepatitis C, where it is used to detect a core protein of the virus. This technique is also very useful in monitoring protein purification and in the cloning of genes.

Co-immunoprecipitation enables the identification of binding partners of a protein

With a monoclonal antibody against a specific protein available, it is also possible to determine the binding partners of that protein under a specific set of conditions. In this technique, known as *co-immunoprecipitation*, the sample of interest—an extract prepared from cultured cells or isolated tissue, for example—is incubated with the specific antibody. Then, agarose beads coated with an antibody-binding protein (such as Protein A from the bacterium *Staphylococcus aureus*) are added to the mixture. Protein A recognizes a portion of the antibody that is separate from the antigen-binding region (the Fc domain, Figure 3.17), and thus does not disrupt the protein-antibody complex. After centrifugation at low speed, the antibody, now bound to the beads, aggregates at the bottom of the tube. Under optimal buffer conditions, the antibody-protein complex will also precipitate any additional proteins that are bound to the original protein (Figure 3.24). Subsequent analysis of the precipitate by SDS-PAGE, followed by either western blot or mass spectrometric fingerprinting (Section 3.3), enables the identification of the binding partners.

FIGURE 3.24 Co-immunoprecipitation. An antibody specific for a particular protein (here, the red oval) is added to an extract isolated from cells or tissue. After an incubation period, agarose beads coated with Protein A are added, and the mixture is subjected to low-speed centrifugation. The beads bind to the antibody-protein complex and precipitate. Any additional proteins that interact with the target protein (here, the yellow rectangle) will also precipitate, and can be identified by SDS-PAGE followed by western blotting or mass spectrometry.

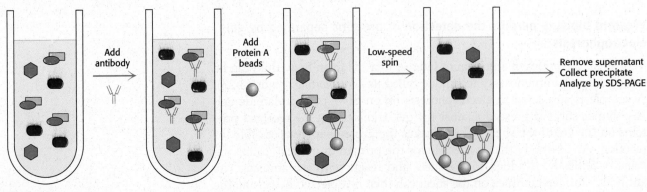

Add antibody

Add Protein A beads

Low-speed spin

Remove supernatant
Collect precipitate
Analyze by SDS-PAGE

Cell or tissue extract

Fluorescent markers make the visualization of proteins in the cell possible

Biochemistry is often performed in test tubes or polyacrylamide gels. However, proteins function in a physiological environment, such as within the cytoplasm of a cell. Fluorescent markers provide a powerful means of examining proteins in their biological context. Cells can be stained with fluorescence-labeled antibodies and examined by *fluorescence micros-copy* to reveal the location of a protein of interest. For example, arrays of parallel bundles are evident in cells stained with antibody specific for actin, a protein that polymerizes into filaments (Figure 3.25). Actin fila-ments are constituents of the cytoskeleton, the internal scaffolding of cells that controls their shape and movement. By tracking protein location, fluorescent markers also provide clues to protein function. For instance, the mineralocorticoid receptor protein binds to steroid hormones, includ-ing cortisol. The receptor was linked to a yellow variant of *green fluorescent protein* (GFP), a naturally fluorescent protein isolated from the jellyfish *Aequorea victoria* (Section 2.6). Fluorescence microscopy revealed that, in the absence of the hormone, the receptor is located in the cytoplasm (Figure 3.26A). On addition of the steroid, the receptor is translocated to the nucleus, where it binds to DNA (Figure 3.26B). These results indicate that the mineralocorticoid receptor protein is a transcription factor that controls gene expression.

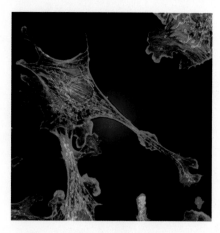

FIGURE 3.25 Actin filaments.
Fluorescence micrograph of a cell shows actin filaments stained green using an antibody specific to actin. [David Becker/ Science Source.]

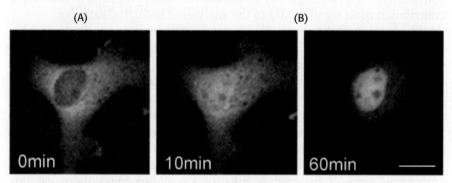

(A)   (B)

0min   10min   60min

FIGURE 3.26 Nuclear localization of a steroid receptor. (A) The mineralocorticoid receptor, made visible by attachment to a yellow variant of GFP, is located predominantly in the cytoplasm of the cultured cell. (B) Subsequent to the addition of corticosterone (a glucocorticoid steroid that also binds to the mineralocorticoid receptor), the receptor moves into the nucleus. [Republished with permission of Society for Neuroscience, from *The Journal of Neuroscience*, Nishi, M. et al., 24, 21, 2004, permission conveyed through Copyright Clearance Center, Inc.]

The highest resolution of fluorescence microscopy is about 0.2 μm (200 nm, or 2000 Å), the wavelength of visible light. Finer spatial resolution can be achieved by electron microscopy if the antibodies are tagged with electron-dense markers. For example, antibodies conjugated to clusters of gold or to ferritin (which has an electron-dense core rich in iron) are highly visible under the electron microscope. *Immunoelectron microscopy* can define the position of antigens to a resolution of 10 nm (100 Å) or finer (Figure 3.27).

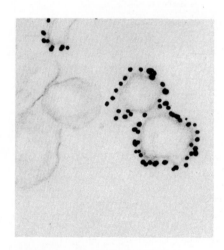

FIGURE 3.27 Immunoelectron microscopy. The opaque particles (150-Å, or 15-nm, diameter) in this electron micrograph are clusters of gold atoms bound to antibody molecules. A gold-labeled antibody against a channel protein (Section 13.4) identifies membrane vesicles at the termini of neurons that contain this protein. [Courtesy of Dr. Peter Sargent.]

3.3 Mass Spectrometry Is a Powerful Technique for the Identification of Peptides and Proteins

In many instances, the study of a particular biological process in its native context is advantageous. For example, if we are interested in a pathway that is localized to the nucleus of a cell, we might conduct studies on an isolated nuclear extract. In these experiments, identification of the proteins present in the sample is often critical. Antibody-based techniques, such as the ELISA method described in the previous section, can be very helpful toward this goal. However, these techniques are limited to the detection of proteins for which an antibody is already available. Mass spectrometry enables the highly precise and sensitive measurement of the atomic composition of a particular molecule, or *analyte,* without prior knowledge of its identity. Originally, this method was relegated to the study of the chemical composition and molecular mass of gases or volatile liquids. However, technological advances in the past two decades have dramatically expanded the utility of mass spectrometry to the study of proteins, even those found at very low concentrations within highly complex mixtures, such as the contents of a particular cell type.

Mass spectrometry enables the highly accurate and sensitive detection of the mass of an analyte. This information can be used to determine the identity and chemical state of the molecule of interest. Mass spectrometers operate by converting analyte molecules into gaseous, charged forms (*gas-phase ions*). Through the application of electrostatic potentials, the ratio of the mass of each ion to its charge (the *mass-to-charge ratio,* or *m/z*) can be measured. Although a wide variety of techniques employed by mass spectrometers are used in current practice, each of them comprises three essential components: the ion source, the mass analyzer, and the detector. Let us consider the first two in greater detail, because improvements in them have contributed most significantly to the analysis of biological samples.

The *ion source* achieves the first critical step in mass spectrometric analysis: conversion of the analyte into gas-phase ions (*ionization*). Until recently, proteins could not be ionized efficiently because of their high molecular weights and low volatility. However, the development of techniques such as *matrix-assisted laser desorption/ionization* (MALDI) and *electrospray ionization* (ESI) has enabled the clearing of this significant hurdle. In MALDI, the analyte is evaporated to dryness in the presence of a volatile, aromatic compound (the *matrix*) that can absorb light at specific wavelengths. A laser pulse tuned to one of these wavelengths excites and vaporizes the matrix, converting some of the analyte into the gas phase. Subsequent gaseous collisions enable the intermolecular transfer of charge, ionizing the analyte. In ESI, a solution of the analyte is passed through an electrically charged nozzle. Droplets of the analyte, now charged, emerge from the nozzle into a chamber of very low pressure, evaporating the solvent and ultimately yielding the ionized analyte.

The newly formed analyte ions then enter the *mass analyzer,* where they are distinguished on the basis of their mass-to-charge ratios. There are a number of different types of mass analyzers. For this discussion, we will consider one of the simplest, the *time-of-flight* (TOF) *mass analyzer,* in which ions are accelerated through an elongated chamber under a fixed electrostatic potential. Given two ions of identical net charge, the smaller ion will require less time to traverse the chamber than will the larger ion. The mass of each ion can be determined by measuring the time required for each ion to pass through the chamber.

The sequential action of the ion source and the mass analyzer enables the highly sensitive measurement of the mass of potentially massive ions, such as those of proteins. Consider an example of a MALDI ion source coupled to a TOF mass analyzer: the MALDI-TOF mass spectrometer (Figure 3.28). Gas-phase ions generated by the MALDI ion source pass directly into the TOF analyzer, where the mass-to-charge ratios are recorded.

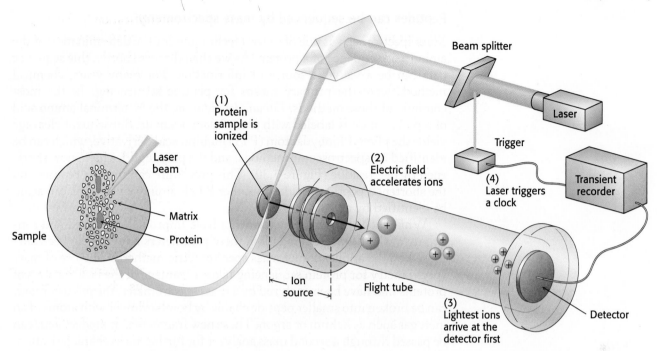

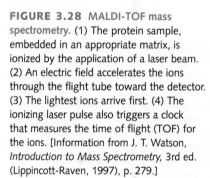

In Figure 3.29, the MALDI-TOF mass spectrum of a mixture of 5 pmol each of insulin and lactoglobulin is shown. The masses determined by MALDI-TOF are 5733.9 and 18,364, respectively. A comparison with the calculated values of 5733.5 and 18,388 reveals that MALDI-TOF is clearly an accurate means of determining protein mass.

In the ionization process, a family of ions, each of the same mass but carrying different total net charges, is formed from a single analyte. Because the mass spectrometer detects ions on the basis of their mass-to-charge ratio, these ions will appear as separate peaks in the mass spectrum. For example, in the mass spectrum of β-lactoglobulin shown in Figure 3.29, peaks near $m/z = 18,364$ (corresponding to the $+1$ charged ion) and $m/z = 9183$ (corresponding to the $+2$ charged ion) are visible (indicated by the blue arrows). Although multiple peaks for the same ion may appear to be a nuisance, they enable the spectrometrist to measure the mass of an analyte ion more than once in a single experiment, improving the overall precision of the calculated result.

FIGURE 3.28 MALDI-TOF mass spectrometry. (1) The protein sample, embedded in an appropriate matrix, is ionized by the application of a laser beam. (2) An electric field accelerates the ions through the flight tube toward the detector. (3) The lightest ions arrive first. (4) The ionizing laser pulse also triggers a clock that measures the time of flight (TOF) for the ions. [Information from J. T. Watson, *Introduction to Mass Spectrometry,* 3rd ed. (Lippincott-Raven, 1997), p. 279.]

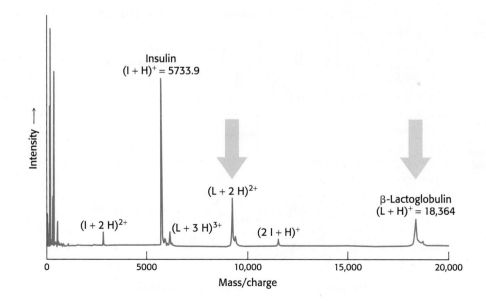

FIGURE 3.29 MALDI-TOF mass spectrum of insulin and **β-lactoglobulin.** A mixture of 5 pmol each of insulin (I) and β-lactoglobulin (L) was ionized by MALDI, which produces predominantly singly charged molecular ions from peptides and proteins—the insulin ion $(I+H)^+$ and the lactoglobulin ion $(L+H)^+$. Molecules with multiple charges, such as those for β-lactoglobulin indicated by the blue arrows, as well as small quantities of a singly charged dimer of insulin $(2\,I+H)^+$ also are produced. [Data from J. T. Watson, *Introduction to Mass Spectrometry,* 3rd ed. (Lippincott-Raven, 1997), p. 282.]

Peptides can be sequenced by mass spectrometry

Mass spectrometry is one of several techniques for the determination of the amino acid sequence of a protein. As we shall discuss shortly, this sequence data can be a valuable source of information. For many years, chemical methods were the primary means for peptide sequencing. In the most common of these methods, *Edman degradation*, the N-terminal amino acid of a polypeptide is labeled with *phenyl isothiocyanate*. Subsequent cleavage yields the phenylthiohydantoin (PTH)-amino acid derivative, which can be identified by spectroscopic methods, and the polypeptide chain, now shortened by one residue (Figure 3.30). This procedure can then be repeated on the shortened peptide, yielding another PTH–amino acid, which can again be identified by chromatography.

While technological advancements have improved the speed and sensitivity of the Edman degradation, these parameters have largely been surpassed by the application of mass spectrometric methods. The use of mass spectrometry for protein sequencing takes advantage of the fact that ions of proteins that have been analyzed by a mass spectrometer, the *precursor ions*, can be broken into smaller peptide chains by bombardment with atoms of an inert gas such as helium or argon. These new fragments, or *product ions*, can be passed through a second mass analyzer for further mass characterization. The utilization of two mass analyzers arranged in this manner is referred to as *tandem mass spectrometry*. Importantly, the product-ion fragments are formed in chemically predictable ways that can provide clues to the amino acid sequence of the precursor ion. For polypeptide analytes, disruption of individual peptide bonds will yield two smaller peptide ions, containing the sequences before and after the cleavage site. Hence, a family of ions can be detected; each ion represents a fragment of the original peptide with one or more amino acids removed from one end (Figure 3.31A). For simplicity, only the carboxyl-terminal peptide fragments are shown in Figure 3.31A.

FIGURE 3.30 The Edman degradation. The labeled amino-terminal residue (PTH–alanine in the first round) can be released without hydrolyzing the rest of the peptide. Hence, the amino-terminal residue of the shortened peptide (Gly-Asp-Phe-Arg-Gly) can be determined in the second round. Three more rounds of the Edman degradation reveal the complete sequence of the original peptide.

EDMAN DEGRADATION

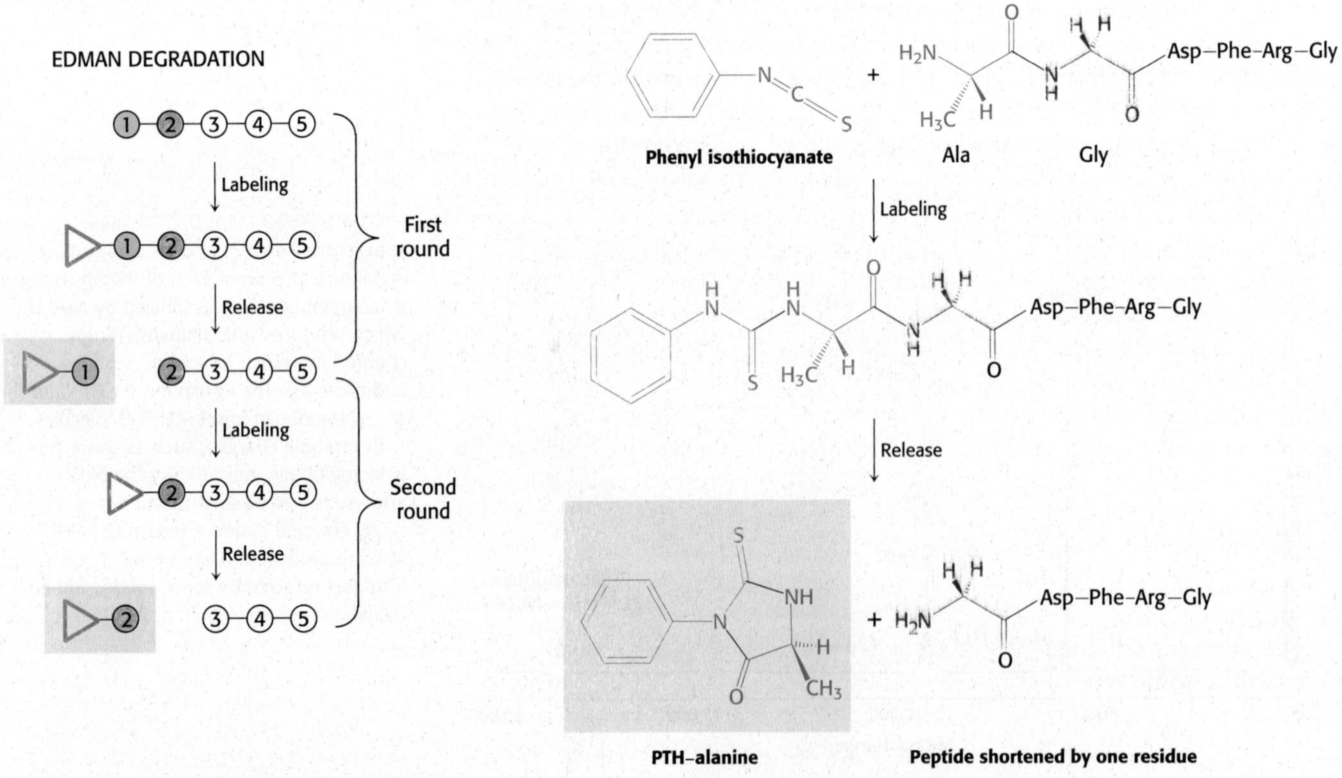

(A)

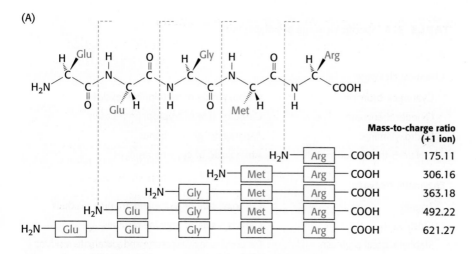

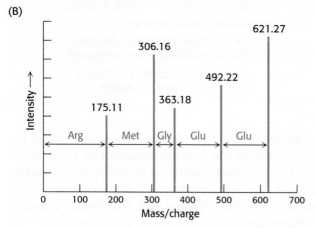

	Mass-to-charge ratio (+1 ion)
H₂N—Arg—COOH	175.11
H₂N—Met—Arg—COOH	306.16
H₂N—Gly—Met—Arg—COOH	363.18
H₂N—Glu—Gly—Met—Arg—COOH	492.22
H₂N—Glu—Glu—Gly—Met—Arg—COOH	621.27

(B)

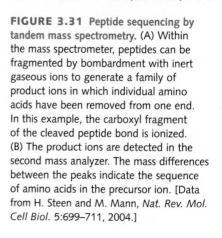

FIGURE 3.31 Peptide sequencing by tandem mass spectrometry. (A) Within the mass spectrometer, peptides can be fragmented by bombardment with inert gaseous ions to generate a family of product ions in which individual amino acids have been removed from one end. In this example, the carboxyl fragment of the cleaved peptide bond is ionized. (B) The product ions are detected in the second mass analyzer. The mass differences between the peaks indicate the sequence of amino acids in the precursor ion. [Data from H. Steen and M. Mann, *Nat. Rev. Mol. Cell Biol.* 5:699–711, 2004.]

Figure 3.31B depicts a representative mass spectrum from a fragmented peptide. The mass differences between the peaks in this fragmentation experiment indicate the amino acid sequence of the precursor peptide ion.

Proteins can be specifically cleaved into small peptides to facilitate analysis

In principle, it should be possible to sequence an entire protein using the Edman degradation or mass spectrometric methods. In practice, the Edman degradation is limited to peptides of 50 residues, because not all peptides in the reaction mixture release the amino acid derivative at each step. For instance, if the efficiency of release for each round were 98%, the proportion of "correct" amino acid released after 60 rounds would be (0.98^{60}), or 0.3—a hopelessly impure mix. Similarly, sequencing of long peptides by mass spectrometry yields a mass spectrum that can be complex and difficult to interpret. This obstacle can be circumvented by cleaving a protein into smaller peptides that can be sequenced. Protein cleavage can be achieved by chemical reagents, such as cyanogen bromide, or proteolytic enzymes, such as trypsin. Table 3.3 gives several other ways of specifically cleaving polypeptide chains. Note that these methods are sequence specific: they disrupt the protein backbone at particular amino acid residues in a predictable manner.

The peptides obtained by specific chemical or enzymatic cleavage are separated by some type of chromatography. The sequence of each purified peptide is then determined by the methods described above. At this point, the amino acid sequences of segments of the protein are known, but the order

TABLE 3.3 Specific cleavage of polypeptides

Reagent	Cleavage site
Chemical cleavage	
Cyanogen bromide	Carboxyl side of methionine residues
O-Iodosobenzoate	Carboxyl side of tryptophan residues
Hydroxylamine	Asparagine–glycine bonds
2-Nitro-5-thiocyanobenzoate	Amino side of cysteine residues
Enzymatic cleavage	
Trypsin	Carboxyl side of lysine and arginine residues
Clostripain	Carboxyl side of arginine residues
Staphylococcal protease	Carboxyl side of aspartate and glutamate residues (glutamate only under certain conditions)
Thrombin	Carboxyl side of arginine
Chymotrypsin	Carboxyl side of tyrosine, tryptophan, phenylalanine, leucine, and methionine
Carboxypeptidase A	Amino side of C-terminal amino acid (not arginine, lysine, or proline)

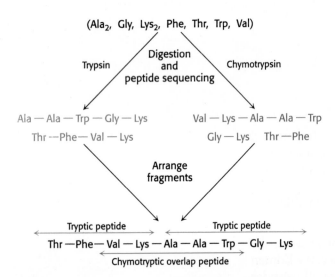

FIGURE 3.32 Overlap peptides. The peptide obtained by chymotryptic digestion overlaps two tryptic peptides, establishing their order.

of these segments is not yet defined. How can we order the peptides to obtain the primary structure of the original protein? The necessary additional information is obtained from *overlap peptides* (Figure 3.32). A second enzyme is used to split the polypeptide chain at different linkages. For example, chymotrypsin cleaves preferentially on the carboxyl side of aromatic and some other bulky nonpolar residues (Chapter 9). Because these chymotryptic peptides overlap two or more tryptic peptides, they can be used to establish the order of the peptides. The entire amino acid sequence of the polypeptide chain is then known.

Additional steps are necessary if the initial protein sample is actually several polypeptide chains. SDS-PAGE under reducing conditions should display the number of chains. Alternatively, the number of distinct N-terminal amino acids could be determined. After a protein has been identified as being made up of two or more polypeptide chains, denaturing agents, such as urea or guanidine hydrochloride, are used to dissociate chains held together by noncovalent bonds. The dissociated chains must be separated from one another before sequence determination can begin. Polypeptide chains linked by disulfide bonds are separated by reduction with thiols such as β-mercaptoethanol or dithiothreitol. To prevent the cysteine residues from recombining, they are alkylated with iodoacetate to form stable *S*-carboxymethyl derivatives (Figure 3.33). Sequencing can then be performed as already described.

Genomic and proteomic methods are complementary

Despite the technological advancements in both chemical and mass spectrometric methods of peptide sequencing, heroic effort is required to elucidate the sequence of large proteins, those with more than 1000 residues. For sequencing such proteins, a complementary experimental approach based on recombinant DNA technology is often more efficient. As will be discussed in Chapter 5, long stretches of DNA can be cloned and sequenced, and the nucleotide sequence can be translated to reveal the amino acid sequence of the protein encoded by the gene (Figure 3.34). Recombinant DNA technology has generated a wealth of amino acid sequence information at a remarkable rate.

Nevertheless, even with the use of the DNA base sequence to determine primary structure, there is still a need to work with isolated proteins. The amino acid sequence deduced by reading the DNA sequence is that of the nascent protein, the direct product of the translational machinery. However, many proteins undergo *posttranslational modifications* after their syntheses. Some have their ends trimmed, and others arise by cleavage of a larger initial polypeptide chain. Cysteine residues in some proteins are oxidized to form disulfide links, connecting either parts within a chain or separate polypeptide chains. Specific side chains of some proteins are altered. Amino acid sequences derived from DNA sequences are rich in information, but they do not disclose these modifications. Chemical analyses of proteins in their mature form are needed to delineate the nature of these changes, which are critical for the biological activities of most proteins. *Thus, genomic and proteomic analyses are complementary approaches to elucidating the structural basis of protein function.*

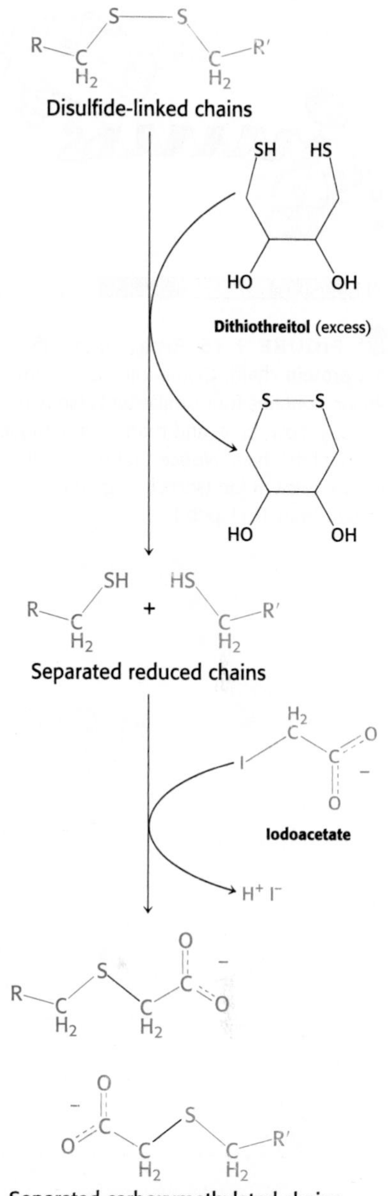

The amino acid sequence of a protein provides valuable information

Regardless of the method used for its determination, the amino acid sequence of a protein can provide the biochemist with a wealth of information as to the protein's structure, function, and history.

1. *The sequence of a protein of interest can be compared with all other known sequences to ascertain whether significant similarities exist.* A search for kinship between a newly sequenced protein and the millions of previously sequenced ones takes only a few seconds on a personal computer (Chapter 6). If the newly isolated protein is a member of an established class of protein, we can begin to infer information about the protein's structure and function. For instance, chymotrypsin and trypsin are members of the serine protease family, a clan of proteolytic enzymes that have a common catalytic mechanism based on a reactive serine residue (Chapter 9). If the sequence of the newly isolated protein shows sequence similarity with trypsin or chymotrypsin, the result suggests that it may be a serine protease.

2. *Comparison of sequences of the same protein in different species yields information about evolutionary pathways.* Genealogical relationships between species can be inferred from sequence differences between their proteins. If we assume that the random mutation rate of proteins over time is constant, then careful sequence comparison of related proteins between two organisms can provide an estimate for when these two evolutionary lines diverged. For example, a comparison of serum albumins found in primates indicates that

FIGURE 3.33 Disulfide-bond reduction. Polypeptides linked by disulfide bonds can be separated by reduction with dithiothreitol followed by alkylation to prevent them from re-forming.

DNA sequence	GGG	TTC	TTG	GGA	GCA	GCA	GGA	AGC	ACT	ATG	GGC	GCA
Amino acid sequence	Gly	Phe	Leu	Gly	Ala	Ala	Gly	Ser	Thr	Met	Gly	Ala

FIGURE 3.34 DNA sequence yields the amino acid sequence. The complete nucleotide sequence of HIV-1 (human immunodeficiency virus), the cause of AIDS (acquired immune deficiency syndrome), was determined within a year after the isolation of the virus. A part of the DNA sequence specified by the RNA genome of the virus is shown here with the corresponding amino acid sequence (deduced from knowledge of the genetic code).

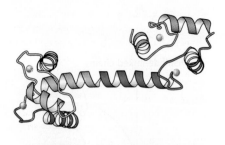

FIGURE 3.35 Repeating motifs in a protein chain. Calmodulin, a calcium sensor, contains four similar units (shown in red, yellow, blue, and orange) in a single polypeptide chain. *Notice* that each unit binds a calcium ion (shown in green). [Drawn from 1CLL.pdb.]

human beings and African apes diverged 5 million years ago, not 30 million years ago as was once thought. Sequence analyses have opened a new perspective on the fossil record and the pathway of human evolution.

3. *Amino acid sequences can be searched for the presence of internal repeats.* Such internal repeats can reveal the history of an individual protein itself. Many proteins apparently have arisen by duplication of primordial genes followed by their diversification. For example, calmodulin, a ubiquitous calcium sensor in eukaryotes, contains four similar calcium-binding modules that arose by gene duplication (Figure 3.35).

4. *Many proteins contain amino acid sequences that serve as signals designating their destinations or controlling their processing.* For example, a protein destined for export from a cell or for location in a membrane contains a *signal sequence*—a stretch of about 20 hydrophobic residues near the amino terminus that directs the protein to the appropriate membrane. Another protein may contain a stretch of amino acids that functions as a *nuclear localization signal,* directing the protein to the nucleus.

5. *Sequence data provide a basis for preparing antibodies specific for a protein of interest.* One or more parts of the amino acid sequence of a protein will elicit an antibody when injected into a mouse or rabbit. These specific antibodies can be very useful in determining the amount of a protein present in solution or in the blood, ascertaining its distribution within a cell, or cloning its gene (Section 3.2).

6. *Amino acid sequences are valuable for making DNA probes that are specific for the genes encoding the corresponding proteins.* Knowledge of a protein's primary structure permits the use of reverse genetics. DNA sequences that correspond to a part of the amino acid sequence can be constructed on the basis of the genetic code. These DNA sequences can be used as probes to isolate the gene encoding the protein so that the entire sequence of the protein can be determined. The gene in turn can provide valuable information about the physiological regulation of the protein. Protein sequencing is an integral part of molecular genetics, just as DNA cloning is central to the analysis of protein structure and function. We will revisit some of these topics in more detail in Chapter 5.

Individual proteins can be identified by mass spectrometry

The combination of mass spectrometry with chromatographic and peptide-cleavage techniques enables highly sensitive protein identification in complex biological mixtures. When a protein is cleaved by chemical or enzymatic methods (Table 3.3), a specific and predictable family of peptide fragments is formed. We learned in Chapter 2 that each protein has a unique, precisely defined amino acid sequence. Hence, the identity of the individual peptides formed from this cleavage reaction—and, importantly, their corresponding masses—is a distinctive signature for that particular protein. Protein cleavage, followed by chromatographic separation and mass spectrometry, enables rapid identification and quantitation of these signatures, even if they are present at very low concentrations. This technique for protein identification is referred to as *peptide mass fingerprinting*.

The speed and sensitivity of mass spectrometry has made this technology critical for the study of proteomics. Let us consider the analysis of the nuclear pore complex from yeast, which facilitates the transport of large molecules into and out of the nucleus. This huge macromolecular complex was purified from yeast cells, then fractionated by HPLC followed by gel

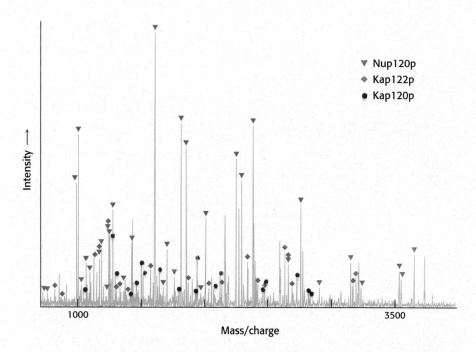

FIGURE 3.36 Proteomic analysis by mass spectrometry. This mass spectrum was obtained by analyzing a trypsin-treated band in a gel derived from a yeast nuclear-pore sample. Many of the peaks were found to match the masses predicted for peptide fragments from three proteins (Nup120p, Kap122p, and Kap120p) within the yeast genome. The band corresponded to an apparent molecular mass of 100 kDa. [Data from M. P. Rout, J. D. Aitchison, A. Suprapto, K. Hjertaas, Y. Zhao, and B. T. Chait, *J. Cell Biol.* 148:635–651, 2000.]

electrophoresis. Individual bands from the gel were isolated, cleaved with trypsin, and analyzed by MALDI-TOF mass spectrometry. The fragments produced were compared with amino acid sequences deduced from the DNA sequence of the yeast genome as shown in Figure 3.36. A total of 174 nuclear pore proteins were identified in this manner. Many of these proteins had not previously been identified as being associated with the nuclear pore despite years of study. Furthermore, mass spectrometric methods are sensitive enough to detect essentially all components of the pore if they are present in the samples used. Thus, a complete list of the components constituting this macromolecular complex could be obtained in a straightforward manner. Proteomic analysis of this type is growing in power as mass spectrometric and biochemical fractionation methods are refined.

3.4 Peptides Can Be Synthesized by Automated Solid-Phase Methods

Peptides of defined sequence can be synthesized to assist in biochemical analysis. These peptides are valuable tools for several purposes.

1. *Synthetic peptides can serve as antigens to stimulate the formation of specific antibodies.* Suppose we want to isolate the protein expressed by a specific gene. Peptides can be synthesized that match the translation of part of the gene's nucleic acid sequence, and antibodies can be generated that target these peptides. These antibodies can then be used to isolate the intact protein or localize it within the cell.

2. *Synthetic peptides can be used to isolate receptors for many hormones and other signal molecules.* For example, white blood cells are attracted to bacteria by formylmethionyl (fMet) peptides released in the breakdown of bacterial proteins. Synthetic formylmethionyl peptides have been useful in identifying the cell-surface receptor for this class of peptide. Moreover, synthetic peptides can be attached to agarose beads to prepare affinity chromatography columns for the purification of receptor proteins that specifically recognize the peptides.

fMet peptide

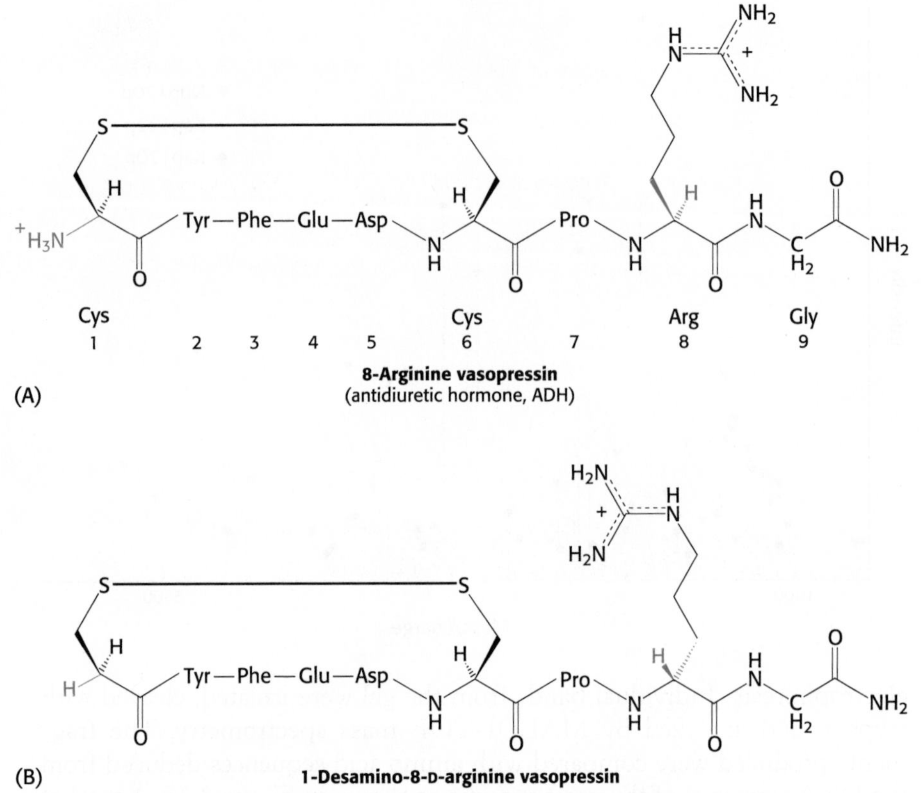

FIGURE 3.37 Vasopressin and a synthetic vasopressin analog. Structural formulas of (A) vasopressin, a peptide hormone that stimulates water resorption, and (B) 1-desamino-8-D-arginine vasopressin, a more stable synthetic analog of this antidiuretic hormone.

(A)

8-Arginine vasopressin
(antidiuretic hormone, ADH)

(B)   **1-Desamino-8-D-arginine vasopressin**

3. *Synthetic peptides can serve as drugs.* Vasopressin is a peptide hormone that stimulates the reabsorption of water in the distal tubules of the kidney, leading to the formation of more-concentrated urine. Patients with diabetes insipidus are deficient in vasopressin (also called *antidiuretic hormone*), and so they excrete large volumes of dilute urine (more than 5 liters per day) and are continually thirsty. This defect can be treated by administering 1-desamino-8-D-arginine vasopressin, a synthetic analog of the missing hormone (Figure 3.37). This synthetic peptide is degraded in vivo much more slowly than vasopressin and does not increase blood pressure.

4. Finally, *studying synthetic peptides can help define the rules governing the three-dimensional structure of proteins.* We can ask whether a particular sequence by itself tends to fold into an α helix, a β strand, or a hairpin turn, or behaves as a random coil. The peptides created for such studies can incorporate amino acids not normally found in proteins, allowing more variation in chemical structure than is possible with the use of only 20 amino acids.

How are these peptides constructed? The amino group of one amino acid is linked to the carboxyl group of another. However, a unique product is formed only if a single amino group and a single carboxyl group are available for reaction. Therefore, it is necessary to block some groups and to activate others to prevent unwanted reactions. First, the carboxyl-terminal amino acid is attached to an insoluble resin by its carboxyl group, effectively protecting it from further peptide-bond-forming reactions (Figure 3.38). The α-amino group of this amino acid is blocked with a protecting group such as a *tert*-butyloxycarbonyl (*t*-Boc) group. The *t*-Boc protecting group of this amino acid is then removed with trifluoroacetic acid.

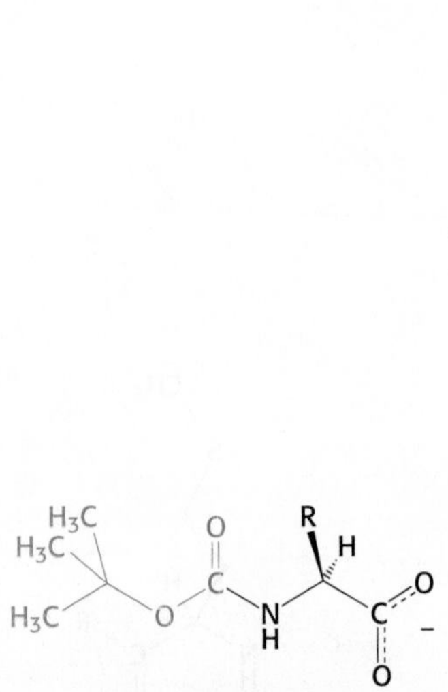

t-**Butyloxycarbonyl amino acid**
(*t*-**Boc amino acid**)

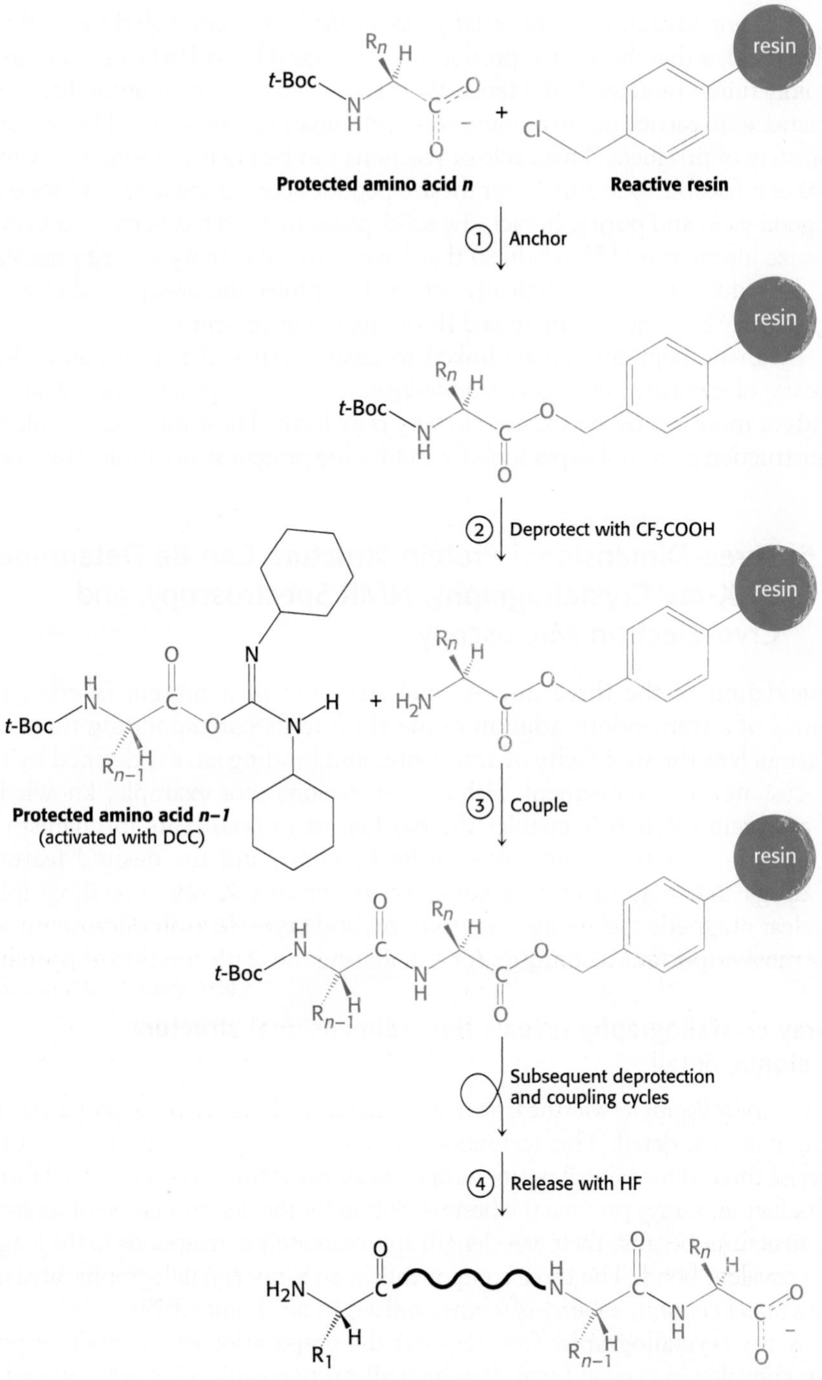

FIGURE 3.38 Solid-phase peptide synthesis. The sequence of steps in solid-phase synthesis is: (1) anchoring of the C-terminal amino acid to a solid resin, (2) deprotection of the amino terminus, and (3) coupling of the free amino terminus with the DCC-activated carboxyl group of the next amino acid. Steps 2 and 3 are repeated for each added amino acid. Finally, in step 4, the completed peptide is released from the resin.

    The next amino acid (in the protected t-Boc form) and *dicyclohexylcarbodiimide* (DCC) are added together. At this stage, only the carboxyl group of the incoming amino acid and the amino group of the resin-bound amino acid are free to form a peptide bond. DCC reacts with the carboxyl group of the incoming amino acid, activating it for the peptide-bond-forming reaction. After the peptide bond has formed, excess reagents are washed away, leaving the desired dipeptide product attached to the beads. Additional amino acids are linked by the same sequence of reactions. At the end of the synthesis, the peptide is released from the beads by the addition of hydrofluoric acid (HF), which cleaves the carboxyl ester anchor without disrupting peptide bonds. Protecting groups on potentially reactive side chains, such as that of lysine, also are removed at this time.

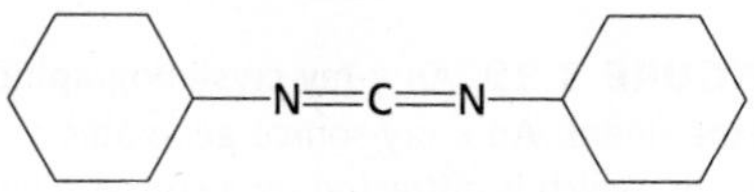

Dicyclohexylcarbodiimide (DCC)

A major advantage of this *solid-phase method,* first developed by R. Bruce Merrifield, is that the desired product at each stage is bound to beads that can be rapidly filtered and washed. Hence, there is no need to purify intermediates. All reactions are carried out in a single vessel, eliminating losses caused by repeated transfers of products. This cycle of reactions can be readily automated, which makes it feasible to routinely synthesize peptides containing about 50 residues in good yield and purity. In fact, the solid-phase method has been used to synthesize interferons (155 residues) that have antiviral activity and ribonuclease (124 residues) that is catalytically active. The protecting groups and cleavage agents may be varied for increased flexibility or convenience.

Synthetic peptides can be linked to create even longer molecules. With the use of specially developed *peptide-ligation* methods, proteins of 100 amino acids or more can by synthesized in very pure form. These methods enable the construction of even sharper tools for examining protein structure and function.

3.5 Three-Dimensional Protein Structure Can Be Determined by X-ray Crystallography, NMR Spectroscopy, and Cryo-Electron Microscopy

Elucidation of the three-dimensional structure of a protein is often the source of a tremendous amount of insight into its corresponding function, inasmuch as the specificity of active sites and binding sites is defined by the precise atomic arrangement within these regions. For example, knowledge of a protein's structure enables the biochemist to predict its mechanism of action, the effects of mutations on its function, and the desired features of drugs that may inhibit or augment its activity. X-ray crystallography, nuclear magnetic resonance spectroscopy, and cryo-electron microscopy are the most important techniques for elucidating the conformation of proteins.

X-ray crystallography reveals three-dimensional structure in atomic detail

X-ray crystallography was the first method developed to determine protein structure in atomic detail. This technique provides the clearest visualization of the precise three-dimensional positions of most atoms within a protein. Of all forms of radiation, x-rays provide the best resolution for the determination of molecular structures because their wavelength approximately corresponds to the length of a covalent bond. The three components in an x-ray crystallographic analysis are a *protein crystal,* a *source of x-rays,* and a *detector* (Figure 3.39).

X-ray crystallography first requires the preparation of a protein or protein complex in crystal form, in which all protein molecules are oriented in a fixed, repeated arrangement with respect to one another. Slowly adding ammonium sulfate or another salt to a concentrated solution of protein to reduce its solubility favors the formation of highly ordered crystals—the process of salting out discussed on page 72. For example, myoglobin crystallizes in 3 M ammonium sulfate. Protein crystallization can be challenging: a concentrated solution of highly pure material is required and it is often difficult to predict which experimental conditions will yield the most-effective crystals. Methods for screening many different crystallization conditions using a small amount of protein sample have been developed. Typically, hundreds of conditions must be tested to obtain crystals fully suitable for crystallographic studies. Nevertheless, increasingly large and complex proteins have been crystallized. For example, poliovirus, an 8500-kDa assembly of 240 protein subunits surrounding an RNA core, has been crystallized

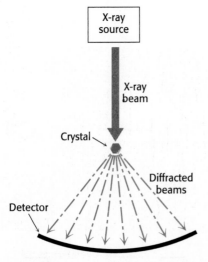

FIGURE 3.39 An x-ray crystallographic experiment. An x-ray source generates a beam, which is diffracted by a crystal. The resulting diffraction pattern is collected on a detector.

and its structure solved by x-ray methods. Crucially, proteins frequently crystallize in their biologically active configuration. Enzyme crystals may display catalytic activity if the crystals are suffused with substrate.

After a suitably pure crystal of protein has been obtained, a source of x-rays is required. A beam of x-rays of wavelength 1.54 Å is produced by accelerating electrons against a copper target. Equipment appropriate for generating x-rays in this manner is often available in laboratories. Alternatively, x-rays can be produced by *synchrotron radiation,* the acceleration of electrons in circular orbits at speeds close to the speed of light. Synchrotron-generated x-ray beams are much more intense than those generated by electrons hitting copper. The higher intensity enables the acquisition of high-quality data from smaller crystals over a shorter exposure time. Several facilities throughout the world generate synchrotron radiation, such as the Advanced Photon Source at Argonne National Laboratory outside Chicago and the Photon Factory in Tsukuba City, Japan.

When a narrow beam of x-rays is directed at the protein crystal, most of the beam passes directly through the crystal while a small part is scattered in various directions. These scattered, or *diffracted,* x-rays can be detected by x-ray film or by a solid-state electronic detector. The scattering pattern provides abundant information about protein structure. The basic physical principles underlying the technique are:

1. *Electrons scatter x-rays.* The amplitude of the wave scattered by an atom is proportional to its number of electrons. Thus, a carbon atom scatters six times as strongly as a hydrogen atom does.

2. *The scattered waves recombine.* Each diffracted beam comprises waves scattered by each atom in the crystal. The scattered waves reinforce one another at the film or detector if they are in phase (in step) there, and they cancel one another if they are out of phase.

3. *The way in which the scattered waves recombine depends only on the atomic arrangement.* The protein crystal is mounted and positioned in a precise orientation with respect to the x-ray beam and the film. The crystal is rotated so that the beam can strike the crystal from many directions. This rotational motion results in an x-ray photograph consisting of a regular array of spots called *reflections.* The x-ray photograph shown in Figure 3.40 is a two-dimensional section through a three-dimensional array of 72,000 reflections. The intensities and positions of these reflections are the basic experimental data of an x-ray crystallographic analysis. Each reflection is formed from a wave with an amplitude proportional to the square root of the observed intensity of the spot. Each wave also has a *phase*—that is, the timing of its crests and troughs relative to those of other waves. Additional experiments or calculations must be performed to determine the phases corresponding to each reflection.

The next step is to reconstruct an image of the protein from the observed reflections. In light microscopy or electron microscopy, the diffracted beams are focused by lenses to directly form an image. However, appropriate lenses for focusing x-rays do not exist. Instead, the image is formed by applying a mathematical relation called a *Fourier transform* to the measured amplitudes and calculated phases of every observed reflection. The image obtained is referred to as

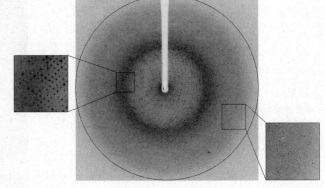

FIGURE 3.40 An x-ray diffraction pattern. A protein crystal diffracts x-rays to produce a pattern of spots, or reflections, on the detector surface. The white silhouette in the center of the image is from a beam stop which protects the detector from the intense, undiffracted x-rays. [S. Lansky et al., *Acta Cryst.* F69:430–434, 2013, Fig. 2. © 2013 IUCr.]

(A)

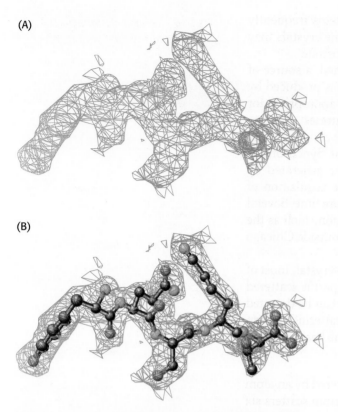

(B)

FIGURE 3.41 Interpretation of an electron-density map. (A) A segment of an electron-density map is drawn as a three-dimensional contour plot, in which the regions inside the "cage" represent the regions of highest electron density. (B) A model of the protein is built into this map so as to maximize the placement of atoms within this density. [Drawn from 1FCH.pdb.]

the *electron-density map*. This map is a three-dimensional graphic representation of where the electrons are most densely localized and it is used to determine the positions of the atoms in the crystallized molecule (Figure 3.41). Critical to the interpretation of the map is its *resolution*, which is established by the number of scattered intensities used in the Fourier transform. The fidelity of the image depends on this resolution. A resolution of 6 Å reveals the course of the polypeptide chain but few other structural details. The reason is that polypeptide chains pack together so that their centers are between 5 Å and 10 Å apart. Maps at higher resolution are needed to delineate groups of atoms, which lie between 2.8 Å and 4.0 Å apart, and individual atoms, which are between 1.0 Å and 1.5 Å apart (Figure 3.42). The ultimate resolution of an x-ray analysis is determined by the degree of perfection of the crystal. For proteins, this limiting resolution is often about 2 Å; however, in exceptional cases, resolutions of 1.0 Å have been obtained.

Nuclear magnetic resonance spectroscopy can reveal the structures of proteins in solution

X-ray crystallography is the most powerful method for determining protein structures. However, some proteins do not readily crystallize. Furthermore, nonphysiological conditions (such as pH and added salts) are often required for successful crystallization. In addition, crystallized proteins adopt conformations that may be influenced by constraints imposed by the crystalline environment. However, since most proteins function in solution under physiological conditions, greater insight into protein function may be gained by structural determination free from crystallization constraints. *Nuclear magnetic resonance* (NMR) *spectroscopy* is unique in being able to reveal the atomic structure of macromolecules *in solution*, provided that highly concentrated solutions (~1 mM, or 15 mg ml^{-1} for a 15-kDa

FIGURE 3.42 Resolution affects the quality of the electron density map. The electron density maps of a tyrosine residue at four different resolution levels (1.0 Å, 2.0 Å, 2.7 Å, and 3.0 Å) are shown. At the lower resolution levels (2.7 Å and 3.0 Å), only a group of atoms corresponding to the side chain is visible, whereas at the highest resolution (1.0 Å), individual atoms within the side chain are distinguishable. [Data from www.rcsb.org/pdb/101/static101 .do?p=education_discussion/Looking-at -Structures/resolution.html.]

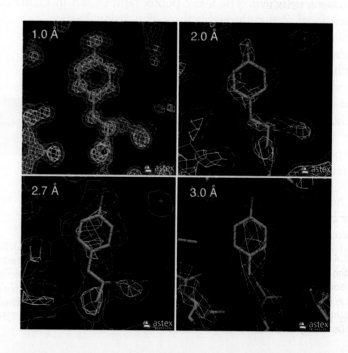

protein) can be obtained. This technique depends on the fact that certain atomic nuclei are intrinsically magnetic. Only a limited number of isotopes display this property, called *spin,* and those most important to biochemistry are listed in Table 3.4. The simplest example is the hydrogen nucleus (^{1}H), which is a proton. The spinning of a proton generates a magnetic moment. This moment can take either of two orientations, or spin states (called α and β), when an external magnetic field is applied (Figure 3.43). The energy difference between these states is proportional to the strength of the imposed magnetic field. The α state has a slightly lower energy because it is aligned with this applied field. Hence, in a given population of nuclei, slightly more will occupy the α state (by a factor of the order of 1.00001 in a typical experiment). A spinning proton in an α state can be raised to an excited state (β state) by applying a pulse of electromagnetic radiation (a radio-frequency, or RF, pulse), provided that the frequency corresponds to the energy difference between the α and the β states. In these circumstances, the spin will change from α to β; in other words, *resonance* will be obtained.

These properties can be used to examine the chemical surroundings of the hydrogen nucleus. The flow of electrons around a magnetic nucleus generates a small, local magnetic field that opposes the applied field. The degree of such shielding depends on the surrounding electron density. Consequently, nuclei in different environments will change states, or resonate, at slightly different field strengths or radiation frequencies. A resonance spectrum for a molecule is obtained by keeping the magnetic field constant and varying the frequency of the electromagnetic radiation. The nuclei of the perturbed sample absorb electromagnetic radiation at a measurable frequency. The different frequencies, termed *chemical shifts,* are expressed in fractional units δ (parts per million, or ppm) relative to the shifts of a standard compound, such as a water-soluble derivative of tetramethylsilane, that is added with the sample. For example, a —CH$_3$ proton typically exhibits a chemical shift (δ) of 1 ppm, compared with a chemical shift of 7 ppm for an aromatic proton. The chemical shifts of most protons in protein molecules fall between 0 and 9 ppm (Figure 3.44). Most protons in many proteins can be resolved by using this technique of *one-dimensional NMR.* With this information, we can then deduce changes to a particular chemical group under different conditions, such as the conformational change of a protein from a disordered structure to an α helix in response to a change in pH.

TABLE 3.4 Biologically important nuclei giving NMR signals

Nucleus	Natural abundance (% by weight of the element)
^{1}H	99.984
^{2}H	0.016
^{13}C	1.108
^{14}N	99.635
^{15}N	0.365
^{17}O	0.037
^{23}Na	100.0
^{25}Mg	10.05
^{31}P	100.0
^{35}Cl	75.4
^{39}K	93.1

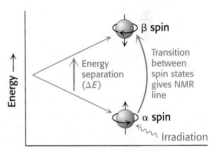

FIGURE 3.43 Basis of NMR spectroscopy. The energies of the two orientations of a nucleus of spin $\frac{1}{2}$ (such as ^{31}P and ^{1}H) depend on the strength of the applied magnetic field. Absorption of electromagnetic radiation of appropriate frequency induces a transition from the lower to the upper level.

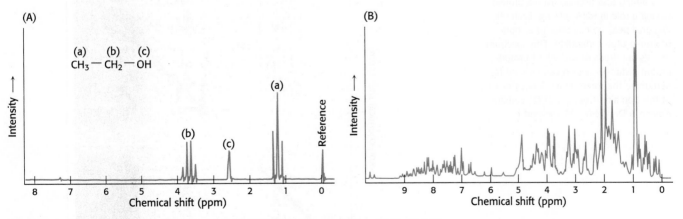

FIGURE 3.44 One-dimensional NMR spectra. (A) ^{1}H-NMR spectrum of ethanol (CH$_3$CH$_2$OH) shows that the chemical shifts for the hydrogen are clearly resolved. (B) ^{1}H-NMR spectrum of a 55 amino acid fragment of a protein having a role in RNA splicing shows a greater degree of complexity. A large number of peaks are present and many overlap. [(A) Data from C. Branden and J. Tooze, *Introduction to Protein Structure* (Garland, 1991), p. 280; (B) courtesy of Dr. Barbara Amann and Dr. Wesley McDermott.]

FIGURE 3.45 The nuclear Overhauser effect. The nuclear Overhauser effect (NOE) identifies pairs of protons that are in close proximity. (A) Schematic representation of a polypeptide chain highlighting five particular protons. Protons 2 and 5 are in close proximity (~4 Å apart), whereas other pairs are farther apart. (B) A highly simplified NOESY spectrum. The diagonal shows five peaks corresponding to the five protons in part A. The peak above the diagonal and the symmetrically related one below reveal that proton 2 is close to proton 5.

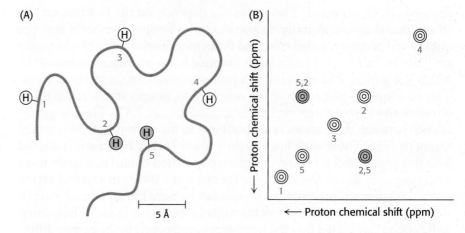

We can garner even more information by examining how the spins on different protons affect their neighbors. By inducing a transient magnetization in a sample through the application of a radio-frequency pulse, we can alter the spin on one nucleus and examine the effect on the spin of a neighboring nucleus. Especially revealing is a *two-dimensional spectrum obtained by nuclear Overhauser enhancement spectroscopy* (NOESY), *which graphically displays pairs of protons that are in close proximity,* even if they are not close together in the primary structure. The basis for this technique is the *nuclear Overhauser effect* (NOE), an interaction between nuclei that is proportional to the inverse sixth power of the distance between them. Magnetization is transferred from an excited nucleus to an unexcited one if the two nuclei are less than about 5 Å apart (Figure 3.45A). In other words, the effect provides a means of detecting the location of atoms relative to one another in the three-dimensional structure of the protein. The peaks that lie along the diagonal of a NOESY spectrum (shown in white in Figure 3.45B) correspond to those present in a one-dimensional NMR experiment. The peaks apart from the diagonal (shown in red in Figure 3.45B), referred to as *off-diagonal peaks* or *cross-peaks,* provide crucial new information: *they identify pairs of protons that are less than 5 Å apart.* A two-dimensional NOESY spectrum for a protein comprising 55 amino acids is shown in Figure 3.46. The large

FIGURE 3.46 Detecting short proton–proton distances. A NOESY spectrum for a 55 amino acid domain from a protein having a role in RNA splicing. Each off-diagonal peak corresponds to a short proton–proton separation. This spectrum reveals hundreds of such short proton–proton distances, which can be used to determine the three-dimensional structure of this domain. [Courtesy of Dr. Barbara Amann & Dr. Wesley McDermott.]

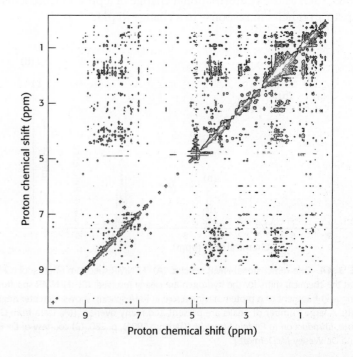

(A)

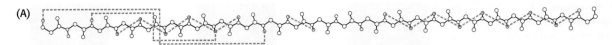

(B)

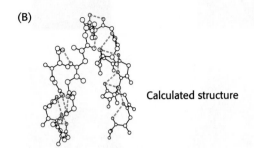

Calculated structure

FIGURE 3.47 Structures calculated on the basis of NMR constraints. (A) NOESY observations show that protons (connected by dotted red lines) are close to one another in space. (B) A three-dimensional structure calculated with these proton pairs constrained to be close together.

number of off-diagonal peaks reveals short proton–proton distances. The three-dimensional structure of a protein can be reconstructed with the use of such proximity relations. Structures are calculated such that proton pairs identified by NOESY cross-peaks must be separated by less than 5 Å in the three-dimensional structure (Figure 3.47). If a sufficient number of distance constraints are applied, the three-dimensional structure can nearly be determined uniquely.

In practice, a family of related structures is generated by NMR spectroscopy for three reasons (Figure 3.48). First, not enough constraints may be experimentally accessible to fully specify the structure. Second, the distances obtained from analysis of the NOESY spectrum are only approximate. Finally, the experimental observations are made not on single molecules but on a large number of molecules in solution that may have slightly different structures at any given moment. Thus, the family of structures generated from NMR structure analysis indicates the range of conformations for the protein in solution. At present, structural determination by NMR spectroscopy is generally limited to proteins less than 50 kDa, but its resolving power is certain to increase. The power of NMR has been greatly enhanced by the ability of recombinant DNA technology to produce proteins labeled uniformly or at specific sites with ^{13}C, ^{15}N, and ^{2}H.

Cryo-electron microscopy is an emerging method of protein structure determination

As we shall see throughout this book, the structures determined by x-ray crystallography and NMR spectroscopy have illuminated much of our understanding of protein function. However, certain proteins, especially those that form large macromolecular complexes or are embedded within the lipid membrane, represent a unique challenge. While crystallographic methods have been developed to enable structure determination of such proteins, these techniques can be quite complex, and are not always reliable. An additional method for protein structure determination, *cryo-electron microscopy* (cryo-EM), has emerged as a viable alternative.

In order to perform cryo-EM, a thin layer of the protein solution is prepared in a fine grid then frozen very quickly, trapping the molecules in an ensemble of orientations. The sample is then placed in a transmission electron microscope under vacuum conditions and exposed to an incident electron beam. Each protein interacts with the beam to produce a two-dimensional projection on the image capture device, or detector. Many projections are detected, each capturing a molecule in a different orientation (Figure 3.49A). Using a process called single-particle analysis, computers use these projections to build a three-dimensional representation of the protein (Figure 3.49B).

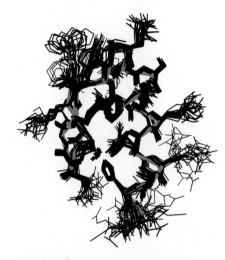

FIGURE 3.48 A family of structures. A set of 25 structures for a 28 amino acid domain from a zinc-finger-DNA-binding protein. The red line traces the average course of the protein backbone. Each of these structures is consistent with hundreds of constraints derived from NMR experiments. The differences between the individual structures are due to a combination of imperfections in the experimental data and the dynamic nature of proteins in solution. [Courtesy of Dr. Barbara Amann.]

FIGURE 3.49 Projections of a multimeric protein captured by cryo-EM. The projections detected in a cryo-EM experiment represent two-dimensional images of the protein trapped in various orientations. Three such projections of the tetrameric protein TRPV1 are shown in (A). Collection of many such projections enables the reconstruction of a three-dimensional model of the protein, as depicted in (B). [(A) Reprinted by permission from Macmillan Publishers Ltd. *Images from M. Liao, et al. Nature,* 504, 107–112, 2013, Fig. 1c; (B) Drawn from 3J5P.pdb.]

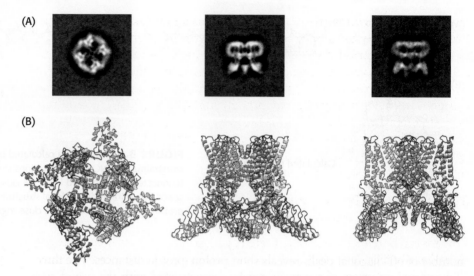

With improvements in image quality and sample preparation, cryo-EM is delivering structures at resolutions approaching 3 Å or better. At this level, details of atomic groups emerge. For example, cryo-EM was used to determine the structure of TRPV1, the tetrameric transmembrane protein depicted in Figure 3.49, in the presence of small molecules that serve to activate the protein. The resolution of this structure was sufficient to allow the identification of the binding sites for these compounds. In addition, cryo-EM enables the visualization of very large molecular complexes. Consider the spliceosome, a complex of 4 oligonucleotide chains and over 30 protein subunits that plays an important role in the processing of eukaryotic messenger RNA (Chapter 30). Although the structural determination of this massive complex has eluded x-ray crystallographic methods, it has been solved to nearly 3.5 Å resolution by cryo-EM.

The structures of over 135,000 proteins had been elucidated by x-ray crystallography, NMR spectroscopy, and cryo-EM by the end of 2017. Several new structures are now determined each day. The coordinates are collected at the Protein Data Bank (www.pdb.org), and the structures can be accessed for visualization and analysis. Knowledge of the detailed molecular architecture of proteins has been a source of insight into how proteins recognize and bind other molecules, how they function as enzymes, how they fold, and how they have evolved. This extraordinarily rich harvest is continuing at a rapid pace and is greatly influencing the entire field of biochemistry as well as other biological and physical sciences.

SUMMARY

The rapid progress in gene sequencing has advanced another goal of biochemistry—elucidation of the proteome. The proteome is the complete set of proteins expressed and includes information about how they are modified, how they function, and how they interact with other molecules.

3.1 The Purification of Proteins Is an Essential First Step in Understanding Their Function

Proteins can be separated from one another and from other molecules on the basis of such characteristics as solubility, size, charge, and binding affinity. SDS–polyacrylamide gel electrophoresis separates the polypeptide chains of proteins under denaturing conditions largely according to mass. Proteins can also be separated electrophoretically on the basis of net charge by isoelectric focusing in a pH gradient. Ultracentrifugation and gel-filtration chromatography resolve proteins according to size, whereas ion-exchange chromatography separates them mainly on the basis of net charge. The high affinity of many proteins for specific chemical groups is exploited in affinity chromatography, in which proteins bind to columns containing beads bearing covalently linked substrates, inhibitors, or other specifically recognized groups. The mass of a protein can be determined by sedimentation-equilibrium measurements.

3.2 Immunology Provides Important Techniques with Which to Investigate Proteins

Proteins can be detected and quantitated by highly specific antibodies; monoclonal antibodies are especially useful because they are homogeneous. Enzyme-linked immunosorbent assays and western blots of SDS–polyacrylamide gels are used extensively. Proteins can also be localized within cells by immunofluorescence microscopy and immunoelectron microscopy.

3.3 Mass Spectrometry Is a Powerful Technique for the Identification of Peptides and Proteins

Techniques such as matrix-assisted laser desorption/ionization (MALDI) and electrospray ionization (ESI) allow the generation of ions of proteins and peptides in the gas phase. The mass of such protein ions can be determined with great accuracy and precision. Masses determined by these techniques act as protein name tags because the mass of a protein or peptide is precisely determined by its amino acid composition and, hence, by its sequence. In addition to chemical methods, such as the Edman degradation, tandem mass spectrometry enables the rapid and highly accurate sequencing of peptides. These sequences are rich in information concerning the kinship of proteins, their evolutionary relationships, and diseases produced by mutations. Knowledge of a sequence provides valuable clues to conformation and function. Mass spectrometric techniques are central to proteomics because they make it possible to analyze the constituents of large macromolecular assemblies or other collections of proteins.

3.4 Peptides Can Be Synthesized by Automated Solid-Phase Methods

Polypeptide chains can be synthesized by automated solid-phase methods in which the carboxyl end of the growing chain is linked to an insoluble support. The carboxyl group of the incoming amino acid is activated by dicyclohexylcarbodiimide and joined to the amino group of the growing chain. Synthetic peptides can serve as drugs and as antigens to stimulate the formation of specific antibodies. They can also be sources of insight into the relation between amino acid sequence and conformation.

3.5 Three-Dimensional Protein Structure Can Be Determined by X-ray Crystallography, NMR Spectroscopy, and Cryo-Electron Microscopy

X-ray crystallography and nuclear magnetic resonance spectroscopy have greatly enriched our understanding of how proteins fold, recognize other molecules, and catalyze chemical reactions. X-ray crystallography is possible because electrons scatter x-rays. The diffraction pattern produced can be analyzed to reveal the arrangement of atoms in a protein. The three-dimensional structures of tens of thousands of proteins are now known in atomic detail. Nuclear magnetic resonance spectroscopy reveals the structure and dynamics of proteins in solution. The chemical shift of nuclei depends on their local environment. Furthermore, the spins of neighboring nuclei interact with each other in ways that provide definitive structural information. This information can be used to determine complete three-dimensional structures of proteins. Cryo-election microscopy has emerged as a third structural determination method, especially for very large protein complexes.

APPENDIX

Critical thinking and problem-solving are essential processes in developing a complete and tangible understanding of biochemistry. In Problem-Solving Strategy Appendices, we present different ways to think about problems and derive solutions. Just as concepts in biochemistry require different approaches for comprehension, different kinds of problems require different strategies.

Problem-Solving Strategies

PROBLEM: *Size estimate.* The relative electrophoretic mobilities of a 30-kDa protein and a 92-kDa protein used as standards on an SDS–polyacrylamide gel are 0.80 and 0.41, respectively. What is the apparent mass of a protein having a mobility of 0.62 on this gel?

SOLUTION: Recall from p. 77 and Figure 3.10 that the relative electrophoretic mobility of proteins is proportional to the *logarithm* of their individual masses.

For protein A (the smaller protein): log(Mass) = log(30 kDa) = 1.48, and mobility = 0.80

For protein B (the larger protein): log(Mass) = log(92 kDa) = 1.96, and mobility = 0.41

With two points defined, we now determine the equation of the line connecting these two points. The slope of the line generated from these data would be:

$$m = \frac{y_1 - y_2}{x_1 - x_2} = \frac{1.48 - 1.96}{0.80 - 0.41} = -1.23$$

To determine the full equation, we need only complete the equation with this slope and one of the points.

$$m = \frac{y - y_1}{x - x_1}$$

$$-1.23 = \frac{y - 1.48}{x - 0.80}$$

$$y = -1.23x + 2.46$$

With the equation of our line, we can now insert the electrophoretic mobility of our known protein, 0.62:

$$y = -1.23(0.62) + 2.46$$

$$y = 1.70$$

Remember, this represents the log(Mass) of our protein! We need only compute the inverse logarithm of 1.70—that is, $10^{(1.70)}$—to determine the mass of our protein: 50 kDa.

KEY TERMS

proteome (p. 70)

assay (p. 71)

specific activity (p. 71)

homogenate (p. 71)

salting out (p. 72)

dialysis (p. 73)

gel-filtration chromatography (p. 73)

ion-exchange chromatography (p. 74)

cation exchange (p. 74)

anion exchange (p. 74)

affinity chromatography (p. 74)

high-performance liquid chromatography (HPLC) (p. 75)

gel electrophoresis (p. 75)

isoelectric point (p. 77)

isoelectric focusing (p. 77)

two-dimensional electrophoresis (p. 78)

sedimentation coefficient (Svedberg unit, S) (p. 80)

antibody (p. 83)

antigen (p. 83)

antigenic determinant (epitope) (p. 83)

polyclonal antibody (p. 84)

monoclonal antibody (p. 85)

hybridoma (p. 86)

enzyme-linked immunosorbent assay (ELISA) (p. 86)

western blotting (p. 87)

co-immunoprecipitation (p. 88)

fluorescence microscopy (p. 89)

green fluorescent protein (GFP) (p. 89)

matrix-assisted laser desorption/ionization (MALDI) (p. 90)

electrospray ionization (ESI) (p. 90)

time-of-flight (TOF) mass analyzer (p. 90)

Edman degradation (p. 92)

phenyl isothiocyanate (p. 92)

tandem mass spectrometry (p. 92)

overlap peptide (p. 94)

peptide mass fingerprinting (p. 96)

solid-phase method (p. 100)

x-ray crystallography (p. 100)

Fourier transform (p. 101)

electron-density map (p. 102)

nuclear magnetic resonance (NMR) spectroscopy (p. 102)

chemical shift (p. 103)

cryo-electron microscopy (p. 105)

PROBLEMS

As you work on these problems, you might find it helpful to visit 🅰 SaplingPlus to see animations of

Animation 1   Gel-Filtration Chromatography

Animation 2   Affinity Chromatography

Animation 3   SDS-PAGE (Gel Electrophoresis)

Animation 4   Isoelectric Focusing

Animation 5   Two-dimensional Electrophoresis

Animation 6   Western Blotting (Immunoblotting)

Animation 7   MALDI-TOF Mass Spectroscopy

Animation 8   Tandem Mass Spectroscopy

1. *Valuable reagents.* The following reagents are often used in protein chemistry:

CNBr   Trypsin
Urea   Performic acid
Mercaptoethanol 6 N HCl
Chymotrypsin   Phenyl isothiocyanate

Which one is the best suited for accomplishing each of the following tasks?

(a) Determination of the amino acid sequence of a small peptide.

(b) Reversible denaturation of a protein devoid of disulfide bonds. Which additional reagent would you need if disulfide bonds were present?

(c) Hydrolysis of peptide bonds on the carboxyl side of aromatic residues.

(d) Cleavage of peptide bonds on the carboxyl side of methionines.

(e) Hydrolysis of peptide bonds on the carboxyl side of lysine and arginine residues.

2. *The only constant is change.* Explain how two different cell types from the same organism will have identical genomes but may have vastly divergent proteomes.

3. *Crafting a new breakpoint.* Ethyleneimine reacts with cysteine side chains in proteins to form S-aminoethyl derivatives. The peptide bonds on the carboxyl side of these modified cysteine residues are susceptible to hydrolysis by trypsin. Why?

4. *Spectrometry.* The absorbance A of a solution is defined as

$$A = \log_{10}(I_0/I)$$

in which I_0 is the incident-light intensity and I is the transmitted-light intensity. The absorbance is related to the molar absorption coefficient (extinction coefficient) ε (in $M^{-1}\,cm^{-1}$), concentration c (in M), and path length l (in cm) by

$$A = \varepsilon l c$$

The absorption coefficient of myoglobin at 580 nm is $15{,}000\ M^{-1}\,cm^{-1}$. What is the absorbance of a $1\ mg\ ml^{-1}$ solution across a 1-cm path? What percentage of the incident light is transmitted by this solution?

5. *It's in the bag.* Suppose that you precipitate a protein with $1\ M\ (NH_4)_2SO_4$ and that you wish to reduce the concentration of the $(NH_4)_2SO_4$. You take 1 ml of your sample and dialyze it in 1000 ml of buffer. At the end of dialysis, what is the concentration of $(NH_4)_2SO_4$ in your sample? How could you further lower the $(NH_4)_2SO_4$ concentration?

6. *Too much or not enough.* Why do proteins precipitate at high salt concentrations? Although many proteins precipitate at high salt concentrations, some proteins require salt to dissolve in water. Explain why some proteins require salt to dissolve.

7. *A slow mover.* Tropomyosin, a 70-kDa muscle protein, sediments more slowly than does hemoglobin (65 kDa). Their sedimentation coefficients are 2.6S and 4.31S, respectively. Which structural feature of tropomyosin accounts for its slow sedimentation?

8. *Sedimenting spheres.* What is the dependence of the sedimentation coefficient s of a spherical protein on its mass? How much more rapidly does an 80-kDa protein sediment than does a 40-kDa protein? ✅❸

9. *Frequently used in shampoos.* The detergent sodium dodecyl sulfate (SDS) denatures proteins. Suggest how SDS destroys protein structure.

10. *Unexpected migration.* Some proteins migrate anomalously in SDS-PAGE gels. For instance, the molecular weight determined from an SDS-PAGE gel is sometimes very different from the molecular weight determined from the amino acid sequence. Suggest an explanation for this discrepancy. ✓❸

11. *Sorting cells.* Fluorescence-activated cell sorting (FACS) is a powerful technique for separating cells according to their content of particular molecules. For example, a fluorescence-labeled antibody specific for a cell-surface protein can be used to detect cells containing such a molecule. Suppose that you want to isolate cells that possess a receptor enabling them to detect bacterial degradation products. However, you do not yet have an antibody directed against this receptor. Which fluorescence-labeled molecule would you prepare to identify such cells?

12. *Column choice.* (a) The octapeptide AVGWRVKS was digested with the enzyme trypsin. Which method would be most appropriate for separating the products: ion-exchange or gel-filtration chromatography? Explain. (b) Suppose that the peptide was digested with chymotrypsin. What would be the optimal separation technique? Explain. ✓❷

13. *Power(ful) tools.* Monoclonal antibodies can be conjugated to an insoluble support by chemical methods. Explain how these antibody-bound beads can be exploited for protein purification. ✓❹

14. *Assay development.* You wish to isolate an enzyme from its native source and need a method for measuring its activity throughout the purification. However, neither the substrate nor the product of the enzyme-catalyzed reaction can be detected by spectroscopy. You discover that the product of the reaction is highly antigenic when injected into mice. Propose a strategy to develop a suitable assay for this enzyme. ✓❶

15. *Making more enzyme?* In the course of purifying an enzyme, a researcher performs a purification step that results in an *increase* in the total activity to a value greater than that present in the original crude extract. Explain how the amount of total activity might increase. ✓❶

16. *Divide and conquer.* The determination of the mass of a protein by mass spectrometry often does not allow its unique identification among possible proteins within a complete proteome, but determination of the masses of all fragments produced by digestion with trypsin almost always allows unique identification. Explain. ✓❸

17. *Know your limits.* Which two amino acids are indistinguishable in peptide sequencing by the tandem mass spectrometry method described in this chapter and why? ✓❺

18. *Protein purification problem.* Complete the following table. ✓❶

Purification procedure	Total protein (mg)	Total activity (units)	Specific activity (units mg⁻¹)	Purification level	Yield (%)
Crude extract	20,000	4,000,000		1	100
(NH₄)₂SO₄ precipitation	5000	3,000,000			
DEAE-cellulose chromatography	1500	1,000,000			
Gel-filtration chromatography	500	750,000			
Affinity chromatography	45	675,000			

19. *Part of the mix.* Your frustrated colleague hands you a mixture of four proteins with the following properties:

	Isoelectric point (pI)	Molecular weight (in kDa)
Protein A	4.1	80
Protein B	9.0	81
Protein C	8.8	37
Protein D	3.9	172

(a) Propose a method for the isolation of Protein B from the other proteins. (b) If Protein B also carried a His tag at its N-terminus, how could you revise your method? ✓❷

20. *The challenge of flexibility.* Structures of proteins comprising domains separated by flexible linker regions can be quite difficult to solve by x-ray crystallographic methods. Why might this be the case? What are possible experimental approaches to circumvent this barrier? ✓❻

Chapter Integration Problems

21. *Quaternary structure.* A protein was purified to homogeneity. Determination of the mass by gel-filtration chromatography yields 60 kDa. Chromatography in the presence of 6 M urea yields a 30-kDa species. When the chromatography is repeated in the presence of 6 M urea and 10 mM β-mercaptoethanol, a single molecular species of 15 kDa results. Describe the structure of the molecule.

22. *Helix–coil transitions.* (a) NMR measurements have shown that poly-L-lysine is a random coil at pH 7 but becomes a helix as the pH is raised above 10. Account for this pH-dependent conformational transition. (b) Predict the pH dependence of the helix–coil transition of poly-L-glutamate.

23. *Peptide mass determination.* You have isolated a protein from the bacterium *E. coli* and seek to confirm its identity by trypsin digestion and mass spectrometry. Determination of the masses of several peptide fragments has enabled you to deduce the identity of the protein. However, there is a discrepancy with one of the peptide fragments, which you believe should have the sequence MLNSFK and an $(M+H)^+$ value of 739.38. In your experiments, you repeatedly obtain an $(M+H)^+$ value of 767.38. What is the cause of this discrepancy and what does it tell you about the region of the protein from which this peptide is derived?

24. *Peptides on a chip.* Large numbers of different peptides can be synthesized in a small area on a solid support. This high-density array can then be probed with a fluorescence-labeled protein to find out which peptides are recognized. The binding of an antibody to an array of 1024 different peptides occupying a total area the size of a thumbnail is shown in the adjoining illustration. How would you synthesize such a peptide array? (Hint: Use light instead of acid to deprotect the terminal amino group in each round of synthesis.)

Fluorescence scan of an array of 1024 peptides in a 1.6-cm² area. Each synthesis site is a 400-μm square. A fluorescently labeled monoclonal antibody was added to the array to identify peptides that are recognized. The height and color of each square denote the fluorescence intensity. [Information from S. P. A. Fodor et al., *Science* 251(1991): 767.]

25. *Exchange rate.* The amide hydrogen atoms of peptide bonds within proteins can exchange with protons in the solvent. In general, amide hydrogen atoms in buried regions of proteins and protein complexes exchange more slowly than those on the solvent-accessible surface do. Determination of these rates can be used to explore the protein-folding reaction, probe the tertiary structure of proteins, and identify the regions of protein–protein interfaces. These exchange reactions can be followed by studying the behavior of the protein in solvent that has been labeled with deuterium (^2H), a stable isotope of hydrogen. What two methods described in this chapter could be readily applied to the study of hydrogen–deuterium exchange rates in proteins?

Data Interpretation Problems

26. *Protein sequencing 1.* Determine the sequence of hexapeptide on the basis of the following data. Note: When the sequence is not known, a comma separates the amino acids (Table 3.3).

Amino acid composition: (2R, A, S, V, Y)

N-terminal analysis of the hexapeptide: A

Trypsin digestion: (R, A, V) and (R, S, Y)

Carboxypeptidase digestion: No digestion.

Chymotrypsin digestion: (A, R, V, Y) and (R, S)

27. *Protein sequencing 2.* Determine the sequence of a peptide consisting of 14 amino acids on the basis of the following data.

Amino acid composition: (4S, 2L, F, G, I, K, M, T, W, Y)

N-terminal analysis: S

Carboxypeptidase digestion: L

Trypsin digestion: (3S, 2L, F, I, M, T, W) (G, K, S, Y)

Chymotrypsin digestion: (F, I, S) (G, K, L) (L, S) (M, T) (S, W) (S, Y)

N-terminal analysis of (F, I, S) peptide: S

Cyanogen bromide treatment: (2S, F, G, I, K, L, M*, T, Y) (2S, L, W)

M*, methionine detected as homoserine.

28. *Applications of two-dimensional electrophoresis.* Performic acid cleaves the disulfide linkage of cystine and converts the sulfhydryl groups into cysteic acid residues, which are then no longer capable of disulfide-bond formation.

Cystine

Performic acid ⟶

Cysteic acid

Consider the following experiment: You suspect that a protein containing three cysteine residues has a single disulfide bond. You digest the protein with trypsin and subject the mixture to electrophoresis along one end of a sheet of paper. After treating the paper with performic acid, you subject the sheet to electrophoresis in the perpendicular direction and stain it with a reagent that detects proteins. How would the paper appear if the protein did not contain any disulfide bonds? If the protein contained a single disulfide bond? Propose an experiment to identify which cysteine residues form the disulfide bond.

Think ▶ Pair ▶ Share Problems

29. Polyacrylamide gels can be prepared in the presence of varying percentages of acrylamide, leading to different pore sizes. Why might this variability be useful?

30. Propose several strategies for the purification of an antibody out of antiserum obtained after antigen inoculation.

31. Explain why it is difficult to draw conclusions about the conformational changes of a protein based on a single crystal structure.

DNA, RNA, and the Flow of Genetic Information

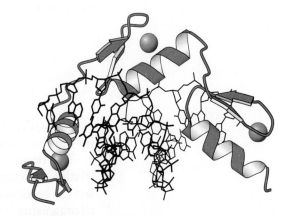

Family resemblance, very evident in this photograph of three sisters, results from having genes in common. Genes must be expressed to exert an effect, and proteins regulate such expression. One such regulatory protein, a zinc-finger protein (zinc ion is blue, protein is red), is shown bound to a control region of DNA (black). [(Left) © Francesca Cesari/LUZ/Redux; (Right) Drawn from 1AAY. pdb.]

✓ LEARNING OBJECTIVES

By the end of this chapter, you should be able to:

1. Distinguish between nucleosides and nucleotides.

2. Identify the bases of DNA and RNA.

3. Distinguish between DNA and RNA.

4. Explain how DNA is replicated.

5. Explain how information flows from DNA to protein.

6. Identify some key differences between bacterial and eukaryotic genes.

OUTLINE

4.1 A Nucleic Acid Consists of Four Kinds of Bases Linked to a Sugar–Phosphate Backbone

4.2 A Pair of Nucleic Acid Strands with Complementary Sequences Can Form a Double-Helical Structure

4.3 The Double Helix Facilitates the Accurate Transmission of Hereditary Information

4.4 DNA Is Replicated by Polymerases That Take Instructions from Templates

4.5 Gene Expression Is the Transformation of DNA Information into Functional Molecules

4.6 Amino Acids Are Encoded by Groups of Three Bases Starting from a Fixed Point

4.7 Most Eukaryotic Genes Are Mosaics of Introns and Exons

DNA and RNA are long linear polymers, called nucleic acids, that carry information in a form that can be passed from one generation to the next. These macromolecules consist of a large number of linked nucleotides, each composed of a sugar, a phosphate, and a base. Sugars linked by phosphates form a common backbone that plays a structural role, whereas the sequence of bases along a nucleic acid strand carries genetic information. The DNA molecule has the form of a *double helix*, a helical structure consisting of two complementary nucleic acid strands. Each strand serves as the template for the other in DNA replication. The genes of all cells and many viruses are made of DNA.

Certain genes specify the kinds of proteins that are made by cells, but DNA is not the direct template for protein synthesis. Rather, a DNA strand is copied into a class of RNA molecules called messenger RNA (mRNA), the information-carrying intermediates in protein synthesis. This process of transcription is followed by translation, the synthesis of proteins according to instructions given by mRNA templates. Information processing in all cells is quite complex. The scheme that underlies information processing at the level of gene expression was first proposed by Francis Crick in 1958.

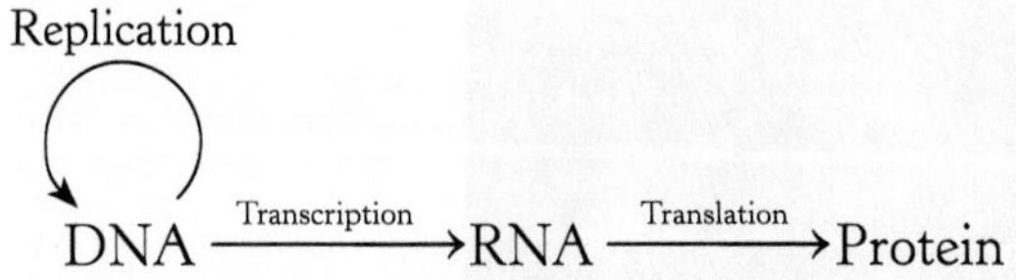

$$\text{DNA} \xrightarrow{\text{Transcription}} \text{RNA} \xrightarrow{\text{Translation}} \text{Protein}$$

Replication

Crick called this scheme the *central dogma*. The basic tenets of this dogma are true, but, as we will see later, this scheme is not as simple as depicted.

This flow of information depends on the genetic code, which defines the relation between the sequence of bases in DNA (or its mRNA transcript) and the sequence of amino acids in a protein. The code is nearly the same in all organisms: a sequence of three bases, called a codon, specifies an amino acid. There is another step in the expression of most eukaryotic genes, which are mosaics of nucleic acid sequences called introns and exons. Both are transcribed, but before translation takes place, introns are cut out of newly synthesized RNA molecules, leaving mature RNA molecules with continuous exons. The existence of introns and exons has crucial implications for the evolution of proteins.

4.1 A Nucleic Acid Consists of Four Kinds of Bases Linked to a Sugar–Phosphate Backbone

The nucleic acids DNA and RNA are well suited to function as the carriers of genetic information by virtue of their covalent structures. These macromolecules are *linear polymers* built from similar units connected end to end (Figure 4.1). Each monomer unit within the polymer is a *nucleotide*. A single nucleotide unit consists of three components: a sugar, a phosphate, and one of four bases. *The sequence of bases in the polymer uniquely characterizes a nucleic acid and constitutes a form of linear information*—information analogous to the letters that spell a person's name.

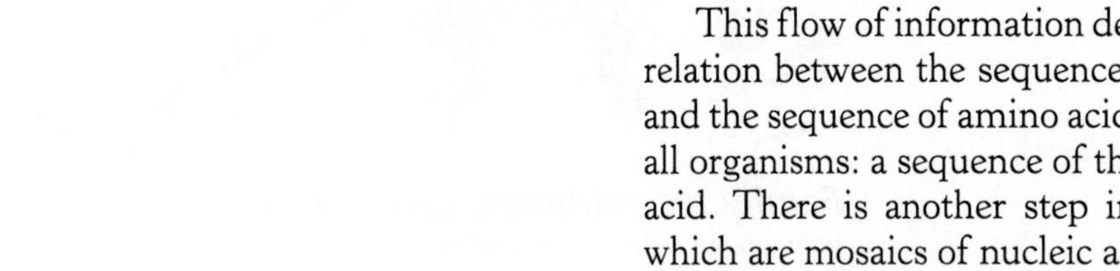

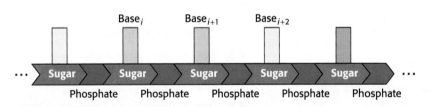

FIGURE 4.1 Polymeric structure of nucleic acids.

RNA and DNA differ in the sugar component and one of the bases

The sugar in *deoxyribonucleic acid* (DNA) is *deoxyribose*. The prefix deoxy indicates that the 2′-carbon atom of the sugar lacks the oxygen atom that is linked to the 2′-carbon atom of *ribose*, as shown in Figure 4.2. Note that sugar carbons are numbered with primes to differentiate them from atoms in the bases. The sugars in both nucleic acids are linked to one another by phosphodiester bridges. Specifically, the 3′-hydroxyl (3′-OH) group of the sugar moiety of one nucleotide is esterified to a phosphate group, which is, in turn, joined to the 5′-hydroxyl group of the adjacent sugar. The chain of sugars linked by phosphodiester bridges is referred to as the *backbone* of the nucleic acid (Figure 4.3). Whereas the backbone is constant in a nucleic acid, the bases vary from one monomer to the next. Two of the bases of DNA are derivatives of *purine*—adenine (A) and guanine (G)—and two of *pyrimidine*—cytosine (C) and thymine (T), as shown in Figure 4.4.

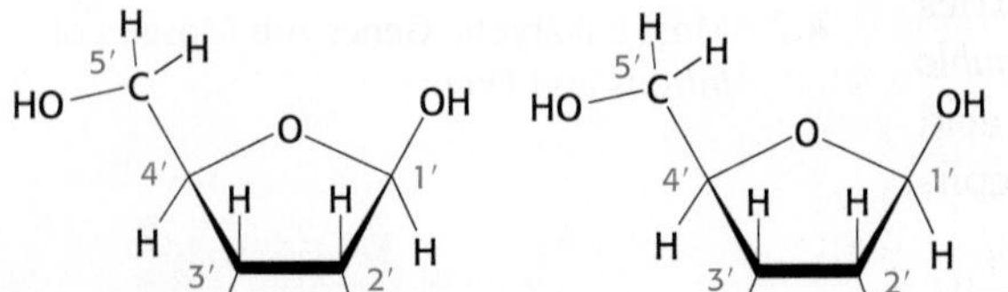

FIGURE 4.2 Ribose and deoxyribose. Atoms in sugar units are numbered with primes to distinguish them from atoms in bases (see Figure 4.4).

Ribose

Deoxyribose

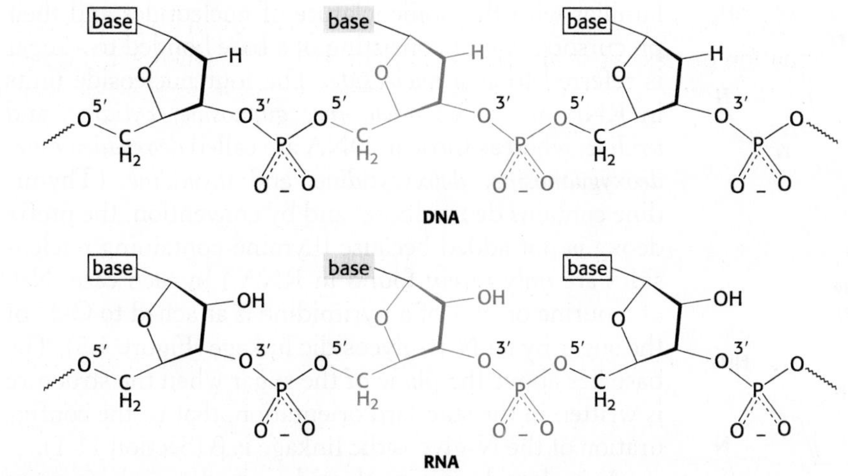

FIGURE 4.3 Backbones of DNA and RNA. The sugar-phosphate backbones of these nucleic acids are formed by 3'-to-5' phosphodiester linkages. A sugar unit is highlighted in red and a phosphate group in blue.

Ribonucleic acid (RNA), like DNA, is a long, unbranched polymer consisting of nucleotides joined by 3'-to-5' phosphodiester linkages (Figure 4.3). The covalent structure of RNA differs from that of DNA in two respects. First, the sugar units in RNA are riboses rather than deoxyriboses. Ribose contains a 2'-hydroxyl group not present in deoxyribose (Figure 4.2). Second, one of the four major bases in RNA is uracil (U) instead of thymine (T) (Figure 4.4).

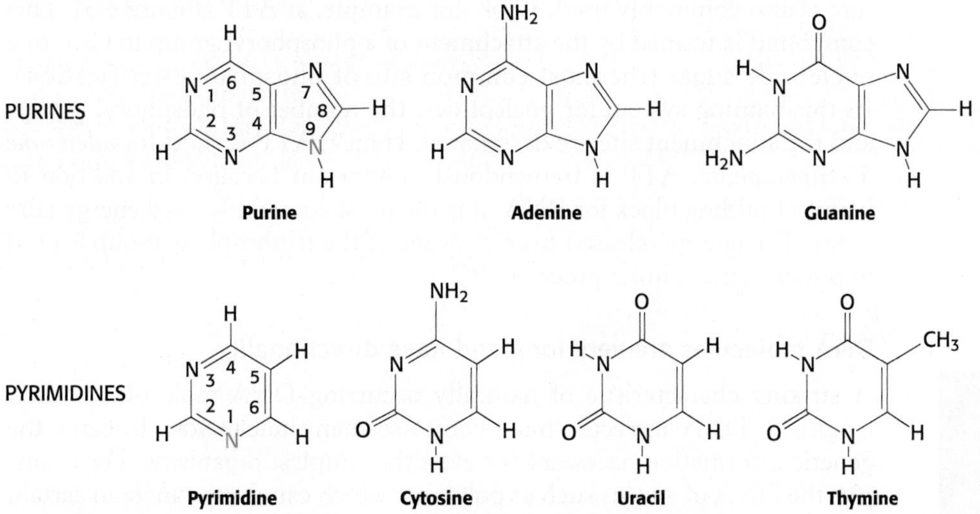

FIGURE 4.4 Purines and pyrimidines. Atoms within bases are numbered without primes. Uracil is present in RNA instead of thymine. N-9 in purines and N-1 in pyrimidines form a bond with the sugar (see Figure 4.5).

Note that each phosphodiester bridge has a negative charge (Figure 4.3). This negative charge repels nucleophilic species such as hydroxide ions, which are capable of hydrolytic attack on the phosphate backbone. This resistance is crucial for maintaining the integrity of information stored in nucleic acids. The absence of the 2'-hydroxyl group in DNA further increases its resistance to hydrolysis. The greater stability of DNA probably accounts for its use rather than RNA as the hereditary material in all modern cells and in many viruses.

Nucleotides are the monomeric units of nucleic acids

The building blocks of nucleic acids and the precursors of these building blocks play many other roles throughout the cell—for instance, as energy currency and as molecular signals. Consequently, it is important to be

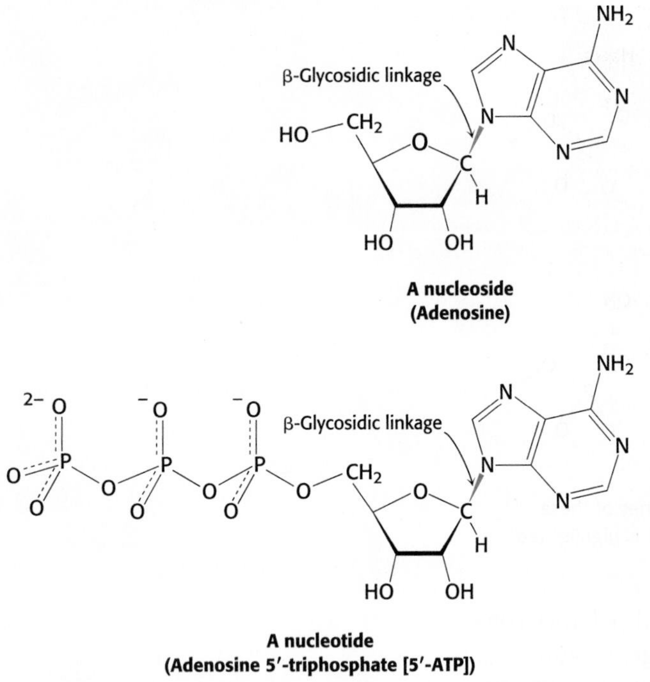

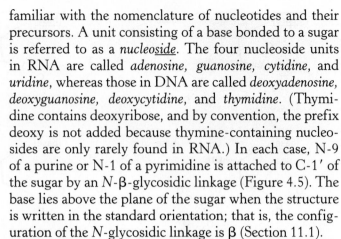

A nucleoside
(Adenosine)

A nucleotide
(Adenosine 5′-triphosphate [5′-ATP])

FIGURE 4.5 A nucleoside and a nucleotide.

familiar with the nomenclature of nucleotides and their precursors. A unit consisting of a base bonded to a sugar is referred to as a *nucleoside*. The four nucleoside units in RNA are called *adenosine, guanosine, cytidine,* and *uridine,* whereas those in DNA are called *deoxyadenosine, deoxyguanosine, deoxycytidine,* and *thymidine.* (Thymidine contains deoxyribose, and by convention, the prefix deoxy is not added because thymine-containing nucleosides are only rarely found in RNA.) In each case, N-9 of a purine or N-1 of a pyrimidine is attached to C-1′ of the sugar by an *N*-β-glycosidic linkage (Figure 4.5). The base lies above the plane of the sugar when the structure is written in the standard orientation; that is, the configuration of the *N*-glycosidic linkage is β (Section 11.1).

A nucleo<u>tide</u> is a nucleoside joined to one or more phosphoryl groups by an ester linkage (Figure 4.5). *Nucleoside triphosphates, nucleosides joined to three phosphoryl groups, are the precursors that form RNA and DNA.* The four nucleotide units that comprise DNA are nucleoside monophosphates called *deoxyadenylate, deoxyguanylate, deoxycytidylate,* and *thymidylate.* Note that a pyrophosphate is released when the nucleotides are linked (p. 125). Similarly, the most common nucleotides that link to form RNA are nucleoside monophosphates *adenylate, guanylate, cytidylate,* and *uridylate.*

This nomenclature does not describe the number of phosphoryl groups or the site of attachment to carbon of the ribose. A more precise nomenclature is also commonly used. Look, for example, at ATP (Figure 4.5). This compound is formed by the attachment of a phosphoryl group to C-5′ of a nucleoside sugar (the most common site of phosphate esterification). In this naming system for nucleotides, the number of phosphoryl groups and the attachment site are designated. Hence, ATP is short for *adenosine 5′-triphosphate.* ATP is tremendously important because, in addition to being a building block for RNA, it is the most commonly used energy currency. The energy released from cleavage of the triphosphate group is used to power many cellular processes (Chapter 15).

DNA molecules are very long and have directionality

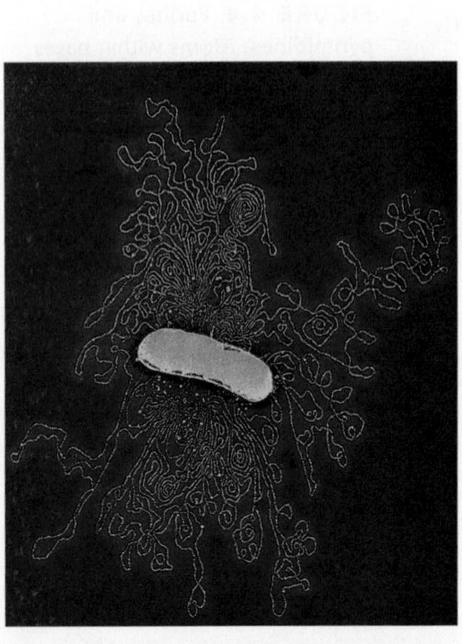

A striking characteristic of naturally occurring DNA molecules is their length. A DNA molecule must comprise many nucleotides to carry the genetic information necessary for even the simplest organisms. For example, the DNA of a virus such as polyoma, which can cause cancer in certain organisms, consists of two paired strands of DNA, each 5100 nucleotides in length. The *E. coli* genome is a single DNA molecule consisting of two strands of 4.6 million nucleotides each (Figure 4.6).

The DNA molecules of higher organisms can be much larger. The human genome comprises approximately 3 billion nucleotides in each strand of DNA, divided among 24 distinct molecules of DNA called chromosomes (22 autosomal chromosomes plus the X and Y sex chromosomes) of different sizes. One of the largest known DNA molecules is found in the Indian muntjac, an Asiatic deer; its genome is nearly as large as the human genome but is distributed on only 3 chromosomes (Figure 4.7). The largest of these chromosomes has two strands of more than 1 billion nucleotides each. If such a DNA molecule could be fully extended, it would stretch more than 1 foot in length. Some plants contain even larger DNA molecules.

FIGURE 4.6 Electron micrograph of part of the *E. coli* genome. [Dr. Gopal Murti/ Science Source.]

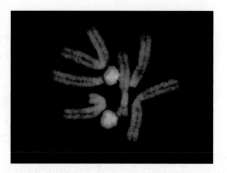

FIGURE 4.7 The Indian muntjac and its chromosomes. Cells from a female Indian muntjac (left) contain three pairs of very large chromosomes (right; stained orange). The image also shows a pair of human chromosomes (stained green) for comparison. [(Left) Super Prin/Shutterstock; (Right) Reprinted by permission from Macmillan Publishers Ltd: Nature Genetics, J-Y Lee, M. Koi, E. J. Stanbridge, M. Oshimura, A. T. Kumamoto, & A. P. Feinberg, Simple purification of human chromosomes to homogeneity using muntjac hybrid cells, vol. 7, p.30, ©1994.]

In spite of the length of DNA molecules, sequences of DNA must be examined to look for similarities in genes or to reveal mutations in a given gene (Chapter 6). Writing the cumbersome chemical structures would be impractical, so scientists have adopted the use of abbreviations. The abbreviated notations pApCpG denote a trinucleotide of DNA consisting of the building blocks deoxyadenylate monophosphate, deoxycytidylate monophosphate, and deoxyguanylate monophosphate linked by two phosphodiester bridges, where "p" denotes a phosphoryl group (Figure 4.8). Note that, like a polypeptide (Section 2.2), *a DNA chain has directionality,* commonly called polarity. One end of the chain has a free 5′-OH group or a 5′-OH group attached to a phosphoryl group, and the other end has a free 3′-OH group, neither of which is linked to another nucleotide. However, when presenting very long sequences of nucleotides, including the phosphates is of no value, given that it is the sequence of bases that represents the genetic information. It is important to remember that even if only the bases are presented, *the base sequence is written in the 5′-to-3′ direction.* Thus, ACG indicates that the unlinked 5′-OH group is on deoxyadenylate, whereas the unlinked 3′-OH group is on deoxyguanylate. Because of this polarity, ACG and GCA correspond to different compounds.

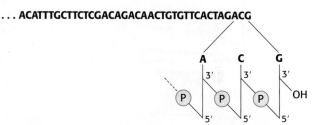

FIGURE 4.8 Structure of a nucleic acid strand. A simplified depiction of a nucleic acid (compare with Figure 4.3). The strand has a 5′ end, which is usually attached to a phosphoryl group, and a 3′ end, which is usually a free hydroxyl group.

4.2 A Pair of Nucleic Acid Strands with Complementary Sequences Can Form a Double-Helical Structure

As discussed in Chapter 1, the covalent structure of nucleic acids accounts for their ability to carry information in the form of a sequence of bases along a nucleic acid strand. The bases on the two separate nucleic acid strands form *specific base pairs* in such a way that a helical structure is formed. The double-helical structure of DNA facilitates the *replication* of the genetic material—that is, the generation of two copies of a nucleic acid from one.

The double helix is stabilized by hydrogen bonds and van der Waals interactions

The ability of nucleic acids to form specific base pairs was discovered in the course of studies directed at determining the three-dimensional structure of DNA. Maurice Wilkins and Rosalind Franklin obtained x-ray diffraction photographs of fibers of DNA (Figure 4.9). The characteristics of

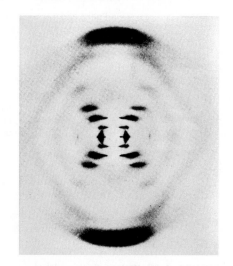

FIGURE 4.9 X-ray diffraction photograph of a hydrated DNA fiber. When crystals of a biomolecule are irradiated with x-rays, the x-rays are diffracted and these diffracted x-rays are seen as a series of spots, called reflections, on a screen behind the crystal. The structure of the molecule can be determined by the pattern of the reflections (Section 3.5). In regard to DNA crystals, the central cross is diagnostic of a helical structure. The strong arcs on the meridian arise from the stack of nucleotide bases, which are 3.4 Å apart. [Omikron/Science Source.]

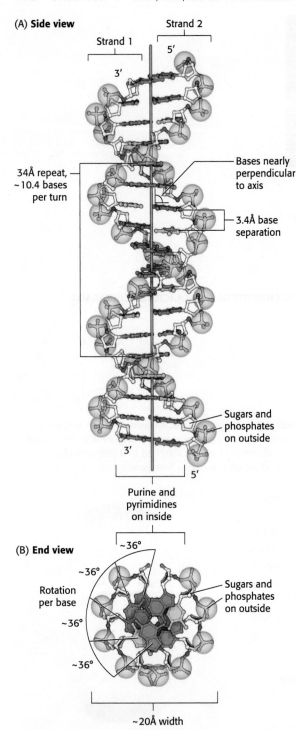

(A) Side view

Strand 1

Strand 2

5′

3′

34Å repeat, ~10.4 bases per turn

Bases nearly perpendicular to axis

3.4Å base separation

Sugars and phosphates on outside

3′

5′

3′

Purine and pyrimidines on inside

(B) End view

~36°

~36°

Rotation per base

Sugars and phosphates on outside

~36°

~36°

~20Å width

FIGURE 4.10 Watson–Crick model of double-helical DNA. (A) Side view. Adjacent bases are separated by 3.4 Å. The structure repeats along the helical axis (vertical) at intervals of 34 Å, which corresponds to approximately 10 nucleotides on each chain. (B) End view, looking down the helix axis, reveals a rotation of 36° per base and shows that the bases are stacked on top of one another [Source: J. L. Tymoczko, J. Berg, and L. Stryer, *Biochemistry: A Short Course*, 2nd ed. (W. H. Freeman and Company, 2013), Fig. 33.11.].

these diffraction patterns indicated that DNA is formed of two strands that wind in a regular helical structure. From these data and others, James Watson and Francis Crick deduced a structural model for DNA that accounted for the diffraction pattern and was the source of some remarkable insights into the functional properties of nucleic acids (Figure 4.10).

The features of the Watson–Crick model of DNA deduced from the diffraction patterns are:

1. Two helical polynucleotide strands are coiled around a common axis with a right-handed screw sense (p. 43). The strands are antiparallel, meaning that they have opposite directionality.

2. The sugar–phosphate backbones are on the outside and the purine and pyrimidine bases lie on the inside of the helix.

3. The bases are nearly perpendicular to the helix axis, and adjacent bases are separated by approximately 3.4 Å. The helical structure repeats on the order of every 34 Å, with about 10.4 bases per turn of helix. There is a rotation of nearly 36 degrees per base (360 degrees per full turn/10.4 bases per turn).

4. The diameter of the helix is about 20 Å.

How is such a regular structure able to accommodate an arbitrary sequence of bases, given the different sizes and shapes of the purines and pyrimidines? In attempting to answer this question, Watson and Crick discovered that guanine can be paired with cytosine and adenine with thymine to form base pairs that have essentially the same shape (Figure 4.11). These base pairs are held together by specific hydrogen bonds, which, although weak (4–21 kJ mol^{-1}, or 1–5 kcal mol^{-1}), stabilize the helix because of their large numbers in a DNA molecule. These *base-pairing rules* account for the observation that the ratios of adenine to thymine and of guanine to cytosine are nearly the same in all species studied, whereas the adenine-to-guanine ratio varies considerably (Table 4.1).

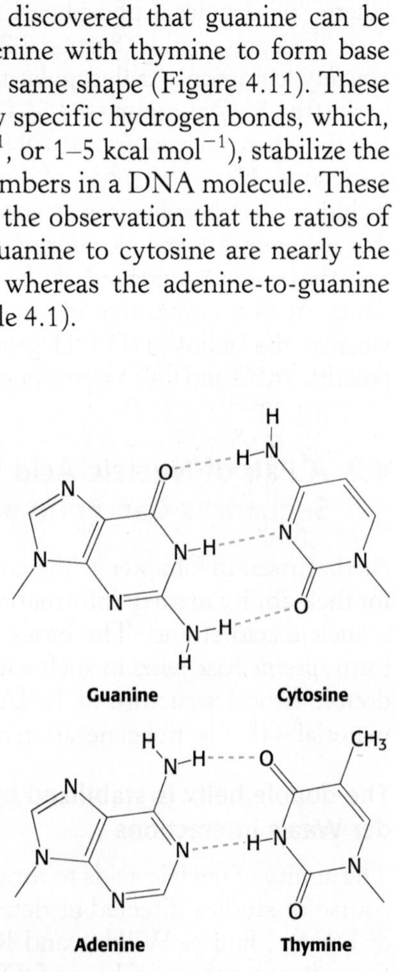

Guanine   **Cytosine**

Adenine   **Thymine**

FIGURE 4.11 Structures of the base pairs proposed by Watson and Crick.

TABLE 4.1 Base compositions experimentally determined for a variety of organisms

Organism	A : T	G : C	A : G
Human being	1.00	1.00	1.56
Salmon	1.02	1.02	1.43
Wheat	1.00	0.97	1.22
Yeast	1.03	1.02	1.67
Escherichia coli	1.09	0.99	1.05
Serratia marcescens	0.95	0.86	0.70

Base stacking
(van der Waal interactions)

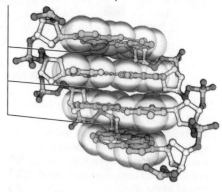

FIGURE 4.12 A side view of DNA. Base pairs are stacked nearly one on top of another in the double helix. The stacked bases interact via van der Waals forces. Such stacking forces help stabilize the double helix. [Source: J. L. Tymoczko, J. Berg, and L. Stryer, *Biochemistry: A Short Course*, 2nd ed. (W. H. Freeman and Company, 2013), Fig. 33.13.]

Inside the helix, the bases are essentially stacked one on top of another (Figure 4.10B). The stacking of base pairs contributes to the stability of the double helix in two ways. First, the formation of the double helix is facilitated by the hydrophobic effect (p. 9). The hydrophobic bases cluster in the interior of the helix away from the surrounding water, whereas the more polar surfaces are exposed to water. This arrangement is reminiscent of protein folding, where hydrophobic amino acids are in the protein's interior and the hydrophilic amino acids are on the exterior (Section 2.4). Second, the stacked base pairs attract one another through van der Waals forces (p. 8), appropriately referred to as *stacking forces,* further contributing to stabilization of the helix (Figure 4.12). The energy associated with a single van der Waals interaction is quite small, typically from 2 to 4 kJ mol^{-1} (0.5–1.0 kcal mol^{-1}). In the double helix, however, a large number of atoms are in van der Waals contact, and the net effect, summed over these atom pairs, is substantial. In addition, base stacking in DNA is favored by the conformations of the somewhat rigid five-membered rings of the backbone sugars.

DNA can assume a variety of structural forms

Watson and Crick based their model (known as the *B-DNA helix*) on x-ray diffraction patterns of highly hydrated DNA fibers, which provided information about properties of the double helix that are averaged over its constituent residues. Under physiological conditions, most DNA is in the B form. X-ray diffraction studies of less-hydrated DNA fibers revealed a different form called *A-DNA*. Like B-DNA, A-DNA is a right-handed double helix made up of anti-parallel strands held together by Watson–Crick base-pairing. The A-form helix is wider and shorter than the B-form helix, and its base pairs are tilted rather than perpendicular to the helix axis (Figure 4.13).

If the A-form helix were simply a property of dehydrated DNA, it would be of little significance. However, double-stranded regions of RNA and at least some RNA–DNA hybrids adopt a double-helical form very similar to that of A-DNA. What is the biochemical basis for differences between the two forms of DNA? Many of the structural differences between B-DNA and A-DNA arise from different puckerings of their

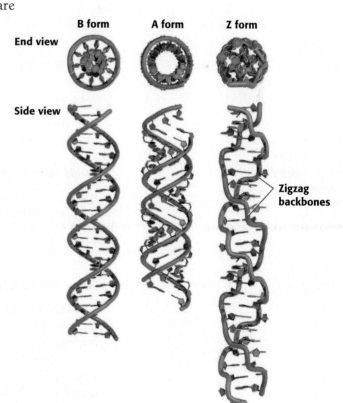

FIGURE 4.13 B-form, A-form, and Z-form of DNA. Models of B-form and A-form DNA depict their right-handed helical structures. The B-form helix is longer and narrower than the A-form helix. Z-DNA is a left-handed helix in which the phosphoryl groups zigzag along the backbone [Drawn from 1BNA.pdb, 1DNZ.pdb, and 131D.pdb.]

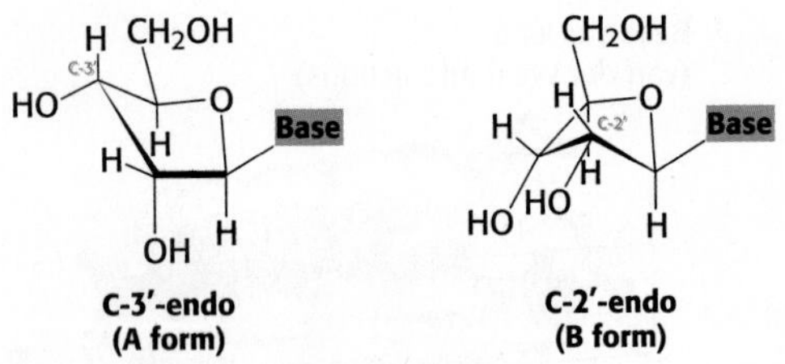

FIGURE 4.14 Sugar pucker. In A-form DNA, the C-3′ carbon atom lies above the approximate plane defined by the four other sugar nonhydrogen atoms (called C-3′ endo). In B-form DNA, each deoxyribose is in a C-2′-endo conformation, in which C-2′ lies out of the plane.

ribose units (Figure 4.14). In A-DNA, C-3′ lies out of the plane (a conformation referred to as C-3′endo) formed by the other four atoms of the ring; in B-DNA, C-2′ lies out of the plane (a conformation called C-2′endo). The C-3′-endo puckering in A-DNA leads to an 11-degree tilting of the base pairs away from perpendicular to the helix. RNA helices are further induced to take the A-DNA form because of steric hindrance from the 2′-hydroxyl group: the 2′-oxygen atom would be too close to three atoms of the adjoining phosphoryl group and to one atom in the next base. In an A-form helix, in contrast, the 2′-oxygen atom projects outward, away from other atoms. The phosphoryl and other groups in the A-form helix bind fewer H_2O molecules than do those in B-DNA. Hence, dehydration favors the A form.

A third type of double helix is *left-handed*, in contrast with the *right-handed* screw sense of the A and B helices. Furthermore, the phosphoryl groups in the backbone are *zigzagged;* hence, this form of DNA is called *Z-DNA* (Figure 4.13).

Although the biological role of Z-DNA is still under investigation, Z-DNA-binding proteins have been isolated, one of which is required for viral pathogenesis of poxviruses, including variola, the agent of smallpox. The existence of Z-DNA shows that DNA is a flexible, dynamic molecule whose parameters are not as fixed as depictions suggest. The properties of A-, B-, and Z-DNA are compared in Table 4.2.

TABLE 4.2 Comparison of A-, B-, and Z-DNA

	A	B	Z
Shape	Broadest	Intermediate	Narrowest
Rise per base pair	2.3 Å	3.4 Å	3.8 Å
Helix diameter	~26 Å	~20 Å	~18 Å
Screw sense	Right-handed	Right-handed	Left-handed
Glycosidic bond*	*anti*	*anti*	Alternating *anti* and *syn*
Base pairs per turn of helix	11	10.4	12
Pitch per turn of helix	25.3 Å	35.4 Å	45.6 Å
Tilt of base pairs from perpendicular to helix axis	19 degrees	1 degree	9 degrees

*Syn and anti refer to the orientation of the N-glycosidic bond between the base and deoxyribose. In the *anti* orientation, the base extends away from the deoxyribose. In the *syn* orientation, the base is above the deoxyribose. Pyrimidines can be in *anti* orientations only, whereas purines can be *anti* or *syn*.

Some DNA molecules are circular and supercoiled

The DNA molecules in human chromosomes are linear. However, electron microscopic and other studies have shown that intact DNA molecules from bacteria and archaea are *circular* (Figure 4.15A). DNA molecules inside cells necessarily have a very compact shape. Note that the *E. coli* chromosome, fully extended, would be about 1000 times as long as the greatest diameter of the bacterium.

A closed DNA molecule has a property unique to circular DNA. The axis of the double helix can itself be twisted or supercoiled into a *superhelix* (Figure 4.15B). A circular DNA molecule without any superhelical turns is known as a *relaxed molecule*. Supercoiling is biologically important for two reasons. First, *a supercoiled DNA molecule is more compact than its relaxed*

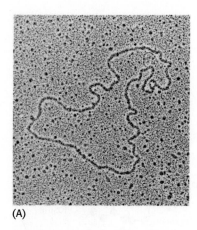

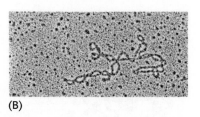

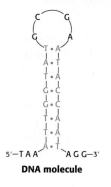

(A)

(B)

FIGURE 4.15 Electron micrographs of circular DNA from mitochondria. (A) Relaxed form. (B) Supercoiled form [Courtesy of Dr. David Clayton.]

counterpart. Second, *supercoiling may hinder or favor the capacity of the double helix to unwind and thereby affect the interactions between DNA and other molecules.* These topological features of DNA will be considered further in Chapter 29.

Single-stranded nucleic acids can adopt elaborate structures

Single-stranded nucleic acids often fold back on themselves to form well-defined structures. Such structures are especially prominent in RNA and RNA-containing complexes such as the ribosome—a large complex of RNAs and proteins on which proteins are synthesized.

The simplest and most-common structural motif formed is a *stem-loop,* created when two complementary sequences within a single strand come together to form double-helical structures (Figure 4.16). In many cases, these double helices are made up entirely of Watson–Crick base pairs. In other cases, however, the structures include mismatched base pairs or unmatched bases that bulge out from the helix. Such mismatches destabilize the local structure but introduce deviations from the standard double-helical structure that can be important for higher-order folding and for function (Figure 4.17).

Single-stranded nucleic acids can adopt complex structures through the interaction of more widely separated bases. Often, three or more bases interact to stabilize these structures. In such cases, hydrogen-bond donors and acceptors that do not participate in Watson–Crick base pairs participate in hydrogen bonds to form nonstandard pairings (Figure 4.17B). Metal ions such as magnesium ion (Mg^{2+}) often assist in the stabilization of these more elaborate structures. These complex structures allow RNA to perform a host of functions that the double-stranded DNA molecule cannot. Indeed, the complexity of some RNA molecules rivals that of proteins, and these RNA molecules perform a number of functions that had formerly been thought the exclusive domain of proteins.

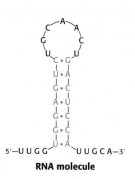

FIGURE 4.16 Stem-loop structures. Stem-loop structures can be formed from single-stranded DNA and RNA molecules.

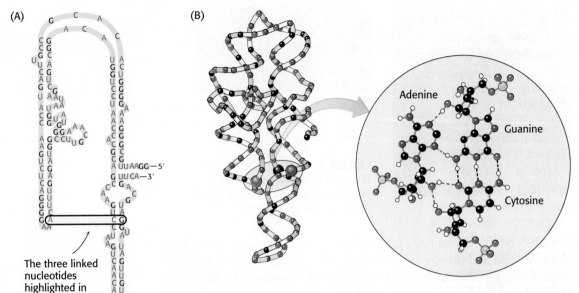

The three linked nucleotides highlighted in part B

FIGURE 4.17 Complex structure of an RNA molecule. A single-stranded RNA molecule can fold back on itself to form a complex structure. (A) The nucleotide sequence showing Watson–Crick base pairs and other nonstandard base pairings in stem-loop structures. (B) The three-dimensional structure and one important long-range interaction between three bases. In the three-dimensional structure at the left, cytidine nucleotides are shown in blue, adenosine in red, guanosine in black, and uridine in green. In the detailed projection, hydrogen bonds within the Watson–Crick base pair are shown as dashed black lines; additional hydrogen bonds are shown as dashed green lines.

4.3 The Double Helix Facilitates the Accurate Transmission of Hereditary Information

The double-helical model of DNA proposed by Watson and Crick and, in particular, the specific base pairing of adenine with thymine and guanine with cytosine, immediately suggested to them how the genetic material might replicate. *The sequence of bases of one strand of the double helix precisely determines the sequence of the other strand.* A guanine on one strand is always paired with a cytosine on the other strand, and so on. Thus, separation of a double helix into its two component strands would yield two single-stranded templates onto which new double helices could be constructed, each of which would have the same sequence of bases as the parent double helix. The hypothesis was that as DNA is replicated, one of the strands of each daughter DNA molecule is newly synthesized, whereas the other is passed unchanged from the parent DNA molecule. This distribution of parental atoms is called *semiconservative replication.*

Differences in DNA density established the validity of the semiconservative replication hypothesis

Matthew Meselson and Franklin Stahl carried out a critical test of this hypothesis in 1958. They labeled the parent DNA with ^{15}N, a heavy isotope of nitrogen, to make it denser than ordinary DNA. The labeled DNA was generated by growing *E. coli* for many generations in a medium that contained $^{15}NH_4Cl$ as the sole nitrogen source. After the incorporation of heavy nitrogen was complete, the bacteria were abruptly transferred to a medium that contained only ^{14}N, the ordinary isotope of nitrogen. The question asked was: What is the distribution of ^{14}N and ^{15}N in the DNA molecules after successive rounds of replication?

The distribution of ^{14}N and ^{15}N was revealed by the technique of *density-gradient equilibrium sedimentation*. A small amount of the *E. coli* DNA was dissolved in a concentrated solution of cesium chloride. This solution was centrifuged until it was nearly at equilibrium. At that point, the opposing processes of sedimentation and diffusion created a stable density gradient in the concentration of cesium chloride across the centrifuge cell. The DNA molecules in this density gradient were driven by centrifugal force into the region where the solution's density was equal to their own. The DNA yielded a narrow band or bands that were detected by absorption of ultraviolet light by the DNA. A mixture of ^{14}N DNA and ^{15}N DNA molecules gave clearly separate bands because they differ in density by about 1% (Figure 4.18).

In their experiment, Meselson and Stahl extracted DNA from the bacteria at various times after the bacteria were transferred from ^{15}N-containing to ^{14}N-containing medium and subjected the samples to density-gradient equilibrium sedimentation. In Figure 4.19, all of the DNA is labeled with ^{15}N at the start of the experiment (generation 0). After one generation, all of the DNA is still in a single band, but the band has shifted. The density of this band (call it hybrid DNA) is precisely halfway between the densities of the ^{14}N DNA and ^{15}N DNA bands. After two generations, there are equal amounts of two bands of DNA. One is hybrid DNA, and the other is ^{14}N DNA. After three rounds of replication, the positions of the two bands remain unchanged, but there is three times the amount of ^{14}N DNA as there is hybrid DNA.

The absence of ^{15}N DNA indicated that parental DNA was not preserved as an intact unit after replication. The absence of ^{14}N DNA in generation 1 indicated that all of the daughter DNA derived some of their atoms from the parent DNA. This proportion had to be half because the density of the hybrid DNA band was halfway between the densities of the ^{14}N DNA and ^{15}N DNA bands.

Meselson and Stahl concluded from these incisive experiments that replication was semiconservative, meaning that each new double helix contains a parent strand and a newly synthesized strand. Their results agreed perfectly with the Watson–Crick model for DNA replication (Figure 4.20).

The double helix can be reversibly melted

During DNA replication and transcription, the two strands of the double helix must be separated from each other, at least in a local region. The two strands of a DNA helix readily come apart when the hydrogen bonds between base pairs are disrupted. In the laboratory, the double helix can be disrupted by

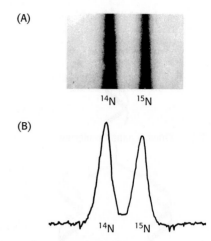

FIGURE 4.18 Resolution of ^{14}N DNA and ^{15}N DNA by density-gradient centrifugation. (A) Ultraviolet-absorption photograph of a special centrifuged tube showing the two distinct bands of DNA. (B) Densitometric tracing of the absorption photograph. [(A) M. Meselson and F. W. Stahl. *Proc. Natl. Acad. Sci. U.S.A.* 44(1958):671.]

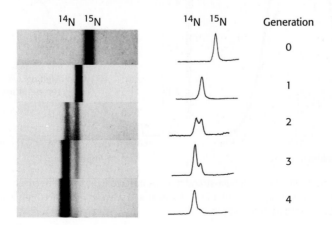

FIGURE 4.19 Detection of semiconservative replication of *E. coli* DNA by density-gradient centrifugation. The position of a band of DNA depends on its content of ^{14}N and ^{15}N. [From M. Meselson and F. W. Stahl. *Proc. Natl. Acad. Sci. U.S.A.* 44(1958):671.]

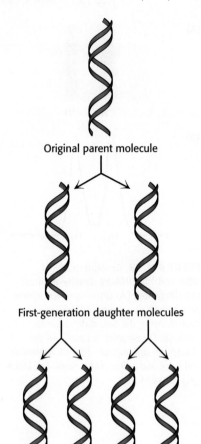

FIGURE 4.20 Diagram of semiconservative replication. Original parental DNA is shown in blue and newly synthesized DNA in red. [Information from M. Meselson and F. W. Stahl, *Proc. Natl. Acad. Sci. U.S.A.* 44: 671–682, 1958.]

Original parent molecule

First-generation daughter molecules

Second-generation daughter molecules

heating a solution of DNA or by adding acid or alkali to ionize its bases. The dissociation of the double helix is called *melting* because it occurs abruptly at a certain temperature. The *melting temperature* (T_m) of DNA is defined as the temperature at which half the helical structure is lost. Inside cells, however, the double helix is not melted by the addition of heat. Instead, proteins called *helicases* use chemical energy (from ATP) to disrupt the helix (Chapter 29).

Stacked bases in nucleic acids absorb less ultraviolet light at a wavelength of 260 nm than do unstacked bases, an effect called *hypochromism* (Figure 4.21A). Thus, as a sample of DNA is heated, the strands separate, increasing the proportion of single-stranded DNA, which is monitored by measuring the increase in absorption of 260 nm light (Figure 4.21B).

Separated complementary strands of nucleic acids spontaneously reassociate to form a double helix when the temperature is lowered below T_m. This renaturation process is sometimes called *annealing*. The facility with which double helices can be melted and then reassociated is crucial for the biological functions of nucleic acids. The ability to melt and reanneal DNA reversibly in the laboratory provides a powerful tool for investigating sequence similarity. We will return to this important technique in Chapter 5.

Unusual circular DNA exists in the eukaryotic nucleus

The discipline of biochemistry has made enormous strides over the past century in explaining the chemical basis of living systems. However, one of the most exciting aspects of biochemistry is that there are still many discoveries to be made. For example, extrachromosomal circular DNA (eccDNA) has recently been discovered in the nuclei of eukaryotic cells. As discussed on page 120, circular DNA has long been known as a component of bacterial cells, but until recently, such DNA was unheard of in nuclei. What is the role of eccDNA? That, of course, is under intense investigation, but the circular DNA may help the cells to specialize, for better or for worse. For instance, more than half of human cancer cells contain eccDNA, and the circles carry genes required for tumor growth. Normal heart cells have eccDNA with genes for the muscle protein titin, which is responsible for muscle plasticity. The discovery of eccDNA opens a new area of research and a new way of looking at eukaryotic DNA and the flexibility of the eukaryotic genome.

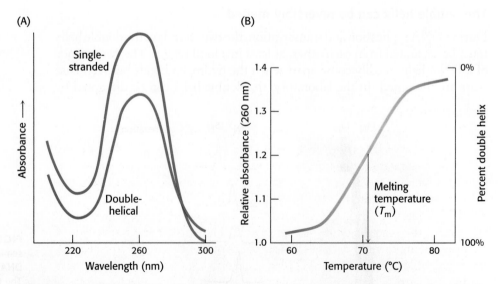

FIGURE 4.21 Hypochromism. (A) Single-stranded DNA absorbs light more effectively than does double-helical DNA. (B) The absorbance of a DNA solution at a wavelength of 260 nm increases when the double helix is melted into single strands.

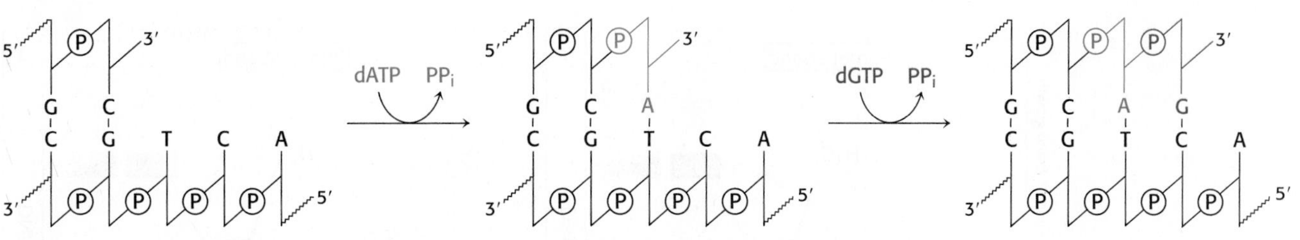

FIGURE 4.22 Polymerization reaction catalyzed by DNA polymerases.

4.4 DNA Is Replicated by Polymerases That Take Instructions from Templates

We now turn to the molecular mechanism of DNA replication. The full replication machinery in a cell comprises more than 20 proteins engaged in intricate and coordinated interplay. The primary catalytic components of the replication machinery are enzymes, called *DNA polymerases*, that promote the formation of the bonds joining units of the DNA backbone. *E. coli* has a number of DNA polymerases that participate in DNA replication and repair (Chapter 29).

DNA polymerase catalyzes phosphodiester-bridge formation

DNA polymerases catalyze the step-by-step addition of deoxyribonucleotide units to a DNA strand (Figure 4.22). The reaction catalyzed, in its simplest form, is

$$(\text{DNA})_n + \text{dNTP} \rightleftharpoons (\text{DNA})_{n+1} + \text{PP}_i$$

Where dNTP stands for any deoxyribonucleotide and PP_i is a pyrophosphate ion.

DNA synthesis has the following characteristics:

1. *The reaction requires all four activated precursors*—that is, the deoxynucleoside 5′-triphosphates dATP, dGTP, dCTP, and TTP—as well as Mg^{2+} ion.

2. *The new DNA strand is assembled directly on a preexisting DNA template.* DNA polymerases catalyze the formation of a phosphodiester linkage efficiently only if the base on the incoming nucleoside triphosphate is complementary to the base on the *template* strand. Thus, DNA polymerase is a *template-directed enzyme* that synthesizes a product with a base sequence complementary to that of the template.

3. *DNA polymerases require a primer to begin synthesis.* A *primer* strand having a free 3′-OH group must be already bound to the template strand. The chain-elongation reaction catalyzed by DNA polymerases is a nucleophilic attack by the 3′-OH terminus of the growing strand on the innermost phosphorus atom of the deoxynucleoside triphosphate (Figure 4.23). A phosphodiester bridge is formed and pyrophosphate is released. At this point, the reaction is readily reversible. The subsequent hydrolysis of pyrophosphate to yield two ions of orthophosphate (P_i) by *pyrophosphatase*, an irreversible reaction under cellular conditions, helps drive the polymerization forward. This is an example of coupled reactions, where a second reaction provides the energy to drive the first reaction forward, a common occurrence in biochemical pathways (Section 15.1).

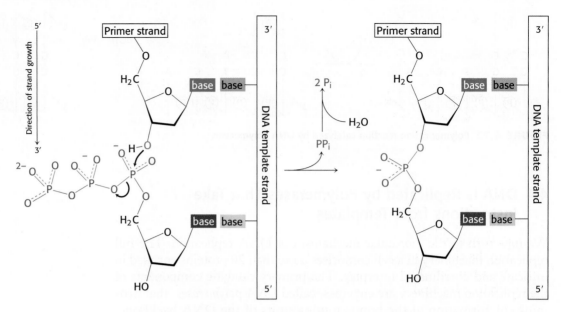

FIGURE 4.23 Strand-elongation reaction. DNA polymerases catalyze the formation of a phosphodiester bridge. [Source: J. L. Tymoczko, J. Berg, and L. Stryer, *Biochemistry: A Short Course*, 2nd ed. (W. H. Freeman and Company, 2013), Fig. 34.2.]

4. *Elongation of the DNA chain proceeds in the 5'-to-3' direction.*

5. *Many DNA polymerases are able to correct mistakes in DNA by removing mismatched nucleotides.* These polymerases have a distinct nuclease activity that allows them to excise incorrect bases by a separate reaction. This nuclease activity contributes to the remarkably high fidelity of DNA replication, which has an error rate of less than 10^{-8} per base pair.

The genes of some viruses are made of RNA

Genes in all cellular organisms are made of DNA. The same is true for some viruses but, for others, the genetic material is RNA. Viruses are genetic elements enclosed in protein coats that can move from one cell to another but are not capable of independent growth. A well-studied example of an RNA virus is the tobacco mosaic virus, which infects the leaves of tobacco plants. This virus consists of a single strand of RNA (6390 nucleotides) surrounded by a protein coat of 2130 identical subunits. An RNA polymerase that takes direction from an RNA template, called an *RNA-directed RNA polymerase,* copies the viral RNA. The infected cells die because of virus-instigated programmed cell death; in essence, the virus instructs the cell to commit suicide. Cell death results in discoloration in the tobacco leaf in a variegated pattern, hence the name mosaic virus.

Another important class of RNA virus comprises the *retroviruses,* so called because the genetic information flows from RNA to DNA rather than from DNA to RNA. This class includes human immunodeficiency virus 1 (HIV-1), the cause of acquired immunodeficiency syndrome (AIDS), as well as a number of RNA viruses that produce tumors in susceptible animals. Retrovirus particles contain two copies of a single-stranded RNA molecule. On entering the cell, the RNA is copied into DNA through the action of a viral enzyme called *reverse transcriptase,*

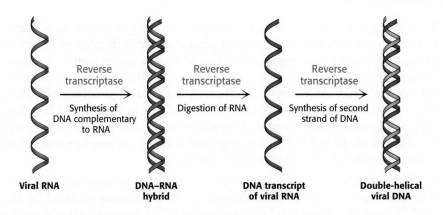

FIGURE 4.24 Flow of information from RNA to DNA in retroviruses. The RNA genome of a retrovirus is converted into DNA by reverse transcriptase, an enzyme brought into the cell by the infecting virus particle. Reverse transcriptase possesses several activities and catalyzes the synthesis of a complementary DNA strand, the digestion of the RNA, and the subsequent synthesis of the DNA strand.

which acts as both a polymerase and an RNase (Figure 4.24). The resulting double-helical DNA version of the viral genome can become incorporated into the chromosomal DNA of the host and is replicated along with the normal cellular DNA. At a later time, the integrated viral genome is expressed to form viral RNA and viral proteins, which assemble into new virus particles.

4.5 Gene Expression Is the Transformation of DNA Information into Functional Molecules

The information stored as DNA becomes useful when it is expressed in the production of RNA and proteins. This rich and complex topic is the subject of several chapters later in this book, but here we introduce the basics of gene expression. DNA can be thought of as archival information, stored and manipulated judiciously to minimize damage (mutations). It is expressed in two steps. First, an RNA copy is made that encodes directions for protein synthesis. This messenger RNA can be thought of as a photocopy of the original information: it can be made in multiple copies, used, and then disposed of. Second, the information in messenger RNA is translated to synthesize functional proteins. Other types of RNA molecules exist to facilitate this translation.

Several kinds of RNA play key roles in gene expression

Scientists used to believe that RNA played a passive role in gene expression, as a mere conveyor of information. However, recent investigations have shown that RNA plays a variety of roles, from catalysis to regulation. Cells contain several kinds of RNA that are involved in gene expression (Table 4.3):

TABLE 4.3 RNA molecules in *E. coli*

Type	Relative amount (%)	Sedimentation coefficient (S)	Mass (kDa)	Number of nucleotides
Ribosomal RNA (rRNA)	80	23	1.2×10^3	3700
		16	0.55×10^3	1700
		5	3.6×10^1	120
Transfer RNA (tRNA)	15	4	2.5×10^1	75
Messenger RNA (mRNA)	5		Heterogeneous	

1. *Messenger RNA* (mRNA) is the template for protein synthesis, or *translation*. An mRNA molecule may be produced for each gene or group of genes that is to be expressed in bacteria, whereas a distinct mRNA is produced for each gene in eukaryotes. Consequently, mRNA is a heterogeneous class of molecules. In bacteria, the average length of an mRNA molecule is about 1.2 kilobases (kb); mRNA of all organisms has structural features, such as stem-loop structures, that regulate the efficiency of translation and the lifetime of the mRNA. Such structural features are more prominent in eukaryotic mRNA (Chapters 30, 32, and 33).

2. *Transfer RNA* (tRNA) carries amino acids in an activated form to the ribosome for peptide-bond formation, in a sequence dictated by the mRNA template. There is at least one kind of tRNA for each of the 20 amino acids. Transfer RNA consists of about 75 nucleotides (having a mass of about 25 kDa).

3. *Ribosomal RNA* (rRNA) is the major component of ribosomes (Chapter 31). In bacteria, there are three kinds of rRNA, called 23S, 16S, and 5S RNA because of their sedimentation behavior (p. 80). One molecule of each of these species of rRNA is present in each ribosome. Ribosomal RNA was once believed to play only a structural role in ribosomes. We now know that rRNA is the actual catalyst for protein synthesis.

Ribosomal RNA is the most abundant of these three types of RNA. Transfer RNA comes next, followed by messenger RNA, which constitutes only 5% of the total RNA. Eukaryotic cells contain additional small RNA molecules that play a variety of roles including the regulation of gene expression, processing of RNA, and the synthesis of proteins. We will examine these small RNAs in later chapters. In this chapter, we will consider rRNA, mRNA, and tRNA.

All cellular RNA is synthesized by RNA polymerases

The synthesis of RNA from a DNA template is called *transcription* and is catalyzed by the enzyme *RNA polymerase* (Figure 4.25). RNA polymerase

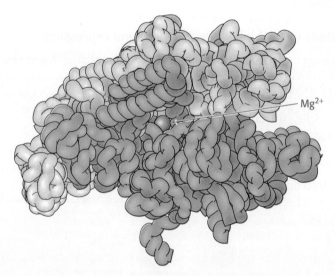

FIGURE 4.25 RNA Polymerase. This large enzyme comprises many subunits, including β (red) and β′ (yellow), which form a "claw" that holds the DNA to be transcribed. *Notice* that the active site includes a Mg^{2+} ion (green) at the center of the structure. The curved tubes making up the protein in the image represent the backbone of the polypeptide chain. [Drawn from 1L9Z.pdb.]

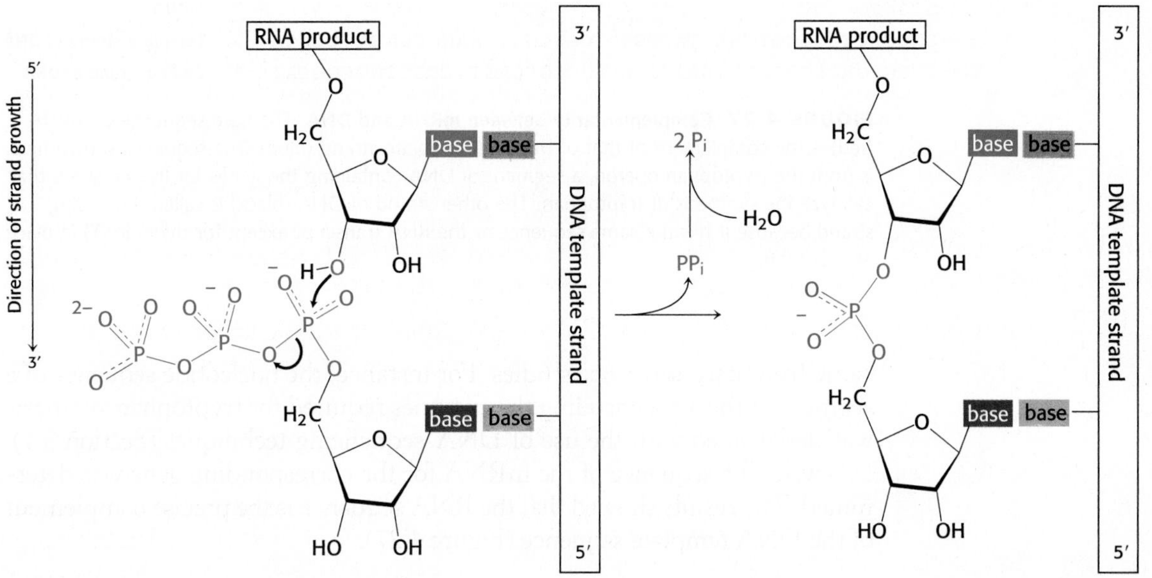

FIGURE 4.26 Transcription mechanism of the chain-elongation reaction catalyzed by RNA polymerase. [Source: J. L. Tymoczko, J. Berg, and L. Stryer, *Biochemistry: A Short Course*, 2nd ed. (W. H. Freeman and Company, 2013), Fig. 36.3.]

catalyzes the initiation and elongation of RNA chains. The reaction catalyzed by this enzyme is

$$(\text{RNA})_{n \text{ residues}} + \text{ribonucleoside triphosphate} \rightleftharpoons (\text{RNA})_{n+1 \text{ residues}} + \text{PP}_i$$

RNA polymerase requires the following components:

1. *A template.* The preferred template is *double-stranded DNA*. Single-stranded DNA also can serve as a template. RNA, whether single or double stranded, is not an effective template; nor are RNA–DNA hybrids.

2. *Activated precursors.* All four *ribonucleoside triphosphates*—ATP, GTP, UTP, and CTP—are required.

3. *A divalent metal ion.* Either Mg^{2+} or Mn^{2+} is effective.

The synthesis of RNA is like that of DNA in several respects (Figure 4.26). First, the direction of synthesis is $5' \rightarrow 3'$. Second, the mechanism of elongation is similar: the $3'$-OH group at the terminus of the growing chain makes a nucleophilic attack on the innermost phosphoryl group of the incoming nucleoside triphosphate. Third, the synthesis is driven forward by the hydrolysis of pyrophosphate. In contrast with DNA polymerase, however, RNA polymerase does not require a primer. In addition, the ability of RNA polymerase to correct mistakes is not as extensive as that of DNA polymerase.

All three types of cellular RNA—mRNA, tRNA, and rRNA—are synthesized in *E. coli* by the same RNA polymerase according to instructions given by a DNA template. In mammalian cells, there is a division of labor among several different kinds of RNA polymerases. We shall return to these RNA polymerases in Chapter 30.

RNA polymerases take instructions from DNA templates

RNA polymerase, like the DNA polymerases described earlier, takes instructions from a DNA template. The earliest evidence was the finding that the *base composition* of newly synthesized RNA is the complement of that of the DNA template strand (the plus or coding strand), as exemplified by the RNA synthesized from a template of single-stranded DNA from the ϕX174 virus (Table 4.4). The strongest evidence for the fidelity of transcription

TABLE 4.4 Base composition (percentage) of RNA synthesized from a viral DNA template

DNA template (plus, or coding, strand of ϕX174)		RNA product	
A	25	U	25
T	33	A	32
G	24	C	23
C	18	G	20

5′—GCGGCGACGCGCAGUUAAUCCCACAGCCGCCAGUUCCGCUGGCGGCAU—3′　**mRNA**
3′—CGCCGCTGCGCGTCAATTAGGGTGTCGGCGGTCAAGGCGACCGCCGTA—5′　**Template strand of DNA**
5′—GCGGCGACGCGCAGTTAATCCCACAGCCGCCAGTTCCGCTGGCGGCAT—3′　**Coding strand of DNA**

FIGURE 4.27 Complementarity between mRNA and DNA. The base sequence of mRNA (red) is the complement of that of the DNA template strand (blue). The sequence shown here is from the tryptophan operon, a segment of DNA containing the genes for five enzymes that catalyze the synthesis of tryptophan. The other strand of DNA (black) is called the coding strand because it has the same sequence as the RNA transcript except for thymine (T) in place of uracil (U).

came from base-sequence studies. For instance, the nucleotide sequence of a segment of the gene encoding the enzymes required for tryptophan synthesis was determined with the use of DNA-sequencing techniques (Section 5.1). Likewise, the sequence of the mRNA for the corresponding gene was determined. The results showed that the RNA sequence is the precise complement of the DNA template sequence (Figure 4.27).

Transcription begins near promoter sites and ends at terminator sites

RNA polymerase must detect and transcribe discrete genes from within large stretches of DNA. What marks the beginning of the unit to be transcribed? DNA templates contain regions called *promoter sites* that specifically bind RNA polymerase and determine where transcription begins. In bacteria, two sequences on the 5′ (upstream) side of the first nucleotide to be transcribed function as promoter sites (Figure 4.28A). One of them, called the *Pribnow box*, has the consensus sequence TATAAT and is centered at −10 (10 nucleotides on the 5′ side of the first nucleotide transcribed, which is denoted by +1). The other, called the −35 *region*, has the consensus sequence TTGACA. The first nucleotide transcribed is usually a purine.

Eukaryotic genes encoding proteins have promoter sites with a TATAAA consensus sequence, called a *TATA box* or a *Hogness box,* centered at about −25 (Figure 4.28B). Many eukaryotic promoters also have a *CAAT box*

Consensus sequence

Not all base sequences of promoter sites are identical. However, they do possess common features, which can be represented by an idealized consensus sequence. Each base in the consensus sequence TATAAT is found in most bacterial promoters. Nearly all promoter sequences differ from this consensus sequence at only one or two bases.

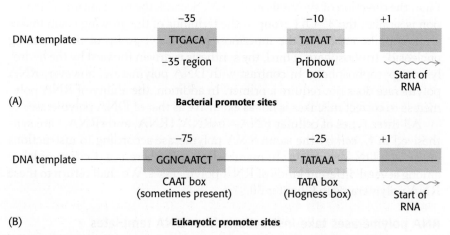

FIGURE 4.28 Promoter sites for transcription in (A) bacteria and (B) eukaryotes. Consensus sequences are shown. The first nucleotide to be transcribed is numbered +1. The adjacent nucleotide on the 5′ side is numbered −1. The sequences shown are those of the coding strand of DNA.

with a GGN*CAAT*CT consensus sequence centered at about −75. The transcription of eukaryotic genes is further stimulated by *enhancer sequences,* which can be quite distant (as many as several kilobases) from the start site, on either its 5′ or its 3′ side.

In *E. coli*, RNA polymerase proceeds along the DNA template, transcribing one of its strands until it synthesizes a terminator sequence. This sequence encodes a termination signal, which is a *base-paired hairpin* on the newly synthesized RNA molecule (Figure 4.29). This hairpin is formed by base-pairing of self-complementary sequences that are rich in G and C. Nascent RNA spontaneously dissociates from RNA polymerase when this hairpin is followed by a string of U residues. Alternatively, RNA synthesis can be terminated by the action of *rho*, a protein. Less is known about the termination of transcription in eukaryotes. A more detailed discussion of the initiation and termination of transcription will be given in Chapter 30. The important point now is that *discrete start and stop signals for transcription are encoded in the DNA template.*

In eukaryotes, the messenger RNA is modified after transcription (Figure 4.30). A "cap" structure, a guanosine nucleotide attached to the mRNA with an unusual 5′-5′ triphosphate linkage, is attached to the 5′ end, and a sequence of adenylates, the poly(A) tail, is added to the 3′ end. These modifications will be presented in detail in Chapter 30.

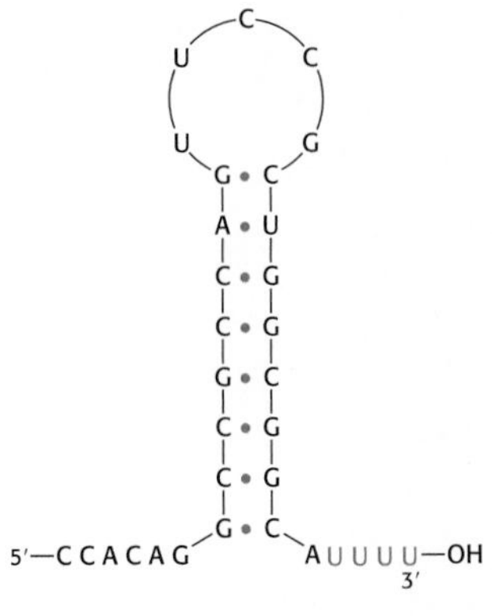

FIGURE 4.29 Base sequence of the 3′ end of an mRNA transcript in *E. coli*. A stable hairpin structure is followed by a sequence of uridine (U) residues.

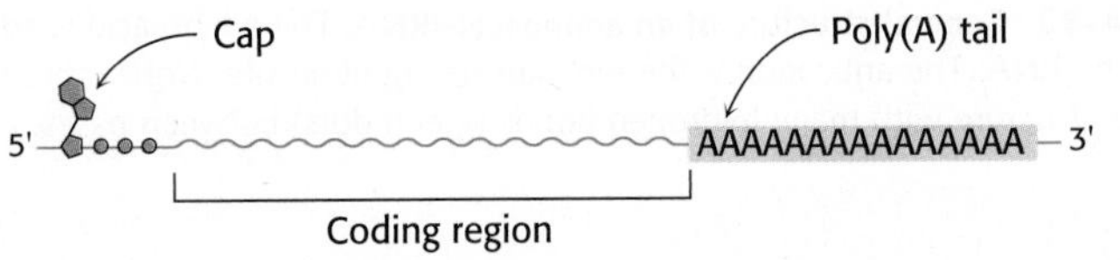

FIGURE 4.30 Modification of mRNA. Messenger RNA in eukaryotes is modified after transcription. A nucleotide "cap" structure is added to the 5′ end, and a poly(A) tail is added at the 3′ end.

Transfer RNAs are the adaptor molecules in protein synthesis

We have seen that mRNA is the template for protein synthesis. How then does it direct amino acids to become joined in the correct sequence to form a protein? Specific adaptor RNA molecules called transfer RNAs bring the amino acids to the mRNA. The structure and reactions of these remarkable molecules will be considered in detail in Chapter 31. For the moment, it suffices to note that tRNAs contain an *amino acid-attachment site* and a *template-recognition site*. A tRNA molecule carries a specific amino acid in an activated form to the ribosome. The carboxyl group of this amino acid is esterified to the 3′- or 2′-hydroxyl group of the ribose unit of an adenylate at the 3′ end of the tRNA molecule. The adenylate is always preceded by two cytidylates to form the CCA arm of the tRNA (Figure 4.31). The joining of an amino acid to a tRNA molecule to form an *aminoacyl-tRNA* is catalyzed by a specific enzyme called an *aminoacyl-tRNA synthetase*. This esterification reaction is driven by ATP cleavage. There is at least one specific synthetase for each of the 20 amino acids. The template-recognition site on tRNA is a sequence of three bases called an *anticodon* (Figure 4.32). The anticodon on tRNA recognizes a complementary sequence of three bases, called a *codon*, on mRNA.

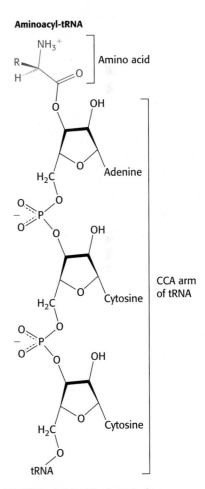

FIGURE 4.31 Attachment of an amino acid to a tRNA molecule. The amino acid (shown in blue) is esterified to the 3′-hydroxyl group of the terminal adenylate of tRNA. [Source: J. L. Tymoczko, J. Berg, and L. Stryer, *Biochemistry: A Short Course*, 2nd ed. (W. H. Freeman and Company, 2013), Fig. 39.3.]

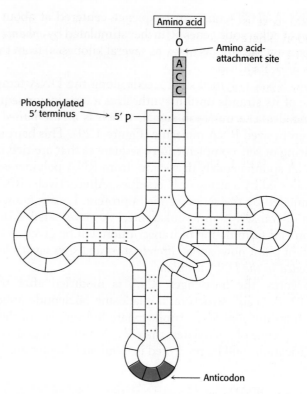

FIGURE 4.32 General structure of an aminoacyl-tRNA. The amino acid is attached at the 3′ end of the RNA. The anticodon is the template-recognition site. *Notice* that the tRNA has a cloverleaf structure with many hydrogen bonds (green dots) between bases.

4.6 Amino Acids Are Encoded by Groups of Three Bases Starting from a Fixed Point

The *genetic code* is the relation between the sequence of bases in DNA (or its RNA transcripts) and the sequence of amino acids in proteins. Numerous experiments established the following features of the genetic code by 1961:

1. *Three nucleotides encode an amino acid.* Proteins are built from a basic set of 20 amino acids, but there are only four bases. Simple calculations show that a minimum of three bases is required to encode at least 20 amino acids. Genetic experiments showed that *an amino acid is in fact encoded by a group of three bases, or codon.*

2. *The code is nonoverlapping.* Consider a base sequence ABCDEF. In an overlapping code, ABC specifies the first amino acid, BCD the next, CDE the next, and so on. In a nonoverlapping code, ABC designates the first amino acid, DEF the second, and so forth. Genetic experiments again established the code to be nonoverlapping.

3. *The code has no punctuation.* In principle, one base (denoted as Q) might serve as a "comma" between groups of three bases.

. . . QABCQDEFQGHIQJKLQ . . .

However, it is not the case. Rather, *the sequence of bases is read sequentially from a fixed starting point*, without punctuation.

4. *The code has directionality.* The code is read from the 5′ end of the messenger RNA to its 3′ end.

5. *The genetic code is degenerate.* Most amino acids are encoded by more than one codon. There are 64 possible base triplets and only 20 amino acids, and in fact 61 of the 64 possible triplets specify particular amino acids. Three triplets (called *stop codons*) designate the termination of translation. Thus, *for most amino acids, there is more than one code word.*

Major features of the genetic code

All 64 codons have been deciphered (Table 4.5). Because the code is highly degenerate, only tryptophan and methionine are encoded by just one triplet each. Each of the other 18 amino acids is encoded by two or more. Indeed, leucine, arginine, and serine are specified by six codons each.

Codons that specify the same amino acid are called *synonyms*. For example, CAU and CAC are synonyms for histidine. Note that synonyms are not distributed haphazardly throughout the genetic code. In Table 4.5, an amino acid specified by two or more synonyms occupies a single box (unless it is specified by more than four synonyms). The amino acids in a box are specified by codons that have the same first two bases but differ in the third base, as exemplified by GUU, GUC, GUA, and GUG. Thus, *most synonyms differ only in the last base of the triplet.* Inspection of the code shows that XYC and XYU always encode the same amino acid, and XYG and XYA usually encode the same amino acid as well. The structural basis for these equivalences of codons becomes evident when we consider the nature of the anticodons of tRNA molecules (Section 31.2).

What is the biological significance of the extensive degeneracy of the genetic code? If the code were not degenerate, 20 codons would designate amino acids and 44 would lead to chain termination. The probability of

---

TABLE 4.5 The genetic code

First position (5′ end)	Second position				Third position (3′ end)
	U	C	A	G	
U	Phe	Ser	Tyr	Cys	U
	Phe	Ser	Tyr	Cys	C
	Leu	Ser	Stop	Stop	A
	Leu	Ser	Stop	Trp	G
C	Leu	Pro	His	Arg	U
	Leu	Pro	His	Arg	C
	Leu	Pro	Gln	Arg	A
	Leu	Pro	Gln	Arg	G
A	Ile	Thr	Asn	Ser	U
	Ile	Thr	Asn	Ser	C
	Ile	Thr	Lys	Arg	A
	Met	Thr	Lys	Arg	G
G	Val	Ala	Asp	Gly	U
	Val	Ala	Asp	Gly	C
	Val	Ala	Glu	Gly	A
	Val	Ala	Glu	Gly	G

Note: This table identifies the amino acid encoded by each triplet. For example, the codon 5′-AUG-3′ on mRNA specifies methionine, whereas CAU specifies histidine. UAA, UAG, and UGA are termination signals. AUG is part of the initiation signal, in addition to coding for internal methionine residues.

mutating to chain termination would therefore be much higher with a nondegenerate code. Chain-termination mutations usually lead to inactive proteins, whereas substitutions of one amino acid for another are usually rather harmless. Moreover, the code is constructed such that a change in any single nucleotide base of a codon results in a synonym or an amino acid with similar chemical properties. Thus, *degeneracy minimizes the deleterious effects of mutations*.

As we will see many times in our study of biochemistry, often things are not as straightforward as they first appear. Recent research has revealed a nonrandom use of synonymous codons in genes in different organisms, a phenomenon called *codon bias*. In other words, different organisms prefer different sets of synonymous codons. The biochemical benefit of codon bias is not yet clearly established, but codon bias may help regulate translation.

Messenger RNA contains start and stop signals for protein synthesis

Messenger RNA is translated into proteins on *ribosomes*—large molecular complexes assembled from proteins and ribosomal RNA. How is mRNA interpreted by the translation apparatus? The start signal for protein synthesis is complex in bacteria. Polypeptide chains in bacteria start with a modified amino acid—namely, formylmethionine (fMet). A specific tRNA, the initiator tRNA, carries fMet. This fMet-tRNA recognizes the codon AUG. However, AUG is also the codon for an internal methionine residue. Hence, the signal for the first amino acid in a bacterial polypeptide chain must be more complex than that for all subsequent ones. *AUG is only part of the initiation signal* (Figure 4.33). In bacteria, the initiating AUG codon is preceded several nucleotides away by a purine-rich sequence, called the *Shine–Dalgarno sequence* (named after John Shine and Lynn Dalgarno, who first described the sequence), that base-pairs with a complementary sequence in a ribosomal RNA molecule (Section 31.3). In eukaryotes, the AUG closest to the 5′ end of an mRNA molecule is usually the start signal for protein synthesis. This particular AUG is read by an initiator tRNA conjugated to methionine. After the initiator AUG has been located, the *reading frame* is established—groups of three nonoverlapping nucleotides are defined, beginning with the initiator AUG codon.

As already mentioned, *UAA, UAG, and UGA designate chain termination*. These codons are read not by tRNA molecules but rather by specific proteins called *release factors* (Section 31.3). Binding of a release factor to the ribosome releases the newly synthesized protein.

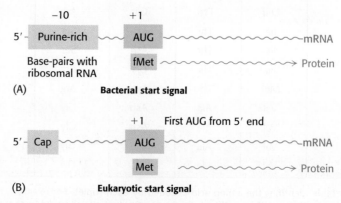

FIGURE 4.33 Initiation of protein synthesis. Start signals are required for the initiation of protein synthesis in (A) bacteria and (B) eukaryotes.

The genetic code is nearly universal

Most organisms use the same genetic code. This universality accounts for the fact that human proteins, such as insulin, can be synthesized in the bacterium *E. coli* and harvested from it for the treatment of diabetes. However, genome-sequencing studies have shown that not all genomes are translated by the same code. Ciliated protozoa, for example, differ from most organisms in that UAA and UAG are read as codons for amino acids rather than as stop signals; UGA is their sole termination signal. The first variations in the genetic code were found in mitochondria from a number of species, including human beings (Table 4.6). The genetic code of mitochondria can differ from that of the rest of the cell because mitochondrial DNA encodes a distinct set of transfer RNAs, adaptor molecules that recognize the alternative codons. Thus, the genetic code is nearly but not absolutely universal.

Why has the code remained nearly invariant through billions of years of evolution, from bacteria to human beings? A mutation that altered the reading of mRNA would change the amino acid sequence of most, if not all, proteins synthesized by that particular organism. Many of these changes would undoubtedly be deleterious, and so there would be strong selection against a mutation with such pervasive consequences.

TABLE 4.6 Distinctive codons of human mitochondria

Codon	Standard Code	Mitochondrial code
UGA	Stop	Trp
UGG	Trp	Trp
AUA	Ile	Met
AUG	Met	Met
AGA	Arg	Stop
AGG	Arg	Stop

4.7 Most Eukaryotic Genes Are Mosaics of Introns and Exons

In bacteria, polypeptide chains are encoded by a continuous array of triplet codons in DNA. For many years, genes in higher organisms were assumed to be organized in the same manner. This view was unexpectedly shattered in 1977, when investigators discovered that most eukaryotic genes are *discontinuous*. The mosaic nature of eukaryotic genes was revealed by electron microscopic studies of hybrids formed between mRNA and a segment of DNA containing the corresponding gene (Figure 4.34). For example, the gene for the β chain of hemoglobin is interrupted within its

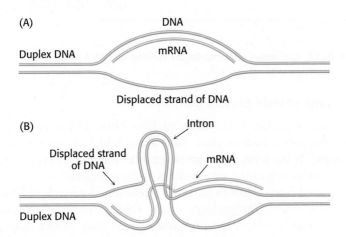

FIGURE 4.34 Detection of introns by electron microscopy. An mRNA molecule (shown in red) is hybridized to genomic DNA containing the corresponding gene. (A) A single loop of single-stranded DNA (shown in blue) is seen if the gene is continuous. (B) Two loops of single-stranded DNA (blue) and a loop of double-stranded DNA (blue and green) are seen if the gene contains an intron. Additional loops are evident if more than one intron is present.

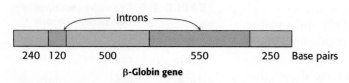

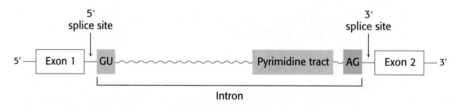

FIGURE 4.35 Structure of the β-globin gene. The figure shows only the exons and introns, and does not include regulatory regions or transcribed but not translated regions of the gene.

amino acid-coding sequence by a long stretch of 550 noncoding base pairs and a short one of 120 base pairs. Thus, the *β-globin gene is split into three coding sequences* (Figure 4.35). Noncoding regions are called *introns* (for *intervening* sequences), whereas coding regions are called *exons* (for *expressed* sequences). The average human gene has 8 introns, and some have more than 100. Intron size ranges from 50 to 10,000 nucleotides.

RNA processing generates mature RNA

At what stage in gene expression are introns removed? Newly synthesized RNA molecules (pre-mRNA or primary transcript) isolated from nuclei are much larger than the mRNA molecules derived from them; in regard to β-globin RNA, the former consists of approximately 1600 nucleotides and the latter approximately 900 nucleotides, which includes nontranslated regions of the mRNA. In fact, the primary transcript of the β-globin gene contains two regions that are not present in the mRNA. *These regions in the primary transcript are excised, and the coding sequences are simultaneously linked by a precise splicing complex to form the mature mRNA* (Figure 4.36). A common feature in the expression of discontinuous, or split, genes is that their exons are ordered in the same sequence in mRNA as in DNA. Thus, the codons in split genes, like continuous genes, are in the same linear order as the amino acids in the polypeptide products.

Splicing is a complex operation that is carried out by *spliceosomes*, which are assemblies of proteins and small nuclear RNA molecules (snRNA). RNA plays the catalytic role (Section 30.3). Spliceosomes recognize signals in the nascent RNA that specify the splice sites. *Introns nearly always begin with GU and end with an AG that is preceded by a pyrimidine-rich tract* (Figure 4.37). *This consensus sequence is part of the signal for splicing.*

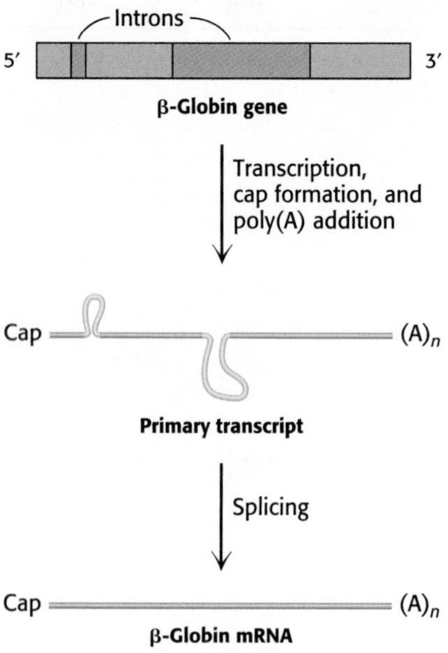

FIGURE 4.36 Transcription and processing of the β-globin gene. The gene is transcribed to yield the primary transcript, which is modified by cap and poly(A) addition. The introns in the primary RNA transcript are removed to form the mRNA.

FIGURE 4.37 Consensus sequence for the splicing of mRNA precursors.

Many exons encode protein domains

Most genes of higher eukaryotes, such as birds and mammals, are split. Lower eukaryotes, such as yeast, have a much higher proportion of continuous genes. In bacteria, split genes are extremely rare. Have introns been inserted into genes in the evolution of higher organisms? Or have introns been removed from genes to form the streamlined genomes of bacteria and simple eukaryotes? Comparisons of the DNA sequences of genes encoding evolutionarily conserved proteins suggest that *introns were present in ancestral genes and were lost in the evolution of organisms that have become optimized for very rapid growth, such as bacteria.* The positions of introns in some genes are at least 1 billion years old. Furthermore, a common mechanism of splicing developed before the divergence of fungi, plants, and vertebrates, as shown by the finding that mammalian cell extracts can splice yeast RNA.

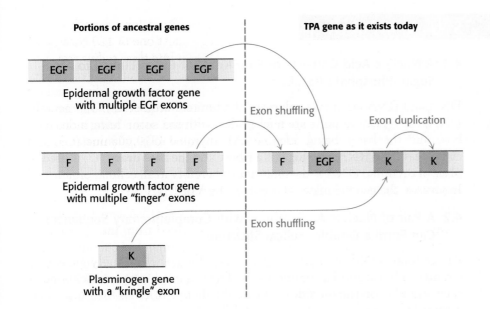

FIGURE 4.38 The tissue plasminogen activator (TPA) gene was generated by exon shuffling. The gene for TPA encodes an enzyme that functions in hemostasis (Section 10.4). This gene consists of 4 exons, one (F) derived from the fibronectin gene which encodes an extracellular matrix protein, one from the epidermal growth factor gene (EGF), and two from the plasminogen gene (K, Section 10.4), the substrate of the TPA protein. The K domain appears to have arrived by exon shuffling and then been duplicated to generate the TPA gene that exists today. [Information from: www.ehu.es/ehusfera/genetica/2012/10/02/demostracion-molecular-de-microevolucion/.]

What advantages might split genes confer? *Many exons encode discrete structural and functional domains of proteins.* An attractive hypothesis is that new *proteins arose in evolution by the rearrangement of exons encoding discrete structural elements, binding sites, and catalytic sites,* a process called *exon shuffling.* Because it preserves functional units but allows them to interact in new ways, exon shuffling is a rapid and efficient means of generating novel genes. Figure 4.38 shows the composition of a gene that was formed in part by exon shuffling. DNA can break and recombine in introns with no deleterious effect on encoded proteins. In contrast, the exchange of sequences within different exons usually leads to loss of protein function.

Another advantage of split genes is the potential for generating a series of related proteins by *alternative splicing* of the primary transcript. For example, a precursor of an antibody-producing cell forms an antibody that is anchored in the cell's plasma membrane (Figure 4.39). The attached antibody recognizes a specific foreign antigen, an event that leads to cell differentiation and proliferation. The activated antibody-producing cells then splice their primary transcript in an alternative manner to form soluble antibody molecules that are secreted rather than retained on the cell surface. *Alternative splicing is a facile means of forming a set of proteins that are variations of a basic motif without requiring a gene for each protein.* Because of alternative splicing, the proteome is more diverse than the genome in eukaryotes.

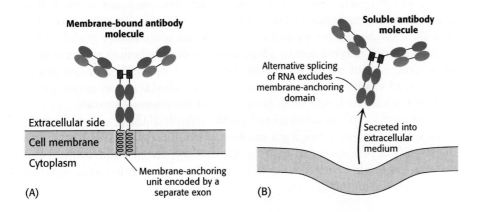

FIGURE 4.39 Alternative splicing. Alternative splicing generates mRNAs that are templates for different forms of a protein: (A) a membrane-bound antibody on the surface of a lymphocyte and (B) its soluble counterpart, exported from the cell. The membrane-bound antibody is anchored to the plasma membrane by a helical segment (highlighted in yellow) that is encoded by its own exon.

SUMMARY

4.1 A Nucleic Acid Consists of Four Kinds of Bases Linked to a Sugar–Phosphate Backbone

DNA and RNA are linear polymers of a limited number of monomers. In DNA, the repeating units are nucleotides, with the sugar being a deoxyribose and the bases being adenine (A), thymine (T), guanine (G), and cytosine (C). In RNA, the sugar is a ribose and the base uracil (U) is used in place of thymine. DNA is the molecule of heredity in all cellular organisms. In viruses, the genetic material is either DNA or RNA.

4.2 A Pair of Nucleic Acid Strands with Complementary Sequences Can Form a Double-Helical Structure

All cellular DNA consists of two very long, helical polynucleotide strands coiled around a common axis. The sugar–phosphate backbone of each strand is on the outside of the double helix, whereas the purine and pyrimidine bases are on the inside stabilized by stacking forces. The two strands are held together by hydrogen bonds between pairs of bases: adenine is always paired with thymine, and guanine is always paired with cytosine. Hence, one strand of a double helix is the complement of the other. The two strands of the double helix run in opposite directions. Genetic information is encoded in the precise sequence of bases along a strand.

DNA is a structurally dynamic molecule that can exist in a variety of helical forms: A-DNA, B-DNA (the classic Watson–Crick double helix), and Z-DNA. In A-, B-, and Z-DNA, two antiparallel chains are held together by Watson–Crick base pairs and stacking interactions between bases in the same strand. A- and B-DNA are right-handed helices. In B-DNA, the base pairs are nearly perpendicular to the helix axis. Z-DNA is a left-handed helix. Most of the DNA in a cell is in the B-form.

Double-stranded DNA can also wrap around itself to form a supercoiled structure. The supercoiling of DNA has two important consequences. Supercoiling compacts the DNA and, because supercoiled DNA is partly unwound, it is more accessible for interactions with other biomolecules.

Single-stranded nucleic acids, most notably RNA, can form complicated three-dimensional structures that may contain extensive double-helical regions that arise from the folding of the chain into hairpins.

4.3 The Double Helix Facilitates the Accurate Transmission of Hereditary Information

The structural nature of the double helix readily accounts for the accurate replication of genetic material because the sequence of bases in one strand determines the sequence of bases in the other strand. In replication, the strands of the helix separate and a new strand complementary to each of the original strands is synthesized. Thus, two new double helices are generated, each composed of one strand from the original molecule and one newly synthesized strand. This mode of replication is called semiconservative replication because each new helix retains one of the parental strands.

In order for replication to take place, the strands of the double helix must be separated. In vitro, heating a solution of double-helical DNA separates the strands, a process called melting. On cooling, the strands reanneal and re-form the double helix. In the cell, special proteins temporarily separate the strands during replication.

4.4 DNA Is Replicated by Polymerases That Take Instructions from Templates

In the replication of DNA, the two strands of a double helix unwind and separate as new strands are synthesized. Each parent strand, with the help of a primer, acts as a template for the formation of a new complementary strand. The replication of DNA is a complex process carried out by many proteins, including several DNA polymerases. The activated precursors in the synthesis of DNA are the four deoxyribonucleoside 5′-triphosphates. The new strand is synthesized in the 5′ → 3′ direction by a nucleophilic attack by the 3′-hydroxyl terminus of the primer strand on the innermost phosphorus atom of the incoming deoxyribonucleoside triphosphate. Most important, DNA polymerases catalyze the formation of a phosphodiester linkage only if the base on the incoming nucleotide is complementary to the base on the template strand. In other words, DNA polymerases are template-directed enzymes. The genes of some viruses, such as tobacco mosaic virus, are made of single-stranded RNA. An RNA-directed RNA polymerase mediates the replication of this viral RNA. Retroviruses, exemplified by HIV-1, have a single-stranded RNA genome that undergoes reverse transcription into double-stranded DNA by reverse transcriptase, an RNA-directed DNA polymerase.

4.5 Gene Expression Is the Transformation of DNA Information into Functional Molecules

The flow of genetic information in normal cells is from DNA to RNA to protein. The synthesis of RNA from a DNA template is called transcription, whereas the synthesis of a protein from an RNA template is termed translation. Cells contain several kinds of RNA, among which are messenger RNA (mRNA), transfer RNA (tRNA), and ribosomal RNA (rRNA), which vary in size from 75 to more than 5000 nucleotides. All cellular RNA is synthesized by RNA polymerases according to instructions given by DNA templates. The activated intermediates are ribonucleoside triphosphates and the direction of synthesis, like that of DNA, is 5′ → 3′. RNA polymerase differs from DNA polymerase in not requiring a primer.

4.6 Amino Acids Are Encoded by Groups of Three Bases Starting from a Fixed Point

The genetic code is the relation between the sequence of bases in DNA (or its RNA transcript) and the sequence of amino acids in proteins. Amino acids are encoded by groups of three bases (called codons) starting from a fixed point. Sixty-one of the 64 codons specify particular amino acids, whereas the other 3 codons (UAA, UAG, and UGA) are signals for chain termination. Thus, for most amino acids, there is more than one code word. That is, the code is degenerate. The genetic code is nearly the same in all organisms. Natural mRNAs contain start and stop signals for translation, just as genes do for directing where transcription begins and ends.

4.7 Most Eukaryotic Genes Are Mosaics of Introns and Exons

Most genes in higher eukaryotes are discontinuous. Coding sequences, called exons, in these split genes are separated by noncoding sequences, called introns, which are removed in the conversion of the primary transcript into mRNA and other functional mature RNA molecules. Split genes, like continuous genes, are colinear with their polypeptide products. A striking feature of many exons is that they encode functional domains in proteins. New proteins probably arose in the course of evolution by the shuffling of exons. Introns may have been present in primordial genes but were lost in the evolution of such fast-growing organisms as bacteria and yeast.

APPENDIX

Problem-Solving Strategies

PROBLEM: The graph below shows the effect of salt concentration on melting temperature of bacterial DNA. How does salt concentration affect the melting temperature of DNA? Account for this effect.

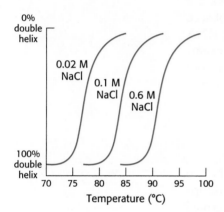

SOLUTION: Analyzing data is fundamental to understanding the results of scientific research, whether the research is published in scientific journals or a newspaper. A common means of displaying scientific results is with a graph using the Cartesian coordinate system, as in the problem.

The first step to understanding the graph is to determine what the coordinates, the x-axis and the y-axis, represent. Recall from previous courses that the y-axis (ordinate) displays the dependent variable or the *results* of an experiment, while the x-axis (abscissa) shows the *independent variable* or the experimental parameter that is manipulated.

▶ **What are the dependent and independent variables in the graph?**

The x-axis shows increasing temperature (°C), while the y-axis shows DNA melting as a decrease in the percent of double helical DNA. Three DNA-melting curves are shown.

To interpret any data, it is crucial to be familiar with the experimental protocol.

▶ **How is the degree of DNA melting determined?**

See Figure 4.21. Single-stranded DNA absorbs more light of 260 nm than does double-stranded DNA. Thus, as the temperature is raised, double-stranded DNA melting is determined by an increase in the absorbance of 260 nm.

Note once again that there are three melting curves.

▶ **In addition to temperature, what other experimental parameter is varied?**

The salt concentration or ionic strength of the DNA solution.

▶ **What is the effect of increasing salt concentration on the melting temperature of DNA?**

As the salt concentration increases, the melting temperature of the DNA also increases.

Now comes the key question:

▶ **Why does increasing salt concentration increase the melting temperature of DNA?**

Sometimes asking a question in a different way might help to more readily reveal the answer.

▶ **Why do higher salt concentrations stabilize the DNA double helix?**

Or

▶ **What aspect of the structure of double-stranded DNA destabilizes the helix in the absence of salt?**

Look at the structures of DNA shown in Figure 4.10 and note the location of the phosphate components of the backbone. Look also at the right-hand half of Figure 4.23. Note that the phosphate has a negative charge. That is, the backbone of the DNA strands is strings of negative charges in close proximity.

▶ **What is the result of like charges, be they phosphates or magnets, being in close proximity?**

Like charges repeal each other (you knew this from the time you started playing with refrigerator magnets as a child). In the case of DNA, the charge repulsion of the phosphates makes the double helix unstable.

▶ **What would happen to the negative charges on the DNA backbone in the presence of the positive charge on a sodium ion (Na⁺)?**

The negative charges on the phosphates would be neutralized.

▶ **Explain the result of the experiment shown in the figure.**

Within the parameters of the experiment, increasing sodium chloride concentration leads to more stable double helices, as measured by the increase in the melting temperature. The simplest explanation is that the sodium reduces charge repulsion in a concentration-dependent manner.

KEY TERMS

double helix (p. 113)

nucleotide (p. 114)

deoxyribonucleic acid (DNA) (p. 114)

deoxyribose (p. 114)

ribose (p. 114)

purine (p. 114)

pyrimidine (p. 114)

ribonucleic acid (RNA) (p. 115)

nucleoside (p. 116)

B-DNA (p. 119)

A-DNA (p. 119)

Z-DNA (p. 120)

semiconservative replication (p. 122)

DNA polymerase (p. 125)

template (p. 125)

primer (p. 125)

reverse transcriptase (p. 126)

messenger RNA (mRNA) (p. 128)

translation (p. 128)

transfer RNA (tRNA) (p. 128)

ribosomal RNA (rRNA) (p. 128)

transcription (p. 128)

RNA polymerase (p. 128)

promoter site (p. 130)

anticodon (p. 131)

codon (p. 131)

genetic code (p. 132)

codon bias (p. 134)

ribosome (p. 134)

Shine–Dalgarno sequence (p. 134)

intron (p. 136)

exon (p. 136)

splicing (p. 136)

spliceosome (p. 136)

exon shuffling (p. 137)

alternative splicing (p. 137)

PROBLEMS

1. *A t instead of an s?* Differentiate between a nucleoside and a nucleotide. ✓❶

2. *A lovely pair.* What is a Watson–Crick base pair?

3. *Chargaff rules!* Biochemist Erwin Chargaff was the first to note that, in DNA, [A] = [T] and [G] = [C], equalities now called Chargaff's rule. Using this rule, determine the percentages of all the bases in DNA that is 20% thymine. ✓❷ ✓❸

4. *But not always.* A single strand of RNA is 20% U. What can you predict about the percentages of the remaining bases? ✓❷ ✓❸

5. *Complements.* Write the complementary sequence (in the standard 5′ → 3′ notation) for (a) GATCAA, (b) TCGAAC, (c) ACGCGT, and (d) TACCAT.

6. *Compositional constraint.* The composition (in mole-fraction units) of one of the strands of a double-helical DNA molecule is [A] = 0.30 and [G] = 0.24. (a) What can you say about [T] and [C] for the same strand? (b) What can you say about [A], [G], [T], and [C] of the complementary strand?

7. *Size matters.* Why are GC and AT the only base pairs permissible in the double helix?

8. *Strong, but not strong enough.* Why does heat denature, or melt, DNA in solution?

9. *Uniqueness.* The human genome contains 3 billion nucleotides arranged in a vast array of sequences. What is the minimum length of a DNA sequence that will, in all probability, appear only once in the human genome? You need consider only one strand and may assume that all four nucleotides have the same probability of appearance.

10. *Coming and going.* What does it mean to say that the DNA strands in a double helix have opposite directionality or polarity?

11. *All for one.* If the forces—hydrogen bonds and stacking forces—holding a helix together are weak, why is it difficult to disrupt a double helix?

12. *Overcharged.* DNA in the form of a double helix must be associated with cations, usually Mg^{2+}. Why is this requirement the case?

13. *Not quite from A to Z.* Identify the three forms that a double helix can assume and describe some key differences.

14. *Lost DNA.* The DNA of a deletion mutant of λ bacteriophage has a length of 15 μm instead of the 17 μm found in the wild type phage. How many base pairs are missing from this mutant?

15. *Axial ratio.* What is the axial ratio (length:diameter) of a DNA molecule 20 μm long?

16. *Guide and starting point.* Define template and primer as they relate to DNA synthesis. ✓❹

17. *An unseen pattern.* What result would Meselson and Stahl have obtained if the replication of DNA were conservative (i.e., the parental double helix stayed together)? Give the expected distribution of DNA molecules after 1.0 and 2.0 generations for conservative replication. ✓❹

18. *Which way?* Explain, on the basis of nucleotide structure, why DNA synthesis proceeds in the 5′-to-3′ direction. ✓❹

19. *Tagging DNA.*

(a) Suppose that you want to radioactively label DNA but not RNA in dividing and growing bacterial cells. Which radioactive molecule would you add to the culture medium? ✓④

(b) Suppose that you want to prepare DNA in which the backbone phosphorus atoms are uniformly labeled with ^{32}P. Which precursors should be added to a solution containing DNA polymerase and primed template DNA? Specify the position of radioactive atoms in these precursors.

20. *Finding a template.* A solution contains DNA polymerase and the Mg^{2+} salts of dATP, dGTP, dCTP, and TTP. The following DNA molecules are added to aliquots of this solution. Which of them would lead to DNA synthesis? Explain your answer. ✓④

(a) A single-stranded closed circle containing 1000 nucleotide units.

(b) A double-stranded closed circle containing 1000 nucleotide pairs.

(c) A single-stranded closed circle of 1000 nucleotides base-paired to a linear strand of 500 nucleotides with a free 3′-OH terminus.

(d) A double-stranded linear molecule of 1000 nucleotide pairs with a free 3′-OH group at each end.

21. *Retrograde.* What is a retrovirus and how does information flow for a retrovirus differ from that for the infected cell?

22. *The right start.* Suppose that you want to assay reverse transcriptase activity. If polyriboadenylate is the template in the assay, what should you use as the primer? Which radioactive nucleotide should you use to follow chain elongation?

23. *Essential degradation.* Reverse transcriptase has ribonuclease activity as well as polymerase activity. What is the role of its ribonuclease activity?

24. *Virus hunting.* You have purified a virus that infects turnip leaves. Treatment of a sample with phenol removes viral proteins. Application of the residual material to scraped leaves results in the formation of progeny virus particles. You infer that the infectious substance is a nucleic acid. Propose a simple and highly sensitive means of determining whether the infectious nucleic acid is DNA or RNA. ✓③

25. *Mutagenic consequences.* Spontaneous deamination of cytosine bases in DNA takes place at low but measurable frequency. Cytosine is converted into uracil by loss of its amino group. After this conversion, which base pair occupies this position in each of the daughter strands resulting from one round of replication? Two rounds of replication?

26. *Information content.*

(a) How many different 8-mer sequences of DNA are there? (Hint: There are 16 possible dinucleotides and 64 possible

trinucleotides.) We can quantify the information-carrying capacity of nucleic acids in the following way. Each position can be one of four bases, corresponding to two bits of information ($2^2 = 4$). Thus, a chain of 5100 nucleotides corresponds to $2 \times 5100 = 10,200$ bits, or 1275 bytes (1 byte = 8 bits).

(b) How many bits of information are stored in an 8-mer DNA sequence? In the *E. coli* genome? In the human genome?

(c) Compare each of these values with the amount of information that can be stored on a 2 gigabyte flash drive.

27. *Key polymerases.* Compare DNA polymerase and RNA polymerase from *E. coli* in regard to each of the following features: (a) activated precursors, (b) direction of chain elongation, (c) conservation of the template, and (d) need for a primer. ✓⑤

28. *Different strands.* Explain the difference between the coding strand and the template strand in DNA. ✓④

29. *Family resemblance.* Differentiate among mRNA, rRNA and tRNA. ✓⑤

30. *A code you can live by.* What are the key characteristics of the genetic code? ✓⑤

31. *Encoded sequences.*

(a) Write the sequence of the mRNA molecule synthesized from a DNA template strand having the following sequence: ✓⑤

$$5′–ATCGTACCGTTA–3′$$

(b) What amino acid sequence is encoded by the following base sequence of an mRNA molecule? Assume that the reading frame starts at the 5′ end.

$$5′–UUGCCUAGUGAUUGGAUG–3′$$

(c) What is the sequence of the polypeptide formed on addition of poly(UUAC) to a cell-free protein-synthesizing system?

32. *A tougher chain.* RNA is readily hydrolyzed by alkali, whereas DNA is not. Why? ✓③

33. *A picture is worth a thousand words.* Write a reaction sequence showing why RNA is more susceptible to nucleophilic attack than DNA. ✓③

34. *Flowing information.* What is meant by the phrase gene expression?

35. *We can all agree on that.* What is a consensus sequence?

36. *A potent blocker.* Cordycepin (3′-deoxyadenosine) is an adenosine analog. When converted into cordycepin 5′-triphosphate, it inhibits RNA synthesis. How does cordycepin 5′-triphosphate block the synthesis of RNA? ✓⑤

37. *Sometimes it is not so bad.* What is meant by the degeneracy of the genetic code? ✓⑤

38. *In fact, it can be good.* What is the biological benefit of a degenerate genetic code? ✓⑤

39. *To bring together as associates.* Match the components in the right-hand column with the appropriate process in the left-hand column. Terms in the left-hand column may be used more than once. ✓⑤

(a) Replication _____
(b) Transcription _____
(c) Translation _____

1. RNA polymerase
2. DNA polymerase
3. Ribosome
4. dNTP
5. tRNA
6. NTP
7. mRNA
8. primer
9. rRNA
10. promoter

40. *A lively contest.* Match the components in the right-hand column with the appropriate process in the left-hand column. ✓⑥

(a) fMet _____
(b) Shine–Dalgarno _____
(c) introns _____
(d) exons _____
(e) pre-mRNA _____
(f) mRNA _____
(g) spliceosome _____

1. continuous message
2. removed during processing
3. the first of many amino acids
4. joins exons
5. joined to make the final message
6. locates the start site
7. discontinuous message

41. *Two from one.* Synthetic RNA molecules of defined sequence were instrumental in deciphering the genetic code. Their synthesis first required the synthesis of DNA molecules to serve as templates. HarGobind Khorana synthesized, by organic-chemical methods, two complementary deoxyribonucleotide oliogmers each with nine residues: d(TAC)₃ and d(GTA)₃. Partly overlapping duplexes that formed on mixing these oligonucleotides then served as templates for the synthesis by DNA polymerase of long, repeating double-helical DNA strands. The next step was to obtain long polyribonucleotide chains with a sequence complementary to only one of the two DNA strands. How did Khorana obtain only poly(UAC)? Only poly(GUA)?

42. *Triple entendre.* The terminal portion of an RNA transcript of a region of T4 phage DNA contains the sequence 5′-AAAUGAGGA-3′. The entire sequence encodes three different polypeptides. What amino acid(s) will terminate the three different polypeptides? ✓⑤

43. *A new translation.* A transfer RNA with a UGU anticodon is enzymatically conjugated to ¹⁴C-labeled cysteine. The cysteine unit is then chemically modified to alanine (with the use of Raney nickel, which removes the sulfur atom of cysteine). The altered aminoacyl-tRNA is added to a protein-synthesizing system containing normal components except for this tRNA. The mRNA added to this mixture contains the following sequence:

$$5'-UUUUGCCAUGUUUGUGCU-3'$$

What is the sequence of the corresponding radiolabeled peptide? ✓⑤

44. *A tricky exchange.* Define exon shuffling and explain why its occurrence might be an evolutionary advantage.

45. *From one, many.* Explain why alternative splicing increases the coding capacity of the genome.

46. *The unity of life.* What is the significance of the fact that human mRNA can be accurately translated in *E. coli*? ✓⑥

Chapter Integration Problems

47. *Back to the bench.* A protein chemist told a molecular geneticist that he had found a new mutant hemoglobin in which aspartate replaced lysine. The molecular geneticist expressed surprise and sent his friend scurrying back to the laboratory. (a) Why did the molecular geneticist doubt the reported amino acid substitution? (b) Which amino acid substitutions would have been more palatable to the molecular geneticist?

48. *Eons apart.* The amino acid sequences of a yeast protein and a human protein having the same function are found to be 60% identical. However, the corresponding DNA sequences are only 45% identical. Account for this differing degree of identity.

Data Interpretation Problems

49. *3 is greater than 2.* The adjoining illustration graphs the relation between the percentage of GC base pairs in DNA and the melting temperature. Suggest a plausible explanation for these results.

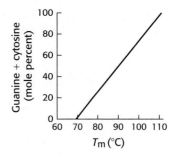

[Data from R. J. Britten and D. E. Kohne, *Science* 161:529–540, 1968.]

50. *Blast from the past.* The illustration below is a graph called a C₀t curve (pronounced "cot"). The *y*-axis shows the percentage of DNA that is double stranded. The *x*-axis is the product of the concentration of DNA and the time required for the double-stranded molecules to form. Explain why the mixture of poly(A) and poly(U) and the three DNAs shown vary in the C_0t value required to completely anneal. MS2 and T4 are bacterial viruses (bacteriophages) with genome sizes of 3569 and 168,903 bp, respectively. The *E. coli* genome is 4.6×10^6 bp.

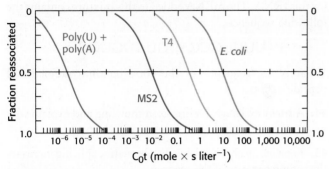

[Data from J. Marmur and P. Doty, *J. Mol. Biol.* 5:120, 1962.]

Think ▶ Pair ▶ Share Problems

51. What properties of DNA make it especially well suited to serve as the genetic material?

52. With regard to the problem in the Problem-Solving example, what other experiments would you design to determine what factors might affect DNA double helix stability?

Exploring Genes and Genomes

Processes such as the development from a caterpillar into a butterfly entail dramatic changes in patterns of gene expression. The expression levels of thousands of genes can be monitored through the use of DNA arrays. At the right, a DNA microarray reveals the expression levels of more than 12,000 human genes; the brightness and color of each spot indicates a change in the expression level of the corresponding gene. [(Left) Cathy Keifer/istockphoto.com. (Right) © Agilent Technologies, Inc. 2013. Reproduced with permission. Courtesy of Agilent Technologies, Inc.]

✓ LEARNING OBJECTIVES

By the end of this chapter, you should be able to:

1. Explain how restriction enzymes work and why they are so important to recombinant DNA technology.

2. Describe several methods commonly used for the sequencing of DNA.

3. List the fundamental steps of the polymerase chain reaction.

4. Describe the various types of vectors used in DNA cloning. Explain the difference between a cloning vector and an expression vector.

5. Describe some of the more commonly used methods to introduce mutations into DNA.

6. Explain how genes can be altered in living organisms, and describe some of the applications of these techniques.

OUTLINE

5.1 The Exploration of Genes Relies on Key Tools

5.2 Recombinant DNA Technology Has Revolutionized All Aspects of Biology

5.3 Complete Genomes Have Been Sequenced and Analyzed

5.4 Eukaryotic Genes Can Be Quantitated and Manipulated with Considerable Precision

Since its emergence in the 1970s, recombinant DNA technology has revolutionized biochemistry. The genetic endowment of organisms can now be precisely changed in designed ways. Recombinant DNA technology is the fruit of several decades of basic research on DNA, RNA, and viruses. It depends, first, on having enzymes that can cut, join, and replicate DNA and those that can reverse transcribe RNA. Restriction enzymes cut very long DNA molecules into specific fragments that can be manipulated; DNA ligases join the fragments together. Many kinds of restriction enzymes are available. By applying this assortment cleverly, researchers can treat DNA sequences as modules that can be moved at will from one DNA molecule to another. Thus, recombinant DNA technology is based on the use of enzymes that act on nucleic acids as substrates.

A second foundation is the base-pairing language that allows complementary sequences to recognize and bind to each other. Hybridization with complementary DNA (cDNA) or RNA probes is a sensitive means of detecting specific nucleotide sequences. In recombinant DNA technology, base-pairing is used to construct new combinations of DNA as well as to detect and amplify particular sequences.

Third, powerful methods have been developed for determining the sequence of nucleotides in DNA. These methods have been harnessed to sequence complete genomes: first, small genomes from viruses; then, larger genomes from bacteria; and, finally, eukaryotic genomes, including the 3-billion-base-pair human genome. Scientists can now exploit the enormous information content of these genome sequences to answer a myriad of biological questions.

Lastly, recombinant DNA technology critically depends on our ability to deliver foreign DNA into host organisms. For example, DNA fragments can be inserted into plasmids, where they can be replicated within a short period of time in their bacterial hosts. In addition, viruses efficiently deliver their own DNA (or RNA) into hosts, subverting them either to replicate the viral genome and produce viral proteins or to incorporate viral DNA into the host genome.

These new methods have wide-ranging benefits across a broad spectrum of disciplines, including biotechnology, agriculture, and medicine. Among these benefits is the dramatic expansion of our understanding of human disease. Throughout this chapter, a specific disorder, amyotrophic lateral sclerosis (ALS), will be used to illustrate the effect that recombinant DNA technology has had on our knowledge of disease mechanisms. ALS was first described clinically in 1869 by the French neurologist Jean-Martin Charcot as a fatal neurodegenerative disease of progressive weakening and atrophy of voluntary muscles. ALS is commonly referred to as Lou Gehrig's Disease, for the baseball legend whose career and life were prematurely cut short as a result of this devastating disease. For many years, little progress had been made in the study of the mechanisms underlying ALS. As we shall see, significant advances have been made with the use of research tools facilitated by recombinant DNA technology.

5.1 The Exploration of Genes Relies on Key Tools

The rapid progress in biotechnology—indeed its very existence—is a result of a few key techniques.

1. *Restriction-enzyme analysis.* Restriction enzymes are precise molecular scalpels that allow an investigator to manipulate DNA segments.

2. *Blotting techniques.* Southern and northern blots are used to separate and identify DNA and RNA sequences, respectively. The western blot, which uses antibodies to characterize proteins, was described in Chapter 3.

3. *DNA sequencing.* The precise nucleotide sequence of a molecule of DNA can be determined. Sequencing has yielded a wealth of information concerning gene architecture, the control of gene expression, and protein structure.

4. *Solid-phase synthesis of nucleic acids.* Precise sequences of nucleic acids can be synthesized de novo and used to identify or amplify other nucleic acids.

5. *The polymerase chain reaction* (PCR). The polymerase chain reaction leads to a billionfold amplification of a segment of DNA. One molecule of DNA can be amplified to quantities that permit characterization and manipulation. This powerful technique can be used to detect pathogens and genetic diseases, determine the source of a hair left at the scene of a crime, and resurrect genes from the fossils of extinct organisms.

A final set of techniques relies on the computer, without which it would be impossible to catalog, access, and characterize the abundant information generated by the methods outlined above. Such uses of the computer will be presented in Chapter 6.

Restriction enzymes split DNA into specific fragments

Restriction enzymes, also called *restriction endonucleases*, recognize specific base sequences in double-helical DNA and cleave both strands of that duplex at specific places. To biochemists, these exquisitely precise scalpels are marvelous gifts of nature. They are indispensable for analyzing chromosome structure, sequencing very long DNA molecules, isolating genes, and creating new DNA molecules that can be cloned.

Restriction enzymes are found in a wide variety of prokaryotes. Their biological role is to cleave foreign DNA molecules, providing the host organism with a primitive immune system. Many restriction enzymes recognize specific sequences of four to eight base pairs and hydrolyze a phosphodiester bond in each strand in this region. A striking characteristic of these cleavage sites is that they almost always possess *twofold rotational symmetry*. In other words, the recognized sequence is *palindromic*, or an inverted repeat, and the cleavage sites are symmetrically positioned. For example, the sequence recognized by a restriction enzyme from *Streptomyces achromogenes* is

Cleavage site
↓
5′ C—C—G—C—G—G 3′

3′ G—G—C—G—C—C 5′
↑
Cleavage site   ⎺Symmetry axis

In each strand, the enzyme cleaves the C–G phosphodiester bond on the 3′ side of the symmetry axis. As we shall see in Chapter 9, this symmetry corresponds to that of the structures of the restriction enzymes themselves.

Several thousand restriction enzymes have been purified and characterized. Their names consist of a three-letter abbreviation for the host organism (e.g., Eco for *Escherichia coli,* Hin for *Haemophilus influenzae,* Hae for *Haemophilus aegyptius*) followed by a strain designation (if needed) and a roman numeral (to distinguish multiple enzymes from the same strain). The specificities of several of these enzymes are shown in Figure 5.1.

Restriction enzymes are used to cleave DNA molecules into specific fragments that are more readily analyzed and manipulated than the entire parent molecule. For example, the 5.1-kb circular duplex DNA of the tumor-producing SV40 virus is cleaved at one site by EcoRI, at four sites by HpaI, and at 11 sites by HindIII. A piece of DNA, called a restriction fragment, produced by the action of one restriction enzyme can be

Palindrome

A word, sentence, or verse that reads the same from right to left as it does from left to right.
   Radar
   Senile felines
   Do geese see God?
   Roma tibi subito motibus ibit amor
Derived from the Greek *palindromos,* "running back again."

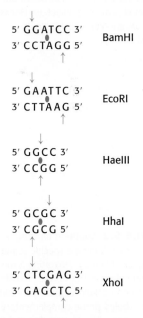

FIGURE 5.1 Specificities of some restriction endonucleases. The sequences that are recognized by these enzymes contain a twofold axis of symmetry. The two strands in these regions are related by a 180-degree rotation about the axis marked by the green symbol. The cleavage sites are denoted by red arrows. The abbreviated name of each restriction enzyme is given at the right of the sequence that it recognizes. Note that the cuts may be staggered or even.

specifically cleaved into smaller fragments by another restriction enzyme. The pattern of such fragments can serve as a *fingerprint* of a DNA molecule, as will be considered shortly. Indeed, complex chromosomes containing hundreds of millions of base pairs can be mapped by using a series of restriction enzymes.

Restriction fragments can be separated by gel electrophoresis and visualized

In Chapter 3, we considered the use of gel electrophoresis to separate protein molecules (Section 3.1). Because the phosphodiester backbone of DNA is highly negatively charged, this technique is also suitable for the separation of nucleic acid fragments. Among the many applications of DNA electrophoresis are the detection of mutations that affect restriction fragment size (such as insertions and deletions) and the isolation, purification, and quantitation of a specific DNA fragment.

For most gels, the shorter the DNA fragment, the farther the migration. Polyacrylamide gels are used to separate, by size, fragments containing as many as 1000 base pairs, whereas more-porous agarose gels are used to resolve mixtures of larger fragments (as large as 20 kilobases, kb). An important feature of these gels is their high resolving power. In certain kinds of gels, fragments differing in length by just one nucleotide out of several hundred can be distinguished. Bands or spots of radioactive DNA in gels can be visualized by autoradiography. Alternatively, a gel can be stained with a dye such as ethidium bromide, which fluoresces an intense orange under irradiation with ultraviolet light when bound to a double-helical DNA molecule (Figure 5.2). A band containing only 10 ng of DNA can be readily seen.

It is often necessary to determine if a particular base sequence is represented in a given DNA sample. For example, one may wish to confirm the presence of a specific mutation in genomic DNA isolated from patients known to be at risk for a particular disease. This specific sequence can be identified by hybridizing it with a labeled complementary DNA strand (Figure 5.3). A mixture of restriction fragments is separated by electrophoresis through an agarose gel, denatured to form single-stranded DNA, and transferred to a membrane composed of nitrocellulose or nylon. The positions of the DNA fragments in the gel are preserved during the transfer. The membrane is then exposed to a ^{32}P-labeled or fluorescently

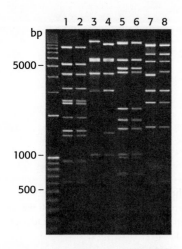

FIGURE 5.2 Gel-electrophoresis pattern of a restriction digest. This gel shows the fragments produced by cleaving DNA from two viral strains (odd- vs. even-numbered lanes) with each of four restriction enzymes. These fragments were made fluorescent by staining the gel with ethidium bromide. [Carr et al., Emerging Infectious Diseases, www.cdc.gov/eid, Vol. 17, No. 8, August 2011.]

FIGURE 5.3 Southern blotting. A DNA fragment containing a specific sequence can be identified by separating a mixture of fragments by electrophoresis, transferring them to nitrocellulose, and hybridizing with a labeled probe complementary to the sequence. The fragment containing the sequence is then visualized by autoradiography or fluorescence imaging.

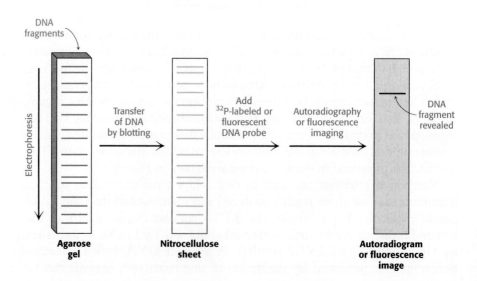

tagged *DNA probe*, a short stretch of single-stranded DNA which contains a known base sequence. The probe hybridizes with a restriction fragment having a complementary sequence, and autoradiography or fluorescence imaging then reveals the position of the restriction-fragment–probe duplex. A particular fragment amid a million others can be readily identified in this way. This powerful technique is named *Southern blotting*, for its inventor Edwin Southern.

In a similar manner, RNA molecules of a specific sequence can also be readily identified. After separation by gel electrophoresis and transfer to nitrocellulose, specific sequences can be detected by DNA probes. This analogous technique for the analysis of RNA has been whimsically termed *northern blotting*. A further play on words accounts for the term *western blotting*, which refers to a technique for detecting a particular protein by staining with specific antibody (Section 3.2).

DNA can be sequenced by controlled termination of replication

The analysis of DNA structure and its role in gene expression has been markedly facilitated by the development of powerful techniques for the *sequencing* of DNA molecules. One of the first and most widely used techniques for DNA sequencing is *controlled termination of replication*, also referred to as the Sanger dideoxy method for its pioneer, Frederick Sanger. The key to this approach is the generation of DNA fragments whose length is determined by the last base in the sequence (Figure 5.4). In the current application of this method, a DNA polymerase is used to make the complement of a particular sequence within a single-stranded DNA molecule. The synthesis is primed by a chemically synthesized fragment that is complementary to a part of the sequence known from other studies. In addition to the four deoxyribonucleoside triphosphates, the reaction mixture contains a small amount of the 2′,3′-*dideoxy analog* of each nucleotide, each carrying a different fluorescent label attached to the base (e.g., a green emitter for termination at A and a red one for termination at T).

2′, 3′-Dideoxy analog

The incorporation of this analog blocks further growth of the new chain because it lacks the 3′-hydroxyl terminus needed to form the next phosphodiester bond. The concentration of the dideoxy analog is low enough that chain termination will take place only occasionally. The polymerase will insert the correct nucleotide sometimes and the dideoxy analog other times, stopping the reaction. For instance, if the dideoxy analog of dATP is present, fragments of various lengths are produced, but all will be terminated by the dideoxy analog. Importantly, this dideoxy analog of dATP will be inserted only where a T was located in the DNA being sequenced. Thus, the fragments of different length will correspond to the positions of T.

The resulting fragments are separated by a technique known as *capillary electrophoresis*, in which the mixture is passed through a very narrow tube

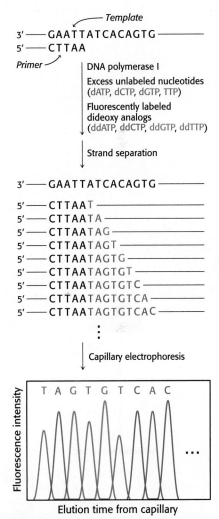

FIGURE 5.4 Fluorescence detection of oligonucleotide fragments produced by the dideoxy method. A sequencing reaction is performed with four chain-terminating dideoxy nucleotides, each labeled with a tag that fluoresces at a different wavelength. The color of each fragment indicates the identity of the last base in the chain. The fragments are separated by size using capillary electrophoresis, and the fluorescence at each of the four wavelengths indicates the sequence of the complement of the original DNA template.

containing a gel matrix at high voltage to achieve efficient separation within a short time. As the DNA fragments emerge from the capillary, they are detected by their fluorescence; the sequence of their colors directly gives the base sequence. Sequences of as many as 1000 bases can be determined in this way. Indeed, automated Sanger sequencing machines can read more than 1 million bases per day.

DNA probes and genes can be synthesized by automated solid-phase methods

DNA strands, like polypeptides (Section 3.4), can be synthesized by the sequential addition of activated monomers to a growing chain that is linked to a solid support. The activated monomers are protected *deoxyribonucleoside 3′-phosphoramidites*. In step 1, the 3′-phosphorus atom of this incoming unit becomes joined to the 5′-oxygen atom of the growing chain to form a *phosphite triester* (Figure 5.5). The 5′-OH group of the activated monomer is unreactive because it is blocked by a dimethoxytrityl (DMT) protecting group, and the 3′-phosphoryl oxygen atom is rendered unreactive by attachment of the β-cyanoethyl (βCE) group. Likewise, amino groups on the purine and pyrimidine bases are blocked.

Coupling is carried out under anhydrous conditions because water reacts with phosphoramidites. In step 2, the phosphite triester (in which P is trivalent) is oxidized by iodine to form a *phosphotriester* (in which P is pentavalent). In step 3, the DMT protecting group on the 5′-OH group of the growing chain is removed by the addition of dichloroacetic acid, which leaves other protecting groups intact. The DNA chain is now elongated by one unit and ready for another cycle of addition. Each cycle takes only about 10 minutes and usually elongates more than 99% of the chains.

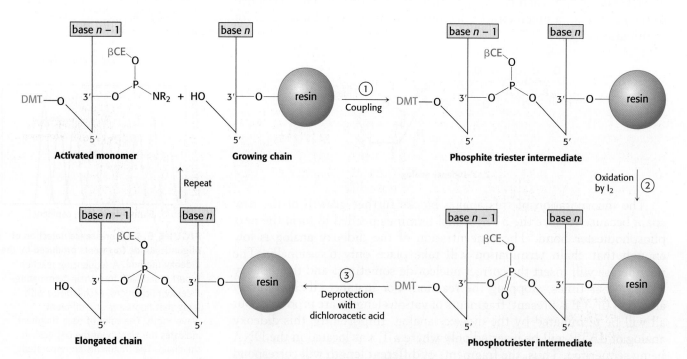

FIGURE 5.5 Solid-phase synthesis of a DNA chain by the phosphite triester method. The activated monomer added to the growing chain is a deoxyribonucleoside 3′-phosphoramidite containing a dimethoxytrityl (DMT) protecting group on its 5′-oxygen atom, a β-cyanoethyl (βCE) protecting group on its 3′-phosphoryl oxygen atom, and a protecting group on the base.

This solid-phase approach is ideal for the synthesis of DNA, as it is for polypeptides, because the desired product stays on the insoluble support until the final release step. All the reactions take place in a single vessel, and excess soluble reagents can be added to drive reactions to completion. At the end of each step, soluble reagents and by-products are washed away from the resin that bears the growing chains. At the end of the synthesis, NH_3 is added to remove all protecting groups and release the oligonucleotide from the solid support. Because elongation is never 100% complete, the new DNA chains are of diverse lengths—the desired chain is the longest one. The sample can be purified by high-performance liquid chromatography or by electrophoresis on polyacrylamide gels. DNA chains of as many as 100 nucleotides can be readily synthesized by this automated method.

The ability to rapidly synthesize DNA chains of any selected sequence opens many experimental avenues. For example, a synthesized oligonucleotide labeled at one end with ^{32}P or a fluorescent tag can be used to search for a complementary sequence in a very long DNA molecule or even in a genome consisting of many chromosomes. The use of labeled oligonucleotides as DNA probes is powerful and general. For example, a DNA probe that can base-pair to a known complementary sequence in a chromosome can serve as the starting point of an exploration of adjacent uncharted DNA. Such a probe can be used as a *primer* to initiate the replication of neighboring DNA by DNA polymerase. An exciting application of the solid-phase approach is the synthesis of new tailor-made genes. New proteins with novel properties can now be produced in abundance by the expression of synthetic genes. Finally, the synthetic scheme heretofore described can be slightly modified for the solid-phase synthesis of RNA oligonucleotides, which can be very powerful reagents for the degradation of specific mRNA molecules in living cells by a technique known as RNA interference (Section 5.4).

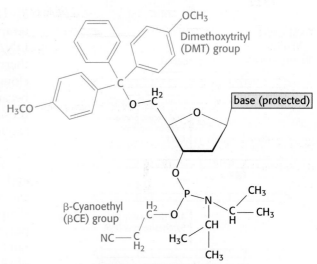

A deoxyribonucleoside 3′-phosphoramidite with DMT and βCE attached

Selected DNA sequences can be greatly amplified by the polymerase chain reaction

In 1984, Kary Mullis devised an ingenious method called the *polymerase chain reaction* (PCR) for amplifying specific DNA sequences. Consider a DNA duplex consisting of a target sequence surrounded by nontarget DNA. Millions of copies of the target sequences can be readily obtained by PCR if the sequences flanking the target are known. PCR is carried out by adding the following components to a solution containing the target sequence: (1) a pair of primers that hybridize with the flanking sequences of the target, (2) all four deoxyribonucleoside triphosphates (dNTPs), and (3) a heat-stable DNA polymerase. A PCR cycle consists of three steps (Figure 5.6):

1. *Strand separation.* The two strands of the parent DNA molecule are separated by heating the solution to 95°C for 15 s.

2. *Hybridization of primers.* The solution is then abruptly cooled to 54°C to allow each primer to hybridize to a DNA strand. One primer hybridizes to the 3′ end of the target on one strand, and the other primer hybridizes to the 3′ end on the complementary target strand. Parent DNA duplexes do not form, because the primers are present in large excess. Primers are typically from 20 to 30 nucleotides long.

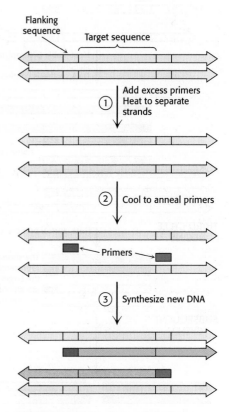

FIGURE 5.6 The first cycle in the polymerase chain reaction (PCR). A cycle consists of three steps: DNA double strand separation, the hybridization of primers, and the extension of primers by DNA synthesis.

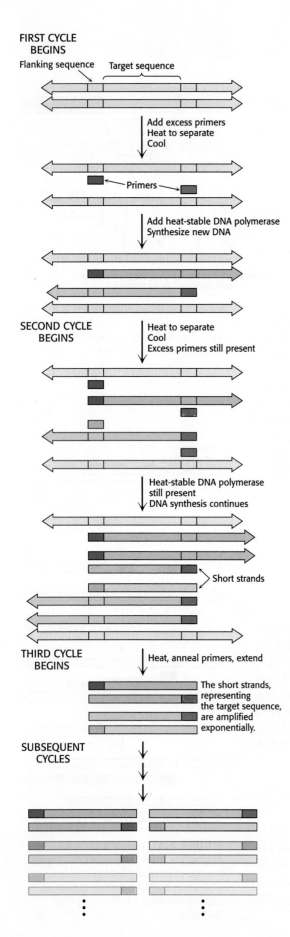

FIRST CYCLE BEGINS

Flanking sequence Target sequence

Add excess primers
Heat to separate
Cool

Primers

Add heat-stable DNA polymerase
Synthesize new DNA

SECOND CYCLE BEGINS

Heat to separate
Cool
Excess primers still present

Heat-stable DNA polymerase still present
DNA synthesis continues

Short strands

THIRD CYCLE BEGINS

Heat, anneal primers, extend

The short strands, representing the target sequence, are amplified exponentially.

SUBSEQUENT CYCLES

3. *DNA synthesis.* The solution is then heated to 72°C, the optimal temperature for heat-stable polymerases. One such enzyme is *Taq* DNA polymerase, which is derived from *Thermus aquaticus,* a thermophilic bacterium that lives in hot springs. The polymerase elongates both primers in the direction of the target sequence because DNA synthesis is in the 5′-to-3′ direction. DNA synthesis takes place on both strands but extends beyond the target sequence in this initial round.

These three steps—strand separation, hybridization of primers, and DNA synthesis—constitute one cycle of the PCR amplification and can be carried out repetitively just by changing the temperature of the reaction mixture. The thermostability of the polymerase makes it feasible to carry out PCR in a closed container; no reagents are added after the first cycle. At the completion of the second cycle, four duplexes containing the target sequence have been generated (Figure 5.7). Of the eight DNA strands comprising these duplexes, two short strands constitute only the target sequence—the sequence including and bounded by the primers. Subsequent cycles will amplify the target sequence exponentially. Ideally, after *n* cycles, the desired sequence is amplified 2^n-fold. The amplification is a millionfold after 20 cycles and a billionfold after 30 cycles, which can be carried out in less than an hour.

Several features of this remarkable method for amplifying DNA are noteworthy. First, the sequence of the target need not be known. All that is required is knowledge of the flanking sequences so that complementary primers can be synthesized. Second, the target can be much larger than the primers. Targets larger than 10 kb have been amplified by PCR. Third, primers do not have to be perfectly matched to flanking sequences to amplify targets. With the use of primers derived from a gene of known sequence, it is possible to search for variations on the theme. In this way, families of genes are being discovered by PCR. Fourth, PCR is highly specific because of the stringency of hybridization at relatively high temperature. *Stringency* is the required closeness of the match between primer and target, which can be controlled by temperature and salt. At high temperatures, only the DNA between hybridized primers is amplified. A gene constituting less than a millionth of the total DNA of a higher organism is accessible by PCR. Fifth, PCR is exquisitely sensitive. A single DNA molecule can be amplified and detected.

PCR is a powerful technique in medical diagnostics, forensics, and studies of molecular evolution

PCR can provide valuable diagnostic information in medicine. Bacteria and viruses can be readily detected with the use of specific primers. For example, PCR can reveal the presence of small amounts of DNA from the human immunodeficiency virus (HIV) in persons who have not yet mounted an immune response

FIGURE 5.7 Multiple cycles of the polymerase chain reaction. The two short double-stranded fragments produced at the end of the third cycle represent the target sequence. Subsequent cycles will amplify the target sequence exponentially and the parent sequence (yellow) arithmetically.

to this pathogen. In these patients, assays designed to detect antibodies against the virus would yield a false negative test result. Finding *Mycobacterium tuberculosis* bacilli in tissue specimens is slow and laborious. With PCR, as few as 10 tubercle bacilli per million human cells can be readily detected. PCR is a promising method for the early detection of certain cancers. This technique can identify mutations of certain growth-control genes, such as the *ras* genes (Chapter 14). The capacity to greatly amplify selected regions of DNA can also be highly informative in monitoring cancer chemotherapy. Tests using PCR can detect when cancerous cells have been eliminated and treatment can be stopped; they can also detect a relapse and the need to immediately resume treatment. PCR is ideal for detecting leukemias caused by chromosomal rearrangements.

In addition, PCR has made an impact on forensics and legal medicine. An individual DNA profile is highly distinctive because many genetic loci are highly variable within a population. For example, variations at one specific location determine a person's HLA type (human leukocyte antigen type; See Section 35.4 on 🍁 SaplingPlus); organ transplants are rejected when the HLA types of the donor and recipient are not sufficiently matched. PCR amplification of multiple genes is being used to establish biological parentage in disputed paternity and immigration cases. Analyses of blood stains and semen samples by PCR have implicated guilt or innocence in numerous assault and rape cases (Figure 5.8). The root of a single shed hair found at a crime scene contains enough DNA for typing by PCR.

DNA is a remarkably stable molecule, particularly when shielded from air, light, and water. Under such circumstances, large fragments of DNA can remain intact for thousands of years or longer. PCR provides an ideal method for amplifying such ancient DNA molecules so that they can be detected and characterized (Section 6.5). PCR can also be used to amplify DNA from microorganisms that have not yet been isolated and cultured. As will be discussed in Chapter 6, sequences from these PCR products can be sources of considerable insight into evolutionary relationships between organisms.

The tools for recombinant DNA technology have been used to identify disease-causing mutations

Let us consider how the techniques just described have been utilized in concert to study ALS, introduced at the beginning of this chapter. Five percent of all patients suffering from ALS have family members who also have been diagnosed with the disease. A heritable disease pattern is indicative of a strong genetic component of disease causation. To identify these disease-causing genetic alterations, researchers identify *polymorphisms* (instances of genetic variation) within an affected family that correlate with the emergence of disease. These polymorphisms may themselves cause disease or be closely linked to the responsible genetic alteration. *Restriction-fragment-length polymorphisms* (RFLPs) are polymorphisms within restriction sites that change the sizes of DNA fragments produced by the appropriate restriction enzyme. Using restriction digests and Southern blots of the DNA from members of ALS-affected families, researchers identified RFLPs that were found preferentially in those family members diagnosed with the disease. For some of these families, strong evidence was obtained for the disease-causing mutation within a specific region of chromosome 21.

After the probable location of one disease-causing gene had been identified, this same research group compared the locations of the

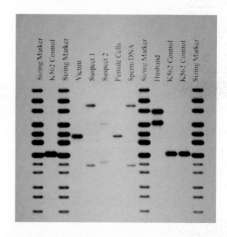

FIGURE 5.8 DNA and forensics. DNA isolated from sperm obtained during the examination of a rape victim was amplified by PCR, then compared with DNA from the victim and three potential suspects—the victim's husband and two additional individuals—using gel electrophoresis and autoradiography. Sperm DNA matched the pattern of Suspect 1, but not that of Suspect 2 or the victim's husband. Sizing marker and K562 lanes refer to control DNA samples. [Martin Shields/Science Source.]

ALS-associated RFLPs with the known sequence of chromosome 21. They noted that this chromosomal locus contains the *SOD1* gene, which encodes the Cu/Zn superoxide dismutase protein SOD1, an enzyme important for the protection of cells against oxidative damage (Section 18.3). PCR amplification of regions of the *SOD1* gene from the DNA of affected family members, followed by Sanger dideoxy sequencing of the targeted fragment, enabled the identification of 11 disease-causing mutations from 13 different families. This work was pivotal for focusing further inquiry into the roles that superoxide dismutase and its corresponding mutant forms play in the pathology of some forms of ALS.

5.2 Recombinant DNA Technology Has Revolutionized All Aspects of Biology

The development of recombinant DNA technology has taken biology from an exclusively analytical science to a synthetic one. New combinations of unrelated genes can be constructed in the laboratory by applying recombinant DNA techniques. These novel combinations can be cloned—amplified many-fold—by introducing them into suitable cells, where they are replicated by the DNA-synthesizing machinery of the host. The inserted genes are often transcribed and translated in their new setting. What is most striking is that the genetic endowment of the host can be permanently altered in a designed way.

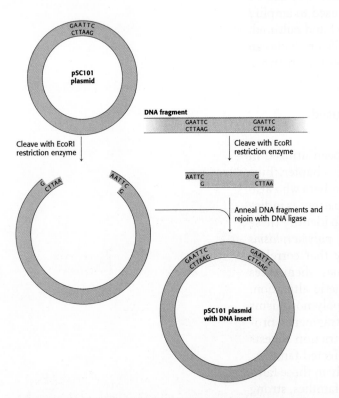

FIGURE 5.9 Joining of DNA molecules by the cohesive-end method. The pSC101 plasmid (blue) is cleaved once by EcoRI to yield a pair of cohesive ends. A fragment of DNA (red), also cleaved with EcoRI, can be ligated to the cut plasmid to form a new plasmid with the DNA fragment inserted.

Restriction enzymes and DNA ligase are key tools in forming recombinant DNA molecules

Let us begin by exploring how novel DNA molecules can be constructed in the laboratory. An essential tool for the manipulation of recombinant DNA is a *vector*, a DNA molecule that can replicate autonomously in an appropriate host organism. Vectors are designed to enable the rapid, covalent insertion of DNA fragments of interest. *Plasmids* (naturally occurring circles of DNA that act as accessory chromosomes in bacteria) and *bacteriophage lambda (λ phage)*, a virus, are choice vectors for cloning in *E. coli*. The vector can be prepared for accepting a new DNA fragment by cleaving it at a single specific site with a restriction enzyme. For example, the plasmid pSC101, a 9.9-kb double-helical circular DNA molecule, is split at a unique site by the EcoRI restriction enzyme. The staggered cuts made by this enzyme produce *complementary single-stranded ends*, which have specific affinity for each other and hence are known as *cohesive* or *sticky ends*. Any DNA fragment can be inserted into this plasmid if it has the same cohesive ends. Such a fragment can be extracted from a larger piece of DNA by using the same restriction enzyme as was used to open the plasmid DNA (Figure 5.9).

The single-stranded ends of the fragment are then complementary to those of the cut plasmid. The DNA fragment and the cut plasmid can be annealed and then joined by *DNA ligase*, which catalyzes the formation of a phosphodiester bond at a break in a DNA chain. DNA ligase requires a free 3′-hydroxyl group and a 5′-phosphoryl group. Furthermore, the chains joined by ligase must be in a double helix. An energy source such as ATP or NAD^+ is required for the joining reaction, as will be discussed in Chapter 29.

What if the target DNA is not naturally flanked by the appropriate restriction sites? How is the fragment cut and annealed to the vector? The cohesive-end method for joining DNA molecules can still be used in these cases by adding a *short, chemically synthesized DNA linker* that can be cleaved by restriction enzymes. First, the linker is covalently joined to the ends of a DNA fragment. For example, the 5′ ends of a decameric linker and a DNA molecule are phosphorylated by polynucleotide kinase and then joined by the ligase from T4 phage (Figure 5.10). This ligase can form a covalent bond between blunt-ended double-helical DNA molecules. Cohesive ends are produced when these terminal extensions are cut by an appropriate restriction enzyme. Thus, *cohesive ends corresponding to a particular restriction enzyme can be added to virtually any DNA molecule.* We see here the fruits of combining enzymatic and synthetic chemical approaches in crafting new DNA molecules.

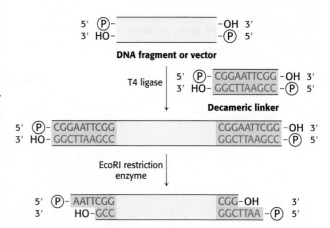

FIGURE 5.10 Formation of cohesive ends. Cohesive ends can be formed by the addition and cleavage of a chemically synthesized linker.

Plasmids and λ phage are choice vectors for DNA cloning in bacteria

Many plasmids and bacteriophages have been ingeniously modified by researchers to enhance the delivery of recombinant DNA molecules into bacteria and to facilitate the genetic selection of bacteria harboring these vectors. As already mentioned, plasmids are circular double-stranded DNA molecules that occur naturally in some bacteria. They range in size from two to several hundred kilobases. Plasmids carry genes for the inactivation of antibiotics, the production of toxins, and the breakdown of natural products. These accessory chromosomes can replicate independently of the host chromosome. In contrast with the host genome, they are dispensable under certain conditions. A bacterial cell may have no plasmids at all or it may house as many as 20 copies of a naturally-occurring plasmid.

Many plasmids have been optimized for a particular experimental task. Some engineered plasmids, for example, can achieve nearly a thousand copies per bacterial cell. One class of plasmids, known as *cloning vectors*, is particularly suitable for the facile insertion and replication of a collection of DNA fragments. These vectors often feature a *polylinker* region that includes many unique restriction sites within its sequence. This polylinker can be cleaved with a variety of restriction enzymes or combinations of enzymes, providing great versatility in the DNA fragments that can be inserted. In addition, these plasmids contain *reporter genes*, which encode easily detectable markers such as antibiotic-resistance enzymes or fluorescent proteins. Creative placement of these reporter genes within these plasmids enables the rapid identification of those vectors that harbor the desired DNA insert.

For example, let us consider the cloning vector pUC18 (Figure 5.11). This plasmid contains three essential components: an origin of replication, so that the plasmid can be replicated when the host bacterium

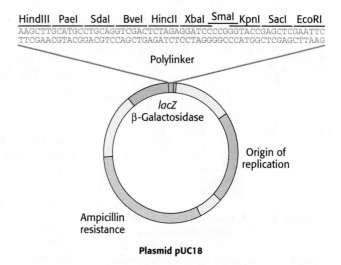

FIGURE 5.11 A polylinker in the plasmid pUC18. In addition to an origin of replication and a gene encoding ampicillin resistance, the plasmid pUC18 includes a polylinker within an essential fragment of the β-galactosidase gene (often called the *lacZ* gene). Insertion of a DNA fragment into one of the many restriction sites within this polylinker can be detected by the absence of β-galactosidase activity.

FIGURE 5.12 Insertional inactivation. Successful insertion of DNA fragments into the polylinker region of pUC18 will result in the disruption of the β-galactosidase gene. Bacterial colonies that harbor such plasmids will no longer convert X-gal into a colored product, and will appear white on the plate.

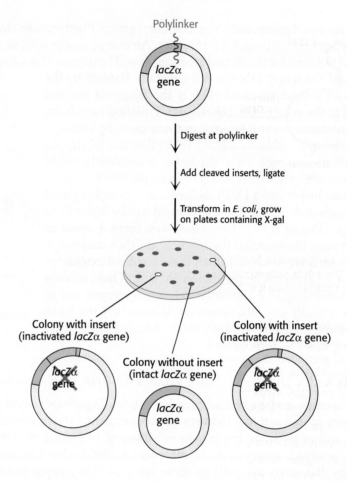

divides; a gene encoding ampicillin resistance, which enables the selection of bacteria that harbor the plasmid; and the *lacZα* gene, which encodes an essential fragment of the enzyme *β-galactosidase*, an enzyme which naturally cleaves the milk sugar lactose (Section 11.2). Insertion of a DNA fragment into the polylinker region disrupts the *lacZα* gene, an effect called *insertional inactivation*. β-Galactosidase also cleaves the synthetic substrate X-gal, releasing a blue dye. Bacterial cells containing a DNA insert at the polylinker will no longer produce the dye in the presence of X-gal, and are readily identified by their white color (Figure 5.12).

Another class of plasmids has been optimized for use as *expression vectors* for the production of large amounts of protein. Many different expression vectors are currently in use. Figure 5.13 depicts the typical elements of a vector designed to express a protein in prokaryotes. In addition to an antibiotic-resistance gene and an origin of replication, it contains promoter and terminator sequences designed to drive the transcription of large amounts of a protein-coding DNA sequence. As in the cloning vectors, these plasmids also contain a polylinker region, which often contain sequences flanking the cloning site that simplify the addition of fusion tags to the protein of interest (Section 3.1), greatly facilitating purification of the overexpressed protein. Many expression vectors also possess transcriptional repressor and operator sequences (Section 31.2), which enable the precise control of when production of the protein will be turned on.

Another widely used vector, λ *phage*, enjoys a choice of lifestyles: this bacteriophage can destroy its host or it can become part of its

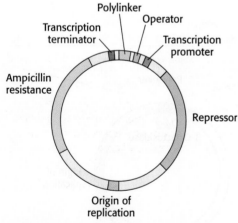

FIGURE 5.13 A prokaryotic expression vector. As with the cloning vector (Figure 5.11), expression vectors contain an origin of replication and an antibiotic-resistance gene. In addition, these plasmids include transcription initiation (promoter) and termination sequences, required for the expression of a protein-coding gene inserted in the polylinker. Many expression plasmids also contain repressor and operator sequences which allow for the control of the timing of protein production.

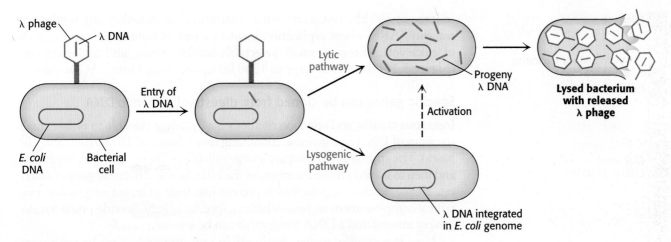

FIGURE 5.14 Alternative infection modes for λ phage. Lambda phage can multiply within a host and lyse it (lytic pathway) or its DNA can become integrated into the host genome (lysogenic pathway), where it is dormant until activated.

host (Figure 5.14). In the *lytic pathway,* viral functions are fully expressed: viral DNA and proteins are quickly produced and packaged into virus particles, leading to the lysis (destruction) of the host cell and the sudden appearance of about 100 progeny virus particles, or *virions.* In the *lysogenic pathway,* the phage DNA becomes inserted into the host-cell genome and can be replicated together with host-cell DNA for many generations, remaining inactive. Certain environmental changes can trigger the expression of this dormant viral DNA, which leads to the formation of progeny viruses and lysis of the host. Large segments of the 48-kb DNA of λ phage are not essential for productive infection and can be replaced by foreign DNA, thus making λ phage an ideal vector.

Mutant λ phages designed for cloning have been constructed. An especially useful one called λgt-λβ contains only two EcoRI cleavage sites instead of the five normally present (Figure 5.15). After cleavage, the middle segment of this λ DNA molecule can be removed. The two remaining pieces of DNA (called arms) have a combined length equal to 72% of a normal genome length. This amount of DNA is too little to be packaged into a λ particle, which can take up only DNA measuring from 78% to 105% of a normal genome. However, a suitably long DNA insert (such as 10 kb)

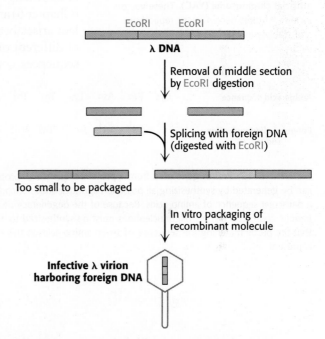

FIGURE 5.15 Mutant λ phage as a cloning vector. The packaging process selects DNA molecules that contain an insert (colored red). DNA molecules that have resealed without an insert are too small to be efficiently packaged.

between the two ends of λ DNA enables such a recombinant DNA molecule (93% of normal length) to be packaged. Nearly all infectious λ particles formed in this way will contain an inserted piece of foreign DNA. Another advantage of using these modified viruses as vectors is that they enter bacteria much more easily than do plasmids. Among the variety of λ mutants that have been constructed for use as cloning vectors, one of them, called a *cosmid,* is essentially a hybrid of λ phage and a plasmid that can serve as a vector for large DNA inserts (as large as 45 kb).

Bacterial and yeast artificial chromosomes

Much larger pieces of DNA can be propagated in *bacterial artificial chromosomes* (BACs) or *yeast artificial chromosomes* (YACs). BACs are highly engineered versions of the *E. coli* fertility (F) factor that can include inserts

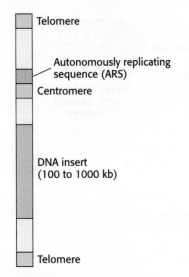

Telomere

Autonomously replicating
sequence (ARS)

Centromere

DNA insert
(100 to 1000 kb)

Telomere

FIGURE 5.16 Diagram of a yeast
artificial chromosome (YAC). These vectors
include features necessary for replication
and stability in yeast cells.

| Amino acid sequence | ... Cys Pro Asn Lys Trp Thr His ... |

Amino acid sequence   ... Cys Pro Asn Lys Trp Thr His ...

Potential oligonucleotide
sequences   TG$_T^C$ CC$_G^C$ AA$_T^C$ AA$_G^A$ TGG AC$_G^C$ CA$_T^C$
      (A above Asn and His; T below Pro and Thr)

FIGURE 5.17 Probes generated from a protein sequence. A probe
can be generated by synthesizing all possible oligonucleotides encoding
a particular sequence of amino acids. Because of the degeneracy of the
genetic code, 256 distinct oligonucleotides must be synthesized to ensure
that the probe matching the sequence of seven amino acids in this example
is present.

as large as 300 kb. YACs contain a centromere, an *autonomously replicating
sequence* (ARS, where replication begins), a pair of telomeres (normal ends
of eukaryotic chromosomes), selectable marker genes, and a cloning site
(Figure 5.16). Inserts as large as 1000 kb can be cloned into YAC vectors.

Specific genes can be cloned from digests of genomic DNA

Ingenious cloning and selection methods have made it possible to isolate small
stretches of DNA in a genome containing more than 3×10^6 kb (i.e., 3×10^9
bases). The approach is to prepare a large collection (*library*) of DNA fragments
and then to identify those members of the collection that have the gene of interest. Hence, to clone a gene that is present just once in an entire genome, two
critical components must be available: a specific oligonucleotide probe for the
gene of interest and a DNA library that can be screened rapidly.

How is a specific probe obtained? In one approach, a probe for a gene
can be prepared if a part of the amino acid sequence of the protein encoded
by the gene is known. Peptide sequencing of a purified protein (Chapter 3)
or knowledge of the sequence of a homologous protein from a related species
(Chapter 6) are two potential sources of such information. However, a problem arises because a single peptide sequence can be encoded by a number
of different oligonucleotides (Figure 5.17). Thus, for this purpose, peptide
sequences containing tryptophan and methionine are preferred, because
these amino acids are specified by a single codon,
whereas other amino acid residues have between two
and six codons (Table 4.5). All the possible DNA
sequences (or their complements) that encode the
targeted peptide sequence are synthesized by the
solid-phase method and made radioactive by phosphorylating their 5′ ends with ^{32}P.

To prepare the DNA library, a sample containing
many copies of total genomic DNA is first mechani-
cally sheared or partly digested by restriction enzymes
into large fragments (Figure 5.18). This process yields

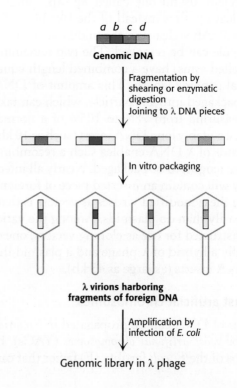

a b c d

Genomic DNA

Fragmentation by
shearing or enzymatic
digestion
Joining to λ DNA pieces

In vitro packaging

**λ virions harboring
fragments of foreign DNA**

Amplification by
infection of *E. coli*

Genomic library in λ phage

FIGURE 5.18 Creation of a genomic
library. A genomic library can be created
from a digest of a whole complex genome.
After fragmentation of the genomic DNA
into overlapping segments, the DNA is
inserted into the λ phage vector (shown
in yellow). Packaging into virions and
amplification by infection in *E. coli* yields a
genomic library.

a nearly random population of overlapping DNA fragments. These fragments are then separated by gel electrophoresis to isolate the set of all fragments that are about 15 kb long. Synthetic linkers are attached to the ends of these fragments, cohesive ends are formed, and the fragments are then inserted into a vector, such as λ phage DNA, prepared with the same cohesive ends. *E. coli* bacteria are then infected with these recombinant phages. These phages replicate themselves and then lyse their bacterial hosts. The resulting lysate contains fragments of human DNA housed in a sufficiently large number of virus particles to ensure that nearly the entire genome is represented. These phages constitute a *genomic library*. Phages can be propagated indefinitely such that the library can be used repeatedly over long periods.

This genomic library is then screened to find the very small number of phages harboring the gene of interest. For the human genome, a calculation shows that a 99% probability of success requires screening about 500,000 clones; hence, a very rapid and efficient screening process is essential. Rapid screening can be accomplished by DNA hybridization.

A dilute suspension of the recombinant phages is first plated on a lawn of bacteria (Figure 5.19). Where each phage particle has landed and infected a bacterium, a *plaque* containing identical phages develops on the plate. A replica of this master plate is then made by applying a sheet of nitrocellulose. Infected bacteria and phage DNA released from lysed cells adhere to the sheet in a pattern of spots corresponding to the plaques. Intact bacteria on this sheet are lysed with sodium hydroxide, which also serves to denature the DNA so that it becomes accessible for hybridization with a ^{32}P-labeled probe. The presence of a specific DNA sequence in a single spot on the replica can be detected by using a radioactive complementary DNA or RNA molecule as a probe. Autoradiography then reveals the positions of spots harboring recombinant DNA. The corresponding plaques are picked out of the intact master plate and grown. A single investigator can readily screen a million clones in a day. This method makes it possible to isolate virtually any gene, provided that a probe is available.

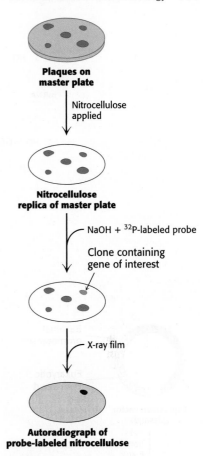

**Plaques on
master plate**

Nitrocellulose
applied

**Nitrocellulose
replica of master plate**

NaOH + ^{32}P-labeled probe

Clone containing
gene of interest

X-ray film

**Autoradiograph of
probe-labeled nitrocellulose**

FIGURE 5.19 Screening a genomic library for a specific gene. Here, a plate is tested for plaques containing gene *a* of Figure 5.18.

Complementary DNA prepared from mRNA can be expressed in host cells

The preparation of eukaryotic DNA libraries presents unique challenges, especially if the researcher is interested primarily in the protein-coding region of a particular gene. Recall that most mammalian genes are mosaics of introns and exons. These interrupted genes cannot be expressed by bacteria, which lack the machinery to splice introns out of the primary transcript. However, this difficulty can be circumvented by causing bacteria to take up recombinant DNA that is complementary to mRNA, where the intronic sequences have been removed.

The key to forming *complementary DNA* is the enzyme *reverse transcriptase*. As discussed in Section 4.4, a retrovirus uses this enzyme to form a DNA–RNA hybrid in replicating its genomic RNA. Reverse transcriptase synthesizes a DNA strand complementary to an RNA template if the transcriptase is provided with a DNA primer that is base-paired to the RNA and contains a free 3′-OH group. We can use a simple sequence of linked thymidine [oligo(T)] residues as the primer. This oligo(T) sequence pairs with the poly(A) sequence at the 3′ end of most eukaryotic mRNA molecules (Section 4.5), as shown in Figure 5.20. The reverse transcriptase then synthesizes the rest of the cDNA strand in the presence of the four deoxyribonucleoside triphosphates (step 1). The template RNA strand of this RNA–DNA hybrid is subsequently hydrolyzed by raising the pH (step 2).

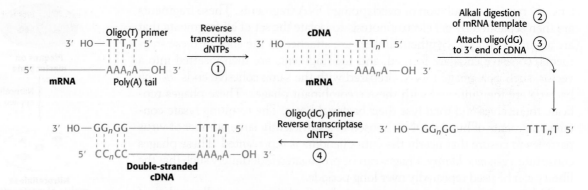

FIGURE 5.20 Formation of a cDNA duplex. A complementary DNA (cDNA) duplex is created from mRNA by (1) use of reverse transcriptase to synthesize a cDNA strand, (2) digestion of the original RNA strand, (3) addition of several G bases to the DNA by terminal transferase, and (4) synthesis of a complementary DNA strand using the newly synthesized cDNA strand as a template.

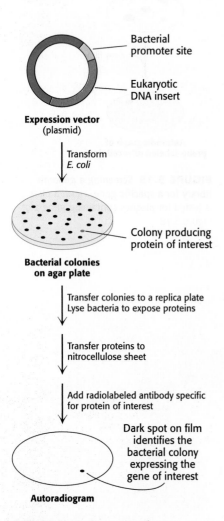

FIGURE 5.21 Screening of cDNA clones. A method of screening for cDNA clones is to identify expressed products by staining with specific antibody.

Unlike RNA, DNA is resistant to alkaline hydrolysis. The single-stranded DNA is converted into double-stranded DNA by creating another primer site. The enzyme *terminal transferase* adds nucleotides—for instance, several residues of dG—to the 3′ end of DNA (step 3). Oligo(dC) can bind to dG residues and prime the synthesis of the second DNA strand (step 4). Synthetic linkers can be added to this double-helical DNA for ligation to a suitable vector. Complementary DNA for all mRNA that a cell contains can be made, inserted into vectors, and then inserted into bacteria. Such a collection is called a *cDNA library*.

Complementary DNA molecules can be inserted into expression vectors to enable the production of the corresponding protein of interest. Clones of cDNA can be screened on the basis of their capacity to direct the synthesis of a foreign protein in bacteria, a technique referred to as *expression cloning*. A labeled antibody specific for the protein of interest can be used to identify colonies of bacteria that express the corresponding protein product (Figure 5.21). As described earlier, spots of bacteria on a replica plate are lysed to release proteins, which bind to an applied nitrocellulose filter. With the addition of labeled antibody specific for the protein of interest, the location of the desired colonies on the master plate can be readily identified. This immunochemical screening approach can be used whenever a protein is expressed and corresponding antibody is available.

Complementary DNA has many applications beyond the generation of genetic libraries. The overproduction and purification of most eukaryotic proteins in prokaryotic cells necessitates the insertion of cDNA into plasmid vectors. For example, proinsulin, a precursor of insulin, is synthesized by bacteria-harboring plasmids that contain DNA complementary to mRNA for proinsulin (Figure 5.22). Indeed, bacteria produce much of the insulin used today by millions of diabetics.

Proteins with new functions can be created through directed changes in DNA

Much has been learned about genes and proteins by analyzing the effects that mutations have on their structure and function. In the classic genetic approach, mutations are generated randomly throughout the genome of a host organism, and those individuals exhibiting a phenotype of interest are

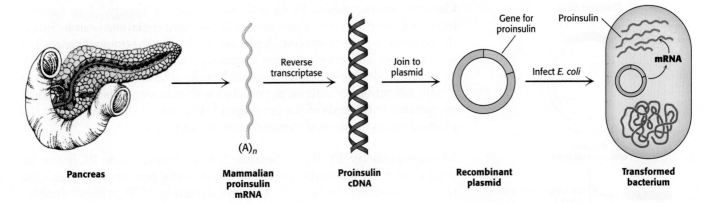

Pancreas → **Mammalian proinsulin mRNA** (A)$_n$ → Reverse transcriptase → **Proinsulin cDNA** → Join to plasmid → **Recombinant plasmid** (Gene for proinsulin) → Infect *E. coli* → **Transformed bacterium** (Proinsulin, mRNA)

FIGURE 5.22 Synthesis of proinsulin by bacteria. Proinsulin, a precursor of insulin, can be synthesized by transformed (genetically altered) clones of *E. coli*. The clones contain the mammalian proinsulin gene.

selected. Analysis of these mutants then reveals which genes are altered, and DNA sequencing identifies the precise nature of the changes. *Recombinant DNA technology makes the creation of specific mutations feasible in vitro.* We can construct new genes with designed properties by making three types of directed changes: *deletions, insertions,* and *substitutions.* A variety of methods can be used to introduce these mutations, including the following examples.

Site-directed mutagenesis. Mutant proteins with single amino acid substitutions can be readily produced by *site-directed mutagenesis* (Figure 5.23). Suppose that we want to replace a particular serine residue with cysteine. This mutation can be made if (1) we have a plasmid containing the gene or cDNA for the protein and (2) we know the base sequence around the site to be altered. If the serine of interest is encoded by TCT, mutation of the central base from C to G yields the TGT codon, which encodes cysteine. This type of mutation is called a *point mutation* because only one base is altered. To introduce this mutation into our plasmid, we prepare an oligonucleotide primer that is complementary to this region of the gene except that it contains TGT instead of TCT. The two strands of the plasmid are separated, and the primer is then annealed to the complementary strand. The mismatch of 1 of 15 base pairs is tolerable if the annealing is carried out at an appropriate temperature. After annealing to the complementary strand, the primer is elongated by DNA polymerase, and the double-stranded circle is closed by adding DNA ligase. Subsequent replication of this duplex yields two kinds of progeny plasmid, half with the original TCT sequence and half with the mutant TGT sequence. Expression of the plasmid containing the new TGT sequence will produce a protein with the desired substitution of cysteine for serine at a unique site. We will encounter many examples of the use of site-directed mutagenesis to precisely alter regulatory regions of genes and to produce proteins with tailor-made features.

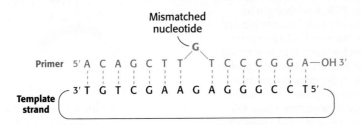

Mismatched nucleotide
G
Primer 5′ A C A G C T T ⌐ T C C C G G A—OH 3′
3′ T G T C G A A G A G G G C C T 5′
Template strand

FIGURE 5.23 Oligonucleotide-directed mutagenesis. A primer containing a mismatched nucleotide is used to produce a desired change in the DNA sequence.

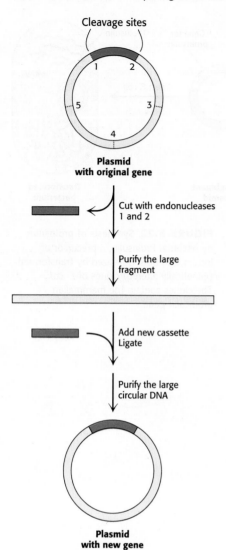

Cleavage sites

**Plasmid
with original gene**

Cut with endonucleases
1 and 2

Purify the large
fragment

Add new cassette
Ligate

Purify the large
circular DNA

**Plasmid
with new gene**

FIGURE 5.24 Cassette mutagenesis.
DNA is cleaved at a pair of unique
restriction sites by two different restriction
endonucleases. A synthetic oligonucleotide
with ends that are complementary to
these sites (the cassette) is then ligated to
the cleaved DNA. The method is highly
versatile because the inserted DNA can
have any desired sequence.

Cassette mutagenesis. In *cassette mutagenesis*, a variety of mutations, including insertions, deletions, and multiple point mutations, can be introduced into the gene of interest. A plasmid harboring the original gene is cut with a pair of restriction enzymes to remove a short segment (Figure 5.24). A synthetic double-stranded oligonucleotide—the *cassette*—carrying the genetic alterations of interest is prepared with cohesive ends that are complementary to the ends of the cut plasmid. Ligation of the cassette into the plasmid yields the desired mutated gene product.

Mutagenesis by PCR. In Section 5.1, we learned how PCR can be used to amplify a specific region of DNA using primers that flank the region of interest. In fact, the creative design of PCR primers enables the introduction of specific insertions, deletions, and substitutions into the amplified sequence. A number of methods have been developed for this purpose. Here, we shall consider one: *inverse PCR* to introduce deletions into plasmid DNA (Figure 5.25). In this approach, primers are designed to flank the sequence to be deleted. However, these primers are oriented in the opposite direction, such that they direct the amplification of the entire plasmid, minus the region to be deleted. If each of the primers contains a 5′ phosphate group, the amplified product can be recircularized with DNA ligase, yielding the desired deletion mutation.

Designer genes. Novel proteins can also be created by splicing together gene segments that encode domains that are not associated in nature. For example, a gene for an antibody can be joined to a gene encoding a toxic protein, yielding a chimeric protein that kills those cells recognized by the antibody. These *immunotoxins* are being evaluated as anticancer agents. Furthermore, noninfectious coat proteins of viruses can be produced in

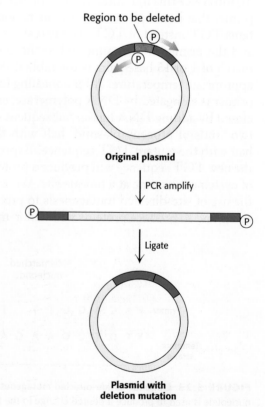

Region to be deleted

Original plasmid

PCR amplify

Ligate

**Plasmid with
deletion mutation**

**FIGURE 5.25 Deletion mutagenesis
by inverse PCR.** A deletion can be
introduced into a plasmid with primers
that flank this region but are oriented
away from the segment to be removed.
PCR amplification yields a linear product
that contains the entire plasmid minus
the unwanted sequence. If the primers
contained a 5′ phosphate group, this
product can be recircularized using DNA
ligase, generating a plasmid with the
desired mutation.

large amounts by recombinant DNA methods. They can serve as *synthetic vaccines* that are safer than conventional vaccines prepared by inactivating pathogenic viruses. A subunit of the hepatitis B virus produced in yeast is proving to be an effective vaccine against this debilitating viral disease. Finally, entirely new genes can be synthesized de novo by the solid-phase method described above. These genes can encode proteins with no known counterparts in nature.

Recombinant methods enable the exploration of the functional effects of disease-causing mutations

The application of recombinant DNA technology to the production of mutated proteins has had a significant effect in the study of ALS. Recall that genetic studies had identified a number of ALS-inducing mutations within the gene encoding Cu/Zn superoxide dismutase. As we shall learn in Section 18.3, SOD1 catalyzes the conversion of the superoxide radical anion into hydrogen peroxide, which, in turn, is converted into molecular oxygen and water by catalase. To study the potential effect of ALS-causing mutations on SOD1 structure and function, the *SOD1* gene was isolated from a human cDNA library by PCR amplification. The amplified fragments containing the gene were then digested by an appropriate restriction enzyme and inserted into a similarly digested plasmid vector. Mutations corresponding to those observed in ALS patients were introduced into these plasmids by site-directed mutagenesis and the protein products were expressed and assayed for their catalytic activity. Surprisingly, these mutations did not significantly alter the enzymatic activity of the corresponding recombinant proteins. These observations have led to the prevailing notion that these mutations impart toxic properties to SOD1. Although the nature of this toxicity is not yet completely understood, one hypothesis is that mutant SOD1 is prone to form toxic aggregates in the cytoplasm of neuronal cells.

5.3 Complete Genomes Have Been Sequenced and Analyzed

The methods just described are extremely effective for the isolation and characterization of fragments of DNA. However, the genomes of organisms ranging from viruses to human beings contain considerably longer sequences, arranged in very specific ways crucial for their integrated functions. Is it possible to sequence complete genomes and analyze them? For small genomes, this sequencing was accomplished soon after DNA-sequencing methods were developed. Sanger and his coworkers determined the complete sequence of the 5386 bases in the genome of the ϕX174 DNA virus in 1977, just a quarter century after Sanger's pioneering elucidation of the amino acid sequence of a protein. This tour de force was followed several years later by the determination of the sequence of human mitochondrial DNA, a double-stranded circular DNA molecule containing 16,569 base pairs. It encodes 2 ribosomal RNAs, 22 transfer RNAs, and 13 proteins. Many other viral genomes were sequenced in subsequent years. However, the genomes of free-living organisms presented a great challenge because even the simplest comprises more than 1 million base pairs. Thus, sequencing projects require both rapid sequencing techniques and efficient methods for assembling many short stretches of 300 to 500 base pairs into a complete sequence.

The genomes of organisms ranging from bacteria to multicellular eukaryotes have been sequenced

With the development of automatic DNA sequencers based on fluorescent dideoxynucleotide chain terminators, high-volume, rapid DNA sequencing became a reality. The genome sequence of the bacterium *Haemophilus influenzae* was determined in 1995 using a "shotgun" approach. The genomic DNA was sheared randomly into fragments that were then sequenced. Computer programs assembled the complete sequence by matching up overlapping regions between fragments. The *H. influenzae* genome comprises 1,830,137 base pairs and encodes approximately 1740 proteins (Figure 5.26). Using similar approaches, as well as more advanced methods described below, investigators have determined the sequences of more than 10,000 bacterial and archaeal species, including key model organisms such as *E. coli, Salmonella typhimurium,* and *Archaeoglobus fulgidus,* as well as pathogenic organisms such as *Yersinia pestis* (bubonic plague) and *Bacillus anthracis* (anthrax).

The first eukaryotic genome to be completely sequenced was that of baker's yeast, *Saccharomyces cerevisiae,* in 1996. This yeast genome comprises approximately 12 million base pairs, distributed on 16 chromosomes, and encodes more than 6000 proteins. This achievement was followed in 1998 by the first complete sequencing of the genome of a multicellular organism, the nematode *Caenorhabditis elegans,* which contains 97 million base pairs. This genome includes more than 19,000 genes. The genomes of many additional organisms widely used in biological and biomedical research have now been sequenced, including those of the fruit fly

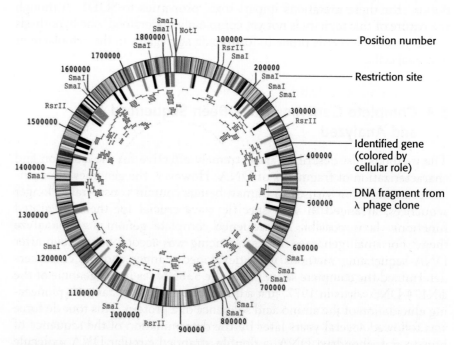

FIGURE 5.26 A complete genome. The diagram depicts the genome of *Haemophilus influenzae,* the first complete genome of a free-living organism to be sequenced. The genome encodes more than 1700 proteins (indicated by the colored bars in the outer circle) and 70 RNA molecules. The likely function of approximately one-half of the proteins was determined by comparisons with sequences of proteins already characterized in other species. Gaps in the final sequence were filled by sequencing genomic DNA inserts from 300 λ phage clones (indicated by the blue bars in the inner circle), many of these overlap one another. [Data from R. D. Fleischmann et al., *Science* 269:496–512, 1995; scan courtesy of The Institute for Genomic Research.]

Drosophila melanogaster, the model plant *Arabidopsis thaliana,* the mouse, the rat, and the dog. Note that even after a genome sequence has been considered complete, some sections, such as the repetitive sequences that make up heterochromatin, may be missing because these DNA sequences are very difficult to manipulate with the use of standard techniques.

The sequence of the human genome has been completed

The ultimate goal of much of genomics research has been the sequencing and analysis of the human genome. Given that the human genome comprises approximately 3 billion base pairs of DNA distributed among 24 chromosomes (Figure 5.27), the challenge of producing a complete sequence was daunting. However, through an organized international effort of academic laboratories and private companies, the human genome progressed from a draft sequence first reported in 2001 to a finished sequence reported in late 2004.

The human genome is a rich source of information about many aspects of humanity, including biochemistry and evolution. Analysis of the genome will continue for many years to come. Developing an inventory of protein-encoding genes is one of the first tasks. At the beginning of the genome-sequencing project, the number of such genes was estimated to be approximately 100,000. With the availability of the completed (but not finished) genome, this estimate was reduced to between 30,000 and 35,000. With the finished sequence, the estimate fell to between 20,000 and 25,000. We will use the estimate of 23,000 throughout this book. The reduction in this estimate is due, in part, to the realization that there are a large number of *pseudogenes,* formerly functional genes that have accumulated mutations such that they no longer produce proteins. For example, more than half of the genomic regions that correspond to olfactory receptors—key molecules responsible for our sense of smell—are pseudogenes (See Section 34.1 on Sapling Plus). The corresponding regions in the genomes of other primates and rodents encode functional olfactory

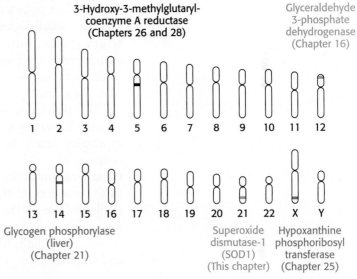

FIGURE 5.27 The human genome. The human genome is arrayed on 46 chromosomes—22 pairs of autosomes and the X and Y sex chromosomes. The locations of several genes associated with important pathways in biochemistry are highlighted.

receptors. Nonetheless, the surprisingly small number of genes belies the complexity of the human proteome. *Many genes encode more than one protein through mechanisms such as alternative splicing of mRNA and posttranslational modifications of proteins.* The different proteins encoded by a single gene often display important variations in functional properties.

The human genome contains a large amount of DNA that does not encode proteins. A great challenge in modern biochemistry and genetics is to elucidate the roles of this noncoding DNA. Much of this DNA is present because of the existence of *mobile genetic elements.* These elements, related to retroviruses (Section 4.4), have inserted themselves throughout the genome over time. Most of these elements have accumulated mutations and are no longer functional. For example, more than 1 million *Alu sequences,* each approximately 300 bases in length, are present in the human genome. *Alu* sequences are examples of *SINES,* short interspersed elements. The human genome also includes nearly 1 million *LINES, long interspersed elements,* DNA sequences that can be as long as 10 kilobases (kb). The roles of these elements as neutral genetic parasites or instruments of genome evolution are under current investigation.

Next-generation sequencing methods enable the rapid determination of a complete genome sequence

Since the introduction of the Sanger dideoxy method in the mid-1970s, significant advances have been made in DNA-sequencing technologies, enabling the readout of progressively longer sequences with higher fidelity and shorter run times. The development of *next-generation sequencing* (NGS) platforms has extended this capability to formerly unforeseen levels. By combining technological breakthroughs in the handling of very small amounts of liquid, high-resolution optics, and computing power, these methods have already made a significant impact on the ability to obtain whole genome sequences rapidly and cheaply (Chapter 1).

Next-generation sequencing refers to a family of technologies, each of which utilizes a unique approach for the determination of a DNA sequence. All of these methods are *highly parallel*: from 1 million to 1 billion DNA fragment sequences are acquired in a single experiment. How are NGS methods capable of attaining such a high number of parallel runs? Individual DNA fragments are amplified by PCR on a solid support—a single bead or a small region of a glass slide—such that clusters of identical DNA fragments are distinguishable by high-resolution imaging. These fragments then serve as templates for DNA polymerase, where the addition of nucleotide triphosphates is converted to a signal that can be detected in a highly sensitive manner.

The technique used to detect individual base incorporation varies among the variety of NGS methods. Let us consider one of these methods in greater detail. In *pyrosequencing*, a single-stranded fragment of DNA tethered to a solid support serves as the sequencing template. A solution containing one of the four deoxyribonucleoside triphosphates is then added to the reaction. If the nucleotide base is not complementary to the next available base on the template strand, no reaction occurs. However, if the nucleotide is incorporated, the released pyrophosphate is coupled to the production of light by the sequential action of the enzymes *ATP sulfurylase* and *luciferase*:

$$PP_i + \text{adenylylsulfate} \xrightleftharpoons{\text{ATP sulfurylase}} ATP + \text{sulfate}$$

$$ATP + \text{luciferin} \xrightleftharpoons{\text{luciferase}} \text{oxyluciferin} + \textbf{light}$$

The emitted light is measured, indicating that a reaction has occurred, thus revealing the identity of the base at that position (Figure 5.28A). If the template contains a run of one particular base, then multiple consecutive additions will take place in a single reaction; the length of this run can be determined by the intensity of the emitted light (Figure 5.28B). At the end of the reaction, the remaining deoxynucleotide and ATP are removed by the enzyme *apyrase*.

$$dNTP \xrightleftharpoons{\text{apyrase}} dNMP + 2P_i$$

$$ATP \xrightleftharpoons{\text{apyrase}} AMP + 2P_i$$

Then, the next nucleotide is added, and the steps are repeated. Nucleotides are added in a defined order, referred to as the *dispensation sequence*, and the pattern of peaks that emerge reveals the sequence of the template strand (Figure 5.28B).

Other sequencing methods are also commonly used. Most can be understood simply by considering the overall reaction of chain elongation catalyzed

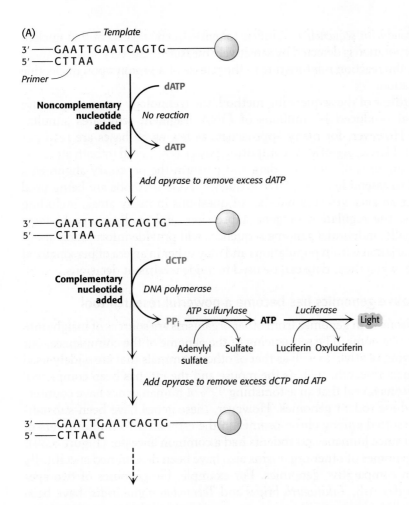

(A)

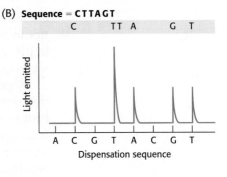

(B)

FIGURE 5.28 Pyrosequencing. (A) The template strand is attached to a solid support and an appropriate primer is added. A single deoxynucleotide is added: if this base is not complementary to the next base in the template strand, then no reaction occurs. However, if the added deoxynucleotide is complementary, pyrophosphate (PP$_i$) is released and coupled to the production of light by the enzymes ATP sulfurylase and luciferin. At the end of each step, apyrase is added to remove remaining nucleotide and ATP (if present), and then the steps are repeated with the next nucleotide. (B) Light is measured at each deoxynucleotide addition. A peak indicates the next base in the sequence and the height of the peak indicates how many of the same deoxynucleotide were incorporated in a row.

by DNA polymerase (Figure 5.29). In the *reversible terminator method*, the four nucleotides are added to the template DNA, with each base tagged with a unique fluorescent label and a reversibly-blocked 3′ end. The blocked end assures that only one phosphodiester linkage will form. Once the nucleotide is incorporated into the growing strand, it is identified by its fluorescent tag, the blocking agent is removed, and the process is repeated. The protocol for

FIGURE 5.29 Detection methods in next-generation sequencing. Measurement of base incorporation in next-generation sequencing methods relies on the detection of the various products of the DNA polymerase reaction. Reversible terminator sequencing measures the nucleotide incorporation in a manner similar to Sanger sequencing, while pyrosequencing and ion semiconductor sequencing detect the release of pyrophosphate and protons, respectively.

ion semiconductor sequencing is similar to pyrosequencing except that nucleotide incorporation is detected by sensitively measuring the very small changes in pH of the reaction mixture due to the release of a proton upon nucleotide incorporation.

Regardless of the sequencing method, the technology exists to quantify the signal produced by millions of DNA fragment templates simultaneously. However, for many approaches, as few as 50 bases are read per fragment. Hence, significant computing power is required to both store the massive amounts of sequence data and perform the necessary alignments required to assemble a completed sequence. NGS methods are being used to answer an ever-growing number of questions in many areas, including genomics, the regulation of gene expression, and evolutionary biology. Additionally, individual genome sequences will provide information about genetic variation within populations and may usher in an era of personalized medicine, when these data can be used to guide treatment decisions.

Comparative genomics has become a powerful research tool

Comparisons with genomes from other organisms are sources of insight into the human genome. The sequencing of the genome of the chimpanzee, our closest living relative, as well as that of other mammals that are widely used in biological research, such as the mouse and the rat, has been completed. Comparisons reveal that an astonishing 99% of human genes have counterparts in these rodent genomes. However, these genes have been substantially reassorted among chromosomes in the estimated 75 million years of evolution since humans and rodents had a common ancestor (Figure 5.30).

The genomes of other organisms also have been determined specifically for use in comparative genomics. For example, the genomes of two species of puffer fish, *Takifuguru bripes* and *Tetraodon nigroviridis*, have been

A puffer fish. [Beth Swanson/Shutterstock.]

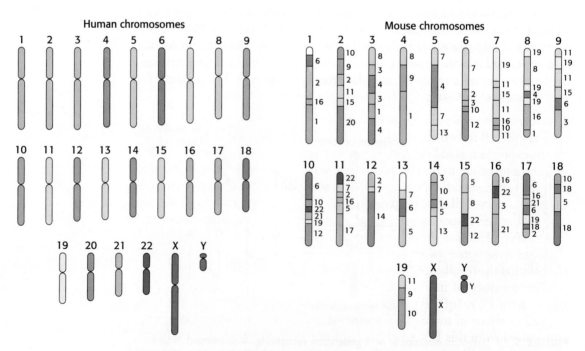

FIGURE 5.30 Genome comparison. A schematic comparison of the human genome and the mouse genome shows reassortment of large chromosomal fragments. The small numbers to the right of the mouse chromosomes indicate the human chromosome to which each region is most closely related.

determined. These genomes were selected because they are very small and lack much of the intergenic DNA present in such abundance in the human genome. The puffer fish genomes include fewer than 400 megabase pairs (Mbp), one-eighth of the number in the human genome, yet the puffer fish and human genomes contain essentially the same number of genes. Comparison of the genomes of these species with that of humans revealed more than 1000 formerly unrecognized human genes. Furthermore, comparison of the two species of puffer fish, which had a common ancestor approximately 25 million years ago, is a source of insight into more-recent events in evolution. Comparative genomics is a powerful tool, both for interpreting the human genome and for understanding major events in the origin of genera and species (Chapter 6).

5.4 Eukaryotic Genes Can Be Quantitated and Manipulated with Considerable Precision

After a gene of interest has been identified, cloned, and sequenced, it is often desirable to understand how that gene and its corresponding protein product function in the context of a whole cell or organism. It is now possible to determine how the expression of a particular gene is regulated, how mutations in the gene affect the function of the corresponding protein product, and how the behavior of an entire cell or model organism is altered by the introduction of mutations within specific genes. Levels of transcription of large families of genes within cells and tissues can be readily quantitated and compared across a range of environmental conditions. Eukaryotic genes can be introduced into bacteria, and the bacteria can be used as factories to produce a desired protein product. DNA can also be introduced into the cells of higher organisms. Genes introduced into animals are valuable tools for examining gene action, and they are the basis of gene therapy. Genes introduced into plants can make the plants resistant to pests, hardy in harsh conditions, or capable of synthesizing greater quantities of essential nutrients. The manipulation of eukaryotic genes holds much promise as a source of medical and agricultural benefits.

Gene-expression levels can be comprehensively examined

Most genes are present in the same quantity in every cell—namely, one copy per haploid cell or two copies per diploid cell. However, the level at which a gene is expressed, as indicated by mRNA quantities, can vary widely, ranging from no expression to hundreds of mRNA copies per cell. Gene-expression patterns vary from cell type to cell type, distinguishing, for example, a muscle cell from a nerve cell. Even within the same cell, gene-expression levels may vary as the cell responds to changes in physiological circumstances. Note that mRNA levels sometimes correlate with the levels of proteins expressed, but this correlation does not always hold. Thus, care must be exercised when interpreting the results of mRNA levels alone.

The quantity of individual mRNA transcripts can be determined by *quantitative PCR* (qPCR), or real-time PCR. RNA is first isolated from the cell or tissue of interest. With the use of reverse transcriptase, cDNA is prepared from this RNA sample. In one qPCR approach, the transcript of interest is PCR amplified with the appropriate primers in the presence of the dye SYBR Green I, which fluoresces brightly when bound to double-stranded DNA. In the initial PCR cycles, not enough duplex is present to allow a detectable fluorescence signal. However, after repeated PCR

(A)

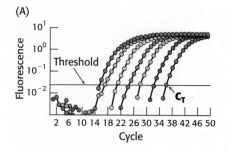

(B)

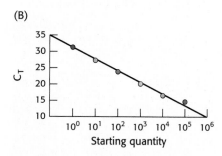

FIGURE 5.31 Quantitative PCR. (A) In qPCR, fluorescence is monitored in the course of PCR amplification to determine C_T, the cycle at which this signal exceeds a defined threshold. Each color represents a different starting quantity of DNA. (B) C_T values are inversely proportional to the number of copies of the original cDNA template. [Data from N. J. Walker, *Science* 296: 557–559, 2002.]

cycles, the fluorescence intensity exceeds the detection threshold and continues to rise as the number of duplexes corresponding to the transcript of interest increases (Figure 5.31). Importantly, the cycle number at which the fluorescence becomes detectable over a defined threshold (or C_T) is inversely proportional to the number of copies of the original template. After the relation between the original copy number and the C_T has been established with the use of a known standard, subsequent qPCR experiments can be used to determine the number of copies of any desired transcript in the original sample, provided the appropriate primers are available.

Although qPCR is a powerful technique for quantitation of a small number of transcripts in any given experiment, we can now use our knowledge of complete genome sequences to investigate an entire *transcriptome*, the pattern and level of expression of all genes in a particular cell or tissue. One of the most powerful methods for this purpose is based on hybridization. Single-stranded oligonucleotides whose sequences correspond to coding regions of the genome are affixed to a solid support such as a microscope slide, creating a *DNA microarray*. Importantly, the position of each sequence within the array is known. mRNA is isolated from the cells of interest (a tumor, for example) as well as a control sample (Figure 5.32). From this mRNA, cDNA is prepared (Section 5.2) in the presence of fluorescent nucleotides using different labels, usually green and red, for the two samples. The samples are combined, separated into single strands, and hybridized to the slide. The relative levels of green and red fluorescence at each spot indicate the differences in expression for each gene. DNA chips have been prepared such that thousands of transcript levels can be assessed in a single experiment. Hence, over several arrays, the differences

FIGURE 5.32 Using DNA microarrays to measure gene expression changes in a tumor. mRNA is isolated from two samples, tumor cells and a control sample. From these transcripts, cDNA is prepared in the presence of a fluorescent nucleotide, with a red label for the tumor sample and a green label for the control sample. The cDNA strands are separated, hybridized to the microarray, and the unbound DNA is washed away. Spots that are red indicate genes that are expressed more highly in the tumor, while the green spots indicate reduced expression relative to control. Spots that are black or yellow indicate comparable expression at either low or high levels, respectively. [Information from D. L. Nelson and M. M. Cox, *Lehninger Principles of Biochemistry,* 6th ed. (W. H. Freeman and Company, 2013).]

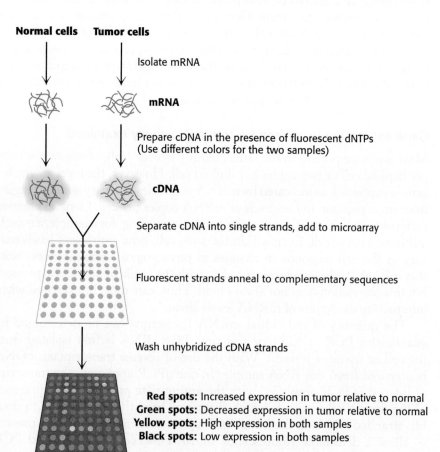

in expression of many genes across a number of different cell types or conditions can be measured (Figure 5.33).

Microarray analyses can be quite informative in the study of gene-expression changes in diseased organisms when compared with their healthy counterparts. As noted earlier, although ALS-causing mutations within the *SOD1* gene have been identified, the mechanism by which the mutant SOD1 protein ultimately leads to motor-neuron loss remains a mystery. Many research groups have used microarray analysis of neuronal cells isolated from humans and mice carrying *SOD1* gene mutations to search for clues into the pathways of disease progression and to suggest potential avenues for treatment. These studies have implicated a variety of biochemical pathways, including immunological activation, handling of oxidative stress, and protein degradation, in the cellular response to the mutant, toxic forms of SOD1.

New genes inserted into eukaryotic cells can be efficiently expressed

Bacteria are ideal hosts for the amplification of DNA molecules. They can also serve as factories for the production of a wide range of prokaryotic and eukaryotic proteins. However, bacteria lack the necessary enzymes to carry out posttranslational modifications such as the specific cleavage of polypeptides and the attachment of carbohydrate units. Thus, many eukaryotic genes can be expressed correctly only in eukaryotic host cells. The introduction of recombinant DNA molecules into cells of higher organisms can also be a source of insight into how their genes are organized and expressed. How are genes turned on and off in embryological development? How does a fertilized egg give rise to an organism with highly differentiated cells that are organized in space and time? These central questions of biology can now be fruitfully approached by expressing foreign genes in mammalian cells.

Recombinant DNA molecules can be introduced into animal cells in several ways. In one method, foreign DNA molecules precipitated by calcium phosphate are taken up by animal cells. A small fraction of the imported DNA becomes stably integrated into the chromosomal DNA. The efficiency of incorporation is low, but the method is useful because it is easy to apply. In another method, DNA is *microinjected* into cells. A fine-tipped glass micropipette containing a solution of foreign DNA is inserted into a nucleus (Figure 5.34). A skilled investigator can inject hundreds of cells per hour. About 2% of injected mouse cells are viable and contain the new gene. In a third method, *viruses* are used to introduce new genes into animal cells. The most effective vectors are *retroviruses*, whose genomes are encoded by RNA and replicate through DNA intermediates. A striking feature of the life cycle of a retrovirus is that the double-helical DNA form of its genome, produced by the action of reverse transcriptase, becomes randomly incorporated into host chromosomal DNA. This DNA version of the viral genome, called *proviral DNA*, can be efficiently expressed by the host cell and replicated along with normal cellular DNA. Retroviruses do not usually kill their hosts. Foreign genes have been efficiently introduced into mammalian cells by infecting them

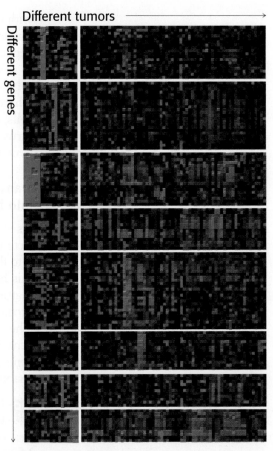

Different tumors ⟶

Different genes

FIGURE 5.33 Gene-expression analysis with the use of microarrays. The expression levels of thousands of genes can be simultaneously analyzed with DNA microarrays. Here, an analysis of 1733 genes in 84 breast tumor samples reveals that the tumors can be divided into distinct classes based on their gene-expression patterns. In this "heat map" representation, each row represents a different gene and each column represents a different breast tumor sample (i.e., a separate microarray experiment). Red corresponds to gene induction and green corresponds to gene repression. [Reprinted by permission from Macmillan Publishers Ltd: *Nature,* 406: 747. C. M. Perou et al., Molecular portraits of human breast tumours. ©2000.]

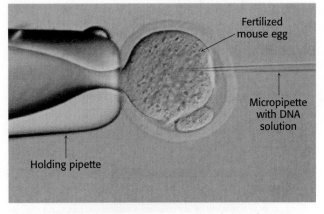

FIGURE 5.34 Microinjection of DNA. Cloned plasmid DNA is being microinjected into the male pronucleus of a fertilized mouse egg. [Marh et al, Hyperactive self-inactivating piggyBac for transposase-enhanced pronuclear microinjection transgenesis, *Proc. Natl. Acad. Sci. U.S.A.,* Vol. 109, no. 47, pp. 19184–19189. Copyright 2012 National Academy of Sciences.]

with vectors derived from the *Moloney murine leukemia virus,* a retrovirus which can accept inserts as long as 6 kb. Some genes introduced by this vector into the genome of a transformed host cell are efficiently expressed.

Two other viral vectors are extensively used. *Vaccinia virus,* a large DNA-containing virus, replicates in the cytoplasm of mammalian cells, where it shuts down host-cell protein synthesis. *Baculovirus* infects insect cells, which can be conveniently cultured. Insect larvae infected with this virus can serve as efficient protein factories. Vectors based on these large-genome viruses have been engineered to express DNA inserts efficiently.

Transgenic animals harbor and express genes introduced into their germ lines

As shown in Figure 5.34, plasmids harboring foreign genes can be microinjected into the male pronucleus of fertilized mouse eggs, which are then inserted into the uterus of a foster-mother mouse. A subset of the resulting embryos in this host will then harbor the foreign gene; these embryos may develop into mature animals. Southern blotting or PCR analysis of DNA isolated from the progeny can be used to determine which offspring carry the introduced gene. These *transgenic mice* are a powerful means of exploring the role of a specific gene in the development, growth, and behavior of an entire organism. Transgenic animals often serve as useful models for a particular disease process, enabling researchers to test the efficacy and safety of a newly developed therapy.

Let us return to our example of ALS. Research groups have generated transgenic mouse lines that express forms of human superoxide dismutase that harbor mutations matching those identified in earlier genetic analyses. Many of these strains exhibit a clinical picture similar to that observed in ALS patients: progressive weakness of voluntary muscles and eventual paralysis, motor-neuron loss, and rapid progression to death. Since their first characterization in 1994, these strains continue to serve as valuable sources of information for the exploration of the mechanism and potential treatment of ALS.

Gene disruption and genome editing provide clues to gene function and opportunities for new therapies

The function of a gene can also be probed by inactivating it and looking for resulting abnormalities. Powerful methods have been developed for accomplishing *gene disruption* (also called *gene knockout*) in organisms such as yeast and mice. These methods rely on the process of *homologous recombination* (Section 29.5), in which two DNA molecules with strong sequence similarity exchange segments. If a region of foreign DNA is flanked by sequences that have high homology to a particular region of genomic DNA, two recombination events will yield the transfer of the foreign DNA into the genome (Figure 5.35). In this manner, specific genes can be targeted if their flanking nucleotide sequences are known.

For example, the gene-knockout approach has been applied to the genes encoding gene-regulatory proteins (also called *transcription factors*) that control the differentiation of muscle cells. When both copies of the gene for the regulatory protein *myogenin* are disrupted, an animal dies at birth because it lacks functional skeletal muscle. Microscopic inspection reveals that the tissues from which muscle normally forms contain precursor cells that have failed to differentiate fully (Figure 5.36A and B). Heterozygous mice

FIGURE 5.35 Gene disruption by homologous recombination. (A) A mutated version of the gene to be disrupted is constructed, maintaining some regions of homology with the normal gene (red). When the foreign mutated gene is introduced into an embryonic stem cell, (B) recombination takes place at regions of homology and (C) the normal (targeted) gene is replaced, or "knocked out," by the foreign gene. The cell is inserted into embryos, and mice lacking the gene (knockout mice) are produced.

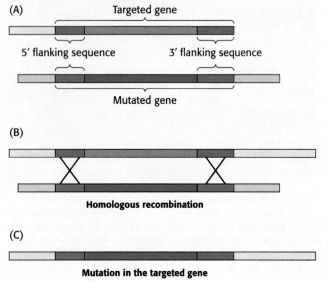

(A)

Targeted gene

5′ flanking sequence   3′ flanking sequence

Mutated gene

(B)

Homologous recombination

(C)

Mutation in the targeted gene

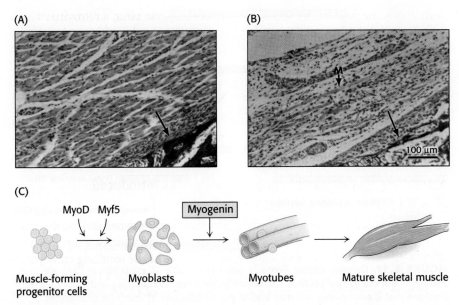

(A)

(B)

100 µm

(C)

MyoD Myf5 | Myogenin

Muscle-forming progenitor cells   Myoblasts   Myotubes   Mature skeletal muscle

FIGURE 5.36 Consequences of gene disruption. Sections of muscle from normal (A) and myogenin-knockout (B) mice, as viewed under the light microscope. The unlabeled arrows in both panels identify comparable sections of the pelvic bone, indicating that similar anatomical regions are depicted. Muscles do not develop properly in mice having both myogenin genes disrupted. A poorly formed muscle fiber in the knockout strain is indicated by the M arrow. (C) The development of mature skeletal muscle from progenitor cells is a highly regulated process involving a number of intermediate cell types and multiple transcription factors. Through the gene-disruption studies in (A) and (B), myogenin was identified as an essential component of this pathway. [(A) and (B) Reprinted by permission from Macmillan Publishers Ltd: *Nature*, v. 364, Hasty, P., Bradley, A., Morris, J. H., Edmondson, D. G., Venuti, J. M., Olson, E. N., Klein, W. H., Muscle deficiency and neonatal death in mice with a targeted mutation in the myogenin gene, pp. 501–506, copyright 1993. (C) Information from S. Hettmer and A. J. Wagers, *Nat. Med.* 16:171–173, 2010, Fig. 1.]

containing one normal myogenin gene and one disrupted gene appear normal, suggesting that a reduced level of myogenin gene expression is still sufficient for normal muscle development. The generation and characterization of this knockout strain provided strong evidence that functional myogenin is essential for proper development of skeletal muscle tissue (Figure 5.36C). Analogous studies have probed the function of many other genes in the context of a living organism. Furthermore, mice harboring gene knockouts can be utilized as animal models for known human genetic diseases, enabling the assessment of new potential therapies for treatment.

Manipulation of genomic DNA using homologous recombination, while a powerful tool, has limitations. Recombination at the desired site can be inefficient and time-consuming. In addition, this method is generally limited to specific model organisms, such as yeast, mice, and fruit flies. Over the past 10 years, new methods for the highly specific modification of genomic DNA, or *genome editing*, have emerged. These approaches rely on the introduction of double-strand breaks at precisely determined sequences within genomic DNA. The resulting cleavage site is repaired by the DNA repair machinery of the host cell in one of two ways (Section 29.5). If no template is provided, the cell will repair the break using a process known as *non-homologous end joining* (NHEJ). However, this process is error-prone, and a variety of insertions or deletions will be introduced into the repair site. It is likely that a subset of these modifications will introduce a premature stop codon into the targeted gene, resulting in a gene knockout. Alternatively, if a DNA fragment containing the desired sequence change is simultaneously introduced with the nucleases, the repair machinery will use this donor template to introduce these changes directly into the genomic sequence, in a process known as *homology directed repair* (HDR) (Figure 5.37).

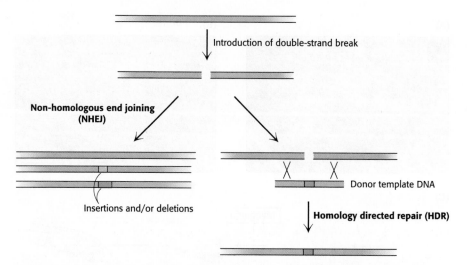

FIGURE 5.37 Genome editing. Precisely generated double-strand breaks in DNA are repaired in the cell by two possible mechanisms. One is non-homologous end joining (NHEJ), a process that is error-prone, and may lead to the introduction of insertions or deletions. Such errors may introduce frameshift mutations that ultimately knock out the entire gene. Alternatively, if a donor template DNA fragment is also provided, the break can be repaired by homology directed repair (HDR), which results in the incorporation of the desired modifications (green) into the targeted gene.

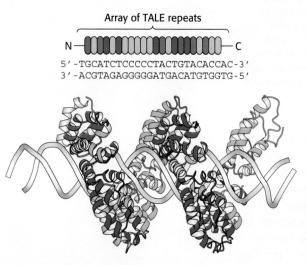

FIGURE 5.38 TALE repeats recognize individual bases in DNA. Each TALE repeat contains 34 amino acids, two of which specify its nucleotide binding partner. In this figure, the identity of these residues is indicated by the color of the repeat. TALE proteins can be designed to uniquely recognize extended oligonucleotide sequences. In this example, a 22 base-pair sequence is bound by a single TALE protein, the bacterial effector PthXo1. [Drawn from 3UGM.pdb.]

How are double-stranded breaks introduced specifically into the gene of interest? In one approach, a sequence-specific nuclease is engineered by fusing the nonspecific nuclease domain of the restriction enzyme FokI to a DNA-binding domain designed to bind to a particular DNA sequence. Two examples of this strategy are in common usage. *Zinc-finger nucleases* (ZFNs) are formed by combining the FokI nuclease domain with the DNA-binding domain that contains a series of zinc finger domains (Section 33.2), small zinc-binding motifs that each recognize a sequence of three base pairs. The preferred DNA-binding sequence can be altered by changing the identity of only four contact residues within each finger. For *transcription activator-like effector nucleases* (TALENs), the DNA-binding domain is comprised of an array of TALE repeats. Each repeat contains 34 amino acids and two α-helices, yet only two of these residues (at positions 12 and 13) are responsible for the unique recognition of a single nucleotide within the double helix (Figure 5.38). Mutation of these residues within an array of repeats enables the recognition of a vast number of possible DNA target sequences with a high degree of specificity. Since FokI functions as a dimer, two ZFN/TALEN constructs are necessary to generate a double-stranded break, one for each strand of DNA (Figure 5.39).

Even greater diversity and potential for customization has been realized with the discovery and application of the *CRISPR-Cas system*. Clustered regularly interspaced palindromic repeats (or CRISPRs) were first discovered in the late 1980s in the genome of the bacterium *E. coli*, then subsequently identified in numerous bacterial and archaeal species. Near these repeats are clusters of genes encoding CRISPR-associated, or *Cas*, proteins, which were predicted to function as enzymes that bind to and cleave DNA. Twenty years later, it was confirmed that these loci constitute a prokaryotic immune system, enabling the cleavage of foreign DNA sequences. Since this discovery, CRISPR-Cas systems have been cleverly adapted to enable customized sequence-specific DNA cleavage.

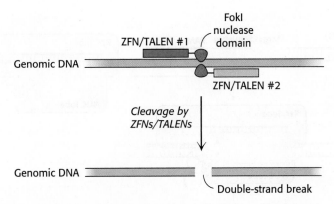

FIGURE 5.39 Introduction of double-strand breaks by ZFNs and TALENs. The nuclease domain of FokI functions as a dimer. Thus, in order to achieve a sequence-specific double-strand break, two ZFNs or TALENs are required, one for each DNA strand.

How do CRISPR-Cas systems achieve sequence-specific double-strand breaks in DNA? Cleavage of target DNA by the CRISPR-Cas system of the bacterium *Streptococcus pyogenes* requires only a single protein, the nuclease Cas9, and an engineered, single-stranded *single guide RNA*, or *sgRNA*. At its 5′ end, the sgRNA contains approximately 20 nucleotides that can be customized to complement the desired target site, followed by multiple stem-loop structures at its 3′ end, which are required for binding to the nuclease (Figure 5.40). Cas9 is a large, 158 kDa protein which contains two lobes: a REC lobe, which binds to the duplex formed between the sgRNA and the target strand of DNA, and a NUC lobe, which contains the two nuclease domains that are responsible for cleavage of the two strands of the target DNA (Figure 5.41A). Cas9 also contains a region that recognizes a short (usually three or four nucleotide) sequence of DNA known as the *protospacer-adjacent motif* (PAM). If the target DNA sequence complementary to the sgRNA is adjacent to a PAM, the Cas9 complex cleaves both target strands using the two nuclease domains of the NUC lobe (Figure 5.41B). As the specificity of this nuclease is driven by a complementary sequence of RNA, the CRISPR-Cas system can be engineered to cleave virtually any DNA sequence, provided it is adjacent to a PAM, without the need for designing new protein recognition domains such as those required for ZFNs or TALENs.

Site-specific nuclease-based genome editing methods have now been applied to a variety of species, including model organisms used in the laboratory (rat, zebrafish, and fruit fly), various forms of livestock (pig, cow), and a number of plants. In addition, their use as therapeutic tools in humans is currently under investigation. For example, a ZFN that inactivates the

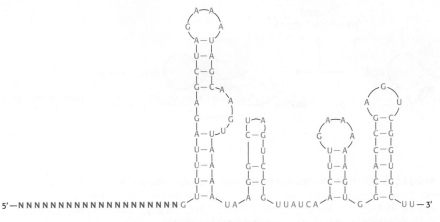

Single-guide RNA (sgRNA)

FIGURE 5.40 Structure of the CRISPR guide RNA. A single guide RNA (sgRNA) contains 20 nucleotides at its 5′ end (blue) which specify the target sequence, followed by multiple stem-loop structures (red) which mediate the interaction with Cas9.

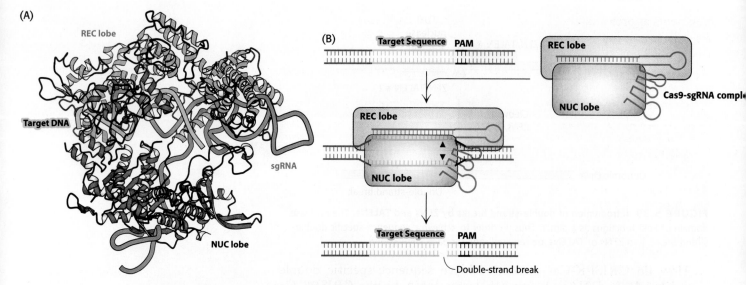

FIGURE 5.41 Cleavage of DNA by the CRISPR-Cas system. (A) The structure of a Cas9-sgRNA complex reveals the two lobes of Cas9: the REC lobe (green) and the NUC lobe (grey). The REC lobe mediates the interaction with the sgRNA, shown in blue and red as in (A). The NUC lobe contains the nuclease domain that will cleave the target DNA (yellow) [Drawn from 4OO8.pdb]. (B) If the target DNA sequence contains an adjacent PAM sequence, the sgRNA-Cas9 complex will bind and cleave both strands of the target, at the sites indicated by the black triangles, yielding a double-strand break.

human CCR5 gene, a coreceptor for cellular invasion of human immunodeficiency virus (HIV), is currently in clinical trials for the treatment of patients infected with HIV.

RNA interference provides an additional tool for disrupting gene expression

An extremely powerful tool for disrupting gene expression was serendipitously discovered in the course of studies that required the introduction of RNA into a cell. The introduction of a specific double-stranded RNA molecule into a cell was found to suppress the transcription of genes that contained sequences present in the double-stranded RNA molecule. Thus, the introduction of a specific RNA molecule can interfere with the expression of a specific gene.

The mechanism of *RNA interference* has been largely established (Figure 5.42). When a double-stranded RNA molecule is introduced into an appropriate cell, the RNA is cleaved by the enzyme Dicer into

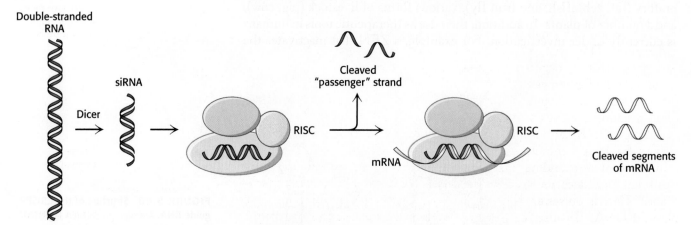

FIGURE 5.42 RNA interference mechanism. A double-stranded RNA molecule is cleaved into 21-bp fragments by the enzyme Dicer to produce siRNAs. These siRNAs are incorporated into the RNA-induced silencing complex (RISC), where the single-stranded RNAs guide the cleavage of mRNAs that contain complementary sequences.

fragments approximately 21 nucleotides in length. Each fragment, termed a small interfering RNA (siRNA), consists of 19 bp of double-stranded RNA and 2 bases of unpaired RNA on each 5′ end. The siRNA is loaded into an assembly of several proteins referred to as the *RNA-induced silencing complex* (RISC), which unwinds the RNA duplex and cleaves one of the strands, the so-called *passenger strand*. The uncleaved single-stranded RNA segment, the *guide strand*, remains incorporated into the enzyme. The fully assembled RISC cleaves mRNA molecules that contain exact complements of the guide-strand sequence. Thus, levels of such mRNA molecules are dramatically reduced. The technique of RNA interference is called *gene knockdown*, because the expression of the gene is reduced but not eliminated, as is the case with gene knockouts.

The machinery necessary for RNA interference is found in many cells. In some organisms such as *C. elegans*, RNA interference is quite efficient. Indeed, RNA interference can be induced simply by feeding *C. elegans* strains of *E. coli* that have been engineered to produce appropriate double-stranded RNA molecules. Although not as efficient in mammalian cells, RNA interference has emerged as a powerful research tool for reducing the expression of specific genes. Moreover, in 2017, an siRNA targeting the protein transthyretin achieved positive results in a Phase III clinical trial for the treatment of familial amyloid polyneuropathy, also referred to as hereditary ATTR amyloidosis.

Tumor-inducing plasmids can be used to introduce new genes into plant cells

The common soil bacterium *Agrobacterium tumefaciens* infects plants and introduces foreign genes into plant cells (Figure 5.43). A lump of tumor tissue called a *crown gall* grows at the site of infection. Crown galls synthesize opines, a group of amino acid derivatives that are metabolized by the infecting bacteria. In essence, the metabolism of the plant cell is diverted to satisfy the highly distinctive appetite of the intruder. *Tumor-inducing plasmids* (Ti plasmids) that are carried by *A. tumefaciens* carry instructions for the switch to the tumor state and the synthesis of opines. A small part of the Ti plasmid becomes integrated into the genome of infected plant cells; this 20-kb segment is called *T-DNA* (transferred DNA; Figure 5.44).

Ti-plasmid derivatives can be used as vectors to deliver foreign genes into plant cells. First, a segment of foreign DNA is inserted into the T-DNA region of a small plasmid through the use of restriction enzymes and ligases. This synthetic plasmid is added to *A. tumefaciens* colonies harboring naturally occurring Ti plasmids. By recombination, Ti plasmids containing the foreign gene are formed. These Ti vectors are valuable tools for exploring the genomes of plant cells and modifying plants to improve their agricultural value and crop yield.

Foreign DNA can also be introduced into plants by applying intense electric fields; a technique called *electroporation* (Figure 5.45). First, the cellulose wall surrounding plant cells is removed by adding cellulase; this treatment produces *protoplasts*, plant cells with exposed plasma membranes. Electric pulses are then applied to a suspension of protoplasts and plasmid DNA. Because high electric fields make membranes transiently permeable to large molecules, plasmid DNA molecules enter the cells. The cell wall is then allowed to reform, and the plant cells are again viable. Maize cells and carrot cells have been stably transformed in this way with the use of plasmid DNA that includes genes for resistance to herbicides. Moreover,

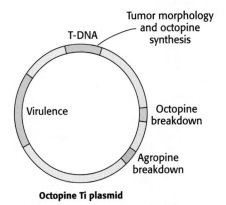

FIGURE 5.43 Tumors in plants. Crown gall, a plant tumor, is caused by a bacterium (*Agrobacterium tumefaciens*) that carries a tumor-inducing plasmid (Ti plasmid). [Matthew A. Escobar, Edwin L. Civerolo, Kristin R. Summerfelt, and Abhaya M. Dandekar. RNAi-mediated oncogene silencing confers resistance to crown gall tumorigenesis. *Proc. Natl. Acad. Sci. U.S.A.* 2001 98 (23) 13437–13442. Copyright 2001 National Academy of Sciences, U.S.A.]

FIGURE 5.44 Ti plasmids. Agrobacteria containing Ti plasmids can deliver foreign genes into some plant cells. [Information from M. Chilton. A vector for introducing new genes into plants. Copyright © 1983 by Scientific American, Inc. All rights reserved.]

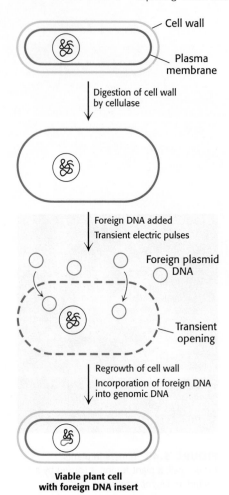

FIGURE 5.45 Electroporation. Foreign DNA can be introduced into plant cells by electroporation, the application of intense electric fields to make their plasma membranes transiently permeable.

the transformed cells efficiently express the plasmid DNA. Electroporation is also an effective means of delivering foreign DNA into animal cells and bacterial cells.

The most effective means of transforming plant cells is through the use of *"gene guns,"* or *bombardment-mediated transformation.* DNA is coated onto 1-μm-diameter tungsten pellets, and these microprojectiles are fired at the target cells with a velocity greater than $400\,\mathrm{m\,s^{-1}}$. Despite its apparent crudeness, this technique is an effective way of transforming plants, especially important crop species such as soybean, corn, wheat, and rice. The gene-gun technique affords an opportunity to develop genetically modified organisms (GMOs) with beneficial characteristics, such as the ability to grow in poor soils, resistance to natural climatic variation, resistance to pests, and the ability to fortify nutritional content. These crops might be most useful in developing countries. However, the use of GMOs is highly controversial, as some fear that their safety risks have not been adequately addressed.

The first GMO to come to market was a tomato characterized by delayed ripening, rendering it ideal for shipment. Pectin is a polysaccharide that gives tomatoes their firmness and is naturally destroyed by the enzyme *polygalacturonase.* As pectin is destroyed, the tomatoes soften, making shipment difficult. DNA was introduced that disrupts the polygalacturonase gene. Less of the enzyme was produced, and the tomatoes stayed fresh longer. However, the tomato's poor taste hindered its commercial success. An especially successful result of the use of Ti plasmid to modify crops is golden rice. Golden rice is a variety of genetically modified rice that contains the genes for β-carotene synthesis, a required precursor for vitamin A synthesis in humans. Consumption of this rice will benefit children and pregnant woman in parts of the world where rice is a dietary staple and vitamin A deficiency is common.

Human gene therapy holds great promise for medicine

The field of *gene therapy* attempts to express specific genes within the human body in such a way that beneficial results are obtained. The gene targeted for expression may be already present or specially introduced. Alternatively, gene therapy may attempt to modify genes containing sequence variations that have harmful consequences. A tremendous amount of research remains to be done before gene therapy becomes practical. Nonetheless, considerable progress has been made. For example, some people lack functional genes for *adenosine deaminase* and succumb to infections if exposed to a normal environment, a condition called *severe combined immunodeficiency* (SCID). Functional genes for this enzyme have been introduced by using gene-therapy vectors based on retroviruses. In 2016, the European Medicines Agency approved the first clinical application of gene therapy for the treatment of SCID patients who do not have a suitable bone marrow donor. This therapy, licensed under the name Strimvelis, involves harvesting bone marrow cells from the patient, isolating the appropriate cell population, and transducing those cells with a retroviral vector that encodes the wild-type adenosine deaminase gene. The treated cells are then returned to the patient's bone marrow. An important hurdle to the approval of this therapy was ensuring the longevity of the clinical effects and eliminating unwanted side effects. Future research promises to transform gene therapy into an important tool for clinical medicine.

SUMMARY

5.1 The Exploration of Genes Relies on Key Tools

The recombinant DNA revolution in biology is rooted in the repertoire of enzymes that act on nucleic acids. Restriction enzymes are a key group among them. These endonucleases recognize specific base sequences in double-helical DNA and cleave both strands of the duplex, forming specific fragments of DNA. These restriction fragments can be separated and displayed by gel electrophoresis. The pattern of these fragments on the gel is a fingerprint of a DNA molecule. A DNA fragment containing a particular sequence can be identified by hybridizing it with a labeled single-stranded DNA probe (Southern blotting).

Rapid sequencing techniques have been developed to further the analysis of DNA molecules. DNA can be sequenced by controlled interruption of replication. The fragments produced are separated by capillary electrophoresis and visualized by fluorescent tags at their 5′ ends.

DNA probes for hybridization reactions, as well as new genes, can be synthesized by the automated solid-phase method. DNA chains as long as 100 nucleotides can be readily synthesized. The polymerase chain reaction makes it possible to greatly amplify specific segments of DNA in vitro. The region amplified is determined by the placement of a pair of primers that are added to the target DNA along with a thermostable DNA polymerase and deoxyribonucleoside triphosphates. The exquisite sensitivity of PCR makes it a choice technique in detecting pathogens and cancer markers, in genotyping, and in amplifying DNA from fossils that are many thousands of years old.

5.2 Recombinant DNA Technology Has Revolutionized All Aspects of Biology

New genes can be constructed in the laboratory, introduced into host cells, and expressed. Novel DNA molecules are made by joining fragments that have complementary cohesive ends produced by the action of a restriction enzyme. DNA ligase seals breaks in DNA chains. Vectors for propagating the DNA include plasmids, λ phage, and bacterial and yeast artificial chromosomes. Specific genes can be cloned from a genomic library with the use of a DNA or RNA probe. Foreign DNA can be expressed after insertion into prokaryotic and eukaryotic cells by the appropriate vector. Specific mutations can be generated in vitro to engineer novel proteins. A mutant protein with a single amino acid substitution can be produced by priming DNA replication with an oligonucleotide encoding the new amino acid. Plasmids can be engineered to permit the facile insertion of a DNA cassette containing any desired mutation. The techniques of protein and nucleic acid chemistry are highly synergistic. Investigators now move back and forth between gene and protein with great facility.

5.3 Complete Genomes Have Been Sequenced and Analyzed

The sequences of many important genomes are known in their entirety. More than 10,000 bacterial and archaeal genomes have been sequenced, including those from key model organisms and important pathogens. The sequence of the human genome has now been completed with nearly full coverage and high precision. Only from 20,000 to 25,000 protein-encoding genes appear to be present in the human genome, a substantially smaller number than earlier estimates. Next-generation sequencing methods have greatly accelerated the pace at which large genomes can be sequenced and analyzed. Comparative genomics has become a powerful tool for analyzing individual genomes and for exploring evolution. Genomewide gene-expression patterns can be examined through the use of DNA microarrays.

5.4 Eukaryotic Genes Can Be Quantitated and Manipulated with Considerable Precision

Changes in gene expression can be readily determined by such techniques as quantitative PCR and hybridization to microarrays. The production of transgenic mice, such as those carrying mutations known to cause ALS in humans, has been a source of considerable insight into disease mechanisms and possible therapies. Genome editing techniques, such as those enabled by the CRISPR/Cas system and ZFNs/TALENs, are greatly expanding our capacity to specifically modify genomes in a wide array of organisms. The functions of particular genes can be investigated by disruption. One method of disrupting the expression of a particular gene is through RNA interference, which depends on the introduction of specific double-stranded RNA molecules into eukaryotic cells. New DNA can be brought into plant cells by the soil bacterium *Agrobacterium tumefaciens*, which harbors Ti plasmids. DNA can also be introduced into plant cells by applying intense electric fields, which render them transiently permeable to very large molecules, or by bombarding them with DNA-coated microparticles. Gene therapy holds great promise for clinical medicine, but many challenges remain.

APPENDIX

Biochemistry in Focus

Improved biofuel production from genetically-engineered algae

Considerable research has been invested in the search for alternative energy sources to fossil fuels. One such alternative is biofuel, a source of energy generated by contemporary biological processes. A number of biofuels have been proposed, including food crops and agricultural biomass. More recently, several species of algae that produce large amounts of lipids have been proposed as the basis for biofuel generation. When subjected to nitrogen starvation, these species can generate high yields of a derivative of fatty acids (Chapter 12), known as fatty acid methyl esters (FAME), a primary component of biofuel. One strain of algae, *Nannochloropsis gaditana*, is a particularly efficient producer of FAME. Its 29 Mbp genome was sequenced in 2012, and the metabolic genes enabling lipid production were identified. Using this sequence information, researchers have attempted to modify this species to engineer a more efficient strain of FAME-producing algae. Recent advancements

in next generation sequencing and genome editing, covered in this chapter, have opened doors to new possibilities in engineering such strains.

One potential hurdle in working with algae for the mass production of biofuel is that nitrogen depletion, while necessary for enhanced lipid production, restricts cell growth. In an effort to circumvent this obstacle, a group of researchers used *RNA-Seq*—a next-generation sequencing method that enables the rapid, high-throughput identification of the entire transcriptome of a cell—to compare the mRNAs expressed in *N. gaditana* before and after nitrogen starvation. By comparing the identified transcripts with annotated sequence databases, these researchers were able to identify 20 transcription factors (Section 33.2), master regulators of gene expression, whose expression was reduced under nitrogen-starved conditions. The researchers reasoned that if the expression of these transcription factors was repressed during lipid production stimulated by nitrogen starvation, then knockout of these genes may improve lipid yields without the need for altering the nitrogen content of their growth media.

In order to test their hypothesis, they used CRISPR-Cas9 technology to generate 18 individual strains where each transcription factor was knocked out (two of the genes could not be knocked out efficiently using this method). Of these strains, only one exhibited enhanced FAME production under nutrient-rich conditions (Figure 5.46). Note that TOC refers to total organic carbon content, and serves as a measure of overall cell growth. The gene disrupted in this strain encodes a transcription factor they named ZnCys. However, a new problem emerged in this experiment: the researchers noted that while FAME production was enhanced in the *ZnCys*-knockout strain ("ZnCys-KO"), the TOC was also reduced, indicating that the growth and viability of this knockout strain was impaired.

To address the cell growth issue, these researchers generated three additional strains: two CRISPR-Cas9-modified strains where the *ZnCys* gene was disrupted in its regulatory

sequences, as opposed to the protein-coding region itself ("*ZnCys*-BASH-3" and "*ZnCys*-BASH-12"), and one strain where they used RNA interference to reduce *ZnCys* expression ("*ZnCys*-RNAi-7"). These strains, which all featured attenuated, but not abolished, *ZnCys* expression, exhibited elevated FAME levels in the presence of nitrogen-containing medium, with only minimally affected TOC levels (Figure 5.47).

This work represents a major step towards understanding how lipid production in biofuel-producing algae strains is regulated. Furthermore, these studies suggest new opportunities for the generation of genetically modified algae strains for optimal biofuel generation.

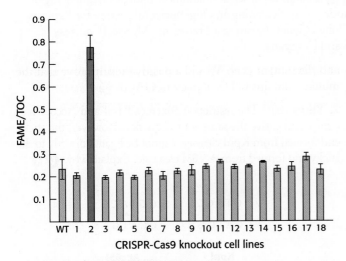

FIGURE 5.46 Knockout *N. chloropsis* cell lines generated by CRISPR-Cas9. Eighteen transcription factor genes were individually disrupted in *N. chloropsis* algae strains using the CRISPR-Cas9 system. Only one strain ("2"), showed a significant increase in FAME production versus the wild-type ("WT") strain. [Information from I. Ajjawi et al., *Nat. Biotech.* 35:647–652, 2017, Fig. S1a.]

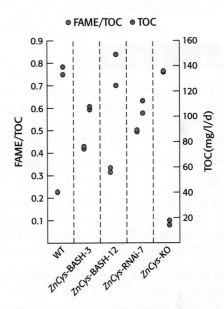

FIGURE 5.47 Attenuation of *ZnCys* expression leads to enhanced lipid production without impairment of cell growth. Attenuation of *ZnCys* expression was achieved by either CRISPR-Cas9-mediated disruption of *ZnCys* regulatory elements ("*ZnCys*-BASH-3" and "*ZnCys*-BASH-12") or by RNA interference ("*ZnCys*-RNAi-7"). Reduction of *ZnCys* expression improved FAME production, but did not exhibit the impaired cell growth (as measured by TOC) shown by the complete knockout strain ("*ZnCys*-KO"). [Information from I. Ajjawi et al., *Nat. Biotech.* 35:647–652, 2017, Fig. 2c.]

KEY TERMS

restriction enzyme (p. 147)

palindrome (p. 147)

DNA probe (p. 149)

Southern blotting (p. 149)

northern blotting (p. 149)

controlled termination of replication (Sanger dideoxy method) (p. 149)

polymerase chain reaction (PCR) (p. 151)

polymorphism (p. 153)

vector (p. 154)

plasmid (p. 154)

bacteriophage lambda (λ phage) (p. 154)

sticky ends (p. 154)

DNA ligase (p. 154)

cloning vector (p. 155)

reporter gene (p. 155)

expression vector (p. 156)

bacterial artificial chromosome (BAC) (p. 157)

yeast artificial chromosome (YAC) (p. 157)

genomic library (p. 159)

complementary DNA (cDNA) (p. 159)

reverse transcriptase (p. 159)

cDNA library (p. 160)

site-directed mutagenesis (p. 161)

cassette mutagenesis (p. 162)

pseudogene (p. 165)

mobile genetic element (p. 165)

short interspersed elements (SINES) (p. 165)

long interspersed elements (LINES) (p. 165)

next-generation sequencing (p. 166)

pyrosequencing (p. 166)

dispensation sequence (p. 166)

quantitative PCR (qPCR) (p. 169)

transcriptome (p. 170)

DNA microarray (gene chip) (p. 170)

transgenic mouse (p. 172)

gene disruption (gene knockout) (p. 172)

genome editing (p. 173)

zinc-finger nuclease (ZFN) (p. 174)

transcription activator-like effector nuclease (TALEN) (p. 174)

CRISPR-Cas system (p. 174)

single guide RNA (sgRNA) (p. 175)

protospacer-adjacent motif (PAM) (p. 175)

RNA interference (p. 176)

RNA-induced silencing complex (RISC) (p. 177)

tumor-inducing plasmid (Ti plasmid) (p. 177)

gene gun (bombardment-mediated transformation) (p. 178)

PROBLEMS

As you work on these problems, you might find it helpful to visit 🐟 **SaplingPlus** to see animations of Dideoxy Sequencing of DNA, Polymerase Chain Reaction, Synthesizing an Oligonucleotide Array, Screening an Oligonucleotide Array for Patterns of Gene Expression, Plasmid Cloning, In Vitro Mutagenesis of Cloned Genes, Creating a Transgenic Mouse. [Courtesy of H. Lodish et al., *Molecular Cell Biology*, 5th ed. (W. H. Freeman and Company, 2004).]

1. *It's not the heat.* Why is *Taq* polymerase especially useful for PCR? ✓❸

2. *The right template.* Ovalbumin is the major protein of egg white. The chicken ovalbumin gene contains eight exons separated by seven introns. Should ovalbumin cDNA or ovalbumin genomic DNA be used to form the protein in *E. coli*? Why?

3. *Cleavage frequency.* The restriction enzyme AluI cleaves at the sequence 5′-AGCT-3′, and NotI cleaves at 5′-GCG-GCCGC-3′. What would be the average distance between cleavage sites for each enzyme on digestion of double-stranded DNA? Assume that the DNA contains equal proportions of A, G, C, and T. ✓❶

4. *Rich or poor?* DNA sequences that are highly enriched in G–C base pairs typically have high melting temperatures. Moreover, once separated, single strands containing these regions can form rigid secondary structures. How might the presence of G–C-rich regions in a DNA template affect PCR amplification? ✓❸

5. *The right cuts.* Suppose that a human genomic library is prepared by exhaustive digestion of human DNA with the EcoRI restriction enzyme. Fragments averaging about 4 kb in length would be generated. Is this procedure suitable for cloning large genes? Why or why not?

6. *A revealing cleavage.* Sickle-cell anemia arises from a mutation in the gene for the β chain of human hemoglobin. The change from GAG to GTG in the mutant eliminates a cleavage site for the restriction enzyme MstII, which recognizes the target sequence CCTGAGG. These findings form the basis of a diagnostic test for the sickle-cell gene. Propose a rapid procedure for distinguishing between the normal and the mutant gene. Would a positive result prove that the mutant contains GTG in place of GAG? ✓❶

7. *Sticky ends?* The restriction enzymes KpnI and Acc65I recognize and cleave the same 6-bp sequence. However, the sticky end formed from KpnI cleavage cannot be ligated directly to the sticky end formed from Acc65I cleavage. Explain why. ✓❶

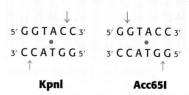

8. *Many melodies from one cassette.* Suppose that you have isolated an enzyme that digests paper pulp and have obtained its cDNA. The goal is to produce a mutant that is effective at high temperature. You have engineered a pair of unique restriction sites in the cDNA that flank a 30-bp coding region. Propose a rapid technique for generating many different mutations in this region. ✓❺

9. *A blessing and a curse.* The power of PCR can also create problems. Suppose someone claims to have isolated dinosaur DNA by using PCR. What questions might you ask to determine if it is indeed dinosaur DNA?

10. *Limited by PAM?* Targeting a DNA sequence for editing by the CRISPR-Cas9 system requires that a PAM sequence be present just downstream of the target sequence. For *S. pyogenes* Cas9 this PAM is 5′-NGG-3′, where N represents any nucleotide. Assuming a fragment of DNA includes equal amounts of A, C, G, and T bases, how far apart would one expect to find these PAMs? ✓❻

11. *Build a better PAM.* Mutants of *S. pyogenes* Cas9 have been developed which recognize different PAM sequences, such as 5'-NGAN-3'. Why might these mutants be useful for genome editing studies? ✓⑥

12. *Questions of accuracy.* The stringency of PCR amplification can be controlled by altering the temperature at which the primers and the target DNA undergo hybridization. How would altering the temperature of hybridization affect the amplification? Suppose that you have a particular yeast gene A and that you wish to see if it has a counterpart in humans. How would controlling the stringency of the hybridization help you? ✓❸

13. *Terra incognita.* PCR is typically used to amplify DNA that lies between two known sequences. Suppose that you want to explore DNA on both sides of a single known sequence. Devise a variation of the usual PCR protocol that would enable you to amplify entirely new genomic terrain.

14. *A puzzling ladder.* A gel pattern displaying PCR products shows four strong bands. The four pieces of DNA have lengths that are approximately in the ratio of 1 : 2 : 3 : 4. The largest band is cut out of the gel, and PCR is repeated with the same primers. Again, a ladder of four bands is evident in the gel. What does this result reveal about the structure of the encoded protein?

15. *Chromosome walking.* Propose a method for isolating a DNA fragment that is adjacent in the genome to a previously isolated DNA fragment. Assume that you have access to a complete library of DNA fragments in a BAC vector but that the sequence of the genome under study has not yet been determined.

16. *Probe design.* Which of the following amino acid sequences would yield the most optimal oligonucleotide probe?

```
Ala-Met-Ser-Leu-Pro-Trp
Gly-Trp-Asp-Met-His-Lys
Cys-Val-Trp-Asn-Lys-Ile
Arg-Ser-Met-Leu-Gln-Asn
```

17. *Man's best friend.* Why might the genomic analysis of dogs be particularly useful for investigating the genes responsible for body size and other physical characteristics?

18. *Of mice and men.* You have identified a gene that is located on human chromosome 20 and wish to identify its location within the mouse genome. On which chromosome would you be most likely to find the mouse counterpart of this gene?

Chapter Integration Problems

19. *Designing primers I.* A successful PCR experiment often depends on designing the correct primers. In particular, the T_m for each primer should be approximately the same. What is the basis of this requirement? ✓❸

20. *Designing primers II.* You wish to amplify a segment of DNA from a plasmid template by PCR with the use of the following primers: 5'-GGATCGATGCTCGCGA-3' and 5'-AGGATCGGGTCGCGAG-3'. Despite repeated attempts, you fail to observe a PCR product of the expected length after electrophoresis on an agarose gel. Instead, you observe a bright smear on the gel with an approximate length of 25 to 30 base pairs. Explain these results. ✓❸

Chapter Integration and Data Interpretation Problem

21. *Any direction but east.* A series of people are found to have difficulty eliminating certain types of drugs from their bloodstreams. The problem has been linked to a gene X, which encodes an enzyme Y. Six people were tested with the use of various techniques of molecular biology. Person A is a normal control, person B is asymptomatic but some of his children have the metabolic problem, and persons C through F display the trait. Tissue samples from each person were obtained. Southern analysis was performed on the DNA after digestion with the restriction enzyme HindIII. Northern analysis of mRNA also was done. In both types of analysis, the gels were probed with labeled X cDNA. Finally, a western blot with an enzyme-linked monoclonal antibody was used to test for the presence of protein Y. The results are shown here. Why is person B without symptoms? Suggest possible defects in the other people.

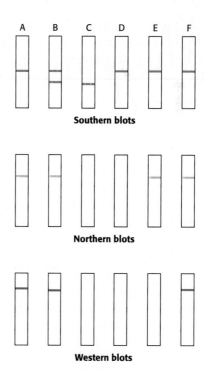

Data Interpretation Problems

22. *DNA diagnostics.* Representations of sequencing chromatograms for variants of the α chain of human hemoglobin are shown here. What is the nature of the amino acid change in each of the variants? The first triplet encodes valine. ✓❷

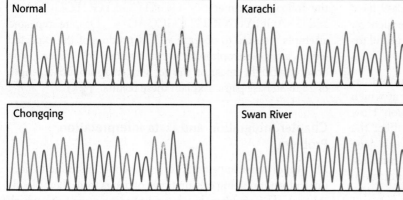

Colors: ddATP, ddCTP, ddGTP, ddTTP

23. *Two peaks.* In the course of studying a gene and its possible mutation in humans, you obtain genomic DNA samples from a collection of persons and PCR amplify a region of interest within this gene. For one of the samples, you obtain the sequencing chromatogram shown here. Provide an explanation for the appearance of these data at position 49 (indicated by the arrow):

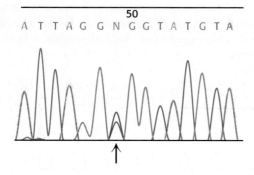

Think ▶ Pair ▶ Share Problem

24. After sequencing the DNA of several affected families, you have discovered a new gene, *BTGS1*, which you suspect plays a role in a rare variant of narcolepsy triggered by the opening of a biochemistry textbook. From its sequence, you suspect the BTGS1 protein functions as a transcription factor, a protein that regulates the expression of other genes. In order to explore this potential mechanism further, you would like to reduce the gene expression—and thus lower the amount of BTGS1 protein—in cultured cells.

What techniques could you use to generate this cell line?

Once successfully accomplished, what are some techniques you could use to determine if the expression of other genes is affected in this cell line?

Propose methods for generating a *BTGS1*-knockout mouse strain.

Exploring Evolution and Bioinformatics

Evolutionary relationships are clear in protein sequences. The close kinship between human beings and chimpanzees, hinted at by the mutual interest shown by Jane Goodall and a chimpanzee in the photograph, is revealed in the amino acid sequences of myoglobin. The human sequence (red) differs from the chimpanzee sequence (blue) in only one amino acid in a protein chain of 153 residues. [(Left) Hugo Van Lawick/National Geographic Creative.]

```
GLSDGEWQLVLNVWGKVEADIPGHGQEVLIRLFKGHPETLEKFDKFKHLKSEDEMKASEDLKKHGATVLTALGGIL-
GLSDGEWQLVLNVWGKVEADIPGHGQEVLIRLFKGHPETLEKFDKFKHLKSEDEMKASEDLKKHGATVLTALGGIL-

KKKGHHEAEIKPLAQSHATKHKIPVKYLEFISECIIQVLHSKHPGDFGADAQGAMNKALELFRKDMASNYKELGFQG
KKKGHHEAEIKPLAQSHATKHKIPVKYLEFISECIIQVLQSKHPGDFGADAQGAMNKALELFRKDMASNYKELGFQG
```

 ## L E A R N I N G   O B J E C T I V E S

By the end of this chapter, you should be able to:

1. Distinguish between homologs, paralogs, and orthologs.

2. Describe how two protein sequences are aligned, and how one can determine whether that alignment is significant.

3. Explain how substitution matrices can be used to improve the possibility of identifying related protein sequences.

4. Distinguish between divergent and convergent evolution.

5. Provide an example of how the evolution of biomolecules can be demonstrated in the laboratory.

O U T L I N E

6.1 Homologs Are Descended from a Common Ancestor

6.2 Statistical Analysis of Sequence Alignments Can Detect Homology

6.3 Examination of Three-Dimensional Structure Enhances Our Understanding of Evolutionary Relationships

6.4 Evolutionary Trees Can Be Constructed on the Basis of Sequence Information

6.5 Modern Techniques Make the Experimental Exploration of Evolution Possible

Like members of a human family, members of molecular families often have features in common. Such family resemblance is most easily detected by comparing three-dimensional structure, the aspect of a molecule most closely linked to function. Consider as an example ribonuclease from cows, which was introduced in our consideration of protein folding (Section 2.6). Comparing structures reveals that the three-dimensional structure of this protein and that of a human ribonuclease are quite similar (Figure 6.1). Although the degree of overlap between these two structures is not unexpected, given their nearly identical biological functions, similarities revealed by other such comparisons are sometimes surprising. For example, angiogenin, a protein that stimulates the growth of new blood vessels, is also structurally similar to ribonuclease—so similar that both angiogenin and ribonuclease are

FIGURE 6.1 Structures of ribonucleases from cows and human beings. Functional similarity follows from structural similarity. [Drawn from 8RAT.pdb and 2RNF.pdb.]

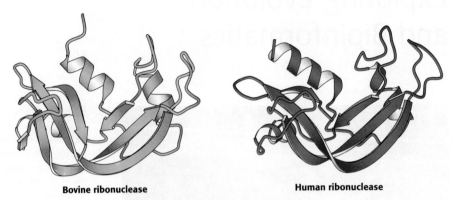

Bovine ribonuclease   Human ribonuclease

Angiogenin

FIGURE 6.2 Structure of angiogenin. The protein angiogenin, identified on the basis of its ability to stimulate blood-vessel growth, is highly similar in three-dimensional structure to ribonuclease. [Drawn from 2ANG.pdb.]

clearly members of the same protein family (Figure 6.2). Angiogenin and ribonuclease must have had a common ancestor at some earlier stage of evolution.

Three-dimensional structures have been determined for only a small proportion of the total number of proteins. In contrast, gene sequences and the corresponding amino acid sequences are available for a great number of proteins, largely owing to the tremendous power of DNA cloning and sequencing techniques, including complete-genome sequencing (Chapter 5). Evolutionary relationships also are clear in amino acid sequences. For example, 35% of the amino acids in corresponding positions are identical in the sequences of bovine ribonuclease and angiogenin. Is this level sufficiently high to ensure an evolutionary relationship? If not, what level is required? In this chapter, we shall examine the methods that are used to compare amino acid sequences and to deduce such evolutionary relationships.

Sequence-comparison methods are powerful tools in modern biochemistry. Sequence databases can be probed for matches to a newly elucidated sequence to identify related molecules. This information can often be a source of considerable insight into the function and mechanism of the newly sequenced molecule. When three-dimensional structures are available, they can be compared to confirm relationships suggested by sequence comparisons and to reveal others that are not readily detected at the level of sequence alone.

By examining the footprints present in modern protein sequences, the biochemist can learn about events in the evolutionary past. Sequence comparisons can often reveal pathways of evolutionary descent and estimated dates of specific evolutionary landmarks. This information can be used to construct evolutionary trees that trace the evolution of a particular protein or nucleic acid, in many cases from Archaea and Bacteria through Eukarya, including human beings. Molecular evolution can also be studied experimentally. In some cases, DNA from fossils can be amplified by PCR methods and sequenced, giving a direct view into the past. In addition, investigators can observe molecular evolution taking place in the laboratory, through experiments based on nucleic acid replication. The results of such studies are revealing much about how evolution proceeds.

6.1 Homologs Are Descended from a Common Ancestor

The exploration of biochemical evolution consists largely of an attempt to determine how proteins, other molecules, and biochemical pathways have been transformed through time. The most fundamental relationship between two entities is *homology*; two molecules are said to be *homologous*

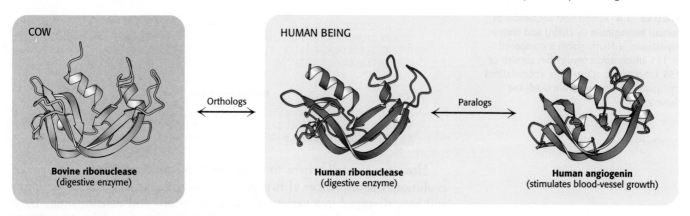

FIGURE 6.3 Two classes of homologs. Homologs that perform identical or very similar functions in different species are called orthologs, whereas homologs that perform different functions within one species are called paralogs.

if they have been derived from a common ancestor. Homologous molecules, or *homologs,* can be divided into two classes (Figure 6.3). *Paralogs* are homologs that are present within one species. Paralogs often differ in their detailed biochemical functions. *Orthologs* are homologs that are present within different species and have very similar or identical functions. Understanding the homology between molecules can reveal the evolutionary history of the molecules as well as information about their function; if a newly sequenced protein is homologous to an already characterized protein, we have a strong indication of the new protein's biochemical function.

How can we tell whether two human proteins are paralogs or whether a yeast protein is the ortholog of a human protein? As will be discussed in Section 6.2, *homology is often detectable by significant similarity in nucleotide or amino acid sequence and almost always revealed in three-dimensional structure.*

6.2 Statistical Analysis of Sequence Alignments Can Detect Homology

A significant sequence similarity between two molecules implies that they are likely to have the same evolutionary origin and, therefore, similar three-dimensional structures, functions, and mechanisms. Both nucleic acid and protein sequences can be compared to detect homology. However, the possibility exists that the observed agreement between any two sequences is solely a product of chance. Because nucleic acids are composed of fewer building blocks than proteins (4 bases versus 20 amino acids), the likelihood of random agreement between two DNA or RNA sequences is significantly greater than that for protein sequences. For this reason, detection of homology between protein sequences is typically far more effective.

To illustrate sequence-comparison methods, let us consider a class of proteins called the *globins.* Myoglobin is a protein that binds oxygen in muscle, whereas hemoglobin is the oxygen-carrying protein in blood (Chapter 7). Both proteins cradle a heme group, an iron-containing organic molecule that binds oxygen. Myoglobin is a monomer, while each human hemoglobin molecule is composed of four heme-containing polypeptide chains, two identical α chains and two identical β chains. Here, we shall consider only the α chain. To examine the similarity between the amino acid sequence of the human α chain and that of human myoglobin (Figure 6.4), we apply a method, referred to as a *sequence alignment,* in which the two sequences are systematically aligned with respect to each other to identify regions of significant overlap.

FIGURE 6.4 Amino acid sequences of human hemoglobin (α chain) and human myoglobin. α-Hemoglobin is composed of 141 amino acids; myoglobin consists of 153 amino acids. (One-letter abbreviations designating amino acids are used; see Table 2.2.)

Human hemoglobin (α chain)

```
VLSPADKTNVKAAWGKVGAHAGEYGAEALERMFLSFPTTKTYFPHFDLSHG
SAQVKGHGKKVADALTNAVAHVDDMPNALSALSDLHAHKLRVDPVNFKLLS
HCLLVTLAAHLPAEFTPAVHASLDKFLASVSTVLTSKYR
```

Human myoglobin

```
GLSDGEWQLVLNVWGKVEADIPGHGQEVLIRLFKGHPETLEKFDKFKHLKS
EDEMKASEDLKKHGATVLTALGGILKKKGHHEAEIKPLAQSHATKHKIPVK
YLEFISECIIQVLQSKHPGDFGADAQGAMNKALELFRKDMASNYKELGFQG
```

How can we tell where to align the two sequences? In the course of evolution, the sequences of two proteins that have an ancestor in common will have diverged in a variety of ways. Insertions and deletions may have occurred at the ends of the proteins or within the functional domains themselves. Individual amino acids may have been mutated to other residues of varying degrees of similarity. To understand how the methods of sequence alignment take these potential sequence variations into account, let us first consider the simplest approach, where we slide one sequence past the other, one amino acid at a time, and count the number of matched residues, or *sequence identities* (Figure 6.5). For α-hemoglobin and myoglobin, the best alignment reveals 23 sequence identities spread throughout the central parts of the sequences.

However, careful examination of all the possible alignments and their scores suggests that important information regarding the relationship between myoglobin and hemoglobin α has been lost with this method. In particular,

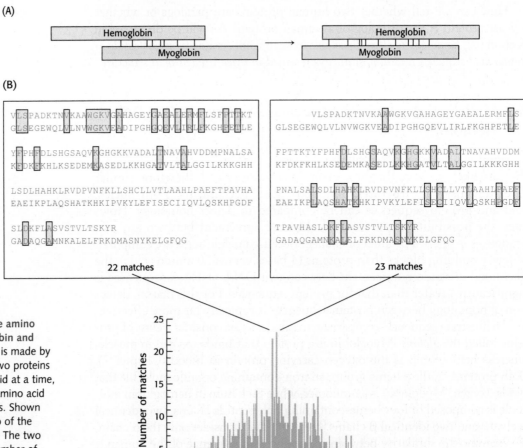

FIGURE 6.5 Comparing the amino acid sequences of α-hemoglobin and myoglobin. (A) A comparison is made by sliding the sequences of the two proteins past each other, one amino acid at a time, and counting the number of amino acid identities between the proteins. Shown are representations of just two of the >100 possible alignments. (B) The two alignments with the largest number of matches are shown above the graph, which plots the matches as a function of alignment.

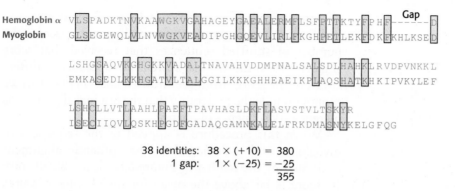

we see that another alignment, featuring 22 identities, is nearly as good (Figure 6.5). This alignment is shifted by six residues relative to the preceding alignment and yields identities that are concentrated toward the amino-terminal end of the sequences. By introducing a *gap* into one of the sequences, the identities found in *both* alignments will be represented (Figure 6.6). The insertion of gaps allows the alignment method to compensate for the insertions or deletions of nucleotides that may have taken place in the gene for one molecule but not the other in the course of evolution.

The use of gaps substantially increases the complexity of sequence alignment because a vast number of possible gaps, varying in both position and length, must be considered throughout each sequence. Moreover, the introduction of an excessive number of gaps can yield an artificially high number of identities. Nevertheless, methods have been developed for the insertion of gaps in the automatic alignment of sequences. These methods use scoring systems to compare different alignments, including penalties for gaps to prevent the insertion of an unreasonable number of them. For example, in one scoring system, each identity between aligned sequences is counted as +10 points, whereas each gap introduced, regardless of size, counts for −25 points. For the alignment shown in Figure 6.6, there are 38 identities (38 × 10 = 380) and 1 gap (1 × −25 = −25), producing a score of (380 + −25 = 355). Overall, there are 38 matched amino acids in an average length of 147 residues; thus, the sequences are 25.9% identical.

Alignments such as the one just described inform us as to the percent identity between two sequences. However, the question remains: how do we interpret this value? For example, we have determined that α-hemoglobin and myoglobin, when allowing for the insertion of gaps, are 25.9% identical. Is this value sufficiently high to suggest a relationship between these two proteins? To answer these questions, we must determine the significance of alignment scores and percent identities.

The statistical significance of alignments can be estimated by shuffling

The sequence similarities in Figure 6.6 appear striking, yet there remains the possibility that a grouping of sequence identities has occurred by chance alone. Because proteins are composed of the same set of 20 amino acid monomers, the alignment of any two unrelated proteins will yield some identities, particularly if we allow the introduction of gaps. Even if two proteins have identical amino acid composition, they may not be linked by evolution. It is the order of the residues within their sequences that implies a relationship between them. Hence, we can assess the significance of our alignment by "shuffling," or randomly rearranging, one of the sequences (Figure 6.7), performing the sequence alignment, and determining a new

THISISTHEAUTHENTICSEQUENCE

| Shuffling

SNUCSNSEATEEITU HEQIH HTTCEI

FIGURE 6.7 The generation of a shuffled sequence.

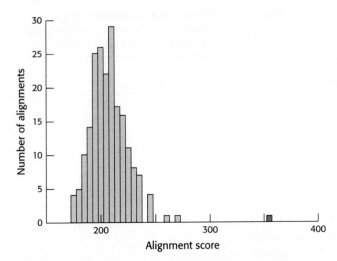

FIGURE 6.8 Statistical comparison of alignment scores. Alignment scores are calculated for many shuffled sequences, and the number of sequences generating a particular score is plotted against the score. The resulting plot is a distribution of alignment scores occurring by chance. The alignment score for unshuffled α-hemoglobin and myoglobin (shown in red) is substantially greater than any of these scores, strongly suggesting that the sequence similarity is significant.

alignment score. This process is repeated many times to yield a histogram showing, for each possible score, the number of shuffled sequences that received that score (Figure 6.8). If the original score is not appreciably different from the scores from the shuffled alignments, then we cannot exclude the possibility that the original alignment is merely a consequence of chance.

When this procedure is applied to the sequences of myoglobin and α-hemoglobin, the authentic alignment (indicated by the red bar in Figure 6.8) clearly stands out. Its score is far above the mean for the alignment scores based on shuffled sequences. The probability that such a deviation occurred by chance alone is approximately 1 in 10^{20}. Thus, we can comfortably conclude that the two sequences are genuinely similar; the simplest explanation for this similarity is that these sequences are homologous—that is, the two molecules have descended from a common ancestor.

Distant evolutionary relationships can be detected through the use of substitution matrices

The scoring scheme described above assigns points only to positions occupied by identical amino acids in the two sequences. No credit is given for any pairing that is not an identity. However, as already discussed, two proteins related by evolution undergo amino acid substitutions as they diverge. A scoring system based solely on amino acid identity cannot account for these changes. To add greater sensitivity to the detection of evolutionary relationships, methods have been developed to compare two amino acids and assess their degree of *similarity*.

Not all substitutions are equivalent. For example, amino acid changes can be classified as structurally conservative or nonconservative. A *conservative substitution* replaces one amino acid with another that is similar in size and chemical properties. Conservative substitutions may have only minor effects on protein structure and often can be tolerated without compromising protein function. In contrast, in a *nonconservative substitution*, an amino acid is replaced by one that is structurally dissimilar. Amino acid changes can also be classified by the fewest number of nucleotide changes necessary to achieve the corresponding amino acid change. Some substitutions arise from the replacement of only a single nucleotide in the gene sequence; whereas others require two or three replacements. Conservative and single-nucleotide substitutions are likely to be more common than are substitutions with more radical effects.

How can we account for the type of substitution when comparing sequences? We can approach this problem by first examining the substitutions that have been observed in proteins known to be evolutionarily related. From an examination of appropriately aligned sequences, substitution matrices have been deduced. A *substitution matrix* describes a scoring system for the replacement of any amino acid with each of the other 19 amino acids. In these matrices, a large positive score corresponds to a substitution that occurs relatively frequently, whereas a large negative score corresponds to a substitution that occurs only rarely. One commonly used substitution matrix, the Blosum-62 (for *Blocks of amino acid substitution matrix*), is illustrated in Figure 6.9. In this depiction, each column in this matrix represents 1 of the 20 amino acids, whereas the position of the single-letter codes

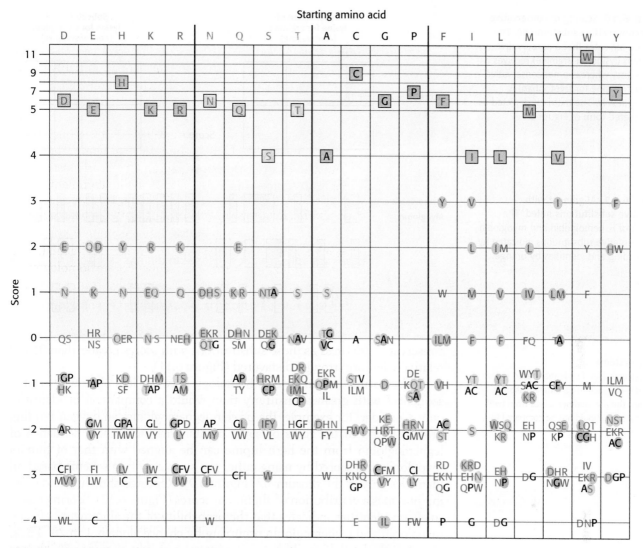

FIGURE 6.9 A graphic view of the Blosum-62. This substitution matrix was derived by examining substitutions within aligned sequence blocks in related proteins. Amino acids are classified into four groups (charged, red; polar, green; large and hydrophobic, blue; other, black). Substitutions that require the change of only a single nucleotide are shaded. Identities are boxed. To find the score for a substitution of, for instance, a Y for an H, you find the Y in the column having H at the top and check the number at the left. In this case, the resulting score is 2.

within each column specifies the score for the corresponding substitution. Notice that scores corresponding to identity (the boxed codes at the top of each column) are not the same for each residue, owing to the fact that less frequently occurring amino acids such as cysteine (C) and tryptophan (W) will align by chance less often than the more common residues. Furthermore, structurally conservative substitutions such as lysine (K) for arginine (R) and isoleucine (I) for valine (V) have relatively high scores, whereas nonconservative substitutions such as lysine for tryptophan result in negative scores (Figure 6.10). When two sequences are compared, each pair of aligned residues is assigned a score based on the matrix. In addition, gap penalties are often assessed. For example, the introduction of a single-residue gap lowers the alignment score by 12 points and the extension of an existing gap costs 2 points per residue. With the use of the Blosum-62 scoring system, the alignment between human α-hemoglobin and human myoglobin shown in Figure 6.6 receives a score of 115. In many regions, most substitutions are

FIGURE 6.10 Scoring of conservative and nonconservative substitutions. The Blosum-62 indicates that a conservative substitution (lysine for arginine) receives a positive score, whereas a nonconservative substitution (lysine for tryptophan) is scored negatively. The matrix is depicted as an abbreviated form of Figure 6.9.

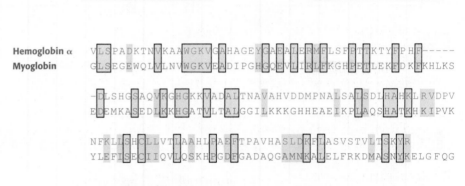

FIGURE 6.11 Alignment with conservative substitutions noted. The alignment of α-hemoglobin and myoglobin with conservative substitutions indicated by yellow shading and identities by orange.

Hemoglobin α   VLSPADKTNVKAAWGKVGAHAGEYGAEALERMFLSFPTTKTYFPHF-----
Myoglobin   GLSEGEWQLVLNVWGKVEADIPGHGQEVLIRLFKGHPETLEKFDKFKHLKS

                -DLSHGSAQVKGHGKKVADALTNAVAHVDDMPNALSALSDLHAHKLRVDPV
                EDEMKASEDLKKHGATVLTALGGILKKKGHHEAEIKPLAQSHATKHKIPVK

                NFKLLSHCLLVTLAAHLPAEFTPAVHASLDKFLASVSTVLTSKYR
                YLEFISECIIQVLQSKHPGDFGADAQGAMNKALELFRKDMASNYKELGFQG

Scoring systems

Scoring systems, like the identity-based system and Blosum-62, each use a different scoring scale. Thus, the raw number scores from two different systems should not be compared to one another.

conservative (defined as those substitutions with scores greater than 0) and relatively few are strongly disfavored (Figure 6.11).

This scoring system detects homology between less obviously related sequences with greater sensitivity than would a comparison of identities only. Consider, for example, the protein leghemoglobin, an oxygen-binding protein found in the roots of some plants. The amino acid sequence of leghemoglobin from the herb lupine can be aligned with that of human myoglobin and scored by using either the simple scoring scheme based on identities only or the Blosum-62 (Figure 6.9). Repeated shuffling and scoring provides a distribution of alignment scores (Figure 6.12). Scoring based solely on identities indicates that the probability of the alignment between myoglobin and leghemoglobin occurring by chance alone is 1 in 20. Thus, although the level of similarity suggests a relationship, there is a 5% chance that the similarity is accidental on the basis of this analysis. In contrast, users of the substitution matrix are able to incorporate the effects of conservative substitutions. From such an analysis, the odds of the alignment occurring by chance are calculated to be approximately 1 in 300. Thus, an analysis performed with the substitution matrix reaches a much firmer conclusion about the evolutionary relationship between these proteins (Figure 6.13).

Experience with sequence analysis has led to the development of simpler rules of thumb. For sequences longer than 100 amino acids, sequence identities greater than 25% are almost certainly not the result of chance alone; such sequences are probably homologous. In contrast, if two sequences are less than 15% identical, their alignment alone is unlikely to indicate

FIGURE 6.12 Alignment of identities only versus the Blosum-62. Repeated shuffling and scoring reveal the significance of sequence alignment for human myoglobin versus lupine leghemoglobin with the use of either (A) the simple, identity-based scoring system or (B) the Blosum-62. The scores for the alignment of the authentic sequences are shown in red. Accounting for amino acid similarity in addition to identity reveals a greater separation between the authentic alignment and the population of shuffled alignments.

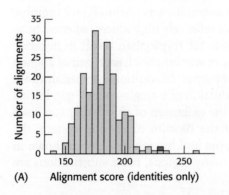

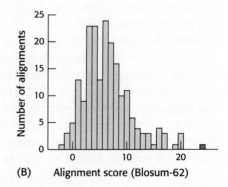

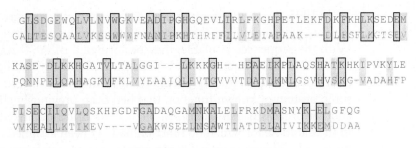

Myoglobin
Leghemoglobin

FIGURE 6.13 Alignment of human myoglobin and lupine leghemoglobin. The use of Blosum-62 yields the alignment shown between human myoglobin and lupine leghemoglobin, illustrating identities (orange boxes) and conservative substitutions (yellow). These sequences are 23% identical.

statistically significant similarity. For sequences that are between 15 and 25% identical, further analysis is necessary to determine the statistical significance of the alignment. It must be emphasized that *the lack of a statistically significant degree of sequence similarity does not rule out homology*. The sequences of many proteins that have descended from common ancestors have diverged to such an extent that the relationship between the proteins can no longer be detected from their sequences alone. As we will see, such homologous proteins can often be identified by examining their three-dimensional structures.

Databases can be searched to identify homologous sequences

When the sequence of a protein is first determined, comparing it with all previously characterized sequences can be a source of tremendous insight into its evolutionary relatives and, hence, its structure and function. Indeed, an extensive sequence comparison is almost always the first analysis performed on a newly elucidated sequence. The sequence-alignment methods just described are used to compare an individual sequence with all members of a database of known sequences.

Database searches for homologous sequences are most often accomplished by using resources available on the Internet at the National Center for Biotechnology Information (www.ncbi.nlm.nih.gov). The procedure used is referred to as a *BLAST* (Basic Local Alignment Search Tool) *search*. An amino acid sequence is typed or pasted into a browser (blast.ncbi.nlm .nih.gov), and a search is performed, most often against a nonredundant database of all known sequences. At the end of 2017, this database included more than 100 million sequences. A BLAST search yields a list of sequence alignments, each accompanied by an estimate giving the likelihood that the alignment occurred by chance (Figure 6.14).

FIGURE 6.14 BLAST search results. Part of the results from a BLAST search of the nonredundant (nr) protein sequence database using the sequence of ribose 5-phosphate isomerase (also called phosphopentose isomerase, Chapter 20) from *E. coli* as a query. Among the thousands of sequences found is the orthologous sequence from humans, and the alignment between these sequences is shown. The number of sequences with this level of similarity expected to be in the database by chance is 2×10^{-25} as shown by the E value (highlighted in red). Because this value is much less than 1, the observed sequence alignment is highly significant.

Name of homologous protein

Homo sapiens ribose 5-phosphate isomerase mRNA complete cds
Sequence ID: AY050633.1 **Length:** 1818 **Number of matches:** 1

Range 1: 252 to 920 — Number of sequences with this similarity expected by chance

Score	Expect	Method	Identities	Positives	Gaps	Frame
113 bits (283)	2e-25	Compositional matrix adjust.	82/224(37%)	118/224(52%)	15/224(6%)	+3

Sequence of queried protein →

```
Query    4   DELKKAVGWAALQ-YVQPGTIVGVGTGSTAAHFIDALGTMKGQIE---GAVSSSDASTEK   59
             +E KK  G AA++ +V+   ++G+G+GST H +  +      Q      + +S + +
Sbjct  252   EEAKKLAGRAAVENHVRNNQVLGIGSGSTIVHAVQRIAERVKQENLNLVCIPTSFQARQL  431
```

Sequence of homologous protein →

```
Query   60   LKSLGIHVFDLNEVDSLGIYVDGADEINGHMQMIKGGGAALTREKIIASVAEKFICIADA  119
             +   G+ + DL+   + + +DGADE++  + +IKGGG  LT+EKI+A  A +FI IAD
Sbjct  432   ILQYGLYLSDLDRHPEIDLAIDGADEVDADLNLIKGGGCLTQEKIVAGYASRFIVIADF  611

Query  120   SKQVDILG---KFPLPVEVIPMARSAVARQLV-KLGGRPEYRQG------VVTDNGNVIL  169
             K   LG    +P+EVIPMA  V+R + K GG E R       VVTDNGN IL
Sbjct  612   RKDSKNLGDQWHKGIPIEVIPMAYVPVSRAVSQKFGGVVELRMAVNKAGPVVTDNGNFIL  791

Query  170   DVHGMEILDPIAMENAINAIPGVVTVGLFANRGADVALIGTPDG  213
             D   +    + AI IPGVV GLF N A+   G DG
Sbjct  792   DWKFDRVHKWSEVNTAIKMIPGVVDTGLFINM-AERVYFGMQDG  920
```

Plus sign = conservative substitution   Letter = identity

In 1995, investigators reported the first complete sequence of the genome of a free-living organism, the bacterium *Haemophilus influenzae* (Figure 5.26). With the sequences available, they performed a BLAST search with each deduced protein sequence. Of 1743 identified protein-coding regions, also referred to as *open reading frames (ORFs)*, 1007 (58%) could be linked to some protein of known function that had been previously characterized in another organism. An additional 347 ORFs could be linked to sequences in the database for which no function had yet been assigned ("hypothetical proteins"). The remaining 389 sequences did not match any sequence present in the database at that time. Thus, investigators were able to identify likely functions for more than half the proteins within this organism solely by sequence comparisons.

6.3 Examination of Three-Dimensional Structure Enhances Our Understanding of Evolutionary Relationships

Sequence comparison is a powerful tool for extending our knowledge of protein function and kinship. However, biomolecules generally function as intricate three-dimensional structures rather than as linear polymers. Mutations occur at the level of sequence, but the effects of the mutations are at the level of function, and function is directly related to tertiary structure. Consequently, to gain a deeper understanding of evolutionary relationships between proteins, we must examine three-dimensional structures, especially in conjunction with sequence information. The techniques of structural determination were presented in Section 3.5.

Tertiary structure is more conserved than primary structure

Because three-dimensional structure is much more closely associated with function than is sequence, tertiary structure is more evolutionarily conserved than is primary structure. This conservation is apparent in the tertiary structures of the globins (Figure 6.15), which are extremely similar even though the similarity between human myoglobin and lupine leghemoglobin is weakly detectable at the sequence level (Figure 6.12) and that between human α-hemoglobin and lupine leghemoglobin is not statistically significant (15% identity). This structural similarity firmly establishes that the framework that binds the heme group and facilitates the reversible binding of oxygen has been conserved over a long evolutionary period.

Anyone aware of the similar biochemical functions of hemoglobin, myoglobin, and leghemoglobin could expect the structural similarities.

FIGURE 6.15
Conservation of three-dimensional structure. The tertiary structures of human hemoglobin (α chain), human myoglobin, and lupine leghemoglobin are conserved. Each heme group contains an iron atom to which oxygen binds. [Drawn from 1HBB.pdb, 1MBD.pdb, and 1GDJ.pdb.]

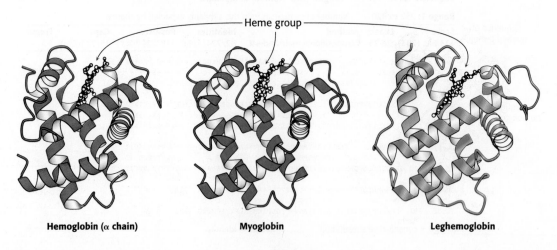

Heme group

Hemoglobin (α chain)　　　　**Myoglobin**　　　　**Leghemoglobin**

In a growing number of other cases, however, a comparison of three-dimensional structures has revealed striking similarities between proteins that were *not* expected to be related, on the basis of their diverse functions. A case in point is the protein actin, a major component of the cytoskeleton (See Section 36.2 on ⊘ SaplingPlus), and heat shock protein 70 (Hsp70), which assists protein folding inside cells. These two proteins were found to be noticeably similar in structure despite only 16% sequence identity (Figure 6.16). On the basis of their three-dimensional structures, actin and Hsp70 are paralogs. The level of structural similarity strongly suggests that, despite their different biological roles in modern organisms, these proteins descended from a common ancestor. As the three-dimensional structures of more proteins are determined, such unexpected kinships are being discovered with increasing frequency. The search for such kinships relies ever more frequently on computer-based searches that are able to compare the three-dimensional structure of any protein with all other known structures.

Knowledge of three-dimensional structures can aid in the evaluation of sequence alignments

The sequence-comparison methods described thus far treat all positions within a sequence equally. However, we have learned from examining families of homologous proteins for which at least one three-dimensional structure is known that regions and residues critical to protein function are more strongly conserved than are other residues. For example, each type of globin contains a bound heme group with an iron atom at its center. A histidine residue that interacts directly with this iron atom (Section 7.1) is conserved in all globins. After we have identified key residues or highly conserved sequences within a family of proteins, we can sometimes identify other family members even when the overall level of sequence similarity is below statistical significance. Thus it may be useful to generate a *sequence template*—a map of conserved residues that are structurally and functionally important and are characteristic of particular families of proteins—which makes it possible to recognize new family members that might be undetectable by other means. Other methods for sequence classification that take advantage of known three-dimensional structures are also available. Still other methods are able to identify conserved residues within a family of homologous proteins, even without a known three-dimensional structure. These methods often use substitution matrices that differ at each position within a family of aligned sequences. Such methods can often detect quite distant evolutionary relationships.

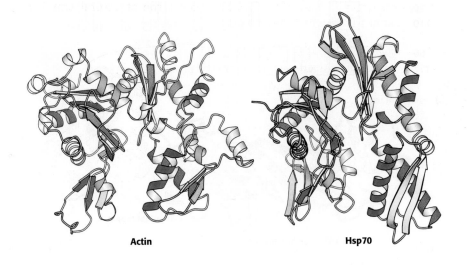

Actin   **Hsp70**

FIGURE 6.16 Structures of actin and a large fragment of heat shock protein 70 (Hsp70). A comparison of the identically colored elements of secondary structure reveals the overall similarity in structure despite the difference in biochemical activities. [Drawn from 1ATN.pdb and 1ATR.pdb.]

Repeated motifs can be detected by aligning sequences with themselves

As discussed in Section 2.4, domains are compact regions of protein structure that can be found in a variety of protein sequences. More than 10% of all proteins contain sets of two or more domains that are similar to one another. Sequence search methods can often detect these internally repeated sequences that have been characterized in other proteins. Often, however, internally repeated sequences do not correspond to previously identified domains. In these cases, their presence can be detected by attempting to align a given sequence with itself. The statistical significance of such repeats can be tested by aligning the regions in question as if these regions were sequences from separate proteins. For the TATA-box-binding protein (Figure 6.17A), a key protein in controlling eukaryotic gene transcription (Section 4.5), such an alignment is highly significant: comparing two 90-residue regions of the protein reveals that 30% of the amino acids are identical (Figure 6.17B). The estimated probability of such an alignment occurring by chance is 1 in 10^{13}. The determination of the three-dimensional structure of the TATA-box-binding protein confirmed the presence of repeated structures; the protein is formed of two nearly identical domains (Figure 6.17C). The evidence is convincing that the gene encoding this protein evolved by duplication of a gene encoding a single domain.

Convergent evolution illustrates common solutions to biochemical challenges

Thus far, we have been exploring proteins derived from common ancestors—that is, through *divergent evolution*. Other cases have been found of proteins that are structurally similar in important ways but are not descended from a common ancestor. How might two unrelated proteins come to resemble each other structurally? Two proteins evolving independently may have

FIGURE 6.17 Sequence alignment of internal repeats. (A) The primary structure of the TATA-box-binding protein. (B) An alignment of the sequences of the two repeats of the TATA-box-binding protein. The amino-terminal repeat is shown in red and the carboxyl-terminal repeat in blue. (C) Structure of the TATA-box-binding protein. The amino-terminal domain is shown in red and the carboxyl-terminal domain in blue. For a different view of this protein, see Figure 30.26. [Drawn from 1VOK.pdb.]

(A) Primary structure of TATA-box-binding protein:

```
  1   MTDQGLEGSN  PVDLSKHPSG  IVPTLQNIVS  TVNLDCKLDL  KAIALQARNA
 51   EYNPKRFAAV  IMRIREPKTT  ALIFASGKMV  CTGAKSEDFS  KMAARKYARI
101   VQKLGFPAKF  KDFKIQNIVG  SCDVKFPIRL  EGLAYSHAAF  SSYEPELFPG
151   LIYRMKVPKI  VLLIFVSGKI  VITGAKMRDE  TYKAFENIYP  VLSEFRKIQQ
```

(B) Self-alignment:

```
  1   MTDQGLEGSNPVDLSKHPS
```

```
 20   GIVPTLQNIVSTVNLDCKLDLKAIALQ-ARNAEYNPKRFAAVIMRIR
110   FKDFKIQNIVGSCDVKFPIRLEGLAYSHAAFSSYEPELFPGLIYRMK
```

```
 66   EPKTTALIFASGKMVCTGAKSEDFSKMAARKYARIVQKLGFPAK
157   VPKIVLLIFVSGKIVITGAKMRDETYKAFENIYPVLSEFRKIQQ
```

(C) N-terminal repeat

C-terminal repeat

TATA-box-binding protein

converged on similar structural features to perform a similar biochemical activity. Perhaps that structure was an especially effective solution to a biochemical problem. The process by which very different evolutionary pathways lead to the same solution is called *convergent evolution*.

An example of convergent evolution is found among the serine proteases. These enzymes, to be considered in more detail in Chapter 9, cleave peptide bonds by hydrolysis. Figure 6.18 shows the structure of the active sites—that is, the sites on the proteins at which the hydrolysis reaction takes place—for two such enzymes, chymotrypsin and subtilisin. These active-site structures are remarkably similar. In each case, a serine residue, a histidine residue, and an aspartate residue are positioned in space in nearly identical arrangements. As we will see, this conserved spatial arrangement is critical for the activity of these enzymes and affords the same mechanistic solution to the problem of peptide bond hydrolysis. At first glance, this similarity might suggest that these proteins are homologous. However, striking differences in the overall structures of these proteins make an evolutionary relationship extremely unlikely (Figure 6.19). Whereas chymotrypsin consists almost entirely of β sheets, subtilisin contains extensive α-helical structure. Moreover, the key serine, histidine, and aspartic acid residues do not even appear in the same order within the two sequences. It is extremely unlikely that two proteins evolving from a common ancestor could have retained similar active-site structures while other aspects of the structure changed so dramatically.

Comparison of RNA sequences can be a source of insight into RNA secondary structures

Homologous RNA sequences can be compared in a manner similar to that already described for protein sequences. Such comparisons can be a source of important insights into evolutionary relationships; in addition, they provide clues to the three-dimensional structure of the RNA itself. As noted in Chapter 4, single-stranded nucleic acid molecules fold back on themselves to form elaborate structures held together by Watson–Crick base-pairing and other noncovalent interactions. In a family of sequences that form similar base-paired structures, base sequences may vary, but base-pairing ability is conserved. Consider, for example, a region from a large RNA molecule present in the ribosomes of all organisms (Figure 6.20). In the region shown, the *E. coli* sequence has a guanine (G) residue in position 9 and a cytosine (C) residue in position 22, whereas the human sequence has uracil (U) in position 9 and adenine (A) in position 22. Examination of the six sequences shown in Figure 6.20 reveals that the bases in positions 9 and 22, as well as several of the neighboring positions, retain the ability to form

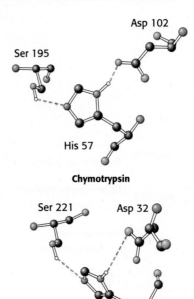

FIGURE 6.18 Convergent evolution of protease active sites. The relative positions of the three key residues shown are nearly identical in the active sites of the serine proteases chymotrypsin and subtilisin.

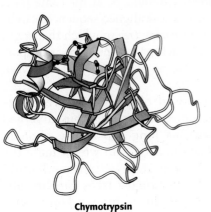

FIGURE 6.19 Structures of mammalian chymotrypsin and bacterial subtilisin. The overall structures are quite dissimilar, in stark contrast with the active sites shown at the top of each structure. The β strands are shown in yellow and the α helices in blue. [Drawn from 1GCT.pdb. and 1SUP.pdb.]

(A)

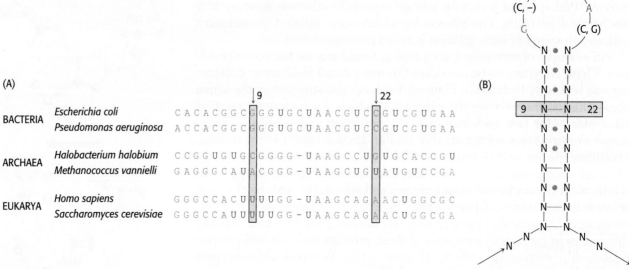

BACTERIA	*Escherichia coli*	C A C A C G G C G G G U G C U A A C G U C C G U C G U G A A
	Pseudomonas aeruginosa	A C C A C G G C G G G U G C U A A C G U C C G U C G U G A A
ARCHAEA	*Halobacterium halobium*	C C G G U G U G C G G G G - U A A G C C U G U G C A C C G U
	Methanococcus vannielli	G A G G G C A U A C G G G - U A A G C U G U A U G U C C G A
EUKARYA	*Homo sapiens*	G G G C C A C U U U U G G - U A A G C A G A A C U G G C G C
	Saccharomyces cerevisiae	G G G C C A U U U U U G G - U A A G C A G A A C U G G C G A

(B)

FIGURE 6.20 Comparison of RNA sequences. (A) A comparison of sequences in a part of ribosomal RNA taken from a variety of species. (B) The implied secondary structure. Green lines indicate positions at which Watson–Crick base-pairing is completely conserved in the sequences shown, whereas dots indicate positions at which Watson–Crick base-pairing is conserved in most cases.

Watson–Crick base pairs even though the identities of the bases in these positions vary. We can deduce that two segments with paired mutations that maintain base-pairing ability are likely to form a double helix. Where sequences are known for several homologous RNA molecules, this type of sequence analysis can often suggest complete secondary structures as well as interactions that do not follow the standard Watson–Crick base pairing rules (Section 4.2). For this particular ribosomal RNA, the subsequent determination of its three-dimensional structure (Section 31.3) confirmed the predicted secondary structure.

6.4 Evolutionary Trees Can Be Constructed on the Basis of Sequence Information

The observation that homology is often manifested as sequence similarity suggests that the evolutionary pathway relating the members of a family of proteins may be deduced by examination of sequence similarity. This approach is based on the notion that sequences that are more similar to one another have had less evolutionary time to diverge than have sequences that are less similar. This method can be illustrated by using the three globin sequences in Figures 6.11 and 6.13, as well as the sequence for the human hemoglobin β chain. These sequences can be aligned with the additional constraint that gaps, if present, should be at the same positions in all of the proteins. These aligned sequences can be used to construct an *evolutionary tree* in which the length of the branch connecting each pair of proteins is proportional to the number of amino acid differences between the sequences (Figure 6.21).

Such comparisons reveal only the relative divergence times—for example, that myoglobin diverged from hemoglobin twice as long ago as the α chain diverged from the β chain. How can we estimate the approximate dates of gene duplications and other evolutionary events? Evolutionary trees can be calibrated by comparing the deduced branch points with

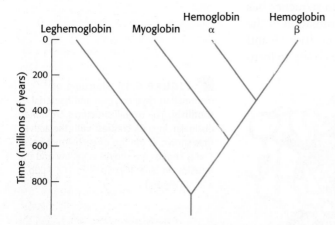

FIGURE 6.21 An evolutionary tree for globins. The branching structure was deduced by sequence comparison, whereas the results of fossil studies provided the overall time scale showing when divergence occurred.

divergence times determined from the fossil record. For example, the duplication leading to the two chains of hemoglobin appears to have occurred 350 million years ago. This estimate is supported by the observation that jawless fish such as the lamprey, which diverged from bony fish approximately 400 million years ago, contain hemoglobin built from a single type of subunit (Figure 6.22). These methods can be applied to both relatively modern and very ancient molecules, such as the ribosomal RNAs that are found in all organisms. Indeed, such an RNA sequence analysis led to the realization that Archaea are a distinct group of organisms that diverged from Bacteria very early in evolutionary history (Figure 1.3).

Horizontal gene transfer events may explain unexpected branches of the evolutionary tree

Evolutionary trees that encompass orthologs of a particular protein across a range of species can lead to unexpected findings. Consider the unicellular red alga *Galdieria sulphuraria*, a remarkable eukaryote that can thrive in extreme environments, including at temperatures up to 56°C, at pH values between 0 and 4, and in the presence of high concentrations of toxic metals. *G. sulphuraria* belongs to the order Cyanidiales, clearly within the eukaryotic branch of the evolutionary tree (Figure 6.23A). However, the complete genome sequence of this organism revealed that nearly 5% of the *G. sulphuraria* ORFs encode proteins that are more closely related to bacterial or archaeal, not eukaryotic, orthologs. Furthermore, the proteins that exhibited these unexpected evolutionary relationships possess functions that are likely to confer a survival advantage in extreme environments, such as the removal of metal ions from inside the cell (Figure 6.23B). One likely explanation for these observations is *horizontal gene transfer*, or *the exchange of DNA between species* that provides a selective advantage to the recipient. Amongst prokaryotes, horizontal gene transfer is a well-characterized and important evolutionary mechanism. For example, as we shall discuss in Chapter 9, exchange of plasmid DNA between bacterial species likely facilitated the acquisition of restriction endonuclease activities. However, recent studies such as those on *G. sulphuraria*, made possible by the expansive growth of complete genome sequence information, suggest that horizontal gene transfer from prokaryotes to eukaryotes, between different domains of life, may also represent evolutionarily significant events.

FIGURE 6.22 The lamprey. A jawless fish whose ancestors diverged from bony fish approximately 400 million years ago, the lamprey has hemoglobin molecules that contain only a single type of polypeptide chain. [Breck P. Kent.]

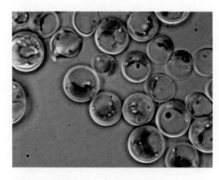

The unicellular red alga, *Galdieria sulphuraria*. [Dr. Gerald Schoenknecht.]

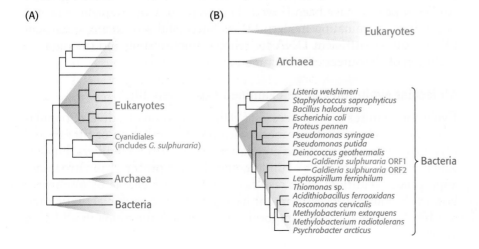

(A)

(B)

Eukaryotes

Archaea

Eukaryotes

Cyanidiales (includes *G. sulphuraria*)

Archaea

Bacteria

Listeria welshimeri
Staphylococcus saprophyticus
Bacillus halodurans
Escherichia coli
Proteus pennen
Pseudomonas syringae
Pseudomonas putida
Deinococcus geothermalis
Galdieria sulphuraria ORF1
Galdieria sulphuraria ORF2
Leptospirillum ferriphilum
Thiomonas sp.
Acidithiobacillus ferrooxidans
Roscomonas cervicalis
Methylobacterium extorquens
Methylobacterium radiotolerans
Psychrobacter arcticus

Bacteria

FIGURE 6.23 Evidence of horizontal gene transfer. (A) The unicellular red alga *Galdieria sulphuraria* belongs to the order Cyanidiales, clearly within the eukaryotic branch of the evolutionary tree. (B) Within the completely sequenced *G. sulphuraria* genome, two ORFs encode proteins involved in transport of arsenate ions across membranes. Alignment of these ORFs against orthologs from a variety of species reveals that these pumps are most closely related to their bacterial counterparts, suggesting that a horizontal gene transfer event occurred during the evolution of this species. [(A) Information from Dr. Gerald Schönknecht; (B) Information from G. Schönknecht et al., *Science* 339:1207–1210, 2013, Fig. 3.]

6.5 Modern Techniques Make the Experimental Exploration of Evolution Possible

Two techniques of biochemistry have made it possible to examine the course of evolution directly and not simply by inference. The polymerase chain reaction (Section 5.1) allows the direct examination of ancient DNA sequences, releasing us, at least in some cases, from the constraints of being able to examine existing genomes from living organisms only. Molecular evolution may also be investigated through the use of *combinatorial chemistry,* the process of producing large populations of molecules en masse and selecting for a biochemical property. This exciting process provides a glimpse into the types of molecules that may have existed very early in evolution.

Ancient DNA can sometimes be amplified and sequenced

The tremendous chemical stability of DNA makes the molecule well suited to its role as the storage site of genetic information. So stable is the molecule that samples of DNA have survived for many thousands of years under appropriate conditions. With the development of PCR and advanced DNA-sequencing methods, such ancient DNA can be amplified and sequenced. This approach was first applied to mitochondrial DNA isolated from a Neanderthal fossil estimated at 38,000 years of age. Comparison of the complete Neanderthal mitochondrial sequence with those from *Homo sapiens* individuals revealed between 201 and 234 substitutions, considerably fewer than the approximately 1500 differences between human beings and chimpanzees over the same region of mitochondrial DNA.

Remarkably, the complete genome sequences of a Neanderthal and a closely related hominin known as a Denisovan have been obtained using DNA isolated from nearly 50,000-year-old fossils. Comparison of these sequences suggests that the common ancestor of modern human beings and Neanderthals lived approximately 570,000 years ago, while the common ancestor between Neanderthals and Denisovans lived nearly 380,000 years ago. An evolutionary tree constructed from these data revealed that the Neanderthal was not an intermediate between chimpanzees and human beings but, instead, was an evolutionary "dead end" that became extinct (Figure 6.24). Further analysis of these sequences has enabled researchers to determine the extent of interbreeding between these groups, elucidate the geographic history of these populations, and make inferences about additional ancestors whose DNA has not yet been sequenced.

A few earlier studies claimed to determine the sequences of far more ancient DNA such as that found in insects trapped in amber, but these studies appear to have been flawed. The source of these sequences turned out to be contaminating modern DNA. Successful sequencing of ancient DNA requires sufficient DNA for reliable amplification and the rigorous exclusion of all sources of contamination.

Molecular evolution can be examined experimentally

Evolution requires three processes: (1) the generation of a diverse population, (2) the selection of members based on some criterion of fitness, and (3) reproduction to enrich the population in these fitter members. Nucleic acid molecules are capable of undergoing all three processes in vitro under appropriate conditions. The results of such studies enable us to glimpse how evolutionary processes might have generated catalytic activities and specific binding abilities—important biochemical functions in all living systems.

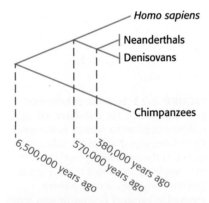

FIGURE 6.24 Placing Neanderthals and Denisovans on an evolutionary tree. Comparison of DNA sequences revealed that neither Neanderthals nor the Denisovans are on the line of direct descent leading to *Homo sapiens* but, instead, branched off earlier and then became extinct.

A diverse population of nucleic acid molecules can be synthesized in the laboratory by the process of combinatorial chemistry, which rapidly produces large populations of a particular type of molecule such as a nucleic acid. A population of molecules of a given size can be generated randomly so that many or all possible sequences are present in the mixture. When an initial population has been generated, it is subjected to a selection process that isolates specific molecules with desired binding or reactivity properties. Finally, molecules that have survived the selection process are replicated through the use of PCR; primers are directed toward specific sequences included at the ends of each member of the population. Errors that occur naturally in the course of the replication process introduce additional variation into the population in each "generation."

Let us consider an application of this approach. Early in evolution, before the emergence of proteins, RNA molecules may have played all major roles in biological catalysis. To understand the properties of potential RNA catalysts, researchers have used the methods heretofore described to create an RNA molecule capable of binding adenosine triphosphate and related nucleotides. An initial population of RNA molecules 169 nucleotides long was created; 120 of the positions differed randomly, with equimolar mixtures of adenine, cytosine, guanine, and uracil. The initial synthetic pool that was used contained approximately 10^{14} RNA molecules. Note that this number is a very small fraction of the total possible pool of random 120-base sequences. From this pool, those molecules that bound to ATP, which had been immobilized on a column, were selected (Figure 6.25).

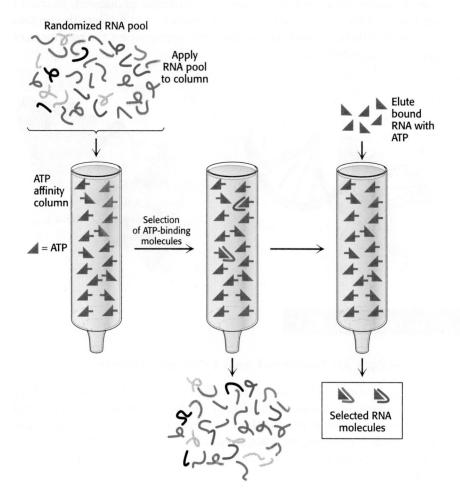

Randomized RNA pool

Apply RNA pool to column

ATP affinity column

⬛ = ATP

Selection of ATP-binding molecules

Elute bound RNA with ATP

Selected RNA molecules

FIGURE 6.25 Evolution in the laboratory. A collection of RNA molecules of random sequences is synthesized by combinatorial chemistry. This collection is selected for the ability to bind ATP by passing the RNA through an ATP affinity column (Section 3.1). The ATP-binding RNA molecules are released from the column by washing with excess ATP and then replicated. The process of selection and replication is then repeated several times. The final RNA products with significant ATP-binding ability are isolated and characterized.

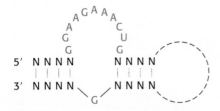

FIGURE 6.26 A conserved secondary structure. The secondary structure shown is common to RNA molecules selected for ATP binding. Bases important for ATP recognition are shown in red. Dashed line indicates a region which varied in length among the various RNA products.

The collection of molecules that bound well to the ATP affinity column was replicated by reverse transcription into DNA, amplification by PCR, and transcription back into RNA. Additional diversity is introduced into the pool by the somewhat error-prone reverse transcriptase, which introduces additional mutations into the population during each cycle. The new population was subjected to additional rounds of selection for ATP-binding activity. After eight generations, members of the selected population were characterized by sequencing. Seventeen different sequences were obtained, 16 of which could form the structure shown in Figure 6.26. Each of these molecules bound ATP with dissociation constants less than 50 μM.

The folded structure of the ATP-binding region from one of these RNAs was determined by nuclear magnetic resonance (NMR) methods (Section 3.5). As expected, this 40-nucleotide molecule is composed of two Watson–Crick base-paired helical regions separated by an 11-nucleotide loop (Figure 6.27A). This loop folds back on itself in an intricate way (Figure 6.27B) to form a deep pocket into which the adenine ring can fit (Figure 6.27C). Thus, a structure had evolved in vitro that was capable of a specific interaction.

Synthetic oligonucleotides that can specifically bind ligands, such as the ATP-binding RNA molecules described above, are referred to as *aptamers*. In addition to their role in understanding molecular evolution, aptamers have shown promise as versatile tools for biotechnology and medicine. They have been developed for diagnostic applications, serving as sensors for ligands ranging from small organic molecules, such as cocaine, to larger proteins, such as thrombin. Several aptamers are also in clinical trials as therapies for diseases ranging from leukemia to diabetes. Macugen (pegaptanib sodium), an aptamer which binds to and inhibits the protein vascular endothelial growth factor, has been approved for the treatment of age-related macular degeneration.

FIGURE 6.27 An evolved ATP-binding RNA molecule. (A) The Watson–Crick base-pairing pattern of an RNA molecule selected to bind adenosine nucleotides. (B) The NMR structure of this RNA molecule reveals the deep pocket into which the ATP molecule is bound. (C) In this surface representation, the ATP molecule has been removed to enable visualization of the pocket. [Drawn from 1RAW.pdb.]

(A)

(B)

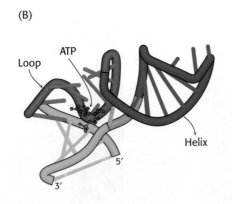

(C)

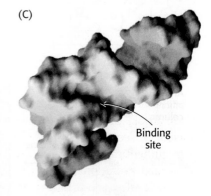

SUMMARY

6.1 Homologs Are Descended from a Common Ancestor

Exploring evolution biochemically often means searching for homology between molecules, because homologous molecules, or homologs, evolved from a common ancestor. Paralogs are homologous molecules that are found in one species and have acquired different functions through evolutionary time. Orthologs are homologous molecules that are found in different species and have similar or identical functions.

6.2 Statistical Analysis of Sequence Alignments Can Detect Homology

Protein and nucleic acid sequences are two of the primary languages of biochemistry. Sequence-alignment methods are the most powerful tools of the evolutionary detective. Sequences can be aligned to maximize their similarity, and the significance of these alignments can be judged by statistical tests. The detection of a statistically significant alignment between two sequences strongly suggests that two sequences are related by divergent evolution from a common ancestor. The use of substitution matrices makes the detection of more-distant evolutionary relationships possible. Any sequence can be used to probe sequence databases to identify related sequences present in the same organism or in other organisms.

6.3 Examination of Three-Dimensional Structure Enhances Our Understanding of Evolutionary Relationships

The evolutionary kinship between proteins may be even more strikingly evident in the conserved three-dimensional structures. The analysis of three-dimensional structure in combination with the analysis of especially conserved sequences has made it possible to determine evolutionary relationships that cannot be detected by other means. Sequence-comparison methods can also be used to detect imperfectly repeated sequences within a protein, indicative of domains that may have been duplicated and linked together over the course of evolution.

6.4 Evolutionary Trees Can Be Constructed on the Basis of Sequence Information

Evolutionary trees can be constructed with the assumption that the number of sequence differences corresponds to the time since the two sequences diverged. Construction of an evolutionary tree based on sequence comparisons revealed approximate times for the gene-duplication events separating myoglobin and hemoglobin as well as the α and β subunits of hemoglobin. Evolutionary trees based on sequences can be compared with those based on fossil records to provide a clearer understanding of the timeline of divergences. Horizontal gene transfer events can appear as unexpected branches on the evolutionary tree.

6.5 Modern Techniques Make the Experimental Exploration of Evolution Possible

The exploration of evolution can also be a laboratory science. In favorable cases, PCR amplification of well-preserved samples allows the determination of nucleotide sequences from extinct organisms. Sequences so determined can help authenticate parts of an evolutionary tree constructed by other means. Molecular evolutionary experiments performed in the test tube can examine how molecules such as ligand-binding RNA molecules might evolve over time.

APPENDIX

Biochemistry in Focus

Using sequence alignments to identify functionally important residues

In this chapter, we have discussed how sequence alignments can be used to determine whether two proteins are related to one another by a common ancestor. In practice, alignment methods, combined with the vast database of known protein sequences, can be versatile and powerful tools to help provide insight into the structure and function of proteins. For example, sequence alignments can be used to identify residues that are critically important to the function of a protein whose structure has not yet been solved.

Let us consider the example of the odorant receptors (ORs; See Section 34.1 on Sapling Plus). OR genes comprise one of the largest mammalian gene superfamilies. Each OR gene encodes a seven-transmembrane-helix (7TM) receptor, one of a large family of proteins that traverse the membrane via seven α-helices (Section 14.1). Each OR is activated by its own unique pattern of ligands, which in this case are odorants. While the structures of several 7TM receptors have been solved, no structure of an OR has been determined. Hence, the atomic details of how odorants, a highly diverse array of compounds, are recognized by their cognate receptors remains of great interest.

In one approach to identify residues important for odorant binding, one research group cleverly used the available OR sequences obtained from the completely sequenced human and mouse genomes. They assumed that residues important for odorant binding should be highly conserved between paired orthologs from both species. Analysis of 218 such ortholog pairs identified 146 positions within the OR sequences that were significantly conserved. However, they then reasoned that these conserved positions would not only contain residues that bind odorant, but would also include amino acids important for preservation of OR structure as well as binding to other proteins. To filter out these residues, the researchers aligned OR sequences from closely related paralogs. Recall that paralogs represent homologs from the same species that have differing functions (here, the recognition of different odorants). In this case, residues involved in protein folding or conserved protein–protein interactions would be conserved between paralogs, but the residues involved in odorant binding would not be conserved. This approach enabled the original list of 146 residues to be reduced to 22.

When mapped onto the known 7TM structure of rhodopsin (see Figure 14.5A), these residues all mapped to a single cluster within three of the helices, and corresponded closely with the location of the retinal ligand site in the rhodopsin structure (Figure 6.28). This odorant-binding-site prediction was enabled by having the completed genomes of human and mouse and the clever application of orthologous and paralogous sequence alignments. Subsequent studies, including site-directed mutagenesis and molecular modeling, have since corroborated and refined these predictions.

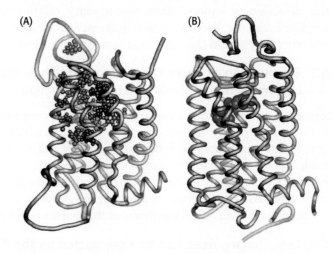

FIGURE 6.28 Modeling of the odorant receptor binding pocket. (A) The 22 residues found to be highly conserved amongst orthologous odorant receptor pairs, but not conserved between paralog pairs, were mapped on the known structure of the GPCR bovine rhodopsin. *Notice* that these residues (colored light green) cluster in one specific region of the receptor that closely aligns with (B) the retinal moiety (colored purple) of rhodopsin. [Republished with permission of John Wiley & Sons, from O. Man, Y. Gilad, and D. Lancet, Protein Sci. 13:240–254, 2004; permission conveyed through Copyright Clearance Center, Inc. Image courtesy Doron Lancet.]

APPENDIX

Problem-Solving Strategies

PROBLEM: *RNA alignment.* Sequences of an RNA fragment from five species have been determined and aligned. Propose a likely secondary structure for these fragments.

```
(1)   UUGGAGAUUCGGUAGAAUCUCCC
(2)   GCCGGGAAUCGACAGAUUCCCCG
(3)   CCCAAGUCCCGGCAGGGACUUAC
(4)   CUCACCUGCCGAUAGGCAGGUCA
(5)   AAUACCACCCGGUAGGGUGGUUC
```

SOLUTION: For a simple alignment over a short RNA fragment such as the one featured in this problem, the trick is to look for two positions that could form a base pair in each of the aligned sequences. Of particular interest in this alignment is the base highlighted in red:

```
(1)   UUGGAGAUUCGGUAGAAUCUCCC
(2)   GCCGGGAAUCGACAGAUUCCCCG
(3)   CCCAAGUCCCGGCAGGGACUUAC
(4)   CUCACCUGCCGAUAGGCAGGUCA
(5)   AAUACCACCCGGUAGGGUGGUUC
```

In all five sequences, this is an absolutely conserved cytosine. In this alignment, there are only two absolutely conserved guanine bases. One is immediately 3′ to this cytosine, and we know from oligonucleotide structure that a base pair between adjacent positions is highly unlikely. The other is the guanine highlighted in blue, which is separated from the cytosine by four nucleotides. An intriguing possibility!

Now, we have an "anchor" from which we can continue looking for additional pairings. In order for this red–blue pairing to form, the RNA fragment would need to fold on itself. Hence, we should look outward from this pair to identify additional pairings.

```
(1)   UUGGAGAUUCGGUAGAAUCUCCC
(2)   GCCGGGAAUCGACAGAUUCCCCG
(3)   CCCAAGUCCCGGCAGGGACUUAC
(4)   CUCACCUGCCGAUAGGCAGGUCA
(5)   AAUACCACCCGGUAGGGUGGUUC
```

Again, we have a conserved pairing. In sequences (1) and (2) the red and blue positions feature a U–A pair, while in sequences (3), (4), and (5), these positions form a C–G pair. If we continue to "walk" along the sequence in this manner, a clear stretch of base-pairing emerges. Moreover, since several of these positions are absolutely conserved, we can include them in our proposed structure (shown here with the original C–G base pair in red/blue):

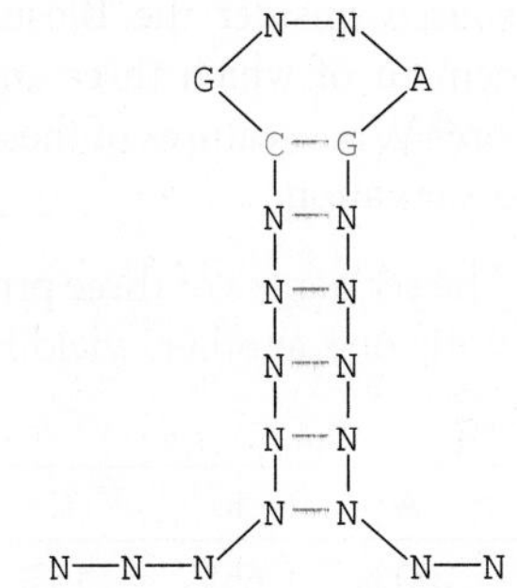

KEY TERMS

homolog (p. 187)

paralog (p. 187)

ortholog (p. 187)

sequence alignment (p. 187)

conservative substitution (p. 190)

nonconservative substitution (p. 190)

substitution matrix (p. 190)

BLAST search (p. 193)

sequence template (p. 195)

divergent evolution (p. 196)

convergent evolution (p. 197)

evolutionary tree (p. 198)

horizontal gene transfer (p. 199)

combinatorial chemistry (p. 200)

aptamers (p. 202)

PROBLEMS

1. *What's the score?* Using the identity-based scoring system (Section 6.2), calculate the score for the following alignment. Do you think these two proteins are related? ✓❷

```
(1) WYLGKITRMDAEVLLKKPTVRDGHFLVTQCESSPGEF
(2) WYFGKITRRESERLLLNPENPRGTFLVRESETTKGAY

    SISVRFGDSVQ-----HFKVLRDQNGKYYLWAVK-FN
    CLSVSDFDNAKGLNVKHYKIRKLDSGGFYITSRTQFS

    SLNELVAYHRTASVSRTHTILLSDMNV
    SSLQQLVAYYSKHADGLCHRLTNV
```

2. *Sequence and structure.* A comparison of the aligned amino acid sequences of two proteins each consisting of 150 amino acids reveals them to be only 8% identical. However, their three-dimensional structures are very similar. Are these two proteins related evolutionarily? Explain. ✓❹

3. *It depends on how you count.* Consider the following two sequence alignments:

```
(1) A-SNLFDIRLIG        (2) ASNLFDIRLI-G
    GSNDFYEVKIMD            GSNDFYEVKIMD
```

Which alignment has a higher score if the identity-based scoring system (Section 6.2) is used? Which alignment has a higher score if the Blosum-62 substitution matrix (Figure 6.9) is used? For the Blosum scoring, you need not apply a gap penalty.

4. *Discovering a new base pair.* Examine the ribosomal RNA sequences in Figure 6.20. In sequences that do not contain Watson–Crick base pairs, what base tends to be paired with G? Propose a structure for your new base pair.

5. *Overwhelmed by numbers.* Suppose that you wish to synthesize a pool of RNA molecules that contain all four bases at each of 40 positions. How much RNA must you have in grams if the pool is to have at least a single molecule of each sequence? The average molecular weight of a nucleotide is 330 g mol^{-1}.

6. *Form follows function.* The three-dimensional structure of biomolecules is more conserved evolutionarily than is sequence. Why? ✓❶

7. *Shuffling.* Using the identity-based scoring system (Section 6.2), calculate the alignment score for the alignment of the following two short sequences: ✓❷

```
(1) ASNFLDKAGK
(2) ATDYLEKAGK
```

Generate a shuffled version of sequence 2 by randomly reordering these 10 amino acids. Align your shuffled sequence with sequence 1 without allowing gaps, and calculate the alignment score between sequence 1 and your shuffled sequence.

8. *Interpreting the score.* Suppose that the sequences of two proteins each consisting of 200 amino acids are aligned and that the percentage of identical residues has been calculated. How would you interpret each of the following results in regard to the possible divergence of the two proteins from a common ancestor? ✓❹

(a) 80%

(b) 50%

(c) 20%

(d) 10%

9. *Particularly unique.* Consider the Blosum-62 matrix in Figure 6.9. Replacement of which three amino acids never yields a positive score? What features of these residues might contribute to this observation?

10. *A set of three.* The sequences of three proteins (A, B, and C) are compared with one another, yielding the following levels of identity:

	A	B	C
A	100%	65%	15%
B	65%	100%	55%
C	15%	55%	100%

Assume that the sequence matches are distributed uniformly along each aligned sequence pair. Would you expect protein A and protein C to have similar three-dimensional structures? Explain.

11. *The more the merrier.* When RNA alignments are used to determine secondary structure, it is advantageous to have many sequences representing a wide variety of species. Why?

12. *To err is human.* You have discovered a mutant form of a thermostable DNA polymerase with significantly reduced fidelity in adding the appropriate nucleotide to the growing DNA strand, compared with wild-type DNA polymerase. How might this mutant be useful in the molecular-evolution experiments described in Section 6.5. ✓⑤

13. *Generation to generation.* When performing a molecular-evolution experiment, such as that described in Section 6.5, why is it important to repeat the selection and replication steps for several generations? ✓⑤

14. *BLAST away.* Using the National Center for Biotechnology Information website (www.ncbi.nlm.nih.gov), find the sequences of the enzyme triose phosphate isomerase from *E. coli* strain K-12 and from *Homo sapiens* (use isoform #1). Use the Global Align tool on the NCBI BLAST website to align these two sequences. What is the percent identity between these two proteins?

Think ▶ Pair ▶ Share Problem

15. In the sequence alignment scoring examples given in this chapter, the penalties imposed by gaps may seem arbitrary. However, these values are important: if the gap penalty is too high, a potential evolutionary relationship between two sequences may be overlooked; if the gap penalty is too low, artificial relationships may be inferred due to the introduction of too many gaps. Can you suggest a method to determine the most appropriate gap penalty? What additional information might you need?

Hemoglobin: Portrait of a Protein in Action

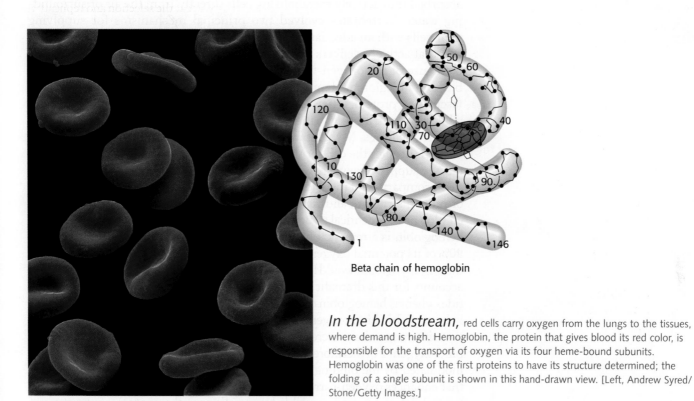

Beta chain of hemoglobin

In the bloodstream, red cells carry oxygen from the lungs to the tissues, where demand is high. Hemoglobin, the protein that gives blood its red color, is responsible for the transport of oxygen via its four heme-bound subunits. Hemoglobin was one of the first proteins to have its structure determined; the folding of a single subunit is shown in this hand-drawn view. [Left, Andrew Syred/Stone/Getty Images.]

 L E A R N I N G   O B J E C T I V E S

By the end of this chapter, you should be able to:

1. Explain how the structure of myoglobin changes upon oxygen binding.

2. Describe the differences in the oxygen-binding properties of hemoglobin and myoglobin.

3. Describe how the cooperative binding of hemoglobin allows for optimal oxygen delivery.

4. Explain the differences between the two models for hemoglobin cooperativity.

5. Identify the key regulators of hemoglobin function.

6. Define the Bohr effect and describe how it explains maximal oxygen delivery to tissues in greatest need.

7. Give several examples of hemoglobin mutations that result in disease.

O U T L I N E

7.1 Binding of Oxygen by Heme Iron

7.2 Hemoglobin Binds Oxygen Cooperatively

7.3 Hydrogen Ions and Carbon Dioxide Promote the Release of Oxygen: The Bohr Effect

7.4 Mutations in Genes Encoding Hemoglobin Subunits Can Result in Disease

The transition from anaerobic to aerobic life was a major step in evolution because it uncovered a rich reservoir of energy. Fifteen times as much energy is extracted from glucose in the presence of oxygen than in its absence. For single-celled and other small organisms, oxygen can be absorbed into actively metabolizing cells directly from the air or surrounding water. Vertebrates evolved two principal mechanisms for supplying their cells with an adequate supply of oxygen. The first is a circulatory system that actively delivers oxygen to cells throughout the body. The second is the use of the oxygen-transport and oxygen-storage proteins, hemoglobin, and myoglobin. Hemoglobin, which is contained in red blood cells, is a fascinating protein, efficiently carrying oxygen from the lungs to the tissues while also contributing to the transport of carbon dioxide and hydrogen ions back to the lungs. Myoglobin, located in muscle, facilitates the diffusion of oxygen through the cell for the generation of cellular energy and provides a reserve supply of oxygen available in time of need.

A comparison of myoglobin and hemoglobin illuminates some key aspects of protein structure and function. These two evolutionarily related proteins employ nearly identical structures for oxygen binding (Chapter 6). However, hemoglobin is a remarkably efficient oxygen carrier, able to use as much as 90% of its potential oxygen-carrying capacity effectively. Under similar conditions, myoglobin would be able to use only 7% of its potential capacity. What accounts for this dramatic difference? Myoglobin exists as a single polypeptide, whereas hemoglobin comprises four polypeptide chains. The four chains in hemoglobin bind oxygen *cooperatively,* meaning that the binding of oxygen to a site in one chain increases the likelihood that the remaining chains will bind oxygen. Furthermore, the oxygen-binding properties of hemoglobin are modulated by the binding of hydrogen ions and carbon dioxide in a manner that enhances oxygen-carrying capacity. Both cooperativity and the response to modulators are made possible by variations in the quaternary structure of hemoglobin when different combinations of molecules are bound.

Hemoglobin and myoglobin have played important roles in the history of biochemistry. They were the first proteins for which three-dimensional structures were determined by x-ray crystallography. Furthermore, the possibility that variations in protein sequence could lead to disease was first proposed and demonstrated for sickle-cell anemia, a blood disease caused by mutation of a single amino acid in one hemoglobin chain. Hemoglobin has been and continues to be a valuable source of knowledge and insight, both in itself and as a prototype for many other proteins that we will encounter throughout our study of biochemistry.

7.1 Binding of Oxygen by Heme Iron

Sperm whale myoglobin was the first protein for which the three-dimensional structure was determined. X-ray crystallographic studies pioneered by John Kendrew revealed the structure of this protein in the 1950s (Figure 7.1). Myoglobin consists largely of α helices that are linked to one another by turns to form a globular structure.

Myoglobin can exist in an oxygen-free form called *deoxymyoglobin* or in a form with an oxygen molecule bound called *oxymyoglobin.* The ability of myoglobin and hemoglobin to bind oxygen depends on the presence of a *heme* molecule. As we shall discuss in Chapter 9, heme is one example of a prosthetic group—a molecule that binds tightly to a protein and is essential for its function.

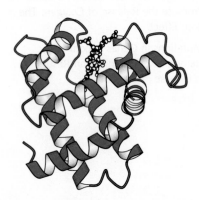

Myoglobin

FIGURE 7.1 Structure of myoglobin. *Notice* that myoglobin consists of a single polypeptide chain, formed of α helices connected by turns, with one oxygen-binding site. [Drawn from 1MBD.pdb.]

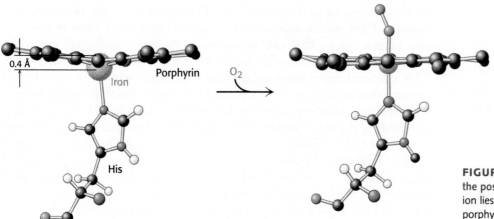

Heme
(Fe-protoporphyrin IX)

The heme group gives muscle and blood their distinctive red color. It consists of an organic component and a central iron atom. The organic component, called *protoporphyrin,* is made up of four pyrrole rings linked by methine bridges to form a tetrapyrrole ring. Four methyl groups, two vinyl groups, and two propionate side chains are attached to the central tetrapyrrole.

The iron atom lies in the center of the protoporphyrin, bonded to the four pyrrole nitrogen atoms. Although the heme-bound iron can be in either the ferrous (Fe^{2+}) or ferric (Fe^{3+}) oxidation state, only the Fe^{2+} state is capable of binding oxygen. The iron ion can form two additional bonds, one on each side of the heme plane. These binding sites are called the fifth and sixth coordination sites. In myoglobin, the fifth coordination site is occupied by the imidazole ring of a histidine residue from the protein. This histidine is referred to as the *proximal histidine.*

Oxygen binding occurs at the sixth coordination site. In deoxymyoglobin, this site remains unoccupied. The iron ion is slightly too large to fit into the well-defined hole within the porphyrin ring; it lies approximately 0.4 Å outside the porphyrin plane (Figure 7.2, left). Binding of the oxygen molecule at the sixth coordination site substantially rearranges the electrons within the iron so that the ion becomes effectively smaller, allowing it to move within the plane of the porphyrin (Figure 7.2, right). Remarkably,

FIGURE 7.2 Oxygen binding changes the position of the iron ion. The iron ion lies slightly outside the plane of the porphyrin in deoxymyoglobin heme (left) but moves into the plane of the heme on oxygenation (right).

the structural changes that take place on oxygen binding were predicted by Linus Pauling on the basis of magnetic measurements in 1936, nearly 25 years before the three-dimensional structures of myoglobin and hemoglobin were elucidated.

Changes in heme electronic structure upon oxygen binding are the basis for functional imaging studies

The change in electronic structure that occurs when the iron ion moves into the plane of the porphyrin is paralleled by alterations in the magnetic properties of hemoglobin; these changes are the basis for *functional magnetic resonance imaging* (fMRI), one of the most powerful methods for examining brain function. Nuclear magnetic resonance techniques detect signals that originate primarily from the protons in water molecules and are altered by the magnetic properties of hemoglobin. With the use of appropriate techniques, images can be generated that reveal differences in the relative amounts of deoxy- and oxyhemoglobin and thus the relative activity of various parts of the brain. When a specific part of the brain is active, blood vessels relax to allow more blood flow to that region. Thus, a more active region of the brain will be richer in oxyhemoglobin.

These noninvasive methods identify areas of the brain that process sensory information. For example, subjects have been imaged while breathing air that either does or does not contain odorants. When odorants are present, fMRI detects an increase in the level of hemoglobin oxygenation (and, hence, of activity) in several regions of the brain (Figure 7.3). These regions are in the primary olfactory cortex, as well as in areas in which secondary processing of olfactory signals presumably takes place. Further analysis reveals the time course of activation of particular regions. Functional MRI shows tremendous potential for mapping regions and pathways engaged in processing sensory information obtained from all the senses. A seemingly incidental aspect of the biochemistry of hemoglobin has enabled observation of the brain in action.

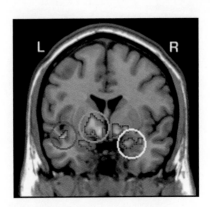

FIGURE 7.3 Functional magnetic resonance imaging of the brain. A functional magnetic resonance image reveals brain response to odorants. The light spots indicate regions of the brain activated by odorants. [Osterbauer, R. et al., "Color of Scents: Chromatic Stimuli Modulate Odor Responses in the Human Brain." *Journal of Neurophysiology* 93 (2005) 3434–3441.]

The structure of myoglobin prevents the release of reactive oxygen species

Oxygen binding to iron in heme is accompanied by the partial transfer of an electron from the ferrous ion to oxygen. In many ways, the structure is best described as a complex between ferric ion (Fe^{3+}) and *superoxide anion* (O_2^-), as illustrated in Figure 7.4. It is crucial that oxygen, when it is released, leaves as dioxygen rather than superoxide, for two important reasons. First, superoxide and other species generated from it are *reactive oxygen species* that can be damaging to many biological materials. Second, release of superoxide would leave the iron ion in the ferric state. This species, termed *metmyoglobin*, does not bind oxygen. Thus, potential oxygen-storage capacity is lost. Features of myoglobin stabilize the oxygen complex such that superoxide is less likely to be released. In particular, the binding pocket of myoglobin includes an additional histidine residue (termed the *distal histidine*) that donates a hydrogen bond to the bound oxygen molecule (Figure 7.5). The superoxide character of the bound oxygen species strengthens this interaction. Thus, *the protein component of myoglobin controls the intrinsic reactivity of heme,* making it more suitable for reversible oxygen binding. The distal histidine may also impair access of carbon monoxide to the heme, which binds tightly to the heme iron with dire consequences (p. 217).

FIGURE 7.4 Iron–oxygen bonding. The interaction between iron and oxygen in myoglobin can be described as a combination of resonance structures, one with Fe^{2+} and dioxygen and another with Fe^{3+} and superoxide ion.

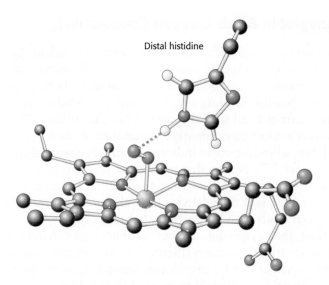

Distal histidine

FIGURE 7.5 Stabilizing bound oxygen. A hydrogen bond (dotted green line) donated by the distal histidine residue to the bound oxygen molecule helps stabilize oxymyoglobin.

Human hemoglobin is an assembly of four myoglobin-like subunits

The three-dimensional structure of hemoglobin from horse heart was solved by Max Perutz shortly after the determination of the myoglobin structure. Since then, the structures of hemoglobins from other species, including humans, have been determined. Hemoglobin consists of four polypeptide chains, two identical α *chains*, and two identical β *chains* (Figure 7.6). Each of the subunits consists of a set of α helices in the same arrangement as the α helices in myoglobin (see Figure 6.15 for a comparison of the structures). The recurring structure is called a *globin fold*. Consistent with this structural similarity, alignment of the amino acid sequences of the α and β chains of human hemoglobin with those of sperm whale myoglobin yields 25% and 24% identity, respectively, and good conservation of key residues such as the proximal and distal histidines. Thus, the α and β chains are related to each other and to myoglobin by divergent evolution (Section 6.2).

The hemoglobin tetramer, referred to as *hemoglobin A* (HbA), is best described as a pair of identical αβ *dimers* ($\alpha_1\beta_1$ and $\alpha_2\beta_2$) that associate to form the tetramer. In deoxyhemoglobin, these αβ dimers are linked by an extensive interface, which includes the carboxyl terminus of each chain. The heme groups are well separated in the tetramer by iron–iron distances ranging from 24 to 40 Å.

(A)

β_1 α_1

α_2 β_2

(B)

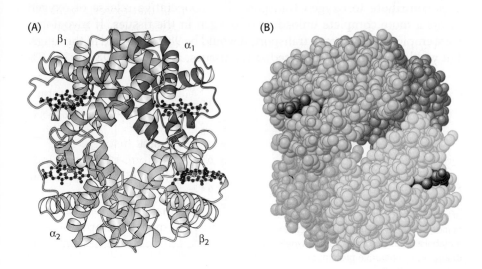

FIGURE 7.6 Quaternary structure of deoxyhemoglobin. Hemoglobin, which is composed of two α chains and two β chains, functions as a pair of αβ dimers. (A) A ribbon diagram. (B) A space-filling model. [Drawn from 1A3N.pdb.]

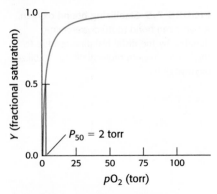

FIGURE 7.7 Oxygen binding by myoglobin. Half the myoglobin molecules have bound oxygen when the oxygen partial pressure is 2 torr.

Torr

A unit of pressure equal to that exerted by a column of mercury 1 mm high at 0°C and standard gravity (1 mm Hg). Named after Evangelista Torricelli (1608–1647), inventor of the mercury barometer.

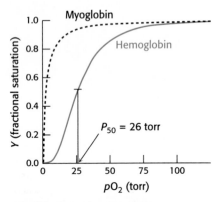

FIGURE 7.8 Oxygen binding by hemoglobin. This curve, obtained for hemoglobin in red blood cells, is shaped somewhat like an "S," indicating that distinct, but interacting, oxygen-binding sites are present in each hemoglobin molecule. Half-saturation for hemoglobin is 26 torr. For comparison, the binding curve for myoglobin is shown as a dashed black curve.

7.2 Hemoglobin Binds Oxygen Cooperatively

We can determine the oxygen-binding properties of each of these proteins by observing its *oxygen-binding curve,* a plot of the *fractional saturation* versus the concentration of oxygen. The fractional saturation, *Y,* is defined as the fraction of possible binding sites that contain bound oxygen. The value of *Y* can range from 0 (all sites empty) to 1 (all sites filled). The concentration of oxygen is most conveniently measured by its *partial pressure,* pO_2. For myoglobin, a binding curve indicating a simple chemical equilibrium is observed (Figure 7.7). Notice that the curve rises sharply as pO_2 increases and then levels off. Half-saturation of the binding sites, referred to as P_{50} (for 50% saturated), is at the relatively low value of 2 torr (mm Hg), indicating that oxygen binds with high affinity to myoglobin.

In contrast, the oxygen-binding curve for hemoglobin in red blood cells shows some remarkable features (Figure 7.8). It does not look like a simple binding curve such as that for myoglobin; instead, it resembles an "S." Such curves are referred to as *sigmoid* because of their S-like shape. In addition, oxygen binding for hemoglobin ($P_{50} = 26$ torr) is significantly weaker than that for myoglobin. Note that this binding curve is derived from hemoglobin in red blood cells.

A sigmoid binding curve indicates that a protein exhibits a special binding behavior. For hemoglobin, this shape suggests that the binding of oxygen at one site within the hemoglobin tetramer increases the likelihood that oxygen binds at the remaining unoccupied sites. Conversely, the unloading of oxygen at one heme facilitates the unloading of oxygen at the others. This sort of binding behavior is referred to as *cooperative,* because the binding reactions at individual sites in each hemoglobin molecule are not independent of one another. We will return to the mechanism of this cooperativity shortly.

What is the physiological significance of the cooperative binding of oxygen by hemoglobin? Oxygen must be transported in the blood from the lungs, where the partial pressure of oxygen is relatively high (approximately 100 torr), to the actively metabolizing tissues, where the partial pressure of oxygen is much lower (typically, 20 torr). Let's consider how the cooperative behavior indicated by the sigmoid curve leads to efficient oxygen transport (Figure 7.9). In the lungs, hemoglobin becomes nearly saturated with oxygen such that 98% of the oxygen-binding sites are occupied. When hemoglobin moves to the tissues and releases O_2, the saturation level drops to 32%. Thus, a total of $98 - 32 = 66\%$ of the potential oxygen-binding sites contribute to oxygen transport. The cooperative release of oxygen favors a more complete unloading of oxygen in the tissues. If myoglobin were employed for oxygen transport, it would be 98% saturated in the lungs, but would remain 91% saturated in the tissues, and so only $98 - 91 = 7\%$

FIGURE 7.9 Cooperativity enhances oxygen delivery by hemoglobin. Because of cooperativity between O_2 binding sites, hemoglobin delivers more O_2 to actively metabolizing tissues than would myoglobin or any noncooperative protein, even one with optimal O_2 affinity.

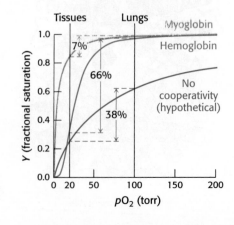

of the sites would contribute to oxygen transport; myoglobin binds oxygen too tightly to be useful in oxygen transport. Nature might have solved this problem by weakening the affinity of myoglobin for oxygen to maximize the difference in saturation between 20 and 100 torr. However, for such a protein, the most oxygen that could be transported from a region in which pO_2 is 100 torr to one in which it is 20 torr is $63 - 25 = 38\%$, as indicated by the blue curve in Figure 7.9. Thus, the cooperative binding and release of oxygen by hemoglobin enables it to deliver nearly 10 times as much oxygen as could be delivered by myoglobin and more than 1.7 times as much as could be delivered by any noncooperative protein.

Closer examination of oxygen concentrations in tissues at rest and during exercise underscores the effectiveness of hemoglobin as an oxygen carrier (Figure 7.10). Under resting conditions, the oxygen concentration in muscle is approximately 40 torr, but during exercise the concentration is reduced to 20 torr. In the decrease from 100 torr in the lungs to 40 torr in resting muscle, the oxygen saturation of hemoglobin is reduced from 98% to 77%, and so $98 - 77 = 21\%$ of the oxygen is released over a drop of 60 torr. In a decrease from 40 torr to 20 torr, the oxygen saturation is reduced from 77% to 32%, corresponding to an oxygen release of 45% over a drop of 20 torr. Thus, because the change in oxygen concentration from rest to exercise corresponds to the steepest part of the oxygen-binding curve, oxygen is effectively delivered to tissues where it is most needed. In Section 7.3, we shall examine other properties of hemoglobin that enhance its physiological responsiveness.

Oxygen binding markedly changes the quaternary structure of hemoglobin

The cooperative binding of oxygen by hemoglobin requires that the binding of oxygen at one site in the hemoglobin tetramer influences the oxygen-binding properties at the other sites. Given the large separation between the iron sites, direct interactions are not possible. Thus, indirect mechanisms for coupling the sites must be at work. These mechanisms are intimately related to the quaternary structure of hemoglobin.

Hemoglobin undergoes substantial changes in quaternary structure on oxygen binding: the $\alpha_1\beta_1$ and $\alpha_2\beta_2$ dimers rotate approximately 15 degrees with respect to one another (Figure 7.11). The dimers themselves are

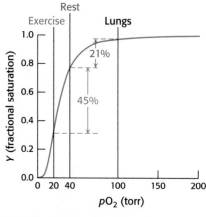

FIGURE 7.10 Responding to exercise. The drop in oxygen concentration from 40 torr in resting tissues to 20 torr in exercising tissues corresponds to the steepest part of the observed oxygen-binding curve. As shown here, hemoglobin is very effective in providing oxygen to exercising tissues.

FIGURE 7.11 Quaternary structural changes on oxygen binding by hemoglobin. *Notice* that, on oxygenation, one $\alpha\beta$ dimer shifts with respect to the other by a rotation of 15 degrees. [Drawn from 1A3N.pdb and 1LFQ.pdb.]

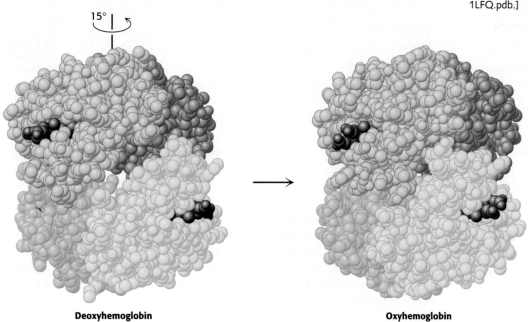

Deoxyhemoglobin   **Oxyhemoglobin**

relatively unchanged, although there are localized conformational shifts. Thus, the interface between the $\alpha_1\beta_1$ and $\alpha_2\beta_2$ dimers is most affected by this structural transition. In particular, the $\alpha_1\beta_1$ and $\alpha_2\beta_2$ dimers are freer to move with respect to one another in the oxygenated state than they are in the deoxygenated state.

The quaternary structure observed in the deoxy form of hemoglobin, *deoxyhemoglobin*, is often referred to as the *T* (for tense) *state* because it is quite constrained by subunit–subunit interactions. The quaternary structure of the fully oxygenated form, *oxyhemoglobin*, is referred to as the *R* (for relaxed) *state*. In light of the observation that the R form of hemoglobin is less constrained, the tense and relaxed designations seem particularly apt. Importantly, in the R state, the oxygen-binding sites are free of strain and are capable of binding oxygen with higher affinity than are the sites in the T state. *By triggering the shift of the hemoglobin tetramer from the T state to the R state, the binding of oxygen to one site increases the binding affinity of other sites.*

Hemoglobin cooperativity can be potentially explained by several models

Two limiting models have been developed to explain the cooperative binding of ligands to a multisubunit assembly such as hemoglobin. In the *concerted model*, also known as the *MWC model* after Jacques Monod, Jeffries Wyman, and Jean-Pierre Changeux, who first proposed it, the overall assembly can exist only in two forms: the T state and the R state. The binding of ligands simply shifts the equilibrium between these two states (Figure 7.12). Thus, as a hemoglobin tetramer binds each oxygen molecule, the probability that the tetramer is in the R state increases. Deoxyhemoglobin tetramers are almost exclusively in the T state. However, the binding of oxygen to one site in the molecule shifts the equilibrium toward the R state. If a molecule assumes the R quaternary structure, the oxygen affinity of its sites increases.

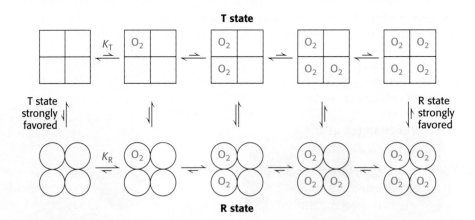

FIGURE 7.12 Concerted model. All molecules exist either in the T state or in the R state. At each level of oxygen loading, an equilibrium exists between the T and R states. The equilibrium shifts from strongly favoring the T state with no oxygen bound to strongly favoring the R state when the molecule is fully loaded with oxygen. The R state has a greater affinity for oxygen than does the T state.

Additional oxygen molecules are now more likely to bind to the three unoccupied sites. Thus, the binding curve for hemoglobin can be seen as a combination of the binding curves that would be observed if all molecules remained in the T state or if all of the molecules were in the R state. As oxygen molecules bind, the hemoglobin tetramers convert from the T state into the R state, yielding the sigmoid binding curve so important for efficient oxygen transport (Figure 7.13).

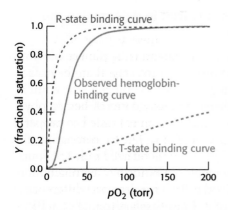

FIGURE 7.13 **T-to-R transition.** The binding curve for hemoglobin can be seen as a combination of the binding curves that would be observed if all molecules remained in the T state or if all of the molecules were in the R state. The sigmoidal curve is observed because molecules convert from the T state into the R state as oxygen molecules bind.

In the concerted model, each tetramer can exist in only two states—the T state and the R state. In an alternative model, the *sequential model,* the binding of a ligand to one site in an assembly increases the binding affinity of neighboring sites without inducing a full conversion from the T into the R state (Figure 7.14).

FIGURE 7.14 **Sequential model.** The binding of a ligand changes the conformation of the subunit to which it binds. This conformational change induces changes in neighboring subunits that increase their affinity for the ligand.

Is the cooperative binding of oxygen by hemoglobin better described by the concerted or the sequential model? Neither model in its pure form fully accounts for the behavior of hemoglobin. Instead, a combined model is required. Hemoglobin behavior is concerted in that the tetramer with three sites occupied by oxygen is almost always in the quaternary structure associated with the R state. The remaining open binding site has an affinity for oxygen more than 20-fold greater than that of fully deoxygenated hemoglobin binding its first oxygen. However, the behavior is not fully concerted, because hemoglobin with oxygen bound to only one of four sites remains primarily in the T-state quaternary structure. Yet, this molecule binds oxygen three times as strongly as does fully deoxygenated hemoglobin, an observation consistent only with a sequential model. These results highlight the fact that the concerted and sequential models represent idealized limiting cases, which real systems may approach but rarely attain.

Structural changes at the heme groups are transmitted to the $\alpha_1\beta_1-\alpha_2\beta_2$ interface

We now examine how oxygen binding at one site is able to shift the equilibrium between the T and R states of the entire hemoglobin tetramer. As in myoglobin, oxygen binding causes each iron atom in hemoglobin to move from outside the plane of the porphyrin into the plane. When the iron atom moves, the proximal histidine residue moves with it. This histidine residue is part of an α helix, which also moves (Figure 7.15). The carboxyl terminal end of this α helix lies in the interface between the two $\alpha\beta$ dimers. The change in position of the carboxyl terminal end of the helix favors the T-to-R transition. Consequently, *the structural transition at the iron ion in one subunit is directly transmitted to the other subunits.* The rearrangement of the dimer interface provides a pathway for communication between subunits, enabling the cooperative binding of oxygen.

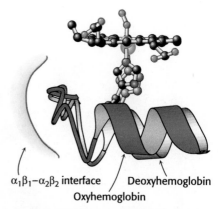

$\alpha_1\beta_1-\alpha_2\beta_2$ interface / Deoxyhemoglobin
Oxyhemoglobin

FIGURE 7.15 Conformational changes in hemoglobin. The movement of the iron ion on oxygenation brings the iron-associated histidine residue toward the porphyrin ring. The associated movement of the histidine-containing α helix alters the interface between the $\alpha\beta$ dimers, instigating other structural changes. For comparison, the deoxyhemoglobin structure is shown in gray behind the oxyhemoglobin structure in red.

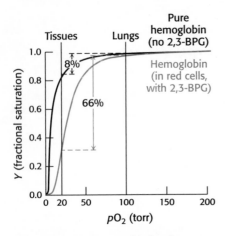

FIGURE 7.16 Oxygen binding by pure hemoglobin compared with hemoglobin in red blood cells. Pure hemoglobin binds oxygen more tightly than does hemoglobin in red blood cells. This difference is due to the presence of 2,3-bisphosphoglycerate (2,3-BPG) in red blood cells.

2,3-Bisphosphoglycerate in red cells is crucial in determining the oxygen affinity of hemoglobin

For hemoglobin to function efficiently, the T state must remain stable until the binding of sufficient oxygen has converted it into the R state. In fact, however, the T state of hemoglobin is highly unstable, pushing the equilibrium so far toward the R state that little oxygen would be released in physiological conditions. Thus, an additional mechanism is needed to properly stabilize the T state. This mechanism was discovered by comparing the oxygen-binding properties of hemoglobin in red blood cells with fully purified hemoglobin (Figure 7.16). Pure hemoglobin binds oxygen much more tightly than does hemoglobin in red blood cells. This dramatic difference is due to the presence within these cells of *2,3-bisphosphoglycerate* (2,3-BPG; also known as 2,3-diphosphoglycerate or 2,3-DPG).

**2,3-Bisphosphoglycerate
(2,3-BPG)**

This highly anionic compound is present in red blood cells at approximately the same concentration as that of hemoglobin (~2 mM). Without 2,3-BPG, hemoglobin would be an extremely inefficient oxygen transporter, releasing only 8% of its cargo in the tissues.

How does 2,3-BPG lower the oxygen affinity of hemoglobin so significantly? Examination of the crystal structure of deoxyhemoglobin in the presence of 2,3-BPG reveals that a single molecule of 2,3-BPG binds in the center of the tetramer, in a pocket present only in the T form (Figure 7.17). On T-to-R transition, this pocket collapses, and 2,3-BPG is released. Thus, in order for the structural transition from T to R to take place, the bonds between hemoglobin and 2,3-BPG must be broken. In the presence of 2,3-BPG, more oxygen-binding sites within the hemoglobin tetramer must be occupied in order to induce the T-to-R transition, and so hemoglobin

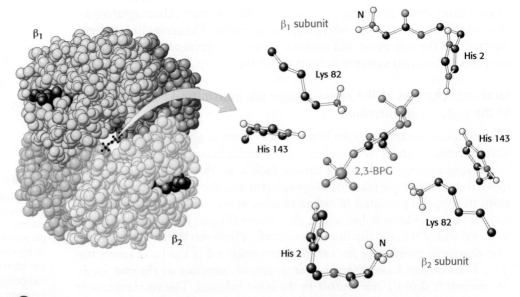

FIGURE 7.17 Mode of binding of 2,3-BPG to human deoxyhemoglobin. 2,3-Bisphosphoglycerate binds to the central cavity of deoxyhemoglobin (left). There, it interacts with three positively charged groups on each β chain (right). [Drawn from 1B86.pdb.]

remains in the lower-affinity T state until higher oxygen concentrations are reached. This mechanism of regulation is remarkable because 2,3-BPG does not in any way resemble oxygen, the molecule on which hemoglobin carries out its primary function. 2,3-BPG is referred to as an *allosteric effector* (from the Greek *allos,* "other," and *stereos,* "structure"). Regulation by a molecule structurally unrelated to oxygen is possible because the allosteric effector binds to a site that is completely distinct from that for oxygen. We will encounter allosteric effects again when we consider enzyme regulation in Chapter 10.

The binding of 2,3-BPG to hemoglobin has other crucial physiological consequences. The globin gene expressed by human fetuses differs from that expressed by adults; *fetal hemoglobin* tetramers include two α chains and two γ chains. The γ chain, a result of gene duplication, is 72% identical in amino acid sequence with the β chain. One noteworthy change is the substitution of a serine residue for His 143 in the β chain, part of the 2,3-BPG-binding site. This change removes two positive charges from the 2,3-BPG-binding site (one from each chain) and reduces the affinity of 2,3-BPG for fetal hemoglobin. Consequently, the oxygen-binding affinity of fetal hemoglobin is higher than that of maternal (adult) hemoglobin (Figure 7.18). This difference in oxygen affinity allows oxygen to be effectively transferred from maternal to fetal red blood cells. We have here an example in which gene duplication and specialization produced a ready solution to a biological challenge—in this case, the transport of oxygen from mother to fetus.

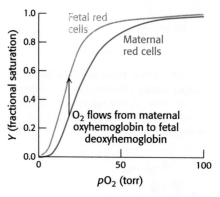

FIGURE 7.18 Oxygen affinity of fetal red blood cells. Fetal red blood cells have a higher oxygen affinity than do maternal red blood cells because fetal hemoglobin does not bind 2,3-BPG as well as maternal hemoglobin does.

Carbon monoxide can disrupt oxygen transport by hemoglobin

Carbon monoxide (CO) is a colorless, odorless gas that binds to hemoglobin at the same site as oxygen, forming a complex termed *carboxyhemoglobin*. Formation of carboxyhemoglobin exerts devastating effects on normal oxygen transport in two ways. First, carbon monoxide binds to hemoglobin about 200-fold more tightly than does oxygen. Even at low partial pressures in the blood, carbon monoxide will displace oxygen from hemoglobin, preventing its delivery. Second, carbon monoxide bound to one site in hemoglobin will shift the oxygen saturation curve of the remaining sites to the left, forcing the tetramer into the R state. This results in an increased affinity for oxygen, preventing its dissociation at tissues.

Exposure to carbon monoxide—from gas appliances and running automobiles, for example—can cause carbon monoxide poisoning, in which patients exhibit nausea, vomiting, lethargy, weakness, and disorientation. One treatment for carbon monoxide poisoning is administration of 100% oxygen, often at pressures greater than atmospheric pressure (this treatment is referred to as *hyperbaric oxygen therapy*). With this therapy, the partial pressure of oxygen in the blood becomes sufficiently high to increase substantially the displacement of carbon monoxide from hemoglobin. Exposure to high concentrations of carbon monoxide, however, can be rapidly fatal: in the United States, approximately 2500 people die each year from carbon monoxide poisoning, about 500 of them from accidental exposures and nearly 2000 by suicide.

7.3 Hydrogen Ions and Carbon Dioxide Promote the Release of Oxygen: The Bohr Effect

We have seen how cooperative release of oxygen from hemoglobin helps deliver oxygen to where it is most needed: tissues exhibiting low oxygen partial pressures. This ability is enhanced by the facility of hemoglobin to respond to other cues in its physiological environment that signal the

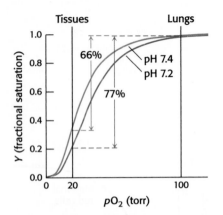

FIGURE 7.19 Effect of pH on the oxygen affinity of hemoglobin. Lowering the pH from 7.4 (red curve) to 7.2 (blue curve) results in the release of O_2 from oxyhemoglobin.

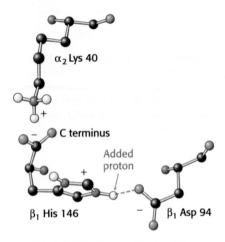

FIGURE 7.20 Chemical basis of the Bohr effect. In deoxyhemoglobin, three amino acid residues form two salt bridges that stabilize the T quaternary structure. The formation of one of the salt bridges depends on the presence of an added proton on histidine β146. The proximity of the negative charge on aspartate β94 in deoxyhemoglobin favors protonation of this histidine. *Notice* that the salt bridge between histidine β146 and aspartate β94 is stabilized by a hydrogen bond (green dashed line).

FIGURE 7.21 Carbon dioxide and pH. Carbon dioxide in the tissues diffuses into red blood cells. Inside a red blood cell, carbon dioxide reacts with water to form carbonic acid, in a reaction catalyzed by the enzyme carbonic anhydrase. Carbonic acid dissociates to form HCO_3^- and H^+, resulting in a drop in pH inside the red cell.

need for oxygen. Rapidly metabolizing tissues, such as contracting muscle, generate large amounts of hydrogen ions and carbon dioxide (Chapter 16). To release oxygen where the need is greatest, hemoglobin has evolved to respond to higher levels of these substances. Like 2,3-BPG, hydrogen ions and carbon dioxide are allosteric effectors of hemoglobin that bind to sites on the molecule that are distinct from the oxygen-binding sites. The regulation of oxygen binding by hydrogen ions and carbon dioxide is called the *Bohr effect* after Christian Bohr, who described this phenomenon in 1904.

The oxygen affinity of hemoglobin decreases as pH decreases from a value of 7.4 (Figure 7.19). Consequently, as hemoglobin moves into a region of lower pH, its tendency to release oxygen increases. For example, transport from the lungs, with pH 7.4 and an oxygen partial pressure of 100 torr, to active muscle, with a pH of 7.2 and an oxygen partial pressure of 20 torr, results in a release of oxygen amounting to 77% of total carrying capacity. Only 66% of the oxygen would be released in the absence of any change in pH. Structural and chemical studies have revealed much about the chemical basis of the Bohr effect. Several chemical groups within the hemoglobin tetramer are important for sensing changes in pH; all of these have pK_a values near pH 7. Consider histidine β146, the residue at the C terminus of the β chain. In deoxyhemoglobin, the terminal carboxylate group of β146 forms an ionic bond, also called a salt bridge, with a lysine residue in the α subunit of the other αβ dimer. This interaction locks the side chain of histidine β146 in a position from which it can participate in a salt bridge with negatively charged aspartate β94 in the same chain, provided that the imidazole group of the histidine residue is protonated (Figure 7.20).

In addition to His β146, the α-amino groups at the amino termini of the α chain and the side chain of histidine α122 also participate in salt bridges in the T state. *The formation of these salt bridges stabilizes the T state, leading to a greater tendency for oxygen to be released.* For example, at high pH, the side chain of histidine β146 is not protonated and the salt bridge does not form. As the pH drops, however, the side chain of histidine β146 becomes protonated, the salt bridge with aspartate β94 forms, and the T state is stabilized.

Carbon dioxide, a neutral species, passes through the red-blood-cell membrane into the cell. This transport is also facilitated by membrane transporters, including proteins associated with Rh blood types. Carbon dioxide stimulates oxygen release by two mechanisms. First, the presence of high concentrations of carbon dioxide leads to a drop in pH within the red blood cell (Figure 7.21). Carbon dioxide reacts with water to form carbonic

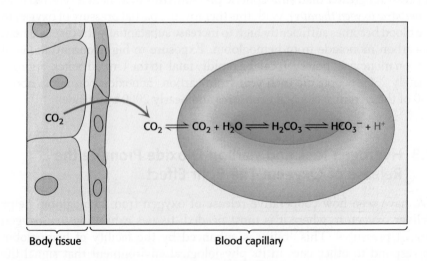

Body tissue   Blood capillary

acid, H_2CO_3. This reaction is accelerated by *carbonic anhydrase,* an enzyme abundant in red blood cells that will be considered extensively in Chapter 9. H_2CO_3 is a moderately strong acid with a pK_a of 3.5. Thus, once formed, carbonic acid dissociates to form bicarbonate ion, HCO_3^- and H^+, resulting in a drop in pH that stabilizes the T state by the mechanism discussed previously.

In the second mechanism, a direct chemical interaction between carbon dioxide and hemoglobin stimulates oxygen release. The effect of carbon dioxide on oxygen affinity can be seen by comparing oxygen-binding curves in the absence and in the presence of carbon dioxide *at a constant pH* (Figure 7.22). In the presence of carbon dioxide at a partial pressure of 40 torr at pH 7.2, the amount of oxygen released approaches 90% of the maximum carrying capacity. Carbon dioxide stabilizes deoxyhemoglobin by reacting with the terminal amino groups to form *carbamate* groups, which are negatively charged, in contrast with the neutral or positive charges on the free amino groups.

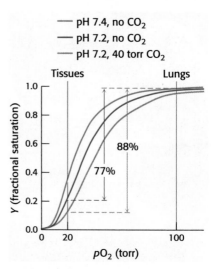

FIGURE 7.22 Carbon dioxide effects. The presence of carbon dioxide decreases the affinity of hemoglobin for oxygen even beyond the effect due to a decrease in pH, resulting in even more efficient oxygen transport from the tissues to the lungs.

Carbamate

The amino termini lie at the interface between the αβ dimers, and these negatively charged carbamate groups participate in salt-bridge interactions that stabilize the T state, favoring the release of oxygen.

Carbamate formation also provides a mechanism for carbon dioxide transport from tissues to the lungs, but it accounts for only about 14% of the total carbon dioxide transport. Most carbon dioxide released from red blood cells is transported to the lungs in the form of HCO_3^- produced from the hydration of carbon dioxide inside the cell (Figure 7.23). Much of the HCO_3^- that is formed leaves the cell through a specific membrane-transport protein that exchanges HCO_3^- from one side of the membrane for Cl^- from the other side. Thus, the serum concentration of HCO_3^- increases. By this means, a large concentration of carbon dioxide is transported from tissues to the lungs in the form of HCO_3^-. In the lungs, this process is reversed: HCO_3^- is converted back into carbon dioxide and exhaled. Thus, carbon dioxide generated by active tissues contributes to a decrease in red-blood-cell pH and, hence, to oxygen release, and is converted into a form that can be transported in the serum and released in the lungs.

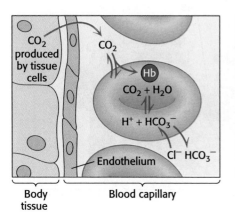

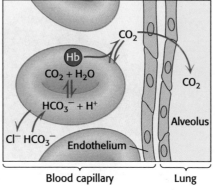

FIGURE 7.23 Transport of CO_2 from tissues to lungs. Most carbon dioxide is transported to the lungs in the form of HCO_3^- produced in red blood cells and then released into the blood plasma. A lesser amount is transported by hemoglobin in the form of an attached carbamate.

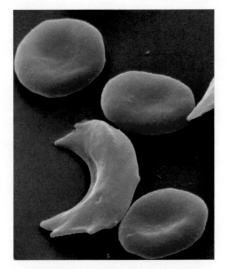

FIGURE 7.24 Sickled red blood cells. A micrograph showing a sickled red blood cell adjacent to normally shaped red blood cells. [Eye of Science/Science Source.]

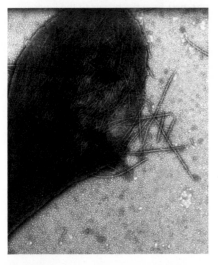

FIGURE 7.25 Sickle-cell hemoglobin fibers. An electron micrograph depicting a ruptured sickled red blood cell with fibers of sickle-cell hemoglobin emerging. [Courtesy of Robert Josephs and Thomas E Wellems University of Chicago.]

7.4 Mutations in Genes Encoding Hemoglobin Subunits Can Result in Disease

In modern times, particularly after the sequencing of the human genome, it is routine to think of genetically encoded variations in protein sequence as a factor in specific diseases. The notion that diseases might be caused by molecular defects was proposed by Linus Pauling in 1949 (four years before Watson and Crick's proposal of the DNA double helix) to explain the blood disease *sickle-cell anemia*. The name of the disorder comes from the abnormal sickle shape of red blood cells deprived of oxygen in people suffering from this disease (Figure 7.24). Pauling proposed that sickle-cell anemia might be caused by a specific variation in the amino acid sequence of one hemoglobin chain. Today, we know that this bold hypothesis is correct. In fact, approximately 7% of the world's population are carriers of some disorder of hemoglobin caused by a variation in its amino acid sequence. In concluding this chapter, we will focus on the two most important of these disorders, sickle-cell anemia and thalassemia.

Sickle-cell anemia results from the aggregation of mutated deoxyhemoglobin molecules

People with sickled red blood cells experience a number of dangerous symptoms. Examination of the contents of these red cells reveals that the hemoglobin molecules have formed large fibrous aggregates (Figure 7.25). These fibers extend across the red blood cells, distorting them so that they clog small capillaries and impair blood flow. In addition, red cells from sickle-cell patients are more adherent to the walls of blood vessels than those from normal individuals, prolonging the opportunity for capillary occlusion. The results may be painful swelling of the extremities and a higher risk of stroke or bacterial infection (due to poor circulation). The sickled red cells also do not remain in circulation as long as normal cells do, leading to anemia.

What is the molecular defect associated with sickle-cell anemia? Vernon Ingram demonstrated in 1956 that a single amino acid substitution in the β chain of hemoglobin is responsible—namely, the replacement of a glutamate residue with valine in position 6. The mutated form is referred to as *hemoglobin S* (HbS). In people with sickle-cell anemia, both alleles of the hemoglobin β-chain gene (HbB) are mutated. The HbS substitution substantially decreases the solubility of deoxyhemoglobin, although it does not markedly alter the properties of oxyhemoglobin.

Examination of the structure of hemoglobin S reveals that the new valine residue lies on the surface of the T-state molecule (Figure 7.26). This new

FIGURE 7.26 Deoxygenated hemoglobin S. The interaction between Val 6 (blue) on a β chain of one hemoglobin molecule and a hydrophobic patch formed by Phe 85 and Leu 88 (gray) on a β chain of another deoxygenated hemoglobin molecule leads to hemoglobin aggregation. The exposed Val 6 residues of other β chains participate in other such interactions in hemoglobin S fibers. [Drawn from 2HBS.pdb.]

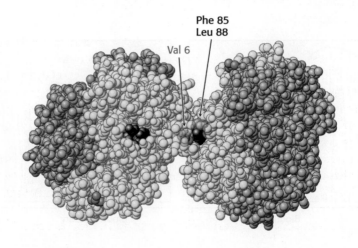

Phe 85
Leu 88

Val 6

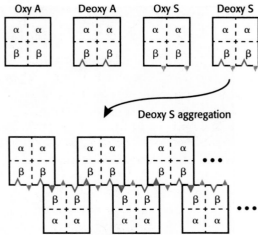

Oxy A   Deoxy A   Oxy S   Deoxy S

Deoxy S aggregation

FIGURE 7.27 The formation of HbS aggregates. The mutation to Val 6 in hemoglobin S is represented by the red triangles, while the hydrophobic patch formed by Phe 85 and Leu 88 in deoxyhemoglobin is represented by the blue indentations. When HbS is in its deoxy form, it exhibits the complementary features necessary for aggregation.

hydrophobic patch interacts with another hydrophobic patch formed by Phe 85 and Leu 88 of the β chain of a neighboring molecule to initiate the aggregation process. More-detailed analysis reveals that a single hemoglobin S fiber is formed from 14 chains of multiple interlinked hemoglobin molecules. Why do these aggregates not form when hemoglobin S is oxygenated? When oxyhemoglobin S is in the R state, residues Phe 85 and Leu 88 on the β chain are largely buried inside the hemoglobin assembly. In the absence of a partner with which to interact, the surface Val residue in position 6 is benign (Figure 7.27).

Approximately 1 in 100 West Africans suffer from sickle-cell anemia. Given the often devastating consequences of the disease, why is the HbS mutation so prevalent in Africa and in some other regions? Recall that both copies of the HbB gene are mutated in people with sickle-cell anemia. Individuals with one copy of the HbB gene and one copy of the HbS gene are said to have *sickle-cell trait* because they can pass the HbS gene to their offspring. While sickle-cell trait is considered a benign condition, rare complications have been identified, including an increased risk of exercise-related death in high-performance athletes. However, people with sickle-cell trait exhibit enhanced resistance to *malaria*, a disease carried by a parasite, *Plasmodium falciparum*, that lives within red blood cells at one stage in its life cycle. The dire effect of malaria on health and reproductive likelihood in historically endemic regions has favored people with sickle-cell trait, increasing the prevalence of the HbS allele (Figure 7.28).

Thalassemia is caused by an imbalanced production of hemoglobin chains

Sickle-cell anemia is caused by the substitution of a single specific amino acid in one hemoglobin chain. *Thalassemia*, the other prevalent inherited disorder of hemoglobin, is caused by the loss or substantial reduction of a single hemoglobin *chain*. The result is low levels of functional hemoglobin and a decreased production of red blood cells, which may lead to anemia, fatigue, pale skin, and spleen and liver malfunction. Thalassemia is a set of related diseases. In α-thalassemia, the α chain of hemoglobin is not produced in sufficient quantity. Consequently, hemoglobin tetramers form that contain only the β chain. These tetramers, referred to as *hemoglobin H* (HbH), bind oxygen with high affinity and no cooperativity. Thus, oxygen release in the tissues is poor. In β-thalassemia, the β chain of hemoglobin is not produced in sufficient quantity. In the absence of β chains,

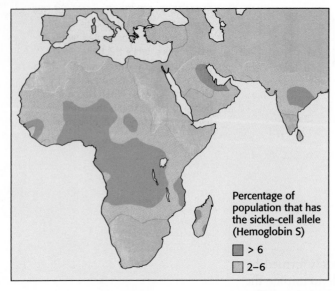
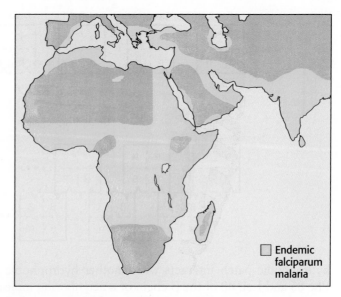

Percentage of
population that has
the sickle-cell allele
(Hemoglobin S)

■ > 6
□ 2–6

□ Endemic
falciparum
malaria

FIGURE 7.28 Sickle-cell trait and malaria. A significant correlation is observed between regions with a high frequency of the HbS allele and regions with a high prevalence of malaria.

the α chains form insoluble aggregates that precipitate inside immature red blood cells. The loss of red blood cells results in anemia. The most severe form of β-thalassemia is called *thalassemia major* or *Cooley anemia*.

Both α- and β-thalassemia are associated with many different genetic variations and display a wide range of clinical severity. The most severe forms of α-thalassemia are usually fatal shortly before or just after birth. However, these forms are relatively rare. An examination of the repertoire of hemoglobin genes in the human genome provides one explanation. Normally, humans have not two but four alleles for the α chain, arranged such that the two genes are located adjacent to each other on one end of each chromosome 16. Thus, the complete loss of α-chain expression requires the disruption of four alleles. β-Thalassemia is more common because humans normally have only two alleles for the β chain, one on each copy of chromosome 11.

The accumulation of free α-hemoglobin chains is prevented

The presence of four genes expressing the α chain, compared with two for the β chain, suggests that the α chain would be produced in excess (given the overly simple assumption that protein expression from each gene is comparable). If this is correct, why doesn't the excess α chain precipitate? One mechanism for maintaining α chains in solution was revealed by the discovery of an 11-kDa protein in red blood cells called *α-hemoglobin stabilizing protein* (AHSP). This protein forms a soluble complex specifically with newly synthesized α-chain monomers. The crystal structure of a complex between AHSP and α-hemoglobin reveals that AHSP binds to the same face of α-hemoglobin as does β-hemoglobin (Figure 7.29). AHSP binds the α chain in both the deoxygenated and oxygenated forms. In the complex with oxygen bound, the distal histidine, rather than the proximal histidine, binds the iron atom.

AHSP serves to bind and ensure the proper folding of α-hemoglobin as it is produced. As β-hemoglobin is expressed, it displaces AHSP because the α-hemoglobin–β-hemoglobin dimer is more stable than the α-hemoglobin–AHSP complex. Thus, AHSP prevents the misfolding, accumulation, and precipitation of free α-hemoglobin. It was originally postulated that

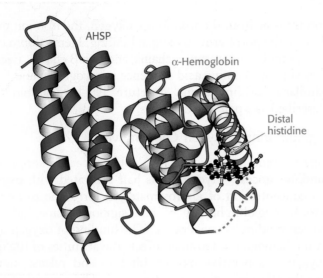

FIGURE 7.29 Stabilizing free α-hemoglobin. The structure of a complex between AHSP and α-hemoglobin is shown. In this complex, the iron atom is bound to oxygen and to the distal histidine. *Notice* that AHSP binds to the same surface of α-hemoglobin as does β-hemoglobin. [Drawn from 1Y01.pdb.]

sequence variation in the gene encoding AHSP may modulate the severity of β-thalassemia, given its important role in preventing an imbalance of hemoglobin subunits. However, demonstration of these effects has proven challenging. Nevertheless, several mutations in α-hemoglobin that appear to disrupt binding to AHSP have been proposed.

Additional globins are encoded in the human genome

In addition to the gene for myoglobin, the two genes for α-hemoglobin, and the one for β-hemoglobin, the human haploid genome contains other globin genes. We have already encountered fetal hemoglobin, which contains the γ chain in place of the β chain. Several other genes encode other hemoglobin subunits that are expressed during development, including the δ chain, the ε chain, and the ζ chain.

Examination of the human genome sequence has revealed two additional globins. Both of these proteins are monomeric proteins, more similar to myoglobin than to hemoglobin. The first, *neuroglobin,* is expressed primarily in the brain and at especially high levels in the retina. Neuroglobin may play a role in protecting neural tissues from hypoxia (insufficient oxygen). The second, *cytoglobin,* is expressed more widely throughout the body. Structural and spectroscopic studies reveal that, in both neuroglobin and cytoglobin, the proximal and the distal histidines are coordinated to the iron atom in the deoxy form. Oxygen binding displaces the distal histidine. The functions of these unique members of the globin family remain an active area of investigation.

SUMMARY

7.1 Binding of Oxygen by Heme Iron

Myoglobin is a largely α-helical protein that binds the prosthetic group heme. Heme consists of protoporphyrin, an organic component with four linked pyrrole rings, and a central iron ion in the Fe^{2+} state. The iron ion is coordinated to the side chain of a histidine residue in myoglobin, referred to as the proximal histidine. One of the oxygen atoms in O_2 binds to an open coordination site on the iron. Because of partial electron transfer from the iron to the oxygen, the iron ion moves into the plane of the porphyrin on oxygen binding. Hemoglobin consists of four polypeptide chains, two α chains, and two β chains. Each of these chains is similar in amino acid sequence to myoglobin and folds into a very similar three-dimensional structure. The hemoglobin tetramer is best described as a pair of αβ dimers.

7.2 Hemoglobin Binds Oxygen Cooperatively

The oxygen-binding curve for myoglobin reveals a simple equilibrium-binding process. Myoglobin is half-saturated with oxygen at an oxygen concentration of approximately 2 torr. The oxygen-binding curve for hemoglobin has an "S"-like (sigmoid) shape, indicating that the oxygen binding is cooperative. The binding of oxygen at one site within the hemoglobin tetramer affects the affinities of the other sites for oxygen. Cooperative oxygen binding and release significantly increase the efficiency of oxygen transport. The amount of the potential oxygen-carrying capacity utilized in transporting oxygen from the lungs (with a partial pressure of oxygen of 100 torr) to tissues (with a partial pressure of oxygen of 20 torr) is 66% compared with 7% if myoglobin had been used as the oxygen carrier.

The quaternary structure of hemoglobin changes on oxygen binding. The structure of deoxyhemoglobin is referred to as the T state. The structure of oxyhemoglobin is referred to as the R state. The two αβ dimers rotate by approximately 15 degrees with respect to one another in the transition from the T to the R state. Cooperative binding can be potentially explained by concerted and sequential models. In the concerted model, each hemoglobin adopts either the T state or the R state; the equilibrium between these two states is determined by the number of occupied oxygen-binding sites. Sequential models allow intermediate structures. Structural changes at the iron sites in response to oxygen binding are transmitted to the interface between αβ dimers, influencing the T-to-R equilibrium.

Red blood cells contain 2,3-bisphosphoglycerate in concentrations approximately equal to that for hemoglobin. 2,3-BPG binds tightly to the T state but not to the R state, stabilizing the T state and lowering the oxygen affinity of hemoglobin. Fetal hemoglobin binds oxygen more tightly than does adult hemoglobin owing to weaker 2,3-BPG binding. This difference allows oxygen transfer from maternal to fetal blood.

7.3 Hydrogen Ions and Carbon Dioxide Promote the Release of Oxygen: The Bohr Effect

The oxygen-binding properties of hemoglobin are markedly affected by pH and by the presence of carbon dioxide, a phenomenon known as the Bohr effect. Increasing the concentration of hydrogen ions—that is, decreasing pH—decreases the oxygen affinity of hemoglobin, owing to the protonation of the amino termini and certain histidine residues. The protonated residues help stabilize the T state. Increasing concentrations of carbon dioxide decrease the oxygen affinity of hemoglobin by two mechanisms. First, carbon dioxide is converted into carbonic acid, which lowers the oxygen affinity of hemoglobin by decreasing the pH inside the red blood cell. Second, carbon dioxide adds to the amino termini of hemoglobin to form carbamates. These negatively charged groups stabilize deoxyhemoglobin through ionic interactions. Because hydrogen ions and carbon dioxide are produced in rapidly metabolizing tissues, the Bohr effect helps deliver oxygen to sites where it is most needed.

7.4 Mutations in Genes Encoding Hemoglobin Subunits Can Result in Disease

Sickle-cell disease is caused by a mutation in the β chain of hemoglobin that substitutes a valine residue for a glutamate residue. As a result, a hydrophobic patch forms on the surface of deoxy (T-state) hemoglobin that leads to the formation of fibrous polymers. These fibers distort red blood cells into sickle shapes. Sickle-cell disease was the first disease to be associated with a change in the amino acid sequence of a protein. Thalassemias are diseases caused by the reduced production of either the α or the β chain, yielding hemoglobin tetramers that contain only one type of hemoglobin chain. Such hemoglobin molecules are characterized by poor oxygen release and low solubility, leading to the destruction of red blood cells in the course of their development. Red blood cell precursors normally produce a slight excess of hemoglobin α chains compared with β chains. To prevent the aggregation of the excess α chains, they produce α-hemoglobin stabilizing protein, which binds specifically to newly synthesized α-chain monomers to form a soluble complex.

APPENDIX

Binding Models Can Be Formulated in Quantitative Terms: The Hill Plot and the Concerted Model

The Hill Plot

A useful way of quantitatively describing cooperative binding processes such as that for hemoglobin was developed by Archibald Hill in 1913. Consider the *hypothetical* equilibrium for a protein X binding a ligand S:

$$X + nS \rightleftharpoons X(S)n \qquad (1)$$

where n is a variable that can take on both integral and fractional values. The parameter n is a measure of the degree of cooperativity in ligand binding. For X = hemoglobin and S = O_2, the maximum value of n is 4. The value of $n = 4$ would apply if oxygen binding by hemoglobin were completely cooperative. If oxygen binding were completely noncooperative, then n would be 1.

Analysis of the equilibrium in equation 1 yields the following expression for the fractional saturation, Y:

$$Y = \frac{[S]^n}{[S]^n + [S_{50}]^n}$$

where $[S_{50}]$ is the concentration at which X is half-saturated. For hemoglobin, this expression becomes

$$Y = \frac{pO_2^n}{pO_2^n + P_{50}^n}$$

where P_{50} is the partial pressure of oxygen at which hemoglobin is half-saturated. This expression can be rearranged to:

$$\frac{Y}{1 - Y} = \frac{pO_2^n}{P_{50}^n}$$

and so

$$\log\left(\frac{Y}{1 - Y}\right) = \log\left(\frac{pO_2^n}{P_{50}^n}\right) = n\log(pO_2) - n\log(P_{50})$$

This equation predicts that a plot of log $(Y/1 - Y)$ versus $\log(pO_2)$, called a *Hill plot*, should be linear with a slope of n.

Hill plots for myoglobin and hemoglobin are shown in Figure 7.30. For myoglobin, the Hill plot is linear with a slope of 1. For hemoglobin, the Hill plot is not completely linear, because the equilibrium on which the Hill plot is based is not entirely correct. However, the plot is approximately linear in the center with a slope of 2.8. The slope, often referred to as the *Hill coefficient*, is a measure of the cooperativity of oxygen binding. The utility of the Hill plot is that it provides a simply derived quantitative assessment of the degree of cooperativity in binding. With the use of the Hill equation and the derived Hill coefficient, a binding curve that closely resembles that for hemoglobin is produced (Figure 7.31).

The Concerted Model

The concerted model can be formulated in quantitative terms. Only four parameters are required: (1) the number of binding sites (assumed to be equivalent) in the protein, (2) the ratio of the concentrations of the T and R states in the absence of bound ligands, (3) the affinity of sites in proteins in the R state for ligand binding, and (4) a measure of how much more tightly subunits in proteins in the R state bind ligands compared with subunits in the T state. The number

Myoglobin

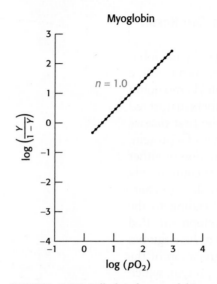

Hemoglobin

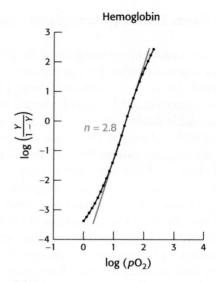

FIGURE 7.30 Hill plots for myoglobin and hemoglobin.

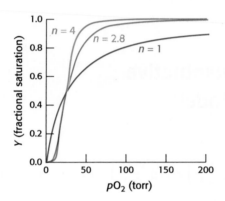

FIGURE 7.31 Oxygen-binding curves for several Hill coefficients. The curve labeled $n = 2.8$ closely resembles the curve for hemoglobin.

of binding sites, n, is usually known from other information. For hemoglobin, $n = 4$. The ratio of the concentrations of the T and R states with no ligands bound, L, is the allosteric constant:

$$L = [T_0]/[R_0]$$

where the subscript refers to the number of ligands bound (in this case, zero). The affinity of subunits in the R state is defined by the dissociation constant for a ligand binding to a single site in the R state, K_R. Similarly, the dissociation constant for a ligand binding to a single site in the T state is K_T. We can define the ratio of these two dissociation constants as

$$c = K_R/K_T$$

This is the measure of how much more tightly a subunit for a protein in the R state binds a ligand compared with a subunit for a protein in the T state. Note that $c < 1$ because K_R and K_T are dissociation constants and tight binding corresponds to a small dissociation constant.

What is the ratio of the concentration of T-state proteins with one ligand bound to the concentration of R-state proteins with one ligand bound? The dissociation constant for a single site in the R state is K_R. For a protein with n sites, there are n possible sites for the first ligand to bind. This statistical factor favors ligand binding compared with a single-site protein. Thus, $[R_1] = n[R_0][S]/K_R$. Similarly, $[T_1] = n[T_0][S]/K_T$. Thus,

$$[T_1]/[R_1] = \frac{n[T_0][S]/K_T}{n[R_0][S]/K_R} = \frac{[T_0]}{[R_0](K_R/K_T)} = cL$$

Similar analysis reveals that, for states with i ligands bound, $[T_i]/[R_i] = c^i L$. In other words, the ratio of the concentrations of the T state to the R state is reduced by a factor of c for each ligand that binds.

Let's define a convenient scale for the concentration of S:

$$\alpha = [S]/K_R$$

This definition is useful because it is the ratio of the concentration of S to the dissociation constant that determines the extent of binding. Using this definition, we see that

$$[R_1] = \frac{n[R_0][S]}{K_R} = n[R_0]\alpha$$

Similarly,

$$[T_1] = \frac{n[T_0][S]}{K_T} = ncL[R_0]\alpha$$

What is the concentration of R-state molecules with two ligands bound? Again, we must consider the statistical factor—that is, the number of ways in which a second ligand can bind to a molecule with one site occupied. The number of ways is $n - 1$. However, because which ligand is the "first" and which is the "second" does not matter, we must divide by a factor of 2. Thus,

$$[R_2] = \frac{\left(\dfrac{n-1}{2}\right)[R_1][S]}{K_R}$$

$$= \left(\frac{n-1}{2}\right)[R_1]\alpha$$

$$= \left(\frac{n-1}{2}\right)(n[R_0]\alpha)\alpha$$

$$= n\left(\frac{n-1}{2}\right)[K_0]\alpha^2$$

We can derive similar equations for the case with i ligands bound and for T states.

We can now calculate the fractional saturation, Y. This is the total concentration of sites with ligands bound divided by the total concentration of potential binding sites. Thus,

$$Y = \frac{([R_1] + [T_1] + 2([R_2]) + [T_2]) + \cdots + n([R_n] + [T_n])}{n([R_0] + [T_0] + [R_1] + [T_1] + \cdots + [R_n] + [T_n])}$$

Substituting into this equation, we find

$$Y = \frac{\begin{array}{l} n[R_0]\alpha + nc[T_0]\alpha + 2(n(n-1)/2)[R_0]\alpha^2 \\ \quad + 2(n(n-1)/2)c^2[T_0]\alpha^2 + \cdots \\ \quad\quad\quad + n[R_0]\alpha^n + nc^n[T^0])\alpha^n \end{array}}{\begin{array}{l} n([R_0] + [T_0] + n[R_0]\alpha + nc[T_0]\alpha + \cdots \\ \quad\quad\quad +[R_0]\alpha^n + c^n[T_0]\alpha^n) \end{array}}$$

Substituting $[T_0] = L[R_0]$ and summing these series yields

$$Y = \frac{\alpha(1+\alpha)^{n-1} + Lc\alpha(1+c\alpha)^{n-1}}{(1+\alpha)^n + L(1+c\alpha)^n}$$

We can now use this equation to fit the observed data for hemoglobin by varying the parameters L, c, and K_R (with $n = 4$). An excellent fit is obtained with $L = 9000$, $c = 0.014$, and $K_R = 2.5$ torr (Figure 7.32).

In addition to the fractional saturation, the concentrations of the species T_0, T_1, T_2, R_2, R_3, and R_4 are shown. The concentrations of all other species are very low. The addition

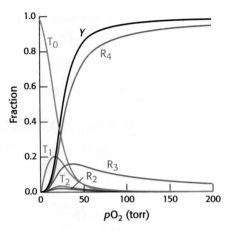

FIGURE 7.32 Modeling oxygen binding with the concerted model. The fractional saturation (γ) as a function pO_2: $L = 9000$, $c = 0.014$, and $K_R = 2.5$ torr. The fraction of molecules in the T state with zero, one, and two oxygen molecules bound (T_0, T_1, and T_2) and the fraction of molecules in the R state with two, three, and four oxygen molecules bound (R_2, R_3, and R_4) are shown. The fractions of molecules in other forms are too low to be shown.

of concentrations is a major difference between the analysis using the Hill equation and this analysis of the concerted model. The Hill equation gives only the fractional saturation, whereas the analysis of the concerted model yields concentrations for all species. In the present case, this analysis yields the expected ratio of T-state proteins to R-state proteins at each stage of binding. This ratio changes from 9000 to 126 to 1.76 to 0.025 to 0.00035 with zero, one, two, three, and four oxygen molecules bound. This ratio provides a quantitative measure of the switching of the population of hemoglobin molecules from the T state to the R state.

The sequential model can also be formulated in quantitative terms. However, the formulation entails many more parameters, and many different sets of parameters often yield similar fits to the experimental data.

APPENDIX

Biochemistry in Focus

A potential antidote for carbon monoxide poisoning?

Neuroglobin has been the subject of significant research investigation as it possesses several unusual qualities. It plays a protective role when blood delivery to the brain is reduced—a clinical situation referred to as *ischemia*. Additionally, the structure of neuroglobin is unique in that the distal histidine (His 64) binds directly to the heme ion and is displaced when oxygen binds (Figure 7.33). Such globins are often referred to as *hexacoordinate globins*. Furthermore, neuroglobin binds ligands such as oxygen and carbon monoxide

quite tightly: for example, human neuroglobin binds O_2 with a P_{50} of 1 torr.

In order to explore the coordination properties of the neuroglobin heme group, researchers have employed site-directed mutagenesis (Section 5.2) to replace the distal histidine side chain with other residues. In the course of these studies, one group made a remarkable discovery: when they mutated His 64 of human neuroglobin to glutamine, and also mutated three surface cysteines to improve solubility, the resulting mutant—"Ngb-H64Q"—exhibited a remarkably high affinity for O_2, with a P_{50} of 0.015 torr! Even more stunning was that Ngb-H64Q bound carbon monoxide 500-fold more tightly than did hemoglobin.

Distal histidine (His 64)

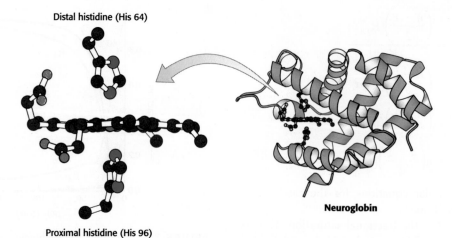

Proximal histidine (His 96)

Neuroglobin

FIGURE 7.33 Structure of neuroglobin. Neuroglobin is a hexacoordinate globin. In the absence of gaseous ligands, its iron atom is coordinated by both the proximal and distal histidine residues. [Drawn from 1OJ6.pdb.]

These data suggested that Ngb-H64Q could potentially serve as an antidote to carbon monoxide poisoning by scavenging CO from carboxyhemoglobin. Such a scavenger should have three essential properties: (1) It should be able to bind CO more tightly than hemoglobin does, (2) it should bind CO more tightly than it does O_2 (so that oxygen in blood does not compete away the CO), and (3) its rate of oxidation to the ferric state should be slow (such that it remains in its active form). These properties were confirmed for Ngb-H64Q. When carboxyhemoglobin was mixed with free Ngb-H64Q, CO transfer was very rapid, with almost complete conversion in less than one minute (Figure 7.34)!

The researchers then applied their new scavenger to the test in an animal model. Mice were given a lethal dose of 3% CO for 4.5 minutes, then switched to room air. At the time that the CO was removed, these mice were given a dose of either Ngb-H64Q, albumin (a protein control), or saline. While the control groups exhibited 10% or worse survival after 40 minutes into the study, Ngb-H64Q yielded a 90% survival rate (Figure 7.35). The discovery of this mutant neuroglobin represents an exciting potential therapeutic option for the treatment of carbon monoxide poisoning that can be rapidly administered by first responders.

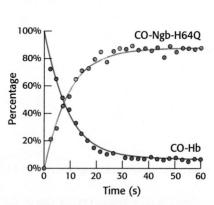

FIGURE 7.34 Ngb-H64Q rapidly scavenges CO from carboxyhemoglobin. In this experiment, carboxyhemoglobin (CO-Hb) was mixed with ligand-free Ngb-H64Q. The CO-Hb species (shown in blue) rapidly decreases while the CO-Ngb-H64Q accumulates. Exchange is nearly complete in less than one minute. [Information from I. Azarov et al., 2016, *Sci. Transl. Med* 8:368ra173, Fig. 2c.]

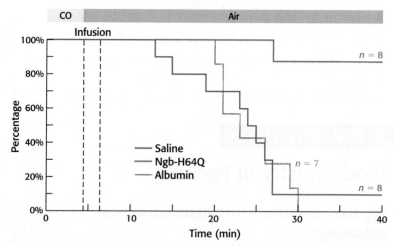

FIGURE 7.35 Ngb-H64Q rescues mice from a fatal carbon monoxide poisoning. This graph represents a survival curve (often referred to as a Kaplan-Meier curve), and indicates the percentage of animals still surviving over a certain period of time. Here, the graph reveals that in the control groups (saline in blue, albumin in red) 10% or fewer animals have survived for 40 minutes after 3% CO inhalation. In contrast, 90% of the animals in the Ngb-H64Q-treated group have survived the duration of the study. [Information from I. Azarov et al., 2016, *Sci. Transl. Med* 8:368ra173, Fig. 5d.]

KEY TERMS

heme (p. 208)

protoporphyrin (p. 209)

proximal histidine (p. 209)

functional magnetic resonance imaging (fMRI) (p. 210)

superoxide anion (p. 210)

metmyoglobin (p. 210)

distal histidine (p. 210)

α chain (p. 211)

β chain (p. 211)

globin fold (p. 211)

αβ dimer (p. 211)

oxygen-binding curve (p. 212)

fractional saturation (p. 212)

partial pressure (p. 212)

sigmoid (p. 212)

cooperative binding (p. 212)

T state (p. 214)

R state (p. 214)

concerted model (MWC model) (p. 214)

sequential model (p. 215)

2,3-bisphosphoglycerate (p. 216)

fetal hemoglobin (p. 217)

carbon monoxide (p. 217)

carboxyhemoglobin (p. 217)

Bohr effect (p. 218)

carbonic anhydrase (p. 219)

carbamate (p. 219)

sickle-cell anemia (p. 220)

hemoglobin S (p. 220)

sickle-cell trait (p. 221)

malaria (p. 221)

thalassemia (p. 221)

hemoglobin H (p. 221)

thalassemia major (Cooley anemia) (p. 222)

α-hemoglobin stabilizing protein (AHSP) (p. 222)

neuroglobin (p. 223)

cytoglobin (p. 223)

Hill plot (p. 225)

Hill coefficient (p. 225)

PROBLEMS

1. *Screening the biosphere.* The first protein structure to have its structure determined was myoglobin from sperm whale. Propose an explanation for the observation that sperm whale muscle is a rich source of this protein. ✓❶

2. *Hemoglobin content.* The average volume of a red blood cell is 87 μm^3. The mean concentration of hemoglobin in red cells is 0.34 g ml^{-1}.

(a) What is the weight of the hemoglobin contained in an average red cell?

(b) How many hemoglobin molecules are there in an average red cell? Assume that the molecular weight of the human hemoglobin tetramer is 65 kDa.

(c) Could the hemoglobin concentration in red cells be much higher than the observed value? (Hint: Suppose that a red cell contained a crystalline array of hemoglobin molecules in a cubic lattice with 65 Å sides.)

3. *Iron content.* How much iron is there in the hemoglobin of a 70-kg adult? Assume that the blood volume is 70 ml kg^{-1} of body weight and that the hemoglobin content of blood is 0.16 g ml^{-1}.

4. *Oxygenating myoglobin.* The myoglobin content of some human muscles is about 8 g kg^{-1}. In sperm whale, the myoglobin content of muscle is about 80 g kg^{-1}.

(a) How much O_2 is bound to myoglobin in human muscle and in sperm whale muscle? Assume that the myoglobin is saturated with O_2 and that the molecular weights of human and sperm whale myoglobin are the same.

(b) The amount of oxygen dissolved in tissue water (in equilibrium with venous blood) at 37°C is about 3.5×10^{-5} M. What is the ratio of oxygen bound to myoglobin to that directly dissolved in the water of sperm whale muscle?

5. *Tuning proton affinity.* The pK_a of an acid depends partly on its environment. Predict the effect of each of the following environmental changes on the pK_a of a glutamic acid side chain.

(a) A lysine side chain is brought into proximity.

(b) The terminal carboxyl group of the protein is brought into proximity.

(c) The glutamic acid side chain is shifted from the outside of the protein to a nonpolar site inside.

6. *Saving grace.* Hemoglobin A inhibits the formation of the long fibers of hemoglobin S and the subsequent sickling of the red cell on deoxygenation. Why does hemoglobin A have this effect?

7. *Carrying a load.* Suppose that you are climbing a high mountain and the oxygen partial pressure in the air is reduced to 75 torr. Estimate the percentage of the oxygen-carrying capacity that will be utilized, assuming that the pH of both tissues and lungs is 7.4 and that the oxygen concentration in the tissues is 20 torr. ✓❸

8. *Bohr for me, not for thee.* Does myoglobin exhibit a Bohr effect? Why or why not? ✓❻

9. *High-altitude adaptation.* After spending a day or more at high altitude (with an oxygen partial pressure of 75 torr), the concentration of 2,3-bisphosphoglycerate (2,3-BPG) in red blood cells increases. What effect would an increased concentration of 2,3-BPG have on the oxygen-binding curve for hemoglobin? Why would this adaptation be beneficial for functioning well at high altitude? ✓ ❸

10. *Blood doping.* Endurance athletes sometimes try an illegal method of blood doping called autologous transfusion. Some blood from the athlete is removed well before competition, and then transfused back into the athlete just before competition.

(a) Why might blood transfusion benefit the athlete?

(b) With time, stored red blood cells become depleted in 2,3-BPG. What might be the consequences of using such blood for a blood transfusion?

11. *I'll take the lobster.* Arthropods such as lobsters have oxygen carriers quite different from hemoglobin. The oxygen-binding sites do not contain heme but, instead, are based on two copper(I) ions. The structural changes that accompany oxygen binding are shown below. How might these changes be used to facilitate cooperative oxygen binding?

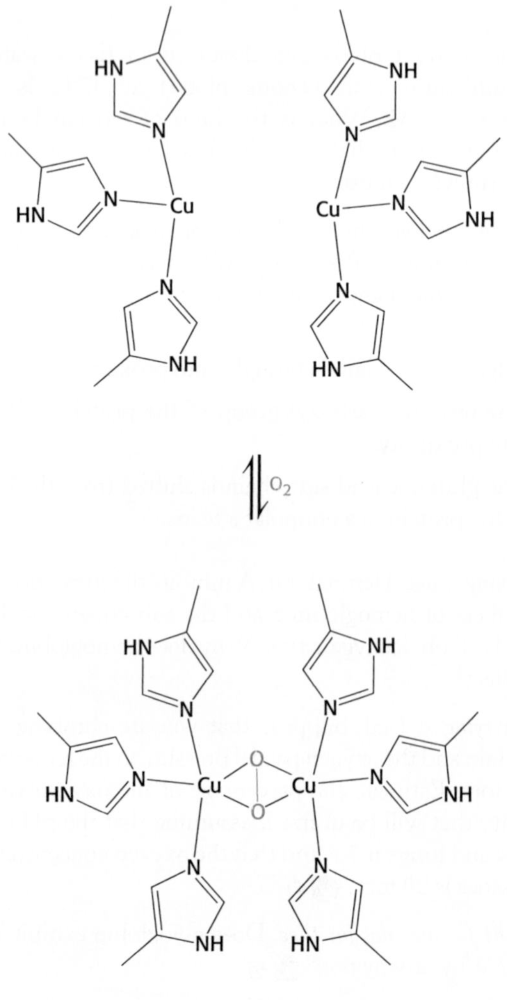

12. *A disconnect.* With the use of site-directed mutagenesis, hemoglobin has been prepared in which the proximal histidine residues in both the α and the β subunits have been replaced by glycine. The imidazole ring from the histidine residue can be replaced by adding free imidazole in solution. Would you expect this modified hemoglobin to show cooperativity in oxygen binding? Why or why not? ✓ ❶

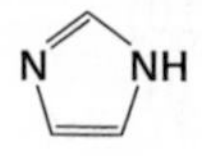

Imidazole

13. *Successful substitution.* Blood cells from some birds do not contain 2,3-bisphosphoglycerate but, instead, contain one of the compounds in parts *a* through *d*, which plays an analogous functional role. Which compound do you think is most likely to play this role? Explain briefly. ✓ ❺

(a)

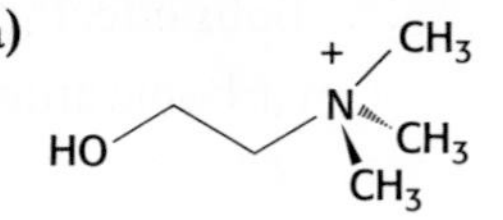

Choline

(b)

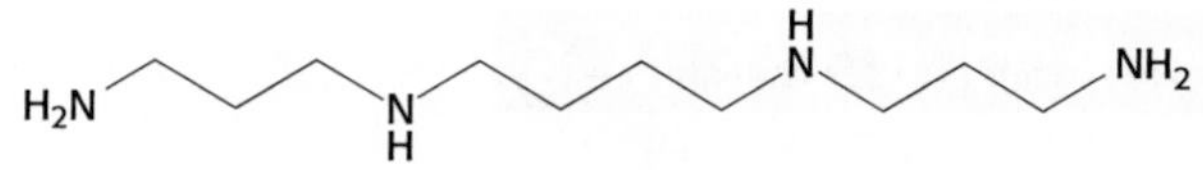

Spermine

(c)

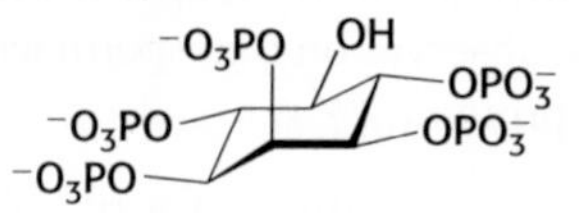

Inositol pentaphosphate

(d)

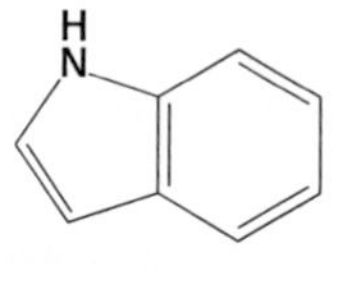

Indole

14. *Theoretical curves.* (a) Using the Hill equation, plot an oxygen-binding curve for a hypothetical two-subunit hemoglobin with $n = 1.8$ and $P_{50} = 10$ torr. (b) Repeat, using the concerted model with $n = 2, L = 1000, c = 0.01$, and $K_R = 1$ torr.

15. *Parasitic effect.* When *P. falciparum* lives inside red blood cells, the metabolism of the parasite tends to release acid. What effect is the presence of acid likely to have on the oxygen-carrying capacity of the red blood cells? On the likelihood that these cells sickle? ✓ ❼

Data Interpretation Problems

16. *Primitive oxygen binding.* Lampreys are primitive organisms whose ancestors diverged from the ancestors of fish and mammals approximately 400 million years ago. Lamprey blood contains a hemoglobin related to mammalian hemoglobin. However, lamprey hemoglobin is monomeric in the oxygenated state. Oxygen-binding data for lamprey hemo-globin are as follows:

pO_2	Y	pO_2	Y	pO_2	Y
0.1	.0060	2.0	.112	50.0	.889
0.2	.0124	3.0	.170	60.0	.905
0.3	.0190	4.0	.227	70.0	.917
0.4	.0245	5.0	.283	80.0	.927
0.5	.0307	7.5	.420	90.0	.935
0.6	.0380	10.0	.500	100	.941
0.7	.0430	15.0	.640	150	.960
0.8	.0481	20.0	.721	200	.970
0.9	.0530	30.0	.812		
1.0	.0591	40.0	.865		

(a) Plot these data to produce an oxygen-binding curve. At what oxygen partial pressure is this hemoglobin half-saturated? On the basis of the appearance of this curve, does oxygen binding seem to be cooperative?

(b) Construct a Hill plot using these data. Does the Hill plot show any evidence for cooperativity? What is the Hill coefficient?

(c) Further studies revealed that lamprey hemoglobin forms oligomers, primarily dimers, in the deoxygenated state. Propose a model to explain any observed cooperativity in oxygen binding by lamprey hemoglobin.

17. *Leaning to the left or to the right.* The illustration below shows several oxygen-dissociation curves. Assume that curve 3 corresponds to hemoglobin with physiological concentrations of CO_2 and 2,3-BPG at pH 7. Which curves represent each of the following perturbations?

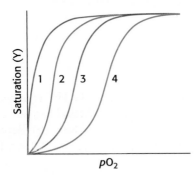

(a) Decrease in CO_2

(b) Increase in 2,3-BPG

(c) Increase in pH

(d) Loss of quaternary structure

Chapter Integration Problems

18. *Location is everything.* 2,3-Bisphosphoglycerate lies in a central cavity within the hemoglobin tetramer, stabilizing the T state. What would be the effect of mutations that placed the BPG-binding site on the surface of hemoglobin?

19. *A therapeutic option.* Hydroxyurea has been shown to increase the expression of fetal hemoglobin in adult red blood cells, by a mechanism that remains unclear. Explain why hydroxyurea can be a useful therapy for patients with sickle-cell anemia.

Think ▸ Pair ▸ Share Problem

20. List all the modulators of oxygen binding by hemoglobin that were discussed in this chapter.

How do these modulators enhance the function of hemoglobin under physiological conditions?

Would you expect any of these modulators to also function similarly in myoglobin? Why or why not?

Enzymes: Basic Concepts and Kinetics

Much of life is motion, whether at the macroscopic level of our daily life or at the molecular level of a cell. Studying motion is what motivated Eadweard James Muybridge to use stop-motion photography to analyze the gallop of a horse in 1878. In biochemistry, kinetics (derived from the Greek *kinesis*, meaning "movement") is used to capture the dynamics of enzyme activity. The energy-rich molecule ATP (right) is a frequent participant in enzyme activity. [(Left) Image Select/Art Resource, NY: (Right) molekuul.be/Alamy.]

✓ LEARNING OBJECTIVES

By the end of this chapter, you should be able to:

1. Describe the relations between the enzyme catalysis of a reaction, the thermodynamics of the reaction, and the formation of the transition state.

2. Explain the relation between the transition state and the active site of an enzyme, and list the characteristics of active sites.

3. Explain what reaction velocity is.

4. Explain how reaction velocity is determined and how reaction velocities are used to characterize enzyme activity.

5. Distinguish between reversible and irreversible inhibitors.

6. Explain how different types of reversible inhibitors can be identified.

7. Describe the advantage of studying enzymes one molecule at a time.

OUTLINE

8.1 Enzymes Are Powerful and Highly Specific Catalysts

8.2 Gibbs Free Energy Is a Useful Thermodynamic Function for Understanding Enzymes

8.3 Enzymes Accelerate Reactions by Facilitating the Formation of the Transition State

8.4 The Michaelis–Menten Model Accounts for the Kinetic Properties of Many Enzymes

8.5 Enzymes Can Be Inhibited by Specific Molecules

8.6 Enzymes Can Be Studied One Molecule at a Time

Enzymes, the catalysts of biological systems, are remarkable molecular devices that determine the patterns of chemical transformations and mediate the transformation of one form of energy into another. Enzymes are also the target of many drugs. For instance, omeprazole is used to inhibit the K^+/H^+ ATPase, the enzyme that acidifies the stomach. In some individuals, the enzyme can be too active, resulting in gastroesophageal reflux disease (GERD) or heartburn, a condition where the stomach acid leaks back to the esophagus.

In addition to being painful, GERD may eventually result in esophageal cancer if left untreated. Omeprazole and similar acting compounds are among the most commonly prescribed drugs. About a quarter of the genes in the human genome encode enzymes, a testament to their importance to life. The most striking characteristics of enzymes are their *catalytic power* and *specificity*. Catalysis takes place at a particular site on the enzyme called the *active site*.

Nearly all known enzymes are proteins. However, proteins do not have an absolute monopoly on catalysis; the discovery of catalytically active RNA molecules provides compelling evidence that RNA was a biocatalyst early in evolution. Proteins as a class of macromolecules are highly effective catalysts for an enormous diversity of chemical reactions because of their capacity *to specifically bind a very wide range of molecules.* By utilizing the full repertoire of intermolecular forces, enzymes bring substrates together in an optimal orientation, the prelude to making and breaking chemical bonds. They catalyze reactions *by stabilizing transition states,* the highest-energy species in reaction pathways. By selectively stabilizing a transition state, an enzyme determines which one of several potential chemical reactions actually takes place.

8.1 Enzymes Are Powerful and Highly Specific Catalysts

Enzymes accelerate reactions by factors of as much as a million or more (Table 8.1). Indeed, most reactions in biological systems do not take place at perceptible rates in the absence of enzymes. Even a reaction as simple as the hydration of carbon dioxide is catalyzed by an enzyme—namely, carbonic anhydrase. The transfer of CO_2 from the tissues to the blood and then to the air in the alveolae of the lungs would be less complete in the absence of this enzyme (p. 221). In fact, carbonic anhydrase is one of the fastest enzymes known. Each enzyme molecule can hydrate 10^6 molecules of CO_2 *per second.* This catalyzed reaction is 10^7 times as fast as the uncatalyzed one. We will consider the mechanism of carbonic anhydrase catalysis in Chapter 9.

TABLE 8.1 Rate enhancement by selected enzymes

Enzyme	Nonenzymatic half-life	Uncatalyzed rate (k_{un} s^{-1})	Catalyzed rate (k_{cat} s^{-1})	Rate enhancement (k_{cat} s^{-1}/k_{un} s^{-1})
OMP decarboxylase	78,000,000 years	2.8×10^{-16}	39	1.4×10^{17}
Staphylococcal nuclease	130,000 years	1.7×10^{-13}	95	5.6×10^{14}
AMP nucleosidase	69,000 years	1.0×10^{-11}	60	6.0×10^{12}
Carboxypeptidase A	7.3 years	3.0×10^{-9}	578	1.9×10^{11}
Ketosteroid isomerase	7 weeks	1.7×10^{-7}	66,000	3.9×10^{11}
Triose phosphate isomerase	1.9 days	4.3×10^{-6}	4300	1.0×10^9
Chorismate mutase	7.4 hours	2.6×10^{-5}	50	1.9×10^6
Carbonic anhydrase	5 seconds	1.3×10^{-1}	1×10^6	7.7×10^6

Abbreviations: OMP, orotidine monophosphate; AMP, adenosine monophosphate.
Source: After A. Radzicka and R. Wolfenden, *Science* 267:90–93, 1995.

Enzymes are highly specific both in the reactions that they catalyze and in their choice of reactants, which are called *substrates*. An enzyme usually catalyzes a single chemical reaction or a set of closely related reactions. Let us consider *proteolytic enzymes* as an example. The biochemical function of these enzymes is to catalyze *proteolysis*, the hydrolysis of a peptide bond.

Peptide   **Carboxyl component**   **Amino component**

Most proteolytic enzymes also catalyze a different but related reaction in vitro—namely, the hydrolysis of an ester bond. Such reactions are more easily monitored than is proteolysis and are useful in experimental investigations of these enzymes.

Ester   **Acid**   **Alcohol**

Proteolytic enzymes differ markedly in their degree of substrate specificity. Papain, which is found in papaya plants, is quite undiscriminating: it will cleave any peptide bond with little regard to the identity of the adjacent side chains. This lack of specificity accounts for its use in meat-tenderizing sauces. The digestive enzyme trypsin, on the other hand, is quite specific and catalyzes the splitting of peptide bonds only on the carboxyl side of lysine and arginine residues (Figure 8.1A). Thrombin, an enzyme that participates in blood clotting (Section 10.4), is even more specific than trypsin. It catalyzes the hydrolysis of Arg–Gly bonds in particular peptide sequences only (Figure 8.1B).

DNA polymerase I, a template-directed enzyme (Section 29.3), is another highly specific catalyst. DNA polymerase adds nucleotides to the strand being synthesized in a sequence determined by the sequence of nucleotides in another DNA strand that serves as a template. DNA polymerase I is remarkably precise in carrying out the instructions given by the template. It inserts the wrong nucleotide into a new DNA strand less than one in a thousand times. *The specificity of an enzyme is due to the precise interaction of the substrate with the enzyme. This precision is a result of the intricate three-dimensional structure of the enzyme protein.*

Many enzymes require cofactors for activity

The catalytic activity of many enzymes depends on the presence of small molecules termed *cofactors*, although the precise role varies with the cofactor and the enzyme. Generally, these cofactors are able to execute chemical reactions that cannot be performed by the standard set of twenty amino acids. An enzyme without its cofactor is referred to as an *apoenzyme;* the complete, catalytically active enzyme is called a *holoenzyme.*

$$\text{Apoenzyme} + \text{cofactor} = \text{holoenzyme}$$

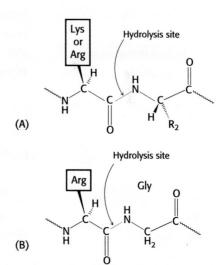

FIGURE 8.1 Enzyme specificity. (A) Trypsin cleaves on the carboxyl side of arginine and lysine residues, whereas (B) thrombin cleaves Arg–Gly bonds in particular sequences only.

Cofactors can be subdivided into two groups: (1) metals and (2) small organic molecules called *coenzymes* (Table 8.2). Often derived from vitamins, coenzymes can be either tightly or loosely bound to the enzyme. Tightly bound coenzymes are called *prosthetic groups*. Loosely associated coenzymes are more like cosubstrates because, like substrates and products, they bind to the enzyme and are released from it. The use of the same coenzyme by a variety of enzymes sets coenzymes apart from normal substrates, however, as does their source in vitamins (Section 15.4). Enzymes that use the same coenzyme usually perform catalysis by similar mechanisms. In Chapter 9, we will examine the importance of metals to enzyme activity and, throughout the book, we will see how coenzymes and their enzyme partners operate in their biochemical context.

Enzymes can transform energy from one form into another

A key activity in all living systems is the conversion of one form of energy into another. For example, in photosynthesis, light energy is converted into chemical-bond energy. In cellular respiration, which takes place in mitochondria, the free energy contained in small molecules derived from food is converted first into the free energy of an ion gradient and then into a different currency—the free energy of adenosine triphosphate. Given their centrality to life, it should come as no surprise that enzymes play vital roles in energy transformation. As we will see, enzymes perform fundamental roles in photosynthesis and cellular respiration. Other enzymes can then use the chemical-bond energy of ATP in diverse ways. For instance, the enzyme myosin converts the energy of ATP into the mechanical energy of contracting muscles (Section 9.4, and Chapter 36 on 🌿 SaplingPlus). Pumps in the membranes of cells and organelles, which can be thought of as enzymes that move substrates rather than chemically alter them, use the energy of ATP to transport molecules and ions across the membrane (Chapter 13). The chemical and electrical gradients resulting from the unequal distribution of these molecules and ions are themselves forms of energy that can be used for a variety of purposes, such as sending nerve impulses.

The molecular mechanisms of these energy-transducing enzymes are being unraveled. We will see in subsequent chapters how unidirectional cycles of discrete steps—binding, chemical transformation, and release—lead to the conversion of one form of energy into another.

TABLE 8.2 Enzyme cofactors

Cofactor	Enzyme
Coenzyme	
Thiamine pyrophosphate	Pyruvate dehydrogenase
Flavin adenine nucleotide	Monoamine oxidase
Nicotinamide adenine dinucleotide	Lactate dehydrogenase
Pyridoxal phosphate	Glycogen phosphorylase
Coenzyme A (CoA)	Acetyl CoA carboxylase
Biotin	Pyruvate carboxylase
5′-Deoxyadenosylcobalamin	Methylmalonyl mutase
Tetrahydrofolate	Thymidylate synthase
Metal	
Zn^{2+}	Carbonic anhydrase
Zn^{2+}	Carboxypeptidase
Mg^{2+}	*Eco*RV
Mg^{2+}	Hexokinase
Ni^{2+}	Urease
Mo	Nitrogenase
Se	Glutathione peroxidase
Mn	Superoxide dismutase
K^+	Acetyl CoA thiolase

8.2 Gibbs Free Energy Is a Useful Thermodynamic Function for Understanding Enzymes

Enzymes speed up the rate of chemical reactions, but the properties of the reaction—whether it can take place at all and the degree to which the enzyme accelerates the reaction—depend on energy differences between reactants and products. *Gibbs free energy (G),* which was touched on in Chapter 1, is a thermodynamic property that is a measure of useful energy, or the energy that is capable of doing work. To understand how enzymes operate, we need to consider only two thermodynamic properties of the reaction: (1)

the free-energy difference (ΔG) between the products and reactants and (2) the energy required to initiate the conversion of reactants into products. The former determines whether the reaction will take place spontaneously, whereas the latter determines the rate of the reaction. Enzymes affect only the latter. Let us review some of the principles of thermodynamics as they apply to enzymes.

The free-energy change provides information about the spontaneity but not the rate of a reaction

As discussed in Chapter 1, the free-energy change of a reaction (ΔG) tells us if the reaction can take place spontaneously:

1. *A reaction can take place spontaneously only if ΔG is negative.* Such reactions are said to be *exergonic*.

2. *A system is at equilibrium and no net change can take place if ΔG is zero.*

3. *A reaction cannot take place spontaneously if ΔG is positive.* An input of free energy is required to drive such a reaction. These reactions are termed *endergonic*.

4. *The ΔG of a reaction depends only on the free energy of the products (the final state) minus the free energy of the reactants (the initial state).* The ΔG of a reaction is independent of the molecular mechanism of the transformation. For example, the ΔG for the oxidation of glucose to CO_2 and H_2O is the same whether it takes place by combustion or by a series of enzyme-catalyzed steps in a cell.

5. *The ΔG provides no information about the rate of a reaction.* A negative ΔG indicates that a reaction *can* take place spontaneously, but it does not signify whether it will proceed at a perceptible rate. As will be discussed shortly (Section 8.3), the rate of a reaction depends on the *free energy of activation* ($\Delta G^{\ddagger}$), which is largely unrelated to the ΔG of the reaction.

The standard free-energy change of a reaction is related to the equilibrium constant

As for any reaction, we need to be able to determine ΔG for an enzyme-catalyzed reaction to know whether the reaction is spontaneous or requires an input of energy. To determine this important thermodynamic parameter, we need to take into account the nature of both the reactants and the products as well as their concentrations.

Consider the reaction

$$A + B \rightleftharpoons C + D$$

The ΔG of this reaction is given by

$$\Delta G = \Delta G^{\circ} + RT\ln\frac{[C][D]}{[A][B]} \tag{1}$$

in which ΔG° is the *standard free-energy change, R* is the gas constant, *T* is the absolute temperature, and [A], [B], [C], and [D] are the molar concentrations (more precisely, the activities) of the reactants. ΔG° is the free-energy change for this reaction under standard conditions—that is, when each of the reactants A, B, C, and D is present at a concentration of 1.0 M (for a gas, the standard state is usually chosen to be 1 atmosphere). Thus, the ΔG of a reaction depends on the *nature* of the reactants (expressed in the ΔG°

Units of energy

A *kilojoule* (kJ) is equal to 1000 J.

A *joule* (J) is the amount of energy needed to apply a 1-newton force over a distance of 1 meter.

A *kilocalorie* (kcal) is equal to 1000 cal.

A *calorie* (cal) is equivalent to the amount of heat required to raise the temperature of 1 gram of water from 14.5°C to 15.5°C. 1 kJ = 0.239 kcal.

TABLE 8.3 Relation between $\Delta G^{\circ\prime}$ and K'_{eq} (at 25°C)

K'_{eq}	$\Delta G^{\circ\prime}$	
	kJ mol^{-1}	kcal mol^{-1}
10^{-5}	28.53	6.82
10^{-4}	22.84	5.46
10^{-3}	17.11	4.09
10^{-2}	11.42	2.73
10^{-1}	5.69	1.36
1	0.00	0.00
10	−5.69	−1.36
10^2	−11.42	−2.73
10^3	−17.11	−4.09
10^4	−22.84	−5.46
10^5	−28.53	−6.82

term of equation 1) and on their *concentrations* (expressed in the logarithmic term of equation 1).

A convention has been adopted to simplify free-energy calculations for biochemical reactions. The standard state is defined as having a pH of 7. Consequently, when H^+ is a reactant, its activity has the value 1 (corresponding to a pH of 7) in equations 1 and 3 (below). The activity of water also is taken to be 1 in these equations. The *standard free-energy change at pH 7*, denoted by the symbol $\Delta G^{\circ\prime}$, will be used throughout this book. The *kilojoule* (abbreviated kJ) and the *kilocalorie* (kcal) will be used as the units of energy. One kilojoule is equivalent to 0.239 kilocalorie.

A simple way to determine $\Delta G^{\circ\prime}$ is to measure the concentrations of reactants and products when the reaction has reached equilibrium. At equilibrium, there is no net change in reactants and products; in essence, the reaction has stopped and $\Delta G = 0$. At equilibrium, equation 1 then becomes

$$0 = \Delta G^{\circ\prime} + RT \ln \frac{[C][D]}{[A][B]} \tag{2}$$

and so

$$\Delta G^{\circ\prime} = -RT \ln \frac{[C][D]}{[A][B]} \tag{3}$$

The equilibrium constant under standard conditions, K'_{eq} is defined as

$$K'_{eq} = \frac{[C][D]}{[A][B]} \tag{4}$$

Substituting equation 4 into equation 3 gives

$$\Delta G^{\circ\prime} = -RT \ln K'_{eq} \tag{5}$$

which can be rearranged to give

$$K'_{eq} = e^{-\Delta G^{\circ\prime}/RT} \tag{6}$$

Substituting $R = 8.315 \times 10^{-3} \, kJ \, mol^{-1} \, deg^{-1}$ and $T = 298 \, K$ (corresponding to 25°C) gives

$$K'_{eq} = e^{-\Delta G^{\circ\prime}/2.47} \tag{7}$$

where $\Delta G^{\circ\prime}$ is here expressed in kilojoules per mole because of the choice of the units for R in equation 7. Thus, the standard free energy and the equilibrium constant of a reaction are related by a simple expression. For example, an equilibrium constant of 10 gives a standard free-energy change of $-5.69 \, kJ \, mol^{-1}$ ($-1.36 \, kcal \, mol^{-1}$) at 25°C (Table 8.3). Note that, for each 10-fold change in the equilibrium constant, the $\Delta G^{\circ\prime}$ changes by $5.69 \, kJ \, mol^{-1}$ ($1.36 \, kcal \, mol^{-1}$).

As an example, let us calculate $\Delta G^{\circ\prime}$ and ΔG for the isomerization of dihydroxyacetone phosphate (DHAP) to glyceraldehyde 3-phosphate (GAP). This reaction takes place in glycolysis (Chapter 16). At equilibrium, the ratio of GAP to DHAP is 0.0475 at 25°C (298 K) and pH 7. Hence, $K'_{eq} = 0.0475$. The standard free-energy change for this reaction is then calculated from equation 5:

$$\Delta G^{\circ\prime} = -RT \ln K'_{eq}$$
$$= -8.315 \times 10^{-3} \times 298 \times \ln (0.0475)$$
$$= +7.53 \, kJ \, mol^{-1} (+1.80 \, kcal \, mol^{-1})$$

Under these conditions, the reaction is endergonic. DHAP will not spontaneously convert into GAP.

Now let us calculate ΔG for this reaction when the initial concentration of DHAP is 2×10^{-4} M and the initial concentration of GAP is 3×10^{-6} M. Substituting these values into equation 1 gives

$$\Delta G = 7.53 \text{ kJ mol}^{-1} - RT \ln \frac{3 \times 10^{-6} \text{M}}{2 \times 10^{-4} \text{M}}$$

$$= 7.53 \text{ kJ mol}^{-1} - (10.42 \text{ kJ mol}^{-1})$$

$$= -2.89 \text{ kJ mol}^{-1} (-0.69 \text{ kcal mol}^{-1})$$

This negative value for the ΔG indicates that the isomerization of DHAP to GAP is exergonic and can take place spontaneously when these species are present at the preceding concentrations. Note that ΔG for this reaction is negative, although $\Delta G^{\circ\prime}$ is positive. *It is important to stress that whether the ΔG for a reaction is larger, smaller, or the same as $\Delta G^{\circ\prime}$ depends on the concentrations of the reactants and products.* The criterion of spontaneity for a reaction is ΔG, not $\Delta G^{\circ\prime}$. This point is important because reactions that are not spontaneous based on $\Delta G^{\circ\prime}$ can be made spontaneous by adjusting the concentrations of reactants and products. This principle is the basis of the coupling of reactions to form metabolic pathways (Chapter 15).

Enzymes alter only the reaction rate and not the reaction equilibrium

Because enzymes are such superb catalysts, it is tempting to ascribe to them powers that they do not have. An enzyme cannot alter the laws of thermodynamics and *consequently cannot alter the equilibrium of a chemical reaction.* Consider an enzyme-catalyzed reaction, the conversion of substrate, S, into product, P. Figure 8.2 shows the rate of product formation with time in the presence and absence of enzyme. Note that the amount of product formed is the same whether or not the enzyme is present but, in the current example, the amount of product formed in seconds when the enzyme is present might take hours (or centuries, see Table 8.1) to form if the enzyme were absent.

Why does the rate of product formation level off with time? The reaction has reached equilibrium. Substrate S is still being converted into product P, but P is being converted into S at a rate such that the amount of P present stays the same.

Let us examine the equilibrium in a more quantitative way. Suppose that, in the absence of enzyme, the forward rate constant (k_F) for the conversion of S into P is 10^{-4} s^{-1} and the reverse rate constant (k_R) for the conversion of P into S is 10^{-6} s^{-1}. The equilibrium constant K is given by the ratio of these rate constants:

$$S \underset{10^{-6} \text{ s}^{-1}}{\overset{10^{-4} \text{ s}^{-1}}{\rightleftarrows}} P$$

$$K = \frac{[\text{P}]}{[\text{S}]} = \frac{k_F}{k_R} = \frac{10^{-4}}{10^{-6}} = 100$$

The equilibrium concentration of P is 100 times that of S, whether or not enzyme is present. However, it might take a very long time to approach this equilibrium without enzyme, whereas equilibrium would be attained rapidly in the presence of a suitable enzyme (Table 8.1). *Enzymes accelerate the attainment of equilibria but do not shift their positions. The equilibrium position is a function only of the free-energy difference between reactants and products.*

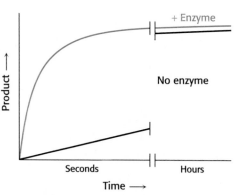

Dihydroxyacetone phosphate (DHAP)

Glyceraldehyde 3-phosphate (GAP)

FIGURE 8.2 Enzymes accelerate the reaction rate. The same equilibrium point is reached but much more quickly in the presence of an enzyme.

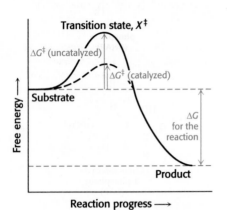

FIGURE 8.3 Enzymes decrease the activation energy. Enzymes accelerate reactions by decreasing $\Delta G^{\ddagger}$, the free energy of activation.

8.3 Enzymes Accelerate Reactions by Facilitating the Formation of the Transition State

The free-energy difference between reactants and products accounts for the equilibrium of the reaction, but enzymes accelerate how quickly this equilibrium is attained. How can we explain the rate enhancement in terms of thermodynamics? To do so, we have to consider not the end points of the reaction but the chemical pathway between the end points.

A chemical reaction of substrate S to form product P goes through a *transition state* $X^{\ddagger}$ that has a higher free energy than does either S or P.

$$S \longrightarrow X^{\ddagger} \longrightarrow P$$

The double dagger denotes the transition state. The transition state is a transitory molecular structure that is no longer the substrate but is not yet the product. The transition state is the least-stable and most-seldom-occupied species along the reaction pathway because it is the one with the highest free energy. The difference in free energy between the transition state and the substrate is called the *Gibbs free energy of activation* or simply the *activation energy*, symbolized by $\Delta G^{\ddagger}$ (Figure 8.3).

$$\Delta G^{\ddagger} = G_X{}^{\ddagger} - G_S$$

Note that the energy of activation, or $\Delta G^{\ddagger}$, does not enter into the final ΔG calculation for the reaction, because the energy required to generate the transition state is released when the transition state forms the product. The activation-energy barrier immediately suggests how an enzyme enhances the reaction rate without altering ΔG of the reaction: enzymes function to lower the activation energy, or, in other words, *enzymes facilitate the formation of the transition state.*

One approach to understanding the increase in reaction rates achieved by enzymes is to assume that the transition state ($X^{\ddagger}$) and the substrate (S) are in equilibrium.

$$S \rightleftharpoons X^{\ddagger} \longrightarrow P$$

The equilibrium constant for the formation of $X^{\ddagger}$ from S is $K^{\ddagger}$ or $[X^{\ddagger}]/[S]$. Using equation 6 (p. 238) and solving for $[X^{\ddagger}]$

$$[X^{\ddagger}] = [S]e^{-\Delta G^{\ddagger}/RT}$$

The rate of the reaction V is equal to the rate of formation of $X^{\ddagger}$ or v multiplied by $[X^{\ddagger}]$, or

$$V = v[X^{\ddagger}] = v[S]e^{-\Delta G^{\ddagger}/RT}$$

Thus the overall rate of the reaction depends on $\Delta G^{\ddagger}$. To be more precise, transition state theory equates v to kT/h.

$$V = v[X^{\ddagger}] = \frac{kT}{h}[S]e^{-\Delta G^{\ddagger}/RT}$$

In this equation, k is Boltzmann's constant, and h is Planck's constant. The value of kT/h at 25°C is $6.6 \times 10^{12}\,\text{s}^{-1}$. Suppose that the free energy of activation is $28.53\,\text{kJ mol}^{-1}$ ($6.82\,\text{kcal mol}^{-1}$). If we were to substitute this value of ΔG in equation 7 (as shown in Table 8.3), this free-energy difference would result when the ratio $[X^{\ddagger}]/[S]$ is 10^{-5}. If we assume for simplicity's sake that $[S] = 1\,\text{M}$, then the reaction rate V is $6.2 \times 10^{7}\,\text{s}^{-1}$. If $\Delta G^{\ddagger}$ were lowered by $5.69\,\text{kJ mol}^{-1}$ ($1.36\,\text{kcal mol}^{-1}$), the ratio $[X^{\ddagger}]/[S]$ would then be 10^{-4}, and the reaction rate would be $6.2 \times 10^{8}\,\text{s}^{-1}$. A decrease of

5.69 kJ mol^{-1} in $\Delta G^{\ddagger}$ yields a 10-fold larger V. A relatively small decrease in $\Delta G^{\ddagger}$ (20% in this particular reaction) results in a much greater increase in V.

Thus, we see the key to how enzymes operate: *enzymes accelerate reactions by decreasing $\Delta G^{\ddagger}$, the activation energy.* The combination of substrate and enzyme creates a reaction pathway whose transition-state energy is lower than that of the reaction in the absence of enzyme (Figure 8.3). Because the activation energy is lower, more molecules have the energy required to reach the transition state. Decreasing the activation barrier is analogous to lowering the height of a high-jump bar; more athletes will be able to clear the bar. *The essence of catalysis is facilitating the formation of the transition state.*

The formation of an enzyme–substrate complex is the first step in enzymatic catalysis

Much of the catalytic power of enzymes comes from their binding to and then altering the structure of the substrate to promote the formation of the transition state. Thus, the first step in catalysis is the formation of an *enzyme–substrate* (ES) complex. Substrates bind to a specific region of the enzyme called the *active site*. Most enzymes are highly selective in the substrates that they bind. Indeed, the catalytic specificity of enzymes depends in part on the specificity of binding.

What is the evidence for the existence of an enzyme–substrate complex?

1. The first clue was the observation that, at a constant concentration of enzyme, the reaction rate increases with increasing substrate concentration until a maximal velocity is reached (Figure 8.4). In contrast, uncatalyzed reactions do not show this saturation effect. *The fact that an enzyme-catalyzed reaction has a maximal velocity suggests the formation of a discrete ES complex.* At a sufficiently high substrate concentration, all the catalytic sites are filled, or saturated, and so the reaction rate cannot increase. Although indirect, the ability to saturate an enzyme with substrate is the most general evidence for the existence of ES complexes.

2. The *spectroscopic characteristics* of many enzymes and substrates change on the formation of an ES complex. These changes are particularly striking if the enzyme contains a colored prosthetic group (problem 45).

3. *X-ray crystallography* has provided high-resolution images of substrates and substrate analogs bound to the active sites of many enzymes (Figure 8.5). In Chapter 9, we will take a close look at several of these complexes.

"I think that enzymes are molecules that are complementary in structure to the activated complexes of the reactions that they catalyze, that is, to the molecular configuration that is intermediate between the reacting substances and the products of reaction for these catalyzed processes. The attraction of the enzyme molecule for the activated complex would thus lead to a decrease in its energy and hence to a decrease in the energy of activation of the reaction and to an increase in the rate of reaction."
—Linus Pauling, Nature 161:707, 1948

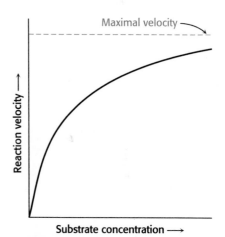

FIGURE 8.4 Reaction velocity versus substrate concentration in an enzyme-catalyzed reaction. An enzyme-catalyzed reaction approaches a maximal velocity.

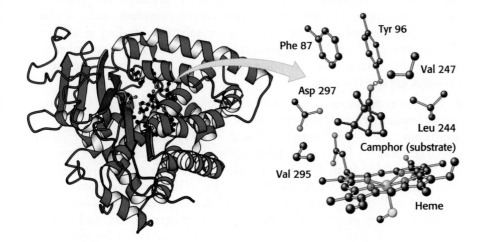

FIGURE 8.5 Structure of an enzyme–substrate complex. (Left) The enzyme cytochrome P450 is illustrated bound to its substrate camphor. (Right) *Notice* that, in the active site, the substrate is surrounded by residues from the enzyme. Note also the presence of a heme cofactor. [Drawn from 2CPP.pdb.]

(A)

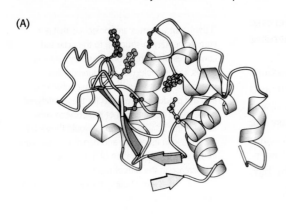

(B)

N ├─────────────────────────────────────┤ C
  1   35   52 62,63   101 108   129

FIGURE 8.6 Active sites may include distant residues.
(A) Ribbon diagram of the enzyme lysozyme with
several components of the active site shown in color.
(B) A schematic representation of the primary structure
of lysozyme shows that the active site is composed of
residues that come from different parts of the polypeptide
chain. [Drawn from 6LYZ.pdb.]

The active sites of enzymes have some common features

The active site of an enzyme is the region that binds the substrates
(and the cofactor, if any). It also contains the amino acid residues
that directly participate in the making and breaking of bonds.
These residues are called the *catalytic groups*. In essence, *the interaction of the enzyme and substrate at the active site promotes the
formation of the transition state*. The active site is the region of the
enzyme that most directly lowers the $\Delta G^{\ddagger}$ of the reaction, thus
providing the rate-enhancement characteristic of enzyme action.
Recall from Chapter 2 that proteins are not rigid structures, but
are flexible and exist in an array of conformations. Thus, the
interaction of the enzyme and substrate at the active site and the
formation of the transition state is a dynamic process. Although
enzymes differ widely in structure, specificity, and mode of catalysis, a number of generalizations concerning their active sites can
be stated:

1. *The active site is a three-dimensional cleft, or crevice,* formed by
groups that come from different parts of the amino acid sequence:
indeed, residues far apart in the amino acid sequence may interact
more strongly than adjacent residues in the sequence, which may
be sterically constrained from interacting with one another. In
lysozyme, the important groups in the active site are contributed by residues numbered 35, 52, 62, 63, 101, and 108 in the sequence of 129 amino
acids (Figure 8.6). Lysozyme, found in a variety of organisms and tissues,
including human tears, degrades the cell walls of some bacteria.

2. *The active site takes up a small part of the total volume of an enzyme.*
Although most of the amino acid residues in an enzyme are not in contact
with the substrate, the cooperative motions of the entire enzyme help to correctly position the catalytic residues at the active site. Experimental attempts
to reduce the size of a catalytically active enzyme show that the minimum
size requires about 100 amino acid residues. In fact, nearly all enzymes are
made up of more than 100 amino acid residues, which gives them a mass
greater than 10 kDa and a diameter of more than 25 Å, suggesting that all
amino acids in the protein, not just those at the active site, are ultimately
required to form a functional enzyme.

3. *Active sites are unique microenvironments.* In all enzymes of known structure, active sites are shaped like a cleft, or crevice, to which the substrates
bind. Water is usually excluded unless it is a reactant. The nonpolar microenvironment of the cleft enhances the binding of substrates as well as catalysis. Nevertheless, the cleft may also contain polar residues, some of which
may acquire special properties essential for substrate binding or catalysis.
The internal positions of these polar residues are biologically crucial exceptions to the general rule that polar residues are located on the surface of
proteins, exposed to water.

4. *Substrates are bound to enzymes by multiple weak attractions.* The noncovalent interactions in ES complexes are much weaker than covalent bonds,
which have energies between -210 and $-460\,\mathrm{kJ\,mol^{-1}}$ (between -50 and
$-110\,\mathrm{kcal\,mol^{-1}}$). In contrast, ES complexes usually have equilibrium constants that range from 10^{-2} to $10^{-8}\,\mathrm{M}$, corresponding to free energies of
interaction ranging from about -13 to $-50\,\mathrm{kJ\,mol^{-1}}$ (from -3 to
$-12\,\mathrm{kcal\,mol^{-1}}$). As discussed in Section 1.3, these weak reversible contacts
are mediated by electrostatic interactions, hydrogen bonds, and van der
Waals forces. Van der Waals forces become significant in binding only when

numerous substrate atoms simultaneously come close to many enzyme atoms through the hydrophobic effect. Hence, the enzyme and substrate should have complementary shapes. The directional character of hydrogen bonds between enzyme and substrate often enforces a high degree of specificity, as seen in the RNA-degrading enzyme ribonuclease (Figure 8.7).

5. *The specificity of binding depends on the precisely defined arrangement of atoms in an active site.* Because the enzyme and the substrate interact by means of short-range forces that require close contact, a substrate must have a matching shape to fit into the site. Emil Fischer proposed the lock-and-key analogy in 1890 (Figure 8.8), which was the model for enzyme–substrate interaction for several decades. We now know that enzymes are flexible and that the shapes of the active sites can be markedly modified by the binding of substrate, a process of dynamic recognition called *induced fit* (Figure 8.9). Moreover, the substrate may bind to only certain conformations of the enzyme, in what is called *conformation selection*. Thus, the mechanism of catalysis is dynamic, involving structural changes with multiple intermediates of both reactants and the enzyme.

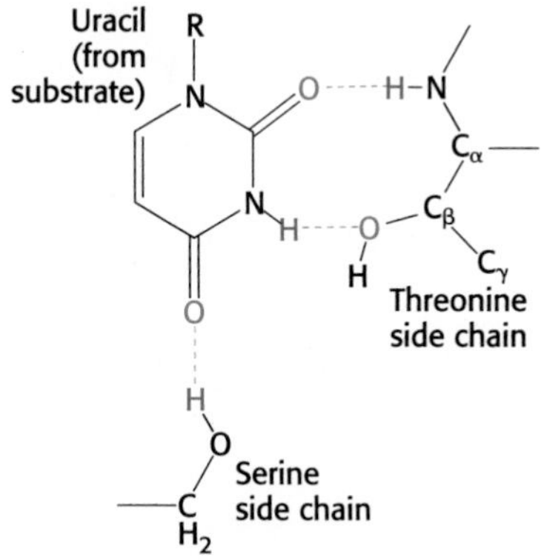

FIGURE 8.7 Hydrogen bonds between an enzyme and substrate. The enzyme ribonuclease forms hydrogen bonds with the uridine component of the substrate. [Information from F. M. Richards, H. W. Wyckoff, and N. Allewell, In *The Neurosciences: Second Study Program,* F. O. Schmidt, Ed. (Rockefeller University Press, 1970), p. 970.]

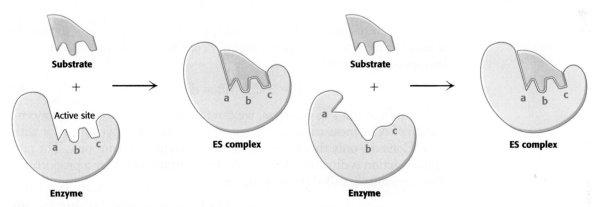

FIGURE 8.8 Lock-and-key model of enzyme–substrate binding. In this model, the active site of the unbound enzyme is complementary in shape to the substrate.

FIGURE 8.9 Induced-fit model of enzyme–substrate binding. In this model, the enzyme changes shape on substrate binding. The active site forms a shape complementary to the substrate only after the substrate has been bound.

The binding energy between enzyme and substrate is important for catalysis

Enzymes lower the activation energy, but where does the energy to lower the activation energy come from? Free energy is released by the formation of a large number of weak interactions between a complementary enzyme and its substrate. The free energy released on binding is called the *binding energy*. Only the correct substrate can participate in most or all of the interactions with the enzyme and thus maximize binding energy, accounting for the exquisite substrate specificity exhibited by many enzymes. Furthermore, *the full complement of such interactions is formed only when the substrate is converted into the transition state.* Thus, the maximal binding energy is released when the enzyme facilitates the formation of the transition state. The energy released by the interaction between the enzyme and the substrate can be thought of as lowering the activation energy. The interaction of the enzyme with the substrate and reaction intermediates is fleeting, with molecular movements resulting in the optimal alignment of functional groups at the active site so that maximum binding energy occurs only between the enzyme and the transition state, the least stable reaction intermediate. However, the transition state is too unstable to exist for long. It collapses to either substrate or product, but which of the two accumulates is determined only by the energy difference between the substrate and the product—that is, by the ΔG of the reaction.

8.4 The Michaelis–Menten Model Accounts for the Kinetic Properties of Many Enzymes

The study of the rates of chemical reactions is called *kinetics,* and the study of the rates of enzyme-catalyzed reactions is called *enzyme kinetics.* A kinetic description of enzyme activity will help us understand how enzymes function. We begin by briefly examining some of the basic principles of reaction kinetics.

Kinetics is the study of reaction rates

What do we mean when we say the "rate" of a chemical reaction? Consider a simple reaction:

$$A \longrightarrow P$$

The rate V is the quantity of A that disappears in a specified unit of time. It is equal to the rate of the appearance of P, or the quantity of P that appears in a specified unit of time.

$$V = -d[A]/dt = d[P]/dt \tag{8}$$

If A is yellow and P is colorless, we can follow the decrease in the concentration of A by measuring the decrease in the intensity of yellow color with time. Consider only the change in the concentration of A for now. The rate of the reaction is directly related to the concentration of A by a proportionality constant, k, called the *rate constant.*

$$V = k[A] \tag{9}$$

Reactions that are directly proportional to the reactant concentration are called *first-order reactions.* First-order rate constants have the units of s^{-1}.

Many important biochemical reactions include two reactants. For example,

$$2A \longrightarrow P$$

or

$$A + B \longrightarrow P$$

They are called *bimolecular reactions* and the corresponding rate equations often take the form

$$V = k[A]^2 \tag{10}$$

and

$$V = k[A][B] \tag{11}$$

The rate constants, called second-order rate constants, have the units $M^{-1} s^{-1}$.

Sometimes, second-order reactions can appear to be first-order reactions. For instance, in reaction 11, if B is present in excess and A is present at low concentrations, the reaction rate will be first order with respect to A and will not appear to depend on the concentration of B. These reactions are called *pseudo-first-order reactions,* and we will see them a number of times in our study of biochemistry.

Interestingly enough, under some conditions, a reaction can be zero order. In these cases, the rate is independent of reactant concentrations. Enzyme-catalyzed reactions can approximate zero-order reactions under some circumstances (p. 247).

The steady-state assumption facilitates a description of enzyme kinetics

The simplest way to investigate the reaction rate is to follow the increase in reaction product as a function of time. First, the extent of product formation is determined as a function of time for a series of substrate concentrations (Figure 8.10A). As expected, in each case, the amount of product formed increases with time, although eventually a time is reached when there is *no net change* in the concentration of S or P. The enzyme is still actively converting substrate into product and vice versa, but the reaction equilibrium has been attained. However, enzyme kinetics is more readily comprehended if we consider only the forward reaction. We can define the rate of catalysis V_0, or the initial rate of catalysis, as the number of moles of product formed per second when the reaction is just beginning—that is, when $t \approx 0$ (Figure 8.10A). These experiments are repeated three to five times with each substrate concentration to insure the accuracy of and assess the variability of the values attained. Next, we plot V_0 versus the substrate concentration [S], assuming a constant amount of enzyme, showing the data points with error bars (Figure 8.10B). Finally, the data points are connected, yielding the results shown in Figure 8.10C. The rate of catalysis rises linearly as substrate concentration increases and then begins to level off and approach a maximum at higher substrate concentrations. For convenience, we will show idealized data without error bars throughout the text, but it is important to keep in mind that in reality, all experiments are repeated multiple times.

In 1913, Leonor Michaelis and Maud Menten proposed a simple model to account for these kinetic characteristics. The critical feature in their treatment is that a specific ES complex is a necessary intermediate in catalysis. The model proposed is

$$E + S \underset{k_{-1}}{\overset{k_1}{\rightleftharpoons}} ES \underset{k_{-2}}{\overset{k_2}{\rightleftharpoons}} E + P$$

An enzyme E combines with substrate S to form an ES complex, with a rate constant k_1. The ES complex has two possible fates. It can dissociate to E and S, with a rate constant k_{-1}, or it can proceed to form product P, with a rate constant k_2. The ES complex can also be reformed from E and P by the reverse reaction with a rate constant k_{-2}. However, as before, we can simplify these reactions by considering the rate of reaction at times close to zero (hence, V_0) when there is negligible product formation and thus no back reaction ($k_{-2}[E][P] \approx 0$).

$$E + S \underset{k_{-1}}{\overset{k_1}{\rightleftharpoons}} ES \overset{k_2}{\longrightarrow} E + P \qquad (12)$$

Thus, for the graph in Figure 8.11, V_0 is determined for each substrate concentration by measuring the rate of product formation at early times before P accumulates (Figure 8.10A).

We want an expression that relates the rate of catalysis to the concentrations of substrate and enzyme and the rates of the individual steps. Our starting point is that the catalytic rate is equal to the product of the concentration of the ES complex and k_2.

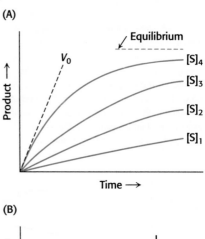

(A)

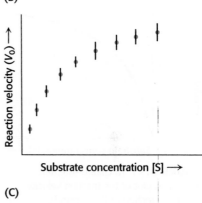

(B)

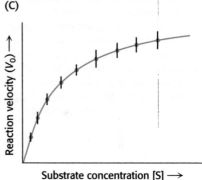

(C)

FIGURE 8.10 Determining the relation between initial velocity and substrate concentration. (A) The amount of product formed at different substrate concentrations is plotted as a function of time. The initial velocity (V_0) for each substrate concentration is determined from the slope of the curve at the beginning of a reaction, when the reverse reaction is insignificant. (B) The values for initial velocity determined in part A are then plotted, with error bars, against substrate concentration. (C) The data points are connected to clearly reveal the relationship of initial velocity to substrate concentration.

In a steady-state system, the concentrations of the intermediates stay the same, even though the concentrations of substrate and products are changing. A sink filled with water that has the tap open just enough to match the loss of water down the drain is in a steady state. The level of the water in the sink never changes even though water is constantly flowing from the faucet through the sink and out through the drain.

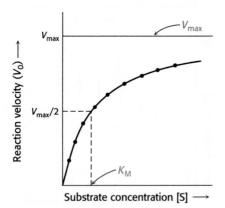

FIGURE 8.11 Michaelis–Menten kinetics. A plot of the reaction velocity (V_0) as a function of the substrate concentration [S] for an enzyme that obeys Michaelis–Menten kinetics shows that the maximal velocity (V_{max}) is approached asymptotically, which is to say, V_{max} will be attained only at an infinite substrate concentration. The Michaelis constant (K_M) is the substrate concentration yielding a velocity of $V_{max}/2$.

$$V_0 = k_2[ES] \tag{13}$$

Now we need to express [ES] in terms of known quantities. The rates of formation and breakdown of ES are given by

$$\text{Rate of formation of ES} = k_1[E][S] \tag{14}$$

$$\text{Rate of breakdown of ES} = (k_{-1}+k_2)[ES] \tag{15}$$

We will use the *steady-state assumption* to simplify matters. In a steady state, the concentrations of intermediates—in this case, [ES]—stay the same even if the concentrations of starting materials and products are changing. This steady state is reached when the rates of formation and breakdown of the ES complex are equal. Setting the right-hand sides of equations 14 and 15 equal gives

$$k_1[E][S] = (k_{-1}+k_2)[ES] \tag{16}$$

By rearranging equation 16, we obtain

$$[E][S]/[ES] = (k_{-1}+k_2)/k_1 \tag{17}$$

Equation 17 can be simplified by defining a new constant, K_M, called the Michaelis constant:

$$K_M = \frac{k_{-1}+k_2}{k_1} \tag{18}$$

Note that K_M has the units of concentration and is independent of enzyme and substrate concentrations. As will be explained, K_M is an important characteristic of enzyme–substrate interactions.

Inserting equation 18 into equation 17 and solving for [ES] yields

$$[ES] = \frac{[E][S]}{K_M} \tag{19}$$

Now let us examine the numerator of equation 19. Because the substrate is usually present at a much higher concentration than that of the enzyme, the concentration of uncombined substrate [S] is very nearly equal to the total substrate concentration. The concentration of uncombined enzyme [E] is equal to the total enzyme concentration $[E]_T$ minus the concentration of the ES complex:

$$[E] = [E]_T - [ES] \tag{20}$$

Substituting this expression for [E] in equation 19 gives

$$[ES] = \frac{([E]_T - [ES])[S]}{K_M} \tag{21}$$

Solving equation 21 for [ES] gives

$$[ES] = \frac{[E]_T[S]/K_M}{1+[S]/K_M} \tag{22}$$

or

$$[ES] = [E]_T\frac{[S]}{[S]+K_M} \tag{23}$$

By substituting this expression for [ES] into equation 13, we obtain

$$V_0 = k_2[E]_T\frac{[S]}{[S]+K_M} \tag{24}$$

The *maximal rate*, V_{max}, is attained when the catalytic sites on the enzyme are saturated with substrate—that is, when $[ES] = [E]_T$. Thus,

$$V_{max} = k_2[E]_T \tag{25}$$

Substituting equation 25 into equation 24 yields the *Michaelis–Menten equation*:

$$V_0 = V_{max}\frac{[S]}{[S]+K_M} \tag{26}$$

This equation accounts for the kinetic data given in Figure 8.11. At very low substrate concentration, when $[S]$ is much less than K_M, $V_0 = (V_{max}/K_M)[S]$; that is, the reaction is first order with the rate directly proportional to the substrate concentration. At high substrate concentration, when $[S]$ is much greater than K_M, $V_0 = V_{max}$; that is, the rate is maximal. The reaction is zero order, independent of substrate concentration.

The significance of K_M is clear when we set $[S] = K_M$ in equation 26. When $[S] = K_M$, then $V_0 = V_{max}/2$. Thus, K_M *is equal to the substrate concentration at which the reaction rate is half its maximal value.* As we will see, K_M is an important characteristic of an enzyme-catalyzed reaction and is significant for its biological function.

Variations in K_M can have physiological consequences

The physiological consequence of K_M is illustrated by the sensitivity of some persons to ethanol. Such persons exhibit facial flushing and rapid heart rate (tachycardia) after ingesting even small amounts of alcohol. In the liver, alcohol dehydrogenase converts ethanol into acetaldehyde.

$$CH_3CH_2OH + NAD^+ \underset{\text{dehydrogenase}}{\overset{\text{Alcohol}}{\rightleftharpoons}} CH_3CHO + NADH + H^+$$

Ethanol   Acetaldehyde

Normally, the acetaldehyde, which is the cause of the symptoms when present at high concentrations, is processed to acetate by aldehyde dehydrogenase.

$$CH_3CHO + NAD^+ + H_2O \underset{\text{dehydrogenase}}{\overset{\text{Aldehyde}}{\rightleftharpoons}} CH_3COO^- + NADH + 2H^+$$

Most people have two forms of the aldehyde dehydrogenase, a low K_M mitochondrial form and a high K_M cytoplasmic form. In susceptible persons, the mitochondrial enzyme is less active owing to the substitution of a single amino acid, and acetaldehyde is processed only by the cytoplasmic enzyme. Because this enzyme has a high K_M, it achieves a high rate of catalysis only at very high concentrations of acetaldehyde. Consequently, less acetaldehyde is converted into acetate; excess acetaldehyde escapes into the blood and accounts for the physiological effects.

K_M and V_{max} values can be determined by several means

K_M is equal to the substrate concentration that yields $V_{max}/2$; however V_{max}, like perfection, is only approached but never attained. How then can we experimentally determine K_M and V_{max}, and how do these parameters enhance our understanding of enzyme-catalyzed reactions? The Michaelis constant, K_M, and the maximal rate, V_{max}, can be readily derived from rates of catalysis measured at a variety of substrate concentrations if an enzyme operates according to the simple scheme given in equation 26. The derivation of K_M and V_{max} is most commonly achieved with the use

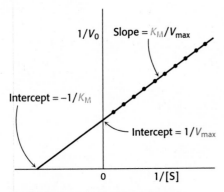

FIGURE 8.12 A double-reciprocal or Lineweaver–Burk plot. A double-reciprocal plot of enzyme kinetics is generated by plotting $1/V_0$ as a function of $1/[S]$. The slope is K_M/V_{max}, the intercept on the vertical axis is $1/V_{max}$, and the intercept on the horizontal axis is $-1/K_M$.

of curve-fitting programs on a computer. However, an older method, although rarely used because the data points at high and low concentrations are weighted differently and thus sensitive to errors, is a source of further insight into the meaning of K_M and V_{max}.

Before the availability of computers, the determination of K_M and V_{max} values required algebraic manipulation of the Michaelis–Menten equation. The Michaelis–Menten equation is transformed into one that gives a straight-line plot that yields values for V_{max} and K_M. Taking the reciprocal of both sides of equation 26 gives

$$\frac{1}{V_0} = \frac{K_M}{V_{max}} \times \frac{1}{S} + \frac{1}{V_{max}} \tag{27}$$

A plot of $1/V_0$ versus $1/[S]$, called a *Lineweaver–Burk* or *double-reciprocal plot*, yields a straight line with a y-intercept of $1/V_{max}$ and a slope of K_M/V_{max} (Figure 8.12). The intercept on the x-axis is $-1/K_M$.

K_M and V_{max} values are important enzyme characteristics

The K_M values of enzymes range widely (Table 8.4). For most enzymes, K_M lies between 10^{-1} and 10^{-7} M. The K_M value for an enzyme depends on the particular substrate and on environmental conditions such as pH, temperature, and ionic strength. The Michaelis constant, K_M, as already noted, is equal to the concentration of substrate at which half the active sites are filled. Thus, K_M provides a measure of the substrate concentration required for significant catalysis to take place. For many enzymes, experimental evidence suggests that the K_M value provides an approximation of the substrate concentration in vivo, which in turn suggests that most enzymes evolved to have a K_M approximately equal to the substrate concentration commonly available. Why might it be beneficial to have a K_M value approximately equal to the commonly available substrate concentration? If the normal concentration of substrate is near K_M, the enzyme will display significant activity and yet the activity will be sensitive to changes in environmental conditions—that is, changes in substrate concentration. At values below K_M, enzymes are very sensitive to changes in substrate concentration but display little activity. At substrate values well above K_M, enzymes have great catalytic activity but are insensitive to changes in substrate concentration. Thus, with the normal substrate concentration being approximately K_M, the enzymes have significant activity ($1/2\,V_{max}$) but are still sensitive to changes in substrate concentration.

Under certain circumstances, K_M reflects the strength of the enzyme–substrate interaction. In equation 18, K_M is defined as $(k_{-1}+k_2)/k_1$. Consider a case in which k_{-1} is much greater than k_2. Under such circumstances, the ES complex dissociates to E and S much more rapidly than product is formed. Under these conditions ($k_{-1} \gg k_2$)

$$K_M \approx \frac{k_{-1}}{k_1} \tag{28}$$

Equation 28 describes the *dissociation constant* of the ES complex.

$$K_{ES} = \frac{[E][S]}{[ES]} = \frac{k_{-1}}{k_1} \tag{29}$$

TABLE 8.4 K_M values of some enzymes

Enzyme	Substrate	K_M (μM)
Chymotrypsin	Acetyl-L-tryptophanamide	5000
Lysozyme	Hexa-N-acetylglucosamine	6
β-Galactosidase	Lactose	4000
Threonine deaminase	Threonine	5000
Carbonic anhydrase	CO_2	8000
Penicillinase	Benzylpenicillin	50
Pyruvate carboxylase	Pyruvate	400
	HCO_3^-	1000
	ATP	60
Arginine-tRNA synthetase	Arginine	3
	tRNA	0.4
	ATP	300

In other words, K_M is equal to the dissociation constant of the ES complex if k_2 is much smaller than k_{-1}. When this condition is met, K_M is a measure of the strength of the ES complex: a high K_M indicates weak binding; a low K_M indicates strong binding. It must be stressed that K_M indicates the affinity of the ES complex only when k_{-1} is much greater than k_2.

The maximal rate, V_{max}, reveals the *turnover number* of an enzyme, which is the number of substrate molecules converted into product by an enzyme molecule in a unit time when the enzyme is fully saturated with substrate. It is equal to the rate constant k_2, which is also called k_{cat}. The maximal rate, V_{max}, reveals the turnover number of an enzyme if the concentration of active sites $[E]_T$ is known, because

$$V_{max} = k_{cat}[E]_T \tag{30}$$

and thus

$$k_{cat} = V_{max}/[E]_T \tag{31}$$

For example, a 10^{-6} M solution of carbonic anhydrase catalyzes the formation of 0.6 M H_2CO_3 per second when the enzyme is fully saturated with substrate. Hence, k_{cat} is $6 \times 10^5 \text{ s}^{-1}$. This turnover number is one of the largest known. Each catalyzed reaction takes place in a time equal to, on average, $1/k_{cat}$, which is 1.7 μs for carbonic anhydrase. The turnover numbers of most enzymes with their physiological substrates range from 1 to 10^4 per second (Table 8.5).

K_M and V_{max} also permit the determination of f_{ES}, the fraction of active sites filled. This relation of f_{ES} to K_M and V_{max} is given by the following equation:

$$f_{ES} = \frac{V}{V_{max}} = \frac{[S]}{[S]+K_M} \tag{32}$$

k_{cat}/K_M is a measure of catalytic efficiency

When the substrate concentration is much greater than K_M, the rate of catalysis is equal to V_{max}, which is a function of k_{cat}, the turnover number, as already described. However, most enzymes are not normally saturated with substrate. Under physiological conditions, the $[S]/K_M$ ratio is typically between 0.01 and 1.0. When $[S] = K_M$, the enzymatic rate is much less than k_{cat} because most of the active sites are unoccupied. Is there a number that characterizes the kinetics of an enzyme under these more typical cellular conditions? Indeed there is, as can be shown by combining equations 13 and 19 to give

$$V_0 = \frac{k_{cat}}{K_M}[E][S] \tag{33}$$

When $[S] \ll K_M$, the concentration of free enzyme $[E]$, is nearly equal to the total concentration of enzyme $[E]_T$; so

$$V_0 = \frac{k_{cat}}{K_M}[S][E]_T \tag{34}$$

Thus, when $[S] \ll K_M$, the enzymatic velocity depends on the values of k_{cat}/K_M, $[S]$, and $[E]_T$. Under these conditions, k_{cat}/K_M is the rate constant for the interaction of S and E. The rate constant k_{cat}/K_M, called the *specificity constant*, is a measure of catalytic efficiency because it takes into account both the rate of catalysis with a particular substrate (k_{cat}) and the nature of the enzyme–substrate interaction (K_M). For instance, by using

TABLE 8.5 Turnover numbers of some enzymes

Enzyme	Turnover number (per second)
Carbonic anhydrase	600,000
3-Ketosteroid isomerase	280,000
Acetylcholinesterase	25,000
Penicillinase	2000
Lactate dehydrogenase	1000
Chymotrypsin	100
DNA polymerase I	15
Tryptophan synthetase	2
Lysozyme	0.5

TABLE 8.6 Substrate preferences of chymotrypsin

Amino acid in ester	Amino acid side chain	k_{cat}/K_M ($s^{-1} M^{-1}$)
Glycine	—H	1.3×10^{-1}
Valine	$\begin{array}{c} CH_2 \\ | \\ -CH \\ | \\ CH_2 \end{array}$	2.0
Norvaline	$-CH_2CH_2CH_3$	3.6×10^2
Norleucine	$-CH_2CH_2CH_2CH_3$	3.0×10^3
Phenylalanine	$-CH_2-\bigcirc$	1.0×10^5

Source: Information from A. Fersht, *Structure and Mechanism in Protein Science: A Guide to Enzyme Catalysis and Protein Folding* (W. H. Freeman and Company, 1999), Table 7.3.

TABLE 8.7 Enzymes for which k_{cat}/K_M is close to the diffusion-controlled rate of encounter

Enzyme	k_{cat}/K_M ($s^{-1} M^{-1}$)
Acetylcholinesterase	1.6×10^8
Carbonic anhydrase	8.3×10^7
Catalase	4×10^7
Crotonase	2.8×10^8
Fumarase	1.6×10^8
Triose phosphate isomerase	2.4×10^8
β-Lactamase	1×10^8
Superoxide dismutase	7×10^9

Source: Information from A. Fersht, *Structure and Mechanism in Protein Science: A Guide to Enzyme Catalysis and Protein Folding* (W. H. Freeman and Company, 1999), Table 4.5.

Circe effect

The utilization of attractive forces to lure a substrate into a site in which it undergoes a transformation of structure, as defined by William P. Jencks, an enzymologist, who coined the term.

A goddess of Greek mythology, Circe lured Odysseus's men to her house and then transformed them into pigs.

k_{cat}/K_M values, we can compare an enzyme's preference for different substrates. Table 8.6 shows the k_{cat}/K_M values for several different substrates of chymotrypsin. Chymotrypsin clearly has a preference for cleaving next to bulky, hydrophobic side chains.

How efficient can an enzyme be? We can approach this question by determining whether there are any physical limits on the value of k_{cat}/K_M. Note that the k_{cat}/K_M *ratio* depends on k_1, k_{-1}, and k_{cat}, as can be shown by substituting for K_M.

$$k_{cat}/K_M = \frac{k_{cat}k_1}{k_{-1}+k_{cat}} = \left(\frac{k_{cat}}{k_{-1}+k_{cat}}\right)k_1 < k_1 \tag{35}$$

Note that the value of k_{cat}/K_M is always less than k_1. Suppose that the rate of formation of product (k_{cat}) is much faster than the rate of dissociation of the ES complex (k_{-1}). The value of k_{cat}/K_M then approaches k_1. Thus, the ultimate limit on the value of k_{cat}/K_M is set by k_1, the rate of formation of the ES complex. *This rate cannot be faster than the diffusion-controlled encounter of an enzyme and its substrate.* Diffusion limits the value of k_1 and so it cannot be higher than between 10^8 and 10^9 $s^{-1} M^{-1}$. Hence, the upper limit on k_{cat}/K_M is between 10^8 and 10^9 $s^{-1} M^{-1}$.

The k_{cat}/K_M ratios of the enzymes superoxide dismutase, acetylcholinesterase, and triose phosphate isomerase are between 10^8 and 10^9 $s^{-1} M^{-1}$. Enzymes that have k_{cat}/K_M ratios at the upper limits have attained *kinetic perfection. Their catalytic velocity is restricted only by the rate at which they encounter substrate in the solution* (Table 8.7). Any further gain in catalytic rate can come only by decreasing the time for diffusion of the substrate into the enzyme's immediate environment. Remember that the active site is only a small part of the total enzyme structure. Yet, for catalytically perfect enzymes, every encounter between enzyme and substrate is productive. In these cases, there may be attractive electrostatic forces on the enzyme that entice the substrate to the active site. These forces are sometimes referred to poetically as *Circe effects*.

The diffusion of a substrate throughout a solution can also be partly overcome by confining substrates and products in the limited volume of a multienzyme complex. Indeed, some series of enzymes are organized into complexes so that the product of one enzyme is very rapidly found by the next enzyme. In effect, products are channeled from one enzyme to the next, much as in an assembly line.

Most biochemical reactions include multiple substrates

Most reactions in biological systems start with two substrates and yield two products. They can be represented by the bisubstrate reaction:

$$A + B \rightleftharpoons P + Q$$

Many such reactions transfer a functional group, such as a phosphoryl or an ammonium group, from one substrate to the other. Those that are oxidation–reduction reactions transfer electrons between substrates. Multiple substrate reactions can be divided into two classes: *sequential* reactions and *double-displacement* reactions.

Sequential reactions. In *sequential reactions*, all substrates must bind to the enzyme before any product is released. Consequently, in a bisubstrate reaction, a *ternary complex* of the enzyme and both substrates forms. Sequential mechanisms are of two types: ordered, in which the substrates bind the enzyme in a defined sequence, and random.

Many enzymes that have NAD^+ or NADH as a substrate exhibit the ordered sequential mechanism. Consider lactate dehydrogenase, an important enzyme in glucose metabolism (Section 16.1). This enzyme reduces pyruvate to lactate while oxidizing NADH to NAD^+.

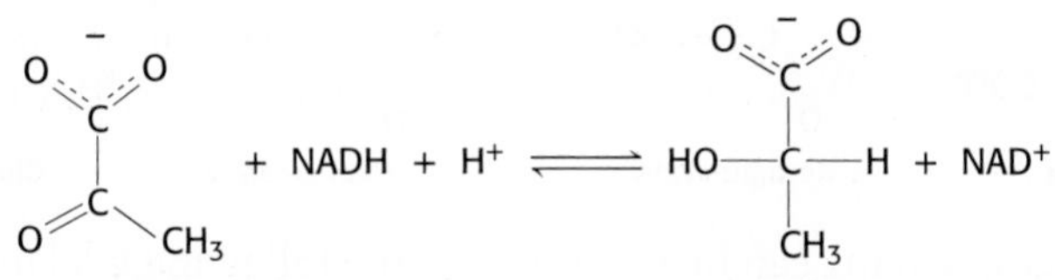

Pyruvate   + NADH + H$^+$ $\rightleftharpoons$ **Lactate** + NAD$^+$

In the ordered sequential mechanism, the coenzyme always binds first and the lactate is always released first. This sequence can be represented by using a notation developed by W. Wallace Cleland:

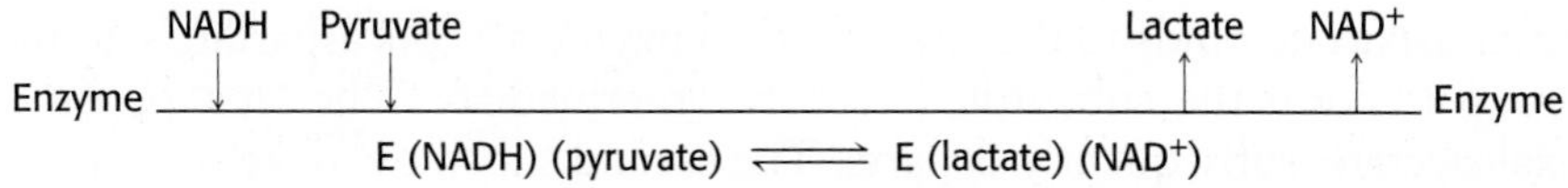

The enzyme exists as a ternary complex consisting of, first, the enzyme and substrates and, after catalysis, the enzyme and products.

In the random sequential mechanism, the order of the addition of substrates and the release of products is random. An example of a random sequential reaction is the formation of phosphocreatine and ADP from creatine and ATP which is catalyzed by creatine kinase (Section 15.2).

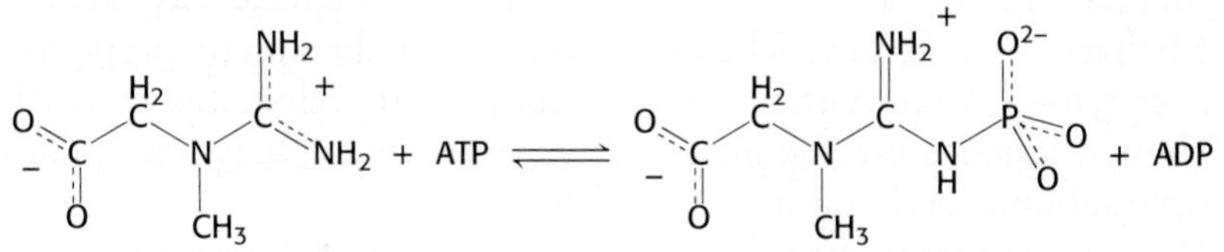

Creatine + ATP $\rightleftharpoons$ **Phosphocreatine** + ADP

Either creatine or ATP may bind first, and either phosphocreatine or ADP may be released first. Phosphocreatine is an important energy source in muscle. Sequential random reactions also can be depicted in the Cleland notation.

Although the order of certain events is random, the reaction still passes through the ternary complexes including, first, substrates and, then, products.

Double-displacement (ping-pong) reactions. In *double-displacement*, or *ping-pong*, *reactions*, one or more products are released before all substrates bind the enzyme. The defining feature of double-displacement reactions is the existence of a *substituted enzyme intermediate*, in which the enzyme is temporarily modified. Reactions that shuttle amino groups between amino acids and α-ketoacids are classic examples of double-displacement mechanisms. The enzyme aspartate aminotransferase catalyzes the transfer of an amino group from aspartate to α-ketoglutarate.

The sequence of events can be portrayed as the following Cleland notation:

After aspartate binds to the enzyme, the enzyme accepts aspartate's amino group to form the substituted enzyme intermediate. The first product, oxaloacetate, subsequently departs. The second substrate, α-ketoglutarate, binds to the enzyme, accepts the amino group from the modified enzyme, and is then released as the final product, glutamate. In the Cleland notation, the substrates appear to bounce on and off the enzyme much as a Ping-Pong ball bounces on a table.

Allosteric enzymes do not obey Michaelis–Menten kinetics

The Michaelis–Menten model has greatly assisted the development of enzymology. Its virtues are simplicity and broad applicability. However, the Michaelis–Menten model cannot account for the kinetic properties of many enzymes. An important group of enzymes that do not obey Michaelis–Menten kinetics are the *allosteric enzymes*. These enzymes consist of multiple subunits and multiple active sites.

Allosteric enzymes often display sigmoidal plots of the reaction velocity V_0 versus substrate concentration [S] (Figure 8.13), rather than the hyperbolic plots predicted by the Michaelis–Menten equation (Figure 8.11). In allosteric enzymes, the binding of substrate to one active site can alter the properties of other active sites in the same enzyme molecule. A possible outcome of this interaction between subunits is that the binding of substrate

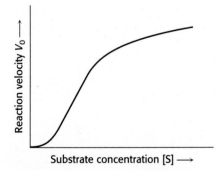

FIGURE 8.13 Kinetics for an allosteric enzyme. Allosteric enzymes display a sigmoidal dependence of reaction velocity on substrate concentration.

becomes *cooperative;* that is, the binding of substrate to one active site facilitates the binding of substrate to the other active sites. Such cooperativity results in a sigmoidal plot of V_0 versus [S]. In addition, the activity of an allosteric enzyme may be altered by regulatory molecules that reversibly bind to specific sites other than the catalytic sites. The catalytic properties of allosteric enzymes can thus be adjusted to meet the immediate needs of a cell. For this reason, allosteric enzymes are key regulators of metabolic pathways (Chapter 10). Recall that we have already met an allosteric protein, hemoglobin, in Chapter 7.

8.5 Enzymes Can Be Inhibited by Specific Molecules

The activity of many enzymes can be inhibited by the binding of specific small molecules and ions. This means of inhibiting enzyme activity serves as a major control mechanism in biological systems, typified by the regulation of allosteric enzymes. In addition, many drugs and toxic agents act by inhibiting enzymes (Chapter 28). This type of enzyme inhibition is not usually the result of evolutionary forces, as it is for allosteric enzymes, but rather due to design of inhibitors by scientists or simple chance discovery of inhibitory molecules. Examining inhibition can be a source of insight into the mechanism of enzyme action: specific inhibitors can often be used to identify residues critical for catalysis. Transition-state analogs are especially potent inhibitors.

Enzyme inhibition can be either irreversible or reversible. An *irreversible inhibitor* dissociates very slowly from its target enzyme because it has become tightly bound to the enzyme, either covalently or noncovalently. Some irreversible inhibitors are important drugs. Penicillin acts by covalently modifying the enzyme transpeptidase, thereby preventing the synthesis of bacterial cell walls and thus killing the bacteria (p. 258). Aspirin acts by covalently modifying the enzyme cyclooxygenase, reducing the synthesis of signaling molecules in inflammation.

Reversible inhibition, in contrast with irreversible inhibition, is characterized by a rapid dissociation of the enzyme–inhibitor complex. In the type of reversible inhibition called *competitive inhibition,* an enzyme can bind substrate (forming an ES complex) or inhibitor (EI) but not both (ESI, enzyme–substrate–inhibitor complex). The competitive inhibitor often resembles the substrate and binds to the active site of the enzyme (Figure 8.14). The substrate is thereby prevented from binding to the same active site. *A competitive inhibitor diminishes the rate of catalysis by reducing the proportion of enzyme molecules bound to a substrate.* At any given inhibitor concentration, competitive inhibition can be relieved by increasing the substrate concentration. Under these conditions, the substrate successfully competes with the inhibitor for the active site. Some competitive inhibitors are useful drugs. One of the earliest examples was the use of sulfanilamide as an antibiotic. Sulfanilamide is an example of a sulfa drug, a sulfur-containing antibiotic. Structurally, sulfanilamide mimics *p*-aminobenzoic acid (PABA), a metabolite required by bacteria for the synthesis of the coenzyme folic acid. Sulfanilamide binds to the enzyme that normally metabolizes PABA and competitively inhibits it, preventing folic acid synthesis. Human beings, unlike bacteria, absorb folic acid from the diet and are thus unaffected by the sulfa drug.

Uncompetitive inhibition is essentially substrate-dependent inhibition in that the inhibitor binds only to the enzyme–substrate complex. The binding site of an uncompetitive inhibitor is created only on interaction of the

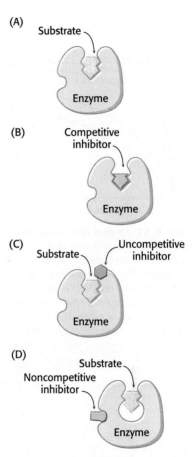

FIGURE 8.14 Distinction between reversible inhibitors. (A) Enzyme–substrate complex; (B) a competitive inhibitor binds at the active site and thus prevents the substrate from binding; (C) an uncompetitive inhibitor binds only to the enzyme–substrate complex; (D) a noncompetitive inhibitor does not prevent the substrate from binding.

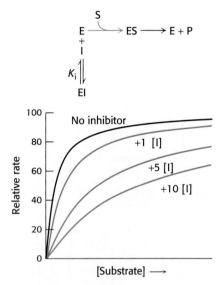

FIGURE 8.15 Kinetics of a competitive inhibitor. As the concentration of a competitive inhibitor increases, higher concentrations of substrate are required to attain a particular reaction velocity. The reaction pathway suggests how sufficiently high concentrations of substrate can completely relieve competitive inhibition.

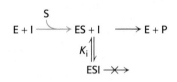

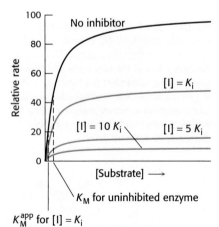

FIGURE 8.16 Kinetics of an uncompetitive inhibitor. The reaction pathway shows that the inhibitor binds only to the enzyme–substrate complex. Consequently, V_{max} cannot be attained, even at high substrate concentrations. The apparent value for K_M is lowered, becoming smaller as more substrate is added.

enzyme and substrate (Figure 8.14C). Uncompetitive inhibition cannot be overcome by the addition of more substrate.

In *noncompetitive inhibition*, the inhibitor and substrate can bind simultaneously to an enzyme molecule at different binding sites (Figure 8.14D). Unlike uncompetitive inhibition, a noncompetitive inhibitor can bind free enzyme or the enzyme–substrate complex. A noncompetitive inhibitor acts by decreasing the concentration of functional enzyme rather than by diminishing the proportion of enzyme molecules that are bound to substrate. The net effect is to decrease the turnover number. Noncompetitive inhibition, like uncompetitive inhibition, cannot be overcome by increasing the substrate concentration. A more complex pattern, called *mixed inhibition*, is produced when a single inhibitor both hinders the binding of substrate and decreases the turnover number of the enzyme.

The different types of reversible inhibitors are kinetically distinguishable

How can we determine whether a reversible inhibitor acts by competitive, uncompetitive, or noncompetitive inhibition? Let us consider only enzymes that exhibit Michaelis–Menten kinetics. Measurements of the rates of catalysis at different concentrations of substrate and inhibitor serve to distinguish the three types of inhibition. In *competitive inhibition*, the inhibitor competes with the substrate for the active site. The dissociation constant for the inhibitor is given by

$$K_i = [\text{E}][\text{I}]/[\text{EI}]$$

The smaller the K_i, the more potent the inhibition. The hallmark of competitive inhibition is that it can be overcome by a sufficiently high concentration substrate (Figure 8.15). The effect of a competitive inhibitor is to increase the apparent value of K_M, meaning that more substrate is needed to obtain the same reaction rate. This new apparent value of K_M, called K_M^{app} is numerically equal to

$$K_M^{app} = K_M(1 + [\text{I}]/K_i)$$

where [I] is the concentration of inhibitor and K_i is the dissociation constant for the enzyme–inhibitor complex. In the presence of a competitive inhibitor, an enzyme will have the same V_{max} as in the absence of an inhibitor. At a sufficiently high concentration, virtually all the active sites are filled with substrate, and the enzyme is fully operative.

Competitive inhibitors are commonly used as drugs. Drugs such as ibuprofen are competitive inhibitors of enzymes that participate in signaling pathways in the inflammatory response. Statins are drugs that reduce high cholesterol levels by competitively inhibiting a key enzyme in cholesterol biosynthesis (Section 26.3).

In *uncompetitive inhibition*, the inhibitor binds only to the ES complex. This enzyme–substrate–inhibitor complex, ESI, does not go on to form any product. Because some unproductive ESI complex will always be present, V_{max} will be lower in the presence of inhibitor than in its absence (Figure 8.16). The uncompetitive inhibitor lowers the apparent value of K_M because the inhibitor binds to ES to form ESI, depleting ES. To maintain the equilibrium between E and ES, more S binds to E, increasing the apparent value of k_1 and thereby reducing the apparent value of K_M (see equation 18). Thus, a lower concentration of S is required to form half of the maximal concentration of ES. The herbicide glyphosate, also known as Roundup, is an uncompetitive inhibitor of an enzyme in the biosynthetic pathway for aromatic amino acids.

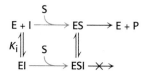

In *noncompetitive inhibition* (Figure 8.17), substrate can bind either to the enzyme or the enzyme–inhibitor complex. However, the enzyme–inhibitor–substrate complex *does not* proceed to form product. In pure noncompetitive inhibition, the K_i for the inhibitor binding to E is the same as for binding to ES complex. The apparent value of V_{max} is decreased to a new value called V_{max}^{app}, whereas the value of K_M is unchanged. The maximal velocity in the presence of a pure noncompetitive inhibitor, V_{max}^{app}, is given by

$$V_{max}^{app} = \frac{V_{max}}{1 + [I]/K_i} \tag{36}$$

Why is V_{max} lowered though K_M remains unchanged? In essence, the inhibitor simply lowers the concentration of functional enzyme. The resulting solution behaves as a more dilute solution of enzyme does. *Noncompetitive inhibition cannot be overcome by increasing the substrate concentration.* Doxycycline, an antibiotic, functions at low concentrations as a noncompetitive inhibitor of a proteolytic enzyme (collagenase). It is used to treat periodontal disease. Some of the toxic effects of lead poisoning may be due to lead's ability to act as a noncompetitive inhibitor of a host of enzymes. Lead reacts with crucial sulfhydryl groups in these enzymes.

Double-reciprocal plots are especially useful for distinguishing between competitive, uncompetitive, and noncompetitive inhibitors. In competitive inhibition, the intercept on the y-axis of the plot of $1/V_0$ versus $1/[S]$ is the same in the presence and in the absence of inhibitor, although the slope is increased (Figure 8.18). The intercept is unchanged because a competitive inhibitor does not alter V_{max}. The increase in the slope of the $1/V_0$ versus $1/[S]$ plot indicates the strength of binding of a competitive inhibitor. In the presence of a competitive inhibitor, equation 27 is replaced by

$$\frac{1}{V_0} = \frac{1}{V_{max}} + \frac{K_M}{V_{max}}\left(1 + \frac{[I]}{K_i}\right)\left(\frac{1}{[S]}\right) \tag{37}$$

In other words, the slope of the plot is increased by the factor $(1 + [I]/K_i)$ in the presence of a competitive inhibitor. Consider an enzyme with a K_M of 10^{-4} M. In the absence of inhibitor, when $V_0 = V_{max}/2$ when $[S] = 10^{-4}$ M. In the presence of a 2×10^{-3} M competitive inhibitor that is bound to the enzyme with a K_i of 10^{-3} M, the apparent K_M (K_M^{app}) will be equal to $K_M(1 + [I]/K_i)$, or 3×10^{-4} M. Substitution of these values into equation 37 gives when $V_0 = V_{max}/4$, when $[S] = 10^{-4}$ M. The presence of the competitive inhibitor thus cuts the reaction rate in half at this substrate concentration.

In uncompetitive inhibition (Figure 8.19), the inhibitor combines only with the enzyme–substrate complex. The equation that describes the double-reciprocal plot for an uncompetitive inhibitor is

$$\frac{1}{V_0} = \frac{K_M}{V_{max}}\frac{1}{[S]} + \frac{1}{V_{max}}\left(1 + \frac{[I]}{K_i}\right) \tag{38}$$

The slope of the line, K_M/V_{max}, is the same as that for the uninhibited enzyme, but the intercept on the y-axis will be increased by $1 + [I]/K_i$. Consequently, the lines in double-reciprocal plots will be parallel.

In pure noncompetitive inhibition (Figure 8.20), the inhibitor can combine with either the enzyme or the enzyme–substrate complex with the same dissociation constant. The value of V_{max} is decreased to the new value V_{max}^{app}, and so the intercept on the vertical axis is increased (equation 36). The new slope, which is equal to K_M/V_{max}^{app}, is larger by the same factor. In contrast with V_{max}, K_M is not affected by pure noncompetitive inhibition.

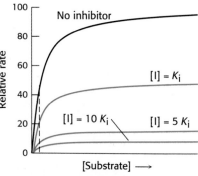

FIGURE 8.17 Kinetics of a noncompetitive inhibitor. The reaction pathway shows that the inhibitor binds both to free enzyme and to an enzyme–substrate complex. Consequently, as with uncompetitive competition, V_{max} cannot be attained. In pure noncompetitive inhibition, K_M remains unchanged, and so the reaction rate increases more slowly at low substrate concentrations than is the case for uncompetitive competition.

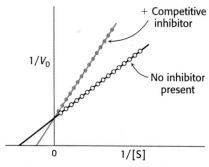

FIGURE 8.18 Competitive inhibition illustrated on a double-reciprocal plot. A double-reciprocal plot of enzyme kinetics in the presence and absence of a competitive inhibitor illustrates that the inhibitor has no effect on V_{max} but increases K_M.

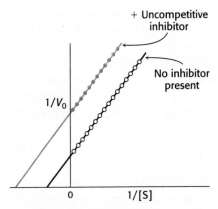

FIGURE 8.19 Uncompetitive inhibition illustrated by a double-reciprocal plot. An uncompetitive inhibitor does not affect the slope of the double-reciprocal plot. V_{max} and K_M are reduced by equivalent amounts.

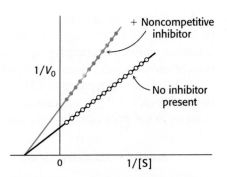

FIGURE 8.20 Noncompetitive inhibition illustrated on a double-reciprocal plot. A double-reciprocal plot of enzyme kinetics in the presence and absence of a pure noncompetitive inhibitor shows that K_M is unaltered and V_{max} is decreased.

Irreversible inhibitors can be used to map the active site

In Chapter 9, we will examine the chemical details of how enzymes function. The first step in obtaining the chemical mechanism of an enzyme is to determine what functional groups are required for enzyme activity. How can we ascertain what these functional groups are? X-ray crystallography of the enzyme bound to its substrate or substrate analog provides one approach. Irreversible inhibitors that covalently bond to the enzyme provide an alternative and often complementary approach: the inhibitors modify the functional groups, which can then be identified. Irreversible inhibitors can be divided into three categories: group-specific reagents, reactive substrate analogs (also called affinity labels), and suicide inhibitors.

Group-specific reagents react with specific side chains of amino acids. An example of a group-specific reagent is diisopropylphosphofluoridate (DIPF). DIPF modifies only 1 of the 28 serine residues in the proteolytic enzyme chymotrypsin and yet inhibits the enzyme, implying that this serine residue is especially reactive. We will see in Chapter 9 that this serine residue is indeed located at the active site. DIPF also revealed a reactive serine residue in acetylcholinesterase, an enzyme important in the transmission of nerve impulses (Figure 8.21). Thus, DIPF and similar compounds

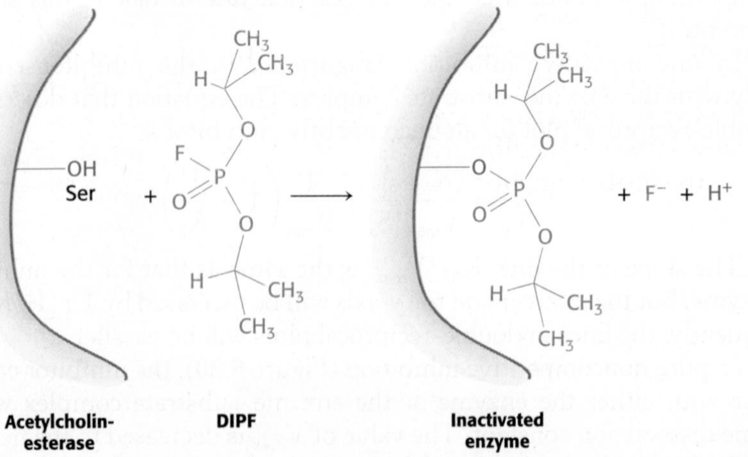

FIGURE 8.21 Enzyme inhibition by diisopropylphosphofluoridate (DIPF), a group-specific reagent. DIPF can inhibit an enzyme by covalently modifying a crucial serine residue.

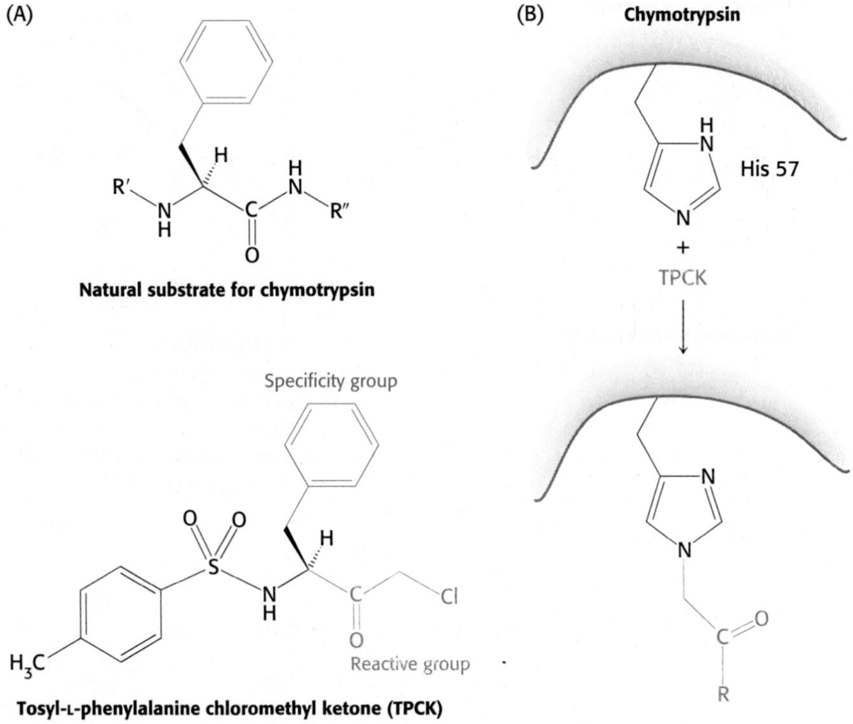

(A)

Natural substrate for chymotrypsin

Specificity group

Tosyl-L-phenylalanine chloromethyl ketone (TPCK)

Reactive group

(B)   **Chymotrypsin**

His 57

+

TPCK

FIGURE 8.22 Affinity labeling.
(A) Tosyl-L-phenylalanine chloromethyl ketone (TPCK) is a reactive analog of the normal substrate for the enzyme chymotrypsin. (B) TPCK binds at the active site of chymotrypsin and modifies an essential histidine residue.

that bind and inactivate acetylcholinesterase are potent nerve gases. Most group-specific reagents do not display the exquisite specificity shown by DIPF. Consequently, more specific means of modifying the active site are required.

Affinity labels, or *reactive substrate analogs,* are molecules that are structurally similar to the substrate for an enzyme and that covalently bind to active-site residues. They are thus more specific for the enzyme's active site than are group-specific reagents. Tosyl-L-phenylalanine chloromethyl ketone (TPCK) is a substrate analog for chymotrypsin (Figure 8.22). TPCK binds at the active site and then reacts irreversibly with a histidine residue at that site, inhibiting the enzyme. The compound 3-bromoacetol phosphate is an affinity label for the enzyme triose phosphate isomerase (TPI). It mimics the normal substrate, dihydroxyacetone phosphate, by binding at the active site; then it covalently modifies the enzyme such that the enzyme is irreversibly inhibited (Figure 8.23).

Suicide inhibitors, or *mechanism-based inhibitors,* are modified substrates that provide the most specific means for modifying an enzyme's active site. The inhibitor binds to the enzyme as a substrate and is initially processed by the normal catalytic mechanism. The mechanism of catalysis then generates a chemically reactive intermediate that inactivates the enzyme through covalent modification. The fact that the enzyme participates in its own irreversible inhibition strongly suggests that the covalently modified

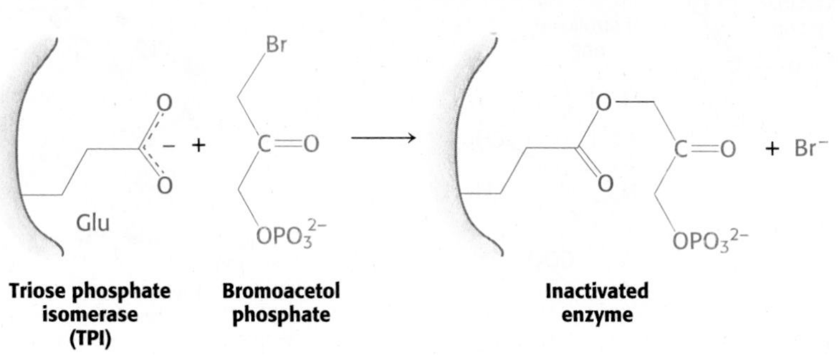

Triose phosphate isomerase (TPI)   **Bromoacetol phosphate**   **Inactivated enzyme**

FIGURE 8.23 Bromoacetol phosphate, an affinity label for triose phosphate isomerase (TPI). Bromoacetol phosphate, an analog of dihydroxyacetone phosphate, binds at the active site of the enzyme and covalently modifies a glutamic acid residue required for enzyme activity.

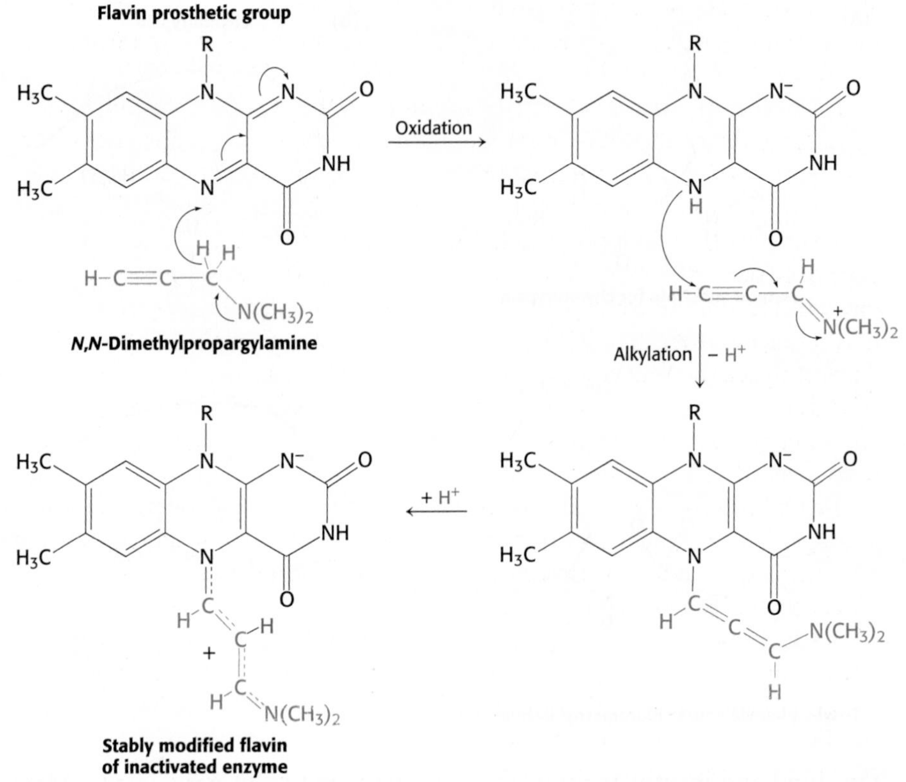

FIGURE 8.24 Mechanism-based (suicide) inhibition. Monoamine oxidase, an enzyme important for neurotransmitter synthesis, requires the cofactor FAD (flavin adenine dinucleotide). *N,N*-Dimethylpropargylamine inhibits monoamine oxidase by covalently modifying the flavin prosthetic group only after the inhibitor has been oxidized. The N-5 flavin adduct is stabilized by the addition of a proton. R represents the remainder of the flavin prosthetic group.

Flavin prosthetic group

N,N-Dimethylpropargylamine

Stably modified flavin of inactivated enzyme

L–Deprenyl (Selegiline)

group on the enzyme is vital for catalysis. An example is *N,N*-dimethylpropargylamine, an inhibitor of the enzyme monoamine oxidase (MAO). A flavin prosthetic group of monoamine oxidase oxidizes the *N,N*-dimethylpropargylamine, which in turn inactivates the enzyme by binding to N-5 of the flavin prosthetic group (Figure 8.24). Monoamine oxidase deaminates neurotransmitters such as dopamine and serotonin, lowering their levels in the brain. Parkinson disease is associated with low levels of dopamine, and depression is associated with low levels of serotonin. *N,N*-Dimethylpropargylamine and L-deprenyl, another suicide inhibitor of monoamine oxidase, are used to treat Parkinson disease and depression.

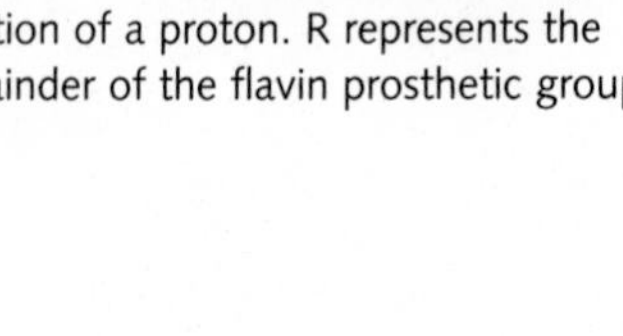

Penicillin irreversibly inactivates a key enzyme in bacterial cell-wall synthesis

Penicillin, the first antibiotic discovered, provides us with another example of a clinically useful suicide inhibitor. Penicillin consists of a thiazolidine ring fused to a *β-lactam* ring to which a variable R group is attached by a peptide bond (Figure 8.25A). In benzylpenicillin, for example, R is a benzyl group (Figure 8.25B). This structure can undergo a variety

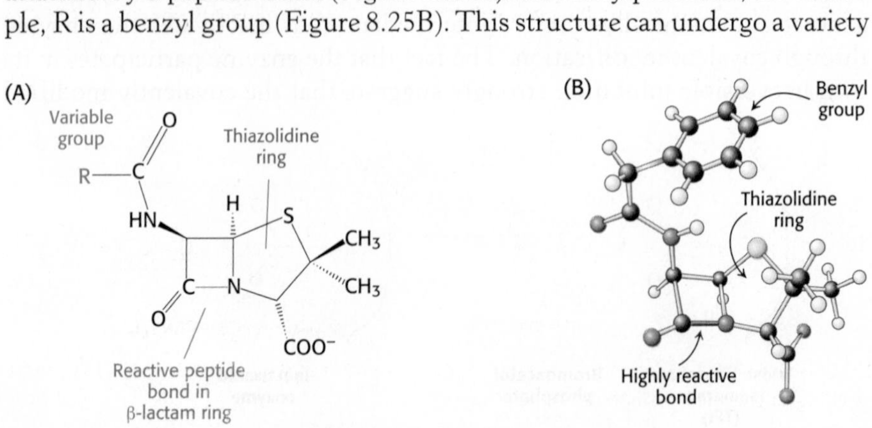

(A)

Variable group

Thiazolidine ring

Reactive peptide bond in β-lactam ring

(B)

Benzyl group

Thiazolidine ring

Highly reactive bond

FIGURE 8.25 The reactive site of penicillin is the peptide bond of its β-lactam ring. (A) Structural formula of penicillin. (B) Representation of benzylpenicillin.

of rearrangements, and, in particular, the β-lactam ring is very labile. Indeed, this instability is closely tied to the antibiotic action of penicillin, as will be evident shortly.

How does penicillin inhibit bacterial growth? Let us consider *Staphylococcus aureus*, the most common cause of staph infections. Penicillin interferes with the synthesis of the *S. aureus* cell wall. The *S. aureus* cell wall is made up of a macromolecule, called a *peptidoglycan* (Figure 8.26), which consists of linear polysaccharide chains that are cross-linked by short peptides (pentaglycines and tetrapeptides). The enormous peptidoglycan molecule confers mechanical support and prevents bacteria from bursting in response to their high internal osmotic pressure. *Glycopeptide transpeptidase* catalyzes the formation of the cross-links that make the peptidoglycan so stable (Figure 8.27). Bacterial cell walls are distinctive in containing D amino acids, which form cross-links by a mechanism different from that used to synthesize proteins.

Penicillin inhibits the cross-linking transpeptidase by the Trojan horse stratagem. The transpeptidase normally forms an *acyl intermediate* with the penultimate D-alanine residue of the D-Ala-D-Ala peptide (Figure 8.28). This covalent acyl-enzyme intermediate then reacts with the amino group of the terminal glycine in another peptide to form the cross-link. Penicillin is welcomed into the active site of the transpeptidase because it mimics the D-Ala-D-Ala moiety of the normal substrate (Figure 8.29). Bound penicillin then forms a covalent bond with a serine residue at the active site of the enzyme. *This penicilloyl-enzyme does not react further. Hence, the transpeptidase is irreversibly inhibited and cell-wall synthesis cannot take place.*

Why is penicillin such an effective inhibitor of the transpeptidase? The highly strained, four-membered β-lactam ring of penicillin makes it especially reactive. On binding to the transpeptidase, the serine residue at the active site attacks the carbonyl carbon atom of the lactam ring to form the penicilloyl-serine derivative (Figure 8.30). Because the peptidase participates in its own inactivation, penicillin acts as a suicide inhibitor.

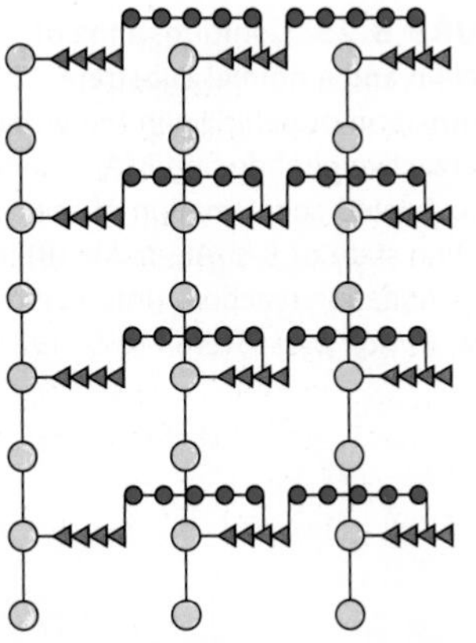

FIGURE 8.26 Schematic representation of the peptidoglycan in *Staphylococcus aureus*. The sugars are shown in yellow, the tetrapeptides in red, and the pentaglycine bridges in blue. The cell wall is a single, enormous, bag-shaped macromolecule because of extensive crosslinking.

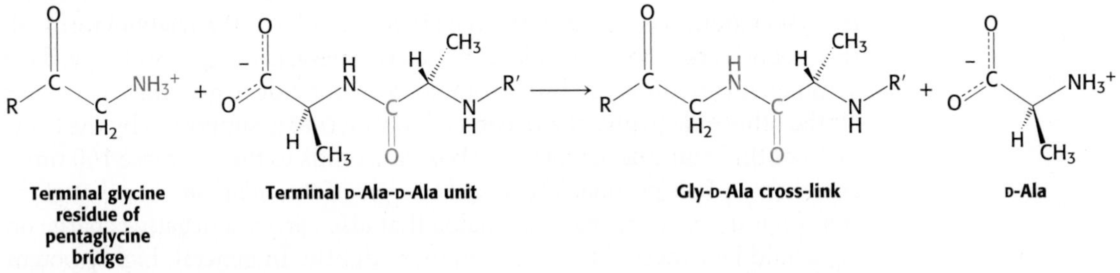

FIGURE 8.27 Formation of cross-links in *S. aureus* peptidoglycan. The terminal amino group of the pentaglycine bridge in the cell wall attacks the peptide bond between two D-alanine residues to form a cross-link.

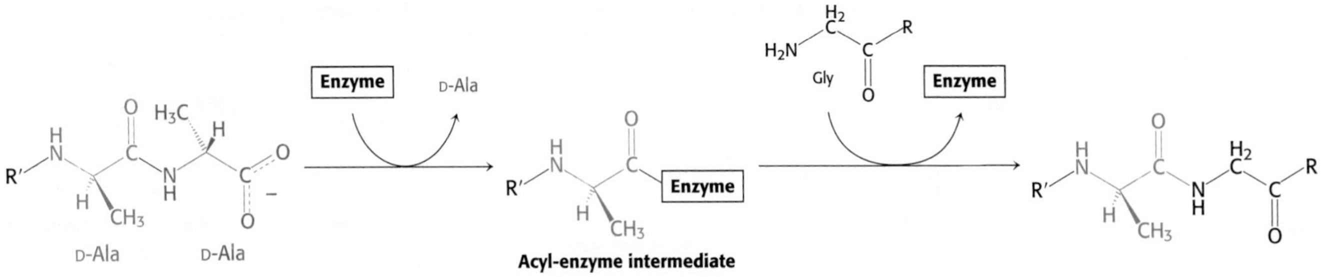

FIGURE 8.28 Transpeptidation reaction. An acyl-enzyme intermediate is formed in the transpeptidation reaction leading to cross-link formation.

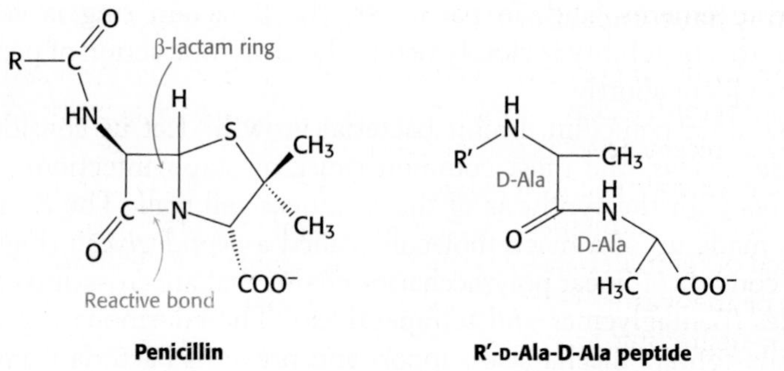

FIGURE 8.29 Conformations of penicillin and a normal substrate. The conformation of penicillin in the vicinity of its reactive peptide bond (A) resembles the postulated conformation of the transition state of R-D-Ala-D-Ala (B) in the transpeptidation reaction. [Information from B. Lee, *J. Mol. Biol.* 61:463–469, 1971.]

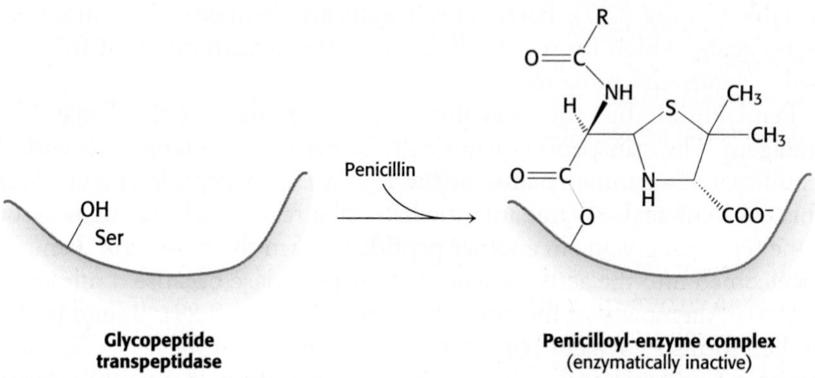

Penicillin

R′-D-Ala-D-Ala peptide

FIGURE 8.30 Formation of a penicilloyl-enzyme derivative. Penicillin reacts irreversibly with the transpeptidase to inactivate the enzyme.

Glycopeptide transpeptidase

Penicilloyl-enzyme complex (enzymatically inactive)

Transition-state analogs are potent inhibitors of enzymes

We turn now to compounds that provide the most intimate views of the catalytic process itself. Linus Pauling proposed in 1948 that compounds resembling the transition state of a catalyzed reaction should be very effective inhibitors of enzymes. These mimics are called *transition-state analogs.* The inhibition of proline racemase is an instructive example. The racemization of proline proceeds through a transition state in which the tetrahedral α-carbon atom has become trigonal (Figure 8.31). In the trigonal form, all three bonds are in the same plane; C_α also carries a net negative charge. This symmetric carbanion can be reprotonated on one side to give the L isomer or on the other side to give the D isomer. This picture is supported by the finding that the inhibitor pyrrole 2-carboxylate binds to the racemase 160 times as tightly as does proline. *The α-carbon atom of this inhibitor, like that of the transition state, is trigonal.* An analog that also carries a negative charge on C_α would be expected to bind even more tightly. In general, highly potent and specific inhibitors of enzymes can be produced by synthesizing compounds that more closely resemble the transition state than the substrate itself. The inhibitory power of transition-state analogs underscores the essence of catalysis: *selective binding of the transition state.*

L-Proline

Planar transition state

D-Proline

Pyrrole 2-carboxylic acid (transition-state analog)

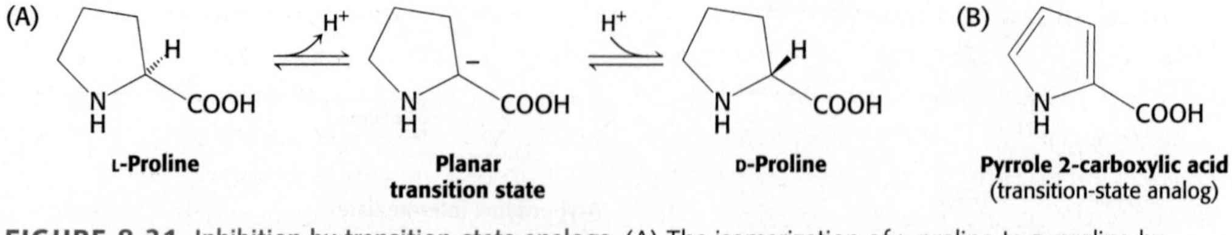

FIGURE 8.31 Inhibition by transition-state analogs. (A) The isomerization of L-proline to D-proline by proline racemase proceeds through a planar transition state in which the α-carbon atom is trigonal rather than tetrahedral. (B) Pyrrole 2-carboxylic acid, a transition-state analog because of its trigonal geometry, is a potent inhibitor of proline racemase.

Enzymes have impact outside the laboratory or clinic

As discussed previously, and as we will see again and again in our study of biochemistry, enzymes play prominent roles in cellular energy metabolism. However, recent developments show that the power of enzymes may be harnessed to generate energy for entire communities, as well as reducing landfill.

Unsorted municipal waste—including paper, food, clothing, and assorted other materials—is treated with a cocktail of enzymes that degrade much of the waste. Treatment with an array of carbohydrate-, lipid-, and protein-degrading enzymes turn much of the waste into a bioliquid of sugars and other biomolecules. The bioliquid is used to fuel the growth of methane-producing bacteria. The methane is harvested and burned to generate electricity. Any waste not degraded by the enzyme cocktail is recycled or incinerated to produce electricity. We are used to thinking that biochemistry research promotes advances in the laboratory and clinics. As this example shows, the power of biochemistry can also be used to attack other societal issues, such as the generation of clean energy.

8.6 Enzymes Can Be Studied One Molecule at a Time

Most experiments that are performed to determine an enzyme characteristic require an enzyme preparation in a buffered solution. Even a few microliters of such a solution will contain millions of enzyme molecules. Much that we have learned about enzymes thus far has come from such experiments, called *ensemble studies*. A basic assumption of ensemble studies is that all of the enzyme molecules are the same or very similar. When we determine an enzymatic property such as the value of K_M in ensemble studies, that value is of necessity an average value of all of the enzyme molecules present. However, as discussed in Chapter 2, we now know that molecular heterogeneity, the ability of a molecule to assume several different structures over time that differ slightly in stability, is an inherent property of all large biomolecules (p. 58). How can we tell if this molecular heterogeneity affects enzyme activity?

By way of example, consider a hypothetical situation. A Martian visits Earth to learn about higher education. The spacecraft hovers high above a university, and our Martian meticulously records how the student population moves about campus. Much information can be gathered from such studies: where students are likely to be at certain times on certain days, which buildings are used when and by how many. Now, suppose our visitor developed a high-magnification camera that could follow one student throughout the day. Such data would provide a much different perspective on college life: What does this student eat? To whom does she talk? How much time does she spend studying? This new *in singulo* method, examining one individual at a time, yields a great deal of new information but also illustrates a potential pitfall of studying individuals, be they students or enzymes: How can we be certain that the student or molecule is representative and not an outlier? This pitfall can be overcome by studying enough individuals to satisfy statistical analysis for validity.

Let us leave our Martian to his observations, and consider a more biochemical situation. Figure 8.32A shows an enzyme that displays molecular heterogeneity, with three active forms that catalyze the same reaction but at different rates. These forms have slightly different stabilities, but thermal noise is sufficient to interconvert the forms. Each form is present as a fraction of the total enzyme population as indicated. If we were to perform an experiment to determine enzyme activity under a particular set of conditions with the use of ensemble methods, we would get a single value, which would represent the average of the heterogeneous assembly (Figure 8.32B). However, were we to perform a sufficient number of single-molecule experiments, we

FIGURE 8.32 Single-molecule studies can reveal molecular heterogeneity. (A) Complex biomolecules, such as enzymes, display molecular heterogeneity. (B) When measuring an enzyme property using ensemble methods, an average value of all of the enzymes present is the result. (C) Single-enzyme studies reveal molecular heterogeneity, with the various forms showing different properties.

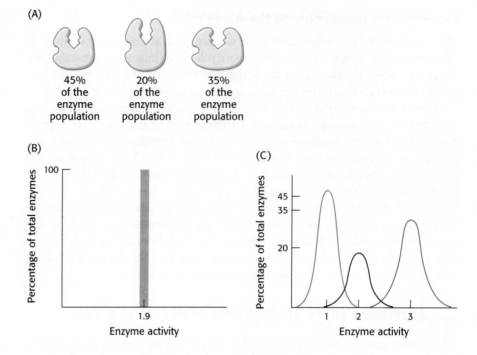

(A)

45% of the enzyme population   20% of the enzyme population   35% of the enzyme population

(B)

(C)

would discover that the enzyme has three different molecular forms with very different activities (Figure 8.32C). Moreover, these different forms would most likely correspond to important biochemical differences.

The development of powerful techniques—such as patch-clamp recording, single-molecule fluorescence, and optical tweezers—has enabled biochemists to look into the workings of individual molecules. We will examine single-molecule studies of membrane channels with the use of patch-clamp recording (Section 13.4), ATP-synthesizing complexes with the use of single-molecule fluorescence (Section 18.4), and molecular motors with the use of an optical trap (See Section 36.2 on **SaplingPlus**). We are now able to observe events at a molecular level that reveal rare or transient structures and fleeting events in a reaction sequence, as well as to measure mechanical forces affecting or generated by an enzyme. Single-molecule studies open a new vista on the function of enzymes in particular and on all large biomolecules in general.

SUMMARY

8.1 Enzymes Are Powerful and Highly Specific Catalysts

Most catalysts in biological systems are enzymes, and nearly all enzymes are proteins. Enzymes are highly specific and have great catalytic power. They can enhance reaction rates by factors of 10^6 or more. Many enzymes require cofactors for activity. Such cofactors can be metal ions or small, vitamin-derived organic molecules called coenzymes.

8.2 Gibbs Free Energy Is a Useful Thermodynamic Function for Understanding Enzymes

Free energy (G) is the most valuable thermodynamic function for understanding the energetics of catalysis. A reaction can take place spontaneously only if the change in free energy (ΔG) is negative. The free-energy change of a reaction that takes place when reactants and products are at unit activity is called the standard free-energy change ($\Delta G°$). Biochemists use $\Delta G°'$, the standard free-energy change at pH 7. Enzymes do not alter reaction equilibria; rather, they increase the rate at which equilibrium is attained.

8.3 Enzymes Accelerate Reactions by Facilitating the Formation of the Transition State

Enzymes serve as catalysts by decreasing the free energy of activation of chemical reactions. Enzymes accelerate reactions by providing a reaction pathway in which the transition state (the highest-energy species) has a lower free energy and hence is more rapidly formed than in the uncatalyzed reaction.

The first step in catalysis is the formation of an enzyme–substrate complex. Substrates are bound to enzymes at active-site clefts from which water is largely excluded when the substrate is bound. The specificity of enzyme–substrate interactions arises mainly from weak reversible contacts mediated by electrostatic interactions, hydrogen bonds, and van der Waals forces, and from the shape of the active site, which rejects molecules that do not have a sufficiently complementary shape. Enzymes facilitate formation of the transition state by a dynamic process in which the substrate binds to specific conformations of the enzyme, accompanied by conformational changes at active sites that result in catalysis.

8.4 The Michaelis–Menten Model Accounts for the Kinetic Properties of Many Enzymes

The kinetic properties of many enzymes are described by the Michaelis–Menten model. In this model, an enzyme (E) combines with a substrate (S) to form an enzyme–substrate (ES) complex, which can proceed to form a product (P) or to dissociate into E and S.

$$E + S \underset{k_{-1}}{\overset{k_1}{\rightleftharpoons}} ES \xrightarrow{k_2} E + P$$

The rate of formation of product V_0 is given by the Michaelis–Menten equation:

$$V_0 = V_{max} \frac{[S]}{[S] + K_M}$$

in which V_{max} is the reaction rate when the enzyme is fully saturated with substrate and K_M, the Michaelis constant, is the substrate concentration at which the reaction rate is half maximal. The maximal rate, V_{max}, is equal to the product of k_2, or k_{cat}, and the total concentration of enzyme. The kinetic constant k_{cat}, called the turnover number, is the number of substrate molecules converted into product per unit time at a single catalytic site when the enzyme is fully saturated with substrate. Turnover numbers for most enzymes are between 1 and 10^4 per second. The ratio of k_{cat}/K_M provides a measure of enzyme efficiency and specificity.

Allosteric enzymes constitute an important class of enzymes whose catalytic activity can be regulated. These enzymes, which do not conform to Michaelis–Menten kinetics, have multiple active sites. These active sites display cooperativity, as evidenced by a sigmoidal dependence of reaction velocity on substrate concentration.

8.5 Enzymes Can Be Inhibited by Specific Molecules

Specific small molecules or ions can inhibit even nonallosteric enzymes. In irreversible inhibition, the inhibitor is covalently linked to the enzyme or bound so tightly that its dissociation from the enzyme is very slow. Covalent inhibitors provide a means of mapping the enzyme's active site. In contrast, reversible inhibition is characterized by a more rapid and less stable interaction between enzyme and inhibitor. A competitive inhibitor prevents the substrate from binding to the active site. It reduces the reaction velocity by diminishing the proportion of enzyme molecules that are bound to substrate. Competitive inhibition can be overcome by raising the substrate concentration. In uncompetitive inhibition, the inhibitor combines only with the enzyme–substrate complex. In noncompetitive inhibition, the inhibitor decreases the turnover number. Uncompetitive and noncompetitive inhibition cannot be overcome by raising the substrate concentration.

The essence of catalysis is selective stabilization of the transition state. Hence, an enzyme binds the transition state more tightly than it binds the substrate. Transition-state analogs are stable compounds that mimic key features of this highest-energy species. They are potent and specific inhibitors of enzymes.

8.6 Enzymes Can Be Studied One Molecule at a Time

Many enzymes are now being studied *in singulo,* at the level of a single molecule. Such studies are important because they yield information that is difficult to obtain in studies of populations of molecules. Single-molecule methods reveal a distribution of enzyme characteristics rather than an average value as is acquired with the use of ensemble methods.

APPENDIX

Enzymes Are Classified on the Basis of the Types of Reactions That They Catalyze

Many enzymes have common names that provide little information about the reactions that they catalyze. For example, a proteolytic enzyme secreted by the pancreas is called trypsin. Most other enzymes are named for their substrates and for the reactions that they catalyze, with the suffix "ase" added. Thus, a peptide hydrolase is an enzyme that hydrolyzes peptide bonds, whereas ATP synthase is an enzyme that synthesizes ATP.

To bring some consistency to the classification of enzymes, in 1964 the International Union of Biochemistry established an Enzyme Commission to develop a nomenclature for enzymes. Reactions were divided into six major groups numbered 1 through 6 (Table 8.8). These groups were subdivided and further subdivided so that a four-number code preceded by the letters *EC* for Enzyme Commission could precisely identify all enzymes.

Consider as an example nucleoside monophosphate (NMP) kinase, an enzyme that we will examine in detail in Section 9.4. It catalyzes the following reaction:

$$ATP + NMP \rightleftharpoons ADP + NDP$$

NMP kinase transfers a phosphoryl group from ATP to any nucleoside monophosphate (NMP) to form a nucleoside diphosphate (NDP) and ADP. Consequently, it is a transferase, or member of group 2. Many groups other than phosphoryl groups, such as sugars and single-carbon units, can be transferred. Transferases that shift a phosphoryl group are designated 2.7. Various functional groups can accept the phosphoryl group. If a phosphate is the acceptor, the transferase is designated 2.7.4. The final number designates the acceptor more precisely. In regard to NMP kinase, a nucleoside monophosphate is the acceptor, and the enzyme's designation is EC 2.7.4.4. Although the common names are used routinely, the classification number is used when the precise identity of the enzyme is not clear from the common name alone.

TABLE 8.8 Six major classes of enzymes

Class	Type of reaction	Example	Chapter
1. Oxidoreductases	Oxidation–reduction	Lactate dehydrogenase	16
2. Transferases	Group transfer	Nucleoside monophosphate kinase (NMP kinase)	9
3. Hydrolases	Hydrolysis reactions (transfer of functional groups to water)	Chymotrypsin	9
4. Lyases	Addition or removal of groups to form double bonds	Fumarase	17
5. Isomerases	Isomerization (intramolecular group transfer)	Triose phosphate isomerase	16
6. Ligases	Ligation of two substrates at the expense of ATP hydrolysis	Aminoacyl-tRNA synthetase	31

APPENDIX Biochemistry in Focus

The effect of temperature on enzyme-catalyzed reactions and the coloring of Siamese cats

As the temperature rises, the rate of most reactions, including enzyme-catalyzed reactions, increases. For most enzymes, there is a temperature at which the increase in catalytic activity ceases and there is a precipitous loss of activity.

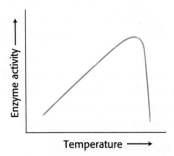

What is the basis of this loss of activity? Recall from Chapter 2 that proteins have a complex three-dimensional structure that is held together by weak bonds. When the temperature increases beyond a certain point, the bonds maintaining the three-dimensional structure are not strong enough to withstand the polypeptide chain's thermal jostling and the protein loses the structure required for activity. The protein is said to be *denatured* (p. 53).

In organisms such as ourselves that maintain a constant body temperature (endotherms), the effect of outside temperature on enzyme activity is minimized. However, in organisms that assume the temperature of the ambient environment (ectotherms), temperature is an important regulator of biochemical and, indeed, biological activity. Lizards, for instance, are most active in warmer temperatures and relatively inactive in cooler temperatures—a behavioral manifestation of biochemical activity. Although endotherms are not as sensitive to ambient temperature as ectotherms, slight tissue temperature alterations are sometimes important. For instance, when athletes "warm up," they are increasing the temperature of their muscles through exertion. This increase in temperature facilitates the biochemistry that will power the exercise.

Which brings us to the coloration of Siamese cats. Interestingly enough, the coloration of Siamese cats can be explained by variation in enzyme sensitivity due to temperature. Coloration is because of the presence of the pigment, melanin, the same pigment that is responsible for human skin color. The first steps in the synthesis of melanin are catalyzed by the enzyme tyrosinase.

Tyrosine → (Tyrosinase) → **Dihydroxyphenylalanine**

→ (Tyrosinase) → **Dopaquinone**

Melanin ←

A deficiency in tyrosinase activity in humans leads to albinism. Now, Siamese cats are often born with very little coloration; sometimes they are even white. As they mature, their extremities—tips of the ears, snout, paws, and end of the tail—develop pigmentation. In many cases, the entire body becomes darker. This suggests that tyrosinase activity is missing at birth, but then returns. A hint as to the nature of the defect in the enzyme comes from the observation that pigmentation develops first at the extremities; that is, the cooler parts of the body. Analysis has established that there is a mutation in the Siamese tyrosinase that results in a loss of activity above 37–39 °C. At lower temperatures—at the extremities—the activity returns. Adult cats have a slightly lower body temperature than newborn kittens, thus allowing pigment formation in adults.

APPENDIX

Problem-Solving Strategies

PROBLEM 1: The velocity of an enzyme is 80% V_{max}. What is the ratio of [S] to K_M under this condition?

SOLUTION: When a question includes the terms V_{max}, [S] and K_M, a good bet is that the solution involves the Michaelis–Menten equation.

▶ **Write out the Michaelis–Menten equation.**

$$V_0 = V_{max}\frac{[S]}{[S] + K_M}$$

The question states the $V_0 = 80\% V_{max}$. That is the only value we are provided with, so we should be able to solve the problem with that value and the equation. First, for convenience, let's convert 80% to 0.8.

▶ **What is the next step?**

Substitute the 0.8 V_{max} for V_0 in the equation.

$$0.8V_{max} = V_{max}\frac{[S]}{[S] + K_M}$$

▶ **Thinking back to algebra, what would be the next step?**

Divide both sides of the equation by V_{max}, which yields

$$0.8 = \frac{[S]}{[S] + K_M}$$

▶ **Now what?**

Simplify the equation so that we arrive at a ratio of $[S]/K_M$.

$$0.8\,[S] + 0.8K_M = [S]$$

$$0.8K_M = [S] - 0.8\,[S]$$

$$0.8K_M = 0.2\,[S]$$

$$4 = \frac{[S]}{K_M}$$

Thus, when [S] is 4-fold greater than K_M, $V_0 = 0.8\ V_{max} = 80\% V_{max}$.

You could confirm this by solving the Michaelis–Menten equation using $[S] = 4K_M$ and solving for V_0 as a fraction of V_{max}.

PROBLEM 2: The graph shows the effect of two different concentrations of a molecule X on an enzyme. How is the molecule affecting enzyme activity?

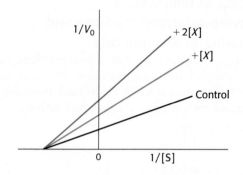

SOLUTION: As with all data interpretation problems, the first step is to make sure that we know what we are looking at. The question concerns enzyme activity, so that should provide some guidance for our initial interpretation.

▶ **What are the axes of the graph?**

The x-axis shows 1/[S], while the y-axis displays $1/V$. Note that these are two reciprocal values.

▶ **What graphical presentation shows enzyme kinetics with double-reciprocal values?**

A Lineweaver–Burk or double reciprocal plot.

▶ **What value is displayed at the point where the graph crosses the y-axis on such a plot? The x-axis?**

The y-axis shows $1/V_{max}$, while the x-axis shows $-1/K_M$.

With this information, we can now determine how X is affecting the enzyme.

▶ **What kinetic values are altered in the presence of X, bearing in mind that the values are presented as reciprocal values?**

Since all of the lines converge on the x-axis, K_M is not altered in the presence of X. However, $1/V_{max}$ increases with increasing amounts of X.

▶ **If $1/V_{max}$ increases in the presence of X, how is V_{max} changing?**

V_{max} must be decreasing.

▶ **Summarize the effects of X on the enzyme's activity. What can you conclude about the nature of the effect of X on enzyme activity?**

K_M is unchanged, but V_{max} is reduced. X must be some type of an inhibitor of the enzyme. In particular, pure noncompetitive inhibitors lower V_{max} without altering K_M.

KEY TERMS

enzyme (p. 234)
substrate (p. 235)
cofactor (p. 235)
apoenzyme (p. 235)
holoenzyme (p. 235)
coenzyme (p. 236)
prosthetic group (p. 236)
free energy (p. 236)
transition state (p. 240)
free energy of activation (p. 237)
active site (p. 241)
induced fit (p. 243)

K_M (the Michaelis constant)
(p. 247)
V_{max} (maximal rate) (p. 247)
Michaelis–Menten equation
(p. 247)
Lineweaver–Burk equation
(double-reciprocal plot) (p. 248)
turnover number (p. 249)
k_{cat}/K_M ratio (the specificity
constant) (p. 249)
sequential reaction (p. 251)

double-displacement (ping-pong)
reaction (p. 252)
allosteric enzyme (p. 252)
competitive inhibition (p. 253)
uncompetitive inhibition (p. 253)
noncompetitive inhibition (p. 254)
group-specific reagent (p. 256)
affinity label (reactive substrate
analog) (p. 257)
mechanism-based (suicide)
inhibition (p. 257)
transition-state analog (p. 260)

PROBLEMS

1. *Raisons d'être*. What are the two properties of enzymes that make them especially useful catalysts?

2. *Partners*. What does an apoenzyme require to become a holoenzyme?

3. *Different partners*. What are the two main types of cofactors?

4. *One a day*. Why are vitamins necessary for good health?

5. *Shared properties*. What are the general characteristics of enzyme active sites? ✓❷

6. *A function of state*. What is the fundamental mechanism by which enzymes enhance the rate of chemical reactions? ✓❷

7. *Nooks and crannies*. What is the structural basis for enzyme specificity? ✓❷

8. *Made for each other*. Match the term with the proper description. ✓❶, ✓❷, ✓❸

(a) Enzyme _____   1. The least-stable reaction intermediate

(b) Substrate _____   2. Site on the enzyme where catalysis takes place

(c) Cofactor _____   3. Enzyme minus its cofactor

(d) Apoenzyme _____   4. Protein catalyst

(e) Holoenzyme _____   5. Function of K'_{eq}

(f) Coenzymes _____   6. Change in enzyme structure

(g) $\Delta G^{\circ\prime}$ _____   7. Reactant in an enzyme-catalyzed reaction

(h) Transition state _____   8. A coenzyme or metal

(i) Active site _____   9. Enzyme plus cofactor

(j) Induced fit _____   10. Small, vitamin-derived organic cofactors

9. *Give with one hand, take with the other*. Why does the activation energy of a reaction not appear in the final ΔG of the reaction? ✓❶, ✓❷

10. *Making progress*. The illustrations below show the reaction-progress curves for two different reactions. Indicate the activation energy as well as the ΔG for each reaction. Which reaction is endergonic? Exergonic? ✓❶

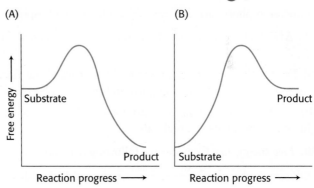

11. *The more things change, the more they stay the same*. Suppose that, in the absence of enzyme, the forward rate constant (k_F) for the conversion of S into P is $10^{-4}\,s^{-1}$ and the reverse rate constant (k_R) for the conversion of P into S is $10^{-6}\,s^{-1}$. ✓❶, ✓❸

$$S \underset{10^{-6}\,s^{-1}}{\overset{10^{-4}\,s^{-1}}{\rightleftharpoons}} P$$

(a) What is the equilibrium for the reaction? What is the $\Delta G^{\circ\prime}$?

(b) Suppose an enzyme enhances the rate of the reaction 100-fold. What are the rate constants for the enzyme-catalyzed reaction? What is the equilibrium constant? The $\Delta G^{\circ\prime}$?

12. *Mountain climbing*. Proteins are thermodynamically unstable. The ΔG of the hydrolysis of proteins is quite negative, yet proteins can be quite stable. Explain this apparent paradox. What does it tell you about protein synthesis?

13. *Protection.* Suggest why the enzyme lysozyme, which degrades cell walls of some bacteria, is present in tears.

14. *Mutual attraction.* What is meant by the term *binding energy*? ✓❷

15. *Catalytically binding.* What is the role of binding energy in enzyme catalysis?

16. *Sticky situation.* What would be the result of an enzyme having a greater binding energy for the substrate than for the transition state? ✓❶, ✓❷

17. *Stability matters.* Transition-state analogs, which can be used as enzyme inhibitors and to generate catalytic antibodies, are often difficult to synthesize. Suggest a reason. ✓❺

18. *Match'em.* Match the K'_{eq} values with the appropriate $\Delta G^{\circ\prime}$ values. ✓❶

K'_{eq}	$\Delta G^{\circ\prime}$ (kJ mol^{-1})
(a) 1	28.53
(b) 10^{-5}	-11.42
(c) 10^4	5.69
(d) 10^2	0
(e) 10^{-1}	-22.84

19. *Free energy!* Assume that you have a solution of 0.1 M glucose 6-phosphate. To this solution, you add the enzyme phosphoglucomutase, which catalyzes the following reaction:

Glucose 6-phosphate $\xrightleftharpoons{\text{Phosphoglucomutase}}$ glucose 1-phosphate

The $\Delta G^{\circ\prime}$ for the reaction is $+7.5$ kJ mol^{-1} ($+1.8$ kcal mol^{-1}). ✓❶

(a) Does the reaction proceed as written? If so, what are the final concentrations of glucose 6-phosphate and glucose 1-phosphate?

(b) Under what cellular conditions could you produce glucose 1-phosphate at a high rate?

20. *Free energy, too!* Consider the following reaction:

Glucose 6-phosphate $\xrightleftharpoons{\text{Phosphoglucomutase}}$ glucose 1-phosphate

After reactant and product were mixed and allowed to reach equilibrium at 25°C, the concentration of each compound was measured:

$$[\text{Glucose 1-phosphate}]_{eq} = 0.01 \text{ M}$$

$$[\text{Glucose 6-phosphate}]_{eq} = 0.19 \text{ M}$$

Calculate K_{eq} and $\Delta G^{\circ\prime}$. ✓❶

21. *Keeping busy.* Many isolated enzymes, if incubated at 37°C, will be denatured. However, if the enzymes are incubated at 37°C, in the presence of substrate, the enzymes are catalytically active. Explain this apparent paradox.

22. *Active yet responsive.* What is the biochemical advantage of having a K_M approximately equal to the substrate concentration normally available to an enzyme?

23. *Affinity or not affinity? That is the question.* The affinity between a protein and a molecule that binds to the protein is frequently expressed in terms of a dissociation constant K_d. ✓❹

Protein + small molecule $\rightleftharpoons$ Protein − small molecule complex

$$K_d = \frac{[\text{protein}][\text{small molecule}]}{[\text{protein} - \text{small molecule complex}]}$$

Does K_M measure the affinity of the enzyme complex? Under what circumstances might K_M approximately equal K_d?

24. *They go together like spaghetti and meatballs.* Match the term with the proper description. ✓❸, ✓❹, ✓❺

(a) Enzyme _____
(b) Kinetics _____
(c) Michaelis–Menten equation _____
(d) Michaelis constant (K_M) _____
(e) Lineweaver–Burk equation _____
(f) Turnover number _____
(g) k_{cat}/K_M _____
(h) Sequential reaction _____
(i) Double-displacement (ping-pong) reaction _____
(j) Allosteric enzyme _____

1. Substrate concentration that yields $1/2 V_{max}$
2. k_2 or k_{cat}
3. Responds to environmental signals
4. The study of reaction rates
5. Describes the kinetics of simple one-substrate reactions
6. A ternary complex is formed
7. Protein catalyst
8. Double-reciprocal plot
9. A measure of enzyme efficiency
10. Includes a substituted enzyme intermediate

25. *Like bread and butter.* Match the term with the description or compound. ✓❻

(a) Competitive inhibition _____
(b) Uncompetitive inhibition _____
(c) Noncompetitive inhibition _____

1. Inhibitor and substrate can bind simultaneously
2. V_{max} remains the same but the K_M^{app} increases
3. Sulfanilamide
4. Binds to the enzyme–substrate complex only
5. Lowers V_{max} and K_M^{app}
6. Roundup
7. K_M remains unchanged but V_{max} is lower
8. Doxycycline
9. Inhibitor binds at the active site

26. *Angry biochemists.* Many biochemists go bananas, and justifiably, when they see a Michaelis–Menten plot like the one shown below. To see why, determine the V_0 as a fraction of V_{max} when the substrate concentration is equal to 10 K_M and 20 K_M. Please control your outrage. ✓❸, ✓❹

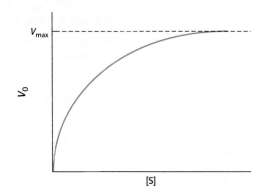

27. *Hydrolytic driving force.* The hydrolysis of pyrophosphate to orthophosphate is important in driving forward biosynthetic reactions such as the synthesis of DNA. This hydrolytic reaction is catalyzed in *E. coli* by a pyrophosphatase that has a mass of 120 kDa and consists of six identical subunits. For this enzyme, a unit of activity is defined as the amount of enzyme that hydrolyzes 10 μmol of pyrophosphate in 15 minutes at 37°C under standard assay conditions. The purified enzyme has a V_{max} of 2800 units per milligram of enzyme. ✓❸, ✓❹

(a) How many moles of substrate are hydrolyzed per second per milligram of enzyme when the substrate concentration is much greater than K_M?

(b) How many moles of active sites are there in 1 mg of enzyme? Assume that each subunit has one active site.

(c) What is the turnover number of the enzyme? Compare this value with others mentioned in this chapter.

28. *Destroying the Trojan horse.* Penicillin is hydrolyzed and thereby rendered inactive by penicillinase (also known as β-lactamase), an enzyme present in some penicillin-resistant bacteria. The mass of this enzyme in *Staphylococcus aureus* is 29.6 kDa. The amount of penicillin hydrolyzed in 1 minute in a 10-ml solution containing 10^{-9} g of purified penicillinase was measured as a function of the concentration of penicillin. Assume that the concentration of penicillin does not change appreciably during the assay. ✓❺

[Penicillin] μM	Amount hydrolyzed (nmol)
1	0.11
3	0.25
5	0.34
10	0.45
30	0.58
50	0.61

(a) Plot V_0 versus [S] and $1/V_0$ versus $1/$[S] for these data. Does penicillinase appear to obey Michaelis–Menten kinetics? If so, what is the value of K_M?

(b) What is the value of V_{max}?

(c) What is the turnover number of penicillinase under these experimental conditions? Assume one active site per enzyme molecule.

29. *Counterpoint.* Penicillinase (β-lactamase) hydrolyzes penicillin. Compare the reaction mechanism of penicillinase to that of glycopeptide transpeptidase. ✓❺, ✓❻

30. *A different mode.* The kinetics of an enzyme is measured as a function of substrate concentration in the presence and absence of 100 μM inhibitor. ✓❺, ✓❻

(a) What are the values of V_{max} and K_M in the presence of this inhibitor?

(b) What type of inhibition is it?

(c) What is the dissociation constant of this inhibitor?

	Velocity (μmol minute^{-1})	
[S] (μM)	No inhibitor	Inhibitor
3	10.4	2.1
5	14.5	2.9
10	22.5	4.5
30	33.8	6.8
90	40.5	8.1

(d) If [S] = 30 μM, what fraction of the enzyme molecules have a bound substrate in the presence and in the absence of 100 μM inhibitor?

31. *Informative inhibition.* What are the four key types of irreversible inhibitors that can be used to study enzyme function? ✓❺

32. *A fresh view.* The plot of $1/V_0$ versus $1/$[S] is sometimes called a Lineweaver–Burk plot. Another way of expressing the kinetic data is to plot V_0 versus $V_0/$[S], which is known as an Eadie–Hofstee plot. ✓❻

(a) Rearrange the Michaelis–Menten equation to give V_0 as a function of $V_0/$[S].

(b) What is the significance of the slope, the *y*-intercept, and the *x*-intercept in a plot of V_0 versus $V_0/$[S]?

(c) Sketch a plot of V_0 versus $V_0/$[S] in the absence of an inhibitor, in the presence of a competitive inhibitor, and in the presence of a noncompetitive inhibitor.

33. *Defining attributes.* What is the defining characteristic for an enzyme catalyzing a sequential reaction? A double-displacement reaction?

34. *Competing substrates.* Suppose that two substrates, A and B, compete for an enzyme. Derive an expression relating the ratio of the rates of utilization of A and B, V_A/V_B, to the concentrations of these substrates and their values of k_{cat} and K_M.

(Hint: Express V_A as a function of k_{cat}/K_M for substrate A, and do the same for V_B.) Is specificity determined by K_M alone?

35. *A tenacious mutant.* Suppose that a mutant enzyme binds a substrate 100 times as tightly as does the native enzyme. What is the effect of this mutation on catalytic rate if the binding of the transition state is unaffected? ✓❷

36. *More Michaelis–Menten.* For an enzyme that follows simple Michaelis–Menten kinetics, what is the value of V_{max} if V_0 is equal to 1 μmol minute^{-1} at 10 K_M? ✓❹

37. *Controlled paralysis.* Succinylcholine is a fast-acting, short-duration muscle relaxant that is used when a tube is inserted into a patient's trachea or when a bronchoscope is used to examine the trachea and bronchi for signs of cancer. Within seconds of the administration of succinylcholine, the patient experiences muscle paralysis and is placed on a respirator while the examination proceeds. Succinylcholine is a competitive inhibitor of acetylcholinesterase, a nervous system enzyme, and this inhibition causes paralysis. However, succinylcholine is hydrolyzed by blood-serum cholinesterase, which shows a broader substrate specificity than does the nervous system enzyme. Paralysis lasts until the succinylcholine is hydrolyzed by the serum cholinesterase, usually several minutes later.

(a) As a safety measure, serum cholinesterase is measured before the examination takes place. Explain why this measurement is good idea.

(b) What would happen to the patient if the serum cholinesterase activity were only 10 units of activity per liter rather than the normal activity of about 80 units?

(c) Some patients have a mutant form of the serum cholinesterase that displays a K_M of 10 mM, rather than the normal 1.4 mM. What will be the effect of this mutation on the patient?

Data Interpretation Problems

38. *A natural attraction, but more complicated.* You have isolated two versions of the same enzyme, a wild type and a mutant differing from the wild type at a single amino acid. Working carefully but expeditiously, you then establish the following kinetic characteristics of the enzymes. ✓❹

	Maximum velocity	K_M
Wild type	100 μmol/min	10 mM
Mutant	1 μmol/min	0.1 mM

(a) With the assumption that the reaction occurs in two steps in which k_{-1} is much larger than k_2, which enzyme has the higher affinity for substrate?

(b) What is the initial velocity of the reaction catalyzed by the wild-type enzyme when the substrate concentration is 10 mM?

(c) Which enzyme alters the equilibrium more in the direction of product?

39. *K_M matters.* The amino acid asparagine is required by cancer cells to proliferate. Treating patients with the enzyme asparaginase is sometimes used as a chemotherapy treatment. Asparaginase hydrolyzes asparagine to aspartate and ammonia. The adjoining illustration shows the Michaelis–Menten curves for two asparaginases from different sources, as well as the concentration of asparagine in the environment (indicated by the arrow). Which enzyme would make a better chemotherapeutic agent? ✓❹

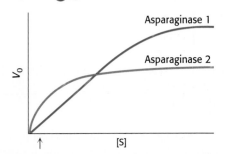

40. *Enzyme specificity.* Catalysis of the cleavage of peptide bonds in small peptides by a proteolytic enzyme is described in the following table.

Substrate	K_M (mM)	k_{cat} (s^{-1})
EMTA↓G	4.0	24
EMTA↓A	1.5	30
EMTA↓F	0.5	18

The arrow indicates the peptide bond cleaved in each case.

(a) If a mixture of these peptides were presented to the enzyme with the concentration of each peptide being the same, which peptide would be digested most rapidly? Most slowly? Briefly explain your reasoning, if any.

(b) The experiment is performed again on another peptide with the following results.

EMTI↓F	9	18

On the basis of these data, suggest the features of the amino acid sequence that dictate the specificity of the enzyme. ✓❹

41. *Varying the enzyme.* For a one-substrate, enzyme-catalyzed reaction, double-reciprocal plots were determined for three different enzyme concentrations. Which of the following three families of curve would you expect to be obtained? Explain. ✓❸

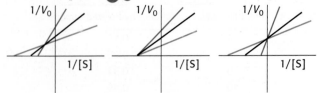

42. *Mental experiment.* Picture in your mind the velocity-versus-substrate concentration curve for a typical Michaelis–Menten enzyme. Now, imagine that the experimental conditions are altered as described below. For each of the conditions described, fill in the table indicating precisely (when possible) the effect on V_{max} and K_M of the imagined Michaelis–Menten enzyme. ✓④, ✓⑥

Experimental condition	V_{max}	K_M
a. Twice as much enzyme is used.		
b. Half as much enzyme is used.		
c. A competitive inhibitor is present.		
d. An uncompetitive inhibitor is present.		
e. A pure noncompetitive inhibitor is present.		

43. *Too much of a good thing.* A simple Michaelis–Menten enzyme, in the absence of any inhibitor, displayed the following kinetic behavior. ✓④, ✓⑤

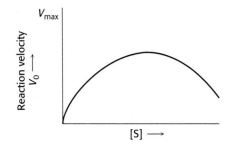

(a) Draw a double-reciprocal plot that corresponds to the velocity-versus-substrate curve.

(b) Suggest a plausible explanation for these kinetic results.

44. *Rate-limiting step.* In the conversion of A into D in the following biochemical pathway, enzymes E_A, E_B, and E_C have the K_M values indicated under each enzyme. If all of the substrates and products are present at a concentration of 10^{-4} M and the enzymes have approximately the same V_{max}, which step will be rate limiting and why? ✓❸

$$A \underset{E_A}{\rightleftharpoons} B \underset{E_B}{\rightleftharpoons} C \underset{E_C}{\rightleftharpoons} D$$
$$K_M = 10^{-2}\text{M} \quad 10^{-4}\text{M} \quad 10^{-4}\text{M}$$

45. *Colored luminosity.* Tryptophan synthetase, a bacterial enzyme that contains a pyridoxal phosphate (PLP) prosthetic group, catalyzes the synthesis of L-tryptophan from L-serine and an indole derivative. The addition of L-serine to the enzyme produces a marked increase in the fluorescence of the PLP group, as the adjoining graph shows. The subsequent addition of indole, the second substrate, reduces this fluorescence to a level even lower than that produced by the enzyme alone. How do these changes in fluorescence support the notion that the enzyme interacts directly with its substrates? ✓❷

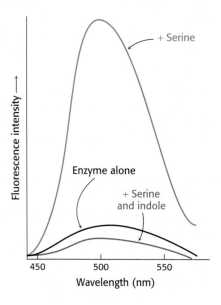

Chapter Integration Problems

46. *Titration experiment.* The effect of pH on the activity of an enzyme was examined. At its active site, the enzyme has an ionizable group that must be negatively charged for substrate binding and catalysis to take place. The ionizable group has a pK_a of 6.0. The substrate is positively charged throughout the pH range of the experiment.

$$E^- + S^+ \rightleftharpoons E^-S^+ \longrightarrow E^- + P^+$$
$$+$$
$$H^+$$
$$\Updownarrow$$
$$EH$$

(a) Draw the V_0-versus-pH curve when the substrate concentration is much greater than the enzyme K_M.

(b) Draw the V_0-versus-pH curve when the substrate concentration is much less than the enzyme K_M.

(c) At which pH will the velocity equal one-half of the maximal velocity attainable under these conditions?

47. *A question of stability.* Pyridoxal phosphate (PLP) is a coenzyme for the enzyme ornithine aminotransferase. The enzyme was purified from cells grown in PLP-deficient media as well as from cells grown in media that contained pyridoxal phosphate. The stability of the two different enzyme preparations was then measured by incubating the enzyme at 37°C for different lengths of time and then assaying for the amount of enzyme activity remaining. The following results were obtained.

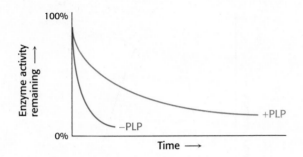

(a) Why does the amount of active enzyme decrease with the time of incubation?

(b) Why does the amount of enzyme from the PLP-deficient cells decline more rapidly?

48. *Not just for enzymes.* Kinetics is useful for studying reactions of all types, not just those catalyzed by enzymes. In Chapters 4 and 5, we learned that DNA could be reversibly melted. When melted double-stranded DNA is allowed to renature, the process can be described as consisting of two steps, a slow second-order reaction followed by a rapid first-order reaction. Explain what is occurring in each step.

Think ▶ Pair ▶ Share Problem

49. Only a few amino acid residues are actually involved in catalysis in enzymes, yet enzymes are constructed of at least 100 amino acids, and often many more. Suggest some functions for the noncatalytic amino acids.

Catalytic Strategies

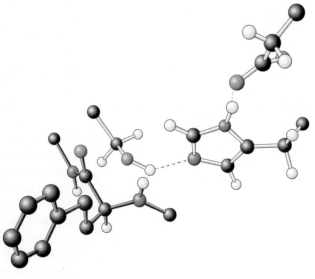

Chess and enzymes have in common the use of strategy, consciously thought out in the game of chess and selected by evolution for the action of an enzyme. The three amino acid residues at the right, denoted by the white bonds, constitute a catalytic triad found in the active site of a class of enzymes that cleave peptide bonds. The substrate, represented by the molecule with the black bonds, is as hopelessly trapped as the king in the photograph of a chess match at the left and is sure to be cleaved. [Wendie Berg.]

✓ LEARNING OBJECTIVES

By the end of this chapter, you should be able to:

1. Discuss four general strategies used by enzymes to accelerate particular reactions.

2. Give examples of specific chemical features of enzyme active sites that facilitate increasing the rates of specific reactions.

3. Give an example of when high absolute rate acceleration is physiologically important and how an enzyme achieves this acceleration.

4. Understand when high specificity is important for an enzyme and how this specificity is achieved.

5. Describe an example of when large conformational changes that occur during an enzymatic reaction cycle are used to drive other processes.

6. Discuss some of the experimental approaches that are used to elucidate enzymatic mechanisms.

OUTLINE

9.1 Proteases Facilitate a Fundamentally Difficult Reaction

9.2 Carbonic Anhydrases Make a Fast Reaction Faster

9.3 Restriction Enzymes Catalyze Highly Specific DNA-Cleavage Reactions

9.4 Myosins Harness Changes in Enzyme Conformation to Couple ATP Hydrolysis to Mechanical Work

What are the sources of the catalytic power and specificity of enzymes? This chapter presents the catalytic strategies used by four classes of enzymes: serine proteases, carbonic anhydrases, restriction endonucleases, and myosins. Each class catalyzes reactions that require the addition of water to a substrate. The mechanisms of these enzymes have been revealed through the use of incisive experimental probes, including the techniques of protein structure determination (Chapter 3) and site-directed mutagenesis (Chapter 5). The mechanisms illustrate many important principles of catalysis. We shall see how these enzymes facilitate the formation of the transition state through the use of binding energy and induced fit as well as additional specific catalytic strategies.

Each of the four classes of enzymes in this chapter illustrates the use of such strategies to solve a different problem. For serine proteases, exemplified by chymotrypsin, the challenge is to promote a reaction that is almost immeasurably slow at neutral pH in the absence of a catalyst. For carbonic anhydrases, the challenge is to achieve a high absolute rate of reaction, suitable for integration with other rapid physiological processes. For restriction endonucleases such as EcoRV, the challenge is to attain a high degree of specificity. Finally, for myosins, the challenge is to utilize the free energy associated with the hydrolysis of adenosine triphosphate (ATP) to drive other processes. Each of the examples selected is a member of a large protein class. For each of these classes, comparison between class members reveals how enzyme active sites have evolved and been refined. Structural and mechanistic comparisons of enzyme action are thus the sources of insight into the evolutionary history of enzymes. In addition, our knowledge of catalytic strategies has been used to develop practical applications, including potent drugs and specific enzyme inhibitors. Finally, although we shall not consider catalytic RNA molecules explicitly in this chapter, the principles also apply to these catalysts.

A few basic catalytic principles are used by many enzymes

In Chapter 8, we learned that enzymatic catalysis begins with substrate binding. The *binding energy* is the free energy released in the formation of a large number of weak interactions between the enzyme and the substrate. The use of this binding energy is the first common strategy used by enzymes. We can envision this binding energy as serving two purposes: it establishes substrate specificity and increases catalytic efficiency. Often only the correct substrate can participate in most or all of the interactions with the enzyme and thus maximize binding energy, accounting for the exquisite substrate specificity exhibited by many enzymes. Furthermore, the full complement of such interactions is formed only when the combination of enzyme and substrate is in the transition state. Thus, interactions between the enzyme and the substrate stabilize the transition state, thereby lowering the free energy of activation. The binding energy can also promote structural changes in both the enzyme and the substrate that facilitate catalysis, a process referred to as *induced fit*.

In addition to the first strategy involving binding energy, enzymes commonly employ one or more of the following four additional strategies to catalyze specific reactions:

1. *Covalent catalysis.* In covalent catalysis, the active site contains a reactive group, usually a powerful nucleophile, that becomes temporarily covalently attached to a part of the substrate in the course of catalysis. The proteolytic enzyme chymotrypsin provides an excellent example of this strategy (Section 9.1).

2. *General acid–base catalysis.* In general acid–base catalysis, a molecule other than water plays the role of a proton donor or acceptor. Chymotrypsin uses a histidine residue as a base catalyst to enhance the nucleophilic power of serine (Section 9.1), whereas a histidine residue in carbonic anhydrase facilitates the removal of a hydrogen ion from a zinc-bound water molecule to generate hydroxide ion (Section 9.2). For myosins, a phosphate group of the ATP substrate serves as a base to promote its own hydrolysis (Section 9.4).

3. *Catalysis by approximation.* Many reactions have two distinct substrates, including all four classes of hydrolases considered in detail in this chapter. In such cases, the reaction rate may be considerably enhanced by bringing the two substrates together along a single binding surface on an enzyme. For example, carbonic anhydrase binds carbon dioxide and water in adjacent sites to facilitate their reaction (Section 9.2).

4. *Metal ion catalysis.* Metal ions can function catalytically in several ways. For instance, a metal ion may facilitate the formation of nucleophiles such as hydroxide ion by direct coordination. A zinc(II) ion serves this purpose in catalysis by carbonic anhydrase (Section 9.2). Alternatively, a metal ion may serve as an electrophile, stabilizing a negative charge on a reaction intermediate. A magnesium(II) ion plays this role in EcoRV (Section 9.3). Finally, a metal ion may serve as a bridge between enzyme and substrate, increasing the binding energy and holding the substrate in a conformation appropriate for catalysis. This strategy is used by myosins (Section 9.4) and, indeed, by essentially all enzymes that utilize ATP as a substrate.

9.1 Proteases Facilitate a Fundamentally Difficult Reaction

Peptide bond hydrolysis is an important process in living systems (Chapter 23). Proteins that have served their purpose must be degraded so that their constituent amino acids can be recycled for the synthesis of new proteins. Proteins ingested in the diet must be broken down into small peptides and amino acids for absorption in the gut. Furthermore, as described in detail in Chapter 10, proteolytic reactions are important in regulating the activity of certain enzymes and other proteins.

Proteases cleave proteins by a hydrolysis reaction—the addition of a molecule of water to a peptide bond:

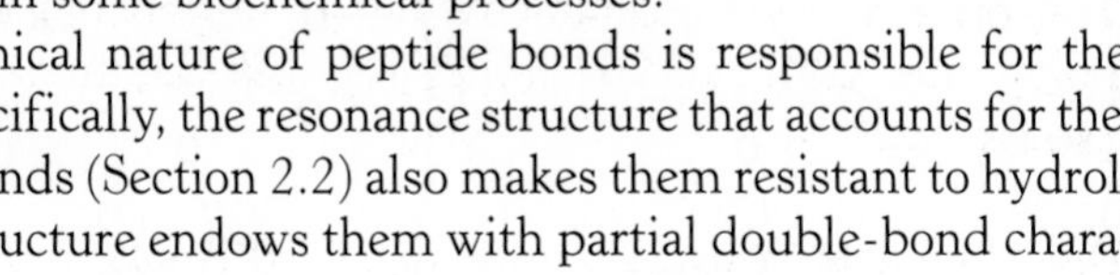

Although the hydrolysis of peptide bonds is thermodynamically favorable, such reactions are extremely slow. In the absence of a catalyst, the half-life for the hydrolysis of a typical peptide at neutral pH is estimated to be between 10 and 1000 years. Yet, peptide bonds must be hydrolyzed within milliseconds in some biochemical processes.

The chemical nature of peptide bonds is responsible for their kinetic stability. Specifically, the resonance structure that accounts for the planarity of peptide bonds (Section 2.2) also makes them resistant to hydrolysis. This resonance structure endows them with partial double-bond character:

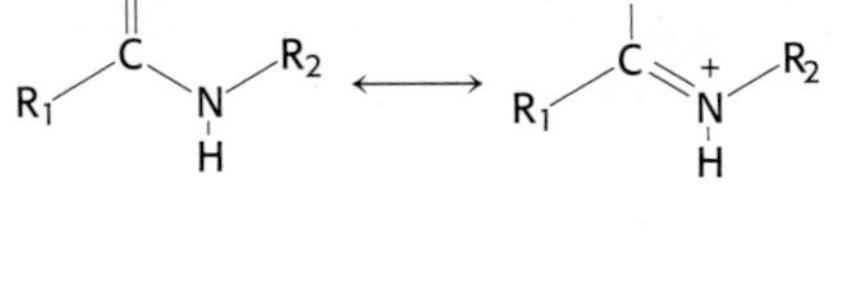

The carbon–nitrogen bond is strengthened by its double-bond character. More importantly, the carbonyl carbon atom is less electrophilic and less susceptible to nucleophilic attack than are the carbonyl carbon atoms in more reactive compounds such as carboxylate esters. Consequently, to promote peptide-bond cleavage, an enzyme must facilitate nucleophilic attack at a normally unreactive carbonyl group.

Chymotrypsin possesses a highly reactive serine residue

A number of proteolytic enzymes participate in the breakdown of proteins in the digestive systems of mammals and other organisms. One such enzyme, chymotrypsin, cleaves peptide bonds selectively on the carboxyl-terminal side of the large hydrophobic amino acids such as tryptophan, tyrosine, phenylalanine, and methionine (Figure 9.1). Chymotrypsin is a good example of the use of *covalent catalysis*. The enzyme employs a powerful nucleophile to attack the unreactive carbonyl carbon atom of the substrate. This nucleophile becomes covalently attached to the substrate briefly in the course of catalysis.

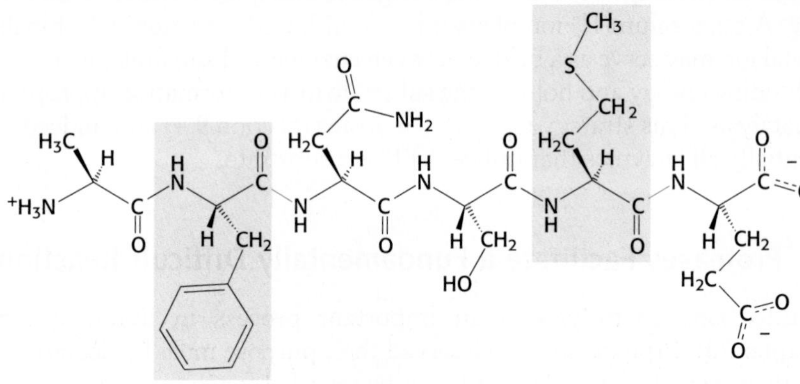

FIGURE 9.1 Specificity of chymotrypsin. Chymotrypsin cleaves proteins on the carboxyl side of aromatic or large hydrophobic amino acids (shaded orange). The bonds cleaved by chymotrypsin are indicated in red.

What is the nucleophile that chymotrypsin employs to attack the substrate carbonyl carbon atom? A clue came from the fact that chymotrypsin contains an extraordinarily reactive serine residue. Chymotrypsin molecules treated with organofluorophosphates such as diisopropylphosphofluoridate (DIPF) lost all activity irreversibly (Figure 9.2). Only a single residue, serine 195, was modified. This *chemical modification reaction* suggested that this unusually reactive serine residue plays a central role in the catalytic mechanism of chymotrypsin.

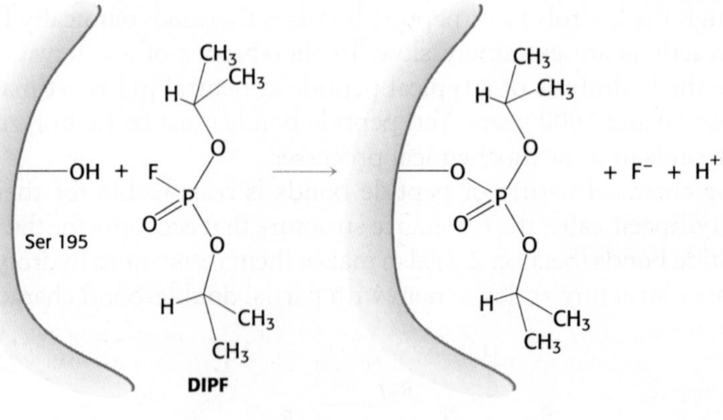

FIGURE 9.2 An unusually reactive serine residue in chymotrypsin. Chymotrypsin is inactivated by treatment with diisopropylphosphofluoridate (DIPF), which reacts only with serine 195 among 28 possible serine residues.

Chymotrypsin action proceeds in two steps linked by a covalently bound intermediate

A study of the kinetics of chymotrypsin provided a second clue to its catalytic mechanism. Enzyme kinetics are often easily monitored by having the enzyme act on a substrate analog that forms a colored product. For chymotrypsin, such a *chromogenic substrate* is *N*-acetyl-L-phenylalanine *p*-nitrophenyl ester. This substrate is an ester rather than an amide, but many proteases will also hydrolyze esters. One of the products formed by chymotrypsin's cleavage of this substrate is *p*-nitrophenolate, which has a yellow color (Figure 9.3). Measurements of the absorbance of light revealed the amount of *p*-nitrophenolate being produced.

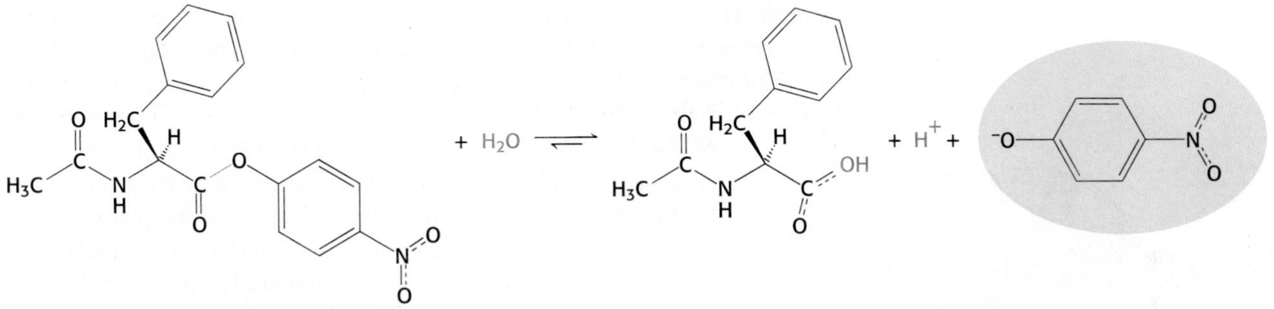

N-Acetyl-L-phenylalanine p-nitrophenyl ester

p-Nitrophenolate

FIGURE 9.3 Chromogenic substrate. *N*-Acetyl-L-phenylalanine *p*-nitrophenyl ester yields a yellow product, *p*-nitrophenolate, on cleavage by chymotrypsin. *p*-Nitrophenolate forms by deprotonation of *p*-nitrophenol at pH 7.

Under steady-state conditions, the cleavage of this substrate obeys Michaelis–Menten kinetics with a K_M of 20 μM and a k_{cat} of 77 s^{-1}. The initial phase of the reaction was examined by using the stopped-flow method, which makes it possible to mix enzyme and substrate and monitor the results within a millisecond. This method revealed an initial rapid burst of colored product, followed by its slower formation as the reaction reached the steady state (Figure 9.4). These results suggest that hydrolysis proceeds in two phases. In the first reaction cycle that takes place immediately after mixing, only the first phase must take place before the colored product is released. In subsequent reaction cycles, both phases must take place. Note that the burst is observed because the first phase is substantially more rapid than the second phase for this substrate.

The two phases are explained by the formation of a covalently bound enzyme–substrate intermediate (Figure 9.5). First, the acyl group of the substrate becomes covalently attached to the enzyme as *p*-nitrophenolate (or an amine if the substrate is an amide rather than an ester) is released. The

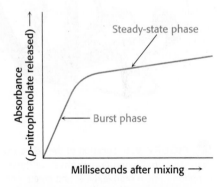

FIGURE 9.4 Kinetics of chymotrypsin catalysis. Two phases are evident in the cleaving of *N*-acetyl-L-phenylalanine *p*-nitrophenyl ester by chymotrypsin: a rapid burst phase (pre-steady-state) and a steady-state phase.

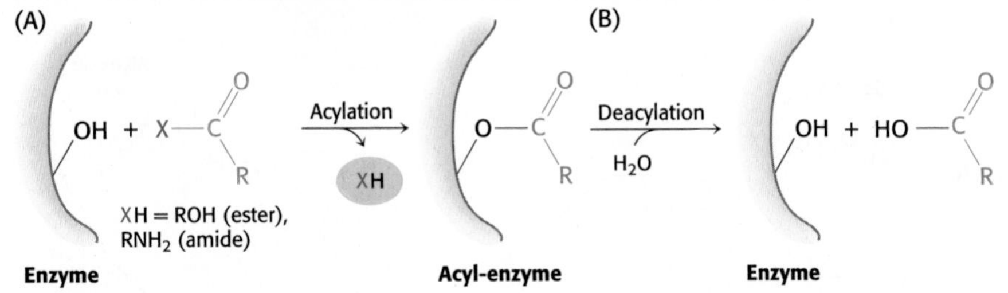

(A)

XH = ROH (ester),
RNH$_2$ (amide)

Enzyme　　　　**Acyl-enzyme**　　　　**Enzyme**

(B)

FIGURE 9.5 Covalent catalysis. Hydrolysis by chymotrypsin takes place in two phases: (A) acylation to form the acyl-enzyme intermediate followed by (B) deacylation to regenerate the free enzyme.

enzyme–acyl group complex is called the *acyl-enzyme intermediate*. Second, the acyl-enzyme intermediate is hydrolyzed to release the carboxylic acid component of the substrate and regenerate the free enzyme. Thus, one molecule of *p*-nitrophenolate is produced rapidly from each enzyme molecule as the acyl-enzyme intermediate is formed. However, it takes longer for the enzyme to be "reset" by the hydrolysis of the acyl-enzyme intermediate, and both phases are required for enzyme turnover.

Serine is part of a catalytic triad that also includes histidine and aspartate

The three-dimensional structure of chymotrypsin revealed that this enzyme is roughly spherical and comprises three polypeptide chains, linked by disulfide bonds. It is synthesized as a single polypeptide, termed *chymotrypsinogen*, which is activated by the proteolytic cleavage of the polypeptide to yield the three chains (Section 10.4). The active site of chymotrypsin, marked by serine 195, lies in a cleft on the surface of the enzyme (Figure 9.6). The structure of the active site explained the special reactivity of serine 195 (Figure 9.7). The side chain of serine 195 is hydrogen bonded to the imidazole ring of histidine 57. The—NH group of this imidazole ring is, in turn, hydrogen bonded to the carboxylate group of aspartate 102. This constellation of residues is referred to as the *catalytic triad*. How does this arrangement of residues lead to the high reactivity of serine 195? The histidine residue serves to position the serine side chain and to polarize its hydroxyl group so that it is poised for deprotonation. In the presence of the substrate, the histidine residue accepts the proton from the serine 195 hydroxyl group. In doing so, the histidine acts as a general base catalyst. The withdrawal of the proton from the hydroxyl group generates an alkoxide ion, which is a much more powerful nucleophile than is an alcohol. The aspartate residue helps orient the histidine residue and make it a better proton acceptor through hydrogen bonding and electrostatic effects.

These observations suggest a mechanism for peptide hydrolysis (Figure 9.8). After substrate binding (step 1), the reaction begins with the oxygen atom of the side chain of serine 195 making a nucleophilic attack on the carbonyl carbon atom of the target peptide bond (step 2). There are now four atoms bonded to the carbonyl carbon, arranged as a tetrahedron, instead of three atoms in a planar arrangement. This inherently unstable *tetrahedral intermediate* bears a formal negative charge on the oxygen atom derived from the carbonyl group. This

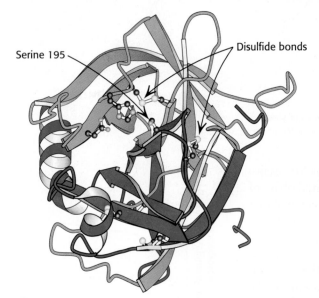

FIGURE 9.6 Location of the active site in chymotrypsin. Chymotrypsin consists of three chains, shown in ribbon form in orange, blue, and green. The side chains of the catalytic triad residues are shown as ball-and-stick representations. *Notice* these side chains, including serine 195, lining the active site in the upper half of the structure. *Also notice* two intrastrand and two interstrand disulfide bonds in various locations throughout the molecule. [Drawn from 1GCT.pdb.]

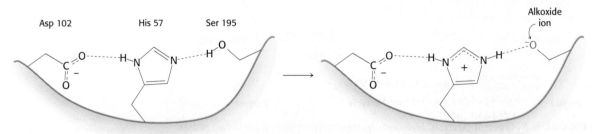

FIGURE 9.7 The catalytic triad. The catalytic triad, shown on the left, converts serine 195 into a potent nucleophile, as illustrated on the right.

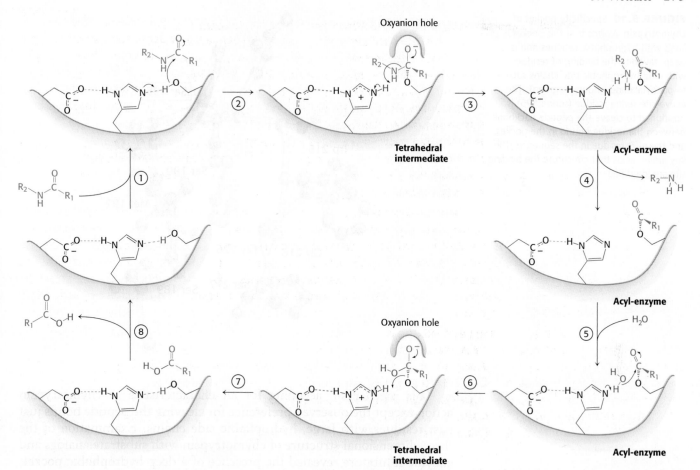

FIGURE 9.8 Peptide hydrolysis by chymotrypsin. The mechanism of peptide hydrolysis illustrates the principles of covalent and acid–base catalysis. The reaction proceeds in eight steps: (1) substrate binding, (2) nucleophilic attack of serine on the peptide carbonyl group, (3) collapse of the tetrahedral intermediate, (4) release of the amine component, (5) water binding, (6) nucleophilic attack of water on the acyl-enzyme intermediate, (7) collapse of the tetrahedral intermediate, and (8) release of the carboxylic acid component. The dashed green lines represent hydrogen bonds.

charge is stabilized by interactions with NH groups from the protein in a site termed the *oxyanion hole* (Figure 9.9). These interactions also help stabilize the transition state that precedes the formation of the tetrahedral intermediate. This tetrahedral intermediate collapses to generate the acyl-enzyme (step 3). This step is facilitated by the transfer of the proton being held by the positively charged histidine residue to the amino group formed by cleavage of the peptide bond. The amine component is now free to depart from the enzyme (step 4), completing the first stage of the hydrolytic reaction—acylation of the enzyme. Such acyl-enzyme intermediates have even been observed using x-ray crystallography by trapping them through adjustment of conditions such as the nature of the substrate, pH, or temperature.

The next stage—deacylation—begins when a water molecule takes the place occupied earlier by the amine component of the substrate (step 5). The ester group of the acyl-enzyme is now hydrolyzed by a process that essentially repeats steps 2 through 4. Now acting as a general acid catalyst, histidine 57 draws a proton away from the water molecule. The resulting OH⁻ ion attacks the carbonyl carbon atom of the acyl group, forming a tetrahedral intermediate (step 6). This structure breaks down to form the carboxylic acid product (step 7). Finally, the release of the carboxylic acid product (step 8) readies the enzyme for another round of catalysis.

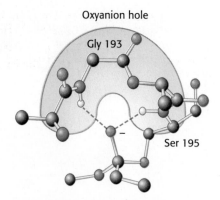

Oxyanion hole

Gly 193

Ser 195

FIGURE 9.9 The oxyanion hole. The structure stabilizes the tetrahedral intermediate of the chymotrypsin reaction. *Notice* that hydrogen bonds (shown in green) link peptide NH groups and the negatively charged oxygen atom of the intermediate.

FIGURE 9.10 Specificity pocket of chymotrypsin. *Notice* that this pocket is lined with hydrophobic residues and is deep, favoring the binding of residues with long hydrophobic side chains such as phenylalanine (shown in green). The active-site serine residue (serine 195) is positioned to cleave the peptide backbone between the residue bound in the pocket and the next residue in the sequence. The key amino acids that constitute the binding site are identified.

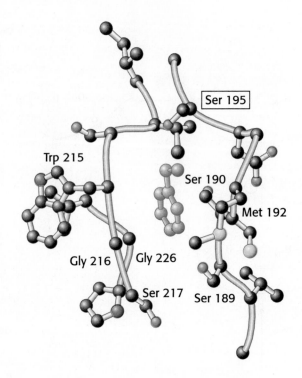

This mechanism accounts for all characteristics of chymotrypsin action except the observed preference for cleaving the peptide bonds just past residues with large, hydrophobic side chains. Examination of the three-dimensional structure of chymotrypsin with substrate analogs and enzyme inhibitors revealed the presence of a deep hydrophobic pocket, called the S_1 pocket, into which the long, uncharged side chains of residues such as phenylalanine and tryptophan can fit. *The binding of an appropriate side chain into this pocket positions the adjacent peptide bond into the active site for cleavage* (Figure 9.10). The specificity of chymotrypsin depends almost entirely on which amino acid is directly on the amino-terminal side of the peptide bond to be cleaved. Other proteases have more-complex specificity patterns. Such enzymes have additional pockets on their surfaces for the recognition of other residues in the substrate. Residues on the amino-terminal side of the scissile bond (the bond to be cleaved) are labeled P_1, P_2, P_3, and so forth, heading away from the scissile bond (Figure 9.11). Likewise, residues on the carboxyl side of the scissile bond are labeled P_1', P_2', P_3', and so forth. The corresponding sites on the enzyme are referred to as S_1, S_2 or S_1', S_2', and so forth.

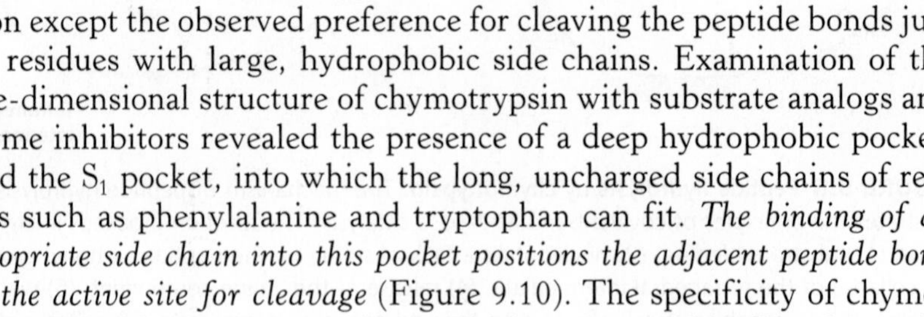

FIGURE 9.11 Specificity nomenclature for protease–substrate interactions. The potential sites of interaction of the substrate with the enzyme are designated P (shown in red), and corresponding binding sites on the enzyme are designated S. The scissile bond (also shown in red) is the reference point.

Catalytic triads are found in other hydrolytic enzymes

Many other peptide-cleaving proteins have subsequently been found to contain catalytic triads similar to that discovered in chymotrypsin. Some, such as trypsin and elastase, are obvious homologs of chymotrypsin. The sequences of these proteins are approximately 40% identical with that of chymotrypsin, and their overall structures are quite similar (Figure 9.12). These proteins operate by mechanisms identical with that of chymotrypsin. However, the three enzymes differ markedly in substrate specificity. Chymotrypsin cleaves at the peptide bond after residues with an aromatic or long, nonpolar side chain. Trypsin cleaves at the peptide bond after residues with long, positively charged side chains—namely, arginine and lysine. Elastase cleaves at the peptide bond after amino acids with small side chains—such as alanine and serine. Comparison of the S_1 pockets of these enzymes reveals that *these different specificities are due to small structural differences*. In trypsin, an aspartate residue (Asp 189) is present at the bottom of the S_1 pocket in place of a serine residue in chymotrypsin. The aspartate residue attracts and stabilizes a positively charged arginine or lysine residue in the substrate. In elastase, two residues at the top of the pocket in chymotrypsin and trypsin are replaced by much bulkier valine residues (Val 190 and Val 216). These residues close off the mouth of the pocket so that only small side chains can enter (Figure 9.13).

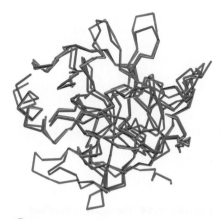

FIGURE 9.12 Structural similarity of trypsin and chymotrypsin. An overlay of the structure of chymotrypsin (red) on that of trypsin (blue) is shown. *Notice* the high degree of similarity. Only α-carbon-atom positions are shown. The mean deviation in position between corresponding α-carbon atoms is quite small, only 1.7 Å. [Drawn from 5PTP.pdb and 1GCT.pdb.]

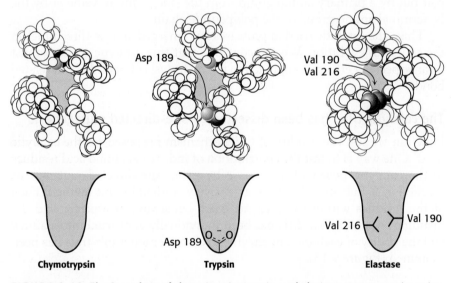

Chymotrypsin   **Trypsin**   **Elastase**

FIGURE 9.13 The S_1 pockets of chymotrypsin, trypsin, and elastase. Certain residues play key roles in determining the specificity of these enzymes. The side chains of these residues, as well as those of the active-site serine residues, are shown in color.

Other members of the chymotrypsin family include a collection of proteins that take part in blood clotting, to be discussed in Chapter 10, as well as the tumor marker protein prostate-specific antigen (PSA). In addition, a wide range of proteases found in bacteria, viruses, and plants belong to this clan.

Other enzymes that are not homologs of chymotrypsin have been found to contain very similar active sites. As noted in Chapter 6, the presence of very similar active sites in these different protein families is a consequence of convergent evolution. Subtilisin, a protease in bacteria such as *Bacillus amyloliquefaciens,* is a particularly well-characterized example. The active site of this enzyme includes both the catalytic triad and the oxyanion hole.

However, one of the NH groups that forms the oxyanion hole comes from the side chain of an asparagine residue rather than from the peptide backbone (Figure 9.14). Subtilisin is the founding member of another large family of proteases that includes representatives from archaea, bacteria, and eukarya.

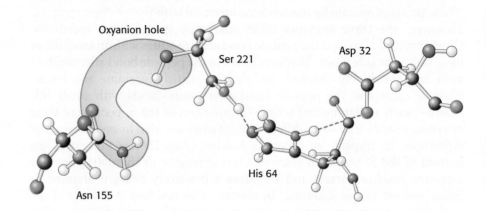

FIGURE 9.14 The catalytic triad and oxyanion hole of subtilisin. *Notice* the two enzyme NH groups (both in the backbone and in the side chain of Asn 155) located in the oxyanion hole. The NH groups will stabilize a negative charge that develops on the peptide bond attacked by nucleophilic serine 221 of the catalytic triad.

Finally, other proteases have been discovered that contain an active-site serine or threonine residue that is activated not by a histidine–aspartate pair but by a primary amino group from the side chain of lysine or by the N-terminal amino group of the polypeptide chain.

Thus, the catalytic triad in proteases has emerged at least three times in the course of evolution. We can conclude that this catalytic strategy must be an especially effective approach to the hydrolysis of peptides and related bonds.

The catalytic triad has been dissected by site-directed mutagenesis

How can we test the validity of the mechanism proposed for the catalytic triad? One way is to test the contribution of individual amino acid residues to the catalytic power of a protease by using site-directed mutagenesis (Section 5.2). Subtilisin has been extensively studied by this method. Each of the residues within the catalytic triad, consisting of aspartic acid 32, histidine 64, and serine 221, has been individually converted into alanine, and the ability of each mutant enzyme to cleave a model substrate has been examined (Figure 9.15).

FIGURE 9.15 Site-directed mutagenesis of subtilisin. Residues of the catalytic triad were mutated to alanine, and the activity of the mutated enzyme was measured. Mutations in any component of the catalytic triad cause a dramatic loss of enzyme activity. Note that the activity is displayed on a logarithmic scale. The mutations are identified as follows: the first letter is the one-letter abbreviation for the amino acid being altered; the number identifies the position of the residue in the primary structure; and the second letter is the one-letter abbreviation for the amino acid replacing the original one. Uncat. refers to the estimated rate for the uncatalyzed reaction.

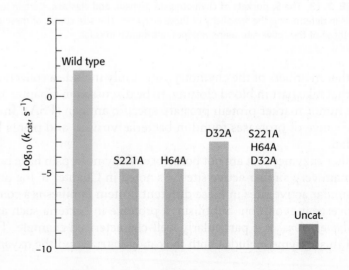

As expected, the conversion of active-site serine 221 into alanine dramatically reduced catalytic power; the value of k_{cat} fell to less than *one millionth* of its value for the wild-type enzyme. The value of K_M was essentially unchanged; its increase by no more than a factor of two indicated that substrate continued to bind normally. The mutation of histidine 64 to alanine reduced catalytic power to a similar degree. The conversion of aspartate 32 into alanine reduced catalytic power by less, although the value of k_{cat} still fell to less than 0.005% of its wild-type value. The simultaneous conversion of all three residues into alanine was no more deleterious than the conversion of serine or histidine alone. These observations support the notion that the catalytic triad and, particularly, the serine–histidine pair, act together to generate a nucleophile of sufficient power to attack the carbonyl carbon atom of a peptide bond. Despite the reduction in their catalytic power, the mutated enzymes still hydrolyze peptides a thousand times as fast as buffer at pH 8.6.

Site-directed mutagenesis also offered a way to probe the importance of the oxyanion hole for catalysis. The mutation of asparagine 155 to glycine eliminated the side-chain NH group from the oxyanion hole of subtilisin. The elimination of the NH group reduced the value of k_{cat} to 0.2% of its wild-type value but increased the value of K_M by only a factor of two. These observations demonstrate that the NH group of the asparagine residue plays a significant role in stabilizing the tetrahedral intermediate and the transition state leading to it.

Cysteine, aspartyl, and metalloproteases are other major classes of peptide-cleaving enzymes

Not all proteases utilize strategies based on activated serine residues. Classes of proteins have been discovered that employ three alternative approaches to peptide-bond hydrolysis (Figure 9.16). These classes are the (1) cysteine proteases, (2) aspartyl proteases, and (3) metalloproteases. In each case, the strategy is to generate a nucleophile that attacks the peptide carbonyl group (Figure 9.17).

The strategy used by the *cysteine proteases* is most similar to that used by the chymotrypsin family. In these enzymes, a cysteine residue, activated by a histidine residue, plays the role of the nucleophile that attacks the peptide bond (Figure 9.17) in a manner quite analogous to that of the serine residue in serine proteases. Because the sulfur atom in cysteine is inherently a better nucleophile than is the oxygen atom in serine, cysteine proteases appear to require only this histidine residue in addition to cysteine and not the full catalytic triad. A well-studied example of these proteins is papain, an enzyme purified from the fruit of the papaya. Mammalian proteases homologous to papain have been discovered, most notably the cathepsins, proteins having a role in the immune system and other systems. The cysteine-based active site arose independently at least twice in the course of evolution; the caspases, enzymes that play a major role in apoptosis (a genetically programmed cell death pathway), have active sites similar to that of papain, but their overall structures are unrelated.

The second class comprises the *aspartyl proteases*. The central feature of the active sites is a pair of aspartic acid residues that act together to allow a water molecule to attack the peptide bond. One aspartic acid residue (in its deprotonated form) activates the attacking water molecule by poising it for deprotonation. The other aspartic acid residue (in its protonated form) polarizes the peptide carbonyl group so that it is more susceptible to attack (Figure 9.17). Members of this class include renin, an enzyme involved in

FIGURE 9.16 Three classes of proteases and their active sites. These examples of a cysteine protease, an aspartyl protease, and a metalloprotease use a histidine-activated cysteine residue, an aspartate-activated water molecule, and a metal-activated water molecule, respectively, as the nucleophile. The two halves of renin are in blue and red to highlight the approximate twofold symmetry of aspartyl proteases. *Notice* how different these active sites are despite the similarity in the reactions they catalyze. [Drawn from 1PPN. pdb.; 1HRN. pdb; 1LND.pdb.]

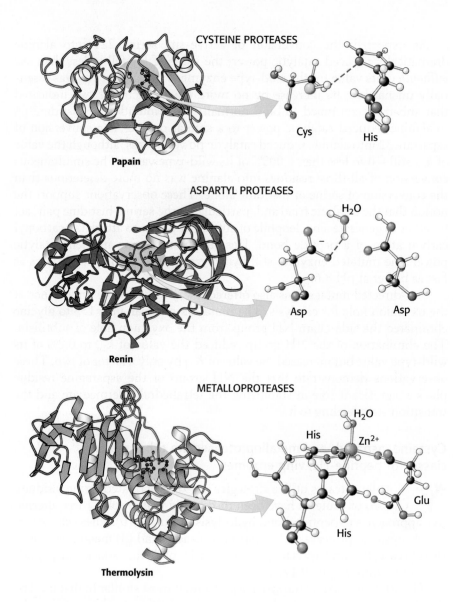

CYSTEINE PROTEASES

Papain

Cys   His

ASPARTYL PROTEASES

H_2O

Renin

Asp   Asp

METALLOPROTEASES

H_2O

His Zn^{2+}

Glu

His

Thermolysin

FIGURE 9.17 The activation strategies for three classes of proteases. The peptide carbonyl group is attacked by (A) a histidine-activated cysteine in the cysteine proteases, (B) an aspartate-activated water molecule in the aspartyl proteases, and (C) a metal-activated water molecule in the metalloproteases. For the metalloproteases, the letter B represents a base (often glutamate) that helps deprotonate the metal-bound water.

the regulation of blood pressure, and the digestive enzyme pepsin. These proteins possess approximate twofold symmetry. A likely scenario is that two copies of a gene for the ancestral enzyme fused to form a single gene that encoded a single-chain enzyme. Each copy of the gene would have contributed an aspartate residue to the active site. The individual chains are now joined to make a single chain in most aspartyl proteases, whereas the proteases present in human immunodeficiency virus (HIV) and other

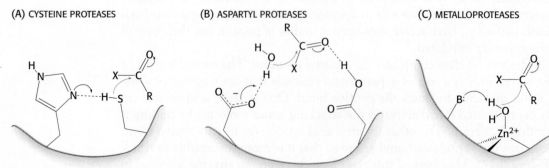

(A) CYSTEINE PROTEASES

(B) ASPARTYL PROTEASES

(C) METALLOPROTEASES

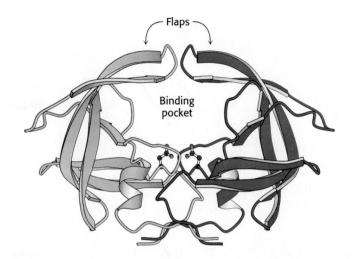

Flaps

Binding
pocket

FIGURE 9.18 HIV protease, a dimeric aspartyl protease. The protease is a dimer of identical subunits, shown in blue and yellow, consisting of 99 amino acids each. *Notice* the placement of active-site aspartic acid residues, one from each chain, which are shown as ball-and-stick structures. The flaps will close down on the binding pocket after substrate has been bound. [Drawn from 3PHV.pdb.]

retroviruses comprise dimers of identical chains (Figure 9.18). This observation is consistent with the idea that larger aspartyl proteases may have evolved by fusion of separate subunits.

The *metalloproteases* constitute the final major class of peptide-cleaving enzymes. The active site of such a protein contains a bound metal ion, almost always zinc, that activates a water molecule to act as a nucleophile to attack the peptide carbonyl group. The bacterial enzyme thermolysin and the digestive enzyme carboxypeptidase A are classic examples of the zinc proteases. Thermolysin, but not carboxypeptidase A, is a member of a large and diverse family of homologous zinc proteases that includes the matrix metalloproteases, enzymes that catalyze the reactions in tissue remodeling and degradation.

In each of these three classes of enzymes, the active site includes features that act to (1) activate a water molecule or another nucleophile, (2) polarize the peptide carbonyl group, and (3) stabilize a tetrahedral intermediate (Figure 9.17).

Protease inhibitors are important drugs

Because of the fundamental biological roles of proteases, these enzymes are important drug targets. For example, captopril, a *protease inhibitor* used to regulate blood pressure, is one of many inhibitors of the angiotensin-converting enzyme (ACE), a metalloprotease. Indinavir (Crixivan), retrovir, and a number of other compounds used in the treatment of AIDS are inhibitors of HIV protease (Figure 9.18), an aspartyl protease. HIV protease cleaves multidomain viral proteins into their active forms; blocking this process completely prevents the virus from being infectious. HIV protease inhibitors, in combination with inhibitors of other key HIV enzymes, have dramatically reduced deaths due to AIDS, assuming that the cost of the treatment could be covered (see Figure 28.21). In many cases, these drugs have converted AIDS from a death sentence to a treatable chronic disease.

Indinavir resembles the peptide substrate of the HIV protease. Indinavir is constructed around an alcohol that mimics the tetrahedral intermediate; other groups are present to bind into the S_2, S_1, S_1', and S_2' recognition sites on the enzyme (Figure 9.19). X-ray crystallographic studies revealed that, in the active site, indinavir adopts a confirmation that approximates the twofold symmetry of the enzyme (Figure 9.20). The active site of HIV protease is covered by two flexible flaps that fold down on top of the bound inhibitor. The OH group of the central alcohol interacts with the two aspartate residues of the active site. In addition, two

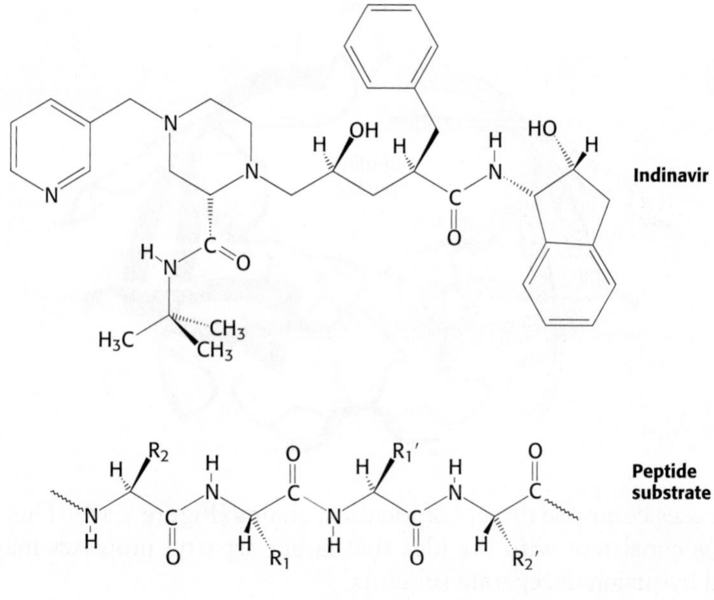

FIGURE 9.19 Indinavir, an HIV protease inhibitor. The structure of indinavir (Crixivan) is shown in comparison with that of a peptide substrate of HIV protease. The scissile bond in the substrate is highlighted in red.

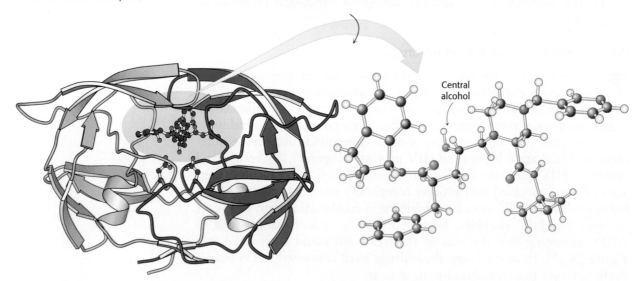

FIGURE 9.20 HIV protease–indinavir complex. (Left) The HIV protease is shown with the inhibitor indinavir bound at the active site. *Notice* the twofold symmetry of the enzyme structure. (Right) The drug has been rotated to reveal its approximately twofold symmetric conformation. [Drawn from 1HSH.pdb.]

carbonyl groups of the inhibitor are hydrogen bonded to a water molecule (not shown in Figure 9.20), which, in turn, is hydrogen bonded to a peptide NH group in each of the flaps. This interaction of the inhibitor with water and the enzyme is not possible within cellular aspartyl proteases such as renin. Thus, the interaction may contribute to the specificity of indinavir for HIV protease. To prevent side effects, protease inhibitors used as drugs must be specific for one enzyme without inhibiting other proteins within the body.

9.2 Carbonic Anhydrases Make a Fast Reaction Faster

Carbon dioxide is a major end product of aerobic metabolism. In mammals, this carbon dioxide is released into the blood and transported to the lungs for exhalation. While in the red blood cells, carbon dioxide reacts with water (Section 7.3). The product of this reaction is a moderately strong acid, carbonic acid ($pK_a = 3.5$), which is converted into bicarbonate ion (HCO_3^-) on the loss of a proton.

$$\underset{O}{\overset{O}{\underset{\|}{\overset{\|}{C}}}} + H_2O \underset{k_{-1}}{\overset{k_1}{\rightleftharpoons}} \underset{\underset{\text{Carbonic}}{\text{acid}}}{HO\overset{\overset{O}{\overset{\|}{C}}}{}OH} \rightleftharpoons \underset{\underset{\text{Bicarbonate}}{\text{ion}}}{HO\overset{\overset{O}{\overset{\|}{C}}}{}\overset{-}{O}} + H^+$$

Even in the absence of a catalyst, this hydration reaction proceeds at a moderately fast pace. At 37°C near neutral pH, the second-order rate constant k_1 is $0.0027 \, M^{-1} \, s^{-1}$. This value corresponds to an effective first-order rate constant of $0.15 \, s^{-1}$ in water ($[H_2O] = 55.5 \, M$). The reverse reaction, the dehydration of HCO_3^-, is even more rapid, with a rate constant of $k_{-1} = 50 \, s^{-1}$. These rate constants correspond to an equilibrium constant of $K_1 = 5.4 \times 10^{-5}$ and a ratio of $[CO_2]$ to $[H_2CO_3]$ of 340:1 at equilibrium.

Carbon dioxide hydration and HCO_3^- dehydration are often coupled to rapid processes, particularly transport processes. Thus, almost all organisms contain enzymes, referred to as *carbonic anhydrases,* that increase the rate of reaction beyond the already reasonable spontaneous rate. For example, carbonic anhydrases dehydrate HCO_3^- in the blood to form CO_2 for exhalation as the blood passes through the lungs. Conversely, they convert CO_2 into HCO_3^- to generate the aqueous humor of the eye and other secretions. Furthermore, both CO_2 and HCO_3^- are substrates and products for a variety of enzymes, and the rapid interconversion of these species may be necessary to ensure appropriate substrate levels. Mutations in some carbonic anhydrases have been found to be associated with osteopetrosis (excessive formation of dense bones accompanied by anemia), defects in kidney function, and mental retardation.

Carbonic anhydrases accelerate CO_2 hydration dramatically. The most-active enzymes hydrate CO_2 at rates as high as $k_{cat} = 10^6 \, s^{-1}$, or a million times a second per enzyme molecule. Fundamental physical processes such as diffusion and proton transfer ordinarily limit the rate of hydration, and so the enzymes employ special strategies to attain such prodigious rates.

Carbonic anhydrase contains a bound zinc ion essential for catalytic activity

Less than 10 years after the discovery of carbonic anhydrase in 1932, this enzyme was found to contain a bound zinc ion. Moreover, the zinc ion appeared to be necessary for catalytic activity. This discovery, remarkable at the time, made carbonic anhydrase the first known zinc-containing enzyme. At present, hundreds of enzymes are known to contain zinc. In fact, more than one-third of all enzymes either contain bound metal ions or require the addition of such ions for activity. Metal ions have several properties that increase chemical reactivity: their positive charges, their ability to form strong yet kinetically labile bonds, and, in some cases, their capacity to be stable in more than one oxidation state. The chemical reactivity of metal ions explains why catalytic strategies that employ metal ions have been adopted throughout evolution.

X-ray crystallographic studies have supplied the most-detailed and direct information about the zinc site in carbonic anhydrase. At least seven carbonic anhydrases, each with its own gene, are present in human beings. They are all clearly homologous, as revealed by substantial sequence identity. Carbonic anhydrase II, a major protein component of red blood cells, has been the most extensively studied (Figure 9.21). It is also one of the most active carbonic anhydrases.

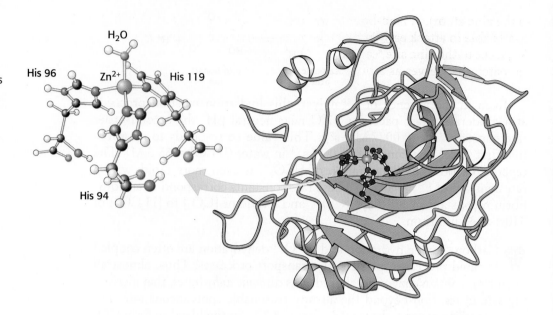

FIGURE 9.21 The structure of human carbonic anhydrase II and its zinc site. (Left) *Notice* that the zinc ion is bound to the imidazole rings of three histidine residues as well as to a water molecule. (Right) *Notice* the location of the zinc site in a cleft near the center of the enzyme. [Drawn from 1CA2.pdb.]

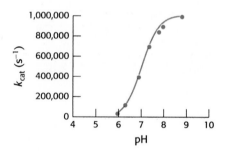

FIGURE 9.22 Effect of pH on carbonic anhydrase activity. Changes in pH alter the rate of carbon dioxide hydration catalyzed by carbonic anhydrase II. The enzyme is maximally active at high pH.

Zinc is found only in the $+2$ state in biological systems. A zinc atom is essentially always bound to four or more ligands; in carbonic anhydrase, three coordination sites are occupied by the imidazole rings of three histidine residues and an additional coordination site is occupied by a water molecule (or hydroxide ion, depending on pH). Because the molecules occupying the coordination sites are neutral, the overall charge on the $Zn(His)_3$ unit remains $+2$.

Catalysis entails zinc activation of a water molecule

How does this zinc complex facilitate carbon dioxide hydration? A major clue comes from the pH profile of enzymatically catalyzed carbon dioxide hydration (Figure 9.22).

At pH 8, the reaction proceeds near its maximal rate. As the pH decreases, the rate of the reaction drops. The midpoint of this transition is near pH 7, suggesting that a group that loses a proton at pH 7 ($pK_a = 7$) plays an important role in the activity of carbonic anhydrase. Moreover, the curve suggests that the deprotonated (high pH) form of this group participates more effectively in catalysis. Although some amino acids, notably histidine, have pK_a values near 7, *a variety of evidence suggests that the group responsible for this transition is not an amino acid but is the zinc-bound water molecule.*

The binding of a water molecule to the positively charged zinc center reduces the pK_a of the water molecule from 15.7 to 7 (Figure 9.23). With the pK_a lowered, the water molecule can more easily lose a proton at neutral pH, generating a substantial concentration of hydroxide ion (bound

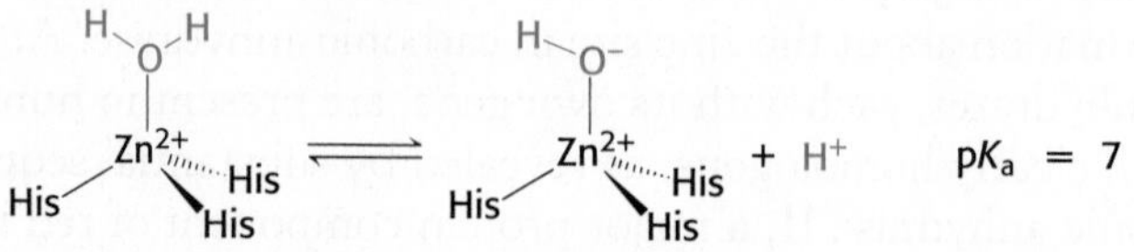

FIGURE 9.23 The pK_a of zinc-bound water. Binding to zinc lowers the pK_a of water from 15.7 to 7.

to the zinc atom). A zinc-bound hydroxide ion (OH^-) is a potent nucleophile able to attack carbon dioxide much more readily than water does. Adjacent to the zinc site, carbonic anhydrase also possesses a hydrophobic patch that serves as a binding site for carbon dioxide (Figure 9.24).

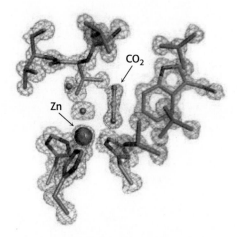

FIGURE 9.24 Carbon dioxide binding site. Crystals of carbonic anhydrase were exposed to carbon dioxide gas at high pressure and low temperature, and x-ray diffraction data were collected. The electron density for carbon dioxide, clearly visible adjacent to the zinc and its bound water, reveals the carbon dioxide binding site. Hydrophobic amino acids are shaded green, and the magenta ones are hydrophilic. [Information from J. F. Domsic et al., *J. Biol. Chem.* 283:30766–30771, 2008.]

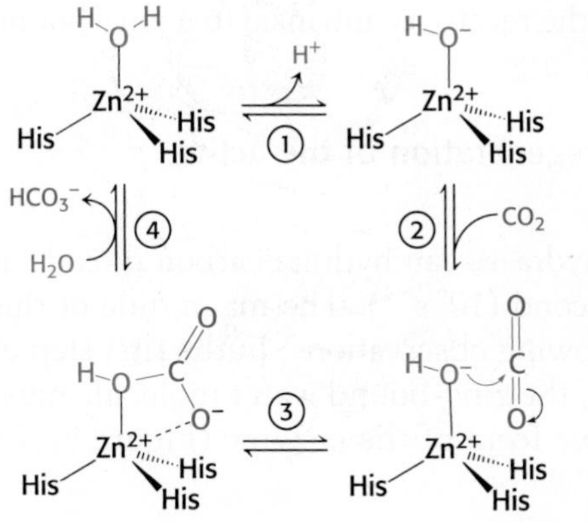

FIGURE 9.25 Mechanism of carbonic anhydrase. The zinc-bound hydroxide mechanism for the hydration of carbon dioxide reveals one aspect of metal ion catalysis. The reaction proceeds in four steps: (1) water deprotonation, (2) carbon dioxide binding, (3) nucleophilic attack by hydroxide on carbon dioxide, and (4) displacement of bicarbonate ion by water.

Based on these observations, a simple mechanism for carbon dioxide hydration can be proposed (Figure 9.25):

1. The zinc ion facilitates the release of a proton from a water molecule, which generates a hydroxide ion.

2. The carbon dioxide substrate binds to the enzyme's active site and is positioned to react with the hydroxide ion.

3. The hydroxide ion attacks the carbon dioxide, converting it into bicarbonate ion, HCO_3^-.

4. The catalytic site is regenerated with the release of HCO_3^- and the binding of another molecule of water.

Thus, the binding of a water molecule to the zinc ion favors the formation of the transition state by facilitating proton release and by positioning the water molecule to be in close proximity to the other reactant.

Studies of a *synthetic analog model system* provide evidence for the mechanism's plausibility. A simple synthetic ligand binds zinc through four nitrogen atoms (compared with three histidine nitrogen atoms in the enzyme), as shown in Figure 9.26. One water molecule remains bound to

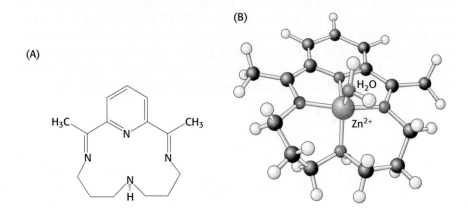

FIGURE 9.26 A synthetic analog model system for carbonic anhydrase. (A) An organic compound, capable of binding zinc, was synthesized as a model for carbonic anhydrase. The zinc complex of this ligand accelerates the hydration of carbon dioxide more than 100-fold under appropriate conditions. (B) The structure of the presumed active complex showing zinc bound to the ligand and to one water molecule.

the zinc ion in the complex. Direct measurements reveal that this water molecule has a pK_a value of 8.7, not as low as the value for the water molecule in carbonic anhydrase but substantially lower than the value for free water. At pH 9.2, this complex accelerates the hydration of carbon dioxide more than 100-fold. Although its rate of catalysis is much less efficient than catalysis by carbonic anhydrase, the model system strongly suggests that the zinc-bound hydroxide mechanism is likely to be correct. Carbonic anhydrases have evolved to employ the reactivity intrinsic to a zinc-bound hydroxide ion as a potent catalyst.

A proton shuttle facilitates rapid regeneration of the active form of the enzyme

As noted earlier, some carbonic anhydrases can hydrate carbon dioxide at rates as high as a million times a second ($10^6 \, s^{-1}$). The magnitude of this rate can be understood from the following observations. In the first step of a carbon dioxide hydration reaction, the zinc-bound water molecule must lose a proton to regenerate the active form of the enzyme (Figure 9.27).

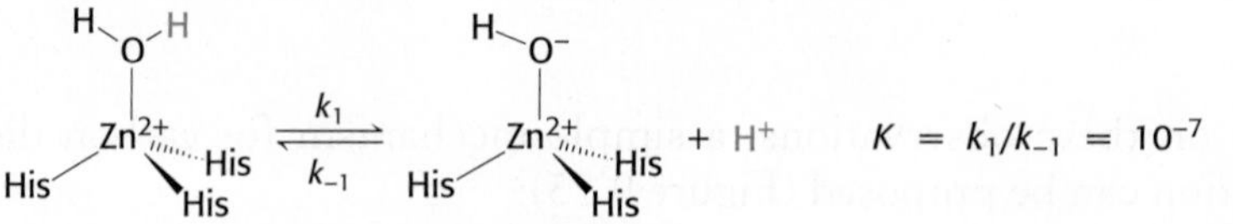

FIGURE 9.27 **Kinetics of water deprotonation.** The kinetics of deprotonation and protonation of the zinc-bound water molecule in carbonic anhydrase.

The rate of the reverse reaction, the protonation of the zinc-bound hydroxide ion, is limited by the rate of proton diffusion. Protons diffuse very rapidly with second-order rate constants near $10^{11} \, M^{-1} \, s^{-1}$. Thus, the backward rate constant k_{-1} must be less than $10^{11} \, M^{-1} \, s^{-1}$. Because the equilibrium constant K is equal to k_1/k_{-1}, the forward rate constant is given by $k_1 = K \times k_{-1}$. Thus, if $k_{-1} \leq 10^{11} \, M^{-1} \, s^{-1}$ and $K = 10^{-7} \, M$ (because $pK_a = 7$), then k_1 must be less than or equal to $10^4 \, s^{-1}$. In other words, the rate of proton diffusion limits the rate of proton release to less than $10^4 \, s^{-1}$ for a group with $pK_a = 7$. However, if carbon dioxide is hydrated at a rate of $10^6 \, s^{-1}$, then every step in the mechanism (Figure 9.25) must take place at least this fast. How is this apparent paradox resolved?

The answer became clear with the realization that *the highest rates of carbon dioxide hydration require the presence of buffer, suggesting that the buffer components participate in the reaction.* The buffer can bind or release protons. The advantage is that, whereas the concentrations of protons and hydroxide ions are limited to $10^{-7} \, M$ at neutral pH, the concentration of buffer components can be much higher, of the order of several millimolar. If the buffer component BH^+ has a pK_a of 7 (matching that for the zinc-bound water molecule), then the equilibrium constant for the reaction in Figure 9.28 is 1. The rate of proton abstraction is given by $k_1' \times [B]$. The second-order rate constants k_1' and k_{-1}' will be limited by buffer diffusion to values less than approximately $10^9 \, M^{-1} \, s^{-1}$.

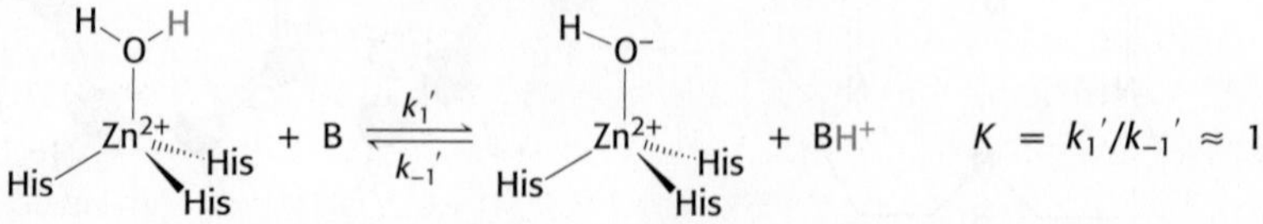

FIGURE 9.28 **The effect of buffer on deprotonation.** The deprotonation of the zinc-bound water molecule in carbonic anhydrase is aided by buffer component B.

Thus, buffer concentrations greater than $[B] = 10^{-3}\,M$ (or 1 mM) may be high enough to support carbon dioxide hydration rates of $10^6\,M^{-1}\,s^{-1}$ because $k_1' \times [B] = (10^9\,M^{-1}\,s^{-1}) \times (10^{-3}\,M) = 10^6\,s^{-1}$. The prediction that the rate increases with increasing buffer concentration has been confirmed experimentally (Figure 9.29).

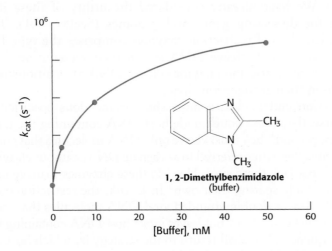

1, 2-Dimethylbenzimidazole
(buffer)

FIGURE 9.29 The effect of buffer concentration on the rate of carbon dioxide hydration. The rate of carbon dioxide hydration increases with the concentration of the buffer 1,2-dimethylbenzimidazole. The buffer enables the enzyme to achieve its high catalytic rates.

The molecular components of many buffers are too large to reach the active site of carbonic anhydrase. Carbonic anhydrase II has evolved a *proton shuttle* to allow buffer components to participate in the reaction from solution. The primary component of this shuttle is histidine 64. This residue transfers protons from the zinc-bound water molecule to the protein surface and then to the buffer (Figure 9.30). Thus, catalytic function has

FIGURE 9.30 Histidine proton shuttle. (1) Histidine 64 abstracts a proton from the zinc-bound water molecule, generating a nucleophilic hydroxide ion and a protonated histidine. (2) The buffer (B) removes a proton from the histidine, regenerating the unprotonated form.

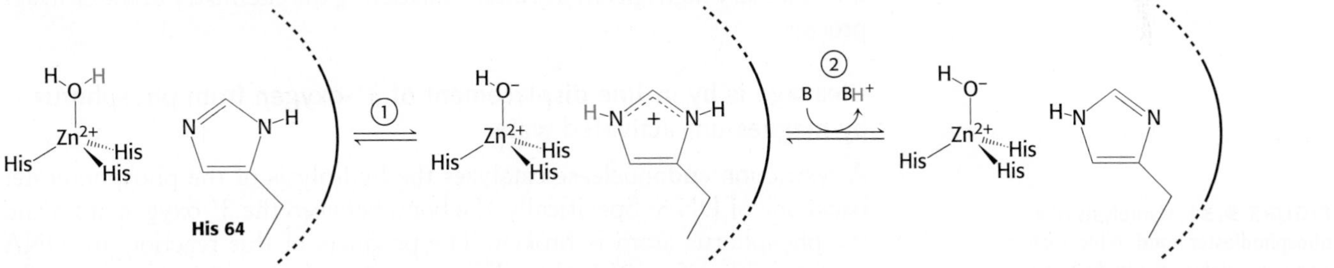

His 64

been enhanced through the evolution of an apparatus for controlling proton transfer from and to the active site. Because protons participate in many biochemical reactions, the manipulation of the proton inventory within active sites is crucial to the function of many enzymes and explains the prominence of acid–base catalysis.

9.3 Restriction Enzymes Catalyze Highly Specific DNA-Cleavage Reactions

We next consider a hydrolytic reaction that results in the cleavage of DNA. Bacteria and archaea have evolved mechanisms to protect themselves from viral infections. Many viruses inject their DNA genomes into cells; once inside, the viral DNA hijacks the cell's machinery to drive the production of viral proteins and, eventually, of progeny virus. Often, a viral infection results in the death of the host cell. A major protective strategy for the host is to use *restriction endonucleases* (restriction enzymes) to degrade the viral DNA on its introduction into a cell. These enzymes recognize particular base sequences, called *recognition sequences* or *recognition sites,* in their target DNA and cleave that DNA at defined positions. We have already considered the utility of these important enzymes for dissecting genes and genomes (Section 5.1). The most well-studied class of restriction enzymes comprises the type II restriction enzymes, which cleave DNA *within* their recognition sequences. Other types of restriction enzymes cleave DNA at positions somewhat distant from their recognition sites.

Restriction endonucleases must show tremendous specificity at two levels. First, they must not degrade host DNA containing the recognition sequences. Second, they must cleave only DNA molecules that contain recognition sites (hereafter referred to as *cognate DNA*) without cleaving DNA molecules that lack these sites. How do these enzymes manage to degrade viral DNA while sparing their own? In *E. coli,* the restriction endonuclease EcoRV cleaves double-stranded viral DNA molecules that contain the sequence 5'-GATATC-3' but leaves intact host DNA containing hundreds of such sequences. We shall return to the strategy by which host cells protect their own DNA at the end of this section.

Restriction enzymes must cleave DNA only at recognition sites, without cleaving at other sites. Suppose that a recognition sequence is six base pairs long. Because there are 4^6, or 4096, sequences having six base pairs, the concentration of sites that must not be cleaved will be approximately 4000-fold higher than the concentration of sites that should be cleaved. Thus, to keep from damaging host-cell DNA, restriction enzymes must cleave cognate DNA molecules much more than 4000 times as efficiently as they cleave nonspecific sites. We shall return to the mechanism used to achieve the necessary high specificity after considering the chemistry of the cleavage process.

Cleavage is by in-line displacement of 3'-oxygen from phosphorus by magnesium-activated water

A restriction endonuclease catalyzes the hydrolysis of the phosphodiester backbone of DNA. Specifically, the bond between the 3'-oxygen atom and the phosphorus atom is broken. The products of this reaction are DNA strands with a free 3'-hydroxyl group and a 5'-phosphoryl group at the cleavage site (Figure 9.31). This reaction proceeds by nucleophilic attack at

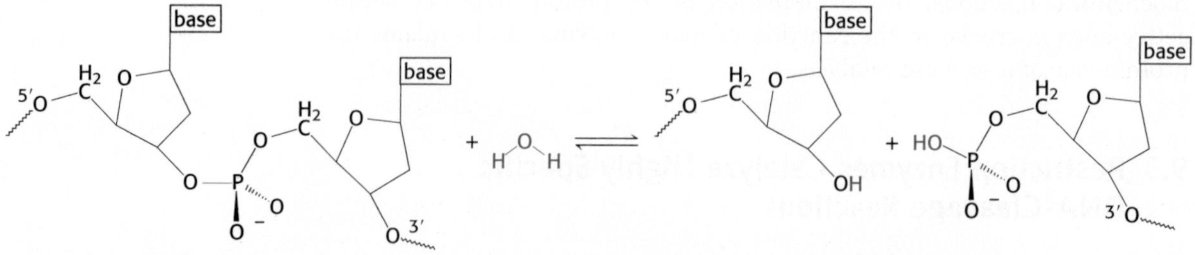

FIGURE 9.31 Hydrolysis of a phosphodiester bond. All restriction enzymes catalyze the hydrolysis of DNA phosphodiester bonds, leaving a phosphoryl group attached to the 5' end. The bond that is cleaved is shown in red.

the phosphorus atom. We will consider two alternative mechanisms, suggested by analogy with the proteases. The restriction endonuclease might cleave DNA by mechanism 1 through a covalent intermediate, employing a potent nucleophile (Nu), or by mechanism 2 through direct hydrolysis:

Mechanism 1 (covalent intermediate)

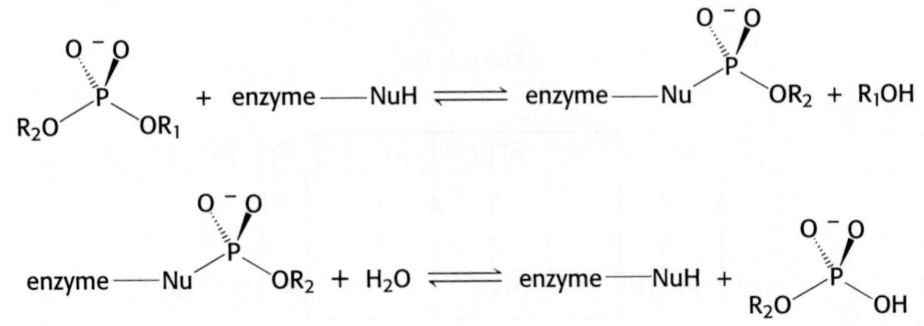

Mechanism 2 (direct hydrolysis)

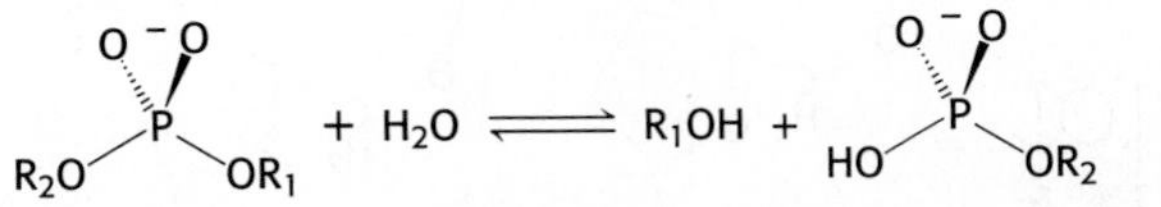

Each mechanism postulates a different nucleophile to attack the phosphorus atom. In either case, each reaction takes place by *in-line displacement:*

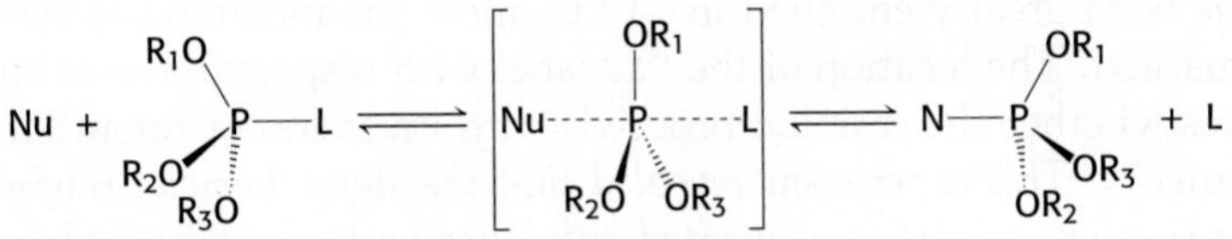

The incoming nucleophile attacks the phosphorus atom, and a pentacoordinate transition state is formed. This species has a trigonal bipyramidal geometry centered at the phosphorus atom, with the incoming nucleophile at one apex of the two pyramids and the group that is displaced (the leaving group, L) at the other apex. Note that the displacement inverts the stereochemical conformation at the tetrahedral phosphorous atom, analogous to the interconversion of the R and S configurations around a tetrahedral carbon center (Section 2.1).

The two mechanisms differ in the number of times that the displacement takes place in the course of the reaction. In the first type of mechanism, a nucleophile in the enzyme (analogous to serine 195 in chymotrypsin) attacks the phosphate group to form a covalent intermediate. In a second step, this intermediate is hydrolyzed to produce the final products. In this case, two displacement reactions take place at the phosphorus atom. Consequently, the stereochemical configuration at the phosphorus atom would be inverted and then inverted again, and the overall configuration would be *retained.* In the second type of mechanism, analogous to that used by the aspartyl-and metalloproteases, an activated water molecule attacks the phosphorus atom directly. In this mechanism, a single displacement reaction takes place at the phosphorus atom. Hence, the stereochemical configuration at the phosphorus atom is *inverted* after cleavage. To determine which mechanism is correct, we examine the stereochemistry at the phosphorus atom after cleavage.

A difficulty is that the stereochemistry is not easily observed, because two of the groups bound to the phosphorus atom are simple oxygen atoms, identical with each other. This difficulty can be circumvented by replacing one oxygen atom with sulfur (producing a species called a phosphorothioate). Let us consider EcoRV endonuclease. This enzyme cleaves the phosphodiester bond between the T and the A at the center of the recognition sequence 5'-GATATC-3'. The first step is to synthesize an appropriate substrate for EcoRV containing phosphorothioates at the sites of cleavage (Figure 9.32). The reaction is then performed in water

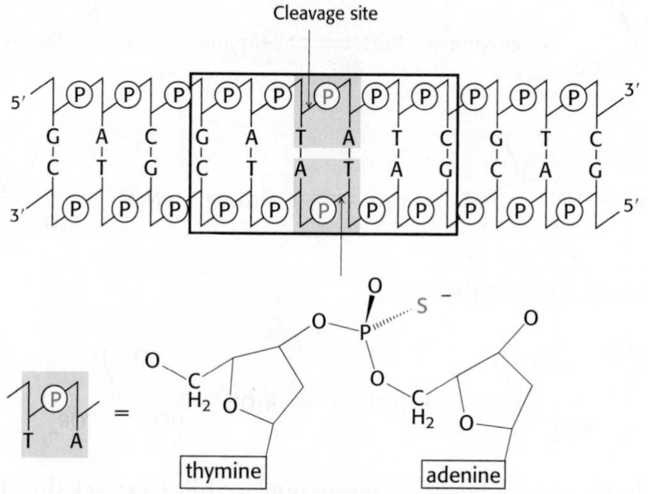

FIGURE 9.32 Labeling with phosphorothioates. Phosphorothioate groups, in which one of the nonbridging oxygen atoms is replaced by a sulfur atom, can be used to label specific sites in the DNA backbone to determine the overall stereochemical course of a displacement reaction. Here, a phosphorothioate is placed at sites that can be cleaved by EcoRV endonuclease.

that has been greatly enriched in ^{18}O to allow the incoming oxygen atom to be marked. The location of the ^{18}O label with respect to the sulfur atom indicates whether the reaction proceeds with inversion or retention of stereochemistry. *This experiment revealed that the stereochemical configuration at the phosphorus atom was inverted only once with cleavage.* This result is consistent with a direct attack by water at the phosphorus atom and rules out the formation of any covalently bound intermediate (Figure 9.33).

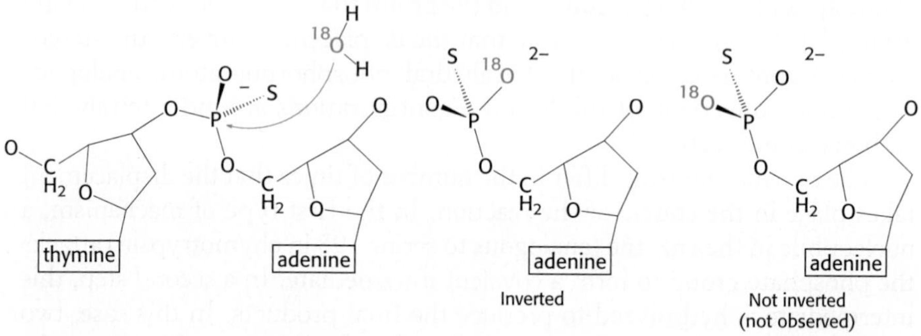

FIGURE 9.33 Stereochemistry of cleaved DNA. Cleavage of DNA by EcoRV endonuclease results in overall inversion of the stereochemical configuration at the phosphorus atom, as indicated by the stereochemistry of the phosphorus atom bound to one bridging oxygen atom, one ^{16}O, one ^{18}O, and one sulfur atom. Two possible products are shown, only one of which is observed, indicating direct attack of water at the phosphorous atom.

Restriction enzymes require magnesium for catalytic activity

Many enzymes that act on phosphate-containing substrates require Mg^{2+} or some other similar divalent cation for activity. One or more Mg^{2+} (or similar) cations are essential to the function of restriction endonucleases. What are the functions of these metal ions?

Direct visualization of the complex between EcoRV endonuclease and cognate DNA molecules in the presence of Mg^{2+} by crystallization has not been possible, because the enzyme cleaves the substrate under these circumstances. Nonetheless, metal ion complexes can be visualized through several approaches. In one approach, crystals of EcoRV endonuclease are prepared bound to oligonucleotides that contain the enzyme's recognition sequence. These crystals are grown in the absence of magnesium to prevent cleavage; after their preparation, the crystals are soaked in solutions containing the metal. Alternatively, crystals have been grown with the use of a mutated form of the enzyme that is less active. Finally, Mg^{2+} can be replaced by metal ions such as Ca^{2+} that bind but do not result in much catalytic activity. In all cases, no cleavage takes place, and so the locations of the metal ion-binding sites are readily determined.

As many as three metal ions have been found to be present per active site. One ion-binding site is occupied in essentially all structures. This metal ion is coordinated to the protein through two aspartate residues and to one of the phosphate-group oxygen atoms near the site of cleavage. This metal ion binds the water molecule that attacks the phosphorus atom, helping to position and activate it in a manner similar to that for the Zn^{2+} ion of carbonic anhydrase (Figure 9.34).

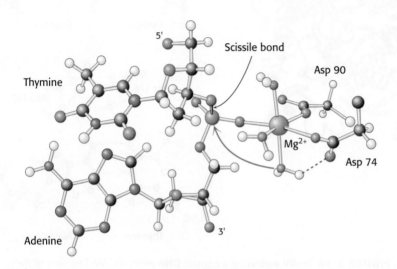

FIGURE 9.34 A magnesium ion-binding site in EcoRV endonuclease. The magnesium ion helps to activate a water molecule and positions it so that it can attack the phosphorus atom.

The complete catalytic apparatus is assembled only within complexes of cognate DNA molecules, ensuring specificity

We now return to the question of specificity, the defining feature of restriction enzymes. The recognition sequences for most restriction endonucleases are *inverted repeats*. This arrangement gives the three-dimensional structure of the recognition site a *twofold rotational symmetry* (Figure 9.35).

The restriction enzymes display a corresponding symmetry: they are dimers whose two subunits are related by twofold rotational symmetry. The matching symmetry of the recognition sequence and the enzyme facilitates

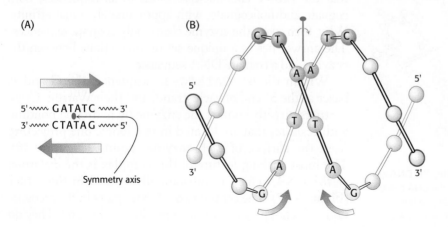

FIGURE 9.35 Structure of the recognition site of EcoRV endonuclease. (A) The sequence of the recognition site, which is symmetric around the axis of rotation designated in green. (B) The inverted repeat within the recognition sequence of EcoRV (and most other restriction endonucleases) endows the DNA site with twofold rotational symmetry.

the recognition of cognate DNA by the enzyme. This similarity in structure has been confirmed by the determination of the structure of the complex between EcoRV endonuclease and DNA fragments containing its recognition sequence (Figure 9.36). The enzyme surrounds the DNA in a tight embrace.

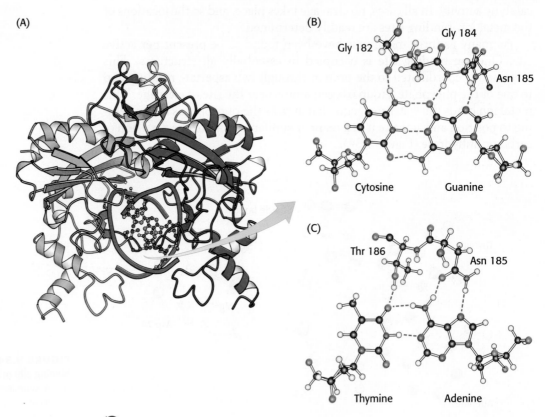

FIGURE 9.36 EcoRV embracing a cognate DNA molecule. (A) This view of the structure of EcoRV endonuclease bound to a cognate DNA fragment is down the helical axis of the DNA. The two protein subunits are in yellow and blue, and the DNA backbone is in red. *Notice* that the twofold axes of the enzyme dimer and the DNA are aligned. One of the DNA-binding loops (in green) of EcoRV endonuclease is shown interacting with the base pairs of its cognate DNA-binding site. Key amino acid residues are shown hydrogen bonding with (B) a CG base pair and (C) an AT base pair. [Drawn from 1RVB.pdb.]

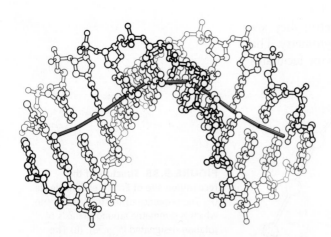

FIGURE 9.37 Distortion of the recognition site. The DNA is represented as a ball-and-stick model. The path of the DNA helical axis, shown in red, is substantially distorted on binding to the enzyme. For the B form of DNA, the axis is straight (not shown).

An enzyme's binding affinity for substrates often determines specificity. Surprisingly, however, binding studies performed in the absence of magnesium have demonstrated that the EcoRV endonuclease binds to all sequences, both cognate and noncognate, with approximately equal affinity. Why, then, does the enzyme cleave only cognate sequences? The answer lies in a unique set of interactions between the enzyme and a cognate DNA sequence.

Within the 5'-GATATC-3' sequence, the G and A bases at the 5' end of each strand, and their Watson–Crick partners directly contact the enzyme by hydrogen bonding with residues that are located in two loops, one projecting from the surface of each enzyme subunit (Figure 9.36). The most striking feature of this complex is the *distortion of the DNA*, which is substantially kinked in the center (Figure 9.37). The central two TA base pairs in the recognition sequence play a key role in producing the kink. They do not make contact with the enzyme but appear to be required because of their ease of distortion. The 5'-TA-3' sequence is known to be among the most easily deformed base pair steps.

The structures of complexes formed with noncognate DNA fragments are strikingly different from those formed with cognate DNA; the noncognate DNA conformation is not substantially distorted (Figure 9.38). *This lack of distortion has important consequences with regard to catalysis. No phosphate is positioned sufficiently close to the active-site aspartate residues to complete a magnesium ion-binding site* (Figure 9.34). Hence, the nonspecific complexes do not bind the magnesium ions and the complete catalytic apparatus is never assembled. The distortion of the substrate and the subsequent binding of the magnesium ion account for the catalytic specificity of more than a million-fold that is observed for EcoRV endonuclease. *Thus, enzyme specificity may be determined by the specificity of enzyme action rather than the specificity of substrate binding.*

We can now see the role of binding energy in this strategy for attaining catalytic specificity. The distorted DNA makes additional contacts with the enzyme, increasing the binding energy. However, the increase in binding energy is canceled by the energetic cost of distorting the DNA from its relaxed conformation (Figure 9.39). Thus, for EcoRV endonuclease, there is little difference in binding affinity for cognate and nonspecific DNA fragments. However, the distortion in the cognate complex dramatically affects catalysis by completing the magnesium ion-binding site. This example illustrates how enzymes can utilize available binding energy to deform substrates and poise them for chemical transformation. Interactions that take place within the distorted substrate complex stabilize the transition state leading to DNA hydrolysis.

Host-cell DNA is protected by the addition of methyl groups to specific bases

How does a host cell harboring a restriction enzyme protect its own DNA? The host DNA is methylated on specific adenine bases within host recognition sequences by other enzymes called *methylases* (Figure 9.40). An endonuclease will not cleave DNA if its recognition sequence is methylated. For each restriction endonuclease, the host cell produces a corresponding methylase that marks the host DNA at the appropriate methylation site. These pairs of enzymes are referred to as *restriction-modification systems*.

The distortion in the DNA explains how methylation blocks catalysis and protects host-cell DNA. The host *E. coli* adds a methyl group to the amino group of the

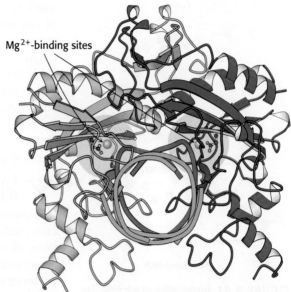

FIGURE 9.38 Nonspecific and cognate DNA within EcoRV endonuclease. A comparison of the positions of the nonspecific (orange) and the cognate DNA (red) within EcoRV. *Notice* that, in the nonspecific complex, the DNA backbone is too far from the enzyme to complete the magnesium ion-binding sites. [Drawn from 1RVB.pdb.]

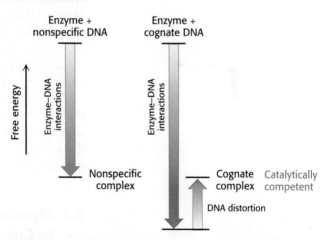

FIGURE 9.39 Greater binding energy of EcoRV endonuclease bound to cognate versus noncognate DNA. The additional interactions between EcoRV endonuclease and cognate DNA increase the binding energy, which can be used to drive DNA distortions necessary for forming a catalytically competent complex.

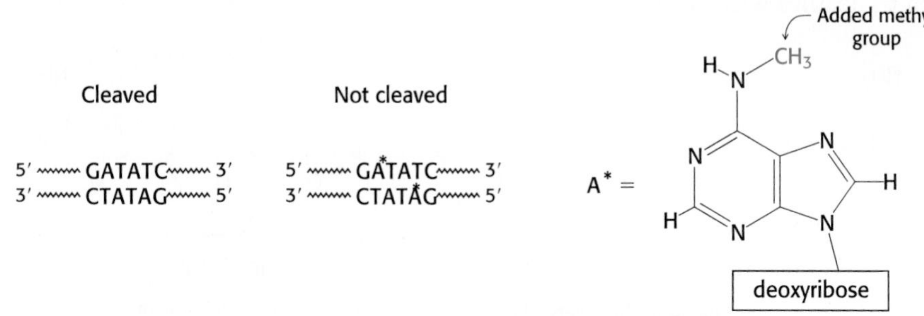

Cleaved   Not cleaved

5' ～～～ GATATC ～～～ 3'   5' ～～～ GA*TATC ～～～ 3'
3' ～～～ CTATAG ～～～ 5'   3' ～～～ CTAT*AG ～～～ 5'

A* =

deoxyribose

Added methyl group

FIGURE 9.40 Protection by methylation. The recognition sequence for EcoRV endonuclease (left) and the sites of methylation (right) in DNA protected from the catalytic action of the enzyme.

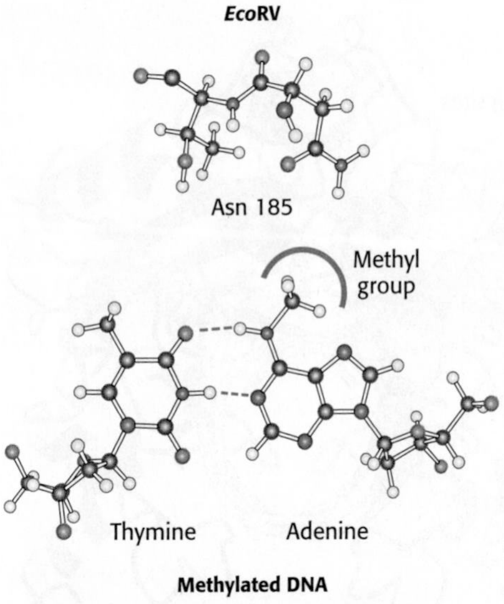

EcoRV

Asn 185

Methyl group

Thymine   Adenine

Methylated DNA

FIGURE 9.41 Methylation of adenine. The methylation of adenine blocks the formation of hydrogen bonds between EcoRV endonuclease and cognate DNA molecules and prevents their hydrolysis.

adenine nucleotide at the 5′ end of the recognition sequence. The presence of the methyl group blocks the formation of a hydrogen bond between the amino group and the side-chain carbonyl group of asparagine 185 (Figure 9.41). This asparagine residue is closely linked to the other amino acids that form specific contacts with the DNA. The absence of the hydrogen bond disrupts other interactions between the enzyme and the DNA substrate, and the distortion necessary for cleavage will not take place.

Type II restriction enzymes have a catalytic core in common and are probably related by horizontal gene transfer

Type II restriction enzymes are prevalent in Archaea and Bacteria. What can we tell of the evolutionary history of these enzymes? Comparison of the amino acid sequences of a variety of type II restriction endonucleases did not reveal significant sequence similarity between most pairs of enzymes. However, a careful examination of three-dimensional structures, taking into account the location of the active sites, revealed the presence of a core structure conserved in the different enzymes.

These observations indicate that many type II restriction enzymes are indeed evolutionarily related. Analyses of the sequences in greater detail suggest that bacteria may have obtained genes encoding these enzymes from other species by *horizontal gene transfer,* the passing of pieces of DNA (such as plasmids) between species that provide a selective advantage in a particular environment. For example, EcoRI (from *E. coli*) and RsrI (from *Rhodobacter sphaeroides*) are 50% identical in sequence over 266 amino acids, clearly indicative of a close evolutionary relationship. However, these species of bacteria are not closely related. Thus, *these species appear to have obtained the gene for these restriction endonucleases from a common source more recently than the time of their evolutionary divergence.* Moreover, the codons used by the gene encoding EcoRI endonuclease to specify given amino acids are strikingly different from the codons used by most *E. coli* genes, which suggests that the gene did not originate in *E. coli.*

Horizontal gene transfer may be a common event. For example, genes that inactivate antibiotics are often transferred, leading to the transmission of antibiotic resistance from one species to another. For restriction-modification systems, protection against viral infections may have favored horizontal gene transfer.

9.4 Myosins Harness Changes in Enzyme Conformation to Couple ATP Hydrolysis to Mechanical Work

The final enzymes that we will consider are the myosins. These enzymes catalyze the hydrolysis of adenosine triphosphate (ATP) to form adenosine diphosphate (ADP) and inorganic phosphate (P_i) and use the energy associated with this thermodynamically favorable reaction to drive the motion of molecules within cells.

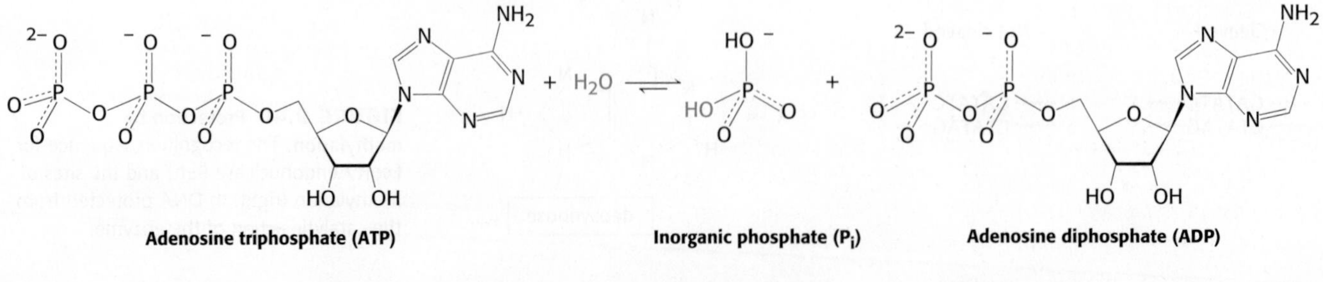

Adenosine triphosphate (ATP)   **Inorganic phosphate (P_i)**   **Adenosine diphosphate (ADP)**

For example, when we lift a book, the energy required comes from ATP hydrolysis catalyzed by myosin in our muscles. Myosins are found in all eukaryotes and the human genome encodes more than 40 different myosins. Myosins generally have elongated structures with globular domains that actually carry out ATP hydrolysis (Figure 9.42). In this chapter, we will focus on the globular ATPase domains, particularly the strategies that allow myosins to hydrolyze ATP in a controlled manner and to use the free energy associated with this reaction to promote substantial conformational changes within the myosin molecule. These conformational changes are amplified by other structures in the elongated myosin molecules to transport proteins or other cargo substantial distances within cells. In Chapter 36 (See Chapter 36 on Sapling Plus), we examine the action of myosins and other molecular-motor proteins in much more detail.

As will be discussed in Chapter 15, ATP is used as the major currency of energy inside cells. Many enzymes use ATP hydrolysis to drive other reactions and processes. In almost all cases, an enzyme that hydrolyzed ATP without any such coupled processes would simply drain the energy reserves of a cell without benefit.

ATP hydrolysis proceeds by the attack of water on the gamma phosphoryl group

In our examination of the mechanism of restriction enzymes, we learned that an activated water molecule performs a nucleophilic attack on phosphorus to cleave the phosphodiester backbone of DNA. The cleavage of ATP by myosins follows an analogous mechanism. To understand the myosin mechanism in more detail, we must first examine the structure of the myosin *ATPase* domain.

The structures of the ATPase domains of several different myosins have been examined. One such domain, that from the soil-living amoeba *Dictyostelium discoideum*, an organism that has been extremely useful for studying cell movement and molecular-motor proteins, has been studied in great detail. The crystal structure of this protein fragment in the absence of nucleotides revealed a single globular domain comprising approximately 750 amino acids. A water-filled pocket is present toward the center of the structure, suggesting a possible nucleotide-binding site. Crystals of this protein were soaked in a solution containing ATP, and the structure was examined again. Remarkably, this structure revealed intact ATP bound in the active site with very little change in the overall structure and without evidence of significant hydrolysis (Figure 9.43). The ATP is also bound to a Mg^{2+} ion.

Kinetic studies of myosins, as well as many other enzymes having ATP or other nucleoside triphosphates as a substrate, reveal that these enzymes are essentially inactive in the absence of divalent metal ions such as magnesium (Mg^{2+}) or manganese (Mn^{2+}) but acquire activity on the addition of these ions. In contrast with the enzymes discussed so far, the metal is not a component of the active site. Rather, nucleotides such as ATP bind these ions, and it is the metal ion–nucleotide complex that is the true substrate for the enzymes. The dissociation constant for the ATP–Mg^{2+} complex is approximately 0.1 mM, and thus, given that intracellular Mg^{2+} concentrations are typically in the millimolar range, essentially all nucleoside triphosphates are present as NTP–Mg^{2+} complexes. *Magnesium or manganese complexes of nucleoside triphosphates are the true substrates for essentially all NTP-dependent enzymes.*

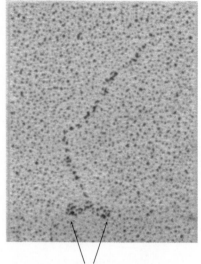

FIGURE 9.42 Elongated structure of muscle myosin. An electron micrograph showing myosin from mammalian muscle. This dimeric protein has an elongated structure with two globular ATPase domains per dimer. [Courtesy of Dr. Paula Flicker, Dr. Theo Walliman, and Dr. Peter Vibert.]

FIGURE 9.43 Myosin–ATP complex structure. An overlay of the structures of the ATPase domain from *Dictyostelium discoideum* myosin with no ligands bound (blue) and the complex of this protein with ATP and magnesium bound (red). *Notice* that the two structures are extremely similar to one another. [Drawn from 1FMV.pdb and 1FMW.pdb].

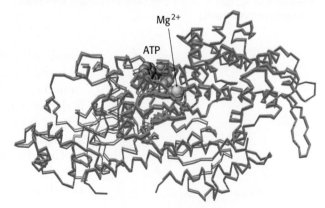

The nucleophilic attack by a water molecule on the γ-phosphoryl group requires some mechanism to activate the water, such as a basic residue or a bound metal ion. Examination of the myosin–ATP complex structure shows no basic residue in an appropriate position and reveals that the bound Mg^{2+} ion is too far away from the phosphoryl group to play this role. These observations suggest why this ATP complex is relatively stable; the enzyme is not in a conformation that is competent to catalyze the reaction. This observation suggests that the domain must undergo a conformational change to catalyze the ATP-hydrolysis reaction.

Formation of the transition state for ATP hydrolysis is associated with a substantial conformational change

The catalytically competent conformation of the myosin ATPase domain must bind and stabilize the transition state of the reaction. In analogy with restriction enzymes, we expect that ATP hydrolysis includes a pentacoordinate transition state.

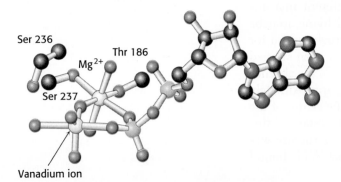

Such pentacoordinate structures based on phosphorus are too unstable to be readily observed. However, transition-state analogs in which other atoms replace phosphorus are more stable. The transition metal vanadium, in particular, forms similar structures. The myosin ATPase domain can be crystallized in the presence of ADP and vanadate, VO_4^{3-}. The result is the formation of a complex that closely matches the expected transition-state structure (Figure 9.44). As expected, the vanadium atom is coordinated to five oxygen atoms, including one oxygen atom from ADP diametrically opposite an oxygen atom that is analogous to the attacking water molecule in the transition state. The Mg^{2+} ion is coordinated to one oxygen atom from the vanadate, one oxygen atom from the ADP, two hydroxyl groups from the enzyme, and two water molecules. In this position, this ion does not appear to play any direct role in activating the attacking water. However, an additional residue from the enzyme, Ser 236, is well positioned to play a role in catalysis (Figure 9.44). In the proposed mechanism of ATP hydrolysis based on this structure, the water molecule attacks the γ-phosphoryl group, with the hydroxyl group of Ser 236 facilitating the transfer of a proton from the attacking water to the hydroxyl group of Ser 236, which, in turn, is deprotonated by one of the oxygen atoms of the γ-phosphoryl group (Figure 9.45). *Thus, in effect, the ATP serves as a base to promote its own hydrolysis.*

FIGURE 9.44 Myosin ATPase transition-state analog. The structure of the transition-state analog formed by treating the myosin ATPase domain with ADP and vanadate (VO_4^{3-}) in the presence of magnesium. *Notice* that the vanadium ion is coordinated to five oxygen atoms including one from ADP. The positions of two residues that bind magnesium as well as Ser 236, a residue that appears to play a direct role in catalysis, are shown. [Drawn from 1VOM.pdb.]

Comparison of the overall structures of the myosin ATPase domain complexed with ATP and with the ADP–vanadate reveals some remarkable differences. Relatively modest structural changes occur in and around the active site. In particular, a stretch of amino acids moves closer to the nucleotide by approximately 2 Å and interacts with the oxygen atom that corresponds to the attacking water molecule. These changes help facilitate the hydrolysis reaction by stabilizing the transition state. However, examination of the overall structure shows even more striking changes.

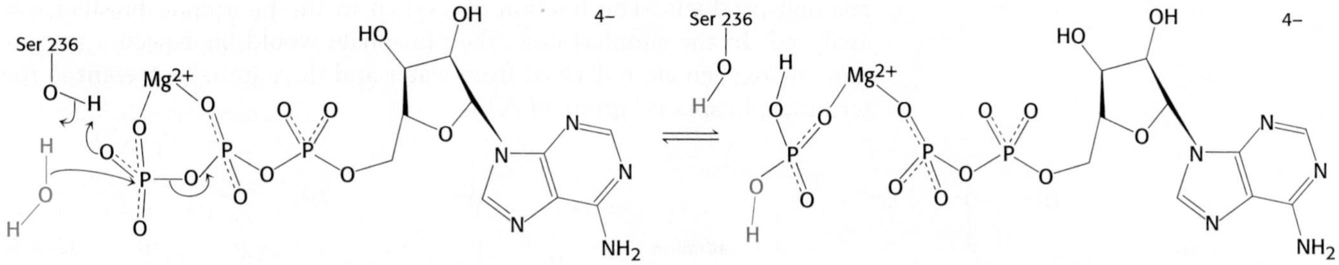

A region comprising approximately 60 amino acids at the carboxyl-terminus of the domain adopts a different configuration in the ADP–vanadate complex, displaced by as much as 25 Å from its position in the ATP complex (Figure 9.46). This displacement tremendously

FIGURE 9.45 Facilitating water attack. The water molecule attacking the γ-phosphoryl group of ATP is deprotonated by the hydroxyl group of Ser 236, which, in turn, is deprotonated by one of the oxygen atoms of the γ-phosphoryl group forming the $H_2PO_4^-$ product.

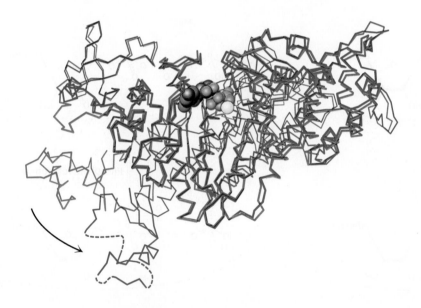

FIGURE 9.46 Myosin conformational changes. A comparison of the overall structures of the myosin ATPase domain with ATP bound (shown in red) and that with the transition-state analog ADP–vanadate (shown in blue). *Notice* the large conformational change of a region at the carboxyl-terminus of the domain (see arrow), some parts of which move as much as 25 Å. [Drawn from 1FMW.pdb and 1VOM.pdb.]

amplifies the relatively subtle changes that take place in the active site. The effect of this motion is amplified even more as this carboxyl-terminal domain is connected to other structures within the elongated structures typical of myosin molecules (Figure 9.42). Thus, the conformation that is capable of promoting the ATP hydrolysis reaction is itself substantially different from other conformational changes that take place in the course of the catalytic cycle.

The altered conformation of myosin persists for a substantial period of time

Myosins are slow enzymes, typically turning over approximately once per second. What steps limit the rate of turnover? In an experiment that was particularly revealing, the hydrolysis of ATP was catalyzed by the myosin ATPase domain from mammalian muscle. The reaction took place in water labeled with ^{18}O to track the incorporation of solvent oxygen into the reaction products. The fraction of oxygen in the phosphate product was analyzed. In the simplest case, the phosphate would be expected to contain one oxygen atom derived from water and three initially present in the terminal phosphoryl group of ATP.

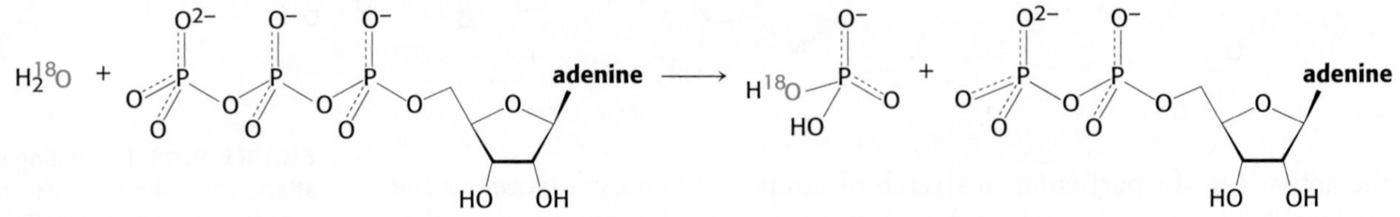

Instead, between two and three of the oxygen atoms in the phosphate were found, on average, to be derived from water. These observations indicate that the ATP hydrolysis reaction within the enzyme active site is reversible. Each molecule of ATP is cleaved to ADP and P_i and then re-formed from these products several times before the products are released from the enzyme (Figure 9.47). At first glance, this observation is startling because

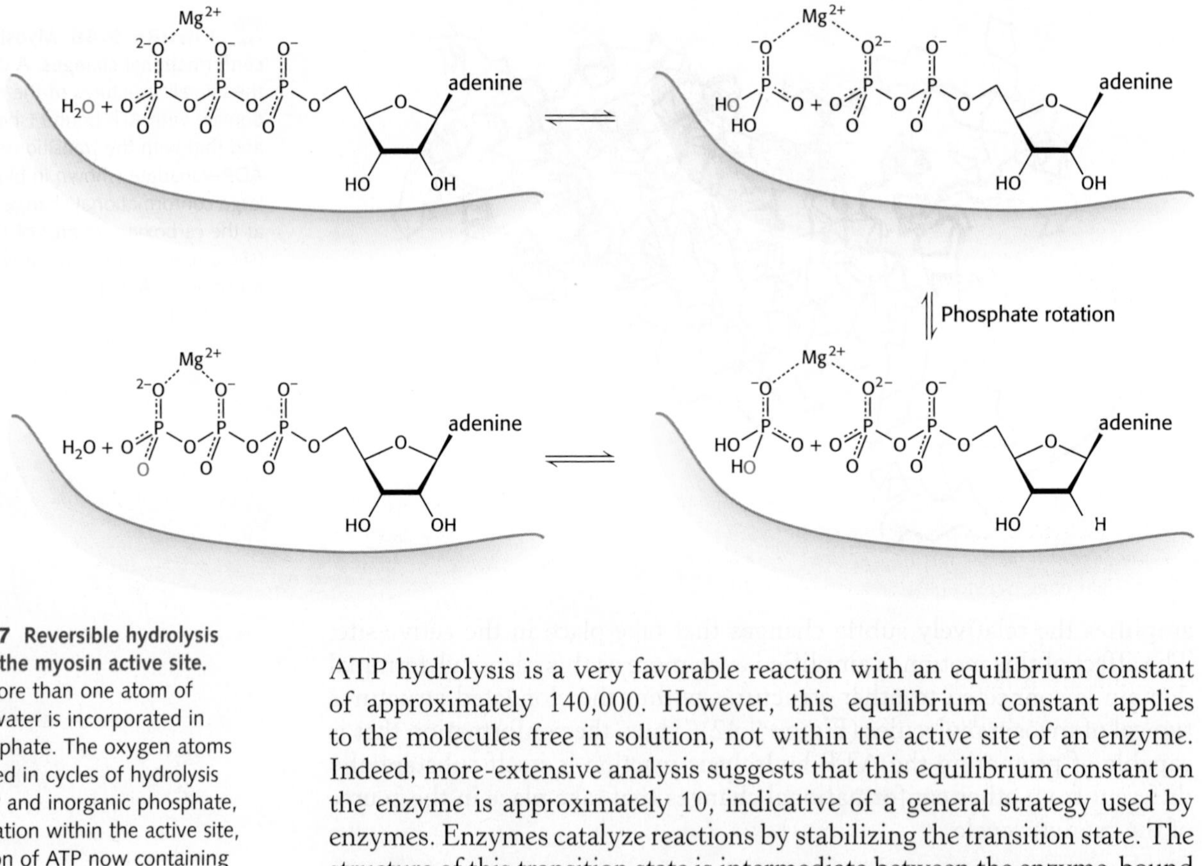

FIGURE 9.47 Reversible hydrolysis of ATP within the myosin active site. For myosin, more than one atom of oxygen from water is incorporated in inorganic phosphate. The oxygen atoms are incorporated in cycles of hydrolysis of ATP to ADP and inorganic phosphate, phosphate rotation within the active site, and reformation of ATP now containing oxygen from water.

ATP hydrolysis is a very favorable reaction with an equilibrium constant of approximately 140,000. However, this equilibrium constant applies to the molecules free in solution, not within the active site of an enzyme. Indeed, more-extensive analysis suggests that this equilibrium constant on the enzyme is approximately 10, indicative of a general strategy used by enzymes. Enzymes catalyze reactions by stabilizing the transition state. The structure of this transition state is intermediate between the enzyme-bound reactants and the enzyme-bound products. Many of the interactions that stabilize the transition state will help equalize the stabilities of the reactants and the products. Thus, *the equilibrium constant between enzyme-bound reactants and products is often close to 1, regardless of the equilibrium constant for the reactants and products free in solution.*

These observations reveal that the hydrolysis of ATP to ADP and P_i is not the rate-limiting step for the reaction catalyzed by myosin. Instead, the release of the products, particularly P_i, from the enzyme is rate limiting. The fact that a conformation of myosin with ATP hydrolyzed but still bound to the enzyme persists for a significant period of time is critical for coupling conformational changes that take place in the course of the reaction to other processes.

Scientists can watch single molecules of myosin move

Myosin molecules function to use the free energy of hydrolysis of ATP to drive macroscopic motion. Myosin molecules move along a filamentous protein termed actin, as we will discuss in more detail in Chapter 36 (See Chapter 36 on 𝒮 **Sapling** Plus). Using a variety of physical methods, scientists have been able to watch *single myosin molecules in action*. For example, a myosin family member termed myosin V can be labeled with fluorescent tags so that it can be localized when fixed on a surface with a precision of less than 15 Å. When this myosin is placed on a surface coated with actin filaments, each molecule remains in a fixed position. However, when ATP is added, each molecule moves along the surface. Tracking individual molecules reveals that each moves in steps of approximately 74 nm, as shown in Figure 9.48. The observation of steps of a fixed size as well as the determination of this step size helps reveal details of the mechanism of action of these tiny molecular motors.

(A)

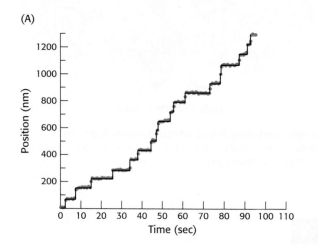

(B)

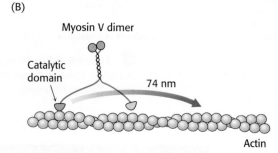

FIGURE 9.48 Single molecule motion. (A) A trace of the position of a single dimeric myosin V molecule as it moves across a surface coated with actin filaments. (B) A model of how the dimeric molecule moves in discrete steps with an average size of 74 ± 5 nm. [Data from A. Yildiz et al., *Science* 300(5628):2061–2065, 2003.]

Myosins are a family of enzymes containing P-loop structures

🜨 X-ray crystallography has yielded the three-dimensional structures of a number of different enzymes that share key structural characteristics and, almost certainly, an evolutionary history with myosin. In particular, a conserved NTP-binding core domain is present. This domain consists of a central β sheet, surrounded on both sides by α helices (Figure 9.49). A characteristic feature of this domain is a loop between the first β strand and the first helix. This loop typically has several glycine residues that are often conserved between more closely related members of this large and diverse family. The loop is often referred to as the *P-loop* because it interacts with phosphoryl groups on the bound nucleotide. P-loop NTPase domains are present in a remarkably wide array of proteins, many of which participate in essential biochemical processes. Examples include ATP synthase, the key enzyme responsible for ATP

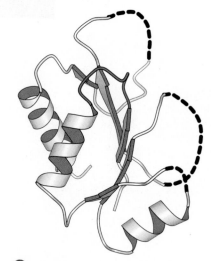

FIGURE 9.49 The core domain of NMP kinases. *Notice* the P-loop shown in green. The dashed lines represent the remainder of the protein structure. [Drawn from 1GKY.pdb.]

generation; signal-transduction proteins such as G proteins; proteins essential for translating mRNA into proteins, such as elongation factor Tu; and DNA- and RNA-unwinding helicases. The wide utility of P-loop NTPase domains is perhaps best explained by their ability to undergo substantial conformational changes on nucleoside triphosphate binding and hydrolysis. We shall encounter these domains throughout the book and shall observe how they function as springs, motors, and clocks. To allow easy recognition of these domains in the book, they will be depicted with the inner surfaces of the ribbons in a ribbon diagram shown in purple and the P-loop shown in green (Figure 9.50).

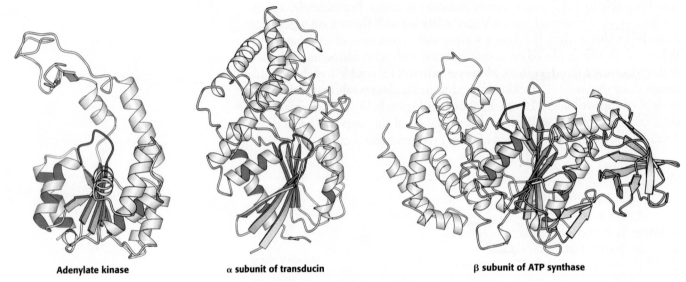

Adenylate kinase   **α subunit of transducin**   **β subunit of ATP synthase**

FIGURE 9.50 Three proteins containing P-loop NTPase domains. *Notice* the conserved domains shown with the inner surfaces of the ribbons in purple and the P-loops in green. [Drawn from 4AKE.pdb; 1TND.pdb; 1BMF.pdb.]

SUMMARY

Enzymes adopt conformations that are structurally and chemically complementary to the transition states of the reactions that they catalyze. Sets of interacting amino acid residues make up sites with the special structural and chemical properties necessary to stabilize the transition state. Enzymes use five basic strategies to form and stabilize the transition state. The first (1) involves the use of binding energy to promote both specificity and catalysis. The others are (2) covalent catalysis, (3) general acid–base catalysis, (4) catalysis by approximation, and (5) metal ion catalysis. The four classes of enzymes examined in this chapter catalyze the addition of water to their substrates but have different requirements for catalytic speed, specificity, and coupling to other processes.

9.1 Proteases Facilitate a Fundamentally Difficult Reaction

The cleavage of peptide bonds by chymotrypsin is initiated by the attack by a serine residue on the peptide carbonyl group. The attacking hydroxyl group is activated by interaction with the imidazole group of a histidine residue, which is, in turn, linked to an aspartate residue. This Ser-His-Asp catalytic triad generates a powerful nucleophile. The product of this initial reaction is a covalent intermediate formed by the enzyme and an acyl group derived from the bound substrate. The hydrolysis of this acyl-enzyme intermediate completes the cleavage process. The tetrahedral intermediates for these reactions have a negative charge on the peptide carbonyl oxygen atom. This negative charge is stabilized by interactions with peptide NH groups in a region on the enzyme termed the oxyanion hole.

Other proteases employ the same catalytic strategy. Some of these proteases, such as trypsin and elastase, are homologs of chymotrypsin. Other proteases, such as subtilisin, contain a very similar catalytic triad that has arisen by convergent evolution. Active-site structures that differ from the catalytic triad are present in a number of other classes of proteases. These classes employ a range of catalytic strategies but, in each case, a nucleophile is generated that is sufficiently powerful to attack the peptide carbonyl group. In some enzymes, the nucleophile is derived from a side chain whereas, in others, an activated water molecule attacks the peptide carbonyl directly.

9.2 Carbonic Anhydrases Make a Fast Reaction Faster

Carbonic anhydrases catalyze the reaction of water with carbon dioxide to generate carbonic acid. The catalysis can be extremely fast: some carbonic anhydrases hydrate carbon dioxide at rates as high as 1 million times per second. A tightly bound zinc ion is a crucial component of the active sites of these enzymes. Each zinc ion binds a water molecule and promotes its deprotonation to generate a hydroxide ion at neutral pH. This hydroxide ion attacks carbon dioxide to form bicarbonate ion, HCO_3^-. Because of the physiological roles of carbon dioxide and bicarbonate ions, speed is of the essence for this enzyme. To overcome limitations imposed by the rate of proton transfer from the zinc-bound water molecule, the most-active carbonic anhydrases have evolved a proton shuttle to transfer protons to a buffer.

9.3 Restriction Enzymes Catalyze Highly Specific DNA-Cleavage Reactions

A high level of substrate specificity is often the key to biological function. Restriction endonucleases that cleave DNA at specific recognition sequences discriminate between molecules that contain these recognition sequences and those that do not. Within the enzyme–substrate complex, the DNA substrate is distorted in a manner that generates a magnesium ion-binding site between the enzyme and DNA. The magnesium ion binds and activates a water molecule, which attacks the phosphodiester backbone.

Some enzymes discriminate between potential substrates by binding them with different affinities. Others may bind many potential substrates but promote chemical reactions efficiently only on specific molecules. Restriction endonucleases such as EcoRV endonuclease employ the latter mechanism. Only molecules containing the proper recognition sequence are distorted in a manner that allows magnesium ion binding and, hence, catalysis. Restriction enzymes are prevented from acting on the DNA of a host cell by the methylation of key sites within its recognition sequences. The added methyl groups block specific interactions between the enzymes and the DNA such that the distortion necessary for cleavage does not take place.

9.4 Myosins Harness Changes in Enzyme Conformation to Couple ATP Hydrolysis to Mechanical Work

Myosins catalyze the hydrolysis of adenosine triphosphate (ATP) to form adenosine diphosphate (ADP) and inorganic phosphate (P_i). The conformations of myosin ATPase domains free of bound nucleotides and with bound ATP are quite similar. Through the use of ADP and vanadate (VO_4^{3-}), an excellent mimic of the transition state for ATP hydrolysis bound to the myosin ATPase domain can be produced. The structure of this complex reveals that dramatic conformational changes take place on formation of this species from the ATP complex. These conformational changes are used to drive substantial motions in molecular motors. The rate of ATP hydrolysis by myosin is relatively low and is limited by the rate of product release from the enzyme. The hydrolysis of ATP to ADP and P_i within the enzyme is reversible with an equilibrium constant of approximately 10, compared with an equilibrium constant of 140,000 for these species free in solution. Myosins are examples of P-loop NTPase enzymes, a large collection of protein families that play key roles in a range of biological processes by virtue of the conformational changes that they undergo with various nucleotides bound.

APPENDIX

Problem-Solving Strategies

Problem-solving strategy 1

PROBLEM: *How many sites?* A researcher has isolated a restriction endonuclease that cleaves at only one particular 10-base-pair site. Would this enzyme be useful in protecting cells from viral infections, given that a typical viral genome is 50,000 base pairs long? Explain.

SOLUTION: This problem focuses on probability.

▶ **What is the significance that the cleavage site is 10-base-pairs long?**

The size of the cleavage site allows us to estimate how frequently, on average, such a site should occur in DNA. Since there are four possible bases at each position, the number of 10-base-pair sites is $4^{10} = 1,048,576$. Thus, each 10-base-pair site should occur approximately once every million base pairs.

▶ **What is the significance of the length of a typical viral genome?**

At 50,000 base pairs, the length given for a typical viral genome is quite small. In particular, 50,000 is much less than 1,048,576. Thus, one would not expect that a typical viral genome would include any particular 10-base-pair site.

Thus, the enzyme would not expect to be useful in protecting the bacterial cells from most viruses.

Repeating this calculation with a more typical 6-base-pair site reveals that there are is $4^6 = 4,096$ possible sites and a typical 50,000-base-pair viral genome would be expected to have several such sites, so that a less specific enzyme would be more effective in protecting the bacterial cells.

Problem-solving strategy 2

PROBLEM: *Labeling strategy.* ATP is added to the myosin ATPase domain in water labeled with ^{18}O. After 50% of the ATP has been hydrolyzed, the remaining ATP is isolated and found to contain ^{18}O. Explain.

SOLUTION: Let's look more closely at the chemical reaction catalyzed by ATPase.

▶ **What is the reaction catalyzed by the ATPase domain of myosin?**

The ATPase domain of myosin catalyzes the hydrolysis of ATP to ADP and inorganic phosphate, P_i, ATP + $H_2O \rightleftharpoons ADP + P_i$.

If only the forward reaction is considered, there is no mechanism for the incorporation of oxygen from water into ATP. Thus, the observation that ^{18}O from water is incorporated into ATP is surprising.

▶ **How could ^{18}O from H_2O be incorporated into ATP?**

When ATP is hydrolyzed, one oxygen atom from water is incorporated into P_i. If the ATP hydrolysis reaction proceeds in the reverse direction, even to a small extent, then the P_i containing ^{18}O might be incorporated into ATP. In solution under normal conditions, the equilibrium lies far to the right, favoring ADP + P_i. However, within the myosin ATPase enzyme active site, the equilibrium is more balanced and both the forward and backward reactions are catalyzed. If the inorganic phosphate is fixed in the enzyme active site, then the oxygen released during the reverse reaction will be the same as the one that entered as water. However, the four oxygens in inorganic phosphate are equivalent, and if the phosphate reorients in the active site, then a different oxygen will be release as water. This allows ^{18}O from water to be incorporated into ATP.

▶ **What is the overall process that accounts for ^{18}O incorporation into ATP?**

Myosin + ATP $\longrightarrow$ Myosin-ATP (ATP binding to myosin)

Myosin-ATP + $H_2^{18}O$ $\longrightarrow$ Myosin-ADP-P_i(^{18}O) (ATP hydrolysis with ^{18}O water)

Myosin-ATP + $H_2^{18}O$ $\longrightarrow$ Myosin-ADP-P_i(^{18}O in different position) (Phosphate reorientation)

Myosin-ADP-P_i(^{18}O) $\longrightarrow$ Myosin-ATP(^{18}O) + $H_2O(^{16}O)$ (ATP synthesis by reaction reversal)

Myosin-ATP(^{18}O) $\longrightarrow$ Myosin + ATP(^{18}O) (Release of ^{18}O-labelled ATP)

KEY TERMS

binding energy (p. 274)
induced fit (p. 274)
covalent catalysis (p. 274)
general acid–base catalysis (p. 275)
catalysis by approximation (p. 275)
metal ion catalysis (p. 275)
chemical modification reaction
  (p. 276)

chromogenic substrate
  (p. 277)
catalytic triad (p. 278)
oxyanion hole (p. 279)
protease inhibitor (p. 285)
proton shuttle (p. 291)
recognition sequence (p. 292)
in-line displacement (p. 293)

methylases (p. 297)
restriction-modification system
  (p. 297)
horizontal gene transfer
  (p. 298)
ATPase (p. 299)
P-loop (p. 303)

PROBLEMS

1. *No burst.* Examination of the cleavage of the amide substrate, A, by chymotrypsin with the use of stopped-flow kinetic methods reveals no burst. The reaction is monitored by noting the color produced by the release of the amino part of the substrate (highlighted in orange). Why is no burst observed? ✓① ✓⑥

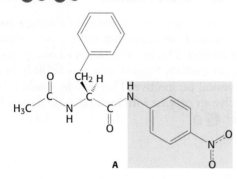

A

2. *Contributing to your own demise.* Consider the subtilisin substrates A and B.

Phe-Ala-Gln-Phe-X   Phe-Ala-His-Phe-X

A   **B**

These substrates are cleaved (between Phe and X) by native subtilisin at essentially the same rate. However, the His 64-to-Ala mutant of subtilisin cleaves substrate B more than 1000-fold as rapidly as it cleaves substrate A. Propose an explanation. ✓②

3. *1 + 1 ≠ 2.* Consider the following argument. In subtilisin, mutation of Ser 221 to Ala results in a 10^6-fold decrease in activity. Mutation of His 64 to Ala results in a similar 10^6-fold decrease. Therefore, simultaneous mutation of Ser 221 to Ala and His 64 to Ala should result in a $10^6 \times 10^6 = 10^{12}$-fold reduction in activity. Is this reduction correct? Why or why not? ✓②

4. *Adding a charge.* In chymotrypsin, a mutant was constructed with Ser 189, which is in the bottom of the substrate-specificity pocket, changed to Asp. What effect would you predict for this Ser 189 → Asp 189 mutation?

5. *Conditional results.* In carbonic anhydrase II, mutation of the proton-shuttle residue His 64 to Ala was expected to result in a decrease in the maximal catalytic rate. However, in buffers such as imidazole with relatively small molecular components, no rate reduction was observed. In buffers with larger molecular components, significant rate reductions were observed. Propose an explanation.

6. *Is faster better?* Restriction endonucleases are, in general, quite slow enzymes with typical turnover numbers of $1\ s^{-1}$. Suppose that endonucleases were faster, with turnover numbers similar to those for carbonic anhydrase ($10^6\ s^{-1}$), such that they act faster than do methylases. Would this increased rate be beneficial to host cells, assuming that the fast enzymes have similar levels of specificity?

7. *Adopting a new gene.* Suppose that one species of bacteria obtained one gene encoding a restriction endonuclease by horizontal gene transfer. Would you expect this acquisition to be beneficial?

8. *Chelation therapy.* Treatment of carbonic anhydrase with high concentrations of the metal chelator EDTA (ethylenediaminetetraacetic acid) results in the loss of enzyme activity. Propose an explanation. ✓❷ ✓❸

9. *An aldehyde inhibitor.* Elastase is specifically inhibited by an aldehyde derivative of one of its substrates:

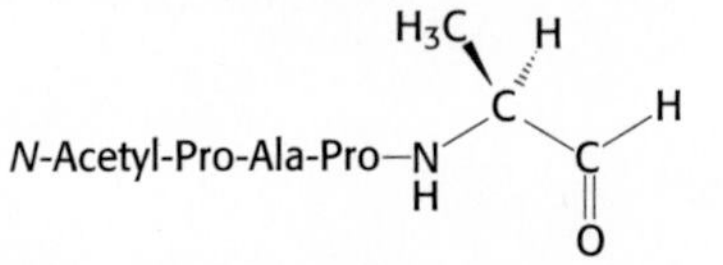

$$\textit{N-Acetyl-Pro-Ala-Pro}-\overset{}{\underset{H}{N}}-\overset{H_3C\quad H}{\underset{}{C}}-\overset{H}{\underset{O}{C}}$$

(a) Which residue in the active site of elastase is most likely to form a covalent bond with this aldehyde?

(b) What type of covalent link would be formed?

10. *Identify the enzyme.* Consider the structure of molecule A. Which enzyme discussed in the chapter do you think molecule A will most effectively inhibit? Explain the basis for your choice. ✓❷

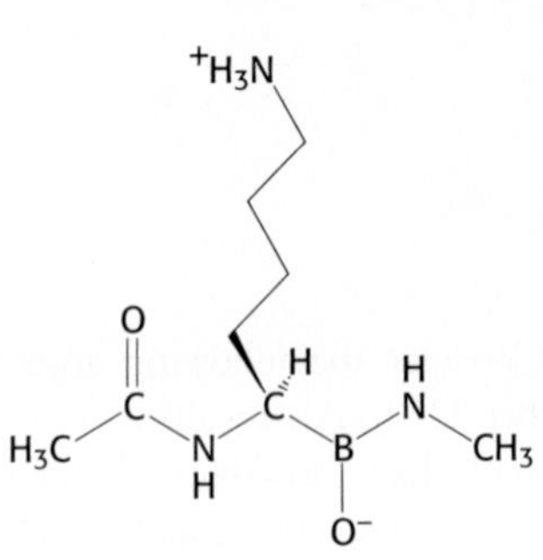

Molecule A

11. *Acid test.* At pH 7.0, carbonic anhydrase exhibits a k_{cat} of 600,000 s^{-1}. Estimate the value expected for k_{cat} at pH 6.0.

12. *Restriction.* To terminate a reaction in which a restriction enzyme cleaves DNA, researchers often add high concentrations of the metal chelator EDTA (ethylenediaminetetraacetic acid). Why does the addition of EDTA terminate the reaction?

13. *Vive la résistance.* Many patients become resistant to HIV protease inhibitors with the passage of time, owing to mutations in the HIV gene that encodes the protease. Mutations are not found in the aspartate residue that interacts with the drugs. Why not? ✓❶ ✓❷

14. *More than one way to skin k_{cat}.* Serine 236 in *Dictyostelium discoideum* myosin has been mutated to alanine. The mutated protein showed modestly reduced ATPase activity. Analysis of the crystal structure of the mutated protein revealed that a water molecule occupied the position of the hydroxyl group of the serine residue in the wild-type protein. Propose a mechanism for the ATPase activity of the mutated enzyme. ✓❶ ✓❷ ✓❻

15. *A power struggle.* The catalytic power of an enzyme can be defined as the ratio of the rate of the enzyme catalyzed reaction to that for the uncatalyzed reaction. Using the information in Figure 9.15 for subtilisin and in Figure 9.22 for carbonic anhydrase, calculate the catalytic powers for these two enzymes.

16. *Wounded but not dead.* How much activity (in terms of relative k_{cat} values) does the version of subtilisin with all three residues in the catalytic triad mutated have compared to uncatalyzed reaction (see Figure 9.15)? Propose an explanation.

Mechanism Problem

17. *Complete the mechanism.* On the basis of the information provided in Figure 9.17, complete the mechanisms for peptide-bond cleavage by (a) a cysteine protease, (b) an aspartyl protease, and (c) a metalloprotease. ✓❶ ✓❷

Think ▶ Pair ▶ Share Problem

18. Most enzymes are quite specific, catalyzing a particular reaction on a set of substrates that are structurally quite similar to one another. Discuss why this is advantageous from a biological perspective. Suggest why this is likely to be true from a chemical perspective. That is, why would it be difficult to evolve an enzyme with high catalytic activity but low specificity? ✓❹

Regulatory Strategies

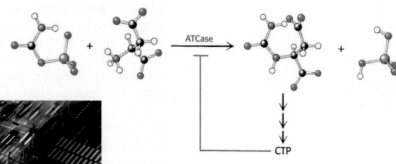

Like motor traffic, metabolic pathways flow more efficiently when regulated by signals. Cytidine triphosphate (CTP), the final product of a multistep pathway, controls flux through the pathway by inhibiting the committed step catalyzed by aspartate transcarbamoylase (ATCase). [Mint Images/Getty Images.]

✓ LEARNING OBJECTIVES

By the end of this chapter, you should be able to:

1. Describe the characteristics of allosteric enzymes and explain how the kinetic curves of such enzymes differ from Michaelis–Menten enzymes.

2. Describe feedback inhibition and how aspartate transcarbamoylase uses it as a regulatory mechanism.

3. Differentiate between homotropic and heterotropic effects and explain how they modify the equilibrium between the T and R forms of an allosteric enzyme.

4. Define isozymes and explain the purpose of isozymes in metabolism.

5. Describe the activity of protein kinases and protein phosphatases.

6. Explain why phosphorylation is an effective means of regulating the activities of target proteins.

7. Define zymogen or proenzyme, and explain why the mechanism of zymogen activation is important physiologically and how it differs from the other control mechanisms discussed in this chapter.

OUTLINE

10.1 Aspartate Transcarbamoylase Is Allosterically Inhibited by the End Product of Its Pathway

10.2 Isozymes Provide a Means of Regulation Specific to Distinct Tissues and Developmental Stages

10.3 Covalent Modification Is a Means of Regulating Enzyme Activity

10.4 Many Enzymes Are Activated by Specific Proteolytic Cleavage

The activity of enzymes must often be regulated so that they function at the proper time and place. This regulation is essential for coordination of the vast array of biochemical processes taking place at any instant in an organism. Enzymatic activity is regulated in five principal ways:

1. *Allosteric control.* Allosteric proteins contain distinct regulatory sites and multiple functional sites. The binding of small signal molecules at regulatory sites controls the activity of these proteins. Moreover, allosteric proteins show the property of *cooperativity:* activity at one functional site affects the activity at others. Proteins displaying allosteric control are thus information transducers: their activity can be modified in response to signal molecules or to information shared among active sites. This chapter examines one of the best-understood allosteric proteins: the enzyme *aspartate transcarbamoylase* (ATCase). Catalysis by aspartate transcarbamoylase of the first step in pyrimidine biosynthesis is inhibited by cytidine triphosphate, the final product of that biosynthesis, in an example of *feedback inhibition*. We have already examined an allosteric protein—hemoglobin, the oxygen transport protein in the blood (Chapter 7).

2. *Multiple forms of enzymes.* Isozymes, or isoenzymes, provide an avenue for varying regulation of the same reaction to meet the specific physiological needs in the particular tissue at a particular time. Isozymes are homologous enzymes within a single organism that catalyze the same reaction but differ slightly in structure and more obviously in K_M and V_{max} values as well as in regulatory properties. Often, isozymes are expressed in a distinct tissue or organelle or at a distinct stage of development.

3. *Reversible covalent modification.* The catalytic properties of many enzymes are markedly altered by the covalent attachment of a modifying group, commonly a phosphoryl group. ATP typically serves as the phosphoryl donor in these reactions, which are catalyzed by *protein kinases*. The removal of phosphoryl groups by hydrolysis is catalyzed by *protein phosphatases*. This chapter considers the structure, specificity, and control of *protein kinase A* (PKA), a ubiquitous eukaryotic enzyme that regulates diverse target proteins.

4. *Proteolytic activation.* The enzymes controlled by some of these regulatory mechanisms cycle between active and inactive states. A different regulatory strategy is used to *irreversibly* convert an inactive enzyme into an active one. Many enzymes are activated by the hydrolysis of a few peptide bonds or even one such bond in inactive precursors called *zymogens* or *proenzymes*. This regulatory mechanism generates digestive enzymes such as chymotrypsin, trypsin, and pepsin. Blood clotting is due to a remarkable cascade of zymogen activations. Active digestive and clotting enzymes are switched off by the irreversible binding of specific inhibitory proteins that are irresistible lures to their molecular prey.

5. *Controlling the amount of enzyme present.* Enzyme activity can also be regulated by adjusting the amount of enzyme present. This important form of regulation usually takes place at the level of transcription. We will consider the control of gene transcription in Chapters 30, 32, and 33.

To begin, we will consider the principles of allostery by examining the enzyme aspartate transcarbamoylase.

10.1 Aspartate Transcarbamoylase Is Allosterically Inhibited by the End Product of Its Pathway

Aspartate transcarbamoylase catalyzes the first step in the biosynthesis of pyrimidines: the condensation of aspartate and carbamoyl phosphate to form *N*-carbamoylaspartate and orthophosphate (Figure 10.1). This reaction is the *committed step* in the pathway, which consists of 10 reactions that will ultimately yield the pyrimidine nucleotides uridine triphosphate (UTP) and cytidine triphosphate (CTP). The committed step means that

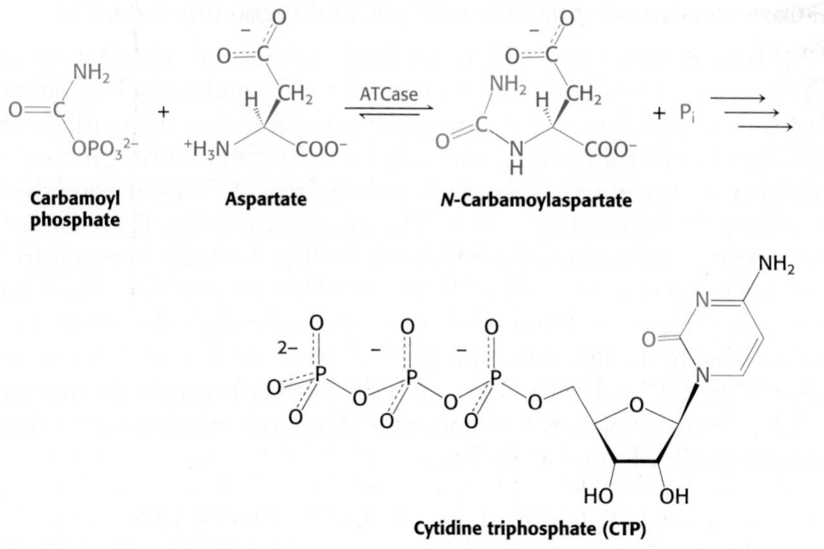

FIGURE 10.1 ATCase reaction.
Aspartate transcarbamoylase catalyzes the committed step, the condensation of aspartate and carbamoyl phosphate to form *N*-carbamoylaspartate, in pyrimidine synthesis.

Carbamoyl phosphate + **Aspartate** $\xrightarrow{\text{ATCase}}$ **N-Carbamoylaspartate** + P_i $\longrightarrow$

Cytidine triphosphate (CTP)

the products of the reaction are committed to ultimate synthesis of the end products of the pathway, in this case UTP and CTP. The committed step, in any pathway, is irreversible under cellular conditions and is the reaction catalyzed by allosteric enzymes. How is ATCase regulated to generate precisely the amount of pyrimidines needed by the cell?

ATCase is inhibited by CTP, the final product of the ATCase-initiated pathway. The rate of the reaction catalyzed by ATCase is fast at low concentrations of CTP but slows as CTP concentration increases (Figure 10.2). Thus, the pathway continues to make new pyrimidines until sufficient quantities of CTP have accumulated. The inhibition of ATCase by CTP is an example of *feedback inhibition,* the inhibition of an enzyme by the end product of the pathway. Feedback inhibition by CTP ensures that *N*-carbamoylaspartate and subsequent intermediates in the pathway are not needlessly formed when pyrimidines are abundant. Another example of feedback inhibition and the deleterious consequences when it goes awry is in Biochemistry in Focus (p. 335).

The inhibitory ability of CTP is remarkable because *CTP is structurally quite different from the substrates of the reaction* (Figure 10.1). Thus CTP must bind to a site distinct from the active site at which substrate binds. Such sites are called *allosteric* or *regulatory sites.* CTP is an example of an *allosteric inhibitor.* In ATCase (but not all allosterically regulated enzymes), the catalytic sites and the regulatory sites are on separate polypeptide chains.

Allosterically regulated enzymes do not follow Michaelis–Menten kinetics

Allosteric enzymes are distinguished by their response to changes in substrate concentration in addition to their susceptibility to regulation by other molecules. Let us examine the rate of product formation as a function of substrate concentration for ATCase (Figure 10.3). The curve differs from that expected for an enzyme that follows Michaelis–Menten kinetics. The observed curve is referred to as sigmoidal because it resembles the letter "S." The vast majority of allosteric enzymes display sigmoidal kinetics. Recall from the discussion of hemoglobin that sigmoidal curves result from cooperation between subunits: the binding of substrate to one active site in a molecule increases the likelihood that substrate will bind to other active sites. To understand the basis of sigmoidal enzyme kinetics and inhibition by CTP, we need to examine the structure of ATCase.

FIGURE 10.2 CTP inhibits ATCase.
Cytidine triphosphate, an end product of the pyrimidine-synthesis pathway, inhibits aspartate transcarbamoylase despite having little structural similarity to reactants or products.

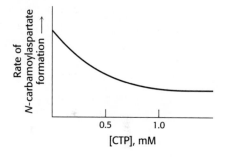

FIGURE 10.3 ATCase displays sigmoidal kinetics. A plot of product formation as a function of substrate concentration produces a sigmoidal curve because the binding of substrate to one active site increases the activity at the other active sites. Thus, the enzyme shows cooperativity.

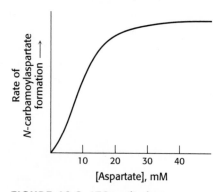

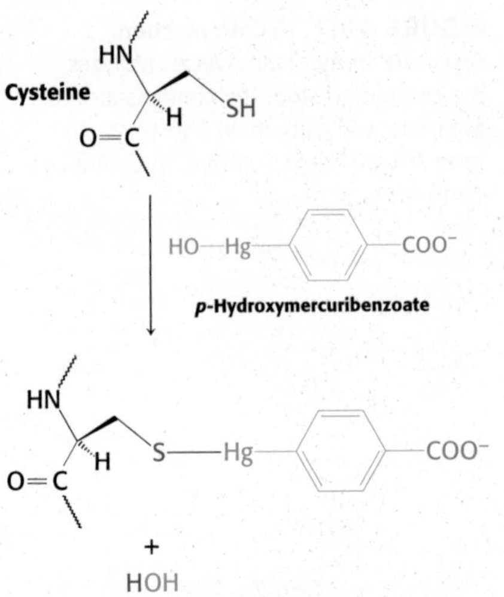

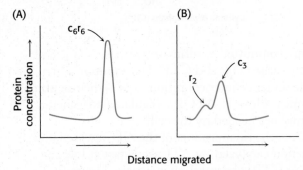

FIGURE 10.4 Modification of cysteine residues. *p*-Hydroxymercuribenzoate reacts with crucial cysteine residues in aspartate transcarbamoylase.

ATCase consists of separable catalytic and regulatory subunits

What is the evidence that ATCase has distinct regulatory and catalytic sites? ATCase can be literally separated into regulatory (r) and catalytic (c) subunits by treatment with a mercurial compound such as *p*-hydroxymercuribenzoate, which reacts with sulfhydryl groups (Figure 10.4). Ultracentrifugation (p. 80) following treatment with mercurials revealed that ATCase is composed of two kinds of subunits (Figure 10.5). The subunits can be readily separated by ion-exchange chromatography because they differ markedly in charge (p. 73) or by centrifugation in a sucrose density gradient because they also differ in size (p. 81). These size differences are manifested in the sedimentation coefficients: that of the native enzyme is 11.6S, whereas those of the dissociated subunits are 2.8S and 5.8S. The attached *p*-mercuribenzoate groups can be removed from the separated subunits by adding an excess of mercaptoethanol, providing isolated subunits for study.

FIGURE 10.5

Ultracentrifugation studies of ATCase. Sedimentation velocity patterns of (A) native ATCase and (B) the enzyme after treatment with *p*-hydroxymercuribenzoate show that the enzyme can be dissociated into regulatory (r) and catalytic (c) subunits. [Data from J. C. Gerhart and H. K. Schachman, *Biochemistry* 4:1054–1062, 1965.]

The larger subunit is the *catalytic subunit*. This subunit has catalytic activity, but when tested by itself, it displays the hyperbolic kinetics of Michaelis–Menten enzymes rather than sigmoidal kinetics. Furthermore, the isolated catalytic subunit is unresponsive to CTP. The isolated smaller subunit can bind CTP but has no catalytic activity. Hence, that subunit is the *regulatory subunit*. The catalytic subunit (c_3) consists of three chains (34 kDa each), and the regulatory subunit (r_2) consists of two chains (17 kDa each). The catalytic and regulatory subunits combine rapidly when they are mixed. The resulting complex has the same structure, c_6r_6, as the native enzyme: two catalytic trimers and three regulatory dimers.

$$2\,c_3 + 3\,r_2 \longrightarrow c_6r_6$$

Most strikingly, the reconstituted enzyme has the same allosteric and kinetic properties as those of the native enzyme. Thus, ATCase is composed of discrete catalytic and regulatory subunits, and *the interaction of the subunits in the native enzyme produces its regulatory and catalytic properties.* The fact that the enzyme can be separated into isolated catalytic and regulatory subunits, which can be reconstituted back to the functional enzyme, allows for a variety of experiments to characterize the allosteric properties of the enzyme (problems 33 and 34).

Allosteric interactions in ATCase are mediated by large changes in quaternary structure

What are the subunit interactions that account for the properties of ATCase? Significant clues have been provided by the three-dimensional structure of various forms of ATCase. Two catalytic trimers are stacked one on top of the other, linked by three dimers of the regulatory chains (Figure 10.6). There are significant contacts between the catalytic and

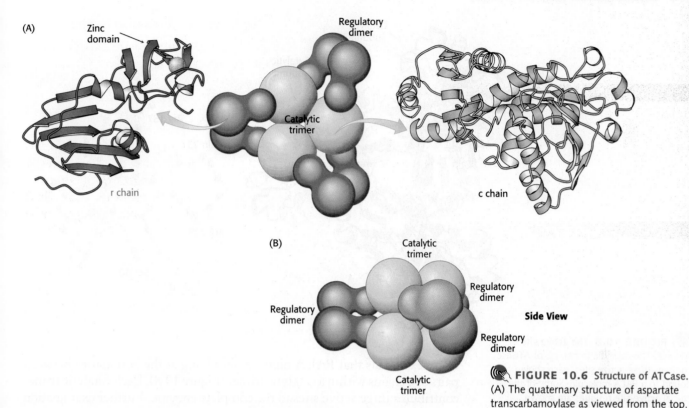

(A) Zinc domain

r chain

Regulatory dimer

Catalytic trimer

c chain

(B)

Catalytic trimer

Regulatory dimer

Regulatory dimer

Regulatory dimer

Side View

Catalytic trimer

FIGURE 10.6 Structure of ATCase. (A) The quaternary structure of aspartate transcarbamoylase as viewed from the top. The drawing in the center is a simplified representation of the relations between subunits. A single catalytic trimer [catalytic (c) chains, shown in yellow] is visible; in this view, the second trimer is hidden below the visible one. *Notice* that each r chain interacts with a c chain through the zinc domain. (B) A side view of the complex. [Drawn from 1RAI.pdb.]

the regulatory subunits: each r chain within a regulatory dimer interacts with a c chain within a catalytic trimer. The c chain makes contact with a structural domain in the r chain that is stabilized by a zinc ion bound to four cysteine residues. The zinc ion is critical for the interaction of the r chain with the c chain. The mercurial compound *p*-hydroxymercuribenzoate is able to dissociate the catalytic and regulatory subunits because mercury binds strongly to the cysteine residues, displacing the zinc and preventing interaction with the c chain.

To locate the active sites, the enzyme is crystallized in the presence of *N*-(phosphonacetyl)-L-aspartate (PALA), a bisubstrate analog (an analog of the two substrates) that resembles an intermediate along the pathway of catalysis (Figure 10.7). PALA is a potent competitive inhibitor of ATCase that binds to and blocks the active sites. The structure of the ATCase–PALA

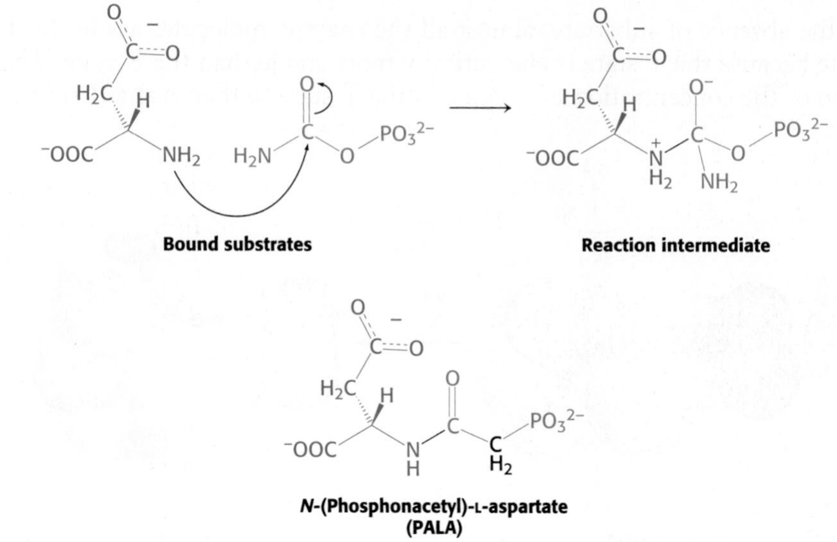

Bound substrates

Reaction intermediate

N-(Phosphonacetyl)-L-aspartate
(PALA)

FIGURE 10.7 PALA, a bisubstrate analog. (Top) Nucleophilic attack by the amino group of aspartate on the carbonyl carbon atom of carbamoyl phosphate generates an intermediate on the pathway to the formation of *N*-carbamoylaspartate. (Bottom) *N*-(Phosphonacetyl)-L-aspartate (PALA) is an analog of the reaction intermediate and a potent competitive inhibitor of aspartate transcarbamoylase.

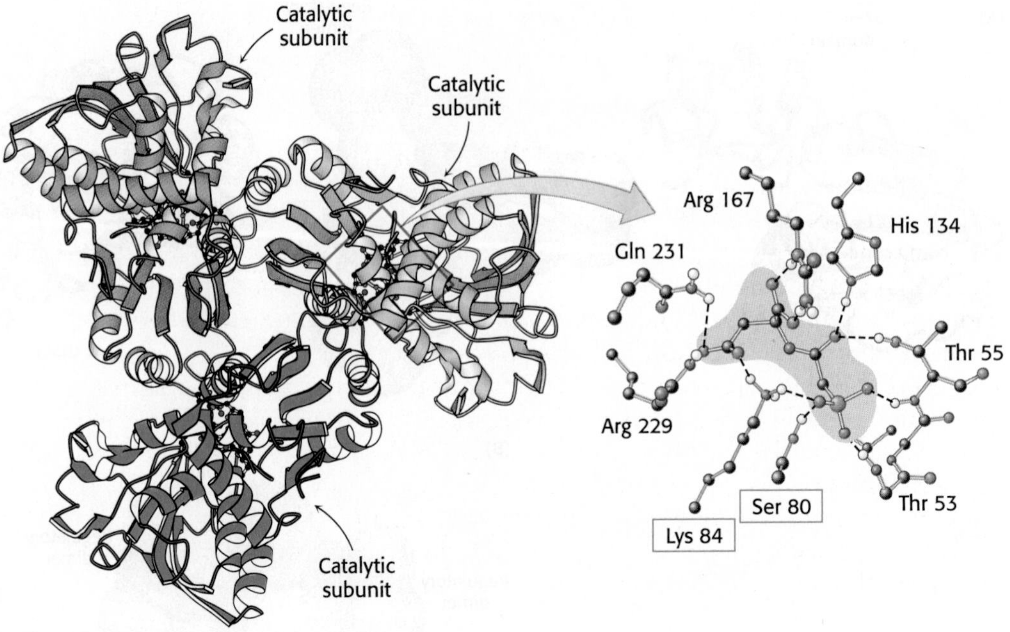

Catalytic subunit

Catalytic subunit

Arg 167

His 134

Gln 231

Thr 55

Arg 229

Ser 80

Thr 53

Lys 84

Catalytic subunit

FIGURE 10.8 The active site of ATCase. The catalytic trimer, c_3, of ATCase contains three active sites, each shown bound to a PALA molecule. Some of the crucial active-site residues are shown binding to the inhibitor PALA (shaded gray) via hydrogen bonds (black dotted lines). *Notice* that an active site is composed mainly of residues from one c chain (yellow bonds), but an adjacent c chain also contributes important residues (green bonds and boxed in green). [Drawn from 8ATC.pdb.]

complex reveals that PALA binds at sites lying at the boundaries between pairs of c chains within a catalytic trimer (Figure 10.8). Each catalytic trimer contributes three active sites to the complete enzyme. Further examination of the ATCase–PALA complex reveals a remarkable change in quaternary structure on binding of PALA. The two catalytic trimers move 12 Å farther apart and rotate approximately 10 degrees about their common threefold axis of symmetry. Moreover, the regulatory dimers rotate approximately 15 degrees to accommodate this motion (Figure 10.9). The enzyme literally expands on PALA binding. In essence, ATCase has two distinct quaternary forms: one that predominates in the absence of substrate or substrate analogs and another that predominates when substrates or analogs are bound. We call these forms the T (for tense) state and the R (for relaxed) state, respectively, as we did for the two quaternary states of hemoglobin.

How can we explain the enzyme's sigmoidal kinetics in light of the structural observations? Like hemoglobin, the enzyme exists in an equilibrium between the T state and the R state.

$$R \rightleftharpoons T$$

In the absence of substrate, almost all the enzyme molecules are in the T state because the T state is energetically more stable than the R state. The ratio of the concentration of enzyme in the T state to that in the R state is

FIGURE 10.9 The T-to-R state transition in ATCase. Aspartate transcarbamoylase exists in two conformations: a compact, relatively inactive form called the tense (T) state and an expanded form called the relaxed (R) state. *Notice* that the structure of ATCase changes dramatically in the transition from the T state to the R State. PALA binding stabilizes the R state.

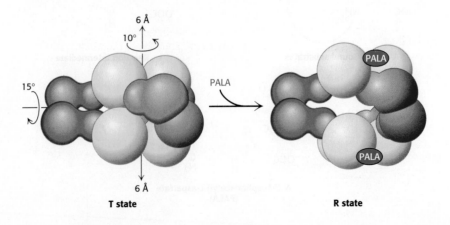

T state

R state

called the allosteric coefficient (L). For most allosteric enzymes, L is on the order of 10^2 to 10^3.

$$L = \frac{T}{R}$$

The T state has a low affinity for substrate and hence exhibits low catalytic activity. The occasional binding of a substrate molecule to one active site in an enzyme increases the likelihood that the entire enzyme shifts to the R state with its higher binding affinity for substrate. The addition of more substrate has two effects. First, it increases the probability that each enzyme molecule will bind at least one substrate molecule. Second, it increases the average number of substrate molecules bound to each enzyme. The presence of additional substrate will increase the fraction of enzyme molecules in the more active R state because *the position of the equilibrium depends on the number of active sites that are occupied by substrate.* We considered this property, called *cooperativity* because the subunits cooperate with one another, when we discussed the sigmoidal oxygen-binding curve of hemoglobin in Chapter 7. The effects of substrates on allosteric enzymes are referred to as *homotropic effects* (from the Greek *homós,* "same").

This mechanism for allosteric regulation is referred to as the *concerted model* because the change in the enzyme is "all or none"; the entire enzyme is converted from T into R, affecting all of the catalytic sites equally. In contrast, the *sequential model* assumes that the binding of ligand to one site on the complex can affect neighboring sites without causing all subunits to undergo the T-to-R transition (p. 214). Although the concerted model explains the behavior of ATCase well, most other allosteric enzymes have features of both models.

The sigmoidal curve for ATCase can be pictured as a composite of two Michaelis–Menten curves, one corresponding to the less-active T state and the other to the more-active R state. At low concentrations of substrate, the curve closely resembles that of the T state enzyme. As the substrate concentration is increased, the curve progressively shifts to resemble that of the R state enzyme (Figure 10.10).

What is the biochemical advantage of sigmoidal kinetics? Allosteric enzymes transition from a less active state to a more active state within a narrow range of substrate concentration. The benefit of this behavior is illustrated in Figure 10.11, which compares the kinetics of a Michaelis–Menten enzyme (blue curve) to that of an allosteric enzyme (red curve). In this example, the Michaelis–Menten enzyme requires an approximately 27-fold increase in substrate concentration to increase V_o from $0.1\,V_{max}$ to

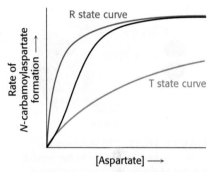

FIGURE 10.10 Basis for the sigmoidal curve. The generation of the sigmoidal curve by the property of cooperativity can be understood by imagining an allosteric enzyme as a mixture of two Michaelis–Menten enzymes, one with a high value of K_M that corresponds to the T state and another with a low value of K_M that corresponds to the R state. As the concentration of substrate is increased, the equilibrium shifts from the T state to the R state, which results in a steep rise in activity with respect to substrate concentration.

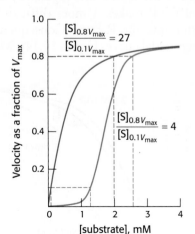

FIGURE 10.11 Allosteric enzymes display threshold effects. As the T-to-R transition occurs, the velocity increases over a narrower range of substrate concentration for an allosteric enzyme (red curve) than for a Michaelis–Menten enzyme (blue curve).

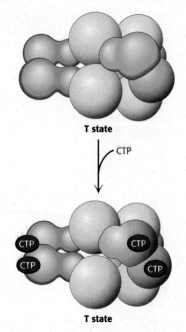

T state

↓ CTP

T state

FIGURE 10.12 CTP stabilizes the T state. The binding of CTP to the regulatory subunit of aspartate transcarbamoylase stabilizes the T state.

$0.8\ V_{\max}$. In contrast, the allosteric enzyme requires only about a fourfold increase in substrate concentration to attain the same increase in velocity. *The activity of allosteric enzymes is more sensitive to changes in substrate concentration near K_M than are Michaelis–Menten enzymes with the same V_{max}.* This sensitivity is called a *threshold effect*: below a certain substrate concentration, there is little enzyme activity. However, after the threshold has been reached, enzyme activity increases rapidly. In other words, much like an "on/off" switch, cooperativity ensures that most of the enzyme is either on (R state) or off (T state). The vast majority of allosteric enzymes display sigmoidal kinetics.

Allosteric regulators modulate the T-to-R equilibrium

We now turn our attention to the effects of pyrimidine nucleotides on ATCase activity. As noted earlier, CTP inhibits the action of ATCase. X-ray studies of ATCase in the presence of CTP reveal (1) that the enzyme is in the T state when bound to CTP and (2) that a binding site for this nucleotide exists in each regulatory chain in a domain that does not interact with the catalytic subunit (Figure 10.12). Each active site is more than 50 Å from the nearest CTP-binding site. The question naturally arises, How can CTP inhibit the catalytic activity of the enzyme when it does not interact with the catalytic chain?

The quaternary structural changes observed on substrate-analog binding suggest a mechanism for inhibition by CTP (Figure 10.13). *The binding*

FIGURE 10.13 The R state and the T state are in equilibrium. Even in the absence of any substrate or regulators, aspartate transcarbamoylase exists in equilibrium between the R and the T states. Under these conditions, the T state is favored by a factor of approximately 200.

T state
(less active)

R state
(more active)

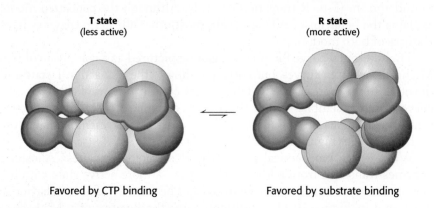

Favored by CTP binding   Favored by substrate binding

of the inhibitor CTP to the T state shifts the T-to-R equilibrium in favor of the T state, decreasing net enzyme activity. CTP increases the allosteric coefficient from 200 in its absence to 1250 when all of the regulatory sites are occupied by CTP. The binding of CTP makes it more difficult for substrate binding to convert the enzyme into the R state. Consequently, CTP increases the initial phase of the sigmoidal curve (Figure 10.14). More

FIGURE 10.14 Effect of CTP and ATP on ATCase kinetics. CTP stabilizes the T state of aspartate transcarbamoylase, making it more difficult for substrate binding to convert the enzyme into the R state. As a result, the curve is shifted to the right, as shown in red. ATP is an allosteric activator of aspartate transcarbamoylase because it stabilizes the R state, making it easier for substrate to bind. As a result, the curve is shifted to the left, as shown in blue.

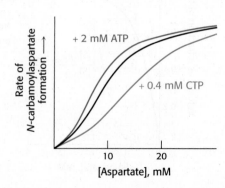

substrate is required to attain a given reaction rate. UTP, the immediate precursor to CTP, also regulates ATCase. While unable to inhibit the enzyme alone, UTP synergistically inhibits ATCase in the presence of CTP.

Interestingly, ATP, too, is an allosteric effector of ATCase, binding to the same site as CTP. However, ATP binding stabilizes the R state, lowering the allosteric coefficient from 200 to 70 and thus *increasing* the reaction rate at a given aspartate concentration (Figure 10.14). In the presence of sufficient ATP, the kinetic profile shows a less-pronounced sigmoidal behavior. Because ATP and CTP bind at the same site, high levels of ATP prevent CTP from inhibiting the enzyme. The effects of nonsubstrate molecules on allosteric enzymes (such as those of CTP and ATP on ATCase) are referred to as *heterotropic effects* (from the Greek *héteros*, "different"). *Substrates generate the sigmoidal curve (homotropic effects), whereas regulators shift the K_M (heterotropic effects). Note, however, that both types of effect are generated by altering the T/R ratio.*

The increase in ATCase activity in response to increased ATP concentration has two potential physiological ramifications. First, high ATP concentration signals a high concentration of purine nucleotides in the cell; the increase in ATCase activity will tend to balance the purine and pyrimidine pools. Second, a high concentration of ATP indicates that energy is available for mRNA synthesis and DNA replication and leads to the synthesis of pyrimidines needed for these processes.

10.2 Isozymes Provide a Means of Regulation Specific to Distinct Tissues and Developmental Stages

Isozymes, or *isoenzymes*, are enzymes that differ in amino acid sequence yet catalyze the same reaction. Typically, these enzymes display different kinetic parameters, such as K_M, or respond to different regulatory molecules. They are encoded by different genes, which usually arise through gene duplication and divergence. Isozymes can often be distinguished from one another by physical properties such as electrophoretic mobility. *Isoform* is a more generic term used when the protein in question is not an enzyme.

The existence of isozymes permits the fine-tuning of metabolism to meet the needs of a given tissue or developmental stage. Consider the example of lactate dehydrogenase (LDH), an enzyme that catalyzes a step in anaerobic glucose metabolism and glucose synthesis. Human beings have two isozymic polypeptide chains for this enzyme: the H isozyme is highly expressed in heart muscle and the M isozyme is expressed in skeletal muscle. The amino acid sequences are 75% identical. Each functional enzyme is tetrameric, and many different combinations of the two isozymic polypeptide chains are possible. The H_4 isozyme, found in the heart, has a higher affinity for substrates than does the M_4 isozyme. In addition, high levels of pyruvate allosterically inhibit the H_4 but not the M_4 isozyme. The other combinations, such as H_3M, have intermediate properties. We will consider these isozymes in their biological context in Chapter 16.

The M_4 isozyme functions optimally in the anaerobic environment of hardworking skeletal muscle, whereas the H_4 isozyme does so in the aerobic environment of heart muscle. Indeed, the proportions of these isozymes change throughout the development of the rat heart as the tissue switches from an anaerobic environment to an aerobic one (Figure 10.15A). Figure 10.15B shows the distribution of tissue-specific forms of lactate dehydrogenase in adult rat tissues. *Essentially all of the enzymes that we will encounter in later chapters, including allosteric enzymes, exist in isozymic forms.*

(A)

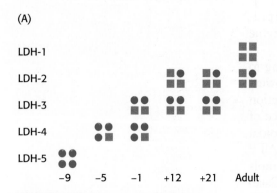

LDH-1

LDH-2

LDH-3

LDH-4

LDH-5

-9   -5   -1   +12   +21   Adult

(B)

	Heart	Kidney	Red blood cell	Brain	Leukocyte	Muscle	Liver
H_4	▬	▬	▬	▬	▬	—	—
H_3M	▬	▬	▬	▬	▬	—	—
H_2M_2	—	▬	—	▬	▬	▬	—
HM_3	—	—	—	—	—	—	—
M_4	—	—	—	—	—	▬	▬

FIGURE 10.15 Isozymes of lactate dehydrogenase. (A) The rat heart lactate dehydrogenase (LDH) isozyme profile changes in the course of development. The H isozyme is represented by squares and the M isozyme by circles. The negative and positive numbers denote the days before and after birth, respectively.
(B) LDH isozyme content varies by tissue. The thickness of the green bars represents the relative amounts of the isozymes.
[(A) Data from W.-H. Li, *Molecular Evolution* (Sinauer, 1997), p. 283; (B) after K. Urich, *Comparative Animal Biochemistry* (Springer Verlag, 1990), p. 542.]

Acetylated lysine

The appearance of some isozymes in the blood is a sign of tissue damage, useful for clinical diagnosis. For instance, an increase in serum levels of H_4 relative to H_3M is an indication that a myocardial infarction, or heart attack, has damaged heart-muscle cells, leading to the release of cellular material.

10.3 Covalent Modification Is a Means of Regulating Enzyme Activity

The covalent attachment of a molecule to an enzyme or other protein can modify its activity. In these instances, a donor molecule provides the functional moiety being attached. Most covalent modifications are reversible. Phosphorylation and dephosphorylation are common means of covalent modification, which we discuss in detail below. The attachment of acetyl groups to lysine residues by acetyltransferases and their removal by deacetylases are another example. Histones—proteins that are packaged with DNA into chromosomes—are extensively acetylated and deacetylated in vivo on lysine residues (Section 33.3). More heavily acetylated histones are associated with genes that are being actively transcribed. Although protein acetylation was originally discovered as a modification of histones, we now know that it is a major means of regulation, with more than 2000 different proteins in mammalian cells regulated by acetylation. Protein acetylation appears to be especially important in the regulation of metabolism. The acetyltransferase and deacetylase enzymes are themselves regulated by phosphorylation, showing that the covalent modification of a protein can be controlled by the covalent modification of the modifying enzymes.

In some cases, the covalent modification is not reversible. The irreversible attachment of a lipid group causes some proteins in signal-transduction pathways, such as Ras (a GTPase) and Src (a protein tyrosine kinase), to become affixed to the cytoplasmic face of the plasma membrane. Fixed in this location, the proteins are better able to receive and transmit information that is being passed along their signaling pathways (Chapter 14). Mutations in both Ras and Src are seen in a wide array of cancers. The attachment of the small protein ubiquitin can signal that a protein is to be destroyed, the ultimate means of regulation (Chapter 23). The protein cyclin must be ubiquitinated and destroyed before a cell can enter anaphase and proceed through the cell cycle.

Virtually all the metabolic processes that we will examine are regulated in part by covalent modification. Indeed, the allosteric properties of many enzymes are altered by covalent modification. Table 10.1 lists some common covalent modifications.

TABLE 10.1 Common covalent modifications of protein activity

Modification	Donor molecule	Example of modified protein	Protein function
Phosphorylation	ATP	Glycogen phosphorylase	Glucose homeostasis; energy transduction
Acetylation	Acetyl CoA	Histones	DNA packing; transcription
Myristoylation	Myristoyl CoA	Src	Signal transduction
ADP ribosylation	NAD^+	RNA polymerase	Transcription
Farnesylation	Farnesyl pyrophosphate	Ras	Signal transduction
γ-Carboxylation	HCO_3^-	Thrombin	Blood clotting
Sulfation	3'-Phosphoadenosine-5'-phosphosulfate	Fibrinogen	Blood-clot formation
Ubiquitination	Ubiquitin	Cyclin	Control of cell cycle

Kinases and phosphatases control the extent of protein phosphorylation

We will see phosphorylation used as a regulatory mechanism in virtually every metabolic process in eukaryotic cells. Indeed, as much as 30% of eukaryotic proteins are phosphorylated. The enzymes catalyzing phosphorylation reactions are called *protein kinases*. These enzymes constitute one of the largest protein families known: there are more than 500 homologous protein kinases in human beings. This multiplicity of enzymes allows regulation to be fine-tuned according to a specific tissue, time, or substrate.

ATP is the most common donor of phosphoryl groups. The terminal (γ) phosphoryl group of ATP is transferred to a specific amino acid of the acceptor protein or enzyme. In eukaryotes, the acceptor residue is commonly one of the three containing a hydroxyl group in its side chain. Transfers to *serine* and *threonine* residues are handled by one class of protein kinases and transfers to *tyrosine* residues by another. Tyrosine kinases, which are unique to multicellular organisms, play pivotal roles in growth regulation, and mutations in these enzymes are commonly observed in cancer cells (Chapter 14).

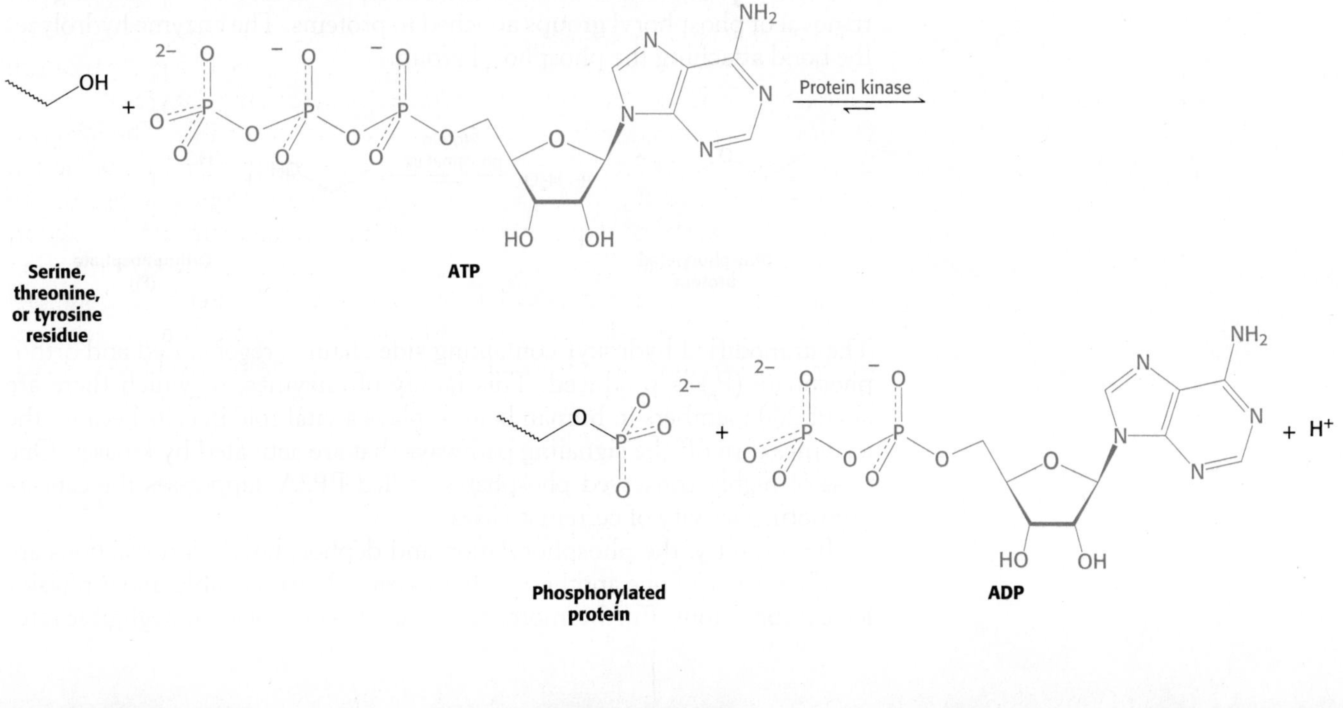

Serine, threonine, or tyrosine residue

ATP

Phosphorylated protein

ADP

TABLE 10.2 Examples of serine and threonine kinases and their activating signals

Signal	Enzyme
Cyclic nucleotides	Cyclic AMP-dependent protein kinase Cyclic GMP-dependent protein kinase
Ca^{2+} and calmodulin	Ca^{2+}–calmodulin protein kinase Phosphorylase kinase or glycogen synthase kinase 2
AMP	AMP-activated kinase
Diacylglycerol	Protein kinase C
Metabolic intermediates and other "local" effectors	Many target-specific enzymes, such as pyruvate dehydrogenase kinase and branched-chain ketoacid dehydrogenase kinase

Source: Information from D. Fell, *Understanding the Control of Metabolism* (Portland Press, 1997), Table 7.2.

Table 10.2 lists a few of the known serine and threonine protein kinases. Proteins that undergo protein-phosphorylation reactions are located inside cells, where the phosphoryl-group donor ATP is abundant. Proteins that are entirely extracellular are not regulated by reversible phosphorylation.

Protein kinases vary in their degree of specificity. *Dedicated protein kinases* phosphorylate a single protein or several closely related ones. *Multifunctional protein kinases* modify many different targets; they have a wide reach and can coordinate diverse processes. Comparisons of amino acid sequences of many phosphorylation sites show that a multifunctional kinase recognizes related sequences. For example, the *consensus sequence* recognized by protein kinase A is Arg-Arg-X-*Ser*-Z or Arg-Arg-X-*Thr*-Z, in which X is a small residue, Z is a large hydrophobic one, and *Ser* or *Thr* is the site of phosphorylation. However, this sequence is not absolutely required. Lysine, for example, can substitute for one of the arginine residues but with some loss of affinity. Thus, *the primary determinant of specificity is the amino acid sequence surrounding the serine or threonine phosphorylation site.* However, distant residues can contribute to specificity. For instance, a change in protein conformation can open or close access to a possible phosphorylation site.

Protein phosphatases reverse the effects of kinases by catalyzing the removal of phosphoryl groups attached to proteins. The enzyme hydrolyzes the bond attaching the phosphoryl group.

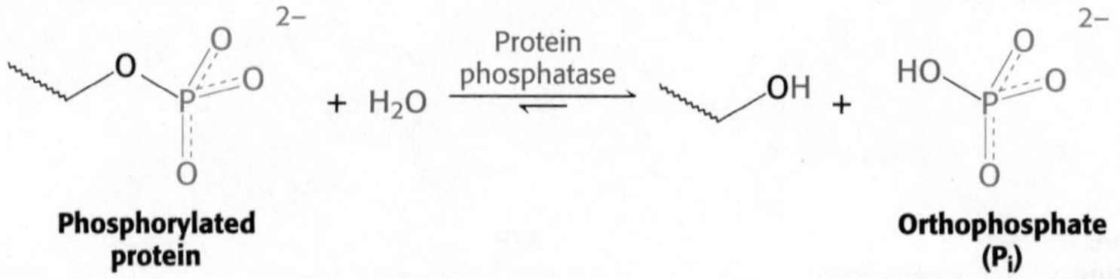

Phosphorylated protein + H_2O → (Protein phosphatase) → —OH + **Orthophosphate (P_i)**

The unmodified hydroxyl-containing side chain is regenerated and orthophosphate (P_i) is produced. This family of enzymes, of which there are about 200 members in human beings, plays a vital role in cells because the enzymes turn off the signaling pathways that are activated by kinases. One class of highly conserved phosphatase called PP2A suppresses the cancer-promoting activity of certain kinases.

Importantly, the phosphorylation and dephosphorylation reactions are not the reverse of one another; each is essentially irreversible under physiological conditions. Furthermore, both reactions take place at negligible rates in the absence of enzymes. Thus, phosphorylation of a protein substrate will take place only through the action of a specific protein kinase and at the expense of ATP cleavage, and dephosphorylation will take place only through the action of a phosphatase. The result is that target proteins cycle between unphosphorylated and phosphorylated forms. The rate of cycling between the phosphorylated and the unphosphorylated states depends on the relative activities of the specific kinases and phosphatases.

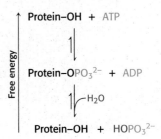

Phosphorylation is a highly effective means of regulating the activities of target proteins

Phosphorylation is a common covalent modification of proteins in all forms of life, which leads to the question: What makes protein phosphorylation so valuable in regulating protein function that its use is ubiquitous? Phosphorylation is a highly effective means of controlling the activity of proteins for several reasons:

1. *The free energy of phosphorylation is large.* Of the -50 kJ mol^{-1} ($-12 \text{ kcal mol}^{-1}$) provided by ATP, about half is consumed in making phosphorylation irreversible; the other half is conserved in the phosphorylated protein. A free-energy change of 5.69 kJ mol^{-1} ($1.36 \text{ kcal mol}^{-1}$) corresponds to a factor of 10 in an equilibrium constant. Hence, phosphorylation can change the conformational equilibrium between different functional states by a large factor, of the order of 10^4. In essence, the energy expenditure allows for a stark shift from one conformation to another.

2. *A phosphoryl group adds two negative charges to a protein.* These new charges may disrupt electrostatic interactions present in the unmodified protein and allow new electrostatic interactions to be formed. Such structural changes can markedly alter substrate binding and catalytic activity.

3. *A phosphoryl group can form three or more hydrogen bonds.* The tetrahedral geometry of a phosphoryl group makes these bonds highly directional, allowing for specific interactions with hydrogen-bond donors.

4. *Phosphorylation and dephosphorylation can take place in less than a second or over a span of hours.* The kinetics can be adjusted to meet the timing needs of a physiological process.

5. *Phosphorylation often evokes highly amplified effects.* A single activated kinase can phosphorylate hundreds of target proteins in a short interval. If the target protein is an enzyme, it can in turn transform a large number of substrate molecules.

6. *ATP is the cellular energy currency* (Chapter 15). The use of this compound as a phosphoryl-group donor links the energy status of the cell to the regulation of metabolism.

Cyclic AMP activates protein kinase A by altering the quaternary structure

Let us examine a specific protein kinase that helps animals cope with stressful situations. The "fight-or-flight" response is common to many animals presented with a dangerous or exciting situation. Muscle becomes primed for action. This priming is the result of the activity of a particular protein kinase. In this case, the hormone epinephrine (adrenaline) triggers the formation of cyclic AMP (cAMP), an intracellular messenger formed by the cyclization of ATP. Cyclic AMP subsequently activates a key enzyme: *protein kinase A* (PKA). The kinase alters the activities of target proteins by

Cyclic adenosine monophosphate (cAMP)

phosphorylating specific serine or threonine residues. The striking finding is that *most effects of cAMP in eukaryotic cells are achieved through the activation of PKA by cAMP.*

PKA provides a clear example of the integration of allosteric regulation and phosphorylation. PKA is activated by cAMP concentrations near 10 nM. The quaternary structure of PKA is reminiscent of that of ATCase. Like that enzyme, PKA in muscle consists of two kinds of subunits: a 49-kDa regulatory (R) subunit and a 38-kDa catalytic (C) subunit. In the absence of cAMP, the regulatory and catalytic subunits form an R_2C_2 complex that is enzymatically inactive (Figure 10.16). The binding of two molecules of cAMP to each of the regulatory subunits leads to the dissociation of R_2C_2 into an R_2 subunit and two C subunits. These free catalytic subunits are enzymatically active. Thus, *the binding of cAMP to the regulatory subunit relieves its inhibition of the catalytic subunit.* PKA, like most other kinases, exists in isozymic forms for fine-tuning regulation to meet the needs of a specific cell or developmental stage. In mammals, four isoforms of the R subunit and three of the C subunit are encoded in the genome.

How does the binding of cAMP activate the kinase? Each R chain contains the sequence Arg-Arg-Gly-*Ala*-Ile, which matches the consensus sequence for phosphorylation except for the presence of alanine in place of serine. In the R_2C_2 complex, this *pseudosubstrate sequence* of R occupies the catalytic site of C, thereby preventing the entry of protein substrates (Figure 10.16). The binding of cAMP to the R chains allosterically moves the pseudosubstrate sequences out of the catalytic sites. The released C chains are then free to bind and phosphorylate substrate proteins. Interestingly, the cAMP-binding domain of the R subunit is highly conserved and found in all organisms.

Mutations in protein kinase A can cause Cushing's syndrome

Cushing's syndrome, a collection of diseases resulting from excess cortisol secretion by the adrenal cortex, is a metabolic disorder characterized by a variety of symptoms such as muscle weakness, thinning skin that is easily bruised, and osteoporosis. Cortisol, a steroid hormone (Section 26.4), has a number of physiological effects including stimulation of glucose synthesis, suppression of the immune response, and inhibition of bone growth. The most common cause of Cushing's syndrome, called Cushing's disease, is a tumor of the pituitary gland that overstimulates cortisol secretion by the adrenal cortex. Recent work shows that a mutation that renders protein kinase A constitutively active also results in the syndrome. In these patients, the catalytic subunit of the enzyme is altered so that it no longer binds

FIGURE 10.16 Regulation of protein kinase A. The binding of four molecules of cAMP activates protein kinase A by dissociating the inhibited holoenzyme (R_2C_2) into a regulatory subunit (R_2) and two catalytically active subunits (C). Each R chain includes cAMP-binding domains and a pseudosubstrate sequence.

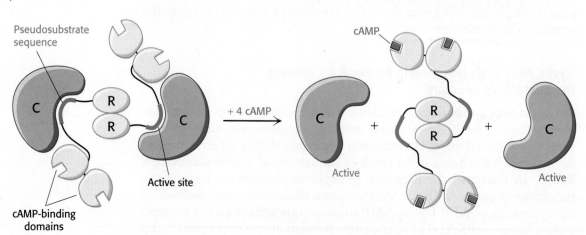

the regulatory subunit. Thus, the enzyme is active even in the absence of cAMP with the ultimate result being unregulated secretion of cortisol.

Exercise modifies the phosphorylation of many proteins

As we will see more clearly in later chapters, exercise is fundamental to sustaining good health, in part by regulating energy metabolism and maintaining whole-body insulin sensitivity. The *phosphoproteome* is the name given to all proteins that are modified by phosphorylation and phosphoproteomics is the study of such proteins. Given the prominence of phosphorylation in biological regulation, it should come as no surprise that exercise modifies the phosphoproteome. Recent research has shown that exercise results in the phosphorylation of more than 1000 potential phosphorylation sites on almost 600 different proteins. Protein kinase A and AMP-activated kinase (AMPK, p. 894) as well as other kinases catalyze these modifications. These modifications alter a number of biological functions, most notably an increase in the ability to process fuels aerobically (Chapters 17 and 18). Determining the functions of the modified proteins will surely keep exercise biochemists engaged for many years.

10.4 Many Enzymes Are Activated by Specific Proteolytic Cleavage

We turn now to a different mechanism of enzyme regulation. While many enzymes acquire full enzymatic activity as they spontaneously fold into their characteristic three-dimensional forms, the folded forms of other enzymes are inactive until one or a few specific peptide bonds are cleaved. The inactive precursor is called a *zymogen* or a *proenzyme*. Since an energy source such as ATP is not needed for specific proteolytic cleavage, even extracellular proteins can be activated by this means. This contrasts with reversible enzyme regulation by phosphorylation, which requires ATP. Another noteworthy difference is that proteolytic activation, unlike allosteric control and reversible covalent modification, is irreversible, taking place just once in the life of an enzyme molecule.

Specific proteolysis is a common means of activating enzymes and other proteins in biological systems. For example:

1. The *digestive enzymes* that hydrolyze foodstuffs are synthesized as zymogens in the stomach and pancreas (Table 10.3).

2. *Blood clotting* is mediated by a cascade of proteolytic activations that ensures a rapid and amplified response to trauma.

3. Some protein hormones are synthesized as inactive precursors. For example, *insulin* is derived from *proinsulin* by proteolytic removal of a peptide.

4. The fibrous protein *collagen,* the major constituent of skin and bone, is derived from *procollagen,* a soluble precursor.

TABLE 10.3 Gastric and pancreatic zymogens

Site of synthesis	Zymogen	Active enzyme
Stomach	Pepsinogen	Pepsin
Pancreas	Chymotrypsinogen	Chymotrypsin
Pancreas	Trypsinogen	Trypsin
Pancreas	Procarboxypeptidase	Carboxypeptidase

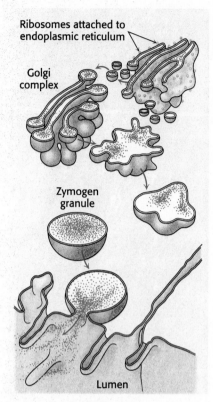

FIGURE 10.17 Secretion of zymogens by an acinar cell of the pancreas. Zymogens are synthesized on ribosomes attached to the endoplasmic reticulum. They are subsequently processed in the Golgi apparatus and packaged into zymogen or secretory granules. With the proper signal, the granules fuse with the plasma membrane, discharging their contents into the lumen of the pancreatic ducts. Cell cytoplasm is depicted as pale green. Membranes and lumen are shown as dark green.

5. Many *developmental processes* are controlled by the activation of zymogens. For example, in the metamorphosis of a tadpole into a frog, large amounts of collagen are resorbed from the tail in the course of a few days. Likewise, much collagen is broken down in a mammalian uterus after delivery. The conversion of *procollagenase* into *collagenase,* the active protease responsible for collagen breakdown, is precisely timed in these remodeling processes.

6. *Programmed cell death,* or *apoptosis,* is mediated by proteolytic enzymes called *caspases,* which are synthesized in precursor form as *procaspases.* When activated by various signals, caspases function to cause cell death in most organisms, ranging from *C. elegans* to human beings. Apoptosis provides a means of sculpting the shapes of body parts in the course of development and is a means of eliminating damaged or infected cells.

We next examine the activation and control of zymogens, using enzymes responsible for digestion and blood-clot formation as examples.

Chymotrypsinogen is activated by specific cleavage of a single peptide bond

Chymotrypsin is a digestive enzyme that hydrolyzes proteins. Chymotrypsin, a member of a large family of serine proteases whose mechanism of action was described in detail in Chapter 9, specifically cleaves peptide bonds on the carboxyl side of amino acid residues with large, hydrophobic R groups (Table 8.6). Its inactive precursor, *chymotrypsinogen,* is synthesized in the pancreas, as are several other zymogens and digestive enzymes. Indeed, the pancreas is one of the most active organs in synthesizing and secreting proteins. The enzymes and zymogens are synthesized in the acinar cells of the pancreas and stored inside membrane-bounded granules (Figure 10.17). The zymogen granules accumulate at the apex of the acinar cell; when the cell is stimulated by a hormonal signal or a nerve impulse, the contents of the granules are released into a duct leading into the duodenum.

Chymotrypsinogen, a single polypeptide chain consisting of 245 amino acid residues, is virtually devoid of enzymatic activity. It is converted into a fully active enzyme when the peptide bond joining arginine 15 and isoleucine 16 is cleaved by trypsin (Figure 10.18). The resulting active enzyme, called π-chymotrypsin, then acts on other π-chymotrypsin

FIGURE 10.18 Proteolytic activation of chymotrypsinogen. The three chains of α-chymotrypsin are linked by two interchain disulfide bonds (A to B and B to C). The approximate positions of both inter- and intrachain disulfide bonds are shown.

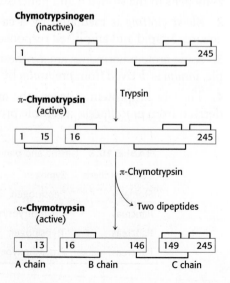

molecules by removing two dipeptides to yield α-chymotrypsin, the stable form of the enzyme. The three resulting chains in α-chymotrypsin remain linked to one another by two interchain disulfide bonds. The striking feature of this activation process is that *cleavage of a single specific peptide bond transforms the protein from a catalytically inactive form into one that is fully active.*

Proteolytic activation of chymotrypsinogen leads to the formation of a substrate-binding site

How does cleavage of a single peptide bond activate the zymogen? The cleavage of the peptide bond between amino acids 15 and 16 triggers key conformational changes, which were revealed by the elucidation of the three-dimensional structure of chymotrypsinogen.

1. The newly formed *amino-terminal group of isoleucine 16 turns inward and forms an ionic bond with aspartate 194 in the interior of the chymotrypsin molecule* (Figure 10.19).

2. This electrostatic interaction triggers a number of conformational changes. Methionine 192 moves from a deeply buried position in the zymogen to the surface of the active enzyme, and residues 187 and 193 move farther apart from each other. These changes result in the formation of the *substrate-specificity site* for aromatic and bulky nonpolar groups. One side of this site is made up of residues 189 through 192 (Figure 9.10). *This cavity for binding part of the substrate is not fully formed in the zymogen.*

3. The tetrahedral transition state generated by chymotrypsin has an oxyanion (a negatively charged carbonyl oxygen atom) that is stabilized by hydrogen bonds with two NH groups of the main chain of the enzyme (Figure 9.9). One of these NH groups is not appropriately located in chymotrypsinogen and so the site stabilizing the oxyanion (the oxyanion hole, p. 279) is incomplete in the zymogen.

4. The conformational changes elsewhere in the molecule are very small. Thus, *the switching on of enzymatic activity in a protein can be accomplished by discrete, highly localized conformational changes that are triggered by the hydrolysis of a single peptide bond.*

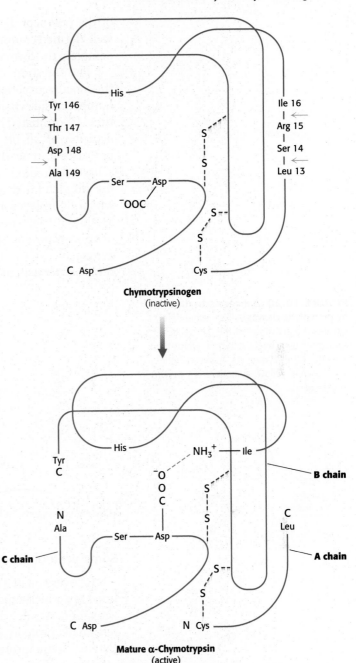

Chymotrypsinogen
(inactive)

Mature α-Chymotrypsin
(active)

FIGURE 10.19 Conformations of chymotrypsinogen and chymotrypsin. The electrostatic interaction between the α-amino group of isoleucine 16 and the carboxylate of aspartate 194, essential for the structure of active chymotrypsin, is possible only after cleavage of the peptide bond between isoleucine and arginine in chymotrypsinogen. [Information from Gregory A. Petsko and Dagmar Ringe, *Protein Structure and Function* (Sinauer, 2003), p. 3–16, Figure 3-31.]

The generation of trypsin from trypsinogen leads to the activation of other zymogens

The structural changes accompanying the activation of *trypsinogen*, the precursor of the proteolytic enzyme *trypsin*, another serine protease, are different from those in the activation of chymotrypsinogen. Four regions of the polypeptide are very flexible in the zymogen, whereas they have a well-defined conformation in trypsin. The resulting structural changes also complete the formation of the oxyanion hole.

The digestion of proteins and other molecules in the duodenum requires the concurrent action of several enzymes, because each is specific for a limited number of side chains. Thus, the zymogens must be switched on at the same time. Coordinated control is achieved by the action of *trypsin as the common activator of all the pancreatic zymogens*—trypsinogen, chymotrypsinogen, proelastase, procarboxypeptidase, and prolipase, the inactive precursor of a lipid-degrading enzyme. To produce active trypsin, the cells that line the duodenum display a membrane-embedded enzyme, *enteropeptidase,* which hydrolyzes a unique lysine–isoleucine peptide bond in trypsinogen as the zymogen enters the duodenum from the pancreas. The small amount of trypsin produced in this way activates more trypsinogen and the other zymogens (Figure 10.20). Thus, *the formation of trypsin by enteropeptidase is the master activation step.*

FIGURE 10.20 Zymogen activation by proteolytic cleavage. Enteropeptidase initiates the activation of the pancreatic zymogens by activating trypsin, which then activates other zymogens. Active enzymes are shown in yellow; zymogens are shown in orange.

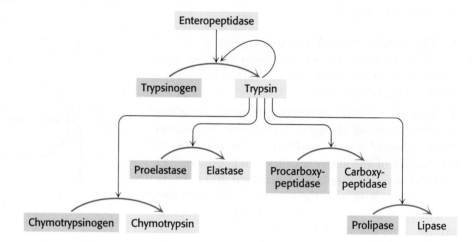

Some proteolytic enzymes have specific inhibitors

The conversion of a zymogen into a protease by cleavage of a single peptide bond is a precise means of switching on enzymatic activity. However, this activation step is irreversible, and so a different mechanism is needed to terminate proteolysis. Specific protease inhibitors accomplish this task. *Serpins, serine protease inhibitors,* are an example of one such family of inhibitors. For example, *pancreatic trypsin inhibitor,* a 6-kDa protein, inhibits trypsin by binding very tightly to its active site. The dissociation constant of the complex is 0.1 pM, which corresponds to a standard free energy of binding of about -75 kJ mol^{-1} ($-18 \text{ kcal mol}^{-1}$). In contrast with nearly all known protein assemblies, this complex is not dissociated into its constituent chains by treatment with denaturing agents such as 8 M urea or 6 M guanidine hydrochloride (p. 52).

The reason for the exceptional stability of the complex is that pancreatic trypsin inhibitor is a very effective substrate analog. X-ray analyses show that the inhibitor lies in the active site of the enzyme, positioned such that the side chain of lysine 15 of this inhibitor interacts with the aspartate side chain in the specificity pocket of trypsin. In addition, there are many hydrogen bonds between the main chain of trypsin and that of its inhibitor. Furthermore, the carbonyl group of lysine 15 and the surrounding atoms of the inhibitor fit snugly in the active site of the enzyme. Comparison of the structure of the inhibitor bound to the enzyme with that of the free inhibitor

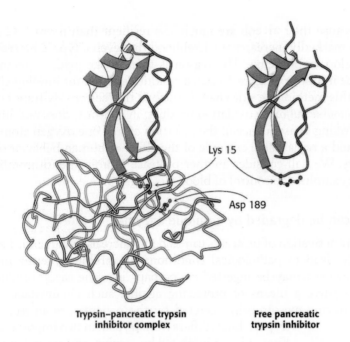

Lys 15

Asp 189

Trypsin–pancreatic trypsin inhibitor complex

Free pancreatic trypsin inhibitor

FIGURE 10.21 Interaction of trypsin with its inhibitor. Structure of a complex of trypsin (yellow) and pancreatic trypsin inhibitor (red). *Notice* that lysine 15 of the inhibitor penetrates into the active site of the enzyme. There it forms an ionic bond with aspartate 189 in the active site. *Also notice* that bound inhibitor and the free inhibitor are almost identical in structure. [Drawn from 1BPI.pdb.]

reveals that *the structure is essentially unchanged on binding to the enzyme* (Figure 10.21). Thus, the inhibitor is preorganized into a structure that is highly complementary to the enzyme's active site. Indeed, the peptide bond between lysine 15 and alanine 16 in pancreatic trypsin inhibitor is cleaved but at a very slow rate: the half-life of the trypsin–inhibitor complex is several months. In essence, the inhibitor is a substrate, but its intrinsic structure is so nicely complementary to the enzyme's active site that it binds very tightly, rarely progressing to the transition state and is turned over slowly.

Why does trypsin inhibitor exist? Recall that trypsin activates other zymogens. Consequently, preventing even small amounts of trypsin from initiating the cascade while the zymogens are still in the pancreas or pancreatic ducts is vital. Trypsin inhibitor binds to any prematurely activated trypsin molecules in the pancreas or pancreatic ducts. This inhibition prevents severe damage to those tissues, which could lead to acute pancreatitis.

Pancreatic trypsin inhibitor is not the only important protease inhibitor. A 53-kDa plasma protein, α_1-*antitrypsin* (also called α_1-*antiproteinase*), protects tissues from digestion by elastase, a secretory product of neutrophils (white blood cells that engulf bacteria). Antielastase would be a more accurate name for this inhibitor, because it blocks *elastase* much more effectively than it blocks trypsin. Like pancreatic trypsin inhibitor, α_1-antitrypsin blocks the action of target enzymes by binding nearly irreversibly to their active sites. Genetic disorders leading to a deficiency of α_1-antitrypsin illustrate the physiological importance of this inhibitor. For example, the substitution of lysine for glutamate at residue 53 in the type Z mutant slows the secretion of this inhibitor from liver cells. Serum levels of the inhibitor are about 15% of normal in people homozygous for this defect. The consequence is that excess elastase destroys alveolar walls in the lungs by digesting elastic fibers and other connective-tissue proteins.

The resulting clinical condition is called *emphysema* (also known as *chronic obstructive pulmonary disease* [COPD]). People with emphysema must breathe much harder than normal people to exchange the same volume

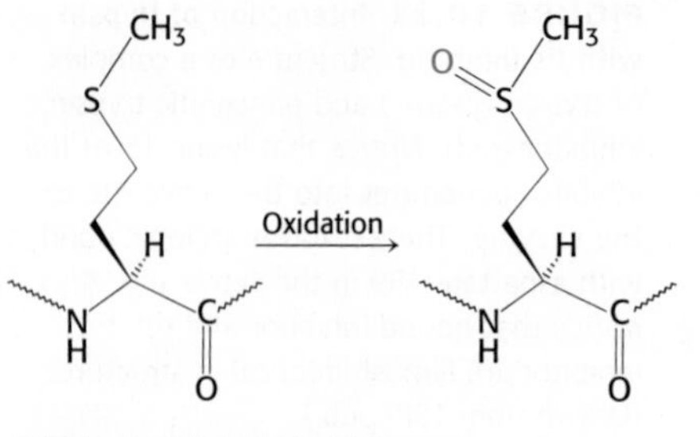

FIGURE 10.22 Oxidation of methionine to methionine sulfoxide.

of air because their alveoli are much less resilient than normal. Cigarette smoking markedly increases the likelihood that even a type Z heterozygote will develop emphysema. The reason is that smoke oxidizes methionine 358 of the inhibitor (Figure 10.22), a residue essential for binding elastase. Indeed, this methionine side chain is the bait that selectively traps elastase. The *methionine sulfoxide* oxidation product, in contrast, does not lure elastase, a striking consequence of the insertion of just one oxygen atom into a protein and a remarkable example of the effect of human behavior on biochemistry. We will consider another protease inhibitor, antithrombin III, when we examine the control of blood clotting.

Serpins can be degraded by a unique enzyme

Premature activation of proteases can lead to pancreatitis, but excess serpins might also lead to pathological conditions by inhibiting serine protease activity. Serpins may be ingested or, perhaps in some cases, produced in excess. Is there a means of protecting against such circumstances? The human enzyme mesotrypsin, secreted by the pancreas, is an isozyme of trypsin, but it differs from trypsin in two important ways. First, it is not inhibited by serpins, and second, it rapidly digests the inhibitor. Interestingly, mutation experiments demonstrate that changing only four amino acids in trypsin renders the mutated enzyme resistant to serpins and confer the catalytic capacity observed in mesotrypsin. Although mesotrypsin is thought to protect against excess serpins, its precise role remains to be discovered.

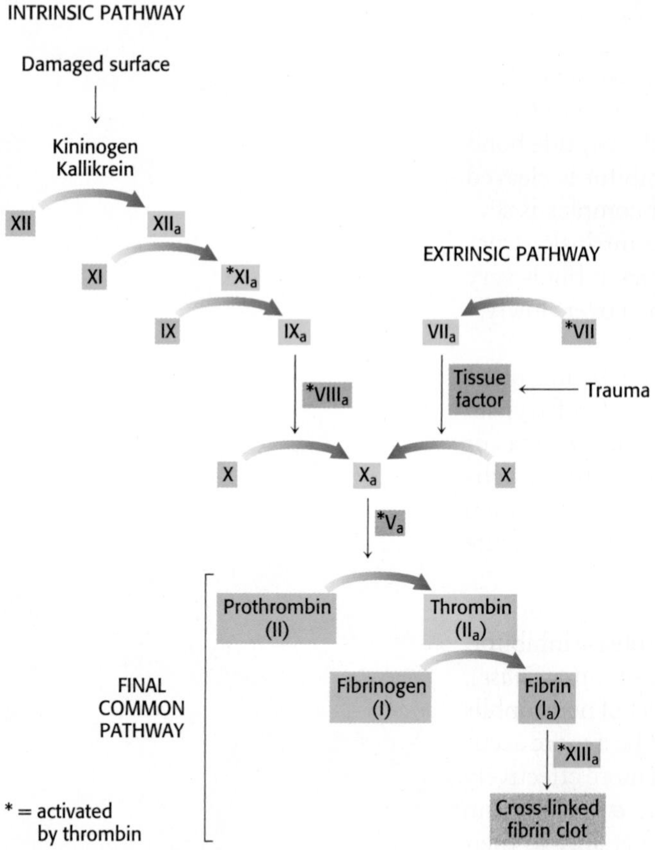

Blood clotting is accomplished by a cascade of zymogen activations

Enzymatic cascades are often employed in biochemical systems to achieve a rapid response. In a cascade, an initial signal institutes a series of steps, each of which is catalyzed by an enzyme. At each step, the signal is amplified. For instance, if a signal molecule activates an enzyme that in turn activates 10 enzymes and each of the 10 enzymes in turn activates 10 additional enzymes, after four steps the original signal will have been amplified 10,000-fold. *Hemostasis*, the process of blood clot formation and dissolution, requires a *cascade of zymogen activations:* the activated form of one clotting factor catalyzes the activation of the next (Figure 10.23). Thus, very small amounts of the initial factors suffice to trigger the cascade, ensuring a rapid response to trauma.

Two means of initiating blood clotting have been described, the *intrinsic pathway* and the *extrinsic pathway.* The intrinsic clotting pathway is activated by exposure of anionic surfaces upon rupture of the endothelial lining of the blood vessels. The extrinsic pathway, which appears to be most crucial in blood clotting, is initiated when trauma exposes *tissue factor* (TF), an integral membrane glycoprotein. Upon exposure to the blood, tissue factor binds to factor VII to activate factor X. Both the intrinsic and extrinsic pathways lead to the activation of factor X (a serine protease), which in turn converts

FIGURE 10.23 Blood-clotting cascade. A fibrin clot is formed by the interplay of the intrinsic, extrinsic, and final common pathways. The intrinsic pathway begins with the activation of factor XII (Hageman factor) by contact with abnormal surfaces produced by injury. The extrinsic pathway is triggered by trauma, which releases tissue factor (TF). TF forms a complex with VII, which initiates a cascade-activating thrombin. Inactive forms of clotting factors are shown in red; their activated counterparts (indicated by the subscript "a") are in yellow. Stimulatory proteins that are not themselves enzymes are shown in blue boxes. A striking feature of this process is that the activated form of one clotting factor catalyzes the activation of the next factor.

prothrombin into *thrombin*, the key protease in clotting. Thrombin then amplifies the clotting process by activating enzymes and factors that lead to the generation of yet more thrombin, an example of positive feedback. Note that the active forms of the clotting factors are designated with a subscript "a," whereas factors that are enzymes or enzyme cofactors activated by thrombin are designated with an asterisk.

Prothrombin must bind to Ca^{2+} to be converted to thrombin

Thrombin is synthesized as a zymogen called prothrombin. The inactive molecule is comprised of four major domains, with the serine protease domain at its carboxyl terminus (Figure 10.24). The first domain, called the *gla domain*, is rich in γ-carboxyglutamate residues (abbreviation gla), and the second and third domains are called *kringle domains* (named after a Danish pastry that they resemble). These three domains work in concert to keep prothrombin in an inactive form. Moreover, because it is rich in γ-carboxyglutamate, the gla

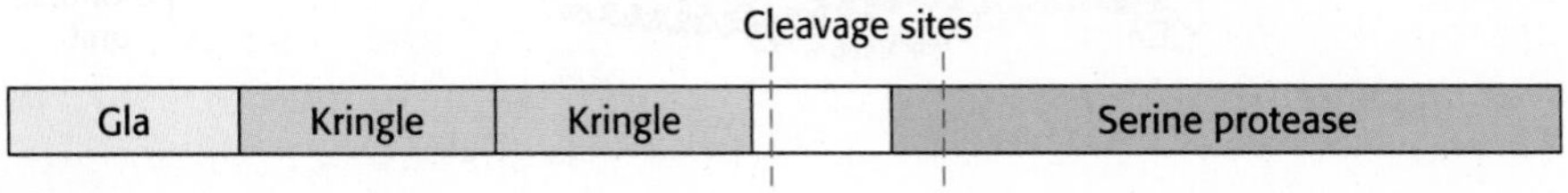

Cleavage sites

Gla	Kringle	Kringle		Serine protease

FIGURE 10.24 Modular structure of prothrombin. Cleavage of two peptide bonds yields thrombin. All the γ-carboxyglutamate residues are in the gla domain.

domain is able to bind Ca^{2+} (Figure 10.25). What is the effect of this binding? The binding of Ca^{2+} by prothrombin anchors the zymogen to phospholipid membranes derived from blood platelets after injury. This binding is crucial because it brings prothrombin into close proximity to two clotting proteins, factor X_a and factor V_a (a stimulatory protein), that catalyze its conversion into thrombin. Factor X_a cleaves the bond between arginine 274 and threonine 275 to release a fragment containing the first three domains. Factor X_a also cleaves the bond between arginine 323 and isoleucine 324 to yield active thrombin.

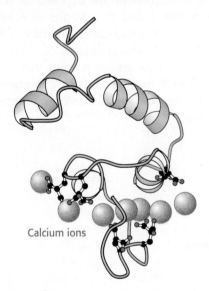

FIGURE 10.25 The calcium-binding region of prothrombin. Prothrombin binds calcium ions with the carboxyl groups of the modified amino acid γ-carboxyglutamate (red). [Drawn from 2PF2.pdb.]

Calcium ions

Fibrinogen is converted by thrombin into a fibrin clot

The best-characterized part of the clotting process is the final step in the cascade: the conversion of *fibrinogen* into fibrin by thrombin. Fibrinogen is a large glycoprotein composed of three nonidential chains, Aα, Bβ, and γ and is found in the blood plasma of all vertebrates. The overall composition of this 340-kDa protein is $(A\alpha_2B\beta_2\,\gamma_2)$. Three globular units are connected by two rods, and the rod regions are triple-stranded α-helical coiled coils, a recurring motif in proteins (Section 2.3; Figure 10.26A). Thrombin cleaves four *arginine–glycine peptide bonds* in the central globular region of fibrinogen (Figure 10.26B). On cleavage, an A peptide of 18 residues is released from each of the two Aα chains, as is a B peptide of 20 residues from each of the two Bβ chains. These A and B peptides are called *fibrinopeptides* (Figure 10.27). A fibrinogen molecule devoid of these fibrinopeptides is called a *fibrin monomer* and has the subunit structure $(\alpha\beta\gamma)_2$.

γ Globular unit

β Globular unit

Central globular unit

β Globular unit

γ Globular unit

Fibrinogen

FIGURE 10.26 Structure of a fibrinogen molecule. (A) A ribbon diagram. The two rod regions are α-helical coiled coils, connected to the three globular regions. (B) A schematic representation showing the positions of the fibrinopeptides (green and purple) in the central globular region. [Part A drawn from 1M1J.pdb.]

Upon thrombin cleavage, amino acid sequences are exposed in the central globular unit that interact with the γ and β subunits of other monomers. Polymerization occurs as more fibrin monomers interact with one another (Figure 10.27). Thus, analogous to the activation of chymotrypsinogen, peptide-bond cleavage exposes new amino termini that can participate in specific interactions. The newly formed "soft clot" is stabilized by the formation of amide bonds between the side chains of lysine and glutamine residues in different monomers.

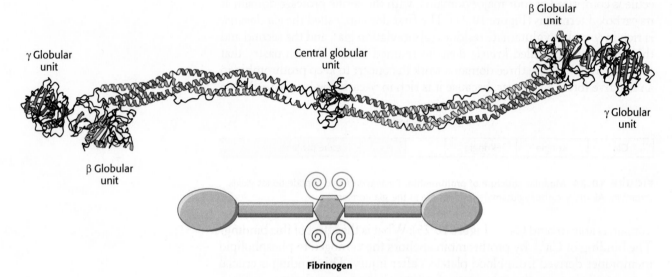

Glutamine   **Lysine**   Transglutaminase

Cross-link $+\ NH_4^+$

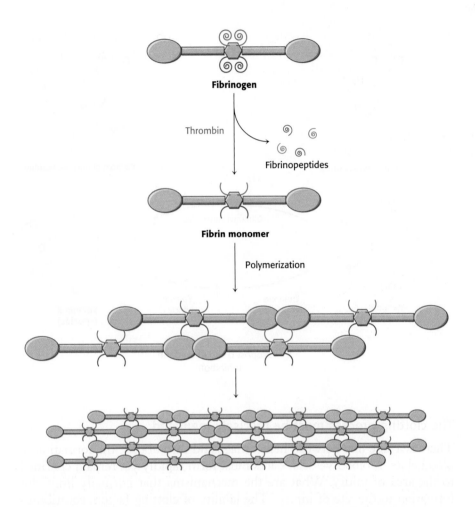

Fibrinogen

Thrombin

Fibrinopeptides

Fibrin monomer

Polymerization

This cross-linking reaction is catalyzed by *transglutaminase (factor XIII$_a$)*, which itself is activated from the protransglutaminase form by thrombin.

Vitamin K is required for the formation of γ-carboxyglutamate

Vitamin K (Figure 10.28) has been known for many years to be essential for the synthesis of prothrombin and several other clotting factors. Indeed, it is called vitamin K because a deficiency in this vitamin results in defective blood *k*oagulation (Scandinavian spelling). After ingestion, vitamin K is reduced to a dihydro derivative that is required by γ-glutamyl carboxy-lase to convert the first 10 glutamate residues in the amino-terminal region of prothrombin into γ-carboxyglutamate (Figure 10.29). Recall that γ-carboxy-glutamate, a strong chelator of Ca^{2+}, is required for the activation of pro-thrombin (p. 329). *Dicoumarol*, which is found in spoiled sweet clover, causes a fatal hemorrhagic disease in cattle fed on this hay. Cows fed dicoumarol synthesize an abnormal prothrombin that does not bind Ca^{2+}, in contrast with normal prothrombin. Dicoumarol was the first *anticoagulant* used to prevent thromboses in patients prone to clot formation. However, it is seldom used now because of poor absorption and gastrointestinal side effects. *Warfa-rin*, another vitamin K antagonist sold under the name of coumadin, is com-monly administered as an anticoagulant. Warfarin appears to inhibit the epoxide reductase and quinone reductase that are required to regenerate the dihydro derivative of vitamin K (Figure 10.29). Dicoumarol, warfarin, and their chemical derivatives serve as effective rat poisons.

Vitamin K

Dicoumarol

Warfarin

FIGURE 10.28 Structures of vitamin K and two antagonists, dicoumarol and warfarin.

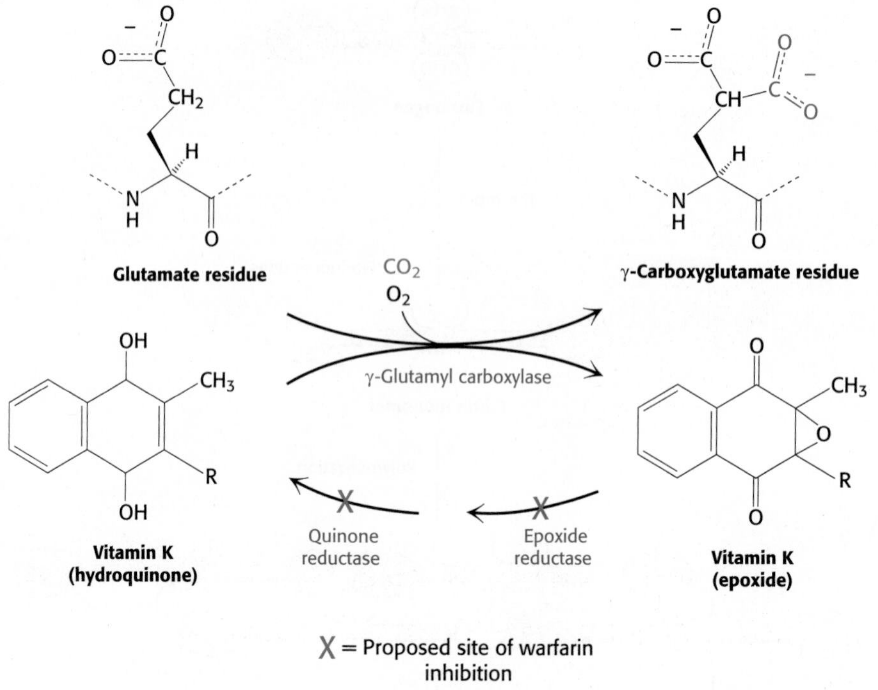

FIGURE 10.29 Synthesis of γ-carboxyglutamate by γ-glutamyl carboxylase. The formation of γ-carboxyglutamate requires the hydroquinone derivative of vitamin K, which is regenerated from the epoxide derivative by the sequential action of epoxide reductase and quinone reductase, both of which are inhibited by warfarin.

X = Proposed site of warfarin inhibition

The clotting process must be precisely regulated

There is a fine line between hemorrhage and thrombosis, the formation of blood clots in blood vessels. Clots must form rapidly yet remain confined to the area of injury. What are the mechanisms that normally limit clot formation to the site of injury? The lability of clotting factors contributes significantly to the control of clotting. Activated factors are short-lived because they are diluted by blood flow, removed by the liver, and degraded by proteases. For example, the stimulatory protein factors V_a and VIII_a are digested by protein C, a protease that is switched on by the action of thrombin. *Thus, thrombin has a dual function: it catalyzes the formation of fibrin and it initiates the deactivation of the clotting cascade.*

Specific inhibitors of clotting factors are also critical in the termination of clotting. For instance, *tissue factor pathway inhibitor* (TFPI) inhibits the complex of TF–VII_a–X_a that activates thrombin. Another key inhibitor is *antithrombin III*, a member of the serpin family of protease inhibitors (p. 326) that forms an irreversible inhibitory complex with thrombin. Antithrombin III resembles α_1-antitrypsin except that it inhibits thrombin much more strongly than it inhibits elastase (Figure 10.21). Antithrombin III also blocks other serine proteases in the clotting cascade—namely, factors XII_a, XI_a, IX_a, and X_a. The inhibitory action of antithrombin III is enhanced by the glycosaminoglycan *heparin*, a negatively charged polysaccharide (Section 11.3) found in mast cells near the walls of blood vessels and on the surfaces of endothelial cells (Figure 10.30). Heparin acts as an anticoagulant by increasing the rate of formation of irreversible complexes between antithrombin III and the serine protease clotting factors.

The importance of the ratio of thrombin to antithrombin is illustrated in the case of a 14-year-old boy who died of a bleeding disorder because of a mutation in his α_1-antitrypsin, which normally inhibits elastase. Methionine 358 in α_1-antitrypsin's binding pocket for elastase was replaced by arginine, resulting in a change in specificity from an elastase inhibitor to a thrombin

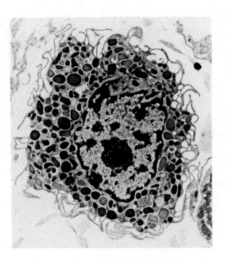

FIGURE 10.30 Electron micrograph of a mast cell. Heparin and other molecules in the dense granules, surrounding the cell nucleus, are released into the extracellular space when the cell is triggered to secrete. [Courtesy of Lynne Mercer.]

inhibitor. Activity of α_1-antitrypsin normally increases markedly after injury to counteract excess elastase arising from stimulated neutrophils. The mutant α_1-antitrypsin caused the patient's thrombin activity to drop to such a low level that hemorrhage ensued. We see here a striking example of how a change of a single residue in a protein can dramatically alter specificity and an example of the critical importance of having the right amount of a protease inhibitor.

Antithrombin limits the extent of clot formation, but what happens to the clots themselves? Clots are not permanent structures but are designed to dissolve when the structural integrity of damaged areas is restored. Fibrin is degraded by *plasmin*, a serine protease that hydrolyzes peptide bonds in the coiled-coil regions. Plasmin molecules can diffuse through aqueous channels in the porous fibrin clot to cut the accessible connector rods. Plasmin is formed by the proteolytic activation of *plasminogen*, an inactive precursor that has a high affinity for the fibrin clots. This conversion is carried out by *tissue-type plasminogen activator* (TPA), a 72-kDa protein that has a domain structure closely related to that of prothrombin (Figure 10.31; compare

Fibrin binding	Kringle	Kringle		Serine protease

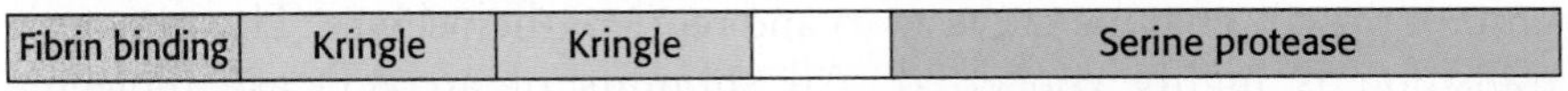

FIGURE 10.31 Modular structure of tissue-type plasminogen activator (TPA).

with Figure 10.24). However, a domain that targets TPA to fibrin clots replaces the membrane-targeting gla domain of prothrombin. The TPA bound to fibrin clots swiftly activates adhering plasminogen. In contrast, TPA activates free plasminogen very slowly. The gene for TPA has been cloned and expressed in cultured mammalian cells. TPA administered at the onset of a heart attack or a stroke caused by a blood clot increases the likelihood of survival without physical or cognitive disabilities (Figure 10.32).

Hemophilia revealed an early step in clotting

Some important breakthroughs in the elucidation of clotting pathways have come from studies of patients with bleeding disorders. *Classic hemophilia*, or *hemophilia A*, is the best-known clotting defect. This disorder is genetically transmitted as a sex-linked recessive characteristic.

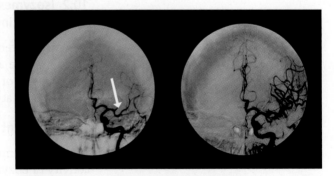

FIGURE 10.32 The effect of tissue-type plasminogen activator. Angiographic images demonstrate the effect of TPA administration. The left image shows an occluded cerebral artery (arrow) prior to TPA injection. The right image, made several hours after injection, reveals the restoration of blood flow to the cerebral artery. [Medical Body Scans/Science Source.]

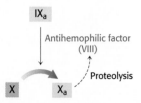

FIGURE 10.33 Action of antihemophilic factor. Antihemophilic factor (Factor VIII) stimulates the activation of factor X by factor IX$_a$. Interestingly, the activity of factor VIII is markedly increased by limited proteolysis by thrombin. This positive feedback amplifies the clotting signal and accelerates clot formation after a threshold has been reached.

In classic hemophilia, factor VIII (antihemophilic factor) of the intrinsic pathway is missing or has markedly reduced activity. Although factor VIII is not itself a protease, it markedly stimulates the activation of factor X, the final protease of the intrinsic pathway, by factor IX$_a$, a serine protease (Figure 10.33). Thus, activation of the intrinsic pathway is severely impaired in classic hemophilia.

In the past, hemophiliacs were treated with transfusions of a concentrated plasma fraction containing factor VIII. This therapy carried the risk of infection. Indeed, many hemophiliacs contracted hepatitis and AIDS. A safer source of factor VIII was urgently needed. With the use of biochemical purification and recombinant DNA techniques, the gene for factor VIII was isolated and expressed in cells grown in culture. Recombinant factor VIII purified from these cells has largely replaced plasma concentrates in treating hemophilia.

SUMMARY

10.1 Aspartate Transcarbamoylase Is Allosterically Inhibited by the End Product of Its Pathway

Allosteric proteins constitute an important class of proteins whose biological activity can be regulated. Specific regulatory molecules can modulate the activity of allosteric proteins by binding to distinct regulatory sites, separate from the active sites. These proteins have multiple functional sites, which display cooperativity as evidenced by a sigmoidal dependence of function on substrate concentration. Aspartate transcarbamoylase (ATCase), one of the best-understood allosteric enzymes, catalyzes the synthesis of N-carbamoylaspartate, the first intermediate in the synthesis of pyrimidines. ATCase is feedback inhibited by CTP, the final product of the pathway, and is stimulated by ATP. ATCase consists of separable catalytic (c_3) subunits (which bind the substrates) and regulatory (r_2) subunits (which bind CTP and ATP). The inhibitory effect of CTP, the stimulatory action of ATP, and the cooperative binding of substrates are mediated by large changes in quaternary structure. On binding substrates, the c_3 subunits of the c_6r_6 enzyme move apart and reorient themselves. This allosteric transition is highly concerted. All subunits of an ATCase molecule simultaneously interconvert from the T (low-affinity for substrate) to the R (high-affinity) state.

10.2 Isozymes Provide a Means of Regulation Specific to Distinct Tissues and Developmental Stages

Isozymes differ in structural characteristics but catalyze the same reaction. They provide a means of fine-tuning metabolism to meet the needs of a given tissue or developmental stage. Isozymes are encoded by separate genes, which are often the result of gene-duplication events.

10.3 Covalent Modification Is a Means of Regulating Enzyme Activity

The covalent modification of proteins is a potent means of controlling the activity of enzymes and other proteins. Phosphorylation is a common type of reversible covalent modification. Signals can be highly amplified by phosphorylation because a single protein kinase can act on many target molecules. The regulatory actions of protein kinases are reversed by protein phosphatases, which catalyze the hydrolysis of attached phosphoryl groups.

Cyclic AMP serves as an intracellular messenger in the transduction of many hormonal and sensory stimuli. Cyclic AMP switches on protein kinase A, a major multifunctional kinase, by binding to the regulatory subunit of the enzyme, thereby releasing the active catalytic subunits of PKA. In the absence of cAMP, the catalytic sites of PKA are occupied by pseudosubstrate sequences of the regulatory subunit.

10.4 Many Enzymes Are Activated by Specific Proteolytic Cleavage

The activation of an enzyme by the proteolytic cleavage of one or a few peptide bonds is a recurring control mechanism seen in processes as diverse as the activation of digestive enzymes and blood clotting. The inactive precursor is a zymogen (proenzyme). Trypsinogen is activated by enteropeptidase or trypsin, and trypsin then activates a host of other zymogens, leading to the digestion of foodstuffs. For instance, trypsin converts chymotrypsinogen, a zymogen, into active chymotrypsin by hydrolyzing a single peptide bond.

A striking feature of the clotting process is that it is accomplished by a cascade of zymogen conversions, in which the activated form of one clotting factor catalyzes the activation of the next precursor. Many of the activated clotting factors are serine proteases. In the final step of clot formation, fibrinogen, a highly soluble molecule in the plasma, is converted by thrombin into fibrin by the hydrolysis of four arginine–glycine bonds. The resulting fibrin monomers spontaneously form long, insoluble fibers called fibrin. Zymogen activation is also essential in the lysis of clots. Plasminogen is converted into plasmin by tissue-type plasminogen activator (TPA). Plasmin, a serine protease, then cleaves fibrin. Although zymogen activation is irreversible, specific inhibitors of some proteases inactivate the enzymes by binding in the active site. The irreversible protein inhibitor antithrombin III holds blood clotting in check in the clotting cascade.

APPENDIX

Biochemistry in Focus

Phosphoribosylpyrophosphate synthetase-induced gout

Gout is a joint disease in which an excess of urate crystallizes in the fluid and lining of the joints, resulting in painful inflammation when cells of the immune system engulf the sodium urate crystals. Urate is a final product of the degradation of purines, the base components of the adenine and guanine nucleotides of DNA and RNA.

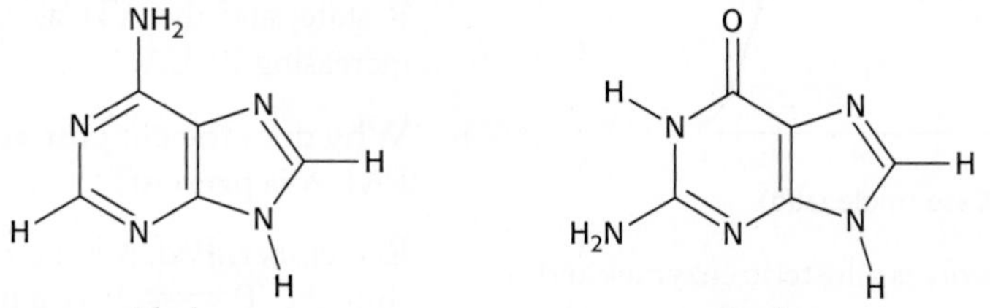

Adenine

Guanine

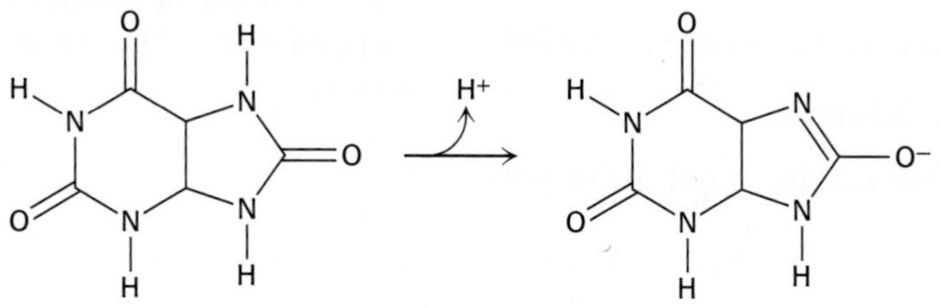

Uric acid

Urate

Although gout can be caused by a number of metabolic deficiencies (Chapter 25), one cause is a mutation in an important enzyme in purine synthesis, phosphoribosylpyrophosphate synthetase (PRS). PRS catalyzes the first step in the metabolic pathway leading to purine synthesis, the synthesis of phosphoribosylpyrophosphate (PRPP).

$$\text{Ribose 5-phosphate} + \text{ATP} \xrightarrow{\substack{\text{Phosphoribosylpyrophosphate} \\ \text{synthetase}}}$$

$$\text{5-phosphoribosyl-1-pyrophosphate} + \text{AMP}$$

Given its location as the first step in the metabolic pathway leading to purine synthesis, we can hypothesize that PRS is an allosteric enzyme that would be inhibited by the purine products of the pathway. Genetic studies have established that a mutation in the gene encoding PRS can result in gout.

What alterations in PRS activity would lead to an overproduction of purines? Two possibilities come to mind. One is a mutation in the regulatory site that renders the enzyme insensitive to feedback inhibition. Another is a mutation that would result in increased enzyme activity, most likely a mutation in the promoter regions of the gene (p. 130) that results in increased enzyme synthesis. In fact, both types of mutations are observed. In certain people, the regulatory site has undergone a mutation that renders PRS insensitive to feedback inhibition. The catalytic activity of PRS is unaffected. In other cases, the increase in enzyme quantity results in greater purine synthesis. In both circumstances, a glut of purine nucleotides results, which are converted into urate. The excess urate accumulates and causes gout.

APPENDIX

Problem-Solving Strategies

Problem-Solving Strategy 1

PROBLEM: Recall that phosphonacetyl-L-aspartate (PALA) is a potent inhibitor of ATCase because it mimics the two physiological substrates. However, in the presence of substrates, low concentrations of this unreactive bisubstrate analog *increase* the reaction velocity. On the addition of PALA, the reaction rate increases until an average of three molecules of PALA are bound per molecule of enzyme. This maximal velocity is 17-fold greater than it is in the absence of PALA. The reaction rate then decreases to nearly zero on the addition of three more molecules of PALA per molecule of enzyme. Why do low concentrations of PALA activate ATCase?

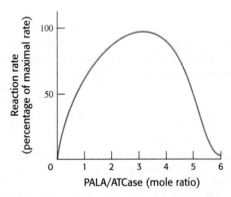

SOLUTION: When a question involves allosteric enzymes and kinetics, it is always a good bet to review cooperativity and the T ⇌ R equilibrium.

▶ **What is cooperativity in terms of allosteric enzymes?**

▶ **What is the T ⇌ R equilibrium?**

Cooperativity means that the activity at one active site affects the activity at others.

ATCase exists in two forms: a T state that is less active and an R state that is catalytically active. The equilibrium favors the T state.

The question asks about the effect of PALA on ATCase activity.

▶ **What is PALA?**

PALA is a bisubstrate analog and a competitive inhibitor of ATCase.

Now the crux of the problem. The question states that, up to a point, increasing amounts of PALA enhance the rate of the reaction for a given substrate concentration.

▶ **How could a competitive inhibitor enhance the rate of the ATCase reaction?**

The binding of PALA switches ATCase from the T to the R state because PALA acts as a substrate analog. PALA disrupts the T ⇌ R equilibrium in favor of R. Thus, the limited amount of substrate can more readily bind the R state, and the ATCase activity initially increases with increasing PALA.

▶ **Why does the enzyme activity eventually fall as more PALA is present?**

Remember, PALA is a competitive inhibitor. PALA will shift the T ⇌ R equilibrium in favor of R, but once the inhibitor occupies more and more active sites, the real substrate will be unable to bind the enzyme. When PALA occupies all of the active sites, the catalytic activity falls to zero.

Problem-Solving Strategy 2

PROBLEM: Examine the metabolic pathway shown below. Which of the enzymes, identified as "e" with a numeric subscript, is likely to be the allosteric enzyme that controls the synthesis of G?

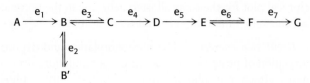

SOLUTION: Allosteric enzymes catalyze the committed step in metabolic pathways.

▶ **What is meant by the committed step?**

The committed step is the reaction that, once it occurs, causes all of the subsequent reactions of the pathway to occur.

▶ **What characterizes the committed step in terms of free energy change?**

The committed step is irreversible under cellular conditions.

▶ **Examining the metabolic pathway above, what reactions are irreversible?**

The reactions catalyzed by e_1, e_4, e_5 and e_7.

▶ **The enzyme e_1 is unlikely to catalyze the committed step. Explain why.**

The product of e_1, B, has two possible fates, conversion to C or B′. Thus, the product is not committed to the synthesis of G.

▶ **Of the remaining enzymes that catalyze irreversible reactions, which is most likely to be the allosteric enzyme.**

The first enzyme that catalyzes a reaction that commits the pathway to the synthesis of G is enzyme e_4. This enzyme is most likely the allosteric enzyme.

KEY TERMS

cooperativity (p. 310)

committed step (p. 310)

feedback (end-product) inhibition (p. 310)

allosteric (regulatory) site (p. 311)

homotropic effect (p. 315)

concerted model (p. 315)

sequential model (p. 315)

heterotropic effect (p. 317)

isozyme (isoenzyme) (isoform) (p. 317)

protein kinase (p. 319)

consensus sequence (p. 320)

protein phosphatase (p. 320)

protein kinase A (PKA) (p. 321)

pseudosubstrate sequence (p. 322)

phosphoproteome (p. 323)

zymogen (proenzyme) (p. 323)

enzymatic cascade (p. 328)

hemostasis (p. 328)

intrinsic pathway (p. 328)

extrinsic pathway (p. 328)

PROBLEMS

1. *Context please.* The allosteric properties of aspartate transcarbamoylase have been discussed in detail in this chapter. What is the function of aspartate transcarbamoylase? ✓ 1, ✓ 2

2. *Like Bonnie and Clyde.* Match the term with the description.

(a) Cooperativity _____

(b) Isozyme _____

(c) Concerted model _____

(d) Sequential model _____

(e) Protein kinase _____

(f) Homotropic effect _____

(g) Feedback inhibition _____

1. End-product regulation

2. Activated by cAMP

3. Different conformations allowed

4. Different primary structure, catalyze same reaction

5. Active sites collaborate

6. Average of many

7. All subunits in the same conformation.

(h) Heterotropic effect _____

(i) Consensus sequence _____

(j) Protein kinase A _____

8. Effect of substrates

9. Effect of regulatory molecules

10. Requires ATP

3. *To tell the truth.* Suppose you have a dimeric allosteric enzyme that follows the concerted model. Which of the following statements are true? ✓ 3

(a) The equilibrium between the T state and R state favors the T state.

(b) The enzyme can exist as a RR dimer.

(c) The enzyme can exist as an RT dimer.

(d) The RR form of the enzyme is most active.

4. *Activity profile.* A histidine residue in the active site of aspartate transcarbamoylase is thought to be important in stabilizing the transition state of the bound substrates. Predict the pH dependence of the catalytic rate, assuming that this interaction is essential and dominates the pH-activity profile of the enzyme. (See equations on p. 15.) ✓❶

5. *Knowing when to say when.* What is feedback inhibition? Why is it a useful property? ✓❷

6. *Knowing when to get going.* What is the biochemical rationale for ATP serving as a positive regulator of ATCase? ✓❶, ✓❸

7. *No T.* What would be the effect of a mutation in an allosteric enzyme that resulted in a T/R ratio of 0? ✓❸

8. *Turned upside down.* An allosteric enzyme that follows the concerted model has a T/R ratio of 300 in the absence of substrate. Suppose that a mutation reversed the ratio. How would this mutation affect the relation between the rate of the reaction and the substrate concentration? ✓❸

9. *Partners.* As shown in Figure 10.2, CTP inhibits ATCase; however, the inhibition is not complete. Can you suggest another molecule that might enhance the inhibition of ATCase? Hint: See Figure 25.2. ✓❶

10. *RT equilibrium.* Differentiate between homotropic and heterotropic effectors. ✓❸

11. *Because it's an enzyme.* X-ray crystallographic studies of ATCase in the R form required the use of the bisubstrate analog PALA. Why was this analog, a competitive inhibitor, used instead of the actual substrates? ✓❶

12. *Allosteric switching.* A substrate binds 100 times as tightly to the R state of an allosteric enzyme as to its T state. Assume that the concerted (MWC) model applies to this enzyme. (See equations for the Concerted Model in the Appendix to Chapter 7.)

(a) By what factor does the binding of one substrate molecule per enzyme molecule alter the ratio of the concentrations of enzyme molecules in the R and T states?

(b) Suppose that L, the ratio of [T] to [R] in the absence of substrate, is 10^7 and that the enzyme contains four binding sites for substrate. What is the ratio of enzyme molecules in the R state to those in the T state in the presence of saturating amounts of substrate, assuming that the concerted model is obeyed? ✓❸

13. *Allosteric transition.* Consider an allosteric protein that obeys the concerted model. Suppose that the ratio of T to R formed in the absence of ligand is 10^5, $K_T = 2$ mM, and $K_R = 5$ μM. The protein contains four binding sites for ligand. What is the fraction of molecules in the R form when 0, 1, 2, 3, and 4 ligands are bound? (See equations for the Concerted Model in the Appendix to Chapter 7.) ✓❸

14. *Negative cooperativity.* You have isolated a dimeric enzyme that contains two identical active sites. The binding of substrate to one active site decreases the substrate affinity of the other active site. Can the concerted model account for this negative cooperativity? Hint: See Section 7.2.

15. *A new view of cooperativity.* Draw a double-reciprocal plot for a typical Michaelis–Menten enzyme and an allosteric enzyme that have the same V_{max} and K_M. Draw a double reciprocal plot for the same allosteric enzyme in the presence of an allosteric inhibitor and an allosteric stimulator. ✓❶

16. *Regulation energetics.* The phosphorylation and dephosphorylation of proteins is a vital means of regulation. Protein kinases attach phosphoryl groups, whereas only a phosphatase will remove the phosphoryl group from the target protein. What is the energy cost of this means of covalent regulation? ✓❺

17. *Vive la différence.* What is an isozyme? ✓❹

18. *Fine-tuning biochemistry.* What is the advantage for an organism to have isozymic forms of an enzyme? ✓❹

19. *Making matches.*

(a) ATCase _____	1. Protein phosphorylation catalyst
(b) T state _____	2. Required to modify glutamate
(c) R state _____	3. Activates a particular kinase
(d) Phosphorylation _____	4. Proenzyme
(e) Kinase _____	5. Activates trypsin
(f) Phosphatase _____	6. Common covalent modification
(g) cAMP _____	7. Inhibited by CTP
(h) Zymogen _____	8. Less-active state of an allosteric protein
(i) Enteropeptidase _____	9. Initiates extrinsic pathway
(j) Vitamin K _____	10. Forms fibrin
(k) Thrombin _____	11. More-active state of an allosteric protein
(l) Tissue factor _____	12. Removes phosphates

20. *Powering change.* Phosphorylation is a common covalent modification of proteins in all forms of life. What energetic advantages accrue from the use of ATP as the phosphoryl donor? ✓❺, ✓❻

21. *No going back.* What is the key difference between regulation by covalent modification and specific proteolytic cleavage? ✓❻, ✓❼

22. *Zymogen activation.* When very low concentrations of pepsinogen are added to acidic media, how does the half-time for activation depend on zymogen concentration? ✓❼

23. *No protein shakes advised.* Predict the physiological effects of a mutation that resulted in a deficiency of enteropeptidase.

24. *A revealing assay.* Suppose that you have just examined a young boy with a bleeding disorder highly suggestive of classic hemophilia (factor VIII deficiency). Because of the late hour, the laboratory that carries out specialized coagulation assays is closed. However, you happen to have a sample of blood from a classic hemophiliac whom you admitted to the hospital an hour earlier. What is the simplest and most rapid test that you can perform to determine whether your present patient also is deficient in factor VIII activity?

25. *Counterpoint.* The synthesis of factor X, like that of pro-thrombin, requires vitamin K. Factor X also contains γ-carboxyglutamate residues in its amino-terminal region. However, activated factor X, in contrast with thrombin, retains this region of the molecule. What is a likely functional consequence of this difference between the two activated species?

26. *A discerning inhibitor.* Antithrombin III forms an irre-versible complex with thrombin but not with prothrombin. What is the most likely reason for this difference in reactivity?

27. *Drug design.* A drug company has decided to use recom-binant DNA methods to prepare a modified α_1-antitrypsin that will be more resistant to oxidation than is the naturally occurring inhibitor. Which single amino acid substitution would you recommend?

28. *Blood must flow.* Why is inappropriate blood-clot forma-tion dangerous?

29. *Hemostasis.* Thrombin functions in both coagulation and fibrinolysis. Explain.

30. *Dissolution row.* What is tissue-type plasminogen activa-tor and what is its role in preventing heart attacks?

31. *Joining together.* What differentiates a soft clot from a mature clot?

Data Interpretation Problems

32. *Distinguishing between models.* The following graph shows the fraction of an allosteric enzyme in the R state (f_R) and the fraction of active sites bound to substrate (Y) as a function of substrate concentration. Which model, the concerted or sequential, best explains these results? ✅❶

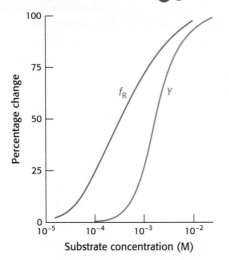

33. *Reporting live from ATCase 1.* ATCase underwent a reac-tion with tetranitromethane to form a colored nitrotyrosine group ($\lambda_{max} = 430$ nm) in each of its catalytic chains. The absorption by this reporter group depends on its immediate environment. An essential lysine residue at each catalytic site also was modified to block the binding of substrate. Cata-lytic trimers from this doubly modified enzyme were then combined with native trimers to form a hybrid enzyme. The absorption by the nitrotyrosine group was measured on addi-tion of the substrate analog succinate. What is the significance of the alteration in the absorbance at 430 nm? ✅❶

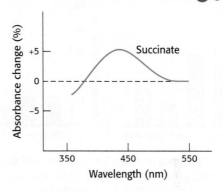

34. *Reporting live from ATCase 2.* A different ATCase hybrid was constructed to test the effects of allosteric activators and inhibitors. Normal regulatory subunits were combined with nitrotyrosine-containing catalytic subunits. The addition of ATP in the absence of substrate increased the absorbance at 430 nm, the same change elicited by the addition of succi-nate (see the graph in Problem 33). Conversely, CTP in the absence of substrate decreased the absorbance at 430 nm. What is the significance of the changes in absorption of the reporter groups? ✅❶

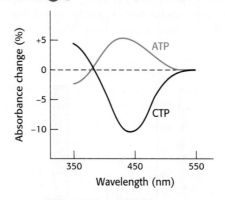

35. *Conviviality and PKA.* Recent studies have suggested that protein kinase A may be important in establishing behaviors in many organisms, including humans. One study investi-gated the role of PKA in locust behavior. Certain species of locust live solitary lives until crowded, at which point they become gregarious—they prefer the crowded life. ✅❺

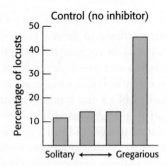

Control (no inhibitor)

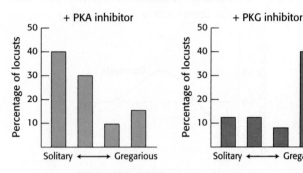

+ PKA inhibitor   + PKG inhibitor

Locusts were grouped together for one hour, and then allowed to stay with the group or move away. Prior to crowding, some insects were injected with a PKA inhibitor, a cyclic GMP-dependent kinase inhibitor or no inhibitor, as indicated. The results are shown above.

(a) What is the response of the control group to crowding?

(b) What is the result if the insects are first treated with PKA inhibitor? PKG inhibitor?

(c) What was the purpose of the experiment with the PKG inhibitor?

(d) What do these results suggest about the role of PKA in the transition from a solitary to a gregarious lifestyle?

The above experiments were repeated on a different species of insect that is always gregarious. The results are shown below.

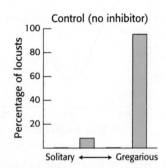

Control (no inhibitor)

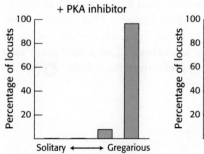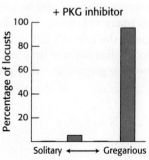

+ PKA inhibitor   + PKG inhibitor

(e) What do these results suggest about the role of PKA in insects that are always gregarious?

Chapter Integration Problems

36. *Repeating heptads.* Each of the three types of fibrin chains contains repeating heptapeptide units (*abcdefg*) in which residues *a* and *d* are hydrophobic. Propose a reason for this regularity.

37. *Density matters.* The sedimentation value of aspartate transcarbamoylase decreases when the enzyme switches to the R state. On the basis of the allosteric properties of the enzyme, explain why the sedimentation value decreases.

38. *Too tight a grip.* Trypsin cleaves proteins on the carboxyl side of lysine. Trypsin inhibitor has a lysine residue, and binds to trypsin, yet it is not a substrate. Explain.

39. *Apparently not the four-leaf variety.* Cows that graze on spoiled sweet clover, which contains dicoumarol, die from hemorrhagic disease. The cause of death is defective pro-thrombin. However, the amino acid composition of the defective prothrombin is identical to that of normal prothrombin. What is the mechanism of action of dicoumarol? Why are the amino acid compositions of the defective and normal pro-thrombin the same?

Mechanism Problems

40. *Aspartate transcarbamoylase.* Write the mechanism (in detail) for the conversion of aspartate and carbamoyl phosphate into *N*-carbamoylaspartate. Include a role for the histidine residue present in the active site. ✔❶

41. *Protein kinases.* Write a mechanism (in detail) for phosphorylation of a serine residue by ATP catalyzed by protein kinase. ✔❺

Think ▶ Pair ▶ Share Problems

42. Why are allosteric enzymes crucial for the control of metabolism?

43. If isolated regulatory subunits and catalytic subunits of ATCase are mixed, the native enzyme is reconstituted. What is the biological significance of the observation?

Carbohydrates

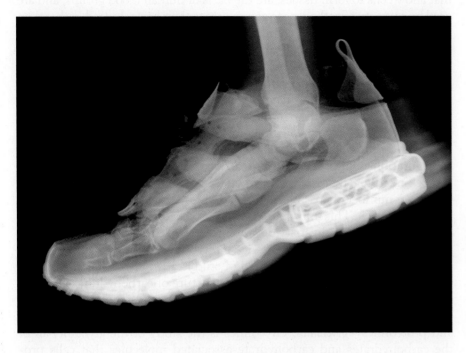

Carbohydrates are important fuel molecules, but they play many other biochemical roles, including protection against high-impact forces. The cartilage of a runner's foot cushions the impact of each step she takes. A key component of cartilage are molecules called glycosaminoglycans, large polymers made up of many repeats of dimers such as the pair shown above. [Nick Veasey/Getty Images.]

LEARNING OBJECTIVES

By the end of this chapter you should be able to:

1. Describe the main roles of carbohydrates in nature.

2. Define carbohydrate and monosaccharide.

3. Describe how simple carbohydrates are linked to form complex carbohydrates.

4. Explain how carbohydrates are linked to proteins.

5. Describe the three main classes of glycoproteins and explain their biochemical roles.

6. Define what lectins are and outline their biochemical functions.

OUTLINE

11.1 Monosaccharides Are the Simplest Carbohydrates

11.2 Monosaccharides Are Linked to Form Complex Carbohydrates

11.3 Carbohydrates Can Be Linked to Proteins to Form Glycoproteins

11.4 Lectins Are Specific Carbohydrate-Binding Proteins

For years, the study of carbohydrates was considered less exciting than many if not most topics of biochemistry. Carbohydrates were recognized as important fuels and structural components but were thought to be peripheral to most key activities of the cell. In essence, they were considered the underlying girders and fuel for a magnificent piece of biochemical architecture. This view has changed dramatically in recent years. We have learned that cells of all organisms are coated in a dense and complex coat of carbohydrates, frequently attached to lipids and proteins. Secreted proteins are often extensively decorated with carbohydrates essential to a protein's function. Indeed, the attachment of a carbohydrate to a protein is the most common post-translational modification of proteins. The extracellular matrix in higher eukaryotes—the environment in which the cells live—is rich in secreted carbohydrates central to cell survival and cell-to-cell communication. Carbohydrates, carbohydrate-containing proteins, and specific carbohydrate-binding proteins are required for interactions that allow cells to form tissues, are the basis of human blood groups, and are used by a variety of pathogens to gain access to their hosts. Indeed, rather than being mere infrastructure components, carbohydrates supply details and enhancements to the biochemical architecture of the cell, helping to define the functionality and uniqueness of the cell.

A key property of carbohydrates that allows their many functions is the tremendous *structural diversity* possible within this class of molecules. Monosaccharides are the simplest carbohydrates. Monosaccharides are small molecules—typically containing from three to nine carbon atoms bound to hydroxyl groups—that vary in size and in the stereochemical configuration at one or more of the carbon centers. These monosaccharides can be linked together to form a large variety of oligosaccharide structures. The sheer number of possible oligosaccharides makes this class of molecules information rich. This information, when attached to proteins, can augment the already immense diversity of proteins.

The realization of the importance of carbohydrates to so many aspects of biochemistry has spawned a field of study called *glycobiology*. Glycobiology is the study of the synthesis and structure of carbohydrates and how carbohydrates are attached to and recognized by other molecules such as proteins. Along with a new field comes a new "omics" to join genomics and proteomics—*glycomics*. Glycomics is the study of the glycome, all of the carbohydrates and carbohydrate-associated molecules that cells produce. Like the proteome, the glycome is dynamic, depending on cellular and environmental conditions. Unraveling oligosaccharide structures and elucidating the effects of their attachment to other molecules constitutes a tremendous challenge in the field of biochemistry.

11.1 Monosaccharides Are the Simplest Carbohydrates

Carbohydrates are carbon-based molecules that are rich in hydroxyl groups. Indeed, the empirical formula for many carbohydrates is $(CH_2O)_n$—literally, a carbon hydrate. Simple carbohydrates are called *monosaccharides*. These simple sugars serve not only as fuel molecules but also as fundamental constituents of living systems. For instance, DNA has a backbone consisting of alternating phosphoryl groups and deoxyribose, a cyclic five-carbon sugar.

Monosaccharides are aldehydes or ketones that have two or more hydroxyl groups. The smallest monosaccharides, composed of three carbon atoms, are dihydroxyacetone and D- and L-glyceraldehyde.

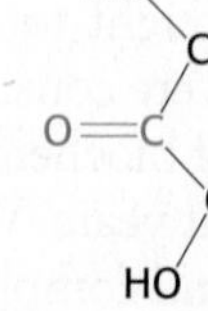

Dihydroxyacetone
(a ketose)

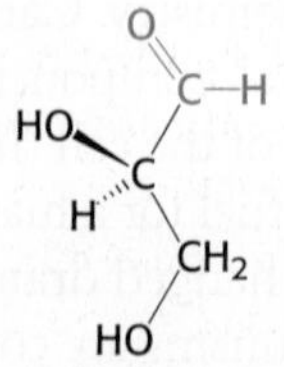

D-Glyceraldehyde
(an aldose)

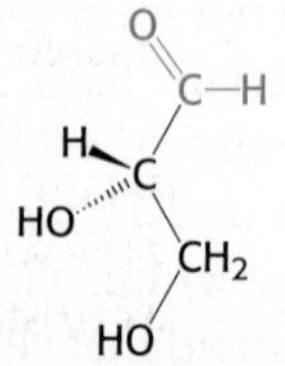

L-Glyceraldehyde
(an aldose)

Dihydroxyacetone is called a *ketose* because it contains a keto group (in red above), whereas glyceraldehyde is called an *aldose* because it contains an aldehyde group (also in red). They are referred to as *trioses* (tri- for three, referring to the three carbon atoms that they contain). Similarly, simple monosaccharides with four, five, six, and seven carbon atoms are called *tetroses, pentoses, hexoses,* and *heptoses,* respectively. Perhaps the monosaccharides of which we are most aware are the hexoses, such as glucose and fructose. Glucose is an essential energy source for virtually all forms of life. Fructose is commonly used as a sweetener that is converted into glucose derivatives inside the cell.

Carbohydrates can exist in a dazzling variety of isomeric forms (Figure 11.1). Dihydroxyacetone and glyceraldehyde are *constitutional isomers* because they have identical molecular formulas but differ in how the atoms are ordered. *Stereoisomers* are isomers that differ in spatial arrangement. Recall from the discussion of amino acids (p. 31) that stereoisomers are designated as having either D or L configuration. Glyceraldehyde has a single asymmetric carbon atom and, thus, there are two stereoisomers of this sugar: D-glyceraldehyde and L-glyceraldehyde. These molecules are a type of stereoisomer called *enantiomers,* which are mirror images of each other. Most vertebrate monosaccharides have the D configuration. According to convention, the D and L isomers are determined by the configuration of the asymmetric carbon atom farthest from the aldehyde or keto group. Dihydroxyacetone is the only monosaccharide without at least one asymmetric carbon atom.

Monosaccharides made up of more than three carbon atoms have multiple asymmetric carbons, and so they can exist not only as enantiomers but also as *diastereoisomers,* isomers that are not mirror images of each other. The number of possible stereoisomers equals 2^n, where n is the number of asymmetric carbon atoms. Thus, a six-carbon aldose with 4 asymmetric carbon atoms can exist as 16 possible diastereoisomers, of which glucose is one such isomer.

FIGURE 11.1 Isomeric forms of carbohydrates.

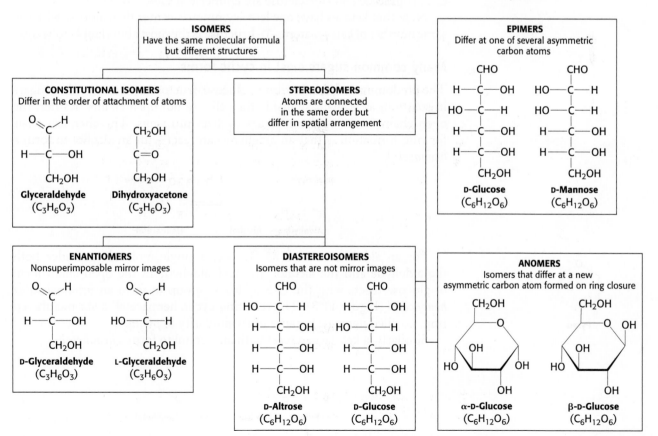

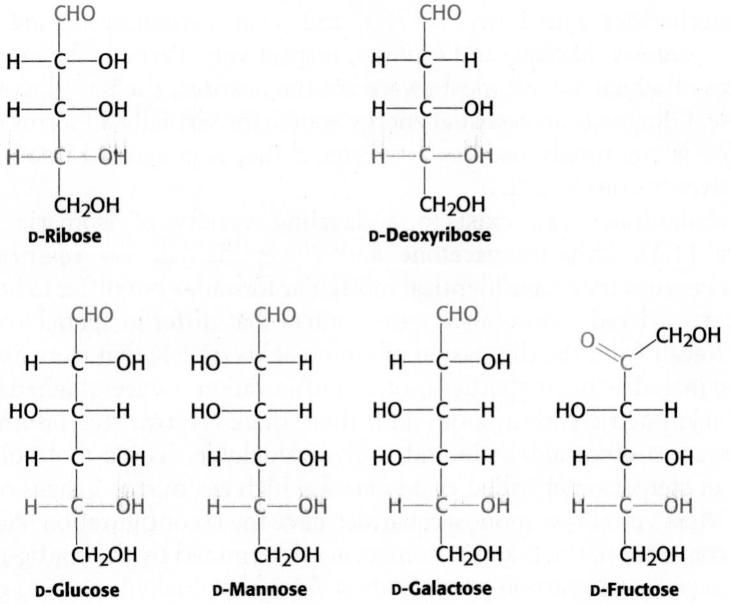

FIGURE 11.2 Common monosaccharides. Aldoses contain an aldehyde group (shown in blue), whereas ketoses, such as fructose, contain a keto group (shown in red). The asymmetric carbon atom (shown in green) farthest from the aldehyde or keto group designates the structures as being in the D configuration.

CHO
H—C—OH
H—C—OH
H—C—OH
CH₂OH
D-Ribose

CHO
H—C—H
H—C—OH
H—C—OH
CH₂OH
D-Deoxyribose

CHO
H—C—OH
HO—C—H
H—C—OH
H—C—OH
CH₂OH
D-Glucose

CHO
HO—C—H
HO—C—H
H—C—OH
H—C—OH
CH₂OH
D-Mannose

CHO
H—C—OH
HO—C—H
HO—C—H
H—C—OH
CH₂OH
D-Galactose

O=C—CH₂OH
HO—C—H
H—C—OH
H—C—OH
CH₂OH
D-Fructose

Figure 11.2 shows the common sugars that we will see most frequently in our study of biochemistry. D-Ribose, the carbohydrate component of RNA, is a five-carbon aldose, as is deoxyribose, the monosaccharide component of deoxynucleotides. D-Glucose, D-mannose, and D-galactose are abundant six-carbon aldoses. Note that D-glucose and D-mannose differ in configuration only at C-2, the carbon atom in the second position. Sugars that are diastereoisomers differing in configuration at only a single asymmetric center are *epimers*. Thus, D-glucose and D-mannose are epimeric at C-2; D-glucose and D-galactose are epimeric at C-4.

Note that ketoses have one less asymmetric center than aldoses with the same number of carbon atoms. D-Fructose is the most abundant ketohexose.

Many common sugars exist in cyclic forms

The predominant forms of ribose, glucose, fructose, and many other sugars in solution, as is the case inside the cell, are not open chains. Rather, the open-chain forms of these sugars cyclize into rings. The chemical basis for ring formation is that an aldehyde can react with an alcohol to form a *hemiacetal*.

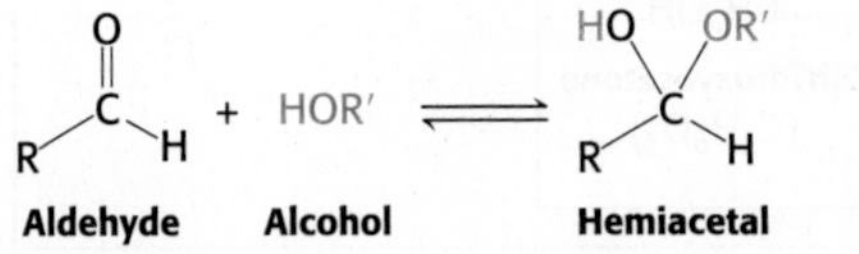

O
‖
R—C—H   +   HOR′   ⇌   R—C(OH)(OR′)—H
Aldehyde   **Alcohol**   **Hemiacetal**

For an aldohexose such as glucose, a single molecule provides both the aldehyde and the alcohol: the C-1 aldehyde in the open-chain form of glucose reacts with the C-5 hydroxyl group to form an *intramolecular hemiacetal* (Figure 11.3). The resulting cyclic hemiacetal, a six-membered ring, is called *pyranose* because of its similarity to *pyran*.

Similarly, a ketone can react with an alcohol to form a *hemiketal*.

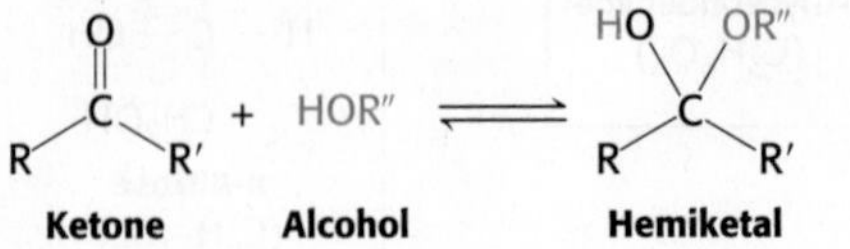

Pyran

O
‖
R—C—R′   +   HOR″   ⇌   R—C(OH)(OR″)—R′
Ketone   **Alcohol**   **Hemiketal**

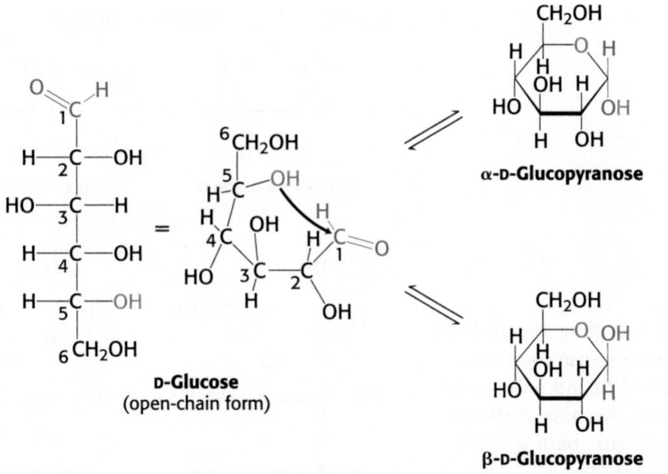

FIGURE 11.3 Pyranose formation. The open-chain form of glucose cyclizes when the C-5 hydroxyl group attacks the oxygen atom of the C-1 aldehyde group to form an intramolecular hemiacetal. Two anomeric forms, designated α and β, can result.

The C-2 keto group in the open-chain form of a ketohexose, such as fructose, can form an *intramolecular hemiketal* by reacting with either the C-6 hydroxyl group to form a six-membered cyclic hemiketal or the C-5 hydroxyl group to form a five-membered cyclic hemiketal (Figure 11.4). The five-membered ring is called a *furanose* because of its similarity to *furan*.

Furan

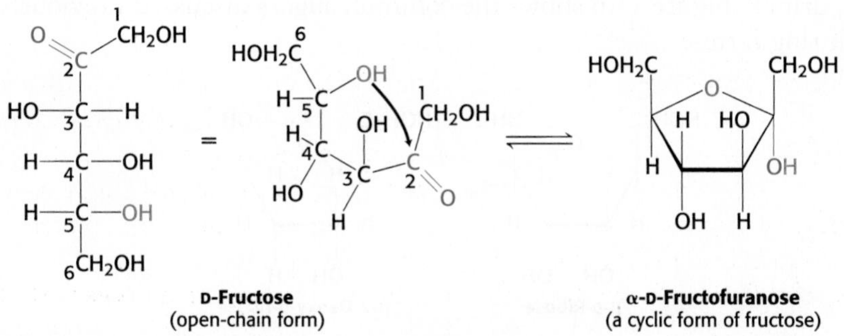

FIGURE 11.4 Furanose formation. The open-chain form of fructose cyclizes to a five-membered ring when the C-5 hydroxyl group attacks the C-2 ketone to form an intramolecular hemiketal. Two anomers are possible, but only the α anomer is shown.

The depictions of glucopyranose (glucose) and fructofuranose (fructose) shown in Figures 11.3 and 11.4 are *Haworth projections*. In such projections, the carbon atoms in the ring are not written out. The approximate plane of the ring is perpendicular to the plane of the paper, with the heavy line on the ring projecting toward the reader.

We have seen that carbohydrates can contain many asymmetric carbon atoms. An additional asymmetric center is created when a cyclic hemiacetal is formed, generating yet another diastereoisomeric form of sugars called *anomers*. In glucose, C-1 (the carbonyl carbon atom in the open-chain form) becomes an asymmetric center. Thus, two ring structures can be formed: α-D-glucopyranose and β-D-glucopyranose (Figure 11.3). For D sugars drawn as Haworth projections in the standard orientation as shown in Figure 11.3, the *designation α means that the hydroxyl group attached to C-1 is on the opposite side of the ring as C-6; β means that the hydroxyl group is on the same side of the ring as C-6*. The C-1 carbon atom is called the *anomeric carbon atom*. An equilibrium mixture of glucose contains approximately one-third α anomer, two-thirds β anomer, and <1% of the open-chain form.

The furanose-ring form of fructose also has anomeric forms, in which α and β refer to the hydroxyl groups attached to C-2, the anomeric carbon atom (Figure 11.4). Fructose forms both pyranose and furanose rings. The pyranose form predominates in fructose that is free in solution, and the furanose form predominates in many fructose derivatives (Figure 11.5).

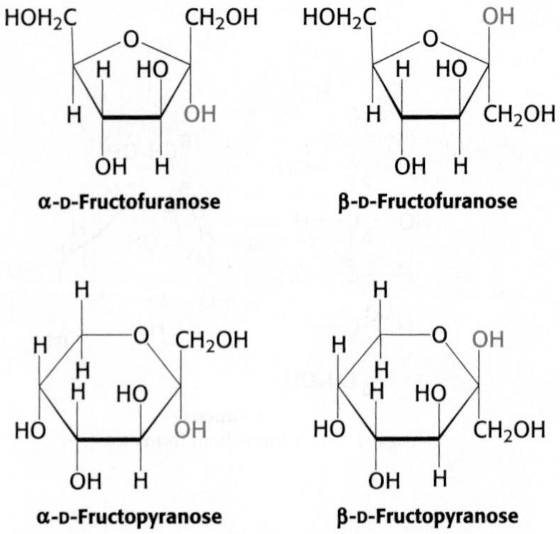

α-D-Fructofuranose

β-D-Fructofuranose

FIGURE 11.5 Ring structures of fructose. Fructose can form both five-membered furanose (top) and six-membered pyranose (bottom) rings. In each case, both α and β anomers are possible.

α-D-Fructopyranose

β-D-Fructopyranose

β-D-Fructopyranose, found in honey, is one of the sweetest chemicals known. The β-D-fructofuranose form is not nearly as sweet. Heating converts β-fructopyranose into the β-fructofuranose form, reducing the sweetness of the solution. For this reason, corn syrup with a high concentration of fructose in the β-D-pyranose form is used as a sweetener in cold, but not hot, drinks. Figure 11.6 shows the common sugars discussed previously in their ring forms.

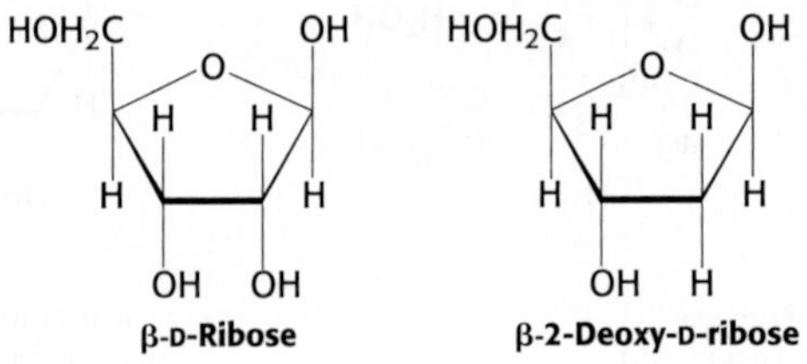

β-D-Ribose

β-2-Deoxy-D-ribose

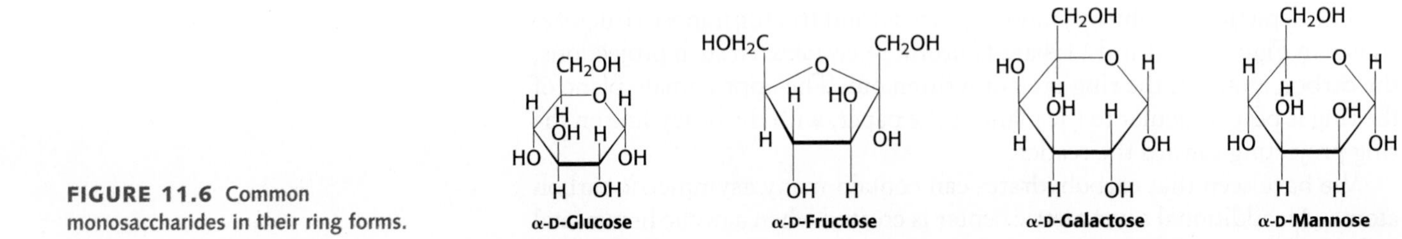

α-D-Glucose

α-D-Fructose

α-D-Galactose

α-D-Mannose

FIGURE 11.6 Common monosaccharides in their ring forms.

Pyranose and furanose rings can assume different conformations

The six-membered pyranose ring is not planar because of the tetrahedral geometry of its saturated carbon atoms. Instead, pyranose rings adopt two classes of conformations, termed chair and boat because of the resemblance to these objects (Figure 11.7). In the chair form, the substituents on the ring carbon atoms have two orientations: axial and equatorial. *Axial* bonds are nearly perpendicular to the average plane of the ring, whereas *equatorial* bonds are nearly parallel to this plane. Axial substituents sterically hinder each other if they emerge on the same side of the ring (e.g., 1,3-diaxial groups). In contrast, equatorial substituents are less crowded. The chair form of β-D-glucopyranose predominates because all axial positions are occupied by hydrogen atoms. The bulkier —OH and —CH₂OH groups emerge at the less-hindered periphery. The boat form of glucose is disfavored because it is quite sterically hindered.

Steric hindrance

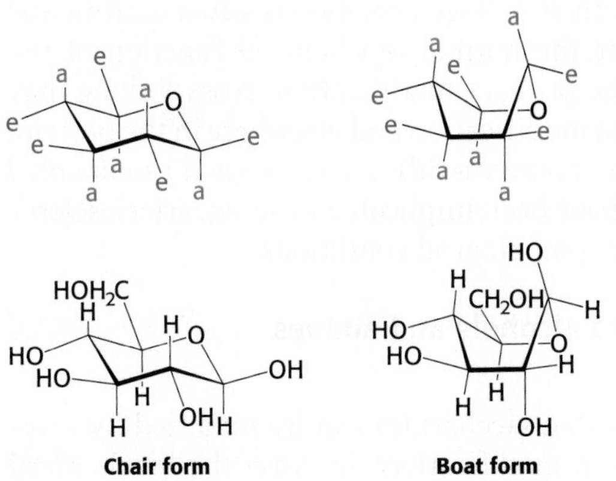

FIGURE 11.7 Chair and boat forms of β-D-glucose. The chair form is more stable because hydrogen atoms occupy all of the axial positions, resulting in less steric hindrance. Abbreviations: a, axial; e, equatorial.

Chair form   **Boat form**

Furanose rings, like pyranose rings, are not planar. They can be puckered so that four atoms are nearly coplanar and the fifth is about 0.5 Å away from this plane (Figure 11.8). This conformation is called an *envelope form* because the structure resembles an opened envelope with the back flap raised. In the ribose moiety of most biomolecules, either C-2 or C-3 is out of the plane on the same side as C-5. These conformations are called C-2-endo and C-3-endo, respectively.

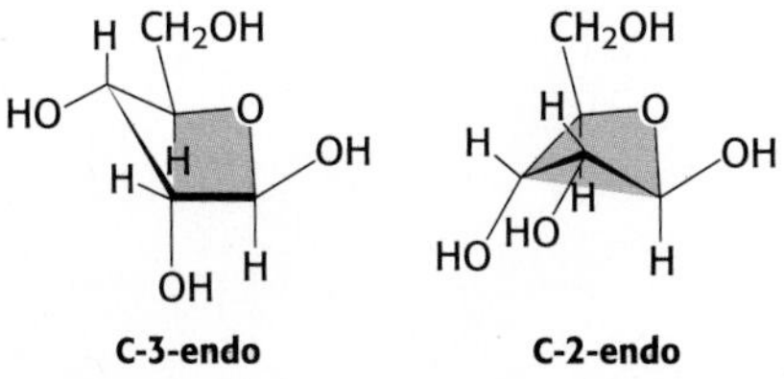

Glucose is a reducing sugar

Because the α and β isomers of glucose are in an equilibrium that passes through the open-chain form, glucose has some of the chemical properties of free aldehydes, such as the ability to react with oxidizing agents. For example, glucose can react with cupric ion (Cu^{2+}), reducing it to cuprous ion (Cu^+), while being oxidized to gluconic acid.

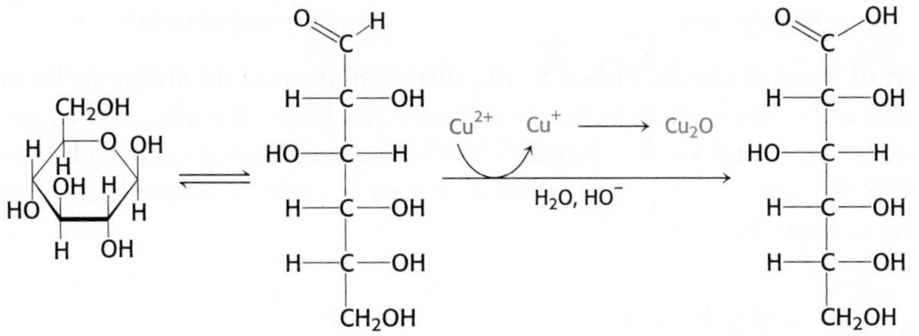

FIGURE 11.8 Envelope conformations of β-D-ribose. The C-3-endo and C-2-endo forms of β-D-ribose are shown. The color indicates the four atoms that lie approximately in a plane.

C-3-endo   C-2-endo

Solutions of cupric ion (known as Fehling's solution) provide a simple test for the presence of sugars such as glucose. Sugars that react are called *reducing sugars*; those that do not are called *nonreducing sugars*. Reducing sugars can often nonspecifically react with a free amino group, usually at a lysine or arginine residue, to form a stable covalent bond. For instance, as a reducing sugar, glucose reacts with hemoglobin to form glycosylated hemoglobin (hemoglobin A1c). Monitoring changes in the amount of glycosylated hemoglobin is an especially useful means of assessing the effectiveness of treatments for diabetes mellitus, a condition characterized by high levels of blood glucose (Section 27.3). Because the glycosylated hemoglobin remains in circulation, the amount of the modified hemoglobin corresponds to the long-term regulation—over several months—of glucose levels. In nondiabetic individuals, less than 6% of the hemoglobin is glycosylated, whereas, in uncontrolled diabetics, almost 10% of the hemoglobin is glycosylated. Although the glycosylation of hemoglobin has no effect on oxygen binding and is thus benign, similar reducing reactions with other proteins such as collagen (p. 49) are often detrimental because the glycosylations alter the normal biochemical function of the modified proteins. Following the primary modification, cross-linking may occur between the site of the first modification and elsewhere in the protein, further compromising function. These modifications, known as *advanced glycation end products* (AGEs), have been implicated in aging, arteriosclerosis, and diabetes, as well as other pathological conditions.

Monosaccharides are joined to alcohols and amines through glycosidic bonds

The biochemical properties of monosaccharides can by modified by reaction with other molecules. These modifications increase the biochemical versatility of carbohydrates, enabling them to serve as signal molecules or facilitating their metabolism. Three common reactants are alcohols, amines, and phosphates. A bond formed between the anomeric carbon atom of a carbohydrate and the oxygen atom of an alcohol is called a *glycosidic bond*—specifically, an O-*glycosidic bond*. O-Glycosidic bonds are prominent when carbohydrates are linked together to form long polymers and when they are attached to proteins (Figure 11.9). In addition, the anomeric carbon

(A)

(B)

α-ᴅ-Methylglucose

Adenosine monophosphate

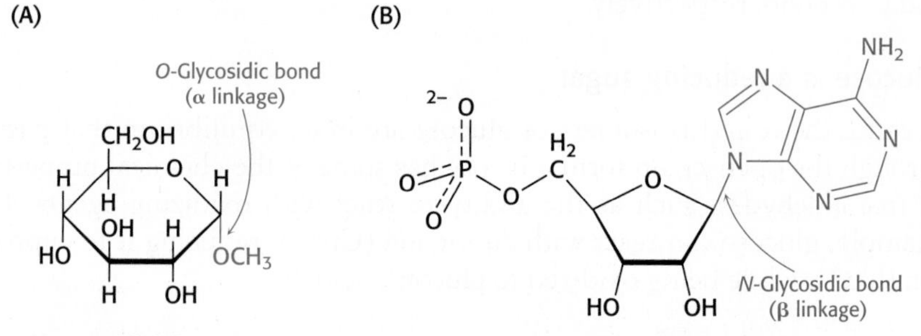

FIGURE 11.9 *O*- and *N*-Glycosidic linkages. (A) An *O*-glycosidic bond links glucose to a methyl group in α-ᴅ-methylglucose. (B) An *N*-glycosidic bond joins ribose to the base adenine in adenosine monophosphate.

atom of a sugar can be linked to the nitrogen atom of an amine to form an N-*glycosidic bond*, such as when nitrogenous bases are attached to ribose units to form nucleosides (Figure 11.9B). Carbohydrates can also be modified by the attachment of functional groups to carbons other than the anomeric carbon (Figure 11.10).

β-ʟ-Fucose
(Fuc)

β-ᴅ-Acetylgalactosamine
(GalNAc)

β-ᴅ-Acetylglucosamine
(GlcNAc)

Sialic acid (Sia)
(*N*-Acetylneuraminate)

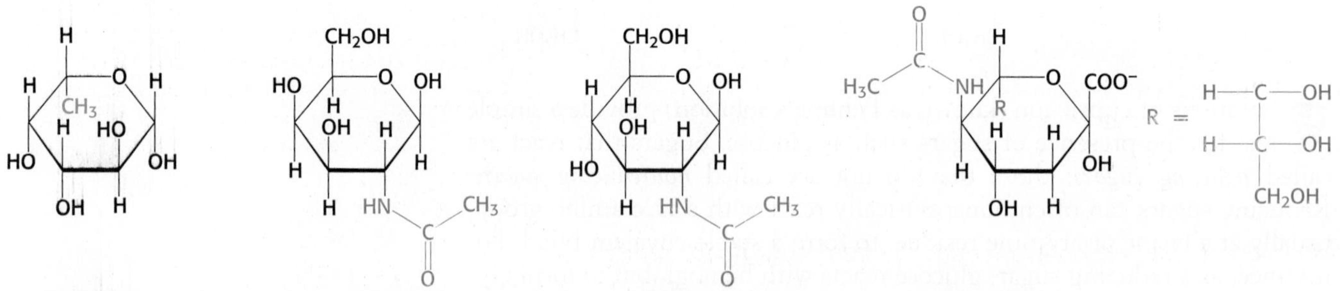

FIGURE 11.10 Modified monosaccharides. Carbohydrates can be modified by the addition of substituents (shown in red) other than hydroxyl groups. Such modified carbohydrates are often expressed on cell surfaces as parts of glycoproteins and glycolipids.

Phosphorylated sugars are key intermediates in energy generation and biosyntheses

One sugar modification deserves special note because of its prominence in metabolism. The addition of phosphoryl groups is a common modification of sugars. For instance, the first step in the breakdown of glucose to obtain energy is its conversion into glucose 6-phosphate. Several subsequent intermediates in this metabolic pathway, such as dihydroxyacetone phosphate and glyceraldehyde 3-phosphate, are phosphorylated sugars.

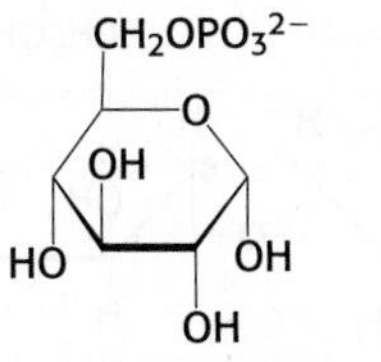

Glucose 6-phosphate (G-6P) 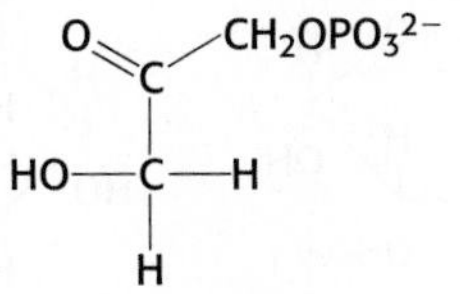 **Dihydroxyacetone phosphate (DHAP)** 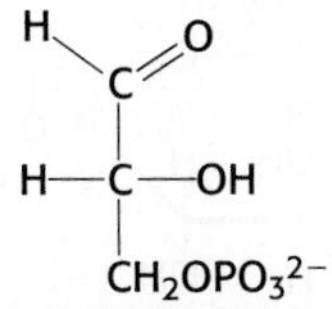 **Glyceraldehyde 3-phosphate (GAP)**

Phosphorylation makes sugars anionic; the negative charge not only prevents these sugars from spontaneously leaving the cell by crossing lipid-bilayer membranes but also prevents them from interacting with transporters of the unmodified sugar. Moreover, phosphorylation *creates reactive intermediates* that will more readily undergo metabolism. For example, a multiply phosphorylated derivative of ribose plays key roles in the biosyntheses of purine and pyrimidine nucleotides (Chapter 25).

11.2 Monosaccharides Are Linked to Form Complex Carbohydrates

Because sugars contain hydroxyl groups, glycosidic bonds can join one monosaccharide to another. *Oligosaccharides* are built by the linkage of two or more monosaccharides by *O*-glycosidic bonds (Figure 11.11). In the disaccharide maltose, for example, two D-glucose residues are joined by a glycosidic linkage between the α-anomeric form of C-1 on one sugar and the hydroxyl oxygen atom on C-4 of the adjacent sugar. Such a linkage is called an α-1,4-glycosidic bond. Just as proteins have a directionality defined by the amino and carboxyl termini, oligosaccharides have a directionality defined by their reducing and nonreducing ends. The carbohydrate unit at the reducing end has a free anomeric carbon atom that has reducing activity because it can form the open-chain form, as discussed earlier (p. 347). By convention, this end of the oligosaccharide is still called the reducing end even when it is bound to another molecule such as a protein and thus no longer has reducing properties.

The fact that monosaccharides have multiple hydroxyl groups means that many different glycosidic linkages are possible. For example, consider three monosaccharides: glucose, mannose, and galactose. These molecules can be linked together in the laboratory to form more than 12,000 structures differing in the order of the monosaccharides and the hydroxyl groups participating in the glycosidic linkages. In this section, we will look at some of the most common oligosaccharides found in nature.

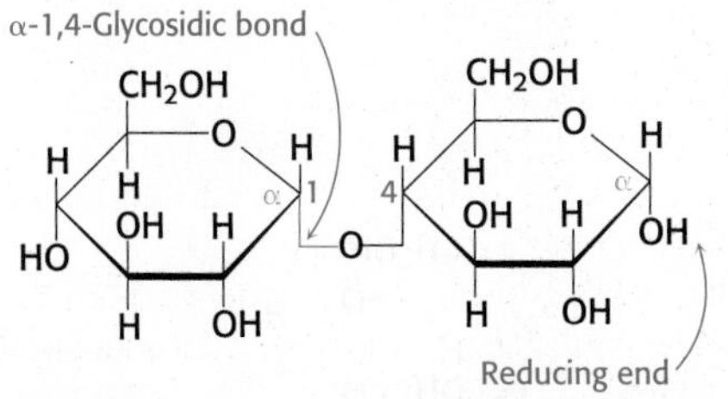

Sucrose, lactose, and maltose are the common disaccharides

A *disaccharide* consists of two sugars joined by an *O*-glycosidic bond. Three abundant disaccharides that we encounter frequently are sucrose, lactose, and maltose (Figure 11.12). *Sucrose* (common table sugar) is obtained commercially from sugar cane or sugar beets. The anomeric carbon atoms of a glucose unit and a fructose unit are joined in this disaccharide; the configuration of this glycosidic linkage is α for glucose and β for fructose. Sucrose can be cleaved into its component monosaccharides by the enzyme *sucrase*, also called *invertase*. *Lactose*, the disaccharide of milk, consists of galactose joined to glucose by a β-1,4-glycosidic linkage. Lactose is hydrolyzed to these monosaccharides by *lactase* in human beings and by *β-galactosidase* in bacteria. In *maltose*, two glucose units are joined by an α-1,4-glycosidic

FIGURE 11.11 Maltose, a disaccharide. Two molecules of glucose are linked by an α-1,4-glycosidic bond to form the disaccharide maltose. The angles in the bonds to the central oxygen atom do not denote carbon atoms. The angles are added only for ease of illustration. The glucose molecule on the right is capable of assuming the open-chain form, which can act as a reducing agent. The glucose molecule on the left cannot assume the open-chain form, because the C-1 carbon atom is bound to another molecule.

Invertase

Invertase was the enzyme Lenore Michaelis and Maud Menten investigated in their classic study, Die Kinetik der Invertinwirkung [(The Kinetics of Invertase) *Biochem. Z.* 49: 333–369, 1913]. In this manuscript, they established that the rate of the invertase reaction is proportional to the concentration of the enzyme–substrate complex as predicted by the Michaelis–Menten equation.

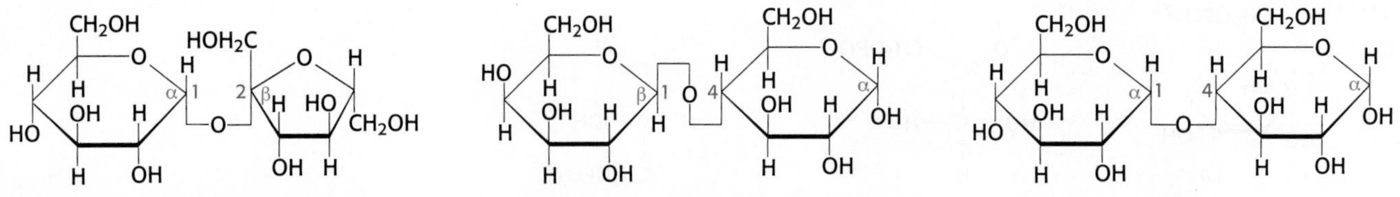

Sucrose
(α-D-Glucopyranosyl-(1→2)-β-D-fructofuranose)

Lactose
(β-D-Galactopyranosyl-(1→4)-α-D-glucopyranose)

Maltose
(α-D-Glucopyranosyl-(1→4)-α-D-glucopyranose)

FIGURE 11.12 Common disaccharides. Sucrose, lactose, and maltose are common dietary components. As in Figure 11.11, the angles in the bonds to the central oxygen atoms do not denote carbon atoms.

linkage. Maltose comes from the hydrolysis of large polymeric oligosaccharides such as starch and glycogen and is in turn hydrolyzed to glucose by *maltase* (α-glucosidase). Maltase will also degrade oligosaccharides linked by α-1,4-glycosidic linkages. Sucrase, lactase, and maltase are located on the outer surfaces of epithelial cells lining the small intestine. The cleavage products of sucrose, lactose, and maltose can be further processed to provide energy in the form of ATP. (See Biochemistry in Focus for a further discussion of maltase.)

Glycogen and starch are storage forms of glucose

Glucose is an important energy source in virtually all life forms. However, free glucose molecules cannot be stored because in high concentrations glucose will disturb the osmotic balance of the cell, potentially resulting in cell death. The solution is to store glucose as units in a large polymer, which is not osmotically active.

Large polymeric oligosaccharides, formed by the linkage of multiple monosaccharides, are called *polysaccharides* and play vital roles in energy storage and in maintaining the structural integrity of an organism. If all of the monosaccharide units in a polysaccharide are the same, the polymer is called a *homopolymer*. The most common homopolymer in animal cells is *glycogen*, the storage form of glucose. Glycogen is present in most of our tissues but is most abundant in muscle and liver. As will be considered in detail in Chapter 21, glycogen is a large, branched polymer of glucose residues. Most of the glucose units in glycogen are linked by α-1,4-glycosidic bonds. The branches are formed by α-1,6-glycosidic bonds, present about once in 12 units (Figure 11.13).

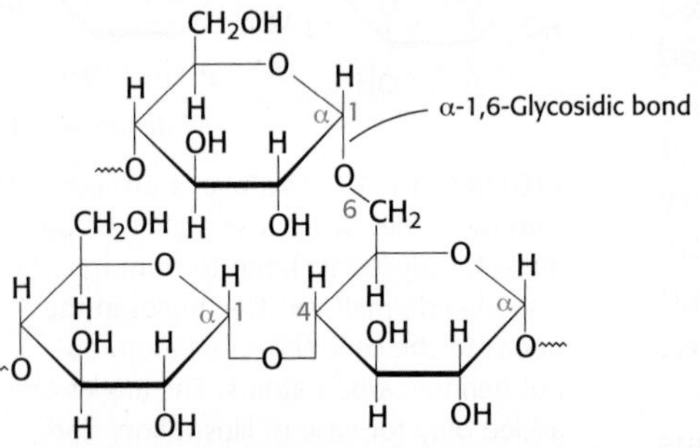

The nutritional reservoir in plants is the homopolymer *starch*, of which there are two forms. *Amylose*, the unbranched type of starch, consists of glucose residues in α-1,4 linkage. *Amylopectin*, the branched form, has about 1 α-1,6 linkage per 30 α-1,4 linkages, in similar fashion to glycogen except for its lower degree of branching. More than half the carbohydrate ingested by human beings is starch found in wheat, potatoes, and rice, to name just a few sources. Amylopectin, amylose, and glycogen are rapidly hydrolyzed by α-*amylase*, an enzyme secreted by the salivary glands and the pancreas.

α-1,6-Glycosidic bond

FIGURE 11.13 Branch point in glycogen. Two chains of glucose molecules joined by α-1,4-glycosidic bonds are linked by an α-1,6-glycosidic bond to create a branch point. Such an α-1,6-glycosidic bond forms at approximately every 12 glucose units, making glycogen a highly branched molecule.

Cellulose, a structural component of plants, is made of chains of glucose

Cellulose, the other major polysaccharide of glucose found in plants, serves a structural rather than a nutritional role as an important component of the plant cell wall. *Cellulose is among the most abundant organic compounds in the biosphere.* Some Some 10^{15} kg of cellulose is synthesized and degraded on Earth each year, an amount 1000 times as great as the combined weight of the human race. Cellulose is an unbranched polymer of glucose

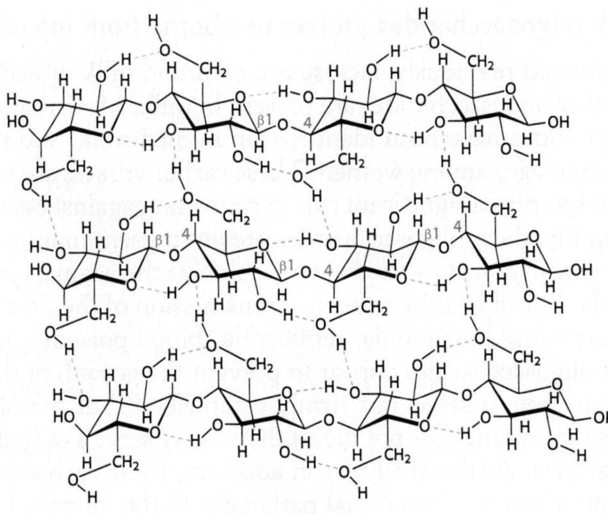

Cellulose
(β-1,4 linkages)

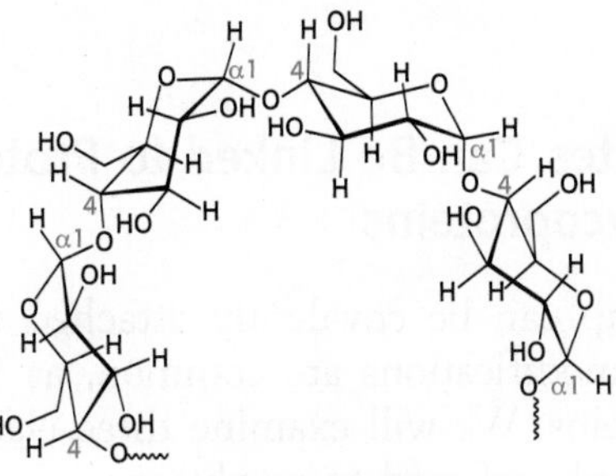

Starch and glycogen
(α-1,4 linkages)

FIGURE 11.14 Glycosidic bonds determine polysaccharide structure. The β-1,4 linkages favor straight chains, which are optimal for structural purposes. The α-1,4 linkages favor bent structures, which are more suitable for storage.

residues joined by β-1,4 linkages, in contrast with the α-1,4 linkage seen in starch and glycogen. This simple difference in stereochemistry yields two molecules with very different properties and biological functions. The β configuration allows cellulose to form very long, straight chains. Fibrils of cellulose are formed by parallel chains that interact with one another through hydrogen bonds, generating a rigid, supportive structure (Figure 11.14, top). The straight chains formed by β linkages are optimal for the construction of fibers having a high tensile strength. The α-1,4 linkages in glycogen and starch produce a very different molecular architecture: a hollow helix is formed instead of a straight chain (Figure 11.14, bottom). The hollow helix formed by α linkages is well suited to the formation of a more compact, accessible store of sugar. Although mammals lack cellulases and therefore cannot digest wood and vegetable fibers, cellulose and other plant fibers are still an important constituent of the mammalian diet as a component of dietary fiber. Soluble fiber such as *pectin* (polygalacturonic acid) slows the movement of food through the gastrointestinal tract, allowing improved digestion and the absorption of nutrients. Insoluble fibers, such as cellulose, increase the rate at which digestion products pass through the large intestine. This increase in rate can minimize exposure to toxins in the diet.

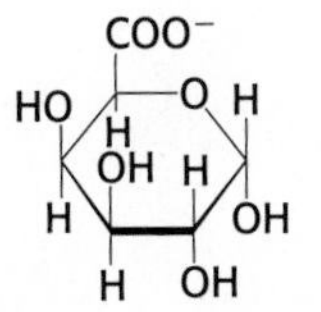

We have considered only homopolymers of glucose. However, given the variety of different monosaccharides that can be put together in any number of arrangements, the number of possible polysaccharides is huge. We will consider some of these polysaccharides shortly.

Galacturonic acid

Human milk oligosaccharides protect newborns from infection

As mentioned previously, lactose is a common milk disaccharide, but many other oligosaccharides are present in milk. More than 150 different oligosaccharides have been identified in human milk, and the amount and composition vary among women. These carbohydrates are not digested by the infant, but play a significant role in protecting against bacterial infection. Importantly, these oligosaccharides are not present in infant formula.

Certain types of *Streptococcus* bacteria colonize the vaginal epithelium of approximately 20% of healthy women. Transmission of the *Streptococcus* to the infant may cause pneumonia, septicemia (blood poisoning), and meningitis. Milk oligosaccharides appear to prevent the growth of the bacteria. How the protection arises is not firmly established, but the oligosaccharides, because the infant does not digest them, may serve as a fuel source for beneficial bacteria. Alternatively, or in addition, these carbohydrates may prevent the attachment of microbial pathogens to the intestinal wall of the newborn. The oligosaccharides appear in the urine of breast-fed infants. Efforts are underway to explore the therapeutic potential of these interesting carbohydrates.

11.3 Carbohydrates Can Be Linked to Proteins to Form Glycoproteins

A carbohydrate group can be covalently attached to a protein to form a *glycoprotein*. Such modifications are common, as 50% of the proteome consists of glycoproteins. We will examine three classes of glycoproteins. The first class is simply referred to as glycoproteins. In glycoproteins of this class, the protein constituent is the largest component by weight. This versatile class plays a variety of biochemical roles. Many glycoproteins are components of cell membranes, where they take part in processes such as cell adhesion and the binding of sperm to eggs. Other glycoproteins are formed by linking carbohydrates to soluble proteins. Many of the proteins secreted from cells are glycosylated, or modified by the attachment of carbohydrates, including most proteins present in the serum component of blood.

The second class of glycoproteins comprises the *proteoglycans*. The protein component of proteoglycans is conjugated to a particular type of polysaccharide called a *glycosaminoglycan*. Carbohydrates make up a much larger percentage by weight of the proteoglycan compared with simple glycoproteins. Proteoglycans function as structural components and lubricants.

Mucins, or *mucoproteins*, are, like proteoglycans, predominantly carbohydrate. *N*-Acetylgalactosamine is usually the carbohydrate moiety bound to the protein in mucins. *N*-Acetylgalactosamine is an example of an *amino sugar*, so named because an amino group replaces a hydroxyl group. Mucins, a key component of mucus, serve as lubricants.

Glycosylation greatly increases the complexity of the proteome. A given protein with several potential glycosylation sites can have many different glycosylated forms, called *glycoforms*. Like protein isoforms (Section 10.2), each is typically generated only in a specific cell type or developmental stage.

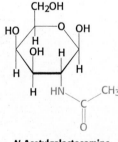

N-Acetylgalactosamine (GalNAc)

Carbohydrates can be linked to proteins through asparagine (*N*-linked) or through serine or threonine (*O*-linked) residues

Sugars in glycoproteins are attached either to the amide nitrogen atom in the side chain of asparagine (termed an N-*linkage*) or to the oxygen atom in the side chain of serine or threonine (termed an O-*linkage*), as shown

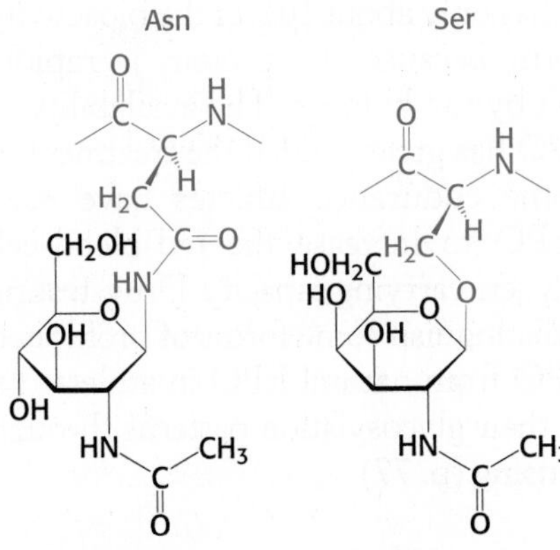

FIGURE 11.15 Glycosidic bonds between proteins and carbohydrates. A glycosidic bond links a carbohydrate to the side chain of asparagine (N-linked) or to the side chain of serine or threonine (O-linked). The glycosidic bonds are shown in blue.

N-linked GlcNAc   **O-linked GalNAc**

in Figure 11.15. An asparagine residue can accept an oligosaccharide only if the residue is part of an Asn-X-Ser or Asn-X-Thr sequence, in which X can be any residue, except proline. However, not all potential sites are glycosylated. Which sites are glycosylated depends on other aspects of the protein structure and on the cell type in which the protein is expressed. All N-linked oligosaccharides have in common a pentasaccharide core consisting of three mannose and two N-acetylglucosamine residues. Additional sugars are attached to this core to form the great variety of oligosaccharide patterns found in glycoproteins (Figure 11.16).

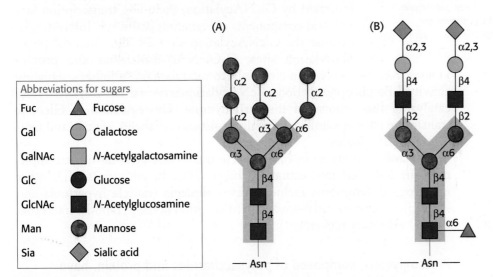

Abbreviations for sugars		
Fuc	▲	Fucose
Gal	●	Galactose
GalNAc	■	N-Acetylgalactosamine
Glc	●	Glucose
GlcNAc	■	N-Acetylglucosamine
Man	●	Mannose
Sia	◆	Sialic acid

FIGURE 11.16 N-linked oligosaccharides. A pentasaccharide core (shaded gray) is common to all N-linked oligosaccharides and serves as the foundation for a wide variety of N-linked oligosaccharides, two of which are illustrated: (A) high-mannose type; (B) complex type. The carbohydrate structures shown are depicted symbolically by employing a scheme that is becoming widely used.

The glycoprotein erythropoietin is a vital hormone

Let us look at a glycoprotein present in the blood serum, whose cloned recombinant form has dramatically improved treatment for anemia, particularly that induced by cancer chemotherapy. The glycoprotein hormone *erythropoietin* (EPO) is secreted by the kidneys and stimulates the production of red blood cells. EPO is composed of 165 amino acids and is N-glycosylated at three asparagine residues and O-glycosylated on a serine residue (Figure 11.17). The mature EPO is 40% carbohydrate by weight, and glycosylation enhances the stability of the protein in the blood.

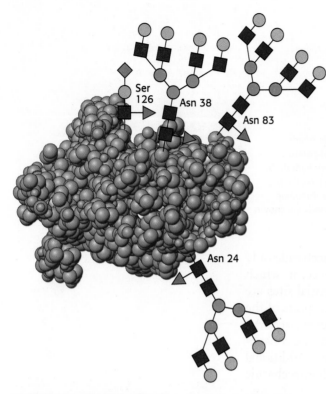

FIGURE 11.17 Oligosaccharides attached to erythropoietin. Erythropoietin has oligosaccharides linked to three asparagine residues and one serine residue. The structures shown are approximately to scale. See Figure 11.16 for the carbohydrate key. [Drawn from 1BUY.pdf.]

Unglycosylated protein has only about 10% of the bioactivity of the glycosylated form because the protein is rapidly removed from the blood by the kidneys. The availability of recombinant human EPO has greatly aided the treatment of anemias. However, some endurance athletes have used recombinant human EPO to increase the red-blood-cell count and hence their oxygen-carrying capacity. Drug-testing laboratories are able to distinguish some forms of prohibited human recombinant EPO from natural EPO in athletes by detecting differences in their glycosylation patterns through the use of isoelectric focusing (p. 77).

Glycosylation functions in nutrient sensing

An especially important glycosylation reaction is the covalent attachment of a single *N*-acetylglucosamine (GlcNAc) to serine or threonine residues of cytoplasmic, nuclear, and mitochondrial proteins. The reaction, one of the most common post-translational modifications, is catalyzed by *O-GlcNAc transferase*, a highly conserved enzyme found in all multicellular animals. The concentration of GlcNAc reflects the active metabolism of carbohydrates, amino acids and fats, indicating that nutrients are abundant (Figure 11.18). More than 4000 proteins are modified by GlcNAcylation, including transcription factors and components of signaling pathways. Interestingly, because the GlcNAcylation sites are also potential phosphorylation sites, *O*-GlcNAc transferase and protein kinases may be involved in cross talk to modulate one another's signaling activity. Like phosphorylation, GlcNAcylation is reversible, with *GlcNAcase* catalyzing the removal of the carbohydrate. Dysregulation of GlcNAc transferase has been linked to insulin resistance, diabetes, cancer, and neurological pathologies.

Recently, mutations in the GlcNAc transferase gene have been implicated in X-linked intellectual disability (ID). In addition to ID, other neurological symptoms include spastic diplegia (muscle hypertonia and spasticity) and pyramidal syndrome (disruption of the pyramidal tracts that control voluntary movement).

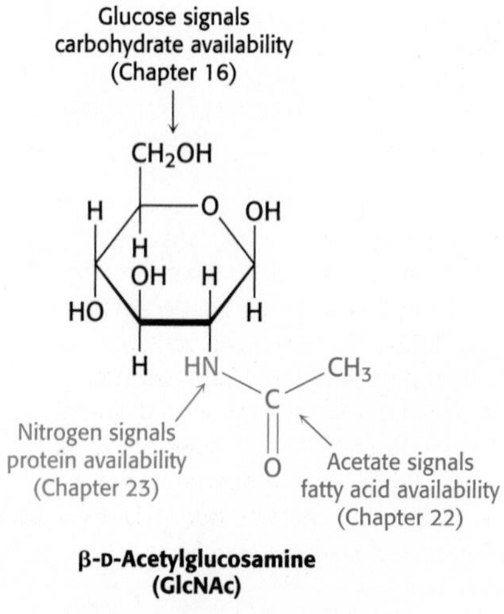

Proteoglycans, composed of polysaccharides and protein, have important structural roles

As stated earlier, proteoglycans are proteins attached to glycosaminoglycans. The glycosaminoglycan makes up as much as 95% of the biomolecule by weight, and so the proteoglycan resembles a polysaccharide more than a protein. Proteoglycans not only function as lubricants and structural components in connective tissue but also mediate the adhesion of cells to the extracellular matrix and bind factors that regulate cell proliferation.

The properties of proteoglycans are determined primarily by the glycosaminoglycan component. Many glycosaminoglycans are made of repeating units of disaccharides containing a derivative of an amino sugar, either glucosamine or galactosamine (Figure 11.19). At least one of the two sugars in the repeating unit has a *negatively charged carboxylate or sulfate group*. The major glycosaminoglycans in animals are chondroitin sulfate,

Glucose signals
carbohydrate availability
(Chapter 16)

Nitrogen signals
protein availability
(Chapter 23)

Acetate signals
fatty acid availability
(Chapter 22)

**β-D-Acetylglucosamine
(GlcNAc)**

FIGURE 11.18 Glycosylation as a nutrient sensor. *N*-Acetylglucosamine is attached to proteins when nutrients are abundant.

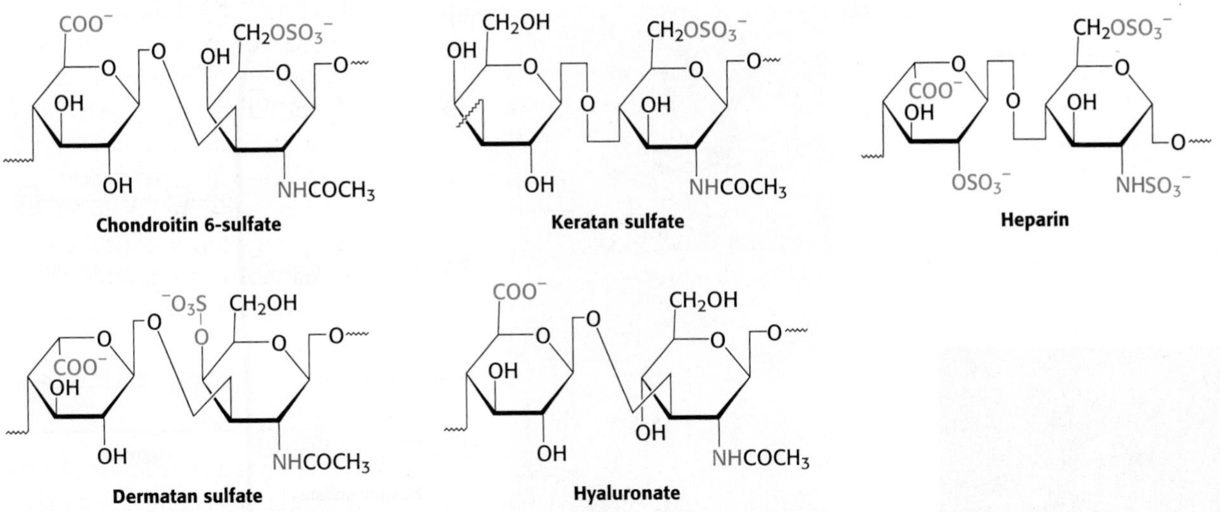

Chondroitin 6-sulfate

Keratan sulfate

Heparin

Dermatan sulfate

Hyaluronate

FIGURE 11.19 Repeating units in glycosaminoglycans. Structural formulas for five repeating units of important glycosaminoglycans illustrate the variety of modifications and linkages that is possible. Amino groups are shown in blue and negatively charged groups in red. Hydrogen atoms have been omitted for clarity. For each molecule shown, the right-hand hexose is a glucosamine derivative.

keratan sulfate, heparin, dermatan sulfate, and hyaluronate. Recall that heparin acts as an anticoagulant to assist the termination of blood clotting (p. 332). *Mucopolysaccharidoses* are a collection of diseases, such as Hurler disease, that result from the inability to degrade glycosaminoglycans (Figure 11.20). Although precise clinical features vary with the disease, all mucopolysaccharidosis result in skeletal deformities and reduced life expectancies.

Proteoglycans are important components of cartilage

Among the best-characterized members of this diverse class is the proteoglycan in the extracellular matrix of cartilage. The proteoglycan *aggrecan* and the protein *collagen* are key components of cartilage. The triple helix of collagen (p. 49) provides structure and tensile strength, whereas aggrecan serves as a shock absorber. The protein component of aggrecan is a large molecule composed of 2397 amino acids. The protein has three globular domains, and the site of glycosaminoglycan attachment is the extended region between globular domains 2 and 3 (Figure 11.21). This linear region contains highly repetitive amino acid sequences, which are sites for the attachment of keratan sulfate and chondroitin sulfate. Many molecules of aggrecan are in turn noncovalently bound through the first globular domain to a very long filament formed by linking together molecules of the glycosaminoglycan hyaluronate (Figure 11.21). Water is bound to the glycosaminoglycans, attracted by the many negative charges. Aggrecan can cushion compressive forces because the adsorbed water enables it to spring back after having been deformed. When pressure is exerted, as when the foot hits the ground while walking, water is squeezed from the glycosaminoglycan, cushioning the impact. When the pressure is released, the water rebinds. *Osteoarthritis*, the most common form of arthritis, results when water is lost from proteoglycan with aging. Other forms of arthritis can result from the proteolytic degradation of aggrecan and collagen in the cartilage.

In addition to being a key component of structural tissues, glycosaminoglycans are common throughout the biosphere. Chitin is a glycosaminoglycan

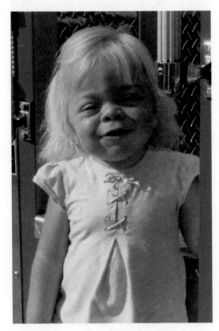

FIGURE 11.20 Hurler disease. Formerly called gargoylism, Hurler disease is a mucopolysaccharidosis having symptoms that include wide nostrils, a depressed nasal bridge, thick lips and earlobes, and irregular teeth. In Hurler disease, glycosaminoglycans cannot be degraded. The excess of these molecules are stored in the soft tissue of the facial regions, resulting in the characteristic facial features. [Courtesy of National MPS Society.]

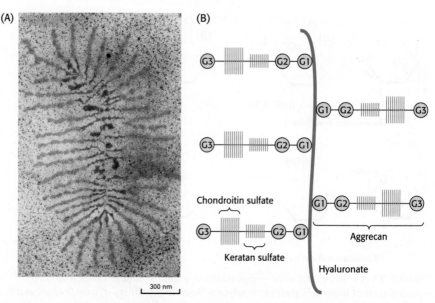

FIGURE 11.21 Structure of proteoglycan from cartilage. (A) Electron micrograph of a proteoglycan from cartilage (with false color added). Proteoglycan monomers emerge laterally at regular intervals from opposite sides of a central filament of hyaluronate. (B) Schematic representation. G = globular domain. [Courtesy of Dr. Lawrence Rosenberg and Joseph A. Buckwalter, see: J.A. Buckwalter and L. Rosenberg. "Structural Changes during Development in Bovine Fetal Epiphyseal Cartilage." *Collagen and Related Research*, 3(1983): 489–504.]

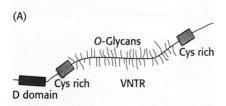

FIGURE 11.22 Chitin, a glycosaminoglycan, is present in insect wings and the exoskeleton. Glycosaminoglycans are components of the exoskeletons of insects, crustaceans, and arachnids. [FLPA/Alamy.]

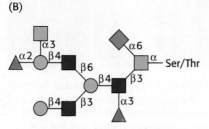

FIGURE 11.23 Mucin structure. (A) A schematic representation of a mucoprotein. The VNTR region is highly glycosylated, forcing the molecule into an extended conformation. The Cys-rich domains and the D domain facilitate the polymerization of many such molecules. (B) An example of an oligosaccharide that is bound to the VNTR region of the protein. See Figure 11.16 for the carbohydrate key. [Information from A. Varki et al. (Eds.), *Essentials of Glycobiology*, 2nd ed. (Cold Spring Harbor Press, 2009), pp. 117, 118.]

found in the exoskeleton of insects, crustaceans, and arachnids and, next to cellulose, is the second most abundant polysaccharide in nature (Figure 11.22). Cephalopods such as squid use their razor sharp beaks, which are made of extensively cross-linked chitin, to disable and consume prey.

Mucins are glycoprotein components of mucus

A third class of glycoproteins is the *mucins* (mucoproteins). In mucins, the protein component is extensively glycosylated at serine or threonine residues by *N*-acetylgalactosamine (Figure 11.10). Mucins are capable of forming large polymeric structures and are common in mucous secretions. These glycoproteins are synthesized by specialized cells in the tracheobronchial, gastrointestinal, and genitourinary tracts. Mucins are abundant in saliva, where they function as lubricants.

A model of a mucin is shown in Figure 11.23A. The defining feature of the mucins is a region of the protein backbone termed the *variable number of tandem repeats* (VNTR) region, which is rich in serine and threonine residues that are *O*-glycosylated. Indeed, the carbohydrate moiety can account for as much as 80% of the molecule by weight. A number of core carbohydrate structures are conjugated to the protein component of mucin. Figure 11.23B shows one such structure.

Mucins adhere to epithelial cells and act as a protective barrier; they also hydrate the underlying cells. In addition to protecting cells from environmental insults, such as stomach acid, inhaled chemicals in the lungs, and bacterial infections, mucins have roles in fertilization, the immune response, and cell adhesion. Mucins are overexpressed in bronchitis and cystic fibrosis, and the overexpression of mucins is characteristic of adenocarcinomas—cancers of the glandular cells of epithelial origin.

Chitin can be processed to a molecule with a variety of uses

We are all aware, in one way or another, of the many uses of the polysaccharide cellulose. It is a major constituent of paper, bioadhesives, and the clothes you are wearing. Less obvious to most of us are uses of the next most abundant polysaccharide chitin. Estimates are that 10,000 tons of chitin could be recovered from the processing of shells of marine crustaceans, chitin that will otherwise become environmental waste.

Processed chitin, chitosan, already has a variety of uses. For instance, it is used as a carrier to assist in drug delivery, as surgical stitches, as a component of skin, hair, and oral care products, and as a stabilizing and thickening component for food products. Recent research shows that chitosan can be used as a component of an adhesive that strongly adheres to wet tissues, allowing the adhesive to be used as a surgical dressing. There are several ways to process chitin, which is water-insoluble, to water-soluble chitosan. We will examine the enzymatic process.

First, the shells are washed and ground to a fine powder, which is then exposed to lactic-acid generating bacteria (Chapter 16). The lactic acid removes minerals from the chitin. Following demineralization, the chitin is treated with protease-secreting bacteria to degrade the protein component of chitin. Next, acetone treatment removes the color by dissolving carotenoids (Chapter 26). Finally, chitin is treated with chitin deacetylase to remove acetyl groups, generating the more versatile chitosan. Here we have a wonderful example of the use of biochemistry in biotechnology to convert environmental waste into a variety of useful materials.

Protein glycosylation takes place in the lumen of the endoplasmic reticulum and in the Golgi complex

The major pathway for protein glycosylation takes place inside the lumen of the *endoplasmic reticulum* (ER) and in the *Golgi complex*, organelles that play central roles in protein trafficking (Figure 11.24). The protein is synthesized by ribosomes attached to the cytoplasmic face of the ER membrane, and the peptide chain is inserted into the lumen of the ER (Section 31.6). The *N*-linked glycosylation begins in the ER and continues in the Golgi complex, whereas the *O*-linked glycosylation takes place exclusively in the Golgi complex.

A large oligosaccharide destined for attachment to the asparagine residue of a protein is assembled on *dolichol phosphate*, a specialized lipid molecule located in the ER membrane and containing about 20 isoprene (C_5) units.

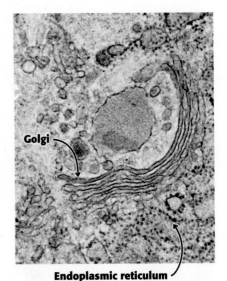

FIGURE 11.24 Golgi complex and endoplasmic reticulum. The electron micrograph shows the Golgi complex and adjacent endoplasmic reticulum. The black dots on the cytoplasmic surface of the ER membrane are ribosomes. [Courtesy of Lynne Mercer.]

$n = 15–19$

Dolichol phosphate

Isoprene

The terminal phosphate group of the dolichol phosphate is the site of attachment of the oligosaccharide. This activated (energy-rich) form of the oligosaccharide is subsequently transferred to a specific asparagine residue of the growing polypeptide chain by an enzyme located on the lumenal side of the ER.

Proteins in the lumen of the ER and in the ER membrane are transported to the Golgi complex, which is a stack of flattened membranous sacs. *Carbohydrate units of glycoproteins are altered and elaborated in the Golgi complex.* The O-linked sugar units are fashioned there, and the N-linked sugars, arriving from the ER as a component of a glycoprotein, are modified in many different ways. *The Golgi complex is the major sorting center of the cell.* Proteins proceed from the Golgi complex to lysosomes, secretory granules, or the plasma membrane, according to signals encoded within their amino acid sequences and three-dimensional structures (Figure 11.25).

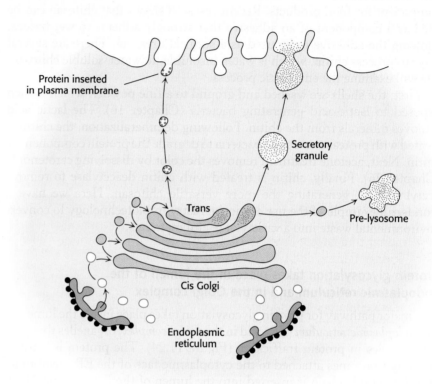

Protein inserted
in plasma membrane

Secretory
granule

Trans

Pre-lysosome

Cis Golgi

Endoplasmic
reticulum

FIGURE 11.25 Golgi complex as sorting center. The Golgi complex is the sorting center in the targeting of proteins to lysosomes, secretory vesicles, and the plasma membrane. The cis face of the Golgi complex receives vesicles from the endoplasmic reticulum, and the trans face sends a different set of vesicles to target sites. Vesicles also transfer proteins from one compartment of the Golgi complex to another. [Courtesy of Dr. Marilyn Farquhar.]

Specific enzymes are responsible for oligosaccharide assembly

How are the complex carbohydrates formed, be they unconjugated molecules such as glycogen or components of glycoproteins? Complex carbohydrates are synthesized through the action of specific enzymes, *glycosyltransferases*, which catalyze the formation of glycosidic bonds. Given the diversity of known glycosidic linkages, many different enzymes are required. Indeed, glycosyltransferases account for 1% to 2% of gene products in all organisms examined.

While dolichol phosphate-linked oligosaccharides are substrates for some glycosyltransferases, the most common carbohydrate donors for glycosyltransferases are activated sugar nucleotides, such as UDP-glucose (UDP is the abbreviation for uridine diphosphate) (Figure 11.26). The attachment of a nucleotide to enhance the energy content of a molecule is a common strategy in biosynthesis that we will see many times in our study of biochemistry. The acceptor substrates for glycosyltransferases are quite varied and include carbohydrates, serine, threonine, and asparagine residues of proteins, lipids, and even nucleic acids.

Blood groups are based on protein glycosylation patterns

The human ABO blood groups illustrate the effects of glycosyltransferases on the formation of glycoproteins. Each blood group is designated by the presence of one of the three different carbohydrates, termed A, B, or O, attached to glycoproteins and glycolipids on the surfaces of red blood cells (Figure 11.27). These structures have in common an oligosaccharide foundation called the O (or sometimes H) antigen. The A and B antigens differ from the O antigen by the addition of one extra monosaccharide, either *N*-acetylgalactosamine (for A) or galactose (for B) through an α-1,3 linkage to a galactose moiety of the O antigen.

Specific glycosyltransferases add the extra monosaccharide to the O antigen. Each person inherits the gene for one glycosyltransferase of this type from each parent. The type A transferase specifically adds *N*-acetylgalactosamine, whereas the type B transferase adds galactose. These enzymes are identical in all but 4 of 354 positions. In the AB blood group, both enzymes are present. The O phenotype results if both enzymes are absent.

These structures have important implications for blood transfusions and other transplantation procedures. If an antigen not normally present in a person is introduced, the person's immune system recognizes it as foreign. Red-blood-cell lysis occurs rapidly, leading to a severe drop in blood pressure (hypotension), shock, kidney failure, and death from circulatory collapse.

Why are different blood types present in the human population? Suppose that a pathogenic organism such as a parasite expresses on its cell surface a carbohydrate antigen similar to one of the blood-group antigens. This antigen may not be readily detected as foreign in a person whose blood type matches the parasite antigen, and the parasite will flourish. However, other people with different blood types will be protected. Hence, there will be selective pressure on human beings to vary blood type to prevent parasitic mimicry and a corresponding selective pressure on parasites to enhance mimicry. The constant "arms race" between pathogenic microorganisms and human beings drives the evolution of diversity of surface antigens within the human population.

Errors in glycosylation can result in pathological conditions

As we saw with EPO, protein function is often compromised when proteins are improperly glycosylated. These errors in glycosylation may result in disease. For instance, certain types of muscular dystrophy can be traced to improper glycosylation of dystroglycan, a membrane protein that links the extracellular matrix with the cytoskeleton. Indeed, an entire family of severe inherited human diseases called *congenital disorders of glycosylation* has been identified. These pathological conditions reveal the importance of proper modification of proteins by carbohydrates and their derivatives.

An especially clear example of the role of glycosylation is provided by *I-cell disease* (also called *mucolipidosis II*), a lysosomal storage disease. Normally, a carbohydrate marker directs certain digestive enzymes from

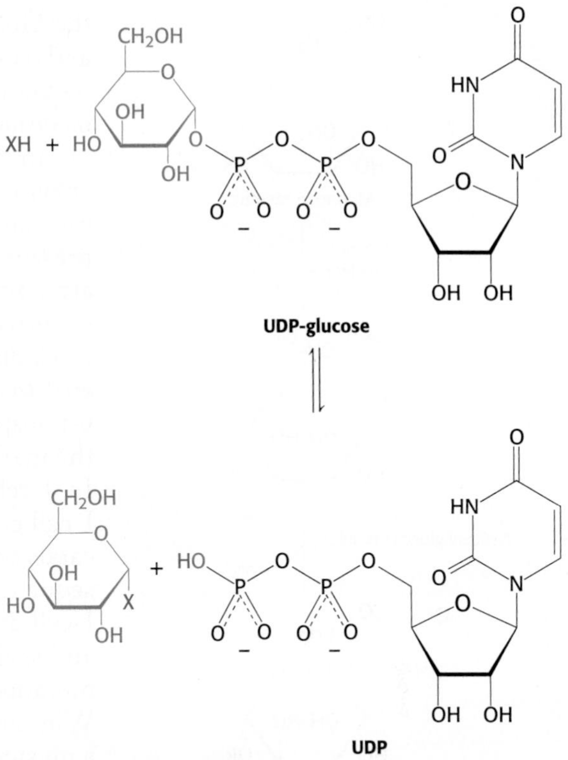

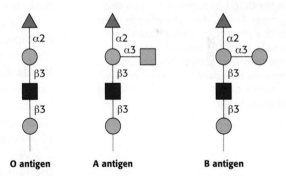

FIGURE 11.26 General form of a glycosyltransferase reaction. The sugar to be added comes from a sugar nucleotide—in this case, UDP-glucose. The acceptor, designated X in this illustration, can be one of a variety of biomolecules, including other carbohydrates or proteins.

O antigen   **A antigen**   **B antigen**

FIGURE 11.27 Structures of A, B, and O oligosaccharide antigens. Different glycosylation patterns account for the different human blood groups.

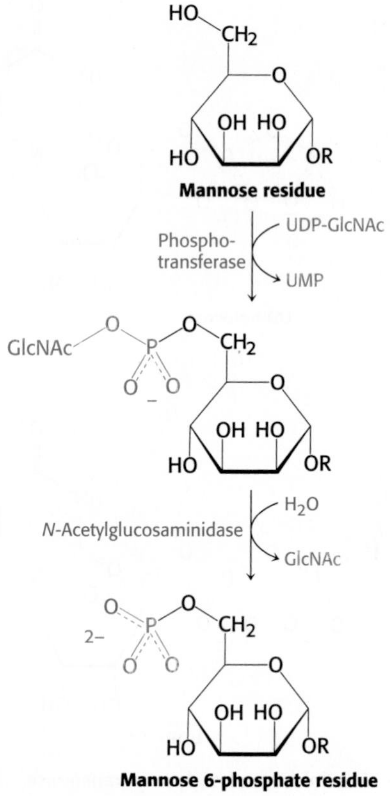

Mannose residue

Phosphotransferase

UDP-GlcNAc

UMP

GlcNAc

Mannose 6-phosphate residue

N-Acetylglucosaminidase

H₂O

GlcNAc

2−

FIGURE 11.28 Formation of a mannose 6-phosphate marker. A glycoprotein destined for delivery to lysosomes acquires a phosphate marker in the Golgi compartment in a two-step process. First, GlcNAc phosphotransferase adds a phospho-*N*-acetylglucosamine unit to the 6-OH group of a mannose, and then an *N*-acetylglucosaminidase removes the added sugar to generate a mannose 6-phosphate residue in the core oligosaccharide.

the Golgi complex to lysosomes. *Lysosomes* are organelles that degrade and recycle damaged cellular components or material brought into the cell by endocytosis. In patients with I-cell disease, lysosomes contain large *inclusions* of undigested glycosaminoglycans and glycolipids—hence the "I" in the name of the disease. These inclusions are present because the enzymes normally responsible for the degradation of glycosaminoglycans are missing from affected lysosomes. Remarkably, the enzymes are present at very high levels in the blood and urine. Thus, active enzymes are synthesized, but, in the absence of appropriate glycosylation, they are exported instead of being sequestered in lysosomes. In other words, *in I-cell disease, a whole series of enzymes are incorrectly addressed and delivered to the wrong location.* Normally, these enzymes contain a mannose 6-phosphate residue as a component of an *N*-oligosaccharide that serves as the marker directing the enzymes from the Golgi complex to lysosomes. In I-cell disease, however, the attached mannose lacks a phosphate. I-cell patients are deficient in the N-*acetylglucosamine phosphotransferase* catalyzing the first step in the addition of the phosphoryl group; the consequence is the mistargeting of eight essential enzymes (Figure 11.28). I-cell disease causes the patient to suffer severe psychomotor retardation and skeletal deformities, similar to those in Hurler disease. Remarkably, mutations in the phosphotransferase have also been linked to stuttering. Why some mutations cause stuttering while others cause I-cell disease is a mystery.

Oligosaccharides can be "sequenced"

How is it possible to determine the structure of a glycoprotein—the oligosaccharide structures and their points of attachment? Most approaches make use of enzymes that cleave oligosaccharides at specific types of linkages.

The first step is to detach the oligosaccharide from the protein. For example, *N*-linked oligosaccharides can be released from proteins by an enzyme such as *peptide N-glycosidase F,* which cleaves the *N*-glycosidic bonds linking the oligosaccharide to the protein. The oligosaccharides can then be isolated and analyzed. Matrix-assisted laser desorption/ ionization/time-of-flight (MALDI-TOF) or other mass spectrometric techniques (Section 3.3) provide the mass of an oligosaccharide fragment. However, many possible oligosaccharide structures are consistent with a given mass. More complete information can be obtained by cleaving the oligosaccharide with enzymes of varying specificities. For example, *β-1,4-galactosidase* cleaves β-glycosidic bonds exclusively at galactose residues. The products can again be analyzed by mass spectrometry (Figure 11.29). The repetition of this process with the use of an array of enzymes of different specificity will eventually reveal the structure of the oligosaccharide.

Proteases applied to glycoproteins can reveal the points of oligosaccharide attachment. Cleavage by a specific protease yields a characteristic pattern of peptide fragments that can be analyzed chromatographically. Fragments attached to oligosaccharides can be picked out because their chromatographic properties will change on glycosidase treatment. Mass spectrometric analysis or direct peptide sequencing can reveal the identity of the peptide in question and, with additional effort, the exact site of oligosaccharide attachment.

While the sequencing of the human genome is complete, the characterization of the much more complex proteome, including the biological roles of glycosylated proteins, presents a challenge to biochemistry.

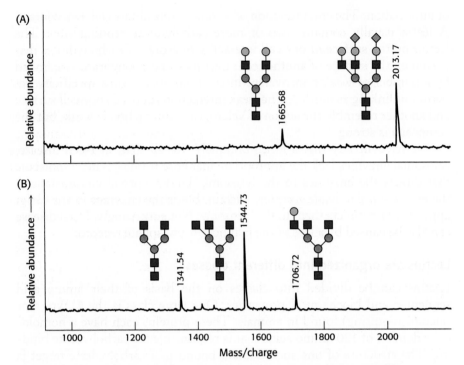

FIGURE 11.29 Mass spectrometric "sequencing" of oligosaccharides. Carbohydrate-cleaving enzymes were used to release and specifically cleave the oligosaccharide component of the glycoprotein fetuin. Parts A and B show the masses obtained with MALDI-TOF spectrometry as well as the corresponding structures of the oligosaccharide-digestion products: (A) digestion with peptide *N*-glycosidase F (to release the oligosaccharide from the protein) and neuraminidase (which cleaves at sialic acid residues); (B) digestion with peptide *N*-glycosidase F, neuraminidase, and β-1,4-galactosidase. Knowledge of the enzyme specificities and the masses of the products permits the characterization of the oligosaccharide. See Figure 11.16 for the carbohydrate key. [Data from A. Varki, R. D. Cummings, J. D. Esko, H. H. Freeze, G. W. Hart, and J. Marth (Eds.), *Essentials of Glycobiology* (Cold Spring Harbor Laboratory Press, 1999), p. 596.]

11.4 Lectins Are Specific Carbohydrate-Binding Proteins

The diversity and complexity of the carbohydrate units and the variety of ways in which they can be joined in oligosaccharides and polysaccharides attest to their biochemical importance. Nature does not construct complex patterns when simple ones suffice. Why all this intricacy and diversity? It is now clear that these carbohydrate structures are the recognition sites for a special class of proteins. Such proteins, termed *glycan-binding proteins*, bind specific carbohydrate structures on neighboring cell surfaces. Originally discovered in plants, glycan-binding proteins are ubiquitous, and no living organisms have been found that lack these key proteins. We will focus on a particular class of glycan-binding proteins termed *lectins* (from Latin *legere*, "to select"). The interaction of lectins with their carbohydrate partners is another example of carbohydrates being information-rich molecules that guide many biological processes. The diverse carbohydrate structures displayed on cell surfaces are well-suited to serving as sites of interaction between cells and their environments. Interestingly, the partners for lectin binding are often the carbohydrate moiety of glycoproteins.

Lectins promote interactions between cells and within cells

Cell–cell contact is a vital interaction in a host of biochemical functions, ranging from building a tissue from isolated cells to facilitating the transmission of information. The chief function of lectins is to facilitate cell–cell contact. A lectin usually contains two or more carbohydrate-binding sites. The lectins on the surface of one cell interact with arrays of carbohydrates displayed on the surface of another cell. Lectins and carbohydrates are linked by a number of weak noncovalent interactions that ensure specificity yet permit unlinking as needed. The weak interactions between one cell surface and another resemble the action of Velcro; each interaction is weak, but the composite is strong.

We have already met a lectin obliquely. Recall that, in I-cell disease, lysosomal enzymes lack the appropriate mannose 6-phosphate, a molecule that directs the enzymes to the lysosome. Under normal circumstances, the *mannose 6-phosphate receptor*, a lectin, binds the enzymes in the Golgi apparatus and directs them to the lysosome. Not surprisingly, I-cell disease can also be caused by a loss of the mannose 6-phosphate receptor.

Lectins are organized into different classes

Lectins can be divided into classes on the basis of their amino acid sequences and biochemical properties. One large class is the C type (for *calcium-requiring*) found in animals. These proteins each have a homologous domain of 120 amino acids that is responsible for carbohydrate binding. The structure of one such domain bound to a carbohydrate target is shown in Figure 11.30.

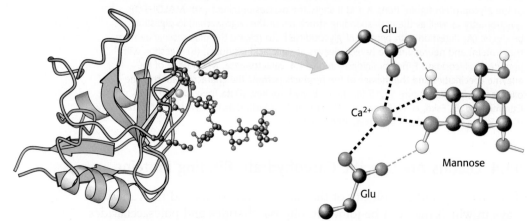

**FIGURE 11.30
Structure of a carbohydrate-binding domain of an animal C-type lectin.** *Notice* that a calcium ion links a mannose residue to the lectin. Selected interactions are shown, with some hydrogen atoms omitted for clarity. [Drawn from 2MSC.pdb.]

A calcium ion on the protein acts as a bridge between the protein and the sugar through direct interactions with sugar OH groups. In addition, two glutamate residues in the protein bind to both the calcium ion and the sugar, and other protein side chains form hydrogen bonds with other OH groups on the carbohydrate. The carbohydrate-binding specificity of a particular lectin is determined by the amino acid residues that bind the carbohydrate. C-type lectins function in a variety of cellular activities, including receptor-mediated endocytosis, a process by which soluble molecules are bound to the cell surface and subsequently internalized (Section 26.3), and cell–cell recognition.

Proteins termed *selectins* are members of the C-type family. Selectins bind immune-system cells to sites of injury in the inflammatory response. The L, E, and P forms of selectins bind specifically to carbohydrates on *lymph*-node vessels, *endothelium*, or activated blood *platelets*, respectively. New therapeutic agents that control inflammation may emerge from a deeper understanding of how selectins bind and distinguish different carbohydrates. *L*-Selectin, originally thought to participate only in the immune response, is produced by embryos when they are ready to attach to the endometrium of the mother's uterus. For a short period of time, the endometrial cells present an oligosaccharide on the cell surface. When the embryo attaches through lectins, the attachment activates signal pathways in the endometrium to make implantation of the embryo possible.

Another large class of lectins comprises the *L*-lectins. These lectins are especially rich in the seeds of leguminous plants, and many of the initial biochemical characterizations of lectins were performed on this readily available lectin. Although the exact role of lectins in plants is unclear, they can serve as potent insecticides. Other *L*-type lectins, such as *calnexin* and *calreticulin,* are prominent chaperones in the eukaryotic endoplasmic reticulum. Recall that chaperones are proteins that facilitate the folding of other proteins (p. 53).

Influenza virus binds to sialic acid residues

Many pathogens gain entry into specific host cells by adhering to cell-surface carbohydrates. For example, influenza virus recognizes sialic acid residues linked to galactose residues that are present on cell-surface glycoproteins. The viral protein that binds to these sugars is a lectin called *hemagglutinin* (Figure 11.31A).

After binding hemagglutinin, the virus is engulfed by the cell and begins to replicate. To exit the cell, a process essentially the reverse of viral entry occurs (Figure 11.31B). Viral assembly results in the budding of the viral particle from the cell. Upon complete assembly, the viral particle is still attached to sialic acid residues of the cell membrane by hemagglutinin on the surface of the new virions. Another viral protein, neuraminidase (sialidase), cleaves the glycosidic bonds between the sialic acid residues and the rest of the cellular glycoprotein, freeing the virus to infect new cells, and thus spreading the infection throughout the respiratory tract. Inhibitors of this enzyme such as oseltamivir (Tamiflu) and zanamivir (Relenza) are important anti-influenza agents.

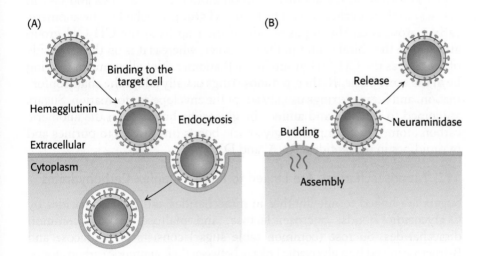

(A) Binding to the target cell

Hemagglutinin

Endocytosis

Extracellular

Cytoplasm

(B) Release

Budding

Neuraminidase

Assembly

FIGURE 11.31 Viral receptors. (A) Influenza virus targets cells by binding to sialic acid residues located at the termini of oligosaccharides present on cell-surface glycoproteins and glycolipids. These carbohydrates are bound by the lectin hemagglutinin, one of the major proteins expressed on the surface of the virus. (B) When viral replication is complete and the viral particle buds from the cell, the other major viral-surface protein, neuraminidase, cleaves oligosaccharide chains to release the viral particle.

Viral hemagglutinin's carbohydrate-binding specificity may play an important role in species specificity of infection and ease of transmission. For instance, avian influenza H5N1 (bird flu) is especially lethal and is readily spread from bird to bird. Although human beings can be infected by this virus, infection is rare and human-to-human transmission is rarer still.

The biochemical basis of these characteristics is that the avian virus hemagglutinin recognizes a different carbohydrate sequence from that recognized in human influenza. Although human beings have the sequence to which the avian virus binds, it is located deep in the lungs. Infection by the avian virus is thus difficult, and, when it does occur, the avian virus is not readily transmitted by sneezing or coughing.

Plasmodium falciparum, the parasitic protozoan that causes malaria, also relies on glycan binding to infect and colonize its host. Glycan-binding proteins of the parasitic form initially injected by the mosquito bind to the glycosaminoglycan heparin sulfate on the liver, initiating the parasite's entry into the cell. On exiting from the liver later in its life cycle, the parasite invades red blood cells by using another glycan-binding protein to bind to the carbohydrate moiety of *glycophorin*, a prominent membrane glycoprotein in red blood cells. Developing means to disrupt the carbohydrate interactions between pathogens and host cells may prove to be clinically useful.

SUMMARY

11.1 Monosaccharides Are the Simplest Carbohydrates

Carbohydrates are aldoses or ketoses that are rich in hydroxyl groups. An aldose is a carbohydrate with an aldehyde group (as in glyceraldehyde and glucose), whereas a ketose contains a keto group (as in dihydroxyacetone and fructose). A sugar belongs to the D series if the absolute configuration of its asymmetric carbon atom farthest from the aldehyde or keto group is the same as that of D-glyceraldehyde. Most naturally occurring sugars belong to the D series. The C-1 aldehyde in the open-chain form of glucose reacts with the C-5 hydroxyl group to form a six-membered pyranose ring. The C-2 keto group in the open-chain form of fructose reacts with the C-5 hydroxyl group to form a five-membered furanose ring. Pentoses such as ribose and deoxyribose also form furanose rings. An additional asymmetric center is formed at the anomeric carbon atom (C-1 in aldoses and C-2 in ketoses) in these cyclizations. The hydroxyl group attached to the anomeric carbon atom is on the opposite side of the ring from the CH_2OH group attached to the chiral center in the α anomer, whereas it is on the same side of the ring as the CH_2OH group in the β anomer. Not all atoms in the ring lie in the same plane. Rather, pyranose rings usually adopt the chair conformation, and furanose rings usually adopt the envelope conformation. Sugars are joined to alcohols and amines by glycosidic bonds from the anomeric carbon atom. For example, *N*-glycosidic bonds link sugars to purines and pyrimidines in nucleotides, RNA, and DNA.

11.2 Monosaccharides Are Linked to Form Complex Carbohydrates

Sugars are linked to one another in disaccharides and polysaccharides by *O*-glycosidic bonds. Sucrose, lactose, and maltose are the common disaccharides. Sucrose (common table sugar) consists of α-glucose and β-fructose joined by a glycosidic linkage between their anomeric carbon atoms. Lactose (in milk) consists of galactose joined to glucose by a β-1,4 linkage. Maltose (from starch) consists of two glucoses joined by an α-1,4 linkage. Starch is a polymeric form of glucose in plants and glycogen serves a similar role in animals. Most of the glucose units in starch and glycogen are in α-1,4 linkage. Cellulose, the major structural polymer of plant cell walls, consists of glucose units joined by β-1,4 linkages. These β linkages give rise to long straight chains that form fibrils with high tensile strength. In contrast, the α linkages in starch and glycogen lead to open helices, in keeping with their roles as mobilizable energy stores.

11.3 Carbohydrates Can Be Linked to Proteins to Form Glycoproteins

Carbohydrates are commonly conjugated to proteins. If the protein component is predominant, the conjugate of protein and carbohydrate is called a glycoprotein. Most secreted proteins, such as the signal molecule erythropoietin, are glycoproteins. Glycoproteins are also prominent on the external surface of the plasma membrane. Proteins bearing covalently linked glycosaminoglycans are proteoglycans. Glycosaminoglycans are polymers of repeating disaccharides. One of the units in each repeat is a derivative of glucosamine or galactosamine. These highly anionic carbohydrates have a high density of carboxylate or sulfate groups. Proteoglycans are found in the extracellular matrices of animals and are key components of cartilage. Mucoproteins, like proteoglycans, are predominantly carbohydrate by weight. The protein component is heavily O-glycosylated with N-acetylgalactosamine joining the oligosaccharide to the protein. Mucoproteins serve as lubricants.

Glycosyltransferases link the oligosaccharide units on proteins either to the side-chain oxygen atom of a serine or threonine residue or to the side-chain amide nitrogen atom of an asparagine residue. Protein glycosylation takes place in the lumen of the endoplasmic reticulum. The N-linked oligosaccharides are synthesized on dolichol phosphate and subsequently transferred to the protein acceptor. Additional sugars are attached in the Golgi complex to form diverse patterns.

11.4 Lectins Are Specific Carbohydrate-Binding Proteins

Carbohydrates on cell surfaces are recognized by proteins called lectins. In animals, the interplay of lectins and their sugar targets guides cell–cell contact. Lectins also play a role inside the cell, such as directing proteins to their proper cellular locations and assisting in protein folding. The viral protein hemagglutinin on the surface of the influenza virus recognizes sialic acid residues on the surfaces of cells invaded by the virus.

APPENDIX

Biochemistry in Focus

α-Glucosidase (maltase) inhibitors can help to maintain blood glucose homeostasis

Maintaining the proper concentration of glucose in the blood (3.9–5.5 mM)—glucose homeostasis—is vital since glucose is reactive and can form advanced glycation products. Moreover, elevated blood glucose (hyperglycemia) is characteristic of diabetes, a serious condition in which insulin, a key hormone in regulating glucose homeostasis is either absent (type 1 diabetes) or ineffective (type 2 diabetes). We will examine the biochemical basis of diabetes many times in our study of biochemistry.

When the body cannot achieve glucose homeostasis on its own, pharmacological intervention is required, as with the administration of insulin or other drugs that control blood glucose. What enzymes might be appropriate targets for drug intervention? In fact, there are several, but one discussed in this chapter is maltase or α-glucosidase (p. 350). Following a meal, starch and glycogen are initially degraded by α-amylase secreted by the salivary glands and pancreas (p. 350). The oligosaccharides generated by α-amylase are further digested by α-glucosidase. Two drugs commonly used to inhibit α-glucosidase are acarbose (Precose) and miglitol (Glyset).

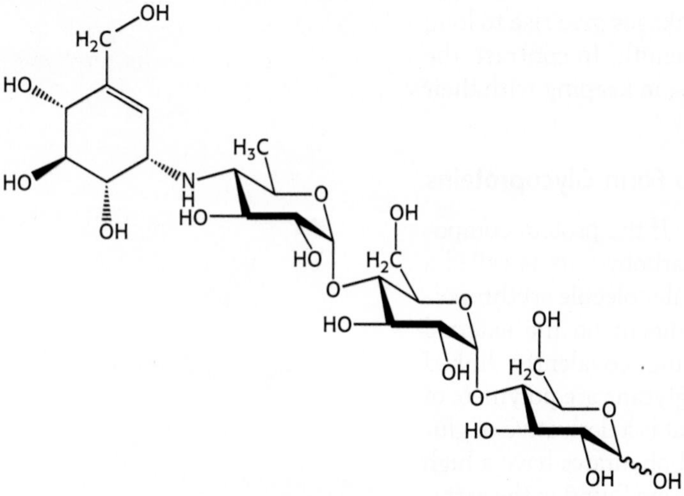

Acarbose (Precose)

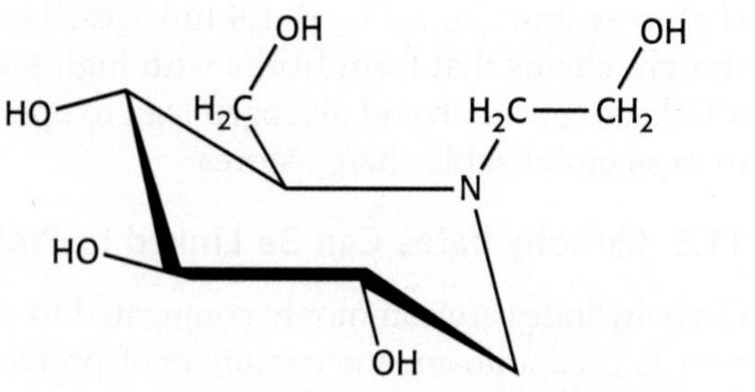

Miglitol (Glyset)

Both acarbose and miglitol are competitive inhibitors (Section 8.5) of α-glucosidase and are usually ingested at the start of a meal to reduce postprandial (after a meal) glucose absorption. Both drugs are used in the treatment of type 2 diabetes to reduce glucose absorption. The drugs may also be used to treat type 1 diabetes if insulin administration alone does not result in glucose homeostasis.

APPENDIX

Problem-Solving Strategies
Problem-Solving 1

PROBLEM: You may remember from organic chemistry that there are a variety of methods to methylate the hydroxyl groups of carbohydrates, that is, replace the hydrogen on all of the hydroxyl groups with a methyl group. You may also remember that polysaccharides can be hydrolyzed by acid to yield monosaccharides. Assume that you have a method that can identify the different methylated monosaccharides. Using these two memories and one assumption, devise an experiment to determine the number of branch points in glycogen or amylopectin.

SOLUTION: The first step is to make sure that we know what a branch point looks like.

▶ **Draw an example of an α-1, 6 glycosidic bond at the branch point. Include two α-1, 4 glycosidic bonds and include a reducing end. Label the two types of bonds and the reducing end.**

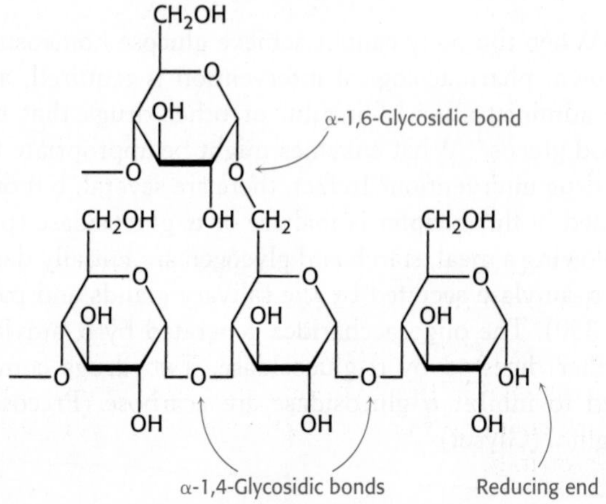

Now, draw the same structure fully methylated.

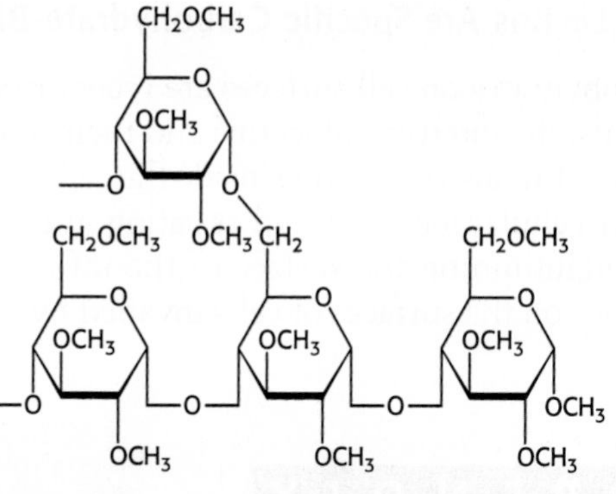

▶ **What products will result from the complete hydrolysis of methylated polysaccharide to monosaccharides?**

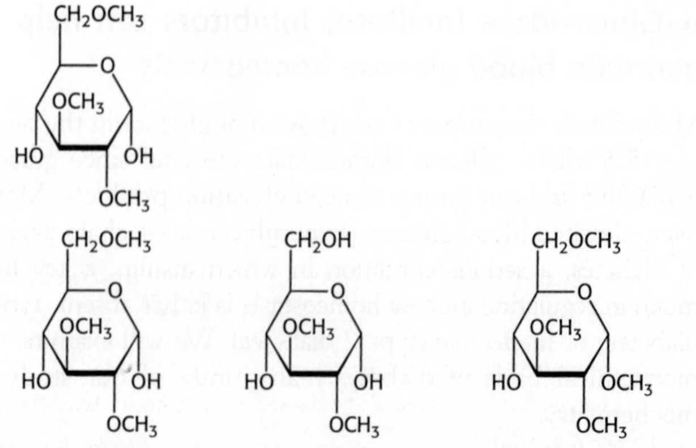

2,3,6-*O*-Trimethylglucose   2,3-*O*-Dimethylglucose   1,2,3,6-*O*-Tetramethylglucose

Examine the resulting monosaccharides.

▶ **How do the methylation patterns differ?**

There are three methylation patterns: 2, 3, 6-tri-O-methylglucose, 2, 3-di-O-methylglucose and 1, 2, 3, 6-tetra-O-methylglucose.

▶ **How can you use this information to determine the amount of branching?**

Because the hydroxyl on carbon 6 is involved in the α-1, 6-glycosidic bond, it will not become methylated. Thus, the amount of 2, 3-di-O-methylglucose corresponds to the number of branch points. Because the reducing end has a free hydroxyl, the number of reducing ends corresponds to the amount of 1, 2, 3, 6-tetra-O-methylglucose.

Here is something to think about.

▶ **How could you use the same experiment to determine the number of nonreducing ends?**

Problem-Solving 2

Data interpretation

PROBLEM: Zika is a mosquito-borne virus. If a pregnant woman is infected with Zika, the virus can cross the placental barrier and cause severe fetal anomalies, most notably microcephaly. Like many viruses, Zika gains entry to the cell by binding to glycosaminoglycans. In the case of Zika, the envelope protein E initiates cell contact. Experiments have established that the E protein binds strongly to pharmaceutical heparin.

Let's examine data in the form of two graphs and see if we can interpret the data.

DATASET: The left bar in the figure below shows protein E bound to heparin-coated plastic. The right bar shows protein E bound to heparin-coated plastic in the presence of excess soluble heparin. The error bars are standard deviations from triplicate experiments.

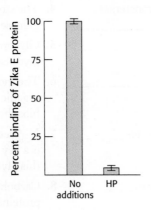

▶ **Why would the amount of protein E bound to immobilized heparin fall in the presence of soluble heparin?**

If protein E binds to the soluble heparin, it is unable to bind to the immobilized heparin. In essence, the soluble heparin outcompetes the immobilized heparin for protein E binding sites.

▶ **How might such a "competition assay" be used to determine the biological glycosaminoglycan that binds to protein E?**

Glycosaminoglycans could be isolated from Zika-sensitive tissues and then tested in the competition assay to determine if they inhibit protein E binding to the immobilized heparin.

DATASET: Various glycosaminoglycans were extracted from human placentas. These glycosaminoglycans were then examined for their ability to inhibit protein E binding to heparin in the competition assay, as shown in the graph below.

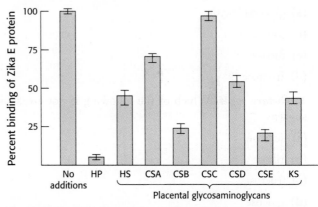

Data from: Kim, S. Y., Zhao, J., Liu, X., Fraser, K., Lin, L., et al. Interaction of Zika virus envelope protein with glycosaminoglycans. *Biochemistry* 56:1151–1162, 2017.

▶ **Which of the glycosaminoglycans are unlikely to bind to protein E in vivo? Explain your answer.**

Sometimes interpreting data is a judgement call. CSC barely inhibits protein E binding, so it is unlikely to bind protein E in vivo. HS, CSA, CSD, and KS also seem unlikely biological partners for protein E.

▶ **Which of the glycosaminoglycans show promise as likely binding sites of protein E?**

CSB and CSE substantially inhibit protein E binding to heparin. Of the glycosaminoglycans tested, these two appear to be the most likely natural binding site(s) for protein E.

KEY TERMS

glycobiology (p. 342)

glycomics (p. 342)

monosaccharide (p. 342)

ketose (p. 342)

aldose (p. 342)

constitutional isomer (p. 343)

stereoisomer (p. 343)

enantiomer (p. 343)

diastereoisomer (p. 343)

epimer (p. 344)

hemiacetal (p. 344)

pyranose (p. 344)

hemiketal (p. 344)

furanose (p. 345)

anomer (p. 345)

reducing sugar (p. 347)

nonreducing sugar (p. 347)

advanced glycation end product (AGE) (p. 348)

glycosidic bond (p. 348)

oligosaccharide (p. 349)

disaccharide (p. 349)

polysaccharide (p. 350)

glycogen (p. 350)

starch (p. 350)

cellulose (p. 350)

glycoprotein (p. 352)

proteoglycan (p. 352)

glycosaminoglycan (p. 352)

mucin (mucoprotein) (p. 352)

glycoform (p. 352)

endoplasmic reticulum (p. 357)

Golgi complex (p. 357)

dolichol phosphate (p. 357)

glycosyltransferase (p. 358)

glycan-binding protein (p. 361)

lectin (p. 361)

selectin (p. 362)

PROBLEMS

1. *Word origin.* Account for the origin of the term *carbohydrate.* ✓❷

2. *A ketose.* Which of the following in not an aldose?

(a) glyceraldehyde

(b) glucose

(c) fucose

(d) fructose

3. *Dietary sugar?* Which of the following is not a reducing sugar?

(a) sucrose

(b) ribose

(c) fructose

(d) glucose

4. *Ham on rye.* Which of the following is found in glycogen and starch?

(a) fructose

(b) β-D-glucose

(c) mannose

(d) α-D-glucose

5. *Diversity.* How many different oligosaccharides can be made by linking one glucose, one mannose, and one galactose? Assume that each sugar is in its pyranose form. Compare this number with the number of tripeptides that can be made from three different amino acids. ✓❸

6. *They go together like a horse and carriage.* Match each term with its description.

(a) Enantiomers _____

(b) Cellulose _____

(c) Lectins _____

(d) Glycosyltransferases _____

(e) Epimers _____

(f) Starch _____

(g) Carbohydrates _____

(h) Proteoglycan _____

(i) Mucoprotein _____

(j) Glycogen _____

1. Has the molecular formula of $(CH_2O)_n$

2. Monosaccharides that differ at a single asymmetric carbon atom

3. The storage form of glucose in animals

4. The storage form of glucose in plants

5. Glycoprotein containing glycosaminoglycans

6. The most abundant organic molecule in the biosphere

7. *N*-Acetylgalactosamine is a key component of this glycoprotein

8. Carbohydrate-binding proteins

9. Enzymes that synthesize oligosaccharides

10. Stereoisomers that are mirror images of each other

7. *Couples.* Indicate whether each of the following pairs of sugars consists of anomers, epimers, or an aldose–ketose pair: ✓❷

(a) D-glyceraldehyde and dihydroxyacetone

(b) D-glucose and D-mannose

(c) D-glucose and D-fructose

(d) α-D-glucose and β-D-glucose

(e) D-ribose and D-ribulose

(f) D-galactose and D-glucose

8. *Carbons and carbonyls.* To which classes of sugars do the monosaccharides shown here belong?

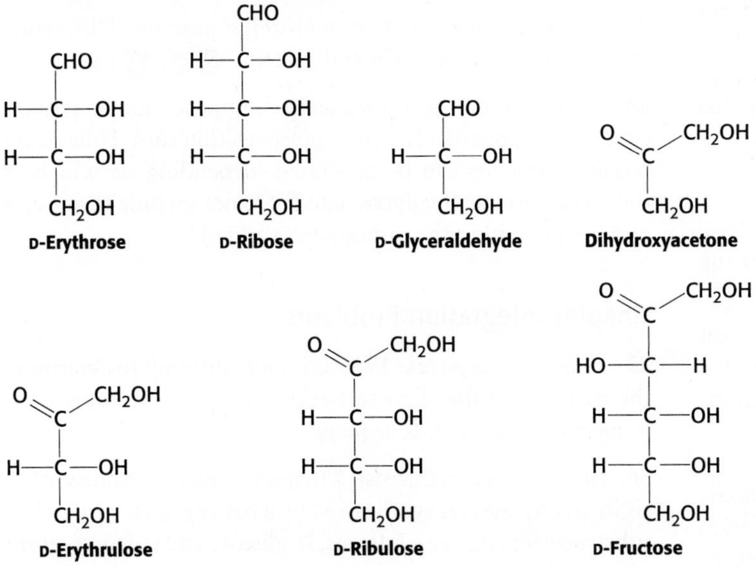

D-Erythrose   D-Ribose   D-Glyceraldehyde   Dihydroxyacetone

D-Erythrulose   D-Ribulose   D-Fructose

9. *Chemical cousins.* Although an aldose with 4 asymmetric carbon atoms is capable of forming 16 diastereoisomers, only 8 of the isomers are commonly observed, including glucose. They are listed below with their structural relation to glucose. Using the structure of glucose as a reference, draw the structures.

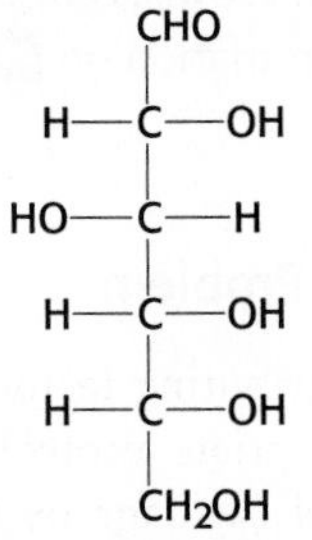

D-Glucose

(a) D-Allose: Epimeric at C-3

(b) D-Altrose: Isomeric at C-2 and C-3

(c) D-Mannose: Epimeric at C-2

(d) D-Gulose: Isomeric at C-3 and C-4

(e) D-Idose: Isomeric at C-2, C-3, and C-4

(f) D-Galactose: Epimeric at C-4

(g) D-Talose: Isomeric at C-2 and C-4

10. *An art project.* Draw the structure of the disaccharide α-glucosyl-(1 → 6)-galactose in the β anomeric form. ✓❸

11. *Mutarotation.* The specific rotations of the α and β anomers of D-glucose are +112 degrees and +18.7 degrees,

respectively. Specific rotation, $[\alpha]_D$, is defined as the observed rotation of light of wavelength 589 nm (the D line of a sodium lamp) passing through 10 cm of a 1 g ml^{-1} solution of a sample. When a crystalline sample of α-D-glucose is dissolved in water, the specific rotation decreases from 112 degrees to an equilibrium value of 52.7 degrees. On the basis of this result, what are the proportions of the α and β anomers at equilibrium? Assume that the concentration of the open-chain form is negligible.

12. *Telltale marker.* Glucose reacts slowly with hemoglobin and other proteins to form covalent compounds. Why is glucose reactive? What is the nature of the adduct formed?

13. *Periodate cleavage.* Compounds containing hydroxyl groups on adjacent carbon atoms undergo carbon–carbon bond cleavage when treated with periodate ion (IO_4^-). How can this reaction be used to distinguish between derivatives of pyranose and furanose?

14. *Sugar lineup.* Identify the following four sugars.

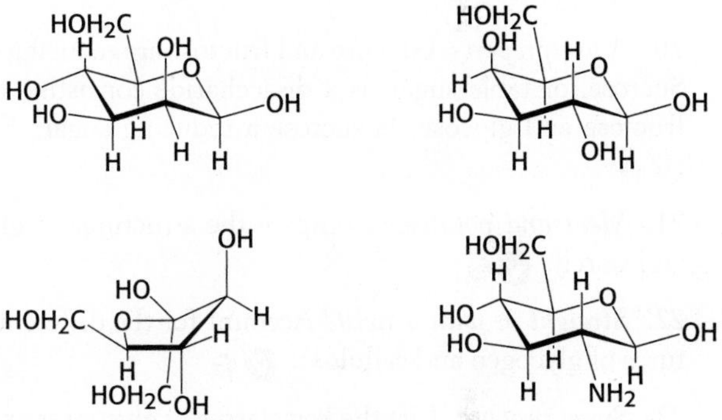

15. *Cellular glue.* A trisaccharide unit of a cell-surface glycoprotein is postulated to play a critical role in mediating cell–cell adhesion in a particular tissue. Design a simple experiment to test this hypothesis.

16. *Component parts.* Raffinose is a trisaccharide and a minor constituent in sugar beets. ✓❸

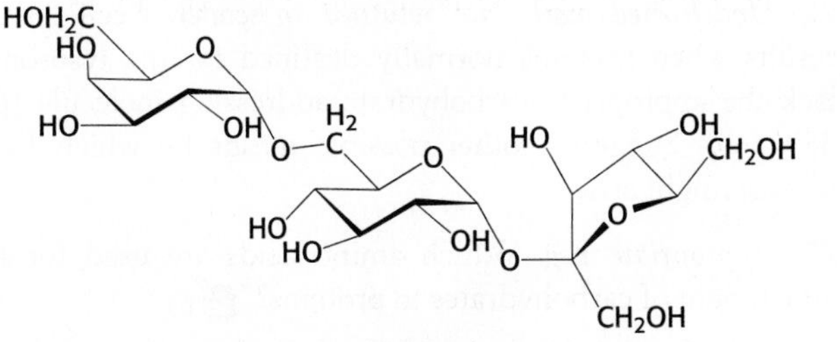

Raffinose

(a) Is raffinose a reducing sugar? Explain.

(b) What are the monosaccharides that compose raffinose?

(c) β-Galactosidase is an enzyme that will remove galactose residues from an oligosaccharide. What are the products of β-galactosidase treatment of raffinose?

17. *Anomeric differences.* α-D-Mannose is a sweet-tasting sugar. β-D-Mannose, on the other hand, tastes bitter. A pure solution of α-D-mannose loses its sweet taste with time as it is converted into the β anomer. Draw the β anomer and explain how it is formed from the α anomer.

18. *A taste of honey.* Fructose in its β-D-pyranose form accounts for the powerful sweetness of honey. The β-D-furanose form, although sweet, is not as sweet as the pyranose form. The furanose form is the more stable form. Draw the two forms and explain why it may not always be wise to cook with honey.

19. *Making ends meet.*

(a) Compare the number of reducing ends to nonreducing ends in a molecule of glycogen.

(b) As we will see in Chapter 21, glycogen is an important fuel-storage form that is rapidly mobilized. At which end—the reducing or nonreducing—would you expect most metabolism to take place?

20. *A lost property.* Glucose and fructose are reducing sugars. Sucrose, or table sugar, is a disaccharide consisting of both fructose and glucose. Is sucrose a reducing sugar? Explain. ✓❸

21. *Meat and potatoes.* Compare the structures of glycogen and starch. ✓❸

22. *Straight or with a twist?* Account for the different structures of glycogen and cellulose. ✓❸

23. *Sweet proteins.* List the key classes of glycoproteins, their defining characteristics, and their biological functions. ✓❶, ✓❺

24. *Life extender.* What is the function of the carbohydrate moiety that is attached to EPO? ✓❶

25. *Cushioning.* What is the role of the glycosaminoglycan in the cushioning provided by cartilage? ✓❶, ✓❺

26. *Undelivered mail. Not returned to sender.* I-cell disease results when proteins normally destined for the lysosomes lack the appropriate carbohydrate-addressing molecule (pp. 359–360). Suggest another possible means by which I-cell disease might arise.

27. *Appropriate peg.* Which amino acids are used for the attachment of carbohydrates to proteins? ✓❹

28. *From one, many.* What is meant by a glycoform?

29. *Ome.* What is meant by the glycome?

30. *Exponential expansion?* Compare the amount of information inherent in the genome, the proteome, and the glycome.

31. *Attachments.* Suppose, one Sunday afternoon, you are relaxing by reading amino acid sequences of various proteins. Being a bit hungry, you are also thinking of sweet snacks. Combining your interests, you wonder whether you can detect *N*-glycosylation sites by simply looking at amino acid sequence. Your roommate, who is taking a biochemistry course, says, "You sure can, to some degree at least, and here's how." What did your roommate say to explain? ✓❹

32. *Locks and keys.* What does the fact that all organisms contain lectins suggest about the role of carbohydrates? ✓❻

33. *Carbohydrates—not just for breakfast anymore.* Differentiate between a glycoprotein and a lectin. ✓❺, ✓❻

34. *Carbohydrates and proteomics.* Suppose that a protein contains six potential *N*-linked glycosylation sites. How many possible proteins can be generated, depending on which of these sites is actually glycosylated? Do not include the effects of diversity within the carbohydrate added.

Chapter Integration Problems

35. *Like a jigsaw puzzle.* Why is it more difficult to determine the structure of the oligosaccharides compared to amino acid sequences or nucleotide sequences?

36. *Stereospecificity.* Sucrose, a major product of photosynthesis in green leaves, is synthesized by a battery of enzymes. The substrates for sucrose synthesis, D-glucose and D-fructose, are a mixture of α and β anomers as well as acyclic compounds in solution. Nonetheless, sucrose consists of α-D-glucose linked by its carbon-1 atom to the carbon-2 atom of β-D-fructose. How can the specificity of sucrose be explained in light of the potential substrates?

37. *Specific recognition.* How might the technique of affinity chromatography (see animation on 🔁 **Sapling Plus**) be used to purify lectins? ✓❻

Data Interpretation Problem

38. *Sore joints.* A contributing factor to the development of arthritis is the inappropriate proteolytic destruction of the aggrecan component of cartilage by the proteolytic enzyme aggrecanase. The immune-system signal molecule interleukin 2 (IL-2) activates aggrecanase; in fact, IL-2 blockers are sometimes used to treat arthritis. Studies were undertaken to determine whether inhibitors of aggrecanase could counteract the effects of IL-2. Pieces of cartilage were incubated in media with various additions and the amount of aggrecan destruction was measured as a function of time.

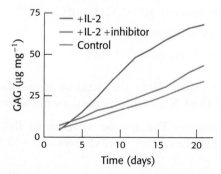

(a) Aggrecan degradation was measured by the release of glycosaminoglycan. What is the rationale for this assay?

(b) Why might glycosaminoglycan release not indicate aggrecan degradation?

(c) What is the purpose of the control (cartilage incubated with no additions)?

(d) What is the effect of adding IL-2 to the system?

(e) What is the response when an aggrecanase inhibitor is added in addition to IL-2?

(f) Why is there some aggrecan destruction in the control with the passage of time?

Think ▶ Pair ▶ Share Problem

39. Describe the various roles that carbohydrates play in nature and explain the biochemical basis for this versatility.

Lipids and Cell Membranes

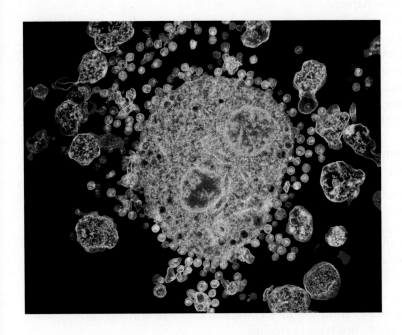

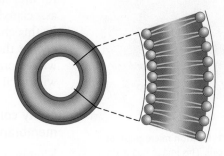

HIV particles (pink) exit an infected cell by membrane budding. Cellular membranes are highly dynamic structures that spontaneously self-assemble. Driven by hydrophobic interactions, as shown in the diagram at right, the fatty acid tails of membrane lipids pack together (yellow), while the polar heads (blue) remain exposed on the surfaces. [NIBSC/Science Source.]

✔ LEARNING OBJECTIVES

By the end of this chapter, you should be able to:

1. Describe the structure of a fatty acid and explain how it is named.

2. Define the three most common types of membrane lipids.

3. Explain how membrane lipids self-assemble into bimolecular sheets.

4. Distinguish between integral and peripheral membrane proteins.

5. Describe how proteins interact with membranes and how membrane-spanning motifs within proteins may be identified.

6. Explain the properties of membrane fluidity and its dependence upon lipid composition.

7. Explain the importance of membranes in eukaryotic cells.

OUTLINE

12.1 Fatty Acids Are Key Constituents of Lipids

12.2 There Are Three Common Types of Membrane Lipids

12.3 Phospholipids and Glycolipids Readily Form Bimolecular Sheets in Aqueous Media

12.4 Proteins Carry Out Most Membrane Processes

12.5 Lipids and Many Membrane Proteins Diffuse Rapidly in the Plane of the Membrane

12.6 Eukaryotic Cells Contain Compartments Bounded by Internal Membranes

The boundaries of all cells are defined by *biological membranes* (Figure 12.1), dynamic structures in which proteins float in a sea of lipids. The lipid component prevents molecules generated inside the cell from leaking out and unwanted molecules from diffusing in, while the protein components act as transport systems that allow the cell to take up specific molecules and remove unwanted ones. Such transport systems confer on membranes the important property of *selective permeability*. We will consider these transport systems in greater detail in the next chapter.

In addition to an external cell membrane (called the *plasma membrane*), eukaryotic cells also contain internal membranes that form the boundaries of organelles such as mitochondria, chloroplasts, peroxisomes, and lysosomes. Functional specialization in the course of evolution has been closely linked to the formation of such compartments. Specific systems have evolved to allow the targeting of selected proteins into or through

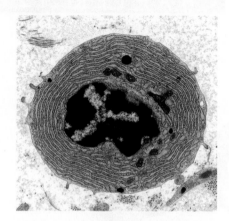

FIGURE 12.1 Electron micrograph of a white blood cell. This image has been colored to indicate the distinct boundary of the cell, formed by its plasma membrane. [Steve Gschmeissner/Science Source.]

particular internal membranes and, hence, into specific organelles. External and internal membranes share essential properties; these features are the subject of this chapter.

Biological membranes serve several additional functions indispensable for life, such as energy storage and information transduction. The proteins associated with the membrane define these functions for any given cell. In this chapter, we will examine the properties of membrane proteins that enable them to exist in the hydrophobic environment of the membrane while connecting two hydrophilic environments. In the next chapter, we will discuss the functions of these proteins.

Many common features underlie the diversity of biological membranes

Membranes are as diverse in structure as they are in function. However, they do have in common a number of important attributes:

1. Membranes are *sheetlike structures*, only two molecules thick, that form *closed boundaries* between different compartments. The thickness of most membranes is between 60 Å (6 nm) and 100 Å (10 nm).

2. Membranes consist mainly of *lipids* and *proteins*. The mass ratio of lipids to proteins ranges from 1:4 to 4:1. Membranes also contain *carbohydrates* that are linked to lipids and proteins.

3. Membrane lipids are small molecules that have both *hydrophilic* and *hydrophobic* moieties. These lipids spontaneously form *closed bimolecular sheets* in aqueous media. These *lipid bilayers* are barriers to the flow of polar molecules.

4. *Specific proteins mediate distinctive functions of membranes.* Proteins serve as pumps, channels, receptors, energy transducers, and enzymes. Membrane proteins are embedded in lipid bilayers, which create suitable environments for their action.

5. Membranes are *noncovalent assemblies*. The constituent protein and lipid molecules are held together by many noncovalent interactions, which act cooperatively.

6. Membranes are *asymmetric*. The two faces of biological membranes always differ from each other.

7. Membranes are *fluid structures*. Lipid molecules diffuse rapidly in the plane of the membrane, as do proteins, unless they are anchored by specific interactions. In contrast, lipid molecules and proteins do not readily rotate across the membrane. Membranes can be regarded as *two-dimensional solutions of oriented proteins and lipids*.

8. Most cell membranes are *electrically polarized*, such that the inside is negative [typically −60 millivolts (mV)]. Membrane potential plays a key role in transport, energy conversion, and excitability (Chapter 13).

12.1 Fatty Acids Are Key Constituents of Lipids

The hydrophobic properties of lipids are essential to their ability to form membranes. Most lipids owe their hydrophobic properties to one component, their fatty acids.

Fatty acid names are based on their parent hydrocarbons

Fatty acids are long hydrocarbon chains of various lengths and degrees of unsaturation that terminate with carboxylic acid groups. The systematic name for a fatty acid is derived from the name of its parent hydrocarbon by the substitution of *oic* for the final *e*. For example, the C_{18} saturated fatty acid is called *octadecanoic acid* because the parent hydrocarbon is octadecane. A C_{18} fatty acid with one double bond is called *octadecenoic* acid; with two double bonds, *octadecadienoic* acid; and with three double bonds, *octadecatrienoic* acid. The notation 18:0 denotes a C_{18} fatty acid with no double bonds, whereas 18:2 signifies that there are two double bonds. The structures of the ionized forms of two common fatty acids—palmitic acid (16:0) and oleic acid (18:1)—are shown in Figure 12.2.

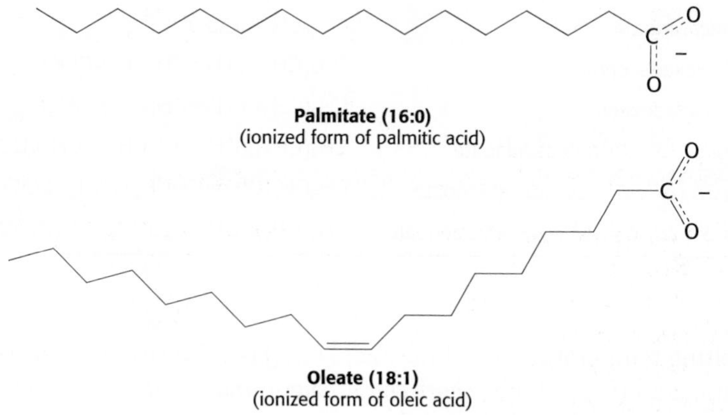

Palmitate (16:0)
(ionized form of palmitic acid)

Oleate (18:1)
(ionized form of oleic acid)

FIGURE 12.2 Structures of two fatty acids. Palmitate is a 16-carbon, saturated fatty acid (16:0), while oleate is an 18-carbon fatty acid with a single cis double bond (18:1).

Fatty acid carbon atoms are numbered starting at the carboxyl terminus, as shown in the margin. Carbon atoms 2 and 3 are often referred to as α and β, respectively. The methyl carbon atom at the distal end of the chain is called the *ω-carbon atom*. The position of a double bond is represented by the symbol Δ followed by a superscript number. For example, *cis*-Δ^9 means that there is a cis double bond between carbon atoms 9 and 10; *trans*-Δ^2 means that there is a trans double bond between carbon atoms 2 and 3. Alternatively, the position of a double bond can be denoted by counting from the distal end, with the ω-carbon atom (the methyl carbon) as number 1. An ω-3 fatty acid, for example, has the structure shown in the margin. Fatty acids are ionized at physiological pH, and so it is appropriate to refer to them according to their carboxylate form: for example, palmitate or hexadecanoate.

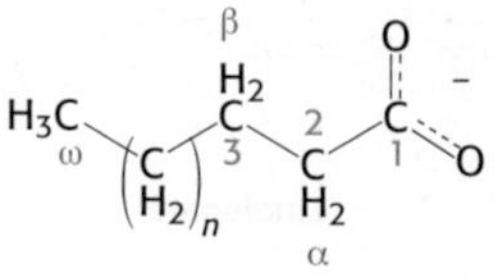

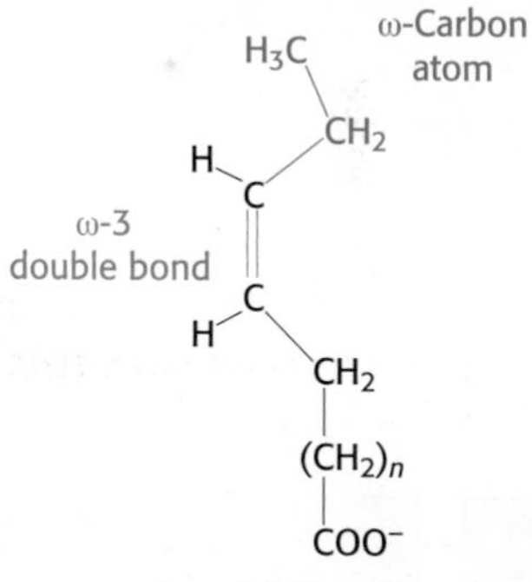

An ω-3 fatty acid

Fatty acids vary in chain length and degree of unsaturation

Fatty acids in biological systems usually contain an even number of carbon atoms, typically between 14 and 24 (Table 12.1). The 16- and 18-carbon fatty acids are most common. The dominance of fatty acid chains containing an even number of carbon atoms reflects the manner in which fatty acids are biosynthesized (Chapter 26). The hydrocarbon chain is almost invariably unbranched in animal fatty acids. The alkyl chain may be saturated or it may contain one or more double bonds. The configuration of the double bonds in most unsaturated fatty acids is cis. The double bonds in polyunsaturated fatty acids are separated by at least one methylene group.

The properties of fatty acids and of lipids derived from them are markedly dependent on chain length and degree of saturation. Unsaturated fatty acids have lower melting points than do saturated fatty acids of the same length. For example, the melting point of stearic acid is 69.6°C, whereas that of oleic acid (which contains one cis double bond) is 13.4°C. The melting points of polyunsaturated fatty acids of the C_{18} series are even lower. Chain length also affects the melting point, as illustrated by the fact that

TABLE 12.1 Some naturally occurring fatty acids in animals

Number of carbons	Number of double bonds	Common name	Systematic name	Formula
12	0	Laurate	n-Dodecanoate	$CH_3(CH_2)_{10}COO^-$
14	0	Myristate	n-Tetradecanoate	$CH_3(CH_2)_{12}COO^-$
16	0	Palmitate	n-Hexadecanoate	$CH_3(CH_2)_{14}COO^-$
18	0	Stearate	n-Octadecanoate	$CH_3(CH_2)_{16}COO^-$
20	0	Arachidate	n-Eicosanoate	$CH_3(CH_2)_{18}COO^-$
22	0	Behenate	n-Docosanoate	$CH_3(CH_2)_{20}COO^-$
24	0	Lignocerate	n-Tetracosanoate	$CH_3(CH_2)_{22}COO^-$
16	1	Palmitoleate	cis-Δ^9-Hexadecenoate	$CH_3(CH_2)_5CH{=}CH(CH_2)_7COO^-$
18	1	Oleate	cis-Δ^9-Octadecenoate	$CH_3(CH_2)_7CH{=}CH(CH_2)_7COO^-$
18	2	Linoleate	cis, cis-Δ^9, Δ^{12}-Octadecadienoate	$CH_3(CH_2)_4(CH{=}CHCH_2)_2(CH_2)_6COO^-$
18	3	Linolenate	all-cis-$\Delta^9, \Delta^{12}, \Delta^{15}$-Octadecatrienoate	$CH_3CH_2(CH{=}CHCH_2)_3(CH_2)_6COO^-$
20	4	Arachidonate	all-cis $\Delta^5, \Delta^8, \Delta^{11}, \Delta^{14}$-Eicosatetraenoate	$CH_3(CH_2)_4(CH{=}CHCH_2)_4(CH_2)_2COO^-$

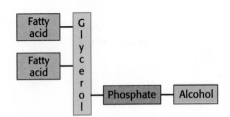

Linolenate

the melting temperature of palmitic acid (C_{16}) is 6.5 degrees lower than that of stearic acid (C_{18}). Thus, *short chain length and unsaturation enhance the fluidity of fatty acids and of their derivatives.*

12.2 There Are Three Common Types of Membrane Lipids

By definition, *lipids are water-insoluble biomolecules that are highly soluble in organic solvents such as chloroform.* Lipids have a variety of biological roles: they serve as fuel molecules, highly concentrated energy stores, signal molecules and messengers in signal-transduction pathways, and components of membranes. The first three roles of lipids will be considered in later chapters. Here, our focus is on lipids as membrane constituents. The three major kinds of membrane lipids are *phospholipids, glycolipids,* and *cholesterol.* We begin with lipids found in eukaryotes and bacteria. The lipids in archaea are distinct, although they have many features related to membrane formation in common with lipids of other organisms.

Phospholipids are the major class of membrane lipids

Phospholipids are abundant in all biological membranes. A phospholipid molecule is constructed from four components:

- one or more fatty acids,
- a platform to which the fatty acids are attached,
- a phosphate, and
- an alcohol attached to the phosphate (Figure 12.3).

The fatty acid components provide a hydrophobic barrier, whereas the remainder of the molecule has hydrophilic properties that enable interaction with the aqueous environment.

The platform on which phospholipids are built may be *glycerol,* a three-carbon alcohol, or *sphingosine,* a more complex alcohol. Phospholipids derived from glycerol are called *phosphoglycerides.* A phosphoglyceride consists of a glycerol backbone to which are attached two fatty acid chains and a phosphorylated alcohol.

FIGURE 12.3 Schematic structure of a phospholipid.

In phosphoglycerides, the hydroxyl groups at C-1 and C-2 of glycerol are esterified to the carboxyl groups of the two fatty acid chains. The C-3 hydroxyl group of the glycerol backbone is esterified to phosphoric acid. When no further additions are made, the resulting compound is *phosphatidate (diacylglycerol 3-phosphate)*, the simplest phosphoglyceride. Only small amounts of phosphatidate are present in membranes. However, the molecule is a key intermediate in the biosynthesis of the other phosphoglycerides (Section 26.1). The absolute configuration of the glycerol 3-phosphate moiety of membrane lipids is shown in Figure 12.4.

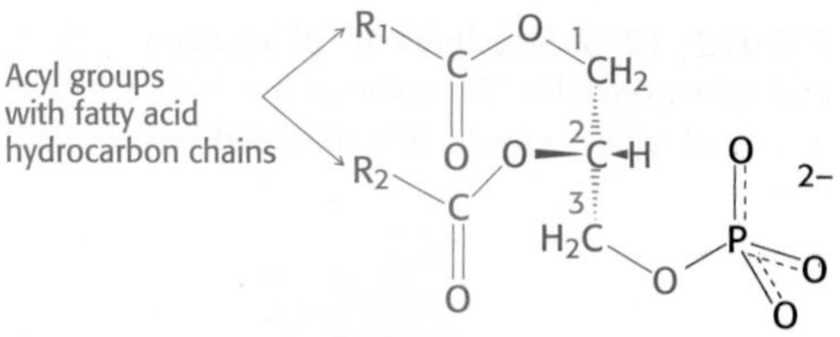

Phosphatidate
(Diacylglycerol 3-phosphate)

FIGURE 12.4 Structure of phosphatidate (diacylglycerol 3-phosphate). The absolute configuration of the center carbon (C-2) is shown.

The major phosphoglycerides are derived from phosphatidate by the formation of an ester bond between the phosphate group of phosphatidate and the hydroxyl group of one of several alcohols. The common alcohol moieties of phosphoglycerides are the amino acid serine, ethanolamine, choline, glycerol, and inositol.

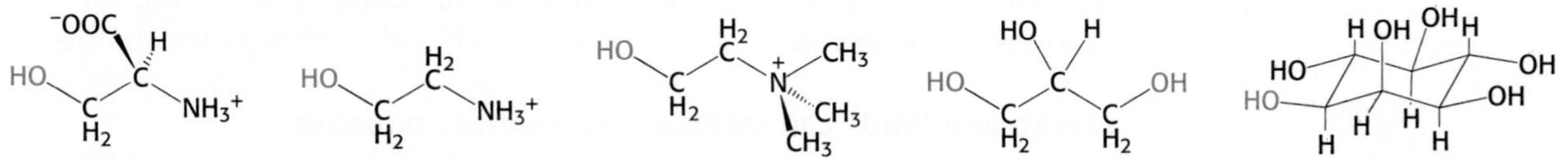

| Serine | Ethanolamine | Choline | Glycerol | Inositol |

The structural formulas of phosphatidylcholine and the other principal phosphoglycerides—namely, phosphatidylethanolamine, phosphatidylserine, phosphatidylinositol, and diphosphatidylglycerol—are given in Figure 12.5.

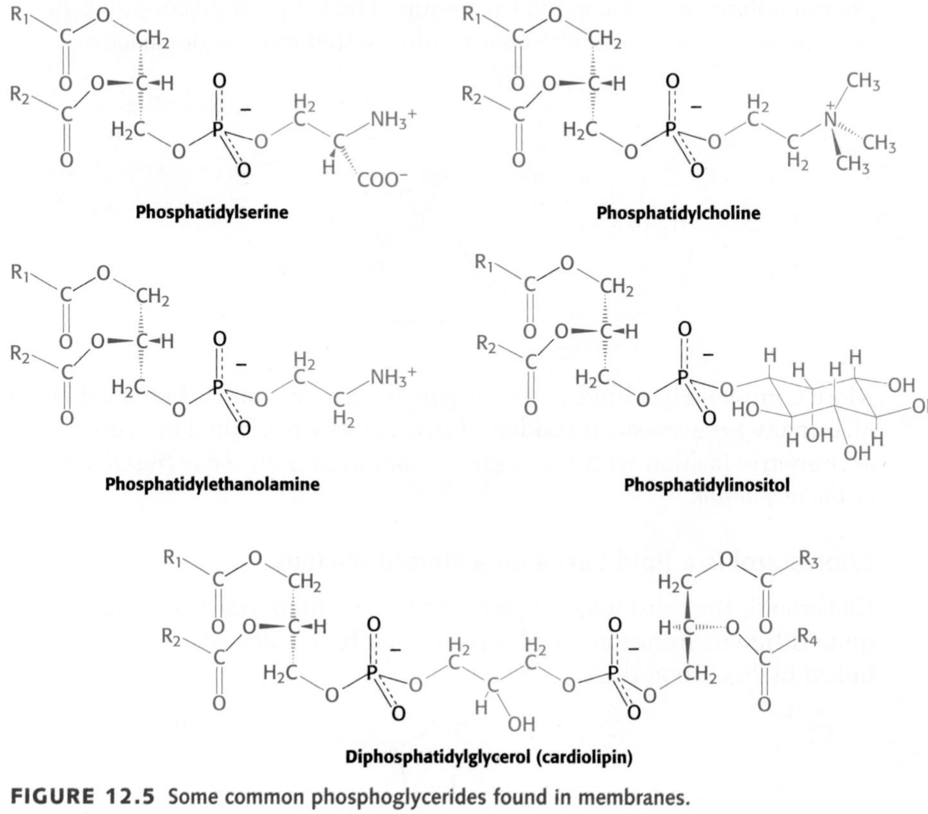

Phosphatidylserine

Phosphatidylcholine

Phosphatidylethanolamine

Phosphatidylinositol

Diphosphatidylglycerol (cardiolipin)

FIGURE 12.5 Some common phosphoglycerides found in membranes.

Sphingomyelin is a phospholipid found in membranes that is not derived from glycerol. Instead, the backbone in sphingomyelin is *sphingosine*, an amino alcohol that contains a long, unsaturated hydrocarbon chain

Sphingosine

Sphingomyelin

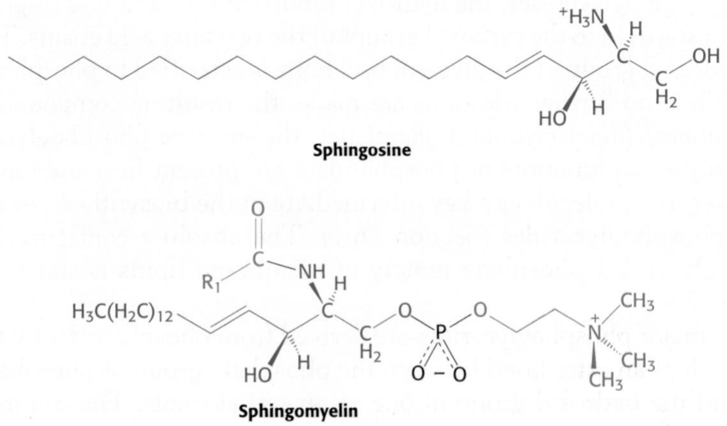

(Figure 12.6). In sphingomyelin, the amino group of the sphingosine backbone is linked to a fatty acid by an amide bond. In addition, the primary hydroxyl group of sphingosine is esterified to phosphorylcholine.

Membrane lipids can include carbohydrate moieties

The second major class of membrane lipids, *glycolipids,* are *sugar-containing lipids.* Like sphingomyelin, the glycolipids in animal cells are derived from sphingosine. The amino group of the sphingosine backbone is acylated by a fatty acid, as in sphingomyelin. Glycolipids differ from sphingomyelin in the identity of the unit that is linked to the primary hydroxyl group of the sphingosine backbone. In glycolipids, one or more sugars (rather than phosphorylcholine) are attached to this group. The simplest glycolipid, called a *cerebroside,* contains a single sugar residue, either glucose or galactose.

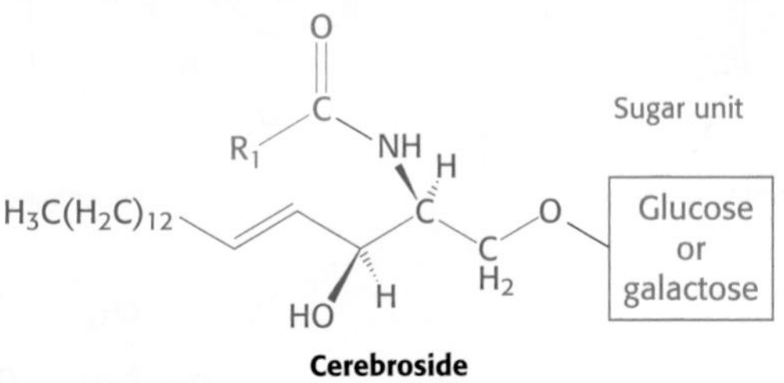

Cerebroside
(a glycolipid)

More complex glycolipids, such as *gangliosides,* contain a branched chain of as many as seven sugar residues. Glycolipids are oriented in a completely asymmetric fashion with the *sugar residues always on the extracellular side of the membrane.*

Cholesterol is a lipid based on a steroid nucleus

Cholesterol, the third major type of membrane lipid, has a structure that is quite different from that of phospholipids. It is a steroid, built from four linked hydrocarbon rings.

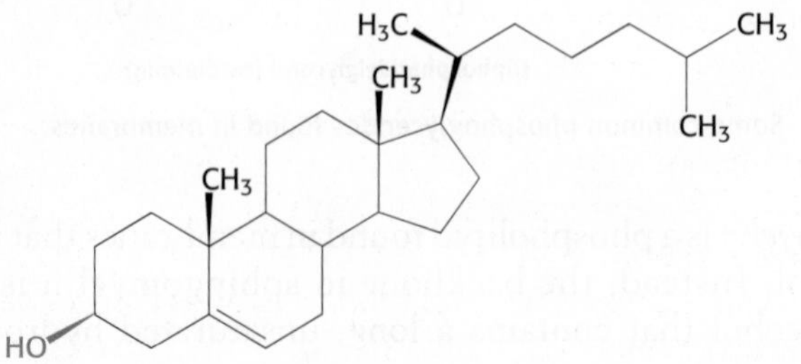

Cholesterol

A hydrocarbon tail is linked to the steroid at one end, and a hydroxyl group is attached at the other end. In membranes, the orientation of the molecule is parallel to the fatty acid chains of the phospholipids, and the hydroxyl group interacts with the nearby phospholipid head groups. Cholesterol is absent from prokaryotes but is found to varying degrees in virtually all animal membranes. It constitutes almost 25% of the membrane lipids in certain nerve cells but is essentially absent from some intracellular membranes.

Archaeal membranes are built from ether lipids with branched chains

The membranes of archaea differ in composition from those of eukaryotes or bacteria in three important ways. Two of these differences clearly relate to the hostile living conditions of many archaea (Figure 12.7).

1. The nonpolar chains are joined to a glycerol backbone by ether rather than ester linkages. The ether linkage is more resistant to hydrolysis.

2. The alkyl chains are branched rather than linear. They are built up from repeats of a fully saturated five-carbon fragment. These branched, saturated hydrocarbons are more resistant to oxidation than the unbranched chains of eukaryotic and bacterial membrane lipids. The ability of archaeal lipids to resist hydrolysis and oxidation may help these organisms to withstand the extreme conditions, such as high temperature, low pH, or high salt concentration, under which some of these archaea grow.

3. The stereochemistry of the central glycerol is inverted compared with that shown in Figure 12.4.

FIGURE 12.7 An archaeon and its environment. Archaea can thrive in habitats as harsh as a volcanic vent. Here, the archaea form an orange mat surrounded by yellow sulfurous deposits. [Images & Volcans/Science Source]

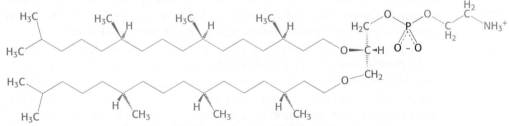

Membrane lipid from the archaeon *Methanococcus jannaschii*

A membrane lipid is an amphipathic molecule containing a hydrophilic and a hydrophobic moiety

The repertoire of membrane lipids is extensive. However, these lipids possess a critical common structural theme: *membrane lipids are amphipathic molecules* (amphiphilic molecules), that is, they contain both a *hydrophilic* and a *hydrophobic* moiety.

Let's look at a model of a phosphoglyceride, such as phosphatidylcholine. Its overall shape is roughly rectangular (Figure 12.8A). The two hydrophobic fatty acid chains are approximately parallel to each other, whereas the hydrophilic phosphorylcholine moiety points in the opposite direction. Sphingomyelin has a similar conformation, as does the archaeal lipid depicted. Therefore, the following shorthand has been adopted to represent these membrane lipids: the hydrophilic unit, also called the *polar head group,* is represented by a circle, and the hydrocarbon tails are depicted by straight or wavy lines (Figure 12.8B).

(A)

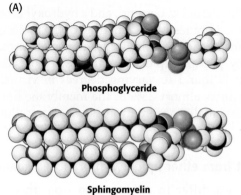

Phosphoglyceride

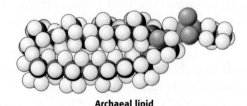

Sphingomyelin

Archaeal lipid

(B)

Shorthand depiction

FIGURE 12.8 Representations of membrane lipids.
(A) Space-filling models of a phosphoglyceride, sphingomyelin, and an archaeal lipid show their shapes and distribution of hydrophilic and hydrophobic moieties. (B) A shorthand depiction of a membrane lipid.

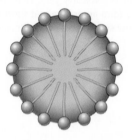

FIGURE 12.9 Diagram of a section of a micelle. Ionized fatty acids readily form such structures, but most phospholipids do not.

FIGURE 12.10 Diagram of a section of a bilayer membrane.

12.3 Phospholipids and Glycolipids Readily Form Bimolecular Sheets in Aqueous Media

What properties enable phospholipids to form membranes? *Membrane formation is a consequence of the amphipathic nature of the molecules.* Their polar head groups favor contact with water, whereas their hydrocarbon tails interact with one another in preference to water. How can molecules with these preferences arrange themselves in aqueous solutions? One way is to form a globular structure called a *micelle.* The polar head groups form the outside surface of the micelle, which is surrounded by water, and the hydrocarbon tails are sequestered inside, interacting with one another (Figure 12.9).

Alternatively, the strongly opposed preferences of the hydrophilic and hydrophobic moieties of membrane lipids can be satisfied by forming a *lipid bilayer,* composed of two lipid sheets (Figure 12.10). A lipid bilayer is also called a *bimolecular sheet.* The hydrophobic tails of each individual sheet interact with one another, forming a hydrophobic interior that acts as a permeability barrier. The hydrophilic head groups interact with the aqueous medium on each side of the bilayer. The two opposing sheets are called leaflets.

The favored structure for most phospholipids and glycolipids in aqueous media is a bimolecular sheet rather than a micelle. The reason is that the two fatty acid chains of a phospholipid or a glycolipid are too bulky to fit into the interior of a micelle. In contrast, salts of fatty acids (such as sodium palmitate, a constituent of soap) readily form micelles because they contain only one chain. *The formation of bilayers instead of micelles by phospholipids is of critical biological importance.* A micelle is a limited structure, usually less than 200 Å (20 nm) in diameter. In contrast, a bimolecular sheet can extend to macroscopic dimensions, as much as a millimeter (10^7 Å, or 10^6 nm) or more. Phospholipids and related molecules are important membrane constituents because they readily form extensive bimolecular sheets.

Lipid bilayers form spontaneously by a *self-assembly process.* In other words, the structure of a bimolecular sheet is inherent in the structure of the constituent lipid molecules. The growth of lipid bilayers from phospholipids is rapid and spontaneous in water. *Hydrophobic interactions are the major driving force for the formation of lipid bilayers.* Recall that hydrophobic interactions also play a dominant role in the stacking of bases in nucleic acids and in the folding of proteins (Sections 1.3 and 2.4). Water molecules are released from the hydrocarbon tails of membrane lipids as these tails become sequestered in the nonpolar interior of the bilayer. Furthermore, *van der Waals attractive forces between the hydrocarbon tails favor close packing of the tails.* Finally, there are *electrostatic and hydrogen-bonding attractions between the polar head groups and water molecules.* Thus, lipid bilayers are stabilized by the full array of forces that mediate molecular interactions in biological systems. Because lipid bilayers are held together by many *reinforcing, noncovalent interactions (predominantly hydrophobic)*, they are *cooperative structures.* These hydrophobic interactions have three significant biological consequences: (1) lipid bilayers have an inherent tendency to be *extensive;* (2) lipid bilayers will tend to *close on themselves* so that there are no edges with exposed hydrocarbon chains, and so they form compartments; and (3) lipid bilayers are *self-sealing* because a hole in a bilayer is energetically unfavorable.

Lipid vesicles can be formed from phospholipids

The propensity of phospholipids to form membranes has been used to create an important experimental and clinical tool. *Lipid vesicles,* or *liposomes,* are aqueous compartments enclosed by a lipid bilayer (Figure 12.11). These structures can be used to study membrane permeability or to deliver chemicals to cells. Liposomes are formed by suspending a suitable lipid, such as phosphatidylcholine, in an aqueous medium, and then *sonicating* (i.e., agitating by high-frequency sound waves) to give a dispersion of closed vesicles that are quite uniform in size. Vesicles formed by this method are nearly spherical and have a diameter of about 500 Å (50 nm). Larger vesicles (of the order of 1 μm or 10^4 Å in diameter) can be prepared by slowly evaporating the organic solvent from a suspension of phospholipid in a mixed-solvent system.

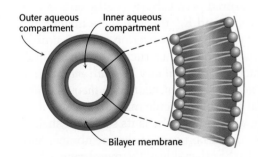

FIGURE 12.11 Liposome. A liposome, or lipid vesicle, is a small aqueous compartment surrounded by a lipid bilayer.

Ions or molecules can be trapped in the aqueous compartments of lipid vesicles by forming the vesicles in the presence of these substances (Figure 12.12). For example, 500-Å-diameter vesicles formed in a 0.1 M glycine solution will trap about 2000 molecules of glycine in each inner aqueous compartment. These glycine-containing vesicles can be separated from the surrounding solution of glycine by dialysis or by gel-filtration chromatography (Section 3.1). The permeability of the bilayer membrane to glycine can then be determined by measuring the rate of efflux of glycine from the inner compartment of the vesicle to the ambient solution. Liposomes can be formed with specific membrane proteins embedded in them by solubilizing the proteins in the presence of detergents and then adding

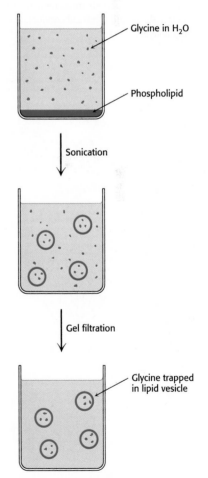

FIGURE 12.12 Preparation of glycine-containing liposomes. Liposomes containing glycine are formed by the sonication of phospholipids in the presence of glycine. Free glycine is removed by gel filtration.

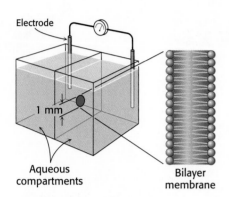

FIGURE 12.13 Experimental arrangement for the study of a planar bilayer membrane. A bilayer membrane is formed across a 1-mm hole in a septum that separates two aqueous compartments. This arrangement permits measurements of the permeability and electrical conductance of lipid bilayers.

them to the phospholipids from which liposomes will be formed. Protein–liposome complexes provide valuable experimental tools for examining a range of membrane-protein functions.

Therapeutic applications for liposomes are currently under active investigation. For example, liposomes containing drugs or DNA can be injected into patients. These liposomes fuse with the plasma membrane of many kinds of cells, introducing into the cells the molecules they contain. Drug delivery with liposomes often lessens its toxicity. Less of the drug is distributed to normal tissues because long-circulating liposomes concentrate in regions of increased blood circulation, such as solid tumors and sites of inflammation. Moreover, the selective fusion of lipid vesicles with particular kinds of cells is a promising means of controlling the delivery of drugs to target cells.

Another well-defined synthetic membrane is a *planar bilayer membrane*. This structure can be formed across a 1-mm hole in a partition between two aqueous compartments by dipping a fine paintbrush into a membrane-forming solution, such as phosphatidylcholine in decane, and stroking the tip of the brush across the hole. The lipid film across the hole thins spontaneously into a lipid bilayer. The electrical conduction properties of this macroscopic bilayer membrane are readily studied by inserting electrodes into each aqueous compartment (Figure 12.13). For example, the permeability of the membrane to ions is determined by measuring the current across the membrane as a function of the applied voltage.

Lipid bilayers are highly impermeable to ions and most polar molecules

Permeability studies of lipid vesicles and electrical-conductance measurements of planar bilayers have shown that *lipid bilayer membranes have a very low permeability for ions and most polar molecules*. Water is a conspicuous exception to this generalization; it traverses such membranes relatively easily because of its low molecular weight, high concentration, and lack of a complete charge. The range of measured permeability coefficients is very wide (Figure 12.14). For example, Na^+ and K^+ traverse these membranes 10^9 times as slowly as does H_2O. Tryptophan, a zwitterion at pH 7, crosses the membrane 10^3 times as slowly as does indole, a structurally related molecule that lacks ionic groups. In fact, *the permeability of small molecules is correlated with their solubility in a nonpolar solvent relative to their solubility in water*. This relation suggests that a small molecule might traverse a lipid bilayer membrane in the following way: first, it sheds its solvation shell of water; then, it is dissolved in the hydrocarbon core of the membrane; and, finally, it diffuses through this core to the other side of the membrane, where it becomes resolvated by water. An ion such as Na^+ traverses membranes

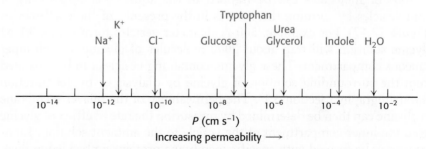

FIGURE 12.14 Permeability coefficients (*P*) of ions and molecules in a lipid bilayer. The ability of molecules to cross a lipid bilayer spans a wide range of values.

very slowly because the replacement of its coordination shell of polar water molecules by nonpolar interactions with the membrane interior is highly unfavorable energetically.

12.4 Proteins Carry Out Most Membrane Processes

We now turn to membrane proteins, which are responsible for most of the dynamic processes carried out by membranes. Membrane lipids form a permeability barrier and thereby establish compartments, whereas *specific proteins mediate nearly all other membrane functions*. In particular, proteins transport chemicals and information across a membrane. Membrane lipids create the appropriate environment for the action of such proteins.

Membranes differ in their protein content. *Myelin*, a membrane that serves as an electrical insulator around certain nerve fibers, has a low content of protein (18%). Relatively pure lipids are well suited for insulation. In contrast, the plasma membranes, or exterior membranes, of most other cells are much more metabolically active. They contain many pumps, channels, receptors, and enzymes. The protein content of these plasma membranes is typically 50%. Energy-transduction membranes, such as the internal membranes of mitochondria and chloroplasts, have the highest content of protein, around 75%.

By serving as an insulator, myelin plays a critical role in enabling the rapid transmission of nerve impulses, or action potentials (Section 13.4). Its structure is quite unique: the cells that generate myelin—oligodendrocytes in the brain, Schwann cells in the periphery—wrap their plasma membranes multiple times around the axon (Figure 12.15) or the part of the neuron that conducts the electrical signal. Significant myelination of neurons in the brain occurs during infancy but persists throughout adolescence, a reminder that the brain is an actively developing organ throughout childhood. The importance of myelination is further underscored by the existence of demyelinating diseases, such as *multiple sclerosis*, where myelin assembly is impaired or existing myelin is damaged.

The protein components of a membrane can be readily visualized by *SDS–polyacrylamide gel electrophoresis*. As we discussed in Section 3.1, the

(A)

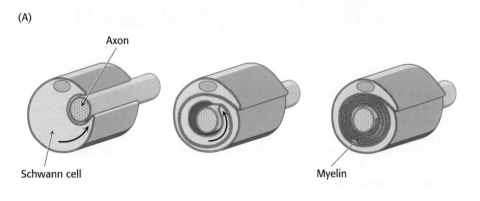

(B)

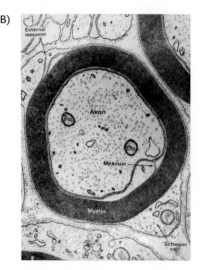

FIGURE 12.15 Myelination of a neuron. (A) Myelin is comprised of the lipid-rich plasma membrane of a Schwann cell in peripheral nerves or the oligodendrocyte in the brain that has wrapped multiple times around the axon, or the portion of the neuron that conducts electrical impulses. [https://biologydictionary.net/myelin-sheath] (B) The myelin sheath around an axon is readily apparent by electron microscopy. [Don W Fawcett/Getty Images]

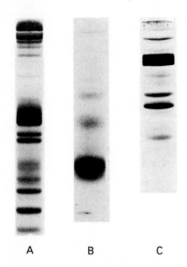

FIGURE 12.16 SDS–acrylamide gel patterns of membrane proteins. (A) The plasma membrane of erythrocytes. (B) The photoreceptor membranes of retinal rod cells. (C) The sarcoplasmic reticulum membrane of muscle cells. [Courtesy of Dr. Theodore Steck and Dr. David MacLennan.]

A   B   C

electrophoretic mobility of many proteins in SDS-containing gels depends on the mass rather than on the net charge of the protein. The gel-electrophoresis patterns of three membranes—the plasma membrane of erythrocytes, the photoreceptor membrane of retinal rod cells, and the sarcoplasmic reticulum membrane of muscle—are shown in Figure 12.16. It is evident that each of these three membranes contains many proteins but has a distinct protein composition. In general, *membranes performing different functions contain different repertoires of proteins.*

Proteins associate with the lipid bilayer in a variety of ways

The ease with which a protein can be dissociated from a membrane indicates how intimately it is associated with the membrane. Some membrane proteins can be solubilized by relatively mild means, such as extraction by a solution of high ionic strength (e.g., 1 M NaCl). Other membrane proteins are bound much more tenaciously; they can be solubilized only by using a detergent or an organic solvent. Membrane proteins can be classified as being either *peripheral* or *integral* on the basis of this difference in dissociability (Figure 12.17). *Integral membrane proteins* interact extensively with the hydrocarbon chains of membrane lipids, and they can be released only by agents that compete for these nonpolar interactions. In fact, most integral membrane proteins span the lipid bilayer. In contrast, *peripheral membrane proteins* are bound to membranes primarily by electrostatic and hydrogen-bond interactions with the head groups of lipids. These polar interactions can be disrupted by adding salts or by changing the pH. Many peripheral membrane proteins are bound to the surfaces of integral proteins, on either the cytoplasmic or the extracellular side of the membrane. Others are anchored to the lipid bilayer by a covalently attached hydrophobic chain, such as a fatty acid.

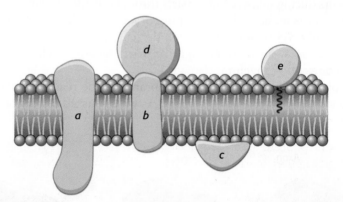

FIGURE 12.17 Integral and peripheral membrane proteins. Integral membrane proteins (*a* and *b*) interact extensively with the hydrocarbon region of the bilayer. Most known integral membrane proteins traverse the lipid bilayer. Peripheral membrane proteins interact with the polar head groups of the lipids (*c*) or bind to the surfaces of integral proteins (*d*). Other proteins are tightly anchored to the membrane by a covalently attached lipid molecule (*e*).

Proteins interact with membranes in a variety of ways

Membrane proteins are more difficult to purify and crystallize than are water-soluble proteins. Nonetheless, researchers using x-ray crystallographic or electron microscopic methods have determined the three-dimensional structures of more than 4000 such proteins at sufficiently high resolution to discern the molecular details. As noted in Chapter 2, membrane proteins differ from soluble proteins in the distribution of hydrophobic and hydrophilic groups. We will consider the structures of three membrane proteins in some detail.

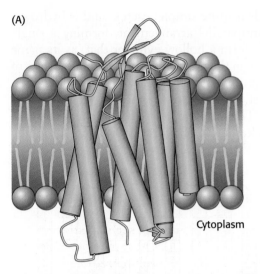

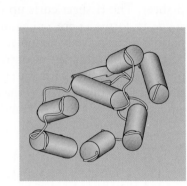

FIGURE 12.18 Structure of bacteriorhodopsin. *Notice* that bacteriorhodopsin consists largely of membrane-spanning α helices (represented by yellow cylinders). (A) View through the membrane bilayer. (B) View from the cytoplasmic side of the membrane. [Drawn from 1BRX.pdb.]

Proteins can span the membrane with alpha helices. The first membrane protein that we consider is the archaeal protein *bacteriorhodopsin,* shown in Figure 12.18. This protein uses light energy to transport protons from inside to outside the cell, generating a proton gradient used to form ATP. Bacteriorhodopsin is built almost entirely of α helices; seven closely packed α helices, nearly perpendicular to the plane of the cell membrane, span its 45-Å width. Examination of the primary structure of bacteriorhodopsin reveals that most of the amino acids in these membrane-spanning α helices are nonpolar and only a very few are charged (Figure 12.19). This distribution of nonpolar amino acids is sensible because these residues are either in contact with the hydrocarbon core of the membrane or with one another. *Membrane-spanning α helices are the most common structural motif in membrane proteins.* As will be considered in Section 12.5, such regions can often be detected by examining amino acid sequence alone.

FIGURE 12.19 Amino acid sequence of bacteriorhodopsin. The seven helical regions are highlighted in yellow and the charged residues in red.

A channel protein can be formed from beta strands. Porin, a protein from the outer membranes of bacteria such as *E. coli* and *Rhodobacter capsulatus,* represents a class of membrane proteins with a completely different type of structure. Structures of this type are built from β strands and contain essentially no α helices (Figure 12.20).

(A)

(B)

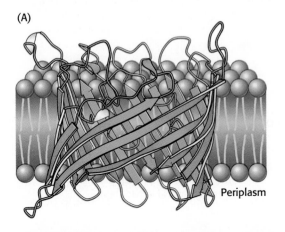

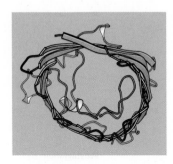

FIGURE 12.20 Structure of bacterial porin (from *Rhodopseudomonas blastica*). *Notice* that this membrane protein is built entirely of β strands. (A) Side view. (B) View from the periplasmic space. Only one monomer of the trimeric protein is shown. [Drawn from 1PRN.pdb.]

The arrangement of β strands is quite simple: each strand is hydrogen bonded to its neighbor in an antiparallel arrangement, forming a single β sheet. The β sheet curls up to form a hollow cylinder that, as its name suggests, forms a pore, or channel, in the membrane. The outside surface of porin is appropriately nonpolar, given that it interacts with the hydrocarbon core of the membrane. In contrast, the inside of the channel is quite hydrophilic and is filled with water. This arrangement of nonpolar and polar surfaces is accomplished by the alternation of hydrophobic and hydrophilic amino acids along each β strand (Figure 12.21).

FIGURE 12.21 Amino acid sequence of a porin. Some membrane proteins, such as porins, are built from β strands that tend to have hydrophobic and hydrophilic amino acids in adjacent positions. The secondary structure of porin from *Rhodopseudomonas blastica* is shown, with the diagonal lines indicating the direction of hydrogen bonding along the β sheet. Hydrophobic residues (F, I, L, M, V, W, and Y) are shown in yellow. These residues tend to lie on the outside of the structure, in contact with the hydrophobic core of the membrane.

Embedding part of a protein in a membrane can link the protein to the membrane surface. The structure of the endoplasmic reticulum membrane-bound enzyme prostaglandin H_2 synthase-1 reveals a rather different role for α helices in protein–membrane associations. This enzyme catalyzes the conversion of arachidonic acid into prostaglandin H_2 in two steps: (1) a cyclooxygenase reaction and (2) a peroxidase reaction (Figure 12.22). Prostaglandin H_2 promotes inflammation and modulates gastric acid secretion. The enzyme that produces prostaglandin H_2 is a homodimer with a rather

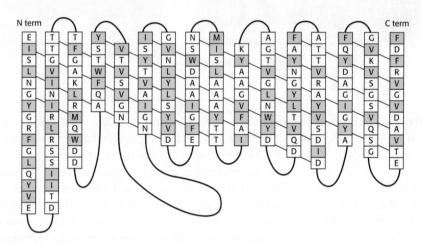

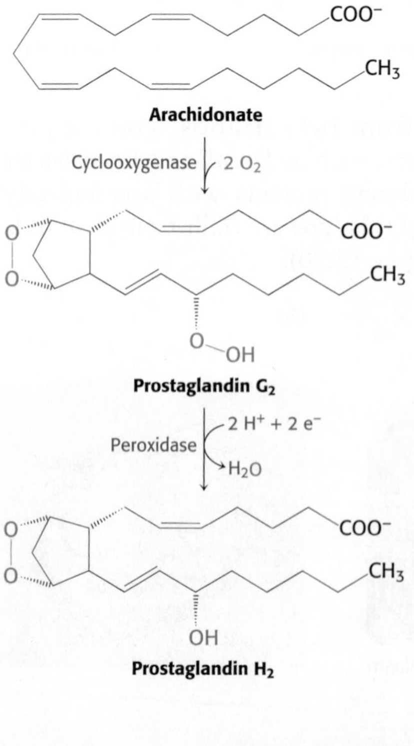

FIGURE 12.22 Formation of prostaglandin H_2. Prostaglandin H_2 synthase-1 catalyzes the formation of prostaglandin H_2 from arachidonic acid in two steps.

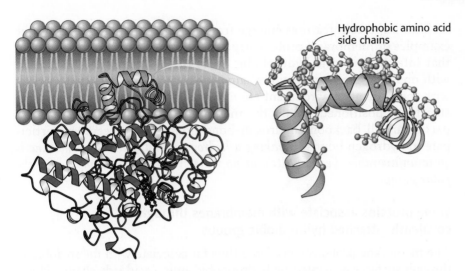

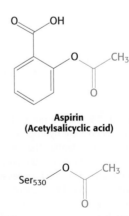

Hydrophobic amino acid side chains

FIGURE 12.23 Attachment of prostaglandin H₂ synthase-1 to the membrane. *Notice* that prostaglandin H₂ synthase-1 is held in the membrane by a set of α helices (orange) coated with hydrophobic side chains. One monomer of the dimeric enzyme is shown. [Drawn from 1PTH.pdb.]

complicated structure consisting primarily of α helices. Unlike bacteriorhodopsin, this protein is not largely embedded in the membrane. Instead, it lies along the outer surface of the membrane, firmly bound by a set of α helices with hydrophobic surfaces that extend from the bottom of the protein into the membrane (Figure 12.23). This linkage is sufficiently strong that only the action of detergents can release the protein from the membrane. Thus, this enzyme is classified as an integral membrane protein, although it does not span the membrane.

The localization of prostaglandin H₂ synthase-1 in the membrane is crucial to its function. The substrate for this enzyme, arachidonic acid, is a hydrophobic molecule generated by the hydrolysis of membrane lipids. Arachidonic acid reaches the active site of the enzyme from the membrane without entering an aqueous environment by traveling through a hydrophobic channel in the protein (Figure 12.24). Indeed, nearly all of us have experienced the importance of this channel: drugs such as aspirin and ibuprofen block the channel and prevent prostaglandin synthesis by inhibiting the cyclooxygenase activity of the synthase. In particular, aspirin acts through the transfer of its acetyl group to a serine residue (Ser 530) that lies along the path to the active site (Figure 12.25). In response to an injury or infection, prostaglandins promote the inflammatory response, which includes swelling, pain, and fever. Cyclooxygenase inhibitors blunt this response, providing pain relief and fever reduction.

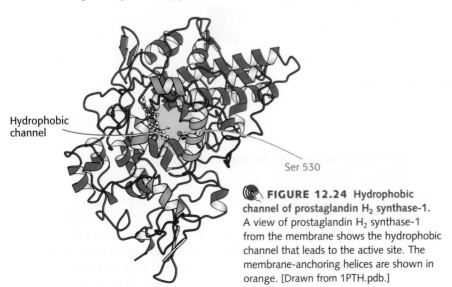

Hydrophobic channel

Ser 530

FIGURE 12.24 Hydrophobic channel of prostaglandin H₂ synthase-1. A view of prostaglandin H₂ synthase-1 from the membrane shows the hydrophobic channel that leads to the active site. The membrane-anchoring helices are shown in orange. [Drawn from 1PTH.pdb.]

Aspirin (Acetylsalicylic acid)

FIGURE 12.25 Aspirin's effects on prostaglandin H₂ synthase-1. Aspirin acts by transferring an acetyl group to a serine residue in prostaglandin H₂ synthase-1.

Two important features emerge from our examination of these three examples of membrane-protein structure. First, the parts of the protein that interact with the hydrophobic parts of the membrane are coated with nonpolar amino acid side chains, whereas those parts that interact with the aqueous environment are much more hydrophilic. Second, the structures positioned within the membrane are quite regular and, in particular, all backbone hydrogen-bond donors and acceptors participate in hydrogen bonds. *Breaking a hydrogen bond within a membrane is quite unfavorable, because little or no water is present to compete for the polar groups.*

Some proteins associate with membranes through covalently attached hydrophobic groups

The membrane proteins considered thus far associate with the membrane through surfaces generated by hydrophobic amino acid side chains. However, even otherwise soluble proteins can associate with membranes if hydrophobic groups are attached to the proteins. Three such groups are shown in Figure 12.26: (1) a palmitoyl group attached to a specific cysteine residue by a thioester bond, (2) a farnesyl group attached to a cysteine residue at the carboxyl terminus, and (3) a glycolipid structure termed a glycosylphosphatidylinositol (GPI) anchor attached to the carboxyl terminus. These modifications are attached by enzyme systems that recognize specific signal sequences near the attachment site.

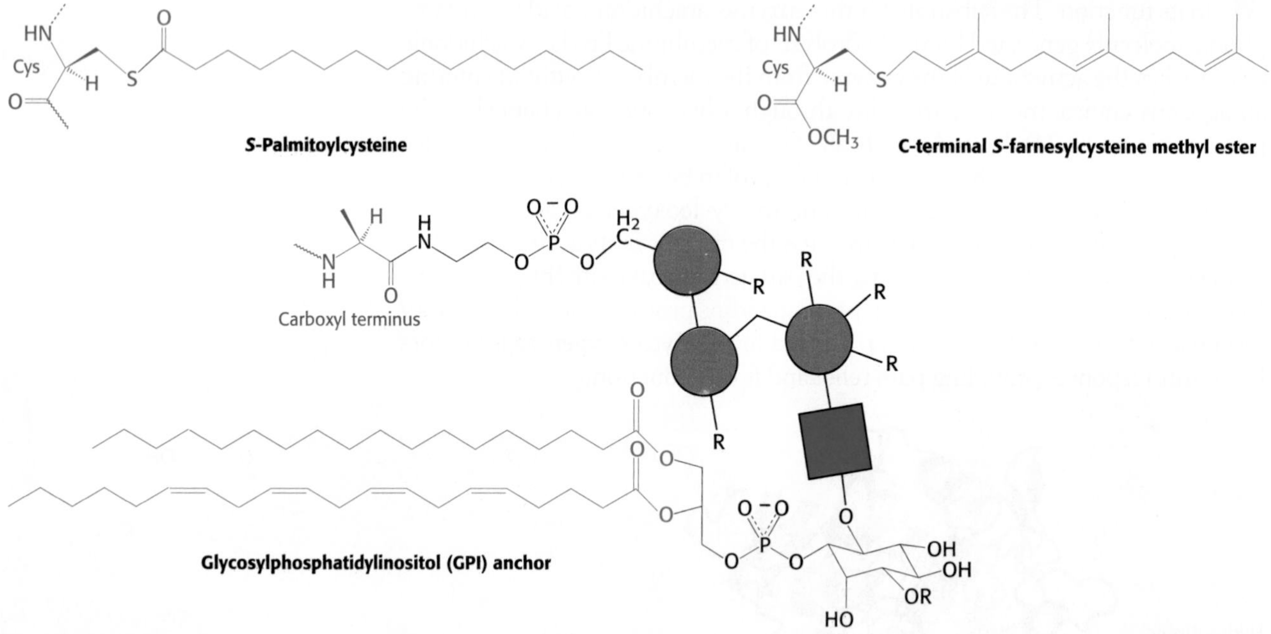

S-Palmitoylcysteine

C-terminal S-farnesylcysteine methyl ester

Carboxyl terminus

Glycosylphosphatidylinositol (GPI) anchor

FIGURE 12.26 Membrane anchors. Membrane anchors are hydrophobic groups that are covalently attached to proteins (in blue) and tether the proteins to the membrane. The green circles and blue square correspond to mannose and β-D-acetylglucosamine (GlcNAc), respectively. R groups represent points of additional modification.

Transmembrane helices can be accurately predicted from amino acid sequences

Many membrane proteins, like bacteriorhodopsin, employ α helices to span the hydrophobic part of a membrane. As noted earlier, most of the residues in these α helices are nonpolar and almost none of them are charged. Can we use this information to identify likely membrane-spanning regions from sequence data alone? One approach to identifying transmembrane helices is to ask whether a postulated helical segment is likely to be more stable in a hydrocarbon environment or in water. Specifically, we want to estimate the free-energy change when a helical segment is transferred from the interior of a membrane to water. Free-energy changes for the transfer of individual amino acid residues from a hydrophobic to an aqueous environment are given in Table 12.2. For example, the transfer of a helix formed entirely of L-arginine residues—a positively charged amino acid—from the interior of a membrane to water would be highly favorable [$-51.5 \text{ kJ mol}^{-1}$ ($-12.3 \text{ kcal mol}^{-1}$) per arginine residue in the helix]. In contrast, the transfer of a helix formed entirely of L-phenylalanine—a hydrophobic amino acid—would be unfavorable [$+15.5 \text{ kJ mol}^{-1}$ ($+3.7 \text{ kcal mol}^{-1}$) per phenylalanine residue in the helix].

The hydrocarbon core of a membrane is typically 30 Å wide, a length that can be traversed by an α helix consisting of 20 residues. We can take the amino acid sequence of a protein and estimate the free-energy change that takes place when a hypothetical α helix formed of residues 1 through 20 is transferred from the membrane interior to water. The same calculation can be made for residues 2 through 21, 3 through 22, and so forth, until we reach the end of the sequence. The span of 20 residues chosen for this calculation is called a *window*. The free-energy change for each window is plotted against the first amino acid at the window to create a *hydropathy plot*. Empirically, a peak of $+84 \text{ kJ mol}^{-1}$ ($+20 \text{ kcal mol}^{-1}$) or more in a hydropathy plot based on a window of 20 residues indicates that a polypeptide segment could be a membrane-spanning α helix. For example, glycophorin—a protein found in the membranes of red blood cells—is predicted by this criterion to have one membrane-spanning helix, in agreement with experimental findings (Figure 12.27). Note, however, that a peak in the hydropathy plot does not prove that a segment is a transmembrane helix. Even soluble proteins may have highly nonpolar regions.

TABLE 12.2 Polarity scale for identifying transmembrane helices

Amino acid residue	Transfer free energy in kJ mol^{-1} (kcal mol^{-1})
Phe	15.5 (3.7)
Met	14.3 (3.4)
Ile	13.0 (3.1)
Leu	11.8 (2.8)
Val	10.9 (2.6)
Cys	8.4 (2.0)
Trp	8.0 (1.9)
Ala	6.7 (1.6)
Thr	5.0 (1.2)
Gly	4.2 (1.0)
Ser	2.5 (0.6)
Pro	−0.8 (−0.2)
Tyr	−2.9 (−0.7)
His	−12.6 (−3.0)
Gln	−17.2 (−4.1)
Asn	−20.2 (−4.8)
Glu	−34.4 (−8.2)
Lys	−37.0 (−8.8)
Asp	−38.6 (−9.2)
Arg	−51.7 (−12.3)

Source: Data from D. M. Engelman, T. A. Steitz, and A. Goldman. *Annu. Rev. Biophys. Biophys. Chem. 15*(1986):21–353.

Note: The free energies are for the transfer of an amino acid residue in an α helix from the membrane interior (assumed to have a dielectric constant of 2) to water.

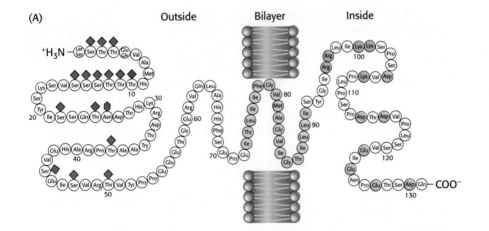

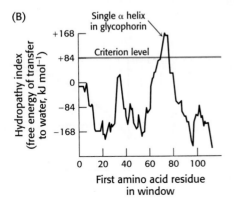

FIGURE 12.27 Locating the membrane-spanning helix of glycophorin. (A) Amino acid sequence and transmembrane disposition of glycophorin A from the red-blood-cell membrane. Fifteen *O*-linked carbohydrate units are shown as diamond shapes, and an *N*-linked unit is shown as a lozenge shape. The hydrophobic residues (yellow) buried in the bilayer form a transmembrane α helix. The carboxyl-terminal part of the molecule, located on the cytoplasmic side of the membrane, is rich in negatively charged (red) and positively charged (blue) residues. (B) Hydropathy plot for glycophorin. The free energy for transferring a helix of 20 residues from the membrane to water is plotted as a function of the position of the first residue of the helix in the sequence of the protein. Peaks of greater than $+84 \text{ kJ mol}^{-1}$ ($+20 \text{ kcal mol}^{-1}$) in hydropathy plots are indicative of potential transmembrane helices. [(A) Information from Dr. Vincent Marchesi; (B) data from D. M. Engelman, T. A. Steitz, and A. Goldman, *Annu. Rev. Biophys. Biophys. Chem.* 15:321–353, 1986. Copyright © 1986 by Annual Reviews, Inc. All rights reserved.]

FIGURE 12.28 Hydropathy plot for porin. No strong peaks are observed for this intrinsic membrane protein, because it is constructed from membrane-spanning β strands rather than α helices.

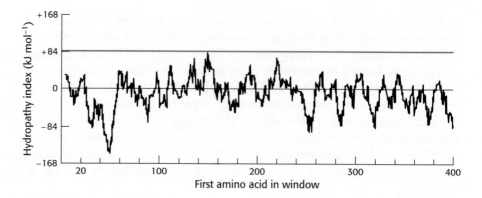

Conversely, some membrane proteins contain membrane-spanning features (such as a set of cylinder-forming β strands) that escape detection by these plots (Figure 12.28).

12.5 Lipids and Many Membrane Proteins Diffuse Rapidly in the Plane of the Membrane

Biological membranes are not rigid, static structures. On the contrary, lipids and many membrane proteins are constantly in lateral motion, a process called *lateral diffusion*. The rapid lateral movement of membrane proteins has been visualized by means of fluorescence microscopy using the technique of *fluorescence recovery after photobleaching* (FRAP; Figure 12.29). First, a cell-surface component is specifically labeled with a fluorescent chromophore. A small region of the cell surface (~3 μm²) is viewed through a fluorescence microscope. The fluorescent molecules in this region are then destroyed (bleached) by a very intense light pulse from a laser, as indicated by the pale spot in Figure 12.29B. The fluorescence of this region is subsequently monitored as a function of time by using a light level sufficiently low to prevent further bleaching. If the labeled component is mobile, bleached molecules leave and unbleached molecules enter the illuminated region, resulting in an increase in the fluorescence intensity. The rate of recovery of fluorescence depends on the lateral mobility of the fluorescence-labeled component, which can be expressed in terms of a diffusion coefficient, D. The average distance S traversed in time t depends on D according to the expression

$$S = (4Dt)^{1/2}$$

The diffusion coefficient of lipids in a variety of membranes is about $1\ \mu m^2\ s^{-1}$. Thus, a phospholipid molecule diffuses an average distance of $2\ \mu m$ in 1 s. This rate means that *a lipid molecule can travel from one end of*

FIGURE 12.29 Fluorescence recovery after photobleaching (FRAP) technique. (A) The cell surface fluoresces because of a labeled surface component. (B) The fluorescent molecules of a small part of the surface are bleached by an intense light pulse. (C) The fluorescence intensity recovers as bleached molecules diffuse out of the region and unbleached molecules diffuse into it. (D) The rate of recovery depends on the diffusion coefficient.

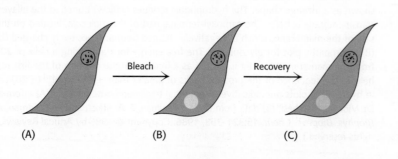

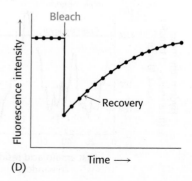

a bacterium to the other in a second. The magnitude of the observed diffusion coefficient indicates that the viscosity of the membrane is about 100 times that of water, rather like that of olive oil.

In contrast, proteins vary markedly in their lateral mobility. *Some proteins are nearly as mobile as lipids, whereas others are virtually immobile.* For example, the photoreceptor protein rhodopsin (See Section 34.3 on 🅢 **Sapling**Plus), a very mobile protein, has a diffusion coefficient of $0.4\ \mu m^2\ s^{-1}$. The rapid movement of rhodopsin is essential for fast signaling. At the other extreme is fibronectin, a peripheral glycoprotein that interacts with the extracellular matrix. For fibronectin, D is less than $10^{-4}\mu m^2\ s^{-1}$. Fibronectin has a very low mobility because it is anchored to actin filaments on the inside of the plasma membrane through *integrin,* a transmembrane protein that links the extracellular matrix to the cytoskeleton.

The fluid mosaic model allows lateral movement but not rotation through the membrane

On the basis of the mobility of proteins in membranes, in 1972 S. Jonathan Singer and Garth Nicolson proposed a *fluid mosaic model* to describe the overall organization of biological membranes. The essence of their model is that *membranes are two-dimensional solutions of oriented lipids and globular proteins.* The lipid bilayer has a dual role: it is both a *solvent* for integral membrane proteins and a *permeability barrier.* Membrane proteins are free to diffuse laterally in the lipid matrix unless restricted by special interactions.

Although the lateral diffusion of membrane components can be rapid, the spontaneous rotation of lipids from one face of a membrane to the other is a very slow process. The transition of a molecule from one membrane surface to the other is called *transverse diffusion* or *flip-flop* (Figure 12.30). The flip-flop of phospholipid molecules in phosphatidylcholine vesicles has been directly measured by electron spin resonance techniques, which show that *a phospholipid molecule flip-flops once in several hours.* Thus, a phospholipid molecule takes about 10^9 times as long to flip-flop across a membrane as it takes to diffuse a distance of 50 Å in the lateral direction. The free-energy barriers to flip-flopping are even larger for protein molecules than for lipids because proteins have more-extensive polar regions. In fact, the flip-flop of a protein molecule has not been observed. Hence, *membrane asymmetry can be preserved for long periods.*

Membrane fluidity is controlled by fatty acid composition and cholesterol content

Many membrane processes, such as transport or signal transduction, depend on the fluidity of the membrane lipids, which in turn depends on the properties of fatty acid chains. Fatty acid chains in membrane bilayers can exist in an ordered, rigid state or in a relatively disordered, fluid state. The transition from the rigid to the fluid state takes place abruptly as the temperature is raised above T_m, the melting temperature (Figure 12.31). *This transition temperature depends on the length of the fatty acid chains and on their degree of unsaturation* (Table 12.3). The presence of saturated fatty acid residues favors the rigid state because their straight hydrocarbon chains interact very favorably with one another. On the other hand, *a cis double bond produces a bend in the hydrocarbon chain. This bend interferes with a highly ordered packing of fatty acid chains, and so* T_m *is lowered* (Figure 12.32). The length of the fatty acid chain also affects the transition temperature. Long hydrocarbon chains interact more strongly than do short ones. Specifically, each additional $-CH_2-$ group makes a favorable

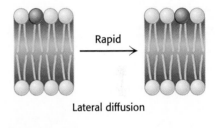

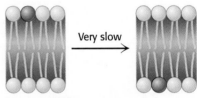

FIGURE 12.30 Lipid movement in membranes. Lateral diffusion of lipids is much more rapid than transverse diffusion (flip-flop).

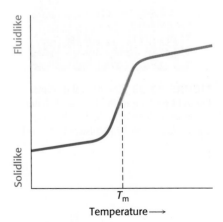

FIGURE 12.31 The phase-transition, or melting, temperature (T_m) for a phospholipid membrane. As the temperature is raised, the phospholipid membrane changes from a packed, ordered state to a more random one.

TABLE 12.3 The melting temperature of phosphatidylcholine containing different pairs of identical fatty acid chains

Number of carbons	Number of double bonds	Fatty acid		
		Common name	Systematic name	T_m (°C)
22	0	Behenate	n-Docosanote	75
18	0	Stearate	n-Octadecanoate	58
16	0	Palmitate	n-Hexadecanoate	41
14	0	Myristate	n-Tetradecanoate	24
18	1	Oleate	cis-Δ^9-Octadecenoate	−22

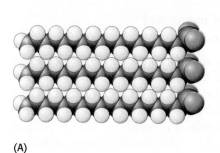

(A)   (B)

FIGURE 12.32 Packing of fatty acid chains in a membrane. The highly ordered packing of fatty acid chains is disrupted by the presence of cis double bonds. The space-filling models show the packing of (A) three molecules of stearate (C₁₈, saturated) and (B) a molecule of oleate (C₁₈, unsaturated) between two molecules of stearate.

contribution of about $-2\,\text{kJ mol}^{-1}$ ($-0.5\,\text{kcal mol}^{-1}$) to the free energy of interaction of two adjacent hydrocarbon chains.

Bacteria regulate the fluidity of their membranes by varying the number of double bonds and the length of their fatty acid chains. For example, the ratio of saturated to unsaturated fatty acid chains in the *E. coli* membrane decreases from 1.6 to 1.0 as the growth temperature is lowered from 42°C to 27°C. This decrease in the proportion of saturated residues prevents the membrane from becoming too rigid at the lower temperature.

In animals, cholesterol is the key regulator of membrane fluidity. Cholesterol contains a bulky steroid nucleus with a hydroxyl group at one end and a flexible hydrocarbon tail at the other end. Cholesterol inserts into bilayers with its long axis perpendicular to the plane of the membrane. The hydroxyl group of cholesterol forms a hydrogen bond with a carbonyl oxygen atom of a phospholipid head group, whereas the hydrocarbon tail of cholesterol is located in the nonpolar core of the bilayer. The different shape of cholesterol compared with that of phospholipids disrupts the regular interactions between fatty acid chains (Figure 12.33).

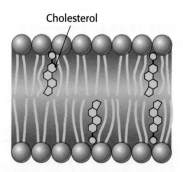

Cholesterol

FIGURE 12.33 Cholesterol disrupts the tight packing of the fatty acid chains. [Information from S. L. Wolfe, *Molecular and Cellular Biology* (Wadsworth, 1993).]

Lipid rafts are highly dynamic complexes formed between cholesterol and specific lipids

In addition to its nonspecific effects on membrane fluidity, cholesterol can form specific complexes with lipids that contain the sphingosine backbone, including sphingomyelin and certain glycolipids, and with GPI-anchored proteins. These complexes concentrate within small (10–200 nm) and highly dynamic regions within membranes. The resulting structures are often referred to as *lipid rafts.* One result of these interactions is the *moderation of membrane fluidity,* making membranes less fluid but at the same time less subject to phase transitions. The presence of lipid rafts thus represents a modification of the original fluid mosaic model for biological membranes. Although their small size and dynamic nature have made them very difficult to study, it appears that lipid rafts may play a role in concentrating proteins that participate in signal transduction pathways and may also serve to regulate membrane curvature and budding.

All biological membranes are asymmetric

Membranes are structurally and functionally asymmetric. The outer and inner surfaces of *all known biological membranes have different components and different enzymatic activities*. A clear-cut example is the pump that regulates the concentration of Na^+ and K^+ ions in cells (Figure 12.34). This transport protein is located in the plasma membrane of nearly all cells in higher organisms. The Na^+–K^+ pump is oriented so that it pumps Na^+ out of the cell and K^+ into it. Furthermore, ATP must be on the inside of the cell to drive the pump. Ouabain, a specific inhibitor of the pump, is effective only if it is located outside. We shall consider the mechanism of this important and fascinating family of pumps in Chapter 13.

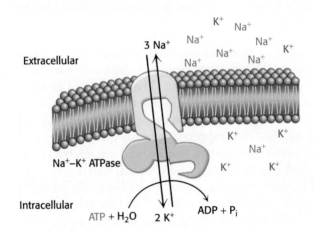

FIGURE 12.34
Asymmetry of the Na^+–K^+ transport system in plasma membranes. The Na^+–K^+ transport system pumps Na^+ out of the cell and K^+ into the cell by hydrolyzing ATP on the intracellular side of the membrane.

Membrane proteins have a unique orientation because, after synthesis, they are inserted into the membrane in an asymmetric manner. This absolute asymmetry is preserved because membrane proteins do not rotate from one side of the membrane to the other and because *membranes are always synthesized by the growth of preexisting membranes*. Lipids, too, are asymmetrically distributed between the two membrane leaflets. For example, in the red-blood-cell membrane, sphingomyelin and phosphatidylcholine are preferentially located in the outer leaflet of the bilayer, whereas phosphatidylethanolamine and phosphatidylserine are located mainly in the inner leaflet. Large amounts of cholesterol are present in both leaflets.

12.6 Eukaryotic Cells Contain Compartments Bounded by Internal Membranes

Thus far, we have considered only the plasma membrane of cells. Some bacteria and archaea have only this single membrane, surrounded by a thick cell wall. Other bacteria, such as *E. coli,* have two membranes separated by a cell wall (made of proteins, peptides, and carbohydrates) lying between them (Figure 12.35). The inner membrane acts as the permeability barrier, and the outer membrane and the cell wall provide additional protection.

(A)

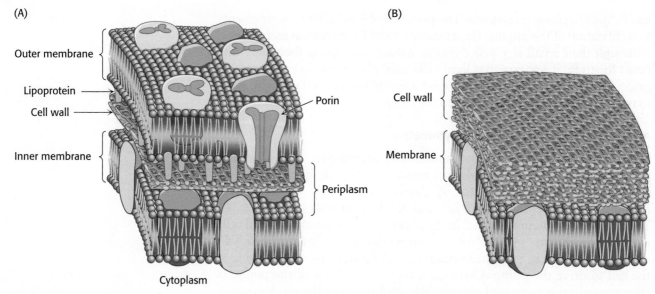

FIGURE 12.35 Cell membranes of prokaryotes. A schematic view of the membrane of bacterial cells surrounded by (A) two membranes or (B) one membrane.

The outer membrane is quite permeable to small molecules, owing to the presence of porins. The region between the two membranes containing the cell wall is called the *periplasm*.

The two types of bacterial membranes depicted in Figure 12.35 can be distinguished microscopically using a technique known as *Gram staining*, named for its inventor Hans Christian Gram. The dye crystal violet is added to a fixed sample of bacteria, followed by iodine to trap the dye in the cell. Then, alcohol is used to wash out the dye. In those cells with a thick cell wall, as shown in Figure 12.35B, the trapped dye binds quite tightly and does not wash out. These bacteria stain dark purple, and are referred to as *Gram positive* (Figure 12.36A). In those cells where the cell wall is very thin, the dye washes out quickly; these bacteria appear pink (due to the addition of a second stain at the conclusion of the experiment) and are referred to as *Gram negative* (Figure 12.36B). Gram staining is an important method for the initial classification of a bacterial sample, particularly in fluid samples from an infected patient, where rapid identification of bacterial type is critical for the selection of antibiotic treatment.

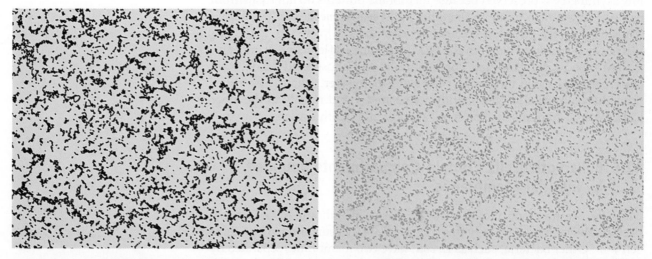

FIGURE 12.36 Gram staining of bacteria. (A) Gram-positive bacteria retain the crystal violet stain in their thick cell walls. (B) Gram-negative bacteria have a thinner cell wall. During the wash step, they lose the crystal violet stain, and appear pink. [(A) Richard J Green/Getty Images (B) Richard J Green/Getty Images]

Eukaryotic cells, with the exception of plant cells, do not have cell walls, and their cell membranes consist of a single lipid bilayer. In plant cells, the cell wall is on the outside of the plasma membrane. Eukaryotic cells are distinguished from prokaryotic cells by the presence of membranes inside the cell that form internal compartments. For example, peroxisomes, organelles that play a major role in the oxidation of fatty acids for energy conversion, are defined by a single membrane. Mitochondria, the organelles in which ATP is synthesized, are surrounded by two membranes. As in the case for a bacterium, the outer membrane is quite permeable to small molecules, whereas the inner membrane is not. Indeed, considerable evidence now indicates that mitochondria evolved from bacteria by *endosymbiosis* (Section 18.1). The nucleus is also surrounded by a double membrane, the *nuclear envelope*, that consists of a set of closed membranes that come together at structures called *nuclear pores* (Figure 12.37). These pores regulate transport into and out of the nucleus. The nuclear envelope is linked to another membrane-defined structure, the *endoplasmic reticulum*, which plays a host of cellular roles, including drug detoxification and the modification of proteins for secretion. Thus, a eukaryotic cell contains interacting compartments, and transport into and out of these compartments is essential to many biochemical processes.

Membranes must be able to separate or join together so that cells and compartments may take up, transport, and release molecules. Many cells take up molecules through the process of *receptor-mediated endocytosis*. Here, a protein or larger complex initially binds to a receptor on the cell surface. After the receptor is bound, specialized proteins act to cause the membrane in this region to invaginate. One of these specialized proteins is *clathrin*, which polymerizes into a lattice network around the growing membrane bud, often referred to as a *clathrin-coated pit* (Figure 12.38).

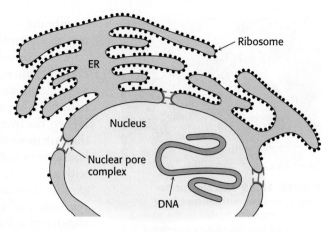

FIGURE 12.37 Nuclear envelope. The nuclear envelope is a double membrane connected to another membrane system of eukaryotes, the endoplasmic reticulum. [Information from E. C. Schirmer and L. Gerace, *Genome Biol.* 3(4):1008.1–1008.4, 2002, reviews, Fig.1.]

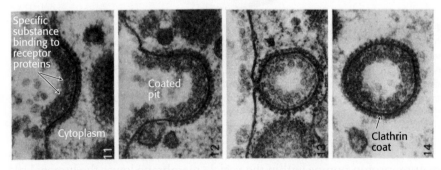

FIGURE 12.38 Vesicle formation by receptor-mediated endocytosis. Receptor binding on the surface of the cell induces the membrane to invaginate, with the assistance of specialized intracellular proteins such as clathrin. The process results in the formation of a vesicle within the cell. [Reproduced with permission: MM Perry, and AB Gilbert Yolk transport in the ovarian follicle of the hen (Gallus domesticus): lipoprotein-like particles at the periphery of the oocyte in the rapid growth phase *J. Cell Sci.* 39: 266.]

The invaginated membrane eventually breaks off and fuses to form a *vesicle.* Various hormones, transport proteins, and antibodies employ receptor-mediated endocytosis to gain entry into a cell. A less-advantageous consequence is that this pathway is available to viruses and toxins as a means of invading cells. The reverse process—the fusion of a vesicle to a

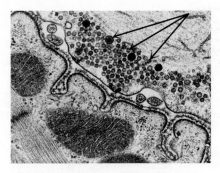

FIGURE 12.39 Neurotransmitter release. Neurotransmitter-containing synaptic vesicles (indicated by the arrows) are arrayed near the plasma membrane of a nerve cell. Synaptic vesicles fuse with the plasma membrane, releasing the neurotransmitter into the synaptic cleft. [Don W. Fawcett/T. Reese/Science Source.]

membrane—is a key step in the release of neurotransmitters from a neuron into the synaptic cleft (Figure 12.39).

Let's consider one example of receptor-mediated endocytosis. Iron is a critical element for the function and structure of many proteins, including hemoglobin and myoglobin (Chapter 7). However, free iron ions are highly toxic to cells, owing to their ability to catalyze the formation of free radicals. Hence, the transport of iron atoms from the digestive tract to the cells where they are most needed must be tightly controlled. In the bloodstream, iron is bound very tightly by the protein *transferrin*, which can bind two Fe^{3+} ions with a dissociation constant of 10^{-23} M at neutral pH. Cells requiring iron express the *transferrin receptor* in their plasma membranes (Section 33.4). Formation of a complex between the transferrin receptor and iron-bound transferrin initiates receptor-mediated endocytosis, internalizing these complexes within vesicles called *endosomes* (Figure 12.40). As the endosomes

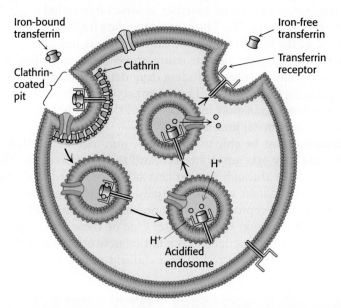

FIGURE 12.40 The transferrin receptor cycle. Iron-bound transferrin binds to the transferrin receptor (TfR) on the surface of cells. Receptor-mediated endocytosis occurs, leading to the formation of a vesicle called an endosome. As the lumen of the endosome is acidified by the action of proton pumps, iron is released from transferrin, passes through channels in the membrane, and is utilized by the cell. The complex between iron-free transferrin and the transferrin receptor is returned to the plasma membrane for another cycle. [Information from L. Zecca et al., *Nat. Rev. Neurosci.* 5:863–873, 2004, Fig.1.]

mature, proton pumps within the vesicle membrane lower the lumenal pH to about 5.5. Under these conditions, the affinity of iron ions for transferrin is reduced; these ions are released and are free to pass through channels in the endosomal membranes into the cytoplasm. The iron-free transferrin complex is recycled to the plasma membrane, where transferrin is released back into the bloodstream and the transferrin receptor can participate in another uptake cycle.

Although budding and fusion appear deceptively simple, the structures of the intermediates in these processes and the detailed mechanisms remain ongoing areas of investigation. These processes must be highly specific in order to prevent incorrect membrane fusion events. *SNARE* (soluble *N*-ethylmaleimide-sensitive-factor attachment protein receptor) *proteins*

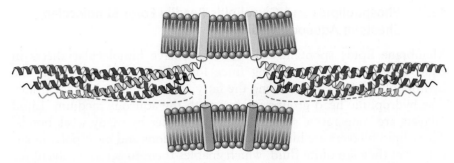

FIGURE 12.41 SNARE complexes initiate membrane fusion. The SNARE protein synaptobrevin (yellow) from one membrane forms a tight four-helical bundle with the corresponding SNARE proteins syntaxin-1 (blue) and SNAP25 (red) from a second membrane. The complex brings the membranes close together, initiating the fusion event. [Drawn from 1SFC.pdb.]

from both membranes help draw appropriate lipid bilayers together by forming tightly coiled four-helical bundles (Figure 12.41). Once these membranes are in close apposition, the fusion process can proceed. SNARE proteins, encoded by gene families in all eukaryotic cells, largely determine the compartment with which a vesicle will fuse. The specificity of membrane fusion ensures the orderly trafficking of membrane vesicles and their cargos through eukaryotic cells.

SUMMARY

Biological membranes are sheetlike structures, typically from 60 to 100 Å thick, that are composed of protein and lipid molecules held together by noncovalent interactions. Membranes are highly selective permeability barriers. They create closed compartments, which may be entire cells or organelles within a cell. Proteins in membranes regulate the molecular and ionic compositions of these compartments. Membranes also control the flow of information between cells.

12.1 Fatty Acids Are Key Constituents of Lipids

Fatty acids are hydrocarbon chains of various lengths and degrees of unsaturation that terminate with a carboxylic acid group. The fatty acid chains in membranes usually contain between 14 and 24 carbon atoms; they may be saturated or unsaturated. Short chain length and unsaturation enhance the fluidity of fatty acids and their derivatives by lowering the melting temperature.

12.2 There Are Three Common Types of Membrane Lipids

The major types of membrane lipids are phospholipids, glycolipids, and cholesterol. Phosphoglycerides, a type of phospholipid, consist of a glycerol backbone, two fatty acid chains, and a phosphorylated alcohol. Phosphatidylcholine, phosphatidylserine, and phosphatidylethanolamine are major phosphoglycerides. Sphingomyelin, a different type of phospholipid, contains a sphingosine backbone instead of glycerol. Glycolipids are sugar-containing lipids derived from sphingosine. Cholesterol, which modulates membrane fluidity, is constructed from a steroid nucleus. A common feature of these membrane lipids is that they are amphipathic molecules, having one hydrophobic and one hydrophilic end.

12.3 Phospholipids and Glycolipids Readily Form Bimolecular Sheets in Aqueous Media

Membrane lipids spontaneously form extensive bimolecular sheets in aqueous solutions. The driving force for membrane formation is the hydrophobic interactions among the fatty acid tails of membrane lipids. The hydrophilic head groups interact with the aqueous medium. Lipid bilayers are cooperative structures, held together by many weak bonds. These lipid bilayers are highly impermeable to ions and most polar molecules, yet they are quite fluid, which enables them to act as a solvent for membrane proteins.

12.4 Proteins Carry Out Most Membrane Processes

Specific proteins mediate distinctive membrane functions such as transport, communication, and energy transduction. Many integral membrane proteins span the lipid bilayer, whereas others are only partly embedded in the membrane. Peripheral membrane proteins are bound to membrane surfaces by electrostatic and hydrogen-bond interactions. Membrane-spanning proteins have regular structures, including β strands, although the α helix is the most common membrane-spanning structure. Sequences of 20 consecutive nonpolar amino acids can be diagnostic of a membrane-spanning α helical region of a protein.

12.5 Lipids and Many Membrane Proteins Diffuse Rapidly in the Plane of the Membrane

Membranes are structurally and functionally asymmetric, as exemplified by the restriction of sugar residues to the external surface of mammalian plasma membranes. Membranes are dynamic structures in which proteins and lipids diffuse rapidly in the plane of the membrane (lateral diffusion), unless restricted by special interactions. In contrast, the rotation of lipids from one face of a membrane to the other (transverse diffusion, or flip-flop) is usually very slow. Proteins do not rotate across bilayers; hence, membrane asymmetry can be preserved. The degree of fluidity of a membrane depends on the chain length of its lipids and on the extent to which their constituent fatty acids are unsaturated. In animals, cholesterol content also regulates membrane fluidity.

12.6 Eukaryotic Cells Contain Compartments Bounded by Internal Membranes

An extensive array of internal membranes in eukaryotes creates compartments within a cell for distinct biochemical functions. For instance, a double membrane surrounds the nucleus (the location of most of the cell's genetic material) and the mitochondria (the location of most ATP synthesis). A single membrane defines the other internal compartments, such as the endoplasmic reticulum. Receptor-mediated endocytosis enables the formation of intracellular vesicles when ligands bind to their corresponding receptor proteins in the plasma membrane. The reverse process—the fusion of a vesicle to a membrane—is a key step in the release of signaling molecules outside the cell.

APPENDIX

Biochemistry in Focus

The curious case of cardiolipin

You may have noticed the unique structure of one of the phospholipids depicted in Figure 12.5. In diphosphatidylglycerol, the central alcohol unit is itself a glycerol moiety onto which two phosphatidate groups are attached.

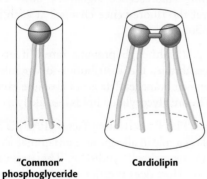

Diphoshatidylglycerol (cardiolipin)

The net result is a phosphoglyceride containing four fatty acyl chains, and a net −2 charge. Owing to its dimeric structure, diphosphatidylglycerol is shaped like a truncated cone, rather than the cylindrical shape of the more typical phospholipid (Figure 12.42).

"Common" phosphoglyceride   Cardiolipin

FIGURE 12.42 The shape of cardiolipin vs. a "common" phosphoglyceride. *Notice* that cardiolipin is shaped like a truncated cone, due to its dimeric diphosphoglycerol head group and four fatty acid tails.

Diphosphatidylglycerol is also known as *cardiolipin*, thus named because it was first identified in the hearts of cows. Cardiolipin is found in the membranes of bacteria and archaea, and is particularly enriched in the inner membrane of mitochondria. The composition of its lipid chains varies by tissue type: in animals, the predominant form of cardiolipin contains four linoleate (18:2) chains. As noted earlier in this chapter, mitochondria are cellular organelles, surrounded by two membranes, that are responsible for the synthesis of cellular ATP. As we shall soon discuss, ATP is generated within inner mitochondrial membrane by a complex of multiple transmembrane proteins known as the respirasome (Section 18.3). Surprisingly, crystal structures of proteins from the respirasome show evidence of highly specific binding sites for cardiolipin in their transmembrane regions. In other words, cardiolipin is not merely a passive component of the lipid "solution," but plays a specific role in the structure and function of these proteins.

The importance of cardiolipin on the function of mitochondria is further demonstrated by individuals with reduced cardiolipin levels due to genetic mutations. Proper synthesis and maintenance of cardiolipin levels requires the enzyme *tafazzin*, which catalyzes the transfer of linoleate chains from phosphatidylcholine to immature cardiolipin, yielding its mature tetralinoleoyl form. Mutations which reduce the catalytic activity of tafazzin result in *Barth syndrome*, which include dilation of the heart chambers, exercise intolerance, and impaired growth. Further investigation of tissue from these patients reveals malformed mitochondria with distorted inner membranes and poorly functioning respirasomes due to improper assembly of the protein complexes.

The importance of cardiolipin on the function of the respirasome provides one example of the interplay between membrane lipids and proteins. The composition of lipid bilayer, and, in fact, individual lipid molecules, can play an integral role in protein function, not merely acting as an organic solvent.

KEY TERMS

fatty acid (p. 396)

phospholipid (p. 376)

sphingosine (p. 376)

phosphoglyceride (p. 376)

sphingomyelin (p. 377)

glycolipid (p. 376)

cerebroside (p. 378)

ganglioside (p. 378)

cholesterol (p. 376)

amphipathic (amphiphilic) molecule (p. 379)

lipid bilayer (p. 374)

liposome (p. 381)

integral membrane protein (p. 384)

peripheral membrane protein (p. 384)

hydropathy plot (p. 389)

lateral diffusion (p. 390)

fluid mosaic model (p. 391)

lipid raft (p. 392)

Gram staining (p. 394)

receptor-mediated endocytosis (p. 395)

clathrin (p. 395)

transferrin (p. 396)

transferrin receptor (p. 396)

endosome (p. 396)

SNARE (soluble *N*-ethylmaleimide-sensitive-factor attachment protein receptor) proteins (p. 396)

PROBLEMS

1. *Population density.* How many phospholipid molecules are there in a 1-μm^2 region of a phospholipid bilayer membrane? Assume that a phospholipid molecule occupies 70 Å^2 of the surface area.

2. *Through the looking-glass.* Phospholipids form lipid bilayers in water. What structure might form if phospholipids were placed in an organic solvent? ✓❸

3. *Lipid diffusion.* What is the average distance traversed by a membrane lipid in 1 μs, 1 ms, and 1 s? Assume a diffusion coefficient of $10^{-8} \text{ cm}^2\text{ s}^{-1}$. ✓❻

4. *Protein diffusion.* The diffusion coefficient, D, of a rigid spherical molecule is given by

$$D = kT/6\pi\eta r$$

in which η is the viscosity of the solvent, r is the radius of the sphere, k is the Boltzman constant (1.38×10^{-16} erg degree^{-1}), and T is the absolute temperature. What is the diffusion coefficient at 37°C of a 100-kDa protein in a membrane that has an effective viscosity of 1 poise (1 poise = 1 erg s^{-1} cm^{-3})? What is the average distance traversed by this protein in 1 μs, 1 ms, and 1 s? Assume that this protein is an unhydrated, rigid sphere with a density of 1.35 g cm^{-3}.

5. *Cold sensitivity.* Some antibiotics act as carriers that bind an ion on one side of a membrane, diffuse through the membrane, and release the ion on the other side. The conductance of a lipid-bilayer membrane containing a carrier antibiotic decreased abruptly when the temperature was lowered from 40°C to 36°C. In contrast, there was little change in conductance of the same bilayer membrane when it contained a channel-forming antibiotic. Why?

6. *Melting point 1.* Explain why oleic acid (18 carbons, one cis bond) has a lower melting point than stearic acid, which has the same number of carbon atoms but is saturated. How would you expect the melting point of *trans*-oleic acid to compare with that of *cis*-oleic acid? Why might most unsaturated fatty acids in phospholipids be in the cis rather than the trans conformation? ✓❻

7. *Melting point 2.* Explain why the melting point of palmitic acid (C$_{16}$) is 6.5 degrees lower than that of stearic acid (C$_{18}$). ✓❻

8. *A sound diet.* Small mammalian hibernators can withstand body temperatures of 0° to 5°C without injury. However, the body fats of most mammals have melting temperatures of approximately 25°C. Predict how the composition of the body fat of hibernators might differ from that of their nonhibernating cousins. ✓❻

9. *Flip-flop 1.* The transverse diffusion of phospholipids in a bilayer membrane was investigated by using a fluorescently labeled analog of phosphatidylserine called NBD-PS.

NBD-phosphatidylserine (NBD-PS)

The fluorescence signal of NBD-PS is quenched when exposed to sodium dithionite, a reducing agent that is not membrane permeable.

Lipid vesicles containing phosphatidylserine (98%) and NBD-PS (2%) were prepared by sonication and purified. Within a few minutes of the addition of sodium dithionite, the fluorescence signal of these vesicles decreased to ~45% of its initial value. Immediately adding a second addition of sodium dithionite yielded no change in the fluorescence signal. However, if the vesicles were allowed to incubate for 6.5 hours, a third addition of sodium dithionite decreased the remaining fluorescence signal by 50%. How would you interpret the fluorescence changes at each addition of sodium dithionite?

10. *Flip-flop 2.* Although proteins rarely if ever flip-flop across a membrane, the distribution of membrane lipids between the membrane leaflets is not absolute except for glycolipids. Why are glycosylated lipids less likely to flip-flop?

11. *Linkages.* Platelet-activating factor (PAF) is a phospholipid that plays a role in allergic and inflammatory responses, as well as in toxic shock syndrome. The structure of PAF is shown here. How does it differ from the structures of the phospholipids discussed in this chapter? ✓❷

Platelet-activating factor (PAF)

12. *A question of competition.* Would a homopolymer of alanine be more likely to form an α helix in water or in a hydrophobic medium? Explain.

13. *A false positive.* Hydropathy plot analysis of your protein of interest reveals a single, prominent hydrophobic peak. However, you later discover that this protein is soluble and not membrane associated. Explain how the hydropathy plot may have been misleading. ✓❺

14. *Maintaining fluidity.* A culture of bacteria growing at 37°C was shifted to 25°C. How would you expect this shift to alter the fatty acid composition of the membrane phospholipids? Explain. ✓❻

15. *Let me count the ways.* Each intracellular fusion of a vesicle with a membrane requires a SNARE protein on the vesicle (called the v-SNARE) and a SNARE protein on the target membrane (called the t-SNARE). Assume that a genome encodes 21 members of the v-SNARE family and 7 members of the t-SNARE family. With the assumption of no specificity, how many potential v-SNARE–t-SNARE interactions could take place? ✓❼

Data Interpretation Problems

16. *Cholesterol effects.* The red curve on the following graph shows the fluidity of the fatty acids of a phospholipid bilayer as a function of temperature. The blue curve shows the fluidity in the presence of cholesterol.

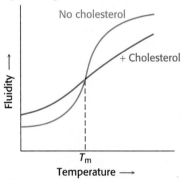

(a) What is the effect of cholesterol?

(b) Why might this effect be biologically important?

17. *Hydropathy plots.* On the basis of the following hydropathy plots for three proteins (A–C), predict which would be membrane proteins. What are the ambiguities with respect to using such plots to determine if a protein is a membrane protein? ✓❺

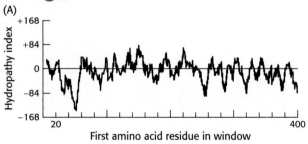

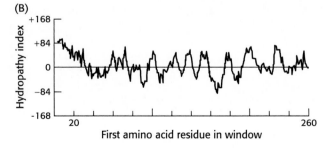

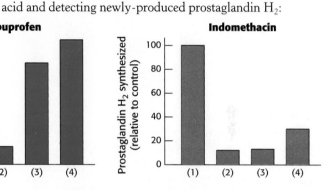

18. *Not all inhibitors are equal.* Ibuprofen and indomethacin are clinically important inhibitors of prostaglandin H₂ synthase-1. Cells expressing this enzyme were incubated under the following conditions, after which the activity of the enzyme was measured by adding radiolabeled arachidonic acid and detecting newly-produced prostaglandin H₂:

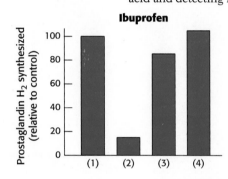

(1) 40 min *without* inhibitor (control)

(2) 40 min with inhibitor

(3) 40 min with inhibitor, after which the cells were resuspended in medium *without* inhibitor

(4) 40 min with inhibitor, after which the cells were resuspended in medium *without* inhibitor and incubated for an additional 30 min.

(a) Provide a hypothesis explaining the different results for these two inhibitors.

(b) How would these results look if aspirin were tested in a similar fashion?

Chapter Integration Problem

19. *The proper environment.* An understanding of the structure and function of membrane proteins has lagged behind that of other proteins. The primary reason is that membrane proteins are more difficult to purify and crystallize. Why might this be the case?

Think ▶ Pair ▶ Share Problem

20. In this chapter, we covered two seemingly contradictory facts about membrane lipids:

  (1) The diffusion ("flip-flop") of membrane lipids from one leaflet to the other is very slow.

  (2) The lipids in biological membranes are asymmetrically distributed.

Propose some mechanisms by which the cell can establish and maintain asymmetric membranes.

Membrane Channels and Pumps

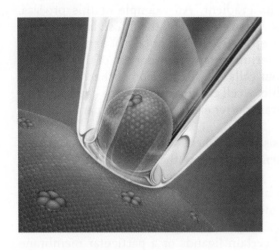

CHAPTER 13

← Closed

← Open

The flow of ions through a single membrane channel (channels are shown in red in the illustration at the left) can be detected by the patch-clamp technique, which records current changes as the channel alternates between open and closed states. [(Left) Illustration by Dana Burns-Pizer (Right) Courtesy of Dr. Mauricio Montal.]

✓ LEARNING OBJECTIVES

By the end of this chapter, you should be able to:

1. Distinguish between passive and active transport.

2. Calculate the free energy stored in a concentration or electrochemical gradient.

3. Define primary and secondary active transport.

4. Compare the mechanisms of the P-type ATPase pump and the ABC transporter pump.

5. Explain how channels are able to rapidly move ions across membranes, and how they achieve ion selectivity.

6. Describe two mechanisms by which ion channels are gated.

7. Explain the steps that lead to the formation of an action potential.

OUTLINE

13.1 The Transport of Molecules Across a Membrane May Be Active or Passive

13.2 Two Families of Membrane Proteins Use ATP Hydrolysis to Pump Ions and Molecules Across Membranes

13.3 Lactose Permease Is an Archetype of Secondary Transporters That Use One Concentration Gradient to Power the Formation of Another

13.4 Specific Channels Can Rapidly Transport Ions Across Membranes

13.5 Gap Junctions Allow Ions and Small Molecules to Flow Between Communicating Cells

13.6 Specific Channels Increase the Permeability of Some Membranes to Water

The lipid bilayer of biological membranes is intrinsically impermeable to ions and polar molecules, yet these species must be able to cross these membranes for normal cell function. Permeability is conferred by three classes of membrane proteins: *pumps*, *carriers*, and *channels*. Pumps use a source of free energy such as ATP hydrolysis or light absorption to drive the thermodynamically uphill transport of ions or molecules. Pump action is an example of *active transport*. Carriers mediate the transport of ions and small molecules across the membrane without consumption of ATP. Channels provide a membrane pore through which ions can flow very rapidly in a thermodynamically downhill direction. The action of channels illustrates *passive transport*, or *facilitated diffusion*.

Pumps are energy transducers in that they convert one form of free energy into another. Two examples of *ATP-driven pumps*—P-type ATPases and the ATP-binding cassette (ABC) transporters—undergo conformational changes on ATP binding and hydrolysis that cause a bound ion to be transported across the membrane. The free energy of

ATP hydrolysis is used to drive the movement of ions against their concentration gradients, a process referred to as *primary active transport*. In contrast, carriers utilize the gradient of one ion to drive the transport of another molecule against its gradient. An example of this process, termed *secondary active transport*, is mediated by the *E. coli* lactose transporter, a well-studied protein responsible for the uptake of a specific sugar from the environment of a bacterium. Many transporters of this class are present in the membranes of our cells. The expression of these transporters determines which metabolites a cell can import from the environment. Hence, adjusting the level of transporter expression is a primary means of controlling metabolism.

Pumps can establish persistent gradients of particular ions across membranes. Specific *ion channels* can allow these ions to flow rapidly across membranes down these gradients. These channels are among the most fascinating molecules in biochemistry in their ability to allow some ions to flow freely through a membrane while blocking the flow of even closely related species. The opening, or gating, of these channels can be controlled by the presence of certain ligands or a particular membrane voltage. Gated ion channels are central to the functioning of our nervous systems, acting as elaborately switched wires that allow the rapid flow of current.

Finally, a different class of channel, the cell-to-cell channel, or *gap junction*, allows the flow of metabolites or ions *between cells*. For example, gap junctions are responsible for synchronizing muscle-cell contraction in the beating heart.

The expression of transporters largely defines the metabolic activities of a given cell type

Each cell type expresses a specific set of transporters in its plasma membrane. This collection of expressed transporters is important because it largely determines the ionic composition inside cells and the compounds that can be taken up from the extracellular environment. In some sense, the specific array of transporters defines the cell's characteristics because a cell can execute only those biochemical reactions for which it has taken up the necessary substrates.

An example from glucose metabolism illustrates this point. As we will see in Chapter 16, tissues differ in their ability to employ different molecules as energy sources. Which tissues can utilize glucose is largely governed by the expression of members of the GLUT family of homologous glucose transporters. For example, GLUT3 is the primary glucose transporter expressed on the plasma membrane of neurons. This transporter binds glucose relatively tightly so that these cells have first call on glucose when it is present at relatively low concentrations. We will encounter many such examples of the critical role that transporter expression plays in the control and integration of metabolism.

13.1 The Transport of Molecules Across a Membrane May Be Active or Passive

We will first consider some general principles of membrane transport. Two factors determine whether a molecule will cross a membrane: (1) the permeability of the molecule in a lipid bilayer and (2) the availability of an energy source.

Many molecules require protein transporters to cross membranes

As stated in Chapter 12, some molecules can pass through cell membranes because they dissolve in the lipid bilayer. Such molecules are called *lipophilic molecules*. The steroid hormones provide a physiological example. These cholesterol relatives can pass through a membrane, but what determines the direction in which they will move? Such molecules will pass through a membrane down their concentration gradient in a process called *simple diffusion*. In accord with the Second Law of Thermodynamics, molecules spontaneously move from a region of higher concentration to one of lower concentration.

Matters become more complicated when the molecule is highly polar. For example, sodium ions are present at 143 mM outside a typical cell and at 14 mM inside the cell. However, sodium does not freely enter the cell, because the charged ion cannot pass through the hydrophobic membrane interior. In some circumstances, as during a nerve impulse, sodium ions must enter the cell. How are they able to do so? Sodium ions pass through specific channels in the hydrophobic barrier formed by membrane proteins. This means of crossing the membrane is called *facilitated diffusion* because the diffusion across the membrane is facilitated by the channel. It is also called *passive transport* because the energy driving the ion movement originates from the ion gradient itself, without any contribution by the transport system. Channels, like enzymes, display substrate specificity in that they facilitate the transport of some ions, but not other, even closely related, ions.

How is the sodium gradient established in the first place? In this case, sodium must move, or be pumped, *against* a concentration gradient. Because moving the ion from a low concentration to a higher concentration results in a decrease in entropy, it requires an input of free energy. Protein transporters embedded in the membrane are capable of using an energy source to move the molecule up a concentration gradient. Because an input of energy from another source is required, this means of crossing the membrane is called *active transport*.

Free energy stored in concentration gradients can be quantified

An unequal distribution of molecules is an energy-rich condition because free energy is minimized when all concentrations are equal. Consequently, to attain such an unequal distribution of molecules requires an input of free energy. How can we quantify the amount of energy required to generate a concentration gradient (Figure 13.1)? Consider an *uncharged* solute molecule. The free-energy change in transporting this species from side 1, where it is present at a concentration of c_1, to side 2, where it is present at concentration c_2, is

$$\Delta G = RT \ln (c_2/c_1) \qquad (1)$$

where R is the gas constant (8.315×10^{-3} kJ mol^{-1} deg^{-1}, or 1.987×10^{-3} kcal mol^{-1} deg^{-1}) and T is the temperature in kelvins. A transport process must be active when ΔG is positive, whereas it can be passive when ΔG is negative. For example, consider the transport of an uncharged molecule from $c_1 = 10^{-3}$ M to $c_2 = 10^{-1}$ M at 25°C (298 K).

$$\Delta G = RT \ln (10^{-1}/10^{-3})$$
$$= (8.315 \times 10^{-3}) \times 298 \times \ln (10^2)$$
$$= +11.4 \text{ kJ mol}^{-1}(+2.7 \text{ kcal mol}^{-1})$$

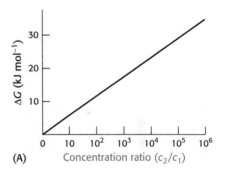

(A) Concentration ratio (c_2/c_1)

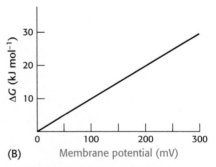

(B) Membrane potential (mV)

FIGURE 13.1 Free energy and transport. The free-energy change in transporting (A) an uncharged solute from a compartment at concentration c_1 to one at c_2 and (B) a singly charged species across a membrane to the side having the same charge as that of the transported ion. *Notice* that the free-energy change imposed by a membrane potential of 59 mV is equivalent to that imposed by a concentration ratio of 10 for a singly charged ion at 25°C.

The ΔG is $+11.4\,\text{kJ mol}^{-1}$ ($+2.7\,\text{kcal mol}^{-1}$), indicating that this transport process requires an input of free energy.

For a *charged* species, the unequal distribution across the membrane generates an electrical potential that also must be considered because the ions will be repelled by like charges. The sum of the concentration and electrical terms is called the *electrochemical potential* or *membrane potential*. The free-energy change is then given by

$$\Delta G = RT \ln (c_2/c_1) + ZF\Delta V \qquad (2)$$

in which Z is the electrical charge of the transported species, ΔV is the potential in volts across the membrane, and F is the Faraday constant ($96.5\,\text{kJ V}^{-1}\,\text{mol}^{-1}$, or $23.1\,\text{kcal V}^{-1}\,\text{mol}^{-1}$).

13.2 Two Families of Membrane Proteins Use ATP Hydrolysis to Pump Ions and Molecules Across Membranes

The extracellular fluid of animal cells has a salt concentration similar to that of seawater. However, cells must control their intracellular salt concentrations to facilitate specific processes, such as signal transduction and action potential propagation, and prevent unfavorable interactions with high concentrations of ions such as Ca^{2+}. For instance, most animal cells contain a high concentration of K^+ and a low concentration of Na^+ relative to the external medium. These ionic gradients are generated by a specific transport system, an enzyme that is called the Na^+–K^+ *pump* or the Na^+–K^+ *ATPase*. The hydrolysis of ATP by the pump provides the energy needed for the active transport of Na^+ out of the cell and K^+ into the cell, generating the gradients. The pump is called the Na^+–K^+ ATPase because the hydrolysis of ATP takes place only when Na^+ and K^+ are present. This ATPase, like all such enzymes, requires Mg^{2+}.

The change in free energy accompanying the transport of Na^+ and K^+ can be calculated. Suppose that the concentrations of Na^+ outside and inside the cell are 143 and 14 mM, respectively, and the corresponding values for K^+ are 4 and 157 mM. At a membrane potential of -50 mV and a temperature of $37°C$, we can use equation 2 (see above) to determine the free-energy change in transporting 3 mol of Na^+ out of the cell and 2 mol of K^+ into the cell:

For the transport of Na^+ ions:

$$\Delta G = RT \ln (c_2/c_1) + ZF\Delta V$$

$$\Delta G = (8.315 \times 10^{-3}\,\text{kJ mol}^{-1}\,\text{deg}^{-1}) \times (310\,\text{deg}) \times \ln (143/14)$$
$$+ (1) \times (96.5\,\text{kJ mol}^{-1}\,\text{V}^{-1}) \times (0.05\,\text{V})$$

$$\Delta G = 10.82\,\text{kJ mol}^{-1}$$

For the transport of K^+ ions:

$$\Delta G = RT \ln (c_2/c_1) + ZF\Delta V$$

$$\Delta G = (8.315 \times 10^{-3}\,\text{kJ mol}^{-1}\,\text{deg}^{-1}) \times (310\,\text{deg}) \times \ln (157/4)$$
$$+ (1) \times (96.5\,\text{kJ mol}^{-1}\,\text{V}^{-1}) \times (-0.05\,\text{V})$$

$$\Delta G = 4.64\,\text{kJ mol}^{-1}$$

Taking into account the transport of 3 Na^+ ions and 2 K^+ ions for every transport cycle, the free energy required to operate the pump is:

$$\Delta G = 3 \times (10.82\,\text{kJ mol}^{-1}) + 2 \times (4.64\,\text{kJ mol}^{-1})$$

$$\Delta G = 41.7\,\text{kJ mol}^{-1}$$

Note that the chemical term is positive (unfavorable) for both ions, but the electrical term is favorable for K^+ (ΔV is negative) and unfavorable for Na^+ (ΔV is positive) as the membrane has net negative charge on its inner surface. Under typical cellular conditions, the hydrolysis of a single ATP molecule per transport cycle provides sufficient free energy, about $-50\,kJ\,mol^{-1}$ ($-12\,kcal\,mol^{-1}$) to drive the uphill transport of these ions. The active transport of Na^+ and K^+ is of great physiological significance. Indeed, more than a third of the ATP consumed by a resting animal is used to pump these ions. The Na^+–K^+ gradient in animal cells controls cell volume, renders neurons and muscle cells electrically excitable, and drives the active transport of sugars and amino acids.

The purification of other ion pumps has revealed a large family of evolutionarily related ion pumps including proteins from bacteria, archaea, and all eukaryotes. Each of these pumps is specific for a particular ion or set of ions. Two are of particular interest: the *sarcoplasmic reticulum Ca^{2+} ATPase* (or *SERCA*) transports Ca^{2+} out of the cytoplasm and into the sarcoplasmic reticulum of muscle cells, and the *gastric H^+–K^+ ATPase* is the enzyme responsible for pumping sufficient protons into the stomach to lower the pH to 1.0. These enzymes and the hundreds of known homologs, including the Na^+–K^+ ATPase, are referred to as *P-type ATPases* because they form a key phosphorylated intermediate. In the formation of this intermediate, a phosphoryl group from ATP is linked to the side chain of a specific conserved aspartate residue in the ATPase to form phosphorylaspartate.

Phosphorylaspartate

P-type ATPases couple phosphorylation and conformational changes to pump calcium ions across membranes

Membrane pumps function by mechanisms that are simple in principle but often complex in detail. Fundamentally, each pump protein can exist in two principal conformational states, one with ion-binding sites open to one side of the membrane and the other with ion-binding sites open to the other side (Figure 13.2). To pump ions in a single direction across a membrane, the free energy of ATP hydrolysis must be coupled to the interconversion between these conformational states.

We will consider the structural and mechanistic features of P-type ATPases by examining SERCA. The properties of this P-type ATPase have been established in great detail by relying on crystal structures of the pump in five different states. This enzyme, which constitutes 80% of the total protein in the sarcoplasmic reticulum membrane, plays an important role in relaxation of contracted muscle. Muscle contraction is triggered by an abrupt rise in the cytoplasmic calcium ion level. Subsequent muscle relaxation depends on the rapid removal of Ca^{2+} from the cytoplasm into the sarcoplasmic reticulum, a specialized compartment for Ca^{2+} storage, by SERCA. This pump maintains a Ca^{2+} concentration of approximately

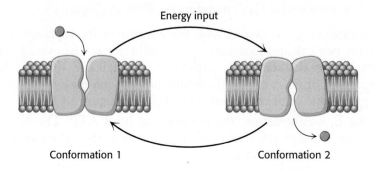

Energy input

Conformation 1 Conformation 2

FIGURE 13.2 Pump action. A simple scheme for the pumping of a molecule across a membrane. The pump interconverts between two conformational states, each with a binding site accessible to a different side of the membrane.

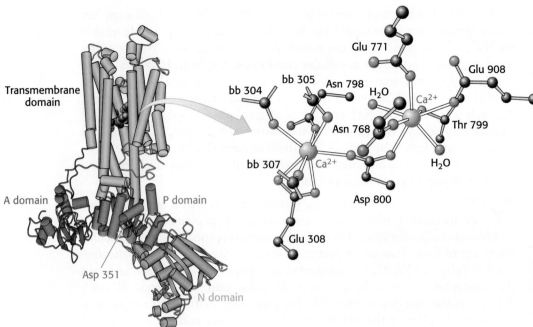

FIGURE 13.3 Calcium-pump structure. The overall structure of the SERCA P-type ATPase. *Notice* the two calcium ions (green) that lie in the center of the transmembrane domain. A conserved aspartate residue (Asp 351) that binds a phosphoryl group lies in the P domain. The designation bb refers to backbone carbonyl groups. [Drawn from 1SU4.pdb.]

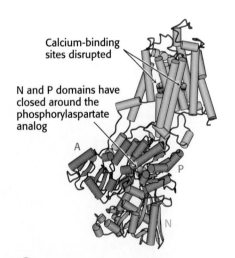

FIGURE 13.4 Conformational changes associated with calcium pumping. This structure was determined in the absence of bound calcium but with a phosphorylaspartate analog present in the P domain. *Notice* how different this structure is from the calcium-bound form shown in Figure 13.3: both the transmembrane part (yellow) and the A, P, and N domains have substantially rearranged. [Drawn from 1WPG.pdb.]

0.1 μM in the cytoplasm, compared to nearly 1.5 mM in the sarcoplasmic reticulum.

The first structure of SERCA to be determined had Ca^{2+} bound, but no nucleotides present (Figure 13.3). SERCA is a single 110-kDa polypeptide with a transmembrane domain consisting of 10 α helices. The transmembrane domain includes sites for binding two calcium ions. Each calcium ion is coordinated to seven oxygen atoms coming from a combination of side-chain glutamate, aspartate, threonine, and asparagine residues, backbone carbonyl groups, and water molecules. A large cytoplasmic headpiece constitutes nearly half the molecular weight of the protein and consists of three distinct domains, each with a distinct function. One domain (N) binds the ATP *nucleotide*, another (P) accepts the *phosphoryl* group on a conserved aspartate residue, and the third (A) serves as an *actuator*, linking changes in the N and P domains to the transmembrane part of the enzyme.

SERCA is a remarkably dynamic protein. For example, the structure of SERCA without bound Ca^{2+}, but with a phosphorylaspartate analog present in the P domain, is shown in Figure 13.4. The N and P domains are now closed around the phosphorylaspartate analog, and the A domain has rotated substantially relative to its position in SERCA with Ca^{2+} bound and without the phosphoryl analog. Furthermore, the transmembrane part of the enzyme has rearranged significantly and the well-organized Ca^{2+}-binding sites are disrupted. These sites are now accessible from the side of the membrane opposite the N, P, and A domains.

The structural results can be combined with other studies to construct a detailed mechanism for Ca^{2+} pumping by SERCA (Figure 13.5):

1. The catalytic cycle begins with the enzyme in its unphosphorylated state with two calcium ions bound. We will refer to the overall enzyme conformation in this state as E_1; with Ca^{2+} bound, it is E_1-$(Ca^{2+})_2$. In this conformation, SERCA can bind calcium ions only on the cytoplasmic side of the membrane. This conformation is shown in Figure 13.3.

2. In the E_1 conformation, the enzyme can bind ATP. The N, P, and A domains undergo substantial rearrangement as they close around the bound ATP, but there is no substantial conformational change in the transmembrane domain. The calcium ions are now trapped inside the enzyme.

3. The phosphoryl group is then transferred from ATP to Asp 351.

4. Upon ADP release, the enzyme again changes its overall conformation, including the membrane domain this time. This new conformation is referred to as E_2 or E_2-P in its phosphorylated form. The process of interconverting the E_1 and E_2 conformations is sometimes referred to as *eversion*.

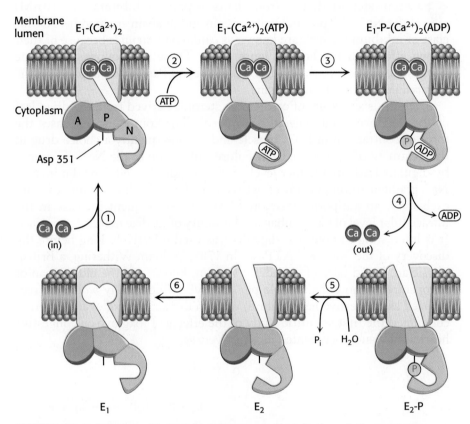

FIGURE 13.5 Pumping calcium. Ca^{2+} ATPase transports Ca^{2+} through the membrane by a mechanism that includes (1) Ca^{2+} binding from the cytoplasm, (2) ATP binding, (3) ATP cleavage with the transfer of a phosphoryl group to Asp 351 on the enzyme, (4) ADP release and eversion of the enzyme to release Ca^{2+} on the opposite side of the membrane, (5) hydrolysis of the phosphorylaspartate residue, and (6) eversion to prepare for the binding of Ca^{2+} from the cytoplasm.

In the E_2-P conformation, the Ca^{2+}-binding sites become disrupted and the calcium ions are released to the side of the membrane opposite that at which they entered; ion transport has been achieved. This conformation is shown in Figure 13.4.

5. The phosphorylaspartate residue is hydrolyzed to release inorganic phosphate.

6. With the release of phosphate, the interactions stabilizing the E_2 conformation are lost, and the enzyme everts to the E_1 conformation.

The binding of two calcium ions from the cytoplasmic side of the membrane completes the cycle.

This mechanism likely applies to other P-type ATPases. For example, Na^+–K^+ ATPase is an $\alpha_2\beta_2$ tetramer. Its α subunit is homologous to SERCA and includes a key aspartate residue analogous to Asp 351. The β subunit does not directly take part in ion transport. A mechanism analogous to that shown in Figure 13.5 applies, with three Na^+ ions binding from the inside of the cell to the E_1 conformation and two K^+ ions binding from outside the cell to the E_2 conformation.

Digitalis specifically inhibits the Na^+–K^+ pump by blocking its dephosphorylation

Certain steroids derived from plants are potent inhibitors ($K_i \approx 10$ nM) of the Na^+–K^+ pump. Digitoxigenin and ouabain are members of this class of inhibitors, which are known as *cardiotonic steroids* because of their strong effects on the heart (Figure 13.6). These compounds inhibit the dephosphorylation of the E_2-P form of the ATPase when applied on the *extracellular* face of the membrane.

Digitalis is a mixture of cardiotonic steroids derived from the dried leaf of the foxglove plant (*Digitalis purpurea*). The compound increases the force of contraction of heart muscle and is consequently a choice drug in the treatment of congestive heart failure. Inhibition of the Na^+–K^+ pump by digitalis leads to a higher level of Na^+ inside the cell. The diminished Na^+ gradient results in slower extrusion of Ca^{2+} by the sodium–calcium exchanger, an antiporter (Section 13.3). The subsequent increase in the intracellular level of Ca^{2+} enhances the ability of cardiac muscle to contract. It is interesting to note that digitalis was used effectively long before the discovery of the Na^+–K^+ ATPase. In 1785, William Withering, a British physician, heard tales of an elderly woman, known as "the old woman of Shropshire," who cured people of "dropsy" (which today would be recognized as congestive heart failure) with an extract of foxglove. Withering conducted the first scientific study of the effects of foxglove on congestive heart failure and documented its effectiveness.

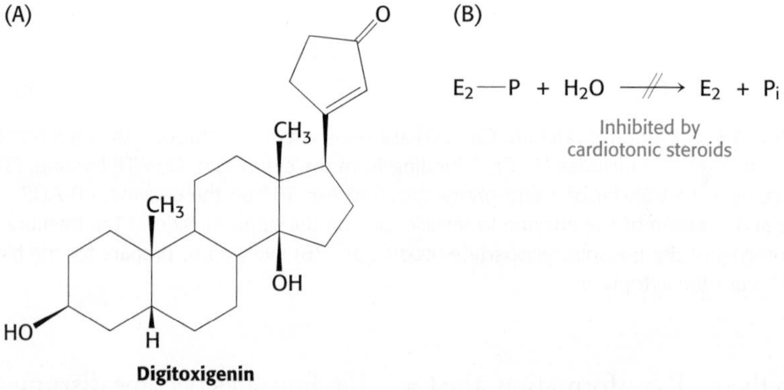

Foxglove (*Digitalis purpurea*) is the source of digitalis, one of the most widely used drugs. [Roger Hall/Shutterstock]

(A)

Digitoxigenin

(B)

$$E_2\text{—}P + H_2O \xrightarrow{\;\;\;/\!\!/\;\;\;} E_2 + P_i$$

Inhibited by cardiotonic steroids

FIGURE 13.6 Digitoxigenin. Cardiotonic steroids such as digitoxigenin shown in (A) inhibit the Na^+–K^+ pump by blocking the dephosphorylation of E_2-P (B).

P-type ATPases are evolutionarily conserved and play a wide range of roles

Analysis of the complete yeast genome revealed the presence of 16 proteins that clearly belong to the P-type ATPase family. More-detailed sequence analysis suggests that 2 of these proteins transport H^+ ions, 2 transport Ca^{2+}, 3 transport Na^+, and 2 transport metals such as Cu^{2+}. In addition, 5 members of this family appear to participate in the transport of phospholipids with amino acid head groups. These 5 proteins help maintain membrane asymmetry by transporting lipids such as phosphatidylserine from the outer to the inner leaflet of the bilayer membrane.

Such enzymes have been termed "flippases." Remarkably, the human genome encodes 70 P-type ATPases. All members of this protein family employ the same fundamental mechanism: The free energy of ATP hydrolysis drives membrane transport by means of conformational changes, which are induced by the addition and removal of a phosphoryl group at an analogous aspartate site in each protein.

Multidrug resistance highlights a family of membrane pumps with ATP-binding cassette domains

Studies of human disease revealed another large and important family of active-transport proteins, with structures and mechanisms quite different from those of the P-type ATPase family. These pumps were identified from studies on tumor cells in culture that developed resistance to drugs that had been initially quite toxic to the cells. Remarkably, the development of resistance to one drug had made the cells less sensitive to a range of other compounds, a phenomenon known as *multidrug resistance*. In a significant discovery, the onset of multidrug resistance was found to correlate with the expression and activity of a membrane protein with an apparent molecular mass of 170 kDa. This protein acts as an ATP-dependent pump that extrudes a wide range of small molecules from cells that express it. The protein is called the *multidrug-resistance* (MDR) *protein* or *P-glycoprotein* ("glyco" because it includes a carbohydrate moiety). Thus, when cells are exposed to a drug, the MDR pumps the drug out of the cell before the drug can exert its effects.

Analysis of the amino acid sequences of MDR and homologous proteins revealed a common architecture (Figure 13.7A). Each protein comprises four domains: two membrane-spanning domains and two ATP-binding domains. The ATP-binding domains of these proteins are called *ATP-binding cassettes* (ABCs) and are homologous to domains in a large family of transport proteins in bacteria and archaea. Transporters that include these domains are called *ABC transporters*. With 79 members, the ABC transporters are the largest single family identified in the *E. coli* genome. The human genome includes more than 150 ABC transporter genes.

The ABC proteins are members of the P-loop NTPase superfamily (Section 9.4). The three-dimensional structures of several members of the ABC transporter family have now been determined, including that of the bacterial lipid transporter MsbA. In contrast with the eukaryotic MDR protein, this protein is a dimer of 62-kDa chains: the amino-terminal half of each protein contains the membrane-spanning domain, and the carboxyl-terminal half contains the ATP-binding cassette (Figure 13.7B). Prokaryotic ABC proteins are often made up of multiple subunits, such as a dimer of identical chains, or as a heterotetramer of two membrane-spanning domain subunits and two ATP-binding-cassette subunits. The consolidation of the enzymatic activities of several polypeptide chains in prokaryotes to a single chain in eukaryotes is a theme that we will see again. The two ATP-binding cassettes are in contact, but they do not interact strongly in the absence of bound ATP (Figure 13.8A). On the basis of several solved structures of ABC transporters, combined with data from other experiments, a mechanism for active transport by these proteins has been developed (Figure 13.9):

1. The catalytic cycle begins with the transporter free of both ATP and substrate. While the distance between the ATP-binding cassettes in this form may vary with the individual transporter, the substrate binding region of the transporter faces inward.

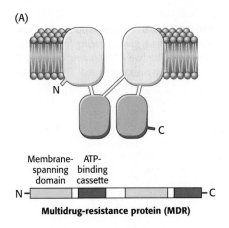

(A)

Membrane-spanning domain | ATP-binding cassette

Multidrug-resistance protein (MDR)

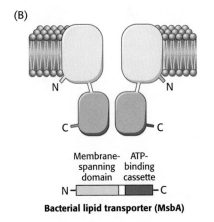

(B)

Membrane-spanning domain | ATP-binding cassette

Bacterial lipid transporter (MsbA)

FIGURE 13.7 Domain arrangement of ABC transporters. ABC transporters are a large family of homologous proteins composed of two transmembrane domains and two ATP-binding domains called ATP-binding cassettes (ABCs). (A) The multidrug-resistance protein is a single polypeptide chain containing all four domains, whereas (B) the bacterial lipid transporter MsbA consists of a dimer of two identical chains, containing one of each domain.

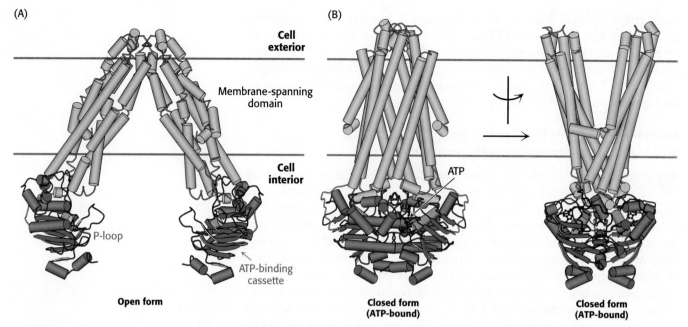

FIGURE 13.8 ABC transporter structure. Two structures of the bacterial lipid transporter MsbA, a representative ABC transporter: (A) The nucleotide-free, inward-facing form and (B) the ATP-bound, outward-facing form is shown in two views (rotated by 90 degrees). In both structures, the two ATP-binding cassettes (blue) are related to the P-loop NTPases and, like them, contain P-loops (green). The α helix adjacent to the P-loop is shown in red. The gray lines indicate the extent of the plasma membrane. [Drawn from 3B5W and 3B60.pdb.]

2. Substrate enters the central cavity of the transporter from inside the cell. Substrate binding induces conformational changes in the ATP-binding cassettes that increase their affinity for ATP.

3. ATP binds to the ATP-binding cassettes, changing their conformations so that the two domains interact strongly with one another. The close interaction of the ABCs reorients the transmembrane helices such that the substrate binding site is now facing outside the cell (Figure 13.8B).

4. The outward facing conformation of the transporter has reduced affinity for the substrate, enabling the release of the substrate on the opposite face of the membrane.

5. The hydrolysis of ATP and the release of ADP and inorganic phosphate reset the transporter for another cycle.

FIGURE 13.9 ABC transporter mechanism. The mechanism includes the following steps: (1) opening of the channel toward the inside of the cell, (2) substrate binding and conformational changes in the ATP-binding cassettes, (3) ATP binding and opening of the channel to the opposite face of the membrane, (4) release of the substrate to the outside of the cell, and (5) ATP hydrolysis to reset the transporter to its initial state.

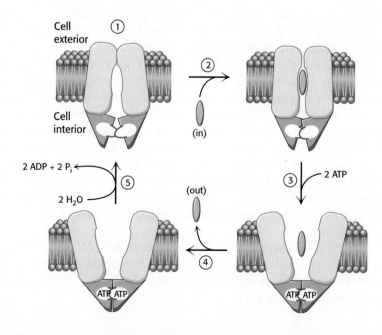

Whereas eukaryotic ABC transporters generally act to export molecules from inside the cell, prokaryotic ABC transporters often act to import specific molecules from *outside* the cell. A specific binding protein acts in concert with the bacterial ABC transporter, delivering the substrate to the transporter and stimulating ATP hydrolysis inside the cell. These binding proteins are present in the periplasm, the compartment between the two membranes that surround some bacterial cells (Figure 12.35A).

Thus, ABC transporters use a substantially different mechanism from the P-type ATPases to couple the ATP hydrolysis reaction to conformational changes. Nonetheless, the net result is the same: The transporters are converted from one conformation capable of binding substrate from one side of the membrane to another that releases the substrate on the other side.

13.3 Lactose Permease Is an Archetype of Secondary Transporters That Use One Concentration Gradient to Power the Formation of Another

Carriers are proteins that transport ions or molecules across the membrane without hydrolysis of ATP. The mechanism of carriers involves both large conformational changes and the interaction of the protein with only a few molecules per transport cycle, limiting the maximum rate at which transport can occur. Although carriers cannot mediate primary active transport—owing to their inability to hydrolyze ATP—they can couple the thermodynamically unfavorable flow of one species of ion or molecule *up* a concentration gradient to the favorable flow of a different species *down* a concentration gradient, a process referred to as secondary active transport. Carriers that move ions or molecules "uphill" by this means are termed *secondary transporters* or *cotransporters*. These proteins can be classified as either *antiporters* or *symporters*. Antiporters couple the downhill flow of one species to the uphill flow of another in the *opposite direction* across the membrane; symporters use the flow of one species to drive the flow of a different species in the *same direction* across the membrane. *Uniporters*, another class of carriers, are able to transport a specific species in either direction governed only by concentrations of that species on either side of the membrane (Figure 13.10).

Secondary transporters are ancient molecular machines, common today in bacteria and archaea as well as in eukaryotes. For example, approximately 160 (of around 4000) proteins encoded by the *E. coli* genome are secondary transporters. Sequence comparison and hydropathy analysis suggest that members of the largest family have 12 transmembrane helices that appear to have arisen by duplication and fusion of a membrane protein containing 6 transmembrane helices. Included in this family is the *lactose permease* of *E. coli*. This symporter uses the H^+ gradient across the *E. coli* membrane (outside has higher H^+ concentration) generated by the oxidation of fuel

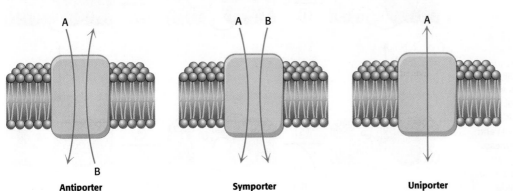

FIGURE 13.10 Antiporters, symporters, and uniporters. Secondary transporters can transport two substrates in opposite directions (antiporters), two substrates in the same direction (symporters), or one substrate in either direction (uniporter).

Antiporter   **Symporter**   **Uniporter**

FIGURE 13.11 Structure of lactose permease with a bound lactose analog. The amino-terminal half of the protein is shown in blue and the carboxyl-terminal half in red. (A) Side view. The gray lines indicate the extent of the plasma membrane. (B) Bottom view (from inside the cell). *Notice* that the structure consists of two halves that surround the sugar and are linked to one another by only a single stretch of polypeptide. [Drawn from 1PV7.pdb.]

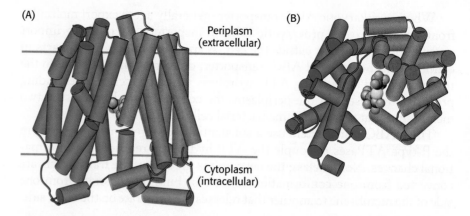

molecules to drive the uptake of lactose and other sugars against a concentration gradient. This transporter has been extensively studied for many decades and is a useful archetype for this family.

The structure of lactose permease has been determined (Figure 13.11). As expected from the sequence analysis, the protein consists of two halves, each of which comprises six membrane-spanning α helices. The two halves are well separated and are joined by a single polypeptide. In this structure, a sugar molecule lies in a pocket in the center of the protein and is accessible from a path that leads from the interior of the cell. On the basis of this structure and a wide range of other experiments, a mechanism for symporter action has been developed. This mechanism (Figure 13.12) has many features similar to those for P-type ATPases and ABC transporters:

1. The cycle begins with the two halves oriented so that the opening to the binding pocket faces outside the cell, in a conformation different from that observed in the structures solved to date. A proton from outside the cell binds to a residue in the permease, quite possibly Glu 325.

2. In the protonated form, the permease binds lactose from outside the cell.

3. The structure everts to the form observed in the crystal structure (Figure 13.11).

4. The permease releases lactose to the inside of the cell.

5. The permease releases a proton to the inside of the cell.

6. The permease everts to complete the cycle.

The site of protonation likely changes in the course of this cycle.

FIGURE 13.12 Lactose permease mechanism. The mechanism begins with the permease open to the outside of the cell (upper left). The permease binds a proton from the outside of the cell (1) and then binds its substrate (2). The permease everts (3) and then releases its substrate (4) and a proton (5) to the inside of the cell. It then everts (6) to complete the cycle.

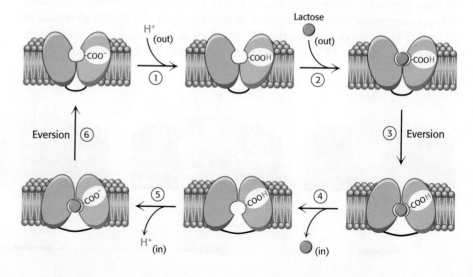

It is believed that this eversion mechanism applies to all classes of secondary transporters, which resemble the lactose permease in overall architecture.

13.4 Specific Channels Can Rapidly Transport Ions Across Membranes

Pumps and carriers can move ions and substrates across the membrane at rates approaching several thousand molecules per second. Other membrane proteins, the passive-transport systems called *ion channels*, are capable of ion-transport rates that are more than 1000 times as fast. These rates of transport through ion channels are close to rates expected for ions diffusing freely through aqueous solution. Yet ion channels are not simply tubes that span membranes through which ions can rapidly flow. Instead, they are highly sophisticated molecular machines that respond to chemical and physical changes in their environments and undergo precisely timed conformational changes.

Action potentials are mediated by transient changes in Na$^+$ and K$^+$ permeability

One of the most important manifestations of ion-channel action is the nerve impulse, which is the fundamental means of communication in the nervous system. A *nerve impulse* is an electrical signal produced by the flow of ions across the plasma membrane of a neuron. The interior of a neuron, like that of most other cells, contains a high concentration of K$^+$ and a low concentration of Na$^+$. These ionic gradients are generated by the Na$^+$–K$^+$ ATPase. The cell membrane has an electrical potential determined by the ratio of the internal to the external concentration of ions. In the resting state, the membrane potential is typically -60 mV. A nerve impulse, or *action potential*, is generated when the membrane potential is depolarized beyond a critical threshold value (e.g., from -60 to -40 mV). The membrane potential becomes positive within about a millisecond and attains a value of about $+30$ mV before turning negative again (repolarization). This amplified depolarization is propagated along the nerve terminal (Figure 13.13).

Ingenious experiments carried out by Alan Hodgkin and Andrew Huxley revealed that action potentials arise from large, transient changes in the permeability of the axon membrane to Na$^+$ and K$^+$ ions. Depolarization of the membrane beyond the threshold level leads to an increase in permeability to Na$^+$. Sodium ions begin to flow into the cell because of the large electrochemical gradient across the plasma membrane. The entry of Na$^+$ further depolarizes the membrane, leading to an added increase in Na$^+$ permeability. This positive feedback yields the very rapid and large change in membrane potential described above and shown in Figure 13.13.

The membrane spontaneously becomes less permeable to Na$^+$ and more permeable to K$^+$. Consequently, K$^+$ flows outward, and so the membrane potential returns to a negative value. The resting level of -60 mV is restored in a few milliseconds as the K$^+$ conductance decreases to the value characteristic of the unstimulated state. The wave of depolarization followed by repolarization moves rapidly along a nerve cell. The propagation of these waves allows a touch at the tip of your toe to be detected in your brain in a few milliseconds.

This model for the action potential postulated the existence of ion channels specific for Na$^+$ and K$^+$. These channels must open in response to changes in membrane potential and then close after having remained open

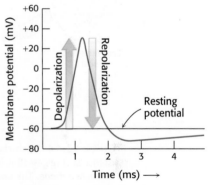

FIGURE 13.13 Action potential. Signals are sent along neurons by the transient depolarization and repolarization of the membrane.

for a brief period of time. This bold hypothesis predicted the existence of molecules with a well-defined set of properties long before tools existed for their direct detection and characterization.

Patch-clamp conductance measurements reveal the activities of single channels

Direct evidence for the existence of these channels was elucidated using the *patch-clamp technique*, which was introduced by Erwin Neher and Bert Sakmann in 1976. This powerful method enables the measurement of the ion conductance through a small patch of cell membrane. A clean glass pipette with a tip diameter of about 1 μm is pressed against an intact cell to form a seal (Figure 13.14). Slight suction leads to the formation of a very

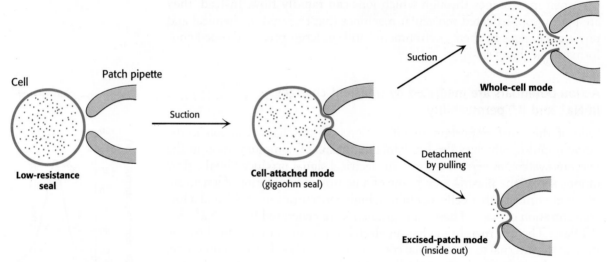

FIGURE 13.14 Patch-clamp modes. The patch-clamp technique for monitoring channel activity is highly versatile. A high-resistance seal (gigaseal) is formed between the pipette and a small patch of plasma membrane. This configuration is called *cell-attached mode*. The breaking of the membrane patch by increased suction produces a low-resistance pathway between the pipette and the interior of the cell. The activity of the channels in the entire plasma membrane can be monitored in this *whole-cell mode*. To prepare a membrane in the *excised-patch mode*, the pipette is pulled away from the cell. A piece of plasma membrane with its cytoplasmic side now facing the medium is monitored by the patch pipette.

FIGURE 13.15 Observing single channels. (A) The results of a patch-clamp experiment revealing the small amount of current, measured in picoamperes (pA, 10^{-12} amperes) passing through a single ion channel. The downward spikes indicate transitions between closed and open states. (B) Closer inspection of one of the spikes in (A) reveals the length of time the channel is in the open state.

tight seal so that the resistance between the inside of the pipette and the bathing solution is many gigaohms (1 gigaohm is equal to 10^9 ohms). Thus, a gigaohm seal (called a *gigaseal*) ensures that an electric current flowing through the pipette is identical with the current flowing through the membrane covered by the pipette. The gigaseal makes possible high-resolution current measurements while a known voltage is applied across the membrane. Remarkably, the flow of ions through a single channel and transitions between the open and the closed states of a channel can be monitored with a time resolution of microseconds (Figure 13.15). Furthermore, the activity

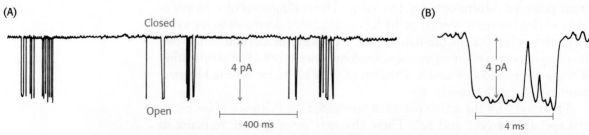

of a channel in its native membrane environment, even in an intact cell, can be directly observed. Patch-clamp methods provided one of the first views of single biomolecules in action. Subsequently, other methods for observing single molecules were invented, opening new vistas on biochemistry at its most fundamental level.

The structure of a potassium ion channel is an archetype for many ion-channel structures

With the existence of ion channels firmly established by patch-clamp methods, scientists sought to identify the molecules that form ion channels. The Na^+ channel was first purified from the electric organ of the electric eel, which is a rich source of the protein forming this channel. The channel was purified on the basis of its ability to bind tetrodotoxin, a neurotoxin from the puffer fish that binds to Na^+ channels very tightly ($K_i \approx 1$ nM). The lethal dose of this poison for an adult human being is about 10 ng.

The isolated Na^+ channel is a single 260-kDa chain (Figure 13.16). Cloning and sequencing of cDNAs encoding Na^+ channels revealed that the channel contains four internal repeats, each having a similar amino acid sequence, suggesting that gene duplication and divergence have produced the gene for this channel. Hydrophobicity profiles indicate that each repeat contains five hydrophobic segments (S1, S2, S3, S5, and S6). Each repeat also contains a highly positively charged S4 segment; positively charged arginine or lysine residues are present at nearly every third residue. It was proposed that segments S1 through S6 are membrane-spanning α helices, while the positively charged residues in S4 act as the voltage sensors of the channel.

Tetrodotoxin

FIGURE 13.16 Sequence relations of ion channels. Like colors indicate structurally similar regions of the sodium, calcium, and potassium channels. Each of these channels exhibits approximate fourfold symmetry, either within one chain (sodium, calcium channels) or by forming tetramers (potassium channels).

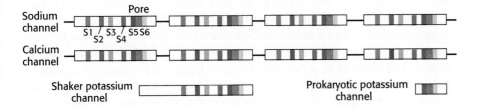

The purification of K^+ channels proved to be much more difficult because of their low abundance and the lack of known high-affinity ligands comparable to tetrodotoxin. The breakthrough came in studies of mutant fruit flies that shake violently when anesthetized with ether. The mapping and cloning of the gene, termed *Shaker*, responsible for this defect revealed the amino acid sequence encoded by a K^+-channel gene. The *Shaker* gene encodes a 70-kDa protein that contains sequences corresponding to segments S1 through S6 in one of the repeated units of the Na^+ channel. Thus, a K^+-channel subunit is homologous to one of the repeated units of Na^+ channels. Consistent with this homology, four Shaker polypeptides come together to form a functional channel. Bacterial K^+ channels have also been discovered that contain only the two membrane-spanning regions corresponding to segments S5 and S6. This and other information suggested that S5 and S6, including the region between them, form the actual pore in the K^+ channel. Segments S1 through S4 contain the apparatus that opens the pore. The sequence relations between these ion channels are summarized in Figure 13.16.

In 1998, Roderick MacKinnon and coworkers determined the structure of a K^+ channel from the bacterium *Streptomyces lividans* by x-ray

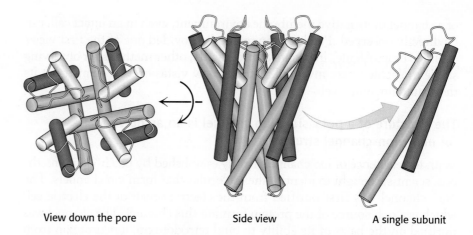

FIGURE 13.17 Structure of the potassium ion channel. The K⁺ channel, composed of four identical subunits, is cone shaped, with the larger opening facing the inside of the cell (center). A view down the pore, looking toward the outside of the cell, shows the relations of the individual subunits (left). One of the four identical subunits of the pore is illustrated at the right, with the pore-forming region shown in gray. [Drawn from 1K4C.pdb.]

View down the pore   Side view   A single subunit

crystallography. This channel contains only the pore-forming segments S5 and S6. As expected, the K⁺ channel is a tetramer of identical subunits, each of which includes two membrane-spanning α helices (Figure 13.17). The four subunits come together to form a pore in the shape of a cone that runs through the center of the structure.

The structure of the potassium ion channel reveals the basis of ion specificity

The structure presented in Figure 13.17 probably represents the K⁺ channel in a closed form. Nonetheless, it suggests how the channel is able to exclude all but K⁺ ions. Beginning from the inside of the cell, the pore starts with a diameter of approximately 10 Å and then constricts to a smaller cavity with a diameter of 8 Å. Both the opening to the outside and the central cavity of the pore are filled with water, and a K⁺ ion can fit in the pore without losing its shell of bound water molecules. Approximately two-thirds of the way through the membrane, the pore becomes more constricted (3-Å diameter). At that point, any K⁺ ions must give up their water molecules and interact directly with groups from the protein. The channel structure effectively reduces the thickness of the membrane from 34 Å to 12 Å by allowing the solvated ions to penetrate into the membrane before the ions must directly interact with the channel (Figure 13.18).

For K⁺ ions to relinquish their water molecules, other polar interactions must replace those with water. The restricted part of the pore is built from residues contributed by the two transmembrane α helices. In particular, a five-amino-acid stretch within this region functions as the *selectivity filter*

FIGURE 13.18 Path through a channel. A potassium ion entering the K⁺ channel can pass a distance of 22 Å into the membrane while remaining solvated with water (blue). At this point, the pore diameter narrows to 3 Å (yellow), and potassium ions must shed their water and interact with carbonyl groups (red) of the pore amino acids.

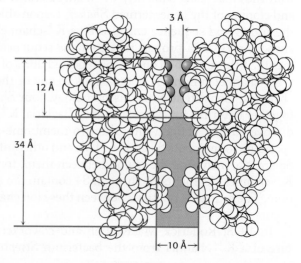

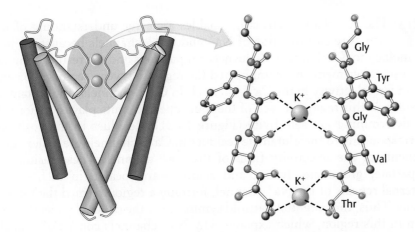

that determines the preference for K^+ over other ions (Figure 13.19). The stretch has the sequence Thr-Val-Gly-Tyr-Gly (TVGYG), and is nearly completely conserved in all K^+ channels. The region of the strand containing the conserved sequence lies in an extended conformation and is oriented such that the peptide carbonyl groups are directed into the channel, in good position to interact with the potassium ions.

Potassium ion channels are 100-fold more permeable to K^+ than to Na^+. How is this high degree of selectivity achieved? Ions having a radius larger than 1.5 Å cannot pass into the narrow diameter (3 Å) of the selectivity filter of the K^+ channel. However, a bare Na^+ is small enough (Table 13.1) to pass through the pore. Indeed, the ionic radius of Na^+ is substantially smaller than that of K^+. How then is Na^+ rejected?

The key point is that the free-energy costs of dehydrating these ions are considerable [Na^+, 301 kJ mol⁻¹ (72 kcal mol⁻¹), and K^+, 230 kJ mol⁻¹ (55 kcal mol⁻¹)]. *The channel pays the cost of dehydrating K^+ by providing optimal compensating interactions with the carbonyl oxygen atoms lining the selectivity filter.* Careful studies of the potassium channel, enabled by the determination of its three-dimensional structure, have revealed that the interior of the pore is a highly dynamic, fluid environment. The favorable interactions between the carbonyl oxygen atoms, which carry a partial negative charge, with the cation are balanced by the repulsion of these oxygen atoms from one another. For this channel, the ideal balance is achieved with K^+, but not with Na^+ (Figure 13.20). Hence, sodium ions are rejected because the higher energetic cost of dehydrating them would not be recovered.

TABLE 13.1 Properties of alkali cations

Ion	Ionic radius (Å)	Hydration free energy in kJ mol⁻¹ (kcal mol⁻¹)
Li^+	0.60	−410 (−98)
Na^+	0.95	−301 (−72)
K^+	1.33	−230 (−55)
Rb^+	1.48	−213 (−51)
Cs^+	1.69	−197 (−47)

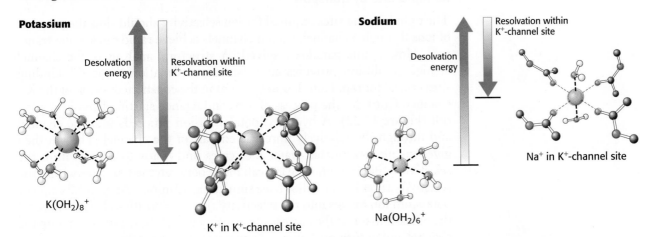

Potassium   Desolvation energy   Resolution within K⁺-channel site

$K(OH_2)_8{}^+$

K^+ in K⁺-channel site

Sodium   Desolvation energy   Resolution within K⁺-channel site

$Na(OH_2)_6{}^+$

Na^+ in K⁺-channel site

FIGURE 13.20 Energetic basis of ion selectivity. The energy cost of dehydrating a potassium ion is compensated by favorable interactions with the selectivity filter. Because a sodium ion is too small to interact favorably with the selectivity filter, the free energy of desolvation cannot be compensated and the sodium ion does not pass through the channel.

The K⁺ channel structure enables a clearer understanding of the structure and function of Na^+ and Ca^{2+} channels because of their homology to K⁺ channels. Sequence comparisons and the results of mutagenesis experiments have implicated the region between segments S5 and S6 in ion selectivity in the Ca^{2+} channel. In Ca^{2+} channels, one glutamate residue of this region in each of the four repeated units plays a major role in determining ion selectivity (Figure 13.21). Residues in the positions corresponding to the glutamate residues in Ca^{2+} channels are major components of the selectivity filter of the Na^+ channel. These residues—aspartate, glutamate, lysine, and alanine—are located in each of the internal repeats of the Na^+ channel, forming a region termed the DEKA locus. Thus, the potential fourfold symmetry of the channel is clearly broken in this region, which explains why Na^+ channels consist of a single large polypeptide chain rather than a noncovalent assembly of four identical subunits. The preference of the Na^+ channel for Na^+ over K⁺ depends on ionic radius; the diameter of the pore determined by these residues and others is sufficiently restricted that small ions such as Na^+ and Li^+ can pass through the channel, but larger ions such as K⁺ are significantly hindered.

FIGURE 13.21 Selectivity filters for calcium and sodium channels. The pores of eukaryotic calcium and sodium channels are comprised of single polypeptide chains. In these views, we are looking down the pore; for simplicity, only the S5 (blue), S6 (yellow), and pore-forming (white) regions are drawn. *Notice* that the selectivity filter for the calcium channel contains four glutamate (E) residues, while the filter for the sodium channel contains four different residues (the DEKA locus): aspartate (D), glutamate (E), lysine (K), and alanine (A). [Drawn from 5GJV.pdb and 5X0M.pdb]

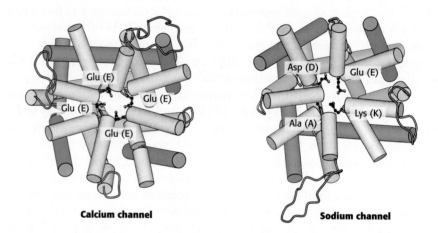

Calcium channel

Sodium channel

The structure of the potassium ion channel explains its rapid rate of transport

The tight binding sites required for ion selectivity should slow the progress of ions through a channel, yet ion channels achieve rapid rates of ion transport. How is this paradox resolved? A structural analysis of the channel at high resolution provides an appealing explanation. Four K⁺-binding sites crucial for rapid ion flow are present in the constricted region of the K⁺ channel. Consider the process of ion conductance starting from inside the cell (Figure 13.22). A hydrated potassium ion proceeds into the channel and through the relatively unrestricted part of the channel. The ion then gives up its coordinated water molecules and binds to a site within the selectivity-filter region. The ion can move between the four sites within the selectivity filter because they have similar ion affinities. As each subsequent potassium ion moves into the selectivity filter, its positive charge will repel the potassium ion at the nearest site, causing it to shift to a site farther up the channel and in turn push upward any potassium ion already bound to a site farther up. Thus, each ion that binds anew favors the release of an ion from the other side of the channel. This multiple-binding-site mechanism solves the paradox of high ion selectivity and rapid flow.

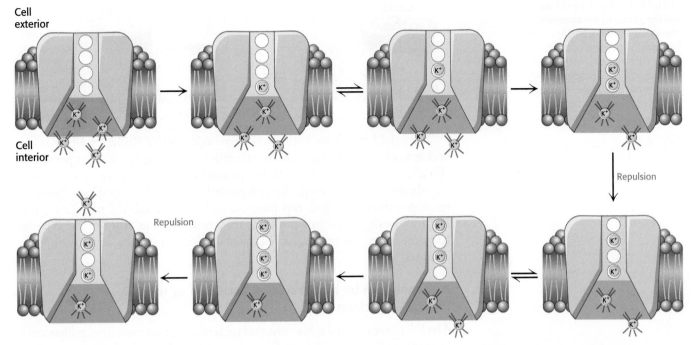

Cell exterior

Cell interior

Repulsion

Repulsion

FIGURE 13.22 Model for K$^+$-channel ion transport. The selectivity filter has four binding sites. Hydrated potassium ions can enter these sites, one at a time, losing their hydration shells. When two ions occupy adjacent sites, electrostatic repulsion forces them apart. Thus, as ions enter the channel from one side, other ions are pushed out the other side.

Voltage gating requires substantial conformational changes in specific ion-channel domains

Some Na$^+$ and K$^+$ channels are gated by membrane potential; that is, they change conformation to a highly conducting form in response to changes in voltage across the membrane. As already noted, these *voltage-gated channels* include segments S1 through S4 in addition to the pore itself formed by S5 and S6. The structure of a voltage-gated K$^+$ channel from *Aeropyrum pernix* has been determined by x-ray crystallography (Figure 13.23). The segments S1 through S4 form domains, termed "paddles," that extend from the core of the channel. These paddles include the segment S4, the voltage sensor itself. Segment S4 forms an α helix lined with positively charged residues. In contrast with expectations, segments S1 through S4 are not enclosed within the protein but, instead, are positioned to lie in the membrane itself.

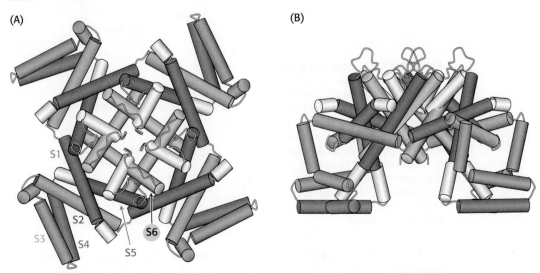

(A)

S1

S2

S3   S4

S5

S6

(B)

FIGURE 13.23 Structure of a voltage-gated potassium channel. (A) A view looking down through the pore. (B) A side view. *Notice* that the positively charged S4 region (red) lies on the outside of the structure at the bottom of the pore. [Drawn from 1ORQ.pdb.]

FIGURE 13.24 A model for voltage gating of ion channels. The voltage-sensing paddles lie in the "down" position below the closed channel (left). Membrane depolarization pulls these paddles through the membrane. The motion pulls the base of the channel apart, opening the channel (right).

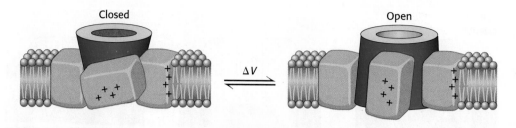

A model for voltage gating has been proposed by Roderick MacKinnon and coworkers on the basis of this structure and a range of other experiments (Figure 13.24). In the closed state, the paddles lie in a "down" position. On membrane depolarization, the cytoplasmic side of the membrane becomes more positively charged, repelling the paddles through the membrane into an "up" position. In this position, they pull the four sides of the base on the pore apart, increasing access to the selectivity filter and opening the channel.

A channel can be inactivated by occlusion of the pore: the ball-and-chain model

The K^+ channel and the Na^+ channel undergo inactivation within milliseconds of opening (Figure 13.25A). A first clue to the mechanism of inactivation came from exposing the cytoplasmic side of either channel to the protease trypsin. Cleavage by trypsin produced trimmed channels that stayed persistently open after depolarization, suggesting that a flexible (i.e., accessible to protease) region of the protein was responsible for inactivation. Furthermore, a mutant Shaker channel lacking 42 amino acids near the amino terminus opened in response to depolarization but did not inactivate (Figure 13.25B). Remarkably, inactivation was restored by adding a synthetic peptide corresponding to the first 20 residues of the native channel (Figure 13.25C).

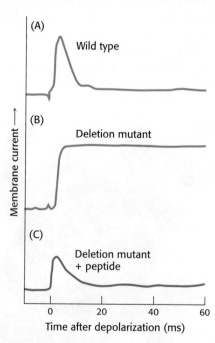

FIGURE 13.25 Inactivation of the potassium ion channel. The amino-terminal region of the K^+ chain is critical for inactivation. (A) The wild-type Shaker K^+ channel displays rapid inactivation after opening. (B) A mutant channel lacking residues 6 through 46 does not inactivate. (C) Inactivation can be restored by adding a peptide consisting of residues 1 through 20 at a concentration of 100 μM. [Data from W. N. Zagotta, T. Hoshi, and R. W. Aldrich, *Science* 250:568–571, 1990.]

These experiments strongly support the *ball-and-chain model* for channel inactivation that had been proposed years earlier (Figure 13.26). According to this model, the first 20 residues of the K^+ channel form a cytoplasmic unit (the *ball*) that is attached to a flexible segment of the

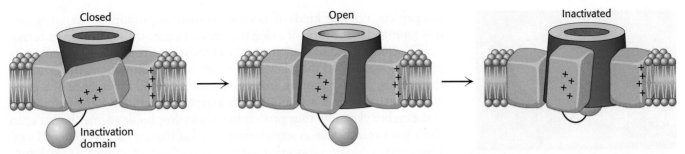

Closed   Open   Inactivated

Inactivation domain

FIGURE 13.26 Ball-and-chain model for channel inactivation. The inactivation domain, or "ball" (gray), is tethered to the channel by a flexible "chain." In the closed state, the ball is located in the cytoplasm. Depolarization opens the channel and creates a binding site for the positively charged ball in the mouth of the pore. Movement of the ball into this site inactivates the channel by occluding it.

polypeptide (the *chain*). When the channel is closed, the ball rotates freely in the aqueous solution. When the channel opens, the ball quickly finds a complementary site in the open pore and occludes it. Hence, the channel opens for only a brief interval before it undergoes inactivation by occlusion. Shortening the chain speeds inactivation because the ball finds its target more quickly. Conversely, lengthening the chain slows inactivation. Thus, the duration of the open state can be controlled by the length and flexibility of the tether. In some senses, the "ball" domains, which include substantial regions of positive charge, can be thought of as large, tethered cations that are pulled into the open channel but get stuck and block further ion conductance.

The acetylcholine receptor is an archetype for ligand-gated ion channels

Nerve impulses are communicated across synapses by small, diffusible molecules called *neurotransmitters*. One neurotransmitter is *acetylcholine*. The presynaptic membrane of a synapse is separated from the postsynaptic membrane by a gap of about 50 nm called the *synaptic cleft*. The arrival of a nerve impulse at the end of an axon leads to the synchronous export of the contents of some 300 membrane-bound compartments, or vesicles, of acetylcholine into the cleft (Figure 13.27). The binding of acetylcholine to the postsynaptic membrane markedly changes its ionic permeability, triggering an action potential. Acetylcholine opens a single kind of cation channel, called the *acetylcholine receptor,* which is almost equally permeable to Na^+ and K^+.

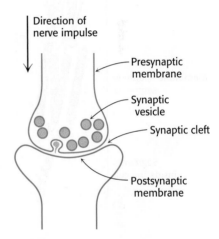

Direction of nerve impulse

Presynaptic membrane

Synaptic vesicle

Synaptic cleft

Postsynaptic membrane

FIGURE 13.27 Schematic representation of a synapse.

Acetylcholine

The acetylcholine receptor is the best-understood *ligand-gated channel.* This type of channel is gated not by voltage but by the presence of specific ligands. The binding of acetylcholine to the channel is followed by its transient opening. The electric organ of *Torpedo marmorata*, an electric ray, is a choice source of acetylcholine receptors for study because its electroplaxes (voltage-generating cells) are very rich in postsynaptic membranes that respond to this neurotransmitter. The receptor is very densely packed in these membranes ($\sim 20,000 \ \mu m^{-2}$). The acetylcholine receptor of the electric organ has been solubilized by adding a nonionic detergent to a postsynaptic membrane preparation and purified by affinity chromatography on a column bearing covalently attached cobratoxin, a small protein toxin from snakes that has a high affinity for acetylcholine receptors. With the use of techniques presented in Chapter 3, the 268-kDa receptor was identified

The torpedo (*Torpedo marmorata*, also known as the electric ray) has an electric organ, rich in acetylcholine receptors, that can deliver a shock of as much as 200 V for approximately 1 s. [Reinhard Dirscherl/Getty Images]

as a pentamer of four kinds of membrane-spanning subunits—α_2, β, γ, and δ—arranged in the form of a ring that creates a pore through the membrane.

The cloning and sequencing of the cDNAs for the four kinds of subunits (50–58 kDa) showed that they have clearly similar sequences; the genes for the α, β, γ, and δ subunits arose by duplication and divergence of a common ancestral gene. Each subunit has a large extracellular domain, followed at the carboxyl end by four predominantly hydrophobic segments that span the bilayer membrane. Acetylcholine binds at the α–γ and α–δ interfaces. Using cryo-EM and x-ray crystallographic methods (Section 3.5), the structure of purified acetylcholine receptors has been determined. The receptor has approximate fivefold symmetry, in harmony with the similarity of its five constituent subunits (Figure 13.28).

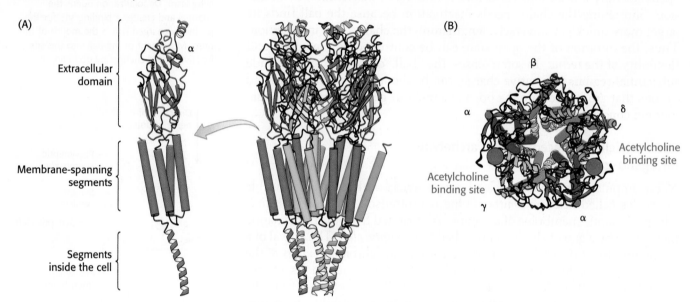

FIGURE 13.28 Structure of the acetylcholine receptor. A model for the structure of the acetylcholine receptor deduced from high-resolution electron microscopic studies reveals that each subunit consists of a large extracellular domain consisting primarily of β strands, four membrane-spanning α helices, and a final α helix inside the cell. (A) A side view shows the pentameric receptor with each subunit type in a different color. One copy of the α subunit is shown in isolation. (B) A view down the channel from outside the cell. The binding sites for the acetylcholine ligand are indicated in green. [Drawn from 2BG9.pdb.]

What is the basis of channel opening? A high-resolution answer to this question has remained elusive, but the details of this mechanism are becoming apparent. Cryo-EM images of the receptor in the open and closed states have been determined, albeit at low resolution. These structures indicate that the binding of acetylcholine to the extracellular domain ultimately leads to the straightening of the α-helices from the α and δ subunits that line the pore (Figure 13.29). In the closed state, the region of narrowest diameter in the pore (the "gate") is located about midway through the membrane. This region is lined with nonpolar residues, which cannot form favorable interactions with K^+ and Na^+ ions. However, once the helices are straightened, the region of narrowest diameter shifts closer to the inner membrane leaflet. This region is lined with polar residues and can conduct K^+ and Na^+ ions freely.

Action potentials integrate the activities of several ion channels working in concert

To see how ligand-gated and voltage-gated channels work together to generate a sophisticated physiological response, we now revisit the action

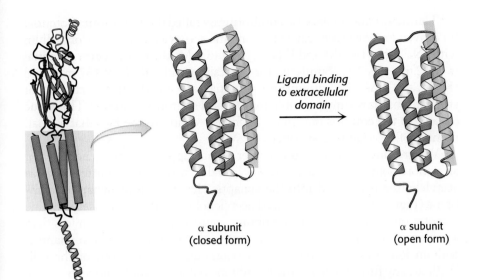

FIGURE 13.29 Opening the acetylcholine receptor. Binding of acetylcholine to the extracellular region of the receptor leads to a series of conformational changes that are transmitted to the pore-lining helices. *Notice* that upon ligand binding, the pore-lining helix of the α subunit straightens (green bar). This subtle structural adjustment changes the gating properties of the pore, enabling cations to pass through. [Drawn from 4AQ5.pdb and 4AQ9.pdb.]

Ligand binding to extracellular domain

α subunit
(closed form)

α subunit
(open form)

potential introduced at the beginning of this section. First, we need to introduce the concept of *equilibrium potential*. Suppose that a membrane separates two solutions that contain different concentrations of some cation X^+, as well as an equivalent amount of anions to balance the charge in each solution (Figure 13.30). Let $[X^+]_{in}$ be the concentration of X^+ on one side of the membrane (corresponding to the inside of a cell) and $[X^+]_{out}$ be the concentration of X^+ on the other side (corresponding to the outside of a cell). Suppose that an ion channel opens that allows X^+ to move across the membrane. What will happen? It seems clear that X^+ will move through the channel from the side with the higher concentration to the side with the lower concentration. However, positive charges will start to accumulate on the side with the lower concentration, making it more difficult to move each additional positively charged ion. An equilibrium will be achieved when the driving force due to the concentration gradient is balanced by the electrostatic force resisting the motion of an additional charge. In these circumstances, the membrane potential is given by the *Nernst equation*:

$$V_{eq} = -(RT/zF) \ln([X]_{in}/[X]_{out})$$

where R is the gas constant and F is the Faraday constant ($96.5 \text{ kJ V}^{-1} \text{ mol}^{-1}$, or $23.1 \text{ kcal V}^{-1} \text{ mol}^{-1}$) and z is the charge on the ion X (e.g., $+1$ for X^+).

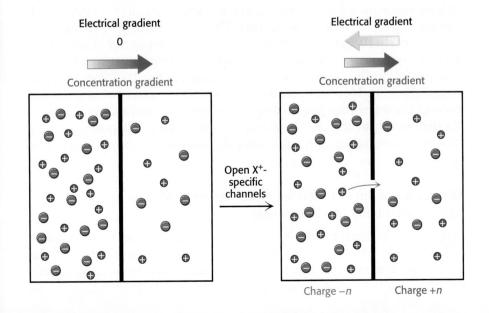

Electrical gradient

0

Concentration gradient

Electrical gradient

Concentration gradient

Open X^+-specific channels

Charge $-n$ Charge $+n$

FIGURE 13.30 Equilibrium potential. The membrane potential reaches equilibrium when the driving force due to the concentration gradient is exactly balanced by the opposing force due to the repulsion of like charges.

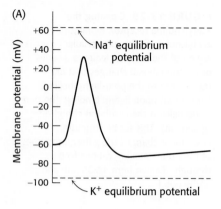

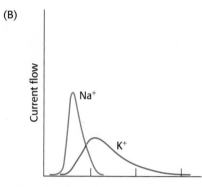

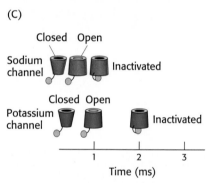

FIGURE 13.31 Action-potential mechanism. (A) On the initiation of an action potential, the membrane potential moves from the resting potential upward toward the Na^+ equilibrium potential and then downward toward the K^+ equilibrium potential. (B) The currents through the Na^+ and K^+ channels underlying the action potential. (C) The states of the Na^+ and K^+ channels during the action potential.

The membrane potential at equilibrium is called the equilibrium potential for a given ion at a given concentration ratio across a membrane. For sodium with $[Na^+]_{in} = 14$ mM and $[Na^+]_{out} = 143$ mM, the equilibrium potential is $+62$ mV at 37°C. Similarly, for potassium with $[K^+]_{in} = 157$ mM and $[K^+]_{out} = 4$ mM, the equilibrium potential is -98 mV. In the absence of stimulation, the resting potential for a typical neuron is -60 mV. This value is close to the equilibrium potential for K^+ owing to the fact that a small number of K^+ channels are open.

We are now prepared to consider what happens in the generation of an action potential (Figure 13.31). Initially, a neurotransmitter such as acetylcholine is released into the synaptic cleft from a presynaptic membrane (Figure 13.27). The released acetylcholine binds to the acetylcholine receptor on the postsynaptic membrane, causing it to open within less than a millisecond. The acetylcholine receptor is a nonspecific cation channel. Sodium ions flow into the cell and potassium ions flow out of the cell. Without any further events, the membrane potential would move to a value corresponding to the average of the equilibrium potentials for Na^+ and K^+, approximately -20 mV. However, as the membrane potential approaches -40 mV, the voltage-sensing paddles of Na^+ channels are pulled into the membrane, opening the Na^+ channels. With these channels open, sodium ions flow rapidly into the cell and the membrane potential rises rapidly toward the Na^+ equilibrium potential (Figure 13.31B, red curve). The voltage-sensing paddles of K^+ channels also are pulled into the membrane by the changed membrane potential, but more slowly than Na^+ channel paddles. Nonetheless, after approximately 1 ms, many K^+ channels start to open. At the same time, inactivation "ball" domains plug the open Na^+ channels, decreasing the Na^+ current. The acetylcholine receptors that initiated these events are also inactivated on this time scale. With the Na^+ channels inactivated and only the K^+ channels open, the membrane potential drops rapidly toward the K^+ equilibrium potential (Figure 13.31B, blue curve). The open K^+ channels are susceptible to inactivation by their "ball" domains, and these K^+ currents, too, are blocked. With the membrane potential returned to close to its initial value, the inactivation domains are released and the channels return to their original closed states. These events propagate along the neuron as the depolarization of the membrane opens channels in nearby patches of membrane.

How much current actually flows across the membrane over the course of an action potential? Consider that a typical nerve cell contains 100 Na^+ channels per square micrometer. At a membrane potential of $+20$ mV, each channel conducts 10^7 ions per second. Thus, in a period of 1 millisecond, approximately 10^5 ions flow through each square micrometer of membrane surface. Assuming a cell volume of 10^4 μm^3 and a surface area of 10^4 μm^2, this rate of ion flow corresponds to an increase in the Na^+ concentration of less than 1%. How can this be? A robust action potential is generated because the membrane potential is very sensitive to even a slight change in the distribution of charge. This sensitivity makes the action potential a very efficient means of signaling over long distances and with rapid repetition rates.

Disruption of ion channels by mutations or chemicals can be potentially life-threatening

The generation of an action potential requires the precise coordination of gating events of a collection of ion channels. Perturbation of this timing can have devastating effects. For example, the rhythmic generation of action potentials by the heart is absolutely essential to maintain delivery of oxygenated blood to the peripheral tissues. *Long QT syndrome* (LQTS) is a genetic disorder in which the recovery of the action potential from its peak potential to the resting equilibrium potential is delayed. The term "QT" refers to a specific feature of the cardiac electrical activity pattern as measured by electrocardiography. LQTS can lead to brief losses of consciousness (syncope), disruption of normal cardiac rhythm (arrhythmia), and sudden death. The most common mutations identified in LQTS patients inactivate K^+ channels or prevent the proper trafficking of these channels to the plasma membrane. The resulting loss in potassium permeability slows the repolarization of the membrane and delays the induction of the subsequent cardiac contraction, rendering the cardiac tissue susceptible to arrhythmias.

Prolongation of the cardiac action potential in this manner can also be induced by a number of therapeutic drugs. In particular, the K^+ channel hERG (for human *ether-a-go-go*-related gene, named for its ortholog in *Drosophila melanogaster*) is highly susceptible to interactions with certain drugs. The hydrophobic regions of these drugs can block hERG by binding to two nonconserved aromatic residues on the internal surface of the channel cavity. In addition, this cavity is predicted to be wider than other K^+ channels because of the absence of a conserved Pro-X-Pro motif within the S6 hydrophobic segment. Inhibition of hERG by these drugs can lead to an increased risk of cardiac arrhythmias and sudden death. Accordingly, a number of these agents, such as the antihistamine terfenadine, have been withdrawn from the market. Screening for the inhibition of hERG is now a critical safety hurdle for the pharmaceutical advancement of a molecule to an approved drug.

13.5 Gap Junctions Allow Ions and Small Molecules to Flow Between Communicating Cells

The ion channels we have considered thus far have narrow pores and are moderately to highly selective in the ions they allow to pass through them. They are closed in the resting state and have short lifetimes in the open state, typically a millisecond, that enable them to transmit frequent neural signals. Let's turn now to a channel with a very different role. *Gap junctions*, also known as *cell-to-cell channels*, serve as passageways between the interiors of contiguous cells. Gap junctions are clustered in discrete regions of the plasma membranes of apposed cells. Electron micrographs of sheets of gap junctions show them tightly packed in a regular hexagonal array (Figure 13.32). An approximately 20-Å central hole, the lumen of the channel, is prominent in each gap junction. These channels span the intervening space, or gap, between apposed cells (hence, the name "gap junction"). The width of the gap between the cytoplasms of the two cells is about 35 Å.

Small hydrophilic molecules as well as ions can pass through gap junctions. The pore size of the junctions was determined by microinjecting a series of fluorescent molecules into cells and observing their passage into adjoining cells. All polar molecules with a mass of less than about 1 kDa can readily pass through these cell-to-cell channels. Thus, *inorganic ions and most metabolites (e.g., sugars, amino acids, and nucleotides) can flow between the interiors of cells joined by gap junctions.* In contrast, proteins, nucleic acids, and polysaccharides are too large to traverse these channels. *Gap junctions are important for intercellular communication.* Cells in some

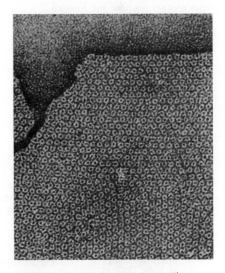

FIGURE 13.32 Gap junctions. This electron micrograph shows a sheet of isolated gap junctions. The cylindrical connexons form a hexagonal lattice having a unit-cell length of 85 Å. The densely stained central hole has a diameter of about 20 Å. [Don W. Fawcett/Science Source.]

excitable tissues, such as heart muscle, are coupled by the rapid flow of ions through these junctions, which ensures a speedy and synchronous response to stimuli. Gap junctions are also essential for the nourishment of cells that are distant from blood vessels, as in lens and bone. Moreover, communicating channels are important in development and differentiation. For example, the quiescent (calm) uterus transforms to a forcefully contracting organ at the onset of labor; the formation of functional gap junctions at that time creates a syncytium of muscle cells that contract in synchrony.

A cell-to-cell channel is made of 12 molecules of *connexin,* one of a family of transmembrane proteins with molecular masses ranging from 30 to 42 kDa. Each connexin molecule contains four membrane-spanning helices (Figure 13.33A). Six connexin molecules are hexagonally arrayed to form a half-channel, called a *connexon* or *hemichannel.* Two connexons join end to end in the intercellular space to form a functional channel between the communicating cells (Figure 13.33B). Each connexon adopts a funnel-shape: At the cytoplasmic face, the inner diameter of the channel is 35 Å, while at its innermost point, the pore narrows to a diameter of 14 Å (Figure 13.33C). Cell-to-cell channels differ from other membrane channels in three respects:

1. They traverse *two* membranes rather than one;

2. They connect cytoplasm to cytoplasm, rather than to the extracellular space or the lumen of an organelle; and

3. The connexons forming a channel are synthesized by different cells.

Gap junctions form readily when cells are brought together. A cell-to-cell channel, once formed, tends to stay open for seconds to minutes. They are closed by high concentrations of calcium ion and by low pH. *The closing of gap junctions by Ca^{2+} and H^+ serves to seal normal cells from injured or dying neighbors.* Gap junctions are also controlled by membrane potential and by hormone-induced phosphorylation.

The human genome encodes 21 distinct connexins. Different members of this family are expressed in different tissues. For example, connexin 26 is expressed in key tissues in the ear. Mutations in this connexin are associated with hereditary deafness. The mechanistic basis for this deafness appears to be insufficient transport of ions or second-messenger molecules, such as inositol trisphosphate, between sensory cells.

FIGURE 13.33 Structure of a gap junction. (A) Six connexins join to form a connexon, or hemichannel, within the plasma membrane (yellow). A single connexin monomer is highlighted in red. The extracellular region of one connexon binds to the same region of a connexon from another cell (orange), forming a complete gap junction. (B) Schematic view of the gap junction, oriented in the same direction as in (A). (C) A bottom-up view looking through the pore of a gap junction. This perspective is visualized in Figure 13.31. [(A) and (C) Drawn from 2ZW3.pdb; (B) Information from Dr. Werner Loewenstein.]

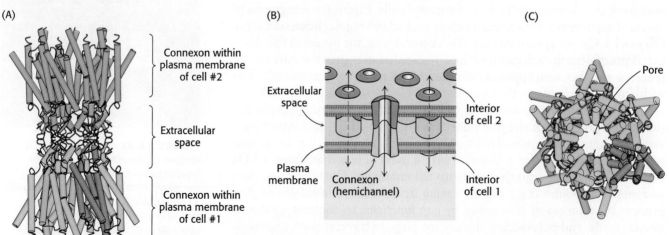

(A)

Connexon within plasma membrane of cell #2

Extracellular space

Connexon within plasma membrane of cell #1

(B)

Extracellular space

Plasma membrane

Connexon (hemichannel)

Interior of cell 2

Interior of cell 1

(C)

Pore

13.6 Specific Channels Increase the Permeability of Some Membranes to Water

One more important class of channels does not take part in ion transport at all. Instead, these channels increase the rate at which water flows through membranes. As noted in Section 12.3, membranes are reasonably permeable to water. Why, then, are water-specific channels required? In certain tissues, in some circumstances, rapid water transport through membranes is necessary. In the kidney, for example, water must be rapidly reabsorbed into the bloodstream after filtration. Similarly, in the secretion of saliva and tears, water must flow quickly through membranes. These observations suggested the existence of specific water channels, but initially the channels could not be identified.

The channels (now called *aquaporins*) were discovered serendipitously. Peter Agre noticed a protein present at high levels in red-blood-cell membranes that had been missed because the protein does not stain well with Coomassie blue. In addition to red blood cells, this protein was found in large quantities in tissues such as the kidney and the cornea, precisely the tissues thought to contain water channels. On the basis of this observation, further studies were designed, revealing that this 24-kDa membrane protein is, indeed, a water channel.

The structure of aquaporin has been determined (Figure 13.34). The protein consists of six membrane-spanning α helices. Two loops containing hydrophilic residues line the actual channel. Water molecules pass through in single file at a rate of 10^6 molecules per second. Importantly, specific positively charged residues toward the center of the channel prevent the transport of protons through aquaporin. Thus, aquaporin channels will not disrupt proton gradients, which play fundamental roles in energy transduction, as we will see in Chapter 18. Remarkably, the aquaporins are channels that have evolved specifically to conduct uncharged substrates.

(A)

(B)

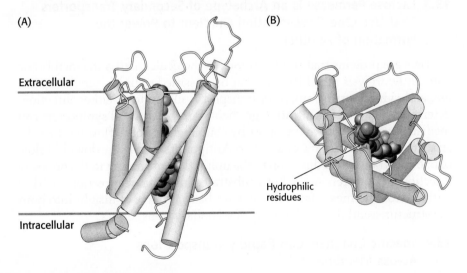

Extracellular

Intracellular

Hydrophilic residues

FIGURE 13.34 Structure of aquaporin. The structure of aquaporin viewed (A) from the side and (B) from the top. *Notice* the hydrophilic residues (shown as space-filling models) that line the water channel. The gray lines in (A) indicate the extent of the plasma membrane. [Drawn from 1J4N.pdb.]

SUMMARY

13.1 The Transport of Molecules Across a Membrane May Be Active or Passive

For a net movement of molecules across a membrane, two features are required: (1) the molecule must be able to cross a hydrophobic barrier and (2) an energy source must power the movement. Lipophilic molecules can pass through a membrane's hydrophobic interior by simple diffusion. These molecules will move down their concentration gradients. Polar or charged molecules require proteins to form passages through the hydrophobic barrier. Passive transport or facilitated diffusion takes place when an ion or polar molecule moves down its concentration gradient. If a molecule moves against a concentration gradient, an external energy source is required; this movement is referred to as active transport and results in the generation of concentration gradients. The electrochemical potential measures the combined ability of a concentration gradient and an uneven distribution of charge to drive species across a membrane.

13.2 Two Families of Membrane Proteins Use ATP Hydrolysis to Pump Ions and Molecules Across Membranes

Active transport is often carried out at the expense of ATP hydrolysis. P-type ATPases pump ions against a concentration gradient and become transiently phosphorylated on an aspartic acid residue during the transport cycle. These ATPases, which include the sarcoplasmic reticulum Ca^{2+} ATPase and the Na^{+}–K^{+} ATPase, are integral membrane proteins with conserved structures and catalytic mechanisms. Another family of ATP-dependent pumps, the ABC transporter proteins, contain ATP-binding cassette domains. Each pump includes four major domains: two domains span the membrane and two others contain ABC P-loop ATPase structures. These pumps are not phosphorylated during pumping; rather, they use the energy of ATP binding and hydrolysis to drive conformational changes that result in the transport of specific substrates across membranes. The multidrug-resistance proteins are ABC transporters that confer resistance on cancer cells by pumping chemotherapeutic drugs out of a cancer cell before the drugs can exert their effects.

13.3 Lactose Permease is an Archetype of Secondary Transporters That Use One Concentration Gradient to Power the Formation of Another

Carriers are proteins that transport ions or molecules across the membrane without hydrolysis of ATP. They can be classified as uniporters, antiporters, and symporters. Uniporters transport a substrate in either direction, determined by the concentration gradient. Antiporters and symporters can mediate secondary active transport by coupling the uphill flow of one substrate to the downhill flow of another. Antiporters couple the downhill flow of one substrate in one direction to the uphill flow of another in the opposite direction. Symporters move both substrates in the same direction. Studies of the lactose permease from *E. coli* have been a source of insight into both the structures and the mechanisms of secondary transporters.

13.4 Specific Channels Can Rapidly Transport Ions Across Membranes

Ion channels allow the rapid movement of ions across the hydrophobic barrier of the membrane. The activity of individual ion-channel molecules can be observed by using patch-clamp techniques. Many ion channels have a common structural framework. In regard to K^{+} channels, hydrated potassium ions must transiently lose their coordinated water molecules as they move to the narrowest part of the channel, termed the selectivity filter. In the selectivity filter, peptide carbonyl groups coordinate the ions. Rapid ion flow through the selectivity filter is facilitated by ion–ion repulsion, with one ion pushing the next ion through the channel. Some ion channels are voltage gated: changes in membrane potential induce conformational changes that open these channels. Many channels spontaneously inactivate after having been open for a short period of time. In some cases, inactivation is due to the binding of a domain of the channel termed the "ball" in the mouth of the channel to block it. Other channels, typified by the acetylcholine receptor, are opened or closed by the binding of ligands. Ligand-gated and voltage-gated channels work in concert to generate action potentials. Inherited mutations or drugs that interfere with the ion channels that produce the action potential can result in potentially life-threatening conditions.

13.5 Gap Junctions Allow Ions and Small Molecules to Flow Between Communicating Cells

In contrast with many channels, which connect the cell interior with the environment, gap junctions, or cell-to-cell channels, serve to connect the interiors of contiguous groups of cells. A cell-to-cell channel is composed of 12 molecules of connexin, which associate to form two 6-membered connexons.

13.6 Specific Channels Increase the Permeability of Some Membranes to Water

Some tissues contain proteins that increase the permeability of membranes to water. Each water-channel-forming protein, termed an aquaporin, consists of six membrane-spanning α helices and a central channel lined with hydrophilic residues that allow water molecules to pass in single file. Aquaporins do not transport protons.

APPENDIX

Biochemistry in Focus

Setting the pace is more than funny business

The heart utilizes coordinated changes in the membrane potential to facilitate the efficient muscular contractions necessary to pump blood throughout the body effectively. However, the heart can beat spontaneously, without any input from the rest of the body. Recall that in our discussion of the action potential, we used the example of a neuron stimulated by acetylcholine. This ligand binds to the acetylcholine receptor, opening a nonspecific cation channel, leading to membrane depolarization and the initiation of an action potential. How does the heart initiate an action potential without such an input?

The answer lies within a collection of cardiac cells known as the pacemaker. While there are several pacemakers in the human heart, the most important of these is the *sinoatrial (SA) node*, located on the posterior wall of the right atrium. The cells of the SA node spontaneously generate action potentials at a rate of about 100 per second. Researchers have known for many years that these cells possess an unusual capacity for mixed Na^+/K^+ conductance when their membranes are hyperpolarized. This membrane permeability has been referred to as the *funny current* (I_f). At rest, this current is active, leading to a gradual depolarization of the membrane to a level that triggers the initiation of an action potential. During the action potential, this current shuts off. However, once the action potential is complete and the membrane returns to its resting potential, the funny current turns back on, repeating the cycle. The action potential propagates from the SA node to the rest of the cardiac tissue in a highly organized manner, stimulating muscle contraction and the movement of blood.

The channels responsible for the funny current have been identified. The *hyperpolarization-activated cyclic nucleotide-gated (HCN) channels* comprise four isoforms. Functional channels are composed of tetramers of these isoforms, either all of the same type (homotetramers) or of a combination of types (heterotetramers). HCN4 is the most prominently expressed in the SA node. These channels are unusual in that they are opened by hyperpolarization, and close when the membrane is depolarized. Moreover, these channels are responsive to changing levels of intracellular cyclic AMP (cAMP), as expected: while the heart can beat spontaneously, neuronal and hormonal input can affect this rate to accommodate the body's changing energetic needs.

Mutations in the *HCN4* gene have been identified in patients with *sick sinus syndrome*, a condition where patients exhibit an abnormally low heart rate (*bradycardia*) that may manifest with symptoms of dizziness, fainting, headaches, and fatigue. In one family, the identified mutation resulted in a greater hyperpolarization required to achieve HCN4 opening (Figure 13.35) when these channels were expressed in cultured cells. In a patient carrying this mutation, fewer HCN4 channels are open at a given resting potential, leading to a slower depolarization and a reduced heart rate.

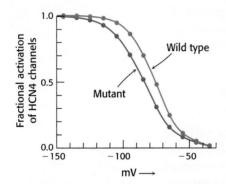

FIGURE 13.35 Opening properties of an HCN4 mutant. This graph shows the fraction of channels open at a given membrane potential. Since these data are from an HCN4 channel, the channel is open (high fractional activation) at hyperpolarized membrane potentials, and closes as the membrane depolarizes. *Notice* that for the mutant HCN4 channels (blue circles) the channel requires a more hyperpolarized state to be open to the same extent as the wild-type channels (red circles). [R. Milanese, M. Baruscotti, T. Gnecchi-Ruscone, and D. DiFrancesco, *New Eng. J. Med.* 354:151–157, 2006, Fig. 3A.]

APPENDIX

Problem-Solving Strategies

PROBLEM: Some animal cells take up glucose by a symporter powered by the simultaneous entry of Na^+. The entry of Na^+ provides a free-energy input of 10.8 kJ mol^{-1} ($2.6 \text{ kcal mol}^{-1}$) under typical cellular conditions (external $[Na^+] = 143$ mM, internal $[Na^+] = 14$ mM, and membrane potential $= -50$ mV). How large a concentration gradient of glucose at 37°C can be generated by this free-energy input?

SOLUTION: Since glucose is uncharged, we can use equation 1 from Section 13.1:

$$\Delta G = RT \ln (c_2/c_1)$$

The problem gives us the free-energy input from the Na^+ gradient (although you should be able to work this out as well, using equation 2), so we need only rearrange this equation to determine the concentration gradient. Recall that for this equation, temperature must be expressed in kelvins:

$$\ln (c_2/c_1) = \frac{\Delta G}{RT}$$

$$\ln (c_2/c_1) = \frac{(10.8 \text{ kJ mol}^{-1})}{(8.315 \times 10^{-3} \text{ kJ mol}^{-1} \text{ deg}^{-1})(310 \text{ K})}$$

$$\ln (c_2/c_1) = 4.19$$

$$c_2/c_1 = e^{(4.19)} = 66$$

Hence, the free energy generated by the Na^+ gradient could power the formation of a 66-fold glucose gradient across the membrane.

KEY TERMS

pump (p. 403)

carrier (p. 403)

channel (p. 403)

active transport (p. 403)

facilitated diffusion (passive transport) (p. 403)

ATP-driven pump (p. 403)

primary active transport (p. 404)

secondary active transport (p. 404)

simple diffusion (p. 405)

electrochemical potential (membrane potential) (p. 406)

Na^+–K^+ pump (Na^+–K^+ ATPase) (p. 406)

sarcoplasmic reticulum Ca^{2+} ATPase (SERCA) (p. 407)

gastric H^+–K^+ ATPase (p. 407)

P-type ATPase (p. 407)

eversion (p. 409)

cardiotonic steroid (p. 410)

digitalis (p. 410)

multidrug resistance (p. 411)

multidrug-resistance (MDR) protein (P-glycoprotein) (p. 411)

ATP-binding cassette (ABC) domain (p. 411)

ABC transporter (p. 411)

secondary transporter (cotransporter) (p. 413)

antiporter (p. 413)

symporter (p. 413)

uniporter (p. 413)

lactose permease (p. 413)

ion channel (p. 404)

nerve impulse (p. 415)

action potential (p. 415)

patch-clamp technique (p. 416)

gigaseal (p. 416)

selectivity filter (p. 418)

voltage-gated channel (p. 421)

ball-and-chain model (p. 422)

neurotransmitter (p. 423)

acetylcholine (p. 423)

synaptic cleft (p. 423)

acetylcholine receptor (p. 423)

ligand-gated channel (p. 423)

equilibrium potential (p. 425)

Nernst equation (p. 425)

long QT syndrome (LQTS) (p. 427)

gap junction (cell-to-cell channels) (p. 427)

connexin (p. 428)

connexon (hemichannel) (p. 428)

aquaporin (p. 429)

PROBLEMS

1. *A helping hand.* Differentiate between simple diffusion and facilitated diffusion.

2. *Powering movement.* What are two forms of energy that can power active transport? ✅❶

3. *Carriers.* Name the three types of carrier proteins. Which of these can mediate secondary active transport? ✅❸

4. *The price of extrusion.* What is the free-energy cost at 25°C of pumping Ca^{2+} out of a cell when the cytoplasmic concentration is 0.4 μM, the extracellular concentration is 1.5 mM, and the membrane potential is −60 mV? ✅❷

5. *Equilibrium potentials.* For a typical mammalian cell, the intracellular and extracellular concentrations of the chloride ion (Cl^-) are 4 μM and 150 mM, respectively. For the calcium ion (Ca^{2+}), the intracellular and extracellular concentrations are 0.2 μM and 1.8 mM, respectively. Calculate the equilibrium potentials at 37°C for these two ions. ✅❷

6. *Variations on a theme.* Write a detailed mechanism for transport by the Na^+–K^+ ATPase based on analogy with the mechanism of the Ca^{2+} ATPase shown in Figure 13.5. ✅❹

7. *Pumping protons.* Design an experiment to show that the action of lactose permease can be reversed in vitro to pump protons.

8. *Opening channels.* Differentiate between ligand-gated and voltage-gated channels. ✅❻

9. *Different directions.* The K^+ channel and the Na^+ channel have similar structures and are arranged in the same orientation in the cell membrane. Yet the Na^+ channel allows sodium ions to flow into the cell and the K^+ channel allows potassium ions to flow out of the cell. Explain. ✅❺

10. *Differing mechanisms.* Distinguish the mechanisms by which uniporters and channels transport ions or molecules across the membrane. ✅❶, ✅❸

11. *Short circuit.* Carbonyl cyanide 4-(trifluoromethoxy) phenylhydrazone (FCCP) is a proton ionophore: it enables protons to pass freely through membranes. Treatment of *E. coli* with FCCP prevents the accumulation of lactose in these cells. Explain.

12. *Working together.* The human genome contains more than 20 connexin-encoding genes. Several of these genes are expressed in high levels in the heart. Why are connexins so highly expressed in cardiac tissue?

13. *Structure–activity relations.* On the basis of the structure of tetrodotoxin, propose a mechanism by which the toxin inhibits Na^+ flow through the Na^+ channel. ✅❺

14. *Hot stuff.* When SERCA is incubated with [γ-^{32}P] ATP (a form of ATP in which the terminal phosphate is labeled with radioactive ^{32}P) and calcium at 0°C for 20 seconds and analyzed by gel electrophoresis, a radioactive band is observed at the molecular weight corresponding to full-length SERCA. Why is a labeled band observed? Would you expect a similar band if you were performing a similar assay, with a suitable substrate, for the MDR protein? ✅❹

15. *A dangerous snail.* Cone snails are carnivores that inject a powerful set of toxins into their prey, leading to rapid paralysis. Many of these toxins are found to bind to specific ion-channel proteins. Why are such molecules so toxic? How might such toxins be useful for biochemical studies?

16. *Pause for effect.* Immediately after the repolarization phase of an action potential, the neuronal membrane is temporarily unable to respond to the stimulation of a second action potential, a phenomenon referred to as the *refractory period*. What is the mechanistic basis for the refractory period? ✅❼

17. *Only a few.* Why do only a small number of sodium ions need to flow through the Na^+ channel to change the membrane potential significantly?

18. *More than one mechanism.* How might a mutation in a cardiac voltage-dependent sodium channel cause long QT syndrome?

19. *Mechanosensitive channels.* Many species contain ion channels that respond to mechanical stimuli. On the basis of the properties of other ion channels, would you expect the flow of ions through a single open mechanosensitive channel to increase in response to an appropriate stimulus? Why or why not?

20. *Concerted opening.* Suppose that a channel obeys the concerted allosteric model (MWC model, Section 7.2). The binding of ligand to the R state (the open form) is 20 times as tight as that to the T state (the closed form). In the absence of ligand, the ratio of closed-to-open channels is 10^5. If the channel is a tetramer, what is the fraction of open channels when 1, 2, 3, and 4 ligands are bound?

21. *Respiratory paralysis.* The neurotransmitter acetylcholine is degraded by a specific enzyme that is inactivated by tabun, sarin, and parathion. On the basis of the structures below, propose a possible source for their lethal actions.

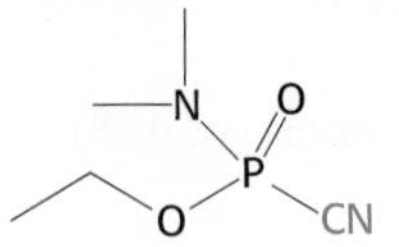

Tabun

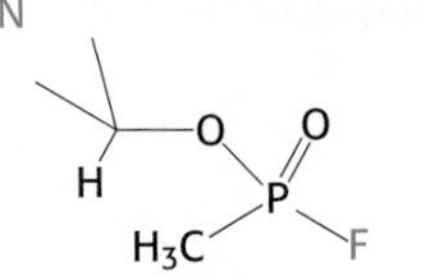

Sarin

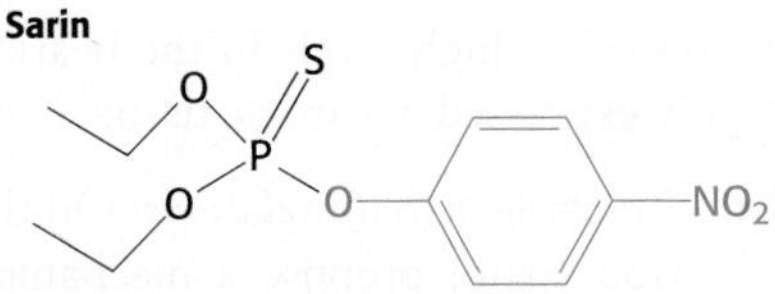

Parathion

22. *Ligand-induced channel opening.* The ratio of open-to-closed forms of the acetylcholine receptor channel containing zero, one, and two bound acetylcholine molecules is 5×10^{-6}, 1.2×10^{-3}, and 14, respectively.

(a) By what factor is the open-to-closed ratio increased by the binding of the first acetylcholine molecule? The second acetylcholine molecule?

(b) What are the corresponding free-energy contributions to channel opening at 25°C? ✓②

(c) Can the allosteric transition be accounted for by the MWC concerted model (Section 7.2)?

23. *Frog poison.* Batrachotoxin (BTX) is a steroidal alkaloid from the skin of *Phyllobates terribilis*, a poisonous Colombian frog (the source of the poison used on blowgun darts). In the presence of BTX, Na^+ channels in an excised patch stay persistently open when the membrane is depolarized. They close when the membrane is repolarized. Which transition is blocked by BTX?

24. *Valium target.* γ-Aminobutyric acid (GABA) opens channels that are specific for chloride ions. The $GABA_A$ receptor channel is pharmacologically important because it is the target of diazepam (Valium), which is used to diminish anxiety.

(a) The extracellular concentration of Cl^- is 123 mM and the intracellular concentration is 4 mM. In which direction does Cl^- flow through an open channel when the membrane potential is in the -60 mV to $+30$ mV range?

(b) What is the effect of Cl^--channel opening on the excitability of a neuron?

(c) The hydropathy profile of the $GABA_A$ receptor resembles that of the acetylcholine receptor. Predict the number of subunits in this Cl^- channel.

25. *Understanding SERCA.* To study the mechanism of SERCA, you prepare membrane vesicles containing this protein oriented such that its ATP binding site is on the outer surface of the vesicle. To measure pump activity, you use an assay that detects the formation of inorganic phosphate in the medium. When you add calcium and ATP to the medium, you observe phosphate production for only a short period of time. Only after the addition of calcimycin, a molecule that makes membranes selectively permeable to calcium, do you observe sustained phosphate production. Explain. ✓④

Chapter Integration Problem

26. *Speed and efficiency matter.* Acetylcholine is rapidly destroyed by the enzyme acetylcholinesterase. This enzyme, which has a turnover number of 25,000 per second, has attained catalytic perfection with a k_{cat}/K_M of $2 \times 10^8 \ M^{-1}s^{-1}$. Why is the efficiency of this enzyme physiologically crucial?

Mechanism Problem

27. *Remembrance of mechanisms past.* Acetylcholinesterase converts acetylcholine into acetate and choline. Like serine proteases, acetylcholinesterase is inhibited by DIPF. Propose a catalytic mechanism for acetylcholine digestion by acetylcholinesterase. Show the reaction as chemical structures.

Data Interpretation Problems

28. *Tarantula toxin.* Acid sensing is associated with pain, tasting, and other biological activities (See Chapter 34 on 🅂 Sapling Plus). Acid sensing is carried out by a ligand-gated channel that permits Na^+ influx in response to H^+. This family of acid-sensitive ion channels (ASICs) includes a number of members. Psalmotoxin 1 (PcTX1), a venom from the tarantula, inhibits some members of this family. The following electrophysiological recordings of cells containing several members of the ASIC family were made in the presence of the toxin at a concentration of 10 nM. The channels were opened by changing the pH from 7.4 to the indicated values. The PcTX1 was present for a short time (indicated by the black bar above the recordings below), after which time it was rapidly washed from the system.

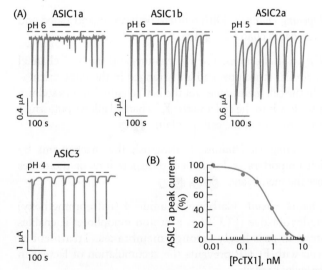

(A) Electrophysiological recordings of cells exposed to tarantula toxin. (B) Plot of peak current of a cell containing the ASIC1a protein versus the toxin concentration. [Data from P. Escoubas et al., *J. Biol. Chem.* 275:25116–25121, 2000.]

(a) Which member of the ASIC family—ASIC1a, ASIC1b, ASIC2a, or ASIC3—is most sensitive to the toxin?

(b) Is the effect of the toxin reversible? Explain.

(c) What concentration of PcTX1 yields 50% inhibition of the sensitive channel?

29. *Channel problems 1.* A number of pathological conditions result from mutations in the acetylcholine receptor channel. One such mutation in the β subunit, βV266M, causes muscle weakness and rapid fatigue. An investigation of the acetylcholine-generated currents through the acetylcholine receptor channel for both a control and a patient yielded the following results.

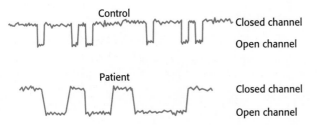

What is the effect of the mutation on channel function? Suggest some possible biochemical explanations for the effect.

30. *Channel problems 2.* The acetylcholine receptor channel can also undergo mutation leading to fast-channel syndrome (FCS), with clinical manifestations similar to those of slow-channel syndrome (problem 29). What would the recordings of ion movement look like in this syndrome? Suggest a biochemical explanation.

31. *Transport differences.* The rate of transport of two molecules, indole and glucose, across a cell membrane is shown below. What are the differences between the transport mechanisms of the two molecules? Suppose that ouabain inhibited the transport of glucose. What would this inhibition suggest about the mechanism of transport?

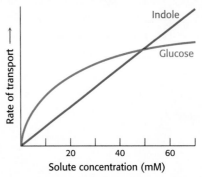

Think ▶ Pair ▶ Share Problem

32. Pumps, carriers (secondary transporters), and channels all function to translocate molecules across the plasma membrane barrier. However, they have very different properties. Compare these three protein classes according to the following characteristics:

- Transport type: primary vs. secondary
- Energy source
- Types of molecules transported
- Kinetics of translocation: fast vs. slow
- Presence of gating or inactivation mechanisms?

Signal-Transduction Pathways

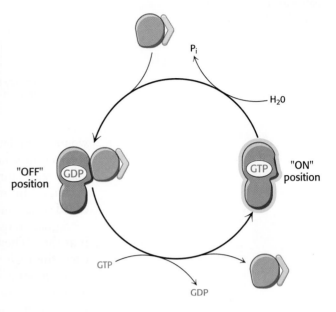

Signal-transduction circuits in biological systems have molecular on–off switches that, like those in a computer chip (above), transmit information when "on." Common among these circuits are those including G proteins (right), which transmit a signal when bound to GTP and are silent when bound to GDP. [(Left) TEK IMAGE/Getty Images.]

✔ LEARNING OBJECTIVES

By the end of this chapter, you should be able to:

1. Describe the essential features of a signal-transduction circuit.

2. Define what is meant by a second messenger.

3. Compare and contrast how the β-adrenergic receptor, the insulin receptor, and the epidermal growth factor receptors transmit a ligand binding event across the plasma membrane to initiate an intracellular signaling cascade.

4. Describe some of the protein domains that are commonly found within signaling proteins.

5. Give examples of how mutations in signaling proteins can result in cancer.

OUTLINE

14.1 Epinephrine and Angiotensin II Signaling: Heterotrimeric G Proteins Transmit Signals and Reset Themselves

14.2 Insulin Signaling: Phosphorylation Cascades Are Central to Many Signal-Transduction Processes

14.3 EGF Signaling: Signal-Transduction Systems Are Poised to Respond

14.4 Many Elements Recur with Variation in Different Signal-Transduction Pathways

14.5 Defects in Signal-Transduction Pathways Can Lead to Cancer and Other Diseases

A cell is highly responsive to specific chemicals in its environment: it may adjust its metabolism or alter gene-expression patterns on sensing the presence of these molecules. In multicellular organisms, these chemical signals are crucial to coordinating physiological responses. Three examples of molecular signals that stimulate a physiological response are epinephrine (sometimes called adrenaline), insulin, and epidermal growth factor (EGF; Figure 14.1). When a mammal is threatened, its adrenal glands release the hormone epinephrine, which stimulates the mobilization of energy stores and leads to improved cardiac function. After a full meal, the β cells in the pancreas release insulin, which stimulates a host of physiological responses, including the uptake of glucose from the bloodstream and its storage as glycogen. The release of EGF in response to a

FIGURE 14.1 Three signal-transduction pathways. The binding of signaling molecules to their receptors initiates pathways that lead to important physiological responses.

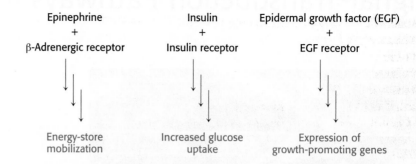

wound stimulates specific cells to grow and divide. In all these cases, the cell receives information that a certain molecule within its environment is present above some threshold concentration. The chain of events that converts the message "this molecule is present" into the ultimate physiological response is called *signal transduction*.

Signal-transduction pathways often comprise many components and branches. They can thus be immensely complicated and confusing. However, the logic of signal transduction can be simplified by examining the common strategies and classes of molecules that recur in these pathways. These principles are introduced here because signal-transduction pathways affect essentially all of the metabolic pathways we will be exploring throughout the rest of the book.

Signal transduction depends on molecular circuits

Signal-transduction pathways follow a broadly similar course that can be viewed as a molecular circuit (Figure 14.2). All such circuits contain certain key steps:

1. *Release of the primary messenger ("signal")*. A stimulus such as a wound or digested meal triggers the release of the signal molecule, also called the *primary messenger*.

2. *Reception of the primary messenger*. Most signal molecules do not enter cells. Instead, proteins in the cell membrane act as *receptors* that convert signal molecule binding on the cell surface to a structural change within the cell's interior. Receptors span the cell membrane and thus have both extracellular and intracellular components. A binding site on the extracellular side specifically recognizes the signal molecule (often referred to as the *ligand*). Such binding sites are analogous to enzyme active sites except that no catalysis takes place within them. The interaction of the ligand and the receptor alters the tertiary or quaternary structure of the receptor so as to induce a structural change on the intracellular side.

3. *Delivery of the message inside the cell by the second messenger*. Other small molecules, called *second messengers*, are used to relay information from receptor–ligand complexes. Second messengers are intracellular molecules that change in concentration in response to environmental signals and mediate the next step in the molecular information circuit. Some particularly important second messengers are cyclic AMP (cAMP) and cyclic GMP (cGMP), calcium ion, inositol 1,4,5-trisphosphate (IP_3), and diacylglycerol (DAG; Figure 14.3).

The use of second messengers has several consequences. First, the signal may be amplified significantly: Only a small number of receptor molecules may be activated by the direct binding of signal molecules, but each activated receptor molecule can lead to the generation of many second messengers. Thus, *a low concentration of signal in the environment, even as little as a single molecule, can yield a large intracellular signal and response*.

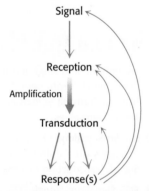

FIGURE 14.2 Principles of signal transduction. An environmental signal is first received by interaction with a cellular component, most often a cell-surface receptor. The information that the signal has arrived is then converted into other chemical forms, or *transduced*. Typically, the transduction process comprises many steps. The signal is often amplified before evoking a response. Feedback pathways regulate the entire signaling process.

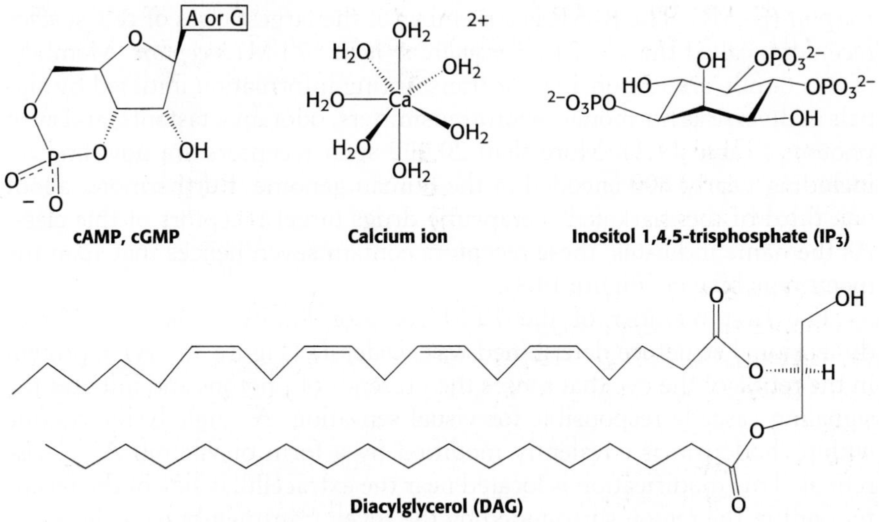

FIGURE 14.3 Common second messengers. Second messengers are intracellular molecules that change in concentration in response to environmental signals. That change in concentration conveys information inside the cell.

cAMP, cGMP   **Calcium ion**   **Inositol 1,4,5-trisphosphate (IP$_3$)**

Diacylglycerol (DAG)

Second, these messengers are often free to diffuse to other cellular compartments where they can influence processes throughout the cell. Third, the use of common second messengers in multiple signaling pathways creates both opportunities and potential problems. Input from several signaling pathways, often called *cross talk,* may alter the concentration of a common second messenger. Cross talk permits more finely tuned regulation of cell activity than would the action of individual independent pathways. However, inappropriate cross talk can result in the misinterpretation of changes in second-messenger concentration.

4. *Activation of effectors that directly alter the physiological response.* The ultimate effect of the signal pathway is to activate (or inhibit) the pumps, channels, enzymes, and transcription factors that directly control metabolic pathways, gene expression, and the permeability of membranes to specific ions.

5. *Termination of the signal.* After a cell has completed its response to a signal, the signaling process must be terminated or the cell loses its responsiveness to new signals. Moreover, signaling processes that fail to terminate properly can have highly undesirable consequences. As we will see, many cancers are associated with signal-transduction processes that are not properly terminated, especially processes that control cell growth.

In this chapter, we will examine components of the three signal-transduction pathways shown in Figure 14.1. In doing so, we will see several classes of adaptor domains present in signal-transduction proteins. These domains usually recognize specific classes of molecules and help transfer information from one protein to another. The components described in the context of these three pathways recur in many other signal-transduction pathways; bear in mind that the specific examples are representative of many such pathways.

14.1 Epinephrine and Angiotensin II Signaling: Heterotrimeric G Proteins Transmit Signals and Reset Themselves

Epinephrine is a hormone secreted by the adrenal glands of mammals in response to internal and external stressors. It exerts a wide range of effects—referred to as the *fight-or-flight response*—to help organisms anticipate the need for rapid muscular activity, including acceleration of heart rate, dilation of the smooth muscle of the airways, and initiation of the breakdown of glycogen (Section 21.3) and fatty acids (Section 22.2). Epinephrine signaling begins with ligand binding to a protein called the *β-adrenergic*

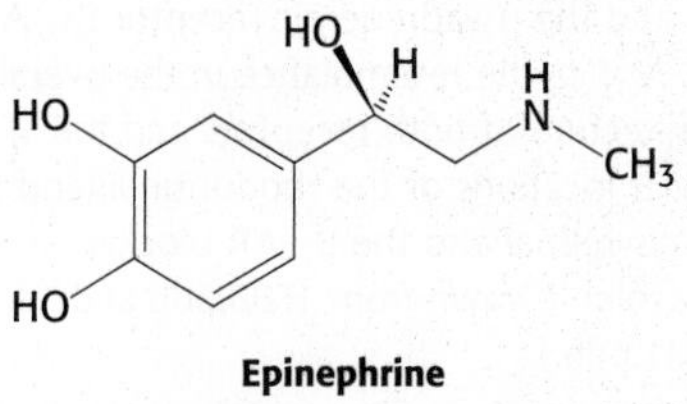

Epinephrine

TABLE 14.1 Biological functions mediated by 7TM receptors

Hormone action
Hormone secretion
Neurotransmission
Chemotaxis
Exocytosis
Control of blood pressure
Embryogenesis
Cell growth and differentiation
Development
Smell
Taste
Vision
Viral infection

Information from J. S. Gutkind, *J. Biol. Chem.* 273:1839–1842, 1998.

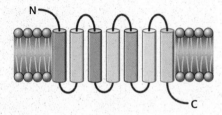

FIGURE 14.4 The 7TM receptor. Schematic representation of a 7TM receptor showing its passage through the membrane seven times.

receptor (β-AR). The β-AR is a member of the largest class of cell-surface receptors, called the *seven-transmembrane-helix* (7TM) *receptors*. Members of this family are responsible for transmitting information initiated by signals as diverse as hormones, neurotransmitters, odorants, tastants, and even photons (Table 14.1). More than 20,000 such receptors are now known, including nearly 800 encoded in the human genome. Furthermore, about one-third of the marketed therapeutic drugs target receptors of this class. As the name indicates, these receptors contain seven helices that span the membrane bilayer (Figure 14.4).

The first member of the 7TM receptor family to have its three-dimensional structure determined was *rhodopsin* (Figure 14.5A), a protein in the retina of the eye that senses the presence of photons and initiates the signaling cascade responsible for visual sensation. A single lysine residue within rhodopsin is covalently modified by a form of vitamin A, 11-*cis*-retinal. This modification is located near the extracellular side of the receptor, within the region surrounded by the seven transmembrane helices. As Section 34.3 discusses in greater detail (See Section 34.3 on ⌘ SaplingPlus), exposure to light induces the isomerization of 11-*cis*-retinal to its all-*trans* form, producing a structural change in the receptor that results in the initiation of an action potential, which is ultimately interpreted by the brain as visual stimulus.

In 2007, the first three-dimensional structure of the β₂ subtype of the human adrenergic receptor (β₂-AR) bound to an inhibitor was solved by x-ray crystallography. This inhibitor, carazolol, competes with epinephrine for binding to the β₂-AR, much in the same way that competitive inhibitors act at enzyme active sites (Section 8.5). The structure of the β₂-AR revealed considerable similarities with that of rhodopsin, particularly with respect to the locations of 11-*cis*-retinal in rhodopsin and the binding site for carazolol (Figure 14.5B).

Ligand binding to 7TM receptors leads to the activation of heterotrimeric G proteins

What is the next step in the pathway? The conformational change in the receptor's cytoplasmic domain activates a protein called a *G protein*, named

FIGURE 14.5 Structures of rhodopsin and the β₂-adrenergic receptor. Three-dimensional structure of rhodopsin (A) and the β₂-adrenergic receptor (β₂-AR). (B). *Notice* the resemblance in the overall architecture of both receptors and the similar locations of the rhodopsin ligand 11-*cis*-retinal and the β₂-AR blocker carazolol. [Drawn from 1F88.pdb and 2RH1.pdb.]

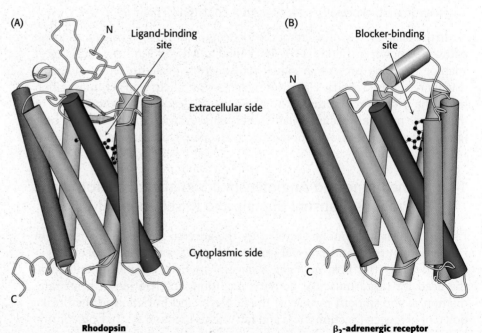

Rhodopsin   **β₂-adrenergic receptor**

for the fact that it binds guanyl nucleotides. The activated G protein stimulates the activity of adenylate cyclase, an enzyme that catalyzes the conversion of ATP into cAMP. The G protein and adenylate cyclase remain attached to the membrane, whereas cAMP, a second messenger, can travel throughout the cell carrying the signal originally brought by the binding of epinephrine. Figure 14.6 provides a broad overview of these steps.

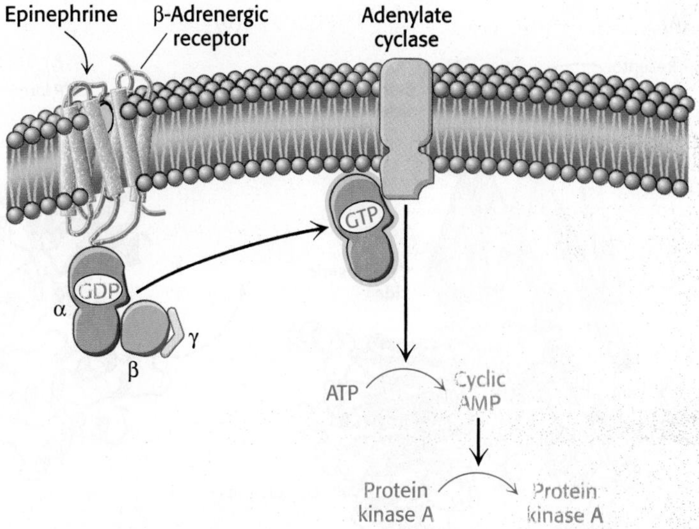

Let us consider the role of the G protein in this signaling pathway in more detail. In its inactive state, the G protein is bound to GDP. In this form, the G protein exists as a heterotrimer consisting of α, β, and γ subunits; the α subunit (referred to as G_α) binds the nucleotide (Figure 14.7). The α subunit is a member of the P-loop NTPase family (Section 9.4), and the P-loop participates in nucleotide binding. The α and γ subunits are usually anchored to the membrane by covalently attached fatty acids. *The role of the hormone-bound receptor is to catalyze the exchange of GTP for bound GDP.* The interaction between the hormone–β_2-AR complex and the heterotrimeric G protein was illustrated in molecular detail when the crystal structure of this complex was determined by Brian Kobilka and coworkers in 2011. In this structure, a synthetic *agonist*, or small molecule that activates a receptor, was used to induce the active conformation of the β_2-AR. Agonist binding results in the movement of two transmembrane helices, yielding an extensive interaction surface for the G_α subunit of the heterotrimer (Figure 14.8A). When bound to the receptor, the nucleotide-binding site of G_α opens substantially, enabling the displacement of GDP by GTP (Figure 14.8B).

FIGURE 14.6 Activation of protein kinase A by a G-protein pathway. Hormone binding to a 7TM receptor initiates a signal-transduction pathway that acts through a G protein and cAMP to activate protein kinase A.

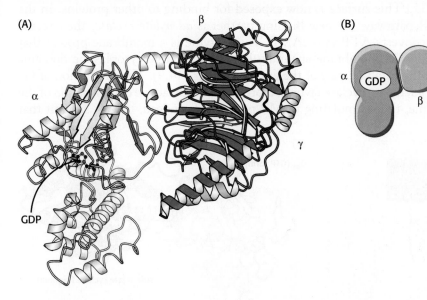

FIGURE 14.7 A heterotrimeric G protein. (A) A ribbon diagram shows the relation between the three subunits. In this complex, the α subunit (gray and purple) is bound to GDP. *Notice* that GDP is bound in a pocket close to the surface at which the α subunit interacts with the $\beta\gamma$ dimer. (B) A schematic representation of the heterotrimeric G protein. [Drawn from 1GOT.pdb.]

On GTP binding, the α subunit simultaneously dissociates from the $\beta\gamma$ dimer ($G_{\beta\gamma}$), transmitting the signal that the receptor has bound its ligand. *A single hormone–receptor complex can stimulate nucleotide exchange in many G-protein heterotrimers.* Thus, hundreds of G_α molecules are converted from their GDP form into their GTP form for each bound molecule of hormone, giving an amplified response. Because they signal through G proteins, 7TM receptors are often called *G-protein-coupled receptors* (GPCRs).

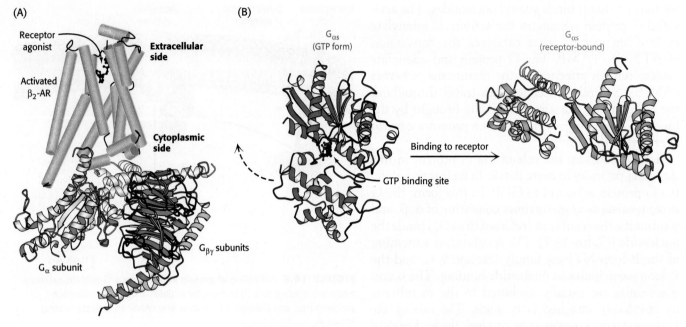

FIGURE 14.8 The complex between the activated β₂-AR and a heterotrimeric G protein. (A) When the β₂-AR (green) binds a receptor agonist, the cytoplasmic face of the receptor forms an interaction surface with the Gₐ subunit of a heterotrimeric G protein. (B) The interaction with the activated receptor leads to a substantial conformational change in the Gₐ protein, in which the GTP binding site is opened, enabling nucleotide exchange. In this figure, Gₐ in its GTP form is shown in purple and in its receptor-bound form is shown in blue. [Drawn from 3SN6.pdb and 1AZT.pdb.]

Activated G proteins transmit signals by binding to other proteins

In the GTP form, the surface of G_α that had been bound to $G_{\beta\gamma}$ has changed its conformation from the GDP form so that it no longer has a high affinity for $G_{\beta\gamma}$. This surface is now exposed for binding to other proteins. In the β-AR pathway, the new binding partner is *adenylate cyclase,* the enzyme that converts ATP into cAMP. This enzyme is a membrane protein that contains 12 membrane-spanning helices; two large cytoplasmic domains form the catalytic part of the enzyme (Figure 14.9). The interaction of G_α with adenylate cyclase favors a more catalytically active conformation of the enzyme, thus stimulating cAMP production. Indeed, the G_α subunit that

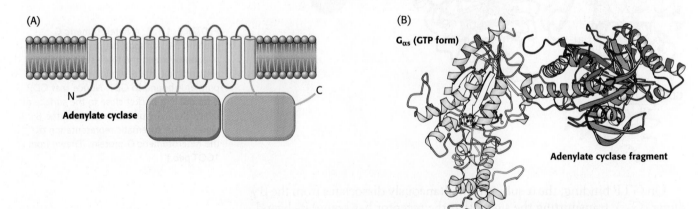

FIGURE 14.9 Adenylate cyclase activation. (A) Adenylate cyclase is a membrane protein with two large intracellular domains that contain the catalytic apparatus. (B) The structure of a complex between Gₐ in its GTP form bound to a catalytic fragment from adenylate cyclase. *Notice* that the surface of Gₐ that had been bound to the βγ dimer (Figure 14.7) now binds adenylate cyclase. [Drawn from 1AZS.pdb.]

participates in the β-AR pathway is called $G_{\alpha s}$ ("s" stands for stimulatory). *The net result is that the binding of epinephrine to the receptor on the cell surface increases the rate of cAMP production inside the cell.* The generation of cAMP by adenylate cyclase provides a second level of amplification because each activated adenylate cyclase can convert many molecules of ATP into cAMP.

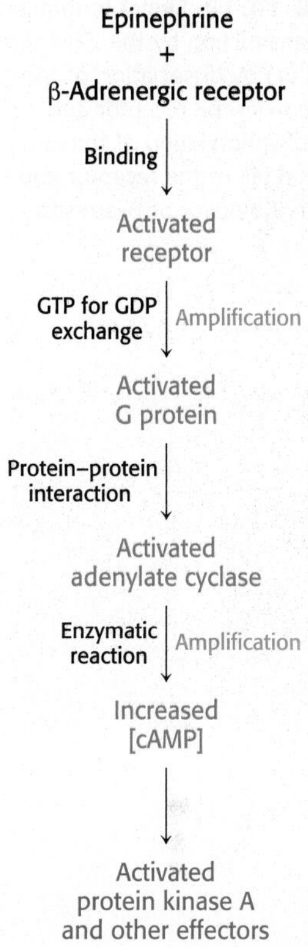

Cyclic AMP stimulates the phosphorylation of many target proteins by activating protein kinase A

The increased concentration of cAMP can affect a wide range of cellular processes. In the muscle, cAMP stimulates the production of ATP for muscle contraction. In other cell types, cAMP enhances the degradation of storage fuels, increases the secretion of acid by the gastric mucosa, leads to the dispersion of melanin pigment granules, diminishes the aggregation of blood platelets, and induces the opening of chloride channels. How does cAMP influence so many cellular processes? *Most effects of cAMP in eukaryotic cells are mediated by the activation of a single protein kinase.* This key enzyme is *protein kinase A* (PKA).

As described earlier, PKA consists of two regulatory (R) chains and two catalytic (C) chains (R_2C_2; Figure 10.16). In the absence of cAMP, the R_2C_2 complex is catalytically inactive. The binding of cAMP to the regulatory chains releases the catalytic chains, which are catalytically active on their own. Activated PKA then phosphorylates specific serine and threonine residues in many targets to alter their activity. For instance, PKA phosphorylates two enzymes that lead to the breakdown of glycogen, the polymeric store of glucose, and the inhibition of further glycogen synthesis (Section 21.3). In addition, *PKA stimulates the expression of specific genes* by phosphorylating a transcriptional activator called the cAMP response element binding (CREB) protein. This activity of PKA illustrates that signal-transduction pathways can extend into the nucleus to alter gene expression.

The signal-transduction pathway initiated by epinephrine is summarized in Figure 14.10.

G proteins spontaneously reset themselves through GTP hydrolysis

How is the signal initiated by epinephrine switched off? G_α *subunits have intrinsic GTPase activity,* which is used to hydrolyze bound GTP to GDP and P_i. This hydrolysis reaction is slow, however, requiring from seconds to minutes. Thus, the GTP form of G_α is able to activate downstream components of the signal-transduction pathway before it is deactivated by GTP hydrolysis. In essence, *the bound GTP acts as a built-in clock that spontaneously resets the G_α subunit after a short time period.* After GTP hydrolysis and the release of P_i, the GDP-bound form of G_α then reassociates with $G_{\beta\gamma}$ to re-form the inactive heterotrimeric protein (Figure 14.11).

FIGURE 14.10 Epinephrine signaling pathway. The binding of epinephrine to the β-adrenergic receptor initiates the signal-transduction pathway. The process in each step is indicated (in black) at the left of each arrow. Steps that have the potential for signal amplification are indicated at the right in green.

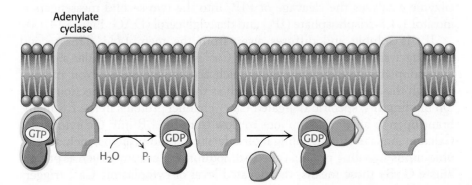

FIGURE 14.11 Resetting G_α. On hydrolysis of the bound GTP by the intrinsic GTPase activity of G_α, G_α reassociates with the βγ dimer to form the heterotrimeric G protein, thereby terminating the activation of adenylate cyclase.

FIGURE 14.12 Signal termination. Signal transduction by the 7TM receptor is halted (1) by dissociation of the signal molecule from the receptor and (2) by phosphorylation of the cytoplasmic C-terminal tail of the receptor and the subsequent binding of β-arrestin.

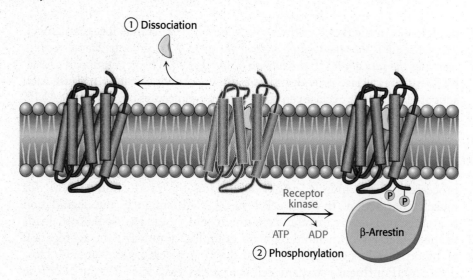

The hormone-bound activated receptor must be reset as well to prevent the continuous activation of G proteins. This resetting is accomplished by two processes (Figure 14.12). First, the hormone dissociates, returning the receptor to its initial inactive state. The likelihood that the receptor remains in its unbound state depends on the extracellular concentration of hormone. Second, the signaling cascade initiated by the hormone–receptor complex activates a kinase that phosphorylates serine and threonine residues in the carboxyl-terminal tail of the receptor. These phosphorylation events result in the deactivation of the receptor. In the example under consideration, *β-adrenergic-receptor kinase* (also called *G-protein receptor kinase 2, or GRK2*) phosphorylates the carboxyl-terminal tail of the hormone–receptor complex but not the unoccupied receptor. Finally, the molecule *β-arrestin* binds to the phosphorylated receptor and further diminishes its ability to activate G proteins.

Some 7TM receptors activate the phosphoinositide cascade

We now turn to another common second-messenger cascade, also employing a 7TM receptor, that is used by many hormones to evoke a variety of responses. The *phosphoinositide cascade*, like the cAMP cascade, converts extracellular signals into intracellular ones. The intracellular second messengers formed by activation of this pathway arise from the cleavage of *phosphatidylinositol 4,5-bisphosphate* (PIP$_2$), a phospholipid present in cell membranes. An example of a signaling pathway based on the phosphoinositide cascade is the one triggered by the receptor for angiotensin II, a peptide hormone that controls blood pressure.

Each type of 7TM receptor signals through a distinct G protein. Whereas the β-adrenergic receptor activates the G protein G$_{\alpha s}$, the angiotensin II receptor activates a G protein called G$_{\alpha q}$. In its GTP form, G$_{\alpha q}$ binds to and activates the β isoform of the enzyme *phospholipase C*. This enzyme catalyzes the cleavage of PIP$_2$ into the two second messengers—inositol 1,4,5-trisphosphate (IP$_3$) and diacylglycerol (DAG; Figure 14.13).

IP$_3$ is soluble and diffuses away from the membrane. This second messenger causes the rapid release of Ca^{2+} from the intracellular stores in the endoplasmic reticulum (ER), which accumulates a reservoir of Ca^{2+} through the action of transporters such as the Ca^{2+} ATPase (Section 13.2). On binding IP$_3$, specific IP$_3$-gated Ca^{2+}–channel proteins in the ER membrane open to allow calcium ions to flow from the ER into the cytoplasm. Calcium ion is itself a signaling molecule: it can bind proteins, including an ubiquitous signaling protein called calmodulin and enzymes such as protein kinase C. By these means, the elevated level of cytoplasmic Ca^{2+} triggers

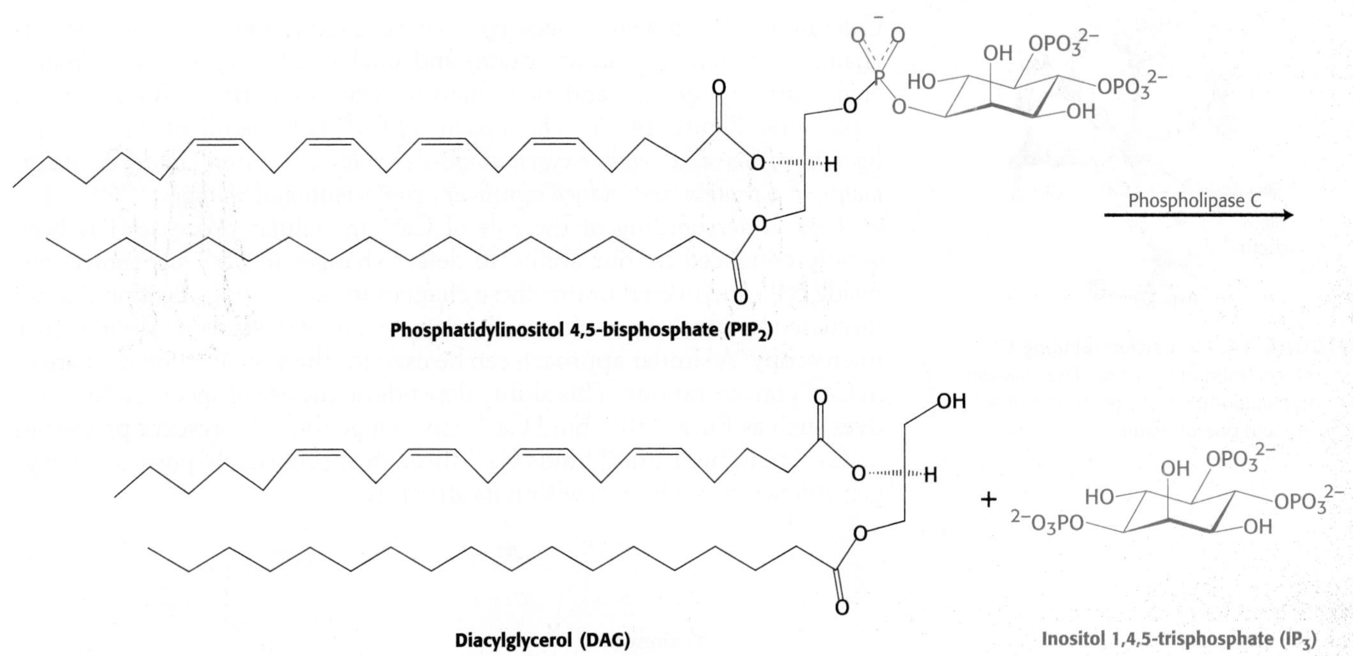

Phosphatidylinositol 4,5-bisphosphate (PIP$_2$)

Phospholipase C →

Diacylglycerol (DAG)

+

Inositol 1,4,5-trisphosphate (IP$_3$)

FIGURE 14.13 Phospholipase C reaction. Phospholipase C cleaves the membrane lipid phosphatidylinositol 4,5-bisphosphate (PIP$_2$) into two second messengers: diacylglycerol (DAG), which remains in the membrane, and inositol 1,4,5-trisphosphate (IP$_3$), which diffuses away from the membrane.

processes such as smooth-muscle contraction, glycogen breakdown, and vesicle release.

DAG remains in the plasma membrane. There, it activates *protein kinase C* (PKC), a protein kinase that phosphorylates serine and threonine residues in many target proteins. To bind DAG, the specialized DAG-binding domains of this kinase require bound calcium. Note that DAG and IP$_3$ work in tandem: IP$_3$ increases the Ca^{2+} concentration, and Ca^{2+} facilitates the DAG-mediated activation of protein kinase C. The phosphoinositide cascade is summarized in Figure 14.14. Both IP$_3$ and DAG act transiently because they are converted into other species by phosphorylation or other processes.

Calcium ion is a widely used second messenger

Calcium ion participates in many signaling processes in addition to the phosphoinositide cascade. Several properties of this ion account for its widespread use as an intracellular messenger. First, fleeting changes in Ca^{2+} concentration are readily detected. At steady state, intracellular levels of Ca^{2+} must be kept low to prevent the precipitation of carboxylated and phosphorylated compounds, which form poorly soluble salts with Ca^{2+}. Transport systems extrude Ca^{2+} from the cytoplasm, maintaining the cytoplasmic concentration of Ca^{2+} at approximately 100 nM—several orders of magnitude lower than that of the extracellular medium (Section 13.2). Given this low steady-state level, transient increases in Ca^{2+} concentration produced by signaling events can be readily sensed.

A second property of Ca^{2+} that makes it a highly suitable intracellular messenger is its ability to bind tightly to proteins and induce substantial structural rearrangements.

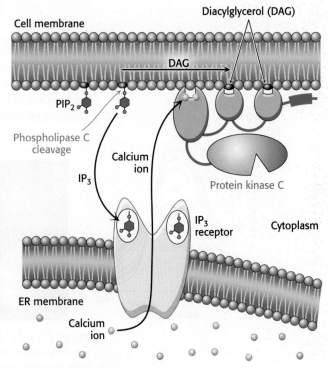

FIGURE 14.14 Phosphoinositide cascade. The cleavage of PIP$_2$ into DAG and IP$_3$ results in the release of calcium ions (owing to the opening of the IP$_3$ receptor ion channels) and the activation of protein kinase C (owing to the binding of protein kinase C to free DAG in the membrane). Calcium ions bind to protein kinase C and help facilitate its activation.

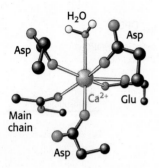

FIGURE 14.15 Calcium-binding site. In one common mode of binding, calcium is coordinated to six oxygen atoms of a protein and one of water.

Calcium ions bind well to negatively charged oxygen atoms (from the side chains of glutamate and aspartate) and uncharged oxygen atoms (main-chain carbonyl groups and side-chain oxygen atoms from glutamine and asparagine; Figure 14.15). *The capacity of Ca^{2+} to be coordinated to multiple ligands—from six to eight oxygen atoms—enables it to cross-link different segments of a protein and induce significant conformational changes.*

Our understanding of the role of Ca^{2+} in cellular processes has been greatly enhanced by our ability to detect changes in Ca^{2+} concentrations inside cells and even monitor these changes in real time. In Section 3.2, we discussed how intact proteins can be detected in live cells using fluorescence microscopy. A similar approach can be used for the visualization of changes in Ca^{2+} concentrations. This ability depends on the use of specially designed dyes such as Fura-2 that bind Ca^{2+} and change their fluorescent properties on Ca^{2+} binding. Fura-2 binds Ca^{2+} through appropriately positioned oxygen atoms (shown in red) within its structure.

Fura-2

When such a dye is introduced into cells, changes in available Ca^{2+} concentration can be monitored with microscopes capable of detecting changes in fluorescence (Figure 14.16). Probes for sensing other second messengers such as cAMP also have been developed. These *molecular-imaging agents* are greatly enhancing our understanding of signal-transduction processes.

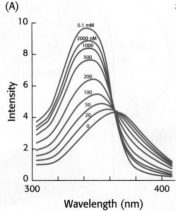

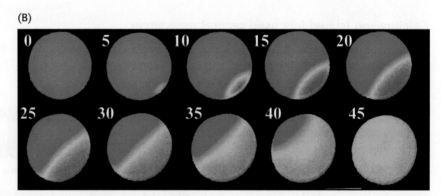

FIGURE 14.16 Calcium imaging. (A) The fluorescence spectra of the calcium-binding dye Fura-2 can be used to measure available calcium ion concentrations in solution and in cells. (B) A series of images show Ca^{2+} spreading across an egg cell following fertilization by sperm. These images were obtained through the use of Fura-2. The images are false colored: orange represents high Ca^{2+} concentrations, and green represents low Ca^{2+} concentrations. [(A) Information from S. J. Lippard and J. M. Berg, *Principles of Bioinorganic Chemistry* (University Science Books, 1994), p. 193; (B) Republished with permission of Company of Biologists, from Exploring the mechanism of action of the sperm-triggered calcium-wave pacemaker in ascidian zygotes, Carroll, M. et al., 2003 *J Cell Sci* 116, 4997–5004; permission conveyed through Copyright Clearance Center, Inc. Courtesy Alex McDougall.]

Calcium ion often activates the regulatory protein calmodulin

Calmodulin (CaM), a 17-kDa protein with four Ca^{2+}-binding sites, serves as a calcium sensor in nearly all eukaryotic cells. *At cytoplasmic concentrations above about 500 nM, Ca^{2+} binds to and activates calmodulin.* Calmodulin is a member of the *EF-hand protein family.* The *EF hand* is a Ca^{2+}-binding motif that consists of a helix, a loop, and a second helix. This motif, originally discovered in the protein parvalbumin, was named the EF hand because the two key helices designated E and F in parvalbumin are positioned like the forefinger and thumb of the right hand (Figure 14.17). These two helices and the intervening loop form the Ca^{2+}-binding motif. Seven oxygen atoms are coordinated to each Ca^{2+}, six from the protein and one from a bound water molecule. Calmodulin is made up of four EF-hand motifs, each of which can bind a single Ca^{2+} ion.

The binding of Ca^{2+} to calmodulin induces substantial conformational changes in its EF hands, exposing hydrophobic surfaces that can be used to bind other proteins. Using its two sets of two EF hands, calmodulin clamps down around specific regions of target proteins—usually exposed α helices with appropriately positioned hydrophobic and charged groups (Figure 14.18). The Ca^{2+}–calmodulin complex stimulates a wide variety of enzymes, pumps, and other target proteins by inducing structural rearrangements in these binding partners. An especially noteworthy set of targets are several *calmodulin-dependent protein kinases* (CaM kinases) that phosphorylate many different proteins and regulate fuel metabolism, ionic permeability, neurotransmitter synthesis, and neurotransmitter release. We see here a recurring theme in signal-transduction pathways: the concentration of a second messenger is increased (in this case, Ca^{2+}); the signal is sensed by a second-messenger-binding protein (in this case, calmodulin); and the second-messenger-binding protein acts to generate changes in enzymes (in this case, calmodulin-dependent kinases) that control effectors.

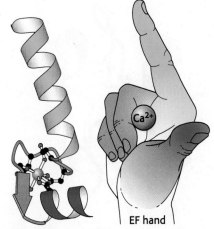

FIGURE 14.17 EF hand. Formed by a helix-loop-helix unit, an EF hand is a binding site for Ca^{2+} in many calcium-sensing proteins. Here, the E helix is yellow, the F helix is blue, and calcium is represented by the green sphere. *Note* that the calcium ion is bound in a loop connecting two nearly perpendicular helices. [Drawn from 1CLL.pdb.]

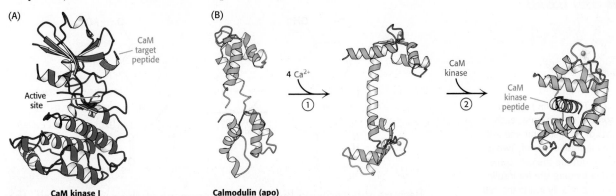

FIGURE 14.18 Calmodulin binds to α helices. (A) An α helix (purple) in CaM kinase I is a target for calmodulin. (B) On Ca^{2+} binding to the apo, or calcium-free, form of calmodulin (1), the two halves of calmodulin clamp down around the target helix (2), binding it through hydrophobic and ionic interactions. In CaM kinase I, this interaction allows the enzyme to adopt an active conformation. [Drawn from 1A06, 1CFD, 1CLL, and 1CM1.pdb.]

14.2 Insulin Signaling: Phosphorylation Cascades Are Central to Many Signal-Transduction Processes

The signaling pathways we have examined so far have activated a protein kinase as a downstream component of the pathway. We now turn to a class of signal-transduction pathways that are *initiated by receptors that include protein kinases as part of their structures.* The activation of these protein kinases sets in motion other processes that ultimately modify the effectors of these pathways.

An example is the signal-transduction pathway initiated by *insulin,* the hormone released in response to increased blood-glucose levels after a meal.

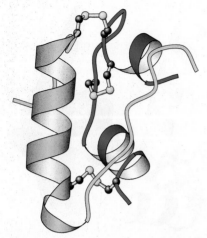

FIGURE 14.19 Insulin structure. *Notice* that insulin consists of two chains (shown in blue and yellow) linked by two interchain disulfide bonds. The α chain (blue) also has an intrachain disulfide bond. [Drawn from 1B2F.pdb.]

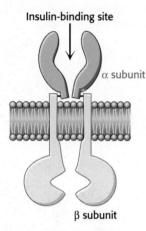

FIGURE 14.20 The insulin receptor. The receptor consists of two units, each of which consists of an α subunit and a β subunit linked by a disulfide bond. Two α subunits, which lie outside the cell, come together to form a binding site for insulin. Each β subunit lies primarily inside the cell and includes a protein kinase domain.

In all of its detail, this multifaceted pathway is quite complex. Hence, we will focus solely on the major branch, which leads to the mobilization of glucose transporters to the cell surface. These transporters allow the cell to take up the glucose that is plentiful in the bloodstream after a meal.

The insulin receptor is a dimer that closes around a bound insulin molecule

Insulin is a peptide hormone, consisting of two chains that are linked by three disulfide bonds (Figure 14.19). Its receptor has a quite different structure from that of the β-AR. The *insulin receptor* is a dimer of two identical units. Each unit consists of one α chain and one β chain linked to one another by a single disulfide bond (Figure 14.20). Each α subunit lies completely outside the cell, whereas each β subunit lies primarily inside the cell, spanning the membrane with a single transmembrane segment. The two α subunits move together to form a binding site for a single insulin molecule—a surprising occurrence because two different surfaces on the insulin molecule must interact with the two identical insulin-receptor chains. The moving together of the dimeric units in the presence of an insulin molecule sets the signaling pathway in motion. *The closing up of an oligomeric receptor or the oligomerization of monomeric receptors around a bound ligand is a strategy used by many receptors to initiate a signal, particularly by those containing a protein kinase.*

Each β subunit consists primarily of a protein kinase domain, homologous to protein kinase A. However, this kinase differs from protein kinase A in two important ways. First, the insulin-receptor kinase is a *tyrosine kinase*; that is, it catalyzes the transfer of a phosphoryl group from ATP to the hydroxyl group of tyrosine, rather than serine or threonine.

$$\text{Tyrosine} \xrightarrow[\text{Tyrosine kinase}]{\text{ATP} \quad \text{ADP}} \text{Phosphotyrosine}$$

Because this tyrosine kinase is a component of the receptor itself, the insulin receptor is referred to as a *receptor tyrosine kinase*. Second, the insulin receptor kinase is in an inactive conformation when the domain is not covalently modified. The kinase is rendered inactive by the position of an unstructured loop (called the *activation loop*) that lies in the center of the structure.

Insulin binding results in the cross-phosphorylation and activation of the insulin receptor

When the two α subunits move together to surround an insulin molecule, the two protein kinase domains on the inside of the cell also are drawn together. It is important to note that as they come together, the flexible activation loop of one kinase subunit is able to fit into the active site of the other kinase subunit within the dimer. With the two β subunits forced together, the kinase domains catalyze the addition of phosphoryl groups from ATP to tyrosine residues in the activation loops. When these tyrosine residues are

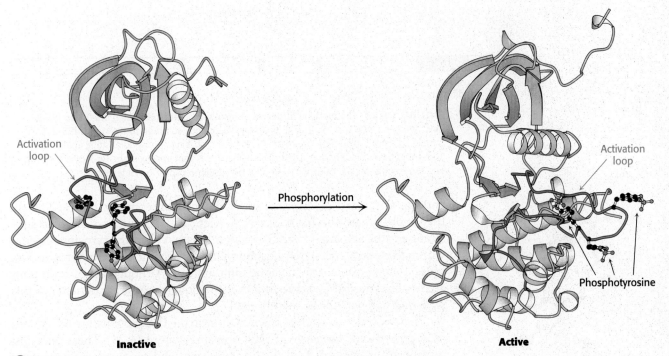

Activation loop

Phosphorylation

Activation loop

Phosphotyrosine

Inactive

Active

FIGURE 14.21 Activation of the insulin receptor by phosphorylation. The activation loop is shown in red in this model of the protein kinase domain of the β subunit of the insulin receptor. The unphosphorylated structure on the left is not catalytically active. *Notice* that, when three tyrosine residues in the activation loop are phosphorylated, the activation loop swings across the structure and the kinase structure adopts a more compact conformation. This conformation is catalytically active. [Drawn from 1IRK.pdb and 1IR3.pdb.]

phosphorylated, a striking conformational change takes place (Figure 14.21). The rearrangement of the activation loop converts the kinase into an active conformation. Thus, *insulin binding on the outside of the cell results in the activation of a membrane-associated kinase within the cell.*

The activated insulin-receptor kinase initiates a kinase cascade

On phosphorylation, the insulin-receptor tyrosine kinase is activated. Because the two units of the receptor are held in close proximity to one another, additional sites within the receptor are also phosphorylated. These phosphorylated sites act as docking sites for other substrates, including a class of molecules referred to as *insulin-receptor substrates* (IRS; Figure 14.22). IRS-1 and IRS-2 are two homologous proteins with a common modular structure (Figure 14.23). The amino-terminal part includes a *pleckstrin homology domain*, which binds phosphoinositide, and a phosphotyrosine-binding domain. These domains act together to anchor the IRS protein to the insulin receptor

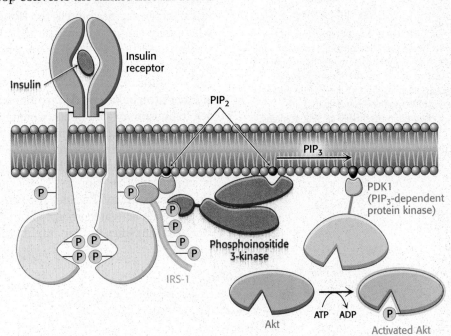

Insulin

Insulin receptor

PIP₂

PIP₃

PDK1 (PIP₃-dependent protein kinase)

Phosphoinositide 3-kinase

IRS-1

Akt

ATP   ADP

Activated Akt

FIGURE 14.22 Insulin signaling. The binding of insulin results in the cross-phosphorylation and activation of the insulin receptor. Phosphorylated sites on the receptor act as binding sites for insulin-receptor substrates such as IRS-1. The lipid kinase phosphoinositide 3-kinase binds to phosphorylated sites on IRS-1 through its regulatory domain, then converts PIP₂ into PIP₃. Binding to PIP₃ activates PIP₃-dependent protein kinase (PDK1), which phosphorylates and activates kinases such as Akt1. Activated Akt1 can then diffuse throughout the cell to continue the signal-transduction pathway.

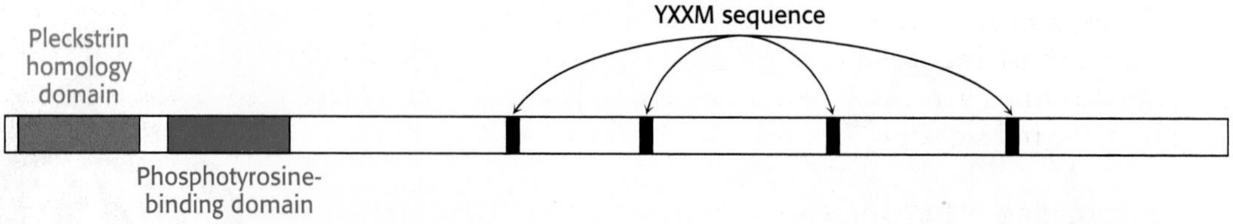

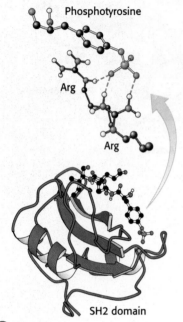

FIGURE 14.23 The modular structure of insulin-receptor substrates IRS-1 and IRS-2. This schematic view represents the amino acid sequence common to IRS-1 and IRS-2. Each protein contains a pleckstrin homology domain (which binds phosphoinositide lipids), a phosphotyrosine-binding domain, and four sequences that approximate Tyr-X-X-Met (YXXM). The four sequences are phosphorylated by the insulin-receptor tyrosine kinase.

FIGURE 14.24 Structure of the SH2 domain. The domain is shown bound to a phosphotyrosine-containing peptide. *Notice* at the top that the negatively charged phosphotyrosine residue interacts with two arginine residues that are conserved in essentially all SH2 domains. [Drawn from 1SPS.pdb.]

and the associated membrane. Each IRS protein contains four sequences that approximate the form Tyr-X-X-Met. These sequences are also substrates for the activated insulin-receptor kinase. When the tyrosine residues within these sequences are phosphorylated to become phosphotyrosine residues, IRS molecules can act as *adaptor proteins*. Although adaptor proteins are not enzymes, they serve to tether the downstream components of this signaling pathway to the membrane.

Phosphotyrosine residues, such as those in the IRS proteins, are recognized most often by *Src homology 2* (SH2) domains (Figure 14.24). These domains, present in many signal-transduction proteins, bind to stretches of polypeptide that contain phosphotyrosine residues. Each specific SH2 domain shows a binding preference for phosphotyrosine in a particular sequence context. Which proteins contain SH2 domains that bind to phosphotyrosine-containing sequences in the IRS proteins? The most important of them are in a class of lipid kinases, called *phosphoinositide 3-kinases* (PI3Ks), that add a phosphoryl group to the 3-position of inositol in phosphatidylinositol 4,5-bisphosphate (PIP_2; Figure 14.25). These enzymes are dimers of 110-kDa catalytic subunits and 85-kDa regulatory subunits. Through SH2 domains in the regulatory subunits, these enzymes bind to the IRS proteins and are drawn to the membrane where they can phosphorylate PIP_2 to form phosphatidylinositol 3,4,5-trisphosphate (PIP_3). PIP_3, in turn, activates a protein kinase, PDK1, by virtue of a pleckstrin homology domain present in this kinase that is specific for PIP_3 (Figure 14.22). The activated PDK1 phosphorylates and activates Akt, another protein kinase. Akt is not membrane anchored; it moves through the cell to phosphorylate targets, which include components that control the trafficking of the glucose receptor GLUT4 to the cell surface as well as enzymes that stimulate glycogen synthesis (Section 21.4).

The cascade initiated by the binding of insulin to the insulin receptor is summarized in Figure 14.26. The signal is amplified at several stages along this pathway. Because the activated insulin receptor itself is a protein kinase, each activated receptor can phosphorylate multiple IRS molecules. Activated enzymes further amplify the signal in at least two of the subsequent

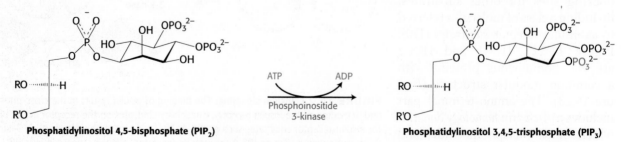

Phosphatidylinositol 4,5-bisphosphate (PIP₂)

ATP   ADP

Phosphoinositide
3-kinase

Phosphatidylinositol 3,4,5-trisphosphate (PIP₃)

FIGURE 14.25 Action of a lipid kinase in insulin signaling. Phosphorylated IRS-1 and IRS-2 activate the enzyme phosphoinositide 3-kinase, an enzyme that converts PIP_2 into PIP_3.

steps. Thus, a small increase in the concentration of circulating insulin can produce a robust intracellular response. Note that although the insulin pathway described here may seem complicated, it is substantially less elaborate than the full signaling network initiated by insulin.

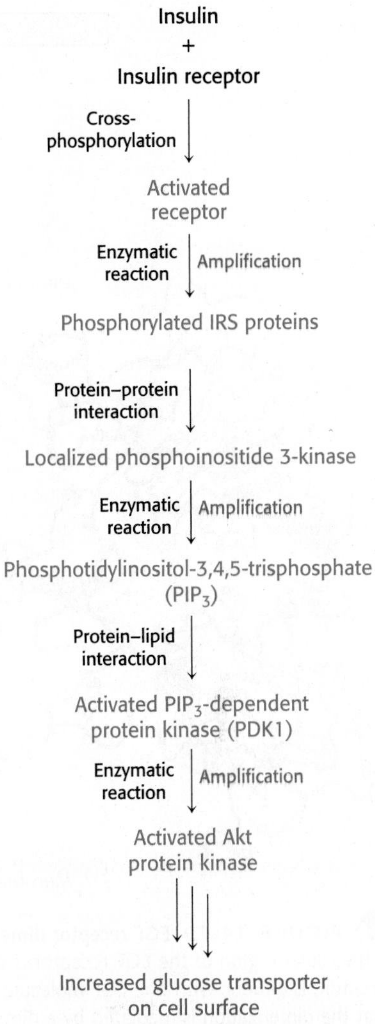

Insulin signaling is terminated by the action of phosphatases

We have seen that the activated G protein promotes its own inactivation by the catalytic release of a phosphoryl group from GTP. In contrast, proteins phosphorylated on serine, threonine, or tyrosine residues are extremely stable kinetically. Specific enzymes, called protein phosphatases, are required to hydrolyze these phosphorylated proteins and return them to their initial states. Similarly, lipid phosphatases are required to remove phosphoryl groups from inositol lipids that had been activated by lipid kinases. In insulin signaling, three classes of enzymes are of particular importance in shutting off the signaling pathway:

1. Protein tyrosine phosphatases that remove phosphoryl groups from tyrosine residues on the insulin receptor and the IRS adaptor proteins;

2. Lipid phosphatases that hydrolyze PIP_3 to PIP_2, and;

3. Protein serine phosphatases that remove phosphoryl groups from activated protein kinases such as Akt.

Many of these phosphatases are activated or recruited as part of the response to insulin. Thus, the binding of the initial signal sets the stage for the eventual termination of the response.

14.3 EGF Signaling: Signal-Transduction Pathways Are Poised to Respond

Our consideration of the signal-transduction cascades initiated by epinephrine and insulin included examples of how components of signal-transduction pathways are poised for action, ready to be activated by minor modifications. For example, G-protein subunits require only the binding of GTP in exchange for GDP to transmit a signal. This exchange reaction is thermodynamically favorable, but it is quite slow in the absence of an appropriate activated 7TM receptor. Similarly, the tyrosine kinase domains of the dimeric insulin receptor are ready for phosphorylation and activation but require insulin bound between two α subunits to draw the activation loop of one tyrosine kinase into the active site of a partner tyrosine kinase to initiate the signaling cascade.

Next, we examine a signal-transduction pathway that reveals another clear example of how these signaling cascades are poised to respond. This pathway is activated by the signal molecule *epidermal growth factor* (EGF). Like that of the insulin receptor, the initiator of this pathway is a receptor tyrosine kinase. Both the extracellular and the intracellular domains of this receptor are ready for action, held in check only by a specific structure that prevents receptors from coming together. Furthermore, in the EGF pathway, we will encounter several additional classes of signaling components that participate in many other signaling networks.

EGF binding results in the dimerization of the EGF receptor

Epidermal growth factor is a 6-kDa polypeptide that stimulates the growth of epidermal and epithelial cells (Figure 14.27). The *EGF receptor* (EGFR), like the insulin receptor, is a dimer of two identical subunits. Each subunit contains an intracellular protein tyrosine kinase domain that participates in

Insulin
+
Insulin receptor

Cross-
phosphorylation

Activated
receptor

Enzymatic | Amplification
reaction

Phosphorylated IRS proteins

Protein–protein
interaction

Localized phosphoinositide 3-kinase

Enzymatic | Amplification
reaction

Phosphotidylinositol-3,4,5-trisphosphate
(PIP_3)

Protein–lipid
interaction

Activated PIP_3-dependent
protein kinase (PDK1)

Enzymatic | Amplification
reaction

Activated Akt
protein kinase

Increased glucose transporter
on cell surface

FIGURE 14.26 Insulin signaling pathway. Key steps in the signal-transduction pathway initiated by the binding of insulin to the insulin receptor.

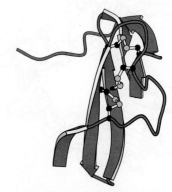

Epidermal growth factor (EGF)

FIGURE 14.27 Structure of epidermal growth factor. *Notice* that three intrachain disulfide bonds stabilize the compact three-dimensional structure of the growth factor. [Drawn from 1EGF.pdb.]

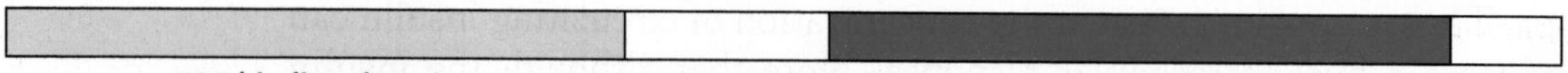

EGF-binding domain	Transmembrane helix	Kinase domain	C-terminal tail (tyrosine-rich)

FIGURE 14.28 Modular structure of the EGF receptor. This schematic view of the amino acid sequence of the EGF receptor shows the EGF-binding domain that lies outside the cell, a single transmembrane helix-forming region, the intracellular tyrosine kinase domain, and the tyrosine-rich domain at the carboxyl terminus.

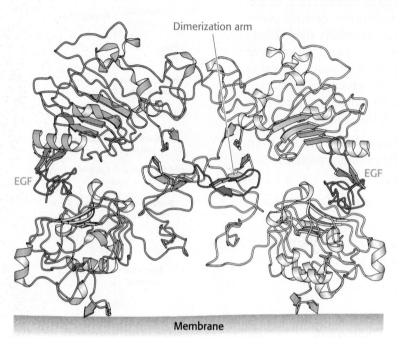

Dimerization arm

EGF

EGF

Membrane

FIGURE 14.29 EGF receptor dimerization. The structure of the extracellular region of the EGF receptor is shown bound to EGF. *Notice* that the structure is dimeric with one EGF molecule bound to each receptor molecule and that the dimerization is mediated by a dimerization arm (red) that extends from each receptor molecule. [Drawn from 1IVO.pdb.]

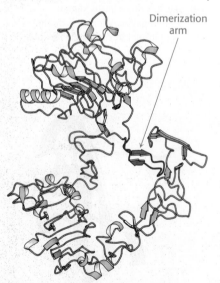

Dimerization arm

FIGURE 14.30 Structure of the unactivated EGF receptor. The extracellular domain of the EGF receptor is shown in the absence of bound EGF. *Notice* that the dimerization arm (red) is bound to a part of the receptor that makes it unavailable for interaction with another receptor molecule. [Drawn from 1NQL.pdb.]

cross-phosphorylation reactions (Figure 14.28). Unlike those of the insulin receptor, however, these units exist as monomers until they bind EGF. Moreover, each EGF receptor monomer binds a single molecule of EGF in its extracellular domain (Figure 14.29). Thus, the dimer binds two ligand molecules, in contrast with the insulin-receptor dimer, which binds only one ligand. Note that each EGF molecule lies far away from the dimer interface. This interface includes a so-called *dimerization arm* from each monomer that reaches out and inserts into a binding pocket on the other monomer.

Although this structure nicely reveals the interactions that support the formation of a receptor dimer favoring cross-phosphorylation, it raises another question: Why doesn't the receptor dimerize and signal in the absence of EGF? This question has been addressed by examining the structure of the EGF receptor in the absence of bound ligand (Figure 14.30). This structure is, indeed, monomeric, and each monomer is in a conformation that is quite different from that observed in the ligand-bound dimer. In particular, the dimerization arm binds to a domain *within the same monomer* that holds the receptor in a closed configuration. In essence, the receptor is poised in a spring-loaded conformation held in position by the contact between the interaction loop and another part of the structure, ready to bind ligand and change into a conformation active for dimerization and signaling.

This observation suggests that a receptor that exists in the extended conformation even in the absence of bound ligand would be constitutively active. Remarkably, such a receptor exists. This receptor, HER2, is approximately 50% identical in amino acid sequence with the EGF receptor and has the same domain structure. HER2 does not bind any known ligand, yet crystallographic studies reveal that it adopts an extended structure very similar to that observed for the ligand-bound EGF receptor. Under normal conditions, HER2 forms heterodimers with the EGF receptor and other members of the EGF receptor family and participates in cross-phosphorylation reactions with these receptors. HER2 is overexpressed in some cancers, presumably contributing to tumor growth by forming homodimers that signal even in the absence of ligand. We will return to HER2 when we consider approaches to cancer treatment based on knowledge of signaling pathways (Section 14.5).

The EGF receptor undergoes phosphorylation of its carboxyl-terminal tail

Like the insulin receptor, the EGF receptor undergoes cross-phosphorylation of one unit by another unit within a dimer. However, unlike that of the insulin receptor, the site of this phosphorylation is not within the activation loop of the kinase, but rather in a region that lies on the C-terminal side of the kinase domain. As many as five tyrosine residues in this region are phosphorylated. The dimerization of the EGF receptor brings the C-terminal region on one receptor into the active site of its partner's kinase. The kinase itself is in an active conformation without phosphorylation, revealing again how this signaling system is poised to respond.

EGF signaling leads to the activation of Ras, a small G protein

The phosphotyrosines on the EGF receptors act as docking sites for SH2 domains on other proteins. The intracellular signaling cascade begins with the binding of *Grb2*, a key adaptor protein that contains one SH2 domain and two *Src homology 3* (SH3) domains. On phosphorylation of the receptor, the SH2 domain of Grb2 binds to the phosphotyrosine residues of the receptor tyrosine kinase. Through its two SH3 domains, Grb2 then binds polyproline-rich polypeptides within a protein called *Sos*. Sos, in turn, binds to *Ras* and activates it. A very prominent signal-transduction component, Ras is a member of a class of proteins called the *small G proteins*. Like the G proteins described in Section 14.1, the small G proteins contain bound GDP in their unactivated forms. Sos opens up the nucleotide-binding pocket of Ras, allowing GDP to escape and GTP to enter in its place. Because of its effect on Ras, Sos is referred to as a *guanine-nucleotide-exchange factor* (GEF). Thus, the binding of EGF to its receptor leads to the conversion of Ras into its GTP form through the intermediacy of Grb2 and Sos (Figure 14.31).

Activated Ras initiates a protein kinase cascade

Ras changes conformation when it is transformed from its GDP into its GTP form. In the GTP form, Ras binds other proteins, including a protein kinase termed *Raf*. When bound to Ras, Raf undergoes a conformational change that activates the Raf protein kinase domain. Both Ras and Raf are anchored to the membrane through covalently bound lipid modifications. Activated Raf then phosphorylates other proteins, including protein kinases termed MEKs. In turn, MEKs activate kinases called *extracellular signal-regulated kinases* (ERKs). ERKs then phosphorylate numerous substrates, including transcription factors in the nucleus as well as other protein kinases. The complete flow of information from the arrival of EGF at the cell surface to changes in gene expression is summarized in Figure 14.32.

Small G proteins, or small GTPases, constitute a large superfamily of proteins—grouped into subfamilies called Ras, Rho, Arf, Rab, and Ran—that play a major role in a host of cell functions including growth, differentiation, cell motility, cytokinesis (the separation of two cells during division), and the transport of materials throughout the cell (Table 14.2). As with the heterotrimeric G proteins, the small G proteins cycle between an active GTP-bound form and an inactive GDP-bound form. They differ from the heterotrimeric G proteins in being smaller (20–25 kDa versus 30–35 kDa) and monomeric. Nonetheless, the two families are related by divergent evolution, and small G proteins have many key mechanistic and structural motifs in common with the G_α subunit of the heterotrimeric G proteins.

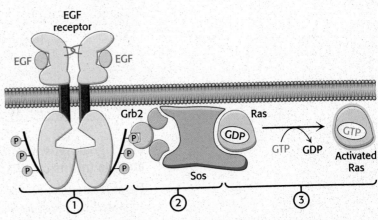

FIGURE 14.31 Ras activation mechanism. The dimerization of the EGF receptor due to EGF binding leads to: (1) the phosphorylation of the C-terminal tails of the receptor, (2) the subsequent recruitment of Grb2 and Sos, and (3) the exchange of GTP for GDP in Ras. This signal-transduction pathway results in the conversion of Ras into its activated GTP-bound form.

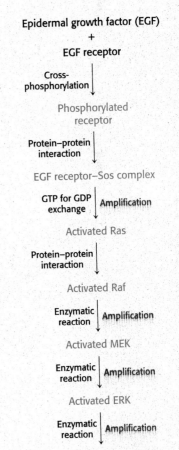

FIGURE 14.32 EGF signaling pathway. The key steps in the pathway initiated by EGF binding to the EGF receptor. A kinase cascade leads to the phosphorylation of transcription factors and concomitant changes in gene expression.

TABLE 14.2 Ras superfamily of GTPases

Subfamily	Function
Ras	Regulates cell growth through serine–threonine protein kinases
Rho	Reorganizes cytoskeleton through serine–threonine protein kinases
Arf	Activates the ADP-ribosyltransferase of the cholera toxin A subunit; regulates vesicular trafficking pathways; activates phospholipase D
Rab	Plays a key role in secretory and endocytotic pathways
Ran	Functions in the transport of RNA and protein into and out of the nucleus

EGF signaling is terminated by protein phosphatases and the intrinsic GTPase activity of Ras

Because so many components of the EGF signal-transduction pathway are activated by phosphorylation, we can expect protein phosphatases to play key roles in the termination of EGF signaling. Indeed, crucial phosphatases remove phosphoryl groups from tyrosine residues on the EGF receptor and from serine, threonine, and tyrosine residues in the protein kinases that participate in the signaling cascade. The signaling process itself sets in motion the events that activate many of these phosphatases. Consequently, signal activation also initiates signal termination.

Like the G proteins activated by 7TM receptors, Ras possesses intrinsic GTPase activity. Thus, the activated GTP form of Ras spontaneously converts into the inactive GDP form. The rate of conversion can be accelerated in the presence of *GTPase-activating proteins* (GAPs), which interact with small G proteins in the GTP form and facilitate GTP hydrolysis. Thus, the lifetime of activated Ras is regulated by accessory proteins in the cell. The GTPase activity of Ras is crucial for shutting off signals leading to cell growth, and so it is not surprising that mutations in Ras are found in many types of cancer, as we will discuss in Section 14.5.

14.4 Many Elements Recur with Variation in Different Signal-Transduction Pathways

We can begin to make sense of the complexity of signal-transduction pathways by taking note of several common themes that have appeared consistently in the pathways described in this chapter and underlie many additional signaling pathways not considered herein.

1. *Protein kinases are central to many signal-transduction pathways.* Protein kinases are central to all three signal-transduction pathways described in this chapter. In the epinephrine-initiated pathway, cAMP-dependent protein kinase (PKA) lies at the end of the pathway, transducing information represented by an increase in cAMP concentration into covalent modifications that alter the activity of key metabolic enzymes. In the insulin- and EGF-initiated pathways, the receptors themselves are protein kinases and several additional protein kinases participate downstream in the pathways. Signal amplification due to protein kinase cascades is a feature common to all three pathways. Although not presented in this chapter, protein kinases often phosphorylate multiple substrates and are thus able to generate a diversity of responses.

2. *Second messengers participate in many signal-transduction pathways.* We have encountered several second messengers, including cAMP, Ca^{2+}, IP_3, and the lipid DAG. Because second messengers are generated by enzymes or by the opening of ion channels, their concentrations can be tremendously amplified compared with the signals that lead to their generation. Specialized proteins sense the concentrations of these second messengers and continue the flow of information along signal-transduction pathways. The second messengers that we have seen recur in many additional signal-transduction pathways. For example, in a consideration of the sensory systems in Chapter 34 (See Chapter 34 on Sapling Plus), we will see how Ca^{2+}-based signaling and cyclic nucleotide-based signaling play key roles in vision and olfaction.

3. *Specialized domains that mediate specific interactions are present in many signaling proteins.* The "wiring" of many signal-transduction pathways is based on distinct protein domains that mediate the interactions between protein components of a particular signaling cascade. We have encountered several of them, including pleckstrin homology domains, which facilitate protein interactions with the lipid PIP_3; SH2 domains, which mediate interactions with polypeptides containing phosphorylated tyrosine residues; and SH3 domains, which interact with peptide sequences that contain multiple proline residues. Many other such domain families exist. In many cases, individual members of each domain family have unique features that allow them to bind to their targets only within a particular sequence context, making them specific for a given signaling pathway and avoiding unwanted cross-talk. *Signal-transduction pathways have evolved in large part by the incorporation of DNA fragments encoding these domains into genes encoding pathway components.*

The presence of these domains is tremendously helpful to scientists trying to unravel signal-transduction pathways. When a protein in a signal-transduction pathway is identified, its amino acid sequence can be analyzed for the presence of these specialized domains by the methods described in Chapter 6. If one or more domains of known function is found, it is often possible to develop clear hypotheses about potential binding partners and signal-transduction mechanisms.

14.5 Defects in Signal-Transduction Pathways Can Lead to Cancer and Other Diseases

In light of their complexity, it comes as no surprise that signal-transduction pathways occasionally fail, leading to disease states. Cancer, a set of diseases characterized by uncontrolled or inappropriate cell growth, is strongly associated with defects in signal-transduction proteins. Indeed, the study of cancer—particularly cancers caused by certain viruses—has contributed greatly to our understanding of signal-transduction proteins and pathways.

For example, Rous sarcoma virus is a retrovirus that causes sarcoma (a cancer of tissues of mesodermal origin such as muscle or connective tissue) in chickens. In addition to the genes necessary for viral replication, this virus carries a gene termed v-*src*. The v-*src* gene is an *oncogene*; it leads to the generation of cancer-like characteristics in susceptible cell types. The protein encoded by the v-*src* gene, v-Src, is a protein tyrosine kinase that includes SH2 and SH3 domains. The v-Src protein is similar in amino acid sequence to a protein normally found in chicken-muscle cells referred to as c-Src (for cellular Src; Figure 14.33A). The c-*src* gene does not induce cell transformation and is termed a *proto-oncogene*, referring to the fact that this gene, when mutated, can be converted into

(A)

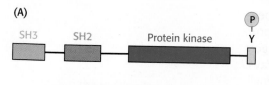

(B)

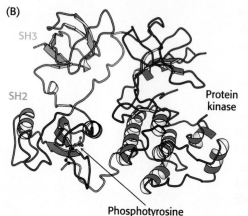

Phosphotyrosine

FIGURE 14.33 Src structure. (A) Cellular Src includes an SH3 domain, an SH2 domain, a protein kinase domain, and a carboxyl-terminal tail that includes a key tyrosine residue. (B) Structure of c-Src in an inactive form with the key tyrosine residue phosphorylated. *Notice* how the three domains work together to keep the enzyme in an inactive conformation: the phosphotyrosine residue is bound in the SH2 domain and the linker between the SH2 domain and the protein kinase domain is bound by the SH3 domain. [Drawn from 2PTK.pdb.]

an oncogene. The protein it encodes is a signal-transduction protein that regulates cell growth.

Why is the biological activity of the v-Src protein so different from that of c-Src? c-Src contains a key tyrosine residue near its C-terminal end that, when phosphorylated, is bound intramolecularly by the upstream SH2 domain (Figure 14.33B). This interaction maintains the kinase domain in an inactive conformation. However, in v-Src, the C-terminal 19 amino acids of c-Src are replaced by a completely different stretch of 11 amino acids that lacks this critical tyrosine residue. Thus, v-Src is always active and can promote unregulated cell growth. Since the discovery of Src, many other mutated protein kinases have been identified as oncogenes.

The gene encoding Ras, a component of the EGF-initiated pathway, is one of the genes most commonly mutated in human tumors. Mammalian cells contain three 21-kDa Ras proteins (H-, K-, and N-Ras), each of which cycles between inactive GDP and active GTP forms. The most common mutations in tumors lead to a loss of the ability to hydrolyze GTP. Thus, the Ras protein is trapped in the "on" position and continues to stimulate cell growth, even in the absence of a continuing signal.

Other genes can contribute to cancer development only when both copies of the gene normally present in a cell are deleted or otherwise damaged. Such genes are called *tumor-suppressor genes*. For example, genes for some of the phosphatases that participate in the termination of EGF signaling are tumor suppressors. Without any functional phosphatase present, EGF signaling persists once initiated, stimulating inappropriate cell growth.

Monoclonal antibodies can be used to inhibit signal-transduction pathways activated in tumors

Mutated or overexpressed receptor tyrosine kinases are frequently observed in tumors. For instance, the epidermal-growth-factor receptor (EGFR) is overexpressed in some human epithelial cancers, including breast, ovarian, and colorectal cancer. Because some small amount of the receptor can dimerize and activate the signaling pathway even without binding to EGF, overexpression of the receptor increases the likelihood that a "grow and divide" signal will be inappropriately sent to the cell. This understanding of cancer-related signal-transduction pathways has led to a therapeutic approach that targets the EGFR. The strategy is to produce monoclonal antibodies to the extracellular domains of the offending receptors. One such antibody, cetuximab (Erbitux), has effectively targeted the EGFR in colorectal cancers. Cetuximab inhibits the EGFR by competing with EGF for the binding site on the receptor. Because the antibody sterically blocks the change in conformation that exposes the dimerization arm, the antibody itself cannot induce dimerization. The result is that the EGFR-controlled pathway is not initiated.

Cetuximab is not the only monoclonal antibody that has been developed to target a receptor tyrosine kinase. Trastuzumab (Herceptin) inhibits another EGFR family member, HER2, that is overexpressed in approximately 30% of breast cancers. Recall that this protein can signal even in the absence of ligand, so it is especially likely that overexpression will stimulate cell proliferation. Breast-cancer patients are now being screened for HER2 overexpression and treated with Herceptin as appropriate. Thus, this cancer treatment is tailored to the genetic characteristics of the tumor.

Protein kinase inhibitors can be effective anticancer drugs

The widespread occurrence of overactive protein kinases in cancer cells suggests that molecules that inhibit these enzymes might act as antitumor agents. For example, more than 90% of patients with chronic myelogenous leukemia (CML) show a specific chromosomal defect in cancer cells (Figure 14.34). The translocation of genetic material between chromosomes 9 and 22 causes the c-*abl* gene, which encodes a tyrosine kinase of the Src family, to be inserted into the *bcr* gene on chromosome 22. The result is the production of a fusion protein called Bcr-Abl that consists primarily of sequences for the c-Abl kinase. However, the *bcr-abl* gene is not regulated appropriately; it is expressed at higher levels than that of the gene encoding the normal c-Abl kinase, stimulating a growth-promoting pathway. Because of this overexpression, leukemia cells express a unique target for chemotherapy. A specific inhibitor of the Bcr-Abl kinase, Gleevec (STI-571, imatinib mesylate), has proved to be a highly effective treatment for patients suffering from CML (Figure 14.35). This approach to cancer chemotherapy is fundamentally distinct from most approaches, which target all rapidly growing cells, including normal ones. Because Gleevec targets tumor cells specifically, side effects caused by the impairment of normal dividing cells can be minimized. *Thus, our understanding of signal-transduction pathways has enabled conceptually new disease treatment strategies.*

Cholera and whooping cough are the result of altered G-protein activity

Although defects in signal-transduction pathways have been most extensively studied in the context of cancer, such defects are important in many other diseases. Cholera and whooping cough are two pathologies of the G-protein-dependent signal pathways. Let us first consider the mechanism of action of the cholera toxin, secreted by the intestinal bacterium *Vibrio cholerae*. Cholera is a potentially life-threatening, acute diarrheal disease transmitted through contaminated water and food. It causes the voluminous secretion of electrolytes and fluids from the intestines of infected persons. The cholera toxin, also called *choleragen*, is a protein composed of two functional units—a β subunit that binds to G_{M1} gangliosides (Section 26.1) of the intestinal epithelium and a catalytic A subunit that enters the cell. The A subunit catalyzes the covalent modification of a $G_{\alpha s}$ protein: the α subunit is modified by the attachment of an ADP-ribose to an arginine residue. This modification stabilizes the GTP-bound form of $G_{\alpha s}$, trapping the molecule in its active conformation. The active G protein, in turn, continuously activates protein kinase A. PKA opens a chloride channel and inhibits sodium absorption by the $Na^+–H^+$ exchanger by phosphorylating both the channel and the exchanger. The net result of the phosphorylation is an excessive loss of NaCl and the loss of large amounts of water into the intestine. Patients suffering from cholera may pass as much as twice their body weight in fluid in 4 to 6 days. Treatment consists of rehydration with a glucose–electrolyte solution.

Whereas cholera is a result of a G protein trapped in the active conformation, causing the signal-transduction pathway to be persistently stimulated, pertussis is a result of the opposite situation. Pertussis toxin is secreted by *Bordetella pertussis*, the bacterium responsible for whooping cough. Like choleragen, pertussis toxin adds an ADP-ribose moiety to a G_α subunit. However, in this case, the ADP-ribose group is added to a $G_{\alpha i}$ protein, a G_α subunit that inhibits adenylate cyclase, closes Ca^{2+} channels, and opens K^+ channels. The effect of this modification is to prevent binding of the heterotrimeric G_i protein to its receptor, trapping it in the "off" conformation. The pulmonary symptoms have not yet been traced to a particular target of the $G_{\alpha i}$ protein.

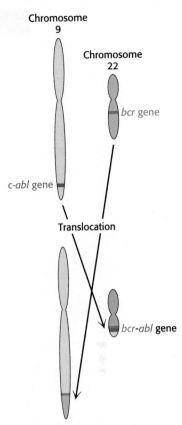

FIGURE 14.34 Formation of the *bcr-abl* gene by translocation. In chronic myelogenous leukemia, parts of chromosomes 9 and 22 are reciprocally exchanged, causing the *bcr* and *abl* genes to fuse. The protein kinase encoded by the *bcr-abl* gene is expressed at higher levels in tumor cells than is the c-*abl* gene in normal cells.

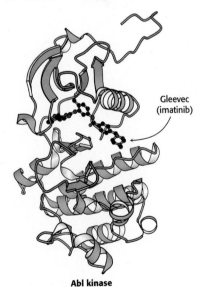

FIGURE 14.35 Inhibition of Abl kinase by Gleevec (imatinib). The structure of the kinase domain of Abl with Gleevec bound is shown. *Notice* that the inhibitor occupies the active site of the kinase (compare, for example, with the location of the active site in the structure of CaM kinase I in Figure 14.18A). [Drawn from 1IEP.pdb]

SUMMARY

In humans and other multicellular organisms, specific signal molecules are released from cells in one organ and are sensed by cells in other organs throughout the body. The message initiated by an extracellular ligand is converted into specific changes in metabolism or gene expression by means of often complex networks referred to as signal-transduction pathways. These pathways amplify the initial signal and lead to changes in the properties of specific effector molecules.

14.1 Epinephrine and Angiotensin II Signaling: Heterotrimeric G Proteins Transmit Signals and Reset Themselves

Epinephrine binds to a cell-surface protein called the β-adrenergic receptor. This receptor is a member of the seven-transmembrane-helix receptor family, so named because each receptor has seven α helices that span the cell membrane. When epinephrine binds to the β-adrenergic receptor on the outside of the cell, the receptor undergoes a conformational change that is sensed inside the cell by a signaling protein termed a heterotrimeric G protein. The α subunit of the G protein exchanges a bound GDP molecule for GTP and concomitantly releases the heterodimer consisting of the β and γ subunits. The α subunit in the GTP form then binds to adenylate cyclase and activates it, leading to an increase in the concentration of the second messenger cyclic AMP. This increase in cyclic AMP concentration, in turn, activates protein kinase A. Other 7TM receptors also signal through heterotrimeric G proteins, although these pathways often include enzymes other than adenylate cyclase. For example, the angiotensin II receptor signals through the phosphoinositide pathway, starting with the activation of phospholipase C, which cleaves a membrane lipid to produce two secondary messengers, diacylglycerol and inositol 1,4,5-trisphosphate. An increased IP_3 concentration leads to the release of calcium ion, another important second messenger, into the cell. G-protein signaling is terminated by the hydrolysis of the bound GTP to GDP.

14.2 Insulin Signaling: Phosphorylation Cascades Are Central to Many Signal-Transduction Processes

Protein kinases are key components in many signal-transduction pathways, including some for which the protein kinase is an integral component of the initial receptor. An example of such a receptor is the membrane tyrosine kinase bound by insulin. Insulin binding causes one subunit within the dimeric receptor to phosphorylate specific tyrosine residues in the other subunit. The resulting conformational changes dramatically increase the kinase activity of the receptor. The activated receptor kinase initiates a signaling cascade that includes both lipid kinases and protein kinases. This cascade eventually leads to the mobilization of glucose transporters to the cell surface, increasing glucose uptake. Insulin signaling is terminated through the action of phosphatases.

14.3 EGF Signaling: Signal-Transduction Systems Are Poised to Respond

Only minor modifications are necessary to transform many signal-transduction proteins from their inactive into their active forms. Epidermal growth factor also signals through a receptor tyrosine kinase. EGF binding induces a conformational change that allows receptor dimerization and cross-phosphorylation. The phosphorylated receptor binds adaptor proteins that mediate the activation of Ras, a small G protein. Activated Ras initiates a protein kinase cascade that eventually leads to the phosphorylation of transcription factors and changes in gene expression. EGF signaling is terminated by the action of phosphatases and the hydrolysis of GTP by Ras.

14.4 Many Elements Recur with Variation in Different Signal-Transduction Pathways

Protein kinases are components of many signal-transduction pathways, both as components of receptors and in other roles. Second messengers, including cyclic nucleotides, calcium, and lipid derivatives, are common in many signaling pathways. The changes in the concentrations of second messengers are often much larger than the changes associated with the initial signal owing to amplification along the pathway. Small domains that recognize phosphotyrosine residues or specific lipids are present in many signaling proteins and are essential to determining the specificity of interactions.

14.5 Defects in Signal-Transduction Pathways Can Lead to Cancer and Other Diseases

Genes encoding components of signal-transduction pathways that control cell growth are often mutated in cancer. Some genes can be mutated to forms called oncogenes that are active regardless of appropriate signals. Monoclonal antibodies directed against cell-surface receptors that participate in signaling have been developed for use in cancer treatment. Our understanding of the molecular basis of cancer has enabled the development of anticancer drugs directed against specific targets, such as the specific kinase inhibitor Gleevec.

APPENDIX

Biochemistry in Focus

Gases get in on the signaling game

In this chapter, we have seen that second messengers play a critical role in transmission of signals throughout the cell. These molecules come in a variety of shapes and sizes, ranging from simple cations (Ca^{2+}), to cyclic nucleotides (cAMP), to lipids (DAG). Remarkably, gaseous molecules have also been identified as second messengers. For example, nitric oxide (NO) is an important signaling molecule in many vertebrate systems, with effects on skeletal and smooth muscle, the heart, and the brain. In particular, NO is a potent *vasodilator*, a molecule that causes relaxation of the smooth muscles of blood vessels, leading to an opening of these vessels. As we will see in Section 24.4, NO is produced by a group of enzymes called the nitric oxide synthases (NOSs), which use arginine and O_2 to generate NO and citrulline. NO possesses unique properties as a second messenger; it can diffuse across biological membranes and is short-lived. Thus, NO can influence signaling not only in the cell that generates it, but within neighboring cells as well.

How is the presence of NO sensed? Inside the cell, NO binds to a heterodimeric protein called *soluble guanylate cyclase* (sGC). Upon binding, the enzyme activity of sGC is enhanced, resulting in the formation of cGMP from GTP. cGMP binds to and activates protein kinase G (PKG), which can, in turn, phosphorylate a number of intracellular targets, including several that enhance calcium removal from the cytosol. Importantly, PKG also phosphorylates and activates phosphodiesterase 5 (PDE), an enzyme which cleaves cGMP into GMP. Hence, as we have seen throughout this chapter, the NO signaling circuit activates the mechanisms for its eventual termination.

The identification of sGC has led to the development of a new class of therapeutics: sGC activators. These molecules bind to sGC and dramatically stimulate catalytic activity, even the absence of NO. As such, they are potent vasodilators. One of these molecules, riociguat, has been approved for the treatment of pulmonary arterial hypertension (PAH), a condition in which the arterial pressure of the vessels of the lungs is elevated.

Potential signaling roles for other gaseous second messengers, including carbon monoxide (CO) and hydrogen sulfide (H_2S), have been identified. Along with NO, these molecules are often referred to as *gasotransmitters*, and are the focus of exciting new research into previously unappreciated signal-transduction pathways and the discovery of new therapeutics.

KEY TERMS

primary messenger (p. 438)

ligand (p. 438)

second messenger (p. 438)

cross talk (p. 439)

β-adrenergic receptor (β-AR) (p. 439)

seven-transmembrane-helix (7TM) receptor (p. 440)

rhodopsin (p. 440)

G protein (p. 440)

agonist (p. 441)

G-protein-coupled receptor (GPCR) (p. 441)

adenylate cyclase (p. 442)

protein kinase A (PKA) (p. 443)

β-adrenergic receptor kinase (p. 444)

phosphoinositide cascade (p. 444)

phosphatidylinositol 4,5-bisphosphate (PIP$_2$) (p. 444)

phospholipase C (p. 444)

protein kinase C (PKC) (p. 445)

calmodulin (CaM) (p. 446)

EF hand (p. 447)

calmodulin-dependent protein kinase (CaM kinase) (p. 446)

insulin (p. 447)

insulin receptor (p. 448)

tyrosine kinase (p. 448)

receptor tyrosine kinase (p. 448)

insulin-receptor substrate (IRS) (p. 449)

pleckstrin homology domain (p. 449)

adaptor protein (p. 450)

Src homology 2 (SH2) domain (p. 450)

phosphoinositide 3-kinases (PI3Ks) (p. 450)

epidermal growth factor (EGF) (p. 451)

EGF receptor (EGFR) (p. 451)

dimerization arm (p. 452)

Src homology 3 (SH3) domain (p. 453)

Ras (p. 453)

small G protein (p. 453)

guanine-nucleotide-exchange factor (GEF) (p. 453)

extracellular signal-regulated kinase (ERK) (p. 453)

GTPase-activating protein (GAP) (p. 454)

oncogene (p. 455)

proto-oncogene (p. 455)

tumor-suppressor gene (p. 456)

PROBLEMS

1. *Active mutants.* Some protein kinases are inactive unless they are phosphorylated on key serine or threonine residues. In some cases, active enzymes can be generated by mutating these serine or threonine residues to aspartate. Explain.

2. *In the pocket.* SH2 domains bind phosphotyrosine residues in deep pockets on their surfaces. Would you expect SH2 domains to bind phosphoserine and phosphothreonine with high affinity? Why or why not? ✔❹

3. *On–off.* Why is the GTPase activity of G proteins crucial to the proper functioning of a cell? Propose a theory as to why G proteins have not evolved to catalyze GTP hydrolysis more efficiently.

4. *Vive la différence.* Why is the fact that a monomeric hormone binds simultaneously to two identical receptor molecules, thus promoting the formation of a dimer of the receptor, considered remarkable?

5. *Antibodies mimicking hormones.* Antibodies have two identical antigen-binding sites. Remarkably, antibodies to the extracellular parts of growth-factor receptors often lead to the same cellular effects as does exposure to growth factors. Explain this observation. ✔❸

6. *Facile exchange.* A mutated form of the α subunit of the heterotrimeric G protein has been identified; this form readily exchanges nucleotides even in the absence of an activated

receptor. What would be the effect on a signaling pathway containing the mutated α subunit? ✔❺

7. *Making connections.* Suppose you were investigating a newly discovered growth-factor signal-transduction pathway. You found that, if you added GTPγS, a nonhydrolyzable analog of GTP, the duration of the hormonal response increased. What can you conclude?

8. *Diffusion rates.* Usually, rates of diffusion vary inversely with molecular weights; so, smaller molecules diffuse faster than do larger ones. In cells, however, calcium ion diffuses more slowly than does cAMP. Propose a possible explanation. ✔❷

9. *Negativity abounds.* Fura-2, as depicted on p. 446, is not effective for the study of calcium levels in intact, living cells. Considering the Fura-2 structure, why is it ineffective?

10. *Awash with glucose.* Glucose is mobilized for ATP generation in muscle in response to epinephrine, which activates G$_{αs}$. Cyclic AMP phosphodiesterase is an enzyme that converts cAMP into AMP. How would inhibitors of cAMP phosphodiesterase affect glucose mobilization in muscle?

11. *Getting it started.* The insulin receptor, on dimerization, cross-phosphorylates the activation loop of the other receptor molecule, leading to activation of the kinase. Propose how this phosphorylation event can take place if the kinase starts in an inactive conformation. ✔❸

12. *Many defects.* Considerable effort has been directed toward determining the genes in which sequence variation contributes to the development of type 2 diabetes. Approximately 800 genes have been implicated. Propose an explanation for this observation.

13. *Growth-factor signaling.* Human growth hormone binds to a cell-surface membrane protein that is not a receptor tyrosine kinase. The intracellular domain of the receptor can bind other proteins inside the cell. Furthermore, studies indicate that the receptor is monomeric in the absence of hormone but dimerizes on hormone binding. Propose a possible mechanism for growth-hormone signaling.

14. *Receptor truncation.* You prepare a cell line that overexpresses a mutant form of EGFR in which the entire intracellular region of the receptor has been deleted. Predict the effect of overexpression of this construct on EGF signaling in this cell line.

15. *Hybrid.* Suppose that, through genetic manipulations, a chimeric receptor is produced that consists of the extracellular domain of the insulin receptor and the transmembrane and intracellular domains of the EGF receptor. Cells expressing this receptor are exposed to insulin, and the level of phosphorylation of the chimeric receptor is examined. What would you expect to observe and why? What would you expect to observe if these cells were exposed to EGF?

16. *Total amplification.* Suppose that each β-adrenergic receptor bound to epinephrine converts 100 molecules of $G_{\alpha s}$ into their GTP forms and that each molecule of activated adenylate cyclase produces 1000 molecules of cAMP per second. With the assumption of a full response, how many molecules of cAMP will be produced in 1 s after the formation of a single complex between epinephrine and the β-adrenergic receptor? ✔️❶

Chapter Integration Problems

17. *Nerve-growth-factor pathway.* Nerve-growth factor (NGF) binds to a protein tyrosine kinase receptor. The amount of diacylglycerol in the plasma membrane increases in cells expressing this receptor when treated with NGF. Propose a simple signaling pathway and identify the isoform of any participating enzymes. Would you expect the concentrations of any other common second messengers to increase on NGF treatment?

18. *Redundancy.* Because of the high degree of genetic variability in tumors, typically no single anticancer therapy is universally effective for all patients, even within a given tumor type. Hence, it is often desirable to inhibit a particular pathway at more than one point in the signaling cascade. In addition to the EGFR-directed monoclonal antibody cetuximab, propose alternative strategies for targeting the EGF signaling pathway for antitumor drug development. ✔️❺

Mechanism Problems

19. *Distant relatives.* The structure of adenylate cyclase is similar to the structures of some types of DNA polymerases, suggesting that these enzymes derived from a common ancestor. Compare the reactions catalyzed by these two enzymes. In what ways are they similar?

20. *Kinase inhibitors as drugs.* Functional and structural analysis indicates that Gleevec is an ATP-competitive inhibitor of the Bcr-Abl kinase. In fact, many kinase inhibitors under investigation or currently marketed as drugs are ATP competitive. Can you suggest a potential drawback of drugs that utilize this particular mechanism of action?

Data Interpretation Problems

21. *Establishing specificity.* You wish to determine the hormone-binding specificity of a newly identified membrane receptor. Three different hormones, X, Y, and Z, were mixed with the receptor in separate experiments, and the percentage of binding capacity of the receptor was determined as a function of hormone concentration, as shown in graph A.

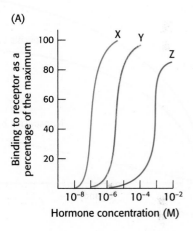

(a) What concentrations of each hormone yield 50% maximal binding?

(b) Which hormone shows the highest binding affinity for the receptor?

You next wish to determine whether the hormone–receptor complex stimulates the adenylate cyclase cascade. To do so, you measure adenylate cyclase activity as a function of hormone concentration, as shown in graph B.

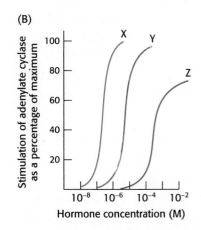

(c) What is the relationship between the binding affinity of the hormone–receptor complex and the ability of the hormone to enhance adenylate cyclase activity? What can you conclude about the mechanism of action of the hormone–receptor complex?

(d) Suggest experiments that would determine whether a $G_{\alpha s}$ protein is a component of the signal-transduction pathway.

22. *Binding issues.* A scientist wishes to determine the number of receptors specific for a ligand X, which he has in both radioactive and nonradioactive form. In one experiment, he adds increasing amounts of radioactive X and measures how much of it is bound to the cells. The result is shown as total binding in the following graph. Next, he performs the same experiment, except that he includes a several hundred-fold excess of nonradioactive X. This result is shown as nonspecific binding. The difference between the two curves is the specific binding.

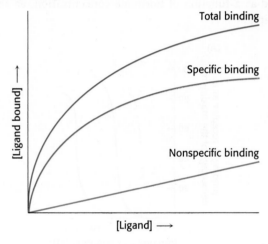

(a) Why is the total binding not an accurate representation of the number of receptors on the cell surface?

(b) What is the purpose of performing the experiment in the presence of excess nonradioactive ligand?

(c) What is the significance of the fact that specific binding attains a plateau?

23. *Counting receptors.* With the use of experiments such as those described in problems 21 and 22, the number of receptors in the cell membrane can be calculated. Suppose that the specific activity of the ligand is 10^{12} cpm per millimole and that the maximal specific binding is 10^4 cpm per milligram of membrane protein. There are 10^{10} cells per milligram of membrane protein. Assume that one ligand binds per receptor. Calculate the number of receptor molecules present per cell.

Think ▶ Pair ▶ Share Problem

24. In Section 14.5, we discussed the importance of a phosphorylated tyrosine residue in c-Src in maintaining the kinase in an inactive state. In humans, this residue is Tyr 527. One of the kinases that phosphorylates Tyr 527 is the creatively named C-terminal Src kinase (Csk). Additionally, one of the phosphatases that targets phospho-Tyr 527 is protein tyrosine phosphatase 1B (PTP1B).

(a) Do you expect Csk to behave as a tumor-suppressor gene or a proto-oncogene? Why?

(b) Do you expect PTP1B to behave as a tumor-suppressor gene or a proto-oncogene? Why?

Metabolism: Basic Concepts and Design

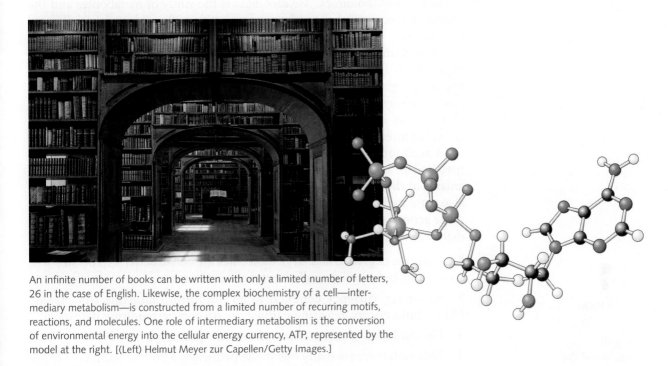

An infinite number of books can be written with only a limited number of letters, 26 in the case of English. Likewise, the complex biochemistry of a cell—intermediary metabolism—is constructed from a limited number of recurring motifs, reactions, and molecules. One role of intermediary metabolism is the conversion of environmental energy into the cellular energy currency, ATP, represented by the model at the right. [(Left) Helmut Meyer zur Capellen/Getty Images.]

 LEARNING OBJECTIVES

By the end of this chapter, you should be able to:

1. Explain what is meant by intermediate metabolism.

2. Identify the factors that make ATP an energy-rich molecule.

3. Explain how ATP can power reactions that would otherwise not take place.

4. Describe the relationship between the oxidation state of a carbon molecule and its usefulness as a fuel.

5. Describe the recurring motifs in metabolic pathways.

OUTLINE

15.1 Metabolism Is Composed of Many Coupled, Interconnecting Reactions

15.2 ATP Is the Universal Currency of Free Energy in Biological Systems

15.3 The Oxidation of Carbon Fuels Is an Important Source of Cellular Energy

15.4 Metabolic Pathways Contain Many Recurring Motifs

The concepts of conformation and dynamics developed in Part I—especially those dealing with the specificity and catalytic power of enzymes, the regulation of their catalytic activity, and the transport of molecules and ions across membranes—are fundamental tools for understanding biochemistry. In this chapter, we will begin to learn the "alphabet" of metabolism. Like the letters in a language alphabet that allow an infinite number of books, these limited reactions and cofactors can be combined to perform a vast array of biochemical feats. The concepts from Part I and the biochemical alphabet of this chapter enable us to now ask questions fundamental to biochemistry:

1. How does a cell extract energy and reducing power from its environment?

2. How does a cell synthesize the building blocks of its macromolecules and then the macromolecules themselves?

These processes are carried out by a highly integrated network of chemical reactions collectively known as *metabolism* or *intermediary metabolism*. Advances in analytical techniques such as mass spectrometry (Section 3.3) have greatly facilitated the study of metabolism and generated yet another "omics"—metabolomics. *Metabolomics* is the study of metabolites and the chemical reactions they undergo during physiological and pathological conditions.

More than a thousand chemical reactions take place in even as simple an organism as *Escherichia coli*. The array of reactions may seem overwhelming at first glance. However, closer scrutiny reveals that metabolism has a *coherent design containing many common motifs*. These motifs include the use of an energy currency and the repeated appearance of a limited number of activated intermediates. In fact, a group of about 100 molecules play central roles in all forms of life. Furthermore, although the number of reactions in metabolism is large, the number of *kinds* of reactions is small and the mechanisms of these reactions are usually quite simple. Metabolic pathways are also regulated in common ways. The purpose of this chapter is to introduce some general principles of metabolism to provide a foundation for the more detailed studies to follow. These principles are:

1. Fuels are degraded and large molecules are constructed step by step in a series of linked reactions called *metabolic pathways*.

2. An energy currency common to all life forms, *adenosine triphosphate (ATP)*, links energy-releasing pathways with energy-requiring pathways.

3. The oxidation of carbon fuels powers the formation of ATP.

4. Although there are many metabolic pathways, a limited number of types of reactions and particular intermediates are common to several pathways.

5. Metabolic pathways are highly regulated.

6. The enzymes involved in metabolism are organized into large complexes. The formation of metabolic enzymes into complexes increases efficiency by facilitating substrate and product movement between the individual enzymes of the complex. Moreover, some pathways produce or consume unstable or even toxic intermediates. Organization of the enzymes allows efficient processing of these products. We will investigate these complexes as we move through our study of biochemistry.

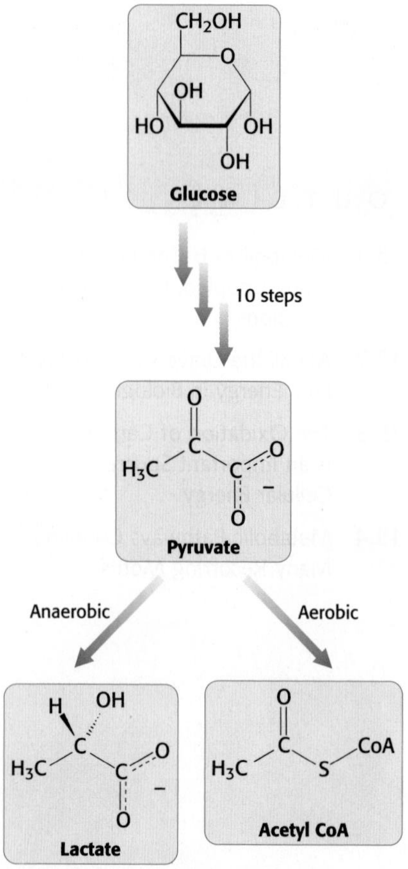

15.1 Metabolism Is Composed of Many Coupled, Interconnecting Reactions

Living organisms require a continual input of free energy for three major purposes: (1) the performance of mechanical work in muscle contraction and cellular movements, (2) the active transport of molecules and ions, and (3) the synthesis of macromolecules and other biomolecules from simple precursors. The free energy used in these processes, which maintain an organism in a state that is far from equilibrium, is derived from the environment. Photosynthetic organisms, or *phototrophs*, obtain this energy by trapping sunlight, whereas *chemotrophs*, which include animals, obtain energy through the oxidation of foodstuffs generated by phototrophs.

Metabolism consists of energy-yielding and energy-requiring reactions

Metabolism is essentially a sequence of chemical reactions that begins with a particular molecule and results in the formation of some other molecule or molecules in a carefully defined fashion (Figure 15.1). There are many

FIGURE 15.1 Glucose metabolism. Glucose is metabolized to pyruvate in 10 linked reactions. Under anaerobic conditions, pyruvate is metabolized to lactate and, under aerobic conditions, to acetyl CoA. The glucose-derived carbons of acetyl CoA are subsequently oxidized to CO_2.

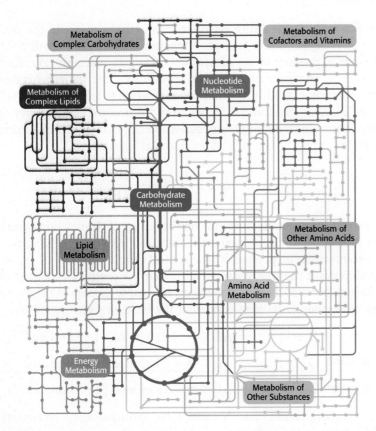

FIGURE 15.2 Metabolic pathways. Each node represents a specific metabolite. [From the Kyoto Encyclopedia of Genes and Genomes (www.genome.ad.jp/kegg).]

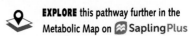

EXPLORE this pathway further in the Metabolic Map on SaplingPlus

such defined pathways in the cell (Figure 15.2), and we will examine a few of them in some detail later. These pathways are interdependent, and their activity is coordinated by exquisitely sensitive means of communication in which allosteric enzymes are predominant (Section 10.1). We considered the principles of this communication in Chapter 14.

We can divide metabolic pathways into two broad classes: (1) those that convert energy from fuels into biologically useful forms and (2) those that require inputs of energy to proceed. Although this division is often imprecise, it is nonetheless a useful distinction in an examination of metabolism. Those reactions that transform fuels into cellular energy are called *catabolic reactions* or, more generally, *catabolism*.

$$\text{Fuel (carbohydrates, fats)} \xrightarrow{\text{Catabolism}} CO_2 + H_2O + \text{useful energy}$$

Those reactions that require energy—such as the synthesis of glucose, fats, or DNA—are called *anabolic reactions* or *anabolism*. The useful forms of energy that are produced in catabolism are employed in anabolism to generate complex structures from simple ones, or energy-rich states from energy-poor ones.

$$\text{Useful energy} + \text{simple precursors} \xrightarrow{\text{Anabolism}} \text{complex molecules}$$

Some pathways can be either anabolic or catabolic, depending on the energy conditions in the cell. These pathways are referred to as *amphibolic pathways*.

An important general principle of metabolism is that *biosynthetic and degradative pathways are almost always distinct*. This separation is necessary for energetic reasons, as will be evident in subsequent chapters. It also facilitates the control of metabolism.

A thermodynamically unfavorable reaction can be driven by a favorable reaction

How are specific pathways constructed from individual reactions? A pathway must satisfy minimally two criteria: (1) the individual reactions must be *specific,* and (2) the entire set of reactions that constitute the pathway must be *thermodynamically favored.* A reaction that is specific will yield only one particular product or set of products from its reactants. As discussed in Chapter 8, enzymes provide this specificity. The thermodynamics of metabolism is most readily approached in relation to free energy, which was discussed in Chapters 1 and 8. A reaction can occur spontaneously only if ΔG, the change in free energy, is negative. Recall that ΔG for the formation of products C and D from substrates A and B is given by

$$\Delta G = \Delta G^{\circ\prime} + RT \ln\frac{[C][D]}{[A][B]}$$

Thus, the ΔG of a reaction depends on the *nature* of the reactants and products (expressed by the $\Delta G^{\circ\prime}$ term, the standard free-energy change) and on their *concentrations* (expressed by the second term).

An important thermodynamic fact is that *the overall free-energy change for a chemically coupled series of reactions is equal to the sum of the free-energy changes of the individual steps.* Consider the following reactions:

A $\rightleftharpoons$ B + C	$\Delta G^{\circ\prime} = +21 \text{ kJ mol}^{-1} (+5 \text{ kcal mol}^{-1})$
B $\rightleftharpoons$ D	$\Delta G^{\circ\prime} = -34 \text{ kJ mol}^{-1} (-8 \text{ kcal mol}^{-1})$
A $\rightleftharpoons$ C + D	$\Delta G^{\circ\prime} = -13 \text{ kJ mol}^{-1} (-3 \text{ kcal mol}^{-1})$

Under standard conditions, A cannot be spontaneously converted into B and C, because $\Delta G^{\circ\prime}$ is positive. However, the conversion of B into D under standard conditions is thermodynamically feasible. Because free-energy changes are additive, the conversion of A into C and D has a $\Delta G^{\circ\prime}$ of $-13 \text{ kJ mol}^{-1} (-3 \text{ kcal mol}^{-1})$, which means that it can occur spontaneously under standard conditions. Thus, *a thermodynamically unfavorable reaction can be driven by a thermodynamically favorable reaction to which it is coupled.* In this example, the reactions are coupled by the shared chemical intermediate B. Metabolic pathways are formed by the coupling of enzyme-catalyzed reactions such that the overall free energy of the pathway is negative.

15.2 ATP Is the Universal Currency of Free Energy in Biological Systems

Just as commerce is facilitated by the use of a common currency, the commerce of the cell—metabolism—is facilitated by the use of a common energy currency, *adenosine triphosphate* (ATP). Part of the free energy derived from the oxidation of foodstuffs and from light is transformed into this highly accessible molecule, which acts as the free-energy donor in most energy-requiring processes such as motion, active transport, and biosynthesis. Indeed, most of catabolism consists of reactions that extract energy from fuels such as carbohydrates and fats and convert it into ATP.

ATP hydrolysis is exergonic

ATP is a nucleotide consisting of adenine, a ribose, and a triphosphate unit (Figure 15.3). The active form of ATP is usually a complex of ATP with Mg^{2+} or Mn^{2+}. In considering the role of ATP as an energy carrier, we can focus on its triphosphate moiety. *ATP is an energy-rich molecule because its triphosphate unit contains two phosphoanhydride bonds.* A large amount of free energy is liberated when ATP is hydrolyzed to adenosine diphosphate

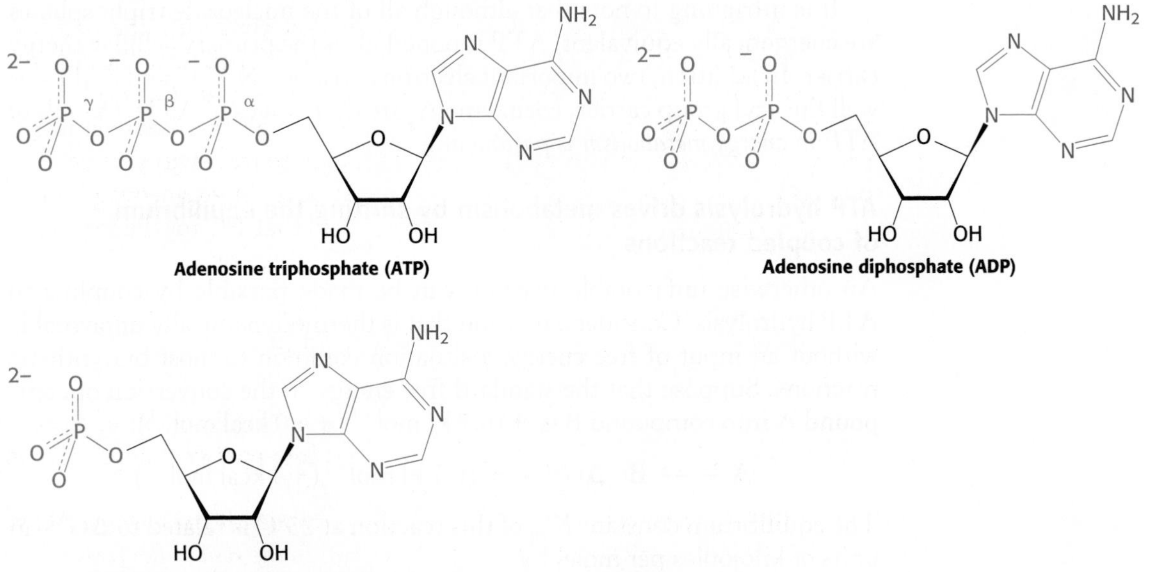

FIGURE 15.3 Structures of ATP, ADP, and AMP. These adenylates consist of adenine (blue), a ribose (black), and a tri-, di-, or monophosphate unit (red). The innermost phosphorus atom of ATP is designated P_α, the middle one P_β, and the outermost one P_γ.

(ADP) and orthophosphate (P_i), or when ATP is hydrolyzed to adenosine monophosphate (AMP) and pyrophosphate (PP_i).

$$ATP + H_2O \rightleftharpoons ADP + Pi$$
$$\Delta G^{\circ\prime} = -30.5 \text{ kJ mol}^{-1} (-7.3 \text{ kcal mol}^{-1})$$

$$ATP + H_2O \rightleftharpoons AMP + PPi$$
$$\Delta G^{\circ\prime} = -45.6 \text{ kJ mol}^{-1} (-10.9 \text{ kcal mol}^{-1})$$

The precise ΔG for these reactions depends on the ionic strength of the medium and on the concentrations of Mg^{2+} and other metal ions (problems 26 and 37). Under typical cellular concentrations, the ΔG for these hydrolyses is approximately $-50 \text{ kJ mol}^{-1} (-12 \text{ kcal mol}^{-1})$.

The free energy liberated in the hydrolysis of ATP is harnessed to drive reactions that require an input of free energy, such as muscle contraction. In turn, ATP is formed from ADP and P_i when fuel molecules are oxidized in chemotrophs or when light is trapped by phototrophs. *This ATP–ADP cycle is the fundamental mode of energy exchange in biological systems.*

Some biosynthetic reactions are driven by the hydrolysis of other nucleoside triphosphates—namely, guanosine triphosphate (GTP), uridine triphosphate (UTP), and cytidine triphosphate (CTP). The diphosphate forms of these nucleotides are denoted by GDP, UDP, and CDP, and the monophosphate forms are denoted by GMP, UMP, and CMP. Enzymes catalyze the transfer of the terminal phosphoryl group from one nucleotide to another. The phosphorylation of nucleoside monophosphates is catalyzed by a family of *nucleoside monophosphate kinases,* as discussed in Section 9.4. The phosphorylation of nucleoside diphosphates is catalyzed by *nucleoside diphosphate kinase,* an enzyme with broad specificity.

$$\underset{\substack{\text{Nucleoside} \\ \text{monophosphate}}}{NMP} + ATP \overset{\substack{\text{Nucleoside monophosphate} \\ \text{kinase}}}{\rightleftharpoons} NDP + ADP$$

$$\underset{\substack{\text{Nucleoside} \\ \text{diphosphate}}}{NDP} + ATP \overset{\substack{\text{Nucleoside diphosphate} \\ \text{kinase}}}{\rightleftharpoons} NTP + ADP$$

It is intriguing to note that although all of the nucleoside triphosphates are energetically equivalent, ATP is nonetheless the primary cellular energy carrier. In addition, two important electron carriers—NAD^+ and FAD—as well the acyl group carrier, coenzyme A, are derivatives of ATP. *The role of ATP in energy metabolism is paramount.*

ATP hydrolysis drives metabolism by shifting the equilibrium of coupled reactions

An otherwise unfavorable reaction can be made possible by coupling to ATP hydrolysis. Consider a reaction that is thermodynamically unfavorable without an input of free energy, a situation common to most biosynthetic reactions. Suppose that the standard free energy of the conversion of compound A into compound B is $+16.7 \text{ kJ mol}^{-1} (+4.0 \text{ kcal mol}^{-1})$:

$$A \rightleftharpoons B \quad \Delta G^{\circ\prime} = +16.7 \text{ kJ mol}^{-1} (+4 \text{ kcal mol}^{-1})$$

The equilibrium constant K'_{eq} of this reaction at 25°C is related to $\Delta G^{\circ\prime}$ (in units of kilojoules per mole) by

$$K'_{eq} = [B]_{eq}/[A]_{eq} = e^{-\Delta G^{\circ\prime}/2.47} = 1.15 \times 10^{-3}$$

Thus, net conversion of A into B cannot take place when the molar ratio of B to A is equal to or greater than 1.15×10^{-3}. However, A can be converted into B under these conditions if the reaction is coupled to the hydrolysis of ATP. Under standard conditions, the $\Delta G^{\circ\prime}$ of hydrolysis is approximately $-30.5 \text{ kJ mol}^{-1} (-7.3 \text{ kcal mol}^{-1})$. The new overall reaction is

$$A + ATP + H_2O \rightleftharpoons B + ADP + P_i$$

$$\Delta G^{\circ\prime} = -13.8 \text{ kJ mol}^{-1} (-3.3 \text{ kcal mol}^{-1})$$

Its free-energy change of $-13.8 \text{ kJ mol}^{-1} (-3.3 \text{ kcal mol}^{-1})$ is the sum of the value of $\Delta G^{\circ\prime}$ for the conversion of A into B $[+16.7 \text{ kJ mol}^{-1} (+4.0 \text{ kcal mol}^{-1})]$ and the value of $\Delta G^{\circ\prime}$ for the hydrolysis of ATP $[-30.5 \text{ kJ mol}^{-1} (-7.3 \text{ kcal mol}^{-1})]$. At pH 7, the equilibrium constant of this coupled reaction is

$$K'_{eq} = \frac{[B]_{eq}}{[A]_{eq}} \times \frac{[ADP]_{eq}[P_i]_{eq}}{[ATP]_{eq}} = e^{13.8/2.47} = 2.67 \times 10^2$$

At equilibrium, the ratio of [B] to [A] is given by

$$\frac{[B]_{eq}}{[A]_{eq}} = K'_{eq} \frac{[ATP]_{eq}}{[ADP]_{eq}[P_i]_{eq}}$$

which means that the hydrolysis of ATP enables A to be converted into B until the [B]/[A] ratio reaches a value of 2.67×10^2. This equilibrium ratio is strikingly different from the value of 1.15×10^{-3} for the reaction A $\longrightarrow$ B in the absence of ATP hydrolysis. In other words, coupling the hydrolysis of ATP with the conversion of A into B under standard conditions has changed the equilibrium ratio of B to A by a factor of about 10^5. If we were to use the ΔG of hydrolysis of ATP under cellular conditions $[-50.2 \text{ kJ mol}^{-1} (-12 \text{ kcal mol}^{-1})]$ in our calculations instead of $\Delta G^{\circ\prime}$, the change in the equilibrium ratio would be even more dramatic, on the order of 10^8.

We see here the thermodynamic essence of ATP's action as an *energy-coupling agent.* Cells maintain ATP levels by using oxidizable substrates or light as sources of free energy for synthesizing the molecule. In the cell, the hydrolysis of an ATP molecule in a coupled reaction then changes the equilibrium ratio of products to reactants by a very large factor, of the order of 10^8. More generally, the hydrolysis of n ATP molecules changes the equilibrium ratio of a coupled reaction (or sequence of reactions) by a factor of 10^{8n}. For example, the hydrolysis of three ATP molecules in a coupled reaction changes the equilibrium ratio by a factor of 10^{24}. Thus, *a thermodynamically unfavorable reaction sequence can be converted into a favorable one by coupling it to the hydrolysis of a sufficient number of ATP molecules in a new reaction.* It should also be emphasized that A and B in the preceding coupled reaction may be interpreted very generally, not only as different chemical species. For example, A and B may represent activated and unactivated conformations of a protein that is activated by phosphorylation with ATP. Through such changes in protein conformation, molecular motors such as myosin, kinesin, and dynein convert the chemical energy of ATP into mechanical energy (See Chapter 35 on Sapling Plus). Indeed, this conversion is the basis of muscle contraction.

Alternatively, A and B may refer to the concentrations of an ion or molecule on the outside and inside of a cell, as in the active transport of a nutrient. The active transport of Na^+ and K^+ across membranes is driven by the phosphorylation of the sodium–potassium pump by ATP and its subsequent dephosphorylation (Section 13.2).

The high phosphoryl potential of ATP results from structural differences between ATP and its hydrolysis products

What makes ATP an efficient phosphoryl-group donor? Let us compare the standard free energy of hydrolysis of ATP with that of a phosphate ester, such as glycerol 3-phosphate:

$$ATP + H_2O \rightleftharpoons ADP + P_i$$

$$\Delta G^{\circ\prime} = -30.5 \text{ kJ mol}^{-1}(-7.3 \text{ kcal mol}^{-1})$$

$$\text{Glycerol 3-phosphate} + H_2O \rightleftharpoons \text{glycerol} + P_i$$

$$\Delta G^{\circ\prime} = -9.2 \text{ kJ mol}^{-1}(-2.2 \text{ kcal mol}^{-1})$$

The magnitude of $\Delta G^{\circ\prime}$ for the hydrolysis of glycerol 3-phosphate is much smaller than that of ATP, which means that ATP has a stronger tendency to transfer its terminal phosphoryl group to water than does glycerol 3-phosphate. In other words, ATP has a higher *phosphoryl-transfer potential (phosphoryl-group-transfer potential)* than does glycerol 3-phosphate.

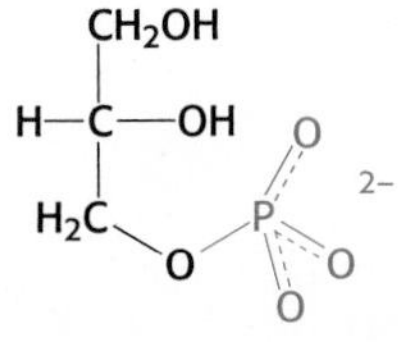

The high phosphoryl-transfer potential of ATP can be explained by features of the ATP structure. Because $\Delta G^{\circ\prime}$ depends on the *difference* in free energies of the products and reactants, we need to examine the structures of both ATP and its hydrolysis products, ADP and P_i, to answer this question. Four factors are important: *resonance stabilization, electrostatic repulsion, increase in entropy,* and *stabilization due to hydration.*

1. *Resonance stabilization.* Orthophosphate (P_i), one of the products of ATP hydrolysis, has greater resonance stabilization than do any of the ATP phosphoryl groups. Orthophosphate has a number of resonance forms of similar energy (Figure 15.4), whereas the γ phosphoryl group

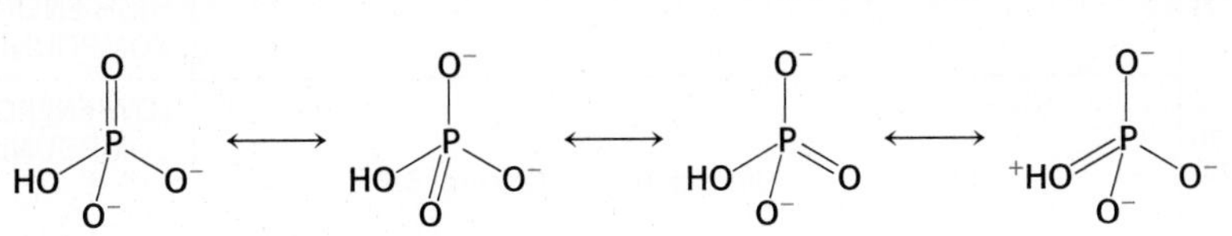

Glycerol 3-phosphate

FIGURE 15.4 Resonance structures of orthophosphate.

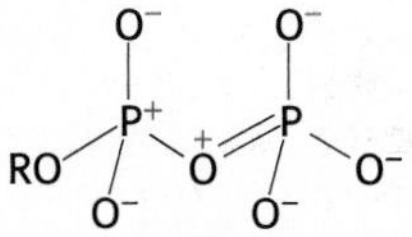

FIGURE 15.5 Improbable resonance structure. The structure contributes little to the terminal part of ATP because two positive charges are placed adjacent to each other.

of ATP has a smaller number. Forms like that shown in Figure 15.6 are unfavorable because a positively charged oxygen atom is adjacent to a positively charged phosphorus atom—an electrostatically unfavorable juxtaposition.

2. *Electrostatic repulsion.* At pH 7, the triphosphate unit of ATP carries about four negative charges. These charges repel one another because they are in close proximity. The repulsion between them is reduced when ATP is hydrolyzed.

3. *Increase in entropy.* The entropy of the products of ATP hydrolysis is greater, in that there are now two molecules instead of a single ATP molecule. We disregard the molecule of water used to hydrolyze the ATP; given the high concentration (55.5 M), there is effectively no change in the concentration of water during the reaction.

4. *Stabilization due to hydration.* Water binds to ADP and P_i, stabilizing these molecules, and rendering the reverse reaction—the synthesis of ATP—more unfavorable.

ATP is often called a high-energy phosphate compound, and its phosphoanhydride bonds are referred to as high-energy bonds. Indeed, a "squiggle" ($\sim P$) is often used to indicate such a bond. Nonetheless, there is nothing special about the bonds themselves. *They are high-energy bonds in the sense that much free energy is released when they are hydrolyzed*, for the reasons listed above.

Phosphoryl-transfer potential is an important form of cellular energy transformation

The standard free energies of hydrolysis provide a convenient means of comparing the phosphoryl-transfer potential of phosphorylated compounds. Such comparisons reveal that ATP is not the only compound with a high phosphoryl-transfer potential. In fact, some compounds in biological systems have a higher phosphoryl-transfer potential than that of ATP. These compounds include phosphoenolpyruvate (PEP), 1,3-bisphosphoglycerate (1,3-BPG), and creatine phosphate (Figure 15.6). Thus, PEP can

FIGURE 15.6 Compounds with high phosphoryl-transfer potential. The role of ATP as the cellular energy currency is illustrated by its relation to other phosphorylated compounds. ATP has a phosphoryl-transfer potential that is intermediate among the biologically important phosphorylated molecules. High-phosphoryl-transfer-potential compounds (1,3-BPG, PEP, and creatine phosphate) derived from the metabolism of fuel molecules are used to power ATP synthesis. In turn, ATP donates a phosphoryl group to other biomolecules to facilitate their metabolism. [Data from D. L. Nelson and M. M. Cox, *Lehninger Principles of Biochemistry*, 5th ed. (W. H. Freeman and Company, 2009), Fig. 13-19.]

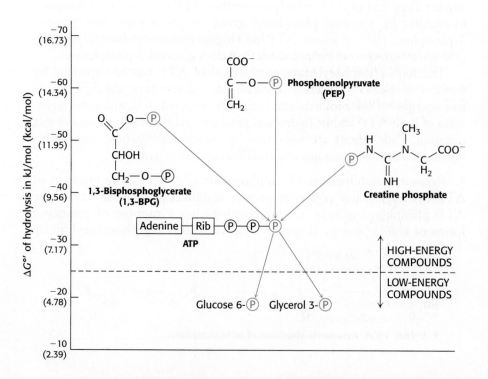

TABLE 15.1 Standard free energies of hydrolysis of some phosphorylated compounds

Compound	kJ mol^{-1}	kcal mol^{-1}
Phosphoenolpyruvate	−61.9	−14.8
1,3-Bisphosphoglycerate	−49.4	−11.8
Creatine phosphate	−43.1	−10.3
ATP (to ADP)	−30.5	−7.3
Glucose 1-phosphate	−20.9	−5.0
Pyrophosphate	−19.3	−4.6
Glucose 6-phosphate	−13.8	−3.3
Glycerol 3-phosphate	−9.2	−2.2

transfer its phosphoryl group to ADP to form ATP. Indeed, this transfer is one of the ways in which ATP is generated in the breakdown of sugars (Chapter 16). It is significant that ATP has a phosphoryl-transfer potential that is intermediate among the biologically important phosphorylated molecules (Table 15.1). *This intermediate position enables ATP to function efficiently as a carrier of phosphoryl groups.*

The amount of ATP in muscle suffices to sustain contractile activity for less than a second. Creatine phosphate in vertebrate muscle serves as a reservoir of high-potential phosphoryl groups that can be readily transferred to ADP. Indeed, we use creatine phosphate to regenerate ATP from ADP every time we exercise strenuously. This reaction is catalyzed by *creatine kinase.*

$$\text{Creatine phosphate} + \text{ADP} \overset{\substack{\text{Creatine} \\ \text{kinase}}}{\rightleftharpoons} \text{ATP} + \text{creatine}$$

At pH 7, the standard free energy of hydrolysis of creatine phosphate is −43.1 kJ mol^{-1} (−10.3 kcal mol^{-1}), compared with −30.5 kJ mol^{-1} (−7.3 kcal mol^{-1}) for ATP. Hence, the standard free-energy change in forming ATP from creatine phosphate is −12.6 kJ mol^{-1} (−3.0 kcal mol^{-1}), which corresponds to an equilibrium constant of 162.

$$K'_{eq} = \frac{[\text{ATP}][\text{creatine}]}{[\text{ADP}][\text{creatine phosphate}]} = e^{-\Delta G^{\circ\prime}/2.47} = e^{12.6/2.47} = 162$$

In resting muscle, typical concentrations of these metabolites are [ATP] = 4 mM, [ADP] = 0.013 mM, [creatine phosphate] = 25 mM, and [creatine] = 13 mM. Because of its abundance and high phosphoryl-transfer potential relative to that of ATP, creatine phosphate is a highly effective phosphoryl buffer. Indeed, creatine phosphate is the major source of phosphoryl groups for ATP regeneration for a runner during the first 4 seconds of a 100-meter sprint. The fact that creatine phosphate can replenish ATP pools is the basis of the use of creatine as a dietary supplement by athletes in sports requiring short bursts of intense activity. After the creatine phosphate pool is depleted, ATP must be generated through metabolism (Figure 15.7).

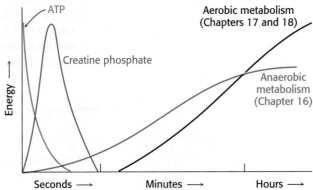

FIGURE 15.7 Sources of ATP during exercise. In the initial seconds, exercise is powered by existing high-phosphoryl-transfer compounds (ATP and creatine phosphate). Subsequently, the ATP must be regenerated by metabolic pathways.

 EXPLORE this pathway further in the Metabolic Map on *Sapling*Plus

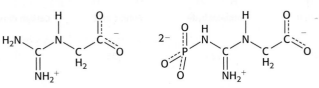

Creatine   **Creatine phosphate**

ATP may have roles other than in energy and signal transduction

The ATP concentration inside the cell is typically between 2 and 8 mM. Interestingly, this is 10-to 100-fold higher than the K_M for most enzymes that use ATP as a substrate. Why would a cell maintain such high concentrations of ATP?

Recent research suggests that ATP may function as a biological *hydrotrope*. Hydrotropes are amphipathic molecules, with a hydrophobic component and a hydrophilic component, but unlike fatty acids or lipids, the hydrophobic component is too small to self-aggregate (Section 12.1). In the case of ATP, the adenine moiety is hydrophobic while the rest of the molecule is hydrophilic.

The cytoplasm has a protein concentration of greater than 100 mg/ml; maintaining protein solubility at such concentrations is problematic. At intracellular concentrations, ATP prevents the formation of protein aggregates and dissolves those that form, implying that another role of ATP is maintaining protein solubility. Interestingly, the concentration of ATP falls with age for reasons under investigation. This decrease could cause, in part, protein aggregation leading to neurodegenerative and other diseases. Reconciling this new proposed role of ATP with its more established roles will be a fascinating area of investigation.

15.3 The Oxidation of Carbon Fuels Is an Important Source of Cellular Energy

ATP serves as the principal *immediate donor of free energy* in biological systems rather than as a long-term storage form of free energy. In a typical cell, an ATP molecule is consumed within a minute of its formation. Although the total quantity of ATP in the body is limited to approximately 100 g, *the turnover of this small quantity of ATP is very high.* For example, a resting human being consumes about 40 kg (88 pounds) of ATP in 24 hours. During strenuous exertion, the rate of utilization of ATP may be as high as 0.5 kg/minute. For a 2-hour run, 60 kg (132 pounds) of ATP is utilized. Clearly, having mechanisms for regenerating ATP is vital. Motion, active transport, signal amplification, and biosynthesis can take place only if ATP is continually regenerated from ADP (Figure 15.8). *The generation of ATP is one of the primary roles of catabolism.* The carbon in fuel molecules—such as glucose and fats—is oxidized to CO_2. The resulting electrons are captured and used to regenerate ATP from ADP and P_i.

In aerobic organisms, the ultimate electron acceptor in the oxidation of carbon is O_2 and the oxidation product is CO_2. Consequently, the more reduced a carbon is to begin with, the more free energy is released by its oxidation. Figure 15.9 shows the $\Delta G^{\circ\prime}$ of oxidation for one-carbon compounds.

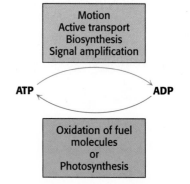

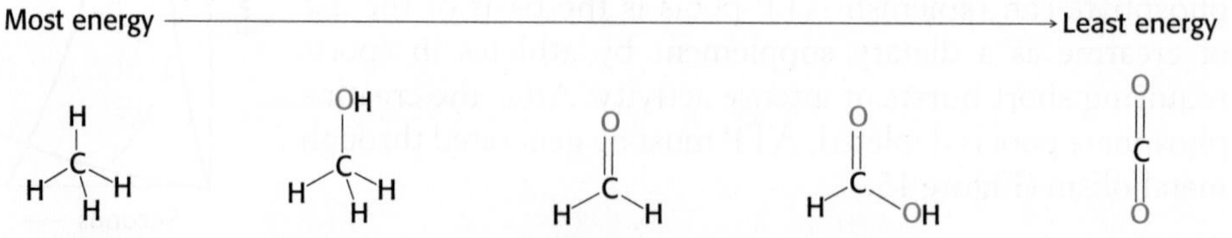

FIGURE 15.8 ATP–ADP cycle. This cycle is the fundamental mode of energy exchange in biological systems.

FIGURE 15.9 Free energy of oxidation of single-carbon compounds.

		Most energy ⟶ Least energy			
		Methane	**Methanol**	**Formaldehyde**	**Formic acid**	**Carbon dioxide**
$\Delta G^{\prime}_{oxidation}$ (kJ mol^{-1})		−820	−703	−523	−285	0
$\Delta G^{\prime}_{oxidation}$ (kcal mol^{-1})		−196	−168	−125	−68	0

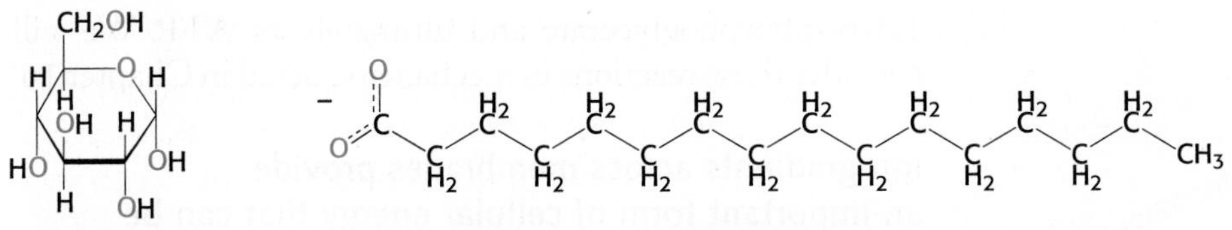

Glucose   Saturated fatty acid

FIGURE 15.10 Prominent fuels. Fats are a more efficient fuel source than carbohydrates such as glucose because the carbon in fats is more reduced.

Fuel molecules are more complex (Figure 15.10) than the single-carbon compounds depicted in Figure 15.9. Nevertheless, oxidation of these fuels takes place one carbon at a time. The carbon-oxidation energy is used in some cases to create a compound with high phosphoryl-transfer potential and in other cases to create an ion gradient. In either case, the end point is the formation of ATP.

Compounds with high phosphoryl-transfer potential can couple carbon oxidation to ATP synthesis

How is the energy released from the oxidation of a carbon compound converted into ATP? As an example, consider glyceraldehyde 3-phosphate (shown in the margin), which is a metabolite of glucose formed in the oxidation of that sugar. The C-1 carbon (shown in red) is at the aldehyde-oxidation level and is not in its most oxidized state. Oxidation of the aldehyde to an acid will release energy.

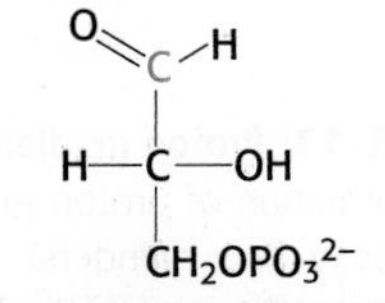

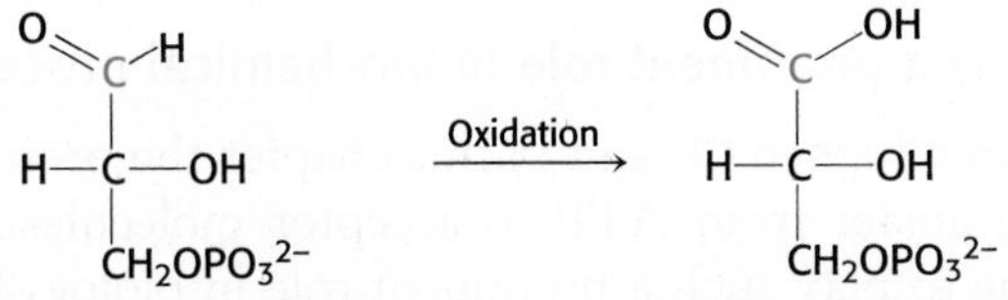

Glyceraldehyde 3-phosphate (GAP)

Glyceraldehyde 3-phosphate →(Oxidation)→ 3-Phosphoglyceric acid

However, the oxidation does not take place directly. Instead, the carbon oxidation generates an acyl phosphate, 1,3-bisphosphoglycerate. The electrons released are captured by NAD^+, which we will consider shortly.

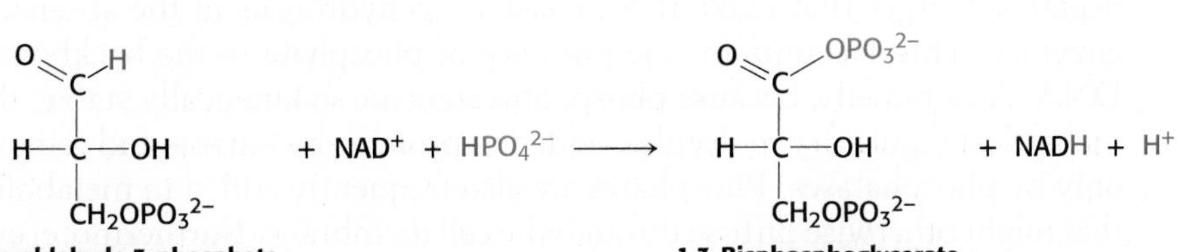

Glyceraldehyde 3-phosphate (GAP) $+ NAD^+ + HPO_4^{2-} \longrightarrow$ 1,3-Bisphosphoglycerate (1,3-BPG) $+ NADH + H^+$

For reasons similar to those discussed for ATP, 1,3-bisphosphoglycerate has a high phosphoryl-transfer potential that is, in fact, greater than that of ATP. Thus, the hydrolysis of 1,3-BPG can be coupled to the synthesis of ATP.

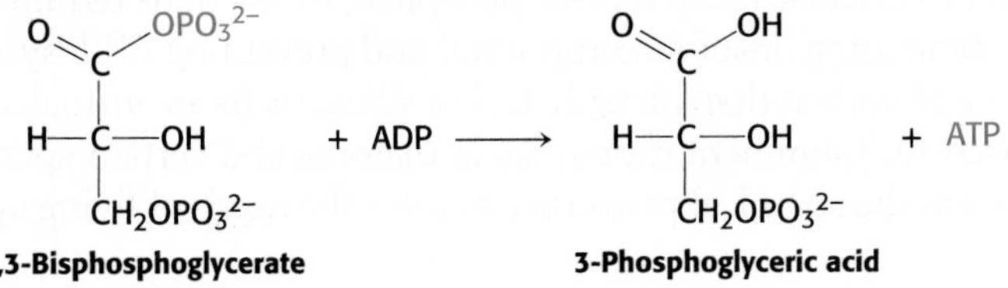

1,3-Bisphosphoglycerate $+ ADP \longrightarrow$ 3-Phosphoglyceric acid $+ ATP$

The energy of oxidation is initially trapped as a high-phosphoryl-transfer-potential compound and then used to form ATP. The oxidation energy of a carbon atom is transformed into phosphoryl-transfer potential, first as

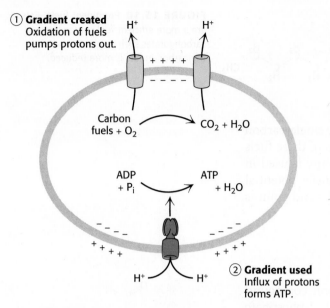

① **Gradient created**
Oxidation of fuels pumps protons out.

Carbon fuels + O_2 → CO_2 + H_2O

ADP + P_i → ATP + H_2O

H^+ H^+

② **Gradient used**
Influx of protons forms ATP.

FIGURE 15.11 Proton gradients. The oxidation of fuels can power the formation of proton gradients by the action of specific proton pumps (yellow cylinders). These proton gradients can in turn drive the synthesis of ATP when the protons flow through an ATP-synthesizing enzyme (red complex).

1,3-bisphosphoglycerate and ultimately as ATP. We will consider these reactions in mechanistic detail in Chapter 16.

Ion gradients across membranes provide an important form of cellular energy that can be coupled to ATP synthesis

As described in Chapter 13, electrochemical potential is an effective means of storing free energy. In fact, the electrochemical potential of *ion gradients across membranes*—produced by the oxidation of fuel molecules or by photosynthesis—ultimately powers the synthesis of most of the ATP in cells. In general, ion gradients are versatile means of coupling thermodynamically unfavorable reactions to favorable ones. In animals, *proton gradients* generated by the oxidation of carbon fuels account for more than 90% of ATP generation (Figure 15.11). This process is called *oxidative phosphorylation* (Chapter 18). ATP hydrolysis can then be used to form ion gradients of different types and functions. The electrochemical potential of a Na^+ gradient, for example, can be tapped to pump Ca^{2+} out of cells or to transport nutrients such as sugars and amino acids into cells.

Phosphates play a prominent role in biochemical processes

We have seen in Chapter 14, and in this chapter the prominence of phosphoryl group transfer from ATP to acceptor molecules. How is it that phosphate came to play such a prominent role in biology? Phosphate and its esters have several characteristics that render it useful for biochemical systems. First, phosphate esters have an important chemical "betwixt and between" characteristic being *thermodynamically unstable* while being *kinetically stable*. Phosphate esters are thus molecules whose energy release can be manipulated by enzymes. The stability of phosphate esters is due to the negative charges that make them resistant to hydrolysis in the absence of enzymes. This accounts for the presence of phosphate in the backbone of DNA. Additionally, because phosphate esters are so kinetically stable, they make ideal regulatory molecules, added to proteins by kinases and removed only by phosphatases. Phosphates are also frequently added to metabolites that might otherwise diffuse through the cell membrane. Furthermore, even when transporters exist for unphosphorylated forms of a metabolite, the addition of a phosphate changes the geometry and polarity of the molecules so that they no longer fit in the binding sites of the transporters.

No other ions have the chemical characteristics of phosphate. Citrate is not sufficiently charged to prevent hydrolysis. Arsenate forms esters that are unstable and susceptible to spontaneous hydrolysis. Indeed, arsenate is poisonous to cells because it can replace phosphate in reactions required for ATP synthesis, generating unstable compounds and preventing ATP synthesis. Silicate is more abundant than phosphate, but silicate salts are virtually insoluble, and are used for biomineralization, as in diatoms and certain sponges. Only phosphate has the chemical properties to meet the needs of living systems.

Energy from foodstuffs is extracted in three stages

Let us take an overall view of the processes of energy conversion in higher organisms before considering them in detail in subsequent chapters. Hans

Krebs described three stages in the generation of energy from the oxidation of foodstuffs (Figure 15.12).

In the first stage, large molecules in food are broken down into smaller units in the process of *digestion*. Proteins are hydrolyzed to their 20 different amino acids, polysaccharides are hydrolyzed to simple sugars such as glucose, and lipids are hydrolyzed to glycerol and fatty acids. The degradation products are then absorbed by the cells of the intestine and distributed throughout the body. This stage is strictly a preparation stage; no useful energy is captured in this phase.

In the second stage, these numerous small molecules are degraded to a few simple units that play a central role in metabolism. In fact, most of them—sugars, fatty acids, glycerol, and several amino acids—are converted into the acetyl unit of acetyl CoA. Some ATP is generated in this stage, but the amount is small compared with that obtained in the third stage.

In the third stage, ATP is produced from the complete oxidation of the acetyl unit of acetyl CoA. The third stage consists of the citric acid cycle and oxidative phosphorylation, which are *the final common pathways in the oxidation of fuel molecules.* Acetyl CoA brings acetyl units into the citric acid cycle [also called the tricarboxylic acid (TCA) cycle or Krebs cycle], where they are completely oxidized to CO_2. Four pairs of electrons are transferred (three to NAD^+ and one to FAD) for each acetyl group that is oxidized. Then, a proton gradient is generated as electrons flow from the reduced forms of these carriers to O_2. This gradient is used to synthesize ATP (Chapters 17 and 18).

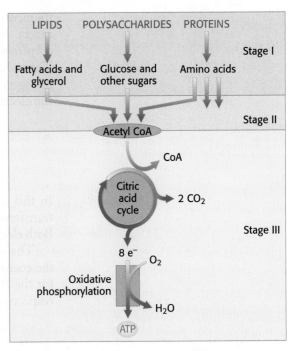

FIGURE 15.12 Stages of catabolism. The extraction of energy from fuels can be divided into three stages.

15.4 Metabolic Pathways Contain Many Recurring Motifs

At first glance, metabolism appears intimidating because of the sheer number of reactants and reactions. Nevertheless, there are unifying themes that make the comprehension of this complexity more manageable. These unifying themes include common metabolites, reactions, and regulatory schemes that stem from a common evolutionary heritage.

Activated carriers exemplify the modular design and economy of metabolism

We have seen that phosphoryl transfer can be used to drive otherwise endergonic reactions, alter the conformation energy of a protein, or serve as a signal to alter the activity of a protein. The phosphoryl-group donor in all of these reactions is ATP. In other words, *ATP is an activated carrier of phosphoryl groups because phosphoryl transfer from ATP is an exergonic process.* The use of activated carriers is a recurring motif in biochemistry, and we will consider several such carriers here. Many such activated carriers function as coenzymes (Section 8.1):

1. *Activated carriers of electrons for fuel oxidation.* In aerobic organisms, the ultimate electron acceptor in the oxidation of fuel molecules is O_2. However, electrons are not transferred directly to O_2. Instead, fuel molecules transfer electrons to special carriers, which are either *pyridine nucleotides* or *flavins*. The reduced forms of these carriers then transfer their high-potential electrons to O_2.

Nicotinamide adenine dinucleotide is a major electron carrier in the oxidation of fuel molecules (Figure 15.13). The reactive part of NAD^+ is

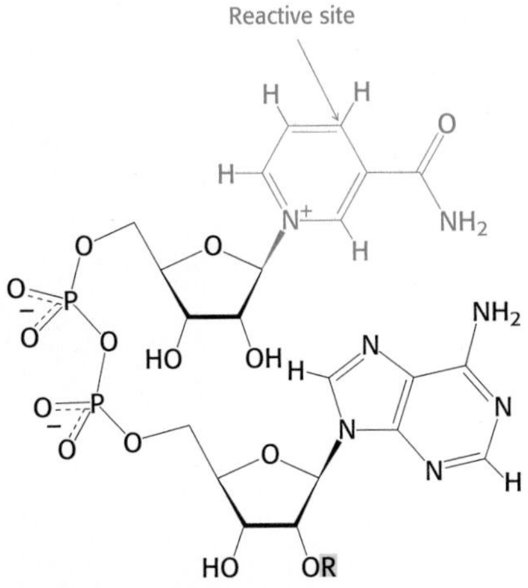

FIGURE 15.13 Structures of the oxidized forms of nicotinamide-derived electron carriers. Nicotinamide adenine dinucleotide (NAD^+) and nicotinamide adenine dinucleotide phosphate ($NADP^+$) are prominent carriers of high-energy electrons. In NAD^+, R = H; in $NADP^+$, R = PO_3^{2-}.

its nicotinamide ring, a pyridine derivative synthesized from the vitamin niacin. *In the oxidation of a substrate, the nicotinamide ring of NAD$^+$ accepts a hydrogen ion and two electrons, which are equivalent to a hydride ion (H:$^-$).* The reduced form of this carrier is called *NADH*. In the oxidized form, the nitrogen atom carries a positive charge, as indicated by NAD$^+$. NAD$^+$ is the electron acceptor in many reactions of the type

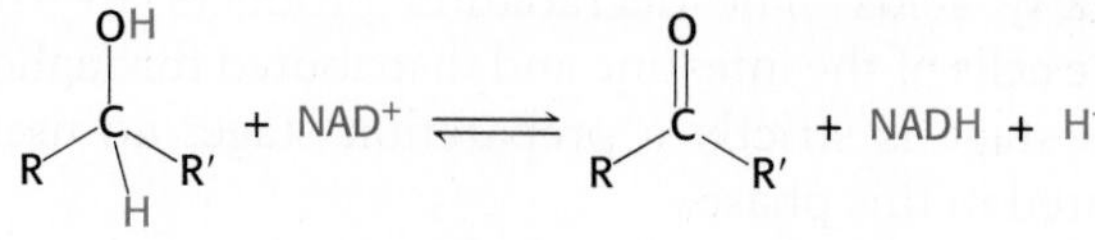

In this dehydrogenation, one hydrogen atom of the substrate is directly transferred to NAD$^+$, whereas the other appears in the solvent as a proton. Both electrons lost by the substrate are transferred to the nicotinamide ring.

The other major electron carrier in the oxidation of fuel molecules is the coenzyme *flavin adenine dinucleotide* (Figure 15.14). The abbreviations for the oxidized and reduced forms of this carrier are FAD and FADH$_2$, respectively. FAD is the electron acceptor in reactions of this type:

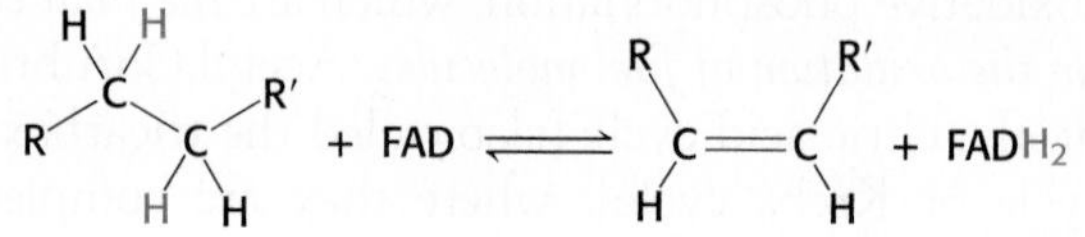

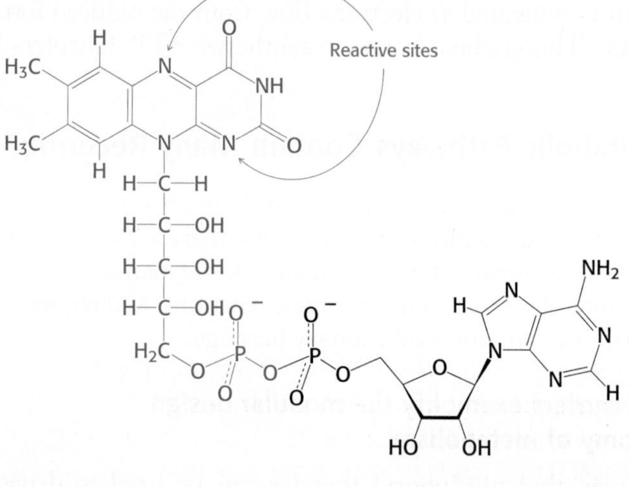

FIGURE 15.14 Structure of the oxidized form of flavin adenine dinucleotide (FAD). This electron carrier consists of a flavin mononucleotide (FMN) unit (shown in blue) and an AMP unit (shown in black).

The reactive part of FAD is its isoalloxazine ring, a derivative of the vitamin riboflavin (Figure 15.15). FAD, like NAD$^+$, can accept two electrons. In doing so, FAD, unlike NAD$^+$, takes up two protons. These carriers of high-potential electrons as well as flavin mononucleotide (FMN)—an electron carrier similar to FAD but lacking the adenine nucleotide—will be considered further in Chapter 18.

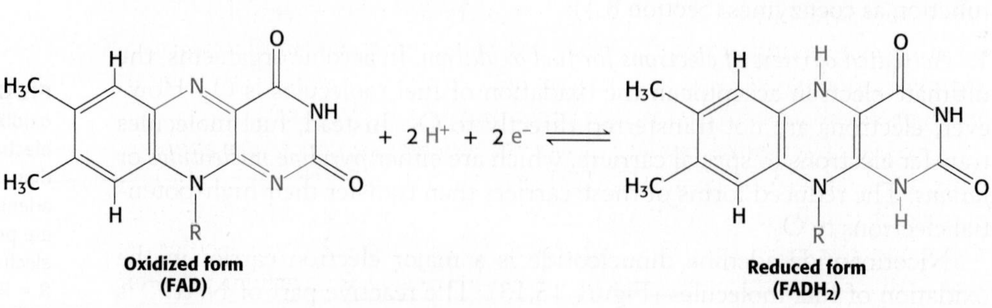

FIGURE 15.15 Structures of the reactive components of FAD and FADH$_2$. The electrons and protons are carried by the isoalloxazine ring component of FAD and FADH$_2$.

Oxidized form (FAD)

Reduced form (FADH$_2$)

2. *An activated carrier of electrons for reductive biosynthesis.* High-potential electrons are required in most biosyntheses because the precursors are more oxidized than the products. Hence, reducing power is needed in addition to ATP. For example, in the biosynthesis of fatty acids, a keto group is reduced to a methylene group in several steps. This sequence of reactions requires an input of four electrons.

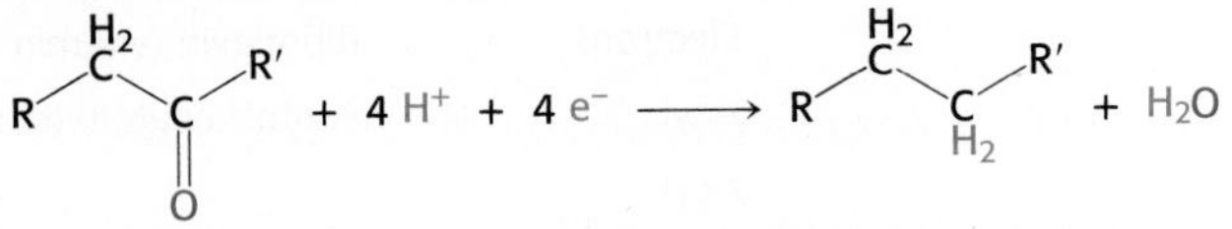

The electron donor in most reductive biosyntheses is NADPH, the reduced form of nicotinamide adenine dinucleotide phosphate ($NADP^+$; Figure 15.13). NADPH differs from NADH in that the $2'$-hydroxyl group of its adenosine moiety is esterified with phosphate. NADPH carries electrons in the same way as NADH. However, *NADPH is used almost exclusively for reductive biosyntheses, whereas NADH is used primarily for the generation of ATP.* The extra phosphoryl group on NADPH is a tag that enables enzymes to distinguish between high-potential electrons to be used in anabolism and those to be used in catabolism.

3. *An activated carrier of two-carbon fragments.* Coenzyme A, another central molecule in metabolism, is a carrier of acyl groups derived from the vitamin pantothenate (Figure 15.16). Acyl groups are important constituents both in catabolism, as in the oxidation of fatty acids, and in anabolism,

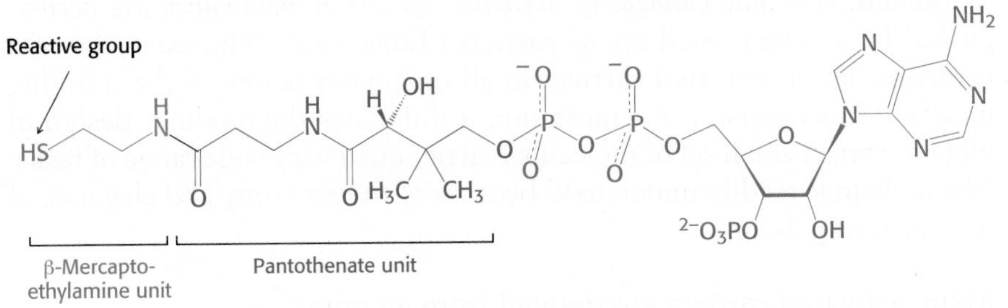

Reactive group

β-Mercapto-ethylamine unit

Pantothenate unit

FIGURE 15.16 Structure of coenzyme A (CoA-SH).

as in the synthesis of membrane lipids. The terminal sulfhydryl group in CoA is the reactive site. Acyl groups are linked to CoA by thioester bonds. The resulting derivative is called an *acyl CoA*. An acyl group often linked to CoA is the acetyl unit; this derivative is called *acetyl CoA*. The $\Delta G^{\circ\prime}$ for the hydrolysis of acetyl CoA has a large negative value:

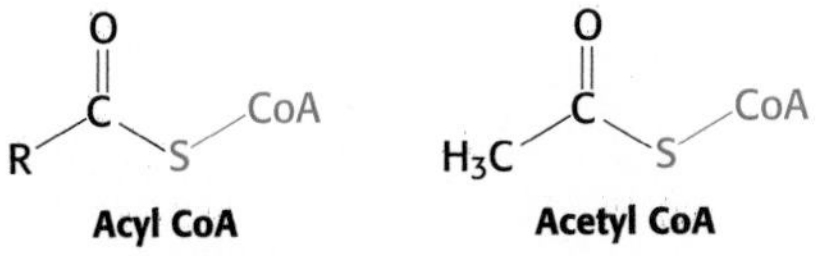

$$\text{Acetyl CoA} + H_2O \rightleftharpoons \text{acetate} + \text{CoA} + H^+$$

$$\Delta G^{\circ\prime} = -31.4 \text{ kJ mol}^{-1} (-7.5 \text{ kcal mol}^{-1})$$

A thioester is thermodynamically more unstable than an oxygen ester because the electrons of the $C{=}O$ bond cannot form resonance structures with the $C{-}S$ bond that are as stable as those that they can form with the $C{-}O$ bond. Consequently, *acetyl CoA has a high acetyl-group-transfer potential because transfer of the acetyl group is exergonic.* Acetyl CoA carries an activated acetyl group, just as ATP carries an activated phosphoryl group.

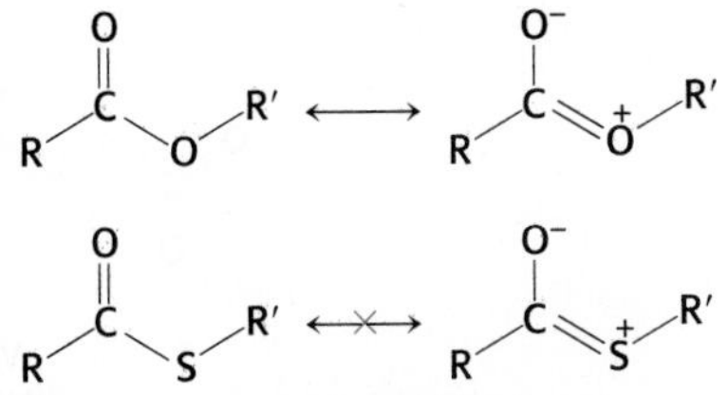

Acyl CoA

Acetyl CoA

Oxygen esters are stabilized by resonance structures not available to thioesters.

The use of activated carriers illustrates two key aspects of metabolism. First, NADH, NADPH, and $FADH_2$ react slowly with O_2 in the absence of a catalyst. Likewise, ATP and acetyl CoA are hydrolyzed slowly (over many hours or even days) in the absence of a catalyst. These molecules are kinetically quite stable in the face of a large thermodynamic driving force for

TABLE 15.2 Some activated carriers in metabolism

Carrier molecule in activated form	Group carried	Vitamin precursor
ATP	Phosphoryl	
NADH and NADPH	Electrons	Nicotinate (niacin) (vitamin B_3)
$FADH_2$	Electrons	Riboflavin (vitamin B_2)
$FMNH_2$	Electrons	Riboflavin (vitamin B_2)
Coenzyme A	Acyl	Pantothenate (vitamin B_5)
Lipoamide	Acyl	
Thiamine pyrophosphate	Aldehyde	Thiamine (vitamin B_1)
Biotin	CO_2	Biotin (vitamin B_7)
Tetrahydrofolate	One-carbon units	Folate (vitamin B_9)
S-Adenosylmethionine	Methyl	
Uridine diphosphate glucose	Glucose	
Cytidine diphosphate diacylglycerol	Phosphatidate	
Nucleoside triphosphates	Nucleotides	

Note: Many of the activated carriers are coenzymes that are derived from water-soluble vitamins.

reaction with O_2 (in regard to the electron carriers) and H_2O (for ATP and acetyl CoA). *The kinetic stability of these molecules in the absence of specific catalysts is essential for their biological function because it enables enzymes to control the flow of free energy and reducing power.*

Second, *most interchanges of activated groups in metabolism are accomplished by a rather small set of carriers* (Table 15.2). The existence of a recurring set of activated carriers in all organisms is one of the unifying motifs of biochemistry. Furthermore, it illustrates the modular design of metabolism. A small set of molecules carries out a very wide range of tasks. Metabolism is readily understood because of the economy and elegance of its underlying design.

Many activated carriers are derived from vitamins

Almost all the activated carriers that act as coenzymes are derived from *vitamins*. Vitamins are organic molecules that are needed in small amounts in the diets of some higher animals. Table 15.3 lists the vitamins that act as coenzymes and Figure 15.17 shows the structures of some of them. This series of vitamins is known as the vitamin B group. In all cases, the vitamin must be modified before it can serve its function. We have already touched

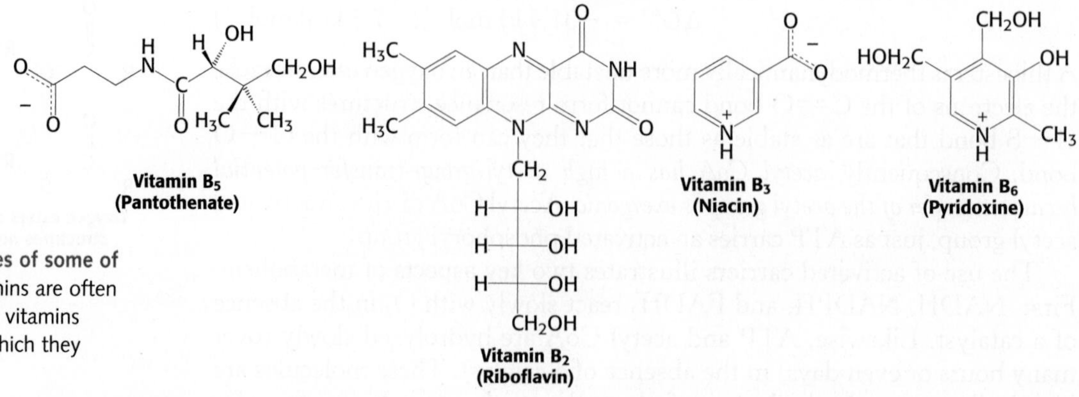

FIGURE 15.17 Structures of some of the B vitamins. These vitamins are often referred to as water-soluble vitamins because of the ease with which they dissolve in water.

TABLE 15.3 The B vitamins

Vitamin	Coenzyme	Typical reaction Type	Consequences of Deficiency
Thiamine (B_1)	Thiamine pyrophosphate	Aldehyde transfer	Beriberi (weight loss, heart problems, neurological dysfunction)
Riboflavin (B_2)	Flavin adenine dinucleotide (FAD)	Oxidation–reduction	Cheliosis and angular stomatitis (lesions of the mouth), dermatitis
Pyridoxine (B_6)	Pyridoxal phosphate	Group transfer to or from amino acids	Depression, confusion, convulsions
Nicotinic acid (niacin) (B_3)	Nicotinamide adenine dinucleotide (NAD^+)	Oxidation–reduction	Pellagra (dermatitis, depression, diarrhea)
Pantothenic acid (B_5)	Coenzyme A	Acyl-group transfer	Hypertension
Biotin (B_7)	Biotin–lysine adducts (biocytin)	ATP-dependent carboxylation and carboxyl-group transfer	Rash about the eyebrows, muscle pain, fatigue (rare)
Folic acid (B_9)	Tetrahydrofolate	Transfer of one-carbon components; thymine synthesis	Anemia, neural-tube defects in development
B_{12}	5'-Deoxyadenosyl-cobalamin	Transfer of methyl groups; intramolecular rearrangements	Anemia, pernicious anemia, methylmalonic acidosis

on the roles of niacin, riboflavin, and pantothenate. We will see these three and the other B vitamins many times in our study of biochemistry.

Vitamins serve the same roles in nearly all forms of life, but higher animals lost the capacity to synthesize them in the course of evolution. For instance, whereas *E. coli* can thrive on glucose and organic salts, human beings require at least 12 vitamins in their diet. The biosynthetic pathways for vitamins can be complex; thus, it is biologically more efficient to ingest vitamins than to synthesize the enzymes required to construct them from simple molecules. This efficiency comes at the cost of dependence on other organisms for chemicals essential for life. Indeed, vitamin deficiency can generate diseases in all organisms requiring these molecules (Tables 15.3 and 15.4).

TABLE 15.4 Noncoenzyme vitamins

Vitamin	Function	Deficiency
A	Roles in vision, growth, Reproduction	Night blindness, cornea damage, damage to respiratory and gastrointestinal tract
C (ascorbic acid)	Antioxidant	Scurvy (swollen and bleeding gums, subdermal hemorrhaging)
D	Regulation of calcium and phosphate metabolism	Rickets (children): skeletal deformities, impaired growth Osteomalacia (adults): soft, bending bones
E	Antioxidant	Lesions in muscles and nerves (rare)
K	Blood coagulation	Subdermal hemorrhaging

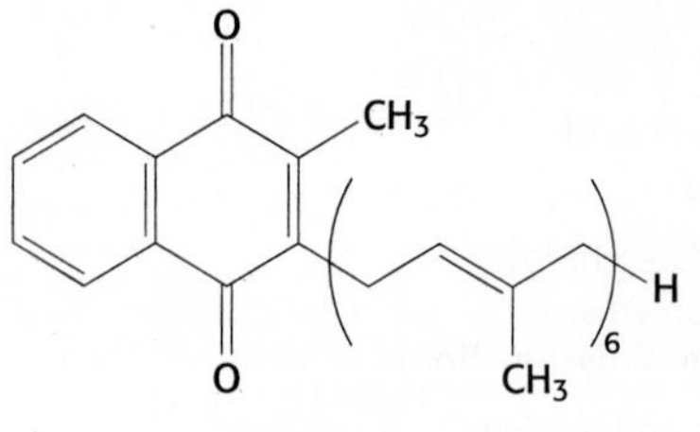

Vitamin K

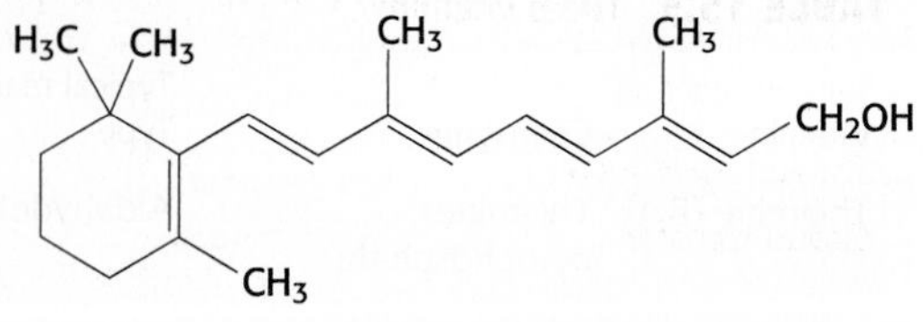

Vitamin A (Retinol)

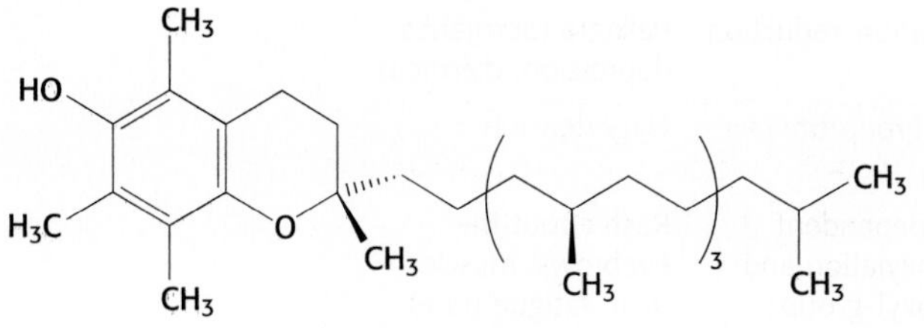

Vitamin E (α-Tocopherol)

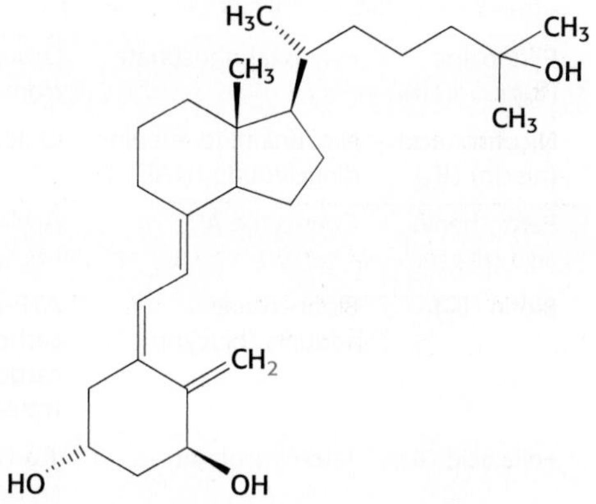

1,25-Dihydroxyvitamin D₃ (Calcitriol)

FIGURE 15.18 Structures of some vitamins that do not function as coenzymes. These vitamins are often called the fat-soluble vitamins because of their hydrophobic nature.

Not all vitamins function as coenzymes. Vitamins designated by the letters A, C, D, E, and K (Figure 15.18 and Table 15.4) have a diverse array of functions.

- Vitamin A (retinol) is the precursor of retinal, the light-sensitive group in rhodopsin and other visual pigments (See Section 34.3 on 🔁 SaplingPlus), and retinoic acid, an important signaling molecule. A deficiency of this vitamin leads to night blindness. In addition, young animals require vitamin A for growth.
- Vitamin C, or ascorbate, acts as an antioxidant. A deficiency in vitamin C results in the formation of unstable collagen molecules and is the cause of scurvy, a disease characterized by skin lesions and blood-vessel fragility (Section 27.6).
- Vitamin D is a precursor to a hormone that regulates the metabolism of calcium and phosphorus. A deficiency in vitamin D impairs bone formation in growing animals.
- Vitamin E (α-tocopherol) deficiency causes a variety of neuromuscular pathologies. This vitamin inactivates reactive oxygen species such as hydroxyl radicals before they can oxidize unsaturated membrane lipids, damaging cell structures.
- Vitamin K is required for normal blood clotting.

We have already met vitamin K in its biochemical context (Section 10.4), and will meet the other noncoenzyme vitamins later in our study of biochemistry.

Key reactions are reiterated throughout metabolism

Just as there is an economy of design in the use of activated carriers, there is also an economy of design in biochemical reactions. The thousands of metabolic reactions, bewildering at first in their variety, can be subdivided into just six types (Table 15.5). These six types of reactions correspond to the six major classes of enzymes discussed in the Appendix to Chapter 8. Specific reactions of each type appear repeatedly, reducing the number of reactions that a student needs to learn.

TABLE 15.5 Types of chemical reactions in metabolism

Type of reaction	Description
Oxidation–reduction	Electron transfer
Group transfer	Transfer of a functional group from one molecule to another
Hydrolytic	Cleavage of bonds by the addition of water
Carbon bond cleavage by means other than hydrolysis or oxidation	Two substrates yielding one product or vice versa. When H_2O or CO_2 are a product, a double bond is formed.
Isomerization	Rearrangement of atoms to form isomers
Ligation requiring ATP cleavage	Formation of covalent bonds (i.e., carbon–carbon bonds)

1. *Oxidation–reduction reactions* are essential components of many pathways. Useful energy is often derived from the oxidation of carbon compounds. Consider the following two reactions:

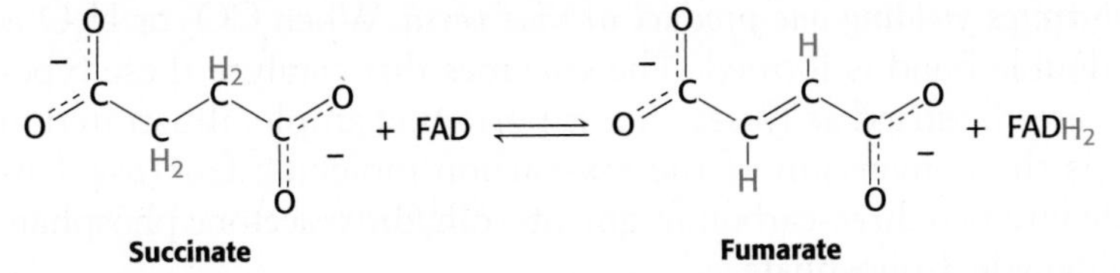

Succinate + FAD $\rightleftharpoons$ **Fumarate** + $FADH_2$ (1)

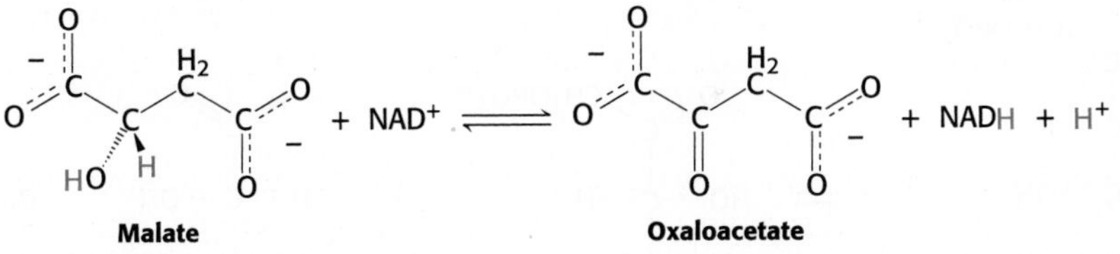

Malate + NAD^+ $\rightleftharpoons$ **Oxaloacetate** + NADH + H^+ (2)

These two oxidation–reduction reactions are components of the citric acid cycle (Chapter 17), which completely oxidizes the activated two-carbon fragment of acetyl CoA to two molecules of CO_2. In reaction 1, $FADH_2$ carries the electrons, whereas, in reaction 2, electrons are carried by NADH.

2. *Group-transfer reactions* play a variety of roles. Reaction 3 is representative of such a reaction. A phosphoryl group is transferred from the activated phosphoryl-group carrier, ATP, to glucose, the initial step in glycolysis, a key pathway for extracting energy from glucose (Chapter 16). This reaction traps glucose in the cell so that further catabolism can take place.

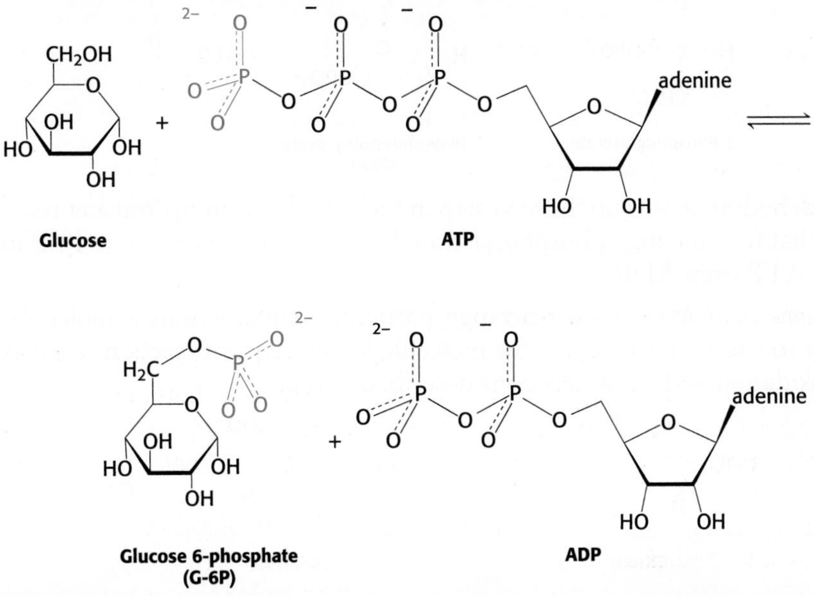

Glucose + **ATP** $\rightleftharpoons$ **Glucose 6-phosphate (G-6P)** + **ADP**   (3)

As stated earlier, group-transfer reactions are used to synthesize ATP. We also saw examples of their use in signaling pathways (Chapter 14).

3. *Hydrolytic reactions* cleave bonds by the addition of water. Hydrolysis is a common means of degrading large molecules, either to facilitate further metabolism or to reuse some of the components for biosynthetic purposes. Proteins are digested by hydrolytic cleavage (Chapters 9 and 10). Reaction 4 illustrates the hydrolysis of a peptide to yield two smaller peptides.

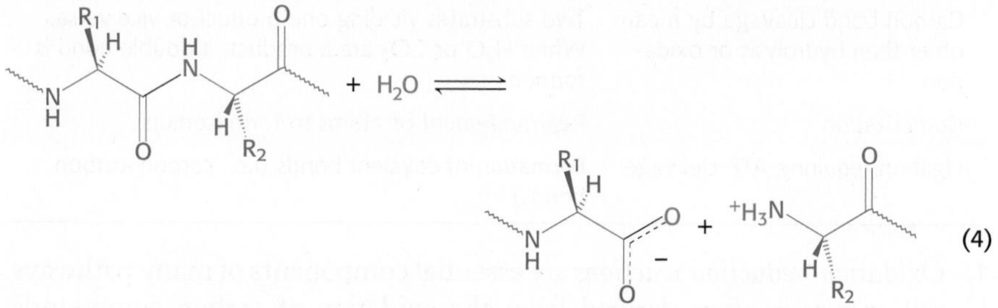

$$\text{(4)}$$

4. *Carbon bonds can be cleaved by means other than hydrolysis or oxidation,* with two substrates yielding one product or vice versa. When CO_2 or H_2O is released, a double bond is formed. The enzymes that catalyze these types of reactions are classified as *lyases*. An important example, illustrated in reaction 5, is the conversion of the six-carbon molecule fructose 1,6-bisphosphate into two three-carbon fragments: dihydroxyacetone phosphate and glyceraldehyde 3-phosphate.

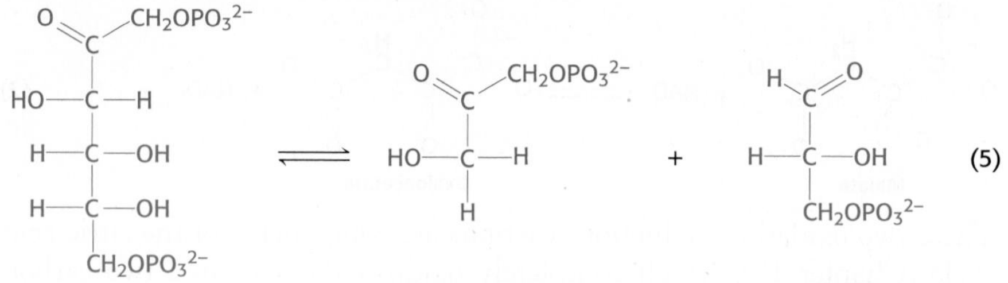

$$\text{(5)}$$

Fructose 1,6-bisphosphate　　　　**Dihydroxyacetone phosphate**　　**Glyceraldehyde 3-phosphate**
(F-1,6-BP)　　　　　　　　　　　**(DHAP)**　　　　　　　　　　　　**(GAP)**

This reaction is a critical step in glycolysis (Chapter 16). Dehydrations to form double bonds, such as the formation of phosphoenolpyruvate (Figure 15.6) from 2-phosphoglycerate (reaction 6), are important reactions of this type.

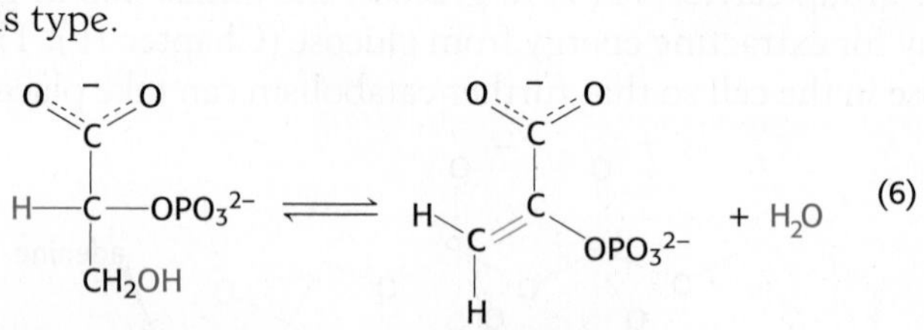

$$\text{(6)}$$

2-Phosphoglycerate　　　　　**Phosphoenolpyruvate**
　　　　　　　　　　　　　　　　(PEP)

The dehydration sets up the next step in the pathway, a group-transfer reaction that uses the high phosphoryl-transfer potential of the product PEP to form ATP from ADP.

5. *Isomerization reactions* rearrange particular atoms within a molecule. Their role is often to prepare the molecule for subsequent reactions such as the oxidation–reduction reactions described in type 1.

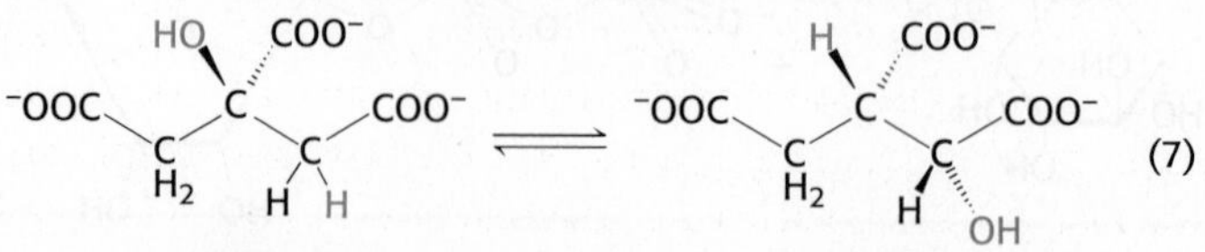

$$\text{(7)}$$

Citrate　　　　　　　　　　　　**Isocitrate**

Reaction 7 is, again, a component of the citric acid cycle. This isomerization prepares the molecule for subsequent oxidation and decarboxylation reactions by moving the hydroxyl group of citrate from a tertiary to a secondary position.

6. *Ligation reactions* form bonds by using free energy from ATP cleavage. Reaction 8 illustrates the ATP-dependent formation of a carbon–carbon bond, necessary to combine smaller molecules to form larger ones. Oxaloacetate is formed from pyruvate and CO_2.

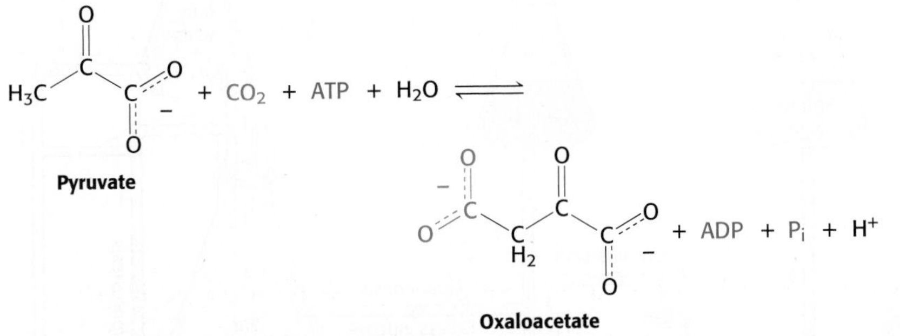

Pyruvate

$+ CO_2 + ATP + H_2O \rightleftharpoons$

Oxaloacetate

$+ ADP + P_i + H^+$ (8)

The oxaloacetate can be used in the citric acid cycle, or converted into glucose or amino acids such as aspartic acid.

These six fundamental reaction types are the basis of metabolism. Remember, all six types can proceed in either direction, depending on the standard free energy for the specific reaction and the concentrations of the reactants and products inside the cell. An effective way to learn is to look for commonalities in the diverse metabolic pathways we will be examining. There is a chemical logic that, when exposed, renders the complexity of the chemistry of living systems more manageable and reveals its elegance.

Metabolic processes are regulated in three principal ways

It is evident that the complex network of metabolic reactions must be rigorously regulated. The levels of available nutrients must be monitored and the activity of metabolic pathways must be altered and integrated to create *homeostasis*, a stable biochemical environment. At the same time, metabolic control must be flexible, able to adjust metabolic activity to the constantly changing external environments of cells. Figure 15.19 illustrates the nutrient pools and their connections that must be monitored and regulated. Metabolism is regulated through control of (1) *the amounts of enzymes,* (2) *their catalytic activities,* and (3) *the accessibility of substrates.*

Controlling the amounts of enzymes. The amount of a particular enzyme depends on both its rate of synthesis and its rate of degradation. The level of many enzymes is adjusted by a change in the *rate of transcription* of the genes encoding them (Chapters 30, 32, and 33). In *E. coli,* for example, the presence of lactose induces within minutes a more than 50-fold increase in the rate of synthesis of β-galactosidase, the enzyme required for the breakdown of this disaccharide.

Controlling catalytic activity. The catalytic activity of enzymes is controlled in several ways. *Allosteric control* is especially important. For example, the first reaction in many biosynthetic pathways is allosterically inhibited by the ultimate product of the pathway. A well-understood example of *feedback inhibition* is the inhibition of aspartate transcarbamoylase by cytidine triphosphate (Section 10.1). This type of control can be almost instantaneous. Another recurring mechanism is *reversible covalent modification* (Section 10.3). For example, glycogen phosphorylase, the enzyme

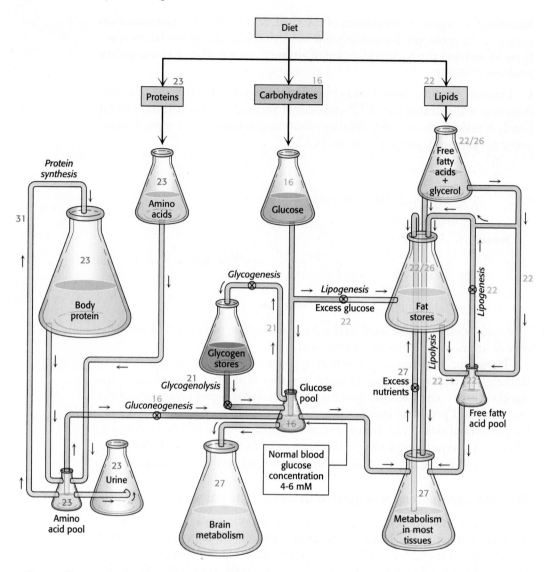

Fats ▢ Carbohydrates ▢ Glycogen ▢ Amino acids

FIGURE 15.19 Homeostasis. Maintaining a constant cellular environment requires complex metabolic regulation that coordinates the use of nutrient pools. The numbers indicate the chapters in which the topics are discussed. [Information from D. U. Silverthorn, *Human Physiology: An Integrated Approach*, 3rd ed. (Pearson, 2004), Figure 22-2.]

catalyzing the breakdown of glycogen, a storage form of sugar, is activated by the phosphorylation of a particular serine residue when glucose is scarce (Section 21.1).

Hormones coordinate metabolic relations between different tissues, often by regulating the reversible modification of key enzymes. For instance, the hormone epinephrine triggers a signal-transduction cascade in muscle, resulting in the phosphorylation and activation of key enzymes and leading to the rapid degradation of glycogen to glucose, which is then used to supply ATP for muscle contraction. As described in Chapter 14, many hormones act through intracellular messengers, such as cyclic AMP and calcium ion, that coordinate the activities of many target proteins.

Many reactions in metabolism are controlled by the *energy status* of the cell. One index of the energy status is the *energy charge*, which is proportional to the mole fraction of ATP plus half the mole fraction of ADP, given that ATP contains two anhydride bonds, whereas ADP contains one. Hence, the energy charge is defined as

$$\text{Energy charge} = \frac{[\text{ATP}] + \frac{1}{2}[\text{ADP}]}{[\text{ATP}] + [\text{ADP}] + [\text{AMP}]}$$

The energy charge can have a value ranging from 0 (all AMP) to 1 (all ATP). *ATP-generating (catabolic) pathways are inhibited by a high-energy charge, whereas ATP-utilizing (anabolic) pathways are stimulated by a high-energy charge.* In plots of the reaction rates of such pathways versus the energy charge, the curves are steep near an energy charge of 0.9, where they usually intersect (Figure 15.20). It is evident that the control of these pathways has evolved to maintain the energy charge within rather narrow limits. In other words, *the energy charge, like the pH of a cell, is buffered.* The energy charge of most cells ranges from 0.90 to 0.95, but can fall to less than 0.7 in muscle during high-intensity exercise. An alternative index of the energy status is the *phosphorylation potential*, which is defined as

$$\text{Phosphorylation potential} = \frac{[\text{ATP}]}{[\text{ADP}] + [\text{P}_i]}$$

The phosphorylation potential, in contrast with the energy charge, depends on the concentration of P_i and is directly related to the free-energy storage available from ATP.

Controlling the accessibility of substrates. Controlling the *availability of substrates* is another means of regulating metabolism in all organisms. For instance, glucose breakdown can take place in many cells only if insulin is present to promote the entry of glucose into the cell. In eukaryotes, metabolic regulation and flexibility are enhanced by compartmentalization. The transfer of substrates from one compartment of a cell to another can serve as a control point. For example, fatty acid oxidation takes place in mitochondria, whereas fatty acid synthesis takes place in the cytoplasm. *Compartmentalization segregates opposed reactions.*

Aspects of metabolism may have evolved from an RNA world

How did the complex pathways that constitute metabolism evolve? The current thinking is that RNA was an early biomolecule that dominated metabolism, serving both as a catalyst and an information storage molecule. This hypothetical time is called the RNA world.

Why do activated carriers such as ATP, NADH, FADH_2, and coenzyme A contain adenosine diphosphate units? A possible explanation is that these molecules evolved from the early RNA catalysts. Non-RNA units such as the isoalloxazine ring may have been recruited to serve as efficient carriers of activated electrons and chemical units, a function not readily performed by RNA itself. We can picture the adenine ring of FADH_2 binding to a uracil unit in a niche of an RNA enzyme (ribozyme) by base-pairing, whereas the isoalloxazine ring protrudes and functions as an electron carrier. When the more versatile proteins replaced RNA as the major catalysts, the ribonucleotide coenzymes stayed essentially unchanged because they were already well-suited to their metabolic roles. The nicotinamide unit of NADH, for example, can readily transfer electrons irrespective of whether the adenine unit interacts with a base in an RNA enzyme or with amino acid residues in a protein enzyme. With the advent of protein enzymes, these important cofactors evolved as free molecules without losing the adenosine diphosphate vestige of their RNA-world ancestry. That molecules and motifs of metabolism are common to all forms of life testifies to their common origin and to the retention of functioning modules through billions

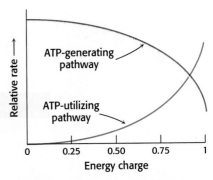

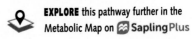

FIGURE 15.20 Energy charge regulates metabolism. When the energy charge is high, ATP inhibits the relative rates of a typical ATP-generating (catabolic) pathway and stimulates the typical ATP-utilizing (anabolic) pathway.

EXPLORE this pathway further in the Metabolic Map on **Sapling** Plus of years of evolution. Our understanding of metabolism, like that of other biological processes, is enriched by inquiry into how these beautifully integrated patterns of reactions came into being.

SUMMARY

All cells transform energy. They extract energy from their environment and use this energy to convert simple molecules into cellular components.

15.1 Metabolism Is Composed of Many Coupled, Interconnecting Reactions

The process of energy transduction takes place through metabolism, a highly integrated network of chemical reactions. Metabolism can be subdivided into catabolism (reactions employed to extract energy from fuels) and anabolism (reactions that use this energy for biosynthesis). The most valuable thermodynamic concept for understanding bioenergetics is free energy. A reaction can occur spontaneously only if the change in free energy (ΔG) is negative. A thermodynamically unfavorable reaction can be driven by a thermodynamically favorable one, which is the hydrolysis of ATP in many cases.

15.2 ATP Is the Universal Currency of Free Energy in Biological Systems

The energy derived from catabolism is transformed into adenosine triphosphate. ATP hydrolysis is exergonic and the energy released can be used to power cellular processes, including motion, active transport, and biosynthesis. Under cellular conditions, the hydrolysis of ATP shifts the equilibrium of a coupled reaction by a factor of 10^8. ATP, the universal currency of energy in biological systems, is an energy-rich molecule because it contains two phosphoanhydride bonds.

15.3 The Oxidation of Carbon Fuels Is an Important Source of Cellular Energy

ATP formation is coupled to the oxidation of carbon fuels, either directly or through the formation of ion gradients. Photosynthetic organisms can use light to generate such gradients. ATP is consumed to drive endergonic reactions and in signal-transduction processes. The extraction of energy from foodstuffs by aerobic organisms comprises three stages. In the first stage, large molecules are broken down into smaller ones, such as amino acids, sugars, and fatty acids. In the second stage, these small molecules are degraded to a few simple units, such as acetyl CoA, that have widespread roles in metabolism. The third stage of metabolism is the citric acid cycle and oxidative phosphorylation, in which ATP is generated as electrons flow to O_2, the ultimate electron acceptor, and fuels are completely oxidized to CO_2.

15.4 Metabolic Pathways Contain Many Recurring Motifs

Metabolism is characterized by common motifs. A small number of recurring activated carriers, such as ATP, NADH, and acetyl CoA, transfer activated groups in many metabolic pathways. NADPH, which carries two electrons at a high potential, provides reducing power for reductive biosynthesis of cell components. Many activated carriers are derived from vitamins—small organic molecules required in the diets of many higher organisms. Moreover, key reaction types are used repeatedly in metabolic pathways.

Metabolism is regulated in a variety of ways. The amounts of some critical enzymes are controlled by regulation of the rate of synthesis and degradation. In addition, the catalytic activities of many enzymes are regulated by allosteric interactions and by covalent modification. The movement of many substrates into cells and subcellular compartments also is controlled. The energy charge, which depends on the relative amounts of ATP, ADP, and AMP, plays a role in metabolic regulation. A high energy charge inhibits ATP-generating (catabolic) pathways, while it stimulates ATP-utilizing (anabolic) pathways.

APPENDIX

Problem-Solving Strategies

PROBLEM: Creatine phosphate is used as a phosphoryl donor for ATP synthesis in muscle.

a. What enzyme catalyzes the synthesis of creatine phosphate? Write out the reaction.

b. For a skeletal muscle at rest, the following metabolites are present at the indicated concentrations.

$$[ATP] = 4\,mM; ADP = 0.013\,mM$$

$$[creatine\ phosphate] = 25\,mM; [creatine] = 13\,mM$$

Calculate the ΔG for the creatine kinase reaction for a muscle at rest.

SOLUTION: Part (a) to this problem is straightforward. You either remember the enzyme and the reaction or you look it up on page 473.

$$\text{Creatine phosphate} + \text{ADP} \underset{}{\overset{\text{Creatine kinase}}{\rightleftharpoons}} \text{ATP} + \text{creatine}$$

Now, to attack part (b).

▶ **What formula do we need to calculate ΔG?**

$$\Delta G = \Delta G^{\circ\prime} + RT\ln\frac{[C][D]}{[A][B]}$$

This formula can be located on page 468.

The values necessary for the right-hand term are provided, but we need to know $\Delta G^{\circ\prime}$ for the kinase reaction.

▶ **How do we determine $\Delta G^{\circ\prime}$?**

We need to dissect the kinase reaction into its two component reactions and look up the $\Delta G^{\circ\prime}$ values for each of the component reactions. Remember, the $\Delta G^{\circ\prime}$ values are additive.

▶ **What are the two component reactions of the creatine kinase reaction?**

$$\text{ADP} + \text{P}_i \longrightarrow \text{ATP} + \text{H}_2\text{O}$$

$$\text{Creatine phosphate} + \text{H}_2\text{O} \longrightarrow \text{creatine} + \text{P}_i$$

Using Table 15.1 in the text, we can determine the $\Delta G^{\circ\prime}$ values for both reactions.

For ATP synthesis, $\Delta G^{\circ\prime} = +30.5\,kJ/mol$. For creatine phosphate hydrolysis, $\Delta G^{\circ\prime} = -43.1\,kJ/mol$.

▶ **What is the $\Delta G^{\circ\prime}$ kinase reaction?**

$\Delta G^{\circ\prime}$ values are additive, so the $\Delta G^{\circ\prime}$ for the kinase reaction is $-43.1\,kJ/mol + 30.5\,kJ/mol = -12.6\,kJ/mol$.

Now, having determined $\Delta G^{\circ\prime}$, we can use the equation above and the values provided to determine ΔG.

$$\Delta G = -12.6\,kJ/mol + RT\frac{[ATP][creatine]}{[ADP][creatine\ phosphate]}$$

$$= -12.9\,kJ/mol + (0.0083)(298)\frac{[4\,mM][13\,mM]}{[0.03\,mM][creatine\ 25\,mM]}$$

$$= 12.6\,kJ/mol + (2.47)\ln(160)$$

$$= 0.1\,kJ/mol$$

Nothing to write home about, but remember, it is a resting muscle.

KEY TERMS

metabolism or intermediary metabolism (p. 464)

metabolomics (p. 464)

phototroph (p. 464)

chemotroph (p. 464)

catabolism (p. 465)

anabolism (p. 465)

amphibolic pathway (p. 465)

adenosine triphosphate (ATP) (p. 466)

phosphoryl-transfer potential (p. 469)

activated carrier (p. 475)

vitamin (p. 478)

oxidation–reduction reaction (p. 481)

group-transfer reaction (p. 481)

hydrolytic reaction (p. 482)

cleavage of carbon bonds by means other than hydrolysis or oxidation (p. 482)

lyase (p. 482)

isomerization reaction (p. 482)

ligation reaction (p. 483)

energy charge (p. 484)

phosphorylation potential (p. 485)

PROBLEMS

1. *Complex patterns.* What is meant by *intermediary metabolism?* ✓❶

2. *Opposites.* Differentiate between anabolism and catabolism. ✓❶

3. *Self-sufficient versus dependent.* Distinguish between a phototroph and a chemotroph.

4. *Recycling.* Describe the ATP–ADP cycle and its role in biological systems. ✓❷

5. *Hans Krebs' idea.* What are the three major stages in the energy extraction from foodstuffs? Assess the contribution of each stage to ATP synthesis.

6. *Graffiti.* While walking to biochemistry class with a friend, you see the following graffiti spray painted on the wall of the science building: "When a system is in equilibrium, the Gibbs free energy is maximum." You are disgusted, not only at the vandalism, but at the ignorance of the vandal. Your friend asks you to explain.

7. *Why bother to eat?* What are the three primary uses for cellular energy?

8. *Match 'em.* ✓❶

1. Cellular energy currency   **(a)** NAD^+
_____

2. Anabolic electron carrier _____   **(b)** Coenzyme A

3. Phototroph _____   **(c)** Precursor to coenzymes

4. Catabolic electron carrier _____   **(d)** Energy-yielding reactions

5. Oxidation-reduction _____   **(e)** Energy-requiring reactions

6. Activated carrier of two _____   **(f)** ATP carbon fragments

7. Vitamin _____   **(g)** Transfers electrons

8. Anabolism _____   **(h)** $NADP^+$

9. Amphibolic reaction _____   **(i)** Converts light energy to chemical energy

10. Catabolism _____   **(j)** Used in anabolism and catabolism

9. *Charges.* In vivo, ATP is usually bound to magnesium or manganese ions. Why is this the case? ✓❷

10. *Energy to burn.* What factors account for the high phosphoryl-transfer potential of nucleoside triphosphates? ✓❷

11. *Back in time.* Account for the fact that ATP, and not another nucleoside triphosphate, is the cellular energy currency.

12. *Currency issues.* Why does it make good sense to have a single nucleotide, ATP, function as the cellular energy currency?

13. *Environmental conditions.* The standard free energy of hydrolysis for ATP is $-30.5 \text{ kJ mol}^{-1}$ ($-7.3 \text{ kcal mol}^{-1}$).

$$ATP + H_2O \rightleftharpoons ADP + P_i$$

How might physiological conditions, such as rest, intense exercise, or an alteration of intracellular ionic environment, alter the free energy of hydrolysis? ✓❷

14. *Brute force?* Metabolic pathways frequently contain reactions with positive standard free-energy values, yet the reactions still take place. How is this possible? ✓❸

15. *Energy flow.* What is the direction of each of the following reactions when the reactants are initially present in equimolar amounts? Use the data given in Table 15.1. ✓❸

(a) $ATP + \text{creatine} \rightleftharpoons \text{creatine phosphate} + ADP$

(b) $ATP + \text{glycerol} \rightleftharpoons \text{glycerol 3-phosphate} + ADP$

(c) $ATP + \text{pyruvate} \rightleftharpoons \text{phosphoenolpyruvate} + ADP$

(d) $ATP + \text{glucose} \rightleftharpoons \text{glucose 6-phosphate} + ADP$

16. *A proper inference.* What information do the $\Delta G^{\circ\prime}$ data given in Table 15.1 provide about the relative rates of hydrolysis of pyrophosphate and acetyl phosphate?

17. *A potent donor.* Consider the following reaction:

$$ATP + \text{pyruvate} \rightleftharpoons \text{phosphoenolpyruvate} + ADP$$

(a) Calculate $\Delta G^{\circ\prime}$ and K'_{eq} at 25°C for this reaction by using the data given in Table 15.1.

(b) What is the equilibrium ratio of pyruvate to phosphoenolpyruvate if the ratio of ATP to ADP is 10? ✓❸

18. *Isomeric equilibrium.* Using the information in Table 15.1, calculate $\Delta G^{\circ\prime}$ for the isomerization of glucose 6-phosphate to glucose 1-phosphate. What is the equilibrium ratio of glucose 6-phosphate to glucose 1-phosphate at 25°C?

19. *Activated acetate.* The formation of acetyl CoA from acetate is an ATP-driven reaction:

$$\text{Acetate} + ATP + CoA \rightleftharpoons \text{acetyl CoA} + AMP + PP_i$$ ✓❸

(a) Calculate $\Delta G^{\circ\prime}$ for this reaction by using data given in this chapter.

(b) The PP_i formed in the preceding reaction is rapidly hydrolyzed in vivo because of the ubiquity of inorganic pyrophosphatase. The $\Delta G^{\circ\prime}$ for the hydrolysis of PP_i is $-19.2\ \text{kJ mol}^{-1}$ ($-4.6\ \text{kcal mol}^{-1}$). Calculate the $\Delta G^{\circ\prime}$ for the overall reaction, including pyrophosphate hydrolysis. What effect does the hydrolysis of PP_i have on the formation of acetyl CoA?

20. *Acid strength.* The pK_a of an acid is a measure of its proton-group-transfer potential.

(a) Derive a relation between $\Delta G^{\circ\prime}$ and pK_a.

(b) What is the $\Delta G^{\circ\prime}$ for the ionization of acetic acid, which has a pK_a of 4.8?

21. *Raison d'être.* The muscles of some invertebrates are rich in arginine phosphate (phosphoarginine). Propose a function for this amino acid derivative.

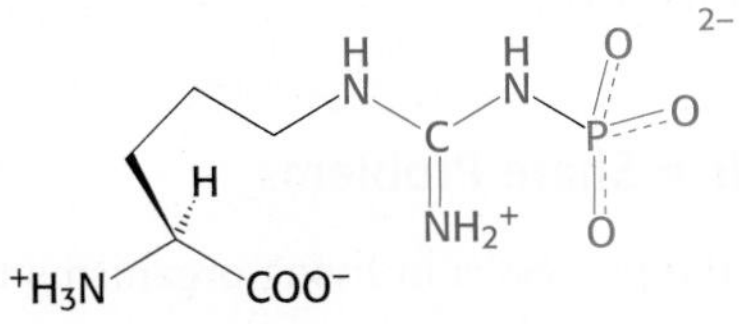

Arginine phosphate

22. *Recurring motif.* What is the structural feature common to ATP, FAD, NAD^+, and CoA? ✓⑤

23. *Ergogenic help or hindrance?* Creatine is a popular, but untested, dietary supplement.

(a) What is the biochemical rationale for the use of creatine?

(b) What type of exercise would most benefit from creatine supplementation? ✓③

24. *Standard conditions versus real life 1.* The enzyme aldolase catalyzes the following reaction in the glycolytic pathway:

Fructose 1, 6-bisphosphate $\xrightleftharpoons{\text{Aldolase}}$

dihydroxyacetone phosphate +

glyceraldehyde 3-phosphate

The $\Delta G^{\circ\prime}$ for the reaction is $+23.8\ \text{kJ mol}^{-1}$ ($+5.7$ kcal mol^{-1}), whereas the ΔG in the cell is $-1.3\ \text{kJ mol}^{-1}$ ($-0.3\ \text{kcal mol}^{-1}$). Calculate the ratio of reactants to products under equilibrium and intracellular conditions. Using your results, explain how the reaction can be endergonic under standard conditions and exergonic under intracellular conditions.

25. *Standard conditions versus real life 2.* On page 470, we showed that a reaction, A $\rightleftharpoons$ B, with a $\Delta G^{\circ\prime} = +16.7$ kJ mol^{-1} ($+4.0\ \text{kcal mol}^{-1}$) has an K'_{eq} of 1.15×10^{-3}. The K'_{eq} is increased to 2.67×10^2 if the reaction is coupled to ATP hydrolysis under standard conditions. The ATP-generating system of cells maintains the $[ATP]/[ADP][P_i]$ ratio at a high level, typically of the order of $500\ \text{M}^{-1}$. Calculate the ratio of B/A under cellular conditions.

26. *Not all alike.* The concentrations of ATP, ADP, and P_i differ with cell type. Consequently, the release of free energy with the hydrolysis of ATP will vary with cell type. Using the following table, calculate the ΔG for the hydrolysis of ATP in liver, muscle, and brain cells. In which cell type is the free energy of ATP hydrolysis most negative? ✓②

	ATP (mM)	ADP (mM)	P_i (mM)
Liver	3.5	1.8	5.0
Muscle	8.0	0.9	8.0
Brain	2.6	0.7	2.7

27. *Oxidation issues.* Examine the pairs of molecules and identify the more-reduced molecule in each pair. ✓④

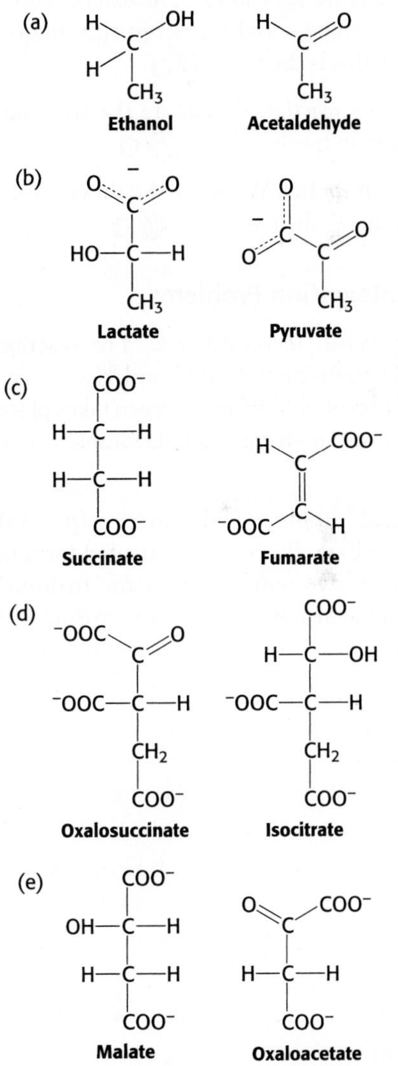

28. *Running downhill.* Glycolysis is a series of 10 linked reactions that convert one molecule of glucose into two molecules of pyruvate with the concomitant synthesis of two molecules of ATP (Chapter 16). The $\Delta G^{\circ\prime}$ for this set of reactions is $-35.6\ \text{kJ mol}^{-1}$ ($-8.5\ \text{kcal mol}^{-1}$), whereas the ΔG, the real-life conditions, is $-90\ \text{kJ mol}^{-1}$ ($-22\ \text{kcal mol}^{-1}$). Explain why the free-energy release is so much greater under intracellular conditions than under standard conditions.

29. *Outsourcing.* Outsourcing, a common business practice, is contracting with another business to perform a particular function. Higher organisms were the original outsourcers, frequently depending on lower organisms to perform key biochemical functions. Give an example from this chapter of biochemical outsourcing. ✅⑤

30. *Breakdown products.* Digestion, the breakdown of complex biochemicals into simpler molecules, is the first stage in the extraction of energy from food, but no useful energy is acquired during this stage. Why is digestion considered a stage in energy extraction? ✅⑤

31. *High-energy electrons.* What are the activated electron carriers for catabolism? For anabolism?

32. *Fewer alternative states.* Thioesters, common in biochemistry, are more unstable (energy-rich) than oxygen esters. Explain why this is the case. ✅④

33. *Classifying reactions.* What are the six common types of reactions seen in biochemistry? ✅⑤

34. *Staying in control.* What are the three principal means of controlling metabolic reactions? ✅⑤

Chapter Integration Problems

35. *Kinetic versus thermodynamic.* The reaction of NADH with oxygen to produce NAD^+ and H_2O is very exergonic, yet the reaction of NADH and oxygen takes place very slowly. Why does a thermodynamically favorable reaction not occur rapidly?

36. *Activated sulfate.* Fibrinogen (p. 331) contains tyrosine-O-sulfate. Propose an activated form of sulfate that could react in vivo with the aromatic hydroxyl group of a tyrosine residue in a protein to form tyrosine-O-sulfate.

Data Interpretation Problem

37. *Opposites attract.* The following graph shows how the ΔG for the hydrolysis of ATP varies as a function of the Mg^{2+} concentration ($pMg = -\log[Mg^{2+}]$). ✅②

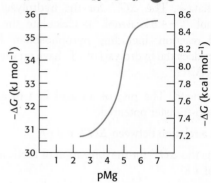

(a) How does decreasing $[Mg^{2+}]$ affect the ΔG of hydrolysis for ATP?

(b) Explain this effect.

Think ▶ Pair ▶ Share Problems

38. Identify the processes in living organisms that require a constant input of free energy.

39. Explain the significance of the free-energy change of reactions 1DG2 and the relationship of DG to DG89, the equilibrium constant and the concentration of reactants and products.

Glycolysis and Gluconeogenesis

Usain Bolt sprints to a win in the 200-meter finals at the Olympics in Rio de Janerio in 2016. Glucose metabolism can generate the ATP to power muscle contraction. During a sprint, when the ATP needs outpace oxygen delivery, as would be the case for Bolt, glucose is metabolized to lactate. When oxygen delivery is adequate, glucose is metabolized more efficiently to carbon dioxide and water. [Odd Andersen/Getty Images.]

A. Low O_2
(last seconds of a sprint)

B. Normal
(long, slow run)

LEARNING OBJECTIVES

By the end of this chapter, you should be able to:

1. Describe how ATP is generated in glycolysis.

2. Explain why the regeneration of NAD^+ is crucial to fermentations.

3. Describe how gluconeogenesis is powered in the cell.

4. Describe the coordinated regulation of glycolysis and gluconeogenesis.

OUTLINE

16.1 Glycolysis Is an Energy-Conversion Pathway in Many Organisms

16.2 The Glycolytic Pathway Is Tightly Controlled

16.3 Glucose Can Be Synthesized from Noncarbohydrate Precursors

16.4 Gluconeogenesis and Glycolysis Are Reciprocally Regulated

The first metabolic pathway that we encounter is *glycolysis*, an ancient pathway employed by a host of organisms. *Glycolysis is the sequence of reactions that metabolizes one molecule of glucose to two molecules of pyruvate with the concomitant net production of two molecules of ATP.* This process is anaerobic, meaning that it does not require O_2, because it evolved before substantial amounts of oxygen accumulated in the atmosphere. Pyruvate can be further processed anaerobically to lactate (*lactic acid fermentation*) or ethanol (*alcoholic fermentation*). Under aerobic conditions, pyruvate can be completely oxidized to CO_2, generating much more ATP, as will be described in Chapters 17 and 18. Figure 16.1 shows some possible fates of pyruvate produced by glycolysis.

Because glucose is such a precious fuel, metabolic products, such as pyruvate and lactate, are salvaged to synthesize glucose in the process

Glycolysis
Derived from the Greek stem *glyk-*, "sweet," and the word *lysis*, "dissolution."

FIGURE 16.1 Some fates of glucose.

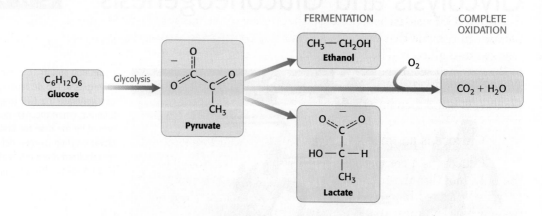

of *gluconeogenesis*. Although glycolysis and gluconeogenesis have some enzymes in common, the two pathways are not simply the reverse of each other. In particular, the highly exergonic, irreversible steps of glycolysis are bypassed in gluconeogenesis. The two pathways are reciprocally regulated so that glycolysis and gluconeogenesis do not take place simultaneously in the same cell to a significant extent.

Our understanding of glucose metabolism, especially glycolysis, has a rich history. Indeed, the development of biochemistry and the delineation of glycolysis went hand in hand. A key discovery was made by Hans and Eduard Buchner in 1897, quite by accident. The Buchners were interested in manufacturing cell-free extracts of yeast for possible therapeutic use. These extracts had to be preserved without the use of antiseptics such as phenol, and so they decided to try sucrose, a commonly used preservative in kitchen chemistry. They obtained a startling result: sucrose was rapidly fermented into alcohol by the yeast juice. The significance of this finding was immense. *The Buchners demonstrated for the first time that fermentation could take place outside living cells.* The accepted view of their day, asserted by Louis Pasteur in 1860, was that fermentation is inextricably tied to living cells. The chance discovery by the Buchners refuted this dogma and opened the door to modern biochemistry. The Buchners' discovery inspired the search for the biochemicals that catalyze the conversion of sucrose into alcohol. *The study of metabolism became the study of chemistry.*

Studies of muscle extracts then showed that many of the reactions of lactic acid fermentation were the same as those of alcoholic fermentation. *This exciting discovery revealed an underlying unity in biochemistry.* The complete glycolytic pathway was elucidated by 1940. Glycolysis is also known as the *Embden–Meyerhof pathway,* named after two pioneers of research on glycolysis.

Glucose is generated from dietary carbohydrates

In our diets, we typically consume a generous amount of starch and a smaller amount of glycogen. These complex carbohydrates must be converted into simpler carbohydrates for absorption by the intestine and transport in the blood. Starch and glycogen are digested primarily by the pancreatic enzyme *α-amylase* and to a lesser extent by salivary α-amylase. Amylase cleaves the α-1,4 bonds of starch and glycogen, but not the α-1,6 bonds. The products are the di- and trisaccharides maltose and maltotriose. The material not digestible because of the α-1,6 bonds is called the *limit dextrin.*

α-Glucosidase (maltase) digests maltotriose and any other oligosaccharides that may have escaped digestion by the amylase; it also cleaves maltose into two glucose molecules. *α-Dextrinase* further digests the limit

Enzyme
A term coined by Friedrich Wilhelm Kühne in 1878 to designate catalytically active substances that had formerly been called ferments. Derived from the Greek words *en,* "in," and *zyme,* "leaven."

dextrin. α-Glucosidase is located on the surface of the intestinal cells, as is *sucrase,* an enzyme that degrades the sucrose contributed by vegetables to fructose and glucose. The enzyme *lactase* is responsible for degrading the milk sugar lactose into glucose and galactose and is also found on the surface of intestinal cells. The monosaccharides are transported into the cells lining the intestine and then into the bloodstream.

Glucose is an important fuel for most organisms

Glucose is a common and important fuel. In mammals, glucose is the only fuel that the brain uses under nonstarvation conditions and the only fuel that red blood cells can use at all. Indeed, almost all organisms use glucose, and most that do, process it in a similar fashion. Recall from Chapter 11 that there are many carbohydrates. Why is glucose instead of some other monosaccharide such a prominent fuel? We can speculate on the reasons. First, glucose is one of several monosaccharides formed from formaldehyde under prebiotic conditions, and so it may have been available as a fuel source for primitive biochemical systems. Second, glucose has a low tendency, relative to other monosaccharides, to nonenzymatically glycosylate proteins. In their open-chain forms, monosaccharides contain carbonyl groups that can react with the amino groups of proteins to form Schiff bases, which rearrange to form a more stable amino–ketone linkage (p. 347). Such nonspecifically modified proteins often do not function effectively. Glucose has a strong tendency to exist in the ring conformation and, consequently, relatively little tendency to modify proteins. Recall that all the hydroxyl groups in the ring conformation of β-glucose are equatorial, contributing to the sugar's high relative stability (p. 346).

16.1 Glycolysis Is an Energy-Conversion Pathway in Many Organisms

We now begin our consideration of the glycolytic pathway. This pathway is common to virtually all cells, both prokaryotic and eukaryotic. In eukaryotic cells, glycolytic enzymes are organized in supramolecular complexes located in the cytoplasm.

The enzymes of glycolysis are associated with one another

Evidence is accumulating that the enzymes of glycolysis are organized into complexes. For example, in yeast, the glycolytic enzymes are associated with the mitochondria, while in mammalian erythrocytes (red blood cells), the enzymes are found bound to the inner surface of the cell membrane. Indeed, some level of organization appears to occur in all cell types. This arrangement increases enzyme efficiency by facilitating movement of substrates and products between enzymes—a process called substrate channeling—and prevents the release of any toxic intermediates. We will see that the organization of metabolic pathways into large complexes is a common occurrence inside the cell.

Glycolysis can be divided into two parts

This pathway can be thought of as comprising two stages (Figure 16.2). Stage 1 is the trapping and preparation phase. No ATP is generated in this stage. In stage 1, glucose is converted into fructose 1,6-bisphosphate in three steps: a phosphorylation, an isomerization, and a second

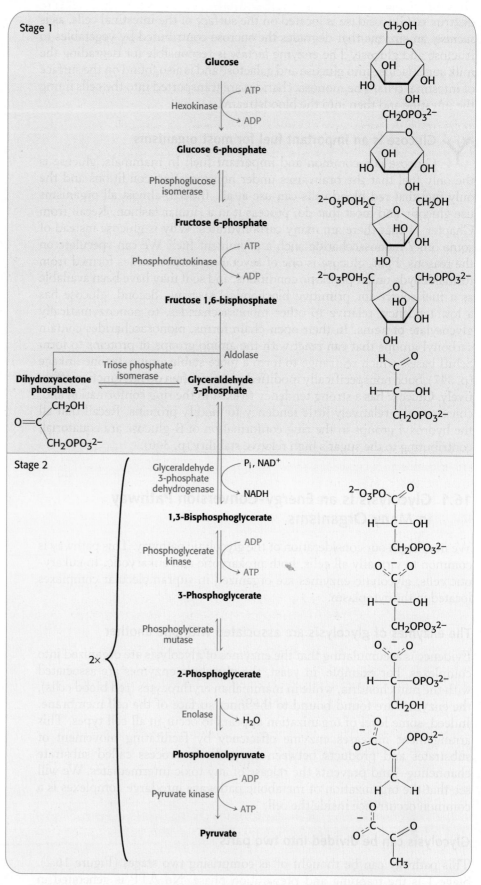

FIGURE 16.2 Stages of glycolysis. The glycolytic pathway can be divided into two stages: (1) glucose is trapped, destabilized, and cleaved into two interconvertible three-carbon molecules generated by cleavage of six-carbon fructose; and (2) ATP is generated.

phosphorylation reaction. *The strategy of these initial steps in glycolysis is to trap the glucose in the cell and form a compound that can be readily cleaved into phosphorylated three-carbon units.* Stage 1 is completed with the cleavage of the fructose 1,6-bisphosphate into two three-carbon fragments. These resulting three-carbon units are readily interconvertible. In stage 2, ATP is harvested when the three-carbon fragments are oxidized to pyruvate.

Hexokinase traps glucose in the cell and begins glycolysis

Glucose enters cells through specific transport proteins (p. 516) and has one principal fate: *it is phosphorylated by ATP to form glucose 6-phosphate.* This step is notable for several reasons. Glucose 6-phosphate cannot pass through the membrane because of the negative charges on the phosphoryl groups, and it is not a substrate for glucose transporters. Also, the addition of the phosphoryl group facilitates the eventual metabolism of glucose into three-carbon molecules with high phosphoryl-transfer potential. The transfer of the phosphoryl group from ATP to the hydroxyl group on carbon 6 of glucose is catalyzed by *hexokinase.*

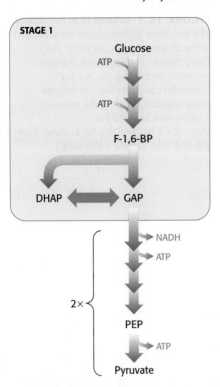

First stage of glycolysis. The first stage of glycolysis begins with the phosphorylation of glucose by hexokinase and ends with the isomerization of dihydroxyacetone phosphate to glyceraldehyde 3-phosphate.

Phosphoryl transfer is a fundamental reaction in biochemistry. *Kinases are enzymes that catalyze the transfer of a phosphoryl group from ATP to an acceptor.* Hexokinase, then, catalyzes the transfer of a phosphoryl group from ATP to a variety of six-carbon sugars (*hexoses*), such as glucose and mannose. *Hexokinase, like adenylate kinase (Section 9.4) and all other kinases, requires Mg^{2+} (or another divalent metal ion such as Mn^{2+}) for activity.* The divalent metal ion forms a complex with ATP.

X-ray crystallographic studies of yeast hexokinase revealed that the binding of glucose induces a large conformational change in the enzyme. Hexokinase consists of two lobes, which move toward each other when glucose is bound (Figure 16.3). On glucose binding, one lobe rotates 12 degrees with respect to the other, resulting in movements of the polypeptide backbone of as much as 8 Å. The cleft between the lobes closes, and the bound glucose becomes surrounded by protein, except for the hydroxyl group of carbon 6, which will accept the phosphoryl group from ATP. The closing of the cleft in hexokinase is a striking example of the role *of induced fit* in enzyme action (Section 8.3).

The glucose-induced structural changes are significant in two respects. First, the environment around the glucose becomes more nonpolar, which favors reaction between the hydrophilic hydroxyl group of glucose and the terminal phosphoryl group of ATP. Second, the conformational changes enable the kinase to discriminate against H_2O as a substrate. The closing of the cleft keeps water molecules away from the active site. If hexokinase were rigid, a molecule of H_2O occupying the binding site for the —CH_2OH of glucose could attack the γ phosphoryl group of ATP, forming ADP and P_i. In other words, a rigid kinase would likely also be an ATPase. It is interesting to note that other kinases taking part in glycolysis—phosphofructokinase, phosphoglycerate kinase, and pyruvate kinase—also contain clefts between

FIGURE 16.3 Induced fit in hexokinase. The two lobes of hexokinase are separated in the absence of glucose (left). The conformation of hexokinase changes markedly on binding glucose (right). *Notice* that two lobes of the enzyme come together, creating the necessary environment for catalysis. [After RSCB Protein Data Bank; drawn from yhx and 1hkg by Adam Steinberg.]

lobes that close when substrate is bound, although the structures of these enzymes are different in other regards. *Substrate-induced cleft closing is a general feature of kinases.* Recall that protein kinase A also undergoes similar structural changes (p. 321).

Fructose 1,6-bisphosphate is generated from glucose 6-phosphate

A crucial step toward completion of the first phase of glycolysis—the formation of fructose 1,6-bisphosphate—is the *isomerization of glucose 6-phosphate to fructose 6-phosphate.* Recall that the open-chain form of glucose has an aldehyde group at carbon 1, whereas the open-chain form of fructose has a keto group at carbon 2. Thus, the isomerization of glucose 6-phosphate to fructose 6-phosphate is a *conversion of an aldose into a ketose.* The reaction catalyzed by *phosphoglucose isomerase* takes several steps because both glucose 6-phosphate and fructose 6-phosphate are present primarily in the cyclic forms. The enzyme must first open the six-membered ring of glucose 6-phosphate, catalyze the isomerization, and then promote the formation of the five-membered ring of fructose 6-phosphate.

Glucose 6-phosphate (G-6P) ⇌ **Glucose 6-phosphate** (open-chain form) ⇌ **Fructose 6-phosphate** (open-chain form) ⇌ **Fructose 6-phosphate (F-6P)**

A second phosphorylation reaction follows the isomerization step. *Fructose 6-phosphate is phosphorylated at the expense of ATP to fructose 1,6-bisphosphate* (F-1,6-BP). The prefix *bis-* in bisphosphate means that two separate monophosphoryl groups are present, whereas the prefix *di-* in diphosphate (as in adenosine diphosphate) means that two phosphoryl groups are present and are connected by an anhydride bond.

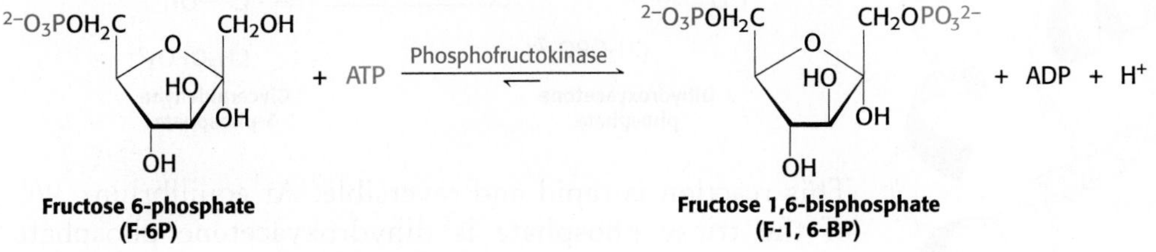

This reaction is catalyzed by *phosphofructokinase* (PFK), an allosteric enzyme that sets the pace of glycolysis. As we will learn, this enzyme plays a central role in the metabolism of many molecules in all parts of the body.

What is the biochemical rationale for the isomerization of glucose 6-phosphate to fructose 6-phosphate and its subsequent phosphorylation to form fructose 1,6-bisphosphate? Had the aldol cleavage taken place in the aldose glucose, a two-carbon and a four-carbon fragment would have resulted. Two different metabolic pathways—one to process the two-carbon fragment and one for the four-carbon fragment—would have been required to extract energy. Phosphorylation of the fructose 6-phosphate to fructose 1,6-bisphosphate prevents the reformation of glucose 6-phosphate. As shown below, aldol cleavage of fructose 1,6-bisphosphate yields two phosphorylated interconvertible three-carbon fragments that will be oxidized in the later steps of glycolysis to capture energy in the form of ATP.

The six-carbon sugar is cleaved into two three-carbon fragments

The newly formed fructose 1,6-bisphosphate is cleaved into *glyceraldehyde 3-phosphate* (GAP) and *dihydroxyacetone phosphate* (DHAP), completing stage 1 of glycolysis. The products of the remaining steps in glycolysis consist of three-carbon units rather than six-carbon units. This reaction, which is readily reversible, is catalyzed by *aldolase*. This enzyme derives its name from the nature of the reverse reaction, an aldol condensation.

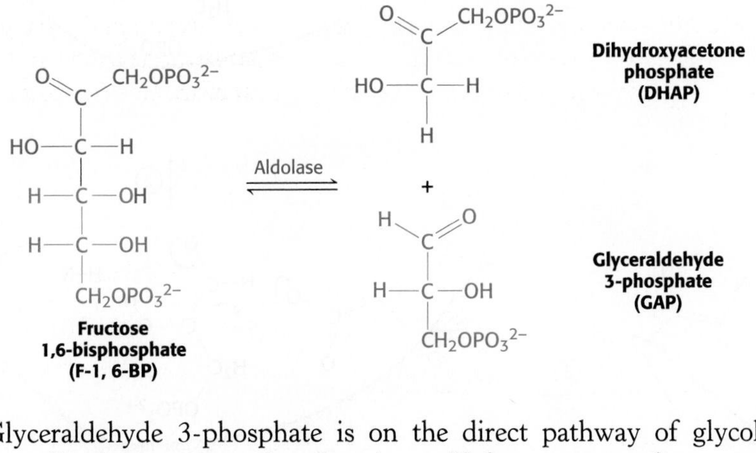

Glyceraldehyde 3-phosphate is on the direct pathway of glycolysis, whereas dihydroxyacetone phosphate is not. Unless a means exists to convert dihydroxyacetone phosphate into glyceraldehyde 3-phosphate, a three-carbon fragment useful for generating ATP will be lost. These compounds are isomers that can be readily interconverted: dihydroxyacetone phosphate is a ketose, whereas glyceraldehyde 3-phosphate is an aldose. The isomerization of these three-carbon phosphorylated sugars is catalyzed by *triose phosphate isomerase* (TPI, sometimes abbreviated TIM; Figure 16.4).

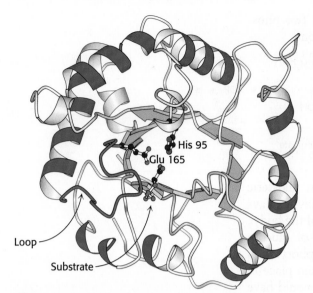

FIGURE 16.4 Structure of triose phosphate isomerase. This enzyme consists of a central core of eight parallel β strands (orange) surrounded by eight α helices (blue). This structural motif, called an αβ barrel, is also found in the glycolytic enzymes aldolase, enolase, and pyruvate kinase. *Notice* that histidine 95 and glutamate 165, essential components of the active site of triose phosphate isomerase, are located in the barrel. A loop (red) closes off the active site on substrate binding. [Drawn from 2YPI.pdb.]

This reaction is rapid and reversible. At equilibrium, 96% of the triose phosphate is dihydroxyacetone phosphate. However, the reaction proceeds readily from dihydroxyacetone phosphate to glyceraldehyde 3-phosphate because the subsequent reactions of glycolysis remove this product. Triose phosphate isomerase deficiency, a rare condition, is the only glycolytic enzymopathy that is lethal. This deficiency is characterized by severe hemolytic anemia and neurodegeneration.

Mechanism: Triose phosphate isomerase salvages a three-carbon fragment

Much is known about the catalytic mechanism of triose phosphate isomerase. TPI catalyzes the transfer of a hydrogen atom from carbon 1 to carbon 2, an intramolecular oxidation–reduction. This isomerization of a ketose into an aldose proceeds through an *enediol intermediate* (Figure 16.5).

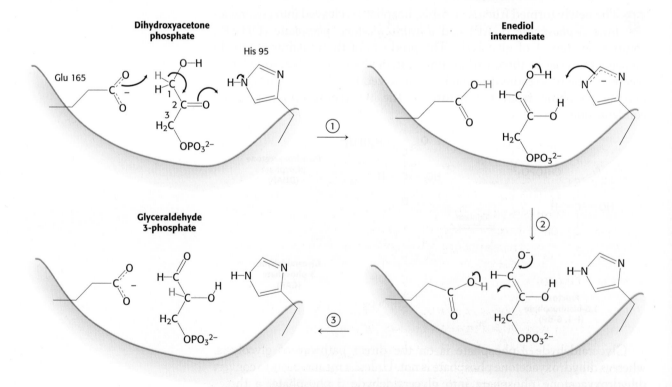

FIGURE 16.5 Catalytic mechanism of triose phosphate isomerase. (1) Glutamate 165 acts as a general base by abstracting a proton (H$^+$) from carbon 1. Histidine 95, acting as a general acid, donates a proton to the oxygen atom bonded to carbon 2, forming the enediol intermediate. (2) Glutamic acid, now acting as a general acid, donates a proton to C-2 while histidine removes a proton from the OH group of C-1. (3) The product is formed, and glutamate and histidine are returned to their ionized and neutral forms, respectively.

X-ray crystallographic and other studies showed that glutamate 165 plays the role of a general acid–base catalyst: it abstracts a proton (H^+) from carbon 1 and then donates it to carbon 2. However, the carboxylate group of glutamate 165 by itself is not basic enough to pull a proton away from a carbon atom adjacent to a carbonyl group. Histidine 95 assists catalysis by donating a proton to stabilize the negative charge that develops on the C-2 carbonyl group.

Two features of this enzyme are noteworthy. First, TPI displays great catalytic prowess. It accelerates isomerization by a factor of 10^{10} compared with the rate obtained with a simple base catalyst such as acetate ion. Indeed, the k_{cat}/K_M ratio for the isomerization of glyceraldehyde 3-phosphate is $2 \times 10^8\,M^{-1}\,s^{-1}$, which is close to the diffusion-controlled limit. In other words, catalysis takes place every time that enzyme and substrate meet. The diffusion-controlled encounter of substrate and enzyme is thus the rate-limiting step in catalysis. TPI is an example of a *kinetically perfect enzyme* (Section 8.4). Second, TPI suppresses an undesired side reaction, the decomposition of the enediol intermediate into methyl glyoxal and orthophosphate.

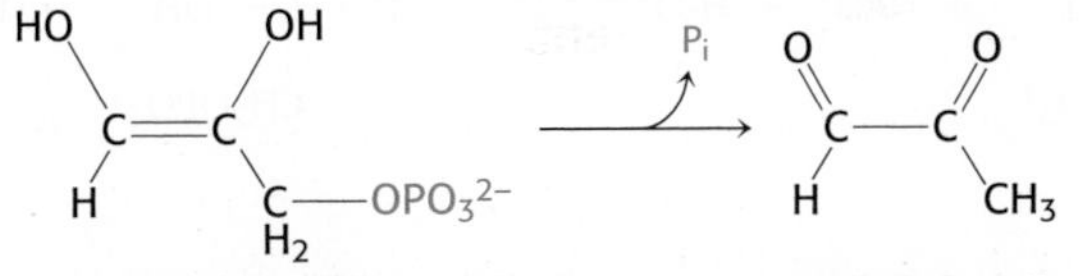

Enediol intermediate   **Methyl glyoxal**

In solution, this physiologically useless reaction is 100 times as fast as isomerization. Moreover, methyl glyoxal is a highly reactive compound that can modify the structure and function of a variety of biomolecules, including proteins and DNA. The reaction of methyl glyoxal with a biomolecule is an example of deleterious reactions called advanced glycation end products, discussed earlier (AGEs, Section 11.1). Hence, TPI must prevent the enediol from leaving the enzyme. This labile intermediate is trapped in the active site by the movement of a loop of 10 residues (Figure 16.4). This loop serves as a lid on the active site, shutting it when the enediol is present and reopening it when isomerization is completed. *We see here a striking example of one means of preventing an undesirable alternative reaction: the active site is kept closed until the desirable reaction takes place.*

Thus, two molecules of glyceraldehyde 3-phosphate are formed from one molecule of fructose 1,6-bisphosphate by the sequential action of aldolase and triose phosphate isomerase. The economy of metabolism is evident in this reaction sequence. The isomerase funnels dihydroxyacetone phosphate into the main glycolytic pathway; a separate set of reactions is not needed. See Biochemistry in Focus for a discussion of triose phosphate isomerase deficiency.

The oxidation of an aldehyde to an acid powers the formation of a compound with high phosphoryl-transfer potential

The preceding steps in glycolysis have transformed one molecule of glucose into two molecules of glyceraldehyde 3-phosphate, but no energy has yet been extracted. On the contrary, thus far, two molecules of ATP have been invested. We come now to the second stage of glycolysis—a series of steps that harvest some of the energy contained in glyceraldehyde 3-phosphate as ATP. The initial reaction in this sequence is the *conversion of glyceraldehyde 3-phosphate into 1,3-bisphosphoglycerate* (1,3-BPG), a reaction catalyzed by *glyceraldehyde 3-phosphate dehydrogenase.*

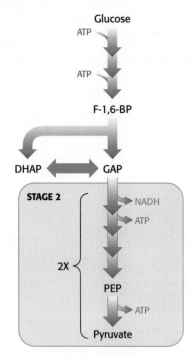

Second stage of glycolysis. The oxidation of three-carbon fragments yields ATP.

1,3-Bisphosphoglycerate is an acyl phosphate, which is a mixed anhydride of phosphoric acid and a carboxylic acid. Such compounds have a high phosphoryl-transfer potential; one of its phosphoryl groups is transferred to ADP in the next step in glycolysis.

The reaction catalyzed by glyceraldehyde 3-phosphate dehydrogenase can be viewed as the sum of two processes: the *oxidation* of the aldehyde to a carboxylic acid by NAD^+ and the *joining* of the carboxylic acid and orthophosphate to form the acyl-phosphate product.

The first reaction is thermodynamically quite favorable, with a standard free-energy change, $\Delta G^{\circ\prime}$, of approximately -50 kJ mol^{-1} ($-12 \text{ kcal mol}^{-1}$), whereas the second reaction is quite unfavorable, with a standard free-energy change of the same magnitude but the opposite sign. If these two reactions simply took place in succession, the second reaction would have a very large activation energy and thus not take place at a biologically significant rate. These two processes *must be coupled* so the favorable aldehyde oxidation can be used to drive the formation of the acyl phosphate. How are these reactions coupled? *The key is an intermediate, formed as a result of the aldehyde oxidation, that is linked to the enzyme by a thioester bond.* Thioesters are high-energy compounds found in many biochemical pathways (Section 15.4). This intermediate reacts with orthophosphate to form the high-energy compound 1,3-bisphosphoglycerate.

The thioester intermediate is higher in free energy than the free carboxylic acid is. The favorable oxidation and unfavorable phosphorylation reactions are coupled by the thioester intermediate, which preserves much of the free energy released in the oxidation reaction. We see here the *use of a covalent enzyme-bound intermediate as a mechanism of energy coupling.* A free-energy profile of the glyceraldehyde 3-phosphate dehydrogenase reaction, compared with a hypothetical process in which the reaction proceeds without this intermediate, reveals how this intermediate allows a favorable process to drive an unfavorable one (Figure 16.6).

Mechanism: Phosphorylation is coupled to the oxidation of glyceraldehyde 3-phosphate by a thioester intermediate

The active site of glyceraldehyde 3-phosphate dehydrogenase includes a reactive cysteine residue, as well as NAD^+ and a crucial histidine (Figure 16.7).

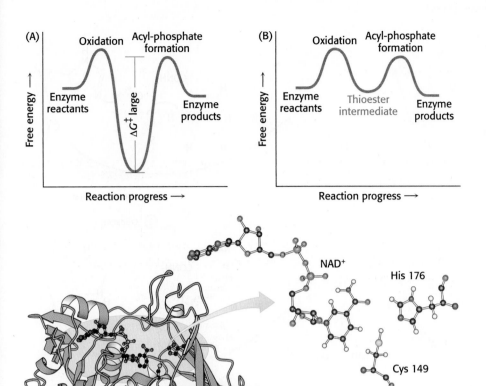

FIGURE 16.6 Free-energy profiles for glyceraldehyde oxidation followed by acyl-phosphate formation. (A) A hypothetical case with no coupling between the two processes. The second step must have a large activation barrier, making the reaction very slow. (B) The actual case with the two reactions coupled through a thioester intermediate.

FIGURE 16.7 Structure of glyceraldehyde 3-phosphate dehydrogenase. *Notice* that the active site includes a cysteine residue and a histidine residue adjacent to a bound NAD$^+$ molecule. The sulfur atom of cysteine will link with the substrate to form a transitory thioester intermediate. [Drawn from 1GAD.pdb.]

Let's consider in detail how these components cooperate in the reaction mechanism (Figure 16.8). In step 1, the aldehyde substrate reacts with the sulfhydryl group of cysteine 149 on the enzyme to form a hemithioacetal. Step 2 is the *transfer of a hydride ion to a molecule of* NAD$^+$ *that is tightly bound to the enzyme and is adjacent to the cysteine residue.* This reaction is favored by the deprotonation of the hemithioacetal by histidine 176. The products of this reaction are the reduced coenzyme NADH and a thioester intermediate. *This thioester intermediate has a free energy close to that of the reactants* (Figure 16.6). In step 3, the NADH formed from the aldehyde oxidation leaves the enzyme and is replaced by a second molecule of NAD$^+$. This step is important because the positive charge on NAD$^+$ polarizes the thioester intermediate to facilitate the attack by orthophosphate. In step 4, orthophosphate attacks the thioester to form 1,3-BPG and free the cysteine residue. This example illustrates the essence of energy transformations and of metabolism itself: energy released by carbon oxidation is converted into high phosphoryl-transfer potential.

ATP is formed by phosphoryl transfer from 1,3-bisphosphoglycerate

1,3-Bisphosphoglycerate is an energy-rich molecule with a greater phosphoryl-transfer potential than that of ATP (Section 15.2). Thus, 1,3-BPG can be used to power the synthesis of ATP from ADP. *Phosphoglycerate kinase* catalyzes the transfer of the phosphoryl group from the acyl phosphate of 1,3-bisphosphoglycerate to ADP; ATP and 3-phosphoglycerate are the products.

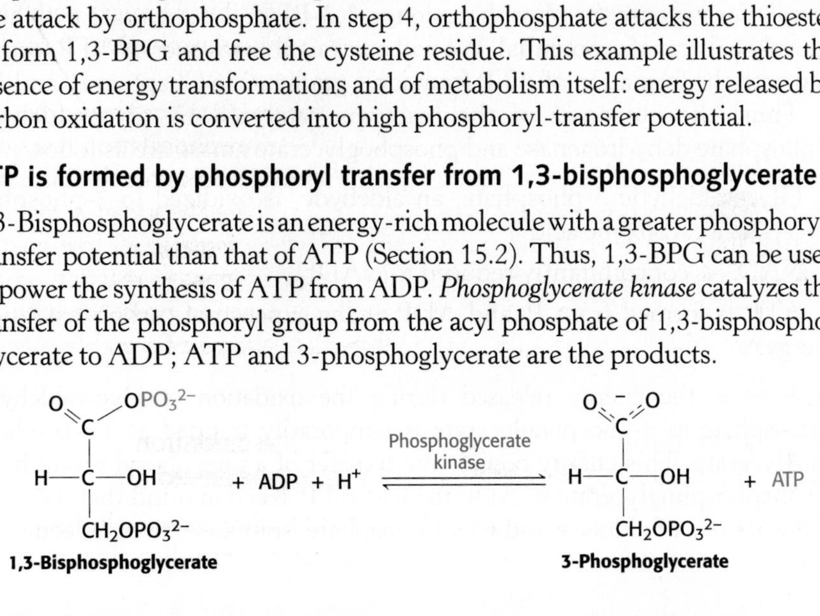

1,3-Bisphosphoglycerate + ADP + H$^+$ $\rightleftharpoons$ **3-Phosphoglycerate** + ATP (Phosphoglycerate kinase)

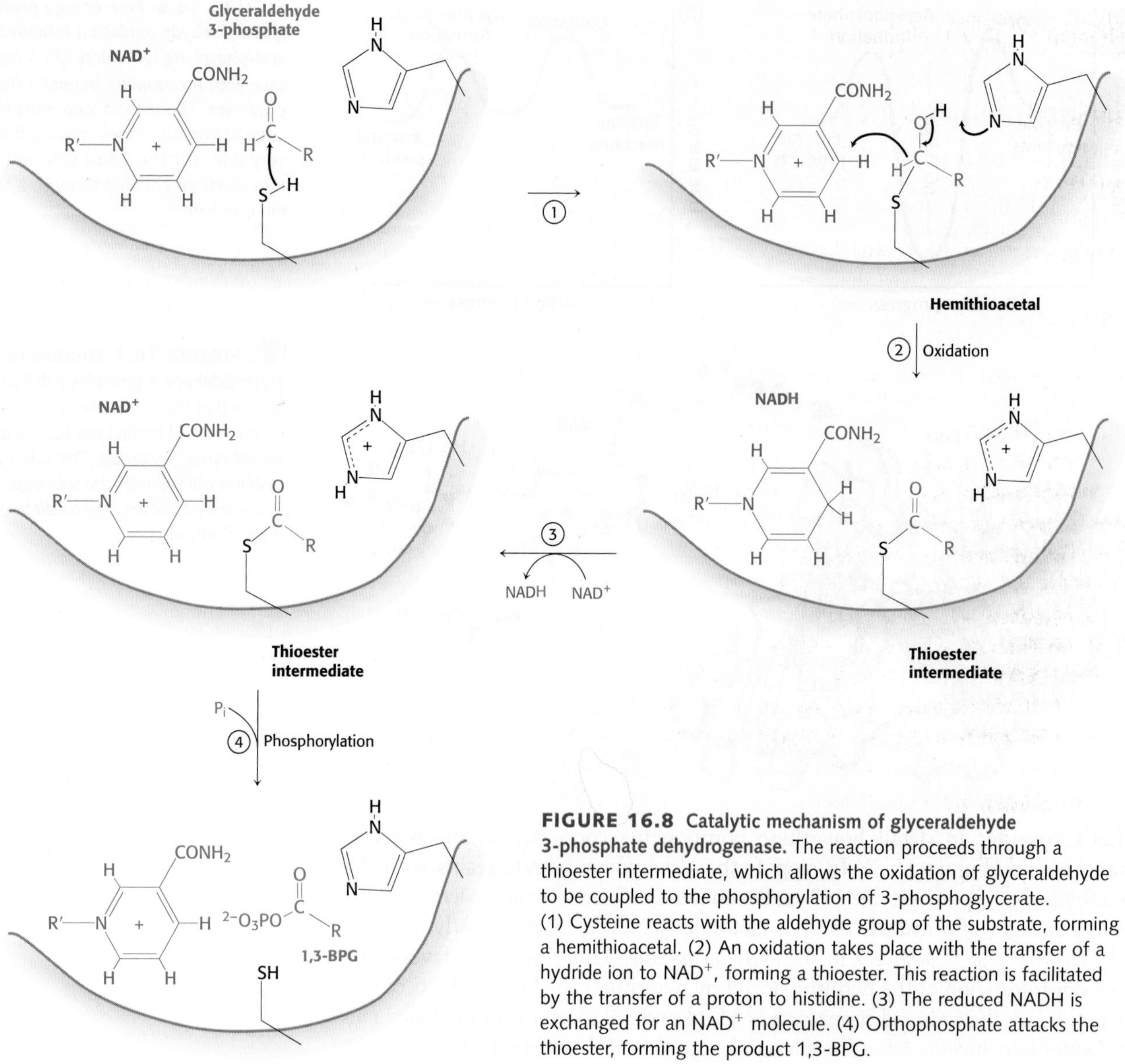

FIGURE 16.8 Catalytic mechanism of glyceraldehyde 3-phosphate dehydrogenase. The reaction proceeds through a thioester intermediate, which allows the oxidation of glyceraldehyde to be coupled to the phosphorylation of 3-phosphoglycerate. (1) Cysteine reacts with the aldehyde group of the substrate, forming a hemithioacetal. (2) An oxidation takes place with the transfer of a hydride ion to NAD^+, forming a thioester. This reaction is facilitated by the transfer of a proton to histidine. (3) The reduced NADH is exchanged for an NAD^+ molecule. (4) Orthophosphate attacks the thioester, forming the product 1,3-BPG.

The formation of ATP in this manner is referred to as *substrate-level phosphorylation* because the phosphate donor, 1,3-BPG, is a substrate with high phosphoryl-transfer potential. We will contrast this manner of ATP formation with the formation of ATP from ionic gradients in Chapters 18 and 19.

Thus, the outcomes of the reactions catalyzed by glyceraldehyde 3-phosphate dehydrogenase and phosphoglycerate kinase are as follows:

1. Glyceraldehyde 3-phosphate, an aldehyde, is oxidized to 3-phosphoglycerate, a carboxylic acid.

2. NAD^+ is concomitantly reduced to NADH.

3. ATP is formed from P_i and ADP at the expense of carbon-oxidation energy.

In essence, the energy released during the oxidation of glyceraldehyde 3-phosphate to 3-phosphoglycerate is temporarily trapped as 1,3-bisphosphoglycerate. This energy powers the transfer of a phosphoryl group from 1,3-bisphosphoglycerate to ADP to yield ATP. Keep in mind that, because of the actions of aldolase and triose phosphate isomerase, two molecules of glyceraldehyde 3-phosphate were formed and hence two molecules of ATP were generated. These ATP molecules make up for the two molecules of ATP consumed in the first stage of glycolysis.

Additional ATP is generated with the formation of pyruvate

In the remaining steps of glycolysis, 3-phosphoglycerate is converted into pyruvate, and a second molecule of ATP is formed from ADP.

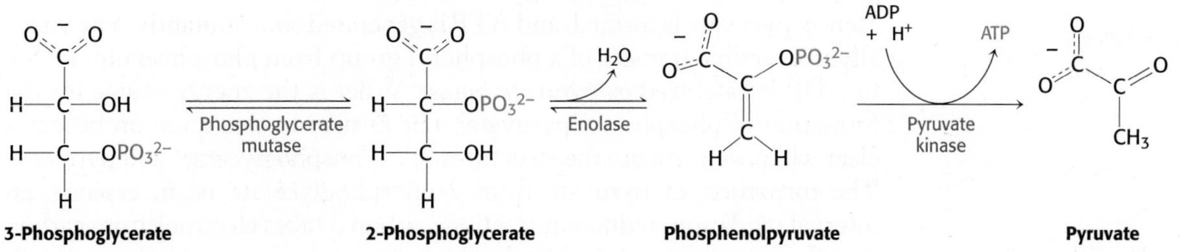

3-Phosphoglycerate   **2-Phosphoglycerate**   **Phosphenolpyruvate**   **Pyruvate**

The first reaction is a rearrangement. The position of the phosphoryl group shifts in the *conversion of 3-phosphoglycerate into 2-phosphoglycerate*, a reaction catalyzed by *phosphoglycerate mutase*. In general, a *mutase* is an enzyme that catalyzes the intramolecular shift of a chemical group, such as a phosphoryl group. The phosphoglycerate mutase reaction has an interesting mechanism: the phosphoryl group is not simply moved from one carbon atom to another. This enzyme requires catalytic amounts of 2,3-bisphosphoglycerate (2,3-BPG) to maintain an active-site histidine residue in a phosphorylated form. This phosphoryl group is transferred to 3-phosphoglycerate to reform 2,3-bisphosphoglycerate.

Enz-His-phosphate + 3-phosphoglycerate $\rightleftharpoons$

Enz-His + 2,3-bisphosphoglycerate

The mutase then functions as a phosphatase: it converts 2,3-bisphosphoglycerate into 2-phosphoglycerate. The mutase retains the phosphoryl group to regenerate the modified histidine.

Enz-His + 2,3-bisphosphoglycerate $\rightleftharpoons$

Enz-His-phosphate + 2-phosphoglycerate

The sum of these reactions yields the mutase reaction:

3-Phosphoglycerate $\rightleftharpoons$ 2-phosphoglycerate

In the next reaction, the dehydration of 2-phosphoglycerate introduces a double bond, creating an *enol*. *Enolase* catalyzes this formation of the enol phosphate *phosphoenolpyruvate* (PEP). This dehydration markedly elevates the transfer potential of the phosphoryl group. An *enol phosphate* has a high phosphoryl-transfer potential, whereas the phosphate ester of an ordinary alcohol, such as 2-phosphoglycerate, has a low one. The $\Delta G^{\circ\prime}$ of the hydrolysis of a phosphate ester of an ordinary alcohol is $-13\,\mathrm{kJ\,mol^{-1}}\,(-3\,\mathrm{kcal\,mol^{-1}})$, whereas that of phosphoenolpyruvate is $-62\,\mathrm{kJ\,mol^{-1}}\,(-15\,\mathrm{kcal\,mol^{-1}})$.

Why does phosphoenolpyruvate have such a high phosphoryl-transfer potential? The phosphoryl group traps the molecule in its unstable enol form. When the phosphoryl group has been donated to ATP, the enol undergoes a conversion into the more stable ketone—namely, pyruvate.

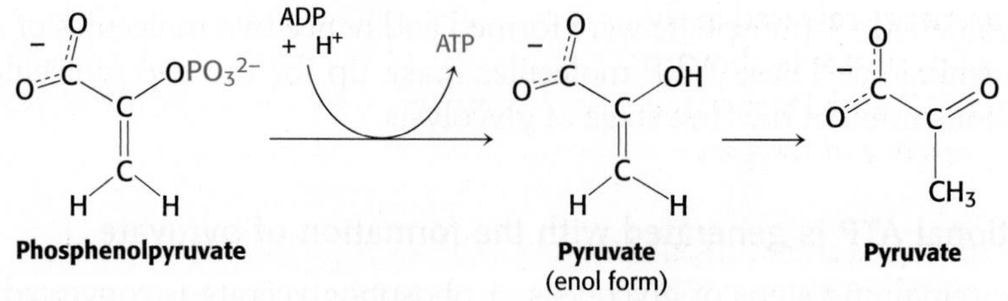

Thus, *the high phosphoryl-transfer potential of phosphoenolpyruvate arises primarily from the large driving force of the subsequent enol–ketone conversion.* Hence, pyruvate is formed, and ATP is generated concomitantly. The virtually irreversible transfer of a phosphoryl group from phosphoenolpyruvate to ADP is catalyzed *by pyruvate kinase.* What is the energy source for the formation of phosphoenolpyruvate? The answer to this question becomes clear when we compare the structures of 2-phosphoglycerate and pyruvate. The formation of pyruvate from 2-phosphoglycerate is, in essence, an internal oxidation–reduction reaction; carbon 3 takes electrons from carbon 2 in the conversion of 2-phosphoglycerate into pyruvate. Compared with 2-phosphoglycerate, C-3 is more reduced in pyruvate, whereas C-2 is more oxidized. Once again, carbon oxidation powers the synthesis of a compound with high phosphoryl-transfer potential, phosphoenolpyruvate here and 1,3-bisphosphoglycerate earlier, which allows the synthesis of ATP.

Because the molecules of ATP used in forming fructose 1,6-bisphosphate have already been regenerated, the two molecules of ATP generated from phosphoenolpyruvate are "profit."

Two ATP molecules are formed in the conversion of glucose into pyruvate

The net reaction in the transformation of glucose into pyruvate is

$$\text{Glucose} + 2\,P_i + 2\,\text{ADP} + 2\,\text{NAD}^+ \longrightarrow$$

$$2\,\text{pyruvate} + 2\,\text{ATP} + 2\,\text{NADH} + 2\,\text{H}^+ + 2\,\text{H}_2\text{O}$$

Thus, *two molecules of ATP are generated in the conversion of glucose into two molecules of pyruvate.* The reactions of glycolysis are summarized in Table 16.1.

TABLE 16.1 Reactions of glycolysis

Step	Reaction
1	Glucose + ATP ⟶ glucose 6-phosphate + ADP + H$^+$
2	Glucose 6-phosphate ⇌ fructose 6-phosphate
3	Fructose 6-phosphate + ATP ⟶ fructose 1,6-bisphosphate + ADP + H$^+$
4	Fructose 1,6-bisphosphate ⇌ dihydroxyacetone phosphate + glyceraldehyde 3-phosphate
5	Dihydroxyacetone phosphate ⇌ glyceraldehyde 3-phosphate
6	Glyceraldehyde 3-phosphate + P$_i$ + NAD$^+$ ⇌ 1,3-bisphosphoglycerate + NADH + H$^+$
7	1,3-Bisphosphoglycerate + ADP ⇌ 3-phosphoglycerate + ATP
8	3-Phosphoglycerate ⇌ 2-phosphoglycerate
9	2-Phosphoglycerate ⇌ phosphoenolpyruvate + H$_2$O
10	Phosphoenolpyruvate + ADP + H$^+$ ⟶ pyruvate + ATP

Note: ΔG, the actual free-energy change, has been calculated from $\Delta G^{\circ\prime}$ and known concentrations of reactants under typical physiological conditions. Glycolysis can proceed only if the ΔG values of all reactions are negative. The small positive ΔG values of three of the above reactions indicate that the concentrations of metabolites in vivo in cells undergoing glycolysis are not precisely known.

The energy released in the anaerobic conversion of glucose into two molecules of pyruvate is about $-90\,kJ\,mol^{-1}$ ($-22\,kcal\,mol^{-1}$). We shall see in Chapters 17 and 18 that much more energy can be released from glucose in the presence of oxygen.

NAD^+ is regenerated from the metabolism of pyruvate

The conversion of glucose into two molecules of pyruvate has resulted in the net synthesis of ATP. However, an energy-converting pathway that stops at pyruvate will not proceed for long, because redox balance has not been maintained. As we have seen, the activity of glyceraldehyde 3-phosphate dehydrogenase, in addition to generating a compound with high phosphoryl-transfer potential, reduces NAD^+ to NADH. In the cell, there are limited amounts of NAD^+, which is derived from the vitamin niacin (B_3), a dietary requirement for human beings. Consequently, NAD^+ must be regenerated for glycolysis to proceed. Thus, the final process in the pathway is the regeneration of NAD^+ through the metabolism of pyruvate.

The sequence of reactions from glucose to pyruvate is similar in most organisms and most types of cells. In contrast, the fate of pyruvate is variable. Three reactions of pyruvate are of primary importance: conversion into ethanol, lactate, or carbon dioxide (Figure 16.9). The first two reactions are fermentations that take place in the absence of oxygen. A *fermentation* is an ATP-generating process in which organic compounds act both as the donor and as the acceptor of electrons. In the presence of oxygen—the most common situation in multicellular organisms and in many unicellular ones—pyruvate is metabolized to carbon dioxide and water through the citric acid cycle and the electron-transport chain, with oxygen serving as the final electron acceptor. We now take a closer look at these three possible fates of pyruvate.

1. *Ethanol* is formed from pyruvate in yeast and several other microorganisms. The first step is the decarboxylation of pyruvate. This reaction is catalyzed by *pyruvate decarboxylase,* which requires the coenzyme thiamine pyrophosphate that is derived from the vitamin thiamine (B_1). The second step is the

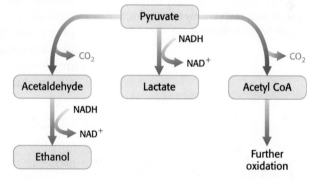

FIGURE 16.9 Diverse fates of pyruvate. Ethanol and lactate can be formed by reactions that include NADH. Alternatively, a two-carbon unit from pyruvate can be coupled to coenzyme A (Chapter 17) to form acetyl CoA.

Enzyme	Reaction type	$\Delta G^{\circ\prime}$ in kJ mol^{-1} (kcal mol^{-1})	ΔG in kJ mol^{-1} (kcal mol^{-1})
Hexokinase	Phosphoryl transfer	-16.7 (-4.0)	-33.5 (-8.0)
Phosphoglucose isomerase	Isomerization	$+1.7$ ($+0.4$)	-2.5 (-0.6)
Phosphofructokinase	Phosphoryl transfer	-14.2 (-3.4)	-22.2 (-5.3)
Aldolase	Aldol cleavage	$+23.8$ ($+5.7$)	-1.3 (-0.3)
Triose phosphate isomerase	Isomerization	$+7.5$ ($+1.8$)	$+2.5$ ($+0.6$)
Glyceraldehyde 3-phosphate dehydrogenase	Phosphorylation coupled to oxidation	$+6.3$ ($+1.5$)	-1.7 (-0.4)
Phosphoglycerate kinase	Phosphoryl transfer	-18.8 (-4.5)	$+1.3$ ($+0.3$)
Phosphoglycerate mutase	Phosphoryl shift	$+4.6$ ($+1.1$)	$+0.8$ ($+0.2$)
Enolase	Dehydration	$+1.7$ ($+0.4$)	-3.3 (-0.8)
Pyruvate kinase	Phosphoryl transfer	-31.4 (-7.5)	-16.7 (-4.0)

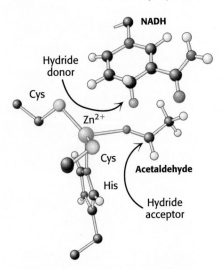

FIGURE 16.10 Active site of alcohol dehydrogenase. The active site contains a zinc ion bound to two cysteine residues and one histidine residue. *Notice* that the zinc ion binds the acetaldehyde substrate through its oxygen atom, polarizing the substrate so that it more easily accepts a hydride from NADH. Only the nicotinamide ring of NADH is shown.

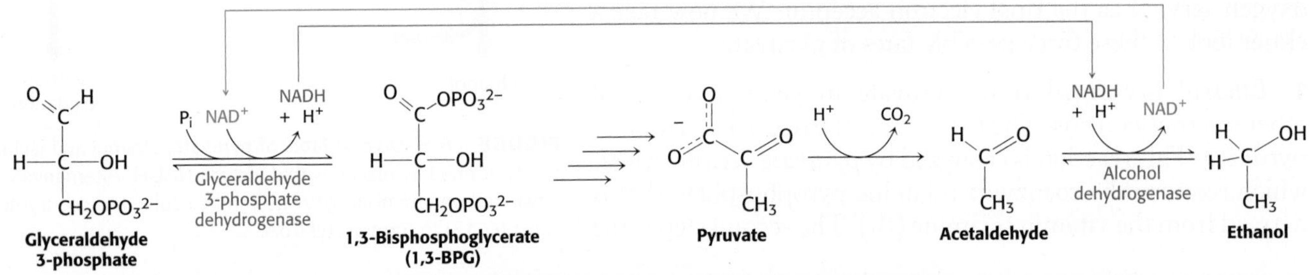

reduction of acetaldehyde to ethanol by NADH, in a reaction catalyzed by *alcohol dehydrogenase.* This reaction regenerates NAD^+.

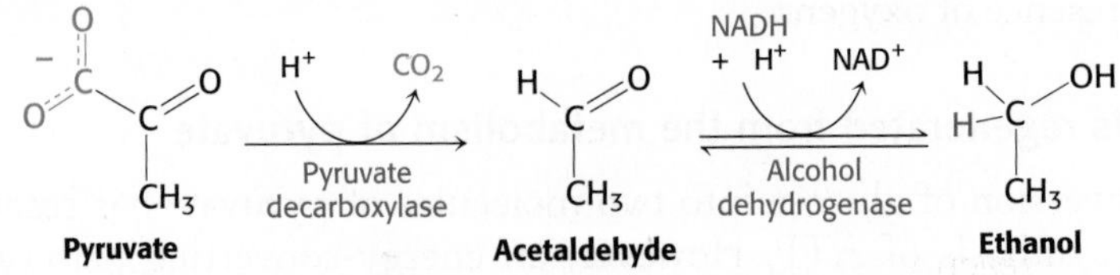

The active site of alcohol dehydrogenase contains a zinc ion that is coordinated to the sulfur atoms of two cysteine residues and a nitrogen atom of histidine (Figure 16.10). This zinc ion polarizes the carbonyl group of the substrate to favor the transfer of a hydride from NADH.

The conversion of glucose into ethanol is an example *of alcoholic fermentation.* The net result of this anaerobic process is

$$Glucose + 2\,P_i + 2\,ADP + 2\,H^+ \longrightarrow$$

$$2\,ethanol + 2\,CO_2 + 2\,ATP + 2\,H_2O$$

Note that NAD^+ and NADH do not appear in this equation, even though they are crucial for the overall process. NADH generated by the oxidation of glyceraldehyde 3-phosphate is consumed in the reduction of acetaldehyde to ethanol. Thus, *there is no net oxidation–reduction in the conversion of glucose into ethanol* (Figure 16.11). The ethanol formed in alcoholic fermentation provides a key ingredient for brewing and winemaking.

FIGURE 16.11 Maintaining redox balance. The NADH produced by the glyceraldehyde 3-phosphate dehydrogenase reaction must be reoxidized to NAD^+ for the glycolytic pathway to continue. In alcoholic fermentation, alcohol dehydrogenase oxidizes NADH and generates ethanol. In lactic acid fermentation (not shown), lactate dehydrogenase oxidizes NADH while generating lactic acid.

2. *Lactate* is formed from pyruvate in a variety of microorganisms in a process called *lactic acid fermentation.* Certain types of skeletal muscles in most animals can also function anaerobically for short periods. For example, a specific type of muscle fiber, called fast twitch or type IIb fibers, performs short bursts of intense exercise. The ATP needs rise faster than the ability of the body to provide oxygen to the muscle. The muscle functions anaerobically until fatigue sets in, which is caused, in part, by lactate buildup. Indeed, the pH of resting type IIb muscle fibers, which is about 7.0, may fall to as low as 6.3 during the bout of exercise. The drop in pH inhibits phosphofructokinase (p. 512). A lactate/H^+ symporter allows the exit of lactate from the muscle cell. The reduction of pyruvate by NADH to form lactate is catalyzed by *lactate dehydrogenase.*

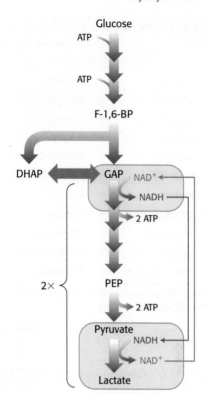

Regeneration of NAD$^+$.

The overall reaction in the conversion of glucose into lactate is

$$\text{Glucose} + 2\,P_i + 2\,\text{ADP} \longrightarrow 2\,\text{lactate} + 2\,\text{ATP} + 2\,H_2O$$

As in alcoholic fermentation, there is no net oxidation–reduction. The NADH formed in the oxidation of glyceraldehyde 3-phosphate is consumed in the reduction of pyruvate. *The regeneration of* NAD$^+$ *in the reduction of pyruvate to lactate or ethanol sustains the continued process of glycolysis under anaerobic conditions.*

3. Only a fraction of the energy of glucose is released in its anaerobic conversion into ethanol or lactate. Much more energy can be extracted aerobically by means of the citric acid cycle and the electron-transport chain. The entry point to this oxidative pathway is *acetyl coenzyme A* (acetyl CoA), which is formed inside mitochondria by the oxidative decarboxylation of pyruvate.

$$\text{Pyruvate} + \text{NAD}^+ + \text{CoA} \longrightarrow \text{acetyl CoA} + CO_2 + \text{NADH} + H^+$$

This reaction, which is catalyzed by the pyruvate dehydrogenase complex, will be considered in detail in Chapter 17. The NAD$^+$ required for this reaction and for the oxidation of glyceraldehyde 3-phosphate is regenerated when NADH ultimately transfers its electrons to O_2 through the electron-transport chain in mitochondria.

Fermentations provide usable energy in the absence of oxygen

Fermentations yield only a fraction of the energy available from the complete combustion of glucose. Why is a relatively inefficient metabolic pathway so extensively used? The fundamental reason is that oxygen is not required. The ability to survive without oxygen affords a host of living accommodations such as soils, deep water, and skin pores. Some organisms, called *obligate anaerobes,* cannot survive in the presence of O_2, a highly reactive compound. The bacterium *Clostridium perfringens,* the cause of gangrene, is an example of an obligate anaerobe. Other pathogenic obligate anaerobes are listed in Table 16.2. Some organisms, such as yeast, are *facultative anaerobes* that metabolize glucose aerobically when oxygen is present and perform fermentation when oxygen is absent.

TABLE 16.2 Examples of pathogenic obligate anaerobes

Bacterium	Result of infection
Clostridium tetani	Tetanus (lockjaw)
Clostridium botulinum	Botulism (an especially severe type of food poisoning)
Clostridium perfringens	Gas gangrene (gas is produced as an end point of the fermentation, distorting and destroying the tissue)
Bartonella hensela	Cat scratch fever (flu-like symptoms)
Bacteroides fragilis	Abdominal, pelvic, pulmonary, and blood infections

TABLE 16.3 Starting and ending points of various fermentations

Glucose	→	Lactate
Lactate	→	Acetate
Glucose	→	Ethanol
Ethanol	→	Acetate
Arginine	→	Carbon Dioxide
Pyrimidines	→	Carbon Dioxide
Purines	→	Formate
Ethylene glycol	→	Acetate
Threonine	→	Propionate
Leucine	→	2-Alkylacetate
Phenylalanine	→	Propionate

Note: The products of some fermentations are the substrates for others.

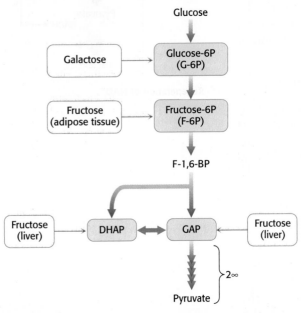

FIGURE 16.12 Entry points in glycolysis for fructose and galactose.

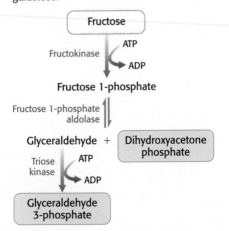

FIGURE 16.13 Fructose metabolism. Fructose enters the glycolytic pathway in the liver through the fructose 1-phosphate pathway.

Many food products, including sour cream, yogurt, various cheeses, beer, wine, and sauerkraut, result from fermentation. Yogurt is produced by the fermentation of lactose in milk to lactate by a mixed culture of *Lactobacillus acidophilus* and *Streptococcus thermophilus*. Sour cream begins with pasteurized light cream, which is fermented to lactate by *Streptococcus lactis*. The lactate is further fermented to ketones and aldehydes by *Leuconostoc citrovorum*. The second fermentation adds to the taste and aroma of sour cream. Yeast, *Saccharomyces cerevisiae*, ferments carbohydrates to ethanol and carbon dioxide, providing some of the ingredients for an array of alcohol beverages. Although we have considered only lactic acid and alcoholic fermentation, microorganisms are capable of generating a wide array of molecules as end points of fermentation (Table 16.3).

Fructose is converted into glycolytic intermediates by fructokinase

Although glucose is the most widely used monosaccharide, others also are important fuels. Let us consider how fructose is funneled into the glycolytic pathway (Figure 16.12). There are no catabolic pathways for metabolizing fructose, and so the strategy is to convert this sugar into a metabolite of glucose.

The main site of fructose metabolism is the liver, using *the fructose 1-phosphate pathway* (Figure 16.13). The first step is the phosphorylation of *fructose* to *fructose 1-phosphate* by *fructokinase*. Fructose 1-phosphate is then split into *glyceraldehyde* and *dihydroxyacetone phosphate*, an intermediate in glycolysis. This aldol cleavage is catalyzed by a specific *fructose 1-phosphate aldolase*. Glyceraldehyde is then phosphorylated to *glyceraldehyde 3-phosphate*, a glycolytic intermediate, by *triose kinase*. In other tissues, such as adipose tissue, *fructose can be phosphorylated to fructose 6-phosphate by hexokinase*.

Excessive fructose consumption can lead to pathological conditions

Fructose, a commonly used sweetener, is a component of sucrose and high fructose corn syrup (which contains approximately 55% fructose and 45% glucose). Epidemiological as well as clinical studies have linked excessive fructose consumption to fatty liver, insulin insensitivity, and obesity. These conditions may eventually result in type 2 diabetes (Chapter 27). Studies have shown that these disorders are not necessarily the result of simple excess energy consumption, but rather how fructose is processed by the liver. What aspects of liver fructose metabolism are the contributing factors then? Note that, as shown in Figure 16.13, the actions of fructokinase and triose kinase bypass the most important regulatory step in glycolysis, the phosphofructokinase-catalyzed reaction. The fructose-derived glyceraldehyde 3-phosphate and dihydroxyacetone phosphate are processed by glycolysis to pyruvate and subsequently to acetyl CoA in an unregulated fashion. As we will see in Chapter 22, this excess acetyl CoA is converted to fatty acids, which can be transported to adipose tissue, resulting in obesity. The liver also begins to accumulate fatty acids, resulting in fatty liver. The activity of the fructokinase and triose kinase can deplete the liver of ATP and inorganic phosphate, compromising liver function. We will return to the topic of obesity and caloric homeostasis in Chapter 27.

Galactose is converted into glucose 6-phosphate

Like fructose, *galactose* is an abundant sugar common in dairy products that must be converted into metabolites of glucose (Figure 16.12). Galactose is converted into *glucose 6-phosphate* in four steps. The first reaction in the *galactose–glucose interconversion pathway* is the phosphorylation of galactose to galactose 1-phosphate by *galactokinase*.

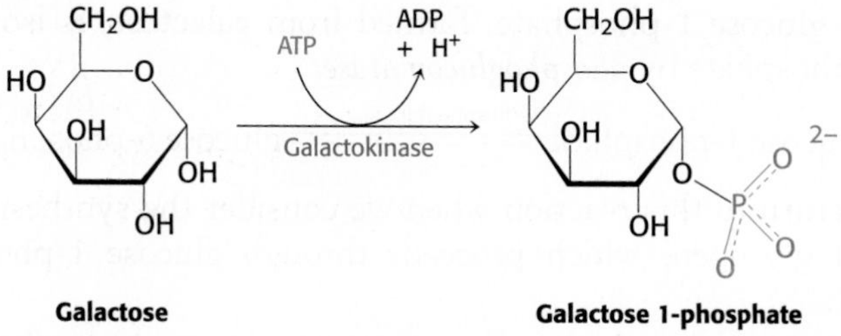

Galactose 1-phosphate then acquires a uridyl group from uridine diphosphate glucose (UDP-glucose), an activated intermediate in the synthesis of carbohydrates (p. 358 and Section 21.4).

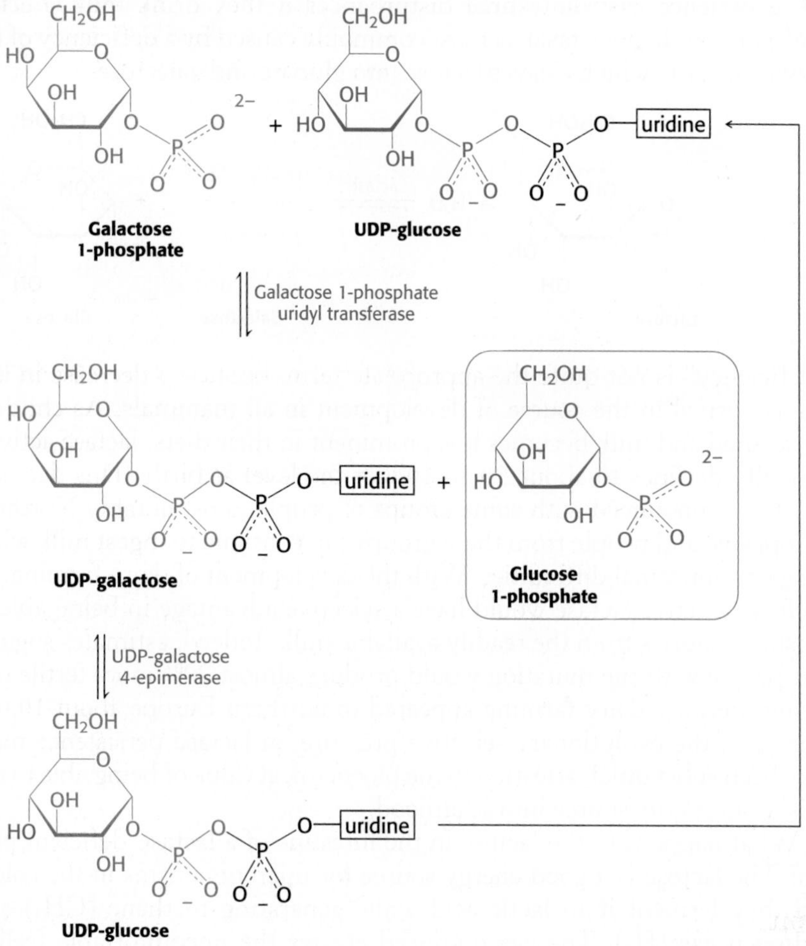

The products of this reaction, which is catalyzed by *galactose 1-phosphate uridyl transferase*, are UDP-galactose and glucose 1-phosphate. The galactose moiety of UDP-galactose is then epimerized to glucose. The configuration of the hydroxyl group at carbon 4 is inverted by *UDP-galactose 4-epimerase*.

The sum of the reactions catalyzed by galactokinase, the transferase, and the epimerase is

$$\text{Galactose} + \text{ATP} \longrightarrow \text{glucose 1-phosphate} + \text{ADP} + \text{H}^+$$

Note that UDP-glucose is not consumed in the conversion of galactose into glucose, because it is regenerated from UDP-galactose by the epimerase. This reaction is reversible, and the product of the reverse direction also is important. *The conversion of UDP-glucose into UDP-galactose is essential for the synthesis of galactosyl residues in complex polysaccharides and glycoproteins if the amount of galactose in the diet is inadequate to meet these needs.*

Finally, glucose 1-phosphate, formed from galactose, is isomerized to glucose 6-phosphate by *phosphoglucomutase.*

$$\text{Glucose 1-phosphate} \underset{\text{Phosphoglucomutase}}{\rightleftharpoons} \text{glucose 6-phosphate}$$

We shall return to this reaction when we consider the synthesis and degradation of glycogen, which proceeds through glucose 1-phosphate, in Chapter 21.

Many adults are intolerant of milk because they are deficient in lactase

Many adults are unable to metabolize the milk sugar lactose and experience gastrointestinal disturbances if they drink milk. *Lactose intolerance,* or hypolactasia, is most commonly caused by a deficiency of the enzyme lactase, which cleaves lactose into glucose and galactose.

Lactose + H₂O $\rightleftharpoons$ (Lactase) **Galactose** + **Glucose**

"Deficiency" is not quite the appropriate term, because a decrease in lactase is normal in the course of development in all mammals. As children are weaned and milk becomes less prominent in their diets, lactase activity normally declines to about 5% to 10% of the level at birth. This decrease is not as pronounced with some groups of people, most notably Northern Europeans, and people from these groups can continue to ingest milk without gastrointestinal difficulties. With the development of dairy farming, an adult with active lactase would have a selective advantage in being able to consume calories from the readily available milk. Indeed, estimates suggest that people with the mutation would produce almost 20% more fertile offspring. Because dairy farming appeared in northern Europe about 10,000 years ago, the evolutionary selective pressure on lactase persistence must have been substantial, attesting to the biochemical value of being able to use milk as an energy source into adulthood.

What happens to the lactose in the intestine of a lactase-deficient person? The lactose is a good energy source for microorganisms in the colon, and they ferment it to lactic acid while generating methane (CH_4) and hydrogen gas (H_2). The gas produced creates the uncomfortable feeling of gut distension and the annoying problem of flatulence. The lactate produced by the microorganisms is osmotically active and draws water into the intestine, as does any undigested lactose, resulting in diarrhea. If severe enough, the gas and diarrhea hinder the absorption of other nutrients such as fats and proteins. The simplest treatment is to avoid the consumption of products containing much lactose. Alternatively, the enzyme lactase can be ingested with milk products.

Scanning electron micrograph of *Lactobacillus.* The anaerobic bacterium *Lactobacillus* is shown here. As suggested by its name, this genus of bacteria ferments glucose into lactic acid and is widely used in the food industry. *Lactobacillus* is also a component of the normal human bacterial flora of the urogenital tract where, because of its ability to generate an acidic environment, it prevents the growth of harmful bacteria. [SPL/Science Source.]

Galactose is highly toxic if the transferase is missing

Less common than lactose intolerance are disorders that interfere with the metabolism of galactose. The disruption of galactose metabolism is referred to as *galactosemia*. The most common form, called classic galactosemia, is an inherited deficiency in galactose 1-phosphate uridyl transferase activity. Afflicted infants fail to thrive. They vomit or have diarrhea after consuming milk, and enlargement of the liver and jaundice are common, sometimes progressing to cirrhosis. Cataracts will form, and lethargy and retarded mental development also are common. The blood-galactose level is markedly elevated, and galactose is found in the urine. The absence of the transferase in red blood cells is a definitive diagnostic criterion.

The most common treatment is to remove galactose (and lactose) from the diet. An enigma of galactosemia is that, although elimination of galactose from the diet prevents liver disease and cataract development, the majority of patients still suffer from central nervous system malfunction, most commonly a delayed acquisition of language skills. Female patients also display ovarian failure.

Cataract formation is better understood. A cataract is the clouding of the normally clear lens of the eye (Figure 16.14). If the transferase is not active in the lens of the eye, the presence of aldose reductase causes the accumulating galactose to be reduced to galactitol.

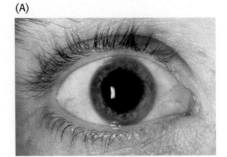

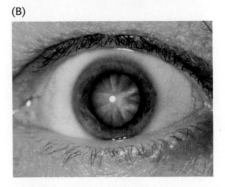

FIGURE 16.14 Cataracts are evident as the clouding of the lens. (A) A healthy eye. (B) An eye with a cataract. [(A) Tim Mainiero/Shutterstock; (B) SPL/Science Source.]

Galactose → (NADPH + H⁺ → NADP⁺, Aldose reductase) → Galactitol

Galactitol is poorly metabolized and accumulates in the lens. Water will diffuse into the lens to maintain osmotic balance, triggering the formation of cataracts. In fact, there is a high incidence of cataract formation with age in populations that consume substantial amounts of milk into adulthood.

16.2 The Glycolytic Pathway Is Tightly Controlled

The glycolytic pathway has a dual role: it degrades glucose to generate ATP and it provides building blocks for biosynthetic reactions. The rate of conversion of glucose into pyruvate is regulated to meet these two major cellular needs. *In metabolic pathways, enzymes catalyzing essentially irreversible reactions are potential sites of control.* In glycolysis, the reactions catalyzed by hexokinase, phosphofructokinase, and pyruvate kinase are virtually irreversible, and each of them serves as a control site. These enzymes become more active or less so in response to the reversible binding of allosteric effectors or to covalent modification. In addition, the amounts of these important enzymes are varied by the regulation of transcription to meet changing metabolic needs. The time required for allosteric control, regulation by phosphorylation, and transcriptional control is measured typically in milliseconds, seconds, and hours, respectively. We will consider the control of glycolysis in two different tissues—skeletal muscle and liver.

Glycolysis in muscle is regulated to meet the need for ATP

Glycolysis in skeletal muscle provides ATP primarily to power contraction. Consequently, *the primary control of muscle glycolysis is the energy charge of the cell*—the ratio of ATP to AMP. Let's examine how each of the key regulatory enzymes responds to changes in the amounts of ATP and AMP present in the cell.

Phosphofructokinase. *Phosphofructokinase is the most important control site in the mammalian glycolytic pathway* (Figure 16.15). High levels of ATP allosterically inhibit the enzyme (a 340-kDa tetramer). ATP binds to a specific regulatory site that is distinct from the catalytic site. The binding of ATP lowers the enzyme's affinity for fructose 6-phosphate. Thus, a high concentration of ATP converts the hyperbolic binding curve of fructose 6-phosphate into a sigmoidal one (Figure 16.16). AMP reverses the inhibitory action of ATP, and so *the activity of the enzyme increases when the ATP/AMP ratio is lowered*. In other words, *glycolysis is stimulated as the energy charge falls*. A decrease in pH also inhibits phosphofructokinase activity by augmenting the inhibitory effect of ATP. The pH might fall when fast-twitch muscle is functioning anaerobically, producing excessive quantities of lactic acid. The inhibitory effect protects the muscle from damage that would result from the accumulation of too much acid.

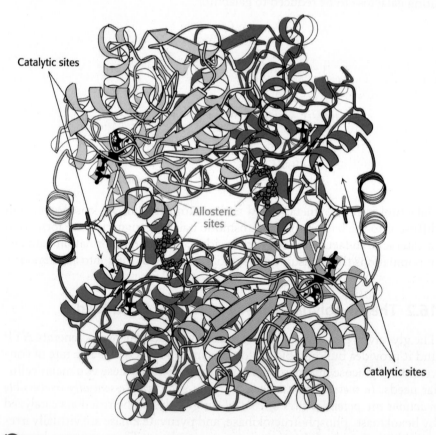

FIGURE 16.15 Structure of phosphofructokinase. The structure of phosphofructokinase from *E. coli* comprises a tetramer of four identical subunits. *Notice* the separation of the catalytic and allosteric sites. Each subunit of the human liver enzyme consists of two domains that are similar to the *E. coli* enzyme. [Drawn from 1PFK.pdb.]

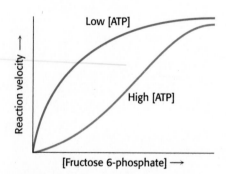

FIGURE 16.16 Allosteric regulation of phosphofructokinase. A high level of ATP inhibits the enzyme by decreasing its affinity for fructose 6-phosphate.

Why is AMP and not ADP the positive regulator of phosphofructokinase? When ATP is being utilized rapidly, the enzyme *adenylate kinase* (Section 9.4) can form ATP from ADP by the following reaction:

$$ADP + ADP \rightleftharpoons ATP + AMP$$

Thus, some ATP is salvaged from ADP, and AMP becomes the signal for the low-energy state. Moreover, the use of AMP as an allosteric regulator provides an especially sensitive control. We can understand why by considering the following. First, the total adenylate pool ([ATP], [ADP], [AMP]) in a cell is constant over the short term. Second, the concentration of ATP is greater than that of ADP, and the concentration of ADP is, in turn, greater than that of AMP. Consequently, small percentage changes in [ATP] result in larger percentage changes in the concentrations of the other adenylate nucleotides. This magnification of small changes in [ATP] to larger changes in [AMP] leads to tighter control by increasing the range of sensitivity of phosphofructokinase (problem 49).

Hexokinase. Phosphofructokinase is the most prominent regulatory enzyme in glycolysis, but it is not the only one. Hexokinase, the enzyme catalyzing the first step of glycolysis, is inhibited by its product, glucose 6-phosphate. High concentrations of this molecule signal that the cell no longer requires glucose for energy or for the synthesis of glycogen, a storage form of glucose (Chapter 21), and the glucose will be left in the blood. A rise in glucose 6-phosphate concentration is a means by which phosphofructokinase communicates with hexokinase. When phosphofructokinase is inactive, the concentration of fructose 6-phosphate rises. In turn, the level of glucose 6-phosphate rises because it is in equilibrium with fructose 6-phosphate. Hence, *the inhibition of phosphofructokinase leads to the inhibition of hexokinase.*

Why is phosphofructokinase rather than hexokinase the pacemaker of glycolysis? The reason becomes evident on noting that glucose 6-phosphate is not solely a glycolytic intermediate. In muscle, glucose 6-phosphate can also be converted into glycogen. The first irreversible reaction unique to the glycolytic pathway, the *committed step* (Section 10.1), is the phosphorylation of fructose 6-phosphate to fructose 1,6-bisphosphate. Thus, it is highly appropriate for phosphofructokinase to be the primary control site in glycolysis. In general, *the enzyme catalyzing the committed step in a metabolic sequence is the most important control element in the pathway.*

Pyruvate kinase. Pyruvate kinase, the enzyme catalyzing the third irreversible step in glycolysis, controls the outflow from this pathway. This final step yields ATP and pyruvate, a central metabolic intermediate that can be oxidized further or used as a building block. ATP allosterically inhibits pyruvate kinase to slow glycolysis when the energy charge is high. When the pace of glycolysis increases, fructose 1,6-bisphosphate—the product of the preceding irreversible step in glycolysis—activates the kinase to enable it to keep pace with the oncoming high flux of intermediates. Such a process is called *feedforward stimulation.* A summary of the regulation of glycolysis in resting and active muscle is shown in Figure 16.17.

The regulation of glycolysis in the liver illustrates the biochemical versatility of the liver

The liver has more diverse biochemical functions than does muscle. Significantly, the liver maintains blood-glucose concentration: it stores glucose as glycogen when glucose is plentiful, and it releases glucose when supplies are low. It also uses glucose to generate reducing power for biosynthesis (Section 20.3) as well as to synthesize a host of biochemicals. So, although the liver has many of the regulatory features of muscle glycolysis, the regulation of glycolysis in the liver is more complex.

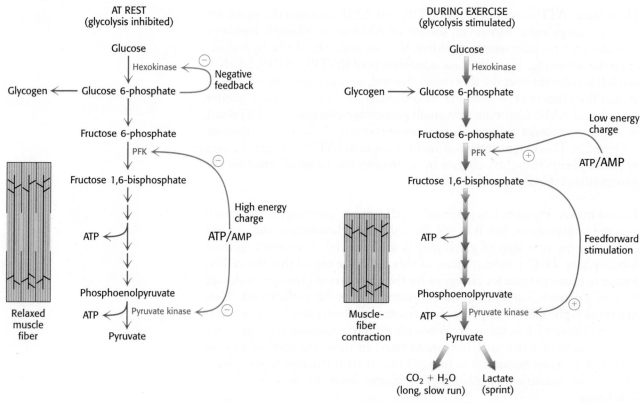

FIGURE 16.17 Regulation of glycolysis in muscle. At rest (left), glycolysis is not very active (thin arrows). The high concentration of ATP inhibits phosphofructokinase (PFK), pyruvate kinase, and hexokinase. Glucose 6-phosphate is converted into glycogen (Chapter 21). During exercise (right), the decrease in the ATP/AMP ratio resulting from muscle contraction activates phosphofructokinase and hence glycolysis. The flux down the pathway is increased, as represented by the thick arrows.

Phosphofructokinase. Liver phosphofructokinase can be regulated by ATP as in muscle, but such regulation is not as important since the liver does not experience the sudden ATP needs that a contracting muscle does. Likewise, low pH is not an important metabolic signal for the liver enzyme, because lactate is not normally produced in the liver. Indeed, as we will see, lactate is converted into glucose in the liver.

Glycolysis in the liver furnishes carbon skeletons for biosyntheses, and so a signal indicating whether building blocks are abundant or scarce should also regulate phosphofructokinase. In the liver, *phosphofructokinase is inhibited by citrate,* an early intermediate in the citric acid cycle (Chapter 17). A high level of citrate in the cytoplasm means that biosynthetic precursors are abundant, and so there is no need to degrade additional glucose for this purpose. Citrate inhibits phosphofructokinase by enhancing the inhibitory effect of ATP.

The key means by which glycolysis in the liver responds to changes in blood glucose is through the signal molecule *fructose 2,6-bisphosphate* (F-2,6-BP), a potent activator of phosphofructokinase. In the liver, the concentration of fructose 6-phosphate rises when blood-glucose concentration is high, and the abundance of fructose 6-phosphate accelerates the synthesis of F-2,6-BP (Figure 16.18). Hence, *an abundance of fructose 6-phosphate leads to a higher concentration of F-2,6-BP.* The binding of fructose 2,6-bisphosphate increases the affinity of phosphofructokinase for fructose 6-phosphate and diminishes the inhibitory effect of ATP (Figure 16.19). Glycolysis is thus accelerated when glucose is abundant. We will explore

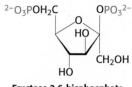

Fructose 2,6-bisphosphate (F-2,6-BP)

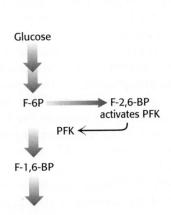

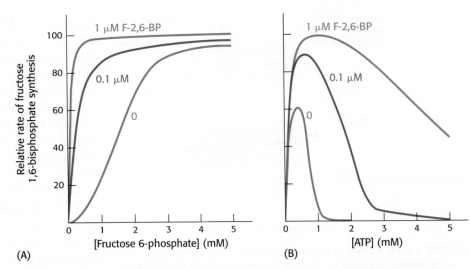

FIGURE 16.18 Regulation of phosphofructokinase by fructose 2,6-bisphosphate. In high concentrations, fructose 6-phosphate (F-6P) activates the enzyme phosphofructokinase (PFK) through an intermediary, fructose 2,6-bisphosphate (F-2,6-BP).

FIGURE 16.19 Activation of phosphofructokinase by fructose 2,6-bisphosphate. (A) The sigmoidal dependence of velocity on substrate concentration becomes hyperbolic in the presence of 1 μM fructose 2,6-bisphosphate. (B) ATP, acting as a substrate, initially stimulates the reaction. As the concentration of ATP increases, it acts as an allosteric inhibitor. The inhibitory effect of ATP is reversed by fructose 2,6-bisphosphate. [Data from E. Van Schaftingen, M. F. Jett, L. Hue, and H. G. Hers, *Proc. Natl. Acad. Sci. U.S.A.* 78:3483–3486, 1981.]

the synthesis and degradation of this important regulatory molecule after we have considered gluconeogenesis.

Hexokinase and glucokinase. The hexokinase reaction in the liver is controlled as in the muscle. However, the liver, in keeping with its role as monitor of blood-glucose levels, possesses another specialized isozyme of hexokinase, called *glucokinase*, which is not inhibited by glucose 6-phosphate. The role of glucokinase is to provide glucose 6-phosphate for the synthesis of glycogen and for the formation of fatty acids (Section 22.1). Remarkably, glucokinase displays the sigmoidal kinetics characteristic of an allosteric enzyme even though it functions as a monomer. Glucokinase phosphorylates glucose only when glucose is abundant because the affinity of glucokinase for glucose is about 50-fold lower than that of hexokinase. Moreover, when glucose concentration is low, glucokinase is inhibited by the liver-specific *glucokinase regulatory protein* (GKRP), which sequesters the kinase in the nucleus until the glucose concentration increases. The low affinity of glucokinase for glucose gives the brain and muscles first call on glucose when its supply is limited, and it ensures that glucose will not be wasted when it is abundant. Drugs that activate liver glucokinase or disrupt its interaction with GKRP are being evaluated as a treatment for type 2 or insulin-insensitive diabetes. Glucokinase is also present in the β cells of the pancreas, where the increased formation of glucose 6-phosphate by glucokinase when blood-glucose levels are elevated leads to the secretion of the hormone insulin. Insulin signals the need to remove glucose from the blood for storage as glycogen or conversion into fat.

Pyruvate kinase. Several isozymic forms of pyruvate kinase (a tetramer of 57-kDa subunits) encoded by different genes are present in mammals: The L type predominates in the liver, and the M type in muscle and the brain. The L and M forms of pyruvate kinase have many properties in common. Indeed, the liver enzyme behaves much like the muscle enzyme with regard to allosteric regulation except that the liver enzyme is also inhibited by alanine (synthesized in one step from pyruvate), a signal that building blocks

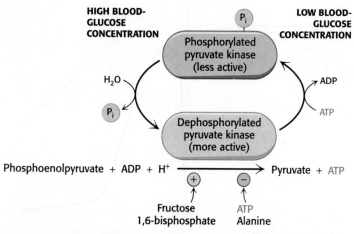

FIGURE 16.20 Control of the catalytic activity of pyruvate kinase. Pyruvate kinase is regulated by allosteric effectors and covalent modification. Fructose 1,6-bisphosphate allosterically stimulates the enzyme, while ATP and alanine are allosteric inhibitors. Glucagon, secreted in response to low blood glucose, promotes phosphorylation and inhibition of the enzyme. When blood glucose concentration is adequate, the enzyme is dephosphorylated and activated.

are available. Moreover, the isozymic forms differ in their susceptibility to covalent modification. The catalytic properties of the L form—but not of the M form—are also controlled by reversible phosphorylation (Figure 16.20). When the blood-glucose level is low, the glucagon-triggered cyclic AMP cascade (p. 526) leads to the phosphorylation of pyruvate kinase, which diminishes its activity. This hormone-triggered phosphorylation prevents the liver from consuming glucose when it is more urgently needed by the brain and muscle. We see here a clear-cut example of how isoenzymes contribute to the metabolic diversity of different organs. We will return to the control of glycolysis after considering gluconeogenesis.

A family of transporters enables glucose to enter and leave animal cells

Several glucose transporters mediate the thermodynamically downhill movement of glucose across the plasma membranes of animal cells. Each member of this protein family, named GLUT1 to GLUT5, consists of a single polypeptide chain about 500 residues long (Table 16.4). Each glucose transporter has a 12-transmembrane-helix structure similar to that of lactose permease (Section 13.3).

TABLE 16.4 Family of glucose transporters

Name	Tissue location	K_M	Comments
GLUT1	All mammalian tissues	1 mM	Basal glucose uptake
GLUT2	Liver and pancreatic β cells	15–20 mM	In the pancreas, plays a role in the regulation of insulin In the liver, removes excess glucose from the blood
GLUT3	All mammalian tissues	1 mM	Basal glucose uptake
GLUT4	Muscle and fat cells	5 mM	Amount in muscle plasma membrane increases with endurance training
GLUT5	Small intestine	—	Primarily a fructose transporter

The members of this family have distinctive roles:

1. GLUT1 and GLUT3, present in nearly all mammalian cells, are responsible for basal glucose uptake. Their K_M value for glucose is about 1 mM, significantly less than the normal serum-glucose level, which typically ranges from 4 mM to 8 mM. Hence, GLUT1 and GLUT3 continually transport glucose into cells at an essentially constant rate.

2. GLUT2, present in liver and pancreatic β cells, is distinctive in having a very high K_M value for glucose (15–20 mM). Hence, glucose enters these tissues at a biologically significant rate only when there is much glucose in the blood. The pancreas can sense the glucose level and accordingly adjust the rate of insulin secretion. The high K_M value of GLUT2 also ensures that glucose rapidly enters liver cells only in times of plenty.

3. GLUT4, which has a K_M value of 5 mM, transports glucose into muscle and fat cells. The number of GLUT4 transporters in the plasma membrane increases rapidly in the presence of insulin, which signals the fed state. Hence, insulin promotes the uptake of glucose by muscle and fat. Endurance exercise training increases the amount of this transporter present in muscle membranes.

4. GLUT5, present in the small intestine, functions primarily as a fructose transporter.

Aerobic glycolysis is a property of rapidly growing cells

Tumors have been known for decades to display enhanced rates of glucose uptake and glycolysis. Indeed, rapidly growing tumor cells will metabolize glucose to lactate even in the presence of oxygen, a process called *aerobic glycolysis* or the *Warburg effect*, after Otto Warburg, the biochemist who first noted this characteristic of cancer cells in the 1920s. In fact, tumors with a high glucose uptake are particularly aggressive, and the cancer is likely to have a poor prognosis. A nonmetabolizable glucose analog, $2\text{-}^{18}\text{F-2-D-deoxyglucose}$, detectable by a combination of positron emission tomography (PET) and computer-aided tomography (CAT), easily visualizes tumors and allows monitoring of treatment effectiveness (Figure 16.21).

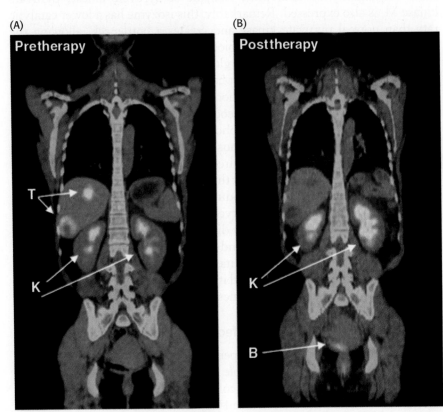

FIGURE 16.21 Tumors can be visualized with $2\text{-}^{18}\text{F-2-D-deoxyglucose}$ (FDG) and positron emission tomography. (A) A nonmetabolizable glucose analog infused into a patient and detected by a combination of positron emission and computer-aided tomography reveals the presence of a malignant tumor (T). (B) After 4 weeks of treatment with a tyrosine kinase inhibitor (Section 14.5), the tumor shows no uptake of FDG, indicating decreased metabolism. Excess FDG, which is excreted in the urine, also visualizes the kidneys (K) and bladder (B). [Courtesy of A. D. Van den Abbeele, MD, Dana-Farber Cancer Institute, Boston.]

What selective advantage does aerobic glycolysis offer the tumor over the energetically more efficient oxidative phosphorylation? Research is being actively pursued to answer the question, but we can speculate on the benefits. First, aerobic glycolysis generates lactic acid that is then secreted. Acidification of the tumor environment has been shown to facilitate tumor invasion. Moreover, lactate impairs the activation of $CD8^+$ T and NK immune system cells (See Section 34.4 on 🔓 **SaplingPlus**) that normally attack the tumor. However, even leukemias perform aerobic glycolysis, and leukemia is not an invasive cancer. Second, and perhaps more importantly, the increased uptake of glucose and formation of glucose 6-phosphate provides substrates for another metabolic pathway—the pentose phosphate pathway (Chapter 20)—that generates biosynthetic reducing power, NADPH. Finally, cancer cells grow more rapidly than the blood vessels that nourish them; thus, as solid tumors grow, the oxygen concentration in their environment falls. In other words, they begin to experience *hypoxia*, a deficiency of oxygen. The use of aerobic glycolysis reduces the dependence of cell growth on oxygen.

What biochemical alterations facilitate the switch to aerobic glycolysis? Again, the answers are not complete, but changes in gene expression of isozymic forms of two glycolytic enzymes may be crucial. Tumor cells express an isozyme of hexokinase that binds to mitochondria. There, the enzyme has ready access to any ATP generated by oxidative phosphorylation and is not susceptible to feedback inhibition by its product, glucose 6-phosphate. More importantly, an embryonic isozyme of pyruvate kinase, pyruvate kinase M, is also expressed. Remarkably, this isozyme has a lower catalytic rate than normal pyruvate kinase and creates a bottleneck, allowing the use of glycolytic intermediates for biosynthetic processes required for cell proliferation. As mentioned above, glucose 6-phosphate can be processed to yield NADPH. Fructose 6-phosphate can be converted into N-acetylglucosamine, a substrate for the modification of many cellular proteins (p. 354), and a component of proteoglycans (p. 354). Dihydroxyacetone phosphate is a precursor for phospholipids, which are required for new membrane synthesis (p. 735). 3-Phosphoglycerate can be converted into glycine, serine, and coenzymes required for nucleotide synthesis. The need for biosynthetic precursors is greater than the need for ATP, suggesting that even glycolysis at a reduced rate produces sufficient ATP to allow cell proliferation. Although originally observed in cancer cells, the Warburg effect is also seen in noncancerous, rapidly dividing cells.

In addition to glucose, cancer and other rapidly dividing cells require large amounts of the amino acid glutamine. Glutamine serves as a carbon source for a number of reactions, and it also supplies the nitrogen for nucleotide, N-acetylglucosamine, and nonessential amino acid synthesis.

Cancer and endurance training affect glycolysis in a similar fashion

The hypoxia that some tumors experience with rapid growth activates a transcription factor, *hypoxia-inducible transcription factor* (HIF-1). HIF-1 increases the expression of most glycolytic enzymes and the glucose transporters GLUT1 and GLUT3 (Table 16.5). These adaptations by the cancer cells enable a tumor to survive until blood vessels can grow. HIF-1 also increases the expression of signal molecules, such as vascular endothelial growth factor (VEGF), that facilitate the growth of blood vessels that will provide nutrients to the cells (Figure 16.22). Without new blood vessels, a tumor would cease to grow and either die or remain harmlessly small. Efforts are underway to develop drugs that inhibit the growth of blood vessels in tumors. Indeed, bevacizumab, a monoclonal antibody that binds to VEGF and prevents activation of angiogenesis, has been approved for treatment of glioblastoma. Glioblastomas are fast-growing cancers of the central nervous system derived from glial cells.

Interestingly, anaerobic exercise training—forcing muscles to rely on lactic acid fermentation for ATP production—also activates HIF-1,

TABLE 16.5 Proteins in glucose metabolism encoded by genes regulated by hypoxia-inducible factor

GLUT1
GLUT3
Hexokinase
Phosphofructokinase
Aldolase
Glyceraldehyde 3-phosphate dehydrogenase
Phosphoglycerate kinase
Enolase
Pyruvate kinase

producing the same effects as those seen in the tumor—enhanced ability to generate ATP anaerobically and a stimulation of blood-vessel growth. These biochemical effects account for the improved athletic performance that results from training and demonstrate how behavior can affect biochemistry. Other signals from sustained muscle contraction trigger muscle mitochondrial biogenesis, allowing for more efficient aerobic energy generation and forestalling the need to resort to lactic acid fermentation for ATP synthesis (Chapter 27).

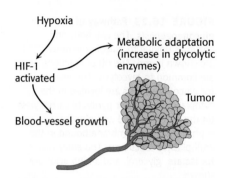

FIGURE 16.22 Alteration of gene expression in tumors owing to hypoxia. The hypoxic conditions inside a tumor mass lead to the activation of the hypoxia-inducible transcription factor (HIF-1), which induces metabolic adaptation (an increase in glycolytic enzymes) and activates angiogenic factors that stimulate the growth of new blood vessels. [Information from C. V. Dang and G. L. Semenza, *Trends Biochem. Sci.* 24:68–72, 1999.]

16.3 Glucose Can Be Synthesized from Noncarbohydrate Precursors

We now turn to the *synthesis of glucose from noncarbohydrate precursors,* a process called *gluconeogenesis.* Maintaining levels of glucose is important because the brain depends on glucose as its primary fuel and red blood cells use glucose as their only fuel. The daily glucose requirement of the brain in a typical adult human being is about 120 g, which accounts for most of the 160 g of glucose needed daily by the whole body. The amount of glucose present in body fluids is about 20 g, and that readily available from glycogen is approximately 190 g. Thus, the direct glucose reserves are sufficient to meet glucose needs for about a day. Gluconeogenesis is especially important during a longer period of fasting or starvation (Section 27.5).

The gluconeogenic pathway converts pyruvate into glucose. Noncarbohydrate precursors of glucose are first converted into pyruvate or enter the pathway at later intermediates such as oxaloacetate and dihydroxyacetone phosphate (Figure 16.23). The major noncarbohydrate precursors are *lactate, amino acids,* and *glycerol.* Lactate is formed by active skeletal muscle when the rate of glycolysis exceeds the rate of oxidative metabolism. Lactate is readily converted into pyruvate by the action of lactate dehydrogenase (p. 506). Amino acids are derived from proteins in the diet and, during starvation, from the breakdown of proteins in skeletal muscle (Section 23.1). The hydrolysis of triacylglycerols (Section 22.2) in fat cells yields glycerol and fatty acids. Glycerol is a precursor of glucose, but animals cannot convert fatty acids into glucose, for reasons that will be given later. Glycerol may enter either the gluconeogenic or the glycolytic pathway at dihydroxyacetone phosphate.

The major site of gluconeogenesis is the *liver,* with a small amount also taking place in the *kidney. Gluconeogenesis in the liver and kidney helps to maintain the glucose concentration in the blood so that the brain and muscle can extract sufficient glucose from it to meet their metabolic demands.*

Gluconeogenesis is not a reversal of glycolysis

In glycolysis, glucose is converted into pyruvate; in gluconeogenesis, pyruvate is converted into glucose. However, *gluconeogenesis is not a reversal of*

FIGURE 16.23 Pathway of gluconeogenesis. The reactions and enzymes unique to gluconeogenesis are shown in red. The other reactions are common to glycolysis. The enzymes for gluconeogenesis are located in the cytoplasm, except for pyruvate carboxylase (in the mitochondria) and glucose 6-phosphatase (membrane bound in the endoplasmic reticulum). The entry points for lactate, glycerol, and amino acids are shown.

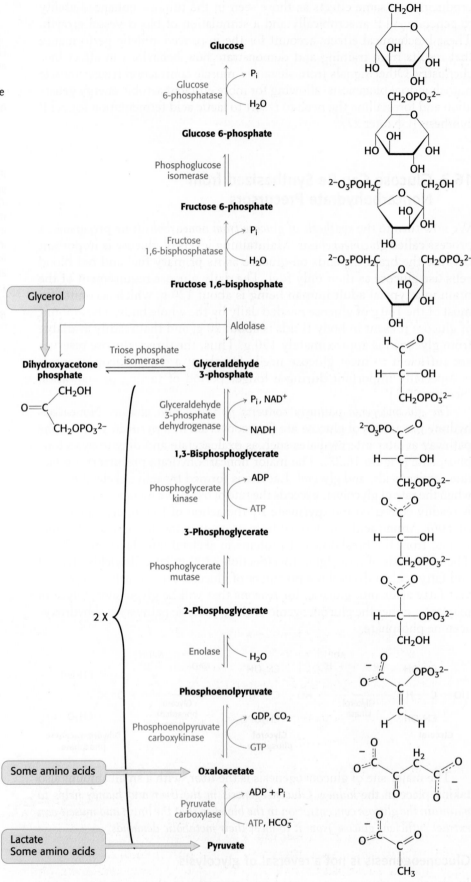

glycolysis. Several reactions must differ because the equilibrium of glycolysis lies far on the side of pyruvate formation. The actual free-energy change for the formation of pyruvate from glucose is about -90 kJ mol^{-1} (-22 kcal mol^{-1}) under typical cellular conditions. Most of the decrease in free energy in glycolysis takes place in the three essentially irreversible steps catalyzed by hexokinase, phosphofructokinase, and pyruvate kinase.

$$\text{Glucose} + \text{ATP} \xrightarrow{\text{Hexokinase}} \text{Glucose 6-phosphate} + \text{ADP}$$

$$\Delta G = -33 \text{ kJ mol}^{-1} \, (-8.0 \text{ kcal mol}^{-1})$$

$$\text{Fructose 6-phosphate} + \text{ATP} \xrightarrow{\text{Phosphofructokinase}}$$

$$\text{fructose 1,6-bisphosphate} + \text{ADP}$$

$$\Delta G = -22 \text{ kJ mol}^{-1} \, (-5.3 \text{ kcal mol}^{-1})$$

$$\text{Phosphoenolpyruvate} + \text{ADP} \xrightarrow{\text{Pyruvate kinase}} \text{pyruvate} + \text{ATP}$$

$$\Delta G = -17 \text{ kJ mol}^{-1} \, (-4.0 \text{ kcal mol}^{-1})$$

In gluconeogenesis, these virtually irreversible reactions of glycolysis must be bypassed.

The conversion of pyruvate into phosphoenolpyruvate begins with the formation of oxaloacetate

The first step in gluconeogenesis is the carboxylation of pyruvate to form oxaloacetate at the expense of a molecule of ATP, a reaction catalyzed by *pyruvate carboxylase.* This reaction occurs in the mitochondria.

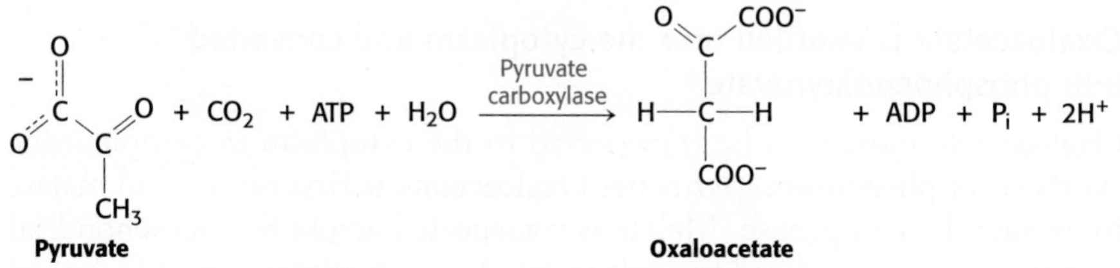

Pyruvate + CO_2 + ATP + H_2O $\xrightarrow{\text{Pyruvate carboxylase}}$ **Oxaloacetate** + ADP + P_i + $2H^+$

Pyruvate carboxylase requires *biotin,* a covalently attached prosthetic group, which serves as the *carrier of activated* CO_2. The carboxylate group of biotin is linked to the ε-amino group of a specific lysine residue by an amide bond (Figure 16.24). Recall that, in aqueous solutions, CO_2 exists primarily as HCO_3^- with the aid of carbonic anhydrase (Section 9.2).

The carboxylation of pyruvate takes place in three stages:

$$HCO_3^- + \text{ATP} \rightleftharpoons HOCO_2 - PO_3^{2-} + \text{ADP}$$

$$\text{Biotin-enzyme} + HOCO_2 - PO_3^{2-} \longrightarrow CO_2 - \text{biotin-enzyme} + P_i$$

$$CO_2 - \text{biotin-enzyme} + \text{pyruvate} \rightleftharpoons \text{biotin-enzyme} + \text{oxaloacetate}$$

Pyruvate carboxylase functions as a tetramer composed of four identical subunits, and each subunit consists of four domains (Figure 16.25). The biotin carboxylase domain (BC) catalyzes the formation of carboxyphosphate and the subsequent attachment of CO_2 to the second domain, the biotin carboxyl carrier protein (BCCP)—the site of the covalently attached biotin. Once bound to CO_2, BCCP leaves the biotin carboxylase active site

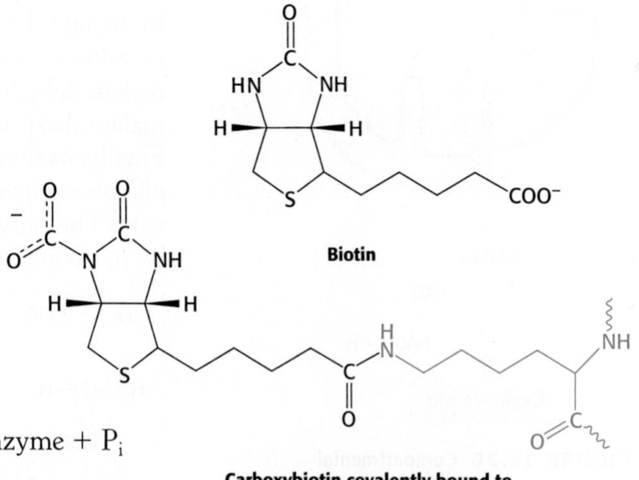

Biotin

Carboxybiotin covalently bound to ε-amino group of a lysine

FIGURE 16.24 Structure of biotin and carboxybiotin.

FIGURE 16.25 A subunit of pyruvate carboxylase. Biotin, covalently attached to the BCCP, transports CO_2 from the BC active site to the PC active site of an adjacent subunit. (BC) *biotin carboxylase;* (BCCP) *biotin carboxyl carrier protein;* (CT) *carboxyl transferase;* (PT) *pyruvate carboxylase tetramerization domain.* [Information from G. Lasso, L. P. C. Yu, D. Gil, S. Xiang, L. Tong, and M. Valle, *Structure* 18:1300–1310, 2010.]

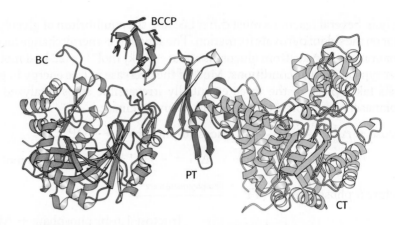

and swings almost the entire length of the subunit (≈ 75 Å) to the active site of the carboxyl transferase domain (CT), which transfers the CO_2 to pyruvate to form oxaloacetate. BCCP in one subunit interacts with the active sites on an adjacent subunit. The fourth domain (PT) facilitates the formation of the tetramer and is the binding site for acetyl CoA, a required allosteric activator.

How acetyl CoA facilitates the carboxylase reaction is under investigation. Recent research suggests that by binding to the enzyme, acetyl CoA enhances conformational communication between the BC and CT domains that enables the movement of the BCCP domain between two catalytic sites. The allosteric activation of pyruvate carboxylase by acetyl CoA is an important physiological control mechanism that will be discussed in Section 17.4. See Biochemistry in Focus 2 for a discussion of pyruvate carboxylase deficiency.

Oxaloacetate is shuttled into the cytoplasm and converted into phosphoenolpyruvate

Oxaloacetate must thus be transported to the cytoplasm to complete the synthesis of phosphoenolpyruvate. Oxaloacetate is first reduced to malate by malate dehydrogenase. Malate is transported across the mitochondrial membrane and reoxidized to oxaloacetate by a cytoplasmic NAD^+-linked malate dehydrogenase (Figure 16.26). The formation of oxaloacetate from malate also provides NADH for use in subsequent steps in gluconeogenesis. Finally, oxaloacetate is simultaneously *decarboxylated* and *phosphorylated* by *phosphoenolpyruvate carboxykinase* (PEPCK) to generate phosphoenolpyruvate. The phosphoryl donor is GTP. The CO_2 that was added to pyruvate by pyruvate carboxylase comes off in this step.

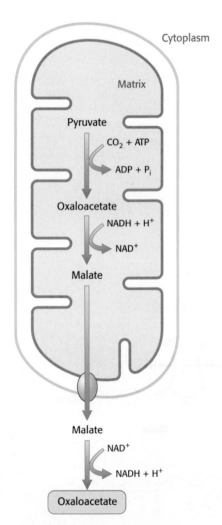

FIGURE 16.26 Compartmental cooperation. Oxaloacetate used in the cytoplasm for gluconeogenesis is formed in the mitochondrial matrix by the carboxylation of pyruvate. Oxaloacetate leaves the mitochondrion by a specific transport system (oval structure in the mitochondrial membranes) in the form of malate, which is reoxidized to oxaloacetate in the cytoplasm.

The sum of the reactions catalyzed by pyruvate carboxylase and phosphoenolpyruvate carboxykinase is

$$\text{Pyruvate} + \text{ATP} + \text{GTP} + \text{H}_2\text{O} \longrightarrow$$

$$\text{phosphoenolpyruvate} + \text{ADP} + \text{GDP} + \text{P}_i + 2\,\text{H}^+$$

This pair of reactions bypasses the irreversible reaction catalyzed by pyruvate kinase in glycolysis.

Why is a carboxylation and a decarboxylation required to form phosphoenolpyruvate from pyruvate? Recall that, in glycolysis, the presence of a phosphoryl group traps the unstable enol isomer of pyruvate as phosphoenolpyruvate (p. 503). However, the addition of a phosphoryl group to pyruvate is a highly unfavorable reaction: the $\Delta G^{\circ}{}'$ of the reverse of the glycolytic reaction catalyzed by pyruvate kinase is $+31\ \mathrm{kJ\ mol^{-1}}\ (+7.5\ \mathrm{kcal\ mol^{-1}})$. In gluconeogenesis, the use of the carboxylation and decarboxylation steps results in a much more favorable $\Delta G^{\circ}{}'$. The formation of phosphoenolpyruvate from pyruvate in the gluconeogenic pathway has a $\Delta G^{\circ}{}'$ of $+\ 0.8\ \mathrm{kJ\ mol^{-1}}\ (+0.2\ \mathrm{kcal\ mol^{-1}})$. A molecule of ATP is used to power the addition of a molecule of CO_2 to pyruvate in the carboxylation step. That CO_2 is then removed to power the formation of phosphoenolpyruvate in the decarboxylation step. *Decarboxylations often drive reactions that are otherwise highly endergonic.* This metabolic motif is used in the citric acid cycle (Chapter 17), the pentose phosphate pathway (Chapter 20), and fatty acid synthesis (Section 22.4).

The conversion of fructose 1,6-bisphosphate into fructose 6-phosphate and orthophosphate is an irreversible step

On formation, phosphoenolpyruvate is metabolized by the enzymes of glycolysis but in the reverse direction. These reactions are near equilibrium under intracellular conditions; so, when conditions favor gluconeogenesis, the reverse reactions will take place until the next irreversible step is reached. This step is the hydrolysis of fructose 1,6-bisphosphate to fructose 6-phosphate and P_i.

$$\text{Fructose 1,6-bisphosphate} + H_2O \xrightarrow{\text{Fructose 1,6-bisphosphatase}}$$

$$\text{fructose 6-phosphate} + P_i$$

The enzyme responsible for this step is *fructose 1,6-bisphosphatase*. Like its glycolytic counterpart, it is an allosteric enzyme that participates in the regulation of gluconeogenesis. We will return to its regulatory properties later in the chapter.

The generation of free glucose is an important control point

The fructose 6-phosphate generated by fructose 1,6-bisphosphatase is readily converted into glucose 6-phosphate. The final step in the generation of free glucose takes place primarily in the liver, a tissue whose metabolic duty is to maintain adequate glucose concentration in the blood for use by other tissues. Free glucose is not formed in the cytoplasm. Rather, glucose 6-phosphate is transported into the lumen of the endoplasmic reticulum, where it is hydrolyzed to glucose by *glucose 6-phosphatase,* which is bound to the ER membrane (Figure 16.27). Glucose and P_i are then shuttled back to the cytoplasm by a pair of transporters. In most tissues, gluconeogenesis ends with the formation of glucose 6-phosphate. Free glucose is not generated because most tissues lack glucose 6-phosphatase. Rather, the glucose 6-phosphate is commonly converted into glycogen, the storage form of glucose (Chapter 21).

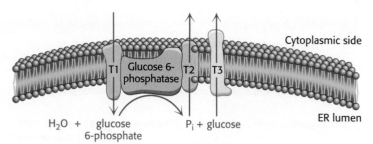

FIGURE 16.27 Generation of glucose from glucose 6-phosphate. Several endoplasmic reticulum (ER) proteins play a role in the generation of glucose from glucose 6-phosphate. T1 transports glucose 6-phosphate into the lumen of the ER, whereas T2 and T3 transport P_i and glucose, respectively, back into the cytoplasm. [Information from A. Buchell and I. D. Waddel, *Biochem. Biophys. Acta* 1092:129–137, 1991.]

Six high-transfer-potential phosphoryl groups are spent in synthesizing glucose from pyruvate

The formation of glucose from pyruvate is energetically unfavorable unless it is coupled to reactions that are favorable. Compare the stoichiometry of gluconeogenesis with that of the reverse of glycolysis.

The stoichiometry of gluconeogenesis is

$$2\,\text{Pyruvate} + 4\,\text{ATP} + 2\,\text{GTP} + 2\,\text{NADH} + 6\,H_2O \longrightarrow$$

$$\text{glucose} + 4\,\text{ADP} + 2\,\text{GDP} + 6\,P_i + 2\,\text{NAD}^+ + 2\,H^+$$

$$\Delta G^{\circ\prime} = -48\,\text{kJ mol}^{-1}\,(-11\,\text{kcal mol}^{-1})$$

In contrast, the stoichiometry for the reversal of glycolysis is

$$2\,\text{Pyruvate} + 2\,\text{ATP} + \text{NADH} + 2\,H_2O \longrightarrow$$

$$\text{glucose} + 2\,\text{ADP} + 2\,P_i + 2\,\text{NAD}^+ + 2\,H^+$$

$$\Delta G^{\circ\prime} = +90\,\text{kJ mol}^{-1}\,(+22\,\text{kcal mol}^{-1})$$

Note that *six* nucleoside triphosphate molecules are hydrolyzed to synthesize glucose from pyruvate in gluconeogenesis, whereas only *two* molecules of ATP are generated in glycolysis in the conversion of glucose into pyruvate. Thus, the extra cost of gluconeogenesis is four high-phosphoryl-transfer-potential molecules for each molecule of glucose synthesized from pyruvate. The four additional molecules having high phosphoryl-transfer potential are needed to turn an energetically unfavorable process (the reversal of glycolysis) into a favorable one (gluconeogenesis). Here we have a clear example of the coupling of reactions: NTP hydrolysis is used to power an energetically unfavorable reaction. The reactions of gluconeogenesis are summarized in Table 16.6.

TABLE 16.6 Reactions of gluconeogenesis

Step	Reaction	Enzyme	Reaction type	$\Delta G^{\circ\prime}$ in kJ mol^{-1} (kcal mol^{-1})
1.	Pyruvate + CO_2 + ATP + H_2O $\longrightarrow$ oxaloacetate + ADP + P_i + 2H$^+$	Pyruvate carboxylase	Phosphoryl transfer	−4.2
2.	Oxaloacetate + GTP $\rightleftharpoons$ phosphoenolpyruvate + GDP + CO_2	Phosphoenolpyruvate carboxykinase	Phosphoryl-coupled oxidation	+5.8
3.	Phosphoenolpyruvate + H_2O $\rightleftharpoons$ 2-phosphoglycerate	Enolase	Hydration	−3.4
4.	2-Phosphoglycerate $\rightleftharpoons$ 3-phosphoglycerate	Phosphoglycerate mutase	Phosphoryl shift	−9.2
5.	3-Phosphoglycerate + ATP $\rightleftharpoons$ 1,3-bisphosphoglycerate + ADP	Phosphoglycerate kinase	Phosphoryl transfer	+37.6
6.	1,3-Bisphosphoglycerate + NADH + H$^+$ $\rightleftharpoons$ glyceraldehyde 3-phosphate + NAD$^+$ + P_i	Glyceraldehyde 3-phosphate dehydrogenase	Phosphoryl-coupled reduction	−12.6
7.	Glyceraldehyde 3-phosphate $\rightleftharpoons$ dihydroxyacetone phosphate	Triose phosphate isomerase	Isomerization	−7.6
8.	Glyceraldehyde 3-phosphate + dihydroxyacetone phosphate $\rightleftharpoons$ fructose 1,6-bisphosphate	Aldolase	Aldol addition	−23.9
9.	Fructose 1,6-bisphosphate + H_2O $\longrightarrow$ fructose 6-phosphate + P_i	Fructose 1,6-bisphosphatase	Phosphoryl transfer	−16.3
10.	Fructose 6-phosphate $\rightleftharpoons$ glucose 6-phosphate	Phosphoglucose isomerase	Isomerization	−1.7
11.	Glucose 6-phosphatase + H_2O $\longrightarrow$ glucose + P_i	Glucose 6-phosphatase	Phosphoryl transfer	−12.1

16.4 Gluconeogenesis and Glycolysis Are Reciprocally Regulated

Gluconeogenesis and glycolysis are coordinated so that, within a cell, one pathway is relatively inactive while the other is highly active. If both sets of reactions were highly active at the same time, the net result would be the hydrolysis of four nucleoside triphosphates (two ATP molecules plus two GTP molecules) per reaction cycle. Both glycolysis and gluconeogenesis are highly exergonic under cellular conditions, and so there is no thermodynamic barrier to such simultaneous activity. However, the *amounts* and *activities* of the distinctive enzymes of each pathway are controlled so that both pathways are not highly active at the same time. The rate of glycolysis is also determined by the concentration of glucose, and the rate of gluconeogenesis by the concentrations of lactate and other precursors of glucose. The basic premise of the reciprocal regulation is that, when energy is needed or glycolytic intermediates are needed for biosynthesis, glycolysis will predominate. When there is a surplus of energy and glucose precursors, gluconeogenesis will take over.

Energy charge determines whether glycolysis or gluconeogenesis will be most active

The first important regulation site is the interconversion of fructose 6-phosphate and fructose 1,6-bisphosphate (Figure 16.28). Consider first a situation in which energy is needed, as in active muscle. In this case, the concentration of AMP is high. Under this condition, AMP stimulates phosphofructokinase but inhibits fructose 1,6-bisphosphatase. Thus, glycolysis is turned on and gluconeogenesis is inhibited. Conversely, high levels of

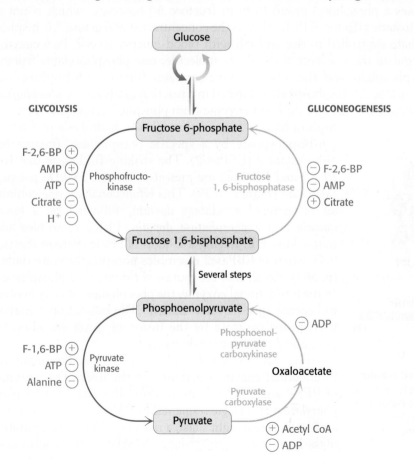

FIGURE 16.28 Reciprocal regulation of gluconeogenesis and glycolysis in the liver. The concentration of fructose 2,6-bisphosphate is high in the fed state and low in starvation. Another important control is the inhibition of pyruvate kinase by phosphorylation during starvation.

ATP and citrate indicate that the energy charge is high and that biosynthetic intermediates are abundant. ATP and citrate inhibit phosphofructokinase, whereas citrate activates fructose 1,6-bisphosphatase. Under these conditions, glycolysis is nearly switched off and gluconeogenesis is promoted. Why does citrate take part in this regulatory scheme? As we will see in Chapter 17, citrate reports on the status of the citric acid cycle, the primary pathway for oxidizing fuels in the presence of oxygen. High levels of citrate indicate an energy-rich situation and the presence of precursors for biosynthesis.

Glycolysis and gluconeogenesis are also reciprocally regulated at the interconversion of phosphoenolpyruvate and pyruvate in the liver. The glycolytic enzyme pyruvate kinase is inhibited by allosteric effectors ATP and alanine, which signal that the energy charge is high and that building blocks are abundant. Conversely, pyruvate carboxylase, which catalyzes the first step in gluconeogenesis from pyruvate, is inhibited by ADP. Likewise, ADP inhibits phosphoenolpyruvate carboxykinase. Pyruvate carboxylase is activated by acetyl CoA, which, like citrate, indicates that the citric acid cycle is producing energy and biosynthetic intermediates (Chapter 17). Hence, gluconeogenesis is favored when the cell is rich in biosynthetic precursors and ATP.

The balance between glycolysis and gluconeogenesis in the liver is sensitive to blood-glucose concentration

In the liver, rates of glycolysis and gluconeogenesis are adjusted to maintain the blood-glucose concentration. *The signal molecule fructose 2,6-bisphosphate strongly stimulates phosphofructokinase* (PFK) *and inhibits fructose 1,6-bisphosphatase.* When blood glucose is low, fructose 2,6-bisphosphate loses a phosphoryl group to form fructose 6-phosphate, which is not an allosteric effector of PFK. How is the concentration of fructose 2,6-bisphosphate controlled to rise and fall with blood-glucose levels? Two enzymes regulate the concentration of this molecule: one phosphorylates fructose 6-phosphate and the other dephosphorylates fructose 2,6-bisphosphate. Fructose 2,6-bisphosphate is formed in a reaction catalyzed by *phosphofructokinase 2* (PFK2), a different enzyme from phosphofructokinase. Fructose 6-phosphate is formed through the hydrolysis of fructose 2,6-bisphosphate by a specific phosphatase, *fructose bisphosphatase 2* (FBPase2). The striking finding is that *both PFK2 and FBPase2 are present in a single 55-kDa polypeptide chain* (Figure 16.29). This *bifunctional enzyme* contains an N-terminal *regulatory domain,* followed by a *kinase domain* and a *phosphatase domain.* PFK2 resembles adenylate kinase in having a P-loop NTPase domain (Section 9.4), whereas FBPase2 resembles phosphoglycerate mutase (p. 503). Recall that the mutase is essentially a phosphatase. In the bifunctional enzyme, the phosphatase activity evolved to become specific for F-2,6-BP. The bifunctional enzyme itself probably arose by the fusion of genes encoding the kinase and phosphatase domains.

What controls whether PFK2 or FBPase2 dominates the bifunctional enzyme's activities in the liver? The activities of PFK2 and FBPase2 are reciprocally controlled by *phosphorylation of a single serine residue.* When glucose is scarce, such as during a night's fast, a rise in the blood concentration of the hormone glucagon triggers a cyclic AMP signal cascade

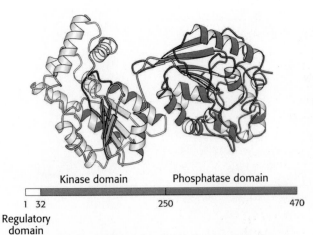

FIGURE 16.29 Domain structure of the bifunctional enzyme phosphofructokinase 2. The kinase domain (purple) is fused to the phosphatase domain (red). The kinase domain is a P-loop NTP hydrolase domain, as indicated by the purple shading (Section 9.4). The bar represents the amino acid sequence of the enzyme. [Drawn from 1BIF.pdb.]

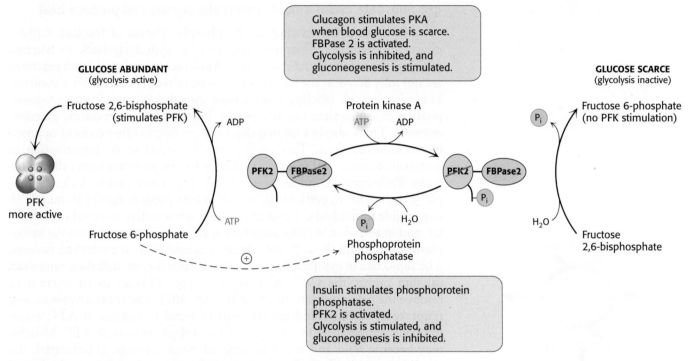

FIGURE 16.30 Control of the synthesis and degradation of fructose 2,6-bisphosphate.
A low blood-glucose concentration as signaled by glucagon leads to the phosphorylation of
the bifunctional enzyme and hence to a lower concentration of fructose 2,6-bisphosphate,
slowing glycolysis. Insulin accelerates the formation of fructose 2,6-bisphosphate by
facilitating the dephosphorylation of the bifunctional enzyme.

(Section 14.1), leading to the phosphorylation of this bifunctional enzyme
by protein kinase A (Figure 16.30). This covalent modification activates
FBPase2 and inhibits PFK2, lowering the level of F-2,6-BP. Gluconeogenesis predominates. Glucose formed by the liver under these conditions
is essential for the viability of the brain. Glucagon stimulation of protein
kinase A also inactivates pyruvate kinase in the liver (p. 516).

Conversely, when the blood-glucose concentration is high, such as after
a meal, gluconeogenesis is not needed. Insulin is secreted and initiates a
signal pathway that activates a protein phosphatase, which removes the
phosphoryl group from the bifunctional enzyme. This covalent modification activates PFK2 and inhibits FBPase2. The resulting rise in the level of
F-2,6-BP accelerates glycolysis. In liver, a key role of glycolysis is to generate metabolites for biosynthesis. The coordinated control of glycolysis and
gluconeogenesis is facilitated by the location of the kinase and phosphatase
domains on the same polypeptide chain as the regulatory domain.

The hormones insulin and glucagon also regulate the amounts of essential enzymes. These hormones alter gene expression primarily by changing
the rate of transcription. Insulin levels rise subsequent to eating, when
there is plenty of glucose for glycolysis. To encourage glycolysis, insulin
stimulates the expression of phosphofructokinase, pyruvate kinase, and the
bifunctional enzyme that makes and degrades F-2,6-BP. Glucagon rises
during fasting, when gluconeogenesis is needed to replace scarce glucose. To
encourage gluconeogenesis, glucagon inhibits the expression of the three regulated glycolytic enzymes and stimulates instead the production of two key
gluconeogenic enzymes, phosphoenolpyruvate carboxykinase and fructose
1,6-bisphosphatase. Transcriptional control in eukaryotes is much slower
than allosteric control, taking hours or days instead of seconds to minutes.

Substrate cycles amplify metabolic signals and produce heat

A pair of reactions such as the phosphorylation of fructose 6-phosphate to fructose 1,6-bisphosphate and its hydrolysis back to fructose 6-phosphate is called a *substrate cycle*. As already mentioned, both reactions are not fully active at the same time because of reciprocal allosteric controls. However, isotope-labeling studies have shown that some fructose 6-phosphate is phosphorylated to fructose 1,6-bisphosphate even during gluconeogenesis. There also is a limited degree of cycling in other pairs of opposed irreversible reactions. This cycling was regarded as an imperfection in metabolic control, and so substrate cycles have sometimes been called *futile cycles*. Pathological conditions may result from futile cycles. A clear example is malignant hyperthermia, which occurs predominantly in muscle in susceptible individuals. These individuals are sensitive to certain anesthetics, and in response to these anesthetics, calcium is released from the sarcoplasmic reticulum through the calcium channel in an uncontrolled fashion. The rapid rise in cytoplasmic calcium activates the sarcoplasmic reticulum calcium pump, the Ca^{2+} ATPase, a P-type ATPase, in an attempt to remove the calcium from the cytoplasm (p. 407). The constant release and reuptake of calcium, and the consequent rapid hydrolysis of ATP, raises the body temperature to 44°C (111°F) and depletes muscle ATP. Muscles may become rigid and suffer substantial tissue damage. If untreated, the condition may be fatal. Treatment includes removal of the anesthetic and treatment with dantrolene, an inhibitor of the calcium channel.

Despite such extraordinary circumstances, substrate cycles now seem likely to be biologically important. One possibility is that *substrate cycles amplify metabolic signals.* Suppose that the rate of conversion of A into B is 100 and of B into A is 90, giving an initial net flux of 10. Assume that an allosteric effector increases the A → B rate by 20% to 120 and reciprocally decreases the B → A rate by 20% to 72. The new net flux is 48, and so a 20% change in the rates of the opposing reactions has led to a 380% increase in the net flux. In the example shown in Figure 16.31, this amplification is made possible by the rapid hydrolysis of ATP. The flux down the glycolytic pathway has been suggested to increase as much as 1000-fold at the initiation of intense exercise. Because the allosteric activation of enzymes alone seems unlikely to explain this increased flux, the existence of substrate cycles may partly account for the rapid rise in the rate of glycolysis.

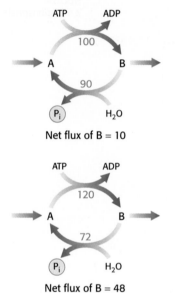

FIGURE 16.31 Substrate cycle. This ATP-driven cycle operates at two different rates. A small change in the rates of the two opposing reactions results in a large change in the *net* flux of product B.

Lactate and alanine formed by contracting muscle are used by other organs

Lactate produced by active skeletal muscle and erythrocytes is a source of energy for other organs. Erythrocytes lack mitochondria and can never oxidize glucose completely. In contracting fast-twitch skeletal muscle fibers during vigorous exercise, the rate at which glycolysis produces pyruvate exceeds the rate at which the citric acid cycle oxidizes it. In these cells, lactate dehydrogenase reduces excess pyruvate to lactate to restore redox balance (p. 506). However, lactate is a dead end in metabolism. It must be converted back into pyruvate before it can be metabolized. Both pyruvate and lactate diffuse out of these cells through carriers into the blood. *In contracting skeletal muscle, the formation and release of lactate lets the muscle generate ATP in the absence of oxygen and shifts the burden of metabolizing lactate from muscle to other organs.* The pyruvate and lactate in the bloodstream have two fates. In one fate, the plasma membranes of some cells—particularly cells in cardiac muscle and slow-twitch (type 1) skeletal muscle—contain carriers that make the cells highly permeable to lactate and pyruvate. These molecules diffuse from the blood into such permeable cells. Once inside these well-oxygenated cells, lactate can be reverted back to pyruvate and metabolized through the citric acid cycle and oxidative phosphorylation to generate ATP. The use of lactate in place of glucose by these cells makes more circulating glucose available to the active muscle cells. In the other fate, excess lactate enters the liver and is converted first into pyruvate and then into glucose by the gluconeogenic pathway. *Contracting skeletal muscle supplies lactate to the liver, which uses it to synthesize and release glucose. Thus, the liver restores the level of glucose necessary for active muscle cells, which derive ATP from the glycolytic conversion of glucose into lactate. These reactions constitute the Cori cycle* (Figure 16.32).

Studies have shown that alanine, like lactate, is a major precursor of glucose in the liver. The alanine is generated in muscle when the carbon skeletons of some amino acids are used as fuels. The nitrogens from these amino acids are transferred to pyruvate to form alanine; the reverse reaction takes place in the liver. This process also helps maintain nitrogen balance. The interplay between glycolysis and gluconeogenesis is summarized in Figure 16.33, which shows how these pathways help meet the energy needs of different cell types.

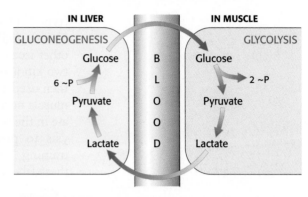
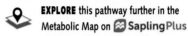

FIGURE 16.32 **The Cori cycle.** Lactate formed by active muscle is converted into glucose by the liver. This cycle shifts part of the metabolic burden of active muscle to the liver. The symbol, ~P represents nucleoside triphosphates.

EXPLORE this pathway further in the Metabolic Map on SaplingPlus

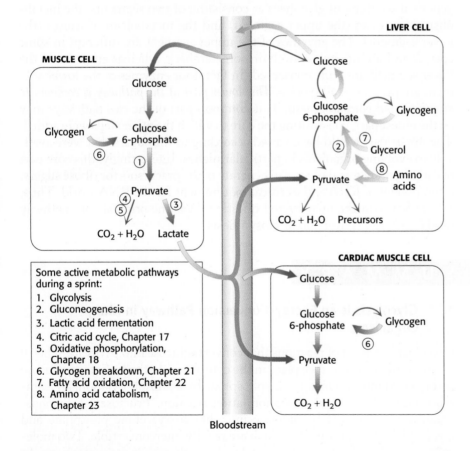

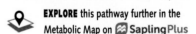
Bloodstream

FIGURE 16.33 PATHWAY INTEGRATION: Cooperation between glycolysis and gluconeogenesis during a sprint. Glycolysis and gluconeogenesis are coordinated, in a tissue-specific fashion, to ensure the energy needs of all cells are met. Consider a sprinter. In skeletal leg muscle, glucose will be metabolized aerobically to CO_2 and H_2O or, more likely (thick arrows) during a sprint, anaerobically to lactate. In cardiac muscle, the lactate can be converted into pyruvate and used as a fuel, along with glucose, to power the heartbeats to keep the sprinter's blood flowing. Gluconeogenesis, a primary function of the liver, will be taking place rapidly (thick arrows) to ensure that enough glucose is present in the blood for skeletal and cardiac muscle, as well as for other tissues. Glycogen, glycerol, and amino acids are other sources of energy that we will learn about in later chapters.

EXPLORE this pathway further in the Metabolic Map on SaplingPlus

Isozymic forms of lactate dehydrogenase in different tissues catalyze the interconversions of pyruvate and lactate (Section 10.2). Lactate dehydrogenase is a tetramer of two kinds of 35-kDa subunits encoded by similar genes: the H type predominates in the heart, and the homologous M type in skeletal muscle and the liver. These subunits associate to form five types of tetramers: H_4, H_3M_1, H_2M_2, H_1M_3, and M_4. The H_4 isozyme (type 1) has higher affinity for substrates than that of the M_4 isozyme (type 5) and, unlike M_4, is allosterically inhibited by high levels of pyruvate. The other isozymes have intermediate properties, depending on the ratio of the two kinds of chains. The H_4 isozyme oxidizes lactate to pyruvate, which is then used as a fuel by the heart through aerobic metabolism. Indeed, heart muscle never functions anaerobically. In contrast, M_4 is optimized to operate in the reverse direction, to convert pyruvate into lactate to allow glycolysis to proceed under anaerobic conditions. Interestingly, endurance training increases the H_4 isozyme in slow-twitch muscle fibers, enabling these fibers to use the lactate generated by other fiber types.

Glycolysis and gluconeogenesis are evolutionarily intertwined

The metabolism of glucose has ancient origins. Organisms living in the early biosphere depended on the anaerobic generation of energy until significant amounts of oxygen began to accumulate 2 billion years ago. Glycolytic enzymes were most likely derived independently rather than by gene duplication, because glycolytic enzymes with similar properties do not have similar amino acid sequences. Although there are four kinases and two isomerases in the pathway, sequence and structural comparisons suggest that these sets of enzymes are not related to one another by divergent evolution.

We can speculate on the relationship between glycolysis and gluconeogenesis if we think of glycolysis as consisting of two segments: the metabolism of hexoses (the upper segment) and the metabolism of trioses (the lower segment). The enzymes of the upper segment are different in some species and are missing entirely in some archaea, whereas enzymes of the lower segment are quite conserved. In fact, four enzymes of the lower segment are present in all species. *This lower part of the pathway is common to glycolysis and gluconeogenesis.* This common part of the two pathways may be the oldest part, constituting the core to which the other steps were added. The upper part would have varied according to the sugars that were available to evolving organisms in particular niches. Interestingly, this core part of carbohydrate metabolism can generate triose precursors for ribose sugars, a component of RNA and a critical requirement for the RNA world. Thus, we are left with the unanswered question: Was the original core pathway used for energy conversion or biosynthesis?

SUMMARY

16.1 Glycolysis Is an Energy-Conversion Pathway in Many Organisms

Glycolysis is the set of reactions that converts glucose into pyruvate. The 10 reactions of glycolysis take place in the cytoplasm. In the first stage, glucose is converted into fructose 1,6-bisphosphate by a phosphorylation, an isomerization, and a second phosphorylation reaction. Fructose 1,6-bisphosphate is then cleaved by aldolase into dihydroxyacetone phosphate and glyceraldehyde 3-phosphate, which are readily interconvertible. Two molecules of ATP are consumed per molecule of glucose in these reactions. In the second stage, ATP is generated. Glyceraldehyde 3-phosphate is oxidized and phosphorylated to form 1,3-bisphosphoglycerate, an acyl phosphate with a high phosphoryl-transfer potential. This molecule transfers a phosphoryl group to ADP to form ATP and 3-phosphoglycerate. A phosphoryl shift and a dehydration form phosphoenolpyruvate, a second intermediate with a high phosphoryl-transfer potential. Another molecule of

ATP is generated as phosphoenolpyruvate is converted into pyruvate. There is a net gain of two molecules of ATP in the formation of two molecules of pyruvate from one molecule of glucose.

The electron acceptor in the oxidation of glyceraldehyde 3-phosphate is NAD^+, which must be regenerated for glycolysis to continue. In aerobic organisms, the NADH formed in glycolysis transfers its electrons to O_2 through the electron-transport chain, which thereby regenerates NAD^+. Under anaerobic conditions and in some microorganisms, NAD^+ is regenerated by the reduction of pyruvate to lactate. In other microorganisms, NAD^+ is regenerated by the reduction of pyruvate to ethanol. These two processes are examples of fermentations.

16.2 The Glycolytic Pathway Is Tightly Controlled

The glycolytic pathway has a dual role: it degrades glucose to generate ATP, and it provides building blocks for the synthesis of cellular components. The rate of conversion of glucose into pyruvate is regulated to meet these two major cellular needs. Under physiological conditions, the reactions of glycolysis are readily reversible except for those catalyzed by hexokinase, phosphofructokinase, and pyruvate kinase. Phosphofructokinase—the most important control element in glycolysis—is inhibited by high levels of ATP and citrate, and it is activated by AMP and fructose 2,6-bisphosphate. In the liver, this bisphosphate signals that glucose is abundant. Hence, phosphofructokinase is active when either energy or building blocks are needed. Hexokinase is inhibited by glucose 6-phosphate, which accumulates when phosphofructokinase is inactive. ATP and alanine allosterically inhibit pyruvate kinase, the other control site, and fructose 1,6-bisphosphate activates the enzyme. Consequently, pyruvate kinase is maximally active when the energy charge is low and glycolytic intermediates accumulate.

16.3 Glucose Can Be Synthesized from Noncarbohydrate Precursors

Gluconeogenesis, which occurs primarily in the liver, is the synthesis of glucose from noncarbohydrate sources, such as lactate, amino acids, glycerol, and alanine produced from pyruvate by active skeletal muscle. Several of the reactions that convert pyruvate into glucose are common to glycolysis. Gluconeogenesis, however, requires four new reactions to bypass the essential irreversibility of three reactions in glycolysis. In two of the new reactions, pyruvate is carboxylated in mitochondria to oxaloacetate, which in turn is decarboxylated and phosphorylated in the cytoplasm to phosphoenolpyruvate. ATP and GTP are consumed in these reactions, which are catalyzed by pyruvate carboxylase and phosphoenolpyruvate carboxykinase, respectively. The other distinctive reactions of gluconeogenesis are the hydrolyses of fructose 1,6-bisphosphate and glucose 6-phosphate, which are catalyzed by specific phosphatases.

16.4 Gluconeogenesis and Glycolysis Are Reciprocally Regulated

Gluconeogenesis and glycolysis are reciprocally regulated so that one pathway is relatively inactive while the other is highly active. Phosphofructokinase and fructose 1,6-bisphosphatase are key control points. Fructose 2,6-bisphosphate, an intracellular signal molecule present at higher levels when glucose is abundant, activates glycolysis and inhibits gluconeogenesis by regulating these enzymes. Pyruvate kinase and pyruvate carboxylase are regulated by other effectors so that both are not maximally active at the same time. Allosteric regulation and reversible phosphorylation, which are rapid, are complemented by transcriptional control, which takes place in hours or days.

APPENDIX Biochemistry in Focus

Biochemistry in Focus 1

Triose phosphate isomerase deficiency (TPID)

TPID is a multisystem disorder presenting in early childhood. Symptoms include congenital hemolytic anemia (deficiency of red cells due to premature destruction of the cell that is present at birth) and progressive neuromuscular disorder, including cardiomyopathy (inflammation and damage to the heart muscle). In severe cases, TPID can lead to death in early childhood. Dihydroxyacetone phosphate accumulates in cells, especially red blood cells. TPID is rare, limiting our understanding of the disorder. The relationship between the biochemical defect and the clinical features remains to be established. Let's generate a hypothesis regarding the pathologies of TPID.

First of all, recall that central nervous system and red blood cells rely on glucose metabolism exclusively for energy. Thus, any disruption in glycolysis would impact these tissues. Also, muscles can use glucose, although many rely on fats as a fuel. This observation accounts for the nature of the symptoms: without energy, neuromuscular function would be compromised. Let's look at how glycolysis would be affected by a deficiency in isomerase activity. The enzyme catalyzes the conversion of dihydroxyacetone phosphate to glyceraldehyde 3-phosphate, which will be further metabolized to yield ATP.

If the isomerase activity is missing, then half of the carbons of glucose cannot be metabolized to yield ATP. At the very least, ATP generation would be compromised. However, research suggests that disruption of energy metabolism is not the cause of the more serious consequences of TPID. Can there be other effects resulting from TPID?

Let's look at the enzyme reaction and its place in glycolysis.

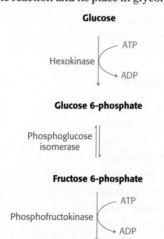

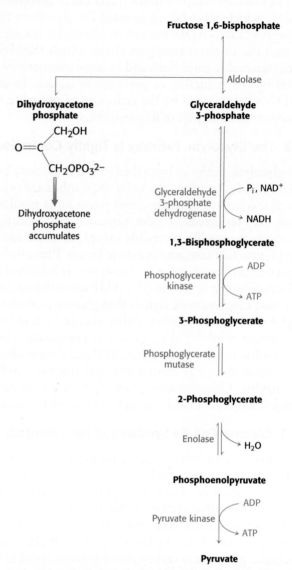

Even if the cells would try to compensate for lack of isomerase activity by processing more glucose, there will an unavoidable buildup of dihydroxyacetone phosphate. Could this buildup have deleterious consequences?

Strong evidence exists that dihydroxyacetone phosphate can be converted into methylglyoxal.

Methylglyoxal

Methylglyoxal, a highly reactive molecule, covalently binds to available amino groups on proteins, yielding advanced glycation end products (AGE) (p. 347). These modifications inhibit protein function. Extensive loss of protein function would then contribute to the pathologies observed in TPID, including early death. AGE has also been implicated in aging, arteriosclerosis (thickening and hardening of artery walls), and diabetes. Clearly establishing the causes of TPID pathologies awaits further research.

Biochemistry in Focus 2

Pyruvate carboxylase deficiency (PCD)

PCD is another rare disorder of carbohydrate metabolism. Several types vary in the seriousness of clinical outcomes, but all are characterized to some extent by hypoglycemia (low concentration of blood glucose) and lactic acidosis (excessive lactic acid in the blood). Symptoms include lethargy and seizures. In severe forms of PCD, death occurs in the first few months of life.

Together, let's try to understand the cause of the two primary biochemical characteristics: hypoglycemia and lactic acidosis. Recall that brain, red blood cells, and muscles use glucose as fuel. Lack of glucose would explain the symptoms described above, as the functioning of the neuromuscular system would be impaired.

First, let's identify the reaction catalyzed by pyruvate carboxylase and then determine its role in glucose metabolism.

Recall that pyruvate carboxylase is a key regulatory enzyme in gluconeogenesis, which occurs primarily in the liver. Pyruvate carboxylase, in cooperation with the next enzyme in the pathway, phosphoenolpyruvate carboxykinase, synthesizes the high-phosphoryl potential compound, phosphoenolpyruvate.

The sum of the reactions is:

Pyruvate + ATP + GTP + H_2O $\longrightarrow$
  phosphoenolpyruvate + ADP + GDP + P_i + $2H^+$

These two reactions bypass the irreversible reaction catalyzed by pyruvate kinase in glycolysis. Under normal circumstances, phosphoenolpyruvate would be metabolized to glucose.

We can surmise that the lack of the pyruvate carboxylase would account for the hypoglycemia: the liver cannot serve its role in maintaining adequate blood glucose concentration for tissues that depend on glucose. Now, how do we account for the lactic acidosis? Red blood cells and many other tissues, such as active muscle, produce lactate as the end product of lactic acid fermentation. The lactate is released into the blood, where it forms lactic acid. An important role of liver is to remove lactic acid from the blood and use it as a gluconeogenic precursor. However, if the lactic acid can't be used to form glucose because of a lack of pyruvate carboxylase, it will remain in the blood leading to a fall in blood pH, resulting in acidosis.

We focused only on the effect of PCD on gluconeogenesis. We will see in later chapters that the oxaloacetate formed by pyruvate carboxylase is an important and versatile molecule, so the lack of carboxylase activity has greater effects than discussed here.

APPENDIX

Problem-Solving Strategies

Problem-Solving Strategies 1.

PROBLEM: Arsenate (AsO_4^{3-}) closely resembles P_i in structure and reactivity. In the reaction catalyzed by glyceraldehyde 3-phosphate dehydrogenase, arsenate can replace phosphate in attacking the energy-rich thioester intermediate. The product of this reaction, 1-arseno-3-phosphoglycerate, is unstable. It and other acyl arsenates are rapidly and spontaneously hydrolyzed. What is the effect of arsenate on ATP generation in a cell?

SOLUTION: One means by which a problem can be solved is by breaking the problem down into a series of questions. This is the approach that we will use here.

▸ **What is the focus of the problem? In other words, what is the problem about?**

The problem explores the effect of arsenate on ATP generation in glycolysis. The question notes that arsenate has a structure similar to that of phosphate. It is possible that the structural similarity is important.

▸ **Compare the structure of arsenate to phosphate.**

Arsenate resembles P_i in structure and reactivity, and P_i is a key component of glycolysis. As stated in the question, arsenate can substitute for P_i in the reactions of glycolysis. In particular, arsenate can serve as a substrate in the reaction catalyzed by glyceraldehyde 3-phosphate dehydrogenase.

▶ **What reaction does this enzyme catalyze in glycolysis?**

▶ **Why is this reaction crucial to ATP synthesis in glycolysis?**

1,3-Bisphosphoglycerate has high phosphoryl potential and can be used to power the synthesis of ATP in the next reaction of glycolysis, a reaction catalyzed by phosphoglycerate kinase.

If glyceraldehyde 3-phosphate dehydrogenase uses arsenate instead of phosphate, the problem states that 1-arseno-3-phosphoglycerate will be formed.

▶ **Referring to the problem, what is the likely fate of 1-arseno-3-phosphoglycerate?**

It will be rapidly and spontaneously hydrolyzed.

▶ **What products will result from the hydrolysis of 1-arseno-3-phosphoglycerate?**

Arsenate and 3-phosphoglycerate.

▶ **3-Phosphoglycerate is a metabolite in glycolysis. Which glycolytic enzyme generates this metabolite?**

Phosphoglycerate kinase (see above).

▶ **What other product is generated by phosphoglycerate kinase?**

ATP.

▶ **If 3-phosphoglycerate is generated by the spontaneous hydrolysis of 1-arseno-3-phosphoglycerate, what is the ATP yield from the phosphoglycerate kinase-catalyzed reaction?**

0.

Thus, use of arsenate instead of phosphate by glyceraldehyde-3 phosphate dehydrogenase eliminates one of the two ATP-synthesizing steps in glycolysis. In fact, it would be unlikely that the arsenate would be used only by glyceraldehyde 3-phosphate dehydrogenase. All enzymes that use phosphate could use arsenate, severely compromising the energy production of the cell, leading to cell damage. If the arsenate concentration were high enough, death would result.

Problem-Solving Strategies 2.

PROBLEM: What would be the effect on an organism's ability to use glucose as an energy source if a mutation inactivated glucose 6-phosphatase in the liver?

SOLUTION: What do we need to know to solve this problem? Two thoughts immediately come to mind:

▶ **What is the reaction catalyzed by glucose 6-phosphatase?**

▶ **What is the role of the liver in maintaining the proper blood glucose concentration?**

The reaction catalyzed by glucose 6-phosphatase is:

$$\text{Glucose 6-phosphate} + H_2O \longrightarrow \text{glucose} + P_i$$

This is the last step in gluconeogenesis, and the free glucose is released into the blood.

An important metabolic role of the liver is the release of glucose into the blood for use by other tissues such as muscle, brain, and red blood cells.

With these two pieces of information, we now can solve the problem:

▶ **What would be the result of a lack of glucose 6-phosphatase activity?**

The blood glucose concentration would fall, resulting in energy deprivation for glucose-dependent tissues.

You might be thinking ahead and wondering, "What would happen to the glucose 6-phosphate synthesized by gluconeogenesis if the glucose 6-phosphatase activity was missing?" Good question! As we will learn in Chapter 21, glycogen is the storage form for glucose. The excess glucose 6-phosphate resulting from a phosphatase deficiency will be converted into glycogen, leading to excess glycogen accumulation in the liver.

Problem-Solving Strategies 3.

PROBLEM: Yeast are facultative anaerobes—they can grow in the absence of oxygen (anaerobically) using alcoholic fermentation or in the presence of oxygen (aerobically) using cellular respiration. Interestingly, yeast cannot live anaerobically using glycerol as their only fuel source. Explain why yeast cannot survive metabolizing glycerol anaerobically.

SOLUTION: This problem asks us to consider glycerol metabolism in yeast under anaerobic conditions. Recall from earlier in this chapter that yeast can generate ATP under anaerobic conditions by processing glucose to ethanol (alcoholic fermentation). According to the problem, yeast cannot survive using glycerol as their sole energy source in the absence of oxygen. How do we prove this?

First, we must determine how glycerol is prepared to enter metabolism.

▶ **What enzymes metabolize glycerol?**

Two enzymes—glycerol kinase and glycerol phosphate dehydrogenase—acting sequentially, convert glycerol into dihydroxyacetone phosphate (DHAP).

Recall that DHAP is readily converted into glyceraldehyde 3-phosphate (GAP). Both DHAP and GAP are gluconeogenic precursors, but only GAP can be metabolized by alcoholic fermentation.

Examine the coupled reactions.

▶ **What reactants are required to generate DHAP?**

The kinase reaction expends an ATP, while the dehydrogenase reaction reduces NAD^+ to NADH.

Now, using Figures 16.1 and 16.11 showing the conversion of pyruvate into ethanol, answer the following questions.

▶ **How many ATPs are synthesized when glycerol is metabolized to ethanol?**

The GAP synthesized from glycerol yields 2 ATP when converted into ethanol, but remember that an ATP was used to phosphorylate glycerol, so the net yield is 1 ATP. The ATP yield is halved. However, the ATP yield is not the major reason for yeast inability to survive.

Examine the use of NAD^+/NADH in the conversion of glycerol into ethanol.

▶ **Is redox balance maintained?**

No. First, consider the use of NAD^+/NADH in the fermentation pathway. Glyceraldehyde 3-phosphate dehydrogenase reduces NAD^+ to NADH, while alcohol dehydrogenase oxidizes NADH to NAD^+. Redox balance is maintained: there is no net change in the NAD^+/NADH ratio. For glycerol to be converted in DHAP and subsequently into GAP, NAD^+ is reduced to NADH, but there is no means to regenerate NAD^+. The pathway stops and the yeast die because of a lack of NAD^+.

KEY TERMS

glycolysis (p. 491)

lactic acid fermentation (p. 491)

alcoholic fermentation (p. 491)

gluconeogenesis (p. 492)

hexokinase (p. 495)

kinase (p. 495)

phosphofructokinase (PFK) (p. 497)

thioester intermediate (p. 500)

substrate-level phosphorylation (p. 502)

pyruvate kinase (p. 504)

fermentation (p. 505)

obligate anaerobe (p. 507)

facultative anaerobe (p. 507)

committed step (p. 513)

feedforward stimulation (p. 514)

glucokinase (p. 515)

aerobic glycolysis (p. 517)

pyruvate carboxylase (p. 521)

biotin (p. 521)

phosphoenolpyruvate carboxykinase (p. 522)

fructose 1,6-bisphosphatase (p. 523)

glucose 6-phosphatase (p. 523)

bifunctional enzyme (p. 526)

substrate cycle (p. 528)

Cori cycle (p. 529)

PROBLEMS

1. *Your choice 1.* The conversion of one molecule of fructose 1,6-bisphosphate into 2 molecules of pyruvate results in the *net* synthesis of:

(a) Two molecules of NADH and two molecules of ATP

(b) Two molecules of NAD^+ and two molecules of ATP

(c) Four molecules of NADH and four molecules of ATP

(d) Two molecules of NADH and four molecules of ATP

2. *Your choice 2.* The conversion of one molecule of glucose into two molecules of lactate results in the *net* synthesis of:

(a) No NADH and two molecules of ATP

(b) No NADH and four molecules of ATP

(c) Two molecules of NADH and four molecules of ATP

(d) Two molecules of NADH and two molecules of ATP

3. *Nothing but the truth.* Which of the following statements is (are) true for a muscle performing lactic acid fermentation?

(a) The process is inhibited by AMP

(b) The process is inhibited by ATP

(c) There is a net synthesis of NADH

(d) The process is exergonic

4. *Like Tarzan and Jane.* Match each term with its description.

(a) aldol cleavage _____   1. enolase

(b) dehydration _____   2. phosphoglycerate mutase

(c) phosphoryl transfer _____   3. triose phosphate isomerase

(d) phosphoryl shift _____   4. glyceraldehyde 3-phosphate dehydrogenase

(e) isomerization _____   5. aldolase

(f) phosphorylation coupled with an oxidation _____   6. pyruvate kinase

5. *Gross versus net.* The gross yield of ATP from the metabolism of glucose to two molecules of pyruvate is four molecules of ATP. However, the net yield is only two molecules of ATP. Why are the gross and net values different? ✓❶

6. *Go together like the owl and the pussycat.* Match each term with its description. ✓❶

(a) Hexokinase _____   1. Forms fructose 1,6-bisphosphate

(b) Phosphoglucose isomerase _____   2. Generates the first high-phosphoryl-transfer-potential compound that is not ATP

(c) Phosphofructokinase _____   3. Converts glucose 6-phosphate into fructose 6-phosphate

(d) Aldolase _____   4. Phosphorylates glucose

(e) Triose phosphate isomerase _____   5. Generates the second molecule of ATP

(f) Glyceraldehyde 3-phosphate dehydrogenase _____   6. Cleaves fructose 1,6-bisphosphate

(g) Phosphoglycerate kinase _____   7. Generates the second high-phosphoryl-transfer-potential compound that is not ATP

(h) Phosphoglycerate mutase _____   8. Catalyzes the interconversion of three-carbon isomers

(i) Enolase _____   9. Converts 3-phosphoglycerate into 2-phosphoglycerate

(j) Pyruvate kinase _____   10. Generates the first molecule of ATP

7. *Who takes? Who gives?* Lactic acid fermentation and alcoholic fermentation are oxidation–reduction reactions. Identify the ultimate electron donor and electron acceptor. ✓❷

8. *ATP yield.* Each of the following molecules is processed by glycolysis to lactate. How much ATP is generated from each molecule? ✓❶

(a) Glucose 6-phosphate

(b) Dihydroxyacetone phosphate

(c) Glyceraldehyde 3-phosphate

(d) Fructose

(e) Sucrose

9. *Enzyme redundancy?* Why is it advantageous for the liver to have both hexokinase and glucokinase to phosphorylate glucose?

10. *Magic?* The interconversion of DHAP and GAP greatly favors the formation of DHAP at equilibrium. Yet the conversion of DHAP by triose phosphate isomerase proceeds readily. Why?

11. *Between two extremes.* What is the role of a thioester in the formation of ATP in glycolysis? ✓❶

12. *Corporate sponsors.* Some of the early research on glycolysis was supported by the brewing industry. Why would the brewing industry be interested in glycolysis? ✓❷

13. *Recommended daily allowance.* The recommended daily allowance for the vitamin niacin is 15 mg per day. How would glycolysis be affected by niacin deficiency?

14. *Who's on first?* Although both hexokinase and phosphofructokinase catalyze irreversible steps in glycolysis and the hexokinase-catalyzed step is first, phosphofructokinase is nonetheless the pacemaker of glycolysis. What does this information tell you about the fate of the glucose 6-phosphate formed by hexokinase?

15. *The tortoise and the hare.* Why is the regulation of phosphofructokinase by energy charge not as important in the liver as it is in muscle?

16. *Like Minneapolis and St. Paul.* Match each term with its description. ✓❸ ✓❹

(a) Lactate _____   1. Generates oxaloacetate

(b) Pyruvate carboxylase _____   2. Readily converted into DHAP

(c) Acetyl CoA _____ 3. Generates a high-phosphoryl-transfer-potential compound

(d) Phosphoenolpyruvate carboxykinase _____ 4. Found predominantly in liver

(e) Glycerol _____ 5. Gluconeogenic counterpart of PFK

(f) Fructose 1,6-bisphosphatase _____ 6. Readily converted into pyruvate

(g) Glucose 6-phosphatase _____ 7. Required for pyruvate carboxylase activity

17. Running in reverse. Why can't the reactions of the glycolytic pathway simply be run in reverse to synthesize glucose? ✓❹

18. Road blocks. What reactions of glycolysis are not readily reversible under intracellular conditions? ✓❹

19. No pickling. Why is it in the muscle's best interest to export lactic acid into the blood during intense exercise? ✓❸

20. Après vous. Why is it physiologically advantageous for the pancreas to use GLUT2, with a high K_M, as the transporter that allows glucose entry into β cells?

21. Bypass. In the liver, fructose can be converted into glyceraldehyde 3-phosphate and dihydroxyacetone phosphate without passing through the phosphofructokinase-regulated reaction. Show the reactions that make this conversion possible. Why might ingesting high levels of fructose have deleterious physiological effects?

22. Match 'em 1. The following sequence is a part of the sequence of reactions in gluconeogenesis.

Pyruvate $\xrightarrow[A]{}$ Oxaloacetate $\xrightarrow[B]{}$ Malate $\xrightarrow[C]{}$

Oxaloacetate $\xrightarrow[D]{}$ Phosphoenolpyruvate

Match the capital letters representing the reaction in the gluconeogenic pathway with parts a, b, c, etc. ✓❸

(a) takes place in mitochondria

(b) takes place in the cytoplasm

(c) produces CO_2

(d) consumes CO_2

(e) requires NADH

(f) produces NADH

(g) requires ATP

(h) requires GTP

(i) requires thiamine

(j) requires biotin

(k) is regulated by acetyl CoA

23. Like Batman and Robin. Match parts a through j with parts 1 through 10.

(a) Glucose 6-phosphate _____ 1. Inhibits phosphofructokinase in the liver

(b) [ATP] < [AMP] _____ 2. Glucokinase

(c) Citrate _____ 3. GLUT2

(d) Low pH _____ 4. Inhibits hexokinase

(e) Fructose 1,6-bisphosphate _____ 5. Inhibits phosphofructokinase in muscle

(f) Fructose 2,6-bisphosphate _____ 6. Inhibits phosphofructokinase

(g) Insulin _____ 7. Stimulates pyruvate kinase

(h) Has a high K_M for glucose _____ 8. Stimulates phosphofructokinase in the liver

(i) Transporter specific to liver and pancreas _____ 9. Causes the insertion of GLUT4 into cell membranes

(j) [ATP] > [AMP] _____ 10. Stimulates phosphofructokinase

24. Trouble ahead. Suppose that a microorganism that was an obligate anaerobe suffered a mutation that resulted in the loss of triose phosphate isomerase activity. How would this loss affect the ATP yield of fermentation? Could such an organism survive? ✓❶

25. Regulation is good. Regulatory enzymes are crucial for the proper functioning and coordination of glycolysis and gluconeogenesis. As a general rule, regulatory enzymes catalyze reactions that are far from equilibrium. Why does this make good biochemical sense?

26. Kitchen chemistry. Sucrose is commonly used to preserve fruits. Why is glucose not suitable for preserving foods?

27. Tracing carbon atoms 1. Glucose labeled with ^{14}C at C-1 is incubated with the glycolytic enzymes and necessary cofactors.

(a) What is the distribution of ^{14}C in the pyruvate that is formed? (Assume that the interconversion of glyceraldehyde 3-phosphate and dihydroxyacetone phosphate is very rapid compared with the subsequent step.)

(b) If the specific activity of the glucose substrate is 10 mCi mmol^{-1} (millicuries per mole, a measure of radioactivity per mole), what is the specific activity of the pyruvate that is formed?

28. Lactic acid fermentation. ✓❷

(a) Write a balanced equation for the conversion of glucose into lactate.

(b) Calculate the standard free-energy change of this reaction by using the data given in Table 16.1 and the fact that $\Delta G°'$ is -25 kJ mol^{-1} (-6 kcal mol^{-1}) for the following reaction:

$$\text{Pyruvate} + \text{NADH} + \text{H}^+ \rightleftharpoons \text{lactate} + \text{NAD}^+$$

What is the free-energy change (ΔG, not $\Delta G°'$) of this reaction when the concentrations of reactants are: glucose, 5 mM; lactate, 0.05 mM; ATP, 2 mM; ADP, 0.2 mM; and P$_i$, 1 mM?

29. *High potential.* What is the equilibrium ratio of phosphoenolpyruvate to pyruvate under standard conditions when [ATP]/[ADP] = 10?

30. *Hexose–triose equilibrium.* What are the equilibrium concentrations of fructose 1,6-bisphosphate, dihydroxyacetone phosphate, and glyceraldehyde 3-phosphate when 1 mM fructose 1,6-bisphosphate is incubated with aldolase under standard conditions?

31. *Double labeling.* 3-Phosphoglycerate labeled uniformly with ^{14}C is incubated with 1,3-BPG labeled with ^{32}P at C-1. What is the radioisotope distribution of the 2,3-BPG that is formed on addition of BPG mutase?

32. *An informative analog.* Xylose has the same structure as that of glucose except that it has a hydrogen atom at G-5 in place of a hydroxymethyl group. The rate of ATP hydrolysis by hexokinase is markedly enhanced by the addition of xylose. Why?

33. *Distinctive sugars.* The intravenous infusion of fructose into healthy volunteers leads to a two- to five-fold increase in the level of lactate in the blood, a far greater increase than that observed after the infusion of the same amount of glucose. ✓❷

(a) Why is glycolysis more rapid after the infusion of fructose?

(b) Fructose has been used in place of glucose for intravenous feeding. Why is this use of fructose unwise?

34. *It is not hard to meet expenses. They are everywhere.* What energetic barrier prevents glycolysis from simply running in reverse to synthesis glucose? What is the energetic cost to overcome this barrier? ✓❸

35. *Match 'em 2.* Indicate which of the conditions listed in the right-hand column increase the activity of the glycolytic and gluconeogenic pathways. ✓❹

(a) glycolysis _____   1. increase in ATP

(b) gluconeogenesis _____   2. increase in AMP

                           3. increase in fructose
                              2,6-bisphosphate

                           4. increase in citrate

                           5. increase in acetyl CoA

                           6. increase in insulin

                           7. increase in glucagon

                           8. fasting

                           9. fed

36. *Waste not, want not.* Why is the conversion of lactic acid from the blood into glucose in the liver in an organism's best interest?

37. *Road blocks bypassed.* How are the irreversible reactions of glycolysis bypassed in gluconeogenesis? ✓❸

38. *Pointlessness averted.* What are the regulatory means that prevent high levels of activity in glycolysis and gluconeogenesis simultaneously? ✓❹

39. *Different needs.* Liver is primarily a gluconeogenic tissue, whereas muscle is primarily glycolytic. Why does this division of labor make good physiological sense?

40. *Never let me go.* Why does the lack of glucose 6-phosphatase activity in the brain and muscle make good physiological sense?

41. *Counting high-energy compounds 1.* How many NTP molecules are required for the synthesis of one molecule of glucose from two molecules of pyruvate? How many NADH molecules? ✓❸

42. *Counting high-energy compounds 2.* How many NTP molecules are required to synthesize glucose from each of the following compounds? ✓❸

(a) Glucose 6-phosphate

(b) Fructose 1,6-bisphosphate

(c) Two molecules of oxaloacetate

(d) Two molecules of dihydroxyacetone phosphate

43. *Lending a hand.* How might enzymes that remove amino groups from alanine and aspartate contribute to gluconeogenesis?

44. *Metabolic mutants.* Predict the effect of each of the following mutations on the pace of glycolysis in liver cells: ✓❹

(a) Loss of the allosteric site for ATP in phosphofructokinase

(b) Loss of the binding site for citrate in phosphofructokinase

(c) Loss of the phosphatase domain of the bifunctional enzyme that controls the level of fructose 2,6-bisphosphate

(d) Loss of the binding site for fructose 1,6-bisphosphate in pyruvate kinase

45. *Yet another metabolic mutant.* What are the likely consequences of a genetic disorder rendering fructose 1,6-bisphosphatase in the liver less sensitive to regulation by fructose 2,6-bisphosphate?

46. *Biotin snatcher.* Avidin, a 70-kDa protein in egg white, has very high affinity for biotin. In fact, it is a highly specific inhibitor of biotin enzymes. Which of the following conversions would be blocked by the addition of avidin to a cell homogenate?

(a) Glucose $\longrightarrow$ pyruvate

(b) Pyruvate $\longrightarrow$ glucose

(c) Oxaloacetate $\longrightarrow$ glucose

(d) Malate $\longrightarrow$ oxaloacetate

(e) Pyruvate $\longrightarrow$ oxaloacetate

(f) Glyceraldehyde 3-phosphate $\longrightarrow$

$\qquad\qquad\qquad\qquad$ fructose 1,6-bisphosphate

47. *Tracing carbon atoms 2.* If cells synthesizing glucose from lactate are exposed to CO_2 labeled with ^{14}C, what will be the distribution of label in the newly synthesized glucose?

48. *Reduce, reuse, recycle.* In the conversion of glucose into two molecules of lactate, the NADH generated earlier in the pathway is oxidized to NAD^+. Why is it not to the cell's advantage to simply make more NAD^+ so that the regeneration would not be necessary? After all, the cell would save much energy because it would no longer need to synthesize lactic acid dehydrogenase. ✅❷

49. *Adenylate kinase again.* Adenylate kinase, an enzyme considered in great detail in Chapter 9, is responsible for interconverting the adenylate nucleotide pool:

$$ADP + ADP \rightleftharpoons ATP + AMP$$

The equilibrium constant for this reaction is close to 1, inasmuch as the number of phosphoanhydride bonds is the same on each side of the equation. Using the equation for the equilibrium constant for this reaction, show why changes in [AMP] are a more effective indicator of the adenylate pool than [ATP].

50. *Working at cross-purposes?* Gluconeogenesis takes place during intense exercise, which seems counterintuitive. Why would an organism synthesize glucose and at the same time use glucose to generate energy? ✅❹

51. *Powering pathways.* Compare the stoichiometries of glycolysis and gluconeogenesis. Recall that the input of one ATP equivalent changes the equilibrium constant of a reaction by a factor of about 10^8 (Section 15.2). By what factor do the additional high-phosphoryl-transfer compounds alter the equilibrium constant of gluconeogenesis? ✅❸

Mechanism Problem

52. *Argument by analogy.* Propose a mechanism for the conversion of glucose 6-phosphate into fructose 6-phosphate by phosphoglucose isomerase based on the mechanism of triose phosphate isomerase.

Chapter Integration Problems

53. *Not just for energy.* People with galactosemia display central nervous system abnormalities even if galactose is eliminated from the diet. The precise reason for it is not known. Suggest a plausible explanation.

54. *Power to the rbc!* Hexokinase in red blood cells has a K_M of approximately 50 μM. Because life is hard enough as it is, let's assume that the hexokinase displays Michaelis–Menten kinetics. What concentration of blood glucose would yield v_o equal to 90% V_{max}? What does this result tell you if normal blood glucose levels range between approximately 3.6 and 6.1 mM?

55. *State function.* Fructose 2,6-bisphosphate is a potent stimulator of phosphofructokinase. Explain how fructose 2,6-bisphosphate might function in the concerted model for allosteric enzymes.

Data Interpretation Problems

56. *Brewing.* The adjoining graph shows results of experiments on the alcoholic fermentation of glucose with the use of yeast extracts. The graph shows the volume of carbon dioxide released (y axis) as a function of time (x axis).

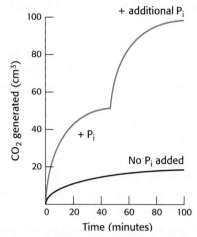

(a) Is measuring the rate of carbon dioxide release a reliable means to measure alcoholic fermentation? Explain.

(b) Why is glucose fermentation dependent on phosphate?

(c) Why is more carbon dioxide generated if the extract is supplemented with phosphate?

(d) During the fermentation, what would the ratio of carbon dioxide generated to phosphate consumed be expected to be?

(e) When fermentation slowed, a hexose bisphosphate accumulated. What is this compound, and why would it accumulate?

57. *Now, that's unusual.* Phosphofructokinase has recently been isolated from the hyperthermophilic archaeon *Pyrococcusfuriosus*. It was subjected to standard biochemical analysis to determine basic catalytic parameters. The processes under study were of the form

Fructose 6-phosphate + $(x - P_i) \longrightarrow$

fructose 1,6-bisphosphate + (x)

The assay measured the increase in fructose 1,6-bisphosphate. Selected results are shown in the adjoining graph.

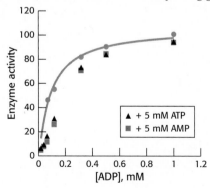

[Data from J. E. Tuininga et al., *J. Biol. Chem.* 274:21023–21028, 1999.]

(a) How does the *P. furiosus* phosphofructokinase differ from the phosphofructokinase considered in this chapter?

(b) What effects do AMP and ATP have on the reaction with ADP?

58. *Cool bees.* In principle, a futile cycle that includes phosphofructokinase and fructose 1,6-bisphosphatase could be used to generate heat. The heat could be used to warm tissues. For instance, certain bumblebees have been reported to use such a futile cycle to warm their flight muscles on cool mornings. Scientists undertook a series of experiments to determine if a number of species of bumblebee use this futile cycle. Their approach was to measure the activity of PFK and F-1,6-BPase in flight muscle.

(a) What was the rationale for comparing the activities of these two enzymes?

(b) The data below show the activities of both enzymes for a variety of bumblebee species (genera *Bombus* and *Psithyrus*). Do these results support the notion that bumblebees use futile cycles to generate heat? Explain.

(c) In which species might futile cycling take place? Explain your reasoning.

(d) Do these results prove that futile cycling does not participate in heat generation?

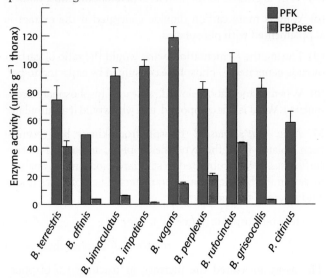

59. *Confused?* The adjoining graph shows muscle phosphofructokinase activity as a function of ATP concentration in the presence of a constant concentration of fructose 6-phosphate. Explain these results and discuss how they relate to the role of phosphofructokinase in glycolysis.

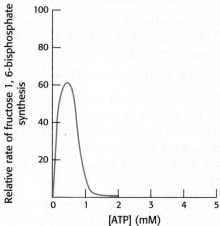

Think ▶ Pair ▶ Share Problems

60. Lactate is a dead-end product in that its sole fate is to be converted back into glucose. Why then do cells bother to form lactate at all?

61. If glucose is such a readily available fuel, why is gluconeogenesis necessary?

The Citric Acid Cycle

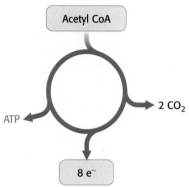

Roundabouts, or traffic circles, function as hubs to facilitate traffic flow. The citric acid cycle is the biochemical hub of the cell, oxidizing carbon fuels, usually in the form of acetyl CoA, as well as serving as a source of precursors for biosynthesis. [Chalermkiat Seedokmai/Getty Images.]

✓ LEARNING OBJECTIVES

By the end of this chapter, you should be able to:

1. Explain why the reaction catalyzed by the pyruvate dehydrogenase complex is a crucial juncture in metabolism.

2. Identify the means by which the pyruvate dehydrogenase complex is regulated.

3. Identify the primary catabolic purpose of the citric acid cycle.

4. Explain the efficiency of using the citric acid cycle to oxidize acetyl CoA.

5. Describe how the citric acid cycle is regulated.

6. Describe the role of the citric acid cycle in anabolism.

7. Identify the biochemical advantages that the glyoxylate cycle provides.

OUTLINE

17.1 The Pyruvate Dehydrogenase Complex Links Glycolysis to the Citric Acid Cycle

17.2 The Citric Acid Cycle Oxidizes Two-Carbon Units

17.3 Entry to the Citric Acid Cycle and Metabolism Through It Are Controlled

17.4 The Citric Acid Cycle Is a Source of Biosynthetic Precursors

17.5 The Glyoxylate Cycle Enables Plants and Bacteria to Grow on Acetate

The metabolism of glucose to pyruvate in glycolysis, an anaerobic process, harvests but a fraction of the ATP available from glucose. Most of the ATP generated in metabolism is provided by the *aerobic* processing of glucose. This process starts with the complete oxidation of glucose derivatives to carbon dioxide. This oxidation takes place in a series of reactions called the *citric acid cycle*, also known as the *tricarboxylic acid* (TCA) *cycle* or the *Krebs cycle*. The citric acid cycle is the *final pathway for the oxidation of fuel molecules*—carbohydrates, fatty acids, and amino acids. Most fuel molecules enter the cycle as *acetyl coenzyme A*.

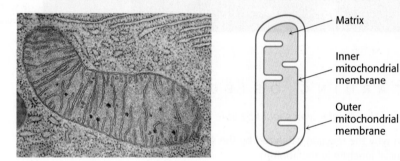

Acetyl coenzyme A (Acetyl CoA)

Under aerobic conditions, the pyruvate generated from glucose is oxidatively decarboxylated to form acetyl CoA. In eukaryotes, the reactions of the citric acid cycle take place in the matrix of the mitochondria (Figure 17.1), in contrast with those of glycolysis, which take place in the cytoplasm.

FIGURE 17.1 Mitochondrion. The double membrane of the mitochondrion is evident in this electron micrograph. The numerous invaginations of the inner mitochondrial membrane are called cristae. The oxidative decarboxylation of pyruvate and the sequence of reactions in the citric acid cycle take place within the matrix. [Omikron/Science Source.]

Matrix

Inner mitochondrial membrane

Outer mitochondrial membrane

The citric acid cycle harvests high-energy electrons

The citric acid cycle is the central metabolic hub of the cell. It is the gateway to the aerobic metabolism of any molecule that can be transformed into an acetyl group or a component of the citric acid cycle. The cycle is also an important source of precursors for the building blocks of many other molecules such as amino acids, nucleotide bases, and porphyrin (the organic component of heme). The citric acid cycle component, oxaloacetate, is also an important precursor to glucose (Section 16.3).

What is the function of the citric acid cycle in transforming fuel molecules into ATP? Recall that fuel molecules are carbon compounds that are capable of being oxidized—that is, of losing electrons (Section 15.3). The citric acid cycle includes a series of oxidation–reduction reactions that result in the oxidation of an acetyl group to two molecules of carbon dioxide. This oxidation generates high-energy electrons that will be used to power the synthesis of ATP. *The catabolic function of the citric acid cycle is the harvesting of high-energy electrons from carbon fuels.*

The overall pattern of the citric acid cycle is shown in Figure 17.2. A four-carbon compound (oxaloacetate) condenses with a two-carbon acetyl unit to yield a six-carbon tricarboxylic acid. The six-carbon compound releases CO_2 twice in two successive oxidative decarboxylations that yield high-energy electrons. A four-carbon compound remains. This four-carbon compound is further processed to regenerate oxaloacetate, which can initiate another round of the cycle. Two carbon atoms enter the cycle as an acetyl unit and two carbon atoms leave the cycle in the form of two molecules of CO_2.

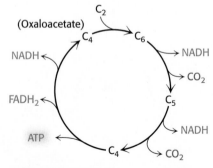

FIGURE 17.2 Overview of the citric acid cycle. The citric acid cycle oxidizes two-carbon units, producing two molecules of CO_2, one molecule of ATP, and high-energy electrons in the form of NADH and $FADH_2$.

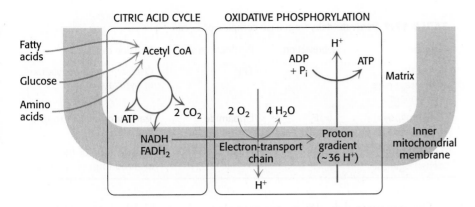

FIGURE 17.3 Cellular respiration. The citric acid cycle constitutes the first stage in cellular respiration, the removal of high-energy electrons from carbon fuels in the form of NADH and FADH$_2$ (blue pathway). These electrons reduce O$_2$ to generate a proton gradient (red pathway), which is used to synthesize ATP (green pathway). The reduction of O$_2$ and the synthesis of ATP constitute oxidative phosphorylation.

Note that the citric acid cycle itself neither generates much ATP nor includes oxygen as a reactant (Figure 17.3). Instead, the citric acid cycle removes electrons from acetyl CoA and uses these electrons to reduce NAD$^+$ and FAD to form NADH and FADH$_2$. Three hydride ions (hence, six electrons) are transferred to three molecules of nicotinamide adenine dinucleotide (NAD$^+$), and one pair of hydrogen atoms (hence, two electrons) is transferred to one molecule of flavin adenine dinucleotide (FAD) each time an acetyl CoA is processed by the cycle. Electrons released in the reoxidation of NADH and FADH$_2$ flow through a series of membrane proteins (referred to as the *electron-transport chain*) to generate a proton gradient across the inner mitochondrial membrane. These protons then flow through ATP synthase to generate ATP from ADP and inorganic phosphate. These electron carriers yield nine molecules of ATP when they are oxidized by O$_2$ in *oxidative phosphorylation* (Chapter 18).

The citric acid cycle, in conjunction with oxidative phosphorylation, provides the preponderance of energy used by aerobic cells—in human beings, greater than 90%. It is highly efficient because the oxidation of a limited number of citric acid cycle molecules can generate large amounts of NADH and FADH$_2$. Note in Figure 17.2 that the four-carbon molecule, oxaloacetate, that initiates the first step in the citric acid cycle is regenerated at the end of one passage through the cycle. Thus, one molecule of oxaloacetate is capable of participating in the oxidation of many acetyl molecules.

17.1 The Pyruvate Dehydrogenase Complex Links Glycolysis to the Citric Acid Cycle

Carbohydrates, most notably glucose, are processed by glycolysis into pyruvate (Chapter 16). Under anaerobic conditions, the pyruvate is converted into lactate or ethanol, depending on the organism. Under aerobic conditions, the pyruvate is transported into mitochondria by a specific carrier protein embedded in the mitochondrial membrane. In the mitochondrial matrix, pyruvate is oxidatively decarboxylated by the *pyruvate dehydrogenase complex* to form acetyl CoA.

$$\text{Pyruvate} + \text{CoA} + \text{NAD}^+ \longrightarrow \text{acetyl CoA} + \text{CO}_2 + \text{NADH} + \text{H}^+$$

This irreversible reaction is the link between glycolysis and the citric acid cycle (Figure 17.4). Note that the pyruvate dehydrogenase complex produces CO$_2$ and captures high-transfer-potential electrons in the form of NADH. Thus, the pyruvate dehydrogenase reaction has many of the key features of the reactions of the citric acid cycle itself.

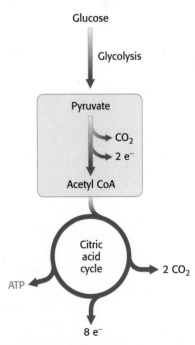

FIGURE 17.4 The link between glycolysis and the citric acid cycle. Pyruvate produced by glycolysis is converted into acetyl CoA, the fuel of the citric acid cycle.

TABLE 17.1 Pyruvate dehydrogenase complex of *E. coli*

Enzyme	Abbreviation	Prosthetic group	Reaction catalyzed
Pyruvate dehydrogenase component	E_1	TPP	Oxidative decarboxylation of pyruvate
Dihydrolipoyl transacetylase	E_2	Lipoamide	Transfer of acetyl group to CoA
Dihydrolipoyl dehydrogenase	E_3	FAD	Regeneration of the oxidized form of lipoamide

The pyruvate dehydrogenase complex, a large, highly integrated complex of three distinct enzymes (Table 17.1), is another example of the organization of enzymes into supramolecular structures (p. 464). Pyruvate dehydrogenase complex is a member of a family of homologous complexes that include the citric acid cycle enzyme α-ketoglutarate dehydrogenase complex (p. 552). These complexes are giant, larger than ribosomes, with molecular masses ranging from 4 million to 10 million kDa. As we will see, their elaborate structures allow groups to travel from one active site to another, connected by tethers to the core of the structure.

Mechanism: The synthesis of acetyl coenzyme A from pyruvate requires three enzymes and five coenzymes

The mechanism of the pyruvate dehydrogenase reaction is wonderfully complex, more so than is suggested by its simple stoichiometry. The reaction requires the participation of the three enzymes of the pyruvate dehydrogenase complex and five coenzymes. The coenzymes *thiamine pyrophosphate* (TPP), *lipoic acid,* and *FAD* serve as catalytic cofactors, and *CoA* and *NAD*$^+$ are stoichiometric cofactors, cofactors that function as substrates.

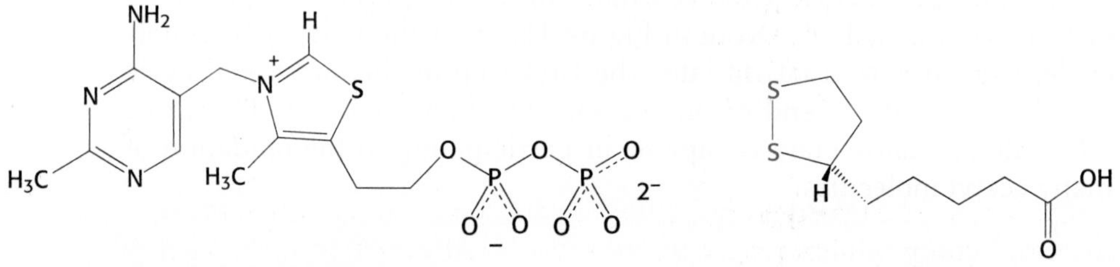

Thiamine pyrophosphate (TPP)   **Lipoic acid**

The conversion of pyruvate into acetyl CoA consists of three steps: decarboxylation, oxidation, and transfer of the resultant acetyl group to CoA. A fourth step is required to regenerate the active enzyme.

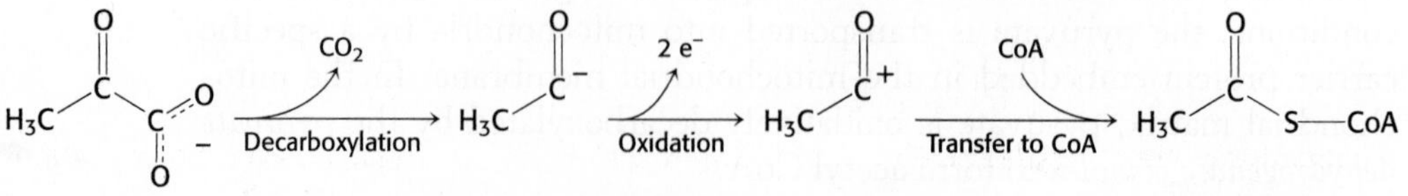

Pyruvate   **Acetyl CoA**

These steps must be coupled to preserve the free energy derived from the decarboxylation step to drive the formation of NADH and acetyl CoA:

1. *Decarboxylation.* Pyruvate combines with TPP and is then decarboxylated to yield hydroxyethyl-TPP (Figure 17.5).

Carbanion of TPP   Pyruvate   Hydroxyethyl-TPP

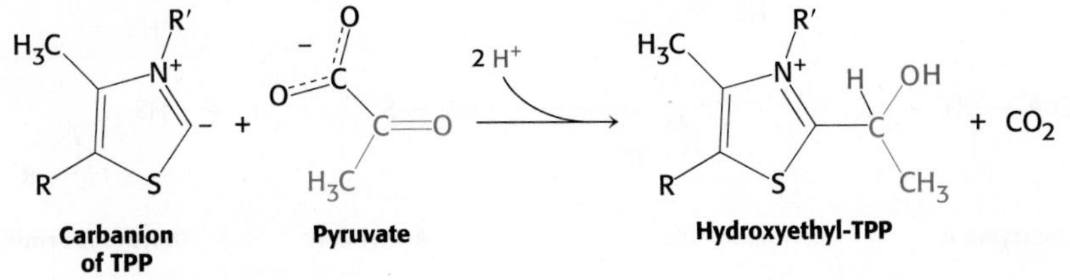

This reaction, the rate-limiting step in acetyl CoA synthesis, is catalyzed by the *pyruvate dehydrogenase component* (E_1) of the multienzyme complex. TPP is the prosthetic group of the pyruvate dehydrogenase component.

TPP   Carbanion   Pyruvate   Addition compound

Resonance forms of hydroxyethyl-TPP   Hydroxyethyl-TPP

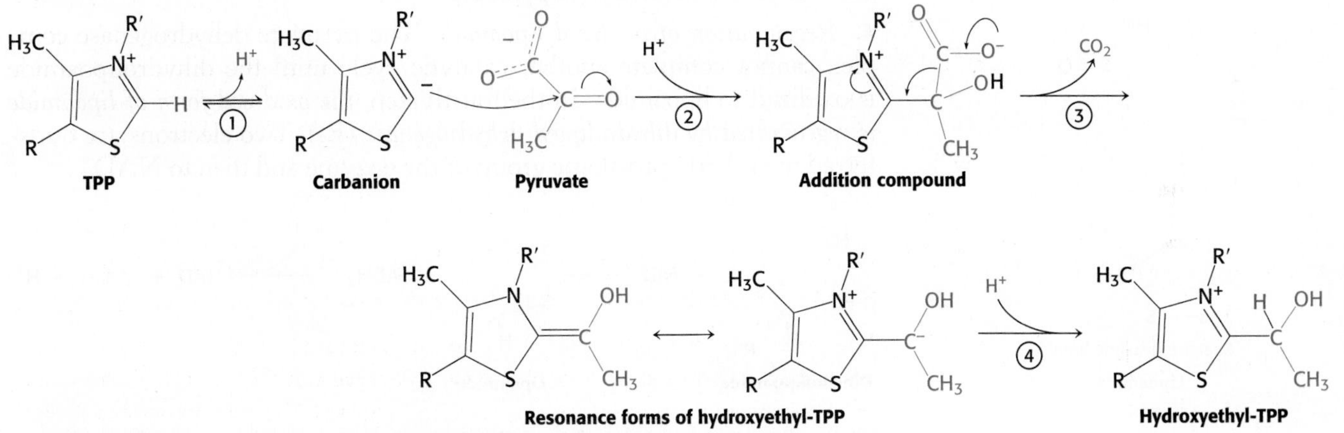

FIGURE 17.5 Mechanism of the E_1 decarboxylation reaction. E_1 is the pyruvate dehydrogenase component of the pyruvate dehydrogenase complex. A key feature of the prosthetic group, TPP, is that the carbon atom between the nitrogen and sulfur atoms in the thiazole ring is much more acidic than most $=C-$ groups, with a pK_a value near 10. (1) This carbon center ionizes to form a *carbanion*. (2) The carbanion readily adds to the carbonyl group of pyruvate. (3) This addition is followed by the decarboxylation of pyruvate. The positively charged ring of TPP acts as an electron sink that stabilizes the negative charge that is transferred to the ring as part of the decarboxylation. (4) Protonation yields hydroxyethyl-TPP.

2. *Oxidation.* The hydroxyethyl group attached to TPP is *oxidized* to form an acetyl group while being simultaneously transferred to lipoamide, a derivative of lipoic acid that is linked to the side chain of a lysine residue by an amide linkage. Note that this transfer results in the formation of an energy-rich thioester bond.

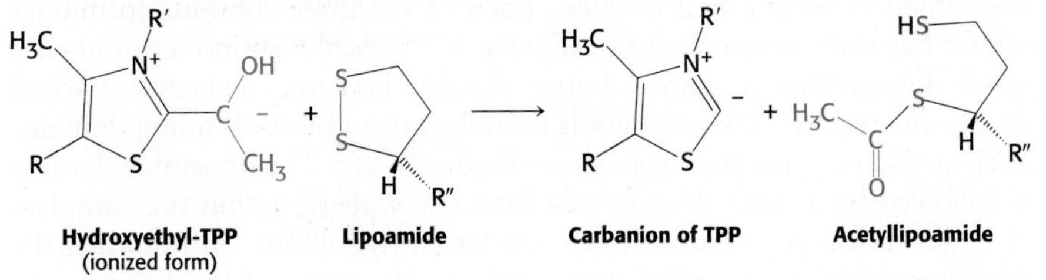

Hydroxyethyl-TPP (ionized form)   Lipoamide   Carbanion of TPP   Acetyllipoamide

The oxidant in this reaction is the disulfide group of lipoamide, which is reduced to its disulfhydryl form. This reaction, also catalyzed by the pyruvate dehydrogenase component E_1, yields *acetyllipoamide*.

3. *Formation of Acetyl CoA.* The acetyl group is transferred from acetyllipoamide to CoA to form acetyl CoA.

Coenzyme A   **Acetyllipoamide**   **Acetyl CoA**   **Dihydrolipoamide**

Dihydrolipoyl transacetylase (E$_2$) catalyzes this reaction. The energy-rich thioester bond is preserved as the acetyl group is transferred to CoA. Recall that CoA serves as a carrier of many activated acyl groups, of which acetyl is the simplest (Section 15.3). *Acetyl CoA, the fuel for the citric acid cycle, has now been generated from pyruvate.*

4. *Regeneration of oxidized lipoamide.* The pyruvate dehydrogenase complex cannot complete another catalytic cycle until the dihydrolipoamide is oxidized to lipoamide. In the fourth step, *the oxidized form of lipoamide is regenerated by dihydrolipoyl dehydrogenase* (E$_3$). Two electrons are transferred to an FAD prosthetic group of the enzyme and then to NAD$^+$.

Dihydrolipoamide   **Lipoamide**

This electron transfer from FAD to NAD$^+$ is unusual because the common role for FAD is to receive electrons from NADH. The electron-transfer potential of FAD is increased by its chemical environment within the enzyme, enabling it to transfer electrons to NAD$^+$. Proteins tightly associated with FAD or flavin mononucleotide (FMN) are called *flavoproteins.*

Flexible linkages allow lipoamide to move between different active sites

The structures and precise composition of the component enzymes of the pyruvate dehydrogenase vary among species. However, there are commonalities.

The core of the complex is formed by 60 molecules of the transacetylase component E$_2$ (Figure 17.6). Transacetylase consists of 20 catalytic trimers assembled to form a hollow cube. Each of the three subunits forming a trimer has three major domains (Figure 17.7). At the amino terminus is a small domain that contains a bound flexible lipoamide cofactor attached to a lysine residue. This domain is homologous to biotin-binding domains such as that of pyruvate carboxylase (Figure 16.25). The lipoamide domain is followed by a small domain that interacts with E$_3$ within the complex. A larger transacetylase domain completes an E$_2$ subunit. Surrounding the core transacetylase is a shell composed of ~45 copies of the E$_1$ and ~10 copies of E$_3$ enzymes. In mammals, E$_1$ is an $\alpha_2\beta_2$ tetramer, and E$_3$ is an $\alpha\beta$ dimer, and this core contains another protein, E$_3$-binding protein (E$_3$-BP), which facilitates the interaction between E$_2$ and E$_3$. If E$_3$-BP is missing, the complex has greatly reduced activity. The gap between the outer shell and

Lysine
side chain

Reactive disulfide bond

Lipoamide

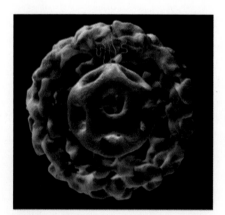

FIGURE 17.6 Structure of the pyruvate dehydrogenase complex from *B. stearothermophilus*. The image of the complex, which was derived from cryo-electron microscopic data (Section 3.5), shows an inner core consisting of the E$_2$ enzyme. The shell surrounding the core consists of E$_1$ and E$_3$ enzymes, although only the E$_1$ enzymes are shown in this structure. Two of the 60 lipoamide arms are shown (red and yellow). [Donald Bliss, National Library of Medicine.]

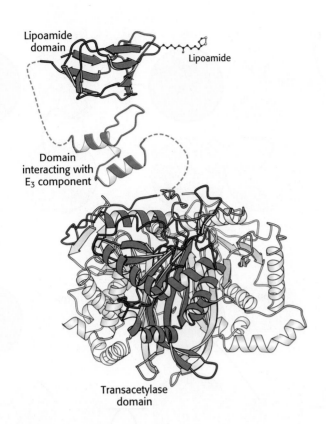

Lipoamide domain

Lipoamide

Domain interacting with E_3 component

Transacetylase domain

FIGURE 17.7 Structure of the transacetylase (E_2) core. The figure shows one subunit of the transacetylase trimer. *Notice* that each subunit consists of three domains: a lipoamide-binding domain, a small domain that interacts with E_3, and a large transacetylase catalytic domain. The catalytic domains interact with one another to form the catalytic trimer. Transacetylase domains of three identical subunits are shown, with one depicted in red and the others in white in the ribbon representation.

the transacetylase core allows the lipoamide arms to visit the various active sites as described below (Figure 17.6).

How do the three distinct active sites work in concert (Figure 17.8)? The key is the long, flexible lipoamide arm of the E_2 subunit, which carries substrate from active site to active site:

1. Pyruvate is decarboxylated at the active site of E_1, forming the hydroxyethyl-TPP intermediate, and CO_2 leaves as the first product. This active site lies deep within the E_1 complex, connected to the enzyme surface by a 20-Å-long hydrophobic channel.

2. E_2 inserts the lipoamide arm of the lipoamide domain into the deep channel in E_1 leading to the active site.

3. E_1 catalyzes the transfer of the acetyl group to the lipoamide. The acetylated arm then leaves E_1 and enters the E_2 cube to visit the active site of E_2, located deep in the cube at the subunit interface.

4. The acetyl moiety is then transferred to CoA, and the second product, acetyl CoA, leaves the cube. The reduced lipoamide arm then swings to the active site of the E_3 flavoprotein.

5. At the E_3 active site, the lipoamide is oxidized by coenzyme FAD. The reactivated lipoamide is ready to begin another reaction cycle.

6. The final product, NADH, is produced with the reoxidation of $FADH_2$ to FAD.

The structural integration of three kinds of enzymes and the long, flexible lipoamide arm make the coordinated catalysis of a complex reaction possible. The proximity of one enzyme to another *increases the overall reaction rate and minimizes side reactions.* All the intermediates in the oxidative decarboxylation of pyruvate remain bound to the complex throughout the reaction sequence and are readily transferred as the flexible arm of E_2 calls on each active site in turn.

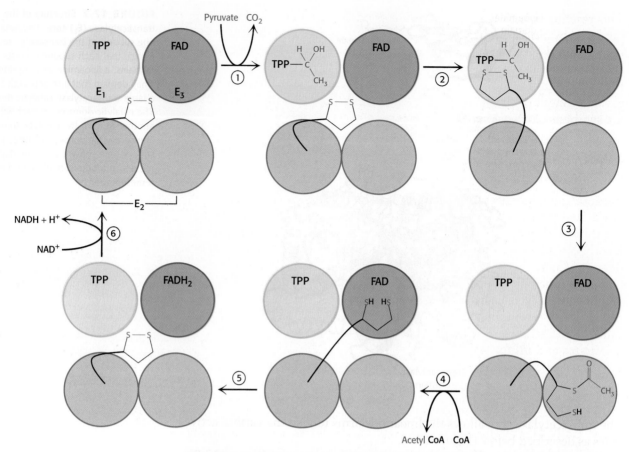

FIGURE 17.8 Reactions of the pyruvate dehydrogenase complex. At the top (left), the enzyme (represented by a yellow, a green, and two red spheres) is unmodified and ready for a catalytic cycle. (1) Pyruvate is decarboxylated to form hydroxyethyl-TPP. (2) The lipoamide arm of E_2 moves into the active site of E_1. (3) E_1 catalyzes the transfer of the two-carbon group to the lipoamide group to form the acetyl–lipoamide complex. (4) E_2 catalyzes the transfer of the acetyl moiety to CoA to form the product acetyl CoA. The dihydrolipoamide arm then swings to the active site of E_3. E_3 catalyzes (5) the oxidation of the dihydrolipoamide and (6) the transfer of the protons and electrons to NAD^+ to complete the reaction cycle.

17.2 The Citric Acid Cycle Oxidizes Two-Carbon Units

The conversion of pyruvate into acetyl CoA by the pyruvate dehydrogenase complex is the link between glycolysis and cellular respiration because *acetyl CoA is the fuel for the citric acid cycle.* Indeed, all fuels are ultimately metabolized to acetyl CoA or components of the citric acid cycle.

Citrate synthase forms citrate from oxaloacetate and acetyl coenzyme A

The citric acid cycle begins with the condensation of a four-carbon unit, oxaloacetate, and a two-carbon unit, the acetyl group of acetyl CoA. Oxaloacetate reacts with acetyl CoA and H_2O to yield citrate and CoA.

This reaction, which is an aldol condensation followed by a hydrolysis, is catalyzed by *citrate synthase*. Oxaloacetate first condenses with acetyl CoA to form *citryl CoA*, a molecule that is energy rich because it contains the thioester bond that originated in acetyl CoA. The hydrolysis of citryl CoA thioester to citrate and CoA drives the overall reaction far in the direction of the synthesis of citrate. In essence, the hydrolysis of the thioester powers the synthesis of a new molecule from two precursors.

Mechanism: The mechanism of citrate synthase prevents undesirable reactions

Because the condensation of acetyl CoA and oxaloacetate initiates the citric acid cycle, it is very important that side reactions, notably the hydrolysis of acetyl CoA to acetate and CoA, be minimized. Let us briefly consider how the citrate synthase prevents the wasteful hydrolysis of acetyl CoA.

Mammalian citrate synthase is a dimer of identical 49-kDa subunits. Each active site is located in a cleft between the large and the small domains of a subunit, adjacent to the subunit interface. X-ray crystallographic studies of citrate synthase and its complexes with several substrates and inhibitors have revealed that the enzyme undergoes large conformational changes in the course of catalysis. Citrate synthase exhibits sequential, ordered kinetics: oxaloacetate binds first, followed by acetyl CoA. The reason for the ordered binding is that *oxaloacetate induces a major structural rearrangement leading to the creation of a binding site for acetyl CoA.* The binding of oxaloacetate converts the open form of the enzyme into a more closed form (Figure 17.9). In each subunit, the small domain rotates 19 degrees relative to the large domain. *Movements as large as 15 Å are produced by the rotation of α helices elicited by quite small shifts of side chains around bound oxaloacetate.* These structural changes create a binding site for acetyl CoA.

Synthase
An enzyme catalyzing a synthetic reaction in which two units are joined usually without the direct participation of ATP (or another nucleoside triphosphate).

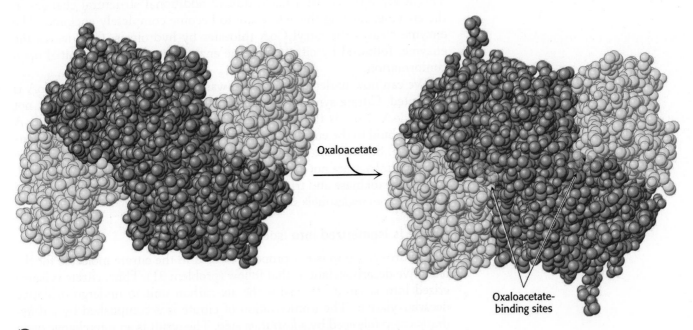

Oxaloacetate

Oxaloacetate-binding sites

FIGURE 17.9 Conformational changes in citrate synthase on binding oxaloacetate. The small domain of each subunit of the homodimer is shown in yellow; the large domains are shown in blue. (Left) Open form of enzyme alone. (Right) Closed form of the oxaloacetate-enzyme complex. [Drawn from 5CSC.pdb and 4CTS.pdb.]

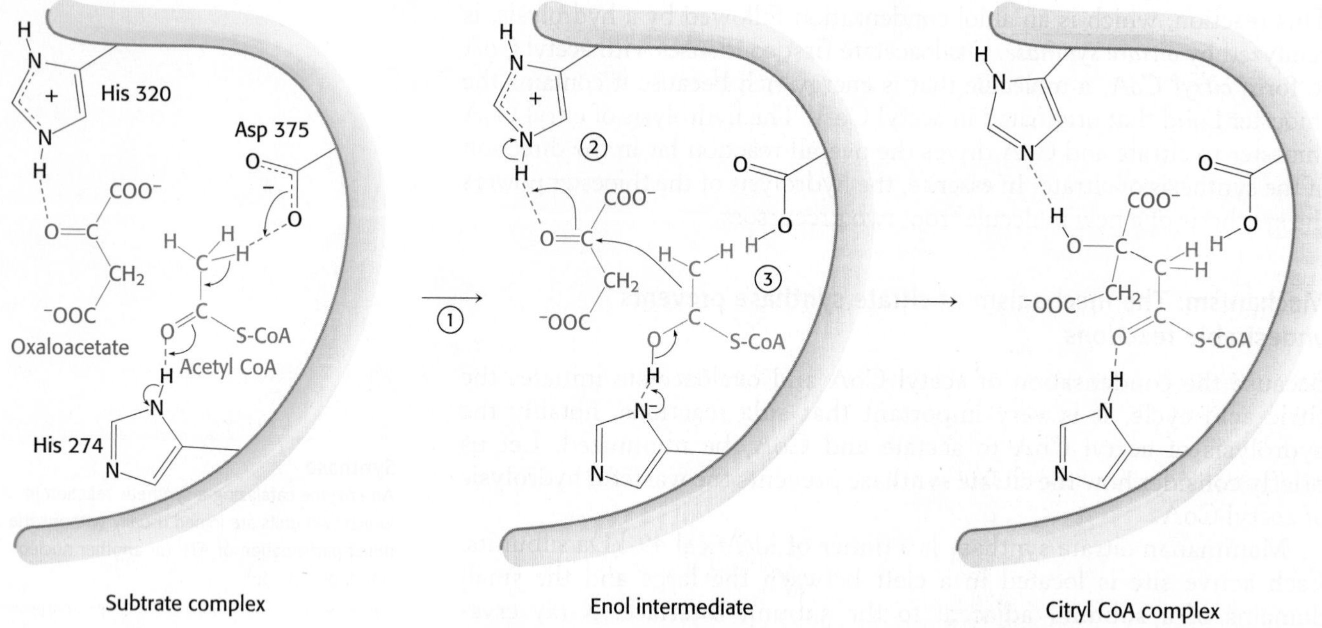

Subtrate complex

Enol intermediate

Citryl CoA complex

FIGURE 17.10 Mechanism of synthesis of citryl CoA by citrate synthase. (1) In the substrate complex (left), His 274 donates a proton to the carbonyl oxygen of acetyl CoA to promote the removal of a methyl proton by Asp 375 to form the enol intermediate (center). (2) Oxaloacetate is activated by the transfer of a proton from His 320 to its carbonyl carbon atom. (3) Simultaneously, the enol of acetyl CoA attacks the carbonyl carbon of oxaloacetate to form a carbon–carbon bond linking acetyl CoA and oxaloacetate. His 274 is reprotonated. Citryl CoA is formed. His 274 participates again as a proton donor to hydrolyze the thioester (not shown), yielding citrate and CoA.

Citrate synthase catalyzes the condensation reaction by bringing the substrates into close proximity, orienting them, and polarizing certain bonds (Figure 17.10). The donation and removal of protons transforms acetyl CoA into an *enol intermediate*. The enol attacks oxaloacetate to form a carbon–carbon double bond linking acetyl CoA and oxaloacetate. The newly formed citryl CoA induces additional structural changes in the enzyme, causing the active site to become completely enclosed. The enzyme cleaves the citryl CoA thioester by hydrolysis. CoA leaves the enzyme, followed by citrate, and the enzyme returns to the initial open conformation.

We can now understand how the wasteful hydrolysis of acetyl CoA is prevented. Citrate synthase is well suited to hydrolyze *citryl* CoA but not *acetyl* CoA. How is this discrimination accomplished? First, acetyl CoA does not bind to the enzyme until oxaloacetate is bound and ready for condensation. Second, the catalytic residues crucial for the hydrolysis of the thioester linkage are not appropriately positioned *until citryl CoA is formed*. As with hexokinase and triose phosphate isomerase (Section 16.1), *induced fit prevents an undesirable side reaction*.

Citrate is isomerized into isocitrate

The hydroxyl group is not properly located in the citrate molecule for the oxidative decarboxylations that follow (problem 31). Thus, citrate is isomerized into isocitrate to enable the six-carbon unit to undergo oxidative decarboxylation. The isomerization of citrate is accomplished by a *dehydration* step followed by a *hydration* step. The result is an interchange of an H and an OH. The enzyme catalyzing both steps is called *aconitase* because cis-*aconitate* is an intermediate.

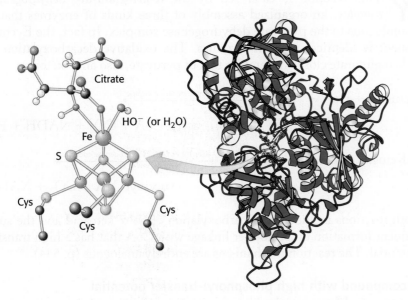

Aconitase is an *iron–sulfur protein*, or *nonheme-iron protein*, in that it contains iron that is not bonded to heme. Rather, its four iron atoms are complexed to four inorganic sulfides and three cysteine sulfur atoms, leaving one iron atom available to bind citrate through one of its COO^- groups and an OH group (Figure 17.11). This Fe-S cluster participates in dehydrating and rehydrating the bound substrate.

FIGURE 17.11 Binding of citrate to the iron–sulfur complex of aconitase. A 4Fe-4S iron–sulfur cluster is a component of the active site of aconitase. *Notice* that one of the iron atoms of the cluster binds to a COO^- group and an OH group of citrate. [Drawn from 1C96.pdb.]

Isocitrate is oxidized and decarboxylated to alpha-ketoglutarate

We come now to the first of four oxidation–reduction reactions in the citric acid cycle. The oxidative decarboxylation of isocitrate is catalyzed by *isocitrate dehydrogenase*.

$$Isocitrate + NAD^+ \longrightarrow \alpha\text{-ketoglutarate} + CO_2 + NADH$$

The intermediate in this reaction is oxalosuccinate, an unstable β-ketoacid. While bound to the enzyme, it loses CO_2 to form α-ketoglutarate.

This oxidation generates the first high-transfer-potential electron carrier, NADH, in the cycle.

Succinyl coenzyme A is formed by the oxidative decarboxylation of alpha-ketoglutarate

The conversion of isocitrate into α-ketoglutarate is followed by a second oxidative decarboxylation reaction, the formation of succinyl CoA from α-ketoglutarate.

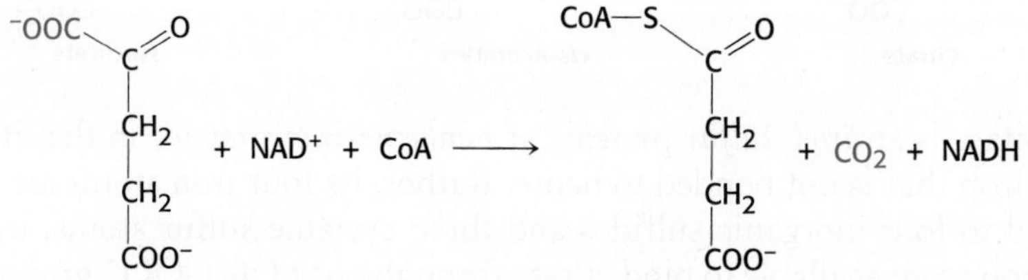

$$\text{α-Ketoglutarate} + NAD^+ + CoA \longrightarrow \text{Succinyl CoA} + CO_2 + NADH$$

This reaction is catalyzed by the *α-ketoglutarate dehydrogenase complex*, an organized assembly of three kinds of enzymes that is homologous to the pyruvate dehydrogenase complex. In fact, the E_3 component is identical in both enzymes. The oxidative decarboxylation of α-ketoglutarate closely resembles that of pyruvate, also an α-ketoacid.

$$\text{Pyruvate} + CoA + NAD^+ \xrightarrow{\text{Pyruvate dehydrogenase complex}} \text{acetyl CoA} + CO_2 + NADH + H^+$$

$$\text{α-Ketoglutarate} + CoA + NAD^+ \xrightarrow{\text{α-Ketoglutarate dehydrogenase complex}} \text{succinyl CoA} + CO_2 + NADH$$

Both reactions include the decarboxylation of an α-ketoacid and the subsequent formation of a thioester linkage with CoA that has a high transfer potential. The reaction mechanisms are entirely analogous (p. 544).

A compound with high phosphoryl-transfer potential is generated from succinyl coenzyme A

Succinyl CoA is an energy-rich thioester compound. The $\Delta G°'$ for the hydrolysis of succinyl CoA is about $-33.5\,\text{kJ mol}^{-1}(-8.0\,\text{kcal mol}^{-1})$, which is comparable to that of ATP ($-30.5\,\text{kJ mol}^{-1}$, or $-7.3\,\text{kcal mol}^{-1}$). In the citrate synthase reaction, the cleavage of the thioester bond powers the synthesis of the six-carbon citrate from the four-carbon oxaloacetate and the two-carbon fragment. *The cleavage of the thioester bond of succinyl CoA is coupled to the phosphorylation of a purine nucleoside diphosphate, usually ADP.* This reaction, which is readily reversible, is catalyzed by *succinyl CoA synthetase* (also called, succinate thiokinase).

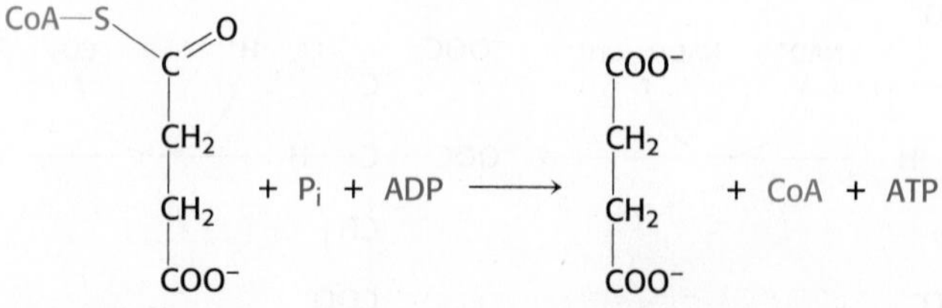

$$\text{Succinyl CoA} + P_i + ADP \longrightarrow \text{Succinate} + CoA + ATP$$

This reaction is the only step in the citric acid cycle that directly yields a compound with high phosphoryl-transfer potential. In mammals, there are two isozymic forms of the enzyme, one specific for ADP and one for GDP.

In tissues that perform large amounts of cellular respiration, such as skeletal and heart muscle, the ADP-requiring isozyme predominates. In tissues that perform many anabolic reactions, such as the liver, the GDP-requiring enzyme is common. The GDP-requiring enzyme is believed to work in reverse of the direction observed in the TCA cycle; that is, GTP is used to power the synthesis of succinyl CoA, which is a precursor for heme synthesis (Section 24.4). The *E. coli* enzyme uses either GDP or ADP as the phosphoryl-group acceptor.

Note that the enzyme *nucleoside diphosphokinase*, which catalyzes the following reaction,

$$GTP + ADP \rightleftharpoons GDP + ATP$$

allows the γ phosphoryl group to be readily transferred from GTP to form ATP, thereby allowing the adjustment of the concentration of GTP or ATP to meet the cell's need.

Mechanism: Succinyl coenzyme A synthetase transforms types of biochemical energy

The mechanism of this reaction is a clear example of an energy transformation: energy inherent in the thioester molecule is transformed into phosphoryl-group-transfer potential (Figure 17.12). The first step is the displacement of coenzyme A by orthophosphate, which generates another energy-rich compound, succinyl phosphate. A histidine residue plays a key role as a moving arm that detaches the phosphoryl group, then swings over to a bound ADP and transfers the group to form ATP. The participation of high-energy compounds in all the steps is attested to by the fact that the reaction is readily reversible: $\Delta G°' = -3.4 \text{ kJ mol}^{-1} (-0.8 \text{ kcal mol}^{-1})$. The formation of ATP at the expense of succinyl CoA is an example of substrate-level phosphorylation (Section 16.1).

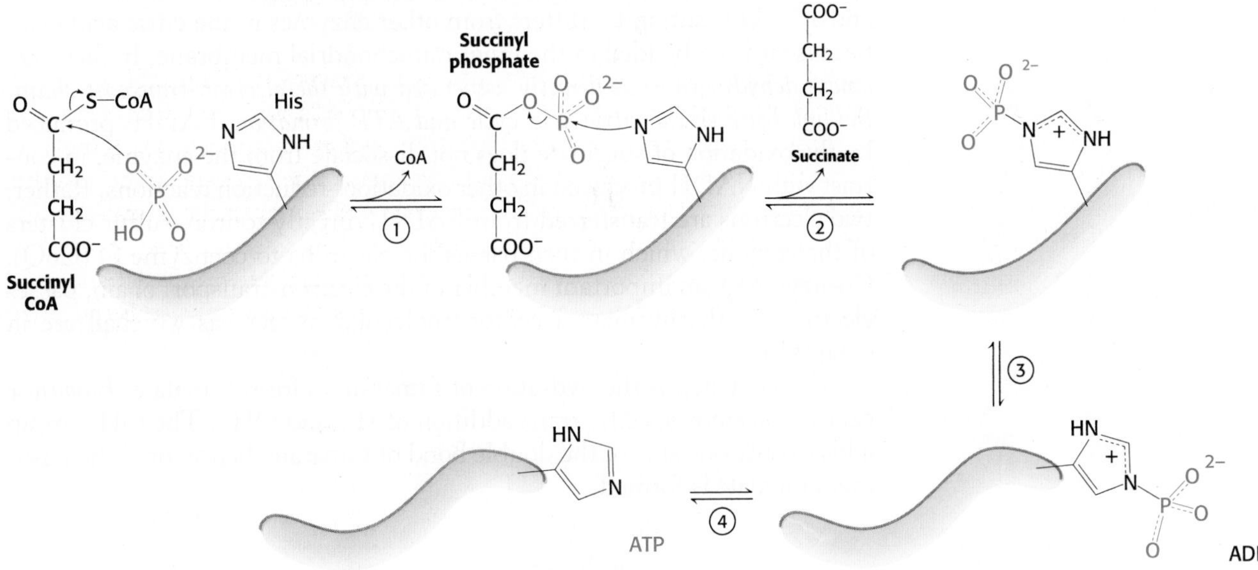

FIGURE 17.12 Reaction mechanism of succinyl CoA synthetase. The reaction proceeds through a phosphorylated enzyme intermediate. (1) Orthophosphate displaces coenzyme A, which generates another energy-rich compound, succinyl phosphate. (2) A histidine residue removes the phosphoryl group with the concomitant generation of succinate and phosphohistidine. (3) The phosphohistidine residue then swings over to a bound ADP, and (4) the phosphoryl group is transferred to form ATP.

Oxaloacetate is regenerated by the oxidation of succinate

Reactions of four-carbon compounds constitute the final stage of the citric acid cycle: the regeneration of oxaloacetate.

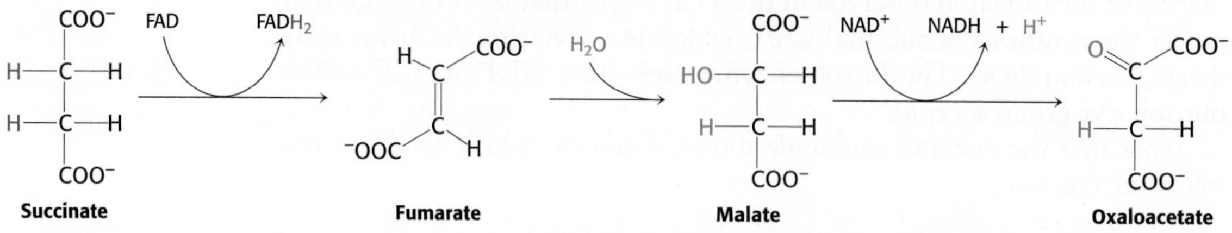

The reactions constitute a metabolic motif that we will see again in fatty acid synthesis and degradation as well as in the degradation of some amino acids. A methylene group (CH_2) is converted into a carbonyl group $C=O$ in three steps: an oxidation, a hydration, and a second oxidation reaction. Oxaloacetate is thereby regenerated for another round of the cycle, and more energy is extracted in the form of $FADH_2$ and NADH.

Succinate is oxidized to fumarate by *succinate dehydrogenase*. The hydrogen acceptor is FAD rather than NAD^+, which is used in the other three oxidation reactions in the cycle. FAD is the hydrogen acceptor in this reaction because the free-energy change is insufficient to reduce NAD^+. FAD is nearly always the electron acceptor in oxidations that remove two hydrogen *atoms* from a substrate. In succinate dehydrogenase, the isoalloxazine ring of FAD is covalently attached to a histidine side chain of the enzyme (denoted E-FAD).

$$\text{E-FAD} + \text{succinate} \rightleftharpoons \text{E-FADH}_2 + \text{fumarate}$$

Succinate dehydrogenase, like aconitase, is an iron–sulfur protein. Indeed, succinate dehydrogenase contains three different kinds of iron–sulfur clusters: 2Fe-2S (two iron atoms bonded to two inorganic sulfides), 3Fe-4S, and 4Fe-4S. Succinate dehydrogenase—which consists of a 70-kDa and a 27-kDa subunit—differs from other enzymes in the citric acid cycle because it is embedded in the inner mitochondrial membrane. In fact, *succinate dehydrogenase is directly associated with the electron-transport chain, the link between the citric acid cycle and ATP formation.* $FADH_2$ produced by the oxidation of succinate does not dissociate from the enzyme, in contrast with NADH produced in other oxidation–reduction reactions. Rather, two electrons are transferred from $FADH_2$ directly to iron–sulfur clusters of the enzyme, which in turn passes the electrons to coenzyme Q (CoQ). Coenzyme Q, an important member of the electron-transport chain, passes electrons to the ultimate acceptor, molecular oxygen, as we shall see in Chapter 18.

The next step is the hydration of fumarate to form L-malate. *Fumarase* catalyzes a stereospecific trans addition of H^+ and OH^-. The OH^- group adds to only one side of the double bond of fumarate; hence, only the L isomer of malate is formed.

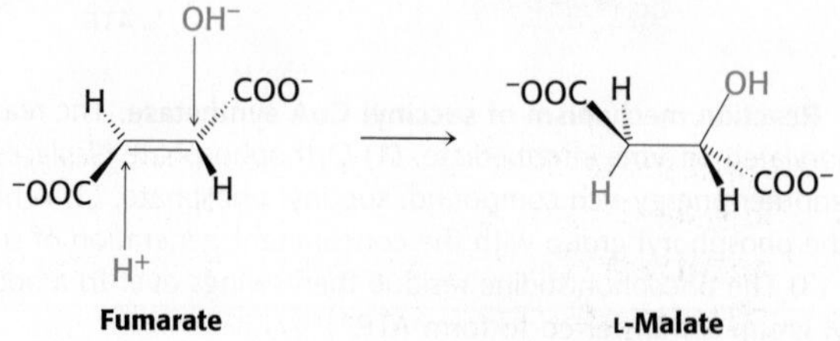

Finally, malate is oxidized to form oxaloacetate. This reaction is catalyzed by *malate dehydrogenase,* and NAD^+ is again the hydrogen acceptor.

$$Malate + NAD^+ \rightleftharpoons oxaloacetate + NADH + H^+$$

The standard free energy for this reaction, unlike that for the other steps in the citric acid cycle, is significantly positive ($\Delta G°' = +29.7 \text{ kJ mol}^{-1}$, or $+7.1 \text{ kcal mol}^{-1}$). The oxidation of malate is driven by the use of the products—oxaloacetate by citrate synthase and NADH by the electron-transport chain.

The citric acid cycle produces high-transfer-potential electrons, ATP, and CO_2

The net reaction of the citric acid cycle is

$$Acetyl CoA + 3 NAD^+ + FAD + ADP + P_i + 2 H_2O \longrightarrow$$

$$2 CO_2 + 3 NADH + FADH_2 + ATP + 2 H^+ + CoA$$

Let us recapitulate the reactions that give this stoichiometry (Figure 17.13 and Table 17.2):

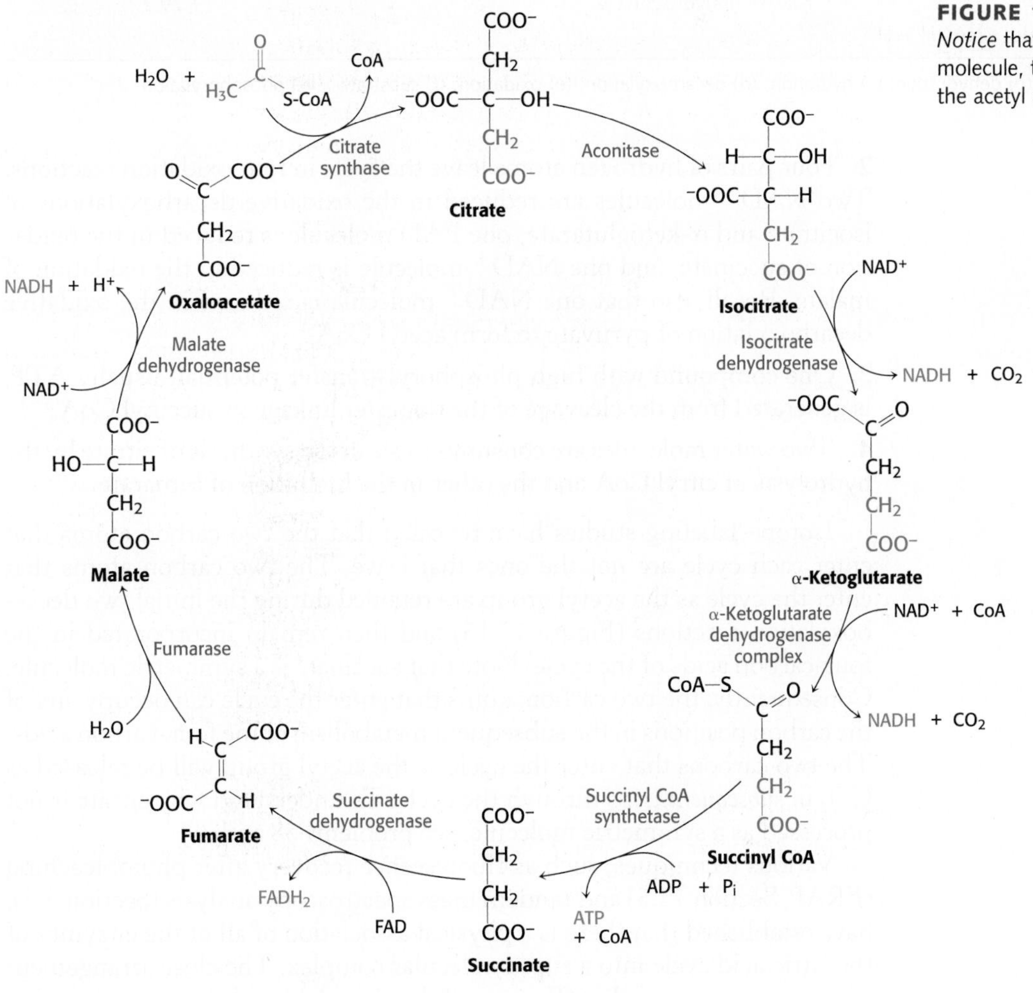

FIGURE 17.13 The citric acid cycle. *Notice* that since succinate is a symmetric molecule, the identity of the carbons from the acetyl unit is lost.

1. Two carbon atoms enter the cycle in the condensation of an acetyl unit (from acetyl CoA) with oxaloacetate. Two carbon atoms leave the cycle in the form of CO_2 in the successive decarboxylations catalyzed by isocitrate dehydrogenase and α-ketoglutarate dehydrogenase.

TABLE 17.2 Citric acid cycle

Step	Reaction	Enzyme	Prosthetic group	Type*	$\Delta G^{\circ\prime}$ kJ mol^{-1}	$\Delta G^{\circ\prime}$ kcal mol^{-1}
1	Acetyl CoA + oxaloacetate + $H_2O \longrightarrow$ citrate + CoA + H^+	Citrate synthase		a	−31.4	−7.5
2a	Citrate $\rightleftharpoons$ cis-aconitate + H_2O	Aconitase	Fe-S	b	+8.4	+2.0
2b	cis-Aconitate + $H_2O \rightleftharpoons$ isocitrate	Aconitase	Fe-S	c	−2.1	−0.5
3	Isocitrate + NAD$^+$ $\rightleftharpoons$ α-ketoglutarate + CO_2 + NADH	Isocitrate dehydrogenase		d + e	−8.4	−2.0
4	α-Ketoglutarate + NAD$^+$ + CoA $\rightleftharpoons$ succinyl CoA + CO_2 + NADH	α-Ketoglutarate dehydrogenase complex	Lipoic acid, FAD, TPP	d + e	−30.1	−7.2
5	Succinyl CoA + P$_i$ + ADP $\rightleftharpoons$ succinate + ATP + CoA	Succinyl CoA synthetase		f	−3.3	−0.8
6	Succinate + FAD (enzyme-bound) $\rightleftharpoons$ fumarate + FADH$_2$ (enzyme-bound)	Succinate dehydrogenase	FAD, Fe-S	e	0	0
7	Fumarate + $H_2O \rightleftharpoons$ L-malate	Fumarase		c	−3.8	−0.9
8	L-Malate + NAD$^+$ $\rightleftharpoons$ oxaloacetate + NADH + H^+	Malate dehydrogenase		E	+29.7	+7.1

*Reaction type: (a) condensation; (b) dehydration; (c) hydration; (d) decarboxylation; (e) oxidation; (f) substrate-level phosphorylation.

2. Four pairs of hydrogen atoms leave the cycle in four oxidation reactions. Two NAD$^+$ molecules are reduced in the oxidative decarboxylations of isocitrate and α-ketoglutarate, one FAD molecule is reduced in the oxidation of succinate, and one NAD$^+$ molecule is reduced in the oxidation of malate. Recall also that one NAD$^+$ molecule is reduced in the oxidative decarboxylation of pyruvate to form acetyl CoA.

3. One compound with high phosphoryl-transfer potential, usually ATP, is generated from the cleavage of the thioester linkage in succinyl CoA.

4. Two water molecules are consumed: one in the synthesis of citrate by the hydrolysis of citryl CoA and the other in the hydration of fumarate.

Isotope-labeling studies have revealed that the two carbon atoms that enter each cycle are not the ones that leave. The two carbon atoms that enter the cycle as the acetyl group are retained during the initial two decarboxylation reactions (Figure 17.13) and then remain incorporated in the four-carbon acids of the cycle. Note that succinate is a symmetric molecule. Consequently, the two carbon atoms that enter the cycle can occupy any of the carbon positions in the subsequent metabolism of the four-carbon acids. The two carbons that enter the cycle as the acetyl group will be released as CO_2 in *subsequent* trips through the cycle. To understand why citrate is not processed as a symmetric molecule, see problems 38 and 39.

Various techniques, such as fluorescence recovery after photobleaching (FRAP, Section 12.5) and tandem mass spectroscopy analysis (Section 3.3), have established that there is a physical association of all of the enzymes of the citric acid cycle into a supramolecular complex. The close arrangement of enzymes enhances the efficiency of the citric acid cycle because a reaction product can pass directly from one active site to the next through connecting channels, a process called substrate channeling.

As will be considered in Chapter 18, the electron-transport chain oxidizes the NADH and FADH$_2$ formed in the citric acid cycle. The transfer of electrons from these carriers to O_2, the ultimate electron acceptor, leads to the generation of a proton gradient across the inner mitochondrial membrane. This proton-motive force then powers the generation of ATP; the net stoichiometry is about 2.5 ATP per NADH, and 1.5 ATP per $FADH_2$. Consequently, nine high-transfer-potential phosphoryl groups are generated when the electron-transport chain oxidizes 3 NADH molecules and 1 $FADH_2$ molecule, and one ATP is directly formed in one round of the citric acid cycle. Thus, one acetyl unit generates approximately 10 molecules of ATP. In dramatic contrast, the anaerobic glycolysis of an entire glucose molecule generates only 2 molecules of ATP (and 2 molecules of lactate).

Recall that molecular oxygen does not participate directly in the citric acid cycle. However, the cycle operates only under aerobic conditions because NAD^+ and FAD can be regenerated in the mitochondrion only by the transfer of electrons to molecular oxygen. *Glycolysis has both an aerobic and an anaerobic mode, whereas the citric acid cycle is strictly aerobic.* Glycolysis can proceed under anaerobic conditions because NAD^+ is regenerated in the conversion of pyruvate into lactate or ethanol.

17.3 Entry to the Citric Acid Cycle and Metabolism Through It Are Controlled

The citric acid cycle is the final common pathway for the aerobic oxidation of fuel molecules. Moreover, as we will see shortly (Section 17.4) and repeatedly elsewhere in our study of biochemistry, the cycle is an important source of building blocks for a host of important biomolecules. As befits its role as the metabolic hub of the cell, entry into the cycle and the rate of the cycle itself are controlled at several stages.

The pyruvate dehydrogenase complex is regulated allosterically and by reversible phosphorylation

As stated earlier, glucose can be formed from pyruvate (Section 16.3). *However, the formation of acetyl CoA from pyruvate is an irreversible step in animals and thus they are unable to convert acetyl CoA back into glucose.* The oxidative decarboxylation of pyruvate to acetyl CoA commits the carbon atoms of glucose to one of two principal fates: oxidation to CO_2 by the citric acid cycle, with the concomitant generation of energy, or incorporation into lipid (Figure 17.14). As expected of an enzyme at a critical branch point in metabolism, the activity of the pyruvate dehydrogenase complex is stringently controlled. High concentrations of reaction products inhibit the reaction: acetyl CoA inhibits the transacetylase component (E_2) by binding directly, whereas NADH inhibits the dihydrolipoyl dehydrogenase (E_3). High concentrations of NADH and acetyl CoA inform the enzyme that the energy needs of the cell have been met or that fatty acids are being degraded to produce acetyl CoA and NADH. In either case, there is no need to metabolize pyruvate to acetyl CoA. This inhibition has the effect of sparing glucose, because most pyruvate is derived from glucose by glycolysis (Section 16.1).

The key means of regulation of the complex in eukaryotes is covalent modification (Figure 17.15). *Phosphorylation of the pyruvate dehydrogenase component (E_1) by pyruvate dehydrogenase kinase (PDK) switches off the activity of the complex.* There are four isozymes of PDK that are expressed in a tissue-specific manner. *Deactivation is reversed by the pyruvate dehydrogenase phosphatase (PDP),* of which there are two isozymic forms. In mammals, the kinase and the phosphatase are associated with the E_2-E_3-BP

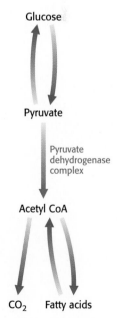

FIGURE 17.14 From glucose to acetyl CoA. The synthesis of acetyl CoA by the pyruvate dehydrogenase complex is a key irreversible step in the metabolism of glucose.

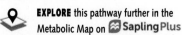

EXPLORE this pathway further in the Metabolic Map on **Sapling**Plus

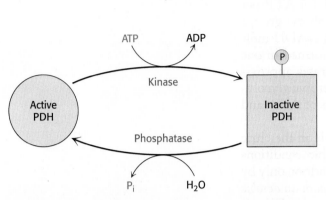

FIGURE 17.15 Regulation of the pyruvate dehydrogenase complex. A specific kinase phosphorylates and inactivates pyruvate dehydrogenase (PDH), and a phosphatase activates the dehydrogenase by removing the phosphoryl group. The kinase and the phosphatase also are highly regulated enzymes.

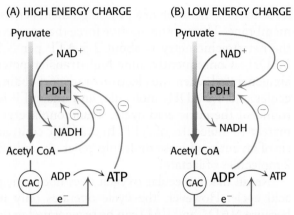

FIGURE 17.16 Response of the pyruvate dehydrogenase complex to the energy charge. The pyruvate dehydrogenase complex is regulated to respond to the energy charge of the cell. (A) The complex is inhibited by its immediate products, NADH and acetyl CoA, as well as by the ultimate product of cellular respiration, ATP. (B) The complex is activated by pyruvate and ADP, which inhibit the kinase that phosphorylates PDH.

complex, again highlighting the structural and mechanistic importance of this core. Both the kinase and the phosphatase are regulated. To see how this regulation works in biological conditions, consider muscle that is becoming active after a period of rest (Figure 17.16). At rest, the muscle will not have significant energy demands. Consequently, the $NADH/NAD^+$, acetyl CoA/ CoA, and ATP/ADP ratios will be high. These high ratios promote phosphorylation and inactivation of the complex by activating PDK. In other words, high concentrations of immediate (acetyl CoA and NADH) and ultimate (ATP) products inhibit the activity. Thus, *pyruvate dehydrogenase is switched off when the energy charge is high.*

As exercise begins, the concentrations of ADP and pyruvate will increase as muscle contraction consumes ATP and glucose is converted into pyruvate to meet the energy demands. Both ADP and pyruvate activate the dehydrogenase by inhibiting the kinase. Moreover, the phosphatase is stimulated by Ca^{2+}, the same signal that initiates muscle contraction. A rise in the cytoplasmic Ca^{2+} level (See Section 36.2 on 🔵 SaplingPlus) elevates the mitochondrial Ca^{2+} level. The rise in mitochondrial Ca^{2+} activates the phosphatase, enhancing pyruvate dehydrogenase activity.

In some tissues, the phosphatase is regulated by hormones. In liver, epinephrine binds to the α-adrenergic receptor to initiate the phosphatidylinositol pathway (Section 14.1), causing an increase in Ca^{2+} concentration that activates the phosphatase. In tissues capable of fatty acid synthesis, such as the liver and adipose tissue, insulin, the hormone that signifies the fed state, stimulates the phosphatase, increasing the conversion of pyruvate into acetyl CoA. Acetyl CoA is the precursor for fatty acid synthesis (Section 22.4). In these tissues, the pyruvate dehydrogenase complex is activated to funnel glucose to pyruvate and then to acetyl CoA and ultimately to fatty acids.

☤ In people with a pyruvate dehydrogenase phosphatase deficiency, pyruvate dehydrogenase is always phosphorylated and thus inactive. Consequently, glucose is processed to lactate rather than acetyl CoA. This condition results in unremitting lactic acidosis—high blood levels of lactic acid. In such an acidic environment, many tissues malfunction, most notably the central nervous system (problem 18). See Biochemistry in Focus

(p. 566) for a discussion of the role of the role of pyruvate dehydrogenase complex in diabetic neuropathy.

The citric acid cycle is controlled at several points

The rate of the citric acid cycle is precisely adjusted to meet an animal cell's needs for ATP (Figure 17.17). The primary control points are the allosteric enzymes isocitrate dehydrogenase and α-ketoglutarate dehydrogenase, the first two enzymes in the cycle to generate high-energy electrons.

The first control site is isocitrate dehydrogenase. The enzyme is allosterically stimulated by ADP, which enhances the enzyme's affinity for substrates. The binding of isocitrate, NAD^+, Mg^{2+}, and ADP is mutually cooperative. In contrast, ATP is inhibitory. The reaction product NADH also inhibits isocitrate dehydrogenase by directly displacing NAD^+. It is important to note that several steps in the cycle require NAD^+ or FAD, which are abundant only when the energy charge is low.

A second control site in the citric acid cycle is α-ketoglutarate dehydrogenase, which catalyzes the rate-limiting step in the citric acid cycle. Some aspects of this enzyme's control are like those of the pyruvate dehydrogenase complex, as might be expected from the homology of the two enzymes. α-Ketoglutarate dehydrogenase is inhibited by succinyl CoA and NADH, the products of the reaction that it catalyzes. In addition, α-ketoglutarate dehydrogenase is inhibited by a high energy charge. Thus, *the rate of the cycle is reduced when the cell has a high level of ATP*. α-Ketoglutarate dehydrogenase deficiency is observed in a number of neurological disorders, including Alzheimer's disease.

The use of isocitrate dehydrogenase and α-ketoglutarate dehydrogenase as control points integrates the citric acid cycle with other pathways and highlights the central role of the citric acid cycle in metabolism. For instance, the inhibition of isocitrate dehydrogenase leads to a buildup of citrate, because the interconversion of isocitrate and citrate is readily reversible under intracellular conditions. Citrate can be transported to the cytoplasm, where it signals phosphofructokinase to halt glycolysis (Section 16.2) and where it can serve as a source of acetyl CoA for fatty acid synthesis (Section 22.4). The α-ketoglutarate that accumulates when α-ketoglutarate dehydrogenase is inhibited can be used as a precursor for several amino acids and the purine bases (Chapter 23 and Chapter 25).

In many bacteria, the funneling of two-carbon fragments into the cycle also is controlled. *The synthesis of citrate from oxaloacetate and acetyl CoA carbon units is an important control point in these organisms.* ATP is an allosteric inhibitor of citrate synthase. The effect of ATP is to increase the value of K_M for acetyl CoA. Thus, as the level of ATP increases, less of this enzyme is bound to acetyl CoA and so less citrate is formed.

Defects in the citric acid cycle contribute to the development of cancer

Four enzymes crucial to cellular respiration are known to contribute to the development of cancer: succinate dehydrogenase, fumarase, pyruvate dehydrogenase kinase, and isocitrate dehydrogenase. Mutations that alter the activity of the first three of these enzymes enhance aerobic glycolysis (Section 16.2). In aerobic glycolysis, cancer cells preferentially metabolize glucose to lactate even in the presence of oxygen. Defects in these enzymes share a common biochemical link: the transcription factor *hypoxia inducible factor 1* (HIF-1).

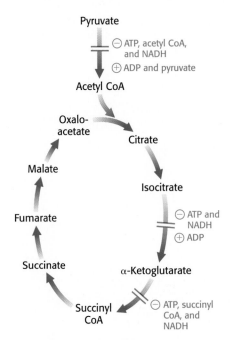

FIGURE 17.17 Control of the citric acid cycle. The citric acid cycle is regulated primarily by the concentration of ATP and NADH. The key control points are the enzymes isocitrate dehydrogenase and α-ketoglutarate dehydrogenase.

Normally, HIF-1 up-regulates the enzymes and transporters that enhance glycolysis only when oxygen concentration falls, a condition called hypoxia. Under normal conditions, HIF-1 is hydroxylated by prolyl hydroxylase 2 and is subsequently destroyed by the proteasome, a large complex of proteolytic enzymes (Section 23.2). The degradation of HIF-1 prevents the stimulation of glycolysis. Prolyl hydroxylase 2 requires α-ketoglutarate, ascorbate (Vitamin C), and oxygen for activity. Thus, when oxygen concentration falls, the prolyl hydroxylase 2 is inactive, HIF-1 is not hydroxylated and not degraded, and the synthesis of proteins required for glycolysis is stimulated. As a result, the rate of glycolysis is increased.

Recent research suggestions that defects in the enzymes of the citric acid cycle may significantly affect the regulation of prolyl hydroxylase 2. When either succinate dehydrogenase or fumarase is defective, succinate and fumarate accumulate in the mitochondria and spill over into the cytoplasm. Both succinate and fumarate are competitive inhibitors of prolyl hydroxylase 2. The inhibition of prolyl hydroxylase 2 results in the stabilization of HIF-1, since HIF-1 is no longer hydroxylated. Lactate, the end product of glycolysis, also appears to inhibit prolyl hydroxylase 2 by interfering with the action of ascorbate. In addition to increasing the amount of the proteins required for glycolysis, HIF-1 also stimulates the production of pyruvate dehydrogenase kinase (PDK). The kinase inhibits the pyruvate dehydrogenase complex, preventing the conversion of pyruvate into acetyl CoA. The pyruvate remains in the cytoplasm, further increasing the rate of aerobic glycolysis. Moreover, mutations in PDK that lead to enhanced activity contribute to increased aerobic glycolysis and the subsequent development of cancer. By enhancing glycolysis and increasing the concentration of lactate, the mutations in PDK result in the inhibition of hydroxylase and the stabilization of HIF-1.

Mutations in isocitrate dehydrogenase result in the generation of an oncogenic metabolite, 2-hydroxyglutarate. The mutant enzyme catalyzes the conversion of isocitrate to α-ketoglutarate, but then reduces α-ketoglutarate to form 2-hydroxyglutarate. 2-Hydroxyglutarate alters the methylation patterns in DNA (Section 33.3) and reduces dependence on growth factors for proliferation. These changes alter gene expression and promote unrestrained cell growth.

2-Hydroxyglutarate

An enzyme in lipid metabolism is hijacked to inhibit pyruvate dehydrogenase activity

A particularly fascinating example of the manipulation of cellular control by cancer cells comes from recent findings on the mitochondrial enzyme acetyl CoA acetyltransferase. Under normal circumstances, this enzyme synthesizes ketone bodies, such as acetoacetate, which are a fuel source for some tissues and an essential fuel source for all tissues under starvation conditions (Section 22.3)

$$2\,\text{Acetyl CoA} + H_2O \xrightarrow[\text{acetyltransferase}]{\text{Acetyl CoA}} \text{acetoacetate} + 2\,\text{CoA}$$

In certain cancers, acetyl CoA acetyltransferase is phosphorylated, which causes the enzyme to form active tetramers. Interestingly, the enzyme then acts as a protein acetyltransferase, adding acetyl groups to pyruvate dehydrogenase and pyruvate dehydrogenase phosphatase. The acetylation inhibits the two enzymes and facilitates the metabolic switch from oxidative phosphorylation to aerobic glycolysis, thereby enhancing the Warburg effect (Section 16.2). In essence, the enzyme is hijacked from its normal role in ketone body formation to further cancer growth, a remarkable occurrence. The appropriated enzyme may provide a unique therapeutic target to inhibit cancer progression.

These observations linking citric acid cycle enzymes to cancer suggest that cancer is also a metabolic disease, not simply a disease of mutant growth factors and cell cycle control proteins. The realization that there is a metabolic component to cancer opens the door to new thinking about the control of cancer. Indeed, preliminary experiments suggest that if cancer cells undergoing aerobic glycolysis are forced by pharmacological manipulation to use oxidative phosphorylation, the cancer cells lose their malignant properties. It is also interesting to note that the citric acid cycle, which has been studied for decades, still has secrets to be revealed by future biochemists.

17.4 The Citric Acid Cycle Is a Source of Biosynthetic Precursors

Thus far, discussion has focused on the citric acid cycle as the *major degradative pathway for the generation of ATP.* As a major metabolic hub of the cell, the citric acid cycle also integrates many of the cell's other metabolic pathways, including those of carbohydrates, fats, amino acids, and porphyrins. The cytoplasmic and mitochondrial pools of citric acid cycle components are interchangeable. *This integration allows the use of citric acid intermediates for biosyntheses* (Figure 17.18). For example, most of the carbon atoms in porphyrins come from *succinyl CoA* in a pathway that occurs both in the cytoplasm and mitochondria. Fats are synthesized in the cytoplasm from mitochondrial citrate. Many of the amino acids, used throughout the cell, are derived from α-*ketoglutarate* and *oxaloacetate.* These biosynthetic processes will be considered in subsequent chapters.

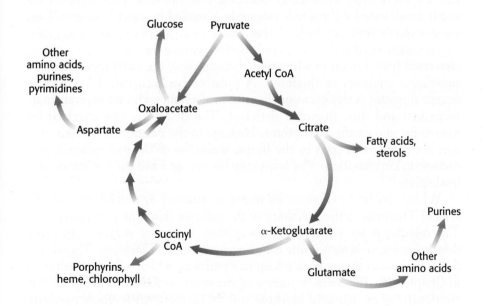

FIGURE 17.18 Biosynthetic roles of the citric acid cycle. Intermediates are drawn off for biosyntheses (shown by red arrows) when the energy needs of the cell are met. Intermediates are replenished by the formation of oxaloacetate from pyruvate.

The citric acid cycle must be capable of being rapidly replenished

An important consideration is that *citric acid cycle intermediates must be replenished if any are drawn off for biosyntheses.* Suppose that much oxaloacetate is converted into amino acids for protein synthesis and, subsequently,

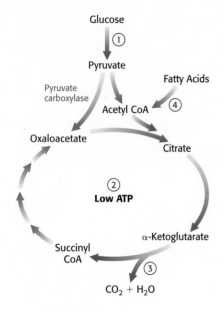

Active pathways

① Glycolysis, Ch. 16
② Citric acid cycle, Ch. 17
③ Oxidative phosphorylation, Ch. 18
④ Fatty acid oxidation, Ch. 22

FIGURE 17.19 PATHWAY INTEGRATION Pathway integration: Metabolic response to exercise after a night's rest. The rate of the citric acid cycle increases during exercise, requiring the replenishment of oxaloacetate and acetyl CoA. Oxaloacetate is replenished by its formation from pyruvate. Acetyl CoA may be produced from the metabolism of both pyruvate and fatty acids.

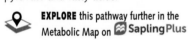 **EXPLORE** this pathway further in the Metabolic Map on 📖 Sapling Plus

Beriberi

A vitamin-deficiency disease first described in 1630 by Jacob Bonitus, a Dutch physician working in Java:

"A certain very troublesome affliction, which attacks men, is called by the inhabitants Beriberi (which means sheep). I believe those, whom this same disease attacks, with their knees shaking and the legs raised up, walk like sheep. It is a kind of paralysis, or rather Tremor: for it penetrates the motion and sensation of the hands and feet indeed sometimes of the whole body."

the energy needs of the cell rise. The citric acid cycle will operate to a reduced extent unless new oxaloacetate is formed because acetyl CoA cannot enter the cycle unless it condenses with oxaloacetate. Even though oxaloacetate is recycled, a minimal level must be maintained to allow the cycle to function.

How is oxaloacetate replenished? Mammals lack the enzymes for the net conversion of acetyl CoA into oxaloacetate or any other citric acid cycle intermediate. Rather, oxaloacetate is formed by the carboxylation of pyruvate, in a reaction catalyzed by the biotin-dependent enzyme *pyruvate carboxylase* (Figure 17.19).

$$\text{Pyruvate} + CO_2 + ATP + H_2O \longrightarrow \text{oxaloacetate} + ADP + P_i + 2H^+$$

Recall that this enzyme plays a crucial role in gluconeogenesis (Section 16.3). It is active only in the presence of acetyl CoA, which signifies the need for more oxaloacetate. If the energy charge is high, oxaloacetate is converted into glucose. If the energy charge is low, oxaloacetate replenishes the citric acid cycle. The synthesis of oxaloacetate by the carboxylation of pyruvate is an example of an *anaplerotic reaction* (*anaplerotic* is of Greek origin, meaning to "fill up"), a reaction that leads to the net synthesis, or replenishment, of pathway components. Note that because the citric acid cycle is a cycle, it can be replenished by the generation of any of the intermediates. Glutamine is an especially important source of citric acid cycle intermediates in rapidly growing cells, including cancer cells. Glutamine is converted into glutamate and then into α-ketoglutarate (Section 23.5).

The disruption of pyruvate metabolism is the cause of beriberi and poisoning by mercury and arsenic

Beriberi, a neurologic and cardiovascular disorder, is caused by a dietary deficiency of thiamine (also called *vitamin B₁*). The disease has been and continues to be a serious health problem in the Far East because rice, the major food, has a rather low content of thiamine. This deficiency is partly ameliorated if the whole rice grain is soaked in water before milling; some of the thiamine in the husk then leaches into the rice kernel. The problem is exacerbated if the rice is polished to remove the outer layer (that is, converted from brown to white rice), because only the outer layer contains significant amounts of thiamine. A form of beriberi, called Wernicke's encephalopathy, is also occasionally seen in alcoholics who are severely malnourished and thus thiamine deficient. The disease is characterized by neurologic and cardiac symptoms. Damage to the peripheral nervous system is expressed as pain in the limbs, weakness of the musculature, and distorted skin sensation. The heart may be enlarged and the cardiac output inadequate.

Which biochemical processes might be affected by a deficiency of thiamine? Thiamine is the precursor of the cofactor thiamine pyrophosphate. *This cofactor is the prosthetic group of three important enzymes: pyruvate dehydrogenase, α-ketoglutarate dehydrogenase, and transketolase.* Transketolase functions in the pentose phosphate pathway, which will be considered in Chapter 20. *The common feature of enzymatic reactions utilizing TPP is the transfer of an activated aldehyde unit.* In beriberi, the levels of pyruvate and α-ketoglutarate in the blood are higher than normal. The increase in the level of pyruvate in the blood is especially pronounced after the ingestion of glucose. A related finding is that the activities of the pyruvate and α-ketoglutarate dehydrogenase complexes in vivo are abnormally low.

The low transketolase activity of red blood cells in beriberi is an easily measured and reliable diagnostic indicator of the disease.

Why does TPP deficiency lead primarily to neurological disorders? The nervous system relies essentially on glucose as its only fuel. The product of glycolysis, pyruvate, can enter the citric acid cycle only through the pyruvate dehydrogenase complex. With that enzyme deactivated, the nervous system has no source of fuel. In contrast, most other tissues can use fats as a source of fuel for the citric acid cycle.

Symptoms similar to those of beriberi appear in organisms exposed to mercury or arsenite (AsO_3^{3-}). Both materials have a high affinity for neighboring (vicinal) sulfhydryls, such as those in the reduced dihydrolipoyl groups of the E_3 component of the pyruvate dehydrogenase complex (Figure 17.20). The binding of mercury or arsenite to the dihydrolipoyl groups inhibits the complex and leads to central nervous system pathologies. The proverbial phrase "mad as a hatter" refers to the strange behavior of poisoned hatmakers who used mercury nitrate to soften and shape animal furs. This form of mercury is absorbed through the skin. Similar symptoms afflicted the early photographers, who used vaporized mercury to create daguerreotypes.

2,3-Dimercaptopropanol (BAL)

Excreted

Dihydrolipoamide from pyruvate dehydrogenase component E₃ **Arsenite** $2 \ H_2O$ **Arsenite chelate on enzyme**   **Restored enzyme**

FIGURE 17.20 Arsenite poisoning. Arsenite inhibits the pyruvate dehydrogenase complex by inactivating the dihydrolipoamide component of the transacetylase. Some sulfhydryl reagents, such as 2,3-dimercaptoethanol, relieve the inhibition by forming a complex with the arsenite that can be excreted.

Treatment for these poisons is the administration of sulfhydryl reagents with adjacent sulfhydryl groups to compete with the dihydrolipoyl residues for binding with the metal ion. The reagent–metal complex is then excreted in the urine. Indeed, 2,3-dimercaptopropanol (Figure 17.20) was developed after World War I as an antidote to lewisite, an arsenic-based chemical weapon. This compound was initially called BAL, for British anti-lewisite.

The citric acid cycle may have evolved from preexisting pathways

How did the citric acid cycle come into being? Although definitive answers are elusive, informed speculation is possible. We can perhaps begin to comprehend how evolution might work at the level of biochemical pathways.

The citric acid cycle was most likely assembled from preexisting reaction pathways. As noted earlier, many of the intermediates formed in the citric

The manuscript proposing the citric acid cycle was submitted for publication to *Nature* but was rejected in June 1937. That same year it was published in *Enzymologia*. Dr. Krebs proudly displayed the rejection letter throughout his career as encouragement for young scientists:

"The editor of NATURE presents his compliments to Dr. H. A. Krebs and regrets that as he has already sufficient letters to fill the correspondence columns of NATURE for seven or eight weeks, it is undesirable to accept further letters at the present time on account of the time delay which must occur in their publication.

If Dr. Krebs does not mind much delay the editor is prepared to keep the letter until the congestion is relieved in the hope of making use of it.

He returns it now, in case Dr. Krebs prefers to submit it for early publication to another periodical."

acid cycle are used in metabolic pathways for amino acids and porphyrins. Thus, compounds such as pyruvate, α-ketoglutarate, and oxaloacetate were likely present early in evolution for biosynthetic purposes. The oxidative decarboxylation of these α-ketoacids is quite favorable thermodynamically and can be used to drive the synthesis of both acyl CoA derivatives and NADH. These reactions almost certainly formed the core of processes that preceded the citric acid cycle evolutionarily. Interestingly, α-ketoglutarate and oxaloacetate can be interconverted by transamination of the respective amino acids by aspartate aminotransferase, another key biosynthetic enzyme. Thus, cycles comprising smaller numbers of intermediates used for a variety of biochemical purposes could have existed before the present form evolved.

17.5 The Glyoxylate Cycle Enables Plants and Bacteria to Grow on Acetate

Acetyl CoA that enters the citric acid cycle has but one fate: oxidation to CO_2 and H_2O. Most organisms thus cannot convert acetyl CoA into glucose. Although oxaloacetate, a key precursor to glucose, is formed in the citric acid cycle, the two decarboxylations that take place before the regeneration of oxaloacetate preclude the *net* conversion of acetyl CoA into glucose.

In plants and in some microorganisms, there is a metabolic pathway that allows the conversion of acetyl CoA generated from fat stores into glucose. This reaction sequence, called the *glyoxylate cycle*, is similar to the citric acid cycle but bypasses the two decarboxylation steps of the cycle. Another important difference is that two molecules of acetyl CoA enter per turn of the glyoxylate cycle, compared with one in the citric acid cycle.

The glyoxylate cycle (Figure 17.21), like the citric acid cycle, begins with the condensation of acetyl CoA and oxaloacetate to form citrate, which is then isomerized to isocitrate. Instead of being decarboxylated, as in the citric acid cycle, isocitrate is cleaved by *isocitrate lyase* into succinate and glyoxylate. The ensuing steps regenerate oxaloacetate from glyoxylate. First, acetyl CoA condenses with glyoxylate to form malate in a reaction catalyzed by *malate synthase,* and then malate is oxidized to oxaloacetate, as in the citric acid cycle. The sum of these reactions is

$$2\,\text{Acetyl CoA} + \text{NAD}^+ + 2\,H_2O \longrightarrow$$

$$\text{succinate} + 2\,\text{CoA} + \text{NADH} + 2\,H^+$$

In plants, these reactions take place in organelles called *glyoxysomes*. This cycle is especially prominent in oil-rich seeds, such as those from sunflowers, cucumbers, and castor beans. Succinate, released midcycle, can be converted into carbohydrates by a combination of the citric acid cycle and gluconeogenesis. The carbohydrates power seedling growth until the cell can begin photosynthesis. Thus, organisms with the glyoxylate cycle gain metabolic versatility because they can use acetyl CoA as a precursor of glucose and other biomolecules. (See Biochemistry in Focus 2 [p. 567] for a discussion of the role of the glyoxylate cycle in tuberculosis.)

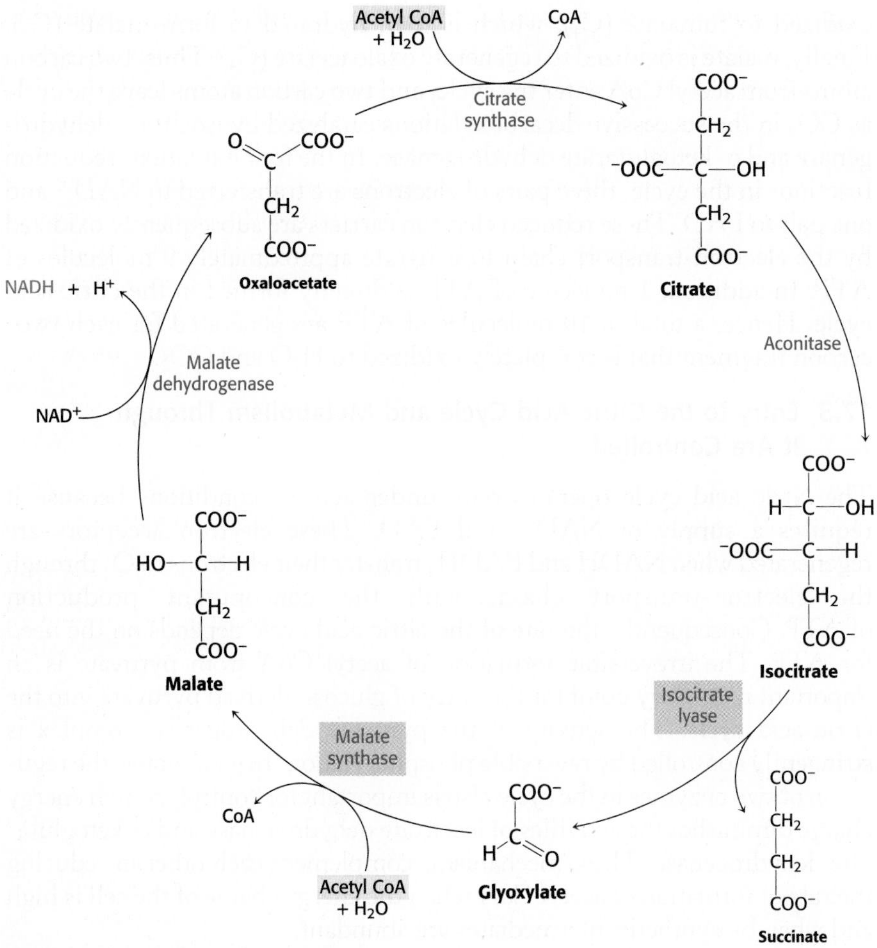

FIGURE 17.21 The glyoxylate cycle. The glyoxylate cycle allows plants and some microorganisms to grow on acetate because the cycle bypasses the decarboxylation steps of the citric acid cycle. The reactions of this cycle are the same as those of the citric acid cycle except for the ones catalyzed by isocitrate lyase and malate synthase, which are boxed in blue.

SUMMARY

The citric acid cycle is the final common pathway for the oxidation of fuel molecules. It also serves as a source of building blocks for biosyntheses.

17.1 The Pyruvate Dehydrogenase Complex Links Glycolysis to the Citric Acid Cycle

Most fuel molecules enter the cycle as acetyl CoA. The link between glycolysis and the citric acid cycle is the oxidative decarboxylation of pyruvate to form acetyl CoA. In eukaryotes, this reaction and those of the cycle take place inside mitochondria, in contrast with glycolysis, which takes place in the cytoplasm.

17.2 The Citric Acid Cycle Oxidizes Two-Carbon Units

The cycle starts with the condensation of oxaloacetate (C_4) and the acetyl unit (C_2) of acetyl CoA to give citrate (C_6), which is isomerized to isocitrate (C_6). Oxidative decarboxylation of this intermediate gives α-ketoglutarate (C_5). The second molecule of carbon dioxide comes off in the next reaction, in which α-ketoglutarate is oxidatively decarboxylated to succinyl CoA (C_4). The thioester bond of succinyl CoA is cleaved by orthophosphate to yield succinate, and an ATP is concomitantly generated. Succinate is oxidized to fumarate (C_4), which is then hydrated to form malate (C_4). Finally, malate is oxidized to regenerate oxaloacetate (C_4). Thus, two carbon atoms from acetyl CoA enter the cycle, and two carbon atoms leave the cycle as CO_2 in the successive decarboxylations catalyzed by isocitrate dehydrogenase and α-ketoglutarate dehydrogenase. In the four oxidation–reduction reactions in the cycle, three pairs of electrons are transferred to NAD^+ and one pair to FAD. These reduced electron carriers are subsequently oxidized by the electron-transport chain to generate approximately 9 molecules of ATP. In addition, 1 molecule of ATP is directly formed in the citric acid cycle. Hence, a total of 10 molecules of ATP are generated for each two-carbon fragment that is completely oxidized to H_2O and CO_2.

17.3 Entry to the Citric Acid Cycle and Metabolism Through It Are Controlled

The citric acid cycle operates only under aerobic conditions because it requires a supply of NAD^+ and FAD. These electron acceptors are regenerated when NADH and $FADH_2$ transfer their electrons to O_2 through the electron-transport chain, with the concomitant production of ATP. Consequently, the rate of the citric acid cycle depends on the need for ATP. The irreversible formation of acetyl CoA from pyruvate is an important regulatory point for the entry of glucose-derived pyruvate into the citric acid cycle. The activity of the pyruvate dehydrogenase complex is stringently controlled by reversible phosphorylation. In eukaryotes, the regulation of two enzymes in the cycle also is important for control. A high energy charge diminishes the activities of isocitrate dehydrogenase and α-ketoglutarate dehydrogenase. These mechanisms complement each other in reducing the rate of formation of acetyl CoA when the energy charge of the cell is high and when biosynthetic intermediates are abundant.

17.4 The Citric Acid Cycle Is a Source of Biosynthetic Precursors

When the cell has adequate energy available, the citric acid cycle can also provide a source of building blocks for a host of important biomolecules, such as nucleotide bases, proteins, and heme groups. This use depletes the cycle of intermediates. When the cycle again needs to metabolize fuel, anaplerotic reactions replenish the cycle intermediates.

17.5 The Glyoxylate Cycle Enables Plants and Bacteria to Grow on Acetate

The glyoxylate cycle enhances the metabolic versatility of many plants and bacteria. This cycle, which uses some of the reactions of the citric acid cycle, enables these organisms to subsist on acetate because it bypasses the two decarboxylation steps of the citric acid cycle.

APPENDIX Biochemistry in Focus

Biochemistry in Focus 1

Diabetic neuropathy may result from inhibition of the pyruvate dehydrogenase complex.

Diabetic neuropathy (DN), a numbness, tingling, or pain in the hands, arm, fingers, toes, feet and legs, is a common complication of both type 1 and type 2 diabetes, affecting approximately 50% of patients. There is no cure for the condition, and treatment relies on painkillers. Recent research in mice and tissue culture suggests that overproduction of lactic acid by cells in the dorsal root ganglion, a part of the nervous system responsible for pain perception, may be a significant contributor to DN. Lactate, produced by some cells of the nervous system, is a common fuel for nerve cells. Neurons import the lactate and convert it to pyruvate for use in cellular respiration. However, a high concentration of lactate is problematic. What is the cause of the increase of lactic acid? Much research needs to be done, but it appears that hyperglycemia (high glucose concentration), the defining feature of diabetes, increases pyruvate dehydrogenase kinase activity in the cells of the dorsal root ganglion. This increase in kinase activity, in turn, leads to the phosphorylation and inhibition of pyruvate dehydrogenase complex. Glycolytically produced pyruvate is them processed to lactate. The overabundance of lactate leads to an increase in acid-sensing nociceptors (pain receptors), a type of G-protein-coupled receptor (Section 14.1), in the dorsal root ganglion. The increased nociceptors result in DN.

Support for the role of lactate in DN comes from a number of experiments. In experimental systems, the lactic acid accumulation can be prevented. Pharmacological inhibition of pyruvate dehydrogenase kinase or lactic acid dehydrogenase or genetically eliminating the kinase gene greatly reduces but does not eliminate DN. The identification of the hyperglycemia-pyruvate dehydrogenase kinase-lactic acid dehydrogenase axis in the dorsal root ganglion as a cause of DN opens the possibility of therapeutic intervention to prevent or ameliorate DN.

Biochemistry in Focus 2

New treatments for tuberculosis may be on the horizon.

Tuberculosis (TB) is one of the leading causes of death throughout the world. In 2016, according to the World Health Organization, 10.4 million people contracted the disease and 1.7 million died, including a quarter of a million children. TB is also the leading cause of death in HIV-positive individuals.

The bacterium responsible, *Mycobacterium tuberculosis*, is transmitted by people with active lung infections by coughing and sneezing. A common treatment for TB is the antibiotic rifampicin, an inhibitor of bacterial RNA synthesis (Chapter 32). However, strains of *M. tuberculosis* are developing resistance to rifampicin, so new treatments are urgently needed.

Some exciting new research suggests a possible treatment that relies on the dependence of the bacteria on the glyoxylate cycle. This dependence is most acute when the bacteria are in a latent state in the lungs (problem 42). Recall that the glyoxylate cycle allows the conversion of fats to glucose. A key enzyme in the cycle is isocitrate lyase, which cleaves isocitrate into glyoxylate and succinate.

In the process of catalysis, a reactive thiolate is formed on Cys_{191} at the active site. The subscript 191 indicates the position of the cysteine in the primary structure.

$$Cys_{191} - S^-$$
Thiolate

A suicide inhibitor or mechanism-based inhibitor (Section 8.2) for the lyase, 2-vinyl-isocitrate, has been synthesized that may lead to an effective treatment for tuberculosis.

2-Vinyl-isocitrate

When the lyase reacts with the inhibitor, succinate is released as in the normal reaction, but a thioether-linked homopyruvyl moiety is covalently linked to Cys_{191}, inhibiting the enzyme.

Homopyruvoyl-modified isocitrate lyase

This may be an important opening into TB therapy because the active site Cys_{191} is conserved in all strains of *M. tuberculosis*,

Isocitrate

Gloxylate

Succinate

which would reduce the likelihood of the development of resistance to the treatment. This research is quite new (2017), but it will be interesting to see if this ultimately leads to an effective medication for TB.

APPENDIX

Problem-Solving Strategies

Problem-solving strategies 1

PROBLEM: Shock is a life-threatening condition that results from a lack of adequate blood flow. Hypovolemic or hemorrhagic shock, the most common variety, is caused by excessive bleeding. Patients in shock often suffer from lactic acidosis owing to a deficiency of O_2.

(a) Why does a lack of blood flow lead to a deficiency of O_2?

(b) Why does a lack of O_2 lead to lactic acid accumulation?

(c) One treatment for shock is to administer dichloroacetate (DCA), which inhibits the kinase associated with the pyruvate dehydrogenase complex. What is the biochemical rationale for this treatment?

SOLUTION: Let's attack (a) by asking another question:

▶ **What is one of the most fundament roles of blood circulation?**

Thinking back to Chapter 7, the blood protein hemoglobin carries O_2 from the lungs to the tissues. A decrease in blood volume would mean less hemoglobin, and thus less O_2 for the tissues. That answers (a). Now, on to part (b).

▶ **What is the role of O_2 in tissues?**

O_2 is required for cellular respiration, which generates 90% of the ATP that a typical tissue requires. We will cover this process in detail in Chapter 18.

▶ **If O_2 is deficient, what is the only means to generate ATP?**

Lactic acid fermentation, as we learned in Chapter 16. So, the answer to (b) is that the tissues of a person in shock resort to lactic acid fermentation in an attempt to generate sufficient ATP in the absence of O_2. Now, let's take on part (c). To answer (c), we need to know the role of the pyruvate dehydrogenase kinase.

▶ **What is the function of the kinase associated with the pyruvate dehydrogenase complex?**

The kinase phosphorylates and inhibits the complex. The kinase, in conjunction with the pyruvate dehydrogenase phosphatase, are the key regulators of the complex.

▶ **What would be the effect of inhibiting the kinase?**

The kinase could not inhibit the pyruvate dehydrogenase complex. Ideally, the complex would then process more pyruvate by cellular respiration, alleviating the potential dangerous lactic acid build up. In fact, most research suggests that treating patients in shock with dichloroacetate is of negligible value.

Problem-solving strategies 2

PROBLEM: Fluoroacetate is a toxic molecule that inhibits the citric acid cycle. When fluoroacetate is added to mitochondria, fluorocitrate builds up.

(a) What step in the citric acid cycle is inhibited by fluoroacetate treatment?

(b) What would be the effect of the inhibition on other intermediates in the citric acid cycle?

SOLUTION: Let's again break this question down into a series of smaller questions, beginning with a simple, but crucial question.

▶ **Fluoroacetate is an analog of what molecule?**

Acetate   **Fluoroacetate**

If we look at the structure of fluoroacetate, it looks like an acetate with fluorine replacing a hydrogen. The fact that fluorocitrate is formed tells us something about the immediate metabolic fate of fluoroacetate. As a hint, think about how acetyl groups enter the citric acid cycle.

▶ **How is it possible for fluoroacetate to enter the citric acid cycle?**

It must first be converted to fluoroacetyl CoA. Good. Now,

▶ **How do you think fluorocitrate can be formed?**

Fluoroacetyl CoA must react with oxaloacetate to form fluorocitrate.
Thinking back to the citric acid cycle,

▶ **What is the normal fate of citrate formed by the condensation of acetyl CoA with oxaloacetate?**

Citrate is converted into isocitrate by the enzyme aconitase. Now, to answer part (a), given what we just remembered:

▶ **What step in the citric acid cycle in inhibited by fluorocitrate?**

The aconitase reaction. It appears the fluorocitrate is not a substrate for aconitase, so it accumulates. Good job.

On to part (b). If the cycle is blocked at the aconitase reaction:

▶ **What would be the effect on the reactions following the aconitase-catalyzed step?**

All the reactions would continue, beginning with the processing of all the isocitrate formed before the introduction of the fluoroacetate. However, no new isocitrate could be generated because of the inhibition of aconitase.

▶ **What citric acid cycle intermediate would be regenerated?**

All the intermediates after the aconitase reaction would be metabolized to oxaloacetate.

▶ **What would be the fate of oxaloacetate if there were sufficient fluoroacetate present?**

It would react with fluoroacetyl CoA until all citric acid cycle components were converted into fluorocitrate. The citric acid cycle would halt, a situation most definitely not compatible with life.

KEY TERMS

citric acid (tricarboxylic acid, TCA; Krebs) cycle (p. 541)

acetyl CoA (p. 541)

oxidative phosphorylation (p. 543)

pyruvate dehydrogenase complex (p. 543)

flavoprotein (p. 546)

citrate synthase (p. 549)

iron–sulfur (nonheme iron) protein (p. 551)

isocitrate dehydrogenase (p. 551)

α-ketoglutarate dehydrogenase (p. 552)

anaplerotic reaction (p. 562)

beriberi (p. 562)

glyoxylate cycle (p. 564)

isocitrate lyase (p. 564)

malate synthase (p. 564)

glyoxysome (p. 564)

PROBLEMS

1. *Essential partners.* Which of the following catalytic cofactors are required by the pyruvate dehydrogenase complex.

(a) NAD^+, lipoic acid, thiamine pyrophosphate

(b) NAD^+, biotin, thiamine pyrophosphate

(c) NAD^+, FAD, lipoic acid

(d) FAD, lipoic acid, thiamine pyrophosphate

2. *Like a sore thumb.* Which of the following is not a component of the citric acid cycle

(a) Succinyl CoA

(b) Propionyl CoA

(c) Malate

(d) Oxalacetate

3. *It is an oxidation after all.* The complete oxidation of one molecule of acetyl CoA by the citric acid cycle results in

(a) Production of one molecule of citrate

(b) Generation of two molecules of ATP

(c) Consumption of one molecule of oxaloacetate

(d) Generation of two molecules of CO_2

4. *An uncanny resemblance.* The pyruvate dehydrogenase complex is mechanistically and structurally similar to

(a) Isocitrate dehydrogenase

(b) α-Ketoglutarate dehydrogenase

(c) Glyceraldehyde 3-phosphate dehydrogenase

(d) Fumarase

5. *Global warming?* Which of the following enzymes catalyzes a reaction that does not generate a molecule of CO_2?

(a) Malate dehydrogenase

(b) Isocitrate dehydrogenase

(c) Pyruvate dehydrogenase

(d) α-ketoglutarate dehydrogenase

6. *A one-way link.* Write the reaction that links glycolysis and the citric acid cycle. What is the enzyme that catalyzes the reaction? ✓❶

7. *Naming names.* What are the five enzymes (including regulatory enzymes) that constitute the pyruvate dehydrogenase complex? Which reactions do they catalyze? ✓❶

8. *Coenzymes.* What coenzymes are required by the pyruvate dehydrogenase complex? What are their roles? ✓❶

9. *More coenzymes.* Distinguish between catalytic coenzymes and stoichiometric coenzymes in the pyruvate dehydrogenase complex.

10. *Waste and fraud?* Figure 17.8 shows the steps in the pyruvate dehydrogenase complex reaction cycle. A key product, acetyl CoA, is released after the fourth step. What is the purpose of the remaining steps? ✓❶

11. *Joined at the hip.* List some of the advantages of organizing the enzymes that catalyze the formation of acetyl CoA from pyruvate into a single large complex. ✓❶

12. *Flow of carbon atoms.* What is the fate of the radioactive label when each of the following compounds is added to a cell extract containing the enzymes and cofactors of the glycolytic pathway, the citric acid cycle, and the pyruvate dehydrogenase complex? (The ^{14}C label is printed in red.) ✓❶

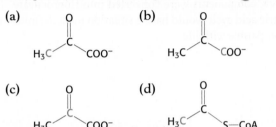

(e) Glucose 6-phosphate labeled at C-1.

13. $C_2 + C_2 \longrightarrow C_4$ ✓❼

(a) Which enzymes are required to get *net synthesis* of oxaloacetate from acetyl CoA?

(b) Write a balanced equation for the net synthesis.

(c) Do mammalian cells contain the requisite enzymes?

14. *Driving force.* What is the $\Delta G°'$ for the complete oxidation of the acetyl unit of acetyl CoA by the citric acid cycle?

15. *Acting catalytically.* The citric acid cycle itself, which is composed of enzymatically catalyzed steps, can be thought of essentially as a supramolecular enzyme. Explain. ✓❹

16. *A potent inhibitor.* Thiamine thiazolone pyrophosphate binds to pyruvate dehydrogenase about 20,000 times as strongly as does thiamine pyrophosphate, and it competitively inhibits the enzyme. Why? ✓❷

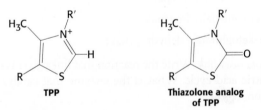

17. *Like Holmes and Watson.* Match each term with its description.

(a) Acetyl CoA _____

(b) Citric acid cycle _____

(c) Pyruvate dehydrogenase complex _____

(d) Thiamine pyrophosphate _____

1. Catalyzes the link between glycolysis and the citric acid cycle

2. Coenzyme required by transacetylase

3. Final product of pyruvate dehydrogenase

4. Catalyzes the formation of acetyl CoA

(e) Lipoic acid _____

(f) Pyruvate dehydrogenase _____

(g) Acetyllipoamide _____

(h) Dihydrolipoyl transacetylase _____

(i) Dihydrolipoyl dehydrogenase _____

(j) Beriberi _____

5. Regenerates active transacetylase

6. Fuel for the citric acid cycle

7. Coenzyme required by pyruvate dehydrogenase

8. Catalyzes the oxidative decarboxylation of pyruvate

9. Due to a deficiency of thiamine

10. Central metabolic hub

18. *Lactic acidosis.* Patients with pyruvate dehydrogenase deficiency show high levels of lactic acid in the blood. However, in some cases, treatment with dichloroacetate (DCA), which inhibits the kinase associated with the pyruvate dehydrogenase complex, lowers lactic acid levels. ✓❷

(a) How does DCA act to stimulate pyruvate dehydrogenase activity?

(b) What does this suggest about pyruvate dehydrogenase activity in patients who respond to DCA?

19. *Energy rich.* What are the thioesters in the reaction catalyzed by PDH complex?

20. *Alternative fates.* Compare the regulation of the pyruvate dehydrogenase complex in muscle and in liver. ✓❷

21. *Mutations.* ✓❷

(a) Predict the effect of a mutation that enhances the activity of the kinase associated with the PDH complex.

(b) Predict the effect of a mutation that reduces the activity of the phosphatase associated with the PDH complex.

22. *Flaking paint, green wallpaper.* Clare Boothe Luce, ambassador to Italy in the 1950s (and Connecticut congressperson, playwright, editor of *Vanity Fair,* and the wife of Henry Luce, founder of *Time* magazine and *Sports Illustrated*) became ill when she was staying at the ambassadorial residence in Rome. The paint on the dining room ceiling, an arsenic-based paint, was flaking; the wallpaper of her bedroom in the ambassadorial residence was colored a mellow green owing to the presence of cupric arsenite in the pigment. Suggest a possible cause of Ambassador Luce's illness. ✓❷, ✓❺

23. *Like Jack and Jill.* Match each enzyme with its description.

(a) Pyruvate dehydrogenase complex _____

(b) Citrate synthase _____

(c) Aconitase _____

(d) Isocitrate dehydrogenase _____

1. Catalyzes the formation of isocitrate

2. Synthesizes succinyl CoA

3. Generates malate

4. Generates ATP

(e) α-Ketoglutarate dehydrogenase _____

(f) Succinyl CoA synthetase
_____

(g) Succinate dehydrogenase
_____

(h) Fumarase
_____

(i) Malate dehydrogenase
_____

(j) Pyruvate carboxylase
_____

5. Converts pyruvate into
acetyl CoA

6. Converts pyruvate into
oxaloacetate

7. Condenses oxaloacetate
and acetyl CoA

8. Catalyzes the formation
of oxaloacetate

9. Synthesizes fumarate

10. Catalyzes the formation
of α-ketoglutarate

24. *A hoax, perhaps?* The citric acid cycle is part of aerobic respiration, but no O_2 is required for the cycle. Explain this paradox. ✓⑤

25. *One of a kind.* How is succinate dehydrogenase unique compared with the other enzymes in the citric acid cycle?

26. *Coupling reactions.* The oxidation of malate by NAD^+ to form oxaloacetate is a highly endergonic reaction under standard conditions $[\Delta G^{\circ\prime} = 29 \text{ kJ mol}^{-1} (7 \text{ kcal mol}^{-1})]$. The reaction proceeds readily under physiological conditions. ✓④

(a) Explain why the reaction proceeds as written under physiological conditions.

(b) Assuming an $[NAD^+]/[NADH]$ ratio of 8 and a pH of 7, what is the lowest [malate]/[oxaloacetate] ratio at which oxaloacetate can be formed from malate?

27. *Synthesizing α-ketoglutarate.* It is possible, with the use of the reactions and enzymes considered in this chapter, to convert pyruvate into α-ketoglutarate without depleting any of the citric acid cycle components. Write a balanced reaction scheme for this conversion, showing cofactors and identifying the required enzymes. ✓⑥

28. *Seven o'clock roadblock.* Malonate is a competitive inhibitor of succinate dehydrogenase. How will the concentrations of citric acid cycle intermediates change immediately after the addition of malonate? Why is malonate not a substrate for succinate dehydrogenase?

$$
\begin{array}{c}
\text{COO}^- \\
| \\
\text{CH}_2 \\
| \\
\text{COO}^-
\end{array}
$$

Malonate

29. *No signal, no activity.* Why is acetyl CoA an especially appropriate activator for pyruvate carboxylase? ✓②, ✓⑤

30. *Power differentials.* As we will see in the next chapter, when NADH reacts with oxygen, 2.5 ATP are generated. When $FADH_2$ reduces oxygen, only 1.5 ATP are generated. Why then does succinate dehydrogenase produce $FADH_2$ and not NADH when succinate is reduced to fumarate?

31. *Back to Orgo (or do you say O chem?).* Before any oxidation can occur in the citric acid cycle, citrate must be isomerized into isocitrate. Why is this the case?

32. *A nod is as good as a wink to a blind horse.* Explain why a GTP molecule, or another nucleoside triphosphate, is energetically equivalent to an ATP molecule in metabolism.

33. *One from two.* The synthesis of citrate from acetyl CoA and oxaloacetate is a biosynthetic reaction. What is the energy source that drives formation of citrate? ✓④

34. *Versatility.* What is the chief benefit of being able to perform the glyoxylate cycle? ✓⑦

Chapter Integration Problems

35. *Fats into glucose?* Fats are usually metabolized into acetyl CoA and then further processed through the citric acid cycle. In Chapter 16, we saw that glucose can be synthesized from oxaloacetate, a citric acid cycle intermediate. Why, then, after a long bout of exercise depletes our carbohydrate stores, do we need to replenish those stores by eating carbohydrates? Why do we not simply replace them by converting fats into carbohydrates?

36. *Alternative fuels.* As we will see (Chapter 22), fatty acid breakdown generates a large amount of acetyl CoA. What will be the effect of fatty acid breakdown on pyruvate dehydrogenase complex activity? On glycolysis? ✓②

Mechanism Problems

37. *Theme and variation.* Propose a reaction mechanism for the condensation of acetyl CoA and glyoxylate in the glyoxylate cycle of plants and bacteria.

38. *Symmetry problems.* In experiments carried out in 1941 to investigate the citric acid cycle, oxaloacetate labeled with ^{14}C in the carboxyl carbon atom farthest from the keto group was introduced to an active preparation of mitochondria.

$$
\begin{array}{c}
\text{O} \diagdown \quad \diagup \text{COO}^- \\
\text{C} \\
| \\
\text{CH}_2 \\
| \\
\text{COO}^-
\end{array}
$$

Oxaloacetate

Analysis of the α-ketoglutarate formed showed that none of the radioactive label had been lost. Decarboxylation of α-ketoglutarate then yielded succinate devoid of radioactivity. All the label was in the released CO_2. Why were the early investigators of the citric acid cycle surprised that *all* the label emerged in the CO_2?

39. *Symmetric molecules reacting asymmetrically.* The interpretation of the experiments described in problem 38 was that citrate (or any other symmetric compound) cannot be an intermediate in the formation of α-ketoglutarate, because of the asymmetric fate of the label. This view seemed compelling until Alexander Ogston incisively pointed out in 1948 that "it is possible that *an asymmetric enzyme which attacks a symmetrical compound can distinguish between its identical groups* [italics added]." For simplicity, consider a molecule in which two hydrogen atoms, a group X and a different group Y, are bonded to a tetrahedral carbon atom as a model for citrate. Explain how a symmetric molecule can react with an enzyme in an asymmetric way.

Data Interpretation Problems

40. *A little goes a long way.* As will become clearer in Chapter 18, the activity of the citric acid cycle can be monitored by measuring the amount of O_2 consumed. The greater the rate of O_2 consumption, the faster the rate of the cycle. Hans Krebs used this assay to investigate the cycle in 1937. He used as his experimental system minced pigeon-breast muscle, which is rich in mitochondria. In one set of experiments, Krebs measured the O_2 consumption in the presence of carbohydrate only and in the presence of carbohydrate and citrate. The results are shown in the following table. ✓⑤

Effect of citrate on oxygen consumption by minced pigeon-breast muscle

	Micromoles of oxygen consumed	
Time (min)	Carbohydrate only	Carbohydrate plus 3 μmol of citrate
10	26	28
60	43	62
90	46	77
150	49	85

(a) How much O_2 would be absorbed if the added citrate were completely oxidized to H_2O and CO_2?

(b) On the basis of your answer to part *a*, what do the results given in the table suggest?

41. *Arsenite poisoning.* The effect of arsenite on the experimental system of problem 40 was then examined. Experimental data (not presented here) showed that the amount of citrate present did not change in the course of the experiment in the absence of arsenite. However, if arsenite was added to the system, different results were obtained, as shown in the following table. ✓⑤

Disappearance of citric acid in pigeon-breast muscle in the presence of arsenite

Micromoles of citrate added	Micromoles of citrate found after 40 minutes	Micromoles of citrate used
22	0.6	21
44	20.0	24
90	56.0	34

(a) What is the effect of arsenite on the disappearance of citrate?

(b) How is the action of arsenite altered by the addition of more citrate?

(c) What do these data suggest about the site of action of arsenite?

42. *Isocitrate lyase and tuberculosis.* The bacterium *Mycobacterium tuberculosis*, the cause of tuberculosis, can invade the lungs and persist in a latent state for years. During this time, the bacteria reside in granulomas—nodular scars containing bacteria and host-cell debris in the center and surrounded by immune cells. The granulomas are lipid-rich, oxygen-poor environments. How these bacteria manage to persist is something of a mystery. The results of research suggest that the glyoxylate cycle is required for the persistence. The following data show the amount of bacteria [presented as colony-forming units (cfu)] in mice lungs in the weeks after an infection. ✓⑦

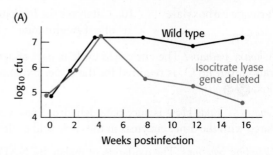

In graph A, the black circles represent the results for wild-type bacteria and the red circles represent the results for bacteria from which the gene for isocitrate lyase was deleted.

(a) What is the effect of the absence of isocitrate lyase?

The techniques described in Chapter 5 were used to reinsert the gene encoding isocitrate lyase into bacteria from which it had previously been deleted. **(See animation on 🅢 Sapling Plus)**

In graph B, black circles represent bacteria into which the gene was reinserted and red circles represent bacteria in which the gene was still missing.

(b) Do these results support those obtained in part *a*?

(c) What is the purpose of the experiment in part *b*?

(d) Why do these bacteria perish in the absence of the glyoxylate cycle?

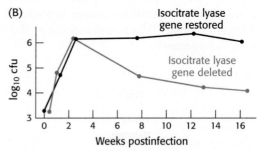

[Data from McKinney et al., *Nature* 406: 735–738, 2000]

Think ▶ Pair ▶ Share Problems

43. The citric acid cycle is often described as the metabolic hub of cellular metabolism. Explain what is meant by this statement.

44. Describe the advantage of regenerating oxaloacetate at the end of each turn of the citric acid cycle.

Oxidative Phosphorylation

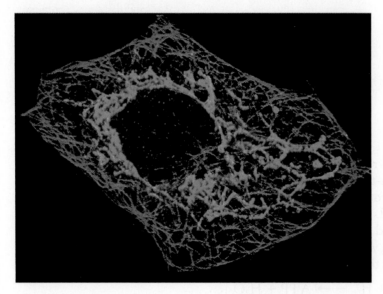

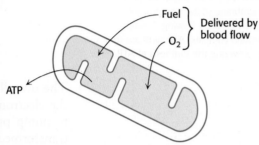

Mitochondria, stained green, form a network inside a fibroblast cell (left). Mitochondria oxidize carbon fuels to form cellular energy in the form of ATP. [(Left) Courtesy of Michael P. Yaffe, Dept. of Biology, University of California at San Diego.]

 LEARNING OBJECTIVES

By the end of this chapter, you should be able to:

1. Describe the key components of the electron-transport chain and how they are arranged.

2. Explain the benefits of having the electron-transport chain located in a membrane.

3. Describe how the proton-motive force is converted into ATP.

4. Identify the ultimate determinant of the rate of cellular respiration.

OUTLINE

18.1 Eukaryotic Oxidative Phosphorylation Takes Place in Mitochondria

18.2 Oxidative Phosphorylation Depends on Electron Transfer

18.3 The Respiratory Chain Consists of Four Complexes: Three Proton Pumps and a Physical Link to the Citric Acid Cycle

18.4 A Proton Gradient Powers the Synthesis of ATP

18.5 Many Shuttles Allow Movement Across Mitochondrial Membranes

18.6 The Regulation of Cellular Respiration Is Governed Primarily by the Need for ATP

The amount of ATP that human beings require to go about their lives is staggering. A sedentary male of 70 kg (154 lbs) requires about 8400 kJ (2000 kcal) for a day's worth of activity. To provide this much energy requires 83 kg of ATP. However, human beings possess only about 250 g of ATP at any given moment. The disparity between the amount of ATP that we have and the amount that we require is compensated by recycling ADP back to ATP. Each ATP molecule is recycled approximately 300 times per day. This recycling takes place primarily through *oxidative phosphorylation*.

We begin our study of oxidative phosphorylation by examining the oxidation–reduction reactions that allow the flow of electrons from NADH and $FADH_2$ to oxygen. The electron flow takes place in four large protein complexes that are embedded in the inner mitochondrial membrane, together called the *respiratory chain* or the *electron-transport chain*.

$$NADH + \tfrac{1}{2}O_2 + H^+ \longrightarrow H_2O + NAD^+$$
$$\Delta G^{\circ\prime} = -220.1 \text{ kJ mol}^{-1}(-52.6 \text{ kcal mol}^{-1})$$

FIGURE 18.1 Overview of oxidative phosphorylation. Oxidation and ATP synthesis are coupled by transmembrane proton fluxes. Electrons from NADH and FADH₂, formed in the TCA cycle, flow through the electron-transport chain to reduce oxygen to water (yellow tube). Some components of the chain pump protons from the mitochondrial matrix to the intermembrane space. The protons return to the matrix by flowing through another protein complex, ATP synthase (red structure), powering the synthesis of ATP.

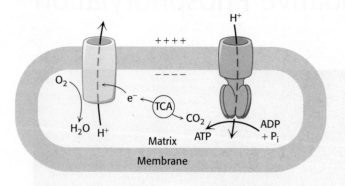

Respiration

An ATP-generating process in which an inorganic compound (such as molecular oxygen) serves as the ultimate electron acceptor. The electron donor can be either an organic compound or an inorganic one.

The overall reaction is exergonic. Importantly, three of the complexes of the electron-transport chain use the energy released by the electron flow to pump protons out of the mitochondrial matrix. In essence, energy is transformed. The resulting unequal distribution of protons generates a pH gradient and a transmembrane electrical potential that creates a *proton-motive force*. ATP is synthesized when protons flow back to the mitochondrial matrix through an enzyme complex.

$$ADP + P_i + H^+ \longrightarrow ATP + H_2O$$
$$\Delta G^{\circ\prime} = +30.5 \text{ kJ mol}^{-1} (+7.3 \text{ kcal mol}^{-1})$$

Thus, *the oxidation of fuels and the phosphorylation of ADP are coupled by a proton gradient across the inner mitochondrial membrane* (Figure 18.1).

Collectively, the generation of high-transfer-potential electrons by the citric acid cycle, their flow through the respiratory chain, and the accompanying synthesis of ATP is called *respiration* or *cellular respiration*.

18.1 Eukaryotic Oxidative Phosphorylation Takes Place in Mitochondria

Recall that a biochemical role of the citric acid cycle, which takes place in the mitochondrial matrix, is the generation of high-energy electrons. It is fitting, therefore, that oxidative phosphorylation, which will convert the energy of these electrons into ATP, also takes place in mitochondria. Mitochondria are oval-shaped organelles, typically about 2 μm in length and 0.5 μm in diameter, about the size of a bacterium.

Mitochondria are bounded by a double membrane

Electron microscopic studies revealed that mitochondria have two membrane systems: an *outer membrane* and an extensive, highly folded *inner membrane*. The inner membrane is folded into a series of internal ridges called *cristae*. Hence, there are two compartments in mitochondria: (1) the *intermembrane space* between the outer and the inner membranes and (2) the *matrix*, which is bounded by the inner membrane (Figure 18.2). The mitochondrial matrix is the site of most of the reactions of the citric acid cycle and fatty acid oxidation. In contrast, oxidative phosphorylation takes place in the inner mitochondrial membrane. The increase in surface area of the inner mitochondrial membrane provided by the cristae creates more sites for oxidative phosphorylation than would be the case with a simple, unfolded membrane. Humans contain an estimated 14,000 m² of inner mitochondrial membrane, which is the approximate equivalent of three football fields in the United States.

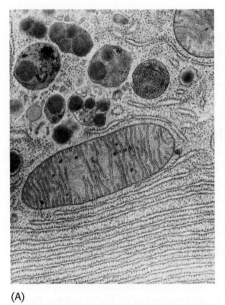

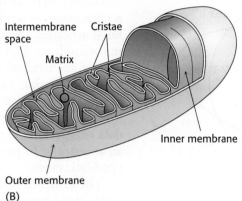

Intermembrane space

Cristae

Matrix

Inner membrane

Outer membrane

(A)   (B)

FIGURE 18.2 Electron micrograph (A) and diagram (B) of a mitochondrion. [(A) Keith R. Porter/Science Source. (B) Information from Wolfe, *Biology of the Cell*, 2e, © 1981 Brooks/Cole, a part of Cengage Learning, Inc. Reproduced by permission www.cengage.com/permission 3.]

The outer membrane is quite permeable to most small molecules and ions because it contains *mitochondrial porin*, a 30- to 35-kDa pore-forming protein also known as VDAC, for *voltage-dependent anion channel*. VDAC, the most prevalent protein in the outer mitochondrial membrane, plays a role in the regulated flux of metabolites—usually anionic species such as phosphate, chloride, organic anions, and the adenine nucleotides—across the outer membrane. In contrast, the inner membrane is impermeable to nearly all ions and polar molecules. A large family of transporters shuttles metabolites such as ATP, pyruvate, and citrate across the inner mitochondrial membrane. The two faces of this membrane will be referred to as the *matrix side* and the *cytoplasmic side* (the latter because it is freely accessible to most small molecules in the cytoplasm). They are also called the *N* and *P* sides, respectively, because the membrane potential is negative on the matrix side and positive on the cytoplasmic side (Figure 18.1).

In bacteria, the electron-driven proton pumps and ATP-synthesizing complexes are located in the cytoplasmic membrane, the inner of two membranes. In many bacteria, the outer membrane, like that of mitochondria, is permeable to most small metabolites because of the presence of porins.

Mitochondria are the result of an endosymbiotic event

Mitochondria are semiautonomous organelles that live in an endosymbiotic relation with the host cell. These organelles contain their own DNA, which encodes a variety of different proteins and RNAs. Mitochondrial DNA is usually portrayed as being circular, but recent research suggests that the mitochondrial DNA of many organisms may be linear. The genomes of mitochondria range broadly in size across species. The mitochondrial genome of the protist *Plasmodium falciparum* consists of fewer than 6000 base pairs (bp), whereas those of some land plants comprise more than 200,000 bp (Figure 18.3). Human mitochondrial DNA comprises 16,569 bp and encodes 13 respiratory-chain proteins as well as the small and large ribosomal RNAs and enough tRNAs to translate all codons. However, mitochondria also contain many proteins encoded by nuclear DNA. Cells that contain mitochondria depend on these organelles for oxidative phosphorylation, and the mitochondria in turn depend on the cell for their very existence. How did this intimate symbiotic relation come to exist?

Rickettsia (a bacterium)

Arabidopsis (a plant)

Plasmodium (a protozoan)

Homo sapiens

FIGURE 18.3 Sizes of mitochondrial genomes. The sizes of three mitochondrial genomes compared with the genome of *Rickettsia*, a relative of the presumed ancestor of all mitochondria. For genomes of more than 60 kbp, the DNA coding region for genes with known function is shown in red.

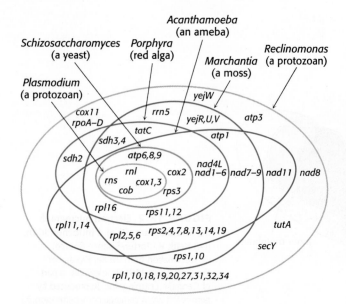

FIGURE 18.4 Overlapping gene complements of mitochondria. The genes present within each oval are those present within the organism represented by the oval. Only rRNA- and protein-coding genes are shown. The genome of *Reclinomonas* contains all the protein-coding genes found in all the sequenced mitochondrial genomes. [Data from M. W. Gray, G. Burger, and B. F. Lang, *Science* 283:1476–1481, 1999.]

An *endosymbiotic event* is thought to have occurred whereby a free-living organism capable of oxidative phosphorylation was engulfed by another cell. The double-membrane, circular DNA (with exceptions) and the mitochondrial-specific transcription and translation machinery all point to this conclusion. Thanks to the rapid accumulation of sequence data for mitochondrial and bacterial genomes, speculation on the origin of the "original" mitochondrion with some authority is now possible. The most mitochondrial-like bacterial genome is that of *Rickettsia prowazekii*, the cause of louse-borne typhus. The genome for this organism is more than 1 million base pairs in size and contains 834 protein-encoding genes. Sequence data suggest that all extant mitochondria are derived from an ancestor of *R. prowazekii* as the result of a single endosymbiotic event.

The evidence that modern mitochondria result from a single event comes from examination of the most bacteria-like mitochondrial genome, that of the protozoan *Reclinomonas americana*. Its genome contains 97 genes, of which 62 specify proteins. The genes encoding these proteins include all of the protein-coding genes found in all of the sequenced mitochondrial genomes (Figure 18.4). Yet, this genome encodes less than 2% of the protein-coding genes in the bacterium *E. coli*. In other words, a small fraction of bacterial genes—2%—is found in all examined mitochondria. How is it possible that all mitochondria have the same 2% of the bacterial genome? It seems unlikely that mitochondrial genomes resulting from several endosymbiotic events could have been independently reduced to the same set of genes found in *R. americana*. What was the fate of the mitochondrial genes no longer harbored by the mitochondria? The reduced genomes of the mitochondria are the result of endosymbiotic gene transfer, in which the mitochondrial genes became part of the nuclear genome. Thus, the original bacterial cell lost DNA, making it incapable of independent living, and the host cell became dependent on the ATP generated by its tenant.

18.2 Oxidative Phosphorylation Depends on Electron Transfer

In Chapter 17, the generation of NADH and $FADH_2$ by the oxidation of acetyl CoA was identified as an important function of the citric acid cycle. In oxidative phosphorylation, electrons from NADH and $FADH_2$ are used to reduce molecular oxygen to water. The highly exergonic reduction is accomplished by a number of electron-transfer reactions, which take place in a set of membrane proteins known as the *electron-transport chain*.

The electron-transfer potential of an electron is measured as redox potential

In oxidative phosphorylation, the *electron-transfer potential* of NADH or $FADH_2$ is converted into the *phosphoryl-transfer potential* of ATP. To better understand this conversion, we need quantitative expressions for these forms of free energy. The measure of phosphoryl-transfer potential is already familiar to us: it is given by $\Delta G^{\circ\prime}$ for the hydrolysis of the activated phosphoryl compound. The corresponding expression for the electron-transfer potential is E_0', the *reduction potential* (also called the *redox potential* or *oxidation–reduction potential*).

Consider a substance that can exist in an oxidized form X and a reduced form X^-. Such a pair is called a *redox couple* and is designated X : X^-. The reduction potential of this couple can be determined by measuring the electromotive force generated by an apparatus called a *sample half-cell* connected to a *standard reference half-cell* (Figure 18.5). The sample half-cell consists of an electrode immersed in a solution of 1 M oxidant (X) and 1 M reductant (X^-). The standard reference half-cell consists of an electrode immersed in a 1 M H^+ solution that is in equilibrium with H_2 gas at 1 atmosphere (1 atm) of pressure. The electrodes are connected to a voltmeter, and an agar bridge allows ions to move from one half-cell to the other, establishing electrical continuity between the half-cells. Electrons then flow from one half-cell to the other through the wire connecting the two half-cells to the voltmeter. If the reaction proceeds in the direction

$$X^- + H^+ \rightarrow X + \tfrac{1}{2} H_2$$

the reactions in the half-cells (referred to as *half-reactions* or *couples*) must be

$$X^- \rightarrow X + e^- \qquad H^+ + e^- \rightarrow \tfrac{1}{2} H_2$$

Thus, electrons flow from the sample half-cell to the standard reference half-cell, and the sample-cell electrode is taken to be negative with respect to the standard-cell electrode. *The reduction potential of the X : X^- couple is the observed voltage at the start of the experiment* (when X, X^-, and H^+ are 1 M with 1 atm of H_2). *The reduction potential of the H^+ : H_2 couple is defined to be 0 volts.* In oxidation–reduction reactions, the donor of electrons, in this case X^-, is called the *reductant* or *reducing agent*, whereas the acceptor of electrons, H^+ here, is called the *oxidant* or *oxidizing agent*.

The meaning of the reduction potential is now evident. A negative reduction potential means that the oxidized form of a substance has lower affinity for electrons than does H_2, as in the preceding example. A positive reduction potential means that the oxidized form of a substance has higher affinity for electrons than does H_2. These comparisons refer to standard conditions—namely, 1 M oxidant, 1 M reductant, 1 M H^+, and 1 atm H_2. Thus, *a strong reducing agent (such as NADH) is poised to donate electrons and has a negative reduction potential, whereas a strong oxidizing agent (such as O_2) is ready to accept electrons and has a positive reduction potential.*

The reduction potentials of many biologically important redox couples are known (Table 18.1). Table 18.1 is like those presented in chemistry textbooks, except that a hydrogen ion concentration of 10^{-7} M (pH 7) instead of 1 M (pH 0) is the standard state adopted by biochemists. This difference is denoted by the prime in E_0'. Recall that the prime in $\Delta G^{\circ\prime}$ denotes a standard free-energy change at pH 7.

The standard free-energy change $\Delta G^{\circ\prime}$ is related to the change in reduction potential $\Delta E_0'$ by

$$\Delta G^{\circ\prime} = -nF\Delta E_0'$$

in which n is the number of electrons transferred, F is a proportionality constant called the *Faraday constant* [96.48 kJ mol^{-1} V^{-1} (23.06 kcal mol^{-1} V^{-1})], $\Delta E_0'$ is in volts, and $\Delta G^{\circ\prime}$ is in kilojoules or kilocalories per mole.

The free-energy change of an oxidation–reduction reaction can be readily calculated from the reduction potentials of the reactants. For example, consider the reduction of pyruvate by NADH, catalyzed by

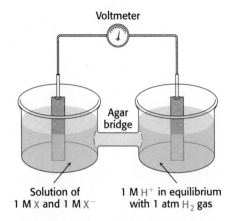

Voltmeter

Agar bridge

Solution of 1 M X and 1 M X^-

1 M H^+ in equilibrium with 1 atm H_2 gas

FIGURE 18.5 Measurement of redox potential. Apparatus for the measurement of the standard oxidation–reduction potential of a redox couple. Electrons flow through the wire connecting the cells, whereas ions flow through the agar bridge.

TABLE 18.1 Standard reduction potentials of some reactions

Oxidant	Reductant	n	E_0' (V)
Succinate $+ CO_2$	α-Ketoglutarate	2	-0.67
Acetate	Acetaldehyde	2	-0.60
Ferredoxin (oxidized)	Ferredoxin (reduced)	1	-0.43
$2 H^+$	H_2	2	-0.42
NAD^+	$NADH + H^+$	2	-0.32
$NADP^+$	$NADPH + H^+$	2	-0.32
Lipoate (oxidized)	Lipoate (reduced)	2	-0.29
Glutathione (oxidized)	Glutathione (reduced)	2	-0.23
FAD	$FADH_2$	2	-0.22
Acetaldehyde	Ethanol	2	-0.20
Pyruvate	Lactate	2	-0.19
$2 H^+$	H_2	2	0.00^1
Fumarate	Succinate	2	$+0.03$
Cytochrome b $(+3)$	Cytochrome b $(+2)$	1	$+0.07$
Dehydroascorbate	Ascorbate	2	$+0.08$
Ubiquinone (oxidized)	Ubiquinone (reduced)	2	$+0.10$
Cytochrome c $(+3)$	Cytochrome c $(+2)$	1	$+0.22$
Fe $(+3)$	Fe $(+2)$	1	$+0.77$
$\frac{1}{2} O_2 + 2 H^+$	H_2O	2	$+0.82$

Note: E_0' is the standard oxidation–reduction potential (pH 7, 25°C) and n is the number of electrons transferred. E_0' refers to the partial reaction written as Oxidant $+ e^- \rightarrow$ reductant.

[1]Standard oxidation − reduction potential at pH = 0. Compare with $E_0' = -0.42$ at pH = 7.

lactate dehydrogenase. Recall that this reaction maintains redox balance in lactic acid fermentation (Figure 16.11).

$$\text{Pyruvate} + \text{NADH} + \text{H}^+ \rightleftharpoons \text{lactate} + \text{NAD}^+ \qquad \text{(A)}$$

The reduction potential of the NAD^+ : NADH couple, or half-reaction, is -0.32 V, whereas that of the pyruvate:lactate couple is -0.19 V. By convention, reduction potentials (as in Table 18.1) refer to partial reactions written as reductions: oxidant $+ e^- \rightarrow$ reductant. Hence,

$$\text{Pyruvate} + 2\,\text{H}^+ + 2\,\text{e}^- \rightarrow \text{lactate} \qquad E_0' = -0.19\,\text{V} \qquad \text{(B)}$$
$$\text{NAD}^+ + \text{H}^+ + 2\,\text{e}^- \rightarrow \text{NADH} \qquad E_0' = -0.32\,\text{V} \qquad \text{(C)}$$

To obtain reaction A from reactions B and C, we need to reverse the direction of reaction C so that NADH appears on the left side of the arrow. In doing so, the sign of E_0' must be changed.

$$\text{Pyruvate} + 2\,\text{H}^+ + 2\,\text{e}^- \rightarrow \text{lactate} \qquad E_0' = -0.19\,\text{V} \qquad \text{(B)}$$
$$\text{NADH} \rightarrow \text{NAD}^+ + \text{H}^+ + 2\,\text{e}^- \qquad E_0' = +0.32\,\text{V} \qquad \text{(D)}$$

For reaction B, the free energy can be calculated with $n = 2$.

$$\Delta G^{\circ\prime} = -2 \times 96.48\,\text{kJ mol}^{-1}\,\text{V}^{-1} \times -0.19\,\text{V}$$
$$= +36.7\,\text{kJ mol}^{-1}(+8.8\,\text{kcal mol}^{-1})$$

Likewise, for reaction D,

$$\Delta G^{\circ\prime} = -2 \times 96.48\,\text{kJ mol}^{-1}\,\text{V}^{-1} \times +0.32\,\text{V}$$
$$= -61.8\,\text{kJ mol}^{-1}(-14.8\,\text{kcal mol}^{-1})$$

Thus, the free energy for reaction A is given by

$$\Delta G^{\circ\prime} = \Delta G^{\circ\prime} \text{ (for reaction B)} + \Delta G^{\circ\prime}\text{(for reaction D)}$$
$$= +36.7 \text{ kJ mol}^{-1} - 61.8 \text{ kJ mol}^{-1}$$
$$= -25.1 \text{ kJ mol}^{-1} (-6.0 \text{ kcal mol}^{-1})$$

Electron flow from NADH to molecular oxygen powers the formation of a proton gradient

The driving force of oxidative phosphorylation is the electron-transfer potential of NADH or $FADH_2$ relative to that of O_2. How much energy is released by the reduction of O_2 with NADH? Let us calculate $\Delta G^{\circ\prime}$ for this reaction. The pertinent half-reactions are

$$\tfrac{1}{2} O_2 + 2 H^+ + 2e^- \rightarrow H_2O \qquad E_0' = +0.82 \text{ V} \qquad \text{(A)}$$
$$NAD^+ + H^+ + 2e^- \rightarrow NADH \qquad E_0' = -0.32 \text{ V} \qquad \text{(B)}$$

The combination of the two half-reactions, as it proceeds in the electron-transport chain, yields

$$\tfrac{1}{2} O_2 + NADH + H^+ \rightarrow H_2O + NAD^+ \qquad \text{(C)}$$

The standard free energy for this reaction is then given by

$$\Delta G^{\circ\prime} = (-2 \times 96.48 \text{ kJ mol}^{-1}V^{-1} \times +0.82 \text{ V})$$
$$+ (-2 \times 96.48 \text{ kJ mol}^{-1}V^{-1} \times +0.32 \text{ V})$$
$$= -158.2 \text{ kJ mol}^{-1} + (-61.9) \text{ kJ mol}^{-1}$$
$$= -220.1 \text{ kJ mol}^{-1} (-52.6 \text{ kcal mol}^{-1})$$

This release of free energy is substantial. Recall that $\Delta G^{\circ\prime}$ for the synthesis of ATP is 30.5 kJ mol^{-1} (7.3 kcal mol^{-1}). The released energy is initially used to generate a proton gradient that is then used for the synthesis of ATP and the transport of metabolites across the mitochondrial membrane.

How can the energy associated with a proton gradient be quantified? Recall from Section 13.1 that the free-energy change for a species moving from one side of a membrane where it is at concentration c_1 to the other side where it is at a concentration c_2 is given by

$$\Delta G = RT \ln(c_2/c_1) + ZF\Delta V$$

in which Z is the electrical charge of the transported species and ΔV is the potential in volts across the membrane. Under typical conditions for the inner mitochondrial membrane, the pH outside is 1.4 units lower than inside [corresponding to ln (c_2/c_1) of 3.2] and the membrane potential is 0.14 V, the outside being positive. Because $Z = +1$ for protons, the free-energy change is $(8.32 \times 10^{-3} \text{ kJ mol}^{-1}\text{K}^{-1} \times 310 \text{ K} \times 3.2) + (+1 \times 96.48 \text{ kJ mol}^{-1}\text{ V}^{-1} \times 0.14 \text{ V}) = 21.8 \text{ kJ mol}^{-1} (5.2 \text{ kcal mol}^{-1})$. Thus, each proton that is transported out of the matrix to the cytoplasmic side corresponds to 21.8 kJ mol^{-1} of free energy.

18.3 The Respiratory Chain Consists of Four Complexes: Three Proton Pumps and a Physical Link to the Citric Acid Cycle

Electrons are transferred from NADH to O_2 through a chain of three large protein complexes called *NADH-Q oxidoreductase, Q-cytochrome c oxidoreductase*, and *cytochrome c oxidase* (Figure 18.6 and Table 18.2). *Electron flow*

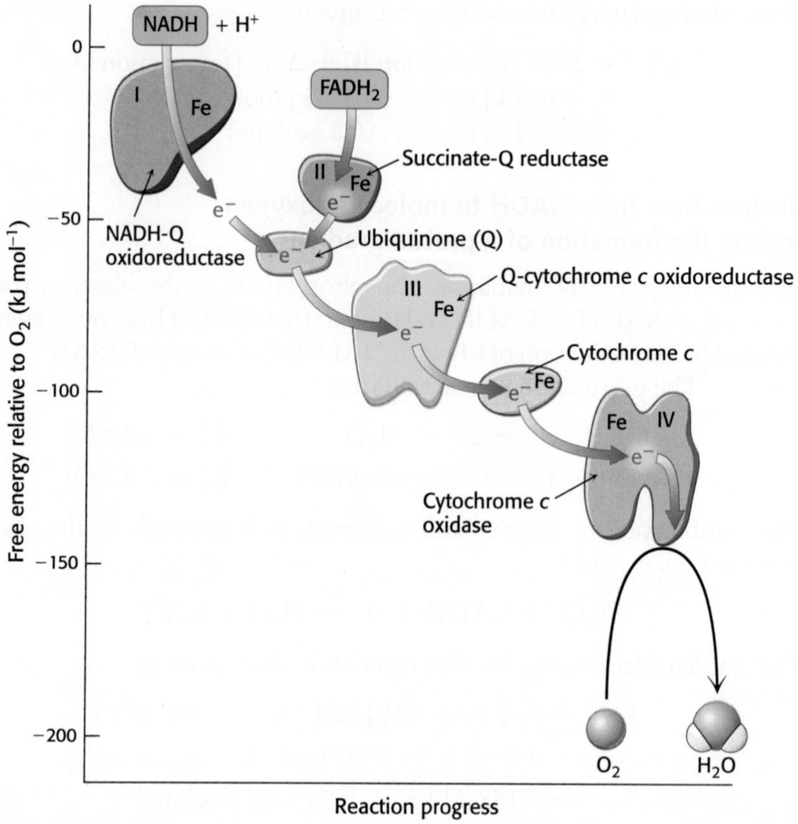

FIGURE 18.6 Components of the electron-transport chain. Electrons flow down an energy gradient from NADH to O_2. The flow is catalyzed by four protein complexes, and the energy released is used to generate a proton gradient. [Information from D. Sadava et al., *Life,* 8th ed. (Sinauer, 2008), p. 150.]

within these transmembrane complexes is highly exergonic and powers the transport of protons across the inner mitochondrial membrane. A fourth large protein complex, called *succinate-Q reductase,* contains the succinate dehydrogenase that generates $FADH_2$ in the citric acid cycle. Electrons from this $FADH_2$ enter the electron-transport chain at Q-cytochrome *c* oxidoreductase. Succinate-Q reductase, in contrast with the other complexes, does not pump protons. NADH-Q oxidoreductase, succinate-Q reductase, Q-cytochrome *c* oxidoreductase, and cytochrome *c* oxidase are also called *Complex I, II, III,* and *IV,* respectively. Complexes I, III, and IV appear to be associated in a supramolecular complex which facilitates the rapid transfer of substrate and prevents the release of reaction intermediates (p. 589).

TABLE 18.2 Components of the mitochondrial electron-transport chain

Enzyme complex	Mass (kDa)	Prosthetic group	Oxidant or reductant		
			Matrix side	Membrane core	Intermembrane side
NADH-Q oxidoreductase	>900	FMN Fe-S	NADH	Q	
Succinate-Q reductase	140	FAD Fe-S	Succinate	Q	
Q-cytochrome *c* oxidoreductase	250	Heme b_H Heme b_L Heme c_1 Fe-S		Q	Cytochrome *c*
Cytochrome *c* oxidase	160	Heme a Heme a_3 Cu_A and Cu_B			Cytochrome *c*

Information from: J. W. DePierre and L. Ernster, *Annu. Rev. Biochem.* 46:215, 1977; Y. Hatefi, *Annu Rev. Biochem.* 54:1015, 1985; and J. E. Walker, *Q. Rev. Biophys.* 25:253, 1992.

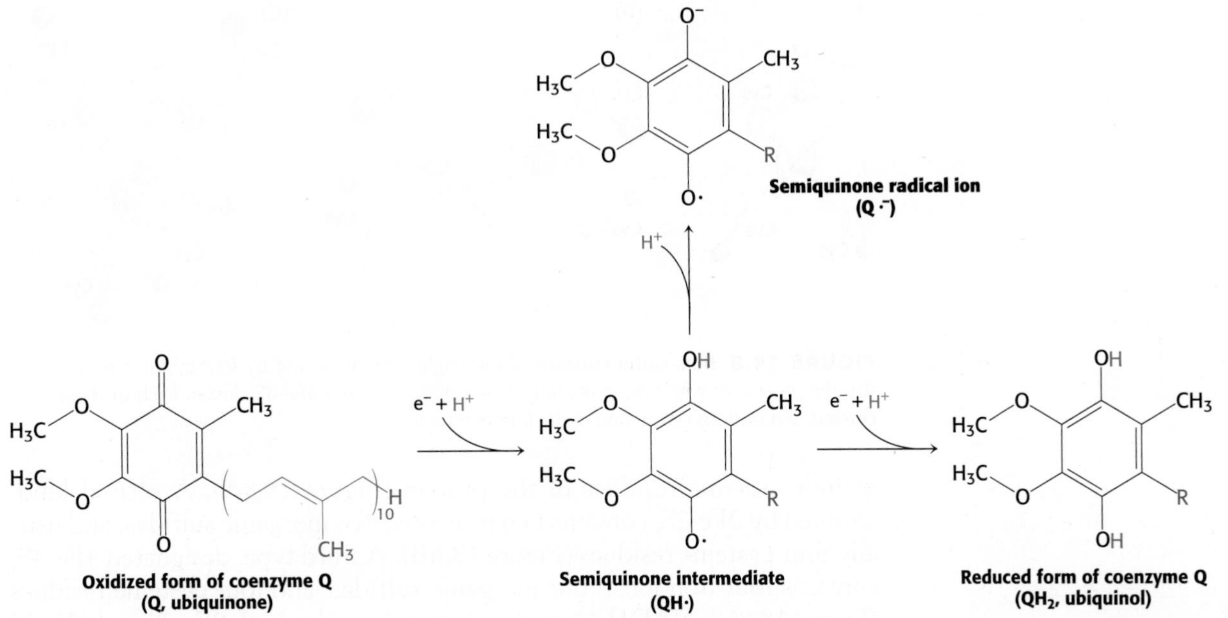

FIGURE 18.7 Oxidation states of quinones. The reduction of ubiquinone (Q) to ubiquinol (QH_2) proceeds through a semiquinone intermediate ($QH\cdot$).

Two special electron carriers ferry the electrons from one complex to the next. The first is *coenzyme Q* (Q), also known as *ubiquinone* because it is a *ubiquitous quinone* in biological systems. Ubiquinone is a hydrophobic quinone that diffuses rapidly within the inner mitochondrial membrane. Electrons are carried from NADH-Q oxidoreductase to Q-cytochrome *c* oxidoreductase, the third complex of the chain, by the reduced form of Q. Electrons from the $FADH_2$ generated by the citric acid cycle are transferred first to ubiquinone and then to the Q-cytochrome *c* oxidoreductase complex.

Coenzyme Q is a quinone derivative with a long tail consisting of five-carbon isoprene units that account for its hydrophobic nature. The number of isoprene units in the tail depends on the species. The most common mammalian form contains 10 isoprene units (coenzyme Q_{10}). For simplicity, the subscript will be omitted from this abbreviation because all varieties function in an identical manner. Quinones can exist in several oxidation states. In the fully oxidized state (Q), coenzyme Q has two keto groups (Figure 18.7). The addition of one electron and one proton results in the semiquinone form ($QH\cdot$). The semiquinone can lose a proton to form a semiquinone radical anion ($Q\cdot^-$). The addition of a second electron and proton to the semiquinone generates ubiquinol (QH_2), the fully reduced form of coenzyme Q, which holds its protons more tightly. Thus, *for quinones, electron-transfer reactions are coupled to proton binding and release,* a property that is key to transmembrane proton transport. Because ubiquinone is soluble in the membrane, a pool of Q and QH_2—the Q *pool*—is thought to exist in the inner mitochondrial membrane, although recent research suggests that the Q pool is confined to the protein complexes of the electron-transport chain.

In contrast with Q, the second special electron carrier is a protein. Cytochrome *c*, a small soluble protein, shuttles electrons from Q-cytochrome *c* oxidoreductase to cytochrome *c* oxidase, the final component in the chain and the one that catalyzes the reduction of O_2.

Iron–sulfur clusters are common components of the electron-transport chain

Iron–sulfur clusters in *iron–sulfur proteins* (also called *nonheme iron proteins*) play a critical role in a wide range of reduction reactions in biological systems. Several types of Fe-S clusters are known (Figure 18.8). In the simplest kind, a single iron ion is tetrahedrally coordinated to the sulfhydryl groups

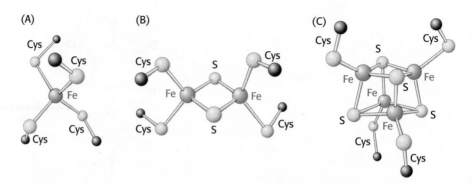

FIGURE 18.8 Iron–sulfur clusters. (A) A single iron ion bound by four cysteine residues. (B) 2Fe-2S cluster with iron ions bridged by sulfide ions. (C) 4Fe-4S cluster. Each of these clusters can undergo oxidation–reduction reactions.

of four cysteine residues of the protein (Figure 18.8A). A second kind, denoted by 2Fe-2S, contains two iron ions, two inorganic sulfides, and usually four cysteine residues (Figure 18.8B). A third type, designated 4Fe-4S, contains four iron ions, four inorganic sulfides, and four cysteine residues (Figure 18.8C). NADH-Q oxidoreductase contains both 2Fe-2S and 4Fe-4S clusters. Iron ions in these Fe-S complexes cycle between Fe^{2+} (reduced) and Fe^{3+} (oxidized) states. Unlike quinones and flavins, iron–sulfur clusters generally undergo oxidation–reduction reactions without releasing or binding protons.

The importance of Fe-S clusters is illustrated by the loss of function of the protein frataxin. Frataxin is a small (14.2-kDa) mitochondrial protein that is crucial for the synthesis of Fe-S clusters. Mutations in frataxin causes Friedreich's ataxia, a disease of varying severity that affects the central and peripheral nervous as well as the heart and skeletal system. Severe cases lead to death in young adult life. The most common mutation is trinucleotide expansion (Section 29.5) in the gene for frataxin.

The high-potential electrons of NADH enter the respiratory chain at NADH-Q oxidoreductase

The electrons of NADH enter the chain at *NADH-Q oxidoreductase* (also called *Complex I* and *NADH dehydrogenase*), an enormous enzyme (>900 kDa) consisting of approximately 45 polypeptide chains organized into 14 core subunits that are found in all species. This proton pump, like that of the other two in the respiratory chain, is encoded by genes residing in both the mitochondria and the nucleus. NADH-Q oxidoreductase is L-shaped, with a horizontal arm lying in the membrane and a vertical arm that projects into the matrix.

The reaction catalyzed by this enzyme appears to be

$$NADH + Q + 5\,H^+_{matrix} \rightarrow NAD^+ + QH_2 + 4\,H^+_{intermembrane\ space}$$

The initial step is the binding of NADH and the transfer of its two high-potential electrons to the *flavin mononucleotide* (FMN) prosthetic group, yielding the reduced form, $FMNH_2$ (Figure 18.9). The electron acceptor of FMN, the isoalloxazine ring, is identical with that of FAD. Electrons are then transferred from $FMNH_2$ to a series of iron–sulfur clusters, the second type of prosthetic group in NADH-Q oxidoreductase.

Recent structural studies have suggested how Complex I acts as a proton pump. What are the structural elements required for proton-pumping? The membrane-embedded part of the complex has four proton half-channels

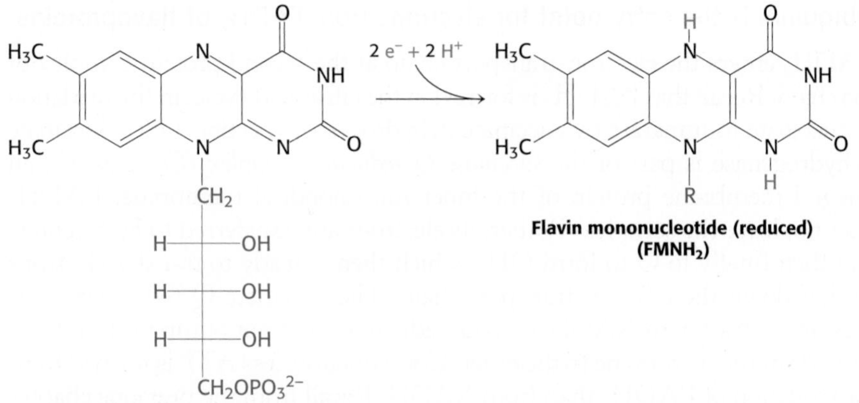

FIGURE 18.9 Oxidation states of flavins.

Flavin mononucleotide (oxidized) (FMN)

$2 e^- + 2 H^+$

Flavin mononucleotide (reduced) (FMNH$_2$)

consisting, in part, of vertical helices. One set of half-channels is exposed to the matrix and the other to the intermembrane space (Figure 18.10). The vertical helices are linked on the matrix side by a long horizontal helix (HL) that connects the matrix half-channels, while the intermembrane space half-channels are joined by a series of β-hairpin-helix connecting elements (βH). An enclosed Q chamber, the site where Q accepts electrons from NADH, exists near the junction of the hydrophilic portion and the membrane-embedded portion. Finally, a hydrophilic funnel connects the Q chamber to a water-lined channel (into which the half channels open) that extends the entire length of the membrane-embedded portion.

How do these structural elements cooperate to pump protons out of the matrix? When Q accepts two electrons from NADH, generating Q^{2-}, the negative charges on Q^{2-} interact electrostatically with negatively charged amino acid residues in the membrane-embedded arm, causing conformational changes in the long horizontal helix and the βH elements. These changes in turn alter the structures of the connected vertical helices that change the pK_a of amino acids, allowing protons from the matrix to first bind to the amino acids, then dissociate into the water-lined channel, and finally enter the inter-membrane space. Thus, *the flow of two electrons from NADH to coenzyme Q through NADH-Q oxidoreductase leads to the pumping of four hydrogen ions out of the matrix of the mitochondrion.* Q^{2-} subsequently takes up two protons from the matrix as it is reduced to QH$_2$. The removal of these protons from the matrix contributes to the formation of the proton-motive force. The QH$_2$ subsequently leaves the enzyme for the Q pool, allowing another reaction cycle to occur.

It is important to note that the citric acid cycle is not the only source of mitochondrial NADH. As we will see in Chapter 22, fatty acid degradation, which also takes place in mitochondria, is another crucial source of NADH for the electron-transport chain. Moreover, electrons from cytoplasmically generated NADH can be transported into mitochondria for use by the electron-transport chain (Section 18.5).

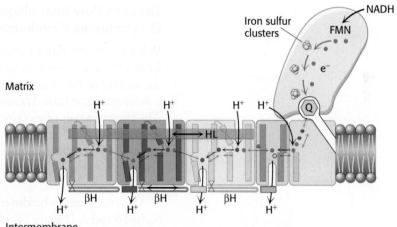

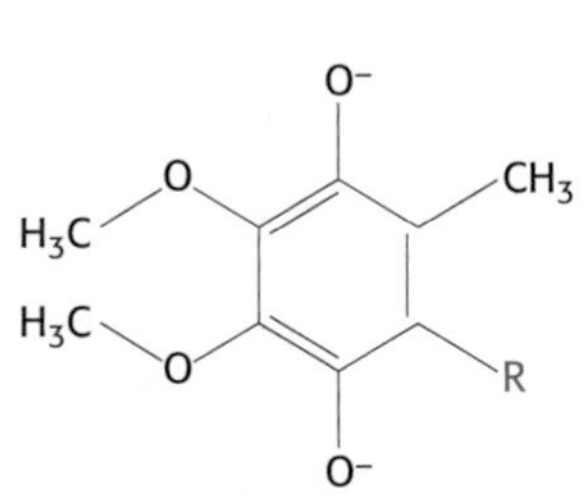

FIGURE 18.10 **Coupled electron–proton transfer reactions through NADH-Q oxidoreductase.** Electrons flow in Complex I from NADH through FMN and a series of iron–sulfur clusters to ubiquinone (Q), forming Q^{2-}. The charges on Q^{2-} are electrostatically transmitted to hydrophilic amino acid residues [shown as red (glutamate) and blue (lysine or histidine) balls] that power the movement of HL and βH components. This movement changes the conformation of the transmembrane helices and results in the transport of four protons out of the mitochondrial matrix. [Information from R. Baradaran et al., *Nature* 494:443–448.]

Reduced ubiquinone (Q^{2-})

Ubiquinol is the entry point for electrons from FADH₂ of flavoproteins

$FADH_2$ enters the electron-transport chain at the second protein complex of the chain. Recall that $FADH_2$ is formed in the citric acid cycle, in the oxidation of succinate to fumarate by succinate dehydrogenase (Section 17.2). Succinate dehydrogenase is part of the *succinate-Q reductase complex (Complex II)*, an integral membrane protein of the inner mitochondrial membrane. $FADH_2$ does not leave the complex. Rather, its electrons are transferred to Fe-S centers and then finally to Q to form QH_2, which then is ready to transfer electrons further down the electron-transport chain. The succinate-Q reductase complex, in contrast with NADH-Q oxidoreductase, does not pump protons from one side of the membrane to the other. Consequently, less ATP is formed from the oxidation of $FADH_2$ than from NADH. Recall from the previous chapter (p. 559) that mutations in succinate dehydrogenase result in an accumulation of succinate that facilitates the development of cancer.

Electrons flow from ubiquinol to cytochrome *c* through Q-cytochrome *c* oxidoreductase

What is the fate of ubiquinol generated by Complexes I and II? The electrons from QH_2 are passed on to *cytochrome* c (Cyt *c*), a water-soluble protein, by the second of the three proton pumps in the respiratory chain, *Q-cytochrome c oxidoreductase* (also known as *Complex III* and as *cytochrome* c *reductase*). The flow of a pair of electrons through this complex leads to the effective net transport of $2\,H^+$ to the intermembrane space, half the yield obtained with NADH-Q reductase because of a smaller thermodynamic driving force.

$$QH_2 + 2\,\text{Cyt}\,c_{ox} + 2\,H^+_{matrix} \rightarrow Q + 2\,\text{Cyt}\,c_{red} + 4\,H^+_{intermembrane\ space}$$

Q-cytochrome *c* oxidoreductase itself contains two types of cytochromes, named *b* and c_1 (Figure 18.11). *A cytochrome is an electron-transferring protein that contains a heme prosthetic group.* The heme prosthetic group in cytochromes *b*, c_1, and *c* is iron-protoporphyrin IX, the same heme present in myoglobin and hemoglobin (Section 7.1). The iron ion of a cytochrome, in contrast to hemoglobin and myoglobin, alternates between a reduced ferrous $(+2)$ state and an oxidized ferric $(+3)$ state during electron transport. The two cytochrome subunits of Q-cytochrome *c* oxidoreductase contain a total of three hemes: two hemes within cytochrome *b*, termed heme b_L (L for low affinity) and heme b_H (H for high affinity), and one heme within cytochrome c_1. These identical hemes have different electron affinities because they are in different polypeptide environments. For example, heme b_L, which is located in a cluster of helices near the intermembrane face of the membrane, has lower affinity for an electron than does heme b_H, which is near the matrix side.

In addition to the hemes, the enzyme contains an iron–sulfur protein with a 2Fe-2S center. This center, termed the *Rieske center*, is unusual in that one of the iron ions is coordinated by two histidine residues rather than two cysteine residues. This coordination stabilizes the center in its reduced form, raising its reduction potential so that it can readily accept electrons from QH_2.

The Q cycle funnels electrons from a two-electron carrier to a one-electron carrier and pumps protons

QH_2 passes two electrons to Q-cytochrome *c* oxidoreductase, but the acceptor of electrons in this complex, cytochrome *c*, can accept only one electron. How does the switch from the two-electron carrier ubiquinol to the one-electron carrier cytochrome *c* take place? The mechanism for the

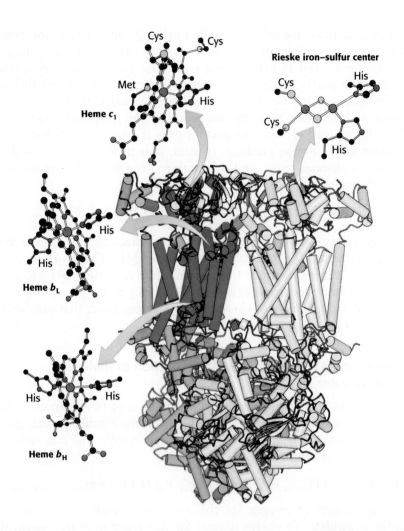

FIGURE 18.11 Structure of Q-cytochrome *c* oxidoreductase. This enzyme is a homodimer with each monomer consisting of 11 distinct polypeptide chains. Some of the more prominent components in one monomer are colored while the other monomer is white. Although each monomer contains the same components, some are identified in one monomer or the other for ease of viewing. *Notice* that the major prosthetic groups, three hemes and a 2Fe-2S cluster, are located either near the edge of the complex bordering the intermembrane space (top) or in the region embedded in the membrane (α helices represented by tubes). They are well-positioned to mediate the electron-transfer reactions between quinones in the membrane and cytochrome *c* in the intermembrane space. [Drawn from 1BCC.pdb.]

coupling of electron transfer from Q to cytochrome *c* to transmembrane proton transport is known as the *Q cycle* (Figure 18.12). Two QH_2 molecules bind to the complex consecutively, each giving up two electrons and two H^+. *These protons are released to the intermembrane space.* The first QH_2 to exit the Q pool binds to the first Q binding site (Q_o), and its two electrons travel through the complex to different destinations. One electron flows, first, to the Rieske 2Fe-2S cluster; then, to cytochrome c_1; and, finally, to a molecule of oxidized cytochrome *c*, converting it into its reduced form. The reduced cytochrome *c* molecule is free to diffuse away from the enzyme to continue down the respiratory chain.

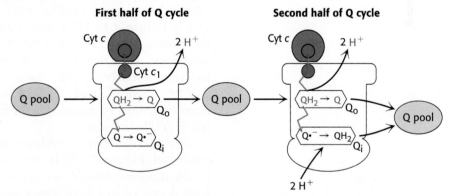

FIGURE 18.12 Q cycle. The Q cycle takes place in Complex III, which is represented in outline form. In the first half of the cycle, two electrons of a bound QH_2 are transferred, one to cytochrome *c* and the other to a bound Q in a second binding site to form the semiquinone radical anion $Q^{\cdot-}$. The newly formed Q dissociates and enters the Q pool. In the second half of the cycle, a second QH_2 also gives up its electrons to Complex III, one to a second molecule of cytochrome *c* and the other to reduce $Q^{\cdot-}$ to QH_2. This second electron transfer results in the uptake of two protons from the matrix. The path of electron transfer is shown in red.

The second electron passes through two heme groups of cytochrome *b* to an oxidized ubiquinone in a second Q binding site (Q_i). The Q in the second binding site is reduced to a semiquinone radical anion ($Q^{\cdot-}$) by the electron from the first QH_2. The now fully oxidized Q leaves the first Q site, free to reenter the Q pool.

A second molecule of QH_2 binds to the Q_o site of Q-cytochrome c oxidoreductase and reacts in the same way as the first. One of the electrons is transferred to cytochrome c and the second electron is transferred to the partly reduced ubiquinone bound in the Q_i binding site. On the addition of the electron from the second QH_2 molecule, this quinone radical anion takes up two protons from the matrix side to form QH_2. *The removal of these two protons from the matrix contributes to the formation of the proton gradient.* In sum, four protons are released into the intermembrane space, and two protons are removed from the mitochondrial matrix.

$$2\,QH_2 + Q + 2\,Cyt\,c_{ox} + 2\,H^+_{matrix} \rightarrow 2\,Q + QH_2 + 2\,Cyt\,c_{red} + 4\,H^+_{intermembrane\ space}$$

In one Q cycle, two QH_2 molecules are oxidized to form two Q molecules, and then one Q molecule is reduced to QH_2. The problem of how to efficiently funnel electrons from a two-electron carrier (QH_2) to a one-electron carrier (cytochrome c) is solved by the Q cycle. The cytochrome b component of the reductase is in essence a recycling device that enables both electrons of QH_2 to be used effectively.

Cytochrome *c* oxidase catalyzes the reduction of molecular oxygen to water

The last of the three proton-pumping assemblies of the respiratory chain is *cytochrome* c *oxidase (Complex IV)*. Cytochrome c oxidase catalyzes the transfer of electrons from the reduced form of cytochrome c to molecular oxygen, the final acceptor.

$$4\,Cyt\,c_{red} + 8\,H^+_{matrix} + O_2 \rightarrow 4\,Cyt\,c_{ox} + 2\,H_2O + 4\,H^+_{intermembrane\ space}$$

The requirement of oxygen for this reaction is what makes "aerobic" organisms aerobic. To obtain oxygen for this reaction is the reason that human beings must breathe. Four electrons are funneled to O_2 to completely reduce it to H_2O, and, concomitantly, protons are pumped from the matrix into the intermembrane space. This reaction is quite thermodynamically favorable. From the reduction potentials in Table 18.1, the standard free-energy change for this reaction is calculated to be $\Delta G^{\circ\prime} = -231.8\ kJ\ mol^{-1}\ (-55.4\ kcal\ mol^{-1})$. As much of this free energy as possible must be captured in the form of a proton gradient for subsequent use in ATP synthesis.

Bovine cytochrome c oxidase is reasonably well understood at the structural level (Figure 18.13). It consists of 13 subunits, 3 of which are encoded by the mitochondrion's own genome. Cytochrome c oxidase contains two *heme A* groups and three *copper ions*, arranged as two copper centers, designated A and B. One center, Cu_A/Cu_A, contains two copper ions linked by two bridging cysteine residues. This center initially accepts electrons from reduced cytochrome c. The remaining copper ion, Cu_B, is bonded to three histidine residues, one of which is modified by covalent linkage to a tyrosine residue. The copper centers alternate between the reduced Cu^+ (cuprous) form and the oxidized Cu^{2+} (cupric) form as they accept and donate electrons.

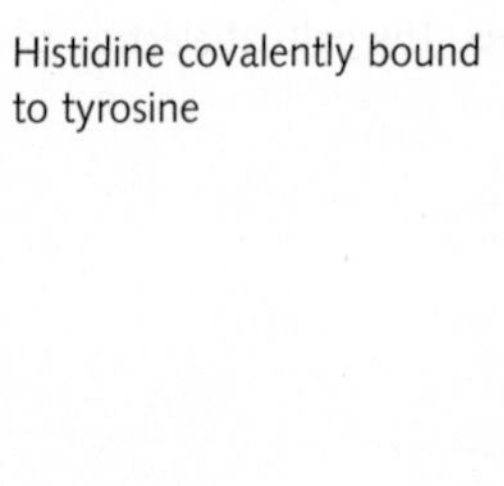

There are two heme A molecules, called *heme* a and *heme* a_3, in cytochrome c oxidase. Heme A differs from the heme in cytochrome c and c_1 in three ways: (1) a formyl group replaces a methyl group, (2) a C_{17}

Histidine covalently bound to tyrosine

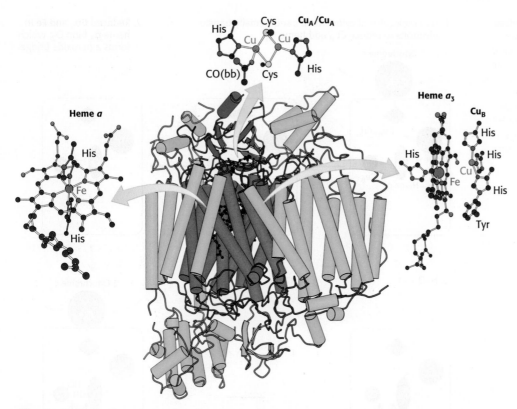

FIGURE 18.13 Structure of cytochrome *c* oxidase. This enzyme consists of 13 polypeptide chains. *Notice* that most of the complex, as well as two major prosthetic groups (heme *a* and heme a_3–Cu_B) are embedded in the membrane (α helices are represented by vertical tubes). Heme a_3–Cu_B is the site of the reduction of oxygen to water. The Cu_A/Cu_A prosthetic group is positioned near the intermembrane space to better accept electrons from cytochrome *c*. CO(bb) is a carbonyl group of the peptide backbone. [Drawn from 2OCC.pdb.]

hydrocarbon chain replaces one of the vinyl groups, and (3) the heme is not covalently attached to the protein.

Heme A

Heme *a* and heme a_3 have distinct redox potentials because they are located in different environments within cytochrome *c* oxidase. An electron flows from cytochrome *c* to Cu_A/Cu_A, to heme *a*, to heme a_3, to Cu_B, and finally to O_2. Heme a_3 and Cu_B are directly adjacent. Together, heme a_3 and Cu_B form the active center at which O_2 is reduced to H_2O.

FIGURE 18.14 Cytochrome *c* oxidase mechanism. The cycle begins and ends with all prosthetic groups in their oxidized forms (shown in blue). Reduced forms are in red. Four cytochrome *c* molecules donate four electrons, which, in allowing the binding and cleavage of an O_2 molecule, also makes possible the import of four H^+ from the matrix to form two molecules of H_2O, which are released from the enzyme to regenerate the initial state.

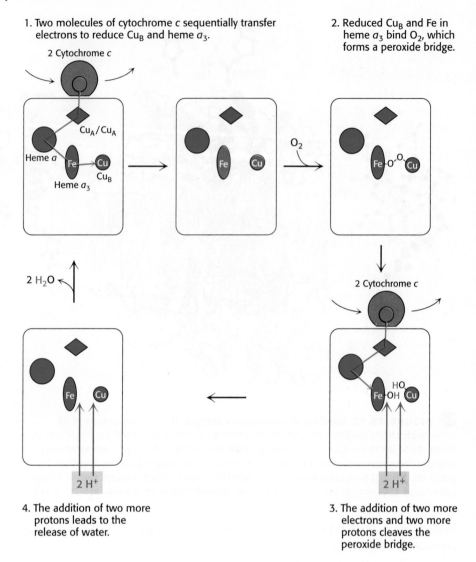

1. Two molecules of cytochrome *c* sequentially transfer electrons to reduce Cu_B and heme a_3.

2. Reduced Cu_B and Fe in heme a_3 bind O_2, which forms a peroxide bridge.

4. The addition of two more protons leads to the release of water.

3. The addition of two more electrons and two more protons cleaves the peroxide bridge.

Four molecules of cytochrome *c* bind consecutively to the enzyme and transfer an electron to reduce one molecule of O_2 to H_2O (Figure 18.14):

1. Electrons from two molecules of reduced cytochrome *c* flow down an electron-transfer pathway within cytochrome *c* oxidase, one stopping at Cu_B and the other at heme a_3. With both centers in the reduced state, they together can now bind an oxygen molecule.

2. As molecular oxygen binds, it abstracts an electron from each of the nearby ions in the active center to form a peroxide (O_2^{2-}) bridge between them (Figure 18.15).

3. Two more molecules of cytochrome *c* bind and release electrons that travel to the active center. The addition of an electron as well as H^+ to each oxygen atom reduces the two ion–oxygen groups to Cu_B^{2+}—OH and Fe^{3+}—OH.

4. Reaction with two more H^+ ions allows the release of two molecules of H_2O and resets the enzyme to its initial, fully oxidized form.

$$4\,\text{Cyt}\,c_{red} + 4\,H^+_{matrix} + O_2 \rightarrow 4\,\text{Cyt}\,c_{ox} + 2\,H_2O$$

The four protons in this reaction come exclusively from the matrix. Thus, the consumption of these four protons contributes directly to the proton gradient. Recall that each proton contributes 21.8 kJ mol^{-1} (5.2 kcal mol^{-1}) to the free energy associated with the proton gradient; so these four protons contribute 87.2 kJ mol^{-1} (20.8 kcal mol^{-1}), an amount substantially less than the free

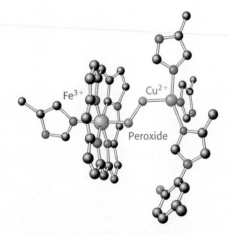

FIGURE 18.15 Peroxide bridge. The oxygen bound to heme a_3 is reduced to peroxide by the presence of Cu_B.

energy available from the reduction of oxygen to water. What is the fate of this missing energy? Remarkably, *cytochrome c oxidase uses this energy to pump four additional protons from the matrix into the intermembrane space in the course of each reaction cycle for a total of eight protons removed from the matrix* (Figure 18.16). The details of how these protons are transported through the protein are still under study. However, two effects contribute to the mechanism. First, charge neutrality tends to be maintained in the interior of proteins. Thus, the addition of an electron to a site inside a protein tends to favor the binding of H^+ to a nearby site. Second, conformational changes take place, particularly around the heme a_3–Cu_B center, in the course of the reaction cycle. Presumably, in one conformation, protons may enter the protein exclusively from the matrix side, whereas, in another, they may exit exclusively to the intermembrane space. Thus, the overall process catalyzed by cytochrome c oxidase is

$$4\,\text{Cyt}\,c_{\text{red}} + 8\,\text{H}^+_{\text{matrix}} + \text{O}_2 \rightarrow 4\,\text{Cyt}\,c_{\text{ox}} + 2\,\text{H}_2\text{O} + 4\,\text{H}^+_{\text{intermembrane space}}$$

Figure 18.17 summarizes the flow of electrons from NADH and FADH$_2$ through the respiratory chain. This series of exergonic reactions is coupled to the pumping of protons from the matrix. As we will see shortly, the energy inherent in the proton gradient will be used to synthesize ATP.

Most of the electron-transport chain is organized into a complex called the respirasome

The portrayal of the electron-transport chain in Figure 18.17 shows the individual components as isolated units. Although fine for illustrative purposes, this depiction is too simplistic. Recent research has established that three of the components of the electron-transport chain are arranged into a large complex called the *respirasome*. The human respirasome consists of 2 copies of Complex I, Complex III, and Complex IV (Figure 18.18). The respirasome forms a circular structure with two copies each of Complex I and Complex IV surrounding two copies of Complex III. Two copies of cytochrome c are located on the surface of Complex III. Although not experimentally established, the structure allows for Complex II to associate in a gap between Complexes I and IV. The structure of the respirasome, which was refractory to x-ray crystallography and nuclear magnetic resonance examination, was solved using the technique of cryo-electron microscopy (cryo-EM) (Section 3.5). The respirasome provides yet another example of multienzyme pathways that are organized into large complexes to enhance efficiency.

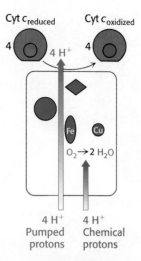

FIGURE 18.16 Proton transport by cytochrome *c* oxidase. Four protons are taken up from the matrix side to reduce one molecule of O$_2$ to two molecules of H$_2$O. These protons are called "chemical protons" because they participate in a clearly defined reaction with O$_2$. Four additional "pumped" protons are transported out of the matrix and released into the intermembrane space in the course of the reaction. The pumped protons double the efficiency of free-energy storage in the form of a proton gradient for this final step in the electron-transport chain.

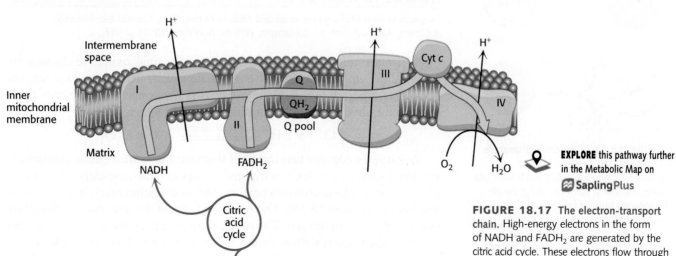

EXPLORE this pathway further in the Metabolic Map on **Sapling**Plus

FIGURE 18.17 The electron-transport chain. High-energy electrons in the form of NADH and FADH$_2$ are generated by the citric acid cycle. These electrons flow through the respiratory chain, which powers proton pumping and results in the reduction of O$_2$.

(A) Side view   Cyt *c*

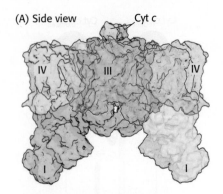

(B) View from top

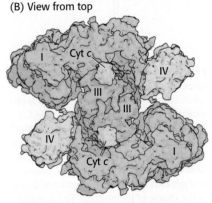

FIGURE 18.18 The respirasome. The structure of the electron-transport chain supercomplex or respirasome has been determined by cryo-electron microscopy. The complex consists of two copies each of Complex I, Complex III and Complex IV. Top and side views are shown, with cytochrome c bound to Complex III.

Dismutation
A reaction in which a single reactant is converted into two different products.

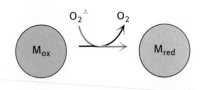

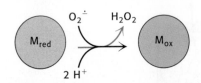

FIGURE 18.19 Superoxide dismutase mechanism. The oxidized form of superoxide dismutase (M_{ox}) reacts with one superoxide ion to form O_2 and generate the reduced form of the enzyme (M_{red}). The reduced form then reacts with a second superoxide and two protons to form hydrogen peroxide and regenerate the oxidized form of the enzyme.

Toxic derivatives of molecular oxygen such as superoxide radicals are scavenged by protective enzymes

As discussed earlier, molecular oxygen is an ideal terminal electron acceptor, because its high affinity for electrons provides a large thermodynamic driving force. However, danger lurks in the reduction of O_2. The transfer of four electrons leads to safe products (two molecules of H_2O), but partial reduction generates hazardous compounds. In particular, *the transfer of a single electron to O_2 forms superoxide ion, whereas the transfer of two electrons yields peroxide.*

$$O_2 \xrightarrow{e^-} O_2^{\cdot -} \xrightarrow{e^-} O_2^{2-}$$
$$\text{Superoxide} \qquad \text{Peroxide}$$
$$\text{ion}$$

Both compounds are potentially destructive. The strategy for the safe reduction of O_2 is clear: *the catalyst does not release partly reduced intermediates.* Cytochrome *c* oxidase meets this crucial criterion by holding O_2 tightly between Fe and Cu ions.

Although cytochrome *c* oxidase and other proteins that reduce O_2 are remarkably successful in not releasing intermediates, small amounts of superoxide anion and hydrogen peroxide are unavoidably formed. Superoxide, hydrogen peroxide, and species that can be generated from them such as the hydroxyl radical ($OH\cdot$) are collectively referred to as *reactive oxygen species* or *ROS*. Oxidative damage caused by ROS has been implicated in the aging process as well as in a growing list of diseases (Table 18.3).

TABLE 18.3 Pathological conditions that may entail free-radical injury

Atherogenesis
Emphysema; bronchitis
Parkinson disease
Duchenne muscular dystrophy
Cervical cancer
Alcoholic liver disease
Diabetes
Acute renal failure
Down syndrome
Retrolental fibroplasia (conversion of the retina into a fibrous mass in premature infants)
Cerebrovascular disorders
Ischemia; reperfusion injury

Information from Michael Lieberman and Allan D. Marks, *Basic Medical Biochemistry: A Clinical Approach*, 4th ed. (Lippincott, Williams & Wilkins, 2012), p. 437.

What are the cellular defense strategies against oxidative damage by ROS? Chief among them is the enzyme *superoxide dismutase*. This enzyme scavenges superoxide radicals by catalyzing the conversion of two of these radicals into hydrogen peroxide and molecular oxygen.

$$2\,O_2^{\cdot -} + 2\,H^+ \xrightarrow{\text{Superoxide dismutase}} O_2 + H_2O_2$$

Eukaryotes contain two forms of this enzyme, a manganese-containing version located in mitochondria and a copper-and-zinc-dependent cytoplasmic form. These enzymes perform the dismutation reaction by a similar mechanism (Figure 18.19). The oxidized form of the enzyme is reduced by superoxide to form oxygen. The reduced form of the enzyme, formed in this reaction, then reacts with a second superoxide ion to form peroxide, which takes up two protons along the reaction path to yield hydrogen peroxide.

The hydrogen peroxide formed by superoxide dismutase and by other processes is scavenged by *catalase*, a ubiquitous heme protein that catalyzes the dismutation of hydrogen peroxide into water and molecular oxygen.

$$2\,H_2O_2 \xrightleftharpoons{\text{Catalase}} O_2 + 2\,H_2O$$

Superoxide dismutase and catalase are remarkably efficient, performing their reactions at or near the diffusion-limited rate (Section 8.4). Glutathione peroxidase also plays a role in scavenging H_2O_2 (Section 20.5). Other cellular defenses against oxidative damage include the antioxidant vitamins, vitamins E and C. Because it is lipophilic, vitamin E is especially useful in protecting membranes from lipid peroxidation.

A long-term benefit of exercise may be to increase the amount of superoxide dismutase in the cell. The elevated aerobic metabolism during exercise causes more ROS to be generated. In response, the cell synthesizes more protective enzymes. The net effect is one of protection, because the increase in superoxide dismutase more effectively protects the cell during periods of rest (problem 54).

Despite the fact that reactive oxygen species are known hazards, recent evidence suggests that, under certain circumstances, the controlled generation of these molecules may be an important component of signal-transduction pathways. For instance, growth factors have been shown to increase ROS levels as part of their signaling pathway, and ROS regulate channels and transcription factors. ROS have been implicated in the control of cell differentiation, the immune response, and autophagy, as well as, other metabolic activities. The dual role of ROS is an excellent example of the wondrous complexity of biochemistry of living systems: even potentially harmful substances can be harnessed to play useful roles.

Electrons can be transferred between groups that are not in contact

How are electrons transferred between electron-carrying groups of the respiratory chain? This question is intriguing because these groups are frequently buried in the interior of a protein in fixed positions and are therefore not directly in contact with one another. Electrons can move through space, even through a vacuum. However, the rate of electron transfer through space falls off rapidly as the electron donor and electron acceptor move apart from each other, decreasing by a factor of 10 for each increase in separation of 0.8 Å. The protein environment provides more-efficient pathways for electron conduction: typically, the rate of electron transfer decreases by a factor of 10 every 1.7 Å (Figure 18.20). For groups in contact, electron-transfer reactions can be quite fast, with rates of approximately $10^{13}\,s^{-1}$. Within proteins in the electron-transport chain, electron-carrying groups are typically separated by 15 Å beyond their van der Waals contact distance. For such separations, we expect electron-transfer rates of approximately $10^4\,s^{-1}$ (i.e., electron transfer in less than 1 ms), assuming that all other factors are optimal. Without the mediation of the protein, an electron transfer over this distance would take approximately 1 day.

The case is more complicated when electrons must be transferred between two distinct proteins, such as when cytochrome *c* accepts electrons from Complex III or passes them on to Complex IV. A series of hydrophobic interactions

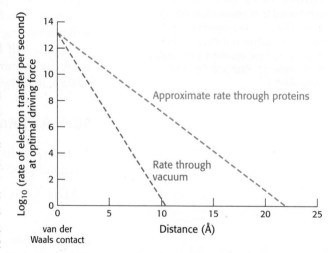

FIGURE 18.20 Distance dependence of electron-transfer rate. The rate of electron transfer decreases as the electron donor and the electron acceptor move apart. In a vacuum, the rate decreases by a factor of 10 for every increase of 0.8 Å. In proteins, the rate decreases more gradually, by a factor of 10 for every increase of 1.7 Å. This rate is only approximate because variations in the structure of the intervening protein medium can affect the rate.

FIGURE 18.21 Conservation of the three-dimensional structure of cytochrome *c*. *Notice* the overall structural similarity of the different molecules from different sources. The side chains are shown for the 21 conserved amino acids as well as the centrally located planar heme. [Drawn from 3CYT.pdb, 3C2C.pdb, and 155C.pdb.]

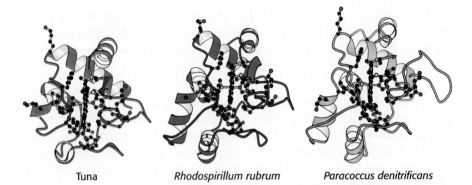

Tuna   *Rhodospirillum rubrum*   *Paracoccus denitrificans*

bring the heme groups of cytochrome *c* and c_1 to within 4.5 Å of each other, with the iron atoms separated by 17.4 Å. This distance could allow cytochrome *c* reduction at a rate of $8.3 \times 10^6\,\text{s}^{-1}$.

The conformation of cytochrome *c* has remained essentially constant for more than a billion years

Cytochrome *c* is present in all organisms having mitochondrial respiratory chains: plants, animals, and eukaryotic microorganisms. This electron carrier evolved more than 1.5 billion years ago, before the divergence of plants and animals. Its function has been conserved throughout this period, as evidenced by the fact that *the cytochrome* c *of any eukaryotic species reacts in vitro with the cytochrome* c *oxidase of any other species tested thus far.* For example, wheat-germ cytochrome *c* reacts with human cytochrome *c* oxidase. Additionally, some bacterial cytochromes, such as cytochrome c_2 from the photosynthetic bacterium *Rhodospirillum rubrum* and cytochrome c_{550} from the denitrifying bacterium *Paracoccus denitrificans*, closely resemble cytochrome *c* from tuna-heart mitochondria (Figure 18.21). This evidence attests to an efficient evolutionary solution to electron transfer bestowed by the structural and functional characteristics of cytochrome *c*.

The resemblance among cytochrome *c* molecules extends to the level of amino acid sequence. Because of the molecule's small size and ubiquity, the amino acid sequences of cytochrome *c* from more than 80 widely ranging eukaryotic species have been determined by direct protein sequencing. The striking finding is that *21 of 104 residues have been invariant for more than one and a half billion years of evolution.* A phylogenetic tree, constructed from the amino acid sequences of cytochrome *c*, reveals the evolutionary relationships between many animal species (Figure 18.22).

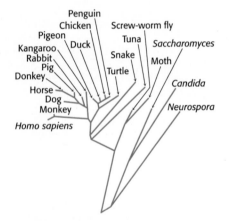

FIGURE 18.22 Evolutionary tree constructed from sequences of cytochrome *c*. Branch lengths are proportional to the number of amino acid changes that are believed to have occurred. [Information from Walter M. Fitch and Emanuel Margoliash.]

18.4 A Proton Gradient Powers the Synthesis of ATP

Thus far, we have considered the flow of electrons from NADH to O_2, an exergonic process.

$$\text{NADH} + \tfrac{1}{2}O_2 + H^+ \rightleftharpoons H_2O + NAD^+$$
$$\Delta G^{\circ\prime} = -220.1\,\text{kJ mol}^{-1}\,(-52.6\,\text{kcal mol}^{-1})$$

Next, we consider how this process is coupled to the synthesis of ATP, an endergonic process.

$$\text{ADP} + P_i + H^+ \rightleftharpoons \text{ATP} + H_2O$$
$$\Delta G^{\circ\prime} = +30.5\,\text{kJ mol}^{-1}\,(+7.3\,\text{kcal mol}^{-1})$$

A molecular assembly in the inner mitochondrial membrane carries out the synthesis of ATP. This enzyme complex was originally called the *mitochondrial*

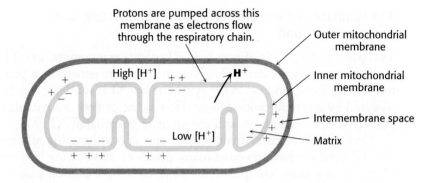

FIGURE 18.23 Chemiosmotic hypothesis. Electron transfer through the respiratory chain leads to the pumping of protons from the matrix to the intermembrane space. The pH gradient and membrane potential constitute a proton-motive force that is used to drive ATP synthesis.

ATPase or F_1F_0 *ATPase* because it was discovered through its catalysis of the reverse reaction, the hydrolysis of ATP. *ATP synthase*, its preferred name, emphasizes its actual role in the mitochondrion. It is also called *Complex V*.

How is the oxidation of NADH coupled to the phosphorylation of ADP? Electron transfer was first suggested to lead to the formation of a covalent high-energy intermediate that serves as a compound having a high phosphoryl-transfer potential, analogous to the generation of ATP by the formation of 1,3-bisphosphoglycerate in glycolysis (Section 16.1). An alternative proposal was that electron transfer aids the formation of an activated protein conformation, which then drives ATP synthesis. The search for such intermediates for several decades proved fruitless.

In 1961, Peter Mitchell suggested a radically different mechanism, *the chemiosmotic hypothesis*. He proposed that electron transport and ATP synthesis are coupled by *a proton gradient across the inner mitochondrial membrane*. In his model, the transfer of electrons through the respiratory chain leads to the pumping of protons from the matrix to the intermembrane space. The H^+ concentration becomes lower in the matrix, and an electric field with the matrix side negative is generated (Figure 18.23). Protons then flow back into the matrix to equalize the distribution. Mitchell's idea was that this flow of protons drives the synthesis of ATP by ATP synthase. The energy-rich unequal distribution of protons is called the *proton-motive force*. The proton-motive force is composed of two components: a chemical gradient and a charge gradient. The chemical gradient for protons can be represented as a pH gradient. The charge gradient is created by the positive charge on the unequally distributed protons forming the chemical gradient. Mitchell proposed that both components power the synthesis of ATP.

Proton-motive force (Δp) = chemical gradient (ΔpH) + charge gradient $(\Delta \psi)$

Mitchell's highly innovative hypothesis that oxidation and phosphorylation are coupled by a proton gradient is now supported by a wealth of evidence. Indeed, electron transport does generate a proton gradient across the inner mitochondrial membrane. The pH outside is 1.4 units lower than inside, and the membrane potential is 0.14 V, the outside being positive. As calculated on page 579, this membrane potential corresponds to a free energy of 21.8 kJ (5.2 kcal) per mole of protons.

An artificial system was created to elegantly demonstrate the basic principle of the chemiosmotic hypothesis. The role of the respiratory chain was played by bacteriorhodopsin, a membrane protein from halobacteria that pumps protons when illuminated. Synthetic vesicles containing bacteriorhodopsin and mitochondrial ATP synthase purified from beef heart were created (Figure 18.24). When the vesicles were exposed to light, ATP was formed. This key experiment clearly showed that *the respiratory chain and ATP synthase are biochemically separate systems, linked only by a proton-motive force.*

Some have argued that, along with the elucidation of the structure of DNA, the discovery that ATP synthesis is powered by a proton gradient is one of the two major advances in biology in the twentieth century. Mitchell's initial postulation of the chemiosmotic theory was not warmly received by all. Efraim Racker, one of the early investigators of ATP synthase, recalls that some thought of Mitchell as a court jester, whose work was of no consequence. Peter Mitchell was awarded the Nobel Prize in chemistry in 1978 for his contributions to understanding oxidative phosphorylation.

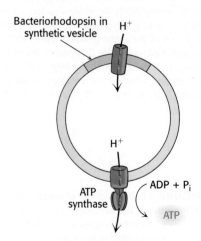

FIGURE 18.24 Testing the chemiosmotic hypothesis. ATP is synthesized when reconstituted membrane vesicles containing bacteriorhodopsin (a light-driven proton pump) and ATP synthase are illuminated.

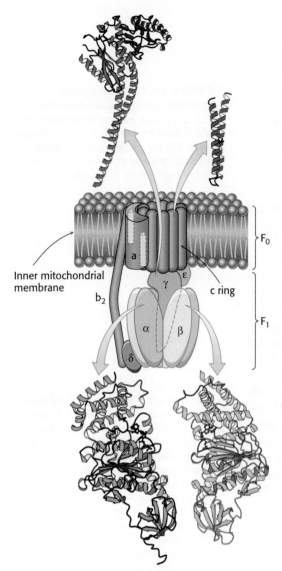

Inner mitochondrial membrane

FIGURE 18.25 Structure of ATP synthase. A schematic structure is shown along with representations of the components for which structures have been determined to high resolution. The P-loop NTPase domains of the α and β subunits are indicated by purple shading. *Notice* that part of the enzyme complex is embedded in the inner mitochondrial membrane, whereas the remainder resides in the matrix. [Drawn from 1E79.pdb and 1C0V.pdb.]

ATP synthase is composed of a proton-conducting unit and a catalytic unit

Two parts of the puzzle of how NADH oxidation is coupled to ATP synthesis are now evident: (1) electron transport generates a proton-motive force, and (2) ATP synthesis by ATP synthase can be powered by a proton-motive force. How is the proton-motive force converted into the high phosphoryl-transfer potential of ATP?

Biochemical, electron microscopic, and crystallographic studies of ATP synthase have revealed many details of its structure (Figure 18.25). It is a large, complex enzyme resembling a ball on a stick. Much of the "stick" part, called the F_0 subunit, is embedded in the inner mitochondrial membrane. The 85-Å-diameter ball, called the F_1 subunit, protrudes into the mitochondrial matrix. The F_1 subunit contains the catalytic activity of the synthase. In fact, isolated F_1 subunits display ATPase activity.

The F_1 subunit consists of five types of polypeptide chains (α_3, β_3, γ, δ, and ε) with the indicated stoichiometry. The α and β subunits, which make up the bulk of the F_1, are arranged alternately in a hexameric ring; they are homologous to one another and are members of the P-loop NTPase family (Section 9.4). Both bind nucleotides but only the β subunits are catalytically active. Just below the α and β subunits is a central stalk consisting of the γ and ε proteins. The γ subunit includes a long helical coiled coil that extends into the center of the $\alpha_3\beta_3$ hexamer. *The γ subunit breaks the symmetry of the $\alpha_3\beta_3$ hexamer: each of the β subunits is distinct by virtue of its interaction with a different face of γ.* Distinguishing the three β subunits is crucial for understanding the mechanism of ATP synthesis.

The F_0 subunit is a hydrophobic segment that spans the inner mitochondrial membrane. F_0 *contains the proton channel of the complex.* This channel consists of a ring comprising from 8 to 14 **c** subunits that are embedded in the membrane. A single **a** subunit binds to the outside of the ring. The F_0 and F_1 subunits are connected in two ways: by the central $\gamma\varepsilon$ stalk and by an exterior column. The exterior column consists of one **a** subunit, two **b** subunits, and the δ subunit.

ATP synthases interact with one another to form dimers, which then associate to form large oligomers of dimers (Figure 18.26). This association stabilizes the individual enzymes to the rotational forces required for catalysis (p. 597) and facilitates

FIGURE 18.26 A dimer of mitochondria ATP synthase. A schematic representation of a dimer of mitochondria ATP synthase is shown embedded in the inner mitochondrial membrane. The dimer, joined by an assortment of proteins, assists in bending the inner mitochondria membrane. The structure of the dimer was determined by cryo-electron microscopy.

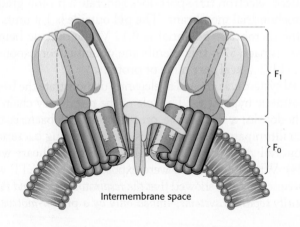

Intermembrane space

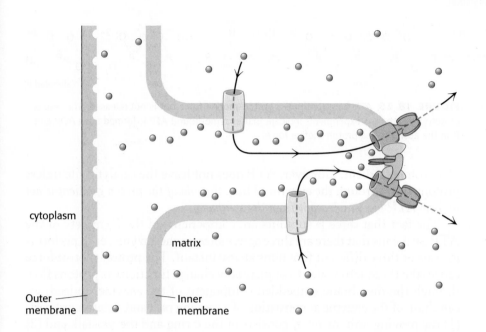

FIGURE 18.27 **ATPase assists in the formation of cristae.** Oligomers of ATP synthase dimers facilitate cristae formation, creating an area where the protons (red balls) are concentrated and have ready access to the F_o portion of the ATP synthase. The electron-transport chain is represented by the yellow cylinders embedded in the inner mitochondrial membrane. [Information from K. M. Davies et al., *Proc. Natl. Acad. Sci. U.S.A.* 108: 14121–14126, 2011.]

the curvature of the inner mitochondrial membrane. The formation of the cristae allows the proton pumps of the electron-transport chain to localize the proton gradient in the vicinity of the synthases, which are located at the tips of the cristae, thereby enhancing the efficiency of ATP synthesis (Figure 18.27).

Proton flow through ATP synthase leads to the release of tightly bound ATP: The binding-change mechanism

ATP synthase catalyzes the formation of ATP from ADP and orthophosphate.

$$ADP^{3-} + HPO_4^{2-} + H^+ \rightleftharpoons ATP^{4-} + H_2O$$

The actual substrates are ADP and ATP complexed with Mg^{2+}, as in all known phosphoryl-transfer reactions with these nucleotides. A terminal oxygen atom of ADP attacks the phosphorus atom of P_i to form a pentacovalent intermediate, which then dissociates into ATP and H_2O (Figure 18.28).

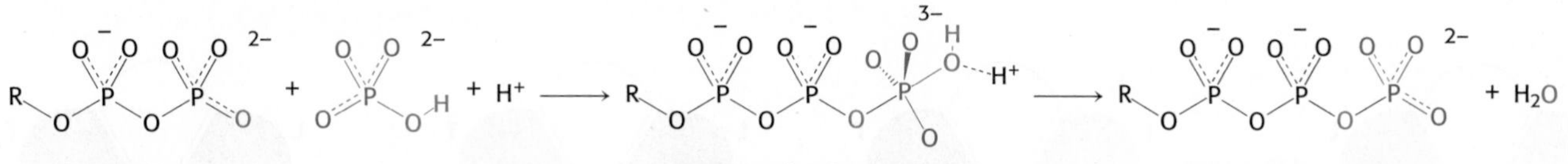

ADP   **P_i**   **Pentacovalent intermediate**   **ATP**

FIGURE 18.28 **ATP-synthesis mechanism.** One of the oxygen atoms of ADP attacks the phosphorus atom of P_i to form a pentacovalent intermediate, which then forms ATP and releases a molecule of H_2O.

How does the flow of protons drive the synthesis of ATP? Isotopic-exchange experiments unexpectedly revealed that *enzyme-bound ATP forms readily in the absence of a proton-motive force.* When ADP and P_i were added to ATP synthase in $H_2^{18}O$, ^{18}O became incorporated into P_i through the synthesis of ATP and its subsequent hydrolysis (Figure 18.29). The rate of incorporation of ^{18}O into P_i showed that about equal amounts of bound ATP and ADP are in equilibrium at the catalytic site, even in the absence

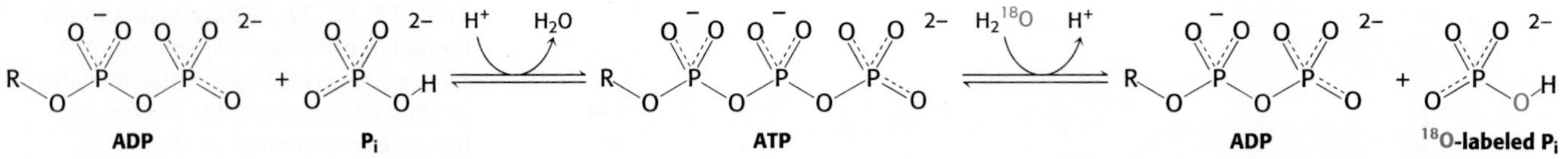

FIGURE 18.29 ATP forms without a proton-motive force but is not released. The results of isotopic-exchange experiments indicate that enzyme-bound ATP is formed from ADP and P_i in the absence of a proton-motive force.

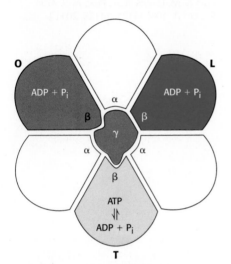

FIGURE 18.30 ATP synthase nucleotide-binding sites are not equivalent. The γ subunit passes through the center of the $\alpha_3\beta_3$ hexamer and makes the nucleotide-binding sites in the β subunits distinct from one another. The β subunits are colored to distinguish them from one another.

Progressive alteration of the forms of the three active sites of ATP synthase

Subunit 1 $L \to T \to O \to L \to T \to O$......
Subunit 2 $O \to L \to T \to O \to L \to T$......
Subunit 3 $T \to O \to L \to T \to O \to L$......

of a proton gradient. However, ATP does not leave the catalytic site unless protons flow through the enzyme. Thus, *the role of the proton gradient is not to form ATP but to release it from the synthase.*

The fact that three β subunits are components of the F_1 moiety of the ATPase means that there are three active sites on the enzyme, each performing one of three different functions at any instant. The proton-motive force causes the three active sites to sequentially change functions as protons flow through the membrane-embedded component of the enzyme. Indeed, we can think of the enzyme as consisting of a moving part and a stationary part: (1) the moving unit, or *rotor,* consists of the **c** ring and the $\gamma\varepsilon$ stalk and (2) the stationary unit, or *stator,* is composed of the remainder of the molecule.

How do the three active sites of ATP synthase respond to the flow of protons? A number of experimental observations suggested a *binding-change mechanism* for proton-driven ATP synthesis. This proposal states that a β subunit can perform each of three sequential steps in the synthesis of ATP by changing conformation. These steps are (1) ADP and P_i binding, (2) ATP synthesis, and (3) ATP release. As already noted, interactions with the γ subunit make the three β subunits structurally distinct (Figure 18.30). At any given moment, one β subunit will be in the L, or loose, conformation. This conformation binds ADP and P_i. A second subunit will be in the T, or tight, conformation. This conformation binds ATP with great avidity, so much so that it will convert bound ADP and P_i into ATP. Both the T and L conformations are sufficiently constrained that they cannot release bound nucleotides. The final subunit will be in the O, or open, form. This form has a more open conformation and can bind or release adenine nucleotides.

The rotation of the γ subunit drives the interconversion of these three forms (Figure 18.31). ADP and P_i bound in the subunit in the T form are transiently combining to form ATP. Suppose that the γ subunit is rotated by 120 degrees in a counterclockwise direction (as viewed from the top). This rotation converts the T-form site into an O-form site with the

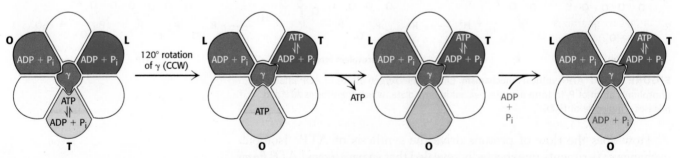

FIGURE 18.31 Binding-change mechanism for ATP synthase. The rotation of the γ subunit interconverts the three β subunits. The subunit in the T (tight) form interconverts ADP and P_i and ATP but does not allow ATP to be released. When the γ subunit is rotated by 120 degrees in a counterclockwise (CCW) direction, the T-form subunit is converted into the O form, allowing ATP release. ADP and P_i can then bind to the O-form subunit. An additional 120-degree rotation (not shown) traps these substrates in an L-form subunit.

nucleotide bound as ATP. Concomitantly, the L-form site is converted into a T-form site, enabling the transformation of an additional ADP and P_i into ATP. The ATP in the O-form site can now depart from the enzyme to be replaced by ADP and P_i. An additional 120-degree rotation converts this O-form site into an L-form site, trapping these substrates. Each subunit progresses from the T to the O to the L form with no two subunits ever present in the same conformational form. This mechanism suggests that ATP can be synthesized and released by driving the rotation of the γ subunit in the appropriate direction.

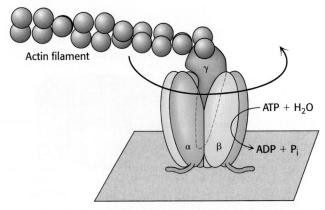

FIGURE 18.32 Direct observation of ATP-driven rotation in ATP synthase. The $\alpha_3\beta_3$ hexamer of ATP synthase is fixed to a surface, with the γ subunit projecting upward and linked to a fluorescently labeled actin filament. The addition and subsequent hydrolysis of ATP result in the counterclockwise rotation of the γ subunit, which can be directly seen under a fluorescence microscope.

Rotational catalysis is the world's smallest molecular motor

Is it possible to observe the proposed rotation directly? Elegant experiments, using single-molecule techniques (Section 8.6), have demonstrated the rotation through the use of a simple experimental system consisting solely of cloned $\alpha_3\beta_3\gamma$ subunits (Figure 18.32). The β subunits were engineered to contain amino-terminal polyhistidine tags, which have a high affinity for nickel ions. This property of the tags allowed the $\alpha_3\beta_3$ assembly to be immobilized on a glass surface that had been coated with nickel ions. The γ subunit was linked to a fluorescently labeled actin filament to provide a long segment that could be observed under a fluorescence microscope. Remarkably, the addition of ATP caused the actin filament to rotate unidirectionally in a counterclockwise direction. *The γ subunit was rotating, driven by the hydrolysis of ATP.* Thus, the catalytic activity of an individual molecule could be observed. The counterclockwise rotation is consistent with the predicted mechanism for hydrolysis because the molecule was viewed from below relative to the view shown in Figure 18.32.

More-detailed analysis in the presence of lower concentrations of ATP revealed that the γ subunit rotates in 120-degree increments. Each increment corresponds to the hydrolysis of a single ATP molecule. In addition, from the results obtained by varying the length of the actin filament and measuring the rate of rotation, the enzyme appears to operate near 100% efficiency; that is, essentially all of the energy released by ATP hydrolysis is converted into rotational motion.

Proton flow around the c ring powers ATP synthesis

The direct observation of rotary motion of the γ subunit is strong evidence for the rotational mechanism for ATP synthesis. The last remaining question is: How does proton flow through F_0 drive the rotation of the γ subunit? The mechanism depends on the structures of the **a** and **c** subunits of F_0 (Figure 18.33). The stationary **a** subunit directly abuts the membrane-spanning ring formed by 8 to 14 **c** subunits. The **a** subunit includes two hydrophilic half-channels that do not span the membrane (Figure 18.33). Thus, protons can pass into either of these channels, but they cannot move completely across the membrane. The **a** subunit is positioned such that each half-channel directly interacts with one **c** subunit.

The structure of the **c** subunit was determined both by NMR methods and by x-ray crystallography. Each polypeptide chain forms a pair of α helices that span the membrane. A glutamic acid (or aspartic acid) residue is found in the middle of one of the helices. If the glutamate is charged (unprotonated), the **c** subunit will not move into the membrane.

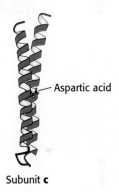

Subunit **c**

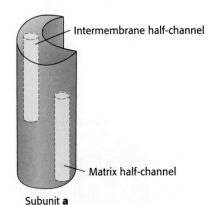

Subunit **a**

FIGURE 18.33 Components of the proton-conducting unit of ATP synthase. The **c** subunit consists of two α helices that span the membrane. In *E. coli*, an aspartic acid residue in one of the helices lies on the center of the membrane. The structure of the **a** subunit has not yet been directly observed, but it appears to include two half-channels that allow protons to enter and pass partway but not completely through the membrane.

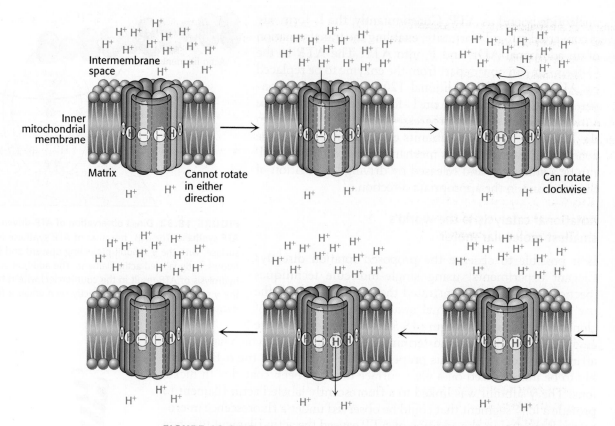

FIGURE 18.34 Proton motion across the membrane drives rotation of the c ring. A proton enters from the intermembrane space into the cytoplasmic half-channel to neutralize the charge on an aspartate residue in a **c** subunit. With this charge neutralized, the **c** ring can rotate clockwise by one **c** subunit, moving an aspartic acid residue out of the membrane into the matrix half-channel. This proton can move into the matrix, resetting the system to its initial state.

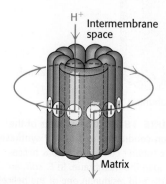

FIGURE 18.35 Proton path through the membrane. Each proton enters the cytoplasmic half-channel, follows a complete rotation of the **c** ring, and exits through the other half-channel into the matrix.

The key to proton movement across the membrane is that, in a proton-rich environment, such as the intermembrane space, a proton will enter a channel and bind the glutamate residue, while the glutamic acid in a proton-poor environment of the other half-channel will release a proton (Figure 18.34). The **c** subunit with the bound proton then moves into the membrane as the ring rotates by one **c** subunit. This rotation brings the newly deprotonated **c** subunit from the matrix half-channel to the proton-rich intermembrane space half-channel, where it can bind a proton. *The movement of protons through the half-channels from the high proton concentration of the intermembrane space to the low proton concentration of the matrix powers the rotation of the **c** ring.* The **a** unit remains stationary as the **c** ring rotates. Each proton that enters the intermembrane space half-channel of the **a** unit moves through the membrane by riding around on the rotating **c** ring to exit through the matrix half-channel into the proton-poor environment of the matrix (Figure 18.35).

How does the rotation of the **c** ring lead to the synthesis of ATP? The **c** ring is tightly linked to the γ and ε subunits. Thus, as the **c** ring turns, the γ and ε subunits are turned inside the $\alpha_3\beta_3$ hexamer unit of F_1. The rotation of the γ subunit in turn promotes the synthesis of ATP through the binding-change mechanism. The exterior column formed by the two **b** chains and the δ subunit prevents the $\alpha_3\beta_3$ hexamer from rotating in sympathy with the **c** ring. Recall that the dimerization and subsequent oligomerization of the synthase in the cristae also stabilizes the enzyme to rotational forces. As stated earlier, the number of **c** subunits in the **c** ring

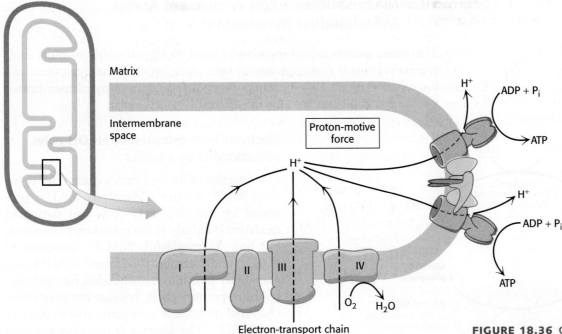

Matrix

Intermembrane space

Proton-motive force

H⁺

Electron-transport chain

I   II   III   IV

O_2 H_2O

H⁺

ADP + P_i

ATP

H⁺

ADP + P_i

ATP

FIGURE 18.36 Overview of oxidative phosphorylation. The electron-transport chain generates a proton gradient, which is used to synthesize ATP.

appears to range between 8 and 14. This number is significant because it determines the number of protons that must be transported to generate a molecule of ATP. Each 360-degree rotation of the γ subunit leads to the synthesis and release of three molecules of ATP. Thus, if there are 10 **c** subunits in the ring (as was observed in a crystal structure of yeast mitochondrial ATP synthase), each ATP generated requires the transport of $10/3 = 3.33$ protons. Recent evidence shows that the **c** rings of all vertebrates are composed of 8 subunits, making vertebrate ATP synthase the most efficient ATP synthase known, with the transport of only 2.7 protons required for ATP synthesis. For simplicity, we will assume that three protons must flow into the matrix for each ATP formed, but we must keep in mind that the true value may differ. As we will see, the electrons from NADH pump enough protons to generate 2.5 molecules of ATP, whereas those from $FADH_2$ yield 1.5 molecules of ATP.

Let us return for a moment to the example with which we began this chapter. If a resting human being requires 85 kg of ATP per day for bodily functions, then 3.3×10^{25} protons must flow through the ATP synthase per day, or 3.9×10^{20} protons per second. Figure 18.36 summarizes the process of oxidative phosphorylation.

ATP synthase and G proteins have several common features

The α and β subunits of ATP synthase are members of the P-loop NTPase family of proteins. In Chapter 14, we learned that the signaling properties of other members of this family, the G proteins, depend on their ability to bind nucleoside triphosphates and diphosphates with great tenacity. They do not exchange nucleotides unless they are stimulated to do so by interaction with other proteins. The binding-change mechanism of ATP synthase is a variation on this theme. The P-loop regions of the β subunits will bind either ADP or ATP (or release ATP), depending on which of three different faces of the γ subunit they interact with. The conformational changes take place in an orderly way, driven by the rotation of the γ subunit.

A little goes a long way

Despite the various molecular machinations and the vast numbers of ATPs synthesized and protons pumped, a resting human being requires surprisingly little power. Approximately 116 watts, the energy output of a typical light bulb, provides enough energy to sustain a resting person.

18.5 Many Shuttles Allow Movement Across Mitochondrial Membranes

The inner mitochondrial membrane must be impermeable to most molecules, yet much exchange has to take place between the cytoplasm and the mitochondria. This exchange is mediated by an array of membrane-spanning transporter proteins (see Figure 13.10).

Electrons from cytoplasmic NADH enter mitochondria by shuttles

One function of the respiratory chain is to regenerate NAD^+ for use in glycolysis. NADH generated by the citric acid cycle and fatty acid oxidation is already in the mitochondrial matrix, but how is cytoplasmic NADH reoxidized to NAD^+ under aerobic conditions? NADH cannot simply pass into mitochondria for oxidation by the respiratory chain, because the inner mitochondrial membrane is impermeable to NADH and NAD^+. The solution is that *electrons from NADH*, rather than NADH itself, are carried across the mitochondrial membrane. One of several means of introducing electrons from NADH into the electron-transport chain is the *glycerol 3-phosphate shuttle* (Figure 18.37). The first step in this shuttle is the transfer of a pair of electrons from NADH to dihydroxyacetone phosphate, a glycolytic intermediate, to form glycerol 3-phosphate. This reaction is catalyzed by a glycerol 3-phosphate dehydrogenase in the cyto-

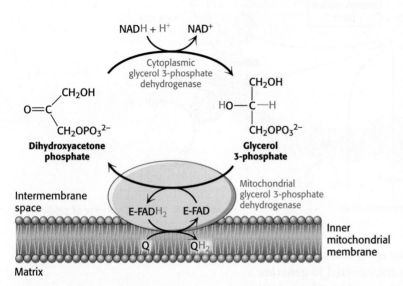

FIGURE 18.37 Glycerol 3-phosphate shuttle. Electrons from NADH can enter the mitochondrial electron-transport chain by being used to reduce dihydroxyacetone phosphate to glycerol 3-phosphate. Glycerol 3-phosphate is reoxidized by electron transfer to an FAD prosthetic group in a membrane-bound glycerol 3-phosphate dehydrogenase. Subsequent electron transfer to Q to form QH_2 allows these electrons to enter the electron-transport chain.

plasm. Glycerol 3-phosphate is reoxidized to dihydroxyacetone phosphate on the outer surface of the inner mitochondrial membrane by a membrane-bound isozyme of glycerol 3-phosphate dehydrogenase. An electron pair from glycerol 3-phosphate is transferred to an FAD prosthetic group in this enzyme to form $FADH_2$. This reaction also regenerates dihydroxyacetone phosphate.

The reduced flavin transfers its electrons to the electron carrier Q, which then enters the respiratory chain as QH_2. *When cytoplasmic NADH transported by the glycerol 3-phosphate shuttle is oxidized by the respiratory chain, 1.5 rather than 2.5 molecules of ATP are formed.* The yield is lower because FAD rather than NAD^+ is the electron acceptor in mitochondrial glycerol 3-phosphate dehydrogenase. The use of FAD enables electrons from cytoplasmic NADH to be transported into mitochondria against an NADH concentration gradient. The price of this transport is one molecule of ATP per two electrons. This glycerol 3-phosphate shuttle is especially prominent in muscle and enables it to sustain a very high rate of oxidative phosphorylation. Indeed, some insects lack lactate dehydrogenase and are completely dependent on the glycerol 3-phosphate shuttle for the regeneration of cytoplasmic NAD^+.

In the heart and liver, electrons from cytoplasmic NADH are brought into mitochondria by the *malate–aspartate shuttle*, which is mediated by two membrane carriers and four enzymes (Figure 18.38). Electrons are transferred from NADH in the cytoplasm to oxaloacetate, forming malate, which traverses the inner mitochondrial membrane in exchange for α-ketoglutarate and is then reoxidized by NAD^+ in the matrix to form NADH in a reaction catalyzed by the citric acid cycle enzyme malate dehydrogenase. The resulting oxaloacetate does not

$$NADH + H^+ + E\text{–}FAD$$
Cytoplasmic   Mitochondrial

$$NAD^+ + E\text{–}FADH_2$$
Cytoplasmic   Mitochondrial

Glycerol 3-phosphate shuttle

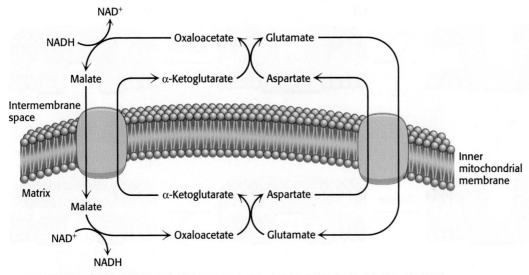

FIGURE 18.38 Malate–aspartate shuttle.

readily cross the inner mitochondrial membrane and so a transamination reaction (Section 23.3) is needed to form aspartate, which can be transported to the cytoplasmic side in exchange for glutamate. Glutamate donates an amino group to oxaloacetate, forming aspartate and α-ketoglutarate. In the cytoplasm, aspartate is then deaminated to form oxaloacetate and the cycle is restarted.

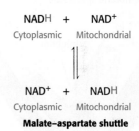

The entry of ADP into mitochondria is coupled to the exit of ATP by ATP-ADP translocase

The major function of oxidative phosphorylation is to generate ATP from ADP. ATP and ADP do not diffuse freely across the inner mitochondrial membrane. How are these highly charged molecules moved across the inner membrane into the cytoplasm? A specific transport protein, *ATP-ADP translocase*, enables these molecules to transverse this permeability barrier. Most important, the flows of ATP and ADP are coupled. *ADP enters the mitochondrial matrix only if ATP exits, and vice versa.* This process is carried out by the translocase, an antiporter:

$$ADP^{3-}_{cytoplasm} + ATP^{4-}_{matrix} \rightarrow ADP^{3-}_{matrix} + ATP^{4-}_{cytoplasm}$$

ATP-ADP translocase, also called the adenine nucleotide translocase (ANT), is highly abundant, constituting about 15% of the protein in the inner mitochondrial membrane. The abundance is a manifestation of the fact that human beings exchange the equivalent of their weight in ATP each day. Despite the fact that mitochondria are the sites of ATP synthesis, both the cytoplasm and the nucleus have more ATP than the mitochondria, a testament to the efficiency of transport.

The 30-kDa translocase contains a single nucleotide-binding site that alternately faces the matrix and the cytoplasmic sides of the membrane (Figure 18.39). ATP and ADP bind to the translocase without Mg^{2+}, and ATP has one more negative charge than that of ADP. Thus, in an actively respiring mitochondrion with a positive membrane potential, ATP transport out of the mitochondrial matrix and ADP transport into the matrix are favored. This ATP-ADP exchange is energetically expensive; about a quarter of the proton-motive force generated by the respiratory chain is consumed by this exchange process. The inhibition of the translocase leads to the subsequent inhibition of cellular respiration as well (p. 608).

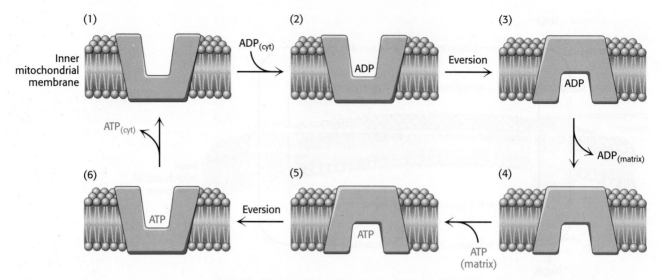

FIGURE 18.39 Mechanism of mitochondrial ATP-ADP translocase. The translocase catalyzes the coupled entry of ADP into the matrix and the exit of ATP from it. The binding of ADP (1) from the cytoplasm favors eversion of the transporter (2) to release ADP into the matrix (3). Subsequent binding of ATP from the matrix to the everted form (4) favors eversion back to the original conformation (5), releasing ATP into the cytoplasm (6).

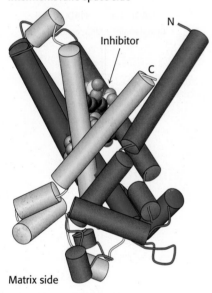

FIGURE 18.40 Structure of mitochondrial transporters. The structure of the ATP-ADP translocase is shown. *Notice* that this structure comprises three similar units (shown in red, blue, and yellow) that come together to form a binding site, here occupied by atractyloside, an inhibitor of this transporter. Other members of the mitochondrial transporter family adopt similar tripartite structures. [Drawn from 1OKC.pdb.]

Mitochondrial transporters for metabolites have a common tripartite structure

Examination of the amino acid sequence of the ATP-ADP translocase revealed that this protein consists of three tandem repeats of a 100-amino-acid module, each of which appears to have two transmembrane segments. This tripartite structure has recently been confirmed by the determination of the three-dimensional structure of this transporter (Figure 18.40). The transmembrane helices form a tepee-like structure with the nucleotide-binding site (marked by a bound inhibitor) lying in the center. Each of the three repeats adopts a similar structure.

ATP-ADP translocase is but one of many mitochondrial transporters for ions and charged metabolites (Figure 18.41). The *phosphate carrier*, which works in concert with ATP-ADP translocase, mediates the electroneutral exchange of $H_2PO_4^-$ for OH^-. The combined action of these two transporters leads to the exchange of cytoplasmic ADP and P_i for matrix ATP at the cost of the influx of one H^+ (owing to the transport of one OH^- out of the matrix). These two transporters, which provide ATP synthase with its substrates, are associated with the synthase to form a large complex called the *ATP synthasome*.

FIGURE 18.41 Mitochondrial transporters. Transporters (also called carriers) are transmembrane proteins that carry specific ions and charged metabolites across the inner mitochondrial membrane.

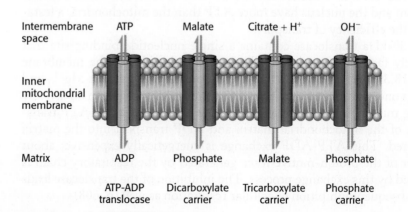

Other homologous carriers also are present in the inner mitochondrial membrane. The dicarboxylate carrier enables malate, succinate, and fumarate to be exported from the mitochondrial matrix in exchange for P_i. The tricarboxylate carrier exchanges citrate and H^+ for malate. In all, more than 40 such carriers are encoded in the human genome. A different type of carrier is responsible for pyruvate transport into the mitochondria. Pyruvate in the cytoplasm enters the mitochondrial membrane by way of a heterodimer composed of two small transmembrane proteins.

18.6 The Regulation of Cellular Respiration Is Governed Primarily by the Need for ATP

Because ATP is the end product of cellular respiration, the ATP needs of the cell are the ultimate determinant of the rate of respiratory pathways and their components.

The complete oxidation of glucose yields about 30 molecules of ATP

We can now estimate how many molecules of ATP are formed when glucose is completely oxidized to CO_2. The number of ATP molecules formed in glycolysis and the citric acid cycle is unequivocally known because it is determined by the stoichiometries of chemical reactions. In contrast, the ATP yield of oxidative phosphorylation is less certain because the stoichiometries of proton pumping, ATP synthesis, and metabolite-transport processes need not be an integer or even have fixed values. As stated earlier, the best current estimates for the number of protons pumped out of the matrix by NADH-Q oxidoreductase, Q-cytochrome c oxidoreductase, and cytochrome c oxidase per electron pair are four, two, and four, respectively. The synthesis of a molecule of ATP is driven by the flow of about three protons through ATP synthase. An additional proton is consumed in transporting ATP from the matrix to the cytoplasm. Hence, about 2.5 molecules of cytoplasmic ATP are generated as a result of the flow of a pair of electrons from NADH to O_2. For electrons that enter at the level of Q-cytochrome c oxidoreductase, such as those from the oxidation of succinate or cytoplasmic NADH transferred by the glycerol-phosphate shuttle, the yield is about 1.5 molecules of ATP per electron pair. Hence, as tallied in Table 18.4, *about 30 molecules of ATP are formed when glucose is completely oxidized to CO_2.* Most of the ATP, 26 of 30 molecules formed, is generated by oxidative phosphorylation. Recall that the anaerobic metabolism of glucose yields only 2 molecules of ATP. One of the effects of endurance exercise, a practice that calls for much ATP for an extended period of time, is to increase the number of mitochondria and blood vessels in muscle and thus increase the extent of ATP generation by oxidative phosphorylation.

The rate of oxidative phosphorylation is determined by the need for ATP

How is the rate of the electron-transport chain controlled? Under most physiological conditions, electron transport is tightly coupled to phosphorylation. *Electrons do not usually flow through the electron-transport chain to O_2 unless ADP is simultaneously phosphorylated to ATP.* When ADP concentration rises, as would be the case in active muscle, the rate of oxidative phosphorylation increases to meet the ATP needs of the muscle. The regulation of the rate of oxidative phosphorylation by the ADP level is called *respiratory control*

TABLE 18.4 ATP yield from the complete oxidation of glucose

Reaction sequence	ATP yield per glucose molecule
Glycolysis: Conversion of glucose into pyruvate (in the cytoplasm)	
Phosphorylation of glucose	−1
Phosphorylation of fructose 6-phosphate	−1
Dephosphorylation of 2 molecules of 1,3-BPG	+2
Dephosphorylation of 2 molecules of phosphoenolpyruvate	+2
2 molecules of NADH are formed in the oxidation of 2 molecules of glyceraldehyde 3-phosphate	
Conversion of pyruvate into acetyl CoA (inside mitochondria)	
2 molecules of NADH are formed	
Citric acid cycle (inside mitochondria)	
2 molecules of adenosine triphosphate are formed from 2 molecules of succinyl CoA	+2
6 molecules of NADH are formed in the oxidation of 2 molecules each of isocitrate, α-ketoglutarate, and malate	
2 molecules of FADH$_2$ are formed in the oxidation of 2 molecules of succinate	
Oxidative phosphorylation (inside mitochondria)	
2 molecules of NADH formed in glycolysis; each yields 1.5 molecules of ATP (assuming transport of NADH by the glycerol 3-phosphate shuttle)	+3
2 molecules of NADH formed in the oxidative decarboxylation of pyruvate; each yields 2.5 molecules of ATP	+5
2 molecules of FADH$_2$ formed in the citric acid cycle; each yields 1.5 molecules of ATP	+3
6 molecules of NADH formed in the citric acid cycle; each yields 2.5 molecules of ATP	+15
Net Yield per Molecule of Glucose	**+30**

Information on the ATP yield of oxidative phosphorylation is from values given in P. C. Hinkle, M. A. Kumar, A. Resetar, and D. L. Harris, *Biochemistry* 30:3576, 1991.

Note: The current value of 30 molecules of ATP per molecule of glucose supersedes the earlier value of 36 molecules of ATP. The stoichiometries of proton pumping, ATP synthesis, and metabolite transport should be regarded as estimates. About 2 more molecules of ATP are formed per molecule of glucose oxidized when the malate–aspartate shuttle rather than the glycerol 3-phosphate shuttle is used.

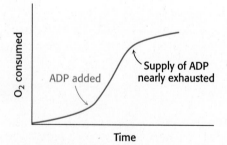

FIGURE 18.42 Respiratory control. Electrons are transferred to O$_2$ only if ADP is concomitantly phosphorylated to ATP.

or *acceptor control*. Experiments on isolated mitochondria demonstrate the importance of ADP level (Figure 18.42). The rate of oxygen consumption by mitochondria increases markedly when ADP is added and then returns to its initial value when the added ADP has been converted into ATP.

The level of ADP likewise affects the rate of the citric acid cycle. At low concentrations of ADP, as in a resting muscle, NADH and FADH$_2$ are not consumed by the electron-transport chain. The citric acid cycle slows because there is less NAD$^+$ and FAD to feed the cycle. As the ADP level rises and oxidative phosphorylation speeds up, NADH and FADH$_2$ are oxidized, and the citric acid cycle becomes more active. *Electrons do not flow from fuel molecules to O$_2$ unless ATP needs to be synthesized.* We see here another example of the regulatory significance of the energy charge. The coordination of the components of cellular respiration as shown in Figure 18.43 makes such regulation possible.

ATP synthase can be regulated

Mitochondria contain an evolutionarily conserved protein, *inhibitory factor 1* (IF1), that specifically inhibits the potential hydrolytic activity of the F$_0$F$_1$ ATP synthase. What is the function of IF1? Consider a circumstance where tissues

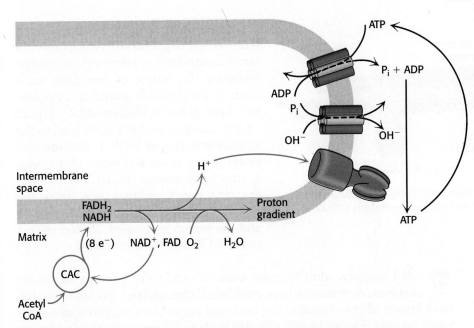

FIGURE 18.43 The integration of cellular respiration. The citric acid cycle (CAC) generates high-energy electrons by oxidizing acetyl CoA. The electrons are used to reduce oxygen to water and in the process develop a proton gradient. The proton gradient powers the synthesis of ATP, which requires the coordination of ATP synthase, the ATP-ADP translocase, and the phosphate carrier.

may be deprived of oxygen (ischemia). Without oxygen as the electron acceptor, the electron-transport chain will be unable to generate the proton-motive force. The ATP in the mitochondria would be hydrolyzed by the synthase, working in reverse (problem 51). The role of IF1 is to prevent the wasteful hydrolysis of ATP by inhibiting the hydrolytic activity of the synthase.

IF1 is overexpressed in many types of cancer. This overexpression plays a role in the induction of the Warburg effect, the switch from oxidative phosphorylation to aerobic glycolysis as the principal means for ATP synthesis (Section 16.2).

Regulated uncoupling leads to the generation of heat

Some organisms possess the ability to uncouple oxidative phosphorylation from ATP synthesis to generate heat. Such uncoupling is a means to maintain body temperature in hibernating animals, in some newborn animals (including human beings), and in many adult mammals, especially those adapted to cold. The skunk cabbage uses an analogous mechanism to heat its floral spikes in early spring, increasing the evaporation of odoriferous molecules that attract insects to fertilize its flowers. In animals, the uncoupling is in *brown adipose tissue* (BAT), which is specialized tissue for the process of *nonshivering thermogenesis*. In contrast, *white adipose tissue* (WAT), which constitutes the bulk of adipose tissue, plays no role in thermogenesis but serves as an energy source and an endocrine gland (Chapters 26 and 27).

Brown adipose tissue is very rich in mitochondria, often called *brown fat mitochondria*. The tissue appears brown from the combination of the greenish-colored cytochromes in the numerous mitochondria and the red hemoglobin present in the extensive blood supply, which helps to carry the heat through the body. The inner mitochondrial membrane of these mitochondria contains a large amount of *uncoupling protein* (UCP-1), or *thermogenin*, a dimer of 33-kDa subunits that resembles ATP-ADP translocase. UCP-1 transports protons from the intermembrane space to the matrix with the assistance of fatty acids. In essence, *UCP-1 generates heat by short-circuiting the mitochondrial proton battery.* The energy of the proton gradient, normally captured as ATP, is released as heat as the protons flow through UCP-1 to the mitochondrial matrix. This dissipative proton

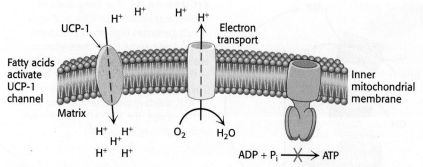

FIGURE 18.44 Action of an uncoupling protein. Uncoupling protein (UCP-1) generates heat by permitting the influx of protons into the mitochondria without the synthesis of ATP.

pathway is activated when the core body temperature begins to fall. In response to a temperature drop, α-adrenergic hormones stimulate the release of free fatty acids from triacylglycerols stored in cytoplasmic lipid granules (Section 22.2) (Figure 18.44). Long-chain fatty acids bind to the cytoplasmic face of UCP-1, and the carboxyl group binds a proton. This causes a structural change in UCP-1 so that the protonated carboxyl now faces the proton-poor environment of the matrix, and the proton is released. Proton release resets UCP-1 to the initial state.

Until recently, adult humans were believed to lack brown fat tissue. However, new studies have established that adults, especially females, have brown adipose tissue in the neck and upper chest regions that is activated by cold (Figure 18.45). Obesity leads to a decrease in brown adipose tissue.

We can witness the effects of a lack of nonshivering thermogenesis by examining pig behavior. The ancestors of pigs are believed to have lost the gene for UCP-1 ~20 million years ago when they inhabited tropical and subtropical environments where they could survive without nonshivering thermogenesis. As the range of pigs expanded, however, the lack of UCP-1 became a liability. Pigs are unusual mammals in that they have large litters and are the only ungulates (hoofed animals) that build nests for birth. These behavioral characteristics appear to be an adaptation to the absence of UCP-1 and, hence, brown fat. Piglets must rely on other means of thermogenesis, such as nesting, large litter size, and shivering.

In addition to UCP-1, two other uncoupling proteins have been identified. UCP-2, which is 56% identical in sequence with UCP-1, is found in a wide variety of tissues. UCP-3 (57% identical with UCP-1 and 73% identical with UCP-2) is localized to skeletal muscle and brown fat. This family of uncoupling proteins, especially UCP-2 and UCP-3, may

FIGURE 18.45 Brown adipose tissue is revealed on exposure to cold. The results of PET–CT scanning show the uptake and distribution of ^{18}F-fluorodeoxyglucose (^{18}F-FDG) in adipose tissue. The patterns of ^{18}F-FDG uptake in the same subject are dramatically different under thermoneutral conditions (A) and after exposure to cold (B). [Republished with permission of American Society for Clinical Investigation, from *J Clin Invest*. 2013;123(8): 3395–3403. doi:10.1172/JCI68993. Permission conveyed through Copyright Clearance Center, Inc.]

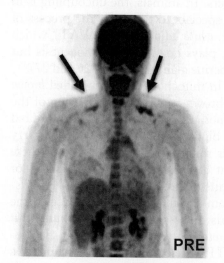

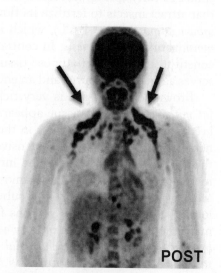

(A) (B)

play a role in energy homeostasis. In fact, the genes for UCP-2 and UCP-3 map to regions of the human and mouse chromosomes that have been linked to obesity, supporting the notion that they function as a means of regulating body weight.

Reintroduction of UCP-1 into pigs may be economically valuable

As discussed above, pigs lack UCP-1. This deficiency has important implications for pig farming. When a piglet is born, it experiences a sudden decrease of its thermal environment of about 15°C. Even with heat lamps to warm the piglets, mortality may be as high as 20% per liter because of hypothermia. Moreover, heating the piglets is a significant cost, accounting for approximately 35% of the energy cost of raising pigs. Furthermore, because of the lack of UCP-1, pigs accumulate fat as a thermal insulator. More fat means there is less lean meat available.

Recently, researchers have inserted the gene for UCP-1 into pig embryos. The resulting pigs are cheaper to raise due to the decrease in energy costs and provide more high quality lean meat. The gene for UCP-1 was inserted into the pig embryos using the CRISPR technique described in Chapter 5. We see here another example of the expanding reach of biochemistry into areas of the economy beyond just the biomedical arena.

Oxidative phosphorylation can be inhibited at many stages

Many potent and lethal poisons exert their effect by inhibiting oxidative phosphorylation at one of a number of different locations (Figure 18.46):

1. *Inhibition of the electron-transport chain. Rotenone,* which is used as a fish and insect poison, and *amytal,* a barbiturate sedative, block electron transfer in NADH-Q oxidoreductase and thereby prevent the utilization of NADH as a substrate. Rotenone, as an electron-transport-chain inhibitor, may play a role, along with genetic susceptibility, in the development of Parkinson disease. In the presence of rotenone and amytal, electron flow resulting from the oxidation of succinate is unimpaired, because these electrons enter through QH_2, beyond the block. *Antimycin A* interferes with electron flow from cytochrome b_H in Q-cytochrome c oxidoreductase. Furthermore, electron flow in cytochrome c oxidase can be blocked by *cyanide* (CN^-), *azide* (N_3^-), and *carbon monoxide* (CO). Cyanide and azide react with the ferric form of heme a_3, whereas carbon monoxide inhibits the ferrous form. Inhibition of the electron-transport chain also inhibits ATP synthesis because the proton-motive force can no longer be generated.

2. *Inhibition of ATP synthase.* Oligomycin, an antibiotic used as an antifungal agent, and dicyclohexylcarbodiimide (DCC) prevent the influx of protons through ATP synthase by binding to the carboxylate group of the **c** subunits required for proton binding. Modification of only one **c** subunit by DCC is sufficient to inhibit the rotation of the entire **c** ring and hence ATP synthesis. If actively respiring mitochondria are exposed to an inhibitor of ATP synthase, the electron-transport chain ceases to operate. This observation clearly illustrates that electron transport and ATP synthesis are normally tightly coupled.

3. *Uncoupling electron transport from ATP synthesis.* The tight coupling of electron transport and phosphorylation in mitochondria can be uncoupled by 2,4-dinitrophenol (DNP) and certain other acidic aromatic compounds. These substances carry protons across the inner mitochondrial membrane, down their concentration gradient. In the presence of these

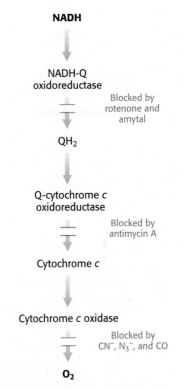

NADH

NADH-Q
oxidoreductase

⎯ Blocked by
rotenone and
amytal

QH_2

Q-cytochrome c
oxidoreductase

⎯ Blocked by
antimycin A

Cytochrome c

Cytochrome c oxidase

⎯ Blocked by
CN^-, N_3^-, and CO

O_2

FIGURE 18.46 Sites of action of some inhibitors of electron transport.

uncouplers, electron transport from NADH to O_2 proceeds in a normal fashion, but ATP is not formed by mitochondrial ATP synthase, because the proton-motive force across the inner mitochondrial membrane is continuously dissipated. This loss of respiratory control leads to increased oxygen consumption and oxidation of NADH. Indeed, in the accidental ingestion of uncouplers, large amounts of metabolic fuels are consumed, but no energy is captured as ATP. Rather, energy is released as heat. DNP is the active ingredient in some herbicides and fungicides. Remarkably, some people consume DNP as a weight-loss drug, despite the fact that the FDA banned its use in 1938. There are also reports that Soviet soldiers were given DNP to keep them warm during the long Russian winters. Chemical uncouplers are nonphysiological, unregulated counterparts of uncoupling proteins.

Drugs are being sought that would function as mild uncouplers, uncouplers not as potentially lethal as DNP, for use in treatment of obesity and related pathologies. Xanthohumol, a prenylated chalcone found in hops and beer, shows promise in this regard. Xanthohumol also scavenges free radicals and is used for treatment of certain types of cancers.

4. *Inhibition of ATP export.* ATP-ADP translocase is specifically inhibited by very low concentrations of *atractyloside* (a plant glycoside) or *bongkrekic acid* (an antibiotic from a mold). Atractyloside binds to the translocase when its nucleotide site faces the intermembrane space, whereas bongkrekic acid binds when this site faces the mitochondrial matrix. Oxidative phosphorylation stops soon after either inhibitor is added, showing that ATP-ADP translocase is essential for maintaining adequate amounts of ADP to accept the energy associated with the proton-motive force.

Mitochondrial diseases are being discovered

The number of diseases that can be attributed to mitochondrial mutations is steadily growing in step with our growing understanding of the biochemistry and genetics of mitochondria. The prevalence of mitochondrial diseases is estimated to be from 10 to 15 per 100,000 people, roughly equivalent to the prevalence of the muscular dystrophies. The first mitochondrial disease to be understood was Leber hereditary optic neuropathy (LHON), a form of blindness that strikes in midlife as a result of mutations in Complex I (see Biochemistry in Focus). Some of these mutations impair NADH utilization, whereas others block electron transfer to Q. Mutations in Complex I are the most frequent cause of mitochondrial diseases. The accumulation of mutations in mitochondrial genes in a span of several decades may contribute to aging, degenerative disorders, and cancer.

A human egg harbors several hundred thousand molecules of mitochondrial DNA, whereas a sperm contributes only a few hundred and thus has little effect on the mitochondrial genotype. Because the maternally inherited mitochondria are present in large numbers and not all of the mitochondria may be affected, the pathologies of mitochondrial mutants can be quite complex. Even within a single family carrying an identical mutation, chance fluctuations in the percentage of mitochondria with the mutation lead to large variations in the nature and severity of the symptoms of the pathological condition as well as the time of onset. As the percentage of defective mitochondria increases, energy-generating capacity diminishes until, at some threshold, the cell can no longer function properly. Defects in cellular respiration are doubly dangerous. Not only does energy transduction decrease, but also the likelihood that reactive oxygen species will be generated increases. Organs that are highly dependent on oxidative phosphorylation, such as the nervous system, the retina, and the heart, are most vulnerable to mutations in mitochondrial DNA.

Mitochondria play a key role in apoptosis

In the course of development or in cases of significant cell damage, individual cells within multicellular organisms undergo *programmed cell death,* or *apoptosis.* Mitochondria act as control centers regulating this process. Although the details have not yet been established, the outer membrane of damaged mitochondria becomes highly permeable, a process referred to as *mitochondrial outer membrane permeabilization* (MOMP). This permeabilization is instigated by a family of proteins (Bcl family) that were initially discovered because of their role in cancer. One of the most potent activators of apoptosis, cytochrome *c,* exits the mitochondria and interacts with apoptotic peptidase-activating factor 1 (APAF-1), which leads to the formation of the *apoptosome.* The apoptosome recruits and activates a proteolytic enzyme called *caspase 9,* a member of the cysteine protease family (Section 9.1) that in turn activates a cascade of other caspases. Each caspase type destroys a particular target, such as the proteins that maintain cell structure. Another target is a protein that inhibits an enzyme that destroys DNA (an enzyme called caspase-activated DNAse or CAD), freeing CAD to cleave the genetic material. This cascade of proteolytic enzymes has been called "death by a thousand tiny cuts."

Power transmission by proton gradients is a central motif of bioenergetics

The main concept presented in this chapter is that mitochondrial electron transfer and ATP synthesis are linked by a transmembrane proton gradient. ATP synthesis in bacteria and chloroplasts also is driven by proton gradients. In fact, proton gradients power a variety of energy-requiring processes such as the active transport of calcium ions by mitochondria, the entry of some amino acids and sugars into bacteria, the rotation of bacterial flagella, and the transfer of electrons from $NADP^+$ to NADPH. Proton gradients can also be used to generate heat, as in nonshivering thermogenesis. It is evident that *proton gradients are a central interconvertible currency of free energy in biological systems* (Figure 18.47). Mitchell noted that the proton-motive force is a marvelously simple and effective store of free energy because it requires only a thin, closed lipid membrane between two aqueous phases.

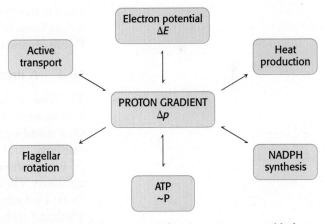

FIGURE 18.47 The proton gradient is an interconvertible form of free energy.

SUMMARY

18.1 Eukaryotic Oxidative Phosphorylation Takes Place in Mitochondria

Mitochondria generate most of the ATP required by aerobic cells through a joint endeavor of the reactions of the citric acid cycle, which take place in the mitochondrial matrix, and oxidative phosphorylation, which takes place in the inner mitochondrial membrane. Mitochondria are descendants of a free-living bacterium that established a symbiotic relation with another cell.

18.2 Oxidative Phosphorylation Depends on Electron Transfer

In oxidative phosphorylation, the synthesis of ATP is coupled to the flow of electrons from NADH or $FADH_2$ to O_2 by a proton gradient across the inner mitochondrial membrane. Electron flow through three asymmetrically oriented transmembrane complexes results in the pumping of protons out of the mitochondrial matrix and the generation of a membrane potential. ATP is synthesized when protons flow back to the matrix through a channel in an ATP-synthesizing complex, called ATP synthase (also known as F_0F_1-ATPase). Oxidative phosphorylation exemplifies a fundamental theme of bioenergetics: the transmission of free energy by proton gradients.

18.3 The Respiratory Chain Consists of Four Complexes: Three Proton Pumps and a Physical Link to the Citric Acid Cycle

The electron carriers in the respiratory assembly of the inner mitochondrial membrane are quinones, flavins, iron–sulfur complexes, heme groups of cytochromes, and copper ions. Electrons from NADH are transferred to the FMN prosthetic group of NADH-Q oxidoreductase (Complex I), the first of four complexes. This oxidoreductase also contains Fe-S centers. The electrons emerge in QH_2, the reduced form of ubiquinone (Q). The citric acid cycle enzyme succinate dehydrogenase is a component of the succinate-Q reductase complex (Complex II), which donates electrons from $FADH_2$ to Q to form QH_2. This hydrophobic carrier transfers its electrons to Q-cytochrome c oxidoreductase (Complex III), a complex that contains cytochromes b and c_1 and an Fe-S center. This complex reduces cytochrome c, a water-soluble peripheral membrane protein. Cytochrome c transfers electrons to cytochrome c oxidase (Complex IV). This complex contains cytochromes a and a_3 and three copper ions. A heme iron ion and a copper ion in this oxidase transfer electrons to O_2, the ultimate acceptor, to form H_2O. Complexes I, III, and IV are organized into a large molecular structure called the respirasome.

18.4 A Proton Gradient Powers the Synthesis of ATP

The flow of electrons through Complexes I, III, and IV leads to the transfer of protons from the matrix side to the cytoplasmic side of the inner mitochondrial membrane. A proton-motive force consisting of a pH gradient (matrix side basic) and a membrane potential (matrix side negative) is generated. The flow of protons back to the matrix side through ATP synthase drives ATP synthesis. The enzyme complex is a molecular motor made of two operational units: a rotating component and a stationary component. The rotation of the γ subunit induces structural changes in the β subunit that result in the synthesis and release of ATP from the enzyme. Proton influx through the **c** ring provides the force for the rotation of the γ subunit.

The flow of two electrons through NADH-Q oxidoreductase, Q-cytochrome c oxidoreductase, and cytochrome c oxidase generates a gradient sufficient to synthesize 1, 0.5, and 1 molecule of ATP, respectively. Hence, 2.5 molecules of ATP are formed per molecule of NADH oxidized in the mitochondrial matrix, whereas only 1.5 molecules of ATP are made per molecule of $FADH_2$ oxidized, because its electrons enter the chain at QH_2, after the first proton-pumping site.

18.5 Many Shuttles Allow Movement Across Mitochondrial Membranes

Mitochondria employ a host of transporters, or carriers, to move molecules across the inner mitochondrial membrane. The electrons of cytoplasmic NADH are transferred into mitochondria by the glycerol phosphate shuttle to form $FADH_2$ from FAD or by the malate–aspartate shuttle to form mitochondrial NADH. The entry of ADP into the mitochondrial matrix is coupled to the exit of ATP by ATP-ADP translocase, a transporter driven by membrane potential.

18.6 The Regulation of Cellular Respiration Is Governed Primarily by the Need for ATP

About 30 molecules of ATP are generated when a molecule of glucose is completely oxidized to CO_2 and H_2O. Electron transport is normally tightly coupled to phosphorylation. NADH and $FADH_2$ are oxidized only if ADP is simultaneously phosphorylated to ATP, a form of regulation called acceptor or respiratory control. Proteins have been identified that uncouple electron transport and ATP synthesis for the generation of heat.

APPENDIX

Biochemistry in Focus

Leber hereditary optic neuropathy can result from defects in Complex I

Leber hereditary optic neuropathy (LHON) is a form of vision loss first described by the German ophthalmologist Theodore Leber in 1871. Vision loss begins in the second or third decade of life and overtime can proceed to blindness. The loss of vision is due to the death of optic neurons and the frequency of occurrence, although not firmly established, appears to be approximately 1 in 40,000 individuals. For unknown reasons, males are affected more than females.

As mentioned in the text (p. 608), LHON is due to mutations in mitochondrial DNA. The vast majority of mitochondria are of maternal origin, and consequently LHON has been referred to as the "Mother's curse."

A number of mutations in the genes encoding Complex I are known to result in LHON. One particular mutation is a simple change of proline to lysine in one of the subunits of Complex I. Genetic manipulation of mice generated a mouse LHON model of the proline to lysine mutation. The LHON mouse model allows a more thorough investigation of the pathology than relying on human tissue.

Mitochondria in the optic nerve of the LHON mice were found to be abnormal in shape and increased in number, characteristics observed in human LHON patients. The increase in number is called compensatory proliferation: the number of mitochondria are increased in an attempt to compensate for decreased function. Mitochondria from LHON mice were isolated to answer a particular question: do the pathologies of LHON result from a decrease in ATP production or an increase in damaging reactive oxygen species?

The first experiment undertaken was to determine the activity of Complex I from LHON mice compared to control (without the mutation) mice as shown in Figure 18.48A. Complex I activity was measured by the ability to oxidize

NADH. Figure 18.48A shows that the LHON Complex I functioned only 70% as well as the Complex I from control mice. There are two possible reasons for the decrease in activity: one is that the LHON Complex I has compromised activity and the other is that the activity is same as that of the control, but there is less Complex I. Control experiments

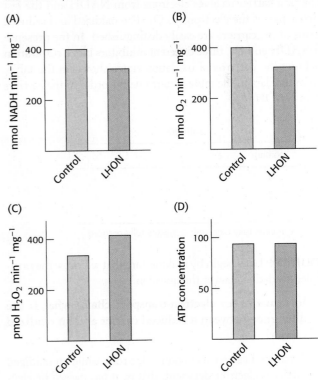

FIGURE 18.48 Complex I mutations and Leber hereditary optic neuropathy. A. The ability of Complex I to oxidize NADH in control and LHON mice. **B.** The reduction of O_2 by mitochondria from control and LHON mice. **C.** The generation of H_2O_2 by mitochondria from control and LHON mice. **D.** The ATP concentration in control and LHON mitochondria.

established that the amount of the complex was the same in both control and LHON mitochondria.

Could it be that, while the activity of LHON Complex I is diminished, the overall activity of the electron-transport chain is unaffected? To address this concern, the capacity of mitochondria to reduce O_2 was compared between the mitochondria from control and LHON mice. Figure 18.48B shows that indeed, the electron-transport chain in LHON mitochondria was again only about 70% as effective as control mitochondria showing that the impairment of Complex I affected the entire electron-transport chain.

The next question to ask is whether the mutation in LHON mice leads to an increase in reactive oxygen generation. Figure 18.48C shows that there is in fact an increase in the production of hydrogen peroxide (H_2O_2), a potent reactive-oxygen species, in LHON mitochondria compared to control mitochondria. This observation is consistent with

the idea that the pathophysiology of LHON is caused by oxidative damage.

But what about ATP production? Could the loss of activity also lead to a decrease in ATP, which might also lead to optic nerve damage? The results shown in Figure 18.48D indicate that ATP synthesizing ability of control and LHON mitochondria are identical, leading support to the idea that reactive oxygen species are the primary cause of damage in LHON.

We should bear in mind that the results described were based on one mutation only, and the results may not apply to other mutations in the Complex I gene. Moreover, the experiments were performed in very complex systems—that is, mitochondria. Even though the experiments were well designed and carried out, it is wise to be a bit skeptical of results obtained in one set of experiments. This is, of course, why it is important that experimental results be replicated by experiments in different laboratories.

APPENDIX

Problem-Solving Strategies

Problem-solving strategies 1

PROBLEM: Four electron carriers—a, b, c, d—are required for electron transport in a bacterial electron-transport chain. The first carrier receives electrons from NADH and the last carrier passes the electrons to O_2. The oxidized and reduced forms of the carriers are easily distinguished. In the presence of NADH and O_2, three different inhibitors block respiration yielding the patterns of oxidation states shown in the table. From the data in the table, determine the order of the carriers from NADH to O_2.

	Carriers			
Inhibitor	a	b	c	d
1	+	+	−	+
2	−	−	−	+
3	+	−	−	+

+ means fully oxidized; − means fully reduced

SOLUTION: Let's begin by making sure that we know the principles needed to answer the question.

▶ **In terms of an electron-transport chain, what is the difference between a reduced carrier and an oxidized carrier?**

A reduced carrier has bound electrons while an oxidized carrier is electron-deficient; that is, it has passed its electrons on without yet binding more electrons.

Now, let's make sure we understand what would happen to the oxidation-reduction state of an electron-transport chain if electron flow is interrupted by an inhibitor. Consider an electron-transport chain NADH → W → X → Y → Z → O_2.

▶ **If an inhibitor prevented electron flow from X to Y, what would be the oxidation-reduction states of the four carriers?**

Carriers "upstream" of the inhibitor would be reduced, while those "downstream" would be oxidized. In the case of our example, W and X would be reduced. They can bind electrons but cannot pass them on. Y and Z would be oxidized. They can transfer electrons but can accept no new electrons because of the inhibitor.

Now with this background, let's tackle the problem.

▶ **What is the site of action of inhibitor 1?**

The carrier c is reduced, so it must not be able to pass on electrons. The other components are oxidized, indicating that they must be downstream of the inhibition. We can't tell what their order is yet, so let's put them in parenthesis. Our preliminary order is c → (a, b, d).

▶ **Where is the site of action of inhibitor 2?**

All carriers are reduced except d. Therefore, the inhibition must be right before d. Our preliminary order is now c → (a,b) → d

Let's see what information inhibitor 3 has to offer.

▶ **What is the site of action of inhibitor 3?**

Both c and b are reduced and a and d are oxidized, so the inhibitor must be acting at b. Combining this with what we already know, the order of the electron-transport chain must be:

$$NADH \rightarrow c \rightarrow b \rightarrow a \rightarrow d \rightarrow O_2.$$

In fact, experiments using inhibitors in a fashion similar to our example here were instrumental in determining the order of the electron carriers in the mitochondrial electron-transport chain (see problem 38).

Problem-solving strategies 2

PROBLEM: 4,6-Dinitro-o-cresol (DNOC) was used as a pesticide until 1991, when it was prohibited because it is extremely toxic to humans.

**4, 6-Dinitro-o-cresol
(DNOC)**

Accidental human exposure to the pesticide has been reported. Symptoms of exposure include a rise in body temperature from 37°C to 39.5°C, profuse sweating, rapid breathing, and a decrease in subcutaneous body fat (fat under the skin). The pesticide was tested on isolated mitochondria to determine the mechanism of action of the pesticide with the results shown in the two graphs in Figure 18.49.

(a) Why is the mitochondrial respiration dependent on ADP?

(b) What is the effect of adding DNOC?

(c) Explain the results of adding potassium cyanide (KCN) in Figure 18.49A?

(d) Now consider Figure 18.49B. Under normal circumstances, what would be the effect of the addition of oligomycin?

(e) What can you conclude about the effect of DNOC?

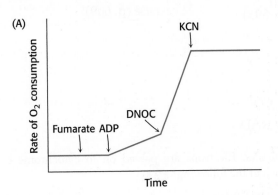

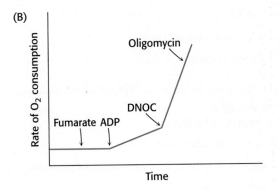

FIGURE 18.49 The impact of DNOC exposure on mitochondria function.

Now for some questions on how DNOC exposure would alter a person's physiology.

(f) Why did the body temperature rise?

(g) Why was body fat absent?

SOLUTION: Before we begin to answer the questions, let's make sure, as always, that we know what we are looking at. The experimental system is isolated mitochondria to which various chemicals are added sequentially. The y-axis shows the rate of O_2 consumption and the x-axis time.

▶ **Why is the measure of the rate of oxygen consumption a means of determining mitochondrial activity?**

Recall that electrons flow down the electron-transport chain, ultimately reducing O_2 to H_2O. Thus, the rate of O_2 consumption measures electron-transport activity and thus mitochondrial activity.

▶ **Examine Figure 18.49A. Explain why the addition of fumarate, a citric acid cycle intermediate, does not result in an increase in O_2 consumption.**

The answer to this question will lead to the answer for question (a). Recall that under normal conditions, electrons will not flow down the electron-transport chain unless ATP is synthesized because otherwise the proton gradient will not be dissipated, an example of acceptor control (Figure 18.42). Thus, election flow as measured by O_2 consumption will occur only if ATP is synthesized from ADP. On to question (b).

▶ **What happens to O_2 consumption upon the addition of DNOC?**

O_2 consumption increases drastically. (This should provide you a hint as to how DNOC acts.) On to question (c).

▶ **What is the effect of adding cyanide to the DNOC-stimulated electron-transport chain?**

Cyanide inhibits Complex IV of the electron-transport chain. Recall that Complex IV reduces O_2 to H_2O. Thus, cyanide completely shuts down the electron-transport chain and O_2 consumption ceases.

Now, let's look at question (d). To answer this question, we need to know how oligomycin acts.

▶ **What is the molecular site of oligomycin action?**

Oligomycin binds to the c ring of ATP synthase inhibiting proton translocation and thus ATP synthesis (p. 607). This leads to question (d).

▶ **Under normal circumstances, that is, in the absence of DNOC, how would inhibition of ATP synthase affect the electron-transport chain?**

Because the electron-transport chain and ATP synthesis are normally coupled, inhibition of ATP synthase should stop the electron-transport chain. Now, we can answer (e).

▶ **How does DNOC act?**

DNOC must be an uncoupler, similar to DNP (p. 607). Recall that uncouplers deplete the proton gradient, allowing the electron-transport chain to operate in the absence of ATP synthesis.

Now, let's approach the physiological questions.

▶ **Why was there a rise in body temperature in the affected individuals?**

In the presence of DNOC, the energy released by the electron-transport chain is released as heat instead of being used to synthesize ATP.

▶ **Why did the affected individuals have reduced body fat?**

All tissues were rapidly metabolizing fat in a futile attempt to synthesize ATP.

KEY TERMS

oxidative phosphorylation (p. 573)

electron-transport chain (p. 573)

cellular respiration (p. 574)

reduction (redox, oxidation–reduction, E_0') potential (p. 577)

coenzyme Q (Q, ubiquinone) (p. 581)

Q pool (p. 581)

iron–sulfur (nonheme iron) protein (p. 581)

NADH-Q oxidoreductase (Complex I) (p. 582)

flavin mononucleotide (FMN) (p. 582)

succinate-Q reductase (Complex II) (p. 584)

cytochrome c (Cyt c) (p. 584)

Q-cytochrome c oxidoreductase (Complex III) (p. 584)

Rieske center (p. 584)

Q cycle (p. 585)

cytochrome c oxidase (Complex IV) (p. 586)

respirasome (p. 589)

superoxide dismutase (p. 590)

catalase (p. 591)

ATP synthase (Complex V, F_1F_0 ATPase) (p. 593)

proton-motive force (p. 593)

glycerol 3-phosphate shuttle (p. 600)

malate–aspartate shuttle (p. 600)

ATP-ADP translocase (adenine nucleotide translocase, ANT) (p. 601)

respiratory (acceptor) control (p. 603)

uncoupling protein (UCP) (p. 605)

programmed cell death (apoptosis) (p. 609)

mitochondrial outer membrane permeabilization (MOMP) (p. 609)

apoptosome (p. 609)

caspase (p. 609)

PROBLEMS

1. *Not a member.* Which of the following electron-transport chain complexes is not a member of the respirasome? ✔❶

(a) Complex I

(b) Complex II

(c) Complex III

(d) Complex IV

2. *Cycle component.* Which of the following electron-transport chain complexes is also a member of the citric acid cycle? ✔❶

(a) Complex I

(b) Complex II

(c) Complex III

(d) Complex IV

3. *The last stop.* The final electron acceptor for the electron-transport chain is which of the following? ✔❶

(a) O_2

(b) Coenzyme Q

(c) CO_2

(d) NAD^+

4. *Cycles.* Electrons are passed on to cytochrome c using which of the following? ✔❶

(a) The Cori cycle

(b) The Q cycle

(c) The glucose-alanine cycle

(d) The Krebs cycle.

5. *May the force be with you.* Which of the following is not true? ✔❷, ✔❸
The proton-motive force:

(a) Requires an intact membrane.

(b) Consists of an unequal distribution of protons.

(c) Used to synthesize ATP.

(d) Is powered by the hydrolysis of ATP.

6. *Breathe or ferment?* Compare fermentation and respiration with respect to electron donors and electron acceptors.

7. *Reference states.* The standard oxidation–reduction potential (E_0') for the reduction of O_2 to H_2O is given as 0.82 V in Table 18.1. However, the value given in textbooks of chemistry is 1.23 V. Account for this difference.

8. *Less energetic electrons.* Why are electrons carried by $FADH_2$ not as energy rich as those carried by NADH? What is the consequence of this difference?

9. *Now prove it.* Calculate the energy released by the reduction of O_2 with $FADH_2$.

10. *Thermodynamic constraint.* Compare the $\Delta G°'$ values for the oxidation of succinate by NAD^+ and by FAD. Use the data given in Table 18.1 to find the E_0' of the $NAD^+ - NADH$ and fumarate-succinate couples, and assume that E_0' for the $FAD - FADH_2$ redox couple is nearly 0.05 V. Why is FAD rather than NAD^+ the electron acceptor in the reaction catalyzed by succinate dehydrogenase?

11. *Give and accept.* Distinguish between an oxidizing agent and a reducing agent.

12. *The benefactor and beneficiary.* Identify the oxidant and the reductant in the following reaction.

$$\text{Pyruvate} + \text{NADH} + H^+ \rightleftharpoons \text{lactate} + NAD^+$$

13. *Six of one, half dozen of the other.* How is the redox potential ($\Delta E_0'$) related to the free-energy change of a reaction ($\Delta G°'$)?

14. *Location, location, location.* Iron is a component of many of the electron carriers of the electron-transport chain. How can it participate in a series of coupled redox reactions if the E_0' value is $+0.77$ V, as seen in Table 18.1? ✓❶

15. *Line up.* Place the following components of the electron-transport chain in their proper order: ✓❶

(a) cytochrome *c*

(b) Q-cytochrome *c* oxidoreductase

(c) NADH-Q reductase

(d) cytochrome *c* oxidase

(e) ubiquinone

16. *Like mac and cheese.* Match each term with its description. ✓❶

(a) Respiration _____

(b) Redox potential _____

(c) Electron-transport chain _____

1. Converts reactive oxygen species into hydrogen peroxide

2. Electron flow from NADH and $FADH_2$ to O_2

3. Facilitates electron flow from FMN to coenzyme Q in Complex I

(d) Flavin mononucleotide (FMN) _____

(e) Iron–sulfur protein _____

(f) Coenzyme Q _____

(g) Cytochrome *c* _____

(h) Q cycle _____

(i) Superoxide dismutase _____

(j) Catalase _____

4. An ATP-generating process in which an inorganic compound serves as the final electron acceptor

5. Measure of the tendency to accept or donate electrons

6. Converts hydrogen peroxide into oxygen and water

7. Funnels electrons from a two-electron carrier to a one-electron carrier

8. Lipid-soluble electron carrier

9. Donates electrons to Complex IV

10. Accepts electrons from NADH in Complex I

17. *Match 'em.* ✓❶

(a) Complex I _____

(b) Complex II _____

(c) Complex III _____

(d) Complex IV _____

(e) Ubiquinone _____

1. Q-cytochrome *c* oxidoreductase

2. Coenzyme Q

3. Succinate-Q reductase

4. NADH-Q oxidoreductase

5. Cytochrome *c* oxidase

18. *Structural considerations.* Explain why coenzyme Q is an effective mobile electron carrier in the electron-transport chain. ✓❷

19. *Inhibitors.* Rotenone inhibits electron flow through NADH-Q oxidoreductase. Antimycin A blocks electron flow between cytochromes *b* and c_1. Cyanide blocks electron flow through cytochrome *c* oxidase to O_2. Predict the relative oxidation–reduction state of each of the following respiratory-chain components in mitochondria that are treated with each of the inhibitors: ✓❶

(a) NAD^+

(b) NADH-Q oxidoreductase

(c) coenzyme Q

(d) cytochrome c_1

(e) cytochrome *c*

(f) cytochrome *a*

20. *Rumored to be a favorite of Elvis.* Amytal is a barbiturate sedative that inhibits electron flow through Complex I. How would the addition of amytal to actively respiring mitochondria affect the relative oxidation–reduction states of the components of the electron-transport chain and the citric acid cycle?

21. *Efficiency.* What is the advantage of having Complexes I, III, and IV associated with one another in the form of a respirasome? ✓❶, ✓❷

22. *Linked In.* What citric acid cycle enzyme is also a component of the electron-transport chain? ✓❷

23. *ROS, not ROUS.* What are the reactive oxygen species and why are they especially dangerous to cells?

24. *Reclaim resources.* Humans have only about 250 g of ATP, but even a couch potato needs about 83 kg of ATP to open a bag of chips and use the remote. How is this discrepancy between requirements and resources reconciled? ✓❸, ✓❹

25. *Energy harvest.* What is the yield of ATP when each of the following substrates is completely oxidized to CO_2 by a mammalian cell homogenate? Assume that glycolysis, the citric acid cycle, and oxidative phosphorylation are fully active.

(a) Pyruvate

(b) Lactate

(c) Fructose 1,6-bisphosphate

(d) Phosphoenolpyruvate

(e) Galactose

(f) Dihydroxyacetone phosphate

26. *Potent poisons.* What is the effect of each of the following inhibitors on electron-transport and ATP formation by the respiratory chain? ✓❶, ✓❸

(a) Azide

(b) Atractyloside

(c) Rotenone

(d) DNP

(e) Carbon monoxide

(f) Antimycin A

27. *A question of coupling.* What is the mechanistic basis for the observation that the inhibitors of ATP synthase also lead to an inhibition of the electron-transport chain? ✓❶, ✓❸, ✓❹

28. *A Brownian ratchet wrench.* What causes the **c** subunits of ATP synthase to rotate? What determines the direction of rotation? ✓❸

29. *An essential residue.* The conduction of protons by the F_0 unit of ATP synthase is blocked by dicyclohexylcarbodiimide, which reacts readily with carboxyl groups. What are the most likely targets of action of this reagent? How might you use site-specific mutagenesis to determine whether this residue is essential for proton conduction? ✓❸, ✓❹

30. *Arsenate again.* Arsenate (AsO_4^{3-}) closely resembles phosphate in structure and reactivity. However, arsenate esters are unstable and are spontaneously hydrolyzed. If arsenate is added to actively respiring mitochondria, what would be the effect on ATP synthesis? On the rate of the electron-transport chain? Briefly explain.

31. *Alternative routes.* The most common metabolic sign of mitochondrial disorders is lactic acidosis. Why?

32. *Connections.* How does the inhibition of ATP-ADP translocase affect the citric acid cycle? Glycolysis?

33. *O_2 consumption.* Oxidative phosphorylation in mitochondria is often monitored by measuring oxygen consumption. When oxidative phosphorylation is proceeding rapidly, the mitochondria will rapidly consume oxygen. If there is little oxidative phosphorylation, only small amounts of oxygen will be used. You are given a suspension of isolated mitochondria and directed to add the following compounds in the order from *a* to *h*. With the addition of each compound, all of the previously added compounds remain present. Predict the effect of each addition on oxygen consumption by the isolated mitochondria. ✓❶, ✓❸, ✓❹

(a) Glucose

(b) ADP + P_i

(c) Citrate

(d) Oligomycin

(e) Succinate

(f) Dinitrophenol

(g) Rotenone

(h) Cyanide

34. *P:O ratios.* The number of molecules of inorganic phosphate incorporated into organic form per atom of oxygen consumed, termed the *P:O ratio*, was frequently used as an index of oxidative phosphorylation.

(a) What is the relation of the P:O ratio to the ratio of the number of protons translocated per electron pair $(H^+/2\,e^-)$ and the ratio of the number of protons needed to synthesize ATP and transport it to the cytoplasm (P/H^+)?

(b) What are the P:O ratios for electrons donated by matrix NADH and by succinate?

35. *Cyanide antidote.* The immediate administration of nitrite is a highly effective treatment for cyanide poisoning. What is the basis for the action of this antidote? (Hint: Nitrite oxidizes ferrohemoglobin to ferrihemoglobin.)

36. *Runaway mitochondria 1.* Suppose that the mitochondria of a patient oxidize NADH irrespective of whether ADP is present. The P:O ratio for oxidative phosphorylation by these mitochondria is less than normal. Predict the likely symptoms of this disorder.

37. *Recycling device.* The cytochrome *b* component of Q-cytochrome *c* oxidoreductase enables both electrons of QH_2 to be effectively utilized in generating a proton-motive force. Cite another recycling device in carbohydrate metabolism that brings a potentially dead-end reaction product back into the mainstream.

38. *Crossover point.* The precise site of action of a respiratory-chain inhibitor can be revealed by the *crossover technique.* Britton Chance devised elegant spectroscopic methods for determining the proportions of the oxidized and reduced forms of each carrier. This determination is feasible because the forms have distinctive absorption spectra, as illustrated in the adjoining graph for cytochrome *c*. You are given a new inhibitor and find that its addition to respiring mitochondria causes the carriers between NADH and QH_2 to become more reduced and those between cytochrome *c* and O_2 to become more oxidized. Where does your inhibitor act? ✓❶, ✓❷

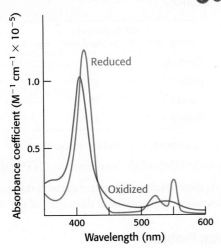

39. *Runaway mitochondria 2.* Years ago, uncouplers were suggested to make wonderful diet drugs. Explain why this idea was proposed and why it was rejected. Why might the producers of antiperspirants be supportive of the idea? ✓❸, ✓❹

40. *Everything is connected.* If actively respiring mitochondria are exposed to an inhibitor of ATP-ADP translocase, the electron-transport chain ceases to operate. Why? ✓❸, ✓❹

41. *Identifying the inhibition.* You are asked to determine whether a chemical is an electron-transport-chain inhibitor or an inhibitor of ATP synthase. Design an experiment to make this determination. ✓❶, ✓❷, ✓❸

42. *To each according to its needs.* It has been noted that the mitochondria of muscle cells often have more cristae than the mitochondria of liver cells. Provide an explanation for this observation. ✓❷

43. *Opposites attract.* An arginine residue (Arg 210) in the **a** subunit of the *E. coli* ATP synthase is near the aspartate residue (Asp 61) in the matrix-side proton channel. How might Arg 210 assist proton flow?

44. *Variable c subunits.* Recall that the number of **c** subunits in the **c** ring appears to range between 8 and 14. This number is significant because it determines the number of protons that must be transported to generate a molecule of ATP. Each 360-degree rotation of the γ subunit leads to the synthesis and release of three molecules of ATP. Thus, if there are 10 **c** subunits in the ring (as was observed in a crystal structure of yeast mitochondrial ATP synthase), each ATP generated requires the transport of $10/3 = 3.33$ protons. How many protons are required to form ATP if the ring has 12 **c** subunits? 14? ✓❸

45. *Counterintuitive.* Under some conditions, mitochondrial ATP synthase has been observed to actually run in reverse.

How would that situation affect the proton-motive force? ✓❸, ✓❹

46. *Etiology? What does that mean?* What does the fact that rotenone appears to increase the susceptibility to Parkinson disease indicate about the etiology of Parkinson disease? ✓❶, ✓❷

47. *Exaggerating the difference.* Why must ATP-ADP translocase (also called adenine nucleotide translocase or ANT) use Mg^{2+}-free forms of ATP and ADP?

48. *Respiratory control.* The rate of oxygen consumption by mitochondria increases markedly when ADP is added and then returns to its initial value when the added ADP has been converted into ATP (Figure 18.42). Why does the rate decrease? ✓❹

49. *Same, but different.* Why is the electroneutral exchange of $H_2PO_4^-$ for OH^- indistinguishable from the electroneutral symport of $H_2PO_4^-$ and H^+?

50. *Multiple uses.* Give an example of the use of the proton-motive force in ways other than for the synthesis of ATP.

Chapter Integration Problems

51. *Just obeying the laws.* Why do isolated F_1 subunits of ATP synthase catalyze ATP hydrolysis? ✓❸

52. *The right location.* Some cytoplasmic kinases, enzymes that phosphorylate substrates at the expense of ATP, bind to voltage-dependent anion channels. What might the advantage of this binding be?

53. *No exchange.* Mice that completely lack ATP-ADP translocase (ANT^-/ANT^-) can be made by using the knockout technique. Remarkably, these mice are viable but have the following pathological conditions: (1) high serum levels of lactate, alanine, and succinate; (2) little electron transport; and (3) a six- to eightfold increase in the level of mitochondrial H_2O_2 compared with that in normal mice. Provide a possible biochemical explanation for each of these conditions.

54. *Maybe you shouldn't take your vitamins.* Exercise is known to increase insulin sensitivity and to ameliorate type 2 diabetes (Chapter 27). Recent research suggests that taking antioxidant vitamins might mitigate the beneficial effects of exercise with respect to ROS protection.

(a) What are the antioxidant vitamins?

(b) How does exercise protect against ROS?

(c) Explain why vitamins might counteract the effects of exercise.

Chapter Integration and Data Interpretation Problem

55. *Monitoring energy sources.* XF technology (Seahorse Bioscience) now allows the measurement of the rate of aerobic respiration and lactic acid fermentation simultaneously in

real time in cultured cells. The extent of aerobic respiration is determined by measuring the oxygen consumption rate (OCR, measured in picomoles of oxygen consumed per minute) while the rate of glycolysis correlates with the extracellular acidification rate [ECAR-milli pH per minute (the changes in pH that occur over time)]. The graph below shows the results of an experiment using the new technology.

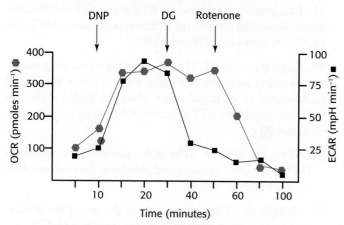

Dinitrophenol (DNP), the glycolysis inhibitor 2-deoxyglucose (DG), and rotenone were added sequentially to cell cultures. ✔️❶, ✔️❷

(a) What is the effect on OCR and ECAR of adding DNP to the cell culture? Explain these results.

(b) Explain the effect of the addition of 2-deoxyglucose.

(c) Explain how 2-deoxyglucose acts as an inhibitor of glycolysis?

(d) Explain the effect of the addition of rotenone.

Data Interpretation Problem

56. *Mitochondrial disease.* A mutation in a mitochondrial gene encoding a component of ATP synthase has been identified. People who have this mutation suffer from muscle weakness, ataxia (a loss of coordination), and retinitis pigmentosa (retinal degeneration). A tissue biopsy was performed on each of three patients having this mutation, and submitochondrial particles were isolated that were capable of succinate-sustained ATP synthesis. First, the activity of the ATP synthase was measured on the addition of succinate and the following results were obtained. ✔️❸, ✔️❹

ATP synthase activity (nmol of ATP formed min^{-1} mg^{-1})	
Controls	3.0
Patient 1	0.25
Patient 2	0.11
Patient 3	0.17

(a) What was the purpose of the addition of succinate?

(b) What is the effect of the mutation on succinate-coupled ATP synthesis?

Next, the ATPase activity of the enzyme was measured by incubating the submitochondrial particles with ATP in the absence of succinate.

ATP hydrolysis (nmol of ATP hydrolyzed min^{-1} mg^{-1})	
Controls	33
Patient 1	30
Patient 2	25
Patient 3	31

(c) Why was succinate omitted from the reaction?

(d) What is the effect of the mutation on ATP hydrolysis?

(e) What do these results, in conjunction with those obtained in the first experiment, tell you about the nature of the mutation?

Mechanism Problem

57. *Chiral clue.* ATPγS, a slowly hydrolyzed analog of ATP, can be used to probe the mechanism of phosphoryl-transfer reactions. Chiral ATPγS has been synthesized containing ^{18}O in a specific γ position and ordinary ^{16}O elsewhere in the molecule. The hydrolysis of this chiral molecule by ATP synthase in ^{17}O-enriched water yields inorganic [$^{16}O,^{17}O,^{18}O$] thiophosphate having the following absolute configuration. In contrast, the hydrolysis of this chiral ATPγS by a calcium-pumping ATPase from muscle gives thiophosphate of the opposite configuration. What is the simplest interpretation of these data? ✔️❸

Think ► Pair ► Share Problems

58. Discuss the functional and structural relationship between the citric acid cycle and the electron-transport chain.

59. Explain the regulation of cellular respiration. Include the roles of glycolysis, the citric acid cycle, the electron-transport chain, ATP synthesis, and the compartmentalization of cellular respiration in the mitochondria.

The Light Reactions
of Photosynthesis

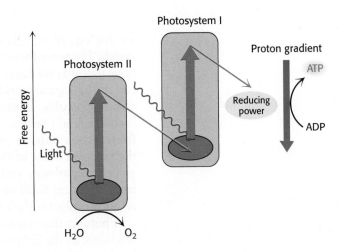

A barley field (left). Throughout the day, the barley converts sunlight into chemical energy with the biochemical process of photosynthesis. Because barley is a major animal feed crop, the sun's energy will soon make its way up the food chain to human beings. [Brian Jannsen/ Alamy.] High-energy electrons in chloroplasts are transported through two photosystems (above). In this transit, which culminates in the generation of reducing power, ATP is synthesized in a manner analogous to mitochondrial ATP synthesis. In contrast with mitochondrial electron transport, however, electrons in chloroplasts are energized by light. [(left) Brian Jannsen/Alamy.]

✓ LEARNING OBJECTIVES

By the end of this chapter, you should be able to:

1. Describe the light reactions.

2. Identify the key products of the light reactions.

3. Explain how redox balance is maintained during the light reactions.

4. Explain how ATP is synthesized in chloroplasts.

5. Describe the function of the light-harvesting complex.

OUTLINE

19.1 Photosynthesis Takes Place in Chloroplasts

19.2 Light Absorption by Chlorophyll Induces Electron Transfer

19.3 Two Photosystems Generate a Proton Gradient and NADPH in Oxygenic Photosynthesis

19.4 A Proton Gradient across the Thylakoid Membrane Drives ATP Synthesis

19.5 Accessory Pigments Funnel Energy into Reaction Centers

19.6 The Ability to Convert Light into Chemical Energy Is Ancient

Every year, Earth is bathed in photons with a total energy content of approximately 10^{24} kJ. By comparison, a major hurricane has 10^{-9} that amount of energy. The source of the energy is, of course, the electromagnetic radiation from the sun. On our planet, there are organisms capable of collecting a fraction, approximately 2%, of this solar energy, and converting it into chemical energy. Green plants are the most obvious of these organisms, although 60% of this conversion is carried out by algae and bacteria. Although only a small amount of the incident energy is captured, it is enough to power all life on Earth. This transformation is perhaps the most important of all of the energy transformations that we will see in our study of biochemistry; without it, life as we know it on our planet simply could not exist.

The process of converting electromagnetic radiation into chemical energy is called *photosynthesis,* which uses light energy to convert carbon dioxide and water into carbohydrates and oxygen.

$$CO_2 + H_2O \xrightarrow{\text{Light}} (CH_2O) + O_2$$

In this equation, CH_2O represents carbohydrate, primarily sucrose and starch. Photosynthetic carbohydrate synthesis is the most common metabolic pathway on Earth. These carbohydrates provide not only the energy to run the biological world, but also the carbon molecules to make a wide array of biomolecules. Photosynthetic organisms are called *autotrophs* (literally, "self-feeders") because they can synthesize chemical fuels such as glucose from carbon dioxide and water by using sunlight as an energy source and then recover some of this energy from the synthesized glucose through the glycolytic pathway and aerobic metabolism. Organisms that obtain energy from chemical fuels only are called *heterotrophs,* which ultimately depend on autotrophs for their fuel.

Photosynthesis is composed of two parts: the light reactions and the dark reactions. In the *light reactions,* light energy is transformed into two forms of biochemical energy with which we are already familiar: reducing power and ATP. The products of the light reactions are then used in the dark reactions to drive the reduction of CO_2 and its conversion into glucose and other sugars. The dark reactions are also called the *Calvin cycle* or *light-independent reactions* and will be discussed in Chapter 20.

Photosynthesis converts light energy into chemical energy

The light reactions of photosynthesis closely resemble the events of oxidative phosphorylation. In Chapters 17 and 18, we learned that cellular respiration is the oxidation of glucose to CO_2 with the reduction of O_2 to water, a process that generates ATP. In photosynthesis, this process must be reversed—reducing CO_2 and oxidizing H_2O to synthesize glucose.

$$\text{Energy} + 6\,H_2O + 6\,CO_2 \xrightarrow{\text{Photosynthesis}} C_6H_{12}O_6 + 6\,O_2$$

$$C_6H_{12}O_6 + 6\,O_2 \xrightarrow[\text{respiration}]{\text{Cellular}} 6\,CO_2 + 6\,H_2O + \text{energy}$$

Although the processes of respiration and photosynthesis are chemically opposite each other, the biochemical principles governing the two processes are nearly identical. The key to both processes is the generation of high-energy electrons. The citric acid cycle oxidizes carbon fuels to CO_2 to generate high-energy electrons. The flow of these high-energy electrons down an electron-transport chain generates a proton-motive force. This proton-motive force is then transduced by ATP synthase to form ATP. To synthesize glucose from CO_2, high-energy electrons are required for two purposes: (1) to provide reducing power in the form of NADPH to reduce CO_2, and (2) to generate ATP to power the reduction. How can high-energy electrons be generated without using a chemical fuel? *Photosynthesis uses energy from light to boost electrons from a low-energy state to a high-energy state.* In the high-energy, unstable state, nearby molecules can abscond with the excited electrons. These electrons are used to produce reducing power, and to generate a proton-motive force across a membrane, which subsequently drives the synthesis of ATP. The reactions that are powered by sunlight are called the *light reactions* (Figure 19.1).

Photosynthetic yield

"If a year's yield of photosynthesis were amassed in the form of sugar cane, it would form a heap over two miles high and with a base 43 square miles."

—Gordon Elliot Fogge
British biologist 1919–2005

If all of this sugar cane were converted into sugar cubes (0.5 inch, or 1.27 cm, on a side) and stacked end to end, the sugar cubes would extend 1.6×10^{10} miles (2.6×10^{10} kilometers) or to the dwarf planet Pluto.

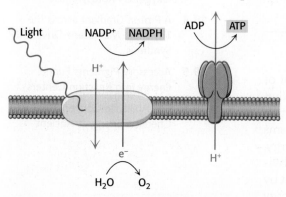

FIGURE 19.1 The light reactions of photosynthesis. Light is absorbed and the energy is used to drive electrons from water to generate NADPH and to drive protons across a membrane. These protons return through ATP synthase (purple structure) to make ATP.

Photosynthesis in green plants is mediated by two kinds of light reactions. Photosystem I generates reducing power in the form of NADPH, but, in the process, becomes electron deficient. Photosystem II oxidizes water and transfers the electrons to replenish the electrons lost by photosystem I. A side product of these reactions is O_2. Electron flow from photosystem II to photosystem I generates the transmembrane proton gradient, augmented by the protons released by the oxidation of water, that drives the synthesis of ATP. In keeping with the similarity of their principles of operation, both processes take place in double-membrane organelles: mitochondria for cellular respiration and chloroplasts for photosynthesis.

Photosynthetic catastrophe

If photosynthesis were to cease, all higher forms of life would be extinct in about 25 years. A milder version of such a catastrophe ended the Cretaceous period 65.1 million years ago when a large asteroid struck the Yucatan Peninsula of Mexico. Enough dust was sent into the atmosphere that photosynthetic capacity was greatly diminished, which apparently led to the disappearance of the dinosaurs and allowed the mammals to rise to prominence.

19.1 Photosynthesis Takes Place in Chloroplasts

Photosynthesis, the means of converting light into chemical energy, takes place in organelles called *chloroplasts*, typically 5 μm long. Like a mitochondrion, a chloroplast has an outer membrane and an inner membrane, with an intervening intermembrane space (Figure 19.2). The inner membrane surrounds a space called the *stroma*, which is the site of the dark reactions

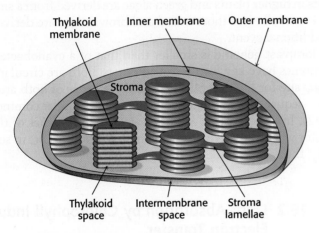

FIGURE 19.2 Diagram of a chloroplast.

of photosynthesis (Section 20.1). In the stroma are membranous structures called *thylakoids*, which are flattened sacs, or discs. The thylakoid sacs are stacked to form a *granum*. Different grana are linked by regions of thylakoid membrane called *stroma lamellae* (Figure 19.3). The thylakoid membranes separate the thylakoid space from the stroma space. Thus, chloroplasts have three different membranes (*outer*, *inner*, and *thylakoid membranes*) and three separate spaces (*intermembrane*, *stroma*, and *thylakoid spaces*). In developing chloroplasts, thylakoids arise from budding of the inner membrane, and so they are analogous to the mitochondrial cristae. Like the mitochondrial cristae, they are the site of coupled oxidation–reduction reactions of the light reactions that generate the proton-motive force.

The primary events of photosynthesis take place in thylakoid membranes

The thylakoid membranes contain the energy-transforming machinery: light-harvesting proteins, reaction centers, electron-transport chains, and ATP synthase. These membranes contain nearly equal amounts of lipids and proteins. The lipid composition is highly distinctive: about 75% of the

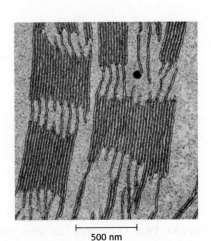

FIGURE 19.3 Electron micrograph of a chloroplast from a spinach leaf. The thylakoid membranes pack together to form grana. [Courtesy of Dr. Kenneth Miller.]

500 nm

total lipids are *galactolipids* and 10% are *sulfolipids*, whereas only 10% are phospholipids. The thylakoid membrane and the inner membrane, like the inner mitochondrial membrane, are impermeable to most molecules and ions. The outer membrane of a chloroplast, like that of a mitochondrion, is highly permeable to small molecules and ions. The stroma contains the soluble enzymes that utilize the NADPH and ATP synthesized by the thylakoids to convert CO_2 into sugar. Plant-leaf cells contain between 1 and 100 chloroplasts, depending on the species, cell type, and growth conditions.

Chloroplasts arose from an endosymbiotic event

Chloroplasts contain their own DNA and the machinery for replicating and expressing it. However, chloroplasts are not autonomous; they also contain many proteins encoded by nuclear DNA. How did the intriguing relation between the cell and its chloroplasts develop? We now believe that, in a manner analogous to the evolution of mitochondria (Section 18.1), chloroplasts are the result of endosymbiotic events in which a photosynthetic microorganism, most likely an ancestor of a cyanobacterium (Figure 19.4), was engulfed by a eukaryotic host. Evidence suggests that chloroplasts in higher plants and green algae are derived from a single endosymbiotic event, whereas those in red and brown algae are derived from at least one additional event.

The chloroplast genome is smaller than that of a cyanobacterium, but the two genomes have key features in common. Both are circular and have a single start site for DNA replication, and the genes of both are arranged in operons—sequences of functionally related genes under common control (Chapter 32). In the course of evolution, many of the genes of the chloroplast ancestor were transferred to the plant cell's nucleus or, in some cases, lost entirely, thus establishing a fully dependent relation.

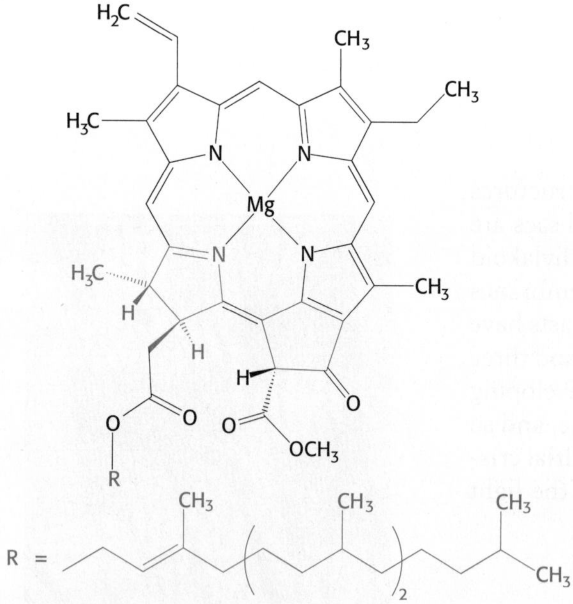

FIGURE 19.4 Cyanobacteria. A colony of the photosynthetic filamentous cyanobacterium *Anabaena* shown at 450 × magnification. Ancestors of these bacteria are thought to have evolved into present-day chloroplasts. [Michael Abbey/Science Source.]

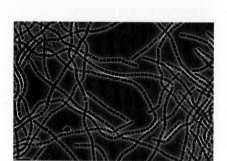

FIGURE 19.5 Chlorophyll. Like heme, chlorophyll *a* is a cyclic tetrapyrrole. One of the pyrrole rings (shown in red) is reduced, and an additional five-carbon ring (shown in blue) is fused to another pyrrole ring. A phytol chain (shown in green) is connected by an ester linkage. Magnesium ion binds at the center of the structure.

19.2 Light Absorption by Chlorophyll Induces Electron Transfer

The trapping of light energy is the key to photosynthesis. The first event is the absorption of light by a photoreceptor molecule. The principal photoreceptor in the chloroplasts of most green plants is the pigment molecule *chlorophyll a*, a substituted tetrapyrrole (Figure 19.5). The four nitrogen atoms of the pyrroles are coordinated to a magnesium ion. Unlike a porphyrin such as heme, chlorophyll has a reduced pyrrole ring and an additional 5-carbon ring fused to one of the pyrrole rings. Another distinctive feature of chlorophyll is the presence of *phytol*, a highly hydrophobic 20-carbon alcohol, esterified to an acid side chain.

Chlorophylls are very effective photoreceptors because they contain networks of conjugated double bonds—alternating single and double bonds. Such compounds are called conjugated *polyenes*. In polyenes, the electrons are not localized to a particular atomic nucleus. Upon the absorption of light energy, the electron transitions from a low-energy molecular orbital to a higher energy orbital. Chlorophylls have very strong absorption bands in the visible region of the spectrum, where the solar output reaching Earth is maximal (Figure 19.6). Chlorophyll *a*'s peak molar extinction coefficient (ε), a measure of a compound's ability to absorb light, is greater than $10^5 \, M^{-1} \, cm^{-1}$, among the highest observed for organic compounds.

What happens when light is absorbed by a pigment molecule? The energy from the light excites an electron from its ground energy level to an excited energy level (Figure 19.7). This high-energy electron can have one of two fates. For most compounds that absorb light, the electron simply returns to the ground state and the absorbed energy is converted into heat. However, if a suitable electron acceptor is nearby, as is the case for chlorophyll in photosynthetic systems, the excited electron can move from the initial molecule to the acceptor (Figure 19.8). A positive charge forms on the initial molecule, owing to the loss of an electron, and a negative charge forms on the acceptor, owing to the gain of an electron. Hence, this process is referred to as *photoinduced charge separation*.

In chloroplasts, the site at which the charge separation takes place within each photosystem is called the *reaction center*. The photosynthetic apparatus is arranged to maximize photoinduced charge separation and minimize an unproductive return of the electron to its ground state. The electron, extracted from its initial site by the absorption of light, now has reducing power: it can reduce other molecules to store the energy originally obtained from light in chemical forms.

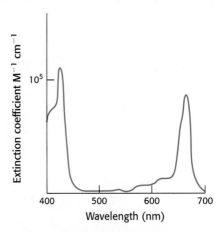

FIGURE 19.6 Light absorption by chlorophyll *a*. Chlorophyll *a* absorbs visible light efficiently, as judged by the extinction coefficient near $10^5 \, M^{-1} \, cm^{-1}$.

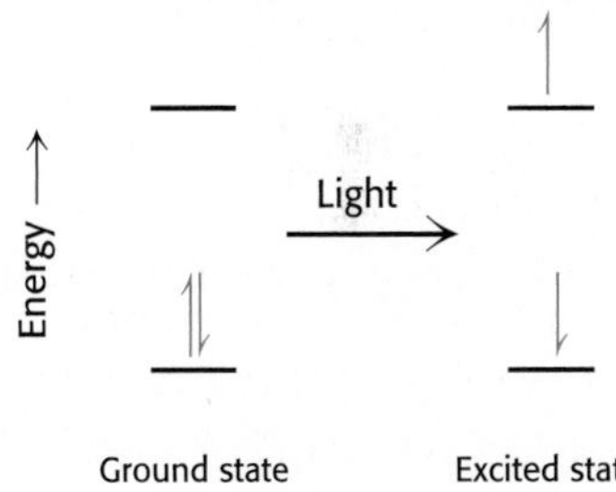

FIGURE 19.7 Light absorption. The absorption of light leads to the excitation of an electron from its ground state to a higher energy level.

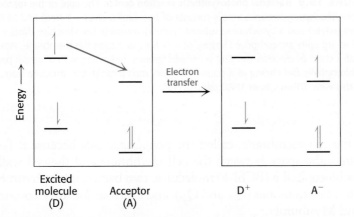

FIGURE 19.8 Photoinduced charge separation. If a suitable electron acceptor is nearby, an electron that has been moved to a high energy level by light absorption can move from the excited molecule to the acceptor.

A special pair of chlorophylls initiate charge separation

Photosynthetic bacteria such as *Rhodopseudomonas viridis* contain a photosynthetic reaction center that has been revealed at atomic resolution. The bacterial reaction center consists of four polypeptides: L (31-kDa, red), M (36-kDa, blue), and H (28-kDa, white) subunits and C, a *c*-type cytochrome with four *c*-type hemes (yellow) (Figure 19.9). *Sequence comparisons and low-resolution structural studies have revealed that the bacterial reaction center is homologous to the more complex plant systems.* Thus, many of our observations of the bacterial system also apply to plant systems.

The L and M subunits form the structural and functional core of the bacterial photosynthetic reaction center (Figure 19.9). Each of these homologous subunits contains five transmembrane helices, in contrast with the H subunit, which has just one. The H subunit lies on the cytoplasmic side of the cell membrane, and the cytochrome subunit lies on the exterior

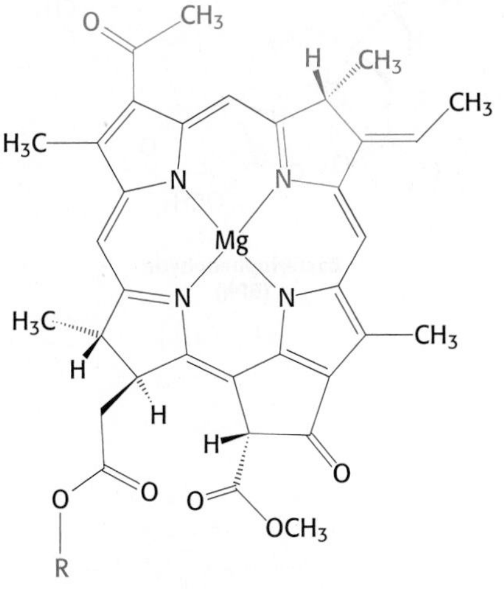

Bacteriochlorophyll *b*
(BChl-*b*)

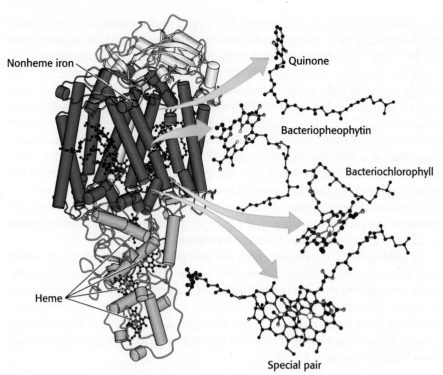

FIGURE 19.9 Bacterial photosynthetic reaction center. The core of the reaction center from *Rhodopseudomonas viridis* consists of two similar chains: L (red) and M (blue). An H chain (white) and a cytochrome subunit (yellow) complete the structure. *Notice* that the L and M subunits are composed largely of α helices that span the membrane. *Also notice* that a chain of electron-carrying prosthetic groups, beginning with a special pair of bacteriochlorophylls and ending at a bound quinone, runs through the structure from bottom to top in this view. [Drawn from 1PRC.pdb.]

Bacteriopheophytin (BPh)

face of the cell membrane, called the *periplasmic side* because it faces the periplasm—the space between the cell membrane and the cell wall. Four bacteriochlorophyll *b* (BChl-*b*) molecules, two bacteriopheophytin *b* (BPh) molecules, two quinones (Q_A and Q_B), and a ferrous ion are associated with the L and M subunits.

Bacteriochlorophylls are photoreceptors similar to chlorophylls, except for the reduction of an additional pyrrole ring and other minor differences that shift their absorption maxima to the near infrared, to wavelengths as long as 1000 nm. *Bacteriopheophytin* is the term for a bacteriochlorophyll that has two protons instead of a magnesium ion at its center.

The reaction begins with light absorption by a pair of BChl-*b* molecules that lie near the periplasmic side of the membrane in the L–M dimer. The pair of BChl-*b* molecules is called the *special pair* because of its fundamental role in photosynthesis. The special pair absorbs light maximally at 960 nm, and, for this reason, is often called *P960* (*P* stands for pigment). After absorbing light, the excited special pair ejects an electron, which is transferred through another BChl-*b* to a bacteriopheophytin (Figure 19.10, steps 1 and 2). This initial charge separation yields a positive charge on the special pair ($P960^+$) and a negative charge on BPh (BPh^-). The electron ejection and transfer take place in less than 10 picoseconds (10^{-12} s).

A nearby electron acceptor, a tightly bound quinone (Q_A), quickly grabs the electron away from BPh^- before the electron has a chance to fall back to the P960 special pair. From Q_A, the electron moves to a more loosely associated quinone, Q_B. The absorption of a second photon and

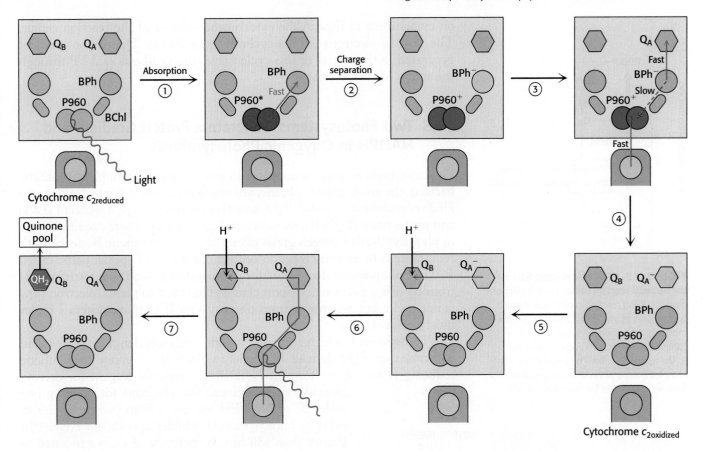

the movement of a second electron from the special pair through the bacteriopheophytin to the quinones completes the two-electron reduction of Q_B from Q to QH_2. Because the Q_B-binding site lies near the cytoplasmic side of the membrane, *two protons are taken up from the cytoplasm, contributing to the development of a proton gradient across the cell membrane* (Figure 19.10, steps 5, 6, and 7).

In their high-energy states, $P960^+$ and BPh^- could undergo charge recombination; that is, the electron on BPh^- could move back to neutralize the positive charge on the special pair. Its return to the special pair would waste a valuable high-energy electron and simply convert the absorbed light energy into heat. How is charge recombination prevented? Two factors in the structure of the reaction center work together to suppress charge recombination nearly completely (Figure 19.10, steps 3 and 4). First, the next electron acceptor (Q_A) is less than 10 Å away from BPh^-, and so the electron is rapidly transferred farther away from the special pair. Second, one of the hemes of the cytochrome subunit is less than 10 Å away from the special pair, and so the positive charge on P960 is neutralized by the transfer of an electron from the reduced cytochrome.

Cyclic electron flow reduces the cytochrome of the reaction center

The cytochrome subunit of the reaction center must regain an electron to complete the cycle. It does so by taking back two electrons from reduced quinone (QH_2). QH_2 first enters the Q pool in the membrane where it is reoxidized to Q by complex bc_1, which is homologous to complex III of the respiratory electron-transport chain. Complex bc_1, transfers the electrons from QH_2 to cytochrome c_2, a water-soluble protein in the periplasm, and in the process pumps protons into the periplasmic space. The electrons now

FIGURE 19.10 Electron chain in the photosynthetic bacterial reaction center. The absorption of light by the special pair (P960) results in the rapid transfer of an electron from this site to a bacteriopheophytin (BPh), creating a photoinduced charge separation (steps 1 and 2). (The asterisk on P960 stands for excited state.) The possible return of the electron from the pheophytin to the oxidized special pair is suppressed by the "hole" in the special pair being refilled with an electron from the cytochrome subunit and the electron from the pheophytin being transferred to a quinone (Q_A) that is farther away from the special pair (steps 3 and 4). Q_A passes the electron to Q_B. The reduction of a quinone (Q_B) on the cytoplasmic side of the membrane results in the uptake of two protons from the cytoplasm (steps 5 and 6). The reduced quinone can move into the quinone pool in the membrane (step 7).

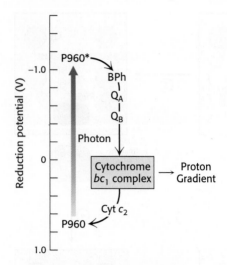

FIGURE 19.11 Cyclic electron flow in the bacterial reaction center. Excited electrons from the P960 reaction center flow through bacteriopheophytin (BPh), a pair of quinone molecules (Q_A and Q_B), cytochrome bc_1 complex, and finally through cytochrome c_2 to the reaction center. The cytochrome bc_1 complex pumps protons as a result of electron flow, which powers the formation of ATP.

on cytochrome c_2 flow to the cytochrome subunit of the reaction center. The flow of electrons is thus cyclic (Figure 19.11). The proton gradient generated in the course of this cycle drives the synthesis of ATP through the action of ATP synthase.

19.3 Two Photosystems Generate a Proton Gradient and NADPH in Oxygenic Photosynthesis

Photosynthesis is more complicated in green plants than in photosynthetic bacteria. In green plants, photosynthesis depends on the interplay of two kinds of membrane-bound, light-sensitive complexes—*photosystem I* (PS I) and *photosystem II* (PS II), as shown in Figure 19.12. There are similarities in photosynthesis between green plants and photosynthetic bacteria. Both require light to energize reaction centers consisting of special pairs, called *P700* for photosystem I and *P680* for photosystem II, and both transfer electrons by using electron-transport chains. However, in plants, electron flow is not cyclic but progresses from photosystem II to photosystem I under most circumstances.

Photosystem I, which responds to light with wavelengths shorter than 700 nm, uses light-derived high-energy electrons to create biosynthetic reducing power in the form of NADPH, a versatile reagent for driving biosynthetic processes. The electrons for creating one molecule of NADPH are taken from two molecules of water by photosystem II, which responds to wavelengths shorter than 680 nm. A molecule of O_2 is generated as a side product of the actions of photosystem II. The electrons travel from photosystem II to photosystem I through cytochrome *bf*, a membrane-bound complex homologous to Complex III in oxidative phosphorylation. Cytochrome *bf* generates a proton gradient across the thylakoid membrane that drives the formation of ATP. Thus, the two photosystems cooperate to produce NADPH and ATP.

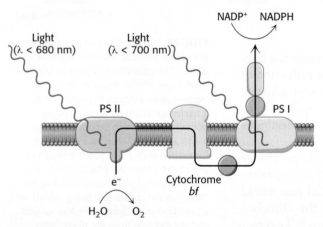

FIGURE 19.12 Two photosystems. The absorption of photons by two distinct photosystems (PS I and PS II) is required for complete electron flow from water to $NADP^+$.

Photosystem II transfers electrons from water to plastoquinone and generates a proton gradient

Photosystem II, an enormous transmembrane assembly of more than 20 subunits, catalyzes the light-driven transfer of electrons from water to plastoquinone. This electron acceptor closely resembles ubiquinone, a component of the mitochondrial electron-transport chain. Plastoquinone cycles between an oxidized form (Q) and a reduced form (QH_2, plastoquinol). The overall reaction catalyzed by photosystem II is

$$2\,Q + 2\,H_2O \xrightarrow{\text{Light}} O_2 + 2\,QH_2$$

The electrons in QH_2 are at a higher redox potential than those in H_2O. Recall that, in oxidative phosphorylation, electrons flow from ubiquinol to an acceptor, O_2, which is at a *lower* potential. Photosystem II drives the reaction in a thermodynamically uphill direction by using the free energy of light.

This reaction is similar to one catalyzed by the bacterial system in that a quinone is converted from its oxidized into its reduced form. Photosystem II is reasonably similar to the bacterial reaction center (Figure 19.13).

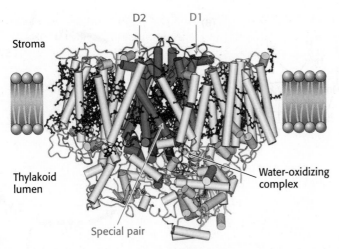

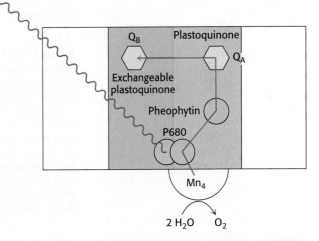

FIGURE 19.13 The structure of photosystem II. The D_1 (red) and D_2 (blue) subunits and the numerous bound chlorophyll molecules (green). *Notice* that the special pair and the water-oxidizing complex lie toward the thylakoid-lumen side of the membrane. [Drawn from 1S5L.pdb.]

The core of the photosystem is formed by D1 and D2, a pair of similar 32-kDa subunits that span the thylakoid membrane. These subunits are homologous to the L and M chains of the bacterial reaction center. Unlike the bacterial system, photosystem II contains a large number of additional subunits that bind more than 30 chlorophyll molecules altogether and increase the efficiency with which light energy is absorbed and transferred to the reaction center (Section 19.5).

The photochemistry of photosystem II begins with excitation of a special pair of chlorophyll molecules that are bound by the D1 and D2 subunits (Figure 19.14). Because the chlorophyll *a* molecules of the special pair absorb light at 680 nm, the special pair is often called *P680*. On excitation, P680 rapidly transfers an electron to a nearby pheophytin. From there, the electron is transferred first to a tightly bound plastoquinone at site Q_A and then to a mobile plastoquinone at site Q_B. This electron flow is entirely analogous to that in the bacterial system. With the arrival of a second electron and the uptake of two protons, the mobile plastoquinone is reduced to QH_2. At this point, the energy of two photons has been safely and efficiently stored in the reducing potential of QH_2.

The major difference between the bacterial system and photosystem II is the source of the electrons that are used to neutralize the positive charge formed on the special pair. $P680^+$, *a very strong oxidant, extracts electrons from water molecules bound at the water-oxidizing complex* (WOC), also called the *manganese center*. The core of this complex includes a calcium ion, four manganese ions, and four water molecules (Figure 19.15A). Manganese was apparently evolutionarily selected for this role because of its ability to exist in multiple oxidation states and to form strong bonds with oxygen-containing species. In its reduced form, the WOC oxidizes two molecules of water to form a single molecule of oxygen. Each time the absorbance of a photon powers the removal of an electron from $P680^+$, the positively charged special pair extracts an electron from a tyrosine residue (often designated Z) of subunit D1 of the WOC, forming a tyrosine

FIGURE 19.14 Electron flow through photosystem II. Light absorption induces electron transfer from P680 down an electron-transfer pathway to an exchangeable plastoquinone. The positive charge on P680 is neutralized by electron flow from water molecules bound at the manganese center.

(A)

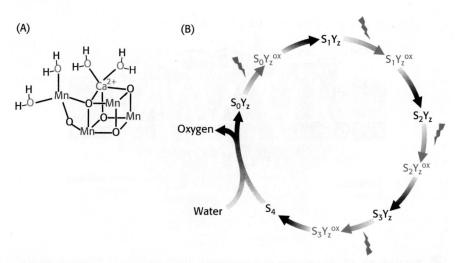

(B)

FIGURE 19.15 The core of the water-oxidizing complex. (A) The deduced core structure of the water-oxidizing complex (WOC), including four manganese ions and one calcium ion, is shown. The valence states of the individual manganese ions are not indicated because of uncertainty about the charge on the individual ions. The center is oxidized, one electron at a time, until two H_2O molecules are oxidized to form a molecule of O_2, which is then released from the complex. **(B)** The absorption of photons by the reaction center generates a tyrosine radical (red arrows), which then extracts electrons from the manganese ions. The structures are designated S_0 to S_4 to indicate the number of electrons that have been removed.

Evolution of oxygen is evident by the generation of bubbles in the aquatic plant *Elodea*. [Colin Milkins/Getty Images.]

radical (Figure 19.15B), The tyrosine radical then removes an electron from a manganese ion. This process occurs four times, with the result that H_2O is oxidized to generate O_2, and H^+. Four photons must be absorbed to extract four electrons from a water molecule (Figure 19.16). The four electrons harvested from water are used to reduce two molecules of Q to QH_2. All

FIGURE 19.16 Four photons are required to generate one oxygen molecule. When dark-adapted chloroplasts are exposed to a brief flash of light, one electron passes through photosystem II. Monitoring the O_2 released after each flash reveals that four flashes are required to generate each O_2 molecule. The peaks in O_2 release are after the 3rd, 7th, and 11th flashes because the dark-adapted chloroplasts start in the S_1 state—that is, the one-electron reduced state.

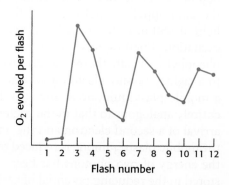

oxygenic phototrophs—the most common type of photosynthetic organism—use the same inorganic core and protein components for capturing the energy of sunlight. A single solution to the biochemical problem of extracting electrons from water evolved billions of years ago, and has been conserved for use under a wide variety of phylogenetic and ecological circumstances.

Photosystem II spans the thylakoid membrane such that the site of quinone reduction is on the side of the stroma, whereas WOC lies in the thylakoid lumen. Thus, the two protons that are taken up with the reduction of Q to QH_2 come from the stroma, and the four protons that are liberated in the course of water oxidation are released into the lumen. This distribution of protons generates a proton gradient across the thylakoid membrane characterized by an excess of protons in the thylakoid lumen compared with the stroma (Figure 19.17).

FIGURE 19.17 Proton-gradient direction. Photosystem II releases protons into the thylakoid lumen and takes them up from the stroma. The result is a pH gradient across the thylakoid membrane with an excess of protons (low pH) inside.

Cytochrome *bf* links photosystem II to photosystem I

Electrons flow from photosystem II to photosystem I through the *cyto-chrome* bf complex. This complex catalyzes the transfer of electrons from plastoquinol (QH_2) to plastocyanin (Pc), a small, soluble copper protein in the thylakoid lumen.

$$QH_2 + 2\,Pc(Cu^{2+}) \rightarrow Q + 2\,Pc(Cu^{+}) + 2\,H^{+}_{\text{thylakoid lumen}}$$

The two protons from plastoquinol are released into the thylakoid lumen. This reaction is reminiscent of that catalyzed by Complex III in oxidative phosphorylation, and most components of the cytochrome *bf* complex are homologous to those of Complex III. The cytochrome *bf* complex includes four subunits: a 23-kDa cytochrome with two *b*-type hemes, a 20-kDa Rieske-type Fe–S protein (p. 584), a 33-kDa cytochrome *f* with a *c*-type cytochrome, and a 17-kDa chain.

This complex catalyzes the reaction by proceeding through the Q cycle (Figure 18.12). In the first half of the Q cycle, plastoquinol (QH_2) is oxidized to plastoquinone (Q), one electron at a time. The electrons from plastoquinol flow through the Fe–S protein to convert oxidized plastocyanin (Pc) into its reduced form.

In the second half of the Q cycle, cytochrome *bf* reduces a molecule of plastoquinone from the Q pool to plastoquinol, taking up two protons from one side of the membrane, and then reoxidizes plastoquinol to release these protons on the other side. The enzyme is oriented so that protons are released into the thylakoid lumen and taken up from the stroma, contributing further to the proton gradient across the thylakoid membrane (Figure 19.18).

Photosystem I uses light energy to generate reduced ferredoxin, a powerful reductant

The final stage of the light reactions is catalyzed by photosystem I, a transmembrane complex consisting of about 15 polypeptide chains and multiple associated proteins and cofactors (Figure 19.19). The core of this system is a pair of similar subunits—psaA (83 kDa, red) and psaB (82 kDa, blue)—which bind 80 chlorophyll molecules as well as other redox factors. These subunits are quite a bit larger than the core subunits of photosystem II and the bacterial reaction center. Nonetheless, they appear to be homologous; the terminal 40% of each subunit is similar to a corresponding subunit of photosystem II. What is termed a *special pair* of chlorophyll *a* molecules lies at the center of the structure and absorb light maximally at 700 nm. This center,

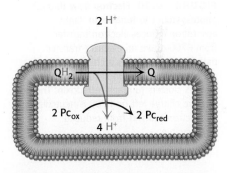

FIGURE 19.18 Cytochrome *bf* contribution to proton gradient. The cytochrome *bf* complex oxidizes QH_2 to Q through the Q cycle. Four protons are released into the thylakoid lumen in each cycle.

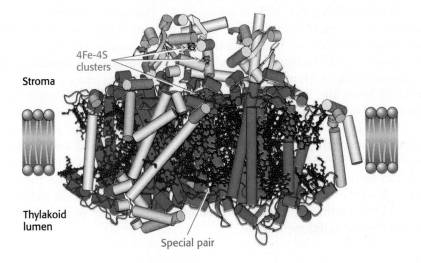

FIGURE 19.19 The structure of photosystem I. The psaA and psaB subunits are shown in red and blue, respectively. *Notice* the numerous bound chlorophyll molecules, shown in green, including the special pair, as well as the iron–sulfur clusters that facilitate electron transfer from the stroma. [Drawn from 1JBO.pdb.]

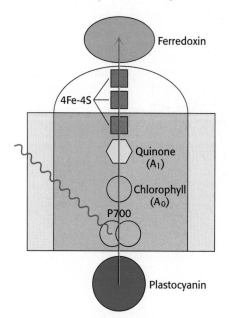

FIGURE 19.20 Electron flow through photosystem I to ferredoxin. Light absorption induces electron transfer from P700 down an electron-transfer pathway that includes a chlorophyll molecule, a quinone molecule, and three 4Fe–4S clusters to reach ferredoxin. The positive charge left on P700 is neutralized by electron transfer from reduced plastocyanin.

called *P700*, initiates photoinduced charge separation (Figure 19.20). The electron travels from P700 down a pathway through chlorophyll at site A_0 and quinone at site A_1 to a set of 4Fe–4S clusters. The next step is the transfer of the electron to ferredoxin (Fd), a soluble protein containing a 2Fe–2S cluster coordinated to four cysteine residues (Figure 19.21). Ferredoxin transfers electrons to $NADP^+$. Meanwhile, $P700^+$ captures an electron from reduced plastocyanin provided by photosystem II to return to P700 so that P700 can be excited again. Thus, the overall reaction catalyzed by photosystem I is a simple one-electron oxidation–reduction reaction.

$$Pc(Cu^+) + Fd_{ox} \xrightarrow{\text{Light}} Pc(Cu^{2+}) + Fd_{red}$$

Given that the reduction potentials for plastocyanin and ferredoxin are $+0.37\,\text{V}$ and $-0.45\,\text{V}$, respectively, the standard free energy for this reaction is $+79.1\,\text{kJ mol}^{-1}$ ($+18.9\,\text{kcal mol}^{-1}$). This uphill reaction is driven by the absorption of a 700-nm photon, which has an energy of $171\,\text{kJ mol}^{-1}$ ($40.9\,\text{kcal mol}^{-1}$).

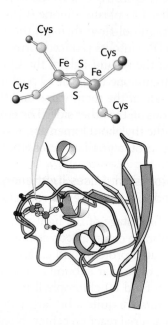

FIGURE 19.21 Structure of ferredoxin. In plants, ferredoxin contains a 2Fe-2S cluster. This protein accepts electrons from photosystem I and carries them to ferredoxin–$NADP^+$ reductase. [Drawn from 1FXA.pdb.]

Ferredoxin–$NADP^+$ reductase converts $NADP^+$ into NADPH

The reduced ferredoxin generated by photosystem I is a strong reductant that is used as an electron source for a variety of reactions, most notably the fixation of N_2 into NH_3 (Section 24.1). However, ferredoxin is not useful for driving many reactions, in part because ferredoxin carries only one available electron. In contrast, NADPH, a two-electron reductant, is a widely used electron donor in biosynthetic processes, including the reactions of the Calvin cycle (Chapter 20). How is reduced ferredoxin used to drive the reduction of $NADP^+$ to NADPH? This reaction is catalyzed by *ferredoxin–$NADP^+$ reductase*, a flavoprotein with an FAD prosthetic group (Figure 19.22A). The bound FAD moiety accepts two electrons and two protons from two molecules of reduced ferredoxin to form $FADH_2$ (Figure 19.22B). The enzyme then transfers a hydride ion (H^-) to $NADP^+$ to form NADPH. This reaction takes place on the stromal side of the membrane. Hence, the uptake of a proton in the reduction of $NADP^+$ further contributes to the generation of the proton gradient across the thylakoid membrane.

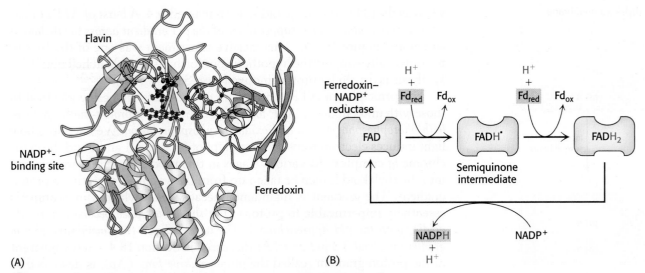

FIGURE 19.22 Structure and function of ferredoxin–NADP⁺ reductase.
(A) Structure of ferredoxin–NADP$^+$ reductase. This enzyme accepts electrons, one at a time, from ferredoxin (shown in orange). (B) Ferredoxin–NADP$^+$ reductase first accepts two electrons, generated by photosystem I, and two protons from the lumen to form two molecules of reduced ferredoxin (Fd). Reduced ferredoxin is then used to form FADH$_2$, which then transfers two electrons and a proton to NADP$^+$ to form NADPH in the stroma. [Drawn from 1EWY.pdb.]

The cooperation between photosystem I and photosystem II creates a flow of electrons from H_2O to NADP$^+$. The pathway of electron flow is called the *Z scheme of photosynthesis* because the redox diagram from P680 to P700* looks like the letter Z (Figure 19.23).

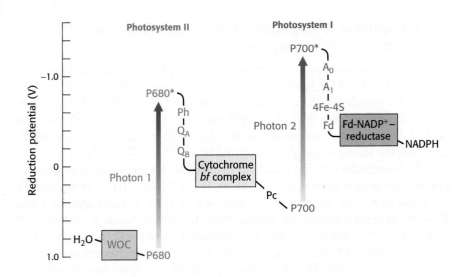

FIGURE 19.23 Pathway of electron flow from H_2O to NADP$^+$ in photosynthesis. This endergonic reaction is made possible by the absorption of light by photosystem II (P680) and photosystem I (P700). Abbreviations: Ph, pheophytin; Q_A and Q_B, plastoquinone-binding proteins; Pc, plastocyanin; A_0 and A_1, acceptors of electrons from P700*; Fd, ferredoxin; WOC, water-oxidizing complex.

19.4 A Proton Gradient across the Thylakoid Membrane Drives ATP Synthesis

In 1966, André Jagendorf showed that chloroplasts synthesize ATP in the dark when an artificial pH gradient is imposed across the thylakoid membrane. To create this transient pH gradient, he soaked chloroplasts in a pH 4 buffer for several hours and then rapidly mixed them with a pH 8 buffer containing ADP and P_i. The pH of the stroma suddenly increased to 8,

Thylakoid membrane

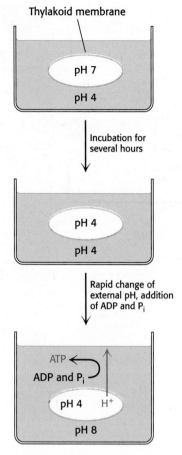

FIGURE 19.24 Jagendorf's demonstration. Chloroplasts synthesize ATP after the imposition of a pH gradient.

whereas the pH of the thylakoid space remained at 4. *A burst of ATP synthesis then accompanied the disappearance of the pH gradient across the thylakoid membrane* (Figure 19.24). This incisive experiment was one of the first to unequivocally support the hypothesis put forth by Peter Mitchell that ATP synthesis is driven by proton-motive force (Section 18.4).

The principles of ATP synthesis in chloroplasts are nearly identical to those with mitochondria. *ATP formation is driven by a proton-motive force in both photophosphorylation and oxidative phosphorylation.* We have seen how light induces electron transfer through photosystems II and I and the cytochrome *bf* complex. At various stages in this process, protons are released into the thylakoid lumen or taken up from the stroma, generating a proton gradient. The gradient is maintained because the thylakoid membrane is essentially impermeable to protons. *The thylakoid space becomes markedly acidic, with the pH approaching 4. The light-induced transmembrane proton gradient is about 3.5 pH units.* As discussed in Section 18.4, energy inherent in the proton gradient, called the *proton-motive force* (Δp), is described as the sum of two components: a charge gradient and a chemical gradient. In chloroplasts, nearly all of Δp arises from the pH gradient, whereas in mitochondria, the contribution from the membrane potential is larger. The reason for this difference is that the thylakoid membrane is quite permeable to Cl^- and Mg^{2+}. The light-induced transfer of H^+ into the thylakoid space is accompanied by the transfer of either Cl^- in the same direction or Mg^{2+} (1 Mg^{2+} per 2 H^+) in the opposite direction. Consequently, electrical neutrality is maintained and no membrane potential is generated. The influx of Mg^{2+} into the stroma plays a role in the regulation of the Calvin cycle (Section 20.2). A pH gradient of 3.5 units across the thylakoid membrane corresponds to a proton-motive force of 0.20 V or a ΔG of $-20.0 \, \text{kJ mol}^{-1} (-4.8 \, \text{kcal mol}^{-1})$.

The ATP synthase of chloroplasts closely resembles those of mitochondria and prokaryotes

The proton-motive force generated by the light reactions is converted into ATP by the *ATP synthase* of chloroplasts, also called the $CF_1 - CF_0$ *complex* (C stands for chloroplast; F for factor). $CF_1 - CF_0$ ATP synthase closely resembles the $F_1 - F_0$ ATP synthase of mitochondria (Section 18.4). CF_0 conducts protons across the thylakoid membrane, whereas CF_1 catalyzes the formation of ATP from ADP and P_i.

CF_0 is embedded in the thylakoid membrane. It consists of four different polypeptide chains known as I (17 kDa), II (16.5 kDa), III (8 kDa), and IV (27 kDa) having an estimated stoichiometry of 1:2:10–14:1. Subunits I and II have sequence similarity to subunit **b** of the mitochondrial F_0 subunit, III corresponds to subunit **c** of the mitochondrial complex, and subunit IV is similar in sequence to subunit **a**. CF_1, the site of ATP synthesis, has a subunit composition $\alpha_3\beta_3\gamma\delta\varepsilon$. The β subunits contain the catalytic sites, similarly to the F_1 subunit of mitochondrial ATP synthase. Remarkably, the β subunits of ATP synthase in corn chloroplasts are more than 60% identical in amino acid sequence with those of human ATP synthase, despite the passage of approximately 1 billion years since the separation of the plant and animal kingdoms.

Note that the membrane orientation of $CF_1 - CF_0$ is reversed compared with that of the mitochondrial ATP synthase (Figure 19.25). However, the functional orientation of the two synthases is identical: protons flow from the lumen through the enzyme to the stroma or matrix where ATP is synthesized. Because CF_1 is on the stromal surface of the thylakoid membrane, the newly synthesized ATP is released directly into the stromal space.

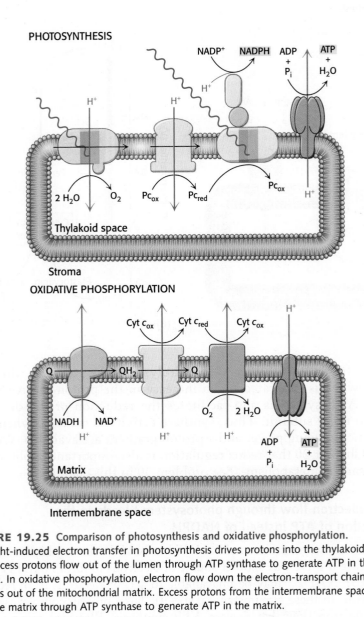

FIGURE 19.25 Comparison of photosynthesis and oxidative phosphorylation.
The light-induced electron transfer in photosynthesis drives protons into the thylakoid lumen. The excess protons flow out of the lumen through ATP synthase to generate ATP in the stroma. In oxidative phosphorylation, electron flow down the electron-transport chain pumps protons out of the mitochondrial matrix. Excess protons from the intermembrane space flow into the matrix through ATP synthase to generate ATP in the matrix.

Likewise, NADPH formed by photosystem I is released into the stromal space. Thus, *ATP and NADPH, the products of the light reactions of photosynthesis, are appropriately positioned for the subsequent dark reactions in which CO_2 is converted into carbohydrate.*

The activity of chloroplast ATP synthase is regulated

The activity of the ATP synthase is sensitive to the redox conditions in the chloroplast. For maximal activity, a specific disulfide bond in the γ subunit must be reduced to two cysteines. The reductant is reduced thioredoxin, which is formed from ferredoxin generated in photosystem I, by ferredoxin–thioredoxin reductase, an iron–sulfur containing enzyme.

$$2 \text{ Reduced ferredoxin} + \text{thioredoxin disulfide} \underset{}{\overset{\text{Ferredoxin-thioredoxin reductase}}{\rightleftharpoons}}$$
$$2 \text{ oxidized ferredoxin} + \text{reduced thioredoxin} + 2 \text{ H}^+$$

Conformational changes in the ε subunit also contribute to synthase regulation. The ε subunit appears to exist in two conformations.

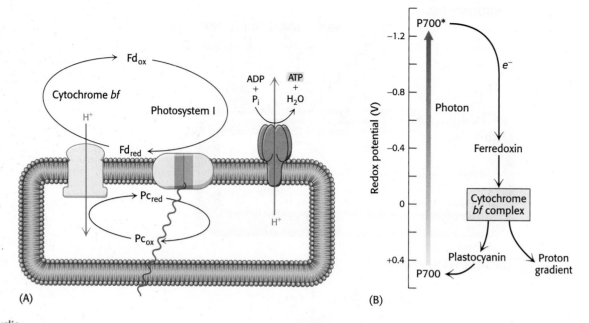

FIGURE 19.26 Cyclic photophosphorylation. (A) In this pathway, electrons from reduced ferredoxin are transferred to cytochrome *bf* rather than to ferredoxin–NADP$^+$ reductase. The flow of electrons through cytochrome *bf* pumps protons into the thylakoid lumen. These protons flow through ATP synthase to generate ATP. Neither NADPH nor O_2 is generated by this pathway. (B) A scheme showing the energetic basis for cyclic photophosphorylation. Abbreviations: Fd, ferredoxin; Pc, plastocyanin.

One conformation inhibits ATP hydrolysis by the synthase, while the other, which is generated by an increase in the proton-motive force, allows ATP synthesis and facilitates the reduction of the disulfide bond in the γ subunit. Thus, synthase activity is maximal when biosynthetic reducing power and a proton gradient are available. We will see in Chapter 20 that redox regulation is also important in photosynthetic carbon metabolism. (See problem 30 in this chapter.)

Cyclic electron flow through photosystem I leads to the production of ATP instead of NADPH

On occasion, when the ratio of NADPH to NADP$^+$ is very high, as might be the case if there was another source of electrons to form NADPH (Section 20.3), NADP$^+$ may be unavailable to accept electrons from reduced ferredoxin. In these circumstances, specific large protein complexes allow cyclic electron flow that powers ATP synthesis. Electrons arising from P700, the reaction center of photosystem I, generate reduced ferredoxin. The electron in reduced ferredoxin is transferred to the cytochrome *bf* complex rather than to NADP$^+$. This electron then flows back through the cytochrome *bf* complex to reduce plastocyanin, which can then be reoxidized by P700$^+$ to complete a cycle. The net outcome of this cyclic flow of electrons is the pumping of protons by the cytochrome *bf* complex. The resulting proton gradient then drives the synthesis of ATP. In this process, called *cyclic photophosphorylation, ATP is generated without the concomitant formation of NADPH* (Figure 19.26). Photosystem II does not participate in cyclic photophosphorylation, and so O_2 is not formed from H_2O.

The absorption of eight photons yields one O_2, two NADPH, and three ATP molecules

We can now estimate the overall stoichiometry for the light reactions. The absorption of four photons by photosystem II generates one molecule of O_2 and releases four protons into the thylakoid lumen. The two molecules of plastoquinol are oxidized by the Q cycle of the cytochrome *bf* complex to release eight protons into the lumen. Finally, the electrons from four molecules of reduced plastocyanin are driven to ferredoxin by the absorption of four additional photons. The four molecules of reduced ferredoxin generate two molecules of NADPH. Thus, the overall reaction is

$$2\,H_2O + 2\,NADP^+ + 10\,H_{stroma}^+ \rightarrow O_2 + 2\,NADPH + 12\,H_{lumen}^+$$

The 12 protons released in the lumen can then flow through ATP synthase. Let's assume that there are 12 subunit III components in CF_0. We expect that 12 protons must pass through CF_0 to complete one full rotation of CF_1. A single rotation generates three molecules of ATP. Given the ratio of 3 ATP for 12 protons, the overall reaction is

$$2\,H_2O + 2\,NADP^+ + 10\,H_{stroma}^+ \longrightarrow O_2 + 2\,NADPH + 12\,H_{lumen}^+$$
$$\underline{3\,ADP^{3-} + 3\,P_i^{2+} + 3\,H^+ + 12\,H_{lumen}^+ \longrightarrow 3\,ATP^{4-} + 3\,H_2O + 12\,H_{stroma}^+}$$
$$2\,NADP^+ + 3\,ADP^{3-} + 3\,P_i^{2-} + H^+ \longrightarrow O_2 + 2\,NADPH + 3\,ATP^{4-} + H_2O$$

Thus, eight photons are required to yield three molecules of ATP (2.7 photons/ATP).

Cyclic photophosphorylation is a somewhat more productive way to synthesize ATP than noncyclic photophosphorylation (the Z scheme). The absorption of four photons by photosystem I leads to the release of eight protons into the lumen by the cytochrome *bf* system. These protons flow through ATP synthase to yield two molecules of ATP. Thus, each two absorbed photons yield one molecule of ATP. No NADPH is produced.

19.5 Accessory Pigments Funnel Energy into Reaction Centers

A light-harvesting system that relied only on the chlorophyll *a* molecules of the special pair would be rather inefficient for two reasons. First, chlorophyll *a* molecules absorb light only at specific wavelengths (Figure 19.6). A large gap is present in the middle of the visible region between approximately 450 and 650 nm. This gap falls right at the peak of the solar spectrum, and so failure to collect this light would constitute a considerable lost opportunity. Second, even on a cloudless day, many photons that can be absorbed by chlorophyll *a* pass through the chloroplast without being absorbed, because the density of chlorophyll *a* molecules in a reaction center is not very great. Accessory pigments, both additional chlorophylls and other classes of molecules, are closely associated with reaction centers. *These pigments absorb light and funnel the energy to the reaction center for conversion into chemical forms.* Accessory pigments prevent the reaction center from sitting idle.

Chlorophyll b and *carotenoids* are important accessory pigments that funnel energy to the reaction center. Chlorophyll *b* differs from chlorophyll *a* in having a formyl group in place of a methyl group. This small difference shifts its two major absorption peaks toward the center of the visible region. In particular, chlorophyll *b* efficiently absorbs light with wavelengths between 450 and 500 nm (Figure 19.27).

Carotenoids are extended polyenes that absorb light between 400 and 500 nm. The carotenoids are responsible for most of the yellow and red colors of fruits and flowers, and they provide the brilliance of fall, when the

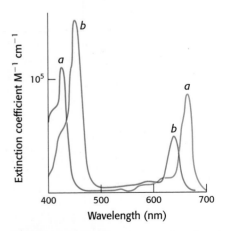

Chlorophyll b

FIGURE 19.27 Absorption spectra of chlorophylls *a* and *b*.

chlorophyll molecules degrade, revealing the carotenoids. Tomatoes are especially rich in lycopene, which accounts for their red color, whereas carrots and pumpkins have abundant β-carotene, accounting for their orange color.

Lycopene

β-Carotene

Resonance energy transfer allows energy to move from the site of initial absorbance to the reaction center

How is energy funneled from accessory pigments to a reaction center? The absorption of a photon does not always lead to electron excitation and transfer. More commonly, excitation energy is transferred from one molecule to a nearby molecule through electromagnetic interactions through space (Figure 19.28). The rate of this process, called *resonance energy transfer*, depends strongly on the distance between the energy-donor and the energy-acceptor molecules; an increase in the distance between the donor and the acceptor by a factor of 2 typically results in a decrease in the energy-transfer rate by a factor of $2^6 = 64$. For reasons of energy conservation, energy transfer must be from a donor in the excited state to an acceptor of equal or lower energy. *The excited state of the special pair of chlorophyll molecules is lower in energy than that of single chlorophyll molecules, allowing reaction centers to trap the energy transferred from other molecules.*

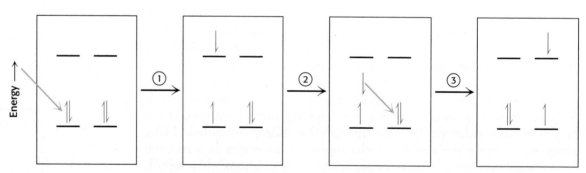

FIGURE 19.28 Resonance energy transfer. (1) An electron can accept energy from electromagnetic radiation of appropriate wavelength and jump to a higher energy state. (2) When the excited electron falls back to its lower energy state, the absorbed energy is released. (3) The released energy can be absorbed by an electron in a nearby molecule, and this electron jumps to a high energy state.

What is the structural relationship between the accessory pigment and the reaction center? The accessory pigments are arranged in numerous *light-harvesting complexes* that completely surround the reaction center (Figure 19.29). The 26-kDa subunit of light-harvesting complex II (LHC-II) is the most abundant membrane protein in chloroplasts. This subunit binds seven chlorophyll *a* molecules, six chlorophyll *b* molecules, and two carotenoid molecules. Similar light-harvesting assemblies exist in photosynthetic bacteria (Figure 19.30).

Chloroplast

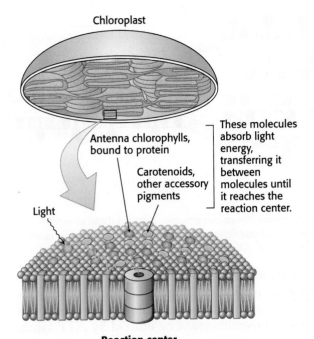

Antenna chlorophylls, bound to protein

Carotenoids, other accessory pigments

These molecules absorb light energy, transferring it between molecules until it reaches the reaction center.

Light

Reaction center
Photochemical reaction here converts the energy of a photon into a separation of charge, initiating electron flow.

FIGURE 19.29 Energy transfer from accessory pigments to reaction centers. Light energy absorbed by accessory chlorophyll molecules or other pigments can be transferred to reaction centers by resonance energy transfer (yellow arrows), where it drives photoinduced charge separation. [Source: D. L. Nelson and M. M. Cox, *Lehninger Principles of Biochemistry*, 5th ed. (W. H. Freeman and Company, 2008), Fig. 19.52.]

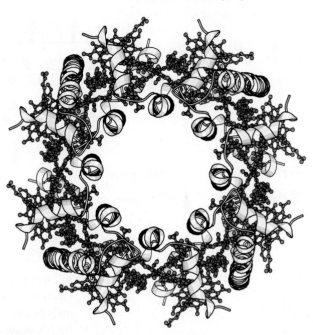

FIGURE 19.30 Structure of a bacterial light-harvesting complex. Eight polypeptides, each of which binds three chlorophyll molecules (green) and a carotenoid molecule (red), surround a central cavity that contains the reaction center (not shown). *Notice* the high concentration of accessory pigments that surround the reaction center. [Drawn from 1LGH.pdb.]

In addition to their role in transferring energy to reaction centers, the carotenoids and other accessory pigments serve a safeguarding function. Carotenoids suppress harmful photochemical reactions, particularly those including oxygen, that can be induced by bright sunlight. The reactive oxygen species damage the cell and eventually the entire plant. To protect against bright sunlight, plants use a mechanism called *nonphotochemical quenching* (NPQ). When NPQ is operating, photons are directed from the light-harvesting complex and the energy of the photons is released as heat. This protection may be especially important in the fall when the primary pigment chlorophyll is being degraded and thus not able to absorb light energy. Plants lacking carotenoids are quickly killed on exposure to light and oxygen.

(See Appendix: Biochemistry in Focus in this chapter for more information about nonphotochemical quenching.)

The components of photosynthesis are highly organized

The intricacy of photosynthesis, seen already in the elaborate interplay of complex components, extends even to the placement of the components in the thylakoid membranes. *Thylakoid membranes of most plants are differentiated into stacked* (appressed) *and unstacked* (nonappressed) *regions* (Figures 19.2 and 19.3). Stacking increases the amount of thylakoid membrane in a given chloroplast volume. Both regions surround a common internal thylakoid space, but only unstacked regions make direct contact with the chloroplast stroma. Stacked and unstacked regions differ in the nature of their photosynthetic assemblies (Figure 19.31). Photosystem I and ATP synthase are located almost exclusively in unstacked regions, whereas photosystem II is present mostly in stacked regions. The cytochrome *bf* complex is found

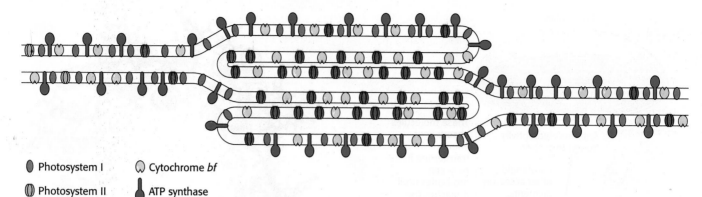

Photosystem I

Cytochrome *bf*

Photosystem II

ATP synthase

FIGURE 19.31 Location of photosynthesis components. Photosynthetic assemblies are differentially distributed in the stacked (appressed) and unstacked (nonappressed) regions of thylakoid membranes. [Information from Dr. Jan M. Anderson and Dr. Bertil Andersson.]

in both regions. Plastoquinone and plastocyanin are the mobile carriers of electrons between assemblies located in different regions of the thylakoid membrane. A common internal thylakoid space enables protons liberated by photosystem II in stacked membranes to be utilized by ATP synthase molecules that are located far away in unstacked membranes.

What is the functional significance of this lateral differentiation of the thylakoid membrane system? The positioning of photosystem I in the unstacked membranes also gives it direct access to the stroma for the reduction of $NADP^+$. ATP synthase is also located in the unstacked region to provide space for its large CF_1 globule and to give access to ADP. In contrast, the tight quarters of the appressed region pose no problem for photosystem II, which interacts with a small polar electron donor (H_2O) and a highly lipid-soluble electron carrier (plastoquinone).

Many herbicides inhibit the light reactions of photosynthesis

Many commercial herbicides kill weeds by interfering with the action of photosystem II or photosystem I. Inhibitors of photosystem II block electron flow, whereas inhibitors of photosystem I divert electrons from the terminal part of this photosystem. Photosystem II inhibitors include urea derivatives such as *diuron* and triazine derivatives such as *atrazine*. These chemicals bind to the Q_B site of the D1 subunit of photosystem II and block the formation of plastoquinol (QH_2).

Paraquat (1,1'-dimethyl-4-4'-bipyridinium) is an inhibitor of photosystem I. Paraquat, a dication, can accept electrons from photosystem I to become a radical. This radical reacts with O_2 to produce reactive oxygen species such as superoxide (O_2^-) and hydroxyl radical (OH•). Such reactive oxygen species react with many biomolecules, including double bonds in membrane lipids, damaging the membrane.

Diuron

Atrazine

19.6 The Ability to Convert Light into Chemical Energy Is Ancient

The ability to convert light energy into chemical energy is a tremendous evolutionary advantage. Geological evidence suggests that oxygenic photosynthesis became important approximately 2 billion years ago. Anoxygenic photosynthetic systems arose much earlier in the 3.5-billion-year history of life on Earth (Table 19.1). The photosynthetic system of the nonsulfur purple bacterium *Rhodopseudomonas viridis* has most features

TABLE 19.1 Major groups of photosynthetic prokaryotes

Bacteria	Photosynthetic electron donor	O_2 use
Green sulphur	H_2, H_2S, S	Anoxygenic
Green nonsulfur	Variety of amino acids and organic acids	Anoxygenic
Purple sulphur	H_2, H_2S, S	Anoxygenic
Purple nonsulfur	Usually organic molecules	Anoxygenic
Cyanobacteria	H_2O	Oxygenic

common to oxygenic photosynthetic systems and clearly predates them. Green sulfur bacteria such as *Chlorobium thiosulfatophilum* carry out a reaction that also seems to have appeared before oxygenic photosynthesis and is even more similar to oxygenic photosynthesis than the photosystem of *R. viridis*. Reduced sulfur species such as H_2S are electron donors in the overall photosynthetic reaction

$$CO_2 + 2H_2S \xrightarrow{\text{Light}} (CH_2O) + 2S + H_2O$$

Nonetheless, photosynthesis did not evolve immediately at the origin of life. No photosynthetic organisms have been discovered in the domain of archaea, implying that photosynthesis evolved in the domain of bacteria after archaea and bacteria diverged from a common ancestor. All domains of life do have electron-transport chains in common, however. As we have seen, components such as the ubiquinone–cytochrome *c* oxidoreductase and cytochrome *bf* family are present in both respiratory and photosynthetic electron-transport chains. These components were the foundations on which light-energy-capturing systems evolved.

Artificial photosynthetic systems may provide clean, renewable energy

As we have learned, photosynthetic organisms use sunlight to oxidize H_2O, producing O_2 and protons used to power ATP synthesis and generate NADPH. Research is currently underway to try to mimic this process in order to provide clean energy. Photovoltaic cells can use light energy to oxidize water, producing O_2 as well as H_2. Hydrogen gas is a fuel that, upon reaction with oxygen, generates energy and only water as a waste product. Major difficulties in creating efficient photovoltaic cells are that the materials required are not durable and often not readily available. Recent work suggests that semiconductors composed of organic–inorganic material show great promise. One such material, perovskite ($CH_3NH_3PbI_3$), is remarkably efficient at capturing sunlight.

Perovskite photovoltaic cells are now being paired with microorganisms to generate a variety of biomolecules, including fuels, in a clean, renewable fashion. One example of such a hybrid-biological-inorganic (HBI) system is the microbial electrosynthesis (MES) reactor (Figure 19.32). Sunlight is converted into an electrical current by the photovoltaic cell. The current flows to the anode, where water is oxidized and the resulting electrons are taken up by the cathode. Bacteria, such as *Ralsontia eutropha* (also used in bioremediation because of its ability to degrade chlorinated hydrocarbons) live in direct contact with the cathode in a biofilm. Biofilms are colonies of

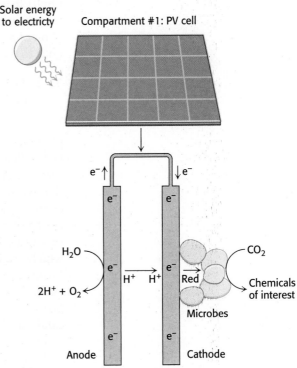

FIGURE 19.32 Artificial photosynthesis. Light is absorbed by a photovoltaic cell. The resulting current flows to the anode to oxidize water, with the electrons then taken up by the cathode. Bacteria in contact with the cathode use the electrons to reduce CO_2 to biomolecules.

bacteria living in a secreted matrix composed of carbohydrates, proteins, and nucleic acids (Section 32.3). The bacteria in the biofilm use the electrons generated by the oxidation of water to reduce CO_2 to industrially important biomolecules. The efficiency of such HBI systems for converting sunlight into chemical energy is near 10%, with the potential for improvement, surpassing the efficiency of natural photosynthesis (2%).

SUMMARY

19.1 Photosynthesis Takes Place in Chloroplasts

The proteins that participate in the light reactions of photosynthesis are located in the thylakoid membranes of chloroplasts. The light reactions result in (1) the creation of reducing power for the production of NADPH, (2) the generation of a transmembrane proton gradient for the formation of ATP, and (3) the production of O_2.

19.2 Light Absorption by Chlorophyll Induces Electron Transfer

Chlorophyll molecules absorb light quite efficiently because they are polyenes. An electron excited to a high-energy state by the absorption of a photon can move to nearby electron acceptors. In photosynthesis, an excited electron leaves a pair of associated chlorophyll molecules known as the special pair. The functional core of photosynthesis, a reaction center from a photosynthetic bacterium, has been studied in great detail. In this system, the electron moves from the special pair (containing bacteriochlorophyll) to a bacteriopheophytin (a bacteriochlorophyll lacking the central magnesium ion) to quinones. The reduction of quinones leads to the generation of a proton gradient, which drives ATP synthesis in a manner analogous to that of oxidative phosphorylation.

19.3 Two Photosystems Generate a Proton Gradient and NADPH in Oxygenic Photosynthesis

Photosynthesis in green plants is mediated by two linked photosystems. In photosystem II, the excitation of a special pair of chlorophyll molecules called P680 leads to electron transfer to plastoquinone in a manner analogous to that of the bacterial reaction center. The electrons are replenished by the extraction of electrons from a water molecule at the water-oxidizing complex, which contains four manganese ions. One molecule of O_2 is generated at this center for each four electrons transferred. The plastoquinol produced at photosystem II is reoxidized by the cytochrome *bf* complex, which transfers the electrons to plastocyanin, a soluble copper protein. From plastocyanin, the electrons enter photosystem I. In photosystem I, the excitation of special pair P700 releases electrons that flow to ferredoxin, a powerful reductant. Ferredoxin–$NADP^+$ reductase, a flavoprotein, then catalyzes the formation of NADPH. A proton gradient is generated as electrons pass through photosystem II, through the cytochrome *bf* complex, and through ferredoxin–$NADP^+$ reductase.

19.4 A Proton Gradient across the Thylakoid Membrane Drives ATP Synthesis

The proton gradient across the thylakoid membrane creates a proton-motive force, used by ATP synthase to form ATP. The ATP synthase of chloroplasts (also called the $CF_0 - CF_1$ complex) closely resembles the

ATP-synthesizing assemblies of bacteria and mitochondria. If the NADPH:$NADP^+$ ratio is high, electrons transferred to ferredoxin by photosystem I can reenter the cytochrome *bf* complex. This process, called cyclic photophosphorylation, leads to the generation of a proton gradient by the cytochrome *bf* complex without the formation of NADPH or O_2.

19.5 Accessory Pigments Funnel Energy into Reaction Centers

Light-harvesting complexes that surround the reaction centers contain additional molecules of chlorophyll *a*, as well as carotenoids and chlorophyll *b* molecules, which absorb light in the center of the visible spectrum. These accessory pigments increase the efficiency of light capture by absorbing light and transferring the energy to reaction centers through resonance energy transfer.

19.6 The Ability to Convert Light into Chemical Energy Is Ancient

The photosystems have structural features in common that suggest a mutual evolutionary origin. Similarities in organization and molecular structure to those of oxidative phosphorylation suggest that the photosynthetic apparatus evolved from an early energy-transduction system.

APPENDIX

Biochemistry in Focus

Increasing the efficiency of photosynthesis will increase crop yields

Nonphotochemical quenching (NPQ) (p. 637) is a means by which plants protect themselves from damaging reactive oxygen species that are generated by intense sunlight—essentially a botanical sunscreen. Understanding NPQ, a complex process consisting of several components, is an active area of research. It is established that NPQ requires a proton gradient across the thylakoid membrane, and in higher photosynthetic organisms such as plants, a particular protein, photosystem II subunit S (PsbS). PsbS, in the presence of a pH gradient, acts in seconds to help to initiate NPQ by interacting with protein components of the light-harvesting complex. This interaction redirects the energy absorbed by the light-harvesting complex from the reaction centers so that the energy is lost as heat. Although quick to be activated, NPQ is slow to turn off, depressing photosynthetic yields to be as much as 30% in some crops. This slow turn off is a target for biochemists hoping to increase the efficiency of photosynthesis and consequently, crop yield.

Why is it necessary to boost photosynthesis? Traditional plant breeding has greatly increased crop yield over the past decades. Additionally, modern farming practices and tools such as fertilizers, pesticides, and irrigation have contributed to increased yield. However, studies suggest that the ability to increase crop yields by traditional means may have reached its limits. And yet, the Food and Agricultural Organization of the United Nations estimates that in 40 years, the world's population will be about 10 billion, an increase of more than 2 billion from the current 7.6 billion. To feed all of these people, agricultural production will need to increase by 50% in the developed world and 400% in the developing world.

To tackle the problem of not enough food, biochemists are trying to increase the efficiency of photosynthesis by reducing the time required to turn off NPQ. Using techniques described in Chapter 5, genes were introduced into the model organism—the tobacco plant—that increased the speed of NPQ shutdown. The result was a 15% increase in plant size. The next step is to modify crops such as wheat, rice, and cowpea—important sources of protein in sub-Saharan Africa.

APPENDIX

Problem-Solving Strategies

PROBLEM: The relative proportions of the components of the proton-motive force that drive ATP synthesis in chloroplasts are different from those that power ATP synthesis in mitochondria. Given that the synthase structure and mechanism of action from both chloroplasts and mitochondria are very similar, explain the basis of this difference.

SOLUTION: As always, let's begin with first principles. The question is about the proton-motive force, so we need to make sure that we understand what the proton-motive force is.

▶ **Write an equation that describes the proton-motive force.**

Proton-motive force (Δp) = chemical gradient (ΔpH) + charge gradient $(\Delta \psi)$

▶ **Now, describe the nature of the two components that make up the proton-motive force.**

The proton-motive force is due to an unequal distribution of H^+ across a proton-impermeable membrane. One component describes the chemical gradient, but since the chemical is a proton, the gradient can be represented as a pH gradient (ΔpH) across the thylakoid membrane of the chloroplast or inner mitochondrial membrane of the mitochondria. The pH is lower (more acidic, more H^+) in the thylakoid lumen and the mitochondrial inner membrane space relative to the stroma or matrix. The second component of the proton-motive force $(\Delta \psi)$ describes the charge difference across the membranes. The charge difference occurs because the chemical forming the gradient, a proton, bears a positive charge. The thylakoid lumen and

inner membrane space of the mitochondria are positive relative to the stroma or matrix.

The question states that although the proton-motive force drives ATP synthesis in both chloroplasts and mitochondria, the proportion of each component is different. Let's think about this.

▶ **Is the thylakoid membrane permeable to H^+?**

No, the thylakoid membrane is not permeable to H^+. This tells us the ΔpH is at least partially responsible for the proton-motive force in chloroplasts.

▶ **Is the thylakoid membrane permeable to ions other than H^+?**

Yes! Recall (p. 632) that the thylakoid membrane is quite permeable to Cl^- and Mg^{2+}. Thus, as H^+ is pumped into the thylakoid lumen, Mg^{2+} (1 Mg^{2+} per 2 H^+) moves out of the lumen into the stroma. Alternatively, Cl^- accompanies H^+ into the lumen from the stroma. In either case, no charge gradient is established, thus, $\Delta \psi \approx 0$. So,

▶ **What components of the proton-motive force power ATP synthesis in chloroplasts?**

In chloroplasts, the proton-motive force consists entirely of the pH gradient, ΔpH. In contrast, in mitochondria, both the pH and the charge gradient contribute to the proton-motive force.

KEY TERMS

light reactions (p. 620)

chloroplast (p. 621)

stroma (p. 621)

thylakoid (p. 621)

granum (p. 621)

chlorophyll *a* (p. 622)

photoinduced charge separation (p. 623)

reaction center (p. 623)

special pair (p. 624)

P960 (p. 624)

photosystem I (PS I) (p. 626)

photosystem II (PS II) (p. 626)

P680 (p. 626)

water-oxidizing complex (WOC) (manganese center) (p. 627)

cytochrome *bf* (p. 629)

P700 (p. 630)

Z scheme of photosynthesis (p. 631)

proton-motive force (p. 632)

ATP synthase ($CF_1 - CF_0$ complex) (p. 632)

cyclic photophosphorylation (p. 634)

carotenoid (p. 635)

light-harvesting complex (p. 636)

nonphotochemical quenching (p. 637)

PROBLEMS

1. *Odd man out.* Which of the following is not a component of chloroplasts?

(a) Thylakoid membrane

(b) Stromal lamellae

(c) Cristae

(d) Granum

2. *Donations are good.* Under normal circumstances, photosystem I donates electrons to:

(a) Photosystem II

(b) ATP synthase

(c) P700

(d) $NADP^+$

3. *Dehydration.* The water-oxidizing complex provides electrons to:

(a) P680

(b) P450

(c) P700

(d) Cytochrome *c*

4. *Either or.* Cyclic photophosphorylation allows the generation of _____ without the synthesis of _____.

(a) ATP, NADPH

(b) NADPH, ATP

(c) ATP, NADH

(d) ATP, O_2

5. *RET or FRET.* Resonance energy transfer:

(a) Results in separation of charge.

(b) Powers ATP synthesis.

(c) Allows energy flux in the light-harvesting complex.

(d) Does not exist.

6. *A crucial prereq.* Human beings do not produce energy by photosynthesis, yet this process is critical to our survival. Explain.

7. *The accounting.* What is the overall reaction for the light reactions of photosynthesis? ✓❶

8. *Like a fife and drum.* Match each term with its description.

(a) Light reactions _____

(b) Chloroplasts _____

(c) Reaction center _____

(d) Chlorophyll _____

(e) Light-harvesting complex _____

(f) Photosystem I _____

(g) Photosystem II _____

(h) Cytochrome *bf* complex _____

(i) Water-oxidizing complex _____

(j) ATP synthase _____

1. Uses resonance energy transfer to reach the reaction center

2. Powers the formation of NADPH

3. Pumps protons

4. Site of photoinduced charge separation

5. Cellular location of photosynthesis

6. CF_1–CF_0 complex

7. Generates ATP, NADPH, and O_2

8. Site of oxygen generation

9. Transfers electrons from H_2O to P680

10. Primary photosynthetic pigment

9. *A single wavelength.* Photosynthesis can be measured by measuring the rate of oxygen production. When plants are exposed to light of wavelength 680 nm, more oxygen is evolved than if the plants are exposed to light of 700 nm. Explain. ✓❶

10. *Combining wavelengths.* If plants described in problem 9 are illuminated by a combination of light of 680 nm and 700 nm, the oxygen production exceeds that of either wavelength alone. Explain. ✓❶

11. *Complementary powers.* Photosystem I produces a powerful reductant, whereas photosystem II produces a powerful oxidant. Identify the reductant and oxidant and describe their roles. ✓❷, ✓❸

12. *If a little is good.* What is the advantage of having an extensive set of thylakoid membranes in the chloroplasts? ✓❶, ✓❷

13. *Cooperation.* Explain how light-harvesting complexes enhance the efficiency of photosynthesis. ✓❺

14. *One thing leads to another.* What is the ultimate electron acceptor in photosynthesis? The ultimate electron donor? What powers the electron flow between the donor and the acceptor? ✓❸

15. *Environmentally appropriate.* Chlorophyll is a hydrophobic molecule. Why is this property crucial for the function of chlorophyll?

16. *Proton origins.* What are the various sources of protons that contribute to the generation of a proton gradient in chloroplasts? ✓❹

17. *Efficiency matters.* What fraction of the energy of 700-nm light absorbed by photosystem I is trapped as high-energy electrons?

18. *That's not right.* Explain the defect or defects in the hypothetical scheme for the light reactions of photosynthesis depicted here. ✓❸

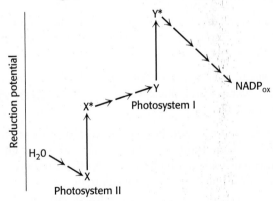

19. *Electron transfer.* Calculate the $\Delta E_0'$ and $\Delta G^{\circ\prime}$ for the reduction of $NADP^+$ by ferredoxin. Use data given in Table 18.1.

20. *To boldly go.* (a) It can be argued that, if life were to exist elsewhere in the universe, it would require some process like photosynthesis. Why is this argument reasonable? (b) If the starship *Enterprise* were to land on a distant planet and find no measurable oxygen in the atmosphere, could the crew conclude that photosynthesis is not taking place? ✓❶

21. Weed killer 1. Dichlorophenyldimethylurea (DCMU), a herbicide, interferes with photophosphorylation and O_2 evolution. However, it does not block O_2 evolution in the presence of an artificial electron acceptor such as ferricyanide. Propose a site for the inhibitory action of DCMU.

22. Weed killer 2. Predict the effect of the herbicide dichlorophenyldimethylurea (DCMU) on a plant's ability to perform cyclic photophosphorylation.

23. Infrared harvest. Consider the relation between the energy of a photon and its wavelength.

a. Some bacteria are able to harvest 1000-nm light. What is the energy (in kilojoules or kilocalories) of a mole (also called an *einstein*) of 1000-nm photons?

b. What is the maximum increase in redox potential that can be induced by a 1000-nm photon?

c. What is the minimum number of 1000-nm photons needed to form ATP from ADP and P_i? Assume a ΔG of 50 kJ mol^{-1} (12 kcal mol^{-1}) for the phosphorylation reaction.

24. Missing acceptors. Suppose that a bacterial reaction center containing only the special pair and the quinones has been prepared. Given the separation of 22 Å between the special pair and the closest quinone, estimate the rate of electron transfer between the excited special pair and this quinone.

25. Close approach. Suppose that energy transfer between two chlorophyll *a* molecules separated by 10 Å takes place in 10 picoseconds. Suppose that this distance is increased to 20 Å with all other factors remaining the same. How long would energy transfer take?

Chapter Integration Problems

26. Functional equivalents. What structural feature of mitochondria corresponds to the thylakoid membranes? ✓❶, ✓❷, ✓❸, ✓❹

27. Compare and contrast. Compare and contrast oxidative phosphorylation and photosynthesis. ✓❶, ✓❷, ✓❸, ✓❹

28. Energy accounts. On page 654, the balance sheet for the cost of the photosynthetically powered synthesis of glucose is presented. Eighteen molecules of ATP are required. Yet, when glucose undergoes combustion in cellular respiration, 30 molecules of ATP are produced. Account for the difference.

Mechanism Problem

29. Hill reaction. In 1939, Robert Hill discovered that chloroplasts evolve O_2 when they are illuminated in the presence of an artificial electron acceptor such as ferricyanide

$[Fe^{3+}(CN)_6]^{3-}$. Ferricyanide is reduced to ferrocyanide $[Fe^{2+}(CN)_6]^{4-}$ in this process. No NADPH or reduced plastocyanin is produced. Propose a mechanism for the Hill reaction.

Data Interpretation and Chapter Integration Problems

30. The same, but different. The $\alpha_3\beta_3\gamma$ complex of mitochondrial or chloroplast ATP synthase functions as an ATPase in vitro. The chloroplast enzyme (both synthase and ATPase activity) is sensitive to redox control, whereas the mitochondrial enzyme is not. To determine where the enzymes differ, a segment of the mitochondrial γ subunit was removed and replaced with the equivalent segment from the chloroplast γ subunit. The ATPase activity of the modified enzyme was then measured as a function of redox conditions. Dithiothreitol (DTT) is a reducing agent. ✓❹

a. What is the redox regulator of the chloroplast ATP synthase in vivo? The adjoining graph shows the ATPase activity of modified and control enzymes under various redox conditions.

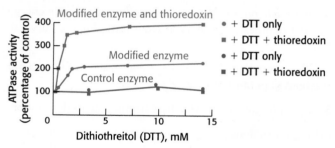

b. What is the effect of increasing the reducing power of the reaction mixture for the control and the modified enzymes?

c. What is the effect of the addition of thioredoxin? How do these results differ from those in the presence of DTT alone? Suggest a possible explanation for the difference.

d. Did the researchers succeed in identifying the region of the γ subunit responsible for redox regulation?

e. What is the biological rationale of regulation by high concentrations of reducing agents?

f. What amino acids in the γ subunit are most likely affected by the reducing conditions?

g. What experiments might confirm your answer to part f?

Think ▶ Pair ▶ Share Problem

31. Albert Szent-Györgyi, Nobel Prize–winning biochemist, once said, "Life is nothing but an electron looking for a place to rest." Explain how this pithy statement applies to photosynthesis and cellular respiration.

The Calvin Cycle and the Pentose Phosphate Pathway

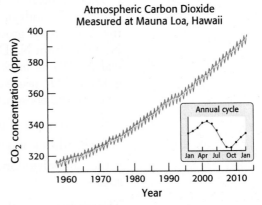

Atmospheric Carbon Dioxide Measured at Mauna Loa, Hawaii

Atmospheric carbon dioxide measurements at Mauna Loa, Hawaii. These measurements show the increase in atmospheric carbon dioxide concentration since 1960. The sawtooth appearance of the curve is the result of annual cycles resulting from seasonal variation in CO_2 fixation by the Calvin cycle in terrestrial plants (see insert). Much of this fixation takes place in rain forests, which account for approximately 50% of terrestrial fixation. [(Left) stillfx/AGE Fotostock. (above) Data from http://www.esrl.noaa.gov/gmd/ccgg/trends]

✓ LEARNING OBJECTIVES

By the end of this chapter, you should be able to:

1. Explain the function of the Calvin cycle.

2. Describe how the light reactions and the Calvin cycle are coordinated.

3. Identify the two stages of the pentose phosphate pathway, and explain how the pathway is coordinated with glycolysis and gluconeogenesis.

4. Identify the enzyme that controls the oxidative phase of the pentose phosphate pathway.

OUTLINE

20.1 The Calvin Cycle Synthesizes Hexoses from Carbon Dioxide and Water

20.2 The Activity of the Calvin Cycle Depends on Environmental Conditions

20.3 The Pentose Phosphate Pathway Generates NADPH and Synthesizes Five-Carbon Sugars

20.4 The Metabolism of Glucose 6-phosphate by the Pentose Phosphate Pathway Is Coordinated with Glycolysis

20.5 Glucose 6-phosphate Dehydrogenase Plays a Key Role in Protection Against Reactive Oxygen Species

Photosynthesis proceeds in two parts: the light reactions and the dark reactions. The light reactions, discussed in Chapter 19, transform light energy into ATP and biosynthetic reducing power, NADPH. Recall that these reactions involve the thylakoid membranes of the chloroplasts (Figure 19.2). The dark reactions, which occur in the stroma of the chloroplasts, use the ATP and NADPH produced by the light reactions to reduce carbon atoms from their fully oxidized state as carbon dioxide to a more reduced state as a hexose. Carbon dioxide is thereby trapped in a form that is useful for many processes and most especially as a fuel. Together, *the light reactions and dark reactions of photosynthesis cooperate to transform light energy into carbon fuel*. The dark reactions are called either the *Calvin–Benson cycle,* after Melvin Calvin and Andrew Benson, the biochemists who elucidated the pathway, or simply the Calvin cycle. The components of the Calvin cycle are called the *dark reactions* because, in contrast with the light reactions, these reactions do not directly depend on the presence of light.

The second half of this chapter examines a pathway common to all organisms, known variously as the *pentose phosphate pathway*, the *hexose monophosphate pathway*, the *phosphogluconate pathway*, or the *pentose shunt*. The pathway provides a means by which glucose can be oxidized to generate NADPH, *the currency of readily available reducing power in cells*. The phosphoryl group on the 2′-hydroxyl group of one of the ribose units of NADPH distinguishes NADPH from NADH. *There is a fundamental distinction between NADPH and NADH in biochemistry: NADH is oxidized by the respiratory chain to generate ATP, whereas NADPH serves as a reductant in biosynthetic processes.* The pentose phosphate pathway can also be used for:

- the catabolism of pentose sugars from the diet,
- the synthesis of pentose sugars for nucleotide biosynthesis, and
- the catabolism and synthesis of less common four- and seven-carbon sugars.

The pentose phosphate pathway and the Calvin cycle have in common several enzymes and intermediates that attest to an evolutionary kinship. Like glycolysis and gluconeogenesis, these pathways are mirror images of each other: the Calvin cycle uses NADPH to reduce carbon dioxide to generate hexoses, whereas the pentose phosphate pathway oxidizes glucose to carbon dioxide to generate NADPH.

20.1 The Calvin Cycle Synthesizes Hexoses from Carbon Dioxide and Water

As stated in Chapter 16, glucose can be formed from noncarbohydrate precursors, such as lactate and amino acids, by gluconeogenesis. The energy powering gluconeogenesis ultimately comes from previous catabolism of carbon fuels. In contrast, photosynthetic organisms can use the Calvin cycle to synthesize glucose from carbon dioxide gas and water, by using sunlight as an energy source. The Calvin cycle introduces into life all the carbon atoms that will be used as fuel and as the carbon backbones of biomolecules. Photosynthetic organisms are called *autotrophs* (literally, *self-feeders*) because they can convert sunlight into chemical energy, which they subsequently use to power their biosynthetic processes. Organisms that obtain energy from chemical fuels only are called *heterotrophs*, and such organisms ultimately depend on autotrophs for their fuel.

The Calvin cycle comprises three stages (Figure 20.1):

1. The fixation of CO_2 to ribulose 1,5-bisphosphate to form two molecules of 3-phosphoglycerate;

2. The reduction of 3-phosphoglycerate to form hexose sugars; and

3. The regeneration of ribulose 1,5-bisphosphate so that more CO_2 can be fixed.

This set of reactions takes place in the stroma of chloroplasts—the photosynthetic organelles (Figure 19.2).

Carbon dioxide reacts with ribulose 1,5-bisphosphate to form two molecules of 3-phosphoglycerate

The first step in the Calvin cycle is the fixation of CO_2. This begins with the conversion of ribulose 1,5-bisphosphate into a highly reactive enediolate intermediate. The CO_2 molecule condenses with the enediolate

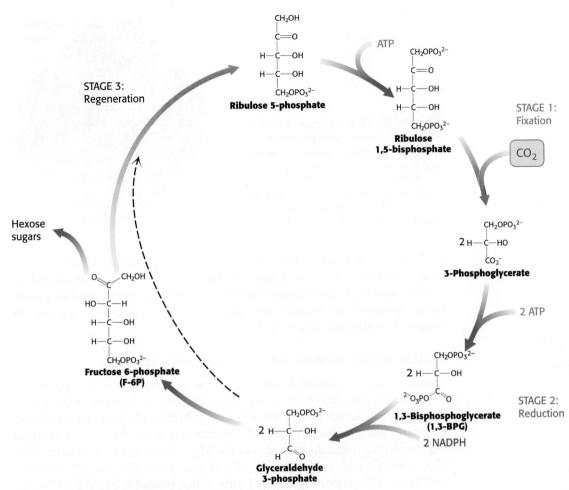

FIGURE 20.1 Calvin cycle. The Calvin cycle consists of three stages. Stage 1 is the fixation of carbon by the carboxylation of ribulose 1,5-bisphosphate. Stage 2 is the reduction of the fixed carbon to begin the synthesis of hexose. Stage 3 is the regeneration of the starting compound ribulose 1,5-bisphosphate.

intermediate to form an unstable six-carbon compound, which is rapidly hydrolyzed to two molecules of 3-phosphoglycerate.

This highly exergonic reaction $[\Delta G^{\circ\prime} = -51.9\,\mathrm{kJ\,mol^{-1}}(-12.4\,\mathrm{kcal\,mol^{-1}})]$ is catalyzed by *ribulose 1,5-bisphosphate carboxylase/oxygenase* (usually called *rubisco*), an enzyme located on the stromal surface of the thylakoid membranes of chloroplasts. This important reaction is the rate-limiting step in hexose synthesis. In plants and green algae, rubisco consists of eight large (L, 55-kDa) subunits and eight small (S, 15-kDa) ones (Figure 20.2). Each L chain contains a catalytic site and a regulatory site. The S chains enhance the catalytic activity of the L chains. This enzyme

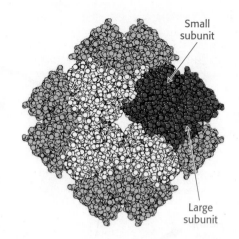

FIGURE 20.2 Structure of rubisco. The enzyme ribulose 1,5-bisphosphate carboxylase/oxygenase (rubisco) comprises eight large subunits (one shown in red and the others in yellow) and eight small subunits (one shown in blue and the others in white). The active sites lie in the large subunits. [Drawn from 1RXO.pdb.]

is abundant in chloroplasts, accounting for approximately 30% of the total leaf protein in some plants. In fact, rubisco is the most abundant enzyme and perhaps the most abundant protein in the entire biosphere. Large amounts are present because rubisco is a sluggish enzyme; its maximal catalytic rate is only $3\,s^{-1}$.

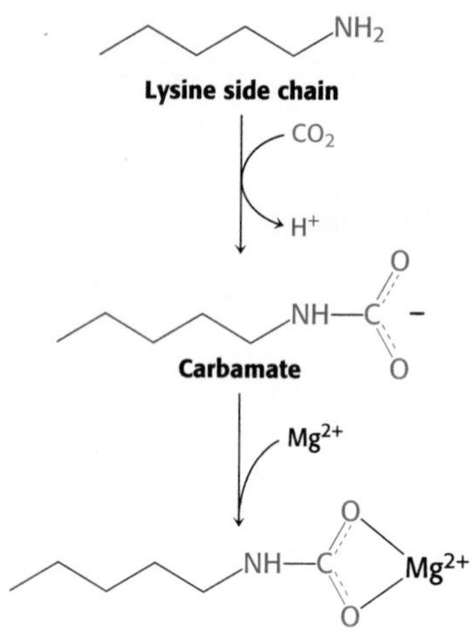

Rubisco activity depends on magnesium and carbamate

Rubisco requires a bound divalent metal ion for activity, usually magnesium ion. Like the zinc ion in the active site of carbonic anhydrase (Section 9.2), this metal ion serves to activate a bound substrate molecule by stabilizing a negative charge. Interestingly, a CO_2 molecule other than the substrate is required to complete the assembly of the Mg^{2+}-binding site in rubisco. This CO_2 molecule adds to the uncharged ε-amino group of lysine 201 to form a *carbamate*. This negatively charged adduct then binds the Mg^{2+} ion.

The metal center plays a key role in binding ribulose 1,5-bisphosphate and activating it so that it reacts with CO_2 (Figure 20.3). Ribulose 1,5-bisphosphate binds to Mg^{2+} through its keto group and an adjacent

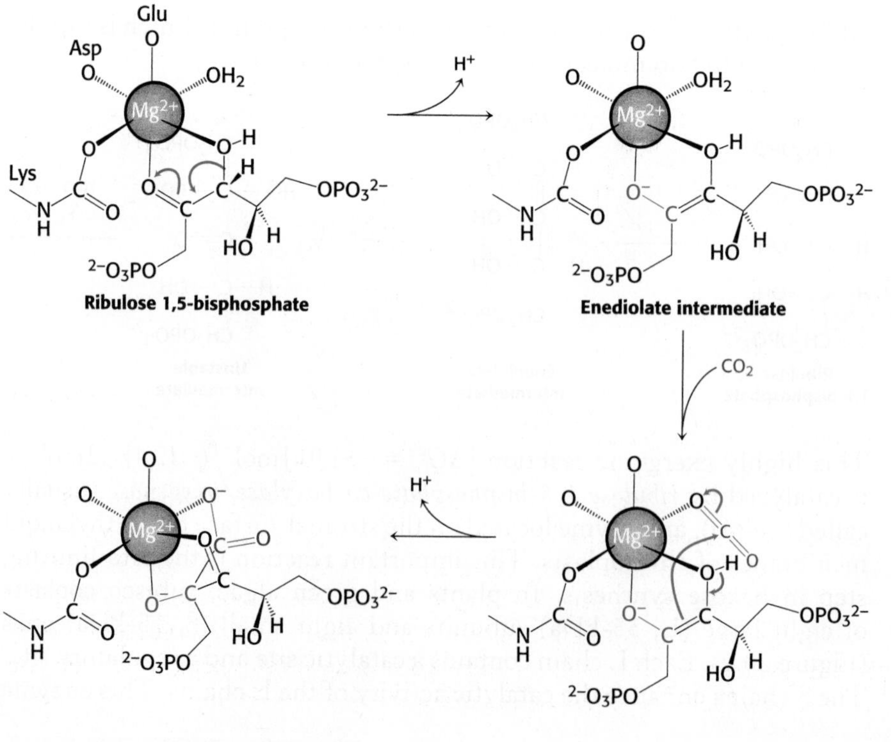

Lysine side chain

Carbamate

FIGURE 20.3 Role of the magnesium ion in the rubisco mechanism. Ribulose 1,5-bisphosphate binds to a magnesium ion that is linked to rubisco through a glutamate residue, an aspartate residue, and the lysine carbamate. The coordinated ribulose 1,5-bisphosphate gives up a proton to form a reactive enediolate species that reacts with CO_2 to form a new carbon–carbon bond.

Ribulose 1,5-bisphosphate

Enediolate intermediate

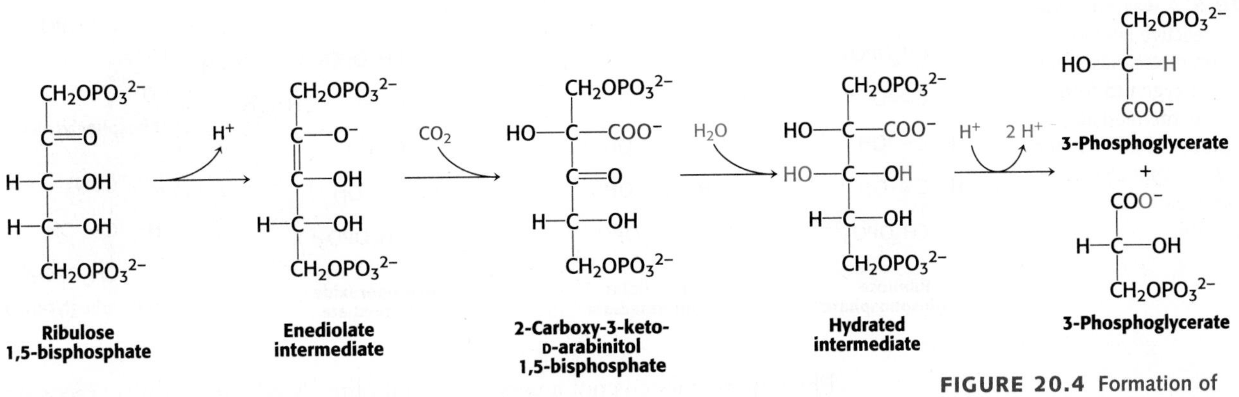

The overall reaction scheme shows the conversion of Ribulose 1,5-bisphosphate through an Enediolate intermediate, 2-Carboxy-3-keto-D-arabinitol 1,5-bisphosphate, and a Hydrated intermediate to form two molecules of 3-Phosphoglycerate.

FIGURE 20.4 Formation of 3-phosphoglycerate. The overall pathway for the conversion of ribulose 1,5-bisphosphate and CO_2 into two molecules of 3-phosphoglycerate. Although the free species are shown, these steps take place on the magnesium ion.

hydroxyl group. This complex is readily deprotonated to form an enediolate intermediate. This reactive species, analogous to the zinc-hydroxide species in carbonic anhydrase, couples with CO_2, forming the new carbon–carbon bond. The resulting 2-carboxy-3-keto-D-arabinitol 1,5-bisphosphate is coordinated to the Mg^{2+} ion through three groups, including the newly formed carboxylate. A molecule of H_2O is then added to this β-ketoacid to form an intermediate that cleaves to form two molecules of 3-phosphoglycerate (Figure 20.4).

Rubisco activase is essential for rubisco activity

In the absence of CO_2, rubisco binds to its substrate, ribulose 1,5-bisphosphate, so tightly that its enzyme activity is inhibited. The enzyme *rubisco activase* uses ATP to induce structural changes in rubisco that allow release of the bound substrate and formation of the required carbamate. The ATP requirement of rubisco activase coordinates rubisco activity with the light reactions.

The activase contains a P-loop domain (Section 9.4) and assembles into a hexameric ring-like structure. The enzyme binds to the C-terminal end of the polypeptide chain that forms the large subunit. Using the energy of ATP hydrolysis, the C-terminal is transiently pulled in to the central pore of the activase. This releases the bound ribulose 1,5-bisphosphate and activates rubisco. Activase is a member of the large AAA family of ATPases that usually assemble into ring-shaped oligomeric structures and use the ATP hydrolysis to power conformational changes in their substrates. AAA ATPases play a variety of roles in biological systems. They are crucial for protein degradation (Chapter 23) and DNA replication (Chapter 29), and also function as molecular motors (See Chapter 36 on SaplingPlus), and indeed rubisco activase is often referred to as motor protein.

Rubisco also catalyzes a wasteful oxygenase reaction: Catalytic imperfection

The reactive enediolate intermediate generated on the Mg^{2+} ion sometimes reacts with O_2 instead of CO_2; thus, rubisco also catalyzes a deleterious oxygenase reaction. The products of this reaction are *phosphoglycolate* and *3-phosphoglycerate* (Figure 20.5). The rate of the carboxylase reaction is four times that of the oxygenase reaction under normal atmospheric conditions at 25°C; the stromal concentration of CO_2 is then 10 μM and that of O_2 is 250 μM. The oxygenase reaction, like the carboxylase reaction, requires that lysine 201 be in the carbamate form. Because this carbamate forms only in the presence of CO_2, rubisco is prevented from catalyzing the oxygenase reaction exclusively when CO_2 is absent.

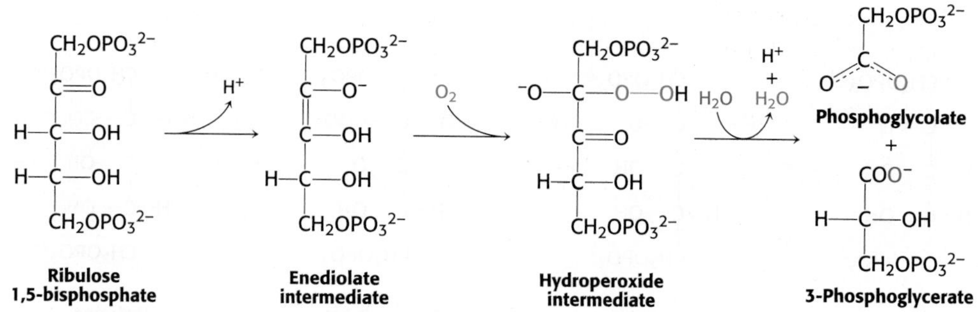

FIGURE 20.5 A wasteful side reaction. The reactive enediolate intermediate on rubisco also reacts with molecular oxygen to form a hydroperoxide intermediate, which then proceeds to form one molecule of 3-phosphoglycerate and one molecule of phosphoglycolate.

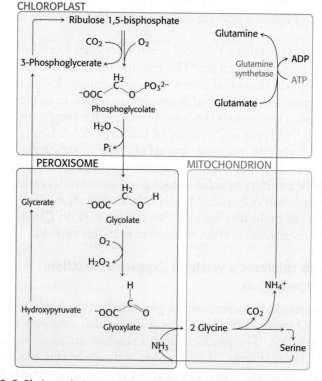

Ribulose 1,5-bisphosphate → Enediolate intermediate → Hydroperoxide intermediate → 3-Phosphoglycerate + Phosphoglycolate

Phosphoglycolate is not a versatile metabolite. A salvage pathway recovers part of its carbon skeleton (Figure 20.6). A specific phosphatase converts phosphoglycolate into *glycolate*, which enters *peroxisomes* (also called *microbodies*; Figure 20.7). Glycolate is then oxidized to *glyoxylate* by glycolate oxidase, an enzyme with a flavin mononucleotide prosthetic group. The H_2O_2 produced in this reaction is cleaved by catalase to H_2O and O_2. Transamination of glyoxylate then yields *glycine*, which then enters the mitochondria. Two glycine molecules can unite to form serine with the release of CO_2 and ammonium ion (NH_4^+). The ammonium ion, used in the synthesis of nitrogen-containing compounds, is salvaged by a glutamine synthetase reaction (Figure 20.6 and Section 23.3). Serine reenters the peroxisome, where it donates its ammonia to glyoxylate and is subsequently converted back into 3-phosphoglycerate.

This salvage pathway serves to recycle three of the four carbon atoms of two molecules of glycolate. However, one carbon atom is lost as CO_2.

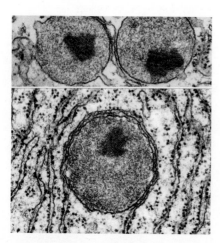

FIGURE 20.7 Electron micrograph of three peroxisomes. The peroxisomes are surrounded by rough endoplasmic reticulum. [Don W. Fawcett/Science Source.]

FIGURE 20.6 Photorespiratory reactions. Phosphoglycolate is formed as a product of the oxygenase reaction in chloroplasts. After dephosphorylation, glycolate is transported into peroxisomes where it is converted into glyoxylate and then glycine. In mitochondria, two glycines are converted into serine, after losing a carbon as CO_2 and ammonium ion. The serine is converted back into 3-phosphoglycerate and the ammonium ion is salvaged in chloroplasts.

This process is called *photorespiration* because O_2 is consumed and CO_2 is released. Photorespiration is wasteful because organic carbon is converted into CO_2 without the production of ATP, NADPH, or another energy-rich metabolite. Although evolutionary processes have presumably enhanced the preference of rubisco for carboxylation—for instance, the rubisco of higher plants is eight-fold as specific for carboxylation as that of photosynthetic bacteria—photorespiration still accounts for the loss of up to 25% of the carbon fixed.

Much research has been undertaken to generate recombinant forms of rubisco that display reduced oxygenase activity, but all such attempts have failed. This raises the following question: what is the biochemical basis of this inefficiency? Structural studies show that when the reactive enediolate intermediate is formed, loops close over the active site to protect the enediolate. A channel to the environment is maintained to allow access by CO_2. However, like CO_2, O_2 is a linear molecule that also fits the channel. In essence, the problem lies not with the enzyme but in the unremarkable structure of CO_2. CO_2 lacks any chemical features that would allow discrimination between it and other gases such as O_2, and thus the oxygenase activity is an inevitable failing of the enzyme. Another possibility exists, however. The oxygenase activity may not be an imperfection of the enzyme, but rather an imperfection in our understanding. Perhaps the oxygenase activity performs a biochemically important role that we have not yet discovered.

Hexose phosphates are made from phosphoglycerate, and ribulose 1,5-bisphosphate is regenerated

The 3-phosphoglycerate product of rubisco is next converted into fructose 6-phosphate, which readily isomerizes to glucose 1-phosphate and glucose 6-phosphate. The mixture of the three phosphorylated hexoses is called the *hexose monophosphate pool*. The steps in this conversion (Figure 20.8) are like those of the gluconeogenic pathway (Figure 16.24), except that glyceraldehyde 3-phosphate dehydrogenase in chloroplasts—which generates glyceraldehyde 3-phosphate (GAP)—is specific for NADPH rather than NADH. These reactions and that catalyzed by rubisco bring CO_2 to the level of a hexose, converting CO_2 into a chemical fuel at the expense of NADPH and ATP generated from the light reactions.

The third phase of the Calvin cycle is the regeneration of ribulose 1,5-bisphosphate, the acceptor of CO_2 in the first step. The problem is to construct a five-carbon sugar from six-carbon and three-carbon sugars. A transketolase and an aldolase play the major role in the rearrangement of the carbon atoms. The *transketolase*, which we will see again in the pentose phosphate pathway, requires the coenzyme thiamine pyrophosphate (TPP) to transfer a two-carbon unit $(CO-CH_2OH)$ from a ketose to an aldose.

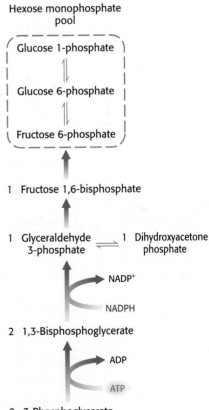

Hexose monophosphate pool

The transketolase converts fructose 6-phosphate and glyceraldehyde 3-phosphate into erythrose 4-phosphate and xylulose 5-phosphate.

Aldolase, which we have already encountered in glycolysis (Section 16.1), catalyzes an aldol condensation between dihydroxyacetone

FIGURE 20.8 Hexose phosphate formation. 3-Phosphoglycerate is converted into fructose 6-phosphate in a pathway parallel to that of gluconeogenesis.

phosphate (DHAP) and an aldehyde. This enzyme is highly specific for dihydroxyacetone phosphate, but it accepts a wide variety of aldehydes.

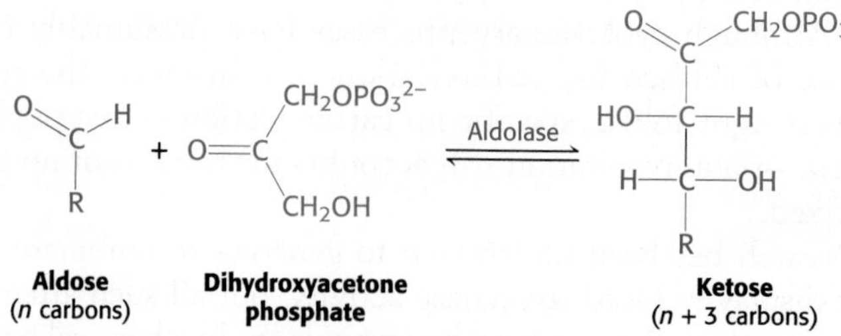

Aldose
(*n* carbons)   **Dihydroxyacetone phosphate**   **Ketose**
(*n* + 3 carbons)

Aldolase generates sedoheptulose 1,7-bisphosphate from erythrose 4-phosphate and dihydroxyacetone phosphate. A phosphatase removes a phosphate from sedoheptulose 1,7-bisphosphosphate to form sedoheptulose 7-phosphate. Transketolase again enters the action to combine sedoheptulose 7-phosphate with another molecule of glyceraldehyde 3-phosphate to form the five-carbon sugar ribose 5-phosphate as well as a second molecule of xylulose 5-phosphate (Figure 20.9).

Fructose 6-phosphate   **Glyceraldehyde 3-phosphate**   **Erythrose 4-phosphate**   **Xylulose 5-phosphate**

Erythrose 4-phosphate   **Dihydroxyacetone phosphate**   **Sedoheptulose 1,7-bisphosphate**   **Sedoheptulose 7-phosphate**

Sedoheptulose 7-phosphate   **Glyceraldehyde 3-phosphate**   **Ribose 5-phosphate**   **Xylulose 5-phosphate**

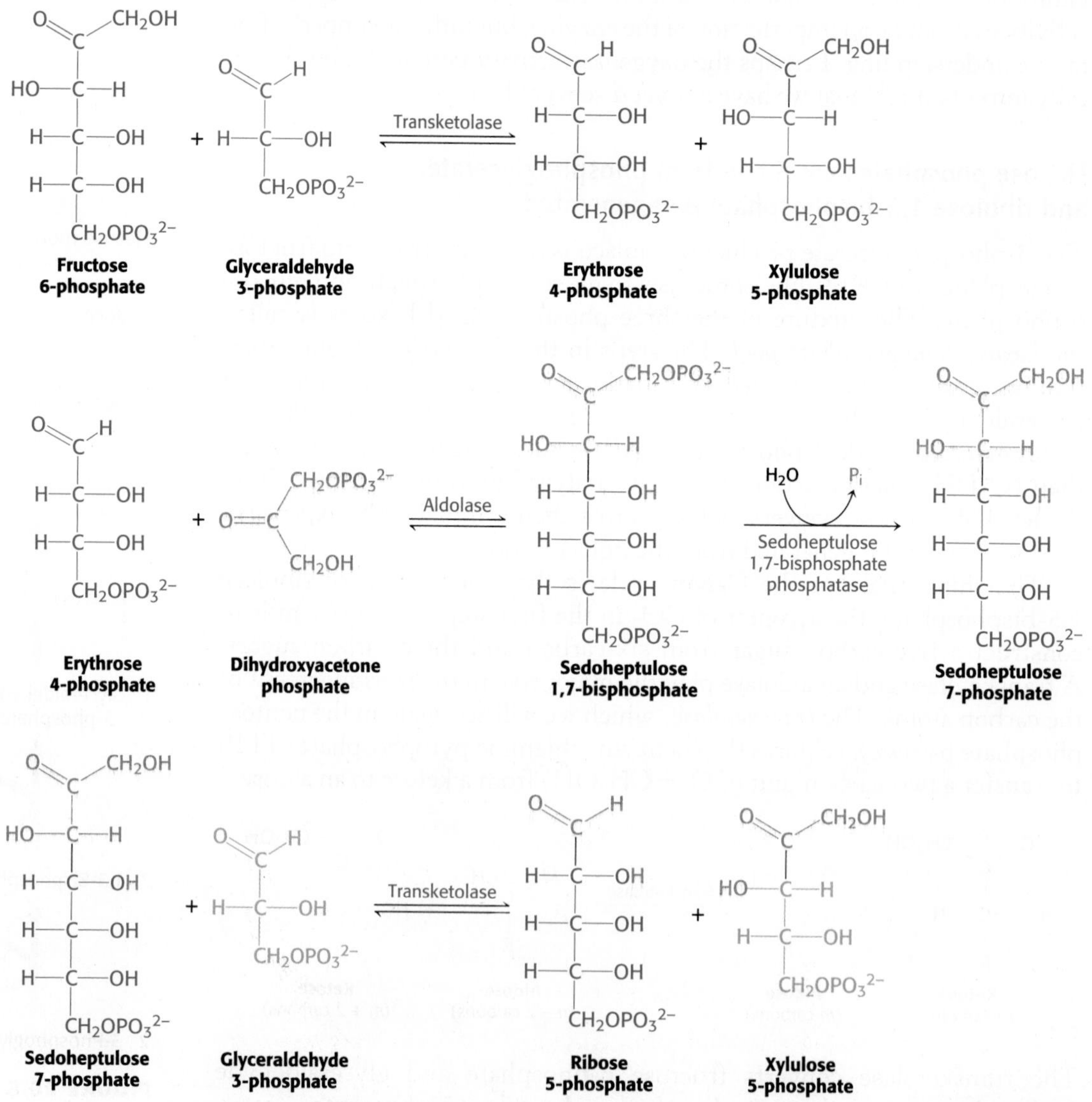

FIGURE 20.9 Formation of five-carbon sugars. First, transketolase converts a six-carbon sugar and a three-carbon sugar into a four-carbon sugar and a five-carbon sugar. Then, aldolase combines the four-carbon product and a three-carbon sugar to form a seven-carbon sugar. Finally, this seven-carbon sugar reacts with another three-carbon sugar to form two additional five-carbon sugars.

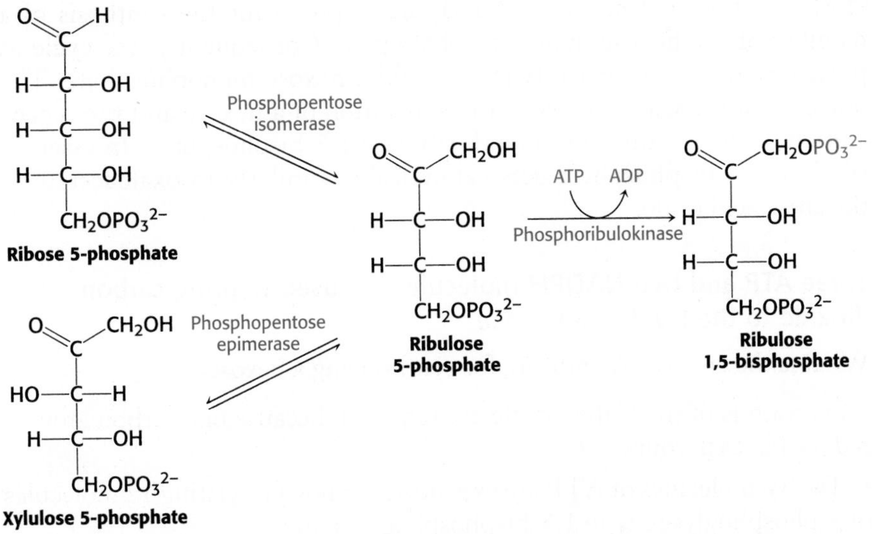

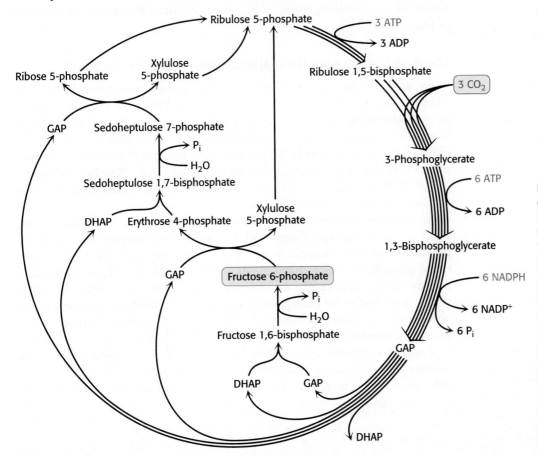

FIGURE 20.10 Regeneration of ribulose 1,5-bisphosphate. Both ribose 5-phosphate and xylulose 5-phosphate are converted into ribulose 5-phosphate, which is then phosphorylated to complete the regeneration of ribulose 1,5-bisphosphate.

Finally, ribose 5-phosphate is converted into ribulose 5-phosphate by *phosphopentose isomerase,* whereas the two molecules of xylulose 5-phosphate are converted into ribulose 5-phosphate by *phosphopentose epimerase.* Ribulose 5-phosphate is converted into ribulose 1,5-bisphosphate through the action of *phosphoribulokinase* (Figure 20.10). The sum of these reactions shown in Figures 20.9 and 20.10 is:

Fructose 6-phosphate + 2 glyceraldehyde 3-phosphate
  + dihydroxyacetone phosphate + 3 ATP $\longrightarrow$
  3 ribulose 1,5-bisphosphate + 3 ADP

Figure 20.11 presents the required reactions with the proper stoichiometry to convert three molecules of CO_2 into one molecule of DHAP.

FIGURE 20.11 Calvin cycle. The diagram shows the reactions necessary with the correct stoichiometry to convert three molecules of CO_2 into one molecule of dihydroxyacetone phosphate (DHAP). The cycle is not as simple as presented in Figure 20.1; rather, it entails many reactions that lead ultimately to the synthesis of glucose and the regeneration of ribulose 1,5-bisphosphate. [Information from J. R. Bowyer and R. C. Leegood. "Photosynthesis," in *Plant Biochemistry,* P. M. Dey and J. B. Harborne, Eds. (Academic Press, 1997), p. 85.]

However, two molecules of DHAP are required for the synthesis of a member of the hexose monophosphate pool. Consequently, the cycle as presented must take place twice to yield a hexose monophosphate. The outcome of the Calvin cycle is the generation of a hexose and the regeneration of the starting compound, ribulose 1,5-bisphosphate. In essence, ribulose 1,5-bisphosphate acts catalytically, similarly to oxaloacetate in the citric acid cycle.

Three ATP and two NADPH molecules are used to bring carbon dioxide to the level of a hexose

What is the energy expenditure for synthesizing a hexose?

• Six rounds of the Calvin cycle are required, because one carbon atom is reduced in each round.

• Twelve molecules of ATP are expended in phosphorylating 12 molecules of 3-phosphoglycerate to 1,3-bisphosphoglycerate.

• 12 molecules of NADPH are consumed in reducing 12 molecules of 1,3-bisphosphoglycerate to glyceraldehyde 3-phosphate.

• An additional six molecules of ATP are spent in regenerating ribulose 1,5-bisphosphate.

We can now write a balanced equation for the net reaction of the Calvin cycle:

$$6\,CO_2 + 18\,ATP + 12\,NADPH + 12\,H_2O \longrightarrow$$
$$C_6H_{12}O_6 + 18\,ADP + 18\,P_i + 12\,NADP^+ + 6\,H^+$$

Thus, three molecules of ATP and two molecules of NADPH are consumed in incorporating a single CO_2 molecule into a hexose such as glucose or fructose.

Starch and sucrose are the major carbohydrate stores in plants

What are the fates of the members of the hexose monophosphate pool? These molecules are used in a variety of ways, but there are two primary roles. Plants contain two major storage forms of sugar: *starch* and *sucrose*. Starch, like its animal counterpart glycogen, is a polymer of glucose residues, but it is less branched than glycogen because it contains a smaller proportion of α-1,6-glycosidic linkages (Section 11.2). Another difference is that ADP-glucose, not UDP-glucose, is the activated precursor. Starch is synthesized and stored in chloroplasts.

In contrast, sucrose (common table sugar), a disaccharide, is synthesized in the cytoplasm. Plants lack the ability to transport hexose phosphates across the chloroplast membrane, but they are able to transport triose phosphates from chloroplasts to the cytoplasm. Triose phosphate intermediates such as glyceraldehyde 3-phosphate cross into the cytoplasm in exchange for phosphate through the action of an abundant triose phosphate-phosphate antiporter. Fructose 6-phosphate formed from triose phosphates joins the glucose unit of UDP-glucose to form sucrose 6-phosphate (Figure 20.12). The hydrolysis of the phosphate ester by sucrose phosphatase yields sucrose, a readily transportable and mobilizable sugar that is stored in many plant cells, such as those in sugar beets and sugar cane.

Fructose 6-phosphate + **UDP-glucose**

Triose phosphates (from chloroplasts) →

Sucrose 6-phosphate synthase

Sucrose 6-phosphate + **UDP**

FIGURE 20.12 Synthesis of sucrose. Sucrose 6-phosphate is formed by the reaction between fructose 6-phosphate and the activated intermediate uridine diphosphate glucose (UDP-glucose). Sucrose phosphatase subsequently generates free sucrose (not shown).

20.2 The Activity of the Calvin Cycle Depends on Environmental Conditions

How do the light reactions communicate with the dark reactions to regulate this crucial process of fixing CO_2 into biomolecules? *The principal means of regulation is alteration of the stromal environment by the light reactions.* The light reactions lead to an increase in stromal pH (a decrease in the H^+ concentration), as well as an increase in the stromal concentrations of Mg^{2+}, NADPH, and reduced ferredoxin—changes that contribute to the activation of certain Calvin-cycle enzymes, located in the stroma (Figure 20.13).

Rubisco is activated by light-driven changes in proton and magnesium ion concentrations

As stated earlier, the rate-limiting step in the Calvin cycle is the carboxylation of ribulose 1,5-bisphosphate to form two molecules of 3-phosphoglycerate. *The activity of rubisco increases markedly on illumination because light facilitates the carbamate formation necessary for enzyme activity (p. 648).* In the stroma, the pH increases from 7 to 8, and the level of Mg^{2+} rises. Both effects are consequences of the light-driven pumping of protons into the thylakoid space. Mg^{2+} ions from the thylakoid space are released into the stroma to compensate for the influx of protons. Carbamate formation is favored at alkaline pH. CO_2 adds to a deprotonated form of lysine 201 of rubisco, and Mg^{2+} ion binds to the carbamate to generate the active form of the enzyme. Thus, light leads not only to the generation of ATP and NADPH, but also to regulatory signals that stimulate CO_2 fixation.

Thioredoxin plays a key role in regulating the Calvin cycle

As we saw in Chapter 19 (Figure 19.23), light-driven reactions lead to electron transfer from water to ferredoxin and, eventually, to NADPH in photosystem I. The presence of reduced ferredoxin and NADPH are good signals

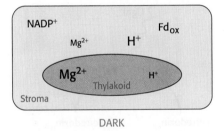

DARK

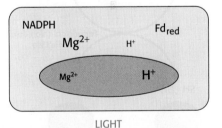

LIGHT

FIGURE 20.13 Light regulation of the Calvin cycle. The light reactions of photosynthesis transfer electrons out of the thylakoid lumen into the stroma and transfer protons from the stroma into the thylakoid lumen. As a consequence of these processes, the concentrations of NADPH, reduced ferredoxin (Fd), and Mg^{2+} in the stroma are higher in the light than in the dark. The stromal pH is also increased (the concentration of H^+ is lowered) in the light as a result of proton pumping from the stroma to the thylakoid lumen. Each of these concentration changes helps couple the Calvin cycle reactions to the light reactions.

FIGURE 20.14 Thioredoxin. The oxidized form of thioredoxin contains a disulfide bond. When thioredoxin is reduced by reduced ferredoxin, the disulfide bond is converted into two free sulfhydryl groups. Reduced thioredoxin can cleave disulfide bonds in enzymes, activating certain Calvin cycle enzymes and inactivating some degradative enzymes. [Drawn from 1F9M.pdb.]

TABLE 20.1 Enzymes regulated by thioredoxin

Enzyme	Pathway
Rubisco	Carbon fixation in the Calvin cycle
Fructose 1,6-bisphosphatase	Gluconeogenesis
Glyceraldehyde 3-phosphate dehydrogenase	Calvin cycle, gluconeogenesis, glycolysis
Sedoheptulose 1,7-bisphosphatase	Calvin cycle
Glucose 6-phosphate dehydrogenase	Pentose phosphate pathway
Phenylalanine ammonia lyase	Lignin synthesis
Phosphoribulokinase	Calvin cycle
$NADP^+$-malate dehydrogenase	C_4 pathway
CF_1–CF_0 ATP synthase	The light reactions

that conditions are right for biosynthesis. One way in which this information is conveyed to biosynthetic enzymes is by *thioredoxin*, a 12-kDa protein containing neighboring cysteine residues that cycle between a reduced sulfhydryl and an oxidized disulfide form (Figure 20.14). The reduced form of thioredoxin activates many biosynthetic enzymes, including *the chloroplast ATP synthase* (p. 633), by reducing disulfide bridges that control their activity. Reduced thioredoxin also inhibits several degradative enzymes by the same means (Table 20.1). In chloroplasts, oxidized thioredoxin is reduced by ferredoxin in a reaction catalyzed by *ferredoxin–thioredoxin reductase* (p. 633). Thus, *the activities of the light and dark reactions of photosynthesis are coordinated through electron transfer from reduced ferredoxin to thioredoxin and then to component enzymes containing regulatory disulfide bonds* (Figure 20.15). We shall return to thioredoxin when we consider the reduction of ribonucleotides (Section 25.3).

NADPH is a signal molecule that activates two biosynthetic enzymes, phosphoribulokinase and glyceraldehyde 3-phosphate dehydrogenase. In the dark, these enzymes are inhibited by association with an 8.5-kDa protein called *CP12*, an intrinsically disordered protein (Section 2.6). NADPH disrupts this association by promoting the formation of two disulfide bonds in CP12, leading to the release of the active enzymes.

The C_4 pathway of tropical plants accelerates photosynthesis by concentrating carbon dioxide

The oxygenase activity of rubisco presents a biochemical challenge to tropical plants because the oxygenase activity increases more rapidly with temperature than does the carboxylase activity. How, then, do plants that grow in hot climates, such as sugar cane, prevent very high rates of wasteful photorespiration? Their solution to this problem is to achieve a high local concentration of CO_2 at the site of the Calvin cycle in their photosynthetic cells. The essence of this process, elucidated by Marshall Davidson Hatch and C. Roger Slack, is that *four-carbon (C_4) compounds such as oxaloacetate and malate carry CO_2 from mesophyll cells, which are in contact with air, to bundle-sheath cells, which are the major sites of photosynthesis* (Figure 20.16). The decarboxylation of the four-carbon compound in a bundle-sheath cell maintains a high concentration of CO_2 at the site of the Calvin cycle. The three-carbon product returns to the mesophyll cell for another round of carboxylation. This metabolic pathway is called the *C_4 pathway* or the *Hatch–Slack pathway*. Corn or maize is a C_4 plant of tremendous economic and agricultural importance, accounting for approximately 20% of human

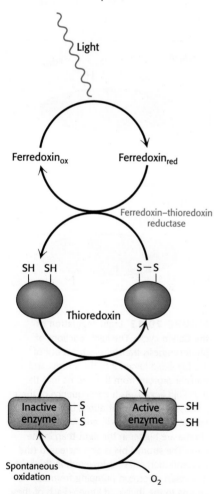

FIGURE 20.15 Enzyme activation by thioredoxin. Reduced thioredoxin activates certain Calvin cycle enzymes by cleaving regulatory disulfide bonds.

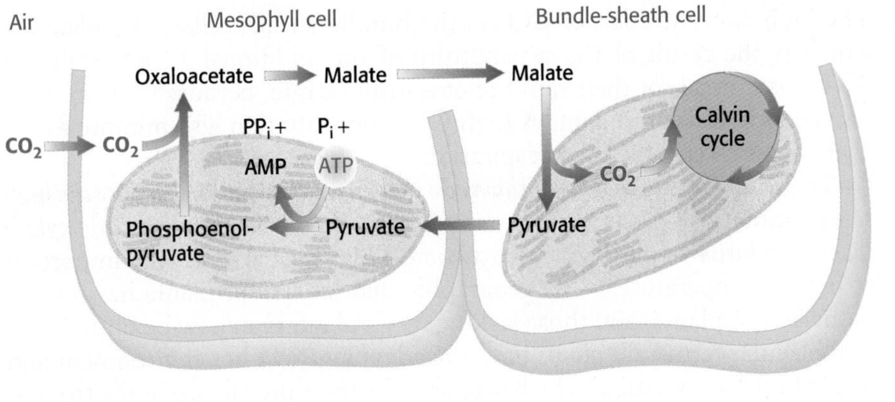

Air   Mesophyll cell   Bundle-sheath cell

Oxaloacetate ⟹ Malate ⟹ Malate

CO_2 ⟶ CO_2 Calvin cycle

$PP_i +$ $P_i +$
AMP   ATP CO_2

Phosphoenol-   Pyruvate ⟵ Pyruvate
pyruvate

FIGURE 20.16 C_4 pathway. Carbon dioxide is concentrated in bundle-sheath cells by the expenditure of ATP in mesophyll cells.

nutrition worldwide. Corn was domesticated by selective breeding from the Mexican grass teosinte about 9000 years ago. Although tropical plants employ the C_4 pathway, grasses and sedges (flowering plants resembling grasses, such as the water chestnut) are the most common users of the C_4 pathway.

The C_4 pathway for the transport of CO_2 starts in a mesophyll cell with the condensation of CO_2 in the form of bicarbonate and phosphoenolpyruvate to form *oxaloacetate* in a reaction catalyzed by *phosphoenolpyruvate carboxylase*. Oxaloacetate is converted into malate by an $NADP^+$-linked malate dehydrogenase. Malate enters the bundle-sheath cell and is oxidatively decarboxylated within the chloroplasts by an $NADP^+$-linked malate dehydrogenase. The released CO_2 enters the Calvin cycle in the usual way by condensing with ribulose 1,5-bisphosphate. Pyruvate formed in this decarboxylation reaction returns to the mesophyll cell. Finally, phosphoenolpyruvate is formed from pyruvate by *pyruvate-P_i dikinase*.

The net reaction of this C_4 pathway is

$$CO_2 \text{ (in mesophyll cell)} + ATP + 2\,H_2O \longrightarrow$$
$$CO_2 \text{ (in bundle-sheath cell)} + AMP + 2\,P_i + 2\,H^+$$

Thus, *the energetic equivalent of two ATP molecules is consumed in transporting CO_2 to the chloroplasts of the bundle-sheath cells.* Note that six molecules of CO_2 are required to synthesize glucose so the C_4 pathway requires an extra 12 ATP when compared to the C_3 pathway. In essence, this process is active transport: the pumping of CO_2 into the bundle-sheath cell is driven by the hydrolysis of one molecule of ATP to one molecule of AMP and two molecules of orthophosphate. The CO_2 concentration can be 20-fold as great in the bundle-sheath cells as in the mesophyll cells. Interestingly, phosphoenolpyruvate carboxylase has a higher affinity for CO_2 (as bicarbonate) than does rubisco. C_4 plants use this affinity difference to their advantage. The high CO_2 affinity of phosphoenolpyruvate carboxylase means rubisco can be adequately supplied with CO_2 while the stomata—the openings in the leaves that allow gas exchange (Figure 20.17)—do not have to open as completely in the heat of the tropical day, thereby preventing water loss.

When the C_4 pathway and the Calvin cycle operate together, the net reaction is

$$6\,CO_2 + 30\,ATP + 12\,NADPH + 24\,H_2O \longrightarrow$$
$$C_6H_{12}O_6 + 30\,ADP + 30\,P_i + 12\,NADP^+ + 18\,H^+$$

Note that 30 molecules of ATP are consumed per hexose molecule formed when the C_4 pathway delivers CO_2 to the Calvin cycle, in contrast with 18 molecules of ATP per hexose molecule in the absence of the C_4 pathway.

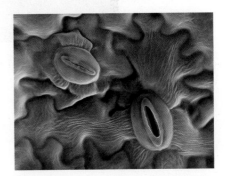

FIGURE 20.17 Scanning electron micrograph of an open stoma and a closed stoma. [Power and Syred/Science Source.]

The high concentration of CO_2 in the bundle-sheath cells of C_4 plants—which is the result of the expenditure of the additional 12 molecules of ATP—is critical for their rapid photosynthetic rate, because CO_2 is limiting when light is abundant. A high CO_2 concentration also minimizes the energy loss caused by photorespiration.

Tropical plants with a C_4 pathway do little photorespiration because the high concentration of CO_2 in their bundle-sheath cells accelerates the carboxylase reaction relative to the oxygenase reaction. This effect is especially important at higher temperatures. The geographic distribution of plants having this pathway (C_4 plants) and those lacking it (C_3 plants) can now be understood in molecular terms. C_4 *plants* have the advantage in a hot environment and under high illumination, which accounts for their prevalence in the tropics. C_3 *plants,* which consume only 18 molecules of ATP per hexose molecule formed in the absence of photorespiration (compared with 30 molecules of ATP for C_4 plants), are more efficient at temperatures lower than about 28°C, and so they predominate in temperate environments (Figure 20.18). Although C_4 plants comprise only 3% of all flowering plants, they account for 25% of all terrestrial carbon fixation because of their minimization of photorespiration.

A   B

FIGURE 20.18 C_3 and C_4 plants. (A) C_3 plants, such as trees, account for approximately 95% of plant species. (B) Corn is a C_4 plant of tremendous agricultural importance. [(A) Audrey Ustuzhanih/FeaturePics; (B) Elena Elisseeva/FeaturePics.]

Rubisco is present in bacteria, eukaryotes, and even archaea, although other photosynthetic components have not been found in archaea. Thus, rubisco emerged early in evolution, when the atmosphere was rich in CO_2 and almost devoid of O_2. The enzyme was not originally selected to operate in an environment like the present one, which is almost devoid of CO_2 and rich in O_2. Photorespiration became significant about 600 million years ago, when the CO_2 concentration fell to present levels. The C_4 plants arose approximately 35 million years ago, but became ecologically significant about 7 million years ago. C_4 plants are a clear example of convergent evolution. C_4 plants evolved independently from C_3 plants many times, and are found in distantly related species, accounting for the fact that none of the enzymes are unique to C_4 plants, suggesting that this pathway made use of already existing enzymes. Interestingly, the use of the C_4 pathway in trees is rare—an observation that is not yet understood.

Crassulacean acid metabolism permits growth in arid ecosystems

Many plants, including some that grow in hot, dry climates, keep the stomata of their leaves closed in the heat of the day to prevent water loss (see Figure 20.17). As a consequence, CO_2 cannot be absorbed during the daylight hours, when it is needed for glucose synthesis. Rather, CO_2 enters the leaf when the stomata open at the cooler temperatures of night. To store the CO_2 until it can be used during the day, such plants make use of an adaptation called *crassulacean acid metabolism* (CAM), named after the genus *Crassulacea* (the succulents). Carbon dioxide is fixed by the C_4 pathway into malate, which is stored in vacuoles. During the day, malate is decarboxylated and the CO_2 becomes available to the Calvin cycle. In contrast with C_4 plants, CAM plants separate CO_2 accumulation from CO_2 utilization temporally rather than spatially.

Although CAM plants do prevent water loss, the use of malate as the sole source of CO_2 comes at a metabolic cost. Because the malate storage is limited, CAM plants cannot generate CO_2 as rapidly as it can be imported by C_3 and C_4 plants. Consequently, the growth rate of CAM plants is slower than that of C_3 and C_4 plants. The saguaro cactus, a CAM plant that can live up to 200 years and reach a height of 60 feet, can take 15 years to grow to only 1 foot in height (Figure 20.19).

FIGURE 20.19 Desert plants. Because of crassulacean acid metabolism, cacti are well suited to life in the desert. [yenwen/Getty Images.]

20.3 The Pentose Phosphate Pathway Generates NADPH and Synthesizes Five-Carbon Sugars

Photosynthetic organisms can use the light reactions for generation of some NADPH. Another pathway, present in all organisms, meets the NADPH needs of nonphotosynthetic organisms and of the nonphotosynthetic tissues in plants. The *pentose phosphate pathway*, which occurs in the cytoplasm, is a crucial source of NADPH for use in reductive biosynthesis (Table 20.2) as well as for protection against oxidative stress. This pathway consists of two phases (Figure 20.20): (1) the oxidative generation of NADPH and (2) the nonoxidative interconversion of sugars.

In the oxidative phase, NADPH is generated when glucose 6-phosphate is oxidized to ribulose 5-phosphate.

$$\text{Glucose 6-phosphate} + 2\,NADP^+ + H_2O \longrightarrow$$
$$\text{ribulose 5-phosphate} + 2\,NADPH + 2\,H^+ + CO_2$$

In the nonoxidative phase, the pathway catalyzes the interconversion of three-, four-, five-, six-, and seven-carbon sugars in a series of nonoxidative reactions. Excess five-carbon sugars may be converted into intermediates of the glycolytic pathway. All these reactions take place in the cytoplasm. These interconversions rely on the same reactions that lead to the regeneration of ribulose 1,5-bisphosphate in the Calvin cycle.

TABLE 20.2 Pathways requiring NADPH
Synthesis
Fatty acid biosynthesis
Cholesterol biosynthesis
Neurotransmitter biosynthesis
Nucleotide biosynthesis
Detoxification
Reduction of oxidized glutathione
Cytochrome P450 monooxygenases

Two molecules of NADPH are generated in the conversion of glucose 6-phosphate into ribulose 5-phosphate

The oxidative phase of the pentose phosphate pathway starts with the dehydrogenation of glucose 6-phosphate at carbon 1, a reaction catalyzed by *glucose 6-phosphate dehydrogenase* (Figure 20.21). This enzyme is highly specific for $NADP^+$; the K_M for NAD^+ is about a thousand times as great as that for

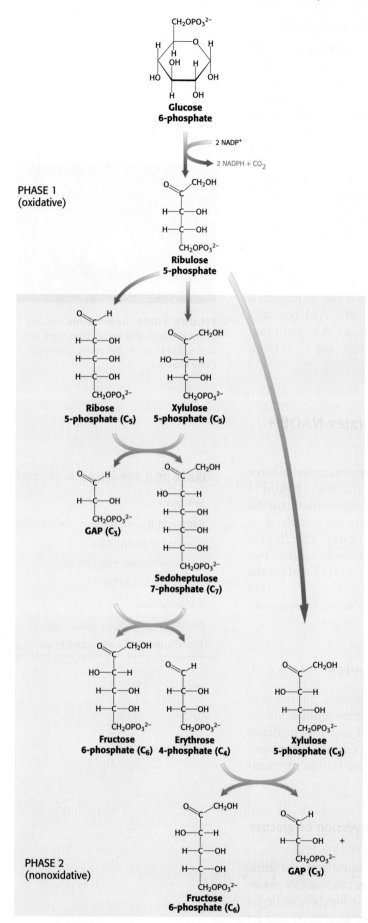

PHASE 1 (oxidative)

2 NADP⁺

2 NADPH + CO₂

PHASE 2 (nonoxidative)

FIGURE 20.20 Pentose phosphate pathway. The pathway consists of (1) an oxidative phase that generates NADPH and (2) a nonoxidative phase that interconverts phosphorylated sugars.

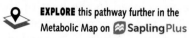

EXPLORE this pathway further in the Metabolic Map on 🍁 SaplingPlus

$NADP^+$. The product is *6-phosphoglucono-δ-lactone*, which is an intramolecular ester between the C-1 carboxyl group and the C-5 hydroxyl group. The next step is the hydrolysis of 6-phosphoglucono-δ-lactone by a specific *lactonase* to give *6-phosphogluconate*. This six-carbon sugar acid is then oxidatively decarboxylated by *6-phosphogluconate dehydrogenase* to yield *ribulose 5-phosphate*. $NADP^+$ is again the electron acceptor.

The pentose phosphate pathway and glycolysis are linked by transketolase and transaldolase

The preceding reactions yield two molecules of NADPH and one molecule of ribulose 5-phosphate for each molecule of glucose 6-phosphate oxidized. The ribulose 5-phosphate is subsequently isomerized to ribose 5-phosphate by phosphopentose isomerase.

Phosphopentose isomerase

Ribulose 5-phosphate   **Ribose 5-phosphate**

Ribose 5-phosphate and its derivatives are components of RNA and DNA, as well as of ATP, NADH, FAD, and coenzyme A. Although ribose 5-phosphate is a precursor to many biomolecules, many cells need NADPH for reductive biosyntheses much more than they need ribose 5-phosphate for incorporation into nucleotides. For instance, adipose tissue, the liver, and mammary glands require large amounts of NADPH for fatty acid synthesis (Chapter 22). In these cases, ribose 5-phosphate is converted into the glycolytic intermediates glyceraldehyde 3-phosphate and fructose 6-phosphate by *transketolase* and *transaldolase*. *These enzymes create a reversible link between the pentose phosphate pathway and glycolysis by catalyzing these three successive reactions.*

$$C_5 + C_5 \xrightleftharpoons{\text{Transketolase}} C_3 + C_7$$

$$C_3 + C_7 \xrightleftharpoons{\text{Transaldolase}} C_6 + C_4$$

$$C_4 + C_5 \xrightleftharpoons{\text{Transketolase}} C_6 + C_3$$

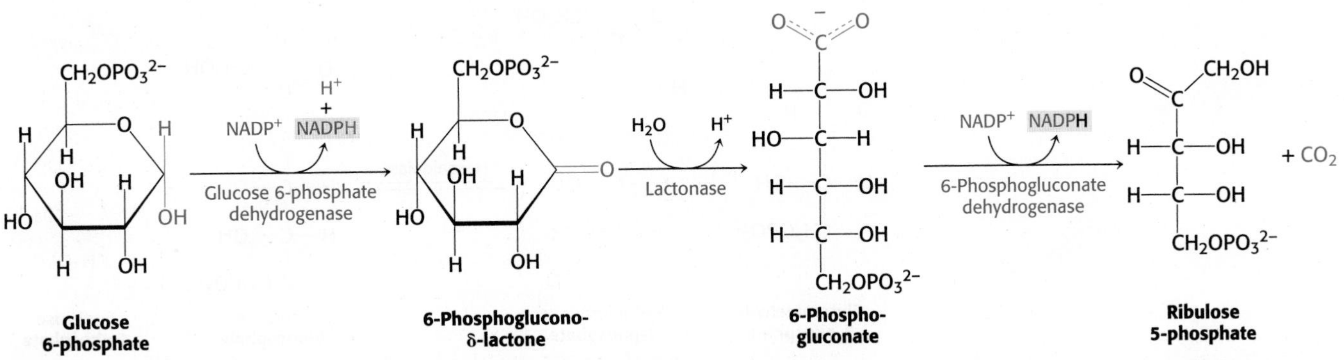

FIGURE 20.21 Oxidative phase of the pentose phosphate pathway. Glucose 6-phosphate is oxidized to 6-phosphoglucono-δ-lactone to generate one molecule of NADPH. The lactone product is hydrolyzed to 6-phosphogluconate, which is oxidatively decarboxylated to ribulose 5-phosphate with the generation of a second molecule of NADPH.

EXPLORE this pathway further in the Metabolic Map on **SaplingPlus**

The net result of these reactions is the *formation of two hexoses and one triose from three pentoses*:

$$3\,C_5 \rightleftharpoons 2\,C_6 + C_3$$

The first of the three reactions linking the pentose phosphate pathway and glycolysis is the formation of *glyceraldehyde 3-phosphate* and *sedoheptulose 7-phosphate* from two pentoses.

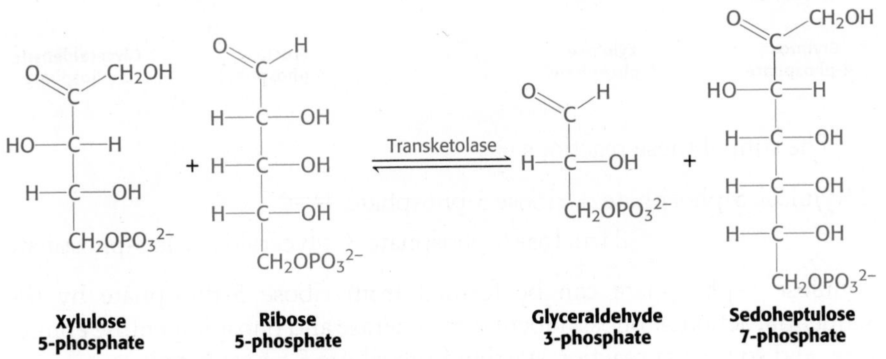

The donor of the two-carbon unit in this reaction is xylulose 5-phosphate, an epimer of ribulose 5-phosphate. A ketose is a substrate of transketolase only if its hydroxyl group at C-3 has the configuration of xylulose rather than ribulose. Ribulose 5-phosphate is converted into the appropriate epimer for the transketolase reaction by *phosphopentose epimerase* in the reverse reaction of that which takes place in the Calvin cycle.

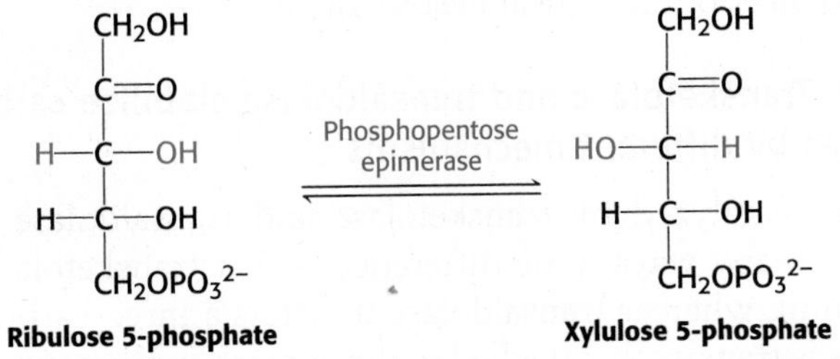

Glyceraldehyde 3-phosphate and sedoheptulose 7-phosphate generated by the transketolase then react to form *fructose 6-phosphate* and *erythrose 4-phosphate*.

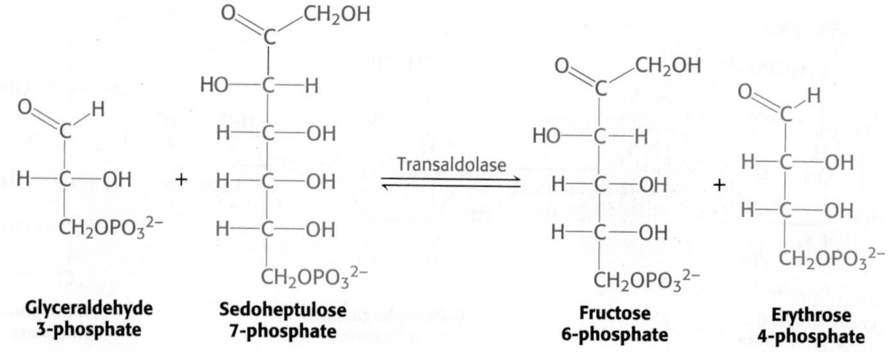

Glyceraldehyde 3-phosphate + **Sedoheptulose 7-phosphate** ⇌ (Transaldolase) **Fructose 6-phosphate** + **Erythrose 4-phosphate**

This synthesis of a four-carbon sugar and a six-carbon sugar is catalyzed by *transaldolase*.

In the third reaction, transketolase catalyzes the synthesis of *fructose 6-phosphate* and *glyceraldehyde 3-phosphate* from erythrose 4-phosphate and xylulose 5-phosphate.

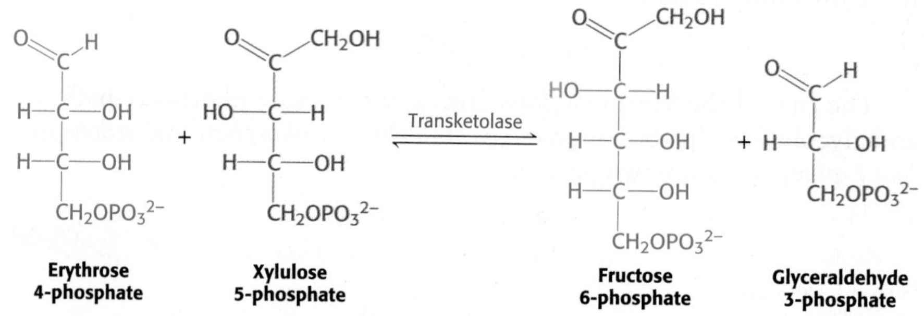

Erythrose 4-phosphate + **Xylulose 5-phosphate** ⇌ (Transketolase) **Fructose 6-phosphate** + **Glyceraldehyde 3-phosphate**

The sum of these reactions is

2 Xylulose 5-phosphate + ribose 5-phosphate ⇌

2 fructose 6-phosphate + glyceraldehyde 3-phosphate

Xylulose 5-phosphate can be formed from ribose 5-phosphate by the sequential action of phosphopentose isomerase and phosphopentose epimerase, and so the net reaction starting from ribose 5-phosphate is

3 Ribose 5-phosphate ⇌

2 fructose 6-phosphate + glyceraldehyde 3-phosphate

Thus, *excess ribose 5-phosphate formed by the pentose phosphate pathway can be completely converted into glycolytic intermediates.* Moreover, any ribose ingested in the diet can be processed into glycolytic intermediates by this pathway. It is evident that the carbon skeletons of sugars can be extensively rearranged to meet physiological needs (Table 20.3).

Mechanism: Transketolase and transaldolase stabilize carbanionic intermediates by different mechanisms

The reactions catalyzed by transketolase and transaldolase are distinct, yet similar in many ways. One difference is that transketolase transfers a two-carbon unit, whereas transaldolase transfers a three-carbon unit. Each of these units is transiently attached to the enzyme in the course of the reaction and thus, these enzymes are examples of double displacement reactions (Section 8.4).

TABLE 20.3 Pentose phosphate pathway

Reaction	Enzyme
Oxidative phase	
Glucose 6-phosphate + NADP$^+$ ⟶ 6-phosphoglucono-δ-lactone + NADPH + H$^+$	Glucose 6-phosphate dehydrogenase
6-Phosphoglucono-δ-lactone + H$_2$O ⟶ 6-phosphogluconate + H$^+$	Lactonase
6-Phosphogluconate + NADP$^+$ → ribulose 5-phosphate + CO$_2$ + NADPH + H$^+$	6-Phosphogluconate dehydrogenase
Nonoxidative phase	
Ribulose 5-phosphate ⇌ ribose 5-phosphate	Phosphopentose isomerase
Ribulose 5-phosphate ⇌ xylulose 5-phosphate	Phosphopentose epimerase
Xylulose 5-phosphate + ribose 5-phosphate ⇌ sedoheptulose 7-phosphate + glyceraldehyde 3-phosphate	Transketolase
Sedoheptulose 7-phosphate + glyceraldehyde 3-phosphate ⇌ fructose 6-phosphate + erythrose 4-phosphate	Transaldolase
Xylulose 5-phosphate + erythrose 4-phosphate ⇌ fructose 6-phosphate + glyceraldehyde 3-phosphate	Transketolase

Transketolase reaction. Transketolase contains a tightly bound thiamine pyrophosphate as its prosthetic group. The enzyme transfers a two-carbon glycoaldehyde from a ketose donor to an aldose acceptor. The site of the addition of the two-carbon unit is the thiazole ring of TPP. Transketolase is homologous to the E_1 subunit of the pyruvate dehydrogenase complex (Section 17.1) and the reaction mechanism is similar (Figure 20.22).

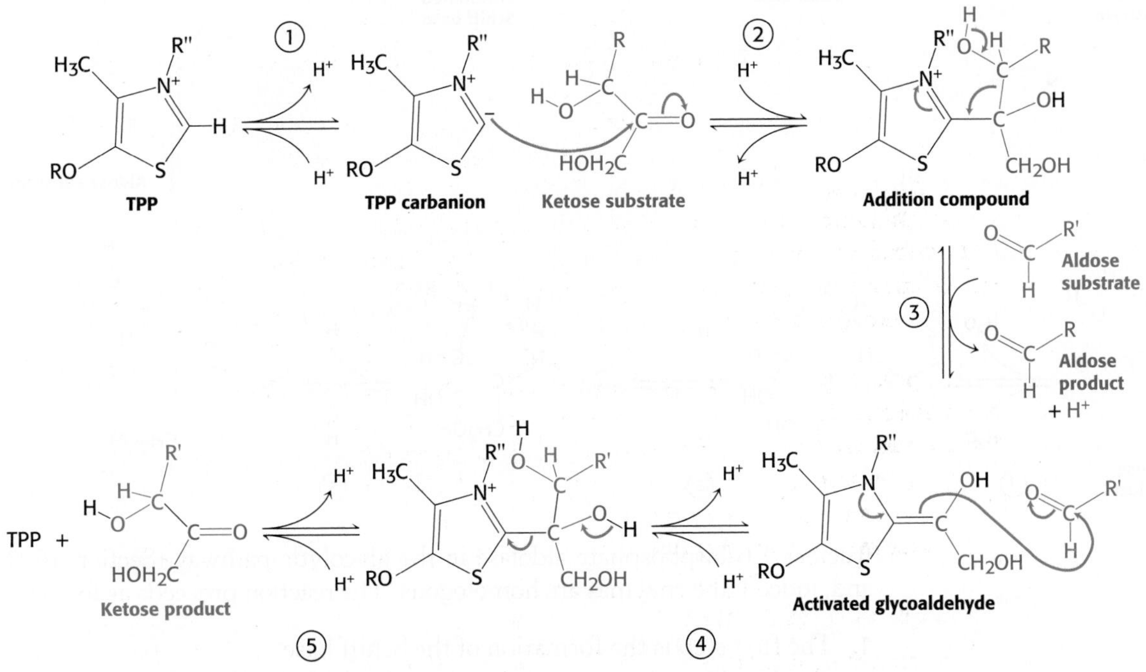

FIGURE 20.22 Transketolase mechanism. (1) Thiamine pyrophosphate (TPP) ionizes to form a carbanion. (2) The carbanion of TPP attacks the ketose substrate. (3) Cleavage of a carbon–carbon bond frees the aldose product and leaves a two-carbon fragment joined to TPP. (4) This activated glycoaldehyde intermediate attacks the aldose substrate to form a new carbon–carbon bond. (5) The ketose product is released, freeing the TPP for the next reaction cycle.

The reaction proceeds as follows:

1. The C-2 carbon atom of bound TPP readily ionizes to give a *carbanion*.

2. The negatively charged carbon atom of this reactive intermediate attacks the carbonyl group of the ketose substrate.

3. The resulting addition compound releases the aldose product to yield an *activated glycoaldehyde unit*.

4. The positively charged nitrogen atom in the thiazole ring acts as an *electron sink* in the development of this activated intermediate. The carbonyl group of a suitable aldose acceptor then condenses with the activated glycoaldehyde unit to form a new ketose.

5. The ketose is then released from the enzyme.

Transaldolase reaction. Transaldolase transfers a three-carbon *dihydroxyacetone* unit from a ketose donor to an aldose acceptor. Transaldolase, in contrast to transketolase, does not contain a prosthetic group. Rather, *a Schiff base is formed between the carbonyl group of the ketose substrate and the ε-amino group of a lysine residue at the active site of the enzyme* (Figure 20.23). This kind of covalent enzyme–substrate intermediate is like that formed in

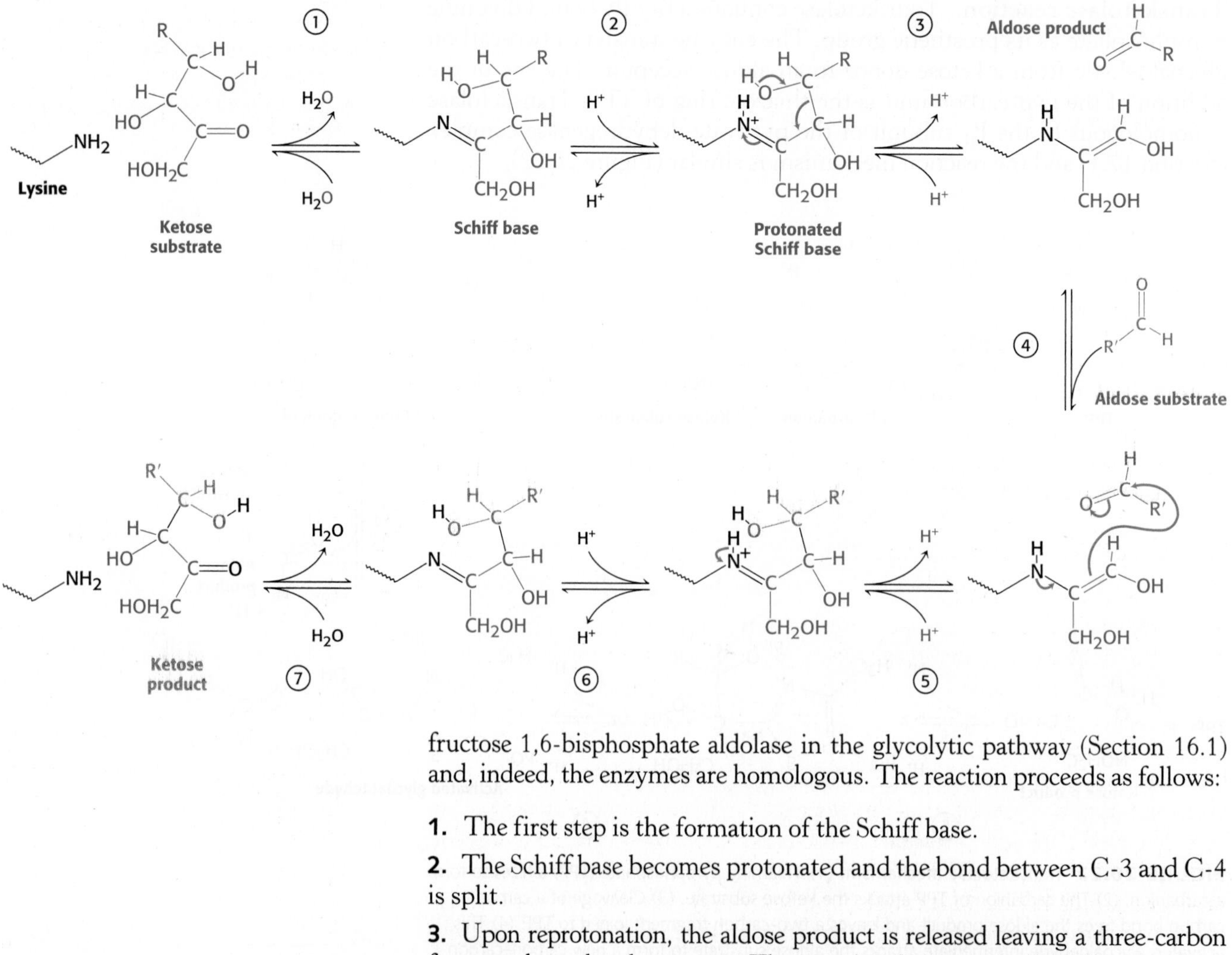

FIGURE 20.23 Transaldolase mechanism. (1) The reaction begins with the formation of a Schiff base between a lysine residue in transaldolase and the ketose substrate. (2) Protonation of the Schiff base occurs. (3) Deprotonation leads to the release of the aldose product, leaving a three-carbon fragment attached to the lysine residue. (4) This intermediate adds to the aldose substrate. (5) Protonation occurs, forming a new carbon–carbon bond. (6) Subsequent deprotonation and (7) hydrolysis of the Schiff base releases the ketose product from the lysine side chain, completing the reaction cycle.

fructose 1,6-bisphosphate aldolase in the glycolytic pathway (Section 16.1) and, indeed, the enzymes are homologous. The reaction proceeds as follows:

1. The first step is the formation of the Schiff base.

2. The Schiff base becomes protonated and the bond between C-3 and C-4 is split.

3. Upon reprotonation, the aldose product is released leaving a three-carbon fragment bound to the enzyme. The negative charge on the Schiff-base carbanion moiety is stabilized by resonance (Figure 20.24). The positively charged nitrogen atom of the protonated Schiff base acts as an electron sink.

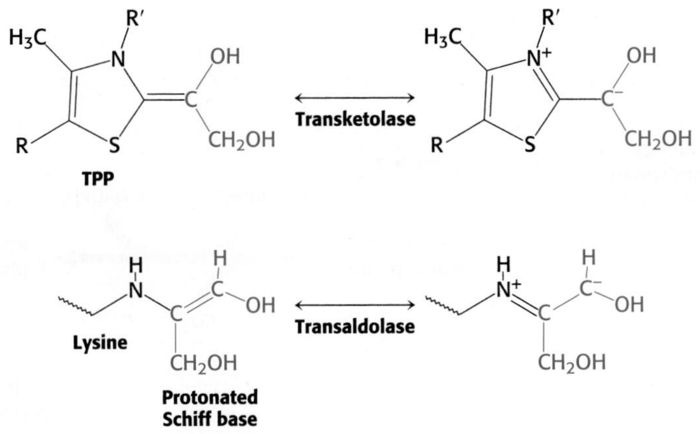

FIGURE 20.24 Carbanion intermediates. For transketolase and transaldolase, a carbanion intermediate is stabilized by resonance. In transketolase, TPP stabilizes this intermediate; in transaldolase, a protonated Schiff base plays this role.

4. The Schiff-base adduct is stable until a suitable aldose becomes bound.

5. The dihydroxyacetone moiety then reacts with the carbonyl group of the aldose. Protonation allows the formation of a new carbon–carbon bond.

6. Subsequent deprotonation occurs.

7. Following deprotonation, hydrolysis of the Schiff base releases the ketose product.

The nitrogen atom of the protonated Schiff base plays the same role in transaldolase as the thiazole-ring nitrogen atom does in transketolase. In each enzyme, a group within an intermediate reacts like a carbanion in attacking a carbonyl group to form a new carbon–carbon bond. In each case, the charge on the carbanion is stabilized by resonance (Figure 20.24).

A **Schiff base**, named after the Italian chemist Hugo Schiff (1834–1915), is a compound with the general structure $R_2C = NR'$. These compounds can be considered a sub-class of imines that are typically formed by the condensation of a ketone or aldehyde with a primary amine.

20.4 The Metabolism of Glucose 6-Phosphate by the Pentose Phosphate Pathway Is Coordinated with Glycolysis

Glucose 6-phosphate is metabolized by both the glycolytic pathway (Chapter 16) and the pentose phosphate pathway. How is the processing of this important metabolite partitioned between these two metabolic routes? The cytoplasmic concentration of $NADP^+$ plays a key role in determining the fate of glucose 6-phosphate.

The rate of the oxidative phase of the pentose phosphate pathway is controlled by the level of $NADP^+$

The first reaction in the oxidative branch of the pentose phosphate pathway—the dehydrogenation of glucose 6-phosphate—is essentially irreversible. In fact, this reaction is rate limiting under physiological conditions and serves as the control site for the oxidative branch of the pathway. The most important regulatory factor is the level of $NADP^+$. Low levels of $NADP^+$ limit the dehydrogenation of glucose 6-phosphate because it is needed as the electron acceptor. The effect of low levels of $NADP^+$ is intensified by the fact that NADPH competes with $NADP^+$ in binding to the enzyme. The ratio of $NADP^+$ to NADPH in the cytoplasm of a liver cell from a well-fed rat is about 0.014. The marked effect of the $NADP^+$ level on the rate of the oxidative phase ensures that NADPH is not generated unless the supply needed for reductive biosyntheses or protection against oxidative stress is low. The nonoxidative phase of the pentose phosphate pathway is controlled primarily by the availability of substrates.

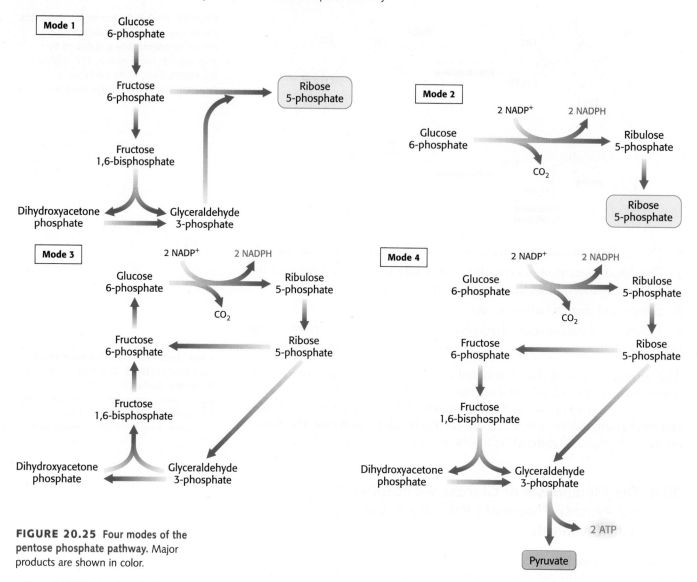

FIGURE 20.25 Four modes of the pentose phosphate pathway. Major products are shown in color.

 EXPLORE this pathway further in the Metabolic Map on **Sapling** Plus

The flow of glucose 6-phosphate depends on the need for NADPH, ribose 5-phosphate, and ATP

We can grasp the intricate interplay between glycolysis and the pentose phosphate pathway by examining the metabolism of glucose 6-phosphate in four different metabolic situations (Figure 20.25).

Mode 1. *Much more ribose 5-phosphate than NADPH is required.* For example, rapidly dividing cells need ribose 5-phosphate for the synthesis of nucleotide precursors of DNA. Most of the glucose 6-phosphate is converted into fructose 6-phosphate and glyceraldehyde 3-phosphate by the glycolytic pathway. Transaldolase and transketolase then convert two molecules of fructose 6-phosphate and one molecule of glyceraldehyde 3-phosphate into three molecules of ribose 5-phosphate by a reversal of the reactions described earlier. The stoichiometry of mode 1 is

$$5 \text{ Glucose 6-phosphate} + \text{ATP} \rightarrow 6 \text{ ribose 5-phosphate} + \text{ADP} + 2 \text{ H}^+$$

Mode 2. *The needs for NADPH and for ribose 5-phosphate are balanced.* Under these conditions, glucose 6-phosphate is processed to one molecule of ribulose 5-phosphate while generating two molecules of NADPH. Ribulose 5-phosphate is then converted into ribose 5-phosphate. The stoichiometry of mode 2 is

$$\text{Glucose 6-phosphate} + 2\,NADP^+ + H_2O \longrightarrow$$
$$\text{ribose 5-phosphate} + 2\,NADPH + 2\,H^+ + CO_2$$

Mode 3. *Much more NADPH than ribose 5-phosphate is required.* For example, adipose tissue requires a high level of NADPH for the synthesis of fatty acids (Table 20.4). In this case, glucose 6-phosphate is completely oxidized to CO_2. Three groups of reactions are active in this situation. First, the oxidative phase of the pentose phosphate pathway forms two molecules of NADPH and one molecule of ribulose 5-phosphate. Then, ribulose

TABLE 20.4 Tissues with active pentose phosphate pathways

Tissue	Function
Adrenal gland	Steroid synthesis
Liver	Fatty acid and cholesterol synthesis
Testes	Steroid synthesis
Adipose tissue	Fatty acid synthesis
Ovary	Steroid synthesis
Mammary gland	Fatty acid synthesis
Red blood cells	Maintenance of reduced glutathione

5-phosphate is converted into fructose 6-phosphate and glyceraldehyde 3-phosphate by transketolase and transaldolase. Finally, glucose 6-phosphate is resynthesized from fructose 6-phosphate and glyceraldehyde 3-phosphate by the gluconeogenic pathway. The stoichiometries of these three sets of reactions are

$$6\,\text{Glucose 6-phosphate} + 12\,NADP^+ + 6\,H_2O \longrightarrow$$
$$6\,\text{Ribose 5-phosphate} + 12\,NADPH + 12\,H^+ + 6\,CO_2$$

$$6\,\text{Ribose 5-phosphate} \longrightarrow$$
$$4\,\text{Fructose 6-phosphate} + 2\,\text{glyceraldehyde 3-phosphate}$$

$$4\,\text{Fructose 6-phosphate} + 2\,\text{glyceraldehyde 3-phosphate} + H_2O \longrightarrow$$
$$5\text{-Glucose 6-phosphate} + P_i$$

The sum of the mode 3 reactions is

$$\text{Glucose 6-phosphate} + 12\,NADP^+ + 7\,H_2O \longrightarrow$$
$$6\,CO_2 + 12\,NADPH + 12\,H^+ + P_i$$

Thus, *the equivalent of glucose 6-phosphate can be completely oxidized to CO_2 with the concomitant generation of NADPH.* In essence, ribose 5-phosphate produced by the pentose phosphate pathway is recycled into glucose 6-phosphate by transketolase, transaldolase, and some of the enzymes of the gluconeogenic pathway.

Mode 4. *Both NADPH and ATP are required.* Alternatively, ribulose 5-phosphate formed from glucose 6-phosphate can be converted into pyruvate. Fructose 6-phosphate and glyceraldehyde 3-phosphate derived from ribose 5-phosphate enter the glycolytic pathway rather than reverting to glucose 6-phosphate. In this mode, *ATP and NADPH are concomitantly generated, and five of the six carbons of glucose 6-phosphate emerge in pyruvate.*

$$3 \text{ Glucose 6-phosphate} + 6 \text{ NADP}^+ + 5 \text{ NAD}^+ + 5 \text{ P}_i + 8 \text{ ADP} \longrightarrow$$
$$5 \text{ pyruvate} + 3 \text{ CO}_2 + 6 \text{ NADPH} + 5 \text{ NADH}$$
$$+ 8 \text{ ATP} + 2 \text{ H}_2\text{O} + 8 \text{ H}^+$$

Pyruvate formed by these reactions can be oxidized to generate more ATP, or it can be used as a building block in a variety of biosyntheses.

The pentose phosphate pathway is required for rapid cell growth

Rapidly dividing cells, such as cancer cells, require ribose 5-phosphate for nucleic acid synthesis and NADPH for fatty acid synthesis, which in turn is required to form membrane lipids (Section 26.1). Recall that rapidly dividing cells switch to aerobic glycolysis to meet their ATP needs (Section 16.2). Glucose 6-phosphate and glycolytic intermediates are then used to generate NADPH and ribose 5-phosphate using the nonoxidative phase of the pentose phosphate pathway. The diversion of glycolytic intermediates into the nonoxidative phase is facilitated by the expression of the gene for a pyruvate kinase isozyme, PKM. PKM has a low catalytic activity, creating a bottleneck in the glycolytic pathway. Glycolytic intermediates accumulate and are then used by the pentose phosphate pathway to synthesize NADPH and ribose 5-phosphate. The shunting of phosphorylated intermediates into the nonoxidative phase of the pentose phosphate pathway is further enabled by the inhibition of triose phosphate isomerase by phosphoenolpyruvate, the substrate of PKM.

Through the looking-glass: The Calvin cycle and the pentose phosphate pathway are mirror images

The complexities of the Calvin cycle and the pentose phosphate pathway are easier to comprehend if we consider them as functional mirror images of each other. The Calvin cycle begins with the fixation of CO_2 and proceeds to use NADPH in the synthesis of glucose. The pentose phosphate pathway begins with the oxidation of a glucose-derived carbon atom to CO_2 and concomitantly generates NADPH. The regeneration phase of the Calvin cycle converts C_6 and C_3 molecules back into the starting material—the C_5 molecule ribulose 1,5-bisphosphate. The pentose phosphate pathway converts a C_5 molecule, ribulose 5-phosphate, into C_6 and C_3 intermediates of the glycolytic pathway. Not surprisingly, in photosynthetic organisms, many enzymes are common to the two pathways. We see the economy of evolution: the use of identical enzymes for similar reactions with different ends.

20.5 Glucose 6-Phosphate Dehydrogenase Plays a Key Role in Protection Against Reactive Oxygen Species

The NADPH generated by the pentose phosphate pathway plays a vital role in protecting the cells from reactive oxygen species (ROS). Reactive oxygen species generated in oxidative metabolism inflict damage on all classes of macromolecules and can ultimately lead to cell death. Indeed, ROS are implicated in a number of human diseases, such as diabetes (Table 18.3). Reduced *glutathione* (GSH), a tripeptide with a free sulfhydryl group, combats oxidative stress by reducing ROS to harmless forms.

Its task accomplished, the glutathione is now in the oxidized form (GSSG) and must be reduced to regenerate GSH. The reducing power is supplied by the NADPH generated by glucose 6-phosphate dehydrogenase in the pentose phosphate pathway. Indeed, cells with reduced levels of glucose 6-phosphate dehydrogenase are especially sensitive to oxidative stress.

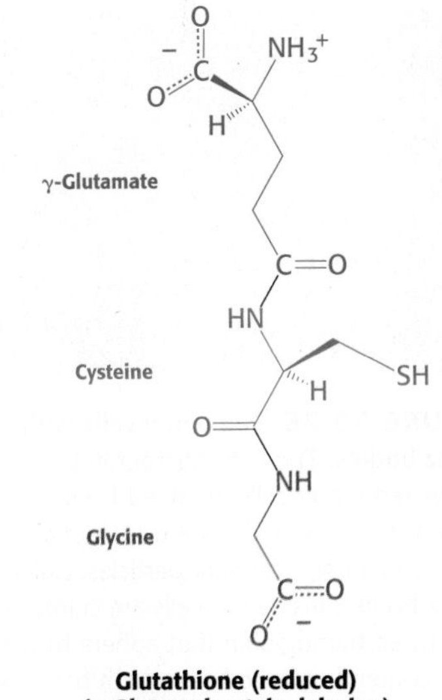

Glucose 6-phosphate dehydrogenase deficiency causes a drug-induced hemolytic anemia

The importance of the pentose phosphate pathway is highlighted by some people's anomalous responses to certain drugs. For instance, pamaquine, the first synthetic antimalarial drug introduced in 1926, was associated with the appearance of severe and mysterious ailments. Most patients tolerated the drug well, but a few developed severe symptoms within a few days after therapy was started. Their urine turned black, jaundice developed, and the hemoglobin content of the blood dropped sharply. In some cases, massive destruction of red blood cells caused death.

This drug-induced *hemolytic anemia* was shown 30 years later to be caused by a *deficiency of glucose 6-phosphate dehydrogenase*, the enzyme catalyzing the first step in the oxidative branch of the pentose phosphate pathway. The result is a dearth of NADPH in all cells, but this deficiency is most acute in red blood cells because they lack mitochondria and have no alternative means of generating reducing power. This defect, which is inherited on the X chromosome, is the most common condition that results from an enzyme malfunction, affecting hundreds of millions of people.

Pamaquine sensitivity is not simply a historical oddity about malaria treatment many decades ago. Primaquine, an antimalarial closely related to pamaquine, is widely used in malaria-prone regions of the world. Vicine, a pyrimidine glycoside of fava beans (which are consumed in countries surrounding the Mediterranean), can also induce hemolysis. People deficient in glucose 6-phosphate dehydrogenase suffer hemolysis from eating fava beans or inhaling the pollen of the fava flowers, a response called *favism*. How can we explain hemolysis caused by pamaquine, primaquine, and vicine biochemically? These chemicals are oxidative agents that generate peroxides, reactive oxygen species that can damage membranes as well as other biomolecules. Peroxides are usually eliminated by the enzyme *glutathione peroxidase*, which uses reduced glutathione as a reducing agent.

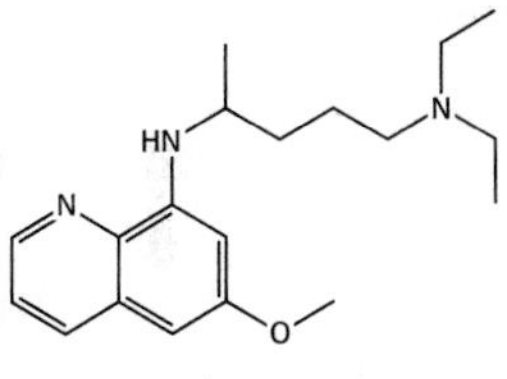

Glutathione (reduced)
(γ-Glutamylcysteinylglycine)

Vicia faba. The Mediterranean plant *Vicia faba* is a source of fava beans that contain the pyrimidine glycoside vicine. [FLPA/Richard Becker/AGE Fotostock.]

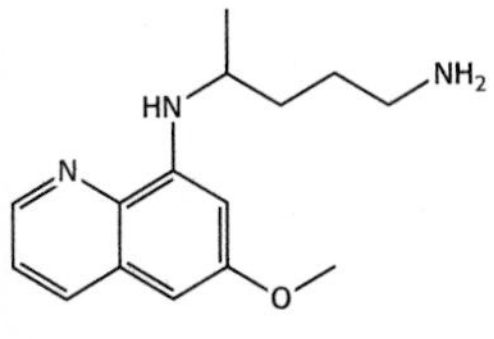

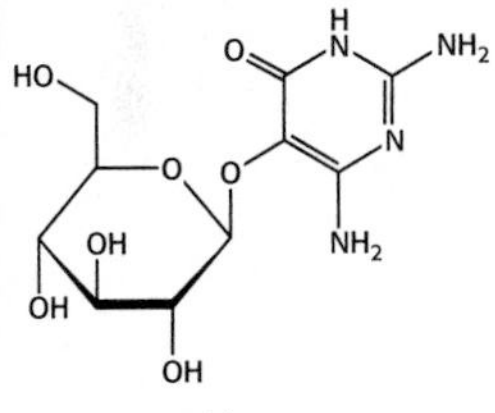

$$2\,GSH + ROOH \xrightarrow{\text{Glutathione peroxidase}} GSSG + H_2O + ROH$$

The major role of NADPH in red blood cells is to reduce the disulfide form of glutathione to the sulfhydryl form. The enzyme that catalyzes the regeneration of reduced glutathione is *glutathione reductase.*

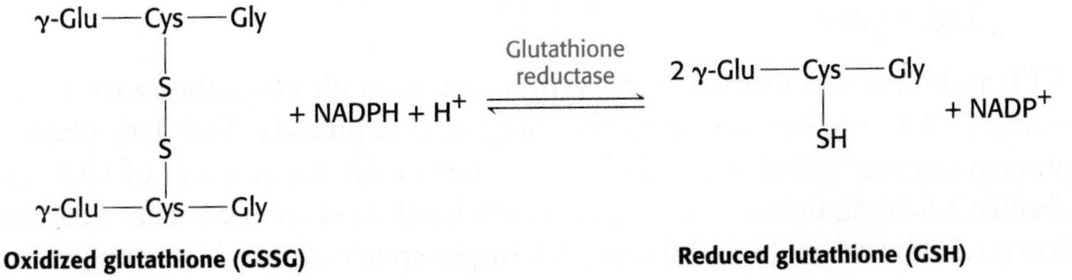

Oxidized glutathione (GSSG)

Reduced glutathione (GSH)

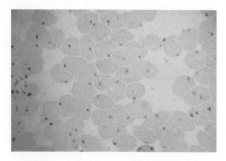

FIGURE 20.26 Red blood cells with Heinz bodies. The light micrograph shows red blood cells obtained from a person deficient in glucose 6-phosphate dehydrogenase. The dark particles, called Heinz bodies, inside the cells are clumps of denatured hemoglobin that adhere to the plasma membrane and stain with basic dyes. Red blood cells in such people are highly susceptible to oxidative damage. [CNRI/ Science Source.]

Red blood cells with a lowered level of reduced glutathione are more susceptible to hemolysis. In the absence of glucose 6-phosphate dehydrogenase, peroxides continue to damage membranes because no NADPH is being produced to restore reduced glutathione. Thus, the answer to our question is that *glucose 6-phosphate dehydrogenase is required to maintain reduced glutathione levels to protect against oxidative stress.* In the absence of oxidative stress, however, the deficiency is quite benign. The sensitivity to oxidative agents of people having this dehydrogenase deficiency also clearly demonstrates that *atypical reactions to drugs may have a genetic basis.*

Reduced glutathione is also essential for maintaining the normal structure of red blood cells by maintaining the structure of hemoglobin. The reduced form of glutathione serves as a sulfhydryl buffer that keeps the residues of hemoglobin in the reduced sulfhydryl form. Without adequate levels of reduced glutathione, the hemoglobin sulfhydryl groups can no longer be maintained in the reduced form. Hemoglobin molecules then cross-link with one another to form aggregates called *Heinz bodies* on cell membranes (Figure 20.26). Membranes damaged by Heinz bodies and reactive oxygen species become deformed, and the cell is likely to undergo lysis.

A deficiency of glucose 6-phosphate dehydrogenase confers an evolutionary advantage in some circumstances

The incidence of the most common form of glucose 6-phosphate dehydrogenase deficiency, characterized by a 10-fold reduction in enzymatic activity in red blood cells, is 11% among Americans of African heritage. This high frequency suggests that the deficiency may be advantageous under certain environmental conditions. Indeed, *glucose 6-phosphate dehydrogenase deficiency protects against falciparum malaria.* The parasites causing this disease require NADPH for growth. Moreover, infection by the parasites induces oxidative stress in the infected cell. Because the pentose phosphate pathway is compromised, the cell and parasite die from oxidative damage. Thus, glucose 6-phosphate dehydrogenase deficiency is a mechanism of protection against malaria, which accounts for its high frequency in malaria-prone regions of the world. We see here once again the interplay of heredity and environment in the effect of drugs and diet on physiology.

The ability of glucose 6-phosphate dehydrogenase deficiency to protect against malaria does, however, create a public health conundrum. Primaquine is a commonly used and highly effective antimalarial drug. However, indiscriminate use of primaquine causes hemolysis in individuals deficient in glucose 6-phosphate dehydrogenase. A solution to this problem may be in the offing, as recent work shows that an anti-malaria vaccine may be within reach.

SUMMARY

20.1 The Calvin Cycle Synthesizes Hexoses from Carbon Dioxide and Water

ATP and NADPH formed in the light reactions of photosynthesis are used to convert CO_2 into hexoses and other organic compounds. The dark phase of photosynthesis, called the *Calvin cycle,* starts with the reaction of CO_2 and ribulose 1,5-bisphosphate to form two molecules of 3-phosphoglycerate. This reaction is catalyzed by rubisco (ribulose 1,5-bisphosphate carboxylase/oxygenase).

The steps in the conversion of 3-phosphoglycerate into fructose 6-phosphate and glucose 6-phosphate are like those of gluconeogenesis, except that glyceraldehyde 3-phosphate dehydrogenase in chloroplasts is specific for NADPH rather than NADH. Ribulose 1,5-bisphosphate is regenerated from fructose 6-phosphate, glyceraldehyde 3-phosphate, and dihydroxyacetone phosphate by a complex series of reactions. Several of the steps in the regeneration of ribulose 1,5-bisphosphate are like those of the pentose phosphate pathway. Three molecules of ATP and two molecules of NADPH are consumed for each molecule of CO_2 converted into a hexose. Starch in chloroplasts and sucrose in the cytoplasm are the major carbohydrate stores in plants.

Rubisco also catalyzes a competing oxygenase reaction, which produces phosphoglycolate and 3-phosphoglycerate. The recycling of phosphoglycolate leads to the release of CO_2 and further consumption of O_2 in a process called photorespiration.

20.2 The Activity of the Calvin Cycle Depends on Environmental Conditions

Reduced thioredoxin formed by the light-driven transfer of electrons from ferredoxin activates enzymes of the Calvin cycle by reducing disulfide bridges. The light-induced increase in pH and Mg^{2+} levels of the stroma are important in stimulating the carboxylation of ribulose 1,5-bisphosphate by rubisco. Photorespiration is minimized in tropical plants, which have an accessory pathway—the C_4 pathway—for concentrating CO_2 at the site of the Calvin cycle. This pathway enables tropical plants to take advantage of high levels of light and minimize the oxygenation of ribulose 1,5-bisphosphate. Other plants use crassulacean acid metabolism (CAM) to prevent dehydration. In CAM plants, the C_4 pathway is active during the night, when the plant exchanges gases with the air. During the day, gas exchange is eliminated and CO_2 is generated from malate stored in vacuoles.

20.3 The Pentose Phosphate Pathway Generates NADPH and Synthesizes Five-Carbon Sugars

Whereas the Calvin cycle is present only in photosynthetic organisms, the pentose phosphate pathway is present in all organisms. The pentose phosphate pathway consists of two phases: an oxidative phase and a nonoxidative phase. The oxidative phase generates NADPH and ribulose 5-phosphate in the cytoplasm. NADPH is used in reductive biosyntheses. The oxidative phase starts with the dehydrogenation of glucose 6-phosphate to form a lactone, which is hydrolyzed to give 6-phosphogluconate and then oxidatively decarboxylated to yield ribulose 5-phosphate. $NADP^+$ is the electron acceptor in both of these oxidations. The nonoxidative phase begins with the isomerization of ribulose 5-phosphate (a ketose) to ribose 5-phosphate (an aldose). Ribose 5-phosphate is converted into glyceraldehyde 3-phosphate and fructose 6-phosphate by *transketolase* and *transaldolase*. These two enzymes create a reversible link between the pentose phosphate pathway and glycolysis. Xylulose 5-phosphate, sedoheptulose 7-phosphate, and erythrose 4-phosphate are intermediates in these interconversions. When NADPH needs are paramount, fructose 6-phosphate and glyceraldehyde 3-phosphate from the nonoxidative phase are used to synthesize glucose 6-phosphate for metabolism in the oxidative phase. In this way, 12 molecules of NADPH can be generated for each molecule of glucose 6-phosphate that is completely oxidized to CO_2.

20.4 The Metabolism of Glucose 6-Phosphate by the Pentose Phosphate Pathway Is Coordinated with Glycolysis

Only the nonoxidative branch of the pathway is significantly active when much more ribose 5-phosphate than NADPH must be synthesized. Under these conditions, fructose 6-phosphate and glyceraldehyde 3-phosphate (formed by the glycolytic pathway) are converted into ribose 5-phosphate without the formation of NADPH. Alternatively, ribose 5-phosphate formed by the oxidative branch can be converted into pyruvate through fructose 6-phosphate and glyceraldehyde 3-phosphate. In this mode, ATP and NADPH are generated, and five of the six carbons of glucose 6-phosphate emerge in pyruvate. The interplay of the glycolytic and pentose phosphate pathways enables the levels of NADPH, ATP, and building blocks such as ribose 5-phosphate and pyruvate to be continuously adjusted to meet cellular needs.

20.5 Glucose 6-Phosphate Dehydrogenase Plays a Key Role in Protection Against Reactive Oxygen Species

NADPH generated by glucose 6-phosphate dehydrogenase maintains the appropriate levels of reduced glutathione required to combat oxidative stress and ensures the proper reducing environment in the cell. Cells with diminished glucose 6-phosphate dehydrogenase activity are especially sensitive to oxidative stress.

APPENDIX   Biochemistry in Focus

Biochemistry in Focus 1
Phosphoenolpyruvate carboxylase allows insight into ancient ecosystems

In addition to having a higher affinity for CO_2 than rubisco, phosphoenolpyruvate carboxylase (PEPC) also incorporates the carbon isotope $^{13}CO_2$ more readily than rubisco. This difference means that C_4 plants and C_3 plants have different $^{13}C/^{12}C$ ratios or different isotopic signatures. These signatures can be detected by mass spectrometry (Section 3.3). Detection of the different signatures has been used by ecologists and evolutionary biologists to explore ancient ecosystems. For example, examination of $^{13}C/^{12}C$ ratio in fossils showed that C_4 plants rapidly expanded and became the dominant plant in prairies and savannahs 10 to 6 million years ago. The isotopic signature even allows the examination of the diets of ancient animals. Fossils of animals that grazed on C_4 plants will have a different isotopic signature than those that fed on C_3 plants. One of the most interesting and exciting developments of modern biological research is the breakdown of barriers between disciplines. A biochemical peculiarity of PEPC—the incorporation of $^{13}CO_2$—can be exploited by ecologists and evolutionary biologists to explore the history of ecosystems.

Biochemistry in Focus 2
Hummingbirds and the pentose phosphate pathway

The evolution of oxygenic photosynthetic organisms allowed the evolution of multicellular organisms. However, as discussed in Chapter 18 and the current chapter, O_2 can be readily converted into reactive oxygen species (ROS), which can cause oxidation damage to most biomolecules. Any aerobic animal that is highly active faces the problem of damage by ROS, and the problem increases with increasing activity. Consider the hummingbird. The flight muscles constitute more than 25% of the birds' weight, the wings flap up to 200 times per second, the heart beats more than 1200 times a minute, and the birds feed while airborne.

[kojihirano/Getty Images.]

This amazing level of activity requires muscles that are capable of maintaining a high metabolic rate for an extended period of time. How do the hummingbirds cope with the ROS generation that is a consequence of high O_2 demand? One means is the nature of the fuel they extract from the flower. The nectar is rich in carbohydrates, the preferred fuel for high intensity activity (Chapter 27). When organisms are using carbohydrates only as a fuel, the respiratory quotient—the ratio of CO_2 produced to O_2 consumed—is 1, as shown in the equation:

$$C_6H_{12}O_6 + 6\,O_2 \xrightarrow[\text{respiration}]{\text{cellular}} 6\,CO_2 + 6\,H_2O \qquad RQ = CO_2/O_2 = 1$$

As we will see in Chapter 27, the RQ of fats and proteins is much lower.

The second advantage of using carbohydrate as fuel is revealed when the RQ of hummingbirds is measured. It is actually greater than 1. What could be the source of the extra CO_2? Recall that the oxidative phase of the pentose phosphate pathway generates NADPH, which protects against ROS (p. 668). In doing this, the oxidative phase also produces CO_2, accounting for the RQ greater than 1:

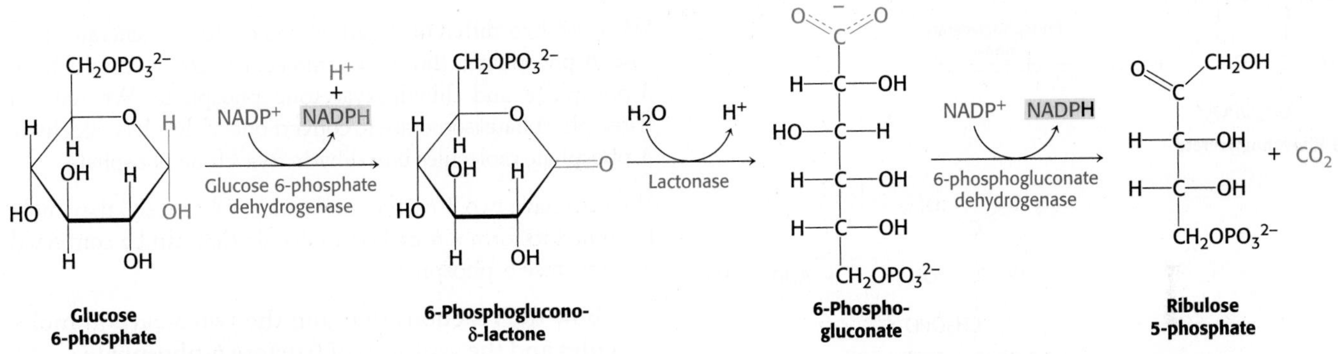

This reveals that, by virtue of a high flux through the pentose phosphate pathway (both its oxidative and nonoxidative phases), hummingbirds are using carbohydrate-rich nectar to produce ATP to power muscle activity and also to protect against ROS. Human athletes who consume carbohydrates during intense, extended exercise may also use the pentose phosphate pathway for ROS protection.

APPENDIX

Problem-Solving Strategies

PROBLEM: Fructose 6-phosphate is a member of the hexose monophosphate pool found in photosynthetic plants.

(a) What are the other members of the pool?
(b) What is the source of the 3-phosphoglycerate in photosynthetic plants?
(c) Outline a pathway for the synthesis of fructose 6-phosphate from 3-phosphoglycerate.

SOLUTION: We don't have to be plant biochemists to come up with various hexose monophosphates; we just need to remember our glucose metabolism from Chapter 17, as hexose monophosphates are prominent.

▶ **What are the other members of the hexose monophosphate pool?**

In addition to fructose 6-phosphate, we have seen glucose 6-phosphate and glucose 1-phosphate. Recall that these hexose monophosphates are readily interconvertible.

▶ **What is the source of the 3-phosphoglycerate in photosynthetic plants?**

This is an important question to be able to answer because it essentially asks: What is the reaction that introduces carbon from the atmosphere into the biosphere? The answer is the reaction catalyzed by rubisco:

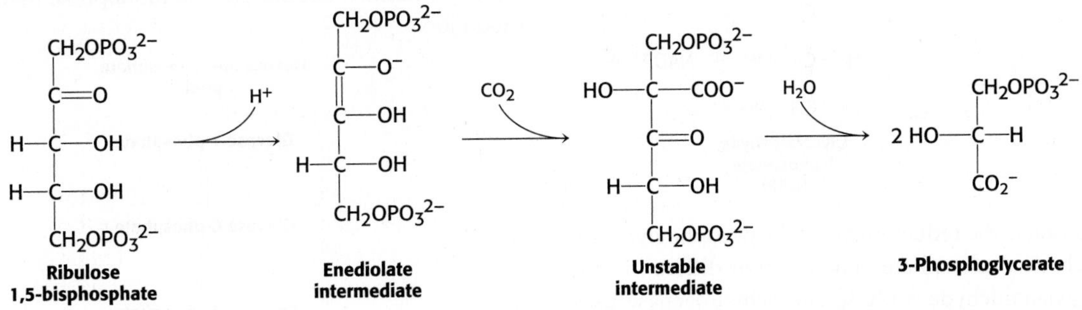

▶ **Outline a pathway for the synthesis of fructose 6-phosphate from 3-phosphoglycerate.**

Now, if you recall, we have dealt with this question before. So, prior to answering the above question, let's ask another.

▶ **What pathway is responsible for the synthesis of glucose from simple precursors?**

Gluconeogenesis. All we need to do then is to examine gluconeogenesis, and make any plant-specific adaptations necessary.

▶ **What is the first step in the conversion of 3-phosphoglycerate into the fructose 6-phosphate?**

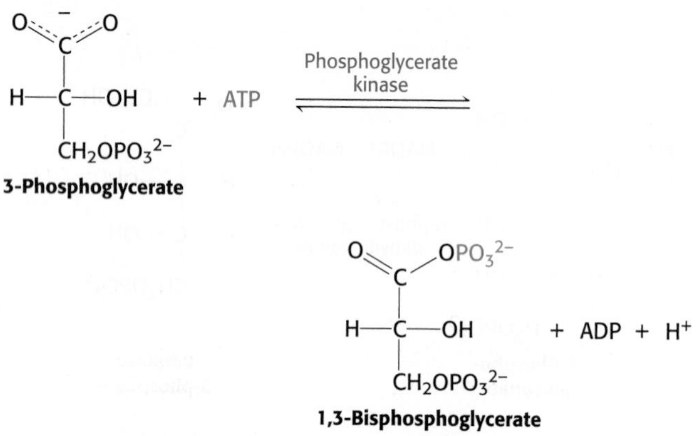

O=C—O⁻ ... (structure)

H—C—OH + ATP $\xrightarrow{\text{Phosphoglycerate kinase}}$

CH₂OPO₃²⁻

3-Phosphoglycerate

O=C—OPO₃²⁻

H—C—OH + ADP + H⁺

CH₂OPO₃²⁻

1,3-Bisphosphoglycerate

The kinase reaction generates the high-energy-phosphoryl-transfer-potential compound 1,3-bisphosphoglycerate. Just to make sure we are clear on everything:

▶ **What is the source of the ATP for the kinase reaction?**

These reactions are occurring in the chloroplast, so there is ready access to ATP produced by the light reactions.

▶ **Using gluconeogenesis as our guide, what is the next reaction in the synthesis of fructose 6-phosphate and how does it differ from the reaction of gluconeogenesis?**

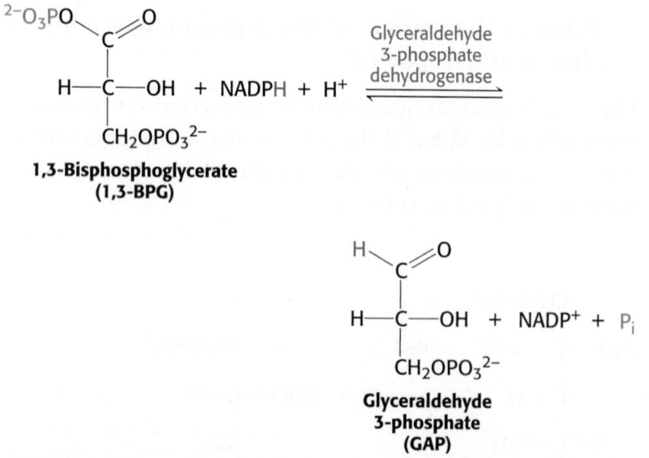

²⁻O₃PO—C=O

H—C—OH + NADPH + H⁺ $\xrightarrow{\text{Glyceraldehyde 3-phosphate dehydrogenase}}$

CH₂OPO₃²⁻

1,3-Bisphosphoglycerate (1,3-BPG)

H—C=O

H—C—OH + NADP⁺ + Pᵢ

CH₂OPO₃²⁻

Glyceraldehyde 3-phosphate (GAP)

The next reaction is the reduction of 1,3-bisphosphoglycerate to glyceraldehyde 3-phosphate. The reaction differs in that chloroplast glyceraldehyde 3-phosphate dehydrogenase uses NADPH instead of NADH. Remember, the NADPH is also

a product of the light reactions. Keep in mind that fructose 6-phosphate is composed of six carbons, while the reactions we have been working with as substrates and products have three carbon molecules. We will thus need to double the reactions above. Let's say we have done that:

▶ **What is the next reaction that must occur?**

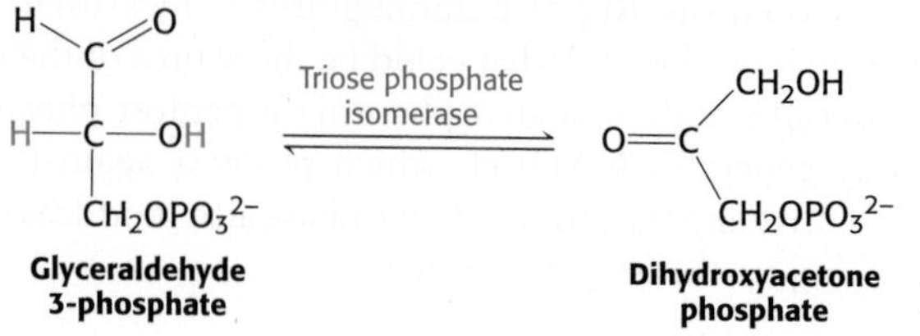

H—C=O

H—C—OH $\xrightarrow{\text{Triose phosphate isomerase}}$ O=C—CH₂OH

CH₂OPO₃²⁻ ... CH₂OPO₃²⁻

Glyceraldehyde 3-phosphate ... **Dihydroxyacetone phosphate**

We need two different 3-carbon molecules to generate fructose 6-phosphate; those two molecules are glyceraldehyde 3-phosphate and dihydroxyacetone phosphate. We rely on triose phosphate isomerase to convert one of the glyceraldehyde 3-phosphate molecules into dihydroxyacetone phosphate.

We now have two 3-carbon molecules. Obviously, they must be joined to form a 6-carbon molecule that can be converted into fructose 6-phosphate.

▶ **Show the reactions that join the two 3-carbon molecules and the synthesis of fructose 6-phosphate.**

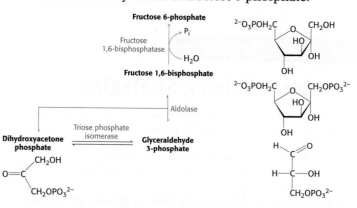

Aldolase joins dihydroxyacetone phosphate and glyceraldehyde 3-phosphate to form fructose 1, 6-bisphosphate, and fructose 1,6-bisphosphatase removes a phosphate to finally yield fructose 6-phosphate. Remember, fructose 6-phosphate is a member of the hexose monophosphate pool, all of the components of which are readily interchangeable. So, we have synthesized not one hexose monophosphate, but three. Good job.

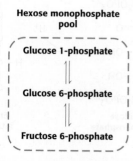

Hexose monophosphate pool

Glucose 1-phosphate
⇅
Glucose 6-phosphate
⇅
Fructose 6-phosphate

KEY TERMS

Calvin cycle (dark reactions) (p. 645)

autotroph (p. 646)

heterotroph (p. 646)

rubisco (ribulose 1,5-bisphosphate carboxylase/oxygenase) (p. 646)

peroxisome (microbody) (p. 650)

photorespiration (p. 651)

hexose monophosphate pool (p. 651)

transketolase (p. 651)

aldolase (p. 651)

starch (p. 654)

sucrose (p. 654)

thioredoxin (p. 655)

C_4 pathway (Hatch-Slack pathway) (p. 656)

C_4 plant (p. 658)

C_3 plant (p. 658)

crassulacean acid metabolism (CAM) (p. 659)

pentose phosphate pathway (p. 646)

glucose 6-phosphate dehydrogenase (p. 656)

glutathione (p. 659)

PROBLEMS

1. *Not the Calvin Klein Cycle.* The Calvin Cycle: ✓①

(a) Cannot occur in the light.

(b) Is independent of the light reactions.

(c) Is a requirement for photosynthetic plants but has no effect on the rest of the biosphere.

(d) Is dependent on the light reactions.

2. *Fixated on CO_2.* The fixation of CO_2 by plants into organic compounds: ✓①

(a) Requires NADPH.

(b) Requires Acetyl CoA.

(c) Generates ATP.

(d) Occurs in the cytoplasm.

3. *Nor the Calvin and Hobbes Cycle.* Why is the Calvin cycle crucial to the functioning of all life forms? ✓①

4. *Be nice to plants.* Differentiate between autotrophs and heterotrophs.

5. *Cabalistic reactions?* Why are the reactions of the Calvin cycle sometimes referred to as the dark reactions? Do they take place only at night or are they grim, secret reactions? ✓①

6. *Compare and contrast.* Identify the similarities and differences between the Krebs cycle and the Calvin cycle. ✓①

7. *Light and dark talk.* Rubisco requires a molecule of CO_2 covalently bound to lysine 201 for catalytic activity. The carboxylation of rubisco is favored by high pH and high Mg^{2+} concentration in the stroma. Why does it make good physiological sense for these conditions to favor rubisco carboxylation? ✓①, ✓②

8. *Labeling experiments.* When Melvin Calvin performed his initial experiments on carbon fixation, he exposed algae to radioactive carbon dioxide. After 5 seconds, only a single organic compound contained radioactivity but, after 60 seconds, many compounds had incorporated radioactivity. (a) What compound initially contained the radioactivity? (b) What compounds contained radioactivity after 60 seconds? ✓①

9. *Three-part harmony.* The Calvin cycle can be thought of as occurring in three parts or stages. Describe the stages. ✓①

10. *Not always to the swiftest.* Suggest a reason why rubisco might be the most abundant enzyme in the world. ✓①

11. *A requirement.* In an atmosphere devoid of CO_2 but rich in O_2, the oxygenase activity of rubisco disappears. Why? ✓①, ✓②

12. *Reduce locally.* Glyceraldehyde 3-phosphate dehydrogenase in chloroplasts uses NADPH to participate in the synthesis of glucose. In gluconeogenesis in the cytoplasm, the isozyme of the dehydrogenase uses NADH. Why is it advantageous for the chloroplast enzyme to use NADPH? ✓①

13. *Total eclipse.* An illuminated suspension of the green alga *Chlorella* is actively carrying out photosynthesis. Suppose that the light is suddenly switched off. How would the levels of 3-phosphoglycerate and ribulose 1,5-bisphosphate change in the next minute? ✓①

14. *CO_2 deprivation.* An illuminated suspension of *Chlorella* is actively carrying out photosynthesis in the presence of 1% CO_2. The concentration of CO_2 is abruptly reduced to 0.003%. What effect would this reduction have on the levels of 3-phosphoglycerate and ribulose 1,5-bisphosphate in the next minute? ✓①

15. *Salvage operation.* Write a balanced equation for the transamination of glyoxylate to yield glycine.

16. *Dog days of August.* Before the days of pampered lawns, most homeowners practiced horticultural Darwinism. A result was that the lush lawns of early summer would often convert into robust cultures of crabgrass in the dog days of August. Provide a possible biochemical explanation for this transition.

17. *Is it hot in here, or is it just me?* Why is the C_4 pathway valuable for tropical plants?

18. *No free lunch.* Explain why maintaining a high concentration of CO_2 in the bundle-sheath cells of C_4 plants is an example of active transport. How much ATP is required per CO_2 to maintain a high concentration of CO_2 in the bundle-sheath cells of C_4 plants?

19. *Breathing pictures?* What is photorespiration, what is its cause, and why is it believed to be wasteful?

20. *Global warming.* C_3 plants are most common in higher latitudes and become less common at latitudes near the equator. The reverse is true of C_4 plants. How might global warming affect this distribution?

21. *Cost effective?* C_3 plants require 18 molecules of ATP to synthesize 1 molecule of glucose. C_4 plants, on the other hand, require 30 molecules of ATP to synthesize 1 molecule of glucose. Why would any plant use C_4 metabolism instead of C_3 metabolism given that C_3 metabolism is so much more efficient?

22. *Communication.* What are the light-dependent changes in the stroma that regulate the Calvin cycle? ✓②

23. *Match 'em.* Match each term with its description.

(a) Calvin cycle _____
(b) Rubisco _____

(c) Carbamate _____
(d) Starch _____

(e) Sucrose _____
(f) Amylose _____
(g) Amylopectin _____

(h) C_3 plants _____

(i) C_4 plants _____
(j) Stomata _____

1. Catalyzes CO_2 fixation
2. Storage form of carbohydrates
3. α-1,4 linkages only
4. 3-Phosphoglycerate is formed after carbon fixation
5. The dark reactions
6. Includes α-1,6 linkages
7. Required for rubisco activity
8. Carbon fixation results in oxaloacetate formation
9. Allow exchange of gases
10. Transport form of carbohydrates

24. *PPP ID please.* Which of the following is not a component of the pentose phosphate pathway? ✓③
(a) NADPH
(b) CO_2
(c) Ribose 5-phosphate
(d) Phosphoenolpyruvate
(e) Erythrose 4-phosphate

25. *Who is in control?* Which of the following enzymes is the regulatory enzyme in the oxidative phase of the pentose phosphate pathway? ✓③, ✓④
(a) 6-Phosphogluconate dehydrogenase
(b) Glyceraldehyde 3-phosphate dehydrogenase

(c) Glucose 6-phosphate dehydrogenase
(d) Glucose 6-phosphatase
(e) Lactonase

26. *Phase shift.* The pentose phosphate pathway is composed of two distinct phases. What are the two phases and what are their roles? ✓③

27. *Linked in.* Describe how the pentose phosphate pathway and glycolysis are linked by transaldolase and transketolase. ✓③

28. *Biochemical taxonomy.* Match the phrases with the reactions or structures below.

(a) Identify 6-phosphoglucono-δ-lactone. _____
(b) Which reactions produce NADPH? _____
(c) Identify ribulose 5-phosphate. _____
(d) What reaction generates CO_2? _____
(e) Identify 6-phosphogluconate. _____
(f) Which reaction is catalyzed by phosphopentose isomerase? _____
(g) Identify ribose 5-phosphate. _____
(h) Which reaction is catalyzed by lactonase? _____
(i) Identify glucose 6-phosphate. _____
(j) Which reaction is catalyzed by 6-phosphogluconate dehydrogenase? _____
(k) Which reaction is catalyzed by glucose 6-phosphate dehydrogenase? _____

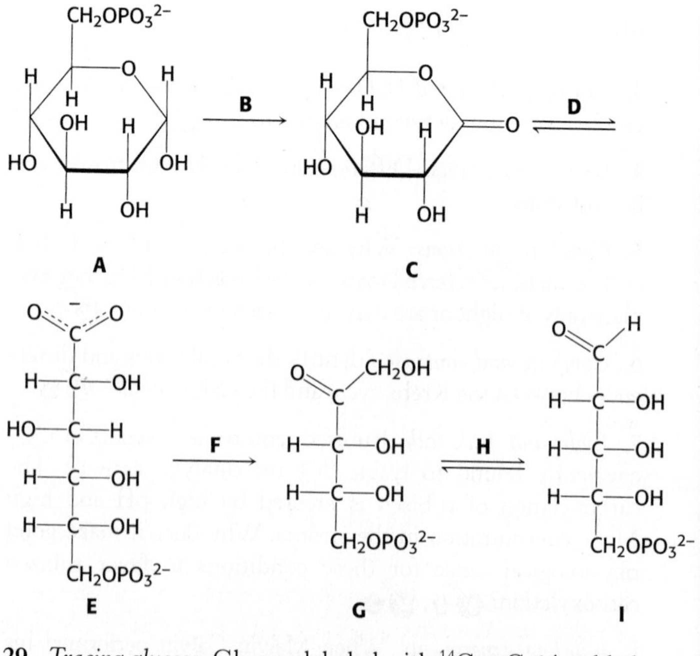

29. *Tracing glucose.* Glucose labeled with ^{14}C at C-6 is added to a solution containing the enzymes and cofactors of the oxidative phase of the pentose phosphate pathway. What is the fate of the radioactive label? ✓③

30. *Recurring decarboxylations.* Which reaction in the citric acid cycle is most analogous to the oxidative decarboxylation of 6-phosphogluconate to ribulose 5-phosphate? What kind of enzyme-bound intermediate is formed in both reactions?

31. *Synthetic stoichiometries.* What is the stoichiometry of the synthesis of (a) ribose 5-phosphate from glucose 6-phosphate without the concomitant generation of NADPH? (b) NADPH from glucose 6-phosphate without the concomitant formation of pentose sugars? ✓❸

32. *Offal or awful?* Liver and other organ meats contain large quantities of nucleic acids. In the course of digestion, RNA is hydrolyzed to ribose, among other chemicals. Explain how ribose can be used as a fuel. ✓❸

33. *A required ATP.* The metabolism of glucose 6-phosphate into ribose 5-phosphate by the joint efforts of the pentose phosphate pathway and glycolysis can be summarized by the following equation.

5 Glucose 6-phosphate + ATP $\longrightarrow$

6 ribose 5-phosphate + ADP

Which reaction requires the ATP? ✓❸

34. *No respiration.* Glucose is normally completely oxidized to CO_2 in the mitochondria. In what circumstance can glucose be completely oxidized to CO_2 in the cytoplasm? ✓❸, ✓❹

35. *Watch your diet, doctor.* The noted psychiatrist Hannibal Lecter once remarked to FBI Agent Clarice Starling that he enjoyed liver with some fava beans and a nice Chianti. Why might this diet be dangerous for some people? ✓❹

36. *No redundancy.* Why do deficiencies in glucose 6-phosphate dehydrogenase frequently present as anemia? ✓❹

37. *Damage control.* What is the role of glutathione in protection against damage by reactive peroxides? Why is the pentose phosphate pathway crucial to this protection?

38. *Reductive power.* What ratio of NADPH to $NADP^+$ is required to sustain [GSH] = 10 mM and [GSSG] = 1 mM? Use the redox potentials given in Table 18.1.

Mechanism Problems

39. *An alternative approach.* The mechanisms of some aldolases do not include Schiff-base intermediates. Instead, these enzymes require bound metal ions. Propose such a mechanism for the conversion of dihydroxyacetone phosphate and glyceraldehyde 3-phosphate into fructose 1,6-bisphosphate.

40. *A recurring intermediate.* Phosphopentose isomerase interconverts the aldose ribose 5-phosphate and the ketose ribulose 5-phosphate. Propose a mechanism.

Chapter Integration Problems

41. *Catching carbons.* Radioactive-labeling experiments can yield estimates of how much glucose 6-phosphate is metabolized by the pentose phosphate pathway and how much is metabolized by the combined action of glycolysis and the citric acid cycle. Suppose that you have samples of two different tissues as well as two radioactively labeled glucose samples, one with glucose labeled with ^{14}C at C-1 and the other with glucose labeled with ^{14}C at C-6. Design an experiment that would enable you to determine the relative activity of the aerobic metabolism of glucose compared with metabolism by the pentose phosphate pathway. ✓❸

42. *You do what you can do.* Red blood cells lack mitochondria. These cells process glucose to lactate, but they also generate CO_2. What is the purpose of producing lactate? How can red blood cells generate CO_2 if they lack mitochondria?

43. *Better to focus on one.* Rubisco catalyzes both a carboxylation reaction and a wasteful oxygenase reaction. Below are the kinetic parameters for the two reactions.

$K_M^{CO_2}$ (μM)	$K_M^{O_2}$ (μM)	$k_{cat}^{CO_2}$ (s^{-1})	$k_{cat}^{O_2}$ (s^{-1})
10	500	3	2

(a) Determine the values of $k_{cat}^{CO_2}/K_M^{CO_2}$ and $k_{cat}^{O_2}/K_M^{O_2}$ as $s^{-1}M^{-1}$.

(b) In light of the k_{cat}/K_M values for the two reactions, why does the oxygenation reaction occur?

44. *Photosynthetic efficiency.* Use the following information to estimate the efficiency of photosynthesis.

The $\Delta G°'$ for the reduction of CO_2 to the level of hexose is $+ 477$ kJ mol^{-1} ($+114$ kcal mol^{-1}).

A mole of 600-nm photons has an energy content of 199 kJ (47.6 kcal).

Assume that the proton gradient generated in producing the required NADPH is sufficient to drive the synthesis of the required ATP.

45. *A violation of the First Law?* The complete combustion of glucose to CO_2 and H_2O yields 30 ATP, as shown in Table 18.4. However, the synthesis of glucose requires only 18 ATP. How is it possible that glucose synthesis from CO_2 and H_2O requires only 18 ATP, but combustion to CO_2 and H_2O yields 30 ATP? Is it a violation of the First Law of Thermodynamics, or perhaps a miracle?

Data Interpretation Problem

46. *Deciding between 3 and 4.* Graph A shows the photosynthetic activity of two species of plant—one a C_4 plant and the other a C_3 plant—as a function of leaf temperature.

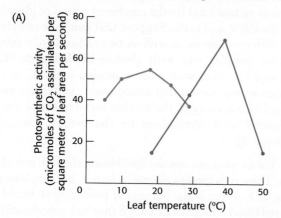

(A)

(a) Which data were most likely generated by the C_4 plant and which by the C_3 plant? Explain.

(b) Suggest some possible explanations for why the photosynthetic activity falls at higher temperatures.

Graph B illustrates how the photosynthetic activity of C_3 and C_4 plants varies with CO_2 concentration when temperature (30°C) and light intensity (high) are constant.

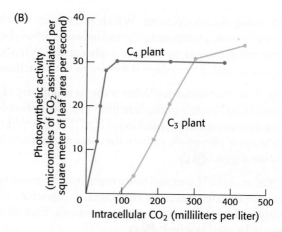

(B)

(c) Why can C_4 plants thrive at CO_2 concentrations that do not support the growth of C_3 plants?

(d) Suggest a plausible explanation why C_3 plants continue to increase photosynthetic activity at higher CO_2 concentrations, whereas C_4 plants reach a plateau.

Think ▸ Pair ▸ Share Problems

47. Explain the differences between the light reactions and the dark reactions of photosynthesis and describe the relationship between these two sets of reactions.

48. Discuss the similarities and differences between the Calvin cycle and the pentose phosphate pathway.

Glycogen Metabolism

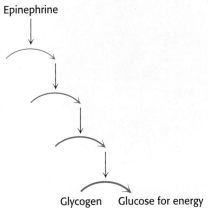

Signaling cascades lead to the mobilization of glycogen to produce glucose, an energy source for cyclists.
[(Left) Jonathan Devich/Epic Images.]

✔ LEARNING OBJECTIVES

By the end of this chapter, you should be able to:

1. List and describe the steps of glycogen breakdown and identify the enzymes required.

2. Explain the regulation of glycogen breakdown.

3. Describe the steps of glycogen synthesis and identify the enzymes required.

4. Explain the regulation of glycogen synthesis.

5. Describe how glycogen degradation and synthesis are coordinated.

6. Compare the different roles of glycogen metabolism in liver and muscle.

OUTLINE

21.1 Glycogen Breakdown Requires the Interplay of Several Enzymes

21.2 Phosphorylase Is Regulated by Allosteric Interactions and Reversible Phosphorylation

21.3 Epinephrine and Glucagon Signal the Need for Glycogen Breakdown

21.4 Glycogen Is Synthesized and Degraded by Different Pathways

21.5 Glycogen Breakdown and Synthesis Are Reciprocally Regulated

Glucose is an important fuel and, as we will see, a key precursor for the biosynthesis of many molecules. However, glucose cannot be stored, because high concentrations of glucose disrupt the osmotic balance of the cell, which would cause cell damage or death. How can adequate stores of glucose be maintained without damaging the cell? The solution to this problem is to store glucose as a nonosmotically active polymer called *glycogen*.

Glycogen is a *readily mobilized storage form of glucose*. It is a very large, branched polymer of glucose residues that can be broken down to yield glucose molecules when energy is needed (Figure 21.1). A glycogen molecule has approximately 12 layers of glucose molecules and can be as large as 40 nm, containing approximately 55,000 glucose residues. Most of the glucose residues in glycogen are linked by α-1,4-glycosidic bonds (Figure 21.2). Branches at about every 12 residues are created by α-1,6-glycosidic bonds. Recall that α-glycosidic linkages form open

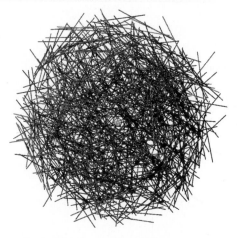

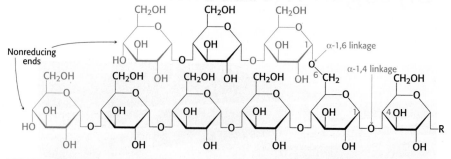

FIGURE 21.2 Glycogen structure. In this structure of two outer branches of a glycogen molecule, the residues at the nonreducing ends are shown in red and the residue that starts a branch is shown in green. The rest of the glycogen molecule is represented by R.

FIGURE 21.1 Glycogen. At the core of the glycogen molecule is the protein glycogenin (yellow). Each line represents glucose molecules joined by α-1,4-glycosidic linkages. The nonreducing ends of the glycogen molecule form the surface of the glycogen granule. Degradation takes place at this surface. [Information from R. Melendez et al., *Biophys. J.* 77:1327–1332, 1999.]

helical polymers, whereas β linkages produce nearly straight strands that form structural fibrils, as in cellulose (Figure 11.14).

Glycogen is not as reduced as fatty acids are and consequently is not as energy rich. Why isn't all excess fuel stored as fatty acids rather than as glycogen? The controlled release of glucose from glycogen maintains blood-glucose concentration between meals. The circulating blood keeps the brain supplied with glucose, which is virtually the only fuel used by the brain, except during prolonged starvation (Chapter 27). Moreover, the readily mobilized glucose from glycogen is a good source of energy for sudden, strenuous activity. Unlike fatty acids, the released glucose can be metabolized in the absence of oxygen and can thus supply energy for anaerobic activity (Section 16.1).

Glycogen is present in bacteria, archaea, and eukaryotes. Recall that plants store glucose as starch, a similar chemical. Thus, storing energy as glucose polymers is common to all forms of life. In humans, most tissues have some glycogen, although the two major sites of glycogen storage are the liver and skeletal muscle. The concentration of glycogen is higher in the liver than in muscle (10% versus 2% by weight), but more glycogen is stored in skeletal muscle overall because of muscle's much greater mass. Glycogen is present in the cytoplasm, with the molecule appearing as granules (Figure 21.3). In the liver, glycogen synthesis and degradation are regulated to maintain the blood-glucose concentration as required to meet the needs of the organism as a whole. In contrast, in muscle, these processes are regulated to meet the energy needs of the muscle itself.

Glycogen metabolism is the regulated release and storage of glucose

Glycogen degradation and synthesis are simple biochemical processes. Glycogen degradation consists of three steps: (1) the release of glucose 1-phosphate from glycogen, (2) the remodeling of the glycogen substrate to permit further degradation, and (3) the conversion of glucose 1-phosphate into glucose 6-phosphate for further metabolism. The glucose 6-phosphate derived from the breakdown of glycogen has three possible fates (Figure 21.4): (1) it can be metabolized by glycolysis, (2) it can be converted into free glucose for release into the bloodstream, and (3) it can be processed by the pentose phosphate pathway to yield NADPH and ribose derivatives. The conversion of glycogen into free glucose occurs mainly in the liver.

Glycogen synthesis, which takes place when glucose is abundant, requires an activated form of glucose, uridine diphosphate glucose (UDP-glucose), formed by the reaction of UTP and glucose 1-phosphate. As is the case for glycogen degradation, the glycogen molecule must be remodeled for continued synthesis.

Glycogen granules

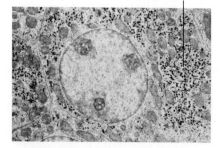

FIGURE 21.3 Electron micrograph of a liver cell. The dark particles in the cytoplasm are glycogen granules. [Courtesy of Dr. George Palade/Yale University, Harvey Cushing/John Hay Whitney Medical Library.]

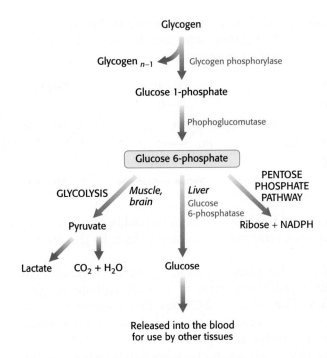

FIGURE 21.4 Fates of glucose 6-phosphate. Glucose 6-phosphate derived from glycogen can be (1) used as a fuel for anaerobic or aerobic metabolism as in, for instance, muscle; (2) converted into free glucose in the liver and subsequently released into the blood; or (3) processed by the pentose phosphate pathway to generate NADPH and ribose in a variety of tissues.

The regulation of glycogen degradation and synthesis is complex, in part because all of the enzymes involved in glycogen metabolism and its regulation are associated with the glycogen particle. Several enzymes taking part in glycogen metabolism allosterically respond to metabolites that signal the energy needs of the cell. *Through these allosteric responses, enzyme activity is adjusted to meet the needs of the cell.* In addition, hormones may initiate signal cascades that lead to the reversible phosphorylation of enzymes, which alters their catalytic rates. *Regulation by hormones adjusts glycogen metabolism to meet the needs of the entire organism.*

21.1 Glycogen Breakdown Requires the Interplay of Several Enzymes

The efficient breakdown of glycogen to provide glucose 6-phosphate for further metabolism requires four enzyme activities: one to degrade glycogen, two to remodel glycogen so that it can remain a substrate for degradation, and one to convert the product of glycogen breakdown into a form suitable for further metabolism. We will examine each of these activities in turn.

Phosphorylase catalyzes the phosphorolytic cleavage of glycogen to release glucose 1-phosphate

Glycogen phosphorylase, the key enzyme in glycogen breakdown, cleaves its substrate by the addition of orthophosphate (P_i) to yield *glucose 1-phosphate.* The cleavage of a bond by the addition of orthophosphate is referred to as *phosphorolysis.*

$$\text{Glycogen} + P_i \rightleftharpoons \text{glucose 1-phosphate} + \text{glycogen}$$
$$(n \text{ residues}) \qquad\qquad\qquad\qquad (n-1 \text{ residues})$$

Phosphorylase catalyzes the sequential removal of glucosyl residues from the nonreducing ends of the glycogen molecule (the ends with a free OH group on carbon 4). Orthophosphate splits the glycosidic linkage between

C-1 of the terminal residue and C-4 of the adjacent one. Specifically, it cleaves the bond between the C-1 carbon atom and the glycosidic oxygen atom, and the α configuration at C-1 is retained.

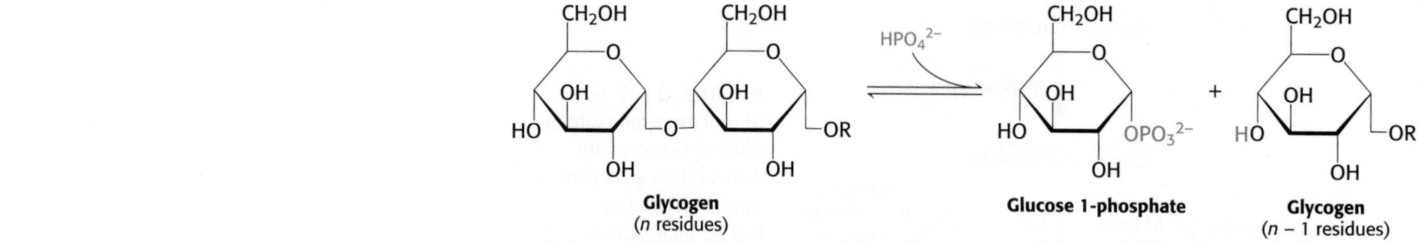

Glycogen (n residues)   **Glucose 1-phosphate**   **Glycogen** (n − 1 residues)

Glucose 1-phosphate released from glycogen can be readily converted into glucose 6-phosphate, an important metabolic intermediate, by the enzyme phosphoglucomutase.

The reaction catalyzed by phosphorylase is readily reversible in vitro. At pH 6.8, the equilibrium ratio of orthophosphate to glucose 1-phosphate is 3.6. The value of $\Delta G^{\circ\prime}$ for this reaction is small because a glycosidic bond is replaced by a phosphoryl ester bond that has a nearly equal transfer potential. However, phosphorolysis proceeds far in the direction of glycogen breakdown in vivo because the $[P_i]/[$glucose 1-phosphate$]$ ratio is usually greater than 100, substantially favoring phosphorolysis. We see here an example of how the cell can alter the free-energy change to favor a reaction's occurrence by altering the ratio of substrate and product.

The phosphorolytic cleavage of glycogen is energetically advantageous because the released sugar is already phosphorylated. In contrast, a hydrolytic cleavage would yield glucose, which would then have to be phosphorylated at the expense of a molecule of ATP to enter the glycolytic pathway. An additional advantage of phosphorolytic cleavage for muscle cells is that no transporters exist for glucose 1-phosphate, which is negatively charged under physiological conditions, and so it cannot be transported or diffuse out of the cell.

Mechanism: Pyridoxal phosphate participates in the phosphorolytic cleavage of glycogen

The special challenge faced by phosphorylase is to cleave glycogen phosphorolytically rather than hydrolytically to save the ATP required to phosphorylate free glucose. Thus, water must be excluded from the active site. Phosphorylase is a dimer of two identical 97-kDa subunits. Each subunit is compactly folded into an *amino-terminal domain* (480 residues) containing a *glycogen-binding site* and a *carboxyl-terminal domain* (360 residues; Figure 21.5). The catalytic site in each subunit is located in a deep crevice formed by residues from both domains. Substrates bind synergistically, causing the crevice to narrow, thereby excluding water. What is the mechanistic basis of the phosphorolytic cleavage of glycogen?

Several clues suggest a mechanism. First, both the glycogen substrate and the glucose 1-phosphate product have an α configuration at C-1. A direct attack by phosphate on C-1 of a sugar would invert the configuration at this carbon atom because the reaction would proceed through a pentacovalent transition state. The observation that the glucose 1-phosphate formed has an α rather than a β configuration suggests that an even number

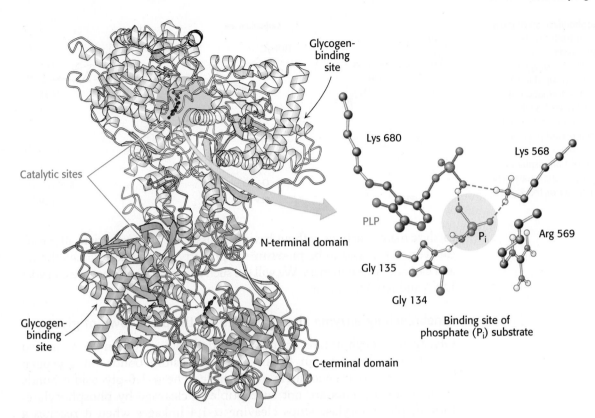

FIGURE 21.5 Structure of glycogen phosphorylase. This enzyme forms a homodimer: one subunit is shown in white and the other in yellow. Each catalytic site includes a pyridoxal phosphate (PLP) group, linked to lysine 680 of the enzyme. The binding site for the phosphate (P$_i$) substrate is shown. *Notice* that the catalytic site lies between the C-terminal domain and the glycogen-binding site. A narrow crevice, which binds four or five glucose units of glycogen, connects the two sites. The separation of the sites allows the catalytic site to phosphorolyze several glucose units before the enzyme must rebind the glycogen substrate. [Drawn from 1NOI.pdb.]

of steps (most simply, two) is required. A likely explanation for these results is that a *carbonium ion intermediate* is formed from the glucose residue.

A second clue to the catalytic mechanism of phosphorylase is its requirement for the coenzyme *pyridoxal phosphate* (PLP), a derivative of pyridoxine (vitamin B$_6$, Section 15.4). The aldehyde group of this coenzyme forms a Schiff-base linkage with a specific lysine side chain of the enzyme (Figure 21.6). Structural studies indicate that the reacting orthophosphate group takes a position between the 5′-phosphate group of PLP and the glycogen substrate (Figure 21.7). *The 5′-phosphate group of PLP acts in tandem with orthophosphate by serving as a proton donor and then as a proton acceptor (i.e., as a general acid–base catalyst).* Orthophosphate (in the HPO$_4^{2-}$ form) donates a proton to the oxygen atom attached to carbon 4 of the departing glycogen chain and simultaneously acquires a proton from PLP. The carbonium ion intermediate formed in this step is then attacked by orthophosphate to form α-glucose 1-phosphate, with the concomitant return of a proton to pyridoxal phosphate.

The glycogen-binding site is 30 Å away from the catalytic site (Figure 21.5), but it is connected to the catalytic site by a narrow crevice able to accommodate four or five glucose units. The large separation between the binding site and the catalytic site enables the enzyme to phosphorolyze many residues without having to dissociate and reassociate after each catalytic cycle. An enzyme that can

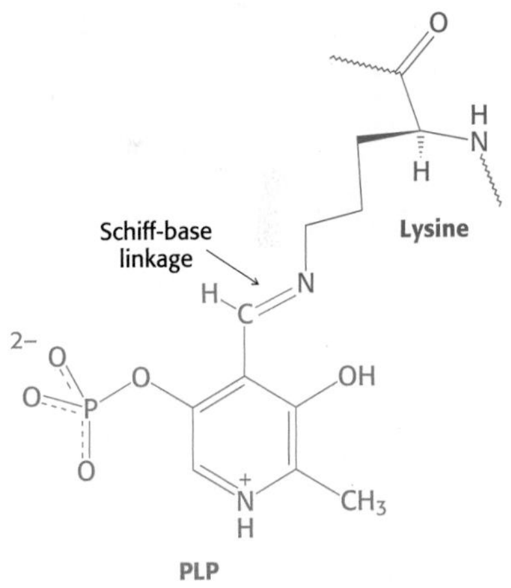

FIGURE 21.6 PLP–Schiff-base linkage. A pyridoxal phosphate (PLP) group (red) forms a Schiff base with a lysine residue (blue) at the active site of phosphorylase, where it functions as a general acid–base catalyst.

A Schiff base, also called an *imine,* is a compound containing a carbon–nitrogen double bond with the nitrogen atom bonded to an organic compound, not a hydrogen atom. A Schiff base is formed by the reaction of a primary amine with an aldehyde or ketone.

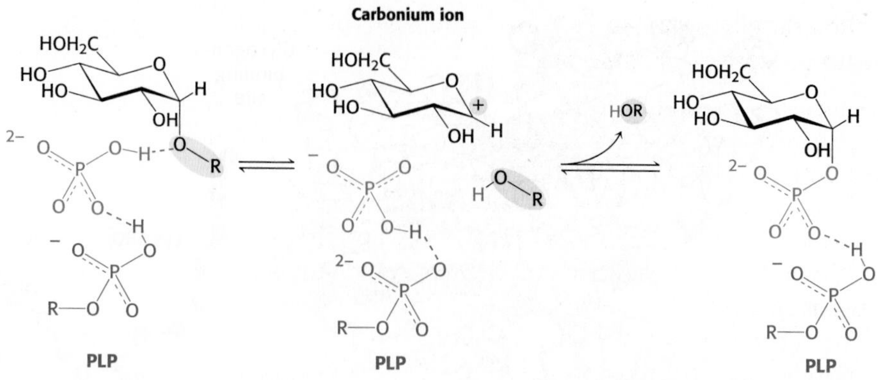

FIGURE 21.7 Phosphorylase mechanism. A bound HPO_4^{2-} group (red) favors the cleavage of the glycosidic bond by donating a proton to the C-4 oxygen of the departing glycosyl group (black). This reaction results in the formation of a carbonium ion and is favored by the transfer of a proton from the protonated phosphate group of the bound pyridoxal phosphate (PLP) group (blue). The carbonium ion and the orthophosphate combine to form glucose 1-phosphate. The 'R' represents the remainder of the PLP coenzyme.

catalyze many reactions without having to dissociate and reassociate after each catalytic step is said to be *processive*—a property of enzymes that synthesize and degrade large polymers. We will see such enzymes again when we consider DNA and RNA synthesis.

A debranching enzyme also is needed for the breakdown of glycogen

Glycogen phosphorylase acting alone degrades glycogen to a limited extent. The enzyme can break α-1,4-glycosidic bonds on glycogen branches but soon encounters an obstacle. The α-1,6-glycosidic bonds at the branch points are not susceptible to cleavage by phosphorylase. Indeed, phosphorylase stops cleaving α-1,4 linkages when it reaches a terminal residue four residues away from a branch point. Because about 1 in 12 residues is branched, cleavage by the phosphorylase alone would come to a halt after the release of eight glucose molecules per branch.

How can the remainder of the glycogen molecule be mobilized for use as a fuel? Two additional enzymes, a *transferase* and *α-1,6-glucosidase*, remodel the glycogen for continued degradation by the phosphorylase (Figure 21.8). *The transferase shifts a block of three glucosyl residues from one outer branch to another.* This transfer exposes a single glucose residue joined by an α-1,6-glycosidic linkage. α-1,6-glucosidase, also known as the debranching enzyme, hydrolyzes the α-1,6-glycosidic bond.

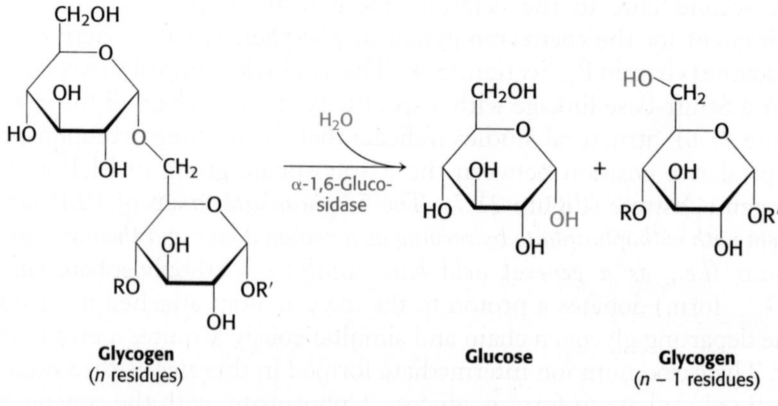

Glycogen (*n* residues)   **Glucose**   **Glycogen** (*n* − 1 residues)

A free glucose molecule is released and then phosphorylated by the glyco-lytic enzyme hexokinase, if the glucose is to be processed by glycolysis or the pentose phosphate pathway. Thus, the transferase and α-1,6-glucosidase convert the branched structure into a linear one, which paves the way for further cleavage by phosphorylase. In eukaryotes, the transferase and the α-1,6-glucosidase activities are present in a single 160-kDa polypeptide chain, providing yet another example of a bifunctional enzyme (Figure 16.29).

Phosphoglucomutase converts glucose 1-phosphate into glucose 6-phosphate

Glucose 1-phosphate formed in the phosphorolytic cleavage of glycogen must be converted into glucose 6-phosphate to enter the metabolic mainstream. This shift of a phosphoryl group is catalyzed by *phosphoglucomutase*. Recall that this enzyme is also used in galactose metabolism (Section 16.1). To effect this shift, the enzyme exchanges a phosphoryl group with the substrate (Figure 21.9). The catalytic site of an active mutase molecule contains a phosphorylated serine residue. The phosphoryl group is transferred from the serine residue to the C-6 hydroxyl group of glucose 1-phosphate to form glucose 1,6-bisphosphate. The C-1 phosphoryl group of this intermediate is then shuttled to the same serine residue, resulting in the formation of glucose 6-phosphate and the regeneration of the phosphoenzyme.

These reactions are like those of *phosphoglycerate mutase*, a glycolytic enzyme (Section 16.1). The role of glucose 1,6-bisphosphate in the interconversion of the phosphoglucoses is like that of 2,3-bisphosphoglycerate (2,3-BPG) in the interconversion of 2-phosphoglycerate and 3-phosphoglycerate in glycolysis. A phosphoenzyme intermediate participates in both reactions.

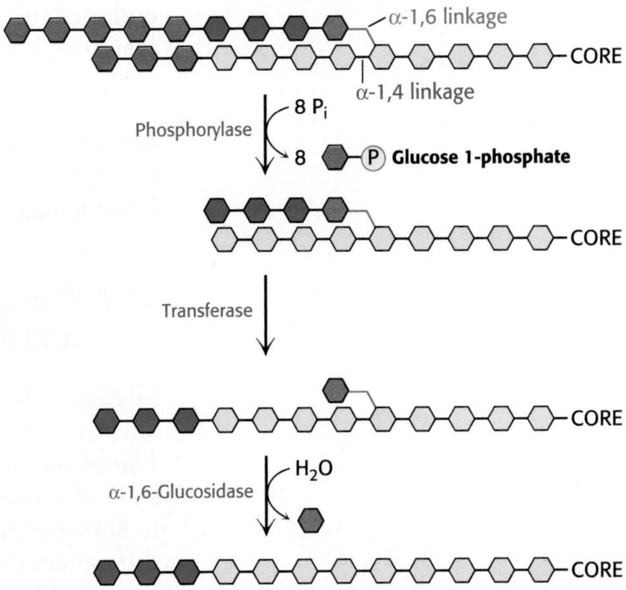

FIGURE 21.8 Glycogen remodeling. First, α-1,4-glycosidic bonds on each branch are cleaved by phosphorylase, leaving four residues along each branch. The transferase shifts a block of three glucosyl residues from one outer branch to the other. In this reaction, the α-1,4-glycosidic link between the blue and the green residues is broken and a new α-1,4 link between the blue and the yellow residues is formed. The green residue is then removed by α-1,6-glucosidase, leaving a linear chain with all α-1,4 linkages, suitable for further cleavage by phosphorylase.

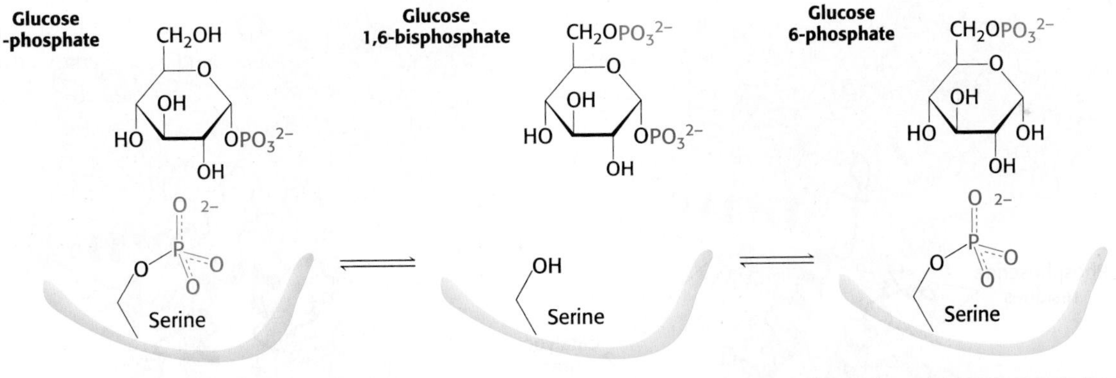

FIGURE 21.9 Reaction catalyzed by phosphoglucomutase. A phosphoryl group is transferred from the enzyme to the substrate, and a different phosphoryl group is transferred back to restore the enzyme to its initial state.

The liver contains glucose 6-phosphatase, a hydrolytic enzyme absent from muscle

A major function of the liver is to maintain a nearly constant concentration of glucose in the blood. The liver releases glucose into the blood between meals and during muscular activity. The released glucose is taken up primarily by the brain, skeletal muscle, and red blood cells. However, the phosphorylated glucose produced by glycogen breakdown is not transported out of cells. The liver contains a hydrolytic enzyme, *glucose 6-phosphatase*, that converts glucose 6-phosphate to glucose, which can leave the liver. This enzyme cleaves the phosphoryl group to form free glucose and orthophosphate. This glucose 6-phosphatase is the same enzyme that releases free glucose at the conclusion of gluconeogenesis. It is located on the lumenal side of the smooth endoplasmic reticulum membrane. Recall that glucose 6-phosphate is transported into the endoplasmic reticulum; glucose and orthophosphate formed by hydrolysis are then shuttled back into the cytoplasm (Section 16.3).

$$\text{Glucose 6-phosphate} + H_2O \longrightarrow \text{glucose} + P_i$$

Glucose 6-phosphatase is absent from most other tissues. Muscle tissues retain glucose 6-phosphate for the generation of ATP. In contrast, glucose is not a major fuel for the liver.

21.2 Phosphorylase Is Regulated by Allosteric Interactions and Reversible Phosphorylation

Glycogen degradation is precisely controlled by multiple interlocking mechanisms. The focus of this control is the enzyme glycogen phosphorylase. *Phosphorylase is regulated by several allosteric effectors that signal the energy state of the cell as well as by reversible phosphorylation, which is responsive to hormones such as insulin, epinephrine, and glucagon.* We will examine the differences in the control of two isozymic forms of glycogen phosphorylase: one specific to liver and one specific to skeletal muscle. These differences are because *the liver maintains glucose homeostasis of the organism as a whole, whereas the muscle uses glucose to produce energy for itself.*

Liver phosphorylase produces glucose for use by other tissues

The dimeric phosphorylase exists in two interconvertible forms: a *usually active* phosphorylated *a* form and a *usually inactive* unphosphorylated *b* form (Figure 21.10). Each of these two forms exists in equilibrium between

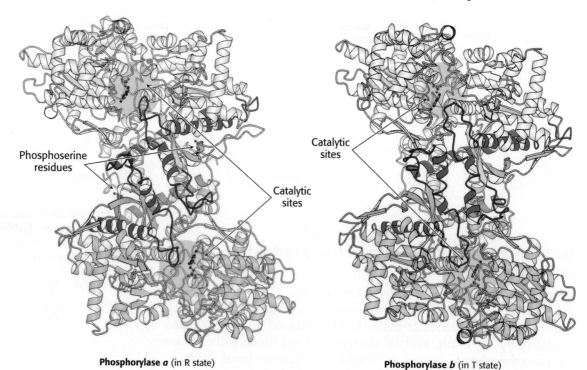

Phosphorylase *a* (in R state)

Phosphorylase *b* (in T state)

FIGURE 21.10 Structures of phosphorylase *a* and phosphorylase *b*. Phosphorylase *a* is phosphorylated on serine 14 of each subunit. This modification favors the structure of the more-active R state. One subunit is shown in white, with helices and loops important for regulation shown in blue and red. The other subunit is shown in yellow, with the regulatory structures shown in orange and green. Phosphorylase *b* is not phosphorylated and exists predominantly in the T state. *Notice* that the catalytic sites are partly occluded in the T state. [Drawn from 1GPA.pdb and 1NOJ.pdb.]

an active relaxed (R) state and a much less active tense (T) state, but the equilibrium for phosphorylase *a* favors the active R state, whereas the equilibrium for phosphorylase *b* favors the less-active T state (Figure 21.11). The role of glycogen degradation in the liver is to form glucose for export to other tissues when the blood-glucose concentration is low. Consequently, we can think of the default state of liver phosphorylase as being the *a* form: glucose is to be generated unless the enzyme is signaled otherwise. The liver phosphorylase *a* form thus exhibits the most responsive R ↔ T transition (Figure 21.12). The binding of glucose to the active site shifts the *a* form from the active R state to the less-active T state. In essence, the enzyme reverts to the low-activity T state only when it detects the presence of sufficient glucose. If glucose is present in the diet, there is no need to degrade glycogen.

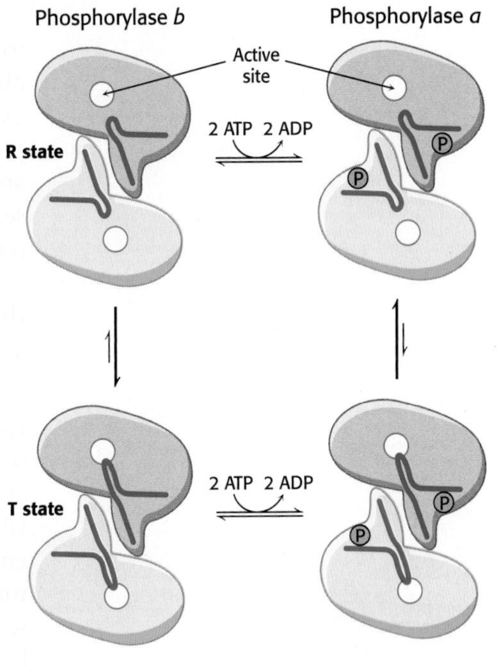

FIGURE 21.11 Phosphorylase regulation. Both phosphorylase *b* and phosphorylase *a* exist as equilibria between an active R state and a less-active T state. Phosphorylase *b* is usually inactive because the equilibrium favors the T state. Phosphorylase *a* is usually active because the equilibrium favors the R state. Regulatory structures are shown in blue and green.

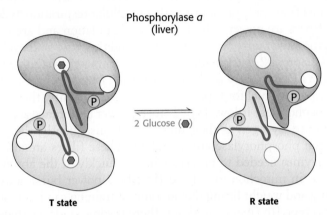

FIGURE 21.12 Allosteric regulation of liver phosphorylase. The binding of glucose to phosphorylase *a* shifts the equilibrium to the T state and inactivates the enzyme. Thus, glycogen is not mobilized when glucose is already abundant.

Muscle phosphorylase is regulated by the intracellular energy charge

In contrast to the liver isozyme, the default state of muscle phosphorylase is the *b* form, owing to the fact that, for muscle, phosphorylase must be active primarily during muscle contraction. Muscle phosphorylase *b* is activated by the presence of high concentrations of AMP, which binds to a nucleotide-binding site and stabilizes the conformation of phosphorylase *b* in the active R state (Figure 21.13). Thus, when a muscle contracts and ATP is converted into AMP by the sequential action of myosin (Section 9.4) and adenylate kinase (Section 16.2), the phosphorylase is signaled to degrade glycogen. ATP acts as a negative allosteric effector by competing with AMP. Thus, *the transition of phosphorylase* b *between the active R state and the less-active T state is controlled by the energy charge of the muscle cell.* If ATP is unavailable, glucose 6-phosphate may bind at the ATP-binding site and stabilize the less-active state of phosphorylase *b*, an example of feedback inhibition. In resting muscle, *phosphorylase* b *is inactive because of the inhibitory effects of ATP and glucose 6-phosphate.* In contrast, *phosphorylase* a *is fully active,* regardless of the concentration of AMP, ATP, and glucose 6-phosphate.

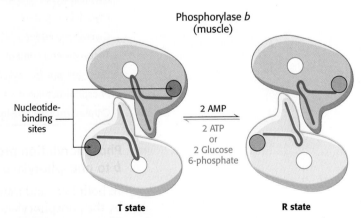

FIGURE 21.13 Allosteric regulation of muscle phosphorylase. A low energy charge, represented by high concentrations of AMP, favors the transition to the R state. ATP and glucose 6-phosphate stabilize the T state.

Unlike the enzyme in muscle, the liver phosphorylase is insensitive to regulation by AMP because the liver does not undergo the dramatic changes in energy charge seen in a contracting muscle. We see here a clear example of the use of isozymes to establish the tissue-specific biochemical properties of the liver and muscle. In human beings, liver phosphorylase and muscle phosphorylase are approximately 90% identical in amino acid sequence, yet the 10% difference results in subtle but important shifts in the regulation of the two enzymes.

Biochemical characteristics of muscle fiber types differ

Not only do the biochemical needs of liver and muscle differ with respect to glycogen metabolism, but the biochemical needs of different muscle fiber types also vary. Skeletal muscle consists of three different fiber types: type I, or slow-twitch muscle; type IIb (also called type IIx), or fast-twitch fibers; and type IIa fibers that have properties intermediate between the other two fiber types. Type I fibers rely predominantly on cellular respiration to derive energy. These fibers are powered by fatty acid degradation and are rich in mitochondria, the site of fatty acid degradation and the citric acid cycle. As we will see in Chapter 22, fatty acids are an excellent energy storage form, but generating ATP from fatty acids is slower than from glycogen. Glycogen is not an important fuel for type I fibers, and consequently the amount of glycogen phosphorylase is low. Type I fibers power endurance activities. Type IIb fibers use glycogen as their main fuel. Consequently, the glycogen and glycogen phosphorylase are abundant. These fibers are also rich in glycolytic enzymes needed to process glucose quickly in the absence of oxygen and poor in mitochondria. Type IIb fibers power burst activities such as sprinting and weight lifting. No amount of training can interconvert type I fibers and type IIb fibers. However, there is some evidence that type IIa fibers are "trainable"; that is, endurance training enhances the oxidative capacity of type IIa fibers while burst activity training enhances the glycolytic capacity. Table 21.1 shows the biochemical profile of the fiber types.

TABLE 21.1 Biochemical characteristics of muscle fiber types

Characteristic	Type I	Type IIa	Type IIb
Fatigue resistance	High	Intermediate	Low
Mitochondrial density	High	Intermediate	Low
Metabolic type	Oxidative	Oxidative/glycolytic	Glycolytic
Myoglobin content	High	Intermediate	Low
Glycogen content	Low	Intermediate	High
Triacylglycerol content	High	Intermediate	Low
Glycogen phosphorylase activity	Low	Intermediate	High
Phosphofructokinase activity	Low	Intermediate	High
Citrate synthase activity	High	Intermediate	Low

Phosphorylation promotes the conversion of phosphorylase *b* to phosphorylase *a*

In both liver and muscle, phosphorylase *b* is converted into phosphorylase *a* by the phosphorylation of a single serine residue (serine 14) in each subunit. This conversion is initiated by hormones. Low blood concentration of glucose leads to the secretion of the hormone glucagon. The rise in glucagon concentration results in phosphorylation of the enzyme, converting it to the phosphorylase *a* form in liver. Emotions like the excitement of exercise or fear will cause the concentration of the hormone epinephrine to increase. Epinephrine binds to receptors in muscle (and in liver), again inducing the phosphorylation of phosphorylase *b* to phosphorylase *a*. The regulatory enzyme *phosphorylase kinase* catalyzes this covalent modification.

Comparison of the structures of phosphorylase *a* in the R state and phosphorylase *b* in the T state reveals that subtle structural changes at the subunit interfaces are transmitted to the active sites (Figure 21.10). The transition from the T state (the prevalent state of phosphorylase *b*) to the R state (the prevalent state of phosphorylase *a*) entails a 10-degree rotation around the twofold axis of the dimer. Most important, this transition is associated with structural changes in α helices that move a loop out of the active site of each subunit. Thus, the T state is less active because the catalytic site is partly blocked. In the R state, the catalytic site is more accessible and a binding site for orthophosphate is well organized.

Phosphorylase kinase is activated by phosphorylation and calcium ions

Phosphorylase kinase converts phosphorylase *b* into the *a* form by attaching a phosphoryl group. The subunit composition of phosphorylase kinase in skeletal muscle is $(\alpha\beta\gamma\delta)_4$, and the mass of this very large protein is 1300 kDa. The enzyme consists of two $(\alpha\beta\gamma\delta)_2$ lobes that are joined by a β_4 bridge that is the core of the enzyme and serves as a scaffold for the remaining subunits. The γ subunit contains the active site, while all of the remaining subunits ($\approx 90\%$ by mass) play regulatory roles. The δ subunit is the Ca^{2+} binding protein *calmodulin,* a calcium sensor that stimulates many enzymes in eukaryotes (Section 14.1). The α and β subunits are targets of protein kinase A. The β subunit is phosphorylated first, followed by the phosphorylation of the α subunit.

Activation of phosphorylase kinase is initiated when Ca^{2+} binds to the δ subunit. This mode of activation of the kinase is especially noteworthy in muscle, where contraction is triggered by the release of Ca^{2+} from the sarcoplasmic reticulum (Figure 21.14). Maximal activation is achieved with the phosphorylation of the β and α subunits of the Ca^{2+}-bound kinase. The stimulation of phosphorylase kinase is one step in a signal-transduction cascade initiated by signal molecules such as glucagon and epinephrine.

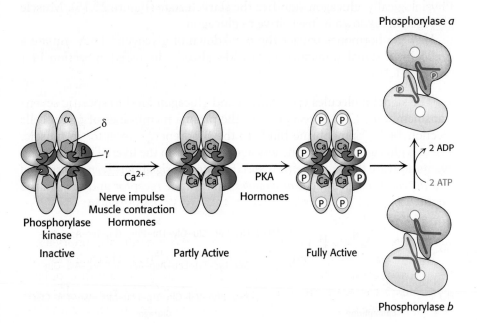

FIGURE 21.14 Activation of phosphorylase kinase. Phosphorylase kinase, an $(\alpha\beta\gamma\delta)_4$ assembly, is partly activated by Ca^{2+} binding to the δ subunits. Activation is maximal when the β and α subunits are phosphorylated in response to hormonal signals. When active, the enzyme converts phosphorylase *b* into phosphorylase *a*.

An isozymic form of glycogen phosphorylase exists in the brain

Recently an isozymic form of glycogen phosphorylase was identified in the brain. Brain glycogen phosphorylase is distinct from the liver and muscle forms, but is more similar to the muscle enzyme than the liver enzyme. Like muscle phosphorylase, brain phosphorylase is stimulated by AMP, but in addition, is regulated by a redox switch, two cysteines that form a disulfide bond. Reactive oxygen species, notably H_2O_2 acting as a signal molecule, lead to the formation of the disulfide bond. The disulfide bond prevents AMP activation of the enzyme, but does not alter its regulation by phosphorylation. Neurons express only the brain isozyme, but astrocytes, cells that facilitate neuron function, express both brain and muscle isoforms.

Brain glycogen phosphorylase may be a target of dithiocarbamates, organosulfur compounds with a variety of uses in industry and agriculture. One such dithiocarbamate, the pesticide thiram, has been linked to neurotoxic poisoning. Dithiocarbamates have broad reactivity, but recent research suggests that thiram disrupts the redox switch of brain phosphorylase, decreasing enzyme activity. The lack of efficient glycogen mobilization in the brain may account for some of the neurological toxicity.

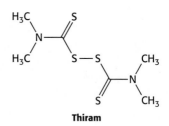

Thiram

21.3 Epinephrine and Glucagon Signal the Need for Glycogen Breakdown

Protein kinase A activates phosphorylase kinase, which in turn activates glycogen phosphorylase. What activates protein kinase A? What is the signal that ultimately triggers an increase in glycogen breakdown?

G proteins transmit the signal for the initiation of glycogen breakdown

Several hormones greatly affect glycogen metabolism. Glucagon and epinephrine trigger the breakdown of glycogen. Muscular activity or its anticipation leads to the release of *epinephrine (adrenaline)*, a catecholamine derived from tyrosine, from the adrenal medulla. Epinephrine markedly stimulates glycogen breakdown in muscle and, to a lesser extent, in the liver. The liver is more responsive to *glucagon*, a polypeptide hormone secreted by the α cells of the pancreas when the blood-sugar concentration is low. Physiologically, glucagon signifies the starved state (Figure 21.15). Muscle glycogen breakdown is insensitive to glucagon.

How do hormones trigger the breakdown of glycogen? They initiate a cyclic AMP signal-transduction cascade, already discussed in Section 14.1 (Figure 21.16).

1. The signal molecules epinephrine and glucagon bind to specific seven-transmembrane (7TM) receptors in the plasma membranes of target cells (Section 14.1). Epinephrine binds to the β-adrenergic receptor in muscle, whereas glucagon binds to the glucagon receptor in the liver. These binding events activate the G_s protein. *A specific external signal has been transmitted into the cell through structural changes,* first in the receptor and then in the G protein.

Epinephrine

$$
\begin{array}{cc}
 & 5 \qquad\qquad 10 \\
^+H_3N\text{--His--Ser--Glu--Gly--Thr--Phe--Thr--Ser--Asp--Tyr--} \\
 & 15 \qquad\qquad 20 \\
\text{--Ser--Lys--Tyr--Leu--Asp--Ser--Arg--Arg--Ala--Gln--} \\
 & 25 \qquad\qquad 29 \\
\text{--Asp--Phe--Val--Gln--Trp--Leu--Met--Asn--Thr--COO}^-
\end{array}
$$

Glucagon

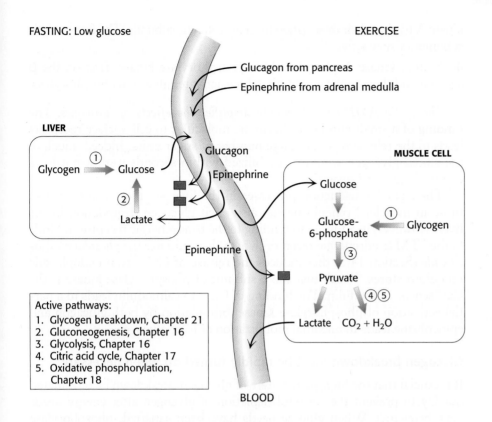

FASTING: Low glucose

EXERCISE

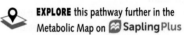

FIGURE 21.15 Pathway integration: Hormonal control of glycogen breakdown. Glucagon stimulates liver-glycogen breakdown when blood glucose is low. Epinephrine enhances glycogen breakdown in muscle and the liver to provide fuel for muscle contraction.

EXPLORE this pathway further in the Metabolic Map on 🟢 Sapling Plus

2. The GTP-bound subunit of G_s activates the transmembrane protein adenylate cyclase. This enzyme catalyzes the formation of the second messenger cyclic AMP (cAMP) from ATP.

3. The elevated cytoplasmic concentration of cAMP activates *protein kinase A* (Section 10.3). The binding of cAMP to inhibitory regulatory subunits of protein

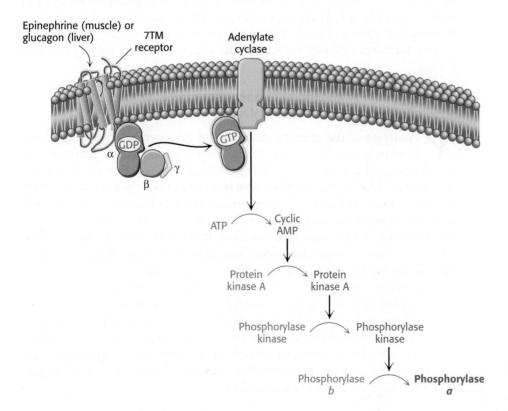

FIGURE 21.16 Regulatory cascade for glycogen breakdown. Glycogen degradation is stimulated by hormone binding to 7TM receptors. Hormone binding initiates a G-protein-dependent signal-transduction pathway that results in the phosphorylation and activation of glycogen phosphorylase. The inactive forms of the enzymes are shown in red and the active forms in blue.

kinase A triggers their dissociation from the catalytic subunits. The free catalytic subunits are now active.

4. Protein kinase A phosphorylates phosphorylase kinase, first on the β subunit and then on the α subunit, which activates glycogen phosphorylase.

The cyclic AMP cascade highly amplifies the effects of hormones. The binding of a small number of hormone molecules to cell-surface receptors leads to the release of a very large number of sugar units. Indeed, much of the stored glycogen would be mobilized within seconds were it not for a counterregulatory system.

The signal-transduction processes in the liver are more complex than those in muscle. Epinephrine can also elicit glycogen degradation in the liver. However, in addition to binding to the β-adrenergic receptor, it binds to the 7TM α-adrenergic receptor, which then initiates the *phosphoinositide cascade* (Section 14.2) that induces the release of Ca^{2+} from endoplasmic reticulum stores. Recall that the δ subunit of phosphorylase kinase is the Ca^{2+} sensor calmodulin. The binding of Ca^{2+} to calmodulin leads to a partial activation of phosphorylase kinase. Stimulation by both glucagon and epinephrine leads to maximal mobilization of liver glycogen.

Glycogen breakdown must be rapidly turned off when necessary

It is crucial that the high-gain system of glycogen breakdown be terminated quickly to prevent the wasteful depletion of glycogen after energy needs have been met. When glucose needs have been satisfied, phosphorylase kinase and glycogen phosphorylase are dephosphorylated and inactivated. Simultaneously, glycogen synthesis is activated.

The signal-transduction pathway leading to the activation of glycogen phosphorylase is shut down automatically when secretion of the initiating hormone ceases. The inherent GTPase activity of the G protein converts the bound GTP into inactive GDP, and phosphodiesterases always present in the cell convert cyclic AMP into AMP. Protein phosphatase 1 (PP1) removes the phosphoryl groups from phosphorylase kinase, thereby inactivating the enzyme. Finally, PP1 also removes the phosphoryl group from glycogen phosphorylase, converting the enzyme into the usually inactive *b* form.

The regulation of glycogen phosphorylase became more sophisticated as the enzyme evolved

Analyses of the primary structures of glycogen phosphorylase from human beings, rats, *Dictyostelium* (slime mold), yeast, potatoes, and *E. coli* have enabled inferences to be made about the evolution of this important enzyme. The 16 residues that come into contact with glucose at the active site are identical in nearly all the enzymes. There is more variation but still substantial conservation of the 15 residues at the pyridoxal phosphate-binding site. Likewise, the glycogen-binding site is well conserved in all the enzymes. The high degree of similarity among these three sites shows that the catalytic mechanism has been maintained throughout evolution.

Differences arise, however, when we compare the regulatory sites. The simplest type of regulation would be feedback inhibition by glucose 6-phosphate. Indeed, the glucose 6-phosphate regulatory site is highly conserved among most of the phosphorylases. The crucial amino acid residues that participate in regulation by phosphorylation and nucleotide binding are well conserved only in the mammalian enzymes. Thus, this level of regulation was a later evolutionary acquisition.

High Gain

Gain is the ratio of the output signal to the input signal. In the case of glycogen metabolism, high gain means that a few hormone signals can rapidly generate large amounts of glucose 6-phosphate.

21.4 Glycogen Synthesis Requires Several Enzymes and Uridine Diphosphate Glucose

As with glycolysis and gluconeogenesis, related biosynthetic and degradative pathways rarely operate by precisely the same reactions in the forward and reverse directions. Glycogen metabolism provided the first known example of this important principle. *Separate pathways afford much greater flexibility, both in energetics and in control.*

Glycogen is synthesized by a pathway that utilizes *uridine diphosphate glucose* (UDP-glucose) as the activated glucose donor.

$$\text{Synthesis: Glycogen}_n + \text{UDP-glucose} \longrightarrow \text{glycogen}_{n+1} + \text{UDP}$$
$$\text{Degradation: Glycogen}_{n+1} + P_i \longrightarrow \text{glycogen}_n + \text{glucose 1-phosphate}$$

**Uridine diphosphate glucose
(UDP-glucose)**

UDP-glucose is an activated form of glucose

UDP-glucose, the glucose donor in the biosynthesis of glycogen, is an *activated form of glucose,* just as ATP and acetyl CoA are activated forms of orthophosphate and acetate, respectively. The C-1 carbon atom of the glucosyl unit of UDP-glucose is activated because its hydroxyl group is esterified to the diphosphate moiety of UDP.

UDP-glucose is synthesized from glucose 1-phosphate and uridine triphosphate (UTP) in a reaction catalyzed by *UDP-glucose pyrophosphorylase.* This reaction liberates the outer two phosphoryl residues of UTP as pyrophosphate.

Glucose 1-phosphate   **UTP**   **UDP-glucose**

This reaction is readily reversible. However, pyrophosphate is rapidly hydrolyzed in vivo to orthophosphate by an inorganic pyrophosphatase. The essentially irreversible hydrolysis of pyrophosphate drives the synthesis of UDP-glucose.

$$\text{Glucose 1-phosphate} + \text{UTP} \rightleftharpoons \text{UDP-glucose} + PP_i$$
$$PP_i + H_2O \longrightarrow 2\,P_i$$
$$\overline{\text{Glucose 1-phosphate} + \text{UTP} + H_2O \longrightarrow \text{UDP-glucose} + 2\,P_i}$$

The synthesis of UDP-glucose exemplifies another recurring theme in biochemistry: *Many biosynthetic reactions are driven by the hydrolysis of pyrophosphate.*

Glycogen synthase catalyzes the transfer of glucose from UDP-glucose to a growing chain

New glucosyl units are added to the nonreducing terminal residues of glycogen. The activated glucosyl unit of UDP-glucose is transferred to the hydroxyl group at C-4 of a terminal residue to form an α-1,4-glycosidic linkage. UDP is displaced by the terminal hydroxyl group of the growing glycogen molecule. This reaction is catalyzed by *glycogen synthase, the key regulatory enzyme in glycogen synthesis.* Humans have two isozymes of glycogen synthase: one is specific to the liver while the other is expressed in muscle and other tissues.

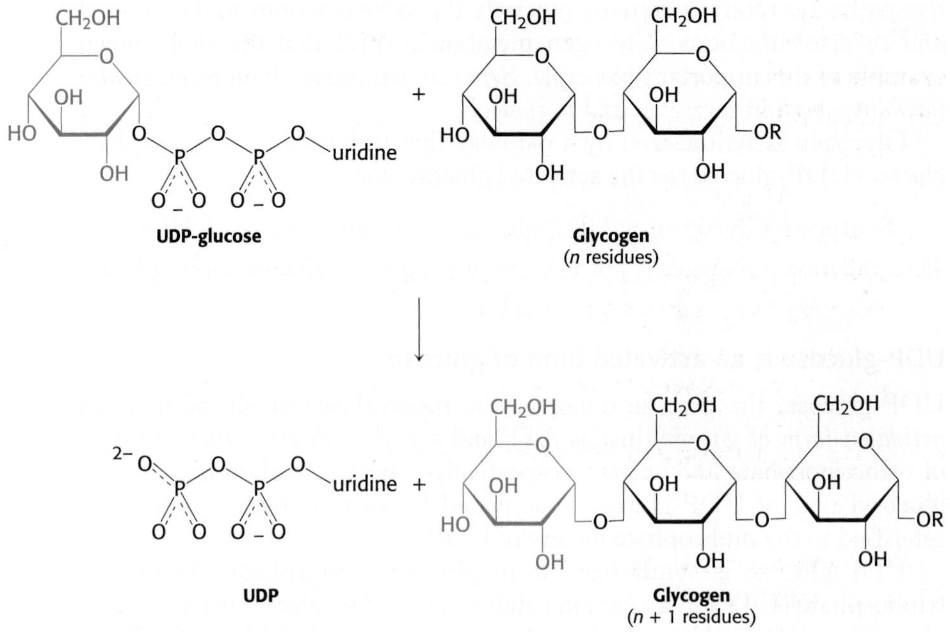

UDP-glucose

Glycogen
(*n* residues)

UDP

Glycogen
(*n* + 1 residues)

Glycogen synthase, a member of the large glycosyltransferase family (Section 11.3), can add glucosyl residues only to a polysaccharide chain already containing at least four residues. Thus, glycogen synthesis requires a *primer*. This priming function is carried out by *glycogenin,* a Mn^{2+}-requiring glycosyltransferase composed of two identical 37-kDa subunits. Each subunit of glycogenin catalyzes the formation of α-1,4-glucose polymers, 10 to 20 glucosyl units in length, which are synthesized stepwise directly on the phenolic hydroxyl group of a specific tyrosine residue in each glycogenin subunit. UDP-glucose is the donor in this autoglycosylation. At this point, glycogen synthase takes over to extend the glycogen molecule. Thus, every glycogen molecule has a glycogenin molecule at its core (Figure 21.1).

Despite no detectable sequence similarity, structural studies have revealed that glycogen synthase is homologous to glycogen phosphorylase. The binding site for UDP-glucose in glycogen synthase corresponds in position to the pyridoxal phosphate in glycogen phosphorylase.

A branching enzyme forms α-1,6 linkages

Glycogen synthase catalyzes only the synthesis of α-1,4 linkages. Another enzyme is required to form the α-1,6 linkages that make glycogen a branched polymer. Branching takes place after a number of glucosyl residues are joined in α-1,4 linkages by glycogen synthase (Figure 21.17). A branch is created by the breaking of an α-1,4

α-1,4 linkage

CORE

UDP-glucose + glycogen synthase

CORE

Branching enzyme

α-1,6 linkage

CORE

Synthase extends both nonreducing ends followed by more branching

FIGURE 21.17 Branching reaction. The branching enzyme removes an oligosaccharide of approximately seven residues from the nonreducing end and creates an internal α-1,6 linkage.

link and the formation of an α-1,6 link. A block of residues, typically 7 in number, is transferred to a more interior site. The *branching enzyme* that catalyzes this reaction requires that the block of 7 or so residues must include the nonreducing terminus, and must come from a chain at least 11 residues long. In addition, the new branch point must be at least 4 residues away from a preexisting one.

Branching is important because it increases the solubility of glycogen. Furthermore, branching creates a large number of terminal residues, the sites of action of glycogen phosphorylase and synthase (Figure 21.18). Thus, *branching increases the rate of glycogen synthesis and degradation.*

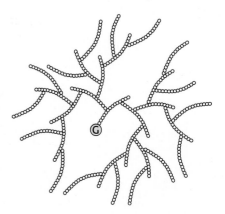

FIGURE 21.18 Cross section of a glycogen molecule. The component labeled G is glycogenin.

Glycogen synthase is the key regulatory enzyme in glycogen synthesis

Glycogen synthase, like phosphorylase, exists in two forms: an active nonphosphorylated *a* form and a usually inactive phosphorylated *b* form. Again, like the phosphorylase, the interconversion of the two forms is regulated by covalent modification mediated by hormones. However, the key means of regulating glycogen synthase is by allosteric regulation of the phosphorylated form of the enzyme, glycogen synthase *b*. Glucose 6-phosphate is a powerful activator of the enzyme, stabilizing the R state of the enzyme relative to the T state.

The covalent modification of glycogen synthase appears to play more of a fine-tuning role. The synthase is phosphorylated at multiple sites by several protein kinases—notably, *glycogen synthase kinase* (GSK), which is under the control of insulin (p. 698), and *protein kinase A*. The function of the multiple phosphorylation sites is still under investigation. Note that *phosphorylation has opposite effects on the enzymatic activities of glycogen synthase and glycogen phosphorylase.*

Glycogen is an efficient storage form of glucose

What is the cost of converting glucose 6-phosphate into glycogen and back into glucose 6-phosphate? The pertinent reactions have already been described, except for reaction 5 below, which is the regeneration of UTP. ATP phosphorylates UDP in a reaction catalyzed by *nucleoside diphosphokinase.*

$$\text{Glucose 6-phosphate} \longrightarrow \text{glucose 1-phosphate} \tag{1}$$

$$\text{Glucose 1-phosphate} + \text{UTP} \longrightarrow \text{UDP-glucose} + \text{PP}_i \tag{2}$$

$$\text{PP}_i + \text{H}_2\text{O} \longrightarrow 2\,\text{P}_i \tag{3}$$

$$\text{UDP-glucose} + \text{glycogen}_n \longrightarrow \text{glycogen}_{n+1} + \text{UDP} \tag{4}$$

$$\text{UDP} + \text{ATP} \longrightarrow \text{UTP} + \text{ADP} \tag{5}$$

Sum: Glucose 6-phosphate + ATP + glycogen$_n$ + H$_2$O $\longrightarrow$

$$\text{glycogen}_{n+1} + \text{ADP} + 2\,\text{P}_i$$

Thus, 1 molecule of ATP is hydrolyzed to incorporate glucose 6-phosphate into glycogen. The energy yield from the breakdown of glycogen is highly efficient. About 90% of the residues are phosphorolytically cleaved to glucose 1-phosphate, which is converted at no cost into glucose 6-phosphate. The other residues are branch residues, which are hydrolytically cleaved. One molecule of ATP is then used to phosphorylate each of these glucose molecules to glucose 6-phosphate. The complete oxidation of glucose 6-phosphate yields about 31 molecules of ATP, and storage consumes slightly more than 1 molecule of ATP per molecule of glucose 6-phosphate; so *the overall efficiency of storage is nearly 97%.*

21.5 Glycogen Breakdown and Synthesis Are Reciprocally Regulated

An important control mechanism prevents glycogen from being synthesized at the same time as it is being broken down. *The same glucagon- and epinephrine-triggered cAMP cascades that initiate glycogen breakdown in the liver and muscle, respectively, also shut off glycogen synthesis. Glucagon and epinephrine control both glycogen breakdown and glycogen synthesis through protein kinase A* (Figure 21.19). Recall that protein kinase A adds a phosphoryl group to phosphorylase kinase, activating that enzyme and initiating glycogen breakdown. Glycogen synthase kinase and protein kinase A add phosphoryl groups to glycogen synthase, but this phosphorylation leads to a *decrease* in enzymatic activity. In this way, glycogen breakdown and synthesis are reciprocally regulated. How is the enzymatic activity reversed so that glycogen breakdown halts and glycogen synthesis begins?

DURING EXERCISE OR FASTING

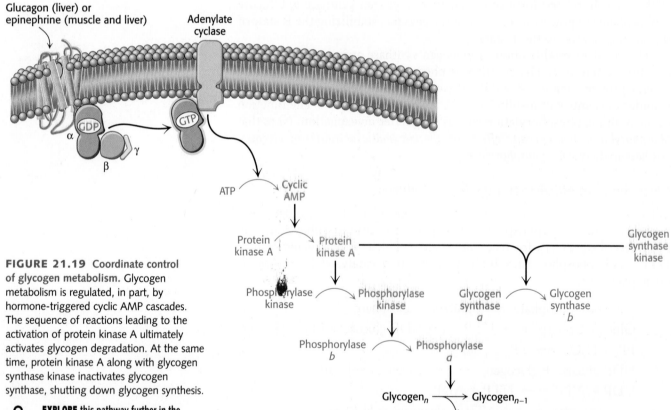

FIGURE 21.19 Coordinate control of glycogen metabolism. Glycogen metabolism is regulated, in part, by hormone-triggered cyclic AMP cascades. The sequence of reactions leading to the activation of protein kinase A ultimately activates glycogen degradation. At the same time, protein kinase A along with glycogen synthase kinase inactivates glycogen synthase, shutting down glycogen synthesis.

 EXPLORE this pathway further in the Metabolic Map on 🍁 SaplingPlus

Protein phosphatase 1 reverses the regulatory effects of kinases on glycogen metabolism

After a bout of exercise, muscle must shift from a glycogen-degrading mode to one of glycogen replenishment. A first step in this metabolic task is to shut down the phosphorylated proteins that stimulate glycogen breakdown. This task is accomplished by *protein phosphatases* that catalyze the hydrolysis of phosphorylated serine and threonine residues in proteins. *Protein phosphatase 1 plays key roles in regulating glycogen metabolism* (Figure 21.20). Protein phosphatase 1 (PP1) inactivates phosphorylase *a* and phosphorylase kinase

AFTER A MEAL OR WHEN AT REST

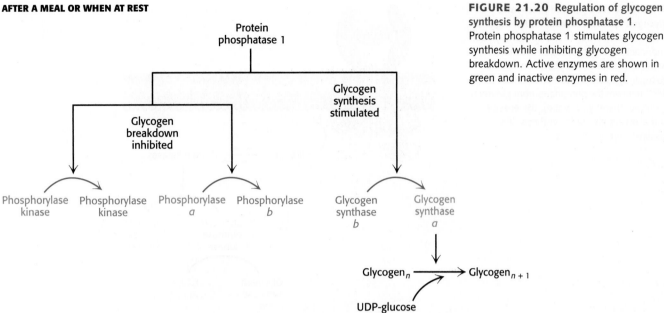

FIGURE 21.20 Regulation of glycogen synthesis by protein phosphatase 1. Protein phosphatase 1 stimulates glycogen synthesis while inhibiting glycogen breakdown. Active enzymes are shown in green and inactive enzymes in red.

by dephosphorylating them. PP1 decreases the rate of glycogen breakdown; it reverses the effects of the phosphorylation cascade. Moreover, *PP1 also removes phosphoryl groups from glycogen synthase b to convert it into the more active glycogen synthase a form.* Here, PP1 also accelerates glycogen synthesis. PP1 is yet another molecular device for coordinating carbohydrate storage. Interestingly, recent research suggests that glycogen phosphorylase is also regulated by acetylation (Section 10.3). Acetylation not only inhibits the enzyme, but also enhances dephosphorylation by promoting the interaction with PP1. Acetylation is stimulated by glucose and insulin, but inhibited by glucagon. It is fascinating that glycogen phosphorylase, one of the first allosteric enzymes identified and one of the most studied enzymes, is still revealing secrets.

The catalytic subunit of PP1 is a 37-kDa single-domain protein. This subunit is usually bound to one of a family of regulatory subunits with masses of approximately 120 kDa. In skeletal muscle and heart, the most prevalent regulatory subunit is called G_M, whereas in the liver, the most prevalent subunit is G_L. These regulatory subunits have modular structures with domains that participate in interactions with glycogen, with the catalytic subunit, and with target enzymes. Thus, *these regulatory subunits act as scaffolds, bringing together the phosphatase and its substrates on the glycogen particle.*

What prevents the phosphatase activity of PP1 from always inhibiting glycogen degradation? When glycogen degradation is called for, epinephrine or glucagon initiates the cAMP cascade that activates protein kinase A (Figure 21.21). Protein kinase A reduces the activity of PP1 by two mechanisms. First, in muscle, G_M is phosphorylated in the domain responsible for binding the catalytic subunit. The catalytic subunit is released from glycogen and from its substrates and its phosphatase activity is greatly

FIGURE 21.21 Regulation of protein phosphatase 1. Protein phosphatase 1 (PP1) regulation in muscle takes place in two steps. Phosphorylation of G_M by protein kinase A dissociates the catalytic subunit from its substrates in the glycogen particle and reduces its activity. Phosphorylation of the inhibitor subunit by protein kinase A and its subsequent binding by the phosphatase completely inactivates the catalytic subunit of PP1.

DURING EXERCISE OR FASTING

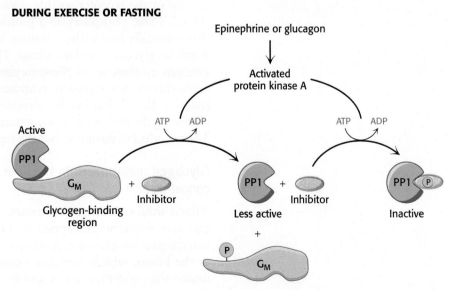

FIGURE 21.22 Insulin inactivates glycogen synthase kinase. Insulin triggers a cascade that leads to the phosphorylation and inactivation of glycogen synthase kinase and prevents the phosphorylation of glycogen synthase. Protein phosphatase 1 (PP1) removes the phosphates from glycogen synthase, thereby activating the enzyme and allowing glycogen synthesis. IRS, insulin-receptor substrate.

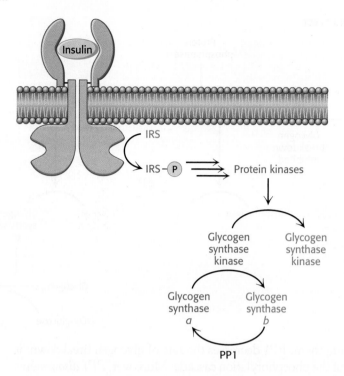

reduced. Second, almost all tissues contain small proteins (Inhibitor in Figure 21.21) that, when phosphorylated, bind to the catalytic subunit of PP1 and inhibit it. Thus, when glycogen degradation is switched on by cAMP, the accompanying phosphorylation of these inhibitors keeps phosphorylase in its active *a* form and glycogen synthase in its inactive *b* form.

Insulin stimulates glycogen synthesis by inactivating glycogen synthase kinase

After exercise, people often consume carbohydrate-rich foods to restock their glycogen stores. How is glycogen synthesis stimulated? When blood-glucose concentration is high, *insulin stimulates the synthesis of glycogen by inactivating glycogen synthase kinase,* one of the enzymes that maintains glycogen synthase in its phosphorylated, inactive state (Figure 21.22). The first step in the action of insulin is its binding to its receptor, a tyrosine kinase receptor in the plasma membrane (Section 14.2). The binding of insulin stimulates the tyrosine kinase activity of the receptor so that it phosphorylates insulin-receptor substrates (IRSs). These phosphorylated proteins trigger signal-transduction pathways that eventually lead to the activation of protein kinases that phosphorylate and inactivate glycogen synthase kinase. The inactive kinase can no longer maintain glycogen synthase in its phosphorylated, inactive state. Protein phosphatase 1 dephosphorylates glycogen synthase, activating it, and restoring glycogen reserves. Recall that insulin increases the amount of glucose in the cell by increasing the number of glucose transporters in the membrane (Section 16.2). The net effect of insulin is thus the replenishment of glycogen stores.

Glycogen metabolism in the liver regulates the blood-glucose concentration

After a meal rich in carbohydrates, blood-glucose concentration rises, and glycogen synthesis is stepped up in the liver. Although insulin is the primary signal for glycogen synthesis, another is the concentration of glucose in the blood, which normally ranges from about 4.4–6.7 mM. The liver senses the concentration of glucose in the blood and takes up or releases glucose accordingly. Experiments performed in rodents established that the amount of liver phosphorylase *a* decreases rapidly when glucose is infused (Figure 21.23). After a lag period, the amount of glycogen synthase *a* increases, which results in glycogen synthesis. *Phosphorylase* a *is the glucose sensor in liver cells,* facilitating the switch from degradation to synthesis. How does phosphorylase perform its sensor function?

Phosphorylase *a* and PP1 are localized to the glycogen particle by interactions with the G_L subunit of PP1. The binding of glucose to phosphorylase *a* shifts its allosteric equilibrium from the active R form to the inactive T form (Figure 21.12). This conformational change *renders the phosphoryl group on serine 14 a substrate for protein phosphatase 1.* PP1 binds tightly to phosphorylase *a* only when the phosphorylase is in the R state, but PP1 is inactive when bound. When glucose induces the transition to the T form, PP1 and the phosphorylase dissociate from each other and the glycogen particle, and PP1 becomes active, converting phosphorylase *a* to the *b* form (Figure 21.11). Recall that the R ↔ T transition of muscle phosphorylase *a* is unaffected by glucose and is thus unaffected by the rise in blood-glucose concentration (Section 21.2).

How does glucose binding to glycogen phosphorylase stimulate glycogen synthesis? As mentioned above, the conversion of *a* into *b* is accompanied by the *release of PP1, which is then free to activate glycogen synthase* (Figure 21.24). The removal of the phosphoryl group of inactive glycogen synthase *b* converts it into the active *a* form. What accounts for the lag between termination of glycogen degradation and the beginnings of glycogen synthesis (Figure 21.23)? There are about 10 phosphorylase *a* molecules per molecule of phosphatase. Consequently, *the activity of glycogen synthase begins to increase only after most of phosphorylase* a *is converted into* b. The lag between the decrease in glycogen degradation and the increase in glycogen synthesis prevents the two pathways from operating simultaneously. This remarkable glucose-sensing system depends on three key elements: (1) communication within phosphorylase between the allosteric site for glucose and the serine phosphate, (2) the use of PP1 to inactivate phosphorylase and activate glycogen synthase, and (3) the binding of the phosphatase to phosphorylase *a* to prevent the premature activation of glycogen synthase.

Efforts are underway to develop drugs that disrupt the interaction of liver phosphorylase with the G_L subunit as a treatment for type 2 diabetes (Section 27.3). One experimental drug binds at a unique site on the enzyme that, synergistically with glucose, stabilizes the T state of

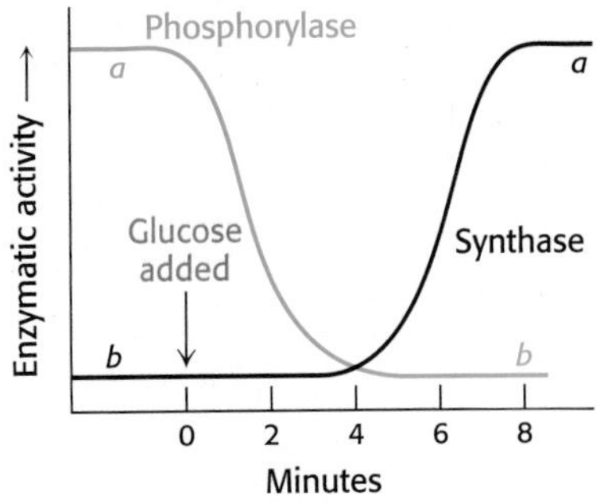

FIGURE 21.23 Blood glucose regulates liver-glycogen metabolism. The infusion of glucose into the bloodstream leads to the inactivation of phosphorylase, followed by the activation of glycogen synthase, in the liver. [Data from W. Stalmans, H. De Wulf, L. Hue, and H.-G. Hers, *Eur. J. Biochem.* 41:117–134, 1974.]

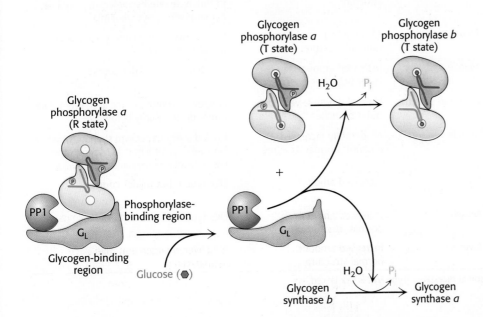

Glycogen phosphorylase *a* (R state)

PP1

G_L

Glycogen-binding region

Phosphorylase-binding region

Glucose (⬤)

Glycogen phosphorylase *a* (T state)

H_2O P_i

Glycogen phosphorylase *b* (T state)

+

PP1

G_L

Glycogen synthase *b* H_2O P_i Glycogen synthase *a*

FIGURE 21.24 Glucose regulation of liver-glycogen metabolism. After a carbohydrate rich meal, glucose binds to and inhibits glycogen phosphorylase *a* in the liver, facilitating the formation of the T state of phosphorylase *a*. The T state of phosphorylase *a* does not bind protein phosphate 1 (PP1), leading to the dissociation of PP1 from glycogen phosphorylase *a*. The free PP1 is no longer inhibited and dephosphorylates glycogen phosphorylase *a* and glycogen synthase *b*, leading to the inactivation of glycogen breakdown and the activation of glycogen synthesis.

phosphorylase and thereby enhances the transition to glycogen synthesis, as described earlier. Type 2 diabetes is characterized by excess blood glucose. Hence, disrupting the association of phosphorylase with the G_L would render it a substrate for PP1. Glycogen degradation would decrease and glucose release into the blood would be inhibited.

A biochemical understanding of glycogen-storage diseases is possible

Edgar von Gierke described the first glycogen-storage disease in 1929. A patient with this disease has a distended abdomen caused by a massive enlargement of the liver. There is a pronounced hypoglycemia between meals. Furthermore, the blood-glucose concentration does not rise on administration of epinephrine and glucagon. An infant with this glycogen-storage disease may have convulsions because of the low blood-glucose concentration.

The enzymatic defect in von Gierke disease was elucidated in 1952 by Carl and Gerty Cori. They found that *glucose 6-phosphatase is missing from the liver of a patient with this disease* (Figure 21.4). This finding was the first demonstration of an inherited deficiency of a liver enzyme. The glycogen in the liver is normal in structure but is present in abnormally large amounts. The absence of glucose 6-phosphatase in the liver causes hypoglycemia because glucose cannot be formed from glucose 6-phosphate. This phosphorylated sugar does not leave the liver, because it cannot cross the plasma membrane. The presence of excess glucose 6-phosphate triggers an increase in glycolysis in the liver, leading to high concentrations of lactate and pyruvate in the blood. Patients who have von Gierke disease also have an increased dependence on fat metabolism. This disease can also be produced by a mutation in the gene that encodes the *glucose 6-phosphate transporter*. Recall that glucose 6-phosphate must be transported into the lumen of the endoplasmic reticulum to be hydrolyzed by phosphatase (Section 16.3). Mutations in the other three essential proteins of this system can likewise lead to von Gierke disease.

Seven other glycogen-storage diseases have been characterized (Table 21.2). In Pompe disease (type II), lysosomes become engorged with glycogen

TABLE 21.2 Glycogen-storage diseases

Type	Defective enzyme	Organ affected	Glycogen in the affected organ	Clinical features
I Von Gierke	Glucose 6-phosphatase or transport system	Liver and kidney	Increased amount; normal structure.	Massive enlargement of the liver; failure to thrive; severe hypoglycemia, ketosis, hyperuricemia, hyperlipemia.
II Pompe	α-1,4-glucosidase (lysosomal)	All organs	Massive increase in amount; normal structure.	Cardiorespiratory failure causes death, usually before age 2.
III Cori	α-1,6-glucosidase (debranching enzyme)	Muscle and liver	Increased amount; short outer branches.	Like type I, but milder course.
IV Andersen	Branching enzyme (α-1,4 → α-1,6)	Liver and spleen	Normal amount; very long outer branches.	Progressive cirrhosis of the liver; liver failure causes death, usually before age 2.
V McArdle	Phosphorylase	Muscle	Moderately increased amount; normal structure.	Limited ability to perform strenuous exercise because of painful muscle cramps; otherwise patient is normal and well developed.
VI Hers	Phosphorylase	Liver	Increased amount.	Like type I, but milder course.
VII	Phosphofructokinase	Muscle	Increased amount; normal structure.	Like type V.
VIII	Phosphorylase kinase	Liver	Increased amount; normal structure.	Mild liver enlargement; mild hypoglycemia.

Note: Types I through VII are inherited as autosomal recessives. Type VIII is sex linked.

because they lack α-1,4-glucosidase, a hydrolytic enzyme confined to these organelles (Figure 21.25). Carl and Gerty Cori also elucidated the biochemical defect in another glycogen-storage disease (type III) which cannot be distinguished from von Gierke disease (type I) by physical examination alone. There are several subtypes of glycogen-storage disease III. In type IIIa and IIIc disease, the structure of both liver and muscle glycogen is abnormal and the amount is markedly increased. Types IIIb and IIId are liver specific. Regardless of subtype, the outer branches of the glycogen are very short. *Patients having this type lack the debranching enzyme (α-1,6-glucosidase),* and so only the outermost branches of glycogen can be effectively utilized. Thus, only a small fraction of this abnormal glycogen is functionally active as an accessible store of glucose.

A defect in glycogen metabolism confined to muscle is found in McArdle disease (type V). *Muscle phosphorylase activity is absent,* and a patient's capacity to perform strenuous exercise is limited because of painful muscle cramps. McArdle disease is examined further in the Biochemistry in Focus Appendix at the end of this chapter. The patient is otherwise normal and well developed. Thus, effective utilization of muscle glycogen is not essential for life.

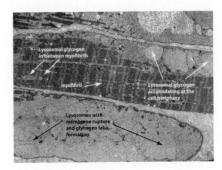

FIGURE 21.25 Electron micrograph of muscle cell from a patient with Pompe disease. Glycogen-engorged lysosomes are seen throughout the cell, including in the myofibrils. As the disease progresses, lysosomes may rupture, releasing large amounts of glycogen into the cytoplasm. Such accumulations of cytoplasmic glycogen are called glycogen lakes. [Reproduced with permission of the author, from B.L. Thurberg et al., Characterization of pre- and post-treatment pathology after enzyme replacement therapy for Pompe disease. *Lab. Invest.* 86(12):1208–1220, 2006.]

SUMMARY

Glycogen, a readily mobilized fuel store, is a branched polymer of glucose residues. Most of the glucose units in glycogen are linked by α-1,4-glycosidic bonds. At about every 12th residue, a branch is created by an α-1,6-glycosidic bond. Glycogen is present in large amounts in muscle cells and in liver cells, where it is stored in the cytoplasm in the form of hydrated granules.

21.1 Glycogen Breakdown Requires the Interplay of Several Enzymes

Most of the glycogen molecule is degraded to glucose 1-phosphate by the action of glycogen phosphorylase, the key enzyme in glycogen breakdown. The glycosidic linkage between C-1 of a terminal residue and C-4 of the adjacent one is split by orthophosphate to give glucose 1-phosphate, which can be reversibly converted into glucose 6-phosphate. Branch points are degraded by the concerted action of an oligosaccharide transferase and an α-1,6-glucosidase.

21.2 Phosphorylase Is Regulated by Allosteric Interactions and Reversible Phosphorylation

Phosphorylase *b*, which is usually inactive, is converted into active phosphorylase *a* by the phosphorylation of a single serine residue in each subunit. This reaction is catalyzed by phosphorylase kinase. The *a* form in the liver is inhibited by glucose. In liver, phosphorylase is activated to liberate glucose for export to other organs, such as skeletal muscle and the brain. The *b* form in muscle can be activated by the binding of AMP, an effect counteracted by ATP and glucose 6-phosphate. In contrast to liver, muscle phosphorylase is activated to generate glucose for use inside the cell as a fuel for contractile activity.

21.3 Epinephrine and Glucagon Signal the Need for Glycogen Breakdown

Epinephrine and glucagon stimulate glycogen breakdown through specific 7TM receptors. Muscle is the primary target of epinephrine, whereas the liver is responsive to glucagon. Both signal molecules initiate a kinase cascade that leads to the activation of phosphorylase kinase, which in turn converts glycogen phosphorylase *b* to the phosphorylated *a* form.

21.4 Glycogen Is Synthesized and Degraded by Different Pathways

The pathway for glycogen synthesis differs from that for glycogen breakdown. UDP-glucose, the activated intermediate in glycogen synthesis, is formed from glucose 1-phosphate and UTP. Glycogen synthase catalyzes the transfer of glucose from UDP-glucose to the C-4 hydroxyl group of a terminal residue in the growing glycogen molecule. Synthesis is primed by glycogenin, an autoglycosylating protein that contains a covalently attached oligosaccharide unit on a specific tyrosine residue. A branching enzyme converts some of the α-1,4 linkages into α-1,6 linkages to increase the number of ends so that glycogen can be made and degraded more rapidly.

21.5 Glycogen Breakdown and Synthesis Are Reciprocally Regulated

Glycogen synthesis and degradation are coordinated by several amplifying reaction cascades. Epinephrine and glucagon stimulate glycogen breakdown and inhibit its synthesis by increasing the cytoplasmic concentration of cyclic AMP, which activates protein kinase A. Protein kinase A activates glycogen breakdown by attaching a phosphate to phosphorylase kinase and inhibits glycogen synthesis by phosphorylating glycogen synthase. Glycogen synthase kinase also inhibits synthesis by phosphorylating the synthase.

The glycogen-mobilizing actions of protein kinase A are reversed by protein phosphatase 1, which is regulated by several hormones. Epinephrine inhibits this phosphatase by blocking its attachment to glycogen molecules and by turning on an inhibitor. Insulin, in contrast, triggers a cascade that phosphorylates and inactivates glycogen synthase kinase. Hence, glycogen synthesis is decreased by epinephrine and increased by insulin. Glycogen synthase and phosphorylase are also regulated by noncovalent allosteric interactions. In fact, phosphorylase is a key part of the glucose-sensing system of liver cells. Glycogen metabolism exemplifies the power and precision of reversible phosphorylation in regulating biological processes.

APPENDIX

Biochemistry in Focus

McArdle disease results from a lack of skeletal muscle glycogen phosphorylase

McArdle disease or Type V glycogen storage disease is a rare condition affecting approximately 1 in 100,000 people. Affected individuals experience painful muscle fatigue and have burgundy-colored urine following an attempt at vigorous exercise. The coloration is due to rhabdomyolsis (rapid breakdown of skeletal muscle), which results in myoglobinuria (myoglobin in the blood and urine). Resting and moderate exercise, such as walking at a relaxed pace, do not induce the symptoms. Let's examine the biochemical basis for these symptoms and explore some other biochemical characteristics of the disorder.

During vigorous exercise, especially the initial few minutes, skeletal muscle contractions are powered by glycogen mobilization by glycogen phosphorylase. If skeletal muscle is made to work vigorously without an adequate energy supply, rhabdomyolysis will occur and the damaged muscle tissue will leak myoglobin into the blood, which will be passed in the urine. Recall the heme group gives globins their red color, and its presence in the urine accounts for the burgundy color. In an unaffected individual, intracellular and blood pH fall as lactic acid is produced during aerobic glycolysis (p. 506). Because McArdle patients cannot mobilize glucose, lactic acid will not be produced and there will be no fall in pH. Interestingly, the pH rises in these individuals. What is the cause of this? Remember that an immediate source of ATP in skeletal muscle is the phosphorylation of ATP at the expense of creatine phosphate.

Creatine phosphate + ADP

Creatine kinase

Creatine + ATP

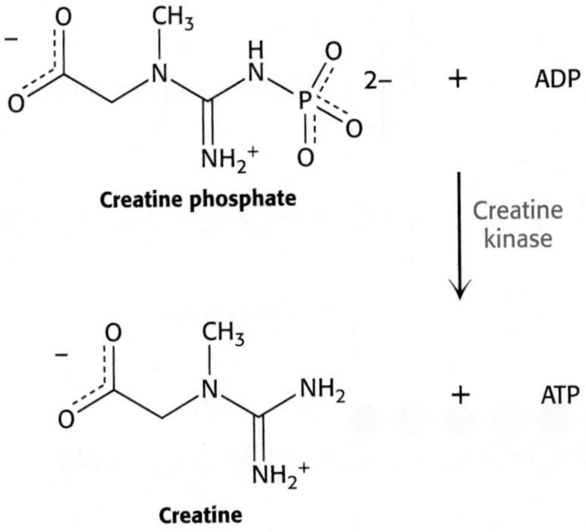

In a futile attempt to power exercise, skeletal muscles of an affected individual rapidly synthesize ATP at the expense of creatine phosphate. However, the guanidinium group of creatine is a strong base, leading to an increase in pH.

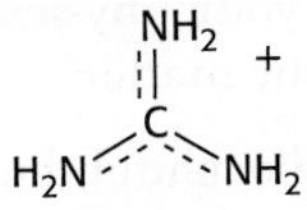

Guanidium

As exercise continues, muscle cells will switch from aerobic glycolysis to cellular respiration. To facilitate this change, there is an increase in citric acid cycle intermediates in unaffected individuals, an increase that is not observed in McArdle patients. Why is the increase absent in McArdle patients? Recall that a key reaction to supplement citric acid cycle components is catalyzed by pyruvate carboxylase.

$$\text{Pyruvate} + \text{CO}_2 + \text{ATP} \xrightarrow{\text{Pyruvate carboxylase}}$$
$$\text{oxaloacetate} + \text{ADP} + \text{P}_i + 2\text{H}^+$$

Oxaloacetate then enters the citric acid cycle. The most common source of pyruvate in skeletal muscle is from the metabolism of glucose, generated by glycogen mobilization. But McArdle patients cannot generate the glucose and thus cannot supplement the citric acid cycle components.

Finally, why don't affected individuals experience symptoms at rest or when performing leisurely exercise? As we will see in the next chapter, under these conditions, skeletal muscle derives most of its energy from fatty acid oxidation rather than from glycogen mobilization.

APPENDIX

Problem-Solving Strategies

PROBLEM: Imagine yourself to be a med student trying to get research experience. The attending physician has a patient with a liver disorder that might be due to a defect in glycogen metabolism. She gives you a sample of glycogen from the patient and tells you to incubate the sample in a test tube with orthophosphate, glycogen phosphorylase, the oligosaccharide transferase, and the debranching enzyme (α-1,6-glucosidase), all of which were stolen from the biochemist's lab down the hall. You find that the ratio of glucose 1-phosphate to glucose formed in this mixture is 100. What do you tell the physician is the most likely enzymatic deficiency in this patient?

SOLUTION: The first item on the agenda is to make sure we know the function of each of the chemicals in the test tube.

▶ **What are the roles of each of the contents of the test tube?**

Phosphate is a substrate for glycogen phosphorylase that allows the phosphorolytic cleavage of glycogen to yield glucose 1-phosphate. The transferase remodels the glycogen by shifting several glucose residues near a branch point from one outer branch to another. The α-1, 6-glucosidase cleaves the α-1, 6-linkage to yield a free glucose. Here is Figure 21.8 again.

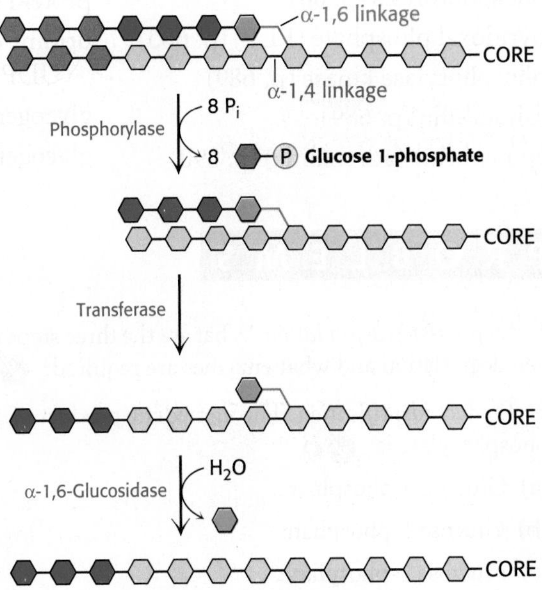

▶ **What is the one piece of data about the glycogen degradation included in the question?**

The ratio of glucose 1-phosphate to free glucose is about 100.

▶ **What aspect of glycogen structure does free glucose represent?**

The only time that free glucose is generated is when α-1, 6-glucosidase cleaves a glucose at a branch point. Thus, each free glucose represents a former branch point.

▶ **Thinking about the discussion of glycogen structure in the chapter, how common are branch points in normal glycogen?**

Branch points occur about once every 12 glucose residues. In other words, when normal glycogen is treated as above, the ratio of glucose 1-phosphate to free glucose should be about 12.

▶ **What does the difference between the glucose 1-phosphate to free glucose ratio in the patient compared to the 12 to 1 ratio normally seen suggest about the structure of glycogen in the patient?**

The data suggest that there are too few branch points in the patient's glycogen.

▶ **How are branch points normally introduced into glycogen?**

Glycogen synthase only extends polymers of glucose. Branches are introduced by the branching enzyme. Figure 21.17 (shown here) shows how branches are added.

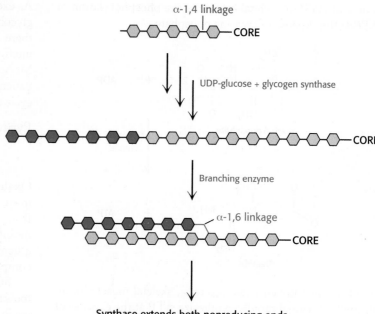

Synthase extends both nonreducing ends followed by more branching

▶ **What do you tell your physician when she asks for your thoughts on the matter?**

You try to appear thoughtful, but enthusiastic, and tell her that it appears that the patient is suffering from a deficiency in the branching enzyme.

KEY TERMS

glycogen phosphorylase (p. 681)
phosphorolysis (p. 681)
pyridoxal phosphate (PLP) (p. 683)
phosphorylase kinase (p. 689)
calmodulin (p. 689)
epinephrine (adrenaline) (p. 690)

glucagon (p. 690)
protein kinase A (PKA) (p. 691)
uridine diphosphate glucose (UDP-glucose) (p. 693)
glycogen synthase (p. 693)
glycogenin (p. 694)

glycogen synthase kinase (p. 695)
protein phosphatase 1 (PP1) (p. 696)
insulin (p. 698)

PROBLEMS

1. *Step-by-step degradation.* What are the three steps in glycogen degradation and what enzymes are required? ✔❶

2. *Product identification.* The immediate product of glycogen phosphorylase is: ✔❶

(a) Glucose 6-phosphate.

(b) Glucose 1-phosphate.

(c) Fructose 1-phosphate.

(d) Glucose 1,6-bisphosphate.

3. *Reaction type.* Glycogen phosphorylase degrades glycogen by: ✔❶

(a) An hydrolysis reaction.

(b) An oxidation-reduction reaction.

(c) A phosphorolysis reaction.

(d) A transferase reaction.

4. *Teamwork.* Which of the following enzymes are not directly involved in glycogen degradation? ✔❶

(a) Phosphoglucose isomerase

(b) α-1,6-Glucosidase

(c) Glycogen phosphorylase

(d) Transferase

5. *Activated glucose.* Which of the following is a substrate for glycogen synthase? ✓❸

(a) UTP-glucose

(b) Glucose 1-phosphate

(c) CDP-glucose

(d) UDP-glucose

6. *Inappropriate.* Which of the following enzymes is not required for glycogen synthesis? ✓❸

(a) Glycogenin

(b) Branching enzyme

(c) Glycogen synthase

(d) α-1,6-Glucosidase

7. *Match 'em.* Match each term with its description.

(a) Glycogen phosphorylase ____

(b) Phosphorolysis ____

(c) Transferase ____

(d) α-1,6-glucosidase ____

(e) Phosphoglucomutase ____

(f) Phosphorylase kinase ____

(g) Protein kinase A ____

(h) Calmodulin ____

(i) Epinephrine ____

(j) Glucagon ____

1. Calcium-binding subunit of phosphorylase kinase

2. Activates glycogen phosphorylase

3. Removal of a glucose residue by the addition of phosphate

4. Stimulates glycogen breakdown in muscle

5. Liberates a free glucose residue

6. Shifts the location of several glucose residues

7. Stimulates glycogen breakdown in the liver

8. Catalyzes phosphorolytic cleavage

9. Prepares glucose 1-phosphate for glycolysis

10. Phosphorylates phosphorylase kinase

8. *Choice is good.* Glycogen is not as reduced as fatty acids are and consequently not as energy rich. Why do animals store any energy as glycogen? Why not convert all excess fuel into fatty acids?

9. *If a little is good, a lot is better.* α-Amylose is an unbranched glucose polymer. Why would this polymer not be as effective a storage form of glucose as glycogen?

10. *Speed matters.* Liver and skeletal muscle have different isozymic forms of glycogen phosphorylase that differ in a number of ways as discussed in the text. One notable difference is that the V_{max} of muscle phosphorylase is considerably greater than the V_{max} of liver phosphorylase. Explain why this difference makes good physiological sense. ✓❻

11. *Dare to be different.* Compare the allosteric regulation of phosphorylase in the liver and in muscle, and explain the significance of the difference. ✓❷, ✓❻

12. *A thumb on the balance.* The reaction catalyzed by phosphorylase is readily reversible in vitro. At pH 6.8, the equilibrium ratio of orthophosphate to glucose 1-phosphate is 3.6. The value of $\Delta G^{\circ\prime}$ for this reaction is small because a glycosidic bond is replaced by a phosphoryl ester bond that has a nearly equal transfer potential. However, phosphorolysis proceeds far in the direction of glycogen breakdown in vivo. Suggest one means by which the reaction can be made irreversible in vivo. ✓❶

13. *Excessive storage.* Suggest an explanation for the fact that the amount of glycogen in type I glycogen-storage disease (von Gierke disease) is increased.

14. *Recouping an essential phosphoryl.* The phosphoryl group on phosphoglucomutase is slowly lost by hydrolysis. Propose a mechanism that utilizes a known catalytic intermediate for restoring this essential phosphoryl group. How might this phosphoryl donor be formed?

15. *Not all absences are equal.* Hers disease results from an absence of liver glycogen phosphorylase and may result in serious illness. In McArdle disease, muscle glycogen phosphorylase is absent. Although exercise is difficult for patients suffering from McArdle disease, the disease is rarely life threatening. Account for the different manifestations of the absence of glycogen phosphorylase in the two tissues. What does the existence of these two different diseases indicate about the genetic nature of the phosphorylase? ✓❻

16. *Hydrophobia.* Why is water excluded from the active site of phosphorylase? Predict the effect of a mutation that allows water molecules to enter. ✓❶

17. *Removing all traces.* In human liver extracts, the catalytic activity of glycogenin was detectable only after treatment with α-amylase, an enzyme the hydrolyzes α-1,4-bonds in starch and glycogen. Why was α-amylase necessary to reveal the glycogenin activity?

18. *Two in one.* A single polypeptide chain houses the transferase and debranching enzymes. Cite a potential advantage of this arrangement. ✓❸

19. *How did they do that?* A strain of mice has been developed that lacks the enzyme phosphorylase kinase. Yet, after strenuous exercise, the glycogen stores of a mouse of this strain are depleted. Explain how this depletion is possible.

20. *An appropriate inhibitor.* What is the rationale for the inhibition of muscle glycogen phosphorylase by glucose 6-phosphate when glucose 1-phosphate is the product of the phosphorylase reaction?

21. *Passing along the information.* Outline the signal-transduction cascade for glycogen degradation in muscle. ✓❷

22. *Slammin' on the brakes* There must be a way to shut down glycogen breakdown quickly to prevent the wasteful depletion of glycogen after energy needs have been met. What mechanisms are employed to turn off glycogen breakdown?

23. *Diametrically opposed.* Phosphorylation has opposite effects on glycogen synthesis and breakdown. What is the advantage of its having opposing effects? ✓⑤

24. *Feeling depleted.* Glycogen depletion resulting from intense, extensive exercise can lead to exhaustion and the inability to continue exercising. Some people also experience dizziness, an inability to concentrate, and a loss of muscle control. Account for these symptoms.

25. *Everyone had a job to do.* What accounts for the fact that liver phosphorylase is a glucose sensor, whereas muscle phosphorylase is not? ✓⑥

26. *Yin and Yang.* Match the terms on the left with the descriptions on the right.

(a) UDP-glucose _____

(b) UDP-glucose pyrophosphorylase _____

(c) Glycogen synthase _____

(d) Glycogenin _____

(e) Branching enzyme _____

(f) Glucose 6-phosphate _____

(g) Glycogen synthase kinase _____

(h) Protein phosphatase 1 _____

(i) Insulin _____

(j) Glycogen phosphorylase *a* _____

1. Has glucose 1-phosphate as one of its substrates

2. Potent activator of glycogen synthase *b*

3. Glucose sensor in the liver

4. Activated substrate for glycogen synthesis

5. Synthesizes α-1,4 linkages between glucose molecules

6. Leads to the inactivation of glycogen synthase kinase

7. Synthesizes α-1,6 linkages between glucose molecules

8. Catalyzes the formation of glycogen synthase *b*

9. Catalyzes the formation of glycogen synthase *a*

10. Provides the primer for glycogen synthesis

27. *Team effort.* What enzymes are required for the synthesis of a glycogen particle starting from glucose 6-phosphate? ✓❸

28. *Force it forward.* The following reaction accounts for the synthesis of UDP-glucose. This reaction is readily reversible. How is it made irreversible in vivo?

Glucose 1-phosphate + UTP ⇌ UDP-glucose + PP$_i$

29. *If you insist.* Why does activation of the phosphorylated *b* form of glycogen synthase by high concentrations of glucose 6-phosphate make good biochemical sense? ✓④

30. *An ATP saved is an ATP earned.* The complete oxidation of glucose 6-phosphate derived from free glucose yields 30 molecules ATP, whereas the complete oxidation of glucose 6-phosphate derived from glycogen yields 31 molecules of ATP. Account for this difference.

31. *Dual roles.* Phosphoglucomutase is crucial for glycogen breakdown as well as for glycogen synthesis. Explain the role of this enzyme in each of the two processes. ✓⑤

32. *Working at cross-purposes.* Write a balanced equation showing the effect of simultaneous activation of glycogen phosphorylase and glycogen synthase. Include the reactions catalyzed by phosphoglucomutase and UDP-glucose pyrophosphorylase. ✓⑤

33. *Achieving immortality.* Glycogen synthase requires a primer. A primer was formerly thought to be provided when the existing glycogen granules are divided between the daughter cells produced by cell division. In other words, parts of the original glycogen molecule were simply passed from generation to generation. Would this strategy have been successful in passing glycogen stores from generation to generation? How are new glycogen molecules now known to be synthesized?

34. *Synthesis signal.* How does insulin stimulate glycogen synthesis? ✓④

Mechanism Problem

35. *Family resemblance.* Enzymes of the α-amylase family (problem 17) catalyze a reaction by forming a covalent intermediate to a conserved aspartate residue. Propose mechanisms for the two enzymes catalyzing steps in glycogen debranching on the basis of their potential membership in the α-amylase family.

Chapter Integration Problems

36. *Double activation.* What pathway in addition to the cAMP-induced signal transduction pathway is used in the liver to maximize glycogen breakdown? ✓❸

37. *Carbohydrate conversion.* Write a balanced equation for the formation of glycogen from galactose.

38. *Working together.* What enzymes are required for the liver to release glucose into the blood when an organism is asleep and fasting?

39. *A shattering experience.* Crystals of phosphorylase *a* grown in the presence of glucose shatter when a substrate such as glucose 1-phosphate is added. Why?

40. *I know I've seen that face before.* UDP-glucose is the activated form of glucose used in glycogen synthesis. However, we have previously met other similar activated forms of carbohydrate in our consideration of metabolism. What other UDP-carbohydrate have we encountered in our study of biochemistry?

41. *Same symptoms, different cause.* Suggest another mutation in glucose metabolism that causes symptoms similar to those of von Gierke disease.

Data Interpretation Problems

42. *An authentic replica.* Experiments were performed in which serine (S) 14 of glycogen phosphorylase was replaced by glutamate (E). The V_{max} of the mutant enzyme was then compared with the wild type phosphorylase in both the *a* and the *b* form.

V_{max} μmoles of glucose 1-PO$_4$ released min^{-1} mg^{-1}	
Wild type phosphorylase *b*	25 ± 0.4
Wild type phosphorylase *a*	100 ± 5
S to E mutant	60 ± 3

(a) Explain the results obtained with the mutant.

(b) Predict the effect of substituting aspartic acid for the serine.

43. *Glycogen isolation 1.* The liver is a major storage site for glycogen. Purified from two samples of human liver, glycogen was either treated or not treated with α-amylase and subsequently analyzed by SDS-PAGE and western blotting with the use of antibodies to glycogenin (See animations on 🌐 **Sapling**Plus). The results are presented in the adjoining illustration below.

(a) Why are no proteins visible in the lanes without amylase treatment?

(b) What is the effect of treating the samples with α-amylase? Explain the results.

(c) List other proteins that you might expect to be associated with glycogen. Why are other proteins not visible?

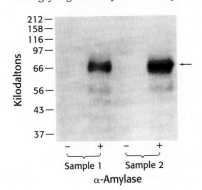

[Courtesy of Dr. Peter J. Roach, Indiana University School of Medicine.]

44. *Glycogen isolation 2.* The gene for glycogenin was transfected into a cell line that normally stores only small amounts of glycogen. The cells were then manipulated according to the following protocol, and glycogen was isolated and analyzed by SDS-PAGE and western blotting by using an antibody to glycogenin with and without α-amylase treatment (See animation on 🌐 **Sapling**Plus). The results are presented in the illustration below.

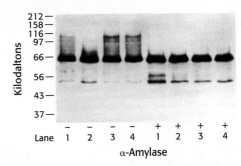

[Courtesy of Dr. Peter J. Roach, Indiana University School of Medicine.]

The protocol: Cells cultured in growth medium and 25 mM glucose (lane 1) were switched to medium containing no glucose for 24 hours (lane 2). Glucose-starved cells were refed with medium containing 25 mM glucose for 1 hour (lane 3) or 3 hours (lane 4). Samples (12 μg of protein) were either treated or not treated with α-amylase, as indicated, before being loaded on the gel.

(a) Why did the western analysis produce a "smear"—that is, the high-molecular-weight staining in lane 1($-$)?

(b) What is the significance of the decrease in high-molecular-weight staining in lane 2($-$)?

(c) What is the significance of the difference between lanes 2($-$) and 3($-$)?

(d) Suggest a plausible reason why there is essentially no difference between lanes 3($-$) and 4($-$).

(e) Why are the bands at 66 kDa the same in the lanes treated with amylase, despite the fact that the cells were treated differently?

Think ▶ Pair ▶ Share Problems

45. Describe the structure of glycogen and explain how the structure facilitates glycogen metabolism.

46. Describe the reciprocal regulation of glycogen catabolism and anabolism in liver. Explain why such reciprocal regulation is required for proper glucose homeostasis.

Fatty Acid Metabolism

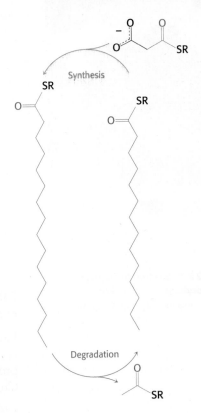

Fats provide efficient means for storing energy for later use. (Right) The processes of fatty acid synthesis (preparation for energy storage) and fatty acid degradation (preparation for energy use) are, in many ways, the reverse of each other. (Above) Many mammals, such as this house mouse, hibernate over the long winter months. Although their metabolism slows during hibernation, the energy needs of the animal must still be met. Fatty acid degradation is a key energy source for this need. [Mark Sisson/FLPA/Science Source.]

✓ LEARNING OBJECTIVES

By the end of this chapter, you should be able to:

1. Identify the repeated steps of fatty acid degradation.

2. Describe ketone bodies and their role in metabolism.

3. Explain how fatty acids are synthesized.

4. Explain how fatty acid metabolism is regulated.

OUTLINE

22.1 Triacylglycerols Are Highly Concentrated Energy Stores

22.2 The Use of Fatty Acids as Fuel Requires Three Stages of Processing

22.3 Unsaturated and Odd-Chain Fatty Acids Require Additional Steps for Degradation

22.4 Ketone Bodies Are a Fuel Source Derived from Fats

22.5 Fatty Acids Are Synthesized by Fatty Acid Synthase

22.6 The Elongation and Unsaturation of Fatty Acids Are Accomplished by Accessory Enzyme Systems

22.7 Acetyl CoA Carboxylase Plays a Key Role in Controlling Fatty Acid Metabolism

We turn now from the metabolism of carbohydrates to that of fatty acids. A fatty acid contains a long hydrocarbon chain and a terminal carboxylate group. Fatty acids have four major physiological roles. First, *fatty acids are fuel molecules.* They are stored as *triacylglycerols* (also called *neutral fats* or *triglycerides*), which are uncharged esters of fatty acids with glycerol. Triacylglycerols are stored mainly in adipose tissue, composed of cells called adipocytes (Figure 22.1). Fatty acids mobilized from triacylglycerols are oxidized to meet the energy needs of a cell or organism. During rest or moderate exercise, such as walking, fatty acids are our

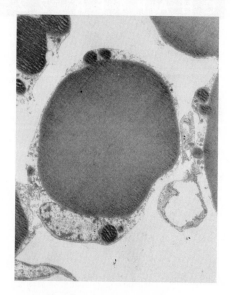

FIGURE 22.1 Electron micrograph of an adipocyte. A small band of cytoplasm surrounds the large deposit of triacylglycerols. [Biophoto Associates/Science Source.]

primary source of energy. Second, *fatty acids are building blocks of phospholipids and glycolipids*. These amphipathic molecules are important components of biological membranes, as discussed in Chapter 12. Third, many proteins are modified by the *covalent attachment of fatty acids, which targets the proteins to membrane locations* (Section 12.4). Fourth, *fatty acid derivatives serve as hormones and intracellular messengers*. In this chapter, we focus on the degradation and synthesis of fatty acids.

A triacylglycerol

Fatty acid degradation and synthesis mirror each other in their chemical reactions

Fatty acid degradation and synthesis consist of four steps that are the reverse of each other in their basic chemistry. Degradation is an oxidative process that converts a fatty acid into a set of activated acetyl units (acetyl CoA) that can be processed by the citric acid cycle (Figure 22.2).

- An activated fatty acid is oxidized to introduce a double bond.
- The double bond is hydrated to introduce a hydroxyl group.
- The alcohol is oxidized to a ketone.
- Finally, the fatty acid is cleaved by coenzyme A to yield acetyl CoA and a fatty acid chain two carbons shorter.

If the fatty acid has an even number of carbon atoms and is saturated, the process is simply repeated until the fatty acid is completely converted into acetyl CoA units.

Fatty acid synthesis is essentially the reverse of this process. The process starts with the individual units to be assembled—in this case, with an activated acyl group (most simply, an acetyl unit) and a malonyl unit (Figure 22.2). The malonyl unit condenses with the acetyl unit to form a four-carbon fragment. To produce the required hydrocarbon chain, the carbonyl group is reduced to a methylene group in three steps: a reduction, a dehydration, and another reduction, exactly the opposite of degradation. The product of the reduction is butyryl CoA. Another activated malonyl group condenses with the butyryl unit, and the process is repeated until a C_{16} or shorter fatty acid is synthesized.

22.1 Triacylglycerols Are Highly Concentrated Energy Stores

Triacylglycerols are highly concentrated stores of metabolic energy because they are *reduced* and *anhydrous* (containing no water). The yield from the complete oxidation of fatty acids is about $38 \, \text{kJ g}^{-1}$ ($9 \, \text{kcal g}^{-1}$), in contrast with about $17 \, \text{kJ g}^{-1}$ ($4 \, \text{kcal g}^{-1}$) for carbohydrates and proteins. The basis of this large difference in caloric yield is that fatty acids are much more reduced than carbohydrates or proteins. Furthermore, triacylglycerols are

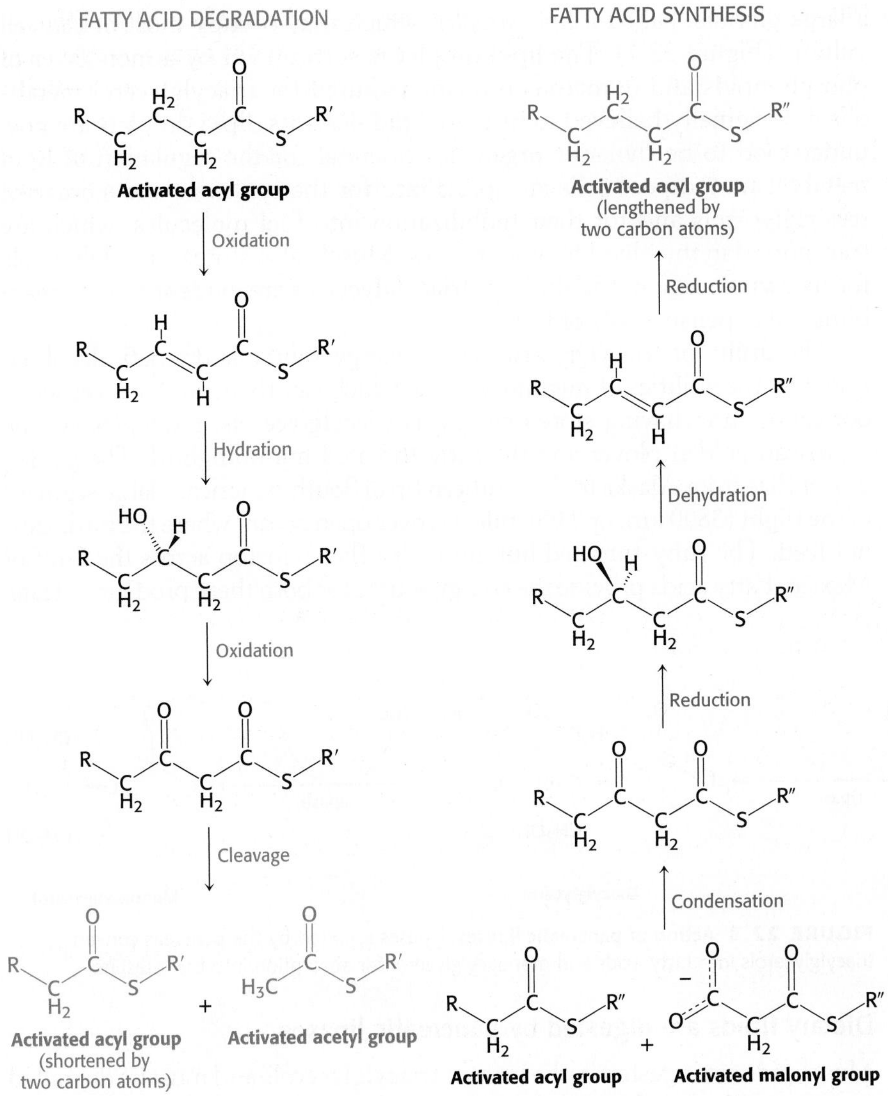

FATTY ACID DEGRADATION

Activated acyl group

Oxidation

Hydration

Oxidation

Cleavage

Activated acyl group
(shortened by
two carbon atoms)

+

Activated acetyl group

FATTY ACID SYNTHESIS

Activated acyl group
(lengthened by
two carbon atoms)

Reduction

Dehydration

Reduction

Condensation

Activated acyl group

+

Activated malonyl group

FIGURE 22.2 Steps in fatty acid degradation and synthesis. The two processes are in many ways mirror images of each other.

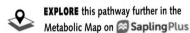

EXPLORE this pathway further in the Metabolic Map on **Sapling**Plus

nonpolar, and so they are stored in a nearly anhydrous form, whereas much more polar carbohydrates are more highly hydrated. In fact, 1 g of dry glycogen binds about 2 g of water. Consequently, *a gram of nearly anhydrous fat stores 6.75 times as much energy as a gram of hydrated glycogen,* which is likely the reason that triacylglycerols rather than glycogen were selected in evolution as the major energy reservoir.

Consider a typical 70-kg man, who has fuel reserves of 420,000 kJ (100,000 kcal) in triacylglycerols, 100,000 kJ (24,000 kcal) in protein (mostly in muscle), 2500 kJ (600 kcal) in glycogen, and 170 kJ (40 kcal) in glucose. Triacylglycerols constitute about 11 kg of his total body weight. If this amount of energy were stored in glycogen, his total body weight would be 64 kg greater. The glycogen and glucose stores provide enough energy to sustain physiological function for about 24 hours, whereas the triacylglycerol stores allow survival for several weeks.

In mammals, the major site of triacylglycerol accumulation is the cytoplasm of *adipose cells (fat cells)*. This fuel-rich tissue is located throughout the body, notably under the skin (subcutaneous fat) and surrounding the internal organs (visceral fat). Droplets of triacylglycerol coalesce to form a large globule, called a *lipid droplet*, which may occupy most of the cell volume (Figure 22.1). The lipid droplet is surrounded by a monolayer of phospholipids and numerous proteins required for triacylglycerol metabolism. Originally believed to be inert-lipid deposits, lipid droplets are now understood to be dynamic organelles essential for the regulation of lipid metabolism. Adipose cells are specialized for the synthesis and storage of triacylglycerols and for their mobilization into fuel molecules, which are transported in the blood to other tissues. Muscle also stores triacylglycerols for its own energy needs. Indeed, triacylglycerols are evident as the "marbling" of expensive cuts of beef.

The utility of triacylglycerols as an energy source is dramatically illustrated by the abilities of migratory birds, which can fly great distances without eating after having stored energy as triacylglycerols. Examples are the American golden plover and the ruby-throated hummingbird. The golden plover flies from Alaska to the southern tip of South America; a large segment of the flight (3800 km, or 2400 miles) is over open ocean, where the birds cannot feed. The ruby-throated hummingbird flies nonstop across the Gulf of Mexico. Fatty acids provide the energy source for both these prodigious feats.

Triacylglycerols fuel the long migration flights of the American golden plover (*Pluvialis dominica*). [Jim Zipp/Science Source.]

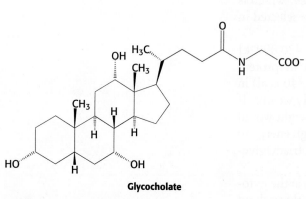

Triacylglycerol   **Diacylglycerol**   **Monoacylglycerol**

FIGURE 22.3 Action of pancreatic lipases. Lipases secreted by the pancreas convert triacylglycerols into fatty acids and monoacylglycerol for absorption into the intestine.

Dietary lipids are digested by pancreatic lipases

Most lipids are ingested in the form of triacylglycerols and must be degraded to fatty acids for absorption across the intestinal epithelium. Intestinal enzymes called *lipases,* secreted by the pancreas, degrade triacylglycerols to free fatty acids and monoacylglycerol (Figure 22.3). Lipids present a special problem because, unlike carbohydrates and proteins, these molecules are not soluble in water. They exit the stomach as an emulsion—particles with a triacylglycerol core surrounded by cholesterol and cholesterol esters. How are the lipids made accessible to the lipases, which are in aqueous solution?

First, the particles are coated with *bile acids* (Figure 22.4), amphipathic molecules synthesized from cholesterol in the liver and secreted from the gallbladder. The ester bonds of each lipid are oriented toward the surface of the bile salt-coated particle, rendering the bond more accessible to digestion by lipases in aqueous solution. However, in this form, the particles are still not substrates for digestion. The protein colipase (another pancreatic secretory product) must bind the lipase to the particle to permit lipid degradation. The final digestion products are carried in micelles to the intestinal epithelium where they are transported across the plasma membrane (Figure 22.5). If the production of bile salts is inadequate due to liver disease, large amounts of fats (as much as 30 g day^{-1}) are excreted in the feces. This condition is referred to as *steatorrhea,* after stearic acid, a common fatty acid.

Glycocholate

FIGURE 22.4 Glycocholate. Bile acids, such as glycocholate, facilitate lipid digestion in the intestine.

Dietary lipids are transported in chylomicrons

In the intestinal mucosal cells, the triacylglycerols are resynthesized from fatty acids and monoacylglycerols and then packaged into lipoprotein transport particles called *chylomicrons*, stable particles approximately 2000 Å (200 nm) in diameter (Figure 22.5). These particles are composed mainly of triacylglycerols, with apolipoprotein B-48 (apo B-48) as the main protein component. Protein constituents of lipoprotein particles are called *apolipoproteins*. Chylomicrons also transport fat-soluble vitamins and cholesterol.

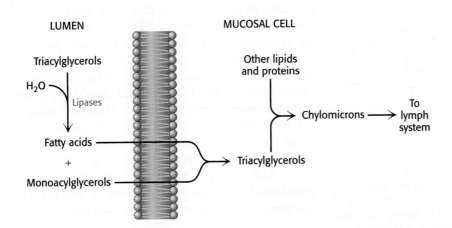

FIGURE 22.5 Chylomicron formation. Free fatty acids and monoacylglycerols are absorbed by intestinal epithelial cells. Triacylglycerols are resynthesized and packaged with other lipids and apolipoprotein B-48 to form chylomicrons, which are then released into the lymph system.

The chylomicrons are released into the lymph system and then into the blood. These particles bind to membrane-bound lipases, primarily at adipose tissue and muscle, where the triacylglycerols are once again degraded into free fatty acids and monoacylglycerols for transport into the tissue. The triacylglycerols are then resynthesized inside the cell and stored. In the muscle, they can be oxidized to provide energy.

22.2 The Use of Fatty Acids as Fuel Requires Three Stages of Processing

Tissues throughout the body gain access to the lipid energy reserves stored in adipose tissue through three stages of processing. First, the lipids must be mobilized. In this process, triacylglycerols are degraded to fatty acids and glycerol, which are released from the adipose tissue and transported to the energy-requiring tissues. Second, at these tissues, the fatty acids must be activated and transported into mitochondria for degradation. Third, the fatty acids are broken down in a step-by-step fashion into acetyl CoA, which is then processed in the citric acid cycle.

Triacylglycerols are hydrolyzed by hormone-stimulated lipases

Consider someone who has just awakened from a night's sleep and begins a bout of exercise. Glycogen stores will be low, but lipids are readily available. How are these lipid stores mobilized?

Before fats can be used as fuels, the triacylglycerol storage form must be hydrolyzed to yield isolated fatty acids. This reaction is catalyzed by hormonally controlled lipases. Under the physiological conditions

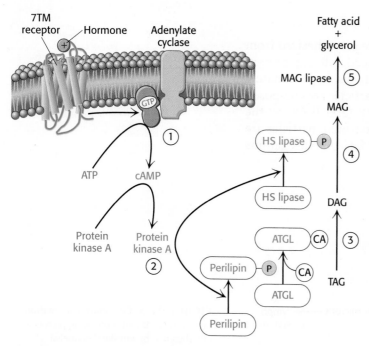

FIGURE 22.6 Mobilization of triacylglycerols. Triacylglycerols in adipose tissue are converted into free fatty acids in response to hormonal signals. 1. The hormones activate protein kinase A through the cAMP cascade. 2. Protein kinase A phosphorylates perilipin, resulting in the restructuring of the lipid droplet and release of the coactivator of ATGL, thereby activating ATGL. 3. ATGL converts triacylglycerol into diacylglycerol. 4. Hormone-sensitive lipase releases a fatty acid from diacylglycerol, generating monoacylglycerol. 5. Monoacylglycerol lipase completes the mobilization process. Abbreviations: 7TM, seven transmembrane receptor; ATGL, adipose triglyceride lipase; CA, coactivator; HS lipase, hormone-sensitive lipase; MAG lipase, monoacylglycerol lipase; DAG, diacylglycerol; TAG, triacylglycerol.

facing an early-morning runner, glucagon and epinephrine will be present. In adipose tissue, these hormones trigger 7 TM receptors that activate adenylate cyclase (Section 14.1). The increased level of cyclic AMP then stimulates protein kinase A, which phosphorylates two key proteins: *perilipin*, a fat-droplet-associated protein, and hormone-sensitive lipase (Figure 22.6). The phosphorylation of perilipin has two crucial effects. First, it restructures the fat droplet so the triacylglycerols are more accessible to the mobilization. Second, the phosphorylation of perilipin triggers the release of a coactivator for *adipose triglyceride lipase* (ATGL). Once bound to the coactivator, ATGL initiates the mobilization of triacylglycerols by releasing a fatty acid from triacylglycerol, forming diacylglycerol. Diacylglycerol is converted into a free fatty acid and monoacylglycerol by the hormone-sensitive lipase. Finally, a monoacylglycerol lipase completes the mobilization of fatty acids with the production of a free fatty acid and glycerol. Thus, *epinephrine and glucagon induce lipolysis.* Although their role in muscle is not as firmly established, these hormones probably also regulate the use of triacylglycerol stores in that tissue.

If the coactivator required by ATGL is missing or defective, a rare condition (incidence unknown) called Chanarin-Dorfman syndrome results. Fats accumulate throughout the body because they cannot be released by ATGL. Other symptoms include dry skin (ichthyosis), enlarged liver, muscle weakness, and mild cognitive disability.

Although adipocytes are the primary site of triacylglycerol metabolism, the liver plays a crucial role in lipid metabolism, as we will examine in more detail in Chapters 26 and 27. Hepatocytes are the most common type of liver cells, and these cells function in all aspects of lipid metabolism, including import, synthesis, storage, and secretion of lipids. These processes are responsive to diet and energy needs of the liver and other tissues. Recent research suggests that the epinephrine/cAMP pathway also regulates lipolysis from lipid droplets in cultured hepatocytes. Interestingly, this signaling pathway is disrupted by ethanol, and this disruption may play a role in the development of fatty liver, a risk factor for obesity and diabetes. See Biochemistry in Focus for further discussion on the ethanol metabolism in the liver.

Free fatty acids and glycerol are released into the blood

Fatty acids are not soluble in aqueous solutions. In order to reach tissues that require fatty acids, the released fatty acids bind to the blood protein albumin, which delivers them to tissues in need of fuel.

Glycerol formed by lipolysis is absorbed by the liver and phosphorylated. It is then oxidized to dihydroxyacetone phosphate, which is isomerized to glyceraldehyde 3-phosphate. This molecule is an intermediate in both the glycolytic and the gluconeogenic pathways.

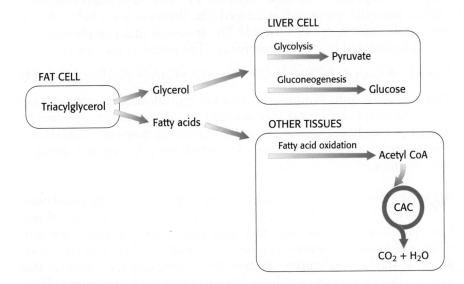

Hence, glycerol can be converted into pyruvate or glucose in the liver, which contains the appropriate enzymes (Figure 22.7). The reverse process can take place by the reduction of dihydroxyacetone phosphate to glycerol 3-phosphate. Hydrolysis by a phosphatase then gives glycerol. Thus, glycerol and glycolytic intermediates are readily interconvertible.

LIVER CELL

Glycolysis → Pyruvate

Gluconeogenesis → Glucose

FAT CELL

Triacylglycerol

Glycerol

Fatty acids

OTHER TISSUES

Fatty acid oxidation → Acetyl CoA

CAC

$CO_2 + H_2O$

FIGURE 22.7 Lipolysis generates fatty acids and glycerol. The fatty acids are used as fuel by many tissues. The liver processes glycerol by either the glycolytic or the gluconeogenic pathway, depending on its metabolic circumstances. Abbreviation: CAC, citric acid cycle.

Fatty acids are linked to coenzyme A before they are oxidized

Fatty acids separate from the albumin in the blood stream and diffuse across the cell membrane with the assistance of fatty acid transport protein. In the cell, fatty acids are shuttled about bound to a fatty-acid-binding-protein.

Fatty acid oxidation occurs in the mitochondrial matrix, but in order to enter the mitochondria, the fatty acids are first activated through the formation of a thioester linkage to coenzyme A. Adenosine triphosphate (ATP) drives the formation of the thioester linkage between the carboxyl group of a fatty acid and the sulfhydryl group of coenzyme A. This activation reaction takes place on the outer mitochondrial membrane, where it is catalyzed by *acyl CoA synthetase* (also called *fatty acid thiokinase*).

Acyl CoA synthetase accomplishes the activation of a fatty acid in two steps. First, the fatty acid reacts with ATP to form an *acyl adenylate*. In this mixed anhydride, the carboxyl group of a fatty acid is bonded to the phosphoryl group of AMP. The other two phosphoryl groups of the ATP substrate are released as pyrophosphate. In the second step, the sulfhydryl group of coenzyme A attacks the acyl adenylate, which is tightly bound to the enzyme, to form acyl CoA and AMP.

Acyl adenylate

$$\text{Fatty acid} + \text{ATP} \rightleftharpoons \text{Acyl adenylate} + \text{PP}_i \qquad (1)$$

$$\text{Acyl (AMP)} + \text{HS—CoA} \rightleftharpoons \text{Acyl CoA} + \text{AMP} \qquad (2)$$

These partial reactions are freely reversible. In fact, the equilibrium constant for the sum of these reactions is close to 1. One high-transfer-potential compound is cleaved (between PP$_i$ and AMP) and one high-transfer-potential compound is formed (the thioester acyl CoA). How is the overall reaction driven forward? The answer is that pyrophosphate is rapidly hydrolyzed by a pyrophosphatase. The complete reaction is

$$\text{RCOO}^- + \text{CoA} + \text{ATP} + \text{H}_2\text{O} \longrightarrow \text{RCO-CoA} + \text{AMP} + 2\,\text{P}_i + 2\,\text{H}^+$$

This reaction is quite favorable because the equivalent of two molecules of ATP is hydrolyzed, whereas only one high-transfer-potential compound is formed. We see here another example of a recurring theme in biochemistry: *Many biosynthetic reactions are made irreversible by the hydrolysis of inorganic pyrophosphate.*

Another motif recurs in this activation reaction. The enzyme-bound acyl adenylate intermediate is not unique to the synthesis of acyl CoA. *Acyl adenylates are frequently formed when carboxyl groups are activated in biochemical reactions.* Amino acids are activated for protein synthesis by a similar mechanism (Section 31.2), although the enzymes that catalyze this process are not homologous to acyl CoA synthetase. Thus, *activation by adenylation recurs in part because of convergent evolution.*

Carnitine carries long-chain activated fatty acids into the mitochondrial matrix

Fatty acids are activated on the outer mitochondrial membrane, whereas they are oxidized in the mitochondrial matrix. A special transport mechanism is needed to carry activated long-chain fatty acids across the inner mitochondrial membrane. These fatty acids must be conjugated to *carnitine*, an alcohol with both a positive and a negative charge (a zwitterion). The acyl group is transferred from the sulfur atom of coenzyme A to the hydroxyl group of carnitine to form *acyl carnitine*. This reaction is catalyzed by *carnitine acyltransferase I*, also called *carnitine palmitoyl transferase I*, which is bound to the outer mitochondrial membrane.

Acyl CoA + **Carnitine** $\rightleftharpoons$ **Acyl carnitine** + **HS—CoA**

Acyl carnitine is then shuttled across the inner mitochondrial membrane by a translocase (Figure 22.8). The acyl group is transferred back to coenzyme A on the matrix side of the membrane. This reaction, which is catalyzed

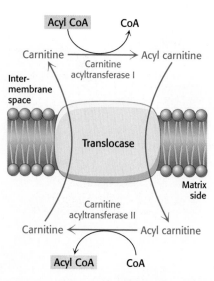

FIGURE 22.8 Acyl carnitine translocase. The entry of acyl carnitine into the mitochondrial matrix is mediated by a translocase. Carnitine returns to the cytoplasmic side of the inner mitochondrial membrane in exchange for acyl carnitine.

by *carnitine acyltransferase II (carnitine palmitoyl transferase II)*, is simply the reverse of the reaction that takes place in the cytoplasm. The reaction is thermodynamically feasible because of the zwitterionic nature of carnitine. The O-acyl link in carnitine has a high group-transfer potential, apparently because, being zwitterions, carnitine and its esters are solvated differently from most other alcohols and their esters. Finally, the translocase returns carnitine to the cytoplasmic side in exchange for an incoming acyl carnitine.

A number of diseases have been traced to a deficiency of carnitine, the transferase, or the translocase. Inability to synthesize carnitine may be a contributing factor to the development of autism in males. The symptoms of carnitine deficiency range from mild muscle cramping to severe weakness and even death. In general, muscle, kidney, and heart are the tissues primarily impaired. Muscle weakness during prolonged exercise is a symptom of a deficiency of carnitine acyltransferases because muscle relies on fatty acids as a long-term source of energy. Medium-chain ($C_8 - C_{10}$) fatty acids are oxidized normally in these patients because these fatty acids can enter the mitochondria, to some degree, in the absence of carnitine. These diseases illustrate that *the impaired flow of a metabolite from one compartment of a cell to another can lead to a pathological condition.*

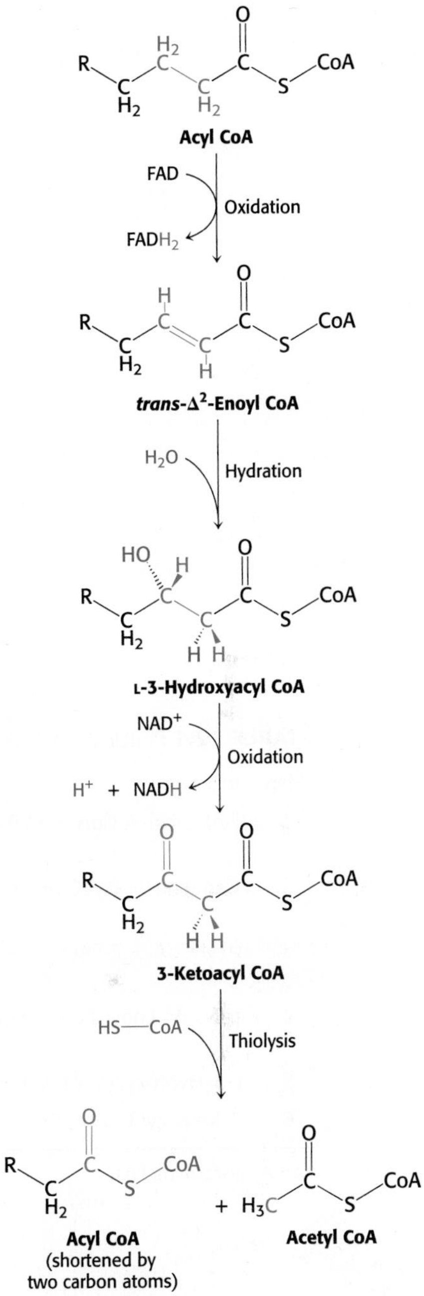

Acetyl CoA, NADH, and FADH₂ are generated in each round of fatty acid oxidation

A saturated acyl CoA is degraded by a recurring sequence of four reactions: oxidation by flavin adenine dinucleotide (FAD), hydration, oxidation by NAD^+, and thiolysis by coenzyme A (Figure 22.9). The fatty acid chain is shortened by two carbon atoms as a result of these reactions and $FADH_2$, NADH, and acetyl CoA are generated. Because oxidation takes place at the β carbon atom, this series of reactions is called the *β-oxidation pathway*.

The first reaction in each round of degradation is the *oxidation* of acyl CoA by an *acyl CoA dehydrogenase* to give an enoyl CoA with a *trans* double bond between C-2 and C-3.

$$\text{Acyl CoA} + \text{E-FAD} \longrightarrow \textit{trans-}\Delta^2\text{-enoyl CoA} + \text{E-FADH}_2$$

As in the dehydrogenation of succinate in the citric acid cycle, FAD rather than NAD^+ is the electron acceptor because the ΔG for this reaction is insufficient to drive the reduction of NAD^+. Electrons from the $FADH_2$ prosthetic group of the reduced acyl CoA dehydrogenase are transferred to a second flavoprotein called *electron-transferring flavoprotein (ETF)*. In turn, ETF donates electrons to *ETF: ubiquinone reductase*, an iron–sulfur protein. Ubiquinone is thereby reduced to ubiquinol, which delivers its high-potential electrons to the second proton-pumping site of the respiratory chain (Section 18.3). Consequently, 1.5 molecules of ATP are generated per molecule of $FADH_2$ formed in this dehydrogenation step, as in the oxidation of succinate to fumarate.

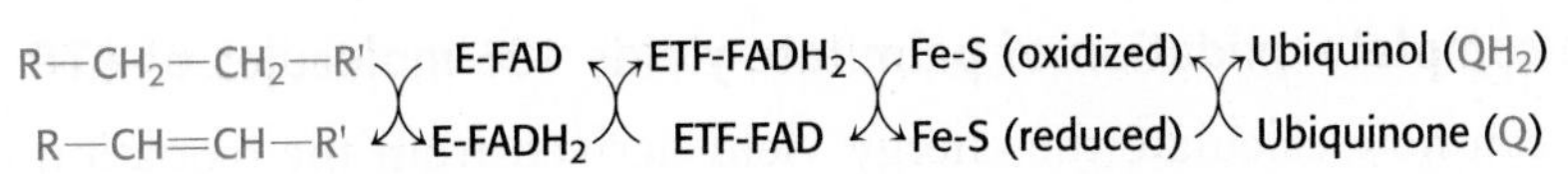

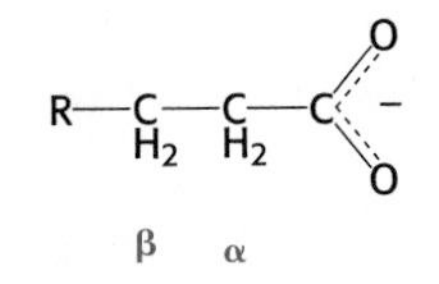

R—CH₂—CH₂—R′   E-FAD   ETF-FADH₂   Fe-S (oxidized)   Ubiquinol (QH₂)

R—CH=CH—R′   E-FADH₂   ETF-FAD   Fe-S (reduced)   Ubiquinone (Q)

The next step is the *hydration* of the double bond between C-2 and C-3 by *enoyl CoA hydratase*.

$$\textit{trans-}\Delta^2\text{-Enoyl CoA} + \text{H}_2\text{O} \longrightarrow \text{L-3-hydroxyacyl CoA}$$

Acyl CoA

FAD — Oxidation — FADH₂

trans-Δ²-Enoyl CoA

H₂O — Hydration

L-3-Hydroxyacyl CoA

NAD⁺ — Oxidation — H⁺ + NADH

3-Ketoacyl CoA

HS—CoA — Thiolysis

Acyl CoA (shortened by two carbon atoms) + **Acetyl CoA**

FIGURE 22.9 Reaction sequence for the degradation of fatty acids. Fatty acids are degraded by the repetition of a four-reaction sequence consisting of oxidation, hydration, oxidation, and thiolysis.

EXPLORE this pathway further in the Metabolic Map on 🐾 Sapling Plus

The hydration of enoyl CoA is stereospecific. Only the L isomer of 3-hydroxyacyl CoA is formed when the $trans$-Δ^2 double bond is hydrated. The enzyme also hydrates a cis-Δ^2 double bond, but the product then is the D isomer. We shall return to this point shortly in considering how unsaturated fatty acids are oxidized.

The hydration of enoyl CoA is a prelude to the second *oxidation* reaction, which converts the hydroxyl group at C-3 into a keto group and generates NADH. This oxidation is catalyzed by *L-3-hydroxyacyl CoA dehydrogenase*, which is specific for the L isomer of the hydroxyacyl substrate.

$$\text{L-3-Hydroxyacyl CoA} + \text{NAD}^+ \rightleftharpoons \text{3-ketoacyl CoA} + \text{NADH} + \text{H}^+$$

The preceding reactions have oxidized the methylene group at C-3 to a keto group. The final step is the *cleavage* of 3-ketoacyl CoA by the thiol group of a second molecule of coenzyme A, which yields acetyl CoA and an acyl CoA shortened by two carbon atoms. This thiolytic cleavage is catalyzed by β-*ketothiolase*.

$$\begin{array}{cc} \text{3-Ketoacyl CoA} + \text{HS-CoA} \rightleftharpoons & \text{acetyl CoA} + \text{acyl CoA} \\ (n \text{ carbons}) & (n - 2 \text{ carbons}) \end{array}$$

Table 22.1 summarizes the reactions in fatty acid oxidation.

TABLE 22.1 Principal reactions in fatty acid oxidation

Step	Reaction	Enzyme
1	Fatty acid + CoA + ATP $\rightleftharpoons$ acyl CoA + AMP + PP$_i$	Acyl CoA synthetase (also called fatty acid thiokinase and fatty acid:CoA ligase)*
2	Carnitine + acyl CoA $\rightleftharpoons$ acyl carnitine + CoA	Carnitine acyltransferase (also called carnitine palmitoyl transferase)
3	Acyl CoA + E-FAD $\longrightarrow$ $trans$-Δ^2-enoyl CoA + E-FADH$_2$	Acyl CoA dehydrogenases (several isozymes having different chain-length specificity)
4	$trans$-Δ^2-Enoyl CoA + H$_2$O $\rightleftharpoons$ L-3-hydroxyacyl CoA	Enoyl CoA hydratase (also called crotonase or 3-hydroxyacyl CoA hydrolyase)
5	L-3-Hydroxyacyl CoA + NAD$^+$ $\rightleftharpoons$ 3-ketoacyl CoA + NADH + H$^+$	L-3-Hydroxyacyl CoA dehydrogenase
6	3-Ketoacyl CoA + CoA $\rightleftharpoons$ acetyl CoA + acyl CoA (shortened by C$_2$)	β-Ketothiolase (also called thiolase)

*An AMP-forming ligase.

The shortened acyl CoA then undergoes another cycle of oxidation, starting with the reaction catalyzed by acyl CoA dehydrogenase (Figure 22.10). Fatty acid chains containing from 12 to 18 carbon atoms are oxidized by the long-chain acyl CoA dehydrogenase. The medium-chain acyl CoA dehydrogenase oxidizes fatty acid chains having from 4 to 14 carbons, whereas the short-chain acyl CoA dehydrogenase acts only on 4- and 6-carbon fatty acid chains. In contrast, β-ketothiolase, hydroxyacyl dehydrogenase, and enoyl CoA hydratase act on fatty acid molecules of almost any length.

The complete oxidation of palmitate yields 106 molecules of ATP

We can now calculate the energy yield derived from the oxidation of a fatty acid. In each reaction cycle, an acyl CoA is shortened by two carbon atoms, and one molecule each of FADH$_2$, NADH, and acetyl CoA are formed.

$$\text{C}_n\text{-acyl CoA} + \text{FAD} + \text{NAD}^+ + \text{H}_2\text{O} + \text{CoA} \longrightarrow$$
$$\text{C}_{n-2}\text{-acyl CoA} + \text{FADH}_2 + \text{NADH} + \text{acetyl CoA} + \text{H}^+$$

The degradation of palmitoyl CoA (C_{16}-acyl CoA) requires seven reaction cycles. In the seventh cycle, the C_4-ketoacyl CoA is thiolyzed to two molecules of acetyl CoA. Hence, the stoichiometry of the oxidation of palmitoyl CoA is

$$\text{Palmitoyl CoA} + 7\,\text{FAD} + 7\,\text{NAD}^+ + 7\,\text{CoA} + 7\,\text{H}_2\text{O} \longrightarrow$$
$$8\,\text{acetyl CoA} + 7\,\text{FADH}_2 + 7\,\text{NADH} + 7\,\text{H}^+$$

Approximately 2.5 molecules of ATP are generated when the respiratory chain oxidizes each of these NADH molecules, whereas 1.5 molecules of ATP are formed for each $FADH_2$ because their electrons enter the chain at the level of ubiquinol. Recall that the oxidation of acetyl CoA by the citric acid cycle yields 10 molecules of ATP. Hence, the number of ATP molecules formed in the oxidation of palmitoyl CoA is 10.5 from the seven $FADH_2$, 17.5 from the seven NADH, and 80 from the eight acetyl CoA molecules, which gives a total of 108. The equivalent of 2 molecules of ATP is consumed in the activation of palmitate, in which ATP is split into AMP and 2 molecules of orthophosphate. Thus, *the complete oxidation of a molecule of palmitate yields 106 molecules of ATP.*

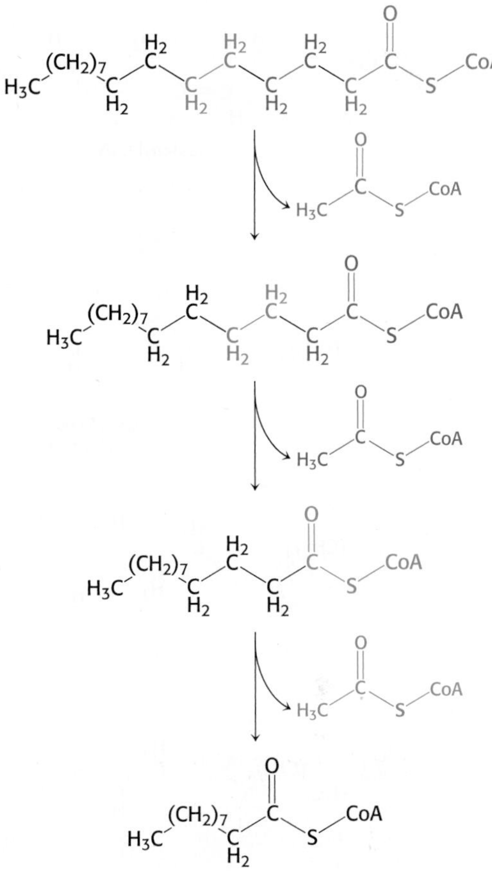

22.3 Unsaturated and Odd-Chain Fatty Acids Require Additional Steps for Degradation

The β-oxidation pathway accomplishes the complete degradation of saturated fatty acids having an even number of carbon atoms. Most fatty acids have such structures because of their mode of synthesis (to be addressed later in this chapter). However, not all fatty acids are so simple. The oxidation of fatty acids containing double bonds requires additional steps, as does the oxidation of fatty acids containing an odd number of carbon atoms.

An isomerase and a reductase are required for the oxidation of unsaturated fatty acids

The oxidation of unsaturated fatty acids presents some difficulties, yet many such fatty acids are available in the diet. Most of the reactions are the same as those for saturated fatty acids. In fact, only two additional enzymes—an isomerase and a reductase—are needed to degrade a wide range of unsaturated fatty acids.

Consider the oxidation of palmitoleate (Figure 22.11). This C_{16} unsaturated fatty acid, which has one double bond between C-9 and C-10, is activated and transported across the inner mitochondrial membrane in the same way as saturated fatty acids are. Palmitoleoyl CoA then undergoes three cycles of degradation, which are carried out by the same enzymes as those in the oxidation of saturated fatty acids. However, the *cis*-Δ^3-enoyl CoA formed in the third round is not a substrate for acyl CoA dehydrogenase. The presence of a double bond between C-3 and C-4 prevents the formation of another double bond between C-2 and C-3. This impasse is resolved by a new reaction that shifts the position and configuration of the *cis*-Δ^3 double bond. *cis*-Δ^3-*Enoyl CoA isomerase* converts this double bond into a *trans*-Δ^2 double bond. The double bond is now between C-2 and C-3. The subsequent reactions are those of the saturated fatty acid oxidation pathway, in which the *trans*-Δ^2-enoyl CoA is a regular substrate.

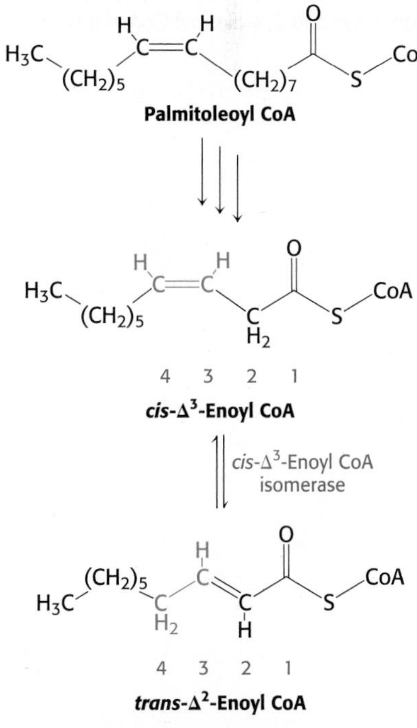

FIGURE 22.10 First three rounds in the degradation of palmitate. Two-carbon units are sequentially removed from the carboxyl end of the fatty acid.

Palmitoleoyl CoA

cis-Δ^3-**Enoyl CoA**

cis-Δ^3-Enoyl CoA isomerase

trans-Δ^2-**Enoyl CoA**

FIGURE 22.11 The degradation of a monounsaturated fatty acid. *Cis*-Δ^3-Enoyl CoA isomerase allows continued β-oxidation of fatty acids with a single double bond.

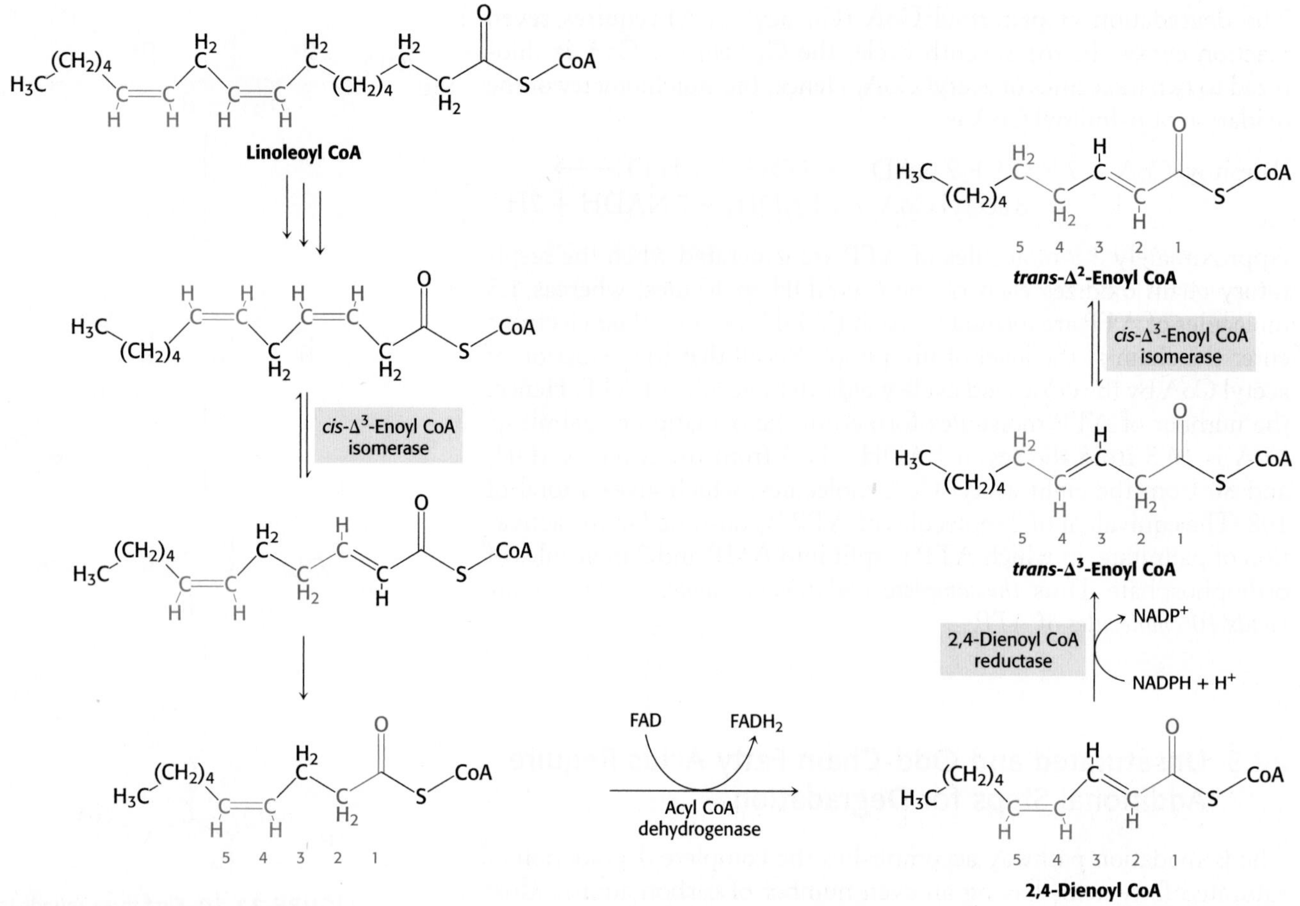

FIGURE 22.12 Oxidation of linoleoyl CoA. The complete oxidation of the diunsaturated fatty acid linoleate is facilitated by the activity of enoyl CoA isomerase and 2,4-dienoyl CoA reductase.

Human beings require polyunsaturated fatty acids (p. 737), which have multiple double bonds, as important precursors for signal molecules. Excess polyunsaturated fatty acids are degraded by β-oxidation. However, another problem arises with the oxidation of polyunsaturated fatty acids. Consider linoleate, a C_{18} polyunsaturated fatty acid with cis-Δ^9 and cis-Δ^{12} double bonds (Figure 22.12). The cis-Δ^3 double bond (between carbons 3 and 4) formed after three rounds of β-oxidation is converted into a $trans$-Δ^2 double bond (between carbons 2 and 3) by the aforementioned isomerase. The acyl CoA produced by another round of β-oxidation contains a cis-Δ^4 (between carbons 4 and 5) double bond. Dehydrogenation of this species by acyl CoA dehydrogenase yields a *2,4-dienoyl intermediate* (double bond between carbons 2 and 3 and carbons 4 and 5), which is not a substrate for the next enzyme in the β-oxidation pathway. This impasse is circumvented by *2,4-dienoyl CoA reductase*, an enzyme that uses NADPH to reduce the 2,4-dienoyl intermediate to $trans$-Δ^3-enoyl CoA. cis-Δ^3-Enoyl CoA isomerase then converts $trans$-Δ^3-enoyl CoA into the $trans$-Δ^2 form, a customary intermediate in the β-oxidation pathway. These catalytic strategies are elegant and economical. Only two extra enzymes are needed for the oxidation of *any* polyunsaturated fatty acid. *Odd-numbered double bonds are handled by the isomerase, and even-numbered ones by the reductase and the isomerase.*

Odd-chain fatty acids yield propionyl CoA in the final thiolysis step

Fatty acids having an odd number of carbon atoms are minor species. They are oxidized in the same way as fatty acids having an even number, except that propionyl CoA and acetyl CoA, rather than two molecules of acetyl CoA, are produced in the final round of degradation. The activated three-carbon unit in propionyl CoA enters the citric acid cycle after it has been converted into succinyl CoA.

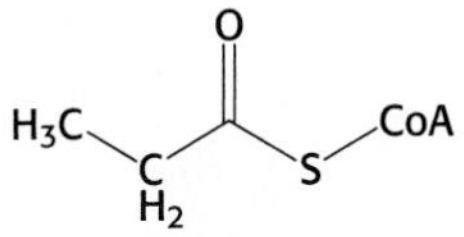

The pathway from propionyl CoA to succinyl CoA is especially interesting because it entails a rearrangement that requires *vitamin B_{12}* (also known as *cobalamin*). Propionyl CoA is carboxylated at the expense of the hydrolysis of a molecule of ATP to yield the D isomer of methylmalonyl CoA (Figure 22.13). This carboxylation reaction is catalyzed by *propionyl CoA carboxylase*, a biotin enzyme that has a catalytic mechanism like that of the homologous enzyme pyruvate carboxylase (Section 16.3). The D isomer of methylmalonyl CoA is racemized to the L isomer, the substrate for a mutase that converts it into *succinyl CoA* by an *intramolecular rearrangement*. The —CO—S—CoA group migrates from C-2 to a methyl group in exchange for a hydrogen atom. This very unusual isomerization is catalyzed by *methylmalonyl CoA mutase*, which contains a derivative of cobalamin as its coenzyme.

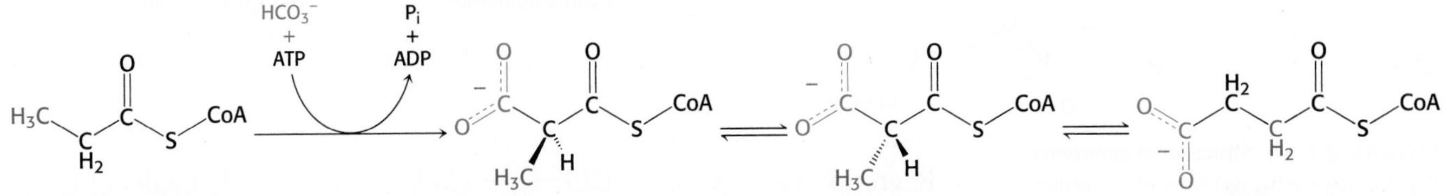

Propionyl CoA

FIGURE 22.13 Conversion of propionyl CoA into succinyl CoA. Propionyl CoA, generated from fatty acids with an odd number of carbons as well as some amino acids, is converted into the citric acid cycle intermediate succinyl CoA.

Vitamin B_{12} contains a corrin ring and a cobalt atom

Cobalamin enzymes, which are present in most organisms, catalyze three types of reactions: (1) *intramolecular rearrangements*; (2) *methylations*, as in the synthesis of methionine (Section 24.2); and (3) the *reduction of ribonucleotides to deoxyribonucleotides* (Section 25.3). In mammals, only two reactions are known to require coenzyme B_{12}. The conversion of L-methylmalonyl CoA into succinyl CoA is one, and the formation of methionine by the methylation of homocysteine is the other (Section 24.2). The latter reaction is especially important because methionine is required for the generation of coenzymes that participate in the synthesis of purines and thymine, which are needed for nucleic acid synthesis.

The core of cobalamin consists of a *corrin ring with a central cobalt atom* (Figure 22.14). The corrin ring, like a porphyrin, has *four pyrrole units*. Two of them are directly bonded to each other, whereas the others are joined by methine bridges, as in porphyrins. The corrin ring is more reduced than that of porphyrins and the substituents are different. A cobalt atom is bonded to the four pyrrole nitrogens. The fifth substituent linked to the cobalt atom is a derivative of *dimethylbenzimidazole* that contains ribose 3-phosphate and aminoisopropanol. One of the nitrogen atoms of dimethylbenzimidazole is linked to the cobalt atom. In coenzyme B_{12}, *the sixth substituent* linked to the cobalt atom is a 5'-*deoxyadenosyl unit* or a methyl group. This position can also be occupied by a cyano group. Cyanocobalamin is the form of the coenzyme administered to treat B_{12} deficiency. In all of these compounds, the cobalt is in the $+3$ oxidation state.

Mechanism: Methylmalonyl CoA mutase catalyzes a rearrangement to form succinyl CoA

The rearrangement reactions catalyzed by coenzyme B_{12} are exchanges of two groups attached to adjacent carbon atoms of the substrate (Figure 22.15). A hydrogen atom migrates from one carbon atom to the next, and an

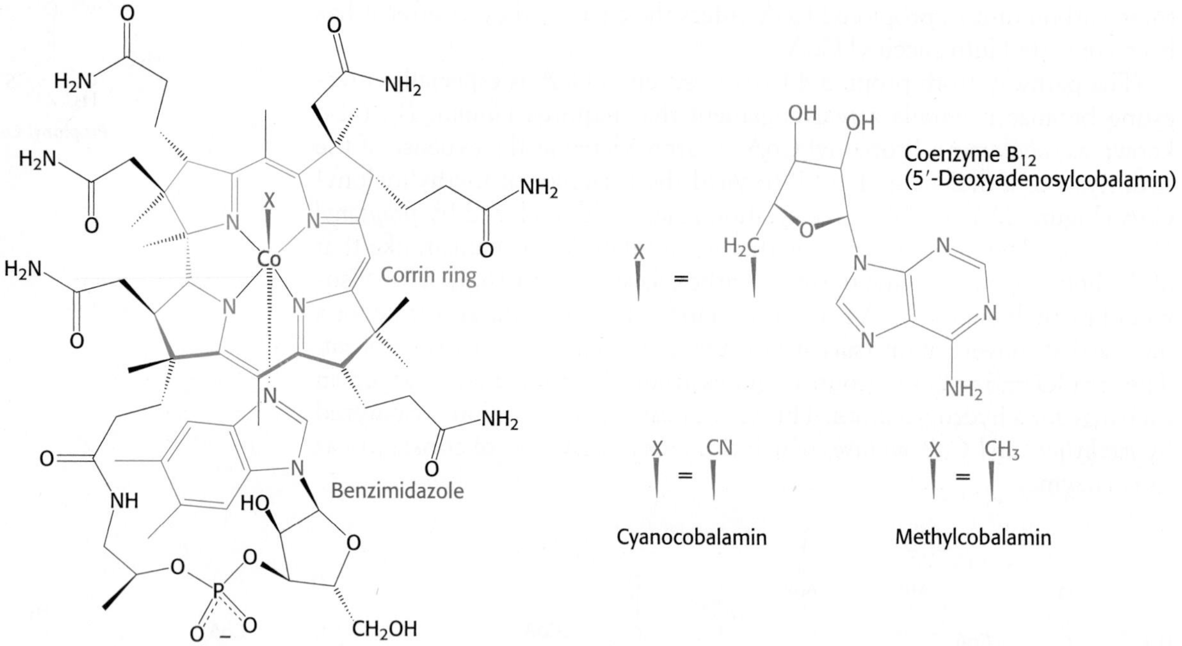

FIGURE 22.14 Structure of coenzyme B₁₂. Coenzyme B_{12} is a class of molecules that vary, depending on the component designated X in the left-hand structure. 5′-Deoxyadenosylcobalamin is the form of the coenzyme in methylmalonyl mutase. Substitution of methyl and cyano groups for X creates methylcobalamin and cyanocobalamin, respectively.

R group (such as the —CO—S—CoA group of methylmalonyl CoA) concomitantly moves in the reverse direction. The first step in these intra-molecular rearrangements is the cleavage of the carbon–cobalt bond of 5′-deoxyadenosylcobalamin to generate the Co^{2+} form of the coenzyme and a 5′-deoxyadenosyl radical, —CH₂· (Figure 22.16). In this *homolytic cleavage reaction*, one electron of the Co–C bond stays with Co (reducing it from the +3 to the +2 oxidation state), whereas the other electron stays with the carbon atom, generating a free radical. In contrast, nearly all other cleavage reactions in biological systems are *heterolytic:* an electron *pair* is transferred to one of the two atoms that were bonded together.

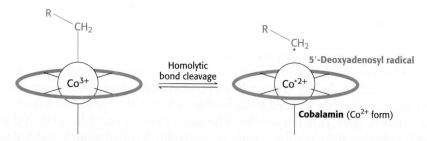

FIGURE 22.15 Rearrangement reaction catalyzed by cobalamin enzymes. The R group can be an amino group, a hydroxyl group, or a substituted carbon.

FIGURE 22.16 Formation of a 5′-deoxyadenosyl radical. The methylmalonyl CoA mutase reaction begins with the homolytic cleavage of the bond joining Co^{3+} of coenzyme B_{12} to a carbon atom of the ribose of the adenosine moiety of the enzyme. The cleavage generates a 5′-deoxyadenosyl radical and leads to the reduction of Co^{3+} to Co^{2+}. The letter R represents the 5′-deoxyadenosyl component of the coenzyme, and the green oval represents the remainder of the coenzyme.

What is the role of this very unusual —CH₂· radical? This highly reactive species abstracts a *hydrogen atom* from the substrate to form 5′-deoxyadenosine and a substrate radical (Figure 22.17). This substrate radical spontaneously rearranges: the carbonyl CoA group migrates to the position formerly occupied by H on the neighboring carbon atom to produce a different radical. This product radical abstracts a hydrogen atom from the methyl group of 5′-deoxyadenosine to complete the rearrangement and return the deoxyadenosyl unit to the radical form. *The role of coenzyme B_{12} in such intramolecular migrations is to serve as a source of free radicals for the abstraction of hydrogen atoms.*

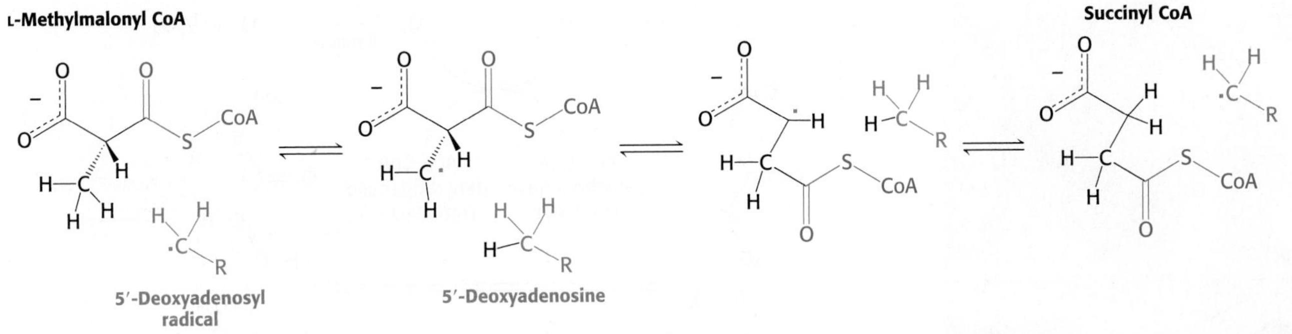

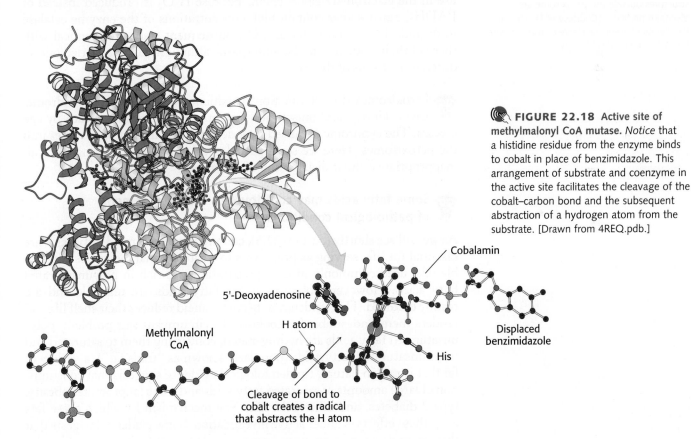

L-Methylmalonyl CoA

Succinyl CoA

5'-Deoxyadenosyl
radical

5'-Deoxyadenosine

FIGURE 22.17 Formation of succinyl CoA by a rearrangement reaction. A free radical abstracts a hydrogen atom in the rearrangement of methylmalonyl CoA to succinyl CoA.

An essential property of coenzyme B_{12} is the weakness of its cobalt–carbon bond, which is readily cleaved to generate a radical. To facilitate the cleavage of this bond, enzymes such as methylmalonyl CoA mutase displace the benzimidazole group from the cobalamin and bind to the cobalt atom through a histidine residue (Figure 22.18). The steric crowding around the cobalt–carbon bond within the corrin ring system contributes to the bond weakness.

FIGURE 22.18 Active site of methylmalonyl CoA mutase. *Notice* that a histidine residue from the enzyme binds to cobalt in place of benzimidazole. This arrangement of substrate and coenzyme in the active site facilitates the cleavage of the cobalt–carbon bond and the subsequent abstraction of a hydrogen atom from the substrate. [Drawn from 4REQ.pdb.]

Cobalamin

5'-Deoxyadenosine

H atom

Methylmalonyl
CoA

Displaced
benzimidazole

His

Cleavage of bond to
cobalt creates a radical
that abstracts the H atom

Fatty acids are also oxidized in peroxisomes

Although most fatty acid oxidation takes place in mitochondria, oxidation of long chain and branched fatty acids takes place in cellular organelles called *peroxisomes* (Figure 22.19) (problems 17 and 45). These organelles are small membrane-bounded compartments that are present in the cells of most eukaryotes. Fatty acid oxidation in these organelles, which halts at octanoyl CoA, serves to shorten very long chains (C_{26}) to make them better substrates of β-oxidation in mitochondria. Peroxisomal oxidation differs from β-oxidation in the initial dehydrogenation reaction (Figure 22.20). In peroxisomes, acyl CoA dehydrogenase—a flavoprotein—transfers electrons

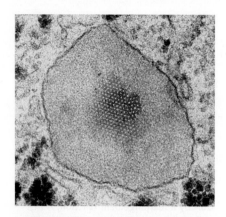

FIGURE 22.19 Electron micrograph of a peroxisome in a liver cell. A crystal of urate oxidase, a peroxisomal enzyme, is present inside at the center of the organelle. The peroxisome is bounded by a single bilayer membrane. The dark granular structures outside the peroxisome are glycogen particles. [Courtesy of Dr. George Palade/Yale University, Harvey Cushing/John Hay Whitney Medical Library.]

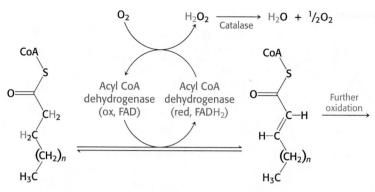

FIGURE 22.20 Initiation of peroxisomal fatty acid degradation. The first dehydrogenation in the degradation of fatty acids in peroxisomes requires a flavoprotein dehydrogenase that transfers electrons from its $FADH_2$ moiety to O_2 to yield H_2O_2.

from the substrate to $FADH_2$ and then to O_2 to yield H_2O_2. In mitochondrial β-oxidation, the high-energy electrons would be captured as $FADH_2$ for use in the electron-transport chain. Because H_2O_2 is produced instead of $FADH_2$, peroxisomes contain high concentrations of the enzyme catalase to degrade H_2O_2 into water and O_2. Subsequent steps are identical with those of their mitochondrial counterparts, although they are carried out by different isoforms of the enzymes.

Peroxisomes do not function in patients with Zellweger syndrome. Liver, kidney, and muscle abnormalities usually lead to death by age 6 years. The syndrome is caused by a defect in the import of enzymes into the peroxisomes. Here we see a pathological condition resulting from an inappropriate cellular distribution of enzymes.

Some fatty acids may contribute to the development of pathological conditions

As we will see shortly (Section 22.5), certain polyunsaturated fatty acids are essential for life, serving as precursors to various signal molecules. Vegetable oils, used commonly in food preparation, are rich in polyunsaturated fatty acids. However, polyunsaturated fatty acids are unstable and are readily oxidized. This tendency to become rancid reduces their shelf life and renders them undesirable for cooking. To circumvent this problem, polyunsaturated fatty acids are hydrogenated, converting them to saturated and trans unsaturated fatty acids (popularly known as "trans fat"), a variety of fat that is rare in nature. Epidemiological evidence suggests that consumption of large amounts of saturated fatty acids and trans fat promotes obesity, type 2 diabetes, and atherosclerosis. The mechanism by which these fats exert these effects is under active investigation. Some evidence suggests that they promote an inflammatory response and may mute the action of insulin and other hormones (Section 27.3).

22.4 Ketone Bodies Are a Fuel Source Derived from Fats

The acetyl CoA formed in fatty acid oxidation enters the citric acid cycle only if fat and carbohydrate degradation are appropriately balanced. Acetyl CoA must combine with oxaloacetate to gain entry to the citric acid cycle. The availability of oxaloacetate, however, depends on an adequate supply of carbohydrate. Recall that oxaloacetate is normally formed from pyruvate, the product of glucose degradation in glycolysis, by pyruvate carboxylase (Section 16.3). If carbohydrate is unavailable or improperly utilized, the concentration of oxaloacetate is lowered and acetyl CoA cannot enter the citric acid cycle. This dependency is the molecular basis of the adage that *fats burn in the flame of carbohydrates*.

In fasting or diabetes, oxaloacetate is consumed to form glucose by the gluconeogenic pathway (Section 16.3) and hence is unavailable for condensation with acetyl CoA. Under these conditions, acetyl CoA is diverted to the formation of acetoacetate and D-3-hydroxybutyrate. Acetoacetate, D-3-hydroxybutyrate, and acetone are often referred to as *ketone bodies*. Abnormally high levels of ketone bodies are present in the blood of untreated diabetics.

Acetoacetate is formed from acetyl CoA in three steps (Figure 22.21). Two molecules of acetyl CoA condense to form acetoacetyl CoA. This reaction, which is catalyzed by thiolase, is the reverse of the thiolysis step in the oxidation of fatty acids. Acetoacetyl CoA then reacts with acetyl CoA and water to give 3-hydroxy-3-methylglutaryl CoA (HMG-CoA) and CoA. This condensation resembles the one catalyzed by citrate synthase (Section 17.2). This reaction, which has a favorable equilibrium owing to the hydrolysis of a thioester linkage, compensates for the unfavorable equilibrium in the formation of acetoacetyl CoA. 3-Hydroxy-3-methylglutaryl CoA is then cleaved to acetyl CoA and acetoacetate. The sum of these reactions is

$$2 \text{ Acetyl CoA} + H_2O \longrightarrow \text{acetoacetate} + 2\text{CoA} + H^+$$

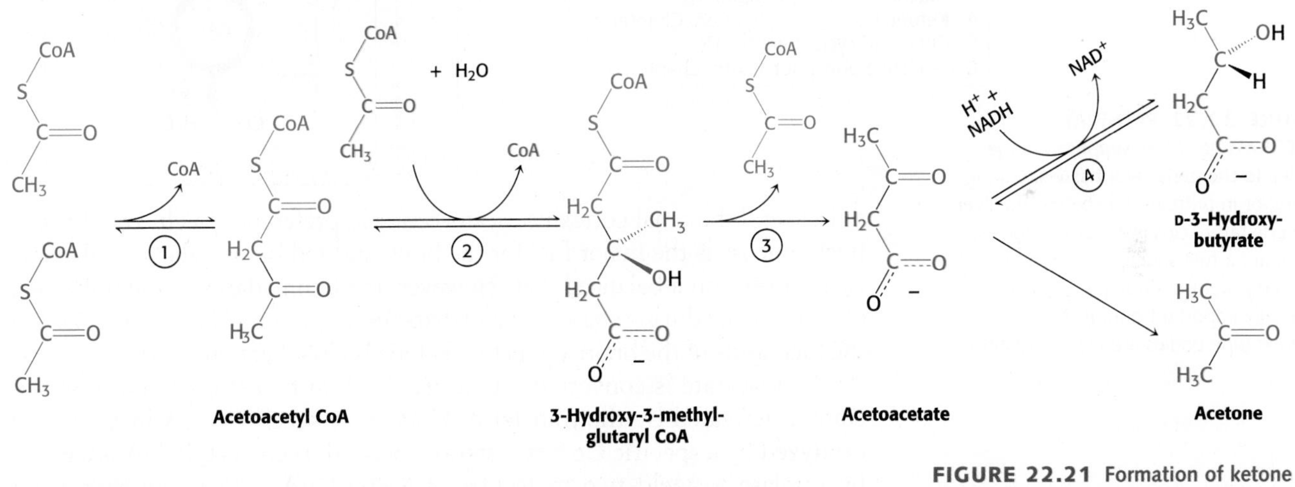

Acetoacetyl CoA   **3-Hydroxy-3-methyl-glutaryl CoA**   **Acetoacetate**   **Acetone**

FIGURE 22.21 Formation of ketone bodies. The ketone bodies—acetoacetate, D-3-hydroxybutyrate, and acetone— are formed from acetyl CoA primarily in the liver. Enzymes catalyzing these reactions are (1) 3-ketothiolase, (2) hydroxymethylglutaryl CoA synthase, (3) hydroxymethylglutaryl CoA cleavage enzyme, and (4) D-3-hydroxybutyrate dehydrogenase. Acetoacetate spontaneously decarboxylates to form acetone.

D-3-Hydroxybutyrate is formed by the reduction of acetoacetate in the mitochondrial matrix by D-3-hydroxybutyrate dehydrogenase. The ratio of hydroxybutyrate to acetoacetate depends on the $NADH/NAD^+$ ratio inside mitochondria.

Because it is a β-ketoacid, acetoacetate also undergoes a slow, spontaneous decarboxylation to acetone. The odor of acetone may be detected in the breath of a person who has a high level of acetoacetate in the blood. Under starvation conditions, the acetone may be captured to synthesize glucose.

Ketone bodies are a major fuel in some tissues

The liver is the major site of the production of acetoacetate and 3-hydroxybutyrate. These molecules are carried from the liver mitochondria into the blood by transport proteins and are conveyed to other tissues such as heart and kidney (Figure 22.22). Acetoacetate and 3-hydroxybutyrate are normal fuels of respiration and are quantitatively important as sources of energy. Indeed, heart

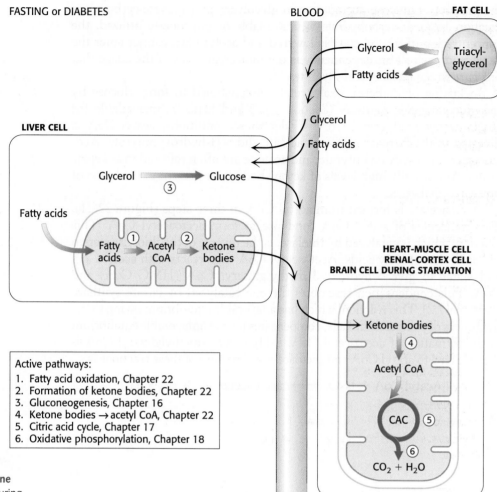

FASTING or DIABETES

FAT CELL

Active pathways:
1. Fatty acid oxidation, Chapter 22
2. Formation of ketone bodies, Chapter 22
3. Gluconeogenesis, Chapter 16
4. Ketone bodies → acetyl CoA, Chapter 22
5. Citric acid cycle, Chapter 17
6. Oxidative phosphorylation, Chapter 18

FIGURE 22.22 PATHWAY INTEGRATION: Liver supplies ketone bodies to the peripheral tissues. During fasting or in untreated diabetics, the liver converts fatty acids into ketone bodies, which are a fuel source for a number of tissues. Ketone-body production is especially important during starvation, when ketone bodies are the predominant fuel.

muscle and the renal cortex use acetoacetate in preference to glucose. In contrast, glucose is the major fuel for the brain and red blood cells in well-nourished people on a balanced diet. However, the brain adapts to the utilization of acetoacetate during starvation and diabetes. In prolonged starvation, 75% of the fuel needs of the brain are met by ketone bodies (Section 27.5).

Acetoacetate is converted into acetyl CoA in two steps. First, acetoacetate is activated by the transfer of CoA from succinyl CoA in a reaction catalyzed by a specific CoA transferase. Second, acetoacetyl CoA is cleaved by thiolase to yield two molecules of acetyl CoA, which can then enter the citric acid cycle (Figure 22.23). The liver has acetoacetate available to supply to other organs because it lacks this particular CoA transferase. 3-Hydroxybutyrate requires an additional step to yield acetyl CoA. It is first oxidized to produce acetoacetate, which is processed as described up to this point, and NADH for use in oxidative phosphorylation.

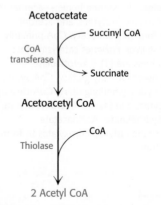

FIGURE 22.23 Utilization of acetoacetate as a fuel. Acetoacetate can be converted into two molecules of acetyl CoA, which then enter the citric acid cycle.

D-3-Hydroxybutyrate   Acetoacetate

Ketone bodies can be regarded as a water-soluble, transportable form of acetyl units. Fatty acids are released by adipose tissue and converted into acetyl units by the liver, which then exports them as acetoacetate. As might be expected, acetoacetate also has a regulatory role. *A high concentration of acetoacetate in the blood signifies an abundance of acetyl units and leads to a decrease in the rate of lipolysis in adipose tissue.*

A high concentration of ketone bodies in the blood, the result of certain pathological conditions, can be life threatening. The most common of these conditions is diabetic ketosis in patients with insulin-dependent diabetes mellitus. These patients are unable to produce insulin. As stated earlier, this hormone, normally released after meals, signals tissues to take up glucose. In addition, it curtails fatty acid mobilization by adipose tissue. The absence of insulin has two major biochemical consequences (Figure 22.24). First, the liver cannot absorb glucose and consequently cannot provide oxaloacetate to process fatty acid-derived acetyl CoA. Second, adipose cells continue to release fatty acids into the bloodstream; these fatty acids are then taken up by the liver and converted into ketone bodies. The liver thus produces large amounts of ketone bodies, which are moderately strong acids. The result is severe acidosis. The decrease in pH impairs tissue function, most importantly in the central nervous system.

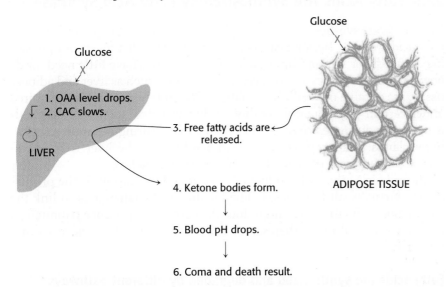

FIGURE 22.24 Diabetic ketosis results when insulin is absent. In the absence of insulin, fats are released from adipose tissue, and glucose cannot be absorbed by the liver or adipose tissue. The liver degrades the fatty acids by β-oxidation but cannot process the acetyl CoA, because of a lack of glucose-derived oxaloacetate (OAA). Excess ketone bodies are formed and released into the blood.

Interestingly, diets that promote ketone-body formation, called *ketogenic diets,* are frequently used as a therapeutic option for children with drug-resistant epilepsy. Ketogenic diets are rich in fats and low in carbohydrates, with adequate amounts of protein. In essence, the body is forced into starvation mode, where fats and ketone bodies become the main fuel source (Section 27.5). Remarkably, recent research in mice suggests that ketogenic diets alter the intestinal flora—the microbiome—and it is this alteration in the microbiome that is responsible for the therapeutic effects of the diet. The altered microbiome in some fashion regulates the levels of certain neurotransmitters, thereby reducing seizures.

Recent research in mice has also established that ketogenic diets extend lifespan, improve memory, and maintain long-term health. It will be interesting to learn if these effects apply to human beings, although gathering such data from humans is much more difficult than experiments with mice.

Animals cannot convert fatty acids into glucose

A typical human being has far greater fat stores than glycogen stores. However, glycogen is necessary to fuel very active muscle, as well as the brain, which normally uses only glucose as a fuel. When glycogen stores are low, why can't the body make use of fat stores and convert fatty acids into glucose? Because *animals are unable to accomplish the net synthesis of glucose from fatty acids.* Specifically, acetyl CoA cannot be converted into pyruvate or oxaloacetate in animals. Recall that the reaction that generates acetyl CoA from pyruvate is irreversible (Section 17.1). The two carbon atoms of the acetyl group of acetyl CoA enter the citric acid cycle, but two carbon atoms leave the cycle in the decarboxylations catalyzed by isocitrate dehydrogenase and α-ketoglutarate dehydrogenase. Consequently, oxaloacetate is regenerated, but it is not formed de novo when the acetyl unit of acetyl CoA is oxidized by the citric acid cycle. In essence, two carbon atoms enter the cycle as an acetyl group, but two carbons leave the cycle as CO_2 before oxaloacetate is generated. As a result, no net synthesis of oxaloacetate is possible. In contrast, plants have two additional enzymes enabling them to convert the carbon atoms of acetyl CoA into oxaloacetate (Section 17.5).

22.5 Fatty Acids Are Synthesized by Fatty Acid Synthase

Fatty acids are synthesized by a complex of enzymes that together are called *fatty acid synthase.* Because eating a typical Western diet meets our physiological needs for fats and lipids, adult human beings have little need for de novo fatty acid synthesis. However, many tissues, such as liver and adipose tissue, are capable of synthesizing fatty acids, and this synthesis is required under certain physiological conditions. For instance, fatty acid synthesis is necessary during embryonic development and during lactation in mammary glands. Inappropriate fatty acid synthesis in the liver from excess caloric consumption or alcohol contributes to liver failure.

Acetyl CoA, the end product of fatty acid degradation, is the precursor for virtually all fatty acids. The biochemical challenge is to link the two carbon units together and reduce the carbons to produce palmitate, a C_{16} fatty acid. Palmitate then serves as a precursor for the variety of other fatty acids.

Fatty acids are synthesized and degraded by different pathways

Although fatty acid synthesis is the reversal of the degradative pathway in regard to basic chemical reactions, the synthetic and degradative pathways are different mechanistically, again exemplifying the principle that *synthetic and degradative pathways are almost always distinct.* Some important differences between the pathways are as follows:

1. Synthesis takes place in the *cytoplasm,* in contrast with degradation, which takes place primarily in the mitochondrial matrix.

2. Intermediates in fatty acid synthesis are covalently linked to the sulfhydryl groups of an *acyl carrier protein* (ACP), whereas intermediates in fatty acid breakdown are covalently attached to the sulfhydryl group of coenzyme A.

3. The enzymes of fatty acid synthesis in higher organisms are joined in a *single polypeptide chain* called *fatty acid synthase.* In contrast, the degradative enzymes are not linked covalently.

4. The growing fatty acid chain is elongated by the *sequential addition of two-carbon units* derived from acetyl CoA. The activated donor of two-carbon units in the elongation step is *malonyl ACP.* The elongation reaction is driven by the release of CO_2.

5. The reductant in fatty acid synthesis is *NADPH*, whereas the oxidants in fatty acid degradation are NAD^+ and *FAD*.

6. The isomeric form of the hydroxyacyl intermediate in degradation is L, while the D form is used in synthesis.

The formation of malonyl CoA is the committed step in fatty acid synthesis

Fatty acid synthesis starts with the carboxylation of acetyl CoA to *malonyl CoA*. This irreversible reaction is the committed step in fatty acid synthesis.

The synthesis of malonyl CoA is catalyzed by the cytoplasmic enzyme *acetyl CoA carboxylase 1*, which contains a biotin prosthetic group. The carboxyl group of biotin is covalently attached to the ε amino group of a lysine residue, as in pyruvate carboxylase (Figure 16.24) and propionyl CoA carboxylase (p. 721). As with these other enzymes, a carboxybiotin intermediate is formed at the expense of the hydrolysis of a molecule of ATP. The activated CO_2 group in this intermediate is then transferred to acetyl CoA to form malonyl CoA.

$$\text{Biotin-enzyme} + \text{ATP} + \text{HCO}_3^- \rightleftharpoons \text{CO}_2\text{-biotin-enzyme} + \text{ADP} + \text{P}_i$$

$$\text{CO}_2\text{-biotin-enzyme} + \text{acetyl CoA} \longrightarrow \text{malonyl CoA} + \text{biotin-enzyme}$$

Acetyl CoA carboxylase 2, an isozyme of carboxylase 1 located in the mitochondria, is the essential regulatory enzyme for fatty acid metabolism.

Intermediates in fatty acid synthesis are attached to an acyl carrier protein

The intermediates in fatty acid synthesis are linked to an acyl carrier protein. Specifically, they are linked to the sulfhydryl terminus of a phosphopantetheine group. In the degradation of fatty acids, this unit is present as part of coenzyme A, whereas, in their synthesis, it is attached to a serine residue of the acyl carrier protein (Figure 22.25). Thus, ACP—a single polypeptide chain of 77 residues—can be regarded as a giant prosthetic group, a "macro CoA."

Fatty acid synthesis consists of a series of condensation, reduction, dehydration, and reduction reactions

The enzyme system that catalyzes the synthesis of saturated long-chain fatty acids from acetyl CoA, malonyl CoA, and NADPH is called *fatty acid synthase*. The synthase is actually a complex of distinct enzymes.

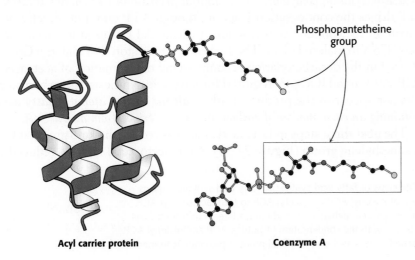

Phosphopantetheine group

Acyl carrier protein

Coenzyme A

FIGURE 22.25 Phosphopantetheine. Both acyl carrier protein and coenzyme A include phosphopantetheine as their reactive units. [Drawn from 1ACP.pdb.]

TABLE 22.2 Principal reactions in fatty acid synthesis in bacteria

Step	Reaction	Enzyme
1	Acetyl CoA + HCO_3^- + ATP $\longrightarrow$ malonyl CoA + ADP + P_i + H^+	Acetyl CoA carboxylase
2	Acetyl CoA + ACP $\rightleftharpoons$ acetyl ACP + CoA	Acetyl transacylase
3	Malonyl CoA + ACP $\rightleftharpoons$ malonyl ACP + CoA	Malonyl transacylase
4	Acetyl ACP + malonyl ACP $\longrightarrow$ acetoacetyl ACP + ACP + CO_2	β-Ketoacyl synthase
5	Acetoacetyl ACP + NADPH + H^+ $\rightleftharpoons$ D-3-hydroxybutyryl ACP + $NADP^+$	β-Ketoacyl reductase
6	D-3-Hydroxybutyryl ACP $\rightleftharpoons$ crotonyl ACP + H_2O	3-Hydroxyacyl dehydratase
7	Crotonyl ACP + NADPH + H^+ $\longrightarrow$ butyryl ACP + $NADP^+$	Enoyl reductase

The fatty acid synthase complex in bacteria is readily dissociated into individual enzymes when the cells are broken apart. The availability of these isolated enzymes has helped biochemists elucidate the steps in fatty acid synthesis (Table 22.2). In fact, the reactions leading to fatty acid synthesis in higher organisms are very much like those of bacteria.

The elongation phase of fatty acid synthesis starts with the formation of acetyl ACP and malonyl ACP. *Acetyl transacylase* and *malonyl transacylase* catalyze these reactions.

$$\text{Acetyl CoA} + \text{ACP} \rightleftharpoons \text{acetyl ACP} + \text{CoA}$$
$$\text{Malonyl CoA} + \text{ACP} \rightleftharpoons \text{malonyl ACP} + \text{CoA}$$

Malonyl transacylase is highly specific, whereas acetyl transacylase can transfer acyl groups other than the acetyl unit, though at a much slower rate. The synthesis of fatty acids with an odd number of carbon atoms starts with propionyl ACP, which is formed from propionyl CoA by acetyl transacylase.

Acetyl ACP and malonyl ACP react to form acetoacetyl ACP (Figure 22.26). The *β-ketoacyl synthase*, also called the condensing enzyme, catalyzes this condensation reaction.

$$\text{Acetyl ACP} + \text{malonyl ACP} \longrightarrow \text{acetoacetyl ACP} + \text{ACP} + CO_2$$

In the condensation reaction, a four-carbon unit is formed from a two-carbon unit and a three-carbon unit, and CO_2 is released. Why is the four-carbon unit not formed from two 2-carbon units—say, two molecules of acetyl ACP? The answer is that the equilibrium for the synthesis of acetoacetyl ACP from two molecules of acetyl ACP is highly unfavorable. In contrast, *the equilibrium is favorable if malonyl ACP is a reactant because its decarboxylation contributes a substantial decrease in free energy.* In effect, ATP drives the condensation reaction, though ATP does not directly participate in the condensation reaction. Instead, ATP is used to carboxylate acetyl CoA to malonyl CoA. The free energy thus stored in malonyl CoA is released in the decarboxylation accompanying the formation of acetoacetyl ACP. Although HCO_3^- is required for fatty acid synthesis, its carbon atom does not appear in the product. Rather, *all the carbon atoms of fatty acids containing an even number of carbon atoms are derived from acetyl CoA.*

The next three steps in fatty acid synthesis reduce the keto group at C-3 to a methylene group (Figure 22.26). First, acetoacetyl ACP is reduced to

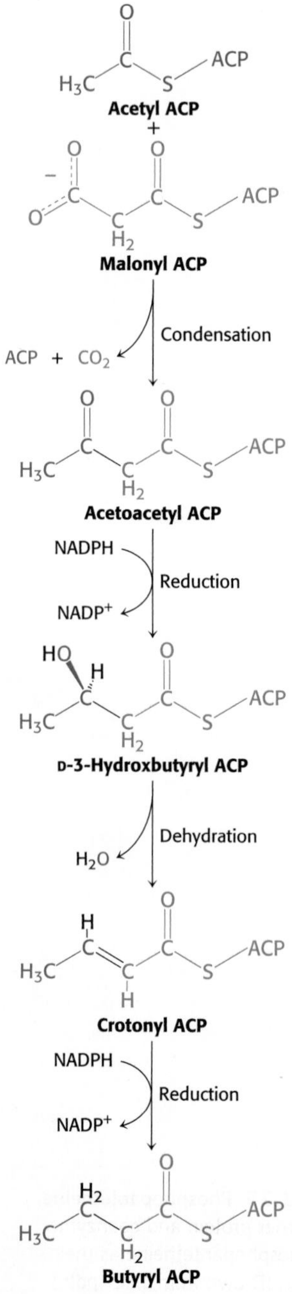

FIGURE 22.26 The steps of fatty acid synthesis. Fatty acid synthesis begins with the condensation of malonyl ACP and acetyl ACP to form acetoacetyl ACP. Acetoacetyl ACP is then reduced, dehydrated, and reduced again to form butyryl ACP. Another cycle begins with the condensation of butyryl ACP and malonyl ACP. The sequence of reactions is repeated until the final product palmitate is formed.

D-3-hydroxybutyryl ACP by β-ketoacyl reductase. This reaction differs from the corresponding one in fatty acid degradation in two respects: (1) the D rather than the L isomer is formed; and (2) NADPH is the reducing agent, whereas NAD^+ is the oxidizing agent in β-oxidation. This difference exemplifies the general principle that *NADPH is consumed in biosynthetic reactions, whereas NADH is generated in energy-yielding reactions.* Then, D-3-hydroxybutyryl ACP is *dehydrated* to form crotonyl ACP, which is a *trans*-Δ^2-enoyl ACP by 3-hydroxyacyl dehydratase. The final step in the cycle *reduces* crotonyl ACP to butyryl ACP. NADPH is again the reductant, whereas FAD is the oxidant in the corresponding reaction in β-oxidation. The bacterial enzyme that catalyzes this step, *enoyl reductase*, can be inhibited by *triclosan,* a broad-spectrum antibacterial agent that is added to a variety of products such as toothpaste, soaps, and skin creams. These last three reactions—a reduction, a dehydration, and a second reduction—convert acetoacetyl ACP into butyryl ACP, which completes the first elongation cycle.

In the second round of fatty acid synthesis, butyryl ACP condenses with malonyl ACP to form a C_6-β-ketoacyl ACP. This reaction is like the one in the first round, in which acetyl ACP condenses with malonyl ACP to form a C_4-β-ketoacyl ACP. Reduction, dehydration, and a second reduction convert the C_6-β-ketoacyl ACP into a C_6-acyl ACP, which is ready for a third round of elongation. The elongation cycles continue until C_{16}-acyl ACP is formed. This intermediate is a good substrate for a thioesterase that hydrolyzes C_{16}-acyl ACP to yield palmitate and ACP. *The thioesterase acts as a ruler to determine fatty acid chain length.* The synthesis of longer-chain fatty acids is discussed in Section 22.6.

Fatty acids are synthesized by a multifunctional enzyme complex in animals

Although the basic biochemical reactions in fatty acid synthesis are very similar in *E. coli* and eukaryotes, the structure of the synthase varies considerably. The component enzymes of animal fatty acid synthases, in contrast with those of *E. coli* and plants, are linked in a large polypeptide chain.

The structure of a large part of the mammalian fatty acid synthase has recently been determined, with the acyl carrier protein and thioesterase remaining to be resolved. The enzyme is a dimer of identical 270-kd subunits. Each chain contains all of the active sites required for activity, as well as an acyl carrier protein tethered to the complex (Figure 22.27). Despite the fact that each chain possesses all of the enzymes required for fatty acid synthesis, the monomers are not active. A dimer is required.

The two component chains interact such that the enzyme activities are partitioned into two distinct compartments, a lower body and an upper body joined by a waist. The lower body contains the β-ketosynthase, while the legs comprise the malonylacetyl transferase domains which are joined to the body by a linker domain. The body and the legs catalyze the condensing reactions. The upper body contains dehydratase and enoyl reductase, and the arms contain the ketoreductase domains. The upper body and the arms catalyze the modification reactions, the reduction and dehydration activities that result in the saturated fatty acid product. Interestingly, the mammalian fatty acid synthase has one active site—malonyl/acetyl transacylase—that adds both acetyl CoA and malonyl CoA. In contrast, most other fatty acid synthases have two separate enzyme activities—one

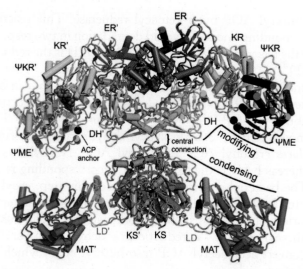

FIGURE 22.27 Structure of the mammalian fatty acid synthase. The complex consists of two parts; the body of the lower part is the β-ketosynthase (KS), while the malonylacetyl transferase (MAT) domains comprise the legs. These structures are joined by a linker domain (LD). The lower part is connected at the waist to the upper part, which consists of dehydratase (DH) and enoyl reductase (ER), which together comprise the upper body. The ketoreductase (KR) domains form the arms. The noncatalytic domains are ψKR and ψME. Bound NADP⁺ cofactors are shown as blue spheres and the attachment site for the acyl carrier protein and thioesterase is shown as a black sphere. The domains of the second chain are indicated by the abbreviation followed by a prime. [Republished with permission of American Assn for the Advancement of Science, from The Crystal Structure of a Mammalian Fatty Acid Synthase, Maier et al., *Science*, Vol. 321, September 5, 2008, pgs. 1315–1322, Figure 1a. © 2008. Permission conveyed through Copyright Clearance Center, Inc. Image courtesy Maier and Ban, ETH Zurich.]

for acetyl CoA and one for malonyl CoA. The enzyme also contains noncatalytic domains, designated by symbol ψ. The ψKR resembles the ketoreductase domain and the ψME domain is homologous to methyltransferase enzymes. Both the ψKR and ψME domains play a role in maintaining enzyme structure.

Let us consider one catalytic cycle of the fatty acid synthase complex (Figure 22.28). An elongation cycle begins when malonyl/acetyl transacylase (MAT) moves an acetyl unit from coenzyme A to the acyl carrier protein (ACP). β-Ketosynthase (β-KS) accepts the acetyl unit, which forms a thioester with a cysteine residue at the β-KS active site. The vacant ACP is reloaded by MAT, this time with a malonyl moiety. Malonyl ACP visits the active site of β-KS where the condensation of the two 2-carbon fragments takes place on the ACP with the concomitant release of CO_2. The selecting and condensing process concludes with the β-ketoacyl product attached to the ACP.

The loaded ACP then sequentially visits the active sites of the modification compartment of the enzyme. Here, the β-keto group of the substrate is reduced to —OH, dehydrated, and finally reduced to yield the saturated acyl product, still attached to the ACP. With the completion of the modification process, the reduced product is transferred to the β-KS while the ACP accepts another malonyl unit. Condensation takes place and is followed by another modification cycle. The process is repeated until the thioesterase releases the final C_{16} palmitic acid product.

Many eukaryotic multienzyme complexes are multifunctional proteins in which different enzymes are linked covalently. Multifunctional enzymes such as fatty acid synthase seem likely to have arisen in eukaryotic evolution by fusion of the individual genes of evolutionary ancestors.

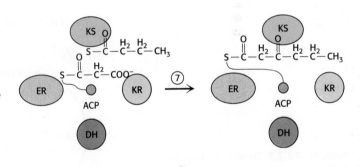

FIGURE 22.28 A catalytic cycle of mammalian fatty acid synthase. The cycle begins when MAT (not shown) attaches an acetyl unit to ACP. (1) ACP delivers the acetyl unit to KS, and MAT then attaches a malonyl unit to ACP. (2) ACP visits KS again, which condenses the acetyl and malonyl units to form the β-ketoacyl product, attached to the ACP. (3) ACP delivers the β-ketoacyl product to the KR enzyme, which reduces the keto group to an alcohol. (4) The β-hydroxyl product then visits DH, which introduces a double bond with the loss of water. (5) The enoyl product is delivered to the ER enzyme, where the double bond is reduced. (6) ACP hands the reduced product to KS and is recharged with malonyl CoA by MAT. (7) KS condenses the two molecules on ACP, which is now ready to begin another cycle. See Figure 22.27 for abbreviations.

The synthesis of palmitate requires 8 molecules of acetyl CoA, 14 molecules of NADPH, and 7 molecules of ATP

The stoichiometry of the synthesis of palmitate is

$$\text{Acetyl CoA} + 7\,\text{malonyl CoA} + 14\,\text{NADPH} + 20\,\text{H}^+ \longrightarrow$$
$$\text{palmitate} + 7\,\text{CO}_2 + 14\,\text{NADP}^+ + 8\,\text{CoA} + 6\,\text{H}_2\text{O}$$

The equation for the synthesis of the malonyl CoA used in the preceding reaction is

$$7\,\text{Acetyl CoA} + 7\,\text{CO}_2 + 7\,\text{ATP} \longrightarrow$$
$$7\,\text{malonyl CoA} + 7\,\text{ADP} + 7\,\text{P}_i + 14\,\text{H}^+$$

Hence, the overall stoichiometry for the synthesis of palmitate is

$$8\,\text{Acetyl CoA} + 7\,\text{ATP} + 14\,\text{NADPH} + 6\,\text{H}^+ \longrightarrow$$
$$\text{palmitate} + 14\,\text{NADP}^+ + 8\,\text{CoA} + 6\,\text{H}_2\text{O} + 7\,\text{ADP} + 7\,\text{P}_i$$

Citrate carries acetyl groups from mitochondria to the cytoplasm for fatty acid synthesis

Fatty acids are synthesized in the cytoplasm, whereas acetyl CoA is formed from pyruvate in mitochondria. Hence, acetyl CoA must be transferred from mitochondria to the cytoplasm for fatty acid synthesis.

Mitochondria, however, are not readily permeable to acetyl CoA. Recall that carnitine carries only long-chain fatty acids. *The barrier to acetyl CoA is bypassed by citrate, which carries acetyl groups across the inner mitochondrial membrane.* Citrate is formed in the mitochondrial matrix by the condensation of acetyl CoA with oxaloacetate (Figure 22.29). When present at high levels, citrate is transported to the cytoplasm, where it is cleaved by *ATP-citrate lyase*.

$$\text{Citrate} + \text{ATP} + \text{CoA} + \text{H}_2\text{O} \longrightarrow$$
$$\text{acetyl CoA} + \text{ADP} + \text{P}_i + \text{oxaloacetate}$$

This reaction occurs in three steps: (1) The formation of a phospho-enzyme with the donation of a phosphoryl group from ATP; (2) binding of citrate and CoA followed by the formation of citroyl CoA and release of the phosphate; and (3) cleavage of citroyl CoA to yield acetyl CoA and oxaloacetate. As we will see shortly (Section 22.6), citrate stimulates acetyl CoA carboxylase, the enzyme that regulates fatty acid metabolism. Recall also, the presence of citrate in the cytoplasm inhibits phosphofructokinase, the enzyme that controls the glycolytic pathway.

Lyases
Enzymes catalyzing the cleavage of C—C, C—O, or C—N bonds by elimination. A double bond is formed in these reactions.

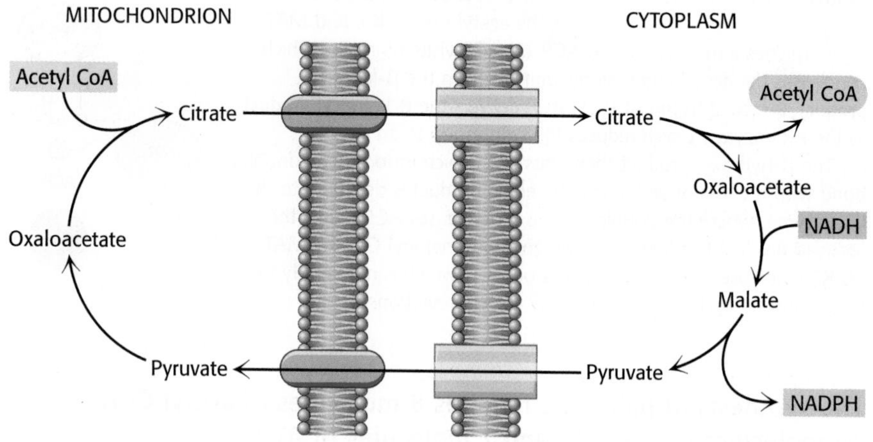

FIGURE 22.29 Transfer of acetyl CoA to the cytoplasm. Acetyl CoA is transferred from mitochondria to the cytoplasm, and the reducing potential of NADH is concomitantly converted into that of NADPH by this series of reactions.

ATP-citrate lyase is stimulated by insulin, which initiates a signal transduction pathway that ultimately results in the phosphorylation and activation of the lyase by protein kinase B (also called Akt).

Several sources supply NADPH for fatty acid synthesis

Oxaloacetate formed in the transfer of acetyl groups to the cytoplasm must now be returned to the mitochondria. The inner mitochondrial membrane is impermeable to oxaloacetate. Hence, a series of bypass reactions are needed. These reactions also generate much of the NADPH needed for fatty acid synthesis. First, oxaloacetate is reduced to malate by NADH. This reaction is catalyzed by a *malate dehydrogenase* in the cytoplasm.

$$\text{Oxaloacetate} + \text{NADH} + \text{H}^+ \rightleftharpoons \text{malate} + \text{NAD}^+$$

Second, malate is oxidatively decarboxylated by an *NADP$^+$-linked malate enzyme* (also called *malic enzyme*).

$$\text{Malate} + \text{NADP}^+ \longrightarrow \text{pyruvate} + \text{CO}_2 + \text{NADPH}$$

The pyruvate formed in this reaction readily enters mitochondria, where it is carboxylated to oxaloacetate by pyruvate carboxylase.

$$Pyruvate + CO_2 + ATP + H_2O \longrightarrow oxaloacetate + ADP + P_i + 2\,H^+$$

The sum of these three reactions is

$$NADP^+ + NADH + ATP + H_2O \longrightarrow$$
$$NADPH + NAD^+ + ADP + P_i + H^+$$

Thus, *one molecule of NADPH is generated for each molecule of acetyl CoA that is transferred from mitochondria to the cytoplasm.* Hence, eight molecules of NADPH are formed when eight molecules of acetyl CoA are transferred to the cytoplasm for the synthesis of palmitate. *The additional six molecules of NADPH required for this process come from the pentose phosphate pathway* (Section 20.3).

The accumulation of the precursors for fatty acid synthesis is a wonderful example of the coordinated use of multiple pathways. The citric acid cycle, transport of oxaloacetate from the mitochondria, and pentose phosphate pathway provide the carbon atoms and reducing power, whereas glycolysis and oxidative phosphorylation provide the ATP to meet the needs for fatty acid synthesis (Figure 22.30).

Fatty acid metabolism is altered in tumor cells

 We have previously seen that cancer cells alter glucose metabolism to meet the needs of rapid cell growth. Cancer cells must also increase fatty acid synthesis for use as signal molecules as well as for incorporation into membrane phospholipids. Many of the enzymes of fatty acid synthesis are overexpressed in most human cancers, and this expression is correlated with tumor malignancy. Recall that normal cells do little de novo fatty acid synthesis, relying instead on dietary intake to meet their fatty needs.

The dependence of de novo fatty acid synthesis provides possible therapeutic targets to inhibit cancer cell growth. Inhibition of β-ketoacyl ACP synthase—the enzyme that catalyzes the condensation step of fatty acid synthesis—does indeed inhibit phospholipid synthesis and subsequent cell growth in some cancers, apparently by inducing apoptosis (p. 611). However, another startling observation was made: *Mice treated with inhibitors of the β-ketoacyl ACP synthase showed remarkable weight loss* because they ate less. Thus, fatty acid synthase inhibitors are exciting candidates both as antitumor and as anti-obesity drugs.

Acetyl CoA carboxylase is also being investigated as a possible target for inhibiting cancer cell growth. Inhibition of the carboxylase in prostate and breast cancer cell lines induces apoptosis in the cancer cells, and yet is without effect in normal cells (problem 59). Understanding the alteration of fatty acid metabolism in cancer cells is a developing area of research that holds promise of generating new cancer therapies.

Triacylglycerols may become an important renewable energy source

 Work is underway to develop efficient means of generating triacylglycerols for use as renewable biodiesel fuel. Although the details of the

FIGURE 22.30 PATHWAY INTEGRATION: Fatty acid synthesis. Fatty acid synthesis requires the cooperation of various metabolic pathways located in different cellular compartments.

EXPLORE this pathway further in the Metabolic Map on **SaplingPlus**

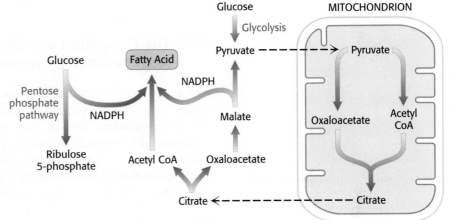

processes are quite complex, we will take an overview of the essential steps. First, CO_2, CO, and H_2 (collectively called syngas because they can be used to make synthetic natural gas) are captured from municipal waste or generated from other waste sources. Next, these gases are fed to acetogenic bacteria, an anaerobic species of bacteria that are ubiquitous in nature. Acetogenic bacteria contain a complex pathway, called the Wood-Ljungdahl pathway, that can synthesize acetate from syngas. In this pathway—which requires the folic acid, a carrier of single carbon molecules (Section 24.2)—CO_2 and CO are reduced to a methyl group. The methyl group then reacts with CO and coenzyme A to form acetyl CoA. A key enzyme in the pathway provides us with another example of a bifunctional enzyme. Carbon monoxide dehydrogenase/acetyl CoA synthase is a nickel-iron sulfur enzyme. The dehydrogenase component reduces CO_2 to CO and the synthase joins the CO with the methyl group and coenzyme A to form acetyl CoA. The acetate can be harvested from the bacteria then used as a carbon source for oleaginous yeast, which can synthesize and accumulate triacylglycerols to 20 to 70% of their cell mass, depending on culture conditions. Using large fermenters and complex algorithms to control bacterial and yeast growth, triacylglycerols can be synthesized in sufficient quantities for biodiesel and other renewable products.

22.6 The Elongation and Unsaturation of Fatty Acids Are Accomplished by Accessory Enzyme Systems

The major product of the fatty acid synthase is palmitate. In eukaryotes, longer fatty acids are formed by elongation reactions catalyzed by enzymes on the cytoplasmic face of the *endoplasmic reticulum membrane*. These reactions add two-carbon units sequentially to the carboxyl ends of both saturated and unsaturated fatty acyl CoA substrates. Malonyl CoA is the two-carbon donor in the elongation of fatty acyl CoAs. Again, condensation is driven by the decarboxylation of malonyl CoA.

Membrane-bound enzymes generate unsaturated fatty acids

Endoplasmic reticulum systems also introduce double bonds into long-chain acyl CoAs. For example, in the conversion of stearoyl CoA into oleoyl CoA, a *cis*-Δ^9 double bond is inserted by an oxidase that employs *molecular oxygen* and *NADH* (or *NADPH*).

$$\text{Stearoyl CoA} + \text{NADH} + \text{H}^+ + \text{O}_2 \longrightarrow$$
$$\text{oleoyl CoA} + \text{NAD}^+ + 2\,\text{H}_2\text{O}$$

This reaction is catalyzed by a complex of three membrane-bound proteins: *NADH-cytochrome b_5 reductase, cytochrome b_5, and stearoyl CoA desaturase* (Figure 22.31). First, electrons are transferred from NADH to the FAD moiety of NADH-cytochrome b_5 reductase. The heme iron atom of

FIGURE 22.31 Electron-transport chain in the desaturation of fatty acids.

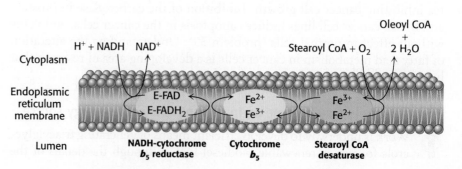

cytochrome b_5 is then reduced to the Fe^{2+} state. The nonheme iron atom of the desaturase is subsequently converted into the Fe^{2+} state, which enables it to interact with O_2 and the saturated fatty acyl CoA substrate. A double bond is formed and two molecules of H_2O are released. Two electrons come from NADH and two from the single bond of the fatty acyl substrate.

A variety of unsaturated fatty acids can be formed from oleate by a combination of elongation and desaturation reactions. For example, oleate can be elongated to a 20:1 *cis*-Δ^{11} fatty acid. Alternatively, a second double bond can be inserted to yield an 18:2 *cis*-Δ^6, Δ^9 fatty acid. Similarly, palmitate (16:0) can be oxidized to palmitoleate (16:1 *cis*-Δ^9), which can then be elongated to *cis*-vaccenate (18:1 *cis*-Δ^{11}).

Unsaturated fatty acids in mammals are derived from either palmitoleate (16:1), oleate (18:1), linoleate (18:2), or linolenate (18:3). The number of carbon atoms from the ω end of a derived unsaturated fatty acid to the nearest double bond identifies its precursor.

Mammals lack the enzymes to introduce double bonds at carbon atoms beyond C-9 in the fatty acid chain. Hence, mammals cannot synthesize linoleate (18:2 *cis*-Δ^9, Δ^{12}) and linolenate (18:3 *cis*-Δ^9, Δ^{12}, Δ^{15}). Linoleate and linolenate are the two essential fatty acids. The term *essential* means that they must be supplied in the diet because they are required by an organism and cannot be synthesized by the organism itself. Linoleate and linolenate furnished by the diet are the starting points for the synthesis of a variety of other unsaturated fatty acids.

Precursor	Formula
Linolenate (ω-3)	$CH_3{-}(CH_2)_2{=}CH{-}R$
Linoleate (ω-6)	$CH_3{-}(CH_2)_5{=}CH{-}R$
Palmitoleate (ω-7)	$CH_3{-}(CH_2)_6{=}CH{-}R$
Oleate (ω-9)	$CH_3{-}(CH_2)_8{=}CH{-}R$

Eicosanoid hormones are derived from polyunsaturated fatty acids

Arachidonate, a 20:4 fatty acid derived from linoleate, is the major precursor of several classes of signal molecules: prostaglandins, prostacyclins, thromboxanes, and leukotrienes (Figure 22.32).

A *prostaglandin is a 20-carbon fatty acid containing a 5-carbon ring* (Figure 22.33). This basic compound is modified by reductases and isomerases to yield nine major classes of prostaglandins, designated PGA through PGI; a subscript denotes the number of carbon–carbon double bonds outside the ring. Prostaglandins with two double bonds, such as PGE_2, are derived from arachidonate; the other two double bonds of this precursor are lost in forming a 5-membered ring. *Prostacyclin* and *thromboxanes* are related compounds that arise from a nascent prostaglandin. They are generated by *prostacyclin synthase* and *thromboxane synthase*, respectively. Alternatively, arachidonate can be converted into *leukotrienes* by the action of *lipoxygenase*. Leukotrienes, first found in leukocytes, contain three conjugated double bonds—hence the name. Prostaglandins, prostacyclin, thromboxanes, and leukotrienes are called *eicosanoids* (from the Greek *eikosi*, "twenty") because they contain 20 carbon atoms.

Prostaglandins and other eicosanoids are *local hormones* because they are short-lived. They alter the activities both of the cells in which they are synthesized and of adjoining cells by binding to 7TM receptors. Their effects may vary from one cell type to another, in contrast with the more-uniform actions of global hormones such as insulin and glucagon. Prostaglandins stimulate inflammation, regulate blood flow to particular organs, control ion transport across membranes, modulate synaptic transmission, and induce sleep.

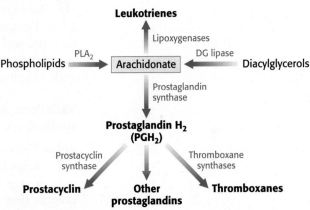

FIGURE 22.32 Arachidonate is the major precursor of eicosanoid hormones. Prostaglandin synthase catalyzes the first step in a pathway leading to prostaglandins, prostacyclins, and thromboxanes. Lipoxygenase catalyzes the initial step in a pathway leading to leukotrienes.

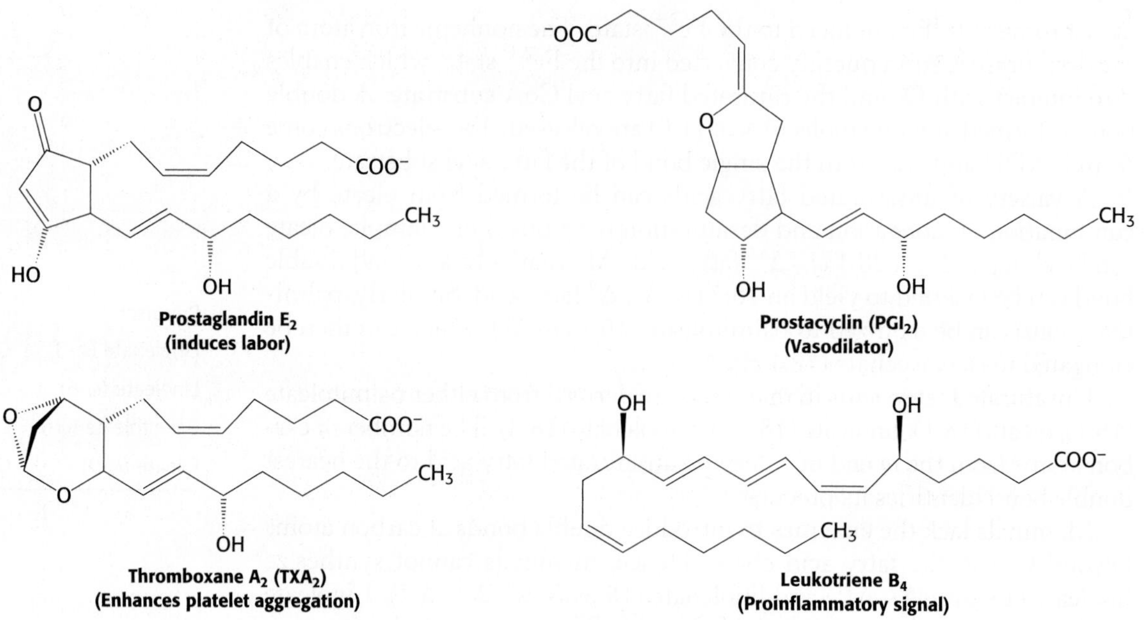

Prostaglandin E₂
(induces labor)

Prostacyclin (PGI₂)
(Vasodilator)

Thromboxane A₂ (TXA₂)
(Enhances platelet aggregation)

Leukotriene B₄
(Proinflammatory signal)

FIGURE 22.33 Structures of several eicosanoids.

Recall that aspirin (acetylsalicylate) blocks access to the active site of the enzyme that converts arachidonate into prostaglandin H_2 (Section 12.3). Because arachidonate is the precursor of other prostaglandins, prostacyclin, and thromboxanes, blocking this step interferes with many signaling pathways. Aspirin's ability to obstruct these pathways accounts for its wide-ranging effects on inflammation, fever, pain, and blood clotting.

Variations on a theme: Polyketide and nonribosomal peptide synthetases resemble fatty acid synthase

The mammalian multifunctional fatty acid synthase is a member of a large family of complex enzymes termed *megasynthases* that participate in step-by-step synthetic pathways. Two important classes of compounds that are synthesized by such enzymes are the *polyketides* and the *nonribosomal peptides*. These classes of compounds provide a variety of useful drugs, including antibiotics, immunosuppressants, antifungal agents, and anticancer drugs. The antibiotic erythromycin is an example of a polyketide, whereas penicillin (Section 8.5) is a nonribosomal peptide.

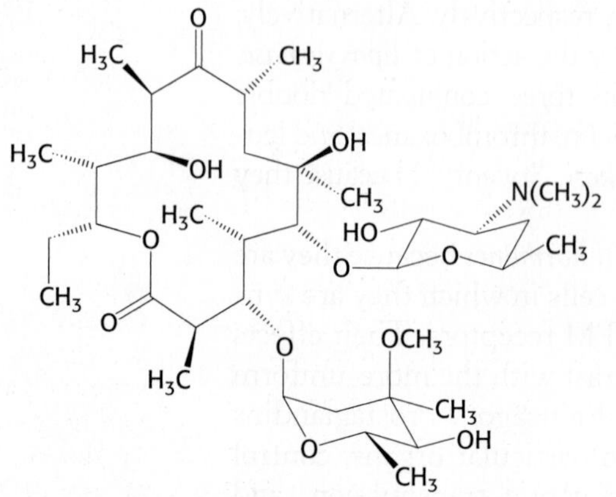

Erythromycin

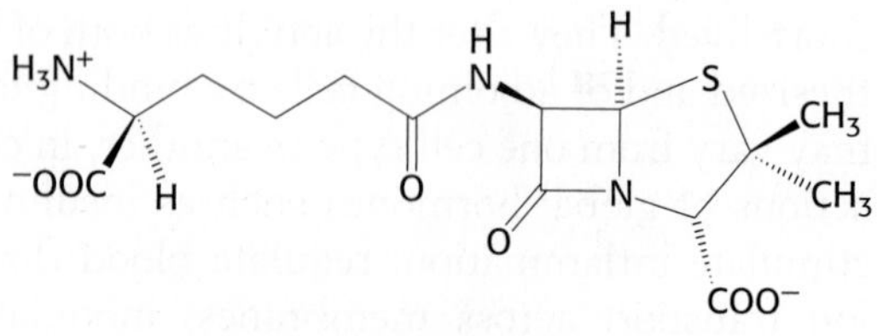

Penicillin

Research is underway to learn how to manipulate the pathways and the enzymes of polyketide and nonribosomal peptide synthesis in order to generate new therapeutics.

22.7 Acetyl CoA Carboxylase Plays a Key Role in Controlling Fatty Acid Metabolism

Fatty acid metabolism is stringently controlled so that synthesis and degradation are highly responsive to physiological needs. Fatty acid synthesis is maximal when carbohydrates and energy are plentiful and when fatty acids are scarce. *Acetyl CoA carboxylase 1 and 2 play essential roles in regulating fatty acid synthesis and degradation.* Recall that this enzyme catalyzes the committed step in fatty acid synthesis: the production of malonyl CoA (the activated two-carbon donor). This important enzyme is subject to both local and hormonal regulation. We will examine each of these levels of regulation in turn.

Acetyl CoA carboxylase is regulated by conditions in the cell

Acetyl CoA carboxylase 1 responds to changes in its immediate environment. *Acetyl CoA carboxylase 1 is switched off by phosphorylation and activated by dephosphorylation* (Figure 22.34). *AMP-activated protein kinase (AMPK)* converts the carboxylase into an inactive form by modifying three serine residues. AMPK is essentially a fuel gauge; it is activated by AMP and inhibited by ATP.

The carboxylase is also allosterically stimulated by citrate. The level of citrate is high when both acetyl CoA and ATP are abundant, signifying that raw materials and energy are available for fatty acid synthesis. Citrate acts in an unusual manner on inactive acetyl CoA carboxylase 1, which exists as isolated inactive dimers. Citrate facilitates the polymerization of the inactive dimers into active filaments (Figure 22.35). However, polymerization by citrate alone requires supraphysiological concentrations. In the cell, citrate-induced polymerization is facilitated by the protein MIG12, which greatly reduces the amount of citrate required. Polymerization can partly reverse the inhibition produced by phosphorylation (Figure 22.36). The stimulatory effect of citrate on the carboxylase is counteracted by *palmitoyl CoA*, which is abundant when there is an excess of fatty acids. Palmitoyl CoA causes the filaments to disassemble into the inactive subunits. Palmitoyl CoA also inhibits the translocase that transports citrate from mitochondria to the cytoplasm, as well as glucose 6-phosphate dehydrogenase, which generates NADPH in the pentose phosphate pathway.

The isozyme acetyl CoA carboxylase 2, located in the mitochondria, plays a role in the regulation of fatty acid degradation. Malonyl CoA, the product of the carboxylase reaction, is present at a high level when fuel molecules are abundant. *Malonyl CoA inhibits carnitine acyltransferase I, preventing the entry of fatty acyl CoAs into the mitochondrial matrix in times of plenty.* Malonyl CoA is an especially effective inhibitor of carnitine acyltransferase I in heart and muscle, tissues that have little fatty acid synthesis capacity of their own. In these tissues, acetyl CoA carboxylase 2 may be a purely regulatory enzyme. Acetyl CoA carboxylase 2 is also phosphorylated and inhibited by AMP-activated kinase. The reduction in the amount of mitochondrial malonyl CoA following the inhibition of AMP-activated kinase allows fatty acid transport into the mitochondria for β-oxidation.

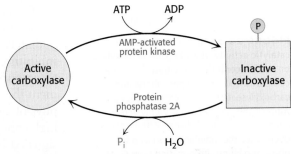

FIGURE 22.34 Control of acetyl CoA carboxylase. Acetyl CoA carboxylase is inhibited by phosphorylation.

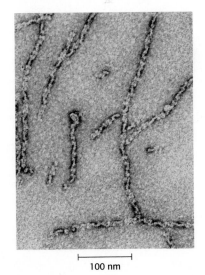

FIGURE 22.35 Filaments of acetyl CoA carboxylase. The electron micrograph shows the enzymatically active filamentous form of acetyl CoA carboxylase from chicken liver. The inactive form is a dimer of 265-kDa subunits. [Courtesy of Dr. M. Daniel Lane.]

(A)

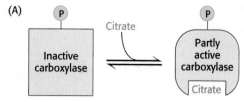

(B)

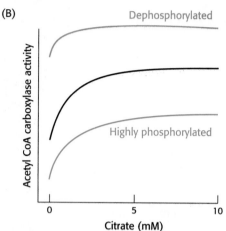

FIGURE 22.36 Dependence of the catalytic activity of acetyl CoA carboxylase on the concentration of citrate. (A) Citrate can partly activate the phosphorylated carboxylase. (B) The dephosphorylated form of the carboxylase is highly active even when citrate is absent. Citrate partly overcomes the inhibition produced by phosphorylation. [Information from G. M. Mabrouk, I. M. Helmy, K. G. Thampy, and S. J. Wakil. *J. Biol. Chem.* 265:6330–6338, 1990.]

Thus, AMPK inhibits fatty acid synthesis while stimulating fatty acid oxidation. Activators of AMPK are being investigated to treat nonalcoholic fatty liver disease (NAFLD), a condition due to overnutrition where fats accumulate in the liver. Untreated NAFLD is often a precursor to type 2 diabetes (Section 27.3).

Acetyl CoA carboxylase is regulated by a variety of hormones

Acetyl CoA carboxylase is controlled by the hormones glucagon, epinephrine, and insulin, which denote the overall energy status of the organism. *Insulin stimulates fatty acid synthesis by activating the carboxylase, whereas glucagon and epinephrine have the reverse effect.*

Regulation by glucagon and epinephrine. Consider, again, a person who has just awakened from a night's sleep and begins a bout of exercise. As mentioned, glycogen stores will be low, but lipids are readily available for mobilization.

As stated earlier, the hormones glucagon and epinephrine, present under conditions of fasting and exercise, will stimulate the mobilization of fatty acids from triacylglycerols in fat cells, which will be released into the blood, and probably from muscle cells, where they will be used immediately as fuel. These same hormones will inhibit fatty acid synthesis by inhibiting acetyl CoA carboxylase. Although the exact mechanism by which these hormones exert their effects is not known, the net result is to augment the inhibition by the AMP-activated kinase. This result makes sound physiological sense: when the energy level of the cell is low, as signified by a high concentration of AMP, and the energy level of the organism is low, as signaled by glucagon, fats should not be synthesized. Epinephrine, which signals the need for immediate energy, enhances this effect. Hence, *these catabolic hormones switch off fatty acid synthesis by keeping the carboxylase in the inactive phosphorylated state.*

Regulation by insulin. Now consider the situation after the exercise has ended and the runner has had a meal. In this case, the hormone insulin inhibits the mobilization of fatty acids and stimulates their accumulation as triacylglycerols by muscle and adipose tissue. Insulin also stimulates fatty acid synthesis by activating acetyl CoA carboxylase. Insulin activates the carboxylase by enhancing the phosphorylation and inactivation of AMPK by protein kinase B. Insulin also promotes the activity of a protein phosphatase that dephosphorylates and activates acetyl CoA carboxylase. Thus, the signal molecules glucagon, epinephrine, and insulin act in concert on triacylglycerol metabolism and acetyl CoA carboxylase to carefully regulate the utilization and storage of fatty acids.

Response to diet. *Long-term control is mediated by changes in the rates of synthesis and degradation of the enzymes participating in fatty acid synthesis.* Animals that have fasted and are then fed high-carbohydrate, low-fat diets show marked increases in their amounts of acetyl CoA carboxylase and fatty acid synthase within a few days. This type of regulation is known as *adaptive control.* This regulation, which is mediated both by insulin and by glucose, is at the level of gene transcription.

AMP-activated protein kinase is a key regulator of metabolism

As we have already seen, AMPK inhibits fatty acid synthesis while simultaneously stimulating fatty acid oxidation. However, this ubiquitous trimeric enzyme ($\alpha\beta\gamma$), which exists in several isozymic forms, regulates a number of other metabolic processes. As in fatty acid metabolism, AMPK in general activates ATP-generating pathways and inhibits pathways that require ATP. Thus, AMPK stimulates glucose uptake and mitochondria biogenesis, while inhibiting cholesterol synthesis and protein synthesis. In some immune cells AMPK moderates the inflammatory response. AMPK also helps initiate nonshivering thermogenesis in brown adipose tissue (p. 607). Finally, its importance to growth is illustrated by the observation that its absence is lethal in mouse embryogenesis, suggesting that AMPK is crucial for early development. Interestingly, the isozymic forms of this key enzyme are themselves regulated by phosphorylation by a number of kinases. Understanding the intricacies of this enzyme will surely keep biochemists engaged for years to come.

SUMMARY

22.1 Triacylglycerols Are Highly Concentrated Energy Stores

Fatty acids are physiologically important as (1) fuel molecules, (2) components of phospholipids and glycolipids, (3) hydrophobic modifiers of proteins, and (4) hormones and intracellular messengers. They are stored in adipose tissue as triacylglycerols (neutral fat).

22.2 The Use of Fatty Acids as Fuel Requires Three Stages of Processing

Triacylglycerols can be mobilized by the hydrolytic action of lipases that are under hormonal control. Glucagon and epinephrine stimulate triacylglycerol breakdown by activating the lipase. Insulin, in contrast, inhibits lipolysis. Fatty acids are activated to acyl CoAs, transported across the inner mitochondrial membrane by carnitine, and degraded in the mitochondrial matrix by a recurring sequence of four reactions: oxidation by FAD, hydration, oxidation by NAD^+, and thiolysis by coenzyme A. The $FADH_2$ and NADH formed in the oxidation steps transfer their electrons to O_2 by means of the respiratory chain, whereas the acetyl CoA formed in the thiolysis step normally enters the citric acid cycle by condensing with oxaloacetate.

22.3 Unsaturated and Odd-Chain Fatty Acids Require Additional Steps for Degradation

Fatty acids that contain double bonds or odd numbers of carbon atoms require ancillary steps to be degraded. An isomerase and a reductase are required for the oxidation of unsaturated fatty acids, whereas propionyl CoA derived from chains with odd numbers of carbon atoms requires a vitamin B_{12}-dependent enzyme to be converted into succinyl CoA.

22.4 Ketone Bodies Are a Fuel Source Derived from Fats

Ketone bodies are an important fuel source for some tissues. The primary ketone bodies—acetoacetate and β-hydroxybutyrate—are formed in the liver by condensation of acetyl CoA units. Mammals are unable to convert fatty acids into glucose because they lack a pathway for the net production of oxaloacetate, pyruvate, or other gluconeogenic intermediates from acetyl CoA.

22.5 Fatty Acids Are Synthesized by Fatty Acid Synthase

Fatty acids are synthesized in the cytoplasm by a different pathway from that of β-oxidation. Fatty acid synthase is the enzyme complex responsible for fatty acid synthesis. Synthesis starts with the carboxylation of acetyl CoA to malonyl CoA, the committed step. This ATP-driven reaction is catalyzed by acetyl CoA carboxylase, a biotin enzyme. The intermediates in fatty acid synthesis are linked to an acyl carrier protein. Acetyl ACP is formed from acetyl CoA, and malonyl ACP is formed from malonyl CoA. Acetyl ACP and malonyl ACP condense to form acetoacetyl ACP, a reaction driven by the release of CO_2 from the activated malonyl unit. A reduction, a dehydration, and a second reduction follow. NADPH is the reductant in these steps. The butyryl ACP formed in this way is ready for a second round of elongation, starting with the addition of a two-carbon unit from malonyl ACP. Seven rounds of elongation yield palmitoyl ACP, which is hydrolyzed to palmitate. In higher organisms, the enzymes catalyzing fatty acid synthesis are covalently linked in a multifunctional enzyme complex. A reaction cycle based on the formation and cleavage of citrate carries acetyl groups from mitochondria to the cytoplasm. NADPH needed for synthesis is generated in the transfer of reducing equivalents from mitochondria by the combined action of cytoplasmic malate dehydrogenase and malic enzyme and by the pentose phosphate pathway.

22.6 The Elongation and Unsaturation of Fatty Acids Are Accomplished by Accessory Enzyme Systems

Fatty acids are elongated and desaturated by enzyme systems in the endoplasmic reticulum membrane. Desaturation requires NADH and O_2 and is carried out by a complex consisting of a flavoprotein, a cytochrome, and a nonheme iron protein. Mammals lack the enzymes to introduce double bonds distal to C-9, and so they require linoleate and linolenate in their diets.

Arachidonate, an essential precursor of prostaglandins and other signal molecules, is derived from linoleate. This 20:4 polyunsaturated fatty acid is the precursor of several classes of signal molecules—prostaglandins, prostacyclins, thromboxanes, and leukotrienes—that act as messengers and local hormones because of their transience. They are called eicosanoids because they contain 20 carbon atoms. Aspirin (acetylsalicylate), an anti-inflammatory and antithrombotic drug, irreversibly blocks the synthesis of these eicosanoids.

22.7 Acetyl CoA Carboxylase Plays a Key Role in Controlling Fatty Acid Metabolism

Fatty acid synthesis and degradation are reciprocally regulated so that both are not simultaneously active. Acetyl CoA carboxylase, the essential control site, is phosphorylated and inactivated by AMP-activated kinase. The phosphorylation is reversed by a protein phosphatase. Citrate, which signals an abundance of building blocks and energy, partly reverses the inhibition by phosphorylation. Carboxylase activity is stimulated by insulin and inhibited by glucagon and epinephrine. In times of plenty, fatty acyl CoAs do not enter the mitochondrial matrix, because malonyl CoA inhibits carnitine acyltransferase I.

APPENDIX

Biochemistry in Focus

Ethanol consumption results in triacylglycerol accumulation in the liver

Ethanol has been a part of the human diet for centuries, partly because of its intoxicating effects and partly because alcoholic beverages provided a safe means of hydration when pure water was scarce. Indeed, throughout the world, only water and tea are consumed more frequently than beer. However, ethanol consumption in excess can result in a number of health problems, most notably liver damage. What is the biochemical basis of these health problems?

Ethanol cannot be excreted and must be metabolized, primarily by the liver. There are several pathways for the metabolism of ethanol (Section 27.6). One pathway consists of two steps. The first step, catalyzed by alcohol dehydrogenase, takes place in the cytoplasm:

$$CH_3CH_2OH + NAD^+ \xrightarrow{\text{Alcohol dehydrogenase}} CH_3CHO + NADH + H^+$$
Ethanol $\qquad\qquad\qquad$ Acetaldehyde

The second step, catalyzed by aldehyde dehydrogenase, takes place in mitochondria.

$$CH_3CHO + NAD^+ + H_2O \xrightarrow{\text{Aldehyde dehydrogenase}} CH_3COO^- + NADH + H^+$$
Acetaldehyde $\qquad\qquad\qquad$ Acetate

Note that ethanol consumption leads to an accumulation of NADH. This high concentration of NADH inhibits gluconeogenesis by preventing the oxidation of lactate to pyruvate. In fact, the high concentration of NADH will cause the reverse reaction to predominate: lactate will accumulate. The consequences may be hypoglycemia (low concentration of blood glucose) and lactic acidosis.

The NADH glut also inhibits fatty acid oxidation. The metabolic purpose of fatty acid oxidation is to generate NADH for ATP generation by oxidative phosphorylation. However, an alcohol consumer's NADH needs are met by ethanol metabolism. In fact, the excess NADH signals that conditions are right for fatty acid synthesis. The excess NADH can be converted to NADPH—the reducing power for fatty acid synthesis—by the combined action of cytoplasmic malate dehydrogenase and NADP$^+$-linked malate enzyme (p. 734).

$$\text{Oxaloacetate} + NADH + H^+ \rightleftharpoons \text{malate} + NAD^+$$
malate dehydrogenase

$$\text{Malate} + NADP^+ \longrightarrow \text{pyruvate} + CO_2 + NADPH$$
NADP$^+$-linked malate enzyme

Moreover, the acetate generated by aldehyde dehydrogenase can be readily converted into acetyl CoA, another substrate for fatty acid synthesis. Thus, ethanol consumption encourages an accumulation of triacylglycerols in the liver, leading to a condition known as fatty liver. Fatty liver may lead to obesity and type 2 diabetes (Section 27.3).

Recent research provides evidence that ethanol also inhibits the β-adrenergic/cAMP pathway in the liver (p. 714). Hepatocytes were exposed to ethanol for several days and then the pathway was stimulated by the β-adrenergic agonist isoproterenol. An agonist is a drug that simulates the action of a signal molecule, in this case epinephrine. The figure below shows the activity of protein kinase A (PKA) in hepatocytes after several days' exposure to ethanol.

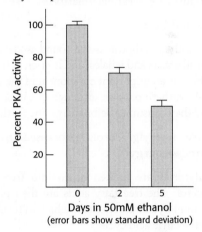

[Schott, M. B., Rasineni, K., Weller, S. G., Schulze, R. J., Sletten, A. C., Casey, C. A., and McNiven, M. A. 2017. β-Adrenergic induction of lipolysis in hepatocytes is inhibited by ethanol exposure. *J. Biol. Chem.* 292: 11815–11828.]

The kinase activity was determined by measuring the phosphorylation of PKA substrates. After 2 days of ethanol exposure, PKA activity fell to 70% of the control value (not exposed to ethanol) and after 5 days, activity fell to 50% of control. Ethanol appears to be inhibiting PKA activity, but might there be an alternate explanation? Might exposure to ethanol inhibit the transcription and translation of PKA and the other components of the pathway? Control experiments determined that the PKA and the other components were present in amounts equal to that of the cells not exposed to ethanol. Thus, in addition to promoting fatty acid synthesis, ethanol also may prevent fatty acid mobilization by the liver by inhibiting control of mobilization, furthering the development of fatty liver. The mechanism by which ethanol inhibits the pathway remains to be determined. Because biochemical systems are so complex, we should always be a bit cautious in giving too much weight to new results. In this case, the studies were performed on rat hepatocytes in culture, so we do not know whether results apply to whole animals including humans.

APPENDIX

Problem-Solving Strategies

Problem-Solving Strategies 1

PROBLEM: Suppose that, for some bizarre reason, you decided to exist on a diet of whale and seal blubber, exclusively.

(a) How would a lack of carbohydrates affect your ability to utilize fats?

(b) What would your breath smell like?

(c) One of your best friends, after trying unsuccessfully to convince you to abandon this diet, makes you promise to consume a healthy dose of odd-chain fatty acids. Does your friend have your best interests at heart? Explain.

SOLUTION: Not clear why you would want to do this, but it is your choice. Just to be clear, let's clarify:

▶ **What is blubber?**

Blubber is thick, adipose tissue found under the skin of creatures like seals and whales, and it's rich in triacylglycerols. Blubber is an important component of the diet of Inuit and other northern peoples, although of course it is not the only food they consume. Before we go any further, let's ask:

▶ **How are triacylglycerols processed to generate biochemical energy?**

Triacylglycerols are broken down into free fatty acids and glycerol. The free fatty acids are then processed by β-oxidation to yield acetyl CoA, which in turn is metabolized by the citric acid cycle.

Now, let's look at part (a).

▶ **How would a lack of carbohydrates affect the processing of fats by β-oxidation?**

Let's answer that question by asking several others:

▶ **What reaction brings acetyl CoA into the citric acid cycle?**

Acetyl CoA condenses with oxaloacetate to form citrate in a reaction catalyzed by citrate synthase. Since we are doing much β-oxidation, we will need plenty of oxaloacetate.

▶ **What is a likely source of oxaloacetate?**

Oxaloacetate is commonly formed by the ATP-powered carboxylation of pyruvate in a reaction catalyzed by pyruvate carboxylase.

▶ **What is a common source of pyruvate?**

Exactly! The metabolism of glucose to pyruvate by glycolysis! So, we would need at least some glucose to provide pyruvate for the synthesis of oxaloacetate for the processing of acetyl CoA generated by the metabolism of your

blubber. This is the basis for the old adage "Fats burn in the flame of carbohydrates." (By the way, how *do* you prepare blubber for your meal?)

▶ **What is the result of a lack of oxaloacetate?**

Acetyl CoA accumulates and begins to form ketone bodies, one of which is acetoacetate. Now, about part (b).

▶ **Acetoacetate undergoes a spontaneous decarboxylation to produce what?**

Acetoacetate undergoes a spontaneous decarboxylation to produce acetone. Your breath would smell like acetone!

On to part (c). Why would your friend want you to make sure your blubber has odd-chain fatty acids? Odd-chain fatty acids are processed by β-oxidation, but the final cleavage product is different.

▶ **How does the final cleavage product of odd-chain fatty acids differ from even-chain fatty acids undergoing β-oxidation?**

The final cleavage product is propionyl CoA instead of acetyl CoA.

▶ **What is a potential fate of propionyl CoA?**

Propionyl CoA can be converted into succinyl CoA, a citric acid cycle intermediate (p. 720) in what is essentially an anaplerotic reaction. The succinyl CoA can be metabolized to oxaloacetate, which will alleviate the buildup of ketone bodies and the bad breath caused by acetone. So, your friend did have your best interests at heart. Buy your friend a chocolate éclair for being such a good friend.

Problem-Solving Strategies 2

PROBLEM: Experiments on fatty acid synthesis in the late 1950s discovered that bicarbonate was required for the synthesis of palmitoyl CoA. Interestingly, the amount of palmitoyl CoA synthesized exceeded the amount of bicarbonate required. If ^{14}C-bicarbonate was used in the experiment, fatty acid synthesis occurred, but interestingly, no ^{14}C was incorporated into the palmitoyl CoA. Explain these observations on the basis of your understanding of fatty acid synthesis.

SOLUTION: First of all, remember that bicarbonate is the hydrated form of carbon dioxide (CO_2). Now, I may not be the sharpest tool in the shed, but even I think the first question to ask is this:

▶ **What is the role of bicarbonate in fatty acid synthesis?**

Bicarbonate is used to form malonyl CoA, the activated precursor for fatty acid synthesis, in a two-step reaction catalyzed by acetyl CoA carboxylase 1.

$$\text{Biotin-enzyme} + \text{ATP} + \text{HCO}_3^- \rightleftharpoons$$
$$\text{CO}_2\text{-biotin-enzyme} + \text{ADP} + \text{P}_i$$

$$\text{CO}_2\text{-biotin-enzyme} + \text{acetyl CoA} \longrightarrow$$
$$\text{malonyl CoA} + \text{biotin-enzyme} \quad (1)$$

▶ **What is the role of malonyl CoA in fatty acid synthesis?**

Malonyl CoA, after being ligated to acyl carrier protein (ACP), condenses with acetyl ACP and subsequent acyl ACP to extend the growing fatty acid by two carbons at a time.

$$\text{Acetyl ACP} + \text{malonyl ACP} \rightarrow$$
$$\text{acetoacetyl ACP} + \text{ACP} + \text{CO}_2 \quad (2)$$

See Figure 22.26.
Okay. Now to the question:

▶ **Why did the amount of palmitoyl CoA exceed the amount of bicarbonate added?**

Examine equations (1) and (2).

▶ **What is the fate of the carbon dioxide added in equation 1?**

It is released when the malonyl CoA condenses with the growing chain, as shown in equation 2. In other words, a single carbon dioxide molecule can facilitate the condensation of many two-carbon units to the growing fatty acid chain.

▶ **When malonyl CoA is synthesized with ^{14}C-bicarbonate, why is none of the ^{14}C found in the newly synthesized fatty acid?**

Because the CO_2 is added and then released. In essence, the bicarbonate, or CO_2, is acting catalytically in facilitating fatty acid synthesis, but it is never being incorporated into the fatty acid.

KEY TERMS

triacylglycerol (neutral fat, triglyceride) (p. 709)
lipid droplet (p. 712)
acyl adenylate (p. 715)
carnitine (p. 716)
β-oxidation pathway (p. 717)
vitamin B$_{12}$ (cobalamin) (p. 721)

peroxisome (p. 723)
ketone body (p. 725)
acyl carrier protein (ACP) (p. 728)
fatty acid synthase (p. 728)
malonyl CoA (p. 729)
acetyl CoA carboxylase 1 and 2 (p. 729)

ATP-citrate lyase (p. 734)
arachidonate (p. 737)
prostaglandin (p. 737)
eicosanoid (p. 737)
polyketide (p. 738)
AMP-activated protein kinase (AMPK) (p. 739)

PROBLEMS

1. *Match 'em.* Match each term with its description.

(a) Triacylglycerol _____
(b) Perilipin _____
(c) Adipose triglyceride lipase _____
(d) Glucagon _____
(e) Acyl CoA synthetase _____
(f) Carnitine _____
(g) β-Oxidation pathway _____
(h) Enoyl CoA isomerase _____
(i) 2,4-Dienoyl CoA reductase _____
(j) Methylmalonyl CoA mutase _____
(k) Ketone body _____

1. The enzyme that initiates lipid degradation
2. Activates fatty acids for degradation
3. Converts a *cis*-Δ^3 double bond into a *trans*-Δ^2 double bond
4. Reduces 2,4-dienoyl intermediate to *trans*-Δ^3-enoyl CoA
5. Storage form of fats
6. Required for entry into mitochondria
7. Requires vitamin B$_{12}$
8. Acetoacetate
9. Means by which fatty acids are degraded
10. Stimulates lipolysis
11. Lipid-droplet-associated protein

2. *After lipolysis.* Write a balanced equation for the conversion of glycerol into pyruvate. Which enzymes are required in addition to those of the glycolytic pathway?

3. *Forms of energy.* The partial reactions leading to the synthesis of acyl CoA (equations 1 and 2, p. 716) are freely reversible. The equilibrium constant for the sum of these reactions is close to 1, meaning that the energy levels of the reactants and products are about equal, even though a molecule of ATP has been hydrolyzed. Explain why these reactions are readily reversible.

4. *Activation fee.* The reaction for the activation of fatty acids before degradation is:

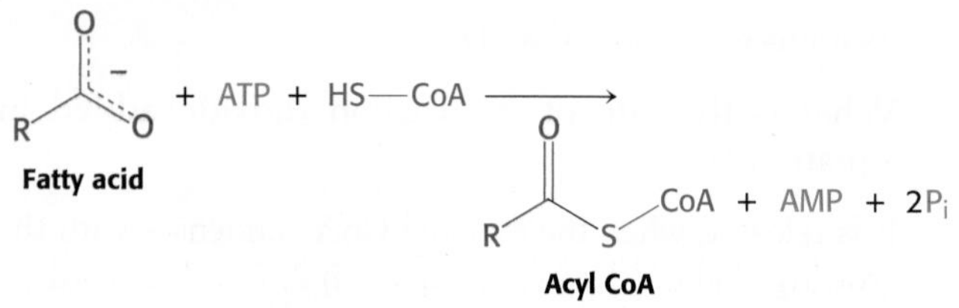

Fatty acid

$+ ATP + HS—CoA \longrightarrow$

Acyl CoA

$CoA + AMP + 2P_i$

This reaction is quite favorable because the equivalent of two molecules of ATP is hydrolyzed. Explain why, from a biochemical bookkeeping point of view, the equivalent of two molecules of ATP is used despite the fact that the left side of the equation has only one molecule of ATP. ✓❶

5. *Proper sequence.* Place the following list of reactions or relevant locations in the β-oxidation of fatty acids in the proper order. ✓❶

(a) Reaction with carnitine

(b) Fatty acid in the cytoplasm

(c) Activation of fatty acid by joining to CoA

(d) Hydration

(e) NAD$^+$-linked oxidation

(f) Thiolysis

(g) Acyl CoA in mitochondrion

(h) FAD-linked oxidation

6. *Remembrance of reactions past.* We have encountered reactions similar to the oxidation, hydration, and oxidation reactions of fatty acid degradation earlier in our study of biochemistry. What other pathway employs this set of reactions? ✓❶

7. *A phantom acetyl CoA?* In the equation for fatty acid degradation shown here, only seven molecules of CoA are required to yield eight molecules of acetyl CoA. How is this difference possible?

Palmitoyl CoA + 7 FAD + 7 NAD$^+$
$\qquad$ + 7 CoA + 7 H$_2$O $\longrightarrow$
$\qquad\qquad$ 8 acetyl CoA + 7 FADH$_2$ + 7 NADH + 7 H$^+$

8. *Comparing yields.* Compare the ATP yields from palmitic acid and palmitoleic acid. ✓❶

9. *Counting ATPs 1.* What is the ATP yield for the complete oxidation of C$_{17}$ (heptadecanoic) fatty acid? Assume that the propionyl CoA ultimately yields oxaloacetate in the citric acid cycle.

10. *Sweet temptation.* Stearic acid is a C$_{18}$ fatty acid component of chocolate. Suppose you had a depressing day and decided to settle matters by gorging on chocolate. How much ATP would you derive from the complete oxidation of stearic acid to CO$_2$? ✓❶

11. *The best storage form.* Compare the ATP yield from the complete oxidation of glucose, a six-carbon carbohydrate, and hexanoic acid, a six-carbon fatty acid. Hexanoic acid is also called *caprioic acid* and is responsible for the "aroma" of goats. Why are fats better fuels than carbohydrates? ✓❶

12. *From fatty acid to ketone body.* Write a balanced equation for the conversion of stearate into acetoacetate. ✓❷

13. *Generous, but not to a fault.* Liver is the primary site of ketone-body synthesis. However, ketone bodies are not used by the liver but are released for other tissues to use. The liver does gain energy in the process of synthesizing and releasing ketone bodies. Calculate the number of molecules of ATP generated by the liver in the conversion of palmitate, a C$_{16}$ fatty acid, into acetoacetate. ✓❷

14. *Counting ATPs 2.* How much energy is attained with the complete oxidation of the ketone body D-3-hydroxybutyrate? ✓❷

15. *Another view.* Why might someone argue that the answer to problem 14 is wrong? ✓❷

16. *An accurate adage.* An old biochemistry adage is that *fats burn in the flame of carbohydrates.* What is the molecular basis of this adage?

17. *Refsum disease.* Phytanic acid is a branched-chain fatty acid component of chlorophyll and is a significant component of milk. In susceptible people, phytanic acid can accumulate, leading to neurological problems. This syndrome is called Refsum disease or phytanic acid storage disease. ✓❶

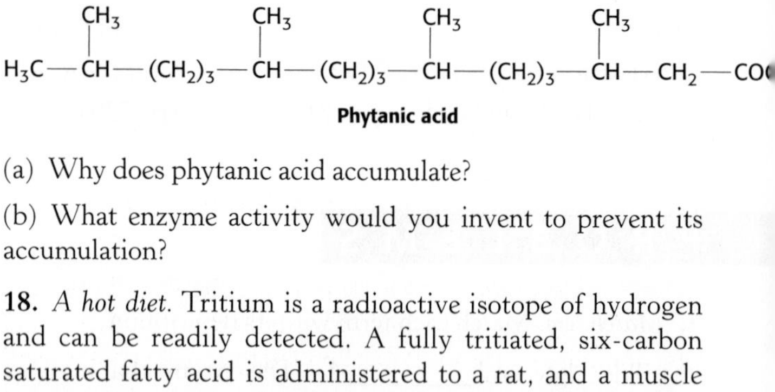

Phytanic acid

(a) Why does phytanic acid accumulate?

(b) What enzyme activity would you invent to prevent its accumulation?

18. *A hot diet.* Tritium is a radioactive isotope of hydrogen and can be readily detected. A fully tritiated, six-carbon saturated fatty acid is administered to a rat, and a muscle biopsy of the rat is taken by concerned, sensitive, and discreet technical assistants. These assistants carefully isolate all of the acetyl CoA generated from the β-oxidation of the radioactive fatty acid and remove the CoA to form acetate. What will be the overall tritium-to-carbon ratio of the isolated acetate? ✓❶

19. *Finding triacylglycerols in all the wrong places.* Insulin-dependent diabetes is often accompanied by high levels of triacylglycerols in the blood. Suggest a biochemical explanation.

20. *A pledge to move forward.* What is the committed step in fatty acid synthesis, and which enzyme catalyzes the step? ✓❸

21. *Match 'em.* Match the terms with the descriptions.

(a) ATP-citrate lyase _____

(b) Malic enzyme _____

(c) Malonyl CoA _____

(d) Acetyl CoA carboxylase _____

(e) Acyl carrier protein _____

(f) β-Ketoacyl synthase _____

(g) Palmitate _____

(h) Eicosanoids _____

(i) Arachidonate _____

(j) AMP-activated protein kinase _____

1. Helps to generate NADPH from NADH
2. Inactivates acetyl CoA carboxylase
3. Molecule on which fatty acids are synthesized
4. A precursor of prostaglandins
5. Activated acetyl CoA
6. The end product of fatty acid synthase
7. Fatty acids containing 20 carbon atoms
8. Catalyzes the committed step in fatty acid synthesis
9. Catalyzes the reaction of acetyl ACP and malonyl ACP
10. Generates cytoplasmic acetyl CoA

22. *Counterpoint.* Compare and contrast fatty acid oxidation and synthesis with respect to

(a) site of the process

(b) acyl carrier

(c) reductants and oxidants

(d) stereochemistry of the intermediates

(e) direction of synthesis or degradation

(f) organization of the enzyme system

23. *The fizz in fatty acid synthesis.* During the initial in vitro studies on fatty acid synthesis, experiments were performed to determine which buffer would result in optimal activity. Bicarbonate buffer proved far superior to phosphate buffer. The researchers could not initially account for this result. From your perspective of many decades later, explain why the result is not surprising. ✓❸

24. *A supple synthesis.* Myristate, a saturated C_{14} fatty acid, is used as an emollient for cosmetics and topical medicinal preparations. Write a balanced equation for the synthesis of myristate. ✓❸

25. *The cost of cleanliness.* Lauric acid is a 12-carbon fatty acid with no double bonds. The sodium salt of lauric acid (sodium laurate) is a common detergent used in a variety of products, including laundry detergent, shampoo, and toothpaste. How many molecules of ATP and NADPH are required to synthesize lauric acid? ✓❸

26. *Proper organization.* Arrange the following steps in fatty acid synthesis in their proper order. ✓❸

(a) Dehydration

(b) Condensation

(c) Release of a C_{16} fatty acid

(d) Reduction of a carbonyl group

(e) Formation of malonyl ACP

27. *No access to assets.* What would be the effect on fatty acid synthesis of a mutation in ATP-citrate lyase that reduces the enzyme's activity? Explain. ✓❸

28. *The truth and nothing but.* True or False. If false, explain. ✓❸

(a) Biotin is required for fatty acid synthase activity.

(b) The condensation reaction in fatty acid synthesis is powered by the decarboxylation of malonyl CoA.

(c) Fatty acid synthesis does not depend on ATP.

(d) Palmitate is the end product of fatty acid synthase.

(e) All of the enzyme activities required for fatty acid synthesis in mammals are contained in a single polypeptide chain.

(f) Fatty acid synthase in mammals is active as a monomer.

(g) The fatty acid arachidonate is a precursor for signal molecules.

(h) Acetyl CoA carboxylase is inhibited by citrate.

29. *Odd fat out.* Suggest how fatty acids with odd numbers of carbons are synthesized. ✓❸

30. *Labels.* Suppose that you had an in vitro fatty acid–synthesizing system with all of the enzymes and cofactors required for fatty acid synthesis except for acetyl CoA. To this system, you added acetyl CoA that contained radioactive hydrogen (^{3}H, tritium) and carbon 14 (^{14}C) as shown here.

The ratio of ^{3}H/^{14}C is 3. What would the ^{3}H/^{14}C ratio be after the synthesis of palmitic acid (C_{16}) with the use of the radioactive acetyl CoA? ✓❸

31. *A tight embrace.* Avidin, a glycoprotein found in eggs, has a high affinity for biotin. Avidin can bind biotin and prevent its use by the body. How might a diet rich in raw eggs affect fatty acid synthesis? What will be the effect on fatty acid synthesis of a diet rich in cooked eggs? Explain. ✓❸

32. *Alpha or omega?* Only one acetyl CoA molecule is used directly in fatty acid synthesis. Identify the carbon atoms in palmitic acid that were donated by acetyl CoA. ✓❸

33. *Now you see it, now you don't.* Although HCO_3^- is required for fatty acid synthesis, its carbon atom does not appear in the product. Explain how this omission is possible. ✓❸

34. *It is all about communication.* Why is citrate an appropriate inhibitor of phosphofructokinase?

35. *Tracing carbon atoms.* Consider a cell extract that actively synthesizes palmitate. Suppose that a fatty acid synthase in this preparation forms one molecule of palmitate in about 5 minutes. A large amount of malonyl CoA labeled with ^{14}C in each carbon atom of its malonyl unit is suddenly added to this system, and fatty acid synthesis is stopped a minute later by altering the pH. The fatty acids are analyzed for radioactivity. Which carbon atom of the palmitate formed by this system is more radioactive—C-1 or C-14? ✓❸

36. *An unaccepting mutant.* The serine residues in acetyl CoA carboxylase that are the target of the AMP-activated protein kinase are mutated to alanine. What is a likely consequence of this mutation? ✓④

37. *Sources.* For each of the following unsaturated fatty acids, indicate whether the biosynthetic precursor in animals is palmitoleate, oleate, linoleate, or linolenate.

(a) 18:1 *cis*-Δ^{11}

(b) 18:3 *cis*-Δ^6, Δ^9, Δ^{12}

(c) 20:2 *cis*-Δ^{11}, Δ^{14}

(d) 20:3 *cis*-Δ^5, Δ^8, Δ^{11}

(e) 22:1 *cis*-Δ^{13}

(f) 22:6 *cis*-Δ^4, Δ^7, Δ^{10}, Δ^{13}, Δ^{16}, Δ^{19}

38. *Driven by decarboxylation.* What is the role of decarboxylation in fatty acid synthesis? Name another key reaction in a metabolic pathway that employs this mechanistic motif. ✓③

39. *Kinase surfeit.* Suppose that a promoter mutation leads to the overproduction of protein kinase A in adipose cells. How might fatty acid metabolism be altered by this mutation? ✓④

40. *Blocked assets.* The presence of a fuel molecule in the cytoplasm does not ensure that the fuel molecule can be effectively used. Give two examples of how impaired transport of metabolites between compartments leads to disease.

41. *Elegant inversion.* Peroxisomes have an alternative pathway for oxidizing polyunsaturated fatty acids. They contain a hydratase that converts D-3-hydroxyacyl CoA into *trans*-Δ^2-enoyl CoA. How can this enzyme be used to oxidize CoAs containing a *cis* double bond at an even-numbered carbon atom (e.g., the *cis*-Δ^{12} double bond of linoleate)?

42. *Arm in arm.* Many eukaryotic enzymes, such as fatty acid synthase, are multifunctional proteins with multiple active sites covalently linked. What is the advantage of this arrangement? ✓③

43. *Covalent catastrophe.* What is a potential disadvantage of having many catalytic sites together on one very long polypeptide chain? ✓③

44. *Missing acyl CoA dehydrogenases.* A number of genetic deficiencies in acyl CoA dehydrogenases have been described. This deficiency presents early in life after a period of fasting. Symptoms include vomiting, lethargy, and sometimes coma. Not only is the blood concentration of glucose low (hypoglycemia), but starvation-induced ketosis is absent. Provide a biochemical explanation for these last two observations.

45. *Effects of clofibrate.* High blood levels of triacylglycerides are associated with heart attacks and strokes. Clofibrate, a drug that increases the activity of peroxisomes, is sometimes used to treat patients with such a condition. What is the biochemical basis for this treatment?

46. *A different kind of enzyme.* Figure 22.36 shows the response of acetyl CoA carboxylase to varying amounts of citrate. Explain this effect in light of the allosteric effects that citrate has on the enzyme. Predict the effects of increasing concentrations of palmitoyl CoA. ✓④

Mechanism Problems

47. *Variation on a theme.* Thiolase is homologous in structure to the condensing enzyme. On the basis of this observation, propose a mechanism for the cleavage of 3-ketoacyl CoA by CoA.

48. *Two plus three to make four.* Propose a reaction mechanism for the condensation of an acetyl unit with a malonyl unit to form an acetoacetyl unit in fatty acid synthesis. ✓③

Chapter Integration Problems

49. *Too tired to exercise.* Explain why people with a hereditary deficiency of carnitine acyltransferase II have muscle weakness. Why are the symptoms more severe during fasting? ✓①

50. *Buff and lean.* It has been suggested that if you are interested in losing body fat, the best time to do strenuous aerobic exercise is in the morning immediately after waking up; that is, after fasting. Don't eat breakfast before exercising, but have a cup of caffeinated coffee. Caffeine is an inhibitor of cAMP phosphodiesterase. Explain why this suggestion might work biochemically. ✓④

51. *NADH to NADPH.* What are the three reactions that allow the conversion of cytoplasmic NADH into NADPH? What enzymes are required? Show the sum of the three reactions. ✓③

52. *Six of one, half a dozen of the other.* People who consume little fat but excess carbohydrates can still become obese. How is this result possible?

53. *Familiars.* One of the reactions critical for the generation of cytoplasmic NADPH from cytoplasmic NADH is also important in gluconeogenesis. What is the reaction and what is the immediate fate of the product of the reaction in gluconeogenesis?

54. *Fats to glycogen.* An animal is fed stearic acid that is radioactively labeled with [^{14}C] carbon. A liver biopsy reveals the presence of ^{14}C-labeled glycogen. How is this finding possible in light of the fact that animals cannot convert fats into carbohydrates?

55. *Enzyme deficit.* Pyruvate dehydrogenase deficiency, a severe condition with a poor prognosis, results in neurological dysfunction and unremitting lactic acidosis. One treatment is placing the patients on a low-carbohydrate/high-fat diet.

(a) Explain why the deficiency would be characterized by neurological dysfunction.

(b) Why does lactic acidosis result?

(c) Explain the rationale for the low-carbohydrate/high-fat diet.

56. *Right enzymes. Wrong location.* Zwellweger syndrome is caused by a defect in enzyme importation into the peroxisomes. What disease of glycosylation is caused by inappropriate cellular distribution of enzymes?

Data Interpretation Problem

57. *Mutant enzyme.* Carnitine palmitoyl transferase I (CPTI) catalyzes the conversion of long-chain acyl CoA into acyl carnitine, a prerequisite for transport into mitochondria and subsequent degradation. A mutant enzyme was constructed with a single amino acid change at position 3 of glutamic acid for alanine. Graphs A through C show data from studies performed to identify the effect of the mutation. ✅❶

(a) What is the effect of the mutation on enzyme activity when the concentration of carnitine is varied (Graph A)? What are the K_M and V_{max} values for the wild-type and mutant enzymes?

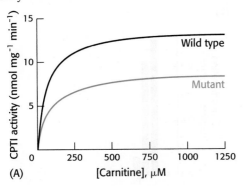

(A)

(b) What is the effect when the experiment is repeated with varying concentrations of palmitoyl CoA (Graph B)? What are the K_M and V_{max} values for the wild-type and mutant enzymes?

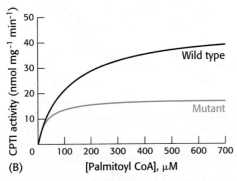

(B)

(c) Graph C shows the inhibitory effect of malonyl CoA on the wild-type and mutant enzymes. Which enzyme is more sensitive to malonyl CoA inhibition?

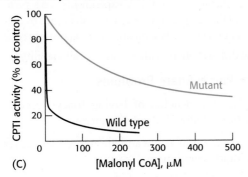

(C)

(d) Suppose that the concentration of palmitoyl CoA = 100 μM, that of carnitine = 100 μM, and that of malonyl CoA = 10 μM. Under these conditions, what is the most prominent effect of the mutation on the properties of the enzyme?

(e) What can you conclude about the role of glutamate 3 in CPTI function?

58. *Partners.* Colipase binds to lipid particles in the intestine and facilitates the subsequent binding and activation of pancreatic lipase. The lipase then hydrolyzes the lipids so they can be absorbed by intestinal cells. Experiments were undertaken using site-directed mutagenesis (Section 5.2) to determine which amino acids in the colipase facilitated lipase binding and which activated the lipase. A portion of these experiments is presented below.

Figure A shows the results of two different amino acid substitutions on the ability of colipase to facilitate the binding of lipase to the lipid particles. In both panels, the blue circles show the control (no amino acid changes) data and the red circles show the data with the altered amino acid. The first letter represents the normal amino acid, the number of its position in the amino acid sequence, and the second letter represents the replacement amino acid.

(a) What is the amino acid alteration in the upper panel?

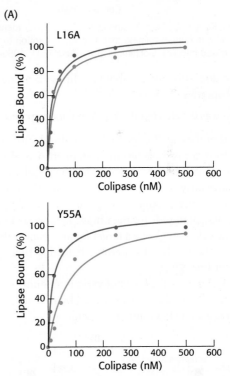

(A)

(b) How does this mutation change the ability of colipase to bind the lipase to the lipid particle?

(c) What is the amino acid substitution in the lower panel?

(d) What is the effect of this substitution on the ability of the colipase to bind the lipase to the lipid particle?

Next, the activity of the bound lipase was examined, as shown in Figure B.

(e) What was the effect of the L16A mutation on lipase activity?

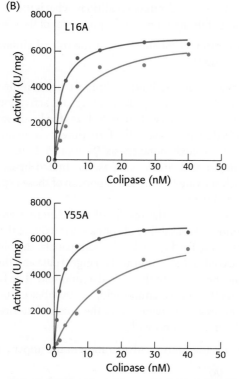

[Data from: Ross, L. E., Xiao, X., and Lowe, M. E. 2013. Identification of amino acids in human colipase that mediate adsorption to lipid emulsions and mixed micelles. Biochim. Biophys. Acta 1831: 1052–1059.]

(f) What does this suggest about the lipase-binding and activation properties of the colipase?

(g) What was the effect of the Y55A mutation on lipase activity?

59. *Nip it in the bud.* Soraphen A is a natural antifungal agent isolated from a myxobacterium. Soraphen A is also a potent inhibitor of acetyl CoA carboxylase, the regulatory enzyme that initiates fatty acid synthesis. As discussed earlier (p. 735), cancer cells must produce a large amount of fatty acids to generate phospholipids for membrane synthesis. Thus, acetyl CoA carboxylase might be a target for anti-cancer drugs. Following are the results of experiments testing this hypothesis on a cancer line. ✓④

Figure A shows the effect of various amounts of soraphen A on fatty acid synthesis as measured by the incorporation of radioactive acetate (^{14}C) into fatty acids.

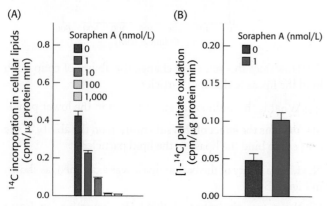

(a) How did soraphen A alter fatty acid synthesis? How was synthesis affected by increasing concentrations of soraphen A?

Figure B shows the results obtained when fatty acid oxidation was measured, as the release of radioactive CO_2 from added radioactive (^{14}C) palmitate, in the presence of soraphen A.

(b) How did soraphen A alter fatty acid oxidation?

(c) Explain the results obtained in B in light of the fact that soraphen A inhibits acetyl CoA carboxylase.

More experiments were undertaken to assess whether carboxylase inhibition did indeed prevent phospholipid synthesis. Again, cells were grown in the presence of radioactive acetate, phospholipids were subsequently isolated, and the amount of ^{14}C incorporated into the phospholipid was determined. The effect of soraphen A on phospholipid synthesis is shown in Figure C.

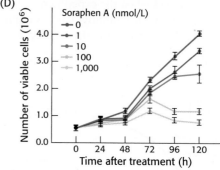

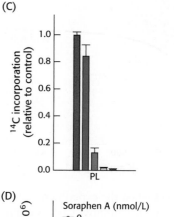

[Data from: Beckers, A., Organe, S., Timmermans, L., Scheys, K., Peeters, A., Brusselmans, K., Verhoeven, G., and Swinnen, J. V. 2007. Chemical inhibition of acetyl-CoA carboxylase induces growth arrest and cytotoxicity selectively in cancer cells. Cancer Res. 67: 8180–8187.]

(d) Did the inhibition of carboxylase by the drug alter phospholipid synthesis?

(e) How might phospholipid synthesis affect cell viability?

Finally, as shown in Figure D, experiments were performed to determine if the drug inhibits cancer growth in vitro by determining the number of cancer cells surviving upon exposure to soraphen A.

(f) How did soraphen A affect cancer cell viability?

Think ▶ Pair ▶ Share Problems

60. Explain the benefits of having triacylglycerols be the primary form of stored metabolic energy.

61. Outline the similarities and differences between fatty acid synthesis and degradation.

62. Discuss the regulation of fatty acid metabolism by the isozymic forms of acetyl CoA carboxylase. Explain the reciprocal control of fatty acid synthesis and degradation and the various means by which this control is achieved.

Protein Turnover and Amino Acid Catabolism

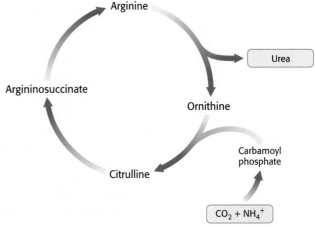

This fourteenth-century hand-colored woodcut from Germany depicts a wheel that classifies urine samples according to their color and consistency. In the middle of the wheel, a doctor inspects a patient's urine by sight, smell, and taste. The vials on the wheel aided physicians in diagnosing diseases. A key component of urine is urea, which is formed from the amino groups released during the metabolism of amino acids. [(Left) Ulrich Pinder. Epiphanie Medicorum. Speculum videndi urinas hominum. Clavis aperiendi portas pulsuum. Berillus discernendi causas & differentias febrium. Nuremberg: 1506. Rosenwald Collection. Rare Book and Special Collections Division, Library of Congress (128.2).]

✓ LEARNING OBJECTIVES

By the end of this chapter, you should be able to:

1. Explain the importance of the regulation of protein turnover.

2. Identify the role of ubiquitin and describe the enzymes required for ubiquitination.

3. Explain the function of the proteasome.

4. Describe the fate of nitrogen that is removed when amino acids are used as fuels.

5. Explain how the carbon skeletons of the amino acids are metabolized after nitrogen removal.

6. Identify metabolic errors in amino acid degradation.

The digestion of dietary proteins in the intestine and the degradation of proteins within the cell provide a steady supply of amino acids to the cell. Many cellular proteins are constantly degraded and resynthesized in response to changing metabolic demands. Others are misfolded or become damaged and they, too, must be degraded. Unneeded or damaged proteins are marked for destruction by the covalent attachment of chains of a small protein called ubiquitin and then degraded by a large, ATP-dependent complex called the *proteasome. The primary use of amino acids provided through degradation or digestion is as building blocks for the synthesis of proteins and other nitrogenous compounds such as nucleotide bases.*

OUTLINE

23.1 Proteins Are Degraded to Amino Acids

23.2 Protein Turnover Is Tightly Regulated

23.3 The First Step in Amino Acid Degradation Is the Removal of Nitrogen

23.4 Ammonium Ion Is Converted into Urea in Most Terrestrial Vertebrates

23.5 Carbon Atoms of Degraded Amino Acids Emerge as Major Metabolic Intermediates

23.6 Inborn Errors of Metabolism Can Disrupt Amino Acid Degradation

Amino acids in excess of those needed for biosynthesis can neither be stored, in contrast with fatty acids and glucose, nor excreted. Rather, surplus amino acids are used as metabolic fuel. *The α-amino group is removed, and the resulting carbon skeleton is converted into a major metabolic intermediate.* Most of the amino groups harvested from surplus amino acids are converted into urea through the *urea cycle,* and their carbon skeletons are transformed into acetyl CoA, acetoacetyl CoA, pyruvate, or one of the intermediates of the citric acid cycle. The carbon skeletons are converted into glucose, glycogen, and fats.

Several coenzymes play key roles in amino acid degradation; foremost among them is *pyridoxal phosphate.* This coenzyme forms Schiff-base intermediates, which are a type of aldimine, that allow α-amino groups to be shuttled between amino acids and ketoacids. We will consider several genetic errors of amino acid degradation that lead to brain damage and cognitive disabilities unless remedial action is initiated soon after birth. The study of amino acid metabolism is especially rewarding because it is rich in connections between basic biochemistry and clinical medicine.

23.1 Proteins Are Degraded to Amino Acids

Dietary protein is a vital source of amino acids. Especially important dietary proteins are those containing the essential amino acids—amino acids that cannot be synthesized and must be acquired in the diet (Table 23.1). Proteins ingested in the diet are digested into amino acids or small peptides that can be absorbed by the intestine and transported in the blood. Another crucial source of amino acids is the degradation of cellular proteins.

The digestion of dietary proteins begins in the stomach and is completed in the intestine

Protein digestion begins in the stomach, where the acidic environment denatures proteins into random coils. Denatured proteins are more accessible as substrates for proteolysis than are native proteins. The primary proteolytic enzyme of the stomach is *pepsin,* a nonspecific protease that, remarkably, is maximally active at pH 2. Thus, pepsin can function in the highly acidic environment of the stomach that disables other proteins.

The partly digested proteins then move from the acidic environment of the stomach to the beginning of the small intestine. The low pH of the food as well as the polypeptide products of pepsin digestion stimulate the release of hormones that promote the secretion from the pancreas of sodium bicarbonate ($NaHCO_3$), which neutralizes the pH of the food, and a variety of pancreatic proteolytic enzymes. Recall that these enzymes are secreted as inactive zymogens that are then converted into active enzymes (Sections 9.1 and 10.4). The battery of enzymes displays a wide array of specificity, and so the substrates are degraded into free amino acids as well as di- and tripeptides. Digestion is further enhanced by proteolytic enzymes, such as aminopeptidase N, that are located in the plasma membrane of the intestinal cells. Aminopeptidases digest proteins from the amino-terminal end. Single amino acids, as well as di- and tripeptides, are transported into the intestinal cells.

At least seven different transporters exist, each specific to a different group of amino acids. A number of inherited disorders result from mutations in these transporters. For example, Hartnup disease, a rare disorder

TABLE 23.1 Essential amino acids in human beings

Histidine
Isoleucine
Leucine
Lysine
Methionine
Phenylalanine
Threonine
Tryptophan
Valine

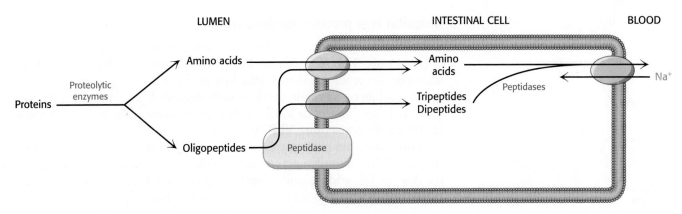

FIGURE 23.1 Digestion and absorption of proteins. Protein digestion is primarily a result of the activity of enzymes secreted by the pancreas. Aminopeptidases associated with the intestinal epithelium further digest proteins. The amino acids and di- and tripeptides are absorbed into the intestinal cells by specific transporters (green and orange ovals). Free amino acids are then released into the blood by transporters (red oval) for use by other tissues.

characterized by rashes, ataxia (lack of muscle control), delayed mental development, and diarrhea, results from a defect in the transporter for tryptophan and other nonpolar amino acids. The absorbed amino acids are subsequently released into the blood by a number of Na^+–amino acid antiporters for use by other tissues (Figure 23.1).

Cellular proteins are degraded at different rates

Protein turnover—the degradation and resynthesis of proteins—takes place constantly in cells. Although some proteins are very stable, many proteins are short lived, particularly those that participate in metabolic regulation. These proteins can be quickly degraded to activate or shut down a signaling pathway. In addition, cells must eliminate damaged proteins. A significant proportion of newly synthesized protein molecules are defective because of errors in translation or misfolding. Even proteins that are normal when first synthesized may undergo oxidative damage or be altered in other ways with the passage of time. These proteins must be removed before they accumulate and aggregate. Indeed, a number of pathological conditions, such as certain forms of Parkinson disease and Huntington disease, are associated with protein aggregation.

The half-lives of proteins range over several orders of magnitude. Ornithine decarboxylase, at approximately 11 minutes, has one of the shortest half-lives of any mammalian protein. This enzyme participates in the synthesis of polyamines, which are cellular cations essential for growth and differentiation. The life of hemoglobin, however, is limited only by the life of the red blood cell, and the lens protein crystallin is limited by the life of the organism.

23.2 Protein Turnover Is Tightly Regulated

How can a cell distinguish proteins that should be degraded? *Ubiquitin* (Ub), a small (76 amino acids) protein present in all eukaryotic cells, is a tag that marks proteins for destruction (Figure 23.2). Ubiquitin is the cellular equivalent of the "black spot" of Robert Louis Stevenson's *Treasure Island*—the signal for death.

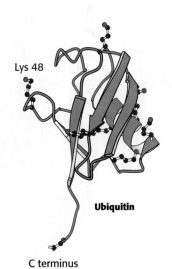

FIGURE 23.2 Structure of ubiquitin. *Notice* that ubiquitin has an extended carboxyl terminus, which is activated and linked to proteins targeted for destruction. Lysine residues, including lysine 48, the major site for linking additional ubiquitin molecules, are shown as ball-and-stick models. [Drawn from 1UBI.pdb.]

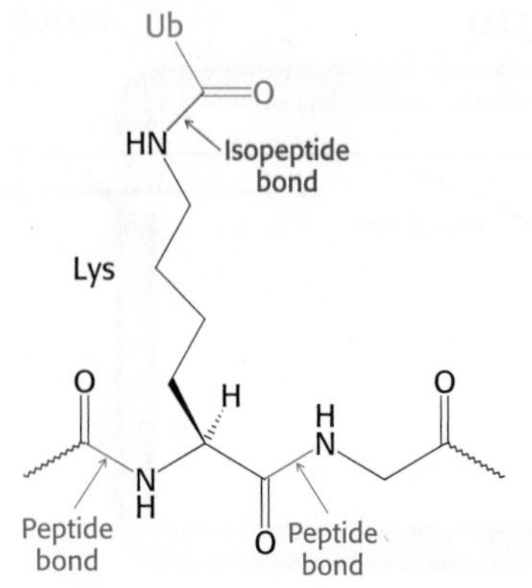

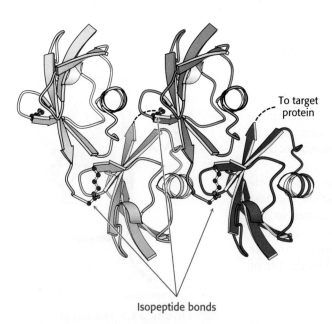

Ubiquitin tags proteins for destruction

Ubiquitin is highly conserved in eukaryotes: yeast and human ubiquitin differ at only 3 of 76 residues. The carboxyl-terminal glycine residue of ubiquitin becomes covalently attached to the ε-amino groups of several lysine residues on a protein destined to be degraded. The energy for the formation of these *isopeptide bonds* (*iso* because ε- rather than α-amino groups are targeted) comes from ATP hydrolysis.

Three enzymes participate in the attachment of ubiquitin to a protein (Figure 23.3): ubiquitin-activating enzyme, or E1; ubiquitin-conjugating enzyme, or E2; and ubiquitin–protein ligase, or E3. First, the C-terminal carboxylate group of ubiquitin becomes linked to a sulfhydryl group of E1

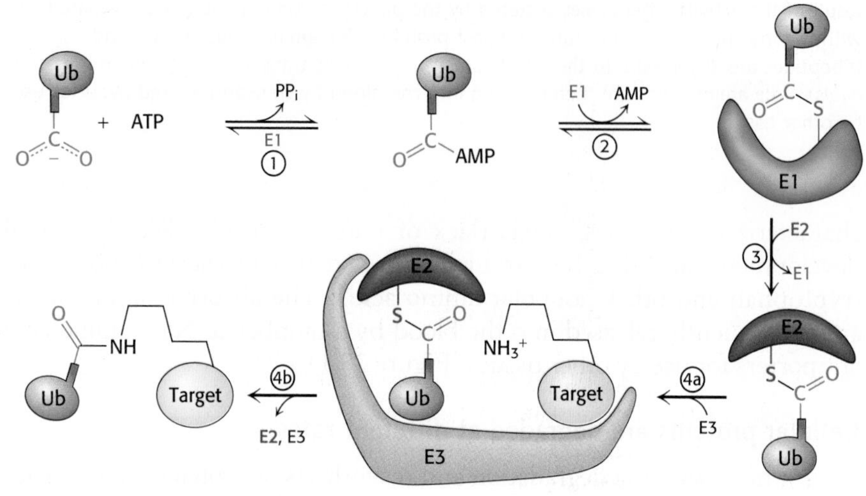

FIGURE 23.3 Ubiquitin conjugation. The ubiquitin-activating enzyme E1 adenylates ubiquitin (Ub) (1) and transfers the ubiquitin to one of its own cysteine residues (2). Ubiquitin is then transferred to a cysteine residue in the ubiquitin-conjugating enzyme E2 by the E2 enzyme (3). Finally, the ubiquitin–protein ligase E3 transfers the ubiquitin to a lysine residue on the target protein (4a and 4b).

FIGURE 23.4 Structure of tetraubiquitin. Four ubiquitin molecules are linked by isopeptide bonds. *Notice* that each isopeptide bond is formed by the linkage of the carboxylate group at the end of the extended C terminus with the ε-amino group of a lysine residue. Dashed lines indicate the positions of the extended C-termini that were not observed in the crystal structure. This unit is the primary signal for degradation when linked to a target protein. [Drawn from 1TBE.pdb]

by a thioester bond. This ATP-driven reaction is reminiscent of fatty acid activation (Section 22.2). In this reaction, an acyl adenylate is formed at the C-terminal carboxylate of ubiquitin with the release of pyrophosphate, and the ubiquitin is subsequently transferred to a sulfhydryl group of a key cysteine residue in E1. The activated ubiquitin is then shuttled to a sulfhydryl group of E2, a reaction catalyzed by E2 itself. Finally, E3 catalyzes the transfer of ubiquitin from E2 to an ε-amino group on the target protein. The ubiquitination reaction is processive: E3 remains bound to the target proteins and generates a chain of ubiquitin molecules by linking the ε-amino group of lysine residue 48 of one ubiquitin molecule to the terminal carboxylate of another. A chain of four or more ubiquitin molecules is especially effective in signaling the need for degradation (Figure 23.4).

What determines whether a protein becomes ubiquitinated? A specific sequence of amino acids, termed a *degron*, indicates that a protein should be degraded. One such signal appears to be quite simple. *The half-life of a cytoplasmic protein is determined to a large extent by its amino-terminal residue* (Table 23.2). This dependency is referred to as the *N-terminal rule* or the *N-terminal degron*. A yeast protein with methionine at its N terminus typically has a half-life of more than 20 hours, whereas one with arginine at this position has a half-life of about 2 minutes. A highly destabilizing N-terminal residue such as arginine or leucine favors rapid ubiquitination, whereas a stabilizing residue such as methionine or proline does not. Why is the N-terminal rule only *apparently* simple? Sometimes the N-terminal degron is exposed only after the protein is proteolytically cleaved. Such degrons are called *pro-N-terminal degrons*, in analogy to proenzymes (Section 10.4), because the protein must be cleaved to expose the signal. In other cases, the destabilizing amino acid is added to the protein after the protein is synthesized. Other modifications, notably N-terminal acetylation, can also activate a degron. Additional degrons thought to identify proteins for degradation include *cyclin destruction boxes*, which are amino acid sequences that mark cell-cycle proteins for destruction, and *PEST sequences*, which contain the amino acid sequence proline (P, single-letter abbreviation), glutamic acid (E), serine (S), and threonine (T).

E3 enzymes are the readers of N-terminal residues. Although most eukaryotes have only one or a small number of distinct E1 enzymes, all eukaryotes have many distinct E2 and E3 enzymes. Moreover, there appears to be only a single family of evolutionarily related E2 proteins but three distinct families of E3 proteins, all together consisting of hundreds of members. Indeed, the E3 family is one of the largest gene families in human beings. The diversity of target proteins that must be tagged for destruction requires a large number of E3 proteins as readers.

Three examples demonstrate the importance of E3 proteins to normal cell function. Proteins that are not broken down owing to a defective E3 may accumulate to create a disease of protein aggregation such as juvenile or early-onset Parkinson disease. A defect in another member of the E3 family causes Angelman syndrome, a severe neurological disorder characterized by an unusually happy disposition, cognitive disability, absence of speech, uncoordinated movement, and hyperactivity. Fascinatingly, if the same ligase is overexpressed, the result is autism. Inappropriate protein turnover also can lead to cancer. For example, human papillomavirus (HPV) encodes a protein that activates a specific E3 enzyme. The enzyme ubiquitinates the tumor suppressor p53 and other proteins that control DNA repair, which are then destroyed. The activation of this E3 enzyme is observed in more than 90% of cervical carcinomas. Thus, the inappropriate marking of key regulatory proteins for destruction can trigger further events, leading to tumor formation.

It is important to note that the role of ubiquitin is much broader than merely marking proteins for destruction. Although we have focused on protein degradation, monoubiquitination also regulates proteins involved in DNA repair, chromatin remodeling, and protein kinase activation, among other biochemical processes.

The proteasome digests the ubiquitin-tagged proteins

If ubiquitin is the mark of death, what is the executioner? *A large protease complex called the proteasome or the 26S proteasome digests the ubiquitinated proteins.* This ATP-driven multisubunit protease spares ubiquitin, which is then recycled. The 26S proteasome is a complex of two components: a 20S catalytic unit and a 19S regulatory unit.

The 20S unit is composed of 28 subunits, encoded by 14 genes, arranged into 4 heteroheptomeric rings stacked to form a structure resembling a

TABLE 23.2 Dependence of the half-lives of cytoplasmic yeast proteins on the identity of their amino-terminal residues

Highly stabilizing residues ($t_{1/2} > 20$ hours)

Ala	Cys	Gly	Met
Pro	Ser	Thr	Val

Intrinsically destabilizing residues ($t_{1/2} = 2$ to 30 minutes)

Arg	His	Ile	Leu
Lys	Phe	Trp	Tyr

Destabilizing residues after chemical modification ($t_{1/2} = 3$ to 30 minutes)

Asn	Asp	Gln	Glu

Data from J. W. Tobias, T. E. Schrader, G. Rocap, and A. Varshavsky. *Science* 254(1991):1374–1377.

FIGURE 23.5 20S proteasome. The 20S proteasome comprises 28 homologous subunits (α-type, red; β-type, blue), arranged in four rings of 7 subunits each. Some of the β-type subunits (right) include protease active sites at their amino termini. [Subunit drawn from 1RYP.pdb.]

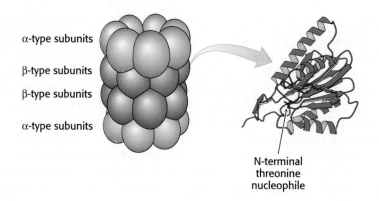

α-type subunits

β-type subunits

β-type subunits

α-type subunits

N-terminal threonine nucleophile

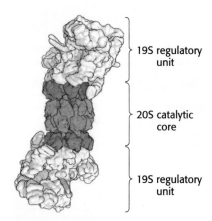

19S regulatory unit

20S catalytic core

19S regulatory unit

FIGURE 23.6 26S proteasome. A 19S regulatory subunit is attached to each end of the 20S catalytic unit.

barrel (Figure 23.5). The outer two rings of the barrel are made up of α-type subunits and the inner two rings of β-type subunits. The 20S catalytic core is a sealed barrel. *Access to its interior is controlled by a 19S regulatory unit,* itself a 700-kDa complex made up of 19 subunits. Two such 19S complexes bind to the 20S proteasome core, one at each end, to form the complete 26S proteasome (Figure 23.6). The 19S regulatory unit has three functions. First, components of the 19S unit are ubiquitin receptors that bind specifically to polyubiquitin chains, thereby ensuring that only ubiquitinated proteins are degraded. Second, an isopeptidase in the 19S unit cleaves off intact ubiquitin molecules from the proteins so that they can be reused. Finally, the doomed protein is unfolded and directed into the catalytic core. Key components of the 19S complex are six ATPases of a type called the AAA class (ATPase *a*ssociated with various cellular *a*ctivities). ATP hydrolysis assists the 19S complex to unfold the substrate and induce conformational changes in the 20S catalytic core so that the substrate can be passed into the center of the complex.

The proteolytic active sites are sequestered in the interior of the barrel to protect potential substrates until they are directed into the barrel. There are three types of active sites in the β subunits, each with a different specificity, but all employ an N-terminal threonine. The hydroxyl group of the threonine residue is converted into a nucleophile that attacks the carbonyl groups of peptide bonds to form acyl-enzyme intermediates. Substrates are degraded in a processive manner without the release of degradation intermediates, until the substrate is reduced to peptides ranging in length from seven to nine residues. These peptide products are released from the proteasome and further degraded by other cellular proteases to yield individual amino acids. Thus, the ubiquitination pathway and the proteasome cooperate to degrade unwanted proteins. Figure 23.7 presents an overview of the fates of amino acids following proteasomal digestion.

The ubiquitin pathway and the proteasome have prokaryotic counterparts

Both the ubiquitin pathway and the proteasome appear to be present in all eukaryotes. Homologs of the proteasome are also found in some prokaryotes. The proteasomes of some archaea are quite similar in overall structure to their eukaryotic counterparts and similarly have 28 subunits (Figure 23.8). In the archaeal proteasome, however, all α outer-ring subunits and all β inner-ring subunits are identical; in eukaryotes, each α or β subunit is one of seven different isoforms. This specialization provides distinct substrate specificity.

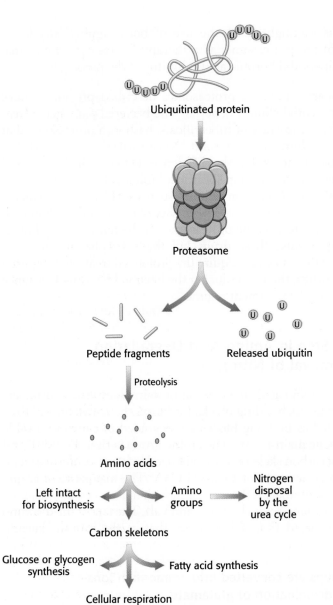

FIGURE 23.7 The proteasome and other proteases generate free amino acids. Ubiquitinated proteins are processed to peptide fragments. Ubiquitin is removed and recycled prior to protein degradation. The peptide fragments are further digested to yield free amino acids, which can be used for biosynthetic reactions, most notably protein synthesis. Alternatively, the amino group can be removed and processed to urea (p. 764) and the carbon skeleton can be used to synthesize carbohydrate or fats or used directly as a fuel for cellular respiration.

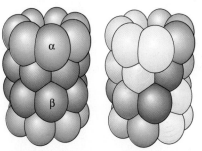

Archaeal proteasome   Eukaryotic proteasome

FIGURE 23.8 Proteasome evolution. The archaeal proteasome consists of 14 identical α subunits and 14 identical β subunits. In the eukaryotic proteasome, gene duplication and specialization has led to 7 distinct subunits of each type. The overall architecture of the proteasome is conserved.

Protein degradation can be used to regulate biological function

Table 23.3 lists a number of physiological processes that are controlled at least in part by protein degradation through the ubiquitin–proteasome pathway. In each case, the proteins being degraded are regulatory proteins. Consider, for example, control of the inflammatory response. A transcription factor called *NF-κB* (NF for nuclear factor) initiates the expression of a number of the genes that take part in this response. This factor is itself activated by the degradation of an attached inhibitory protein, *I-κB* (I for inhibitor). In response to inflammatory signals that bind to membrane-bound receptors, I-κB is phosphorylated at two serine residues, creating an E3 binding site. The binding of E3 leads to the ubiquitination and degradation of I-κB, unleashing NF-κB. The liberated transcription factor migrates from the cytoplasm to the nucleus to stimulate the transcription of the target genes. The NF-κB–I-κB system illustrates the interplay of several key regulatory motifs: receptor-mediated signal transduction, phosphorylation, compartmentalization, controlled and specific degradation, and selective gene expression. The importance of the ubiquitin–proteasome system for the regulation of

TABLE 23.3 Processes regulated by protein degradation

Gene transcription

Cell-cycle progression

Organ formation

Circadian rhythms

Inflammatory response

Tumor suppression

Cholesterol metabolism

Antigen processing

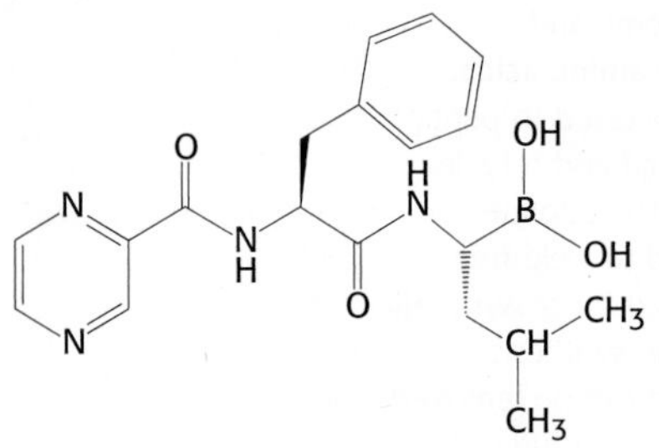

Bortezomib
(a dipeptidyl boronic acid)

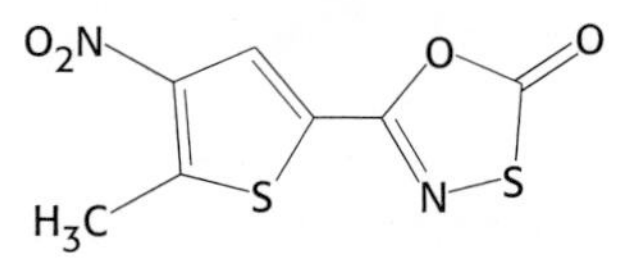

HT1171
[5-(2-methyl-3-nitrothiophen-2-yl)-
1,3,4-oxathiazol-2-one]

gene expression is highlighted by the use of bortezomib (Velcade), a potent inhibitor of the proteasome, as a therapy for multiple myeloma. Bortezomib is a dipeptidyl boronic acid inhibitor of the proteasome.

The evolutionary studies of proteasomes described previously have also yielded potential clinical benefits. The bacterial pathogen *Mycobacterium tuberculosis*, the cause of tuberculosis, harbors a proteasome that is very similar to the human counterpart. Nevertheless, recent work has shown that it is possible to exploit the differences between the human and the bacterial proteasomes to develop specific inhibitors of the *M. tuberculosis* complex. Oxathiazol-2-one compounds such as HT1171 are suicide inhibitors (Section 8.5) of the proteolytic activity of the *M. tuberculosis* proteasome, but have no effect on the proteasomes of the human host. This is especially exciting because these drugs kill the nonreplicating form of *M. tuberculosis*, and thus may not require the prolonged treatment required with conventional drugs, thereby reducing the likelihood of drug resistance due to interruption of the treatment regime.

23.3 The First Step in Amino Acid Degradation Is the Removal of Nitrogen

What is the fate of amino acids released on protein digestion or turnover? The first call is for use as building blocks for biosynthetic reactions. However, any not needed as building blocks are degraded to compounds able to enter the metabolic mainstream. The amino group is first removed, and then the remaining carbon skeleton is metabolized to glucose, one of several citric acid cycle intermediates, or to acetyl CoA. The major site of amino acid degradation in mammals is the liver, although muscles readily degrade the branched-chain amino acids (Leu, Ile, and Val). The fate of the α-amino group will be considered first, followed by that of the carbon skeleton (Section 23.5).

Alpha-amino groups are converted into ammonium ions by the oxidative deamination of glutamate

The α-amino group of many amino acids is transferred to α-ketoglutarate to form *glutamate*, which is then oxidatively deaminated to yield ammonium ion (NH_4^+).

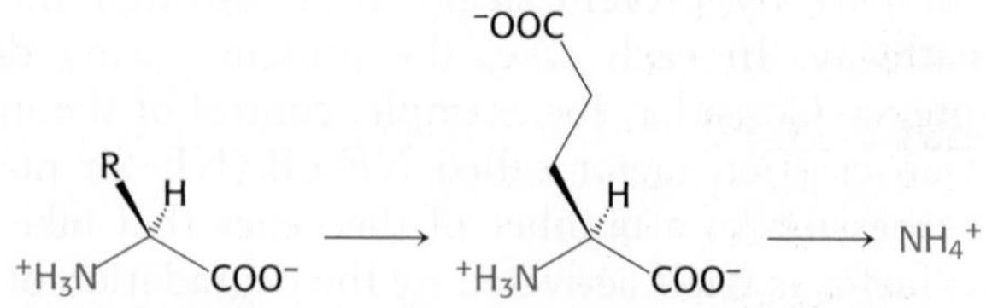

Aminotransferases catalyze the transfer of an α-amino group from an α-amino acid to an α-ketoacid.

These enzymes, also called *transaminases*, generally funnel α-amino groups from a variety of amino acids to α-ketoglutarate for conversion into NH_4^+.

Aspartate aminotransferase, one of the most important of these enzymes, catalyzes the transfer of the amino group of aspartate to α-ketoglutarate.

$$\text{Asparate} + \alpha\text{-ketoglutarate} \rightleftharpoons \text{oxaloacetate} + \text{glutamate}$$

Alanine aminotransferase catalyzes the transfer of the amino group of alanine to α-ketoglutarate.

$$\text{Alanine} + \alpha\text{-ketoglutarate} \rightleftharpoons \text{pyruvate} + \text{glutamate}$$

Transamination reactions are reversible and can thus be used to synthesize amino acids from α-ketoacids, as we shall see in Chapter 24.

The nitrogen atom in glutamate is converted into free ammonium ion by oxidative deamination, a reaction catalyzed by *glutamate dehydrogenase*. This enzyme is unusual in being able to utilize either NAD^+ or $NADP^+$, at least in some species. The reaction proceeds by dehydrogenation of the C—N bond, followed by hydrolysis of the resulting ketimine.

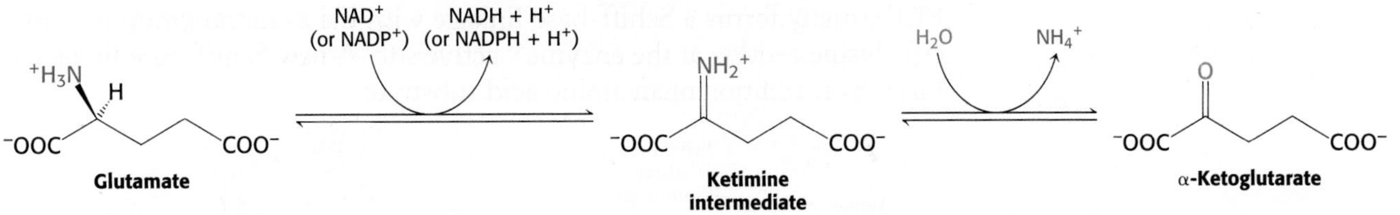

This reaction equilibrium constant is close to 1 in the liver, so the direction of the reaction is determined by the concentrations of reactants and products. Normally, the reaction is driven forward by the rapid removal of ammonium ion. Glutamate dehydrogenase, essentially a liver-specific enzyme, is located in mitochondria, as are some of the other enzymes required for the production of urea. This compartmentalization sequesters free ammonium ion, which is toxic.

In mammals, but not in other organisms, glutamate dehydrogenase is allosterically inhibited by GTP and stimulated by ADP. These nucleotides exert their regulatory effects in a unique manner. An *abortive complex* is formed on the enzyme when a product is replaced by substrate before the reaction is complete. For instance, the enzyme bound to glutamate and NAD(P)H is an abortive complex. GTP facilitates the formation of such complexes while ADP destabilizes the complexes.

The sum of the reactions catalyzed by aminotransferases and glutamate dehydrogenase is

$$\alpha\text{-Amino acid} + NAD^+ + H_2O \rightleftharpoons \alpha\text{-ketoacid} + NH_4^+ + NADH + H^+$$
$$\text{(or } NADP^+) \qquad\qquad\qquad \text{(or } NADPH)$$

In most terrestrial vertebrates, NH_4^+ is converted into urea, which is excreted.

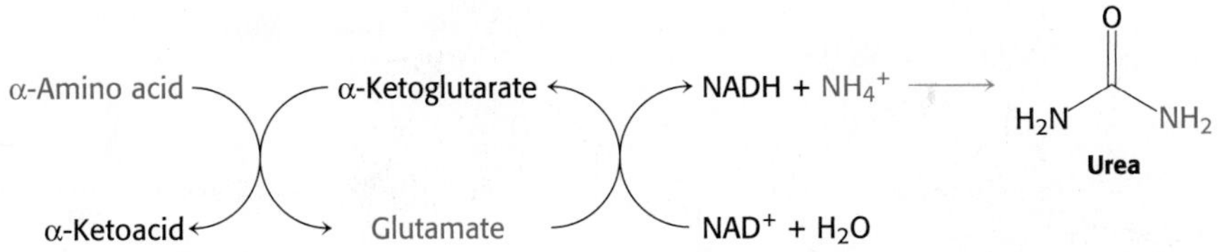

Mechanism: Pyridoxal phosphate forms Schiff-base intermediates in aminotransferases

All aminotransferases contain the prosthetic group *pyridoxal phosphate (PLP)*, which is derived from *pyridoxine (vitamin B_6)*. Pyridoxal phosphate includes a pyridine ring that is slightly basic to which is attached an OH

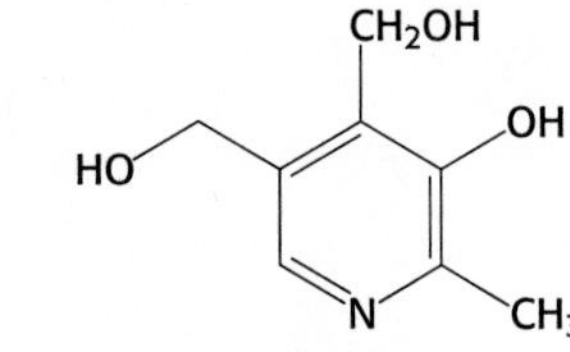

Pyridoxine
(vitamin B₆)

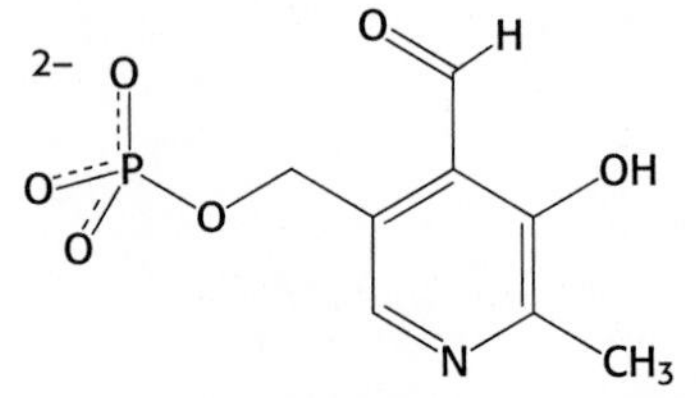

Pyridoxal phosphate
(PLP)

group that is slightly acidic. Thus, pyridoxal phosphate derivatives can form a stable tautomeric form in which the pyridine nitrogen atom is protonated and, hence, positively charged, while the OH group loses a proton and hence is negatively charged, forming a phenolate.

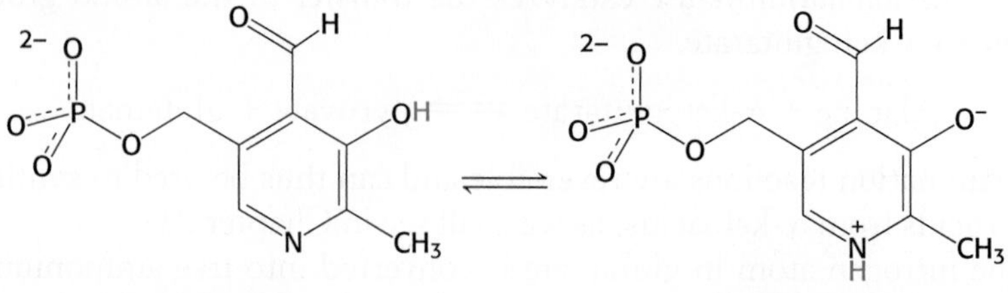

PLP (protonated)   PLP (phenolate)

The most important functional group on PLP is the aldehyde. This group forms covalent Schiff-base intermediates (p. 685) with amino acid substrates. Indeed, even in the absence of substrate, the aldehyde group of PLP usually forms a Schiff-base linkage with the ε-amino group of a specific lysine residue at the enzyme's active site. A new Schiff-base linkage is formed on addition of an amino acid substrate.

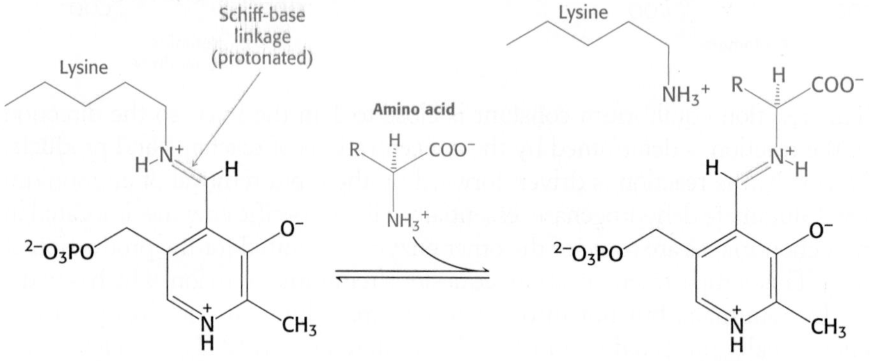

Internal aldimine   External aldimine

The α-amino group of the amino acid substrate displaces the ε-amino group of the active-site lysine residue. In other words, an *internal* aldimine becomes an *external* aldimine. The amino acid–PLP Schiff base that is formed remains tightly bound to the enzyme by multiple noncovalent interactions. The Schiff-base linkage often accepts a proton at the nitrogen of the pyridine ring, with the positive charge stabilized by interaction with the negatively charged phenolate group of PLP.

The Schiff base between the amino acid substrate and PLP, the external *aldimine*, loses a proton from the α-carbon atom of the amino acid to form a *quinonoid* intermediate (Figure 23.9). Reprotonation of this intermediate at

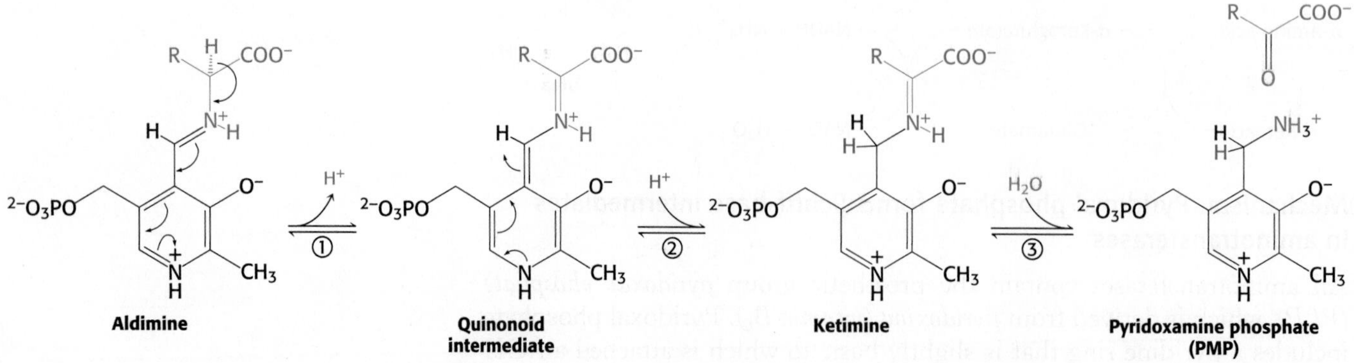

FIGURE 23.9 Transamination mechanism. (1) The external aldimine loses a proton to form a quinonoid intermediate. (2) Reprotonation of this intermediate at the aldehyde carbon atom yields a ketimine. (3) This intermediate is hydrolyzed to generate the α-ketoacid product and pyridoxamine phosphate.

Aldimine   Quinonoid intermediate   Ketimine   Pyridoxamine phosphate (PMP)

the aldehyde carbon atom yields a *ketimine*. The ketimine is then hydrolyzed to an α-ketoacid and *pyridoxamine phosphate (PMP)*. These steps constitute half of the transamination reaction.

$$\text{Amino acid}_1 + \text{E-PLP} \rightleftharpoons \alpha\text{-ketoacid}_1 + \text{E-PMP}$$

The second half takes place by the reverse of the preceding pathway. A second α-ketoacid reacts with the enzyme–pyridoxamine phosphate complex (E-PMP) to yield a second amino acid and regenerate the enzyme–pyridoxal phosphate complex (E-PLP).

$$\alpha\text{-Ketoacid}_2 + \text{E-PMP} \rightleftharpoons \text{amono acid}_2 + \text{E-PLP}$$

The sum of these partial reactions is

$$\text{Amino acid}_1 + \alpha\text{-ketoacid} \rightleftharpoons \text{amino acid}_2 + \alpha\text{-ketoacid}_1$$

Pyridoxamine phosphate
(PMP)

Aspartate aminotransferase is an archetypal pyridoxal-dependent transaminase

The mitochondrial enzyme aspartate aminotransferase provides an especially well studied example of PLP as a coenzyme for transamination reactions (Figure 23.10). X-ray crystallographic studies provided detailed views of how PLP and substrates are bound and confirmed much of the proposed catalytic mechanism. Each of the identical 45-kDa subunits of this dimer consists of a large domain and a small one. PLP is bound to the large domain, in a pocket near the domain interface. In the absence of substrate, the aldehyde group of PLP is in a Schiff-base linkage with lysine 258, as expected. Adjacent to the coenzyme's binding site is a conserved arginine residue that interacts with the α-carboxylate group of the amino acid substrate, helping to orient the substrate appropriately in the active site. A base is necessary to remove a proton from the α-carbon group of the amino acid and to transfer it to the aldehyde carbon atom of PLP (Figure 23.9, steps 1 and 2). The lysine amino group that was initially in Schiff-base linkage with PLP appears to serve as the proton donor and acceptor.

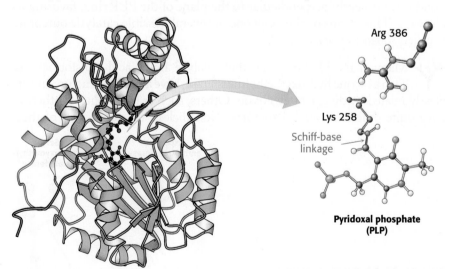

FIGURE 23.10 Aspartate aminotransferase. The active site of this prototypical PLP-dependent enzyme includes pyridoxal phosphate attached to the enzyme by a Schiff-base linkage with lysine 258. An arginine residue in the active site helps orient substrates by binding to their α-carboxylate groups. Only one of the enzyme's two subunits is shown. [Drawn from 1AAW.pdb.]

Blood levels of aminotransferases serve a diagnostic function

The presence of alanine and aspartate aminotransferase in the blood is an indication of liver damage. Liver damage can occur for a number of reasons, including viral hepatitis, long-term excessive alcohol consumption, and reaction to drugs such as acetaminophen (Section 28.1). Under these

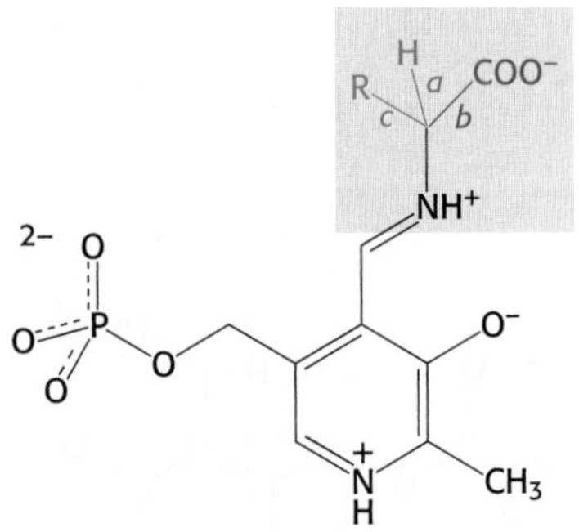

FIGURE 23.11 Bond cleavage by PLP enzymes. Pyridoxal phosphate enzymes labilize one of three bonds at the α-carbon atom of an amino acid substrate. For example, bond *a* is labilized by aminotransferases, bond *b* by decarboxylases, and bond *c* by aldolases (such as threonine aldolases). PLP enzymes also catalyze reactions at the β- and γ-carbon atoms of amino acids.

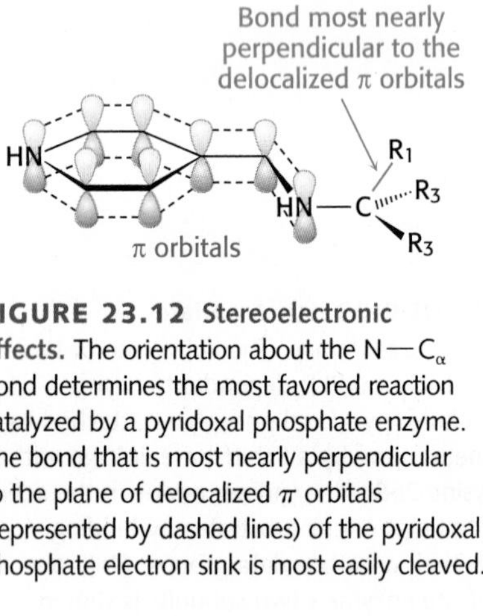

FIGURE 23.12 Stereoelectronic effects. The orientation about the N—C$_\alpha$ bond determines the most favored reaction catalyzed by a pyridoxal phosphate enzyme. The bond that is most nearly perpendicular to the plane of delocalized π orbitals (represented by dashed lines) of the pyridoxal phosphate electron sink is most easily cleaved.

conditions, liver cell membranes are damaged and some cellular proteins, including the aminotransferases, leak into the blood. Normal blood values for alanine and aspartate aminotransferase activity are 5–30 units/l and 40–125 units/l, respectively. Depending on the extent of liver damage, the values will reach 200–300 units/l.

Pyridoxal phosphate enzymes catalyze a wide array of reactions

Transamination is just one of a wide range of amino acid transformations that are catalyzed by PLP enzymes. The other reactions catalyzed by PLP enzymes at the α-carbon atom of amino acids are decarboxylations, deaminations, racemizations, and aldol cleavages (Figure 23.11). In addition, PLP enzymes catalyze elimination and replacement reactions at the β-carbon atom (e.g., tryptophan synthetase in the synthesis of tryptophan) and the γ-carbon atom (e.g., cystathionine β-synthase in the synthesis of cysteine) of amino acid substrates. Three common features of PLP catalysis underlie these diverse reactions.

1. A Schiff base is formed by the amino acid substrate (the amine component) and PLP (the carbonyl component).

2. The protonated form of PLP acts as an *electron sink* to stabilize catalytic intermediates that are negatively charged. Electrons from these intermediates are attracted to the positive charge on the ring nitrogen atom. In other words, PLP is an *electrophilic catalyst*.

3. The product Schiff base is cleaved at the completion of the reaction.

How does an enzyme selectively break a particular one of three bonds at the α-carbon atom of an amino acid substrate? An important principle is that *the bond being broken must be perpendicular to the π orbitals of the electron sink* (Figure 23.12). An aminotransferase, for example, binds the amino acid substrate so that the C$_\alpha$—H bond is perpendicular to the PLP ring (Figure 23.13). In serine hydroxymethyltransferase, the enzyme that converts serine into glycine, the N—C$_\alpha$ bond is rotated so that the C$_\alpha$—C$_\beta$ bond is most nearly perpendicular to the plane of the PLP ring, favoring its cleavage. This means of choosing one of several possible catalytic outcomes is called *stereoelectronic control*.

Many of the PLP enzymes that catalyze amino acid transformations, such as serine hydroxymethyltransferase, have a similar structure and are clearly related by divergent evolution. Others, such as tryptophan synthetase, have quite different overall structures. Nonetheless, the active sites of these

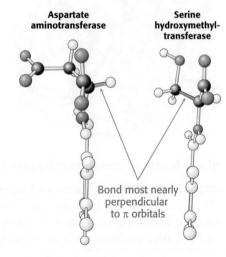

Aspartate aminotransferase

Serine hydroxymethyltransferase

Bond most nearly perpendicular to π orbitals

FIGURE 23.13 Reaction choice. In aspartate aminotransferase, the C$_\alpha$—H bond is most nearly perpendicular to the π-orbital system and is cleaved. In serine hydroxymethyltransferase, a small rotation about the N—C$_\alpha$ bond places the C$_\alpha$—C$_\beta$ bond perpendicular to the π system, favoring its cleavage.

enzymes are remarkably similar to that of aspartate aminotransferase, revealing the effects of convergent evolution.

Serine and threonine can be directly deaminated

The α-amino groups of serine and threonine can be directly converted into NH_4^+ without first being transferred to α-ketoglutarate. These direct deaminations are catalyzed by *serine dehydratase* and *threonine dehydratase*, in which PLP is the prosthetic group.

$$\text{Serine} \longrightarrow \text{pyruvate} + NH_4^+$$
$$\text{Threonine} \longrightarrow \alpha\text{-ketobutyrate} + NH_4^+$$

These enzymes are called *dehydratases* because *dehydration precedes deamination*. Serine loses a hydrogen ion from its α-carbon atom and a hydroxide ion group from its β-carbon atom to yield aminoacrylate. This unstable compound reacts with H_2O to give pyruvate and NH_4^+. Thus, the presence of a hydroxyl group attached to the β-carbon atom in each of these amino acids permits the direct deamination.

Peripheral tissues transport nitrogen to the liver

Although most amino acid degradation takes place in the liver, other tissues can degrade amino acids. For instance, muscle uses branched-chain amino acids as a source of fuel during prolonged exercise and fasting. How is the nitrogen processed in these other tissues? As in the liver, the first step is the removal of the nitrogen from the amino acid. However, muscle lacks the enzymes of the urea cycle, and so the nitrogen must be released in a nontoxic form that can be absorbed by the liver and converted into urea.

Nitrogen is transported from muscle to the liver in two principal transport forms. Glutamate is formed by transamination reactions, but the nitrogen is then transferred to pyruvate to form alanine, which is released into the blood (Figure 23.14). The liver takes up the alanine and converts it back into pyruvate by transamination. The pyruvate can be used for gluconeogenesis and the amino group eventually appears as urea. This transport is referred to as the *glucose–alanine cycle*. It is reminiscent of the Cori cycle discussed earlier (Figure 16.32). However, in contrast with the Cori cycle, pyruvate is not reduced to lactate by NADH, and thus more high-energy electrons are available for oxidative phosphorylation in the muscle.

FIGURE 23.14 PATHWAY INTEGRATION: The glucose–alanine cycle. During prolonged exercise and fasting, muscle uses branched-chain amino acids as fuel. The nitrogen removed is transferred (through glutamate) to alanine, which is released into the blood stream. In the liver, alanine is taken up and converted into pyruvate for the subsequent synthesis of glucose.

 EXPLORE this pathway further in the Metabolic Map on 🌀 Sapling Plus

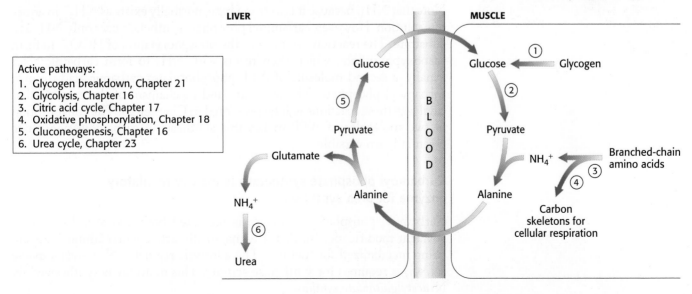

Active pathways:
1. Glycogen breakdown, Chapter 21
2. Glycolysis, Chapter 16
3. Citric acid cycle, Chapter 17
4. Oxidative phosphorylation, Chapter 18
5. Gluconeogenesis, Chapter 16
6. Urea cycle, Chapter 23

Glutamine is also a key transport form of nitrogen. Glutamine synthetase catalyzes the synthesis of glutamine from glutamate and NH_4^+ in an ATP-dependent reaction:

$$NH_4^+ + \text{glutamate} + \text{ATP} \xrightarrow{\text{Glutamine synthetase}} \text{glutamine} + \text{ADP} + P_i$$

The nitrogens of glutamine can be converted into urea in the liver.

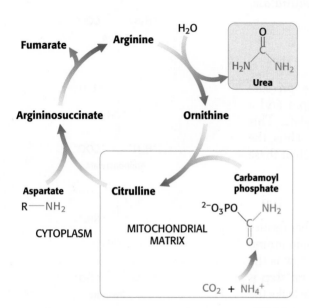

FIGURE 23.15 The urea cycle.

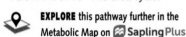

EXPLORE this pathway further in the Metabolic Map on **Sapling**Plus

23.4 Ammonium Ion Is Converted into Urea in Most Terrestrial Vertebrates

Some of the NH_4^+ formed in the breakdown of amino acids is consumed in the biosynthesis of nitrogen compounds. In most terrestrial vertebrates, the excess NH_4^+ is converted into *urea* and then excreted. Such organisms are referred to as *ureotelic*.

In terrestrial vertebrates, urea is synthesized by the *urea cycle* (Figure 23.15). One of the nitrogen atoms of urea is transferred from an amino acid, aspartate. The other nitrogen atom is derived directly from free NH_4^+, and the carbon atom comes from HCO_3^- (derived by the hydration of CO_2; Section 9.2).

$$NH_4^+ \longrightarrow$$

Urea

The urea cycle begins with the formation of carbamoyl phosphate

The urea cycle begins with the coupling of free NH_3 with HCO_3^- to form carbamoyl phosphate, the committed reaction of the urea cycle, which is catalyzed by *carbamoyl phosphate synthetase I*. Carbamoyl phosphate is a simple molecule, but its synthesis is complex, requiring three steps.

Note that NH_3, because it is a strong base, normally exists as NH_4^+ in aqueous solution. However, carbamoyl phosphate synthetase uses only NH_3 as a substrate. The reaction begins with the phosphorylation of HCO_3^- to form carboxyphosphate, which then reacts with NH_3 to form carbamic acid. Finally, a second molecule of ATP phosphorylates carbamic acid to form carbamoyl phosphate. The structure and mechanism of the enzyme that catalyzes these reactions will be presented in Chapter 25. The consumption of two molecules of ATP makes this synthesis of carbamoyl phosphate essentially irreversible.

Carbamoyl phosphate synthetase is the key regulatory enzyme for urea synthesis

Carbamoyl phosphate synthetase is regulated both allosterically and by covalent modification so that it is maximally active when amino acids are being metabolized for fuel use. The allosteric regulator N-*acetylglutamate* (NAG) is required for synthetase activity. This molecule is synthesized by N-*acetylglutamate synthase*.

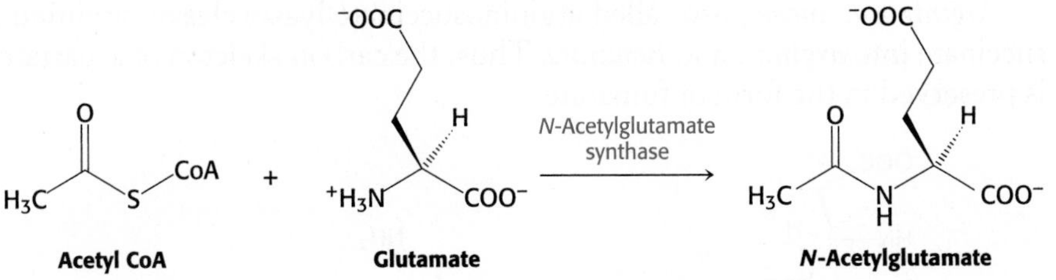

Acetyl CoA + **Glutamate** →(N-Acetylglutamate synthase)→ **N-Acetylglutamate**

N-acetylglutamate synthase is itself activated by arginine. Thus, NAG is synthesized when amino acids, as represented by arginine and glutamate, are readily available, and carbamoyl phosphate synthetase is then activated to process the generated ammonia. When ammonia is not being generated, the synthetase is inhibited by acetylation. A rise in the mitochondrial NAD^+, indicative of an energy-poor state, stimulates a deacetylase that removes the acetyl group, activating the synthetase and readying the enzyme for processing ammonia from protein degradation. The control of the acetylation of the synthetase is not yet clear.

Carbamoyl phosphate reacts with ornithine to begin the urea cycle

The carbamoyl group of carbamoyl phosphate has a high transfer potential because of its anhydride linkage. The carbamoyl group is transferred to *ornithine* to form *citrulline*, in a reaction catalyzed by *ornithine transcarbamoylase*.

Ornithine + **Carbamoyl phosphate** →(Ornithine transcarbamoylase, P_i)→ **Citrulline**

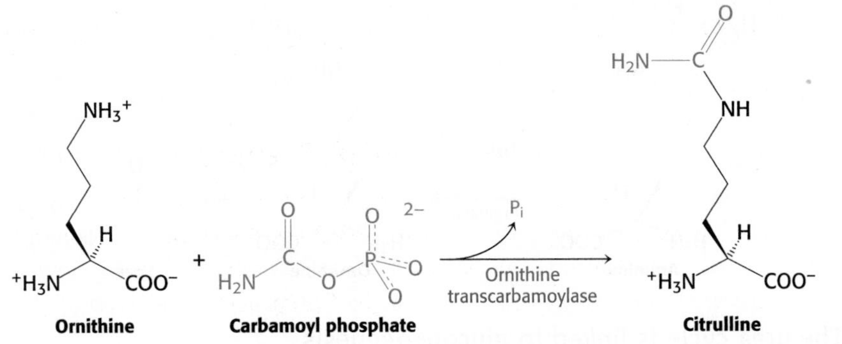

Ornithine and citrulline are amino acids, but they are not used as building blocks of proteins. The formation of NH_4^+ by glutamate dehydrogenase, its incorporation into carbamoyl phosphate as NH_3, and the subsequent synthesis of citrulline take place in the mitochondrial matrix. In contrast, the next three reactions of the urea cycle, which lead to the formation of urea, take place in the cytoplasm.

Citrulline is transported to the cytoplasm, where it condenses with aspartate, the donor of the second amino group of urea. This synthesis of *argininosuccinate*, catalyzed by *argininosuccinate synthetase*, is driven by the cleavage of ATP into AMP and pyrophosphate and by the subsequent hydrolysis of pyrophosphate.

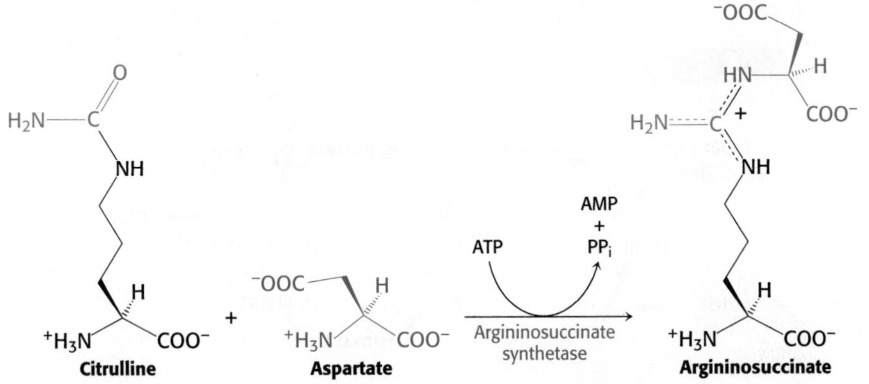

Citrulline + **Aspartate** →(Argininosuccinate synthetase, ATP → AMP + PP_i)→ **Argininosuccinate**

Argininosuccinase (also called argininosuccinate lyase) cleaves argininosuccinate into *arginine* and *fumarate*. Thus, the carbon skeleton of aspartate is preserved in the form of fumarate.

Finally, arginine is hydrolyzed to generate urea and ornithine in a reaction catalyzed by *arginase*. Ornithine is then transported back into the mitochondrion to begin another cycle. The urea is excreted. Indeed, human beings excrete about 10 kg (22 pounds) of urea per year.

In ancient Rome, urine was a valuable commodity. Vessels were placed on street corners for passersby to urinate into. Bacteria would degrade the urea, releasing ammonium ion, which would act as a bleach to brighten togas.

The urea cycle is linked to gluconeogenesis

The stoichiometry of urea synthesis is

$$CO_2 + NH_4^+ + 3ATP + aspartate + 2H_2O \longrightarrow$$
$$urea + 2ADP + 2P_i + AMP + PP_i + fumarate$$

Pyrophosphate is rapidly hydrolyzed, and so the equivalent of four molecules of ATP are consumed in these reactions to synthesize one molecule of urea. The synthesis of fumarate by the urea cycle is important because it is a precursor for glucose synthesis (Figure 23.16). Fumarate is hydrated to malate, which is in turn oxidized to oxaloacetate. Oxaloacetate can be converted into glucose by gluconeogenesis or transaminated to aspartate.

FIGURE 23.16 Metabolic integration of nitrogen metabolism. The urea cycle, gluconeogenesis, and the transamination of oxaloacetate are linked by fumarate and aspartate.

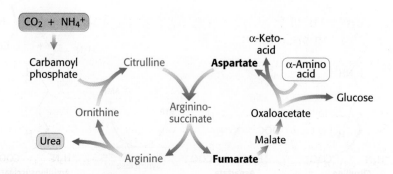

Urea-cycle enzymes are evolutionarily related to enzymes in other metabolic pathways

Carbamoyl phosphate synthetase generates carbamoyl phosphate for both the urea cycle and the first step in pyrimidine biosynthesis (Section 25.1). In mammals, two distinct isozymes of the enzyme are present. Carbamoyl phosphate synthetase II, used in pyrimidine biosynthesis, differs in two important ways from its urea-cycle counterpart. First, this enzyme utilizes glutamine as a nitrogen source rather than NH_3. The side-chain amide of glutamine is hydrolyzed within one domain of the enzyme, and the ammonia generated moves through a tunnel in the enzyme to a second active site, where it reacts with carboxyphosphate. Second, this enzyme is part of a large complex that catalyzes several steps in pyrimidine biosynthesis (Section 25.1). Interestingly, the domain in which glutamine hydrolysis takes place is largely preserved in the urea-cycle enzyme, although that domain is catalytically inactive. This site binds *N*-acetylglutamate, an allosteric activator of the enzyme. *A catalytic site in one isozyme has been adapted to act as an allosteric site in another isozyme having a different physiological role.*

Can we find homologs for the other enzymes in the urea cycle? Ornithine transcarbamoylase is homologous to aspartate transcarbamoylase, which catalyzes the first step in pyrimidine biosynthesis, and the structures of their catalytic subunits are quite similar (Figure 23.17). Thus, two consecutive steps in the pyrimidine biosynthetic pathway were adapted for urea synthesis. The next step in the urea cycle is the addition of aspartate to citrulline to form argininosuccinate, and the subsequent step is the removal of fumarate. These two steps together accomplish the net addition of an amino group to citrulline to form arginine. Remarkably, these steps are analogous to two consecutive steps in the purine biosynthetic pathway (Section 25.2).

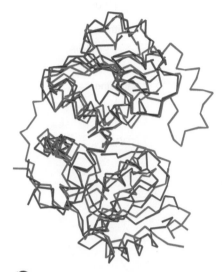

FIGURE 23.17 Homologous enzymes. The structure of the catalytic subunit of ornithine transcarbamoylase (blue) is quite similar to that of the catalytic subunit of aspartate transcarbamoylase (red), indicating that these two enzymes are homologs. [Drawn from 1AKM.pdb and 1RAI.pdb.]

Aspartate — **Fumarate**

P-ribose → P-ribose → P-ribose

The enzymes that catalyze these steps are homologous to argininosuccinate synthetase and argininosuccinase, respectively. Thus, four of the five enzymes in the urea cycle were adapted from enzymes taking part in nucleotide biosynthesis. The remaining enzyme, arginase, appears to be an ancient enzyme found in all domains of life.

Inherited defects of the urea cycle cause hyperammonemia and can lead to brain damage

The synthesis of urea in the liver is the major route for the removal of NH_4^+. Urea cycle disorders occur with a prevalence of about 1 in 15,000. A blockage of carbamoyl phosphate synthesis or of any of the four steps of the urea cycle has devastating consequences because there is no alternative pathway for the synthesis of urea. *All defects in the urea cycle lead to an elevated level of NH_4^+ in the blood (hyperammonemia).* Some of these genetic defects become evident a day or two after birth, when the afflicted infant becomes lethargic and

Arginine
(excess supplied)

Urea

Ornithine

Carbamoyl
phosphate

Citrulline

Aspartate

Argininosuccinate
(excreted)

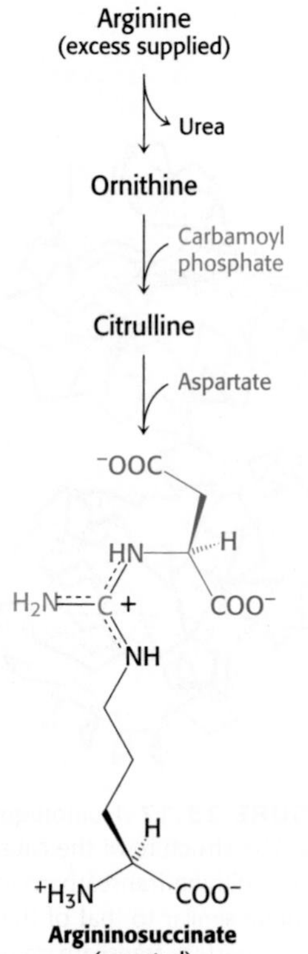

FIGURE 23.18 Treatment of argininosuccinase deficiency. Argininosuccinase deficiency can be managed by supplementing the diet with arginine. Nitrogen is excreted in the form of argininosuccinate.

vomits periodically. Coma and irreversible brain damage may soon follow, a condition called *hepatic encephalopathy*. Why are high levels of NH_4^+ toxic? The answer to this question is not yet known. Recent work, however, suggests that NH_4^+ may inappropriately activate a sodium-potassium-chloride cotransporter. This activation disrupts the osmotic balance of the nerve cell, causing swelling that damages the cell and results in neurological disorders.

Ingenious strategies for coping with deficiencies in urea synthesis have been devised on the basis of a thorough understanding of the underlying biochemistry. Consider, for example, *argininosuccinase deficiency*. This defect can be partly bypassed by *providing a surplus of arginine in the diet and restricting the total protein intake*. In the liver, arginine is split into urea and ornithine, which then reacts with carbamoyl phosphate to form citrulline (Figure 23.18). This urea-cycle intermediate condenses with aspartate to yield argininosuccinate, which is then excreted. Note that two nitrogen atoms—one from carbamoyl phosphate and the other from aspartate—are eliminated from the body per molecule of arginine provided in the diet. In essence, *argininosuccinate substitutes for urea in carrying nitrogen out of the body*.

The treatment of *carbamoyl phosphate synthetase deficiency* or *ornithine transcarbamoylase deficiency* illustrates a different strategy for circumventing a metabolic block. Citrulline and argininosuccinate cannot be used to dispose of nitrogen atoms because their formation is impaired. Under these conditions, excess nitrogen accumulates in glycine and glutamine. The challenge then is to rid the body of the nitrogen accumulating in these two amino acids. That goal is accomplished by supplementing a protein-restricted diet with *large amounts of benzoate and phenylacetate*. Benzoate is activated to benzoyl CoA, which reacts with glycine to form hippurate, which is excreted (Figure 23.19). Likewise, phenylacetate is activated to phenylacetyl CoA, which reacts with glutamine to form phenylacetylglutamine, which is also excreted. These conjugates substitute for urea in the disposal of nitrogen. Thus, *latent biochemical pathways can be activated to partly bypass a genetic defect*.

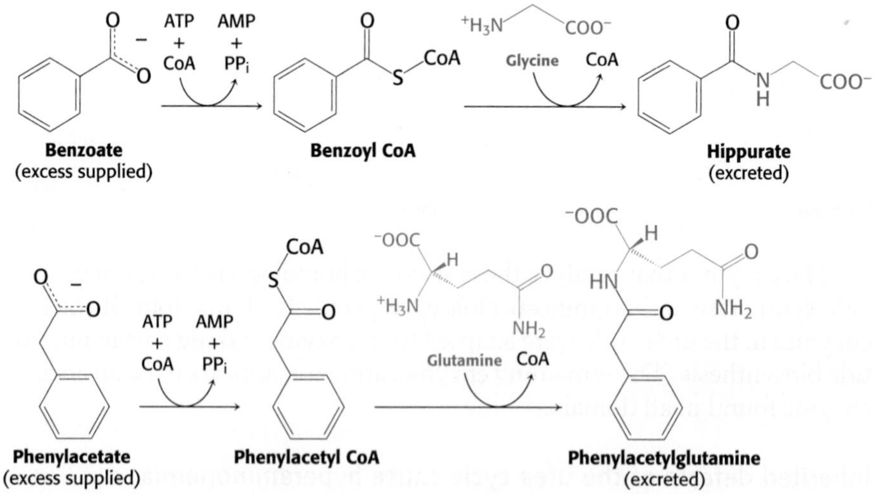

FIGURE 23.19 Treatment of carbamoyl phosphate synthetase and ornithine transcarbamoylase deficiencies. Both deficiencies can be treated by supplementing the diet with benzoate and phenylacetate. Nitrogen is excreted in the form of hippurate and phenylacetylglutamine.

ATP   AMP
  +   +
CoA PP_i

Benzoate
(excess supplied)

Benzoyl CoA

Glycine   CoA

Hippurate
(excreted)

ATP   AMP
  +   +
CoA PP_i

Glutamine   CoA

Phenylacetate
(excess supplied)

Phenylacetyl CoA

Phenylacetylglutamine
(excreted)

Urea is not the only means of disposing of excess nitrogen

As stated earlier, most terrestrial vertebrates are ureotelic; they excrete excess nitrogen as urea. However, urea is not the only excretable form of nitrogen. *Ammoniotelic organisms, such as aquatic vertebrates and invertebrates, release nitrogen as NH_4^+* and rely on the aqueous environment to dilute this toxic substance. Interestingly, lungfish, which are normally ammoniotelic, become ureotelic in time of drought, when they live out of the water.

Both ureotelic and ammoniotelic organisms depend on sufficient water, to varying degrees, for nitrogen excretion. *In contrast, uricotelic organisms, such as birds and reptiles, secrete nitrogen as the purine uric acid.* Uric acid is secreted as an almost solid slurry requiring little water. The secretion of uric acid also has the advantage of removing four atoms of nitrogen per molecule. The pathway for nitrogen excretion developed in the course of evolution clearly depends on the habitat of the organism.

Uric acid

23.5 Carbon Atoms of Degraded Amino Acids Emerge as Major Metabolic Intermediates

We now turn to the fates of the carbon skeletons of amino acids after the removal of the α-amino group. *The strategy of amino acid degradation is to transform the carbon skeletons into major metabolic intermediates that can be converted into glucose or oxidized by the citric acid cycle.* The conversion pathways range from extremely simple to quite complex. In fact, some amino acids can be degraded by alternate pathways. For instance, threonine can be metabolized to succinyl CoA or pyruvate. In all cases, however, the carbon skeletons of the diverse set of 20 fundamental amino acids are funneled into only seven molecules: *pyruvate, acetyl CoA, acetoacetyl CoA, α-ketoglutarate, succinyl CoA, fumarate,* and *oxaloacetate.* We see here an example of the remarkable economy of metabolic conversions.

Amino acids that are degraded to acetyl CoA or acetoacetyl CoA are termed *ketogenic amino acids* because they can give rise to ketone bodies or fatty acids. Amino acids that are degraded to pyruvate, α-ketoglutarate, succinyl CoA, fumarate, or oxaloacetate are termed *glucogenic amino acids.* Oxaloacetate, generated from pyruvate and other citric acid cycle intermediates, can be converted into phosphoenolpyruvate and then into glucose (Section 16.3). Recall that mammals lack a pathway for the net synthesis of glucose from acetyl CoA or acetoacetyl CoA.

Of the basic set of 20 amino acids, only leucine and lysine are solely ketogenic (Figure 23.20). Isoleucine, phenylalanine, tryptophan, and tyrosine

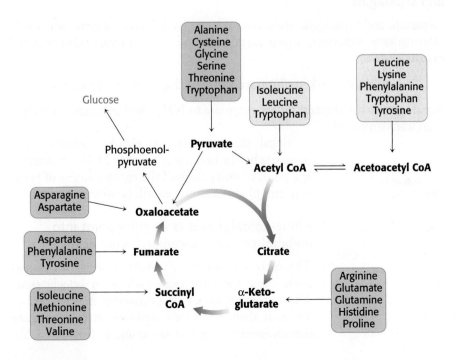

FIGURE 23.20 Fates of the carbon skeletons of amino acids. Glucogenic amino acids are shaded red, and ketogenic amino acids are shaded yellow. Several amino acids are both glucogenic and ketogenic.

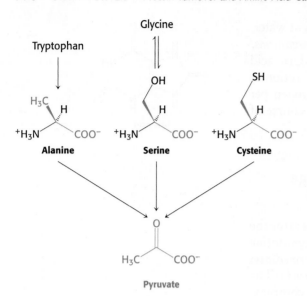

FIGURE 23.21 Pyruvate formation from amino acids. Pyruvate is the point of entry for alanine, serine, cysteine, glycine and tryptophan.

are both ketogenic and glucogenic. Some of their carbon atoms emerge in acetyl CoA or acetoacetyl CoA, whereas others appear in potential precursors of glucose. The other 14 amino acids are classed as solely glucogenic. We will identify the degradation pathways by the entry point into metabolism.

Pyruvate is an entry point into metabolism for a number of amino acids

Pyruvate is the entry point of the three-carbon amino acids—alanine, serine, and cysteine—into the metabolic mainstream (Figure 23.21). The transamination of alanine directly yields pyruvate.

$$\text{Alanine} + \alpha\text{-ketoglutarate} \rightleftharpoons \text{pyruvate} + \text{glutamate}$$

As mentioned earlier in the chapter, glutamate is then oxidatively deaminated, yielding NH_4^+ and regenerating α-ketoglutarate. The sum of these reactions is

$$\text{Alanine} + \text{NAD(P)}^+ + H_2O \rightleftharpoons \text{pyruvate} + NH_4^+ + \text{NAD(P)H} + H^+$$

Another simple reaction in the degradation of amino acids is the *deamination of serine to pyruvate* by *serine dehydratase* (p. 763).

$$\text{Serine} \longrightarrow \text{pyruvate} + NH_4^+$$

Cysteine can be converted into pyruvate by several pathways, with its sulfur atom emerging in H_2S, SCN^-, or SO_3^{2-}.

The carbon atoms of three other amino acids can be converted into pyruvate. *Glycine* can be converted into serine by the enzymatic addition of a hydroxymethyl group or it can be cleaved to give CO_2, NH_4^+, and an activated one-carbon unit. *Threonine* can give rise to pyruvate through the intermediate 2-amino-3-ketobutyrate. Three carbon atoms of *tryptophan* can emerge in alanine, which can be converted into pyruvate.

Oxaloacetate is an entry point into metabolism for aspartate and asparagine

Aspartate and asparagine are converted into oxaloacetate, a citric acid cycle intermediate. *Aspartate*, a four-carbon amino acid, is directly *transaminated to oxaloacetate.*

$$\text{Aspartate} + \alpha\text{-ketoglutarate} \rightleftharpoons \text{oxaloacetate} + \text{glutamate}$$

Asparagine is hydrolyzed by *asparaginase* to NH_4^+ and aspartate, which is then transaminated.

Recall that aspartate can also be converted into *fumarate* by the urea cycle (Figure 23.16). Fumarate is a point of entry for half the carbon atoms of tyrosine and phenylalanine, as will be discussed shortly.

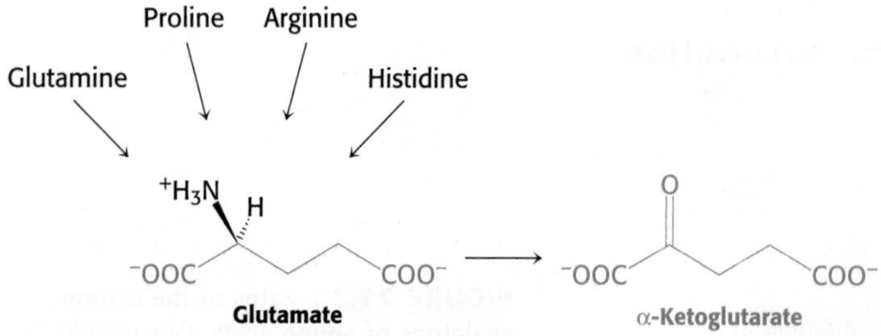

Alpha-ketoglutarate is an entry point into metabolism for five-carbon amino acids

The carbon skeletons of several five-carbon amino acids enter the citric acid cycle at α-ketoglutarate. These amino acids are first converted into *glutamate*, which is then oxidatively deaminated by glutamate dehydrogenase to yield α-ketoglutarate (Figure 23.22).

FIGURE 23.22 α-Ketoglutarate formation from amino acids. α-Ketoglutarate is the point of entry of several five-carbon amino acids that are first converted into glutamate.

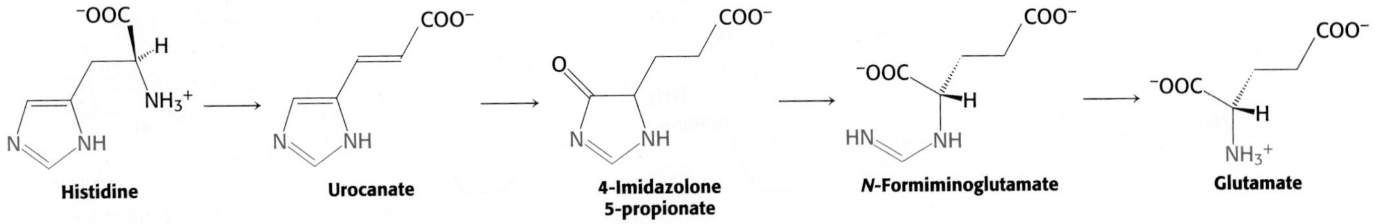

FIGURE 23.23 Histidine degradation. Conversion of histidine into glutamate.

Histidine is converted into 4-imidazolone 5-propionate (Figure 23.23). The amide bond in the ring of this intermediate is hydrolyzed to the N-formimino derivative of glutamate, which is then converted into glutamate by the transfer of its formimino group to tetrahydrofolate, a carrier of activated one-carbon units (Figure 24.9).

Glutamine is hydrolyzed to glutamate and NH_4^+ by *glutaminase*. *Proline* and *arginine* are each converted into glutamate γ-semialdehyde, which is then oxidized to glutamate (Figure 23.24).

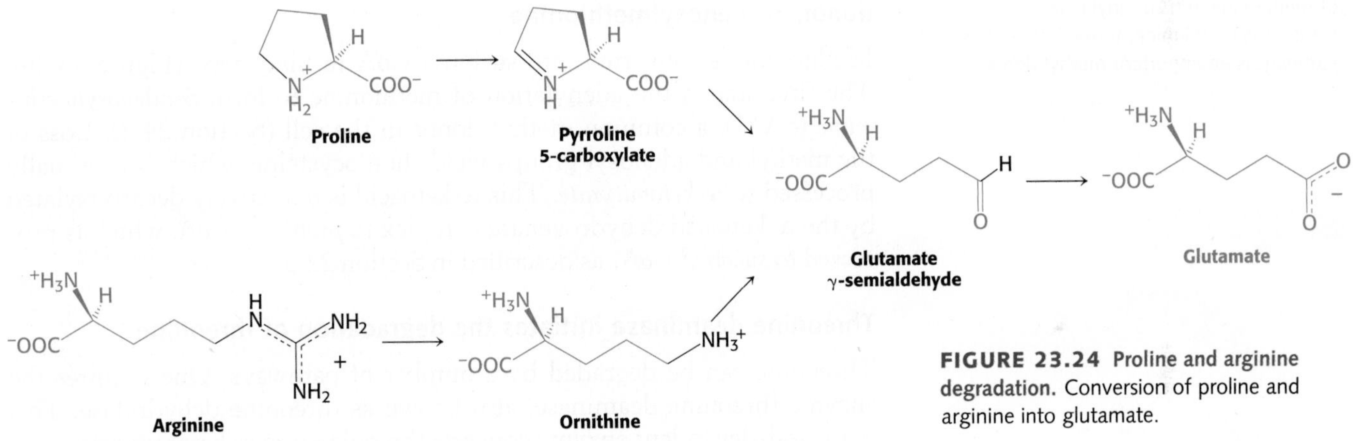

FIGURE 23.24 Proline and arginine degradation. Conversion of proline and arginine into glutamate.

Succinyl coenzyme A is a point of entry for several amino acids

Succinyl CoA is a point of entry for some of the carbon atoms of methionine, isoleucine, threonine, and valine. Propionyl CoA and then methylmalonyl CoA are intermediates in the breakdown of these four amino acids (Figure 23.25). This pathway from propionyl CoA to succinyl CoA is also used in the oxidation of fatty acids that have an odd number of carbon atoms. The mechanism for the interconversion of propionyl CoA and methylmalonyl CoA was presented in Section 22.3. (See Biochemistry in Focus at the end of this chapter.)

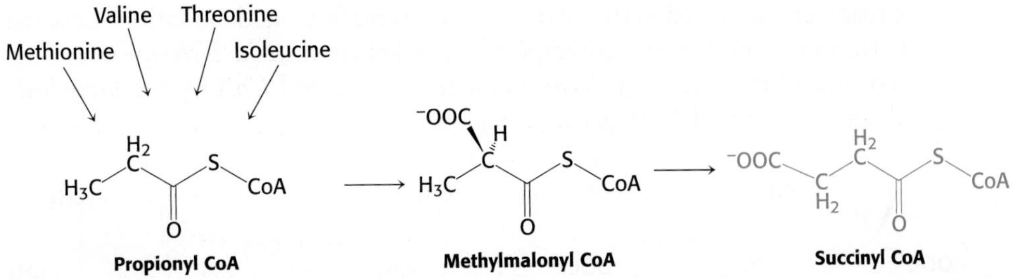

FIGURE 23.25 Succinyl CoA formation. Conversion of methionine, isoleucine, threonine, and valine into succinyl CoA.

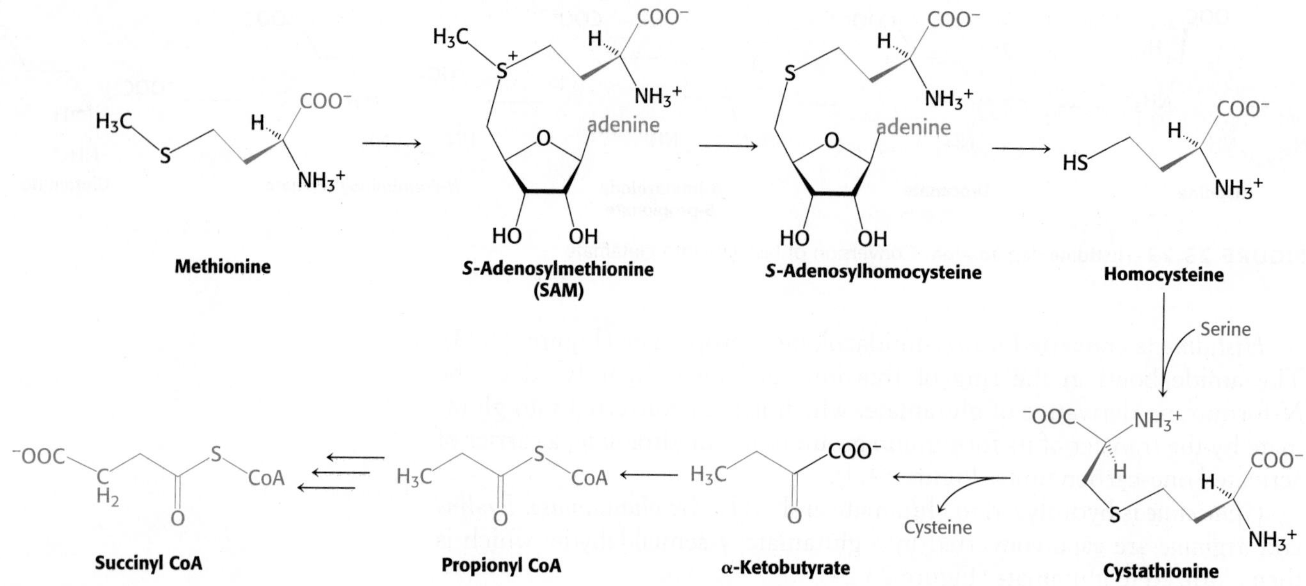

FIGURE 23.26 Methionine metabolism. The pathway for the conversion of methionine into succinyl CoA. S-Adenosylmethionine, formed along this pathway, is an important methyl donor.

Methionine degradation requires the formation of a key methyl donor, S-adenosylmethionine

Methionine is converted into succinyl CoA in nine steps (Figure 23.26). The first step is the adenylation of methionine to form S-*adenosylmethionine* (SAM), a common methyl donor in the cell (Section 24.2). Loss of the methyl and adenosyl groups yields homocysteine, which is eventually processed to α-*ketobutyrate*. This α-ketoacid is oxidatively decarboxylated by the α-ketoacid dehydrogenase complex to *propionyl CoA*, which is processed to *succinyl CoA,* as described in Section 22.3.

Threonine deaminase initiates the degradation of threonine

Threonine can be degraded by a number of pathways. One requires the enzyme threonine deaminase, also known as threonine dehydratase. This pyridoxal-dependent enzyme converts threonine into α-ketobutyrate.

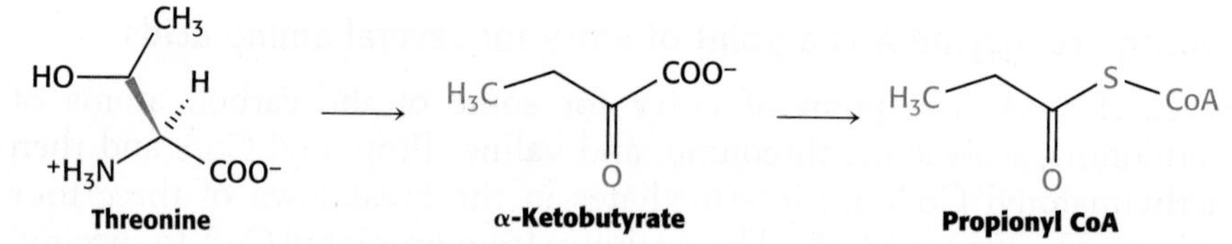

α-Ketobutyrate then forms propionyl CoA, which is subsequently metabolized to succinyl CoA as shown in Figure 23.25.

The branched-chain amino acids yield acetyl CoA, acetoacetate, or propionyl CoA

The branched-chain amino acids are degraded by reactions that we have already encountered in the citric acid cycle and fatty acid oxidation. Leucine is transaminated to the corresponding α-ketoacid, α-*ketoisocaproate*. This α-ketoacid is *oxidatively decarboxylated to isovaleryl CoA by the branched-chain α-ketoacid dehydrogenase complex.*

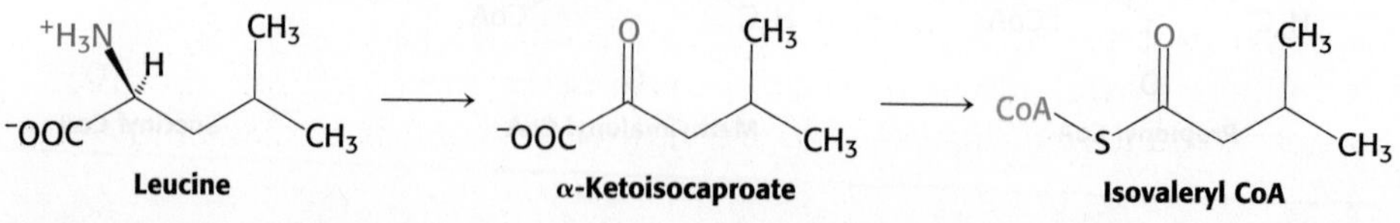

The α-ketoacids of valine and isoleucine, the other two branched-chain aliphatic amino acids, also are substrates (as is α-ketobutyrate derived from methionine). The oxidative decarboxylation of these α-ketoacids is analogous to that of pyruvate to acetyl CoA and of α-ketoglutarate to succinyl CoA. The branched-chain α-ketoacid dehydrogenase, a multienzyme complex, is a homolog of pyruvate dehydrogenase complex (Section 17.1) and α-ketoglutarate dehydrogenase complex (Section 17.2). Indeed, the E3 components of these enzymes, which regenerate the oxidized form of lipoamide, are identical.

The isovaleryl CoA derived from leucine is *dehydrogenated* to yield β-*methylcrotonyl CoA*. This oxidation is catalyzed by *isovaleryl CoA dehydrogenase*. The hydrogen acceptor is FAD, as in the analogous reaction in fatty acid oxidation that is catalyzed by acyl CoA dehydrogenase. β-*Methylglutaconyl CoA* is then formed by the *carboxylation* of β-methylcrotonyl CoA at the expense of the hydrolysis of a molecule of ATP. As might be expected, the carboxylation mechanism of β-methylcrotonyl CoA carboxylase is similar to that of pyruvate carboxylase and acetyl CoA carboxylase.

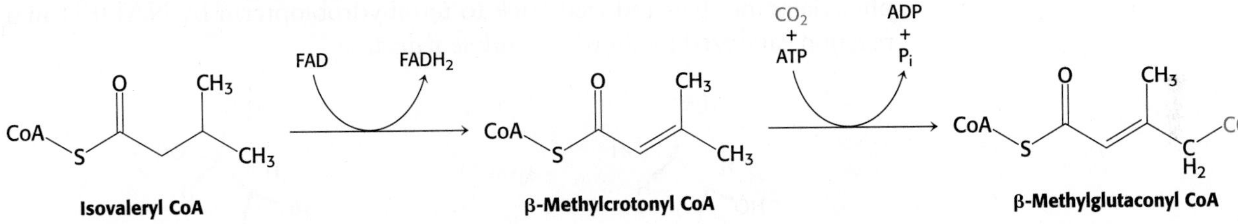

β-Methylglutaconyl CoA is then *hydrated* to form *3-hydroxy-3-methylglutaryl CoA*, which is cleaved into *acetyl CoA* and *acetoacetate*. This reaction has already been discussed in regard to the formation of ketone bodies from fatty acids (Section 22.3).

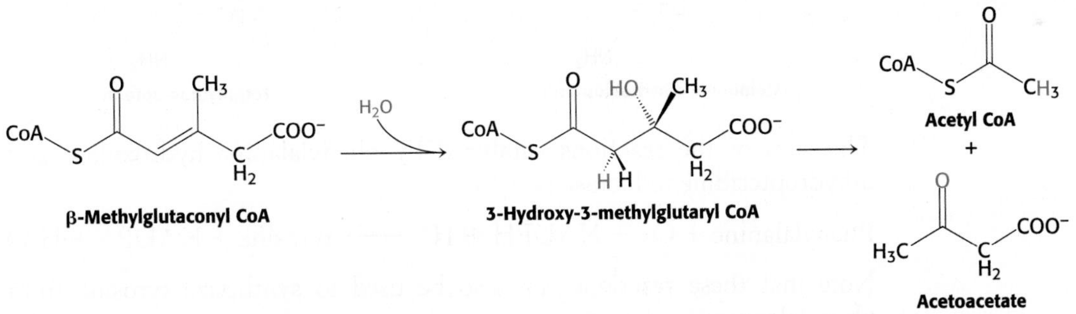

The degradative pathways of valine and isoleucine resemble that of leucine. After transamination and oxidative decarboxylation to yield a CoA derivative, the subsequent reactions are like those of fatty acid oxidation. Isoleucine yields acetyl CoA and propionyl CoA, whereas valine yields CO_2 and propionyl CoA. The degradation of leucine, valine, and isoleucine validate a point made earlier (Chapter 15): the number of reactions in metabolism is large, but the number of *kinds* of reactions is relatively small. The degradation of leucine, valine, and isoleucine provides a striking illustration of the underlying simplicity and elegance of metabolism.

Oxygenases are required for the degradation of aromatic amino acids

The degradation of the aromatic amino acids yields the common intermediates acetoacetate, fumarate, and pyruvate. The degradation pathway is not as straightforward as that of the amino acids previously discussed. For the aromatic amino acids, *molecular oxygen is used to break the aromatic ring.*

The degradation of phenylalanine begins with its hydroxylation to tyrosine, a reaction catalyzed by *phenylalanine hydroxylase*. This enzyme is called a *monooxygenase* (or *mixed-function oxygenase*) because *one atom of* O_2 *appears in the product and the other in* H_2O.

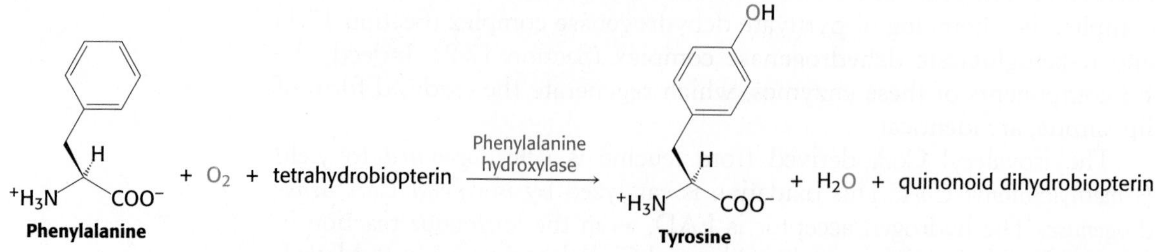

The reductant here is *tetrahydrobiopterin*, an electron carrier that has not been previously discussed and is derived from the cofactor *biopterin*. Because biopterin is synthesized in the body, it is not a vitamin. The quinonoid form of dihydrobiopterin is produced in the hydroxylation of phenylalanine. It is reduced back to tetrahydrobiopterin by NADPH in a reaction catalyzed by *dihydropteridine reductase*.

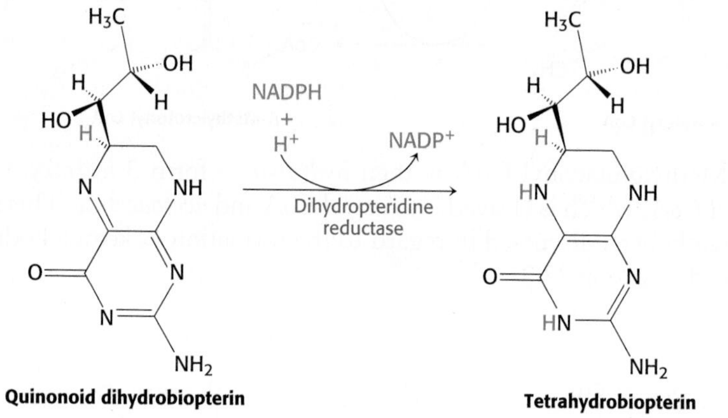

The sum of the reactions catalyzed by phenylalanine hydroxylase and dihydropteridine reductase is

$$\text{Phenylalanine} + O_2 + \text{NADPH} + H^+ \longrightarrow \text{tyrosine} + \text{NADP}^+ + H_2O$$

Note that these reactions can also be used to synthesize tyrosine from phenylalanine.

The next step in the degradation of phenylalanine and tyrosine is the transamination of tyrosine to p-*hydroxyphenylpyruvate* (Figure 23.27). This α-ketoacid then reacts with O_2 to form *homogentisate*. The enzyme catalyzing this complex reaction, p-*hydroxyphenylpyruvate hydroxylase*, is

FIGURE 23.27 Phenylalanine and tyrosine degradation. The pathway for the conversion of phenylalanine into acetoacetate and fumarate.

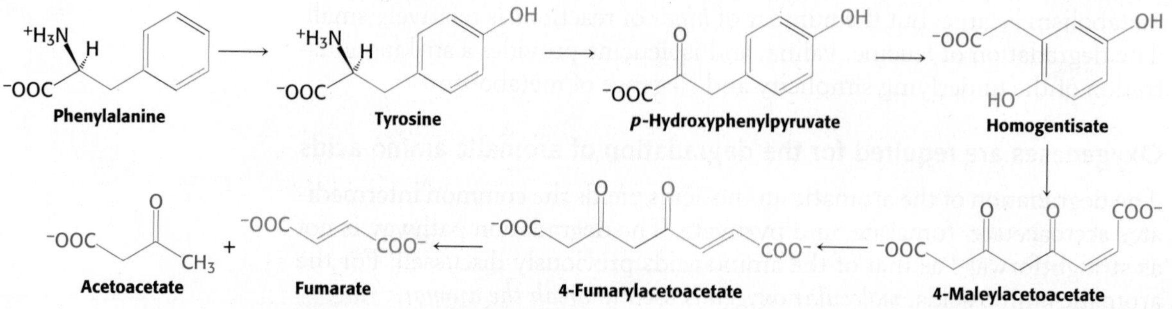

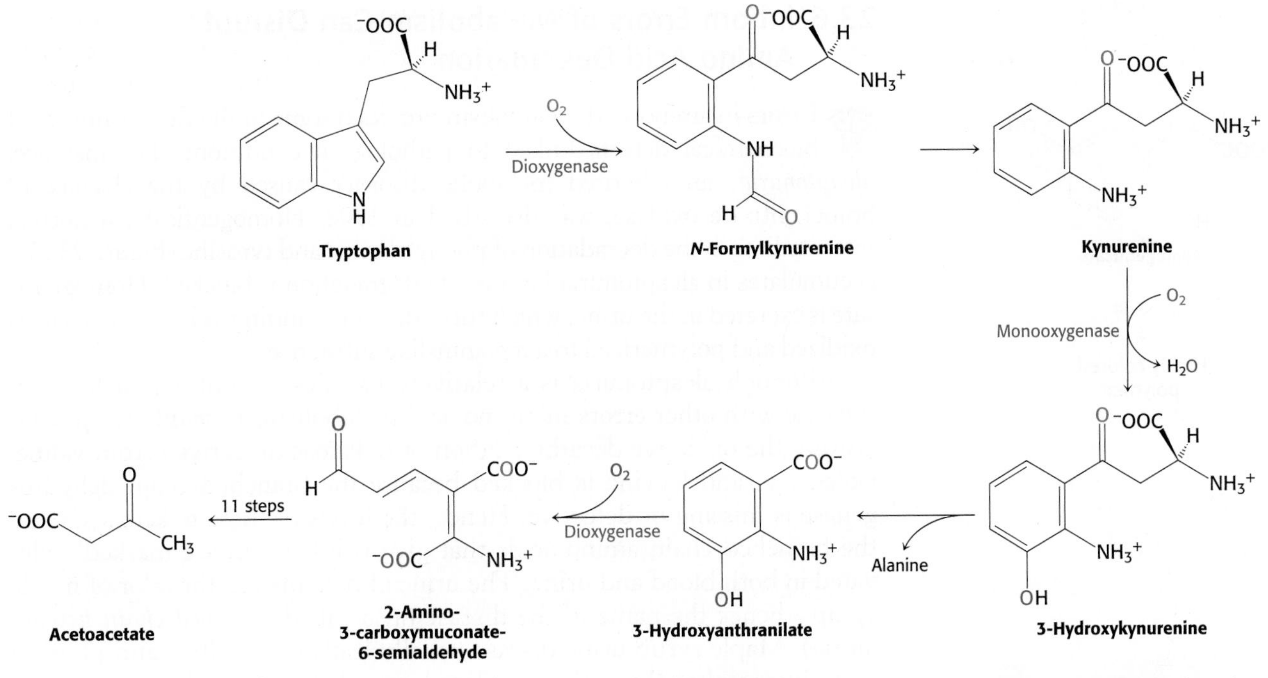

FIGURE 23.28 Tryptophan degradation. The pathway for the conversion of tryptophan into alanine and acetoacetate.

called a *dioxygenase* because *both atoms of O₂* become incorporated into the product, one on the ring and one in the carboxyl group. The aromatic ring of homogentisate is then cleaved by O_2, which yields 4-maleylacetoacetate. This reaction is catalyzed by *homogentisate oxidase*, another dioxygenase. 4-Maleylacetoacetate is then isomerized to *4-fumarylacetoacetate* by an enzyme that uses glutathione as a cofactor. Finally, 4-fumarylacetoacetate is hydrolyzed to *fumarate* and *acetoacetate*.

Tryptophan degradation requires several oxygenases (Figure 23.28). Tryptophan 2,3-dioxygenase cleaves the pyrrole ring, and kynurenine 3-monooxygenase hydroxylates the remaining benzene ring, a reaction similar to the hydroxylation of phenylalanine to form tyrosine. Alanine is removed and the 3-hydroxyanthranilate is cleaved by another dioxygenase and subsequently processed to acetoacetyl CoA. *Nearly all cleavages of aromatic rings in biological systems are catalyzed by dioxygenases.* The active sites of these enzymes contain iron that is not part of heme or an iron–sulfur cluster.

Protein metabolism helps to power the flight of migratory birds

Earlier we learned that fats are the primary fuel for migratory birds that fly great distances without feeding (Section 22.1). Proteins are also degraded to contribute fuel for these impressive feats of endurance. Why would proteins be degraded? There are several likely reasons. First, since the birds are not feeding, proteins involved with digestion and transport of nutrients from the gut are not needed and thus can be degraded. Second, gluconeogenesis is required to provide glucose for the central nervous system, and the glucogenic amino acids will meet this need. Moreover, some amino acid carbon skeletons can replenish citric acid cycle intermediates to facilitate fat metabolism. Third, the reduction in body mass resulting from protein degradation will reduce the energy cost of flight. There is one final benefit. In addition to not feeding during flight, the migratory birds cannot hydrate, but yet they still require water. While the catabolism of fats yields little water—0.029 g H_2O kJ^{-1}—protein provides fivefold more water than fats—0.155 g H_2O kJ^{-1}—and may be a vital source of water for migratory flights. It is fascinating to realize the biochemical challenges that migratory birds must overcome.

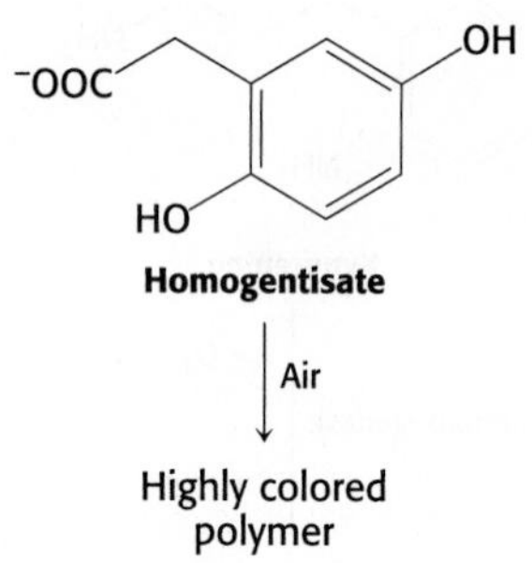

23.6 Inborn Errors of Metabolism Can Disrupt Amino Acid Degradation

Errors in amino acid metabolism provided some of the first examples of biochemical defects linked to pathological conditions. For instance, *alcaptonuria,* an inherited metabolic disorder caused by the absence of homogentisate oxidase, was described in 1902. Homogentisate, a normal intermediate in the degradation of phenylalanine and tyrosine (Figure 23.27), accumulates in alcaptonuria because its degradation is blocked. Homogentisate is excreted in the urine, which turns dark on standing as homogentisate is oxidized and polymerized to a melanin-like substance.

Although alcaptonuria is a relatively harmless condition, such is not the case with other errors in amino acid metabolism. In *maple syrup urine disease,* the oxidative decarboxylation of α-ketoacids derived from valine, isoleucine, and leucine is blocked because the branched-chain dehydrogenase is missing or defective. Hence, the levels of these α-ketoacids and the branched-chain amino acids that give rise to them are markedly elevated in both blood and urine. The urine of patients has the odor of maple syrup—hence the name of the disease (also called *branched-chain ketoaciduria*). Maple syrup urine disease usually leads to cognitive and physical disabilities unless the patient is placed on a diet low in valine, isoleucine, and leucine early in life. The disease can be readily detected in newborns by screening urine samples with 2,4-dinitrophenylhydrazine, which reacts with α-ketoacids to form 2,4-dinitrophenylhydrazone derivatives. A definitive diagnosis can be made by mass spectrometry. Table 23.4 lists some other diseases of amino acid metabolism.

Homogentisate

Air

Highly colored polymer

2,4-Dinitrophenyl-hydrazine

α-Ketoacid

2,4-Dinitrophenylhydrazone derivative

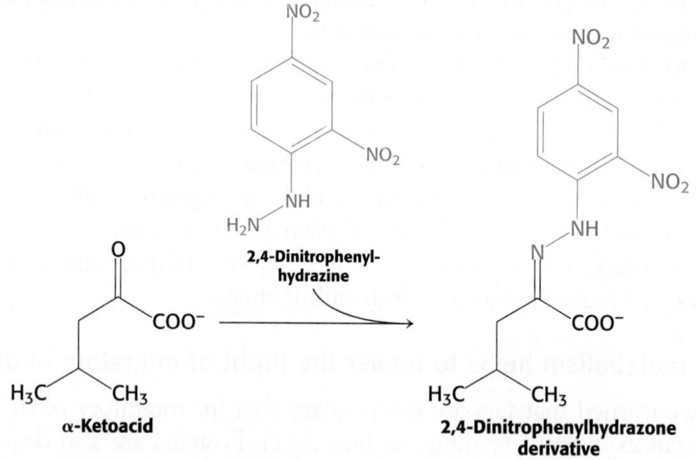

TABLE 23.4 Inborn errors of amino acid metabolism

Disease	Enzyme deficiency	Symptoms
Citrullinema	Arginosuccinate lyase	Lethargy, seizures, reduced muscle tension
Tyrosinemia	Various enzymes of tyrosine degradation	Weakness, liver damage, cognitive disability
Albinism	Tyrosinase	Absence of pigmentation
Homocystinuria	Cystathionine β-synthase	Scoliosis, muscle weakness, cognitive disability, thin blond hair
Hyperlysinemia	α-Aminoadipic semialdehyde dehydrogenase	Seizures, cognitive disability, lack of muscle tone, ataxia

Phenylketonuria is one of the most common metabolic disorders

Phenylketonuria is perhaps the best known of the diseases of amino acid metabolism. Phenylketonuria, which occurs with a prevalence of 1 in 10,000 births, is caused by an *absence or deficiency of phenylalanine hydroxylase* or, more rarely, of its tetrahydrobiopterin cofactor. *Phenylalanine accumulates in all body fluids because it cannot be converted into tyrosine.* Normally, three-quarters of phenylalanine molecules are converted into tyrosine, and the other quarter become incorporated into proteins. Because the major outflow pathway is blocked in phenylketonuria, the blood level of phenylalanine is typically at least 20-fold as high as in normal people. Minor fates of phenylalanine in normal people, such as the formation of phenylpyruvate, become major fates in phenylketonurics. Indeed, the initial description of phenylketonuria in 1934 was made by observing the reaction of phenylpyruvate in the urine of phenylketonurics with $FeCl_3$, which turns the urine olive green.

Almost all untreated phenylketonurics have severe cognitive disabilities. The brain weight of these people is below normal, myelination of their nerves is defective, and their reflexes are hyperactive. The life expectancy of untreated phenylketonurics is drastically shortened. Half die by age 20 and three-quarters by age 30. Phenylketonurics appear normal at birth but are severely defective by age 1 if untreated. The therapy for phenylketonuria is a *low-phenylalanine diet* supplemented with tyrosine, because tyrosine is normally synthesized from phenylalanine. The aim is to provide just enough phenylalanine to meet the needs for growth and replacement. Proteins that have a low content of phenylalanine, such as casein from milk, are hydrolyzed and phenylalanine is removed by adsorption. A low-phenylalanine diet must be started very soon after birth to prevent irreversible brain damage. In one study, the average IQ of phenylketonurics treated within a few weeks after birth was 93; a control group treated starting at age 1 had an average IQ of 53.

Early diagnosis of phenylketonuria is essential and has been accomplished by mass screening programs of all babies born in the United States and Canada. The phenylalanine level in the blood is the preferred diagnostic criterion because it is more sensitive and reliable than the $FeCl_3$ test. Prenatal diagnosis of phenylketonuria with DNA probes has become feasible because the gene has been cloned and the exact locations of many mutations have been discovered in the protein.

Determining the basis of the neurological symptoms of phenylketonuria is an active area of research

The biochemical basis of the cognitive disabilities is not firmly established, but one hypothesis suggests that the lack of hydroxylase reduces the amount of tyrosine, an important precursor to neurotransmitters such as dopamine. Moreover, high concentrations of phenylalanine prevent amino acid transport of any tyrosine present as well as tryptophan, a precursor to the neurotransmitter serotonin, into the brain. Because all three of the amino acids are transported by the same carrier, phenylalanine will saturate the carrier, preventing access to tyrosine and tryptophan. Lack of these amino acids will impair protein synthesis in the brain. Finally, high blood levels of phenylalanine result in higher levels of phenylalanine in the brain, and evidence suggests this elevated concentration inhibits glycolysis at pyruvate kinase, disrupts myelination of nerve fibers, and reduces the synthesis of several neurotransmitters.

Phenylalanine

α-Ketoacid

α-Amino acid

Phenylpyruvate

SUMMARY

23.1 Proteins Are Degraded to Amino Acids

Dietary protein is digested in the intestine, producing amino acids that are transported throughout the body. Cellular proteins are degraded at widely variable rates, ranging from minutes to the life of the organism.

23.2 Protein Turnover Is Tightly Regulated

The turnover of cellular proteins is a regulated process requiring complex enzyme systems. Proteins to be degraded are conjugated with ubiquitin, a small conserved protein, in a reaction driven by ATP hydrolysis. The ubiquitin-conjugating system is composed of three distinct enzymes. A large, barrel-shaped complex called the proteasome digests the ubiquitinated proteins. The proteasome also requires ATP hydrolysis to function. The resulting amino acids provide a source of precursors for protein, nucleotide bases, and other nitrogenous compounds.

23.3 The First Step in Amino Acid Degradation Is the Removal of Nitrogen

Surplus amino acids are used as building blocks and as metabolic fuel. The first step in their degradation is the removal of their α-amino groups by transamination to α-ketoacids. Pyridoxal phosphate is the coenzyme in all aminotransferases and in many other enzymes catalyzing amino acid transformations. The α-amino group funnels into α-ketoglutarate to form glutamate, which is then oxidatively deaminated by glutamate dehydrogenase to give NH_4^+ and α-ketoglutarate. NAD^+ or $NADP^+$ is the electron acceptor in this reaction.

23.4 Ammonium Ion Is Converted into Urea in Most Terrestrial Vertebrates

The first step in the synthesis of urea is the formation of carbamoyl phosphate, which is synthesized from HCO_3^-, NH_3, and two molecules of ATP by carbamoyl phosphate synthetase. Ornithine is then carbamoylated to citrulline by ornithine transcarbamoylase. These two reactions take place in mitochondria. Citrulline leaves the mitochondrion and condenses with aspartate to form argininosuccinate, which is cleaved into arginine and fumarate. The other nitrogen atom of urea comes from aspartate. Urea is formed by the hydrolysis of arginine, which also regenerates ornithine.

23.5 Carbon Atoms of Degraded Amino Acids Emerge as Major Metabolic Intermediates

The carbon atoms of degraded amino acids are converted into pyruvate, acetyl CoA, acetoacetate, or an intermediate of the citric acid cycle. Most amino acids are solely glucogenic, two are solely ketogenic, and a few are both ketogenic and glucogenic. Alanine, serine, cysteine, glycine, threonine, and tryptophan are degraded to pyruvate. Asparagine and aspartate are converted into oxaloacetate. α-Ketoglutarate is the point of entry for glutamate and four amino acids (glutamine, histidine, proline, and arginine) that can be converted into glutamate. Succinyl CoA is the point of entry for some of the carbon atoms of three amino acids (methionine, isoleucine, threonine, and valine) that are degraded through the intermediate methylmalonyl CoA. Leucine is degraded to acetoacetate and acetyl CoA. The breakdown of valine and isoleucine is like that of leucine. Their α-ketoacid derivatives are oxidatively decarboxylated by the branched-chain α-ketoacid dehydrogenase.

The rings of aromatic amino acids are degraded by oxygenases. Phenylalanine hydroxylase, a monooxygenase, uses tetrahydrobiopterin as the reductant. One of the oxygen atoms of O_2 emerges in tyrosine and the other in water. Subsequent steps in the degradation of these aromatic amino acids are catalyzed by dioxygenases, which catalyze the insertion of both atoms of O_2 into organic products. Four of the carbon atoms of phenylalanine and tyrosine are converted into fumarate, and four emerge in acetoacetate.

23.6 Inborn Errors of Metabolism Can Disrupt Amino Acid Degradation

Errors in amino acid metabolism were sources of some of the first insights into the correlation between pathology and biochemistry. Phenylketonuria is the best known of the many hereditary errors of amino acid metabolism. This condition results from the accumulation of high levels of phenylalanine in the body fluids. This accumulation leads to cognitive disabilities unless the afflicted are placed on low-phenylalanine diets immediately after birth.

APPENDIX

Biochemistry in Focus

Methylmalonic acidemia results from an inborn error of metabolism

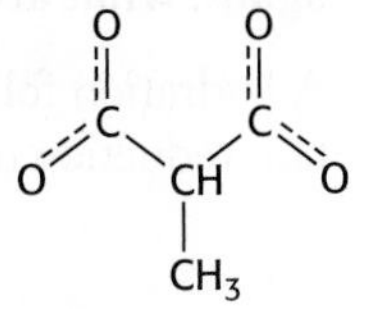

Methylmalonic acidemia (also known as methylmalonic acidosis) is an inherited disorder, occurring about 1 in every 75,000 births. In this disorder, the blood concentration of methylmalonic acid may be dangerously high. The clinical spectrum of the disorder is quite wide, ranging from benign to death in infancy. Affected infants experience vomiting, hypotonia (muscle weakness), lethargy and

Methylmalonic acid

an overall failure to thrive. Without treatment, methylmalonic acidemia can be fatal.

Do you remember where we first encountered methylmalonic acid? Recall from Chapter 22 (p. 720) that the degradation of odd chain fatty acids yields propionyl CoA, which can be converted into the citric acid cycle intermediate succinyl CoA. Methylmalonic acid is an intermediate in this conversion. In particular, the vitamin B_{12}-dependent enzyme methylmalonyl CoA mutase converts methylmalonyl CoA to succinyl CoA. If the mutase activity is low or absent, methylmalonyl CoA accumulates and is converted into methylmalonic acid, which is released into the blood.

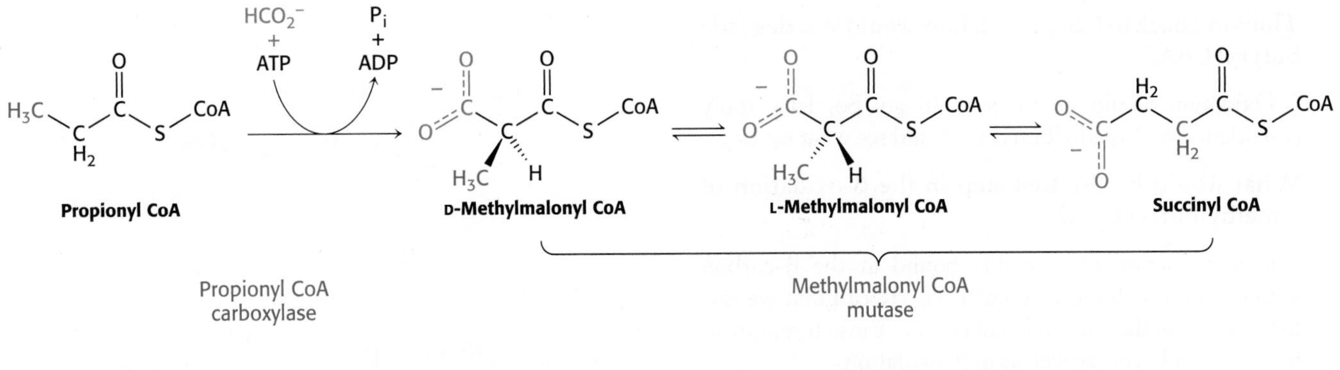

Propionyl CoA → D-Methylmalonyl CoA → L-Methylmalonyl CoA → Succinyl CoA

Propionyl CoA carboxylase

Methylmalonyl CoA mutase

As mentioned in Chapter 22, odd chain fatty acids are rare. Consequently, methylmalonic acidemia is unlikely to result simply from the metabolism of odd chain fatty acids. There must be another dietary source of propionyl CoA. What could this source be?

Recall from this chapter that four essential amino acids—methionine, valine, isoleucine and threonine—have propionyl CoA as an intermediate (Figure 23.25). These four amino acids are the main source of the methymalonic acid. Treatment then of the condition is a diet low in the offending amino acids with just enough to meet the needs of protein synthesis.

Can you suggest the nature of the metabolic error that results in methylmalonic acidemia? In fact, there are two types of causes.

In the first case, the mutase itself is absent or mutated so as to be inactive. In the second case, the supply of vitamin B_{12} is inadequate. This later circumstance can be treated by the administration of the vitamin, mitigating the need for a restricted protein diet.

APPENDIX

Problem-Solving Strategies

Problem-Solving Strategies 1

PROBLEM: Isoleucine is said to be both a ketogenic and glucogenic amino acid. Starting from 2-methylbutryl CoA, a degradation intermediate of isoleucine, prove that it is both ketogenic and glucogenic.

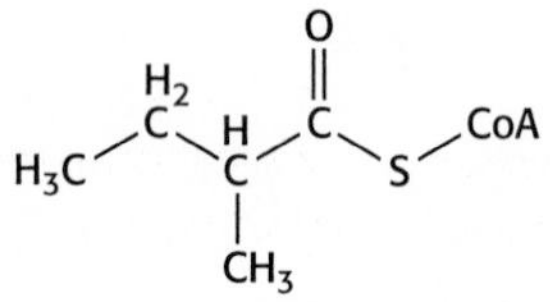

2-Methylbutyryl CoA

SOLUTION: Okay, let's begin by thinking about the structure of 2-methylbutryl CoA. We have seen similar molecules before.

▶ **What general type of biochemical does this look like?**

It looks like a fatty acyl CoA, except that there is a methyl group where a hydrogen would be in a fatty acyl CoA. To prove this, let's look at the molecule with the substitution of a hydrogen for the methyl group, butyryl CoA.

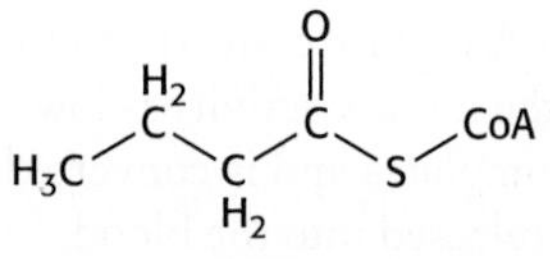

Butyryl CoA

▶ **Thinking back to Chapter 22, how would you degrade butyryl CoA?**

β-Oxidation would be the way to go. So, let's apply β-oxidation to 2-methylbutryl CoA and see what we get.

▶ **What would be the first step in the β-oxidation of 2-methylbutryl CoA?**

The introduction of a double bound at the β-carbon accompanied with reduction of FAD. Recall that we saw the reaction in the conversion of succinate into fumarate in the citric acid cycle as well as in β-oxidation.

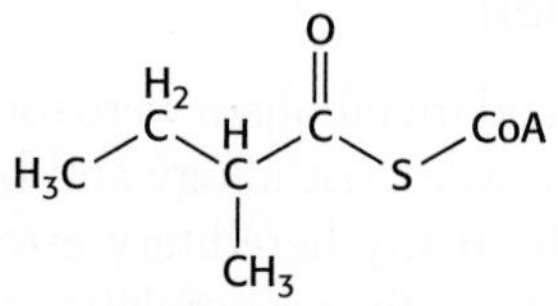

2-Methylbutyryl CoA

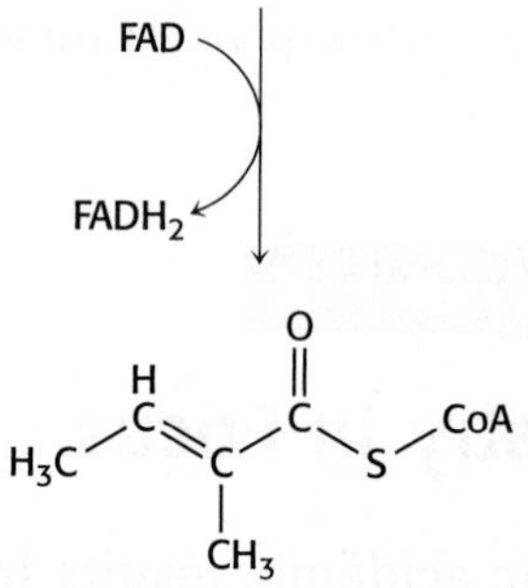

▶ **Using the β-oxidation (or citric acid cycle) motif again, what are the next two reactions?**

A hydration followed by an oxidation, with the concomitant reduction of NAD^+.

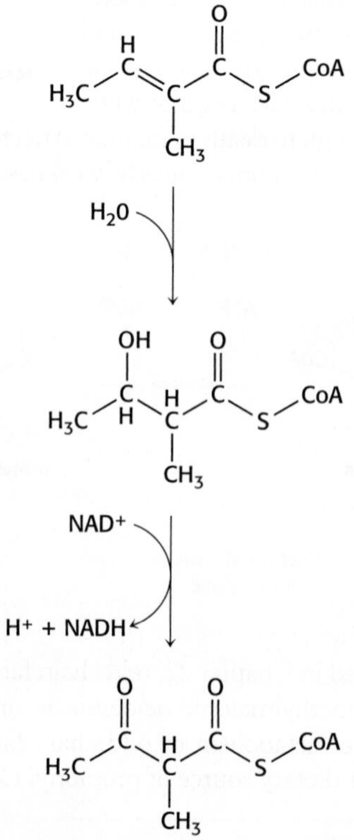

Do you need a prompt as to what is the next reaction? Probably not, but let's do it any way.

▶ **What is the next step in the degradation pathway, again using β-oxidation as a guide?**

Thiolysis by addition of CoA.

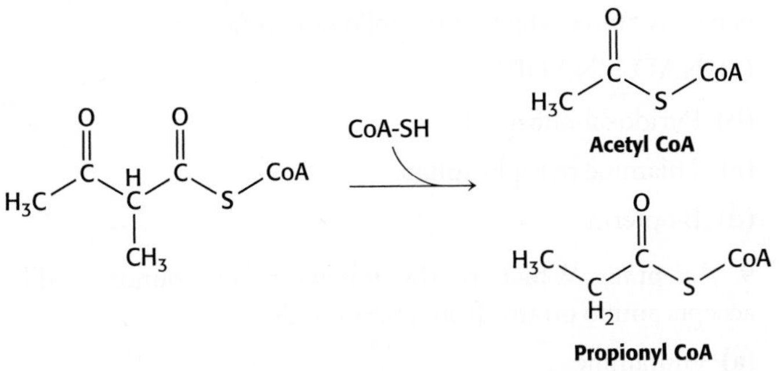

Thiolysis yields acetyl CoA. Acetyl CoA is a precursor for fatty acid synthesis, accounting for the fact that isoleucine is a ketogenic amino acid. But is it a glucogenic amino acid?

▶ **What is the metabolic fate of propionyl CoA?**

As we saw in Section 22.3, Section 23.5 and in the Biochemistry in Focus for this chapter, propionyl CoA is converted into the citric acid intermediate succinyl CoA, which can be metabolized to oxaloacetate. Oxaloacetate is a gluconeogenic precursor (Section 16.3). So, yes, isoleucine is both a ketogenic and glucogenic amino acid. QED as the Romans used to say.

Problem-Solving Strategies 2

PROBLEM: Muscles use branched chain amino acids (BCAA) as a fuel. However, this presents something of a problem, since muscles do not possess the urea cycle to metabolize the nitrogen that must be removed from the BCAA before they can be used as a fuel. Explain how muscles have resolved this conundrum.

SOLUTION: Let's break this down into a series of small questions, starting with a very straight-forward question.

▶ **What are the branched chain amino acids?**
Valine, leucine, and isoleucine.

In order to use BCAA as a fuel, the nitrogen must first be removed.

▶ **How is nitrogen removed from the BCAA in muscle?**

As we saw in the chapter, nitrogen is transferred to α-ketoglutarate by aminotransferases to form glutamate. Although some glutamate is found in the blood, it is not the common transport form of nitrogen in the blood. Rather, nitrogen is transferred from glutamate to form another amino acid.

▶ **What is the fate of the nitrogen on the newly formed glutamate?**

The nitrogen is transferred to pyruvate to form alanine, which is then released into the blood. Recall that this transport cycle is called the glucose-alanine cycle (p. 763). Nitrogen can also be transferred in the blood as glutamine, formed from glutamate by glutamine synthetase.

▶ **What is the fate of the alanine and glutamine in the blood?**

These amino acids are removed from the blood by the liver. The nitrogen is transferred to α-ketoglutarate to form glutamate by an aminotransferase and subsequently used to form urea. The pyruvate formed by the deamination of alanine can be used to form glucose, accounting for the name of the glucose-alanine cycle.

▶ **Identify another cycle where the metabolic burden of muscle is shifted to the liver?**

Using the Cori cycle (Section 16.4), the liver removes lactate, produced by lactic acid fermentation in active muscle, from the blood and converts it into glucose.

KEY TERMS

ubiquitin (p. 751)

degron (p. 754)

proteasome (p. 751)

aminotransferase (transaminase) (p. 758)

glutamate dehydrogenase (p. 759)

pyridoxal phosphate (PLP) (p. 759)

pyridoxamine phosphate (PMP) (p. 760)

glucose–alanine cycle (p. 763)

urea cycle (p. 764)

carbamoyl phosphate synthetase (p. 764)

N-acetylglutamate (p. 764)

ketogenic amino acid (p. 769)

glucogenic amino acid (p. 769)

biopterin (p. 774)

phenylketonuria (p. 777)

PROBLEMS

1. *Getting exposure.* Proteins are denatured by acid in the stomach. This denaturation makes them better substrates for proteolysis. Explain why this is the case.

2. *Targeting for destruction.* What are the steps required to attach ubiquitin to a target protein? ✓❶, ✓❷

3. *Not the dating service.* Match the description on the right with the term on the left.

(a) Pepsin _____	**1.** Requires an adenylate intermediate
(b) N-terminal rule _____	**2.** Marks a protein for destruction
(c) Ubiquitin _____	**3.** 19S regulatory subunit
(d) PEST sequence _____	**4.** Determines half-life of a protein
(e) Threonine nucleophiles _____	**5.** 20S core
(f) ATP-dependent protein unfolding _____	**6.** Substrate for ligase
(g) Proteasome _____	**7.** Stomach proteolytic enzyme
(h) Ubiquitin-activating enzyme _____	**8.** Recognizes protein to be degraded
(i) Ubiquitin-conjugating enzyme _____	**9.** Protein-degrading machine
(j) Ubiquitin-ligase _____	**10.** Pro-Glu-Ser-Thr

4. *Wasted energy?* Protein hydrolysis is an exergonic process, yet the 26S proteasome is dependent on ATP hydrolysis for activity. ✓❶, ✓❸

(a) Explain why ATP hydrolysis is required by the 26S proteasome.

(b) Small peptides can be hydrolyzed without the expenditure of ATP. How does this information concur with your answer to part *a*?

5. *The benefits of specialization.* The archaeal proteasome contains 14 identical active β subunits, whereas the eukaryotic proteasome has two sets of 7 distinct β subunits. What are the potential benefits of having several distinct active subunits? ✓❶, ✓❸

6. *Propose a structure.* The 19S subunit of the proteasome contains six subunits that are members of the AAA ATPase family. Other members of this large family are associated into homohexamers with sixfold symmetry. Propose a structure for the AAA ATPases within the 19S proteasome. How might you test and refine your prediction?

7. *Required partner.* Aminotransferases require which of the following cofactors:

(a) $NAD^+/NADP^+$

(b) Pyridoxal phosphate

(c) Thiamine pyrophosphate

(d) Biopterin

8. *Butch Cassidy and the Sundance Kid.* Glutamate dehydrogenase requires which of the following cofactors:

(a) $NAD^+/NADP^+$

(b) Pyridoxal phosphate

(c) Thiamine pyrophosphate

(d) Biopterin

9. *Accepting.* Which of the following compounds readily accepts amino groups from amino acids?

(a) Glutamine

(b) Isocitrate

(c) Malate

(d) α-Ketoglutarate

10. *Donations welcome.* The immediate donors of the nitrogen atoms of urea are:

(a) Aspartate and glutamate

(b) Glutamate and carbamoyl phosphate

(c) Aspartate and carbamoyl phosphate

(d) Glutamine and aspartate

11. *Traveler.* Which compound of the urea cycle is synthesized in the mitochondria and transported into the cytoplasm?

(a) Ornithine

(b) Citrulline

(c) Arginiosuccinate

(d) Arginine

12. *Keto counterparts.* Name the α-ketoacid that is formed by the transamination of each of the following amino acids: ✓❹, ✓❺

(a) Alanine   **(d)** Leucine

(b) Aspartate   **(e)** Phenylalanine

(c) Glutamate   **(f)** Tyrosine

13. *A versatile building block.*

(a) Write a balanced equation for the conversion of aspartate into glucose through the intermediate oxaloacetate. Which coenzymes participate in this transformation?

(b) Write a balanced equation for the conversion of aspartate into oxaloacetate through the intermediate fumarate.

14. *Effective electron sinks.* Pyridoxal phosphate stabilizes carbanionic intermediates by serving as an electron sink. Which other prosthetic group catalyzes reactions in this way?

15. *Cooperation.* How do aminotransferases and glutamate dehydrogenase cooperate in the metabolism of the amino group of amino acids? ✓④

16. *Taking away the nitrogen.* What amino acids yield citric acid cycle components and glycolysis intermediates when deaminated? ✓④, ✓⑤

17. *One reaction only.* What amino acids can be deaminated directly? ✓④

18. *Useful products.* What are the common features of the breakdown products of the carbon skeletons of amino acids? ✓⑤

19. *Helping hand.* Propose a role for the positively charged guanidinium nitrogen atom in the cleavage of argininosuccinate into arginine and fumarate.

20. *Nitrogen sources.* What are the immediate biochemical sources for the two nitrogen atoms in urea? ✓④

21. *Counterparts.* Match the biochemical on the right with the property on the left. ✓④

(a) Formed from NH_4^+ _____   1. Aspartate
(b) Hydrolyzed to yield urea _____   2. Urea
(c) A second source of nitrogen _____ 3. Ornithine
(d) Reacts with aspartate _____   4. Carbamoyl phosphate
(e) Cleavage yields fumarate _____   5. Arginine
(f) Accepts the first nitrogen _____ 6. Citrulline
(g) Final product _____   7. Arginosuccinate

22. *Line up.* Identify structures A–D, and place them in the order that they appear in the urea cycle. ✓④

(A)

(B)

(C)

(D)

23. *Completing the cycle.* Four high-transfer-potential phosphoryl groups are consumed in the synthesis of urea according to the stoichiometry given on page 766. In this reaction, aspartate is converted into fumarate. Suppose that fumarate is converted into oxaloacetate. What is the resulting stoichiometry of urea synthesis? How many high-transfer-potential phosphoryl groups are spent?

24. *A good bet.* A friend bets you a bazillion dollars that you can't prove that the urea cycle is linked to the citric acid cycle and other metabolic pathways. Can you collect?

25. *Inhibitor design.* Compound A has been synthesized as a potential inhibitor of a urea-cycle enzyme. Which enzyme do you think compound A might inhibit? Explain your answer.

Compound A

26. *Ammonia toxicity.* Glutamate is an important neurotransmitter whose levels must be carefully regulated in the brain. Explain how a high concentration of ammonia might disrupt this regulation. How might a high concentration of ammonia alter the citric acid cycle?

27. *A precise diagnosis.* The urine of an infant gives a positive reaction with 2,4-dinitrophenylhydrazine. Mass spectrometry shows abnormally high blood levels of pyruvate, α-ketoglutarate, and the α-ketoacids of valine, isoleucine, leucine and threonine. Identify a likely molecular defect and propose a definitive test of your diagnosis. ✓⑥

28. *Therapeutic design.* How would you treat an infant who is deficient in argininosuccinate synthetase? Which molecules would carry nitrogen out of the body? ✓⑥

29. *Damaged liver.* As we will see later (Chapter 27), liver damage (cirrhosis) often results in ammonia poisoning. Explain why this is the case. ✓④

30. *Argininosuccinic aciduria.* Argininosuccinic aciduria is a condition that results when the urea-cycle enzyme argininosuccinase is deficient. Argininosuccinate is present in the blood and urine. Suggest how this condition might be treated while still removing nitrogen from the body. ✓⑥

31. *Sweet hazard.* Why should phenylketonurics avoid using aspartame, an artificial sweetener? (Hint: Aspartame is L-aspartyl-L-phenylalanine methyl ester.) ✓⑥

32. *Déjà vu.* N-Acetylglutamate is required as a cofactor in the synthesis of carbamoyl phosphate. How is N-acetylglutamate synthesized from glutamate?

33. *Negative nitrogen balance.* A deficiency of even one amino acid results in a negative nitrogen balance. In this state, more protein is degraded than is synthesized, and so more nitrogen is excreted than is ingested. Why would protein be degraded if one amino acid were missing?

34. *Precursors.* Differentiate between ketogenic amino acids and glucogenic amino acids. ✓⑤

35. *A sleight of hand.* The end products of tryptophan degradation are acetyl CoA and acetoacetyl CoA, yet tryptophan is a gluconeogenic amino acid in animals. Explain. ✓⑤

36. *Closely related.* Pyruvate dehydrogenase complex and α-ketoglutarate dehydrogenase complex are huge enzymes consisting of three discrete enzymatic activities. Which amino acids require a related enzyme complex, and what is the name of the enzyme?

37. *Supply lines.* The carbon skeletons of the 20 common amino acids can be degraded into a limited number of end products. What are the end products and in what metabolic pathway are they commonly found? ✓⑤

Mechanism Problems

38. *Serine dehydratase.* Write out a complete mechanism for the conversion of serine into aminoacrylate catalyzed by serine dehydratase.

39. *Serine racemase.* The nervous system contains a substantial amount of D-serine, which is generated from L-serine by serine racemase, a PLP-dependent enzyme. Propose a mechanism for this reaction. What is the equilibrium constant for the reaction L-serine ⇌ D-serine?

Chapter Integration Problems

40. *Multiple substrates.* In Chapter 8, we learned that there are two types of bisubstrate reactions, sequential and double-displacement. Which type characterizes the action of aminotransferases? Explain your answer.

41. *Double duty.* Degradation signals are commonly located in protein regions that also facilitate protein–protein interactions. Explain why this coexistence of two functions in the same domain might be useful. ✓❶, ✓❷

42. *Fuel choice.* Within a few days after a fast begins, nitrogen excretion accelerates to a higher-than-normal level. After a few weeks, the rate of nitrogen excretion falls to a lower level and continues at this low rate. However, after the fat stores have been depleted, nitrogen excretion rises to a high level. ✓❹, ✓❺

(a) What events trigger the initial surge of nitrogen excretion?

(b) Why does nitrogen excretion fall after several weeks of fasting?

(c) Explain the increase in nitrogen excretion when the lipid stores have been depleted.

43. *A serious situation.* Pyruvate carboxylase deficiency is a fatal disorder. Patients with pyruvate carboxylase deficiency sometimes display some or all of the following symptoms: lactic acidosis, hyperammonemia (excess NH_4^+ in the blood), hypoglycemia, and demyelination of the regions of

the brain due to insufficient lipid synthesis. Provide a possible biochemical rationale for each of these observations.

44. *Many roles.* Pyridoxal phosphate is an important coenzyme in transamination reactions. We have seen this coenzyme before, in glycogen metabolism. Which enzyme in glycogen metabolism requires pyridoxal phosphate and what role does the coenzyme play in this enzyme?

45. *Enough cycles to have a race.* The glucose–alanine cycle is reminiscent of the Cori cycle, but the glucose–alanine cycle can be said to be more energy efficient. Explain why this is so. ✓❹

Data Interpretation Problem

46. *Another helping hand.* In eukaryotes, the 20S proteasome component in conjunction with the 19S component degrades ubiquitinated proteins with the hydrolysis of a molecule of ATP. Archaea lack ubiquitin and the 26S proteasome but do contain a 20S proteasome. Some archaea also contain an ATPase that is homologous to the ATPases of the eukaryotic 19S component. This archaeal ATPase activity was isolated as a 650-kDa complex (called PAN) from the archaeon *Thermoplasma,* and experiments were performed to determine if PAN could enhance the activity of the 20S proteasome from *Thermoplasma* as well as other 20S proteasomes.

Protein degradation was measured as a function of time and in the presence of various combinations of components. Graph A shows the results. ✓❶, ✓❷, ✓❸

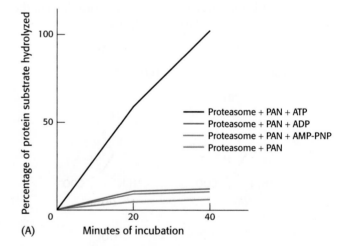

(A)

AMP-PNP

(a) What is the effect of PAN on archaeal proteasome activity in the absence of nucleotides?

(b) What is the nucleotide requirement for protein digestion?

(c) What evidence suggests that ATP hydrolysis, and not just the presence of ATP, is required for digestion?

A similar experiment was performed with a small peptide as a substrate for the proteasome instead of a protein. The results obtained are shown in graph B.

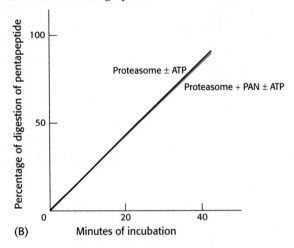

(B)

(d) How do the requirements for peptide digestion differ from those of protein digestion?

(e) Suggest some reasons for the difference.

The ability of PAN from the archaeon *Thermoplasma* to support protein degradation by the 20S proteasomes from the archaeon *Methanosarcina* and rabbit muscle was then examined.

Percentage of digestion of protein substrate
(Source of the 20S proteasome)

Additions	Thermoplasma	Methanosarcina	Rabbit muscle
None	11	10	10
PAN	8	8	8
PAN + ATP	100	40	30
PAN + ADP	12	9	10

[Data from P. Zwickl, D. Ng, K. M. Woo, H.-P. Klenk, and A. L. Goldberg. An archaebacterial ATPase, homologous to ATPase in the eukaryotic 26S proteasome, activates protein breakdown by 20S proteasomes. *J. Biol. Chem.* 274(1999): 26008–26014.]

(f) Can the *Thermoplasma* PAN augment protein digestion by the proteasomes from other organisms?

(g) What is the significance of the stimulation of rabbit muscle proteasome by *Thermoplasma* PAN?

Think ▶ Pair ▶ Share Problems

47. Identify the means by which protein turnover is regulated.

48. Describe the common features of amino acid degradation.

49. Account for the fact that the symptoms of methylmalonic acidemia can be so variable, from benign to fatal.

The Biosynthesis of Amino Acids

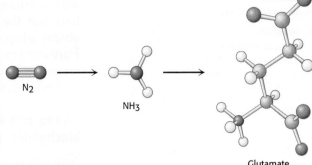

Nitrogen is a key component of amino acids. The atmosphere is rich in nitrogen gas (N_2), a very unreactive molecule. Certain organisms, such as bacteria that live in the root nodules of yellow clover, can convert nitrogen gas into ammonia (NH_3), which can then be used to synthesize, first, glutamate and then other amino acids. [(Left) Hugh Spencer/Science Source.]

✓ LEARNING OBJECTIVES

By the end of this chapter, you should be able to:

1. Explain the centrality of nitrogen fixation to life and describe how atmospheric nitrogen is converted into biologically useful forms of nitrogen.

2. Identify the sources of carbon atoms for amino acid synthesis.

3. Describe the role of feedback inhibition in controlling amino acid synthesis.

4. Identify important biomolecules that are derived from amino acids.

OUTLINE

24.1 Nitrogen Fixation: Microorganisms Use ATP and a Powerful Reductant to Reduce Atmospheric Nitrogen to Ammonia

24.2 Amino Acids Are Made from Intermediates of the Citric Acid Cycle and Other Major Pathways

24.3 Feedback Inhibition Regulates Amino Acid Biosynthesis

24.4 Amino Acids Are Precursors of Many Biomolecules

The assembly of biological macromolecules, including proteins and nucleic acids, requires the generation of appropriate starting materials. We have already considered the assembly of carbohydrates in discussions of the Calvin cycle, the pentose phosphate pathway (Chapter 20), and glycogen synthesis (Chapter 21). The present chapter and the next two examine the assembly of the other important building blocks—namely, amino acids, nucleotides, and lipids.

The pathways for the biosynthesis of these molecules are extremely ancient, going back to the last common ancestor of all living things. Indeed, these pathways probably predate many of the pathways of energy transduction discussed in Part II and may have provided key selective advantages in early evolution. Many of the intermediates in energy-yielding pathways play a role in biosynthesis as well. These common intermediates allow efficient interplay between energy-yielding (catabolic) and energy-requiring biosynthetic (anabolic) pathways. Thus, cells are able to balance the degradation of compounds for energy mobilization and the synthesis of starting materials for macromolecular construction.

Anabolism

Biosynthetic processes.

Catabolism

Degradative processes.

Derived from the Greek *ana*, "up"; *kata*, "down"; *ballein*, "to throw."

We begin our consideration of biosynthesis with amino acids—the building blocks of proteins and the nitrogen source for many other important molecules, including nucleotides, neurotransmitters, and prosthetic groups such as porphyrins. Amino acid biosynthesis is intimately connected with nutrition because many higher organisms, including human beings, have lost the ability to synthesize some amino acids and must therefore obtain adequate quantities of these essential amino acids in their diets. Furthermore, because some amino acid biosynthetic enzymes are absent in mammals but present in plants and microorganisms, they are useful targets for herbicides and antibiotics.

Amino acid synthesis requires solutions to three key biochemical problems

Nitrogen is an essential component of amino acids. Earth has an abundant supply of nitrogen, but it is primarily in the form of atmospheric nitrogen gas (N_2), a remarkably inert molecule. Thus, a fundamental problem for biological systems is to obtain nitrogen in a more usable form. This problem is solved by certain microorganisms capable of reducing the inert $N \equiv N$ molecule of nitrogen gas to two molecules of ammonia in one of the most amazing reactions in biochemistry. Nitrogen in the form of ammonia is the source of nitrogen for all the amino acids. The carbon backbones come from the glycolytic pathway, the pentose phosphate pathway, or the citric acid cycle.

In amino acid production, we encounter another important problem in biosynthesis—namely, stereochemical control. Because all amino acids except glycine are chiral, biosynthetic pathways must generate the correct isomer with high fidelity. In each of the 19 pathways for the generation of chiral amino acids, the stereochemistry at the α-carbon atom is established by a transamination reaction that includes pyridoxal phosphate (PLP). Almost all the aminotransferases that catalyze these reactions descend from a common ancestor, illustrating once again that effective solutions to biochemical problems are retained throughout evolution.

Finally, how are the proper concentrations of amino acids maintained inside the cell? Biosynthetic pathways are often highly regulated such that building blocks are synthesized only when supplies are low. Frequently, a high concentration of the final product of a pathway inhibits the activity of allosteric enzymes that function early in the pathway to control the committed step. These enzymes are similar in functional properties to aspartate transcarbamoylase and its regulators (Section 10.1). Feedback and allosteric mechanisms ensure that all 20 amino acids are maintained in sufficient amounts for protein synthesis and other processes.

24.1 Nitrogen Fixation: Microorganisms Use ATP and a Powerful Reductant to Reduce Atmospheric Nitrogen to Ammonia

One of the most challenging biochemical problems organisms face is obtaining nitrogen in a usable form. The nitrogen in amino acids, purines, pyrimidines, and other biomolecules ultimately comes from atmospheric nitrogen, N_2. The extremely strong $N \equiv N$ bond, which has a bond energy of $940 \, \text{kJ mol}^{-1}$ ($225 \, \text{kcal mol}^{-1}$), is highly resistant to chemical attack. Indeed, Antoine Lavoisier named nitrogen gas "azote," from Greek words meaning "without life," because it is so unreactive. Nevertheless, the conversion of nitrogen and hydrogen to form ammonia is thermodynamically favorable; the reaction is difficult kinetically because of the activation energy required to form intermediates along the reaction pathway.

Although higher organisms are unable to fix nitrogen, this conversion is carried out by some bacteria and archaea. The biosynthetic process starts with the reduction of N_2 to NH_3 (ammonia), a process called *nitrogen fixation*. Symbiotic *Rhizobium* bacteria invade the roots of leguminous plants and form root nodules in which they fix nitrogen, supplying both the bacteria and the plants. The importance of nitrogen fixation by *diazotrophic (nitrogen-fixing) microorganisms* to the metabolism of all higher eukaryotes cannot be overstated: the amount of N_2 fixed by these species has been estimated to be 10^{11} kilograms per year, about 60% of Earth's newly fixed nitrogen. Lightning and ultraviolet radiation fix another 15%; the other 25% is fixed by industrial processes. The industrial process for nitrogen fixation devised by Fritz Haber in 1910 is still being used in fertilizer factories.

$$N_2 + 3\,H_2 \rightleftharpoons 2\,NH_3$$

The fixation of N_2 is typically carried out by mixing it with H_2 gas over an iron catalyst at about 500°C and a pressure of 300 atmospheres.

To meet the kinetic challenge, the biological process of nitrogen fixation requires a complex enzyme with multiple redox centers. The *nitrogenase complex*, which carries out this fundamental transformation, consists of two proteins: a *reductase* (also called the iron protein or Fe protein), which provides electrons with high reducing power, and *nitrogenase* (also called the molybdenum–iron protein or MoFe protein), which uses these electrons to reduce N_2 to NH_3. The transfer of electrons from the reductase to the nitrogenase component is coupled to the hydrolysis of ATP by the reductase (Figure 24.1).

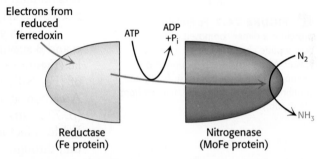

FIGURE 24.1 Nitrogen fixation. Electrons flow from ferredoxin to the reductase (iron protein, or Fe protein) to nitrogenase (molybdenum–iron protein, or MoFe protein) to reduce nitrogen to ammonia. ATP hydrolysis within the reductase drives conformational changes necessary for the efficient transfer of electrons.

In principle, the reduction of N_2 to NH_3 is a six-electron process.

$$N_2 + 6\,e^- + 6\,H^+ \longrightarrow 2\,NH_3$$

However, the biological reaction always generates at least 1 mol of H_2 in addition to 2 mol of NH_3 for each mol of $N \equiv N$. Hence, an input of two additional electrons is required.

$$N_2 + 8\,e^- + 8\,H^+ \longrightarrow 2\,NH_3 + H_2$$

In most nitrogen-fixing microorganisms, *the eight high-potential electrons come from reduced ferredoxin,* generated by oxidative processes. Two molecules of ATP are hydrolyzed for each electron transferred. Thus, *at least 16 molecules of ATP are hydrolyzed for each molecule of N_2 reduced.*

$$N_2 + 8\,e^- + 8\,H^+ + 16\,ATP + 16\,H_2O \longrightarrow$$
$$2\,NH_3 + H_2 + 16\,ADP + 16\,P_i$$

Recall that O_2 is required for oxidative phosphorylation to generate the ATP necessary for nitrogen fixation. However, the nitrogenase complex is exquisitely sensitive to inactivation by O_2. To allow ATP synthesis and nitrogenase to function simultaneously, leguminous plants maintain a very low concentration of free O_2 in their root nodules, the location of the nitrogenase. This is accomplished by binding O_2 to *leghemoglobin,* a homolog of hemoglobin (Figure 6.15).

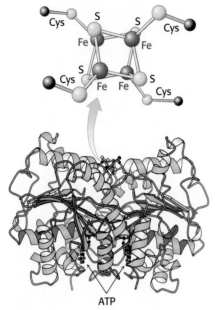

FIGURE 24.2 Fe Protein. This protein is a dimer composed of two polypeptide chains linked by a 4Fe–4S cluster. *Notice* that each monomer is a member of the P-loop NTPase family and contains an ATP-binding site. [Drawn from 1N₂C.pdb.]

The iron–molybdenum cofactor of nitrogenase binds and reduces atmospheric nitrogen

Both the reductase and the nitrogenase components of the complex are *iron–sulfur proteins,* in which iron is bonded to the sulfur atom of a cysteine residue and to inorganic sulfide. Recall that iron–sulfur clusters act as electron carriers (Section 18.3). The *reductase* is a dimer of identical 30-kDa subunits bridged by a 4Fe–4S cluster (Figure 24.2).

The role of the reductase is to transfer electrons from a suitable donor, such as reduced ferredoxin, to the nitrogenase component. The 4Fe–4S cluster carries the electrons, one at a time, to nitrogenase. The binding and hydrolysis of ATP triggers a conformational change that moves the reductase closer to the nitrogenase component, whence it is able to transfer its electron to the center of nitrogen reduction. The structure of the ATP-binding region reveals it to be a member of the P-loop NTPase family (Section 9.4) that is clearly related to the nucleotide-binding regions found in G proteins and related proteins. Thus, we see another example of how this domain has been recruited in evolution because of its ability to couple nucleoside triphosphate hydrolysis to conformational changes.

The nitrogenase component is an $\alpha_2\beta_2$ tetramer (240-kDa), in which the α and β subunits are homologous to each other and structurally quite similar (Figure 24.3). The nitrogenase requires the FeMo cofactor, which consists of $[Fe_4–S_3]$ and $[Mo–Fe_3–S_3]$ subclusters joined by three disulfide bonds. A carbon atom (the interstitial carbon), donated by S-adenosylmethionine (p. 722), sits at the interstices of the iron atoms of the FeMo cofactor. The FeMo cofactor is also coordinated to a homocitrate moiety and to the α subunit through one histidine residue and one cysteinate residue.

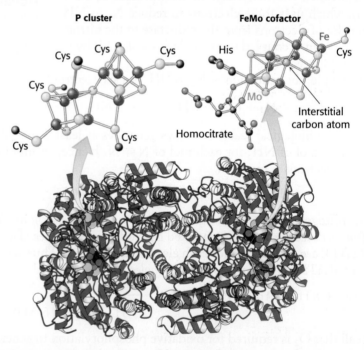

FIGURE 24.3 MoFe protein. This protein is a heterotetramer composed of two α subunits (red) and two β subunits (blue). *Notice* that the protein contains two copies each of two types of clusters: P clusters and FeMo cofactors. Each P cluster contains eight iron atoms (green) and seven sulfides linked to the protein by six cysteine residues. Each FeMo cofactor contains one molybdenum atom, seven iron atoms, nine sulfides, the interstitial carbon atom, and a homocitrate, and is linked to the protein by one cysteinate residue and one histidine residue. [Drawn from 1M1N.pdb.]

Electrons from the reductase enter at the *P clusters,* which are located at the α–β interface. The role of the P clusters is to store electrons until they can be used productively to reduce nitrogen at the FeMo cofactor. *The FeMo cofactor is the site of nitrogen fixation.* One face of the FeMo cofactor is likely to be the site of nitrogen reduction. The electron-transfer reactions from the P cluster take place in concert with the binding of hydrogen ions to nitrogen as it is reduced. Further studies are under way to elucidate the mechanism of this remarkable reaction.

Ammonium ion is assimilated into an amino acid through glutamate and glutamine

The NH_3 generated by the nitrogenase complex is a base, and becomes NH_4^+ in aqueous solutions. The next step in the assimilation of nitrogen into biomolecules is the entry of NH_4^+ into amino acids. The amino acids *glutamate* and *glutamine* play pivotal roles in this regard, acting as nitrogen donors for most amino acids. The α-amino group of most amino acids comes from the α-amino group of glutamate by transamination (Section 23.3). Glutamine, the other major nitrogen donor, contributes its side-chain nitrogen atom in the biosynthesis of a wide range of important compounds, including the amino acids tryptophan and histidine.

Glutamate is synthesized from NH_4^+ and α-ketoglutarate, a citric acid cycle intermediate, by the action of *glutamate dehydrogenase.* We have already encountered this enzyme in the degradation of amino acids (Section 23.3). Recall that NAD^+ is the oxidant in catabolism, whereas NADPH is the reductant in biosyntheses. Glutamate dehydrogenase is unusual in that it does not discriminate between NADH and NADPH, at least in some species.

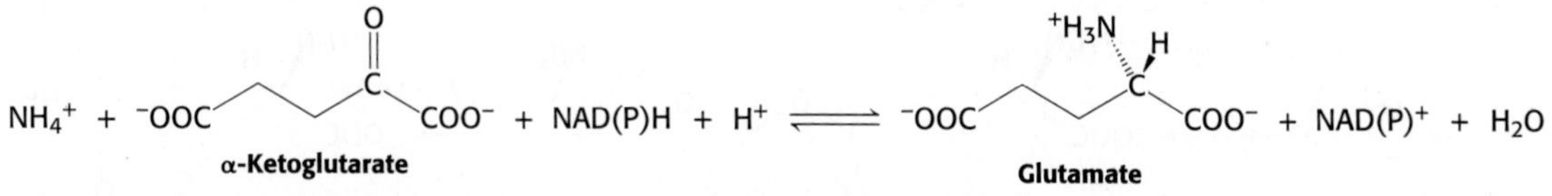

The reaction proceeds in two steps. First, a Schiff base forms between ammonium ion and α-ketoglutarate. The formation of a Schiff base between an amine and a carbonyl compound is a key reaction that takes place at many stages of amino acid biosynthesis and degradation.

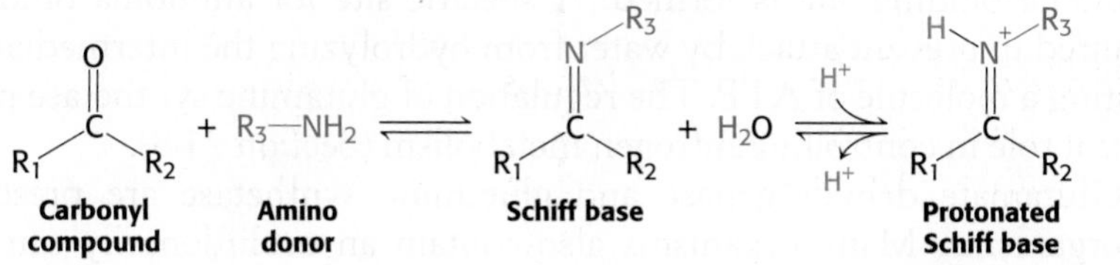

Schiff bases are easily protonated. In the second step, the protonated Schiff base is reduced by the transfer of a hydride ion from NAD(P)H to form glutamate.

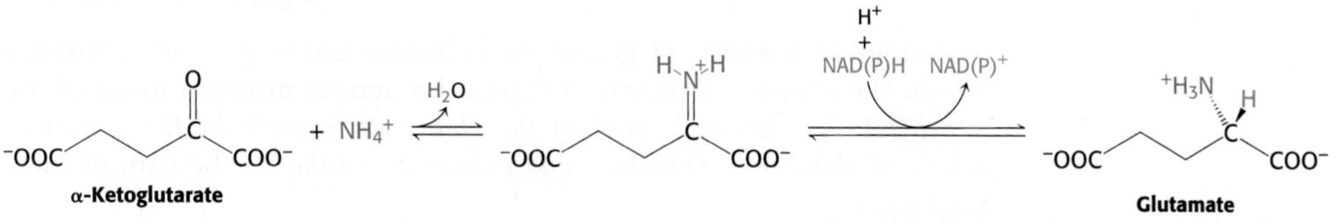

This reaction is crucial because it establishes the stereochemistry of the α-carbon atom (*S* absolute configuration) in glutamate. The enzyme binds

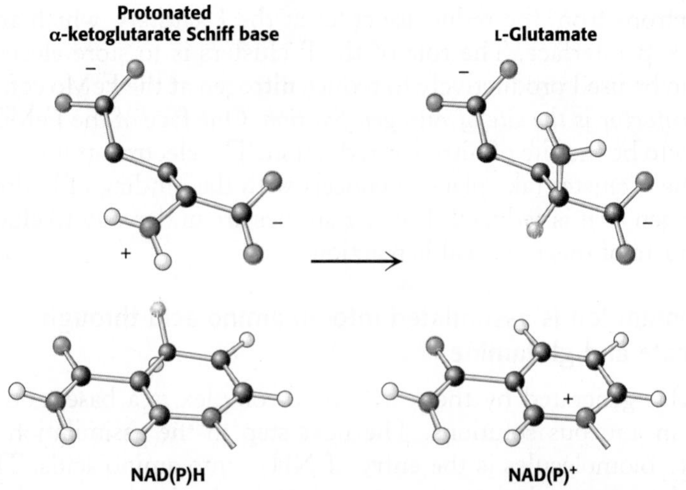

FIGURE 24.4 Establishment of chirality. In the active site of glutamate dehydrogenase, hydride transfer (green) from NAD(P)H to a specific face of the achiral protonated Schiff base of α-ketoglutarate establishes the L configuration of glutamate.

Protonated α-ketoglutarate Schiff base → L-Glutamate

NAD(P)H   NAD(P)$^+$

the α-ketoglutarate substrate in such a way that hydride transferred from NAD(P)H is added to form the L isomer of glutamate (Figure 24.4). As we shall see, this stereochemistry is established for other amino acids by transamination reactions that rely on pyridoxal phosphate. Recall that establishing the correct stereochemistry is one of the fundamental challenges of nitrogen incorporation into biomolecules.

A second ammonium ion is incorporated into glutamate to form glutamine by the action of *glutamine synthetase*. This amidation is driven by the hydrolysis of ATP. ATP participates directly in the reaction by phosphorylating the side chain of glutamate to form an acylphosphate intermediate, which then reacts with ammonia to form glutamine.

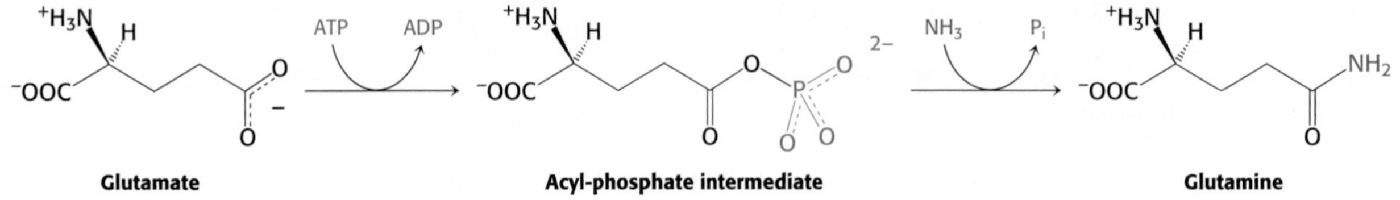

Glutamate   Acyl-phosphate intermediate   Glutamine

NH_4^+ initially binds to the enzyme, but upon formation of the acylphosphate intermediate, ammonium ion is deprotonated as a high-affinity ammonia-binding site is formed. A specific site for ammonia binding is required to prevent attack by water from hydrolyzing the intermediate and wasting a molecule of ATP. The regulation of glutamine synthetase plays a critical role in controlling nitrogen metabolism (Section 24.3).

Glutamate dehydrogenase and glutamine synthetase are present in all organisms. Many organisms also contain an evolutionarily unrelated enzyme, *glutamate synthase*, which catalyzes the reductive amination of α-ketoglutarate to glutamate. Glutamine is the nitrogen donor.

$$\alpha\text{-Ketoglutarate} + \text{glutamine} + \text{NADPH} + H^+ \rightleftharpoons$$
$$2\,\text{glutamate} + \text{NADP}^+$$

The side-chain amide of glutamine is hydrolyzed to generate ammonia within the enzyme, a recurring theme throughout nitrogen metabolism. *When NH_4^+ is limiting, most of the glutamate is made by the sequential action of glutamine synthetase and glutamate synthase.* The sum of these reactions is

$$NH_4^+ + \alpha\text{-ketoglutarate} + \text{NADPH} + \text{ATP} \longrightarrow$$
$$\text{glutamate} + \text{NADP}^+ + \text{ADP} + P_i$$

Note that this stoichiometry differs from that of the glutamate dehydrogenase reaction in that ATP is hydrolyzed. Why do organisms sometimes use this more expensive pathway? The answer is that the value of K_M of glutamate dehydrogenase for NH_4^+ is high (~ 1 mM), and so this enzyme is not saturated when NH_4^+ is limiting. In contrast, glutamine synthetase has very low K_M for NH_4^+. Thus, ATP hydrolysis is required to capture ammonia when it is scarce.

24.2 Amino Acids Are Made from Intermediates of the Citric Acid Cycle and Other Major Pathways

Thus far, we have considered the conversion of N_2 into NH_4^+ and the assimilation of NH_4^+ into glutamate and glutamine. We turn now to the biosynthesis of the other amino acids, the majority of which obtain their nitrogen from glutamate or glutamine. The pathways for the biosynthesis of amino acids are diverse. However, they have an important common feature: *their carbon skeletons come from intermediates of glycolysis, the pentose phosphate pathway, or the citric acid cycle.* On the basis of these starting materials, amino acids can be grouped into six biosynthetic families (Figure 24.5).

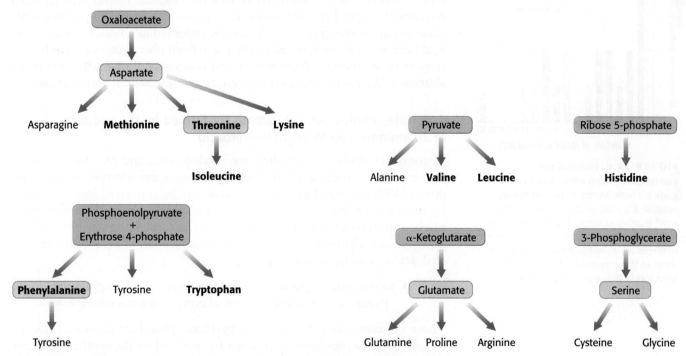

FIGURE 24.5 Biosynthetic families of amino acids in bacteria and plants. Major metabolic precursors are shaded blue. Amino acids that give rise to other amino acids are shaded yellow. Essential amino acids are in boldface type.

Human beings can synthesize some amino acids but must obtain others from their diet

Most microorganisms, such as *E. coli,* can synthesize the entire basic set of 20 amino acids, whereas human beings cannot make 9 of them. The amino acids that *must* be supplied in the diet are called *essential amino acids,* whereas the others, which can be synthesized if dietary content is insufficient, are termed *nonessential amino acids* (Table 24.1). These designations refer to the needs of an organism under a particular set of conditions.

TABLE 24.1 Basic set of 20 amino acids

Nonessential	Essential
Alanine	Histidine
Arginine	Isoleucine
Asparagine	Leucine
Aspartate	Lysine
Cysteine	Methionine
Glutamate	Phenylalanine
Glutamine	Threonine
Glycine	Tryptophan
Proline	Valine
Serine	
Tyrosine	

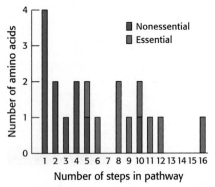

FIGURE 24.6 Essential and nonessential amino acids. Some amino acids are nonessential to human beings because they can be biosynthesized in a small number of steps. Those amino acids requiring a large number of steps for their synthesis are essential in the diet because some of the enzymes for these steps have been lost in the course of evolution.

For example, enough arginine is synthesized by the urea cycle to meet the needs of an adult, but perhaps not those of a growing child. Likewise, tyrosine is sometimes designated as an essential amino acid, but is not if phenylalanine is present in adequate amounts. In mammals, tyrosine can be synthesized from phenylalanine in one step.

Phenylalanine → (Phenylalanine hydroxylase) → **Tyrosine**

A deficiency of even one amino acid results in a *negative nitrogen balance*. In this state, more protein is degraded than is synthesized, and so more nitrogen is excreted than is ingested.

The nonessential amino acids are synthesized by quite simple reactions, whereas the pathways for the formation of the essential amino acids are quite complex. For example, the nonessential amino acids *alanine* and *aspartate* are synthesized in a single step from pyruvate and oxaloacetate, respectively. In contrast, the pathways for the essential amino acids require from 5 to 16 steps (Figure 24.6). The sole exception to this pattern is arginine, inasmuch as the synthesis of this nonessential amino acid de novo requires 10 steps. Typically, however, it is made in only 3 steps from ornithine as part of the urea cycle. Tyrosine, classified as a nonessential amino acid because it can be synthesized in 1 step from phenylalanine, requires 10 steps to be synthesized from scratch and is essential if phenylalanine is not abundant. We begin with the biosynthesis of nonessential amino acids.

Aspartate, alanine, and glutamate are formed by the addition of an amino group to an alpha-ketoacid

Three α-ketoacids—α-ketoglutarate, oxaloacetate, and pyruvate—can be converted into amino acids in one step through the addition of an amino group. We have seen that α-ketoglutarate can be converted into glutamate by reductive amination (Section 23.3). The amino group from glutamate can be transferred to other α-ketoacids by transamination reactions. Thus, aspartate and alanine can be made from the addition of an amino group to oxaloacetate and pyruvate, respectively.

$$\text{Oxaloacetate} + \text{glutamate} \rightleftharpoons \text{aspartate} + \alpha\text{-ketoglutarate}$$
$$\text{Pyruvate} + \text{glutamate} \rightleftharpoons \text{alanine} + \alpha\text{-ketoglutarate}$$

These reactions are carried out by *pyridoxal phosphate-dependent aminotransferases*. Transamination reactions are required for the synthesis of most amino acids.

In Section 23.3, we considered the mechanism of aminotransferases as applied to the metabolism of amino acids. Let us review the aminotransferase mechanism as it operates in the *biosynthesis* of amino acids (Figure 23.10). The reaction pathway begins with *pyridoxal phosphate* (PLP) in a Schiff-base linkage with lysine at the aminotransferase active site, forming an internal aldimine (Figure 24.7). An amino group is transferred from glutamate to form pyridoxamine phosphate (PMP), the actual amino donor, in a multistep process. PMP then reacts with an incoming α-ketoacid to form a ketimine. Proton loss forms a quinonoid intermediate that then accepts a proton at a different site to form an external aldimine. The newly formed amino acid is released with the concomitant formation of the internal aldimine.

FIGURE 24.7 Amino acid biosynthesis by transamination. (1) Within an aminotransferase, the internal aldimine is converted into pyridoxamine phosphate (PMP) by reaction with glutamate in a multistep process not shown. (2) PMP then reacts with an α-ketoacid to generate a ketimine. (3) This intermediate is converted into a quinonoid intermediate (4), which in turn yields an external aldimine. (5) The aldimine is cleaved to release the newly formed amino acid to complete the cycle.

A common step determines the chirality of all amino acids

Aspartate aminotransferase is the prototype of a large family of PLP-dependent enzymes. Comparisons of amino acid sequences as well as several three-dimensional structures reveal that almost all aminotransferases having roles in amino acid biosynthesis are related to aspartate aminotransferase by divergent evolution. An examination of the aligned amino acid sequences reveals that two residues are completely conserved. These residues are the lysine residue that forms the Schiff base with the PLP cofactor (lysine 258 in aspartate aminotransferase) and an arginine residue that interacts with the α-carboxylate group of the ketoacid (Figure 23.10).

An essential step in the transamination reaction is the protonation of the quinonoid intermediate to form the external aldimine. *The chirality of the amino acid formed is determined by the direction from which this proton is added to the quinonoid form* (Figure 24.8). The interaction between the conserved arginine residue and the α-carboxylate group helps orient the substrate so that the lysine residue transfers a proton to the bottom face of the quinonoid intermediate, generating an aldimine with an L configuration at the C_α center.

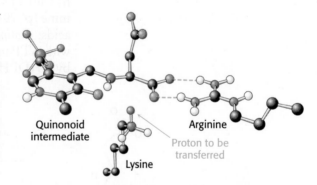

FIGURE 24.8 Stereochemistry of proton addition. In an aminotransferase active site, the addition of a proton from the lysine residue to the bottom face of the quinonoid intermediate determines the L configuration of the amino acid product. The conserved arginine residue interacts with the α-carboxylate group and helps establish the appropriate geometry of the quinonoid intermediate.

The formation of asparagine from aspartate requires an adenylated intermediate

The formation of asparagine from aspartate is chemically analogous to the formation of glutamine from glutamate. Both transformations are amidation reactions and both are driven by the hydrolysis of ATP. The actual reactions are different, however. In bacteria, the reaction for asparagine synthesis is

$$\text{Aspartate} + \text{NH}_4^+ + \text{ATP} \longrightarrow \text{asparagine} + \text{AMP} + \text{PP}_i + \text{H}^+$$

Thus, the products of ATP hydrolysis are AMP and PP_i rather than ADP and P_i. Aspartate is activated by adenylation rather than by phosphorylation.

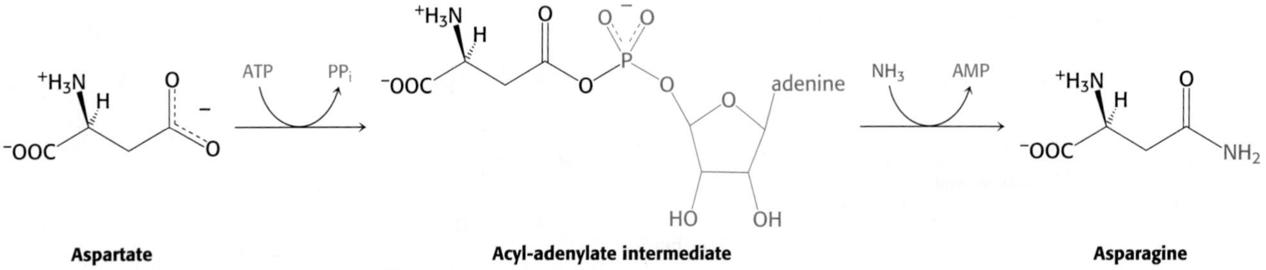

Aspartate   **Acyl-adenylate intermediate**   **Asparagine**

We have encountered this mode of activation in fatty acid degradation and will see it again in lipid and protein synthesis.

In mammals, the nitrogen donor for asparagine is glutamine rather than ammonia as in bacteria. Ammonia is generated by hydrolysis of the side chain of glutamine and directly transferred to activated aspartate, bound in the active site. An advantage is that the cell is not directly exposed to NH_4^+, which is toxic at high levels to human beings and other mammals. *The use of glutamine hydrolysis as a mechanism for generating ammonia for use within the same enzyme is a motif common throughout biosynthetic pathways.*

Glutamate is the precursor of glutamine, proline, and arginine

The synthesis of glutamate by the reductive amination of α-ketoglutarate has already been discussed, as has the conversion of glutamate into glutamine (p. 791). Glutamate is the precursor of two other nonessential amino acids: *proline* and *arginine*. First, the γ-carboxyl group of glutamate reacts with ATP to form an acyl phosphate. This mixed anhydride is then reduced by NADPH to an aldehyde.

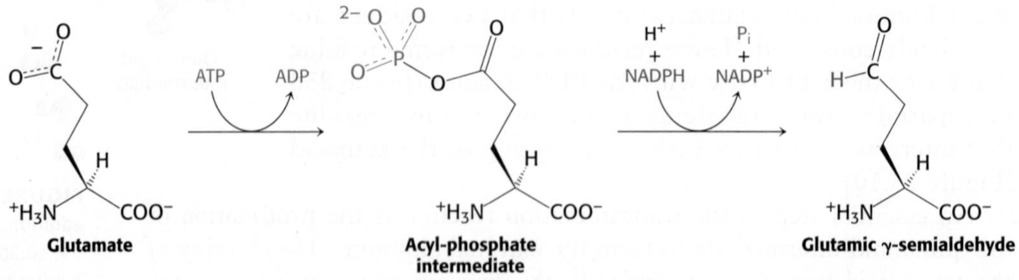

Glutamate   **Acyl-phosphate intermediate**   **Glutamic γ-semialdehyde**

Glutamic γ-semialdehyde cyclizes with a loss of H_2O in a nonenzymatic process to give Δ^1-pyrroline 5-carboxylate, which is reduced by NADPH to proline. Alternatively, the semialdehyde can be transaminated to ornithine, which is converted in several steps into arginine in the urea cycle (Figure 23.16).

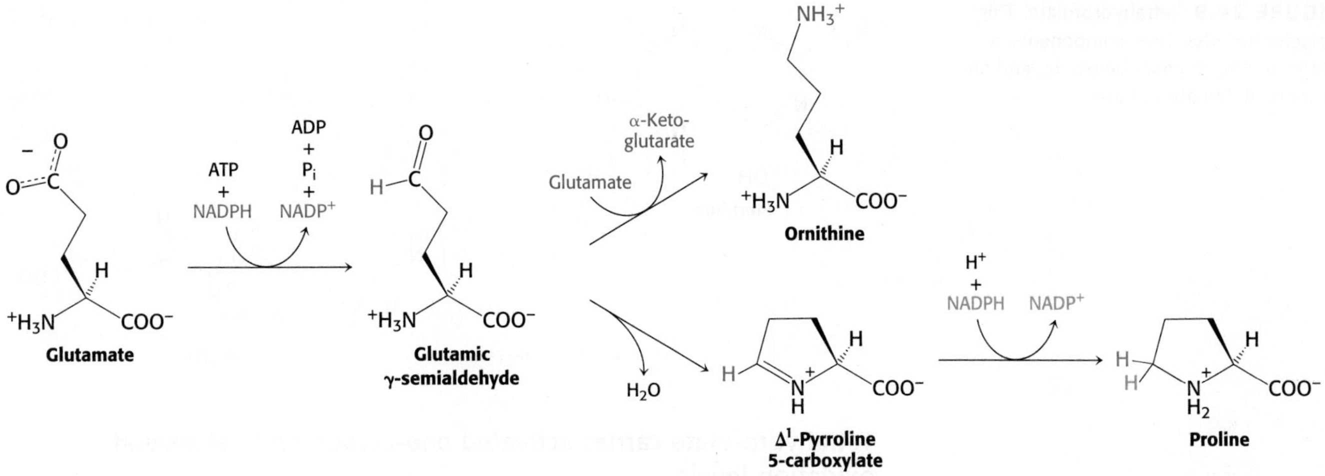

3-Phosphoglycerate is the precursor of serine, cysteine, and glycine

Serine is synthesized from 3-phosphoglycerate, an intermediate in glycolysis. The first step is an oxidation to 3-phosphohydroxypyruvate. This α-ketoacid is transaminated to 3-phosphoserine, which is then hydrolyzed to serine.

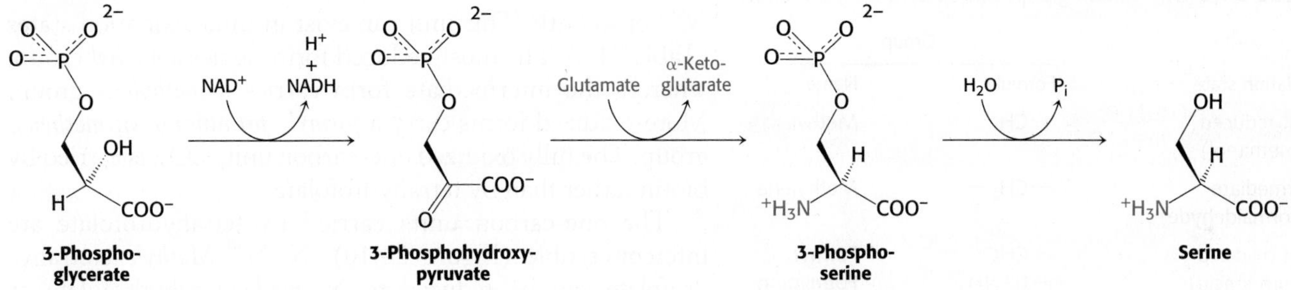

Serine is the precursor of *cysteine* and *glycine*. As we shall see, the conversion of serine into cysteine requires the substitution of a sulfur atom derived from methionine for the side-chain oxygen atom. In the formation of glycine, the side-chain methylene group of serine is transferred to *tetrahydrofolate*, a carrier of one-carbon units that will be discussed shortly (Figure 24.9).

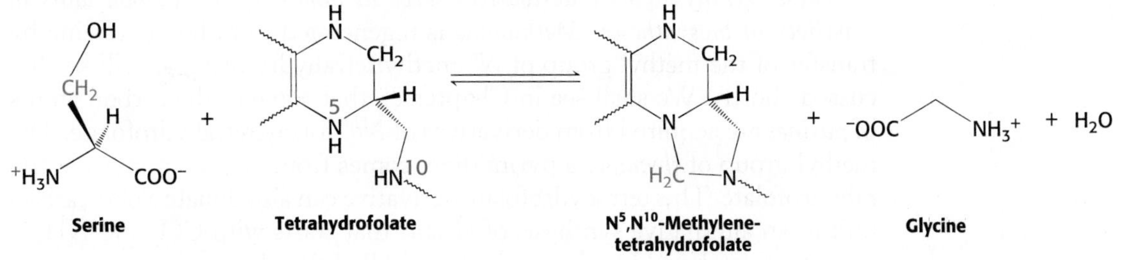

This interconversion is catalyzed by *serine hydroxymethyltransferase*, a PLP enzyme that is homologous to aspartate aminotransferase. The formation of the Schiff base of serine renders the bond between its α- and β-carbon atoms susceptible to cleavage, enabling the transfer of the β-carbon to tetrahydrofolate and producing the Schiff base of glycine.

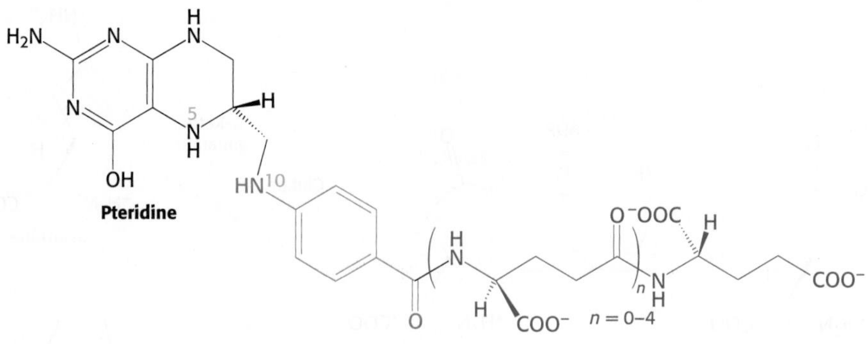

FIGURE 24.9 Tetrahydrofolate. This cofactor includes three components: a pteridine ring, *p*-aminobenzoate, and one or more glutamate residues.

Pteridine   *p*-Aminobenzoate   Glutamate

Tetrahydrofolate carries activated one-carbon units at several oxidation levels

Tetrahydrofolate (also called *tetrahydropteroylglutamate*) is a highly versatile carrier of activated one-carbon units. This cofactor consists of three groups: a substituted pteridine, *p*-aminobenzoate, and a chain of one or more glutamate residues (Figure 24.9). Mammals can synthesize the pteridine ring, but they are unable to conjugate it to the other two units. They obtain tetrahydrofolate from their diets or from microorganisms in their intestinal tracts.

The one-carbon group carried by tetrahydrofolate is bonded to its N-5 or N-10 nitrogen atom (denoted as N^5 and N^{10}) or to both. This unit can exist in three oxidation states (Table 24.2). The most-reduced form carries a *methyl* group, whereas the intermediate form carries a *methylene* group. More-oxidized forms carry a *formyl*, *formimino*, or *methenyl* group. The fully oxidized one-carbon unit, CO_2, is carried by biotin rather than by tetrahydrofolate.

The one-carbon units carried by tetrahydrofolate are interconvertible (Figure 24.10). N^5,N^{10}-*Methylenetetrahydrofolate* can be reduced to N^5-*methyltetrahydrofolate* or oxidized to N^5,N^{10}-*methenyltetrahydrofolate*. N^5,N^{10}-*Methenyltetrahydrofolate* can be converted into N^5-*formiminotetrahydrofolate* or N^{10}-*formyltetrahydrofolate*, both of which are at the same oxidation level. N^{10}-*Formyltetrahydrofolate* can also be synthesized from tetrahydrofolate, formate, and ATP. N^5-*Formyltetrahydrofolate* can be reversibly isomerized to N^{10}-*formyltetrahydrofolate* or it can be converted into N^5,N^{10}-*methenyltetrahydrofolate*.

These tetrahydrofolate derivatives serve as donors of one-carbon units in a variety of biosyntheses. Methionine is regenerated from homocysteine by transfer of the methyl group of N^5-methyltetrahydrofolate, as will be discussed shortly. We shall see in Chapter 25 that some of the carbon atoms of *purines* are acquired from derivatives of N^{10}-formyltetrahydrofolate. The methyl group of *thymine*, a pyrimidine, comes from N^5,N^{10}-methylenetetrahydrofolate. This tetrahydrofolate derivative can also donate a one-carbon unit in an alternative synthesis of *glycine* that starts with CO_2 and NH_4^+, a reaction catalyzed by *glycine synthase* (called the *glycine cleavage enzyme* when it operates in the reverse direction).

$$CO_2 + NH_4^+ + N^5,N^{10}\text{-methylenetetrahydrofolate} + NADH \rightleftharpoons$$
$$\text{glycine} + \text{tetrahydofolate} + NAD^+$$

Thus, one-carbon units at each of the three oxidation levels are utilized in biosyntheses. Furthermore, *tetrahydrofolate serves as an acceptor of*

TABLE 24.2 One-carbon groups carried by tetrahydrofolate

Oxidation state	Group	
	Formula	Name
Most reduced (= methanol)	$-CH_3$	Methyl
Intermediate (= formaldehyde)	$-CH_2-$	Methylene
Most oxidized (= formic acid)	$-CHO$	Formyl
	$-CHNH$	Formimino
	$-CH=$	Methenyl

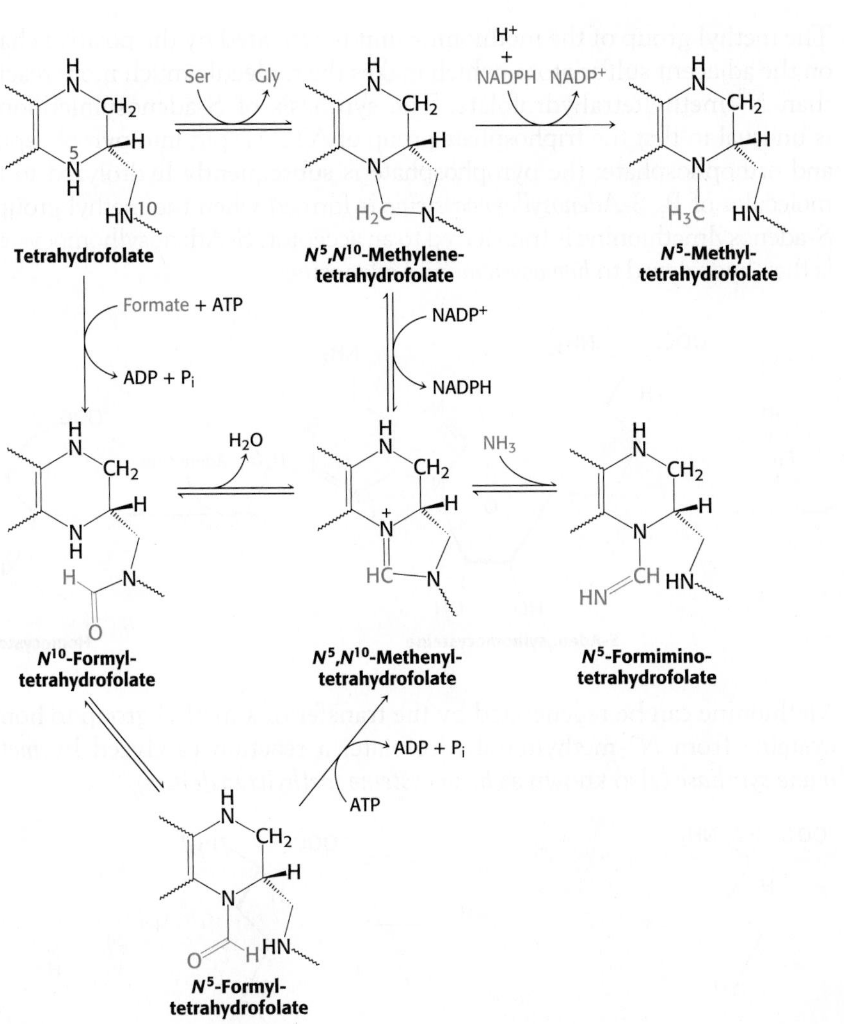

FIGURE 24.10 Conversions of one-carbon units attached to tetrahydrofolate.

one-carbon units in degradative reactions. The major source of one-carbon units is the facile conversion of serine into glycine by serine hydroxymethyltransferase (p. 797), which yields N^5,N^{10}-methylenetetrahydrofolate. Serine can be derived from 3-phosphoglycerate, and so *this pathway enables one-carbon units to be formed de novo from carbohydrates.*

S-Adenosylmethionine is the major donor of methyl groups

Tetrahydrofolate can carry a methyl group on its N-5 atom, but its transfer potential is not sufficiently high for most biosynthetic methylations. Rather, the activated methyl donor is usually *S-adenosylmethionine* (SAM), which is synthesized by the transfer of an adenosyl group from ATP to the sulfur atom of methionine.

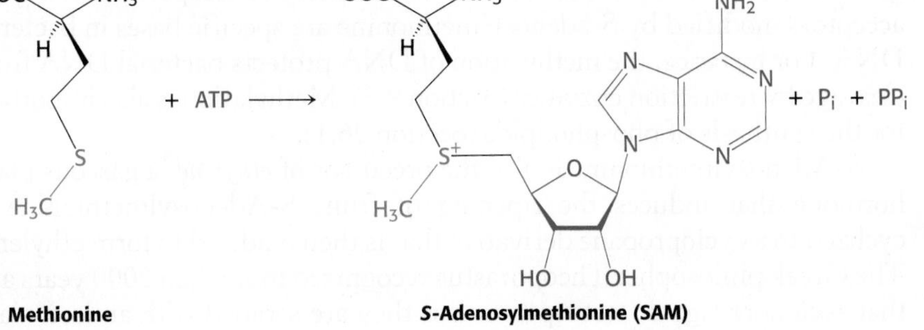

The methyl group of the methionine unit is activated by the positive charge on the adjacent sulfur atom, which makes the molecule much more reactive than N^5-methyltetrahydrofolate. The synthesis of S-adenosylmethionine is unusual in that the triphosphate group of ATP is split into pyrophosphate and orthophosphate; the pyrophosphate is subsequently hydrolyzed to two molecules of P_i. S-*Adenosylhomocysteine* is formed when the methyl group of S-adenosylmethionine is transferred to an acceptor. S-Adenosylhomocysteine is then hydrolyzed to *homocysteine* and adenosine.

S-Adenosylmethionine (SAM) → **S-Adenosylhomocysteine** → **Homocysteine**

Methionine can be regenerated by the transfer of a methyl group to homocysteine from N^5-methyltetrahydrofolate, a reaction catalyzed by *methionine synthase* (also known as *homocysteine methyltransferase*).

Homocysteine + **N^5-Methyl-tetrahydrofolate** → **Methionine** + **Tetrahydrofolate**

The coenzyme that mediates this transfer of a methyl group is *methylcobalamin,* derived from vitamin B_{12}. In fact, this reaction and the rearrangement of L-methylmalonyl CoA to succinyl CoA (p. 721), catalyzed by a homologous enzyme, are the only two B_{12}-dependent reactions known to take place in mammals. Another enzyme that converts homocysteine into methionine without vitamin B_{12} also is present in many organisms.

These reactions constitute the *activated methyl cycle* (Figure 24.11). Methyl groups enter the cycle in the conversion of homocysteine into methionine and are then made highly reactive by the addition of an adenosyl group, which makes the sulfur atoms positively charged and the methyl groups much more electrophilic. The high transfer potential of the S-methyl group enables it to be transferred to a wide variety of acceptors. Among the acceptors modified by S-adenosylmethionine are specific bases in bacterial DNA. For instance, the methylation of DNA protects bacterial DNA from cleavage by restriction enzymes (Section 9.3). Methylation is also important for the synthesis of phospholipids (Section 26.1).

S-Adenosylmethionine is also the precursor of *ethylene,* a gaseous plant hormone that induces the ripening of fruit. S-Adenosylmethionine is cyclized to a cyclopropane derivative that is then oxidized to form ethylene. The Greek philosopher Theophrastus recognized more than 2000 years ago that sycamore figs do not ripen unless they are scraped with an iron claw.

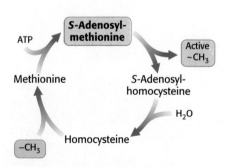

FIGURE 24.11 Activated methyl cycle. The methyl group of methionine is activated by the formation of S-adenosylmethionine.

The reason is now known: *Wounding triggers ethylene production, which in turn induces ripening.*

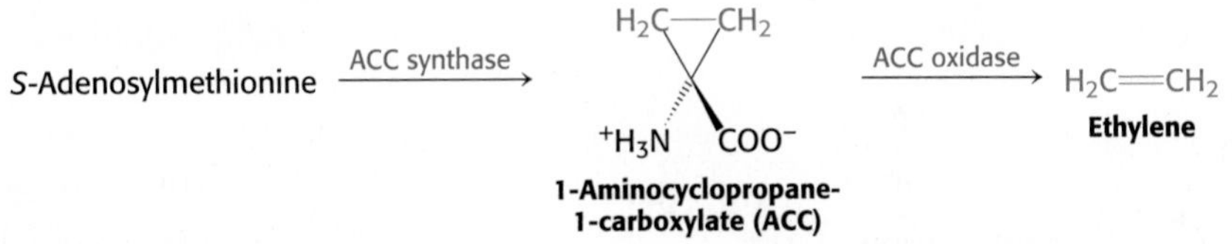

S-Adenosylmethionine is subsequently regenerated from 5′-methyladenosine in five steps.

Cysteine is synthesized from serine and homocysteine

In addition to being a precursor of methionine in the activated methyl cycle, homocysteine is an intermediate in the synthesis of cysteine. Serine and homocysteine condense to form *cystathionine.* This reaction is catalyzed by *cystathionine β-synthase.* Cystathionine is then deaminated and cleaved to cysteine and α-ketobutyrate by *cystathionine γ-lyase* or *cystathionase.* Both of these enzymes utilize PLP and are homologous to aspartate aminotransferase. The net reaction is

$$\text{Homocysteine + serine} \rightleftharpoons \text{cysteine} + \alpha\text{-ketobutyrate} + NH_4^+$$

Note that the sulfur atom of cysteine is derived from homocysteine, whereas the carbon skeleton comes from serine.

High homocysteine levels correlate with vascular disease

People with elevated serum levels of homocysteine (homocysteinemia) or the disulfide-linked dimer homocystine (homocystinuria) have an unusually high risk for coronary heart disease and arteriosclerosis. The most common genetic cause of high homocysteine levels is a mutation within the gene encoding cystathionine β-synthase. High levels of homocysteine appear to damage cells lining blood vessels and to increase the growth of vascular smooth muscle. The amino acid raises oxidative stress as well and has also been implicated in the development of type 2 diabetes (Section 27.3). The molecular basis of homocysteine's action has not been clearly identified, but may result from stimulation of the inflammatory response. Vitamin treatments are sometimes effective in reducing homocysteine levels in some people. Treatment with vitamins maximizes the activity of the two major metabolic pathways processing homocysteine. Pyridoxal phosphate, a vitamin B_6 derivative, is necessary for the activity of cystathionine β-synthase, which converts homocysteine into cystathione; tetrahydrofolate, as well as vitamin B_{12}, supports the methylation of homocysteine to methionine.

Shikimate and chorismate are intermediates in the biosynthesis of aromatic amino acids

We turn now to the biosynthesis of essential amino acids. These amino acids are synthesized by plants and microorganisms, and those in the human diet are ultimately derived primarily from plants. The essential amino acids are formed by much more complex routes than are the nonessential amino acids. The pathways for the synthesis of aromatic amino acids in bacteria have been selected for discussion here because they are well understood and exemplify recurring mechanistic motifs.

Phenylalanine, tyrosine, and tryptophan are synthesized by a common pathway in *E. coli* (Figure 24.12). The initial step is the condensation of phosphoenolpyruvate (a glycolytic intermediate) with erythrose 4-phosphate (a pentose phosphate pathway intermediate). The resulting seven-carbon

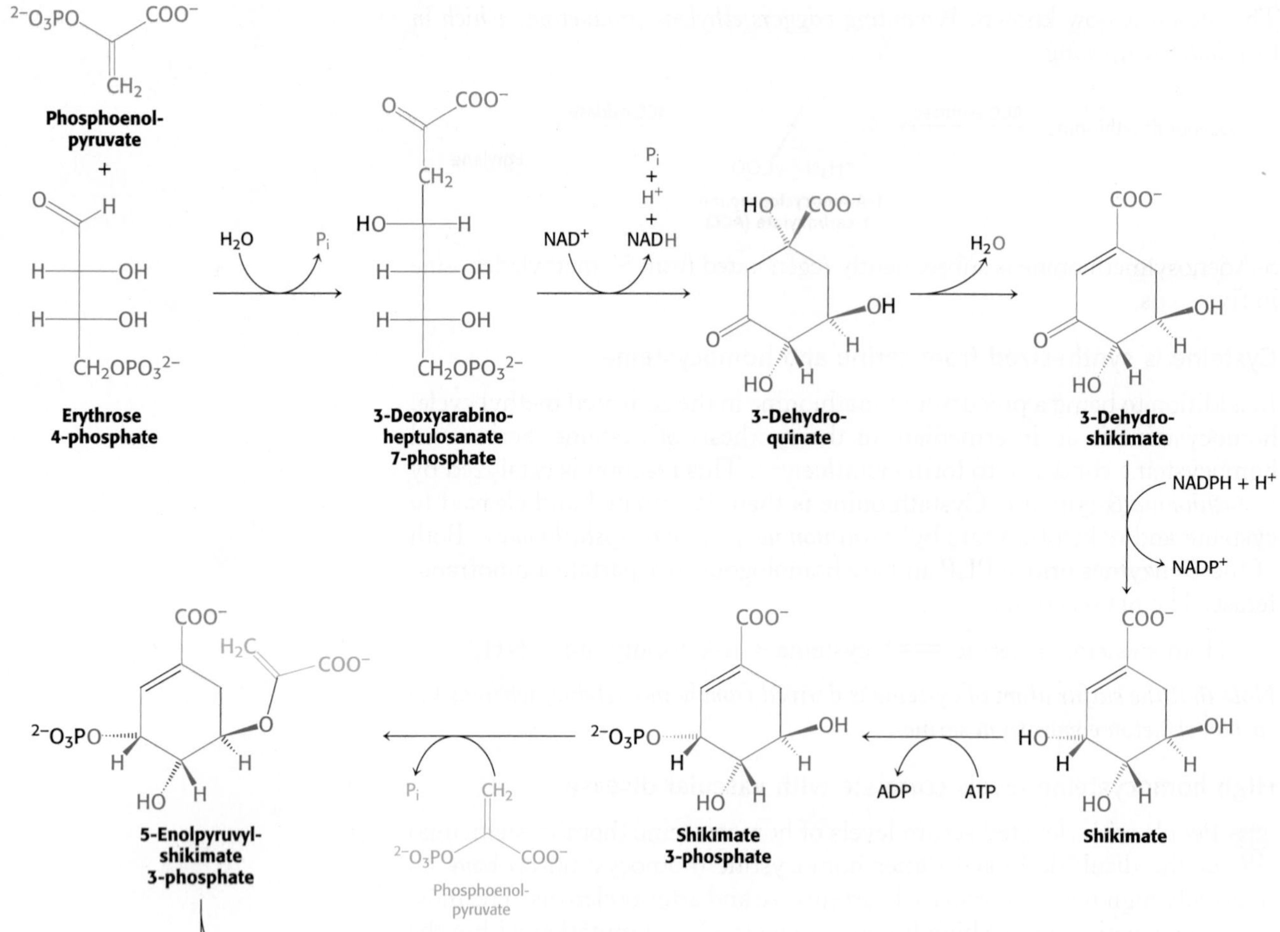

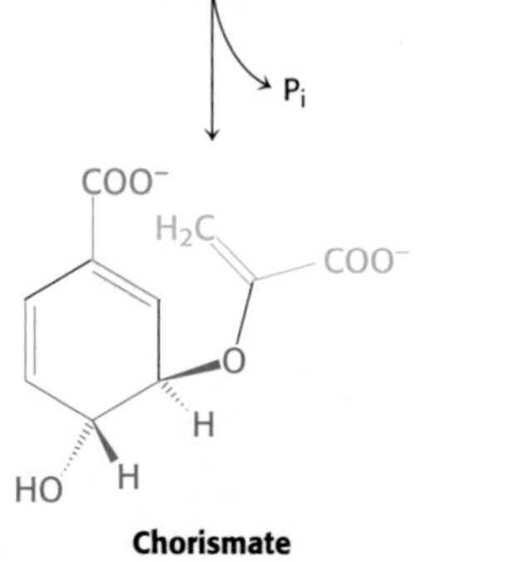

FIGURE 24.12 Pathway to chorismate. Chorismate is an intermediate in the biosynthesis of phenylalanine, tyrosine, and tryptophan.

open-chain sugar is oxidized, loses its phosphoryl group, and cyclizes to 3-dehydroquinate. Dehydration then yields 3-dehydroshikimate, which is reduced by NADPH to shikimate. The phosphorylation of shikimate by ATP gives shikimate 3-phosphate, which condenses with a second molecule of phosphoenolpyruvate. The resulting 5-enolpyruvyl intermediate loses its phosphoryl group, yielding chorismate, the common precursor of all three aromatic amino acids. The importance of this pathway is revealed by the effectiveness of glyphosate (commercially known as Roundup), a broad-spectrum herbicide. This compound is an uncompetitive inhibitor of the enzyme that produces 5-enolpyruvylshikimate 3-phosphate. It blocks aromatic amino acid biosynthesis in plants but is fairly nontoxic in animals because they lack the enzyme. Recently, there have been claims that glyphosate may cause cancer, although most scientific evidence does not support the claims.

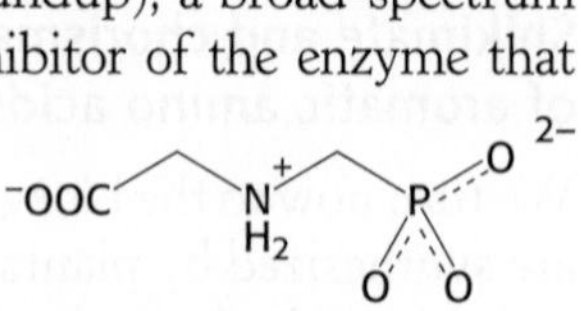

Glyphosate
(Roundup)

The pathway bifurcates at chorismate. Let us first follow the *prephenate branch* (Figure 24.13). A mutase converts chorismate into prephenate, the immediate precursor of the aromatic ring of phenylalanine and tyrosine. This fascinating conversion is a rare example of an electrocyclic reaction in biochemistry, mechanistically similar to the well-known Diels–Alder reaction in organic chemistry. Dehydration and decarboxylation yield *phenylpyruvate*. Alternatively, prephenate can be oxidatively decarboxylated to

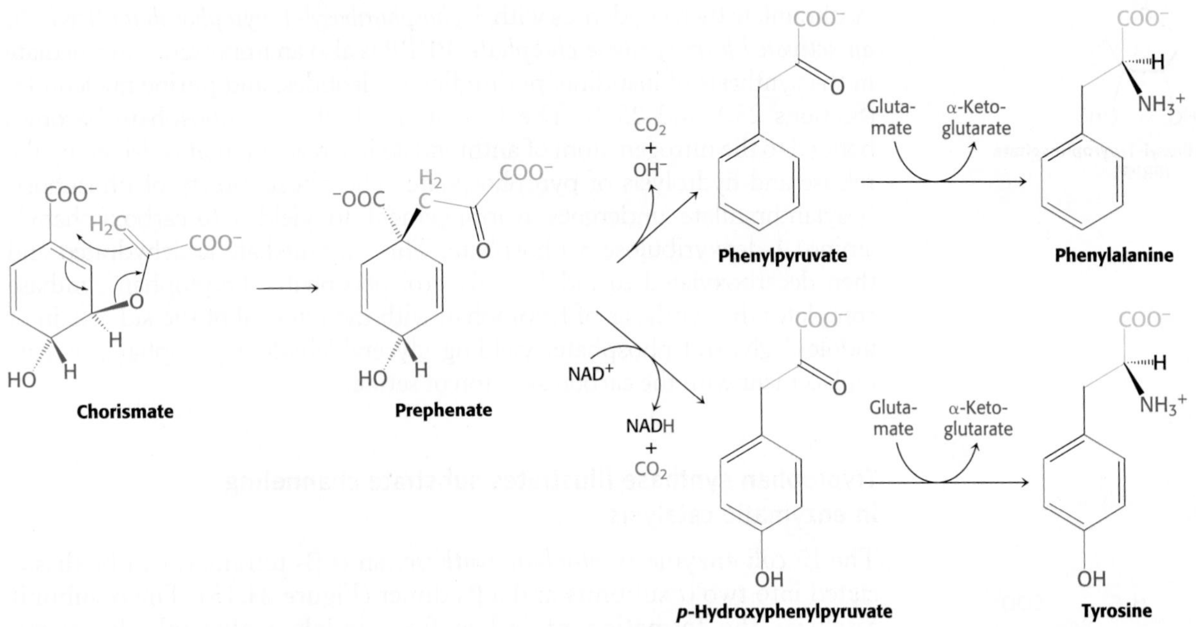

FIGURE 24.13 Synthesis of phenylalanine and tyrosine. Chorismate can be converted into prephenate, which is subsequently converted into phenylalanine and tyrosine.

p-*hydroxyphenylpyruvate*. These α-ketoacids are then transaminated to form *phenylalanine* and *tyrosine*.

The branch starting with *anthranilate* leads to the synthesis of tryptophan (Figure 24.14). Chorismate acquires an amino group derived from the hydrolysis of the side chain of glutamine and releases pyruvate to form anthranilate.

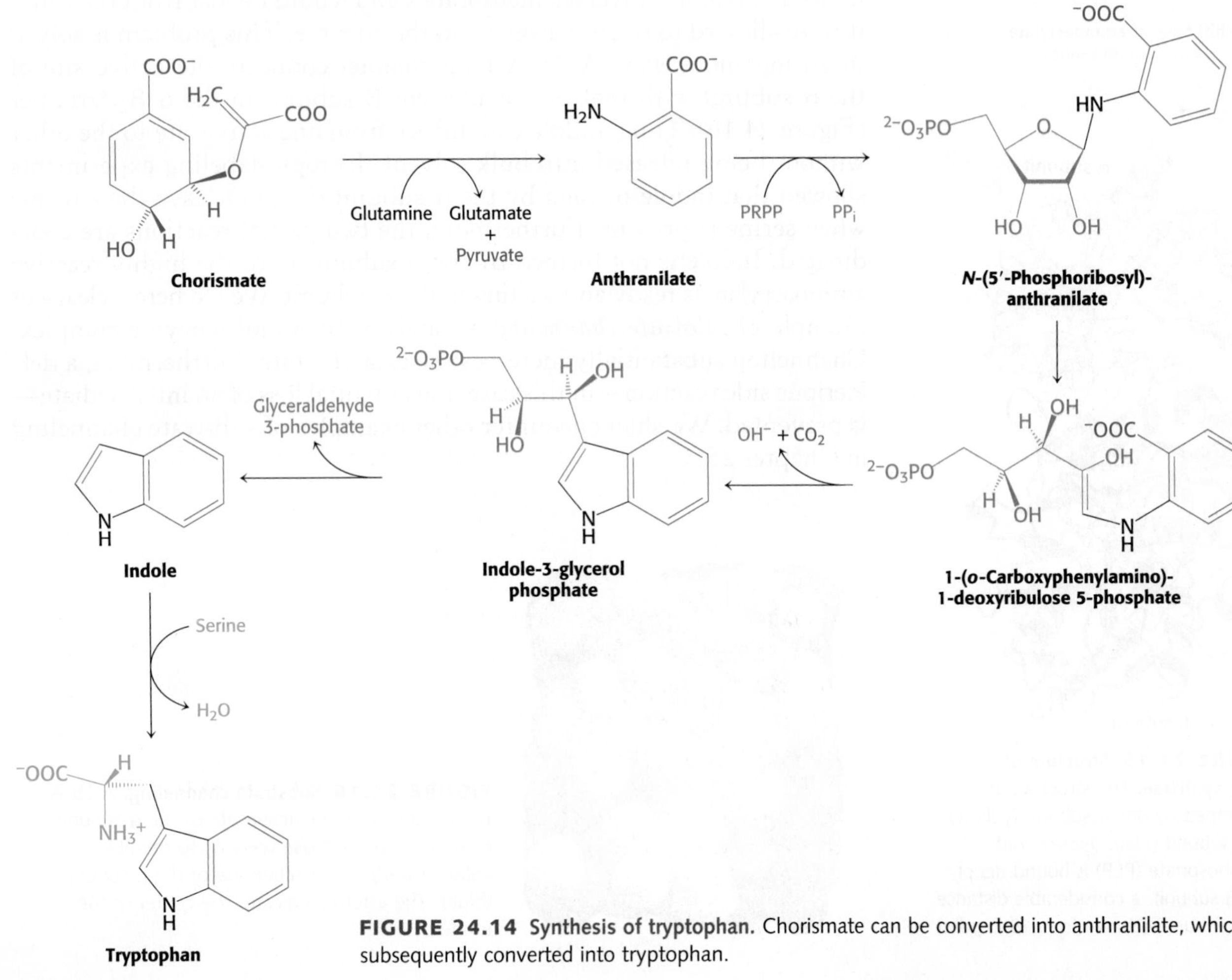

FIGURE 24.14 Synthesis of tryptophan. Chorismate can be converted into anthranilate, which is subsequently converted into tryptophan.

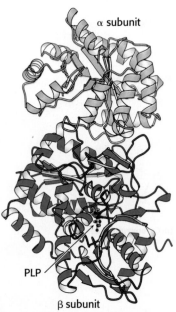

5-Phosphoribosyl-1-pyrophosphate
(PRPP)

Indole

Schiff base of aminoacrylate
(derived from serine)

α subunit

PLP

β subunit

FIGURE 24.15 **Structure of**
tryptophan synthase. The structure of the
complex formed by one α subunit (yellow)
and one β subunit (blue). *Notice* that
pyridoxal phosphate (PLP) is bound deeply
inside the β subunit, a considerable distance
from the α subunit. [Drawn from 1BKS.pdb.]

Anthranilate then condenses with *5-phosphoribosyl-1-pyrophosphate (PRPP)*, *an activated form of ribose phosphate*. PRPP is also an important intermediate in the synthesis of histidine, pyrimidine nucleotides, and purine nucleotides (Sections 25.1 and 25.2). The C-1 atom of ribose 5-phosphate becomes bonded to the nitrogen atom of anthranilate in a reaction that is driven by the release and hydrolysis of pyrophosphate. The ribose moiety of phosphoribosylanthranilate undergoes rearrangement to yield 1-(*o*-carboxyphenylamino)-1-deoxyribulose 5-phosphate. This intermediate is dehydrated and then decarboxylated to indole-3-glycerol phosphate. Tryptophan synthase completes the synthesis of tryptophan with the removal of the side chain of indole-3-glycerol phosphate, yielding glyceraldehyde 3-phosphate, and its replacement with the carbon skeleton of serine.

Tryptophan synthase illustrates substrate channeling in enzymatic catalysis

The *E. coli* enzyme *tryptophan synthase*, an $\alpha_2\beta_2$ tetramer, can be dissociated into two α subunits and a β_2 dimer (Figure 24.15). The α subunit catalyzes the formation of indole from indole-3-glycerol phosphate, whereas each β subunit has a PLP-containing active site that catalyzes the condensation of indole and serine to form tryptophan. Serine forms a Schiff base with this PLP, which is then dehydrated to give the *Schiff base of aminoacrylate*. This reactive intermediate is attacked by indole to give tryptophan.

The synthesis of tryptophan poses a challenge. Indole, a hydrophobic molecule, readily traverses membranes and would be lost from the cell if it were allowed to diffuse away from the enzyme. This problem is solved in an ingenious way. A 25-Å-long channel connects the active site of the α subunit with that of the adjacent β subunit in the $\alpha_2\beta_2$ tetramer (Figure 24.16). Thus, indole can diffuse from one active site to the other without being released into bulk solvent. Isotopic-labeling experiments showed that indole formed by the α subunit does not leave the enzyme when serine is present. Furthermore, the two partial reactions are coordinated. Indole is not formed by the α subunit until the highly reactive aminoacrylate is ready and waiting in the β subunit. We see here a clear-cut example of *substrate channeling* in catalysis by a multienzyme complex. Channeling substantially increases the catalytic rate. Furthermore, a deleterious side reaction—in this case, the potential loss of an intermediate—is prevented. We shall encounter other examples of substrate channeling in Chapter 25.

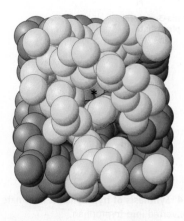

FIGURE 24.16 Substrate channeling. A 25-Å
tunnel runs from the active site of the α subunit
of tryptophan synthase (yellow) to the PLP
cofactor (red) in the active site of the β subunit
(blue). The asterisk indicates the center of the
tunnel.

24.3 Feedback Inhibition Regulates Amino Acid Biosynthesis

The rate of synthesis of amino acids depends mainly on the *amounts* of the biosynthetic enzymes and on their *activities*. We now consider the control of enzymatic activity. The regulation of enzyme synthesis in eukaryotes will be discussed in Chapter 33.

In a biosynthetic pathway, the first irreversible reaction, called the *committed step*, is usually an important regulatory site. The final product of the pathway (Z) often inhibits the enzyme that catalyzes the committed step (A → B).

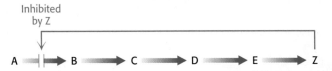

This kind of control is essential for the conservation of building blocks and metabolic energy. Consider the biosynthesis of serine (p. 797). The committed step in this pathway is the oxidation of 3-phosphoglycerate, catalyzed by the enzyme *3-phosphoglycerate dehydrogenase*. The *E. coli* enzyme is a tetramer of four identical subunits, each comprising a catalytic domain and a serine-binding regulatory domain (Figure 24.17). The binding of serine to a regulatory site reduces the value of V_{max} for the enzyme; an enzyme bound to four molecules of serine is essentially inactive. Thus, if serine is abundant in the cell, the enzyme activity is inhibited, and so 3-phosphoglycerate, a key building block that can be used for other processes, is not wasted.

Branched pathways require sophisticated regulation

The regulation of branched pathways is more complicated because the concentration of two products must be accounted for. In fact, several intricate feedback mechanisms have been found in branched biosynthetic pathways.

Feedback inhibition and activation. Two pathways with a common initial step may each be inhibited by its own product and activated by the product of the other pathway. Consider, for example, the biosynthesis of the branched chain amino acids valine, leucine, and isoleucine. A common intermediate, hydroxyethyl thiamine pyrophosphate (hydroxyethyl-TPP; Section 17.1), initiates the pathways leading to all three of these amino acids. Hydroxyethyl-TPP reacts with α-ketobutyrate in the initial step for the synthesis of isoleucine. Alternatively, hydroxyethyl-TPP reacts with pyruvate in the committed step for the pathways leading to valine and leucine. Thus, the relative concentrations of α-ketobutyrate and pyruvate determine how much isoleucine is produced compared with valine and leucine. *Threonine deaminase*, the PLP enzyme that catalyzes the formation of α-ketobutyrate, is allosterically inhibited by isoleucine (Figure 24.18). This enzyme is also allosterically activated by valine. Thus, this enzyme is inhibited by the end product of the pathway that it initiates and is activated by the end product of a competitive pathway. This mechanism balances the amounts of different amino acids that are synthesized.

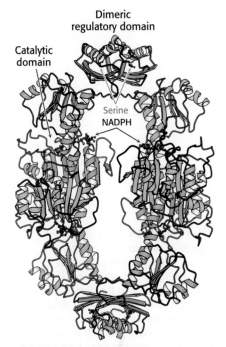

FIGURE 24.17 Structure of 3-phosphoglycerate dehydrogenase. This enzyme, which catalyzes the committed step in the serine biosynthetic pathway, is inhibited by serine. *Notice* the two serine-binding dimeric regulatory domains—one at the top and the other at the bottom of the structure. [Drawn from 1PSD.pdb.]

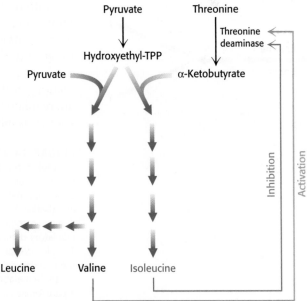

FIGURE 24.18 Regulation of threonine deaminase. Threonine is converted into α-ketobutyrate in the committed step, leading to the synthesis of isoleucine. The enzyme that catalyzes this step, threonine deaminase, is inhibited by isoleucine and activated by valine, the product of a parallel pathway.

The regulatory domain in threonine deaminase is very similar in structure to the regulatory domain in 3-phosphoglycerate dehydrogenase (Figure 24.19). In the latter enzyme, regulatory domains of two subunits interact to form a dimeric serine-binding regulatory unit, and so the tetrameric enzyme contains two such regulatory units. Each unit is capable of binding two serine molecules. In threonine deaminase, the two regulatory domains are fused into a single unit with two differentiated amino acid-binding sites, one for isoleucine and the other for valine. Sequence analysis shows that similar regulatory domains are present in other amino acid biosynthetic enzymes. *The similarities suggest that feedback-inhibition processes may have evolved by the linkage of specific regulatory domains to the catalytic domains of biosynthetic enzymes.*

FIGURE 24.19 A recurring regulatory domain. The regulatory domain formed by two subunits of 3-phosphoglycerate dehydrogenase is structurally related to the single-chain regulatory domain of threonine deaminase. *Notice* that both structures have four α helices and eight β strands in similar locations. Sequence analyses have revealed this amino acid-binding regulatory domain to be present in other enzymes as well. [Drawn from 1PSD and 1TDJ.pdb.]

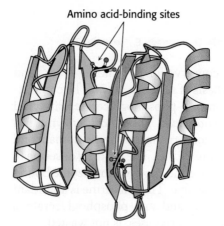

Amino acid-binding sites

Dimeric regulatory domain of 3-phosphoglycerate dehydrogenase

Single-chain regulatory domain of threonine deaminase

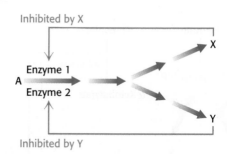

Inhibited by X

Enzyme 1

A

Enzyme 2

X

Y

Inhibited by Y

Enzyme multiplicity. The committed step can be catalyzed by two or more isozymes, enzymes with essentially identical catalytic mechanisms, but different regulatory properties (Section 10.2). For example, the phosphorylation of aspartate is the committed step in the biosynthesis of threonine, methionine, and lysine. Three distinct aspartokinases, which evolved by gene duplication, catalyze this reaction in *E. coli* (Figure 24.20). The catalytic domains of these enzymes show approximately 30% sequence identity. Although the mechanisms of catalysis are the same, their activities are regulated differently: one enzyme is not subject to feedback inhibition, another is inhibited by threonine, and the third is inhibited by lysine.

FIGURE 24.20 Domain structures of three aspartokinases. Each catalyzes the committed step in the biosynthesis of a different amino acid: (top) methionine, (middle) threonine, and (bottom) lysine. They have a catalytic domain in common (red) but differ in their regulatory domains (yellow and orange). The blue domain represents another enzyme (homoserine dehydrogenase) involved in aspartate metabolism. Thus, the top two aspartokinases are bifunctional enzymes.

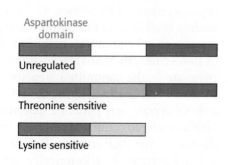

Aspartokinase domain

Unregulated

Threonine sensitive

Lysine sensitive

Cumulative feedback inhibition. A common step is partly inhibited by multiple final products, each acting independently. The regulation of glutamine synthetase in *E. coli* is a striking example of cumulative feedback inhibition. Recall that glutamine is synthesized from glutamate, NH_4^+, and ATP.

Glutamine synthetase consists of 12 identical 50-kDa subunits arranged in two hexagonal rings that face each other. This enzyme regulates the flow of nitrogen and hence plays a key role in controlling bacterial metabolism. The amide group of glutamine is a source of nitrogen in the biosyntheses of a variety of compounds, such as tryptophan, histidine, carbamoyl phosphate, glucosamine 6-phosphate, cytidine triphosphate, and adenosine monophosphate. Glutamine synthetase is cumulatively inhibited by each of these final products of glutamine metabolism, as well as by alanine and glycine. *In cumulative inhibition, each inhibitor can reduce the activity of the enzyme, even when other inhibitors are bound at saturating levels.* The enzymatic activity of glutamine synthetase is switched off almost completely when all final products are bound to the enzyme.

The sensitivity of glutamine synthetase to allosteric regulation is altered by covalent modification

The activity of glutamine synthetase is also controlled by *reversible covalent modification*—the attachment of an *AMP unit* by a phosphodiester linkage to the hydroxyl group of a specific tyrosine residue in each subunit. *This adenylylated enzyme is less active and more susceptible to cumulative feedback inhibition than is the deadenylylated form.* The covalently attached AMP unit is removed from the adenylylated enzyme by phosphorolysis.

The adenylylation and phosphorolysis reactions are catalyzed by the same enzyme, *adenylyl transferase*. The adenylyl transferase is composed of two homologous halves, one half catalyzing the adenylylation reaction and the other half the phosphorolytic deadenylylation reaction. What determines whether an AMP unit is added or removed? The specificity of adenylyl transferase is controlled by a *regulatory protein* P_{II}, a trimeric protein that can exist in two forms, unmodified (P_{II}) or covalently bound to UMP (P_{II}-UMP). The complex of P_{II} and adenylyl transferase catalyzes the attachment of an AMP unit to glutamine synthetase, which reduces its activity. Conversely, the complex of P_{II}-UMP and adenylyl transferase removes AMP from the adenylylated enzyme (Figure 24.21).

This scheme of regulation immediately raises the question, how is the modification of P_{II} controlled? P_{II} is converted into P_{II}-UMP by the attachment of uridine monophosphate to a specific tyrosine

Tyrosine residue modified by adenylylation

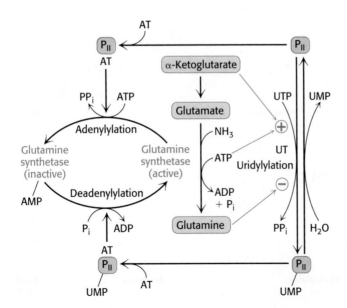

FIGURE 24.21 Covalent regulation of glutamine synthetase. Adenylyl transferase (AT), in association with the regulatory protein P_{II}, adenylylates and inactivates the synthetase. When associated with P_{II} bound to UMP, AT deadenylylates the synthetase, thereby activating the enzyme. Uridylyl transferase (UT), the enzyme that modifies P_{II}, is allosterically regulated by α-ketoglutarate, ATP, and glutamine. [Information from D. L. Nelson and M. M. Cox, *Lehninger Principles of Biochemistry* 7th ed. (W. H. Freeman and Company, 2013), Fig. 22.9.]

residue (Figure 24.21). This reaction, which is catalyzed by *uridylyl transferase,* is stimulated by ATP and α-ketoglutarate, whereas it is inhibited by glutamine. In turn, the UMP units on P_{II} are removed by hydrolysis, a reaction promoted by glutamine and inhibited by α-ketoglutarate. These opposing catalytic activities are present on a single polypeptide chain and are controlled so that the enzyme does not simultaneously catalyze uridylylation and hydrolysis. In essence, if glutamine is present, the covalent modification system favors adenylylation and inactivation of glutamine synthetase. If glutamine is absent, as indicated by the presence of its precursors α-ketoglutarate and ATP, the control system results in the deadenylylation and activation of the synthetase.

The integration of nitrogen metabolism in a cell requires that a large number of input signals be detected and processed. In addition, the regulatory protein P_{II} also participates in regulating the transcription of genes for glutamine synthetase and other enzymes taking part in nitrogen metabolism. The evolution of covalent regulation superimposed on feedback inhibition provided many more regulatory sites and made possible a finer tuning of the flow of nitrogen in the cell. We have previously seen such a dual regulatory format in the regulation of glycogen metabolism (Section 21.5).

24.4 Amino Acids Are Precursors of Many Biomolecules

In addition to being the building blocks of proteins and peptides, amino acids serve as precursors of many kinds of small molecules that have important and diverse biological roles. Let us briefly survey some of the biomolecules that are derived from amino acids (Figure 24.22).

Purines and *pyrimidines* are derived largely from amino acids. The biosynthesis of these precursors of DNA, RNA, and numerous coenzymes will be discussed in detail in Chapter 25. The reactive terminus of *sphingosine,* an intermediate in the synthesis of sphingolipids, comes from serine. *Histamine,* a potent vasodilator, is derived from histidine by decarboxylation. Tyrosine is a precursor of *thyroxine* (tetraiodothyronine, a hormone that modulates metabolism), *epinephrine* (adrenaline), and *melanin* (a complex polymeric molecule responsible for skin pigmentation). The neurotransmitter *serotonin* (5-hydroxytryptamine) and the *nicotinamide ring* of NAD^+

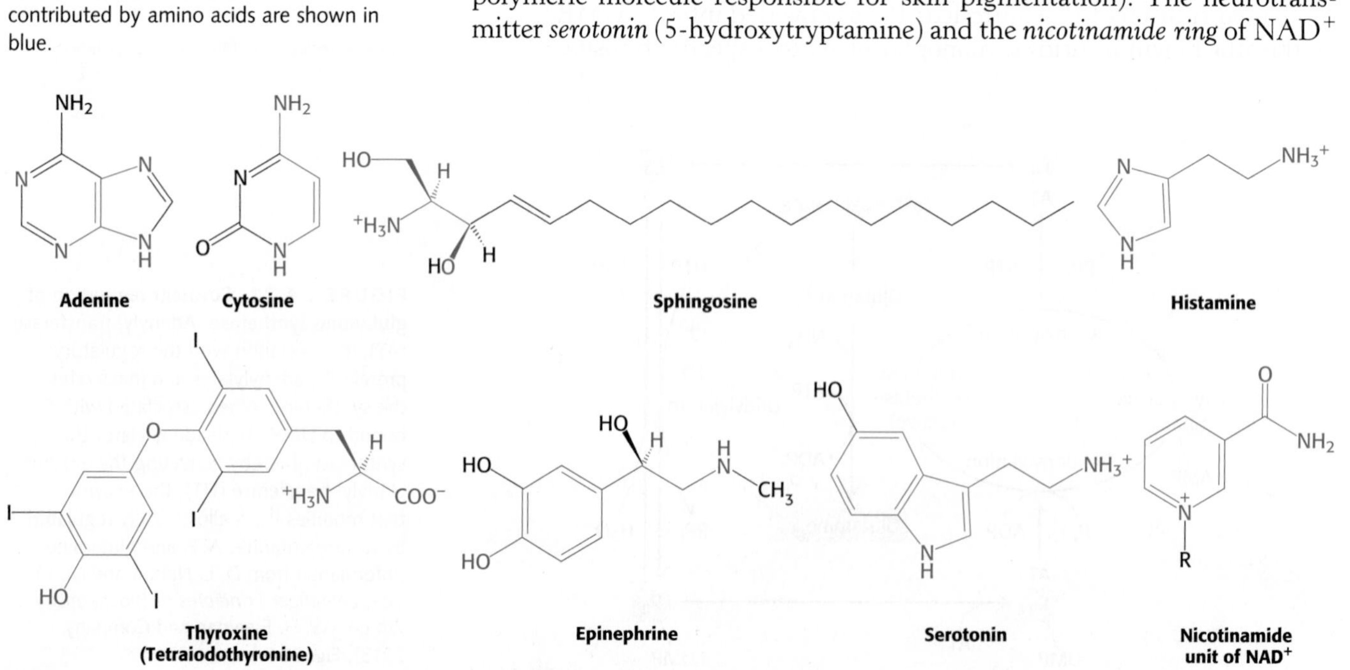

FIGURE 24.22 Selected biomolecules derived from amino acids. The atoms contributed by amino acids are shown in blue.

Adenine   Cytosine   Sphingosine   Histamine

Thyroxine (Tetraiodothyronine)   Epinephrine   Serotonin   Nicotinamide unit of NAD$^+$

are synthesized from tryptophan. Let us now consider in more detail three particularly important biochemicals derived from amino acids. (See Biochemistry in Focus for a discussion of melanin synthesis.)

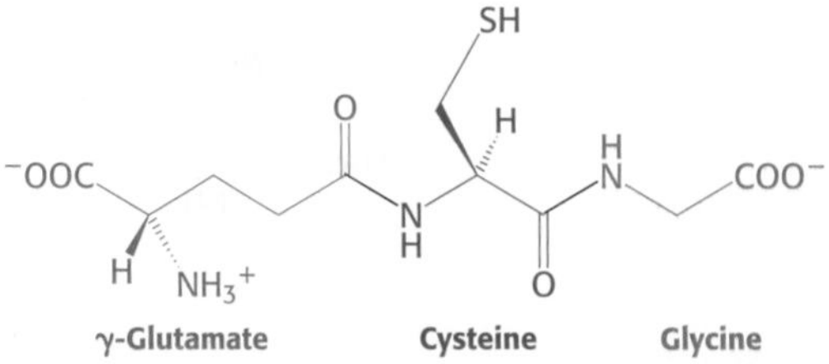

Glutathione, a gamma-glutamyl peptide, serves as a sulfhydryl buffer and an antioxidant

Glutathione, a tripeptide containing a sulfhydryl group, is a highly distinctive amino acid derivative with several important roles (Figure 24.23). For example, glutathione, present at high levels (~5 mM) in animal cells, protects red blood cells from oxidative damage by serving as a sulfhydryl buffer (Section 20.5). It cycles between a reduced thiol form (GSH) and an oxidized form (GSSG) in which two tripeptides are linked by a disulfide bond.

$$2\,GSH + RO\!-\!OH \rightleftharpoons GSSG + H_2O + ROH$$

GSSG is reduced to GSH by *glutathione reductase*, a flavoprotein that uses NADPH as the electron source. The ratio of GSH to GSSG in most cells is greater than 500. *Glutathione plays a key role in detoxification by reacting with hydrogen peroxide and organic peroxides, the harmful by-products of aerobic life* (Section 20.5).

Glutathione peroxidase, the enzyme catalyzing the reaction with peroxides, is remarkable in having a modified amino acid containing a *selenium* (Se) atom (Figure 24.24). Specifically, its active site contains the selenium analog of cysteine, in which selenium has replaced sulfur. The selenolate (E-Se$^-$) form of this residue reduces the peroxide substrate to an alcohol and is in turn oxidized to selenenic acid (E-SeOH). Glutathione then comes into action by forming a selenosulfide adduct (E-Se-S-G). A second molecule of glutathione then regenerates the active form of the enzyme by attacking the selenosulfide to form oxidized glutathione (Figure 24.25).

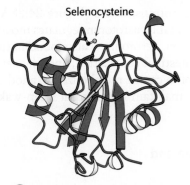

FIGURE 24.24 Structure of glutathione peroxidase. This enzyme, which has a role in peroxide detoxification, contains a selenocysteine residue in its active site. [Drawn from 1GP1.pdb.]

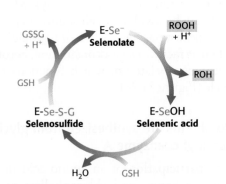

FIGURE 24.25 Catalytic cycle of glutathione peroxidase. [Information from O. Epp, R. Ladenstein, and A. Wendel. *Eur. J. Biochem.* 133(1983):51–69.]

Nitric oxide, a short-lived signal molecule, is formed from arginine

Nitric oxide (NO) is an important messenger in many vertebrate signal-transduction processes, identified first as a relaxing factor in the cardiovascular system. It is now known to have a variety of roles not only in the cardiovascular system, but also in the immune and nervous systems (Table 24.3). NO has also been shown to stimulate mitochondrial biogenesis.

FIGURE 24.23 Glutathione. This tripeptide consists of a cysteine residue flanked by a glycine residue and a glutamate residue that is linked to cysteine by an isopeptide bond between glutamate's side-chain carboxylate group and cysteine's amino group.

^{15}N labeling: A pioneer's account
"Myself as a Guinea Pig
 . . . in 1944, I undertook, together with David Rittenberg, an investigation on the turnover of blood proteins of man. To this end I synthesized 66 g of glycine labeled with 35 percent ^{15}N at a cost of $1000 for the ^{15}N. On 12 February 1945, I started the ingestion of the labeled glycine. Since we did not know the effect of relatively large doses of the stable isotope of nitrogen and since we believed that the maximum incorporation into the proteins could be achieved by the administration of glycine in some continual manner, I ingested 1 g samples of glycine at hourly intervals for the next 66 hours . . . . At stated intervals, blood was withdrawn and after proper preparation the ^{15}N concentrations of different blood proteins were determined."

—David Shemin
Bioessays 10(1989):30

TABLE 24.3 Some functions of nitric oxide
Reduces blood pressure by relaxing vascular muscle
Increases blood flow to the kidney
Acts as a neurotransmitter to increase cerebral blood flow
Dilates pulmonary vessels
Mediates erectile function
Controls peristalsis of the gastrointestinal tract
Regulates inflammation

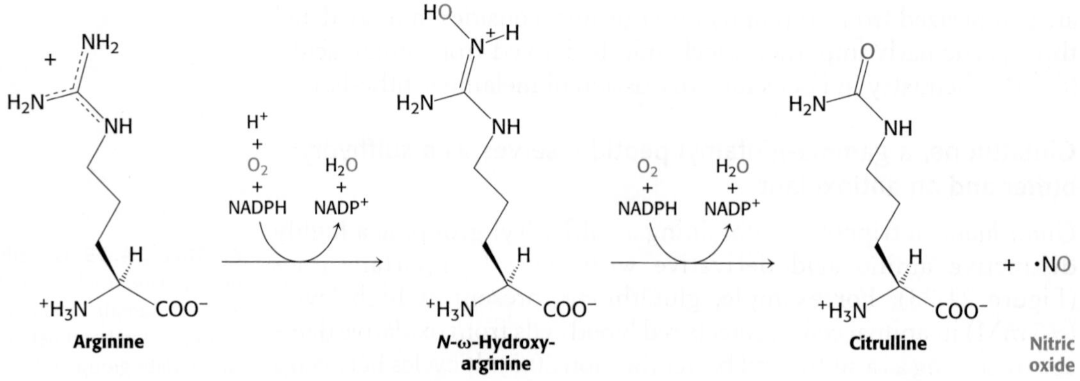

FIGURE 24.26 Formation of nitric oxide. NO is generated by the oxidation of arginine.

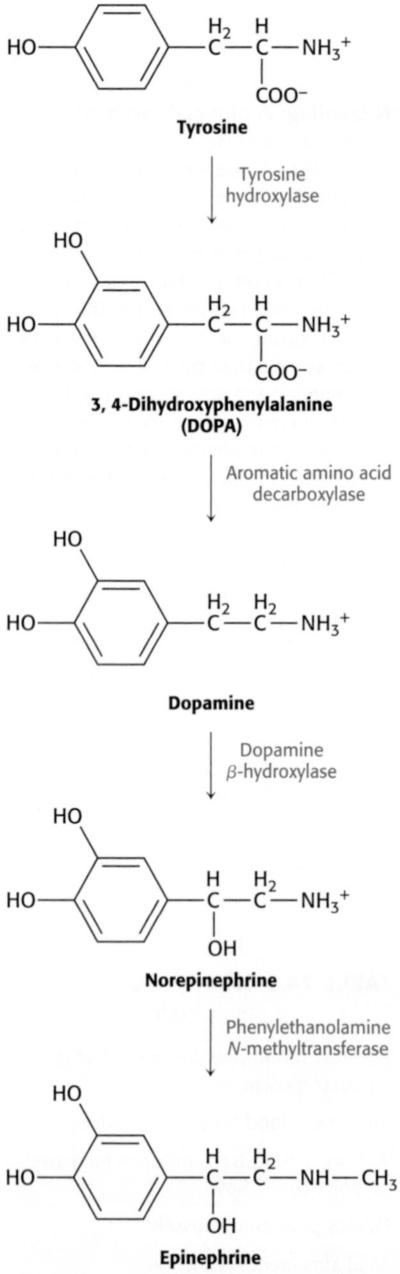

FIGURE 24.27 Tyrosine is a precursor for the catecholamines. The catecholamines—dopamine, norepinephrine, and epinephrine—function as stress-signal molecules.

This free-radical gas is produced endogenously from *arginine* in a complex reaction that is catalyzed by *nitric oxide synthase*. NADPH and O_2 are required for the synthesis of nitric oxide (Figure 24.26). Nitric oxide acts by binding to and activating soluble guanylate cyclase, an important enzyme in signal transduction (See Section 34.3 on SaplingPlus). This enzyme is homologous to adenylate cyclase but includes a heme-containing domain that binds NO.

Amino acids are precursors for a number of neurotransmitters

Catecholamines, signal molecules with a variety of roles, are derived from tyrosine (Figure 24.27). Norepinephrine and dopamine are neurotransmitters, and epinephrine is familiar to us for the regulation of fuel use, stimulating glycogen breakdown and lipid mobilization. In general, the catecholamines are associated with stress and play a role in the "fight or flight" response.

Tryptophan is a precursor for two neurotransmitters (Figure 24.28A). Serotonin functions in several ways. Most notably, serotonin regulates moods and may be involved in alleviating depression. Serotonin is also found in the gastrointestinal tract, where it regulates intestinal contraction, and platelets, where it facilitates vasoconstriction. Serotonin can be metabolized to another hormone, melatonin, which functions to maintain our natural sleep-wake cycle (Figure 24.28B).

Porphyrins are synthesized from glycine and succinyl coenzyme A

The participation of an amino acid in the biosynthesis of the porphyrin rings of hemes and chlorophylls was first revealed by isotope-labeling experiments carried out by David Shemin and his colleagues. In 1945, they showed that the nitrogen atoms of heme were labeled after the feeding of [^{15}N] glycine to human subjects, whereas the ingestion of [^{15}N]glutamate resulted in very little labeling (p. 809).

Experiments using ^{14}C, which had just become available, revealed that 8 of the carbon atoms of heme in nucleated duck erythrocytes are derived from the α-carbon atom of glycine and none from the carboxyl carbon atom. Subsequent studies demonstrated that the other 26 carbon atoms of heme can arise from acetate. Moreover, the ^{14}C in methyl-labeled acetate emerged in 24 of these carbon atoms, whereas the ^{14}C in carboxyl-labeled acetate appeared only in the other 2 (Figure 24.29).

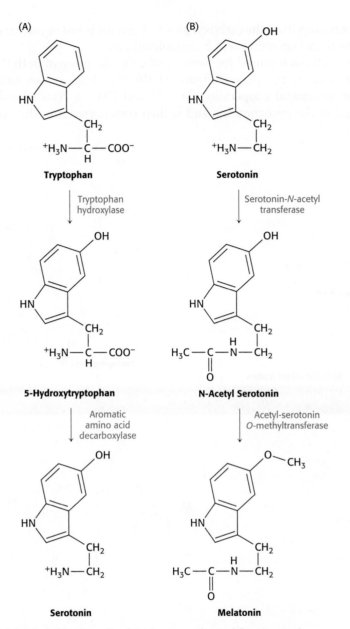

FIGURE 24.28 Serotonin and melatonin are synthesized from tryptophan.

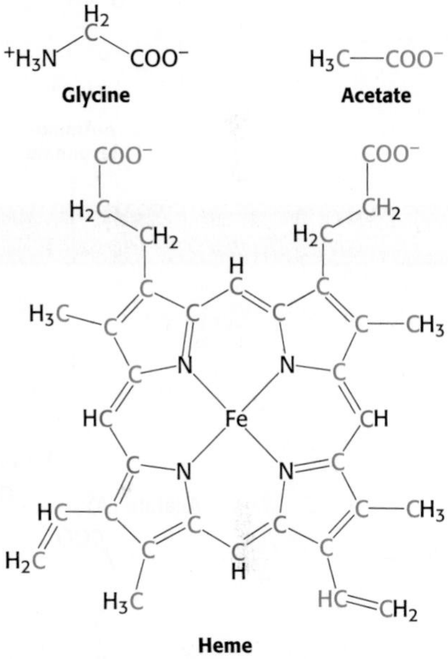

FIGURE 24.29 Heme labeling. The origins of atoms in heme revealed by the results of isotopic labeling studies.

This highly distinctive labeling pattern suggested that acetate is converted to succinyl-CoA through enzymes from the citric acid cycle (Section 17.2) and that a heme precursor is formed by the condensation of glycine with succinyl-CoA. Indeed, *the first step in the biosynthesis of porphyrins in mammals is the condensation of glycine and succinyl CoA to form δ-aminolevulinate.*

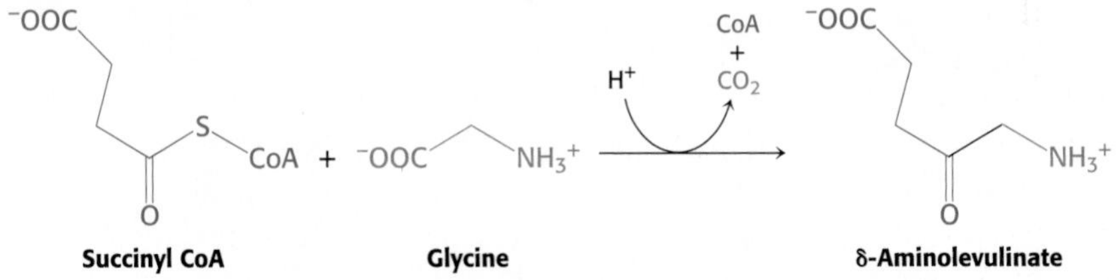

This reaction is catalyzed by *δ-aminolevulinate synthase*, a PLP enzyme present in mitochondria. Consistent with the labeling studies described earlier, the carbon atom from the carboxyl group of glycine is lost as carbon dioxide, while the α-carbon remains in δ-aminolevulinate.

The reactions required for heme synthesis take place in both the mitochondria and the cytoplasm (Figure 24.30), revealing another example of intercompartmental cooperation (p. 522 and 734). δ-Aminolevulinate is generated in the mitochondria and is then transported into the cytoplasm

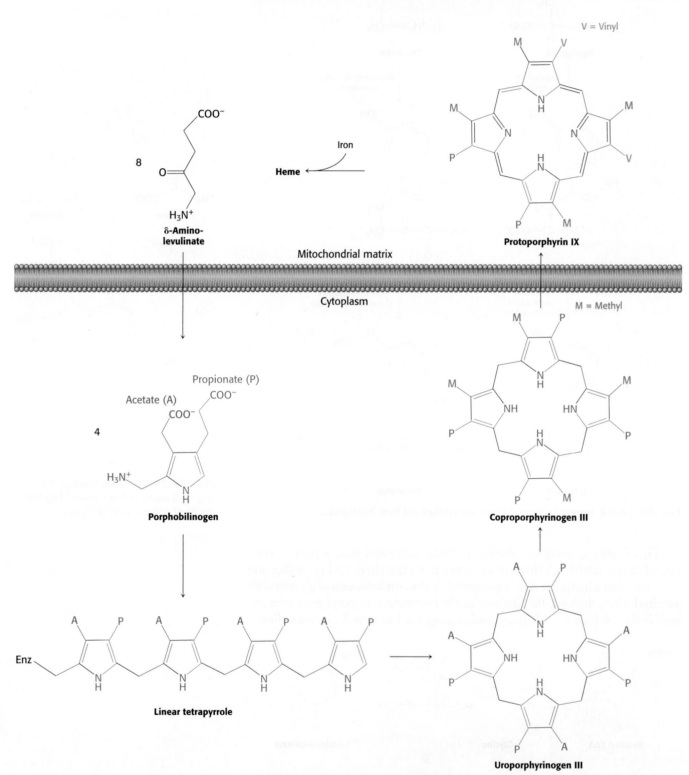

FIGURE 24.30 Heme biosynthetic pathway. The pathway for the formation of heme starts with eight molecules of δ-aminolevulinate and involves inter-compartmental cooperation.

where two molecules of δ-aminolevulinate condense to form *porpho-bilinogen*, the next intermediate. Four molecules of porphobilinogen then condense head to tail to form a linear *tetrapyrrole* in a reaction catalyzed by *porphobilinogen deaminase*. The enzyme-bound linear tetrapyrrole then cyclizes to form *uroporphyrinogen III*, which has an asymmetric arrangement of side chains. This reaction requires a *cosynthase*. In the presence of synthase alone, uroporphyrinogen I, the nonphysiological symmetric isomer, is produced. Uroporphyrinogen III is also a key intermediate in the synthesis of vitamin B_{12} by bacteria and that of chlorophyll by bacteria and plants (Figure 24.30).

The porphyrin skeleton is now formed. Subsequent reactions alter the side chains and the degree of saturation of the porphyrin ring (Figure 24.30). *Coproporphyrinogen III* is formed by the decarboxylation of the acetate side chains and then transported into the mitochondria. There, the desaturation of the porphyrin ring and the conversion of two of the propionate side chains into vinyl groups yield *protoporphyrin IX*. The chelation of iron finally gives *heme*, the prosthetic group of proteins such as myoglobin, hemoglobin, catalase, peroxidase, and cytochrome *c*. The insertion of the *ferrous* form of iron is catalyzed by *ferrochelatase*. Iron is transported in the plasma by *transferrin*, a protein that binds two ferric ions, and is stored in tissues inside molecules of *ferritin* (Section 33.4).

Studies with ^{15}N-labeled glycine revealed that the normal human erythrocyte has a life span of about 120 days. The first step in the degradation of the heme group is the cleavage of its α-methine bridge to form the green pigment *biliverdin*, a linear tetrapyrrole. This reaction is catalyzed by *heme oxygenase*. The central methine bridge of biliverdin is then reduced by *biliverdin reductase* to form *bilirubin*, a red pigment (Figure 24.31). The changing color of a bruise is a highly graphic indicator of these degradative reactions.

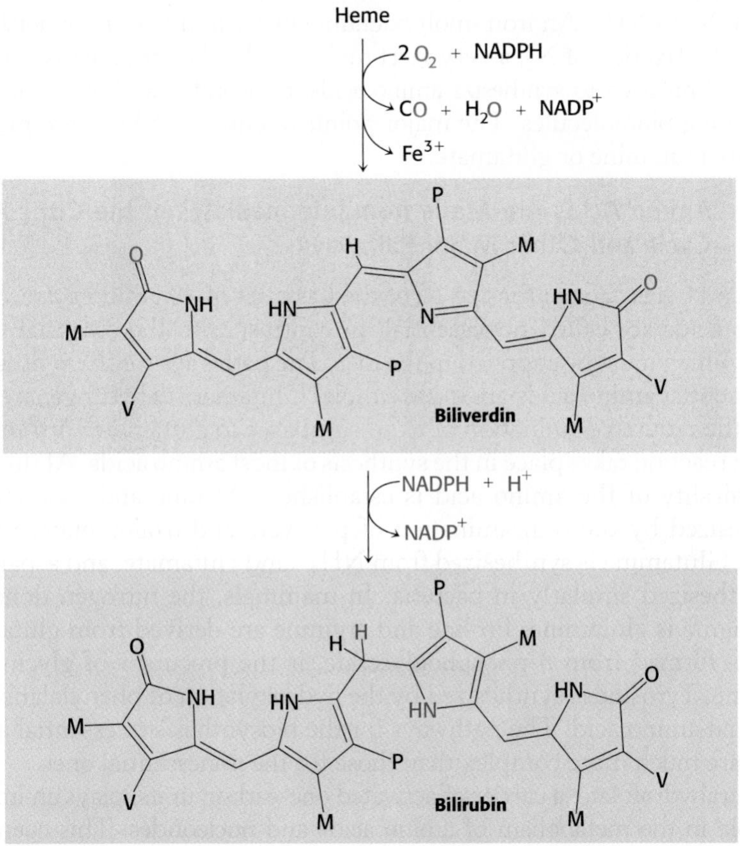

FIGURE 24.31 Heme degradation. The formation of the heme-degradation products biliverdin and bilirubin is responsible for the color of bruises. Abbreviations: M, methyl; V, vinyl.

Porphyrins accumulate in some inherited disorders of porphyrin metabolism

Porphyrias are inherited or acquired disorders caused by a deficiency of enzymes in the heme biosynthetic pathway. Porphyrin is synthesized in both the erythroblasts and the liver, and either one may be the site of a disorder. *Congenital erythropoietic porphyria,* for example, prematurely destroys erythrocytes. This disease results from insufficient cosynthase. In this porphyria, the synthesis of the required amount of uroporphyrinogen III is accompanied by the formation of very large quantities of uroporphyrinogen I, the useless symmetric isomer. Uroporphyrin I, coproporphyrin I, and other symmetric derivatives also accumulate. The urine of patients having this disease is red because of the excretion of large amounts of uroporphyrin I. Their teeth exhibit a strong red fluorescence under ultraviolet light because of the deposition of porphyrins. Furthermore, their skin is usually very sensitive to light because photoexcited porphyrins are quite reactive. *Acute intermittent porphyria* is the most prevalent of the porphyrias affecting the liver. This porphyria is characterized by the overproduction of porphobilinogen and δ-aminolevulinate, which results in severe abdominal pain and neurological dysfunction. The "madness" of George III, king of England during the American Revolution, may have been due to this porphyria.

SUMMARY

24.1 Nitrogen Fixation: Microorganisms Use ATP and a Powerful Reductant to Reduce Atmospheric Nitrogen to Ammonia

Microorganisms use ATP and reduced ferredoxin, a powerful reductant, to reduce N_2 to NH_3. An iron–molybdenum cluster in nitrogenase deftly catalyzes the fixation of N_2, a very inert molecule. Higher organisms consume the fixed nitrogen to synthesize amino acids, nucleotides, and other nitrogen-containing biomolecules. The major points of entry of NH_4^+ into metabolism are glutamine or glutamate.

24.2 Amino Acids Are Made from Intermediates of the Citric Acid Cycle and Other Major Pathways

Human beings can synthesize 11 of the basic set of 20 amino acids. These amino acids are called nonessential, in contrast with the essential amino acids, which must be supplied in the diet. The pathways for the synthesis of nonessential amino acids are quite simple. Glutamate dehydrogenase catalyzes the reductive amination of α-ketoglutarate to glutamate. A transamination reaction takes place in the synthesis of most amino acids. At this step, the chirality of the amino acid is established. Alanine and aspartate are synthesized by the transamination of pyruvate and oxaloacetate, respectively. Glutamine is synthesized from NH_4^+ and glutamate, and asparagine is synthesized similarly in bacteria. In mammals, the nitrogen donor for asparagine is glutamine. Proline and arginine are derived from glutamate. Serine, formed from 3-phosphoglycerate, is the precursor of glycine and cysteine. Tyrosine is synthesized by the hydroxylation of phenylalanine, an essential amino acid. The pathways for the biosynthesis of essential amino acids are much more complex than those for the nonessential ones.

Tetrahydrofolate, a carrier of activated one-carbon units, plays an important role in the metabolism of amino acids and nucleotides. This coenzyme carries one-carbon units at three oxidation states, which are interconvertible: most reduced—methyl; intermediate—methylene; and most oxidized—formyl, formimino, and methenyl. The major donor of activated methyl groups is S-adenosylmethionine, which is synthesized by the transfer of an adenosyl group from ATP to the sulfur atom of methionine. S-Adenosylhomocysteine is formed when the activated methyl group is transferred to an acceptor. It is hydrolyzed to adenosine and homocysteine, and the latter is then methylated to methionine to complete the activated methyl cycle.

24.3   Feedback Inhibition Regulates Amino Acid Biosynthesis

Most of the pathways of amino acid biosynthesis are regulated by feedback inhibition in which the committed step is allosterically inhibited by the final product. The regulation of branched pathways requires extensive interaction among the branches that includes both negative and positive regulation. The regulation of glutamine synthetase in *E. coli* is a striking demonstration of cumulative feedback inhibition and of control by a cascade of reversible covalent modifications.

24.4   Amino Acids Are Precursors of Many Biomolecules

Amino acids are precursors of a variety of biomolecules. Glutathione (γ-Glu-Cys-Gly) serves as a sulfhydryl buffer and detoxifying agent. Glutathione peroxidase, a selenoenzyme, catalyzes the reduction of hydrogen peroxide and organic peroxides by glutathione. Nitric oxide, a short-lived messenger, is formed from arginine. Porphyrins are synthesized from glycine and succinyl CoA, which condense to give δ-aminolevulinate. Two molecules of this intermediate become linked to form porphobilinogen. Four molecules of porphobilinogen combine to form a linear tetrapyrrole, which cyclizes to uroporphyrinogen III. Oxidation and side-chain modifications lead to the synthesis of protoporphyrin IX, which acquires an iron atom to form heme.

A P P E N D I X

Biochemistry in Focus

Tyrosine is a precursor for human pigments

In Biochemistry in Focus in Chapter 8, we discussed the temperature-sensitive mutation in the enzyme tyrosinase that causes the coloration patterns in Siamese cats. In humans, the lack of tyrosinase results in a specific type of albinism called oculocutaneous albinism, type 1. The characteristics of this variety of albinism are milky white skin, white hair, and blue eyes. Because the melanin pigments play a role in the retina and in optic nerve development, these individuals also suffer from a variety of vision problems.

Tyrosinase, a copper-requiring enzyme, catalyzes the first two steps in melanin synthesis, and evidence suggests it may catalyze yet another step.

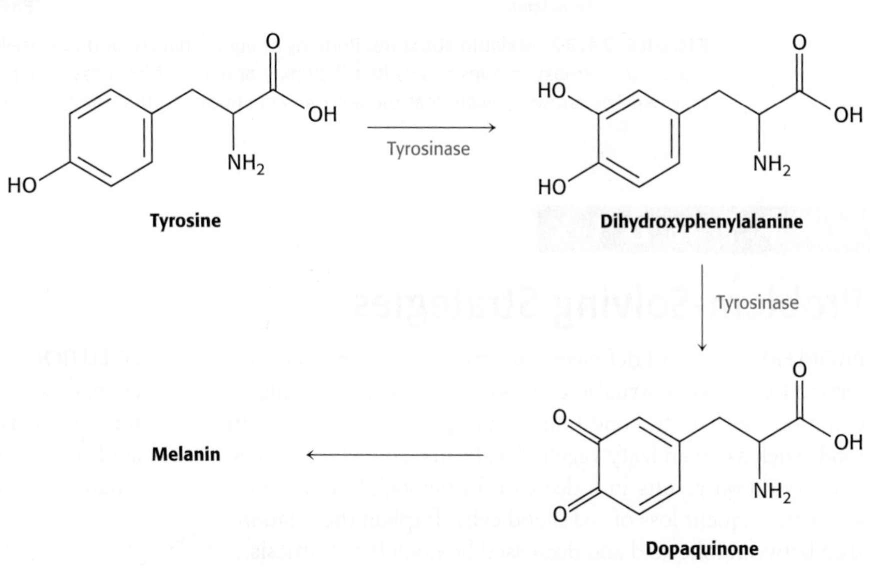

Tyrosine → (Tyrosinase) → **Dihydroxyphenylalanine** → (Tyrosinase) → **Dopaquinone** → **Melanin**

There are two types of melanin, eumelanin and pheom-elanin, and two varieties of eumelanin, brown and black. Pheomelanin yields a pinkish hue, and is found primarily at the lips, nipples and genitals. Variations in the amounts of the two eumelanins and pheomelanin account for the variations in skin and hair color.

The eumelanins are large, highly cross-linked polymers of 5, 6-dihydroxyindole and 5, 6-dihydroxyindole-2-carboxylic acid.

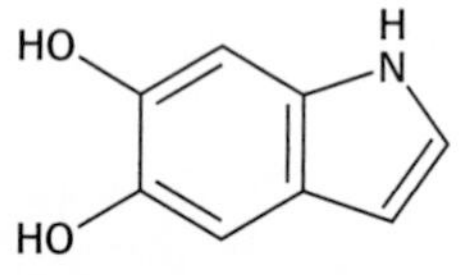

5, 6-Dihydroxyindole

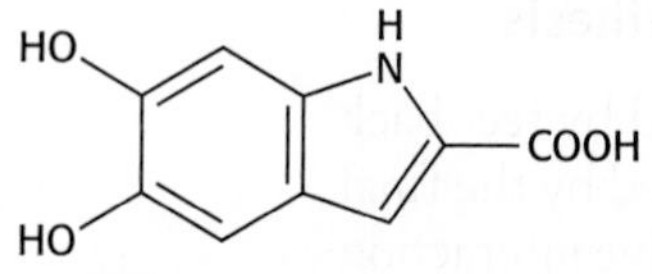

5, 6-Dihydroxyindole-2-carboxylic acid

Brown and black eumelanin are thought to differ in the bonding patterns of the polymers. A portion of a eumelanin polymer is shown in Figure 24.32A.

Pheomelanin is formed from two metabolites of tyrosine, benzothiozine and benzothiazole, as well as the amino acid cysteine.

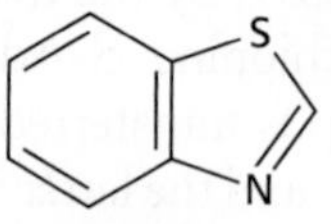

Benzothiazole

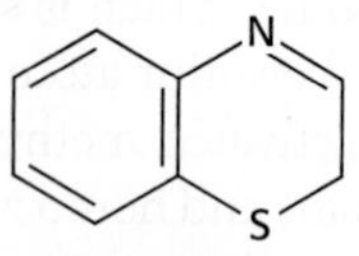

Benzothiazine

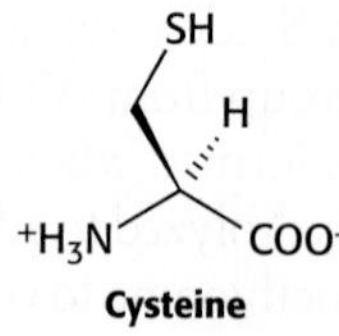

Cysteine

A portion of the structure of pheomelanin is shown in Figure 24.32B. In addition to providing us with the delightful palette of skin and hair color, melanins also protect us by absorbing ultraviolet radiation. People who live or whose ancestors lived near the equator have large amounts of eumelanin in their skins. Exposure to sunlight increases eumelanin production as a protective response. This is, of course, the process of tanning. Sunlight also increases the amount of vitamin D (Section 26.4) produced by skin cells. It is well established, however, that too much sunlight overwhelms the protection provided by eumela-nins and results in skin damage and skin cancer.

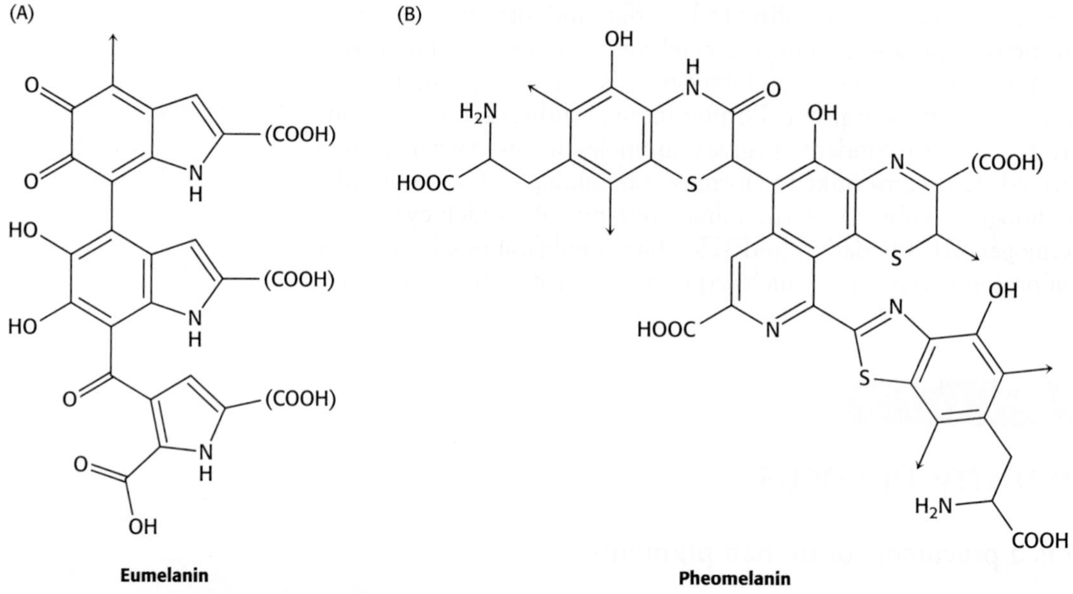

FIGURE 24.32 Melanin structure. Portions of eumelanin (A) and pheomelanin (B) are shown. The parenthesis around the carboxyl groups means that they may or may not be present, depending on their specific location in the polymer. The arrows indicate that the polymer continues and that the figure represents only a portion of the polymer.

APPENDIX

Problem-Solving Strategies

PROBLEM: Folic acid deficiency anemia is a condition charac-terized by pale skin, irritability, loss of appetite, and fatigue. A common cause of the condition is an improper diet, lacking in foods such as green leafy vegetables, beans, and whole grains. The condition results in a decrease in hemoglobin synthesis and a subsequent loss of red blood cells. Explain the relation-ship between folic acid and decreased hemoglobin synthesis.

SOLUTION: First of all, keep in mind what material was covered in the chapter. Second, note that the question focuses on two key biomolecules, hemoglobin and folic acid. Let's start by dissecting the big question into several smaller ones about these two biomolecules.

▶ **What components make up a hemoglobin molecule?**

Two α-globin proteins, two β-globin proteins and a heme group (Chapter 7).

    This chapter does not focus on proteins, so let's focus our attention on the heme group.

▶ **What are the two constituents of heme?**

Iron and protoporphyrin IX. Again, keeping in mind the material in the chapter, let's take a closer look at protoporphyrin. Examine the synthesis of protoporphyrin IX on pages 810–813.

▶ **Summoning all of your acquired biochemical knowledge, what do you think is the key step in the synthesis of protoporphyrin IX?**

Most likely the first step, the synthesis of δ-aminolevulinate. Note that all of the remaining steps are simply condensations of derivatives of δ-aminolevulinate.

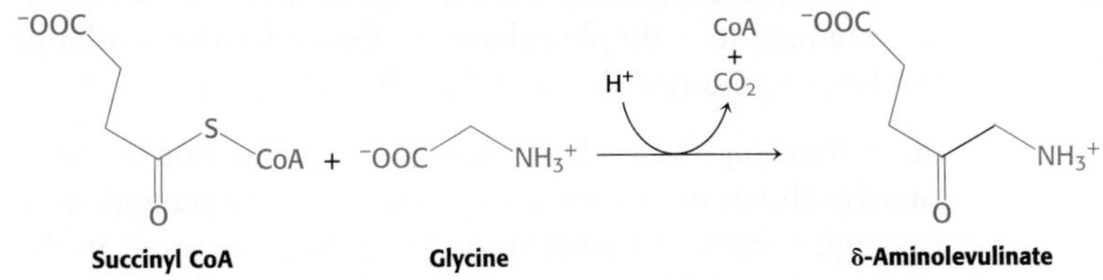

| Succinyl CoA | Glycine | δ-Aminolevulinate |

Good. Now let's do some thinking about the other biomolecule in the question.

▶ **What is the biochemical role of folic acid?**

It is converted into the versatile coenzyme tetrahydrofolate, which plays a role in the synthesis of many amino acids. Think back to the synthesis of protoporphyrin IX.

▶ **Are there any amino acids involved in the synthesis of protoporphyrin IX?**

Exactly one—glycine. I bet you can guess that the next question we are going to ask is:

▶ **Is tetrahydrofolate involved in the synthesis of glycine?**

Bingo! Indeed, it is.

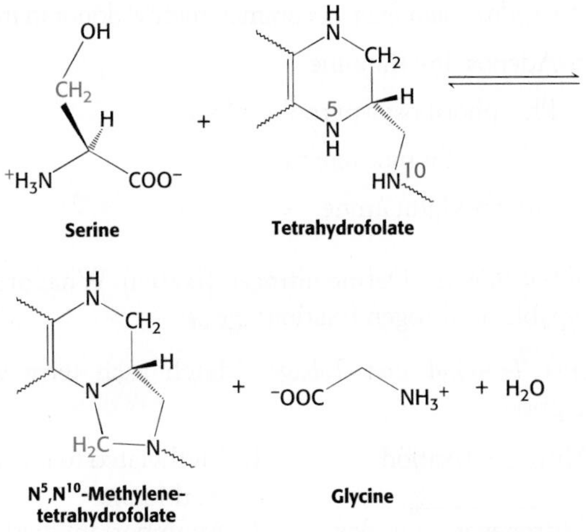

| Serine | Tetrahydrofolate |

| N⁵,N¹⁰-Methylene-tetrahydrofolate | Glycine |

The lack of glycine due to a lack of tetrahydrofolate results in a decrease in heme and thus hemoglobin. The decrease in hemoglobin causes the anemia. Now go eat some spinach. And if there is any chance you will be pregnant soon, take folic acid supplements. You will have to wait until the next chapter to find out why that is case.

KEY TERMS

nitrogen fixation (p. 789)

nitrogenase complex (p. 789)

essential amino acids (p. 793)

nonessential amino acids (p. 793)

pyridoxal phosphate (p. 794)

tetrahydrofolate (p. 797)

S-adenosylmethionine (SAM) (p. 799)

activated methyl cycle (p. 800)

substrate channeling (p. 804)

committed step (p. 805)

enzyme multiplicity (p. 806)

cumulative feedback inhibition (p. 806)

glutathione (p. 809)

nitric oxide (NO) (p. 809)

porphyria (p. 814)

PROBLEMS

1. *Cooperation.* The two enzyme activities required for nitrogen fixation are:

(a) Nitrogenase

(b) Glutamine synthetase

(c) Reductase

(d) Glutamate dehydrogenase

2. *Making donations.* The electron donor for nitrogen fixation is:

(a) NADPH

(b) P680

(c) NADH

(d) Ferredoxin

3. *Amine donor.* The principal amine group donors in amino acid synthesis are

(a) Asparagine

(b) Glutamine

(c) Glutamate

(d) Lysine

4. *The methyl donor man.* A common methyl donor in the cell is:

(a) *S*-Adenosylmethionine

(b) 5-Phosphoribosyl-pyrophosphate

(c) *S*-Adenosylhomocycteine

(d) *S*-Adenosylglutamine

5. *Out of thin air.* Define nitrogen fixation. What organisms are capable of nitrogen fixation? ✓❶

6. *Like Trinidad and Tobago.* Match each term with its description.

(a) Nitrogen fixation
_____

(b) Nitrogenase complex
_____

(c) Glutamate
_____

(d) Essential amino acids
_____

(e) Nonessential amino acids _____

(f) Aminotransferase
_____

(g) Pyridoxal phosphate
_____

(h) Tetrahydrofolate
_____

(i) *S*-Adenosylmethionine
_____

(j) Homocysteine
_____

1. Methylated to form methionine
2. An important methyl donor
3. Coenzyme required by aminotransferases
4. Conversion of N_2 into NH_3
5. A carrier of various one-carbon units
6. Amino acids that are dietary requirements
7. Amino acids that are readily synthesized
8. Responsible for nitrogen fixation
9. Transfers amino groups between ketoacids
10. A common amino group donor

7. *Teamwork.* Identify the two components of the nitrogenase complex and describe their specific tasks. ✓❶

8. *The fix is in.* "The mechanistic complexity of nitrogenase is necessary because nitrogen fixation is a thermodynamically unfavorable process." True or false? Explain. ✓❶

9. *Siphoning resources.* Nitrogen-fixing bacteria on the roots of some plants can consume as much as 20% of the ATP produced by their host—consumption that does not seem very beneficial to the plant. Explain why this loss of valuable resources is tolerated and what the bacteria are doing with the ATP. ✓❶

10. *From few, many.* What are the seven precursors of the 20 amino acids? ✓❷

11. *Vital, in the truest sense.* Why are certain amino acids defined as essential for human beings? ✓❷

12. *From sugar to amino acid.* Write a balanced equation for the synthesis of alanine from glucose.

13. *From air to blood.* What are the intermediates in the flow of nitrogen from N_2 to heme? ✓❷, ✓❹

14. *Common component.* What cofactor is required by all aminotransferases? ✓❷

15. *Here, hold this.* In this chapter, we considered three different cofactors/cosubstrates that act as carriers of one-carbon units. Name them.

16. *One-carbon transfers.* Which derivative of folate is a reactant in the conversion of (a) glycine into serine? (b) homocysteine into methionine?

17. *Telltale tag.* In the reaction catalyzed by glutamine synthetase, an oxygen atom is transferred from the side chain of glutamate to orthophosphate, as shown by the results of ^{18}O-labeling studies. Account for this finding. ✓❷

18. *Telltale tag, redux.* In contrast to the production of glutamine by glutamine synthetase (problem 17), the generation of asparagine from ^{18}O-labeled aspartate does not result in the transfer of an ^{18}O atom to orthophosphate. In what molecule do you expect to find one of the ^{18}O atoms? ✓❷

19. *Therapeutic glycine.* Isovaleric acidemia is an inherited disorder of leucine metabolism caused by a deficiency of isovaleryl CoA dehydrogenase. Many infants having this disease die in the first month of life. The administration of large amounts of glycine sometimes leads to marked clinical improvement. Propose a mechanism for the therapeutic action of glycine. ✓❷

20. *Lending a hand.* The atoms from tryptophan shaded below are derived from two other amino acids. Name them. ✓❷

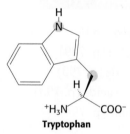

Tryptophan

21. *Deprived bacteria.* Blue-green algae (cyanobacteria) form heterocysts when deprived of ammonia and nitrate. In this form, the cyanobacteria lack nuclei and are attached to adjacent vegetative cells. Heterocysts have photosystem I activity but are entirely devoid of photosystem II activity. What is their role? ✓❶

22. *Cysteine and cystine.* Most cytoplasmic proteins lack disulfide bonds, whereas extracellular proteins usually contain them. Why?

23. *Through the looking-glass.* Suppose that aspartate aminotransferase was chemically synthesized with the use of D-amino acids only. What products would you expect if this

mirror-image enzyme were treated with (a) L-aspartate and α-ketoglutarate; (b) D-aspartate and α-ketoglutarate?

24. To and fro. The synthesis of δ-aminolevulinate takes place in the mitochondrial matrix, whereas the formation of porphobilinogen takes place in the cytoplasm. Propose a reason for the mitochondrial location of the first step in heme synthesis. ✓❹

25. Direct synthesis. Which of the 20 amino acids can be synthesized directly from a common metabolic intermediate by a transamination reaction? ✓❷

26. Alternative route to proline. Certain species of bacteria possess an enzyme, ornithine cyclodeaminase, that can catalyze the conversion of L-ornithine into L-proline in a single catalytic cycle.

Ornithine   **Proline**

The enzyme *lysine* cyclodeaminase has also been identified. Predict the product of the reaction catalyzed by lysine cyclodeaminase.

27. Lines of communication. For the following example of a branched pathway, propose a feedback inhibition scheme that would result in the production of equal amounts of Y and Z. ✓❸

28. Cumulative feedback inhibition. Consider the branched pathway in problem 27. The first common step (A ⟶ B) is partly inhibited by both of the final products, each acting independently of the other. Suppose that a high level of Y alone decreased the rate of the A ⟶ B step from 100 to 60 s^{-1} and that a high level of Z alone decreased the rate from 100 to 40 s^{-1}. What would the rate be in the presence of high levels of both Y and Z? ✓❸

29. Recovered activity. Free sulfhydryl groups can be alkylated with 2-bromoethylamine to the corresponding thioether.

Researchers prepared a mutant form of aspartate aminotransferase in which lysine 258 was replaced by cysteine (Lys258Cys). This mutant protein has no observable catalytic

2-Bromoethylamine

activity. However, treatment of Lys258Cys with 2-bromoethylamine yielded a protein with ∼7% activity relative to the wild-type enzyme. Explain why alkylation recovered some enzyme activity.

Mechanism Problems

30. Ethylene formation. Propose a mechanism for the conversion of *S*-adenosylmethionine into 1-aminocyclopropane-1-carboxylate (ACC) by ACC synthase, a PLP enzyme. What is the other product?

31. Mirror-image serine. Brain tissue contains substantial amounts of D-serine, which acts as a neurotransmitter. D-serine is generated from L-serine by serine racemase, a PLP enzyme. Propose a mechanism for the interconversion of L- and D-serine. What is the equilibrium constant for the reaction L-serine ⇌ D-serine?

32. An unusual amino acid. Elongation factor-2 (eEF-2), a protein taking part in translation, contains a histidine residue that is modified posttranslationally in several steps to a complex side chain known as diphthamide. An intermediate along this pathway is referred to as diphthine. ✓❹

(a) Labeling experiments indicate that the diphthine intermediate is formed by the modification of histidine with four molecules of *S*-adenosylmethionine (indicated by the four colors). Propose a mechanism for the formation of diphthine.

(b) The final conversion of diphthine into diphthamide is known to be ATP dependent. Propose two possible mechanisms for the final amidation step.

Histidine

Diphthine   **Diphthamide**

Chapter Integration Problems

33. *Here, hold this again.* In this chapter, we considered three different cofactors/cosubstrates that act as carriers of one-carbon units (problem 15). Now, name another carrier of one carbon units that we have encountered previously.

34. *Connections.* How might increased synthesis of aspartate and glutamate affect energy production in a cell? How would the cell respond to such an effect?

35. *Protection required.* Suppose that a mutation in bacteria resulted in the diminished activity of methionine adenosyl-transferase, the enzyme responsible for the synthesis of SAM from methionine and ATP. Predict how this diminished activity might affect the stability of the mutated bacteria's DNA.

36. *Heme biosynthesis.* Shemin and coworkers used acetate-labeling experiments to conclude that succinyl-CoA is a key intermediate in the biosynthesis of heme. Identify the intermediates in the conversion of acetate into succinyl-CoA. ✓④

37. *Comparing K_M.* Glutamate dehydrogenase (p. 791) and glutamine synthetase (p. 792) are present in all organisms. Many organisms also contain another enzyme, *glutamate synthase*, which catalyzes the reductive amination of α-ketoglutarate with the use of glutamine as the nitrogen donor.

α-Ketoglutarate + glutamine + NADPH + H$^+$

$$\xrightarrow[\text{synthase}]{\text{Glutamate}} 2 \text{ glutamate} + \text{NADP}^+$$

The side-chain amide of glutamine is hydrolyzed to generate ammonia within the enzyme. When NH$_4^+$ is limiting, most of the glutamate is made by the sequential action of glutamine synthetase and glutamate synthase. The sum of these reactions is

NH$_4^+$ + α-ketoglutarate + NADPH + ATP $\longrightarrow$
glutamate + NADP$^+$ + ADP + P$_i$

Note that this stoichiometry differs from that of the glutamate dehydrogenase reaction in that ATP is hydrolyzed. Why do some organisms sometimes use this more expensive pathway? (Hint: The K_M value for NH$_4^+$ of glutamate dehydrogenase is higher than that of glutamine synthetase.)

Chapter Integration and Data Interpretation Problem

38. *Light effects.* The adjoining graph shows the concentration of several free amino acids in light- and dark-adapted plants.

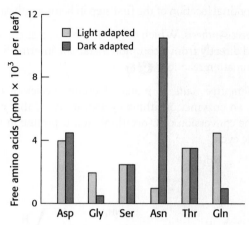

[Data from B.B. Buchanan, W. Gruissem, and R.L. Jones, *Biochemistry and Molecular Biology of Plants* (American Society of Plant Physiology, 2000), Fig. 8.3, p. 363.]

(a) Of the amino acids shown, which are most affected by light–dark adaptation?

(b) Suggest a plausible biochemical explanation for the difference observed.

(c) White asparagus, a culinary delicacy, is the result of growing asparagus plants in the dark. What chemical might you think enhances the taste of white asparagus?

Think ▶ Pair ▶ Share Problems

39. Explain the process of nitrogen fixation and why it is crucial to all life on earth.

40. Why is advantageous for farmers to alternate yearly planting of leguminous and non-leguminous crops in their fields?

Nucleotide Biosynthesis

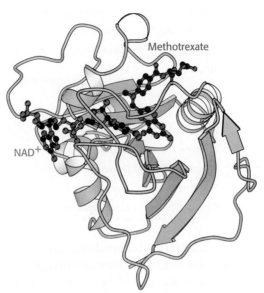

Methotrexate

NAD$^+$

Charles V, Holy Roman Emperor (1519–1558), who also ruled as Charles I, king of Spain (1516–1556), was one of Europe's most powerful rulers. Under his reign, the Aztec and Inca empires of the New World fell to the Spanish. Charles V also suffered from severe gout, a pathological condition caused by the accumulation of the purine urate in the joints. This painting by Titian shows the emperor's swollen left hand, which he rests gingerly in his lap. Recent analysis of a part of his finger, saved as a religious relic, confirms the diagnosis of gout. Xanthine oxidase (left) is the enzyme responsible for the synthesis of urate in the process of purine degradation. [(Far left) Erich Lessing/Art Resource, NY.]

✓ LEARNING OBJECTIVES

By the end of this chapter, you should be able to:

1. Describe how pyrimidine nucleotides are synthesized.

2. Describe the pathway for purine nucleotide synthesis.

3. Explain how deoxyribonucleotides are formed.

4. Name the regulatory steps in nucleotide synthesis.

5. Identify the pathological conditions resulting from impaired nucleotide synthesis.

OUTLINE

25.1 The Pyrimidine Ring Is Assembled de Novo or Recovered by Salvage Pathways

25.2 Purine Bases Can Be Synthesized de Novo or Recycled by Salvage Pathways

25.3 Deoxyribonucleotides Are Synthesized by the Reduction of Ribonucleotides Through a Radical Mechanism

25.4 Key Steps in Nucleotide Biosynthesis Are Regulated by Feedback Inhibition

25.5 Disruptions in Nucleotide Metabolism Can Cause Pathological Conditions

Nucleotides are key biomolecules required for a variety of life processes. First, nucleotides are the *activated precursors of nucleic acids*, necessary for the replication of the genome and the transcription of the genetic information into RNA. Second, an adenine nucleotide, ATP, is *the universal currency of energy*. A guanine nucleotide, GTP, also serves as an energy source for a more select group of biological processes. Third, nucleotide derivatives such as UDP-glucose *participate in biosynthetic processes*, for example, the formation of glycogen. Fourth, nucleotides are *essential components of signal-transduction pathways*. Cyclic nucleotides such as cyclic AMP and cyclic GMP are second messengers that transmit signals both within and between cells. Furthermore, ATP acts as the donor of phosphoryl groups transferred by protein kinases in a variety of signaling pathways and, in some cases, ATP is secreted as a signal molecule.

In this chapter, we continue along the path begun in Chapter 24, which described the incorporation of nitrogen into amino acids from inorganic sources such as nitrogen gas. The amino acids glycine and aspartate are the scaffolds on which the ring systems present in nucleotides are assembled. Furthermore, aspartate and the side chain of glutamine serve as sources of NH$_2$ groups in the formation of nucleotides.

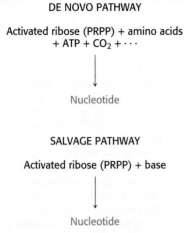

DE NOVO PATHWAY

Activated ribose (PRPP) + amino acids
+ ATP + CO_2 + $\cdots$

↓

Nucleotide

SALVAGE PATHWAY

Activated ribose (PRPP) + base

↓

Nucleotide

FIGURE 25.1 De novo and salvage pathways. In de novo synthesis, the base itself is synthesized from simpler starting materials, including amino acids. ATP hydrolysis is required for de novo synthesis. In a salvage pathway, a base is reattached to a ribose, activated in the form of 5-phosphoribosyl-1-pyrophosphate (PRPP).

TABLE 25.1 Nomenclature of bases, nucleosides, and nucleotides

RNA		
Base	Ribonucleoside	Ribonucleotide (5′-monophosphate)
Adenine (A)	Adenosine	Adenylate (AMP)
Guanine (G)	Guanosine	Guanylate (GMP)
Uracil (U)	Uridine	Uridylate (UMP)
Cytosine (C)	Cytidine	Cytidylate (CMP)
DNA		
Base	Deoxyribonucleoside	Deoxyribonucleotide (5′-monophosphate)
Adenine (A)	Deoxyadenosine	Deoxyadenylate (dAMP)
Guanine (G)	Deoxyguanosine	Deoxyguanylate (dGMP)
Thymine (T)	Thymidine	Thymidylate (TMP)
Cytosine (C)	Deoxycytidine	Deoxycytidylate (dCMP)

Nucleotide biosynthetic pathways are tremendously important as intervention points for therapeutic agents. Many of the most widely used drugs in the treatment of cancer block steps in nucleotide biosynthesis, particularly steps in the synthesis of DNA precursors.

Nucleotides can be synthesized by de novo or salvage pathways

The pathways for the biosynthesis of nucleotides fall into two classes: *de novo* pathways and *salvage* pathways (Figure 25.1). In de novo (from scratch) pathways, the nucleotide bases are assembled from simpler compounds. The framework for a *pyrimidine* base is assembled first and then attached to ribose. In contrast, the framework for a *purine* base is synthesized piece by piece directly onto a ribose-based structure. These pathways each comprise a small number of elementary reactions that are repeated with variations to generate different nucleotides, as might be expected for pathways that appeared very early in evolution. In salvage pathways, preformed bases are recovered and reconnected to a ribose unit.

De novo pathways lead to the synthesis of *ribo*nucleotides. However, DNA is built from *deoxyribo*nucleotides. Consistent with the notion that RNA preceded DNA in the course of evolution, all deoxyribonucleotides are synthesized from the corresponding ribonucleotides. The deoxyribose sugar is generated by the reduction of ribose within a fully formed nucleotide. Furthermore, the methyl group that distinguishes the thymine of DNA from the uracil of RNA is added at the last step in the pathway.

The nomenclature of nucleotides and their constituent units was presented in Chapter 4. Recall that a *nucleoside* is a purine or pyrimidine base linked to a sugar and that a *nucleotide* is a phosphate ester of a nucleoside. The names of the major bases of RNA and DNA, and of their nucleoside and nucleotide derivatives, are given in Table 25.1.

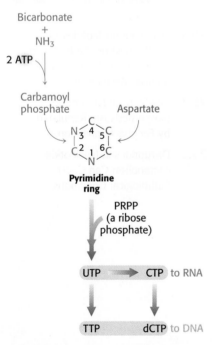

Bicarbonate
+
NH_3

2 ATP

↓

Carbamoyl
phosphate   Aspartate

Pyrimidine
ring

PRPP
(a ribose
phosphate)

↓

UTP → CTP   to RNA

↓   ↓

TTP   dCTP   to DNA

FIGURE 25.2 De novo pathway for pyrimidine nucleotide synthesis. The C-2 and N-3 atoms in the pyrimidine ring come from carbamoyl phosphate, whereas the other atoms of the ring come from aspartate.

25.1 The Pyrimidine Ring Is Assembled de Novo or Recovered by Salvage Pathways

In de novo synthesis of pyrimidines, the ring is synthesized first and then it is attached to a ribose phosphate to form a *pyrimidine nucleotide* (Figure 25.2). Pyrimidine rings are assembled from bicarbonate, aspartate, and ammonia. Although an ammonia molecule already present in solution can be used, the ammonia is usually produced from the hydrolysis of the side chain of glutamine.

Bicarbonate and other oxygenated carbon compounds are activated by phosphorylation

The first step in de novo pyrimidine biosynthesis is the synthesis of *carbamoyl phosphate* from bicarbonate and ammonia in a multistep process, requiring the cleavage of two molecules of ATP. This reaction is catalyzed by *carbamoyl phosphate synthetase II* (CPS II). Recall that carbamoyl phosphate synthetase I facilitates ammonia incorporation into urea (Section 23.4). Carbamoyl phosphate synthetase II is a dimer composed of a small subunit that hydrolyzes glutamine to form NH_3 and a large subunit that completes the synthesis of carbamoyl phosphate. Analysis of the structure of the large subunit reveals two homologous domains, each of which catalyzes an ATP-dependent step (Figure 25.3).

In the first step, bicarbonate is phosphorylated by ATP to form carboxyphosphate and ADP. Ammonia then reacts with carboxyphosphate to form carbamic acid and inorganic phosphate.

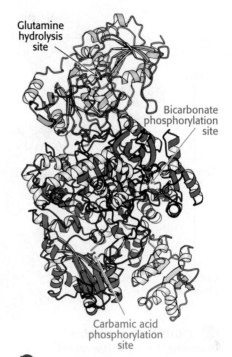

FIGURE 25.3 Structure of carbamoyl phosphate synthetase II. *Notice* that the enzyme contains sites for three reactions. This enzyme consists of two chains. The smaller chain (yellow) contains a site for glutamine hydrolysis to generate ammonia. The larger chain includes two ATP-grasp domains (blue and red). In one ATP-grasp domain (blue), bicarbonate is phosphorylated to carboxyphosphate, which then reacts with ammonia to generate carbamic acid. In the other ATP-grasp domain, the carbamic acid is phosphorylated to produce carbamoyl phosphate. [Drawn from 1JDB.pdb.]

Bicarbonate → Carboxyphosphate → Carbamic acid

The active site for this reaction lies in a domain formed by the amino-terminal third of CPS. This domain forms a structure called an *ATP-grasp fold,* which surrounds ATP and holds it in an orientation suitable for nucleophilic attack at the γ phosphoryl group. Proteins containing ATP-grasp folds catalyze the formation of carbon–nitrogen bonds through acylphosphate intermediates. Such ATP-grasp folds are widely used in nucleotide biosynthesis.

In the second step catalyzed by carbamoyl phosphate synthetase II, carbamic acid is phosphorylated by another molecule of ATP to form carbamoyl phosphate.

Carbamic acid → Carbamoyl phosphate

This reaction takes place in a second ATP-grasp domain within the enzyme. The active sites leading to carbamic acid formation and carbamoyl phosphate formation are very similar, revealing that this enzyme evolved by a gene duplication event. Indeed, duplication of a gene encoding an ATP-grasp domain followed by specialization was central to the evolution of nucleotide biosynthetic processes (p. 827).

The side chain of glutamine can be hydrolyzed to generate ammonia

Glutamine is the primary source of ammonia for carbamoyl phosphate synthetase II. In this case, the small subunit of the enzyme hydrolyzes glutamine to form ammonia and glutamate. The active site of the glutamine-hydrolyzing component contains a catalytic dyad comprising a cysteine and a histidine residue. Such a catalytic dyad, reminiscent of the active site of cysteine proteases (Figure 9.16), is conserved in a family of amidotransferases, including CTP synthetase and GMP synthetase.

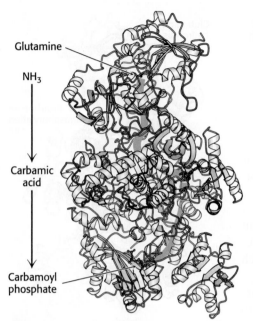

Glutamine

NH₃

Carbamic acid

Carbamoyl phosphate

FIGURE 25.4 Substrate channeling. The three active sites of carbamoyl phosphate synthetase II are linked by a channel (yellow) through which intermediates pass. Glutamine enters one active site, and carbamoyl phosphate, which includes the nitrogen atom from the glutamine side chain, leaves another 80 Å away. [Drawn from 1JDB.pdb.]

Intermediates can move between active sites by channeling

Carbamoyl phosphate synthetase II contains three different active sites (Figure 25.3), separated from one another by a total of 80 Å. Intermediates generated at one site move to the next without leaving the enzyme. These intermediates move within the enzyme by means of substrate channeling, similar to the process described for tryptophan synthetase (Figure 25.4; also Figure 24.16). The ammonia generated in the glutamine-hydrolysis active site travels 45 Å through a channel within the enzyme to reach the site at which carboxyphosphate has been generated. The carbamic acid generated at this site diffuses an additional 35 Å through an extension of the channel to reach the site at which carbamoyl phosphate is generated. This channeling serves two roles: (1) intermediates generated at one active site are captured with no loss caused by diffusion and (2) labile intermediates, such as carboxyphosphate and carbamic acid (which decompose in less than 1 s at pH 7), are protected from hydrolysis. We will see additional examples of substrate channeling later in this chapter.

Orotate acquires a ribose ring from PRPP to form a pyrimidine nucleotide and is converted into uridylate

Carbamoyl phosphate reacts with aspartate to form carbamoylaspartate in a reaction catalyzed by *aspartate transcarbamoylase* (Section 10.1). Carbamoylaspartate then cyclizes to form dihydroorotate, which is then oxidized by NAD^+ to form orotate.

Carbamoyl phosphate → (Aspartate, P_i) → **Carbamoylaspartate** → (H^+, H_2O) → **Dihydroorotate** → (NAD^+, NADH + H^+) → **Orotate**

In mammals, the enzymes that form orotate are part of a single large polypeptide chain called CAD, for *c*arbamoyl phosphate synthase, *a*spartate transcarbamoylase and *d*ihydroorotase.

At this stage, orotate couples to ribose, in the form of *5-phosphoribosyl-1-pyrophosphate* (PRPP), a form of ribose activated to accept nucleotide bases. *5-Phosphoribosyl-1-pyrophosphate synthetase* synthesizes PRPP by adding a pyrophosphate from ATP to ribose-5-phosphate, which is formed by the pentose phosphate pathway.

Ribose 5-phosphate → (ATP, AMP, PRPP synthetase) → **PRPP**

Orotate reacts with PRPP to form *orotidylate*, a pyrimidine nucleotide. This reaction is driven by the hydrolysis of pyrophosphate. The enzyme that catalyzes this addition, *orotate phosphoribosyltransferase,* is homologous to a number of other phosphoribosyltransferases that add different groups to PRPP to form the other nucleotides.

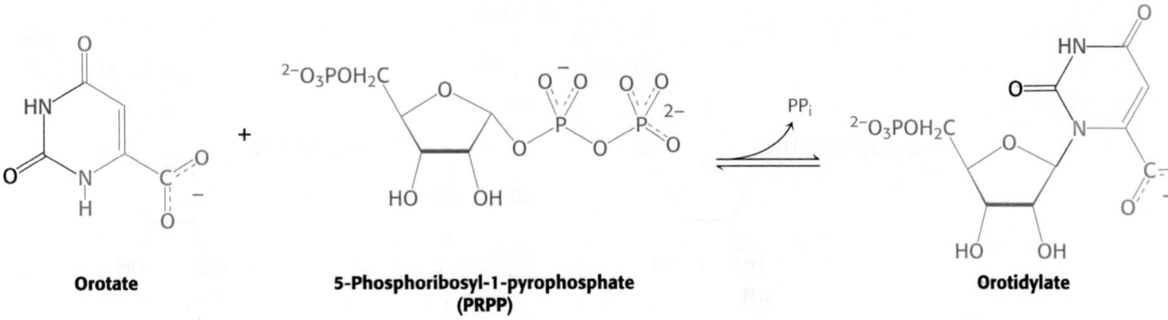

Orotate + 5-Phosphoribosyl-1-pyrophosphate (PRPP) → Orotidylate

Orotidylate is then decarboxylated to form *uridylate* (UMP), a major pyrimidine nucleotide that is a precursor to RNA. This reaction is catalyzed by *orotidylate decarboxylase*, also called orotidine-5′ phosphate decarboxylase.

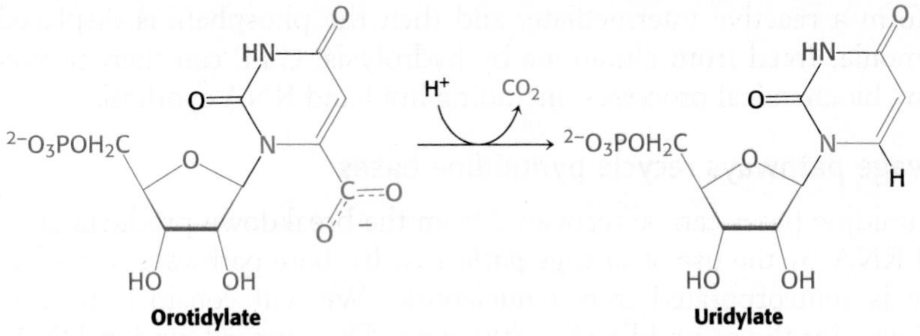

Orotidylate → Uridylate

Orotidylate decarboxylase is one of the most proficient enzymes known. In its absence, decarboxylation is extremely slow and is estimated to take place once every 78 million years; with the enzyme present, it takes place approximately once per second, a rate enhancement of 10^{17}-fold. The phosphoribosyltransferase and decarboxylase activities are located on the same polypeptide chain, providing another example of a bifunctional enzyme. The bifunctional enzyme is called *uridine monophosphate synthetase*.

Nucleotide mono-, di-, and triphosphates are interconvertible

How is the other major pyrimidine ribonucleotide, cytidine, formed? It is synthesized from the uracil base of UMP, but the synthesis can take place only after UMP has been converted into UTP. Recall that the diphosphates and triphosphates are the active forms of nucleotides in biosynthesis and energy conversions. Nucleoside monophosphates are converted into nucleoside triphosphates in stages. First, nucleoside monophosphates are converted into diphosphates by specific *nucleoside monophosphate kinases* that utilize ATP as the phosphoryl-group donor. For example, UMP is phosphorylated to UDP by *UMP kinase*.

$$UMP + ATP \rightleftharpoons UDP + ADP$$

Nucleoside diphosphates and triphosphates are interconverted by *nucleoside diphosphate kinase*, an enzyme that has broad specificity, in contrast with the monophosphate kinases. X and Y represent any of several ribonucleosides or even deoxyribonucleosides:

$$XDP + YTP \rightleftharpoons XTP + YDP$$

CTP is formed by amination of UTP

After uridine triphosphate has been formed, it can be transformed into *cytidine triphosphate* by the replacement of a carbonyl group by an amino group, a reaction catalyzed by *cytidine triphosphate synthetase*.

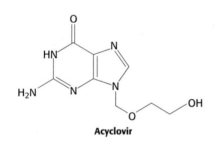

Like the synthesis of carbamoyl phosphate, this reaction requires ATP and uses glutamine as the source of the amino group. The reaction proceeds through an analogous mechanism in which the O-4 atom is phosphorylated to form a reactive intermediate, and then the phosphate is displaced by ammonia, freed from glutamine by hydrolysis. CTP can then be used in many biochemical processes, including lipid and RNA synthesis.

Salvage pathways recycle pyrimidine bases

Pyrimidine bases can be recovered from the breakdown products of DNA and RNA by the use of *salvage pathways*. In these pathways, a preformed base is reincorporated into a nucleotide. We will consider the salvage pathway for the pyrimidine base thymine. Thymine is found in DNA and base-pairs with adenine in the DNA double helix. Thymine released from degraded DNA is salvaged in two steps. First, thymine is converted into the nucleoside thymidine by *thymidine phosphorylase*.

$$\text{Thymine} + \text{deoxyribose-1-phosphate} \rightleftharpoons \text{thymidine} + P_i$$

Thymidine is then converted into a nucleotide by *thymidine kinase*.

$$\text{Thymidine} + \text{ATP} \rightleftharpoons \text{TMP} + \text{ADP}$$

Viral thymidine kinase differs from the mammalian enzyme and thus provides a therapeutic target. For instance, herpes simplex infections are treated with acyclovir (acycloguanosine), which viral thymidine kinase converts into acyclovir monophosphate by adding a phosphate to the hydroxyl group of acyclovir. Viral thymidine kinase binds acyclovir more than 200 times as tightly as cellular thymidine kinase, accounting for its effect on infected cells only. Acyclovir monophosphate is phosphorylated by cellular enzymes to yield acyclovir triphosphate. Acyclovir triphosphate competes with dGTP for DNA polymerase. Once incorporated into viral DNA, it acts as a chain terminator since it lacks a 3′-hydroxyl required for chain extension. As we will see shortly, thymidine kinase also plays a role in the de novo synthesis of thymidylate.

25.2 Purine Bases Can Be Synthesized de Novo or Recycled by Salvage Pathways

Like pyrimidine nucleotides, *purine nucleotides* can be synthesized de novo or by a salvage pathway. When synthesized de novo, purine synthesis begins with simple starting materials such as amino acids and bicarbonate (Figure 25.5). Unlike the bases of pyrimidines, the purine bases are assembled already attached to the ribose ring. Alternatively, purine bases, released by the hydrolytic degradation of nucleic acids and nucleotides, can be salvaged and recycled. Purine salvage pathways are especially noted for the energy that they save and the remarkable effects of their absence (p. 840).

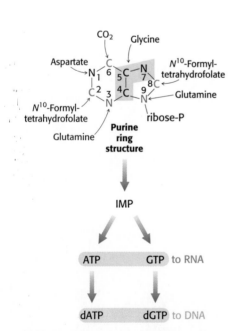

FIGURE 25.5 De novo pathway for purine nucleotide synthesis. The origins of the atoms in the purine ring are indicated.

The purine ring system is assembled on ribose phosphate

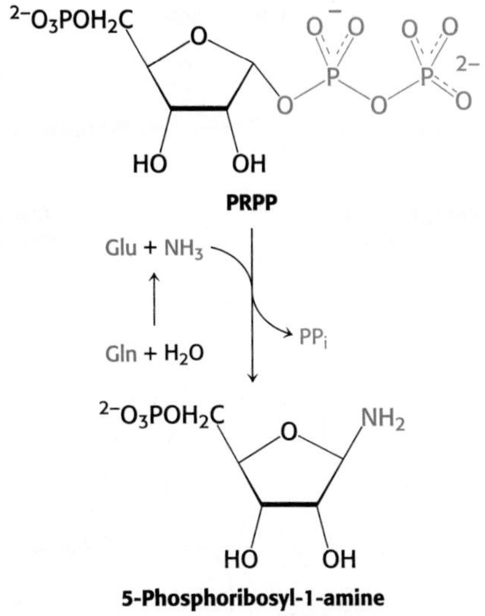

De novo purine biosynthesis, like pyrimidine biosynthesis, requires PRPP but, for purines, PRPP provides the foundation on which the bases are constructed step by step. The initial step is the displacement of pyrophosphate by ammonia, rather than by a preassembled base, to produce *5-phosphoribosyl-1-amine*, with the amine in the β configuration.

Glutamine phosphoribosyl amidotransferase catalyzes this reaction, which is the committed step in purine regulation. This enzyme comprises two domains: the first is homologous to the phosphoribosyltransferases in purine salvage pathways (p. 830), whereas the second produces ammonia from glutamine by hydrolysis. However, this glutamine-hydrolysis domain is distinct from the domain that performs the same function in carbamoyl phosphate synthetase II. In glutamine phosphoribosyl amidotransferase, a cysteine residue located at the amino terminus facilitates glutamine hydrolysis. To prevent wasteful hydrolysis of either substrate, the amidotransferase assumes the active configuration only on binding of both PRPP and glutamine. As is the case with carbamoyl phosphate synthetase II, the ammonia generated at the glutamine-hydrolysis active site passes through a channel to reach PRPP without being released into solution.

PRPP

5-Phosphoribosyl-1-amine

The purine ring is assembled by successive steps of activation by phosphorylation followed by displacement

Nine additional steps are required to assemble the purine ring. Remarkably, the first six steps are analogous reactions. Most of these steps are catalyzed by enzymes with ATP-grasp domains that are homologous to those in carbamoyl phosphate synthetase. *Each step consists of the activation of a carbon-bound oxygen atom (typically a carbonyl oxygen atom) by phosphorylation, followed by the displacement of the phosphoryl group by ammonia or an amine group acting as a nucleophile* (Nu).

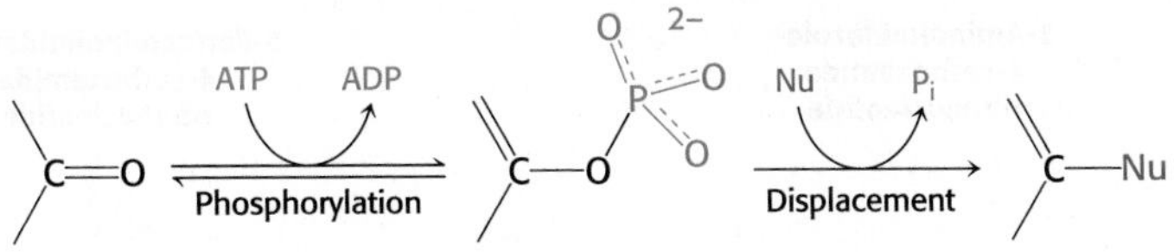

De novo purine biosynthesis proceeds as shown in Figure 25.6. Table 25.2 lists the enzymes that catalyze each step of the reaction.

1. The carboxylate group of a glycine residue is activated by phosphorylation and then coupled to the amino group of phosphoribosylamine. A new amide bond is formed, and the amino group of glycine is free to act as a nucleophile in the next step.

TABLE 25.2 The enzymes of de novo purine synthesis

Step	Enzyme
1	Glycinamide ribonucleotide (GAR) synthetase
2	GAR transformylase
3	Formylglycinamidine synthase
4	Aminoimidazole ribonucleotide synthetase
5	Carboxyaminoimidazole ribonucleotide synthetase
6	Succinylaminoimidazole carboxamide ribonucleotide synthetase
7	Adenylosuccinate lyase
8	Aminoimidazole carboxamide ribonucleotide transformylase
9	Inosine monophosphate cyclohydrolyase

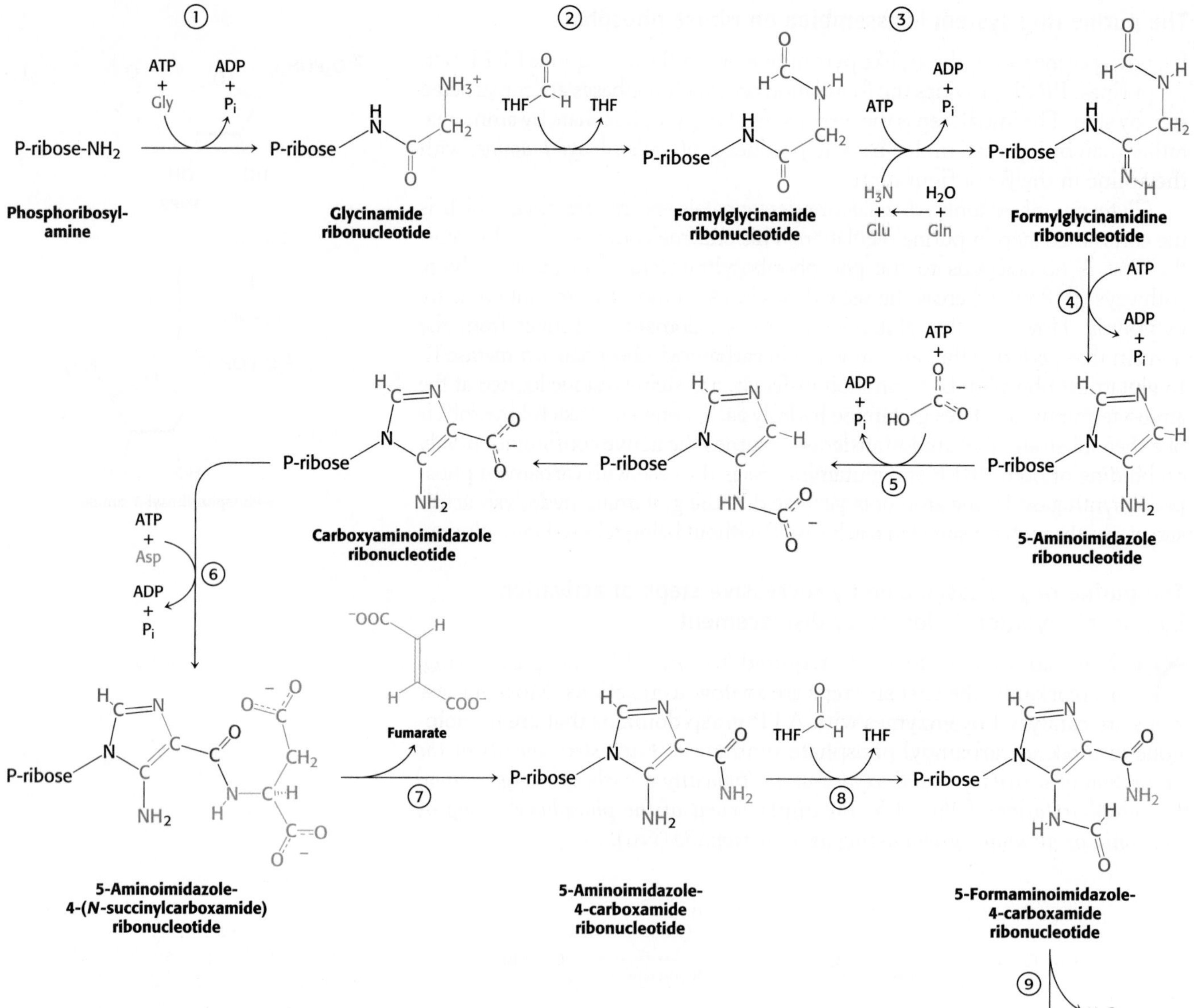

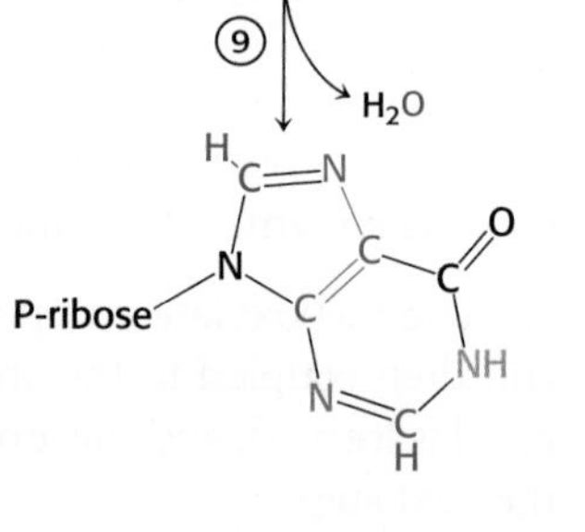

FIGURE 25.6 De novo purine biosynthesis. (1) Glycine is coupled to the amino group of phosphoribosylamine. (2) N^{10}-Formyltetrahydrofolate (THF) transfers a formyl group to the amino group of the glycine residue. (3) The inner amide group is phosphorylated and converted into an amidine by the addition of ammonia derived from glutamine. (4) An intramolecular coupling reaction forms the five-membered imidazole ring. (5) Bicarbonate adds first to the exocyclic amino group and then to a carbon atom of the imidazole ring. (6) The imidazole carboxylate is phosphorylated, and the phosphate is displaced by the amino group of aspartate. (7) Fumarate is released. (8) A second formyl group is donated from N^{10}-formyltetrahydrofolate (THF). (9) Cyclization completes the synthesis of inosinate, a purine nucleotide.

2. N^{10}-formyltetrahydrofolate donates a formyl moiety to this amino group to form formylglycinamide ribonucleotide.

3. The inner carbonyl group is activated by phosphorylation and then converted into an amidine by the addition of ammonia derived from glutamine.

4. The product of this reaction, formylglycinamidine ribonucleotide, cyclizes to form the five-membered imidazole ring found in purines. Although this cyclization is likely to be favorable thermodynamically, a molecule of ATP is consumed to ensure irreversibility. The familiar pattern is repeated: a phosphoryl group from the ATP molecule activates the carbonyl group and is displaced by the nitrogen atom attached to the ribose molecule. Cyclization is thus an intramolecular reaction in which the nucleophile and phosphate-activated carbon atom are present within the same molecule.

In higher eukaryotes, the enzymes catalyzing steps 1, 2, and 4 (Table 25.2) are components of a single polypeptide chain.

5. Bicarbonate is activated by phosphorylation and then attacked by the exocyclic amino group. The product of the reaction in step 5 rearranges to transfer the carboxylate group to the imidazole ring. Interestingly, mammals do not require ATP for this step; bicarbonate apparently attaches directly to the exocyclic amino group and is then transferred to the imidazole ring.

6. The imidazole carboxylate group is phosphorylated again and the phosphate group is displaced by the amino group of aspartate. Once again, in higher eukaryotes, the enzymes catalyzing steps 5 and 6 (Table 25.2) share a single polypeptide chain.

7. Fumarate, an intermediate in the citric acid cycle, is eliminated, leaving the nitrogen atom from aspartate joined to the imidazole ring. The use of aspartate as an amino-group donor and the concomitant release of fumarate are reminiscent of the conversion of citrulline into arginine in the urea cycle, and these steps are catalyzed by homologous enzymes in the two pathways (Section 23.4).

8. A formyl group from N^{10}-formyltetrahydrofolate is added to this nitrogen atom to form a final 5-formaminoimidazole-4-carboxamide ribonucleotide.

9. 5-Formaminoimidazole-4-carboxamide ribonucleotide cyclizes with the loss of water to form inosinate.

Many of the intermediates in the de novo purine biosynthesis pathway degrade rapidly in water. Their instability in water suggests that the product of one enzyme must be channeled directly to the next enzyme along the pathway. Recent evidence shows that the enzymes do indeed form complexes when purine synthesis is required (p. 830).

AMP and GMP are formed from IMP

A few steps convert inosinate into either AMP or GMP (Figure 25.7). *Adenylate* is synthesized from inosinate by the substitution of an amino group for the carbonyl oxygen atom at C-6. Again, the addition of aspartate followed by the elimination of fumarate contributes the amino group. GTP, rather than ATP, is the phosphoryl-group donor in the synthesis of the adenylosuccinate intermediate from inosinate and aspartate. In accord with the use of GTP, the enzyme that promotes this conversion, *adenylosuccinate synthetase*, is structurally related to the G-protein family and does not contain an ATP-grasp domain.

FIGURE 25.7 Generating AMP and GMP. Inosinate is the precursor of AMP and GMP. AMP is formed by the addition of aspartate followed by the release of fumarate. GMP is generated by the addition of water, dehydrogenation by NAD^+, and the replacement of the carbonyl oxygen atom by $-NH_2$ derived by the hydrolysis of glutamine.

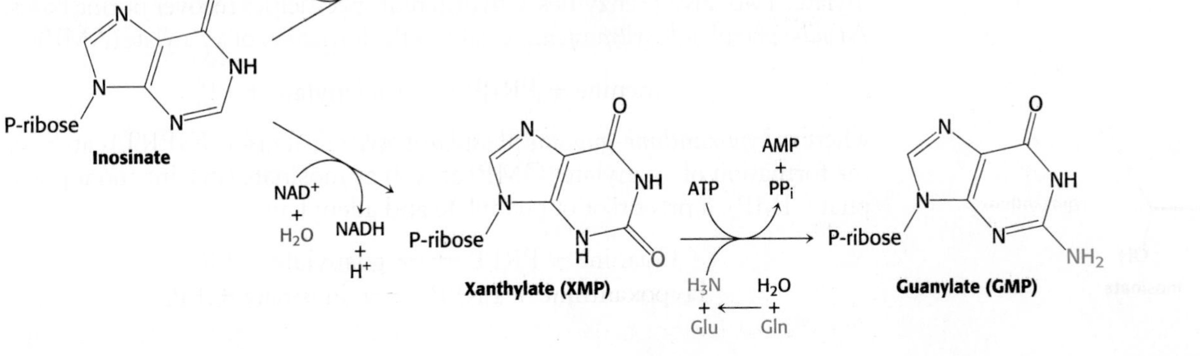

Guanylate is synthesized by the oxidation of inosinate to xanthylate (XMP), followed by the incorporation of an amino group at C-2. NAD^+ is the hydrogen acceptor in the oxidation of inosinate. Xanthylate is activated by the transfer of an AMP group (rather than a phosphoryl group) from ATP to the oxygen atom in the newly formed carbonyl group. Ammonia, generated by the hydrolysis of glutamine, then displaces the AMP group to form guanylate, in a reaction catalyzed by *GMP synthetase*. Note that the synthesis of adenylate requires GTP, whereas the synthesis of guanylate requires ATP. This reciprocal use of nucleotides by the pathways creates an important regulatory opportunity (Section 25.4).

Enzymes of the purine synthesis pathway associate with one another in vivo

Biochemists believe that the enzymes of many metabolic pathways, such as glycolysis and the citric acid cycle, are physically associated with one another. Such associations would increase the efficiency of pathways by facilitating the movement of the product of one enzyme to the active site of the next enzyme in the pathway. The evidence for such associations comes primarily from experiments in which one component of a pathway, carefully isolated from the cell, is found to be bound to other components of the pathway. However, these observations raise the question, do enzymes associate with one another in vivo or do they spuriously associate during the isolation procedure? Recent in vivo evidence shows that the enzymes of the purine synthesis pathway associate with one another when purine synthesis is required. Various enzymes of the pathway were fused with the green fluorescent protein (GFP) (Figure 2.65) and transfected into cells. When cells were grown in the presence of purine, the GFP was spread diffusely throughout the cytoplasm (Figure 25.8A). When the cells were switched to growth media without purines, purine synthesis began and the enzymes became associated with one another, forming complexes dubbed *purinosomes* (Figure 25.8B). The experiments were repeated with other enzymes of the purine synthesis pathway bearing the GFP, and the results were the same: purine synthesis occurs when the enzymes form the purinosomes. What actually causes complex formation? While the results are not yet established, it appears that several G-protein coupled receptors, including those responding to epinephrine as well as ATP and ADP (purinergic receptors), instigate complex formation, while human creatine kinase II (hCK2), responding to the presence of purines, causes disassembly of the purinosome.

Salvage pathways economize intracellular energy expenditure

As we have seen, the de novo synthesis of purines requires a substantial investment of ATP. Purine salvage pathways provide a more economical means of generating purines. Free purine bases, derived from the turnover of nucleotides or from the diet, can be attached to PRPP to form purine nucleoside monophosphates, in a reaction analogous to the formation of orotidylate. Two salvage enzymes with different specificities recover purine bases. *Adenine phosphoribosyltransferase* catalyzes the formation of adenylate (AMP):

$$\text{Adenine} + \text{PRPP} \longrightarrow \text{adenylate} + \text{PP}_i$$

whereas *hypoxanthine-guanine phosphoribosyltransferase* (HGPRT) catalyzes the formation of guanylate (GMP) as well as inosinate (inosine monophosphate, IMP), a precursor of guanylate and adenylate.

$$\text{Guanine} + \text{PRPP} \longrightarrow \text{guanylate} + \text{PP}_i$$
$$\text{Hypoxanthine} + \text{PRPP} \longrightarrow \text{inosinate} + \text{PP}_i$$

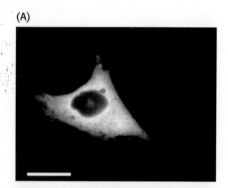

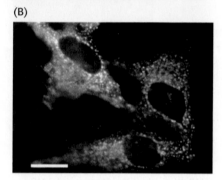

FIGURE 25.8 Formation of purinosomes. A gene construct encoding a fusion protein consisting of formylglycinamidine synthase and GFP was transfected into and expressed in Hela cells, a human cell line. (A) In the presence of purines (the absence of purine synthesis), the GFP was seen as a diffuse stain throughout the cytoplasm. (B) When the cells were shifted to a purine-free medium, purinosomes formed, seen as cytoplasmic granules, and purine synthesis occurred. White scale bars, 10 μm [From An, S., Kumar, R., Sheets, E. D., and Benkovic, S. J. 2008. Reversible compartmentalization of de novo purine biosynthetic complexes in living cells. *Science* 320:103–106. Reprinted with permission from AAAS.]

Inosinate
Hypoxanthine

25.3 Deoxyribonucleotides Are Synthesized by the Reduction of Ribonucleotides Through a Radical Mechanism

We turn now to the synthesis of deoxyribonucleotides. These precursors of DNA are formed by the reduction of ribonucleotides; specifically, the 2'-hydroxyl group on the ribose moiety is replaced by a hydrogen atom. The substrates are ribonucleoside diphosphates, and the ultimate reductant is NADPH. The enzyme *ribonucleotide reductase* is responsible for the reduction reaction for all four ribonucleotides. The ribonucleotide reductases of different organisms are a remarkably diverse set of enzymes. Yet detailed studies have revealed that they have a common reaction mechanism, and their three-dimensional structural features indicate that these enzymes are homologous. We will focus on the best understood of these enzymes, that of *E. coli* living aerobically.

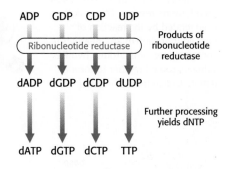

Mechanism: A tyrosyl radical is critical to the action of ribonucleotide reductase

The ribonucleotide reductase of *E. coli* consists of two subunits: R1 (an 87-kDa dimer) and R2 (a 43-kDa dimer). The R1 subunit contains the active site as well as two allosteric control sites (Section 25.4). This subunit includes three conserved cysteine residues and a glutamate residue, all four of which participate in the reduction of ribose to deoxyribose (Figure 25.9). The R2 subunit's role in catalysis is to generate a free radical in each of its two chains. Each R2 chain contains a *tyrosyl radical* with an unpaired electron delocalized onto its aromatic ring, generated by a

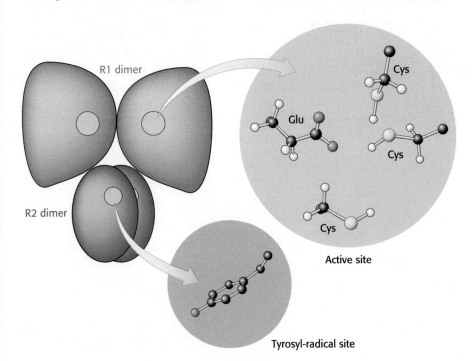

Active site

Tyrosyl-radical site

FIGURE 25.9 Ribonucleotide reductase. Ribonucleotide reductase reduces ribonucleotides to deoxyribonucleotides in its active site, which contains three key cysteine residues and one glutamate residue. Each R2 subunit contains a tyrosyl radical that accepts an electron from one of the cysteine residues in the active site to initiate the reduction reaction. Two R1 subunits come together to form a dimer as do two R2 subunits.

nearby *iron center* consisting of two ferric (Fe^{3+}) ions bridged by an oxide (O^{2-}) ion (Figure 25.10). This very unusual free radical is remarkably stable, with a half-life of 4 days at 4°C. By contrast, free tyrosine radicals in solution have microsecond lifetimes.

In the synthesis of a deoxyribonucleotide, the OH bonded to C-2' of the ribose ring is replaced by H, with retention of the configuration at the C-2' carbon atom (Figure 25.11).

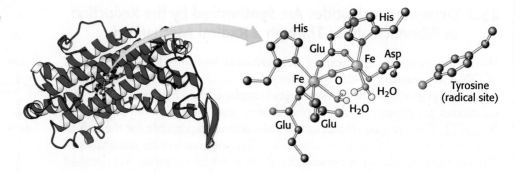

FIGURE 25.10 Ribonucleotide reductase R2 subunit. The R2 subunit contains a stable free radical on a tyrosine residue. This radical is generated by the reaction of oxygen (not shown) at a nearby site containing two iron atoms. Two R2 subunits come together to form a dimer. [Drawn from 1RIB.pdb.]

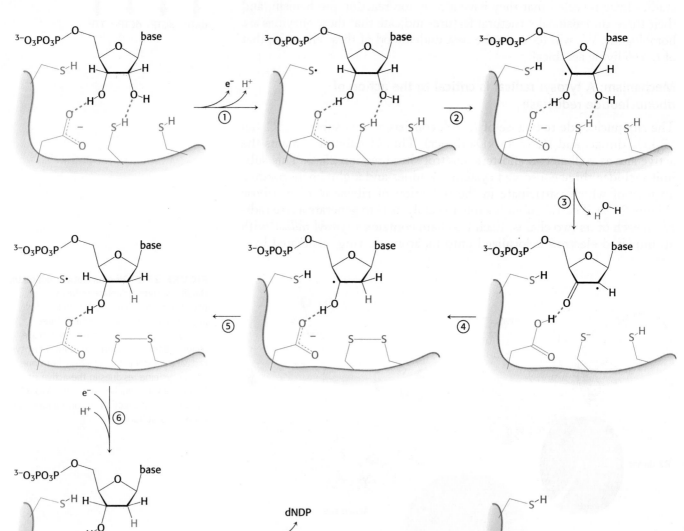

FIGURE 25.11 Ribonucleotide reductase mechanism. (1) An electron is transferred from a cysteine residue on R1 to a tyrosine radical on R2, generating a highly reactive cysteine thiyl radical. (2) This radical abstracts a hydrogen atom from C-3' of the ribose unit. (3) The radical at C-3' releases OH^- from the C-2' carbon atom. Combined with a proton from a second cysteine residue, the OH^- is eliminated as water. (4) A hydride ion is transferred from a third cysteine residue with the concomitant formation of a disulfide bond. (5) The C-3' radical recaptures the originally abstracted hydrogen atom. (6) An electron is transferred from R2 to reduce the thiyl radical, which also accepts a proton. (7) The deoxyribonucleotide leaves R1. (8) The disulfide formed in the active site is reduced to begin another cycle.

1. The reaction begins with the transfer of an electron from a cysteine residue on R1 to the tyrosyl radical on R2. The loss of an electron generates a highly reactive *cysteine thiyl radical* within the active site of R1.

2. This radical then abstracts a hydrogen atom from C-3′ of the ribose unit, generating a radical at that carbon atom.

3. The radical at C-3′ promotes the release of the OH^- from the C-2′ carbon atom. Protonated by a second cysteine residue, the departing OH^- leaves as a water molecule.

4. A hydride ion (a proton with two electrons) is then transferred from a third cysteine residue to complete the reduction of the position, form a disulfide bond, and regenerate a radical.

5. This C-3′ radical recaptures the same hydrogen atom originally abstracted by the first cysteine residue, and the deoxyribonucleotide is free to leave the enzyme.

6. R2 provides an electron to reduce the thiyl radical.

7. The newly synthesized deoxyribonucleotide exits the active site.

8. The disulfide bond generated in the enzyme's active site is reduced to regenerate the active enzyme.

The electrons for this reduction come from NADPH, but not directly. One carrier of reducing power linking NADPH with the reductase is *thioredoxin*, a 12-kDa protein with two exposed cysteine residues near each other. These sulfhydryls are oxidized to a disulfide in the reaction catalyzed by ribonucleotide reductase itself. In turn, reduced thioredoxin is regenerated by electron flow from NADPH. This reaction is catalyzed by *thioredoxin reductase*, a flavoprotein. Electrons flow from NADPH to bound FAD of the reductase, to the disulfide of oxidized thioredoxin, and then to ribonucleotide reductase and finally to the ribose unit.

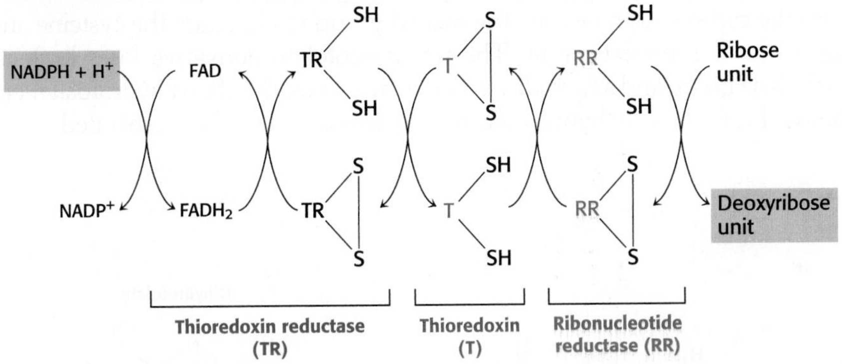

Stable radicals other than tyrosyl radical are employed by other ribonucleotide reductases

Ribonucleotide reductases that do not contain tyrosyl radicals have been characterized in other organisms. Instead, these enzymes contain other stable radicals that are generated by other processes. For example, in one class of reductases, the coenzyme adenosylcobalamin (vitamin B_{12}) is the radical source (Section 22.3). Despite differences in the stable radical employed, the active sites of these enzymes are similar to that of the *E. coli* ribonucleotide reductase, and they appear to act by the same mechanism, based on the exceptional reactivity of cysteine radicals. Thus, these enzymes have a common ancestor but evolved a range of mechanisms for generating stable radical species that function well under different growth conditions.

The primordial enzymes appear to have been inactivated by oxygen, whereas enzymes such as the *E. coli* enzyme make use of oxygen to generate the initial tyrosyl radical. Note that the reduction of ribonucleotides to deoxyribonucleotides is a difficult reaction chemically, likely to require a sophisticated catalyst. The existence of a common protein enzyme framework for this process strongly suggests that proteins joined the RNA world before the evolution of DNA as a stable storage form for genetic information.

Thymidylate is formed by the methylation of deoxyuridylate

Uracil, produced by the pyrimidine synthesis pathway, is not a component of DNA. Rather, DNA contains *thymine*, a methylated analog of uracil. Another step is required to generate thymidylate from uracil. *Thymidylate synthase* catalyzes this finishing touch: deoxyuridylate (dUMP) is methylated to thymidylate (TMP). Recall that thymidylate synthase also functions in the thymine salvage pathway (p. 826). As will be described in Chapter 29, the methylation of this nucleotide marks sites of DNA damage for repair and, hence, helps preserve the integrity of the genetic information stored in DNA. The methyl donor in this reaction is N^5, N^{10}-methylenetetrahydrofolate rather than *S*-adenosylmethionine (Section 24.2).

The methyl group becomes attached to the C-5 atom within the aromatic ring of dUMP, but this carbon atom is not a good nucleophile and cannot itself attack the appropriate group on the methyl donor. Thymidylate synthase promotes methylation by adding a thiolate from a cysteine side chain to this ring to generate a nucleophilic species that can attack the methylene group of N^5, N^{10}-methylenetetrahydrofolate (Figure 25.12). This methylene group, in turn, is activated by distortions imposed by the enzyme that favor opening the five-membered ring. The activated dUMP's attack on the methylene group forms the new carbon–carbon bond. The intermediate formed is then converted into product: a hydride ion is transferred from the tetrahydrofolate ring to transform the methylene group into a methyl group, and a proton is abstracted from the carbon atom bearing the methyl group to eliminate the cysteine and regenerate the aromatic ring. The tetrahydrofolate derivative loses both its methylene group and a hydride ion and, hence, is oxidized to dihydrofolate. For the synthesis of more thymidylate, tetrahydrofolate must be regenerated.

FIGURE 25.12 Thymidylate synthesis. Thymidylate synthase catalyzes the addition of a methyl group (derived from N^5, N^{10}-methylenetetrahydrofolate) to dUMP to form TMP. The addition of a thiolate from the enzyme activates dUMP. Opening the five-membered ring of the THF derivative prepares the methylene group for nucleophilic attack by the activated dUMP. The reaction is completed by the transfer of a hydride ion to form dihydrofolate.

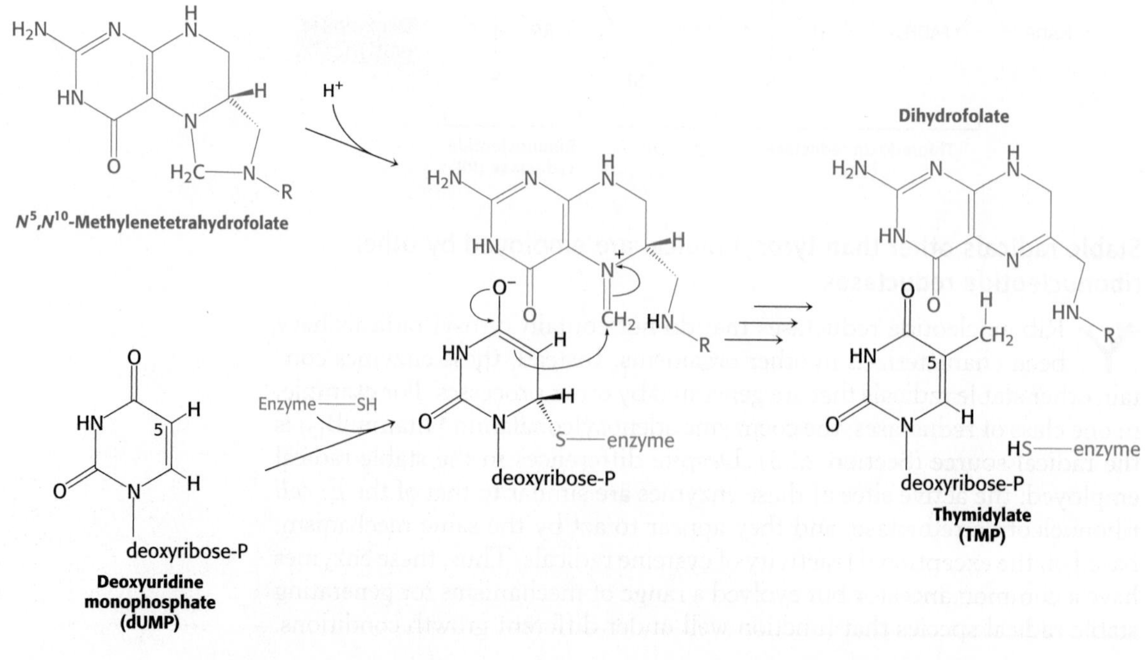

Dihydrofolate reductase catalyzes the regeneration of tetrahydrofolate, a one-carbon carrier

Tetrahydrofolate is regenerated from the dihydrofolate that is produced in the synthesis of thymidylate. This regeneration is accomplished by *dihydrofolate reductase* with the use of NADPH as the reductant.

Dihydrofolate + NADPH + H$^+$ ⟶ **Tetrahydrofolate** + NADP$^+$

A hydride ion is directly transferred from the nicotinamide ring of NADPH to the pteridine ring of dihydrofolate. The bound dihydrofolate and NADPH are held in close proximity to facilitate the hydride transfer.

Several valuable anticancer drugs block the synthesis of thymidylate

Rapidly dividing cells require an abundant supply of thymidylate for the synthesis of DNA. The vulnerability of these cells to the inhibition of TMP synthesis has been exploited in the treatment of cancer. Thymidylate synthase and dihydrofolate reductase are choice targets of chemotherapy (Figure 25.13).

Fluorouracil, an anticancer drug, is converted in vivo into *fluorodeoxyuridylate* (F-dUMP). This analog of dUMP irreversibly inhibits thymidylate synthase after acting as a normal substrate through part of the catalytic cycle. Recall that the formation of TMP requires the removal of a proton (H$^+$) from C-5 of the bound nucleotide (Figure 25.12). However, the enzyme cannot abstract F$^+$ from F-dUMP, and so catalysis is blocked at the stage of the covalent complex formed by F-dUMP, methylenetetrahydrofolate, and the sulfhydryl group of the enzyme (Figure 25.14). We see here an example of *suicide inhibition*, in which an enzyme converts a substrate into a reactive inhibitor that halts the enzyme's catalytic activity (Section 8.5).

The synthesis of TMP can also be blocked by inhibiting the regeneration of tetrahydrofolate. Analogs of dihydrofolate,

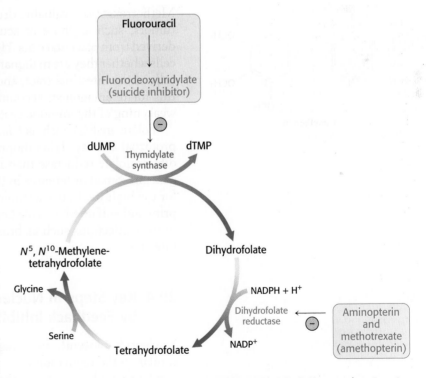

FIGURE 25.13 Anticancer drug targets. Thymidylate synthase and dihydrofolate reductase are choice targets in cancer chemotherapy because the generation of large quantities of precursors for DNA synthesis is required for rapidly dividing cancer cells.

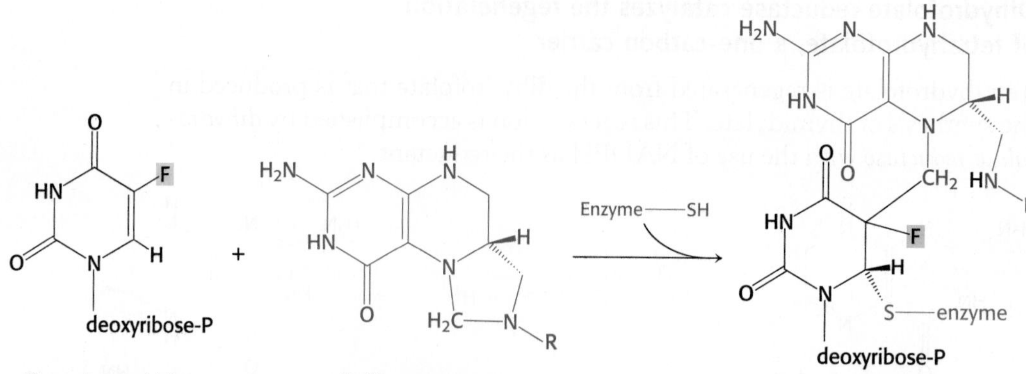

FIGURE 25.14 Suicide inhibition. Fluorodeoxyuridylate (generated from fluorouracil) traps thymidylate synthase in a form that cannot proceed down the reaction pathway.

Fluorodeoxyuridylate   + N^5,N^{10}-Methylene-tetrahydrofolate   →   Stable adduct

deoxyribose-P

deoxyribose-P

Enzyme—SH

such as *aminopterin* and *methotrexate* (amethopterin), are potent competitive inhibitors ($K_i < 1$ nM) of dihydrofolate reductase.

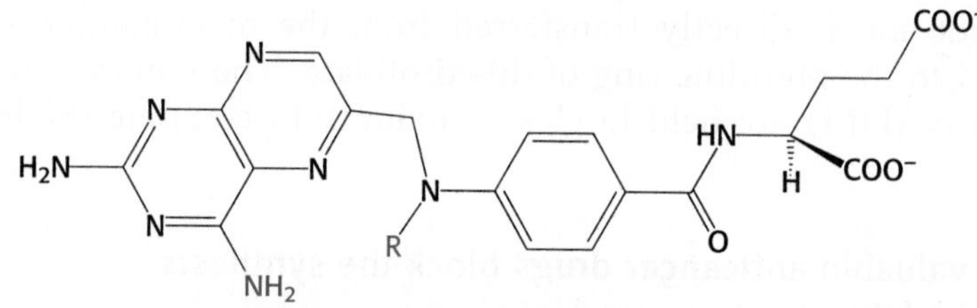

Aminopterin (R = H) or methotrexate (R = CH₃)

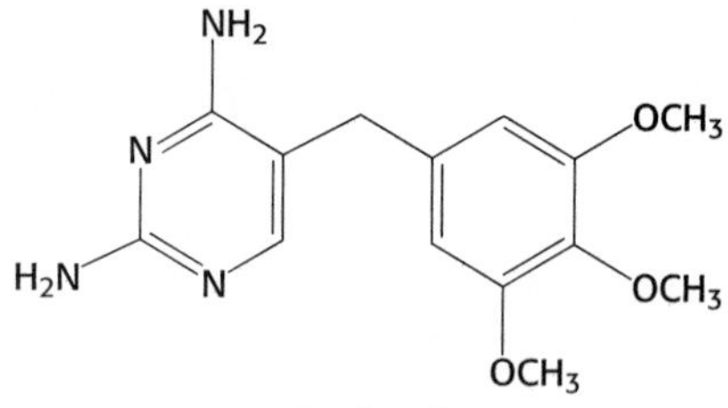

Methotrexate is a valuable drug in the treatment of many rapidly growing tumors, such as those in acute leukemia and choriocarcinoma, a cancer derived from placental cells. However, methotrexate kills rapidly replicating cells whether they are malignant or not. Stem cells in bone marrow, epithelial cells of the intestinal tract, and hair follicles are vulnerable to the action of this folate antagonist, accounting for its toxic side effects, which include weakening of the immune system, nausea, and hair loss.

Folate analogs such as *trimethoprim* have potent antibacterial and antiprotozoal activity. Trimethoprim binds 10^5-fold less tightly to mammalian dihydrofolate reductase than it does to reductases of susceptible microorganisms. Small differences in the active-site clefts of these enzymes account for the highly selective antimicrobial action. The combination of trimethoprim and sulfamethoxazole (an inhibitor of folate synthesis) is widely used to treat infections such as bronchitis, traveler's diarrhea, and urinary tract infections.

Trimethoprim

25.4 Key Steps in Nucleotide Biosynthesis Are Regulated by Feedback Inhibition

Nucleotide biosynthesis is regulated by feedback inhibition in a manner similar to the regulation of amino acid biosynthesis (Section 24.3). These regulatory pathways ensure that the various nucleotides are produced in the required quantities.

Pyrimidine biosynthesis is regulated by aspartate transcarbamoylase

Aspartate transcarbamoylase, one of the key enzymes for the regulation of pyrimidine biosynthesis in bacteria, was described in detail in Chapter 10.

Recall that *ATCase is inhibited by CTP, the final product of pyrimidine biosynthesis,* and stimulated by ATP.

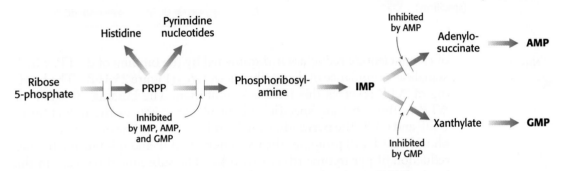

The synthesis of purine nucleotides is controlled by feedback inhibition at several sites

The regulatory scheme for purine nucleotides is more complex than that for pyrimidine nucleotides (Figure 25.15).

FIGURE 25.15 Control of purine biosynthesis. Feedback inhibition controls both the overall rate of purine biosynthesis and the balance between AMP and GMP production.

1. The committed step in purine nucleotide biosynthesis is the conversion of PRPP into phosphoribosylamine by *glutamine phosphoribosyl amido-transferase.* This important enzyme is feedback-inhibited by many purine ribonucleotides. It is noteworthy that AMP and GMP, the final products of the pathway, synergistically inhibit the amidotransferase.

2. Inosinate is the branch point in the synthesis of AMP and GMP. *The reactions leading away from inosinate are sites of feedback inhibition.* AMP inhibits the conversion of inosinate into adenylosuccinate, its immediate precursor. Similarly, GMP inhibits the conversion of inosinate into xanthylate, its immediate precursor.

3. As already noted, GTP is a substrate in the synthesis of AMP, whereas ATP is a substrate in the synthesis of GMP (Figure 25.7). This *reciprocal substrate relation* tends to balance the synthesis of adenine and guanine ribonucleotides.

Note that the synthesis of PRPP by PRPP synthetase is highly regulated even though it is not the committed step in purine synthesis. Mutations have been identified in PRPP synthetase that result in a loss of allosteric response to nucleotides without any effect on catalytic activity of the enzyme. A consequence of this mutation is an overabundance of purine nucleotides that can result in gout, a pathological condition discussed in Section 25.5.

The synthesis of deoxyribonucleotides is controlled by the regulation of ribonucleotide reductase

The reduction of ribonucleotides to deoxyribonucleotides is precisely controlled by allosteric interactions. Each polypeptide of the R1 subunit of the aerobic *E. coli* ribonucleotide reductase contains two allosteric sites: one of them controls the *overall activity* of the enzyme, and the other regulates *substrate specificity* (Figure 25.16A). The overall catalytic activity

FIGURE 25.16 Regulation of ribonucleotide reductase. (A) Each subunit in the R1 dimer contains two allosteric sites in addition to the active site. One site regulates the overall activity and the other site regulates substrate specificity. (B) The patterns of regulation with regard to different nucleoside diphosphates demonstrated by ribonucleotide reductase.

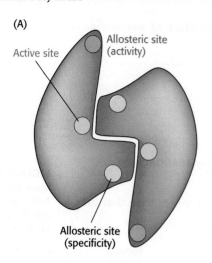

(A)

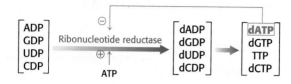

(B) Regulation of overall activity

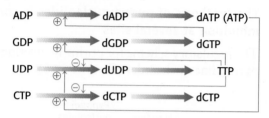

Regulation of substrate specificity

of ribonucleotide reductase is diminished by the binding of dATP, which signals an abundance of deoxyribonucleotides (Figure 25.16B). The binding of ATP reverses this feedback inhibition. The binding of dATP or ATP to the substrate-specificity control site enhances the reduction of UDP and CDP, the pyrimidine nucleotides. The binding of thymidine triphosphate (TTP) promotes the reduction of GDP and inhibits the further reduction of pyrimidine ribonucleotides. The subsequent increase in the level of dGTP stimulates the reduction of ATP to dATP. This complex pattern of regulation supplies the appropriate balance of the four deoxyribonucleotides needed for the synthesis of DNA.

Ribonucleotide reductase is an attractive target for cancer therapy and a number of clinically approved anticancer drugs mimic substrates and regulators of the enzyme. The pyrimidine analog gemcitabine, once converted to the diphosphate form in vivo, becomes a suicide inhibitor of the reductase used to treat advanced pancreatic cancer. The purine analogs clofarabine and cladribine, upon conversion to their triphosphate forms in vivo, are dATP analogs and act as allosteric inhibitors of the enzyme. Clofarabine is used to treat pediatric acute myeloid leukemia while cladribine is effective against some forms of chronic lymphoid leukemia.

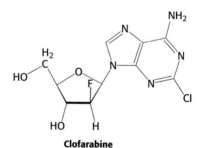

Gemcitabine

Clofarabine

Cladribine

25.5 Disruptions in Nucleotide Metabolism Can Cause Pathological Conditions

Nucleotides are vital to a host of biochemical processes. It is not surprising, then, that disruption of nucleotide metabolism would have a variety of physiological effects. The nucleotides of a cell undergo continual turnover. Nucleotides are hydrolytically degraded to nucleosides by *nucleotidases*. The phosphorolytic cleavage of nucleosides to free bases and ribose 1-phosphate (or deoxyribose 1-phosphate) is catalyzed by *nucleoside phosphorylases*. Ribose 1-phosphate is isomerized by *phosphoribomutase* to ribose 5-phosphate, a substrate in the synthesis of PRPP. Some of the bases are reused to form nucleotides by salvage pathways. Others are degraded to products that are excreted (Figure 25.17). A deficiency of an enzyme can disrupt these pathways, leading to a pathological condition.

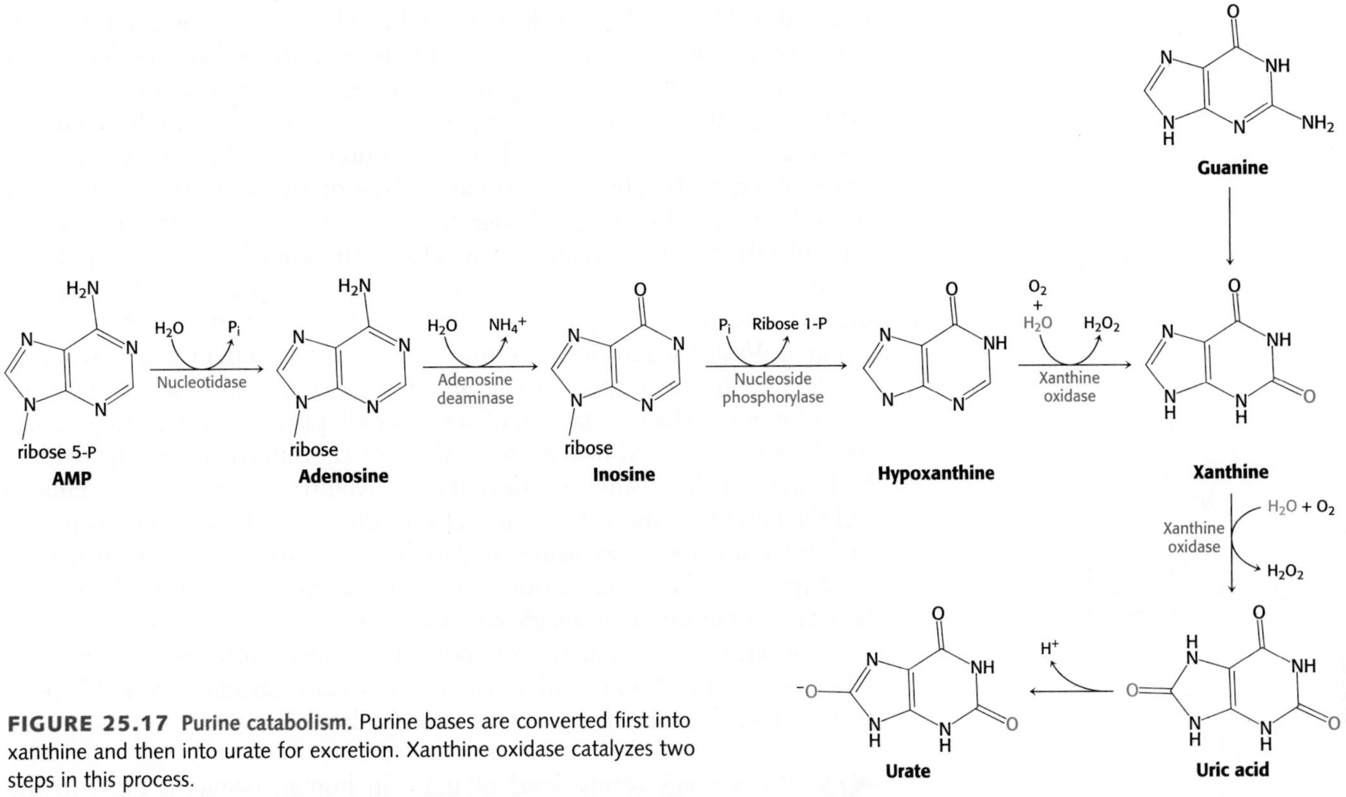

FIGURE 25.17 Purine catabolism. Purine bases are converted first into xanthine and then into urate for excretion. Xanthine oxidase catalyzes two steps in this process.

The loss of adenosine deaminase activity results in severe combined immunodeficiency

The pathway for the degradation of AMP includes an extra step because adenosine is not a substrate for nucleoside phosphorylase. First, the phosphate is removed by a nucleotidase to yield the nucleoside adenosine (Figure 25.17). In the extra step, adenosine is deaminated by *adenosine deaminase* to form inosine.

A deficiency in adenosine deaminase activity is associated with some forms of *severe combined immunodeficiency* (SCID), an immunological disorder. Persons with the disorder have acute recurring infections, often leading to death at an early age. SCID is characterized by a loss of T cells, which are crucial to the immune response (See Section 35.5 on 🅢 **Sapling**Plus). Although the biochemical basis of the disorder is not clearly established, a lack of adenosine deaminase results in an increase of 50 to 100 times the normal level of dATP, which inhibits ribonucleotide reductase and, consequently, DNA synthesis. Moreover, adenosine itself is a powerful signal molecule with a role in a number of regulatory pathways. Disruption in the levels of adenosine may also be deleterious. SCID is often called the "bubble boy disease" because its treatment may include complete isolation of the patient from the environment. Adenosine deaminase deficiency has been successfully treated by gene therapy.

Gout is induced by high serum levels of urate

Inosine generated by adenosine deaminase is subsequently metabolized by *nucleoside phosphorylase* to hypoxanthine. *Xanthine oxidase*, a molybdenum- and iron-containing flavoprotein, oxidizes hypoxanthine to *xanthine* and then to *uric acid*. Molecular oxygen, the oxidant in both reactions, is reduced to H_2O_2, which is decomposed to H_2O and O_2 by catalase. Uric acid loses a proton at physiological pH to form *urate*. *In human beings, urate is the final product of purine degradation and is excreted in the urine.*

High serum levels of urate (hyperuricemia) induce the painful joint disease *gout*. In this disease, the sodium salt of urate crystallizes in the fluid and lining of the joints. The small joint at the base of the big toe is a common site for sodium urate buildup, although the salt accumulates at other joints also. Painful inflammation results when cells of the immune system engulf the sodium urate crystals. The kidneys, too, may be damaged by the deposition of urate crystals. Gout is a common medical problem, affecting 1% of the population of Western countries. It is nine times as common in men as in women.

Administration of *allopurinol*, an analog of hypoxanthine, is one treatment for gout. The mechanism of action of allopurinol is interesting: it acts *first as a substrate and then as an inhibitor* of xanthine oxidase. The oxidase hydroxylates allopurinol to *alloxanthine (oxipurinol)*, which then remains tightly bound to the active site. The binding of alloxanthine keeps the molybdenum atom of xanthine oxidase in the +4 oxidation state instead of it returning to the +6 oxidation state as in a normal catalytic cycle. We see here another example of *suicide inhibition.*

The synthesis of urate from hypoxanthine and xanthine decreases soon after the administration of allopurinol. The serum concentrations of hypoxanthine and xanthine rise, and that of urate drops.

Allopurinol

The average serum level of urate in human beings is close to the solubility limit and is higher than levels found in other primates. What is the selective advantage of a urate level so high that it teeters on the brink of gout in many people? It turns out that urate has a markedly beneficial action. Urate is a highly effective scavenger of reactive oxygen species. Indeed, urate is about as effective as ascorbate (vitamin C) as an antioxidant. The increased level of urate in human beings may protect against reactive oxygen species that are implicated in a host of pathological conditions (Table 18.3).

Lesch–Nyhan syndrome is a dramatic consequence of mutations in a salvage-pathway enzyme

Mutations in genes that encode nucleotide biosynthetic enzymes can reduce levels of needed nucleotides and can lead to an accumulation of intermediates. A nearly total absence of hypoxanthine-guanine phosphoribosyltransferase (HGPRT) has unexpected and devastating consequences. The most striking expression of this inborn error of metabolism, called the *Lesch–Nyhan syndrome*, is *compulsive self-destructive behavior*. At age 2 or 3, children with this disease begin to bite their fingers and lips and will chew them off if unrestrained. These children also behave aggressively toward others. *Cognitive disabilities* and *spasticity* are other characteristics of the Lesch–Nyhan syndrome. Elevated levels of urate in the serum lead to the formation of kidney stones early in life, followed by the symptoms of gout years later. The disease is inherited as a sex-linked recessive disorder.

What is the connection between the absence of HGPRT activity and the behavioral characteristics of Lesch-Nyhan syndrome? The answer is not clear, but it is possible to speculate. The brain has limited capacity for de novo purine synthesis. Consequently, lack of HGPRT results in a deficiency of purine nucleotides. ATP and ADP, formed from inosinate, are especially important in the brain as signal molecules. These nucleotides bind to and activate G-protein coupled receptors that regulate the dopamine secreting neurons. Thus, the lack of HGPRT results in an imbalance of key neurotransmitters. Moreover, the guanosine nucleotides required for G-protein function may also be in short supply. The Lesch–Nyhan syndrome demonstrates that the salvage pathway for the synthesis of IMP and GMP is not of minor importance. Moreover, the Lesch–Nyhan syndrome reveals that *abnormal behavior such as self-mutilation and extreme hostility can be caused by the absence of a single enzyme.* Psychiatry will no doubt benefit from the unraveling of the molecular basis of such mental disorders.

Folic acid deficiency promotes birth defects such as spina bifida

Spina bifida is one of a class of birth defects characterized by the incomplete or incorrect formation of the neural tube early in development. In the United States, the prevalence of *neural-tube defects* is approximately 1 case per 1000 births. A variety of studies have demonstrated that the prevalence of neural-tube defects is reduced by as much as 70% when women take folic acid as a dietary supplement before and during the first trimester of pregnancy. One hypothesis is that more folate derivatives are needed for the synthesis of DNA precursors when cell division is frequent and substantial amounts of DNA must be synthesized.

SUMMARY

25.1 The Pyrimidine Ring Is Assembled de Novo or Recovered by Salvage Pathways

The pyrimidine ring is assembled first and then linked to ribose phosphate to form a pyrimidine nucleotide. 5-Phosphoribosyl-1-pyrophosphate is the donor of the ribose phosphate moiety. The synthesis of the pyrimidine ring starts with the formation of carbamoylaspartate from carbamoyl phosphate and aspartate, a reaction catalyzed by aspartate transcarbamoylase. Dehydration, cyclization, and oxidation yield orotate, which reacts with PRPP to give orotidylate. Decarboxylation of this pyrimidine nucleotide yields UMP. CTP is then formed by the amination of UTP.

25.2 Purine Bases Can Be Synthesized de Novo or Recycled by Salvage Pathways

The purine ring is assembled from a variety of precursors: glutamine, glycine, aspartate, N^{10}-formyltetrahydrofolate, and CO_2. The committed step in the de novo synthesis of purine nucleotides is the formation of 5-phosphoribosylamine from PRPP and glutamine. The purine ring is assembled on ribose phosphate, in contrast with the de novo synthesis of pyrimidine nucleotides. The addition of glycine, followed by formylation, amination, and ring closure, yields 5-aminoimidazole ribonucleotide. This intermediate contains the completed five-membered ring of the purine skeleton. The addition of CO_2, the nitrogen atom of aspartate, and a formyl group, followed by ring closure, yields inosinate, a purine ribonucleotide. AMP and GMP are formed from IMP. Purine ribonucleotides can also be synthesized by a salvage pathway in which a preformed base reacts directly with PRPP.

25.3 Deoxyribonucleotides Are Synthesized by the Reduction of Ribonucleotides Through a Radical Mechanism

Deoxyribonucleotides, the precursors of DNA, are formed in *E. coli* by the reduction of ribonucleoside diphosphates. These conversions are catalyzed by ribonucleotide reductase. Electrons are transferred from NADPH to sulfhydryl groups at the active sites of this enzyme by thioredoxin. A tyrosyl free radical generated by an iron center in the reductase initiates a radical reaction on the sugar, leading to the exchange of H for OH at C-2′. TMP is formed by the methylation of dUMP. The donor of a methylene group and a hydride in this reaction is N^5, N^{10}-methylenetetrahydrofolate, which is converted into dihydrofolate. Tetrahydrofolate is regenerated by the reduction of dihydrofolate by NADPH. Dihydrofolate reductase, which catalyzes this reaction, is inhibited by folate analogs such as aminopterin and methotrexate. These compounds and fluorouracil, an inhibitor of thymidylate synthase, are used as anticancer drugs.

25.4 Key Steps in Nucleotide Biosynthesis Are Regulated by Feedback Inhibition

Pyrimidine biosynthesis in *E. coli* is regulated by the feedback inhibition of aspartate transcarbamoylase, the enzyme that catalyzes the committed step. CTP inhibits and ATP stimulates this enzyme. The feedback inhibition of glutamine-PRPP amidotransferase by purine nucleotides is important in regulating their biosynthesis. Disruption of the regulation and activity of ribonucleotide reductase by drugs is an effective chemotherapy for some types of cancer.

25.5 Disruptions in Nucleotide Metabolism Can Cause Pathological Conditions

Severe combined immunodeficiency results from the absence of adenosine deaminase, an enzyme in the purine degradation pathway. Purines are degraded to urate in human beings. Gout, a disease that affects joints and leads to arthritis, is associated with an excessive accumulation of urate. The Lesch–Nyhan syndrome, a genetic disease characterized by self-mutilation, cognitive disabilites, and gout, is caused by the absence of hypoxanthine-guanine phosphoribosyltransferase. This enzyme is essential for the synthesis of purine nucleotides by the salvage pathway. Neural-tube defects are more frequent when a pregnant woman is deficient in folate derivatives early in pregnancy, possibly because of the important role of these derivatives in the synthesis of DNA precursors.

APPENDIX

Biochemistry in Focus

Uridine plays a role in caloric homeostasis

Uridine, a pyrimidine nucleoside, and its metabolites play a variety of important roles in the cell.

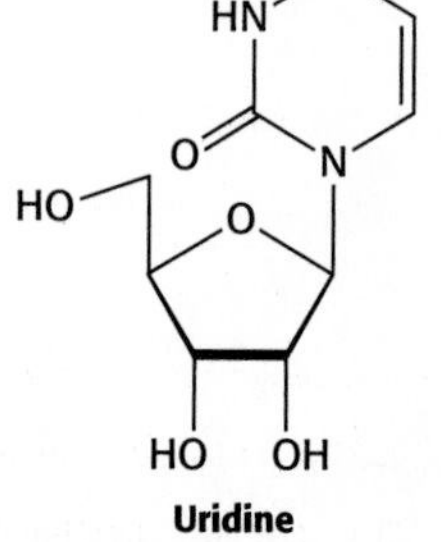

Uridine

It is a component of RNA (Chapter 30), a precursor to thymidine nucleotides (p. 834), is required for glycogen synthesis (p. 695) and the glycosylation of proteins (p. 358) and lipids (Section 26.1). As we will see in Chapter 28, it also plays a crucial role in the detoxification of xenobiotics, chemicals foreign to an organism. Recent research points to a new role of uridine as a signal molecule in the maintenance of caloric homeostasis. Caloric homeostasis is the ability to maintain adequate but not excessive energy stores. We will examine this topic in more detail in Chapter 27.

The concentration of uridine in the blood of rodents and humans is dependent on the nutritional status of the organism—whether it is fasting or fed. The physiological effects of uridine are dependent on complex interactions between adipose (fat) tissue and the liver. Adipocytes release uridine into the blood during fasting and it is thought to act on the brain to increase appetite. Uridine may also be responsible for the small drop in body temperature that is a characteristic of fasting and an overall decrease in the metabolic rate of the organism during fasting.

Upon refeeding, uridine is cleared from the blood by the liver. In the liver, some of the uridine is transported to the gallbladder, where it becomes a component of bile. Bile is a fluid containing biochemicals that assists with the digestion

and absorption of dietary fats (Section 26.4). Upon refeeding, bile is released into the intestine from the gallbladder. Uridine, in some fashion, stimulates the uptake of glucose from the intestine.

This new role of uridine is complex, and much research needs to be done to understand how uridine actually functions. However, it is fascinating to see how many important biochemical activities require uridine and its metabolites.

APPENDIX

Problem-Solving Strategies

PROBLEM: Suppose you have cells growing in culture that are capable of de novo synthesis of purine nucleotides. Now imagine that these cells have the proper concentration of AMP but the concentration of GMP is low.

(a) Explain the regulatory processes of purine synthesis that would result in an increase in the concentration of GMP.

(b) What would be the result of having adequate concentrations of both AMP and GMP?

SOLUTION: Let's look at (a) first. It should be reasonably clear that this question is about the regulation of purine nucleotide synthesis, and, more specifically how AMP and GMP contribute to the regulation.

▶ **Outline the regulatory scheme for purine synthesis.**

The regulatory scheme is shown in Figure 25.15, but we know how busy you are, so to save you some time, we show it below.

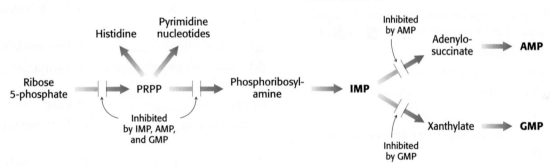

Examine the regulatory pathway.

▶ **What would be the effect of adequate AMP concentration?**

This is a bit tricky. You could say that AMP would inhibit the production of PRPP and phosphoribosylamine, limiting the amount of IMP synthesized. However, that inhibition would not increase the concentration of GMP, and yet the cell requires adequate amounts of both AMP and GMP.

▶ **How could AMP inhibition of a pathway lead to an increase in GMP?**

AMP also inhibits the synthesis of adenylosuccinate from IMP, a step on the pathway to AMP synthesis. In

other words, there is no need to make AMP if there is an adequate amount already present. Consequently, any IMP formed would be directed to GMP synthesis.

What about part (b)?

▶ **What would be the effect of adequate concentration of AMP and GMP?**

AMP and GMP would inhibit their own synthesis from IMP. As a result, the concentration of IMP would rise. The elevated concentration of AMP, GMP, and IMP would inhibit the production of PRPP and phosphoribosylamine, effectively shutting down purine nucleotide synthesis.

KEY TERMS

pyrimidine nucleotide (p. 822)

carbamoyl phosphate synthetase II (CPS II) (p. 823)

ATP-grasp fold (p. 823)

5-phosphoribosyl-1-pyrophosphate (PRPP) (p. 824)

orotidylate (p. 824)

salvage pathway (p. 826)

purine nucleotide (p. 826)

glutamine phosphoribosyl amidotransferase (p. 827)
ribonucleotide reductase (p. 831)
thymidylate synthase (p. 834)

dihydrofolate reductase (p. 835)
severe combined immunodeficiency (SCID) (p. 839)
gout (p. 840)

Lesch–Nyhan syndrome (p. 840)
spina bifida (p. 841)
neural-tube defect (p. 841)

PROBLEMS

1. *Contribution.* Which of the following pairs of molecules form the pyrimidine ring? ✓❶

(a) Glutamate and carbamoyl phosphate

(b) Aspartate and glutamine

(c) Aspartate and carbamoyl phosphate

(d) Glycine and carbamoyl phosphate

2. *No contribution.* Which of the following molecules do NOT contribute atoms to purine synthesis? ✓❷

(a) Glutamine

(b) Aspartate

(c) Glycine

(d) Lysine

3. *Activated sugar.* The phosphorylated 5-carbon monosaccharide required for nucleotide synthesis is: ✓❶, ✓❷

(a) Ribulose 1, 5-bisphosphate

(b) 5-phosphoribosyl-1-pyrophosphate

(c) Xylulose 5-phosphate

(d) Ribose 5-phosphate

4. *At the controls.* The key regulatory enzyme in pyrimidine synthesis is: ✓❹

(a) Aspartate transcarbamoylase.

(b) Glutamine phosphoribosyl amidotransferase.

(c) Uridine monophosphate synthetase.

(d) Orotidylate decarboxylase

5. *Pathway management.* The key regulatory enzyme is purine synthesis is: ✓❹

(a) Glycinamide ribonucleotide synthetase.

(b) Adenylosuccinate lyase.

(c) Glutamine phosphoribosyl amidotransferase.

(d) Carbamoyl phosphate synthetase.

6. *From the beginning or extract and save and reuse.* Differentiate between the de novo synthesis of nucleotides and salvage-pathway synthesis.

7. *Finding their roots 1.* Identify the source of the atoms in the pyrimidine ring. ✓❶

8. *Finding their roots 2.* Identify the source of the atoms in the purine ring. ✓❷

9. *Multifaceted.* List some of the biochemical roles played by nucleotides.

10. *An s instead of a t?* Differentiate between a nucleoside and a nucleotide.

11. *Associate 'em.*

(a) Excessive urate _____   1. Spina bifida

(b) Lack of adenosine deaminase _____   2. Precursor to both ATP and GTP

(c) Lack of HGPRT _____   3. Purine

(d) Carbamoyl phosphate _____   4. Deoxynucleotide synthesis

(e) Inosinate _____   5. UTP

(f) Ribonucleotide reductase _____   6. Lesch-Nyhan disease

(g) Lack of folic acid _____   7. Immunodeficiency

(h) Glutamine phosphoribosyl transferase _____   8. Pyrimidine

(i) Single ring _____   9. Gout

(j) Bicyclic ring _____   10. First step in pyrimidine synthesis

(k) Precursor to CTP _____   11. Committed step in purine synthesis

12. *Safe passage.* What is substrate channeling? How does it affect enzyme efficiency?

13. *Activated ribose phosphate.* Write a balanced equation for the synthesis of PRPP from glucose through the oxidative branch of the pentose phosphate pathway.

14. *Making a pyrimidine.* Write a balanced equation for the synthesis of orotate from glutamine, CO_2, and aspartate. ✓❶

15. *Identifying the donor.* What is the activated reactant in the biosynthesis of each of the following compounds?

(a) Phosphoribosylamine

(c) Orotidylate (from orotate)

(b) Carbamoylaspartate

(d) Phosphoribosylanthranilate

16. *Inhibiting purine biosynthesis.* Amidotransferases are inhibited by the antibiotic azaserine (*O*-diazoacetyl-L-serine), which is an analog of glutamine. ✓❷

Azaserine

Which intermediates in purine biosynthesis would accumulate in cells treated with azaserine?

17. *The price of methylation.* Write a balanced equation for the synthesis of TMP from dUMP that is coupled to the conversion of serine into glycine. ✓❸

18. *Sulfa action.* Bacterial growth is inhibited by sulfanilamide and related sulfa drugs, and there is a concomitant accumulation of 5-aminoimidazole-4-carboxamide ribonucleotide. This inhibition is reversed by the addition of *p*-aminobenzoate.

Sulfanilamide

Propose a mechanism for the inhibitory effect of sulfanilamide.

19. *HAT medium.* Mutant cells unable to synthesize nucleotides by salvage pathways are very useful tools in molecular and cell biology. Suppose that cell A lacks thymidine kinase, the enzyme catalyzing the phosphorylation of thymidine to thymidylate, and that cell B lacks hypoxanthine-guanine phosphoribosyl transferase.

(a) Cell A and cell B do not proliferate in a HAT medium containing *hypoxanthine*, *aminopterin* or *amethopterin* (methotrexate), and *thymine*. However, cell C, formed by the fusion of cells A and B, grows in this medium. Why?

(b) Suppose that you want to introduce foreign genes into cell A. Devise a simple means of distinguishing between cells that have taken up foreign DNA and those that have not.

20. *Vitamin requirement.* What is the role of folate in pyrimidine synthesis and what is the consequence of a folate deficiency during development? ✓❶

21. *Bringing equilibrium.* What is the reciprocal substrate relation in the synthesis of ATP and GTP? ✓❹

22. *Find the label.* Suppose that cells are grown on amino acids that have all been labeled at the α carbons with ^{13}C. Identify the atoms in cytosine and guanine that will be labeled with ^{13}C. ✓❶, ✓❷

23. *Need a map?* Describe the path for the synthesis of TTP from UTP. ✓❸

24. *Adjunct therapy.* Allopurinol is sometimes given to patients with acute leukemia who are being treated with anticancer drugs. Why is allopurinol used?

25. *A hobbled enzyme.* Both side-chain oxygen atoms of aspartate 27 at the active site of dihydrofolate reductase form hydrogen bonds with the pteridine ring of folates. The importance of this interaction was assessed by studying two mutants at this position, Asn 27 and Ser 27. The dissociation constant of methotrexate was 0.07 nM for the wild type, 1.9 nM for the Asn 27 mutant, and 210 nM for the Ser 27 mutant, at 25°C. Calculate the standard free energy of the binding of methotrexate by these three proteins. What is the decrease in binding energy resulting from each mutation? ✓❸

26. *Needed supplies.* Why are cancer cells especially sensitive to inhibitors of TMP synthesis? ✓❸

27. *Alternative routes.* Orotic aciduria results when one of the enzyme activities of UMP synthetase is absent. This syndrome is characterized by large amounts of orotic acid in the blood and urine, megaloblastic anemia (characterized by large, immature, and dysfunctional red blood cells) and retarded growth. Suggest a possible treatment for this condition. ✓❺

28. *Correcting deficiencies.* Suppose that a person is found who is deficient in an enzyme required for IMP synthesis. How might this person be treated? ✓❺

29. *Labeled nitrogen.* Purine biosynthesis is allowed to take place in the presence of [^{15}N] aspartate, and the newly synthesized GTP and ATP are isolated. What positions are labeled in the two nucleotides? ✓❷

30. *On the trail of carbons.* Tissue culture cells were incubated with glutamine labeled with ^{15}N in the amide group. Subsequently, IMP was isolated and found to contain some ^{15}N. Which atoms in IMP were labeled? ✓❷

31. *Mechanism of action.* What is the biochemical basis of allopurinol treatment for gout? ✓❺

32. *Changed inhibitor.* Xanthine oxidase treated with allopurinol results in the formation of a new compound that is an extremely potent inhibitor of the enzyme. Propose a structure for this compound. ✓❺

33. *Calculate the ATP footprint.* How many molecules of ATP are required to synthesize one molecule of CTP from scratch? ✓❶

34. *Blockages.* What intermediate in purine synthesis will accumulate if a strain of bacteria is lacking each of the following biochemicals? ✓❷

(a) Aspartate

(b) Tetrahydrofolate

(c) Glycine

(d) Glutamine

Mechanism Problems

35. *The same and not the same.* Write out mechanisms for the conversion of phosphoribosylamine into glycinamide ribonucleotide and of xanthylate into guanylate. ✓❷

36. *Closing the ring.* Propose a mechanism for the conversion of 5-formamidoimidazole-4-carboxamide ribonucleotide into inosinate.

Chapter Integration Problems

37. *Different strokes.* Human beings contain two different carbamoyl phosphate synthetase enzymes. One uses glutamine as a substrate, whereas the other uses ammonia. What are the functions of these two enzymes?

38. *A generous donor.* What major biosynthetic reactions utilize PRPP?

39. *They're everywhere!* Nucleotides play a variety of roles in the cell. Give an example of a nucleotide that acts in each of the following roles or processes.

(a) Second messenger

(b) Phosphoryl-group transfer

(c) Activation of carbohydrates

(d) Activation of acetyl groups

(e) Transfer of electrons

(f) DNA sequencing

(g) Chemotherapy

(h) Allosteric effector

40. *Pernicious anemia.* Purine biosynthesis is impaired by vitamin B_{12} deficiency. Why? How might fatty acid and amino acid metabolism also be affected by a vitamin B_{12} deficiency? ✓❷

41. *Folate deficiency.* Suppose someone was suffering from a folate deficiency. What cells would you think might be most affected? Symptoms may include diarrhea and anemia.

42. *Hyperuricemia.* Many patients with glucose 6-phosphatase deficiency have high serum levels of urate. Hyperuricemia can be induced in normal people by the ingestion of alcohol or by strenuous exercise. Propose a common mechanism that accounts for these findings. ✓❺

43. *Labeled carbon.* Succinate uniformly labeled with ^{14}C is added to cells actively engaged in pyrimidine biosynthesis. Propose a mechanism by which carbon atoms from succinate could be incorporated into a pyrimidine. At what positions is the pyrimidine labeled? ✓❶

44. *Something funny going on here.* Cells were incubated with glucose labeled with ^{14}C in carbon 2, shown in red in the structure below. Later, uracil was isolated and found to contain ^{14}C in carbons 4 and 6. Account for this labeling pattern. ✓❶

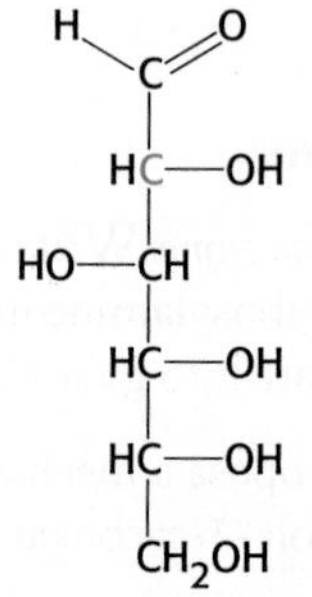

45. *Side effects.* Azathioprine is an immunosuppressant drug used in kidney transplants. In vivo, azathioprine is metabolized to the hypoxanthine analog 6-mercaptopurine, which is then converted into 6-mercaptopurine ribose monophosphate, the active form of the drug. 6-Mercaptopurine ribose monophosphate also inhibits de novo purine synthesis, reducing uric acid levels in the blood and urine. However, when administered to Lesch–Nyhan patients, there was no effect on uric acid levels. Explain why this is the case. ✓❷, ✓❺

46. *Exercising muscle.* Some interesting reactions take place in muscle tissue to facilitate the generation of ATP for contraction. In muscle contraction, ATP is converted into ADP. Adenylate kinase converts two molecules of ADP into a molecule of ATP and AMP. ✓❷

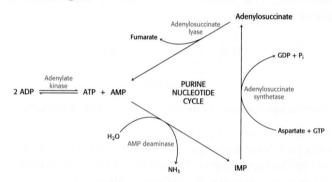

(a) Why is this reaction beneficial to contracting muscle?

(b) Why is the equilibrium for the adenylate kinase approximately equal to 1?

Muscle can metabolize AMP by using the purine nucleotide cycle. The initial step in this cycle, catalyzed by AMP deaminase, is the conversion of AMP into IMP.

(c) Why might the deamination of AMP facilitate ATP formation in muscle?

(d) How does the purine nucleotide cycle assist the aerobic generation of ATP?

47. *A common step.* What three reactions transfer an amino group from aspartate to yield the aminated product and fumarate?

48. *Your pet duck.* You suspect that your pet duck has gout. Why should you think twice before administering a dose of allopurinol-laced bread? ✓❺

Data Interpretation Problem

49. *Not all absences are equal.* In order to better understand the nature of the defects in HGPRT in patients suffering from Lesch–Nyhan disease, cell lines from two Lesch–Nyhan patients (LND 1 and LND 2) as well as a normal individual were established. HGPRT was then purified from each of the cell lines using the techniques examined in Chapter 3 (see animation on 🅢 Sapling Plus). Using the purified enzyme, the basic kinetics of the three enzymes were compared, as shown in Figure A. In each case, the same amount of purified enzyme was used in the analysis. ✓❺

(a) Describe the results shown in the figure.

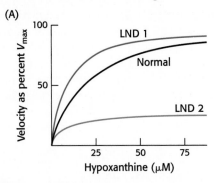

(b) Suggest some possible explanations for the appearance of Lesch–Nyhan disease even in the presence of apparently normal enzyme.

(c) The stability of the HGPRT was tested to determine if the LND 1 enzyme was more labile. The same amounts of the normal and LND 1 enzyme were incubated at 37°C in the absence of any substrates or products. At various times, a portion of the enzyme was removed and assayed for enzyme activity. The results are shown in Figure B. Explain these results.

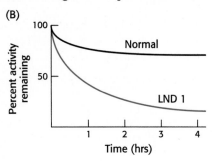

Think ▸ Pair ▸ Problems

50. Describe how the concentrations of the four deoxyribonucleotides are balanced by ribonucleotide reductase.

51. Explain the biochemical defect that leads to Lesch-Nyhan syndrome and suggest how the defect might cause the behavioral symptoms characteristic of the disorder.

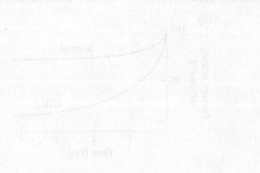

The Biosynthesis of Membrane Lipids and Steroids

Fats such as the triacylglycerol molecule (below) are widely used to store excess energy for later use and to fulfill other purposes, illustrated by the insulating blubber of whales. The natural tendency of fats to exist in nearly water-free forms makes these molecules well suited to these roles. [© Francois Gohier/ardea.com.]

 LEARNING OBJECTIVES

By the end of this chapter, you should be able to:

1. Describe the relation between triacylglycerol synthesis and phospholipid synthesis.

2. Describe the three stages of cholesterol synthesis.

3. Identify the regulatory processes in cholesterol synthesis.

4. Identify the important molecules synthesized from cholesterol precursors and from cholesterol.

OUTLINE

26.1 Phosphatidate Is a Common Intermediate in the Synthesis of Phospholipids and Triacylglycerols

26.2 Cholesterol Is Synthesized from Acetyl Coenzyme A in Three Stages

26.3 The Complex Regulation of Cholesterol Biosynthesis Takes Place at Several Levels

26.4 Important Biochemicals Are Synthesized from Cholesterol and Isoprene

This chapter examines the biosynthesis of three important components of biological membranes—phospholipids, sphingolipids, and cholesterol—which were introduced in Chapter 12. Triacylglycerols also are considered here because the pathway for their synthesis overlaps that of phospholipids. Cholesterol is of interest both as a membrane component and as a precursor of potent signal molecules, such as the steroid hormones progesterone, testosterone, estradiol, and cortisol.

The transport and uptake of cholesterol vividly illustrate a recurring mechanism for the entry of metabolites and signal molecules into cells. Cholesterol is transported in blood by the low-density lipoprotein (LDL) and taken up into cells by the LDL receptor on the cell surface. The LDL receptor is absent in people with *familial hypercholesterolemia*, a genetic disease. People lacking the receptor have markedly elevated cholesterol levels in the blood and cholesterol deposits on blood vessels, and they are prone to childhood heart attacks. Indeed, cholesterol is implicated in the development of atherosclerosis in people who do not have genetic defects. Thus, the regulation of cholesterol synthesis and transport can be a source of especially clear insight into the role that our understanding of biochemistry plays in medicine.

26.1 Phosphatidate Is a Common Intermediate in the Synthesis of Phospholipids and Triacylglycerols

Lipid synthesis requires the coordinated action of gluconeogenesis and fatty acid metabolism, as illustrated in Figure 26.1. The common step in the synthesis of both phospholipids for membranes and triacylglycerols for energy storage is the formation of *phosphatidate* (diacylglycerol 3-phosphate). In mammalian cells, phosphatidate is synthesized in the endoplasmic reticulum and the outer mitochondrial membrane. The pathway begins with *glycerol 3-phosphate,* which is formed primarily by the reduction of dihydroxyacetone phosphate (DHAP) synthesized by the gluconeogenic pathway, and to a lesser extent by the phosphorylation of glycerol. The addition of two fatty acids to glycerol-3-phosphate yields phosphatidate. First, acyl coenzyme A contributes a fatty acid chain to form *lysophosphatidate* and, then, a second acyl CoA contributes a fatty acid chain to yield phosphatidate.

Glycerol 3-phosphate   **Lysophosphatidate**   **Phosphatidate**

These acylations are catalyzed by *glycerol phosphate acyltransferase.* In most phosphatidates, the fatty acid chain attached to the C-1 atom is saturated,

FIGURE 26.1 PATHWAY INTEGRATION. Sources of intermediates in the synthesis of triacylglycerols and phospholipids. Phosphatidate, synthesized from dihydroxyacetone phosphate (DHAP) produced in gluconeogenesis and fatty acids, can be further processed to produce triacylglycerol or phospholipids. Phospholipids and other membrane lipids are continuously produced in all cells.

 EXPLORE this pathway further in the Metabolic Map on **Sapling**Plus

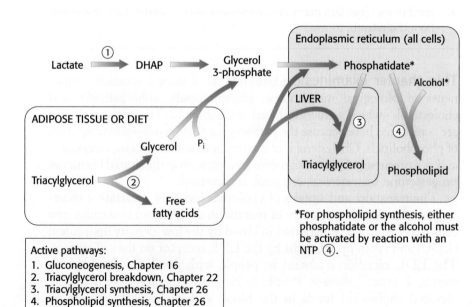

Active pathways:
1. Gluconeogenesis, Chapter 16
2. Triacylglycerol breakdown, Chapter 22
3. Triacylglycerol synthesis, Chapter 26
4. Phospholipid synthesis, Chapter 26

*For phospholipid synthesis, either phosphatidate or the alcohol must be activated by reaction with an NTP ④.

whereas the one attached to the C-2 atom is unsaturated. Phosphatidate can also be synthesized from *diacylglycerol* (DAG), in what is essentially a salvage pathway, by the action of *diacylglycerol kinase*:

$$\text{Diacylglycerol} + \text{ATP} \longrightarrow \text{phosphatidate} + \text{ADP}$$

The phospholipid and triacylglycerol pathways diverge at phosphatidate. In the synthesis of triacylglycerols, a key enzyme in the regulation of lipid synthesis, *phosphatidic acid phosphatase*, hydrolyzes phosphatidate to give a diacylglycerol. This intermediate is acylated to a *triacylglycerol* through the addition of a third fatty acid chain in a reaction that is catalyzed by *diglyceride acyltransferase*. Both enzymes are associated in a *triacylglycerol synthetase complex* that is bound to the endoplasmic reticulum membrane.

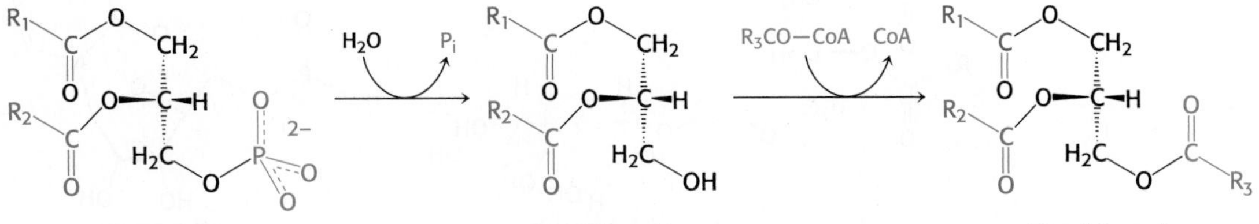

Phosphatidate **Diacylglycerol (DAG)** **Triacylglycerol**

The liver is the primary site of triacylglycerol synthesis. From the liver, the triacylglycerols are transported to the muscles for energy conversion or to the adipose cells for storage.

The synthesis of phospholipids requires an activated intermediate

Membrane-lipid synthesis continues in the endoplasmic reticulum and in the Golgi apparatus. *Phospholipid* synthesis requires the combination of a diacylglycerol with an alcohol. As in most anabolic reactions, one of the components must be activated. In this case, either the diacylglycerol or the alcohol may be activated, depending on the source of the reactants.

The synthesis of some phospholipids begins with the reaction of phosphatidate with cytidine triphosphate (CTP) to form the activated diacylglycerol, *cytidine diphosphodiacylglycerol* (CDP-diacylglycerol). This reaction, like those of many biosyntheses, is driven forward by the hydrolysis of pyrophosphate.

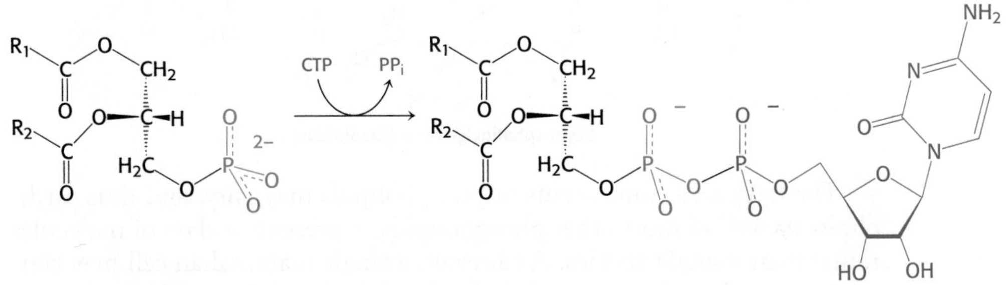

Phosphatidate **CDP-diacylglycerol**

The activated phosphatidyl unit then reacts with the hydroxyl group of an alcohol to form a phosphodiester linkage. If the alcohol is inositol, the products are *phosphatidylinositol* and cytidine monophosphate (CMP).

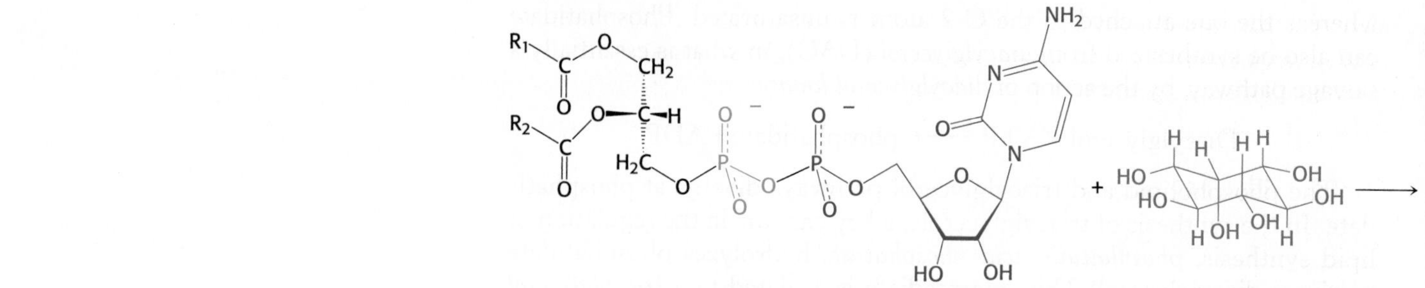

CDP-diacylglycerol + **Inositol** ⟶

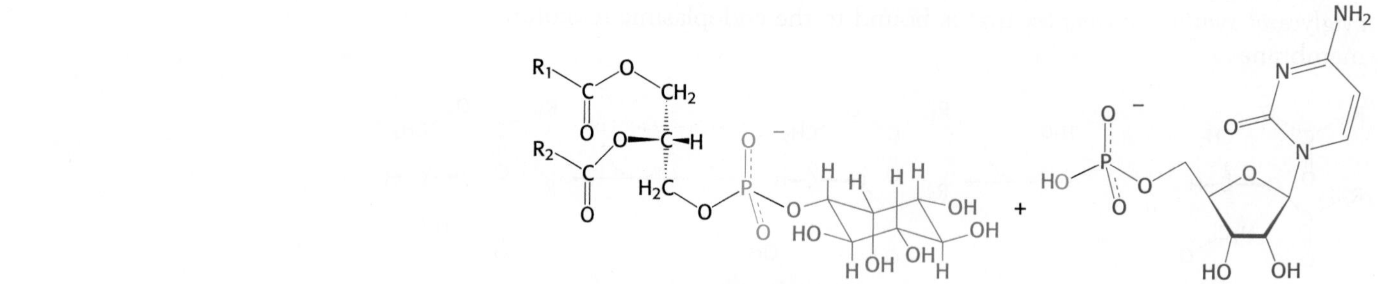

Phosphatidylinositol + **CMP**

Subsequent phosphorylations catalyzed by specific kinases lead to the synthesis of *phosphatidylinositol 4,5-bisphosphate,* the precursor of two intracellular messengers—diacylglycerol and inositol 1,4,5-trisphosphate (Section 14.2). If the alcohol is phosphatidylglycerol, the products are *diphosphatidylglycerol (cardiolipin)* and CMP. In eukaryotes, cardiolipin is synthesized in the mitochondria and is located exclusively in inner mitochondrial membranes. Cardiolipin plays an important role in the organization of the protein components of oxidative phosphorylation. For example, it is required for the full activity of cytochrome *c* oxidase (Section 18.3). Remarkably, very recent research shows that in nerve cells, cardiolipin facilitates the proper folding of α-synuclein. Misfolding of α-synuclein causes the formation of aggregates that are toxic to nerve cells. These α-synuclein deposits are believed to play a role in the development of Parkinson's disease.

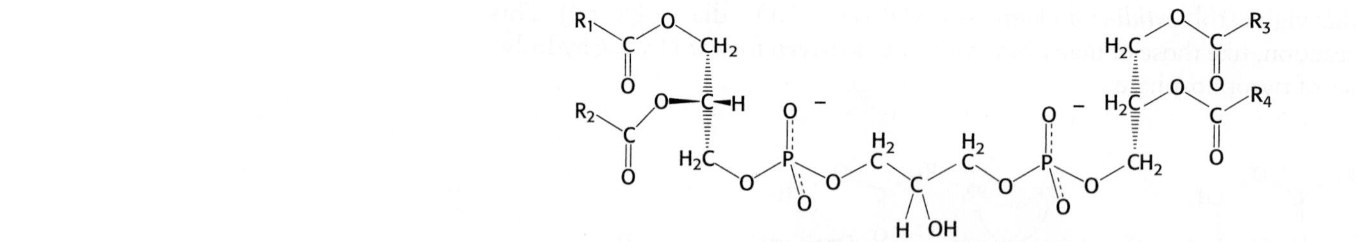

Diphosphatidylglycerol (Cardiolipin)

The fatty acid components of phospholipids may vary, and thus cardiolipin, as well as most other phospholipids, represents a class of molecules rather than a single species. As a result, a single mammalian cell may contain thousands of distinct phospholipids. Phosphatidylinositol is unusual in that it has a nearly fixed fatty acid composition. Stearic acid usually occupies the C-1 position and arachidonic acid the C-2 position.

Some phospholipids are synthesized from an activated alcohol

Phosphatidylethanolamine, the major phospholipid of the inner leaflet of cell membranes, is synthesized from the alcohol ethanolamine. To activate the alcohol, ethanolamine is phosphorylated by ATP to form the precursor, *phosphorylethanolamine.* This precursor then reacts with CTP to form the activated alcohol, *CDP-ethanolamine.* The phosphorylethanolamine unit of CDP-ethanolamine is transferred to a diacylglycerol to form *phosphatidylethanolamine.*

Phosphatidylcholine is an abundant phospholipid

The most common phospholipid in mammals is phosphatidylcholine, comprising approximately 50% of the membrane mass. Dietary choline is activated in a series of reactions analogous to those in the activation of ethanolamine. CTP-phosphocholine cytidylyltransferase (CCT) catalyzes the formation of CDP-choline, the rate-limiting step in phosphatidylcholine synthesis. CCT is an amphitropic enzyme—a class of enzymes whose regulator ligand is the membrane itself. A portion of the enzyme, normally associated with the membrane, detects a fall in phosphatidylcholine as an alteration in the physical properties of the membrane. When this occurs, another portion of the enzyme is inserted into the membrane, leading to enzyme activation. Indeed, the k_{cat}/K_M value (Section 8.4) increases three orders of magnitude upon activation, resulting in the restoration of phosphatidylcholine levels.

Earlier (Section 22.4), we examined how cancer cells increase lipogenesis to meet the fatty acid needs for membrane synthesis. Evidence is accumulating that CCT is specifically activated in some cancers to generate the required phosphocholine.

The liver also possesses an enzyme, *phosphatidylethanolamine methyltransferase,* which synthesizes phosphatidylcholine from phosphatidylethanolamine when dietary choline is insufficient, a testament to the importance of this vital phospholipid. The amino group of this phosphatidylethanolamine is methylated three times to form *phosphatidylcholine.* S*-Adenosylmethionine* is the methyl donor (Section 24.2).

Phosphatidyl-ethanolamine

Phosphatidyl-choline

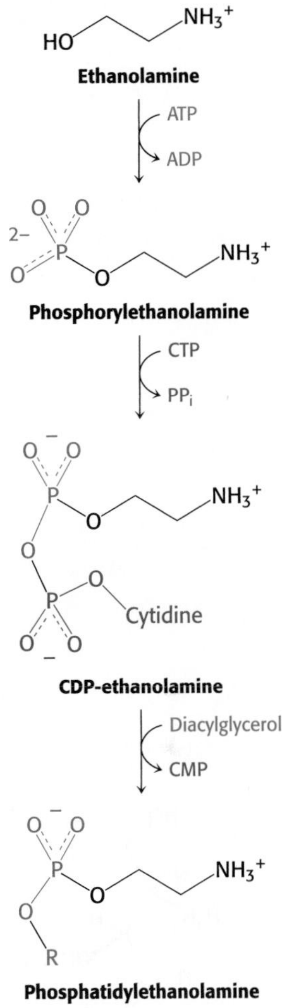

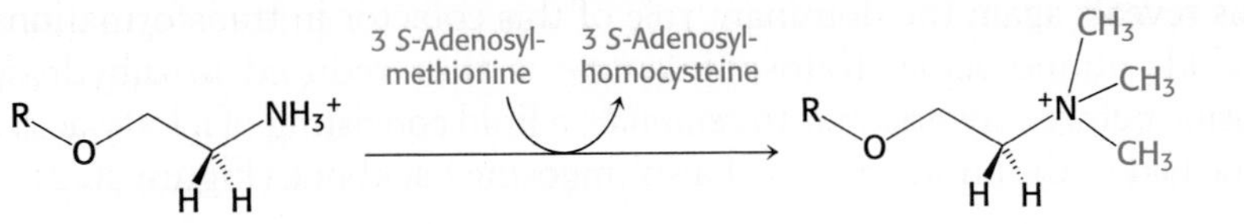

Thus, phosphatidylcholine can be produced by two distinct pathways in mammals, ensuring that this phospholipid can be synthesized even if the components for one pathway are in limited supply.

Excess choline is implicated in the development of heart disease

Choline is a popular dietary supplement, believed by some to enhance liver and neuronal function. Although the effectiveness of choline supplements is not established, the dangers of excess choline consumption are becoming clear. Gut bacteria convert excess choline into trimethylamine (TMA), a gas that smells like rotten fish, and the liver converts the absorbed TMA into trimethylamine-*N*-oxide (TMAO). TMAO stimulates cholesterol uptake by macrophages, a process that can result in atherosclerosis (p. 867). Foods rich in phosphatidylcholine, such as red meats and dairy products, may also result in TMAO production.

Trimethylamine

Trimethylamine-*N*-oxide

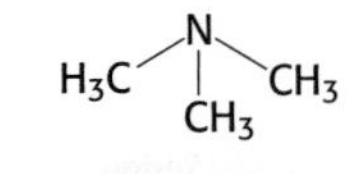

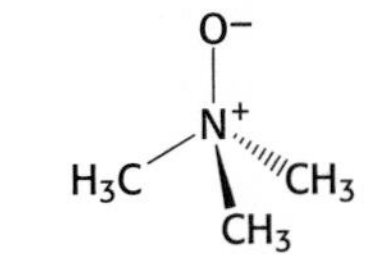

Base-exchange reactions can generate phospholipids

Phosphatidylserine makes up 10% of the phospholipids in mammals. This phospholipid is synthesized in a *base-exchange reaction* of serine with phosphatidylcholine or phosphatidylethanolamine. In the reaction, serine replaces choline or ethanolamine.

$$\text{Phosphatidylcholine} + \text{serine} \longrightarrow \text{choline} + \text{phosphatidylserine}$$

$$\text{Phosphatidylethanolamine} + \text{serine} \longrightarrow$$
$$\text{ethanolamine} + \text{phosphatidylserine}$$

Phosphatidylserine is normally located in the inner leaflet of the plasma membrane bilayer but is moved to the outer leaflet in apoptosis (Section 18.6). There, it serves to attract phagocytes to consume the cell remnants after apoptosis is complete. Phosphatidylserine is translocated from one side of the membrane to the other by an ATP-binding cassette translocase (Section 13.2).

Note that a cytidine nucleotide plays the same role in the synthesis of these phosphoglycerides as a uridine nucleotide does in the formation of glycogen (Section 21.4). In all of these biosyntheses, an activated intermediate (UDP-glucose, CDP-diacylglycerol, or CDP-alcohol) is formed from a phosphorylated substrate (glucose 1-phosphate, phosphatidate, or a phosphorylalcohol) and a nucleoside triphosphate (UTP or CTP). The activated intermediate then reacts with a hydroxyl group (the terminus of glycogen, an alcohol, or a diacylglycerol).

Sphingolipids are synthesized from ceramide

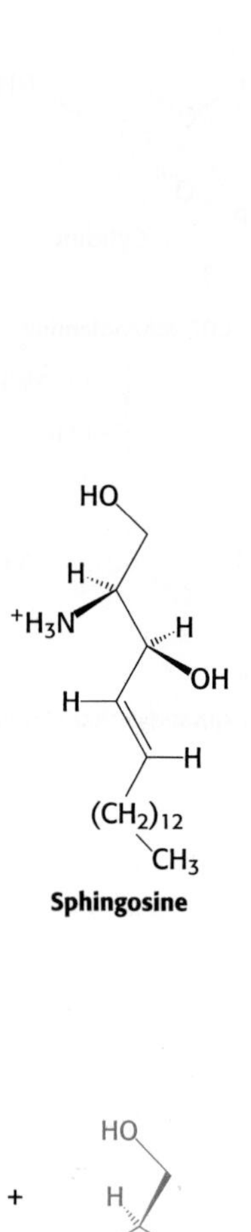

We now turn from glycerol-based phospholipids to another class of membrane lipid—the *sphingolipids*. These lipids are found in the plasma membranes of all eukaryotic cells, although the concentration is highest in the cells of the central nervous system. The backbone of a sphingolipid is *sphingosine*, rather than glycerol. Palmitoyl CoA and serine condense to form 3-ketosphinganine. The serine–palmitoyl transferase catalyzing this reaction is the rate-limiting step in the pathway and requires pyridoxal phosphate; this reveals again the dominant role of this cofactor in transformations that include amino acids. Ketosphinganine is then reduced to dihydrosphingosine before conversion into *ceramide*, a lipid consisting of a fatty acid chain attached to the amino group of a sphingosine backbone (Figure 26.2).

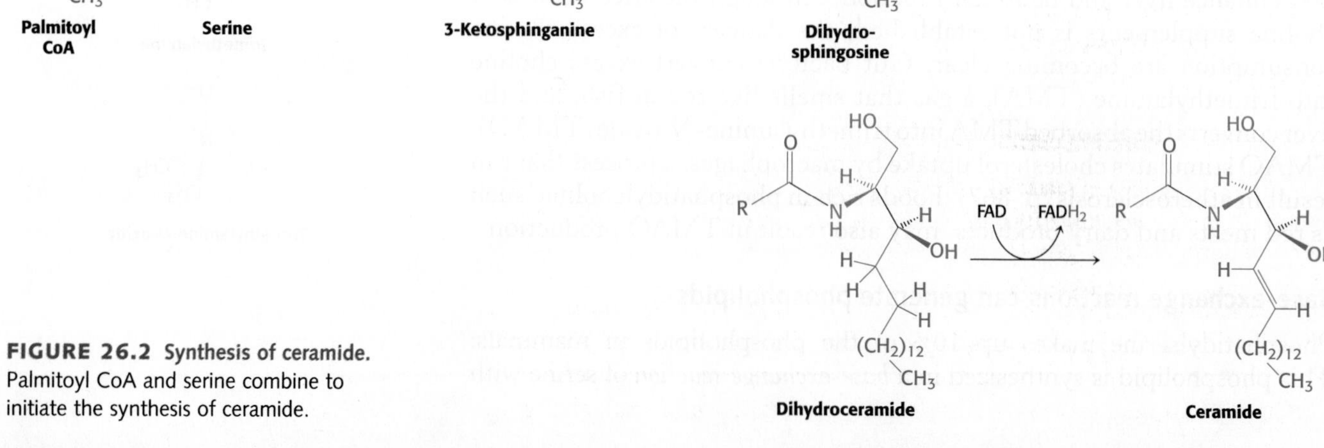

Sphingosine

FIGURE 26.2 Synthesis of ceramide. Palmitoyl CoA and serine combine to initiate the synthesis of ceramide.

In all sphingolipids, the amino group of ceramide is acylated. The terminal hydroxyl group also is substituted (Figure 26.3). In *sphingomyelin*, a component of the myelin sheath covering many nerve fibers, the substituent is phosphoryl-choline, which comes from phosphatidylcholine. In a *cerebroside*, the substituent is glucose or galactose. UDP-glucose or UDP-galactose is the sugar donor.

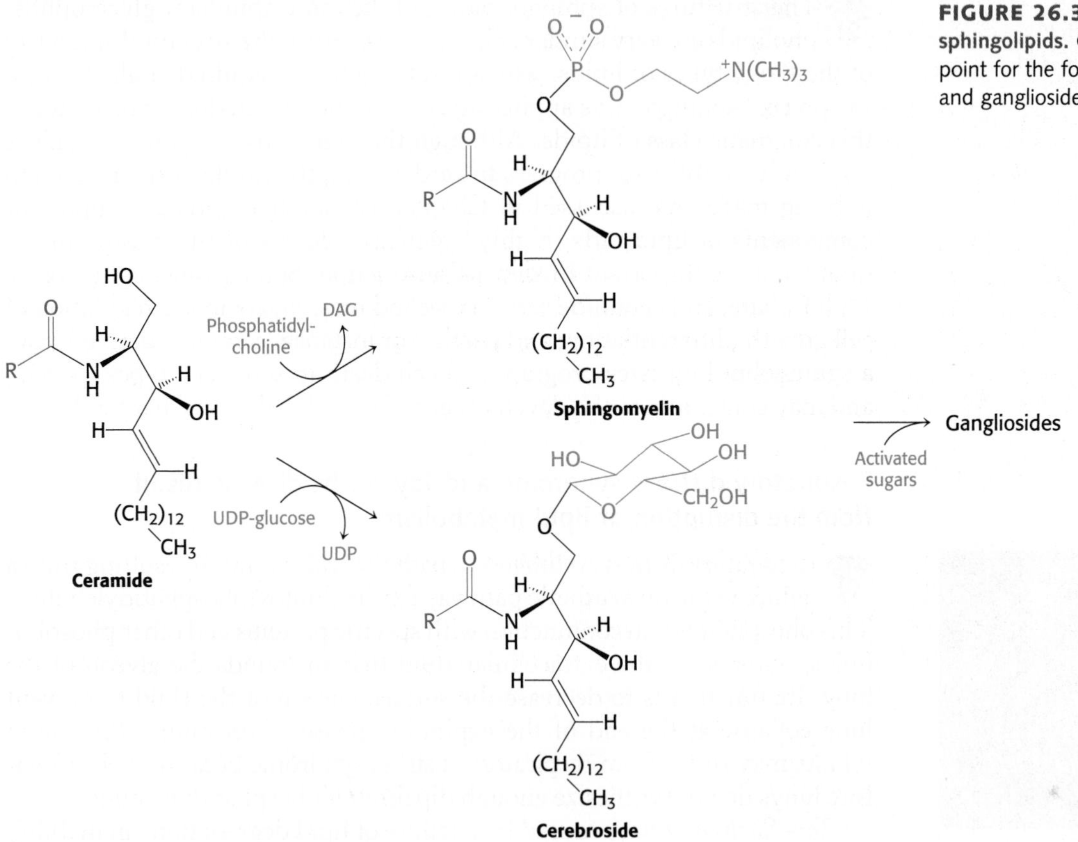

Sphingomyelin

Ceramide

Cerebroside

Gangliosides

Activated sugars

FIGURE 26.3 Synthesis of sphingolipids. Ceramide is the starting point for the formation of sphingomyelin and gangliosides.

Gangliosides are carbohydrate-rich sphingolipids that contain acidic sugars

Gangliosides are the most complex sphingolipids. In a *ganglioside*, an *oligosaccharide chain* is linked to the terminal hydroxyl group of ceramide by a glucose residue (Figure 26.4). This oligosaccharide chain contains at least one acidic sugar, either N-*acetylneuraminate* or N-*glycolylneuraminate*. These acidic sugars are called *sialic acids*. Their nine-carbon backbones are synthesized from phosphoenolpyruvate (a three-carbon unit) and N-acetylmannosamine 6-phosphate (a six-carbon unit).

Gangliosides are synthesized by the ordered, step-by-step addition of sugar residues to ceramide. The synthesis of these complex lipids requires the activated sugars UDP-glucose, UDP-galactose, and UDP-N-acetyl-galactosamine, as well as the CMP derivative of N-acetylneuraminate. CMP-N-acetylneuraminate is synthesized from CTP and N-acetylneur-aminate. The sugar composition of the resulting ganglioside is determined by the specificity of the glycosyltransferases in the cell. Almost 200 different gangliosides have been characterized (see Figure 26.4 for the composition of ganglioside G_{M1}).

Ganglioside-binding by the cholera toxin is the first step in the development of cholera, a pathological condition characterized by severe diarrhea (Section 14.5). Enterotoxigenic *E. coli*, the most common cause of diarrhea, including traveler's diarrhea, produces a toxin that also gains

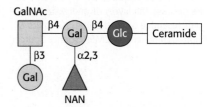

FIGURE 26.4 Ganglioside G_{M1}. This ganglioside consists of five monosaccharides linked to ceramide: one glucose (Glc) molecule, two galactose (Gal) molecules, one N-acetylgalactosamine (GalNAc) molecule, and one N-acetylneuraminate (NAN) molecule.

R₂ = H, *N*-acetylneuraminate

R₂ = OH, *N*-glycolylneuraminate

FIGURE 26.5 Lysosome with lipids. An electron micrograph of a lysosome engorged with lipids. Such lysosomes are sometimes described as being "onion-skin" like because the layers of undigested lipids resemble a sliced onion. [Graphics & Photography Service/CMSP/Science Source.]

access to the cell by first binding to gangliosides. Gangliosides are also crucial for binding immune-system cells to sites of injury in the inflammatory response.

Sphingolipids confer diversity on lipid structure and function

The structures of sphingolipids and the more abundant glycerophospholipids are very similar (Figure 12.8). Given the structural similarity of these two types of lipids, why are sphingolipids required at all? Indeed, the prefix "sphingo" was applied to capture the "sphinxlike" properties of this enigmatic class of lipids. Although the precise role of sphingolipids is not firmly established, progress toward solving the riddle of their function is being made. As discussed in Chapter 12, sphingolipids are important components of lipid rafts, highly organized regions of the plasma membrane that are important in signal transduction. Sphingosine, sphingosine 1-phosphate, and ceramide serve as second messengers in the regulation of cell growth, differentiation, and death. For instance, ceramide derived from a sphingolipid initiates programmed cell death in some cell types (p. 857) and may contribute to the development of type 2 diabetes (Chapter 27).

Respiratory distress syndrome and Tay–Sachs disease result from the disruption of lipid metabolism

Respiratory distress syndrome is a pathological condition resulting from a failure in the biosynthetic pathway for dipalmitoylphosphatidylcholine. This phospholipid, in conjunction with specific proteins and other phospholipids, is found in the extracellular fluid that surrounds the alveoli of the lung. Its function is to decrease the surface tension of the fluid to prevent lung collapse at the end of the expiration phase of breathing. Premature infants may suffer from respiratory distress syndrome because their immature lungs do not synthesize enough dipalmitoylphosphatidylcholine.

Tay–Sachs disease is caused by a failure of lipid degradation: an inability to degrade gangliosides. Gangliosides are found in highest concentration in the nervous system, particularly in gray matter, where they constitute 6% of the lipids. Gangliosides are normally degraded inside lysosomes by the sequential removal of their terminal sugars, but in Tay–Sachs disease this degradation does not take place. As a consequence, neurons become significantly swollen with lipid-filled lysosomes (Figure 26.5). An affected infant displays weakness and retarded psychomotor skills before 1 year of age. The child is demented and blind by age 2 and usually dies before age 3.

The ganglioside content of the brain of an infant with Tay–Sachs disease is greatly elevated. *The concentration of ganglioside G$_{M2}$ is many times higher than normal because its terminal N-acetylgalactosamine residue is removed very slowly or not at all.* The missing or deficient enzyme is a specific β-N-*acetylhexosaminidase*.

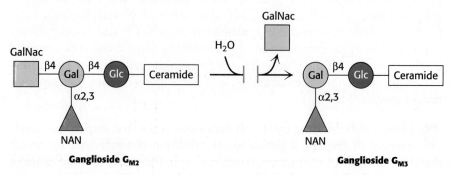

Tay–Sachs disease can be diagnosed in the course of fetal development. Cells obtained from either the villi of the placenta (chorionic villus sampling) or from the amniotic fluid (amniocentesis) are tested for the presence of the defective gene. Tay–Sachs disease was especially prominent among Ashkenazi Jews (descendants of Jews from central and eastern Europe). A genetic testing program, initiated in the early 1970s upon the development of a simple blood test to identify carriers, has virtually eliminated the disease in the population.

Ceramide metabolism stimulates tumor growth

Ceramide is a precursor for sphingomyelin, cerebroside, and ganglio-sides. Ceramide itself, however, induces programmed cell death or apoptosis (Section 18.6). Recall that cancer cells require all types of lipids for membrane formation (p. 853 and Section 22.4). How do cancer cells prevent ceramide-induced cell death? It appears that these cells destroy the apoptotic signal by converting it into a pro-mitotic signal. *Ceramidase* removes the fatty acid from the amino group of ceramide, generating sphingosine. Sphingosine is then converted into sphingosine 1-phosphate by sphingosine kinase, which stimulates cell division.

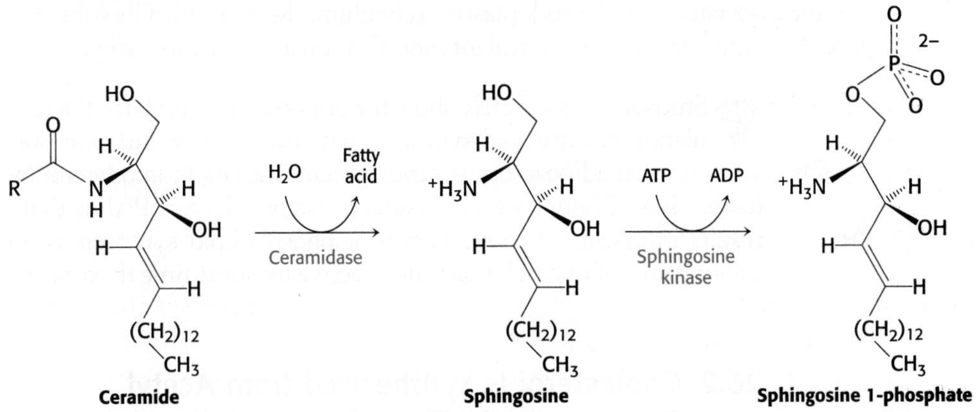

Thus, cancer cells convert a potentially lethal signal molecule into one that promotes tumor growth. Efforts are underway to develop inhibitors of ceramidase for use as chemotherapeutic agents.

Phosphatidic acid phosphatase is a key regulatory enzyme in lipid metabolism

Although the details of the regulation of lipid synthesis remain to be elucidated, evidence suggests that *phosphatidic acid phosphatase* (PAP), working in concert with diacylglycerol kinase, plays a key role in the regulation of lipid synthesis. PAP, also called lipin 1 in mammals, controls the extent to which triacylglycerols are synthesized relative to phospholipids and regulates the type of phospholipid synthesized (Figure 26.6). For instance, when PAP activity is high, phosphatidate is dephosphorylated and diacylglycerol is produced, which can react with the appropriate activated alcohols to yield phosphatidylethanolamine, phosphatidylserine, or phosphatidylcholine. Diacylglycerol can also be converted into triacylglycerols. Evidence suggests that the formation of triacylglycerols may act as a fatty acid buffer. This buffering helps to regulate the levels of diacylglycerol and sphingolipids, which serve signaling functions.

When PAP activity is lower, phosphatidate is used as a precursor for different phospholipids, such as phosphatidylinositol and cardiolipin.

FIGURE 26.6 Regulation of lipid synthesis. Phosphatidic acid phosphatase is the key regulatory enzyme in lipid synthesis. When active, PAP generates diacylglycerol (DAG), which can react with activated alcohols to form phospholipids or with fatty acyl CoA to form triacylglycerols. When PAP is inactive, phosphatidate is converted into CMP-DAG for the synthesis of different phospholipids. PAP also controls the amount of DAG and phosphatidate, both of which function as second messengers.

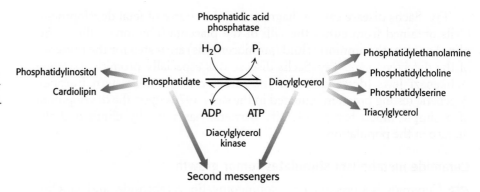

 EXPLORE this pathway further in the Metabolic Map on 🅜 Sapling Plus

Moreover, phosphatidate is a signal molecule itself. Phosphatidate regulates the growth of endoplasmic reticulum and nuclear membranes and acts as a cofactor that stimulates the expression of genes in phospholipid synthesis.

What are the signal molecules that regulate the activity of PAP? CDP-diacylglycerol, phosphatidylinositol, and cardiolipin enhance PAP activity, and sphingosine and dihydrosphingosine inhibit it. The enzyme also undergoes extensive phosphorylation and dephosphorylation. When phosphorylated, the enzyme resides in the cytoplasm. Upon dephosphorylation, the enzyme associates with the endoplasmic reticulum, the location of its substrate phosphatidate. Details of the covalent modification are under investigation.

Studies in mice clearly show the importance of PAP for the regulation of fatty acid synthesis. The loss of PAP function prevents normal adipose-tissue development, leading to lipodystrophy (severe loss of body fat) and insulin resistance. Excess PAP activity results in obesity. The regulation of phospholipid synthesis is an exciting area of research that will be active for some time to come.

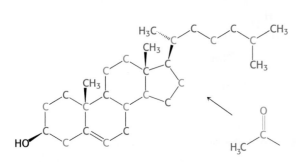

FIGURE 26.7 Labeling of cholesterol. Isotope-labeling experiments reveal the source of carbon atoms in cholesterol synthesized from acetate labeled in its methyl group (blue) or carboxylate atom (red).

Cholesterol

"Cholesterol is the most highly decorated small molecule in biology. Thirteen Nobel Prizes have been awarded to scientists who devoted major parts of their careers to cholesterol. Ever since it was isolated from gallstones in 1784, cholesterol has exerted an almost hypnotic fascination for scientists from the most diverse areas of science and medicine. . . . Cholesterol is a Janus-faced molecule. The very property that makes it useful in cell membranes, namely its absolute insolubility in water, also makes it lethal."

—Michael Brown and Joseph Goldstein, on the occasion of their receipt of a Nobel Prize for elucidating the control of blood levels of cholesterol.
Nobel Lectures (1985) © The Nobel Foundation, 1985

26.2 Cholesterol Is Synthesized from Acetyl Coenzyme A in Three Stages

We now turn our attention to the synthesis of the fundamental lipid *cholesterol*. This steroid modulates the fluidity of animal cell membranes (Section 12.5) and is the precursor of steroid hormones such as progesterone, testosterone, estradiol, and cortisol. *All 27 carbon atoms of cholesterol are derived from acetyl CoA in a three-stage synthetic process* (Figure 26.7).

1. Stage one is the synthesis of isopentenyl pyrophosphate, an activated isoprene unit that is the key building block of cholesterol.

2. Stage two is the condensation of six molecules of isopentenyl pyrophosphate to form squalene.

3. In stage three, squalene cyclizes and the tetracyclic product is subsequently converted into cholesterol.

The first stage takes place in the cytoplasm, and the next two in the endoplasmic reticulum.

The synthesis of mevalonate, which is activated as isopentenyl pyrophosphate, initiates the synthesis of cholesterol

The first stage in the synthesis of cholesterol is the formation of isopentenyl pyrophosphate from acetyl CoA. This set of reactions starts with the formation

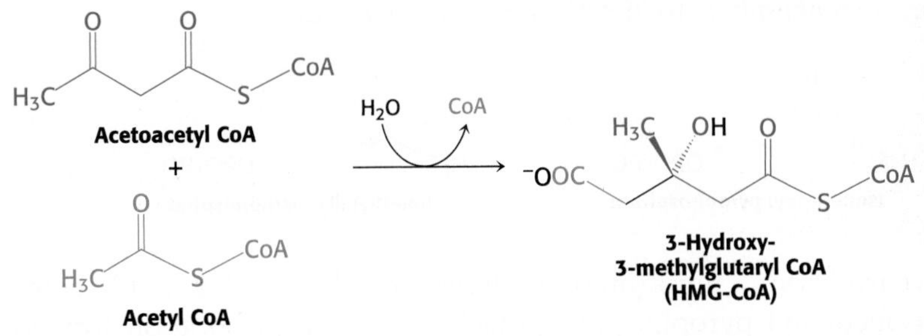

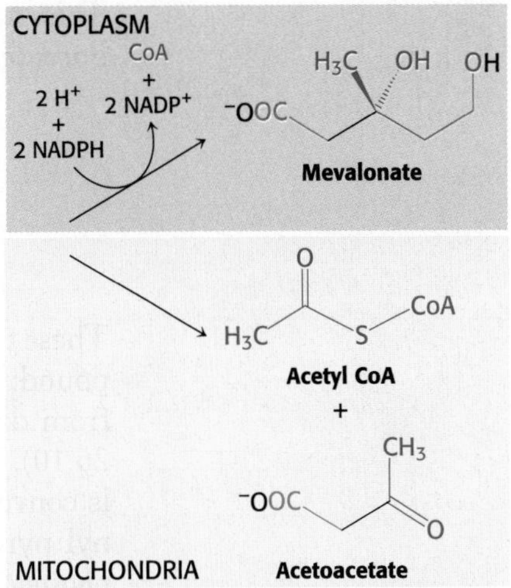

FIGURE 26.8 Fates of 3-hydroxy-3-methylglutaryl CoA. In the cytoplasm, HMG-CoA is converted into mevalonate. In mitochondria, it is converted into acetyl CoA and acetoacetate.

of 3-hydroxy-3-methylglutaryl CoA (HMG CoA) from acetyl CoA and acetoacetyl CoA. This intermediate is reduced to *mevalonate* for the synthesis of cholesterol (Figure 26.8). Recall that, alternatively, 3-hydroxy-3-methylglutaryl CoA may be generated in the mitochondria and processed to form ketone bodies, which are subsequently secreted to provide fuel for other tissues, notably the brain under starvation conditions (Section 22.3).

The synthesis of mevalonate is the committed step in cholesterol formation. The enzyme catalyzing this irreversible step, *3-hydroxy-3-methylglutaryl CoA reductase* (HMG-CoA reductase), is an important control site in cholesterol biosynthesis, as will be discussed shortly.

3-Hydroxy-3-methylglutaryl CoA + 2 NADPH + 2 H$^+$ $\longrightarrow$
$$\text{mevalonate} + 2\,\text{NADP}^+ + \text{CoA}$$

HMG-CoA reductase is an integral membrane protein in the endoplasmic reticulum.

Mevalonate is converted into *3-isopentenyl pyrophosphate* in three consecutive reactions requiring ATP (Figure 26.9). In the last step, the release of CO_2 yields isopentenyl pyrophosphate, an activated isoprene unit that is a key building block for many important biomolecules throughout the kingdoms of life (Section 26.4).

FIGURE 26.9 Synthesis of isopentenyl pyrophosphate. This activated intermediate is formed from mevalonate in three steps requiring ATP, followed by a decarboxylation.

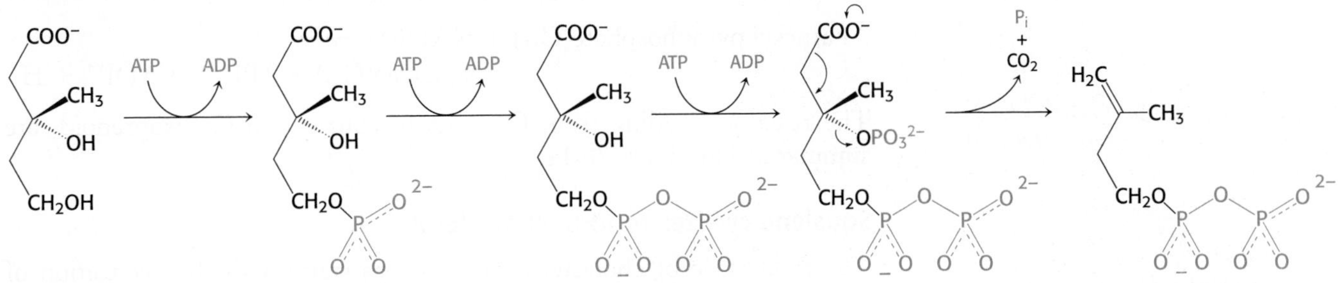

Mevalonate   5-Phospho-mevalonate   5-Pyrophospho-mevalonate   3-Isopentenyl pyrophosphate

Squalene (C$_{30}$) is synthesized from six molecules of isopentenyl pyrophosphate (C$_5$)

Squalene is synthesized from isopentenyl pyrophosphate by the reaction sequence

$$C_5 \longrightarrow C_{10} \longrightarrow C_{15} \longrightarrow C_{30}$$

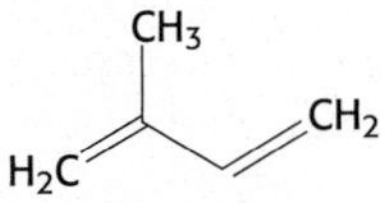

Isoprene

This stage in the synthesis of cholesterol starts with the isomerization of *isopentenyl pyrophosphate* to *dimethylallyl pyrophosphate*.

Isopentenyl pyrophosphate   **Dimethylallyl pyrophosphate**

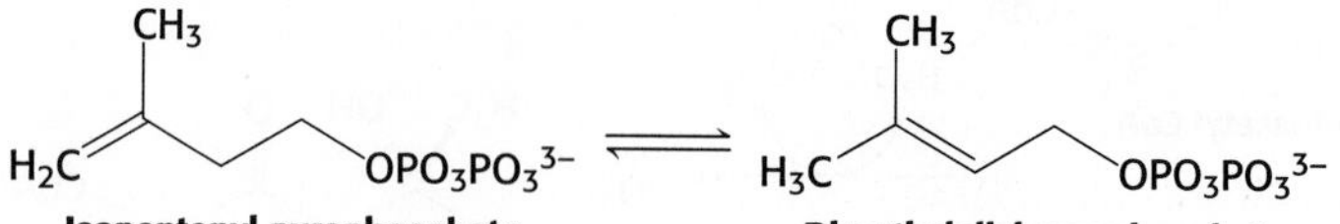

These two isomeric C_5 units (one of each type) condense to form a C_{10} compound: isopentenyl pyrophosphate attacks an allylic carbocation formed from dimethylallyl pyrophosphate to yield *geranyl pyrophosphate* (Figure 26.10). The same kind of reaction takes place again: geranyl pyrophosphate is converted into an allylic carbonium ion, which is attacked by isopentenyl pyrophosphate. The resulting C_{15} compound is called *farnesyl pyrophosphate*. The same enzyme, *geranyl transferase*, catalyzes each of these condensations.

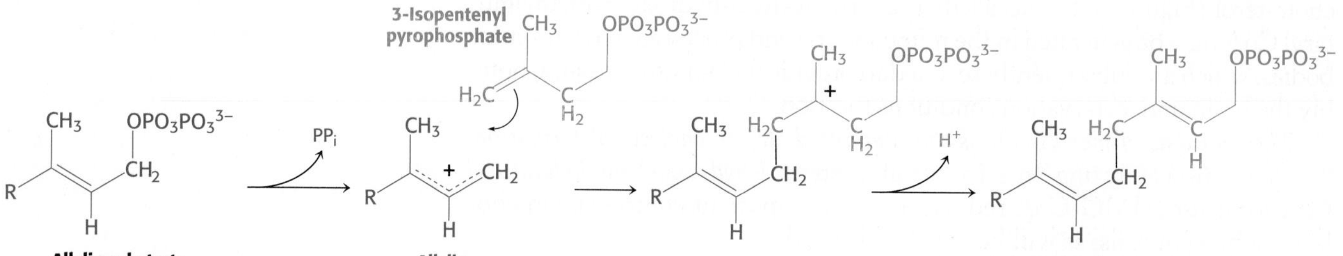

FIGURE 26.10 Condensation mechanism in cholesterol synthesis. The isomeric C_5 units dimethylallyl pyrophosphate and isopentenyl pyrophosphate join to form geranyl pyrophosphate. The same mechanism is used to add an additional isopentenyl pyrophosphate to form farnesyl pyrophosphate.

The last step in the synthesis of *squalene* is a reductive tail-to-tail condensation of two molecules of farnesyl pyrophosphate catalyzed by the endoplasmic reticulum enzyme *squalene synthase*.

$$2 \text{ Farnesyl pyrophosphate}(C_{15}) + \text{NADPH} \longrightarrow$$
$$\text{squalene}(C_{30}) + 2 \text{ PP}_i + \text{NADP}^+ + \text{H}^+$$

The reactions leading from C_5 units to squalene, a C_{30} isoprenoid, are summarized in Figure 26.11.

Squalene cyclizes to form cholesterol

The final stage of cholesterol biosynthesis starts with the cyclization of squalene (Figure 26.12). Squalene is first activated by conversion into squalene epoxide (2,3-oxidosqualene) in a reaction that uses O_2 and NADPH. Squalene epoxide is then cyclized to *lanosterol* by *oxidosqualene cyclase*. The enzyme holds squalene epoxide in an appropriate conformation and initiates the reaction by protonating the epoxide oxygen. The carbocation formed spontaneously rearranges to produce lanosterol. Lanosterol is converted into cholesterol in a multistep process by the removal of three methyl groups, the reduction of one double bond by NADPH, and the migration of the other double bond (Figure 26.13).

26.3 The Complex Regulation of Cholesterol Biosynthesis Takes Place at Several Levels

Cholesterol can be obtained from the diet or it can be synthesized de novo. Cholesterol biosynthesis is one of the most highly regulated metabolic pathways known. Biosynthetic rates may vary several hundred-fold, depending on how much cholesterol is consumed in the diet. An adult on a low-cholesterol diet typically synthesizes about 800 mg of cholesterol per day. The liver is the major site of cholesterol synthesis in mammals, although the intestine also forms significant amounts. The rate of cholesterol formation by these organs is highly responsive to the cellular level of cholesterol. *This feedback regulation is mediated primarily by changes in the amount and activity of 3-hydroxy-3-meth-ylglutaryl CoA reductase.* As described earlier (p. 859), this enzyme catalyzes the formation of mevalonate, the committed step in cholesterol biosynthesis. HMG CoA reductase is controlled in multiple ways:

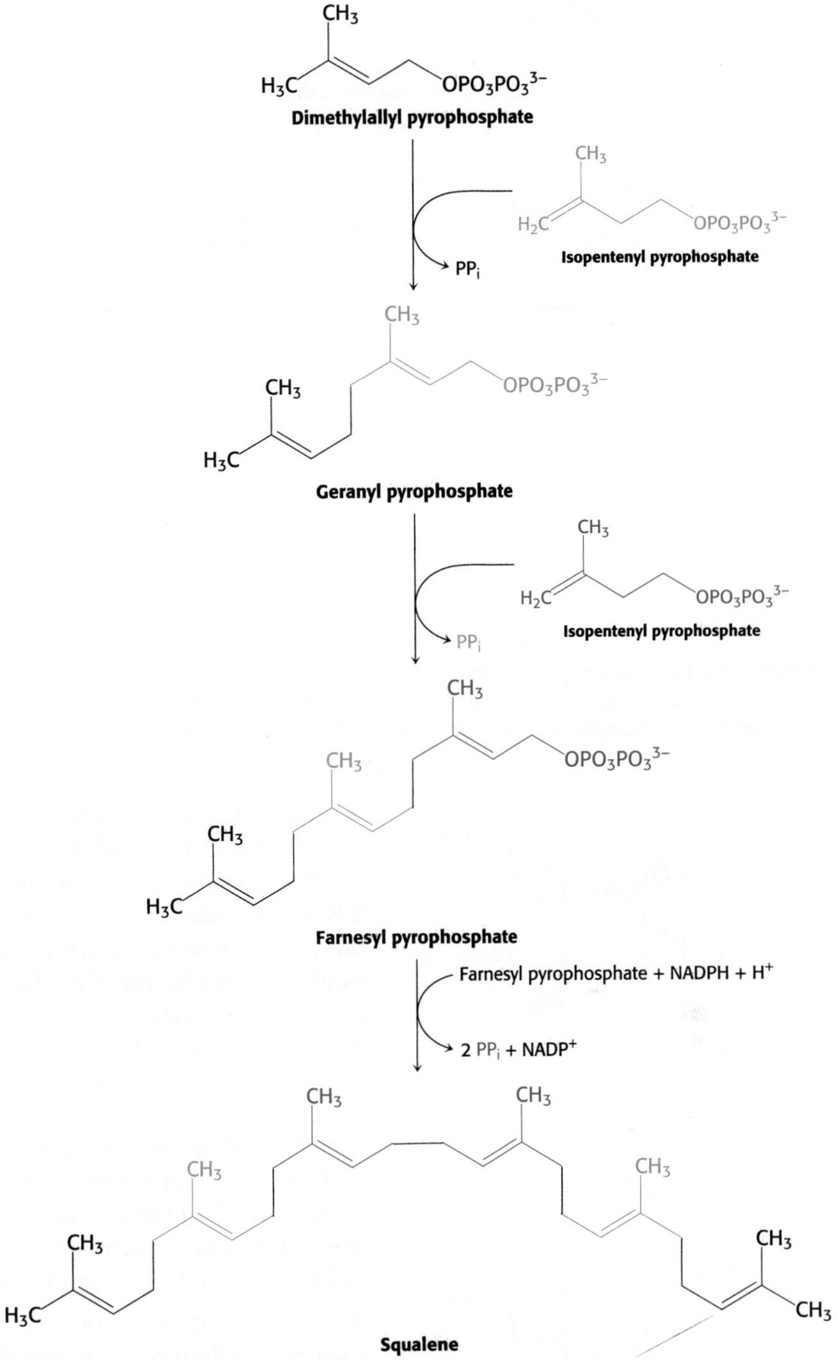

1. The rate of *synthesis of reductase mRNA* is controlled by the *sterol regulatory element binding protein* (SREBP), a member of a family of transcription factors that regulate the proteins required for lipid synthesis. This transcription factor binds to a short DNA sequence called the *sterol regulatory element* (SRE) on the 5′ side of the reductase gene when cholesterol levels are low, enhancing transcription of the gene. In its inactive state, the SREBP resides in the endoplasmic reticulum membrane, where it is associated with the SREBP cleavage activating protein (SCAP), an integral membrane protein. SCAP is the cholesterol sensor. When cholesterol levels fall, SCAP escorts SREBP in small membrane vesicles to the Golgi complex, where it is released from the membrane by two specific proteolytic cleavages (Figure 26.14). The first cleavage frees a fragment of SREBP from SCAP, whereas the second cleavage releases the regulatory domain from the membrane. The released protein migrates to the nucleus and binds the SRE of the HMG-CoA reductase gene, as well as several other genes in the cholesterol biosynthetic pathway, to enhance transcription. When cholesterol levels rise, the proteolytic release of the SREBP is blocked, and the SREBP in the nucleus

FIGURE 26.11 Squalene synthesis. One molecule of dimethyallyl pyrophosphate and two molecules of isopentenyl pyrophosphate condense to form farnesyl pyrophosphate. The tail-to-tail coupling of two molecules of farnesyl pyrophosphate yields squalene.

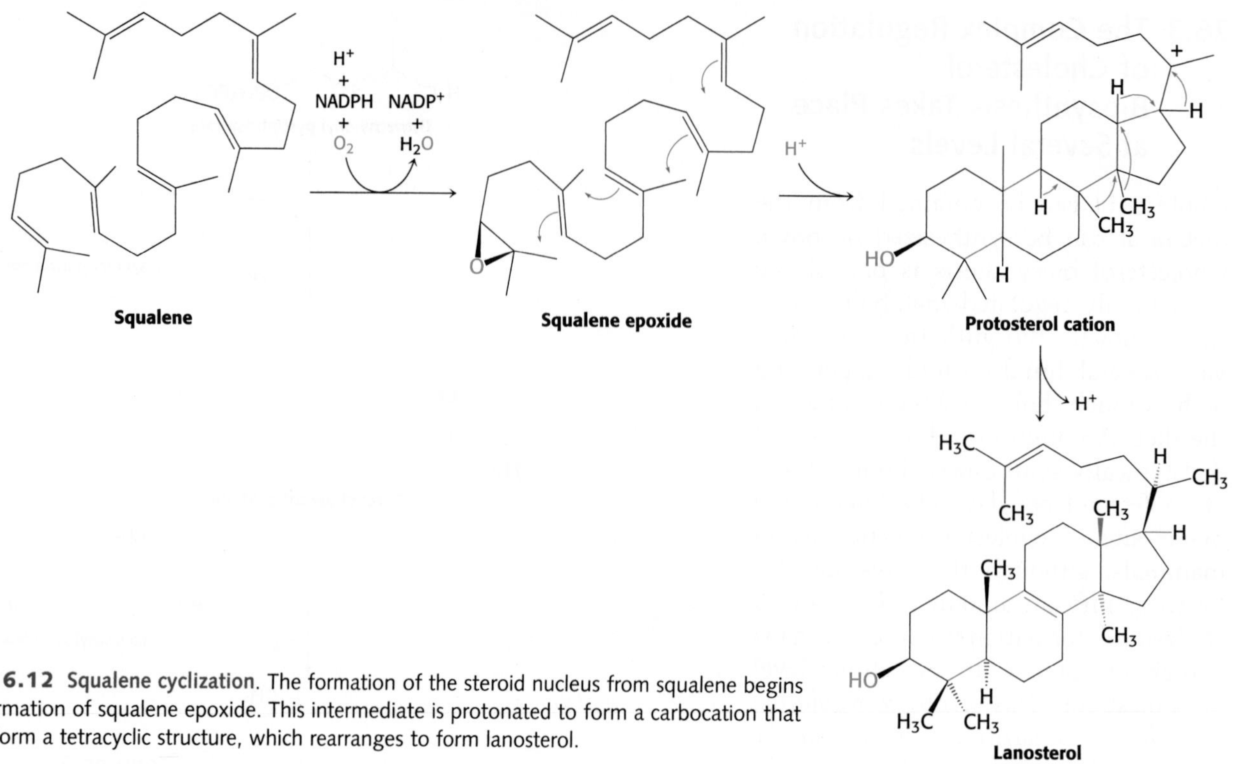

FIGURE 26.12 Squalene cyclization. The formation of the steroid nucleus from squalene begins with the formation of squalene epoxide. This intermediate is protonated to form a carbocation that cyclizes to form a tetracyclic structure, which rearranges to form lanosterol.

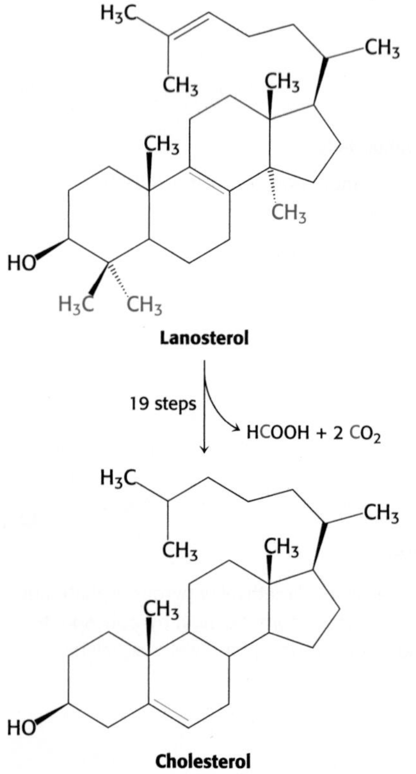

FIGURE 26.13 Cholesterol formation. Lanosterol is converted into cholesterol in a complex process.

is rapidly degraded by proteasomes located in the nucleus. These two events halt the transcription of genes of the cholesterol biosynthetic pathways.

What is the molecular mechanism that retains SCAP–SREBP in the ER when cholesterol is present but allows movement to the Golgi complex when cholesterol concentration is low? When cholesterol is low, SCAP binds to vesicular proteins that facilitate the transport of SCAP–SREBP to the Golgi apparatus, as previously described. When cholesterol is present, SCAP binds cholesterol, which causes a structural change in SCAP so that it binds to another endoplasmic reticulum protein called Insig (*insulin-induced gene*) (Figure 26.15). Insig is the anchor that retains SCAP and thus SREBP in the endoplasmic reticulum in the presence of cholesterol. The interactions between SCAP and Insig can also be forged when Insig binds 25-hydroxycholesterol, a metabolite of cholesterol. Thus, two distinct steroid–protein interactions prevent the inappropriate movement of SCAP–SREBP to the Golgi complex.

2. The rate of *translation of reductase mRNA* is inhibited by nonsterol metabolites derived from mevalonate.

3. The *degradation of the reductase* is stringently controlled. The enzyme is bipartite: its cytoplasmic domain carries out catalysis and *its membrane domain senses signals that lead to its degradation.* The membrane domain may undergo structural changes in *response to increasing concentrations of sterols such as lanosterol and 25-hydroxycholesterol.* Under these conditions, the reductase appears to bind to a subset of Insigs that are also associated with the ubiquitinating enzymes (Figure 26.16). The reductase is polyubiquitinated and subsequently extracted from the membrane in a process that requires gerenylgeraniol. The extracted reductase is then degraded by the proteasome (Section 23.2). The combined regulation at the levels of transcription, translation, and degradation can alter the amount of enzyme in the cell more than 200-fold.

4. *Phosphorylation decreases the activity of the reductase.* This enzyme, like acetyl CoA carboxylase (which catalyzes the committed step in fatty

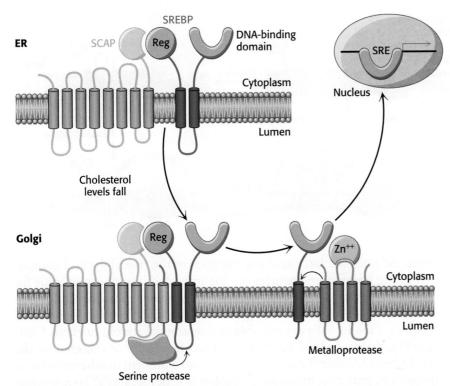

FIGURE 26.14 The SREBP pathway. SREBP resides in the endoplasmic reticulum (ER), where it is bound to SCAP by its regulatory (Reg) domain. When cholesterol levels fall, SCAP and SREBP move to the Golgi complex, where SREBP undergoes successive proteolytic cleavages by a serine protease and a metalloprotease. The released DNA-binding domain moves to the nucleus to alter gene expression. [Information from an illustration provided by Dr. Michael Brown and Dr. Joseph Goldstein.]

acid synthesis, Section 22.5), is switched off by an AMP-activated protein kinase. Thus, cholesterol synthesis ceases when the ATP level is low.

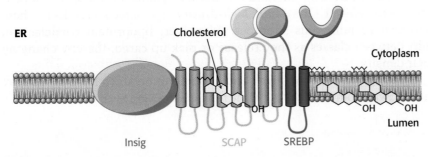

FIGURE 26.15 Insig regulates SCAP–SREBP movement. In the presence of cholesterol, Insig interacts with SCAP–SREBP and prevents the activation of SREBP. Cholesterol binding to SCAP or 25-hydroxycholesterol binding to Insig facilitates the interaction of Insig and SCAP, retaining SCAP–SREBP in the endoplasmic reticulum. [Information from M. S. Brown and J. L. Goldstein. Cholesterol feedback: From Schoenheimer's bottle to Scap's MELADL, *J. Lipid Res.* 50:S15–S27, 2009.]

Lipoproteins transport cholesterol and triacylglycerols throughout the organism

Cholesterol and triacylglycerols are transported in body fluids in the form of *lipoprotein particles*, which are important for a number of reasons. First, lipoprotein particles are the means by which triacylglycerols are delivered to tissues, from the intestine or liver, for use as fuel or for

FIGURE 26.16 Insig facilitates the degradation of HMG-CoA reductase. In the presence of sterols, a subclass of Insig associated with ubiquitinating enzymes binds HMG-CoA reductase. This interaction results in the ubiquitination of the enzyme. This modification and the presence of geranylgeraniol results in extraction of the enzyme from the membrane and degradation by the proteasome. [Information from R. A. DeBose-Boyd. Feedback regulation of cholesterol synthesis: Sterol-accelerated ubiquitination and degradation of HMG CoA reductase, *Cell Res.* 18:609–621, 2008.]

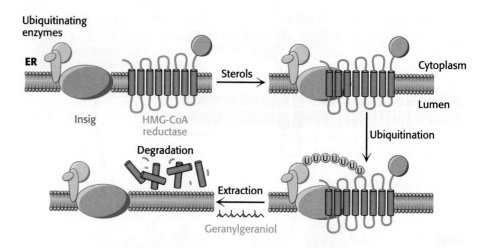

storage. Second, the fatty acid constituents of the triacylglycerol components of the lipoprotein particles are incorporated into phospholipids for membrane synthesis. Likewise, cholesterol is a vital component of membranes and is a precursor to the powerful signal molecules—the steroid hormones. Finally, cells are not able to degrade the steroid nucleus. Consequently, the cholesterol must be used biochemically or excreted by the liver. Excess cholesterol plays a role in the development of atherosclerosis. Lipoprotein particles function in cholesterol homeostasis, transporting the molecule from sites of synthesis to sites of use, and finally to the liver for excretion.

Each lipoprotein particle consists of a core of hydrophobic lipids surrounded by a shell of more-polar lipids and proteins. The protein components of these macromolecular aggregates, called apoproteins, have two roles: *they solubilize hydrophobic lipids and contain cell-targeting signals.* Apolipoproteins are synthesized and secreted by the liver and the intestine. Lipoprotein particles are classified according to increasing density (Table 26.1): *chylomicrons, chylomicron remnants, very low-density lipoproteins* (VLDLs), *intermediate-density lipoproteins* (IDLs), *low-density lipoproteins* (LDLs), and *high-density lipoproteins* (HDLs). These classes have numerous subtypes. Moreover, lipoprotein particles can shift between classes as they release or pick up cargo, thereby changing their density.

Triacylglycerols, cholesterol, and other lipids obtained from the diet are carried away from the intestine in the form of large *chylomicrons* (Section 22.1).

TABLE 26.1 Properties of plasma lipoproteins

Plasma lipoproteins	Density (g ml⁻¹)	Diameter (nm)	Apolipoprotein	Physiological role	Composition (%)				
					TAG	CE	C	PL	P
Chylomicron	<0.95	75–1200	B-48, C, E	Dietary fat transport	86	3	1	8	2
Very low-density lipoprotein	0.95–1.006	30–80	B-100, C, E	Endogenous fat transport	52	14	7	18	8
Intermediate-density lipoprotein	1.006–1.019	15–35	B-100, E	LDL precursor	38	30	8	23	11
Low-density lipoprotein	1.019–1.063	18–25	B-100	Cholesterol transport	10	38	8	22	21
High-density lipoprotein	1.063–1.21	7.5–20	A	Reverse cholesterol transport	5–10	14–21	3–7	19–29	33–57

Abbreviations: TAG, triacylglycerol; CE, cholesteryl ester; C, free cholesterol; PL, phospholipid; P, protein.

These particles have a very low density because triacylglycerols constitute about 90% of their content. Apolipoprotein B-48 (apo B-48), a large protein (240 kDa), forms an amphipathic spherical shell around the fat globule; the external face of this shell is hydrophilic. The triacylglycerols in chylomicrons are released through hydrolysis by *lipoprotein lipases*. These enzymes are located on the lining of blood vessels in muscle and other tissues that use fatty acids as fuels or in the synthesis of lipids. The liver then takes up the cholesterol-rich residues, known as *chylomicron remnants*.

Lipoprotein particles are also crucial for the transport of lipids from the liver, which is a major site of triacylglycerol and cholesterol synthesis, to other tissues in the body (Figure 26.17). Triacylglycerols and cholesterol in excess of the liver's own needs are exported into the blood in the form of very low-density lipoproteins. These particles are stabilized by two apolipoproteins—apo B-100 (513 kDa) and apo E (34 kDa). Triacylglycerols in very low-density lipoproteins, as in chylomicrons, are hydrolyzed by lipases on capillary surfaces, with the released fatty acids being taken up by the muscle and other tissues. The resulting remnants, which are rich in cholesteryl esters, are called *intermediate-density lipoproteins*. These particles have two fates. Half of them are taken up by the liver for processing, and half are converted into low-density lipoprotein by the removal of more triacylglycerol by tissue lipases that absorb the released fatty acids. Interestingly, apo B-100—one of the largest proteins known—is a longer version of apo B-48, the protein component of chylomicrons. Both apo B proteins are encoded by the same gene and produced from the same initial RNA transcript. In the intestine, RNA editing (Section 30.3) modifies the transcript to generate the mRNA for apo B-48, the truncated form.

Low-density lipoprotein is the major carrier of cholesterol in blood (Figure 26.18). It contains a core of some 1500 cholesterol molecules esterified to fatty acids; the most common fatty acid chain in these esters is linoleate, a polyunsaturated fatty acid. A shell of phospholipids and unesterified cholesterol molecules surrounds this highly hydrophobic core. The shell also contains a single copy of apo B-100, which is recognized by target cells. *The role of LDL is to transport cholesterol to peripheral tissues and regulate de novo cholesterol synthesis at these sites*, as described on page 861. A different purpose is served by *high-density lipoprotein*, which picks up cholesterol released into the plasma from dying cells and from membranes undergoing turnover and delivers the cholesterol to the liver for excretion. An acyltransferase in HDL esterifies these cholesterols, which are then returned by HDL to the liver (Figure 26.19).

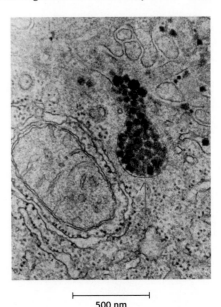

├─────── 500 nm ───────┤

FIGURE 26.17 Site of cholesterol synthesis. Electron micrograph of a part of a liver cell actively engaged in the synthesis and secretion of very low-density lipoprotein (VLDL). The arrow points to a vesicle that is releasing its content of VLDL particles. [Courtesy of Dr. George Palade/ Yale University, Harvey Cushing/John Hay Whitney Medical Library.]

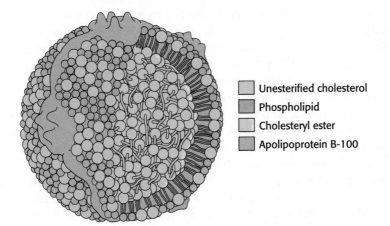

■ Unesterified cholesterol
■ Phospholipid
■ Cholesteryl ester
■ Apolipoprotein B-100

FIGURE 26.18 Schematic model of low-density lipoprotein. The LDL particle is approximately 22 nm (220 Å) in diameter.

FIGURE 26.19 An overview of lipoprotein particle metabolism. Fatty acids are abbreviated FFA. [Information from J. G. Hardman (Ed.), L. L. Limbird (Ed.), and A. G. Gilman (Consult. Ed.), *Goodman and Gilman's The Pharmacological Basis of Therapeutics,* 10th ed. (McGraw-Hill, 2001), p. 975, Fig. 36.1.]

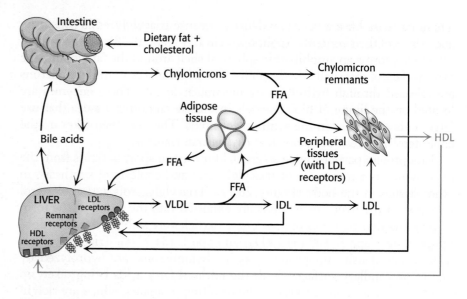

Low-density lipoproteins play a central role in cholesterol metabolism

Cholesterol metabolism must be precisely regulated to prevent atherosclerosis. The mode of control in the liver, the primary site of cholesterol synthesis, has already been considered: dietary cholesterol reduces the activity and amount of 3-hydroxy-3-methylglutaryl CoA reductase, the enzyme catalyzing the committed step. In general, cells outside the liver and intestine obtain cholesterol from the plasma rather than synthesizing it de novo. Specifically, *their primary source of cholesterol is the low-density lipoprotein.* The process of LDL uptake, called *receptor-mediated endocytosis,* serves as a paradigm for the uptake of many molecules (Figure 26.20).

Endocytosis begins when apolipoprotein B-100 on the surface of an LDL particle binds to a specific receptor protein on the plasma membrane of nonhepatic cells. The receptors for LDL are localized in specific regions called *coated pits,* which contain a protein called *clathrin.* The receptor–LDL complex is then internalized by *endocytosis;* that is, the plasma membrane in the vicinity of the complex invaginates and then fuses to form an endocytic vesicle called an *endosome.* The endosome is acidified by an ATP-dependent proton pump homologous to the Na^+/K^+ ATPase (Section 13.2), which causes the receptor to release its cargo.

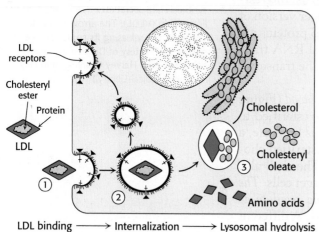

FIGURE 26.20 Receptor-mediated endocytosis. The process of receptor-mediated endocytosis is illustrated for the cholesterol-carrying complex, low-density lipoprotein (LDL): (1) LDL binds to a specific receptor, the LDL receptor; (2) this complex invaginates to form an endosome; (3) after separation from its receptor, the LDL-containing vesicle fuses with a lysosome, leading to the degradation of the LDL and the release of the cholesterol.

Some of the receptor is returned to the cell membrane in a recycling vesicle, while a portion is degraded along with its cargo (p. 869). The round-trip time for a receptor is about 10 minutes; in its lifetime of about a day, each receptor may bring hundreds of LDL particles into the cell. The vesicles containing LDL, and in some cases the receptor, subsequently fuse with *lysosomes,* acidic vesicles that carry a wide array of degradative enzymes. The protein component of LDL is hydrolyzed to free amino acids. The cholesteryl esters in LDL are hydrolyzed by a lysosomal acid lipase. *The released unesterified cholesterol can then be used for membrane biosynthesis.* Alternatively, it can be *reesterified for storage inside the cell.* In fact, free cholesterol activates *acyl CoA:cholesterol acyltransferase* (ACAT), the enzyme catalyzing this reaction. Reesterified cholesterol contains mainly oleate and palmitoleate, which are monounsaturated fatty acids, in contrast with the cholesteryl esters in LDL, which are rich in linoleate, a polyunsaturated fatty acid (Table 12.1). It is imperative that the cholesterol be reesterified. High concentrations of unesterified cholesterol disrupt the integrity of cell membranes.

The synthesis of the LDL receptor is itself subject to feedback regulation. Studies of cultured fibroblasts show that, *when cholesterol is abundant inside the cell, new LDL receptors are not synthesized, and so the uptake of additional cholesterol from plasma LDL is blocked.* The gene for the LDL receptor, like that for the reductase, is regulated by SREBP, which binds to a sterol regulatory element that controls the rate of mRNA synthesis.

The absence of the LDL receptor leads to hypercholesterolemia and atherosclerosis

The pioneering studies of *familial hypercholesterolemia* by Michael Brown and Joseph Goldstein revealed the physiological importance of the LDL receptor. The total concentration of cholesterol and LDL in the blood plasma is markedly elevated in this genetic disorder, which results from a mutation at a single autosomal locus. The cholesterol level in the plasma of homozygotes is typically 680 mg dl^{-1}, compared with 300 mg dl^{-1} in heterozygotes (clinical assay results are often expressed in milligrams per deciliter, which is equal to milligrams per 100 milliliters). A value of <200 mg dl^{-1} is regarded as desirable, but many people have higher levels. *In familial hypercholesterolemia, cholesterol is deposited in various tissues because of the high concentration of LDL cholesterol in the plasma.* Nodules of cholesterol called *xanthomas* are prominent in skin and tendons. LDL accumulates under the endothelial cells lining the blood vessels. Of particular concern is the oxidation of the excess LDL to form oxidized LDL (oxLDL), which can instigate the inflammatory response by the immune system, a response that has been implicated in the development of cardiovascular disease. The oxLDL is taken up by immune system cells called macrophages, which become engorged to form foam cells. These foam cells become trapped in the walls of the blood vessels and contribute to the formation of atherosclerotic plaques that cause arterial narrowing and lead to heart attacks (Figure 26.21). In fact, *most homozygotes die of coronary artery disease in childhood.* The disease in heterozygotes (1 in 500 people) has a milder and more variable clinical course.

The molecular defect in most cases of familial hypercholesterolemia is an absence or deficiency of functional receptors for LDL. Receptor mutations that disrupt each of the stages in the endocytotic pathway have been identified.

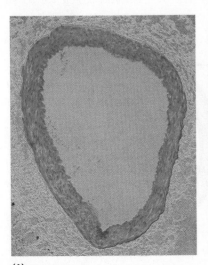 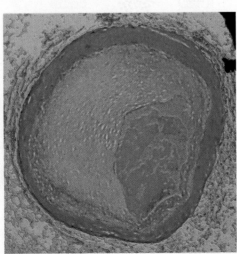

(A)   (B)

FIGURE 26.21 The effects of excess cholesterol. Cross section of (A) a normal artery and (B) an artery blocked by a cholesterol-rich plaque. [SPL/Science Source.]

Homozygotes have almost no functional receptors for LDL, whereas heterozygotes have about half the normal number. Consequently, the entry of LDL into liver and other cells is impaired, leading to an increased level of LDL in the blood plasma. Furthermore, less IDL enters liver cells because IDL entry, too, is mediated by the LDL receptor. Consequently, IDL stays in the blood longer in familial hypercholesterolemia, and more of it is converted into LDL than in normal people. All deleterious consequences of an absence or deficiency of the LDL receptor can be attributed to the ensuing elevated level of LDL cholesterol in the blood.

Mutations in the LDL receptor prevent LDL release and result in receptor destruction

One class of mutations that results in familial hypercholesterolemia generates receptors that are reluctant to give up the LDL cargo. Let us begin by examining the makeup of the receptor. The human LDL receptor is a 160-kDa glycoprotein, which is composed of six different types of domains (Figure 26.22A). The amino-terminal region of the receptor—the site of LDL binding— consists of seven homologous *LDL receptor type-A* (LA) domains, with domains 4 and 5 most important for LDL binding. A second type of domain is homologous to one found in the epidermal growth factor (EGF). This domain is repeated three times, and in between the second and third repeat is a propeller structure consisting of six blade-like domains. This part of the receptor is crucial in releasing LDL in the endosome. The fourth domain, which is very rich in serine and threonine residues, contains *O*-linked sugars. These oligosaccharides function as struts to keep the receptor extended from the membrane so that the LDL-binding domain is accessible to LDL. The fifth type of domain consists of 22 hydrophobic residues that span the plasma membrane. The sixth and final domain consists of 50 residues and emerges on the cytoplasmic

FIGURE 26.22 LDL receptor releases LDL in the endosome. (A) A schematic representation of the domain structure of the LDL receptor. (B) In the endosome, the receptor converts from the open structure into the closed structure, resulting in the release of the LDL into the endosome. [Information from I. D. Campbell, *Biochem. Soc. Trans.* 31(pt. 6p):1107–1114, 2003, Fig. 1A.]

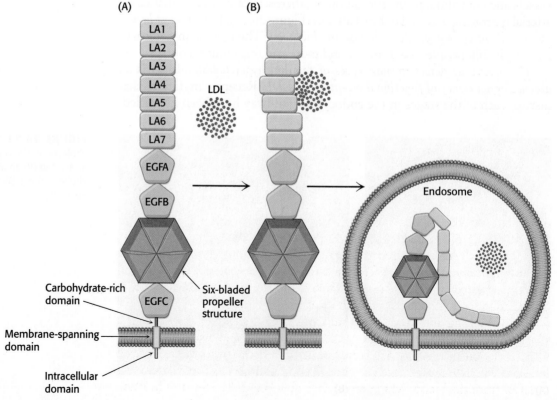

side of the membrane, where it controls the interaction of the receptor with coated pits and participates in endocytosis.

How does the LDL receptor relinquish its cargo on entering the endosome? The receptor exists in two interconvertible states: an extended or open state, capable of binding LDL, and a closed state that results in release of the LDL in the endosome. The receptor maintains the open state while in the plasma membrane, on binding LDL, and throughout its journey to the endosome. The conversion from the open state into the closed state takes place on exposure to the acidic environment of the endosome (Figure 26.22B). Three contiguous modules LA7, EGFA, and EGFB rigidly position the propeller module to facilitate displacement of the LDL as the closed state is formed. Under neutral pH, aspartate residues of the propeller blades form hydrogen bonds that hold each blade to the rest of the propeller structure. Exposure to the low-pH environment of the endosome causes the propeller-like structures to interact with the LDL-binding domain. This interaction displaces the LDL, which is then digested by the lysosome. The receptor is often returned to the plasma membrane to again bind LDL. The importance of this process is highlighted by the fact that more than half of the point mutations that result in familial hypercholesterolemia are due to disruptions in the interconversion between the open state and the closed state. These mutations result in a failure to release the LDL cargo and loss of the receptor by degradation.

Inability to transport cholesterol from the lysosome causes Niemann-Pick disease

Niemann-Pick diseases are a group of lipid storage disorders of varying severity. One fatal variety is caused by the accumulation of cholesterol in lysosomes that results in multiple organ failure. Upon release from the LDL receptor, cholesterol is immediately bound to Niemann-Pick C2 (NPC2) protein, a soluble protein in the lumen of the lysosome. Cholesterol-laden NPC2 then docks with Niemann-Pick C1 (NPC1) protein. NPC1, which is embedded in the lysosomal membrane by 13 transmembrane helices, accepts the cholesterol of NPC2, and transfers the cholesterol to the lysosomal membrane itself. The cholesterol is subsequently moved to the endoplasmic reticulum and the plasma membrane. Most mutations (~95%) causing Niemann-Pick disease are found in NPC1. Interestingly, Ebola virus binding to NPC1 is required for infection.

Cycling of the LDL receptor is regulated

PCSK9 (proprotein convertase subtilisin/kexin type) is a protease, secreted by the liver, which plays a crucial role in the regulation of cycling of the LDL receptor. Despite the fact that PCSK9 is a protease, enzymatic activity of the protein is not required for cycling regulation. PCSK9 in the blood binds to the EGFA domain of the receptor (Figure 26.22). PCSK9 locks the receptor in the open conformation even in the acidic conditions of the endosome. Failure to adopt the closed conformation prevents the receptor from returning to the plasma membrane, and it is degraded in the lysosome along with its cargo.

Individuals having a mutation that reduces the amount of PCSK9 in the blood have greatly reduced levels of LDL in the blood and display an almost 90% reduction in the rate of cardiovascular disease. Presumably, reduced levels of PCSK9 allow more receptor cycling and more efficient removal of LDL from the blood. Much research is now being directed at inhibiting

PCSK9-mediated receptor degradation in individuals with high cholesterol levels. Phase 3 trials (Section 28.4) are underway, testing the efficacy of a monoclonal antibody to PCSK9. Monoclonal antibody treatments are expensive, so the search continues for a chemical inhibitor of the protein.

HDL appears to protect against atherosclerosis

Although the events that result in atherosclerosis take place rapidly in familial hypercholesterolemia, a similar sequence of events takes place in people who develop atherosclerosis over decades. In particular, the formation of foam cells and plaques are especially hazardous occurrences. HDL and its role in returning cholesterol to the liver appear to be important in mitigating these life-threatening circumstances.

HDL has a number of antiatherogenic properties, including the inhibition of LDL oxidation. However, HDL's best-characterized property is the removal of cholesterol from cells, especially macrophages. Earlier, we learned that HDL retrieves cholesterol from other tissues in the body to return the cholesterol to the liver for excretion as bile or in the feces. This transport, called *reverse cholesterol transport*, is especially important in regard to macrophages. Indeed, when the transport fails, macrophages become foam cells and facilitate the formation of plaques. Macrophages that collect cholesterol from LDL normally transport the cholesterol to HDL particles. The more HDL, the more readily this transport takes place and the less likely that the macrophages will develop into foam cells. Presumably, this robust reverse cholesterol transport accounts for the observation that higher HDL levels confer protection against atherosclerosis.

The importance of reverse cholesterol transport is illustrated by the occurrence of mutations that inactivate a cholesterol-transport protein in endothelial cells and macrophages, ABCA1 (ATP-binding cassette transporter, subfamily A1) (Figure 13.7). Loss of activity of cholesterol-transport protein ABCA1 results in a very rare condition called *Tangier disease*, which is characterized by HDL deficiency, accumulation of cholesterol in macrophages, and premature atherosclerosis. Under normal conditions, the apoprotein component of HDL, apoA-I, binds to ABCA1 to facilitate LDL transport. Moreover, the interaction between apoA-I and ABCA1 initiates a signal transduction pathway in the endothelial cells that inhibits the inflammatory response.

Until recently, high levels of HDL-bound cholesterol ("good cholesterol") relative to LDL-bound cholesterol ("bad cholesterol") were believed to protect against cardiovascular disease. This belief was based on epidemiological studies. However, a number of recent clinical trials revealed that increased levels of HDL-bound cholesterol had no protective effects at all. These studies do not discount the protective effects of HDL alone, but they illustrate the danger of equating free HDL and cholesterol-bound HDL.

The clinical management of cholesterol levels can be understood at a biochemical level

Homozygous familial hypercholesterolemia can be treated only by a liver transplant. A more generally applicable therapy is available for heterozygotes and others with high levels of cholesterol. *The goal is to reduce the amount of cholesterol in the blood by stimulating the synthesis of more than the customary number of LDL receptors.* We have already observed that the production of LDL receptors is controlled by the cell's need for cholesterol.

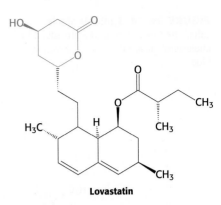

The therapeutic strategy is to deprive the cell of ready sources of cholesterol. When cholesterol is required, the amount of mRNA for the LDL receptor rises and more receptors are found on the cell surface. This state can be induced by a two-pronged approach. First, the reabsorption of bile salts (cholesterol derivatives that promote the absorption of dietary cholesterol and dietary fats) from the intestine is inhibited. Second, de novo synthesis of cholesterol is blocked.

The reabsorption of bile is impeded by oral administration of positively charged polymers, such as cholestyramine, that bind negatively charged bile salts and are not themselves absorbed. Cholesterol synthesis can be effectively blocked by a class of compounds called *statins*. A well-known example of such a compound is lovastatin, which is also called mevacor (Figure 26.23). These compounds are potent competitive inhibitors ($K_i = 1$ nM) of HMG-CoA reductase, the essential control point in the biosynthetic pathway. Plasma cholesterol levels decrease by 50% in many patients given both lovastatin and inhibitors of bile-salt reabsorption. Lovastatin and other inhibitors of HMG-CoA reductase are widely used to lower the plasma-cholesterol level in people who have atherosclerosis, which is the leading cause of death in industrialized societies. Preliminary studies suggest that reducing levels of PCSK9 (p. 869) and HMG-CoA reductase activity may be an especially effective means of reducing cholesterol levels. The development of statins as effective drugs is further described in Chapter 28.

26.4 Important Biochemicals Are Synthesized from Cholesterol and Isoprene

Although cholesterol is well known in its own right as a contributor to the development of heart disease, metabolites of cholesterol—the steroid hormones—also appear in the news frequently. Indeed, steroid-hormone abuse seems to be as prominent in the sports pages as any athlete is. In addition to steroid hormones, cholesterol is a precursor for two other important molecules: bile salts and vitamin D. We begin with a look at the bile salts, molecules crucial for the uptake of lipids in the diet.

Bile salts. *Bile salts* are polar derivatives of cholesterol. These compounds are highly effective *detergents* because they contain both polar and nonpolar regions. Bile salts are synthesized in the liver, stored and concentrated in the gallbladder, and then released into the small intestine. Bile salts, the major constituent of bile, *solubilize dietary lipids*. Solubilization increases the effective surface area of lipids with two consequences: (1) more surface area is exposed to the digestive action of lipases and (2) lipids are more readily absorbed by the intestine. Bile salts are also the major breakdown products of cholesterol. The bile salts glycocholate, the primary bile salt, and taurocholate are shown in Figure 26.24.

In addition to bile salts, bile is composed of cholesterol, phospholipids, and the breakdown products of heme, bilirubin, and biliverdin (Section 24.4). If too much cholesterol is present in the bile, it will precipitate to form gallbladder stones (cholelithiasis). These stones may block bile secretion and inflame the gallbladder, a condition called cholecystitis. Symptoms include pain in the upper right abdomen, especially after a fatty meal, and nausea. If need be, the gall bladder is removed, and bile flows from the liver through the bile duct directly into the intestine.

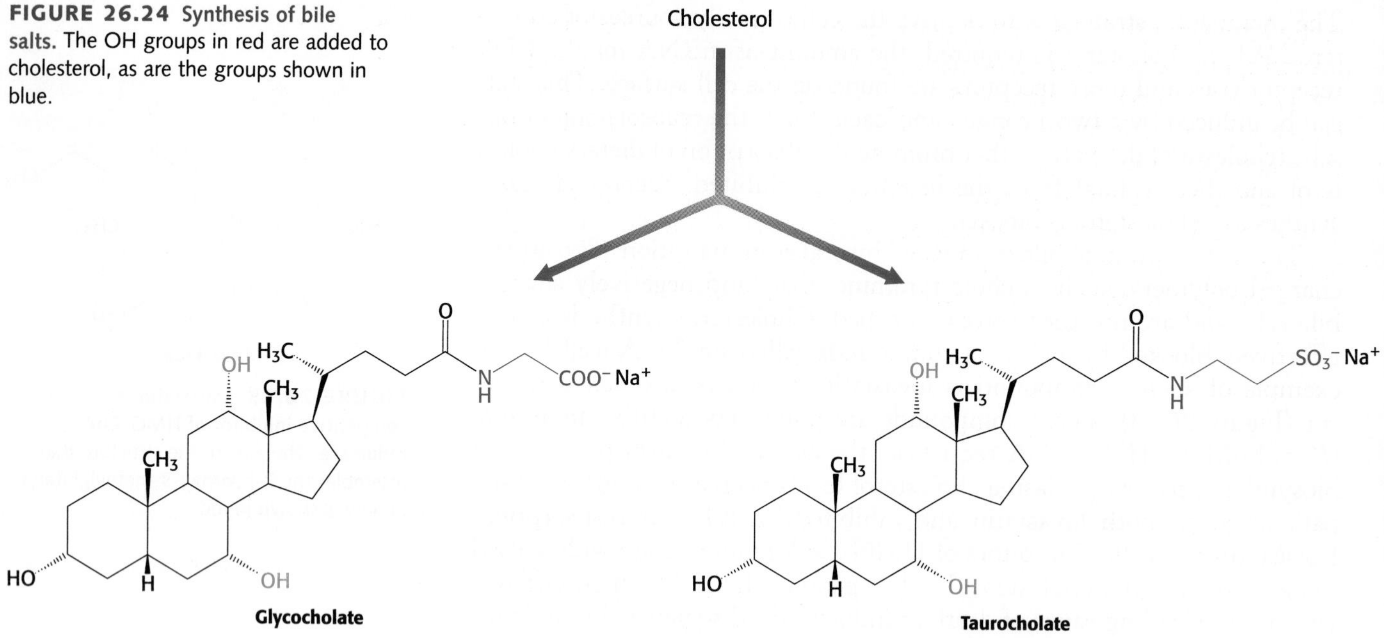

FIGURE 26.24 Synthesis of bile salts. The OH groups in red are added to cholesterol, as are the groups shown in blue.

Cholesterol

Glycocholate

Taurocholate

Steroid hormones. Cholesterol is the precursor of the five major classes of *steroid hormones:* progestogens, glucocorticoids, mineralocorticoids, androgens, and estrogens (Figure 26.25). These hormones are powerful signal molecules that regulate a host of organismal functions. *Progesterone, a progestogen,* prepares the lining of the uterus for the implantation of an ovum. Progesterone is also essential for the maintenance of pregnancy by preventing premature uterine contractions. *Androgens* (such as *testosterone*) are responsible for the development of male secondary sex characteristics, whereas *estrogens* (such as *estradiol*) are required for the development of female secondary sex characteristics. Estrogens, along with progesterone, also participate in the ovarian cycle. *Glucocorticoids* (such as *cortisol*) promote gluconeogenesis and glycogen synthesis, enhance the degradation of fat and protein, and inhibit the inflammatory response. They enable animals to respond to stress; indeed, the absence of glucocorticoids can be fatal. *Mineralocorticoids* (primarily *aldosterone*) act on the distal tubules of the kidney to increase the reabsorption of Na^+ and the excretion of K^+ and H^+, which leads to an increase in blood volume and blood pressure. The major sites of synthesis of these classes of hormones are the corpus luteum, for progestogens; the testes, for androgens; the ovaries, for estrogens; and the adrenal cortex, for glucocorticoids and mineralocorticoids.

Steroid hormones bind to and activate receptor molecules that serve as transcription factors to regulate gene expression (Section 33.2). These small similar molecules are able to have greatly differing effects because the slight structural variations among them allow interactions with specific receptor molecules.

Letters identify the steroid rings and numbers identify the carbon atoms

Carbon atoms in steroids are numbered, as shown for cholesterol in Figure 26.26. The rings in steroids are denoted by the letters A, B, C, and D. Cholesterol contains two angular

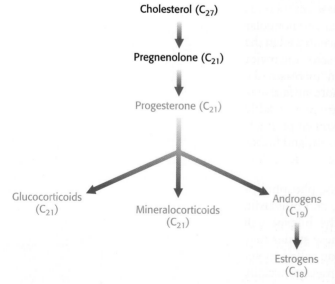

Cholesterol (C$_{27}$)

Pregnenolone (C$_{21}$)

Progesterone (C$_{21}$)

Glucocorticoids (C$_{21}$)

Mineralocorticoids (C$_{21}$)

Androgens (C$_{19}$)

Estrogens (C$_{18}$)

FIGURE 26.25 Biosynthetic relations of classes of steroid hormones and cholesterol.

methyl groups: the C-19 methyl group is attached to C-10, and the C-18 methyl group is attached to C-13. The C-18 and C-19 methyl groups of cholesterol lie *above* the plane containing the four rings. A substituent that is above the plane is termed *β oriented*, whereas a substituent that is below the plane is *α oriented*.

If a hydrogen atom is attached to C-5, it can be either α or β oriented. The A and B steroid rings are fused in a *trans* conformation if the C-5 hydrogen is α oriented, and *cis* if it is β oriented. The absence of a Greek letter for the C-5 hydrogen atom on the steroid nucleus implies a trans fusion.

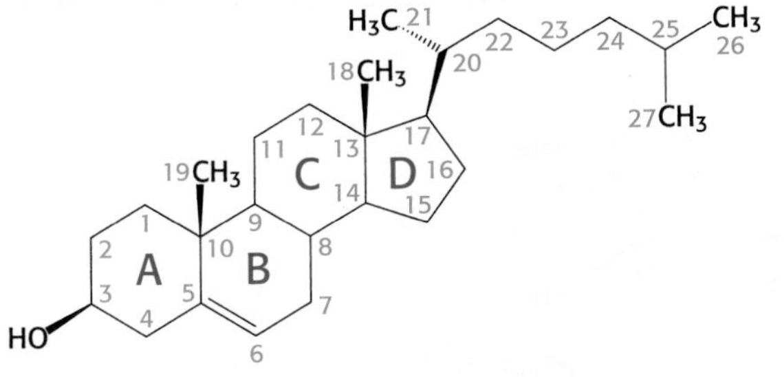

FIGURE 26.26 Cholesterol carbon numbering. The numbering scheme for the carbon atoms in cholesterol and other steroids.

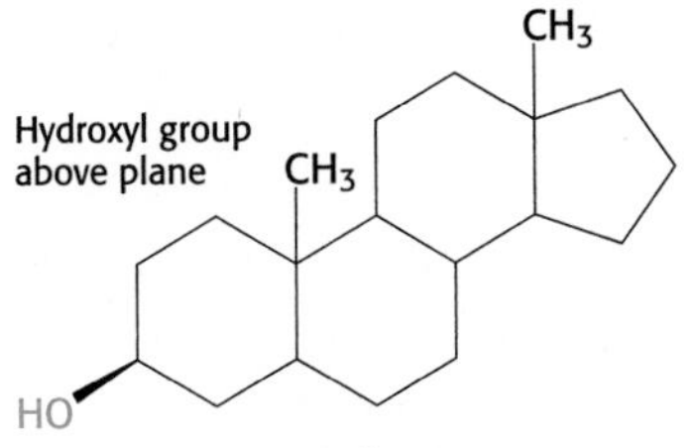

3β-Hydroxy

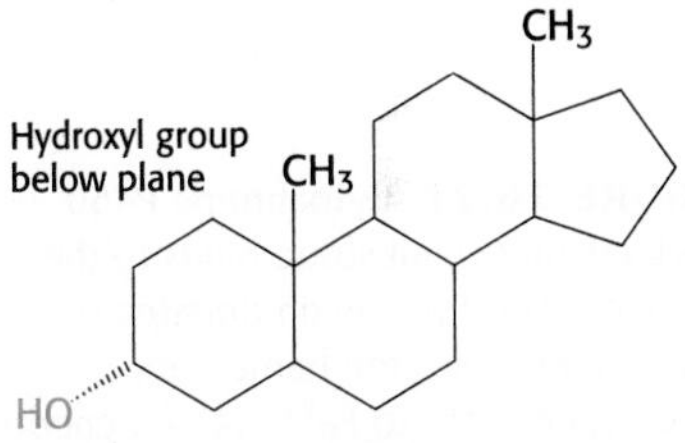

3α-Hydroxy

The C-5 hydrogen atom is α oriented in all steroid hormones that contain a hydrogen atom in that position. In contrast, bile salts have a β-oriented hydrogen atom at C-5. Thus, *a cis fusion is characteristic of the bile salts, whereas a trans fusion is characteristic of all steroid hormones that possess a hydrogen atom at C-5.* A trans fusion yields a nearly planar structure, whereas a cis fusion gives a buckled structure.

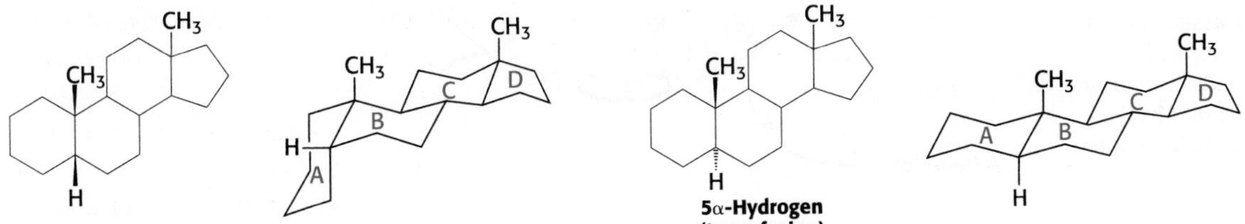

5β-Hydrogen (cis fusion)

5α-Hydrogen (trans fusion)

Steroids are hydroxylated by cytochrome P450 monooxygenases that use NADPH and O₂

The addition of OH groups plays an important role in the synthesis of cholesterol from squalene and in the conversion of cholesterol into steroid hormones and bile salts. All these hydroxylations require NADPH and O_2. The oxygen atom of the incorporated hydroxyl group comes from O_2 rather than from H_2O. Whereas one oxygen atom of the O_2 molecule goes into the substrate, the other is reduced to water. The enzymes catalyzing these reactions are called *monooxygenases* (or *mixed-function oxygenases*). Recall that a monooxygenase also participates in the hydroxylation of aromatic amino acids (Section 23.5).

$$\text{RH} + O_2 + \text{NADPH} + H^+ \longrightarrow \text{ROH} + H_2O + \text{NADP}^+$$

Hydroxylation requires the activation of oxygen. In the synthesis of steroid hormones and bile salts, activation is accomplished by members of the *cytochrome P450* family, a family of cytochromes that absorb light maximally at 450 nm when complexed in vitro with exogenous carbon monoxide. These membrane-anchored proteins (~50 kDa) contain a heme prosthetic group. Oxygen is activated through its binding to the iron atom in the heme group.

Because the hydroxylation reactions promoted by P450 enzymes are oxidation reactions, it is at first glance surprising that they also consume the reductant NADPH. NADPH transfers its high-potential electrons to a flavoprotein, which transfers them, one at a time, to *adrenodoxin*, a non-heme iron protein. Adrenodoxin transfers one electron to reduce the ferric (Fe^{3+}) form of P450 to the ferrous (Fe^{2+}) form (Figure 26.27).

Without the addition of this electron, P450 will not bind oxygen. Recall that only the ferrous form (Fe^{2+}) of myoglobin and hemoglobin binds oxygen (Section 7.1). The binding of O_2 to the heme is followed by the acceptance of a second electron from adrenodoxin. The acceptance of this second electron leads to cleavage of the O—O bond. One of the oxygen atoms is

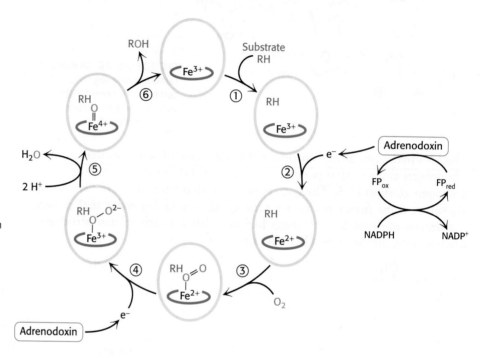

FIGURE 26.27 Cytochrome P450 mechanism. (1) Substrate binds to the enzyme. (2) Adrenodoxin donates an electron, reducing the heme iron. (3) Oxygen binds to Fe^{2+}. (4) Adrenodoxin donates a second electron. (5) The bond between the oxygen atoms is cleaved, a molecule of water is released, and an $Fe^{4+} = O$ intermediate is formed. (6) The $Fe^{4+} = O$ intermediate forms the hydroxylated product and returns the iron to the Fe^{3+} state, ready for another reaction cycle.

then protonated and released as water. The remaining oxygen atom forms a highly reactive ferryl (Fe^{4+}) Fe=O intermediate. This intermediate abstracts a hydrogen atom from the substrate RH to form R•. This transient free radical captures the OH group from the iron atom to form ROH, the hydroxylated product, returning the iron atom to the ferric state.

The cytochrome P450 system is widespread and performs a protective function

The cytochrome P450 system, which in mammals is located primarily in the smooth endoplasmic reticulum of the liver and small intestine, is also important in the *detoxification of foreign substances* (xenobiotic compounds). For example, the hydroxylation of phenobarbital, a barbiturate, *increases its solubility* and *facilitates its excretion*. Likewise, polycyclic aromatic hydrocarbons that are ingested by drinking contaminated water are hydroxylated by P450, providing sites for conjugation with highly polar units (e.g., glucuronate or sulfate) that markedly increase the solubility of the modified aromatic molecule. One of the most relevant functions of the cytochrome P450 system to human beings is its role in metabolizing drugs such as caffeine and ibuprofen (Chapter 28). Some members of the cytochrome P450 system also metabolize ethanol (Section 27.6). The duration of action of many medications depends on their rate of inactivation by the

P450 system. Despite its general protective role in the removal of foreign chemicals, the action of the P450 system is not always beneficial. *Some of the most powerful carcinogens are generated from harmless compounds by the P450 system in vivo in the process of metabolic activation* (Figure 29.34). In plants, the cytochrome P450 system plays a role in the synthesis of toxic compounds as well as the pigments of flowers.

Earlier (Section 9.1), we examined the use of protease inhibitors to treat HIV infection. These inhibitors, in conjunction with inhibitors of other HIV enzymes, have drastically reduced deaths due to AIDS. The effectiveness of these drugs is sometimes compromised because P450 enzymes, such as cytochrome P450-3A4, inactivate them. The drug ritonavir was developed in the search for more effective inhibitors. Interestingly, ritonavir, although a protease inhibitor, is also a potent inhibitor of cytochrome P450-3A4. This serendipitous "side-effect" allows the clinician to administer low doses of ritonavir in combination with other protease inhibitors, effectively increasing the concentrations of these drugs while allowing a reduction in dosage and dosing frequency.

Pregnenolone, a precursor of many other steroids, is formed from cholesterol by cleavage of its side chain

Steroid hormones contain 21 or fewer carbon atoms, whereas cholesterol contains 27. Thus, the first stage in the synthesis of steroid hormones is the removal of a six-carbon unit from the side chain of cholesterol to form *pregnenolone.* The side chain of cholesterol is hydroxylated at C-20 and then at C-22, and the bond between these carbon atoms is subsequently cleaved by *desmolase.* Three molecules of NADPH and three molecules of O_2 are consumed in this six-electron oxidation.

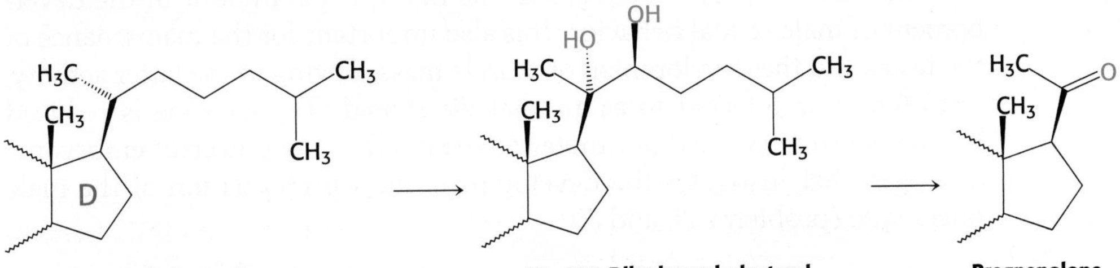

Cholesterol   20α,22β-Dihydroxycholesterol   Pregnenolone

Progesterone and corticosteroids are synthesized from pregnenolone

Progesterone is synthesized from pregnenolone in two steps. The 3-hydroxyl group of pregnenolone is oxidized to a 3-keto group, and the Δ^5 double bond is isomerized to a Δ^4 double bond (Figure 26.28). *Cortisol,* the major glucocorticoid, is synthesized from progesterone by hydroxylations at C-11, C-17, and C-21; C-17 must be hydroxylated before C-21 is hydroxylated, whereas C-11 can be hydroxylated at any stage. The enzymes catalyzing these hydroxylations are highly specific. The initial step in the synthesis of *aldosterone,* the major mineralocorticoid, is the hydroxylation of progesterone at C-21. The resulting deoxycorticosterone is hydroxylated at C-11. The oxidation of the C-18 angular methyl group to an aldehyde then yields aldosterone.

Androgens and estrogens are synthesized from pregnenolone

Androgens and estrogens also are synthesized from pregnenolone through the intermediate progesterone. Androgens contain 19 carbon atoms. The

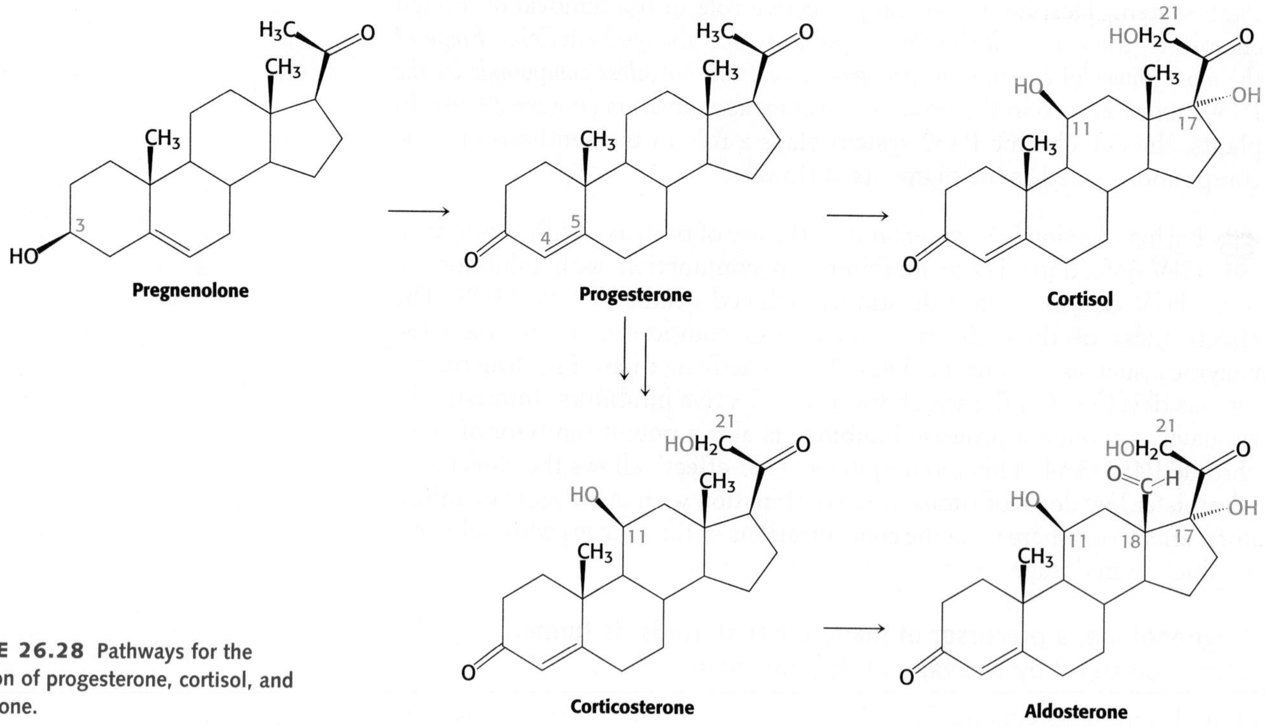

FIGURE 26.28 Pathways for the formation of progesterone, cortisol, and aldosterone.

synthesis of androgens starts with the hydroxylation of progesterone at C-17 (Figure 26.29). The side chain consisting of C-20 and C-21 is then cleaved to yield *androstenedione*, an androgen. *Testosterone,* another androgen, is formed by the reduction of the 17-keto group of androstenedione. Testosterone, through its actions in the brain, is paramount in the development of male sexual behavior. It is also important for the maintenance of the testes and the development of muscle mass. Owing to the latter activity, testosterone is referred to as an *anabolic steroid.* Testosterone is reduced by *5α-reductase* to yield *dihydrotestosterone* (DHT), a powerful embryonic androgen that instigates the development and differentiation of the male phenotype (problems 26 and 29).

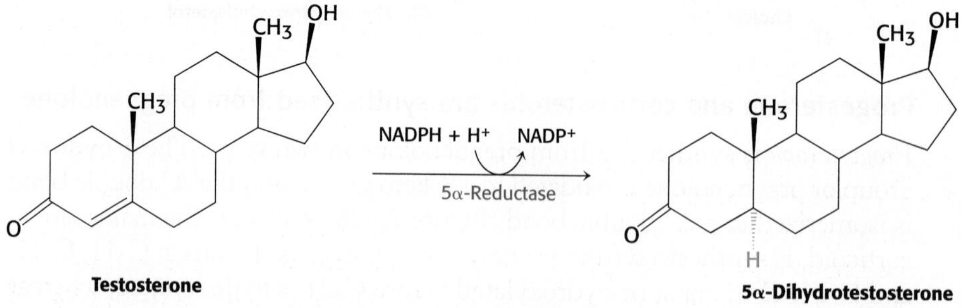

Estrogens are synthesized from androgens by the loss of the C-19 angular methyl group and the formation of an aromatic A ring. *Estrone,* an estrogen, is derived from androstenedione, whereas *estradiol,* the biologically most potent estrogen, is formed from testosterone. Estradiol can also be formed from estrone. The formation of the aromatic A ring is catalyzed by the P450 enzyme *aromatase.*

Because breast and ovarian cancers frequently depend on estrogens for growth, aromatase inhibitors are often used as a treatment for these cancers. Anastrozole is a competitive inhibitor of the enzyme, whereas

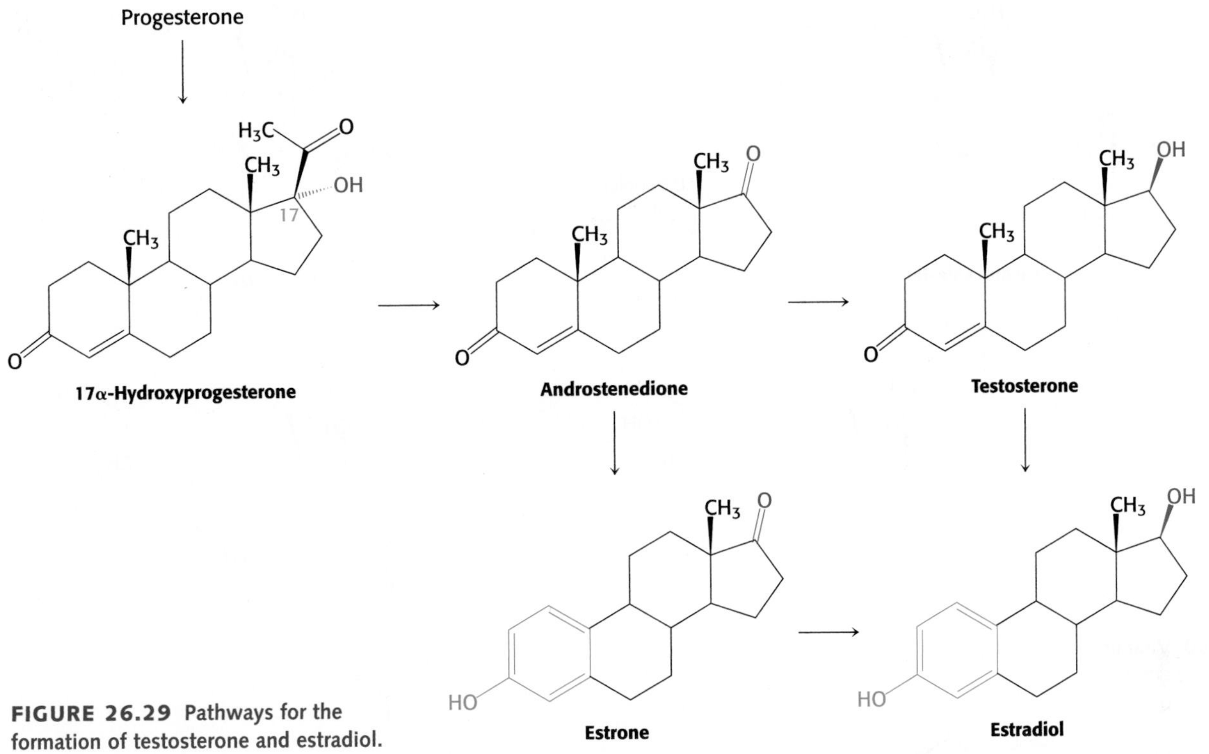

FIGURE 26.29 Pathways for the formation of testosterone and estradiol.

exemestane is a suicide inhibitor that covalently modifies and inactivates the enzyme.

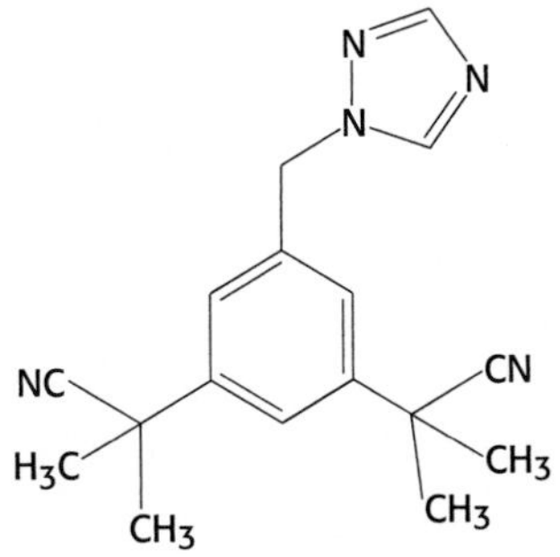

Vitamin D is derived from cholesterol by the ring-splitting activity of light

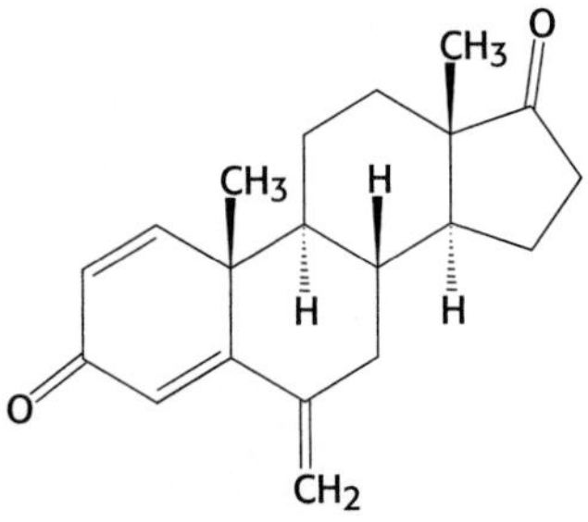

Cholesterol is also the precursor of vitamin D, which plays an essential role in the control of calcium and phosphorus metabolism. *7-Dehydrocholesterol (provitamin D₃) is photolyzed by the ultraviolet light of sunlight to previtamin D₃, which spontaneously isomerizes to vitamin D₃* (Figure 26.30). Vitamin D₃ (cholecalciferol) is converted into *calcitriol* (1,25-dihydroxycholecalciferol), the active hormone, by hydroxylation reactions in the liver and kidneys. Although not a steroid, vitamin D acts in an analogous fashion. It binds to a receptor, structurally similar to the steroid receptors, to form a complex that functions as a transcription factor, regulating gene expression.

Vitamin D deficiency in childhood produces *rickets,* a disease characterized by inadequate calcification of cartilage and bone. Rickets was so common in 17th-century England that it was called the "children's disease of the English." The 7-dehydrocholesterol in the skin of these children was not photolyzed to previtamin D₃, because there was little sunlight for many months of the year. Furthermore, their diets provided little vitamin D, because most naturally occurring foods have a low content of this vitamin. Fish-liver oils are a notable exception. Cod-liver oil, abhorred by generations of children because of its unpleasant taste, was used in the past as a rich source of vitamin D. Today, the most reliable dietary sources of vitamin D are fortified foods. Milk, for example, is fortified to a level of 400 international units per quart (10 μg per quart). The recommended daily intake of vitamin D is 200 international units until age 50, after which it increases with age. In adults, vitamin D deficiency leads to softening and weakening of bones, a condition called *osteomalacia.* The occurrence of osteomalacia in Muslim women who are clothed so

Anastrozole

Exemestane

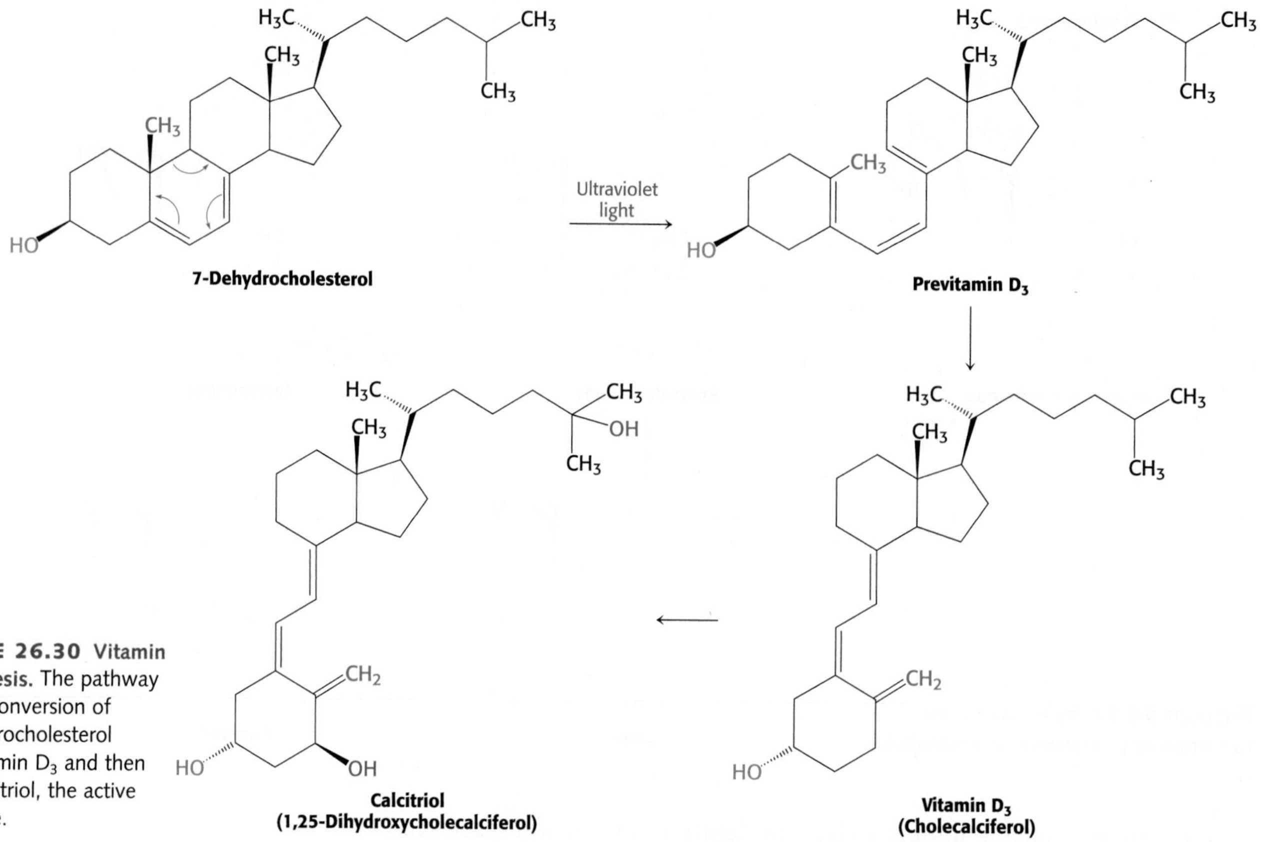

FIGURE 26.30 Vitamin D synthesis. The pathway for the conversion of 7-dehydrocholesterol into vitamin D_3 and then into calcitriol, the active hormone.

that only their eyes are exposed to sunlight is a striking reminder that vitamin D is needed by adults as well as by children.

Research over the past few years indicates that vitamin D may play a much larger biochemical role than simply the regulation of bone metabolism. Muscle seems to be a target of vitamin D action. In muscle, vitamin D appears to affect a number of biochemical processes with the net effect being enhanced muscle performance. Studies also suggest that vitamin D prevents cardiovascular disease, reduces the incidence of a variety of cancers, and protects against autoimmune diseases including diabetes. Moreover, vitamin D deficiency appears to be more common than previously thought. In the United States, 75% of Blacks and many Hispanics and Asians have insufficient blood levels of vitamin D. This recent research on vitamin D shows again the dynamic nature of biochemical investigations. Vitamin D, a chemical whose biochemical role was believed to be well established, now offers new frontiers of biomedical research.

Five-carbon units are joined to form a wide variety of biomolecules

The synthesis of squalene (C_{30}) from isopentenyl pyrophosphate (C_5) exemplifies a fundamental mechanism for the assembly of carbon skeletons of biomolecules. *A remarkable array of compounds are formed from isopentenyl pyrophosphate, the basic five-carbon building block.* The fragrances of many plants arise from volatile C_{10} and C_{15} compounds, which are called *terpenes*. For example, myrcene ($C_{10}H_{16}$) from bay leaves consists of two isoprene units, as does limonene ($C_{10}H_{15}$) from lemon oil (Figure 26.31). Zingiberene ($C_{15}H_{24}$), from the oil of ginger, is made up of three isoprene units. Some terpenes, such as geraniol from geraniums and menthol from peppermint oil, are alcohols, and others, such as citro-nellal, are aldehydes. *Natural rubber* is a linear polymer of *cis*-isoprene units.

We have already encountered several molecules that contain isoprenoid side chains. The C_{30} hydrocar-bon *side chain of vitamin K*, a key molecule in clotting (p. 331), is built from six C_5 units. *Coenzyme Q_{10}* in the mitochondrial respiratory chain (p. 581) has a side chain made up of 10 isoprene units. Yet another example is the *phytol side chain of chlorophyll* (p. 622), which is formed from four isoprene units. Recall also that proteins are targeted to membranes by the covalent attachment of a modified farnesyl (C_{15}) unit to their carboxyl-terminal cysteine residue (p. 388).

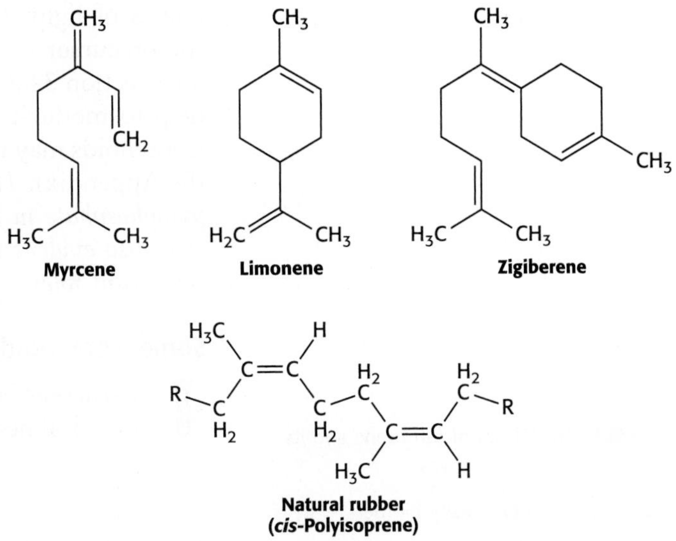

Isoprenoids can delight by their color as well as by their fragrance. Indeed, isoprenoids can be regarded as sensual molecules! The color of tomatoes and carrots comes from *carotenoids*, specifically from *lycopene* and β-carotene, respectively. These compounds absorb light because they contain extended networks of single and double bonds—that is, they are *polyenes*. Their C_{40} carbon skeletons are built by the successive addition of C_5 units to form *geranylgeranyl pyrophosphate, a C_{20} intermediate*, which then condenses tail-to-tail with another molecule of geranylgeranyl pyrophosphate. This biosynthetic pathway is like that of squalene except that C_{20} rather than C_{15} units are assembled and condensed.

$$C_5 \longrightarrow C_{10} \longrightarrow C_{15} \longrightarrow C_{30} \quad \text{(Squalene)}$$
$$C_5 \longrightarrow C_{10} \longrightarrow C_{15} \longrightarrow C_{20} \longrightarrow C_{40} \quad \text{(Phytoene)}$$

Phytoene, the C_{40} condensation product, is dehydrogenated to yield lycopene. Cyclization of both ends of lycopene gives β-carotene (Figure 26.32). Carotenoids serve as light-harvesting molecules in photosynthetic assemblies and also play a role in protecting bacteria from the deleterious

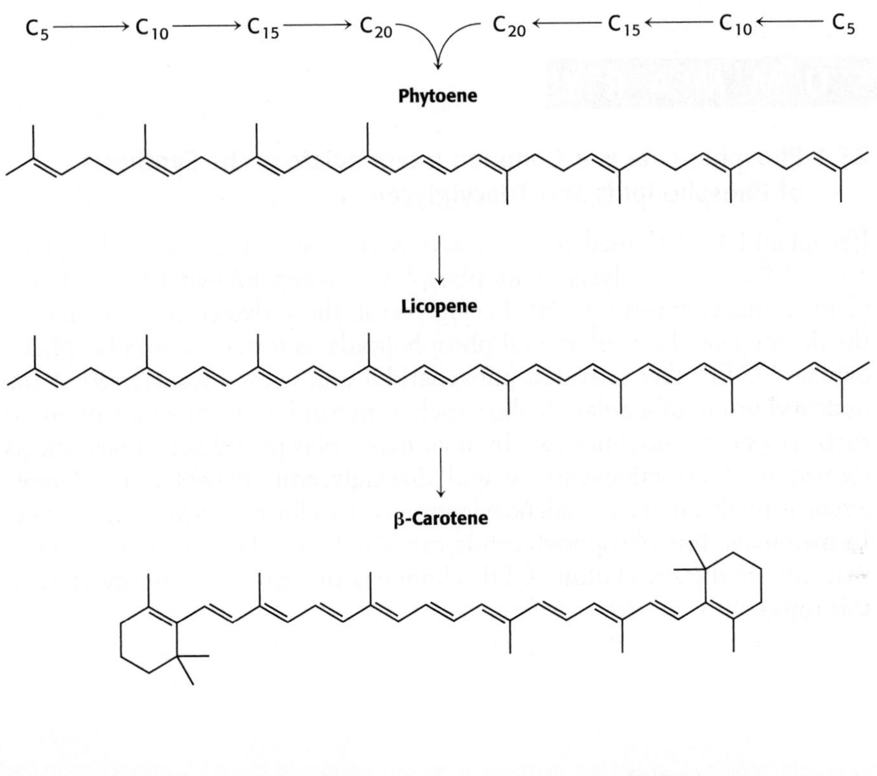

FIGURE 26.31 Formulas for some isoprenoids.

Perfumes, colors and sounds echo one another
—Charles Baudelaire
French poet 1821–1867
Correspondances

FIGURE 26.32 Synthesis of C_{40} isoprenoids.

effects of light. Carotenoids are also essential for vision. β-Carotene is the precursor of retinal, the chromophore in all known visual pigments (See Section 34.3 in **Sapling**Plus). There is also evidence that isoprenoids help to modulate the inflammatory response, and a decrease in certain isoprenoids may result in inflammation (see Problem-Solving Strategies in the Appendix). *These examples illustrate the fundamental role of isopentenyl pyrophosphate in the assembly of extended carbon skeletons of biomolecules. It is also evident that isoprenoids are ubiquitous in nature and have diverse significant roles.*

Some isoprenoids have industrial applications

 Farnesene is an isoprenoid that has potential to meet a wide variety of practical needs (Table 26.2).

TABLE 26.2 Uses of farnesene and its derivatives

As a high-energy density biofuel
As sealants and adhesives
As solvents and lubricants
To make automobile tires that yield improved gas mileage and wet road grip
As a component of cosmetics
To make flavors and fragrances

Farnesene

However, to meet these potential uses in an economical manner, large amounts of farnesene must be produced.

To solve the problem of cheap and readily accessible farnesene, scientists at the biotechnology company Amyris started with the model organism *Saccharomyces cerevisiae* —baker's yeast. Then, using the genetic techniques discussed in Chapter 5 and shown in the animation (See Chapter 5 animated techniques in **Sapling**Plus), they introduced genes for four enzymes that were not native to yeast. These genes enabled the yeast to now ferment sugar cane syrup to farnesene, and then to secrete the farnesene. The secreted farnesene is 93% pure and distillation enhances the purity to 98%. This process can be scaled up to industrial levels, thus generating a versatile, renewable biomolecule with a variety of applications. This example shows that to do useful, important, and exciting biochemistry you do not have to be at a university or medical school. Indeed, the extent of the usefulness of biochemistry are limited only by one's imagination.

SUMMARY

26.1 Phosphatidate Is a Common Intermediate in the Synthesis of Phospholipids and Triacylglycerols

Phosphatidate is formed by successive acylations of glycerol 3-phosphate by acyl CoA. Hydrolysis of its phosphoryl group followed by acylation yields a triacylglycerol. CDP-diacylglycerol, the activated intermediate in the de novo synthesis of several phospholipids, is formed from phosphatidate and CTP. The activated phosphatidyl unit is then transferred to the hydroxyl group of a polar alcohol, such as inositol, to form a phospholipid such as phosphatidylinositol. In mammals, phosphatidylethanolamine is formed by CDP-ethanolamine and diacylglycerol. Phosphatidylethanolamine is methylated by *S*-adenosylmethionine to form phosphatidylcholine. In mammals, this phosphoglyceride can also be synthesized by a pathway that utilizes dietary choline. CDP-choline is the activated intermediate in this route.

Sphingolipids are synthesized from ceramide, which is formed by the acylation of sphingosine. Gangliosides are sphingolipids that contain an oligosaccharide unit having at least one residue of N-acetylneuraminate or a related sialic acid. They are synthesized by the step-by-step addition of activated sugars, such as UDP-glucose, to ceramide.

26.2 Cholesterol Is Synthesized from Acetyl Coenzyme A in Three Stages

Cholesterol is a steroid component of animal membranes and a precursor of steroid hormones. The committed step in its synthesis is the formation of mevalonate from 3-hydroxy-3-methylglutaryl CoA (derived from acetyl CoA and acetoacetyl CoA). Mevalonate is converted into isopentenyl pyrophosphate (C_5), which condenses with its isomer, dimethylallyl pyrophosphate (C_5), to form geranyl pyrophosphate (C_{10}). The addition of a second molecule of isopentenyl pyrophosphate yields farnesyl pyrophosphate (C_{15}), which condenses with itself to form squalene (C_{30}). This intermediate cyclizes to lanosterol (C_{30}), which is modified to yield cholesterol (C_{27}).

26.3 The Complex Regulation of Cholesterol Biosynthesis Takes Place at Several Levels

In the liver, cholesterol synthesis is regulated by changes in the amount and activity of 3-hydroxy-3-methylglutaryl CoA reductase. Transcription of the gene, translation of the mRNA, and degradation of the enzyme are stringently controlled. In addition, the activity of the reductase is regulated by phosphorylation.

Triacylglycerols exported by the intestine are carried by chylomicrons and then hydrolyzed by lipases lining the capillaries of target tissues. Cholesterol and other lipids in excess of those needed by the liver are exported in the form of very low-density lipoprotein. After delivering its content of triacylglycerols to adipose tissue and other peripheral tissue, VLDL is converted into intermediate-density lipoprotein and then into low-density lipoprotein. IDL and LDL carry cholesteryl esters, primarily cholesteryl linoleate. Liver and peripheral tissue cells take up LDL by receptor-mediated endocytosis. The LDL receptor, a protein spanning the plasma membrane of the target cell, binds LDL and mediates its entry into the cell. Absence of the LDL receptor in the homozygous form of familial hypercholesterolemia leads to a markedly elevated plasma level of LDL cholesterol and the deposition of cholesterol on blood-vessel walls, which in turn may result in childhood heart attacks. High-density lipoproteins transport cholesterol from the peripheral tissues to the liver.

26.4 Important Biochemicals Are Synthesized from Cholesterol and Isoprene

In addition to bile salts, which facilitate the digestion of lipids, five major classes of steroid hormones are derived from cholesterol: progestogens, glucocorticoids, mineralocorticoids, androgens, and estrogens. Hydroxylations by P450 monooxygenases that use NADPH and O_2 play an important role in the synthesis of steroid hormones and bile salts from cholesterol. P450 enzymes, a large superfamily, also participate in the detoxification of drugs and other foreign substances.

Pregnenolone (C_{21}) is an essential intermediate in the synthesis of steroids. This steroid is formed by scission of the side chain of cholesterol. Progesterone (C_{21}), synthesized from pregnenolone, is the precursor of cortisol and aldosterone. Hydroxylation of progesterone and cleavage of its side chain yields androstenedione, an androgen (C_{19}). Estrogens (C_{18}) are synthesized from androgens by the loss of an angular methyl group and the formation of an aromatic A ring. Vitamin D, which is important in the control of calcium and phosphorus metabolism, is formed from a derivative of cholesterol by the action of light.

A remarkable array of biomolecules in addition to cholesterol and its derivatives are synthesized from isopentenyl pyrophosphate, the basic five-carbon building block. The hydrocarbon side chains of vitamin K, coenzyme Q, and chlorophyll are extended chains constructed from this activated C_5 unit. Many proteins are targeted to membranes by prenyl groups that are derived from this activated intermediate.

APPENDIX

Biochemistry in Focus

Excess ceramides may cause insulin insensitivity

We are aware of the role that excess cholesterol plays in cardiovascular disease (p. 870). We also learned that ceramides may promote tumor growth. Recent research suggests that ceramide might be the "new cholesterol" in that it promotes a number of metabolic diseases, including insulin resistance. As we will learn in Chapter 27, insulin resistance—the inability to respond to insulin—often results in type 2 diabetes.

A hallmark of obesity is that lipids accumulate in tissues that are not suited for lipid storage, such as liver and muscle. Many types of lipids build up with excess caloric consumption, but ceramides are among the most deleterious since they have been implicated in the development of insulin resistance. Ceramides are generally not found in the diet, but rather are synthesized from dietary fat, notably saturated fatty acids. In obese individuals, excess ceramides are found in muscle, the β-cells of the pancreas, and liver.

How might a high ceramide concentration result in insulin resistance? Although the details are far from clear, ceramide is thought to act in a number of ways. Ceramide stimulates fatty acid import, which leads to lipid accumulation. Moreover, ceramide activates a protein phosphatase that inhibits protein kinases that are necessary for insulin action. In other words, in the presence of excess ceramide, cells become insulin resistant. The inability of insulin to mobilize lipids leads to yet more lipid accumulation in muscle and liver.

Interestingly, a hormone secreted by adipose tissue, adiponectin (Section 27.2), has been shown in mice to lower cellular and blood levels of ceramides. Adiponectin acts by binding to a membrane bound receptor in target tissues. Adiponectin binding to its receptor activates a ceramidase activity that is inherent in the receptor. The ceramidase converts ceramides to sphingosine thus blocking the pathological effects of ceramide.

APPENDIX

Problem-Solving Strategies

PROBLEM: Mevalonate kinase deficiency, also called mevalonic aciduria, is a rare disorder caused by a mutation in gene encoding the enzyme mevalonate kinase. Mevalonate kinase catalyzes the formation of 5-phosphomevalonate from mevalonate. Mevalonate kinase deficiency is characterized by anatomical birth defects, psychomotor impairment, ataxia, seizures, enlarged liver, and inflamed lymph nodes. High concentrations of mevalonate are found in the blood and urine. The spectrum of pathologies varies with the amount of residual enzyme activity.

(a) Explain why excess mevalonate is present in the blood and urine?
(b) What would be the effect of the enzyme deficiency on cholesterol synthesis?
(c) How might the level of HMG-CoA reductase activity be altered by the enzyme deficiency?

(d) Suggest what might be the actual cause of the various pathologies characterizing mevalonate kinase deficiency.

SOLUTION: As usual, we will break the question down into a series of smaller ones.

▶ **Mevalonate is a component of which biochemical pathway?**

If you don't remember, take a look at Figure 26.9. Upon doing so, you will see that mevalonate is an early intermediate in the synthesis of cholesterol and isoprenoids.

Now, summoning all of your biochemical acumen, answer the following question.

▶ **What reaction does mevalonate kinase (the "star" of the question we are dissecting) catalyze?**

With your level of acumen, you would write that it phosphorylates mevalonate to form 5-phosphomevalonate. Or you might just write the equation.

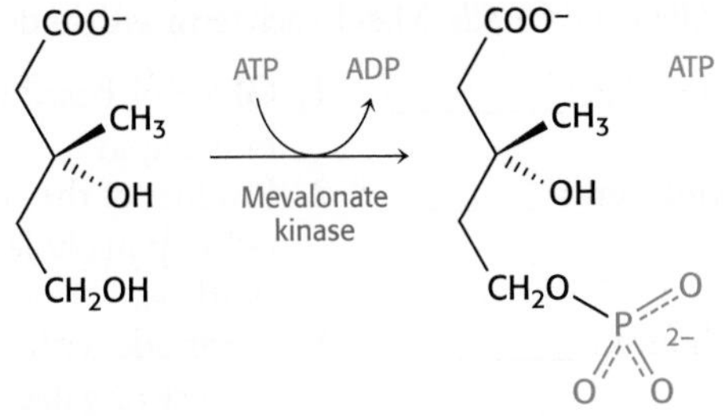

Mevalonate   **5-Phosphpo-mevalonate**

Now that we have established some parameters, let's tackle part (a).

▶ **Why is there an excess of mevalonate in blood and urine?**

If the mevalonate cannot be metabolized, it will accumulate and eventually leave the tissues and appear in the blood and urine.

▶ **How would the absence of 5-phosphomevalonate impact cholesterol and isoprenoid synthesis?**

5-Phosphomevalonate is a fundamental building block for cholesterol and isoprenoids. No 5-phosphomevalonate = no cholesterol and no isoprenoids.

Now, let's get started on (c).

▶ **What is the role of HMG CoA reductase in cholesterol and isoprenoid synthesis?**

HMG CoA reductase catalyzes the rate-limiting step in cholesterol biosynthesis.

Think back to the principles of cholesterol regulation.

▶ **If the concentration of cholesterol is low, how might a cell that requires cholesterol respond?**

The response would be to increase the activity and amount of HMG CoA reductase to increase the concentration of mevalonate. However, this tactic would be to no avail, since, in the absence of kinase, more mevalonate would simply be released from the tissues.

What about part (d)?

▶ **The lack of mevalonate kinase will alter the concentrations of a number of biomolecules. How might these altered concentrations account for the symptoms of mevalonate kinase deficiency?**

This is an interesting question, which does not yet have a definitive answer. The most obvious answer would be too much mevalonate and/or too little cholesterol. However, recall that many isoprenoids will also be missing because of the deficiency. Recent work implies that lack of specific isoprenoids may be the cause of the pathologies. Of course, the answer might also be "all of the above."

KEY TERMS

phosphatidate (p. 850)

triacylglycerol (p. 851)

phospholipid (p. 851)

cytidine diphosphodiacylglycerol (CDP-diacylglycerol) (p. 851)

sphingolipid (p. 854)

ceramide (p. 854)

cerebroside (p. 855)

ganglioside (p. 855)

cholesterol (p. 858)

mevalonate (p. 859)

3-hydroxy-3-methylglutaryl CoA reductase (HMG-CoA reductase) (p. 859)

3-isopentenyl pyrophosphate (p. 859)

sterol regulatory element binding protein (SREBP) (p. 861)

lipoprotein particles (p. 863)

low-density lipoprotein (LDL) (p. 864)

high-density lipoprotein (HDL) (p. 864)

receptor-mediated endocytosis (p. 866)

reverse cholesterol transport (p. 870)

bile salt (p. 871)

steroid hormone (p. 872)

PROBLEMS

1. *Building block.* All glycerol-containing phospholipids are derived from: ✓❶

(a) Ceramide

(b) Ganglioside

(c) Mevalonate

(d) Phosphatidate

2. *Human head with a lion's body.* Sphingolipids are synthesized from:

(a) Ceramide

(b) Phosphatidylserine

(c) Diacylglycerol

(d) Isoprene

3. *Sam I Am.* The synthesis of phosphatidylcholine from phosphatidylethanolamine requires: ✓❶

(a) Tetrahydrofolate

(b) Glycine

(c) Choline

(d) S-Adenosylmethionine

4. *In its entirety.* Cholesterol is synthesized entirely from: ✓❷

(a) Oxaloacetate

(b) Glycerol 3-phosphate

(c) Acetyl CoA

(d) Serine

5. *One of the four humors.* Bile salts are synthesized from: ✓❹

(a) Cholesterol

(b) Prostaglandin

(c) Triacylglycerol

(d) Ceramide

6. *Different roles.* Describe the roles of glycerol 3-phosphate, phosphatidate, and diacylglycerol in triacylglycerol and phospholipid synthesis. ✓❶

7. *Needed supplies.* How is the glycerol 3-phosphate required for phosphatidate synthesis generated? ✓❶

8. *Making fat.* Write a balanced equation for the synthesis of a triacylglycerol, starting from glycerol and fatty acids.

9. *Making a phospholipid.* Write a balanced equation for the synthesis of phosphatidylethanolamine by the de novo pathway, starting from ethanolamine, glycerol, and fatty acids.

10. *ATP needs.* How many high-phosphoryl-transfer-potential molecules are required to synthesize phosphatidylethanolamine from ethanolamine and diacylglycerol? Assume that the ethanolamine is the activated component. ✓❶

11. *Identifying differences.* Differentiate among sphingomyelin, a cerebroside, and a ganglioside.

12. *Let's count the ways.* There may be 50 ways to leave your lover, but, in principle, there are only three ways to make a glycerol-based phospholipid. Describe the three pathways.

13. *Activated donors.* What is the activated reactant in each of the following biosyntheses?

(a) Phosphatidylinositol from inositol

(b) Phosphatidylethanolamine from ethanolamine

(c) Ceramide from sphingosine

(d) Sphingomyelin from ceramide

(e) Cerebroside from ceramide

(f) Ganglioside G_{M1} from ganglioside G_{M2}

(g) Farnesyl pyrophosphate from geranyl pyrophosphate

14. *No DAG, no TAG.* What would be the effect of a mutation that decreased the activity of phosphatidic acid phosphatase? ✓❶

15. *Like Wilbur and Orville.* Match each term with its description.

(a) Phosphatidate _____

(b) Triacylglycerol _____

(c) Phospholipid _____

(d) Sphingolipid _____
(e) Cerebroside _____

(f) Ganglioside _____

(g) Cholesterol _____
(h) Mevalonate _____

(i) Lipoprotein particle
_____

(j) Steroid hormone
_____

1. Glycerol-based membrane lipid
2. Product of the committed step in cholesterol synthesis
3. Ceramide with either glucose or galactose attached
4. Storage form of fatty acids
5. Squalene is a precursor to this molecule
6. Transports cholesterol and lipids
7. Derived from cholesterol
8. Precursor to both phospholipids and triacylglycerols
9. Formed from ceramide by the attachment of phosphocholine
10. Ceramide with multiple carbohydrates attached

16. *The Law of Three Stages.* What are the three stages required for the synthesis of cholesterol?

17. *Many regulations to follow.* Outline the mechanisms of the regulation of cholesterol biosynthesis. ✓❷

18. *Telltale labels.* What is the distribution of isotopic labeling in cholesterol synthesized from each of the following precursors? ✓❷

(a) Mevalonate labeled with ^{14}C in its carboxyl carbon atom

(b) Malonyl CoA labeled with ^{14}C in its carboxyl carbon atom

19. *Too much, too soon.* What is familial hypercholesterolemia and what are its causes? ✓❸

20. *Familial hypercholesterolemia.* Several classes of LDL-receptor mutations have been identified as causes of this disease. Suppose that you have been given the following: cells from patients with different mutations; an antibody specific for the LDL receptor that can be seen with an electron microscope; and access to an electron microscope. What differences in antibody distribution might you expect to find in the cells from different patients? ✓❸

21. *Breakfast conversation.* You and a friend are eating breakfast together. While eating, your friend is reading the back of her cereal box and comes across the following statement: "Cholesterol plays beneficial roles in your body, making cells, hormones, and tissues." Knowing that you are taking biochemistry, she asks if the statement makes sense. What do you reply?

22. *A good thing.* What are statins? What is their pharmacological function? ✓❸

23. *Too much of a good thing.* Would the development of a "super statin" that inhibited all HMG CoA reductase activity be a useful drug? Explain.

24. *RNA editing.* A shortened version (apo B-48) of apolipoprotein B is formed by the intestine, whereas the full-length protein (apo B-100) is synthesized by the liver. A glutamine codon (CAA) is changed into a stop codon. Propose a simple mechanism for this change.

25. *A means of entry.* Describe the process of receptor-mediated endocytosis by using LDL as an example. ✓❸

26. *Inspiration for drug design.* Some actions of androgens are mediated by dihydrotestosterone, which is formed by the reduction of testosterone. This finishing touch is catalyzed by an NADPH-dependent 5α-reductase (p. 876). Chromosomal XY males with a genetic deficiency of this reductase are born with a male internal urogenital tract but predominantly female external genitalia. These people are usually reared as girls. At puberty, they masculinize because the testosterone level rises. The testes of these reductase-deficient men are normal, whereas their prostate glands remain small. How might this information be used to design a drug to treat *benign prostatic hypertrophy*, a common consequence of the normal aging process in men? A majority of men older than age 55 have some degree of prostatic enlargement, which often leads to urinary obstruction. ✓❹

27. *Drug idiosyncrasies.* Debrisoquine, a β-adrenergic blocking agent, has been used to treat hypertension. The optimal dose varies greatly (20–400 mg daily) in a population of patients. The urine of most patients taking the drug contains a high level of 4-hydroxydebrisoquine. However, those most sensitive to the drug (about 8% of the group studied) excrete debrisoquine and very little of the 4-hydroxy derivative. Propose a molecular basis for this drug idiosyncrasy. Why should caution be exercised in giving other drugs to patients who are very sensitive to debrisoquine?

Debrisoquine

28. *Removal of odorants.* Many odorant molecules are highly hydrophobic and concentrate within the olfactory epithelium. They would give a persistent signal independent of their concentration in the environment if they were not rapidly modified. Propose a mechanism for converting hydrophobic odorants into water-soluble derivatives that can be rapidly eliminated. ✓❹

29. *Development difficulties.* Propecia (finasteride) is a synthetic steroid that functions as a competitive and specific inhibitor of 5α-reductase (p. 876), the enzyme responsible for the synthesis of dihydrotestosterone from testosterone. ✓❹

Finasteride

In addition to its use as a treatment for benign prostatic hypertrophy (problem 26), finasteride is also widely used to retard the development of male pattern hair loss. Pregnant women are advised to avoid handling this drug. Why is it vitally important that pregnant women avoid contact with Propecia?

30. *Lifestyle consequences.* Human beings and the plant *Arabidopsis* evolved from the same distant ancestor possessing a small number of cytochrome P450 genes. Human beings have approximately 50 such genes, whereas *Arabidopsis* has more than 250 of them. Propose a role for the large number of P450 isozymes in plants.

31. *Personalized medicine.* The cytochrome P450 system metabolizes many medicinally useful drugs. Although all human beings have the same number of P450 genes, individual polymorphisms exist that alter the specificity and efficiency of the proteins encoded by the genes. How could knowledge of individual polymorphisms be useful clinically?

32. *Honeybee crisis.* In 2006, there was a sudden, unexplained die-off of honeybee colonies throughout the United States. The die-off was economically significant because one-third of the human diet comes from insect-pollinated plants, and honeybees are responsible for 80% of the pollination. In October of 2006, the sequence of the honeybee genome was reported. Interestingly, the genome was found to contain far fewer cytochrome P450 genes than do the genomes of other insects. Suggest how the die-off and the scarcity of P450 genes may be related.

33. *Missing enzyme.* Congenital adrenal hyperplasia is a life-threatening condition that results from a deficiency in the P450 enzyme steroid 21-hydroxylase. This enzyme catalyzes the first step in the conversion of progesterone into cortisol and aldosterone (Figures 26.28 and 26.29). A characteristic of congenital adrenal hyperplasia is an increase in sex hormone production. Explain why this is the case. ✓❹

34. *Let the sun shine in.* At a biochemical level, vitamin D functions like a steroid hormone (Chapter 32). Therefore, it is sometimes referred to as an honorary steroid. Why is vitamin D not an actual steroid? ✓❹

Mechanism Problems

35. *An interfering phosphate.* In the course of the overall reaction catalyzed by HMG-CoA reductase, a histidine residue protonates a coenzyme A thiolate, CoA– S$^-$, generated in an earlier step. ✔️❸

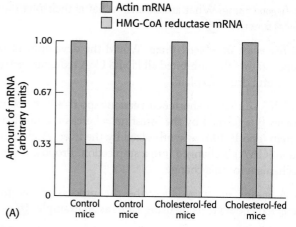

A nearby serine residue can be phosphorylated by AMP-activated kinase, which results in a loss of activity. Propose an explanation for why phosphorylation of the serine residue inhibits enzyme activity.

36. *Demethylation.* Methyl amines are often demethylated by cytochrome P450 enzymes. Propose a mechanism for the formation of methylamine from dimethylamine catalyzed by cytochrome P450. What is the other product?

Chapter Integration Problems

37. *Similarities.* Compare the role of CTP in phosphoglyceride synthesis with the role of UTP in glycogen synthesis.

38. *Hold on tight or you might be thrown to the cytoplasm.* Many proteins are modified by the covalent attachment of a farnesyl (C$_{15}$) or a geranylgeranyl (C$_{20}$) unit to the carboxyl-terminal cysteine residue of the protein. Suggest why this modification might occur. ✔️❹

39. *Fork in the road.* 3-Hydroxy-3-methylglutaryl CoA is on the pathway for cholesterol biosynthesis. It is also a component of another pathway. Name the pathway. What determines which pathway 3-hydroxy-3-methylglutaryl CoA follows?

40. *Requires a club membership.* How is methionine metabolism related to the synthesis of phosphatidylcholine?

41. *Drug resistance.* Dichlorodiphenyltrichloroethane (DDT) is a potent insecticide rarely used today because of its effects on other forms of life. In insects, DDT disrupts sodium channel function, leading to eventual death. Mosquitoes have developed resistance to DDT and other insecticides that function in a similar fashion. Suggest two means by which DDT resistance might develop.

42. *ATP requirements.* Explain how cholesterol synthesis depends on the activity of ATP-citrate lyase.

Data Interpretation and Chapter Integration Problems

43. *Cholesterol feeding.* Mice were divided into four groups, two of which were fed a normal diet and two of which were fed a cholesterol-rich diet. HMG-CoA reductase mRNA and protein from liver were then isolated and quantified. Graph A shows the results of the mRNA isolation. ✔️❸

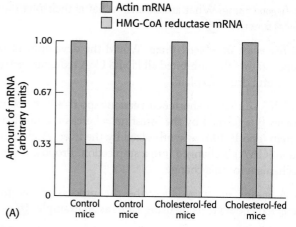

(a) What is the effect of cholesterol feeding on the amount of HMG-CoA reductase mRNA?

(b) What is the purpose of also isolating the mRNA for the protein actin, which is not under the control of the sterol regulatory element?

HMG-CoA reductase protein was isolated by precipitation with a monoclonal antibody to HMG-CoA reductase. The amount of HMG-CoA protein in each group is shown in graph B.

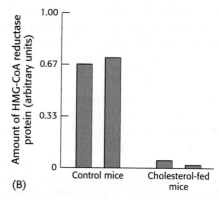

(c) What is the effect of the cholesterol diet on the amount of HMG-CoA reductase protein?

(d) Why is this result surprising in light of the results in graph A?

(e) Suggest possible explanations for the results in graph B.

44. *Gaucher disease.* Gaucher disease, the most common lysosomal storage disease in humans, is caused by mutations in the gene encoding glucocerebrosidase (GCase), the lysosomal enzyme that degrades glucocerebrosides. The disease has a variety of symptoms depending on the severity of the disease and the person affected. Identical twins can display very different levels of severity. Symptoms include bone pain, enlarged liver, excessive fatigue, and cognitive disabilities.

In order to better understand the disease, researchers undertook a series of experiments to characterize the nature of the defect in the enzyme. Cells were obtained from a person without the disease (control) as well as from a patient with the disease (GD). The cells were grown in culture, the enzyme was isolated, and enzyme activity of 10 μg of each sample was measured (Figure A).

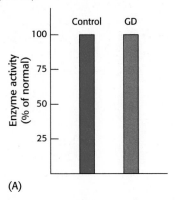

(A)

Data from J. Lu, *Proc. Natl. Acad. Sci. U.S.A.* 107:21665–21670, 2012.

(a) What do the results tell about the catalytic activity of the enzyme from the GD cells? Why were these results surprising to the researchers?

Again, cells were grown in culture and cell extracts were made from the control and GD samples. Care was taken to ensure that the same number of cells were used in both extractions. A western blot with antibodies to GCase was performed on the cell extracts, with the results shown in Figure B.

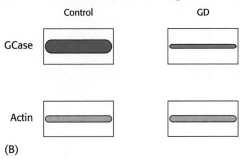

(B)

Data from J. Lu, *Proc. Natl. Acad. Sci. U.S.A.* 107:21665–21670, 2012.

(b) Provide two possible explanations for the result shown in Figure B. Why was it crucial to ensure that the extract was made from the same number of cells? What was the purpose of including the western blot of actin?

(c) The researchers then measured the amount of mRNA for the enzyme in both the control and GD cells. The amount of mRNA was identical in the two samples. With this information, reinterpret the results shown in Figure B.

Next, cells from the control and GD individuals were grown in the presence and absence of a potent proteasome inhibitor. The amount of GCase was determined using western blots. The results are shown in Figure C.

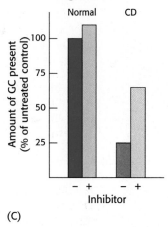

(C)

Data from J. Lu, *Proc. Natl. Acad. Sci. U.S.A.* 107:21665–21670, 2012.

(d) Suggest the nature of the defect in the GD enzyme. What is the significance of the increase in activity of the enzyme from the normal cell observed in the presence of the inhibitor?

Think ▶ Pair ▶ Share Problems

45. Explain the regulation of lipid metabolism.

46. Describe the biological significance of cholesterol.

47. Explain the biochemical importance of the components of the cholesterol biosynthetic pathway.

The Integration of Metabolism

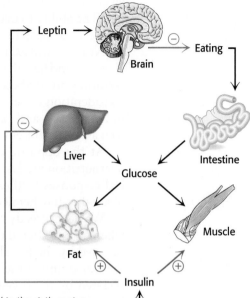

The image above shows a detail of runners on a Greek amphora painted in the sixth century B.C. Athletic feats, as well as others as seemingly simple as the maintenance of blood-glucose levels, require elaborate metabolic integration. The representation at the right shows the organs that have essential roles in the metabolic integration that regulates blood-glucose levels during exercise and at rest. Insulin and leptin (secreted by adipocytes) are two of the hormones that modulate the metabolic pathways of organs throughout the body such that adequate energy is available to meet the demands of living. [(Left) Euphiletos Painter (sixth century B.C.E.), Panathenaic prize amphora, c. 530 B.C.E. Reverse. Terracotta, h. 24 1/2 in. (62.2 cm.), Archaic Greek, Attic. Rogers Fund, 1914 (14.130.12). The Metropolitan Museum of Art, New York, NY, USA. Image copyright © The Metropolitan Museum of Art/Art Resource, NY.]

✓ LEARNING OBJECTIVES

By the end of this chapter, you should be able to:

1. Describe caloric homeostasis and its biochemical importance.

2. Explain the role of the neurotransmitters and hormones in maintaining caloric homeostasis.

3. Differentiate between type 1 and type 2 diabetes and explain how each develops.

4. Explain how exercise alters cellular biochemistry.

5. Describe the biochemical adaptations to food intake and starvation.

6. Explain how ethanol alters metabolism in the liver.

OUTLINE

27.1 Caloric Homeostasis Is a Means of Regulating Body Weight

27.2 The Brain Plays a Key Role in Caloric Homeostasis

27.3 Diabetes Is a Common Metabolic Disease Often Resulting from Obesity

27.4 Exercise Beneficially Alters the Biochemistry of Cells

27.5 Food Intake and Starvation Induce Metabolic Changes

27.6 Ethanol Alters Energy Metabolism in the Liver

We have been examining the biochemistry of metabolism one pathway at a time. We have seen how useful energy is extracted from fuels and used to power biosynthetic reactions and signal-transduction pathways. In Chapters 29 through 33, we will extend our study of biosynthetic reactions to the synthesis of nucleic acids and proteins. Before we do that, however, in this chapter we will take a step back to examine large-scale biochemical interactions that constitute the physiology of the organism.

In keeping with a central theme of life—energy manipulations—we will look at the regulation of energy at the organismal level, which boils down to an apparently simple but actually quite complex question: at the biochemical level, how does an organism know when to eat and when to refrain from eating? The ability to maintain adequate but not excessive energy stores is called *caloric homeostasis* or *energy homeostasis*.

Next, we will examine a significant perturbation of caloric homeostasis—obesity—and how this physiological condition affects insulin action, frequently resulting in diabetes. We then turn our attention to a biochemical consideration of one of the most beneficial activities humans can engage in—exercise—and explore how exercise mitigates the effects of diabetes as well as how different forms of exercise use different sources of fuel.

At the opposite end of the physiological spectrum from obesity and overnutrition are fasting and starvation, and we will examine the biochemical responses to these challenges. The chapter ends with a consideration of another biochemical energy perturbation—excess alcohol consumption.

We have already encountered instances of organismal energy regulation when we considered the actions of insulin and glucagon. Recall that insulin, secreted by the β cells of the pancreas, causes glucose to be removed from the blood and stimulates the synthesis of glycogen and lipids. Glucagon, secreted by the α cells of the pancreas, has effects opposite those of insulin. Glucagon increases the level of blood glucose by stimulating glycogen breakdown and gluconeogenesis. This chapter introduces two other hormones that play key roles in caloric homeostasis. Leptin and adiponectin, secreted by the adipose tissue, work in concert with insulin to regulate caloric homeostasis.

27.1 Caloric Homeostasis Is a Means of Regulating Body Weight

By now in our study of biochemistry, we are well aware of the fact that carbohydrates and lipids are sources of energy. We consume these energy sources as foods, convert the energy into ATP, and use the ATP to power our lives. Like all energy transformations, our energy consumption and expenditure are governed by the laws of thermodynamics. Recall that the First Law of Thermodynamics states that energy can neither be created nor destroyed. Translated into the practical terms of our diets,

$$\text{Energy consumed} = \text{energy expended} + \text{energy stored}$$

This simple equation has severe physiological and health implications: according to the First Law of Thermodynamics, if we consume more energy than we expend, we will become overweight or obese. Obesity is generally defined as a body mass index (BMI) of more than 30 kg m^{-2}, whereas overweight is defined as a BMI of more than 25 kg m^{-2} (Figure 27.1). Recall that excess fat is stored in adipocytes as triacylglycerols. The number of adipocytes becomes fixed in adults, and so obesity results in engorged adipocytes. Indeed, the cells may increase as much as 1000-fold in size.

We are all aware that many of us, especially in the developed world, are becoming obese or have already attained that state. In the United States, obesity has become an epidemic, with nearly 40% of adults and 20% of adolescents classified as such. Obesity is identified as a risk factor in a host of pathological conditions including diabetes mellitus, hypertension, and cardiovascular disease (Table 27.1). The cause of obesity is quite simple in

Height in feet and inches (in cm)

Weight in pounds (in kg)		4′8″ (142)	4′10″ (147)	5′0″ (152)	5′2″ (157)	5′4″ (163)	5′6″ (168)	5′8″ (173)	5′10″ (178)	6′0″ (183)	6′2″ (188)	6′4″ (193)
260	(117.9)	58	54	51	48	45	42	40	37	35	33	32
250	(113.4)	56	52	49	46	43	40	38	36	34	32	30
240	(108.9)	54	50	47	44	41	39	36	34	33	31	29
230	(104.3)	52	48	45	42	39	37	35	33	31	30	28
220	(99.8)	49	46	43	40	38	36	33	32	30	28	27
210	(95.3)	47	44	41	38	36	34	32	30	28	27	26
200	(90.7)	45	42	39	37	34	32	30	29	27	26	24
190	(86.2)	43	40	37	35	33	31	29	27	26	24	23
180	(81.6)	40	38	35	33	31	29	27	26	24	23	22
170	(77.1)	38	36	33	31	29	27	26	24	23	22	21
160	(72.6)	36	33	31	29	27	26	24	23	22	21	19
150	(68.0)	34	31	29	27	26	24	23	22	20	19	18
140	(63.5)	31	29	27	26	24	23	21	20	19	18	17
130	(59.0)	29	27	25	24	22	21	20	19	18	17	16
120	(54.4)	27	25	23	22	21	19	18	17	16	15	15
110	(49.9)	25	23	21	20	19	18	17	16	15	14	13
100	(45.4)	22	21	20	18	17	16	15	14	14	13	12
90	(40.8)	20	19	18	16	15	15	14	13	12	12	11
80	(36.3)	18	17	16	15	14	13	12	11	11	10	10

>30	Obese
25–30	Overweight
18.5–25	Normal
<18.5	Underweight

$$BMI = \frac{weight}{height^2}$$

FIGURE 27.1 Body mass index (BMI). The BMI value for an individual person is a reliable indicator of obesity for most people. [Data from the Centers for Disease Control.]

most cases: more food is consumed than is needed, and the excess calories are stored as fat. We will consider the biochemical basis of pathologies caused by obesity later in the chapter.

Before we undertake a biochemical analysis of the results of overconsumption, let us consider why the obesity epidemic is occurring in the first place. There are several possible, overlapping explanations. The first is a commonly held view that our bodies are programmed to rapidly store excess calories in times of plenty, an evolutionary adaptation from times past when humans were not assured of having ample food, as many of us are today. Consequently, we store calories as if a fast might begin tomorrow, but no such fast arrives. A second possible explanation is that we no longer face the

TABLE 27.1 Health consequences of obesity or being overweight

Coronary heart disease

Type 2 diabetes

Cancers (endometrial, breast, colon, and others)

Hypertension (high blood pressure)

Dyslipidemia (disruption of lipid metabolism, e.g., high cholesterol and triacylglycerols)

Stroke

Liver and gallbladder disease

Sleep apnea and respiratory problems

Osteoarthritis (degeneration of cartilage and underlying bone at a joint)

Gynecological problems (abnormal menses, infertility)

Male infertility problems

Information from: Centers for Disease Control and Prevention website (www.cdc.gov).

risks of predation. Evidence indicates that predation was a common cause of death for our ancestors. An obese individual would more likely have been culled from a group of our ancestors than would a more nimble, lean individual. As the risk of predation declined, leanness became less beneficial.

A third possibility, one currently receiving much attention, is that calorie-dense, highly palatable foods—foods rich in sugar and fats—that are readily accessible to most people in the developed world, act as drugs, triggering the same reward pathways in the brain that are triggered by drugs such as cocaine. These reward pathways can be strong enough to override appetite-suppressing signals. Fourth, a growing body of research suggests that our intestinal microbiome—the bacteria that inhibit our guts—plays a significant role in how we process our food. For instance, germ-free mice do not become obese, even when given unfettered access to a high-fat diet. However, when such mice are exposed to the intestinal flora of obese mice, the mice now become obese even on a normal chow diet. Moreover, the intestinal microbiome of obese mice triggers an inflammatory response that may blunt the effect of signal molecules that normally regulate the desire to eat as well as how the mice process the calories they consume. Many of these results have been extended to humans. Finally, it is evident that individuals respond differently to obesity-inducing environmental conditions and that this difference has a large genetic component. Various studies have indicated the heritability of fat mass to be between 30% and 70%, suggesting that the tendency toward obesity may be highly heritable, making it extra difficult to lose weight.

Regardless of why we may have a propensity to gain weight, this predisposition can be counteracted behaviorally—by eating less and exercising more. It also appears that it is easier to prevent weight gain than it is to lose weight. Clearly, caloric homeostasis is a complicated biological phenomenon. Understanding this phenomenon will engage biomedical research scientists for some time to come.

As disturbing as the obesity epidemic is, an equally intriguing, almost amazing observation is that many people are able to maintain an approximately constant weight throughout adult life, despite consuming tons of food over a lifetime (problems 6 and 7). Although willpower, exercise, and a bathroom scale often play a role in this homeostasis, some biochemical signaling must be taking place to achieve this remarkable physiological feat.

27.2 The Brain Plays a Key Role in Caloric Homeostasis

What makes this remarkable balance of energy input and output possible? As one might imagine, the answer is complicated, entailing many biochemical signals as well as a host of behavioral factors. We will focus on a few key biochemical signals, and divide our discussion into two parts: short-term signals that are active during a meal and long-term signals that report on the overall energy status of the body. These signals originate in the gastrointestinal tract, the β cells of the pancreas, and fat cells. The primary target of these signals is the brain, in particular a group of neurons in a region of the hypothalamus called the arcuate nucleus.

Signals from the gastrointestinal tract induce feelings of satiety

Short-term signals relay feelings of satiety from the gut to various regions of the brain and thus reduce the urge to eat (Figure 27.2). The best-studied short-term signal is cholecystokinin (CCK). *Cholecystokinin* is actually a

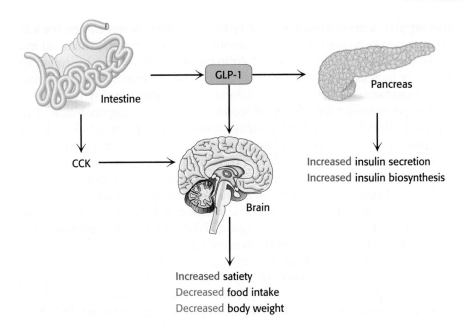

FIGURE 27.2 Satiation signals. Cholecystokinin (CCK) and glucagon-like peptide 1 (GLP-1) are signal molecules that induce feelings of satiety in the brain. CCK is secreted by specialized cells of the small intestine in response to a meal and activates satiation pathways in the brain. GLP-1, secreted by L cells in the intestine, also activates satiation pathways in the brain and potentiates insulin action in the pancreas. [Information from S. C. Wood, *Cell Metab.* 9:489–498, 2009, Fig. 1.]

family of peptide hormones of various lengths (from 8 to 58 amino acids in length, depending on posttranslational processing) secreted into the blood by cells in the duodenum and jejunum regions of the small intestine as a postprandial satiation signal. The CCK binds to the CCK receptor, a G-protein-coupled receptor (Section 14.1) located in various peripheral neurons, that relay signals to the brain. This binding initiates a signal-transduction pathway in the brain that generates a feeling of satiety. CCK also plays an important role in digestion, stimulating the secretion of pancreatic enzymes and bile salts from the gallbladder.

Another important gut signal is *glucagon-like peptide 1* (GLP-1), a hormone of approximately 30 amino acids in length. GLP-1 is secreted by intestinal L cells, hormone-secreting cells located throughout the lining of the gastrointestinal tract. GLP-1 has a variety of effects, all apparently facilitated by binding to its receptor, another G-protein-coupled receptor. Like CCK, GLP-1 induces feelings of satiety that inhibit further eating. GLP-1 also potentiates glucose-induced insulin secretion by the pancreatic β cells while inhibiting glucagon secretion.

Although we have examined only two short-term signals, many others are believed to exist (Table 27.2). Most of the short-term signals thus far identified are appetite suppressants. However, ghrelin, a peptide that is 28 amino acids in length and secreted by the stomach, acts on regions of the hypothalamus to stimulate appetite through its G-protein-coupled receptor. Ghrelin secretion increases before a meal and decreases afterward.

Leptin and insulin regulate long-term control over caloric homeostasis

Two key signal molecules regulate energy homeostasis over the time scale of hours or days: *leptin*, which is secreted by the adipocytes, and *insulin*, which is secreted by the pancreatic β cells. Leptin reports on the status of the triacylglycerol stores, whereas insulin reports on the status of glucose in the blood—in other words, of carbohydrate availability. We will consider leptin first.

Adipose tissue was formerly considered an inert depot of triacylglycerols. However, recent work has shown that adipose tissue is an active endocrine tissue, secreting signal molecules called *adipokines*, such as leptin,

TABLE 27.2 Gastrointestinal peptides that regulate food intake

Appetite-suppressing signals
Cholecystokinin
Glucagon-like peptide 1
Glucagon-like peptide 2
Amylin
Somatostatin
Bombesin
Enterostatin
Apolipoprotein A-IV
Gastric inhibitory peptide
Appetite-enhancing peptides
Ghrelin

Information from: M. H. Stipanuk, ed., *Biochemical, Physiological, Molecular Aspects of Human Nutrition*, 2d ed. (Saunders/Elsevier, 2006), p. 627, Box 22-1.

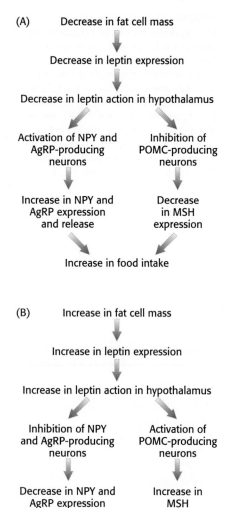

(A)   Decrease in fat cell mass

Decrease in leptin expression

Decrease in leptin action in hypothalamus

Activation of NPY and AgRP-producing neurons

Inhibition of POMC-producing neurons

Increase in NPY and AgRP expression and release

Decrease in MSH expression

Increase in food intake

(B)   Increase in fat cell mass

Increase in leptin expression

Increase in leptin action in hypothalamus

Inhibition of NPY and AgRP-producing neurons

Activation of POMC-producing neurons

Decrease in NPY and AgRP expression and release

Increase in MSH expression

Decrease in food intake

FIGURE 27.3 The effects of leptin in the brain. Leptin is an adipokine secreted by adipose tissue in direct relation to fat mass. (A) When leptin levels fall, as in fasting, appetite-enhancing neuropeptides NPY and AgRP are secreted, whereas the secretion of appetite-suppressing signals such as MSH is inhibited. (B) When fat mass increases, leptin inhibits NPY and AgRP secretion while stimulating the release of appetite-suppressing hormone MSH. [Information from M. H. Stipanuk, *Biochemical, Physiological, & Molecular Aspects of Human Nutrition*, 2d ed. (Saunders/Elsevier, 2006), Fig. 22-2.]

that regulate a host of physiological processes. Leptin is secreted by the adipocytes in direct proportion to the amount of fat present. The more fat in a body, the more leptin is secreted. Leptin binding to its receptor throughout the body increases the sensitivity of muscle and the liver to insulin, stimulates β oxidation of fatty acids, and decreases triacylglycerol synthesis.

Let us consider the effects of leptin in the brain. Leptin binds to its receptor, thereby activating a signal transduction pathway. The leptin receptor is found in various regions of the brain, but particularly in the arcuate nucleus of the hypothalamus. There, one population of neurons expresses appetite-stimulating (orexigenic) peptides, called neuropeptide Y (NPY) and agouti-related peptide (AgRP). Fasting stimulates the production of NPY and AgRP owing to the decrease in leptin levels that results from diminishing adipose tissue (Figure 27.3A). Leptin, on the other hand, inhibits the NPY/AgRP neurons, preventing the release of NPY and AgRP and thus repressing the desire to eat (Figure 27.3B).

The second population of neurons containing leptin receptors expresses a large precursor polypeptide, proopiomelanocortin (POMC). In response to leptin binding to its receptor on POMC neurons, POMC is proteolytically processed to yield a variety of signal molecules, one of which, *melanocyte-stimulating hormone* (MSH), is especially important in this context. MSH, originally discovered as a stimulator of melanocytes (cells that synthesize the pigment melanin), activates appetite-suppressing (anorexigenic) neurons and thus inhibits food consumption. Fasting inhibits MSH activity and thus stimulates eating. AgRP inhibits MSH activity by acting as an antagonist, binding to the MSH receptor but failing to activate the receptor (Figure 27.3). Thus, the net effect of leptin binding to its receptor is the initiation of a complex signal-transduction pathway that ultimately curtails food intake.

Insulin receptors are also present in the hypothalamus, although the mechanism of insulin action in the brain is less clear than that of leptin. Insulin appears to inhibit NPY/AgRP-producing neurons, thus inhibiting food consumption.

Leptin is one of several hormones secreted by adipose tissue

Leptin was the first adipokine discovered because of the dramatic effects of its absence. Researchers discovered a strain of mice called ob/ob mice, which lack leptin and, as a result, are extremely obese. These mice display hyperphagia (overeating), hyperlipidemia (accumulation of triacylglycerols in muscle and liver), and an insensitivity to insulin. Since the discovery of leptin, other adipokines have been detected. For instance, *adiponectin* is another signal molecule produced by the adipocytes. Secretion of adiponectin falls in direct proportion to increases in fat mass. A key function of adiponectin appears to be increasing the sensitivity of the organism to insulin. Both leptin and adiponectin exert their effects through the key regulatory enzyme, AMP-activated protein kinase (AMPK). Recall that this enzyme is active when AMP levels are elevated and ATP levels are diminished, and this activation leads to a decrease in anabolism and an increase in catabolism, most notably an increase in fatty acid oxidation (Section 22.6). In insulin-resistant obese animals such as the ob/ob mice, leptin levels increase while those of adiponectin decrease.

Adipocytes also produce two hormones, *RBP4* (originally discovered as a *retinol binding protein*) and *resistin*, that promote insulin resistance. Although it is unclear why adipocytes secrete hormones that facilitate insulin resistance, a pathological condition, we can speculate on the reason. These signal molecules may help to fine-tune the actions of leptin and adiponectin or perhaps to act as "brakes" on the action of leptin and adiponectin to prevent hypoglycemia in the fasted state. Some evidence indicates that enlarged adipocytes that result from obesity may secrete higher levels of insulin-antagonizing hormones and thus contribute to insulin resistance. Resistin has recently been implicated as a causal factor in the increased incidence of cardiovascular disease in obese individuals.

Leptin resistance may be a contributing factor to obesity

If leptin is produced in proportion to body-fat mass and leptin inhibits eating, why do people become obese? Obese people, in most cases, have both functioning leptin receptors and a high blood concentration of leptin. The failure to respond to the anorexigenic effects of leptin is called *leptin resistance*. What is the basis of leptin resistance?

As for most questions in the exciting area of energy homeostasis, the answer is not well worked out, but recent evidence suggests that a group of proteins called *suppressors of cytokine signaling* (SOCS) may take part. These proteins fine-tune some hormonal systems by inhibiting receptor action. SOCS proteins inhibit receptor signaling by a number of means. Consider, for example, the effect of SOCS proteins on the insulin receptor. Recall that insulin stimulates the autophosphorylation of tyrosine residues on the insulin receptor, which in turn phosphorylates IRS-1, initiating the insulin-signaling pathway (Figure 27.4A). SOCS proteins bind to phosphorylated tyrosine residues on receptors or other members of the signal-transduction pathway, thereby disrupting signal flow and thus altering the cell's biochemical activity (Figure 27.4B). In other cases, the binding of SOCS proteins to components of the signal-transduction pathway may also enhance proteolytic degradation of these components by the proteasome (Section 23.2). Evidence in support of a role for SOCS in leptin resistance comes from mice that have had SOCS selectively deleted from POMC-expressing neurons. These mice display an enhanced sensitivity to leptin and are resistant to weight gain even when fed a high-fat diet. The reason why the activity of SOCS proteins increases, leading to leptin resistance, remains to be determined.

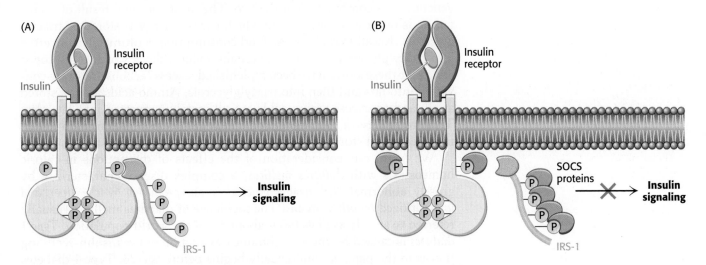

FIGURE 27.4 Suppressors of cytokine signaling (SOCS) regulate receptor function. (A) Insulin binding results in phosphorylation of the receptor and subsequent phosphorylation of IRS-1. These processes initiate the insulin-signaling pathway. (B) SOCS proteins disrupt interactions of components of the insulin-signaling pathway by binding phosphorylated proteins and thereby inhibiting the pathway. The binding of a signal component by SOCS results in proteasomal degradation in some cases. (IRS-1, insulin-receptor substrate 1; SOCS, suppressor of cytokine signaling.)

Leptin resistance can arise by means other than the attenuation of leptin signaling by SOCS. To have systemic effects, leptin must enter the brain. Mutations in proteins that transport leptin into the brain or in the brain leptin receptors could also result in leptin resistance. (See Biochemistry in Focus 1 for a further discussion of leptin and adiponectin.)

Dieting is used to combat obesity

Given the obesity epidemic that we currently face and its associated disorders, much attention has been focused on determining the most effective weight-loss diet. In general, two categories of diet try to help us control our caloric intake—low-carbohydrate diets and low-fat diets. Low-carbohydrate diets usually emphasize protein consumption. Although studies of the effects of diets on humans are immensely complex, data are beginning to accumulate suggesting that low-carbohydrate–high-protein diets *may* be the most effective for losing weight. The exact reasons are not clear, but there are two common hypotheses. First, proteins seem to induce a feeling of satiation more effectively than do fats or carbohydrates. Second, proteins require more energy to digest than do fats or carbohydrates, and the increased energy expenditure contributes to weight loss. For instance, some studies show that a diet that is 30% protein requires almost 30% more energy to digest than that required by a diet that is 10% protein. The mechanisms by which protein-rich diets enhance energy expenditure and feelings of satiation remain to be determined. Regardless of the type of diet, the adage "Eat less, exercise more" always applies.

27.3 Diabetes Is a Common Metabolic Disease Often Resulting from Obesity

Having taken an overview of the regulation of body weight, we now examine the biochemical results when regulation fails because of behavior, genetics, or a combination of the two. The most common result of such a failure is obesity, a condition in which excess energy is stored as triacylglycerols. Recall that all excess food consumption is ultimately converted into triacylglycerols. Humans maintain about a day's worth of glycogen and, after these stores have been replenished, excess carbohydrates are converted into fats and then into triacylglycerols. Amino acids are not stored at all, and so excess amino acids are ultimately converted into fats also. Thus, regardless of the type of food consumed, excess consumption results in increased fat stores.

We begin our consideration of the effects of disruptions in caloric homeostasis with *diabetes mellitus,* a complex disease characterized by grossly abnormal fuel usage: *glucose is overproduced by the liver and underutilized by other organs.* The incidence of diabetes mellitus (usually referred to simply as *diabetes*) is about 5% of the world population. *Type 1 diabetes* is caused by the autoimmune destruction of the insulin-secreting β cells in the pancreas and usually begins before age 20. Type 1 diabetes is also referred to as insulin-dependent diabetes, meaning that the affected person requires the administration of insulin to live. Most diabetics, in contrast, have a normal or even higher level of insulin in their blood, but they are unresponsive to the hormone, a characteristic called *insulin resistance*. This form of the disease, known as *type 2 diabetes*, typically arises later in life than does the insulin-dependent form. Type 2 diabetes accounts for approximately 90% of the diabetes cases throughout the world and is the most common metabolic disease in the world, with about 9% of the world population and 10% of the U.S. population affected. In the United States, it is the leading cause of blindness, kidney failure, and amputation. *Obesity is a significant predisposing factor for the development of type 2 diabetes.*

Insulin initiates a complex signal-transduction pathway in muscle

What is the biochemical basis of insulin resistance? How does insulin resistance lead to failure of the pancreatic β cells that results in type 2 diabetes? How does obesity contribute to this progression? To answer these questions and begin to unravel the mysteries of metabolic disorders, let us examine the mechanism of action of insulin in muscle, the largest tissue target of insulin. Indeed, muscle uses about 85% of the glucose ingested during a meal.

In a normal cell, insulin binds to its receptor, which then autophosphorylates on tyrosine residues, with each subunit of the receptor phosphorylating its partner. Phosphorylation of the receptor generates binding sites for insulin-receptor substrates (IRSs), such as IRS-1 (Figure 27.5). Subsequent phosphorylation of IRS-1 by the tyrosine kinase activity of the insulin receptor engages the insulin-signaling pathway (1). Phosphorylated IRS-1 binds to phosphoinositide 3-kinase (PI3K) and activates it. PI3K catalyzes the conversion of phosphatidylinositol 4,5-bisphosphate (PIP_2) into phosphatidylinositol 3,4,5-trisphosphate (PIP_3), a second messenger (2). (PIP_3) activates the phosphatidylinositol-dependent protein kinase (PDK) (3), which in turn activates several other kinases, most notably Akt (4), also known as protein kinase B (PKB). Akt facilitates the translocation of the glucose transporter (GLUT4)-containing vesicles to the cell membrane, which leads to a more robust absorption of glucose from the blood. Moreover, Akt phosphorylates and inhibits glycogen synthase kinase (GSK3). Recall that GSK3 inhibits glycogen synthase (Section 21.5). Thus, insulin leads to the absorption of glucose from the blood, activation of glycogen synthase, and enhanced glycogen synthesis.

Like all signal pathways, the insulin-signaling cascade must be capable of being turned off. Three different processes contribute to the down-regulation of insulin signaling. First, phosphatases deactivate the insulin receptor and destroy a key second messenger. *Tyrosine phosphatase IB* removes phosphoryl groups from the receptor, thus inactivating it. The second messenger PIP_3 is inactivated by the phosphatase *PTEN* (phosphatase and *tensin* homolog), which dephosphorylates it, forming PIP_2, which itself has no second-messenger properties.

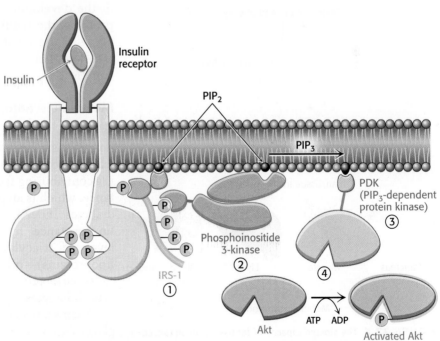

FIGURE 27.5 Insulin signaling. The binding of insulin results in the cross-phosphorylation and activation of the insulin receptor. Phosphorylated sites on the receptor act as binding sites for insulin-receptor substrates such as IRS-1. The lipid kinase phosphoinositide 3-kinase binds to phosphorylated sites on IRS-1 through its regulatory domain and then converts PIP_2 into PIP_3. Binding to PIP_3 activates PIP_3-dependent protein kinase, which phosphorylates and activates kinases such as Akt. Activated Akt can then diffuse throughout the cell and continue the signal-transduction pathway.

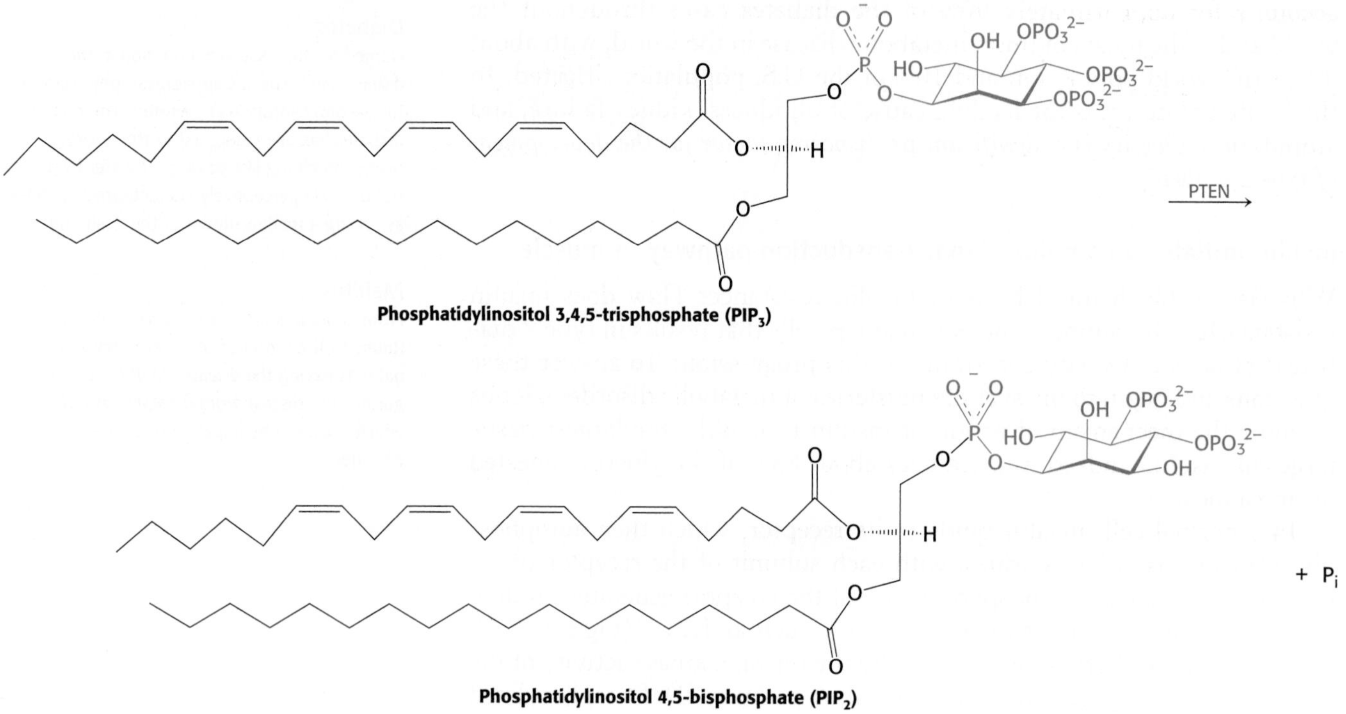

Phosphatidylinositol 3,4,5-trisphosphate (PIP$_3$)

PTEN →

Phosphatidylinositol 4,5-bisphosphate (PIP$_2$)

+ P$_i$

Second, the IRS protein can be inactivated by phosphorylation on serine residues by specific Ser/Thr kinases. These kinases are activated by overnutrition and other stress signals and may play a role in the development of insulin resistance. Finally, SOCS proteins, the regulatory proteins discussed earlier, interact with the insulin receptor and IRS-1 and apparently facilitate their proteolytic degradation by the proteasome complex.

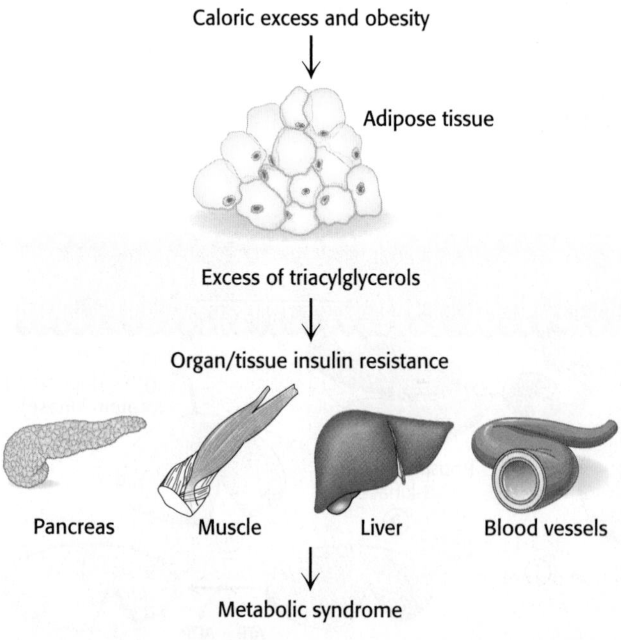

Metabolic syndrome often precedes type 2 diabetes

With our knowledge of the key components of energy homeostasis, let us begin our investigation of the biochemical basis of insulin resistance and type 2 diabetes. Obesity is a contributing factor to the development of insulin resistance, which is an early development on the path to type 2 diabetes. Indeed, a cluster of pathologies—including insulin resistance, hyperglycemia, dyslipidemia (high blood levels of triacylglycerols, cholesterol, and low-density lipoproteins)—often develop together. This clustering, called *metabolic syndrome*, is thought to be a predecessor of type 2 diabetes.

A consequence of obesity is that the amount of triacylglycerols consumed exceeds the adipose tissue's storage capacity. As a result, other tissues begin to accumulate fat, most notably liver, a condition called hepatic steatosis, and muscle (Figure 27.6). For reasons to be presented later in the chapter, this accumulation results in insulin resistance and ultimately in pancreatic failure. We will focus on muscle and the pancreatic β cells.

Caloric excess and obesity

Adipose tissue

Excess of triacylglycerols

Organ/tissue insulin resistance

Pancreas   Muscle   Liver   Blood vessels

Metabolic syndrome

FIGURE 27.6 The storage capacity of fat tissue can be exceeded in obesity. In caloric excess, the storage capacity of adipocytes can be exceeded with deleterious results. The excess fat accumulates in other tissues, resulting in biochemical malfunction of the tissues. When the pancreas, muscle, liver, and cells lining the blood vessels are affected, metabolic syndrome, a condition that often precedes type 2 diabetes, may result. [Information from S. Fröjdö, H. Vidal, and L. Pirola, *Biochim. Biophys. Acta* 1792:83–92, 2009, Fig. 1.]

Excess fatty acids in muscle modify metabolism

We have seen many times the importance of fats as fuels. In regard to obesity, more fats are present than can be processed by muscle. Although the rate of β oxidation increases in response to the high concentration of fats, mitochondria are not capable of processing all of the fatty acids by β oxidation; fatty acids accumulate in the mitochondria and eventually spill over into the cytoplasm. Indeed, the inability to process all of the fatty acids results in their reincorporation into triacylglycerols and the accumulation of fat in the cytoplasm. In the cytoplasm, levels of diacylglycerol and ceramide (a component of sphingolipids) also increase. Diacylglycerol is a second messenger that activates protein kinase C (PKC) (Section 14.1). When active, PKC and other Ser/Thr protein kinases are capable of phosphorylating IRS and reducing the ability of IRS to propagate the insulin signal. Saturated and trans unsaturated fatty acids may also activate kinases that block the insulin signal. Ceramide or its metabolites inhibit glucose uptake and glycogen synthesis, apparently by inhibiting PDK and Akt (p. 882). The result is a diet-induced insulin resistance (Figure 27.7).

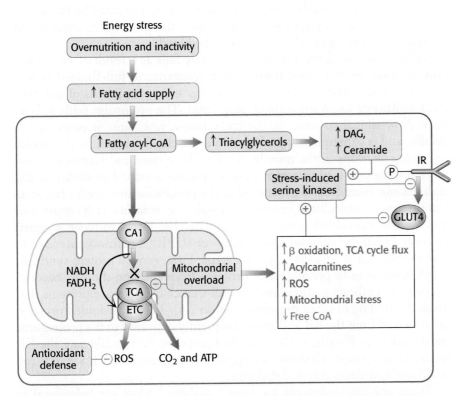

FIGURE 27.7 Excess fat in peripheral tissues can result in insulin insensitivity. Excess fat accumulation in peripheral tissues, most notably muscle, can disrupt some signal-transduction pathways and inappropriately activate others. In particular, diacylglycerols and ceramide activate stress-induced pathways that interfere with insulin signaling, resulting in insulin resistance. (Abbreviations: DAG, diacylglycerol; TGs, triacylglycerols; ROS, reactive oxygen species; CA1, carnitine acyltransferase 1; GLUT4, glucose transporter 4; IR, insulin receptor; ETC, electron-transport chain; TCA, tricarboxylic acid cycle.)

Insulin resistance in muscle facilitates pancreatic failure

What is the effect of overnutrition on the pancreas? This question is important because a primary function of the pancreas is to respond to the presence of glucose in the blood by secreting insulin, a process referred to as *glucose-stimulated insulin secretion* (GSIS). Indeed, the β cell is a virtual

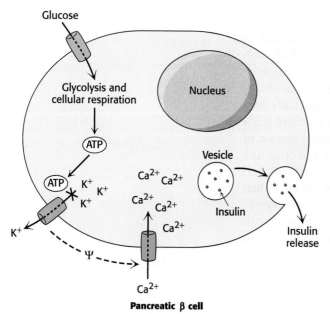

FIGURE 27.8 Insulin release is regulated by ATP. The metabolism of glucose by glycolysis and cellular respiration increases the concentration of ATP, which causes an ATP-sensitive potassium channel to close. The closure of this channel alters the charge across the membrane (ψ) and causes a calcium channel to open. The influx of calcium causes insulin-containing granules to fuse with the plasma membrane, releasing insulin into the blood.

insulin factory. Proinsulin mRNA, which encodes proinsulin, the precursor to insulin, constitutes 20% of the total mRNA in the pancreas. Proinsulin constitutes 50% of the total protein synthesized in the pancreas.

Glucose enters the pancreatic β cells through the glucose transporter GLUT2. Recall that GLUT2 will allow glucose transport only when blood glucose is plentiful, ensuring that insulin is secreted only when glucose is abundant, such as after a meal (Section 16.2). The β cell metabolizes glucose to CO_2 and H_2O in the process of cellular respiration, generating ATP (Chapters 16, 17, and 18). The resulting increase in the ATP/ADP ratio closes an ATP-sensitive K^+ channel that, when open, allows potassium to flow out of the cell (Figure 27.8). The resulting alteration in the cellular ionic environment opens a Ca^{2+} channel. The influx of Ca^{2+} causes insulin-containing secretory vesicles to fuse with the cell membrane and release insulin into the blood. Thus, the increase in energy charge resulting from the metabolism of glucose has been translated by the membrane proteins into a physiological response—the secretion of insulin and the subsequent removal of glucose from the blood.

What aspect of β cell function ultimately fails as a result of overnutrition, causing the transition from insulin resistance to full-fledged type 2 diabetes? Recall that, under normal circumstances, the β cells of the pancreas synthesize large amounts of proinsulin. The proinsulin folds into its three-dimensional structure in the endoplasmic reticulum, is processed to insulin, and is subsequently packaged into vesicles for secretion. As insulin resistance develops in the muscle, the β cells respond by synthesizing yet more insulin in a futile attempt to drive insulin action. The ability of the endoplasmic reticulum to process all of the proinsulin and insulin becomes compromised, a condition known as *endoplasmic reticulum* (ER) *stress*, and unfolded or misfolded proteins accumulate. ER stress initiates a signal pathway called the *unfolded protein response* (UPR), a pathway intended to save the cell. UPR consists of several steps. First, general protein synthesis is inhibited so as to prevent more proteins from entering the ER. Second, chaperone synthesis is stimulated. Recall that chaperones are proteins that assist the folding of other proteins (Section 2.6). Third, misfolded proteins are removed from the ER and are subsequently delivered to the proteasome for destruction. Finally, if the described response fails to alleviate the ER stress, programmed cell death is triggered, which ultimately leads to β cell death and full-fledged type 2 diabetes.

What are the treatments for type 2 diabetes? Most are behavioral in nature. Diabetics are advised to count calories, making sure that energy intake does not exceed energy output; to consume a diet rich in vegetables, fruits, and grains; and to get plenty of aerobic exercise. Note that these guidelines are the same as those for healthy living, even for those not suffering from type 2 diabetes. Treatments specific for type 2 diabetes include the monitoring of blood-glucose concentration so that it is within the target range (normal is 3.6 to 6.1 mM). For those who are not able to maintain proper glucose levels with the behaviors described herein, drug treatments are required. The administration of insulin may be necessary upon pancreatic failure, and treatment with metformin (Glucophage), which activates AMPK, may be effective. AMPK promotes the oxidation of fats while inhibiting fat synthesis and storage. It also stimulates glucose uptake and storage by muscle while inhibiting gluconeogenesis in the liver.

Metabolic derangements in type 1 diabetes result from insulin insufficiency and glucagon excess

We now turn to the more-straightforward type 1 diabetes. In type 1 diabetes, insulin production is insufficient because of autoimmune destruction of the pancreatic β cells. Consequently, the glucagon/insulin ratio is at higher-than-normal levels. In essence, the diabetic person is in biochemical fasting mode despite a high concentration of blood glucose. Because insulin is deficient, *the entry of glucose into adipose and muscle cells is impaired.* The liver becomes stuck in a gluconeogenic and ketogenic state. The gluconeogenic state is characterized by excessive production of glucose. The excessive level of glucagon relative to that of insulin leads to a decrease in the amount of fructose 2,6-bisphosphate (F-2, 6-BP), which stimulates glycolysis and inhibits gluconeogenesis in the liver. Hence, glycolysis is inhibited and gluconeogenesis is stimulated because of the opposite effects of F-2, 6-BP on phosphofructokinase and fructose-1, 6-bisphosphatase (Section 16.4; Figure 27.9). Essentially, the cells' response to a lack of insulin amplifies the amount of glucose in the blood. The high glucagon/insulin ratio in diabetes also promotes glycogen breakdown. Hence, *an excessive amount of glucose is produced by the liver and released into the blood.* Glucose is excreted in the urine (hence the name *mellitus*) when its concentration in the blood exceeds the reabsorptive capacity of the renal tubules. Water accompanies the excreted glucose, and so an untreated diabetic in the acute phase of the disease is hungry and thirsty.

Because carbohydrate utilization is impaired, a lack of insulin leads to the uncontrolled breakdown of lipids and proteins, resulting in the ketogenic state. Large amounts of acetyl CoA are then produced by β oxidation. However, much of the acetyl CoA cannot enter the citric acid cycle because there is insufficient oxaloacetate for the condensation step. Recall that mammals can synthesize oxaloacetate from pyruvate, a product of glycolysis, but not from acetyl CoA; instead, they generate ketone bodies (Section 22.3). *A striking feature of diabetes is the shift in fuel usage from carbohydrates to fats; glucose, more abundant than ever, is spurned.* In high concentrations, ketone bodies overwhelm the kidney's capacity to maintain acid–base balance. The untreated diabetic can go into a coma because of a lowered blood-pH level and dehydration. Interestingly, diabetic ketosis is rarely a problem in type 2 diabetes because insulin is active enough to prevent excessive lipolysis in liver and adipose tissue.

What is the treatment for type 1 diabetes? Insulin injections are required for survival. Likewise, blood-glucose concentration must be monitored. Many of the behaviors recommended for type 2 diabetes apply to type 1: watching calories, exercising, and eating a healthy diet.

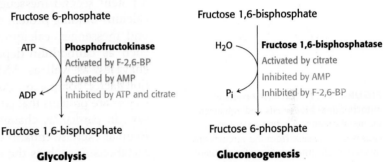

Glycolysis

Fructose 6-phosphate
ATP → **Phosphofructokinase** / Activated by F-2,6-BP / Activated by AMP / Inhibited by ATP and citrate → ADP
Fructose 1,6-bisphosphate

Gluconeogenesis

Fructose 1,6-bisphosphate
H_2O → **Fructose 1,6-bisphosphatase** / Activated by citrate / Inhibited by AMP / Inhibited by F-2,6-BP → P_i
Fructose 6-phosphate

FIGURE 27.9 Regulation of glycolysis and gluconeogenesis. Phosphofructokinase is the key enzyme in the regulation of glycolysis, whereas fructose 1,6-bisphosphatase is the principal enzyme controlling the rate of gluconeogenesis. Note the reciprocal relation between the pathways and the signal molecules.

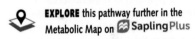

EXPLORE this pathway further in the Metabolic Map on 🟢 Sapling Plus

27.4 Exercise Beneficially Alters the Biochemistry of Cells

Skeletal muscle accounts for ≈40% of total body mass and ≈35% of resting metabolic activity. Moreover, skeletal muscle is the largest target tissue for insulin. Given its biochemical prominence, increased muscle activity—exercise—coupled with a healthy diet is one of the most effective treatments for diabetes as well as a host of other pathological conditions. These include coronary disease, hypertension, depression, age-related muscle wasting (sarcopenia), and a variety of cancers. With regard to diabetes, exercise increases the insulin sensitivity of people who are insulin resistant or type 2 diabetics and is in fact more effective than pharmacological interventions. What is the basis of this beneficial effect? (See Biochemistry in Focus 2 for a further discussion of the biochemical effects of exercise.)

Mitochondrial biogenesis is stimulated by muscular activity

When muscle is stimulated to contract during exercise by receiving nerve impulses from motor neurons, calcium is released from the sarcoplasmic reticulum. Calcium induces muscle contraction, as will be discussed later in the text (See Chapter 36 in 🅿 SaplingPlus). Recall that calcium is also a potent second messenger and frequently works in association with the calcium-binding protein calmodulin (Section 14.1). In its capacity as a second messenger, calcium stimulates various calcium-dependent enzymes, such as calmodulin-dependent protein kinase. The calcium-dependent enzymes, as well as AMPK, subsequently activate particular transcription-factor complexes. As we will see in Chapters 30 and 32, transcription factors are proteins that control gene expression. Two patterns of gene expression, in particular, change in response to regular exercise (Figure 27.10). Regular exercise enhances the production of proteins required for fatty acid metabolism, such as the enzymes of β oxidation. Interestingly, fatty acids themselves function as signal molecules to activate the transcription of enzymes of fatty acid metabolism. Additionally, another set of transcription factors activated by the calcium signal cascade initiates a signal pathway that increases mitochondrial biogenesis. In concert, *the increase in fatty acid oxidizing capability and additional mitochondria allow for the efficient metabolism of fatty acids.* Because an excess of fatty acids results in insulin resistance, as already discussed, efficient metabolism of fatty acids results in *an increase in insulin sensitivity.* Indeed, muscles of well-trained athletes may contain high concentrations of triacylglycerols and still maintain exquisite sensitivity to insulin. These alterations of muscle biochemistry are some of the molecular manifestations of the training effect of exercise.

FIGURE 27.10 Exercise results in mitochondrial biogenesis and enhanced fat metabolism. An action potential causes Ca²⁺ release from the sarcoplasmic reticulum (SR), the muscle-cell equivalent of the endoplasmic reticulum. The Ca²⁺, in addition to instigating muscle contraction, activates nuclear transcription factors that stimulate the expression of specific genes. The products of these genes, in conjunction with the products of mitochondrial genes, are responsible for mitochondrial biogenesis. Fatty acids activate a different set of genes that increase the fatty acid oxidation capability of mitochondria. [Information from D. A. Hood, *J Appl. Physiol.* 90:1137–1157, 2001, Fig. 2.]

Fuel choice during exercise is determined by the intensity and duration of activity

In keeping with our theme of energy use under different physiological conditions, we now examine how fuels are used in different types of exercise. The fuels used in anaerobic exercises—sprinting, for example—differ from those used in aerobic exercises—such as distance running. The selection of fuels during these different forms of exercise illustrates many important facets of energy transduction and metabolic integration. ATP directly powers myosin, the protein immediately responsible for converting chemical energy into movement (Section 9.4, and see Section 36.1, and 36.2 on 🅿 SaplingPlus). However, the amount of ATP in muscle is small. Hence, the power output and, in turn, the velocity of running depend on the rate of ATP production from other fuels. As shown in Table 27.3, *creatine phosphate* (phosphocreatine) can

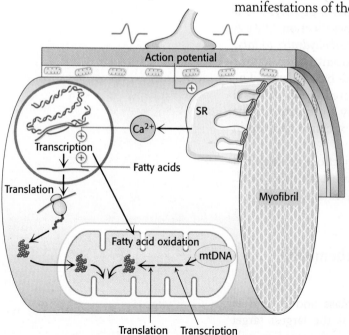

TABLE 27.3 Fuel sources for muscle contraction

Fuel source	Maximal rate of ATP production (mmol s^{-1})	Total ~P available (mmol)
Muscle ATP		223
Creatine phosphate	73.3	446
Conversion of muscle glycogen into lactate	39.1	6700
Conversion of muscle glycogen into CO_2	16.7	84,000
Conversion of liver glycogen into CO_2	6.2	19,000
Conversion of adipose-tissue fatty acids into CO_2	6.7	4,000,000

Note: Fuels stored are estimated for a 70-kg person having a muscle mass of 28 kg.
Data from E. Hultman and R. C. Harris. In *Principles of Exercise Biochemistry*, edited by J. R. Poortmans (Karger, 2004), pp. 78–119.

swiftly transfer its high-potential phosphoryl group to ADP to generate ATP. However, the amount of creatine phosphate, like that of ATP itself, is limited. Creatine phosphate and ATP can power intense muscle contraction for 5 to 6 s. Maximum speed in a sprint can thus be maintained for only 5 to 6 s (Figure 15.7). Thus, the winner in a 100-meter sprint is the runner who both achieves the highest initial velocity and then slows down the least.

During a ~10 second sprint, the ATP level in muscle drops from 5.2 to 3.7 mM, and that of creatine phosphate decreases from 9.1 to 2.6 mM. Anaerobic glycolysis provides fuel to make up for the loss of ATP and creatine phosphate. *A 100-meter sprint is powered by stored ATP, creatine phosphate, and the anaerobic glycolysis of muscle glycogen.* The conversion of muscle glycogen into lactate can generate a good deal more ATP, but the rate is slower than that of phosphoryl-group transfer from creatine phosphate. Because of anaerobic glycolysis, the blood-lactate level is elevated from 1.6 to 8.3 mM. The release of H^+ from the intensely active muscle concomitantly lowers the blood pH from 7.42 to 7.24. Sprinting is powered by fast-twitch muscle fiber (type IIb) specialized for anaerobic glycolysis. Recall that one of the effects of high-intensity training is to increase the amount of lactate transporters in the membranes of slow-twitch fibers (type 1), which remove lactate from the blood and slow the fall in blood pH (Section 16.4). Nonetheless, the pace of a 100-meter sprint cannot be sustained in a 1000-meter run (~132 s) for two reasons. First, creatine phosphate is consumed within a few seconds. Second, the lactate produced would cause acidosis. Thus, alternative fuel sources are needed.

Let us examine the fuel use of a middle-distance runner. The complete oxidation of muscle glycogen to CO_2 by aerobic respiration substantially increases the energy available to power the 1000-meter run, but this aerobic process is a good deal slower than anaerobic glycolysis. Because ATP is produced more slowly by oxidative phosphorylation than by glycolysis (Table 27.3), the runner's pace is necessarily slower than in a 100-meter sprint. The championship velocity for the 1000-meter run is about 7.6 m s^{-1}, compared with approximately 10.4 m s^{-1} for the 100-meter event (Figure 27.11).

The running of a marathon (26 miles 385 yards, or 42,200 meters) requires a different selection of fuels and is characterized by cooperation (depending on the capabilities

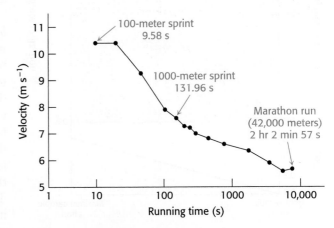

FIGURE 27.11 Dependency of the velocity of running on the duration of the race. The values shown are world track records. [Data from trackandfieldnews.com.]

of the runner) between muscle, liver, and adipose tissue. Liver glycogen complements muscle glycogen as an energy store that can be tapped. Elite marathoners can use the aerobic combustion of glucose to power the entire race, provided they consume fuels during the race. However, for most of us, the total body glycogen stores (103 mol of ATP at best) are insufficient to provide the 150 mol of ATP needed for this grueling event. Much larger quantities of ATP can be obtained by the oxidation of fatty acids derived from the breakdown of *fat in adipose tissue,* but the maximal rate of ATP generation is slower yet than that of glycogen oxidation and is more than 10-fold slower than that with creatine phosphate. Thus, *ATP is generated much more slowly from high-capacity stores than from limited ones,* accounting for the different velocities of anaerobic and aerobic events. Fats are rapidly consumed in activities such as distance running, explaining why extended aerobic exercise is beneficial for people who are insulin resistant.

Is it possible to determine the contribution of each fuel as a function of exercise intensity? The percentage contribution of each fuel can be measured with the use of a respirometer, which measures the respiratory quotient (RQ), the ratio of CO_2 produced to O_2 consumed. Consider the complete combustion of glucose:

$$C_6H_{12}O_6 + 6\,O_2 \longrightarrow 6\,CO_2 + 6\,H_2O$$
Glucose

The RQ for glucose is 1. Now consider the oxidation of a typical fatty acid, palmitate:

$$C_{16}H_{32}O_2 + 23\,O_2 \longrightarrow 16\,CO_2 + 16\,H_2O$$
Palmitate

The RQ for palmitate oxidation is 0.7. Thus, as aerobic exercise intensity increases, the RQ will rise from 0.7 (only fats are used as fuel) to 1.0 (only glucose is used as fuel). Between these values, a mixture of fuels is used (Figure 27.12).

What is the optimal mix of fuels for use during a marathon? As suggested above, this is a complex question that varies with the athlete and level of training. Studies have shown that, when muscle glycogen has been depleted, the power output of the muscle falls to approximately 50% of maximum. Power output decreases despite the fact that ample supplies of fat are available, suggesting that fats can supply only about 50% of maximal aerobic effort. Indeed, depletion of glycogen stores during a race is referred to as "bonking" or "hitting the wall," with the result that the athlete must greatly reduce the pace.

How is an optimal mix of these fuels achieved? *A low blood-sugar level leads to a high glucagon/insulin ratio, which in turn mobilizes fatty acids from adipose tissue.* Fatty acids readily enter muscle, where they are degraded by β oxidation to acetyl CoA and then to CO_2. The elevated acetyl CoA level decreases the activity of the pyruvate dehydrogenase complex to block the conversion of pyruvate into acetyl CoA. Hence, fatty acid oxidation decreases the funneling of glucose into the citric acid cycle and oxidative phosphorylation. Glucose is spared so that just enough remains available at the end of the marathon to increase the pace as the finish line draws near. The simultaneous use of both fuels gives a higher mean velocity than would be attained if glycogen were totally consumed

FIGURE 27.12 An idealized representation of fuel use as a function of aerobic exercise intensity. (A) With increased exercise intensity, the use of fats as fuels falls as the utilization of glucose increases. (B) The respiratory quotient (RQ) measures the alteration in fuel use.

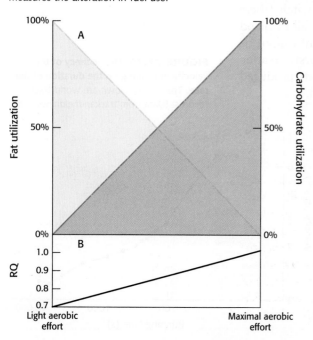

before the start of fatty acid oxidation. It is important to bear in mind that fuel utilization is only one of many factors that determine running ability.

If carbohydrate-rich meals are consumed after glycogen depletion, glycogen stores are rapidly restored. In addition, glycogen synthesis continues during the consumption of carbohydrate-rich meals, increasing glycogen stores far above normal. This phenomenon is called "super compensation" or, more commonly, carbo-loading.

27.5 Food Intake and Starvation Induce Metabolic Changes

Thus far, we have been considering metabolism in the context of excess calorie consumption, as in obesity, or extreme caloric needs, as in exercise. We now look at the opposite physiological condition—lack of calories.

The starved–fed cycle is the physiological response to a fast

We begin with a physiological condition called the *starved–fed cycle*, which we all experience in the hours after an evening meal and through the night's fast. This nightly starved–fed cycle has three stages: the well-fed state after a meal, the early fasting during the night, and the refed state after breakfast. A major goal of the many biochemical alterations in this period is to maintain *glucose homeostasis*—that is, a constant blood-glucose concentration. Maintaining glucose homeostasis is crucial because glucose is normally the only fuel source for the brain. As discussed earlier, the major defect in diabetes is the inability to perform this vital task. The two primary signals regulating the starved–fed cycle are insulin and glucagon.

1. *The well-fed, or postprandial, state.* After we consume and digest an evening meal, glucose and amino acids are transported from the intestine to the blood. The dietary lipids are packaged into chylomicrons and transported to the blood by the lymphatic system. This fed condition leads to the secretion of insulin, which signals the fed state. Insulin stimulates glycogen synthesis in both muscle and the liver, suppresses gluconeogenesis by the liver, and stimulates protein synthesis. Insulin also accelerates glycolysis in the liver, which in turn increases the synthesis of fatty acids.

The liver helps to limit the amount of glucose in the blood during times of plenty by storing it as glycogen so as to be able to release glucose in times of scarcity. How is the excess blood glucose present after a meal removed? The liver is able to trap large quantities of glucose because it possesses an isozyme of hexokinase called *glucokinase*, which converts glucose into glucose 6-phosphate, which cannot be transported out of the cell. Recall that glucokinase has a high K_M value and is thus active only when blood-glucose concentration is high (Section 16.2). Furthermore, glucokinase is not inhibited by glucose 6-phosphate as hexokinase is. Consequently, *the liver forms glucose 6-phosphate more rapidly as the blood-glucose concentration rises. The increase in glucose 6-phosphate, which activates glycogen synthase, coupled with insulin action leads to a buildup of glycogen stores.* The hormonal effects on glycogen synthesis and storage are reinforced by a direct action of glucose itself. *Phosphorylase a is a glucose sensor in addition to being the enzyme that cleaves glycogen.* When the glucose level is high, the binding of glucose to phosphorylase *a* renders the enzyme susceptible to the action of a phosphatase that converts it into phosphorylase *b,* which does not readily degrade glycogen (Figure 21.24). Thus, *glucose allosterically shifts the glycogen system from a degradative to a synthetic mode.*

The high insulin level in the fed state also promotes *the entry of glucose into muscle and adipose tissue*. Insulin stimulates the synthesis of glycogen by muscle as well as by the liver. The entry of glucose into adipose tissue provides glycerol 3-phosphate for the synthesis of triacylglycerols. The action of insulin also extends to amino acid and protein metabolism. Insulin promotes the uptake of branched-chain amino acids (valine, leucine, and isoleucine) by muscle. Indeed, insulin has a general stimulating effect on protein synthesis, which favors a building up of muscle protein. In addition, it inhibits the intracellular degradation of proteins.

2. *The early fasting, or postabsorptive, state.* The blood-glucose concentration begins to drop several hours after a meal, leading to a decrease in insulin secretion and a rise in glucagon secretion by the α cells of the pancreas. The regulation of glucagon secretion is poorly understood, but when glucose is abundant, β cells inhibit glucagon secretion. When glucose concentration falls, the inhibition is relieved and glucagon is secreted by the α cells of the pancreas. Just as insulin signals the fed state, glucagon signals the starved state, serving to mobilize glycogen stores when there is no dietary intake of glucose. *The main target organ of glucagon is the liver.* Glucagon stimulates glycogen breakdown and inhibits glycogen synthesis by triggering the cyclic AMP cascade; this leads to the phosphorylation and activation of phosphorylase and the inhibition of glycogen synthase (Section 21.5). Glucagon also inhibits fatty acid synthesis by diminishing the production of pyruvate and by maintaining the inactive phosphorylated state of acetyl CoA carboxylase. In addition, glucagon stimulates gluconeogenesis in the liver and blocks glycolysis by lowering the level of F-2, 6-BP (Figure 27.9).

All known actions of glucagon are mediated by protein kinases that are activated by cyclic AMP. The activation of the cyclic AMP cascade results in a higher level of phosphorylase *a* activity and a lower level of glycogen synthase *a* activity. Glucagon's effect on this cascade is reinforced by the low concentration of glucose in the blood. The diminished binding of glucose to phosphorylase *a* makes the enzyme less susceptible to the hydrolytic action of the phosphatase. Instead, the phosphatase remains bound to phosphorylase *a*, and so the synthase stays in the inactive phosphorylated form. Consequently, there is a rapid mobilization of glycogen.

The large amount of glucose formed by the hydrolysis of glucose 6-phosphate derived from glycogen is then released from the liver into the blood. The entry of glucose into muscle and adipose tissue decreases in response to a low insulin level. The diminished utilization of glucose by muscle and adipose tissue also contributes to the maintenance of the blood-glucose concentration. The net result of these actions of glucagon is to *markedly increase the release of glucose by the liver*. Both muscle and the liver use fatty acids as fuel when the blood-glucose concentration drops, saving the glucose for use by the brain and red blood cells. Thus, *the blood-glucose concentration is kept at or above 4.4 mM by three major factors: (1) the mobilization of glycogen and the release of glucose by the liver, (2) the release of fatty acids by adipose tissue, and (3) the shift in the fuel used from glucose to fatty acids by muscle.*

What is the result of the depletion of the liver's glycogen stores? Gluconeogenesis from lactate and alanine continues, but this process merely replaces glucose that had already been converted into lactate and alanine by tissues such as muscle and red blood cells. Moreover, the brain oxidizes glucose completely to CO_2 and H_2O. Thus, for the net synthesis of glucose to take place, another source of carbon is required. Glycerol released from adipose tissue on lipolysis provides some of the carbon atoms, with the remaining carbon atoms coming from the hydrolysis of muscle proteins.

3. *The refed state.* What are the biochemical responses to a hearty breakfast? Fat is processed exactly as it is processed in the normal fed state. However, that is not the case for glucose. The liver does not initially absorb glucose from the blood, but, instead, leaves it for the other tissues. Moreover, the liver remains in a gluconeogenic mode. Now, however, the newly synthesized glucose is used to replenish the liver's glycogen stores. As the blood-glucose concentration continues to rise, the liver completes the replenishment of its glycogen stores and begins to process the remaining excess glucose for fatty acid synthesis.

Metabolic adaptations in prolonged starvation minimize protein degradation

Earlier, we considered the metabolic results of overnutrition, a condition becoming all too common in prosperous nations. Let us now examine the opposite extreme. What are the adaptations if fasting is prolonged to the point of starvation, a circumstance affecting nearly a billion people worldwide? A typical well-nourished 70-kg man has fuel reserves totaling about 670,000 kJ (161,000 kcal; Table 27.4). The energy need for a 24-hour period ranges from about 6700 kJ (1600 kcal) to 25,000 kJ (6000 kcal), depending on the extent of activity. Thus, stored fuels suffice to meet caloric needs in starvation for 1 to 3 months. However, the carbohydrate reserves are exhausted in only a day.

TABLE 27.4 Fuel reserves in a typical 70-kg man

Organ	Available energy in kilojoules (kcal)					
	Glucose or glycogen		Triacylglycerols		Mobilizable proteins	
Blood	250	(60)	20	(45)	0	(0)
Liver	1700	(400)	2000	(450)	1700	(400)
Brain	30	(8)	0	(0)	0	(0)
Muscle	5000	(1200)	2000	(450)	100,000	(24,000)
Adipose tissue	330	(80)	560,000	(135,000)	170	(40)

Data from G. F. Cahill, Jr., *Clin. Endocrinol. Metab.* 5:398, 1976.

Even under starvation conditions, the blood-glucose concentration must be maintained above 2.2 mM. *The first priority of metabolism in starvation is to provide sufficient glucose to the brain and other tissues (such as red blood cells) that are absolutely dependent on this fuel.* However, precursors of glucose are not abundant. Most energy is stored in the fatty acyl moieties of triacylglycerols. Moreover, recall that fatty acids cannot be converted into glucose, because acetyl CoA resulting from fatty acid breakdown cannot be transformed into pyruvate (Section 22.3). The glycerol moiety of triacylglycerol can be converted into glucose, but only a limited amount is available. The only other potential source of glucose is the carbon skeletons of amino acids derived from the breakdown of proteins. However, proteins are not stored, and so any breakdown will necessitate a loss of function. Thus, *the second priority of metabolism in starvation is to preserve protein, which is accomplished by shifting the fuel being used from glucose to fatty acids and ketone bodies* (Figure 27.13).

The metabolic changes on the first day of starvation are like those after an overnight fast. The low blood-sugar level leads to decreased insulin

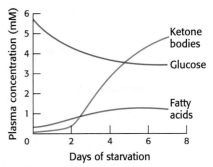

FIGURE 27.13 Fuel choice during starvation. The plasma concentration of fatty acids and ketone bodies increase in starvation, whereas that of glucose decreases.

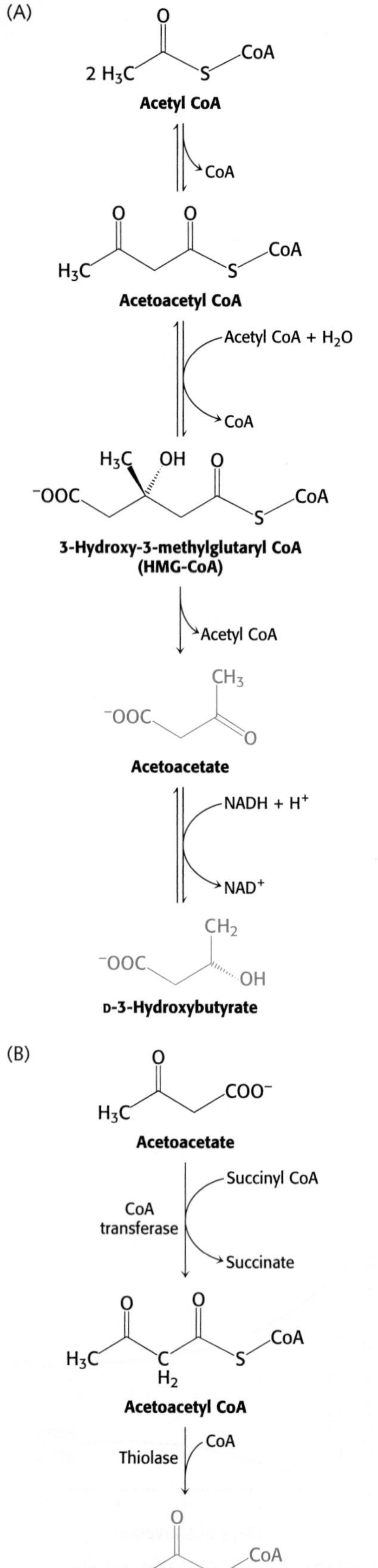

(A)

Acetyl CoA

Acetoacetyl CoA

3-Hydroxy-3-methylglutaryl CoA (HMG-CoA)

Acetoacetate

D-3-Hydroxybutyrate

(B)

Acetoacetate

CoA transferase

Acetoacetyl CoA

Thiolase

Acetyl CoA

FIGURE 27.14 Synthesis of ketone bodies by the liver (A) and entry of ketone bodies into the citric acid cycle (B).

secretion and increased glucagon secretion. *The dominant metabolic processes are the mobilization of triacylglycerols in adipose tissue and gluconeogenesis by the liver.* The liver obtains energy for its own needs by oxidizing fatty acids released from adipose tissue. The concentrations of acetyl CoA and citrate consequently increase, which switches off glycolysis. The uptake of glucose by muscle is markedly diminished because of the low insulin level, whereas fatty acids enter freely. Consequently, *muscle uses no glucose and relies exclusively on fatty acids for fuel.* The β oxidation of fatty acids by muscle halts the conversion of pyruvate into acetyl CoA, because acetyl CoA stimulates the phosphorylation of the pyruvate dehydrogenase complex, which renders it inactive (Section 17.3). Hence, any available pyruvate, lactate, and alanine are exported to the liver for conversion into glucose. Glycerol derived from the cleavage of triacylglycerols is another raw material for the synthesis of glucose by the liver.

Proteolysis also provides carbon skeletons for gluconeogenesis. During starvation, degraded proteins are not replenished and serve as carbon sources for glucose synthesis. Initial sources of protein are those that turn over rapidly, such as proteins of the intestinal epithelium and the secretions of the pancreas. Proteolysis of muscle protein provides some of the three-carbon precursors of glucose. However, survival for most animals depends on being able to move rapidly, which requires a large muscle mass, and so muscle loss must be minimized.

How is the loss of muscle curtailed? After about 3 days of starvation, the liver forms large amounts of the ketone bodies, acetoacetate, and D-3-hydroxybutyrate (Figure 27.14A). Their synthesis from acetyl CoA increases markedly because the citric acid cycle is unable to oxidize all the acetyl units generated by fatty acid degradation. Gluconeogenesis depletes the supply of oxaloacetate, which is essential for the entry of acetyl CoA into the citric acid cycle. Consequently, the liver produces large quantities of ketone bodies, which are released into the blood. At this time, *the brain begins to consume significant amounts of acetoacetate in place of glucose.* After 3 days of starvation, about a quarter of the brain's energy needs are met by ketone bodies (Table 27.5). The heart also uses ketone bodies as fuel.

After several weeks of starvation, ketone bodies become the major fuel of the brain. Acetoacetate is activated by the transfer of CoA from succinyl CoA to give acetoacetyl CoA (Figure 27.14B). Cleavage by thiolase then yields two molecules of acetyl CoA, which enter the citric acid cycle. In essence,

TABLE 27.5 Fuel metabolism in starvation

	Amount formed or consumed in 24 hours (grams)	
Fuel exchanges and consumption	3rd day	40th day
Fuel use by the brain		
Glucose	100	40
Ketone bodies	50	100
All other use of glucose	50	40
Fuel mobilization		
Adipose-tissue lipolysis	180	180
Muscle-protein degradation	75	20
Fuel output of the liver		
Glucose	150	80
Ketone bodies	150	150

ketone bodies are equivalents of fatty acids that are an accessible fuel source for the brain. Only 40 g of glucose is then needed per day for the brain, compared with about 120 g in the first day of starvation (Figure 27.15A). *The effective conversion of fatty acids into ketone bodies by the liver and their use by the brain markedly diminishes the need for glucose. Hence, less muscle is degraded than in the first days of starvation.* The breakdown of 20 g of muscle daily compared with 75 g early in starvation is most important for survival. How are the amino acids released by muscle processed? The amino acids are transported to the liver where nitrogen is removed as urea, gluconeogenic amino acids are metabolized into glucose, and ketogenic amino acids are used as fuel for the liver or processed to ketone bodies.

What happens after depletion of the triacylglycerol stores? The ketone body contribution disappears, and the only source of fuel that remains is protein (Figure 27.15B). Protein degradation accelerates, and death inevitably results from a loss of heart, liver, or kidney function. A person's survival time is mainly determined by the size of the triacylglycerol depot. Examinations of people who have starved themselves to death as a political protest showed that lean individuals succumb after about 70 days whereas obese individuals can survive for 6 or 7 months.

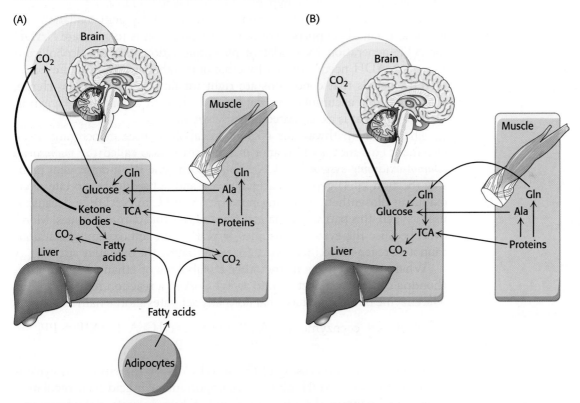

FIGURE 27.15 Fuel utilization during prolonged starvation. A simplified depiction of the relation of adipose tissue, liver, muscle, and brain is shown. (A) Fat mass is adequate. (B) Upon depletion of fat mass. Only the amino acids alanine and glutamine are shown since they are also important for nitrogen transport.

27.6 Ethanol Alters Energy Metabolism in the Liver

Ethanol has been a part of the human diet for centuries, partly because of its intoxicating effects and partly because alcoholic beverages provided a safe means of hydration when pure water was scarce. However, its consumption in excess can result in a number of health problems, most notably liver damage. What is the biochemical basis of these health problems?

Ethanol metabolism leads to an excess of NADH

Ethanol cannot be excreted and must be metabolized, primarily by the liver. This metabolism is accomplished by two pathways. The first pathway comprises two steps. The first step, catalyzed by the enzyme *alcohol dehydrogenase*, takes place in the cytoplasm:

$$CH_3CH_2OH + NAD^+ \xrightarrow{\text{Alcohol dehydrogenase}} CH_3CHO + NADH + H^+$$

Ethanol → Acetaldehyde

The second step, catalyzed by *aldehyde dehydrogenase*, takes place in mitochondria:

$$CH_3CHO + NAD^+ + H_2O \xrightarrow{\text{Aldehyde dehydrogenase}} CH_3COO^- + NADH + H^+$$

Acetaldehyde → Acetate

Note that *ethanol consumption leads to an accumulation of NADH*. This high concentration of NADH inhibits gluconeogenesis by preventing the oxidation of lactate to pyruvate. In fact, the high concentration of NADH will cause the reverse reaction to predominate, and lactate will accumulate. The consequences may be hypoglycemia and lactic acidosis.

The overabundance of NADH also inhibits fatty acid oxidation. An important metabolic purpose of fatty acid oxidation is to generate NADH for ATP generation by oxidative phosphorylation, but an alcohol consumer's NADH needs are met by ethanol metabolism. In fact, the excess NADH signals that conditions are right for fatty acid synthesis. Hence, triacylglycerols accumulate in the liver, leading to a condition known as "fatty liver" that is exacerbated in obese persons.

The second pathway for ethanol metabolism uses cytochrome P450 enzymes. This means of ethanol metabolism is also called the *microsomal ethanol-oxidizing system* (MEOS). This cytochrome P450-dependent pathway (Section 26.4) generates acetaldehyde and subsequently acetate while oxidizing biosynthetic reducing power, NADPH, to $NADP^+$. Because it uses oxygen, this pathway generates free radicals that damage tissues. Moreover, because the system consumes NADPH, the antioxidant glutathione cannot be regenerated (Section 20.5), exacerbating the oxidative stress.

What are the effects of the other metabolites of ethanol? Liver mitochondria can convert acetate into acetyl CoA in a reaction requiring ATP. The enzyme is the thiokinase that normally activates short-chain fatty acids.

$$\text{Acetate} + \text{coenzyme A} + \text{ATP} \longrightarrow \text{acetyl CoA} + \text{AMP} + \text{PP}_i$$
$$\text{PP}_i \longrightarrow 2\,\text{P}_i$$

However, further processing of the acetyl CoA by the citric acid cycle is blocked, because NADH inhibits two important citric acid cycle regulatory enzymes—isocitrate dehydrogenase and α-ketoglutarate dehydrogenase. The accumulation of acetyl CoA has several consequences. First, ketone bodies will form and be released into the blood, aggravating the acidic condition already resulting from the high lactate concentration. The processing of the acetate in the liver becomes inefficient, leading to a buildup of acetaldehyde. This very reactive compound forms covalent bonds with many important functional groups in proteins, impairing protein function. If ethanol is consistently consumed at high levels, the acetaldehyde can significantly damage the liver, eventually leading to cell death.

Ethanol also alters gene transcription, further exacerbating liver damage. Ethanol in some fashion stimulates the synthesis of SREBP (p. 861), a transcription factor that up-regulates the genes that promote fatty acid synthesis. At the same time, acetaldehyde derived from ethanol inactivates another transcription factor, PPARα (peroxisome proliferator-activated receptor), which normally stimulates the genes involved in fatty acid oxidation. The biochemical effects of ethanol consumption can be quite rapid. For instance, fat accumulates in the liver within a few days of moderate alcohol consumption.

Liver damage from excessive ethanol consumption occurs in three stages. The first stage is the aforementioned development of fatty liver. In the second stage—alcoholic hepatitis—groups of cells die and inflammation results. This stage can itself be fatal. In stage three—cirrhosis—fibrous structure and scar tissue are produced around the dead cells. Cirrhosis impairs many of the liver's biochemical functions. The cirrhotic liver is unable to convert ammonia into urea, and blood levels of ammonia rise. Ammonia is toxic to the nervous system and can cause coma and death. Cirrhosis of the liver arises in about 25% of alcoholics, and about 75% of all cases of liver cirrhosis are the result of alcoholism. Viral hepatitis is a nonalcoholic cause of liver cirrhosis.

Excess ethanol consumption disrupts vitamin metabolism

The adverse effects of ethanol are not limited to the metabolism of ethanol itself. Vitamin A (retinol) is converted into retinoic acid, an important signal molecule for growth and development in vertebrates, by the same dehydrogenases that metabolize ethanol. Consequently, this activation does not take place in the presence of ethanol, which acts as a competitive inhibitor. Moreover, the P450 enzymes induced by ethanol inactivate retinoic acid. These disruptions in the retinoic acid signaling pathway are believed to be responsible, at least in part, for fetal alcohol syndrome as well as the development of a variety of cancers.

The disruption of vitamin A metabolism is a direct result of the biochemical changes induced by excess ethanol consumption. Other disruptions in metabolism result from another common characteristic of alcoholics—malnutrition. Alcoholics will frequently drink instead of eating. A dramatic neurological disorder, referred to as *Wernicke–Korsakoff syndrome*, results from insufficient intake of the vitamin thiamine. Symptoms include mental confusion, unsteady gait, and lack of fine motor skills. The symptoms of Wernicke–Korsakoff syndrome are similar to those of beriberi (Section 17.4) because both conditions result from a lack of thiamine. Thiamine is converted into the coenzyme thiamine pyrophosphate, a key constituent of the pyruvate dehydrogenase complex. Recall that this complex links glycolysis with the citric acid cycle. Disruptions in the pyruvate dehydrogenase complex are most evident as neuromuscular disorders because the nervous system is normally dependent on glucose for energy generation.

Alcoholic scurvy is occasionally observed because of an insufficient ingestion of vitamin C. Vitamin C is required for the formation of stable collagen fibers. The symptoms of scurvy include skin lesions and blood-vessel fragility. Most notable are bleeding gums, the loss of teeth, and periodontal infections. Gums are especially sensitive to a lack of vitamin C because the collagen in gums turns over rapidly. What is the biochemical basis for scurvy? Vitamin C is required for the synthesis of 4-hydroxyproline, an amino acid necessary for collagen stability. To form this unusual amino acid, proline residues on the amino side of glycine residues in nascent collagen chains become hydroxylated. One oxygen atom from O_2 becomes attached to C-4 of proline, while the other oxygen atom is taken up by α-ketoglutarate, which is converted into succinate (Figure 27.16). This reaction is catalyzed

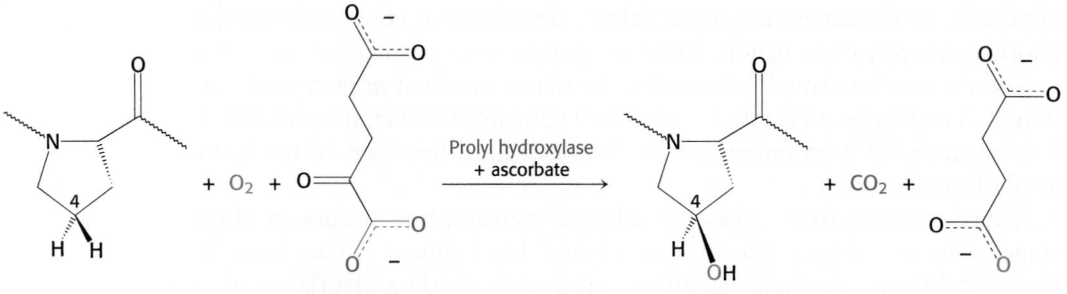

FIGURE 27.16 Formation of 4-hydroxyproline. Proline is hydroxylated at C-4 by the action of prolyl hydroxylase, an enzyme that activates molecular oxygen.

by *prolyl hydroxylase*, a *dioxygenase*, which requires an Fe^{2+} ion to activate O_2. The enzyme also converts α-ketoglutarate into succinate without hydroxylating proline. In this partial reaction, an oxidized iron complex is formed, which inactivates the enzyme. How is the active enzyme regenerated? *Ascorbate (vitamin C)* comes to the rescue by reducing the ferric ion of the inactivated enzyme. In the recovery process, ascorbate is oxidized to dehydroascorbic acid (Figure 27.17). Thus, ascorbate serves here as a specific *antioxidant*. Why does impaired hydroxylation have such devastating consequences? *Collagen synthesized in the absence of ascorbate is less stable than the normal protein.* Hydroxyproline stabilizes the collagen triple helix by forming interstrand hydrogen bonds. The abnormal fibers formed by insufficiently hydroxylated collagen account for the symptoms of scurvy.

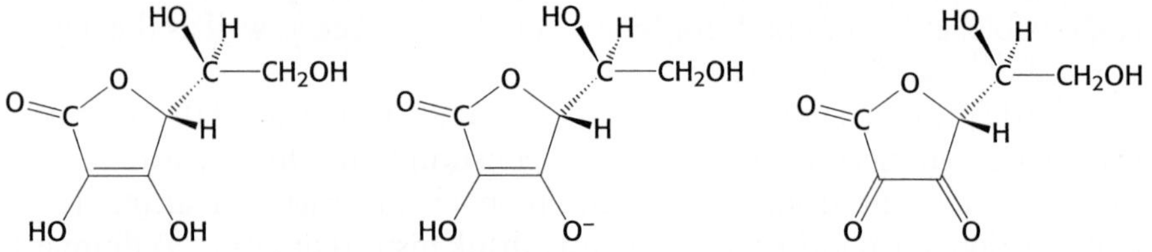

FIGURE 27.17 Forms of ascorbic acid (vitamin C). Ascorbate is the ionized form of vitamin C, and dehydroascorbic acid is the oxidized form of ascorbate.

SUMMARY

27.1 Caloric Homeostasis Is a Means of Regulating Body Weight

Many people are able to maintain a near-constant body weight throughout adult life. This ability is a demonstration of caloric homeostasis, a physiological condition in which energy needs match energy intake. When energy intake is greater than energy needs, weight gain results. In the developed world, obesity is at epidemic proportions and is implicated as a contributing factor in a host of pathological conditions.

27.2 The Brain Plays a Key Role in Caloric Homeostasis

Various signal molecules act on the brain to control appetite. Short-term signals such as CCK and GLP-1 relay satiety signals to the brain while eating is in progress. Long-term signals include leptin and insulin. Leptin, secreted by adipose tissue in direct proportion to adipose-tissue mass, is an indication of fat stores. Leptin inhibits eating. Insulin also works in the brain, signaling carbohydrate availability.

Leptin acts by binding to a receptor in brain neurons, which initiates signal-transduction pathways that reduce appetite. Obesity can develop in individuals with normal amounts of leptin and the leptin receptor, suggesting that such individuals are leptin resistant. Suppressors of cytokine signaling may inhibit leptin signaling, leading to leptin resistance and obesity.

27.3 Diabetes Is a Common Metabolic Disease Often Resulting from Obesity

Diabetes is the most common metabolic disease in the world. Type 1 diabetes results when insulin is absent due to autoimmune destruction of the pancreatic β cells. Type 2 diabetes is characterized by normal or higher levels of insulin, but the target tissues of insulin, notably muscle, do not respond to the hormone—a condition called insulin resistance. Obesity is a significant predisposing factor for type 2 diabetes.

In muscle, excess fats accumulate in an obese individual. These fats are processed to second messengers that activate signal-transduction pathways, which inhibit insulin signaling, leading to insulin resistance. Insulin resistance in target tissues ultimately leads to pancreatic β-cell failure. The pancreas tries to compensate for a lack of insulin action by synthesizing more insulin, resulting in ER stress and subsequent activation of apoptotic pathways that lead to β cell death.

Type 1 diabetes results when insulin is absent due to autoimmune destruction of the pancreatic β cells. The resulting lack of insulin and excess of glucagon leads to an elevated blood-glucose level, the mobilization of triacylglycerols, and excessive ketone-body formation. Accelerated ketone-body formation can lead to acidosis, coma, and death in untreated insulin-dependent diabetics.

27.4 Exercise Beneficially Alters the Biochemistry of Cells

Exercise is a useful prescription for insulin resistance and type 2 diabetes. Muscle activity stimulates mitochondrial biogenesis in a calcium-dependent manner. The increase in the number of mitochondria facilitates fatty acid oxidation in the muscle, resulting in increased insulin sensitivity.

Fuel choice in exercise is determined by the intensity and duration of the bout of exercise. Sprinting and marathon running are powered by different fuels to maximize power output. The 100-meter sprint is powered by stored ATP, creatine phosphate, and anaerobic glycolysis. In contrast, the oxidation of both muscle glycogen and fatty acids derived from adipose tissue is essential in the running of a marathon, a highly aerobic process.

27.5 Food Intake and Starvation Induce Metabolic Changes

Insulin signals the fed state; it stimulates the formation of glycogen and triacylglycerols and the synthesis of proteins. In contrast, glucagon signals a low blood-glucose concentration; it stimulates glycogen breakdown and gluconeogenesis by the liver and triacylglycerol hydrolysis by adipose tissue. After a meal, the rise in the blood-glucose concentration leads to an increased secretion of insulin and a decreased secretion of glucagon. Consequently, glycogen is synthesized in muscle and the liver. When the blood-glucose concentration drops several hours later, glucose is then formed by the degradation of glycogen and by the gluconeogenic pathway, and fatty acids are released by the hydrolysis of triacylglycerols. The liver and muscle then increasingly use fatty acids instead of glucose to meet their own energy needs so that glucose is conserved for use by the brain and the red blood cells.

The metabolic adaptations in starvation serve to minimize protein degradation. Large amounts of ketone bodies are formed by the liver from fatty acids and released into the blood within a few days after the onset of starvation. After several weeks of starvation, ketone bodies become the major fuel of the brain. The diminished need for glucose decreases the rate of muscle breakdown, and so the likelihood of survival is enhanced.

27.6 Ethanol Alters Energy Metabolism in the Liver

The oxidation of ethanol results in an unregulated overproduction of NADH, which has several consequences. A rise in the blood concentration of lactic acid and ketone bodies causes a fall in blood pH, or acidosis. The liver is damaged because the excess NADH causes excessive fat formation as well as the generation of acetaldehyde, a reactive molecule. Alterations in the activity of several transcription factors due to ethanol consumption also contribute to liver pathology. Severe liver damage can result with continued ethanol use.

APPENDIX Biochemistry in Focus

Biochemistry in Focus 1

Adipokines help to regulate fuel metabolism in the liver

As we have seen many times in our study of biochemistry, the liver plays a central role in the regulation of glucose and lipid metabolism. This role is facilitated by communication between adipocytes and liver. The adipokines leptin and adiponectin regulate fuel metabolism in liver (Figure 27.18). Recall that earlier in the text, we stated that leptin secretion increases as fat mass increases, while adiponectin secretion increases as fat mass decreases. Both hormones act to prevent lipid accumulation in the liver and stimulate fatty acid oxidation in a similar fashion. Binding to their respective membrane receptors results in the phosphorylation and activation of AMP-activated protein kinase (AMPK, p. 740). Activated AMPK in turn

phosphorylates and inactivates acetyl CoA carboxylase 1, the principal regulatory enzyme of fatty acid metabolism. With the inhibition of acetyl CoA carboxylase 1, the concentration of its product, malonyl CoA falls. Recall that malonyl CoA inhibits fatty acid oxidation by inhibiting the carnitine acyl transferase, the enzyme that transports fatty acids into the mitochondria to undergo β-oxidation (p. 739). Consequently, β-oxidation of fatty acids is enhanced. Activated AMPK also decreases the expression of transcription factors, such as SREBP, that control genes required for lipogenesis and cholesterol synthesis. The overall effect of leptin and adiponectin on fatty acid metabolism, then, is to decrease synthesis and increase oxidation, preventing lipid accumulation in the liver.

Adiponectin can also ameliorate some to the effects of lipid accumulation. As discussed in the Biochemistry in Focus section of Chapter 26, a consequence of lipid accumulation is an increase in the concentration of ceramide. Ceramide stimulates fatty acid import, which leads to lipid accumulation and inhibits protein kinases required insulin action. In other words, in the presence of excess ceramide, cells become insulin resistant. Upon binding to its receptor, adiponectin activates the ceramidase activity associated with the receptor, which converts ceramide into sphingosine. The decrease in ceramide thus heightens insulin sensitivity. Adiponectin also inhibits gluconeogenesis by mechanisms that remain ill-defined.

The combined action of leptin and adiponectin on the liver results in

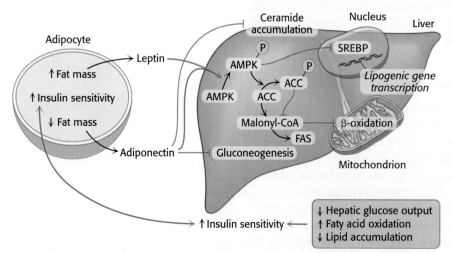

FIGURE 27.18 Adipokines help to maintain systemic lipid and glucose homeostasis.

decreased glucose output, decreased lipid accumulation, and increased fatty acid oxidation in the liver. All of these alter-ations enhance insulin sensitivity for the liver and other tissues including the adipocytes.

Biochemistry in Focus 2

Exercise alters muscle and whole-body metabolism

We have discussed the effects of exercise on biochemical processes a number of times in this text, most recently in the current chapter. In this Biochemistry in Focus feature, we will take an overview of the molecular adaptations to exercise, and then compare the effects of endurance and resistance training on muscle and whole-body biochemistry.

Regardless of the type of exercise, the timing of the metabolic response, in general, is the same (Figure 27.19). The first response to exercise is an alteration in the patterns of gene expression. (A complete discussion of the regulation of gene expression is presented in Chapters 30 and 33.) A variety of transcription factors are activated by a variety of signal molecules, and all of these transcription factors stimulate genes that encode proteins that facilitate exercise. For instance, as discussed on page 902, fatty acids and Ca^{2+} stimulate the transcription of genes that stimulate mitochondrial biogenesis and fatty acid oxidation. Recall also that hypoxia (low oxygen) activates the hypoxia inducible factor (p.518), which stimulates the genes encoding glycolytic enzymes, glucose transporters, and angiogenesis stimulating factors. Remarkably, as shown in Figure 27.18, even a single bout of exercise can stimulate gene expression several-fold, with levels returning to normal in 24 hours. However, it takes repeated bouts of exercise to accumulate enough mRNA to enhance the synthesis of proteins responding to exercise. In addition to the proteins mentioned above, other examples of such proteins include muscle contractile and structural proteins (Sections 9.4, 36.1, and Section 36.2 in 🌀 **Sapling Plus**) and the oxygen-binding proteins hemoglobin and myoglobin (Chapter 7). When these and other proteins accumulate to a sufficient extent, the physiological effects of exercise become apparent. You can run longer and faster or you can do more repetitions with heavier weights.

Table 27.6 shows some of the molecular adaptations and health benefits accruing from endurance and resistance training. For most of us, some combination of the two will result in maximal health benefits. The American Medical Association and the American College of Sports Medicine launched a global initiative several years ago called Exercise is Medicine. As Table 27.6 illustrates, and as we have learned throughout our study of biochemistry, this is indeed the case: exercise *is* medicine.

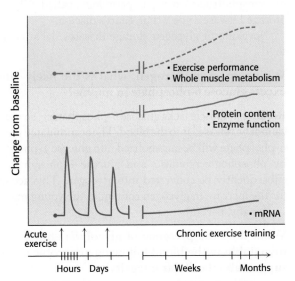

FIGURE 27.19 The molecular adaptation to exercise. Changes as a function of duration of exercise are shown for mRNA synthesis (bottom), protein synthesis (middle), and exercise performance (top). [Figure 1 from Egan, B., and Zierath, J. R. 2013. Exercise metabolism and the molecular regulation of skeletal muscle adaptation. *Cell Metab.* 17:162–184.]

TABLE 27.6 Adaptations and health benefits of aerobic and resistance training

	Aerobic (endurance) training	Resistance (strength) training
Skeletal Muscle morphology and exercise performance		
Muscle growth (hypertrophy)	↔	↑↑↑
Muscle strength and power	↔	↑↑↑
Muscle fiber size	↔↑	↑↑↑
Anaerobic capacity	↑	↑↑
Muscle protein synthesis	↔↑	↑↑↑
Mitochondria protein synthesis	↑↑	↔↑
Lactate tolerance	↑↑	↔↑
Mitochondria density and oxidative capacity	↑↑↑	↔↑
Endurance capacity	↑↑↑	↔↑
Whole-body and metabolic health		
Bone mineral density	↑↑	↑↑
Percent body fat	↓↓	↓
Lean body mass	↔	↑↑
Insulin sensitivity	↑↑	↑↑
Inflammatory makers	↓↓	↓
Resting heart rate	↓↓	↔
Stroke volume*, resting and maximal	↑↑	↔
Blood pressure at rest	↔↓	↔
Cardiovascular risk	↓↓↓	↓
Basal metabolic rate	↑	↑↑

* Volume of the blood pumped from the left ventricle of the heart per beat. ↑ Values increase; ↓ values decrease; ↔ no change: ↔↑ or ↔↓ little or no change.

Table 1 from Egan, B., and Zierath, J. R. 2013. Exercise metabolism and the molecular regulation of skeletal muscle adaptation. *Cell Metab.* 17:162–184.

APPENDIX

Problem-Solving Strategies

PROBLEM: Imagine yourself to be a biochemist at a prestigious biomedical research institution. You are sitting at your desk late one afternoon thinking about the results of the day's experiments. A clinician friend from down the hall comes into your office to ask for some guidance. He has a patient, a young man, who enjoys walking but experiences painful cramps upon intense exercise. Blood tests after an anaerobic treadmill exercise show no increase in lactate but did reveal the presence of glycolytic enzymes and myoglobin in the blood. Blood glucose concentration was normal and epinephrine treatment led to glycogen mobilization. The results of further blood work are shown in Table 27.7.

TABLE 27.7 Blood test results for Problem-Solving Strategies

	Glycogen (mg/gm tissue)	Glucose 6-phosphate (µmol/gm tissue)	Fructose 6-phosphate (µmol/gm tissue)	Fructose 1, 6 bisphosphate (µmol/gm tissue)
Your friend's patient	47	10	3	0.01
Normal values	8–11	0.4–0.7	0.09–0.15	0.7–1.0

Your friend sheepishly admits he didn't pay much attention in his biochemistry class and asks for your help in interpreting the information concerning his patient. You smile, perhaps a bit too condescendingly, and say: "Sure. If you buy me a chocolate éclair later." He agrees and then asks you a series of incisive questions.

(a) What do the results of the anaerobic exercise test suggest is wrong with this poor guy's muscles?

(b) What is the significance of the results of treating the patient with epinephrine?

(c) What do the results of the muscle biopsy reveal to you?

(d) Why is the muscle glycogen higher than normal?

(e) What is the significance of the glycolytic enzymes and myoglobin in the blood?

SOLUTION: Let's proceed with our normal dissection of each question into simpler questions. Consider (a) first.

▶ **What is the normal end product of anaerobic exercise in muscle?**

Lactate.

▶ **What does the absence of lactate in the patient tell you?**

Glycolysis must not be functioning properly. Why this is the case we do not yet know.
    On to (b).

▶ **What is the biochemical function of epinephrine in muscle?**

Epinephrine stimulates glycogen breakdown. The fact that the patient responded normally to epinephrine shows that glycogen degradation is functional. So far, we have established that glycolysis is not functioning well but glycogen breakdown is normal.

    What about part (c)?

▶ **Describe the results of the muscle biopsy.**

Glycogen, glucose 6-phosphate, and fructose 6-phosphate are present in higher-than-normal concentrations, while fructose 1,6-bisphosphate is at a much lower concentration than normal. Note that glucose 6-phosphate, fructose 1,6-bisphosphate, and fructose 1,6-bisphosphate are glycolytic intermediates. The fact that glycogen could be mobilized in response to epinephrine establishes that the excess glycogen is not a result of impaired glycogen degradation. Taken together, these results suggest something is amiss in glycolysis—perhaps an impaired enzyme—that results in glycogen accumulation.

▶ **What glycolytic enzyme appears to be impaired? (If necessary, see Figure 16.2 for assistance.)**

Intermediates upstream of the phosphofructokinase reaction are present in excess, while the one downstream metabolite measured, fructose 1,6-bisphosphate, is deficient. This condition could be explained by a deficiency of phosphofructokinase, a pathology called Glycogen Storage Disease Type VII or Tauri's disease.
    Speaking of glycogen storage diseases, let's move on to (d).

▶ **With respect to glycogen, what is the likely fate of excess glucose 6-phosphate in muscle?**

Recall that muscle lacks glucose 6-phosphatase, so glucose cannot be released into the blood. The accumulated glucose 6-phosphate will be metabolized into glucose 1-phosphate by phosphoglucomutase, and glucose 1-phosphate will subsequently be converted into glycogen. This accounts for the increase in glycogen content in the young man.
    Finally, part (e).

▶ **What does the presence of glycolytic enzymes and myoglobin in the blood following anaerobic exercise suggest about cellular integrity?**

Cells must be severely damaged due to a lack of ATP during intense exercise. Cells die and the cellular contents are released into the blood—a condition called rhabdomyolysis.
    Now, head to the bakery across the street from your lab and have your friend buy you the chocolate éclair.

KEY TERMS

caloric homeostasis (energy homeostasis) (p. 890)
cholecystokinin (CCK) (p. 892)
glucagon-like peptide 1 (GLP-1)
  (p. 893)
leptin (p. 893)
insulin (p. 893)

adiponectin (p. 894)
leptin resistance (p. 895)
type 1 diabetes (p. 896)
insulin resistance (p. 896)
type 2 diabetes (p. 896)
metabolic syndrome (p. 898)

endoplasmic reticulum (ER) stress
  (p. 900)
unfolded protein response (UPR)
  (p. 900)
starved–fed cycle (p. 905)
glucose homeostasis (p. 905)

PROBLEMS

1. *Energy reserves.* Which of the following tissues provide the largest energy supply in a well-nourished human?

(a) Liver

(b) Muscle

(c) Brain

(d) Adipose tissue

2. *Target tissues.* Which of the following tissues are primary targets of insulin?

(a) Muscle

(b) Brain

(c) Adipose tissue

(d) Liver

3. *Brain power.* Under normal circumstances, what is the most commonly used fuel by the brain?

(a) Glycerol

(b) Glucose

(c) Fatty acids

(d) Ketone bodies

4. *No stimulation.* Which of the following pathways is not stimulated by epinephrine?

(a) Glycogen degradation

(b) Mobilization of fatty acids from adipose tissue

(c) Glycogen synthesis

(d) Glycolysis in muscle

5. *Indistinguishable.* Type 1 diabetes is the result of which of the following?

(a) Autoimmune disorder

(b) Metabolic syndrome

(c) Improper diet

(d) Excessive insulin secretion

6. *Weight control.* As disturbing as the obesity epidemic is, an equally intriguing, almost amazing observation is that many people are able to maintain an approximately constant weight throughout adult life. A few simple calculations of a simplified situation illustrates how remarkable this feat is. Consider a 120-pound woman whose weight does not change significantly between the ages of 25 and 65. Let's say that the woman requires 8,400 kJ (2000 kcal) per day^{-1}. For simplicity's sake, let us assume the woman's diet consists predominantly of fatty acids derived from lipids. The energy density of fatty acids is 38 kJ (9 kcal) g^{-1}. How much food has she consumed over the 40 years in question? ✓❶

7. *Spare tire.* Suppose that our test subject from problem 1 gained 55 pounds between the ages of 25 and 65 (a sadly common occurrence), and that her weight at 65 years of age is 175 pounds. Calculate how many excess calories she consumed per day to gain the 55 pounds over 40 years. Assume that our test subject is 5 feet 6 inches tall. What is her BMI? Would she be considered obese at 175 lbs? ✓❶

8. *Depot fat.* Adipose tissue was once only considered a storage site for fat. Why is this view no longer considered correct?

9. *Balancing act.* What is meant by caloric homeostasis? ✓❶

10. *Dynamic duo.* What are the key hormones responsible for maintaining caloric homeostasis? ✓❷

11. *Dual roles.* What two biochemical roles does CCK play? GLP-1? ✓❷

12. *Failure to communicate.* Leptin inhibits eating and is secreted in amounts in direct proportion to body fat. Moreover, obese people have normal amounts of leptin and leptin receptor. Why, then, do people become obese? ✓❷

13. *Many signals.* Match the characteristic (a–i) with the appropriate hormone (1–6). ✓❷

(a) Secreted by adipose tissue

———————

(b) Stimulates liver gluconeogenesis ——————

(c) GPCR pathway

———————

1. leptin

2. adiponectin

3. GLP-1

4. CCK

(d) Satiety signal _____   5. insulin

_____   6. glucagon

(e) Enhances insulin secretion

_____

(f) Secreted by the pancreas during a fast _____

(g) Secreted after a meal

_____

(h) Stimulates glycogen synthesis _____

(i) Missing in type 1 diabetes

_____

14. *A key chemical.* What are the sources of glucose 6-phosphate in liver cells?

15. *Neither option is good.* Differentiate between type 1 and type 2 diabetes. ✓❸

16. *Fighting diabetes.* Leptin is considered an "anti-diabetogenic" hormone. Explain. ✓❷

17. *Metabolic energy and power.* The rate of energy expenditure of a typical 70-kg person at rest is about 70 watts (W), like that of a light bulb.

(a) Express this rate in kilojoules per second and in kilocalories per second.

(b) How many electrons flow through the mitochondrial electron-transport chain per second under these conditions?

(c) Estimate the corresponding rate of ATP production.

(d) The total ATP content of the body is about 50 g. Estimate how often an ATP molecule turns over in a person at rest.

18. *Respiratory quotient (RQ).* This classic metabolic index is defined as the volume of CO_2 released divided by the volume of O_2 consumed. ✓❹

(a) Calculate the RQ values for the complete oxidation of glucose and of tripalmitoylglycerol.

(b) What do RQ measurements reveal about the contributions of different energy sources during intense exercise? (Assume that protein degradation is negligible.)

19. *Camel's hump.* Compare the H_2O yield from the complete oxidation of 1 g of glucose with that of 1 g of tripalmitoylglycerol. Relate these values to the evolutionary selection of the contents of a camel's hump.

20. *Hungry–nourished.* What is meant by the starved–fed cycle? ✓❺

21. *Of course, too much is bad for you.* What are the primary means of processing ethanol? ✓❻

22. *Started out with burgundy, but soon hit the harder stuff.* Describe the three stages of ethanol consumption that lead to liver damage and possibly death. ✓❻

23. *The wages of sin.* How long does a person have to jog to offset the calories obtained from eating 10 macadamia nuts

(75 kJ, or 18 kcal, per nut)? (Assume an incremental power consumption of 400 W.) ✓❶

24. *Sweet hazard.* Ingesting large amounts of glucose immediately before a marathon might seem to be a good way of increasing the fuel stores. However, experienced runners do not ingest glucose before a race. What is the biochemical reason for their avoidance of this potential fuel? (Hint: consider the effect of glucose ingestion on the level of insulin.) ✓❷, ✓❹

25. *Lipodystrophy.* Lipodystrophy is a condition in which an individual lacks adipose tissue. The muscles and liver from such individuals are insulin resistant, and both tissues accumulate large amounts of triacylglycerols (hyperlipidemia). The administration of leptin partly ameliorates this condition. What does it indicate about the relation of adipose tissue to insulin action? ✓❷

26. *Therapeutic target.* What would be the effect of a mutation in the gene for PTP1B (protein tyrosine phosphatase 1B) that inactivated the enzyme in a person who has type 2 diabetes? ✓❸

27. *An effect of diabetes.* Insulin-dependent diabetes is often accompanied by hypertriglyceridemia, which is an excess blood level of triacylglycerols in the form of very low-density lipoproteins (VLDL). Suggest a biochemical explanation. ✓❸

28. *Sharing the wealth.* The hormone glucagon signifies the starved state, yet it inhibits glycolysis in the liver. How does this inhibition of an energy-production pathway benefit the organism? ✓❷

29. *Compartmentation.* Glycolysis takes place in the cytoplasm, whereas fatty acid degradation takes place in mitochondria. What metabolic pathways depend on the interplay of reactions that take place in both compartments?

30. *Kwashiorkor.* The most common form of malnutrition in children in the world, kwashiorkor, is caused by a diet having ample calories but little protein. The high levels of carbohydrate result in high levels of insulin. What is the effect of high levels of insulin on ✓❺

(a) lipid utilization?

(b) protein metabolism?

(c) Children suffering from kwashiorkor often have large distended bellies caused by water from the blood leaking into extracellular spaces. Suggest a biochemical basis for this condition.

31. *One for all, all for one.* How is the metabolism of the liver coordinated with that of skeletal muscle during strenuous exercise? ✓❹

32. *A little help, please?* What is the advantage of converting pyruvate into lactate in skeletal muscle? ✓❹

33. *Fuel choice.* What is the major fuel for resting muscle? What is the major fuel for muscle under strenuous work conditions? ✓❹

34. *Hefty reimbursement.* Endurance athletes sometimes follow the exercise-and-diet plan described here: seven days before an event, do exhaustive exercises so as to all but deplete glycogen stores. For the next 2 to 3 days, consume few carbohydrates and do moderate- to low-intensity exercises. Finally, 3 to 4 days before the event, consume a diet rich in carbohydrates. Explain the benefits of this regime. ✓④

35. *Oxygen deficit.* After light exercise, the oxygen consumed in recovery is approximately equal to the oxygen deficit, which is the amount of additional oxygen that would have been consumed had oxygen consumption reached steady state immediately. How is the oxygen consumed in recovery used? ✓④

36. *Excess post-exercise oxygen consumption.* The oxygen consumed after strenuous exercise stops is significantly greater than the oxygen deficit and is termed *excess post-exercise oxygen consumption* (EPOC). Why is so much more oxygen required after intense exercise? ✓④

37. *Psychotropic effects.* Ethanol is unusual in that it is freely soluble in both water and lipids. Thus, it has access to all regions of the highly vascularized brain. Although the molecular basis of ethanol action in the brain is not clear, ethanol evidently influences a number of neurotransmitter receptors and ion channels. Suggest a biochemical explanation for the diverse effects of ethanol. ✓⑥

38. *Fiber type.* Skeletal muscle has several distinct fiber types. Type I is used primarily for aerobic activity, whereas type IIb is specialized for short, intense bursts of activity. How could you distinguish between these types of muscle fiber if you viewed them with an electron microscope? ✓④

39. *Tour de France.* Cyclists in the Tour de France (more than 2000 miles in 3 weeks) require about 836,000 kJ (200,000 kcal) of energy, or 41,840 kJ (10,000 kcal) day^{-1} (a resting male requires, 8368 kJ, or 2000 kcal, day^{-1}). ✓①

(a) With the assumptions that the energy yield of ATP is about 50.2 kJ (12 kcal) mol^{-1} and that ATP has a molecular weight of 503 g mol^{-1}, how much ATP would be expended by a Tour de France cyclist?

(b) Pure ATP can be purchased at a cost of approximately $150 per gram. How much would it cost to power a cyclist through the Tour de France if the ATP had to be purchased?

40. *Responding to stress.* Why does it make good physiological sense that regular bouts of prolonged exercise will result in mitochondrial biogenesis? ✓④

41. *Burning fats.* How does the activation of AMP-activated protein kinase (AMPK) during aerobic exercise foster the switch to fatty acid oxidation for distance running? ✓④

42. *Saving protein.* What metabolic and hormonal changes account for decreased gluconeogenesis during the first several weeks of starvation in humans? ✓⑤

43. *Did I do that!?* Alcohol consumption on an empty stomach results in some interesting biochemical as well as embarrassing behavioral alterations. We will ignore the latter. Gluconeogenesis falls; there are increases in intracellular ratios of lactate to pyruvate, of glycerol 3-phosphate to

dihydroxyacetone phosphate, of glutamate to α-ketoglutarate, and of D-3-hydroxybutyrate to acetoacetate. Hypoglycemia develops rapidly. Blood pH also falls. Alcohol consumption by a well-fed individual does not lead to hypoglycemia or alteration in blood pH. ✓⑥

(a) Why does ethanol consumption result in the altered ratios?

(b) Why does hypoglycemia and blood acidosis result in the hungry individual?

(c) Why does the well-fed person not experience hypoglycemia?

44. *Skip the candy bar.* Humans can survive longer by total fasting than on a diet consisting entirely of small amounts of carbohydrates. What is going on? ✓⑤

45. *Too much of a good thing.* What is the relation between fatty acid oxidation and insulin resistance in the muscle? ✓③

46. *Aneurin? Really?* Why are the symptoms of beriberi similar to those of Wernicke–Korsakoff syndrome?

47. *Graham crackers and chocolate milk.* You have the opportunity to shadow some physicians at a prestigious medical school in the Midwest. The physicians are examining a 2 ½-year-old boy with the following symptoms: he is unable to walk, his legs are painful to the touch, and x-rays reveal a broken leg bone, his body is covered with a rash, and his gums are bleeding. His mother explains that the condition has been ongoing for about 6 weeks. Your team is baffled. Various ideas are presented but none seems to account for the poor little guy's condition. As the only non-MD participating in the examination, you ask, tentatively of the mom: what does he like to eat? She responds that he is a picky eater and for the past couple of months he has consumed only graham crackers and chocolate milk. A light bulb appears over your head, to the astonishment of all present. What is the matter with the child?

Data Interpretation Problem

48. *Lactate threshold.* The graph shows the relation between blood-lactate levels, oxygen consumption, and heart rate during exercise of increasing intensity. The values for oxygen consumption and heart rate are indicators of the degree of exertion. ✓④

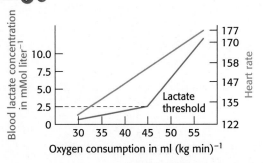

(a) Why is some lactate produced even when exercise is moderate?

(b) Biochemically, what is taking place when the lactate concentration begins to rise rapidly, a point called the lactate threshold?

(c) Endurance athletes will sometimes measure blood-lactate levels during training so that they know their lactate threshold. Then, during events, they will race just at or below their lactate threshold until the late stages of the race. Biochemically, why is this practice wise?

(d) Training can increase the lactate threshold. Explain.

Chapter Integration Problems

49. *Feeling the burn.* Under certain circumstances, the respiratory quotient (RQ value) for an athlete exercising intensely can rise above 1. How is this possible? ✓❹

50. *So many channels. Like cable TV.* Describe the role of ion channels in insulin secretion by the β cells of the pancreas.

Think ▶ Pair ▶ Share Problems

51. Discuss the biochemical processes that contribute to maintaining caloric homeostasis.

52. Describe the biochemical alterations that result in type 2 diabetes.

53. With respect to the patient discussed in Problem-Solving Strategies, explain why the young man was able to perform moderate exercise without discomfort.

Drug Development

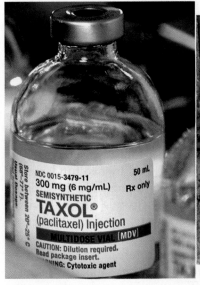

CHAPTER 28

Taxol

Many drugs are based on natural products.

Taxol (above) is a molecule isolated from the bark of the Pacific yew tree (*Taxus brevifolia*) that potently inhibits cell division and is used to treat various forms of cancer. More recently, methods have been developed to synthesize Taxol without destruction of the Pacific yew, significantly reducing the ecological impact of drug preparation. [Credit: Courtesy Bristol-Myers Squibb, Inga Spence/Science Source]

✓ LEARNING OBJECTIVES

By the end of this chapter, you should be able to:

1. Describe two approaches to early drug discovery.

2. Identify the criteria the compounds must meet in order to be developed into drugs.

3. Describe the processes by which the body interacts with drug molecules.

4. Describe how structural information can be used to assist in drug development.

5. Provide examples of how genetics and genomics can be used to progress drug discovery.

6. Distinguish the clinical phases of drug development.

OUTLINE

28.1 Compounds Must Meet Stringent Criteria to Be Developed into Drugs

28.2 Drug Candidates Can Be Discovered by Serendipity, Screening, or Design

28.3 Genomic Analyses Can Aid Drug Discovery

28.4 The Clinical Development of Drugs Proceeds Through Several Phases

The development of drugs is among the most important interfaces between biochemistry and medicine. Throughout the previous chapters in this book, we have encountered a number of types of proteins—enzymes, receptors, and transporters—that serve as targets for a majority of the drugs in clinical use (Figure 28.1). In most cases, drugs act by binding to these proteins and inhibiting or otherwise modulating their activities. Thus, knowledge of these molecules and the pathways in which they participate is crucial to drug development. An effective drug is much more than a potent modulator of its target, however. Drugs must be readily administered to patients, preferably as small tablets taken orally, and must survive within the body

FIGURE 28.1 The targets of drugs in current use. In this pie chart, currently used drugs that modulate human proteins are sorted by target type: enzymes in green, receptors in blue, transporters in red, and other targets in gray. [Data from M. Rask-Andersen et al., *Nat. Rev. Drug Discov.* 10:579–590, 2011.]

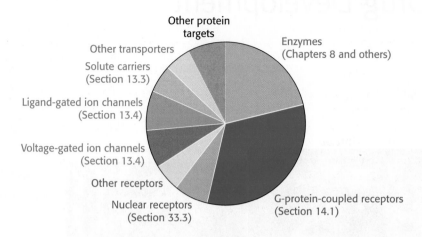

long enough to reach their targets. Furthermore, to prevent unwanted physiological effects, drugs must not modulate the properties of biomolecules other than the intended targets. These requirements tremendously limit the number of compounds that have the potential to be clinically useful drugs.

Initial drug–like molecules are typically discovered by two fundamentally distinct approaches. The *target-based screening* approach first requires the identification of a *target*—a protein or an oligonucleotide whose activity, when altered by a drug, will alter the progression of a human disease process (Figure 28.2A). Before embarking on the time-consuming and costly process of developing a drug against a particular target, researchers must first accumulate evidence that the target exhibits two critical properties. First, the target must be *valid*: there must be sufficient evidence that the target plays an important role in disease progression in humans. Researchers accumulate evidence for target validation from a variety of experiments, such as human genetic analyses, RNA interference, transgenic or knockout animal models (Section 5.4), and studies with *tool compounds*—molecules which may be usable in cellular or animal experiments but not suitable themselves as drugs for humans. Second, the target must be *tractable*, or *druggable*. A tractable target contains structural features, such as ligand-binding pockets, into which drug-like molecules can be identified and optimized. Once the target is identified, assays for its activity can be developed (Section 3.1). Then, libraries of compounds can be tested in this assay; active molecules can be identified and optimized. Many unexpected results may be encountered in this process as the complexity of biological systems reveals itself.

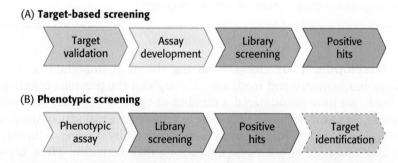

FIGURE 28.2 Two paths to early drug discovery. (A) Target-based screening. A molecular target is identified and validated. An assay for this target's activity is used to screen a compound library. Hits can then be tested for their physiological effects and optimized into drug candidates. (B) Phenotypic screening. A phenotypic assay is developed in a cellular or animal model system. This assay is used to screen a compound library, from which hits can be optimized into drug candidates. These compounds can also be used to identify the molecular target.

Alternatively, drugs may be identified by the *phenotypic screening* approach (Figure 28.2B). In this method, libraries of compounds are screened in an assay designed to detect a desired phenotype in a cellular system or animal model. For example, compounds may be tested for their ability to promote cellular survival or enhance resistance to specific toxic stimuli. In this approach, a biological effect is known before the target is identified. The mode of action of the compound is only later identified after substantial additional work.

In this chapter, we explore the science of pharmacology. We examine a number of case histories that illustrate drug development, including many of its concepts, methods, and challenges. We then see how the concepts and tools from genomics are influencing approaches to drug discovery. The chapter ends with a summary of the clinical trial phases required for drug development.

Pharmacology
The science that deals with the discovery, chemistry, composition, identification, biological and physiological effects, uses, and manufacture of drugs.

28.1 Compounds Must Meet Stringent Criteria to Be Developed into Drugs

Many compounds have significant effects when taken into the body, but only a very small fraction of them have the potential to be useful drugs. A foreign compound, not adapted to its role in the cell through long evolution, must have a range of special properties to function effectively without causing serious harm. Let us consider some of the challenges faced by drug developers.

Drugs must be potent and selective

Most drugs bind to specific proteins, usually receptors or enzymes, within the body. To be effective, a drug should bind a sufficient number of its target proteins when taken at a dose that can be reasonably administered to patients. One factor in determining drug effectiveness is the strength of the interaction between the drug and its target. A molecule that binds to some target protein is often referred to as a *ligand*. A ligand-binding curve is shown in Figure 28.3. Ligand molecules occupy progressively more target-binding sites as ligand concentration increases until essentially all of the available sites are occupied. The tendency of a ligand to bind to its target is measured by the *dissociation constant*, K_d, defined by the expression

$$K_d = [R][L]/[RL]$$

where [R] is the concentration of the free receptor, [L] is the concentration of the free ligand, and [RL] is the concentration of the receptor–ligand complex. The dissociation constant is a measure of the strength of the interaction between the drug candidate and the target; the lower the value, the stronger the interaction. The concentration of free ligand at which one-half of the binding sites are occupied equals the dissociation constant, as long as the concentration of binding sites is substantially less than the dissociation constant.

In many cases, biological assays using living cells or tissues (rather than direct enzyme or binding assays) are employed to examine the potency of drug candidates. For example, the fraction of bacteria killed by a drug might indicate the potency of a potential antibiotic. In these cases, values such as EC_{50} are used. EC_{50} is the concentration of the drug candidate required to elicit 50% of the maximal biological response (Figure 28.4). Similarly, EC_{90} is the concentration required to achieve 90% of the maximal response.

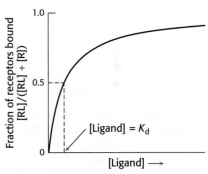

FIGURE 28.3 Ligand binding. The titration of a receptor, R, with a ligand, L, results in the formation of the complex RL. In uncomplicated cases, the binding reaction follows a simple saturation curve. Half of the receptors are bound to ligand when the ligand concentration equals the dissociation constant, K_d, for the RL complex.

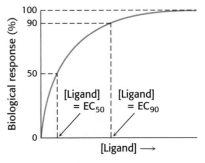

FIGURE 28.4 Effective concentrations. The concentration of a ligand required to elicit a biological response can be quantified in terms of EC_{50}, the concentration required to give 50% of the maximum response, and EC_{90}, the concentration required to give 90% of the maximum response.

In the example of an antibiotic, EC_{90} would be the concentration required to kill 90% of bacteria exposed to the drug. For drug candidates that are inhibitors, the corresponding terms IC_{50} and IC_{90} are often used to describe the concentrations of the inhibitor required to reduce a response by 50% or 90%, relative to its value in the absence of an inhibitor, respectively.

Values such as the IC_{50} and EC_{50} are measures of the *potency* of a drug candidate in modulating the activity of the desired biological target. To prevent unwanted effects, often called *side effects*, ideal drug candidates should also be *selective*. That is, they should not bind biomolecules other than the target to any significant extent. Developing such a drug can be quite challenging, particularly if the drug target is a member of a large family of evolutionarily related proteins. The degree of selectivity can be described in terms of the ratio of the K_d values for the binding of the drug candidate to any other molecules to the K_d value for the binding of the drug candidate to the desired target.

Under physiological conditions, achieving sufficient binding of a target site by a drug can be quite challenging. Most drug targets also bind ligands normally present in tissues; often, the drug and these ligands will compete for binding sites on the target. We encountered this scenario when we considered competitive inhibitors in Chapter 8. Suppose that the drug target is an enzyme and the drug candidate is competitive with respect to the enzyme's natural substrate. The concentration of the drug candidate necessary to inhibit the enzyme will depend on the physiological concentration of the normal substrate (Figure 28.5). Biochemists Yung-Chi Cheng and William Prusoff described the relation between the IC_{50} of an enzyme inhibitor and its *inhibition constant* K_i (analogous to the dissociation constant, K_d, of a ligand). *For a competitive inhibitor,*

$$IC_{50} = K_i(1 + [S]/K_M)$$

This relation, referred to as the *Cheng–Prusoff equation*, demonstrates that the IC_{50} of a competitive inhibitor will depend on the concentration and the Michaelis constant (K_M) for the substrate S. The higher the concentration of the natural substrate, the higher the concentration of drug needed to inhibit the enzyme.

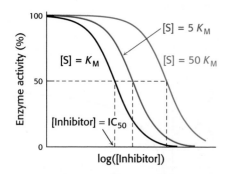

FIGURE 28.5 Inhibitors compete with substrates for enzyme active sites. The measured IC_{50} of a competitive inhibitor to its target enzyme depends on the concentration of substrate present.

Drugs must have suitable properties to reach their targets

Thus far, we have focused on the ability of molecules to act on specific target molecules. However, an effective drug also has other requisite characteristics. It must be easily administered and must reach its target at sufficient concentration to be effective. After a drug molecule has entered the body, it is acted upon by several processes—absorption, distribution, metabolism, and excretion—that will determine the effective concentration of this molecule over time (Figure 28.6). A drug's response to these processes is referred to as its *ADME* (pronounced "add-me") properties.

Administration and absorption. Ideally, a drug can be taken orally as a small tablet. An orally administered active compound must be able to survive the acidic conditions in the gut and then be absorbed through the intestinal epithelium. Thus, the compound must be able to pass through cell membranes at a significant rate. Larger molecules such as proteins cannot be administered orally, because they often cannot survive the acidic conditions in the stomach and, if they do, they are not readily absorbed. Even many small molecules are not absorbed well; they may be too polar and unable to pass through cell membranes easily, for example. The ability to be absorbed

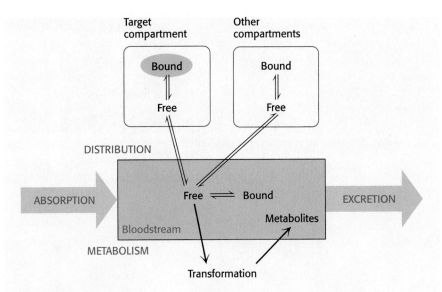

FIGURE 28.6 Absorption, distribution, metabolism, and excretion (ADME). The concentration of a compound at its target site (yellow) is affected by the extents and rates of absorption, distribution, metabolism, and excretion.

is often quantified in terms of the *oral bioavailability*. This quantity is defined as the ratio of the peak concentration of a compound given orally to the peak concentration of the same dose injected directly into the bloodstream. Bioavailability can vary considerably from species to species, and so results from animal studies may be difficult to apply to human beings. Despite this variability, some useful generalizations have been made. One powerful set consists of *Lipinski's rules*.

Lipinski's rules tell us that poor absorption is likely when

1. the molecular weight is greater than 500 daltons.
2. the number of hydrogen-bond donors is greater than 5.
3. the number of hydrogen-bond acceptors is greater than 10.
4. the partition coefficient [measured as $\log(P)$] is greater than 5.

The *partition coefficient* is a way to measure the tendency of a molecule to dissolve in membranes, which correlates with its ability to dissolve in organic solvents. It is determined by allowing a compound to equilibrate between water and an organic phase, *n*-octanol. The $\log(P)$ value is defined as $\log_{10}$ of the ratio of the concentration of a compound in *n*-octanol to the concentration of the compound in water:

$$\log(P) = \log\left(\frac{[\text{compound}]_{\text{n-octanol}}}{[\text{compound}]_{\text{water}}}\right)$$

For example, if the concentration of the compound in the *n*-octanol phase is 100 times that in the aqueous phase, then $\log(P)$ is 2. Although the ability of a drug to partition in organic solvents is ideal, as it implies that the compound can penetrate membranes, a $\log(P)$ value that is too high suggests that the molecule may be poorly soluble in an aqueous environment.

Morphine, for example, satisfies all of Lipinski's rules and has moderate bioavailability (Figure 28.7). A drug that violates one or more of these rules may still have satisfactory bioavailability. Nonetheless, these rules serve as guiding principles for evaluating new drug candidates.

Distribution. Compounds taken up by intestinal epithelial cells can pass into the bloodstream. However, hydrophobic compounds and many others do not freely dissolve in the bloodstream. These compounds bind to proteins, such as albumin (Figure 28.8), that are abundant in the

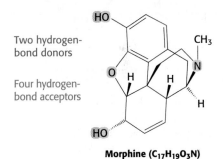

Two hydrogen-bond donors

Four hydrogen-bond acceptors

Morphine ($C_{17}H_{19}O_3N$)

Molecular weight = 285

$\log(P) = 1.27$

FIGURE 28.7 Lipinski's rules applied to morphine. Morphine satisfies all of Lipinski's rules and has an oral bioavailability in human beings of 33%.

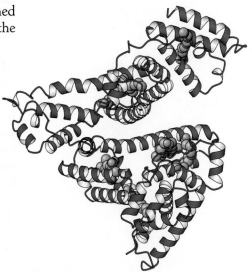

FIGURE 28.8 Structure of the drug carrier human serum albumin. Seven hydrophobic molecules (in red) are shown bound to a single molecule of serum albumin. [Drawn from 1BKE.pdb.]

FIGURE 28.9 Distribution of the drug fluconazole. After having been taken in, compounds distribute themselves to various organs within the body. The distribution of the antifungal agent fluconazole has been monitored through the use of positron emission tomography (PET) scanning. These images were taken of a healthy human volunteer 90 minutes after injection of a dose of 5 mg kg^{-1} of fluconazole containing trace amounts of fluconazole labeled with the positron-emitting isotope ^{18}F. [Reprinted by permission from Macmillan Publishers Ltd: *Nature Reviews Drug Discovery*, Rudin, M., and Weissleder, R., "Molecular imaging in drug discovery and devlopment," 2: 2, p. 123–131, Copyright 2003.]

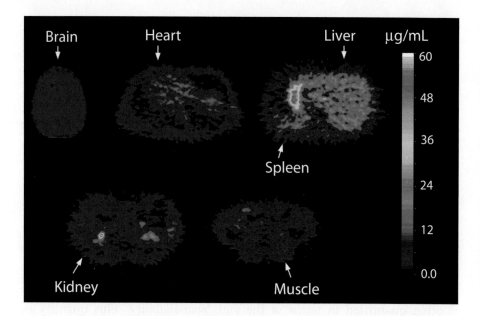

Fluconazole

blood serum. As albumin is present in the circulation at high concentrations (nearly 0.7 mM) and can bind a wide variety of compounds, it can dramatically affect the free concentration of drug in the bloodstream. Moreover, the binding capacity of albumin enables transport of the drug throughout the circulatory system.

When a compound has reached the bloodstream, it is distributed to different fluids and tissues, which are often referred to as *compartments*. Some compounds are highly concentrated in their target compartments, either by binding to the target molecules themselves or by other mechanisms. Other compounds are distributed more widely (Figure 28.9). An effective drug will reach the target compartment in sufficient quantity; the concentration of the compound in the target compartment is reduced whenever the compound is distributed into other compartments.

Some target compartments are particularly hard to reach. Many compounds are excluded from the central nervous system by the *blood–brain barrier*, the tight junctions between endothelial cells that line blood vessels within the brain and spinal cord.

Metabolism. A potential drug molecule must evade the body's defenses against foreign compounds. Many such compounds (called *xenobiotic compounds*) are released from the body in the urine or stool, often after having been metabolized to aid in excretion. This *drug metabolism* poses a considerable threat to drug effectiveness because the concentration of the desired compound decreases as it is metabolized. Thus, a rapidly metabolized compound must be administered more frequently or at higher doses.

Two of the most common pathways in xenobiotic metabolism are *oxidation* and *conjugation*. Oxidation reactions can aid excretion in at least two ways: by increasing water solubility, and thus ease of transport, and by introducing functional groups that participate in subsequent metabolic steps. These reactions are often promoted by cytochrome P450 enzymes in the liver (Section 26.4). The human genome encodes more than 50 different P450 isozymes, many of which participate in xenobiotic metabolism. A typical reaction catalyzed by a P450 isozyme is the hydroxylation of ibuprofen (Figure 28.10).

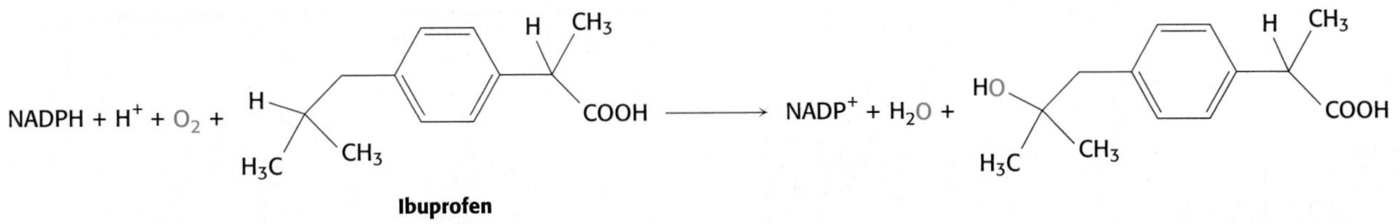

Ibuprofen

FIGURE 28.10 P450 conversion of ibuprofen. Cytochrome P450 isozymes, primarily in the liver, catalyze xenobiotic metabolic reactions such as hydroxylation. The reaction introduces an oxygen atom derived from molecular oxygen.

Conjugation is the addition of particular groups to the xenobiotic compound. Common groups added are glutathione (Section 20.5), glucuronic acid, and sulfate (Figure 28.11). These additions often increase water solubility and provide labels that can be recognized to target excretion. Examples of conjugation include the addition of glutathione to the anticancer drug cyclophosphamide, the addition of glucuronidate to the analgesic morphine, and the addition of a sulfate group to the hair-growth stimulator minoxidil.

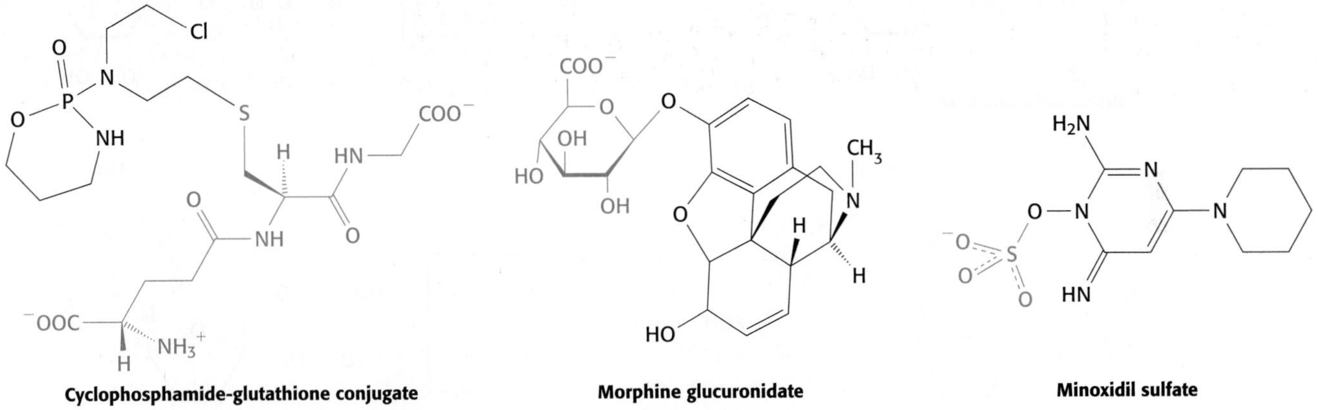

Cyclophosphamide-glutathione conjugate

Morphine glucuronidate

Minoxidil sulfate

Interestingly, the sulfation of minoxidil produces a compound that is more active in stimulating hair growth than is the unmodified compound. Consequently, the metabolic products of a drug, though usually less active than the drug, can sometimes be more active.

Note that an oxidation reaction often precedes conjugation because the oxidation reaction can generate modifications, such as the hydroxyl group, to which groups such as glucuronic acid can be attached. The oxidation reactions of xenobiotic compounds are often referred to as *phase I transformations*, and the conjugation reactions are referred to as *phase II transformations*. These reactions take place primarily in the liver. Because blood flows from the intestine directly to the liver through the portal vein, xenobiotic metabolism often alters drug compounds before they ever reach full circulation. This *first-pass metabolism* can substantially limit the availability of compounds taken orally.

Excretion. After compounds have entered the bloodstream, they can be removed from circulation and excreted from the body by two primary pathways. First, they can be absorbed through the kidneys and excreted in the urine. In this process, the blood passes through *glomeruli*, networks of fine capillaries in the kidney that act as filters. Compounds with molecular weights less than approximately 60,000 pass though the glomeruli. Many of the water molecules, glucose molecules, nucleotides, and other low-molecular-weight compounds that pass through the glomeruli are reabsorbed into the bloodstream, either by transporters that have broad specificities or by the passive transfer of hydrophobic molecules through membranes. Drugs and metabolites that pass through the first filtration step and are not reabsorbed are excreted.

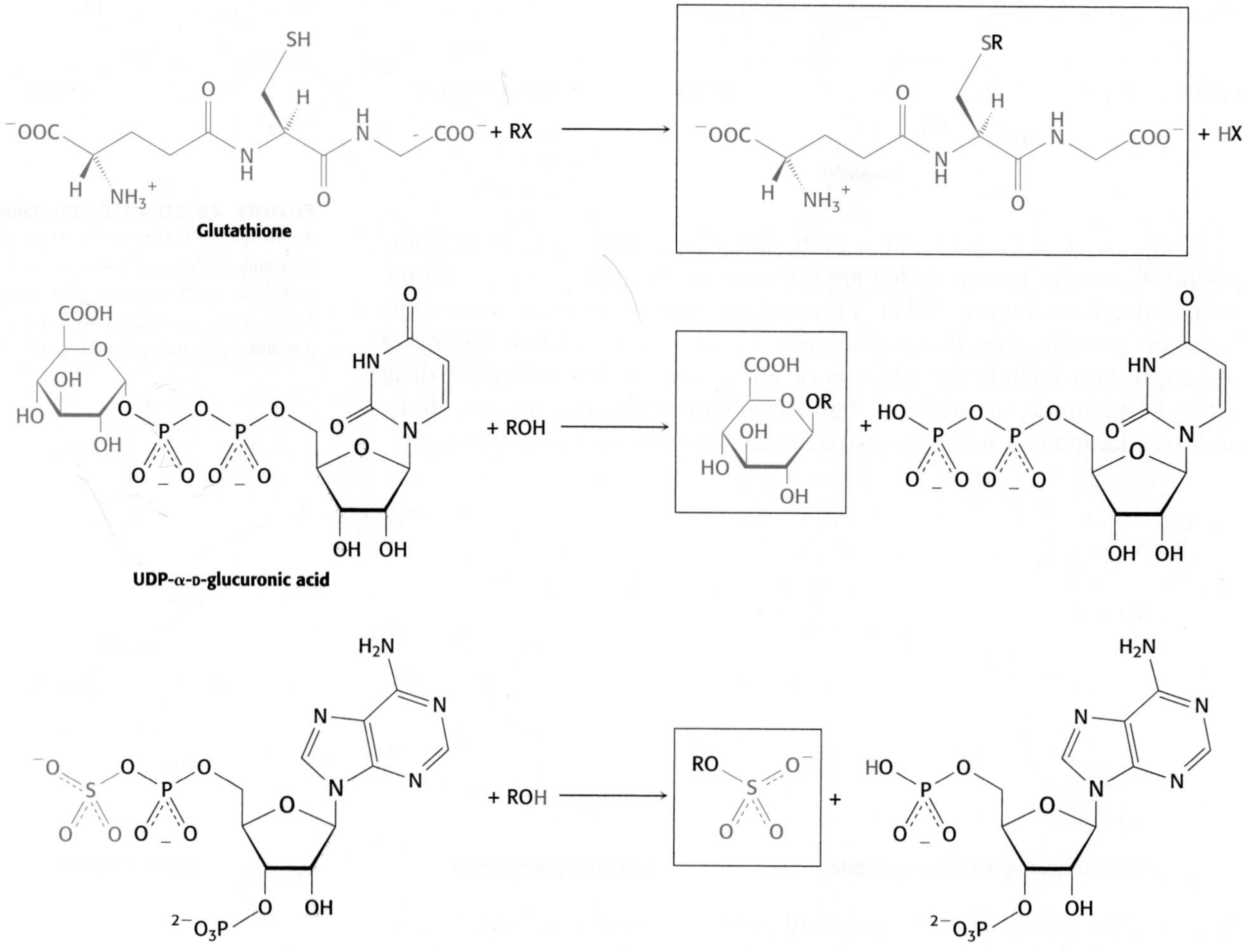

FIGURE 28.11 Conjugation reactions. Compounds that have appropriate groups are often modified by conjugation reactions. Such reactions include the addition of glutathione (top), glucuronic acid (middle), or sulfate (bottom). The conjugated product is shown boxed.

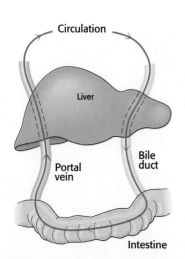

FIGURE 28.12 Enterohepatic cycling. Some drugs can move from the blood circulation to the liver, into the bile, into the intestine, to the liver, and back into circulation. This cycling decreases the rate of drug excretion.

Second, compounds can be actively transported into bile, a process that takes place in the liver. After concentration in the gallbladder, bile flows into the intestine. In the intestine, the drugs and metabolites can be excreted through the stool, reabsorbed into the bloodstream, or further degraded by digestive enzymes. Sometimes, compounds are recycled from the bloodstream into the intestine and back into the bloodstream, a process referred to as *enterohepatic cycling* (Figure 28.12). This process can significantly decrease the rate of excretion of some compounds because they escape from an excretory pathway and reenter the circulation.

The kinetics of compound excretion is often complex. In some cases, a fixed percentage of the remaining compound is excreted over a given period of time (Figure 28.13). This pattern of excretion results in exponential loss of the compound from the bloodstream that can be characterized by a half-life ($t_{1/2}$). The half-life is the fixed period of time required to eliminate 50% of the remaining compound. It is a measure of how long an effective concentration of the compound remains in the system after administration. As such, the half-life is a major factor in determining how often a drug must be taken. A drug with a long half-life might need to be taken only once per day, whereas a drug with a short half-life might need to be taken three or four times per day.

Toxicity can limit drug effectiveness

An effective drug must not be so toxic that it seriously harms the person who takes it. A drug may be toxic for any of several reasons. First, it may modulate the target molecule itself *too* effectively, referred to as *mechanism-based* or *on-target toxicity*. For example, the presence of too much of the anticoagulant drug coumadin can result in dangerous, uncontrolled bleeding and death. Second, the compound may modulate the properties of proteins that are distinct from the target molecule itself, referred to as *off-target toxicity*. Compounds that are directed to one member of a family of enzymes or receptors often bind to other family members. For example, an antiviral drug directed against viral proteases may be toxic if it also inhibits proteases normally present in the body such as those that regulate blood pressure.

A compound may also be toxic if it modulates the activity of a protein unrelated to its intended target. For example, many compounds block ion channels such as the potassium channel hERG (Section 13.4), causing potentially life-threatening disturbances of cardiac rhythm. To prevent cardiac side effects, many compounds are screened for their ability to block such channels.

Finally, even if a compound is not itself toxic, its metabolic by-products may be. Phase I metabolic processes can generate damaging reactive groups in products. An important example is liver toxicity observed with large doses of the common pain reliever acetaminophen (Figure 28.14). A particular cytochrome P450 isozyme oxidizes acetaminophen to *N*-acetyl-*p*-benzoquinone imine. The resulting compound is conjugated to glutathione. With large doses, however, the liver concentration of glutathione drops dramatically, and the liver is no longer able to protect itself from this reactive compound and others. Initial symptoms of excessive acetaminophen include nausea and vomiting. Within 24 to 48 hours, symptoms of liver failure may appear. Acetaminophen poisoning accounts for about 35% of cases of severe liver failure in the United States. A liver transplant is often the only effective treatment.

The toxicity of a drug candidate can be described in terms of the *therapeutic index*. This measure of toxicity is determined through animal tests, usually with mice or rats. The therapeutic index is defined as the ratio of the

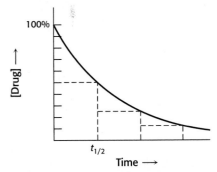

FIGURE 28.13 Half-life of drug excretion. In the case shown, the concentration of a drug in the bloodstream decreases to one-half of its value in a period of time, $t_{1/2}$, referred to as its half-life.

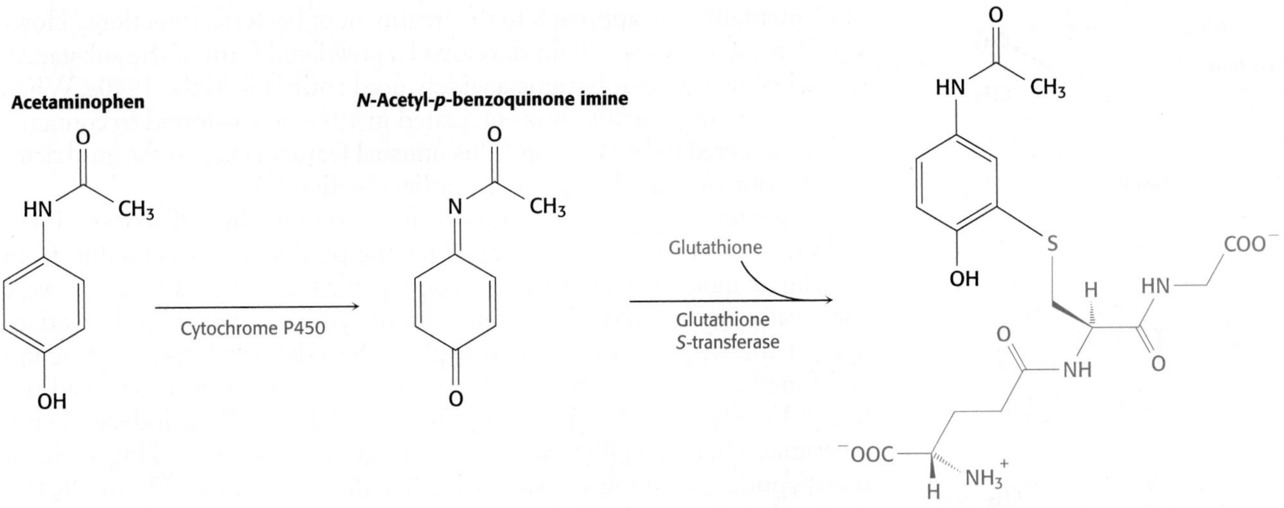

FIGURE 28.14 Acetaminophen toxicity. A minor metabolic product of acetaminophen is *N*-acetyl-*p*-benzoquinone imine. This metabolite is conjugated to glutathione. Large doses of acetaminophen can deplete liver glutathione stores.

dose of a compound that is required to kill one-half of the animals (referred to as the LD_{50} for "lethal dose") to a comparable measure of the effective dose, usually the EC_{50}. So, if the therapeutic index is 1000, then lethality is significant only when 1000 times the effective dose is administered. Analogous indices can provide measures of toxicity less severe than lethality.

Many compounds have favorable properties in vitro, yet fail when administered to a living organism because of difficulties with ADME and toxicity. Expensive and time-consuming animal studies are required to verify that a drug candidate is not toxic, yet differences between animal species in their response can confound decisions about moving forward with a compound toward human studies. One hope is that, with more understanding of the biochemistry of these processes, scientists can develop computer-based models to replace or augment animal tests. Such models would need to accurately predict the fate of a compound inside a living organism from its molecular structure or other properties that are easily measured in the laboratory without the use of animals.

28.2 Drug Candidates Can Be Discovered by Serendipity, Screening, or Design

Traditionally, many drugs were discovered by serendipity, or chance observation. More recently, drugs have been discovered by screening collections of natural products or large compound libraries for molecules that have desired medicinal properties. Alternatively, scientists have designed specific drug candidates by using their knowledge about a preselected molecular target. We will examine several examples of each of these pathways to reveal common principles.

Serendipitous observations can drive drug development

Perhaps the most well-known observation in the history of drug development is Alexander Fleming's chance observation in 1928 that colonies of the bacterium *Staphylococcus aureus* died when they were adjacent to colonies of the mold *Penicillium notatum*. Spores of the mold had landed accidentally on plates growing the bacteria. Fleming soon realized that the mold produced a substance that could kill disease-causing bacteria. This discovery led to a fundamentally new approach to the treatment of bacterial infections. Howard Florey and Ernest Chain developed a powdered form of the substance, termed penicillin, that became a widely used antibiotic in the 1940s. When the structure of penicillin was elucidated in 1945, it was found to contain a four-membered β-lactam ring. This unusual feature is key to the antibacterial function of penicillin, as noted earlier (Section 8.5).

Three steps were crucial to fully capitalize on Fleming's discovery. First, an industrial process was developed for the production of penicillin from *Penicillium* mold on a large scale. Second, penicillin and its derivatives were chemically synthesized. The availability of synthetic penicillin derivatives opened the way for scientists to explore the relations between structure and function. Many such penicillin derivatives remain in widespread use today. Finally, in 1965, Jack Strominger and James Park independently determined that penicillin exerts its antibiotic activity by blocking a critical transpeptidase reaction in bacterial cell-wall biosynthesis (Figure 28.15), as introduced in Section 8.5.

Many other drugs have been discovered by serendipitous observations. Chlorpromazine (Thorazine), a drug used to treat psychosis, was discovered

Penicillin

Chlorpromazine

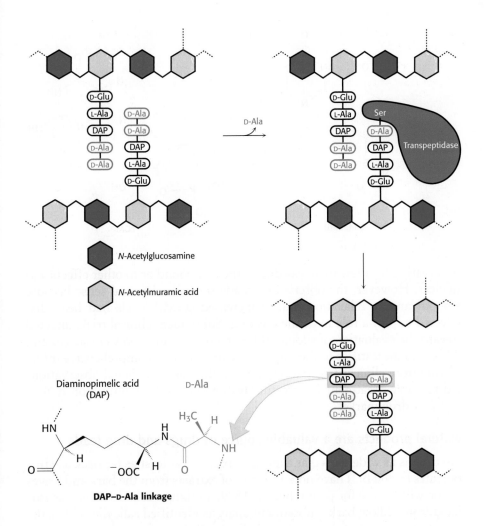

N-Acetylglucosamine

N-Acetylmuramic acid

Diaminopimelic acid
(DAP)

D-Ala

DAP–D-Ala linkage

FIGURE 28.15 Mechanism of cell-wall biosynthesis disrupted by penicillin. A transpeptidase enzyme catalyzes the formation of cross-links between peptidoglycan groups. In the case shown, the transpeptidase catalyzes the linkage of D-alanine at the end of one peptide chain to the amino acid diaminopimelic acid (DAP) on another peptide chain. The diaminopimelic acid linkage (bottom left) is found in Gram-negative bacteria such as *E. coli*. Linkages of glycine-rich peptides are found in Gram-positive bacteria. Penicillin inhibits the action of the transpeptidase, so bacteria exposed to the drug have weak cell walls that are susceptible to lysis.

in the course of investigations directed toward the treatment of shock in surgical patients. In 1952, French surgeon Henri Laborit noticed that, after taking the compound, his patients were remarkably calm. This observation suggested that chlorpromazine could benefit psychiatric patients, and, indeed, the drug has been used for many years to treat patients with schizophrenia and bipolar disorder. The drug does have significant side effects (including movement disorders, sedation, weight gain, and reduced blood pressure upon standing), and its use has been largely superseded by more recently developed drugs.

Chlorpromazine acts by binding to receptors for the neurotransmitter dopamine and blocking them. Dopamine D2 receptors are the targets of many other psychoactive drugs. In the search for drugs with more limited side effects, studies have been performed to correlate drug effects with biochemical parameters such as dissociation constants and binding and release rate constants.

A more recent example of a drug discovered by chance observation is sildenafil (Viagra). This compound was developed as an inhibitor of phosphodiesterase 5 (PDE5), an enzyme that catalyzes the hydrolysis of cGMP to GMP (Figure 28.16). The compound was intended as a treatment for hypertension and angina because cGMP plays a central role in the relaxation of smooth-muscle cells in blood vessels (Figure 28.17). The inhibition of PDE5 was expected to increase the concentration of cGMP by blocking the pathway for its degradation. In the course of early clinical trials in Wales, some men reported unusual penile erections. Whether this chance

Dopamine

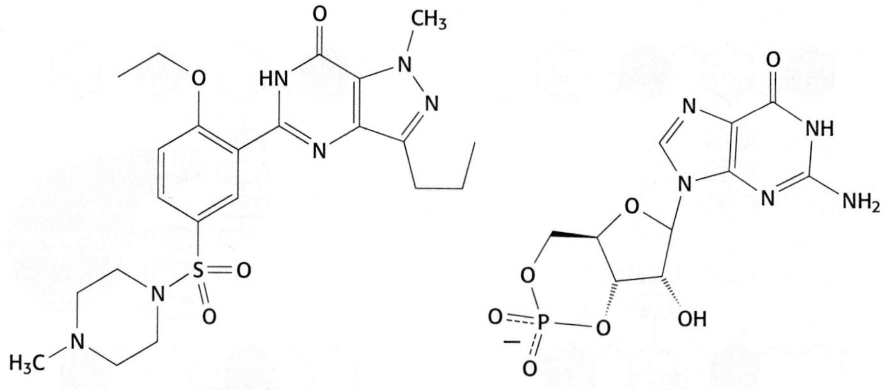

FIGURE 28.16 Sildenafil, a mimic of cGMP. Sildenafil was designed to resemble cGMP, the substrate of phosphodiesterase 5 (PDE5).

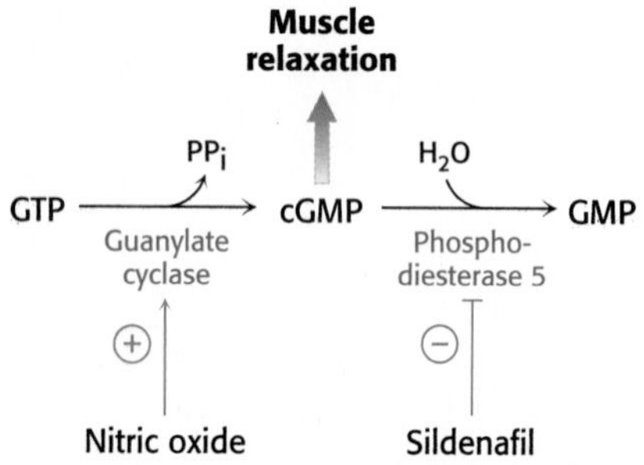

Sildenafil

cGMP

FIGURE 28.17 Muscle-relaxation pathway. Increases in nitric oxide levels stimulate guanylate cyclase, which produces cGMP. The increased cGMP concentration promotes smooth-muscle relaxation. PDE5 hydrolyzes cGMP, which lowers the cGMP concentration. The inhibition of PDE5 by sildenafil maintains elevated levels of cGMP.

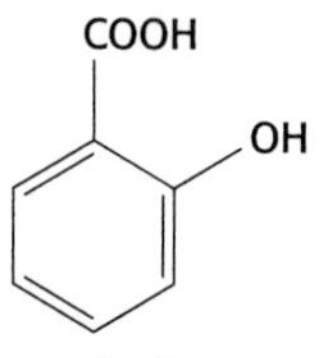

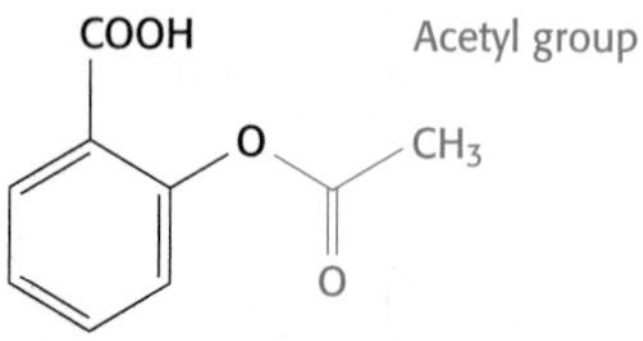

Salicylic acid

Aspirin (acetylsalicylic acid)

observation by a few men was due to the compound or to other effects was unclear. However, the observation made some biochemical sense because smooth-muscle relaxation due to increased cGMP levels had been discovered to play a role in penile erection. Subsequent clinical trials directed toward the evaluation of sildenafil for erectile dysfunction were successful. This account testifies to the importance of collecting comprehensive information from clinical-trial participants. In this case, incidental observations led to a new treatment for erectile dysfunction and a multibillion-dollar-per-year drug market.

Natural products are a valuable source of drugs and drug leads

No drug is as widely used as aspirin. Observers at least as far back as Hippocrates (~400 BC) have noted the use of extracts from the bark and leaves of the willow tree for pain relief. In 1829, a mixture called *salicin* was isolated from willow bark. Subsequent analysis identified salicylic acid as the active component of this mixture. Salicylic acid was formerly used to treat pain, but this compound often irritated the stomach. Several investigators attempted to find a means to neutralize salicylic acid. Felix Hoffmann, a chemist working at the German company Bayer, developed a less irritating derivative by treating salicylic acid with a base and acetyl chloride. This derivative, acetylsalicylic acid, was named *aspirin* from "a" for acetyl chloride, "spir" for *Spiraea ulmaria* (meadowsweet, a flowering plant that also contains salicylic acid), and "in" (a common ending for drugs). Each year, approximately 35,000 tons of aspirin are taken worldwide, nearly the weight of the *Titanic*.

As discussed in Chapter 12, the acetyl group in aspirin is transferred to the side chain of a serine residue that lies along the path to the active site of the cyclooxygenase component of prostaglandin H_2 synthase (Figure 12.24). In this position, the acetyl group blocks access to the active site. Thus, even though aspirin binds in the same pocket on the enzyme as salicylic acid, the acetyl group of aspirin dramatically increases its effectiveness as a drug. This account illustrates the value of screening extracts from plants and other materials that are believed to have medicinal properties for active compounds. The large number of herbal and folk medicines are a treasure trove of new drug leads.

Let us consider another example, which has also had a significant impact on the clinical management of individuals with cardiovascular disease. More than 100 years ago, a fatty, yellowish material was discovered on the arterial walls of patients who had died of vascular disease. The presence of the material was termed *atheroma* from the Greek word for porridge.

This material proved to be cholesterol. The Framingham Heart Study, initiated in 1948, documented a correlation between high blood-cholesterol levels and high mortality rates from heart disease. This observation led to the notion that blocking cholesterol synthesis might lower blood-cholesterol levels and, in turn, lower the risk of heart disease. Initial attempts at blocking cholesterol synthesis focused on steps near the end of the pathway. However, these efforts were abandoned because the accumulation of the insoluble substrate for the inhibited enzyme led to the development of cataracts and other side effects. Investigators eventually identified a more favorable target—namely, the enzyme HMG-CoA reductase (Section 26.2). This enzyme acts on a substrate, HMG-CoA (3-hydroxy-3-methylglutaryl coenzyme A), that does not accumulate, because it is water-soluble and can be utilized by other pathways.

A promising natural product, compactin, was discovered in a screen of compounds from a fermentation broth from *Penicillium citrinum* in a search for antibacterial agents. In some, but not all, animal studies, compactin was found to inhibit HMG-CoA reductase and to lower serum cholesterol levels. In 1982, a new HMG-CoA reductase inhibitor was discovered in a fermentation broth from *Aspergillus cereus*. This compound, now called lovastatin, was found to be structurally very similar to compactin, bearing one additional methyl group.

In clinical trials, lovastatin significantly reduced serum cholesterol levels with few side effects. Most side effects could be prevented by treatment with mevalonate (the product of HMG-CoA reductase), indicating that the side effects were mechanism-based. One notable side effect is muscle pain or weakness (termed *myopathy*), although its cause remains to be fully established. After many studies, the Food and Drug Administration (FDA) approved lovastatin for treating high serum cholesterol levels.

A structurally related HMG-CoA reductase inhibitor was later shown to cause a statistically significant decrease in deaths due to coronary heart disease. This result validated the benefits of lowering serum cholesterol levels. Further mechanistic analysis revealed that the HMG-CoA reductase inhibitor acts not only by lowering the rate of cholesterol biosynthesis, but also by inducing the expression of the low-density lipoprotein (LDL) receptor (Section 26.3). Cells with such receptors remove LDL particles from the bloodstream, and so these particles cannot contribute to atheroma.

Screening libraries of synthetic compounds expands the opportunity for identification of drug leads

Lovastatin and its relatives are either natural products or readily derived from natural products. After the discovery of these compounds, totally synthetic molecules were developed that are more-potent inhibitors of HMG-CoA reductase (Figure 28.18). These compounds were found to be effective at lower dose levels, reducing side effects. The original HMG-CoA reductase inhibitors or their precursors were found by screening libraries of natural products. More recently, drug developers have tried screening large libraries of both natural products and purely synthetic compounds prepared in the course of many drug-development programs. Under favorable circumstances, millions of compounds can be tested in this process, termed *high-throughput screening*. Compounds in these libraries can be synthesized one at a time for testing. An alternative approach is to synthesize a large number of structurally related compounds that differ from one another at only one or a few positions all at once. This approach is often termed *combinatorial chemistry*. Here, compounds are synthesized with the use of

Compactin

Lovastatin

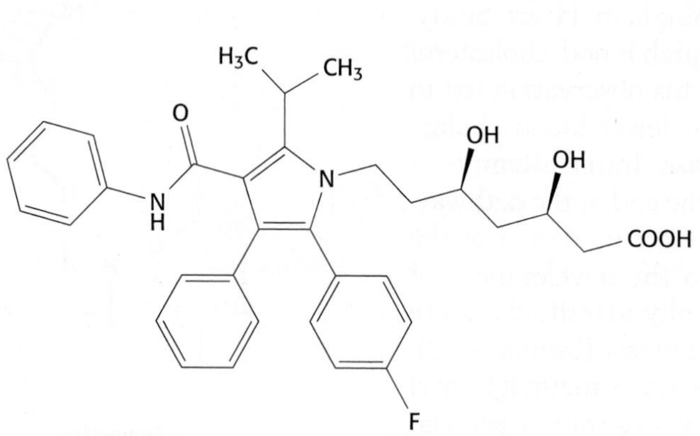

Atorvastatin

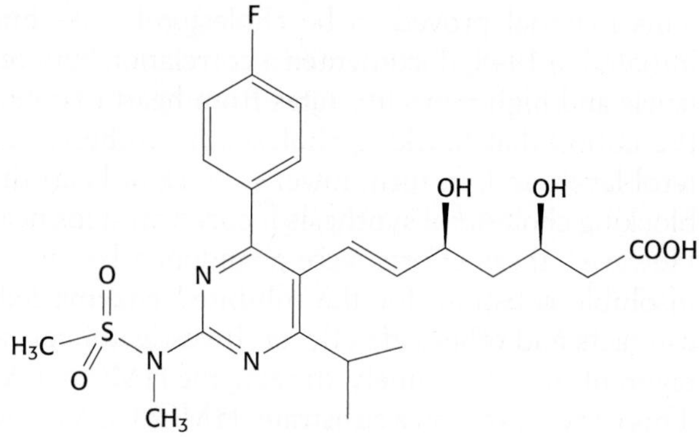

Rosuvastatin

FIGURE 28.18 Synthetic statins.
Atorvastatin (Lipitor) and rosuvastatin
(Crestor) are completely synthetic drugs
that inhibit HMG-CoA reductase.

the same chemical reactions but a variable set of reactants. Suppose that a molecular scaffold is constructed with two reactive sites and that 20 reactants can be used in the first site and 40 reactants can be used in the second site. A total of $20 \times 40 = 800$ possible compounds can be produced.

A key method in combinatorial chemistry is *split-pool synthesis* (Figure 28.19). The method depends on solid-phase synthetic methods, first

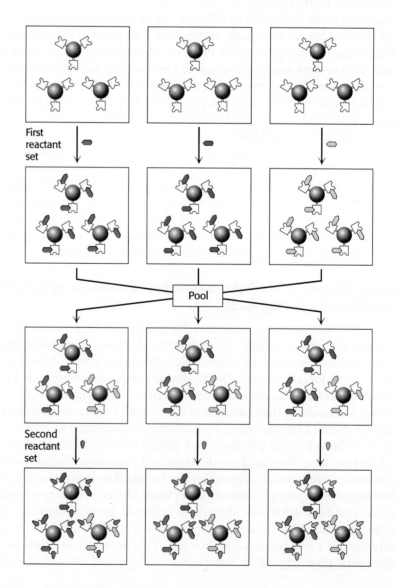

FIGURE 28.19 Split-pool synthesis.
Reactions are performed on beads. Each of
the reactions with the first set of reactants
is performed on a separate set of beads.
The beads are then pooled, mixed, and
split into sets. The second set of reactants
is then added. Many different compounds
will be produced, but all of the compounds
on a single bead will be identical.

developed for the synthesis of peptides (Section 3.5). Compounds are synthesized on small beads. Beads containing an appropriate starting *scaffold* are produced and divided (split) into *n* sets, with *n* corresponding to the number of building blocks to be used at one site. Reactions adding the reactants at the first site are run, and the beads are isolated by filtration. The *n* sets of beads are then combined (pooled), mixed, and split again into *m* sets, with *m* corresponding to the number of reactants to be used at the second site. Reactions adding these *m* reactants are run, and the beads are again isolated. The important result is that each bead contains only one compound, even though the entire library of beads contains many. Furthermore, although only *n* + *m* reactions were run, *n* × *m* compounds are produced. With the preceding values for *n* and *m*, 20 + 40 = 60 reactions produce 20 × 40 = 800 compounds. In some cases, assays can be performed directly with the compounds still attached to the bead to find compounds with desired properties (Figure 28.20). Alternatively, each bead can be isolated and the compound can be cleaved from the bead to produce free compounds for analysis. After an interesting compound has been identified, analytical methods of various types must be used to identify which of the *n* × *m* compounds are present.

Note that the "universe" of drug-like compounds is vast. More than an estimated 10^{40} compounds are possible with molecular weights less than 750. So, even with "large" libraries of millions of compounds, only a tiny fraction of the chemical possibilities is present for study.

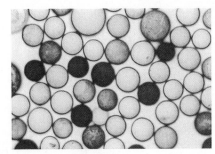

FIGURE 28.20 Screening a library of synthesized carbohydrates. A small combinatorial library of carbohydrates synthesized on the surface of 130-mm beads is screened for carbohydrates that are bound tightly by a lectin from peanuts. Beads that have such carbohydrates are darkly stained through the action of an enzyme linked to the lectin. [From Figure 3 in Liang et al., "Polyvalent binding to carbohydrates immobilized on an insoluble resin," *Proceedings of the National Academy of Sciences, USA*, vol. 94, pp. 10554–10559. Copyright 1997 National Academy of Sciences, USA.]

Drugs can be designed on the basis of three-dimensional structural information about their targets

Many drugs bind to their targets in a manner reminiscent of Emil Fischer's lock and key (Figure 8.8). In principle, we should be able to design a key, given enough knowledge about the shape and chemical composition of the lock. In the idealized case, we could design a small molecule that is complementary in shape and electronic structure to a target protein so that it binds effectively to the targeted site. Despite our ability to determine three-dimensional structures rapidly, the achievement of this goal remains elusive. Designing stable compounds *from scratch* that have the correct shape and properties to fit precisely into a binding site is difficult because predicting the structure that will best fit into a binding site is difficult. Prediction of binding affinity requires a detailed understanding of the interactions between a compound and its binding partner *and* of the interactions between the compound and the solvent when the compound is free in solution.

Nonetheless, *structure-based drug design* has proved to be a powerful tool in drug development. Among its most prominent successes has been the development of drugs that inhibit the protease from the human immunodeficiency virus (HIV; see Section 35.4 on SaplingPlus). Consider the development of the protease inhibitor indinavir (Crixivan; Figure 9.19). Two sets of promising inhibitors that had high potency but poor solubility and bioavailability were discovered. X-ray crystallographic analysis and molecular-modeling findings suggested that a hybrid of these two inhibitor sets might have both high potency and improved bioavailability (Figure 28.21). The synthesized hybrid compound did show improvements but required further optimization. The structural data suggested one point where modifications could be tolerated. A series of compounds were prepared and tested for their ability to inhibit the protease (Figure 28.22). These data are

FIGURE 28.21 Initial design of an HIV protease inhibitor. This compound was designed by combining part of one compound with good inhibition activity but poor solubility (shown in red) with part of another compound with better solubility (shown in blue).

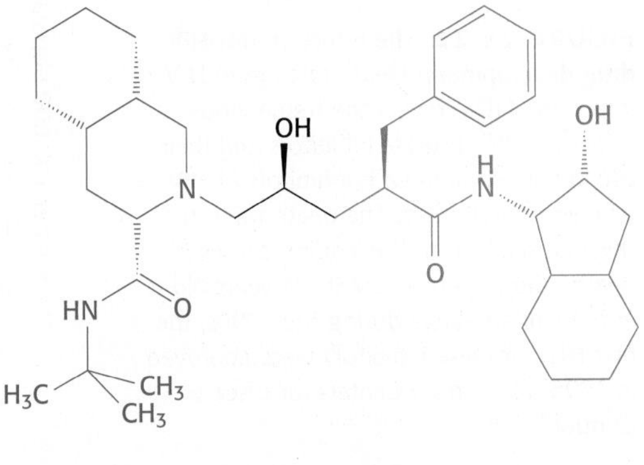

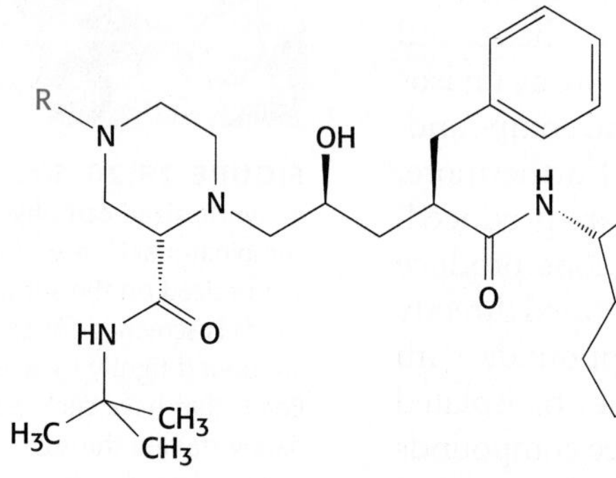

R =	IC$_{50}$ (nM)	log(P)	c_{max} (μM)
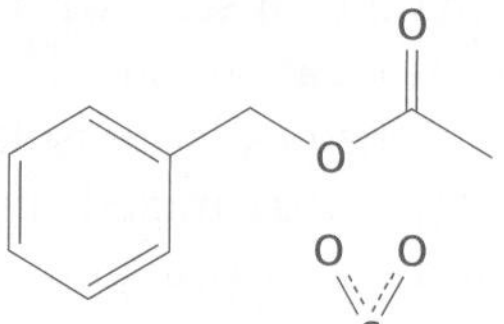	0.4	4.67	< 0.1
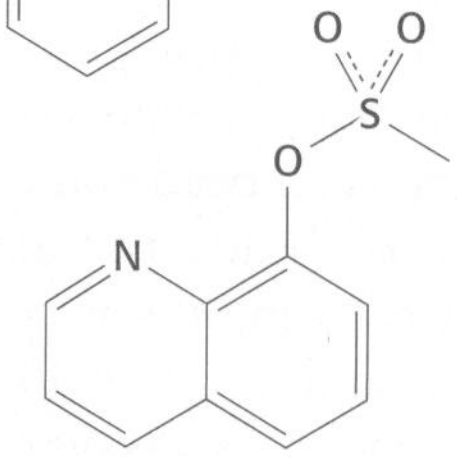	0.01	3.70	< 0.1
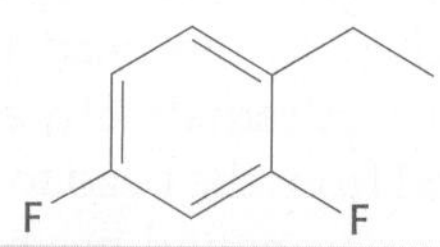	0.3	3.69	0.7
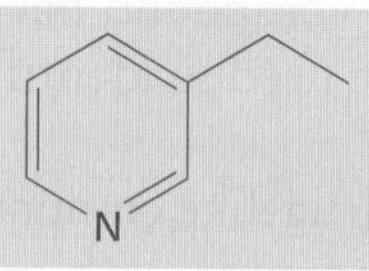	0.6	2.92	11

FIGURE 28.22 Compound optimization. Four compounds are evaluated for characteristics including the IC$_{50}$, log(P), and c_{max} (the maximal concentration of compound present) measured in the serum of dogs. The compound shown at the bottom (highlighted in yellow) has the weakest inhibitory power (measured by IC$_{50}$) but by far the best bioavailability (measured by c_{max}). This compound was selected for further development, leading to the drug indinavir (Crixivan).

a demonstration of a *structure–activity relationship (SAR)*; they provide an opportunity to correlate structure with function and guide design of further molecules. The most active compound showed poor bioavailability, but one of the other compounds (highlighted in yellow in Figure 28.22) showed good bioavailability and acceptable activity. The maximum serum concentration available through oral administration was significantly higher than the levels required to suppress replication of the virus. This drug, as well as other protease inhibitors developed at about the same time, has been used in combination with other drugs to treat AIDS with much more encouraging results than had been obtained previously (Figure 28.23).

Structural information can also be used to optimize the selectivity of drugs for their intended target. As we have discussed earlier, aspirin inhibits the cyclooxygenase activity of prostaglandin H$_2$ synthase. Animal studies suggested that mammals contain not one but two distinct cyclooxygenase enzymes, both of which are targeted by aspirin. The more recently discovered enzyme, cyclooxygenase 2 (COX2), is expressed primarily as part of the inflammatory response, whereas cyclooxygenase 1 (COX1) is expressed

FIGURE 28.23 The effect of anti-HIV drug development. Death rates from HIV infection (AIDS) reveal the tremendous effect of HIV protease inhibitors and their use in combination with inhibitors of HIV reverse transcriptase. The death rates in this graph are from the leading causes of death among persons 24 to 44 years old in the United States during the 1990s; the first HIV protease inhibitors were approved in 1995. [Data from Centers for Disease Control.]

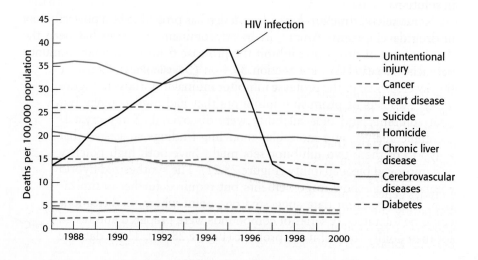

more ubiquitously. These observations suggested that a cyclooxygenase inhibitor that was specific for COX2 might be able to reduce inflammation in conditions such as arthritis without producing the gastric and other side effects associated with aspirin.

The amino acid sequences of COX1 and COX2 were deduced from cDNA cloning studies. These sequences are more than 60% identical, clearly indicating that the enzymes have the same overall structure. Nevertheless, there are some differences in the residues around the aspirin-binding site. X-ray crystallography revealed that an extension of the binding pocket was present in COX2, but absent in COX1. This structural difference suggested a strategy for constructing COX2-specific inhibitors—namely, to synthesize compounds that had a protuberance that would fit into the pocket in the COX2 enzyme. Such compounds were designed and synthesized and then further refined to produce effective drugs familiar as Celebrex and Vioxx (Figure 28.24). Vioxx (rofecoxib) was subsequently withdrawn from the market because some patients experienced adverse effects. Some of these effects may be due to mechanism-based toxicity. Thus, although the development of these drugs is a triumph for structure-based drug design, these outcomes highlight that the inhibition of important enzymes can lead to complex physiological responses.

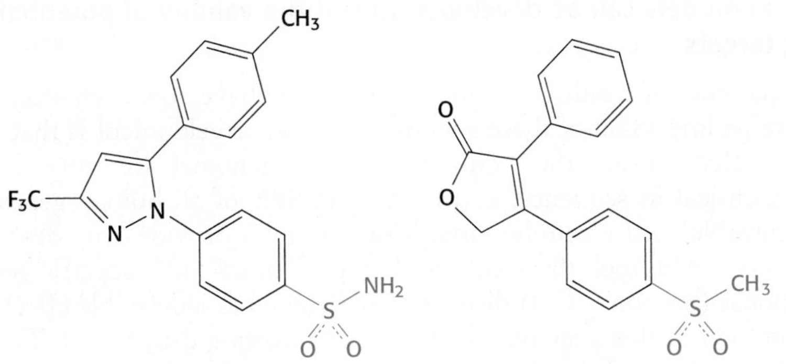

Celecoxib (Celebrex)   **Rofecoxib (Vioxx)**

FIGURE 28.24 COX2-specific inhibitors. These compounds have protuberances (shown in red) that fit into a pocket in the COX2 isozyme but sterically clash with the COX1 isozyme.

28.3 Genomic Analyses Can Aid Drug Discovery

The completion of the sequencing of human and other genomes is a potentially powerful driving force for the development of new drugs. Genomic sequencing and analysis projects have vastly increased our knowledge of the proteins encoded by the human genome. This new source of knowledge may greatly accelerate early stages of the drug-development process or even allow drugs to be tailored to the individual patient.

Potential targets can be identified in the human proteome

The human genome encodes approximately 21,000 proteins, not counting the variation produced by alternative mRNA splicing and posttranslational modifications. Many of these proteins are potential drug targets, in particular those that are enzymes or receptors and have significant biological effects when activated or inhibited. Several large protein families are particularly rich sources of targets. For example, the human genome includes genes for more than 500 protein kinases that can be recognized by comparing the deduced amino acid sequences. Many of these kinases are known to play a role in the progression of a variety of diseases. For example, Bcr-Abl kinase, a dysregulated kinase formed from a specific chromosomal

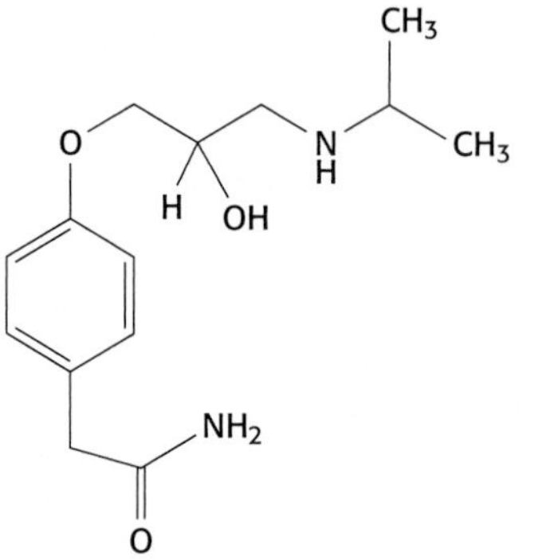

Atenolol

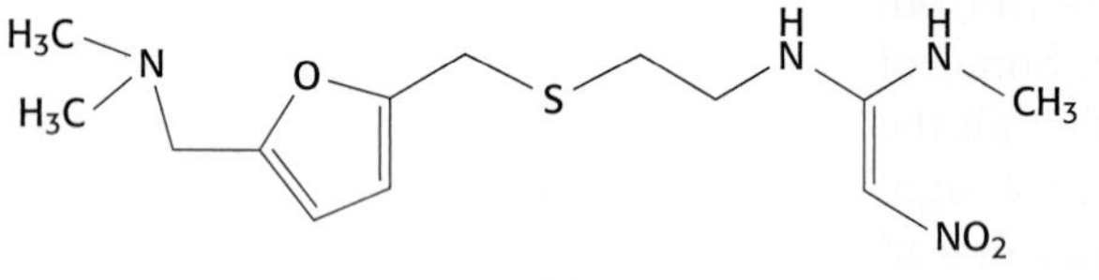

Ranitidine

defect, is known to contribute to certain leukemias and is the target of the drug imatinib mesylate (Gleevec; Section 14.5). Some of the other protein kinases undoubtedly play central roles in particular cancers as well. Similarly, the human genome encodes approximately 800 7TM receptors (Section 14.1), of which approximately 350 are odorant receptors. Many of the remaining 7TM receptors are known or potential drug targets. For example, the beta-blockers, which are widely used to treat hypertension, target the β_1-adrenergic receptor, and the antiulcer medication ranitidine (Zantac) targets the histamine H_2 receptor (a 7TM receptor that participates in the control of gastric acid secretion).

Novel proteins that are not part of large families already supplying drug targets can be more readily identified through the use of genomic information. There are a number of ways to identify proteins that could serve as targets of drug-development programs. One way is to look for changes in expression patterns, protein localization, or posttranslational modifications in cells from disease-afflicted organisms. Another is to perform studies of tissues or cell types in which particular genes are expressed. Such analyses of the human genome should increase the number of actively pursued drug targets.

Animal models can be developed to test the validity of potential drug targets

The genomes of a number of model organisms have now been sequenced. The most important of these genomes for drug development is that of the mouse. Remarkably, the mouse and human genomes are approximately 85% identical in sequence, and more than 98% of all human genes have recognizable mouse counterparts. Mouse studies provide drug developers with a powerful tool: the ability to disrupt ("knock out") specific genes in the mouse (Section 5.4). If disruption of a gene has a desirable effect, then the product of this gene may represent a promising drug target. The utility of this approach has been demonstrated retrospectively. For example, disruption of the gene for the α subunit of the $H^+–K^+$ ATPase, the key protein for secreting acid into the stomach, produces mice with a gastric pH of 6.9. Under similar conditions, their wild-type counterparts produce a gastric pH of 3.2. This protein is the target of the drugs omeprazole (Prilosec) and lansoprazole (Prevacid and Takepron), used for treating gastroesophageal reflux disease.

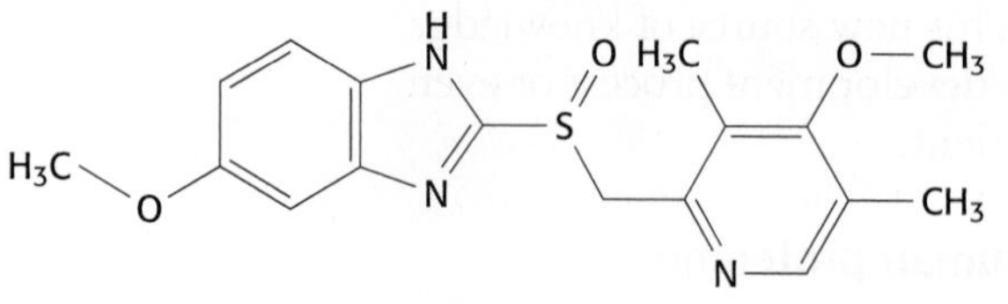

Omeprazole

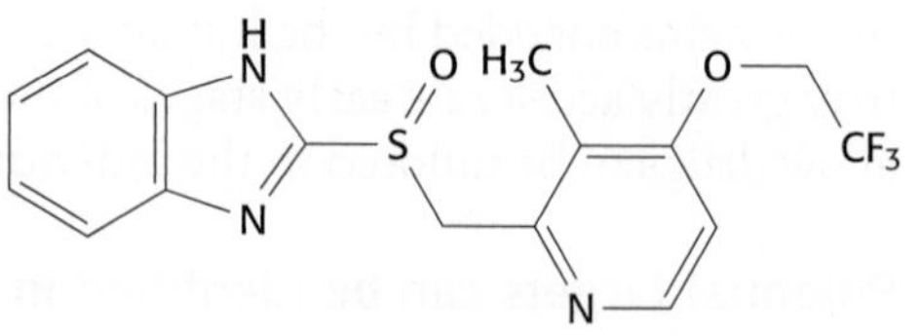

Lansoprazole

Several large-scale efforts are underway to generate thousands of mouse strains, each having a different gene disrupted. The phenotypes of these mice are a good indication of whether the protein encoded by a disrupted gene is a promising drug target. However, these data should be interpreted carefully. For instance, some enzymes possess functions beyond their catalytic activity, such as providing important protein–protein interactions. A knockout of the entire gene would remove all such functions, while a drug acting as an enzyme inhibitor would only impact the catalytic function.

Nevertheless, this approach allows drug developers to evaluate potential targets without any preconceived notions regarding physiological function.

Potential targets can be identified in the genomes of pathogens

Human proteins are not the only important drug targets. Drugs such as penicillin and HIV protease inhibitors act by targeting proteins within a pathogen. The genomes of hundreds of pathogens have now been sequenced, and these genome sequences can be mined for potential targets.

New antibiotics are needed to combat bacteria that are resistant to many existing antibiotics. One approach seeks proteins essential for cell survival that are conserved in a wide range of bacteria. Drugs that inactivate such proteins are expected to be broad-spectrum antibiotics, useful for treating infections from any of a range of different bacteria. One such protein is peptide deformylase, the enzyme that removes formyl groups present at the amino termini of bacterial proteins immediately after translation (Figure 31.19).

Alternatively, a drug may be needed against a specific pathogen. An example of such a pathogen is the strain of coronavirus responsible for severe acute respiratory syndrome (SARS). From late 2002 through early 2003, nearly 8500 individuals contracted SARS. While the initial outbreak occurred in the Guangdong Province of China, it ultimately claimed nearly 800 fatalities worldwide. Within one month of the recognition of this disease, investigators had isolated the virus that causes the syndrome, and, within weeks, completely sequenced its 29,751-base genome. This sequence revealed the presence of a gene encoding a viral protease, Mpro, known to be essential for viral replication from studies of other members of the coronavirus family to which the SARS virus belongs. This structure has opened the possibility for specific antiviral treatments for this virus and related viruses (Figure 28.25).

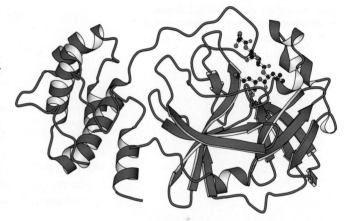

FIGURE 28.25 Emerging drug target. The structure of a protease, Mpro, from the coronavirus that causes SARS (severe acute respiratory syndrome) is shown bound to an inhibitor. This structure was determined less than a year after the identification of the virus. [Drawn from 1P9S.pdb.]

Genetic differences influence individual responses to drugs

Many drugs are not effective in everyone, often because of genetic differences among individuals. Nonresponding individuals may have slight differences in either a drug's target or the proteins taking part in drug transport and metabolism. The goal of the emerging fields of pharmacogenetics and pharmacogenomics is to design drugs that either act more consistently from person to person or are tailored to individuals with particular genotypes.

Drugs such as metoprolol that target the β_1-adrenergic receptor are widely used treatments for hypertension. These drugs are often referred to as *beta-blockers*.

Metoprolol

However, some individuals do not respond well to beta-blockers. Two variants of the gene coding for the β_1-adrenergic receptor are common in the U.S. population. The most common allele has serine in position 49 and arginine in position 389. In some persons, however, glycine replaces one or

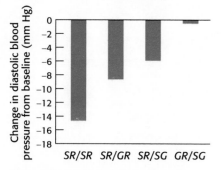

FIGURE 28.26 Phenotype–genotype correlation. Average changes in diastolic blood pressure on treatment with metoprolol. Persons with two copies of the most common allele ($S_{49}R_{389}$) showed significant decreases in blood pressure. Those with one variant allele (GR or SG) showed more modest decreases, and those with two variant alleles (GR/SG) showed no decrease. [Data from J. A. Johnson et al., *Clin. Pharmacol. Ther.* 74:p. 44–52, 2003.]

the other of these residues. In clinical studies, participants with two copies of the most common allele responded well to metoprolol: their daytime diastolic blood pressure was reduced by 14.7 ± 2.9 mm Hg on average. In contrast, participants with one variant allele showed a smaller reduction in blood pressure, and the drug had no significant effect on participants with two variant alleles (Figure 28.26). These observations suggest the potential utility of genotyping individual patients at these positions. One could then predict whether treatment with metoprolol or other beta-blockers is likely to be effective.

Given the importance of ADME and toxicity properties in determining drug efficacy, it is not surprising that variations in proteins participating in drug transport and metabolism can alter a drug's effectiveness. An important example is the use of thiopurine drugs such as 6-thioguanine, 6-mercaptopurine, and azothioprine to treat diseases including leukemia, immune disorders, and inflammatory bowel disease.

6-Thioguanine **6-Mercaptopurine** **Azathioprine**

A minority of patients who are treated with these drugs show signs of toxicity at doses that are well tolerated by most patients. These differences between patients are due to rare variations in the gene encoding the xenobiotic-metabolizing enzyme thiopurine methyltransferase, which adds a methyl group to sulfur atoms.

6-Mercaptopurine + *S*-adenosylmethionine $\rightleftharpoons$ [Thiopurine methyltransferase] + *S*-adenosylhomocysteine + H^+

The variant enzyme is less stable. Patients with these variant enzymes can build up toxic levels of the drugs if appropriate care is not taken. Thus, genetic variability in an enzyme participating in drug metabolism plays a large role in determining the variation in the tolerance of different persons to particular drug levels. Many other drug-metabolism enzymes and drug-transport proteins have been implicated in controlling individual reactions to specific drugs. The identification of the genetic factors will allow a deeper understanding of why some drugs work well in some individuals but poorly in others, and may enable the tailoring of a patient's drug therapy based on his or her genotype.

28.4 The Clinical Development of Drugs Proceeds Through Several Phases

In the United States, the FDA requires demonstration that drug candidates be effective and safe before they may be used in human beings on a large scale. This requirement is particularly true for drug candidates that are to be taken by people who are relatively healthy. More side effects are acceptable for drug candidates intended to treat significantly ill patients such as those with serious forms of cancer, where there are clear, unfavorable consequences for not having an effective treatment.

Clinical trials are time-consuming and expensive

Clinical trials test the effectiveness and potential side effects of a candidate drug before it is approved by the FDA for general use. These trials proceed in at least three phases (Figure 28.27). In Phase I, a small number (typically from 10 to 100) of usually healthy volunteers take the drug for an initial assessment of safety. These volunteers are given a range of doses and are monitored for signs of toxicity. The efficacy of the drug candidate is not specifically evaluated.

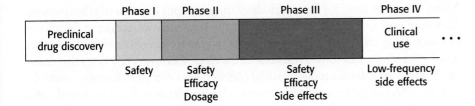

FIGURE 28.27 Clinical trial phases. Clinical trials proceed in phases examining safety and efficacy in increasingly large groups.

In Phase II, the efficacy of the drug candidate is tested in a small number of persons who might benefit from the drug. Further data regarding the drug's safety are obtained. Such trials are often *controlled* and *double-blinded*. In a controlled study, subjects are divided randomly into two groups. Subjects in the treatment group are given the treatment under investigation. Subjects in the control group are given either a *placebo*—that is, a treatment such as sugar pills known to not have intrinsic value—or the best standard treatment available, if withholding treatment altogether would be unethical. In a double-blinded study, neither the subjects nor the researchers know which subjects are in the treatment group and which are in the control group. A double-blinded study prevents bias in the course of the trial. When the trial has been completed, the assignments of the subjects into treatment and control groups are unsealed and the results for the two groups are compared. A variety of doses are often investigated in Phase II trials to determine which doses appear to be free of serious side effects and which doses appear to be effective.

The power of the *placebo effect*—that is, the tendency to perceive improvement in a subject who believes that he or she is receiving a potentially beneficial treatment—should not be underestimated. In a study of arthroscopic surgical treatment for knee pain, for example, subjects who were led to believe that they had received surgery through the use of videotapes and other means showed the same level of improvement, on average, as subjects who actually received the procedure.

In Phase III, similar studies are performed on a larger and more diverse population. This phase is intended to more firmly establish the efficacy of the drug candidate and to detect side effects that may develop in a small percentage of the subjects who receive treatment. Thousands of subjects may participate in a typical Phase III study.

Clinical trials can be extremely costly and time-consuming. Hundreds or thousands of patients must be recruited and monitored for the duration of the trial. Many physicians, nurses, clinical pharmacologists, statisticians, and others participate in the design and execution of the trial. Costs can run from tens of millions to hundreds of millions of dollars. Extensive records must be kept, including documentation of any adverse reactions. These data are compiled and submitted to the FDA. Furthermore, many drugs fail to show efficacy in these larger studies where the patient population is more diverse. The full cost of developing a drug is currently estimated to be more than $800 million.

Even after a drug has been approved and is in use, difficulties can arise. Clinical trials run after a drug has entered the market, referred to as Phase IV or *postmarketing surveillance* studies, are designed to identify low-frequency side effects that may only emerge after widespread or long-term use. As mentioned earlier, rofecoxib (Vioxx) was withdrawn from the market after significant cardiac side effects were detected in Phase IV clinical trials. Such events highlight the necessity for users of any drug to balance beneficial effects against potential risks.

The evolution of drug resistance can limit the utility of drugs for infectious agents and cancer

Many drugs are used for long periods of time without any loss of effectiveness. However, in some cases, particularly for the treatment of cancer or infectious diseases, drug treatments that were initially effective become less so. In other words, the disease becomes resistant to the drug therapy. Why does this resistance develop? Infectious diseases and cancer have a common feature—namely, that an affected person contains many cells (or viruses) that can mutate and reproduce. These conditions are necessary for evolution to take place. Thus, an individual microorganism or cancer cell may, by chance, have a genetic variation that makes it more suitable for growth and reproduction in the presence of the drug. These microorganisms or cells are more fit than others in their population, and they will tend to take over the population. As the selective pressure due to the drug is continually applied, the population of microorganisms or cancer cells will tend to become more and more resistant to the presence of the drug. Note that resistance can develop by a number of mechanisms.

The HIV protease inhibitors discussed earlier provide an important example of the evolution of drug resistance. Retroviruses are very well suited to this sort of evolution because reverse transcriptase carries out replication without a proofreading mechanism. In a genome of approximately 9750 bases, each possible single point mutation is estimated to appear in a virus particle more than 1000 times per day in each infected person. Many multiple mutations also occur. Most of these mutations either have no effect or are detrimental to the virus. However, a few of the mutant virus particles encode proteases that are less susceptible to inhibition by the drug. In the presence of an HIV protease inhibitor, these virus particles will tend to replicate more effectively than does the population at large. With the passage of time, the less susceptible viruses will come to dominate the population and the virus population will become resistant to the drug.

Pathogens may become resistant to antibiotics by completely different mechanisms. Some pathogens contain enzymes that inactivate or degrade specific antibiotics. For example, many bacteria are resistant to β-lactams such as penicillin because they contain β-lactamase enzymes. These enzymes hydrolyze the β-lactam ring and render the drugs inactive.

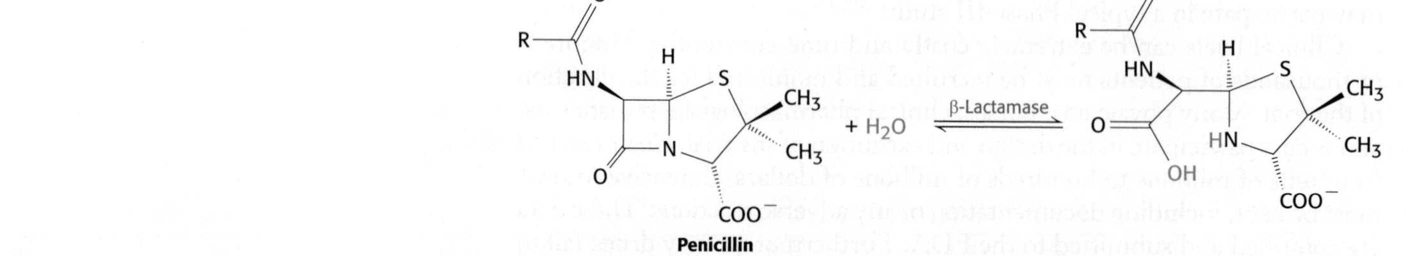

Penicillin

Many of these enzymes are encoded in plasmids, small circular pieces of DNA often carried by bacteria. Many plasmids are readily transferred from one bacterial cell to another, transmitting the capability for antibiotic resistance. Plasmid transfer thus contributes to the spread of antibiotic resistance, a major health-care challenge. On the other hand, plasmids have been harnessed for use in recombinant DNA methods (Section 5.2).

Drug resistance commonly emerges in the course of cancer treatment. Cancer cells are characterized by their ability to grow rapidly without the constraints that apply to normal cells. Many drugs used for cancer chemotherapy inhibit processes that are necessary for this rapid cell growth. However, individual cancer cells may accumulate genetic changes that mitigate the effects of such drugs. These altered cancer cells will tend to grow more rapidly than others and will become dominant within the cancer-cell population. This ability of cancer cells to mutate quickly has been a challenge to one of the major breakthroughs in cancer treatment: the development of inhibitors for proteins specific to cancer cells present in certain leukemias (Section 14.5). For example, tumors become undetectable in patients treated with imatinib mesylate, which is directed against the Bcr-Abl protein kinase. Unfortunately, the tumors of many of the patients treated with imatinib mesylate recur after a period of years. In many of these cases, mutations have altered the Bcr-Abl protein so that it is no longer inhibited by the concentrations of imatinib mesylate used in therapy.

Cancer patients often take multiple drugs concurrently in the course of chemotherapy and, in many cases, cancer cells become simultaneously resistant to many or all of them. This multiple-drug resistance can be due to the proliferation of cancer cells that overexpress a number of ABC transporter proteins that pump drugs out of the cell (Section 13.2). Thus, cancer cells can evolve drug resistance by overexpressing normal human proteins or by modifying proteins responsible for the cancer phenotype.

SUMMARY

28.1 Compounds Must Meet Stringent Criteria to Be Developed into Drugs

Most drugs act by binding to enzymes or receptors and modulating their activities. To be effective, drugs must bind to these targets with high affinity and specificity. However, compounds with the desired affinity and specificity do not necessarily make suitable drugs. Most compounds are poorly absorbed or rapidly excreted from the body or they are modified by metabolic pathways that target foreign compounds. Consequently, when taken orally, these compounds do not reach their targets at appropriate concentrations for a sufficient period of time. A drug's properties related to its absorption, distribution, metabolism, and excretion are called ADME properties. Oral bioavailability is a measure of a drug's ability to be absorbed; it is the ratio of the peak concentration of a compound given orally to the peak concentration of the same dose directly injected. The structure of a compound can affect its bioavailability in complicated ways, but generalizations called Lipinski's rules provide useful guidelines. Drug-metabolism pathways include oxidation by cytochrome P450 enzymes (phase I metabolism) and conjugation to glutathione, glucuronic acid, and sulfate (phase II metabolism). A compound may also not be a useful drug because it is toxic; it may either modulate the target molecule too effectively (mechanism-based toxicity) or bind to proteins other than the target (off-target toxicity). The liver and kidneys play central roles in drug metabolism and excretion.

28.2 Drug Candidates Can Be Discovered by Serendipity, Screening, or Design

Many drugs have been discovered by serendipity—that is, by chance observation. The antibiotic penicillin is produced by a mold that accidentally contaminated a culture dish, killing nearby bacteria. Drugs such as chlorpromazine and sildenafil were discovered to have beneficial, but unexpected, effects on human physiology. The cholesterol-lowering statin drugs were developed after large collections of compounds were screened for potentially interesting activities. Combinatorial chemistry methods have been developed to generate large collections of chemically related yet diverse compounds for screening. In some cases, the three-dimensional structure of a drug target is available and can be used to aid the design of potent and specific inhibitors. Examples of drugs designed in this manner are the HIV protease inhibitor indinavir and cyclooxygenase 2 inhibitors such as celecoxib.

28.3 Genomic Analyses Can Aid Drug Discovery

The human genome encodes approximately 21,000 proteins, and many more if derivatives due to alternative mRNA splicing and post-translational modification are included. The genome sequences can be examined for potential drug targets. Large families of proteins known to participate in key physiological processes such as the protein kinases and 7TM receptors have each yielded several targets for which drugs have been developed. The genomes of model organisms also are useful for drug-development studies. Strains of mice with particular genes disrupted have been useful in validating certain drug targets. The genomes of bacteria, viruses, and parasites encode many potential drug targets that can be exploited owing to their important functions and their differences from human proteins, minimizing the potential for side effects. Genetic differences between individuals can be examined and correlated with differences in responses to drugs, potentially aiding both clinical treatments and drug development.

28.4 The Clinical Development of Drugs Proceeds Through Several Phases

Before compounds can be given to human beings as drugs, they must be extensively tested for safety and efficacy. Clinical trials are performed in stages: first testing safety, then safety and efficacy in a small population, and finally safety and efficacy in a larger population to detect rarer adverse effects. Largely due to the expenses associated with clinical trials, the cost of developing a new drug has been estimated to be more than $800 million. Even when a drug has been approved for use, complications can arise. With regard to infectious diseases and cancer, patients often develop resistance to a drug after it has been administered for a period of time because variants of the disease agent that are less susceptible to the drug arise and replicate, even when the drug is present.

Biochemistry in Focus

Monoclonal antibodies: Expanding the drug developer's toolbox

In this chapter, we have primarily focused on small molecules as drugs. In most cases, these molecules are less than 500 Da in molecular weight (as described by Lipinski's rules), may be orally bioavailable, and bind to specific structural pockets on their targets. In contrast, proteins have emerged as another prominent and rapidly growing class of drugs. These proteins, referred to as *biologics*, present several advantages when compared with small molecules.

The most prominent examples of protein therapeutics are monoclonal antibodies. In Section 3.2, we discussed the hybridoma method for the preparation of monoclonal antibodies that bind tightly to specific ligands. As we shall discuss in Chapter 35 (on ⊘ SaplingPlus), antibody-producing cells have evolved remarkable mechanisms for generating a broad range of recognition diversity. Scientists can take advantage of nature's own combinatorial chemistry to generate numerous distinct antibodies that can bind to a specific ligand. As drugs, monoclonal antibodies have unique advantages over small molecules. For example, they can recognize a broader surface of the target, not limited to the smaller pockets typically bound by small molecule drugs. This larger interaction surface also enables the development of drugs with very high potency. Furthermore, the Fc region of monoclonal antibodies binds to an endogenous salvage pathway receptor, the *neonatal Fc receptor* (FcRn), permitting the drug to persist in an active form for very long half-lives, on the order of weeks. Hence, monoclonal antibodies can be dosed much less frequently (e.g., weekly to monthly) than small-molecule drugs (e.g., daily).

There are some limitations to monoclonal antibodies as drugs. First, as they are proteins, they are unable to withstand the harsh environment of the gastrointestinal tract. Hence, antibodies cannot be dosed orally and are usually administered under the skin (subcutaneously) or directly into blood vessels (intravenously). Additionally, antibodies do not cross membranes readily, so they generally are only usable against secreted or cell-surface targets. Despite these limitations, however, over 60 monoclonal antibodies are currently in clinical use as therapeutics, with hundreds more in clinical development.

How do monoclonal antibodies exert their therapeutic effects? In many cases, they serve as antagonists, blocking receptors from interacting with their corresponding ligands. This was the case for cetuximab (Erbitux), the chemotherapeutic monoclonal antibody that blocks the epidermal growth factor (EGF) receptor from binding EGF (Section 14.5). Alternatively, some monoclonal antibodies bind specific circulating ligands, preventing them from exerting their cellular effects. In other cases, monoclonal antibodies can be used to target specific cell-surface proteins—such as markers specific for tumor cells—for degradation by the host's own immune system.

Another unique example is the *antibody-drug conjugate* (ADC). In this case, a small molecule drug is conjugated to an antibody specific for a particular cell type. Once the antibody-target complex is internalized by the cell, the linker between the antibody and the drug is cleaved, and the free drug can exert its effects. In this manner, the off-target toxicity of a particular drug can be limited by directing its activity to a small subset of cells within the body.

Monoclonal antibodies represent a rapidly growing and exciting drug class. These versatile molecules have been exploited in numerous creative ways to provide clinical benefit for a variety of diseases besides cancer, including inflammatory, infectious, cardiovascular, and ophthalmologic conditions.

KEY TERMS

target-based screening (p. 922)

phenotypic screening (p. 923)

ligand (p. 923)

dissociation constant (K_d) (p. 923)

side effect (p. 924)

inhibition constant (K_i) (p. 924)

Cheng–Prusoff equation (p. 924)

ADME (p. 924)

oral bioavailability (p. 925)

Lipinski's rules (p. 925)

partition coefficient (p. 925)

compartment (p. 926)

blood–brain barrier (p. 926)

xenobiotic compounds (p. 926)

drug metabolism (p. 926)

oxidation (p. 926)

conjugation (p. 926)

phase I transformation (p. 927)

phase II transformation (p. 927)

first-pass metabolism (p. 927)

glomeruli (p. 927)

enterohepatic cycling (p. 928)

mechanism-based toxicity (p. 929)

off-target toxicity (p. 929)

therapeutic index (p. 929)

atheroma (p. 932)

myopathy (p. 933)

high-throughput screening (p. 933)

combinatorial chemistry (p. 933)

split-pool synthesis (p. 934)

structure-based drug design
(p. 935)

structure–activity relationship
(SAR) (p. 936)

placebo (p. 941)
placebo effect (p. 941)

PROBLEMS

1. *Routes to discovery.* For each of the following drugs, indicate whether the physiological effects of the drug were known before or after the target was identified. ✓❶

(a) Penicillin

(b) Sildenafil (Viagra)

(c) Rofecoxib (Vioxx)

(d) Atorvastatin (Lipitor)

(e) Aspirin

(f) Indinavir (Crixivan)

2. *Lipinski's rules.* Which of the following compounds satisfy all of Lipinski's rules? [Log(P) values are given in parentheses.] ✓❷

(a) Atenolol (0.23)

(b) Sildenafil (3.18)

(c) Indinavir (2.78)

3. *Calculating log tables.* Considerable effort has been expended to develop computer programs that can estimate log(P) values entirely on the basis of chemical structure. Why would such programs be useful? ✓❷

4. *An ounce of prevention.* Legislation has been proposed that would require that *N*-acetylcysteine be added to acetaminophen tablets. Speculate about the role of this additive. ✓❸

5. *Clinical-trial design.* Distinguish between Phase I and Phase II clinical trials in regard to number of persons enrolled, the state of health of the subjects, and the goals of the study. ✓❻

6. *Drug interactions.* As noted in this chapter, coumadin can be a very dangerous drug because too much can cause uncontrolled bleeding. Persons taking coumadin must be careful about taking other drugs, particularly those that bind to albumin. Propose a mechanism for this drug–drug interaction.

7. *A bad combination.* Explain why drugs that inhibit P450 enzymes may be particularly dangerous when used in combination with other medications. ✓❸

8. *Mechanistically speaking.* Name one advantage of a noncompetitive inhibitor as a potential drug compared with a competitive inhibitor.

9. *A helping hand.* You have developed a drug that is capable of inhibiting the ABC transporter MDR. Suggest a possible application for this drug in cancer chemotherapy.

10. *Find the target.* Trypanosomes are unicellular parasites that cause sleeping sickness. During one stage of their life cycle, these organisms live in the bloodstream and derive all of their energy from glycolysis, which takes place in a specialized organelle called a glycosome inside the parasite. Propose potential targets for treating sleeping sickness. What are some potential difficulties with your approach?

11. *Knowledge is power.* How might genomic information be helpful for the effective use of imatinib mesylate (Gleevec) in cancer chemotherapy? ✓❺

12. *Multiple targets, same goal.* Sildenafil induces its physiological effects by increasing the intracellular concentrations of cGMP, leading to muscle relaxation. On the basis of the scheme shown in Figure 28.17, identify another approach for increasing cGMP levels with a small molecule.

Mechanism Problem

13. *Variations on a theme.* The metabolism of amphetamine by cytochrome P450 enzymes results in the conversion shown here. Propose a mechanism and indicate any additional products.

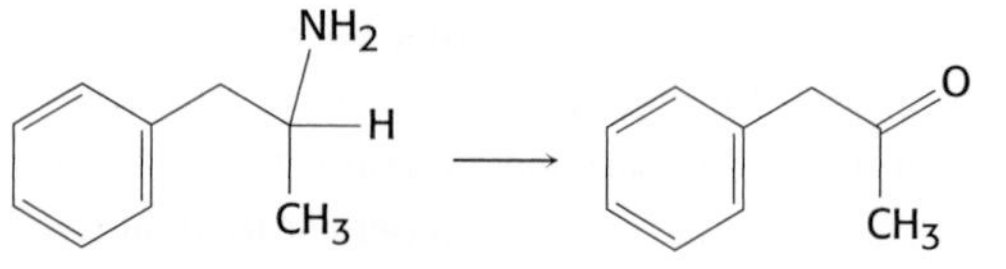

Amphetamine

Data Interpretation Problem

14. *HIV protease inhibitor design.* Compound A is one of a series that were designed to be potent inhibitors of HIV protease.

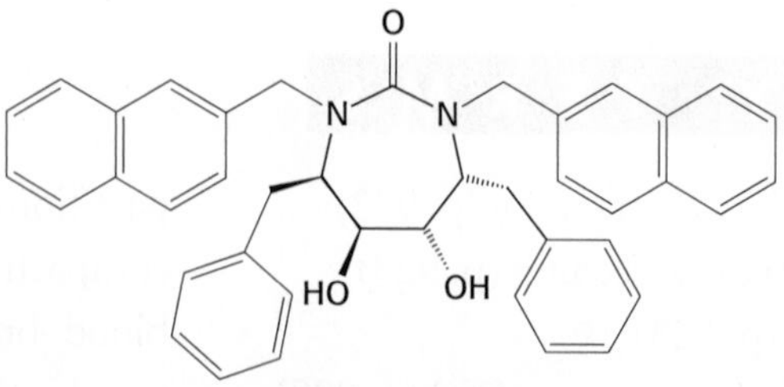

Compound A

Compound A was tested by using two assays: (1) direct inhibition of HIV protease in vitro and (2) inhibition of viral RNA production in HIV-infected cells, a measure of viral replication. The results of these assays are shown below. The HIV protease activity is measured with a substrate peptide present at a concentration equal to its K_M value.

Compound A (nM)	HIV protease activity (arbitrary units)
0	11.2
0.2	9.9
0.4	7.4
0.6	5.6
0.8	4.8
1	4.0
2	2.2
10	0.9
100	0.2

Compound A (nM)	Viral RNA production (arbitrary units)
0	760
1.0	740
2.0	380
3.0	280
4.0	180
5.0	100
10	30
50	20

Estimate the values for the K_I of compound A in the protease-activity assay and for its IC_{50} in the viral-RNA-production assay.

Treating rats with the relatively high oral dose of 20 mg kg^{-1} results in a maximum concentration of compound A of 0.4 μM. On the basis of this value, do you expect compound A to be effective in preventing HIV replication when taken orally?

Think ▸ Pair ▸ Share Problem

15. Let us suppose that you are studying a collection of competitive inhibitors of an exciting new enzyme. The K_M value for the natural substrate is 1 μM. One member of your team presents to you a list of IC_{50} values for these compounds using an assay that employs a substrate concentration of 20 μM. A week later, another member of your team tells you about her assay that requires a substrate concentration of only 1 μM. She presents you a list of IC_{50} values for the same set of compounds, which are consistently about 10-fold lower than those previously determined. Explain why this drop in IC_{50} is not at all surprising.

DNA Replication, Repair, and Recombination

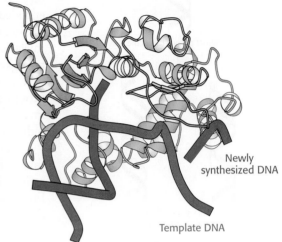

CHAPTER 29

Newly
synthesized DNA

Template DNA

Faithful copying is essential to the storage of genetic
information. With the precision of a careful artisan copying
an exemplar, a DNA polymerase (above) copies DNA strands,
preserving the precise sequence of bases with very few
errors. [(Left) Lou Linwei/Alamy.]

✔ LEARNING OBJECTIVES

By the end of this chapter, you should be able to:

1. Describe some of the challenges associated with faithfully copying and maintaining the genome of an organism.

2. Describe the reactions catalyzed by DNA polymerase, DNA primase, and DNA ligase and the roles of the template, the primer, and Okazaki fragments.

3. Discuss the reaction catalyzed by helicases and key aspects of their mechanisms.

4. Define topoisomers of circular DNA molecules and explain the relationship between twist and writhe in determining the linking number.

5. Describe how different activities of DNA replication are coordinated to accurately copy a complete genome.

6. Compare and contrast prokaryotic and eukaryotic replication.

7. Define telomeres and describe how they are synthesized.

8. Discuss types of DNA damage and how such lesions can be repaired.

9. Discuss the mechanism of DNA recombination and its role in a variety of biological processes.

OUTLINE

29.1 DNA Replication Proceeds by the Polymerization of Deoxyribonucleoside Triphosphates Along a Template

29.2 DNA Unwinding and Supercoiling Are Controlled by Topoisomerases

29.3 DNA Replication Is Highly Coordinated

29.4 Many Types of DNA Damage Can Be Repaired

29.5 DNA Recombination Plays Important Roles in Replication, Repair, and Other Processes

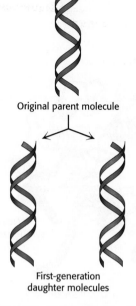

Original parent molecule

First-generation
daughter molecules

FIGURE 29.1 DNA replication. Each strand of one double helix (shown in blue) acts as a template for the synthesis of a new complementary strand (shown in red).

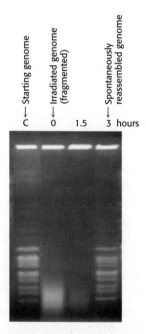

FIGURE 29.3 DNA repair in action. The bacterium *Deinococcus radiodurans* can reassemble its genome over the course of 3 hours after irradiation with gamma rays has fragmented the genome into many pieces. To aid analysis, the genomic DNA samples were digested with a restriction enzyme that cuts at only a few sites within the genome. C = control [Reprinted by permission from Macmilllan Publishers Ltd: Zahradka et al. Reassembly of shattered chromosomes in Deinococcus radiodurans. Nature 443, 569–573 (4 October 2006), © 2006.]

Perhaps the most exciting aspect of the structure of DNA deduced by Watson and Crick was, as expressed in their words, that the "specific pairing we have postulated immediately suggests a possible copying mechanism for the genetic material." A double helix separated into two single strands can be replicated because each strand serves as a template on which its complementary strand can be assembled (Figure 29.1). To preserve the information encoded in DNA through many cell divisions, copying of the genetic information must be extremely faithful. To replicate the human genome without mistakes, an error rate of less than 1 bp per 6×10^9 bp must be achieved. Such remarkable accuracy is achieved through a multilayered system of accurate DNA synthesis (which has an error rate of 1 per 10^3–10^4 bases inserted), proofreading during DNA synthesis (which reduces that error rate to approximately 1 per 10^6–10^7 bp), and post replication mismatch repair (which reduces the error rate to approximately 1 per 10^9–10^{10} bp).

Even after DNA has been initially replicated, the genome is still not safe. Although DNA is remarkably robust, ultraviolet light as well as a range of chemical species can damage DNA, introducing changes in the DNA sequence (mutations) or lesions that can block further DNA replication (Figure 29.2). All organisms contain DNA-repair systems that detect DNA damage and act to preserve the original sequence. Mutations in genes that encode components of DNA-repair systems are key factors in the development of cancer. Among the most potentially devastating types of DNA damage are double-stranded breaks in DNA. With both strands of the double helix broken in a local region, neither strand is intact to act as a template for future DNA synthesis. A mechanism used to repair such lesions relies on DNA recombination—that is, the reassortment of DNA sequences present on two different double helices. In addition to its role in DNA repair, recombination is crucial for the generation of genetic diversity in meiosis. Recombination is also the key to generating a highly diverse repertoire of genes for key molecules in the immune system (See Chapter 35 on 🌐 SaplingPlus).

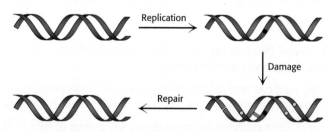

FIGURE 29.2 DNA Replication, damage, and repair. Some errors (shown as a black dot) may arise in the replication processes. Additional defects (shown in yellow), including modified bases, cross-links, and single- and double-strand breaks, are introduced into DNA by subsequent DNA-damaging reactions. Many of the errors are detected and subsequently repaired.

The bacterium *Deinococcus radiodurans* illustrates the extraordinary power of DNA repair systems. This bacterium was discovered in 1956 when scientists were studying the use of high doses of gamma radiation to sterilize canned meat. In some cases, the meat still spoiled due to the growth of a bacterial species that withstood doses of gamma radiation more than 1000 times larger than those that would kill a human being. Each *D. radiodurans* cell contains between 4 and 10 copies of its genome. Even when these bacterial chromosomes are broken into many fragments by the ionizing radiation, they can reassemble and recombine to regenerate the intact genome with essentially no loss of information (Figure 29.3). These cells can also survive extreme desiccation, that is, drying out, much better than other organisms. This ability is believed to be the selective advantage that favored the evolution of this and related species.

29.1 DNA Replication Proceeds by the Polymerization of Deoxyribonucleoside Triphosphates Along a Template

The base sequences of newly synthesized DNA must faithfully match the sequences of parent DNA. To achieve faithful replication, each strand within the parent double helix acts as a *template* for the synthesis of a new DNA strand with a complementary sequence, as discussed in Section 4.3. The building blocks for the synthesis of the new strands are deoxyribonucleoside triphosphates. They are added, one at a time, to the 3′ end of an existing strand of DNA.

Although this reaction is in principle quite simple, it is significantly complicated by specific features of the DNA double helix. First, the two strands of the double helix run in opposite directions. Because DNA strand synthesis always proceeds in the 5′-to-3′ direction, the DNA replication process must have special mechanisms to accommodate the oppositely directed strands. Second, the two strands of the double helix interact with one another in such a way that the bases, key templates for replication, are on the inside of the helix. Thus, the two strands must be separated from each other so as to generate appropriate templates. Finally, the two strands of the double helix wrap around each other. Thus, strand separation also entails the unwinding of the double helix. This unwinding creates supercoils that must themselves be resolved as replication continues. We begin with a consideration of the chemistry that underlies the formation of the phosphodiester backbone of newly synthesized DNA.

DNA polymerases require a template and a primer

DNA polymerases catalyze the formation of polynucleotide chains. Each incoming nucleoside triphosphate first forms an appropriate base pair with a base in the template. Only then does the DNA polymerase link the incoming base with the predecessor in the chain. Thus, *DNA polymerases are template-directed enzymes.*

DNA polymerases add nucleotides to the 3′ end of a polynucleotide chain. The polymerase catalyzes the nucleophilic attack by the 3′-hydroxyl-group terminus of the polynucleotide chain on the α phosphoryl group of the nucleoside triphosphate to be added (Figure 4.25). To initiate this reaction, DNA polymerases require a *primer* with a free 3′-hydroxyl group already base-paired to the template. They cannot start from scratch by adding nucleotides to a free single-stranded DNA template. RNA polymerase, in contrast, can initiate RNA synthesis without a primer, as we shall see in Chapter 30.

All DNA polymerases have structural features in common

The three-dimensional structures of a number of DNA polymerase enzymes are known. The first such structure was elucidated by Tom Steitz and coworkers, who determined the structure of the so-called *Klenow fragment* of DNA polymerase I from *E. coli* (Figure 29.4). This fragment comprises two main parts of the full enzyme, including the polymerase unit. This unit approximates the shape of a right hand with domains that are referred to as the fingers, the thumb, and the palm. In addition to the polymerase, the Klenow fragment includes a domain with 3′ → 5′ *exonuclease* activity that participates in proofreading and correcting the polynucleotide product.

Primer
The initial segment of a polymer that is to be extended on which elongation depends.

Template
A sequence of DNA or RNA that directs the synthesis of a complementary sequence.

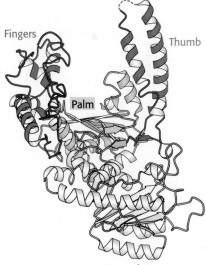

FIGURE 29.4 DNA polymerase structure. The first DNA polymerase structure determined was that of a fragment of *E. coli* DNA polymerase I called the Klenow fragment. *Notice* that, like other DNA polymerases, the polymerase unit resembles a right hand with fingers (blue), palm (yellow), and thumb (red). The Klenow fragment also includes an exonuclease domain that removes incorrect nucleotide bases. [Drawn from 1DPI.pdb.]

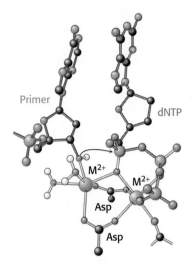

FIGURE 29.5 DNA polymerase mechanism. Two metal ions (typically, Mg^{2+}) participate in the DNA polymerase reaction. One metal ion coordinates the 3'-hydroxyl group of the primer, whereas the other metal ion interacts only with the dNTP. The phosphoryl group of the nucleoside triphosphate bridges between the two metal ions. The hydroxyl group of the primer attacks the phosphoryl group to form a new O—P bond.

DNA polymerases are remarkably similar in overall shape, although they differ substantially in detail. At least five structural classes have been identified; some of them are clearly homologous, whereas others appear to be the products of convergent evolution. In all cases, the finger and thumb domains wrap around DNA and hold it across the enzyme's active site, which comprises residues primarily from the palm domain. Furthermore, all DNA polymerases use similar strategies to catalyze the polymerase reaction, making use of a mechanism in which two metal ions take part.

Two bound metal ions participate in the polymerase reaction

Like all enzymes with nucleoside triphosphate substrates, DNA polymerases require metal ions for activity. Examination of the structures of DNA polymerases with bound substrates and substrate analogs reveals the presence of two metal ions in the active site. One metal ion binds both the deoxynucleoside triphosphate (dNTP) and the 3'-hydroxyl group of the primer, whereas the other interacts only with the dNTP (Figure 29.5). The two metal ions are bridged by the carboxylate groups of two aspartate residues in the palm domain of the polymerase. These side chains hold the metal ions in the proper positions and orientations. The metal ion bound to the primer activates the 3'-hydroxyl group of the primer, facilitating its attack on the α phosphoryl group of the dNTP substrate in the active site. The two metal ions together help stabilize the negative charge that accumulates on the pentacoordinate transition state. The metal ion initially bound to dNTP stabilizes the negative charge on the pyrophosphate product.

The specificity of replication is dictated by complementarity of shape between bases

DNA must be replicated with high fidelity. Each base added to the growing chain should, with high probability, be the Watson–Crick complement of the base in the corresponding position in the template strand. The binding of the dNTP containing the proper base is favored by the formation of a base pair with its partner on the template strand. Although hydrogen bonding contributes to the formation of this base pair, overall shape complementarity is crucial. Studies show that a nucleotide with a base that is very similar in shape to adenine but lacks the ability to form base-pairing hydrogen bonds can still direct the incorporation of thymidine, both in vitro and in vivo (Figure 29.6).

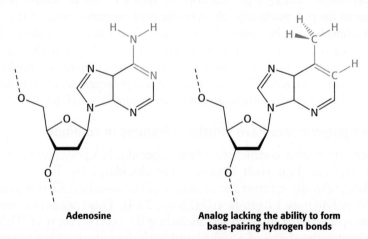

FIGURE 29.6 Shape complementarity. The base analog on the right has the same shape as adenosine, but groups that form hydrogen bonds between base pairs have been replaced by groups (shown in red) not capable of hydrogen bonding. Nonetheless, studies reveal that, when incorporated into the template strand, this analog directs the insertion of thymidine in DNA replication.

An examination of the crystal structures of various DNA polymerases reveals why shape complementarity is so important. First, residues of the enzyme form hydrogen bonds with *the minor-groove side of the base pair in the active site* (Figure 29.7). In the minor groove, hydrogen-bond acceptors are present in the same positions for all Watson–Crick base pairs. These interactions act as a "ruler" that measures whether a properly spaced base pair has formed in the active site.

Second, DNA polymerases close down around the incoming dNTP (Figure 29.8). The binding of a deoxyribonucleoside triphosphate into the active site of a DNA polymerase triggers a conformational change: the finger domain rotates to form a tight pocket into which only a properly shaped base pair will readily fit. Many of the residues lining this pocket are important to ensure the efficiency and fidelity of DNA synthesis. For example, mutation of a conserved tyrosine residue that forms part of the pocket results in a polymerase that is approximately 40 times as error prone as the parent polymerase.

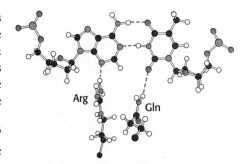

FIGURE 29.7 Minor-groove interactions. DNA polymerases donate two hydrogen bonds to base pairs in the minor groove. Hydrogen-bond acceptors are present in these two positions for all Watson–Crick base pairs, including the A—T base pair shown.

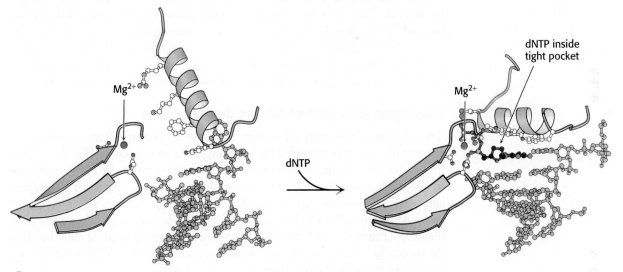

FIGURE 29.8 Shape selectivity. The binding of a deoxyribonucleoside triphosphate (dNTP) to DNA polymerase induces a conformational change, generating a tight pocket for the base pair consisting of the dNTP and its partner on the template strand. Such a conformational change is possible only when the dNTP corresponds to the Watson–Crick partner of the template base. [Drawn from 2BDP.pdb and 1T7P.pdb.]

An RNA primer synthesized by primase enables DNA synthesis to begin

DNA polymerases cannot initiate DNA synthesis without a primer, a section of nucleic acid having a free 3′ end that forms a double helix with the template. How is this primer formed? An important clue came from the observation that RNA synthesis is essential for the initiation of DNA synthesis. In fact, *RNA primes the synthesis of DNA.* An RNA polymerase called *primase* synthesizes a short stretch of RNA (about five nucleotides) that is complementary to one of the template DNA strands (Figure 29.9). Primase, like other RNA polymerases, can initiate synthesis without a primer. After DNA synthesis has been initiated, the short stretch of RNA is removed by hydrolysis and replaced by DNA.

One strand of DNA is made continuously, whereas the other strand is synthesized in fragments

Both strands of parental DNA serve as templates for the synthesis of new DNA. The site of DNA synthesis is called the *replication fork*

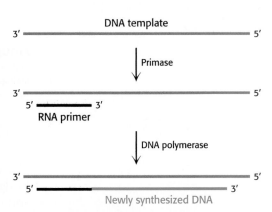

FIGURE 29.9 Priming. DNA replication is primed by a short stretch of RNA that is synthesized by primase, an RNA polymerase. The RNA primer is removed at a later stage of replication.

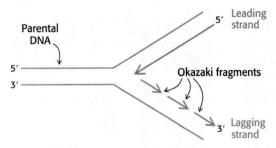

FIGURE 29.10 Okazaki fragments. At a replication fork, both strands are synthesized in the 5′ → 3′ direction. The leading strand is synthesized continuously, whereas the lagging strand is synthesized in short pieces termed Okazaki fragments.

because the complex formed by the newly synthesized daughter helices arising from the parental duplex resembles a two-pronged fork. Recall that the two strands are antiparallel; that is, they run in opposite directions. During DNA replication, both daughter strands appear on cursory examination to grow in the same direction. However, all known DNA polymerases synthesize DNA in the 5′ → 3′ direction but not in the 3′ → 5′ direction. How then does one of the daughter DNA strands appear to grow in the 3′ → 5′ direction?

This dilemma was resolved by Reiji Okazaki, who found that *a significant proportion of newly synthesized DNA exists as small fragments*. These units of about a thousand nucleotides (called *Okazaki fragments*) are present briefly in the vicinity of the replication fork (Figure 29.10).

As replication proceeds, these fragments become covalently joined through the action of the enzyme DNA ligase to form a continuous daughter strand. The other new strand is synthesized continuously. The strand formed from Okazaki fragments is termed the *lagging strand*, whereas the one synthesized without interruption is the *leading strand*. The discontinuous assembly of the lagging strand enables 5′ → 3′ polymerization at the nucleotide level to give rise to overall growth in the 3′ → 5′ direction.

DNA ligase joins ends of DNA in duplex regions

The joining of Okazaki fragments requires an enzyme that catalyzes the joining of the ends of two DNA chains. The existence of circular DNA molecules also points to the existence of such an enzyme. In 1967, scientists in several laboratories simultaneously discovered *DNA ligase. This enzyme catalyzes the formation of a phosphodiester linkage between the 3′-hydroxyl group at the end of one DNA chain and the 5′-phosphoryl group at the end of the other* (Figure 29.11). An energy source is required to drive this thermodynamically uphill reaction. In eukaryotes and archaea, ATP is the energy source. In bacteria, NAD⁺ typically plays this role.

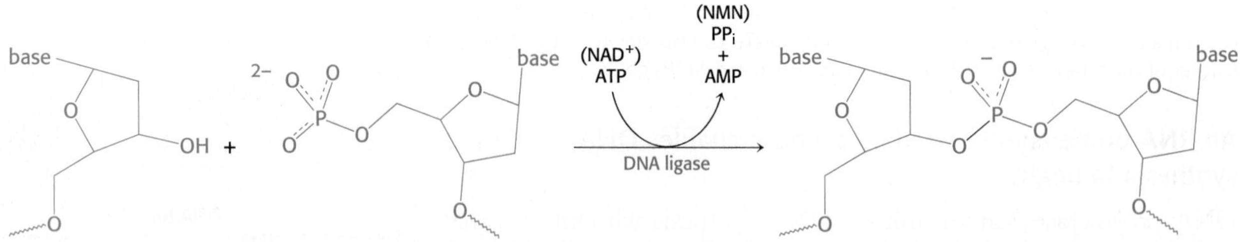

FIGURE 29.11 DNA ligase reaction. DNA ligase catalyzes the joining of one DNA strand with a free 3′-hydroxyl group to another with a free 5′-phosphoryl group. In eukaryotes and archaea, ATP is cleaved to AMP and PPᵢ to drive this reaction. In bacteria, NAD⁺ is cleaved to AMP and nicotinamide mononucleotide (NMN).

DNA ligase cannot link two molecules of single-stranded DNA or circularize single-stranded DNA. Rather, *ligase seals breaks in double-stranded DNA molecules*. The enzyme from *E. coli* ordinarily forms a phosphodiester bridge only if there are at least a few bases of single-stranded DNA on the end of a double-stranded fragment that can come together with those on another fragment to form base pairs. Ligase encoded by T4 bacteriophage can link two blunt-ended double-helical fragments, a capability that is exploited in recombinant DNA technology.

The separation of DNA strands requires specific helicases and ATP hydrolysis

For a double-stranded DNA molecule to replicate, the two strands of the double helix must be separated from each other, at least locally. This separation allows each strand to act as a template on which a new polynucleotide chain can be assembled. Specific enzymes, termed *helicases*, utilize the energy of ATP hydrolysis to power strand separation.

Helicases are a large and diverse family of enzymes taking part in many biological processes. The helicases in DNA replication are typically oligomers containing six subunits that form a ring structure. The structure of one such helicase, that from bacteriophage T7, has been a source of considerable insight into the helicase mechanism (Figure 29.12). Each of the subunits within this hexameric structure has a core structure that includes a P-loop NTPase domain (Figure 9.49). In addition to the P-loop, each subunit has two loops that extend toward the center of the ring structure and interact with DNA. Each subunit interacts closely with its two neighbors within the ring structure. Closer examination of this structure reveals that the ring deviates significantly from sixfold symmetry. This deviation is even more apparent when helicase is crystallized in the presence of the nonhydrolyzable ATP analog AMP-PNP.

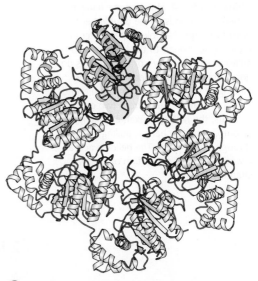

FIGURE 29.12 Helicase structure. The structure of the hexameric helicase from bacteriophage T7. One of the six subunits is shown in yellow with the P-loop NTPase shown in purple. The loops that participate in DNA binding are highlighted by a yellow oval. *Notice* that each subunit interacts closely with its neighbors and that the DNA-binding loops line the hole in the center of the structure. [Drawn from 1E0K.pdb.]

AMP-PNP

The AMP-PNP binds to only four of the six subunits within the ring (Figure 29.13). Furthermore, the four nucleotide-binding sites are not identical but fall into two classes. One class appears to be well positioned to bind ATP but not catalyze its hydrolysis, whereas the other class is more well-suited to catalyze the hydrolysis but not release the hydrolysis products. The classes are analogous to myosin's two different conformations—one for binding ATP and one for hydrolyzing it (Section 9.4). Finally, the six subunits fall into three classes with regard to their orientation with respect to the overall ring structure, with differences in rotation around an axis in the plane of the ring of approximately 30°. These differences in orientation affect the position of the two DNA-binding loops in each subunit.

These observations are consistent with the following mechanism for the helicase (Figure 29.14). Only a single strand of DNA can fit through the center on the ring. This single strand binds to loops on two adjacent subunits, one of which has bound ATP and the other of which has bound ADP + P$_i$. The binding of ATP to the domains that initially had no bound nucleotides leads to a conformational change within the entire hexamer, leading to the release of ADP + P$_i$ from two subunits and the binding of the single-stranded DNA by one of the domains that just bound ATP. This conformational change pulls the DNA through the center of the hexamer. The protein acts as a wedge, forcing the two strands of the double helix apart. This cycle then repeats itself, moving two bases along the DNA strand with each cycle.

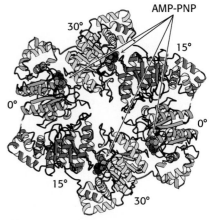

FIGURE 29.13 Helicase asymmetry. The structure of the T7 helicase complexes with the ATP analog AMP-PNP is shown. The three classes of helicase subunits are shown in blue, red, and yellow. The rotation relative to the plane of the hexamer is shown for each subunit. *Notice* that only four of the subunits, those shown in blue and yellow, bind AMP-PNP. [Drawn from 1E0K.pdb.]

FIGURE 29.14 Helicase mechanism. One of the strands of the double helix passes through the hole in the center of the helicase, bound to the loops of two adjacent subunits. Two of the subunits do not contain bound nucleotides. On the binding of ATP to these two subunits and the release of ADP + P_i from two other subunits, the helicase hexamer undergoes a conformational change, pulling the DNA through the helicase. The helicase acts as a wedge to force separation of the two strands of DNA.

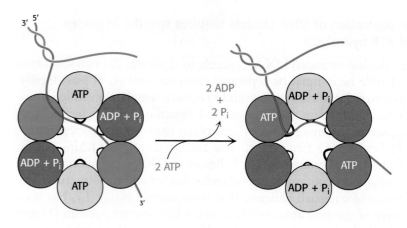

29.2 DNA Unwinding and Supercoiling Are Controlled by Topoisomerases

As a helicase moves along unwinding DNA, the DNA in front of the helicase will become overwound in the absence of other changes. As discussed in Section 4.2, DNA double helices that are torsionally stressed tend to fold up on themselves to form tertiary structures created by *supercoiling*. We will first consider the supercoiling of DNA in quantitative terms and then turn to topoisomerases, enzymes that can directly modulate DNA winding and supercoiling. Supercoiling is most readily understood by considering circular DNA molecules, but it also applies to linear DNA molecules constrained to be in loops by other means. Most DNA molecules inside cells are subject to supercoiling.

Consider a linear 260-bp DNA duplex in the B-DNA form (Figure 29.15A). Because the number of base pairs per turn in an unstressed DNA molecule averages 10.4, this linear DNA molecule has 25 (260/10.4) turns. The ends of this helix can be joined to produce a *relaxed* circular DNA (Figure 29.15B). A different circular DNA can be formed by unwinding the linear duplex by two turns before joining its ends (Figure 29.15C). What is the structural consequence of unwinding before ligation? Two limiting conformations are possible. The DNA can fold into a structure containing 23 turns of B helix and an unwound loop (Figure 29.15D). Alternatively, the double helix can fold up to cross itself. Such crossings are called *supercoils*. In particular, a supercoiled structure with 25 turns of B helix and 2 turns of *right-handed* (termed *negative*) superhelix can be formed (Figure 29.15E).

Supercoiling markedly alters the overall form of DNA. *A supercoiled DNA molecule is more compact than a relaxed DNA molecule of the same length.* Hence, supercoiled DNA moves faster than relaxed DNA when analyzed by centrifugation or electrophoresis. Unwinding will cause supercoiling in circular DNA molecules, whether covalently closed or constrained in closed configurations by other means.

The linking number of DNA, a topological property, determines the degree of supercoiling

Our understanding of the conformation of DNA is enriched by concepts drawn from topology, a branch of mathematics dealing with structural properties that are unchanged by deformations such as stretching and bending. A key topological property of a circular DNA molecule is its *linking number (Lk)*, which is equal to the number of times that a strand of DNA winds in the right-handed direction around the helix axis when the axis

(A)

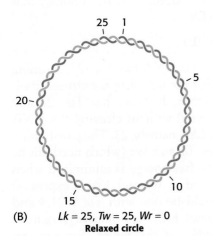

(B) $Lk = 25$, $Tw = 25$, $Wr = 0$
Relaxed circle

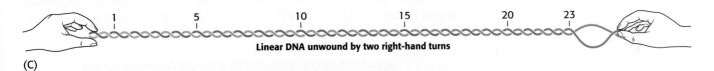

Linear DNA unwound by two right-hand turns

(C)

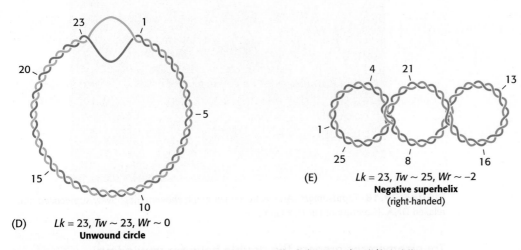

(D) $Lk = 23$, $Tw \sim 23$, $Wr \sim 0$
Unwound circle

(E) $Lk = 23$, $Tw \sim 25$, $Wr \sim -2$
Negative superhelix
(right-handed)

FIGURE 29.15 Linking number. The relations between the linking number (Lk), twisting number (Tw), and writhing number (Wr) of a circular DNA molecule revealed schematically. [Information from W. Saenger, *Principles of Nucleic Acid Structure* (Springer Verlag, 1984), p. 452.]

lies in a plane, as in Figure 29.15A. For the relaxed DNA shown in Figure 29.15B, $Lk = 25$. For the partly unwound molecule shown in part D and the supercoiled one shown in part E, $Lk = 23$ because the linear duplex was unwound two complete turns before closure. Molecules differing only in linking number are *topological isomers*, or *topoisomers*, of one another. *Topoisomers of DNA can be interconverted only by cutting one or both DNA strands and then rejoining them.*

The unwound DNA and supercoiled DNA shown in Figure 29.15D and E are topologically identical but geometrically different. They have the same value of Lk but differ in *twist* (Tw) and *writhe* (Wr). Although the rigorous definitions of twist and writhe are complex, twist is a measure of the helical winding of the DNA strands around each other, whereas writhe is a measure of the coiling of the axis of the double helix—that is, supercoiling. A right-handed coil is assigned a negative number (negative supercoiling) and a left-handed coil is assigned a positive number (positive supercoiling).

Is there a relation between Tw and Wr? Indeed, there is. Topology tells us that the sum of Tw and Wr is equal to Lk.

$$Lk = Tw + Wr$$

In Figure 29.15, the partly unwound circular DNA has $Tw \sim 23$, meaning the helix has 23 turns and $Wr \sim 0$, meaning the helix has not crossed itself to create a supercoil. The supercoiled DNA, however has $Tw \sim 25$ and $Wr \sim -2$. These forms can be interconverted without cleaving the DNA chain because they have the same value of Lk—namely, 23. The partitioning of Lk (which must be an integer) between Tw and Wr (which need not be integers) is determined by energetics. The free energy is minimized when about 70% of the change in Lk is expressed in Wr and 30% is expressed in Tw. Hence, the most stable form would be one with $Tw = 24.4$ and $Wr = -1.4$. Thus, *a lowering of* Lk *causes both right-handed (negative) supercoiling of the DNA axis and unwinding of the duplex.* Topoisomers differing by just 1 in Lk, and consequently by 0.7 in Wr, can be readily separated by agarose gel electrophoresis because their hydrodynamic volumes are quite different; *supercoiling condenses DNA* (Figure 29.16).

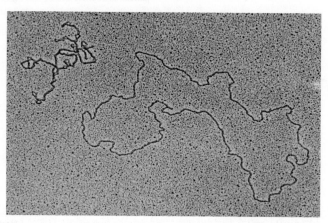

FIGURE 29.16 Topoisomers. An electron micrograph showing negatively supercoiled and relaxed DNA. [Courtesy of Dr. Jack Griffith.]

Topoisomerases prepare the double helix for unwinding

Most naturally occurring DNA molecules are negatively supercoiled. What is the basis for this prevalence? As already stated, negative supercoiling arises from the unwinding or underwinding of the DNA. In essence, negative supercoiling prepares DNA for processes requiring separation of the DNA strands, such as replication. Positive supercoiling condenses DNA as effectively, but it makes strand separation more difficult.

The presence of supercoils in the immediate area of unwinding would, however, make unwinding difficult. Therefore, negative supercoils must be continuously removed, and the DNA relaxed, as the double helix unwinds. Specific enzymes called *topoisomerases* that introduce or eliminate supercoils were discovered by James Wang and Martin Gellert. *Type I topoisomerases* catalyze the relaxation of supercoiled DNA, a thermodynamically favorable process. *Type II topoisomerases* utilize free energy from ATP hydrolysis to add negative supercoils to DNA. Both type I and type II topoisomerases play important roles in DNA replication as well as in transcription and recombination.

These enzymes alter the linking number of DNA by catalyzing a three-step process: (1) the *cleavage* of one or both strands of DNA, (2) the *passage* of a segment of DNA through this break, and (3) the *resealing* of the DNA break. Type I topoisomerases cleave just one strand of DNA, whereas type II enzymes cleave both strands. The two types of enzymes have several common features, including the use of key tyrosine residues to form covalent links to the polynucleotide backbone that is transiently broken.

Type I topoisomerases relax supercoiled structures

The three-dimensional structures of several type I topoisomerases have been determined (Figure 29.17). These structures reveal many features of the reaction mechanism. Human type I topoisomerase comprises four domains, which are arranged around a central cavity having a diameter of 20 Å, just the correct size to accommodate a double-stranded DNA molecule. This cavity also includes a tyrosine residue (Tyr 723), which acts as a nucleophile to cleave the DNA backbone in the course of catalysis.

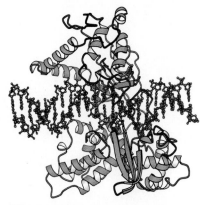

FIGURE 29.17 Structure of topoisomerase I. The structure of a complex between a fragment of human topoisomerase I and DNA is shown. *Notice that DNA lies in a central cavity within the enzyme.* [Drawn from 1EJ9.pdb.]

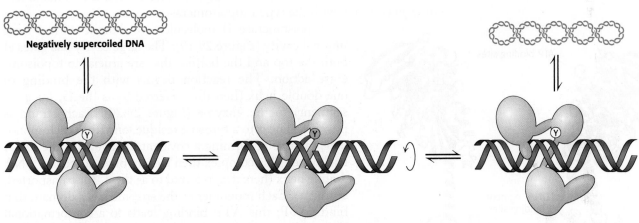

Negatively supercoiled DNA

FIGURE 29.18 Topoisomerase I mechanism. On binding to DNA, topoisomerase I cleaves one strand of the DNA by means of a tyrosine (Y) residue attacking a phosphoryl group. When the strand has been cleaved, it rotates in a controlled manner around the other strand. The reaction is completed by religation of the cleaved strand. This process results in partial or complete relaxation of a supercoiled plasmid.

From analyses of these structures and the results of other studies, the relaxation of negatively supercoiled DNA molecules is known to proceed in the following manner (Figure 29.18). First, the DNA molecule binds inside the cavity of the topoisomerase. The hydroxyl group of tyrosine 723 attacks a phosphoryl group on one strand of the DNA backbone to form a phosphodiester linkage between the enzyme and the DNA, cleaving the DNA and releasing a free 5′-hydroxyl group.

With the backbone of one strand cleaved, the DNA can now rotate around the remaining strand, its movement driven by the release of energy stored because of the supercoiling. The rotation of the DNA unwinds the supercoils. The enzyme controls the rotation so that the unwinding is not rapid. The free hydroxyl group of the DNA attacks the phosphotyrosine residue to reseal the backbone and release tyrosine. The DNA is then free to dissociate from the enzyme. Thus, reversible cleavage of one strand of supercoiled DNA allows controlled rotation to partly relax the supercoils.

Type II topoisomerases can introduce negative supercoils through coupling to ATP hydrolysis

Supercoiling requires an input of energy because a supercoiled molecule, in contrast with its relaxed counterpart, is torsionally stressed. The introduction of an additional supercoil into a 3000-bp plasmid typically requires about 30 kJ mol^{-1} (7 kcal mol^{-1}).

Supercoiling can be catalyzed by type II topoisomerases. These elegant molecular machines couple the binding and hydrolysis of ATP to the directed passage of one DNA double helix through another, temporarily cleaved DNA double helix. These enzymes have several mechanistic features in common with the type I topoisomerases.

Topoisomerase II molecules are dimeric with a large internal cavity (Figure 29.19). The large cavity has gates at both the top and the bottom that are crucial to topoisomerase action. The reaction begins with the binding of one double helix (hereafter referred to as the G, for gate, segment) to the enzyme (Figure 29.20). Each strand is positioned next to a tyrosine residue, one from each monomer, capable of forming a covalent linkage with the DNA backbone. This complex then loosely binds a second DNA double helix (hereafter referred to as the T, for transported, segment). Each monomer of the enzyme has a domain that binds ATP; this ATP binding leads to a conformational change that strongly favors the coming together of the two domains. As these domains come closer together, they trap the bound T segment. This conformational change also forces the separation and cleavage of the two strands of the G segment. Each strand is linked to the enzyme by a tyrosine–phosphodiester bond. Unlike the type I enzymes, the type II topoisomerases hold the DNA tightly so that it cannot rotate. The T segment then passes through the cleaved G segment and into the large central cavity. The ligation of the G segment leads to the release of the T segment through the gate at the bottom of the enzyme. The hydrolysis of ATP and the release of ADP and orthophosphate allow the ATP-binding domains to separate, preparing the enzyme to bind another T segment. The overall process leads to a decrease in the linking number by two.

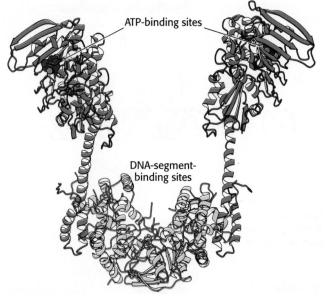

ATP-binding sites

DNA-segment-binding sites

FIGURE 29.19 Structure of topoisomerase II. The dimeric structure of a typical topoisomerase II, that from the archaeon *Sulfolobus shibatae*. *Notice that each half of the enzyme has one domain (shown in yellow) that contains a region for binding a DNA double helix and another domain (shown in green) that contains ATP binding sites.* [Drawn from 2ZBK.pdb.]

The bacterial topoisomerase II (often called DNA gyrase) is the target of several antibiotics that inhibit the prokaryotic enzyme much more than the eukaryotic one. *Novobiocin* blocks the binding of ATP to gyrase. *Nalidixic acid* and *ciprofloxacin*, in contrast, interfere with the breakage and rejoining of DNA chains. These two gyrase inhibitors are widely used to treat urinary tract and other infections including those due to *Bacillus anthracis* (anthrax). *Camptothecin*, an antitumor agent, inhibits

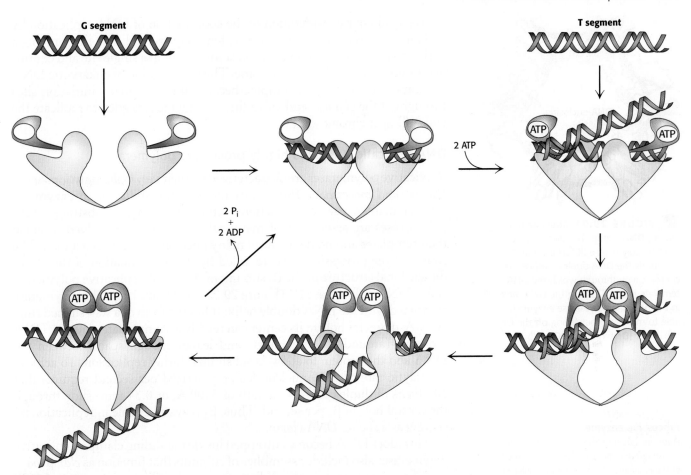

FIGURE 29.20 Topoisomerase II mechanism. Topoisomerase II first binds one DNA duplex termed the G (for gate) segment. The binding of ATP to the two N-terminal domains brings these two domains together. This conformational change leads to the cleavage of both strands of the G segment and the binding of an additional DNA duplex, the T segment. This T segment then moves through the break in the G segment and out the bottom of the enzyme. The hydrolysis of ATP resets the enzyme with the G segment still bound. The orientation of the entering T segment results in the introduction of negative supercoils.

human topoisomerase I by stabilizing the form of the enzyme covalently linked to DNA.

Nalidixic acid

Ciprofloxacin

29.3 DNA Replication Is Highly Coordinated

DNA replication must be very rapid, given the sizes of the genomes and the rates of cell division. The *E. coli* genome contains 4.6 million base pairs and is copied in less than 40 minutes. Thus, 2000 bases are incorporated per second. Enzyme activities must be highly coordinated to replicate entire genomes precisely and rapidly.

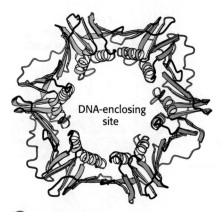

DNA-enclosing site

FIGURE 29.21 Structure of a sliding DNA clamp. The dimeric β subunit of DNA polymerase III forms a ring that surrounds the DNA duplex. *Notice* the central cavity through which the DNA template slides. Clasping the DNA molecule in the ring, the polymerase enzyme is able to move without falling off the DNA substrate. [Drawn from 2POL.pdb.]

Processive enzyme

From the Latin *procedere*, "to go forward."

An enzyme that catalyzes multiple rounds of the elongation or digestion of a polymer while the polymer stays bound. A *distributive* enzyme, in contrast, releases its polymeric substrate between successive catalytic steps.

We begin our consideration of the coordination of DNA replication by looking at *E. coli*, which has been extensively studied. For this organism with a relatively small genome, replication begins at a single site and continues around the circular chromosome. The coordination of eukaryotic DNA replication is much more complex because there are many initiation sites throughout the genome and an additional enzyme is needed to replicate the ends of linear chromosomes.

DNA replication requires highly processive polymerases

Replicative polymerases are characterized by their *very high catalytic potency, fidelity, and processivity. Processivity* refers to the ability of an enzyme to catalyze many consecutive reactions without releasing its substrate. These polymerases are assemblies of many subunits that have evolved to grasp their templates and not let go until many nucleotides have been added. The source of the processivity was revealed by the determination of the three-dimensional structure of the β_2 subunit of the *E. coli* replicative polymerase called DNA polymerase III (Figure 29.21). This unit keeps the polymerase associated with the DNA double helix. It has the form of a star-shaped ring. A 35-Å-diameter hole in its center can readily accommodate a duplex DNA molecule (around 20 Å diameter) and leaves enough space between the DNA and the protein to allow rapid sliding during replication. To achieve a catalytic rate of 1000 nucleotides polymerized per second requires that 100 turns of duplex DNA (a length of 3400 Å, or 0.34 mm) slide through the central hole of β_2 per second. Thus, β_2 plays a key role in replication by serving as a *sliding DNA clamp*.

How does DNA become entrapped inside the sliding clamp? Replicative polymerases also include assemblies of subunits that function as *clamp loaders*. These enzymes grasp the sliding clamp and, utilizing the energy of ATP binding, pull apart one of the interfaces between the two subunits of the sliding clamp. DNA can move through the gap, inserting itself through the central hole. ATP hydrolysis then releases the clamp, which closes around the DNA.

The leading and lagging strands are synthesized in a coordinated fashion

Replicative polymerases such as DNA polymerase III synthesize the leading and lagging strands simultaneously at the replication fork (Figure 29.22). DNA polymerase III begins the synthesis of the leading strand starting from the RNA primer formed by primase. The duplex DNA ahead of the polymerase is unwound by a hexameric helicase called DnaB. Copies of single-stranded-binding protein (SSB) bind to the unwound strands, keeping the strands separated so that both strands can serve as templates. The leading strand is synthesized continuously by polymerase III. Topoisomerase II concurrently introduces right-handed (negative) supercoils to avert a topological crisis.

The mode of synthesis of the lagging strand is necessarily more complex. As mentioned earlier, the lagging strand is synthesized in fragments so that $5' \rightarrow 3'$ polymerization leads to overall growth in the $3' \rightarrow 5'$ direction. Yet the synthesis of the lagging strand is coordinated with the synthesis of the leading strand. How is this coordination accomplished? Examination of the subunit composition of the DNA polymerase III holoenzyme reveals an elegant solution (Figure 29.23). The holoenzyme includes two copies of the polymerase core enzyme, which consists of the DNA polymerase itself (the α subunit); the ε subunit, a $3'$-to-$5'$ proofreading exonuclease; another subunit called θ; and two copies of the dimeric β-subunit sliding clamp.

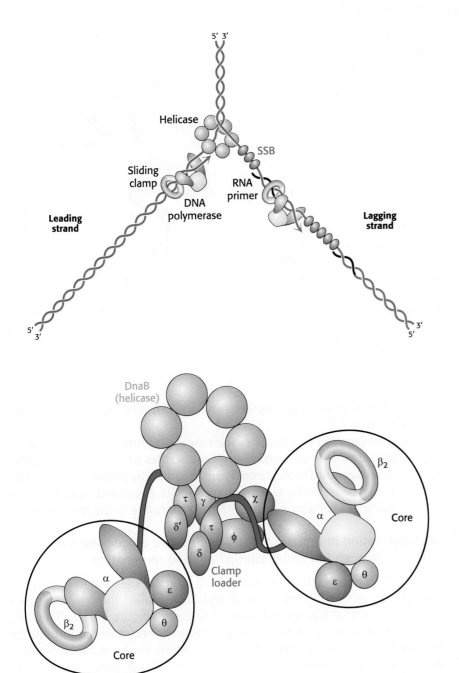

FIGURE 29.22 Replication fork.
A schematic view of the arrangement of DNA polymerase III and associated enzymes and proteins present in the replication of DNA. The helicase separates the two strands of the parent double helix, allowing DNA polymerases to use each strand as a template for DNA synthesis. Abbreviation: SSB, single-stranded-binding protein.

FIGURE 29.23 DNA polymerase holoenzyme. Each holoenzyme consists of two copies of the polymerase core enzyme, which comprises the α, ε, and θ subunits and two copies of the β subunit, linked to a central structure. The central structure includes the clamp-loader complex and the hexameric helicase DnaB.

The core enzymes are linked to a central structure having the subunit composition $\gamma\tau_2\delta\delta'\chi\phi$. The $\gamma\tau_2\delta\delta'$ complex is the clamp loader, and the χ and ϕ subunits interact with the single-stranded-DNA–binding protein. The entire apparatus interacts with the hexameric helicase DnaB. Eukaryotic replicative polymerases have similar, albeit slightly more complicated, subunit compositions and structures.

The lagging-strand template is looped out so that it passes through the polymerase site in one subunit of a dimeric polymerase III in the same direction as that of the leading-strand template in the other subunit, $5' \rightarrow 3'$. DNA polymerase III lets go of the lagging-strand template after adding about 1000 nucleotides by releasing the sliding clamp. A new loop is then formed, a sliding clamp is added, and primase again synthesizes a short stretch of RNA primer to initiate the formation of another Okazaki fragment. This mode of replication has been termed the *trombone model*

FIGURE 29.24 Trombone model. The replication of the leading and lagging strands is coordinated by the looping out of the lagging strand to form a structure that acts somewhat as a trombone slide, growing as the replication fork moves forward. When the polymerase on the lagging strand reaches a region that has been replicated, the sliding clamp is released and a new loop is formed.

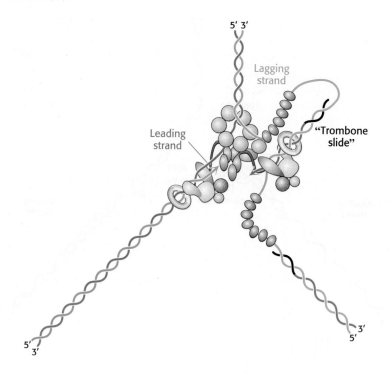

because the size of the loop lengthens and shortens like the slide on a trombone (Figure 29.24).

The gaps between fragments of the nascent lagging strand are filled by DNA polymerase I. This essential enzyme also uses its $5' \rightarrow 3'$ exonuclease activity to remove the RNA primer lying ahead of the polymerase site. The primer cannot be erased by DNA polymerase III, because the enzyme lacks $5' \rightarrow 3'$ editing capability. Finally, DNA ligase connects the fragments.

DNA replication in *Escherichia coli* begins at a unique site and proceeds through initiation, elongation, and termination

In *E. coli*, DNA replication starts at a unique site within the entire 4.6×10^6 bp genome. This *origin of replication*, called the oriC *locus*, is a 245-bp region that has several unusual features (Figure 29.25). The *oriC* locus contains five copies of a sequence that is the preferred binding site for the origin-recognition protein DnaA. In addition, the locus contains a tandem array of 13-bp sequences that are rich in AT base pairs. Several steps are required to prepare for the start of replication:

1. *The binding of DnaA proteins to DNA is the first step in the preparation for replication.* DnaA is a member of the P-loop NTPase family related to the hexameric helicases. Each DnaA monomer comprises an ATPase domain

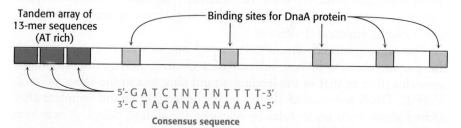

FIGURE 29.25 Origin of replication in *E. coli*. The *oriC* locus has a length of 245 bp. It contains a tandem array of three nearly identical 13-nucleotide sequences (green) and five binding sites (yellow) for the DnaA protein.

linked to a DNA-binding domain at its C-terminus. DnaA molecules are able to bind to each other through their ATPase domains; a group of bound DnaA molecules will break apart on the binding and hydrolysis of ATP. The binding of DnaA molecules to one another signals the start of the preparatory phase, and their breaking apart signals the end of that phase. The DnaA proteins bind to the five high-affinity sites in *oriC* and then come together with DnaA molecules bound to lower-affinity sites to form an oligomer, possibly a cyclic hexamer. The DNA is wrapped around the outside of the DnaA hexamer (Figure 29.26).

2. *Single DNA strands are exposed in the prepriming complex.* With DNA wrapped around a DnaA hexamer, additional proteins are brought into play. The hexameric helicase DnaB is loaded around the DNA with the help of the helicase loader protein DnaC. Local regions of *oriC*, including the AT regions, are unwound and trapped by the single-stranded-DNA–binding protein. The result of this process is the generation of a structure called the *prepriming complex*, which makes single-stranded DNA accessible to other proteins (Figure 29.27). Significantly, the primase, DnaG, is now able to insert the RNA primer.

3. *The polymerase holoenzyme assembles.* The DNA polymerase III holoenzyme assembles on the prepriming complex, initiated by interactions between DnaB and the sliding-clamp subunit of DNA polymerase III. These interactions also trigger ATP hydrolysis within the DnaA subunits, signaling the initiation of DNA replication. The breakup of the DnaA assembly prevents additional rounds of replication from beginning at the replication origin.

Once initiated, DNA replication could continue around the circular chromosome. To efficiently terminate replication, the *E. coli* chromosome includes specific termination sites (or Ter sites) that act as binding sites for a protein termed *termination utilization substance* or Tus (Figure 29.28). The complexes that form when Tus binds to the Ter sites are not symmetrical and this asymmetry is central for its function. When a replication fork, led by the helicase DnaB, reaches a Tus-Ter complex from one direction, it can move past it. However, when it reaches it from the opposite direction, the Tus-Ter complex effectively blocks the replication fork. The positions and orientations of the Ter sites within the *E. coli* genome facilitate efficient termination when one cycle of termination is complete.

DNA synthesis in eukaryotes is initiated at multiple sites

Replication in eukaryotes is mechanistically similar to replication in prokaryotes but is more challenging for a number of reasons. One of them is

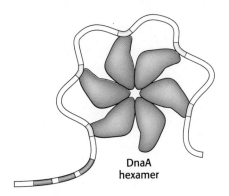

FIGURE 29.26 Assembly of DnaA. Monomers of DnaA bind to their binding sites (shown in yellow) in *oriC* and come together to form a complex structure, possibly the cyclic hexamer shown here. This structure marks the origin of replication and favors DNA strand separation in the AT-rich sites (green).

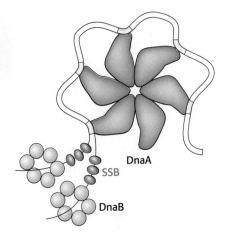

FIGURE 29.27 Prepriming complex. The AT-rich regions are unwound and trapped by the single-stranded-binding protein (SSB). The hexameric DNA helicase DnaB is loaded on each strand. At this stage, the complex is ready for the synthesis of the RNA primers and assembly of the DNA polymerase III holoenzyme.

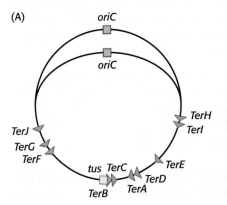

(A)

FIGURE 29.28 Replication termination in E. coli. The E. coli genome includes 10 Ter sites that bind the termination utilization substance (Tus) protein. The Ter-Tus complexes block incoming replication forks in a unidirectional manner (indicated by the orientation of the triangles representing the Ter sites), facilitating termination of replication after a single cycle.

sheer size: *E. coli* must replicate 4.6 million base pairs, whereas a human diploid cell must replicate more than 6 billion base pairs. Second, the genetic information for *E. coli* is contained on 1 chromosome, whereas, in human beings, 23 pairs of chromosomes must be replicated. Finally, whereas the *E. coli* chromosome is circular, human chromosomes are linear. Unless countermeasures are taken, linear chromosomes are subject to shortening with each round of replication.

The first two challenges are met by the use of multiple origins of replication. In human beings, replication requires about 30,000 origins of replication, with each chromosome containing several hundred. Each origin of replication is the starting site for a replication unit, or *replicon*. DNA replication can be monitored by single-molecule methods, revealing bidirectional synthesis from different sites (Figure 29.29). In contrast with *E. coli*, the origins of replication in human beings do not contain regions of sharply defined sequence. Instead, more broadly defined AT-rich sequences are the sites around which the *origin of replication complexes* (ORCs) are assembled.

FIGURE 29.29 Eukaryotic origins of replication. The image shows a single molecule of DNA containing two origins of replication. The origins were identified by labeling new replicated DNA in human cells first with one thymine analog (iodo-deoxyuridine, I-dU) and then another (chloro-deoxyuridine, Cl-dU). DNA molecules from these cells were then extended on a microscope slide and labeled with antibodies to I-dU (green) and Cl-dU (red) to visualize the DNA. This method allows the detection of replication origins as well as determination of the rate of DNA synthesis. [Courtesy Aaron Bensimon. Data from: Conti et al. "Replication Fork Velocities at Adjacent Replication Origins Are Coordinately Modified during DNA Replication in Human Cell." *Molecular Biology of the Cell* 18(2007):3059–3067.]

1. *The assembly of the ORC is the first step in the preparation for replication.* In human beings, the ORC is composed of six different proteins, each homologous to DnaA. These proteins come together to form a hexameric structure analogous to the assembly formed by DnaA.

2. *Licensing factors recruit a helicase that exposes single strands of DNA.* After the ORC has been assembled, additional proteins are recruited, including Cdc6, a homolog of the ORC subunits, and Cdt1. These proteins, in turn, recruit a hexameric helicase with six distinct subunits called Mcm2-7. These proteins, including the helicase, are sometimes called *licensing factors* because they permit the formation of the initiation complex. After the initiation complex has formed, Mcm2-7 separates the parental DNA strands, and the single strands are stabilized by the binding of *replication protein A*, a single-stranded-DNA–binding protein.

3. *Two distinct polymerases are needed to copy a eukaryotic replicon.* An initiator polymerase called *polymerase α* begins replication but is soon replaced by a more processive enzyme. This process is called *polymerase switching* because one polymerase has replaced another. This second enzyme, called *DNA polymerase δ*, is the principal replicative polymerase in eukaryotes (Table 29.1).

Replication begins with the binding of DNA polymerase α. This enzyme includes a primase subunit, used to synthesize the RNA primer, as well as

TABLE 29.1 Some types of DNA polymerases

Name	Function
Prokaryotic Polymerases	
DNA polymerase I	Erases primer and fills in gaps on lagging strand
DNA polymerase II (error-prone polymerase)	DNA repair
DNA polymerase III	Primary enzyme of DNA synthesis
Eukaryotic Polymerases	
DNA polymerase a Primase subunit DNA polymerase unit	Initiator polymerase Synthesizes the RNA primer Adds stretch of about 20 nucleotides to the primer
DNA polymerase β (error-prone polymerase)	DNA repair
DNA polymerase δ	Primary enzyme of DNA synthesis

an active DNA polymerase. After this polymerase has added a stretch of about 20 deoxynucleotides to the primer, another replication protein, called *replication factor C* (RFC), displaces DNA polymerase α. Replication factor C attracts a sliding clamp called *proliferating cell nuclear antigen* (PCNA), which is homologous to the β_2 subunit of *E. coli* polymerase III. The binding of PCNA to DNA polymerase δ renders the enzyme highly processive and suitable for long stretches of replication. Replication continues in both directions from the origin of replication until adjacent replicons meet and fuse. RNA primers are removed and the DNA fragments are ligated by DNA ligase.

The use of multiple origins of replication requires mechanisms for ensuring that each sequence is replicated once and only once. The events of eukaryotic DNA replication are linked to the eukaryotic *cell cycle* (Figure 29.30). The processes of DNA synthesis and cell division are coordinated in the cell cycle so that the replication of all DNA sequences is complete before the cell progresses into the next phase of the cycle. This coordination requires several *checkpoints* that control the progression along the cycle. A family of small proteins termed *cyclins* are synthesized and degraded by proteasomal digestion in the course of the cell cycle. Cyclins act by binding to specific *cyclin-dependent protein kinases* and activating them. One such kinase, cyclin-dependent kinase 2 (cdk2) binds to assemblies at origins of replication and regulates replication through a number of interlocking mechanisms.

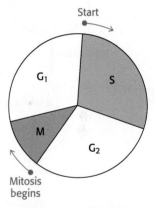

FIGURE 29.30 Eukaryotic cell cycle. DNA replication and cell division must take place in a highly coordinated fashion in eukaryotes. Mitosis (M) takes place only after DNA synthesis (S). Two gaps (G_1 and G_2) in time separate the two processes.

Telomeres are unique structures at the ends of linear chromosomes

Whereas the genomes of essentially all prokaryotes are circular, the chromosomes of human beings and other eukaryotes are linear. The free ends of linear DNA molecules introduce several complications that must be resolved by special enzymes. In particular, complete replication of DNA ends is difficult because polymerases act only in the $5' \rightarrow 3'$ direction. The lagging strand would have an incomplete 5′ end after the removal of the RNA primer. Each round of replication would further shorten the chromosome.

The first clue to how this problem is resolved came from sequence analyses of the ends of chromosomes, which are called *telomeres* (from the Greek *telos*, "an end"). Telomeric DNA contains hundreds of tandem repeats of a six-nucleotide sequence. One of the strands is G rich at the 3′ end, and it is slightly longer than the other strand. In human beings, the repeating G-rich sequence is AGGGTT.

The structure adopted by telomeres has been extensively investigated. Evidence suggests that they may form large duplex loops (Figure 29.31). The single-stranded region at the very end of the structure has been proposed to loop back to form a DNA duplex with another part of the repeated sequence, displacing a part of the original telomeric duplex. This loop-like structure is formed and stabilized by specific telomere-binding proteins. Such structures would nicely mask and protect the end of the chromosome.

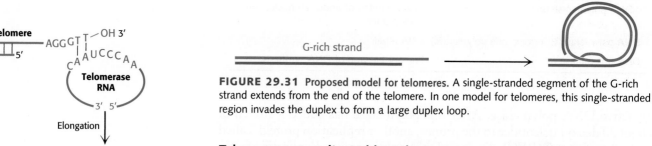

FIGURE 29.31 Proposed model for telomeres. A single-stranded segment of the G-rich strand extends from the end of the telomere. In one model for telomeres, this single-stranded region invades the duplex to form a large duplex loop.

Telomeres are replicated by telomerase, a specialized polymerase that carries its own RNA template

How are the repeated sequences generated? An enzyme, termed *telomerase*, that executes this function has been purified and characterized. When a primer ending in GGTT is added to human telomerase in the presence of deoxynucleoside triphosphates, the sequences GGTTAGGGTT and GGTTAGGGTTAGGGTT, as well as longer products, are generated. Elizabeth Blackburn and Carol Greider discovered that the enzyme adding the repeats contains an RNA molecule that serves as the template for the elongation of the G-rich strand (Figure 29.32). Thus, telomerase carries the information necessary to generate the telomere sequences. The exact number of repeated sequences is not crucial.

Subsequently, a protein component of telomerases also was identified. This component is related to reverse transcriptases, enzymes first discovered in retroviruses that copy RNA into DNA (Section 5.2). Thus, *telomerase is a specialized reverse transcriptase that carries its own template.* Telomerase is generally expressed at high levels only in rapidly growing cells. Thus, telomeres and telomerase can play important roles in cancer-cell biology and in cell aging.

Because cancer cells express high levels of telomerase, whereas most normal cells do not, telomerase is a potential target for anticancer therapy. A variety of approaches for blocking telomerase expression or blocking its activity are under investigation for cancer treatment and prevention.

29.4 Many Types of DNA Damage Can Be Repaired

We have examined how even very large and complex genomes can, in principle, be replicated with considerable fidelity. However, DNA does become damaged, both in the course of replication and through other processes. Damage to DNA can be as simple as the misincorporation of a single base or it can take more-complex forms such as the chemical modification of bases, chemical cross-links between the two strands of the double helix, or

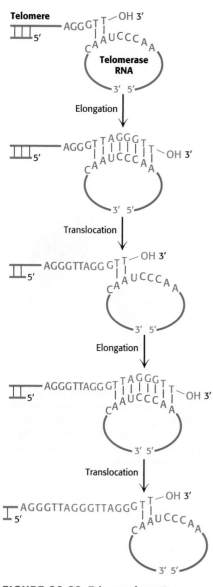

FIGURE 29.32 Telomere formation. Mechanism of synthesis of the G-rich strand of telomeric DNA. The RNA template of telomerase is shown in blue and the nucleotides added to the G-rich strand of the primer are shown in red. [Information from E. H. Blackburn, *Nature* 350:569–573, 1991.]

breaks in one or both of the phosphodiester backbones. The results may be cell death or cell transformation, changes in the DNA sequence that can be inherited by future generations, or blockage of the DNA replication process itself. A variety of DNA-repair systems have evolved that can recognize these defects and, in many cases, restore the DNA molecule to its undamaged form. We begin with some of the sources of DNA damage.

Errors can arise in DNA replication

Errors introduced in the replication process are the simplest source of damage in the double helix. With the addition of each base, there is the possibility that an incorrect base might be incorporated, forming a non-Watson–Crick base pair. These non-Watson–Crick base pairs can locally distort the DNA double helix. Furthermore, such mismatches can be *mutagenic*; that is, they can result in permanent changes in the DNA sequence. When a double helix containing a non-Watson–Crick base pair is replicated, the two daughter double helices will have different sequences because the mismatched base is very likely to pair with its Watson–Crick partner. Errors other than mismatches include insertions, deletions, and breaks in one or both strands. Furthermore, replicative polymerases can stall or even fall off a damaged template entirely. As a consequence, replication of the genome may halt before it is complete.

A variety of mechanisms have evolved to deal with such interruptions, including specialized DNA polymerases that can replicate DNA across many lesions. A drawback is that such polymerases are substantially more error prone than are normal replicative polymerases. Nonetheless, these *translesion* or *error-prone polymerases* allow the completion of a draft sequence of the genome that can be at least partly repaired by DNA-repair processes. DNA recombination (Section 29.5) provides an additional mechanism for salvaging interruptions in DNA replication.

Bases can be damaged by oxidizing agents, alkylating agents, and light

A variety of chemical agents can alter specific bases within DNA after replication is complete. Such *mutagens* include reactive oxygen species such as hydroxyl radical. For example, hydroxyl radical reacts with guanine to form 8-oxoguanine. 8-Oxoguanine is mutagenic because it often pairs with adenine rather than cytosine in DNA replication. Its choice of pairing partner differs from that of guanine because it uses a different edge of the base to form base pairs (Figure 29.33). Deamination is another potentially deleterious process. For example, adenine can be deaminated to form hypoxanthine (Figure 29.34). This process is mutagenic because hypoxanthine pairs with cytosine rather than thymine. Guanine and cytosine also can be deaminated to yield bases that pair differently from the parent base.

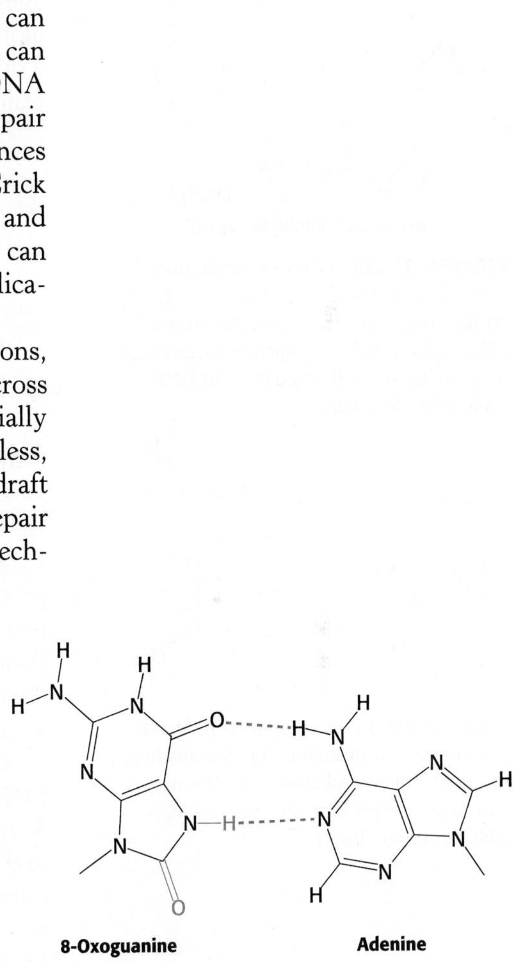

8-Oxoguanine　　　　**Adenine**

FIGURE 29.33 Oxoguanine–adenine base pair. When guanine is oxidized to 8-oxoguanine, the damaged base can form a base pair with adenine through an edge of the base that does not normally participate in base-pair formation.

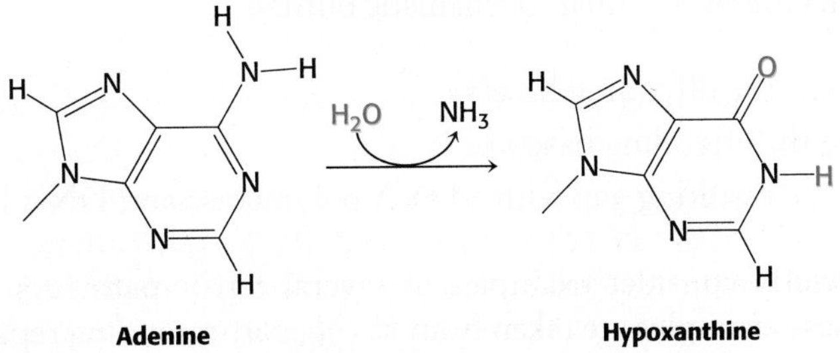

Adenine　　　　　　**Hypoxanthine**

FIGURE 29.34 Adenine deamination. The base adenine can be deaminated to form hypoxanthine. Hypoxanthine forms base pairs with cytosine in a manner similar to that of guanine, and so the deamination reaction can result in mutation.

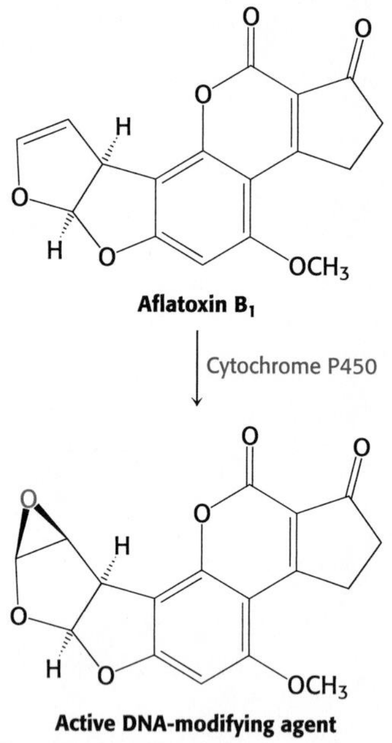

Aflatoxin B₁

Cytochrome P450

Active DNA-modifying agent

FIGURE 29.35 Aflatoxin activation. The compound, produced by molds that grow on peanuts, is activated by cytochrome P450 to form a highly reactive species that modifies bases such as guanine in DNA, leading to mutations.

Psoralen

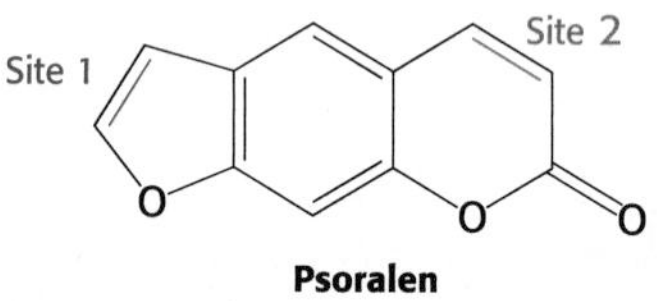

FIGURE 29.37 A cross-linking agent. The compound psoralen and its derivatives can form interstrand cross-links through two reactive sites that can form adducts with nucleotide bases.

In addition to oxidation and deamination, nucleotide bases are subject to alkylation. Electrophilic centers can be attacked by nucleophiles such as N-7 of guanine and adenine to form alkylated adducts. Some compounds are converted into highly active electrophiles through the action of enzymes that normally play a role in detoxification. A striking example is aflatoxin B₁, a compound produced by molds that grow on peanuts and other foods. A cytochrome P450 enzyme (Section 26.4) converts this compound into a highly reactive epoxide (Figure 29.35). This agent reacts with the N-7 atom of guanosine to form a mutagenic adduct that frequently leads to a G–C-to-T–A transversion.

The ultraviolet component of sunlight is a ubiquitous DNA-damaging agent. Its major effect is to covalently link adjacent pyrimidine residues along a DNA strand (Figure 29.36). Such a pyrimidine dimer cannot fit into a double helix, and so replication and gene expression are blocked until the lesion is removed.

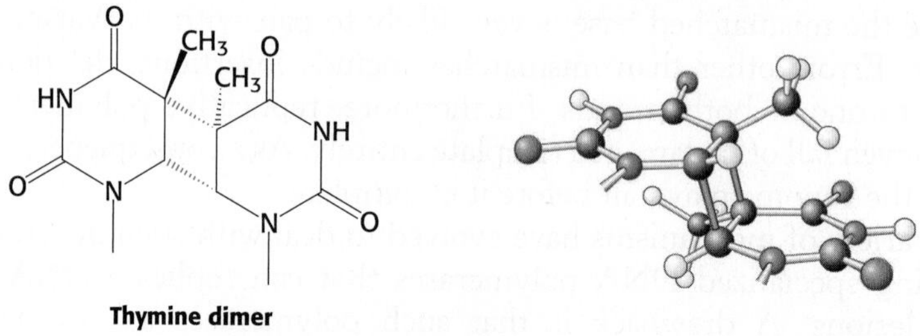

Thymine dimer

FIGURE 29.36 Cross-linked dimer of two thymine bases. Ultraviolet light induces cross-links between adjacent pyrimidines along one strand of DNA.

A thymine dimer is an example of an *intra*strand cross-link because both participating bases are in the same strand of the double helix. Cross-links between bases on opposite strands also can be introduced by various agents. Psoralens are compounds, produced by a number of plants including the common fig, that form such *inter*strand cross-links (Figure 29.37). Interstrand cross-links disrupt replication because they prevent strand separation.

High-energy electromagnetic radiation such as x-rays can damage DNA by producing high concentrations of reactive species in solution. X-ray exposure can induce several types of DNA damage including single- and double-stranded breaks in DNA. This ability to induce such DNA damage led Hermann Muller to discover the mutagenic effects of x-rays in *Drosophila* in 1927. This discovery contributed to the development of *Drosophila* as one of the premier organisms for genetic studies.

DNA damage can be detected and repaired by a variety of systems

To protect the genetic message, a wide range of DNA-repair systems are present in most organisms. Many systems repair DNA by using sequence information from the uncompromised strand. Such single-strand replication systems follow a similar mechanistic outline:

1. Recognize the offending base(s).
2. Remove the offending base(s).
3. Repair the resulting gap with a DNA polymerase and DNA ligase.

We will briefly consider examples of several repair pathways. Although many of these examples are taken from *E. coli*, corresponding repair systems are present in most other organisms, including humans.

The replicative DNA polymerases themselves are able to correct many DNA mismatches produced in the course of replication. For example, the ε subunit of *E. coli* DNA polymerase III functions as a 3′-to-5′ exonuclease. This domain removes mismatched nucleotides from the 3′ end of DNA by hydrolysis. How does the enzyme sense whether a newly added base is correct? As a new strand of DNA is synthesized, it is *proofread*. If an incorrect base is inserted, then DNA synthesis slows down owing to the difficulty of threading a non-Watson–Crick base pair into the polymerase. In addition, the mismatched base is weakly bound and therefore able to fluctuate in position. The delay from the slowdown allows time for these fluctuations to take the newly synthesized strand out of the polymerase active site and into the exonuclease active site (Figure 29.38). There, the DNA is degraded, one nucleotide at a time, until it moves back into the polymerase active site and synthesis continues.

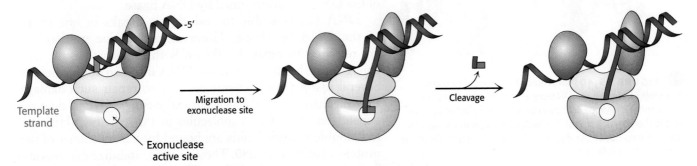

FIGURE 29.38 Proofreading. The growing polynucleotide chain occasionally leaves the polymerase site and migrates to the active site of the exonuclease. There, one or more nucleotides are excised from the newly synthesized chain, removing potentially incorrect bases.

A second mechanism is present in essentially all cells to correct errors made in replication that are not corrected by proofreading (Figure 29.39). *Mismatch-repair* systems consist of at least two proteins, one for detecting the mismatch and the other for recruiting an endonuclease that cleaves the newly synthesized DNA strand close to the lesion to facilitate repair. In *E. coli*, these proteins are called MutS and MutL and the endonuclease is called MutH.

Another mechanism of DNA repair is *direct repair*, one example of which is the photochemical cleavage of pyrimidine dimers. Nearly all cells contain a *photoreactivating enzyme* called *DNA photolyase*. The *E. coli* enzyme, a 35-kDa protein that contains bound N^5,N^{10}-methenyltetrahydrofolate and flavin adenine dinucleotide (FAD) cofactors, binds to the distorted region of DNA. The enzyme uses light energy—specifically, the absorption of a photon by the N^5,N^{10}-methenyltetrahydrofolate coenzyme—to form an excited state that cleaves the dimer into its component bases.

The excision of modified bases such as 3-methyladenine by the *E. coli* enzyme *AlkA* is an example of *base-excision repair*. The binding of this enzyme to damaged DNA flips the affected base out of the DNA double helix and into the active site of the enzyme (Figure 29.40). The enzyme then acts as *a glycosylase*, cleaving the glycosidic bond to release the damaged base. At this stage, the DNA backbone is intact, but a base is missing. This hole is called an *AP site* because it is apurinic (devoid of A or G) or apyrimidinic (devoid of C or T). An *AP endonuclease* recognizes this defect and nicks the backbone adjacent to the missing base. *Deoxyribose phosphodiesterase* excises the residual deoxyribose phosphate unit, and DNA polymerase I inserts an undamaged nucleotide, as dictated by the base on the undamaged complementary strand. Finally, the repaired strand is sealed by DNA ligase.

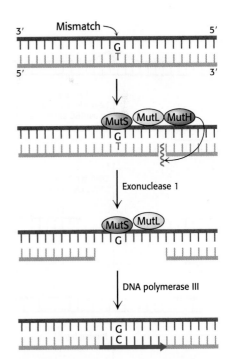

FIGURE 29.39 Mismatch repair. DNA mismatch repair in *E. coli* is initiated by the interplay of MutS, MutL, and MutH proteins. A G—T mismatch is recognized by MutS. MutH cleaves the backbone in the vicinity of the mismatch. A segment of the DNA strand containing the erroneous T is removed by exonuclease I and synthesized anew by DNA polymerase III. [Information from R. F. Service, *Science* 263:1559–1560, 1994.]

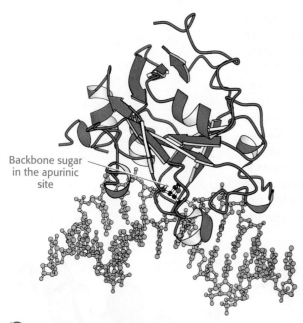

Backbone sugar in the apurinic site

FIGURE 29.40   Structure of DNA-repair enzyme. A complex between the DNA-repair enzyme AlkA and an analog of a DNA molecule missing a purine base (an apurinic site) is shown. *Notice* that the backbone sugar in the apurinic site is flipped out of the double helix into the active site of the enzyme. [Drawn from 1BNK.pdb.]

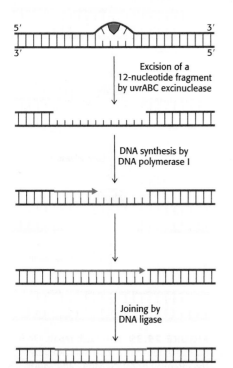

Excision of a 12-nucleotide fragment by uvrABC excinuclease

DNA synthesis by DNA polymerase I

Joining by DNA ligase

FIGURE 29.41   Nucleotide-excision repair. Repair of a region of DNA containing a thymine dimer by the sequential action of a specific excinuclease, a DNA polymerase, and a DNA ligase. The thymine dimer is shown in blue and the new region of DNA is in red. [Information from P. C. Hanawalt, *Endeavour* 31:83, 1982.]

One of the best-understood examples of *nucleotide-excision repair* is utilized for the excision of a pyrimidine dimer. Three enzymatic activities are essential for this repair process in *E. coli* (Figure 29.41). First, an enzyme complex consisting of the proteins encoded by the *uvrABC* genes detects the distortion produced by the DNA damage. The UvrABC enzyme then cuts the damaged DNA strand at two sites, 8 nucleotides away from the damaged site on the 5′ side and 4 nucleotides away on the 3′ side. The 12-residue oligonucleotide excised by this highly specific *excinuclease* (from the Latin *exci*, "to cut out") then diffuses away. DNA polymerase I enters the gap to carry out repair synthesis. The 3′ end of the nicked strand is the primer, and the intact complementary strand is the template. Finally, the 3′ end of the newly synthesized stretch of DNA and the original part of the DNA chain are joined by DNA ligase.

DNA ligase is able to seal simple breaks in one strand of the DNA backbone. However, alternative mechanisms are required to repair breaks on both strands that are close enough together to separate the DNA into two double helices. Several distinct mechanisms are able to repair such damage. One mechanism, *nonhomologous end joining* (NHEJ), does not depend on other DNA molecules in the cell. In NHEJ, the free double-stranded ends are bound by a heterodimer of two proteins, Ku70 and Ku80. These proteins stabilize the ends and mark them for subsequent manipulations. Through mechanisms that are not yet well understood, the Ku70/80 heterodimers act as handles used by other proteins to draw the two double-stranded ends close together so that enzymes can seal the break.

Alternative mechanisms of double-stranded-break repair can operate if an intact stretch of double-stranded DNA with an identical or very similar sequence is present in the cell. These repair processes use homologous recombination, presented in Section 29.5.

As noted in Section 5.4, researchers can now take advantage of double-stranded-break repair mechanisms for executing targeted changes in eukaryotic genomes. Specific engineered nucleases are used to introduce double-strand breaks in particular locations within the genome. The DNA repair machinery in the targeted cells will then repair the break, through either NHEJ to disrupt the gene or homologous recombination-based repair with added double-stranded DNA fragments to introduce more elaborate modifications. For example, mutations associated with human diseases can be specifically introduced into the genomes of model organisms such as mice or zebrafish to examine the consequences.

The presence of thymine instead of uracil in DNA permits the repair of deaminated cytosine

The presence in DNA of thymine rather than uracil, as in RNA, was an enigma for many years. Both bases pair with adenine. The only difference between them is a methyl group in thymine in place of the C-5 hydrogen atom in uracil. Why is a methylated base employed in DNA and not in RNA? The existence of an active repair system to correct the deamination of cytosine provides a convincing solution to this puzzle.

Cytosine in DNA spontaneously deaminates at a perceptible rate to form uracil. The deamination of cytosine is potentially mutagenic because uracil pairs with adenine, and so one of the daughter strands will contain a U–A base pair rather than the original C–G base pair. This mutation is prevented by a repair system that recognizes uracil to be foreign to DNA (Figure 29.42). The repair enzyme, *uracil DNA glycosylase*, is homologous to AlkA. The enzyme hydrolyzes the glycosidic bond between the uracil and deoxyribose moieties but does not attack thymine-containing nucleotides. The AP site generated is repaired to reinsert cytosine. Thus, *the methyl group on thymine is a tag that distinguishes thymine from deaminated cytosine.* If thymine were not used in DNA, uracil correctly in place would be indistinguishable from uracil formed by deamination. The defect would persist unnoticed, and so a C–G base pair would necessarily be mutated to U–A in one of the daughter DNA molecules. This mutation is prevented by a repair system that searches for uracil and leaves thymine alone. *Thymine is used instead of uracil in DNA to enhance the fidelity of the genetic message.*

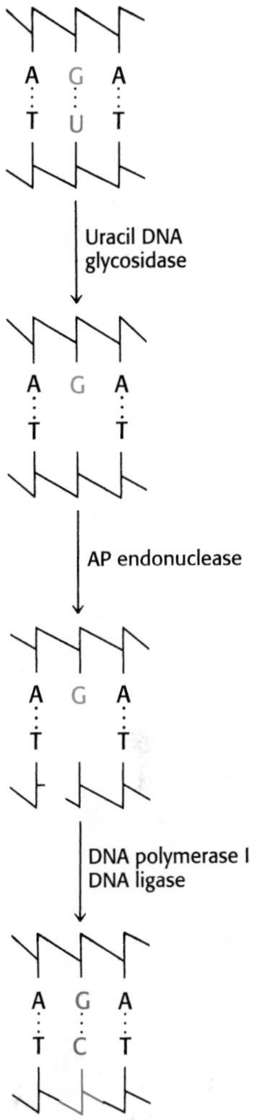

Some genetic diseases are caused by the expansion of repeats of three nucleotides

Some genetic diseases are caused by the presence of DNA sequences that are inherently prone to errors in the course of repair and replication. A particularly important class of such diseases is characterized by the presence of long tandem arrays of repeats of three nucleotides. An example is *Huntington disease*, an autosomal dominant neurological disorder with a variable age of onset. The mutated gene in this disease expresses a protein in the brain called huntingtin, which contains a stretch of consecutive glutamine residues. These glutamine residues are encoded by a tandem array of CAG sequences within the gene. In unaffected persons, this array is between 6 and 31 repeats, whereas, in those with the disease, the array is between 36 and 82 repeats or longer. Moreover, the array tends to become longer from one generation to the next. The consequence is a phenomenon called *anticipation*: the children of an affected parent tend to show symptoms of the disease at an earlier age than did the parent.

The tendency of these *trinucleotide repeats* to expand is explained by the formation of alternative structures in the course of DNA repair. On cleavage of the DNA backbone, part of the array can loop out without disrupting base-pairing outside this region. Then, in replication, DNA polymerase extends this strand through the remainder of the array, leading to an increase in the number of copies of the trinucleotide sequence.

A number of other neurological diseases are characterized by expanding arrays of trinucleotide repeats. How do these long stretches of repeated amino acids cause disease? For huntingtin, it appears that the polyglutamine stretches become increasingly prone to aggregate as their length increases; the additional consequences of such aggregation are still under investigation.

Many cancers are caused by the defective repair of DNA

As described in Chapter 14, cancers are caused by mutations in genes associated with growth control. Defects in DNA-repair systems increase the overall frequency of mutations and, hence, the likelihood of cancer-causing mutations. Indeed, the synergy between studies of mutations that predispose people to cancer and studies of DNA repair in model organisms has been tremendous in revealing the biochemistry of DNA-repair pathways. Genes for DNA-repair proteins are often *tumor suppressor genes*; that is, they suppress tumor development when at least one copy of the gene is free of a deleterious mutation. When both copies of a gene are mutated, however, tumors develop at rates greater than those for the

FIGURE 29.42 Uracil repair. Uridine bases in DNA, formed by the deamination of cytidine, are excised and replaced by cytidine.

TABLE 29.2

Disease	Selected Genes	Repair Pathway
Xeroderma pigmentosum (skin)	XPA, XPB, XPC	Nucleotide excision repair
Lynch syndrome (colon cancer)	MSH2, MLH1	Mismatch repair
Breast and ovarian cancer	BRCA1 and BRCA2	Double-strand break repair
Renal and lung cancer	OGG1	Base excision repair

population at large. People who inherit defects in a single tumor-suppressor allele do not necessarily develop cancer but are susceptible to developing the disease because only the one remaining normal copy of the gene must develop a new defect to further the development of cancer. Table 29.2 lists selected genes that are associated with diseases including cancer and that encode proteins involved with DNA repair processes.

Consider, for example, *xeroderma pigmentosum*, a rare human skin disease. The skin of an affected person is extremely sensitive to sunlight or ultraviolet light. In infancy, severe changes in the skin become evident and worsen with time. The skin becomes dry, and there is a marked atrophy of the dermis. Keratoses appear, the eyelids become scarred, and the cornea ulcerates. Skin cancer usually develops at several sites. Many patients die before age 30 from metastases of these malignant skin tumors. Studies of xeroderma pigmentosum patients have revealed that mutations occur in genes for a number of different proteins. These proteins are components of the human nucleotide-excision-repair pathway, including homologs of the UvrABC subunits.

Defects in other repair systems can increase the frequency of other tumors. For example, *hereditary nonpolyposis colorectal cancer* (HNPCC, or *Lynch syndrome*) results from defective DNA mismatch repair. HNPCC is not rare—as many as 1 in 200 people will develop this form of cancer. Mutations in two genes, called *hMSH2* and *hMLH1*, account for most cases of this hereditary predisposition to cancer. The striking finding is that these genes encode the human counterparts of MutS and MutL of *E. coli*. Mutations in *hMSH2* and *hMLH1* seem likely to allow mutations to accumulate throughout the genome. In time, genes important in controlling cell proliferation become altered, resulting in the onset of cancer.

Not all tumor-suppressor genes are specific to particular types of cancer. *The gene for a protein called p53 is mutated in more than half of all tumors.* The p53 protein helps control the fate of damaged cells. First, it plays a central role in sensing DNA damage, especially double-stranded breaks. Then, after sensing damage, the protein either promotes a DNA-repair pathway or activates the apoptosis pathway, leading to cell death. Most mutations in the p53 gene are sporadic; that is, they occur in somatic cells rather than being inherited. People who inherit a deleterious mutation in one copy of the p53 gene suffer from *Li-Fraumeni syndrome* and have a high probability of developing several types of cancer.

Cancer cells often have two characteristics that make them especially vulnerable to agents that damage DNA molecules. First, they divide frequently, and so their DNA replication pathways are more active than they are in most cells. Second, as already noted, cancer cells often have defects in DNA-repair pathways. Several agents widely used in cancer chemotherapy, including cyclophosphamide and cisplatin, act by damaging DNA. Cancer cells are less able to avoid the effect of the induced damage than are normal cells, providing a therapeutic window for specifically killing cancer cells.

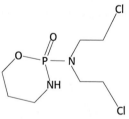

Cyclophosphamide

Cisplatin

Many potential carcinogens can be detected by their mutagenic action on bacteria

Many human cancers are caused by exposure to chemicals that cause mutations. It is important to identify such compounds and to ascertain their potency so that human exposure to them can be minimized. Bruce Ames devised a simple and sensitive test for detecting chemical mutagens. In the *Ames test*, a thin layer of agar containing about 10^9 bacteria of a specially constructed tester strain of *Salmonella* is placed on a petri plate. These bacteria are unable to grow in the absence of histidine, because a mutation is present in one of the genes for the biosynthesis of this amino acid. The addition of a chemical mutagen to the center of the plate results in many new mutations. A small proportion of them reverse the original mutation, and histidine can be synthesized. These *revertants* multiply in the absence of an external source of histidine and appear as discrete colonies after the plate has been incubated at 37°C for 2 days (Figure 29.43). For example, 0.5 µg of 2-aminoanthracene gives 11,000 revertant colonies, compared with only 30 spontaneous revertants in its absence. A series of concentrations of a chemical can be readily tested to generate a dose–response curve.

Some of the tester strains are responsive to *base-pair substitutions*, whereas others detect *deletions or additions of base pairs (frameshifts)*. The sensitivity of these specially designed strains has been enhanced by the genetic deletion of their excision-repair systems. Potential mutagens enter the tester strains easily because the lipopolysaccharide barrier that normally coats the surface of *Salmonella* is incomplete in these strains. A key feature of this detection system is the inclusion of a *mammalian liver homogenate*. Recall that some potential carcinogens such as aflatoxin are converted into their active forms by enzyme systems in the liver or other mammalian tissues. Bacteria lack these enzymes, and so the test plate requires a few milligrams of a liver homogenate to activate this group of mutagens.

The *Salmonella* test is extensively used to help evaluate the mutagenic and carcinogenic risks of a large number of chemicals. This rapid and inexpensive bacterial assay for mutagenicity complements epidemiological surveys and animal tests that are necessarily slower, more laborious, and far more expensive. The *Salmonella* test for mutagenicity is an outgrowth of studies of gene–protein relations in bacteria. It is a striking example of how fundamental research in molecular biology can lead directly to important advances in public health.

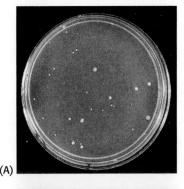

(A)

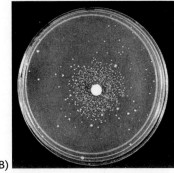

(B)

FIGURE 29.43 Ames test. (A) A petri plate containing about 10^9 *Salmonella* bacteria that cannot synthesize histidine and (B) a petri plate containing a filter-paper disc with a mutagen, which produces a large number of revertants that can synthesize histidine. After 2 days, the revertants appear as rings of colonies around the disc. The small number of visible colonies in plate A are spontaneous revertants. [Reprinted from *Mutation Resarch/Environmental Mutagenesis and Related Subjects*, 31:6, Ames, B.N., et al. Methods for detecting carcinogens and mutagens with the salmonella/mammalian-microsome mutagenicity test, pp. 347–363, Copyright 1975, with permission from Elsevier.]

29.5 DNA Recombination Plays Important Roles in Replication, Repair, and Other Processes

Most processes associated with DNA replication function to copy the genetic message as faithfully as possible. However, several biochemical processes require the *recombination* of genetic material between two DNA molecules. In genetic recombination, two daughter molecules are formed by the exchange of genetic material between two parent molecules (Figure 29.44). Recombination is essential in the following processes.

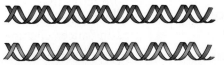 Recombination
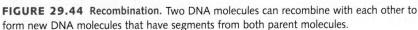

FIGURE 29.44 Recombination. Two DNA molecules can recombine with each other to form new DNA molecules that have segments from both parent molecules.

1. When replication stalls, recombination processes can reset the replication machinery so that replication can continue.

2. Some double-stranded breaks in DNA are repaired by recombination.

3. In meiosis, the limited exchange of genetic material between paired chromosomes provides a simple mechanism for generating genetic diversity in a population.

4. As we shall see in Chapter 35 (See 🔗 **Sapling**Plus), recombination plays a crucial role in generating molecular diversity for antibodies and some other molecules in the immune system.

5. Some viruses employ recombination pathways to integrate their genetic material into the DNA of a host cell.

6. Recombination is used to manipulate genes in, for example, the generation of gene knockout mice and other targeted genome modifications (Section 5.4).

Recombination is most efficient between DNA sequences that are similar in sequence. In homologous recombination, parent DNA duplexes align at regions of sequence similarity, and new DNA molecules are formed by the breakage and joining of homologous segments.

RecA can initiate recombination by promoting strand invasion

In many recombination pathways, a DNA molecule with a free end recombines with a DNA molecule having no free ends available for interaction. DNA molecules with free ends are the common result of double-stranded DNA breaks, but they may also be generated in DNA replication if the replication complex stalls. This type of recombination has been studied extensively in *E. coli*, but it also takes place in other organisms through the action of proteins homologous to those of *E. coli*. Often dozens of proteins participate in the complete recombination process. However, the key protein is RecA, the homolog of which is called Rad51 in human cells. To accomplish the exchange, the single-stranded DNA displaces one of the strands of the double helix (Figure 29.45). The resulting three-stranded structure is called a *displacement loop* or *D-loop*. This process is often referred to as *strand invasion*. Because a free 3′ end is now base-paired to a contiguous strand of DNA, the 3′ end can act as a primer to initiate new DNA synthesis. Strand invasion can initiate many processes, including the repair of double-stranded breaks and the reinitiation of replication after the replication apparatus has come off its template. In the repair of a break, the recombination partner is an intact DNA molecule with an overlapping sequence.

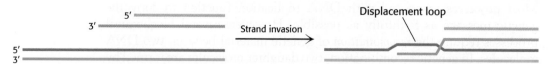

FIGURE 29.45 Strand invasion. This process, promoted by proteins such as RecA, can initiate recombination.

Some recombination reactions proceed through Holliday-junction intermediates

In recombination pathways for meiosis and some other processes, intermediates form that are composed of four polynucleotide chains in a cross-like structure. Intermediates with these cross-like structures are often referred to as *Holliday junctions*, after Robin Holliday, who proposed their role in recombination in 1964. Such intermediates have been characterized by a wide range of techniques, including x-ray crystallography.

Specific enzymes, termed *recombinases*, bind to these structures and resolve them into separated DNA duplexes. The Cre recombinase from bacteriophage P1 has been extensively studied. The mechanism begins with the recombinase binding to the DNA substrates (Figure 29.46).

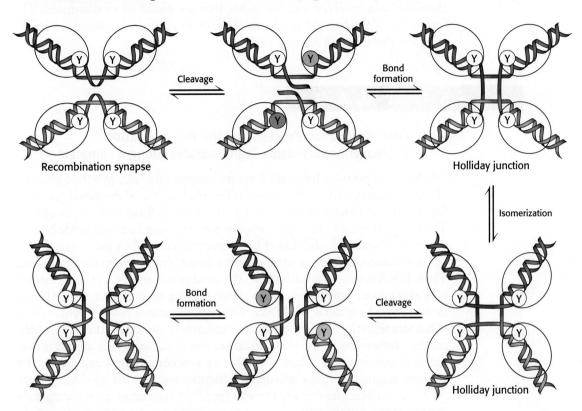

FIGURE 29.46 Recombination mechanism. Recombination begins as two DNA molecules come together to form a recombination synapse. One strand from each duplex is cleaved by transesterification with the 3′ end of each of the cleaved strands linked to a tyrosine residue on the recombinase enzyme. New phosphodiester bonds are formed when a 5′ end of the other cleaved strand in the complex attacks these tyrosine–DNA adducts. After isomerization, these steps are repeated to form the recombined products.

Four molecules of the enzyme and two DNA molecules come together to form a *recombination synapse*. The reaction begins with the cleavage of one strand from each duplex via transesterification reactions with the 3′-phosphoryl groups becoming linked to specific tyrosine residues in the recombinase while the 5′-hydroxyl group of each cleaved strand remains free. The free 5′ ends invade the other duplex and participate in transesterification reactions to form new phosphodiester bonds and free the tyrosine residues. These reactions result in the formation of a Holliday junction. This junction can then isomerize to form a structure in which the polynucleotide chains in the center of the structure are reoriented. From this junction, the processes of strand cleavage and phosphodiester-bond formation repeat. The result is a synapse containing the two recombined duplexes. Dissociation of this complex generates the final recombined products.

Cre catalyzes the formation of Holliday junctions as well as their resolution. In contrast, other proteins bind to Holliday junctions that have already been formed by other processes and resolve them into separate duplexes. In many cases, these proteins also promote the process of branch migration whereby a Holliday junction is moved along the two component double helices. Branch migration can affect which segments of DNA are exchanged in a recombination process.

Recombination can also occur during cell division between different, non-homologous chromosomes. These recombination events, termed translocations, can be harmless or can result in diseases, depending on the chromosomes involved and the genes that are modified or disrupted. The formation of the bcr-abl gene, which leads to chronic myelogenous leukemia, was discussed in Section 14.5.

SUMMARY

29.1 DNA Replication Proceeds by the Polymerization of Deoxyribonucleoside Triphosphates Along a Template

DNA polymerases are template-directed enzymes that catalyze the formation of phosphodiester bonds by nucleophilic attack by the 3'-hydroxyl group on the innermost phosphorus atom of a deoxyribonucleoside 5'-triphosphate. The complementarity of shape between correctly matched nucleotide bases is crucial to ensuring the fidelity of base incorporation. DNA polymerases cannot start chains de novo; a primer with a free 3'-hydroxyl group is required. Thus, DNA synthesis is initiated by the synthesis of an RNA primer, the task of a specialized primase enzyme. After serving as a primer, the RNA is degraded and replaced by DNA. DNA polymerases always synthesize a DNA strand in the 5'-to-3' direction. So that both strands of the double helix can be synthesized in the same direction simultaneously, one strand is synthesized continuously while the other is synthesized in fragments called Okazaki fragments. Gaps between the fragments are sealed by DNA ligases. ATP-driven helicases prepare the way for DNA replication by separating the strands of the double helix.

29.2 DNA Unwinding and Supercoiling Are Controlled by Topoisomerases

A key topological property of DNA is its linking number (Lk), which is defined as the number of times one strand of DNA winds around the other in the right-hand direction when the DNA axis is constrained to lie in a plane. Molecules differing in linking number are topoisomers of one another and can be interconverted only by cutting one or both DNA strands; these reactions are catalyzed by topoisomerases. Changes in linking number generally lead to changes in both the number of turns of double helix and the number of turns of superhelix. Topoisomerase II catalyzes the ATP-driven introduction of negative supercoils, which leads to the compaction of DNA and renders it more susceptible to unwinding. Supercoiled DNA can be relaxed by topoisomerase I or topoisomerase II. Topoisomerase I acts by transiently cleaving one strand of DNA in a double helix, whereas topoisomerase II transiently cleaves both strands simultaneously.

29.3 DNA Replication Is Highly Coordinated

Replicative DNA polymerases are processive; that is, they catalyze the addition of many nucleotides without dissociating from the template. A major contributor to processivity is the DNA sliding clamp, such as the dimeric β subunit of the *E. coli* replicative polymerase. The sliding clamp has a ring structure that encircles the DNA double helix and keeps the enzyme and DNA associated. The DNA polymerase holoenzyme is a large DNA-copying machine formed by two DNA polymerase enzymes, one to act on each template strand, associated with other subunits including a sliding clamp and a clamp loader.

The synthesis of the leading and lagging strands of a double-stranded DNA template is coordinated. As a replicative polymerase moves along a DNA template, the leading strand is copied smoothly while the lagging strand forms loops that change length in the course of the synthesis of each Okazaki fragment. The mode of action is referred to as the trombone model.

DNA replication is initiated at a single site within the *E. coli* genome. A set of specific proteins recognizes this origin of replication and assembles the enzymes needed for DNA synthesis, including a helicase that promotes strand separation. The initiation of replication in eukaryotes is more complex. DNA synthesis is initiated at thousands of sites throughout the genome. Complexes homologous to those in *E. coli*, but more complicated, are assembled at each eukaryotic origin of replication. A special polymerase called telomerase that relies on an RNA template synthesizes specialized structures called telomeres at the ends of linear chromosomes.

29.4 Many Types of DNA Damage Can Be Repaired

A wide variety of DNA damage can occur. For example, mismatched bases may be incorporated in the course of DNA replication or individual bases may be damaged by oxidation or alkylation after DNA replication. Other forms of damage are the formation of cross-links and the introduction of single- or double-stranded breaks in the DNA backbone. Several different repair systems detect and repair DNA damage. Repair begins with the process of proofreading in DNA replication: mismatched bases that were incorporated in the course of synthesis are excised by exonuclease activity present in replicative polymerases. Some DNA lesions such as thymine dimers can be directly reversed through the action of specific enzymes. Other DNA-repair pathways act through the excision of single damaged bases (base-excision repair) or short segments of nucleotides (nucleotide-excision repair). Double-stranded breaks in DNA can be repaired by homologous or non-homologous end-joining processes. Defects in DNA-repair components are associated with susceptibility to many different sorts of cancer. Such defects are a common target of cancer treatments. Many potential carcinogens can be detected by their mutagenic action on bacteria (the Ames test).

29.5 DNA Recombination Plays Important Roles in Replication, Repair, and Other Processes

Recombination is the exchange of segments between two DNA molecules. Recombination is important in some types of DNA repair as well as other processes such as meiosis, the generation of antibody diversity, and the life cycles of some viruses. Some recombination pathways are initiated by strand invasion, in which a single strand at the end of a DNA double helix forms base pairs with one strand of DNA in another double helix and displaces the other strand. A common intermediate formed in other recombination pathways is the Holliday junction, which consists of four strands of DNA that come together to form a cross-like structure. Recombinases promote recombination reactions through the introduction of specific DNA breaks and the formation and resolution of Holliday-junction intermediates.

APPENDIX

Biochemistry in Focus

Identifying amino acids crucial for DNA replication fidelity

The faithful replication of DNA is central to life. How can the fidelity of DNA replication be measured? What are the structural features of DNA polymerases that promote accurate copying of DNA sequences? These questions can be addressed using the three-dimensional structure of DNA polymerases to identify amino acids that might be important for fidelity, site-directed mutagenesis to modify these amino acids, and assays similar to the Ames test (p. 975) to measure the fidelity of DNA replication.

Using the structure of the Klenow fragment (Figure 29.4), residues that are highly conserved among evolutionary related DNA polymerases and that are in the parts of enzyme that would be in close proximity to a DNA substrate were identified and more than 25 were changed, one at a time, to alanine using site-directed mutagenesis. The mutated proteins were expressed and purified and then used to complete the synthesis of a circular DNA substrate containing a single-stranded region. The DNA substrate used was the genome of a bacteriophage, M13, that could be assayed by transfecting the reaction product into *E. coli* bacteria. Functional bacterio-phages would produce plaques when the bacteria were grown on a solid growth medium due to the slower growth of bacteria that had been infected with the bacteriophage. Moreover, the M13 bacteriophage encoded an enzyme that converted a substance in the growth medium to produce a blue color. If the gene encoding this enzyme was mutated during DNA synthesis, then plaques lacking this blue color were produced. Thus, the frequency with which mutated forms of the DNA polymerase introduced errors during the synthesis of the DNA substrate could be measured by counting the frequency at which non-blue-colored plaques were produced. The wild-type enzyme produced mutated bacteriophages at a frequency of 57×10^{-4}. Some mutated forms of the DNA polymerase produced mutated bacteriophages at frequencies that were twofold or more higher than the wild type.

This illustrates how biochemical techniques can be combined to explore the structural features that are crucial in promoting accurate replication of DNA. Variations of this approach can be used to examine mutated forms of human DNA polymerases that occur in cancer cells to determine if the mutations decrease the fidelity of these polymerases and contribute to the genomic instability of the cancer cells.

KEY TERMS

template (p. 951)

DNA polymerase (p. 951)

primer (p. 951)

exonuclease (p. 951)

primase (p. 953)

replication fork (p. 953)

Okazaki fragment (p. 954)

lagging strand (p. 954)

leading strand (p. 954)

DNA ligase (p. 954)

helicase (p. 955)

supercoiling (p. 956)

linking number (p. 956)

topoisomer (p. 957)

twist (p. 957)

writhe (p. 957)

topoisomerase (p. 958)

processivity (p. 962)

sliding DNA clamp (p. 962)

trombone model (p. 963)

origin of replication (p. 964)

origin of replication complex (ORC) (p. 966)

cell cycle (p. 967)

telomere (p. 967)

telomerase (p. 968)

mutagen (p. 969)

mismatch repair (p. 971)

direct repair (p. 971)

base-excision repair (p. 971)

nucleotide-excision repair (p. 972)

nonhomologous end joining (NHEJ) (p. 972)

trinucleotide repeat (p. 973)

tumor suppressor gene (p. 973)

Ames test (p. 975)

RecA (p. 976)

Holliday junction (p. 977)

recombinase (p. 977)

recombination synapse (p. 977)

PROBLEMS

1. *Activated intermediates.* DNA polymerase I, DNA ligase, and topoisomerase I catalyze the formation of phosphodiester bonds. What is the activated intermediate in the linkage reaction catalyzed by each of these enzymes? What is the leaving group? ✔❷

2. *Life in a hot tub.* An archaeon (*Sulfolobus acidocaldarius*) found in acidic hot springs contains a topoisomerase that catalyzes the ATP-driven introduction of positive supercoils into DNA. How might this enzyme be advantageous to this unusual organism?

3. *Which way?* Provide a chemical explanation of why DNA synthesis proceeds in a 5′-to-3′ direction. ✓②

4. *Nucleotide requirement.* DNA replication does not take place in the absence of the ribonucleotides ATP, CTP, GTP, and UTP. Propose an explanation.

5. *Close contact.* Examination of the structure of DNA polymerases bound to nucleotide analogs reveals that conserved residues come within van der Waals contact of C-2′ of the bound nucleotide. What is the potential significance of this interaction?

6. *Molecular motors in replication.*

(a) How fast does template DNA spin (expressed in revolutions per second) at an *E. coli* replication fork? ✓③

(b) What is the velocity of movement (in micrometers per second) of DNA polymerase III holoenzyme relative to the template?

7. *Wound tighter than a drum.* Why would replication come to a halt in the absence of topoisomerase II?

8. *The missing link.* One form of a plasmid shows a twist of $Tw = 48$ and a writhe of $Wr = 3$. ✓④

(a) What is the linking number?

(b) What would the value of writhe be for a form with twist $Tw = 50$ if the linking number is the same as that for the preceding form?

9. *Telomeres and cancer.* Telomerase is not active in most human cells. Some cancer biologists have suggested that activation of the telomerase gene would be a requirement for a cell to become cancerous. Explain why this might be the case. ✓⑦

10. *Backward?* Bacteriophage T7 helicase moves along DNA in the 5′-to-3′ direction. Other helicases have been reported to move in the 3′-to-5′ direction. Is there any fundamental reason why you would expect helicases to move in one direction or the other? ✓③

11. *Nick translation.* Suppose that you wish to make a sample of DNA duplex highly radioactive to use as a DNA probe. You have a DNA endonuclease that cleaves the DNA internally to generate 3′-OH and 5′-phosphoryl groups, intact DNA polymerase I, and radioactive dNTPs. Suggest a means for making the DNA radioactive.

12. *Revealing tracks.* Suppose that replication is initiated in a medium containing *moderately* radioactive tritiated thymine. After a few minutes of incubation, the bacteria are transferred to a medium containing *highly* radioactive tritiated thymidine. Sketch the autoradiographic pattern that would be seen for (a) undirectional replication and (b) bidirectional replication, each from a single origin.

13. *Mutagenic trail.* Suppose that the single-stranded RNA from tobacco mosaic virus was treated with a chemical mutagen, that mutants were obtained having serine or leucine instead of proline at a specific position, and that further treatment of these mutants with the same mutagen yielded phenylalanine at this position.

What are the plausible codon assignments for these four amino acids?

14. *Induced spectrum.* DNA photolyases convert the energy of light in the near-ultraviolet or visible region (300–500 nm) into chemical energy to break the cyclobutane ring of pyrimidine dimers. In the absence of substrate, these photoreactivating enzymes do not absorb light of wavelengths longer than 300 nm. Why is the substrate-induced absorption band advantageous? ✓⑧

15. *Missing telomerase.* Cells lacking telomerase can grow for several cell divisions without obvious defects. However, after more cell divisions, such cells tend to show chromosomes that have fused together. Propose an explanation for the formation of the chromosomes. ✓⑦

16. *I need to unwind.* With the assumption that the energy required to break an average base pair in DNA is 10 kJ mol^{-1} $(2.4 \text{ kcal mol}^{-1})$, estimate the maximum number of base pairs that could be broken per ATP hydrolyzed by a helicase operating under standard conditions. ✓③

17. *Triplet oxidation.* The oxidation of guanine bases in the context of triplet repeats such as CAGCAGCAG can lead to the expansion of the repeat. Explain.

Mechanism Problem

18. *A revealing analog.* AMP-PNP, the β,γ-imido analog of ATP (p. 833), is hydrolyzed very slowly by most ATPases. The addition of AMP-PNP to topoisomerase II and circular DNA leads to the negative supercoiling of a single molecule of DNA per enzyme. DNA remains bound to the enzyme in the presence of this analog. What does this finding reveal about the catalytic mechanism?

Data Interpretation and Chapter Integration Problems

19. *Like a ladder.* Circular DNA from SV40 virus was isolated and subjected to gel electrophoresis (See animation on **SaplingPlus**). The results are shown in lane A (the control) of the adjoining gel patterns.

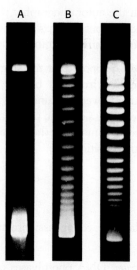

[From W. Keller. PNAS 72(1975):2553.]

(a) Why does the DNA separate in agarose gel electrophoresis? How does the DNA in each band differ?

The DNA was then incubated with topoisomerase I for 5 minutes and again analyzed by gel electrophoresis with the results shown in lane B.

(b) What types of DNA do the various bands represent?

Another sample of DNA was incubated with topoisomerase I for 30 minutes and again analyzed as shown in lane C.

(c) What is the significance of the fact that more of the DNA is in slower-moving forms?

20. *Ames test.* The following illustration shows four petri plates used for the Ames test. A piece of filter paper (white circle in the center of each plate) was soaked in one of four preparations and then placed on a petri plate. The four preparations contained (A) purified water (control), (B) a known mutagen, (C) a chemical whose mutagenicity is under investigation, and (D) the same chemical after treatment with liver homogenate. The number of revertants, visible as colonies on the petri plates, was determined in each case.

(a) What was the purpose of the control plate, which was exposed only to water?

(b) Why was it wise to use a known mutagen in the experimental system?

(c) How would you interpret the results obtained with the experimental compound?

(d) What liver components would you think are responsible for the effects observed in preparation D?

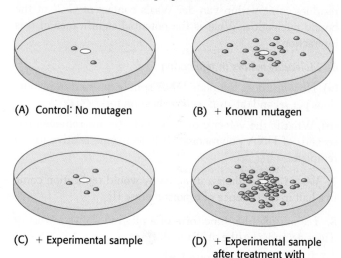

(A) Control: No mutagen

(B) + Known mutagen

(C) + Experimental sample

(D) + Experimental sample after treatment with liver homogenate

Think ▶ Pair ▶ Share Problem

21. Describe various ways in which the sequence of DNA is verified or corrected. Why are there so many pathways?

RNA Synthesis and Processing

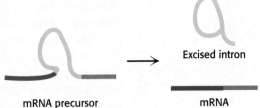

mRNA precursor

Excised intron

mRNA

RNA synthesis is a key step in the expression of genetic information. For eukaryotic cells, the initial RNA transcript (the mRNA precursor) is often spliced, removing introns that do not encode protein sequences. Often, the same pre-mRNA is spliced differently in different cell types or at different developmental stages. In the image at the left, proteins associated with RNA splicing (stained with a fluorescent antibody) highlight regions of the newt genome that are being actively transcribed. [(Left) SPL/Science Source.]

✓ LEARNING OBJECTIVES

By the end of this chapter, you should be able to:

1. Discuss the primary function of RNA polymerases, the reaction that they catalyze and its chemical mechanism.

2. Describe transcription, including the processes of initiation, elongation, and termination.

3. Discuss the regulation of transcription in eukaryotes, including the roles of transcription factors and enhancers.

4. Discuss the roles of RNA polymerases I, II, and III in producing ribosomal, transfer, and messenger RNAs.

5. Describe the process of RNA splicing, including the roles of the spliceosome and self-splicing RNA molecules.

OUTLINE

30.1 RNA Polymerases Catalyze Transcription

30.2 Transcription in Eukaryotes Is Highly Regulated

30.3 The Transcription Products of Eukaryotic Polymerases Are Processed

30.4 The Discovery of Catalytic RNA Was Revealing in Regard to Both Mechanism and Evolution

DNA stores genetic information in a stable form that can be readily replicated. The expression of this genetic information requires its flow from DNA to RNA and, usually, to protein, as was introduced in Chapter 4. This chapter examines RNA synthesis, or *transcription*, which is the process of synthesizing an RNA transcript with the transfer of sequence information from a DNA template. We begin with a discussion of RNA polymerases, the large and complex enzymes that carry out the synthetic process. We then turn to transcription in bacteria and focus on the three stages of transcription: promoter binding and initiation, elongation of the nascent RNA transcript, and termination. We then examine transcription in eukaryotes, focusing on the distinctions between bacterial and eukaryotic transcription.

RNA transcripts in eukaryotes are extensively modified, as exemplified by the capping of the 5′ end of an mRNA precursor and the addition of a long poly(A) tail to its 3′ end. One of the most striking examples of RNA modification is the splicing of mRNA precursors, which is catalyzed by spliceosomes, protein complexes consisting of small nuclear ribonucleoprotein particles (snRNPs). Remarkably, some RNA molecules can splice themselves in the absence of protein. Greatly influencing our view of molecular evolution, this landmark discovery by Thomas Cech and Sidney Altman revealed that RNA molecules can serve as catalysts.

RNA splicing is not merely a curiosity. Many genetic diseases have been associated with mutations that affect RNA splicing. Moreover, the same pre-mRNA can be spliced differently in various cell types, at different stages of development, or in response to other biological signals. In addition, individual bases in some pre-mRNA molecules are changed in a process called RNA editing. One of the biggest surprises of the sequencing of the human genome was that only about 20,000 genes were identified compared with previous estimates of 100,000 or more. The ability of one gene to encode more than one distinct mRNA by alternative splicing and, hence, more than one protein may play a key role in expanding the repertoire of our genomes.

Furthermore, the investigation of different classes of RNA molecules has been one of the most productive areas of biochemical research in recent years. In the next chapter we will explore the long-known ribosomal and transfer RNA molecules that are central to protein synthesis. We have also encountered microRNAs, molecules for which our understanding is still rapidly expanding. Many other classes of RNAs have been discovered more recently, including long non-coding RNAs, the functions of which are still under active investigation.

RNA synthesis comprises three stages: Initiation, elongation, and termination

RNA synthesis is catalyzed by large enzymes called *RNA polymerases*. The basic biochemistry of RNA synthesis is common to all organisms, commonality that has been beautifully illustrated by the three-dimensional structures of representative RNA polymerases from prokaryotes and eukaryotes (Figure 30.1). Despite substantial differences in size and number of polypeptide subunits, the overall structures of these enzymes are quite similar, revealing a common evolutionary origin.

RNA synthesis, like all biological polymerization reactions, takes place in three stages: *initiation, elongation,* and *termination.* RNA polymerases perform multiple functions in this process:

1. They search DNA for initiation sites, also called *promoter sites* or simply *promoters.* For instance, *E. coli* DNA has about 2000 promoter sites in its 4.6×10^6 bp genome.

2. They unwind a short stretch of double-helical DNA to produce single-stranded DNA templates from which the sequence of bases can be easily read out.

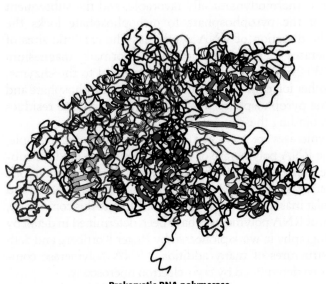

Prokaryotic RNA polymerase

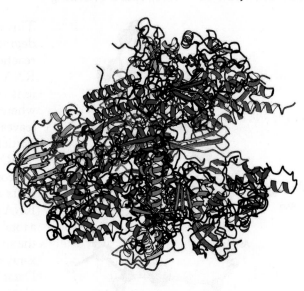

Eukaryotic RNA polymerase

FIGURE 30.1 RNA polymerase structures. The three-dimensional structures of RNA polymerases from a prokaryote *(Thermus aquaticus)* and a eukaryote *(Saccharomyces cerevisiae)*. The two largest subunits for each structure are shown in dark red and dark blue. *Notice* that both structures contain a central metal ion (green) in their active sites, near a large cleft on the right. The similarity of these structures reveals that these enzymes have the same evolutionary origin and have many mechanistic features in common. [Drawn from 1I6V.pdb and 1I6H.pdb.]

3. They select the correct ribonucleoside triphosphate and catalyze the formation of a phosphodiester bond. This process is repeated many times as the enzyme moves along the DNA template. RNA polymerase is completely processive—a transcript is synthesized from start to end by a single RNA polymerase molecule.

4. They detect termination signals that specify where a transcript ends.

5. They interact with activator and repressor proteins that modulate the rate of transcription initiation over a wide range. Gene expression is controlled substantially at the level of transcription, as will be discussed in detail in Chapters 32 and 33.

The chemistry of RNA synthesis is identical for all forms of RNA, including messenger RNAs, transfer RNAs, ribosomal RNAs, and small regulatory RNAs. The basic steps just outlined apply to all forms. Their synthetic processes differ mainly in regulation, posttranscriptional processing, and the specific RNA polymerase that participates.

30.1 RNA Polymerases Catalyze Transcription

The fundamental reaction of RNA synthesis is the formation of a phosphodiester bond. The 3′-hydroxyl group of the last nucleotide in the chain nucleophilically attacks the α phosphoryl group of the incoming nucleoside triphosphate with the concomitant release of a pyrophosphate.

NTP

RNA 5′-N—N—N—N—N   5′-N—N—N—N—N   + PP$_i$
DNA 3′-N—N—N—N—N—N—N—N—N-5′   ⇌   3′-N—N—N—N—N—N—N—N—N-5′

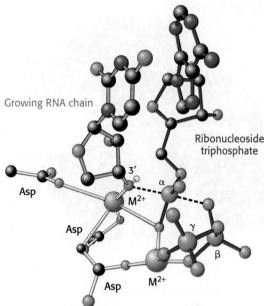

Growing RNA chain

Ribonucleoside triphosphate

Asp

Asp

Asp

FIGURE 30.2 RNA polymerase active site. A model of the transition state for phosphodiester-bond formation in the active site of RNA polymerase. The 3′-hydroxyl group of the growing RNA chain attacks the α-phosphoryl group of the incoming nucleoside triphosphate, resulting in the release of pyrophosphate. This transition state is structurally similar to that in the active site of DNA polymerase (Figure 29.5).

This reaction is thermodynamically favorable, and the subsequent degradation of the pyrophosphate to orthophosphate locks the reaction in the direction of RNA synthesis. The catalytic sites of RNA polymerases include two metal ions, normally magnesium ions (Figure 30.2). One ion remains tightly bound to the enzyme, whereas the other ion comes in with the nucleoside triphosphate and leaves with the pyrophosphate. Three conserved aspartate residues participate in binding these metal ions.

RNA polymerases are very large, complex enzymes. For example, the core of the RNA polymerase of *E. coli* consists of 5 kinds of subunits with the composition $\alpha_2\beta\beta'\omega$ (Table 30.1). A typical eukaryotic RNA polymerase is larger and more complex, having 12 subunits and a total molecular mass of more than 0.5 MDa. Despite this complexity, the structures of RNA polymerases have been determined in detail by x-ray crystallography in work pioneered by Roger Kornberg and Seth Darst. The structures of many additional RNA polymerase complexes have been determined by cryo-electron microscopy.

The polymerization reactions catalyzed by RNA polymerases take place within a complex in DNA termed a *transcription bubble* (Figure 30.3). This complex consists of double-stranded DNA that has been locally unwound in a region of approximately 17 base pairs. The edges of the bases that normally take part in Watson–Crick base pairs are exposed in the unwound region. We will begin with a detailed examination of the elongation process, including the role of the DNA template read by RNA polymerase and the reactions catalyzed by the polymerase, before returning to the more complex processes of initiation and termination.

RNA chains are formed de novo and grow in the 5′-to-3′ direction

Let us begin our examination of transcription by considering the DNA template. The first nucleotide (the start site) of a DNA sequence to be transcribed

TABLE 30.1 Subunits of RNA polymerase from *E. coli*

Subunit	Gene	Number	Mass (kDa)
α	*rpoA*	2	37
β	*rpoB*	1	151
β′	*rpoC*	1	155
ω	*rpoZ*	1	10
σ^{70}	*rpoD*	1	70

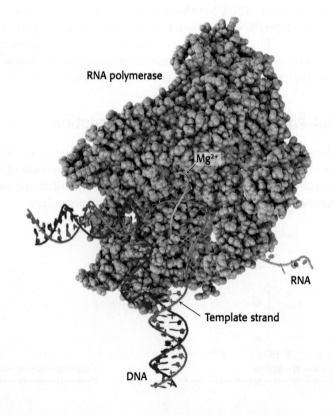

RNA polymerase

Mg²⁺

RNA

Template strand

DNA

FIGURE 30.3 Transcription bubble. RNA polymerase separates a region of the double helix to form a structure called the "transcription bubble." The red (template) and blue (nontemplate) strands of DNA are shown along with the RNA molecule being synthesized (shown in green). The position of the active site magnesium is indicated. The DNA enters from the left and exits at the bottom.

is denoted as +1 and the second one as +2; the nucleotide preceding the start site is denoted as −1. These designations refer to the coding strand of DNA. Recall that the sequence of the *template strand of DNA* is the *complement* of that of the RNA transcript (Figure 30.4). In contrast, the *coding strand of DNA* has the *same* sequence as that of the RNA transcript except for thymine (T) in place of uracil (U). The coding strand is also known as the *sense* (+) *strand*, and the template strand as the *antisense* (−) *strand*.

```
                        |AACGUAGGGUCACAUC...   RNA transcript
...GCATACAACACACC|TTGCATCCCAGTGTAG...   Template, or antisense (–), strand
...CGTATGTTGTGTGG|AACGTAGGGTCACATC...   Coding, or sense (+), strand
                 −1|+1+2
```

FIGURE 30.4 Template and coding strands. The template, or antisense (−), strand is complementary in sequence to the RNA transcript.

In contrast with DNA synthesis, *RNA synthesis can start de novo, without the requirement for a primer*. Most newly synthesized RNA chains carry a highly distinctive tag on the 5′ end: the first base at that end is either *pppG* or *pppA*.

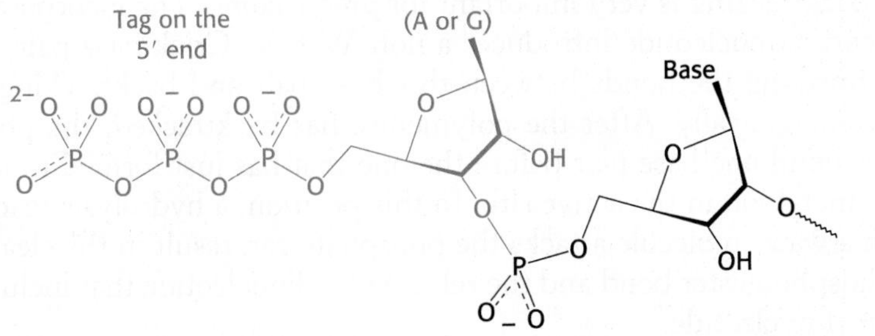

The presence of the triphosphate moiety confirms that RNA synthesis starts at the 5′ end.

The dinucleotide shown above is synthesized by RNA polymerase as part of the complex process of initiation, which will be discussed later in the chapter. After initiation takes place, RNA polymerase elongates the nucleic acid chain in the following manner (Figure 30.5). A ribonucleoside triphosphate binds in the active site of the RNA polymerase, directly adjacent to the growing RNA chain. The incoming ribonucleoside triphosphate forms a Watson–Crick base pair with the template strand. The 3′-hydroxyl group of the growing RNA chain, oriented and activated by the tightly bound metal ion, attacks the α-phosphoryl group to form a new phosphodiester bond, displacing pyrophosphate.

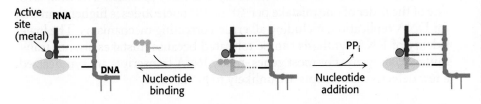

To proceed to the next step, the RNA–DNA hybrid must move relative to the polymerase to bring the 3′ end of the newly added nucleotide into proper position for the next nucleotide to be added (Figure 30.6). This translocation step does not include breaking any bonds between base pairs and is reversible but, once it has taken place, the addition of the next nucleotide, favored by the triphosphate cleavage and pyrophosphate release and cleavage, drives the polymerization reaction forward.

FIGURE 30.5 Elongation mechanism. A ribonucleoside triphosphate binds adjacent to the growing RNA chain and forms a Watson–Crick base pair with a base on the DNA template strand. The 3′-hydroxyl group at the end of the RNA chain attacks the newly bound nucleotide and forms a new phosphodiester bond, releasing pyrophosphate.

FIGURE 30.6 Translocation. After nucleotide addition, the RNA–DNA hybrid can translocate through the RNA polymerase, bringing a new DNA base into position to base-pair with an incoming nucleoside triphosphate.

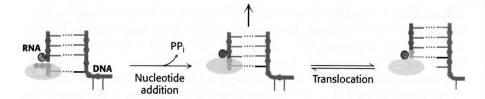

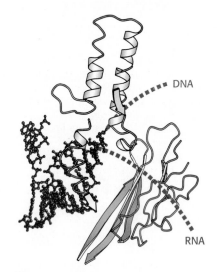

FIGURE 30.7 RNA–DNA hybrid separation. A structure within RNA polymerase forces the separation of the RNA–DNA hybrid. *Notice* that the DNA strand exits in one direction and the RNA product exits in another. [Drawn from 1I6H. pdb.]

The lengths of the RNA–DNA hybrid and of the unwound region of DNA stay rather constant as RNA polymerase moves along the DNA template. The length of the RNA–DNA hybrid is determined by a structure within the enzyme that forces the RNA–DNA hybrid to separate, allowing the RNA chain to exit from the enzyme and the DNA chain to rejoin its DNA partner (Figure 30.7).

RNA polymerases backtrack and correct errors

The RNA–DNA hybrid can also move in the direction opposite that of elongation (Figure 30.8). This backtracking is less favorable energetically than moving forward because it breaks the bonds between a base pair. However, backtracking is very important for *proofreading*. The incorporation of an incorrect nucleotide introduces a non-Watson–Crick base pair. In this case, breaking the bonds between this base pair and backtracking is less costly energetically. After the polymerase has backtracked, the phosphodiester bond one base pair before the one that has just formed is adjacent to the metal ion in the active site. In this position, a hydrolysis reaction in which a water molecule attacks the phosphate can result in the cleavage of the phosphodiester bond and the release of a dinucleotide that includes the incorrect nucleotide.

FIGURE 30.8 Backtracking. The RNA–DNA hybrid can occasionally backtrack within the RNA polymerase. In the backtracked position, hydrolysis can take place, producing a configuration equivalent to that after translocation. Backtracking is more likely if a mismatched base is added, facilitating proofreading.

Studies of single molecules of RNA polymerase have confirmed that the enzymes hesitate and backtrack to correct errors. Furthermore, these proofreading activities are often enhanced by accessory proteins. The final error rate of the order of one mistake per 10^4 or 10^5 nucleotides is higher than that for DNA replication, including all error-correcting mechanisms. The lower fidelity of RNA synthesis can be tolerated because mistakes are not transmitted to progeny. For most genes, many RNA transcripts are synthesized; a few defective transcripts are unlikely to be harmful.

RNA polymerase binds to promoter sites on the DNA template to initiate transcription

The elongation process is common to all organisms. In contrast, the processes of initiation and termination differ substantially in bacteria and eukaryotes. We begin with a discussion of these processes in bacteria, starting with initiation of transcription. The bacterial RNA polymerase discussed earlier with the composition $\alpha_2\beta\beta'\omega$ is referred to as the *core enzyme*.

The inclusion of an additional subunit produces the *holoenzyme* with composition $\alpha_2\beta\beta'\omega\sigma$. The σ subunit helps find sites on DNA where transcription begins, referred to as *promoter sites* or, simply, *promoters*. At these sites, the σ subunit participates in the initiation of RNA synthesis and then dissociates from the rest of the enzyme.

Sequences upstream of the promoter site are important in determining where transcription begins. A striking pattern was evident when the sequences of bacterial promoters were compared. *Two common motifs are present on the upstream side of the transcription start site.* They are known as the −10 sequence and the −35 sequence because they are centered at about 10 and 35 nucleotides upstream of the start site. The region containing these sequences is called the *core promoter*. The −10 and −35 sequences are each 6 bp long. Their *consensus sequences*, deduced from analyses of many promoters (Figure 30.9), are

$$\underset{\text{5'}}{}\overset{-35}{\text{TTGACA}}\cdots\cdots\overset{-10}{\text{TATAAT}}\cdots\overset{+1}{\underset{\text{site}}{\text{Start}}}$$

Promoters differ markedly in their efficacy. Some genes are transcribed frequently—as often as every 2 seconds in *E. coli*. The promoters for these genes are referred to as *strong promoters*. In contrast, other genes are transcribed much less frequently, about once in 10 minutes; the promoters for these genes are *weak promoters*. The −10 and −35 regions of most strong promoters have sequences that correspond closely to the consensus sequences, whereas weak promoters tend to have multiple substitutions at these sites. Indeed, mutation of a single base in either the −10 sequence or the −35 sequence can diminish promoter activity. The distance between these conserved sequences also is important; a separation of 17 nucleotides is optimal. Thus, *the efficiency or strength of a promoter sequence serves to regulate transcription.* Regulatory proteins that bind to specific sequences near promoter sites and interact with RNA polymerase (Chapter 32) also markedly influence the frequency of transcription of many genes.

Outside the core promoter in a subset of highly expressed genes is the *upstream element* (also called the UP element for *upstream element*). This sequence is present from 40 to 60 nucleotides upstream of the transcription start site. The UP element is bound by the α subunit of RNA polymerase and serves to increase the efficiency of transcription by creating an additional interaction site for the polymerase.

Sigma subunits of RNA polymerase recognize promoter sites

To initiate transcription, the $\alpha_2\beta\beta'\omega$ core of RNA polymerase must bind the promoter. However, it is the σ *subunit* that makes this binding possible by enabling *RNA polymerase to recognize promoter sites.* In the presence of the σ subunit, the RNA polymerase binds weakly to the DNA and slides along the double helix until it dissociates or encounters a promoter. The σ subunit recognizes the promoter through several interactions with the nucleotide bases of the promoter DNA. The structure of a bacterial RNA polymerase holoenzyme bound to a promoter site shows the σ subunit interacting with DNA at the −10 and −35 regions essential to promoter recognition (Figure 30.10). Therefore, *the σ subunit is responsible for the specific binding of the RNA polymerase to a promoter site on the template DNA.* The σ subunit is generally released when the nascent RNA chain reaches 9 or 10 nucleotides in length. After its release, it can assist

Transcription
starts here
↓

	5' −10	3'
(A)	G A T A T A A T G A G C A A A	
(B)	C G T A T A A T T T G A C C A	
(C)	G T T A A C T A G T A C G C A	
(D)	G T G A T A C T G A G C A C A	
(E)	G T T T T C A T G C C T C C A	
	T A T A A T	

FIGURE 30.9 Bacterial promoter sequences. A comparison of five sequences from prokaryotic promoters reveals a recurring sequence of TATAAT centered on position −10. The −10 consensus sequence (in red) was deduced from a large number of promoter sequences. The sequences are from the (A) *uvrD*, (B) *uncI*, and (C) *trp* operons of *E. coli*, from (D) λ phage, and from (E) ϕX174 phage.

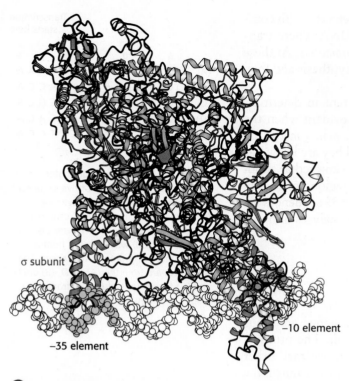

σ subunit

−10 element

−35 element

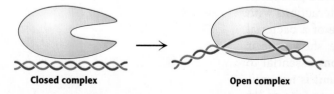 **FIGURE 30.10 RNA polymerase holoenzyme complex.**
Notice that the σ subunit (blue) of the bacterial RNA polymerase
holoenzyme makes sequence-specific contacts with the −10 (red) and
−35 promoter elements (yellow). The active site of the polymerase is
revealed by bound metal ion (purple). [Drawn from 1L9Z.pdb].

initiation by another core enzyme. Thus, the σ *subunit
acts catalytically.*

E. *coli* has seven distinct σ factors for recognizing
several types of promoter sequences in *E. coli* DNA.
The type that recognizes the consensus sequences
described earlier is called σ^{70} because it has a mass of
70 kDa. A different σ factor comes into play when the
temperature is raised abruptly. *E. coli* responds by synthesizing σ^{32}, which recognizes the promoters of *heat-
shock genes*. These promoters exhibit −10 sequences
that are somewhat different from the −10 sequence
for standard promoters (Figure 30.11). The increased
transcription of heat-shock genes leads to the coordinated synthesis of a series of protective proteins. Other
σ factors respond to environmental conditions, such as
nitrogen starvation. These findings demonstrate that *σ
plays the key role in determining where RNA polymerase
initiates transcription.*

Some other bacteria contain a much larger number of σ factors. For example, the genome of the
soil bacterium *Streptomyces coelicolor* encodes more
than 60 σ factors recognized on the basis of their amino
acid sequences. This repertoire allows these cells to
adjust their gene-expression programs to the wide range
of conditions, in regard to nutrients and competing
organisms, that they may experience.

−35   −10

5′~~~T T G A C A~~~~~~~~T A T A A T~~~~3′   Standard promoter

5′~~~T N N C N C N C T T G A A~~~~~~C C C A T N T~~~~3′   Heat-shock promoter

5′~~~C T G G G N A~~~~~~~~T T G C A~~~~3′   Nitrogen-starvation promoter

FIGURE 30.11 Alternative promoter sequences. A comparison of the consensus sequences
of standard, heat-shock, and nitrogen-starvation promoters of *E. coli*. These promoters are
recognized by σ^{70}, σ^{32}, and σ^{54}, respectively.

RNA polymerases must unwind the template double helix for transcription to take place

Although RNA polymerases can search for promoter sites when bound to
double-helical DNA, a segment of the DNA double helix must be unwound
before synthesis can begin. The transition from the *closed promoter complex* (in which DNA is double helical) to the *open promoter complex* (in
which a DNA segment is unwound) is an essential event in transcription
(Figure 30.12). The free energy necessary to break the bonds between
approximately 17 base pairs in the double helix is derived from additional
interactions that are possible when the DNA distorts to
wrap around the RNA polymerase and from interactions
between the single-stranded DNA regions and other parts
of the enzyme. These interactions stabilize the open promoter complex and help pull the template strand into the
active site. The −35 element remains in a double-helical
state, whereas the −10 element is unwound. The stage is
now set for the formation of the first phosphodiester bond
of the new RNA chain.

Closed complex   **Open complex**

FIGURE 30.12 DNA unwinding. The transition from the closed
promoter complex to the open promoter complex requires the
unwinding of approximately 17 base pairs of DNA.

Elongation takes place at transcription bubbles that move along the DNA template

The elongation phase of RNA synthesis begins with the formation of the first phosphodiester bond. Repeated cycles of nucleotide addition can take place at this point. However, until about 10 nucleotides have been added, RNA polymerase sometimes releases the short RNA, which dissociates from the DNA. Once RNA polymerase passes this point, the enzyme stays bound to its template until a termination signal is reached. The region containing RNA polymerase, DNA, and nascent RNA corresponds to the transcription bubble (Figure 30.13). The newly synthesized RNA forms a hybrid helix with the template DNA strand. This RNA–DNA helix is about 8 bp long, which corresponds to nearly one turn of a double helix. The 3′-hydroxyl group of the RNA in this hybrid helix is positioned so that it can attack the α-phosphorus atom of an incoming ribonucleoside triphosphate. The core enzyme also contains a binding site for the coding strand of DNA. About 17 bp of DNA are unwound throughout the elongation phase, as in the initiation phase. The transcription bubble moves a distance of 170 Å (17 nm) in a second, which corresponds to a rate of elongation of about 50 nucleotides per second. Although rapid, it is much slower than the rate of DNA synthesis, which is 800 nucleotides per second.

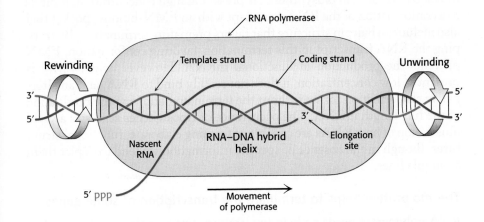

FIGURE 30.13 Transcription bubble. A schematic representation of a transcription bubble in the elongation of an RNA transcript. Duplex DNA is unwound at the forward end of RNA polymerase and rewound at its rear end. The RNA–DNA hybrid rotates during elongation.

Sequences within the newly transcribed RNA signal termination

In bacteria, the termination of transcription is as precisely controlled as its initiation. In the termination phase of transcription, the formation of phosphodiester bonds ceases, the RNA–DNA hybrid dissociates, the unwound region of DNA rewinds, and RNA polymerase releases the DNA. What determines where transcription is terminated? *The transcribed regions of DNA templates contain stop signals.* The simplest one is a *palindromic GC-rich region followed by an AT-rich region.* The RNA transcript of this DNA palindrome is self-complementary (Figure 30.14). Hence, its bases can pair to form a hairpin structure with a stem and loop, a structure favored by its high content of G and C residues. Guanine–cytosine base pairs are more stable than adenine–thymine pairs because of the extra hydrogen bond in the base pair. This stable hairpin is followed by a sequence of four or more uracil residues, which also are crucial for termination. The RNA transcript ends within or just after them.

How does this combination hairpin–oligo(U) structure terminate transcription? First, RNA polymerase likely pauses immediately after it has synthesized a stretch of RNA that folds into a hairpin. Furthermore, the RNA–DNA hybrid helix produced after the hairpin is unstable because

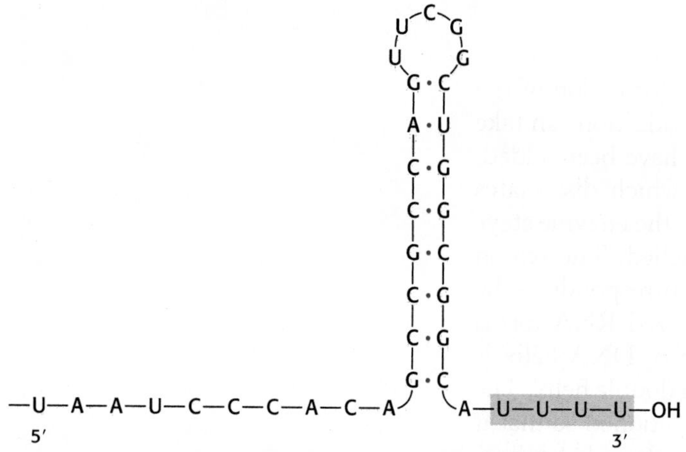

FIGURE 30.14 Termination signal. A termination signal found at the 3′ end of an mRNA transcript consists of a series of bases that form a stable stem-loop structure and a series of U residues.

FIGURE 30.15 Riboswitch. (A) The 5′-end of an mRNA that encodes proteins engaged in the production of flavin mononucleotide (FMN) folds to form a structure that is stabilized by binding FMN. This structure includes a terminator that leads to premature termination of the mRNA. At lower concentrations of FMN, an alternative structure that lacks the terminator is formed, leading to the production of full-length mRNA. (B) The three-dimensional structure of a related FMN-binding riboswitch bound to FMN. The blue and yellow stretches correspond to regions highlighted in the same colors in part A. *Notice* how the yellow strand contacts the bound FMN, stabilizing the structure. [Drawn from 3F2Q.pdb].

its rU–dA base pairs are the weakest of the four kinds. Hence, the pause in transcription caused by the hairpin permits the weakly bound *nascent RNA to dissociate from the DNA template* and then from the enzyme. The solitary DNA template strand rejoins its partner to re-form the DNA duplex, and the transcription bubble closes.

Some messenger RNAs directly sense metabolite concentrations

As we shall explore in Chapters 32 and 33, the expression of many genes is controlled in response to the concentrations of metabolites and signaling molecules within cells. One set of control mechanisms depends on the remarkable ability of some mRNA molecules to form special secondary structures, some of which are capable of directly binding small molecules. These structures are termed *riboswitches*. Consider a riboswitch that controls the synthesis of genes that participate in the biosynthesis of riboflavin in *Bacillus subtilis* (Figure 30.15). When flavin mononucleotide (FMN), a key intermediate in riboflavin biosynthesis, is present at high concentration, it binds to a conformation of the RNA transcript with an FMN-binding pocket that also includes a hairpin structure that favors premature termination. By trapping the RNA transcript in this termination-favoring conformation, FMN prevents the production of functional mRNA. However, when FMN is present at low concentration, it does not readily bind to RNA, and an alternative conformation forms without the terminator, allowing the production of the full-length mRNA. The occurrence of riboswitches serves as a vivid illustration of how RNAs are capable of forming elaborate, functional structures, though in the absence of specific information we tend to depict them as simple lines.

The *rho* protein helps to terminate the transcription of some genes

RNA polymerase needs no help to terminate transcription at a hairpin followed by several U residues. At other sites, however, termination requires the participation of an additional factor. This discovery was prompted by the observation that some RNA molecules synthesized in vitro by RNA

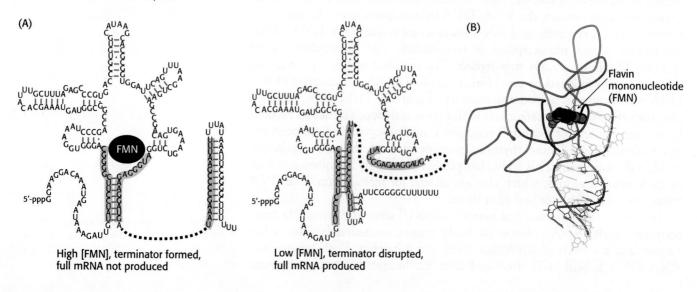

(A) High [FMN], terminator formed, full mRNA not produced

Low [FMN], terminator disrupted, full mRNA produced

(B) Flavin mononucleotide (FMN)

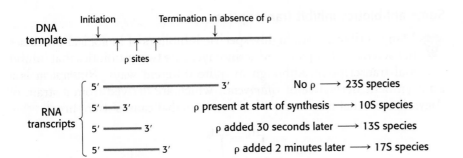

FIGURE 30.16 Effect of ρ protein on the size of RNA transcripts.

polymerase acting alone are *longer* than those made in vivo. The missing factor, a protein that caused the correct termination, was isolated and named *rho* (ρ). Additional information about the action of ρ was obtained by adding this termination factor to an incubation mixture at various times after the initiation of RNA synthesis (Figure 30.16). RNAs with sedimentation coefficients of 10S, 13S, and 17S were obtained when ρ was added at initiation, a few seconds after initiation, and 2 minutes after initiation, respectively. If no ρ was added, transcription yielded a 23S RNA product. It is evident that the template contains at least three termination sites that respond to ρ (yielding 10S, 13S, and 17S RNA) and one termination site that does not (yielding 23S RNA). Thus, specific termination at a site producing 23S RNA can take place in the absence of ρ. However, ρ detects additional termination signals that are not recognized by RNA polymerase alone.

How does ρ provoke the termination of RNA synthesis? *A key clue is the finding that ρ hydrolyzes ATP in the presence of single-stranded RNA but not in the presence of DNA or duplex RNA.* Hexameric ρ is a helicase, homologous to the helicases that we encountered in our consideration of DNA replication (Section 29.1). A stretch of nucleotides is bound in such a way that the RNA passes through the center of the structure (Figure 30.17). The ρ protein is brought into action by sequences located in the nascent RNA that are rich in cytosine and poor in guanine. The helicase activity of ρ enables the protein to pull the nascent RNA while pursuing RNA polymerase. When ρ catches RNA polymerase at the transcription bubble, it breaks the RNA–DNA hybrid helix by functioning as an RNA–DNA helicase.

Proteins in addition to ρ may provoke termination. For example, the *nusA protein* enables RNA polymerase in *E. coli* to recognize a characteristic class of termination sites. *A common feature of protein-independent and protein-dependent termination is that the functioning signals lie in newly synthesized RNA rather than in the DNA template.*

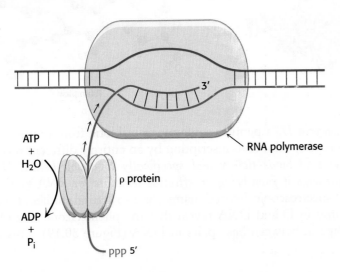

FIGURE 30.17 Mechanism for the termination of transcription by ρ protein. This protein is an ATP-dependent helicase that binds the nascent RNA chain and pulls it away from RNA polymerase and the DNA template. The small arrows represent the movement of ρ protein along the transcript.

Some antibiotics inhibit transcription

Many antibiotics are highly specific inhibitors of biological processes in bacteria. Rifampicin and actinomycin are two antibiotics that inhibit bacterial transcription, although in quite different ways. *Rifampicin* is a semisynthetic derivative of *rifamycins,* which are derived from a strain of *Amycolatopsis* that is related to the bacterium that causes strep throat.

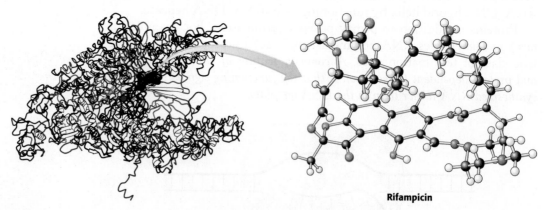

Rifampicin

This antibiotic *specifically inhibits the initiation of RNA synthesis.* Rifampicin interferes with the formation of the first few phosphodiester bonds in the RNA chain. The structure of a complex between a prokaryotic RNA polymerase and rifampicin reveals that the antibiotic blocks the channel into which the RNA–DNA hybrid generated by the enzyme must pass (Figure 30.18). The binding site is 12 Å from the active site itself. Rifampicin can inhibit only the initiation of transcription, not elongation, because the RNA–DNA hybrid present in the enzyme during elongation prevents the antibiotic from binding. The pocket in which rifampicin binds is conserved among bacterial RNA polymerases, but not eukaryotic polymerases, and so rifampicin can be used as an antibiotic in antituberculosis therapy.

Rifampicin

FIGURE 30.18 Antibiotic action.
Rifampicin binds to a pocket in the channel that is normally occupied by the newly formed RNA–DNA hybrid. Thus, the antibiotic blocks elongation after only two or three nucleotides have been added.

Actinomycin D, a peptide-containing antibiotic from a different strain of *Streptomyces,* inhibits transcription by an entirely different mechanism. *Actinomycin D binds tightly and specifically to double-helical DNA and thereby prevents it from being an effective template for RNA synthesis.* The results of spectroscopic, hydrodynamic, and structural studies of complexes of actinomycin D and DNA reveal that the phenoxazone ring of actinomycin slips in between base pairs in DNA (Figure 30.19). This mode of

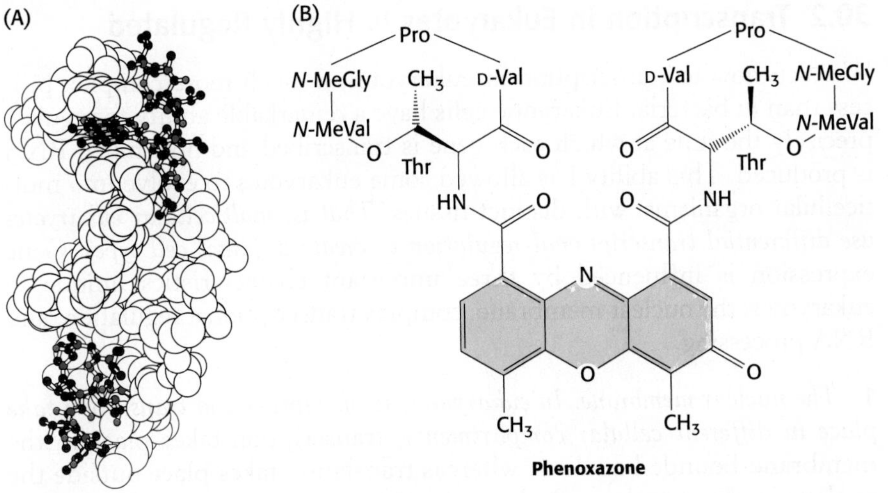

(A)

(B)

Phenoxazone

FIGURE 30.19 Actinomycin–DNA complex structure. (A) The structure of a complex between a DNA duplex (shown as a space-filling model) and actinomycin B (shown as a ball-and-stick model). Two actinomycin B molecules are bound in the complex. (B) The structure of actinomycin B showing the phenoxazone ring. *Notice* in (A) near the top, how the phenoxazone (yellow) slides between base pairs of the DNA. Abbreviation: Me, methyl. [Drawn from 1I3W.pdb]

binding is called *intercalation*. At low concentrations, actinomycin D inhibits transcription without significantly affecting DNA replication or protein synthesis. Hence, *actinomycin D is extensively used as a highly specific inhibitor of the formation of new RNA in both prokaryotic and eukaryotic cells*. Its ability to inhibit the growth of rapidly dividing cells makes it an effective therapeutic agent in the treatment of some cancers.

Precursors of transfer and ribosomal RNA are cleaved and chemically modified after transcription in prokaryotes

In prokaryotes, messenger RNA molecules undergo little or no modification after synthesis by RNA polymerase. Indeed, many mRNA molecules are translated while they are being transcribed. In contrast, *transfer RNA (tRNA) and ribosomal RNA (rRNA) molecules are generated by cleavage and other modifications of nascent RNA chains*. For example, in *E. coli*, the three rRNAs and a tRNA are excised from a single primary RNA transcript that also contains spacer regions (Figure 30.20). Other transcripts contain arrays of several kinds of tRNA or of several copies of the same tRNA. The nucleases that cleave and trim these precursors of rRNA and tRNA are highly precise. *Ribonuclease P* (RNase P), for example, generates the correct 5′ terminus of all tRNA molecules in *E. coli*. Sidney Altman and his coworkers showed that this interesting enzyme contains a catalytically active RNA molecule. *Ribonuclease III* (RNase III) excises 5S, 16S, and 23S rRNA precursors from the primary transcript by cleaving double-helical hairpin regions at specific sites.

A second type of processing is the *addition of nucleotides to the termini of some RNA chains*. For example, CCA, a terminal sequence required for the function of all tRNAs, is added to the 3′ ends of tRNA molecules for which this terminal sequence is not encoded in the DNA. The enzyme that catalyzes the addition of CCA is atypical for an RNA polymerase in that it does not use a DNA template. A third type of processing is the *modification of bases and ribose units* of ribosomal RNAs. In prokaryotes, some bases of rRNA are methylated. Unusual bases are found in all tRNA molecules. They are formed by the enzymatic modification of a standard ribonucleotide in a tRNA precursor. For example, uridylate residues are modified after transcription to form *ribothymidylate* and *pseudouridylate*. These modifications generate diversity, allowing greater structural and functional versatility.

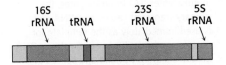

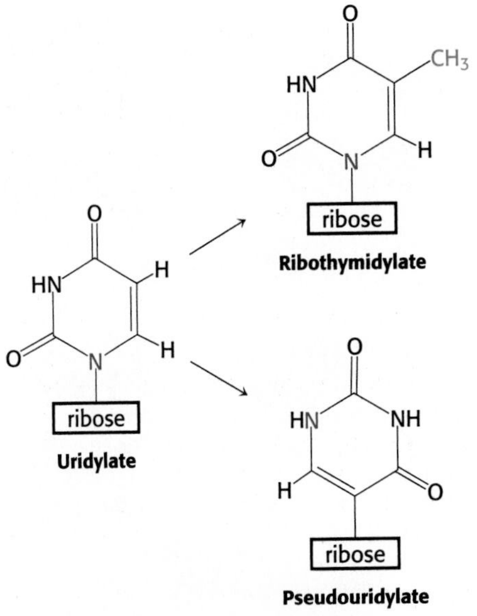

FIGURE 30.20 Primary transcript. Cleavage of this transcript produces 5S, 16S, and 23S rRNA molecules and a tRNA molecule. Spacer regions are shown in yellow.

Uridylate → **Ribothymidylate**

Uridylate → **Pseudouridylate**

30.2 Transcription in Eukaryotes Is Highly Regulated

We turn now to transcription in eukaryotes, a much more complex process than in bacteria. Eukaryotic cells have a remarkable ability to regulate precisely the time at which each gene is transcribed and how much RNA is produced. This ability has allowed some eukaryotes to evolve into multicellular organisms, with distinct tissues. *That is, multicellular eukaryotes use differential transcriptional regulation to create different cell types.* Gene expression is influenced by three important characteristics unique to eukaryotes: the nuclear membrane, complex transcriptional regulation, and RNA processing.

1. *The nuclear membrane. In eukaryotes, transcription and translation take place in different cellular compartments:* transcription takes place in the membrane-bounded nucleus, whereas translation takes place outside the nucleus in the cytoplasm. In bacteria, the two processes are closely coupled (Figure 30.21). Indeed, the translation of bacterial mRNA begins while the transcript is still being synthesized. *The spatial and temporal separation of transcription and translation enables eukaryotes to regulate gene expression in much more intricate ways, contributing to the richness of eukaryotic form and function.*

FIGURE 30.21 Transcription and translation. These two processes are closely coupled in prokaryotes whereas they are spatially and temporally separate in eukaryotes. (A) In prokaryotes, the primary transcript serves as mRNA and is used immediately as the template for protein synthesis. (B) In eukaryotes, mRNA precursors are processed and spliced in the nucleus before being transported to the cytoplasm for translation into protein. [Information from J. Darnell, H. Lodish, and D. Baltimore. *Molecular Cell Biology,* 2nd ed. (Scientific American Books, 1990), p. 230.]

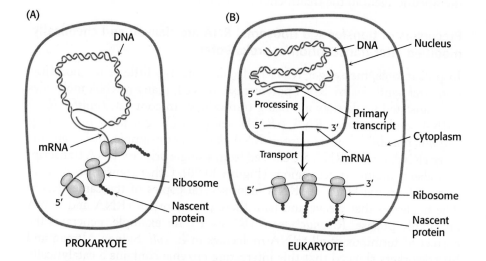

2. *Complex transcriptional regulation.* Like bacteria, eukaryotes rely on conserved sequences in DNA to regulate the initiation of transcription. But bacteria have only three promoter elements (the -10, -35, and UP elements), whereas eukaryotes use a variety of types of promoter elements, each identified by its own conserved sequence. Not all possible types will be present together in the same promoter. *In eukaryotes, elements that regulate transcription can be found at a variety of locations in DNA,* upstream or downstream of the start site and sometimes at distances much farther from the start site than in prokaryotes. For example, enhancer elements located on DNA far from the start site increase the promoter activity of specific genes.

3. *RNA processing.* Although both bacteria and eukaryotes modify RNA, *eukaryotes very extensively process nascent RNA destined to become mRNA.* This processing includes modifications to both ends and, most significantly, splicing out segments of the primary transcript. RNA processing is described in Section 30.3.

Three types of RNA polymerase synthesize RNA in eukaryotic cells

In bacteria, RNA is synthesized by a single kind of polymerase. In contrast, the nucleus of a typical eukaryotic cell contains three types of RNA polymerase differing in template specificity and location in the nucleus (Table 30.2). All these polymerases are large proteins, containing from 8 to 14 subunits and having total molecular masses greater than 500 kDa. *RNA polymerase I* is located in specialized structures within the nucleus called nucleoli, where it transcribes the tandem array of genes for 18S, 5.8S, and 28S rRNA. The other rRNA molecule (5S rRNA) and all the tRNA molecules are synthesized by *RNA polymerase III*, which is located in the nucleoplasm rather than in nucleoli. *RNA polymerase II*, which also is located in the nucleoplasm, synthesizes the precursors of mRNA as well as several small RNA molecules, such as those of the splicing apparatus and many of the precursors to small regulatory RNAs.

TABLE 30.2 Eukaryotic RNA polymerases

Type	Location	Cellular transcripts	Effects of a-amanitin
I	Nucleolus	18S, 5.8S, and 28S rRNA	Insensitive
II	Nucleoplasm	mRNA precursors and snRNA	Strongly inhibited
III	Nucleoplasm	tRNA and 5S rRNA	Inhibited by high concentrations

Although all eukaryotic RNA polymerases are homologous to one another and to prokaryotic RNA polymerases, RNA polymerase II contains a unique *carboxyl-terminal domain* on the 220-kDa subunit called the CTD; this domain is unusual because it contains multiple repeats of a YSPTSPS consensus sequence. The activity of RNA polymerase II is regulated by phosphorylation mainly on the serine residues of the CTD.

Another major distinction among the polymerases lies in their responses to the toxin α-amanitin, a cyclic octapeptide that contains several modified amino acids.

α-Amanitin

Amanita phalloides, also called the *death cap*. [Jacana/Science Source.]

α-Amanitin is produced by the poisonous mushroom *Amanita phalloides*, which is also called the *death cap* or the *destroying angel*. More than a hundred deaths result worldwide each year from the ingestion of poisonous mushrooms. α-Amanitin binds very tightly ($K_d = 10\,\text{nM}$) to RNA polymerase II and thereby blocks the elongation phase of RNA synthesis.

RNA polymerase I promoter

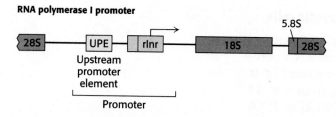

RNA polymerase II promoter

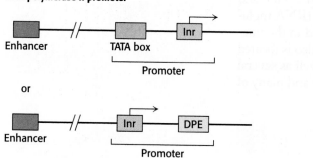

RNA polymerase III promoter

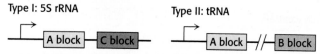

FIGURE 30.22 Common eukaryotic promoter elements. Each eukaryotic RNA polymerase recognizes a set of promoter elements—sequences in DNA that promote transcription. The RNA polymerase I promoter consists of a ribosomal initiator (rInr) and an upstream promoter element (UPE). The RNA polymerase II promoter likewise includes an initiator element (Inr) and may also include either a TATA box or a downstream promoter element (DPE). Separate from the promoter region, enhancer elements bind specific transcription factors. RNA polymerase III promoters consist of conserved sequences that lie within the transcribed genes.

$$5'\ T_{82}\ A_{97}\ T_{93}\ A_{85}\ A_{63}\ A_{88}\ A_{50}\ 3'$$

TATA box

FIGURE 30.23 TATA box. Comparisons of the sequences of more than 100 eukaryotic promoters led to the consensus sequence shown. The subscripts denote the frequency (%) of the base at that position.

Higher concentrations of α-amanitin (1 μM) inhibit polymerase III, whereas polymerase I is insensitive to this toxin. This pattern of sensitivity is highly conserved throughout the animal and plant kingdoms.

Eukaryotic polymerases also differ from each other in the promoters to which they bind. Eukaryotic genes, like prokaryotic genes, require promoters for transcription initiation. Like prokaryotic promoters, eukaryotic promoters consist of conserved sequences that attract the polymerase to the start site. However, eukaryotic promoters differ distinctly in sequence and position, depending on the type of RNA polymerase to which they bind (Figure 30.22).

1. *RNA polymerase I.* The ribosomal DNA (rDNA) transcribed by polymerase I is arranged in several hundred tandem repeats, each containing a copy of each of three rRNA genes. The promoter sequences are located in stretches of DNA separating the genes. At the transcriptional start site lies a TATA-like sequence called the *ribosomal initiator element* (rInr). Farther upstream, 150 to 200 bp from the start site, is the *upstream promoter element* (UPE). Both elements aid transcription by binding proteins that recruit RNA polymerase I.

2. *RNA polymerase II.* Promoters for RNA polymerase II, like prokaryotic promoters, include a set of consensus sequences that define the start site and recruit the polymerase. However, the promoter can contain any combination of a number of possible consensus sequences. Unique to eukaryotes, they also include enhancer elements that can be very distant (more than 1 kb) from the start site.

3. *RNA polymerase III.* Promoters for RNA polymerase III are *within* the transcribed sequence, downstream of the start site. There are two types of intergenic promoters for RNA polymerase III. Type I promoters, found in the 5S rRNA gene, contain two short, conserved sequences known as the A block and the C block. Type II promoters, found in tRNA genes, consist of two 11-bp sequences, the A block and the B block, situated about 15 bp from either end of the gene.

Three common elements can be found in the RNA polymerase II promoter region

RNA polymerase II transcribes all of the protein-coding genes in eukaryotic cells. Promoters for RNA polymerase II, like those for bacterial polymerases, are generally located on the 5′ side of the start site for transcription. Because these sequences are on the *same* molecule of DNA as the genes being transcribed, they are called *cis-acting elements*. The most commonly recognized cis-acting element for genes transcribed by RNA polymerase II is called the *TATA box* on the basis of its consensus sequence (Figure 30.23). The TATA box is usually found between positions −30 and −100. Note that the eukaryotic TATA box closely resembles the prokaryotic −10 sequence (TATAAT) but is farther from the start site. The mutation of a single base in the TATA box markedly impairs promoter activity. Thus, the precise sequence, not just a high content of AT pairs, is essential.

The TATA box is often paired with an *initiator element* (Inr), a sequence found at the transcriptional start site, between positions −3 and +5. This sequence defines the start site because the other promoter elements are at variable distances from that site. Its presence increases transcriptional activity.

A third element, the *downstream core promoter element* (DPE), is commonly found in conjunction with the Inr in transcripts that lack the TATA box. In contrast with the TATA box, the DPE is found downstream of the start site, between positions +28 and +32.

Additional regulatory sequences are located between −40 and −150. Many promoters contain a *CAAT box,* and some contain a *GC box* (Figure 30.24). Constitutive genes (genes that are continuously expressed rather than regulated) tend to have GC boxes in their promoters. The positions of these upstream sequences vary from one promoter to another, in contrast with the quite constant location of the −35 region in prokaryotes. Another difference is that the CAAT box and the GC box can be effective when present on the template (antisense) strand, unlike the −35 region, which must be present on the coding (sense) strand. These differences between prokaryotes and eukaryotes correspond to fundamentally different mechanisms for the recognition of cis-acting elements. The −10 and −35 sequences in prokaryotic promoters are binding sites for RNA polymerase and its associated σ factor. In contrast, the TATA, CAAT, and GC boxes and other cis-acting elements in eukaryotic promoters are recognized by proteins other than by RNA polymerase itself.

The TFIID protein complex initiates the assembly of the active transcription complex

Cis-acting elements constitute only part of the puzzle of eukaryotic gene expression. *Transcription factors* that bind to these elements also are required. For example, RNA polymerase II is guided to the start site by a set of transcription factors known collectively as *TFII* (*TF* stands for transcription factor, and *II* refers to RNA polymerase II). Individual TFII factors are called TFIIA, TFIIB, and so on.

In TATA-box promoters, the key initial event is the recognition of the TATA box by the TATA-box-binding protein (TBP), a 30-kDa component of the 700-kDa TFIID complex (Figure 30.25). In TATA-less promoters, other proteins in the TFIID complex bind the core promoter elements but, because less is known about these interactions, we will consider only the TATA-box–TBP binding interaction. TBP binds 10^5 times as tightly to the TATA box as to nonconsensus sequences; the dissociation constant of the TBP–TATA-box complex is approximately 1 nM. TBP is a saddle-shaped protein consisting of two similar domains (Figure 30.26).

5′ G G N C A A T C T 3′
CAAT box

5′ G G G C G G 3′
GC box

FIGURE 30.24 CAAT box and GC box. Consensus sequences for the CAAT and GC boxes of eukaryotic promoters for mRNA precursors.

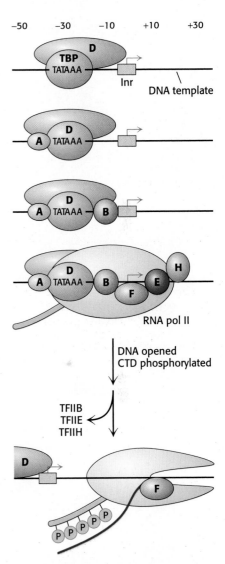

FIGURE 30.25 Transcription initiation. Transcription factors TFIIA, B, D, E, F, and H are essential in initiating transcription by RNA polymerase II. The step-by-step assembly of these general transcription factors begins with the binding of TFIID (purple) to the TATA box. [The TATA-box-binding protein (TBP), a component of TFIID, recognizes the TATA box.] After assembly, TFIIH opens the DNA double helix and phosphorylates the carboxyl-terminal domain (CTD), allowing the polymerase to leave the promoter and begin transcription. The red arrow marks the transcription start site.

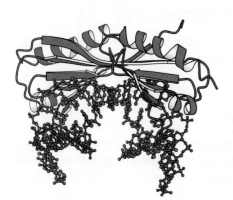

FIGURE 30.26 Complex formed by TATA-box-binding protein and DNA. The saddlelike structure of the protein sits atop a DNA fragment. *Notice* that the DNA is significantly unwound and bent. [Drawn from 1CDW.pdb.]

The TATA box of DNA binds to the concave surface of TBP. This binding induces large conformational changes in the bound DNA. The double helix is substantially unwound to widen its *minor groove*, enabling it to make extensive contact with the antiparallel β strands on the concave side of TBP. Hydrophobic interactions are prominent at this interface. Four phenylalanine residues, for example, are intercalated between base pairs of the TATA box. The flexibility of AT-rich sequences is generally exploited here in bending the DNA. Immediately outside the TATA box, classical B-DNA resumes. The TBP–TATA-box complex is distinctly asymmetric. The asymmetry is crucial for specifying a unique start site and ensuring that transcription proceeds unidirectionally.

TBP bound to the TATA box is the heart of the initiation complex (Figure 30.25). The surface of the TBP saddle provides docking sites for the binding of other components. Additional transcription factors assemble on this nucleus in a defined sequence. TFIIA is recruited, followed by TFIIB; then TFIIF, RNA polymerase II, TFIIE, and TFIIH join the other factors to form a complex called the *basal transcription apparatus*. During the formation of the basal transcription apparatus, the carboxyl-terminal domain (CTD) is unphosphorylated and plays a role in transcription regulation through its binding to an enhancer-associated complex called mediator (Section 33.2). The phosphorylated CTD stabilizes transcription elongation by RNA polymerase II and recruits RNA-processing enzymes that act in the course of elongation. *Phosphorylation of the CTD by TFIIH marks the transition from initiation to elongation.* The importance of the carboxyl-terminal domain is highlighted by the finding that yeast containing mutant polymerase II with fewer than 10 repeats in the CTD are not viable. Most of the factors are released before the polymerase leaves the promoter and can then participate in another round of initiation.

Multiple transcription factors interact with eukaryotic promoters

The basal transcription complex described in the preceding section initiates transcription at a low frequency. Additional transcription factors that bind to other sites are required to achieve a high rate of mRNA synthesis. Their role is to selectively stimulate specific genes. Upstream stimulatory sites in eukaryotic genes are diverse in sequence and variable in position. Their variety suggests that they are recognized by many different specific proteins. Indeed, many transcription factors have been isolated, and their binding sites have been identified by footprinting experiments. For example, *heat-shock transcription factor* (HSTF) is expressed in *Drosophila* after an abrupt increase in temperature. This 93-kDa DNA-binding protein binds to the following consensus sequence:

5′–CNNGAANNTCCNNG–3′

Several copies of this sequence, known as the *heat-shock response element*, are present starting at a site 15 bp upstream of the TATA box.

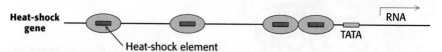

HSTF differs from σ^{32}, a heat-shock protein of *E. coli* (p. 990), in binding directly to response elements in heat-shock promoters rather than first becoming associated with RNA polymerase.

Enhancer sequences can stimulate transcription at start sites thousands of bases away

The activities of many promoters in higher eukaryotes are greatly increased by another type of cis-acting element called an *enhancer*. Enhancer sequences have no promoter activity of their own *yet can exert their stimulatory actions over distances of several thousand base pairs. They can be upstream, downstream, or even in the midst of a transcribed gene.* Moreover, enhancers are effective when present on *either the coding or non-coding DNA strand.*

A particular enhancer is effective only in certain cells. For example, the immunoglobulin enhancer functions in B lymphocytes but not elsewhere. Cancer can result if the relation between genes and enhancers is disrupted. In Burkitt lymphoma and B-cell leukemia, a chromosomal translocation brings the proto-oncogene *myc* (a transcription factor itself) under the control of a powerful immunoglobulin enhancer. The consequent dysregulation of the *myc* gene is believed to play a role in the progression of the cancer.

Transcription factors and other proteins that bind to regulatory sites on DNA can be regarded as passwords that cooperatively open multiple locks, giving RNA polymerase access to specific genes. The discovery of promoters and enhancers has allowed us to gain a better understanding of how genes are selectively expressed in eukaryotic cells. The regulation of eukaryotic gene transcription, discussed in Chapter 33, is the fundamental means of controlling gene expression.

Although bacteria lack TBP, archaea utilize a TBP molecule that is structurally quite similar to the eukaryotic protein. In fact, transcriptional control processes in archaea are, in general, much more similar to those in eukaryotes than are the processes in bacteria. Many components of the eukaryotic transcriptional machinery evolved from those in an ancestor of archaea.

30.3 The Transcription Products of Eukaryotic Polymerases Are Processed

Virtually all the initial products of transcription are further processed in eukaryotes. For example, primary transcripts (pre-mRNA molecules), the products of RNA polymerase II action, acquire a cap at their 5′ ends and a poly(A) tail at their 3′ends. Most importantly, *nearly all mRNA precursors in higher eukaryotes are spliced.* Introns are precisely excised from primary transcripts, and exons are joined to form mature mRNAs with continuous messages. Some mature mRNAs are only a tenth the size of their precursors, which can be as large as 30 kb or more. The pattern of splicing can be regulated in the course of development to generate variations on a theme, such as membrane-bound or secreted forms of antibody molecules. Alternative splicing enlarges the repertoire of proteins in eukaryotes and is one clear illustration of why the proteome is more complex than the genome. The particular processing steps and the factors taking part vary according to the type of RNA polymerase.

RNA polymerase I produces three ribosomal RNAs

Several RNA molecules are key components of ribosomes. RNA polymerase I transcription produces a single precursor (45S in mammals) that encodes

FIGURE 30.27 Processing of eukaryotic pre-rRNA. The mammalian pre-rRNA transcript contains the RNA sequences destined to become the 18S, 5.8S, and 28S rRNAs of the small and large ribosomal subunits. First, nucleotides are modified: small nucleolar ribonucleoproteins methylate specific nucleoside groups and convert selected uridines into pseudouridines (indicated by red lines). Next, the pre-rRNA is cleaved and packaged to form mature ribosomes, in a highly regulated process in which more than 200 proteins take part.

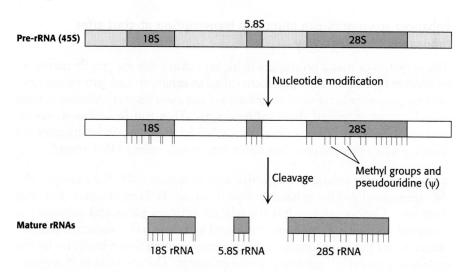

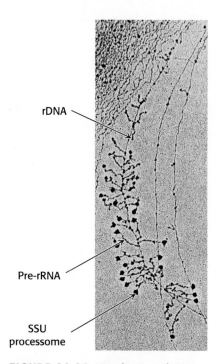

FIGURE 30.28 Visualization of rRNA transcription and processing in eukaryotes. Transcription of rRNA and its assembly into precursor ribosomes can be visualized by electron microscopy. The structures resemble Christmas trees: the trunk is the rDNA and each branch is a pre-rRNA transcript. Transcription starts at the top of the tree, where the shortest transcripts can be seen, and progresses down the rDNA to the end of the gene. The terminal knobs visible at the end of some pre-rRNA transcripts likely correspond to the SSU processome, a large ribonucleoprotein required for processing the pre-rRNA. [Reprinted by permission from Macmillan Publishers Ltd: *Nature* 417: 967–970, F. Dragon et. al. A large nucleolar U3 ribonucleoprotein required for 18S ribosomal RNA biogenesis, © 2002.]

three RNA components of the ribosome: the 18S rRNA, the 28S rRNA, and the 5.8S rRNA (Figure 30.27). The 18S rRNA is the RNA component of the small ribosomal subunit (40S), and the 28S and 5.8S rRNAs are two RNA components of the large ribosomal subunit (60S). The other RNA component of the large ribosomal subunit, the 5S rRNA, is transcribed by RNA polymerase III as a separate transcript.

The cleavage of the precursor into three separate rRNAs is actually the final step in its processing. First, the nucleotides of the pre-rRNA sequences destined for the ribosome undergo extensive modification, on both ribose and base components, directed by many *small nucleolar ribonucleoproteins* (snoRNPs), each of which consists of one snoRNA and several proteins. The pre-rRNA is assembled with ribosomal proteins, as guided by processing factors, in a large ribonucleoprotein. For instance, the small-subunit (SSU) processome is required for 18S rRNA synthesis and can be visualized in electron micrographs as a terminal knob at the 5′ ends of the nascent rRNAs (Figure 30.28). Finally, rRNA cleavage (sometimes coupled with additional processing steps) releases the mature rRNAs assembled with ribosomal proteins as ribosomes. Like those of RNA polymerase I transcription itself, most of these processing steps take place in the cell nucleolus, a nuclear subcompartment.

RNA polymerase III produces transfer RNA

Eukaryotic tRNA transcripts are among the most processed of all RNA polymerase III transcripts. Like those of prokaryotic tRNAs, the 5′ leader is cleaved by RNase P, the 3′ trailer is removed, and CCA is added by the CCA-adding enzyme (Figure 30.29). Eukaryotic tRNAs are also heavily modified on base and ribose moieties; these modifications are important for function. In contrast with prokaryotic tRNAs, many eukaryotic pre-tRNAs are also spliced by an endonuclease and a ligase to remove an intron.

The product of RNA polymerase II, the pre-mRNA transcript, acquires a 5′ cap and a 3′ poly(A) tail

Perhaps the most extensively studied transcription product is the product of RNA polymerase II: most of this RNA will be processed to mRNA. The immediate product of RNA polymerase II is sometimes referred to as precursor-to-messenger RNA, or *pre-mRNA*. Most pre-mRNA molecules

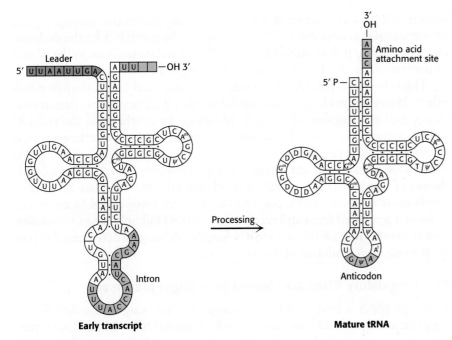

FIGURE 30.29 Transfer RNA precursor processing. The conversion of a yeast tRNA precursor into a mature tRNA requires the removal of a 14-nucleotide intron (yellow), the cleavage of a 5′ leader (green), and the removal of UU and the attachment of CCA at the 3′ end (red). In addition, several bases are modified.

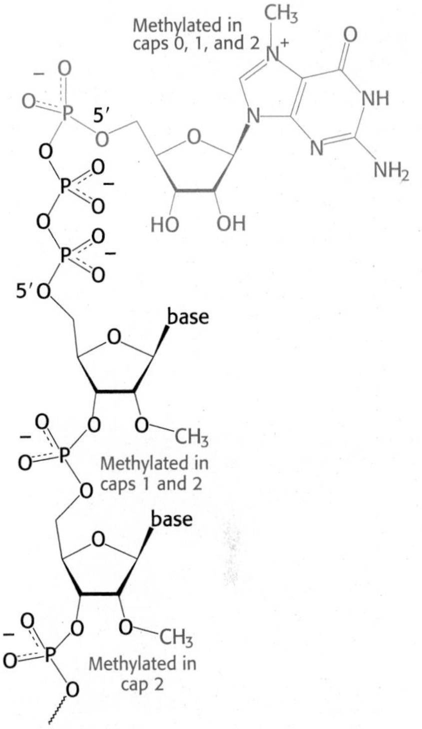

are spliced to remove the introns. Moreover, both the 5′ and the 3′ ends are modified, and both modifications are retained as the pre-mRNA is converted into mRNA.

As in prokaryotes, eukaryotic transcription usually begins with A or G. However, the 5′ triphosphate end of the nascent RNA chain is immediately modified. First, a phosphoryl group is released by hydrolysis. The diphosphate 5′ end then attacks the α-phosphorus atom of GTP to form a very unusual 5′–5′ triphosphate linkage. This distinctive terminus is called a *cap* (Figure 30.30). The N-7 nitrogen of the terminal guanine is then methylated by S-adenosylmethionine to form *cap 0*. The adjacent riboses may be methylated to form *cap 1* or *cap 2*. Transfer RNA and ribosomal RNA molecules, in contrast with messenger RNAs and with small RNAs that participate in splicing, do not have caps. Caps contribute to the stability of mRNAs by protecting their 5′ ends from phosphatases and nucleases. In addition, caps enhance the translation of mRNA by eukaryotic protein-synthesizing systems.

As mentioned earlier, pre-mRNA is also modified at the 3′ end. *Most eukaryotic mRNAs contain a polyadenylate, poly(A), tail at that end,* added after transcription has ended. The DNA template does not encode this poly(A) tail. Indeed, the nucleotide preceding poly(A) is not the last nucleotide to be transcribed. Some primary transcripts contain hundreds of nucleotides beyond the 3′ end of the mature mRNA.

How is the 3′ end of the pre-mRNA given its final form? *Eukaryotic primary transcripts are cleaved by a specific endonuclease that recognizes the sequence AAUAAA* (Figure 30.31). Cleavage does not take place if this sequence or a segment of some 20 nucleotides on its 3′ side is deleted. The presence of internal AAUAAA sequences in some

FIGURE 30.30 Capping the 5′ end. Caps at the 5′ end of eukaryotic mRNA include 7-methylguanylate (red) attached by a triphosphate linkage to the ribose at the 5′ end. None of the riboses are methylated in cap 0, one is methylated in cap 1, and both are methylated in cap 2.

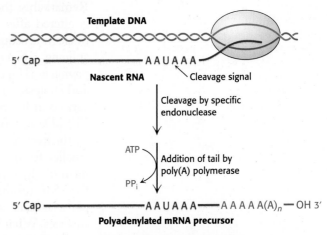

FIGURE 30.31 Polyadenylation of a primary transcript. A specific endonuclease cleaves the RNA downstream of AAUAAA. Poly(A) polymerase then adds about 250 adenylate residues.

mature mRNAs indicates that AAUAAA is only part of the cleavage signal; its context also is important. After cleavage of the pre-RNA by the endonuclease, a *poly(A) polymerase* adds about 250 adenylate residues to the 3′ end of the transcript; ATP is the donor in this reaction.

The role of the poly(A) tail is still not firmly established despite much effort. However, evidence is accumulating that it enhances translation efficiency and the stability of mRNA. Blocking the synthesis of the poly(A) tail by exposure to *3′-deoxyadenosine (cordycepin)* does not interfere with the synthesis of the primary transcript. Messenger RNA devoid of a poly(A) tail can be transported out of the nucleus. However, an mRNA molecule devoid of a poly(A) tail is usually a much less effective template for protein synthesis than is one with a poly(A) tail. Indeed, some mRNAs are stored in an unadenylated form and receive the poly(A) tail only when translation is imminent. The half-life of an mRNA molecule may be determined in part by the rate of degradation of its poly(A) tail.

Small regulatory RNAs are cleaved from larger precursors

Cleavage plays a role in the processing of small single-stranded RNAs (approximately 20–23 nucleotides) called *microRNAs*. MicroRNAs play key roles in gene regulation in eukaryotes, as we shall see in Chapter 33. They are generated from initial transcripts produced by RNA polymerase II and, in some cases, RNA polymerase III. These transcripts fold into hairpin structures that are cleaved by specific nucleases at various stages (Figure 30.32). The final single-stranded RNAs are bound by members of the Argonaute family of proteins where the RNAs help target the regulation of specific genes.

FIGURE 30.32 Small regulatory RNA production. A pathway from a transcription product including a microRNA to the mature microRNA bound to an Argonaute protein. The initial transcription product, a pri-microRNA, is first cleaved to a small double-stranded RNA called a pre-microRNA. One of the strands of the pre-microRNA, the mature microRNA, is then bound by an Argonaute protein.

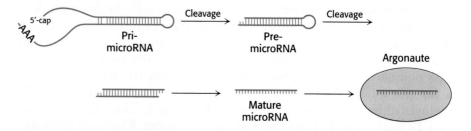

RNA editing changes the proteins encoded by mRNA

Remarkably, the amino acid sequence information encoded by some mRNAs is altered after transcription. *RNA editing* is the term for a change in the nucleotide sequence of RNA after transcription by processes other than RNA splicing. RNA editing is prominent in some systems already discussed. *Apolipoprotein B* (apo B) plays an important role in the transport of triacylglycerols and cholesterol by forming an amphipathic spherical shell around the lipids carried in lipoprotein particles (Section 26.3). Apo B exists in two forms, a 512-kDa *apo B-100* and a 240-kDa *apo B-48*. The larger form, synthesized by the liver, participates in the transport of lipids synthesized in the cell. The smaller form, synthesized by the small intestine, carries dietary fat in the form of chylomicrons. Apo B-48 contains the 2152 N-terminal residues of the 4536-residue apo B-100. This truncated molecule can form lipoprotein particles but cannot bind to the low-density-lipoprotein receptor on cell surfaces. What is the relationship between these two forms of apo B? Experiments revealed that a totally unexpected mechanism for generating diversity is at work: *the changing of the nucleotide sequence of mRNA after its synthesis* (Figure 30.33). *A specific cytidine residue of mRNA is deaminated to uridine, which changes the codon at residue 2153 from CAA (Gln) to UAA (stop).* The deaminase that catalyzes this reaction is present in the small intestine, but not in the liver, and is expressed only at certain developmental stages.

RNA editing is not confined to apolipoprotein B. Glutamate opens cation-specific channels in the vertebrate central nervous system by binding to receptors in postsynaptic membranes. RNA editing changes a single glutamine codon (CAG) in the mRNA for the glutamate receptor to the codon for arginine (read as CGG). The substitution of Arg for Gln in the receptor prevents Ca^{2+}, but not Na^+, from flowing through the channel.

RNA editing is likely much more common than was formerly thought. The chemical reactivity of nucleotide bases, including the susceptibility to deamination that necessitates complex DNA-repair mechanisms, has been harnessed as an engine for generating molecular diversity at the RNA and, hence, protein levels.

In trypanosomes (parasitic protozoans), a different kind of RNA editing markedly changes several mitochondrial mRNAs. Nearly half the uridine residues in these mRNAs are inserted by RNA editing. A *guide RNA molecule* identifies the sequences to be modified, and a *poly(U) tail* on the guide donates uridine residues to the mRNAs undergoing editing. DNA sequences evidently do not always faithfully disclose the sequence of encoded proteins: functionally crucial changes to mRNA can take place.

Sequences at the ends of introns specify splice sites in mRNA precursors

Most genes in higher eukaryotes are composed of exons and introns (Section 4.7). The introns must be excised and the exons must be linked to form the final mRNA in a process called *RNA splicing.* This splicing must be exquisitely sensitive: splicing just one nucleotide upstream or downstream of the intended site would create a one-nucleotide shift, which would alter the reading frame on the 3' side of the splice to give an entirely different amino acid sequence, likely including a premature stop codon. Thus, the correct splice site must be clearly marked. Does a particular sequence denote the splice site? The sequences of thousands of intron–exon junctions within RNA transcripts are known. In eukaryotes from yeast to mammals, these sequences have a common structural motif: *the intron begins with GU and ends with AG.* The consensus sequence at the 5' splice in vertebrates is AGGUAAGU, where the GU is invariant (Figure 30.34). At the 3' end of an intron, the consensus sequence is a stretch of *10 pyrimidines* (U or C; termed the *polypyrimidine tract*), followed by any base and then by C, and ending with the invariant AG. Introns also have an important internal site located between 20 and 50 nucleotides upstream of the 3' splice site; it is called the *branch site* for reasons that will be evident shortly. In yeast, the branch-site sequence is nearly always UACUAAC, whereas, in mammals, a variety of sequences are found.

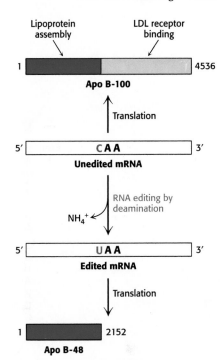

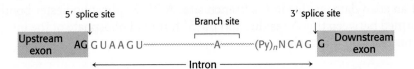

FIGURE 30.34 Splice sites. Consensus sequences for the 5' splice site and the 3' splice site are shown. Py stands for pyrimidine.

FIGURE 30.33 RNA editing. Enzyme-catalyzed deamination of a specific cytidine residue in the mRNA for apolipoprotein B-100 changes a codon for glutamine (CAA) to a stop codon (UAA). Apolipoprotein B-48, a truncated version of the protein lacking the LDL receptor-binding domain, is generated by this posttranscriptional change in the mRNA sequence. [Information from P. Hodges and J. Scott, *Trends Biochem. Sci.* 17:77, 1992.]

The 5' and 3' splice sites and the branch site are essential for determining where splicing takes place. Mutations in each of these three critical regions lead to aberrant splicing. Introns vary in length from 50 to 10,000 nucleotides, and so the splicing machinery may have to find the 3' site several thousand nucleotides away. Specific sequences near the splice sites (in both

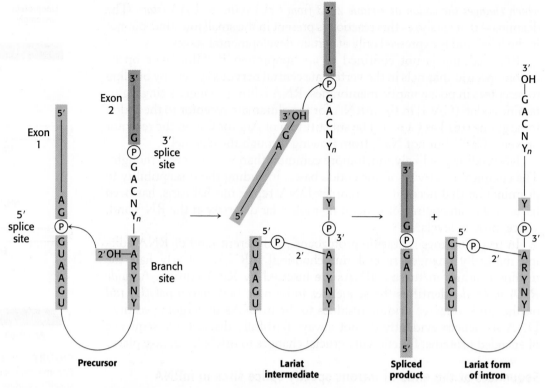

FIGURE 30.35 Splicing mechanism used for mRNA precursors. The upstream (5′) exon is shown in blue, the downstream (3′) exon in green, and the branch site in yellow. Y stands for a pyrimidine nucleotide, R for a purine nucleotide, and N for any nucleotide. The 5′ splice site is attacked by the 2′-OH group of the branch-site adenosine residue. The 3′ splice site is attacked by the newly formed 3′-OH group of the upstream exon. The exons are joined, and the intron is released in the form of a lariat. [Information from P. A. Sharp, *Cell* 42:397–408, 1985.]

the introns and the exons) play an important role in splicing regulation, particularly in designating splice sites when there are many alternatives. Researchers are currently attempting to determine the factors that contribute to splice-site selection for individual mRNAs. Despite our knowledge of splice-site sequences, predicting pre-mRNAs and their protein products from genomic DNA sequence information remains a challenge.

Splicing consists of two sequential transesterification reactions

The splicing of nascent mRNA molecules is a complicated process. It requires the cooperation of several small RNAs and proteins that form a large complex called a *spliceosome*. However, the chemistry of the splicing process is simple. Splicing begins with the cleavage of the phosphodiester bond between the upstream exon (exon 1) and the 5′ end of the intron (Figure 30.35). The attacking group in this reaction is the 2′-OH group of an adenylate residue in the branch site. A 2′–5′ phosphodiester bond is formed between this A residue and the 5′ terminal phosphate of the intron. This reaction is a transesterification.

Note that this adenylate residue is also joined to two other nucleotides by normal 3′–5′ phosphodiester bonds (Figure 30.36). Hence a *branch* is generated at this site, and a *lariat intermediate* is formed.

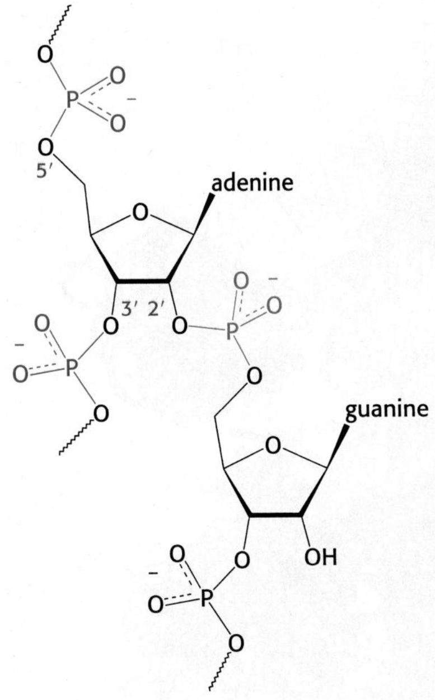

FIGURE 30.36 Splicing branch point. The structure of the branch point in the lariat intermediate in which the adenylate residue is joined to three nucleotides by phosphodiester bonds. The new 2′-to-5′ linkage is shown in red, and the usual 3′-to-5′ linkages are shown in blue.

The 3′-OH terminus of exon 1 then attacks the phosphodiester bond between the intron and exon 2. Exons 1 and 2 become joined, and the intron is released in lariat form. Again, this reaction is a transesterification. Splicing is thus accomplished by two *transesterification reactions* rather than by hydrolysis followed by ligation. The first reaction generates a free 3′-OH group at the 3′ end of exon 1, and the second reaction links this group to the 5′-phosphate of exon 2. *The number of phosphodiester bonds stays the same during these steps,* which is crucial because it allows the splicing reaction itself to proceed without an energy source such as ATP or GTP.

Small nuclear RNAs in spliceosomes catalyze the splicing of mRNA precursors

The nucleus contains many types of small RNA molecules with fewer than 300 nucleotides, referred to as *snRNAs* (small nuclear RNAs). A few of them—designated U1, U2, U4, U5, and U6—are essential for splicing mRNA precursors. The secondary structures of these RNAs are highly conserved in organisms ranging from yeast to human beings. These RNA molecules are associated with specific proteins to form complexes termed *snRNPs* (small nuclear ribonucleoprotein particles); investigators often speak of them as "snurps." Spliceosomes are large (60S) dynamic assemblies composed of snRNPs, hundreds of other proteins called *splicing factors,* and the mRNA precursors being processed (Table 30.3).

TABLE 30.3 Small nuclear ribonucleoprotein particles (snRNPs) in the splicing of mRNA precursors

snRNP	Size of snRNA (nucleotides)	Role
U1	165	Binds the 59 splice site
U2	185	Binds the branch site
U5	116	Binds the 59 splice site and then the 39 splice site
U4	145	Masks the catalytic activity of U6
U6	106	Catalyzes splicing

FIGURE 30.37 Spliceosome structure. The three-dimensional structure of one form of the spliceosome from a yeast determined by high-resolution cryo-electron microscopy. The key RNA components U2, U5, and U6 are shown in red, yellow, and green, respectively. [Drawn from 3JB9.pdb.]

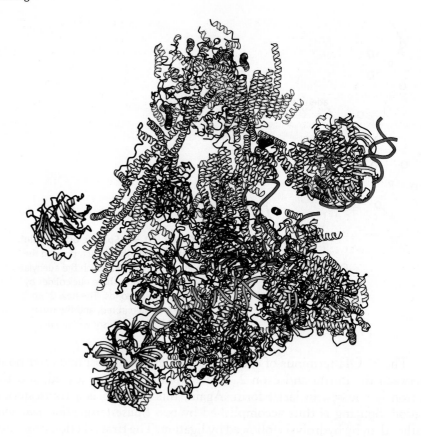

The large and dynamic nature of the spliceosome made the determination of the detailed three-dimensional structure a great challenge. However, with the maturation of cryo-electron microscopy (Section 3.5), the structures of spliceosomes have been determined from several species in a number of different stages of its function (Figure 30.37).

Splicing begins with the recognition of the 5′ splice site by the U1 snRNP (Figure 30.38). U1 snRNA contains a highly conserved six-nucleotide sequence, not covered by protein in the snRNP, that base-pairs to the 5′ splice site of the pre-mRNA. This binding initiates spliceosome assembly on the pre-mRNA molecule.

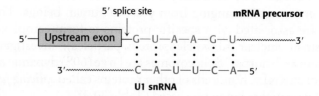

U2 snRNP then binds the branch site in the intron by base-pairing between a highly conserved sequence in U2 snRNA and the pre-mRNA. U2 snRNP binding requires ATP hydrolysis. A preassembled U4-U5-U6 tri-snRNP joins this complex of U1, U2, and the mRNA precursor to form the spliceosome. This association also requires ATP hydrolysis.

A revealing view of the interplay of RNA molecules in this assembly came from examining the pattern of cross-links formed by *psoralen*, a reagent that joins neighboring pyrimidines in base-paired regions on treatment with

FIGURE 30.38 Spliceosome assembly and action. U1 binds the 5′ splice site and U2 binds to the branch point. A preformed U4-U5-U6 complex then joins the assembly to form the complete spliceosome. The U6 snRNA re-folds and binds the 5′ splice site, displacing U1. Extensive interactions between U6 and U2 displace U4. Then, in the first transesterification step, the branch-site adenosine attacks the 5′ splice site, making a lariat intermediate. U5 holds the two exons in close proximity, and the second transesterification takes place, with the 5′ splice-site hydroxyl group attacking the 3′ splice site. These reactions result in the mature spliced mRNA and a lariat form of the intron bound by U2, U5, and U6. [Information from T. Villa, J. A. Pleiss, and C. Guthrie, *Cell* 109:149–152, 2002.]

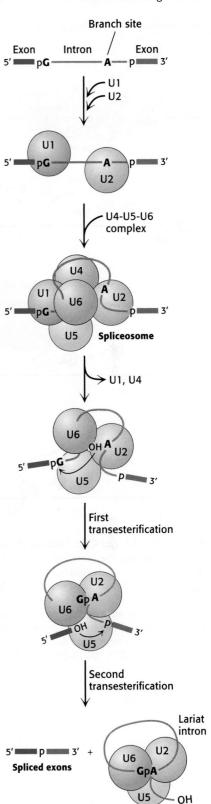

light. These cross-links suggest that splicing takes place in the following way. First, U5 interacts with exon sequences in the 5′ splice site and subsequently with the 3′ exon. Next, U6 disengages from U4 and undergoes an intramolecular rearrangement that permits base-pairing with U2 as well as interaction with the 5′ end of the intron, displacing U1 from the spliceosome. The catalytic center includes two bound magnesium ions bound primarily by phosphate groups from the U6 RNA (Figure 30.39). U4 serves as an inhibitor that masks U6 until the specific splice sites are aligned. These rearrangements result in the first transesterification reaction, cleaving the 5′ exon and generating the lariat intermediate.

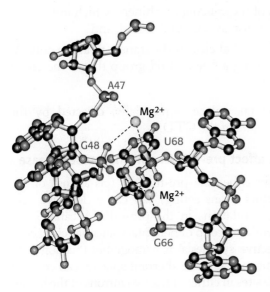

FIGURE 30.39 Splicing catalytic center. The structure of the catalytic center from the splicesome includes two key magnesium ions bound to the U6 RNA. These ions serve to promote the two transesterification reactions that are central to RNA splicing.

Further rearrangements of RNA in the spliceosome facilitate the second transesterification. In these rearrangements, U5 aligns the free 5′ exon with the 3′ exon such that the 3′-hydroxyl group of the 5′ exon is positioned to nucleophilically attack the 3′ splice site to generate the spliced product. U2, U5, and U6 bound to the excised lariat intron are released to complete the splicing reaction. Both transesterification reactions are promoted by the pair of bound magnesium ions, in reactions reminiscent of those for DNA and RNA polymerases.

Many of the steps in the splicing process require ATP hydrolysis. How is the free energy associated with ATP hydrolysis used to power splicing? To achieve the well-ordered rearrangements necessary for splicing, ATP-powered RNA helicases must unwind RNA helices and allow alternative base-pairing arrangements to form. Thus, two features of the splicing process are noteworthy. First, *RNA molecules play key roles in directing the alignment of splice sites and in carrying out catalysis.* Second, *ATP-powered*

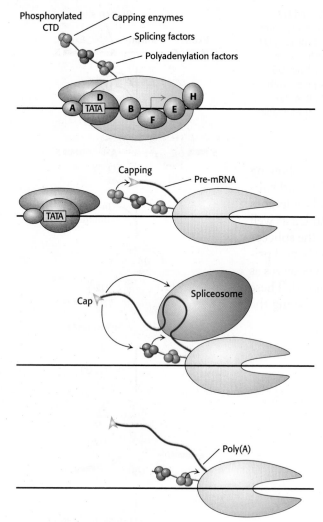

FIGURE 30.40 The CTD: Coupling transcription to pre-mRNA processing. The transcription factor TFIIH phosphorylates the carboxyl-terminal domain (CTD) of RNA polymerase II, signaling the transition from transcription initiation to elongation. The phosphorylated CTD binds factors required for pre-mRNA capping, splicing, and polyadenylation. These proteins are brought in close proximity to their sites of action on the nascent pre-mRNA as it is transcribed during elongation. [Information from P. A. Sharp, *TIBS* 30:279–281, 2005.]

helicases unwind RNA duplex intermediates that facilitate catalysis and induce the release of snRNPs from the mRNA.

Transcription and processing of mRNA are coupled

Although the transcription and processing of mRNAs have been described herein as separate events in gene expression, experimental evidence suggests that the two steps are coordinated by the carboxyl-terminal domain of RNA polymerase II. We have seen that the CTD consists of a unique repeated seven-amino-acid sequence, YSPTSPS. Either S_2 or S_5 or both may be phosphorylated in the various repeats. The phosphorylation state of the CTD is controlled by a number of kinases and phosphatases and leads the CTD to bind many of the proteins having roles in RNA transcription and processing. The CTD contributes to efficient transcription by recruiting these proteins to the pre-mRNA (Figure 30.40), including:

1. Capping enzymes, which methylate the 5′ guanine on the pre-mRNA immediately after transcription begins;

2. Components of the splicing machinery, which initiate the excision of each intron as it is synthesized; and

3. An endonuclease that cleaves the transcript at the poly(A) addition site, creating a free 3′-OH group that is the target for 3′ adenylation.

These events take place sequentially, directed by the phosphorylation state of the CTD.

Mutations that affect pre-mRNA splicing cause disease

Mutations in either the pre-mRNA (cis-acting) or the splicing factors (trans-acting) can cause defective pre-mRNA splicing. Mutations in the pre-mRNA cause some forms of thalassemia, a group of hereditary anemias characterized by the defective synthesis of hemoglobin (Section 7.4). Cis-acting mutations that cause aberrant splicing can occur at the 5′ or 3′ splice sites in either of the two introns of the hemoglobin β chain or in its exons. The mutations usually result in an incorrectly spliced pre-mRNA that, because of a premature stop codon, cannot encode a full-length protein. The defective mRNA is normally degraded rather than translated. Mutations in the 5′ splice site may alter that site such that the splicing machinery cannot recognize it, forcing the machinery to find another 5′ splice site in the intron and introducing the potential for a premature stop codon. Mutations in the intron itself may create a new 5′ splice site; in this case, either one of the two splice sites may be recognized (Figure 30.41). Consequently, some normal protein can be made, and so the disease is less severe. *Mutations affecting splicing have been estimated to cause at least 15% of all genetic diseases.*

Disease-causing mutations may also appear in splicing factors. Retinitis pigmentosa is a disease of acquired blindness, first described in 1857, with an incidence of 1/3500. About 5% of the autosomal dominant form of retinitis pigmentosa is likely due to mutations in the hPrp8 protein, a pre-mRNA splicing factor that is a component of the U4-U5-U6 tri-snRNP. How a mutation in a splicing factor that is present in all cells causes disease

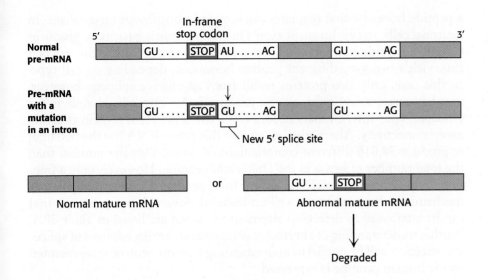

FIGURE 30.41 A splicing mutation that causes thalassemia. An A-to-G mutation within the first intron of the gene for the human hemoglobin β chain creates a new 5′ splice site (GU). Both 5′ splice sites are recognized by the U1 snRNP; thus splicing may sometimes create a normal mature mRNA and an abnormal mature mRNA that contains intron sequences. The normal mature mRNA is translated into a hemoglobin β chain. Because it includes intron sequences, the abnormal mature mRNA now has a premature stop codon and is degraded.

only in the retina is not clear; nevertheless, retinitis pigmentosa is a good example of how mutations that disrupt spliceosome function can cause disease.

Most human pre-mRNAs can be spliced in alternative ways to yield different proteins

Alternative splicing is a widespread mechanism for generating protein diversity. Different combinations of exons from the same gene may be spliced into a mature RNA, producing distinct forms of a protein for specific tissues, developmental stages, or signaling pathways. What controls which splicing sites are selected? The selection is determined by the binding of trans-acting splicing factors to cis-acting sequences in the pre-mRNA. Most alternative splicing leads to changes in the coding sequence, resulting in proteins with different functions. *Alternative splicing provides a powerful mechanism for expanding the versatility of genomic sequences through combinatorial control.* Consider a gene with five positions at which splicing can take place. With the assumption that these alternative splicing pathways can be regulated independently, a total of $2^5 = 32$ different mRNAs can be generated.

Sequencing of the human genome has revealed that most pre-mRNAs are alternatively spliced, leading to a much greater number of proteins than would be predicted from the number of genes. An example of alternative splicing leading to the expression of two different proteins, each in a different tissue, is provided by the gene encoding both calcitonin and calcitonin-gene-related peptide (CGRP; Figure 30.42). In the thyroid gland, the inclusion of exon 4 in one splicing pathway produces calcitonin,

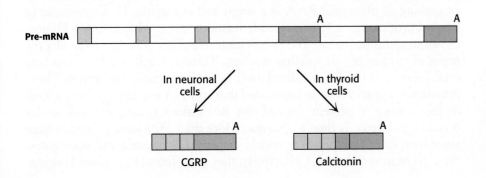

FIGURE 30.42 An example of alternative splicing. In human beings, two very different hormones are produced from a single calcitonin/CGRP pre-mRNA. Alternative splicing produces the mature mRNA for either calcitonin or CGRP (calcitonin-gene-related protein), depending on the cell type in which the gene is expressed. Each alternative transcript incorporates one of two alternative polyadenylation signals (A) present in the pre-mRNA.

a peptide hormone that regulates calcium and phosphorus metabolism. In neuronal cells, the exclusion of exon 4 in another splicing pathway produces CGRP, a peptide hormone that acts as a vasodilator. A single pre-mRNA thus yields two very different peptide hormones, depending on cell type. In this case, only two proteins result from alternative splicing; however, in other cases, many more can be produced. An extreme example is the *Drosophila* pre-mRNA that encodes DSCAM, a neuronal protein affecting axon connectivity. Alternative splicing of this pre-mRNA has the potential to produce 38,016 different combinations of exons, a greater number than the total number of genes in the *Drosophila* genome. However, only a fraction of these potential mRNAs appear to be produced, owing to regulatory mechanisms that are not yet well understood. Several human diseases that can be attributed to defects in alternative splicing are listed in Table 30.4. Further understanding of alternative splicing and the mechanisms of splice-site selection will be crucial to understanding how the proteome represented by the human genome is expressed.

TABLE 30.4 Selected human diseases attributed to defects in alternative splicing

Disorder	Gene or its product
Acute intermittent porphyria	Porphobilinogen deaminase
Breast and ovarian cancer	*BRCA*1
Cystic fibrosis	CFTR
Frontotemporal dementia	τ protein
Hemophilia A	Factor VIII
HGPRT deficiency (Lesch–Nyhan syndrome)	Hypoxanthine-guanine phosphoribosyltransferase
Leigh encephalomyelopathy	Pyruvate dehydrogenase E1α
Severe combined immunodeficiency	Adenosine deaminase
Spinal muscle atrophy	*SMN*1 or *SMN*2

30.4 The Discovery of Catalytic RNA Was Revealing in Regard to Both Mechanism and Evolution

RNAs form a surprisingly versatile class of molecules. As we have seen, splicing is catalyzed largely by RNA molecules, with proteins playing a secondary role. Another enzyme that contains a key RNA component is ribonuclease P, which catalyzes the maturation of tRNA by endonucleolytic cleavage of nucleotides from the 5′ end of the precursor molecule. Finally, as we shall see in Chapter 31, the RNA component of ribosomes is the catalyst that carries out protein synthesis.

The versatility of RNA first became clear from observations of the processing of ribosomal RNA in a single-cell eukaryote. In *Tetrahymena* (a ciliated protozoan), a 414-nucleotide intron is removed from a 6.4-kb precursor to yield the mature 26S rRNA molecule (Figure 30.43). In an elegant series of studies of this splicing reaction, Thomas Cech and his coworkers established that the RNA spliced itself to precisely excise the intron. These remarkable experiments demonstrated that an RNA molecule can *splice itself* in the absence of protein. Indeed, the RNA alone is catalytic and, under certain conditions, is thus a *ribozyme*. More than 1500 similar introns have since been found in species as widely dispersed as bacteria and eukaryotes, though not in vertebrates. Collectively, they are referred to as *group I introns*.

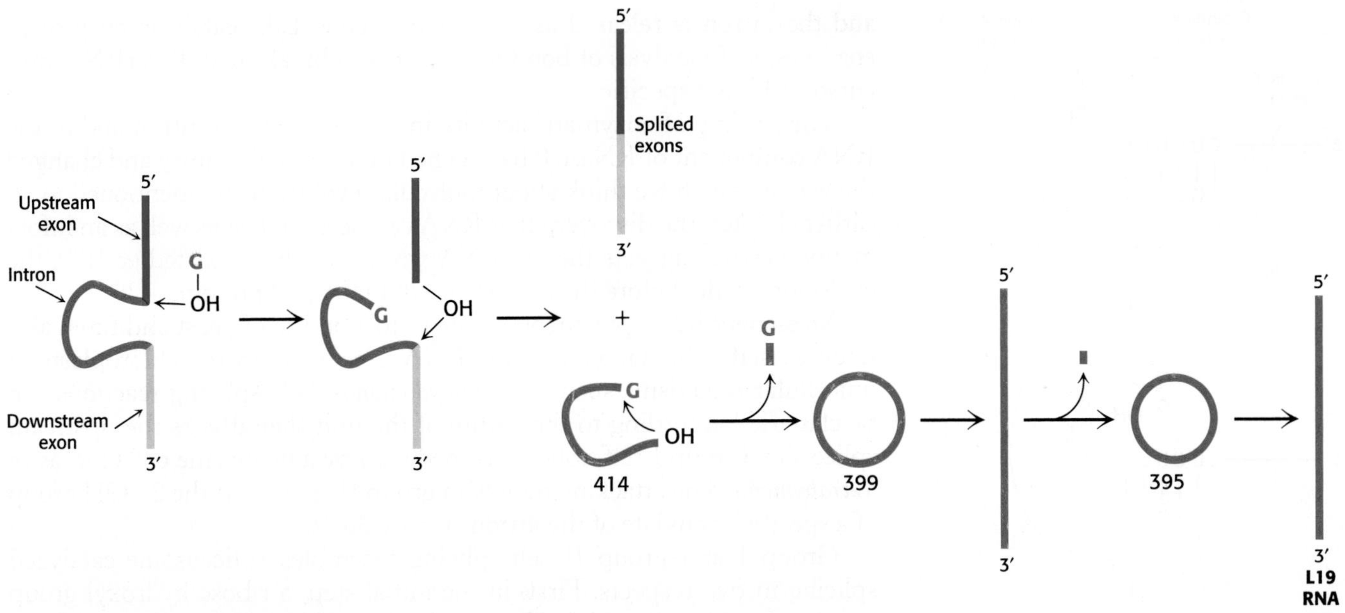

FIGURE 30.43 Self-splicing. A ribosomal RNA precursor from *Tetrahymena*, representative of the group I introns, splices itself in the presence of a guanosine cofactor (G, shown in green). A 414-nucleotide intron (red) is released in the first splicing reaction. This intron then splices itself twice again to produce a linear RNA that has lost a total of 19 nucleotides. This L19 RNA is catalytically active. [Information from T. Cech. RNA as an enzyme. Copyright © 1986 by Scientific American, Inc. All rights reserved.]

The *self-splicing* reaction in the group I intron requires an added guanosine nucleotide. Nucleotides were originally included in the reaction mixture because it was thought that ATP or GTP might be needed as an energy source. Instead, the nucleotides were found to be necessary as cofactors. The required cofactor proved to be a guanosine unit, in the form of guanosine, GMP, GDP, or GTP. G (denoting any one of these species) serves not as an energy source but as an attacking group that becomes transiently incorporated into the RNA (Figure 30.43). G binds to the RNA and then attacks the 5′ splice site to form a phosphodiester bond with the 5′ end of the intron. This transesterification reaction generates a 3′-OH group at the end of the upstream exon. This newly attached 3′-OH group then attacks the 3′ splice site. This second transesterification reaction joins the two exons and leads to the release of the 414-nucleotide intron.

Self-splicing depends on the structural integrity of the RNA precursor. Much of the group I intron is needed for self-splicing. This molecule, like many RNAs, has a folded structure formed by many double-helical stems and loops (Figure 30.44), with a well-defined pocket for binding the guanosine. Examination of the three-dimensional structure of a catalytically active group I intron determined by x-ray crystallography reveals the coordination of magnesium ions in the active site analogous to that observed in protein enzymes such as DNA polymerase.

Analysis of the base sequence of the rRNA precursor suggested that the splice sites are aligned with the catalytic residues by base-pairing between the *internal guide sequence* (IGS) in the intron and the 5′ and 3′ exons (Figure 30.45). The IGS first brings together the guanosine cofactor and the 5′ splice site so that the 3′-OH group of G can nucleophilically attack the phosphorus atom at this splice site. The IGS then holds the downstream exon in position for attack by the newly formed 3′-OH group of the upstream exon. A phosphodiester bond is formed between the two exons,

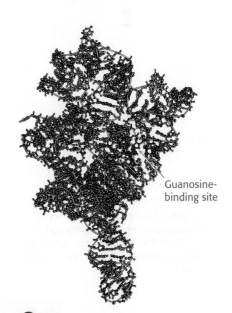

FIGURE 30.44 Structure of a self-splicing intron. The structure of a large fragment of the self-splicing intron from *Tetrahymena* reveals a complex folding pattern of helices and loops. Bases are shown in green, A; yellow, C; purple, G; and orange, U. [Drawn from 1GRZ.pdb].

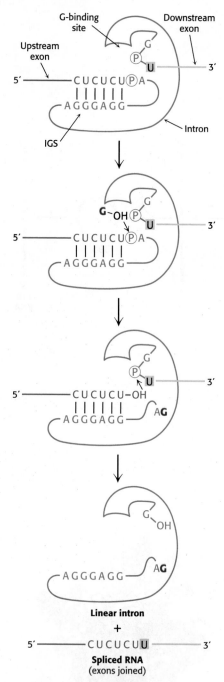

FIGURE 30.45 Self-splicing mechanism. The catalytic mechanism of the group I intron includes a series of transesterification reactions. [Information from T. Cech. RNA as an enzyme. Copyright © 1986 by Scientific American, Inc. All rights reserved.]

and the intron is released as a linear molecule. Like catalysis by protein enzymes, self-catalysis of bond formation and breakage in this rRNA precursor is highly specific.

The finding of enzymatic activity in the self-splicing intron and in the RNA component of RNase P has opened new areas of inquiry and changed the way in which we think about molecular evolution. As mentioned in an earlier chapter, the discovery that RNA can be a catalyst as well as an information carrier suggests that an RNA world may have existed early in the evolution of life, before the appearance of DNA and protein.

Messenger RNA precursors in the mitochondria of yeast and fungi also undergo self-splicing, as do some RNA precursors in the chloroplasts of unicellular organisms such as *Chlamydomonas*. Self-splicing reactions can be classified according to the nature of the unit that attacks the upstream splice site. Group I self-splicing is mediated by a guanosine cofactor, as in *Tetrahymena*. The attacking moiety in group II splicing is the 2'-OH group of a specific adenylate of the intron (Figure 30.46).

Group I and group II self-splicing resembles spliceosome-catalyzed splicing in two respects. First, in the initial step, a ribose hydroxyl group attacks the 5' splice site. The newly formed 3'-OH terminus of the upstream exon then attacks the 3' splice site to form a phosphodiester bond with the downstream exon. Second, both reactions are transesterifications in which the phosphate moieties at each splice site are retained in the products. The number of phosphodiester bonds stays constant. Group II splicing is like the spliceosome-catalyzed splicing of mRNA precursors in several additional ways. The attack at the 5' splice site is carried out by a part of the intron itself (the 2'-OH group of adenosine) rather than by an external cofactor (G). In both cases, the intron is released in the form of a lariat. Moreover, in some instances, the group II intron is transcribed in pieces that assemble through hydrogen bonding to the catalytic intron, in a manner analogous to the assembly of the snRNAs in the spliceosome.

These similarities have led to the suggestion that the spliceosome-catalyzed splicing of mRNA precursors evolved from RNA-catalyzed self-splicing. Group II splicing may well be an intermediate between group I splicing and the splicing in the nuclei of higher eukaryotes. *A major step in this transition was the transfer of catalytic power from the intron itself to other molecules.* The formation of spliceosomes gave genes a new freedom because introns were no longer constrained to provide the catalytic center for splicing. Another advantage of external catalysts for splicing is that they can be more readily regulated. However, it is important to note that similarities do not establish ancestry. The similarities between group II introns and mRNA splicing may be a result of convergent evolution. Perhaps there are only a limited number of ways to carry out efficient, specific intron excision. The determination of whether these similarities stem from ancestry or from chemistry will require expanding our understanding of RNA biochemistry.

SUMMARY

30.1 RNA Polymerases Catalyze Transcription

All cellular RNA molecules are synthesized by RNA polymerases according to instructions given by DNA templates. The activated monomer substrates are ribonucleoside triphosphates. The direction of RNA synthesis is

SELF-SPLICING INTRONS

SPLICEOSOME-CATALYZED SPLICING OF NUCLEAR mRNA

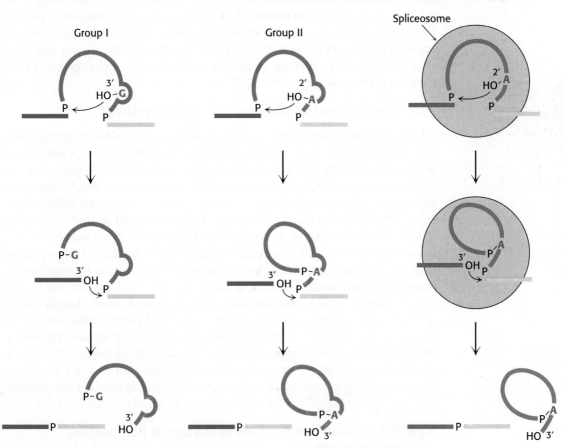

FIGURE 30.46 Comparison of splicing pathways. The exons being joined are shown in blue and yellow and the attacking unit is shown in green. The catalytic site is formed by the intron itself (red) in group I and group II splicing. In contrast, the splicing of nuclear mRNA precursors is catalyzed by snRNAs and their associated proteins in the spliceosome. [Information from P. A. Sharp, *Science* 235:766–771, 1987.]

$5' \rightarrow 3'$, as in DNA synthesis. RNA polymerases, unlike DNA polymerases, do not need a primer.

RNA polymerase in *E. coli* is a multisubunit enzyme. The subunit composition of the ~500-kDa holoenzyme is $\alpha_2\beta\beta'\omega\sigma$ and that of the core enzyme is $\alpha_2\beta\beta'\omega$. Transcription is initiated at promoter sites consisting of two sequences, one centered near -10 and the other near -35; that is, 10 and 35 nucleotides away from the start site in the 5' (upstream) direction. The consensus sequence of the -10 region is TATAAT. The σ subunit enables the holoenzyme to recognize promoter sites. When the growth temperature is raised, *E. coli* expresses a special σ subunit that selectively binds the distinctive promoter of heat-shock genes. RNA polymerase must unwind the template double helix for transcription to take place. Unwinding exposes some 17 bases on the template strand and sets the stage for the formation of the first phosphodiester bond. Newly synthesized RNA chains usually start with pppG or pppA. The σ subunit usually dissociates from the holoenzyme after the initiation of the new chain.

Elongation takes place at transcription bubbles that move along the DNA template at a rate of about 50 nucleotides per second. RNA polymerase occasionally backtracks, a process that can facilitate proofreading of the RNA transcript. The nascent RNA chain contains stop signals that end transcription. One stop signal is an RNA hairpin, which is followed by several U residues. A different stop signal is read by the *rho* protein, an ATPase. Some genes are regulated by riboswitches, structures that form in RNA transcripts and bind specific metabolites. In *E. coli*, precursors of transfer RNA and ribosomal RNA are cleaved and chemically modified after transcription, whereas messenger RNA is used unchanged as a template for protein synthesis.

30.2 Transcription in Eukaryotes Is Highly Regulated

RNA synthesis in eukaryotes takes place in the nucleus, whereas protein synthesis takes place in the cytoplasm. There are three types of RNA polymerase in the nucleus: RNA polymerase I makes ribosomal RNA precursors, II makes messenger RNA precursors, and III makes transfer RNA precursors. Eukaryotic promoters are complex, being composed of several different elements. Promoters for RNA polymerase II may be located on the 5′ side or the 3′ side of the start site for transcription. A common type of eukaryotic promoter consists of a TATA box centered between −30 and −100 and paired with an initiator element. Eukaryotic promoter elements are recognized by proteins called transcription factors rather than by RNA polymerase II. The saddle-shaped TATA-box-binding protein unwinds and sharply bends DNA at TATA-box sequences and serves as a focal point for the assembly of transcription complexes. The TATA-box-binding protein initiates the assembly of the active transcription complex. The activity of many promoters is greatly increased by enhancer sequences that have no promoter activity of their own. Enhancer sequences can act over distances of several kilobases, and they can be located either upstream or downstream of a gene.

30.3 The Transcription Products of Eukaryotic Polymerases Are Processed

The 5′ ends of mRNA precursors become capped and methylated in the course of transcription. A 3′ poly(A) tail is added to most mRNA precursors after the nascent chain has been cleaved by an endonuclease. RNA editing alters the nucleotide sequence of some mRNAs, such as the one for apolipoprotein B.

The splicing of mRNA precursors is carried out by spliceosomes, which consist of small nuclear ribonucleoprotein particles. Splice sites in mRNA precursors are specified by sequences at ends of introns and by branch sites near their 3′ ends. The 2′-OH group of an adenosine residue in the branch site attacks the 5′ splice site to form a lariat intermediate. The newly generated 3′-OH terminus of the upstream exon then attacks the 3′ splice site to become joined to the downstream exon. Splicing thus consists of two transesterification reactions, with the number of phosphodiester bonds remaining constant during reactions. Small nuclear RNAs in spliceosomes catalyze the splicing of mRNA precursors. A pair of magnesium ions bound to U6 snRNA are key components of the active center of spliceosomes.

The events in posttranscriptional processing of mRNA are controlled by the phosphorylation state of the carboxy-terminal domain, part of RNA polymerase II.

30.4 The Discovery of Catalytic RNA Was Revealing in Regard to Both Mechanism and Evolution

Some RNA molecules, such as those containing the group I intron, undergo self-splicing in the absence of protein. A self-modified version of this rRNA intron displays true catalytic activity and is thus a ribozyme. Spliceosome-catalyzed splicing may have evolved from self-splicing. The discovery of catalytic RNA has opened new vistas in our exploration of early stages of molecular evolution and the origins of life.

APPENDIX

Biochemistry in Focus

Discovering enzymes made of RNA

The discovery that the catalysts that accelerate almost all biological reactions were specific protein molecules was one of the foundational discoveries in biochemistry. This led to decades of very productive research directed toward purifying and identifying the protein enzymes that play roles in many fundamental biochemical processes. It was in this context that the laboratory of Tom Cech, a relatively new faculty member in the Department of Chemistry and Biochemistry at the University of Colorado, set out to purify the enzyme(s) involved in splicing a particular class of RNA molecules.

Having purified the unspliced RNA substrate, Cech and Art Zaug, a technician in his laboratory, began to study extracts from the cell nuclei of the organism they were using (*Tetrahymena thermophila*) which was likely to contain the desired enzymes. They found that treating the unspliced

RNA with the nuclear extract did result in splicing. However, they also did a control experiment in which they omitted the extract and found that splicing occurred to the same extent. Their first thought was that they had made a mistake and added nuclear extract to the control sample. However, careful repetition revealed that no mistake had been made and that the RNA splicing reaction was occurring without the extract. The challenge now shifted to understanding this self-splicing reaction. Subsequent work revealed the mechanism for RNA self-splicing discussed in Section 30.4 and opened up an entirely new area of biochemistry.

This example highlights a key aspect of biochemical research. Important discoveries are often made not when results are obtained that match the researchers' expectation but rather when unexpected results are obtained, leading to initial confusion and then a reframing of the problem to be addressed.

KEY TERMS

transcription (p. 983)

RNA polymerase (p. 984)

promoter site (p. 984)

transcription bubble (p. 986)

consensus sequence (p. 989)

sigma (σ) subunit (p. 989)

riboswitch (p. 992)

rho (ρ) protein (p. 993)

carboxy-terminal domain (CTD) (p. 997)

TATA box (p. 998)

transcription factor (p. 999)

enhancer (p. 1001)

small nucleolar ribonucleoprotein (snoRNP) (p. 1002)

pre-mRNA (p. 1002)

5′ cap (p. 1003)

poly(A) tail (p. 1003)

microRNA (p. 1004)

RNA editing (p. 1004)

RNA splicing (p. 1005)

spliceosome (p. 1006)

small nuclear RNA (snRNA) (p. 1007)

small nuclear ribonucleoprotein particle (snRNP) (p. 1007)

alternative splicing (p. 1011)

catalytic RNA (ribozyme) (p. 1012)

self-splicing (p. 1013)

PROBLEMS

1. *Complements.* The sequence of part of an mRNA is

5′-AUGGGGAACAGCAAGAGUGGGGGCCCUGUCCAAGGAG-3′

What is the sequence of the DNA coding strand? Of the DNA template strand?

2. *Checking for errors.* Why is RNA synthesis not as carefully monitored for errors as is DNA synthesis? ✓❶, ✓❷

3. *Speed is not of the essence.* Why is it advantageous for DNA synthesis to be more rapid than RNA synthesis?

4. *Active sites.* The overall structures of RNA polymerase and DNA polymerase are very different, yet their active sites show considerable similarities. What do the similarities suggest about the evolutionary relationship between these two important enzymes? ✓❶

5. *Potent inhibitor.* Heparin inhibits transcription by binding to RNA polymerase. What properties of heparin allow it to bind so effectively to RNA polymerase?

6. *A loose cannon.* Sigma protein by itself does not bind to promoter sites. Predict the effect of a mutation enabling σ

to bind to the -10 region in the absence of other subunits of RNA polymerase. ✓❷, ✓❸

7. *Stuck sigma.* What would be the likely effect of a mutation that prevents σ from dissociating from the RNA polymerase core? ✓❷

8. *Transcription time.* What is the minimum length of time required for the synthesis by *E. coli* polymerase of an mRNA encoding a 100-kDa protein?

9. *Rapid search.* RNA polymerase finds promoter sites very rapidly. The observed rate constant for the binding of the RNA polymerase holoenzyme to promoter sequences is $10^{10}\,M^{-1}s^{-1}$. The rate constant for two macromolecules encountering each other is typically $10^{8}\,M^{-1}s^{-1}$. Propose an explanation for the 100-fold larger rate for a protein finding a particular site along a DNA molecule.

10. *Where to begin?* Identify the likely transcription start site in the following DNA sequence:

5'-GCCGTTGACACCGTTCGGCGATCGATCCGCTATAATGTG
TGGATCCGCTT-39'

3'-CGGCAACTGTGGCAAGCCGCTAGCTAGGCGATATTACAC
ACCTAGGCGAA-5' ✓❷

11. *Between bubbles.* How far apart are transcription bubbles on *E. coli* genes that are being transcribed at a maximal rate?

12. *A revealing bubble.* Consider the synthetic RNA–DNA transcription bubble illustrated here. Refer to the coding DNA strand, the template strand, and the RNA strand as strands 1, 2, and 3, respectively.

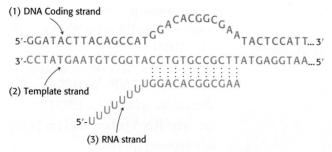

(1) DNA Coding strand
5'-GGATACTTACAGCCAT GGACACGGCGAA TACTCCATT...3'
3'-CCTATGAATGTCGGTACCTGTGCCGCTTATGAGGTAA...5'
(2) Template strand
UGGACACGGCGAA
5'-UUUUUUU
(3) RNA strand

(a) Suppose that strand 3 is labeled with ^{32}P at its 5' end and that polyacrylamide gel electrophoresis is carried out under nondenaturing conditions. Predict the autoradiographic pattern for (i) strand 3 alone, (ii) strands 1 and 3, (iii) strands 2 and 3, (iv) strands 1, 2, and 3, and (v) strands 1, 2, and 3 and core RNA polymerase.

(b) What is the likely effect of rifampicin on RNA synthesis in this system?

(c) Heparin blocks elongation of the RNA primer if it is added to core RNA polymerase before the onset of transcription but not if added after transcription starts. Account for this difference.

(d) Suppose that synthesis is carried out in the presence of ATP, CTP, and UTP. Compare the length of the longest product obtained with that expected when all four ribonucleoside triphosphates are present.

13. *Proofreading marks.* The major products of proofreading by RNA polymerase are dinucleotides rather than mononucleotides. Why?

14. *Abortive cycling.* Di- and trinucleotides are occasionally released from RNA polymerase at the very start of transcription, a process called abortive cycling. This process requires the restart of transcription. Suggest a plausible explanation for abortive cycling.

15. *Polymerase inhibition.* Cordycepin inhibits poly(A) synthesis at low concentrations and RNA synthesis at higher concentrations.

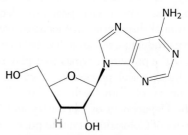

Cordycepin (3'-deoxyadenosine)

(a) What is the basis of inhibition by cordycepin?

(b) Why is poly(A) synthesis more sensitive than the synthesis of other RNAs to the presence of cordycepin?

(c) Does cordycepin need to be modified to exert its effect?

16. *Alternative splicing.* A gene contains eight sites where alternative splicing is possible. Assuming that the splicing pattern at each site is independent of that at all other sites, how many splicing products are possible? ✓❺

17. *Supercoiling.* Negative supercoiling of DNA favors the transcription of genes because it facilitates unwinding. However, not all promoter sites are stimulated by negative supercoiling. The promoter site for topoisomerase II itself is a noteworthy exception. Negative supercoiling decreases the rate of transcription of this gene. Propose a possible mechanism for this effect and suggest a reason why it may occur.

18. *An extra piece.* In one type of mutation leading to a form of thalassemia, the mutation of a single base (G to A) generates a new 3' splice site (blue in the illustration below) akin to the normal one (yellow) but farther upstream.

Normal 3' end of intron

5' CCTATTGGTCTATTTTCCACCCTTAGGCTGCTG 3'

5' CCTATTAGTCTATTTTCCACCCTTAGGCTGCTG 3'

What is the amino acid sequence of the extra segment of protein synthesized in a thalassemic patient having a mutation leading to aberrant splicing? The reading frame after the splice site begins with TCT.

19. *A long-tailed messenger.* Another thalassemic patient had a mutation leading to the production of an mRNA for the β chain of hemoglobin that was 900 nucleotides longer than the normal one. The poly(A) tail of this mutant mRNA was located a few nucleotides after the only AAUAAA sequence in the additional sequence. Propose a mutation that would lead to the production of this altered mRNA.

Mechanism Problem

20. *RNA editing.* Many uridine molecules are inserted into some mitochondrial mRNAs in trypanosomes. The uridine residues come from the poly(U) tail of a donor strand. Nucleoside triphosphates do not participate in this reaction. Propose a reaction mechanism that accounts for these findings. (Hint: Relate RNA editing to RNA splicing.)

Chapter Integration Problems

21. *Proteome complexity.* What processes considered in this chapter make the proteome more complex than the genome? What processes might further enhance this complexity?

22. *Separation technique.* Suggest a means by which you could separate mRNA from the other types of RNA in a eukaryotic cell.

Data Interpretation Problems

23. *Run-off experiment.* Nuclei were isolated from brain, liver, and muscle. The nuclei were then incubated with α-[^{32}P] UTP under conditions that allow RNA synthesis, except that an inhibitor of RNA initiation was present. The radioactive RNA was isolated and annealed to various DNA sequences that had been attached to a gene chip [See animation on ⊘ **Sapling**Plus]. In the adjoining graphs, the intensity of the shading indicates roughly how much mRNA was attached to each DNA sequence.

Liver   Muscle   Brain

(a) Why does the intensity of hybridization differ between genes?

(b) What is the significance of the fact that some of the RNA molecules display different hybridization patterns in different tissues?

(c) Some genes are expressed in all three tissues. What would you guess is the nature of these genes?

(d) Suggest a reason why an initiation inhibitor was included in the reaction mixture.

24. *Christmas trees.* The adjoining autoradiograph depicts several bacterial genes undergoing transcription. Identify the DNA. What are the strands of increasing length? Where is the beginning of transcription? The end of transcription? On the page, what is the direction of RNA synthesis? What can you conclude about the number of enzymes participating in RNA synthesis on a given gene?

[Medical Images RM/Phototake, Inc.]

Think ▶ Pair ▶ Share Problem

25. Compare the biological roles of RNA and DNA. What aspects of RNA structure and chemistry make it so versatile?

Protein Synthesis

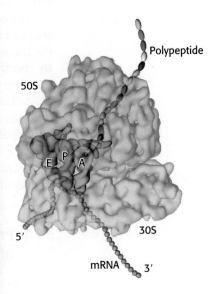

CHAPTER 31

Polypeptide

50S

E P A

5′

30S

mRNA   3′

The ribosome, shown at the right, is a factory for the manufacture of polypeptides. Amino acids are carried into the ribosome, one at a time, connected to transfer RNA molecules. Each amino acid is joined to the growing polypeptide chain, which detaches from the ribosome only after the polypeptide has been completed. This assembly-line approach allows even very long polypeptide chains to be assembled rapidly and with impressive accuracy. [(Left) Mary Evans Picture Library/The Image Works.]

✓ LEARNING OBJECTIVES

By the end of this chapter, you should be able to:

1. Explain how nucleic acid information is translated into amino acid sequence and define the role of aminoacyl tRNA synthetases in this process.

2. Define the role of ribosomes in protein synthesis.

3. Compare and contrast bacterial and eukaryotic protein synthesis.

4. Describe the pathway for the synthesis of secretory and membrane proteins.

OUTLINE

31.1 Protein Synthesis Requires the Translation of Nucleotide Sequences into Amino Acid Sequences

31.2 Aminoacyl Transfer RNA Synthetases Read the Genetic Code

31.3 The Ribosome Is the Site of Protein Synthesis

31.4 Eukaryotic Protein Synthesis Differs from Bacterial Protein Synthesis Primarily in Translation Initiation

31.5 A Variety of Antibiotics and Toxins Can Inhibit Protein Synthesis

31.6 Ribosomes Bound to the Endoplasmic Reticulum Manufacture Secretory and Membrane Proteins

The genes that encode proteins are key components of the genetic information because proteins play most of the functional roles in cells. In Chapters 29 and 30, we examined how DNA is replicated and transcribed into RNA. We now turn to the mechanism of protein synthesis, a process called *translation* because the 4-letter alphabet of nucleic acids is translated into the entirely different 20-letter alphabet of proteins. Translation is a conceptually more complex process than either replication or transcription, both of which take place within the framework of a common base-pairing language. As befits its position linking the nucleic acid and protein languages, the process of protein synthesis critically depends on both nucleic acid and protein factors. Protein synthesis takes place on *ribosomes*—enormous complexes containing three large RNA molecules and more than 50 proteins. Interestingly, *the ribosome is* *a ribozyme;* that is, the RNA components catalyze protein synthesis. This observation strongly supports the notion that life evolved through an RNA world, and the ribosome is a surviving inhabitant of that world.

Transfer RNA molecules (tRNAs) and messenger RNA (mRNA) also are key participants in protein synthesis. The link between amino acids and nucleic acids is first made by enzymes called aminoacyl-tRNA synthetases. By specifically linking a particular amino acid to each tRNA, these enzymes translate the genetic code.

Although RNA is paramount in the process of translation, protein factors also are required for the efficient synthesis of a protein. Protein factors participate in the initiation, elongation, and termination of protein synthesis. This chapter focuses primarily on protein synthesis in bacteria because it illustrates many general principles and is well understood. Some distinctive features of protein synthesis in eukaryotes also are presented.

31.1 Protein Synthesis Requires the Translation of Nucleotide Sequences into Amino Acid Sequences

The basics of protein synthesis are the same across all kingdoms of life—evidence that the protein-synthesis system arose very early in evolution. An mRNA is decoded, or read, in the 5′-to-3′ direction, one codon at a time, and the corresponding protein is synthesized in the amino-to-carboxyl direction by the sequential addition of amino acids to the carboxyl end of the growing peptide chain (Figure 31.1). The amino acids arrive at the growing chain in activated form as aminoacyl-tRNAs, created by joining the carboxyl group of an amino acid to the 3′ end of a tRNA molecule. The linking of an amino acid to its corresponding tRNA is catalyzed by an *aminoacyl-tRNA synthetase*. For each amino acid, there is usually one activating enzyme and at least one kind of tRNA.

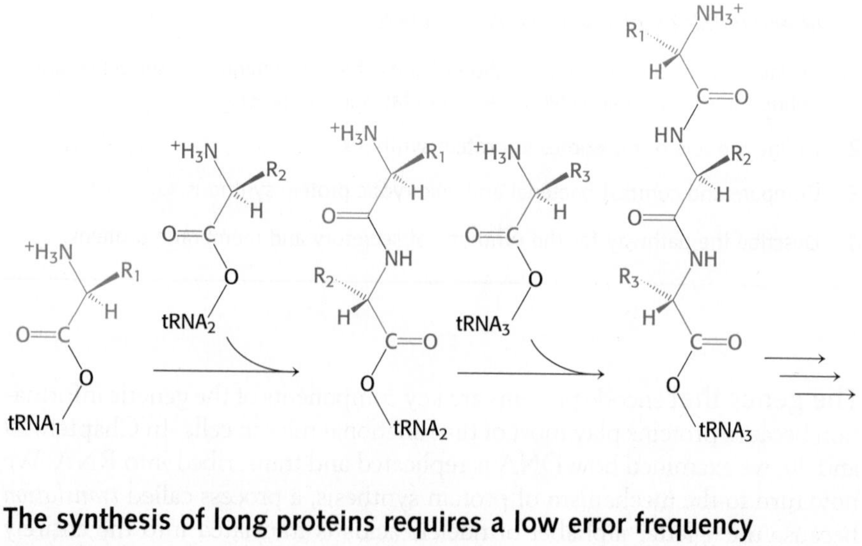

FIGURE 31.1 Polypeptide-chain growth. Proteins are synthesized by the successive addition of amino acids to the carboxyl terminus.

The synthesis of long proteins requires a low error frequency

The process of transcription is analogous to copying, word for word, a page of a book. There is no change of alphabet or vocabulary, so the likelihood of a change in meaning is small. Translating the base sequence of an mRNA molecule into a sequence of amino acids is analogous to translating the page of a book into another language. Translation is a complex process, entailing many steps and dozens of molecules. The potential for error exists at each step. The complexity of translation creates a conflict between two requirements: the process must be both accurate and fast enough to meet a cell's needs. In *E. coli,* translation can take place at a rate of 20 amino acids per second, a truly impressive speed considering the complexity of the process.

How accurate must protein synthesis be? Let us consider error rates. The probability of forming a protein with no errors depends on the number of amino acid residues and on the frequency (ε) of insertion of a wrong amino acid. As Table 31.1 shows, an error frequency of 10^{-2} is intolerable, even for quite small proteins. An ε value of 10^{-3} usually leads to the error-free synthesis of a 300-residue protein (~ 33 kDa) but not of a 1000-residue protein (~ 110 kDa). Thus, the error frequency must not exceed approximately 10^{-4} to produce the larger proteins effectively. Lower error frequencies are conceivable; however, except for the largest proteins, they will not dramatically increase the percentage of proteins with accurate sequences. In addition, such lower error rates are likely to be possible only by a reduction in the rate of protein synthesis because additional time for proofreading is required. *In fact, the observed values of ε are close to 10^{-4}.* An error frequency of about 10^{-4} per amino acid residue was selected in the course of evolution to accurately produce proteins consisting of as many as 1000 amino acids while maintaining a remarkably rapid rate for protein synthesis.

TABLE 31.1 Accuracy of protein synthesis

Frequency of inserting an incorrect amino acid	Probability of synthesizing an error-free protein		
	Number of amino acid residues		
	100	300	1000
10^{-2}	0.366	0.049	0.000
10^{-3}	0.905	0.741	0.368
10^{-4}	0.990	0.970	0.905
10^{-5}	0.999	0.997	0.990

Note: The probability p of forming a protein with no errors depends on n, the number of amino acids, and ε, the frequency of insertion of a wrong amino acid: $p = (1 - \varepsilon)^n$.

Transfer RNA molecules have a common design

The fidelity of protein synthesis requires accurate recognition of three-base *codons* on messenger RNA. Recall that the genetic code relates each amino acid to a three-letter codon (Section 4.6). An amino acid cannot itself recognize a codon. Consequently, an amino acid is attached to a specific tRNA molecule that recognizes the codon by Watson–Crick base-pairing. *Transfer RNA serves as the adapter molecule that binds to a specific codon and brings with it an amino acid for incorporation into the polypeptide chain.*

Consider yeast alanyl-tRNA, so called because it will carry the amino acid alanine. This adapter molecule is a single chain of 76 ribonucleotides (Figure 31.2). The 5′ terminus is phosphorylated (pG), whereas the 3′ terminus has a free hydroxyl group. The *amino acid-attachment site* is the 3′-hydroxyl group of the adenosine residue at the 3′ terminus of the molecule. The sequence 5′-IGC-3′ in the middle of the molecule

Anticodon

Codon

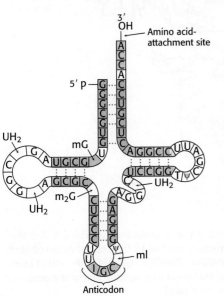

FIGURE 31.2 Alanyl-tRNA sequence. The base sequence of yeast alanyl-tRNA and the deduced cloverleaf secondary structure are shown. Modified nucleosides are abbreviated as follows: methylinosine (mI), dihydrouridine (UH₂), ribothymidine (T), pseudouridine (ψ), methylguanosine (mG), and (dimethylguanosine (m₂G). Inosine (I), another modified nucleoside, is part of the anticodon.

Inosine

5-Methylcytidine (mC)

Dihydrouridine (UH₂)

is the *anticodon*, where I is the purine base inosine. It is complementary to 5'-GCC-3', one of the codons for alanine.

Thousands of tRNA sequences are known. The striking finding is that all of them can be arranged in a cloverleaf pattern in which about half the residues are base-paired (Figure 31.3). Hence, *tRNA molecules have many common structural features.* This finding is not unexpected, because all tRNA molecules must be able to interact in nearly the same way with the ribosomes, mRNAs, and protein factors that participate in translation.

All known transfer RNA molecules have the following features:

1. Each is a single chain containing between *73 and 93 ribonucleotides* (~25 kDa).

2. The molecule is L-shaped (Figure 31.4).

3. They contain *many unusual bases,* typically between 7 and 15 per molecule. Some of these bases are methylated or dimethylated derivatives of A, U, C, and G formed by enzymatic modification of a precursor tRNA. Some methylations prevent the formation of certain base pairs, thereby rendering such bases accessible for interactions with other bases. In addition, methylation imparts a hydrophobic character to some regions of tRNAs, which may be important for their interaction with synthetases and ribosomal proteins. Other modifications alter codon recognition, as will be described shortly.

4. About half the nucleotides in tRNAs are base-paired to form double helices. The four helical regions are arranged to form two apparently continuous segments of double helix. These segments are like A-form DNA, as expected for an RNA helix (Section 4.2). One helix, containing the 5' and 3' ends, runs horizontally in the model shown in Figure 31.5. The other helix, which contains the anticodon and runs vertically in Figure 31.5, forms the other arm of the L.

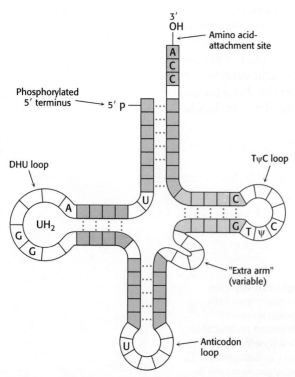

FIGURE 31.3 General structure of tRNA molecules. Comparison of the base sequences of many tRNAs reveals a number of conserved features.

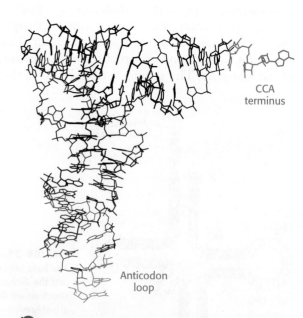

FIGURE 31.4 Transfer RNA structure. *Notice* the L-shaped structure revealed by this skeletal model of yeast phenylalanyl-tRNA. The CCA region is at the end of one arm, and the anticodon loop is at the end of the other. [Drawn from 1EHZ.pdb.]

Five groups of bases are not base-paired in this way: the *3′ CCA terminal region,* which is part of a region called the *acceptor stem;* the *TψC loop,* which acquired its name from the sequence ribothymine-pseudouracil-cytosine; the *"extra arm",* which contains a variable number of residues; *the DHU loop,* which contains several *di*hydrouracil residues; and the *anticodon loop.* Most of the bases in the nonhelical regions participate in hydrogen-bonding interactions, even if the interactions are not like those in Watson–Crick base pairs. The structural diversity generated by this combination of helices and loops containing modified bases ensures that the tRNAs can be uniquely distinguished, though they are structurally similar overall.

5. The 5′ end of a tRNA is phosphorylated. The 5′ terminal residue is usually pG.

6. An activated amino acid is attached to a hydroxyl group of the adenosine residue in the amino acid-attachment site, located at the end of the 3′ CCA component of the acceptor stem (Figure 31.6). This single-stranded region can change conformation in the course of amino acid activation and protein synthesis.

7. The anticodon loop, which is present in a loop near the center of the sequence, is at the other end of the L, making accessible the three bases that constitute the anticodon.

Thus, the architecture of the tRNA molecule is well suited to its role as adaptor: the anticodon is available to interact with an appropriate codon on mRNA while the end that is linked to an activated amino acid is well positioned to participate in peptide-bond formation.

Some transfer RNA molecules recognize more than one codon because of wobble in base-pairing

What are the rules that govern the recognition of a codon by the anticodon of a tRNA? A simple hypothesis is that each of the bases of the codon forms a Watson–Crick type of base pair with a complementary base on the anticodon of the tRNA. The codon and anticodon would then be lined up in an antiparallel fashion. In the diagram in the margin (p. 1026), the prime denotes the complementary base. Thus X and X′ would be either A and U (or U and A) or G and C (or C and G). According to this model, a particular anticodon can recognize only one codon.

The facts are otherwise. As found experimentally, *some tRNA molecules can recognize more than one codon.* For example, the yeast alanyl-tRNA binds to *three* codons: GCU, GCC, and GCA. The first two bases of these codons are the same, whereas the third is different. Could it be that recognition of the third base of a codon is sometimes less discriminating than recognition of the other two? The pattern of degeneracy of the genetic code indicates that it might be so. XYU and XYC always encode the same amino acid; XYA and XYG usually do. These data suggest that the steric criteria might be less stringent for pairing of the third base than for the other two. In other words, there is some steric freedom ("wobble") in the pairing of the third base of the codon.

The *wobble hypothesis* is now firmly established (Table 31.2). The anticodons of tRNAs of known sequence bind to the codons predicted by this hypothesis. For example, the anticodon of yeast alanyl-tRNA is IGC. This tRNA recognizes the codons GCU, GCC, and GCA. Recall that, by convention, nucleotide sequences are written in the 5′ → 3′ direction unless otherwise noted. Hence, I (the 5′ base of this anticodon) pairs with U, C, or A (the 3′ base of the codon), as predicted.

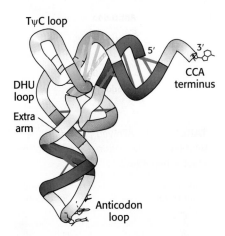

FIGURE 31.5 Helix stacking in tRNA. The four double-stranded regions of the tRNA (Figure 31.3) stack to form an L-shaped structure. [Drawn from 1EHZ.pdb.]

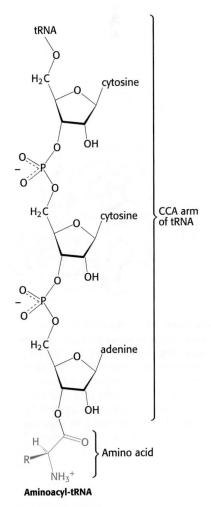

FIGURE 31.6 Aminoacyl-tRNA. Amino acids are coupled to tRNAs through ester linkages to either the 2′- or the 3′-hydroxyl group of the 3′-adenosine residue. A linkage to the 3′-hydroxyl group is shown.

Anticodon

$$-\overset{3'}{X'}-\overset{}{Y'}-\overset{5'}{Z'}-$$

$$-\overset{5'}{X}-\overset{}{Y}-\overset{3'}{Z}-$$

Codon

TABLE 31.2 Allowed pairings at the third base of the codon according to the wobble hypothesis

First base of anticodon	Third base of codon
C	G
A	U
U	A or G
G	U or C
I	U, C, or A

Inosine–cytidine base pair   **Inosine–uridine base pair**   **Inosine–adenosine base pair**

Two generalizations concerning the codon–anticodon interaction can be made:

1. The first two bases of a codon pair in the standard way. Recognition is precise. Hence, *codons that differ in either of their first two bases must be recognized by different tRNAs.* For example, both UUA and CUA encode leucine but are read by different tRNAs.

2. The first base of an anticodon determines whether a particular tRNA molecule reads one, two, or three kinds of codons: C or A (one codon), U or G (two codons), or I (three codons). Thus, *part of the degeneracy of the genetic code arises from imprecision (wobble) in the pairing of the third base of the codon with the first base of the anticodon.* We see here a strong reason for the frequent appearance of inosine, one of the unusual nucleosides, in anticodons. *Inosine maximizes the number of codons that can be read by a particular tRNA molecule.* The inosine bases in tRNA are formed by the deamination of adenosine after the synthesis of the primary transcript.

Why is wobble tolerated in the third position of the codon but not in the first two? This question is answered by considering the interaction of the tRNA with the ribosome. As we will see, ribosomes are huge RNA–protein complexes consisting of two subunits, the 30S and 50S subunits. The 30S subunit has an RNA molecule, the 16S rRNA, that has three universally conserved bases—adenine 1492, adenine 1493, and guanine 530—that form hydrogen bonds on the minor-groove side but only with correctly formed base pairs of the codon–anticodon duplex (Figure 31.7). These interactions check whether Watson–Crick base pairs are present in the first two positions of the codon–anticodon duplex, but not the third position, so more-varied base pairs are tolerated. *Thus, the ribosome plays an active role in decoding the codon–anticodon interactions.*

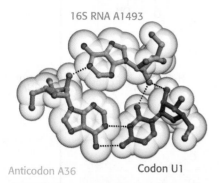

FIGURE 31.7 16S rRNA monitors base-pairing between the codon and the anticodon. Adenine 1493, one of three universally conserved bases in 16S rRNA, forms hydrogen bonds with the bases in both the codon and the anticodon only if the codon and anticodon are correctly paired. [Information from J. M. Ogle and V. Ramakrishnan, *Annu. Rev. Biochem.* 74:129–177, 2005, Fig. 2a.]

31.2 Aminoacyl Transfer RNA Synthetases Read the Genetic Code

Before codon and anticodon meet, the amino acids required for protein synthesis must first be attached to specific tRNA molecules, linkages that are crucial for two reasons. *First, the attachment of a given amino acid to a particular tRNA establishes the genetic code.* When an amino acid has been linked to a tRNA, it will be incorporated into a growing polypeptide chain at a position dictated by the anticodon of the tRNA. *Second, because the formation of a peptide bond between free amino acids is not thermodynamically favorable, the amino acid must first be activated in order for protein synthesis to proceed. The activated intermediates in protein synthesis are amino acid esters,* in which the carboxyl group of an amino acid is linked to either the 2′- or the 3′-hydroxyl group of the ribose unit at the 3′ end of tRNA. An amino acid ester of tRNA is called an *aminoacyl-tRNA* or sometimes a *charged tRNA* (Figure 31.6). For a specific amino acid attached to its cognate tRNA—for instance, threonine—the charged tRNA is designated Thr-tRNA^{Thr}.

Amino acids are first activated by adenylation

The activation reaction is catalyzed by specific *aminoacyl-tRNA synthetases.* The first step is the formation of an *aminoacyl adenylate* from an amino acid and ATP.

$$\text{Amino acid} + \text{ATP} \rightleftharpoons \text{aminoacyl-AMP} + \text{PP}_i$$

This activated species is a mixed anhydride in which the carboxyl group of the amino acid is linked to the phosphoryl group of AMP; hence, it is also known as *aminoacyl-AMP.*

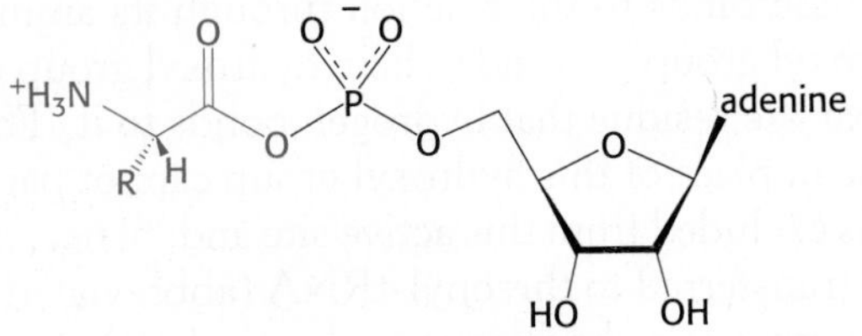

Aminoacyl adenylate

The next step is the transfer of the aminoacyl group of aminoacyl-AMP to a particular tRNA molecule to form *aminoacyl-tRNA.*

$$\text{Aminoacyl-AMP} + \text{tRNA} \rightleftharpoons \text{aminoacyl-tRNA} + \text{AMP}$$

The sum of these activation and transfer steps is

$$\text{Amino acid} + \text{ATP} + \text{tRNA} \rightleftharpoons \text{aminoacyl-tRNA} + \text{AMP} + \text{PP}_i$$

The $\Delta G^{\circ\prime}$ of this reaction is close to 0, because the free energy of hydrolysis of the ester bond of aminoacyl-tRNA is similar to that for the hydrolysis of ATP to AMP and PP$_i$. As we have seen many times, the reaction is driven by the hydrolysis of pyrophosphate. The sum of these three reactions is highly exergonic:

$$\text{Amino acid} + \text{ATP} + \text{tRNA} + \text{H}_2\text{O} \longrightarrow \text{aminoacyl-tRNA} + \text{AMP} + 2\text{P}_i$$

Thus, *the equivalent of two molecules of ATP is consumed in the synthesis of each aminoacyl-tRNA.* One of them is consumed in forming the ester linkage of aminoacyl-tRNA, whereas the other is consumed in driving the reaction forward.

The activation and transfer steps for a particular amino acid are catalyzed by the same aminoacyl-tRNA synthetase. Indeed, *the aminoacyl-AMP intermediate does not dissociate from the synthetase.* Rather, it is tightly bound to the active site of the enzyme by noncovalent interactions.

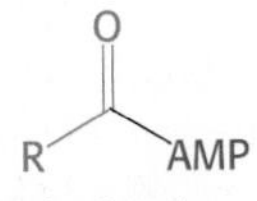

Acyl adenylate intermediate

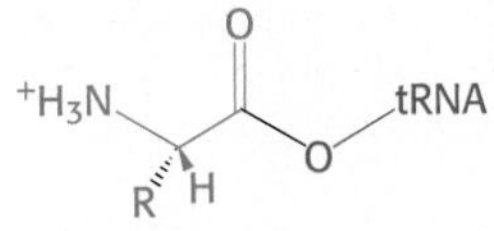

Aminoacyl-tRNA

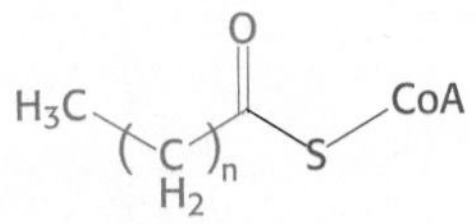

Fatty acyl CoA

Threonine

Valine

Serine

We have already encountered an acyl adenylate intermediate in fatty acid activation (Section 22.2). The major difference between these reactions is that the acceptor of the acyl group is CoA in fatty acid activation and tRNA in amino acid activation. The energetics of these biosyntheses are very similar: both are made irreversible by the hydrolysis of pyrophosphate.

Aminoacyl-tRNA synthetases have highly discriminating amino acid activation sites

Each aminoacyl-tRNA synthetase is highly specific for a given amino acid. Indeed, a synthetase will incorporate the incorrect amino acid only once in 10^4 or 10^5 reactions. How is this level of specificity achieved? Each aminoacyl-tRNA synthetase takes advantage of the properties of its amino acid substrate. Let us consider the challenge faced by threonyl-tRNA synthetase. Threonine is especially similar to two other amino acids—namely, valine and serine. Valine has almost exactly the same shape as that of threonine, except that valine has a methyl group in place of a hydroxyl group. Serine has a hydroxyl group, as does threonine, but lacks the methyl group. How can the threonyl-tRNA synthetase avoid coupling these incorrect amino acids to threonyl-tRNA?

The structure of the amino acid-binding site of threonyl-tRNA synthetase reveals how valine is avoided (Figure 31.8). The synthetase contains a zinc ion, bound to the enzyme by two histidine residues and one cysteine residue. The remaining coordination sites are available for substrate binding. Threonine binds to the zinc ion through its amino group and its side-chain hydroxyl group. The side-chain hydroxyl group is further recognized by an aspartate residue that hydrogen bonds to it. The methyl group present in valine in place of this hydroxyl group cannot participate in these interactions; it is excluded from this active site and, hence, does not become adenylated and transferred to threonyl-tRNA (abbreviated tRNA^{Thr}). The use of a zinc ion appears to be unique to threonyl-tRNA synthetase; other aminoacyl-tRNA synthetases have different strategies for recognizing their cognate amino acids. The carboxylate group of the correctly positioned threonine is available to attack the α phosphoryl group of ATP to form the aminoacyl adenylate.

The zinc site is less able to discriminate against serine because this amino acid does have a hydroxyl group that can bind to the zinc ion. Indeed,

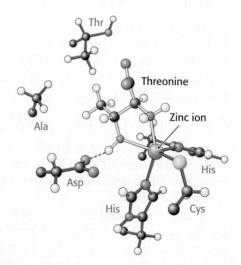

FIGURE 31.8 Active site of threonyl-tRNA synthetase. *Notice* that the amino acid-binding site includes a zinc ion (green ball) that binds threonine through its amino and hydroxyl groups.

with only this mechanism available, threonyl-tRNA synthetase does mistakenly couple serine to threonyl-tRNA at a rate 10^{-2} to 10^{-3} times that for threonine. As noted on page 1023, this error rate is likely to lead to many translation errors. How is a higher level of specificity achieved?

Proofreading by aminoacyl-tRNA synthetases increases the fidelity of protein synthesis

Threonyl-tRNA synthetase can be incubated with tRNA^{Thr} that has been covalently linked with serine (Ser-tRNA^{Thr}); the tRNA has been "mischarged." The reaction is immediate: a rapid hydrolysis of the aminoacyl-tRNA forms serine and free tRNA. In contrast, incubation with correctly charged Thr-tRNA^{Thr} results in no reaction. Thus, threonyl-tRNA synthetase contains an additional functional site that hydrolyzes Ser-tRNA^{Thr} but not Thr-tRNA^{Thr}. This editing site provides an opportunity for the synthetase to correct its mistakes and improve its fidelity to less than one mistake in 10^4. The results of structural and mutagenesis studies revealed that the editing site is more than 20 Å from the activation site (Figure 31.9). This editing site readily accepts and cleaves Ser-tRNA^{Thr} but does not cleave Thr-tRNA^{Thr}. The discrimination of serine from threonine is easy because threonine contains an *extra* methyl group; a site that conforms to the structure of serine will sterically exclude threonine. The structure of the complex between threonyl-tRNA synthetase and its substrate reveals that the aminoacylated CCA can swing out of the activation site and into the editing site (Figure 31.10). Thus, the aminoacyl-tRNA can be edited without dissociating from the synthetase. This proofreading depends on the conformational flexibility of a short stretch of polynucleotide sequence.

Most aminoacyl-tRNA synthetases contain editing sites in addition to activation sites. These complementary pairs of sites function as a *double sieve* to ensure very high fidelity. In general, the acylation site rejects amino acids that are *larger* than the correct one because there is insufficient room for them, whereas the hydrolytic site cleaves activated species that are *smaller* than the correct one.

A few synthetases achieve high accuracy without editing. For example, tyrosyl-tRNA synthetase has no difficulty discriminating between tyrosine and phenylalanine; the hydroxyl group on the tyrosine ring enables tyrosine to bind to the enzyme 10^4 times as strongly as phenylalanine. *Proofreading has been selected in evolution only when fidelity must be enhanced beyond what can be obtained through an initial binding interaction.*

Synthetases recognize various features of transfer RNA molecules

How do synthetases choose their tRNA partners? This enormously important step is the point at which "translation" takes place—at which the correlation between the amino acid and the nucleic acid worlds is made. In a sense, aminoacyl-tRNA synthetases are the only molecules in biology that "know" the genetic code. Their precise recognition of tRNAs is as important for high-fidelity protein synthesis as is the accurate selection of amino acids, and this recognition is sometimes referred to as the "second genetic code." In general, tRNA recognition by the synthetase is different for each synthetase and tRNA pair. Consequently, generalities are difficult to make.

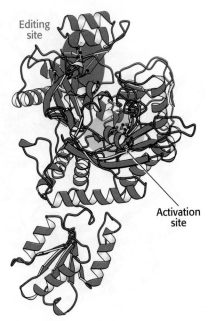

FIGURE 31.9 Editing site. Mutagenesis studies revealed the position of the editing site (shown in green) in threonyl-tRNA synthetase. The activation site is shown in yellow. Only one subunit of the dimeric enzyme is shown here and in subsequent illustrations. [Drawn from 1QF6.pdb.]

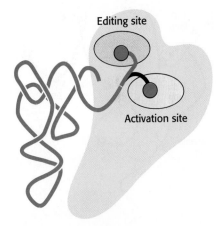

FIGURE 31.10 Editing of aminoacyl-tRNA. The flexible CCA arm of an aminoacyl-tRNA can move the amino acid between the activation site and the editing site. If the amino acid fits well into the editing site, the amino acid is removed by hydrolysis.

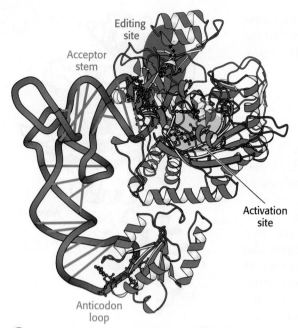

FIGURE 31.11 Threonyl-tRNA synthetase complex. The structure shows the complex between threonyl-tRNA synthetase (blue) and tRNA^Thr (red). *Notice* that the synthetase binds to both the acceptor stem and the anticodon loop. [Drawn from 1QF6.pdb.]

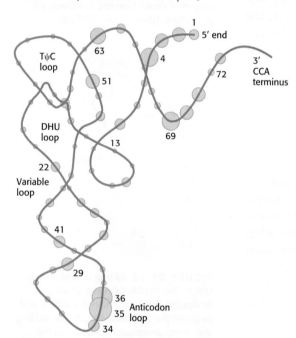

FIGURE 31.12 Recognition sites on tRNA. Circles represent nucleotides, and the sizes of the circles are proportional to the frequency with which they are used as recognition sites by aminoacyl-tRNA synthetases. The numbers indicate the positions of the nucleotides in the base sequence, beginning from the 5′ end of the tRNA molecule. [Information from M. Ibba and D. Söll, *Annu. Rev. Biochem.* 69:617–650, 1981, p. 636.]

Some synthetases recognize their tRNA partners primarily on the basis of their anticodons, although they may also recognize other aspects of tRNA structure that vary among different tRNAs. The most direct evidence comes from crystallographic studies of complexes formed between synthetases and their cognate tRNAs. Consider, for example, the structure of the complex between threonyl-tRNA synthetase and tRNA^Thr (Figure 31.11). As expected, the CCA arm extends into the zinc-containing activation site, where it is well positioned to accept threonine from threonyl adenylate. The enzyme interacts extensively not only with the acceptor stem of the tRNA, but also with the anticodon loop. The interactions with the anticodon loop are particularly revealing. Each base within the sequence 5′-CGU-3′ of the anticodon participates in hydrogen bonds with the enzyme; those with the second two bases (G and U) appear to be more important because the synthetase interacts just as efficiently with the anticodons GGU and UGU. Although interactions between the enzyme and the anticodon are often crucial for correct recognition, Figure 31.12 shows that many aspects of tRNA molecules are recognized by synthetases. Note that many of the recognition sites are loops rich in unusual bases that can provide structural identifiers.

Aminoacyl-tRNA synthetases can be divided into two classes

At least one aminoacyl-tRNA synthetase exists for each amino acid. The diverse sizes, subunit composition, and sequences of these enzymes were bewildering for many years. Could it be that essentially all synthetases evolved independently? The determination of the three-dimensional structures of several synthetases followed by more-refined sequence comparisons revealed that different synthetases are, in fact, related. Specifically, synthetases fall into two classes, termed *class I* and *class II*, each of which includes enzymes specific for 10 of the 20 amino acids (Table 31.3). Intriguingly, synthetases from the two classes bind to different faces of the tRNA molecule (Figure 31.13). The CCA arm of tRNA adopts different conformations to accommodate these interactions; the arm is in the helical conformation observed for free tRNA (Figures 31.4 and 31.5) for class II enzymes and in a hairpin conformation for class I enzymes. These two classes also differ in other ways.

1. Class I enzymes acylate the 2′-hydroxyl group of the terminal adenosine of tRNA, whereas class II enzymes (except the enzyme for Phe-tRNA) acylate the 3′-hydroxyl group.

2. The two classes bind ATP in different conformations.

3. Most class I enzymes are monomeric, whereas most class II enzymes are dimeric.

Why did two distinct classes of aminoacyl-tRNA synthetases evolve? The observation that the two classes bind to distinct faces of tRNA suggests a possibility. Recognition sites on both faces of tRNA may have been required to allow the recognition of 20 different tRNAs.

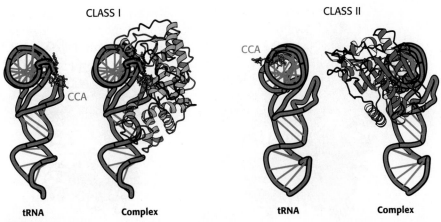

CLASS I

CLASS II

CCA

CCA

tRNA   Complex   tRNA   Complex

FIGURE 31.13 Classes of aminoacyl-tRNA synthetases. *Notice* that class I and class II synthetases recognize different faces of the tRNA molecule. The CCA arm of tRNA adopts different conformations in complexes with the two classes of synthetase. Note that the CCA arm of the tRNA is turned toward the viewer (Figures 31.4 and 31.5). [Drawn from 1EUY.pdb and 1QF6.pdb.]

TABLE 31.3 Classification and subunit structure of aminoacyl-tRNA synthetases in *E. coli*

Class I	Class II
Arg (α)	Ala ($\alpha4$)
Cys (α)	Asn ($\alpha2$)
Gln (α)	Asp ($\alpha2$)
Glu (α)	Gly ($\alpha2\beta2$)
Ile (α)	His ($\alpha2$)
Leu (α)	Lys ($\alpha2$)
Met (α)	Phe ($\alpha2\beta2$)
Trp ($\alpha2$)	Ser ($\alpha2$)
Tyr ($\alpha2$)	Pro ($\alpha2$)
Val (α)	Thr ($\alpha2$)

31.3 The Ribosome Is the Site of Protein Synthesis

We turn now to ribosomes, the molecular machines that coordinate the interplay of aminoacyl-tRNAs, mRNA, and proteins that leads to protein synthesis. An *E. coli* ribosome is a ribonucleoprotein assembly with a mass of about 2500 kDa, a diameter of approximately 250 Å, and a sedimentation coefficient (Section 3.1) of 70S. The 20,000 ribosomes in a bacterial cell constitute nearly a quarter of its mass.

A ribosome can be dissociated into a *large subunit (50S)* and a *small subunit (30S)*. These subunits can be further split into their constituent proteins and RNAs. The 30S subunit contains 21 different proteins (referred to as S1 through S21) and a 16S RNA molecule. The 50S subunit contains 34 different proteins (L1 through L34) and two RNA molecules, a 23S and a 5S species. A ribosome contains one copy of each RNA molecule, two copies each of the L7 and L12 proteins, and one copy of each of the other proteins. The L7 protein is identical with L12 except that its amino terminus is acetylated (Section 10.3). Both the 30S and the 50S subunits can be reconstituted in vitro from their constituent proteins and RNA. *This reconstitution is an outstanding example of the principle that supramolecular complexes can form spontaneously from their macromolecular constituents.*

The structures of both the 30S and the 50S subunits as well as the complete 70S ribosome have been determined (Figure 31.14). The features of these structures are in remarkable agreement with interpretations of less-direct experimental probes. These structures provide an invaluable framework for examining the mechanism of protein synthesis.

Ribosomal RNAs (5S, 16S, and 23S rRNA) play a central role in protein synthesis

The prefix *ribo* in the name *ribosome* is apt because RNA constitutes nearly two-thirds of the mass of these large molecular assemblies. The three RNAs present—5S, 16S, and 23S—are critical for ribosomal architecture and function. They are formed by the cleavage of primary 30S transcripts and further processing. These molecules fold into structures that allow them to form internal base pairs. Their base-pairing patterns were deduced by comparing the nucleotide sequences of many species to detect conserved sequences as

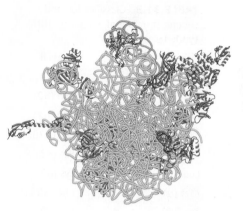

50S subunit

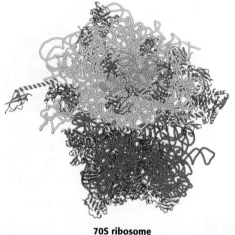

70S ribosome

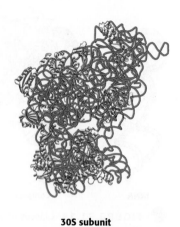

30S subunit

FIGURE 31.14 The ribosome at high resolution. Detailed models of the ribosome based on the results of x-ray crystallographic studies of the 70S ribosome and the 30S and 50S subunits: (left) view of the part of the 50S subunit that interacts with the 30S subunit; (center) side view of the 70S ribosome; (right) view of the part of the 30S subunit that interacts with the 50S subunit. 23S RNA is shown in yellow, 5S RNA in orange, 16S RNA in green, proteins of the 50S subunit in red, and proteins of the 30S subunit in blue. *Notice that the interface between the 50S and the 30S subunits consists entirely of RNA.* [Drawn from 1GIX. pdb and 1GIY.pdb.]

well as conserved base pairings. For instance, the 16S RNA of one species may have a G–C base pair, whereas another may have an A–U base pair, but the location of the base pair is the same in both molecules. Chemical modification and digestion experiments supported the structures deduced from sequence comparisons (Figure 31.15). The striking finding is that across all species, *ribosomal RNAs* (rRNAs) *are folded into defined structures that have many short duplex regions.*

For many years, ribosomal proteins were presumed to orchestrate protein synthesis and ribosomal RNAs were presumed to serve primarily as structural scaffolding. The current view is almost the reverse. The discovery of catalytic RNA (Section 30.4) made biochemists receptive to the possibility that RNA plays a much more active role in ribosomal function. The detailed structures make it clear that the key sites in the ribosome, such as those that catalyze the formation of the peptide bond and interact with mRNA and tRNA, are composed almost entirely of RNA. Contributions from the proteins are minor. The almost inescapable conclusion is that the ribosome initially consisted only of RNA and that the proteins were added later to fine-tune its functional properties. This conclusion has the pleasing consequence of dodging a "chicken and egg" question: How can complex proteins be synthesized if complex proteins are required for protein synthesis?

Ribosomes have three tRNA-binding sites that bridge the 30S and 50S subunits

Three tRNA-binding sites in ribosomes are arranged to allow the formation of peptide bonds between amino acids encoded by the codons on mRNA (Figure 31.16). The mRNA fragment being translated at a given moment is bound within the 30S subunit. Each of the tRNA molecules is in contact with both the 30S subunit and the 50S subunit. At the 30S end, two of the three tRNA molecules are bound to the mRNA through anticodon–codon base pairs. These binding sites are called the A site (for *a*minoacyl) and the P site (for *p*eptidyl). The third tRNA molecule is bound to an adjacent site called the E site (for *e*xit).

The other end of each tRNA molecule, the end without the anticodon, interacts with the 50S subunit. The acceptor stems of the tRNA molecules occupying the A site and the P site converge at a site where a peptide bond is formed. A tunnel connects this site to the back of the ribosome, through which the polypeptide chain passes during synthesis (Figure 31.17).

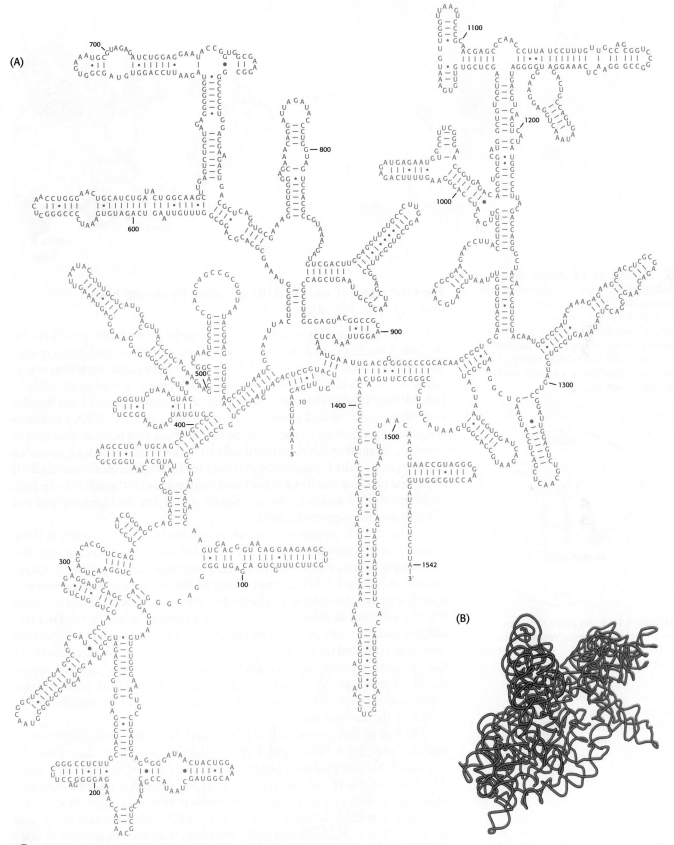

FIGURE 31.15 Ribosomal RNA folding pattern. (A) The secondary structure of 16S ribosomal RNA deduced from sequence comparison and the results of chemical studies. (B) The tertiary structure of 16S RNA determined by x-ray crystallography. [(A) Courtesy of Dr. Bryn Weiser and Dr. Harry Noller; (B) drawn from 1FJG.pdb.]

(A)

(B)

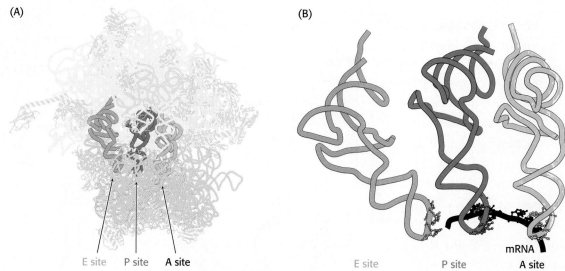

E site   P site   A site

E site   P site   A site

mRNA

FIGURE 31.16 Transfer RNA-binding sites. (A) Three tRNA-binding sites are present on the 70S ribosome. They are called the A (for aminoacyl), P (for peptidyl), and E (for exit) sites. Each tRNA molecule contacts both the 30S and the 50S subunit. (B) The tRNA molecules in sites A and P are base-paired with mRNA. [(B) Drawn from 1JGP. pdb.]

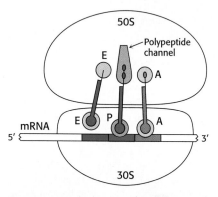

50S

Polypeptide channel

E

A

E   P   A

mRNA

5′   3′

30S

FIGURE 31.17 An active ribosome. This schematic representation shows the relations among the key components of the translation machinery.

The term polycistronic is derived from cistron, a no longer used name for a gene.

The start signal is usually AUG preceded by several bases that pair with 16S rRNA

How does protein synthesis start? The simplest possibility would be for the first 3 nucleotides of each mRNA to serve as the first codon; no special start signal would then be needed. However, experiments show that translation in bacteria does not begin immediately at the 5′ terminus of mRNA. Indeed, the first translated codon is nearly always more than 25 nucleotides away from the 5′ end. Furthermore, in bacteria, many mRNA molecules are *polycistronic*—that is, they encode two or more polypeptide chains. For example, a single mRNA molecule about 7000-nucleotides long specifies five enzymes in the biosynthetic pathway for tryptophan in *E. coli*. Each of these five proteins has its own start and stop signals on the mRNA. In fact, *all known mRNA molecules contain signals that define the beginning and end of each encoded polypeptide chain*.

A clue to the mechanism of initiation was the finding that nearly half the amino-terminal residues of proteins in *E. coli* are methionine. In fact, the initiating codon in mRNA is AUG (methionine) or, less frequently, GUG (valine) or, rarely UUG (leucine). What additional signals are necessary to specify a translation start site? The first step toward answering this question was the isolation of initiator regions from a number of mRNAs. This isolation was accomplished by using ribonuclease to digest mRNA–ribosome complexes (formed under conditions in which protein synthesis could begin but elongation could not take place). As expected, each initiator region usually displays an AUG codon (Figure 31.18). In addition, each initiator region contains a purine-rich sequence centered about 10 nucleotides on the 5′ side of the initiator codon.

The role of this purine-rich region, called the *Shine–Dalgarno sequence*, (named after John Shine and Lynn Dalgarno, who first described the sequence), became evident when the sequence of 16S rRNA was elucidated. The 3′ end of this rRNA component of the 30S subunit contains a sequence of several bases that is complementary to the purine-rich region in the initiator sites of mRNA. Mutagenesis of the CCUCC sequence near the 3′ end of 16S rRNA to ACACA markedly interferes with the recognition of start sites in mRNA. This result and other evidence show that the initiator region of mRNA binds very near the 3′ end of the 16S rRNA. The number of base pairs linking mRNA and 16S rRNA ranges from three to nine. Thus, *two kinds of interactions determine where protein synthesis starts: (1) the pairing of*

```
5'                                                    3'
AGCACGAGGGGAAAUCUGAUGGAACGCUAC    E. coli trpA
UUUGGAUGGAGUGAAACGAUGGCGAUUGCA    E. coli araB
GGUAACCAGGUAACAACCAUGCGAGUGUUG    E. coli thrA
CAAUUCAGGGUGGUGAAUGUGAAACCAGUA    E. coli lacI
AAUCUUGGAGGCUUUUUUUAUGGUUCGUUCU   φX174 phage A protein
UAACUAAGGAUGAAAUGCAUGUCUAAGACA    Qβ phage replicase
UCCUAGGAGGUUUGACCUAUGCGAGCUUUU    R17 phage A protein
AUGUACUAAGGAGGUUGUAUGGAACAACGC    λ phage cro
```

Pairs with   Pairs with
16S rRNA   initiator tRNA

FIGURE 31.18 Initiation sites.
Sequences of mRNA initiation sites for protein synthesis in some bacterial and viral mRNA molecules. Comparison of these sequences reveals some recurring features.

mRNA bases with the 3′ end of 16S rRNA and (2) the pairing of the initiator codon on mRNA with the anticodon of an initiator tRNA molecule.

Bacterial protein synthesis is initiated by formylmethionyl transfer RNA

As stated earlier, methionine is the first amino acid in many *E. coli* proteins. However, the methionine residue found at the amino-terminal end of *E. coli* proteins is usually modified. In fact, *protein synthesis in bacteria starts with the modified amino acid N-formylmethionine (fMet)*. A special tRNA brings formylmethionine to the ribosome to initiate protein synthesis. This *initiator tRNA* (abbreviated as tRNA$_f$) differs from the tRNA that inserts methionine in internal positions (abbreviated as tRNA$_m$). The subscript "f" indicates that methionine attached to the initiator tRNA can be formylated, whereas it cannot be formylated when attached to tRNA$_m$. Although virtually all proteins synthesized in *E. coli* begin with formylmethionine, in approximately one-half of the proteins, *N*-formylmethionine is removed when the nascent chain is 10 amino acids long.

Methionine is linked to these two kinds of tRNAs by the same aminoacyl-tRNA synthetase. A specific enzyme then formylates the amino group of the methionine molecule that is attached to tRNA$_f$ (Figure 31.19). The activated formyl donor in this reaction is N^{10}-formyltetrahydrofolate, a folate derivative that carries activated one-carbon units (Section 24.2). Free methionine and methionyl-tRNA$_m$ are not substrates for this transformylase.

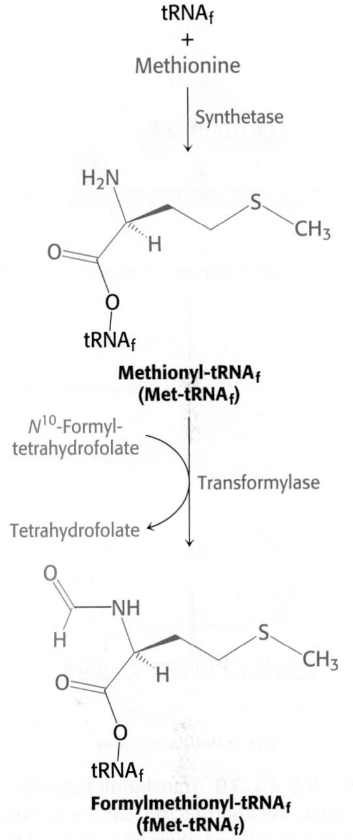

Formylmethionyl-tRNA$_f$ is placed in the P site of the ribosome in the formation of the 70S initiation complex

Messenger RNA and formylmethionyl-tRNA$_f$ must be brought to the ribosome for protein synthesis to begin. How is this task accomplished? Three protein *initiation factors* (IF1, IF2, and IF3) are essential. The 30S ribosomal subunit first forms a complex with IF1 and IF3 (Figure 31.20). The binding of these factors to the 30S subunit prevents it from prematurely joining the 50S subunit to form a dead-end 70S complex, devoid of mRNA and fMet-tRNA$_f$. IF1 binds near the A site and directs the fMet-tRNA$_f$ to the P site. Initiation factor 2, a member of the G-protein family, binds GTP, and the concomitant conformational change enables IF2 to associate with fMet-tRNA$_f$. The IF2–GTP–initiator-tRNA complex binds with mRNA (correctly positioned by the interaction of the Shine–Dalgarno sequence with the 16S rRNA) and the 30S subunit to form the *30S initiation complex*. Structural changes then lead to the ejection of IF1 and IF3. IF2 stimulates

FIGURE 31.19 Formylation of methionyl-tRNA. Initiator tRNA (tRNA$_f$) is first charged with methionine, and then a formyl group is transferred to the methionyl-tRNA$_f$ from N^{10}-formyltetrahydrofolate.

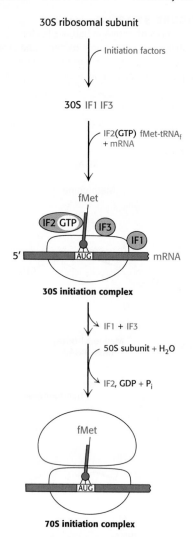

30S ribosomal subunit

Initiation factors

30S IF1 IF3

IF2(GTP) fMet-tRNA$_f$ + mRNA

fMet

5′ **IF2 GTP** **IF3** **IF1** mRNA AUG

30S initiation complex

IF1 + IF3

50S subunit + H$_2$O

IF2, GDP + P$_i$

fMet

AUG

70S initiation complex

FIGURE 31.20 Translation initiation in bacteria. Initiation factors aid the assembly first of the 30S initiation complex and then of the 70S initiation complex.

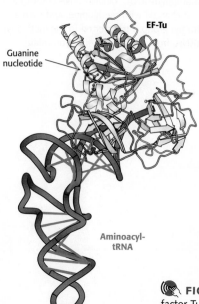

EF-Tu

Guanine nucleotide

Aminoacyl-tRNA

the association of the 50S subunit to the complex. The GTP bound to IF2 is hydrolyzed upon arrival of the 50S subunit, releasing IF2. The result is a *70S initiation complex*. The formation of the 70S initiation complex is the rate-limiting step in protein synthesis.

When the 70S initiation complex has been formed, the ribosome is ready for the elongation phase of protein synthesis. The fMet-tRNA$_f$ molecule occupies the P site on the ribosome, positioned so that its anticodon pairs with the initiating codon on mRNA. The other two sites for tRNA molecules, the A site and the E site, are empty. This interaction establishes the *reading frame* for the translation of the entire mRNA. After the initiator codon has been located, groups of three nonoverlapping nucleotides are defined.

Elongation factors deliver aminoacyl-tRNA to the ribosome

At this point, fMet-tRNA$_f$ occupies the P site, and the A site is vacant. The particular species inserted into the empty A site depends on the mRNA codon in the A site. However, the appropriate aminoacyl-tRNA does not simply leave the synthetase and diffuse to the A site. Rather, it is delivered to the A site in association with a 43-kDa protein called *elongation factor Tu* (EF-Tu), another member of the G-protein family. EF-Tu, the most abundant bacterial protein, binds aminoacyl-tRNA only in its GTP form (Figure 31.21). The binding of EF-Tu to aminoacyl-tRNA serves two functions. First, EF-Tu protects the delicate ester linkage in aminoacyl-tRNA from hydrolysis. Second, EF-Tu contributes to the accuracy of protein synthesis because GTP hydrolysis and expulsion of the EF-Tu-GDP complex from the ribosome occurs only if the pairing between the anticodon and the codon is correct. EF-Tu interacts with the 16S RNA, which monitors the accuracy of the base pairing of position 1 and 2 of the codon (p. 1026). Correct codon recognition induces structural changes in the 30S subunit that move the EF-Tu to a highly conserved site in the 23S RNA in the 50S subunit called the sarcin-ricin loop (SRL). The interaction with the SRL of the 50S subunit activates the GTPase activity of EF-Tu, releasing EF-Tu-GDP from the ribosome. These same structural changes also rotate the aminoacyl-tRNA in the A site so that the amino acid is brought into proximity with the aminoacyl-tRNA in the P site on the 50S subunit, a process called *accommodation*. Accommodation aligns the amino acids for peptide bond formation.

Released EF-Tu is then reset to its GTP form by a second elongation factor, *elongation factor Ts* (Figure 31.22). EF-Ts induces the dissociation of GDP. GTP binds to EF-Tu, and EF-Ts concomitantly departs. It is noteworthy that *EF-Tu does not interact with fMet-tRNA$_f$*. Hence, this initiator tRNA is not delivered to the A site. In contrast, Met-tRNA$_m$, like all other aminoacyl-tRNAs, does bind to EF-Tu. These findings account for the fact that *internal AUG codons are not read by the initiator tRNA*. Conversely, IF2 recognizes fMet-tRNA$_f$ but no other tRNA. The cycle of elongation continues until a termination codon is met.

This GTP–GDP cycle of EF-Tu is reminiscent of those of the heterotrimeric G proteins in signal transduction (Section 14.1) and the Ras proteins in growth control (Section 14.3). This similarity is due to their

FIGURE 31.21 Structure of elongation factor Tu. The structure of a complex between elongation factor Tu (EF-Tu) and an aminoacyl-tRNA. *Notice* the P-loop NTPase domain (purple shading) at the amino-terminal end of EF-Tu. This NTPase domain is similar to those in other G proteins. [Drawn from 1B23. pdb.]

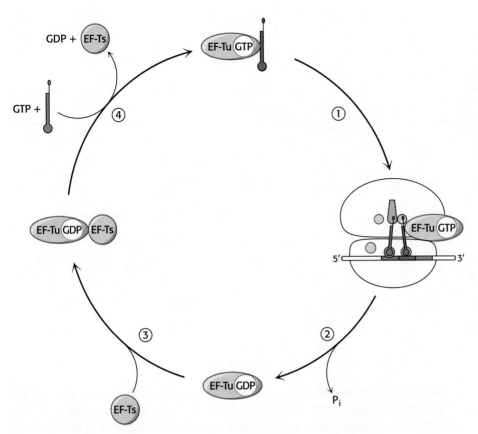

FIGURE 31.22 GTP–GDP cycle of EF-Tu. (1) EF-Tu-GTP binds tRNA and delivers it to the A site on the ribosome. (2) Correct codon recognition stimulates the GTPase activity of EF-Tu, which leaves the ribosome in its GDP form. (3) EF-Ts binds to EF-Tu-GDP. (4) EF-Ts induces release of GDP. EF-Ts departs as another GTP and tRNA bind to EF-Tu-GTP, and the complex is ready for another delivery to the ribosome.

shared evolutionary heritage, seen in the homology of the amino-terminal domain of EF-Tu to the P-loop NTPase domains in the other G proteins. In all these related enzymes, the change in conformation between the GTP and the GDP forms leads to a change in interaction partners. A further similarity is the requirement that an additional protein catalyzes the exchange of GTP for GDP; EF-Ts catalyzes the exchange for EF-Tu, just as an activated receptor does for a heterotrimeric G protein.

Peptidyl transferase catalyzes peptide-bond synthesis

With both the P site and the A site occupied by aminoacyl-tRNA, the stage is set for the formation of a peptide bond: the formylmethionine molecule linked to the initiator tRNA will be transferred to the amino group of the amino acid in the A site. The formation of the peptide bond, one of the most important reactions in life, is a thermodynamically spontaneous reaction catalyzed by a site on the 23S rRNA of the 50S subunit called the *peptidyl transferase center*. This catalytic center is located deep in the 50S subunit near the tunnel that allows the nascent peptide to leave the ribosome.

The ribosome, which enhances the rate of peptide bond synthesis by a factor of 10^7 over the uncatalyzed reaction ($\sim 10^{-4}\,\mathrm{M}^{-1}\,\mathrm{s}^{-1}$), derives much of its catalytic power from *catalysis by proximity and orientation*. The ribosome positions and orients the two substrates so that they are situated to take advantage of the inherent reactivity of an amine group (on the aminoacyl-tRNA in the A site) with an ester (on the initiator tRNA in the P site). The amino group of the aminoacyl-tRNA in the A site, in its unprotonated state, makes a nucleophilic attack on the ester linkage between the initiator tRNA and the formylmethionine molecule in the P site (Figure 31.23A). The nature of the transition state that follows the attack is not established and several models are plausible. One model proposes roles for the 2′OH of

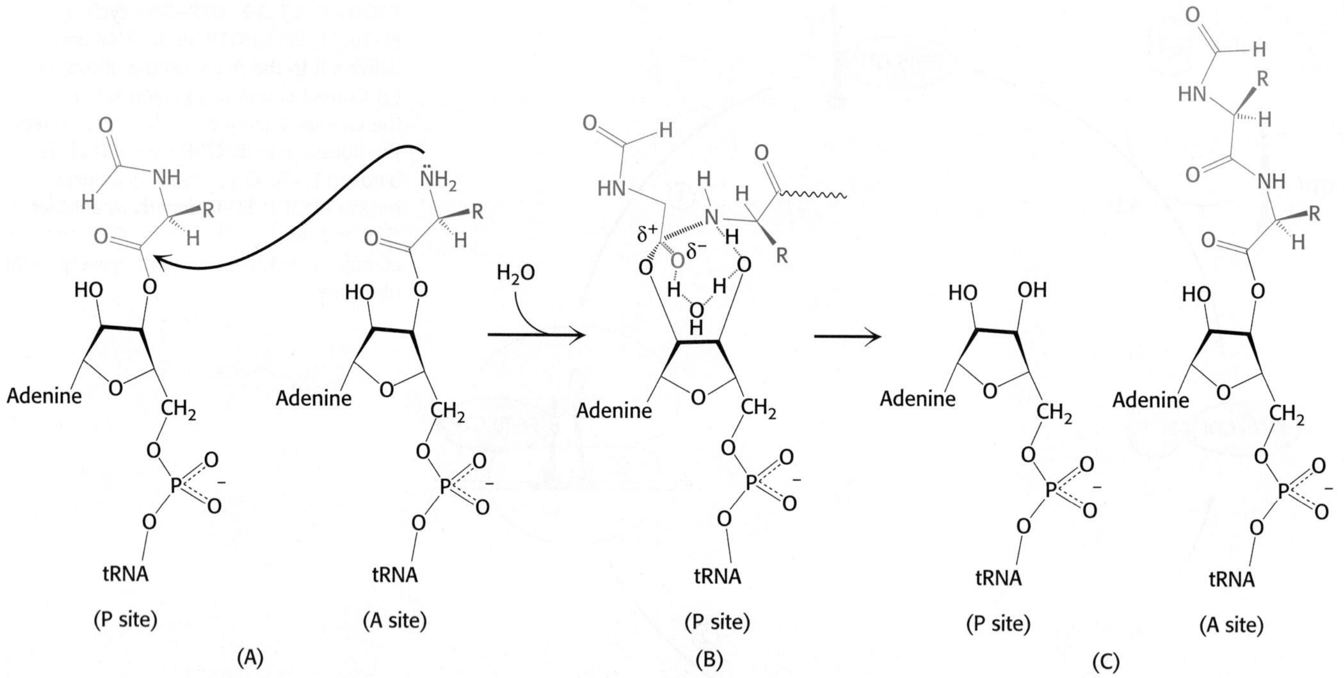

FIGURE 31.23 Peptide-bond formation.
(A) The amino group of the aminoacyl-tRNA attacks the carbonyl group of the ester linkage of the peptidyl-tRNA.
(B) An eight-membered transition state is formed with the addition of a molecule of water. Note: Not all atoms are shown and some bond lengths are exaggerated for clarity. (C) This transition state collapses to form the peptide bond and release the deacylated tRNA.

the adenosine of the tRNA in the P site and a molecule of water at the peptidyl transferase center (Figure 31.23B). The nucleophilic attack of the α-amino group generates an eight-membered transition state in which three protons are shuttled about in a concerted manner. The proton of the attacking amino group hydrogen bonds to the 2′ oxygen of ribose of the tRNA. The hydrogen of 2′OH in turn interacts with the oxygen of the water molecule at the center, which then donates a proton to the carbonyl oxygen. Collapse of the transition state with the formation of the peptide bond allows protonation of the 3′OH of the now empty tRNA in the P site (Figure 31.23C). The stage is now set for translocation and formation of the next peptide bond.

The formation of a peptide bond is followed by the GTP-driven translocation of tRNAs and mRNA

With the formation of the peptide bond, the peptide chain is now attached to the tRNA whose anticodon is in the A site on the 30S subunit. The two subunits rotate with respect to one another, and this structural change places the CCA end of the same tRNA and its peptide in the P site of the large subunit (Figure 31.24). Another aminoacyl-tRNA is delivered to the A site by EF-Tu bound to GTP (EF-Tu is not shown in the figure.) (1). Again, peptide bond synthesis occurs (2). However, protein synthesis cannot continue without the translocation of the mRNA and the tRNAs within the ribosome. *Elongation factor G* (EF-G, also called *translocase*) catalyzes the movement of mRNA, at the expense of GTP hydrolysis, by a distance of three nucleotides. Now, the next codon is positioned in the A site for interaction with the incoming aminoacyl-tRNA-EF-Tu-GTP complex. (3). The peptidyl-tRNA moves out of the A site into the P site on the 30S subunit and at the same time, the deacylated tRNA moves out of the P site into the E site and is subsequently released from the ribosome (4). The movement of the peptidyl-tRNA into the P site shifts the mRNA by one codon, exposing the next codon to be translated in the A site.

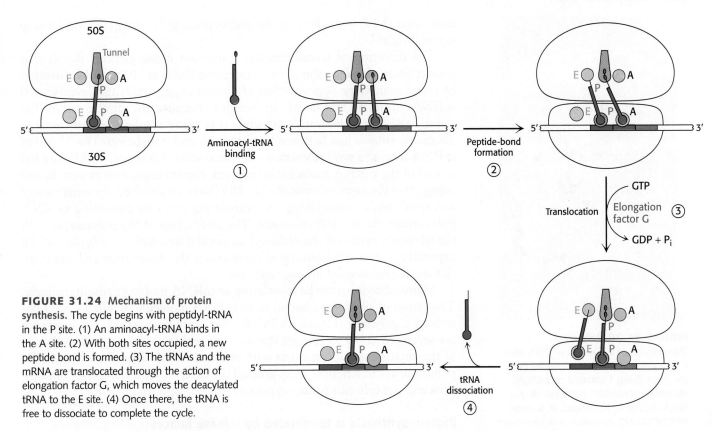

FIGURE 31.24 Mechanism of protein synthesis. The cycle begins with peptidyl-tRNA in the P site. (1) An aminoacyl-tRNA binds in the A site. (2) With both sites occupied, a new peptide bond is formed. (3) The tRNAs and the mRNA are translocated through the action of elongation factor G, which moves the deacylated tRNA to the E site. (4) Once there, the tRNA is free to dissociate to complete the cycle.

The three-dimensional structure of the ribosome undergoes significant change during translocation, and evidence suggests that translocation may result from properties of the ribosome itself. However, EF-G accelerates the process. A possible mechanism for accelerating the translocation process is shown in Figure 31.25. First, EF-G in the GTP form binds to the ribosome near the A site, interacting with the 23S rRNA of the 50S subunit. The binding of EF-G to the ribosome stimulates the GTPase activity of EF-G. On GTP hydrolysis, EF-G undergoes a conformational change that displaces the peptidyl-tRNA in the A site to the P site, which carries the mRNA and the deacylated tRNA with it. The dissociation of EF-G leaves the ribosome ready to accept the next aminoacyl-tRNA into the A site.

Note that *the peptide chain remains in the P site on the 50S subunit throughout this cycle,* growing into the exit tunnel. This cycle is repeated, with mRNA translation taking place in the 5′ → 3′ direction, as new aminoacyl-tRNAs

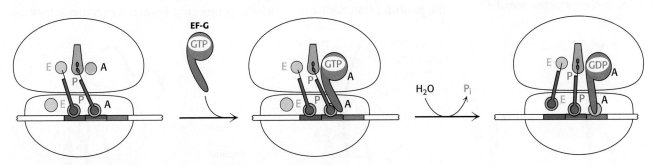

FIGURE 31.25 Translocation mechanism. In the GTP form, EF-G binds to the A site on the 50S subunit. This binding stimulates GTP hydrolysis, inducing a conformational change in EF-G that forces the tRNAs and mRNA to move through the ribosome by a distance corresponding to one codon.

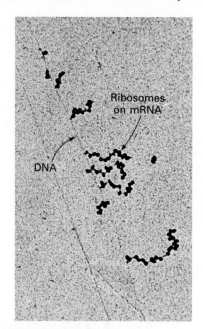

FIGURE 31.26 Polysomes. Transcription of a segment of DNA from *E. coli* generates mRNA molecules that are immediately translated by multiple ribosomes. [From O. L. Miller, Jr., B. A. Hamkalo, and C. A. Thomas, Jr. *Science* 169(1970): 392. Reprinted with permission from AAAS.]

move into the A site, allowing the polypeptide to be elongated until a stop signal is found.

The direction of translation has important consequences. Recall that transcription also is in the 5′ → 3′ direction (Section 30.1). If the direction of translation were opposite that of transcription, only fully synthesized mRNA could be translated. In contrast, because the directions are the same, mRNA can be translated while it is being synthesized. In bacteria, almost no time is lost between transcription and translation. The 5′ end of mRNA interacts with ribosomes very soon after it is made, well before the 3′ end of the mRNA molecule is finished. Recent experiments with *E. coli* using cryo-electron microscopy (p. 105) have established the existence of an *expressome*, a transcribing and translating complex consisting of RNA polymerase and the 60S ribosome. The interaction of the polymerase with the ribosome occurs at the carboxyl terminal domain of the polymerase. *An important feature of bacterial gene expression is that translation and transcription are closely coupled in space and time.*

Many ribosomes can be translating an mRNA molecule simultaneously. The group of ribosomes bound to an mRNA molecule is called a *polyribosome* or a *polysome* (Figure 31.26). Recent work shows that the ribosomes are arranged so as to protect the mRNA and to facilitate easy exchange of the substrates and products with the cytoplasm. The ribosomes in the polysome are in a helical array around the mRNA with the tRNA binding sites and peptide exit tunnel exposed to the cytoplasm.

Protein synthesis is terminated by release factors that read stop codons

The final phase of translation is termination. How does the synthesis of a polypeptide chain come to an end when a stop codon is encountered? No tRNAs with anticodons complementary to the stop codons—UAA, UGA, or UAG—exist in normal cells. Instead, these *stop codons are recognized by proteins called release factors* (RFs). One of these release factors, RF1, recognizes UAA or UAG. A second factor, RF2, recognizes UAA or UGA. A third factor, RF3, another GTPase, catalyzes the removal of RF1 or RF2 from the ribosome upon release of the newly synthesized protein.

RF1 and RF2 are compact proteins that, in eukaryotes, resemble a tRNA molecule. When bound to the ribosome, the proteins unfold to bridge the gap between the stop codon on the mRNA and the peptidyl transferase center on the 50S subunit (Figure 31.27). The RF interacts with the peptidyl transferase center using a loop containing a highly conserved glycine-glycine-glutamine (GGQ) sequence, with the glutamine methylated on the amide nitrogen atom of the R group. This modified glutamine (assisted by the peptidyl transferase) is crucial in promoting a water molecules attack on

FIGURE 31.27 Termination of protein synthesis. A release factor recognizes a stop codon in the A site and stimulates the release of the completed protein from the tRNA in the P site.

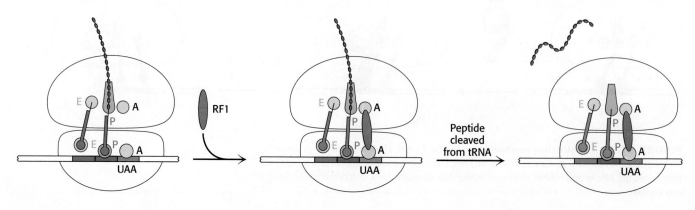

the ester linkage between the tRNA and the polypeptide chain, freeing the polypeptide chain. The detached polypeptide leaves the ribosome. Transfer RNA and messenger RNA remain briefly attached to the 70S ribosome until the entire complex is dissociated through the hydrolysis of GTP in response to the binding of EF-G and another factor, called the *ribosome release factor* (RRF).

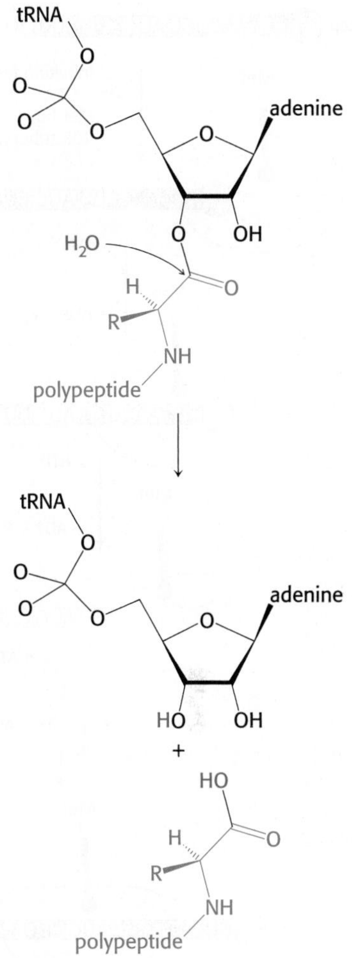

31.4 Eukaryotic Protein Synthesis Differs from Bacterial Protein Synthesis Primarily in Translation Initiation

The basic plan of protein synthesis in eukaryotes and archaea is similar to that in bacteria. The major structural and mechanistic themes recur in all domains of life. However, eukaryotic protein synthesis entails more protein components than does bacterial protein synthesis, and some steps are more intricate. Some noteworthy similarities and differences are as follows:

1. *Ribosomes.* Eukaryotic ribosomes are larger. They consist of a 60S large subunit and a 40S small subunit, which come together to form an 80S particle having a mass of 4200 kDa, compared with 2700 kDa for the bacterial 70S ribosome. The 40S subunit contains an 18S RNA that is homologous to the bacterial 16S RNA. The 60S subunit contains three RNAs: the 5S RNA, which is homologous to the bacterial 5S rRNA; the 28S RNA, which is homologous to the bacterial 23S molecules; and the 5.8S RNA, which is homologous to the 5′ end of the 23S RNA of bacteria.

2. *Initiator tRNA.* In eukaryotes, the initiating amino acid is methionine rather than *N*-formylmethionine. However, as in bacteria, a special tRNA participates in initiation. This aminoacyl-tRNA is called Met-tRNA$_i$ (the subscript "i" stands for initiation).

3. *Initiation.* The initiating codon in eukaryotes is always AUG. Eukaryotes, in contrast with bacteria, do not have a specific purine-rich sequence on the 5′ side to distinguish initiator AUGs from internal ones. Instead, the AUG nearest the 5′ end of mRNA is usually selected as the start site. Eukaryotes utilize many more initiation factors than do bacteria, and their interplay is much more complex. The prefix *eIF* denotes a eukaryotic initiation factor.

Initiation begins with the formation of a ternary complex consisting of the 40S ribosome and Met-tRNA$_i$ in association with eIF-2. The complex is called the 43S preinitiation complex (PIC). The PIC binds to the 5′ end of mRNA and begins searching for an AUG codon by moving step-by-step in the 3′ direction. Initiation factor eIF-4E binds to the 5′-cap of the mRNA (Section 30.3) and facilitates binding of PIC to the mRNA (Figure 31.28). This scanning process is catalyzed by helicases that move along the mRNA powered by ATP hydrolysis. Pairing of the anticodon of Met-tRNA$_i$ with the AUG codon of mRNA signals that the target has been found. In almost all cases, eukaryotic mRNA has only one start site and hence is the template for only a single protein. In contrast, a bacterial mRNA can have multiple Shine–Dalgarno sequences and, hence, start sites, and it can serve as a template for the synthesis of several proteins.

The difference in initiation mechanism between bacteria and eukaryotes is, in part, a consequence of the difference in RNA processing. The 5′ end of mRNA is readily available to ribosomes immediately after transcription in bacteria. In contrast, in eukaryotes pre-mRNA must be processed and transported to the cytoplasm before translation is initiated. The 5′ cap

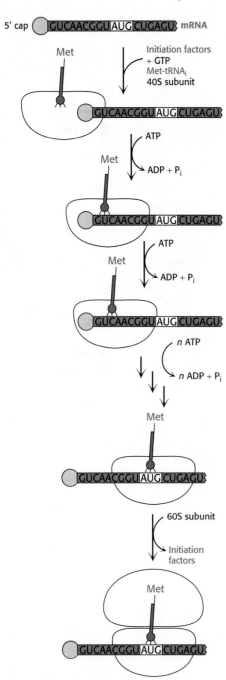

FIGURE 31.28 Eukaryotic translation initiation. In eukaryotes, translation initiation starts with the assembly of a complex on the 5′ cap that includes the 40S subunit and Met-tRNA$_i$. Driven by ATP hydrolysis, this complex scans the mRNA until the first AUG is reached. The 60S subunit is then added to form the 80S initiation complex.

provides an easily recognizable starting point. In addition, the complexity of eukaryotic translation initiation provides another mechanism for regulation of gene expression that we shall explore further in Chapter 33.

Although most eukaryotic mRNA molecules rely on the 5′ cap to initiate protein synthesis, recent work has established that some mRNA molecules can recruit ribosomes for initiation without the use of a 5′-cap and cap-binding proteins. In these mRNAs, highly structured RNA sequences called *internal ribosome entry sites* (IRES) facilitate 40S ribosome binding to the mRNA. IRES were first discovered in the genomes of RNA viruses and have since been found in other viruses, as well as in a subset of cellular mRNA that appears to take part in development and cell stress. The molecular mechanism by which IRES function to initiate protein synthesis remains to be determined.

4. *The Structure of mRNA.* Soon after the PIC binds the mRNA, eIF-4G links eIF-4E to a protein associated with the poly(A) tail, the poly(A)-binding protein 1 (PABP1; Figure 31.29). Cap and tail are thus brought together to form a circle of mRNA. The benefits of circularization of mRNA remain to be determined, but a requirement for circularization may prevent translation of mRNA molecules that have lost their poly(A) tails (Section 30.3).

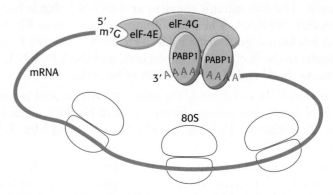

FIGURE 31.29 Protein interactions circularize eukaryotic mRNA. [Information from H. Lodish et al., *Molecular Cell Biology*, 5th ed. (W. H. Freeman and Company, 2004), Fig. 4.31.]

5. *Elongation and Termination.* Eukaryotic elongation factors EF1α and EF1βγ are the counterparts of bacterial EF-Tu and EF-Ts. The GTP form of EF1α delivers aminoacyl-tRNA to the A site of the ribosome, and EF1βγ catalyzes the exchange of GTP for bound GDP. Eukaryotic EF2 mediates GTP-driven translocation in much the same way as does bacterial EF-G. Termination in eukaryotes is carried out by a single release factor, eRF1, compared with two in bacteria. Release factor eIF-3 accelerates the activity of eRF-1.

6. *Organization.* The components of the translation machinery in higher eukaryotes are organized into large complexes associated with the cytoskeleton. This association is believed to facilitate the efficiency of protein synthesis. Recall that organization of elaborate biochemical processes into physical complexes is a recurring theme in biochemistry (the ATP synthasome in Section 18.5 and purine synthesis in Section 25.2).

Mutations in initiation factor 2 cause a curious pathological condition

Mutations in eukaryotic initiation factor 2 result in a rare, mysterious disease, called *vanishing white matter* (VWM) *disease,* in which

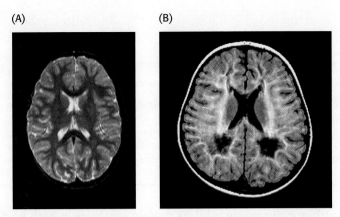

FIGURE 31.30 The effects of vanishing white matter disease. (A) In the normal brain, magnetic resonance imaging (MRI) visualizes the white matter as dark gray. (B) In the diseased brain, this MRI reveals that white matter is replaced by cerebrospinal fluid, seen as white. [(A) Du Cane Medical Imaging Ltd/Science Source (B) Reprinted from *Lancet Neurology*, 5, M.C. van der Knaap, M.S., Pronk, J.C., Scheper, G.C., "Vanishing white matter disease" Fig. B, p. 414 © 2006, with permission from Elsevier.]

nerve cells in the brain disappear and are replaced by cerebrospinal fluid (Figure 31.30). The white matter of the brain consists predominantly of nerve axons that connect the gray matter of the brain to the rest of the body. Death, resulting from fever or extended coma, can occur anywhere from a few years to decades after the onset of the disease. Symptoms usually appear in young children but can develop shortly after birth or even in adulthood. An especially puzzling aspect of the disease is its tissue specificity. A mutation in a biochemical process as fundamental to life as protein-synthesis initiation would be predicted to be lethal or at least to affect all tissues of the body. Diseases such as VWM graphically show that, although much progress has been made in biochemistry, much more research will be required to understand the complexities of health and disease. (See Biochemistry in Focus for a discussion of ribosomes and disease.)

31.5 A Variety of Antibiotics and Toxins Can Inhibit Protein Synthesis

Many chemicals that inhibit various aspects of protein synthesis have been identified. These chemicals are powerful experimental tools and clinically useful drugs.

Some antibiotics inhibit protein synthesis

The differences between eukaryotic and bacterial ribosomes can be exploited for the development of antibiotics (Table 31.4). For example, the antibiotic *streptomycin*, a highly basic trisaccharide, interferes with the binding of fMet-tRNA to ribosomes in bacteria and thereby prevents the correct initiation of protein synthesis. Other *aminoglycoside antibiotics* such as neomycin, kanamycin, and gentamycin interfere with the interaction between tRNA and the 16S rRNA of the 30S subunit of bacterial ribosomes. *Chloramphenicol* acts by inhibiting peptidyl transferase activity. *Erythromycin* binds to the 50S subunit and blocks translocation.

Streptomycin

TABLE 31.4 Antibiotic inhibitors of protein synthesis

Antibiotic	Action
Streptomycin and other aminoglycosides	Inhibit initiation and cause the misreading of mRNA (bacteria)
Tetracycline	Binds to the 30S subunit and inhibits the binding of aminoacyl-tRNAs (bacteria)
Chloramphenicol	Inhibits the peptidyl transferase activity of the 50S ribosomal subunit (bacteria)
Cycloheximide	Inhibits translocation (eukaryotes)
Erythromycin	Binds to the 50S subunit and inhibits translocation (bacteria)
Puromycin	Causes premature chain termination by acting as an analog of aminoacyl-tRNA (bacteria and eukaryotes)

The antibiotic *puromycin* inhibits protein synthesis in both bacteria and eukaryotes by causing nascent polypeptide chains to be released before their synthesis is completed. Puromycin is an analog of the terminal part of aminoacyl-tRNA (Figure 31.31). It binds to the A site on the ribosome and blocks the entry of aminoacyl-tRNA. Furthermore, puromycin contains an α-amino group. This amino group, like the one on aminoacyl-tRNA, forms a peptide bond with the carboxyl group of the growing peptide chain. The product, a peptide having a covalently attached puromycin residue at its carboxyl end, dissociates from the ribosome. No longer used medicinally, puromycin remains an experimental tool for the investigation of protein synthesis. *Cycloheximide*, another antibiotic, blocks translocation in eukaryotic ribosomes, making a useful laboratory tool for blocking protein synthesis in eukaryotic cells.

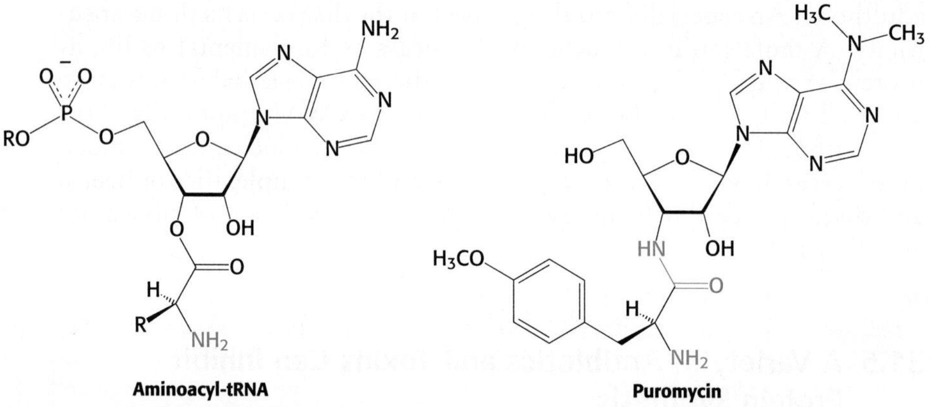

FIGURE 31.31 Antibiotic action of puromycin. Puromycin resembles the aminoacyl terminus of an aminoacyl-tRNA. Its amino group joins the carboxyl group of the growing polypeptide chain to form peptidyl-puromycin that dissociates from the ribosome. Peptidyl-puromycin is stable because puromycin has an amide (shown in red) rather than an ester linkage.

Aminoacyl-tRNA

Puromycin

Diphtheria toxin blocks protein synthesis in eukaryotes by inhibiting translocation

Many antibiotics, harvested from bacteria for medicinal purposes, are inhibitors of bacterial protein synthesis. However, some bacteria produce protein-synthesis inhibitors that inhibit eukaryotic protein synthesis, leading to diseases such as diphtheria, which was a major cause of death in childhood before the advent of effective immunization. Symptoms include painful sore throat, hoarseness, fever, and difficulty breathing. The lethal effects of this disease are due mainly to a protein toxin produced by a phage infecting *Corynebacterium diphtheriae*, a bacterium that grows in the upper respiratory tract of an infected person. A few micrograms of diphtheria toxin is usually lethal in an unimmunized person because it inhibits protein synthesis. The toxin consists of a single polypeptide chain that is cleaved into a 21-kDa A fragment and a 40-kDa B fragment shortly after entering the cell. The role of the B-fragment in the intact protein is to bind to the cell, enabling the toxin to enter the cytoplasm of its target cell. The A fragment catalyzes the covalent modification of EF2, the elongation factor catalyzing translocation in eukaryotic protein synthesis, resulting in protein synthesis inhibition and cell death.

A single A fragment of the toxin in the cytoplasm can kill a cell. Why is it so lethal? EF2 contains *diphthamide,* an unusual amino acid residue that enhances the fidelity of codon shifting during translocation. Diphthamide is formed by a highly conserved complicated pathway that posttranslationally modifies histidine. The A fragment of the diphtheria toxin catalyzes the transfer of the ADP ribose unit of NAD^+ to the diphthamide ring (Figure 31.32). *This ADP ribosylation of a single side chain of EF2 blocks EF2's capacity to carry out the translocation of the growing polypeptide chain.* Protein synthesis ceases, accounting for the remarkable toxicity of diphtheria toxin.

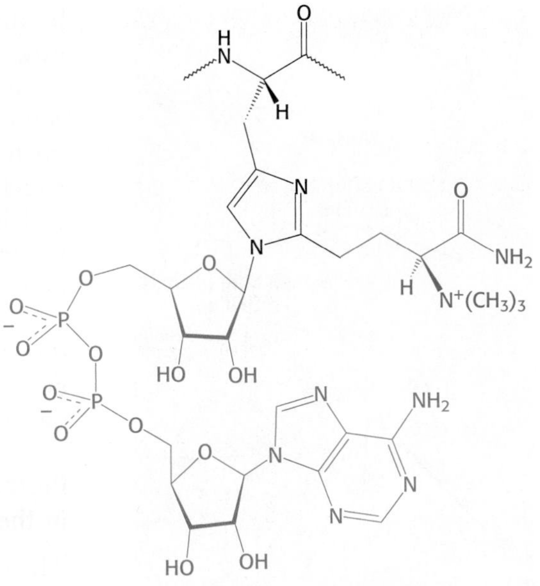

Some toxins modifiy 28S ribosomal RNA

Ricin is a small protein (65 kDa) found in the seeds of the castor oil plant, *Ricinus communis* (Figure 31.33). It is indeed a deadly molecule because as little as 500 µg is lethal for an adult human being, and a single molecule can inhibit all protein synthesis in a cell, resulting in cell death.

Ricin is a heterodimeric protein composed of a catalytic A chain joined by a single disulfide bond to a B chain. The B chain allows the toxin to bind to the target cell, and this binding leads to an endocytotic uptake of the dimer and the eventual release of the A chain into the cytoplasm. The A chain, an N-glycoside hydrolase, cleaves adenine from a particular adenosine nucleotide on the 28S rRNA that is found in all eukaryotic ribosomes. Removal of the adenine base completely inactivates the ribosome by preventing the binding of elongation factors.

In contrast, another toxin, α-sarcin is a ribonuclease rather than a glycoside hydrolase that cleaves a single phosphodiester linkage in the 28S RNA in the same loop where ricin acts. This cleavage completely inhibits protein synthesis. α-Sarcin is a 17kDa protein secreted by filamentous fungi that penetrates cells by interacting with membrane lipids. Little is known about the in vivo toxicity of α-sarcin. Thus, ricin, α-sarcin, and diphtheria toxin all act by inhibiting protein-synthesis elongation; ricin and α-sarcin do so by covalently modifying rRNA, and diphtheria toxin does so by covalently modifying the elongation factor. Interestingly, attempts are underway to modify α-sarcin for use as an antitumor drug.

31.6 Ribosomes Bound to the Endoplasmic Reticulum Manufacture Secretory and Membrane Proteins

Not all newly synthesized proteins are destined to function in the cytoplasm. Eukaryotic cells can direct proteins to internal sites such as mitochondria, the nucleus, and the endoplasmic reticulum, a process called *protein targeting* or *protein sorting.* How is sorting accomplished? There are two general mechanisms by which sorting takes place. In one mechanism, the protein is synthesized in the cytoplasm, and then the completed protein is delivered to its intracellular location posttranslationally. Proteins destined for the nucleus, chloroplasts, mitochondria, and peroxisomes are delivered

ADP-ribose

FIGURE 31.32 Blocking of translocation by diphtheria toxin. Diphtheria toxin blocks protein synthesis in eukaryotes by catalyzing the transfer of an ADP-ribose unit from NAD^+ to diphthamide, a modified amino acid residue in EF2 (translocase). Diphthamide is formed by the posttranslational modification (blue) of a histidine residue.

On September 7, 1978, an agent of the Soviet Union State Security Department (KGB), using a weapon built into an umbrella, embedded a small pellet containing ricin into the thigh of Bulgarian dissident Georgi Markov as he crossed Waterloo Bridge in London. Mr. Markov felt a sharp sting, as if from a bug bite. He died four days later.

FIGURE 31.33 Castor beans. The seeds of castor beans from *Ricinus communis* are a rich source of oils with a wide variety of uses, including the production of biodiesel fuels. The seeds are also rich in the toxin ricin. [Ted Kinsman/Science Source.]

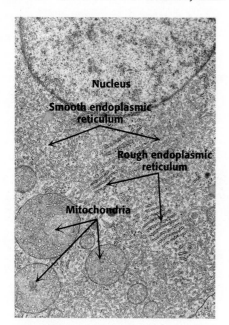

FIGURE 31.34 Ribosomes are bound to the endoplasmic reticulum. In this electron micrograph, ribosomes appear as small black dots binding to the cytoplasmic side of the endoplasmic reticulum to give a rough appearance. In contrast, the smooth endoplasmic reticulum is devoid of ribosomes. [Don W. Fawcett/Science Source.]

by this general process. The other mechanism, termed the *secretory pathway*, directs proteins into the *endoplasmic reticulum* (ER), the extensive membrane system that comprises about half the total membrane of a cell, cotranslationally—that is, while the protein is being synthesized. Approximately 30% of all proteins are sorted by the secretory pathway, including secreted proteins, residents of the ER, the Golgi complex, lysosomes, and integral membrane proteins of these organelles as well as integral plasma-membrane proteins. We will focus our attention on the secretory pathway only.

In eukaryotic cells, a ribosome remains free in the cytoplasm unless it is directed to the ER. The region that binds ribosomes is called the *rough ER* because of its studded appearance in contrast with the *smooth ER*, which is devoid of ribosomes (Figure 31.34).

Protein synthesis begins on ribosomes that are free in the cytoplasm

The synthesis of proteins sorted by the secretory pathway begins on free ribosomes that soon associate with the ER. Free ribosomes that are synthesizing proteins for use in the cell are apparently identical with those attached to the ER. What is the process that directs the ribosome synthesizing a protein destined to enter the ER to bind to the ER?

Signal sequences mark proteins for translocation across the endoplasmic reticulum membrane

The synthesis of proteins destined to leave the cell or become embedded in the plasma membrane begins on a free ribosome but, shortly after synthesis begins, it is halted until the ribosome is directed to the cytoplasmic side of the ER. When the ribosome docks with the ER membrane, protein synthesis begins again. As the newly forming peptide chain exits the ribosome, it is transported, cotranslationally, through the membrane into the lumen of the ER. The translocation consists of four components.

1. *The signal sequence.* The signal sequence is *a sequence of 9 to 12 hydrophobic amino acid residues, sometimes containing positively charged amino acids* (Figure 31.35). This sequence, which adopts an α-helical structure, is usually near the amino terminus of the nascent polypeptide chain. The presence of the signal sequence identifies the nascent peptide as one that must cross the ER membrane. Some signal sequences are maintained in the mature protein, whereas others are cleaved by a *signal peptidase* on the lumenal side of the ER membrane.

		Cleavage site
Human growth hormone	M A T G S R T S L L L A F G L L C L P W L Q E G S A	F P T
Human proinsulin	M A L W M R L L P L L A L L A L W G P D P A A A	F V N
Bovine proalbumin	M K W V T F I S L L L L F S S A Y S	R G V
Mouse antibody H chain	M K V L S L L Y L L T A I P H I M S	D V Q
Chicken lysozyme	M R S L L I L V L C F L P K L A A L G	K V F
Bee promellitin	M K F L V N V A L V F M V V Y I S Y I Y A	A P E
Drosophila glue protein	M K L L V V A V I A C M L I G F A D P A S G	C K D
Zea maize protein 19	M A A K I F C L I M L L G L S A S A A T A	S I F
Yeast invertase	M L L Q A F L F L L A G F A A K I S A	S M T
Human influenza virus A	M K A K L L V L L Y A F V A G	D Q I

FIGURE 31.35 Amino-terminal signal sequences of some eukaryotic secretory and plasma-membrane proteins. The hydrophobic core (yellow) is preceded by basic residues (blue) and followed by a cleavage site (red) for signal peptidase.

2. *The signal-recognition particle (SRP).* The signal-recognition particle recognizes the signal sequence and binds the sequence and the ribosome as soon as the signal sequence exits the ribosome. SRP then shepherds the ribosome and its nascent polypeptide chain to the ER membrane. SRP is a ribonucleoprotein consisting of a 7S RNA and six different proteins (Figure 31.36). One protein, SRP54, is a GTPase that is crucial for SRP function. SRP samples all ribosomes until it locates one exhibiting a signal sequence. After SRP is bound to the signal sequence, interactions between the ribosome and the SRP occlude the elongation-factor-binding site, thereby halting protein synthesis.

3. *The SRP receptor (SR).* The SRP–ribosome complex diffuses to the ER, where SRP binds the SRP receptor, an integral membrane protein consisting of two subunits, SRα and SRβ. SRα is, like SRP54, a GTPase.

4. *The translocon.* The SRP–SR complex delivers the ribosome to the translocation machinery, called the *translocon,* a multisubunit assembly of integral and peripheral membrane proteins. The translocon is a protein-conducting channel. This channel opens when the translocon and ribosome bind to each other. Protein synthesis resumes with the growing polypeptide chain passing through the translocon channel into the lumen of the ER.

The interactions of the components of the translocation machinery are shown in Figure 31.37. Both the SRP54 and the SRα subunits of SR must

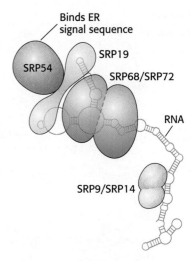

FIGURE 31.36 The signal-recognition particle. The signal-recognition particle (SRP) consists of six proteins and one 300-nucleotide RNA molecule. The RNA has a complex structure with many double-helical stretches punctuated by single-stranded regions, shown as circles. [Information from H. Lodish et al., *Molecular Cell Biology,* 5th ed. (W. H. Freeman and Company, 2004). Data from K. Strub et al., *Mol. Cell Biol.* 11:3949–3959, 1991, and S. High and B. Dobberstein, *J. Cell Biol.* 113:229–233, 1991.]

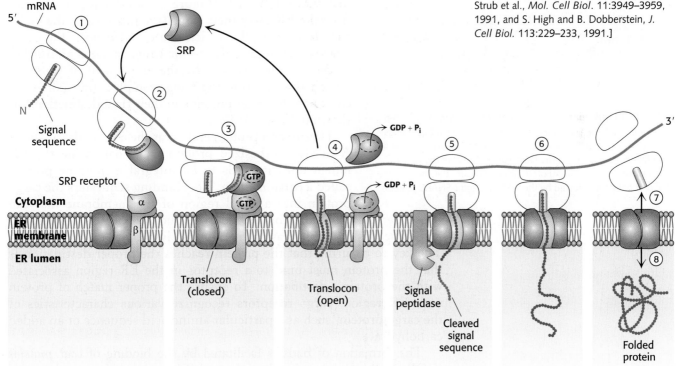

FIGURE 31.37 The SRP targeting cycle. (1) Protein synthesis begins on free ribosomes. (2) After the signal sequence has exited the ribosome, it is bound by the SRP, and protein synthesis halts. (3) The SRP–ribosome complex docks with the SRP receptor in the ER membrane. (4) The SRP and the SRP receptor simultaneously hydrolyze bound GTPs. Protein synthesis resumes and the SRP is free to bind another signal sequence. (5) The signal peptidase may remove the signal sequence as it enters the lumen of the ER. (6) Protein synthesis continues as the protein is synthesized directly into the ER. (7) On completion of protein synthesis, the ribosome is released. (8) The protein tunnel in the translocon closes. [Information from H. Lodish et al., *Molecular Cell Biology,* 5th ed. (W. H. Freeman and Company, 2004), Fig. 16.6.]

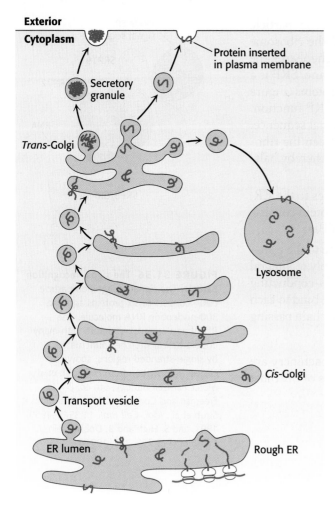

FIGURE 31.38 Protein-sorting pathways. Newly synthesized proteins in the lumen of the ER are collected into membrane buds. These buds pinch off to form transport vesicles. The transport vesicles carry the cargo proteins to the Golgi complex, where the cargo proteins are modified. Transport vesicles then carry the cargo to the final destination as directed by the v-SNARE and t-SNARE proteins.

bind GTP to facilitate the formation of the SRP–SR complex. For the SRP–SR complex to then deliver the ribosome to the translocon, the two GTP molecules—one in SRP and the other in SR—are aligned in what is essentially an active site shared by the two proteins. The formation of the alignment is catalyzed by the 7S RNA of the SRP. After the ribosome has been passed along to the translocon, the GTPs are hydrolyzed, SRP and SR dissociate, and SRP is free to search for another signal sequence to begin the cycle anew. Thus, SRP acts catalytically. The signal peptidase, which is associated with the translocon in the lumen of the ER, removes the signal sequence from most proteins.

Transport vesicles carry cargo proteins to their final destination

As the proteins are synthesized, they fold to form their three-dimensional structures in the lumen of the ER. Some proteins are modified by the attachment of N-linked carbohydrates (Section 11.3). Finally, the proteins must be sorted and transported to their final destinations. Regardless of the destination, the principles of transport are the same. Transport is mediated by *transport vesicles* that bud off the ER (Figure 31.38). Transport vesicles from the ER carry their cargo (the proteins) to the Golgi complex, where the vesicles fuse and deposit the cargo inside the complex. There the cargo proteins are further modified—for instance, by the attachment of O-linked carbohydrates. From the Golgi complex, transport vesicles carry the cargo proteins to their final destinations, as shown in Figure 31.38.

How does a protein end up at the correct destination? A newly synthesized protein will float inside the ER lumen until it binds to an integral membrane protein called a *cargo receptor*. This binding sequesters the cargo protein into a small region of the membrane that can subsequently form a membrane bud. The bud will carry the protein to a specific destination—plasma membrane, lysosome, or cell exterior. The key to ensuring that the protein reaches the proper destination is that the protein must bind to a receptor in the ER region associated with the protein's destination. To ensure the proper match of protein with ER region, cargo receptors recognize various characteristics of the cargo protein, such as a particular amino acid sequence or an added carbohydrate.

The formation of buds is facilitated by the binding of *coat proteins* (COPs) to the cytoplasmic side of the bud. The coat proteins associate with one another to pinch off the vesicle. After the transport vesicle has formed and is released, the coat proteins are shed to reveal another integral protein called *v-SNARE* ("v" for *v*esicle) (Section 12.6). v-SNARE will bind to a particular *t-SNARE* ("t" for *t*arget) in the target membrane. This binding leads to the fusion of the transport vesicle to the target membrane, and the cargo is delivered. Thus, the assignment of identical v-SNARE proteins to the same region of the ER membrane causes an ER region to be associated with a particular destination.

SUMMARY

31.1 Protein Synthesis Requires the Translation of Nucleotide Sequences into Amino Acid Sequences

Protein synthesis is called translation because information present as a nucleic acid sequence is translated into a different language, the sequence of amino acids in a protein. This complex process is mediated by the coordinated interplay of more than a hundred macromolecules, including mRNA, rRNAs, tRNAs, aminoacyl-tRNA synthetases, and protein factors. Given that a protein typically comprises from 100 to 1000 amino acids, the frequency at which an incorrect amino acid is incorporated in the course of protein synthesis must be less than 10^{-4}. Transfer RNAs are the adaptors that make the link between a nucleic acid and an amino acid. These molecules, single chains of about 80 nucleotides, have an L-shaped structure.

31.2 Aminoacyl Transfer RNA Synthetases Read the Genetic Code

Each amino acid is activated and linked to a specific transfer RNA by an enzyme called an aminoacyl-tRNA synthetase. The synthetase links the carboxyl group of an amino acid to the 2′- or 3′-hydroxyl group of the adenosine unit of a CCA sequence at the 3′ end of the tRNA by an ester linkage. There is at least one specific aminoacyl-tRNA synthetase and at least one specific tRNA for each amino acid. A synthetase utilizes both the functional groups and the shape of its cognate amino acid to prevent the attachment of an incorrect amino acid to a tRNA. Some synthetases have a separate active site at which incorrectly linked amino acids are removed by hydrolysis. A synthetase recognizes the anticodon, the acceptor stem, and sometimes other parts of its tRNA substrate. By specifically recognizing both amino acids and tRNAs, aminoacyl-tRNA synthetases implement the instructions of the genetic code.

The codons of messenger RNA recognize the anticodons of transfer RNAs rather than the amino acids attached to the tRNAs. A codon on mRNA forms base pairs with the anticodon of the tRNA. Some tRNAs are recognized by more than one codon because pairing of the third base of a codon is less crucial than that of the other two (the wobble mechanism). There are two evolutionary distinct classes of synthetases, each recognizing 10 amino acids. The two classes recognize opposite faces of tRNA molecules.

31.3 The Ribosome Is the Site of Protein Synthesis

Protein synthesis takes place on ribosomes—ribonucleoprotein particles (about two-thirds RNA and one-third protein) consisting of large and small subunits. In *E. coli,* the 70S ribosome (2500 kDa) is made up of 30S and 50S subunits. The 30S subunit consists of 16S ribosomal RNA and 21 different proteins; the 50S subunit consists of 23S and 5S rRNA and 34 different proteins. The ribosome includes three sites for tRNA binding called the A (aminoacyl) site, the P (peptidyl) site, and the E (exit) site.

Protein synthesis takes place in three phases: initiation, elongation, and termination. In bacteria, mRNA, formylmethionyl-tRNA$_f$ (the special initiator tRNA that recognizes AUG), and a 30S ribosomal subunit come together with the assistance of initiation factors to form a 30S initiation complex. A 50S ribosomal subunit then joins this complex to form a 70S initiation complex, in which fMet-tRNA$_f$ occupies the P site of the ribosome.

Elongation factor Tu delivers the appropriate aminoacyl-tRNA to the ribosome's A site as an EF-Tu–aminoacyl-tRNA–GTP ternary complex. EF-Tu serves both to protect the aminoacyl-tRNA from premature cleavage and to increase the fidelity of protein synthesis by ensuring that the correct anticodon–codon pairing has taken place before hydrolyzing GTP and releasing aminoacyl-tRNA into the A site. A peptide bond is formed when the amino group of the aminoacyl-tRNA nucleophilically attacks the ester linkage of the peptidyl-tRNA. On peptide-bond formation, the tRNAs and mRNA are translocated for the next cycle to begin. The deacylated tRNA moves to the E site and then leaves the ribosome, and the peptidyl-tRNA moves from the A site into the P site. Elongation factor G uses the free energy of GTP hydrolysis to drive translocation. Protein synthesis is terminated by release factors, which recognize the termination codons UAA, UGA, and UAG and cause the hydrolysis of the ester bond between the polypeptide and tRNA.

31.4 Eukaryotic Protein Synthesis Differs from Bacterial Protein Synthesis Primarily in Translation Initiation

The basic plan of protein synthesis in eukaryotes is similar to that in bacteria, but there are some significant differences between them. Eukaryotic ribosomes (80S) consist of a 40S small subunit and a 60S large subunit. The initiating amino acid is again methionine, but it is not formylated. The initiation of protein synthesis is more complex in eukaryotes than in bacteria. In eukaryotes, the AUG closest to the 5′ end of mRNA is nearly always the start site. The 40S ribosome finds this site by binding to the 5′ cap and then scanning the RNA until AUG is reached. The regulation of translation in eukaryotes provides a means for regulating gene expression.

31.5 A Variety of Antibiotics and Toxins Can Inhibit Protein Synthesis

Many clinically important antibiotics function by inhibiting protein synthesis. All steps of protein synthesis are susceptible to inhibition by one antibiotic or another. Diphtheria toxin inhibits protein synthesis by covalently modifying an elongation factor, thereby preventing elongation. Ricin, a toxin from castor beans, inhibits elongation by removing a crucial adenine residue from rRNA. α-Sarcin is a ribonuclease that cleaves a single phosphodiester linkage in the 28S RNA in the same loop where ricin acts. This cleavage completely inhibits protein synthesis.

31.6 Ribosomes Bound to the Endoplasmic Reticulum Manufacture Secretory and Membrane Proteins

Proteins contain signals that determine their ultimate destination. The synthesis of all proteins begins on free ribosomes in the cytoplasm. In eukaryotes, protein synthesis continues in the cytoplasm unless the nascent chain contains a signal sequence that directs the ribosome to the endoplasmic reticulum. Amino-terminal signal sequences consist of a hydrophobic segment of 9 to 12 residues preceded by a positively charged amino acid. The signal-recognition particle, a ribonucleoprotein assembly, recognizes signal sequences and brings ribosomes bearing them to the ER. A GTP–GDP cycle releases the signal sequence from SRP and then detaches SRP from its receptor. The nascent chain is then translocated across the ER membrane. Proteins are transported throughout the cell in transport vesicles.

APPENDIX

Biochemistry in Focus

Selective control of gene expression by ribosomes

Ribosomes were believed to be homogeneous; that is, all ribosomes were the same, regardless of cell type or physiological state. Recent evidence suggests otherwise and points to a more significant role for ribosomes in the regulation of gene expression than as mere (but complicated!) workbenches on which proteins are made.

How might this heterogeneity be manifested? There are a number of ways, but two come to mind immediately. First, the composition of the ribosomes may be altered in a cell-specific manner. Second, nonribosomal proteins may associate with the ribosomes. In both cases, ribosomes with different properties would result. Evidence for the first possibility comes from pathological conditions, called *ribosomopathies*, caused by mutations in specific ribosomal proteins. For instance, mutations in certain ribosomal proteins result in Diamond-Blackfan anemia, which is characterized by bone marrow failure in blood stem-cell differentiation. Although protein synthesis is impaired in the blood stem cells, protein synthesis in other tissues is apparently normal. Another example is isolated congenital asplenia, where the child is born without a spleen, resulting in severely compromised immune function. The condition is caused by a mutation in a small ribosomal protein. How the lack of this protein results in the absence of the spleen in not known. As with Diamond-Blackfan anemia, the ribosomal activity in other tissues appears normal. If ribosomal mutations lead to loss of ribosomal function in a tissue-specific manner, might other physiological alterations in the amount or posttranslational modification of ribosomal proteins regulate ribosomal function in a tissue-specific manner under normal physiological conditions? This question remains to be answered.

Nonribosomal proteins may also control ribosome activity. FMRP (*fragile X mental retardation protein*) is found in the brain of humans and is essential for cognitive development and also for reproductive function in women. FMRP, an RNA-binding protein, binds to the 80S ribosome to inhibit the translation of specific mRNAs. Mutations in FMRP result in cognitive disabilities, ovarian failure, and other pathologies. In another startling example, an isozyme of the glycolytic enzyme pyruvate kinase, pyruvate kinase M (p. 518), binds to ribosome to control the translation of proteins destined for the endoplasmic reticulum.

How else might ribosomes regulate translation? Ribosome concentration varies 3- to 10-fold among different tissues. Low concentration of ribosomes could possibly discriminate against certain mRNAs that do not initiate well. This might be the case if mRNA has a complex three-dimensional structure in the 5′ untranslated region, the part of the mRNA upstream of the initiator codon. Some cases of Diamond-Blackfan anemia may be caused by insufficient ribosome levels, resulting in impaired translation of a specific subset of mRNA. What determines ribosome concentration in different cell types remains unknown. Einstein is reported to have said, "The more I learn, the more I realize how much I don't know." He might as well have been talking about ribosomes. "Ribosomology" is certain to keep biochemists busy for years to come.

APPENDIX

Problem-Solving Strategies

PROBLEM: A portion of the nucleotide sequence of the template strand of DNA from *E. coli* is shown below. This sequence is known to encode the carboxyl terminus of a long protein. Determine the encoded amino acid sequence.

5′-ACCGATTACTTTGCATGG-3′

SOLUTION: The problem provides us with the template strand, but asks us to determine amino acid sequence, which is encoded in the mRNA sequence. Our first question must be:

▶ **What is the relation of the sequence of the template strand of DNA to the nucleotide sequence of the mRNA?**

Recall from Figure 30.4, that the sense strand of the DNA is complementary to the template strand. Recall also that the transcript, or mRNA in our case, has the same sequence as the sense strand, with U replacing T.

▶ **Write out the complementary strand for the sequence given in the problem.**

Remember that the strands of double stranded DNA have opposite polarity.

Template	5′-A	C	C	G	A	T	T	A	C	T	T	T	G	C	A	T	G	G-3′
Sense (coding)	3′-T	G	G	C	T	A	A	T	G	A	A	A	C	G	T	A	C	C-5′

The mRNA will correspond to the sense strand, with U substituted for T.

▶ **In which direction is mRNA translated?**

The direction of translation is $5' \rightarrow 3'$, so to avoid confusion, write out the $5' \rightarrow 3'$ sequence of the mRNA.

5'-CCAUGCAAAGUAAUCGGU-3'

All that we know about this sequence is that it is at the 3' end of the mRNA. The question doesn't give us the reading frame—that is, what are the actual codons. For instance, we can't tell whether the first codon of this sequence is CCA, CAU, or AUG. We need to think of a means of establishing the reading frame.

▶ **What codon is present near the 3' end of all mRNA molecules?**

There must be a stop codon.

▶ **What are the stop codons?**

UAA, UAG, and UGA.

The sequence in question contains a UAA. So, working in the 5' direction from the stop codon, the reading frame is:

5'-C-**CAU-GCA-AAG**-UAA- UCGGU-3'

All you need now is a copy of the genetic code, which is Table 4.5, and you will find that the last three amino acids of this *E. coli* protein are: His-Ala-Lys.

KEY TERMS

translation (p. 1021)

ribosome (p. 1021)

aminoacyl-tRNA synthetase (p. 1022)

codon (p. 1023)

transfer RNA (tRNA) (p. 1023)

anticodon (p. 1024)

wobble hypothesis (p. 1025)

30S subunit (p. 1031)

50S subunit (p. 1031)

Shine–Dalgarno sequence (p. 1034)

initiation factor (p. 1035)

elongation factor Tu (EF-Tu) (p. 1036)

accommodation (p. 1036)

elongation factor Ts (EF-Ts) (p. 1036)

peptidyl transferase center (p. 1037)

elongation factor G (EF-G) (translocase) (p. 1038)

polysome (p. 1040)

release factor (RF) (p. 1040)

signal sequence (p. 1046)

signal peptidase (p. 1046)

signal-recognition particle (SRP) (p. 1047)

SRP receptor (SR) (p. 1047)

translocon (p. 1047)

transport vesicle (p. 1048)

coat proteins (p. 1048)

v-SNARE (p. 1048)

t-SNARE (p. 1048)

PROBLEMS

1. *Not a Klingon.* A codon is:

(a) An alternative name for gene

(b) Three amino acids that encode a nucleotide

(c) Three nucleotides that encode an amino acid

(d) One of three nucleotides that encode and amino acid

2. *tRNA enzyme.* Any given aminoacyl-tRNA synthetase:

(a) Attaches the amino acid to the 5'-end of the tRNA

(b) Always recognizes only one specific tRNA

(c) Recognizes all tRNA molecules

(d) Forms an ester linkage between the amino acid and the tRNA

3. *Initiation.* Bacterial protein synthesis is initiated by:

(a) *S*-adenosylmethionyl tRNA

(b) Methionyl tRNA

(c) N-formylmethionyl tRNA

(d) N^{10}-formyltetrahydrofolate tRNA

4. *Bond former.* Peptide bond formation is catalyzed by:

(a) rRNA

(b) A protein in large ribosomal subunit

(c) A protein in the small ribosomal subunit

(d) Aminoacyl tRNA synthetase

5. *A cluster of ribosomes.* In a polysome, or polyribosome, the polypeptides associated with which ribosomes will be the longest:

(a) They will be the same length since the rate of translation is constant.

(b) Those at the 3'-end of the mRNA.

(c) Those at the 5'-end of the mRNA.

(d) Those in the middle of the mRNA.

6. *Babel fish.* Why is protein synthesis also called translation?

7. *Careful, but not too careful.* Why is it crucial that protein synthesis has an error frequency of 10^{-4}?

8. *Commonalities.* What features are common to all tRNA molecules? ✓❶

9. *The ol' two step.* What two reaction steps are required for the formation of an aminoacyl-tRNA? ✓❶

10. *The same but different.* Why must tRNA molecules have both unique structural features and common structural features? ✓❶

11. *Charge it.* In the context of protein synthesis, what is meant by an activated amino acid? ✓❶

12. *Synthetase mechanism.* The formation of isoleucyl-tRNA proceeds through the reversible formation of an enzyme-bound Ile-AMP intermediate. Predict whether ^{32}P-labeled ATP is formed from ^{32}PP$_i$ when each of the following sets of components is incubated with the specific activating enzyme: ✓❶

(a) ATP and ^{32}PP$_i$

(b) tRNA, ATP, and ^{32}PP$_i$

(c) Isoleucine, ATP, and ^{32}PP$_i$

13. *1 = 2, for sufficiently large values of 1.* The energetic equivalent of two molecules of ATP is used to activate an amino acid, yet only one molecule of ATP is used. Explain.

14. *Sieves.* Using threonyl-tRNA synthetase as an example, account for the specificity of threonyl-tRNA formation. ✓❶

15. *Use all available information.* Suggest a reason why there are two classes of aminoacyl-tRNA synthetases, with each class recognizing a different face of the tRNA. ✓❶

16. *Going wobbly.* Explain how it is possible that some tRNA molecules recognize more than one codon. ✓❶

17. *Light and heavy ribosomes.* Ribosomes were isolated from bacteria grown in a "heavy" medium (^{13}C and ^{15}N) and from bacteria grown in a "light" medium (^{12}C and ^{14}N). These 70S ribosomes were added to an in vitro system engaged in protein synthesis. An aliquot removed several hours later was analyzed by density-gradient centrifugation. How many bands of 70S ribosomes would you expect to see in the density gradient? ✓❷

18. *The price of protein synthesis.* What is the smallest number of molecules of ATP and GTP consumed in the synthesis of a 200-residue protein, starting from amino acids? Assume that the hydrolysis of PP$_i$ is equivalent to the hydrolysis of ATP for this calculation. ✓❶, ✓❷

19. *Correct phasing.* What is meant by the phrase *reading frame*? ✓❷

20. *Suppressing frameshifts.* The insertion of a base in a coding sequence leads to a shift in the reading frame, which in most cases produces a nonfunctional protein. Propose a mutation in a tRNA that might suppress frameshifting. ✓❶, ✓❷

21. *Tagging a ribosomal site.* Design an affinity-labeling reagent for one of the tRNA-binding sites in *E. coli* ribosomes. ✓❷

22. *Viral mutation.* An mRNA transcript of a T7 phage gene contains the base sequence

↓

5′–AACUGCACGA**G**GUAACACAAGAUGGCU–3′

Predict the effect of a mutation that changes the red G to A.

23. *A new translation.* A transfer RNA with a UGU anticodon is enzymatically conjugated to ^{14}C-labeled cysteine. The cysteine unit is then chemically modified to alanine. The altered aminoacyl-tRNA is added to a protein-synthesizing system containing normal components except for this tRNA. The mRNA added to this mixture contains the following sequence:

5′–UUUUGCCAUGUUUGUGCU–3′

What is the sequence of the corresponding radiolabeled peptide?

24. *Two synthetic modes.* Compare and contrast protein synthesis by ribosomes with protein synthesis by the solid-phase method (Section 3.4). ✓❶, ✓❷

25. *Triggered GTP hydrolysis.* Ribosomes markedly accelerate the hydrolysis of GTP bound to the complex of EF-Tu and aminoacyl-tRNA. What is the biological significance of this enhancement of GTPase activity by ribosomes? ✓❷

26. *Blocking translation.* Devise an experimental strategy for switching off the expression of a specific mRNA without changing the gene encoding the protein or the gene's control elements. ✓❷

27. *Directional problem.* Suppose that you have a protein-synthesis system that is synthesizing a protein designated A. Furthermore, you know that protein A has four trypsin-sensitive sites, equally spaced in the protein, that, on digestion with trypsin, yield the peptides A$_1$, A$_2$, A$_3$, A$_4$, and A$_5$. Peptide A$_1$ is the amino-terminal peptide, and A$_5$ is the carboxyl-terminal peptide. Finally, you know that your system requires 4 minutes to synthesize a complete protein A. At $t = 0$, you add all 20 amino acids, each carrying a ^{14}C label. ✓❷

(a) At $t = 1$ minute, you isolate intact protein A from the system, cleave it with trypsin, and isolate the five peptides. Which peptide is most heavily labeled?

(b) At $t = 3$ minutes, what will be the order of the labeling of peptides from greatest to least?

(c) What does this experiment tell you about the direction of protein synthesis?

28. *Translator.* Aminoacyl-tRNA synthetases are the only component of gene expression that decodes the genetic code. Explain. ✓❶

29. *A timing device.* EF-Tu, a member of the G-protein family, plays a crucial role in the elongation process of translation. Suppose that a slowly hydrolyzable analog of GTP were added to an elongating system. What would be the effect on the rate of protein synthesis? ✓❷

30. *Not just RNA.* What are the roles of the protein factors required for protein synthesis? ✓❸

31. *Membrane transport.* What four components are required for the translocation of proteins across the endoplasmic reticulum membrane? ✓❹

32. *Push. Don't pull.* What is the energy source that powers the cotranslational movement of proteins across the endoplasmic reticulum? ✓❹

33. *You have to know where to look.* Bacterial messenger RNAs usually contain many AUG codons. How does the ribosome identify the AUG specifying initiation? ✓❷

34. *Fundamentally the same, yet....* List the differences between bacterial and eukaryotic protein synthesis. ✓❸

35. *Like a border collie.* What is the role of the signal-recognition particle in protein translocation? ✓❹

36. *An assembly line.* Why is the fact that protein synthesis takes place on polysomes advantageous? ✓❷

37. *Match 'em* ✓❶, ✓❷

(a) Initiation _____   1. GTP

(b) Elongation _____   2. AUG

(c) Termination _____   3. fMet

4. RRF

5. IF2

6. Shine–Dalgarno

7. EF-Tu

8. Peptidyl transferase

9. UGA

10. Transformylase

38. *Wasted effort?* Transfer RNA molecules are quite large, given that the anticodon consists of only three nucleotides. What is the purpose of the rest of the tRNA molecule? ✓❶, ✓❷

Mechanism Problem

39. *Evolutionary amino acid choice.* Ornithine is structurally similar to lysine except ornithine's side chain is one methylene group shorter than that of lysine. Attempts to chemically synthesize and isolate ornithinyl-tRNA proved unsuccessful. Propose a mechanistic explanation. (Hint: Six-membered rings are more stable than seven-membered rings.)

Chapter Integration Problems

40. *Svedberg addition.* You are studying your biochemistry notes. In your notebook, you have written:

Bacterial: 30S + 50S = 70S
Eukaryotic: 40S + 60S = 80S

Your friend, who is looking over your shoulder, says: "Wow! I didn't realize biochemists were so bad at math." Explain, politely, why the equations that you wrote are correct. ✓❷

41. *Contrasting modes of elongation.* The two basic mechanisms for the elongation of biomolecules are represented in the adjoining illustration. In type 1, the activating group (X) is released from the growing chain. In type 2, the activating group is released from the incoming unit as it is added to the growing chain. Indicate whether each of the following biosyntheses is by means of a type 1 or a type 2 mechanism:

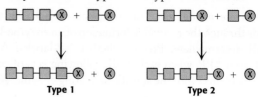

Type 1   **Type 2**

(a) Glycogen synthesis

(b) Fatty acid synthesis

(c) $C_5 \rightarrow C_{10} \rightarrow C_{15}$ in cholesterol synthesis

(d) DNA synthesis

(e) RNA synthesis

(f) Protein synthesis

42. *Enhancing fidelity.* Compare the accuracy of DNA replication, RNA synthesis, and protein synthesis. Which mechanisms are used to ensure the fidelity of each of these processes?

43. *Déjà vu.* Which protein in G-protein cascades plays a role similar to that of elongation factor Ts?

44. *Family resemblance.* Eukaryotic elongation factor 2 is inhibited by ADP ribosylation catalyzed by diphtheria toxin. What other G proteins are sensitive to this mode of inhibition? ✓❸

45. *The exceptional* E. coli. In contrast with *E. coli*, most bacteria do not have a full complement of aminoacyl-tRNA synthetases. For instance, *Helicobacter pylori*, the cause of stomach ulcers, has tRNA^{Gln}, but no Gln-tRNA synthetase. However, glutamine is a common amino acid in *H. pylori* proteins. Suggest a means by which glutamine can be incorporated into proteins in *H. pylori*. (Hint: Glu-tRNA synthetase can misacylate tRNA^{Gln}.)

46. *The final step.* What aspect of primary structure allows the transfer of linear nucleic acid information into the functional three-dimensional structure of proteins?

Data Interpretation Problems

47. *Helicase helper.* The initiation factor eIF-4 displays ATP-dependent RNA helicase activity. Another initiation factor, eIF-4H, has been proposed to assist the action of eIF-4. Graph A shows some of the experimental results from an assay that can measure the activity of eIF-4 helicase in the presence of eIF-4H.

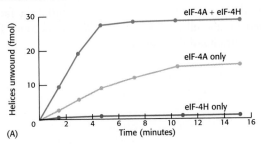

(A)

(a) What are the effects on eIF-4 helicase activity in the presence of eIF-4H?

(b) Why did measuring the helicase activity of eIF-4H alone serve as an important control?

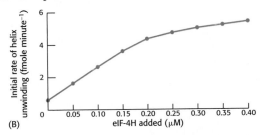

(B)

(c) The initial rate of helicase activity of 0.2 μM of eIF-4 was then measured with varying amounts of eIF-4H (graph B). What ratio of eIF-4H to eIF-4 yielded optimal activity?

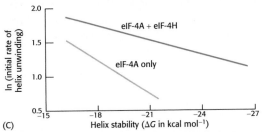

(C)

[Data from N. J. Richter, G. W. Rodgers, Jr., J. O. Hensold, and W. C. Merrick. Further biochemical and kinetic characterization of human eukaryotic initiation factor 4H. *J. Biol. Chem.* 274:35415–35424, 1999.]

(d) Next, the effect of RNA–RNA helix stability on the initial rate of unwinding in the presence and absence of eIF-4H was tested (graph C). How does the effect of eIF-4H vary with helix stability?

(e) How might eIF-4H affect the helicase activity of eIF-4A?

48. *Size separation.* The protein-synthesizing machinery was isolated from eukaryotic cells and briefly treated with a low concentration of RNase. The sample was then subjected to sucrose gradient centrifugation. The gradient was fractionated and the absorbance, or optical density (OD), at 254 nm was recorded for each fraction. The following plot was obtained. ✓❸

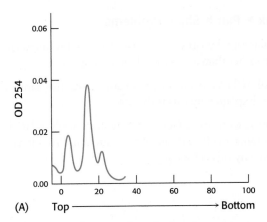

(A)

(a) What do the three peaks of absorbance in graph A represent?

The experiment was repeated except that, this time, the RNase treatment was omitted.

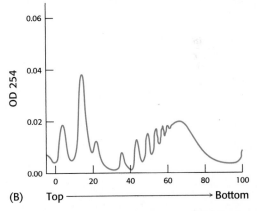

(B)

(b) Why is the centrifugation pattern in graph B more complex? What do the series of peaks near the bottom of the centrifuge tube represent?

Before the isolation of the protein-synthesizing machinery, the cells were grown in low concentrations of oxygen (hypoxic conditions). Again the experiment was repeated without RNase treatment (graph C).

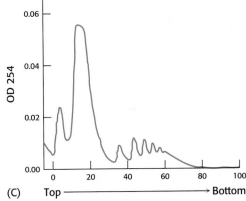

(C)

[Data from M. Koritzinsky et al., *EMBO J.* 25:1114–1125, 2006.]

(c) What is the effect of growing cells under hypoxic conditions?

Think ▸ Pair ▸ Share Problems

49. Suggest why ribosomes exist as two subunits in all forms of life rather than as a single, larger complex.

50. All tRNAs from all organisms have the same overall shape. Explain why this is the case.

51. Some elongation factors are evolutionarily related to the G-proteins involved in signal transduction. Provide a possible reason why this is the case.

52. On page 1042, we discussed vanishing white matter disease, a condition that affects only the brain. The disease is caused by a mutation in eukaryotic initiation factor 2, required by all cells. Present some ideas to account for the tissue specificity of the mutation.

The Control of Gene Expression in Prokaryotes

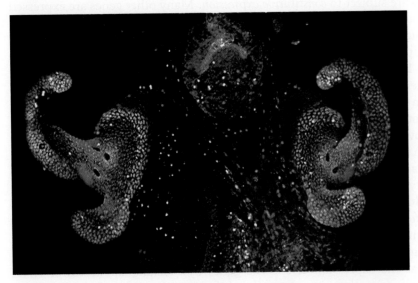

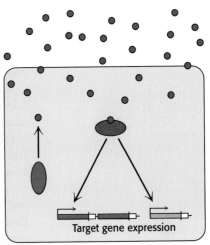

CHAPTER 32

Target gene expression

Bacteria respond to changes in their environments. A micrograph of the light organ of a newly hatched squid (*Euprymna scolopes*) is shown on the left. The light spots are due to colonies of the bacteria *Vibrio fischeri* that live symbiotically within these organs. These bacteria become luminescent when they reach an appropriately high density. The density is sensed by the circuit shown on the right in which each bacterium releases a small molecule into the environment. The molecule is subsequently taken up by other bacterial cells, which start a signaling cascade that stimulates the expression of specific genes. [Spencer V. Nyholm, Eric V. Stabb, Edward G. Ruby, and Margaret J. McFall-Ngai, "Establishment of an animal–bacterial association: Recruiting symbiotic vibrios from the environment." *PNAS* 97(2000): 10231–10235. Copyright 2000 National Academy of Sciences.]

✓ LEARNING OBJECTIVES

By the end of this chapter, you should be able to:

1. Discuss the role of DNA-binding proteins in controlling transcription in prokaryotes.

2. Describe the basic structural features of DNA-binding proteins that allow them to recognize specific DNA sequences.

3. Discuss the concept of an operon and the roles of the various components in the *E. coli lac* operon.

4. Discuss the genetic switch that controls the lytic versus lysogenic path in bacteriophage lambda.

5. Describe the mechanisms by which gene regulation is controlled by changes in the concentration of small molecules including quorum sensing.

6. Discuss posttranslational mechanisms for controlling prokaryotic gene expression including attentuators and riboswitches.

OUTLINE

32.1 Many DNA-Binding Proteins Recognize Specific DNA Sequences

32.2 Prokaryotic DNA-Binding Proteins Bind Specifically to Regulatory Sites in Operons

32.3 Regulatory Circuits Can Result in Switching Between Patterns of Gene Expression

32.4 Gene Expression Can Be Controlled at Posttranscriptional Levels

Even simple prokaryotic cells must respond to changes in their metabolism or in their environments. Much of this response takes place through changes in gene expression. A gene is *expressed* when it is transcribed into RNA and, for most genes, translated into proteins. Genomes comprise thousands of genes. Some of these genes are expressed all the time. These genes are subject to *constitutive expression*. Many other genes are expressed only under some circumstances—that is, under a particular set of physiological conditions. These genes are subject to *regulated expression*. For example, the level of expression of some genes in bacteria may vary more than 1000-fold in response to the supply of nutrients or to environmental challenges.

In this chapter, we will examine gene-regulation mechanisms in prokaryotes, particularly *E. coli*, because many of these processes were first discovered in this organism. In Chapter 33, we will turn to gene-regulation mechanisms in eukaryotes. We shall see both substantial similarities and fundamental differences in comparing gene-regulatory mechanisms of the two types of organisms.

How is gene expression controlled? *Gene activity is controlled first and foremost at the level of transcription.* Whether a gene is transcribed is determined largely by the interplay between specific DNA sequences and certain proteins that bind to these sequences. Most often, these proteins repress the expression of specific genes by blocking the access of RNA polymerase to their promoters. In some cases, however, the proteins can activate the expression of specific genes. We shall learn about several different strategies that allow the coordinated regulation of sets of genes. Some genes are also controlled at stages beyond the level of transcription and we shall examine several mechanisms at these stages. Finally, we will examine several important examples of how gene expression is regulated in response to changes in the concentrations of specific molecules in the environment of prokaryotic cells.

32.1 Many DNA-Binding Proteins Recognize Specific DNA Sequences

How do regulatory systems distinguish the genes that need to be activated or repressed from genes that are constitutive? After all, the DNA sequences of genes themselves do not have any distinguishing features that would allow regulatory systems to recognize them. Instead, gene regulation depends on other sequences in the genome. In prokaryotes, these regulatory sites are close to the region of the DNA that is transcribed. Regulatory sites are usually binding sites for specific DNA-binding proteins, which can stimulate or repress gene expression. These regulatory sites were first identified in *E. coli* in studies of changes in gene expression. In the presence of the sugar lactose, the bacterium starts to express a gene encoding β-galactosidase, an enzyme that can process lactose for use as a carbon and energy source. The sequence of the regulatory site for this gene is shown in Figure 32.1. The nucleotide sequence of this site shows a nearly perfect inverted repeat, indicating that the DNA in this region has an approximate twofold axis of symmetry. Recall that cleavage sites for restriction enzymes such as EcoRV have similar symmetry properties

FIGURE 32.1 Sequence of the *lac* regulatory site. The nucleotide sequence of this regulatory site shows a nearly perfect inverted repeat, corresponding to twofold rotational symmetry in the DNA. Parts of the sequences that are related by this symmetry are shown in the same color.

5'-...T G T G T G G A A T T G T G A G C G G A T A A C A A T T T C A C A C A...3'
3'-...A C A C A C C T T A A C A C T C G C C T A A T G T T A A A G T G T G T...5'

(Section 9.3). Symmetry in such regulatory sites usually corresponds to symmetry in the protein that binds the site. *Symmetry matching is a recurring theme in protein–DNA interactions.*

To understand these protein–DNA interactions in detail, scientists examined the structure of the complex between an oligonucleotide that includes this site and the DNA-binding unit that recognizes it (Figure 32.2). The DNA-binding unit comes from a protein called the *lac repressor*, which represses the expression of the lactose-processing gene. As expected, this DNA-binding unit binds as a dimer, and the twofold axis of symmetry of the dimer matches the symmetry of the DNA. An α helix from each monomer of the protein is inserted into the major groove of the DNA, where amino acid side chains make specific contacts with exposed edges of the base pairs. For example, the side chain of an arginine residue of the protein forms a pair of hydrogen bonds with a guanine residue of the DNA, which would not be possible with any other base. This interaction and similar ones allow the *lac* repressor to bind more tightly to this site than to the wide range of other sites present in the *E. coli* genome.

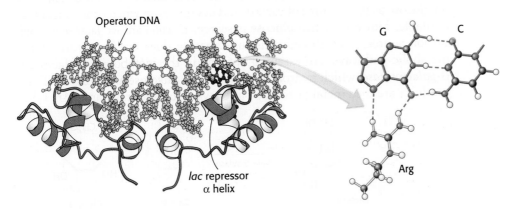

FIGURE 32.2 The *lac* repressor–DNA complex. The DNA-binding domain from a gene-regulatory protein, the *lac* repressor, binds to a DNA fragment containing its preferred binding site (referred to as operator DNA) by inserting an α helix into the major groove of operator DNA. *Notice* that a specific contact forms between an arginine residue of the repressor and a G–C base pair in the binding site. [Drawn from 1EFA.pdb.]

The helix-turn-helix motif is common to many prokaryotic DNA-binding proteins

Are similar strategies utilized by other prokaryotic DNA-binding proteins? The structures of many such proteins have now been determined, and amino acid sequences are known for many more. Strikingly, the DNA-binding surfaces of many, but not all, of these proteins consist of a pair of α helices separated by a tight turn (Figure 32.3). In complexes with

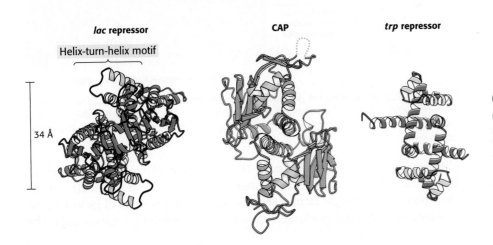

FIGURE 32.3 Helix-turn-helix motif. These structures show three sequence-specific DNA-binding proteins that interact with DNA through a helix-turn-helix motif (highlighted in yellow). *Notice* that, in each case, the helix-turn-helix units within a protein dimer are approximately 34 Å apart, corresponding to one full turn of DNA. [Drawn from 1EFA, 1RUN, and 1TRO.pdb.]

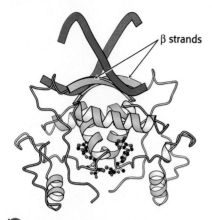

FIGURE 32.4 DNA recognition through β strands. A methionine repressor is shown bound to DNA. *Notice* that residues in β strands, rather than in α helices, participate in the crucial interactions between the protein and the DNA. [Drawn from 1CMA.pdb.]

DNA, the second of these two helices (often called the *recognition helix*) lies in the major groove, where amino acid side chains make contact with the edges of base pairs. In contrast, residues of the first helix participate primarily in contacts with the DNA backbone. *Helix-turn-helix motifs* are present on many proteins that bind DNA as dimers, and thus two of the units will be present, one on each monomer.

Although the helix-turn-helix motif is the most commonly observed DNA-binding unit in prokaryotes, not all regulatory proteins bind DNA through such units. A striking example is provided by the *E. coli* methionine repressor (Figure 32.4). This protein binds DNA through the insertion of a pair of β strands into the major groove.

32.2 Prokaryotic DNA-Binding Proteins Bind Specifically to Regulatory Sites in Operons

A historically important example reveals many common principles of gene regulation by DNA-binding proteins. Bacteria such as *E. coli* usually rely on glucose as their source of carbon and energy, even when other sugars are available. However, when glucose is scarce, *E. coli* can use lactose as their carbon source, even though this disaccharide does not lie on any major metabolic pathways. An essential enzyme in the metabolism of lactose is *β-galactosidase*, which hydrolyzes lactose into galactose and glucose. These products are then metabolized by pathways discussed in Chapter 16.

Lactose **Galactose** **Glucose**

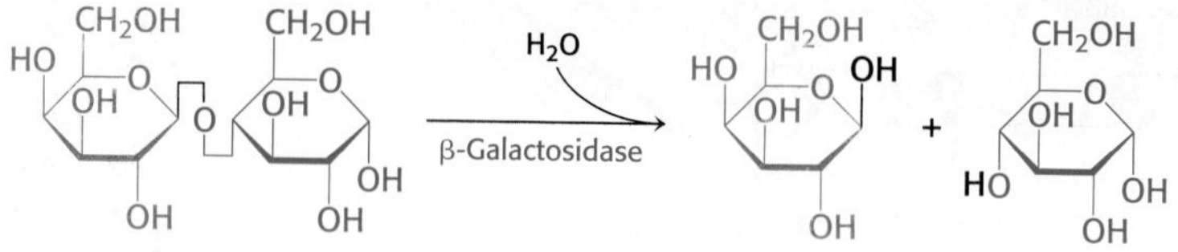

This reaction can be conveniently followed in the laboratory through the use of alternative galactoside substrates that form colored products such as X-Gal (Figure 32.5). An *E. coli* cell growing on a carbon source such as

X-Gal

5,5′-Dibromo-4,4′-dichloro-indigo

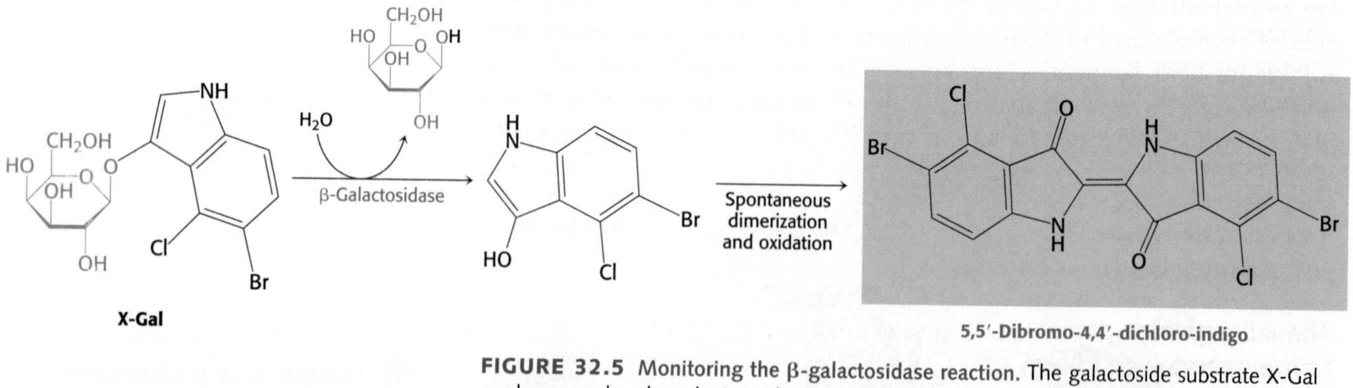

FIGURE 32.5 Monitoring the β-galactosidase reaction. The galactoside substrate X-Gal produces a colored product on cleavage by β-galactosidase. The appearance of this colored product provides a convenient means for monitoring the amount of the enzyme both in vitro and in vivo.

glucose or glycerol contains fewer than 10 molecules of β-galactosidase. In contrast, the same cell will contain several thousand molecules of the enzyme when grown on lactose (Figure 32.6). The presence of lactose in the culture medium induces a large increase in the amount of β-galactosidase by eliciting the synthesis of new enzyme molecules rather than by activating a preexisting but inactive precursor.

A crucial clue to the mechanism of gene regulation was the observation that two other proteins are synthesized in concert with β-galactosidase—namely, *galactoside permease* and *thiogalactoside transacetylase*. The permease is required for the transport of lactose across the bacterial cell membrane (Section 13.3). The transacetylase is not essential for lactose metabolism but appears to play a role in the detoxification of compounds that also may be transported by the permease. Thus, *the expression levels of a set of enzymes that all contribute to the adaptation to a given change in the environment change together*. Such a coordinated unit of gene expression is called an *operon*.

An operon consists of regulatory elements and protein-encoding genes

The parallel regulation of β-galactosidase, the permease, and the transacetylase suggested that the expression of genes encoding these enzymes is controlled by a common mechanism. François Jacob and Jacques Monod proposed the *operon model* to account for this parallel regulation as well as the results of other genetic experiments. The genetic elements of the model are a *regulator gene* that encodes a regulatory protein, a regulatory DNA sequence called *an operator site*, and a *set of structural genes* (Figure 32.7).

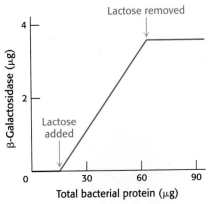

FIGURE 32.6 β-Galactosidase induction. The addition of lactose to an *E. coli* culture causes the production of β-galactosidase to increase from very low amounts to much larger amounts. The increase in the amount of enzyme parallels the increase in the number of cells in the growing culture. β-Galactosidase constitutes 6.6% of the total protein synthesized in the presence of lactose.

FIGURE 32.7 Operons. (A) The general structure of an operon as conceived by Jacob and Monod. (B) The structure of the lactose operon. In addition to the promoter, *p*, in the operon, a second promoter is present in front of the regulator gene, *i*, to drive the synthesis of the regulator.

The regulator gene encodes a repressor protein that binds to the operator site. The binding of the repressor to the operator prevents transcription of the structural genes. The operator and its associated structural genes constitute the operon. For the *lactose (lac) operon*, the *i* gene encodes the repressor, *o* is the operator site, and the *z*, *y*, and *a* genes are the structural genes for β-galactosidase, the permease, and the transacetylase, respectively. The operon also contains a promoter site (denoted by *p*), which directs the RNA polymerase to the correct transcription initiation site. The *z*, *y*, and *a* genes are transcribed to give a single mRNA molecule that encodes all three proteins. An mRNA molecule encoding more than one protein is known as a *polygenic* or *polycistronic* transcript.

The *lac* repressor protein in the absence of lactose binds to the operator and blocks transcription

In the absence of lactose, the lactose operon is repressed. How does the *lac* repressor mediate this repression? The lac repressor exists as a tetramer of 37-kDa subunits with two pairs of subunits coming together to form the DNA-binding unit previously discussed. In the absence of lactose, the repressor binds very tightly and rapidly to the operator. When the *lac* repressor is bound to DNA, the repressor prevents RNA polymerase from transcribing the protein-coding genes inasmuch as the operator site is directly adjacent to and downstream of the promoter site where the repressor would block the progress of RNA polymerase.

How does the *lac* repressor locate the operator site in the *E. coli* chromosome? The *lac* repressor binds 4×10^6 times as strongly to operator DNA as it does to random sites in the genome. This high degree of selectivity allows the repressor to find the operator efficiently even with a large excess (4.6×10^6) of other sites within the *E. coli* genome. The dissociation constant for the repressor–operator complex is approximately 0.1 pM (10^{-13} M). The rate constant for association ($\approx 10^{10}$ M^{-1} s^{-1}) is strikingly high, indicating that the repressor finds the operator primarily by diffusing along a DNA molecule (a one-dimensional search) rather than encountering it from the aqueous medium (a three-dimensional search). This diffusion has been confirmed by studies that monitored the behavior of fluorescently labeled single molecules of *lac* repressor inside living *E. coli* cells.

Inspection of the complete *E. coli* genome sequence reveals two sites within 500 bp of the primary operator site that approximate the sequence of the operator. When one dimeric DNA-binding unit binds to the operator site, the other DNA-binding unit of the *lac* repressor tetramer can bind to one of these sites with similar sequences. The DNA between the two bound sites forms a loop. No other sites that closely match the sequence of the *lac operator* site are present in the rest of the *E. coli* genome sequence. Thus, *the DNA-binding specificity of the* lac *repressor is sufficient to specify two closely related sites within the* E. coli *genome.*

The three-dimensional structure of the *lac* repressor has been determined in various forms. Each monomer consists of a small amino-terminal domain that binds DNA and a larger domain that mediates the formation of the dimeric DNA-binding unit and the tetramer (Figure 32.8). A pair of the amino-terminal domains come together to form the functional DNA-binding unit. Each monomer has a helix-turn-helix unit that interacts with the major groove of the bound DNA.

Ligand binding can induce structural changes in regulatory proteins

In the situation just described, glucose is present and lactose is absent, and the *lac* operon is repressed. How does the presence of lactose trigger the relief of this repression and, hence, the expression of the *lac* operon? Interestingly, lactose itself does not have this effect; rather, *allolactose*, a combination of galactose and glucose with an α-1,6 rather than an α-1,4 linkage, does. Allolactose is thus referred to as the *inducer* of the *lac* operon. Allolactose is a side product of the β-galactosidase reaction and is produced at low levels by the few molecules of β-galactosidase that are present before induction. Some other β-galactosides such as *isopropylthiogalactoside* (IPTG) are potent inducers of β-galactosidase expression, although they are not substrates of the enzyme. IPTG is useful in the laboratory as a tool for inducing gene expression in engineered bacterial strains.

The inducer triggers gene expression by preventing the *lac* repressor from binding the operator. *The inducer binds to the* lac *repressor and thereby greatly reduces the repressor's affinity for operator DNA.* An inducer molecule binds in the center of the large domain within each monomer. This binding leads to subtle conformational changes, which modify the relation between the two small DNA-binding domains (Figure 32.9). These domains can no longer easily contact DNA simultaneously, leading to a dramatic reduction in DNA-binding affinity.

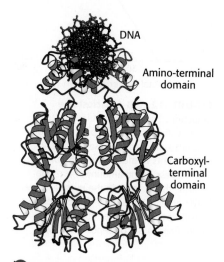

FIGURE 32.8 Structure of the *lac* **repressor.** A *lac* repressor dimer is shown bound to DNA. *Notice* that the amino-terminal domain binds to DNA, whereas the carboxyl-terminal domain forms a separate structure. A part of the structure that mediates the formation of *lac* repressor tetramers is not shown. [Drawn from 1EFA.pdb.]

1,6-Allolactose

Isopropylthiogalactoside (IPTG)

Let us recapitulate the processes that regulate gene expression in the lactose operon (Figure 32.10). In the absence of inducer, the *lac* repressor is bound to DNA in a manner that blocks RNA polymerase from transcribing the *z*, *y*, and *a* genes. Thus, very little β-galactosidase, permease, or transacetylase are produced. The addition of lactose to the environment leads to the formation of allolactose. This inducer binds to the *lac* repressor, leading to conformational changes and the release of DNA by the *lac* repressor. With the operator site unoccupied, RNA polymerase can then transcribe the other *lac* genes and the bacterium will produce the proteins necessary for the efficient use of lactose.

The structure of the large domain of the *lac* repressor is similar to those of a large class of proteins that are present in *E. coli* and other bacteria. This family of homologous proteins binds ligands such as sugars and amino acids at their centers. Remarkably, domains of this family are utilized by eukaryotes in taste proteins and in neurotransmitter receptors, as will be discussed in Chapter 34 (see Chapter 34 on ♨ SaplingPlus).

The operon is a common regulatory unit in prokaryotes

Many other gene-regulatory networks function in ways analogous to those of the *lac* operon. For example, genes taking part in purine and, to a lesser degree, pyrimidine biosynthesis are repressed by the *pur repressor*. This dimeric protein is 31% identical in sequence with the *lac* repressor and has a similar three-dimensional structure. However, the behavior of the *pur* repressor is opposite that of the *lac* repressor: whereas the *lac* repressor is *released* from DNA by binding to a small molecule, *the* pur *repressor binds DNA specifically, blocking transcription, only when bound to a small molecule.* Such a small molecule is called a *corepressor.* For the *pur* repressor, the corepressor can be either guanine or hypoxanthine. The dimeric *pur* repressor binds to inverted-repeat DNA sites of the form 5'-ANGCAANCGNTTNCNT-3', in which the bases shown in boldface type are particularly important. Examination of the *E. coli* genome sequence reveals the presence of more than 20 such sites, regulating 19 operons and including more than 25 genes (Figure 32.11).

Because the DNA-binding sites for these regulatory proteins are short, it is likely that they evolved independently and are not related by divergence from an ancestral regulatory site. Once a ligand-regulated DNA-binding protein is present in a cell, binding sites for the protein may arise by mutation adjacent to additional genes. Binding sites for the *pur* repressor have evolved in the regulatory

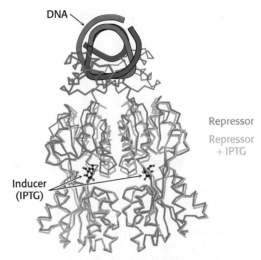

FIGURE 32.9 Effects of IPTG on *lac* repressor structure. The structure of the *lac* repressor bound to the inducer isopropylthiogalactoside (IPTG), shown in orange, is superimposed on the structure of the *lac* repressor bound to DNA, shown in purple. *Notice* that the binding of IPTG induces subtle structural changes but these are largest near the interface to the DNA-binding domains so that the two DNA-binding domains cannot interact effectively with DNA. The DNA-binding domains of the *lac* repressor bound to IPTG are not shown, because these regions are not well ordered in the crystals studied.

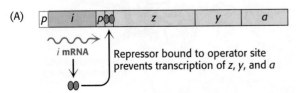

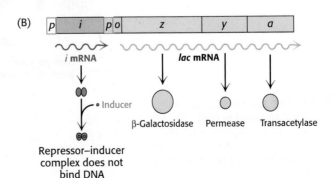

FIGURE 32.10 Induction of the *lac* operon. (A) In the absence of lactose, the *lac* repressor binds DNA and represses transcription from the *lac* operon. (B) Allolactose or another inducer binds to the *lac* repressor, leading to its dissociation from DNA and to the production of *lac* mRNA.

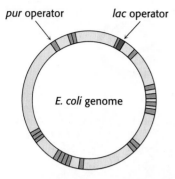

FIGURE 32.11 Binding-site distributions. The *E. coli* genome contains only a single region that closely matches the sequence of the *lac* operator (shown in blue). In contrast, 19 sites match the sequence of the *pur* operator (shown in red). Thus, the *pur* repressor regulates the expression of many more genes than does the *lac* repressor.

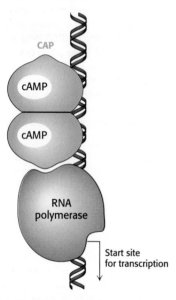

FIGURE 32.12 Binding site for catabolite activator protein (CAP). This protein binds as a dimer to an inverted repeat that is at the position −61 relative to the start site of transcription. The CAP-binding site on DNA is adjacent to the position at which RNA polymerase binds.

FIGURE 32.13 Structure of a dimer of CAP bound to DNA. The residues shown in yellow in each CAP monomer have been implicated in direct interactions with RNA polymerase. [Drawn from 1RUN.pdb.]

regions of a wide range of genes taking part in nucleotide biosynthesis. All such genes can then be regulated in a concerted manner.

The organization of prokaryotic genes into operons is useful for the analysis of completed genome sequences. Sometimes a gene of unknown function is discovered to be part of an operon containing well-characterized genes. Such associations can provide powerful clues to the biochemical and physiological functions of the uncharacterized gene.

Transcription can be stimulated by proteins that contact RNA polymerase

All the DNA-binding proteins discussed thus far function by inhibiting transcription until some environmental condition, such as the presence of lactose, is met. There are also DNA-binding proteins that stimulate transcription. One particularly well-studied example is a protein in *E. coli* that stimulates the expression of catabolic enzymes.

E. coli grown on glucose, a preferred energy source, have very low levels of catabolic enzymes for metabolizing other sugars. Clearly, the synthesis of these enzymes when glucose is abundant would be wasteful. Glucose has an inhibitory effect on the genes encoding these enzymes, an effect called *catabolite repression*. It is due to the fact that glucose lowers the concentration of cyclic AMP in *E. coli*. When the cAMP concentration is high, it stimulates the concerted transcription of many catabolic enzymes by acting through a protein called the *catabolite activator protein (CAP)*, which is also known as the cAMP receptor protein (CRP).

When bound to cAMP, CAP stimulates the transcription of lactose- and arabinose-catabolizing genes. CAP is a sequence-specific DNA-binding protein. Within the *lac* operon, CAP binds to an inverted repeat that is centered near position −61 relative to the start site for transcription (Figure 32.12). This site is approximately 70 base pairs from the operator site. As expected from the symmetry of the binding site, CAP functions as a dimer of identical subunits. CAP binding also bends DNA in a manner that favors interactions with RNA polymerase.

The CAP–cAMP complex stimulates the initiation of transcription by approximately a factor of 50. Energetically favorable contacts between CAP and RNA polymerase increase the likelihood that transcription will be initiated at sites to which the CAP–cAMP complex is bound (Figure 32.13).

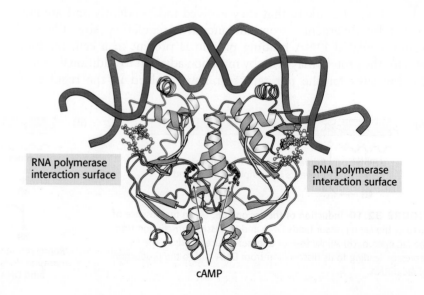

Thus, in regard to the *lac* operon, gene expression is maximal when the binding of allolactose relieves the inhibition by the *lac* repressor and the CAP–cAMP complex stimulates the binding of RNA polymerase.

The *E. coli* genome contains many CAP-binding sites in positions appropriate for interactions with RNA polymerase. Thus, an increase in the cAMP level inside an *E. coli* bacterium results in the formation of CAP–cAMP complexes that bind to many promoters and stimulate the transcription of genes encoding a variety of catabolic enzymes.

32.3 Regulatory Circuits Can Result in Switching Between Patterns of Gene Expression

The study of viruses that infect bacteria has led to significant advances in our understanding of the processes that control gene expression. Again, sequence-specific DNA-binding proteins play key roles in these processes. Investigations of bacteriophage λ have been particularly revealing. We examined the alternative infection modes of λ phage in Chapter 5. In the lytic pathway, most of the genes in the viral genome are transcribed, initiating the production of many virus particles and leading to the eventual lysis of the bacterial cell with the concomitant release of approximately 100 virus particles. In the lysogenic pathway, the viral genome is incorporated into the bacterial DNA where most of the viral genes remain unexpressed, allowing the viral genome to be carried along as the bacteria replicate. Two key proteins and a set of regulatory sequences in the viral genome are responsible for the switch that determines which of these two pathways is followed.

The λ repressor regulates its own expression

The first protein that we shall consider is the λ *repressor*, sometimes known as the λ cI protein. This protein is key because it blocks, either directly or indirectly, the transcription of almost all genes encoded by the virus. The one exception is the gene that encodes the λ repressor itself. The λ repressor consists of an amino-terminal DNA-binding domain and a carboxyl-terminal domain that participates in protein oligomerization (Figure 32.14). This protein binds to a number of key sites in the λ phage genome. The sites of greatest interest for our present discussion are in the so-called right operator (Figure 32.15). This region includes three binding sites for the λ repressor dimer, as well as two promoters within a region of approximately 80 base pairs. One promoter drives the expression of the gene for the λ repressor itself, whereas the other drives the expression of a number of other viral genes.

The λ repressor does not have the same affinity for the three sites; it binds the site O_R1 with the highest affinity. In addition, the binding to adjacent sites is cooperative so that, after a λ repressor dimer has bound at O_R1,

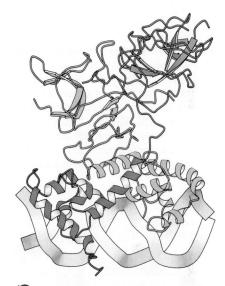

FIGURE 32.14 Structure of the λ repressor bound to DNA. The λ repressor binds to DNA as a dimer. The amino-terminal domain of one subunit is shown in red and the carboxyl-terminal domain is shown in blue. In the other subunit, both domains are shown in yellow. *Notice* how α helices on the amino-terminal domains fit into the major groove of the DNA. [Drawn from 3DBN.pdb.]

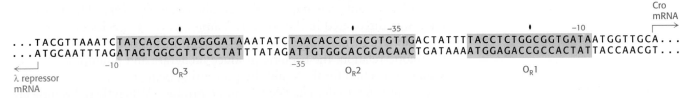

FIGURE 32.15 Sequence of the λ right operator. The three operator sites (O_R1, O_R2, and O_R3) are shaded yellow with their centers indicated. The start sites for the λ repressor mRNA and the Cro mRNA are indicated, as are their −10 and −35 sequences.

(A) LAMBDA REPRESSOR LEVEL LOW
λ repressor gene expression stimulated

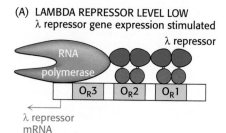

(B) LAMBDA REPRESSOR LEVEL HIGHER
λ repressor gene expression blocked

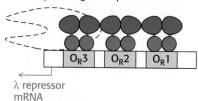

FIGURE 32.16 The λ repressor controls its own synthesis. (A) When λ repressor levels are relatively low, the repressor binds to sites O_R1 and O_R2 and stimulates the transcription of the gene that encodes the λ repressor itself. (B) When λ repressor levels are higher, the repressor also binds to site O_R3, blocking access to its promoter and repressing transcription from this gene.

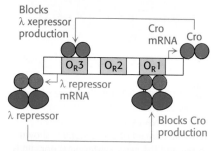

FIGURE 32.17 The λ repressor and Cro form a genetic circuit. The λ repressor blocks the production of Cro by binding most favorably to site O_R1 whereas Cro blocks the production of the λ repressor by binding most favorably to site O_R3. This circuit forms a switch that determines whether the lysogenic or the lytic pathway is followed.

the likelihood that a protein will bind to the adjacent site O_R2 increases by approximately 25-fold. Thus, when λ repressor is present in the cell at moderate concentrations, the most likely configuration has λ repressor bound at O_R1 and O_R2, but not at O_R3. In this configuration, the λ repressor dimer bound at O_R1 blocks access to the promoter on the right side of the operator sites, repressing transcription of the adjacent gene, which encodes a protein termed Cro (*c*ontroller of *r*epressor and *o*thers), while the repressor dimer at O_R2 can be in contact with RNA polymerase and stimulate transcription of the promoter that controls the transcription of the gene that encodes the λ repressor itself.

Thus, the λ repressor stimulates its own production. As the concentration of the λ repressor increases further, an additional repressor dimer can bind to the O_R3 site, blocking the other promoter and repressing the production of additional repressor. Thus, the right operator serves to maintain the λ repressor in a narrow, stable concentration range (Figure 32.16). The λ repressor also blocks other promoters in the λ phage genome so that the repressor is the only phage protein produced, which corresponds to the lysogenic state.

A circuit based on the λ repressor and Cro forms a genetic switch

What stimulates the switch to the lytic pathway? Changes such as DNA damage initiate the cleavage of the λ repressor at a specific bond between the DNA-binding and oligomerization domains. This process is mediated by the *E. coli* RecA protein (Section 29.5). After this cleavage has taken place, the affinity of the λ repressor for DNA is reduced. After the λ repressor is no longer bound to the O_R1 site, the Cro gene can be transcribed. *Cro* is a small protein that binds to the same sites as the λ repressor does, but with a different order of affinity for the three sites in the right operator. In particular, Cro has the highest affinity for O_R3. Cro bound in this site blocks the production of new repressor. The absence of repressor leads to the production of other phage genes, leading to the production of virus particles and the eventual lysis of the host cells. Thus, this genetic circuit acts as a switch with two stable states: (1) repressor high, Cro low, corresponding to the lysogenic state and (2) Cro high, repressor low, corresponding to the lytic state (Figure 32.17). Regulatory circuits with different DNA-binding proteins controlling the expression of each other's genes constitute a common motif for controlling gene expression.

Many prokaryotic cells release chemical signals that regulate gene expression in other cells

Prokaryotic cells have been traditionally viewed as solitary single cells. However, it is becoming increasingly clear that, in many circumstances, prokaryotic cells live in complex communities, interacting with other cells of their own and different species. *These social interactions change the patterns of gene expression within the cells.*

An important type of interaction is called *quorum sensing*. This phenomenon was discovered in the bacteria *Vibrio fischeri*, a species of bacterium that can live inside a specialized light organ in the bobtail squid. In this symbiotic relation, the bacteria produce luciferase and bioluminesce, providing protection for the squid (by preventing being backlit by moonlight) in exchange for a safer place to live and reproduce. When these bacteria are grown in culture at low density, they are not bioluminescent. However, when the cell density reaches a critical level, the gene for luciferase is expressed and the cells bioluminesce. A key observation was that, when *V. fischeri* cells were transferred to a sterile medium in which other *V. fischeri* cells had been grown to high density, the cells became bioluminescent even at low cell density. This experiment revealed that a chemical, subsequently shown to be *N*-3-oxo-hexanoyl homoserine lactone (hereafter AHL for acyl-homoserine lactone), had been released into the medium where it triggered the development of the bioluminescence (Figure 32.18). This compound and other compounds that play similar roles are termed *autoinducers*.

Cells of *V. fischeri* release the autoinducer into their environment and other *V. fischeri* cells take up the chemical. *V. fischeri* cells express a DNA-binding protein LuxR that serves as the receptor for the autoinducer. LuxR comprises two domains, one of which binds AHL and the other of which binds DNA through a helix-turn-helix motif (Figure 32.19). After the AHL concentration inside the cell has increased to an appropriate level, a substantial fraction of the LuxR molecules bind AHL. When bound to AHL, LuxR dimers bind to specific sites on DNA and increase the rate of transcription initiation at specific genes. The target genes include an operon that contains LuxA and LuxB, which together encode the luciferase enzyme, and LuxI, which produces an enzyme that catalyzes the formation of more AHL.

Because each cell produces only a small amount of the autoinducer, this regulatory system allows each *V. fischeri* cell to determine the density of the *V. fischeri* population in its environment—hence the term *quorum sensing* for this process. Studies of other prokaryotic cells are revealing an elaborate chemical language of different autoinducers (as well as autorepressors that repress specific genes). The "words" in this language include other acyl-homoserine lactone molecules with different acyl chain lengths and functionalities, as well as other distinct classes of molecules.

Biofilms are complex communities of prokaryotes

Many species of prokaryotes can be found in specialized structures termed *biofilms* that can form on surfaces. Biofilms are of considerable medical importance because organisms within them are often quite resistant to the immune response of the host as well as to antibiotics. Quorum sensing appears to play a major role in the formation of biofilms, in that cells are able to sense other cells in their environments and to promote the formation of communities with particular compositions. Some genes controlled by quorum-sensing mechanisms promote the formation of specific molecules that serve as scaffolds for the biofilm. An intriguing recent discovery is that many of the organisms that are present in biofilms on or in our bodies (perhaps 95% or more) have not been grown in culture. Through DNA-sequencing methods, we are developing a better census of our microbiome and are working toward understanding the gene-regulatory mechanisms that support these complex communities.

Understanding microbial communities and biofilms is crucial for many areas of health care. Dental plaque is a biofilm, and changes in the component populations of bacteria can influence dental caries and other disorders. Biofilms can also form on surgical

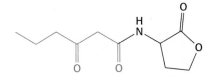

Acyl-homoserine lactone (AHL)

FIGURE 32.18 Autoinducer structure. The structure of the acyl-homoserine lactone *N*-3-oxo-hexanoyl homoserine lactone, the autoinducer from *V. fischeri*. The autoinducers from other bacterial species can have different acyl groups (shown in red).

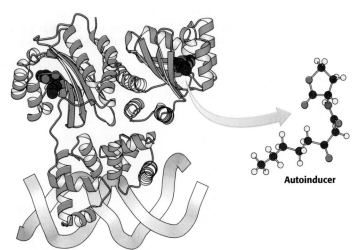

Autoinducer

FIGURE 32.19 Quorum-sensing gene regulator. The structure of a homolog of LuxR (TraR from the bacterium *Agrobacterium tumefaciens*) is shown. *Notice* that the dimeric protein binds to DNA through an α-helical domain, whereas the autoinducer binds to a separate domain.

equipment, and understanding the factors that support and disrupt biofilms can influence choices of materials and sterilization procedures.

32.4 Gene Expression Can Be Controlled at Posttranscriptional Levels

The modulation of the rate of transcription initiation is the most common mechanism of gene regulation. However, other stages of transcription also can be targets for regulation. In addition, the process of translation provides other points of intervention for regulating the level of a protein produced in a cell. In Chapter 30, we considered riboswitches that control transcription termination (Section 30.1). Other riboswitches control gene expression by other mechanisms such as the formation of structures that inhibit translation. Additional mechanisms for posttranscriptional gene regulation have been discovered, one of which will be described here.

Attenuation is a prokaryotic mechanism for regulating transcription through the modulation of nascent RNA secondary structure

A means for regulating transcription in bacteria was discovered by Charles Yanofsky and his colleagues as a result of their studies of the tryptophan operon. This operon encodes five enzymes that convert chorismate into tryptophan. Analysis of the 5′ end of *trp* mRNA revealed the presence of a *leader sequence* of 162 nucleotides before the initiation codon of the first enzyme. The next striking observation was that bacteria produced a transcript consisting of only the first 130 nucleotides when the tryptophan level was high, but they produced a 7000-nucleotide *trp* mRNA, including the entire leader sequence, when tryptophan was scarce. Thus, when tryptophan is plentiful and the biosynthetic enzymes are not needed, transcription is abruptly broken off before any mRNA coding for the enzymes is produced. The site of termination is called the attenuator, and this mode of regulation is called *attenuation*.

Attenuation depends on features at the 5′ end of the mRNA product (Figure 32.20). The first part of the leader sequence encodes a 14-amino-acid

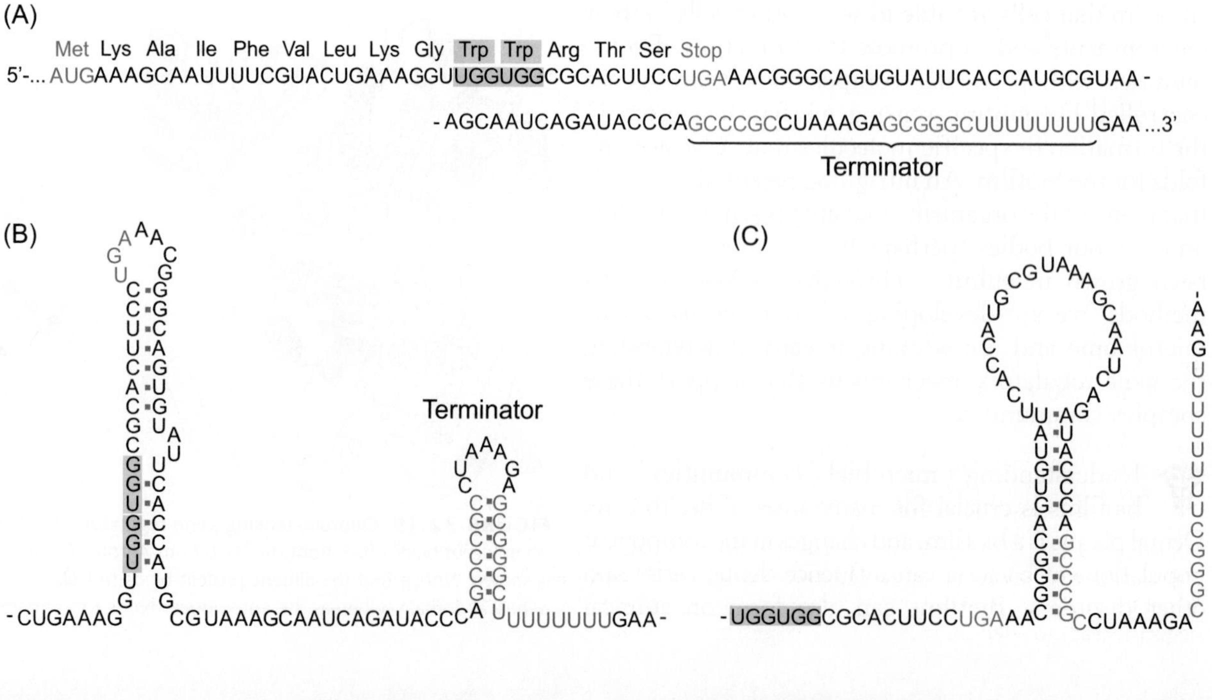
(A)
Met Lys Ala Ile Phe Val Leu Lys Gly Trp Trp Arg Thr Ser Stop
5'-... AUGAAAGCAAUUUUCGUACUGAAAGGUUGGUGGCGCACUUCCUGAAACGGGCAGUGUAUUCACCAUGCGUAA-
-AGCAAUCAGAUACCCAGCCCGCCUAAAGAGCGGGCUUUUUUUUGAA ...3'
Terminator
(B)
Terminator
-CUGAAAG
CGUAAAGCAAUCAGAUACCCA
UUUUUUUUGAA-
(C)
-UGGUGGCGCACUUCCUGAAAC
GCCUAAAGA

FIGURE 32.20 Leader region of *trp* mRNA. (A) The nucleotide sequence of the 5′ end of *trp* mRNA includes a short open reading frame that encodes a peptide comprising 14 amino acids; the leader encodes two tryptophan residues and has an untranslated attenuator region that includes a region that can form a terminator structure (red and blue). (B and C) The attenuator region can adopt either of two distinct stem-loop structures.

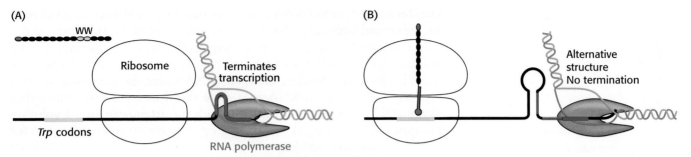

FIGURE 32.21 Attenuation. (A) In the presence of adequate concentrations of tryptophan (and, hence, Trp-tRNA), translation proceeds rapidly and an RNA structure forms that terminates transcription. (B) At low concentrations of tryptophan, translation stalls while awaiting Trp-tRNA, giving time for an alternative RNA structure to form that prevents the formation of the terminator and transcription can proceed.

leader peptide. Following the open reading frame for the peptide is the attenuator, a region of RNA that is capable of forming several alternative structures. Recall that transcription and translation are tightly coupled in bacteria. Thus, the translation of the *trp* mRNA begins soon after the ribosome-binding site has been synthesized.

How does the level of tryptophan alter transcription of the *trp* operon? An important clue was the finding that the 14-amino-acid leader peptide includes two adjacent tryptophan residues. A ribosome is able to translate the leader region of the mRNA product only in the presence of adequate concentrations of tryptophan. When enough tryptophan is present, a stem-loop structure forms in the attenuator region, which leads to the release of RNA polymerase from the DNA (Figure 32.21). However, when tryptophan is scarce, transcription is terminated less frequently. Little tryptophanyl-tRNA is present, and so the ribosome stalls at the tandem UGG codons encoding tryptophan. This delay leaves the adjacent region of the mRNA exposed as transcription continues. An alternative RNA structure that does not function as a terminator is formed, and transcription continues into and through the coding regions for the enzymes. Thus, attenuation provides an elegant means of sensing the supply of tryptophan required for protein synthesis.

Several other operons for the biosynthesis of amino acids in *E. coli* also are regulated by attenuator sites. The leader peptide of each contains an abundance of the amino acid residues of the type synthesized by the operon (Figure 32.22). For example, the leader peptide for the phenylalanine operon includes 7 phenylalanine residues among 15 residues.

(A)
Met - Lys - Arg - Ile - Ser - Thr - Thr - Ile - Thr - Thr - Thr - Ile - Thr - Ile - Thr - Thr -
5′ AUG AAA CGC AUU AGC ACC ACC AUU ACC ACC ACC AUC ACC AUU ACC ACA 3′

(B)
Met - Lys - His - Ile - Pro - Phe - Phe - Phe - Ala - Phe - Phe - Phe - Thr - Phe - Pro - Stop
5′ AUG AAA CAC AUA CCG UUU UUC UUC GCA UUC UUU UUU ACC UUC CCC UGA 3′

(C)
Met - Thr - Arg - Val - Gln - Phe - Lys - His - His - His - His - His - His - His - Pro - Asp -
5′ AUG ACA CGC GUU CAA UUU AAA CAC CAC CAU CAU CAC CAU CAU CCU GAC 3′

FIGURE 32.22 Leader peptide sequences. Amino acid sequences and the corresponding mRNA nucleotide sequences of the (A) threonine operon, (B) phenylalanine operon, and (C) histidine operon. In each case, an abundance of one amino acid in the leader peptide sequence leads to attenuation.

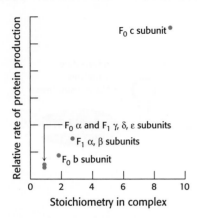

FIGURE 32.23 Translational control of protein stoichiometry. The relative rates of the subunits of *E. coli* ATP synthase produced from a single polycistronic RNA. The rates of production match the stoichiometry of the complex due to differences in rates of translation. [Information from G.-W. Li, D. Burkhardt, C. Gross, and J. S. Weissman, *Cell* 157: 624–635, 2014.]

The threonine operon encodes enzymes required for the synthesis of both threonine and isoleucine; the leader peptide contains 8 threonine and 4 isoleucine residues in a 16-residue sequence. The leader peptide for the histidine operon includes 7 histidine residues in a row. In each case, low levels of the corresponding charged tRNA cause the ribosome to stall, trapping the nascent mRNA in a state that can form a structure that allows RNA polymerase to read through the attenuator site. Evolution has apparently converged on this strategy repeatedly as a mechanism for controlling amino acid biosynthesis.

A remarkable example of posttranslational control of gene expression is provided by ATP synthase (Section 18.4) from *E. coli*. Recall that this protein machine includes eight different types of subunits with a stoichiometry of approximately 1:1:1:1:2:3:3:~10. These proteins are produced from a single polycistronic RNA, yet protein production closely matches the final stoichiometry due to differences in the rates of translation for the different genes (Figure 32.23). This illustrates the interplay between transcription and translation in determining protein expression.

The examples of prokaryotic gene regulation that we have been describing come from bacteria, as opposed to archaea. The transcriptional apparatus present in archaea shares many features with that found in eukaryotes. This commonality is frequently interpreted to suggest that eukaryotes evolved after a cell fusion event in which a bacterial cell was engulfed by an archaeal cell. Nonetheless, the key principles of gene regulation such as the occurrence of operons and the roles of DNA-binding proteins in directly blocking or stimulating RNA polymerase are the same in bacteria and archaea, perhaps because of similar genome sizes. As we shall see in the next chapter, eukaryotic cells with larger genomes and, often, many distinct cell types, utilize quite different strategies.

SUMMARY

32.1 Many DNA-Binding Proteins Recognize Specific DNA Sequences

The regulation of gene expression depends on the interplay between specific sequences within the genome and proteins that bind specifically to these sites. Specific DNA-binding proteins recognize regulatory sites that usually lie adjacent to the genes whose transcription is regulated by these proteins. The proteins of the largest family contain a helix-turn-helix motif. The first helix of this motif inserts into the major groove of DNA and makes specific hydrogen-bonding and other contacts with the edges of the base pairs.

32.2 Prokaryotic DNA-Binding Proteins Bind Specifically to Regulatory Sites in Operons

In prokaryotes, many genes are clustered into operons, which are units of coordinated genetic expression. An operon consists of control sites (an operator and a promoter) and a set of structural genes. In addition, regulator genes encode proteins that interact with the operator and promoter sites to stimulate or inhibit transcription. The treatment of *E. coli* with lactose induces an increase in the production of β-galactosidase and two additional proteins that are encoded in the lactose operon. In the absence of lactose or a similar galactoside inducer, the *lac* repressor protein binds to an operator site on the DNA and blocks transcription. The binding of allolactose, a derivative of lactose, to the *lac* repressor induces a conformational change that leads to dissociation from DNA. RNA polymerase can then move through the operator to transcribe the *lac* operon.

Some proteins activate transcription by directly contacting RNA polymerase. For example, cyclic AMP stimulates the transcription of many catabolic operons by binding to the catabolite activator protein. The binding of the cAMP–CAP complex to a specific site in the promoter region of an inducible catabolic operon enhances the binding of RNA polymerase and the initiation of transcription.

32.3 Regulatory Circuits Can Result in Switching Between Patterns of Gene Expression

The study of bacterial viruses, particularly bacteriophage λ, has revealed key aspects of gene-regulatory networks. Bacteriophage λ can develop by either a lytic or a lysogenic pathway. A key regulatory protein, the λ repressor, regulates its own expression, promoting transcription of the gene that encodes the repressor when repressor levels are low and blocking transcription of the gene when levels are high. This behavior depends on the λ right operator, which includes three sites to which λ repressor dimers can bind. Cooperative binding of the λ repressor to two of the sites stabilizes the state in which two λ repressor dimers are bound. The Cro protein binds to the same sites as does the λ repressor, but with reversed affinities. When Cro is present at sufficient concentrations, it blocks the transcription of the gene for the λ repressor while allowing the transcription of its own gene. Thus, these two proteins and the operator form a genetic switch that can exist in either of two states.

Some prokaryotic species participate in quorum sensing. This process includes the release of chemicals called autoinducers into the medium surrounding the cells. These autoinducers are often, but not always, acyl homoserine lactones. Autoinducers are taken up by surrounding cells. When the autoinducer concentration reaches an appropriate level, it is bound by receptor proteins that activate the expression of genes, including those that promote the synthesis of more autoinducer. These chemically mediated social interactions allow these prokaryotes to change their gene-expression patterns in response to the number of other cells in their environments. Biofilms are complex communities of prokaryotes that are promoted by quorum-sensing mechanisms.

32.4 Gene Expression Can Be Controlled at Posttranscriptional Levels

Gene expression can also be regulated at the level of translation. In prokaryotes, many operons important in amino acid biosynthesis are regulated by attenuation, a process that depends on the formation of alternative structures in mRNA, one of which favors the termination of transcription. Attenuation is mediated by the translation of a leader region of mRNA. A ribosome stalled by the absence of an aminoacyl-tRNA needed to translate the leader mRNA alters the structure of mRNA, allowing RNA polymerase to transcribe the operon beyond the attenuator site.

Biochemistry in Focus

Regulating gene expression through proteolysis

In Section 32.3 we discussed some aspects of how the λ repressor functions to control the life cycle of bacteriophage λ. Specifically, in the lysogenic state, the λ repressor allows the bacteriophage to remain within the bacterial genome without many of the viral genes being expressed. In this state, the viral genome is replicated with the bacterial genome without disruption. However, in times of stress on the bacterial cell, such as exposure to ionizing radiation, the virus activates the lytic pathway, causing expression of many more of the viral genes, replication of the viral DNA, and production of viral particles that are released from the lysed bacterial cell.

The elucidation of this genetic switch was an important step in the history of our understanding of gene regulation mechanisms. A key observation was that during the stress response, the λ repressor is cleaved in a specific position (between amino acid residues 111 and 112) within its sequence. Cleavage is promoted by the bacterial protein RecA, which is produced in response to stress. This cleavage dramatically decreases the affinity of the λ repressor for it specific DNA binding sites. Once the λ repressor has released the DNA, expression of the gene encoding the Cro protein is allowed and the full switch to the lytic state occurs.

Why does cleavage of the λ repressor reduce its affinity for DNA so strikingly? The λ repressor, like many of the gene-regulatory proteins that we have encountered in this chapter, binds to DNA as a dimer. Properly structured dimers facilitate tight binding to DNA because binding of the DNA with one member of the dimer positions the other member so that both monomers can interact productively with the same DNA molecule. The λ repressor consists of two domains: the amino-terminal domain, which interacts with DNA, and the carboxyl-terminal domain, which forms a stable dimer. Cleavage by RecA occurs between these two domains, making the λ repressor dimer much less stable and allowing the monomers of the DNA binding domains to release the DNA independently.

This illustrates a general mechanism for coupling other processes to changes in gene expression. Controlled proteolysis of gene regulatory proteins is a common theme, both in prokaryotes and in eukaryotes. A major challenge is understanding how activation of proteolytic processes within cells targets specific proteins without inducing widespread proteolysis. There are many variations on this theme, including pathways that involve the proteasome (Section 23.2).

KEY TERMS

helix-turn-helix motif (p. 1060)
β-galactosidase (p. 1060)
operon model (p. 1061)
operator site (p. 1061)
repressor (p. 1061)
lac repressor (p. 1061)
lac operator (p. 1062)

inducer (p. 1062)
isopropylthiogalactoside (IPTG) (p. 1062)
pur repressor (p. 1063)
corepressor (p. 1063)
catabolite repression (p. 1064)
catabolite activator protein (CAP) (p. 1064)

λ repressor (p. 1065)
Cro (p. 1066)
quorum sensing (p. 1066)
autoinducer (p. 1067)
biofilm (p. 1067)
attenuation (p. 1068)

PROBLEMS

1. *Missing genes.* Predict the effects of deleting the following regions of DNA:

(a) The gene encoding *lac* repressor
(b) The *lac* operator
(c) The gene encoding CAP ✓❶, ✓❸

2. *Minimal concentration.* Calculate the concentration of *lac* repressor, assuming that one molecule is present per cell. Assume that each *E. coli* cell has a volume of 10^{-12} cm^3. Would you expect the single molecule to be free or bound to DNA?

3. *Counting sites.* Calculate the expected number of times that a given 8-base-pair DNA site should be present in the *E. coli* genome. Assume that all four bases are equally probable. Repeat for a 10-base-pair site and a 12-base-pair site.

4. *The same but not the same.* The *lac* repressor and the *pur* repressor are homologous proteins with very similar three-dimensional structures, yet they have different effects on gene expression. Describe two important ways in which the gene-regulatory properties of these proteins differ.

5. *The opposite direction.* Some compounds called anti-inducers bind to repressors such as the *lac* repressor and inhibit the action of inducers; that is, transcription is repressed and higher concentrations of inducer are required to induce transcription. Propose a mechanism of action for anti-inducers.

6. *Inverted repeats.* Suppose that a nearly perfect 20-base-pair inverted repeat is observed in a DNA sequence. Provide two possible explanations.

7. *Broken operators.* Consider a hypothetical mutation in O_R2 that blocks both λ repressor and Cro binding. How would this mutation affect the likelihood of bacteriophage λ entering the lytic phase?

8. *Promoters.* Compare the −10 and −35 sequences for the λ repressor and Cro genes in the right operator. How many differences are there between these sequences?

9. *Positive and negative feedback.* What is the effect of an increased Cro concentration on the expression of the gene for the λ repressor? Of an increased concentration of λ repressor on the expression of the Cro gene? Of an increased concentration of λ repressor on the expression of the λ repressor gene?

10. *Leaderless.* The mRNA for the λ repressor begins with 5'-AUG-3', which encodes the methionine residue that begins the protein. What is unusual about this beginning? Would it cause the mRNA to translate efficiently or not?

11. *Quorum count.* Suppose you have a series of compounds that you wish to test for the autoinducer activity in *Vibrio fischeri.* Propose a simple assay, assuming that you can grow *V. fischeri* cultures at low cell densities.

12. *Codon utilization.* There are four codons that encode threonine. Consider the leader sequence in Figure 32.22A. What codons are used and with what frequency?

Mechanism Problem

13. *Follow the stereochemistry.* The hydrolysis of lactose is catalyzed by β-galactosidase. Does the overall reaction proceed with retention or inversion of configuration? Given that each step likely proceeds with inversion of configuration, what does the overall change in stereochemistry suggest about the mechanism? A key residue in the reaction has been identified to be Glu 537. Propose an overall mechanism for the hydrolysis of lactose.

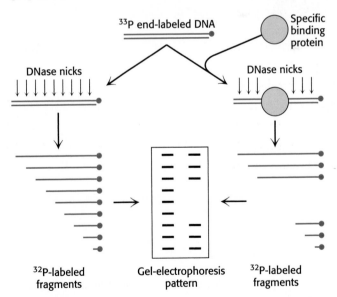

Lactose

Galactose   **Glucose**

Data Interpretation Problem

14. *Leaving tracks.* A powerful method for examining protein–DNA interactions is called DNA footprinting. In this method, a DNA fragment containing a potential binding site is radiolabeled on one end. The labeled DNA is then treated with a DNA-cleaving agent such as DNase I such that each DNA molecule within the population is cut only once. The same cleavage process is carried out in the presence of the DNA-binding protein. The bound protein protects some sites within the DNA from cleavage. The patterns of DNA fragments in the cleaved pool of DNA molecules are then examined by electrophoresis followed by autoradiography (a).

This method is applied to a DNA fragment containing a single binding site for the λ repressor in the presence of different concentrations of the λ repressor. The results are shown below:

λ repressor concentration

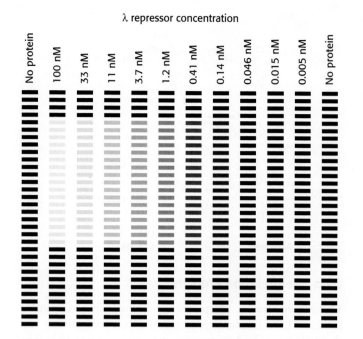

Estimate the dissociation constant for the λ repressor–DNA complex and the standard free energy of binding. ✓❹

Think ▶ Pair ▶ Share Problem

15. Gene expression involves the conversion of information present in a genomic DNA sequence into a functional protein. Give examples of the control of gene expression at different steps along this pathway.

The Control of Gene Expression in Eukaryotes

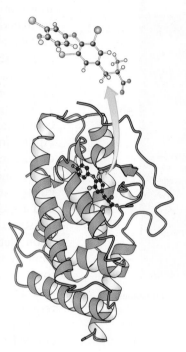

Complex biological processes often require coordinated control of the expression of many genes. The maturation of a tadpole into a frog is largely controlled by thyroid hormone. This hormone regulates gene expression by binding to a protein, the thyroid-hormone receptor, as shown at the right. In response to the hormone's binding, this protein binds to specific DNA sites in the genome and modulates the expression of nearby genes. [(Left) Gary Meszaros/Getty Images.]

 LEARNING OBJECTIVES

By the end of this chapter, you should be able to:

1. Discuss the packaging of DNA into chromatin in eukaryotes including the relationships between the DNA and histone proteins.

2. Describe some of differences between control of gene expression in prokaryotes and in eukaryotes.

3. Describe the structures of some common motifs found in eukaryotic transcription factors.

4. Discuss the role of chromatin and modification of chromatin structure in controlling eukaryotic gene expression.

5. Define epigenetics and discuss the role of modifications to DNA and histone proteins in controlling gene expression and determining cell type.

6. Discuss the family of nuclear hormone receptors and explain how ligand binding to such a receptor can result in changes in gene expression.

7. Give examples of mechanisms of posttranscriptional gene regulation.

OUTLINE

33.1 Eukaryotic DNA Is Organized into Chromatin

33.2 Transcription Factors Bind DNA and Regulate Transcription Initiation

33.3 The Control of Gene Expression Can Require Chromatin Remodeling

33.4 Eukaryotic Gene Expression Can Be Controlled at Posttranscriptional Levels

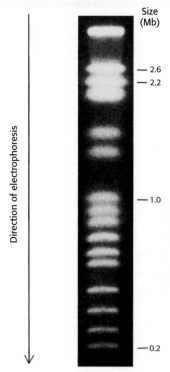

FIGURE 33.1 Yeast chromosomes. Pulsed-field electrophoresis allows the separation of 16 yeast chromosomes. [From G. Chu, D. Wollrath, D. and R.W. Davis, Separation of large DNA molecules by contour-clamped homogeneous electric fields. *Science* 234:1583–1585. 1986. Reprinted with permission from AAAS.]

Megabase

A length of DNA consisting of 10^6 base pairs (if double stranded) or 10^6 bases (if single stranded).

1 Mb = 10^3 kb = 10^6 bases

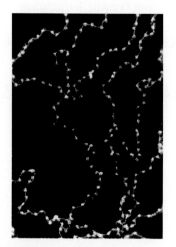

FIGURE 33.2 Chromatin structure. An electron micrograph of chromatin showing its "beads on a string" character. The beads correspond to DNA complexes with specific proteins. [Don. W. Fawcett/Science Source.]

Many of the most important and intriguing features in modern biology and medicine, such as the pathways crucial for the development of multicellular organisms, the changes that distinguish normal cells and cancer cells, and the evolutionary changes leading to new species, entail networks of gene-regulatory pathways. Gene regulation in eukaryotes is significantly more complex than in prokaryotes in several ways. First, the genomes being regulated are significantly larger. The *E. coli* genome consists of a single, circular chromosome containing 4.6 megabases (Mb). This genome encodes approximately 2000 proteins. In comparison, one of the simplest eukaryotes, *Saccharomyces cerevisiae* (baker's yeast), contains 16 chromosomes ranging in size from 0.2 to 2.2 Mb (Figure 33.1). The yeast genome totals 12 Mb and encodes approximately 6000 proteins. The genome within a human cell contains 23 pairs of chromosomes ranging in size from 50 to 250 Mb. Approximately 20,000 protein-coding genes are present within the 3000 Mb of human DNA.

Second, whereas prokaryotic genomic DNA is relatively accessible, eukaryotic DNA is packaged into chromatin, a complex between the DNA and a special set of proteins (Figure 33.2). Although the principles for the construction of chromatin are relatively simple, the chromatin structure for a complete genome is quite complex. Importantly, in a given eukaryotic cell, some genes and their associated regulatory regions are relatively accessible for transcription and regulation, whereas other genes are tightly packaged and are thus rendered inactive. Eukaryotic gene regulation frequently requires the manipulation of chromatin structure.

A manifestation of this complexity is the presence of many different *cell types* in most eukaryotes. A liver cell, a pancreatic cell, and an embryonic stem cell contain the same DNA sequences, yet the subset of genes highly expressed in cells from the pancreas, which secretes digestive enzymes, differs markedly from the subset highly expressed in the liver, the site of lipid transport and energy transduction. Embryonic stem cells do not express any subset of genes at high levels; the most highly expressed genes are "housekeeping" genes involved in the cytoskeleton and processes such as translation (Table 33.1). The existence of stable cell types is due to differences in the *epigenome*, differences in chromatin structure, and covalent modifications of the DNA, not in the DNA sequence itself. Thus, different cell types share the same genome (DNA sequence) but differ in their epigenomes, the packaging and modification of this genome.

In addition, eukaryotic genes are not generally organized into operons. Instead, genes that encode proteins for steps within a given pathway are often spread widely across the genome. This characteristic requires that other mechanisms function to regulate genes in a coordinated way.

Despite these differences, some aspects of gene regulation in eukaryotes are quite similar to those in prokaryotes. In particular, activator and repressor proteins that recognize specific DNA sequences are central to many gene-regulatory processes. In this chapter, we will focus first on chromatin structure. We will then turn to transcription factors—DNA-binding proteins similar in many ways to the prokaryotic proteins that we encountered in the preceding chapter. Key features of eukaryotic promoters, transcription factors, and enhancer sequences that can act far away from transcription start sites were introduced in Section 30.2. Eukaryotic transcription factors can act directly by interacting with the transcriptional machinery or indirectly by influencing chromatin structure. Finally, we examine selected posttranscriptional gene-regulatory

TABLE 33.1 Highly expressed protein-encoding genes of the pancreas, liver, and embryonic stem cells (as percentage of total mRNA pool)

Rank	Proteins expressed in pancreas	%	Proteins expressed in liver	%	Proteins expressed in stem cells	%
1	Procarboxypeptidase A1	7.6	Albumin	3.5	Glyceraldehyde-3-phosphate dehydrogenase	0.7
2	Pancreatic trypsinogen 2	5.5	Apolipoprotein A-I	2.8	Translation elongation factor 1 α1	0.6
3	Chymotrypsinogen	4.4	Apolipoprotein C-I	2.5	Tubulin α	0.5
4	Pancreatic trypsin 1	3.7	Apolipoprotein C-III	2.1	Translationally controlled tumor protein	0.5
5	Elastase IIIB	2.4	ATPase 6/8	1.5	Cyclophilin A	0.4
6	Protease E	1.9	Cytochrome oxidase 3	1.1	Cofilin	0.4
7	Pancreatic lipase	1.9	Cytochrome oxidase 2	1.1	Nucleophosmin	0.3
8	Procarboxypeptidase B	1.7	α_1-Antitrypsin	1.0	Connexin 43	0.3
9	Pancreatic amylase	1.7	Cytochrome oxidase 1	0.9	Phosphoglycerate mutase	0.2
10	Bile-salt-stimulated lipase	1.4	Apolipoprotein E	0.9	Translation elongation factor 1 β2	0.2

Sources: Data for pancreas from V. E. Velculescu, L. Zhang, B. Vogelstein, and K. W. Kinzler, *Science* 270:484–487, 1995. Data for liver from T. Yamashita, S. Hashimoto, S. Kaneko, S. Nagai, N. Toyoda, T. Suzuki, K. Kobayashi, and K. Matsushima, *Biochem. Biophys. Res. Commun.* 269:110–116, 2000. Data for stem cells from M. Richards, S. P. Tan, J. H. Tan, W. K. Chan, and A. Bongso, *Stem Cells* 22:51–64, 2004.

mechanisms, including those based on microRNAs, an important class of gene-regulatory molecules discovered in recent years.

33.1 Eukaryotic DNA Is Organized into Chromatin

Eukaryotic DNA is tightly bound to a group of small basic proteins called *histones*. In fact, histones constitute half the mass of a eukaryotic chromosome. The entire complex of a cell's DNA and associated protein is called *chromatin*. Chromatin compacts and organizes eukaryotic DNA and its presence has dramatic consequences for gene regulation.

Nucleosomes are complexes of DNA and histones

Chromatin is made up of repeating units, each containing 200 bp of DNA and two copies each of four histone proteins H2A, H2B, H3, and H4. Histones have strikingly basic properties because a quarter of the residues in each histone are either arginine or lysine, positively charged amino acids that strongly interact with the negatively charged DNA. The protein complex is called the *histone octamer*. The repeating units of the histone octamer and the associated DNA are known as *nucleosomes*. Chromatin viewed with the electron microscope has the appearance of beads on a string (Figure 33.2); each bead has a diameter of approximately 100 Å. Partial digestion of chromatin with DNase yields the isolated beads. These particles consist of fragments of DNA about 200 bp in length bound to the histone octamer. More-extensive digestion yields a shorter DNA fragment of 145 bp bound to the octamer. The smaller complex formed by the histone octamer and the 145-bp DNA fragment is the *nucleosome core particle*. The DNA connecting core particles in undigested chromatin is called linker DNA. Histone H1 binds, in part, to the linker DNA.

DNA wraps around histone octamers to form nucleosomes

The overall structure of the nucleosome was revealed through electron microscopic and x-ray crystallographic studies pioneered by Aaron Klug and his colleagues. More recently, the three-dimensional structures of reconstituted nucleosome core particles have been determined to higher resolution by x-ray diffraction methods (Figure 33.3). The DNA "string" actually wraps around the histone protein "beads." The four types of histone that make up the protein core are homologous and similar in structure (Figure 33.4). The eight histones in the core are arranged into a $(H3)_2(H4)_2$ tetramer and a pair of H2A–H2B dimers. The tetramer and dimers come together to form a left-handed superhelical ramp around which the DNA wraps. In addition, each histone has an amino-terminal tail that extends out from the core structure. These tails are flexible and contain many lysine and arginine residues. As we shall see, *covalent modifications of these tails play an essential role in regulating gene expression.*

The DNA forms a left-handed superhelix as it wraps around the outside of the histone octamer. The protein core forms contacts with the inner surface of the DNA superhelix at many points, particularly along the phosphodiester backbone and the minor groove. Nucleosomes will form on almost all DNA sites, although some sequences are preferred because the dinucleotide steps are properly spaced to favor bending around the histone core. A histone with a different structure from that of the others, called histone H1, seals off the nucleosome at the location at which the linker DNA enters and leaves. The amino acid sequences of histones, including their amino-terminal tails, are remarkably conserved from yeast to human beings.

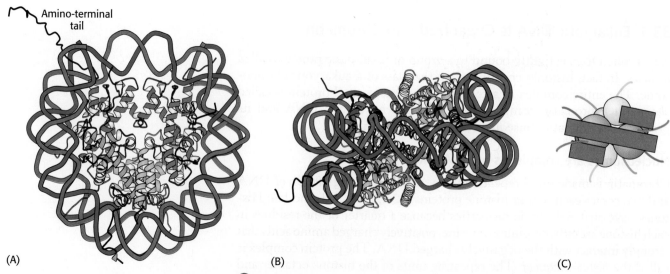

Amino-terminal tail

(A)   (B)   (C)

FIGURE 33.3 Nucleosome core particle. The structure consists of a core of eight histone proteins surrounded by DNA. (A) A view showing the DNA wrapping around the histone core. (B) A 90-degree rotation of the view in part (A). *Notice* that the DNA forms a left-handed superhelix as it wraps around the core. (C) A schematic view. [Drawn from 1AOI.pdb.]

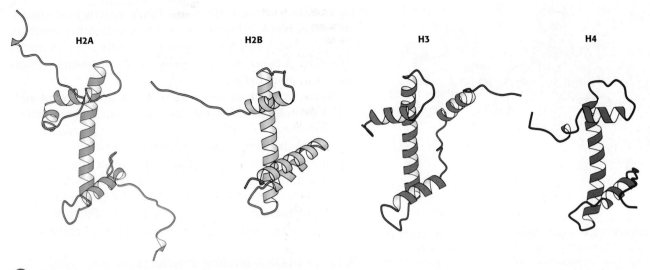

H2A   **H2B**   **H3**   **H4**

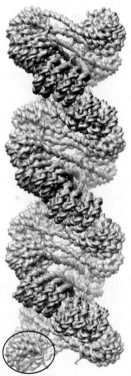

FIGURE 33.4 **Homologous histones.** Histones H2A, H2B, H3, and H4 adopt a similar three-dimensional structure as a consequence of common ancestry. Some parts of the tails at the termini of the proteins are not shown. [Drawn from 1AOI.pdb.]

The winding of DNA around the nucleosome core contributes to the packing of DNA by decreasing its linear extent. An extended 200-bp stretch of DNA would have a length of about 680 Å. Wrapping this DNA around the histone octamer reduces the length to approximately 100 Å along the long dimension of the nucleosome. Thus, the DNA is compacted by a factor of 7. However, human chromosomes in metaphase, which are highly condensed, are compacted by a factor of 10^4. Clearly, the nucleosome is just the first step in DNA compaction. What is the next step? The nucleosomes themselves can stack in helical arrays approximately 300 Å across (Figure 33.5). The folding of these fibers of nucleosomes into loops further compacts DNA.

The wrapping of DNA around the histone octamer as a left-handed helix also stores negative supercoils; if the DNA in a nucleosome is straightened out, the DNA will be underwound. This underwinding is exactly what is needed to separate the two DNA strands during replication and transcription.

33.2 Transcription Factors Bind DNA and Regulate Transcription Initiation

DNA-binding *transcription factors* are key to gene regulation in eukaryotes, just as they are in prokaryotes. However, the roles of eukaryotic transcription factors are different in several ways. First, whereas the DNA-binding sites crucial for the control of gene expression in prokaryotes are usually quite close to promoters, those in eukaryotes can be farther away from promoters and can exert their action at a distance. Second, most prokaryotic genes are regulated by single transcription factors, and multiple genes in a pathway are expressed in a coordinated fashion because such genes are often transcribed as part of a polycistronic mRNA. In eukaryotes, the expression of each gene is typically controlled by multiple transcription factors, and the coordinated expression of different genes depends on having similar transcription-factor-binding sites in each gene in the set. Third, in prokaryotes, transcription factors usually interact directly with RNA polymerase. In eukaryotes, while some transcription

Nucleosome

FIGURE 33.5 **Higher-order chromatin structure.** The structure of a helical array of nucleosomes determined by cryo-electron microscopy revealing a mode of chromatin condensation. The structures in cells are more heterogeneous and dynamic. [From Feng Song et al., Cryo-EM Study of the Chromatin Fiber Reveals a Double Helix Twisted by Tetranucleosomal Units. *Science* 344, 376–380. 2014. Reprinted with permission from AAAS.]

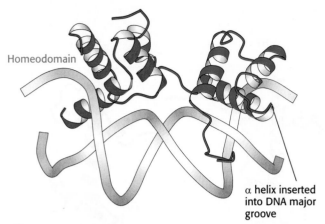

Homeodomain

α helix inserted into DNA major groove

FIGURE 33.6 Homeodomain structure. The structure of a heterodimer formed from two different DNA-binding domains, each based on a homeodomain. *Notice* that each homeodomain has a helix-turn-helix motif with one helix inserted into the major groove of DNA. [Drawn from 1AKH.pdb.]

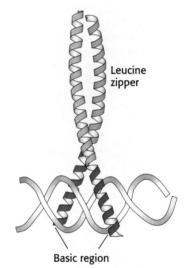

Leucine zipper

Basic region

(📖) **FIGURE 33.7 Basic-leucine zipper.** This heterodimer comprises two basic-leucine zipper proteins. *Notice* that the basic region lies in the major groove of DNA. The leucine-zipper part stabilizes the protein dimer. [Drawn from 1FOS.pdb.]

factors interact directly with RNA polymerase, many others act more indirectly, interacting with other proteins associated with RNA polymerase or modifying chromatin structure. Let us now examine eukaryotic transcription factors in more detail.

Eukaryotic transcription factors usually consist of several domains. The *DNA-binding domain* binds to regulatory sequences that can either be adjacent to the promoter or at some distance from it. Most commonly, transcription factors include additional domains that help activate transcription. When a transcription factor is bound to the DNA, its *activation domain* promotes transcription by interacting with RNA polymerase II, by interacting with other associated proteins, or by modifying the local structure of chromatin.

A range of DNA-binding structures are employed by eukaryotic DNA-binding proteins

The structures of many eukaryotic DNA-binding proteins have been determined and a range of structural motifs have been observed, but we will focus on three that reveal the common features and the diversity of these motifs. The first class of eukaryotic DNA-binding unit that we will consider is the *homeodomain* (Figure 33.6). The structure of the homeodomain and its mode of recognition of DNA are very similar to those of the prokaryotic helix-turn-helix proteins. In eukaryotes, homeodomain proteins often form heterodimeric structures, sometimes with other homeodomain proteins, that recognize asymmetric DNA sequences.

The second class of eukaryotic DNA-binding unit comprises the *basic-leucine zipper (bZip) proteins* (Figure 33.7). This DNA-binding unit consists of a pair of long α helices. The first part of each α helix is a basic region that lies in the major groove of the DNA and makes contacts responsible for DNA-site recognition. The second part of each α helix forms a coiled-coil structure with its partner. Because these units are often stabilized by appropriately spaced leucine residues, these structures are often referred to as *leucine zippers*.

The final class of eukaryotic DNA-binding units that we will consider here are the Cys_2His_2 *zinc-finger domains* (Figure 33.8). A DNA-binding unit of this class comprises tandem sets of small domains, each of which binds a zinc ion through conserved sets of two cysteine and two histidine

(📖) **FIGURE 33.8 Zinc-finger domains.** A DNA-binding domain comprising three Cys_2His_2 zinc-finger domains (shown in yellow, blue, and red) is shown in a complex with DNA. Each zinc-finger domain is stabilized by a bound zinc ion (shown in green) through interactions with two cysteine residues and two histidine residues. *Notice* how the protein wraps around the DNA in the major groove. [Drawn from 1AAY.pdb.]

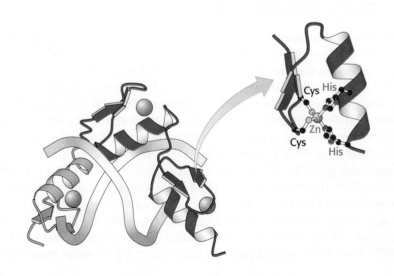

Cys His
Zn
Cys
His

residues. These structures, often called *zinc-finger domains*, form a string that follows the major groove of DNA. An α helix from each domain makes specific contact with the edges of base pairs within the groove. Some proteins contain arrays of 10 or more zinc-finger domains, potentially enabling them to contact long stretches of DNA. The human genome encodes several hundred proteins that contain zinc-finger domains of this class. We will encounter another class of zinc-based DNA-binding domain when we consider nuclear hormone receptors in Section 33.3.

Activation domains interact with other proteins

The activation domains of transcription factors generally recruit other proteins that promote transcription. In some cases, these activation domains interact directly with RNA polymerase II or closely associated proteins. The activation domains act through intermediary proteins that bridge between the transcription factors and the polymerase. An important target of activators is *mediator*, a protein complex of 25 to 30 subunits conserved from yeast to human beings that acts as a bridge between transcription factors and promoter-bound RNA polymerase II (Figure 33.9). Studies using cryo-electron microscopy are revealing the details of mediator structure and its interactions with RNA polymerase II and transcription factors. These structures provide insights into how mediator facilitates the phosphorylation of the carboxyl-terminal domain of RNA polymerase to promote the transition from transcription initiation to elongation.

Activation domains are less conserved than DNA-binding domains. In fact, very little sequence similarity has been found. For example, they may be acidic, hydrophobic, glutamine rich, or proline rich. However, they have certain features in common. First, they are often *redundant;* that is, a part of the activation domain can be deleted without loss of function. Second, they are *modular* and can activate transcription when paired with a variety of DNA-binding domains. Third, they can act *synergistically:* two activation domains acting together create a stronger effect than either acting separately.

We have been considering the case in which gene control increases the expression level of a gene. In many cases, the expression of a gene must be decreased by blocking transcription. The agents in such cases are *transcriptional repressors*. Like activators, transcriptional repressors act in many cases by altering chromatin structure.

Multiple transcription factors interact with eukaryotic regulatory regions

The basal transcription complex described in Chapter 30 initiates transcription at a low frequency. Recall that several general transcription factors join with RNA polymerase II to form the basal transcription complex. Additional transcription factors must bind to other sites that can be near the promoter or quite distant for a gene to achieve a higher rate of mRNA synthesis. In contrast with the regulators of prokaryotic transcription, few eukaryotic transcription factors have any effect on their own. Instead, each factor recruits other proteins to build up large complexes that interact with the transcriptional machinery to activate transcription.

A major advantage of this mode of regulation is that a given regulatory protein can have different effects, depending on what other proteins are present in the same cell. This phenomenon, called *combinatorial control*, is crucial to multicellular organisms that have many different cell types. Even in unicellular eukaryotes such as yeast, combinatorial control allows the generation of distinct cell types.

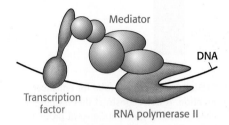

FIGURE 33.9 Mediator. Mediator, a large complex of protein subunits, acts as a bridge between transcription factors bearing activation domains and RNA polymerase II. These interactions help recruit and stabilize RNA polymerase II near specific genes that are then transcribed.

Gene encoding
green fluorescent
protein

4 kilobases
upstream of
Islet-1 including
enhancer

FIGURE 33.10 A probe for enhancer function. A DNA construct that includes a gene encoding green fluorescent protein adjacent to 4 kilobases of the DNA upstream of the start site for the protein Islet-1. If this construct is introduced into cells in which Islet-1 is expressed, then the enhancer and promoter should lead to the production of green fluorescent protein.

Enhancers can stimulate transcription in specific cell types

Transcription factors can often act even if their binding sites lie at a considerable distance from the promoter. These distant regulatory sites are called *enhancers* (Chapter 30). Enhancers function by serving as binding sites for specific transcription factors. An enhancer is effective only in the specific cell types in which appropriate regulatory proteins are expressed. In many cases, these DNA-binding proteins influence transcription initiation by perturbing the local chromatin structure to expose a gene or its regulatory sites rather than by direct interactions with RNA polymerase. This mechanism accounts for the ability of enhancers to act at a distance.

The properties of enhancers are illustrated by studies of the enhancer controlling the gene Islet-1. This gene encodes a homeodomain-containing protein that plays important roles in the developing nervous system. DNA fragments were constructed containing the promoter and enhancer regions associated with the Islet-1 gene connected to a gene expressing green fluorescent protein (GFP; Figure 33.10). When this DNA construct was introduced into zebrafish, an organism useful for imaging studies because of its size and transparency, the expression of green fluorescent protein could be visualized by microscopy. The green fluorescence of GFP was only seen in motor neurons and only during one stage of zebrafish development (Figure 33.11), revealing the power of the enhancer in limiting gene expression to particular cells.

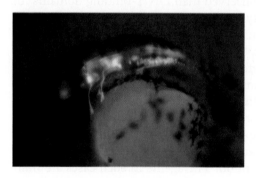

FIGURE 33.11 An experimental demonstration of enhancer function. A microscopic image of a developing zebrafish embryo into which the probe for enhancer function (Figure 33.10) has been introduced. The green fluorescence reveals that green fluorescent protein is expressed only in a subset of cells, shown to be cranial motor neurons. [From Higashijima et al., Visualization of Cranial Motor Neurons in Live Zebrafish. *J. Neurosci.* 20, 206–218. 2000. © Society for Neuroscience.]

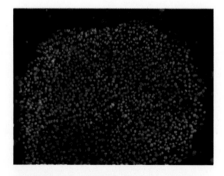

FIGURE 33.12 Induced pluripotent stem cells. A micrograph of human-induced pluripotent stem cells stained green for a transcription factor that is characteristic of pluripotent cells. [Reprinted from *Cell*, K. Takahashi, K. Tanabe, M. Ohnuki, M. Narita, T. Ichisaka, K. Tomoda, and S. Yamanaka, Induction of Pluripotent Stem Cells from Adult Human Fibroblasts by Defined Factors, Vol. 131, 861–872, 2007, Figure 1, part N, with permission from Elsevier. Courtesy Prof. Shinya Yamanaka, Kyoto University.]

Induced pluripotent stem cells can be generated by introducing four transcription factors into differentiated cells

An important application illustrating the power of transcription factors is the development of *induced pluripotent stem* (iPS) *cells*. Pluripotent stem cells have the ability to differentiate into many different cell types on appropriate treatment. Previously isolated cells derived from embryos show a very high degree of pluripotency. Over time, researchers identified dozens of genes in embryonic stem cells that contributed to this pluripotency when expressed. In a remarkable experiment reported for mouse cells in 2006 and human cells in 2007, Shinya Yamanaka demonstrated that just four genes out of this entire set could induce pluripotency in already-differentiated skin cells. Yamanaka introduced genes encoding four transcription factors into skin cells called fibroblasts. The fibroblasts de-differentiated into cells that appeared to have characteristics very nearly identical with those of embryonic stem cells (Figure 33.12).

These iPS cells represent powerful research tools and, potentially, a new class of therapeutic agents. The proposed concept is that a sample of a patient's fibroblasts could be readily isolated and converted into iPS cells.

These iPS cells could be treated to differentiate into a desired cell type that could then be transplanted into the patient. For example, such an approach might be used to restore a particular class of nerve cells that had been depleted by a neurodegenerative disease. Although the field of iPS cell research is still evolving, it holds great promise as a possible approach to treat many common and difficult-to-treat diseases.

33.3 The Control of Gene Expression Can Require Chromatin Remodeling

Early observations suggested that chromatin structure plays a major role in controlling eukaryotic gene expression. DNA that is densely packaged into chromatin is less susceptible to cleavage by the nonspecific DNA-cleaving enzyme DNase I. Regions adjacent to genes that are being transcribed are more sensitive to cleavage by DNase I than are other sites in the genome, suggesting that the DNA in these regions is less compacted than it is elsewhere and more accessible to proteins. In addition, some sites, usually within 1 kb of the start site of an active gene, are exquisitely sensitive to DNase I and other nucleases. These *hypersensitive sites* correspond to regions that have few nucleosomes or contain nucleosomes in an altered conformation. *Hypersensitive sites are cell-type specific and developmentally regulated.* For example, globin genes in the precursors of erythroid cells from 20-hour-old chicken embryos are insensitive to DNase I. However, when hemoglobin synthesis begins at 35 hours, regions adjacent to these genes become highly susceptible to digestion. In tissues such as the brain that produce no hemoglobin, the globin genes remain resistant to DNase I throughout development and into adulthood. These studies suggest that a prerequisite for gene expression is a relaxing of the chromatin structure.

Recent experiments even more clearly revealed the role of chromatin structure in regulating access to DNA-binding sites. Genes required for galactose utilization in yeast are activated by a transcription factor called GAL4, which recognizes DNA-binding sites with two 5′-CGG-3′ sequences on complementary strands separated by 11 base pairs (Figure 33.13). While thousands of GAL4 binding sites of the form 5′-CGG(N)$_{11}$CCG-3′ are present in the yeast genome, only a handful of them regulate genes necessary for galactose metabolism. How is GAL4 targeted to this small fraction of potential binding sites? This question is addressed through the use of a technique called *chromatin immunoprecipitation* (ChIP; Figure 33.14). GAL4 is first cross-linked to its DNA-binding

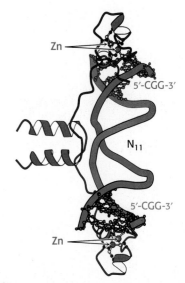

FIGURE 33.13 GAL4 binding sites. The yeast transcription factor GAL4 binds to DNA sequences of the form 5′-CGG(N)$_{11}$CCG-3′. Two zinc-based domains are present in the DNA-binding region of this protein. *Notice* that these domains contact the 5′-CGG-3′ sequences, leaving the center of the site uncontacted. [Drawn from 1D66.pdb.]

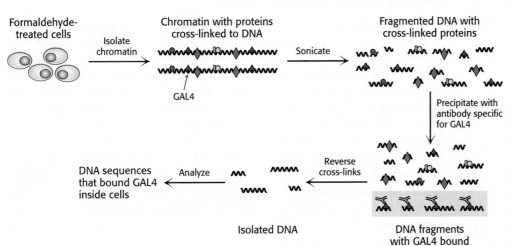

FIGURE 33.14 Chromatin immunoprecipitation. Cells or isolated nuclei are treated with formaldehyde to cross-link proteins to DNA. The cells are then lysed and the DNA is fragmented by sonication. DNA fragments bound to a particular protein are isolated through the use of an antibody specific for that protein. The cross-links are then reversed and the DNA fragments are characterized.

sites in chromatin. The DNA is then fragmented into small pieces, and antibodies to GAL4 are used to isolate the chromatin fragments containing GAL4. The cross-linking is reversed, and the DNA is isolated and characterized. The results of these studies reveal that approximately 10 of the 4000 potential GAL4 sites are occupied by GAL4 when the cells are growing on galactose, while more than 99% of the sites appear to be blocked, presumably by the local chromatin structure. Thus, whereas in prokaryotes all sites appear to be equally accessible, chromatin structure shields a large number of the potential binding sites in eukaryotic cells. GAL4 is thereby prevented from binding to sites that are unimportant in galactose metabolism. These lines of evidence and others reveal that chromatin structure is altered in active genes compared with inactive ones.

The methylation of DNA can alter patterns of gene expression

The degree of methylation of DNA provides another mechanism, in addition to packaging with histones, for inhibiting gene expression inappropriate to a specific cell type. Carbon 5 of cytosine can be methylated by specific methyltransferases. About 70% of the 5′-CpG-3′ sequences (where "p" represents the phosphate residue in the DNA backbone) in mammalian genomes are methylated. However, the distribution of these methylated cytosines varies, depending on the cell type. Consider the β-globin gene. In cells that are actively expressing hemoglobin, the region from approximately 1 kb upstream of the start site to approximately 100 bp downstream of the start site is less methylated than the corresponding region in cells that do not express this gene. The relative absence of 5-methylcytosines near the start site is referred to as *hypomethylation*. The methyl group of 5-methylcytosine protrudes into the major groove where it could easily interfere with the binding of proteins that stimulate transcription.

The distribution of CpG sequences in mammalian genomes is not uniform. Many CpG sequences have been converted into TpG through mutation by the deamination of 5-methylcytosine to thymine. However, sites near the 5′ ends of genes have been maintained because of their role in gene expression. Thus, most genes are found in *CpG islands*, regions of the genome that contain approximately four times as many CpG sequences as does the remainder of the genome.

Steroids and related hydrophobic molecules pass through membranes and bind to DNA-binding receptors

We next look at an example that illustrates how transcription factors can stimulate changes in chromatin structure that affect transcription. We will consider in some detail the system that detects and responds to estrogens. Synthesized and released by the ovaries, *estrogens,* such as estradiol, are cholesterol-derived steroid hormones (Section 26.4). They are required for the development of female secondary sex characteristics and, along with progesterone, participate in the ovarian cycle.

Because they are hydrophobic molecules, estrogens easily diffuse across cell membranes. When inside a cell, estrogens bind to highly specific, soluble receptor proteins. Estrogen receptors are members of a large family of proteins that act as receptors for a wide range of hydrophobic molecules, including other steroid hormones, thyroid hormones, and retinoids.

NH$_2$

CH$_3$

O

H

deoxyribose

5-Methylcytosine

CH$_3$ OH

HO

Estradiol
(an estrogen)

All-*trans*-retinoic acid
(a retinoid)

Thyroxine
(L-3,5,3′,5′-Tetraiodothyronine)
(a thyroid hormone)

The human genome encodes approximately 50 members of this family, often referred to as *nuclear hormone receptors*. The genomes of other multicellular eukaryotes encode similar numbers of nuclear hormone receptors, although they are absent in yeast.

All these receptors have a similar mode of action. On binding of the signal molecule (called, generically, a *ligand*), the ligand–receptor complex modifies the expression of specific genes by binding to control elements in the DNA. Estrogen receptors bind to specific DNA sites (referred to as *estrogen response elements* or EREs) that contain the consensus sequence 5′-**AGGTCAN-NNTGACCT**-3′. As expected from the symmetry of this sequence, an estrogen receptor binds to such sites as a dimer.

A comparison of the amino acid sequences of members of this family reveals two highly conserved domains: a DNA-binding domain and a ligand-binding domain (Figure 33.15). The DNA-binding domain lies toward the center of the molecule and consists of a set of zinc-based domains different from the Cys_2His_2 zinc-finger proteins introduced in Section 33.2. These zinc-based domains bind to specific DNA sequences by virtue of an α helix that lies in the major groove in the specific DNA complexes formed by estrogen receptors.

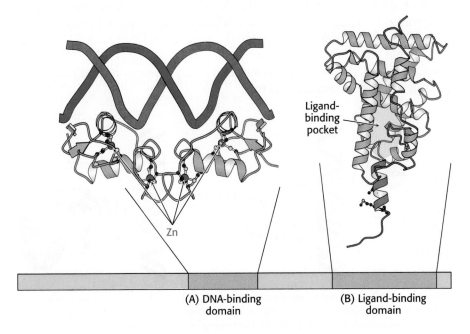

FIGURE 33.15 Structure of two nuclear hormone-receptor domains. Nuclear hormone receptors contain two crucial conserved domains: (A) a DNA-binding domain toward the center of the sequence and (B) a ligand-binding domain toward the carboxyl terminus. The structure of a dimer of the DNA-binding domain bound to DNA is shown, as is one monomer of the normally dimeric ligand-binding domain. [Drawn from 1HCQ and 1LBD.pdb.]

Ligand-binding pocket

Zn

(A) DNA-binding domain

(B) Ligand-binding domain

Nuclear hormone receptors regulate transcription by recruiting coactivators to the transcription complex

The second highly conserved domain of the nuclear receptor proteins lies near the carboxyl terminus and is the ligand-binding site. This domain

FIGURE 33.16 Ligand binding to nuclear hormone receptor. The ligand lies completely surrounded within a pocket in the ligand-binding domain. *Notice* that the last α helix, helix 12 (shown in purple), folds into a groove on the side of the structure on ligand binding. [Drawn from 1LDB and 1ERE.pdb.]

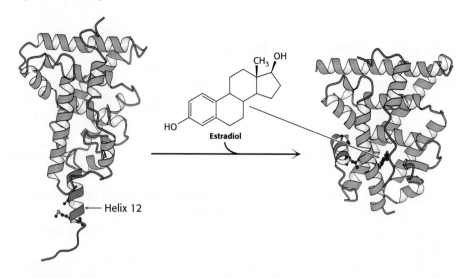

folds into a structure that consists almost entirely of α helices, arranged in three layers. The ligand binds in a hydrophobic pocket that lies in the center of this array of helices (Figure 33.16). This domain changes conformation when it binds its ligand, estrogen. How does ligand binding lead to changes in gene expression? The simplest model would have the binding of ligand alter the DNA-binding properties of the receptor, analogously to the *lac* repressor in prokaryotes. However, experiments with purified nuclear hormone receptors revealed that ligand binding does *not* significantly alter DNA-binding affinity and specificity. Another mechanism is operative.

Investigators sought to determine whether specific proteins might bind to the nuclear hormone receptors only in the presence of ligand. Such searches led to the identification of several related proteins called *coactivators*, such as SRC-1 (*s*teroid *r*eceptor *c*oactivator-1), GRIP-1 (*g*lucocorticoid *r*eceptor *i*nteracting *p*rotein-1), and NcoA-1 (*n*uclear hormone receptor *co*activator-1). These coactivators are referred to as the p160 family because of their size. The binding of ligand to the receptor induces a conformational change that allows the recruitment of a coactivator (Figure 33.17). In many cases, these coactivators are enzymes that catalyze reactions that lead to the modification of chromatin structure.

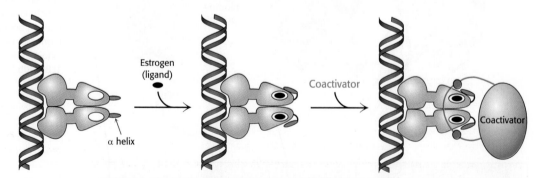

FIGURE 33.17 Coactivator recruitment. The binding of ligand to a nuclear hormone receptor induces a conformational change in the ligand-binding domain. This change in conformation generates favorable sites for the binding of a coactivator.

Steroid-hormone receptors are targets for drugs

Molecules such as estradiol that bind to a receptor and trigger signaling pathways are called *agonists*. Athletes sometimes take natural and synthetic agonists of the androgen receptor, a member of the family of nuclear hormone receptors, because their binding to the androgen receptor stimulates the expression of genes that enhance the development of lean muscle mass.

Androstendione
(a natural androgen)

Dianabol
(methandrostenolone)
(a synthetic androgen)

Referred to as *anabolic steroids*, such compounds used in excess are not without side effects. In men, excessive use leads to a decrease in the secretion of testosterone, to testicular atrophy, and sometimes to breast enlargement (gynecomastia) if some of the excess androgen is converted into estrogen. In women, excess testosterone causes a decrease in ovulation and estrogen secretion; it also causes breast regression and growth of facial hair.

Other molecules bind to nuclear hormone receptors but do not effectively trigger signaling pathways. Such compounds are called *antagonists* and are, in many ways, like competitive inhibitors of enzymes. Some important drugs are antagonists that target the estrogen receptor. For example, *tamoxifen* and *raloxifene* are used in the treatment and prevention of breast cancer, because some breast tumors rely on estrogen-mediated pathways for growth. Because some of these compounds have distinct effects on different forms of the estrogen receptor, they are referred to as *selective estrogen receptor modulators* (SERMs).

Tamoxifen

Raloxifene

The determination of the structures of complexes between the estrogen receptor and these drugs revealed the basis for their antagonist effect (Figure 33.18). Tamoxifen binds to the same site as estradiol does. However, tamoxifen has a group that extends out of the normal ligand-binding pocket, as do other antagonists. These groups block the normal conformational changes induced by estrogen. Tamoxifen blocks the binding of coactivators and thus inhibits the activation of gene expression.

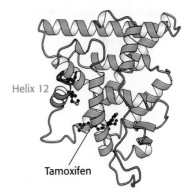

FIGURE 33.18 Estrogen receptor–tamoxifen complex. Tamoxifen binds in the pocket normally occupied by estrogen. However, *notice* that part of the tamoxifen structure extends from this pocket, and so helix 12 cannot pack in its usual position. Instead, this helix blocks the coactivator-binding site. [Drawn from 3ERT.pdb.]

Chromatin structure is modulated through covalent modifications of histone tails

We have seen that nuclear receptors respond to signal molecules by recruiting coactivators. Now we can ask how coactivators modulate transcriptional activity. *These proteins act to loosen the histone complex from the DNA, exposing additional DNA regions to the transcription machinery.*

Much of the effectiveness of coactivators appears to result from their ability to covalently modify the amino-terminal tails of histones as well as regions on other proteins. Some of the p160 coactivators and the proteins that they recruit catalyze the transfer of acetyl groups from acetyl CoA to specific lysine residues in these amino-terminal tails.

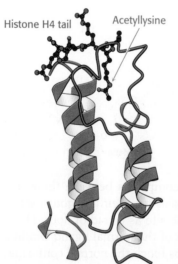

Lysine in histone tail   **Acetyl CoA**

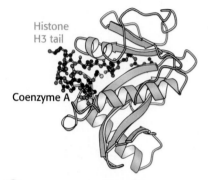

FIGURE 33.19 Structure of histone acetyltransferase. The amino-terminal tail of histone H3 extends into a pocket in which a lysine side chain can accept an acetyl group from acetyl Coenzyme A bound in an adjacent site. [Drawn from 1QSN.pdb.]

Enzymes that catalyze such reactions are called *histone acetyltransferases* (HATs). The histone tails are readily extended so they can fit into the HAT active site and become acetylated (Figure 33.19).

What are the consequences of histone acetylation? Lysine bears a positively charged ammonium group at neutral pH. The addition of an acetyl group generates an uncharged amide group. This change dramatically reduces the affinity of the tail for DNA and modestly decreases the affinity of the entire histone complex for DNA, loosening the histone complex from the DNA.

In addition, the acetylated lysine residues interact with a specific *acetyllysine-binding domain* that is present in many proteins that regulate eukaryotic transcription. This domain, termed a *bromodomain*, comprises approximately 110 amino acids that form a four-helix bundle containing a peptide-binding site at one end (Figure 33.20).

Histone H4 tail   Acetyllysine

FIGURE 33.20 Structure of a bromodomain. This four-helix-bundle domain binds peptides containing acetyllysine. *Notice* that an acetylated peptide from histone H4 is bound in the structure. [Drawn from 1EGI.pdb.]

Bromodomain-containing proteins are components of two large complexes essential for transcription. One is a complex of more than 10 polypeptides that binds to the *TATA-box-binding protein*. Recall that the TATA-box-binding protein is an essential transcription factor for many genes (Section 30.2). Proteins that bind to the TATA-box-binding protein are called *TAFs* (for *TATA-box-binding protein associated factors*). In particular, TAF1 contains a pair of bromodomains near its carboxyl terminus. The two domains are oriented such that each can bind one of two acetyllysine residues at positions 5 and 12 in the histone H4 tail. Thus, *acetylation of the histone tails provides a mechanism for recruiting other components of the transcriptional machinery.*

Bromodomains are also present in some components of large complexes known as *chromatin-remodeling complexes* or *chromatin-remodeling engines.* These complexes, which also contain domains homologous to those of helicases, utilize the free energy of ATP hydrolysis to shift the positions of nucleosomes along the DNA and to induce other conformational changes in chromatin (Figure 33.21). Histone acetylation can lead to a reorganization of the chromatin structure, potentially exposing binding sites for other factors. *Thus, histone acetylation can activate transcription through a combination of three mechanisms: by reducing the affinity of the histones for DNA, by recruiting other components of the transcriptional machinery, and by initiating the remodeling of the chromatin structure.*

FIGURE 33.21 Chromatin remodeling. Eukaryotic gene regulation begins with an activated transcription factor bound to a specific site on DNA. One scheme for the initiation of transcription by RNA polymerase II requires five steps: (1) recruitment of a coactivator, (2) acetylation of lysine residues in the histone tails, (3) binding of a remodeling-engine complex to the acetylated lysine residues, (4) ATP-dependent remodeling of the chromatin structure to expose a binding site for RNA polymerase II or for other factors, and (5) recruitment of RNA polymerase II. Only two subunits are shown for each complex, although the actual complexes are much larger. Other schemes are possible.

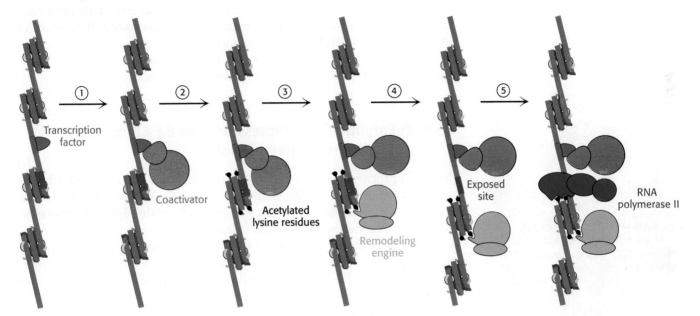

Nuclear hormone receptors also include regions that interact with components of the mediator complex. Thus, two mechanisms of gene regulation can work in concert. Modification of histones and chromatin remodeling can open up regions of chromatin into which the transcription complex can be recruited through protein–protein interactions.

Transcriptional repression can be achieved through histone deacetylation and other modifications

Just as in prokaryotes, some changes in a cell's environment lead to the repression of genes that had been active. The modification of histone tails again plays an important role. However, in repression, a key reaction appears to be the deacetylation of acetylated lysine, catalyzed by specific *histone deacetylase* enzymes.

In many ways, the acetylation and deacetylation of lysine residues in histone tails (and, likely, in other proteins) is analogous to the phosphorylation and dephosphorylation of serine, threonine, and tyrosine residues in other stages of signaling processes. Like the addition of phosphoryl groups, the addition of acetyl groups can induce conformational changes and generate novel binding sites. Without a means of removal of these groups, however,

TABLE 33.2 Selected histone modifications

Modification	Associated with
H3 K9 acetylation	Activation
H3 K27 acetylation	Activation
H4 K16 acetylation	Activation
H3 K4 monomethylation	Activation
H3 K9 trimethylation	Repression
H3 K27 trimethylation	Repression
H3 R17 methylation	Activation
H2B S14 phosphorylation	DNA repair
H2B K120 ubiquitination	Activation

these signaling switches will become stuck in one position and lose their effectiveness. Like phosphatases, deacetylases help reset the switches.

Acetylation is not the only modification of histones and other proteins in gene-regulation processes. The methylation of specific lysine and arginine residues also can be important. Some of the more common modifications are shown in Table 33.2.

The elucidation of the roles of these modifications is a very active area of research at present. The relation between various histone modifications and their roles in controlling gene expression is sometimes referred to as "the histone code" and important generalizations have been elucidated. For example, trimethylation of lysine 27 on histone H3 (H3 K27) is associated with repression of gene expression. This modification is promoted by a protein complex termed *polycomb repressive complex-2* (PRC-2). PRC-2 both binds to histones containing the trimethylated H3 K27 and catalyzes the addition of methyl groups to other H3 K27 substrates. This allows an initial modification to spread to adjacent histone octamers, facilitating gene repression. The development and application of methods that allow probing of histone modifications on a genome-wide scale in a range of cell types has provided large data sets that can be examined to test and refine models of transcription regulation via chromatin modifications.

33.4 Eukaryotic Gene Expression Can Be Controlled at Posttranscriptional Levels

Just as in prokaryotes, gene expression in eukaryotes can be regulated subsequent to transcription. We shall consider two examples. The first is the regulation of genes taking part in iron metabolism through key features in RNA secondary structure, similar in many ways to prokaryotic posttranscriptional regulation (Section 32.4). The second entails an entirely new mechanism, first glimpsed with the discovery of RNA interference. Certain small regulatory RNA molecules allow the regulation of gene expression through interaction with a range of mRNA molecules. Remarkably, this mechanism, discovered only recently, affects the expression of approximately 60% of all human genes.

Genes associated with iron metabolism are translationally regulated in animals

RNA secondary structure plays a role in the regulation of iron metabolism in eukaryotes. Iron is an essential nutrient, required for the synthesis of hemoglobin, cytochromes, and many other proteins. However, excess iron can be quite harmful because, untamed by a suitable protein environment, iron can initiate a range of free-radical reactions that damage proteins, lipids, and nucleic acids. Animals have evolved sophisticated systems for the accumulation of iron in times of scarcity and for the safe storage of excess iron for later use. Key proteins include *transferrin*, a transport protein that carries iron in the serum, *transferrin receptor*, a membrane protein that binds iron-loaded transferrin and initiates its entry into cells, and *ferritin*, an impressively efficient iron-storage protein found primarily in the liver and kidneys. Twenty-four ferritin polypeptides form a nearly spherical shell that encloses as many as 2400 iron atoms, a ratio of one iron atom per amino acid (Figure 33.22).

Ferritin and transferrin-receptor expression levels are reciprocally related in their responses to changes in iron levels. When iron is scarce,

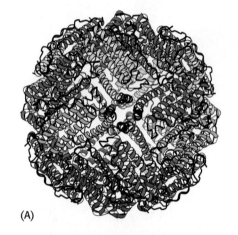

(A)

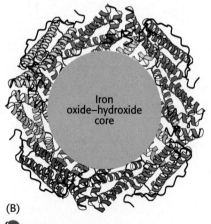

(B)

FIGURE 33.22 Structure of ferritin. (A) Twenty-four ferritin polypeptides form a nearly spherical shell. (B) A cutaway view reveals the core that stores iron as an iron oxide–hydroxide complex [Drawn from 1IES.pdb.]

Iron oxide–hydroxide core

the amount of transferrin receptor increases and little or no new ferritin is synthesized. Interestingly, the extent of mRNA synthesis for these proteins does not change correspondingly. Instead, regulation takes place at the level of translation.

Consider ferritin first. Ferritin mRNA includes a stem-loop structure termed an *iron-response element* (IRE) in its 5′ untranslated region (Figure 33.23). This stem-loop binds a 90-kDa protein, called an *IRE-binding protein* (IRP), that blocks the initiation of translation as the IRE binds on the 5′ side of the coding region. When the iron level increases, the IRP binds iron as a 4Fe-4S cluster. The IRP bound to iron cannot bind RNA, because the binding sites for iron and RNA substantially overlap. Thus, in the presence of iron, ferritin mRNA is released from the IRP and translated to produce ferritin, which sequesters the excess iron.

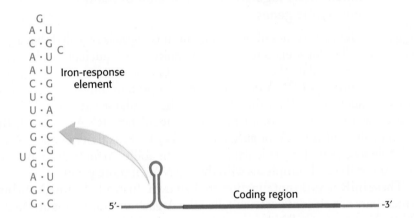

FIGURE 33.23 Iron-response element. Ferritin mRNA includes a stem-loop structure, termed an iron-response element (IRE), in its 5′ untranslated region. The IRE binds a specific protein that blocks the translation of this mRNA under low iron conditions.

An examination of the nucleotide sequence of transferrin-receptor mRNA reveals the presence of several IRE-like regions. However, these regions are located in the 3′ untranslated region rather than in the 5′ untranslated region (Figure 33.24). Under low-iron conditions, IRP binds to these IREs. However, given the location of these binding sites, the transferrin-receptor mRNA can still be translated. What happens when the iron level increases and the IRP no longer binds transferrin-receptor mRNA? Freed from the IRP, transferrin-receptor mRNA is rapidly degraded. Thus, an increase in the cellular iron level leads to the destruction of transferrin-receptor mRNA and, hence, a reduction in the production of transferrin-receptor protein.

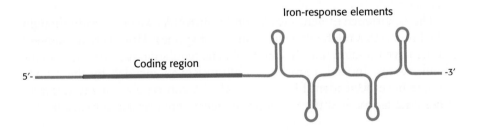

FIGURE 33.24 Transferrin-receptor mRNA. This mRNA has a set of iron-response elements (IREs) in its 3′ untranslated region. The binding of the IRE-binding protein to these elements stabilizes the mRNA but does not interfere with translation.

The purification of the IRP and the cloning of its cDNA were sources of truly remarkable insight into evolution. The IRP was found to be approximately 30% identical in amino acid sequence with the citric acid cycle enzyme aconitase from mitochondria. Further analysis revealed that the IRP is, in fact, an active aconitase enzyme; it is a cytoplasmic aconitase that had been known for a long time, but its function was not well

(A)

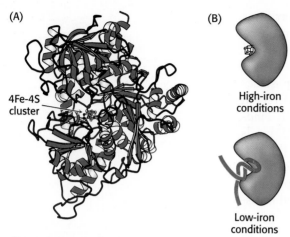

4Fe-4S cluster

(B)

High-iron conditions

Low-iron conditions

FIGURE 33.25 The IRP is an aconitase. (A) Aconitase contains an unstable 4Fe-4S cluster at its center. (B) Under conditions of low iron, the 4Fe-4S cluster dissociates and appropriate RNA molecules can bind in its place. [Drawn from 1C96.pdb.]

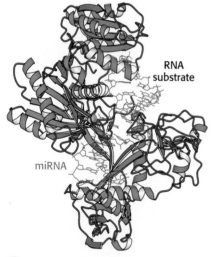

RNA substrate

miRNA

FIGURE 33.26 MicroRNA–Argonaute complex. The miRNA (shown in red) is bound by the Argonaute protein. *Notice* that the miRNA serves as a guide that binds the RNA substrate (shown in gray) through the formation of a double helix. Two magnesium ions are shown in green. [Drawn from 3HK2.pdb.]

understood (Figure 33.25). The iron–sulfur center at the active site of the IRP is rather unstable, and loss of the iron triggers significant changes in protein conformation. Thus, this protein can serve as an iron-sensing factor.

Other mRNAs, including those taking part in heme synthesis, have been found to contain IREs. Thus, genes encoding proteins required for iron metabolism acquired sequences that, when transcribed, provided binding sites for the iron-sensing protein. An environmental signal—the concentration of iron—controls the translation of proteins required for the metabolism of this metal. Thus, mutations in the untranslated region of mRNAs have been selected for beneficial regulation by iron levels.

Small RNAs regulate the expression of many eukaryotic genes

Genetic studies of development in *C. elegans* revealed that a gene called *lin-4* encodes an RNA molecule 61 nucleotides long that can regulate the expression of certain other genes. The 61-nucleotide *lin-4* RNA does not encode a protein, but rather is cleaved into a 22-nucleotide RNA that possesses the regulatory activity. This discovery was the first view of a large class of regulatory RNA molecules that are now called *microRNAs* or *miRNAs*. The key to the activity of microRNAs and their specificity for particular genes is their ability to form Watson–Crick base-pair-stabilized complexes with the mRNAs of those genes.

These miRNAs do not function on their own. Instead, the miRNAs bind to members of a class of proteins called the *Argonaute* family (Figure 33.26). These Argonaute–miRNA complexes can then bind mRNAs that have sequences that are substantially complementary to the miRNAs. Once bound, such an mRNA can be cleaved by the Argonaute–miRNA complex through the action of a magnesium-based active site. Thus, *the miRNAs serve as guide RNAs that determine the specificity of the Argonaute complex* (Figure 33.27). Cleavage of mRNA by the Argonaute–miRNA complex is similar to the mechanism of RNA interference. The small RNA molecules participating in RNAi, however, come from a different source. In RNAi, double-stranded RNAs are cleaved into 21-nucleotide fragments, single-stranded components of which are bound by members of the Argonaute family to form an RNA-induced silencing complex (RISC) that cleaves complementary mRNAs. For miRNAs, the single-stranded RNAs are generated from larger genetically encoded precursors, as described in Chapter 30.

The occurrence of gene regulation by miRNAs was originally thought to be limited to a relatively small number of species. However, subsequent studies have revealed that this mode of gene regulation is nearly ubiquitous in eukaryotes. Indeed, more than 1900 miRNAs encoded by the human genome have been identified. Each miRNA can regulate many different genes because many different target sequences are present in each mRNA.

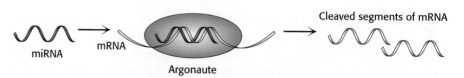

Cleaved segments of mRNA

miRNA   mRNA   Argonaute

FIGURE 33.27 MicroRNA action. MicroRNAs bind to members of the Argonaute family where they target specific mRNA molecules for cleavage.

An estimated 60% of all human genes are regulated by one or more miRNAs. As one example, consider the human miRNA called miR-206. This miRNA down-regulates the expression of one isoform of the estrogen receptor. In addition, it appears to down-regulate the expression of several different coactivators that interact with the estrogen receptor. Thus, this miRNA can mute the influence of estrogen by blocking the estrogen-initiated signaling pathway at several different steps.

The microRNA pathway has tremendous implications for the evolution of gene-regulatory pathways. Most of the target sites for miRNAs are present in the 3'-untranslated regions of mRNAs. These sequences are quite free to mutate because they do not encode proteins and are not required to fold into any specific structures. Thus, in the context of a set of expressed miRNAs, mutations in this region of any particular gene could, in principle, increase or decrease the affinity for one or more miRNAs and, hence, alter the regulation of the gene.

SUMMARY

33.1 Eukaryotic DNA Is Organized into Chromatin

Eukaryotic DNA is tightly bound to basic proteins called histones; the combination is called chromatin. DNA wraps twice around an octamer of core histones to form a nucleosome. The four core histones are homologous and fold into similar structures. Each core histone has an amino-terminal tail rich in lysine and arginine residues. Nucleosomes are the first stage of compaction of eukaryotic DNA. Chromatin blocks access to many potential DNA-binding sites. Changes in chromatin structure play a major role in regulating gene expression.

33.2 Transcription Factors Bind DNA and Regulate Transcription Initiation

Most eukaryotic genes are not expressed unless they are activated by the binding of specific proteins, called transcription factors, to sites on the DNA. These specific DNA-binding proteins interact directly or indirectly with RNA polymerases or their associated proteins. Eukaryotic transcription factors are modular: they consist of separate DNA-binding and activation domains. Important classes of DNA-binding proteins include the homeodomains, the basic-leucine zipper proteins, and Cys_2His_2 zinc-finger proteins. Each of these classes of proteins uses an α helix to make specific contacts with DNA. Activation domains interact with RNA polymerases or their associated factors or with other protein complexes such as mediator. Enhancers are DNA elements that can modulate gene expression from more than 1000 bp away from the start site of transcription. Enhancers are often specific for certain cell types, depending on which DNA-binding proteins are present. The introduction of genes for a specific set of four transcription factors into fibroblasts can cause these cells to dedifferentiate into induced pluripotent stem cells.

33.3 The Control of Gene Expression Can Require Chromatin Remodeling

Chromatin structure is crucial to the control of gene expression; it is more open near the transcription start sites of actively transcribed genes. Steroids such as estrogens bind to eukaryotic transcription factors called nuclear hormone receptors. These proteins are capable of binding DNA whether or not ligands are bound. The binding of ligands induces a conformational change that allows the recruitment of additional proteins called coactivators. Among the most important functions of coactivators is to catalyze the addition of acetyl groups to lysine residues in the tails of histone proteins. Histone acetylation decreases the affinity of the histones for DNA, making additional genes accessible for transcription. In addition, acetylated histones are targets for proteins containing specific binding units called bromodomains. Bromodomains are components of two classes of large complexes: (1) chromatin-remodeling engines and (2) factors associated with RNA polymerase II. These complexes open up sites on chromatin and initiate transcription.

33.4 Eukaryotic Gene Expression Can Be Controlled at Posttranscriptional Levels

Genes encoding proteins that transport and store iron are regulated at the translational level. Iron-response elements, structures that are present in certain mRNAs, are bound by an IRE-binding protein when this protein is not binding iron. Whether the expression of a gene is stimulated or inhibited in response to changes in the iron status of a cell depends on the location of the IRE within the mRNA. The IRE-binding protein is a cytoplasmic aconitase that loses its iron–sulfur center under low-iron conditions. MicroRNAs are specialized RNA molecules encoded as parts of larger RNA precursors. They bind to proteins of the Argonaute family; the bound miRNAs function as guides that help bind specific mRNA molecules that are then cleaved.

APPENDIX

Biochemistry in Focus

A mechanism for consolidating epigenetic modifications

Modification of histone proteins plays a key role in controlling eukaryotic gene expression. An important characteristic of some histone modifications is that they tend to spread, that is, the modification of one histone octamer increases the likelihood that nearby histone octamers will be modified in the same way. This characteristic is crucial since it facilitates the formation of regions or domains with a high frequency of a given modification that represents a robust signal. What is a possible biochemical mechanism for the spreading of modifications? In this chapter, we encountered enzymatic activities that introduce modifications such as histone acetyltransferases and briefly noted other modifications such as lysine methylation, which are introduced by different classes of enzymes. We also noted the existence of proteins that bind to specific modifications such as proteins containing bromodomains that bind specifically to peptides containing the acetyllysine modification.

With this knowledge, we can imagine a mechanism for facilitating the spreading of a modification. Suppose that a pro-

tein or protein complex includes both an enzyme that promotes a particular modification and a binding domain or region that recognizes the same modification, separated by an appropriate distance. Such a complex could then spread the modification since binding of the complex to one modified histone would increase the likelihood that the enzyme active site would productively interact with a nearby, unmodified histone, introducing an additional modification. After the complex releases the initial modified histone, this process could then repeat, spreading the modification to many nearby histones.

A structure recently determined by cryo-electron microscopy provides support for this hypothetical mechanism. The structure comprises the human **polycomb repressive complex-2** (PRC-2) bound to a DNA fragment that includes two nucleosomes. One of the nucleosomes includes trimethylated lysines in residue 27 on histone H3, one of which is bound within a binding site at one end of the protein complex. The other nucleosome has one of its histone H3 lysine 27 residues bound within the active site of an enzyme domain known to function as a lysine methyltransferase. This structure provides a framework to understand modification spreading and for designing other experiments to test and extend this understanding.

KEY TERMS

cell type (p. 1076)

histone (p. 1077)

chromatin (p. 1077)

nucleosome (p. 1077)

nucleosome core particle (p. 1078)

transcription factor (p. 1079)

homeodomain (p. 1080)

basic-leucine zipper (bZip) protein (p. 1080)

Cys$_2$His$_2$ zinc-finger domain (p. 1080)

mediator (p. 1081)

combinatorial control (p. 1081)

enhancer (p. 1082)

induced pluripotent stem (iPS) cell (p. 1082)

hypersensitive site (p. 1083)

chromatin immunoprecipitation (ChIP) (p. 1083)

hypomethylation (p. 1084)

CpG island (p. 1084)

nuclear hormone receptor (p. 1085)

estrogen response element (ERE) (p. 1085)

coactivator (p. 1086)

agonist (p. 1086)

anabolic steroid (p. 1087)

antagonist (p. 1087)

selective estrogen receptor modulator (SERM) (p. 1087)

histone acetyltransferase (HAT) (p. 1088)

acetyllysine-binding domain (p. 1088)

bromodomain (p. 1088)

TATA-box-binding protein associated factor (TAF) (p. 1088)

chromatin-remodeling complex (p. 1089)

histone deacetylase (p. 1089)

polycomb repressive complex-2 (p. 1090)

transferrin (p. 1090)

transferrin receptor (p. 1090)

ferritin (p. 1090)

iron-response element (IRE) (p. 1091)

IRE-binding protein (IRP) (p. 1091)

microRNA (miRNA) (p. 1092)

Argonaute family proteins (p. 1092)

PROBLEMS

1. *Charge neutralization.* Given the histone amino acid sequences illustrated below, estimate the charge of a histone octamer at pH 7. Assume that histidine residues are uncharged at this pH. How does this charge compare with the charge on 150 base pairs of DNA? ✓❶

Histone H2A

MSGRGKQGGKARAKAKTRSSRAGLQFPVGRVHRLLRKGNYSERVGAGAPVYLAAVLEYLTAEILELAGNA

ARDNKKTRIIPRHLQLAIRNDEELNKLLGRVTIAQGGVLPNIQAVLLPKKTESHHKAKGK

Histone H2B

MPEPAKSAPAPKKGSKKAVTKAQKKDGKKRKRSRKESYSVYVYKVLKQVHPDTGISSKAMGIMNSFVNDI

FERIAGEASRLAHYNKRSTITSREIQTAVRLLLPGELAKHAVSEGTKAVTKYTSSK

Histone H3

MARTKQTARKSTGGKAPRKQLATKAARKSAPSTGGVKKPHRYRPGTVALREIRRYQKSTELLIRKLPFQR

LVREIAQDFKTDLRFQSAAIGALQEASEAYLVGLFEDTNLCAIHAKRVTIMPKDIQLARRIRGERA

Histone H4

MSGRGKGGKGLGKGGAKRHRKVLRDNIQGITKPAIRRLARRGGVKRISGLIYEETRGVLKVFLENVIRDA

VTYTEHAKRKTVTAMDVVYALKRQGRTLYGFGG

2. *Chromatin immunoprecipitation.* You have used the technique of chromatin immunoprecipitation to isolate DNA fragments containing a DNA-binding protein of interest. Suppose that you wish to know whether a particular known DNA fragment is present in the isolated mixture. How might you detect its presence?

3. *Around we go.* Assuming that 145 base pairs of DNA wrap around the histone octamer 1.75 times, estimate the radius of the histone octamer. Assume 3.4 Å per base pair and simplify the calculation by assuming that the wrapping is in two rather than three dimensions and neglecting the thickness of the DNA. ✓❶

4. *Nitrogen substitution.* Growth of mammalian cells in the presence of 5-azacytidine results in the activation of some normally inactive genes. Propose an explanation.

5-Azacytidine

5. *A new domain.* A protein domain that recognizes 5-methylcytosine in the context of double-stranded DNA has been characterized. What role might proteins containing such a domain play in regulating gene expression? Where on a double-stranded DNA molecule would you expect such a domain to bind? ✓❸

6. *Hybrid receptor.* Through recombinant DNA methods, a modified steroid hormone receptor was prepared that consists of an estrogen receptor with its ligand-binding domain replaced by the ligand-binding domain from the progesterone receptor. Predict the expected responsiveness of gene expression for cells treated with estrogen or with progesterone. ✓❻

7. *Different modifications.* What is the effect of acetylation of a lysine residue on the charge of a histone protein? Of lysine methylation? ✓④

8. *Transformer.* The following amino acid sequence of one of the four transcription factors is used to generate iPS cells:

HTCDYAGCGKTYTKSSHLKAHLRTHTGEKPYHCDWDGCGWK
FARSDELTRHYRKHTGHRPFQCQKCDRAFSRSDHLALHMKRHF

This transcription factor belongs to one of the three structural classes discussed in Section 33.2. Identify the class. ✓③

9. *Coverage.* What percentage of the DNA sites in yeast are accessible, assuming that the fraction of sites observed for GAL4 is typical? To how many base pairs of the 12-Mb yeast genome does this percentage correspond?

10. *Modifications.* Examination of the histone modifications around a gene reveals an abundance of histone H3 with lysine 27 modified with three methyl groups. Does this suggest that this gene is activated or repressed? How would your answer change if many lysine 27 residues were modified with acetyl groups? ✓⑤

11. *Iron regulation.* What effect would you expect from the addition of an IRE to the 5′ end of a gene that is not normally regulated by iron levels? To the 3′ end? ✓⑦

12. *Predicting microRNA regulation.* Suppose that you have identified an miRNA that has the sequence 5′-GCCUAGC CUUAGCAUUGAUUGG-3′. Propose a strategy for identifying mRNA that might be regulated by this miRNA, given the sequences of all mRNAs encoded by the human genome. ✓⑦

Mechanism Problem

13. *Acetyltransferases.* Propose a mechanism for the transfer of an acetyl group from acetyl CoA to the amino group of lysine.

Data Interpretation Problem

14. *Limited restriction.* The restriction enzyme HpaII is a powerful tool for analyzing DNA methylation. This enzyme cleaves sites of the form 5′-CCGG-3′ but will not cleave such sites if the DNA is methylated on any of the cytosine residues. Genomic DNA from different organisms is treated with HpaII and the results are analyzed by gel electrophoresis (see the adjoining patterns) (See animation on 🌀 SaplingPlus). Provide an explanation for the observed patterns.

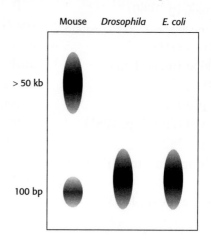

Think ▶ Pair ▶ Share Problem

15. Almost all cells within an animal contain DNA with the same sequence, yet different cells can have very different properties and gene expression patterns. What are the primary mechanisms that facilitate the existence of distinct cell types in eukaryotes?

Chapter 1

1. The hydrogen-bond donors are the NH and NH_2 groups. The hydrogen-bond acceptors are the carbonyl oxygen atoms and those ring nitrogen atoms that are not bonded to hydrogen or to deoxyribose.

2. Interchange the positions of the single and double bonds in the six-membered ring.

3. (a) Ionic interactions; (b) van der Waals interactions.

4. Processes *a* and *b*.

5. $\Delta S_{system} = -661\,J\,mol^{-1}\,K^{-1}(-158\,kcal\,mol^{-1}\,K^{-1})$
$\Delta S_{surroundings} = +842\,J\,mol^{-1}\,K^{-1}(+201\,cal\,mol^{-1}\,K^{-1})$

6. (a) 1.0; (b) 13.0; (c) 1.3; (d) 12.7

7. 2.88

8. 1.96

9. 55.5 M

10. 11.83

11. 447; 0.00050

12. 0.00066 M

13. 6.0

14. 5.53

15. 6.48

16. 7.8

17. 100

18. (a) 5.0; (b) 8.0; (c) 3.16

19. (a) 1.6; (b) 0.51; (c) 0.16

20.

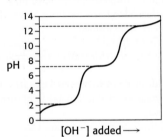

21. (a) No; (b) Around pH 2.

22. 0.1 M sodium acetate solution: 6.34; 6.03; 5.70; 4.75; 0.01 M sodium acetate solution: 5.90; 4.75; 3.38; 1.40.

23. 90 mM acetic acid; 160 mM sodium acetate, 0.18 moles acetic acid; 0.32 moles sodium acetate; 10.81 g acetic acid; 26.25 g sodium acetate.

24. 0.50 moles of acetic acid; 0.32 moles of NaOH; 30.03 g of acetic acid; 12.80 g of NaOH.

25. 250 mM; yes; no; it will also contain 90 mM NaCl.

26. 8.63 g Na_2HPO_4; 4.71 g NaH_2PO_4

27. 7.0; this buffer will not be very useful, because the pH value is far from the pK_a value.

28. (a) MOPS; (b) MES

29. 50 mM

30. Buffer, because the sodium ions will shield the electrostatic repulsion between the phosphate groups.

31. $+1.45\,kJ\,mol^{-1}(+0.35\,kcal\,mol^{-1})$; $+57.9\,kJ\,mol^{-1}$ $(+13.8\,kcal\,mol^{-1})$

32. The hydrophobic effect.

33. There will be approximately 15 million differences.

34. $(20!)/(10!(\times)10!) = 184{,}756$

35. 7.9%

Chapter 2

1. (A) Proline, Pro, P; (B) tyrosine, Tyr, Y; (C) leucine, Leu, L; (D) lysine, Lys, K.

2. (a) C, B, A; (b) D; (c) D, B; (d) B, D; (e) B.

3. (a) 6; (b) 2; (c) 3; (d) 1; (e) 4; (f) 5.

4. (a) Ala; (b) Tyr; (c) Ser; (d) His.

5. Ser, Glu, Tyr, Thr

6. (a) Alanine-glycine-serine; (b) Alanine; (c and d):

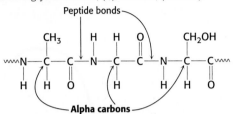

7. At pH 5.5, the net charge is +1:

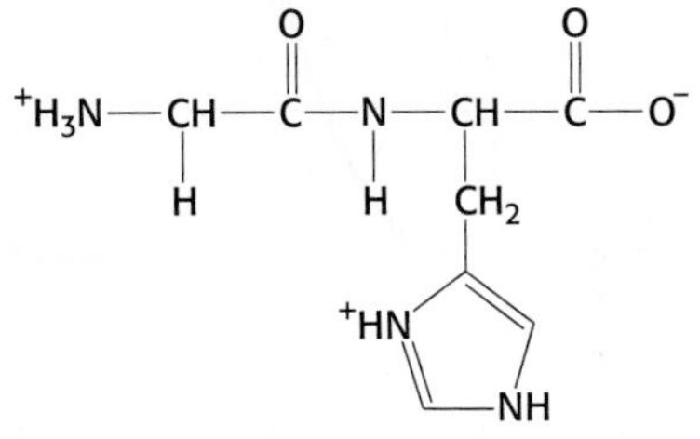

At pH 7.5, the net charge is 0:

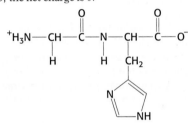

8. There are 20 choices for each of the 50 amino acids: 20^{50}, or 1×10^{65}.

9.

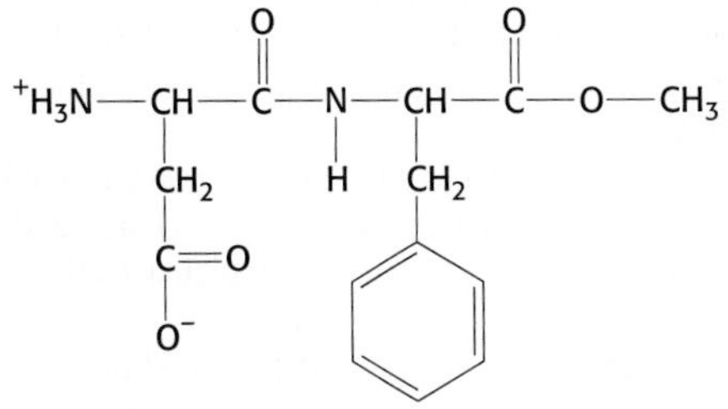

10. The solution would have no net charge at the pH in the midpoint of the zwitterionic species curve (blue), effectively at the midpoint between pK_1 and pK_2:

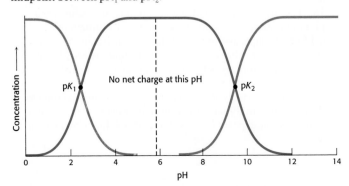

11.

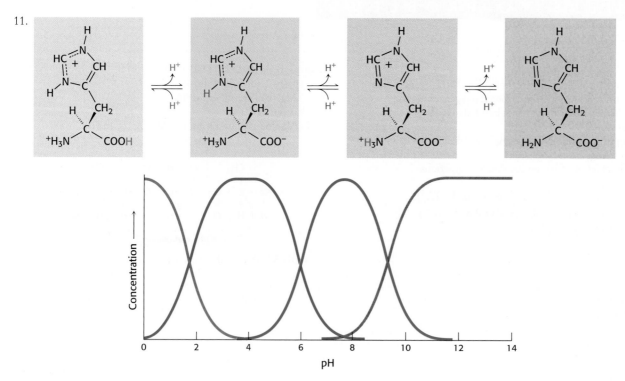

12. The (nitrogen–α carbon–carbonyl carbon) repeating unit.

13. Side chain is the functional group attached to the α-carbon atom of an amino acid.

14. Amino acid composition refers simply to the amino acids that make up the protein. The order is not specified. Amino acid sequence is the same as the primary structure—the sequence of amino acids from the amino terminal to the carboxyl terminal of the protein. Different proteins may have the same amino acid composition, but amino acid sequence identifies a unique protein.

15. (a) Each strand is 35 kDa and hence has about 318 residues (the mean residue mass is 110 daltons). Because the rise per residue in an α helix is 1.5 Å, the length is 477 Å. More precisely, for an α-helical coiled coil, the rise per residue is 1.46 Å; so the length is 464 Å.
(b) Eighteen residues in each strand (40 minus 4 divided by 2) are in a β-sheet conformation. Because the rise per residue is 3.5 Å, the length is 63 Å.

16. The methyl group attached to the β-carbon atom of isoleucine sterically interferes with α-helix formation. In leucine, this methyl group is attached to the γ-carbon atom, which is farther from the main chain and hence does not interfere.

17. Proline and glycine. The cyclic side chain of proline linking the nitrogen and α-carbon atoms limits ϕ to a very narrow range (around −60 degrees). The lack of steric hindrance exhibited by the side chain hydrogen atom of glycine enables this amino acid to access a much greater area of the Ramachandran plot.

18. The first mutation destroys activity because valine occupies more space than alanine does, and so the protein must take a different shape, assuming that this residue lies in the closely packed interior. The second mutation restores activity because of a compensatory reduction of volume; glycine is smaller than isoleucine.

19. Loops invariably are on the surface of proteins, exposed to the environment. Because many proteins exist in aqueous environments, the exposed loops will be hydrophilic to interact with water.

20. The native conformation of insulin is not the thermodynamically most stable form, because it contains two separate chains linked by disulfide bonds. Insulin is formed from proinsulin, a single-chain precursor, that is cleaved to form insulin, a 51-residue molecule, after the disulfide bonds have formed.

21. A segment of the main chain of the protease could hydrogen bond to the main chain of the substrate to form an extended parallel or antiparallel pair of β strands.

22. Glycine has the smallest side chain of any amino acid. Its size is often critical in allowing polypeptide chains to make tight turns or to approach one another closely.

23. Glutamate, aspartate, and the terminal carboxylate can form salt bridges with the guanidinium group of arginine. In addition, this group can be a hydrogen-bond donor to the side chains of glutamine, asparagine, serine, threonine, aspartate, tyrosine, and glutamate and to the main-chain carbonyl group. Histidine can form hydrogen bonds with arginine at pH 7.

24. Disulfide bonds in hair are broken by adding a thiol-containing reagent and applying gentle heat. The hair is curled, and an oxidizing agent is added to re-form disulfide bonds to stabilize the desired shape.

25. Some proteins that span biological membranes are "the exceptions that prove the rule" because they have the reverse distribution of hydrophobic and hydrophilic amino acids. For example, consider *porins*, proteins found in the outer membranes of many bacteria. Membranes are built largely of hydrophobic chains. Thus, porins are covered on the outside largely with hydrophobic residues that interact with the neighboring hydrophobic chains. In contrast, the center of the protein contains many charged and polar amino acids that surround a water-filled channel running through the middle of the protein. Thus, because porins function in hydrophobic environments, they are "inside out" relative to proteins that function in aqueous solution.

26. The amino acids would be hydrophobic in nature. An α helix is especially suited to crossing a membrane because all the amide hydrogen atoms and carbonyl oxygen atoms of the peptide

backbone take part in intrachain hydrogen bonds, thus stabilizing these polar atoms in a hydrophobic environment.

27. This example demonstrates that the pK_a values are affected by the environment. A given amino acid can have a variety of pK_a values, depending on the chemical environment inside the protein.

28. Recall that hemoglobin exists as a tetramer while myoglobin is a monomer. Consequently, the hydrophobic residues on the surface of hemoglobin subunits are probably involved in van der Waals interactions with similar regions on the other subunits, and will be shielded from the aqueous environment by this interaction.

29. A possible explanation is that the severity of the symptoms corresponds to the degree of structural disruption. Hence, substitution of alanine for glycine might result in mild symptoms, but substitution of the much larger tryptophan may result in little or no collagen triple-helix formation.

30. The energy barrier that must be crossed to go from the polymerized state to the hydrolyzed state is large even though the reaction is thermodynamically favorable.

31. Using the Henderson–Hasselbalch equation, we find the ratio of alanine-COOH to alanine-COO⁻ at pH 7 to be 10^{-4}. The ratio of alanine-NH_2 to alanine-NH_3^+, determined in the same fashion, is 10^{-1}. Thus, the ratio of neutral alanine to the zwitterionic species is $10^{-4} \times 10^{-1} = 10^{-5}$.

32. The assignment of absolute configuration requires the assignment of priorities to the four groups connected to a tetrahedral carbon atom. For all amino acids except cysteine, the priorities are: (1) amino group; (2) carbonyl group; (3) side chain; (4) hydrogen. For cysteine, because of the sulfur atom in its side chain, the side chain has a greater priority than does the carbonyl group, leading to the assignment of an R rather than S configuration.

33. ELVISISLIVINGINLASVEGAS

34. No, Pro–X would have the characteristics of any other peptide bond. The steric hindrance in X–Pro arises because the R group of Pro is bonded to the amino group. Hence, in X–Pro, the proline R group is near the R group of X, which would not be the case in Pro–X.

35. A, c; B, e; C, d; D, a; E, b.

36. The reason is that the wrong disulfides formed pairs in urea. There are 105 different ways of pairing eight cysteine molecules to form four disulfides; only one of these combinations is enzymatically active. The 104 wrong pairings have been picturesquely termed "scrambled" ribonuclease.

Chapter 3

1. (a) Phenyl isothiocyanate; (b) urea; β-mercaptoethanol to reduce disulfides; (c) chymotrypsin; (d) CNBr; (e) trypsin.

2. For each cell within an organism, the genome is a fixed property. However, the proteome is dynamic, reflecting different environmental conditions and external stimuli. Two different cell types will likely express different subsets of proteins encoded within the genome.

3. The S-aminoethylcysteine side chain resembles that of lysine. The only difference is a sulfur atom in place of a methylene group.

4. A 1 mg ml⁻¹ solution of myoglobin (17.8 kDa; Table 3.2) corresponds to 5.62×10^{-5} M. The absorbance of a 1-cm path length is 0.84, which corresponds to an I_0/I ratio of 6.96. Hence 14.4% of the incident light is transmitted.

5. The sample was diluted 1000-fold. The concentration after dialysis is thus 0.001 M, or 1 mM. You could reduce the salt concentration by dialyzing your sample, now 1 mM, in more buffer free of $(NH_4)_2SO_4$.

6. If the salt concentration becomes too high, the salt ions interact with the water molecules. Eventually, there will not be enough water molecules to interact with the protein, and the protein will precipitate. If there is lack of salt in a protein solution, the proteins may interact with one another—the positive charges on one protein with the negative charges on another or several others. Such an aggregate becomes too large to be solubilized by water alone. If salt is added, the salt neutralizes the charges on the proteins, preventing protein–protein interactions.

7. Tropomyosin is rod shaped, whereas hemoglobin is approximately spherical.

8. The frictional coefficient, f, and the mass, m, determine s. Specifically, f is proportional to r (see equation 2 on p. 75). Hence, f is proportional to $m^{1/3}$, and so s is proportional to $m^{2/3}$ (see the equation on p. 80). An 80-kDa spherical protein undergoes sedimentation 1.59 times as rapidly as a 40-kDa spherical protein.

9. The long hydrophobic tail on the SDS molecule (see p. 76) disrupts the hydrophobic interactions in the interior of the protein. The protein unfolds, with the hydrophobic R groups now interacting with SDS rather than with one another.

10. The protein may be modified. For instance, asparagine residues in the protein may be modified with carbohydrate units (Section 2.6).

11. A fluorescence-labeled derivative of a bacterial degradation product (e.g., a formylmethionyl peptide) would bind to cells containing the receptor of interest.

12. (a) Trypsin cleaves after arginine (R) and lysine (K), generating AVGWR, VK, and S. Because they differ in size, these products could be separated by molecular exclusion chromatography.
(b) Chymotrypsin, which cleaves after large aliphatic or aromatic R groups, generates two peptides of equal size (AVGW) and (RVKS). Separation based on size would not be effective. The peptide RVKS has two positive charges (R and K), whereas the other peptide is neutral. Therefore, the two products could be separated by ion-exchange chromatography.

13. Antibody molecules bound to a solid support can be used for affinity purification of proteins for which a ligand molecule is unknown or unavailable.

14. If the product of the enzyme-catalyzed reaction is highly antigenic, it may be possible to obtain antibodies to this particular molecule. These antibodies can be used to detect the presence of product by ELISA, providing an assay format suitable for the purification of this enzyme.

15. An inhibitor of the enzyme being purified might have been present and subsequently removed by a purification step. This removal would lead to an apparent increase in the total amount of enzyme present.

16. Many proteins have similar masses but different sequences and different patterns when digested with trypsin. The set of masses of tryptic peptides forms a detailed "fingerprint" of a protein that is very unlikely to appear at random in other proteins regardless of size. (A conceivable analogy is: "Just as similarly sized fingers will give different individual fingerprints, so also similarly sized proteins will give different digestion patterns with trypsin.")

17. Isoleucine and leucine are isomers and, hence, have identical masses. Peptide sequencing by mass spectrometry as described in this chapter is incapable of distinguishing these residues. Further analytical techniques are required to differentiate these residues.

18.

Purification procedure	Total protein (mg)	Total activity (units)	Specific activity (units mg^{-1})	Purification level	Yield (%)
Crude extract	20,000	4,000,000	200	1	100
(NH$_4$)$_2$SO$_4$ precipitation	5000	3,000,000	600	3	75
DEAE-cellulose chromatography	1500	1,000,000	667	3.3	25
Gel-filtration chromatography	500	750,000	1500	7.5	19
Affinity chromatography	45	675,000	15,000	75	17

19. (a) Ion exchange chromatography will remove Proteins A and D, which have substantially lower isoelectric point; then gel filtration chromatography will remove Protein C, which has a lower molecular weight. (b) If Protein B carries a His tag, a single affinity chromatography step with an immobilized nickel(II) column may be sufficient to isolate the desired protein from the others.

20. Protein crystal formation requires the ordered arrangement of identically positioned molecules. Proteins with flexible linkers can introduce disorder into this arrangement and prevent the formation of suitable crystals. A ligand or binding partner may induce an ordered conformation to this linker and could be included in the solution to facilitate crystal growth. Alternatively, the individual domains separated by the linker may be expressed by recombinant methods and their crystal structures solved separately.

21. Treatment with urea will disrupt noncovalent bonds. Thus the original 60-kDa protein must be made of two 30-kDa subunits. When these subunits are treated with urea and β-mercaptoethanol, a single 15-kDa species results, suggesting that disulfide bonds link the 30-kDa subunits.

22. (a) Electrostatic repulsion between positively charged ε-amino groups hinders α-helix formation at pH 7. At pH 10, the side chains become deprotonated, allowing α-helix formation.
(b) Poly-L-glutamate is a random coil at pH 7 and becomes α helical below pH 4.5 because the γ-carboxylate groups become protonated.

23. The difference between the predicted and the observed masses for this fragment equals 28.0, exactly the mass shift that would be expected in a formylated peptide. This peptide is likely formylated at its amino terminus, and corresponds to the most N-terminal fragment of the protein.

24. Light was used to direct the synthesis of these peptides. Each amino acid added to the solid support contained a photolabile protecting group instead of a t-Boc protecting group at its α-amino group. Illumination of selected regions of the solid support led to the release of the protecting group, which exposed the amino groups in these sites to make them reactive. The pattern of masks used in these illuminations and the sequence of reactants define the ultimate products and their locations.

25. Mass spectrometry is highly sensitive and capable of detecting the mass difference between a protein and its deuterated counterpart. Fragmentation techniques can be used to identify the amino acids that retained the isotope label. Alternatively, NMR spectroscopy can be used to detect the isotopically labeled atoms because the deuteron and the proton have very different nuclear-spin properties.

26. First amino acid: A
   Last amino acid: R (not cleaved by carboxypeptidase)
   Sequence of N-terminal tryptic peptide: AVR (tryptic peptide ends in K)
   Sequence of N-terminal chymotryptic peptide: AVRY (chymotryptic peptide ends in Y)
   Sequence: AVRYSR

27. First amino acid: S
   Last amino acid: L
   Cyanogen bromide cleavage: M is 10th position
   C-terminal residues are: (2S,L,W)
   Amino-terminal residues: (G,K,S,Y), tryptic peptide, ends in K
   Amino-terminal sequence: SYGK
   Chymotryptic peptide order: (S,Y), (G,K,L), (F,I,S), (M,T), (S,W), (S,L)
   Sequence: SYGKLSIFTMSWSL

28. If the protein did not contain any disulfide bonds, then the electrophoretic mobility of the trypsin fragments would be the same before and after performic acid treatment: all the fragments would lie along the diagonal of the paper. If one disulfide bond were present, the disulfide-linked trypsin fragments would run as a single peak in the first direction, and then as two separate peaks after performic acid treatment. The result would be two peaks appearing off the diagonal:

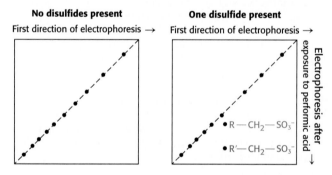

These fragments could then be isolated from the chromatography paper and analyzed by mass spectrometry to determine their amino acid composition and thus identify the cysteines participating in the disulfide bond.

Chapter 4

1. A nucleoside is a base attached to a ribose or deoxyribose sugar. A nucleotide is a nucleoside with one or more phosphoryl groups attached to the ribose or deoxyribose.

2. Hydrogen-bond pairing between the base A and the base T as well as hydrogen-bond pairing between the base G and the base C in DNA.

3. T is always equal to A, and so these two nucleotides constitute 40% of the bases. G is always equal to C, and so the remaining 60% must be 30% G and 30% C.

4. Nothing, because the base-pair rules do not apply to single-stranded nucleic acids.

5. (a) TTGATC; (b) GTTCGA; (c) ACGCGT; (d) ATGGTA.

6. (a) [T] + [C] = 0.46. (b) [T] = 0.30, [C] = 0.24, and [A] + [G] = 0.46.

7. Stable hydrogen bonding occurs only between GC and AT pairs. Moreover, two purines are too large to fit inside the double helix, and two pyrimidines are too small to form base pairs with each other.

8. The thermal energy causes the chains to wiggle about, which disrupts the hydrogen bonds between base pairs and the stacking forces between bases and thereby causes the strands to separate.

9. The probability that any sequence will appear is $1/4^n$, where 4 is the number of nucleotides and n is the length of the sequence. The probability of any 15-base sequence appearing is $1/4^{15}$, or $1/1,073,741,824$. Thus, a 15-nucleotide sequence would be likely to appear approximately three times (3 billion × probability of appearance). The probability of a 16-base sequence appearing is $1/4^{16}$, which is equal to $1/4,294,967,296$. Such a sequence will be unlikely to appear more than once.

10. One end of a nucleic acid polymer ends with a free 5′-hydroxyl group (or a phosphoryl group esterified to the hydroxyl group), and the other end has a free 3′-hydroxyl group. Thus, the ends are different. Two strands of DNA can form a double helix only if the strands are running in different directions—that is, have opposite polarity.

11. Although the individual bonds are weak, the population of thousands to millions of such bonds provides much stability. There is strength in numbers.

12. There would be too much charge repulsion from the negative charges on the phosphoryl groups. These charges must be countered by the addition of cations.

13. The three forms are A-DNA, B-DNA and Z-DNA, with B-DNA being the most common. There are many differences (Table 4.2). Some key differences are: A-DNA and B-DNA are right-handed, whereas Z-DNA is left-handed. A-DNA forms in less-hydrated conditions than does B-DNA. The A form is shorter and wider than the B form.

14. 5.88×10^3 base pairs.

15. The diameter of DNA is 20 Å, and 1 Å = 0.1 nm, so the diameter is 2 nm. Because 1 μm = 10^3 nm, the length is 2×10^4 nm. Thus, the axial ratio is 1×10^4.

16. A template is the sequence of DNA or RNA that directs the synthesis of a complementary sequence. A primer is the initial segment of a polymer that is to be extended during elongation.

17. In conservative replication, after 1.0 generation, half of the molecules would be ^{15}N-^{15}N, the other half ^{14}N-^{14}N. After 2.0 generations, one-quarter of the molecules would be ^{15}N-^{15}N, the other three-quarters ^{14}N-^{14}N. Hybrid ^{14}N-^{15}N molecules would not be observed in conservative replication.

18. The nucleotides used for DNA synthesis have the triphosphate attached to the 5′-hydroxyl group with free 3′-hydroxyl groups. Such nucleotides can be utilized only for 5′-to-3′ DNA synthesis.

19. (a) Tritiated thymine or tritiated thymidine. (b) dATP, dGTP, dCTP, and TTP labeled with ^{32}P in the innermost (α) phosphorus atom.

20. Molecules in parts *a* and *b* would not lead to DNA synthesis, because they lack a 3′-OH group (a primer). The molecule in part *d* has a free 3′-OH group at one end of each strand but no template strand beyond. Only the molecule in part *c* would lead to DNA synthesis.

21. A retrovirus is a virus that has RNA as its genetic material. However, for the information to be expressed, it must first be converted into DNA, a reaction catalyzed by the enzyme reverse transcriptase. Thus, at least initially, information flow is opposite that of a normal cell: RNA → DNA rather than DNA → RNA.

22. A thymidylate oligonucleotide should be used as the primer. The poly(A) template specifies the incorporation of T; hence, radioactive thymidine triphosphate (labeled in the α phosphoryl group) should be used in the assay.

23. The ribonuclease serves to degrade the RNA strand, a necessary step in forming duplex DNA from the RNA–DNA hybrid.

24. Treat one aliquot of the sample with ribonuclease and another with deoxyribonuclease. Test these nuclease-treated samples for infectivity.

25. Deamination changes the original GC base pair into a GU pair. After one round of replication, one daughter duplex will contain a GC pair and the other duplex will contain an AU pair. After two rounds of replication, there will be two GC pairs, one AU pair, and one AT pair.

26. (a) $4^8 = 65,536$. In computer terminology, there are 64K 8-mers of DNA. (b) A bit specifies two bases (say, A and C) and a second bit specifies the other two (G and T). Hence, two bits are needed to specify a single nucleotide (base pair) in DNA. For example, 00, 01, 10, and 11 could encode A, C, G, and T. An 8-mer stores 16 bits ($2^{16} = 65,536$), the *E. coli* genome [4.6×10^6 bp] stores 9.2×10^6 bits, and the human genome (3.0×10^9 bases) stores 6.0×10^9 bits of genetic information. (c) A 2 gigabyte flash drive can hold 2×10^9 bits. A large number of 8-mer sequences could be stored on such a flash drive. The DNA sequence of *E. coli* could be written on a single 2 gigabyte flash drive with room to spare for a lot of spring break photos. We would need three such flash drives to record the entire human genome, although you could purchase a 16 gigabyte flash drive for only $10.00 at Target, leaving a lot of room for photos.

27. (a) Deoxyribonucleoside triphosphates versus ribonucleoside triphosphates. (b) 5′ → 3′ for both. (c) Semiconserved for DNA polymerase I; conserved for RNA polymerase. (d) DNA polymerase I needs a primer, whereas RNA polymerase does not.

28. The template strand has a sequence complementary to that of the RNA transcript. The coding strand has the same sequence as that of the RNA transcript except for thymine (T) in place of uracil (U).

29. Messenger RNA encodes the information that, on translation, yields a protein. Ribosomal RNA is the catalytic component of ribosomes, the molecular complexes that synthesize proteins. Transfer RNA is an adaptor molecule, capable of binding a specific amino acid and recognizing a corresponding codon. Transfer RNAs with attached amino acids are substrates for the ribosome.

30. Three nucleotides encode an amino acid; the code is nonoverlapping; the code has no punctuation; the code exhibits directionality; the code is degenerate.

31. (a) 5′-UAACGGUACGAU-3′
(b) Leu-Pro-Ser-Asp-Trp-Met
(c) Poly(Leu-Leu-Thr-Tyr)

32. The 2′-OH group in RNA acts as an intramolecular nucleophile. In the alkaline hydrolysis of RNA, it forms a 2′–3′-cyclic intermediate.

33.

34. Gene expression is the process of expressing the information of a gene in its functional molecular form. For many genes, the functional information is a protein molecule. Thus, gene expression includes transcription and translation.

35. A nucleotide sequence whose bases represent the most-common, but not necessarily the only, members of the sequence. A consensus sequence can be thought of as the average of many similar sequences.

36. Cordycepin terminates RNA synthesis. An RNA chain containing cordycepin lacks a $3'$-OH group.

37. Degeneracy of the code refers to the fact that most amino acids are encoded by more than one codon.

38. If only 20 of the 64 possible codons encoded amino acids, then a mutation that changed a codon would likely result in a nonsense codon, leading to termination of protein synthesis. With degeneracy, a nucleotide change might yield a synonym or a codon for an amino acid with similar chemical properties.

39. (a) 2, 4, 8; (b) 1, 6, 10; (c) 3, 5, 7, 9.

40. (a) 3; (b) 6; (c) 2; (d) 5; (e) 7; (f) 1; (g) 4.

41. Incubation with RNA polymerase and only UTP, ATP, and CTP led to the synthesis of only poly(UAC). Only poly(GUA) was formed when GTP was used in place of CTP.

42. A peptide terminating with Lys (UGA is a stop codon), another containing -Asn-Glu-, and a third containing -Met-Arg-.

43. Phe-Cys-His-Val-Ala-Ala.

44. Exon shuffling is a molecular process that can lead to the generation of new proteins by the rearrangement of exons within genes. Because many exons encode functional protein domains, exon shuffling is a rapid and efficient means of generating new genes.

45. Alternate splicing allows one gene to code for several different but related proteins.

46. It shows that the genetic code and the biochemical means of interpreting the code are common to even very distantly related life forms. It also testifies to the unity of life: that all life arose from a common ancestor.

47. (a) A codon for lysine cannot be changed to one for aspartate by the mutation of a single nucleotide. (b) Arg, Asn, Gln, Glu, Ile, Met, or Thr.

48. The genetic code is degenerate. Of the 20 amino acids, 18 are specified by more than one codon. Hence, many nucleotide changes (especially in the third base of a codon) do not alter the nature of the encoded amino acid. Mutations leading to an altered amino acid are usually more deleterious than those that do not and hence are subject to more stringent selection.

49. GC base pairs have three hydrogen bonds compared with two for AT base pairs. Thus, the higher content of GC means more hydrogen bonds and greater helix stability.

50. C_0t value essentially corresponds to the complexity of the DNA sequence—in other words, how long it will take for a sequence of DNA to find its complementary strand to form a double helix. The more complex the DNA, the slower it reassociates to make the double-stranded form.

Chapter 5

1. *Taq* polymerase is the DNA polymerase from the thermophilic bacterium that lives in hot springs. Consequently, it is heat stable and can withstand the high temperatures required for PCR without denaturing. Thus, it isn't necessary to add it to the reaction before each new cycle.

2. Ovalbumin cDNA should be used. *E. coli* lacks the machinery to splice the primary transcript arising from genomic DNA.

3. The presence of the AluI sequence would, on average, be $(1/4)^4$, or 1/256, because the likelihood of any base being at any position is one-fourth and there are four positions. By the same reasoning, the

presence of the NotI sequence would be $(1/4)^8$, or 1/65,536. Thus, the average product of digestion by AluI would be 250 base pairs (0.25 kb) in length, whereas that by NotI would be 66,000 base pairs (66 kb) in length.

4. PCR amplification is greatly hindered by the presence of G–C-rich regions within the template. Owing to their high melting temperatures, these templates do not denature easily, thus preventing the initiation of an amplification cycle. In addition, rigid secondary structures prevent the progress of DNA polymerase along the template strand during elongation.

5. No, because most human genes are much longer than 4 kb. A fragment would contain only a small part of a complete gene.

6. Southern blotting of an MstII digest would distinguish between the normal and the mutant genes. The loss of a restriction site would lead to the replacement of two fragments on the Southern blot by a single longer fragment. Such a finding would not prove that GTG replaced GAG; other sequence changes at the restriction site could yield the same result.

7. Although the two enzymes cleave the same recognition site, they each break different bonds within the 6-bp sequence. Cleavage by KpnI yields an overhang on the $3'$ strand, whereas cleavage by Acc65I produces an overhang on the $5'$ strand. These sticky ends do not overlap.

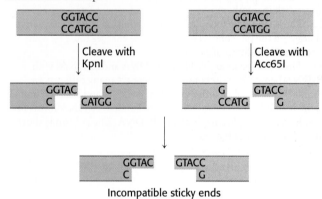

Incompatible sticky ends

8. A simple strategy for generating many mutants is to synthesize a degenerate set of cassettes by using a mixture of activated nucleosides in particular rounds of oligonucleotide synthesis. Suppose that the 30-bp coding region begins with GTT, which encodes valine. If a mixture of all four nucleotides is used in the first and second rounds of synthesis, the resulting oligonucleotides will begin with the sequence XYT (where X and Y denote A, C, G, or T). These 16 different versions of the cassette will encode proteins containing either Phe, Leu, Ile, Val, Ser, Pro, Thr, Ala, Tyr, His, Asn, Asp, Cys, Arg, or Gly at the first position. Likewise, degenerate cassettes can be made in which two or more codons are simultaneously varied.

9. Because PCR can amplify as little as one molecule of DNA, statements claiming the isolation of ancient DNA need to be greeted with some skepticism. The DNA would need to be sequenced. Is it similar to human, bacterial, or fungal DNA? If so, contamination is the likely source of the amplified DNA. Is it similar to that of birds or crocodiles? This sequence similarity would strengthen the case that it is dinosaur DNA because these species are evolutionarily close to dinosaurs.

10. Since the N could be any one of the four bases, the presence of the NGG trinucleotide would occur with a frequency of $\frac{4}{4} \times \frac{1}{4} \times \frac{1}{4}$, or once every 16 bases, on average.

11. Since PAM sequences are required downstream of a particular target site, variants of Cas9 that recognize different PAMs would increase the likelihood that a sgRNA could be designed against a particular target sequence.

12. At high temperatures of hybridization, only very close matches between primer and target would be stable because all (or most) of the bases would need to find partners to stabilize the primer–target helix. As the temperature is lowered, more mismatches would be tolerated; so the amplification is likely to yield genes with less sequence similarity. In regard to the yeast gene, synthesize primers corresponding to the ends of the gene, and then use these primers and human DNA as the target. If nothing is amplified at 54°C, the human gene differs from the yeast gene, but a counterpart may still be present. Repeat the experiment at a lower temperature of hybridization.

13. Digest genomic DNA with a restriction enzyme, and select the fragment that contains the known sequence. Circularize this fragment. Then carry out PCR with the use of a pair of primers that serve as templates for the synthesis of DNA away from the known sequence.

14. The encoded protein contains four repeats of a specific sequence.

15. Use chemical synthesis or the polymerase chain reaction to prepare hybridization probes that are complementary to both ends of the known (previously isolated) DNA fragment. Challenge clones representing the library of DNA fragments with both of the hybridization probes. Select clones that hybridize to one of the probes but not the other; such clones are likely to represent DNA fragments that contain one end of the known fragment along with the adjacent region of the particular chromosome.

16. The codon(s) for each amino acid can be used to determine the number of possible nucleotide sequences that encode each peptide sequence (Table 4.5):

Ala–Met–Ser–Leu–Pro–Trp:
$4 \times 1 \times 6 \times 6 \times 4 \times 1 = 576$ total sequences

Gly–Trp–Asp–Met–His–Lys:
$4 \times 1 \times 2 \times 1 \times 2 \times 2 = 32$ total sequences

Cys–Val–Trp–Asn–Lys–Ile:
$2 \times 4 \times 1 \times 2 \times 2 \times 3 = 96$ total sequences

Arg–Ser–Met–Leu–Gln–Asn:
$6 \times 6 \times 1 \times 6 \times 2 \times 2 = 864$ total sequences

The set of DNA sequences encoding the peptide Gly-Trp-Asp-Met-His-Lys would be most ideal for probe design because it encompasses only 32 total oligonucleotides.

17. Within a single species, individual dogs show enormous variation in body size and substantial diversity in other physical characteristics. Therefore, genomic analysis of individual dogs would provide valuable clues concerning the genes responsible for the diversity within the species.

18. On the basis of the comparative genome map shown in Figure 5.30, the region of greatest overlap with human chromosome 20 can be found on mouse chromosome 2.

19. T_m is the melting temperature of a double-stranded nucleic acid. If the melting temperatures of the primers are too different, the extent of hybridization with the target DNA will differ during the annealing phase, which would result in differential replications of the strands.

20. Careful comparison of the sequences reveals that there is a 7-bp region of complementarity at the 3′ ends of these two primers:

5′-GGATCGATG**CTCGCGA**-3′

3′-**GAGCGCT**GGGCTAGGA-5′

In a PCR experiment, these primers would likely anneal to one another, preventing their interaction with the template DNA. During DNA synthesis by the polymerase, each primer would act as a template for the other primer, leading to the amplification of a 25-bp sequence corresponding to the overlapped primers.

21. A mutation in person B has altered one of the alleles for gene X, leaving the other intact. The fact that the mutated allele is smaller suggests that a deletion has occurred in one copy of the gene. The one functioning copy is transcribed and translated and apparently produces enough protein to render the person asymptomatic.

Person C has only the smaller version of the gene. This gene is neither transcribed (negative northern blot) nor translated (negative western blot).

Person D has a normal-size copy of the gene but no corresponding RNA or protein. There may be a mutation in the promoter region of the gene that prevents transcription.

Person E has a normal-size copy of the gene that is transcribed, but no protein is made, which suggests that a mutation prevents translation. There are a number of possible explanations, including a mutation that introduced a premature stop codon in the mRNA.

Person F has a normal amount of protein but still displays the metabolic problem. This finding suggests that the mutation affects the activity of the protein—for instance, a mutation that compromises the active site of enzyme Y.

22. Chongqing: residue 2, L → R, CTG → CGG
Karachi: residue 5, A → P, GCC → CCC
Swan River: residue 6, D → G, GAC → GGC

23. This particular person is heterozygous for this particular mutation: one allele is wild type, whereas the other carries a point mutation at this position. Both alleles are PCR amplified in this experiment, yielding the "dual peak" appearance on the sequencing chromatogram.

Chapter 6

1. There are 27 identities (highlighted in yellow) and two gaps, for a score of 220. The two sequences are approximately 27% identical. For an alignment of 27% identity over almost 100 residues, it is likely these two proteins are evolutionarily related and structurally similar.

```
WYLGKITRMDAEVLLKKPTVRDGHFLVTQCESSPGEF
WYFGKITRRESERLLLNPENPRGTFLVRESETTKGAY

SISVRFGDSVQ-----HFKVLRDQNGKYYLWAVK-FN
CLSVSDFDNAKGLNVKHYKIRKLDSGGFYITSRTQFS

SLNELVAYHRTASVSRTHTILLSDMNV
SSLQQLVAYYSKHADGLCHRLTNV
```

2. They are likely related by divergent evolution, because three-dimensional structure is more conserved than is sequence identity.

3. (1) Identity score = −25; Blosum score = 14;
(2) identity score = 15; Blosum score = 4.

4. U. One possible structure:

5. There are 4^{40}, or 1.2×10^{24}, different molecules. Each molecule has a mass of 2.2×10^{-20}, because 1 mol of polymer has a mass of 330 g mol$^{-1} \times 40$, and there are 6.02×10^{23} molecules per mole. Therefore, 26.4 kg of RNA would be required.

6. Because three-dimensional structure is much more closely associated with function than is sequence, tertiary structure is more evolutionarily conserved than is primary structure. In other words, protein function is the most important characteristic, and protein function is determined by structure. Thus, the structure must be conserved but not necessarily a specific amino acid sequence.

7. Alignment score of sequences (1) and (2) is $6 \times 10 = 60$. Many answers are possible, depending on the randomly reordered sequence. A possible result is

Shuffled sequence: (2) TKADKAGEYL

Alignment: (1) ASNFLDKAGK
                    (2) TKADKAGEYL

Alignment score is $4 \times 10 = 40$.

8. (a) Almost certainly diverged from a common ancestor. (b) Almost certainly diverged from a common ancestor. (c) May have diverged from a common ancestor, but the sequence alignment may not provide supporting evidence. (d) May have diverged from a common ancestor, but the sequence alignment is unlikely to provide supporting evidence.

9. Replacement of cysteine, glycine, and proline never yields a positive score. Each of these residues exhibits features unlike those of its other 19 counterparts: cysteine is the only amino acid capable of forming disulfide bonds, glycine is the only amino acid without a side chain and is highly flexible, and proline is the only amino acid that is highly constrained through the bonding of its side chain to its amine nitrogen.

10. Protein A is clearly homologous to protein B, given 65% sequence identity, and so A and B are expected to have quite similar three-dimensional structures. Likewise, proteins B and C are clearly homologous, given 55% sequence identity, and so B and C are expected to have quite similar three-dimensional structures. Thus, proteins A and C are likely to have similar three-dimensional structures, even though they are only 15% identical in sequence.

11. To detect pairs of residues with correlated mutations, there must be variability in these sequences. If the alignment is over-represented by closely related organisms, there may not be enough changes in their sequences to allow the identification of potential base-pairing patterns.

12. After RNA molecules have been selected and reverse transcribed, PCR is performed to introduce additional mutations into these strands. The use of this error-prone, thermostable polymerase in the amplification step would enhance the efficiency of this random mutagenesis.

13. The initial pool of RNA molecules used in a molecular-evolution experiment is typically much smaller than the total number of possible sequences. Hence, the best possible RNA sequences will likely not be represented in the initial set of oligonucleotides. Mutagenesis of the initial selected RNA molecules allows for the iterative improvement of these sequences for the desired property.

14. 44% identity.

Chapter 7

1. The whale swims long distances between breaths. A high concentration of myoglobin in the whale muscle maintains a ready supply of oxygen for the muscle between breathing episodes.

2. (a) 2.96×10^{-11} g; (b) 2.74×10^8 molecules; (c) No. There would be 3.17×10^8 hemoglobin molecules in a red cell if they were packed in a cubic crystalline array. Hence, the actual packing density is about 84% of the maximum possible.

3. 2.65 g (or 4.75×10^{-2} mol) of Fe

4. (a) In human beings, 1.44×10^{-2} g (4.49×10^{-4} mol) of O_2 per kilogram of muscle. In sperm whale, 0.144 g (4.49×10^{-3} mol) of O_2 per kilogram. (b) 128:1

5. The pK_a is (a) lowered; (b) raised; and (c) raised.

6. Deoxy Hb A contains a complementary site, and so it can add on to a fiber of deoxy Hb S. The fiber cannot then grow further, because the terminal deoxy Hb A molecule lacks a sticky patch.

7. 62.7% oxygen-carrying capacity

8. Myoglobin does not exhibit a Bohr effect. The interactions responsible for mediating the Bohr effect in hemoglobin are dependent upon a tetrameric structure. Myoglobin is a monomer.

9. A higher concentration of BPG would shift the oxygen-binding curve to the right, causing an increase in P_{50}. The larger value of P_{50} would promote dissociation of oxygen in the tissues

and would thereby increase the percentage of oxygen delivered to the tissues.

10. (a) The transfusion would increase the number of red blood cells, which increases the oxygen-carrying capacity of the blood, allowing more sustained effort. (b) BPG stabilizes the T-state of hemoglobin, which results in a more efficient release of oxygen. If BPG is depleted, the oxygen will not be released even though the red blood cells are carrying more oxygen.

11. Oxygen binding appears to cause the copper ions and their associated histidine ligands to move closer to one another, thereby also moving the helices to which the histidines are attached (in similar fashion to the conformational change in hemoglobin).

12. The modified hemoglobin should not show cooperativity. Although the imidazole in solution will bind to the heme iron (in place of histidine) and will facilitate oxygen binding, the imidazole lacks the crucial connection to the particular α helix that must move so as to transmit the change in conformation.

13. Inositol pentaphosphate (part c) is highly anionic, much like 2,3-bisphosphoglycerate.

14.

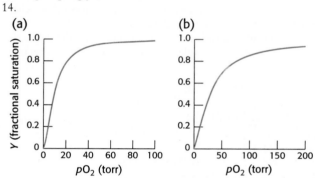

15. Release of acid will lower the pH. A lower pH promotes oxygen dissociation in the tissues. However, the enhanced release of oxygen in the tissues will increase the concentration of deoxy-Hb, thereby increasing the likelihood that the cells will sickle.

16. (a) $Y = 0.5$ when $pO_2 = 10$ torr. The plot of Y versus pO_2 appears to indicate little or no cooperativity.

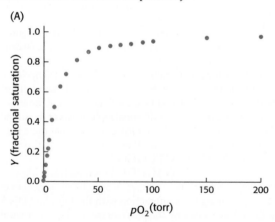

(b) The Hill plot shows slight cooperativity with $n = 1.3$ in the central region.

(B)

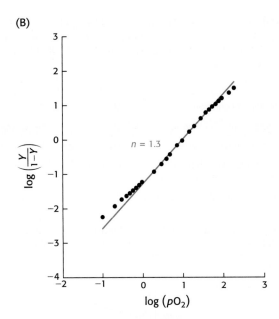

$n = 1.3$

(c) Deoxy dimers of lamprey hemoglobin could have lower affinity for oxygen than do the monomers. If the binding of the first oxygen atom to a dimer causes dissociation of the dimer to give two monomers, then the process would be cooperative. In this mechanism, oxygen binding to each monomer would be easier than binding the first oxygen atom to a deoxy dimer.

17. (a) 2; (b) 4; (c) 2; (d) 1.

18. The electrostatic interactions between BPG and hemoglobin would be weakened by competition with water molecules. The T state would not be stabilized.

19. Fetal hemoglobin contains two α chains and two γ chains. The increase in levels of the γ chain provides an alternative to the mutated β chain in sickle-cell anemia patients. Hence, by increasing fetal hemoglobin expression, hydroxyurea reduces the potential for the formation of insoluble hemoglobin aggregates.

Chapter 8

1. Rate enhancement and substrate specificity.

2. A cofactor.

3. Coenzymes and metals.

4. Vitamins are converted into coenzymes.

5. The active site is a three-dimensional crevice or cleft; it makes up only a small part of the total volume of the enzyme. Active sites have unique microenvironments. A substrate binds to the active site with multiple weak interactions. The specificity of the active site depends on the active site's precise three-dimensional structure.

6. Enzymes facilitate the formation of the transition state.

7. The intricate three-dimensional structure of proteins allows the construction of active sites that will recognize only specific substrates.

8. (a) 4; (b) 7; (c) 8; (d) 3; (e) 9; (f) 10; (g) 5; (h) 1; (i) 2; (j) 6.

9. The energy required to reach the transition state (the activation energy) is returned when the transition state proceeds to product.

10. The product is more stable than the substrate in graph A; so ΔG is negative and the reaction is exergonic. In graph B, the product has more energy than the substrate has; ΔG is positive, meaning that the reaction is endergonic.

11. (a) $K = \dfrac{[\text{P}]}{[\text{S}]} = \dfrac{k_F}{k_R} = \dfrac{10^{-4}}{10^{-6}} = 100$. Using equation 5 in the text, $\Delta G^{\circ\prime} = -11.42 \text{ kJ mol}^{-1}$ ($-2.73 \text{ kcal mol}^{-1}$). (b) $k_F = 10^{-2} \text{ s}^{-1}$

and $k_R = 10^{-4} \text{ s}^{-1}$. The equilibrium constant and $\Delta G^{\circ\prime}$ values are the same for both the uncatalyzed and catalyzed reactions.

12. Protein hydrolysis has a large activation energy. Protein synthesis must require energy to proceed.

13. The enzymes help protect the fluid that surrounds the eyes from bacterial infection.

14. Binding energy is the free energy released when two molecules bind together, such as when an enzyme and a substrate interact.

15. Binding energy is maximized when an enzyme interacts with the transition state, thereby facilitating the formation of the transition state and enhancing the rate of the reaction.

16. There would be no catalytic activity. If the enzyme–substrate complex is more stable than the enzyme–transition-state complex, the transition state would not form and catalysis would not take place.

17. Transition states are very unstable. Consequently, molecules that resemble transition states are themselves likely to be unstable and, hence, difficult to synthesize.

18. (a) 0; (b) $+28.53$; (c) -22.84; (d) -11.42; (e) $+5.69$.

19. (a) $\Delta G^{\circ\prime} = -RT \ln K'_{\text{eq}}$
$+1.8 = -(1.98 \times 10^{-3} \text{ kcal}^{-1} \text{ K}^{-1} \text{ mol}^{-1})(298 \text{ K})$
$(\ln[\text{G1P}]/[\text{G6P}])$
$-3.05 = \ln[\text{G1P}]/[\text{G6P}]$
$+3.05 = \ln[\text{G6P}]/[\text{G1P}]$
$K'^{-1}_{\text{eq}} = 21 \text{ or } K'_{\text{eq}} = 4.8 \times 10^{-2}$

Because $[\text{G6P}]/[\text{G1P}] = 21$, there is 1 molecule of G1P for every 21 molecules of G6P. Because we started with 0.1 M, the [G1P] is $1/22(0.1 \text{ M}) = 0.0045 \text{ M}$ and [G6P] must be $21/22(0.1 \text{ M})$ or 0.096 M. Consequently, the reaction does not proceed as written to a significant extent. (b) Supply G6P at a high rate and remove G1P at a high rate by other reactions. In other words, make sure that the [G6P]/[G1P] is kept large.

20. $K_{\text{eq}} = 19$, $\Delta G^{\circ\prime} = -7.3 \text{ kJ mol}^{-1}$ ($-1.77 \text{ kcal mol}^{-1}$)

21. The three-dimensional structure of an enzyme is stabilized by interactions with the substrate, reaction intermediates, and products. This stabilization minimizes thermal denaturation.

22. At substrate concentrations near the K_M, the enzyme displays significant catalysis yet is sensitive to changes in substrate concentration.

23. No, K_M is not equal to the dissociation constant and thus does not measure affinity because the numerator also contains k_2, the rate constant for the conversion of the enzyme substrate complex into enzyme and product. If, however, k_2 is much smaller than k_{-1}, $K_M \approx K_d$.

24. (a) 7; (b) 4; (c) 5; (d) 1; (e) 8; (f) 2; (g) 9; (h) 6; (i) 10; (j) 3.

25. Competitive inhibition: 2, 3, 9; uncompetitive: 4, 5, 6; noncompetitive: 1, 7, 8.

26. When $[\text{S}] = 10 \, K_M$, $V_0 = 0.91 \, V_{\max}$. When $[\text{S}] = 20 \, K_M$, $V_0 = 0.95 \, V_{\max}$. So, any Michaelis–Menten curves showing that the enzyme actually attains $V_{\max}$ are pernicious lies.

27. (a) 31.1 μmol; (b) 0.05 μmol; (c) 622 s^{-1}, a midrange value for enzymes (Table 8.5).

28. (a) Yes, $K_M = 5.2 \times 10^{-6}$ M; (b) $V_{\max} = 6.8 \times 10^{-10} \text{ mol minute}^{-1}$; (c) 337 s^{-1}.

29. Penicillinase, like glycopeptide transpeptidase, forms an acyl-enzyme intermediate with its substrate but transfers the intermediate to water rather than to the terminal glycine residue of the pentaglycine bridge.

30. (a) $V_{\max}$ is 9.5 μmol minute^{-1}. K_M is 1.1×10^{-5} M, the same as without inhibitor. (b) Noncompetitive. (c) 2.5×10^{-5} M (d) $f_{\text{ES}} = 0.73$, in the presence or absence of this noncompetitive inhibitor.

31. Group-specific reagents; affinity labels; suicide inhibitors; transition-state analogs.

32. (a) $V_0 = V_{max} - (V_0/[S]) K_M$
(b) Slope $= -K_M$, y-intercept $= V_{max}$, x-intercept $= V_{max}/K_M$
(c) An Eadie Hofstee plot.

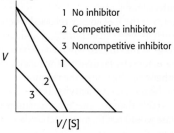

1 No inhibitor
2 Competitive inhibitor
3 Noncompetitive inhibitor

33. Sequential reactions are characterized by the formation of a ternary complex consisting of the enzyme and both substrates. Double-displacement reactions always require the formation of a temporarily substituted enzyme intermediate.

34. The rates of utilization of substrates A and B are given by

$$V_A = \left(\frac{k_{cat}}{K_M}\right)_A [E][A]$$

and

$$V_B = \left(\frac{k_{cat}}{K_M}\right)_B [E][B]$$

Hence, the ratio of these rates is

$$V_A/V_B = \left(\frac{k_{cat}}{K_M}\right)_B [A]/\left(\frac{k_{cat}}{K_M}\right)_A [B]$$

Thus, an enzyme discriminates between competing substrates on the basis of their values of k_{cat}/K_M rather than of K_M alone.

35. The mutation slows the reaction by a factor of 100 because the activation of free energy is increased by $+11.42$ kJ mol^{-1} ($+2.73$ kcal mol^{-1}). Strong binding of the substrate relative to the transition state slows catalysis.

36. 11 μmol minute^{-1}

37. (a) This piece of information is necessary for determining the correct dosage of succinylcholine to administer. (b) The duration of the paralysis depends on the ability of the serum cholinesterase to clear the drug. If there were one-eighth the amount of enzyme activity, paralysis could last eight times as long. (c) K_M is the concentration needed by the enzyme to reach $^1/_2 V_{max}$. Consequently, for a given concentration of substrate, the reaction catalyzed by the enzyme with the lower K_M will have the higher rate. The mutant patient with the higher K_M will clear the drug at a much lower rate.

38. (a) K_M is a measure of affinity *only* if k_2 is rate limiting, which is the case here. Therefore, the lower K_M means higher affinity. The mutant enzyme has a higher affinity. (b) 50 μmol minute^{-1}. 10 mM is K_M, and K_M yields $^1/_2 V_{max}$. V_{max} is 100 μmol minute^{-1}, and so... (c) Enzymes do not alter the equilibrium of the reaction.

39. Enzyme 2. Despite the fact that enzyme 1 has a higher V_{max} than enzyme 2, enzyme 2 shows greater activity at the concentration of the substrate in the environment because enzyme 2 has a lower K_M for the substrate.

40. (a) The most effective means of measuring the efficiency of any enzyme substrate complex is to determine the k_{cat}/K_M values. For the three substrates in question, the respective values of k_{cat}/K_M are: 6, 15, and 36. Thus, the enzyme exhibits a strong preference for cleaving peptide bonds in which the second amino acid is a large hydrophobic amino acid. (b) The k_{cat}/K_M for this substrate is 2. Not very effective. This value suggests that the enzyme prefers to cleave peptide bonds with the following specificity: small R group—large hydrophobic R group.

41. If the total amount of enzyme $[E]_T$ is increased, V_{max} will increase, because $V_{max} = k_2[E]_T$. But $K_M = (k_{-1} + k_2)/k_1$; that is, it is independent of substrate concentration. The middle graph describes this situation.

42.

Experimental condition	V_{max}	K_M
(a) Twice as much enzyme is used	Doubles	No change
(b) Half as much enzyme is used	Half as large	No change
(c) A competitive inhibitor is present	No change	Increases
(d) An uncompetitive inhibitor is present	Decreases	Decreases
(e) A pure noncompetitive is present	Decreases	No change

43. (a)

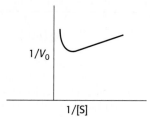

(b) This behavior is substrate inhibition: at high concentrations, the substrate forms unproductive complexes at the active site. The adjoining drawing shows what might happen. Substrate normally binds in a defined orientation, shown in the drawing as red to red and blue to blue. At high concentrations, the substrate may bind at the active site such that the proper orientation is met for each end of the molecule, but two different substrate molecules are binding.

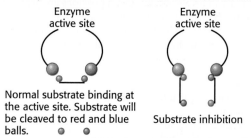

Normal substrate binding at the active site. Substrate will be cleaved to red and blue balls.

Substrate inhibition

44. The first step will be the rate-limiting step. Enzymes E_B and E_C are operating at $^1/_2 V_{max}$, whereas the K_M for enzyme E_A is greater than the substrate concentration. E_A would be operating at approximately $10^{-2}V_{max}$.

45. The fluorescence spectroscopy reveals the existence of an enzyme–serine complex and of an enzyme–serine–indole complex.

46. (a) When $[S^+]$ is much greater than the value of K_M, pH will have a negligible effect on the enzyme because S^+ will interact with E^- as soon as the enzyme becomes available.

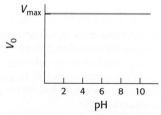

(b) When $[S^+]$ is much less than the value of K_M, the plot of V_0 versus pH becomes essentially a titration curve for the ionizable groups, with enzyme activity being the titration marker. At low pH, the high concentration of H$^+$ will keep the enzyme in the EH form and inactive. As the pH rises, more and more of the enzyme will be in the E⁻ form and active. At high pH (low H⁺), all of the enzyme is E⁻.

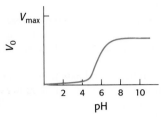

(c) The midpoint on this curve will be the pK_a of the ionizable group, which is stated to be pH 6.

47. (a) Incubating the enzyme at 37°C leads to a denaturation of enzyme structure and a loss of activity. For this reason, most enzymes must be kept cool if they are not actively catalyzing their reactions. (b) The coenzyme apparently helps to stabilize enzyme structure, because enzyme from PLP-deficient cells denatures faster. Cofactors often help stabilize enzyme structure.

48. The slow second-order step occurs when complementary sequences on two different molecules bind with one another. Once the initial sequence alignment occurs, the remainder of the molecule can quickly locate complementary sequences from this nucleation site and reanneal, a first-order reaction.

Chapter 9

1. For the amide substrate, the formation of the acyl-enzyme intermediate is slower than the hydrolysis of the acyl-enzyme intermediate, and so no burst is observed. A burst is observed for ester substrates, because the formation of the acyl-enzyme intermediate is faster, leading to the observed burst.

2. The histidine residue in the substrate can substitute to some extent for the missing histidine residue of the catalytic triad of the mutant enzyme.

3. No. The catalytic triad works as a unit. After this unit has been made ineffective by the mutation of histidine to alanine, the further mutation of serine to alanine should have only a small effect.

4. The substitution corresponds to one of the key differences between trypsin and chymotrypsin, and so trypsin-like specificity (cleavage after lysine and arginine) might be predicted. In fact, additional changes are required to effect this specificity change.

5. Imidazole is apparently small enough to reach the active site of carbonic anhydrase and compensate for the missing histidine. Buffers with large molecular components cannot do so, and the effects of the mutation are more evident.

6. No, because the enzyme would destroy the host DNA before protective methylation could take place.

7. No. The bacteria receiving the enzyme would have their own DNA destroyed because they would likely lack the appropriate protective methylase.

8. EDTA will bind to Zn^{2+} and remove the ion, which is required for enzyme activity, from the enzyme.

9. (a) The aldehyde reacts with the active-site serine. (b) A hemiacetal is formed.

10. Trypsin, because molecule A contains a lysine side chain, which can fit into the S_1 pocket of trypsin.

11. The reaction is expected to be slower by a factor of 10 because the rate depends on the pK_a of the zinc-bound water. $k_{cat} = 60{,}000 \text{ s}^{-1}$.

12. EDTA binds the Mg^{2+} necessary for the enzymatic reaction to proceed.

13. If the aspartate is mutated, the protease is inactive and the virus will not be viable.

14. Water substitutes for the hydroxyl group of serine 236 in mediating proton transfer from the attacking water and the γ-phosphoryl group.

15. For subtilisin, the catalytic power is approximately $30 \text{ s}^{-1}/10^{-8} \text{ s}^{-1} = 3 \times 10^9$. For carbonic anhydrase (at pH 7), the catalytic power is approximately $500{,}000 \text{ s}^{-1}/0.15 \text{ s}^{-1} = 3.3 \times 10^6$. Subtilisin is the more powerful enzyme by this criterion.

16. The triple mutant still catalyzes the reaction by a factor of approximately 1000-fold compared to the uncatalyzed reaction. This could be due to the enzyme binding the substrate and holding it in a conformation that is susceptible to attack by water.

17. (a) Cysteine protease: The same as Figure 9.8, except that cysteine replaces serine in the active site and no aspartate is present.
(b) Aspartyl protease:

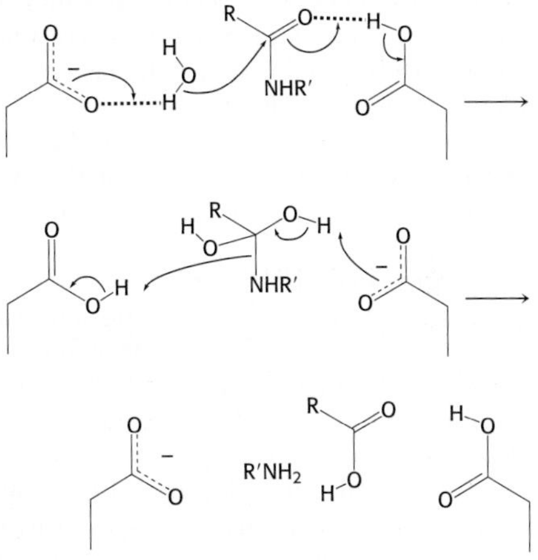

(c) Metalloprotease:

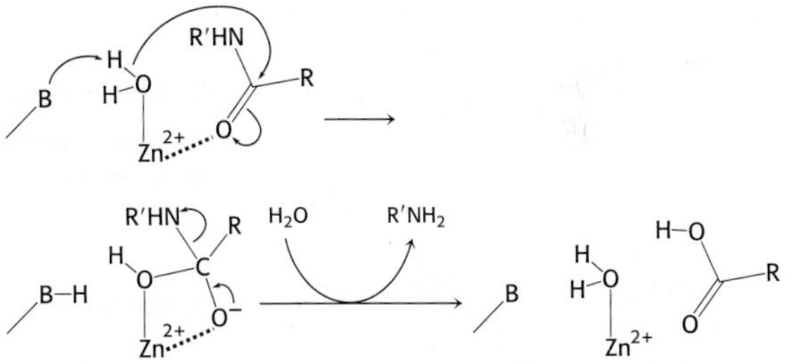

Chapter 10

1. The enzyme catalyzes the first step in the synthesis of pyrimidines. It facilitates the condensation of carbamoyl phosphate and aspartate to form *N*-carbamoylaspartate and inorganic phosphate.

2. (a) 5; (b) 4; (c) 7; (d) 3; (e) 10; (f) 8; (g) 1; (h) 9; (i) 6; (j) 2.

3. a, b, and d are true.

4. The protonated form of histidine probably stabilizes the negatively charged carbonyl oxygen atom of the scissile bond (bond to be broken) in the transition state. Deprotonation would lead to a loss of activity. Hence, the rate is expected to be half maximal at a pH of about 6.5 (the pK_a of an unperturbed histidine side chain in a protein) and to decrease as the pH is raised.

5. The inhibition of an allosteric enzyme by the end product of the pathway controlled by the enzyme. It prevents the production of too much end product and the consumption of substrates when product is not required.

6. High concentrations of ATP might signal two overlapping situations. The high levels of ATP might suggest that some *nucleotides* are available for nucleic acid synthesis, and consequently, UTP and CTP should be synthesized. The high levels of ATP indicate that *energy* is available for nucleic acid synthesis, and so UTP and CTP should be produced.

7. All of the enzyme would be in the R form all of the time. There would be no cooperativity. The kinetics would look like that of a Michaelis–Menten enzyme.

8. The enzyme would show simple Michaelis–Menten kinetics because it is essentially always in the R state.

9. CTP is formed by the addition of an amino group to UTP. Evidence indicates the UTP is also capable of inhibiting ATCase in the presence of CTP.

10. Homotropic effectors are the substrates of allosteric enzymes. Heterotropic effectors are the regulators of allosteric enzymes. Homotropic effectors account for the sigmoidal nature of the velocity versus substrate concentration curve, whereas heterotropic effectors alter the K_M of the curve. Ultimately, both types of effectors work by altering the T/R ratio.

11. If substrates had been used, the enzyme would catalyze the reaction. Intermediates would not accumulate on the enzyme. Consequently, any enzyme that crystallized would have been free of substrates or products.

12. (a) 100. The change in the [R]/[T] ratio on binding one substrate molecule must be the same as the ratio of the substrate affinities of the two forms. (b) 10. The binding of four substrate molecules changes the [R]/[T] by a factor of $100^4 = 10^8$. The ratio in the absence of substrate is 10^{-7}. Hence, the ratio in the fully liganded molecule is $10^8 \times 10^{-7} = 10$.

13. The fraction of molecules in the R form is 10^{-5}, 0.004, 0.615, 0.998, and 1 when 0, 1, 2, 3, and 4 ligands, respectively, are bound.

14. The sequential model can account for negative cooperativity, whereas the concerted model cannot.

15.

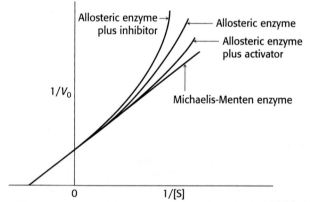

16. The net outcome of the two reactions is the hydrolysis of ATP to ADP and P_i, which has a ΔG of $-50\ kJ\ mol^{-1}$ ($-12\ kcal\ mol^{-1}$) under cellular conditions.

17. Isozymes are homologous enzymes that catalyze the same reaction but have different kinetic or regulatory properties.

18. Although the same reaction may be required in a variety of tissues, the biochemical properties of tissues will differ according to their biological function. Isozymes allow the fine-tuning of catalytic and regulatory properties to meet the specific needs of the tissue.

19. (a) 7; (b) 8; (c) 11; (d) 6; (e) 1; (f) 12; (g) 3; (h) 4; (i) 5; (j) 2; (k) 10; (l) 9.

20. When phosphorylation takes place at the expense of ATP, sufficient energy is expended to dramatically alter the structure and hence activity of a protein. Moreover, because ATP is the cellular energy currency, protein modification is linked to the energy status of the cell.

21. Covalent modification is reversible, whereas proteolytic cleavage is irreversible.

22. Activation is independent of zymogen concentration because the reaction is intramolecular.

23. Although quite rare, cases of enteropeptidase deficiency have been reported. The affected person has diarrhea and fails to thrive because digestion is inadequate. In particular, protein digestion is impaired.

24. Add blood from the second patient to a sample from the first. If the mixture clots, the second patient has a defect different from that of the first. This type of assay is called a complementation test.

25. Activated factor X remains bound to blood-platelet membranes, which accelerates its activation of prothrombin.

26. Antithrombin III is a very slowly hydrolyzed substrate of thrombin. Hence, its interaction with thrombin requires a fully formed active site on the enzyme.

27. Replacing methionine with leucine would be a good choice. Leucine is resistant to oxidation and has nearly the same volume and degree of hydrophobicity as methionine has.

28. Inappropriate clot formation could block arteries in the brain, causing a stroke, or the heart, causing a heart attack.

29. Thrombin catalyzes the hydrolysis of fibrinogen to form active fibrin. But it also has a role in shutting down the cascade by activating protein C, a protease that digests other clotting enzymes V_a and VIII_a.

30. Tissue-type plasminogen activator, or TPA, is a serine protease that leads to the dissolution of blood clots. TPA activates plasminogen that is bound to a fibrin clot, converting it into active plasmin, which then hydrolyzes the fibrin of the clot.

31. A mature clot is stabilized by amide linkages between the side chains of lysine and glutamine that are absent in a soft clot. The linkages are formed by transglutaminase.

32. The simple sequential model predicts that the fraction of catalytic chains in the R state, f_R, is equal to the fraction containing bound substrate, Y. The concerted model, in contrast, predicts that f_R increases more rapidly than Y as the substrate concentration is increased. The change in f_R leads to the change in Y on addition of substrate, as predicted by the concerted model.

33. The binding of succinate to the functional catalytic sites of the native c_3 moiety changed the visible absorption spectrum of nitrotyrosine residues in the *other* c_3 moiety of the hybrid enzyme. Thus, the binding of substrate analog to the active sites of one trimer altered the structure of the other trimer.

34. According to the concerted model, an allosteric activator shifts the conformational equilibrium of all subunits toward the R state, whereas an allosteric inhibitor shifts it toward the T state. Thus, ATP (an allosteric activator) shifted the equilibrium to the R form, resulting in an absorption change similar to that obtained when substrate is bound. CTP had a different effect. Hence, this allosteric inhibitor shifted the equilibrium to the T form. Thus, the concerted model accounts for the ATP-induced and CTP-induced (heterotropic), as well as for the substrate-induced (homotropic), allosteric interactions of ATCase.

35. (a) When crowded, the control group displays gregarious behavior. (b) Inhibition of PKA appears to prevent gregarious behavior, while inhibition of PKG has no effect on the behavior. (c) The effect of PKG inhibition was investigated to establish that the effect seen with PKA inhibition is specific and not just due to the inhibition of any kinase. (d) PKA plays a role in altering behavior of insects. (e) Gregariousness in locusts that are always gregarious is not affected by PKA inhibition, suggesting that PKA may play a role in modifying behavior patterns but not in establishing the patterns.

36. Residues *a* and *d* are located in the interior of an α-helical coiled coil, near the axis of the superhelix. Hydrophobic interactions between these side chains contribute to the stability of the coiled coil.

37. In the R state, ATCase expands and becomes less dense. This decrease in density results in a decrease in the sedimentation value (see the formula on p. 80).

38. The interaction between trypsin and the inhibitor is so stable that the transition state is rarely formed. Recall that maximal binding energy is released when an enzyme binds to the transition state. If the substrate-enzyme interaction is too stable, the transition state rarely forms.

39. Dicoumarol is a competitive inhibitor of γ-glutamyl carboxylase. Consequently, γ-carboxyglutamate, which is required for prothrombin to be converted into thrombin, is not formed. Amino acid composition is determined by hydrolyzing the protein at elevated temperatures in the presence of strong acid. Under these conditions, the carboxyl group of γ-carboxyglutamate is removed from the normal prothrombin, leaving simply glutamate.

40.

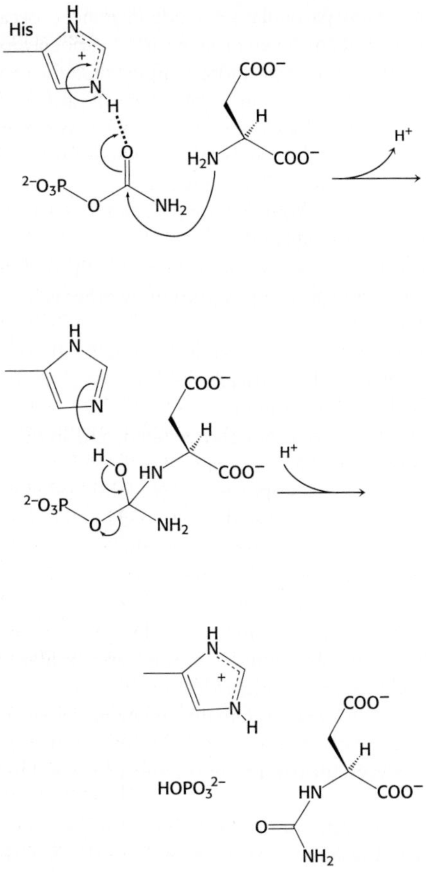

41.

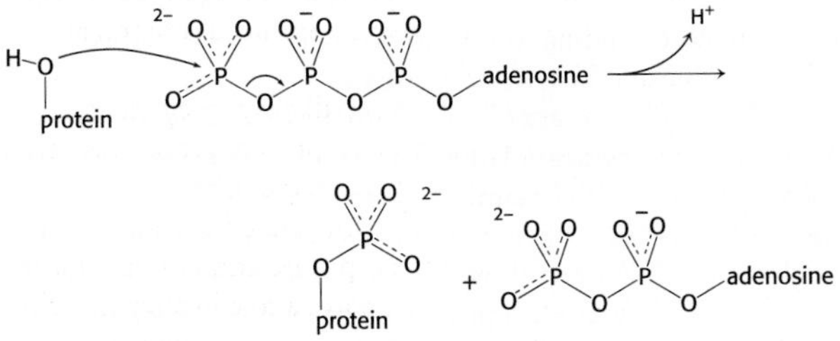

Chapter 11

1. Carbohydrates were originally regarded as *hydrates* of *carbon* because the empirical formula of many of them is $(CH_2O)_n$.

2. (d)

3. (a)

4. (d)

5. Three amino acids can be linked by peptide bonds in only six different ways. However, three different monosaccharides can be linked in a plethora of ways. The monosaccharides can be linked in a linear or branched manner, with α or β linkages, with bonds between C-1 and C-3, between C-1 and C-4, between C-1 and C-6, and so forth. In fact, the three monosaccharides can form 12,288 different trisaccharides.

6. (a) 10; (b) 6; (c) 8; (d) 9; (e) 2; (f) 4; (g) 1; (h) 5; (i) 7; (j) 3.

7. (a) aldose-ketose; (b) epimers; (c) aldose-ketose (d) anomers; (e) aldose-ketose; (f) epimers.

8. Erythrose: tetrose aldose; ribose: pentose aldose; glyceraldehyde: triose aldose; dihydroxyacetone: triose ketose; erythrulose: tetrose ketose; ribulose: pentose ketose; fructose: hexose ketose.

9.

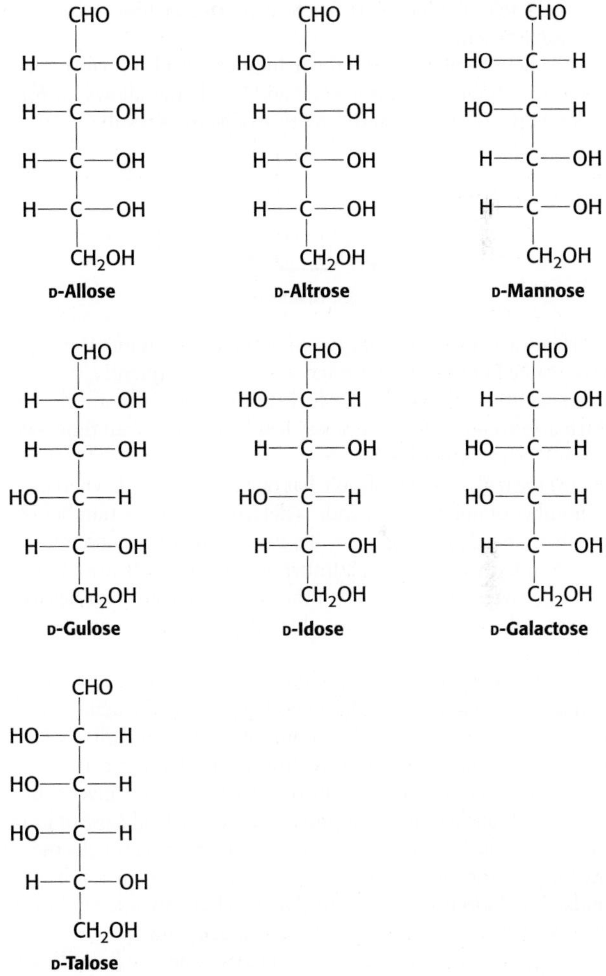

ᴅ-Allose ᴅ-Altrose ᴅ-Mannose

ᴅ-Gulose ᴅ-Idose ᴅ-Galactose

ᴅ-Talose

10.

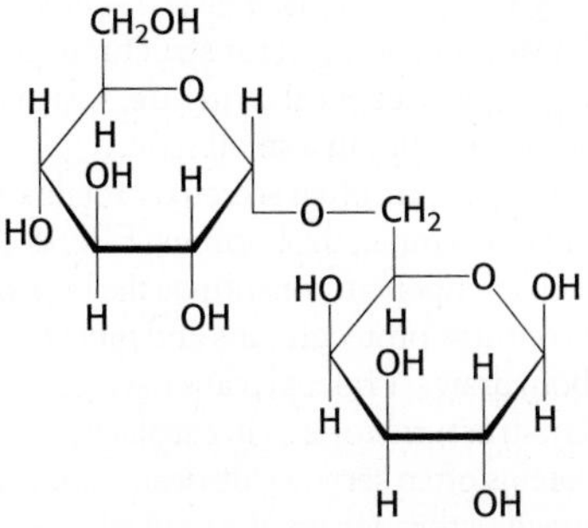

α-Glucosyl-(1 → 6)-galactose

11. The proportion of the α anomer is 0.36, and that of the β anomer is 0.64.

12. Glucose is reactive because of the presence of an aldehyde group in its open-chain form. The aldehyde group slowly condenses with amino groups to form aldimine products of a type called Schiff-base adducts (a compound with the general structure $R_2C = NR'$, where R' is not an H).

13. A pyranoside reacts with two molecules of periodate; formate is one of the products. A furanoside reacts with only one molecule of periodate; formate is not formed.

14. (a) β-D-Mannose; (b) β-D-galactose; (c) β-D-fructose; (d) β-D-glucosamine.

15. The trisaccharide itself should be a competitive inhibitor of cell adhesion if the trisaccharide unit of the glycoprotein is critical for the interaction.

16. (a) Not a reducing sugar; no open-chain forms are possible. (b) D-galactose, D-glucose, D-fructose. (c) D-galactose and sucrose (glucose + fructose).

17. The hemiketal linkage of the α anomer is broken to form the open form. Rotation about the C-1 and C-2 bonds allows the formation of the β anomer, and a mixture of isomers results.

β-D-Mannose

18. Heating converts the very sweet pyranose form into the more-stable but less-sweet furanose form. Consequently, the sweetness of the preparation is difficult to accurately control, which also accounts for why honey loses sweetness with time. See Figure 11.5 for structures.

19. (a) Each glycogen molecule has one reducing end, whereas the number of nonreducing ends is determined by the number of branches, or α-1,6 linkages. (b) Because the number of nonreducing ends greatly exceeds the number of reducing ends in a collection of glycogen molecules, all of the degradation and synthesis of glycogen takes place at the nonreducing ends, thus maximizing the rate of degradation and synthesis.

20. No, sucrose is not a reducing sugar. The anomeric carbon atom acts as the reducing agent in both glucose and fructose but, in sucrose, the anomeric carbon atoms of fructose and glucose are joined by a covalent bond and are thus not available to react.

21. Glycogen is a polymer of glucose linked by α-1,4-glycosidic bonds with branches formed approximately every 12 glucose units by α-1,6-glycosidic bonds. Starch consists of two polymers of glucose. Amylose is a straight-chain polymer formed by α-1,4-glycosidic bonds. Amylopectin is similar to glycogen but amylopectin has fewer branches, one branch per 30 or so glucose units.

22. Cellulose is a linear polymer of glucose joined by β-1,4 linkages. Glycogen is a branched polymer with the main chain being formed by α-1,4-glycosidic bonds. The β-1,4 linkages allow the formation of a linear polymer ideal for structural roles. The α-1,4 linkages of glycogen form a helical structure, which allows the storage of many glucose moieties in a small space.

23. Simple glycoproteins are often secreted proteins and thus play a variety of roles. For example, the hormone EPO is a glycoprotein. Usually, the protein component constitutes the bulk of the glycoprotein by mass. In contrast, proteoglycans and mucoproteins are predominantly carbohydrates. Proteoglycans have glycosaminoglycans attached, and play structural roles as in cartilage and the extracellular matrix. Mucoproteins often serve as lubricants and have multiple carbohydrates attached through an N-acetylgalactosamine moiety.

24. The attachment of the carbohydrate allows the EPO to stay in circulation longer and thus to function for longer periods of time than would a carbohydrate-free EPO.

25. The glycosaminoglycan, because it is heavily charged, binds many water molecules. When cartilage is stressed, such as when your heel hits the ground, the water is released, thus cushioning the impact. When you lift your heel, the water rebinds.

26. The lectin that binds the mannose 6-phosphate might be defective and not recognize a correctly addressed protein.

27. Asparagine, serine, and threonine.

28. Different molecular forms of a glycoprotein that differ in the amount of carbohydrate attached, or the location of attachment, or both.

29. The total collection of carbohydrates synthesized by a cell at particular times and under particular environmental conditions.

30. The genome comprises all of the genes present in an organism. The proteome includes all of the possible protein products and modified proteins that a cell expresses under any particular set of circumstances. The glycome consists of all of the carbohydrates synthesized by the cell under any particular set of circumstances. Because the genome is static, but any given protein can be variously expressed and modified, the proteome is more complex than the genome. The glycome, which includes not only glycoforms of proteins, but also many possible carbohydrate structures, must be even more complex.

31. An asparagine residue can be glycosylated if the residue is part of an Asn-X-Ser or Asn-X-Thr sequence, in which X can be any residue, except proline. Your roommate's hedging is because not all potential sites are glycosylated.

32. It suggests that carbohydrates are on the cell surfaces of all organisms for the purpose of recognition by other cells, organisms, or the environment.

33. A glycoprotein is a protein that is decorated with carbohydrates. A lectin is a protein that specifically recognizes carbohydrates. A lectin can also be a glycoprotein.

34. Each site either is or is not glycosylated, and so there are $2^6 = 64$ possible proteins.

35. The 20 amino acid components of proteins and the 4 nucleotide components of nucleic acids are linked by the same type of bond—the peptide bond in proteins and the 5′-to-3′ phosphodiester linkage in nucleic acids. In contrast, there are many different carbohydrates that can be modified and linked in a variety of ways. Moreover, oligosaccharides can be branched. Finally, many sugars have the same or similar chemical formula, and similar chemical properties, making specific identification and linkage difficult.

36. As discussed in Chapter 9, many enzymes display stereochemical specificity. Clearly, the enzymes of sucrose synthesis are able to distinguish between the isomers of the substrates and link only the correct pair.

37. If the carbohydrate specificity of the lectin is known, an affinity column with the appropriate carbohydrate attached could be prepared. The protein preparation containing the lectin of interest could be passed over the column. The use of this method was indeed how the glucose-binding lectin concanavalin A was purified.

38. (a) Aggrecan is heavily decorated with glycosaminoglycans. If glycosaminoglycans are released into the media, aggrecan must be undergoing degradation. (b) Another enzyme might be present that cleaves glycosaminoglycans from aggrecan without degrading aggrecan. Other experiments not shown established that glycosaminoglycan release is an accurate measure of aggrecan destruction. (c) The control provides a baseline of "background" degradation inherent in the assay. (d) Aggrecan degradation is greatly enhanced. (e) Aggrecan degradation is reduced to the background system. (f) It is an in vitro system in which not all the factors contributing to cartilage stabilization in vivo are present.

Chapter 12

1. 2.86×10^6 molecules, because each leaflet of the bilayer contains 1.43×10^6 molecules.

2. Essentially an "inside-out" membrane. The hydrophilic groups would come together on the interior of the structure, away from the solvent, whereas the hydrocarbon chains would interact with the solvent.

3. 2×10^{-7} cm, 6×10^{-6} cm, and 2×10^{-4} cm.

4. The radius of this molecule is 3.1×10^{-7} cm, and its diffusion coefficient is 7.4×10^{-9} cm^2 s^{-1}. The average distances traversed are 1.7×10^{-7} cm in 1 μs, 5.4×10^{-6} cm in 1 ms, and 1.7×10^{-4} cm in 1 s.

5. The membrane underwent a phase transition from a highly fluid to a nearly frozen state when the temperature was lowered. A carrier can shuttle ions across a membrane only when the bilayer is highly fluid. A channel, in contrast, allows ions to traverse its pore even when the bilayer is quite rigid.

6. The presence of a cis double bond introduces a kink in the fatty acid chain that prevents tight packing and reduces the number of atoms in van der Waals contact. The kink lowers the melting point compared with that of a saturated fatty acid. Trans fatty acids do not have the kink, and so their melting temperatures are higher, more similar to those of saturated fatty acids. Because trans fatty acids have no structural effect, they are rarely observed.

7. Palmitic acid is shorter than stearic acid. So, when the chains pack together, there is less opportunity for van der Waals interaction and the melting point is thus lower than that of the longer stearic acid.

8. Hibernators selectively feed on plants that have a high proportion of polyunsaturated fatty acids with lower melting temperature.

9. The initial decrease in fluorescence with the first addition of sodium dithionite results from the quenching of NBD-PS molecules in the outer leaflet of the bilayer. Sodium dithionite does not traverse the membrane under these experimental conditions; hence, it does not quench the labeled phospholipids in the inner leaflet. A second addition of sodium dithionite has no effect, as the NBD-PS molecules in the outer leaflet remain quenched. However, after a 6.5-hour incubation, about half the NBD-PS has flipped over to the outer leaflet of the bilayer, resulting in the 50% decrease in fluorescence when sodium dithionite is added.

10. The addition of the carbohydrate introduces a significant energy barrier to the flip-flop because a hydrophilic carbohydrate moiety would need to be moved through a hydrophobic environment. This energetic barrier enhances membrane asymmetry.

11. The C_{16} alkyl chain is attached by an ether linkage. The C-2 carbon atom of glycerol has only an acetyl group attached by an ester linkage instead of a fatty acid, as is the case with most phospholipids.

12. In a hydrophobic environment, the formation of intrachain hydrogen bonds stabilizes the amide hydrogen atoms and carbonyl oxygen atoms of the polypeptide chain, and so an α helix forms. In an aqueous environment, these groups are stabilized by interaction with water, and so there is no energetic reason to form an α helix. Thus, the α helix would be more likely to form in a hydrophobic environment.

13. The protein may contain an α helix that passes through the hydrophobic core of the protein. This helix is likely to feature a stretch of hydrophobic amino acids similar to those observed in transmembrane helices.

14. The shift to the lower temperature would decrease fluidity by enhancing the packing of the hydrophobic chains by van der Waals interactions. To prevent this packing, new phospholipids having shorter chains and a greater number of cis double bonds would be synthesized. The shorter chains would reduce the number of van der Waals interactions, and the cis double bonds, which cause the kink in structure, would prevent the packing of the fatty acid tails of the phospholipids.

15. Each of the 21 v-SNARE proteins could interact with each of 7 t-SNARE partners. Multiplication gives the total number of different interacting pairs: $7 \times 21 = 147$ different v-SNARE–t-SNARE pairs.

16. (a) The graph shows that, as temperature increases, the phospholipid bilayer becomes more fluid. T_m is the temperature of the transition from the predominantly less-fluid state to the predominantly more-fluid state. Cholesterol broadens the transition from the less-fluid to the more-fluid state. In essence, cholesterol makes membrane fluidity less sensitive to temperature changes. (b) This effect is important because the presence of cholesterol tends to stabilize membrane fluidity by preventing sharp transitions. Because protein function depends on the proper fluidity of the membrane, cholesterol maintains the proper environment for membrane-protein function.

17. The protein plotted in part c is a transmembrane protein from C. elegans. It spans the membrane with four α helices that are prominently displayed as hydrophobic peaks in the hydropathy plot. Interestingly, the protein plotted in part a also is a membrane protein, a porin. This protein is made primarily of β strands, which lack the prominent hydrophobic window of membrane helices. This example shows that, although hydropathy plots are useful, they are not infallible.

18. (a) Prostaglandin H_2 synthase-1 recovers its activity immediately after removal of ibuprofen, suggesting that this inhibitor dissociates rapidly from the enzyme. In contrast, the enzyme remains significantly inhibited 30 minutes after removal of indomethacin, suggesting that this inhibitor dissociates slowly from its active site. (b) Aspirin covalently modifies prostaglandin H_2 synthase-1, indicating that it would dissociate very slowly (if at all). Hence, one would anticipate that very low activity would be evident in all conditions where inhibitor has been added (columns 2, 3, and 4).

19. To purify any protein, the protein must first be solubilized. For a membrane protein, solubilization usually requires a detergent—hydrophobic molecules that bind to the protein and thus replace the lipid environment of the membrane. If the detergent is removed, the protein aggregates and precipitates from solution. Often, the steps in purification, such as ion-exchange chromatography, are difficult to perform in the presence of sufficient detergent to solubilize the protein. Crystals of appropriate protein–detergent complexes must be generated.

Chapter 13

1. In simple diffusion, the substance in question can diffuse down its concentration gradient through the membrane. In facilitated diffusion, the substance is not lipophilic and cannot directly diffuse through the membrane. A channel or carrier is required to facilitate movement down the gradient.

2. Two forms are (1) ATP hydrolysis and (2) the movement of one molecule down its concentration gradient coupled with the movement of another molecule up its concentration gradient.

3. The three types of carriers are symporters, antiporters, and uniporters. Symporters and antiporters can mediate secondary active transport.

4. The free-energy cost is $+32$ kJ mol^{-1} ($+7.6$ kcal mol^{-1}). The chemical work performed is $+20.4$ kJ mol^{-1} ($+4.9$ kcal mol^{-1}), and the electrical work performed is $+11.5$ kJ mol^{-1} ($+2.8$ kcal mol^{-1}).

5. For chloride, $z = -1$; for calcium $z = +2$. At the concentrations given, the equilibrium potential for chloride is -97 mV and the equilibrium potential for calcium is $+122$ mV.

6. By analogy with the Ca^{2+} ATPase, with three Na^+ ions binding from inside the cell to the E_1 conformation and with two K^+ ions binding from outside the cell to the E_2 conformation, a plausible mechanism is as follows:

(i) A catalytic cycle could begin with the enzyme in its unphosphorylated state (E_1) with three sodium ions bound.
(ii) The E_1 conformation binds ATP. A conformational change traps sodium ions inside the enzyme.
(iii) The phosphoryl group is transferred from ATP to an aspartyl residue.
(iv) On ADP release, the enzyme changes its overall conformation, including the membrane domain. This new conformation (E_2) releases the sodium ions to the side of the membrane opposite that at which they entered and binds two potassium ions from the side where sodium ions are released.
(v) The phosphorylaspartate residue is hydrolyzed to release inorganic phosphate. With the release of phosphate, the interactions stabilizing E_2 are lost, and the enzyme everts to the E_1 conformation. Potassium ions are released to the cytoplasmic side of the membrane. The binding of three sodium ions from the cytoplasmic side of the membrane completes the cycle.

7. Establish a lactose gradient across vesicle membranes that contain properly oriented lactose permease. Initially, the pH should be the same on both sides of the membrane and the lactose concentration should be higher on the "exit" side of lactose permease. As the lactose flows "in reverse" through the permease, down its concentration gradient, whether or not a pH gradient becomes established as the lactose gradient is dissipated can be tested.
8. Ligand-gated channels open in response to the binding of a molecule by the channel, whereas voltage-gated channels open in response to changes in the membrane potential.
9. An ion channel must transport ions in either direction at the same rate. The net flow of ions is determined only by the composition of the solutions on either side of the membrane.
10. Uniporters act as enzymes do; their transport cycles include large conformational changes, and only a few molecules interact with the protein per transport cycle. In contrast, channels, after having opened, provide a pore in the membrane through which many ions may pass. As such, channels mediate transport at a much higher rate than do uniporters.
11. FCCP effectively creates a pore in the bacterial membrane through which protons can pass rapidly. Protons that are pumped out of the bacteria will pass through this pore preferentially (the "path of least resistance"), rather than participate in H^+/lactose symport.
12. Cardiac muscle must contract in a highly coordinated manner in order to pump blood effectively. Numerous gap junctions, abundant in connexin proteins, mediate the orderly cell-to-cell propagation of the action potential through the heart during each beat.
13. The positively charged guanidinium group resembles Na^+ and binds to negatively charged carboxylate groups in the mouth of the channel.
14. SERCA, a P-type ATPase, uses a mechanism by which a covalent phosphorylated intermediate (at an aspartate residue) is formed. At steady state, a subset of the SERCA molecules is trapped in the E_2-P state and, as a result, radiolabeled. The MDR protein is an ABC transporter and does not operate through a phosphorylated intermediate. Hence, a radiolabeled band would not be observed for MDR.
15. The blockage of ion channels inhibits action potentials, leading to loss of nervous function. Like tetrodotoxin, these toxin molecules are useful for isolating and specifically inhibiting particular ion channels.

16. After repolarization, the ball domains of the ion channels engage the channel pore, rendering them inactive for a short period of time. During this time, the channels cannot be reopened until the ball domains disengage and the channel returns to the "closed" state.
17. Because sodium ions are charged and sodium channels carry only sodium ions (but not anions), the accumulation of excess positive charge on one side of the membrane occurs very rapidly. The establishment of a membrane potential in this manner dominates long before any chemical gradients form.
18. A mutation that impairs the ability of the sodium channel to inactivate would prolong the duration of the depolarizing sodium current, thus lengthening the cardiac action potential.
19. No; an external stimulus will control whether ion channels are more likely to be open or closed (in a manner similar to voltage-gated ion channels), but the unit conductance of the open channel will be influenced very little.
20. The ratio of closed-to-open forms of the channel is 10^5, 5000, 250, 12.5, and 0.625 when zero, one, two, three, and four ligands, respectively, are bound. Hence, the fraction of open channels is 1.0×10^{-5}, 2.0×10^{-4}, 4.0×10^{-3}, 7.4×10^{-2}, and 0.62.
21. These organic phosphates inhibit acetylcholinesterase by reacting with the active-site serine residue to form a stable phosphorylated derivative. They cause respiratory paralysis by blocking synaptic transmission at cholinergic synapses.
22. (a) The binding of the first acetylcholine molecule increases the open-to-closed ratio by a factor of 240, and the binding of the second increases it by a factor of 11,700. (b) The free-energy contributions are $+14$ kJ mol^{-1} ($+3.3$ kcal mol^{-1}) and $+23$ kJ mol^{-1} ($+5.6$ kcal mol^{-1}), respectively. (c) No; the MWC model predicts that the binding of each ligand will have the same effect on the open-to-closed ratio.
23. Batrachotoxin blocks the transition from the open to the closed state.
24. (a) Chloride ions flow into the cell. (b) Chloride flux is inhibitory because it hyperpolarizes the membrane. (c) The channel consists of five subunits.
25. After the addition of ATP and calcium, SERCA will pump Ca^{2+} ions into the vesicle. However, the accumulation of Ca^{2+} ions inside the vesicle will rapidly lead to the formation of an electrical gradient that cannot be overcome by ATP hydrolysis. The addition of calcimycin will allow the pumped Ca^{2+} ions to flow back out of the vesicle, dissipating the charge buildup, and enabling the pump to operate continuously.
26. The catalytic prowess of acetylcholinesterase ensures that the duration of the nerve stimulus will be short, preventing sustained nerve stimulation and enabling subsequent impulses to be generated more rapidly.
27. See reaction below.

28. (a) Only ASIC1a is inhibited by the toxin. (b) Yes; when the toxin was removed, the activity of the acid-sensing channel began to be restored. (c) 0.9 nM.

29. This mutation is one of a class of mutations that result in slow-channel syndrome (SCS). The results suggest a defect in channel closing; so, the channel remains open for prolonged periods. Alternatively, the channel may have a higher affinity for acetylcholine than does the control channel.

30. The mutation reduces the affinity of acetylcholine for the receptor. The recordings would show the channel opening only infrequently.

31. Glucose displays a transport curve that suggests the participation of a carrier because the initial rate is high but then levels off at higher concentrations, consistent with saturation of the carrier, which is reminiscent of Michaelis–Menten enzymes (Section 8.4). Indole shows no such saturation phenomenon, which implies that the molecule is lipophilic and simply diffuses across the membrane. Ouabain is a specific inhibitor the Na^+–K^+ pump. If ouabain were to inhibit glucose transport, then a Na^+-glucose cotransporter would be assisting in transport.

Chapter 14

1. The negatively charged glutamate residues mimic the negatively charged phosphoserine or phosphothreonine residues and stabilize the active conformation of the enzyme.

2. No. Phosphoserine and phosphothreonine are considerably shorter than phosphotyrosine.

3. The GTPase activity terminates the signal. Without such activity, after a pathway has been activated, it remains activated and is unresponsive to changes in the initial signal. If the GTPase activity were more efficient, the lifetime of the GTP-bound G_α subunit would be too short to achieve downstream signaling.

4. Two identical receptor molecules must recognize different aspects of the same signal molecule.

5. Growth-factor receptors can be activated by dimerization. If an antibody causes a receptor to dimerize, the signal-transduction pathway in a cell will be activated.

6. The mutated α subunit will always be in the GTP form and, hence, in the active form, which would stimulate its signaling pathway.

7. A G protein is a component of the signal-transduction pathway. GTPγS is not hydrolyzed by the G_α subunit, leading to prolonged activation.

8. Calcium ions diffuse slowly because they bind to many protein surfaces within a cell, impeding their free motion. Cyclic AMP does not bind as frequently, and so it diffuses more rapidly.

9. Fura-2 is a highly negatively charged molecule, with five carboxylate groups. Its charge prevents it from effectively crossing the hydrophobic region of the plasma membrane.

10. $G_{\alpha s}$ stimulates adenylate cyclase, leading to the generation of cAMP. This signal then leads to glucose mobilization (Chapter 21).

If cAMP phosphodiesterase were inhibited, then cAMP levels would remain high even after the termination of the epinephrine signal, and glucose mobilization would continue.

11. If the two kinase domains are forced to be within close proximity of each other, the activation loop of one kinase, in its inactivating conformation, can be displaced by the activation loop of the neighboring kinase, which acts as a substrate for phosphorylation.

12. The full network of pathways initiated by insulin includes a large number of proteins and is substantially more elaborate than indicated in Figure 14.26. Furthermore, many additional proteins take part in the termination of insulin signaling. A defect in any of the proteins in the insulin signaling pathways or in the subsequent termination of the insulin response could potentially cause problems. Therefore, it is not surprising that many different gene defects can cause type 2 diabetes.

13. The binding of growth hormone causes its monomeric receptor to dimerize. The dimeric receptor can then activate a separate tyrosine kinase to which the receptor binds. The signaling pathway can then continue in similar fashion to the pathways that are activated by the insulin receptor or other mammalian EGF receptors.

14. The truncated receptor will dimerize with the full-length monomers on EGF-binding, but cross-phosphorylation cannot take place, because the truncated receptor possesses neither the substrate for the neighboring kinase domain nor its own kinase domain to phosphorylate the C-terminal tail of the other monomer. Hence, these mutant receptors will block normal EGF signaling.

15. Insulin would elicit the response that is normally caused by EGF. Insulin binding will likely stimulate dimerization and phosphorylation of the chimeric receptor and thereby signal the downstream events that are normally triggered by EGF binding. Exposure of these cells to EGF would have no effect.

16. 10^5

17. The formation of diacylglycerol implies the participation of phospholipase C. A simple pathway would entail receptor activation by cross-phosphorylation, followed by the binding of phospholipase C (through its SH2 domains). The participation of phospholipase C indicates that IP_3 would be formed and, hence, calcium concentrations would increase.

18. Other potential drug targets within the EGF signaling cascade include, but are not limited to, the kinase active sites of the EGF receptor, Raf, MEK, or ERK.

19. In the reaction catalyzed by adenylate cyclase, the 3′-OH group nucleophilically attacks the α-phosphorus atom attached to the 5′-OH group, leading to displacement of pyrophosphate. The reaction catalyzed by DNA polymerase is similar except that the 3′-OH group is on a different nucleotide.

20. ATP-competitive inhibitors are likely to act on multiple kinases because every kinase domain contains an ATP-binding site. Hence, these drugs may not be selective for the desired kinase target.

21. (a) $X \approx 10^{-7}\,M; Y \approx 5 \times 10^{-6}\,M; Z \approx 10^{-3}\,M$. (b) Because much less X is required to fill half of the sites, X displays the highest affinity. (c) The binding affinity almost perfectly matches the ability to stimulate adenylate cyclase, suggesting that the hormone receptor complex leads to the stimulation of adenylate cyclase. (d) Try performing the experiment in the presence of antibodies to $G_{\alpha s}$.

22. (a) The total binding does not distinguish binding to a specific receptor from binding to different receptors or from nonspecific binding to the membrane. (b) The rationale is that the receptor will have a high affinity for the ligand. Thus, in the presence of excess nonradioactive ligand, the receptor will bind to nonradioactive ligand. Therefore, any binding of the radioactive ligand must be nonspecific. (c) The plateau suggests that the number of receptor-binding sites in the cell membrane is limited.

23. Number of receptors per cell =

$$\frac{10^4 \, \text{cpm}}{\text{mg of membrane protein}} \times \frac{\text{mg of membrane protein}}{10^{10} \, \text{cells}} \times$$

$$\frac{\text{mmol}}{10^{12} \, \text{cpm}} \times \frac{6.023 \times 10^{20} \, \text{molecules}}{\text{mmol}} = 600$$

Chapter 15

1. The highly integrated biochemical reactions that take place inside the cell.

2. Anabolism is the set of biochemical reactions that use energy to build new molecules and ultimately new cells. Catabolism is the set of biochemical reactions that extract energy from fuel sources or break down biomolecules.

3. Phototrophs derive energy by transforming light into chemical energy in the process of photosynthesis. Chemotrophs derive energy from the oxidation of organic molecules such as glucose and fatty acids.

4. The ATP–ADP cycle is the repeated conversion of ATP to ADP followed by the synthesis of ATP from ADP. The ATP is converted to ADP to power motion, biosynthesis, active transport, and signal amplification. The synthesis of ATP is powered by photosynthesis or the oxidation of fuel molecules.

5. Stage 1 is digestion, where the larger molecules in food are broken into smaller molecules. No ATP is generated. In stage 2, these smaller molecules are converted to key molecules of metabolism. A small amount of ATP is generated in this stage. Stage 3 consists of the complete oxidation of fuel molecules to CO_2 and H_2O. Most of the ATP needs of the cell are met in stage 3.

6. You reply that vandalism is disrespectful and expensive. Part of your tuition money will now have to pay to remove the vandalism. Plus, the fool should know that Gibbs free energy is at a minimum when a system is in equilibrium.

7. Cellular movements and the performance of mechanical work; active transport; biosynthetic reactions.

8. 1. f; 2. h; 3. i; 4. a; 5. g; 6. b; 7. c; 8. e; 9. j; 10. d.

9. These ions neutralize the charges on the ATP and also facilitate interactions with macromolecules that bind ATP.

10. Charge repulsion, resonance stabilization, increase in entropy, and stabilization by hydration.

11. Trick question. The answer is not known. Adenine appears to form more readily under prebiotic conditions, so ATP may have predominated initially.

12. Having only one nucleotide represent the available energy allows the cell to better monitor its energy status.

13. Increasing the concentration of ATP or decreasing the concentration of cellular ADP or P_i (by rapid removal by other reactions, for instance) would make the reaction more exergonic. Likewise, altering the Mg^{2+} concentration could raise or lower the ΔG of the reaction (see Data Interpretation, problem 37).

14. The free-energy changes of the individual steps in a pathway are summed to determine the overall free-energy change of the entire pathway. Consequently, a reaction with a positive free-energy value can be powered to take place if coupled to a sufficiently exergonic reaction.

15. Reactions in parts *a* and *c,* to the left; reactions in parts *b* and *d,* to the right.

16. None whatsoever.

17. $\Delta G^{\circ\prime} = +31.4 \, \text{kJ mol}^{-1} (+7.5 \, \text{kcal mol}^{-1})$ and $K'_{eq} = 3.06 \times 10^{-6}$; (b) 3.28×10^4.

18. $\Delta G^{\circ\prime} = +7.1 \, \text{kJ mol}^{-1} (+1.7 \, \text{kcal mol}^{-1})$. The equilibrium ratio is 17.5.

19. (a) Acetate + CoA + H^+ goes to acetyl CoA + H_2O, $\Delta G^{\circ\prime} = -31.4 \, \text{kJ mol}^{-1} (-7.5 \, \text{kcal mol}^{-1})$. ATP hydrolysis to AMP and PP_i, $\Delta G^{\circ\prime} = -45.6 \, \text{kJ mol}^{-1} (-10.9 \, \text{kcal mol}^{-1})$.

Overall reaction, $\Delta G^{\circ\prime} = -14.2 \, \text{kJ mol}^{-1} (-3.4 \, \text{kcal mol}^{-1})$.
(b) With pyrophosphate hydrolysis, $\Delta G^{\circ\prime} = -33.4 \, \text{kJ mol}^{-1}$ $(-7.98 \, \text{kcal mol}^{-1})$. Pyrophosphate hydrolysis makes the overall reaction even more exergonic.

20. (a) For an acid AH,

$$AH \rightleftharpoons A^- + H^+ \quad K_a = \frac{[A^-][H^+]}{[AH]}$$

The pK_a is defined as $pK_a = -\log_{10} K_a$. $\Delta G^{\circ\prime}$ is the standard free-energy change at pH 7. Thus, $\Delta G^{\circ\prime} = -RT \ln K_a = -2.303 \, RT \log_{10} K_a = +2.303 \, RT \, pK_a$. (b) $\Delta G^{\circ\prime} = +27.32 \, \text{kJ mol}^{-1}$ $(+6.53 \, \text{kcal mol}^{-1})$.

21. Arginine phosphate in invertebrate muscle, like creatine phosphate in vertebrate muscle, serves as a reservoir of high-potential phosphoryl groups. Arginine phosphate maintains a high level of ATP in muscular exertion.

22. An ADP unit.

23. (a) The rationale behind creatine supplementation is that it would be converted into creatine phosphate and thus serve as a rapid means of replenishing ATP after muscle contraction. (b) If creatine supplementation is beneficial, it would affect activities that depend on short bursts of activity; any sustained activity would require ATP generation by fuel metabolism, which, as Figure 15.7 shows, requires more time.

24. Under standard conditions, $\Delta G^{\circ\prime} = -RT \ln$ [products]/ [reactants]. Substituting $+23.8 \, \text{kJ mol}^{-1} (+5.7 \, \text{kcal mol}^{-1})|$ for $\Delta G^{\circ\prime}$ and solving for [products]/[reactants] yields 9.9×10^{-5}. In other words, the forward reaction does not take place to a significant extent. Under intracellular conditions, ΔG is $-1.3 \, \text{kJ mol}^{-1} (-0.3 \, \text{kcal mol}^{-1})$. Using the equation $\Delta G = \Delta G^{\circ\prime} + RT \ln$ [products]/[reactants] and solving for [products]/[reactants] gives a ratio of 5.96×10^{-5}. Thus, a reaction that is endergonic under standard conditions can be converted into an exergonic reaction by maintaining the [products]/[reactants] ratio below the equilibrium value. This conversion is usually attained by using the products in another coupled reaction as soon as they are formed.

25. Under standard conditions,

$$K'_{eq} = \frac{[B]_{eq}}{[A]_{eq}} \times \frac{[ADP]_{eq}[P_i]_{eq}}{[ATP]_{eq}} = 10^{33/1.36} = 2.67 \times 10^2$$

At equilibrium, the ratio of [B] to [A] is given by

$$\frac{[B]_{eq}}{[A]_{eq}} = K'_{eq} \frac{[ATP]_{eq}}{[ADP]_{eq}[P_i]_{eq}}$$

The ATP-generating system of cells maintains the [ATP]/[ADP] $[P_i]$ ratio at a high level, typically about 500 M^{-1}. For this ratio,

$$\frac{[B]_{eq}}{[A]_{eq}} = 2.67 \times 10^2 \times 500 = 1.34 \times 10^5$$

This equilibrium ratio is strikingly different from the value of 1.15×10^{-3} for the reaction A $\rightleftharpoons$ B, in the absence of ATP hydrolysis. In other words, coupling the hydrolysis of ATP with the conversion of A into B has changed the equilibrium ratio of B to A by a factor of about 10^8.

26. Liver: $-45.2 \, \text{kJ mol}^{-1} (-10.8 \, \text{kcal mol}^{-1})$; muscle: $-48.1 \, \text{kJ mol}^{-1} (-11.5 \, \text{kcal mol}^{-1})$; brain: $-48.5 \, \text{kJ mol}^{-1}$ $(-11.6 \, \text{kcal mol}^{-1})$. The ΔG is most negative in brain cells.

27. (a) Ethanol; (b) lactate; (c) succinate; (d) isocitrate; (e) malate.

28. Recall that $\Delta G = \Delta G^{\circ\prime} + RT \ln$ [products]/[reactants]. Altering the ratio of products to reactants will cause ΔG to vary. In glycolysis, the concentrations of the components of the pathway result in a value of ΔG greater than that of $\Delta G^{\circ\prime}$.

29. Higher organisms cannot make vitamins, and thus are dependent on obtaining them from other organisms.

30. Unless the ingested food is converted into molecules capable of being absorbed by the intestine, no energy can ever be extracted by the body.

31. NADH and $FADH_2$ are electron carriers for catabolism; NADPH is the carrier for anabolism.

32. The electrons of the C–O bond cannot form resonance structures with the C–S bond that are as stable as those that they can form with the C–O bond. Thus, the thioester is not stabilized by resonance to the same degree as an oxygen ester is stabilized.

33. Oxidation–reduction reactions; ligation reactions; isomerization reactions; group-transfer reactions; hydrolytic reactions; cleavage of bonds by means other than hydrolysis or oxidation.

34. Controlling the amount of enzymes; controlling enzyme activity; controlling the availability of substrates.

35. Although the reaction is thermodynamically favorable, the reactants are kinetically stable because of the large activation energy. Enzymes lower the activation energy so that reactions take place on time scales required by the cell.

36. The activated form of sulfate in most organisms is 3′-phosphoadenosine-5′-phosphosulfate.

37. (a) As the Mg^{2+} concentration falls, the ΔG of hydrolysis becomes larger and more negative. Note that pMg is a logarithmic plot, and so each number on the x-axis represents a 10-fold change in $[Mg^{2+}]$. (b) Mg^{2+} would bind to the phosphates of ATP and help to mitigate charge repulsion. As the $[Mg^{2+}]$ falls, charge stabilization of ATP would be less, leading to greater charge repulsion and a larger and more negative ΔG of hydrolysis.

Chapter 16

1. (d)
2. (d)
3. (b) and (d)
4. (a) 5; (b) 1; (c) 6; (d) 2 (e) 3; (f) 4.
5. Two molecules of ATP are produced per molecule of glyceraldehyde 3-phosphate and, because two molecules of GAP are produced per molecule of glucose, the total ATP yield is four. However, two molecules of ATP are required to convert glucose into fructose 1,6-bisphosphate. Thus, the net yield is only two molecules of ATP.
6. (a) 4; (b) 3; (c) 1; (d) 6; (e) 8; (f) 2; (g) 10; (h) 9; (i) 7; (j) 5.
7. In both cases, the electron donor is glyceraldehyde 3-phosphate. In lactic acid fermentation, the electron acceptor is pyruvate, converting it into lactate. In alcoholic fermentation, acetaldehyde is the electron acceptor, forming ethanol.
8. (a) 3 ATP; (b) 2 ATP; (c) 2 ATP; (d) 2 ATP; (e) 4 ATP.
9. Glucokinase enables the liver to remove glucose from the blood when hexokinase is saturated, ensuring that glucose is captured for later use.
10. The GAP formed is immediately removed by subsequent reactions, resulting in the conversion of DHAP into GAP by the enzyme.
11. A thioester couples the oxidation of glyceraldehyde 3-phosphate to 3-phosphoglycerate with the formation of 1,3-bisphosphoglycerate. 1,3-Bisphosphoglycerate can subsequently power the formation of ATP.
12. Glycolysis is a component of alcoholic fermentation, the pathway that produces alcohol for beer and wine. The belief was that understanding the biochemical basis of alcohol production might lead to a more efficient means of producing beer.
13. The conversion of glyceraldehyde 3-phosphate into 1,3-bisphosphoglycerate would be impaired. Glycolysis would be less effective.
14. Glucose 6-phosphate must have other fates. Indeed, it can be converted into glycogen (Chapter 21) or be processed to yield reducing power for biosynthesis (Chapter 20).
15. The energy needs of a muscle cell vary widely, from rest to intense exercise. Consequently, the regulation of phosphofructokinase by energy charge is vital. In other tissues, such as the liver,

ATP concentration is less likely to fluctuate and will not be a key regulator of phosphofructokinase.

16. (a) 6; (b) 1; (c) 7; (d) 3; (e) 2; (f) 5; (g) 4.
17. The $\Delta G°′$ for the reverse of glycolysis is $+90$ kJ mol^{-1} ($+22$ kcal mol^{-1}), far too endergonic to take place.
18. The conversion of glucose into glucose 6-phosphate by hexokinase; the conversion of fructose 6-phosphate into fructose 1,6-bisphosphate by phosphofructokinase; the formation of pyruvate from phosphoenolpyruvate by pyruvate kinase.
19. Lactic acid is a moderately strong acid. If it remained in the cell, the pH of the cell would fall, which could lead to the denaturation of muscle protein and result in muscle damage.
20. GLUT2 transports glucose only when the blood concentration of glucose is high, which is precisely the condition in which the β cells of the pancreas secrete insulin.
21. Fructose + ATP → fructose 1-phosphate + ADP: Fructokinase

Fructose 1-phosphate → dihydroxyacetone phosphate + glyceraldehyde: Fructose 1-phosphate aldolase

Glyceraldehyde + ATP → glyceraldehyde 3-phosphate + ADP: Triose kinase

The primary controlling step of glycolysis catalyzed by phosphofructokinase is bypassed by the preceding reactions. Glycolysis will proceed in an unregulated fashion.

22. (a) A, B; (b) C, D; (c) D; (d) A; (e) B; (f) C; (g) A; (h) D; (i) none; (j) A; (k) A.
23. (a) 4; (b) 10; (c) 1; (d) 5; (e) 7; (f) 8; (g) 9; (h) 2; (i) 3; (j) 6.
24. Without triose isomerase, only one of the two three-carbon molecules generated by aldolase could be used to generate ATP. Only two molecules of ATP would result from the metabolism of each glucose. But two molecules of ATP would still be required to form fructose 1,6-bisphosphate, the substrate for aldolase. The net yield of ATP would be zero, a yield incompatible with life.
25. If a reaction were at equilibrium, increasing the rate of the reaction would change nothing, as the reaction would still be at equilibrium. On the other hand, if the reaction were far from equilibrium, increasing the rate of the reaction would increase the amount of product. Consequently, regulatory enzymes must catalyze reactions that are far from equilibrium to be effective.
26. Glucose is reactive because its open-chain form contains an aldehyde group.
27. (a) The label is in the methyl carbon atom of pyruvate. (b) 5 mCi mM^{-1}. The specific activity is halved because the number of moles of product (pyruvate) is twice that of the labeled substrate (glucose).
28. (a) Glucose + 2 P_i + 2 ADP → 2 lactate + 2 ATP. (b) $\Delta G = -114$ kJ mol^{-1} (-27.2 kcal mol^{-1}).
29. 3.06×10^{-5}
30. The equilibrium concentrations of fructose 1,6-bisphosphate, dihydroxyacetone phosphate, and glyceraldehyde 3-phosphate are 7.8×10^{-4} M, 2.2×10^{-4} M, and 2.2×10^{-4} M, respectively.
31. All three carbon atoms of 2,3-BPG are ^{14}C labeled. The phosphorus atom attached to the C-2 hydroxyl group is ^{32}P labeled.
32. Hexokinase has a low ATPase activity in the absence of a sugar because it is in a catalytically inactive conformation. The addition of xylose closes the cleft between the two lobes of the enzyme. However, xylose lacks a hydroxymethyl group, and so it cannot be phosphorylated. Instead, a water molecule at the site normally occupied by the C-6 hydroxymethyl group acts as the acceptor of the phosphoryl group from ATP.
33. (a) The fructose 1-phosphate pathway forms glyceraldehyde 3-phosphate. (b) Phosphofructokinase, a key control enzyme, is bypassed. Unregulated flow through the glycolytic pathway can result in excess fat accumulation and ATP depletion in the liver.

34. The reverse of glycolysis is highly endergonic under cellular conditions. The expenditure of six NTP molecules in gluconeogenesis renders gluconeogenesis exergonic.

35. (a) 2, 3, 6, 9; (b) 1, 4, 5, 7, 8.

36. Lactic acid is capable of being further oxidized and is thus useful energy. The conversion of this acid into glucose saves the carbon atoms for future combustion.

37. In glycolysis, the formation of pyruvate and ATP by pyruvate kinase is irreversible. This step is bypassed by two reactions in gluconeogenesis: (1) the formation of oxaloacetate from pyruvate and CO_2 by pyruvate carboxylase and (2) the formation of phosphoenolpyruvate from oxaloacetate and GTP by phosphoenolpyruvate carboxykinase. The formation of fructose 1,6-bisphosphate by phosphofructokinase is bypassed by fructose 1,6-bisphosphatase in gluconeogenesis, which catalyzes the conversion of fructose 1,6-bisphosphate into fructose 6-phosphate. Finally, the hexokinase-catalyzed formation of glucose 6-phosphate in glycolysis is bypassed by glucose 6-phosphatase, but only in the liver.

38. Reciprocal regulation at the key allosteric enzymes in the two pathways. For instance, PFK is stimulated by fructose 2,6-bisphosphate and AMP. The effect of these signals is opposite that of fructose 1,6-bisphosphatase. If both pathways were operating simultaneously, a futile cycle would result. ATP would be hydrolyzed, yielding only heat.

39. Muscle is likely to produce lactic acid during contraction. Lactic acid is a moderately strong acid and cannot accumulate in muscle or blood. Liver removes the lactic acid from the blood and converts it into glucose. The glucose can be released into the blood or stored as glycogen for later use.

40. Glucose is an important energy source for both tissues and is essentially the only energy source for the brain. Consequently, these tissues should never release glucose. Glucose release is prevented by the absence of glucose 6-phosphatase.

41. 6 NTP (4 ATP and 2 GTP); 2 NADH.

42. (a) None; (b) none; (c) 4 (2 ATP and 2 GTP); (d) none.

43. If the amino groups are removed from alanine and aspartate, the ketoacids pyruvate and oxaloacetate are formed. Both of these molecules are components of the gluconeogenic pathway.

44. (a) Increased; (b) increased; (c) increased; (d) decreased.

45. Fructose 2,6-bisphosphate, present at high concentration when glucose is abundant, normally inhibits gluconeogenesis by blocking fructose 1,6-bisphosphatase. In this genetic disorder, the phosphatase is active irrespective of the glucose level. Hence, substrate cycling is increased. The level of fructose 1,6-bisphosphate is consequently lower than normal. Less pyruvate is formed and thus less ATP is generated.

46. Reactions in parts b and e would be blocked.

47. There will be no labeled carbons. The CO_2 added to pyruvate (formed from the lactate) to form oxaloacetate is lost with the conversion of oxaloacetate into phosphoenolpyruvate.

48. This example illustrates the difference between the *stoichiometric* and the *catalytic* use of a molecule. If cells used NAD^+ stoichiometrically, a new molecule of NAD^+ would be required each time a molecule of lactate was produced. The synthesis of NAD^+ requires ATP. On the other hand, if the NAD^+ that is converted into NADH could be recycled and reused, a small amount of the molecule could regenerate a vast amount of lactate, which is the case in the cell. NAD^+ is regenerated by the oxidation of NADH and reused. NAD^+ is thus used catalytically.

49. Consider the equilibrium equation of adenylate kinase:

$$K_{eq} = [ATP][AMP]/[ADP]^2 \qquad (1)$$

or

$$AMP = K_{eq}[ADP]^2/[ATP] \qquad (2)$$

Recall that $[ATP] > [ADP] > [AMP]$ in the cell. As ATP is utilized, a small decrease in its concentration will result in a larger percentage increase in [ADP] because its concentration is greater than that of ADP. This larger percentage increase in [ADP] will result in an even greater percentage increase in [AMP] because the concentration of AMP is related to the square of [ADP]. In essence, equation 2 shows that monitoring the energy status with AMP magnifies small changes in [ATP], leading to tighter control.

50. The synthesis of glucose during intense exercise provides a good example of interorgan cooperation in higher organisms. When muscle is actively contracting, lactate is produced from glucose by glycolysis. The lactate is released into the blood and absorbed by the liver, where it is converted by gluconeogenesis into glucose. The newly synthesized glucose is then released and taken up by the muscle for energy generation.

51. The input of four additional high-phosphoryl-transfer-potential molecules in gluconeogenesis changes the equilibrium constant by a factor of 10^{32}, which makes the conversion of pyruvate into glucose thermodynamically feasible. Without this energetic input, gluconeogenesis would not take place.

52. The mechanism is analogous to that for triose phosphate isomerase (Figure 16.5). It proceeds through an enediol intermediate. The active site would be expected to have a general base (analogous to Glu 165 in TPI) and a general acid (analogous to His 95 in TPI).

53. Galactose is a component of glycoproteins. Possibly, the absence of galactose leads to the improper formation or function of glycoproteins required in the central nervous system. More generally, the fact that the symptoms arise in the absence of galactose suggests that galactose is required in some fashion.

54. Using the Michaelis–Menten equation to solve for [S] when $K_M = 50 \,\mu M$ and $V_o = 0.9V_{max}$ shows that a substrate concentration of 0.45 mM yields 90% of V_{max}. Under normal conditions, the enzyme is essentially working at V_{max}.

55. Fructose 2,6-bisphosphate stabilizes the R state of the enzyme.

56. (a) Carbon dioxide is a good indicator of the rate of alcoholic fermentation because it is, along with ethanol, a product of alcoholic fermentation. (b) Phosphate is a required substrate for the reaction catalyzed by glyceraldehyde 3-phosphate dehydrogenase. The phosphate is incorporated into 1,3-bisphosphoglycerate. (c) It indicates that the amount of free phosphate must have been limiting. (d) The ratio would be expected to be 1. A phosphate would be consumed for every pyruvate decarboxylated. (e) Lack of phosphate would inhibit glyceraldehyde 3-phosphate dehydrogenase. This inhibition would "back up" glycolysis, resulting in the accumulation of fructose 1,6-bisphosphate.

57. (a) Curiously, the enzyme uses ADP as the phosphoryl donor rather than ATP. (b) Both AMP and ATP behave as competitive inhibitors of ADP, the phosphoryl donor. Apparently, the *P. furiosus* enzyme is not allosterically inhibited by ATP.

58. (a) If both enzymes operated simultaneously, the following reactions would take place:

The net result would be simply:

$$ATP + H_2O \rightarrow ADP + P_i$$

The energy of ATP hydrolysis would be released as heat. (b) Not really. For the cycle to generate heat, both enzymes must be functional at the same time in the same cell. (c) The species *B. terrestris* and *B. rufocinctus* might show some futile cycling because both enzymes are active to a substantial degree. (d) No. These results simply suggest that simultaneous activity of phosphofructokinase and fructose 1,6-bisphosphatase is unlikely to be employed to generate heat in the species shown.

59. ATP initially stimulates PFK activity, as would be expected for a substrate. Higher concentrations of ATP inhibit the enzyme. Although this effect seems counterintuitive for a substrate, recall that the function of glycolysis in muscle is to generate ATP. Consequently, high concentrations of ATP signal that the ATP needs are met and glycolysis should stop. In addition to being a substrate, ATP is an allosteric inhibitor of PFK.

Chapter 17

1. (a)
2. (b)
3. (d)
4. (b)
5. (a)
6. The pyruvate dehydrogenase complex catalyzes the following reaction, linking glycolysis and the citric acid cycle:

$$Pyruvate + CoA + NAD^+ \rightarrow acetyl\ CoA + NADH + H^+ + CO_2$$

7. Pyruvate dehydrogenase catalyzes the decarboxylation of pyruvate and the formation of acetyllipoamide. Dihydrolipoyl transacetylase catalyzes the formation of acetyl CoA. Dihydrolipoyl dehydrogenase catalyzes the reduction of the oxidized lipoic acid. The kinase associated with the complex phosphorylates and inactivates the complex, whereas the phosphatase dephosphorylates and activates the complex.
8. Thiamine pyrophosphate plays a role in the decarboxylation of pyruvate. Lipoic acid (as lipoamide) transfers the acetyl group. Coenzyme A accepts the acetyl group from lipoic acid to form acetyl CoA. FAD accepts the electrons and hydrogen ions when reduced lipoic acid is oxidized. NAD^+ accepts electrons from $FADH_2$.
9. Catalytic coenzymes (TPP, lipoic acid, and FAD) are modified but regenerated in each reaction cycle. Thus, they can play a role in the processing of many molecules of pyruvate. Stoichiometric coenzymes (coenzyme A and NAD^+) are used in only one reaction because they are the components of products of the reaction.
10. The remaining steps regenerate oxidized lipoamide, which is required to begin the next reaction cycle. Moreover, this regeneration results in the production of high-energy electrons in the form of NADH.
11. The advantages are as follows:
   - The reaction is facilitated by having the active sites in proximity.
   - The reactants do not leave the enzyme until the final product is formed.
   - Constraining the reactants minimizes loss due to diffusion and minimizes side reactions.
   - All of the enzymes are present in the correct amounts.
   - Regulation is more efficient because the regulatory enzymes—the kinase and phosphatase—are part of the complex.
12. (a) After one round of the citric acid cycle, the label emerges in C-2 and C-3 of oxaloacetate. (b) The label emerges in CO_2 in the formation of acetyl CoA from pyruvate. (c) After one round of the citric acid cycle, the label emerges in C-1 and C-4 of oxaloacetate. (d) and (e) Same fate as that in part *a*.
13. (a) Isocitrate lyase and malate synthase are required in addition to the enzymes of the citric acid cycle. (b) 2 Acetyl CoA + 2 NAD^+ + FAD + 3 $H_2O \rightarrow$ oxaloacetate + 2 CoA + 2 NADH + $FADH_2$ + 3 H^+. (c) No. Hence, mammals cannot carry out the net synthesis of oxaloacetate from acetyl CoA.
14. $-41.0\ kJ\ mol^{-1}\ (-9.8\ kcal\ mol^{-1})$
15. Enzymes or enzyme complexes are biological catalysts. Recall that a catalyst facilitates a chemical reaction without the catalyst itself being permanently altered. Oxaloacetate can be thought of as a catalyst because it binds to an acetyl group, leads to the oxidative decarboxylation of the two carbon atoms, and is regenerated at the completion of a cycle. In essence, oxaloacetate (and any cycle intermediate) acts as a catalyst.
16. Thiamine thiazolone pyrophosphate is a transition-state analog. The sulfur-containing ring of this analog is uncharged, and so it closely resembles the transition state of the normal coenzyme in thiamine-catalyzed reactions (e.g., the uncharged resonance form of hydroxyethyl-TPP).
17. (a) 6; (b) 10; (c) 1; (d) 7; (e) 2; (f) 8; (g) 3; (h) 4; (i) 5; (j) 9.
18. (a) DCA inhibits pyruvate dehydrogenase kinase. (b) The fact that inhibiting the kinase results in more dehydrogenase activity suggests that there must be some residual activity that is being inhibited by the kinase.
19. Acetyllipoamide and acetyl CoA.
20. In muscle, the acetyl CoA generated by the complex is used for energy generation. Consequently, signals that indicate an energy-rich state (high ratios of ATP/ADP and NADH/NAD^+) inhibit the complex, whereas the reverse conditions stimulate the enzyme. Calcium as the signal for muscle contraction (and, hence, energy need) also stimulates the enzyme. In liver, acetyl CoA derived from pyruvate is used for biosynthetic purposes, such as fatty acid synthesis. Insulin, the hormone denoting the fed state, stimulates the complex.
21. (a) Enhanced kinase activity will result in a decrease in the activity of the PDH complex because phosphorylation by the kinase inhibits the complex. (b) Phosphatase activates the complex by removing a phosphate. If the phosphatase activity is diminished, the activity of the PDH complex also will decrease.
22. She might have been ingesting, in some fashion, the arsenite from the peeling paint or the wallpaper. Also, she might have been breathing arsine gas from the wallpaper, which would be oxidized to arsenite in her body. In any of these circumstances, the arsenite inhibited enzymes that require lipoic acid—notably, the PDH complex.
23. (a) 5; (b) 7; (c) 1; (d) 10; (e) 2; (f) 4; (g) 9; (h) 3; (i) 8; (j) 6.
24. The TCA cycle depends on a steady supply of NAD^+ as an oxidant, generating NADH. O_2 is never directly utilized in the cycle. However, NAD^+ is regenerated via donation of electrons to O_2 by way of the electron transport chain, so eventually a lack of O_2 will cause the cycle to cease due to a lack of NAD^+.
25. Succinate dehydrogenase is the only enzyme in the citric acid cycle that is embedded in the mitochondrial membrane, which makes it associated with the electron-transport chain.
26. (a) The steady-state concentrations of the products are low compared with those of the substrates. (b) The ratio of malate to oxaloacetate must be greater than 1.57×10^4 for oxaloacetate to be formed.
27.

$$Pyruvate + CoA + NAD^+ \xrightarrow{\text{Pyruvate dehydrogenase complex}} acetyl\ CoA + CO_2 + NADH$$

$$Pyruvate + CO_2 + ATP + H_2O \xrightarrow{\text{Pyruvate carboxylase}} oxaloacetate + ADP + P_i + H^+$$

$$Oxaloacetate + acetyl\ CoA + H_2O \xrightarrow{\text{Citrate synthase}} citrate + CoA + H^+$$

$$Citrate \xrightarrow{\text{Aconitase}} isocitrate$$

$$Isocitrate + NAD^+ \xrightarrow{\text{Isocitrate dehydrogenase}} \alpha\text{-ketoglutarate} + CO_2 + NADH$$

Net: 2 Pyruvate + 2 NAD^+ + ATP + $H_2O \longrightarrow$
α-ketoglutarate + CO_2 + ADP + P_i + 2 NADH + 3 H^+

28. Succinate will increase in concentration, followed by α-ketoglutarate and the other intermediates "upstream" of the site of inhibition. Succinate has two methylene groups that are required for the dehydrogenation, whereas malonate has but one.

29. Pyruvate carboxylase should be active only when the acetyl CoA concentration is high. Acetyl CoA might accumulate if the energy needs of the cell are not being met, because of a deficiency of oxaloacetate. Under these conditions the pyruvate carboxylase catalyzes an anaplerotic reaction. Alternatively, acetyl CoA might accumulate because the energy needs of the cell have been met. In this circumstance, pyruvate will be converted back into glucose, and the first step in this conversion is the formation of oxaloacetate.

30. The energy released when succinate is oxidized to fumarate is not sufficient to power the synthesis of NADH but is sufficient to reduce FAD.

31. Citrate is a tertiary alcohol that cannot be oxidized, because oxidation requires a hydrogen atom to be removed from the alcohol and a hydrogen atom to be removed from the carbon atom bonded to the alcohol. No such hydrogen exists in citrate. The isomerization converts the tertiary alcohol into isocitrate, which is a secondary alcohol that can be oxidized.

32. The enzyme nucleoside diphosphokinase transfers a phosphoryl group from GTP (or any nucleoside triphosphate) to ADP according to the reversible reaction:

$$GTP + ADP \rightleftharpoons GDP + ATP$$

33. The reaction is powered by the hydrolysis of a thioester. Acetyl CoA provides the thioester that is converted into citryl CoA. When this thioester is hydrolyzed, citrate is formed in an irreversible reaction.

34. It enables organisms such as plants and bacteria to convert fats, through acetyl CoA, into glucose.

35. We cannot get the net conversion of fats into glucose, because the only means to get the carbon atoms from fats into oxaloacetate, the precursor of glucose, is through the citric acid cycle. However, although two carbon atoms enter the cycle as acetyl CoA, two carbon atoms are lost as CO_2 before oxaloacetate is formed. Thus, although some carbon atoms from fats may end up as carbon atoms in glucose, we cannot obtain a *net* synthesis of glucose from fats.

36. Acetyl CoA will inhibit the complex. Glucose metabolism to pyruvate will be slowed because acetyl CoA is being derived from an alternative source.

37. The enol intermediate of acetyl CoA attacks the carbonyl carbon atom of glyoxylate to form a C—C bond. This reaction is like the condensation of oxaloacetate with the enol intermediate of acetyl CoA in the reaction catalyzed by citrate synthase. Glyoxylate contains a hydrogen atom in place of the $-CH_2COO^-$ group of oxaloacetate; the reactions are otherwise nearly identical.

38. Citrate is a symmetric molecule. Consequently, the investigators assumed that the two $-CH_2COO^-$ groups in it would react identically. Thus, for every citrate molecule undergoing the reactions shown in path 1, they thought that another citrate molecule would react as shown in path 2. If so, then only *half* the label should have emerged in the CO_2.

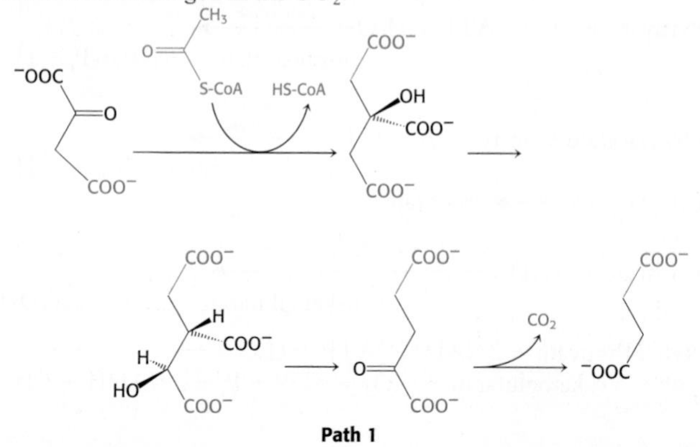

Path 1

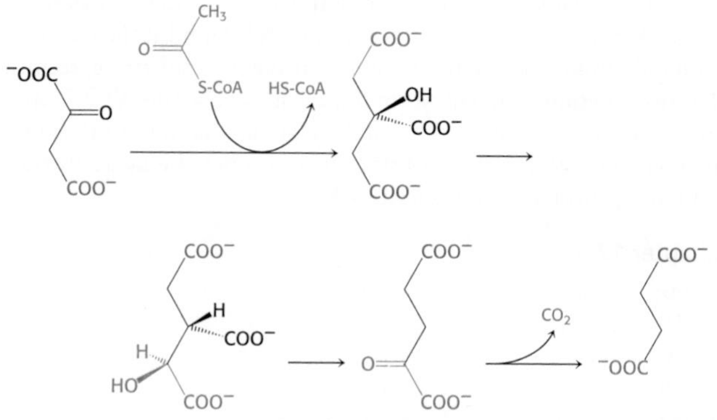

39. Call one hydrogen atom A and the other B. Now suppose that an enzyme binds three groups of this substrate—X, Y, and H—at three complementary sites. The adjoining diagram shows X, Y, and H_A bound to three points on the enzyme. In contrast, X, Y, and H_B cannot be bound to this active site; two of these three groups can be bound, but not all three. Thus, H_A and H_B will have different fates.

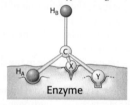

Sterically nonequivalent groups such as H_A and H_B will almost always be distinguished in enzymatic reactions. The essence of the differentiation of these groups is that the enzyme holds the substrate in a specific orientation. Attachment at three points, as depicted in the diagram, is a readily visualized way of achieving a particular orientation of the substrate, but it is not the only means of doing so.

40. (a) The complete oxidation of citrate requires 4.5 μmol of O_2 for every micromole of citrate.

$$C_6H_8O_7 + 4.5\,O_2 \rightarrow 6\,CO_2 + 4\,H_2O$$

Thus, 13.5 μmol of O_2 would be consumed by 3 μmol of citrate. (b) Citrate led to the consumption of far more O_2 than can be accounted for simply by the oxidation of citrate itself. Citrate thus facilitated O_2 consumption.

41. (a) In the absence of arsenite, the amount of citrate remained constant. In its presence, the concentration of citrate fell, suggesting that it was being metabolized. (b) The action of arsenite is not altered. Citrate still disappears. (c) Arsenite is preventing the regeneration of citrate. Recall (p. 563) that arsenite inhibits the pyruvate dehydrogenase complex.

42. (a) The initial infection is unaffected by the absence of isocitrate lyase, but the absence of this enzyme inhibits the latent phase of the infection. (b) Yes. (c) A critic could say that, in the process of deleting the isocitrate lyase gene, some other gene was damaged, and it is the absence of this other gene that prevents latent infection. Reinserting the isocitrate lyase gene into the bacteria from which it had been removed renders the criticism less valid. (d) Isocitrate lyase enables the bacteria to synthesize carbohydrates that are necessary for survival, including carbohydrate components of the cell membrane.

Chapter 18

1. (b)
2. (b)
3. (a)
4. (b)
5. (d)
6. In fermentations, organic compounds are both the donors and the acceptors of electrons. In respiration, the electron donor is

usually an organic compound, whereas the electron acceptor is an inorganic molecule, such as oxygen.

7. Biochemists use E_0', the value at pH 7, whereas chemists use E_0, the value in 1 M H^+. The prime denotes that pH 7 is the standard state.

8. The reduction potential of $FADH_2$ is less than that of NADH (Table 18.1). Consequently, when those electrons are passed along to oxygen, less energy is released. The consequence of the difference is that electron flow from $FADH_2$ to O_2 pumps fewer protons than do the electrons from NADH.

9. The $\Delta G^{\circ\prime}$ for the reduction of oxygen by $FADH_2$ is -200 kJ mol^{-1} (-48 kcal mol^{-1}).

10. $\Delta G^{\circ\prime}$ is $+67$ kJ mol^{-1} ($+16.1$ kcal mol^{-1}) for oxidation by NAD^+ and -3.8 kJ mol^{-1} (-0.92 kcal mol^{-1}) for oxidation by FAD. The oxidation of succinate by NAD^+ is not thermodynamically feasible.

11. An oxidizing agent, or oxidant, accepts electrons in oxidation–reduction reactions. A reducing reagent, or reductant, donates electrons in such reactions.

12. Pyruvate accepts electrons and is thus the oxidant. NADH gives up electrons and is the reductant.

13. $\Delta G^{\circ\prime} = -nF\Delta E_0'$

14. The $\Delta E_0'$ value of iron can be altered by changing the environment of the ion.

15. c, e, b, a, d.

16. (a) 4; (b) 5; (c) 2; (d) 10; (e) 3; (f) 8; (g) 9; (h) 7; (i) 1; (j) 6.

17. (a) 4; (b) 3; (c) 1; (d) 5; (e) 2.

18. The 10 isoprene units render coenzyme Q soluble in the hydrophobic environment of the inner mitochondrial membrane. The two oxygen atoms can reversibly bind two electrons and two protons as the molecule transitions between the quinone form and quinol form.

19. Rotenone: NADH, NADH-Q oxidoreductase will be reduced. The remainder will be oxidized. Antimycin A: NADH, NADH-Q oxidoreductase and coenzyme Q will be reduced. The remainder will be oxidized. Cyanide: All will be reduced.

20. Complex I would be reduced, whereas Complexes II, III, and IV would be oxidized. The citric acid cycle would halt because it has no way to oxidize NADH.

21. The respirasome is another example of the use of supramolecular complexes in biochemistry. Having the three complexes that are proton pumps associated with one another will enhance the efficiency of electron flow from complex to complex, which in turn will cause more-efficient proton pumping.

22. Succinate dehydrogenase is a component of Complex II.

23. Hydroxyl radical (OH ·), hydrogen peroxide (H_2O_2), superoxide ion (O_2^-), and peroxide (O_2^{2-}). These small molecules react with a host of macromolecules—including proteins, nucleotides, and membranes—to disrupt cell structure and function.

24. The ATP is recycled by ATP-generating processes, most notably oxidative phosphorylation.

25. (a) 12.5; (b) 14; (c) 32; (d) 13.5; (e) 30; (f) 16.

26. (a) It blocks electron transport and proton pumping at Complex IV. (b) It blocks electron transport and ATP synthesis by inhibiting the exchange of ATP and ADP across the inner mitochondrial membrane. (c) It blocks electron transport and proton pumping at Complex I. (d) It blocks ATP synthesis without inhibiting electron transport by dissipating the proton gradient. (e) It blocks electron transport and proton pumping at Complex IV. (f) It blocks electron transport and proton pumping at Complex III.

27. If the proton gradient is not dissipated by the influx of protons into a mitochondrion with the generation of ATP, eventually the outside of the mitochondrion develops such a large positive charge that the electron-transport chain can no longer pump protons against the gradient.

28. The subunits are jostled by background thermal energy (Brownian motion). The proton gradient makes clockwise rotation

more likely because that direction results in protons flowing down their concentration gradient.

29. Dicyclohexylcarbodiimide reacts readily with carboxyl groups. Hence, the most likely targets are aspartate and glutamate side chains. In fact, Asp 61 of subunit **c** of *E. coli* F_0 is specifically modified by this reagent. The conversion of Asp 61 into asparagine by site-specific mutagenesis eliminates proton conduction.

30. ATP synthase will from an arsenate (AsO_4^{3-}) anhydride with ADP. Such compounds are unstable and rapidly hydrolyzed. Under these conditions, no ATP synthesis can occur. But the synthase will continue catalyzing the useless reaction. Consequently, the electron-transport system will keep running, most likely at a higher than normal rate. In essence, arsenate is acting as an uncoupler.

31. In the presence of poorly functioning mitochondria, the only means of generating ATP is by anaerobic glycolysis, which will lead to an accumulation of lactic acid in blood.

32. If ADP cannot get into mitochondria, the electron-transport chain will cease to function because there will be no acceptor for the energy. NADH will build up in the matrix. Recall that NADH inhibits some citric acid cycle enzymes and that NAD^+ is required by several citric acid cycle enzymes. Glycolysis will switch to lactic acid fermentation so that the NADH can be reoxidized to NAD^+ by lactate dehydrogenase.

33. (a) No effect; mitochondria cannot metabolize glucose. (b) No effect; no fuel is present to power the synthesis of ATP. (c) The $[O_2]$ falls because citrate is a fuel and ATP can be formed from ADP and P_i. (d) Oxygen consumption stops because oligomycin inhibits ATP synthesis, which is coupled to the activity of the electron-transport chain. (e) No effect, for the reasons given in part *d*. (f) $[O_2]$ falls rapidly because the system is uncoupled and does not require ATP synthesis to lower the proton-motive force. (g) $[O_2]$ falls, though at a lower rate. Rotenone inhibits Complex I, but the presence of succinate will enable electrons to enter at Complex II. (h) Oxygen consumption ceases because Complex IV is inhibited and the entire chain backs up.

34. (a) The P:O ratio is equal to the product of $(H^+/2\,e^-)$ and (P/H^+). Note that the P:O ratio is identical with the P:2 e^- ratio. (b) 2.5 and 1.5, respectively.

35. Cyanide can be lethal because it binds to the ferric form of cytochrome oxidase and thereby inhibits oxidative phosphorylation. Nitrite converts ferrohemoglobin into ferrihemoglobin, which also binds cyanide. Thus, ferrihemoglobin competes with cytochrome *c* oxidase for cyanide. This competition is therapeutically effective because the amount of ferrihemoglobin that can be formed without impairing oxygen transport is much greater than the amount of cytochrome *c* oxidase.

36. Such a defect (called Luft syndrome) was found in a 38-year-old woman who was incapable of performing prolonged physical work. Her basal metabolic rate was more than twice normal, but her thyroid function was normal. A muscle biopsy showed that her mitochondria were highly variable and atypical in structure. Biochemical studies then revealed that oxidation and phosphorylation were not tightly coupled in these mitochondria. In this patient, much of the energy of fuel molecules was converted into heat rather than ATP.

37. Triose phosphate isomerase converts dihydroxyacetone phosphate (a potential dead end) into glyceraldehyde 3-phosphate (a mainstream glycolytic intermediate).

38. This inhibitor (like antimycin A) blocks the reduction of cytochrome c_1 by QH_2, the crossover point.

39. If oxidative phosphorylation were uncoupled, no ATP could be produced. In a futile attempt to generate ATP, much fuel would be consumed. The danger lies in the dose. Too much uncoupling would lead to tissue damage in highly aerobic organs such as the brain and heart, which would have severe consequences for the organism as a whole. The energy that is normally transformed into ATP would be released as heat. To maintain body temperature,

sweating might increase, although the very process of sweating itself depends on ATP.

40. If ATP and ADP cannot exchange between the matrix and the mitochondria, ATP synthase will cease to function because its substrate ADP is absent. The proton gradient will eventually become so large that the energy released by the electron-transport chain will not be great enough to pump protons against the larger-than-normal gradient.

41. Add the inhibitor with and without an uncoupler, and monitor the rate of O_2 consumption. If the O_2 consumption increases again in the presence of inhibitor and uncoupler, the inhibitor must be inhibiting ATP synthase. If the uncoupler has no effect on the inhibition, the inhibitor is inhibiting the electron-transport chain.

42. Presumably, because the muscle has greater energy needs, especially during exercise, it will require more ATP. This requirement means that more sites of oxidative phosphorylation are called for, and these sites can be provided by an increase in the amount of cristae.

43. The arginine residue, with its positive charge, will facilitate proton release from aspartic acid by stabilizing the negatively charged aspartate.

44. 4; 4.7

45. The ATP synthase would pump protons at the expense of ATP hydrolysis, thus maintaining the proton-motive force. The synthase would function as an ATPase. There is some evidence that damaged mitochondria use this tactic to maintain, at least temporarily, the proton-motive force.

46. It suggests that malfunctioning mitochondria may play a role in the development of Parkinson disease. Specifically, it implicates Complex I.

47. The extra negative charge on ATP relative to that on ADP accounts for ATP's more rapid translocation out of the mitochondrial matrix. If the charge differences between ATP and ADP were lessened by the binding of Mg^{2+}, ADP might more readily compete with ATP for transport to the cytoplasm.

48. When all of the available ADP has been converted into ATP, ATP synthase can no longer function. The proton gradient becomes large enough that the energy of the electron-transport chain is not enough to pump against the gradient, and electron transport and, hence, oxygen consumption falls.

49. The effect on the proton gradient is the same in each case.

50. ATP export from the matrix. Phosphate import into the matrix.

51. Recall from the discussion of enzyme-catalyzed reactions that the direction of a reaction is determined by the ΔG difference between substrate and products. An enzyme speeds up the rate of both the forward and the backward reactions. The hydrolysis of ATP is exergonic, and so ATP synthase will enhance the hydrolytic reaction.

52. The cytoplasmic kinases thereby obtain preferential access to the exported ATP.

53. The organic acids in the blood are indications that the mice are deriving a large part of their energy needs through anaerobic glycolysis. Lactate is the end product of anaerobic glycolysis. Alanine is an aminated transport form of pyruvate, which is formed from lactate. Alanine formation plays a role in succinate formation, which is caused by the reduced state of the mitochondria.

The electron-transport chain is slowed because the inner mitochondrial membrane is hyperpolarized. Without ADP to accept the energy of the proton-motive force, the membrane becomes polarized to such an extent that protons can no longer be pumped. The excess H_2O_2 is probably due to the fact that the superoxide radical is present in higher concentration because the oxygen can no longer be effectively reduced.

$$O_2\cdot^- + O_2\cdot^- + 2\,H^+ \rightarrow O_2 + H_2O_2$$

Indeed, these mice display evidence of such oxidative damage.

54. (a) Vitamins C and E. (b) Exercise induces superoxide dismutase, which converts ROS in hydrogen peroxide and oxygen. (c) The answer to this question is not fully established. Two possibilities are (1) the suppression of ROS by vitamins prevents the expression of more superoxide dismutase and (2) some ROS may be signal molecules required to stimulate insulin-sensitivity pathways.

55. (a) DNP is an uncoupler that prevents the use of the proton-motive force for ATP synthesis. Consequently, the rate of oxygen consumption rises (reflecting the rate of the electron transport chain) in a futile attempt to synthesize ATP. Because mitochondrial ATP synthesis in inhibited, the rate of glycolysis, as measured by ECAR, increases in an attempt meet the cells ATP needs. (b) Because glycolysis is now inhibited, no lactic acid will be produced and the rate of extracellular acidification will fall. Because DNP is still present, oxygen consumption will still occur at a high rate. (c) A key step in glycolysis is the isomerization of glucose 6-phosphate to fructose 6-phosphate. 2-Deoxyglucose is incapable of undergoing the isomerization. (d) Rotenone inhibits electron flow through Complex I, the electron transport chain is inhibited and oxygen consumption ceases.

56. (a) Succinate is oxidized by Complex II, and the electrons are used to establish a proton-motive force that powers ATP synthesis. (b) The ability to synthesize ATP is greatly reduced. (c) The goal was to measure ATP hydrolysis. If succinate had been added in the presence of ATP, no reaction would have taken place, because of respiratory control. (d) The mutation has little effect on the ability of the enzyme to catalyze the hydrolysis of ATP. (e) They suggest two things: (1) the mutation did not affect the catalytic site on the enzyme, because ATP synthase is still capable of catalyzing the reverse reaction, and (2) the mutation did not affect the amount of enzyme present, given that the controls and patients had similar amounts of activity.

57. The absolute configuration of thiophosphate indicates that inversion at phosphorus has taken place in the reaction catalyzed by ATP synthase. This result is consistent with an inline phosphoryl-transfer reaction taking place in a single step. The retention of configuration in the Ca^{2+}-ATPase reaction points to two phosphoryl-transfer reactions—inversion by the first and a return to the starting configuration by the second. The Ca^{2+}-ATPase reaction proceeds by a phosphorylated enzyme intermediate.

Chapter 19

1. (c)
2. (d)
3. (a)
4. (a)
5. (c)
6. The oxygen we require is produced by photosynthesis. Moreover, all of the carbon atoms of which we are made, not just carbohydrates, enter the biosphere through the process of photosynthesis.
7. $2\,NADP^+ + 3\,ADP^{3-} + 3\,P_i^{2-} + H^+ \rightarrow O_2 + 2\,NADPH + 3\,ATP^{4-} + H_2O$
8. (a) 7; (b) 5; (c) 4; (d) 10; (e) 1; (f) 2; (g) 9; (h) 3; (i) 8; (j) 6.

9. Photosystem II, in conjunction with the water-oxidizing complex, powers oxygen release. The reaction center of photosystem II absorbs light maximally at 680 nm.

10. Oxygen consumption will be maximal when photosystems I and II are operating cooperatively. Oxygen will be efficiently generated when electrons from photosystem II fill the electron holes in photosystem I, which were generated when the reaction center of photosystem I was illuminated by light of 700 nm.

11. Photosystem I generates reduced ferredoxin, which reduces $NADP^+$ to NADPH, a biosynthetic reducing power. Photosystem II activates the water-oxidizing complex, generating electrons for photosynthesis, and generating protons to form a proton gradient and to reduce ferredoxin and O_2.

12. The light reactions take place on thylakoid membranes. Increasing the membrane surface increases the number of ATP- and NADPH-generating sites.

13. These complexes absorb more light than a reaction center alone can absorb. The light-harvesting complexes funnel light to the reaction centers.

14. $NADP^+$ is the acceptor. H_2O is the donor. Light energy.

15. Chlorophyll is readily inserted into the hydrophobic interior of the thylakoid membranes.

16. Protons released by the oxidation of water; protons pumped into the lumen by the cytochrome *bf* complex; protons removed from the stroma by the reduction of $NADP^+$ and plastoquinone.

17. 700-nm photons have an energy content of $+172 \, kJ \, mol^{-1}$. The absorption of light by photosystem I results in a $\Delta E_0'$ of $-1.0 \, V$. Recall that $\Delta E_0' = -nF\Delta E_0'$ where $F = 96.48 \, kJ \, mol^{-1} \, V^{-1}$. Under standard conditions, the energy change for the electrons is $96.5 \, kJ$. Thus, the efficiency is $96.5/172 = 56\%$.

18. The electron flow from photosystem II to photosystem I is uphill, or exergonic. For this uphill flow, ATP would need to be consumed, defeating the purpose of photosynthesis.

19. $\Delta E_0' = 10.11 \, V$, and $\Delta G^{\circ\prime} = -21.3 \, kJ \, mol^{-1}$ ($-5.1 \, kcal \, mol^{-1}$).

20. (a) All ecosystems require an energy source from outside the system, because the chemical energy sources will ultimately be limited. The photosynthetic conversion of sunlight is one example of such a conversion. (b) Not at all. Spock would point out that chemicals other than water can donate electrons and protons.

21. DCMU inhibits electron transfer in the link between photosystems II and I. O_2 can evolve in the presence of DCMU if an artificial electron acceptor such as ferricyanide can accept electrons from Q.

22. DCMU will have no effect, because it blocks photosystem II, and cyclic photophosphorylation uses photosystem I and the cytochrome *bf* complex.

23. (a) $+120 \, kJ \, einstein^{-1} (+28.7 \, kcal \, einstein^{-1})$; (b) $1.24 \, V$; (c) One 1000-nm photon has the free-energy content of 2.4 molecules of ATP. A minimum of 0.42 photon is needed to drive the synthesis of a molecule of ATP.

24. At this distance, the expected rate is one electron per second.

25. The distance doubles, and so the rate should decrease by a factor of 64 to 640 ps.

26. The cristae.

27. In eukaryotes, both processes take place in specialized organelles. Both depend on high-energy electrons to generate ATP. In oxidative phosphorylation, the high-energy electrons originate in fuels and are extracted as reducing power in the form of NADH. In photosynthesis, the high-energy electrons are generated by light and are captured as reducing power in the form of NADPH. Both processes use redox reactions to generate a proton gradient, and the enzymes that convert the proton gradient into ATP are very similar in both processes. In both systems, electron transport takes place in membranes inside organelles.

28. We need to factor in the NADPH because it is an energy-rich molecule. Recall from Chapter 18 that NADH is worth 2.5 ATP if oxidized by the electron-transport chain; 12 NADPH = 30 ATP. Eighteen molecules of ATP are used directly, and so the equivalent of 48 molecules of ATP are required for the synthesis of glucose.

29. The electrons flow through photosystem II directly to ferricyanide. No other steps are required.

30. (a) Thioredoxin. (b) The control enzyme is unaffected, but the mitochondrial enzyme with part of the chloroplast γ subunit increases activity as the concentration of DTT increases. (c) The increase was even larger when thioredoxin was present. Thioredoxin is the natural reductant for the chloroplast enzyme, and so it presumably operates more efficiently than would DTT, which probably functions to keep the thioredoxin reduced. (d) They seem to have done so. (e) The enzyme is susceptible to control by the redox state. In plant cells, reduced thioredoxin is generated by photosystem I. Thus, the enzyme is active when photosynthesis is taking place. (f) Cysteine. (g) Group-specific modification or site-specific mutagenesis.

Chapter 20

1. (d)

2. (a)

3. The Calvin cycle is the primary means of converting gaseous CO_2 into organic matter—that is, biomolecules. Essentially, every carbon atom in your body passed through rubisco and the Calvin cycle at some time in the past.

4. Autotrophs can use the energy of sunlight, carbon dioxide, and water to synthesize carbohydrates, which can subsequently be used for catabolic or anabolic purposes. Heterotrophs require chemical fuels and are thus ultimately dependent on autotrophs.

5. Nothing grim or secret about these reactions. They are sometimes called the dark reactions because they do not directly depend on light.

6.

Calvin cycle	Krebs cycle
Stroma	Matrix
Carbon chemistry for photosynthesis	Carbon chemistry for oxidative phosphorylation
Fixes CO_2	Releases CO_2
Requires high-energy electrons (NADPH)	Generates high-energy electrons (NADH)
Regenerates starting compound (ribulose 1,5-bisphosphate)	Regenerates starting compound (oxaloacetate)
Requires ATP	Generates ATP
Complex stoichiometry	Simple stoichiometry

7. The stroma will accumulate Mg^{2+} and become alkaline as a consequence of the movement of protons from the stroma to the thylakoid space. Thus, rubisco is primed for activity when the light reactions are providing ATP and NADPH required for carbon fixation and glucose synthesis.

8. (a) 3-Phosphoglycerate. (b) The other members of the Calvin cycle.

9. Stage 1 is the fixation of CO_2 with ribulose 1,5-bisphosphate and the subsequent formation of 3-phosphoglycerate. Stage 2 is the conversion of some of the 3-phosphoglycerate into hexose. Stage 3 is the regeneration of ribulose 1,5-bisphosphate.

10. It catalyzes a crucial reaction, but it is highly inefficient. Consequently, it is required in large amounts to overcome its slow catalysis.

11. Because carbamate forms only in the presence of CO_2, this property prevents rubisco from catalyzing the oxygenase reaction exclusively when CO_2 is absent.

12. Because NADPH is generated in the chloroplasts by the light reactions.

13. The concentration of 3-phosphoglycerate would increase, whereas that of ribulose 1,5-bisphosphate would decrease.

14. The concentration of 3-phosphoglycerate would decrease, whereas that of ribulose 1,5-bisphosphate would increase.

15. Aspartate + glyoxylate → oxaloacetate + glycine

16. The oxygenase activity of rubisco increases with temperature. Crabgrass is a C_4 plant, whereas most grasses lack this capability. Consequently, the crabgrass will thrive at the hottest part of the summer because the C_4 pathway provides an ample supply of CO_2.

17. The C_4 pathway allows the CO_2 concentration to increase at the site of carbon fixation. High concentrations of CO_2 inhibit the oxygenase reaction of rubisco. This inhibition is important for tropical plants because the oxygenase activity increases more rapidly with temperature than does the carboxylase activity.

18. ATP is required to form phosphoenolpyruvate (PEP) from pyruvate. The PEP combines with CO_2 to form oxaloacetate and, subsequently, malate. Two ATP molecules are required because a second ATP molecule is needed to phosphorylate AMP to ADP.

19. Photorespiration is the consumption of oxygen by plants with the production of CO_2, but it does not generate energy. Photorespiration is due to the oxygenase activity of rubisco. It is wasteful because, instead of fixing CO_2 for conversion into hexoses, rubisco is generating CO_2.

20. As global warming progresses, C_4 plants will invade the higher latitudes, and C_3 plants will retreat to cooler regions.

21. C_4 metabolism allows rubisco to function efficiently even when the temperatures are high, which favor oxygenase activity. Moreover, C_4 metabolism allows desert plants to accumulate CO_2 at night when the temperatures are cooler and water evaporation is not a problem.

22. The light reactions lead to an increase in the stromal concentrations of NADPH, reduced ferredoxin, and Mg^{2+}, as well as an increase in pH.

23. (a) 5; (b) 1; (c) 7; (d) 2; (e) 10; (f) 3; (g) 6; (h) 4; (i) 8; (j) 9.

24. (d)

25. (c)

26. The oxidative phase generates NADPH and is irreversible. The nonoxidative phase allows for the interconversion of phosphorylated sugars.

27. The enzymes catalyze the transformation of the five-carbon sugar formed by the oxidative phase of the pentose phosphate pathway into fructose 6-phosphate and glyceraldehyde 3-phosphate—intermediates in glycolysis (and gluconeogenesis).

28. (a) C; (b) B and F; (c) G; (d) F; (e) E; (f) H; (g) I; (h) D; (i) A; (j) F; (k) B.

29. The label emerges at C-5 of ribulose 5-phosphate.

30. Oxidative decarboxylation of isocitrate to α-ketoglutarate. A β-ketoacid intermediate is formed in both reactions.

31. (a) 5 Glucose 6-phosphate + ATP → 6 ribose 5-phosphate + ADP + H^+. (b) Glucose 6-phosphate + 12 $NADP^+$ + 7 H_2O → 6 CO_2 + 12 NADPH + 12 H^+ + P_i.

32. The nonoxidative phase of the pentose phosphate pathway can be used to convert three molecules of ribose 5-phosphate into two molecules of fructose 6-phosphate and one molecule of glyceraldehyde 3-phosphate. These molecules are components of the glycolytic pathway.

33. The conversion of fructose 6-phosphate into fructose 1,6-bisphosphate by phosphofructokinase requires ATP.

34. When much NADPH is required. The oxidative phase of the pentose phosphate pathway is followed by the nonoxidative phase. The resulting fructose 6-phosphate and glyceraldehyde 3-phosphate are used to generate glucose 6-phosphate through gluconeogenesis, and the cycle is repeated until the equivalent of one glucose molecule is oxidized to CO_2.

35. Fava beans contain vicine, a pyrimidine glycoside that can lead to the generation of peroxides—reactive oxygen species that can damage membranes as well as other biomolecules. Glutathione is used to detoxify the ROS. The regeneration of glutathione depends on an adequate supply of NADPH, which is synthesized by the oxidative phase of the pentose phosphate pathway. People with low levels of the dehydrogenase are especially susceptible to vicine toxicity.

36. Because red blood cells do not have mitochondria and the only means to obtain NADPH is through the pentose phosphate pathway. There are biochemical means to convert mitochondrial NADH into cytoplasmic NADPH.

37. Reactive peroxides are a type of reactive oxygen species. The enzyme glutathione peroxidase uses reduced glutathione to neutralize peroxides by converting them into alcohols while generating oxidized glutathione. Reduced glutathione is regenerated by glutathione reductase with the use of NADPH, the product of the oxidative phase of the pentose phosphate pathway.

38. $\Delta E_0'$ for the reduction of glutathione by NADPH is $+0.09$ V. Hence, $\Delta G°'$ is -17.4 kJ mol^{-1} (-4.2 kcal mol^{-1}), which corresponds to an equilibrium constant of 1126. The required [NADPH]/[$NADP^+$] ratio is 8.9×10^{-5}.

39. The enolate form of dihydroxyacetone phosphate would still participate in the reaction, but a divalent metal ion rather than a protonated Schiff base would stabilize the enolate intermediate. The enolate would then add to the aldehyde of glyceraldehyde 3-phosphate.

Dihydroxyacetone phosphate

Glyceraldehyde 3-phosphate

Fructose 1,6-bisphosphate

40.

Ribose 5-phosphate

Enediol intermediate

Ribulose 5-phosphate

41. Incubate an aliquot of a tissue homogenate with glucose labeled with ^{14}C at C-1, and incubate another with glucose labeled with ^{14}C at C-6. Compare the radioactivity of the CO_2 produced

by the two samples. The rationale of this experiment is that only C-1 is decarboxylated by the pentose phosphate pathway, whereas C-1 and C-6 are decarboxylated equally when glucose is metabolized by the glycolytic pathway, the pyruvate dehydrogenase complex, and the citric acid cycle. The reason for the equivalence of C-1 and C-6 in the latter set of reactions is that glyceraldehyde 3-phosphate and dihydroxyacetone phosphate are rapidly inter-converted by triose phosphate isomerase.

42. Lacking mitochondria, red blood cells metabolize glucose to lactate to obtain energy in the form of ATP. The CO_2 results from extensive use of the pentose phosphate pathway coupled with gluconeogenesis. This coupling allows the generation of much NADPH with the complete oxidation of glucose by the oxidative branch of the pentose phosphate pathway.

43. (a) $k_{cat}^{CO_2}/K_M^{CO_2} = 3 \times 10^5\,s^{-1}M^{-1}$
$k_{cat}^{O_2}/K_M^{O_2} = 4 \times 10^3\,s^{-1}M^{-1}$ (b) Despite the fact that the specificity constant for CO_2 as a substrate is much greater than that of O_2, the concentration of O_2 in the atmosphere is higher than that of CO_2, allowing the oxygenation reaction to occur.

44. The reduction of each mole of CO_2 to the level of a hexose requires two moles of NADPH. The reduction of $NADP^+$ is a two-electron process. Hence, the formation of two moles of NADPH requires the pumping of four moles of electrons by photosystem I. The electrons given up by photosystem I are replenished by photosystem II, which needs to absorb an equal number of photons. Hence, eight photons are needed to generate the required NADPH. The energy input of eight moles of photons is +1594 kJ (+381 kcal). Thus, the overall efficiency of photosynthesis under standard conditions is at least 477/1594, or 30%.

45. It is neither a violation nor a miracle. The equation on page 654 requires not only 18 ATP, but also 12 NADPH. These electrons, if transferred to NAD^+ and used in the electron-transport chain, would yield 30 ATP. Thus, the synthesis of glucose requires the equivalent of 48 ATP.

46. (a) The curve on the right in graph A was generated by the C_4 plant. Recall that the oxygenase activity of rubisco increases with temperature more rapidly than does the carboxylase activity. Consequently, at higher temperatures, the C_3 plants would fix less carbon. Because C_4 plants can maintain a higher CO_2 concentration, the rise in temperature is less deleterious. (b) The oxygenase activity will predominate. Additionally, when the temperature rise is very high, the evaporation of water might become a problem. The higher temperatures can begin to damage protein structures as well. (c) The C_4 pathway is a very effective active-transport system for concentrating CO_2, even when environmental concentrations are very low. (d) With the assumption that the plants have approximately the same capability to fix CO_2, the C_4 pathway is apparently the rate-limiting step in C_4 plants.

Chapter 21

1. Step 1 is the release of glucose 1-phosphate from glycogen by glycogen phosphorylase. Step 2 is the formation of glucose 6-phosphate from glucose 1-phosphate, a reaction catalyzed by phosphoglucomutase. Step 3 is the remodeling of the glycogen by the transferase and the α-1,6-glucosidase.

2. (b)
3. (c)
4. (a)
5. (d)
6. (d)
7. (a) 8; (b) 3; (c) 6; (d) 5; (e) 9; (f) 2; (g) 10; (h) 1; (i) 4; (j) 7.
8. Glycogen is an important fuel reserve for several reasons. Unlike fatty acids, the released glucose can provide energy in the absence of oxygen and can thus supply energy for anaerobic activity. Moreover, the controlled breakdown of glycogen and release of glucose

increase the amount of glucose that is available between meals. Hence, glycogen serves as a buffer to maintain blood-glucose concentration. Glycogen's role in maintaining blood-glucose concentration is especially important because glucose is virtually the only fuel used by the brain, except during prolonged starvation. In addition, the glucose from glycogen is readily mobilized and is therefore a good source of energy for sudden, strenuous activity.

9. As an unbranched polymer, α-amylose has only one nonreducing end. Therefore, only one glycogen phosphorylase molecule could degrade each α-amylose molecule. Because glycogen is highly branched, there are many nonreducing ends per molecule. Consequently, many phosphorylase molecules can release many glucose molecules per glycogen molecule providing a rapid source of energy.

10. The ATP needs of the liver are less variable that those of the muscle. For instance, when a runner begins a race, her thigh muscles need ATP to power the contractions. Muscle glycogen breakdown is the key source of energy, especially in the beginning of the race.

11. In muscle, the b form of phosphorylase is activated by AMP. In the liver, the a form is inhibited by glucose. The difference corresponds to the difference in the metabolic role of glycogen in each tissue. Muscle uses glycogen as a fuel for contraction, whereas the liver uses glycogen to maintain blood-glucose concentration.

12. Cells maintain the $[P_i]/[\text{glucose 1-phosphate}]$ ratio at greater than 100, making the phosphorolysis essentially irreversible. We see here an example of how the cell can alter the free-energy change to favor a reaction taking place by altering the ratio of substrate and product.

13. The high concentration of glucose 6-phosphate in von Gierke disease, resulting from the absence of glucose 6-phosphatase or the transporter, shifts the allosteric equilibrium of phosphorylated glycogen synthase toward the active form.

14. The phosphoryl donor is glucose 1,6-bisphosphate. Such a reaction could be catalyzed by a kinase using ATP as a substrate. The reaction is, in fact, catalyzed by phosphoglucokinase.

15. The different manifestations correspond to the different roles of the liver and muscle. Liver glycogen phosphorylase plays a crucial role in the maintenance of blood-glucose concentration. Recall that glucose is the primary fuel for the brain. Muscle glycogen phosphorylase provides glucose only for the muscle and, even then, only when the energy needs of the muscle are high, as during exercise. The fact that there are two different diseases suggests that there are two different isozymic forms of the glycogen phosphorylase—a liver-specific isozyme and a muscle-specific isozyme.

16. Water is excluded from the active site to prevent hydrolysis. The entry of water could lead to the formation of glucose rather than glucose 1-phosphate. A site-specific mutagenesis experiment is revealing in this regard. In phosphorylase, Tyr 573 is hydrogen bonded to the 2'-OH group of a glucose residue. The ratio of glucose 1-phosphate to glucose product is 9000 : 1 for the wild-type enzyme, and 500 : 1 for the Phe 573 mutant. Model building suggests that a water molecule occupies the site normally filled by the phenolic OH group of tyrosine and occasionally attacks the oxocarbonium ion intermediate to form glucose.

17. The amylase activity was necessary to remove all of the glycogen from the glycogenin. Recall that glycogenin synthesizes oligosaccharides of about eight glucose units, and then activity stops. Consequently, if the glucose residues are not removed by extensive amylase treatment, glycogenin will not function.

18. The substrate can be handed directly from the transferase site to the debranching site.

19. During exercise, [ATP] falls and [AMP] rises. Recall that AMP is an allosteric activator of glycogen phosphorylase b. Thus, even in the absence of covalent modification by phosphorylase kinase, glycogen is degraded.

20. Although glucose 1-phosphate is the actual product of the phosphorylase reaction, glucose 6-phosphate is a more versatile molecule with respect to metabolism. Among other fates, glucose 6-phosphate can be processed to yield energy or building blocks. In the liver, glucose 6-phosphate can be converted into glucose and released into the blood.

21. Epinephrine binds to its G-protein-coupled receptor. The resulting structural changes activate a G_α protein, which in turn activates adenylate cyclase. Adenylate cyclase synthesizes cAMP, which activates protein kinase A. Protein kinase A partly activates phosphorylase kinase, which phosphorylates and activates glycogen phosphorylase. The calcium released during muscle contraction also activates the phosphorylase kinase, leading to further stimulation of glycogen phosphorylase.

22. First, the signal-transduction pathway is shut down when the initiating hormone is no longer present. Second, the inherent GTPase activity of the G protein converts the bound GTP into inactive GDP. Third, phosphodiesterases convert cyclic AMP into AMP. Fourth, PP1 removes the phosphoryl group from glycogen phosphorylase, converting the enzyme into the usually inactive b form.

23. It prevents both from operating simultaneously, which would lead to a useless expenditure of energy.

24. All these symptoms suggest central nervous system problems. If exercise is exhaustive enough or the athlete has not prepared well enough or both, liver glycogen also can be depleted. The brain depends on glucose derived from liver glycogen. The symptoms suggest that the brain is not getting enough fuel (hypoglycemia).

25. Liver phosphorylase a is inhibited by glucose, which facilitates the $R \rightarrow T$ transition. This transition releases PP1, which inactivates glycogen breakdown and stimulates glycogen synthesis. Muscle phosphorylase is insensitive to glucose.

26. (a) 4; (b) 1; (c) 5; (d) 10; (e) 7; (f) 2; (g) 8; (h) 9; (i) 6; (j) 3.

27. Phosphoglucomutase, UDP-glucose pyrophosphorylase, pyrophosphatase, glycogenin, glycogen synthase, and branching enzyme.

28. The enzyme pyrophosphatase converts the pyrophosphate into two molecules of inorganic phosphate. This conversion renders the overall reaction irreversible.

$$\text{Glucose 1-phosphate} + \text{UTP} \rightleftharpoons \text{UDP-glucose} + \text{PP}_i$$
$$\underline{\text{PP}_i + \text{H}_2\text{O} \longrightarrow 2\,\text{P}_i}$$
$$\text{Glucose 1-phosphate} + \text{UTP} + \text{H}_2\text{O} \longrightarrow \text{UDP-glucose} + 2\,\text{P}_i$$

29. The presence of high concentrations of glucose 6-phosphate indicates that glucose is abundant and that it is not being used by glycolysis. Therefore, this valuable resource is saved by incorporation into glycogen.

30. Free glucose must be phosphorylated at the expense of a molecule of ATP. Glucose 6-phosphate derived from glycogen is formed by phosphorolytic cleavage, sparing one molecule of ATP. Thus, the net yield of ATP when glycogen-derived glucose is processed to pyruvate is three molecules of ATP compared with two molecules of ATP from free glucose.

31. Breakdown: Phosphoglucomutase converts glucose 1-phosphate, liberated from glycogen breakdown, into glucose 6-phosphate, which can be released as free glucose (liver) or processed in glycolysis (muscle and liver). Synthesis: Converts glucose 6-phosphate into glucose 1-phosphate, which reacts with UTP to form UDP-glucose, the substrate for glycogen synthase.

32. $\text{Glycogen}_n + \text{P}_i \rightarrow \text{glycogen}_{n-1} + \text{glucose 6-phosphate}$

Glucose 6-phosphate $\rightarrow$ glucose 1-phosphate

UTP + glucose 1-phosphate $\rightarrow$ UDP-glucose + $2\,\text{P}_i$

Glycogen_{n-1} + UDP-glucose $\rightarrow$ glycogen_n + UDP

Sum: Glycogen_n + UTP $\rightarrow$ glycogen_n + UDP + P_i

33. In principle, having glycogen be the only primer for the further synthesis of glycogen should be a successful strategy. However, if the glycogen granules were not evenly divided between daughter cells, glycogen stores for future generations of cells might be compromised. Glycogenin synthesizes the primer for glycogen synthase.

34. Insulin binds to its receptor and activates the tyrosine kinase activity of the receptor, which in turn triggers a pathway that activates protein kinases. The kinases phosphorylate and inactivate glycogen synthase kinase. Protein phosphatase 1 then removes the phosphate from glycogen synthase and thereby activates the synthase.

35.

Transferase reaction

α-1,6-Glucosidase reaction

36. In the liver, glucagon stimulates the cAMP-dependent pathway that activates protein kinase A. Epinephrine binds to a 7TM α-adrenergic receptor in the liver plasma membrane, which activates phospholipase C and the phosphoinositide cascade. This activation causes calcium ions to be released from the endoplasmic reticulum, which bind to calmodulin, and further stimulates phosphorylase kinase and glycogen breakdown.

37. $\text{Galactose} + \text{ATP} + \text{UTP} + \text{H}_2\text{O} + \text{glycogen}_n \longrightarrow \text{glycogen}_{n+1} + \text{ADP} + \text{UDP} + 2\,\text{P}_i + \text{H}^+$.

38. Phosphorylase, transferase, glucosidase, phosphoglucomutase, and glucose 6-phosphatase.

39. Glucose is an allosteric inhibitor of phosphorylase a. Hence, crystals grown in its presence are in the T state. The addition of glucose 1-phosphate, a substrate, shifts the R-to-T equilibrium toward the R state. The conformational differences between these states are sufficiently large that the crystal shatters unless it is stabilized by chemical cross-links.

40. Galactose is converted into UDP-galactose to eventually form glucose 6-phosphate.

41. This disease can also be produced by a mutation in the gene that encodes the glucose 6-phosphate transporter. Recall that glucose 6-phosphate must be transported into the lumen of the endoplasmic reticulum to be hydrolyzed by phosphatase. Mutations in the other three essential proteins of this system can likewise lead to von Gierke disease.

42. (a) Apparently, the glutamate, with its negatively charged R group, can mimic to some extent the presence of a phosphoryl group on serine. That the stimulation is not as great is not surprising in that the carboxyl group is smaller and not as charged as the phosphate. (b) Substitution of aspartate would give some stimulation, but because it is smaller than the glutamate, the stimulation would be smaller.

43. (a) Glycogen was too large to enter the gel and, because analysis was by western blot with the use of an antibody specific to glycogenin, we would not expect to see background proteins. (b) α-Amylase degrades glycogen, releasing the protein glycogenin, which can be visualized by a western blot. (c) Glycogen phosphorylase, glycogen synthase, and protein phosphatase 1. These proteins might be visible if the gel were stained for protein, but a western analysis using only an anti-glycogenin antibody reveals the presence of glycogenin only.

44. (a) The smear was due to molecules of glycogenin with increasingly large amounts of glycogen attached to them. (b) In the absence of glucose in the medium, glycogen is metabolized, resulting in a loss of the high-molecular-weight material. (c) Glycogen could have been resynthesized and added to the glycogenin when the cells were fed glucose again. (d) No difference between lanes 3 and 4 suggests that, by 1 hour, the glycogen molecules had attained maximum size in this cell line. Prolonged incubation does not apparently increase the amount of glycogen. (e) α-Amylase removes essentially all of the glycogen, and so only the glycogenin remains.

Chapter 22

1. (a) 5; (b) 11; (c) 1; (d) 10; (e) 2; (f) 6; (g) 9; (h) 3; (i) 4; (j) 7; (k) 8.

2. Glycerol + 2 NAD$^+$ + P$_i$ + ADP ⟶ pyruvate + ATP + H$_2$O + 2 NADH + H$^+$

Glycerol kinase and glycerol phosphate dehydrogenase

3. The ready reversibility is due to the high-energy nature of the thioester in the acyl CoA.

4. To return the AMP to a form that can be phosphorylated by oxidative phosphorylation or substrate-level phosphorylation, another molecule of ATP must be expended in the reaction:

$$\text{ATP} + \text{AMP} \rightleftharpoons 2\,\text{ADP}$$

5. b, c, a, g, h, d, e, f.

6. The citric acid cycle. The reactions that take succinate to oxaloacetate, or the reverse, are similar to those of fatty acid metabolism (Section 17.2).

7. The next-to-last degradation product, acetoacetyl CoA, yields two molecules of acetyl CoA with the thiolysis by only one molecule of CoA.

8. Palmitic acid yields 106 molecules of ATP. Palmitoleic acid has a double bond between carbons C-9 and C-10. When palmitoleic acid is processed in β-oxidation, one of the oxidation steps (to introduce a double bond before the addition of water) will not take place, because a double bond already exists. Thus, FADH$_2$ will not be generated, and palmitoleic acid will yield 1.5 fewer molecules of ATP than palmitic acid, for a total of 104.5 molecules of ATP.

9.

Activation fee to form the acyl CoA	−2 ATP
Seven rounds of yield:	
7 acetyl CoA at 10 ATP/acetyl CoA	+70 ATP
7 NADH at 2.5 ATP/NADH	+17.5 ATP
7 FADH$_2$ at 1.5 ATP/FADH$_2$	+10.5 ATP
Propionyl CoA, which requires an ATP to be converted into succinyl CoA	−1 ATP
Succinyl CoA ⟶ succinate	+1 ATP
Succinate ⟶ fumarate 1 FADH$_2$ FADH$_2$ at 1.5 ATP/FADH$_2$	+1.5 ATP
Fumarate ⟶ malate	
Malate ⟶ oxaloacetate + NADH NADH at 2.5 ATP/NADH	+2.5 ATP
Total	120 ATP

10. You might hate yourself in the morning, but at least you won't have to worry about energy. To form stearoyl CoA requires the equivalent of 2 molecules of ATP.

Stearoyl CoA + 8 FAD + 8 NAD$^+$ + 8 CoA + 8 H$_2$O ⟶
9 acetyl CoA + 8 FADH$_2$ + 8 NADH + 8 H$^+$

9 acetyl CoA at 10 ATP/acetyl CoA	+90 ATP
8 NADH at 2.5 ATP/NADH	+20 ATP
8 FADH$_2$ at 1.5 ATP/FADH$_2$	+12 ATP
Activation fee	−2.0
Total	122 ATP

11. Keep in mind that, in the citric acid cycle, 1 molecule of FADH$_2$ yields 1.5 ATP, 1 molecule of NADH yields 2.5 ATP, and 1 molecule of acetyl CoA yields 10 ATP. Two molecules of ATP are produced when glucose is degraded to 2 molecules of pyruvate. Two molecules of NADH also are produced, but the electrons are transferred to FADH$_2$ to enter the electron transport chain. Each molecule of FADH$_2$ can generate 1.5 ATP. Each molecule of pyruvate will produce 1 molecule of NADH. Each molecule of acetyl CoA generates 3 molecules of NADH, 1 molecule of FADH$_2$, and 1 molecule of ATP. So, we have a total of 10 ATP per acetyl CoA, or 20 for the 2 molecules of acetyl CoA. The total for glucose is 30 ATP. Now, what about hexanoic acid? Caprioic acid is activated to caprioic CoA at the expense of 2 ATP, and so we are 2 ATP in the hole. The first cycle of β-oxidation generates 1 FADH$_2$, 1 NADH, and 1 acetyl CoA. After the acetyl CoA has been run through the citric acid cycle, this step will have generated a total of 14 ATP. The second cycle of β-oxidation generates 1 FADH$_2$ and 1 NADH but 2 acetyl CoA. After the acetyl CoA has been run through the citric acid cycle, this step will have generated a total of 24 ATP. The total is 36 ATP. Thus, the foul-smelling caprioic acid has a net yield of 36 ATP. So, on a per carbon basis, this fat yields 20% more ATP than does glucose, a manifestation of the fact that fats are more reduced than carbohydrates.

12. Stearate + ATP + 13.5 H$_2$O + 8 FAD + 8 NAD$^+$ ⟶ 4.5 acetoacetate + 14.5 H$^+$ + 8 FADH$_2$ + 8 NADH + AMP + 2 P$_i$.

13. Palmitate is activated and then processed by β-oxidation according to the following reactions.

Palmitate + CoA + ATP ⟶ palmitoyl CoA + AMP + 2 P$_i$
Palmitoyl CoA + 7 FAD + 7 NAD$^+$ + 7 CoASH + H$_2$O ⟶
8 acetyl CoA + 7 FADH$_2$ + 7 NADH + 7 H$^+$

The eight molecules of acetyl CoA combine to form four molecules of acetoacetate for release into the blood, and so they do not contribute to the energy yield in the liver. However, the $FADH_2$ and NADH generated in the preparation of acetyl CoA can be processed by oxidative phosphorylation to yield ATP.

$$1.5 \, ATP/FADH_2 \times 7 = 10.5 \, ATP$$
$$2.5 \, /ATP/NADH \times 7 = 17.5 \, ATP$$

The equivalent of 2 ATP were used to form palmitoyl CoA. Thus, 26 ATP were generated for use by the liver.

14. NADH produced with the oxidation to acetoacetate = 2.5 ATP. Acetoacetate is converted into acetoacetyl CoA. Two molecules of acetyl CoA result from the hydrolysis of acetoacetyl CoA, each worth 10 ATP when processed by the citric acid cycle. Total ATP yield is 22.5.

15. Because a molecule of succinyl CoA is used to form acetoacetyl CoA. Succinyl CoA could be used to generate one molecule of ATP, so someone could argue that the yield is 21.5.

16. For fats to be combusted, not only must they be converted into acetyl CoA, but the acetyl CoA must be processed by the citric acid cycle. In order for acetyl CoA to enter the citric acid cycle, there must be a supply of oxaloacetate. Oxaloacetate can be formed by the metabolism of glucose to pyruvate and the subsequent carboxylation of pyruvate to form oxaloacetate.

17. (a)

$$H_3C{-}CH(CH_3){-}(CH_2)_3{-}CH(CH_3){-}(CH_2)_3{-}CH(CH_3){-}(CH_2)_3{-}CH(CH_3){-}CH_2{-}COO^-$$

Phytanic acid

The problem with phytanic acid is that, as it undergoes β-oxidation, we encounter the dreaded pentavalent carbon atom. Because the pentavalent carbon atom does not exist, β-oxidation cannot take place and phytanic acid accumulates.

$$R{-}CH(CH_3){-}CH_2{-}COO^- \longrightarrow$$

$$R{-}C(CH_3){=}CH{-}COO^- \longrightarrow R{-}C(CH_3)(OH){-}CH_2{-}COO^- \longrightarrow$$

$$R{-}C(CH_3)({=}O){-}CH_2{-}COO^-$$

The dreaded pentavalent carbon atom

(b) Removing methyl groups, though theoretically possible, would be time-consuming and lacking in elegance. What would we do with the methyl groups? Our livers solve the problem by using α oxidation.

$$R{-}CH(CH_3){-}CH_2{-}COO^- \longrightarrow R{-}CH(CH_3){-}CH(OH){-}COO^- \longrightarrow$$

$$R{-}CH(CH_3){-}C({=}O){-}COO^- \longrightarrow R{-}CH(CH_3){-}C({=}O){-}O^- + CO_2$$

One round of α oxidation rather than β-oxidation converts phytanic acid into a β-oxidation substrate.

18. The first oxidation removes two tritium atoms. The hydration adds nonradioactive H and OH. The second oxidation removes another tritium atom from the β-carbon atom. Thiolysis removes an acetyl CoA with only one tritium atom; so the tritium-to-carbon ratio is 1:2. This ratio will be the same for two of the acetates. The last one, however, does not undergo oxidation, and so all tritium remains. The ratio for this acetate is 3:2. The ratio for the entire molecule is then 5:6.

19. In the absence of insulin, lipid mobilization will take place to an extent that it overwhelms the ability of the liver to convert the lipids into ketone bodies.

20. The formation of malonyl CoA from acetyl CoA by acetyl CoA carboxylase 1.

21. (a) 10; (b) 1; (c) 5; (d) 8; (e) 3; (f) 9; (g) 6; (h) 7; (i) 4; (j) 2.

22. (a) Oxidation in mitochondria; synthesis in the cytoplasm. (b) Coenzyme A in oxidation; acyl carrier protein for synthesis. (c) FAD and NAD^+ in oxidation; NADPH for synthesis. (d) The L isomer of 3-hydroxyacyl CoA in oxidation; the D isomer in synthesis. (e) From carboxyl to methyl in oxidation; from methyl to carboxyl in synthesis. (f) The enzymes of fatty acid synthesis, but not those of oxidation, are organized in a multienzyme complex.

23. Bicarbonate is required for the synthesis of malonyl CoA from acetyl CoA by acetyl CoA carboxylase.

24. 7 acetyl CoA + 6 ATP + 12 NADPH + 12 H^+ ⟶
myristate + 7 CoA + 6 ADP + 6 P_i + 12 $NADP^+$ + 5H_2O.

25. We will need six acetyl CoA units. One acetyl CoA unit will be used directly to become the two carbon atoms farthest from the acid end. The other five units must be converted into malonyl CoA. The synthesis of each malonyl CoA molecule costs 1 molecule of ATP; so 5 molecules of ATP will be required. Each round of elongation requires 2 molecules of NADPH, 1 molecule to reduce the keto group to an alcohol, and 1 molecule to reduce the double bond. As a result, 10 molecules of NADPH will be required. Therefore, 5 molecules of ATP and 10 molecules of NADPH are required to synthesize lauric acid.

26. e, b, d, a, c.

27. Such a mutation would inhibit fatty acid synthesis because the enzyme cleaves cytoplasmic citrate to yield acetyl CoA for fatty acid synthesis.

28. (a) False. Biotin is required for acetyl CoA carboxylase activity. (b) True. (c) False. ATP is required to synthesize malonyl CoA. (d) True. (e) True. (f) False. Fatty acid synthase is a dimer. (g) True. (h) False. Acetyl CoA carboxylase is stimulated by citrate, which is cleaved to yield its substrate acetyl CoA.

29. Fatty acids with odd numbers of carbon atoms are synthesized starting with propionyl ACP (instead of acetyl ACP), which is formed from propionyl CoA by acetyl transacetylase.

30. All of the labeled carbon atoms will be retained. Because we need 8 acetyl CoA molecules and only 1 carbon atom is labeled in the acetyl group, we will have 8 labeled carbon atoms. The only acetyl CoA used directly will retain 3 tritium atoms. The 7 acetyl CoA molecules used to make malonyl CoA will lose 1 tritium atom on addition of the CO_2 and another one at the dehydration step. Each of the 7 malonyl CoA molecules will retain 1 tritium atom. Therefore, the total retained tritium is 10 atoms. The ratio of tritium to carbon is 1.25.

31. With a diet rich in raw eggs, avidin will inhibit fatty acid synthesis by reducing the amount of biotin required by acetyl CoA carboxylase. Cooking the eggs will denature avidin, and so it will no longer bind biotin.

32. The only acetyl CoA used directly, not in the form of malonyl CoA, provides the two carbon atoms at the ω end of the fatty acid chain. Because palmitic acid is a C_{16} fatty acid, acetyl CoA will have provided carbons 15 and 16.

33. HCO_3^- is attached to acetyl CoA to form malonyl CoA. When malonyl CoA condenses with acetyl CoA to form the four-carbon ketoacyl CoA, the HCO_3^- is lost as CO_2.

34. Phosphofructokinase controls the flux down the glycolytic pathway. Glycolysis functions to generate ATP or building blocks for biosynthesis, depending on the tissue. The presence of citrate in the cytoplasm indicates that those needs are met, and there is no need to metabolize glucose.

35. C-1 is more radioactive.

36. The mutant enzyme will be persistently active because it cannot be inhibited by phosphorylation. Fatty acid synthesis will be abnormally active. Such a mutation might lead to obesity.

37. (a) Palmitoleate; (b) linoleate; (c) linoleate; (d) oleate; (e) oleate; (f) linolenate.

38. Decarboxylation drives the condensation of malonyl ACP and acetyl ACP. In contrast, the condensation of two molecules of acetyl ACP is energetically unfavorable. In gluconeogenesis, decarboxylation drives the formation of phosphoenolpyruvate from oxaloacetate.

39. Fat mobilization in adipocytes is activated by phosphorylation. Hence, overproduction of the cAMP-activated kinase will lead to an accelerated breakdown of triacylglycerols and a depletion of fat stores.

40. Carnitine translocase deficiency and glucose 6-phosphate transporter deficiency.

41. In the fifth round of β-oxidation, *cis*-Δ^2-enoyl CoA is formed. Dehydration by the classic hydratase yields D-3-hydroxyacyl CoA, the wrong isomer for the next enzyme in β-oxidation. This dead end is circumvented by a second hydratase that removes water to give *trans*-Δ^2-enoyl CoA. The addition of water by the classic hydratase then yields L-3-hydroxyacyl CoA, the appropriate isomer. Thus, hydratases of opposite stereospecificities serve to epimerize (invert the configuration of) the 3-hydroxyl group of the acyl CoA intermediate.

42. An advantage of this arrangement is that the synthetic activity of different enzymes is coordinated. In addition, intermediates can be efficiently handed from one active site to another without leaving the assembly. Furthermore, a complex of covalently joined enzymes is more stable than one formed by noncovalent attractions. Each of the component enzymes is recognizably homologous to its bacterial counterpart.

43. The probability of synthesizing an error-free polypeptide chain decreases as the length of the chain increases. A single mistake can make the entire polypeptide ineffective. In contrast, a defective subunit can be spurned in the formation of a noncovalent multienzyme complex; the good subunits are not wasted.

44. The absence of ketone bodies is due to the fact that the liver—the source of ketone bodies in the blood—cannot oxidize fatty acids to produce acetyl CoA. Moreover, because of the impaired fatty acid oxidation, the liver becomes more dependent on glucose as an energy source. This dependency results in a decrease in gluconeogenesis and a drop in blood-glucose concentration, which is exacerbated by the lack of fatty acid oxidation in muscle and a subsequent increase in glucose uptake from the blood.

45. Peroxisomes enhance the degradation of very long chain fatty acids. Consequently, increasing the activity of peroxisomes could help to lower levels of blood triacylglycerides. In fact, clofibrate is rarely used because of serious side effects.

46. Citrate works by facilitating, in cooperation with the protein MIG12, the formation of active filaments from inactive monomers. In essence, it increases the number of active sites available, or the concentration of enzyme. Consequently, its effect is visible as an increase in the value of V_{max}. Allosteric enzymes that alter their V_{max} values in response to regulators are sometimes called V-class enzymes. The more common type of allosteric enzyme, in which K_m is altered, comprises K-class enzymes. Palmitoyl CoA causes depolymerization and thus inactivation.

47. The thiolate anion of CoA attacks the 3-keto group to form a tetrahedral intermediate. This intermediate collapses to form acyl CoA and the enolate anion of acetyl CoA. Protonation of the enolate yields acetyl CoA.

48.

Malonyl-ACP

Acetyl-ACP

Acetoacetyl-ACP

49. Fatty acids cannot be transported into mitochondria for oxidation. The muscles cannot use fats as a fuel. Muscles could use glucose derived from glycogen. However, when glycogen stores are depleted, as after a fast, the effect of the deficiency is especially apparent.

50. After a night's sleep, glycogen stores will be low, but fats will be plentiful. Muscles will burn fat as a fuel. Why the caffeine? Lipid mobilization is stimulated by glucagon and epinephrine, and both of these hormones can work through the cAMP cascade. cAMP stimulates protein kinase A, which stimulates the breakdown of triacylglycerols in the adipose tissue. If cAMP hydrolysis to AMP is inhibited, protein kinase A will be maximally stimulated, fat will be maximally mobilized, and you will look as buff and lean as a svelte waterfowl.

51. The first reaction

$$\text{Oxaloacetate} + \text{NADH} + \text{H}^+ \rightleftharpoons \text{malate} + \text{NAD}^+$$

is catalyzed by cytoplasmic malate dehydrogenase. The next reaction is catalyzed by malic enzyme:

$$\text{Malate} + \text{NADP}^+ \longrightarrow \text{pyruvate} + \text{CO}_2 + \text{NADPH}$$

Finally, oxaloacetate is regenerated from pyruvate by pyruvate carboxylase.

$$\text{Pyruvate} + \text{CO}_2 + \text{ATP} + \text{H}_2\text{O} \longrightarrow$$
$$\text{oxaloacetate} + \text{ADP} + \text{P}_i + 2\,\text{H}^+$$

The sum of the reactions is

$$\text{NADP}^+ + \text{NADH} + \text{ATP} + \text{H}_2\text{O} \longrightarrow$$
$$\text{NADPH} + \text{NAD}^+ + \text{ADP} + \text{P}_i + \text{H}^+$$

52. When glycogen stores are filled, the excess carbohydrates are metabolized to acetyl CoA, which is then converted into fats. Human beings cannot convert fats into carbohydrates, but they can certainly convert carbohydrates into fats.

53. $\text{Pyruvate} + \text{CO}_2 + \text{ATP} + \text{H}_2\text{O} \longrightarrow$
$$\text{oxaloacetate} + \text{ADP} + \text{P}_i + 2\,\text{H}^+$$

In gluconeogenesis, the oxaloacetate is converted into phosphoenolpyruvate by PEP carboxykinase at the expense of a molecule of GTP.

54. A labeled fat can enter the citric acid cycle as acetyl CoA and yield labeled oxaloacetate, which can be converted into glucose and then glycogen. However, oxaloacetate is formed only after two carbon atoms have been lost as CO_2. Consequently, even though oxaloacetate may be labeled, there can be no net synthesis of oxaloacetate and hence no net synthesis of glucose or glycogen.

55. (a) Glucose is the primary fuel used by the brain. Lack of pyruvate dehydrogenase would prevent complete oxidation of glucose-derived pyruvate by cellular respiration. (b) Because ATP cannot be generated aerobically, glucose would be metabolized to lactate to obtain some ATP. (c) Such a diet would generate ketone bodies for use as fuel by the brain.

56. I-cell disease (Section 11.3) results when lysosomal enzymes are secreted instead of imported into the lysosome.

57. (a) The V_{max} is decreased and the K_M is increased. V_{max} (wild type) = 13 nmol minute^{-1} mg^{-1}; K_M (wild type) = 45 μM; V_{max} (mutant) = 8.3 nmol minute^{-1} mg^{-1}; K_M (mutant) = 74 μM. (b) Both the V_{max} and the K_M are decreased. V_{max} (wild type) = 41 nmol minute^{-1} mg^{-1}; K_M (wild type) = 104 μM; V_{max} (mutant) = 23 nmol minute^{-1} mg^{-1}; K_M (mutant) = 69 μM. (c) The wild type is significantly more sensitive to malonyl CoA. (d) With respect to carnitine, the mutant displays approximately 65% of the activity of the wild type; with respect to palmitoyl CoA, approximately 50% activity. On the other hand, 10 μM of malonyl CoA inhibits approximately 80% of the wild type but has essentially no effect on the mutant enzyme. (e) The glutamate appears to play a more prominent role in regulation by malonyl CoA than in catalysis.

58. (a) Leucine is changed to alanine. (b) No change relative to control. (c) Tyrosine to alanine. (d) Greatly reduced lipase-binding compared to the control. (e) Lipase activity was diminished. (f) This suggests they are separate. The L16A mutation allows binding of the lipase but apparently does not allow activation of the lipase. (g) This mutation decreased both binding and activity, suggesting that the tyrosine is involved in both aspects of colipase activity.

59. (a) Soraphen A inhibits fatty acid synthesis in a dose-dependent manner. (b) Fatty acid oxidation is increased in the presence of soraphen A. (c) Recall that acetyl CoA carboxylase 2 synthesizes malonyl CoA to inhibit the transport of fatty acids into the mitochondria, thereby preventing fatty acid oxidation. Soraphen A apparently inhibits both forms of the carboxylase. (d) Phospholipid synthesis was inhibited in a dose-dependent manner. (e) Phospholipids are required for membrane synthesis. (f) Soraphen A inhibits cell proliferation, especially at higher concentrations.

Chapter 23

1. When the proteins are denatured, all of the peptide bonds are accessible to proteolytic enzymes. If the three-dimensional structure of a protein is maintained, access to many peptide bonds is denied to the proteolytic enzymes.

2. First, the ubiquitin-activating enzyme (E1) links ubiquitin to a sulfhydryl group on E1 itself. Next, the ubiquitin is transferred to a cysteine residue on the ubiquitin-conjugating enzyme (E2) by E2. The ubiquitin-protein ligase (E3), using the ubiquitinated E2 as a substrate, transfers the ubiquitin to the target protein.

3. (a) 7; (b) 4; (c) 2; (d) 10; (e) 5; (f) 3; (g) 9; (h) 1; (i) 6; (j) 8.

4. (a) The ATPase activity of the 26S proteasome resides in the 19S subunit. The energy of ATP hydrolysis is used to unfold the substrate, which is too large to enter the catalytic barrel. ATP may also be required for translocation of the substrate into the barrel. (b) Substantiates the answer in part *a*. Because they are small, the peptides do not need to be unfolded. Moreover, small peptides could probably enter all at once and not require translocation.

5. In the eukaryotic proteasome, the distinct β subunits have different substrate specificities, allowing proteins to be more thoroughly degraded.

6. The six subunits probably exist as a heterohexamer. Cross-linking experiments could test the model and help determine which subunits are adjacent to one another.

7. (b)

8. (a)

9. (d)

10. (c)

11. (b)

12. (a) Pyruvate; (b) oxaloacetate; (c) α-ketoglutarate; (d) α-ketoisocaproate; (e) phenylpyruvate; (f) hydroxyphenylpyruvate.

13. (a) Aspartate + α-ketoglutarate + GTP + ATP + 2 H$_2$O + NADH + H$^+$ $\longrightarrow$ $^1/_2$ glucose + glutamate + CO$_2$ + ADP + GDP + NAD$^+$ + 2 P$_i$.
The required coenzymes are pyridoxal phosphate in the transamination reaction and NAD$^+$/NADH in the redox reactions. (b) Aspartate + CO$_2$ + NH$_4^+$ + 3 ATP + NAD$^+$ + 4 H$_2$O $\longrightarrow$ oxaloacetate + urea + 2 ADP + 4 P$_i$ + AMP + NADH + H$^+$.

14. Thiamine pyrophosphate.

15. Aminotransferases transfer the α-amino group to α-ketoglutarate to form glutamate. Glutamate is oxidatively deaminated to form an ammonium ion.

16. Aspartate (oxaloacetate), glutamate (α-ketoglutarate), alanine (pyruvate).

17. Serine and threonine.

18. They are either fuels for the citric acid cycle, components of the citric acid cycle, or molecules that can be converted into a fuel for the citric acid cycle in one step.

19. It acts as an electron sink.

20. Carbamoyl phosphate and aspartate.

21. (a) 4; (b) 5; (c) 1; (d) 6; (e) 7; (f) 3; (g) 2.

22. A, arginine; B, citrulline; C, ornithine; D, arginosuccinate. The order of appearance: C, B, D, A.

23. CO$_2$ + NH$_4^+$ + 3 ATP + NAD$^+$ + aspartate + 3 H$_2$O $\longrightarrow$ urea + 2 ADP + 2 P$_i$ + AMP + PP$_i$ + NADH + H$^+$ + oxaloacetate.
Four high-transfer-potential phosphoryl groups are spent. Note, however, that an NADH is generated if fumarate is converted into oxaloacetate. NADH can generate 2.5 ATP in the electron-transport chain. Taking these ATP into account, only 1.5 high-transfer-potential phosphoryl groups are spent.

24. The synthesis of fumarate by the urea cycle is important because it links the urea cycle and the citric acid cycle. Fumarate is hydrated to malate, which, in turn, is oxidized to oxaloacetate. Oxaloacetate has several possible fates: (1) transamination to aspartate, (2) conversion into glucose by the gluconeogenic pathway, (3) condensation with acetyl CoA to form citrate, or (4) conversion into pyruvate. You can collect.

25. Ornithine transcarbamoylase (Compound A is analogous to PALA; Chapter 10).

26. Ammonia could lead to the amination of α-ketoglutarate, producing a high concentration of glutamate in an unregulated fashion. α-Ketoglutarate for glutamate synthesis could be removed from the citric acid cycle, thereby diminishing the cell's respiration capacity.

27. The mass spectrometric analysis strongly suggests that three enzymes—pyruvate dehydrogenase, α-ketoglutarate dehydrogenase, and the branched-chain α-ketoacid dehydrogenase—are deficient. Most likely, the common E$_3$ component of these enzymes is missing or defective. This proposal could be tested by purifying these three enzymes and assaying their ability to catalyze the regeneration of lipoamide.

28. Benzoate, phenylacetate, and arginine would be given to supplement a protein-restricted diet. Nitrogen would emerge in hippurate, phenylacetylglutamine, and citrulline.

29. The liver is the primary tissue for capturing nitrogen as urea. If the liver is damaged (for instance, by hepatitis or the excessive consumption of alcohol), free ammonia is released into the blood.

30. This defect can be partly bypassed by providing a surplus of arginine in the diet and restricting the total protein intake. In the liver, arginine is split into urea and ornithine, which then reacts

with carbamoyl phosphate to form citrulline. This urea-cycle intermediate condenses with aspartate to yield argininosuccinate, which is then excreted. Note that two nitrogen atoms—one from carbamoyl phosphate and the other from aspartate—are eliminated from the body per molecule of arginine provided in the diet. In essence, argininosuccinate substitutes for urea in carrying nitrogen out of the body. The formation of argininosuccinate removes the nitrogen, and the restriction on protein intake relieves the aciduria.

31. Aspartame, a dipeptide ester (L-aspartyl-L-phenylalanine methyl ester), is hydrolyzed to L-aspartate and L-phenylalanine. High levels of phenylalanine are harmful in phenylketonurics.

32. N-Acetylglutamate is synthesized from acetyl CoA and glutamate. Once again, acetyl CoA serves as an activated acetyl donor. This reaction is catalyzed by N-acetylglutamate synthase.

33. Not all proteins are created equal: some are more important than others. Some proteins would be degraded to provide the missing amino

acid. The nitrogen from the other amino acids would be excreted as urea. Consequently, more nitrogen would be excreted than ingested.

34. The carbon skeletons of ketogenic amino acids can be converted into ketone bodies or fatty acids. Only leucine and lysine are purely ketogenic. Glucogenic amino acids are those whose carbon skeletons can be converted into glucose.

35. As shown in Figure 23.28, alanine, a gluconeogenic amino acid, is released during the metabolism of tryptophan to acetyl CoA and acetoacetyl CoA.

36. The branched-chain amino acids leucine, isoleucine, valine and threonine. The required enzyme is the branched-chain α-keto-acid dehydrogenase complex.

37. Pyruvate (glycolysis and gluconeogenesis), acetyl CoA (citric acid cycle and fatty acid synthesis), acetoacetyl CoA (ketone-body formation), α-ketoglutarate (citric acid cycle), succinyl CoA (citric acid cycle), fumarate (citric acid cycle), and oxaloacetate (citric acid cycle and gluconeogenesis).

38.

39.

The equilibrium constant for the interconversion of L-serine and D-serine is exactly 1.

40. Double-displacement. A substituted enzyme intermediate is formed.

41. Exposure of such a domain suggests that a component of a multiprotein complex has failed to form properly or that

one component has been synthesized in excess. This exposure leads to rapid degradation and the restoration of appropriate stoichiometries.

42. (a) Depletion of glycogen stores. When they are gone, proteins must be degraded to meet the glucose needs of the brain. The resulting amino acids are deaminated, and the nitrogen atoms are

excreted as urea. (b) The brain has adapted to the use of ketone bodies, which are derived from fatty acid catabolism. In other words, the brain is being powered by fatty acid breakdown. (c) When the glycogen and lipid stores are gone, the only available energy source is protein.

43. The precise cause of all of the symptoms is not firmly established, but a likely explanation depends on the centrality of oxaloacetate to metabolism. A lack of pyruvate carboxylase would reduce the amount of oxaloacetate. The lack of oxaloacetate would reduce the activity of the citric acid cycle and so ATP would be generated by lactic acid formation. If the concentration of oxaloacetate is low, aspartate cannot be formed and the urea cycle would be compromised. Oxaloacetate is also required to form citrate, which transports acetyl CoA to the cytoplasm for fatty acid synthesis. Finally, oxaloacetate is required for gluconeogenesis.

44. Glycogen phosphorylase. The coenzyme serves as an acid base catalyst.

45. In the Cori cycle, the carbon atoms are transferred from muscle to liver as lactate. For lactate to be of any use, it must be reduced to pyruvate. This reduction requires high-energy electrons in the form of NADH. When the carbon atoms are transferred as alanine, transamination yields pyruvate directly.

46. (a) Virtually no digestion in the absence of nucleotides. (b) Protein digestion is greatly stimulated by the presence of ATP. (c) AMP-PNP, a nonhydrolyzable analog of ATP, is no more effective than ADP. (d) The proteasome requires neither ATP nor PAN to digest small substrates. (e) PAN and ATP hydrolysis may be required to unfold the peptide and translocate it into the proteasome. (f) Although *Thermoplasma* PAN is not as effective with the other proteasomes, it nonetheless results in threefold to fourfold stimulation of digestion. (g) In light of the fact that the archaea and eukarya diverged several billion years ago, the fact that *Thermoplasma* PAN can stimulate the rabbit muscle proteasome suggests homology not only between the proteasomes, but also between PAN and the 19S subunit (most likely the ATPases) of the mammalian 26S proteasome.

Chapter 24

1. (a) and (c)
2. (d)
3. (b) and (c)
4. (a)
5. Nitrogen fixation is the conversion of atmospheric N_2 into NH_4^+. Diazotrophic (nitrogen-fixing) microorganisms are able to fix nitrogen.
6. (a) 4; (b) 8; (c) 10; (d) 6; (e) 7; (f) 9; (g) 3; (h) 5; (i) 2; (j) 1.
7. The reductase provides electrons with high reducing power, whereas the nitrogenase, which requires ATP hydrolysis, uses the electrons to reduce N_2 to NH_3.
8. False. Nitrogen fixation is thermodynamically favored. ATP expenditure by the nitrogenase is required to make the reaction kinetically possible.
9. The bacteria provide the plant with ammonia by reducing atmospheric nitrogen. This reduction is energetically expensive, and the bacteria use ATP from the plant.
10. Oxaloacetate, pyruvate, ribose-5-phosphate, phosphoenolpyruvate, erythrose-4-phosphate, α-ketoglutarate, and 3-phosphoglycerate.
11. Human beings do not have the biochemical pathways to synthesize certain amino acids from simpler precursors. Consequently, these amino acids are "essential" and must be obtained from the diet.
12. Glucose + 2 ADP + 2 P_i + 2 NAD^+ + 2 glutamate $\longrightarrow$ 2 alanine + 2 α-ketoglutrate + 2 ATP + 2 NADH + 2 H_2O + 2 H^+
13. $N_2 \longrightarrow NH_4^+ \longrightarrow$ glutamate $\longrightarrow$ serine $\longrightarrow$ glycine $\longrightarrow$ δ-aminolevulinate $\longrightarrow$ porphobilinogen $\longrightarrow$ heme.

14. Pyridoxal phosphate (PLP).
15. *S*-Adenosylmethionine, tetrahydrofolate, and methylcobalamin.
16. (a) N^5,N^{10}-Methylenetetrahydrofolate; (b) N^5-methyltetrahydrofolate.
17. γ-Glutamyl phosphate is a likely reaction intermediate.
18. The synthesis of asparagine from aspartate passes through an acyl-adenylate intermediate. One of the products of the reaction will be ^{18}O-labeled AMP.
19. Glycine reacts with the excess isovaleric acid to form isovalerylglycine. This water-soluble conjugate, in contrast with isovaleric acid, is excreted very rapidly by the kidneys. Thus, the administration of glycine results in the removal of isovlaeric acid.
20. The nitrogen atom shaded red is derived from glutamine. The carbon atom shaded blue is derived from serine.
21. They carry out nitrogen fixation. The absence of photosystem II provides an environment in which O_2 is not produced. Recall that the nitrogenase is very rapidly inactivated by O_2.
22. The cytoplasm is a reducing environment, whereas the extracellular milieu is an oxidizing environment.
23. (a) None; (b) D-glutamate and oxaloacetate.
24. Succinyl CoA is formed in the mitochondrial matrix as part of the citric acid cycle.
25. Alanine from pyruvate; aspartate from oxaloacetate; glutamate from α-ketoglutarate.
26. Lysine cyclodeaminase converts L-lysine into the six-membered ring analog of proline, also referred to as L-homoproline or L-pipecolate:

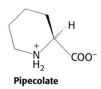

Pipecolate

27. Y could inhibit the C → D step, Z could inhibit the C → F step, and C could inhibit A → B. This scheme is an example of sequential feedback inhibition. Alternatively, Y could inhibit the C → D step, Z could inhibit the C → F step, and the A → B step would be inhibited only in the presence of both Y and Z. This scheme is called concerted feedback inhibition.
28. The rate of the A → B step in the presence of high levels of Y and Z would be 24 s^{-1} (0.6 × 0.4 × 100 s^{-1}).
29. Lysine 258 is absolutely essential for the activity of aspartate aminotransferase, as it is responsible both for the formation of the internal aldimine with the pyridoxal phosphate cofactor and for transferring the proton between the ketimine and quinonoid intermediates. Mutation of this residue to cysteine would be expected to dramatically impair catalysis, as cysteine cannot occupy the same space as lysine and also exhibits differing pK_a properties. Upon treatment with 2-bromoethylamine, however, the resulting thioether now has a shape and pK_a similar to the original lysine side chain. Hence, some catalytic activity is restored.
30. An external aldimine forms with SAM, which is deprotonated to form the quinonoid intermediate. The deprotonated carbon atom attacks the carbon atom adjacent to the sulfur atom to form the cyclopropane ring and release methylthioadenosine, the other product.
31. An external aldimine forms with L-serine, which is deprotonated to form the quinonoid intermediate. This intermediate is reprotonated on its opposite face to form an aldimine with D-serine. This compound is cleaved to release D-serine. The equilibrium constant for a racemization reaction is 1 because the reactant and product are exact mirror images of each other.

32. (a) In the first step, histidine attacks the methylene group from the methionine subgroup of SAM (rather than the usual methyl substituent), resulting in the transfer of an aminocarboxypropyl group. Three subsequent conventional SAM-mediated methylations of the primary amine yield diphthine.

SAM

Diphthine

(b) In this chapter, we have observed two examples of an ATP-dependent conversion of a carboxylate into an amide: glutamine synthetase, which uses an acyl-phosphate intermediate, and asparagine synthetase, which uses an acyl-adenylate intermediate. Either mechanism is possible in formation of diphthamide from diphthine.

33. Biotin.

34. Synthesis from oxaloacetate and α-ketoglutarate would deplete the citric acid cycle, which would decrease ATP production. Anapleurotic reactions would be required to replenish the citric acid cycle.

35. SAM is the donor for DNA methylation reactions that protect a host from digestion by its own restriction enzymes. A lack of SAM would render the bacterial DNA susceptible to digestion by the cell's own restriction enzymes.

36. Acetate $\longrightarrow$ acetyl-CoA $\longrightarrow$ citrate $\longrightarrow$ isocitrate $\longrightarrow$ α-ketoglutarate $\longrightarrow$ succinyl-CoA.

37. The value of K_M of glutamate dehydrogenase for NH_4^+ is high ($\gg 1$ mM), and so this enzyme is not saturated when NH_4^+ is limiting. In contrast, glutamine synthetase has very low K_M for NH_4^+. Thus, ATP hydrolysis is required to capture ammonia when it is scarce.

38. (a) Asparagine is much more abundant in the dark. More glutamine is present in the light. These amino acids show the most dramatic effects. Glycine also is more abundant in the light.

(b) Glutamine is a more metabolically reactive amino acid, used in the synthesis of many other compounds. Consequently, when energy is available as light, glutamine will be preferentially synthesized. Asparagine, which carries more nitrogen per carbon atom and is thus a more efficient means of storing nitrogen when energy is short, is synthesized in the dark. Glycine is more prevalent in the light because of photorespiration. (c) White asparagus has an especially high concentration of asparagine, which accounts for its intense taste. All asparagus has a large amount of asparagine. In fact, as suggested by its name, asparagine was first isolated from asparagus.

Chapter 25

1. (c)

2. (d)

3. (b)

4. (a)

5. (c)

6. In de novo synthesis, the nucleotides are synthesized from simpler precursor compounds, in essence from scratch. In salvage pathways, preformed bases are recovered and attached to riboses.

7. Carbon 2 and nitrogen 3 come from carbamoyl phosphate. Nitrogen 1 and carbons 4, 5, and 6 are derived from aspartate.

8. Nitrogen 1: aspartate; carbon 2: N^{10}-formyltetrahydrofolate; nitrogen 3: glutamine; carbons 4 and 5 and nitrogen 7: glycine; carbon 6: CO_2; carbon 8: N^{10}-formyltetrahydrofolate; nitrogen 9: glutamine.

9. Energy currency: ATP; signal transduction: ATP and GTP; RNA synthesis: ATP, GTP, CTP, and UTP; DNA synthesis: dATP, dCTP, dGTP, and TTP; components of coenzymes: ATP in CoA, FAD, and NAD(P)$^+$; carbohydrate synthesis: UDP-glucose. These are just some of the uses.

10. A nucleoside is a base attached to ribose. A nucleotide is a nucleoside with the ribose bearing one or more phosphates.

11. (a) 9; (b) 7; (c) 6; (d) 10; (e) 2; (f) 4; (g) 1; (h) 11; (i) 8; (j) 3; (k) 5.

12. Substrate channeling is the process whereby the product of one active site moves to become a substrate at another active site without ever leaving the enzyme. A channel connects the active sites. Substrate channeling greatly enhances enzyme efficiency and minimizes the diffusion of a substrate to an active site.

13. Glucose + 2 ATP + 2 NADP$^+$ + H_2O $\longrightarrow$
      PRPP + CO_2 + ADP + AMP + 2 NADPH + 3 H$^+$.

14. Glutamine + aspartate + CO_2 + 2 ATP + NAD$^+$ $\longrightarrow$
      orotate + 2 ADP + 2 P$_i$ + glutamate + NADH + H$^+$.

15. (a, c, and d) PRPP; (b) carbamoyl phosphate.

16. PRPP and formylglycinamide ribonucleotide.

17. dUMP + serine + NADPH + H$^+$ $\longrightarrow$
      TMP + NADP$^+$ + glycine.

18. Sulfanilamide inhibits the synthesis of folate by acting as an analog of p-aminobenzoate, one of the precursors of folate.

19. (a) Cell A cannot grow in a HAT medium, because it cannot synthesize TMP either from thymidine or from dUMP. Cell B cannot grow in this medium, because it cannot synthesize purines by either the de novo pathway or the salvage pathway. Cell C can grow in a HAT medium because it contains active thymidine kinase from cell B (enabling it to phosphorylate thymidine to TMP) and hypoxanthine guanine phosphoribosyltransferase from cell A (enabling it to synthesize purines from hypoxanthine by the salvage pathway). (b) Transform cell A with a plasmid containing foreign genes of interest and a functional thymidine kinase gene. The only cells that will grow in a HAT medium are those that have acquired a thymidylate kinase gene; nearly all of these transformed cells will also contain the other genes on the plasmid.

20. The folate derivative N^5, N^{10}-methylenetetrahydrofolate is required by thymidylate synthase to add a methyl group to dUMP, forming TMP. Insufficient folate could result in spina bifida.

21. The reciprocal substrate relation refers to the fact that AMP synthesis requires GTP, whereas GMP synthesis requires ATP. These requirements tend to balance the synthesis of ATP and GTP.

22. Ring carbon 6 in cytosine will be labeled. In guanine, only carbon 5 will be labeled with ^{13}C.

23. UTP is first converted into UDP. Ribonucleotide reductase generates dUDP. DeoxyUDP is converted to dUMP. Thymidylate synthase generates TMP from dUMP. Monophosphate and diphosphate kinases subsequently form TTP.

24. These patients have a high level of urate because of the breakdown of nucleic acids. Allopurinol prevents the formation of kidney stones and blocks other deleterious consequences of hyperuricemia by preventing the formation of urate.

25. The free energies of binding are -57.7 (wild type), -49.8 (Asn 27), and -38.1 (Ser 27) kJ mol^{-1} (-13.8, -11.9, and -9.1 kcal mol^{-1}, respectively). The loss in binding energy is $+7.9$ kJ mol^{-1} ($+1.9$ kcal mol^{-1}) and $+19.7$ kJ mol^{-1} ($+4.7$ kcal mol^{-1}).

26. By their nature, cancer cells divide rapidly and thus require frequent DNA synthesis. Inhibitors of TMP synthesis will impair DNA synthesis and cancer growth.

27. Uridine and cytidine are administered to bypass the defective enzyme in orotic aciduria.

28. Inosine or hypoxanthine could be administered.

29. N-1 in both cases, and the amine group linked to C-6 in ATP.

30. Nitrogen atoms 3 and 9 in the purine ring.

31. Allopurinol, an analog of hypoxanthine, is a suicide inhibitor of xanthine oxidase.

32. An oxygen atom is added to allopurinol to form alloxanthine.

Alloxanthine

33.

The synthesis of carbamoyl phosphate requires 2 ATP	2 ATP
The formation of PRPP from ribose 5-phosphate yields an AMP*	2 ATP
The conversion of UMP to UTP requires 2 ATP	2 ATP
The conversion of UTP to CTP requires 1 ATP	1 ATP
Total	7 ATP

*Remember that AMP is the equivalent of 2 ATP because an ATP must be expended to generate ADP, the substrate for ATP synthesis.

34. (a) Carboxyaminoimidazole ribonucleotide; (b) glycinamide ribonucleotide; (c) phosphoribosyl amine; (d) formylglycinamide ribonucleotide.

35. The first reaction proceeds by phosphorylation of glycine to form an acyl phosphate followed by nucleophilic attack by the amine of phosphoribosylamine to displace orthophosphate. The second reaction consists of adenylation of the carbonyl group of xanthylate followed by nucleophilic attack by ammonia to displace AMP.

36. The —NH$_2$ group attacks the carbonyl carbon atom to form a tetrahedral intermediate. Removal of a proton leads to the elimination of water to form inosinate.

37. The enzyme that uses ammonia, carbamoyl phosphate synthetase I, forms carbamoyl phosphate for a reaction with ornithine, the first step of the urea cycle. The enzyme that uses glutamine, carbamoyl phosphate synthetase II, generates carbamoyl phosphate for use in the first step of pyrimidine biosynthesis.

38. PRPP is the activated intermediate in the synthesis of phosphoribosylamine in the de novo pathway of purine formation; of purine nucleotides from free bases by the salvage pathway; of orotidylate in the formation of pyrimidines; of nicotinate ribonucleotide; and of phosphoribosylanthranilate in the pathway leading to tryptophan.

39. (a) cAMP; (b) ATP; (c) UDP-glucose; (d) acetyl CoA; (e) NAD$^+$, FAD; (f) dideoxynucleotides; (g) fluorouracil; (h) CTP inhibits ATCase.

40. In vitamin B$_{12}$ deficiency, methyltetrahydrofolate cannot donate its methyl group to homocysteine to regenerate methionine. Because the synthesis of methyltetrahydrofolate is irreversible, the cell's tetrahydrofolate will ultimately be converted into this form. No formyl or methylene tetrahydrofolate will be left for nucleotide synthesis. Vitamin B$_{12}$ is also required to metabolize propionyl CoA generated in the oxidation of odd-chain fatty acids and in the degradation of several amino acids.

41. Because folate is required for nucleotide synthesis, cells that are dividing rapidly would be most readily affected. They would include cells of the intestine, which are constantly replaced, and precursors to blood cells. A lack of intestinal cells and blood cells would account for the symptoms often observed.

42. In patients with glucose 6-phosphatase deficiency, the cytoplasmic level of ATP in the liver falls as a result of increased glycogen synthesis. In all three conditions, AMP rises above normal, and the excess AMP is degraded to urate.

43. Succinate $\longrightarrow$ malate $\longrightarrow$ oxaloacetate by the citric acid cycle. Oxaloacetate $\longrightarrow$ aspartate by transamination, followed by pyrimidine synthesis. Carbons 4, 5, and 6 are labeled.

44. Glucose will most likely be converted into two molecules of pyruvate, one of which will be labeled in the 2 position:

Now consider two common fates of pyruvate—conversion into acetyl CoA and subsequent processing by the citric acid cycle or carboxylation by pyruvate carboxylase to form oxaloacetate. Formation of citrate by condensing the labeled pyruvate with oxaloacetate will yield labeled citrate:

The labeled carbon will be retained through one round of the citric acid cycle but, on the formation of the symmetric succinate, the label will appear in two different positions. Thus, when succinate is metabolized to oxaloacetate, which may be aminated to form aspartate, two carbons will be labeled:

When this aspartate is used to form uracil, the labeled COO⁻ attached to the α-carbon is lost and the other COO⁻ becomes incorporated into uracil as carbon 4.

Suppose, instead, that labeled 2-[^{14}C] pyruvate is carboxylated to form oxaloacetate and processed to form aspartate. In this case, the α-carbon of aspartate bears the label.

When this aspartate is used to synthesize uracil, carbon 6 bears the label:

45. HGPRT, which is nonfunctional in Lesch–Nyhan patients, is required to form 6-mercaptopurine ribose monophosphate. Consequently, de novo purine synthesis continues.

46. (a) Some ATP can be salvaged from the ADP that is being generated. (b) There are equal numbers of high-phosphoryl-transfer-potential groups on each side of the equation. (c) Because the adenylate kinase reaction is at equilibrium, the removal of AMP would lead to the formation of more ATP. (d) Essentially, the cycle serves as an anaplerotic reaction for the generation of the citric acid cycle intermediate fumarate.

47. (i) The formation of 5-aminoididazole-4-carboxamide ribonucleotide from 5-aminoimidazole-4-(N-succinylcarboxamide) ribonucleotide in the synthesis of IMP. (ii) The formation of AMP from adenylosuccinate. (iii) The formation of arginine from argininosuccinate in the urea cycle.

48. Allopurinol is an inhibitor of xanthine oxidase, which is on the pathway for urate synthesis. In your pet duck, this pathway is the means by which excess nitrogen is excreted. If xanthine oxidase were inhibited in your duck, nitrogen could not be excreted, with severe consequences such as the formation of a dead duck.

49. (a) The enzyme from the LND 2 patient showed much less activity than the normal enzyme, suggesting that the defect in the enzyme impaired its catalytic ability. The results for LND 1 are puzzling. The enzyme displays activity similar to the enzyme from the normal cell line, and yet the patient was suffering from Lesch–Nyhan disease. (b) Possible explanations include: there may be an inhibitor in the cells that prevents the enzyme from acting in vivo but is lost during the purification procedure; the enzyme may be degraded more rapidly in vivo than the normal enzyme; the enzyme may be inherently less stable than the normal enzyme. (c) The enzyme from LND 1 lost activity much faster than the normal enzyme, suggesting that the enzyme was structurally unstable and would lose enzyme activity in the cell, accounting for the appearance of the disease

Chapter 26

1. (d)
2. (a)
3. (d)
4. (c)
5. (a)
6. Glycerol 3-phosphate is the foundation for both triacylglycerol and phospholipid synthesis. Glycerol 3-phosphate is acylated twice to form phosphatidate. In triacylglycerol synthesis, the phosphoryl group is removed from glycerol 3-phosphate to form diacylglycerol, which is then acylated to form triacylglycerol. In phospholipid synthesis, phosphatidate commonly reacts with CTP to form CDP-diacylglycerol, which then reacts with an alcohol to form a phospholipid. Alternatively, diacylglycerol may react with a CDP-alcohol to form a phospholipid.

7. Glycerol 3-phosphate is formed primarily by the reduction of dihydroxyacetone phosphate, a gluconeogenic intermediate, and to a lesser extent by the phosphorylation of glycerol.

8. Glycerol + 4 ATP + 3 fatty acids + 4 H$_2$O ⟶ triacylglycerol + ADP + 3 AMP + 7 P$_i$ + 4 H$^+$.

9. Glycerol + 3 ATP + 2 H$_2$O + CTP + ethanolamine → phosphatidylethanolamine + CMP + ADP + 2 AMP + 6 P$_i$ + 3 H$^+$.

10. Three. One molecule of ATP to form phosphorylethanolamine and two molecules of ATP to regenerate CTP from CMP.

11. All are synthesized from ceramide. In sphingomyelin, the terminal hydroxyl group of ceramide is modified with phosphorylcholine. In a cerebroside, the hydroxyl group has a glucose or galactose attached. In a ganglioside, oligosaccharide chains are attached to the hydroxyl group.

12. (i) Activate the diacylglycerol as CDP-DAG. (ii) Activate the alcohol as CDP-alcohol. (iii) Use the base-exchange reaction.

13. (a) CDP-diacylglycerol; (b) CDP-ethanolamine; (c) acyl CoA; (d) phosphatidylcholine; (e) UDP-glucose or UDP-galactose; (f) UDP-galactose; (g) geranyl pyrophosphate.

14. Such mutations are seen in mice. The amount of adipose tissue would decrease severely because diacylglycerol could not be formed. Normally, diacylglycerol is acylated to form triacylglycerols. If there were deficient phosphatidic acid phosphatase activity, no triacylglycerols would form.

15. (a) 8; (b) 4; (c) 1; (d) 9; (e) 3; (f) 10; (g) 5; (h) 2; (i) 6; (j) 7.

16. (i) The synthesis of activated isoprene units (isopentyl pyrophosphate), (ii) the condensation of six of the activated isoprene units to form squalene, and (iii) cyclization of the squalene to form cholesterol.

17. The amount of reductase and its activity control the regulation of cholesterol biosynthesis. Transcriptional control is mediated by SREBP. Translation of the reductase mRNA also is controlled. The reductase itself may undergo regulated proteolysis. Finally, the activity of the reductase is inhibited by phosphorylation by AMP kinase when ATP levels are low.

18. None in either (a) or (b), because the label is lost as CO$_2$.

19. The hallmark of this genetic disease is elevated cholesterol levels in the blood of even young children. The excess cholesterol is taken up by macrophages, which eventually results in plaque formation and heart disease. There are many mutations that cause the disease, but all result in malfunctioning of the LDL receptor.

20. The categories of mutations are:
(i) no receptor is synthesized; (ii) receptors are synthesized but do not reach the plasma membrane, because they lack signals for intracellular transport or do not fold properly; (iii) receptors reach the cell surface, but they fail to bind LDL normally because of a defect in the LDL-binding domain; (iv) receptors reach the cell surface and bind LDL, but they fail to cluster in coated pits because of a defect in their carboxyl-terminal regions.

21. "None of your business" and "I don't talk biochemistry until after breakfast" are appropriate but rude and uninformative answers. A better answer might be: "Although it is true that cholesterol is a precursor to steroid hormones, the rest of the statement is oversimplified. Cholesterol is a component of membranes, and membranes literally define cells, and cells make up tissues. But to say that cholesterol 'makes' cells and tissues is wrong."

22. Statins are competitive inhibitors of HMG-CoA reductase. They are used as drugs to inhibit cholesterol synthesis in patients with high levels of cholesterol.

23. No. Cholesterol is essential for membrane function and as a precursor for bile salts and steroid hormones. The complete lack of cholesterol would be lethal.

24. Deamination of cytidine to uridine changes CAA (Gln) into UAA (stop).

25. The LDL contains apolipoprotein B-100, which binds to an LDL receptor on the cell surface in a region known as a coated pit. On binding, the complex is internalized by endocytosis to form an internal vesicle. The vesicle is separated into two components. One, with the receptor, is transported back to the cell surface and fuses with the membrane, allowing continued use of the receptor. The other vesicle fuses with lysosomes inside the cell. The cholesteryl esters are hydrolyzed, and free cholesterol is made available for cellular use. The LDL protein is hydrolyzed to free amino acids.

26. Benign prostatic hypertrophy can be treated by inhibiting 5α-reductase. Finasteride, the 4-azasteroid analog of dihydrotestosterone, competitively inhibits the reductase but does not act on androgen receptors. Patients taking finasteride have a markedly lower plasma level of dihydrotestosterone and a nearly normal level of testosterone. The prostate gland becomes smaller, whereas testosterone-dependent processes such as fertility, libido, and muscle strength appear to be unaffected.

Finasteride

27. Patients who are most sensitive to debrisoquine have a deficiency of a liver P450 enzyme encoded by a member of the *CYP2* subfamily. This characteristic is inherited as an autosomal recessive trait. The capacity to degrade other drugs may be impaired in people who hydroxylate debrisoquine at a slow rate, because a single P450 enzyme usually handles a broad range of substrates.

28. Many hydrophobic odorants are deactivated by hydroxylation. Molecular oxygen is activated by a cytochrome P450 monooxygenase. NADPH serves as the reductant. One oxygen atom of O_2 goes into the odorant substrate, whereas the other is reduced to water.

29. Recall that dihydrotestosterone is crucial for the development of male characteristics in the embryo. If a pregnant woman were to be exposed to Propecia, the 5α-reductase of the male embryo would be inhibited, which could result in severe developmental abnormalities.

30. The oxygenation reactions catalyzed by the cytochrome P450 family permit greater flexibility in biosynthesis. Because plants are not mobile, they must rely on physical defenses, such as thorns, and chemical defenses, such as toxic alkaloids. The larger P450 array might permit greater biosynthetic versatility.

31. This knowledge might enable clinicians to characterize the likelihood of a patient's having an adverse drug reaction or being susceptible to chemical-induced illnesses. It might also permit a personalized and especially effective drug-treatment regime for diseases such as cancer.

32. The honeybees may be especially sensitive to environmental toxins, including pesticides, because these chemicals are not readily detoxified by honeybees, owing to their minimal P450 system.

33. A deficiency of the hydroxylase will cause a build-up of progesterone, which will then be converted into estradiol and testosterone.

34. The core structure of a steroid is four fused rings: three cyclohexane rings and one cyclopentane ring. In vitamin D, the B ring is split by ultraviolet light.

35. The negatively charged phosphoserine residue interacts with the positively charged protonated histidine residue and decreases its ability to transfer a proton to the thiolate.

His

36. The methyl group is first hydroxylated. The hydroxymethylamine eliminates formaldehyde to form methylamine.

37. Note that a cytidine nucleotide plays the same role in the synthesis of these phosphoglycerides as a uridine nucleotide does in the formation of glycogen (Section 21.4). In all of these biosyntheses, an activated intermediate (UDP-glucose, CDP-diacylglycerol, or CDP-alcohol) is formed from a phosphorylated substrate (glucose 1-phosphate, phosphatidate, or a phosphorylalcohol) and a nucleoside triphosphate (UTP or CTP). The activated intermediate then reacts with a hydroxyl group (the terminus of glycogen, the side chain of serine, or a diacylglycerol).

38. The attachment of isoprenoid side chains confers hydrophobic character. Proteins having such a modification are targeted to membranes.

39. 3-Hydroxy-3-methylglutaryl CoA is also a precursor for ketone-body synthesis. If fuel is needed elsewhere in the body, as might be the case during a fast, 3-hydroxy-3-methylglutaryl CoA is converted into the ketone body acetoacetate. If energy needs are met, the liver will synthesize cholesterol. Remember also that cholesterol synthesis occurs in the cytoplasm while ketone bodies are synthesized in the mitochondria.

40. One way in which phosphatidylcholine can be synthesized is by the addition of three methyl groups to phosphatidylethanolamine. The methyl donor is a modified form of methionine, *S*-adenosylmethionine or SAM (Section 24.2).

41. Mutations could occur in the gene encoding the sodium channel that would prevent the action of DDT. Alternatively, P450 enzyme synthesis could be increased to accelerate metabolism of the insecticide to inactive metabolites. In fact, both types of responses to insecticides have been observed.

42. Citrate is transported out of the mitochondria in times of plenty. ATP-citrate lyase yields acetyl CoA and oxaloacetate. The acetyl CoA can then be used to synthesize cholesterol.

43. (a) There is no effect. (b) Because actin is not controlled by cholesterol, the amount isolated should be the same in both experimental groups; a difference would suggest a problem in the RNA isolation. (c) The presence of cholesterol in the diet dramatically reduces the amount of HMG-CoA reductase protein. (d) A common means of regulating the amount of a protein present is to regulate transcription, which is clearly not the case here. (e) The translation of mRNA could be inhibited, and the protein could be rapidly degraded.

44. (a) The activity of the enzyme from the GD patient is the same as that from the control. This suggests that GD is not caused by a mutation that impairs enzyme activity. (b) There is much less GCase in the GD sample. Because the same number of cells were used in the two samples (confirmed by the identical levels of actin in both samples), we know that the variation in amount of GCase is not due to different quantities of cells. Two possible explanations are that transcription is impaired in the GD sample or that the enzyme in GD is more readily destroyed than the control. (c) Apparently, the enzyme is being synthesized but then degraded. (d) The defect in the enzyme from the GD patient seems to result in more rapid degradation of the enzyme by the proteasome. The fact that the amount of GCase present in the control increased upon proteasomal inhibition suggests that GCase may undergo rapid turnover under normal conditions.

Chapter 27

1. (d)
2. (a), (c), and (d)
3. (b)
4. (c)
5. (a)
6. Over the 40 years under consideration, our test subject will have consumed

$$40 \text{ years} \times 365 \text{ days year}^{-1} \times 8400 \text{ KJ (2000 kcal) day}^{-1}$$
$$= 1.2 \times 10^8 \text{ kJ } (2.9 \times 10^7 \text{ kcals}) \text{ in 40 years.}$$

Thus, over the 40-year span, our subject has ingested

$$1.2 \times 108 \, (2.9 \times 10^7 \text{ kcal})/38 \text{ kJ } (9 \text{ kcal g}^{-1})$$
$$= 3.2 \times 10^6 \text{ g} = 3200 \text{ kg of food,}$$

which is equivalent to more than 6 tons of food!
7. 2.55 pounds = 25 kg = 25,000 g = total weight gain

$$40 \text{ years} \times 365 \text{ days year}^{-1} = 14,600 \text{ days}$$

$$25,000 \text{ g}/14,600 \text{ days} = 1.7 \text{ g day}^{-1}$$

which is equivalent to an extra pat of butter per day. Her BMI is 26.5, and she would be considered overweight but not obese.
8. Adipose tissue is now known to be an active endocrine organ, secreting signal molecules called adipokines.
9. Caloric homeostasis is the condition in which the energy expenditure of an organism is equal to the energy intake.
10. Leptin and insulin.
11. CCK produces a feeling of satiety and stimulates the secretion of digestive enzymes by the pancreas and the secretion of bile salts by the gallbladder. GLP-1 also produces a feeling of satiety; in addition, it potentiates the glucose-induced secretion of insulin by the β cells of the pancreas.
12. Obviously, something is amiss. Although the answer is firmly established, the leptin-signaling pathway appears to be inhibited by suppressors of cytokine signaling, the regulatory proteins.
13. (a) 1, 2; (b) 6; (c) 3, 4; (d) 3, 4; (e) 3; (f) 6; (g) 5; (h) 5; (i) 5.
14. Phosphorylation of dietary glucose after it enters the liver; gluconeogenesis; glycogen breakdown.
15. Type 1 diabetes is due to autoimmune destruction of the insulin-producing cells of the pancreas. Type 1 diabetes is also called insulin-dependent diabetes because affected people require insulin to survive. Type 2 diabetes is characterized by insulin resistance. Insulin is produced, but the tissues that should respond to insulin, such as muscle, do not.
16. Leptin stimulates processes that are impaired in diabetes. For instance, leptin stimulates fatty acid oxidation, inhibits triacylglycerol synthesis, and increases the sensitivity of muscle and the liver to insulin.
17. (a) A watt is equal to 1 joule (J) per second (0.239 calorie per second). Hence, 70 W is equivalent to 0.07 kJ s^{-1} (0.017 kcal s^{-1}). (b) A watt is a current of 1 ampere (A) across a potential of 1 volt (V). For simplicity, let us assume that all the electron flow is from NADH to O_2 (a potential drop of 1.14 V). Hence, the current is 61.4 A, which corresponds to 3.86×10^{20} electrons per second (1 A = 1 coulomb s^{-1} = 6.28×10^{18} charge s^{-1}). (c) About 2.5 molecules of ATP are formed per molecule of NADH oxidized (two electrons). Hence, 1 molecule of ATP is formed per 0.8 electron transferred. A flow of 3.86×10^{20} electrons per second therefore leads to the generation of 4.83×10^{20} molecules of ATP per second, or 0.80 mmol s^{-1}. (d) The molecular weight of ATP is 507. The total body content of ATP of 50 g is equal to 0.099 mol. Hence, ATP turns over about once in 125 seconds when the body is at rest.
18. (a) The stoichiometry of the complete oxidation of glucose is

$$C_6H_{12}O_6 + 6\,O_2 \longrightarrow 6\,CO_2 + 6\,H_2O$$

and that of tripalmitoylglycerol is

$$C_{51}H_{98}O_2 + 72.5\,O_2 \longrightarrow 51\,CO_2 + 49\,H_2O$$

Hence, the RQ values are 1.0 and 0.703, respectively. (b) An RQ value reveals the relative use of carbohydrates and fats as fuels. The RQ of an amateur marathon runner typically decreases from 0.97 to 0.77 in the course of a race. The lowering of the RQ indicates the shift in fuel from carbohydrates to fat.
19. One gram of glucose (molecular weight 180.2) is equal to 5.55 mmol, and one gram of tripalmitoylglycerol (molecular weight 807.3) is equal to 1.24 mmol. The reaction stoichiometries (problem 18) indicate that 6 mol of H_2O is produced per mole of glucose oxidized, and 49 mol of H_2O is produced per mole of tripalmitoylglycerol oxidized. Hence, the H_2O yields per gram of fuel are 33.3 mmol (0.6 g) for glucose and 60.8 mmol (1.09 g) for tripalmitoylglycerol. Thus, complete oxidation of this fat gives 1.82 times as much water as does glucose. Another advantage of triacylglycerols is that they can be stored in essentially anhydrous form, whereas glucose is stored as glycogen, a highly hydrated polymer. A hump consisting mainly of glycogen would be an intolerable burden—far more than the straw that broke the camel's back.
20. The starved–fed cycle is the nightly hormonal cycle that humans experience during sleep and on eating. The cycle maintains adequate amounts of blood glucose. The starved part—sleep—is characterized by increased glucagon secretion and decreased insulin secretion. After a meal, glucagon concentration falls and insulin concentration rises.
21. Ethanol is oxidized to yield acetaldehyde by alcohol dehydrogenase, which is subsequently oxidized to acetate and acetaldehyde. Ethanol is also metabolized to acetaldehyde by the P450 enzymes, with the subsequent depletion of NADPH.
22. First, fatty liver develops owing, in part, to the increased amounts of NADH that inhibit fatty acid oxidation and stimulate fatty acid synthesis. Ethanol alters gene expression, stimulating fatty acid synthesis and inhibiting fatty acid oxidation. Second, alcoholic hepatitis begins, owing to oxidative damage and damage due to excess acetaldehyde that results in cell death. Finally, fibrous tissues form, creating scars that impair blood flow and biochemical function of the liver. Ammonia cannot be converted into urea, and its toxicity leads to coma and death.
23. A typical macadamia nut has a mass of about 2 g. Because it consists mainly of fats (~ 37 kJ g^{-1}, ~ 9 kcal g^{-1}), a nut has a value of about 75 kJ (18 kcal). The ingestion of 10 nuts results in an intake of about 753 kJ (180 kcal). As stated in the answer to problem 17, a power consumption of 1 W corresponds to 1 J s^{-1} (0.239 cal s^{-1}), and so 400-W running requires 0.4 kJ s^{-1} (0.0956 kcal s^{-1}). Hence, a person would have to run 1882 s, or about 31 minutes, to spend the calories provided by 10 nuts.
24. A high blood-glucose level triggers the secretion of insulin, which stimulates removal of glucose from the blood for the synthesis of glycogen and triacylglycerols. A high insulin level would impede the mobilization of fuel reserves during the marathon.
25. A lack of adipose tissue leads to an accumulation of fats in the muscle, with the generation of insulin resistance. The experiment shows that adipokines secreted by the adipose tissue (here, leptin), facilitate in some fashion the action of insulin in muscle.
26. Such a mutation would increase the phosphorylation of the insulin receptor and IRS in muscle and would improve insulin sensitivity. Indeed, PTP1B is an attractive therapeutic target for type 2 diabetes.
27. Lipid mobilization can be so rapid that it exceeds the ability of the liver to oxidize the lipids or convert them into ketone bodies. The excess is re-esterified and released into the blood as VLDLs.
28. A role of the liver is to provide glucose for other tissues. In the liver, glycolysis is used not for energy production but for biosynthetic purposes. Consequently, in the presence of glucagon, liver glycolysis stops so the glucose can be released into the blood.
29. The urea cycle and gluconeogenesis.

30. (a) Insulin inhibits lipid utilization. (b) Insulin stimulates protein synthesis, but there are no amino acids in the children's diet. Moreover, insulin inhibits protein breakdown. Consequently, muscle proteins cannot be broken down and used for the synthesis of essential proteins. (c) Because proteins cannot be synthesized, blood osmolarity is too low. Consequently, fluid leaves the blood. An especially important protein for maintaining blood osmolarity is albumin.

31. During strenuous exercise, muscle converts glucose into pyruvate through glycolysis. Some of the pyruvate is processed by cellular respiration. However, some of it is converted into lactate and released into the blood. The liver takes up the lactate and converts it into glucose through gluconeogenesis. Muscle may process the carbon skeletons of branched-chain amino acids aerobically. The nitrogens of these amino acids are transferred to pyruvate to form alanine, which is released into the blood and taken up by the liver. After the transamination of the amino group to α-ketoglutarate, the resulting pyruvate is converted into glucose. Finally, muscle glycogen may be mobilized, and the released glucose can be used by muscle.

32. This conversion allows muscle to function anaerobically. NAD^+ is regenerated when pyruvate is reduced to lactate, and so energy can continue to be extracted from glucose during strenuous exercise. The liver converts the lactate into glucose.

33. Fatty acids and glucose, respectively.

34. This practice is called carbo-loading. Depleting the glycogen stores will initially cause the muscles to synthesize a large amount of glycogen when dietary carbohydrates are provided, and will lead to the supercompensation of glycogen stores.

35. The oxygen consumption at the end of exercise is used to replenish ATP and creatine phosphate and to oxidize any lactate produced.

36. Oxygen is used in oxidative phosphorylation to resynthesize ATP and creatine phosphate. The liver converts lactate released by the muscle into glucose. Blood must be circulated to return the body temperature to normal, and so the heart cannot return to its resting rate immediately. Hemoglobin must be reoxygenated to replace the oxygen used in exercise. The muscles that power breathing must continue working at the same time as the exercised muscles are returning to resting states. In essence, all the biochemical systems activated in intense exercise need increased oxygen to return to the resting state.

37. Ethanol may replace water that is hydrogen bonded to proteins and membrane surfaces. This alteration of the hydration state of the protein would alter its conformation and hence function. Ethanol may also alter phospholipid packing in membranes. The two effects suggest that integral membrane proteins would be most sensitive to ethanol, as indeed seems to be the case.

38. Cells from the type I fiber would be rich in mitochondria, whereas those of the type II fiber would have few mitochondria.

39. (a) The ATP expended during this race amounts to about 8380 kg, or 18,400 pounds. (b) The cyclist would need about $1,260,000,000 to complete the race. The winner of the Tour de France receives about 600,000 US dollars, not nearly enough to pay for the ATP.

40. Exercise greatly enhances the ATP needs of muscle cells. To more efficiently meet these needs, more mitochondria are synthesized.

41. AMPK activity increases as ATP is used for muscle contraction. AMPK inactivates acetyl CoA carboxylase 1. Recall that malonyl CoA, the product of acetyl CoA carboxylase 1, inhibits transport of fatty acids into the mitochondria. The decrease in muscle malonyl CoA allows fatty acid oxidation in muscle.

42. After the first several days of starvation, most tissues are using fatty acids. Ketone bodies provide much of the brain's energy, decreasing the glucose requirement. The low insulin/high glucagon ration stimulates lipolysis and gluconeogenesis. However, gluconeogenesis will not occur because the increase in fatty acid

oxidation increases acetyl CoA and NADH. The high concentration of NADH inhibits the TCA cycle as well as gluconeogenesis. The acetyl CoA forms ketone bodies.

43. (a) The increase in all of the ratios is due to the NADH glut caused by the metabolism of alcohol. (b) The increased amounts of lactate and D-3-hydroxybutyrate are released into the blood, accounting for the acidosis. (c) Drinking on an empty stomach suggests that glycogen stores are low. Because of the NADH glut, gluconeogenesis cannot occur. Consequently, hypoglycemia results. A well-fed drinker will have glucose in the blood from the meal and thus will not experience hypoglycemia.

44. If glucose is always provided, even in small amounts, the brain will use glucose as a fuel rather than adapting to ketone body use. During the fast, muscle protein will be broken down to meet the brain's glucose needs. This protein degradation will lead to organ failure sooner than if the brain had adapted to ketone body utilization.

45. The inability of muscle mitochondria to process all of the fatty acids produced by overnutrition leads to excessive levels of diacylglycerol and ceramide in the muscle cytoplasm. These second-messenger molecules activate enzymes that impair insulin signaling.

46. Both are due to a lack of thiamine (vitamin B_1). Thiamine, which is sometimes called aneurin, is required most notably for the proper functioning of pyruvate dehydrogenase.

47. The little fellow is suffering from an absence of vitamin C; which is to say, he has scurvy. Fortunately, this is a condition that is easy to rectify.

48. (a) Red blood cells always produce lactate, and fast-twitch or type II muscle fibers (problem 38) also produce a large amount of lactate. (b) At that point, the athlete is beginning to move into anaerobic exercise, in which most energy is produced by anaerobic glycolysis. (c) The lactate threshold is essentially the point at which the athlete switches from aerobic exercise, which can be done for extended periods, to anaerobic exercise, essentially sprinting, which can be done for only short periods. The idea is to race at the extreme of his or her aerobic capacity until the finish line is in sight and then to switch to anaerobic. (d) Training increases the number of blood vessels and of muscle mitochondria. Together, they increase the ability to process glucose aerobically. Consequently, a greater effort can be expended before the switch to anaerobic energy production.

49. Consider the graph of lactate production as a function of effort shown in problem 48. For an athlete racing at her lactate threshold or just below it, the RQ value will be 1. With the finish line in sight, our runner ups her pace so that she is now also processing glucose to lactic acid in addition to processing glucose aerobically; that is, she begins running above her lactate threshold. The lactic acid released into the blood will ionize

$$CH_3COOH \rightleftharpoons CH_3COO^- + H^+$$
Lactic acid

The increase in H^+ will alter the blood buffer system, leading to the formation of carbonic acid

$$H^+ + HCO_3^- \rightleftharpoons H_2CO_3$$
Carbonic acid

The carbonic acid will dissociate into water and carbon dioxide

$$H_2CO_3 \rightleftharpoons CO_2 + H_2O$$

This carbon dioxide will be superimposed on the carbon dioxide generated by the combustion of glucose aerobically, leading to an RQ greater than 1.

50. The increase in ATP resulting from the processing of glucose in the β cells closes a potassium channel. The closing of the potassium channel alters the voltage across the cell membrane, which leads to an opening of a calcium channel. The influx of calcium causes the insulin-containing secretory granules to fuse with the cell membrane, resulting in the release of insulin.

Chapter 28

1. (a) Before; (b) after; (c) after; (d) after; (e) before; (f) after.

2. a) Yes; (b) yes; (c) no (MW > 600).

3. If computer programs could estimate $\log(P)$ values on the basis of chemical structure, then the required laboratory time for drug development could be shortened. The determination of the relative solubilities of pharmaceutical candidates by allowing each compound to equilibrate between water and an organic phase would no longer be necessary.

4. Perhaps N-acetylcysteine would conjugate to some of the N-acetyl-p-benzoquinone imine that is produced by the metabolism of acetaminophen, thereby preventing the depletion of the liver's supply of glutathione.

5. In Phase I clinical trials, approximately 10 to 100 usually healthy volunteers are typically enrolled in a study designed to assess safety. In contrast, a larger number of subjects are enrolled in a typical Phase II trial. Moreover, these persons may benefit from the drug administered. In a Phase II trial, efficacy, dosage, and safety can be assessed.

6. The binding of other drugs to albumin could cause extra coumadin to be released. (Albumin is a general carrier for hydrophobic molecules.)

7. A drug that inhibits a P450 enzyme may dramatically affect the disposition of another drug that is metabolized by that same enzyme. If this inhibited metabolism is not accounted for when dosing, the second drug may reach very high, and sometimes toxic, levels in the blood.

8. Unlike competitive inhibition, noncompetitive inhibition cannot be overcome with additional substrate. Hence, a drug that acts by a noncompetitive mechanism will be unaffected by changing levels of the physiological substrate.

9. An inhibitor of MDR could prevent the efflux of a chemotherapeutic drug from tumor cells. Hence, this type of an inhibitor could be useful in averting resistance to cancer chemotherapy.

10. Agents that inhibit one or more enzymes of the glycolytic pathway could act to deprive trypanosomes of energy and thus be useful for treating sleeping sickness. A difficulty is that glycolysis in the host cells also would be inhibited.

11. Imatinib is an inhibitor of the Bcr-Abl kinase, a mutant kinase present only in tumor cells that have undergone a translocation between chromosomes 9 and 22 (Figure 14.34). Before initiating treatment with imatinib, we could sequence the DNA of the tumor cells and determine (i) whether this translocation has taken place and (ii) whether the sequence of *BCR-ABL* carries any mutations that would render the kinase resistant to imatinib. If the translocation has not taken place or if the gene carries resistance mutations, then imatinib would likely not be an effective treatment for the patient carrying this particular tumor.

12. Sildenafil increases cGMP levels by inhibiting the phosphodiesterase-mediated breakdown of cGMP to GMP. Intracellular cGMP levels can also be increased by activating its synthesis. This activation can be achieved with the use of NO donors (such as sodium nitroprusside and nitroglycerin) or compounds that activate guanylate cyclase activity. Drugs that act by the latter mechanism are currently in clinical trials.

13. A reasonable mechanism would be an oxidative deamination following an overall mechanism similar to that in Figure 28.10, with release of ammonia.

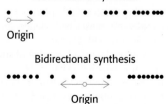

14. $K_I \approx 0.3$ nM. $IC_{50} \approx 2.0$ nM. Yes, compound A should be effective when taken orally because 400 nM is much greater than the estimated values of K_I and IC_{50}.

Chapter 29

1. DNA polymerase I uses deoxyribonucleoside triphosphates; pyrophosphate is the leaving group. DNA ligase uses DNA–adenylate (AMP joined to the $5'$-phosphoryl group) as a reaction partner; AMP is the leaving group. Topoisomerase I uses a DNA–tyrosyl intermediate ($5'$-phosphoryl group linked to the phenolic OH group); the tyrosine residue of the enzyme is the leaving group.

2. Positive supercoiling resists the unwinding of DNA. The melting temperature of DNA increases in proceeding from negatively supercoiled to relaxed to positively supercoiled DNA. Positive supercoiling is probably an adaptation to high temperature.

3. The nucleotides used for DNA synthesis have the triphosphate attached to the $5'$-hydroxyl group with free $3'$-hydroxyl groups. Such nucleotides can be utilized only for $5'$-to-$3'$ DNA synthesis.

4. DNA replication requires RNA primers. Without appropriate ribonucleotides, such primers cannot be synthesized.

5. This close contact prevents the incorporation of ribonucleotides rather than $2'$-deoxyribonucleotides.

6. (a) 96.2 revolutions per second (1000 nucleotides per second divided by 10.4 nucleotides per turn for B-DNA gives 96.2 rps). (b) 0.34 μm s^{-1} (1000 nucleotides per second corresponds to 3400 Å s^{-1} because the axial distance between nucleotides in B-DNA is 3.4 Å).

7. Eventually, the DNA would become so tightly wound that movement of the replication complex would be energetically impossible.

8. (a) Linking number $Lk = Tw + Wr = 48 + 3 = 51$.
(b) If $Tw = 50$, then $Wr = 1$.

9. A hallmark of most cancer cells is prolific cell division, which requires DNA replication. If the telomerase were not activated, the chromosomes would shorten until they became nonfunctional, leading to cell death.

10. No.

11. Treat the DNA briefly with endonuclease to occasionally nick each strand. Add the polymerase with the radioactive dNTPs. At the broken bond, or nick, the polymerase will degrade the existing strand with its $5' \to 3'$ exonuclease activity and replace it with a radioactive complementary copy by using its polymerase activity. This reaction scheme is referred to as "nick translation" because the nick is moved, or translated, along the DNA molecule without ever becoming sealed.

12. If replication were unidirectional, tracks with a low grain density at one end and a high grain density at the other end would be seen. On the other hand, if replication were bidirectional, the middle of a track would have a low density, as shown in the diagram below. For *E. coli*, the grain tracks are denser on both ends than in the middle, indicating that replication is bidirectional.

Unidirectional synthesis

• • •   • •   ••• • •••• •••
○——→
Origin

Bidirectional synthesis

•••••• •   • • •   •• ••••••
←——○——→
Origin

13. Pro (CCC), Ser (UCC), Leu (CUC), and Phe (UUC). Alternatively, the last base of each of these codons could be U.

14. Potentially deleterious side reactions are prevented. The enzyme itself might be damaged by light if it could be activated by light in the absence of bound DNA harboring a pyrimidine dimer.

15. The free DNA ends that appear in the absence of telomeres are repaired by DNA fusion.

16. The free energy of ATP hydrolysis under standard conditions is $-30.5\,\text{kJ mol}^{-1}$ ($-7.3\,\text{kcal mol}^{-1}$). In principle, it could be used to break three base pairs.

17. The oxidation of guanine could lead to DNA repair: DNA strand cleavage could allow looping out of the triplet repeat regions and triplet expansion.

18. The release of DNA topoisomerase II after the enzyme has acted on its DNA substrate requires ATP hydrolysis. Negative supercoiling requires only the binding of ATP, not its hydrolysis.

19. (a) Size; the top is relaxed and the bottom is supercoiled DNA. (b) Topoisomers. (c) The DNA is becoming progressively more unwound, or relaxed, and thus slower moving.

20. (a) It was used to determine the number of spontaneous revertants—that is, the background mutation rate. (b) To firmly establish that the system was working. A known mutagen's failure to produce revertants would indicate that something was wrong with the experimental system. (c) The chemical itself has little mutagenic ability but is apparently activated into a mutagen by the liver homogenate. (d) Cytochrome P450 system.

Chapter 30

1. The sequence of the coding (+ , sense) strand is

5′-ATGGGGAACAGCAAGAGTGGGGCCCTGTCCAAGGAG-3′

and the sequence of the template (− , antisense) strand is

3′-TACCCCTTGTCGTTCTCACCCCGGGACAGGTTCCTC-5′

2. An error will affect only one molecule of mRNA of many synthesized from a gene. In addition, the errors do not become a permanent part of the genomic information.

3. At any given instant, only a fraction of the genome (total DNA) is being transcribed. Consequently, speed is not essential.

4. The active sites are related by convergent evolution.

5. Heparin, a glycosaminoglycan, is highly anionic. Its negative charges, like the phosphodiester bridges of DNA templates, allow it to bind to lysine and arginine residues of RNA polymerase.

6. This mutant σ will competitively inhibit the binding of holoenzyme and prevent the specific initiation of RNA chains at promoter sites.

7. The core enzyme without σ binds more tightly to the DNA template than does the holoenzyme. The retention of σ after chain initiation would make the mutant RNA polymerase slower. The mutant enzyme would also be unlikely to bind alternative σ factors.

8. A 100-kDa protein contains about 910 residues, which are encoded by 2730 nucleotides. At a maximal transcription rate of 50 nucleotides per second, the mRNA would be synthesized in 55 s.

9. The RNA polymerase slides along the DNA rapidly rather than simply diffusing through three-dimensional space.

10. The start site is in red:

5′-GCCGTTGACACCGTTCGGCGATCGATCCGCTAT
AATGTGTGGATCCGCTT-3′

11. Initiation at strong promoters takes place every 2 s. In this interval, 100 nucleotides are transcribed. Hence, centers of transcription bubbles are 34 nm (340 Å) apart.

12. (a) The lowest band on the gel will be that of strand 3 alone (*i*). Band *ii* will be at the same position as band *i* because the RNA is not complementary to the nontemplate strand, whereas band *iii* will be higher because a complex is formed between RNA and the template strand. Band *iv* will be higher than the others because strand 1 is complexed to 2, and strand 2 is complexed to 3. Band *v* is the highest because core polymerase associates with the three strands. (b) None, because rifampicin acts before the formation of the open complex. (c) RNA polymerase is processive. When the template is bound, heparin cannot enter the DNA-binding site.

(d) When GTP is absent, synthesis stops when the first cytosine residue downstream of the bubble is encountered in the template strand. In contrast, with all four nucleoside triphosphates present, synthesis will continue to the end of the template.

13. Backtracking must occur prior to cleavage leading to dinucleotide products.

14. The base-pairing energy of the di- and trinucleotide DNA RNA hybrids formed at the very beginning of transcription is not sufficient to prevent strand separation and loss of product.

15. (a) Because cordycepin lacks a 3′-OH group, it cannot participate in 3′ → 5′ bond formation. (b) Because the poly(A) tail is a long stretch of adenosine nucleotides, the likelihood that a molecule of cordycepin would become incorporated is higher than with most RNA. (c) Yes, it must be converted into cordycepin 5′-triphosphate.

16. There are $2^8 = 256$ possible products.

17. The relationship between the −10 and −35 sequences could be affected by torsional strain. Since topoisomerase II introduces negative supercoils in DNA, this prevents this enzyme from overstimulating the expression of its own gene.

18. Ser-Ile-Phe-His-Pro-Stop

19. A mutation that disrupted the normal AAUAAA recognition sequence for the endonuclease could account for this finding. In fact, a change from U to C in this sequence caused this defect in a thalassemic patient. Cleavage occurred at the AAUAAA 900 nucleotides downstream from this mutant AACAAA site.

20. One possibility is that the 3′ end of the poly(U) donor strand cleaves the phosphodiester bond on the 5′ side of the insertion site. The newly formed 3′ terminus of the acceptor strand then cleaves the poly(U) strand on the 5′ side of the nucleotide that initiated the attack. In other words, a uridine residue could be added by two transesterification reactions. This postulated mechanism is similar to the one in RNA splicing.

21. Alternative splicing, RNA editing. Covalent modification of the proteins subsequent to synthesis.

22. Attach an oligo(dT) or oligo(U) sequence to an inert support to create an affinity column. When RNA is passed through the column, only poly(A)-containing RNA will be retained.

23. (a) Different amounts of RNA are present for the various genes. (b) Although all of the tissues have the same genes, the genes are expressed to different extents in different tissues. (c) These genes are called housekeeping genes—genes that most tissues express. They might include genes for glycolysis or citric acid cycle enzymes. (d) The point of the experiment is to determine which genes are initiated in vivo. The initiation inhibitor is added to prevent initiation at start sites that may have been activated during the isolation of the nuclei.

24. DNA is the single strand that forms the trunk of the tree. Strands of increasing length are RNA molecules; the beginning of transcription is where growing chains are the smallest; the end of transcription is where chain growth stops. Direction is left to right. Many enzymes are actively transcribing each gene.

Chapter 31

1. (c)
2. (d)
3. (c)
4. (a)
5. (b)

6. The Oxford English Dictionary defines translation as the action or process of turning from one language into another. Protein synthesis converts nucleic acid sequence information into amino acid sequence information.

7. An error frequency of 1 incorrect amino acid every 10^4 incorporations allows for the rapid and accurate synthesis of proteins as large as 1000 amino acids. Higher error rates would result in too

many defective proteins. Lower error rates would likely slow the rate of protein synthesis without a significant gain in accuracy.

8. (i) Each is a single chain. (ii) They contain unusual bases. (iii) Approximately half of the bases are base-paired to form double helices. (iv) The 5′ end is phosphorylated and is usually pG. (v) The amino acid is attached to the hydroxyl group of the A residue of the CCA sequence at the 3′ end of the tRNA. (vi) The anticodon is located in a loop near the center of the tRNA sequence. (vii) The molecules are L-shaped.

9. First is the formation of the aminoacyl adenylate, which then reacts with the tRNA to form the aminoacyl-tRNA. Both steps are catalyzed by aminoacyl-tRNA synthetase.

10. Unique features are required so that the aminoacyl-tRNA synthetases can distinguish among the tRNAs and attach the correct amino acid to the proper tRNA. Common features are required because all tRNAs must interact with the same protein-synthesizing machinery.

11. An activated amino acid is one linked to the appropriate tRNA.

12. (a) No; (b) no; (c) yes.

13. The ATP is cleaved to AMP and PP_i. Consequently, a second ATP is required to convert AMP into ADP, the substrate for oxidative phosphorylation.

14. Amino acids larger than the correct amino acid cannot fit into the active site of the tRNA. Smaller but incorrect amino acids that become attached to the tRNA fit into the editing site and are cleaved from the tRNA.

15. Recognition sites on both faces of the tRNAs may be required to uniquely identify the 20 different tRNAs.

16. The first two bases in a codon form Watson–Crick base pairs that are checked for fidelity by bases of the 16S rRNA. The third base is not inspected for accuracy, and so some variation is tolerated.

17. Four bands: light, heavy, a hybrid of light 30S and heavy 50S, and a hybrid of heavy 30S and light 50S.

18. Two hundred molecules of ATP are converted into 200 AMP + 400 P_i to activate the 200 amino acids, which is equivalent to 400 molecules of ATP. One molecule of GTP is required for initiation, and 398 molecules of GTP are needed to form 199 peptide bonds.

19. The reading frame is a set of contiguous, nonoverlapping three-nucleotide codons that begins with a start codon and ends with a stop codon.

20. A mutation caused by the insertion of an extra base can be suppressed by a tRNA that contains a fourth base in its anticodon. For example, UUUC rather than UUU is read as the codon for phenylalanine by a tRNA that contains 3′-AAAG-5′ as its anticodon.

21. One approach is to synthesize a tRNA that is acylated with a reactive amino acid analog. For example, bromoacetyl-phenylalanyl-tRNA is an affinity-labeling reagent for the P site of *E. coli* ribosomes.

22. The sequence GAGGU is complementary to a sequence of five bases at the 3′ end of 16S rRNA and is located several bases upstream of an AUG start codon. Hence, this region is a start signal for protein synthesis. The replacement of G by A would be expected to weaken the interaction of this mRNA with the 16S rRNA and thereby diminish its effectiveness as an initiation signal. In fact, this mutation results in a 10-fold decrease in the rate of synthesis of the protein specified by this mRNA.

23. The peptide would be Phe-Cys-His-Val-Ala-Ala. The codons UGC and UGU encode cysteine but, because the cysteine was modified to alanine, alanine is incorporated in place of cysteine.

24. Proteins are synthesized from the amino to the carboxyl end on ribosomes, whereas they are synthesized in the reverse direction in the solid-phase method. The activated intermediate in ribosomal synthesis is an aminoacyl-tRNA; in the solid-phase method, it is the adduct of the amino acid and dicyclohexylcarbodiimide.

25. GTP is not hydrolyzed until aminoacyl-tRNA is delivered to the A site of the ribosome. An earlier hydrolysis of GTP would be wasteful because EF-Tu–GDP has little affinity for aminoacyl-tRNA.

26. The translation of an mRNA molecule can be blocked by antisense RNA, an RNA molecule with the complementary sequence. The antisense–sense RNA duplex cannot serve as a template for translation; single-stranded mRNA is required. Furthermore, the antisense–sense duplex is degraded by nucleases. Antisense RNA added to the external medium is spontaneously taken up by many cells. A precise quantity can be delivered by microinjection. Alternatively, a plasmid encoding the antisense RNA can be introduced into target cells.

27. (a) A_5. (b) $A_5 > A_4 > A_3 > A_2$. (c) Synthesis is from the amino terminus to the carboxyl terminus.

28. These enzymes convert nucleic acid information into protein information by interpreting the tRNA and linking it to the proper amino acid.

29. The rate would fall because the elongation step requires that the GTP be hydrolyzed before any further elongation can take place.

30. Protein factors modulate the initiation of protein synthesis. The role of IF1 and IF3 is to prevent premature binding of the 30S and 50S ribosomal subunits, whereas IF2 delivers Met-tRNA$_f$ to the ribosome. Protein factors are also required for elongation (EF-G and EF-Tu), for termination (release factors, RFs), and for ribosome dissociation (ribosome release factors, RRFs).

31. The signal sequence, signal-recognition particle (SRP), the SRP receptor, and the translocon.

32. The formation of peptide bonds, which in turn are powered by the hydrolysis of the aminoacyl-tRNAs.

33. The Shine–Dalgarno sequence of the mRNA base-pairs with a part of the 16S rRNA of the 30S subunit, which positions the subunit so that the initiator AUG is recognized.

34.

	Bacteria	Eukaryote
Ribosome size	70S	80S
mRNA	Polycistronic	Not polycistronic
Initiation	Shine–Dalgarno is required	First AUG is used
Protein factors	Required	Many more required
Relation to transcription	Translation can start before transcription is completed	Transcription and translation are spatially separated
First amino acid	fMet	Met

35. The SRP binds to the signal sequence and inhibits further translation. The SRP ushers the inhibited ribosome to the ER, where it interacts with the SRP receptor (SR). The SRP–SR complex binds the translocon and simultaneously hydrolyzes GTP. On GTP hydrolysis, SRP and SR dissociate from each other and from the ribosome. Protein synthesis resumes and the nascent protein is channeled through the translocon.

36. The alternative would be to have a single ribosome translating a single mRNA molecule. The use of polysomes allows more protein synthesis per mRNA molecule in a given period of time and thus the production of more protein.

37. (a) 1, 2, 3, 5, 6, 10; (b) 1, 2, 7, 8; (c) 1, 4, 8, 9.

38. Transfer RNAs have roles in several recognition processes. A tRNA must be recognized by the appropriate aminoacyl-tRNA synthetase, and the tRNA must interact with the ribosome and, in particular, with the peptidyl transferase.

39. The ornithinyl-tRNA can be initially synthesized. However, the side-chain amino group attacks the ester linkage to form a six-membered amide, releasing the tRNA.

40. The S stands for a Svedberg unit, a measure of how fast a particle moves under centrifugal force (Section 3.1). The numbers are not arithmetic values.

41. (a, d, and e) Type 2; (b, c, and f) Type 1.

42. The error rates of DNA, RNA, and protein synthesis are of the order of 10^{-10}, 10^{-5}, and 10^{-4}, respectively, per nucleotide (or amino acid) incorporated. The fidelity of all three processes depends on the precision of base-pairing to the DNA or mRNA template. Few errors are corrected in RNA synthesis. In contrast, the fidelity of DNA synthesis is markedly increased by the $3' \rightarrow 5'$ proofreading nuclease activity and by postreplicative repair. In protein synthesis, the mischarging of some tRNAs is corrected by the hydrolytic action of aminoacyl-tRNA synthetase. Proofreading also takes place when aminoacyl-tRNA occupies the A site on the ribosome; the GTPase activity of EF-Tu sets the pace of this final stage of editing.

43. EF-Ts catalyzes the exchange of GTP for GDP bound to EF-Tu. In G-protein cascades, an activated 7TM receptor catalyzes GTP–GDP exchange in a G protein.

44. The α subunits of G proteins are inhibited by a similar mechanism in cholera and whooping cough (Section 14.5).

45. Glu-tRNA^{Gln} is formed by misacylation. The activated glutamate is subsequently amidated to form Gln-tRNA^{Gln}. Ways in which glutamine is formed from glutamate were discussed in Section 24.2. In regard to *H. pylori*, a specific enzyme, Glu-tRNA^{Gln} amidotransferase, catalyzes the following reaction:

$$\text{Gln} + \text{Glu-tRNA}^{Gln} + \text{ATP} \rightarrow \text{Gln-tRNA}^{Gln} + \text{Glu} + \text{ADP} + P_i$$

Glu-tRNA^{Glu} is not a substrate for the enzyme; so the transferase must also recognize aspects of the structure of tRNA^{Gln}.

46. The fact is that primary structure determines the three-dimensional structure of the protein. Thus, the final phase of information transfer from DNA to RNA to protein synthesis is the folding of the protein into its functional state, which is a function of the primary structure.

47. (a) eIF-4H has two effects: (i) the extent of unwinding is increased and (ii) the rate of unwinding is increased, as indicated by the increased rise in activity at early reaction times. (b) To firmly establish that the effect of eIF-H4 was not due to any inherent helicase activity. (c) Half-maximal activity was achieved at 0.11 μM of eIF-4H. Therefore, maximal stimulation would be achieved at a ratio of 1:1. (d) eIF-4H enhances the rate of unwinding of all helices, but the effect is greater as the helices increase in stability. (e) The results in graph C suggest that eIF-4H increases the processivity.

48. (a) The three peaks represent, from left to right, the 40S ribosomal subunit, the 60S ribosomal subunit, and the 80S ribosome. (b) Not only are ribosomal subunits and the 80S ribosome present, but polysomes of various lengths also are apparent. The individual peaks in the polysome region represent polysomes of discrete length. (c) The treatment significantly inhibited the number of polysomes while increasing the number of free ribosomal subunits. This outcome could be due to inhibited protein-synthesis initiation or inhibited transcription.

Chapter 32

1. (a) and (b) Cells will express β-galactosidase, *lac* permease, and thiogalactoside transacetylase even in the absence of lactose. (c) The levels of catabolic enzymes such as β-galactosidase and arabinose isomerase will remain low even at low levels of glucose.

2. The concentration is $1/(6 \times 10^{23})$ moles per 10^{-15} liter $= 1.7 \times 10^{-9}$ M. Because $K_d = 10^{-13}$ M, the single molecule should be bound to its specific binding site.

3. The number of possible 8-bp sites is $4^8 = 65,536$. In a genome of 4.6×10^6 base pairs, the average site should appear $4.6 \times 10^6/65,536 = 70$ times. Each 10-bp site should appear 4 times. Each 12-bp site should appear 0.27 times (many 12-bp sites will not appear at all).

4. The *lac* repressor does not bind DNA when the repressor is bound to a small molecule (the inducer), whereas the *pur* repressor binds DNA only when the repressor is bound to a small molecule (the corepressor). The *E. coli* genome contains only a single *lac* repressor-binding region, whereas it has many sites for the *pur* repressor.

5. Anti-inducers bind to the conformation of repressors, such as the *lac* repressor, that are capable of binding DNA. They occupy a site that overlaps that for the inducer and, therefore, compete for binding to the repressor.

6. The inverted repeat may be a binding site for a dimeric DNA-binding protein or it may correspond to a stem-loop structure in the encoded RNA.

7. Bacteriophage λ would be more likely to enter the lytic phase because the cooperative binding of the λ repressor to O_R^2 and O_R^1, which supports the lysogenic pathway, would be disrupted.

8.

λ repressor gene	-10 region GATTTA	-35 region TAGATA
Cro gene	-10 region TAATGG	-35 region TTGACT

There are four differences in the -10 region and three differences in the -35 region.

9. Increased Cro concentration reduces the expression of the λ repressor gene. Increased λ repressor concentration reduces the expression of the Cro gene. At low λ repressor concentrations, increased λ repressor concentration increases the expression of the λ repressor gene. At higher λ repressor concentrations, increased λ repressor concentration decreases the expression of the λ repressor gene.

10. Normally, bacterial mRNAs have a leader sequence in which a Shine–Dalgarno sequence precedes the AUG start codon. The absence of a leader would be expected to lead to inefficient translation.

11. Add each compound to a culture of *V. fischeri* at low density and look for the development of luminescence.

12. ACC, 7; ACA, 1; ACU, 0; ACG, 0.

13. Retention. The reaction likely proceeds in two steps: Attack of Glu 537 on the carbon of the galactose moiety (with inversion) followed by attack of water on this carbon (with a second inversion) to release the galactose from the enzyme.

14. The footprint appears to have approximately 50% of its intensity near 3.7 nM so that the dissociation constant is approximately 3.7 nM, corresponding to a standard free energy of binding of -48 kJ/mol (-11 kcal/mol) at T = 298K.

Chapter 33

1. The distribution of charged amino acids is H2A (13 K, 13 R, 2 D, 7 E, charge $= +17$), H2B (20 K, 8 R, 3 D, 7 E, charge $= +18$), H3 (13 K, 18 R, 4 D, 7 E, charge $= +20$), H4 (11 K, 14 R, 3 D, 4 E, charge $= +18$). The total charge of the histone octamer is estimated to be $2 \times (17 + 18 + 20 + 18) = +146$. The total charge on 150 base pairs of DNA is -300. Thus, the histone octamer neutralizes approximately one-half of the charge.

2. The presence of a particular DNA fragment could be detected by hybridization, by PCR, or by direct sequencing.

3. The total length of the DNA is estimated to be 145 bp × 3.4 Å/bp = 493 Å, which represents 1.75 turns or $1.75 \times 2\pi r = 11.0r$. Thus, the radius is estimated to be $r = 493$ Å/11.0 = 44.8 Å.

4. 5-Azacytidine cannot be methylated. Some genes, normally repressed by methylation, will be active. 5-Azacytidine is also an inhibitor of DNA methyltransferase, resulting in lower levels of methylation and, hence, lower levels of gene repression.

5. Proteins containing these domains will be targeted to methylated DNA in repressed promoter regions. They would likely bind in the major groove because that is where the methyl group is located.

6. Gene expression is not expected to respond to the presence of estrogen. However, genes for which expression normally responds to estrogen will respond to the presence of progesterone.

7. The acetylation of lysine will reduce the charge from $+1$ to 0. The methylation of lysine will not reduce the charge.

8. On the basis of the pattern of cysteine and histidine residues, this region appears to contain three zinc-finger domains.

9. $10/4000 = 0.25\%$. 0.25% of 12 Mb = 30 kilobase pairs.

10. Trimethylated-repressed; Acetylated-activated.

11. The addition of an IRE to the 5′ end of the mRNA is expected to block translation in the absence of iron. The addition of an IRE to the 3′ end of the mRNA is not expected to block translation, but it might affect mRNA stability.

12. The sequences of all of the mRNAs would be searched for sequences that are fully or nearly complementary to the sequence of the miRNA. These sequences would be candidates for regulation by this mRNA.

13. The amino group of the lysine residue, formed from the protonated form by a base, attacks the carbonyl group of acetyl CoA to generate a tetrahedral intermediate. This intermediate collapses to form the amide bond and release CoA.

14. In mouse DNA, most of the HpaII sites are methylated and therefore not cut by the enzyme, resulting in large fragments. Some small fragments are produced from CpG islands that are unmethylated. For *Drosophila* and *E. coli* DNA, there is no methylation and all sites are cut.

For Chapters 34-36, please refer to the online Answers to Problems section.

CHAPTER 2

Where to Start

Service, R. F. 2008. Problem solved* (*sort of) (a brief review of protein folding). *Science* 321:784–786.

Doolittle, R. F. 1985. Proteins. *Sci. Am.* 253(4):88–99.

Richards, F. M. 1991. The protein folding problem. *Sci. Am.* 264(1):54–57.

Weber, A. L., and Miller, S. L. 1981. Reasons for the occurrence of the twenty coded protein amino acids. *J. Mol. Evol.* 17:273–284.

Books

Petsko, G. A., and Ringe, D. 2004. *Protein Structure and Function*. New Science Press.

Tanford, C., and Reynolds, J. 2004. *Nature's Robots: A History of Proteins*. Oxford.

Branden, C., and Tooze, J. 1999. *Introduction to Protein Structure* (2nd ed.). Garland.

Creighton, T. E. 1992. *Proteins: Structures and Molecular Principles* (2nd ed.). W. H. Freeman and Company.

Conformation of Proteins

Smock, R. G., and Gierasch, L. M. 2009. Sending signals dynamically. *Science* 324:198–203.

Tokuriki, N., and Tawfik, D. S. 2009. Protein dynamism and evolvability. *Science* 324:203–207.

Pace, C. N., Grimsley, G. R., and Scholtz, J. M. 2009. Protein ionizable groups: pK values and their contribution to protein stability and solubility. *J. Biol. Chem.* 284:13285–13289.

Breslow, R., and Cheng, Z.-L. 2009. On the origin of terrestrial homochirality for nucleosides and amino acids. *Proc. Natl. Acad. Sci. U.S.A.* 106:9144–9146.

Secondary Structure

Shoulders, M. D., and Raines, R. T. 2009. Collagen structure and stability. *Annu. Rev. Biochem.* 78:929–958.

O'Neil, K. T., and DeGrado, W. F. 1990. A thermodynamic scale for the helix-forming tendencies of the commonly occurring amino acids. *Science* 250:646–651.

Zhang, C., and Kim, S. H. 2000. The anatomy of protein beta-sheet topology. *J. Mol. Biol.* 299:1075–1089.

Regan, L. 1994. Protein structure: Born to be beta. *Curr. Biol.* 4:656–658.

Srinivasan, R., and Rose, G. D. 1999. A physical basis for protein secondary structure. *Proc. Natl. Acad. Sci. U.S.A.* 96:14258–14263.

Intrinsically Unstructured Proteins

Galea, C. A., Wang, Y., Sivakolundu, S. G., and Kriwacki, R. W. 2008. Regulation of cell division by intrinsically unstructured proteins: Intrinsic flexibility, modularity, and signaling conduits. *Biochemistry* 47:7598–7609.

Raychaudhuri, S., Dey, S., Bhattacharyya, N. P., and Mukhopadhyay, D. 2009. The role of intrinsically unstructured proteins in neurodegenerative diseases. *PLoS One* 4:e5566.

Tompa, P., and Fuxreiter, M. 2008. Fuzzy complexes: Polymorphism and structural disorder in protein–protein interactions. *Trends Biochem. Sci.* 33:2–8.

Tuinstra, R. L., Peterson, F. C., Kutlesa, E. S., Elgin, S., Kron, M. A., and Volkman, B. F. 2008. Interconversion between two unrelated protein folds in the lymphotactin native state. *Proc. Natl. Acad. Sci. U.S.A.* 105:5057–5062.

Domains

Jin, J., Xie, X., Chen, C., Park, J. G., Stark, C., James, D. A., Olhovsky, M., Lindinger, R., Mao, Y., and Pawson, T. 2009. Eukaryotic protein domains as functional units of cellular evolution. *Sci. Signal.* 2:ra76.

Bennett, M. J., Choe, S., and Eisenberg, D. 1994. Domain swapping: Entangling alliances between proteins. *Proc. Natl. Acad. Sci. U.S.A.* 91:3127–3131.

Bergdoll, M., Eltis, L. D., Cameron, A. D., Dumas, P., and Bolin, J. T. 1998. All in the family: Structural and evolutionary relationships among three modular proteins with diverse functions and variable assembly. *Protein Sci.* 7:1661–1670.

Hopfner, K. P., Kopetzki, E., Kresse, G. B., Bode, W., Huber, R., and Engh, R. A. 1998. New enzyme lineages by subdomain shuffling. *Proc. Natl. Acad. Sci. U.S.A.* 95:9813–9818.

Ponting, C. P., Schultz, J., Copley, R. R., Andrade, M. A., and Bork, P. 2000. Evolution of domain families. *Adv. Protein Chem.* 54:185–244.

Protein Folding

Caughey, B., Baron, G. S., Chesebro, B., and Jeffrey, M. 2009. Getting a grip on prions: Oligomers, amyloids, and pathological membrane interactions. *Annu. Rev. Biochem.* 78:177–204.

Cobb, N. J., and Surewicz, W. K. 2009. Prion diseases and their biochemical mechanisms. *Biochemistry* 48:2574–2585.

Soto, C. 2011. Prion diseases: The end of the controversy? *Trends Biochem. Sci.* 36:151–158.

Daggett, V., and Fersht, A. R. 2003. Is there a unifying mechanism for protein folding? *Trends Biochem. Sci.* 28:18–25.

Selkoe, D. J. 2003. Folding proteins in fatal ways. *Nature* 426:900–904.

Anfinsen, C. B. 1973. Principles that govern the folding of protein chains. *Science* 181:223–230.

Baldwin, R. L., and Rose, G. D. 1999. Is protein folding hierarchic? I. Local structure and peptide folding. *Trends Biochem. Sci.* 24:26–33.

Baldwin, R. L., and Rose, G. D. 1999. Is protein folding hierarchic? II. Folding intermediates and transition states. *Trends Biochem. Sci.* 24:77–83.

Kuhlman, B., Dantas, G., Ireton, G. C., Varani, G., Stoddard, B. L., and Baker, D. 2003. Design of a novel globular protein with atomic-level accuracy. *Science* 302:1364–1368.

Staley, J. P., and Kim, P. S. 1990. Role of a subdomain in the folding of bovine pancreatic trypsin inhibitor. *Nature* 344:685–688.

Covalent Modification of Proteins

Tarrant, M. K., and Cole, P. A. 2009. The chemical biology of protein phosphorylation. *Annu. Rev. Biochem.* 78:797–825.

Krishna, R. G., and Wold, F. 1993. Post-translational modification of proteins. *Adv. Enzymol. Relat. Areas. Mol. Biol.* 67:265–298.

Aletta, J. M., Cimato, T. R., and Ettinger, M. J. 1998. Protein methylation: A signal event in post-translational modification. *Trends Biochem. Sci.* 23:89–91.

Tsien, R. Y. 1998. The green fluorescent protein. *Annu. Rev. Biochem.* 67:509–544.

CHAPTER 3

Where to Start

Sanger, F. 1988. Sequences, sequences, sequences. *Annu. Rev. Biochem.* 57:1–28.

Merrifield, B. 1986. Solid phase synthesis. *Science* 232:341–347.

Hunkapiller, M. W., and Hood, L. E. 1983. Protein sequence analysis: Automated microsequencing. *Science* 219:650–659.

Milstein, C. 1980. Monoclonal antibodies. *Sci. Am.* 243(4):66–74.

Moore, S., and Stein, W. H. 1973. Chemical structures of pancreatic ribonuclease and deoxyribonuclease. *Science* 180:458–464.

Books

Methods in Enzymology. Academic Press.

Wilson, K., and Walker, J. (Eds.). 2010. *Principles and Techniques of Practical Biochemistry* (7th ed.). Cambridge University Press.

Van Holde, K. E., Johnson, W. C., and Ho, P.-S. 1998. *Principles of Physical Biochemistry.* Prentice Hall.

Wilkins, M. R., Williams, K. L., Appel, R. D., and Hochstrasser, D. F. 1997. *Proteome Research: New Frontiers in Functional Genomics (Principles and Practice).* Springer Verlag.

Johnstone, R. A. W. 1996. *Mass Spectroscopy for Chemists and Biochemists* (2nd ed.). Cambridge University Press.

Kyte, J. 1994. *Structure in Protein Chemistry.* Garland.

Creighton, T. E. 1993. *Proteins: Structure and Molecular Properties* (2nd ed.). W. H. Freeman and Company.

Cantor, C. R., and Schimmel, P. R. 1980. *Biophysical Chemistry.* W. H. Freeman and Company.

Protein Purification and Analysis

Blackstock, W. P., and Weir, M. P. 1999. Proteomics: Quantitative and physical mapping of cellular proteins. *Trends Biotechnol.* 17:121–127.

Deutscher, M. (Ed.). 1997. *Guide to Protein Purification.* Academic Press.

Dunn, M. J. 1997. Quantitative two-dimensional gel electrophoresis: From proteins to proteomes. *Biochem. Soc. Trans.* 25:248–254.

Scopes, R. K., and Cantor, C. 1994. *Protein Purification: Principles and Practice* (3rd ed.). Springer Verlag.

Aebersold, R., Pipes, G. D., Wettenhall, R. E., Nika, H., and Hood, L. E. 1990. Covalent attachment of peptides for high sensitivity solid-phase sequence analysis. *Anal. Biochem.* 187:56–65.

Ultracentrifugation and Mass Spectrometry

Steen, H., and Mann, M. 2004. The ABC's (and XYZ's) of peptide sequencing. *Nat. Rev. Mol. Cell Biol.* 5:699–711.

Glish, G. L., and Vachet, R. W. 2003. The basics of mass spectrometry in the twenty-first century. *Nat. Rev. Drug Discov.* 2:140–150.

Li, L., Garden, R. W., and Sweedler, J. V. 2000. Single-cell MALDI: A new tool for direct peptide profiling. *Trends Biotechnol.* 18:151–160.

Yates, J. R., 3d. 1998. Mass spectrometry and the age of the proteome. *J. Mass Spectrom.* 33:1–19.

Pappin, D. J. 1997. Peptide mass fingerprinting using MALDI-TOF mass spectrometry. *Methods Mol. Biol.* 64:165–173.

Schuster, T. M., and Laue, T. M. 1994. *Modern Analytical Ultracentrifugation.* Springer Verlag.

Arnott, D., Shabanowitz, J., and Hunt, D. F. 1993. Mass spectrometry of proteins and peptides: Sensitive and accurate mass measurement and sequence analysis. *Clin. Chem.* 39:2005–2010.

Chait, B. T., and Kent, S. B. H. 1992. Weighing naked proteins: Practical, high-accuracy mass measurement of peptides and proteins. *Science* 257:1885–1894.

Edmonds, C. G., Loo, J. A., Loo, R. R., Udseth, H. R., Barinaga, C. J., and Smith, R. D. 1991. Application of electrospray ionization mass spectrometry and tandem mass spectrometry in combination with capillary electrophoresis for biochemical investigations. *Biochem. Soc. Trans.* 19:943–947.

Jardine, I. 1990. Molecular weight analysis of proteins. *Methods Enzymol.* 193:441–455.

Proteomics

Yates, J. R., 3rd. 2004. Mass spectral analysis in proteomics. *Annu. Rev. Biophys. Biomol. Struct.* 33:297–316.

Weston, A. D., and Hood, L. 2004. Systems biology, proteomics, and the future of health care: Toward predictive, preventative, and personalized medicine. *J. Proteome Res.* 3:179–196.

Pandey, A., and Mann, M. 2000. Proteomics to study genes and genomes. *Nature* 405:837–846.

Dutt, M. J., and Lee, K. H. 2000. Proteomic analysis. *Curr. Opin. Biotechnol.* 11:176–179.

Rout, M. P., Aitchison, J. D., Suprapto, A., Hjertaas, K., Zhao, Y., and Chait, B. T. 2000. The yeast nuclear pore complex: Composition, architecture, and transport mechanism. *J. Cell Biol.* 148:635–651.

X-ray Crystallography, NMR Spectroscopy, and Cryo-electron Microscopy

Rhodes, G. 2006. *Crystallography Made Crystal Clear* (3rd ed.). Elsevier/Academic Press.

Moffat, K. 2003. The frontiers of time-resolved macromolecular crystallography: Movies and chirped X-ray pulses. *Faraday Discuss.* 122:65–88.

Bax, A. 2003. Weak alignment offers new NMR opportunities to study protein structure and dynamics. *Protein Sci.* 12:1–16.

Wery, J. P., and Schevitz, R. W. 1997. New trends in macromolecular x-ray crystallography. *Curr. Opin. Chem. Biol.* 1:365–369.

Glusker, J. P. 1994. X-ray crystallography of proteins. *Methods Biochem. Anal.* 37:1–72.

Clore, G. M., and Gronenborn, A. M. 1991. Structures of larger proteins in solution: Three- and four-dimensional heteronuclear NMR spectroscopy. *Science* 252:1390–1399.

Wüthrich, K. 1989. Protein structure determination in solution by nuclear magnetic resonance spectroscopy. *Science* 243:45–50.

Wüthrich, K. 1986. *NMR of Proteins and Nucleic Acids.* Wiley-Interscience.

Fernandez-Leiro, R., and Scheres, S. H. W. 2016. Unravelling the structures of biological macromolecules by cryo-EM. *Nature* 537:339–346.

Milne, J. L. S., Borgnia, M. J., Bartesaghi, A., Tran, E. E. H., Earl, L. A., Schauder, D. M., Lengyel, J., Pierson, J., Patwardhan, A., and Subramaniam, S. 2013. Cryo-electron microscopy—a primer for the non-microscopist. *FEBS J.* 280:28–45.

Liao, M., Cao, E., Julius, D., and Cheng, Y. 2013. Structure of the TRPV1 ion channel determined by electron cryo-microscopy. *Nature* 504:107–112.

Monoclonal Antibodies and Fluorescent Molecules

Immunology Today. 2000. Volume 21, issue 8.

Tsien, R. Y. 1998. The green fluorescent protein. *Annu. Rev. Biochem.* 67:509–544.

Kendall, J. M., and Badminton, M. N. 1998. *Aequorea victoria* bioluminescence moves into an exciting era. *Trends Biotechnol.* 16:216–234.

Goding, J. W. 1996. *Monoclonal Antibodies: Principles and Practice.* Academic Press.

Köhler, G., and Milstein, C. 1975. Continuous cultures of fused cells secreting antibody of predefined specificity. *Nature* 256:495–497.

Chemical Synthesis of Proteins

Bang, D., Chopra, N., and Kent, S. B. 2004. Total chemical synthesis of crambin. *J. Am. Chem. Soc.* 126:1377–1383.

Dawson, P. E., and Kent, S. B. 2000. Synthesis of native proteins by chemical ligation. *Annu. Rev. Biochem.* 69:923–960.

Mayo, K. H. 2000. Recent advances in the design and construction of synthetic peptides: For the love of basics or just for the technology of it. *Trends Biotechnol.* 18:212–217.

CHAPTER 4

Where to Start

Crick, F. H. C. 1954. The structure of the hereditary material. *Sci. Am.* 191(4):54–61.

Chambon, P. 1981. Split genes. *Sci. Am.* 244(5):60–71.

Watson, J. D., and Crick, F. H. C. 1953. Molecular structure of nucleic acids: A structure for deoxyribose nucleic acid. *Nature* 171:737–738.

Watson, J. D., and Crick, F. H. C. 1953. Genetic implications of the structure of deoxyribonucleic acid. *Nature* 171:964–967.

Meselson, M., and Stahl, F. W. 1958. The replication of DNA in *Escherichia coli. Proc. Natl. Acad. Sci. U.S.A.* 44:671–682.

Books

Bloomfield, V. A., Crothers, D. M., Tinoco, I., and Hearst, J. 2000. *Nucleic Acids: Structures, Properties, and Functions.* University Science Books.

Singer, M., and Berg, P. 1991. *Genes and Genomes: A Changing Perspective.* University Science Books.

Lodish, H., Berk, A., Kaiser, C. A., Krieger, M. Bretscher, A., Ploegh, H., Amon, A., and Scott, M. P. 2016. *Molecular Cell Biology* (8th ed.). W. H. Freeman and Company.

Krebs, J. E, Goldstein, E. S., and Kilpatrick, S. T. 2017. *Lewin's Genes XII* (12th ed.). Jones and Bartlett.

Watson, J. D., Baker, T. A., Bell, S. P., Gann, A., Levine, M., and Losick, R. 2013. *Molecular Biology of the Gene* (7th ed.). Benjamin Cummings.

DNA Structure

Neidle, S. 2007. *Principles of Nucleic Acid Structure.* Academic Press.

Dickerson, R. E., Drew, H. R., Conner, B. N., Wing, R. M., Fratini, A. V., and Kopka, M. L. 1982. The anatomy of A-, B-, and Z-DNA. *Science* 216:475–485.

Sinden, R. R. 1994. *DNA Structure and Function.* Academic Press.

DNA Replication

Lehman, I. R. 2003. Discovery of DNA polymerase. *J. Biol. Chem.* 278:34733–34738.

Hübscher, U., Maga, G., and Spardari, S. 2002. Eukaryotic DNA polymerases. *Annu. Rev. Biochem.* 71:133–163.

Hübscher, U., Nasheuer, H.-P., and Syväoja, J. E. 2000. Eukaryotic DNA polymerases: A growing family. *Trends Biochem. Sci.* 25:143–147.

Brautigam, C. A., and Steitz, T. A. 1998. Structural and functional insights provided by crystal structures of DNA polymerases and their substrate complexes. *Curr. Opin. Struct. Biol.* 8:54–63.

Kornberg, A. 2005. *DNA Replication* (2nd ed.). University Science Books.

Discovery of Messenger RNA

Jacob, F., and Monod, J. 1961. Genetic regulatory mechanisms in the synthesis of proteins. *J. Mol. Biol.* 3:318–356.

Brenner, S., Jacob, F., and Meselson, M. 1961. An unstable intermediate carrying information from genes to ribosomes for protein synthesis. *Nature* 190:576–581.

Hall, B. D., and Spiegelman, S. 1961. Sequence complementarity of T2-DNA and T2-specific RNA. *Proc. Natl. Acad. Sci. U.S.A.* 47:137–146.

Genetic Code

Quax, T. E. F., Claassens, N. J., Söll, D., and van der Oost, J. 2015. Codon bias as a means to fine-tune gene expression. *Mol. Cell* 59:149–161.

Novoa, E. M., and Ribas de Pouplana, L. 2012. Speeding with control: Codon usage, tRNAs, and ribosomes. *Trends Genet.* 28:574–581.

Koonin, E. V., and Novozhilov, A. S. 2009. Origin and evolution of the genetic code: The universal enigma. *IUBMB Life* 61:99–111.

Yarus, M., Caporaso, J. G., and Knight, R. 2005. Origins of the genetic code: The escaped triplet theory. *Annu. Rev. Biochem.* 74:179–198.

Freeland, S. J., and Hurst, L. D. 2004. Evolution encoded. *Sci. Am.* 290(4):84–91.

Crick, F. H. C., Barnett, L., Brenner, S., and Watts-Tobin, R. J. 1961. General nature of the genetic code for proteins. *Nature* 192:1227–1232.

Knight, R. D., Freeland, S. J., and Landweber L. F. 1999. Selection, history and chemistry: The three faces of the genetic code. *Trends Biochem. Sci.* 24(6):241–247.

Introns, Exons, and Split Genes

Liu, M., and Grigoriev, A. 2004. Protein domains correlate strongly with exons in multiple eukaryotic genomes—evidence of exon shuffling? *Trends Genet.* 20:399–403.

Dorit, R. L., Schoenbach, L., and Gilbert, W. 1990. How big is the universe of exons? *Science* 250:1377–1382.

Cochet, M., Gannon, F., Hen, R., Maroteaux, L., Perrin, F., and Chambon, P. 1979. Organization and sequence studies of the 17-piece chicken conalbumin gene. *Nature* 282:567–574.

Tilghman, S. M., Tiemeier, D. C., Seidman, J. G., Peterlin, B. M., Sullivan, M., Maizel, J. V., and Leder, P. 1978. Intervening sequence of DNA identified in the structural portion of a mouse β-globin gene. *Proc. Natl. Acad. Sci. U.S.A.* 75:725–729.

Reminiscences and Historical Accounts

Watson, J. D., Gann, A., and Witkowski, J. (Eds.). 2012. *The Annotated and Illustrated Double Helix*. Simon and Shuster.

Nirenberg, M. 2004. Deciphering the genetic code—A personal account. *Trends Biochem. Sci.* 29:46–54.

Clayton, J., and Dennis, C. (Eds.). 2003. *50 Years of DNA*. Palgrave Macmillan.

Watson, J. D. 1968. *The Double Helix*. Atheneum.

McCarty, M. 1985. *The Transforming Principle: Discovering That Genes Are Made of DNA*. Norton.

Cairns, J., Stent, G. S., and Watson, J. D. 2000. *Phage and the Origins of Molecular Biology*. Cold Spring Harbor Laboratory.

Olby, R. 1974. *The Path to the Double Helix*. University of Washington Press.

Judson, H. F. 1996. *The Eighth Day of Creation*. Cold Spring Harbor Laboratory.

Sayre, A. 2000. *Rosalind Franklin and DNA*. Norton.

CHAPTER 5

Where to Start

Berg, P. 1981. Dissections and reconstructions of genes and chromosomes. *Science* 213:296–303.

Gilbert, W. 1981. DNA sequencing and gene structure. *Science* 214:1305–1312.

Sanger, F. 1981. Determination of nucleotide sequences in DNA. *Science* 214:1205–1210.

Mullis, K. B. 1990. The unusual origin of the polymerase chain reaction. *Sci. Am.* 262(4):56–65.

Books on Recombinant DNA Technology

Watson, J. D., Myers, R. M., Caudy, A. A., and Witkowski, J. 2007. *Recombinant DNA: Genes and Genomes* (3rd ed.). W. H. Freeman and Company.

Grierson, D. (Ed.). 1991. *Plant Genetic Engineering*. Chapman and Hall.

Mullis, K. B., Ferré, F., and Gibbs, R. A. (Eds.). 1994. *The Polymerase Chain Reaction*. Birkhaüser.

Green, M. R., and Sambrook, S. 2014. *Molecular Cloning: A Laboratory Manual* (4th ed.). Cold Spring Harbor Laboratory Press.

Ausubel, F. M., Brent, R., Kingston, R. E., and Moore, D. D. (Eds.). 2002. *Short Protocols in Molecular Biology: A Compendium of Methods from Current Protocols in Molecular Biology*. Wiley.

Birren, B., Green, E. D., Klapholz, S., Myers, R. M., Roskams, J., Riethamn, H., and Hieter, P. (Eds.). 1999. *Genome Analysis* (vols. 1–4). Cold Spring Harbor Laboratory Press.

Methods in Enzymology. Academic Press. [Many volumes in this series deal with recombinant DNA technology.]

DNA Sequencing and Synthesis

Hunkapiller, T., Kaiser, R. J., Koop, B. F., and Hood, L. 1991. Large-scale and automated DNA sequence determination. *Science* 254:59–67.

Sanger, F., Nicklen, S., and Coulson, A. R. 1977. DNA sequencing with chain-terminating inhibitors. *Proc. Natl. Acad. Sci. U.S.A.* 74:5463–5467.

Maxam, A. M., and Gilbert, W. 1977. A new method for sequencing DNA. *Proc. Natl. Acad. Sci. U.S.A.* 74:560–564.

Smith, L. M., Sanders, J. Z., Kaiser, R. J., Hughes, P., Dodd, C., Connell, C. R., Heiner, C., Kent, S. B. H., and Hood, L. E. 1986. Fluorescence detection in automated DNA sequence analysis. *Nature* 321:674–679.

Pease, A. C., Solas, D., Sullivan, E. J., Cronin, M. T., Holmes, C. P., and Fodor, S. P. A. 1994. Light-generated oligonucleotide arrays for rapid DNA sequence analysis. *Proc. Natl. Acad. Sci. U.S.A.* 91:5022–5026.

Venter, J. C., Adams, M. D., Sutton, G. G., Kerlavage, A. R., Smith, H. O., and Hunkapiller, M. 1998. Shotgun sequencing of the human genome. *Science* 280:1540–1542.

Mardis, E. R. 2008. Next-generation DNA sequencing methods. *Annu. Rev. Genomics Hum. Genet.* 9:387–402.

Metzker, M. L. 2010. Sequencing technologies—the next generation. *Nature Rev. Genet.* 11:31–46.

Rothberg, J. M., Hinz, W., Rearick, T. M., Schultz, J., Mileski, W., Davey, M., Leamon, J. H., Johnson, K., Milgrew, M. J., Edwards, M., et al. 2011. An integrated semiconductor device enabling non-optical genome sequencing. *Nature* 475:348–352.

Polymerase Chain Reaction

Arnheim, N., and Erlich, H. 1992. Polymerase chain reaction strategy. *Annu. Rev. Biochem.* 61:131–156.

Kirby, L. T. (Ed.). 1997. *DNA Fingerprinting: An Introduction.* Stockton Press.

Eisenstein, B. I. 1990. The polymerase chain reaction: A new method for using molecular genetics for medical diagnosis. *New Engl. J. Med.* 322:178–183.

Foley, K. P., Leonard, M. W., and Engel, J. D. 1993. Quantitation of RNA using the polymerase chain reaction. *Trends Genet.* 9:380–386.

Pääbo, S. 1993. Ancient DNA. *Sci. Am.* 269(5):86–92.

Hagelberg, E., Gray, I. C., and Jeffreys, A. J. 1991. Identification of the skeletal remains of a murder victim by DNA analysis. *Nature* 352:427–429.

Lawlor, D. A., Dickel, C. D., Hauswirth, W. W., and Parham, P. 1991. Ancient HLA genes from 7500-year-old archaeological remains. *Nature* 349:785–788.

Krings, M., Geisert, H., Schmitz, R. W., Krainitzki, H., and Pääbo, S. 1999. DNA sequence of the mitochondrial hypervariable region II for the Neanderthal type specimen. *Proc. Natl. Acad. Sci. U.S.A.* 96:5581–5585.

Ovchinnikov, I. V., Götherström, A., Romanova, G. P., Kharitonov, V. M., Lidén, K., and Goodwin, W. 2000. Molecular analysis of Neanderthal DNA from the northern Caucasus. *Nature* 404:490–493.

Genome Sequencing

International Human Genome Sequencing Consortium. 2004. Finishing the euchromatic sequence of the human genome. *Nature* 431:931–945.

Lander, E. S., Linton, L. M., Birren, B., Nusbaum, C., Zody, M. C., Baldwin, J., Devon, K., Dewar, K., Doyle, M., FitzHugh, W., et al. 2001. Initial sequencing and analysis of the human genome. *Nature* 409:860–921.

Venter, J. C., Adams, M. D., Myers, E. W., Li, P. W., Mural, R. J., Sutton, G. G., Smith, H. O., Yandell, M., Evans, C. A., Holt, R. A., et al. 2001. The sequence of the human genome. *Science* 291:1304–1351.

Waterston, R. H., Lindblad-Toh, K., Birney, E., Rogers, J., Abril, J. F., Agarwal, P., Agarwala, R., Ainscough, R., Alexandersson, M., An, P., et al. 2002. Initial sequencing and comparative analysis of the mouse genome. *Nature* 420:520–562.

Koonin, E. V. 2003. Comparative genomics, minimal gene-sets and the last universal common ancestor. *Nat. Rev. Microbiol.* 1:127–236.

Gilligan, P., Brenner, S., and Venkatesh, B. 2002. Fugu and human sequence comparison identifies novel human genes and conserved non-coding sequences. *Gene* 294:35–44.

Enard, W., and Pääbo, S. 2004. Comparative primate genomics. *Annu. Rev. Genomics Hum. Genet.* 5:351–378.

Quantitative PCR and DNA Arrays

Duggan, D. J., Bittner, J. M., Chen, Y., Meltzer, P., and Trent, J. M. 1999. Expression profiling using cDNA microarrays. *Nat. Genet.* 21:10–14.

Golub, T. R., Slonim, D. K., Tamayo, P., Huard, C., Gaasenbeek, M., Mesirov, J. P., Coller, H., Loh, M. L., Downing, J. R., Caligiuri, M. A., et al. 1999. Molecular classification of cancer: Class discovery and class prediction by gene expression monitoring. *Science* 286:531–537.

Perou, C. M., Sørlie, T., Eisen, M. B., van de Rijn, M., Jeffery, S. S., Rees, C. A., Pollack, J. R., Ross, D. T., Johnsen, H., Akslen, L. A., et al. 2000. Molecular portraits of human breast tumours. *Nature* 406:747–752.

Walker, N. J. 2002. A technique whose time has come. *Science* 296:557–559.

Manipulation of Eukaryotic Genes

Anderson, W. F. 1992. Human gene therapy. *Science* 256:808–813.

Friedmann, T. 1997. Overcoming the obstacles to gene therapy. *Sci. Am.* 277(6):96–101.

Blaese, R. M. 1997. Gene therapy for cancer. *Sci. Am.* 277(6):111–115.

Brinster, R. L., and Palmiter, R. D. 1986. Introduction of genes into the germ lines of animals. *Harvey Lect.* 80:1–38.

Capecchi, M. R. 1989. Altering the genome by homologous recombination. *Science* 244:1288–1292.

Hasty, P., Bradley, A., Morris, J. H., Edmondson, D. G., Venuti, J. M., Olson, E. N., and Klein, W. H. 1993. Muscle deficiency and neonatal death in mice with a targeted mutation in the myogenin gene. *Nature* 364:501–506.

Parkmann, R., Weinberg, K., Crooks, G., Nolta, J., Kapoor, N., and Kohn, D. 2000. Gene therapy for adenosine deaminase deficiency. *Annu. Rev. Med.* 51:33–47.

Gaj, T., Gersbach, C. A., and Barbas III, C. F. 2013. ZFN, TALEN, and CRISPR/Cas-based methods for genome engineering. *Trends Biotechnol.* 31:397–405.

Doudna, J. A., and Charpentier, E. 2014. The new frontier of genome engineering with CRISPR-Cas9. *Science* 346:1258096.

Sander, J. D., and Joung, J. K. 2014. CRISPR-Cas systems for editing, regulating and targeting genomes. *Nat. Biotech.* 32:347–355.

Nishimasu, H., Ran, F. A., Hsu, P. D., Konermann, S., Shehata, S. I., Dohmae, N., Ishitani, R., Zhang, F., and Nureki, O. 2014. Crystal structure of Cas9 in complex with guide RNA and target DNA. *Cell* 156:935–949.

RNA Interference

Rana, T. M. 2007. Illuminating the silence: Understanding the structure and function of small RNAs. *Nat. Rev. Mol. Cell Biol.* 8:23–36.

Novina, C. D., and Sharp, P. A. 2004. The RNAi revolution. *Nature* 430:161–164.

Hannon, G. J., and Rossi, J. J. 2004. Unlocking the potential of the human genome with RNA interference. *Nature* 431:371–378.

Meister, G., and Tuschl, T. 2004. Mechanisms of gene silencing by double-stranded RNA. *Nature* 431:343–349.

Elbashir, S. M., Harborth, J., Lendeckel, W., Yalcin, A., Weber, K., and Tuschl, T. 2001. Duplexes of 21-nucleotide RNAs mediate RNA interference in cultured mammalian cells. *Nature* 411:494–498.

Fire, A., Xu, S., Montgomery, M. K., Kostas, S. A., Driver, S. E., and Mello, C. C. 1998. Potent and specific genetic interference by double-stranded RNA in *Caenorhabditis elegans*. *Nature* 391:806–811.

Genetic Engineering of Plants

Gasser, C. S., and Fraley, R. T. 1992. Transgenic crops. *Sci. Am.* 266(6):62–69.

Gasser, C. S., and Fraley, R. T. 1989. Genetically engineering plants for crop improvement. *Science* 244:1293–1299.

Shimamoto, K., Terada, R., Izawa, T., and Fujimoto, H. 1989. Fertile transgenic rice plants regenerated from transformed protoplasts. *Nature* 338:274–276.

Chilton, M.-D. 1983. A vector for introducing new genes into plants. *Sci. Am.* 248(6):50–59.

Hansen, G., and Wright, M. S. 1999. Recent advances in the transformation of plants. *Trends Plant Sci.* 4:226–231.

Hammond, J. 1999. Overview: The many uses of transgenic plants. *Curr. Top. Microbiol. Immunol.* 240:1–20.

Finer, J. J., Finer, K. R., and Ponappa, T. 1999. Particle bombardment mediated transformation. *Curr. Top. Microbiol. Immunol.* 240:60–80.

Hiei, Y., Ishida, Y., and Komari, T. 2014. Progress of cereal transformation technology mediated by *Agrobacterium tumefaciens*. *Front. Plant Sci.* 5:628.

Amyotrophic Lateral Sclerosis

Siddique, T., Figlewicz, D. A., Pericak-Vance, M. A., Haines, J. L., Rouleau, G., Jeffers, A. J., Sapp, P., Hung, W.-Y., Bebout, J., McKenna-Yasek, D., et al. 1991. Linkage of a gene causing familial amyotrophic lateral sclerosis to chromosome 21 and evidence of genetic-locus heterogeneity. *New Engl. J. Med.* 324:1381–1384.

Rosen, D. R., Siddique, T., Patterson, D., Figlewicz, D. A., Sapp, P., Hentati, A., Donaldson, D., Goto, J., O'Regan, J. P., Deng, H.-X., et al. 1993. Mutations in Cu/Zn superoxide dismutase gene are associated with familial amyotrophic lateral sclerosis. *Nature* 362:59–62.

Gurney, M. E., Pu, H., Chiu, A. Y., Dal Canto, M. C., Polchow, C. Y., Alexander, D. D., Caliendo, J., Hentati, A., Kwon, Y. W., Deng, H.-X., et al. 1994. Motor neuron degeneration in mice that express a human Cu, Zn superoxide dismutase mutation. *Science* 264:1772–1774.

Borchelt, D. R., Lee, M. K., Slunt, H. S., Guarnieri, M., Xu, Z.-S., Wong, P. C., Brown, R. H., Jr., Price, D. L., Sisodia, S. S., and Cleveland, D. W. 1994. Superoxide dismutase 1 with mutations linked to familial amyotrophic lateral sclerosis possesses significant activity. *Proc. Natl. Acad. Sci. U.S.A.* 91:8292–8296.

Taylor, J. P., Brown, Jr., R. H., and Cleveland, D. W. 2016. Decoding ALS: from genes to mechanism. *Nature* 539:197–206.

Biochemistry in Focus

Ajjawi, et al. 2017. Lipid production in *Nannochloropsis gaditana* is doubled by decreasing expression of a single transcriptional regulator. *Nat. Biotech.* 35:647–652.

Radakovits, R., et al. 2012. Draft genome sequence and genetic transformation of the oleaginous alga *Nannochloropsis gaditana*. *Nat. Commun.* 3:686.

CHAPTER 6

Where to Start

Hogewer, P. 2011. The roots of bioinformatics in theoretical biology. *PLoS Comp. Biol.* 7:e1002021.

Searls, D. B. 2010. The roots of bioinformatics. *PLoS Comp. Biol.* 6: e1000809.

Books

Claverie, J.-M., and Notredame, C. 2003. *Bioinformatics for Dummies*. Wiley.

Pevsner, J. 2003. *Bioinformatics and Functional Genomics*. Wiley-Liss.

Doolittle, R. F. 1987. *Of URFS and ORFS*. University Science Books.

Sequence Alignment

Schaffer, A. A., Aravind, L., Madden, T. L., Shavirin, S., Spouge, J. L., Wolf, Y. I., Koonin, E. V., and Altschul, S. F. 2001. Improving the accuracy of PSI-BLAST protein database searches with composition-based statistics and other refinements. *Nucleic Acids Res.* 29:2994–3005.

Henikoff, S., and Henikoff, J. G. 1992. Amino acid substitution matrices from protein blocks. *Proc. Natl. Acad. Sci. U.S.A.* 89:10915–10919.

Johnson, M. S., and Overington, J. P. 1993. A structural basis for sequence comparisons: An evaluation of scoring methodologies. *J. Mol. Biol.* 233:716–738.

Eddy, S. R. 2004. Where did the BLOSUM62 alignment score matrix come from? *Nat. Biotechnol.* 22:1035–1036.

Aravind, L., and Koonin, E. V. 1999. Gleaning non-trivial structural, functional and evolutionary information about proteins by iterative database searches. *J. Mol. Biol.* 287:1023–1040.

Altschul, S. F., Madden, T. L., Schaffer, A. A., Zhang, J., Zhang, Z., Miller, W., and Lipman, D. J. 1997. Gapped BLAST and PSI-BLAST: A new generation of protein database search programs. *Nucleic Acids Res.* 25:3389–3402.

Structure Comparison

Orengo, C. A., Bray, J. E., Buchan, D. W., Harrison, A., Lee, D., Pearl, F. M., Sillitoe, I., Todd, A. E., and Thornton, J. M. 2002. The CATH protein family database: A resource for structural and functional annotation of genomes. *Proteomics* 2:11–21.

Bashford, D., Chothia, C., and Lesk, A. M. 1987. Determinants of a protein fold: Unique features of the globin amino acid sequences. *J. Mol. Biol.* 196:199–216.

Harutyunyan, E. H., Safonova, T. N., Kuranova, I. P., Popov, A. N., Teplyakov, A. V., Obmolova, G. V., Rusakov, A. A., Vainshtein, B. K., Dodson, G. G., Wilson, J. C., et al. 1995. The structure of deoxy- and oxy-leghaemoglobin from lupin. *J. Mol. Biol.* 251:104–115.

Flaherty, K. M., McKay, D. B., Kabsch, W., and Holmes, K. C. 1991. Similarity of the three-dimensional structures of actin and the ATPase fragment of a 70-kDa heat shock cognate protein. *Proc. Natl. Acad. Sci. U.S.A.* 88:5041–5045.

Murzin, A. G., Brenner, S. E., Hubbard, T., and Chothia, C. 1995. SCOP: A structural classification of proteins database for the investigation of sequences and structures. *J. Mol. Biol.* 247:536–540.

Hadley, C., and Jones, D. T. 1999. A systematic comparison of protein structure classification: SCOP, CATH and FSSP. *Struct. Fold. Des.* 7:1099–1112.

Domain Detection

Marchler-Bauer, A., Anderson, J. B., DeWeese-Scott, C., Fedorova, N. D., Geer, L. Y., He, S., Hurwitz, D. I., Jackson, J. D., Jacobs, A. R., Lanczycki, C. J., et al. 2003. CDD: A curated Entrez database of conserved domain alignments. *Nucleic Acids Res.* 31:383–387.

Ploegman, J. H., Drent, G., Kalk, K. H., and Hol, W. G. 1978. Structure of bovine liver rhodanese I: Structure determination at 2.5 Å resolution and a comparison of the conformation and sequence of its two domains. *J. Mol. Biol.* 123:557–594.

Nikolov, D. B., Hu, S. H., Lin, J., Gasch, A., Hoffmann, A., Horikoshi, M., Chua, N. H., Roeder, R. G., and Burley, S. K. 1992. Crystal structure of TFIID TATA-box binding protein. *Nature* 360:40–46.

Doolittle, R. F. 1995. The multiplicity of domains in proteins. *Annu. Rev. Biochem.* 64:287–314.

Heger, A., and Holm, L. 2000. Rapid automatic detection and alignment of repeats in protein sequences. *Proteins* 41:224–237.

Evolutionary Trees

Wolf, Y. I., Rogozin, I. B., Grishin, N. V., and Koonin, E. V. 2002. Genome trees and the tree of life. *Trends Genet.* 18:472–479.

Doolittle, R. F. 1992. Stein and Moore Award address. Reconstructing history with amino acid sequences. *Protein Sci.* 1:191–200.

Zuckerkandl, E., and Pauling, L. 1965. Molecules as documents of evolutionary history. *J. Theor. Biol.* 8:357–366.

Schönknecht, G., Chen, W.-H., Ternes, C. M., Barbier, G. G., Shrestha, R. P., Stanke, M., Bräutigam, A., Baker, B. J., Banfield, J. F., Garavito, R. M., et al. 2013. Gene transfer from bacteria and archaea facilitated evolution of an extremophilic eukaryote. *Science* 339:1207–1210.

Ancient DNA

Prüfer, K., Racimo, F., Patterson, N., Jay, F., Sankararaman, S., Sawyer, S., Heinze, A., Renaud, G., Sudmant, P. H., de Filippo C., et al. 2014. The complete genome sequence of a Neanderthal from the Altai Mountains. *Nature* 505:43–49.

Meyer, M., Kircher, M., Gansauge, M. T., Li, H., Racimo, F., Mallick, S., Schraiber, J. G., Jay, F., Prüfer, K., de Filippo, C., et al. 2012. A high-coverage genome sequence from an archaic Denisovan individual. *Science* 338:222–226.

Green, R. E., Malaspinas, A.-S., Krause, J., Briggs, A. W., Johnson, P. L. F., Uhler, C., Meyer, M., Good, J. M., Maricic, T., Stenzel, U., et al. 2008. A complete Neandertal mitochondrial genome sequence determined by high-throughput sequencing. *Cell* 134:416–426.

Pääbo, S., Poinar, H., Serre, D., Jaenicke-Despres, V., Hebler, J., Rohland, N., Kuch, M., Krause, J., Vigilant, L., and Hofreiter, M. 2004. Genetic analyses from ancient DNA. *Annu. Rev. Genet.* 38:645–679.

Evolution in the Laboratory

Sassanfar, M., and Szostak, J. W. 1993. An RNA motif that binds ATP. *Nature* 364:550–553.

Gold, L., Polisky, B., Uhlenbeck, O., and Yarus, M. 1995. Diversity of oligonucleotide functions. *Annu. Rev. Biochem.* 64:763–797.

Wilson, D. S., and Szostak, J. W. 1999. In vitro selection of functional nucleic acids. *Annu. Rev. Biochem.* 68:611–647.

Hermann, T., and Patel, D. J. 2000. Adaptive recognition by nucleic acid aptamers. *Science* 287:820–825.

Keefe, A. D., Pai, S., and Ellington, A. 2010. Aptamers as therapeutics. *Nat. Rev. Drug Discov.* 9:537–550.

Radom, F., Jurek, P. M., Mazurek, M. P., Otlewski, J., and Jelen´, F. 2013. Aptamers: Molecules of great potential. *Biotechnol. Adv.* 31:1260–1274.

Websites

The Protein Data Bank (PDB) site is the repository for three-dimensional macromolecular structures. It currently contains more than 100,000 structures. (https://www.wwpdb.org)

National Center for Biotechnology Information (NCBI) contains molecular biological databases and software for analysis. (http://www.ncbi.nlm.nih.gov/)

Biochemistry in Focus

Man, O., Gilad, Y., and Lancet, D. 2004. Prediction of the odorant binding site of olfactory receptor proteins by human–mouse comparisons. *Protein Sci.* 13:240–254.

CHAPTER 7

Where to Start

Changeux, J.-P. 2011. 50th anniversary of the word "Allosteric." *Protein Sci.* 20:1119–1124.

Perutz, M. F. 1978. Hemoglobin structure and respiratory transport. *Sci. Am.* 239(6):92–125.

Perutz, M. F. 1980. Stereochemical mechanism of oxygen transport by haemoglobin. *Proc. R. Soc. Lond. Biol. Sci.* 208:135–162.

Kilmartin, J. V. 1976. Interaction of haemoglobin with protons, CO_2, and 2,3-diphosphoglycerate. *Brit. Med. Bull.* 32:209–222.

Structure

Kendrew, J. C., Bodo, G., Dintzis, H. M., Parrish, R. G., Wyckoff, H., and Phillips, D. C. 1958. A three-dimensional model of the myoglobin molecule obtained by x-ray analysis. *Nature* 181:662–666.

Shaanan, B. 1983. Structure of human oxyhaemoglobin at 2.1 Å resolution. *J. Mol. Biol.* 171:31–59.

Frier, J. A., and Perutz, M. F. 1977. Structure of human foetal deoxyhaemoglobin. *J. Mol. Biol.* 112:97–112.

Perutz, M. F. 1969. Structure and function of hemoglobin. *Harvey Lect.* 63:213–261.

Perutz, M. F. 1962. Relation between structure and sequence of haemoglobin. *Nature* 194:914–917.

Harrington, D. J., Adachi, K., and Royer, W. E., Jr. 1997. The high resolution crystal structure of deoxyhemoglobin S. *J. Mol. Biol.* 272:398–407.

Interaction of Hemoglobin with Allosteric Effectors

Benesch, R., and Beesch, R. E. 1969. Intracellular organic phosphates as regulators of oxygen release by haemoglobin. *Nature* 221:618–622.

Fang, T. Y., Zou, M., Simplaceanu, V., Ho, N. T., and Ho, C. 1999. Assessment of roles of surface histidyl residues in the molecular basis of the Bohr effect and of β 143 histidine in the binding of 2,3-bisphosphoglycerate in human normal adult hemoglobin. *Biochemistry* 38:13423–13432.

Arnone, A. 1992. X-ray diffraction study of binding of 2,3-diphosphoglycerate to human deoxyhaemoglobin. *Nature* 237:146–149.

Models for Cooperativity

Changeux, J.-P. 2012. Allostery and the Monod-Wyman-Changeux model after 50 years. *Annu. Rev. Biophys.* 41:103–133.

Monod, J., Wyman, J., and Changeux, J.-P. 1965. On the nature of allosteric interactions: A plausible model. *J. Mol. Biol.* 12:88–118.

Koshland, D. L., Jr., Nemethy, G., and Filmer, D. 1966. Comparison of experimental binding data and theoretical models in proteins containing subunits. *Biochemistry* 5:365–385.

Ackers, G. K., Doyle, M. L., Myers, D., and Daugherty, M. A. 1992. Molecular code for cooperativity in hemoglobin. *Science* 255:54–63.

Sickle-Cell Anemia and Thalassemia

Herrick, J. B. 1910. Peculiar elongated and sickle-shaped red blood corpuscles in a case of severe anemia. *Arch. Intern. Med.* 6:517–521.

Pauling, L., Itano, H. A., Singer, S. J., and Wells, L. C. 1949. Sickle cell anemia: A molecular disease. *Science* 110:543–548.

Ingram, V. M. 1957. Gene mutation in human hemoglobin: The chemical difference between normal and sickle cell haemoglobin. *Nature* 180:326–328.

Eaton, W. A., and Hofrichter, J. 1990. Sickle cell hemoglobin polymerization. *Adv. Prot. Chem.* 40:63–279.

Weatherall, D. J. 2001. Phenotype genotype relationships in monogenic disease: Lessons from the thalassemias. *Nat. Rev. Genet.* 2:245–255.

Tsaras, G., Owusu-Ansah, A., Boateng, F. O., and Amoateng-Adjepong, Y. 2009. Complications associated with sickle cell trait: A brief narrative review. *Am. J. Med.* 122: 507–512.

Globin-Binding Proteins and Other Globins

Helbo, S., Weber, R. E., and Fago, A. 2013. Expression patterns and adaptive functional diversity of vertebrate myoglobins. *Biochim. Biophys. Acta* 1834:1832–1839.

Kihm, A. J., Kong, Y., Hong, W., Russell, J. E., Rouda, S., Adachi, K., Simon, M. C., Blobel, G. A., and Weiss, M. J. 2002. An abundant erythroid protein that stabilizes free α-haemoglobin. *Nature* 417:758–763.

Feng, L., Zhou, S., Gu, L., Gell, D. A., Mackay, J. P., Weiss, M. J., Gow, A. J., and Shi, Y. 2005. Structure of oxidized α-haemoglobin bound to AHSP reveals a protective mechanism for haem. *Nature* 435:697–701.

Yu, X., Kong, Y., Dore, L. C., Abdulmalik, O., Katein, A. M., Zhou, S., Choi, J. K., Gell, D., Mackay, J. P., Gow, A. J., et al. 2007. An erythroid chaperone that facilitates folding of α-globin subunits for hemoglobin synthesis. *J. Clin. Invest.* 117:1856–1865.

Burmester, T., Haberkamp, M., Mitz, S., Roesner, A., Schmidt, M., Ebner, B., Gerlach, F., Fuchs, C., and Hankeln, T. 2004. Neuroglobin and cytoglobin: Genes, proteins and evolution. *IUBMB Life* 56:703–707.

Hankeln, T., Ebner, B., Fuchs, C., Gerlach, F., Haberkamp, M., Laufs, T. L., Roesner, A., Schmidt, M., Weich, B., Wystub, S., et al. 2005. Neuroglobin and cytoglobin in search of their role in the vertebrate globin family. *J. Inorg. Biochem.* 99:110–119.

Burmester, T., Ebner, B., Weich, B., and Hankeln, T. 2002. Cytoglobin: A novel globin type ubiquitously expressed in vertebrate tissues. *Mol. Biol. Evol.* 19:416–421.

Zhang, C., Wang, C., Deng, M., Li, L., Wang, H., Fan, M., Xu, W., Meng, F., Qian, L., and He, F. 2002. Full-length cDNA cloning of human neuroglobin and tissue expression of rat neuroglobin. *Biochem. Biophys. Res. Commun.* 290:1411–1419.

Biochemistry in Focus

Azarov, I., Wang, L., Rose, J. J., Xu, Q., Huang, X. N., Belanger, A., Wang, Y., Guo, L., Liu, C., Ucer, K. B., McTiernan, C. F., O'Donnell, C. P., Shiva, S., Tejero, J., Kim-Shapiro, D. B., and Gladwin, M. T. 2016. Five-coordinate H64Q neuroglobin as a ligand-trap antidote for carbon monoxide poisoning. *Sci. Transl. Med.* 8:368ra173.

Pesce, A., Dewilde, S., Nardini, M., Moens, L., Ascenzi, P., Hankeln, T., Burmester, T., and Bolognesi, M. 2003. Human brain neuroglobin structure reveals a distinct mode of controlling oxygen affinity. *Structure* 11:1087–1095.

CHAPTER 8

Where to Start

Zalatan, J. G., and Herschlag, D. 2009. The far reaches of enzymology. *Nat. Chem. Biol.* 5:516–520.

Hammes, G. G. 2008. How do enzymes really work? *J. Biol. Chem.* 283:22337–22346.

Koshland, D. E., Jr. 1987. Evolution of catalytic function. *Cold Spring Harbor Symp. Quant. Biol.* 52:1–7.

Jencks, W. P. 1987. Economics of enzyme catalysis. *Cold Spring Harbor Symp. Quant. Biol.* 52:65–73.

Lerner, R. A., and Tramontano, A. 1988. Catalytic antibodies. *Sci. Am.* 258(3):58–70.

Books

Cook, P. F., and Cleland, W. W. 2007. *Enzyme Kinetics and Mechanism*. Garland Press.

Fersht, A. 1999. *Structure and Mechanism in Protein Science: A Guide to Enzyme Catalysis and Protein Folding*. W. H. Freeman and Company.

Walsh, C. 1979. *Enzymatic Reaction Mechanisms*. W. H. Freeman and Company.

Bender, M. L., Bergeron, R. J., and Komiyama, M. 1984. *The Bioorganic Chemistry of Enzymatic Catalysis*. Wiley-Interscience.

Abelson, J. N., and Simon, M. I. (Eds.). 1992. *Methods in Enzymology*. Academic Press.

Friedmann, H. C. (Ed.). 1981. *Benchmark Papers in Biochemistry*, vol. 1, *Enzymes*. Hutchinson Ross.

Transition-State Stabilization, Analogs, and Other Enzyme Inhibitors

Schramm, V. L. 2007. Enzymatic transition state theory and transition state analog design. *J. Biol. Chem.* 282:28297–28300.

Pauling, L. 1948. Nature of forces between large molecules of biological interest. *Nature* 161:707–709.

Leinhard, G. E. 1973. Enzymatic catalysis and transition-state theory. *Science* 180:149–154.

Kraut, J. 1988. How do enzymes work? *Science* 242:533–540.

Waxman, D. J., and Strominger, J. L. 1983. Penicillin-binding proteins and the mechanism of action of β-lactam antibiotics. *Annu. Rev. Biochem.* 52:825–869.

Abraham, E. P. 1981. The β-lactam antibiotics. *Sci. Am.* 244(6):76–86.

Walsh, C. T. 1984. Suicide substrates, mechanism-based enzyme inactivators: Recent developments. *Annu. Rev. Biochem.* 53:493–535.

Enzyme Kinetics and Mechanisms

Hammes, G. G., Benkovic, S. J., and Hammes-Schiffer, S. 2011. Flexibility, Diversity, and Cooperativity: Pillars of Enzyme Catalysis. *Biochemistry* 50:10422–10430.

Johnson, K. A., and Goody, R. S. 2011. The Original Michaelis Constant: Translation of the 1913 Michaelis–Menten Paper. *Biochemistry* 50:8264–8269.

Hammes-Schiller, S., and Benkovic, S. J. 2006. Relating protein motion to catalysis. *Annu. Rev. Biochem.* 75:519–541.

Benkovic, S. J., and Hammes-Schiller, S. 2003. A perspective on enzyme catalysis. *Science* 301:1196–1202.

Hur, S., and Bruice, T. C. 2003. The near attack conformation approach to the study of the chorismate to prephenate reaction. *Proc. Natl. Acad. Sci. U.S.A.* 100:12015–12020.

Miles, E. W., Rhee, S., and Davies, D. R. 1999. The molecular basis of substrate channeling. *J. Biol. Chem.* 274:12193–12196.

Warshel, A. 1998. Electrostatic origin of the catalytic power of enzymes and the role of preorganized active sites. *J. Biol. Chem.* 273:27035–27038.

Cannon, W. R., and Benkovic, S. J. 1999. Solvation, reorganization energy, and biological catalysis. *J. Biol. Chem.* 273:26257–26260.

Cleland, W. W., Frey, P. A., and Gerlt, J. A. 1998. The low barrier hydrogen bond in enzymatic catalysis. *J. Biol. Chem.* 273:25529–25532.

Romesberg, F. E., Santarsiero, B. D., Spiller, B., Yin, J., Barnes, D., Schultz, P. G., and Stevens, R. C. 1998. Structural and kinetic evidence for strain in biological catalysis. *Biochemistry* 37:14404–14409.

Fersht, A. R., Leatherbarrow, R. J., and Wells, T. N. C. 1986. Binding energy and catalysis: A lesson from protein engineering of the tyrosyl-tRNA synthetase. *Trends Biochem. Sci.* 11:321–325.

Jencks, W. P. 1975. Binding energy, specificity, and enzymic catalysis: The Circe effect. *Adv. Enzymol.* 43:219–410.

Knowles, J. R., and Albery, W. J. 1976. Evolution of enzyme function and the development of catalytic efficiency. *Biochemistry* 15:5631–5640.

Single Molecule Studies

Allewell, N. M. 2010. Thematic Minireview Series: Single-molecule Measurements in Biochemistry and Molecular Biology. *J. Biol. Chem.* 285:18959–18983. A series of reviews on single-molecule studies.

Min, W., English, B. P., Lou, G., Cherayil, B. J., Kou, S. C., and Xie, X. S. 2005. Fluctuating Enzymes: Lessons from Single-Molecule Studies. *Acc. Chem. Res.* 38:923–931.

Xie, X. S., and Lu, H. P. 1999. Single-molecule enzymology. *J. Biol. Chem.* 274:15967–15970.

Lu, H. P., Xun, L., and Xie, X. S. 1998. Single-molecule enzymatic dynamics. *Science* 282:1877–1882.

CHAPTER 9

Where to Start

Stroud, R. M. 1974. A family of protein-cutting proteins. *Sci. Am.* 231(1):74–88.

Kraut, J. 1977. Serine proteases: Structure and mechanism of catalysis. *Annu. Rev. Biochem.* 46:331–358.

Lindskog, S. 1997. Structure and mechanism of carbonic anhydrase. *Pharmacol. Ther.* 74:1–20.

Jeltsch, A., Alves, J., Maass, G., and Pingoud, A. 1992. On the catalytic mechanism of EcoRI and EcoRV: A detailed proposal based on biochemical results, structural data and molecular modelling. *FEBS Lett.* 304:4–8.

Bauer, C. B., Holden, H. M., Thoden, J. B., Smith, R., and Rayment, I. 2000. X-ray structures of the apo and MgATP-bound states of *Dictyostelium discoideum* myosin motor domain. *J. Biol. Chem.* 275:38494–38499.

Lolis, E., and Petsko, G. A. 1990. Transition-state analogues in protein crystallography: Probes of the structural source of enzyme catalysis. *Annu. Rev. Biochem.* 59:597–630.

Books

Fersht, A. 1999. *Structure and Mechanism in Protein Science: A Guide to Enzyme Catalysis and Protein Folding.* W. H. Freeman and Company.

Silverman, R. B. 2000. *The Organic Chemistry of Enzyme-Catalyzed Reactions.* Academic Press.

Page, M., and Williams, A. 1997. *Organic and Bio-organic Mechanisms.* Addison Wesley Longman.

Chymotrypsin and Other Serine Proteases

Fastrez, J., and Fersht, A. R. 1973. Demonstration of the acyl-enzyme mechanism for the hydrolysis of peptides and anilides by chymotrypsin. *Biochemistry* 12:2025–2034.

Sigler, P. B., Blow, D. M., Matthews, B. W., and Henderson, R. 1968. Structure of crystalline-chymotrypsin II: A preliminary report including a hypothesis for the activation mechanism. *J. Mol. Biol.* 35:143–164.

Kossiakoff, A. A., and Spencer, S. A. 1981. Direct determination of the protonation states of aspartic acid-102 and histidine-57 in the tetrahedral intermediate of the serine proteases: Neutron structure of trypsin. *Biochemistry* 20:6462–6474.

Carter, P., and Wells, J. A. 1988. Dissecting the catalytic triad of a serine protease. *Nature* 332:564–568.

Carter, P., and Wells, J. A. 1990. Functional interaction among catalytic residues in subtilisin BPN′. *Proteins* 7:335–342.

Koepke, J., Ermler, U., Warkentin, E., Wenzl, G., and Flecker, P. 2000. Crystal structure of cancer chemopreventive Bowman-Birk inhibitor in ternary complex with bovine trypsin at 2.3 Å resolution: Structural basis of Janus-faced serine protease inhibitor specificity. *J. Mol. Biol.* 298:477–491.

Gaboriaud, C., Rossi, V., Bally, I., Arlaud, G. J., and Fontecilla-Camps, J. C. 2000. Crystal structure of the catalytic domain of human complement C1s: A serine protease with a handle. *EMBO J.* 19:1755–1765.

Bachovchin D. A., and Cravatt B. F. 2012. The pharmacological landscape and therapeutic potential of serine hydrolases. *Nature Reviews Drug Discovery* 11:52–68.

Other Proteases

Vega, S., Kang, L. W., Velazquez-Campoy, A., Kiso, Y., Amzel, L. M., and Freire, E. 2004. A structural and thermodynamic escape mechanism from a drug resistant mutation of the HIV-1 protease. *Proteins* 55:594–602.

Kamphuis, I. G., Kalk, K. H., Swarte, M. B., and Drenth, J. 1984. Structure of papain refined at 1.65 Å resolution. *J. Mol. Biol.* 179:233–256.

Kamphuis, I. G., Drenth, J., and Baker, E. N. 1985. Thiol proteases: Comparative studies based on the high-resolution structures of papain and actinidin, and on amino acid sequence information for cathepsins B and H, and stem bromelain. *J. Mol. Biol.* 182:317–329.

Sivaraman, J., Nagler, D. K., Zhang, R., Menard, R., and Cygler, M. 2000. Crystal structure of human procathepsin X: A cysteine protease with the proregion covalently linked to the active site cysteine. *J. Mol. Biol.* 295:939–951.

Davies, D. R. 1990. The structure and function of the aspartic proteinases. *Annu. Rev. Biophys. Biophys. Chem.* 19:189–215.

Dorsey, B. D., Levin, R. B., McDaniel, S. L., Vacca, J. P., Guare, J. P., Darke, P. L., Zugay, J. A., Emini, E. A., Schleif, W. A., Quintero, J. C., et al. 1994. L-735,524:

The design of a potent and orally bio-available HIV protease inhibitor. *J. Med. Chem.* 37:3443–3451.

Chen, Z., Li, Y., Chen, E., Hall, D. L., Darke, P. L., Culberson, C., Shafer, J. A., and Kuo, L. C. 1994. Crystal structure at 1.9-Å resolution of human immunodeficiency virus (HIV) II protease complexed with L-735,524, an orally bioavailable inhibitor of the HIV proteases. *J. Biol. Chem.* 269:26344–26348.

Ollis, D. L., Cheah, E., Cygler, M., Dijkstra, B., Frolow, F., Franken, S. M., Harel, M., Remington, S. J., Silman, I., Schrag, J., et al. 1992. The α/β hydrolase fold. *Protein Eng.* 5:197–211.

Miller, M. 2012. The early years of retroviral protease crystal structures. *Biopolymers* 94:521–529.

Carbonic Anhydrase

Lindskog, S., and Coleman, J. E. 1973. The catalytic mechanism of carbonic anhydrase. *Proc. Natl. Acad. Sci. U.S.A.* 70:2505–2508.

Kannan, K. K., Notstrand, B., Fridborg, K., Lovgren, S., Ohlsson, A., and Petef, M. 1975. Crystal structure of human erythrocyte carbonic anhydrase B: Three-dimensional structure at a nominal 2.2-Å resolution. *Proc. Natl. Acad. Sci. U.S.A.* 72:51–55.

Boriack-Sjodin, P. A., Zeitlin, S., Chen, H. H., Crenshaw, L., Gross, S., Dantanarayana, A., Delgado, P., May, J. A., Dean, T., and Christianson, D. W. 1998. Structural analysis of inhibitor binding to human carbonic anhydrase II. *Protein Sci.* 7:2483–2489.

Wooley, P. 1975. Models for metal ion function in carbonic anhydrase. *Nature* 258:677–682.

Jonsson, B. H., Steiner, H., and Lindskog, S. 1976. Participation of buffer in the catalytic mechanism of carbonic anhydrase. *FEBS Lett.* 64:310–314.

Sly, W. S., and Hu, P. Y. 1995. Human carbonic anhydrases and carbonic anhydrase deficiencies. *Annu. Rev. Biochem.* 64:375–401.

Maren, T. H. 1988. The kinetics of HCO_3^- synthesis related to fluid secretion, pH control, and CO_2, elimination. *Annu. Rev. Physiol.* 50:695–717.

Roy, A., and Taraphder, S. 2010. Role of protein motions on proton transfer pathways in human carbonic anhydrase II. *Biochim. Biophys. Acta* 1804:352–361.

Ozensoy Guler, O., Capasso, C., and Supuran, C. T. 2016. A magnificent enzyme superfamily: carbonic anhydrase, their purification and characterization. *J. Enzyme Inhib. Med. Chem.* 31:689–694.

Taraphder, S., Maupin, C. M., Swanson, J. M., and Voth, G. A. 2016. Coupling protein dynamics with proton transport in human carbonic anhydrase II. *J. Phys. Chem. B.* 120:8389–8404.

Restriction Enzymes

Selvaraj, S., Kono, H., and Sarai, A. 2002. Specificity of protein-DNA recognition revealed by structure-based potentials: Symmetric/asymmetric and cognate/non-cognate binding. *J. Mol. Biol.* 322:907–915.

Winkler, F. K., Banner, D. W., Oefner, C., Tsernoglou, D., Brown, R. S., Heathman, S. P., Bryan, R. K., Martin, P. D., Petratos, K., and Wilson, K. S. 1993. The crystal structure of EcoRV endonuclease and of its complexes with cognate and non-cognate DNA fragments. *EMBO J.* 12:1781–1795.

Kostrewa, D., and Winkler, F. K. 1995. Mg^{2+} binding to the active site of EcoRV endonuclease: A crystallographic study of complexes with substrate and product DNA at 2 Å resolution. *Biochemistry* 34:683–696.

Athanasiadis, A., Vlassi, M., Kotsifaki, D., Tucker, P. A., Wilson, K. S., and Kokkinidis, M. 1994. Crystal structure of PvuII endonuclease reveals extensive structural homologies to EcoRV. *Nat. Struct. Biol.* 1:469–475.

Sam, M. D., and Perona, J. J. 1999. Catalytic roles of divalent metal ions in phosphoryl transfer by EcoRV endonuclease. *Biochemistry* 38:6576–6586.

Jeltsch, A., and Pingoud, A. 1996. Horizontal gene transfer contributes to the wide distribution and evolution of type II restriction-modification systems. *J. Mol. Evol.* 42:91–96.

Advani S., Mishra P., Dubey S., and Thakur S. 2010. Categoric prediction of metal ion mechanisms in the active sites of 17 select type II restriction endonucleases. *Biochem. Biophys. Res. Commun.* 402:177–179.

Horton, N. C., and Perona, J. J. 2000. Crystallographic snapshots along a protein-induced DNA-bending pathway. *Proc. Natl. Acad. Sci. U.S.A.* 97:5729–5734.

Myosins

Grigorenko, B. L., Rogov, A. V., Topol, I. A., Burt, S. K., Martinez, H. M., and Nemukhin, A. V. 2007. Mechanism of the myosin catalyzed hydrolysis of ATP as rationalized by molecular modeling. *Proc. Natl. Acad. Sci. U.S.A.* 104:7057–7061.

Gulick, A. M., Bauer, C. B., Thoden, J. B., and Rayment, I. 1997. X-ray structures of the MgADP, MgATPγ S, and MgAMPPNP complexes of the *Dictyostelium discoideum* myosin motor domain. *Biochemistry* 36:11619–11628.

Kovacs, M., Malnasi-Csizmadia, A., Woolley, R. J., and Bagshaw, C. R. 2002. Analysis of nucleotide binding to *Dictyostelium* myosin II motor domains containing a single tryptophan near the active site. *J. Biol. Chem.* 277:28459–28467.

Kuhlman, P. A., and Bagshaw, C. R. 1998. ATPase kinetics of the *Dictyostelium discoideum* myosin II motor domain. *J. Muscle Res. Cell Motil.* 19:491–504.

Smith, C. A., and Rayment, I. 1996. X-ray structure of the magnesium(II) ADP vanadate complex of the *Dictyostelium discoideum* myosin motor domain to 1.9 Å resolution. *Biochemistry* 35:5404–5417.

Yildiz A., Forkey J. N., McKinney S. A., Ha T., Goldman Y. E., and Selvin P. R. 2003. Myosin V walks hand-over-hand: Single fluorophore imaging with 1.5-nm localization. *Science.* 300:2061–2065.

Houdusse, A., and Sweeney, H. L. 2016. How myosin generates force on actin filaments. *Trends Biochem. Sci.* 41:989–997.

CHAPTER 10

Where to Start

Kyriakis, J. M. 2014. In the beginning, there was protein phosphorylation. *J. Biol. Chem.* 289:9460–9462.

Changeux, J.-P. 2011. 50th anniversary of the word "Allosteric." *Protein Sci.* 20:1119–1124.

Kantrowitz, E. R., and Lipscomb, W. N. 1990. *Escherichia coli* aspartate transcarbamoylase: The molecular basis for a concerted allosteric transition. *Trends Biochem. Sci.* 15:53–59.

Schachman, H. K. 1988. Can a simple model account for the allosteric transition of aspartate transcarbamoylase? *J. Biol. Chem.* 263:18583–18586.

Neurath, H. 1989. Proteolytic processing and physiological regulation. *Trends Biochem. Sci.* 14:268–271.

Bode, W., and Huber, R. 1992. Natural protein proteinase inhibitors and their interaction with proteinases. *Eur. J. Biochem.* 204:433–451.

Aspartate Transcarbamoylase and Allosteric Interactions

Changeux, J.-P., and Christopoulos, A. 2016. Allosteric modulation as a unifying mechanism for receptor function and regulation. *Cell* 166:1084–1102.

Changeux, J.-P. 2012. Allostery and the Monod-Wyman-Changeux Model After 50 Years. *Annu. Rev. Biophys.* 41:103–133.

Peterson, A. W., Cockrell, G. M., and Kantrowitz, E. R. 2012. A second allosteric site in *Escherichia coli* aspartate transcarbamoylase. *Biochemistry* 51:4776–4778.

Rabinowitz, J. D., Hsiao, J. J., Gryncel, K. R., Kantrowitz, E. R., Feng, X.-J., Li, G., and Rabitz H. 2008. Dissecting enzyme regulation by multiple allosteric effectors: Nucleotide regulation of aspartate transcarbamoylase. *Biochemistry* 47:5881–5888.

West, J. M., Tsuruta, H., and Kantrowitz, E. R. 2004. A fluorescent probe-labeled *Escherichia coli* aspartate transcarbamoylase that monitors the allosteric conformation state. *J. Biol. Chem.* 279:945–951.

Endrizzi, J. A., Beernink, P. T., Alber, T., and Schachman, H. K. 2000. Binding of bisubstrate analog promotes large structural changes in the unregulated catalytic trimer of aspartate transcarbamoylase: Implications for allosteric regulation. *Proc. Natl. Acad. Sci. U.S.A.* 97:5077–5082.

Beernink, P. T., Endrizzi, J. A., Alber, T., and Schachman, H. K. 1999. Assessment of the allosteric mechanism of aspartate transcarbamoylase based on the crystalline structure of the unregulated catalytic subunit. *Proc. Natl. Acad. Sci. U.S.A.* 96:5388–5393.

Wales, M. E., Madison, L. L., Glaser, S. S., and Wild, J. R. 1999. Divergent allosteric patterns verify the regulatory paradigm for aspartate transcarbamoylase. *J. Mol. Biol.* 294:1387–1400.

Newell, J. O., Markby, D. W., and Schachman, H. K. 1989. Cooperative binding of the bisubstrate analog *N*-(phosphonacetyl)-L-aspartate to aspartate transcarbamoylase and the heterotropic effects of ATP and CTP. *J. Biol. Chem.* 264:2476–2481.

Stevens, R. C., Gouaux, J. E., and Lipscomb, W. N. 1990. Structural consequences of effector binding to the T state of aspartate carbamoyl-transferase: Crystal structures of the unliganded and ATP- and CTP-complexed enzymes at 2.6-Å resolution. *Biochemistry* 29:7691–7701.

Gouaux, J. E., and Lipscomb, W. N. 1990. Crystal structures of phosphonoacetamide ligated T and phosphonoacetamide and malonate ligated R states of aspartate carbamoyltransferase at 2.8-Å resolution and neutral pH. *Biochemistry* 29:389–402.

Labedan, B., Boyen, A., Baetens, M., Charlier, D., Chen, P., Cunin, R., Durbeco, V., Glansdorff, N., Herve, G., Legrain, C., et al. 1999. The evolutionary history of carbamoyltransferases: A complex set of paralogous genes was already present in the last universal common ancestor. *J. Mol. Evol.* 49:461–473.

Covalent Modification

Hoffman, N. J., Parker, B. L., Chaudhuri, R., Fisher-Wellman, K. H., Kleinert, M., et al. 2015. Global phosphoproteomic analysis of human skeletal muscle reveals a network of exercise-regulated kinases and AMPK substrates. *Cell Metab.* 22:922–935.

Endicott, J. A., Noble, M. E. M., and Johnson, L. N. 2012. The structural basis for control of eukaryotic protein kinases. *Annu. Rev. Biochem.* 81:587–613.

Tarrant, M. K., and Cole, P. A. 2009. The chemical biology of protein phosphorylation *Annu. Rev. Biochem.* 78:797–825.

Guarente, L. 2011. The logic linking protein acetylation and metabolism. *Cell Metab.* 14:151–153.

Guan, K-L., and Xiong, Y. 2011. Regulation of intermediary metabolism by protein acetylation. *Trends Biochem. Sci.* 36:108–116.

Johnson, L. N., and Barford, D. 1993. The effects of phosphorylation on the structure and function of proteins. *Annu. Rev. Biophys. Biomol. Struct.* 22:199–232.

Barford, D., Das, A. K., and Egloff, M. P. 1998. The structure and mechanism of protein phosphatases: Insights into catalysis and regulation. *Annu. Rev. Biophys. Biomol. Struct.* 27:133–164.

Protein Kinase A

Taylor, S. S., Ilouz, R., Zhang, P., and Kornev, A. P. 2012. Assembly of allosteric macromolecular switches: Lessons from PKA. *Nature Rev. Mol. Cell Biol.* 13:646–658.

Zhang, P., Smith-Nguyen, E. V., Keshwani, M. M., Deal, M. S., Kornev, A. P., and Taylor, S. S. 2012. Structure and allostery of the PKA RIIb tetrameric holoenzyme. *Science* 334:712–716.

Taylor, S. S., and Kornev, A. P. 2011. Protein kinases: evolution of dynamic regulatory proteins. *Trends Biochem. Sci.* 36:65–77.

Pearlman, S. M., Serber, Z., and Ferrell Jr., J. E. 2011. A mechanism for the evolution of phosphorylation sites. *Cell* 147:934–946.

Knighton, D. R., Zheng, J. H., TenEyck, L., Xuong, N. H., Taylor, S. S., and Sowadski, J. M. 1991. Structure of a peptide inhibitor bound to the catalytic subunit of cyclic adenosine monophosphate-dependent protein kinase. *Science* 253:414–420.

Zymogen Activation

Artenstein, A. W., and Opal, S. M. 2011. Proprotein convertases in health and disease. *New Engl. J. Med.* 65:2507–2518.

Neurath, H. 1986. The versatility of proteolytic enzymes. *J. Cell. Biochem.* 32:35–49.

Bode, W., and Huber, R. 1986. Crystal structure of pancreatic serine endopeptidases. In *Molecular and Cellular Basis of Digestion* (pp. 213–234), edited by P. Desnuelle, H. Sjostrom, and O. Noren. Elsevier.

James, M. N. 1991. Refined structure of porcine pepsinogen at 1.8 Å resolution. *J. Mol. Biol.* 219:671–692.

Protease Inhibitors

Gooptu, B., Dickens, J. A., and Lomas, D. A. 2014. The molecular and cellular pathology of α_1-antitrypsin deficiency. *Trends Mol. Med.* 20:116–127.

Stockley, R. A., and Turner, A. M. 2014. α-1-Antitrypsin deficiency: clinical variability, assessment, and treatment. *Trends Mol. Med.* 20:104–115.

Alloy, A. P., Kayode, O., Wang, R., Hockla, A., Soares, A. S., and Radisky, E. S. 2015. Mesotrypsin has evolved four unique residues to cleave trypsin inhibitors as substrates. *J. Biol. Chem.* 290:21523–21535.

Gooptu, B., and Lomas, D. A. 2009. Conformational Pathology of the Serpins: Themes, Variations, and Therapeutic Strategies. *Annu. Rev. Biochem.* 78:147–167.

Carp, H., Miller, F., Hoidal, J. R., and Janoff, A. 1982. Potential mechanism of emphysema: α_1-Proteinase inhibitor recovered from lungs of cigarette smokers contains oxidized methionine and has decreased elastase inhibitory capacity. *Proc. Natl. Acad. Sci. U.S.A.* 79:2041–2045.

Owen, M. C., Brennan, S. O., Lewis, J. H., and Carrell, R. W. 1983. Mutation of antitrypsin to antithrombin. *New Engl. J. Med.* 309:694–698.

Clotting Cascade

Kollman, J. M., Pandi, L., Sawaya, M. R., Riley, M., and Doolittle, R. F. 2009. Crystal structure of human fibrinogen. *Biochemistry* 48:3877–3886.

Furie, B., and Furie, B. C. 2008. Mechanisms of thrombus formation. *New Engl. J. Med.* 359:938–949.

Orfeo, T., Brufatto, N., Nesheim, M. E., Xu, H., Butenas, S., and Mann, K. G. 2004. The factor V activation paradox. *J. Biol. Chem.* 279:19580–19591.

Mann, K. G. 2003. Thrombin formation. *Chest* 124:4S–10S.

Rose, T., and Di Cera, E. 2002. Three-dimensional modeling of thrombin–fibrinogen interaction. *J. Biol. Chem.* 277:18875–18880.

Krem, M. M., and Di Cera, E. 2002. Evolution of cascades from embryonic development to blood coagulation. *Trends Biochem. Sci.* 27:67–74.

Fuentes-Prior, P., Iwanaga, Y., Huber, R., Pagila, R., Rumennik, G., Seto, M., Morser, J., Light, D. R., and Bode, W. 2000. Structural basis for the anticoagulant activity of the thrombin–thrombomodulin complex. *Nature* 404:518–525.

Lawn, R. M., and Vehar, G. A. 1986. The molecular genetics of hemophilia. *Sci. Am.* 254(3):48–65.

CHAPTER 11

Where to Start

Glycochemistry and glycobiology. A series of review articles. 2007. *Nature* 446:999–1051.

Maeder, T. 2002. Sweet medicines. *Sci. Am.* 287(1):40–47.

Freeze, H. H. 2013. Understanding human glycosylation disorders. *J. Biol. Chem.* 288:6936–6945.

Coutinho, M. F., Prata M., J., and Alves, S. 2012. Mannose-6-phosphate pathway: A review on its role in lysosomal function and dysfunction. *Mol. Gen. Metab.* 105:542–550.

Books

Varki, A., Cummings, R., Esko, J., Freeze, H., Stanley, P., Hart, G., et al. 2017. *Essentials of Glycobiology* (3rd ed.). Cold Spring Harbor Laboratory Press.

Stick, R. V., and Williams, S. 2008. *Carbohydrates: The Essential Molecules of Life* (2nd ed.). Elsevier Science.

Lindhorst, T. K. 2007. *Essentials of Carbohydrate Chemistry and Biochemistry* (3rd ed.). Wiley-VCH.

Taylor, M. E. 2006. *Introduction to Glycobiology* (2nd ed.). Oxford University Press.

Glycoproteins

Kim, S. Y., Zhao, J., Liu, X., Fraser, K., Lin, L., et al. 2017. Interaction of Zika Virus Envelope Protein with Glycosaminoglycans. *Biochemistry* 56:1151−1162.

Tran, D. T., and Hagen, K. G. T. 2013. Mucin-type O-glycosylation during development. *J. Biol. Chem.* 288:6921–6929.

Gill, D. J., Clausen H., and Bard, F. 2011. Location, location, location: New insights into O-GalNAc protein glycosylation. *Trends Cell Biol.* 21:149–158.

Foley, R. N. 2008. Erythropoietin: Physiology and molecular mechanisms. *Heart Failure Rev.* 13:404–414.

Fisher, J. W. 2003. Erythropoietin: Physiology and pharmacology update. *Exp. Biol. Med.* 228:1–14.

Cheetham, J. C., Smith, D. M., Aoki, K. H., Stevenson, J. L., Hoeffel, T. J., Syed, R. S., Egrie, J., and Harvey, T. S. 1998. NMR structure of human erythropoietin and a comparison with its receptor bound conformation. *Nat. Struct. Biol.* 5:861–866.

Hattrup, C. L., and Gendler, S. J. 2008. Structure and function of the cell surface (tethered) mucins. *Annu. Rev. Physiol.* 70:431–457.

Thorton, D. J., Rousseau, K., and McGuckin, M. A. 2008. Structure and function of mucins in airways mucus. *Annu. Rev. Physiol.* 70:459–486.

Rose, M. C., and Voynow, J. A. 2007. Respiratory tract mucin genes and mucin glycoproteins in health and disease. *Physiol. Rev.* 86:245–278.

Lamoureux, F., Baud'huin, M., Duplomb, L., Heymann, D., and Rédini, F. 2007. Proteoglycans: Key partners in bone cell biology. *Bioessays* 29:758–771.

Carraway, K. L., Funes, M., Workman, H. C., and Sweeney, C. 2007. Contribution of membrane mucins to tumor progression through modulation of cellular growth signaling pathways. *Curr. Top. Dev. Biol.* 78:1–22.

Yan, A., and Lennarz, W. J. 2005. Unraveling the mechanism of protein N-glycosylation. *J. Biol. Chem.* 280:3121–3124.

Pratta, M. A., Yao, W., Decicco, C., Tortorella, M., Liu, R.-Q., Copeland, R. A., Magolda, R., Newton, R. C., Trzaskos, J. M., and Arner, E. C. 2003. Aggrecan protects cartilage collagen from proteolytic cleavage. *J. Biol. Chem.* 278:45539–45545.

Glycosyltransferases

Willems, A. P., Gundogdu, M., Kempers, M. J. E., Giltay, J. C., Pfundt, R. et al. 2017. Mutations in N-acetylglucosamine (O-GlcNAc) transferase in patients with X-linked intellectual disability. *J. Biol. Chem.* 292:12621–12631.

Wells, L. 2013. The O-mannosylation pathway: Glycosyltransferases and proteins implicated in congenital muscular dystrophy. *J. Biol. Chem.* 288:6930–6935.

Vocadlo, D. J. 2012. O-GlcNAc processing enzymes: Catalytic mechanisms, substrate specificity, and enzyme regulation. *Curr. Opin. Chem. Biol.* 16:488–497.

Hurtado-Guerrero, R., and Davies, G. J. 2012. Recent structural and mechanistic insights into post-translational enzymatic glycosylation. *Curr. Opin. Chem. Biol.* 16:479–487.

Hart, G. W., Slawson, C., Ramirez-Correa, G., and Lagerlof, O. 2011. Cross talk between O-GlcNAcylation and phosphorylation: Roles in signaling, transcription, and chronic disease. *Annu. Rev. Biochem.* 80:825–858.

Lazarus, M. B., Nam, Y., Jiang, J., Sliz, P., and Walker, S. 2011. Structure of human O-GlcNAc transferase and its complex with a peptide substrate. *Nature* 469:564–569.

Lee, W.-S., Kang, C., Drayna, D., and Kornfeld, S. 2011. Analysis of mannose 6-phosphate uncovering enzyme mutations associated with persistent stuttering. *J. Biol. Chem.* 286:39786–39793.

Lairson, L. L., Henrissat, B., Davies, G. J., and Withers, S. G. 2008. Glycosyltransferases: Structures, functions and mechanisms. *Annu. Rev. Biochem.* 77:521–555.

Qasba, P. K., Ramakrishnan, B., and Boeggeman, E. 2005. Substrate-induced conformational changes in glycosyltransferases. *Trends Biochem. Sci.* 30:53–62.

Carbohydrate-Binding Proteins

Lin, A. E., Autran, C. A., Szyszka, A., Escajadillo, T., Huang, M. et al. 2017. Human milk oligosaccharides inhibit growth of group B *Streptococcus*. *J. Biol. Chem.* 292:11243–11249.

Gabius, H.-J., André, S., Jiménez-Barbero, J., Romero, A., and Solís, D. 2011. From lectin structure to functional glycomics: Principles of the sugar code. *Trends Biochem. Sci.* 36:298–313.

Wasserman, P. M. 2008. Zona pellucida glycoproteins. *J. Biol. Chem.* 283:24285–24289.

Sharon, N. 2008. Lectins: Past, present and future. *Biochem. Soc. Trans.* 36:1457–1460.

Balzarini, J. 2007. Targeting the glycans of glycoproteins: A novel paradigm for antiviral therapy. *Nat. Rev. Microbiol.* 5:583–597.

Sharon, N. 2007. Lectins: Carbohydrate-specific reagents and biological recognition molecules. *J. Biol. Chem.* 282:2753–2764.

Stevens, J., Blixt, O., Tumpey, T. M., Taubenberger, J. K., Paulson, J. C., and Wilson, I. A. 2006. Structure and receptor specificity of hemagglutinin from an H5N1 influenza virus. *Science* 312:404–409.

Cambi, A., Koopman, M., and Figdor, C. G. 2005. How C-type lectins detect pathogens. *Cell. Microbiol.* 7:481–488.

Clothia, C., and Jones, E. V. 1997. The molecular structure of cell adhesion molecules. *Annu. Rev. Biochem.* 66:823–862.

Bouckaert, J., Hamelryck, T., Wyns, L., and Loris, R. 1999. Novel structures of plant lectins and their complexes with carbohydrates. *Curr. Opin. Struct. Biol.* 9:572–577.

Weis, W. I., and Drickamer, K. 1996. Structural basis of lectin–carbohydrate recognition. *Annu. Rev. Biochem.* 65:441–473.

Carbohydrate Sequencing

Venkataraman, G., Shriver, Z., Raman, R., and Sasisekharan, R. 1999. Sequencing complex polysaccharides. *Science* 286:537–542.

Zhao, Y., Kent, S. B. H., and Chait, B. T. 1997. Rapid, sensitive structure analysis of oligosaccharides. *Proc. Natl. Acad. Sci. U.S.A.* 94:1629–1633.

Rudd, P. M., Guile, G. R., Küster, B., Harvey, D. J., Opdenakker, G., and Dwek, R. A. 1997. Oligosaccharide sequencing technology. *Nature* 388:205–207.

Industrial Uses of Polysaccharides

Li, L., Celiz, A. D., Yang, J., Yang, Q., Wamala, I. et al. 2017. Tough adhesives for diverse wet surfaces. *Science* 357:378–381.

Hamed, I., Özogul, F., and Joe M. Regenstein, J. M. 2016. Industrial applications of crustacean by-products (chitin, chitosan, and chitooligosaccharides): A review. *Trends Food Sci. Techno.* 48:40–50.

CHAPTER 12

Where to Start

De Weer, P. 2000. A century of thinking about cell membranes. *Annu. Rev. Physiol.* 62:919–926.

Bretscher, M. S. 1985. The molecules of the cell membrane. *Sci. Am.* 253(4):100–108.

Unwin, N., and Henderson, R. 1984. The structure of proteins in biological membranes. *Sci. Am.* 250(2):78–94.

Deisenhofer, J., and Michel, H. 1989. The photosynthetic reaction centre from the purple bacterium *Rhodopseudomonas viridis. EMBO J.* 8:2149–2170.

Singer, S. J., and Nicolson, G. L. 1972. The fluid mosaic model of the structure of cell membranes. *Science* 175:720–731.

Jacobson, K., Sheets, E. D., and Simson, R. 1995. Revisiting the fluid mosaic model of membranes. *Science* 268:1441–1442.

Books

Gennis, R. B. 1989. *Biomembranes: Molecular Structure and Function.* Springer Verlag.

Vance, D. E., and Vance, J. E. (Eds.). 2008. *Biochemistry of Lipids, Lipoproteins, and Membranes* (5th ed.). Elsevier.

Lipowsky, R., and Sackmann, E. 1995. *The Structure and Dynamics of Membranes.* Elsevier.

Racker, E. 1985. *Reconstitutions of Transporters, Receptors, and Pathological States.* Academic Press.

Tanford, C. 1980. *The Hydrophobic Effect: Formation of Micelles and Biological Membranes* (2nd ed.). Wiley-Interscience.

Membrane Lipids and Dynamics

Lingwood, D., and Simons, K. 2010. Lipid rafts as a membrane-organizing principle. *Science.* 327:46–50.

Pike, L. J. 2009. The challenge of lipid rafts. *J. Lipid Res.* 50:S323–S328.

Simons, K., and Vaz, W. L. 2004. Model systems, lipid rafts, and cell membranes. *Annu. Rev. Biophys. Biomol. Struct.* 33:269–295.

Anderson, T. G., and McConnell, H. M. 2002. A thermodynamic model for extended complexes of cholesterol and phospholipid. *Biophys. J.* 83:2039–2052.

Saxton, M. J., and Jacobson, K. 1997. Single-particle tracking: Applications to membrane dynamics. *Annu. Rev. Biophys. Biomol. Struct.* 26:373–399.

Bloom, M., Evans, E., and Mouritsen, O. G. 1991. Physical properties of the fluid lipid-bilayer component of cell membranes: A perspective. *Q. Rev. Biophys.* 24:293–397.

Elson, E. L. 1986. Membrane dynamics studied by fluorescence correlation spectroscopy and photobleaching recovery. *Soc. Gen. Physiol. Ser.* 40:367–383.

Zachowski, A., and Devaux, P. F. 1990. Transmembrane movements of lipids. *Experientia* 46:644–656.

Devaux, P. F. 1992. Protein involvement in transmembrane lipid asymmetry. *Annu. Rev. Biophys. Biomol. Struct.* 21:417–439.

Silvius, J. R. 1992. Solubilization and functional reconstitution of biomembrane components. *Annu. Rev. Biophys. Biomol. Struct.* 21:323–348.

Yeagle, P. L., Albert, A. D., Boesze-Battaglia, K., Young, J., and Frye, J. 1990. Cholesterol dynamics in membranes. *Biophys. J.* 57:413–424.

Nagle, J. F., and Tristram-Nagle, S. 2000. Lipid bilayer structure. *Curr. Opin. Struct. Biol.* 10:474–480.

Dowhan, W. 1997. Molecular basis for membrane phospholipid diversity: Why are there so many lipids? *Annu. Rev. Biochem.* 66:199–232.

Huijbregts, R. P. H., de Kroon, A. I. P. M., and de Kruijff, B. 1998. Rapid transmembrane movement of newly synthesized phosphatidylethanolamine across the inner membrane of *Escherichia coli. J. Biol. Chem.* 273:18936–18942.

Structure of Membrane Proteins

Walian, P., Cross, T. A., and Jap, B. K. 2004. Structural genomics of membrane proteins. *Genome Biol.* 5:215.

Werten, P. J., Remigy, H. W., de Groot, B. L., Fotiadis, D., Philippsen, A., Stahlberg, H., Grubmuller, H., and Engel, A. 2002. Progress in the analysis of membrane protein structure and function. *FEBS Lett.* 529:65–72.

Popot, J.-L., and Engleman, D. M. 2000. Helical membrane protein folding, stability and evolution. *Annu. Rev. Biochem.* 69:881–922.

White, S. H., and Wimley, W. C. 1999. Membrane protein folding and stability: Physical principles. *Annu. Rev. Biophys. Biomol. Struct.* 28:319–365.

Marassi, F. M., and Opella, S. J. 1998. NMR structural studies of membrane proteins. *Curr. Opin. Struct. Biol.* 8:640–648.

Lipowsky, R. 1991. The conformation of membranes. *Nature* 349:475–481.

Altenbach, C., Marti, T., Khorana, H. G., and Hubbell, W. L. 1990. Transmembrane protein structure: Spin labeling of bacteriorhodopsin mutants. *Science* 248:1088–1092.

Fasman, G. D., and Gilbert, W. A. 1990. The prediction of transmembrane protein sequences and their conformation: An evaluation. *Trends Biochem. Sci.* 15:89–92.

Jennings, M. L. 1989. Topography of membrane proteins. *Annu. Rev. Biochem.* 58:999–1027.

Engelman, D. M., Steitz, T. A., and Goldman, A. 1986. Identifying non-polar transbilayer helices in amino acid sequences of membrane proteins. *Annu. Rev. Biophys. Biophys. Chem.* 15:321–353.

Udenfriend, S., and Kodukola, K. 1995. How glycosyl-phosphatidyl-inositol-anchored membrane proteins are made. *Annu. Rev. Biochem.* 64:563–591.

Intracellular Membranes

Skehel, J. J., and Wiley, D. C. 2000. Receptor binding and membrane fusion in virus entry: The influenza hemagglutinin. *Annu. Rev. Biochem.* 69:531–569.

Roth, M. G. 1999. Lipid regulators of membrane traffic through the Golgi complex. *Trends Cell Biol.* 9:174–179.

Jahn, R., and Sudhof, T. C. 1999. Membrane fusion and exocytosis. *Annu. Rev. Biochem.* 68:863–911.

Stroud, R. M., and Walter, P. 1999. Signal sequence recognition and protein targeting. *Curr. Opin. Struct. Biol.* 9:754–759.

Teter, S. A., and Klionsky, D. J. 1999. How to get a folded protein across a membrane. *Trends Cell Biol.* 9:428–431.

Hettema, E. H., Distel, B., and Tabak, H. F. 1999. Import of proteins into peroxisomes. *Biochim. Biophys. Acta* 1451:17–34.

Membrane Fusion

Sollner, T. H., and Rothman, J. E. 1996. Molecular machinery mediating vesicle budding, docking and fusion. *Experientia* 52:1021–1025.

Ungar, D., and Hughson, F. M. 2003. SNARE protein structure and function. *Annu. Rev. Cell Dev. Biol.* 19:493–517.

Martens, S., and McMahon, H. T. 2008. Mechanisms of membrane fusion: Disparate players and common principles. *Nat. Rev. Mol. Cell Biol.* 9:543–556.

Biochemistry in Focus

Saric, A., Andreau, K., Armand, A.-S., Møller, I. M., and Petit, P. X. 2016. Barth syndrome: from mitochondrial dysfunctions associated with aberrant production of reactive oxygen species to pluripotent stem cell studies. *Front. Genet.* 6:359.

CHAPTER 13

Where to Start

Lancaster, C. R. 2004. Structural biology: Ion pump in the movies. *Nature* 432:286–287.

Unwin, N. 2003. Structure and action of the nicotinic acetylcholine receptor explored by electron microscopy. *FEBS Lett.* 555:91–95.

Abramson, J., Smirnova, I., Kasho, V., Verner, G., Iwata, S., and Kaback, H. R. 2003. The lactose permease of *Escherichia coli*: Overall structure, the sugar-binding site and the alternating access model for transport. *FEBS Lett.* 555:96–101.

Lienhard, G. E., Slot, J. W., James, D. E., and Mueckler, M. M. 1992. How cells absorb glucose. *Sci. Am.* 266(1):86–91.

King, L. S., Kozono, D., and Agre, P. 2004. From structure to disease: The evolving tale of aquaporin biology. *Nat. Rev. Mol. Cell Biol.* 5:687–698.

Neher, E., and Sakmann, B. 1992. The patch clamp technique. *Sci. Am.* 266(3):28–35.

Sakmann, B. 1992. Elementary steps in synaptic transmission revealed by currents through single ion channels. *Science* 256:503–512.

Books

Ashcroft, F. M. 2000. *Ion Channels and Disease*. Academic Press.

Conn, P. M. (Ed.). 1998. *Ion Channels*, vol. 293, *Methods in Enzymology*. Academic Press.

Aidley, D. J., and Stanfield, P. R. 1996. *Ion Channels: Molecules in Action*. Cambridge University Press.

Hille, B. 2001. *Ionic Channels of Excitable Membranes* (3rd ed.). Sinauer.

Läuger, P. 1991. *Electrogenic Ion Pumps*. Sinauer.

Stein, W. D. 1990. *Channels, Carriers, and Pumps: An Introduction to Membrane Transport*. Academic Press.

Hodgkin, A. 1992. *Chance and Design: Reminiscences of Science in Peace and War*. Cambridge University Press.

P-Type ATPases

Sorensen, T. L., Moller, J. V., and Nissen, P. 2004. Phosphoryl transfer and calcium ion occlusion in the calcium pump. *Science* 304:1672–1675.

Sweadner, K. J., and Donnet, C. 2001. Structural similarities of Na, K-ATPase and SERCA, the Ca^{2+}-ATPase of the sarcoplasmic reticulum. *Biochem. J.* 356:685–704.

Toyoshima, C., and Mizutani, T. 2004. Crystal structure of the calcium pump with a bound ATP analogue. *Nature* 430:529–535.

Toyoshima, C., Nakasako, M., Nomura, H., and Ogawa, H. 2000. Crystal structure of the calcium pump of sarcoplasmic reticulum at 2.6 Å resolution. *Nature* 405:647–655.

Auer, M., Scarborough, G. A., and Kuhlbrandt, W. 1998. Three-dimensional map of the plasma membrane H^+-ATPase in the open conformation. *Nature* 392:840–843.

Axelsen, K. B., and Palmgren, M. G. 1998. Evolution of substrate specificities in the P-type ATPase superfamily. *J. Mol. Evol.* 46:84–101.

Pedersen, P. A., Jorgensen, J. R., and Jorgensen, P. L. 2000. Importance of conserved α-subunit segment [709]GDGVND for Mg^{2+} binding, phosphorylation, energy transduction in Na, K-ATPase. *J. Biol. Chem.* 275:37588–37595.

Blanco, G., and Mercer, R. W. 1998. Isozymes of the Na-K-ATPase: Heterogeneity in structure, diversity in function. *Am. J. Physiol.* 275:F633–F650.

Estes, J. W., and White, P. D. 1965. William Withering and the purple foxglove. *Sci. Am.* 212(6):110–117.

ATP-Binding Cassette Proteins

Locher, K. P. 2009. Structure and mechanism of ATP-binding cassette transporters. *Phil. Trans. R. Soc. B* 364:239–245.

Rees, D. C., Johnson, E., and Lewinson, O. 2009. ABC transporters: The power to change. *Nat. Rev. Mol. Cell Biol.* 10:218–227.

Ward, A., Reyes, C. L., Yu, J., Roth, C. B., and Chang, G. 2007. Flexibility in the ABC transporter MsbA: Alternating access with a twist. *Proc. Natl. Acad. Sci. U.S.A.* 104:19005–19010.

Locher, K. P., Lee, A. T., and Rees, D. C. 2002. The *E. coli* BtuCD structure: A framework for ABC transporter architecture and mechanism. *Science* 296:1091–1098.

Borths, E. L., Locher, K. P., Lee, A. T., and Rees, D. C. 2002. The structure of *Escherichia coli* BtuF and binding to its cognate ATP binding cassette transporter. *Proc. Natl. Acad. Sci. U.S.A.* 99:16642–16647.

Dong, J., Yang, G., and McHaourab, H. S. 2005. Structural basis of energy transduction in the transport cycle of MsbA. *Science* 308:1023–1028.

Akabas, M. H. 2000. Cystic fibrosis transmembrane conductance regulator: Structure and function of an epithelial chloride channel. *J. Biol. Chem.* 275:3729–3732.

Chen, J., Sharma, S., Quiocho, F. A., and Davidson, A. L. 2001. Trapping the transition state of an ATP-binding cassette transporter: Evidence for a concerted mechanism of maltose transport. *Proc. Natl. Acad. Sci. U.S.A.* 98:1525–1530.

Sheppard, D. N., and Welsh, M. J. 1999. Structure and function of the CFTR chloride channel. *Physiol. Rev.* 79:S23–S45.

Chen, Y., and Simon, S. M. 2000. In situ biochemical demonstration that P-glycoprotein is a drug efflux pump with broad specificity. *J. Cell Biol.* 148:863–870.

Saier, M. H., Jr., Paulsen, I. T., Sliwinski, M. K., Pao, S. S., Skurray, R. A., and Nikaido, H. 1998. Evolutionary origins of multidrug and drug-specific efflux pumps in bacteria. *FASEB J.* 12:265–274.

Symporters and Antiporters

Kaback, H. R. 2013. A chemiosmotic mechanism of symport. *Proc. Natl. Acad. Sci. USA* 112:1259–1264.

Abramson, J., Smirnova, I., Kasho, V., Verner, G., Kaback, H. R., and Iwata, S. 2003. Structure and mechanism of the lactose permease of *Escherichia coli*. *Science* 301:610–615.

Philipson, K. D., and Nicoll, D. A. 2000. Sodium-calcium exchange: A molecular perspective. *Annu. Rev. Physiol.* 62:111–133.

Pao, S. S., Paulsen, I. T., and Saier, M. H., Jr. 1998. Major facilitator superfamily. *Microbiol. Mol. Biol. Rev.* 62:1–34.

Wright, E. M., Hirsch, J. R., Loo, D. D., and Zampighi, G. A. 1997. Regulation of Na^+/glucose cotransporters. *J. Exp. Biol.* 200:287–293.

Kaback, H. R., Bibi, E., and Roepe, P. D. 1990. β-Galactoside transport in *E. coli*: A functional dissection of lac permease. *Trends Biochem. Sci.* 8:309–314.

Hilgemann, D. W., Nicoll, D. A., and Philipson, K. D. 1991. Charge movement during Na^+ translocation by native and cloned cardiac Na^+/Ca^{2+} exchanger. *Nature* 352:715–718.

Hediger, M. A., Turk, E., and Wright, E. M. 1989. Homology of the human intestinal Na^+/glucose and *Escherichia coli* Na^+/proline cotransporters. *Proc. Natl. Acad. Sci. U.S.A.* 86:5748–5752.

Ion Channels

Zhou, Y., and MacKinnon, R. 2003. The occupancy of ions in the K1 selectivity filter: Charge balance and coupling of ion binding to a protein conformational change underlie high conduction rates. *J. Mol. Biol.* 333:965–975.

Zhou, Y., Morais-Cabral, J. H., Kaufman, A., and MacKinnon, R. 2001. Chemistry of ion coordination and hydration revealed by a K^+ channel-Fab complex at 2.0 Å resolution. *Nature* 414:43–48.

Jiang, Y., Lee, A., Chen, J., Cadene, M., Chait, B. T., and MacKinnon, R. 2002. The open pore conformation of potassium channels. *Nature* 417:523–526.

Jiang, Y., Lee, A., Chen, J., Ruta, V., Cadene, M., Chait, B. T., and MacKinnon, R. 2003. X-ray structure of a voltage-dependent K^+ channel. *Nature* 423:33–41.

Jiang, Y., Ruta, V., Chen, J., Lee, A., and MacKinnon, R. 2003. The principle of gating charge movement in a voltage-dependent K^+ channel. *Nature* 423:42–48.

Mackinnon, R. 2004. Structural biology: Voltage sensor meets lipid membrane. *Science* 306:1304–1305.

Noskov, S. Y., Bernèche, S., and Roux, B. 2004. Control of ion selectivity in potassium channels by electrostatic and dynamic properties of carbonyl ligands. *Nature* 431:830–834.

Bezanilla, F. 2000. The voltage sensor in voltage-dependent ion channels. *Physiol. Rev.* 80:555–592.

Shieh, C.-C., Coghlan, M., Sullivan, J. P., and Gopalakrishnan, M. 2000. Potassium channels: Molecular defects, diseases, and therapeutic opportunities. *Pharmacol. Rev.* 52:557–594.

Horn, R. 2000. Conversation between voltage sensors and gates of ion channels. *Biochemistry* 39:15653–15658.

Perozo, E., Cortes, D. M., and Cuello, L. G. 1999. Structural rearrangements underlying K^+-channel activation gating. *Science* 285:73–78.

Doyle, D. A., Morais Cabral, J., Pfuetzner, R. A., Kuo, A., Gulbis, J. M., Cohen, S. L., Chait, B. T., and MacKinnon R. 1998. The structure of the potassium channel: Molecular basis of K^+ conduction and selectivity. *Science* 280:69–77.

Marban, E., Yamagishi, T., and Tomaselli, G. F. 1998. Structure and function of the voltage-gated Na^+ channel. *J. Physiol.* 508:647–657.

Miller, R. J. 1992. Voltage-sensitive Ca^{2+} channels. *J. Biol. Chem.* 267:1403–1406.

Stephens, R. F., Guan, W., Zhorov, B. S., and Spafford, J. D. 2015. Selectivity filters and cysteine-rich extracellular loops in voltage-gated sodium, calcium, and NALCN channels. *Front. Physiol.* 6:153.

Catterall, W. A. 1991. Excitation-contraction coupling in vertebrate skeletal muscle: A tale of two calcium channels. *Cell* 64:871–874.

Ligand-Gated Ion Channels

Unwin, N. 2005. Refined structure of the nicotinic acetylcholine receptor at 4 Å resolution. *J. Mol. Biol.* 346:967–989.

Unwin, N., and Fujiyoshi, Y. 2012. Gating movement of acetylcholine receptor caught by plunge-freezing. *J. Mol. Biol.* 422:617–634.

Miyazawa, A., Fujiyoshi, Y., Stowell, M., and Unwin, N. 1999. Nicotinic acetylcholine receptor at 4.6 Å resolution: Transverse tunnels in the channel wall. *J. Mol. Biol.* 288:765–786.

Jiang, Y., Lee, A., Chen, J., Cadene, M., Chait, B. T., and MacKinnon, R. 2002. Crystal structure and mechanism of a calcium-gated potassium channel. *Nature* 417:515–522.

Barrantes, F. J., Antollini, S. S., Blanton, M. P., and Prieto, M. 2000. Topography of the nicotinic acetylcholine receptor membrane-embedded domains. *J. Biol. Chem.* 275:37333–37339.

Cordero-Erausquin, M., Marubio, L. M., Klink, R., and Changeux, J. P. 2000. Nicotinic receptor function: New perspectives from knockout mice. *Trends Pharmacol. Sci.* 21:211–217.

Le Novère, N., and Changeux, J. P. 1995. Molecular evolution of the nicotinic acetylcholine receptor: An example of multigene family in excitable cells. *J. Mol. Evol.* 40:155–172.

Kunishima, N., Shimada, Y., Tsuji, Y., Sato, T., Yamamoto, M., Kumasaka, T., Nakanishi, S., Jingami, H., and Morikawa, K. 2000. Structural basis of glutamate recognition by dimeric metabotropic glutamate receptor. *Nature* 407:971–978.

Betz, H., Kuhse, J., Schmieden, V., Laube, B., Kirsch, J., and Harvey, R. J. 1999. Structure and functions of inhibitory and excitatory glycine receptors. *Ann. N. Y. Acad. Sci.* 868:667–676.

Unwin, N. 1995. Acetylcholine receptor channel imaged in the open state. *Nature* 373:37–43.

Colquhoun, D., and Sakmann, B. 1981. Fluctuations in the microsecond time range of the current through single acetylcholine receptor ion channels. *Nature* 294:464–466.

Long QT Syndrome and hERG

Saenen, J. B., and Vrints, C. J. 2008. Molecular aspects of the congenital and acquired Long QT Syndrome: Clinical implications. *J. Mol. Cell. Cardiol.* 44:633–646.

Zaręba, W. 2007. Drug induced QT prolongation. *Cardiol. J.* 14:523–533.

Fernandez, D., Ghanta, A., Kauffman, G. W., and Sanguinetti, M. C. 2004. Physicochemical features of the hERG channel drug binding site. *J. Biol. Chem.* 279:10120–10127.

Mitcheson, J. S., Chen, J., Lin, M., Culberson, C., and Sanguinetti, M. C. 2000. A structural basis for drug-induced long QT syndrome. *Proc. Natl. Acad. Sci. U.S.A.* 97:12329–12333.

Gap Junctions

Maeda, S., Nakagawa, S., Suga, M., Yamashita, E., Oshima, A., Fujiyoshi, Y., and Tsukihara, T. 2009. Structure of the connexin 26 gap junction channel at 3.5 Å resolution. *Nature* 458:597–604.

Saez, J. C., Berthoud, V. M., Branes, M. C., Martinez, A. D., and Beyer, E. C. 2003. Plasma membrane channels formed by connexins: Their regulation and functions. *Physiol. Rev.* 83:1359–1400.

Revilla, A., Bennett, M. V. L., and Barrio, L. C. 2000. Molecular determinants of membrane potential dependence in vertebrate gap junction channels. *Proc. Natl. Acad. Sci. U.S.A.* 97:14760–14765.

Unger, V. M., Kumar, N. M., Gilula, N. B., and Yeager, M. 1999. Three-dimensional structure of a recombinant gap junction membrane channel. *Science* 283:1176–1180.

Simon, A. M. 1999. Gap junctions: More roles and new structural data. *Trends Cell Biol.* 9:169–170.

Beltramello, M., Piazza, V., Bukauskas, F. F., Pozzan, T., and Mammano, F. 2005. Impaired permeability to Ins(1,4,5) P3 in a mutant connexin underlies recessive hereditary deafness. *Nat. Cell Biol.* 7:63–69.

White, T. W., and Paul, D. L. 1999. Genetic diseases and gene knockouts reveal diverse connexin functions. *Annu. Rev. Physiol.* 61:283–310.

Water Channels

Agre, P., King, L. S., Yasui, M., Guggino, W. B., Ottersen, O. P., Fujiyoshi, Y., Engel, A., and Nielsen, S. 2002. Aquaporin water channels: From atomic structure to clinical medicine. *J. Physiol.* 542:3–16.

Agre, P., and Kozono, D. 2003. Aquaporin water channels: Molecular mechanisms for human diseases. *FEBS Lett.* 555:72–78.

de Groot, B. L., Engel, A., and Grubmuller, H. 2003. The structure of the aquaporin-1 water channel: A comparison between cryo-electron microscopy and X-ray crystallography. *J. Mol. Biol.* 325:485–493.

Biochemistry in Focus

Sartiani, L., Mannaioni, G., Masi, A., Romanelli, M. N., and Cerbai, E. 2017. The hyperpolarization-activated cyclic nucleotide-gated channels: from biophysics to pharmacology of a unique family of ion channels. *Physiol. Rev.* 69:354–395.

Milanese, R., Baruscotti, M., Gnecchi-Ruscone, T., and DiFrancesco, D. 2006. Familial sinus bradycardia associated with a mutation in the cardiac pacemaker channel. *New Eng. J. Med.* 354:151–157.

Ludwig, A., Zong, X., Stieber, J., Hullin, R., Hofman, F., and Biel, M. 1999. Two pacemaker channels from human heart with profoundly different activation kinetics. *EMBO J.* 18:2323–2329.

CHAPTER 14

Where to Start

Scott, J. D., and Pawson, T. 2000. Cell communication: The inside story. *Sci. Am.* 282(6):7279.

Pawson, T. 1995. Protein modules and signalling networks. *Nature* 373:573–580.

Okada, T., Ernst, O. P., Palczewski, K., and Hofmann, K. P. 2001. Activation of rhodopsin: New insights from structural and biochemical studies. *Trends Biochem. Sci.* 26:318–324.

Tsien, R. Y. 1992. Intracellular signal transduction in four dimensions: From molecular design to physiology. *Am. J. Physiol.* 263: C723–C728.

Loewenstein, W. R. 1999. *Touchstone of Life: Molecular Information, Cell Communication, and the Foundations of Life.* Oxford University Press.

G Proteins and 7TM Receptors

Palczewski, K., Kumasaka, T., Hori, T., Behnke, C. A., Motoshima, H., Fox, B. A., Le Trong, I., Teller, D. C., Okada, T., Stenkamp, R. E., et al. 2000. Crystal structure of rhodopsin: A G protein-coupled receptor. *Science* 289:739–745.

Rasmussen, S. G. F., Choi, H.-J., Rosenbaum, D. M., Kobilka, T. S., Thian, F. S., Edwards, P. C., Burghammer,

M., Ratnala, V. R. P., Sanishvili, R., Fischetti, R. F., et al. 2007. Crystal structure of the human β₂ adrenergic G-protein-coupled receptor. *Nature* 450:383–387.

Rosenbaum, D. M., Cherezov, V., Hanson, M. A., Rasmussen, S. G. F., Thian, F. S., Kobilka, T. S., Choi, H.-J., Yao, X.-J., Weis, W. I., Stevens, R. C., et al. 2007. GPCR engineering yields high-resolution structural insights into β₂-adrenergic receptor function. *Science* 318:1266–1273.

Rasmussen, S. G. F., DeVree, B. T., Zou, Y., Kruse, A. C., Chung, K. Y., Kobilka, T. S., Thian, F. S., Chae, P. S., Pardon, E., Calinski, D., et al. 2011. Crystal structure of the β₂ adrenergic receptor–Gs protein complex. *Nature* 477:549–555.

Lefkowitz, R. J. 2000. The superfamily of heptahelical receptors. *Nat. Cell Biol.* 2:E133–E136.

Audet, M., and Bouvier, M. 2012. Restructuring G-protein-coupled receptor activation. *Cell* 151:14–22.

Bourne, H. R., Sanders, D. A., and McCormick, F. 1991. The GTPase superfamily: Conserved structure and molecular mechanism. *Nature* 349:117–127.

Lambright, D. G., Noel, J. P., Hamm, H. E., and Sigler, P. B. 1994. Structural determinants for activation of the α-subunit of a heterotrimeric G protein. *Nature* 369:621–628.

Noel, J. P., Hamm, H. E., and Sigler, P. B. 1993. The 2.2 Å crystal structure of transducin-α complexed with GTPγS. *Nature* 366:654–663.

Sondek, J., Lambright, D. G., Noel, J. P., Hamm, H. E., and Sigler, P. B. 1994. GTPase mechanism of G proteins from the 1.7-Å crystal structure of transducin α-GDP-AlF₄⁻. *Nature* 372:276–279.

Sondek, J., Bohm, A., Lambright, D. G., Hamm, H. E., and Sigler, P. B. 1996. Crystal structure of a G-protein βγ dimer at 2.1 Å resolution. *Nature* 379:369–374.

Wedegaertner, P. B., Wilson, P. T., and Bourne, H. R. 1995. Lipid modifications of trimeric G proteins. *J. Biol. Chem.* 270:503–506.

Farfel, Z., Bourne, H. R., and Iiri, T. 1999. The expanding spectrum of G protein diseases. *New Engl. J. Med.* 340:1012–1020.

Bockaert, J., and Pin, J. P. 1999. Molecular tinkering of G protein-coupled receptors: An evolutionary success. *EMBO J.* 18:1723–1729.

Cyclic AMP Cascade

Hurley, J. H. 1999. Structure, mechanism, and regulation of mammalian adenylyl cyclase. *J. Biol. Chem.* 274:7599–7602.

Kim, C., Xuong, N. H., and Taylor, S. S. 2005. Crystal structure of a complex between the catalytic and regulatory (RI) subunits of PKA. *Science* 307:690–696.

Tesmer, J. J., Sunahara, R. K., Gilman, A. G., and Sprang, S. R. 1997. Crystal structure of the catalytic domains of adenylyl cyclase in a complex with Gₛα-GTPγS. *Science* 278:1907–1916.

Smith, C. M., Radzio-Andzelm, E., Madhusudan, Akamine, P., and Taylor, S. S. 1999. The catalytic subunit of cAMP-dependent protein kinase: Prototype for an extended network of communication. *Prog. Biophys. Mol. Biol.* 71:313–341.

Taylor, S. S., Buechler, J. A., and Yonemoto, W. 1990. cAMP-dependent protein kinase: Framework for a diverse family of regulatory enzymes. *Annu. Rev. Biochem.* 59:971–1005.

Phosphoinositide Cascade

Berridge, M. J., and Irvine, R. F. 1989. Inositol phosphates and cell signalling. *Nature* 341:197–205.

Berridge, M. J. 1993. Inositol trisphosphate and calcium signalling. *Nature* 361:315–325.

Essen, L. O., Perisic, O., Cheung, R., Katan, M., and Williams, R. L. 1996. Crystal structure of a mammalian phosphoinositide-specific phospholipase C δ. *Nature* 380:595–602.

Ferguson, K. M., Lemmon, M. A., Schlessinger, J., and Sigler, P. B. 1995. Structure of the high affinity complex of inositol trisphosphate with a phospholipase C pleckstrin homology domain. *Cell* 83:1037–1046.

Baraldi, E., Carugo, K. D., Hyvonen, M., Surdo, P. L., Riley, A. M., Potter, B. V., O'Brien, R., Ladbury, J. E., and Saraste, M. 1999. Structure of the PH domain from Bruton's tyrosine kinase in complex with inositol 1,3,4,5-tetrakisphosphate. *Struct. Fold. Design* 7:449–460.

Waldo, G. L., Ricks, T. K., Hicks, S. N., Cheever, M. L., Kawano, T., Tsuboi, K., Wang, X., Montell, C., Kozasa, T., Sondek, J., et al. 2010. Kinetic scaffolding mediated by a phospholipase C-β and Gq signaling complex. *Science* 330:974–980.

Calcium

Ikura, M., Clore, G. M., Gronenborn, A. M., Zhu, G., Klee, C. B., and Bax, A. 1992. Solution structure of a calmodulin-target peptide complex by multidimensional NMR. *Science* 256:632–638.

Kuboniwa, H., Tjandra, N., Grzesiek, S., Ren, H., Klee, C. B., and Bax, A. 1995. Solution structure of calcium-free calmodulin. *Nat. Struct. Biol.* 2:768–776.

Grynkiewicz, G., Poenie, M., and Tsien, R. Y. 1985. A new generation of Ca²⁺ indicators with greatly improved fluorescence properties. *J. Biol. Chem.* 260:3440–3450.

Kerr, R., Lev-Ram, V., Baird, G., Vincent, P., Tsien, R. Y., and Schafer, W. R. 2000. Optical imaging of calcium transients in neurons and pharyngeal muscle of *C. elegans*. *Neuron* 26:583–594.

Chin, D., and Means, A. R. 2000. Calmodulin: A prototypical calcium sensor. *Trends Cell Biol.* 10:322–328.

Dawson, A. P. 1997. Calcium signalling: How do IP3 receptors work? *Curr. Biol.* 7:R544–R547.

Protein Kinases, Including Receptor Tyrosine Kinases

Riedel, H., Dull, T. J., Honegger, A. M., Schlessinger, J., and Ullrich, A. 1989. Cytoplasmic domains determine signal specificity, cellular routing characteristics and influence ligand binding of epidermal growth factor and insulin receptors. *EMBO J.* 8:2943–2954.

Taylor, S. S., Knighton, D. R., Zheng, J., Sowadski, J. M., Gibbs, C. S., and Zoller, M. J. 1993. A template for the protein kinase family. *Trends Biochem. Sci.* 18:84–89.

Sicheri, F., Moarefi, I., and Kuriyan, J. 1997. Crystal structure of the Src family tyrosine kinase Hck. *Nature* 385:602–609.

Waksman, G., Shoelson, S. E., Pant, N., Cowburn, D., and Kuriyan, J. 1993. Binding of a high affinity phosphotyrosyl peptide to the Src SH2 domain: Crystal structures of the complexed and peptide-free forms. *Cell* 72:779–790.

Schlessinger, J. 2000. Cell signaling by receptor tyrosine kinases. *Cell* 103:211–225.

Simon, M. A. 2000. Receptor tyrosine kinases: Specific outcomes from general signals. *Cell* 103:13–15.

Robinson, D. R., Wu, Y. M., and Lin, S. F. 2000. The protein tyrosine kinase family of the human genome. *Oncogene* 19:5548–5557.

Hubbard, S. R. 1999. Structural analysis of receptor tyrosine kinases. *Prog. Biophys. Mol. Biol.* 71:343–358.

Carter-Su, C., and Smit, L. S. 1998. Signaling via JAK tyrosine kinases: Growth hormone receptor as a model system. *Recent Prog. Horm. Res.* 53:61–82.

Insulin Signaling Pathway

Khan, A. H., and Pessin, J. E. 2002. Insulin regulation of glucose uptake: A complex interplay of intracellular signalling pathways. *Diabetologia* 45:1475–1483.

Bevan, P. 2001. Insulin signalling. *J. Cell Sci.* 114:1429–1430.

De Meyts, P., and Whittaker, J. 2002. Structural biology of insulin and IGF1 receptors: Implications for drug design. *Nat. Rev. Drug Discov.* 1:769–783.

Dhe-Paganon, S., Ottinger, E. A., Nolte, R. T., Eck, M. J., and Shoelson, S. E. 1999. Crystal structure of the pleckstrin homology-phosphotyrosine binding (PH-PTB) targeting region of insulin receptor substrate 1. *Proc. Natl. Acad. Sci. U.S.A.* 96:8378–8383.

Domin, J., and Waterfield, M. D. 1997. Using structure to define the function of phosphoinositide 3-kinase family members. *FEBS Lett.* 410:91–95.

Hubbard, S. R. 1997. Crystal structure of the activated insulin receptor tyrosine kinase in complex with peptide substrate and ATP analog. *EMBO J.* 16:5572–5581.

Hubbard, S. R., Wei, L., Ellis, L., and Hendrickson, W. A. 1994. Crystal structure of the tyrosine kinase domain of the human insulin receptor. *Nature* 372:746–754.

EGF Signaling Pathway

Burgess, A. W., Cho, H. S., Eigenbrot, C., Ferguson, K. M., Garrett, T. P., Leahy, D. J., Lemmon, M. A., Sliwkowski, M. X., Ward, C. W., and Yokoyama, S. 2003. An open-and-shut case? Recent insights into the activation of EGF/ErbB receptors. *Mol. Cell* 12:541–552.

Cho, H. S., Mason, K., Ramyar, K. X., Stanley, A. M., Gabelli, S. B., Denney, D. W., Jr., and Leahy, D. J. 2003.

Structure of the extracellular region of HER2 alone and in complex with the Herceptin Fab. *Nature* 421:756–760.

Chong, H., Vikis, H. G., and Guan, K. L. 2003. Mechanisms of regulating the Raf kinase family. *Cell. Signal.* 15:463–469.

Stamos, J., Sliwkowski, M. X., and Eigenbrot, C. 2002. Structure of the epidermal growth factor receptor kinase domain alone and in complex with a 4-anilinoquinazoline inhibitor. *J. Biol. Chem.* 277:46265–46272.

Ras

Milburn, M. V., Tong, L., deVos, A. M., Brunger, A., Yamaizumi, Z., Nishimura, S., and Kim, S. H. 1990. Molecular switch for signal transduction: Structural differences between active and inactive forms of protooncogenic Ras proteins. *Science* 247:939–945.

Boriack-Sjodin, P. A., Margarit, S. M., Bar-Sagi, D., and Kuriyan, J. 1998. The structural basis of the activation of Ras by Sos. *Nature* 394:337–343.

Maignan, S., Guilloteau, J. P., Fromage, N., Arnoux, B., Becquart, J., and Ducruix, A. 1995. Crystal structure of the mammalian Grb2 adaptor. *Science* 268:291–293.

Takai, Y., Sasaki, T., and Matozaki, T. 2001. Small GTP-binding proteins. *Physiol. Rev.* 81:153–208.

Cancer

Druker, B. J., Sawyers, C. L., Kantarjian, H., Resta, D. J., Reese, S. F., Ford, J. M., Capdeville, R., and Talpaz, M. 2001. Activity of a specific inhibitor of the BCR-ABL tyrosine kinase in the blast crisis of chronic myeloid leukemia and acute lymphoblastic leukemia with the Philadelphia chromosome. *New Engl. J. Med.* 344:1038–1042.

Vogelstein, B., and Kinzler, K. W. 1993. The multistep nature of cancer. *Trends Genet.* 9:138–141.

Ellis, C. A., and Clark, G. 2000. The importance of being K-Ras. *Cell. Signal.* 12:425–434.

Hanahan, D., and Weinberg, R. A. 2000. The hallmarks of cancer. *Cell* 100:57–70.

McCormick, F. 1999. Signalling networks that cause cancer. *Trends Cell Biol.* 9:M53–M56.

Biochemistry in Focus

Mustafa, A. K., Gadella, M. M., and Snyder, S. H. 2009. Signaling by gasotransmitters. *Sci. Signaling* 2:re2.

Kolluru, G. K., Shen, X., Yuan, S., and Kevil, C. G. 2017. Gasotransmitter heterocellular signaling. *Antiox. Redox Signal.* 26:936–960.

CHAPTER 15

Where to Start

Liu, X., and Locasale, J. W. 2017. Metabolomics: A primer. *Trends Biochem. Sci.* 42:274–284.

Stipanuk, M. H., and Caudill, M. A. (Eds.). 2012. *Biochemical, Physiological, Molecular Aspects of Human Nutrition* (3rd ed.). Elsevier.

McGrane, M. M., Yun, J. S., Patel, Y. M., and Hanson, R. W. 1992. Metabolic control of gene expression: In vivo studies with transgenic mice. *Trends Biochem. Sci.* 17:40–44.

Westheimer, F. H. 1987. Why nature chose phosphates. *Science* 235:1173–1178.

Kamerlin, S. C. L., Sharma, P. K., Prasad, R. B., and Warshel, A. 2013. Why nature really chose phosphate. *Q. Rev. Biophys.* 46:1–132.

Books

Atkins, P., and de Paula, J. 2015. *Physical Chemistry for the Life Sciences* (3rd ed.). Oxford University Press.

Harold, F. M. 1986. *The Vital Force: A Study of Bioenergetics.* W. H. Freeman and Company.

Krebs, H. A., and Kornberg, H. L. 1957. *Energy Transformations in Living Matter.* Springer Verlag.

Nicholls, D. G., and Ferguson, S. J. 2013. *Bioenergetics* (4th ed.). Academic Press.

Frayn, K. N. 2010. *Metabolic Regulation: A Human Perspective* (3rd ed.). Wiley-Blackwell.

Fell, D. 1997. *Understanding the Control of Metabolism.* Portland Press.

Harris, D. A. 1995. *Bioenergetics at a Glance.* Blackwell Scientific.

Von Baeyer, H. C. 1999. *Warmth Disperses and Time Passes: A History of Heat.* Modern Library.

Thermodynamics

Alberty, R. A. 1993. Levels of thermodynamic treatment of biochemical reaction systems. *Biophys. J.* 65:1243–1254.

Alberty, R. A., and Goldberg, R. N. 1992. Standard thermodynamic formation properties for the adenosine $5'$-triphosphate series. *Biochemistry* 31:10610–10615.

Alberty, R. A. 1968. Effect of pH and metal ion concentration on the equilibrium hydrolysis of adenosine triphosphate to adenosine diphosphate. *J. Biol. Chem.* 243:1337–1343.

Goldberg, R. N. 1984. *Compiled Thermodynamic Data Sources for Aqueous and Biochemical Systems: An Annotated Bibliography (1930–1983).* National Bureau of Standards Special Publication 685, U.S. Government Printing Office.

Frey, P. A., and Arabshahi, A. 1995. Standard free energy change for the hydrolysis of the α,β-phosphoanhydride bridge in ATP. *Biochemistry* 34:11307–11310.

Bioenergetics and Metabolism

Patel, A., Malinovska, L., Saha, S., Wang, J., Alberti, S., Krishnan, Y., and Hyman, A. A. 2017. ATP as a biological hydrotrope. *Science* 356:753–756.

Schilling, C. H., Letscher, D., and Palsson, B. O. 2000. Theory for the systemic definition of metabolic pathways and their use in interpreting metabolic function from a pathway-oriented perspective. *J. Theor. Biol.* 203:229–248.

DeCoursey, T. E., and Cherny, V. V. 2000. Common themes and problems of bioenergetics and voltage-gated proton channels. *Biochim. Biophys. Acta* 1458:104–119.

Giersch, C. 2000. Mathematical modelling of metabolism. *Curr. Opin. Plant Biol.* 3:249–253.

Rees, D. C., and Howard, J. B. 1999. Structural bioenergetics and energy transduction mechanisms. *J. Mol. Biol.* 293:343–350.

Regulation of Metabolism

Schmitt, D. L., and An, S. 2017. Spatial organization of metabolic enzyme complexes in cells. *Biochemistry* 56:3184–3196.

Frederick, D. W., Loro, E., Liu, L., Davila, Jr., A., Chellappa, K., et al. 2016. Loss of NAD homeostasis leads to progressive and reversible degeneration of skeletal muscle. *Cell Metab.* 24:269–282.

Kemp, G. J. 2000. Studying metabolic regulation in human muscle. *Biochem. Soc. Trans.* 28:100–103.

Towle, H. C., Kaytor, E. N., and Shih, H. M. 1996. Metabolic regulation of hepatic gene expression. *Biochem. Soc. Trans.* 24:364–368.

Hofmeyr, J. H. 1995. Metabolic regulation: A control analytic perspective. *J. Bioenerg. Biomembr.* 27:479–490.

Historical Aspects

Kalckar, H. M. 1991. 50 years of biological research: From oxidative phosphorylation to energy requiring transport regulation. *Annu. Rev. Biochem.* 60:1–37.

Kalckar, H. M. (Ed.). 1969. *Biological Phosphorylations.* Prentice Hall.

Fruton, J. S. 1972. *Molecules and Life.* Wiley-Interscience.

Lipmann, F. 1971. *Wanderings of a Biochemist.* Wiley-Interscience.

CHAPTER 16

Where to Start

McCracken, A. N., and Edinger, A. L. 2013. Nutrient transporters: The Achilles' heel of anabolism. *Trends Endocrin. Met.* 24:200–208.

Curry, A. 2013. The milk revolution. *Nature* 500:20–22.

Bar-Even, A., Flamholz, A., Noor, E., and Milo, R. 2012. Rethinking glycolysis: On the biochemical logic of metabolic pathways. *Nature Chem. Biol.* 8:509–517.

Ward, P. S., and Thompson, C. B. 2012. Metabolic reprogramming: A cancer hallmark even Warburg did not anticipate. *Cancer Cell* 21:297–308.

Herling, A., König, M., Bulik, S., and Holzhütter, H.-G. 2011. Enzymatic features of the glucose metabolism in tumor cells. *FEBS J.* 278:2436–2459.

Lin, H. V., and Accili, D. 2011. Hormonal regulation of hepatic glucose production in health and disease. *Cell Metab.* 14:9–19.

Hirabayashi, J. 1996. On the origin of elementary hexoses. *Quart. Rev. Biol.* 71:365–380.

Books and Reviews

Tong, L. 2013. Structure and function of biotin-dependent carboxylases. *Cell. Mol. Life Sci.* 70:863–891.

Frayn, K. N. 2010. *Metabolic Regulation: A Human Perspective* (3rd ed.). Wiley-Blackwell.

Fell, D. 1997. *Understanding the Control of Metabolism.* Portland.

Fersht, A. 1999. *Structure and Mechanism in Protein Science: A Guide to Enzyme Catalysis and Protein Folding.* W. H. Freeman and Company.

Poortmans, J. R. (Ed.). 2004. *Principles of Exercise Biochemistry.* Krager.

Structure of Glycolytic and Gluconeogenic Enzymes

Lietzan, A. D., and St. Maurice, M. 2013. A substrate-induced biotin binding pocket in the carboxyltransferase domain of pyruvate carboxylase. *J. Biol. Chem.* 288:19915–19925.

Banaszak, L., Mechin, I., Obmolova, G., Oldham, M., Chang, S. H., Ruiz, T., Radermacher, M., Kopperschläger, G., and Rypniewski, W. 2011. The crystal structures of eukaryotic phosphofructokinases from baker's yeast and rabbit skeletal muscle. *J. Mol. Biol.* 407:284–297.

Lasso, G., Yu, L. P. C., Gil, D., Xiang, S., Tong, L., and Valle, M. 2010. Cryo-EM analysis reveals new insights into the mechanism of action of pyruvate carboxylase. *Structure* 18:1300–1310.

Ferreras, C., Hernández, E. D., Martínez-Costa, O. H., and Aragón, J. J. 2009. Subunit interactions and composition of the fructose 6-phosphate catalytic site and the fructose 2,6-bisphosphate allosteric site of mammalian phosphofructokinase. *J. Biol. Chem.* 284:9124–9131.

Hines, J. K., Chen, X., Nix, J. C., Fromm, H. J., and Honzatko. R. B. 2007. Structures of mammalian and bacterial fructose-1, 6- bisphosphatase reveal the basis for synergism in AMP/fructose-2, 6-bisphosphate inhibition. *J. Biol. Chem.* 282:36121–36131.

Ferreira-da-Silva, F., Pereira, P. J., Gales, L., Roessle, M., Svergun, D. I., Moradas-Ferreira, P., and Damas, A. M. 2006. The crystal and solution structures of glyceraldehyde-3-phosphate dehydrogenase reveal different quaternary structures. *J. Biol. Chem.* 281:33433–33440.

Kim, S.-G., Manes, N. P., El-Maghrabi, M. R., and Lee, Y.-H. 2006. Crystal structure of the hypoxia-inducible form of 6-phosphofructo-2-kinase/fructose-2,6-phosphatase (PFKFB3): A possible target for cancer therapy. *J. Biol. Chem.* 281:2939–2944.

Aleshin, A. E., Kirby, C., Liu, X., Bourenkov, G. P., Bartunik, H. D., Fromm, H. J., and Honzatko, R. B. 2000. Crystal structures of mutant monomeric hexokinase I reveal multiple ADP binding sites and conformational changes relevant to allosteric regulation. *J. Mol. Biol.* 296:1001–1015.

Bernstein, B. E., and Hol, W. G. 1998. Crystal structures of substrates and products bound to the phosphoglycerate kinase active site reveal the catalytic mechanism. *Biochemistry* 37:4429–4436.

Rigden, D. J., Alexeev, D., Phillips, S. E. V., and Fothergill-Gilmore, L. A. 1998. The 2.3 Å X-ray crystal structure of *S. cerevisiae* phosphoglycerate mutase. *J. Mol. Biol.* 276:449–459.

Zhang, E., Brewer, J. M., Minor, W., Carreira, L. A., and Lebioda, L. 1997. Mechanism of enolase: The crystal structure of asymmetric dimer enolase-2-phospho-D-glycerate/enolase-phosphoenolpyruvate at 2.0 Å resolution. *Biochemistry* 36:12526–12534.

Hasemann, C. A., Istvan E. S., Uyeda, K., and Deisenhofer, J. 1996. The crystal structure of the bifunctional enzyme 6-phosphofructo-2-kinase/fructose-2,6-biphosphatase reveals distinct domain homologies. *Structure* 4:1017–1029.

Tari, L. W., Matte, A., Pugazhenthi, U., Goldie, H., and Delbaere, L. T. J. 1996. Snapshot of an enzyme reaction intermediate in the structure of the ATP-Mg^{2+}-oxalate ternary complex of *Escherichia coli* PEP carboxykinase. *Nat. Struct. Biol.* 3:355–363.

Catalytic Mechanisms

Soukri, A., Mougin, A., Corbier, C., Wonacott, A., Branlant, C., and Branlant, G. 1989. Role of the histidine 176 residue in glyceraldehyde-3-phosphate dehydrogenase as probed by site-directed mutagenesis. *Biochemistry* 28:2586–2592.

Bash, P. A., Field, M. J., Davenport, R. C., Petsko, G. A., Ringe, D., and Karplus, M. 1991. Computer simulation and analysis of the reaction pathway of triosephosphate isomerase. *Biochemistry* 30:5826–5832.

Knowles, J. R., and Albery, W. J. 1977. Perfection in enzyme catalysis: The energetics of triosephosphate isomerase. *Acc. Chem. Res.* 10:105–111.

Regulation

Herman, M. A., and Samuel, V. T. 2016. The sweet path to metabolic demise: Fructose and lipid synthesis. *Trends Endo. Metab.* 27:719–730.

Schmitt, D. L., and An, S. 2017. Spatial organization of metabolic enzyme complexes in cells. *Biochemistry* 56:3184−3196.

Casey, A. K., and Miller, B. G. 2016. Kinetic basis of carbohydrate-mediated inhibition of human glucokinase by the glucokinase regulatory protein. *Biochemistry* 55:2899−2902.

Westerhold, L. E., Bridges, L. C., Shaikh, S. R., and Zeczycki, T. N. 2017. Kinetic and thermodynamic analysis of acetyl-CoA activation of *Staphylococcus Aureus* pyruvate carboxylase. *Biochemistry* 56:3492−3506.

Liu, S., Ammirati, M. J., Song, X., Knafels, J. D., Zhang, J., Greasley, S. E., Pfefferkorn, J. A., and Qiu, X. 2012. Insights into mechanism of glucokinase activation: Observation of multiple distinct protein conformations. *J. Biol. Chem.* 287:13598–13610.

Brüser, A., Kirchberger, J., Kloos, M., Sträter, N., and Schöneberg, T. 2012. Functional linkage of adenine nucleotide binding sites in mammalian muscle 6-phosphofructokinase. *J. Biol. Chem.* 287:17546–17553.

Anderka, O., Boyken, J., Aschenbach, U., Batzer, A., Boscheinen, O., and Schmoll, D. 2008. Biophysical characterization of the interaction between hepatic glucokinase and its regulatory protein: Impact of physiological and pharmacological effectors. *J. Biol. Chem.* 283:31333–31340.

Iancu, C. V., Mukund, S., Fromm, H. J., and Honzatko, R. B. 2005. R-state AMP complex reveals initial steps of the quaternary transition of fructose-1,6-bisphosphatase. *J. Biol. Chem.* 280:19737–19745.

Lee, Y. H., Li, Y., Uyeda, K., and Hasemann, C. A. 2003. Tissue-specific structure/function differentiation of the five isoforms of 6-phosphofructo-2-kinase/fructose-2,6-bisphosphatase. *J. Biol. Chem.* 278:523–530.

Gleeson, T. T. 1996. Post-exercise lactate metabolism: A comparative review of sites, pathways, and regulation. *Annu. Rev. Physiol.* 58:556–581.

Jitrapakdee, S., and Wallace, J. C. 1999. Structure, function and regulation of pyruvate carboxylase. *Biochem. J.* 340:1–16.

van de Werve, G., Lange, A., Newgard, C., Mechin, M. C., Li, Y., and Berteloot, A. 2000. New lessons in the regulation of glucose metabolism taught by the glucose 6-phosphatase system. *Eur. J. Biochem.* 267:1533–1549.

Sugar Transporters

Blodgett, D. M., Graybill, C., and Carruthers, A. 2008. Analysis of glucose transporter topology and structural dynamics. *J. Biol. Chem.* 283:36416–36424.

Huang, S., and Czech, M. P. 2007. The GLUT4 glucose transporter. *Cell Metab.* 5:237–252.

Czech, M. P., and Corvera, S. 1999. Signaling mechanisms that regulate glucose transport. *J Biol. Chem.* 274:1865–1868.

Silverman, M. 1991. Structure and function of hexose transporters. *Annu. Rev. Biochem.* 60:757–794.

Thorens, B., Charron, M. J., and Lodish, H. F. 1990. Molecular physiology of glucose transporters. *Diabetes Care* 13:209–218.

Glycolysis and Cancer

Pavlova, N. N., and Thompson, C. B. 2016. The emerging hallmarks of cancer metabolism. *Cell Metab.* 23:27–47.

Olson, K. A., Schell, J. C., and Rutter, J. 2016. Pyruvate and metabolic flexibility: Illuminating a path toward selective cancer therapies. *Trends Biochem. Sci.* 41:219–230.

Morgan, H. P., O'Reilly, F. J., Wear, M. A., O'Neill, J. R., Fothergill-Gilmore, L. A., Hupp, T., and Walkinshaw, M. D. 2013. M2 pyruvate kinase provides a mechanism for nutrient sensing and regulation of cell proliferation. *Proc. Natl. Acad. Sci. U.S.A.* 110:5881–5886.

Schulze, A., and Harris, A. L. 2012. How cancer metabolism is tuned for proliferation and vulnerable to disruption. *Nature* 491:364–373.

Lunt, S. Y., and Vander Heiden, M. G. 2011. Aerobic glycolysis: Meeting the metabolic requirements of cell proliferation. *Annu. Rev. Cell Dev. Biol.* 27:441–64.

Vander Heiden, M. G., Cantley, L. C., and Thompson, C. B. 2009. Understanding the Warburg effect: The metabolic requirements of cell proliferation. *Science* 324:1029–1033.

Mathupala, S. P., Ko, Y. H., and Pedersen, P. L. 2009. Hexokinase-2 bound to mitochondria: Cancer's stygian link to the "Warburg effect" and a pivotal target for effective therapy. *Sem. Cancer Biol.* 19:17–24.

Kroemer, G. K., and Pouyssegur, J. 2008. Tumor cell metabolism: Cancer's Achilles' heel. *Cancer Cell* 12:472–482.

Hsu, P. P., and Sabatini, D. M. 2008. Cancer cell metabolism: Warburg and beyond. *Cell* 134:703–707.

Genetic Diseases

Orosz, F., Oláh, J., and Ovádi, J. 2009. Triosephosphate isomerase deficiency: New insights into an enigmatic disease. *Biochim. Biophys. Acta* 1792:1168–1174.

Scriver, C. R., Beaudet, A. L., Valle, D., Sly, W. S., Childs, B., Kinzler, K., and Vogelstein, B. (Eds.). 2001. *The Metabolic and Molecular Basis of Inherited Disease* (8th ed.). McGraw-Hill.

Evolution

Dandekar, T., Schuster, S., Snel, B., Huynen, M., and Bork, P. 1999. Pathway alignment: Application to the comparative analysis of glycolytic enzymes. *Biochem. J.* 343:115–124.

Heinrich, R., Melendez-Hevia, E., Montero, F., Nuno, J. C., Stephani, A., and Waddell, T. G. 1999. The structural design of glycolysis: An evolutionary approach. *Biochem. Soc. Trans.* 27:294–298.

Walmsley, A. R., Barrett, M. P., Bringaud, F., and Gould, G. W. 1998. Sugar transporters from bacteria, parasites and mammals: Structure-activity relationships. *Trends Biochem. Sci.* 23:476–480.

Maes, D., Zeelen, J. P., Thanki, N., Beaucamp, N., Alvarez, M., Thi, M. H., Backmann, J., Martial, J. A., Wyns, L., Jaenicke, R., et al. 1999. The crystal structure of triosephosphate isomerase (TIM) from *Thermotoga maritima*: A comparative thermostability structural analysis of ten different TIM structures. *Proteins* 37:441–453.

Historical Aspects

Friedmann, H. C. 2004. From *Butyribacterium* to *E. coli*: An essay on unity in biochemistry. *Perspect. Biol. Med.* 47:47–66.

Fruton, J. S. 1999. *Proteins, Enzymes, Genes: The Interplay of Chemistry and Biology.* Yale University Press.

Kalckar, H. M. (Ed.). 1969. *Biological Phosphorylations: Development of Concepts.* Prentice Hall.

CHAPTER 17

Where to Start

Sugden, M. C., and Holness, M. J. 2003. Recent advances in mechanisms regulating glucose oxidation at the level of the pyruvate dehydrogenase complex by PDKs. *Am. J. Physiol. Endocrinol. Metab.* 284:E855–E862.

Owen, O. E., Kalhan, S. C., and Hanson, R. W. 2002. The key role of anaplerosis and cataplerosis for citric acid function. *J. Biol. Chem.* 277:30409–30412.

Pyruvate Dehydrogenase Complex

Patel, K. P., O'Brien, T. W., Subramony, S. H., Shuster, J., and Stacpoole, P. W. 2012. The spectrum of pyruvate dehydrogenase complex deficiency: Clinical, biochemical and genetic features in 371 patients. *Mol. Genet. Metab.* 105:34–43.

Vijayakrishnan, S., Callow, P., Nutley, M. A., Mcgow, D. P., Gilbert, D., Kropholler, P., Cooper, A., Byron, O., and Lindsay, J. G. 2011. Variation in the organization and subunit composition of the mammalian pyruvate dehydrogenase complex E2/E3BP core assembly. *Biochem. J.* 437:565–574.

Vijayakrishnan, S., Kelly, S. M., Gilbert, R. J., Callow, P., Bhella, D., Forsyth, T., Lindsay, J. G., and Byron, O. 2010. Solution structure and characterization of the human pyruvate dehydrogenase complex core assembly. *J. Mol. Biol.* 399:71–93.

Brautigam, C. A., Wynn, R. M., Chuang, J. L., and Chuang, D. T. 2009. Subunit and catalytic component stoichiometries of an in vitro reconstituted human pyruvate dehydrogenase complex. *J. Biol. Chem.* 284:13086–13098.

Lengyel, J. S., Stott, K. M., Wu, X., Brooks, B. R., Balbo, A., et al. 2008. Extended polypeptide linkers establish the spatial architecture of a pyruvate dehydrogenase multienzyme complex. *Structure* 16:93–103

Hiromasa, Y., Fujisawa, T., Aso, Y., and Roche, T. E. 2004. Organization of the cores of the mammalian pyruvate dehydrogenase complex formed by E2 and E2 plus the E3-binding proteins and their capacities to bind the E1 and E3 components. *J. Biol Chem.* 279:6921–6933.

Domingo, G. J., Chauhan, H. J., Lessard, I. A., Fuller, C., and Perham, R. N. 1999. Self-assembly and catalytic activity of the pyruvate dehydrogenase multienzyme complex from *Bacillus stearothermophilus*. *Eur. J. Biochem.* 266:1136–1146.

Structure of Citric Acid Cycle Enzymes

Fraser, M. E., Hayakawa, K., Hume, M. S., Ryan, D. G., and Brownie, E. R. 2006. Interactions of GTP with the ATP-grasp domain of GTP-specific succinyl-CoA synthetase. *J. Biol. Chem.* 281:11058–11065.

Yankovskaya, V., Horsefield, R., Törnroth, S., Luna-Chavez, C., Miyoshi, H., Léger, C., Byrne, B., Cecchini, G., and Iowata, S. 2003. Architecture of succinate dehydrogenase and reactive oxygen species generation. *Science* 299:700–704.

Fraser, M. E., James, M. N., Bridger, W. A., and Wolodko, W. T. 1999. A detailed structural description of *Escherichia coli* succinyl-CoA synthetase. *J. Mol. Biol.* 285:1633–1653. [Published erratum appears in May 7, 1999, issue of *J. Mol. Biol.* 288(3):501.]

Lloyd, S. J., Lauble, H., Prasad, G. S., and Stout, C. D. 1999. The mechanism of aconitase: 1.8 Å resolution crystal structure of the S642A:citrate complex. *Protein Sci.* 8:2655–2662.

Rose, I. A. 1998. How fumarase recycles after the malate → fumarate reaction: Insights into the reaction mechanism. *Biochemistry* 37:17651–17658.

Organization of the Citric Acid Cycle

Lambeth, D. O., Tews, K. N., Adkins, S., Frohlich, D., and Milavetz, B. I. 2004. Expression of two succinyl-CoA

specificities in mammalian tissues. *J. Biol. Chem.* 279:36621–36624.

Velot, C., Mixon, M. B., Teige, M., and Srere, P. A. 1997. Model of a quinary structure between Krebs TCA cycle enzymes: A model for the metabolon. *Biochemistry* 36:14271–14276.

Haggie, P. M., and Brindle, K. M. 1999. Mitochondrial citrate synthase is immobilized in vivo. *J. Biol. Chem.* 274:3941–3945.

Morgunov, I., and Srere, P. A. 1998. Interaction between citrate synthase and malate dehydrogenase: Substrate channeling of oxaloacetate. *J. Biol. Chem.* 273:29540–29544.

Regulation

Shi, Q., Xu, H., Yu, H., Zhang, N., Ye, Y., Estevez, A. G., Deng, H., and Gibson, G. E. 2011. Inactivation and reactivation of the mitochondrial α-ketoglutarate dehydrogenase complex. *J. Biol. Chem.* 286:17640–17648.

Phillips, D., Aponte, A. M., French, S. A., Chess, D. J., and Balaban, R. S. 2009. Succinyl-CoA synthetase is a phosphate target for the activation of mitochondrial metabolism. *Biochemistry* 48:7140–7149.

Taylor, A. B., Hu, G., Hart, P. J., and McAlister-Henn, L. 2008. Allosteric motions in structures of yeast NAD⁺-specific isocitrate dehydrogenase. *J. Biol. Chem.* 283:10872–10880.

Green, T., Grigorian, A., Klyuyeva, A., Tuganova, A., Luo, M., and Popov, K. M. 2008. Structural and functional insights into the molecular mechanisms responsible for the regulation of pyruvate dehydrogenase kinase. *J. Biol. Chem.* 283:15789–15798.

Hiromasa, Y., and Roche, T. E. 2003. Facilitated interaction between the pyruvate dehydrogenase kinase isoform 2 and the dihydrolipoyl acetyltransferases. *J. Biol. Chem.* 278:33681–33693.

Jitrapakdee, S., and Wallace, J. C. 1999. Structure, function and regulation of pyruvate carboxylase. *Biochem. J.* 340:1–16.

The Citric Acid Cycle and Cancer

Fan, J., Lin, R., Xia, S., Chen, D., Elf, S. E., Liu, S., et al. 2016. Tetrameric acetyl-CoA acetyltransferase 1 is important for tumor growth. *Mol. Cell* 64:859–874.

Wang, F., Travins, J., DeLaBarre, B., Penard-Lacronique, V., Schalm, S., Hansen, E., Straley, K., Kernytsky, A., Liu, W., Gliser, C., et al. 2013. Targeted inhibition of mutant IDH2 in leukemia cells induces cellular differentiation. *Science* 340:622–626.

Rohle, D., Popovici-Muller, J., Palaskas, N., Turcan, S., Grommes, C., Campos, C., Tsoi, J., Clark, O., Oldrini, B., Komisopoulou, E., et al. 2013. An inhibitor of mutant IDH1 delays growth and promotes differentiation of glioma cells. *Science* 340:626–630.

Losman, J.-A., Koivunen, P., Lee, S., Schneider, R. K., McMahon, C., Cowley, G. S., Root, D. E., Ebert, B. L., Kaelin, W. G. Jr., et al. 2013. (*R*)-2-Hydroxyglutarate is sufficient to promote leukemogenesis and its effects are reversible. *Science* 339:1621–1625.

Sakai, C., Tomitsuka, T., Esumi, H., Harada, S., and Kita, K. 2012. Mitochondrial fumarate reductase as a target of chemotherapy: From parasites to cancer cells. *Biochim. Biophys. Acta* 1820:643–651.

Xekouki P., and Stratakis, C. A. 2012. Succinate dehydrogenase (SDHx) mutations in pituitary tumors: Could this be a new role for mitochondrial complex II and/or Krebs cycle defects? *Endocr.-Relat. Cancer* 19:C33–C40.

Thompson, C. B. 2009. Metabolic enzymes as oncogenes or tumor suppressors. *New Engl. J. Med.* 360:813–815.

McFate, T., Mohyeldin, A., Lu, H., Thakar, J., Henriques, J., Halim, N. D., Wu, H., Schell, M. J., Tsang, T. M., Teahan, O., Zhou, S., Califano, J. A., Jeoung, M. N., Harris, R. A., and Verma, A. 2008. Pyruvate dehydrogenase complex activity controls metabolic and malignant phenotype in cancer cells. *J. Biol. Chem.* 283:22700–22708.

Gogvadze, V., Orrenius, S., and Zhivotovsky, B. 2008. Mitochondria in cancer cells: What is so special about them? *Trends Cell Biol.* 18:165–173.

Evolutionary Aspects

Meléndez-Hevia, E., Waddell, T. G., and Cascante, M. 1996. The puzzle of the Krebs citric acid cycle: Assembling the pieces of chemically feasible reactions, and opportunism in the design of metabolic pathways in evolution. *J. Mol. Evol.* 43:293–303.

Baldwin, J. E., and Krebs, H. 1981. The evolution of metabolic cycles. *Nature* 291:381–382.

Gest, H. 1987. Evolutionary roots of the citric acid cycle in prokaryotes. *Biochem. Soc. Symp.* 54:3–16.

Weitzman, P. D. J. 1981. Unity and diversity in some bacterial citric acid cycle enzymes. *Adv. Microbiol. Physiol.* 22:185–244.

Discovery of the Citric Acid Cycle

Kornberg, H. 2000. Krebs and his trinity of cycles. *Nat. Rev. Mol. Cell. Biol.* 1:225–228.

Krebs, H. A., and Johnson, W. A. 1937. The role of citric acid in intermediate metabolism in animal tissues. *Enzymologia* 4:148–156.

Krebs, H. A. 1970. The history of the tricarboxylic acid cycle. *Perspect. Biol. Med.* 14:154–170.

Krebs, H. A., and Martin, A. 1981. *Reminiscences and Reflections.* Clarendon Press.

Biochemistry in Focus

Rahman, M. H., Jha, M. K., Kim, J.-H., Nam, Y., Lee, M. G., Younghoon Go, et al. 2016. Pyruvate dehydrogenase kinase-mediated glycolytic metabolic shift in the dorsal root ganglion drives painful diabetic neuropathy. *J. Biol. Chem.* 291:6011–6015.

Pham, T. V., Murkin, A. S., Moynihan, M. M., Harris, L., Tyler, P. C., et al. 2017. Mechanism-based inactivator of isocitrate lyases 1 and 2 from *Mycobacterium tuberculosis*. *Proc. Natl. Acad. Sci. U.S.A.* 114:7617–7622.

CHAPTER 18

Where to Start

Guarente, L. 2008. Mitochondria: A nexus for aging, calorie restriction, and sirtuins? *Cell* 132:171–176.

Wallace, D. C. 2007. Why do we still have a maternally inherited mitochondrial DNA? Insights from evolutionary medicine. *Annu. Rev. Biochem.* 76:781–821.

Hosler, J. P., Ferguson-Miller, S., and Mills, D. A. 2006. Energy transduction: Proton transfer through the respiratory complexes. *Annu. Rev. Biochem.* 75:165–187.

Gray, M. W., Burger, G., and Lang, B. F. 1999. Mitochondrial evolution. *Science* 283:1476–1481.

Shultz, B. E., and Chan, S. I. 2001. Structures and proton-pumping strategies of mitochondrial respiratory enzymes. *Annu. Rev. Biophys. Biomol. Struct.* 30:23–65.

Books

Scheffler, I. E. 2007. *Mitochondria.* Wiley.

Lane, N. 2005. *Power, Sex, Suicide: Mitochondria and the Meaning of Life.* Oxford.

Nicholls, D. G., and Ferguson, S. J. 2013. *Bioenergetics* (4th ed.). Academic Press.

Electron-Transport Chain

Guo, R., Zong, S., Wu, M., Gu, J., and Yang, M. 2017. Architecture of human mitochondrial respiratory megacomplex $I_2III_2IV_2$ *Cell* 170:1247–1257

Fedor, J. G., Jones, A. J. Y., Di Luca, A., Kaila, V. R. I., and Hirst, J. 2017. Correlating kinetic and structural data on ubiquinone binding and reduction by respiratory complex I. *Proc. Natl. Acad. Sci. U.S.A.* 114:12737–12742.

Baradaran, R., Berrisford, J. M., Minhas, G. S., and Sazanov, L. A. 2013. Crystal structure of the entire respiratory complex I. *Nature* 494:443–448.

Lapuente-Brun, E., Moreno-Loshuertos, R., Acín-Pérez, R., Latorre-Pellicer, A., Colás, C., Balsa, E., Perales-Clemente, E., Quirós, P. M., Calvo, E., Rodríguez-Hernández, M. A., et al. 2013. Supercomplex assembly determines electron flux in the mitochondrial electron transport chain. *Science* 340:1567–1570.

Cammack, R. 2012. Iron-sulfur proteins. *The Biochemist* 35:14–17.

Yoshikawa, S., Muramoto, K., and Shinzawa-Itoh, K. 2011. Proton-pumping mechanism of cytochrome *c* oxidase. *Annu. Rev. Biophys.* 40:205–23.

Qin, L., Liu, J., Mills, D. A., Proshlyakov, D. A., Hiser, C., and Ferguson-Miller, S. 2009. Redox-dependent conformational changes in cytochrome *c* oxidase suggest a gating mechanism for proton uptake. *Biochemistry* 48:5121–5130.

Lill, R. 2009. Function and biogenesis of iron–sulphur proteins. *Nature* 460:831–838.

Cooley, C. W., Lee, D.-W., and Daldal, F. 2009. Across membrane communication between the Q_o and Q_i active sites of cytochrome bc_1. *Biochemistry* 48:1888–1899.

Verkhovskaya, M. L., Belevich, N., Euro, L., Wikström, M., and Verkhovsky, M. I. 2008. Real-time electron transfer in respiratory complex I. *Proc. Natl. Acad. Sci. U.S.A.* 105:3763–3767.

Acín-Pérez, R., Fernández-Silva, P., Peleato, M. L., Pérez-Martos, A., and Enriquez, J. A. 2008. Respiratory active mitochondrial super-complexes. *Mol. Cell* 32:529–539.

Kruse, S. E., Watt, W. C., Marcinek, D. J., Kapur, R. P., Schenkman, K. A., and Palmiter, R. D. 2008. Mice with mitochondrial Complex I deficiency develop a fatal encephalomyopathy. *Cell Metab.* 7:312–320.

Sun, F., Huo, X., Zhai, Y., Wang, A., Xu, J., Su, D., Bartlam, M., and Ral, Z. 2005. Crystal structure of mitochondrial respiratory membrane protein complex II. *Cell* 121:1043–1057.

Crofts, A. R. 2004. The cytochrome bc_1 complex: Function in the context of structure. *Annu. Rev. Physiol.* 66:689–733.

Bianchi, C., Genova, M. L., Castelli, G. P., and Lenaz, G. 2004. The mitochondrial respiratory chain is partially organized in a supramolecular complex. *J. Biol. Chem.* 279:36562–36569.

Cecchini, G. 2003. Function and structure of Complex II of the respiratory chain. *Annu. Rev. Biochem.* 72:77–109.

Lange, C., and Hunte, C. 2002. Crystal structure of the yeast cytochrome bc_1 complex with its bound substrate cytochrome c. *Proc. Natl. Acad. Sci. U.S.A.* 99:2800–2805.

ATP Synthase

Guo, H., Bueler, S. A., Rubinstein, J. L. 2017. Atomic model for the dimeric F_0 region of mitochondrial ATP synthase. *Science* 358:936–940.

Toei, M., and Noji, H. 2013. Single-molecule analysis of F_0F_1-ATP synthase inhibited by N, N-dicyclohexylcarbodiimide. *J. Biol. Chem.* 288:25717–25726.

Watt, I. N., Montgomery, M. G., Runswick, M. J., Leslie, A. G. W., and Walker, J. E. 2010. Bioenergetic cost of making an adenosine triphosphate molecule in animal mitochondria. *Proc. Natl. Acad. Sci. U.S.A.* 107:16823–16827.

Wittig, I., and Hermann, S. 2009. Supramolecular organization of ATP synthase and respiratory chain in mitochondrial membranes. *Biochim. Biophys. Acta* 1787:672–680.

Junge, W., Sielaff, H., and Engelbrecht S. 2009. Torque generation and elastic power transmission in the rotary F_0F_1-ATPase. *Nature* 459:364–370.

von Ballmoos, C., Cook, G. M., and Dimroth, P. 2008. Unique rotary ATP synthase and its biological diversity. *Annu. Rev. Biophys.* 37:43–64.

Adachi, K., Oiwa, K., Nishizaka, T., Furuike, S., Noji, H., Itoh, H., Yoshida, M., and Kinosita, K., Jr. 2007. Coupling of rotation and catalysis in F_1-ATPase revealed by single-molecule imaging and manipulation. *Cell* 130:309–321.

Chen, C., Ko, Y., Delannoy, M., Ludtke, S. J., Chiu, W., and Pedersen, P. L. 2004. Mitochondrial ATP synthasome: Three-dimensional structure by electron microscopy of

the ATP synthase in complex formation with the carriers for P_i and ADP/ATP. *J. Biol. Chem.* 279:31761–31768.

Noji, H., and Yoshida, M. 2001. The rotary machine in the cell: ATP synthase. *J. Biol. Chem.* 276:1665–1668.

Yasuda, R., Noji, H., Kinosita, K., Jr., and Yoshida, M. 1998. F_1-ATPase is a highly efficient molecular motor that rotates with discrete 120 degree steps. *Cell* 93:1117–1124.

Noji, H., Yasuda, R., Yoshida, M., and Kinosita, K., Jr., 1997. Direct observation of the rotation of F_1-ATPase. *Nature* 386:299–302.

Tsunoda, S. P., Aggeler, R., Yoshida, M., and Capaldi, R. A. 2001. Rotation of the c subunit oligomer in fully functional $F_1 F_0$ ATP synthase. *Proc. Natl. Acad. Sci. U.S.A.* 987:898–902.

Gibbons, C., Montgomery, M. G., Leslie, A. G. W., and Walker, J. 2000. The structure of the central stalk in F_1-ATPase at 2.4 Å resolution. *Nat. Struct. Biol.* 7:1055–1061.

Sambongi, Y., Iko, Y., Tanabe, M., Omote, H., Iwamoto-Kihara, A., Ueda, I., Yanagida, T., Wada, Y., and Futai, M. 1999. Mechanical rotation of the c subunit oligomer in ATP synthase (F_0F_1): Direct observation. *Science* 286:1722–1724.

Translocators and Channels

Villarroya, F., and Vidal-Puig, A. 2013. Beyond the sympathetic tone: The new brown fat activators. *Cell Metab.* 17:638–643.

Rey, M., Forest, E., and Pelosi, L. 2012. Exploring the conformational dynamics of the bovine ADP/ATP carrier in mitochondria. *Biochemistry* 51:9727–9735.

Divakaruni, A. S., Humphrey, D. M., and Brand, M. D. 2012. Fatty acids change the conformation of uncoupling protein 1 (UCP1). *J. Biol. Chem.* 44:36845–36853.

Fedorenko, A., Lishko, P. V., and Kirichok, Y. 2012. Mechanism of fatty-acid-dependent UCP1 uncoupling in brown fat mitochondria. *Cell* 151:400–413.

van Marken Lichtenbelt, W. D., Vanhommerig, J. W., Smulders, N. M., Drossaerts, J. M., Kemerink, G. J., Bouvy, N. D., Schrauwen, P., and Teule, G. J. 2009. Cold-activated brown adipose tissue in healthy men. *New Engl. J. Med.* 360:1500–1508.

Cypess, A. M., Sanaz Lehman, S., Gethin Williams, G., Tal, I., Rodman, D., Goldfine, A. B., Kuo, F. C., Palmer, E. L., Tseng, Y.-H., Doria, A., et al. 2009. Identification and importance of brown adipose tissue in adult humans. *New Engl. J. Med.* 360:1509–1517.

Virtanen, K. A., Lidell, M. E., Orava, J., Heglind, M., Westergren, R., Niemi, T., Taittonen, M., Laine, J., Savisto, N.-J., Enerbäck, S., et al. 2009. Functional brown adipose tissue in healthy adults. *New Engl. J. Med.* 360:1518–1525.

Bayrhuber, M., Meins, T., Habeck, M., Becker, S., Giller, K., Villinger, S., Vonrhein, C., Griesinger, C., Zweckstetter, M., and Zeth, K. 2008. Structure of the human voltage-dependent anion channel. *Proc. Natl. Acad. Sci. U.S.A.* 105:15370–15375.

Bamber, L., Harding, M., Monné, M., Slotboom, D.-J., and Kunji, E. R. 2007. The yeast mitochondrial ADP/ATP carrier functions as a monomer in mitochondrial membranes. *Proc. Natl. Acad. Sci. U.S.A.* 10:10830–10843.

Pebay-Peyroula, E., Dahout, C., Kahn, R., Trézéguet, V., Lauquin, G. J.-M., and Brandolin, G. 2003. Structure of mitochondrial ADP/ATP carrier in complex with carboxyatractyloside. *Nature* 246:39–44.

Reactive Oxygen Species, Superoxide Dismutase, and Catalase

Banerjee, R. 2017. Introduction to the Thematic Minireview Series: Redox metabolism and signaling. *J. Biol. Chem.* 292:16802–16803. The first in a series of short reviews on the role of reactive oxygens species in metabolism and signaling.

Sena, L. A., and Chandel, N. S. 2012. Physiological roles of mitochondrial reactive oxygen species. *Mol. Cell* 48:158–167.

Forman, H. J., Maiorino, M., and Ursini, F. 2010. Signaling functions of reactive oxygen species. *Biochemistry* 49:835–842.

Murphy, M. P. 2009. How mitochondria produce reactive oxygen species. *Biochem. J.* 417:1–13.

Leitch, J. M., Yick, P. J., and Culotta, V. V. 2009. The right to choose: Multiple pathways for activating copper, zinc superoxide dismutase. *J. Biol. Chem.* 284:24679–24683.

Winterbourn, C. C. 2008. Reconciling the chemistry and biology of reactive oxygen species. *Nat. Chem. Biol.* 4:278–286.

Veal, E. A., Day, A. M., and Morgan, B. A. 2007. Hydrogen peroxide sensing and signaling. *Mol. Cell* 26:1–14.

Stone, J. R., and Yang, S. 2006. Hydrogen peroxide: A signaling messenger. *Antioxid. Redox Signal.* 8:243–270.

Valentine, J. S., Doucette, P. A., and Potter S. Z. 2005. Copper-zinc superoxide dismutase and amyotrophic lateral sclerosis. *Annu. Rev. Biochem.* 74:563–593.

Mitochondrial Diseases

Papa, S., and De Rasmo, D. 2013. Complex I deficiencies in neurological disorders. *Trends Mol. Med.* 19:61–69.

Koopman, W. J. H., Willems, P. H. G. M., and Smeitink, J. A. M. 2012. Monogenic mitochondrial disorders. *New Engl. J. Med.* 366:1132–41.

Lin, C. S., Sharpley, M. S., Fan W., Waymire, K. G., Sadun, A. A., Carelli, V., Ross-Cisneros, F. N., Baciu, P., Sung, E., McManus, M. J., et al. 2012. Mouse mtDNA mutant model of Leber hereditary optic neuropathy. *Proc. Natl. Acad. Sci. U.S.A.* 109:20065–20070.

Mitochondria Disease. 2009. A compendium of nine articles on mitochondrial diseases. *Biochem. Biophys. Acta Mol. Basis Disease* 1792:1095–1167.

Cicchetti, F., Drouin-Ouellet, J., and Gross, R. E. 2009. Environmental toxins and Parkinson's disease: What have we learned from pesticide-induced animal models? *Trends Pharm. Sci.* 30:475–483.

DiMauro, S., and Schon, E. A. 2003. Mitochondrial respiratory-chain disease. *New Engl. J. Med.* 348:2656–2668.

Smeitink, J., van den Heuvel, L., and DiMauro, S. 2001. The genetics and pathology of oxidative phosphorylation. *Nat. Rev. Genet.* 2:342–352.

Industrial Application of Biochemistry

Zheng, Q., Lin, J., Huang, J., Zhang, H., Zhang, R., et al. 2017. Reconstitution of UCP1 using CRISPR/Cas9 in the white adipose tissue of pigs decreases fat deposition and improves thermogenic capacity. *Proc. Natl. Acad. Sci. U.S.A.* E9474–E9482

Apoptosis

Qi, S., Pang, Y., Hu, Q., Liu, Q., Li, H., Zhou, Y., He, T., Liang, Q., Liu, Y., Yuan, X., et al. 2010. Crystal structure of the *Caenorhabditis elegans* apoptosome reveals an octameric assembly of CED-4. *Cell* 141:446–457.

Chan, D. C. 2006. Mitochondria: Dynamic organelles in disease, aging, and development. *Cell* 125:1241–1252.

Green, D. R. 2005. Apoptotic pathways: Ten minutes to dead. *Cell* 121:671–674.

Historical Aspects

Prebble, J., and Weber, B. 2003. *Wandering in the Gardens of the Mind: Peter Mitchell and the Making of Glynn.* Oxford.

Mitchell, P. 1979. Keilin's respiratory chain concept and its chemiosmotic consequences. *Science* 206:1148–1159.

Preeble, J. 2002. Peter Mitchell and the ox phos wars. *Trends Biochem. Sci.* 27:209–212.

Mitchell, P. 1976. Vectorial chemistry and the molecular mechanics of chemiosmotic coupling: Power transmission by proticity. *Biochem. Soc. Trans.* 4:399–430.

Racker, E. 1980. From Pasteur to Mitchell: A hundred years of bioenergetics. *Fed. Proc.* 39:210–215.

Kalckar, H. M. 1991. Fifty years of biological research: From oxidative phosphorylation to energy requiring transport and regulation. *Annu. Rev. Biochem.* 60:1–37.

CHAPTER 19

Where to Start

Bourzac, K. 2017. Solar upgrade. *Nature* 544: S11–S13.

Stolstad, E. 2116. Engineered crops could have it made in the shade. *Science* 354:816.

Zhang, T. 2015. More efficient together. *Science* 350:738–739.

Barber, J., and Andersson, B. 1994. Revealing the blueprint of photosynthesis. *Nature* 370:31–34.

Books and General Reviews

Nelson, N., and Yocum, C. 2006. Structure and functions of photosystems I and II. *Annu. Rev. Plant Biol.* 57:521–565.

Merchant, S., and Sawaya, M. R. 2005. The light reactions: A guide to recent acquisitions for the picture gallery. *Plant Cell* 17:648–663.

Blankenship, R. E. 2009. *Molecular Mechanisms of Photosynthesis.* Wiley-Blackwell.

Nicholls, D. G., and Ferguson, S. J. 2013. *Bioenergetics* (4th ed.). Academic Press.

Electron-Transfer Mechanisms

Beratan, D., and Skourtis, S. 1998. Electron transfer mechanisms. *Curr. Opin. Chem. Biol.* 2:235–243.

Moser, C. C., Keske, J. M., Warncke, K., Farid, R. S., and Dutton, P. L. 1992. Nature of biological electron transfer. *Nature* 355:796–802.

Boxer, S. G. 1990. Mechanisms of long-distance electron transfer in proteins: Lessons from photosynthetic reaction centers. *Annu. Rev. Biophys. Biophys. Chem.* 19:267–299.

Photosystem II

Vinyard, D. J., Ananyev, G. M., and Dismukes, G. C. 2013. Photosystem II: The reaction center of oxygenic photosynthesis. *Annu. Rev. Biochem.* 82:577–606.

Kirchhoff, H., Tremmel, I., Haase, W., and Kubitscheck, U. 2004. Supramolecular photosystem II organization in grana of thylakoid membranes: Evidence for a structured arrangement. *Biochemistry* 43:9204–9213.

Diner, B. A., and Rappaport, F. 2002. Structure, dynamics, and energetics of the primary photochemistry of photosystem II of oxygenic photosynthesis. *Annu. Rev. Plant Biol.* 54:551–580.

Zouni, A., Witt, H. T., Kern, J., Fromme, P., Krauss, N., Saenger, W., and Orth, P. 2001. Crystal structure of photosystem II from *Synechococcus elongatus* at 3.8 Å resolution. *Nature* 409:739–743.

Deisenhofer, J., and Michel, H. 1991. High-resolution structures of photosynthetic reaction centers. *Annu. Rev. Biophys. Biophys. Chem.* 20:247–266.

Oxygen Evolution

Barber, J. 2016. Mn_4Ca Cluster of photosynthetic oxygen-evolving center: Structure, function and evolution. *Biochemistry* 55:5901–5906.

Umena, Y., Kawakami, K., Shen, J.-R., and Kamiya, N. 2011. Crystal structure of oxygen-evolving photosystem II at a resolution of 1.9 Å. *Nature* 473:55–60.

Barber, J. 2008. Crystal structure of the oxygen-evolving complex of photosystem II. *Inorg. Chem.* 47:1700–1710.

Pushkar, Y., Yano, J., Sauer, K., Boussac, A., and Yachandra, V. K. 2008. Structural changes in the Mn_4Ca cluster and the mechanism of photosynthetic water splitting. *Proc. Natl. Acad. Sci. U.S.A.* 105:1879–1884.

Renger, G. 2007. Oxidative photosynthetic water splitting: Energetics, kinetics and mechanism. *Photosynth. Res.* 92:407–425.

Renger, G., and Kühn, P. 2007. Reaction pattern and mechanism of light induced oxidative water splitting in photosynthesis. *Biochim. Biophys. Acta* 1767:458–471.

Photosystem I and Cytochrome *bf*

Schöttler, M. A., Albus, C. A., and Bock, R. 2011. Photosystem I: Its biogenesis and function in higher plants. *J. Plant Physiol.* 168:1452–1461.

Iwai, M., Takizawa, K., Tokutsu, R., Okamuro, A., Takahashi, Y., and Minagawa, J. 2010. Isolation of the elusive supercomplex that drives cyclic electron flow in photosynthesis. *Nature* 464:1210–1214.

Amunts, A., Drory, O., and Nelson, N. 2007. The structure of photosystem I supercomplex at 3.4 Å resolution. *Nature* 447:58–63.

Cramer, W. A., Zhang, H., Yan, J., Kurisu, G., and Smith, J. L. 2004. Evolution of photosynthesis: Time-independent structure of the cytochrome b_6f complex. *Biochemistry* 43:5921–5929.

Kargul, J., Nield, J., and Barber, J. 2003. Three-dimensional reconstruction of a light-harvesting complex I-photosystem I (LHCI-PSI) supercomplex from the green alga *Chlamydomonas reinhardtii*. *J. Biol. Chem.* 278:16135–16141.

Schubert, W. D., Klukas, O., Saenger, W., Witt, H. T., Fromme, P., and Krauss, N. 1998. A common ancestor for oxygenic and anoxygenic photosynthetic systems: A comparison based on the structural model of photosystem I. *J. Mol. Biol.* 280:297–314.

ATP Synthase

Kohzuma, K., Dal Bosco, C., Meurer, J., and Kramer, D. M. 2013. Light- and metabolism-related regulation of the chloroplast ATP synthase has distinct mechanisms and functions. *J. Biol. Chem.* 288:13156–13163.

Vollmar, M., Schlieper, D., Winn, D., Büchner, C., and Groth, G. 2009. Structure of the c14 rotor ring of the proton translocating chloroplast ATP synthase. *J. Biol. Chem.* 284:18228–18235.

Varco-Merth, B., Fromme, R., Wang, M., and Fromme, P. 2008. Crystallization of the c14-rotor of the chloroplast ATP synthase reveals that it contains pigments. *Biochim. Biophys. Acta* 1777:605–612.

Oster, G., and Wang, H. 1999. ATP synthase: Two motors, two fuels. *Structure* 7:R67–R72.

Weber, J., and Senior, A. E. 2000. ATP synthase: What we know about ATP hydrolysis and what we do not know about ATP synthesis. *Biochim. Biophys. Acta* 1458:300–309.

Light-Harvesting Assemblies

Collins, A. M., Qian, P., Tang, Q., Bocian, D. F., Hunter, C. N., Blankenship, R. E. 2010. Light-harvesting antenna system from the phototrophic bacterium *Roseiflexus castenholzii*. *Biochemistry* 49:7524–7531.

Melkozernov, A. N., Barber, J., and Blankenship, R. E. 2006. Light harvesting in photosystem I supercomplexes. *Biochemistry* 45:331–345.

Conroy, M. J., Westerhuis, W. H., Parkes-Loach, P. S., Loach, P. A., Hunter, C. N., and Williamson, M. P. 2000. The solution structure of *Rhodobacter sphaeroides* LH1b reveals two helical domains separated by a more flexible region: Structural consequences for the LH1 complex. *J. Mol. Biol.* 298:83–94.

Koepke, J., Hu, X., Muenke, C., Schulten, K., and Michel, H. 1996. The crystal structure of the light-harvesting complex II (B800–850) from *Rhodospirillum molischianum*. *Structure* 4:581–597.

Grossman, A. R., Bhaya, D., Apt, K. E., and Kehoe, D. M. 1995. Light-harvesting complexes in oxygenic photosynthesis: Diversity, control, and evolution. *Annu. Rev. Genet.* 29:231–288.

Evolution

Hohmann-Marriott, M. F., and Blankenship, R. E. 2011. Evolution of photosynthesis. *Annu. Rev. Plant Biol.* 62:515–48.

Chen, M., and Zhang, Y. 2008. Tracking the molecular evolution of photosynthesis through characterization of atomic contents of the photosynthetic units. *Photosynth. Res.* 97:255–261.

Iverson, T. M. 2006. Evolution and unique bioenergetic mechanisms in oxygenic photosynthesis. *Curr. Opin. Chem. Biol.* 10:91–100.

Cavalier-Smith, T. 2002. Chloroplast evolution: Secondary symbiogenesis and multiple losses. *Curr. Biol.* 12:R62–64.

Nelson, N., and Ben-Shem, A. 2005. The structure of photosystem I and evolution of photosynthesis. *BioEssays* 27:914–922.

Green, B. R. 2001. Was "molecular opportunism" a factor in the evolution of different photosynthetic light-harvesting pigment systems? *Proc. Natl. Acad. Sci. U.S.A.* 98:2119–2121.

Dismukes, G. C., Klimov, V. V., Baranov, S. V., Nozlov, Y. N., Das Gupta, J., and Tyryshkin, A. 2001. The origin of atmospheric oxygen on Earth: The innovation of oxygenic photosynthesis. *Proc. Natl. Acad. Sci. U.S.A.* 98:2170–2175.

Moreira, D., Le Guyader, H., and Phillippe, H. 2000. The origin of red algae and the evolution of chloroplasts. *Nature* 405:69–72.

Cavalier-Smith, T. 2000. Membrane heredity and early chloroplast evolution. *Trends Plant Sci.* 5:174–182.

Industrial Application of Photosynthesis: Artificial Photosynthesis

Liu, C., Colón, B. C., Ziesack, M., Silver, P. A., and Nocera, D. G. 2016. Water splitting–biosynthetic system with CO_2 reduction efficiencies exceeding photosynthesis. *Science* 352:1210–1213.

Nocera, D. G. 2017. Solar fuels and solar chemicals industry. *Acc. Chem. Res.* 50:616–619.

Nangle, S. N., Sakimoto, K. K., Silver, P. A., and Nocera, D. G. 2017. Biological-inorganic hybrid systems as a generalized platform for chemical production. *Curr. Opin. Chem. Biol.* 41:107–113.

Biochemistry in Focus

Correa-Galvis, V., Redekop, P., Guan, K., Griess, A., Truong, T. B., Wakao, S., Niyogi, K. K., and Jahns, P. 2016. Photosystem II subunit PsbS is involved in the induction of LHCSR protein-dependent energy dis-
sipation in *Chlamydomonas reinhardtii*. *J. Biol. Chem.* 291:17478–17487.

Lu, Y., Liu, H., Saer, R., Li, V. L., Zhang, H. et al. 2017. A molecular mechanism for nonphotochemical quenching in cyanobacteria. *Biochemistry* 56:2812–2823.

Gerotto, C., Franchin, C., Arrigoni, G., and Morosinotto, T. 2015. In vivo identification of photosystem II light harvesting complexes interacting with photosystem subunit S. *Plant Physiol.* 168:1747–1761.

CHAPTER 20

Where to Start

Buchanan, B. B., and Wong, J. H. 2013. A conversation with Andrew Benson: reflections on the discovery of the Calvin–Benson cycle. *Photosynth. Res.* 114:207–214.

Ellis, R. J. 2010. Tackling unintelligent design. *Nature* 463:164–165.

Gutteridge, S., and Pierce, J. 2006. A unified theory for the basis of the limitations of the primary reaction of photosynthetic CO_2, fixation: Was Dr. Pangloss right? *Proc. Natl. Acad. Sci. U.S.A.* 103:7203–7204.

Horecker, B. L. 1976. Unravelling the pentose phosphate pathway. In *Reflections on Biochemistry* (pp. 65–72), edited by A. Kornberg, L. Cornudella, B. L. Horecker, and J. Oro. Pergamon.

Levi, P. 1984. Carbon. In *The Periodic Table*. Random House.

Books and General Reviews

Parry, M. A. J., Andralojc, P. J., Mitchell, R. A. C., Madgwick, P. J., and Keys, A. J. 2003. Manipulation of rubisco: The amount, activity, function and regulation. *J. Exp. Bot.* 54:1321–1333.

Spreitzer, R. J., and Salvucci, M. E. 2002. Rubisco: Structure, regulatory interactions, and possibilities for a better enzyme. *Annu. Rev. Plant Biol.* 53:449–475.

Wood, T. 1985. *The Pentose Phosphate Pathway*. Academic Press.

Buchanan, B. B., Gruissem, W., and Jones, R. L. 2000. *Biochemistry and Molecular Biology of Plants*. American Society of Plant Physiologists.

Enzymes and Reaction Mechanisms

Peterson-Forbrook, D. S., Hilton, M. T., Tichacek, L., Henderson, J. N., Bui, H. Q., and Wachter, R. M. 2017. Nucleotide dependence of subunit rearrangements in short-form rubisco activase from spinach. *Biochemistry* 56:4906–4921.

Harrison, D. H., Runquist, J. A., Holub, A., and Miziorko, H. M. 1998. The crystal structure of phosphoribulokinase from *Rhodobacter sphaeroides* reveals a fold similar to that of adenylate kinase. *Biochemistry* 37:5074–5085.

Miziorko, H. M. 2000. Phosphoribulokinase: Current perspectives on the structure/function basis for regulation and catalysis. *Adv. Enzymol. Relat. Areas Mol. Biol.* 74:95–127.

Thorell, S., Gergely, P., Jr., Banki, K., Perl, A., and Schneider, G. 2000. The three-dimensional structure of human transaldolase. *FEBS Lett.* 475:205–208.

Carbon Dioxide Fixation and Rubisco

Satagopan, S., Scott, S. S., Smith, T. G., and Tabita, F. R. 2009. A rubisco mutant that confers growth under a normally "inhibitory" oxygen concentration. *Biochemistry* 48:9076–9083.

Tcherkez, G. G. B., Farquhar, G. D., and Andrews, J. T. 2006. Despite slow catalysis and confused substrate specificity, all ribulose bisphosphate carboxylases may be nearly perfectly optimized. *Proc. Natl. Acad. Sci. U.S.A.* 103:7246–7251.

Sugawara, H., Yamamoto, H., Shibata, N., Inoue, T., Okada, S., Miyake, C., Yokota, A., and Kai, Y. 1999. Crystal structure of carboxylase reaction-oriented ribulose 1,5-bisphosphate carboxylase/oxygenase from a thermophilic red alga, *Galdieria partita. J. Biol. Chem.* 274:15655–15661.

Hansen, S., Vollan, V. B., Hough, E., and Andersen, K. 1999. The crystal structure of rubisco from *Alcaligenes eutrophus* reveals a novel central eight-stranded β-barrel formed by β-strands from four subunits. *J. Mol. Biol.* 288:609–621.

Knight, S., Andersson, I., and Branden, C. I. 1990. Crystallographic analysis of ribulose 1,5-bisphosphate carboxylase from spinach at 2.4 Å resolution: Subunit interactions and active site. *J. Mol. Biol.* 215:113–160.

Taylor, T. C., and Andersson, I. 1997. The structure of the complex between rubisco and its natural substrate ribulose 1,5-bisphosphate. *J. Mol. Biol.* 265:432–444.

Cleland, W. W., Andrews, T. J., Gutteridge, S., Hartman, F. C., and Lorimer, G. H. 1998. Mechanism of rubisco: The carbamate as general base. *Chem. Rev.* 98:549–561.

Buchanan, B. B. 1992. Carbon dioxide assimilation in oxygenic and anoxygenic photosynthesis. *Photosynth. Res.* 33:147–162.

Hatch, M. D. 1987. C_4 photosynthesis: A unique blend of modified biochemistry, anatomy, and ultrastructure. *Biochim. Biophys. Acta* 895:81–106.

Regulation

Keown, J. R., Griffin, M. D. W., Mertens, H. D. T., and Pearce, F. G. 2013. Small oligomers of ribulose-bisphosphate carboxylase/oxygenase (rubisco) activase are required for biological activity. *J. Biol. Chem.* 288:20607–20615.

Carmo-Silva, A. E., and Salvucci, M. E. 2013. The regulatory properties of rubisco activase differ among species and affect photosynthetic induction during light transitions. *Plant Physiol.* 161:1645–1655.

Gontero, B., and Maberly, S. C. 2012. An intrinsically disordered protein, CP12: Jack of all trades and master of the Calvin cycle. *Biochem. Soc. Trans.* 40:995–999.

Stotz, M., Mueller-Cajar, O., Ciniawsky, S., Wendler, P., Hartl, F.-U., Bracher, A., and Hayer-Hartl, M. 2011.

Structure of green-type Rubisco activase from tobacco. *Nature Struct. Mol. Biol.* 18:1366–1370.

Lebreton, S., Andreescu, S., Graciet, E., and Gontero, B. 2006. Mapping of the interaction site of CP12 with glyceraldehyde-3-phosphate dehydrogenase from *Chlamydomonas reinhardtii*. Functional consequences for glyceraldehyde-3-phosphate dehydrogenase. *FEBS J.* 273:3358–3369.

Graciet, E., Lebreton, S., and Gontero, B. 2004. The emergence of new regulatory mechanisms in the Benson-Calvin pathway via protein-protein interactions: A glyceraldehyde-3-phosphate dehydrogenase/CP12/phosphoribulokinase complex. *J. Exp. Bot.* 55:1245–1254.

Balmer, Y., Koller, A., del Val, G., Manieri, W., Schürmann, P., and Buchanan, B. B. 2003. Proteomics gives insight into the regulatory function of chloroplast thioredoxins. *Proc. Natl. Acad. Sci. U.S.A.* 100:370–375.

Wedel, N., Soll, J., and Paap, B. K. 1997. CP12 provides a new mode of light regulation of Calvin cycle activity in higher plants. *Proc. Natl. Acad. Sci. U.S.A.* 94:10479–10484.

Avilan, L., Lebreton, S., and Gontero, B. 2000. Thioredoxin activation of phosphoribulokinase in a bi-enzyme complex from *Chlamydomonas reinhardtii* chloroplasts. *J. Biol. Chem.* 275:9447–9451.

Irihimovitch, V., and Shapira, M. 2000. Glutathione redox potential modulated by reactive oxygen species regulates translation of rubisco large subunit in the chloroplast. *J. Biol. Chem.* 275:16289–16295.

Glucose 6-Phosphate Dehydrogenase

Howes, R. E., Piel, F. B., Patil, A. P., Nyangiri, O. A., Gething, P. W., Dewi, M., Hogg, M. M., Battle, K. E., Padilla, C. D., Baird, et al. 2012. G6PD deficiency prevalence and estimates of affected populations in malaria endemic countries: A geostatistical model-based map. *PLoS Med.* 9:e1001339.

Wang, X.-T., and Engel, P. C. 2009. Clinical mutants of human glucose 6-phosphate dehydrogenase: Impairment of $NADP^+$ binding affects both folding and stability. *Biochim. Biophys. Acta* 1792:804–809.

Au, S. W., Gover, S., Lam, V. M., and Adams, M. J. 2000. Human glucose-6-phosphate dehydrogenase: The crystal structure reveals a structural NADP(+) molecule and provides insights into enzyme deficiency. *Struct. Fold. Des.* 8:293–303.

Salvemini, F., Franze, A., Iervolino, A., Filosa, S., Salzano, S., and Ursini, M. V. 1999. Enhanced glutathione levels and oxidoresistance mediated by increased glucose-6-phosphate dehydrogenase expression. *J. Biol. Chem.* 274:2750–2757.

Tian, W. N., Braunstein, L. D., Apse, K., Pang, J., Rose, M., Tian, X., and Stanton, R. C. 1999. Importance of glucose-6-phosphate dehydrogenase activity in cell death. *Am. J. Physiol.* 276:C1121–C1131.

Tian, W. N., Braunstein, L. D., Pang, J., Stuhlmeier, K. M., Xi, Q. C., Tian, X., and Stanton, R. C. 1998. Importance of glucose-6-phosphate dehydrogenase activity for cell growth. *J. Biol. Chem.* 273:10609–10617.

Ursini, M. V., Parrella, A., Rosa, G., Salzano, S., and Martini, G. 1997. Enhanced expression of glucose-6-phosphate dehydrogenase in human cells sustaining oxidative stress. *Biochem. J.* 323:801–806.

Evolution

Williams, B. P., Aubry S., and Hibberd, J. M. 2012. Molecular evolution of genes recruited into C_4 photosynthesis. *Trends Plant Sci.* 4:213–220.

Sage, R. F., Sage, T. L., and Kocacinar, F. 2012. Photorespiration and the evolution of C_4 photosynthesis. *Annu. Rev. Plant Biol.* 63:19–47.

Deschamps, P., Haferkamp, I., d'Hulst, C., Neuhaus, H. E., and Ball, S. G. 2008. The relocation of starch metabolism to chloroplasts: When, why and how. *Trends Plant Sci.* 13:574–582.

Coy, J. F., Dubel, S., Kioschis, P., Thomas, K., Micklem, G., Delius, H., and Poustka, A. 1996. Molecular cloning of tissue-specific transcripts of a transketolase-related gene: Implications for the evolution of new vertebrate genes. *Genomics* 32:309–316.

Schenk, G., Layfield, R., Candy, J. M., Duggleby, R. G., and Nixon, P. F. 1997. Molecular evolutionary analysis of the thiamine-diphosphate-dependent enzyme, transketolase. *J. Mol. Evol.* 44:552–572.

Notaro, R., Afolayan, A., and Luzzatto, L. 2000. Human mutations in glucose 6-phosphate dehydrogenase reflect evolutionary history. *FASEB J.* 14:485–494.

Wedel, N., and Soll, J. 1998. Evolutionary conserved light regulation of Calvin cycle activity by NADPH-mediated reversible phosphoribulokinase/CP12/glyceraldehyde-3-phosphate dehydrogenase complex dissociation. *Proc. Natl. Acad. Sci. U.S.A.* 95:9699–9704.

Martin, W., and Schnarrenberger, C. 1997. The evolution of the Calvin cycle from prokaryotic to eukaryotic chromosomes: A case study of functional redundancy in ancient pathways through endosymbiosis. *Curr. Genet.* 32:1–18.

Ku, M. S., Kano-Murakami, Y., and Matsuoka, M. 1996. Evolution and expression of C4 photosynthesis genes. *Plant Physiol.* 111:949–957.

Pereto, J. G., Velasco, A. M., Becerra, A., and Lazcano, A. 1999. Comparative biochemistry of CO_2 fixation and the evolution of autotrophy. *Int. Microbiol.* 2:3–10.

Biochemistry in Focus 2

Levin, E., Lopez-Martinez, G., Fane, G., and Davidowitz, G. 2017. Hawkmoths use nectar sugar to reduce oxidative damage from flight. *Science* 355:733–735.

CHAPTER 21

Where to Start

Fisher, E. H. 2013. Cellular regulation by protein phosphorylation. *Biochem. Biophys. Res. Commun.* 430:865–867.

Greenberg, C. C., Jurczak, M. J., Danos, A. M., and Brady, M. J. 2006. Glycogen branches out: New perspectives on the role of glycogen metabolism in the integration of metabolic pathways. *Am. J. Physiol. Endocrinol. Metab.* 291:E1–E8.

Books and General Reviews

Roach, P. J, Depaoli-Roach, A. A., Hurley, T. D., and Tagliabracci, V. S. 2012. Glycogen and its metabolism: Some new developments and old themes. *Biochem. J.* 441: 763–787.

Palm, D. C., Rohwer J. M., and Hofmeyr, J.-H. S. 2013. Regulation of glycogen synthase from mammalian skeletal muscle: A unifying view of allosteric and covalent regulation. *FEBS J.* 280:2–27.

Agius, L. 2008. Glucokinase and molecular aspects of liver glycogen metabolism. *Biochem. J.* 414:1–18.

Structural Studies

Mathieu, C., de la Sierra-Gallay, I. L., Duval, R., Xu, X., Cocaign, A., Legerc, T., et al. 2016. Insights into brain glycogen metabolism: The structure of human brain glycogen phosphorylase. *J. Biol Chem.* 291:18072–18083.

Nadeau, O. W., Lane, L. A., Xu, D., Sage, J., Priddy, T. S., Artigues, A., Villar, M. T., Yang, Q., Robinson, C. V., Zhang, Y. et al. 2012. Structure and location of the regulatory β subunits in the $(\alpha\beta\gamma\delta)_4$ phosphorylase kinase complex. *J. Biol. Chem.* 287:36651–36661.

Horcajada, C., Guinovart, J. J., Fita, I., and Ferrer, J. C. 2006. Crystal structure of an archaeal glycogen synthase: Insights into oligomerization and substrate binding of eukaryotic glycogen synthases. *J. Biol. Chem.* 281:2923–2931.

Buschiazzo, A., Ugalde, J. E., Guerin, M. E., Shepard, W., Ugalde, R. A., and Alzari, P. M. 2004. Crystal structure of glycogen synthase: Homologous enzymes catalyze glycogen synthesis and degradation. *EMBO J.* 23:3196–3205.

Gibbons, B. J., Roach, P. J., and Hurley, T. D. 2002. Crystal structure of the autocatalytic initiator of glycogen biosynthesis, glycogenin. *J. Mol. Biol.* 319:463–477.

Priming of Glycogen Synthesis

Lomako, J., Lomako, W. M., and Whelan, W. J. 2004. Glycogenin: The primer for mammalian and yeast glycogen synthesis. *Biochim. Biophys. Acta* 1673:45–55.

Lin, A., Mu, J., Yang, J., and Roach, P. J. 1999. Self-glucosylation of glycogenin, the initiator of glycogen biosynthesis, involves an inter-subunit reaction. *Arch. Biochem. Biophys.* 363:163–170.

Catalytic Mechanisms

Mathieu, C., Bui, L.-C., Petit, E., Haddad, I., Agbulut, O., Vinh, J., Dupret, J.-M., and Rodrigues-Lima, F. 2017. Molecular mechanisms of allosteric inhibition of brain glycogen phosphorylase by neurotoxic dithiocarbamate chemicals. *J. Biol. Chem.* 292:1603–1612.

Skamnaki, V. T., Owen, D. J., Noble, M. E., Lowe, E. D., Lowe, G., Oikonomakos, N. G., and Johnson, L. N. 1999. Catalytic mechanism of phosphorylase kinase probed by mutational studies. *Biochemistry* 38:14718–14730.

Buchbinder, J. L., and Fletterick, R. J. 1996. Role of the active site gate of glycogen phosphorylase in allosteric inhibition and substrate binding. *J. Biol. Chem.* 271:22305–22309.

Regulation of Glycogen Metabolism

Zhang, T., Wang, S., Lin, Y., Xu, W., Ye, D., Xiong, Y., Zhao, S., and Guan, K.-L. 2012. Acetylation negatively regulates glycogen phosphorylase by recruiting protein phosphatase 1. *Cell Metab.* 15:75–87.

Díaz, A., Martínez-Pons, C., Fita, I., Ferrer, J. C., and Guinovart, J. J. 2011. Processivity and subcellular localization of glycogen synthase depend on a non-catalytic high affinity glycogen-binding site. *J. Biol. Chem.* 286:18505–18514.

Bouskila, M., Hunter, R. W., Ibrahim, A. D. F., Delattre, L., Peggie, M., van Diepen, J. A., Voshol, P. J., Jensen, J., and Sakamoto, K. 2010. Allosteric regulation of glycogen synthase controls glycogen synthesis in muscle. *Cell Metab.*12:456–466.

Boulatnikov, I. G., Peters, J. L., Nadeau, O. W., Sage, J. M., Daniels, P. J., Kumar, P., Walsh, D. A., and Carlson, G. M. 2009. Expressed phosphorylase *b* kinase and its αγδ subcomplex as regulatory models for the rabbit skeletal muscle holoenzyme. *Biochemistry* 48:10183–10191.

Ros, S., García-Rocha, M., Domínguez, J., Ferrer, J. C., and Guinovart, J. J. 2009. Control of liver glycogen synthase activity and intracellular distribution by phosphorylation. *J. Biol. Chem.* 284:6370–6378.

Danos, A. M., Osmanovic, S., and Brady, M. J. 2009. Differential regulation of glycogenolysis by mutant protein phosphatase-1 glycogen-targeting subunits. *J. Biol. Chem.* 284:19544–19553.

Pautsch, A., Stadler, N., Wissdorf, O., Langkopf, E., Moreth, M., and Streicher, R. 2008. Molecular recognition of the protein phosphatase 1 glycogen targeting subunit by glycogen phosphorylase. *J. Biol. Chem.* 283:8913–8918.

Jope, R. S., and Johnson, G. V. W. 2004. The glamour and gloom of glycogen synthase kinase-3. *Trends Biochem. Sci.* 29:95–102.

Doble, B. W., and Woodgett, J. R. 2003. GSK-3: Tricks of the trade for a multi-tasking kinase. *J. Cell Sci.* 116:1175–1186.

Pederson, B. A., Cheng, C., Wilson, W. A., and Roach, P. J. 2000. Regulation of glycogen synthase: Identification of residues involved in regulation by the allosteric ligand glucose-6-P and by phosphorylation. *J. Biol. Chem.* 275:27753–27761.

Melendez, R., Melendez-Hevia, E., and Canela, E. I. 1999. The fractal structure of glycogen: A clever solution to optimize cell metabolism. *Biophys. J.* 77:1327–1332.

Franch, J., Aslesen, R., and Jensen, J. 1999. Regulation of glycogen synthesis in rat skeletal muscle after glycogen-depleting contractile activity: Effects of adrenaline on glycogen synthesis and activation of glycogen synthase and glycogen phosphorylase. *Biochem. J.* 344:231–235.

Aggen, J. B., Nairn, A. C., and Chamberlin, R. 2000. Regulation of protein phosphatase-1. *Chem. Biol.* 7: R13–R23.

Egloff, M. P., Johnson, D. F., Moorhead, G., Cohen, P. T., Cohen, P., and Barford, D. 1997. Structural basis for the recognition of regulatory subunits by the catalytic subunit of protein phosphatase 1. *EMBO J.* 16:1876–1887.

Wu, J., Liu, J., Thompson, I., Oliver, C. J., Shenolikar, S., and Brautigan, D. L. 1998. A conserved domain for glycogen binding in protein phosphatase-1 targeting subunits. *FEBS Lett.* 439:185–191.

Genetic Diseases

Nyhan, W. L., Barshop, B. A., and Ozand, P. T. 2005. *Atlas of Metabolic Diseases*. (2nd ed., pp. 373–408). Hodder Arnold.

Chen, Y.-T. 2001. Glycogen storage diseases. In *The Metabolic and Molecular Bases of Inherited Diseases* (8th ed., pp. 1521–1552), edited by C. R. Scriver., W. S. Sly, B. Childs, A. L. Beaudet, D. Valle, K. W. Kinzler, and B. Vogelstein. McGraw-Hill.

Burchell, A., and Waddell, I. D. 1991. The molecular basis of the hepatic microsomal glucose-6-phosphatase system. *Biochim. Biophys. Acta* 1092:129–137.

Lei, K. J., Shelley, L. L., Pan, C. J., Sidbury, J. B., and Chou, J. Y. 1993. Mutations in the glucose-6-phosphatase gene that cause glycogen storage disease type Ia. *Science* 262:580–583.

Ross, B. D., Radda, G. K., Gadian, D. G., Rocker, G., Esiri, M., and Falconer-Smith, J. 1981. Examination of a case of suspected McArdle's syndrome by ^{31}P NMR. *New Engl. J. Med.* 304:1338–1342.

Evolution

Holm, L., and Sander, C. 1995. Evolutionary link between glycogen phosphorylase and a DNA modifying enzyme. *EMBO J.* 14:1287–1293.

Hudson, J. W., Golding, G. B., and Crerar, M. M. 1993. Evolution of allosteric control in glycogen phosphorylase. *J. Mol. Biol.* 234:700–721.

Rath, V. L., and Fletterick, R. J. 1994. Parallel evolution in two homologues of phosphorylase. *Nat. Struct. Biol.* 1:681–690.

Melendez, R., Melendez-Hevia, E., and Cascante, M. 1997. How did glycogen structure evolve to satisfy the requirement for rapid mobilization of glucose? A problem of physical constraints in structure building. *J. Mol. Evol.* 45:446–455.

Rath, V. L., Lin, K., Hwang, P. K., and Fletterick, R. J. 1996. The evolution of an allosteric site in phosphorylase. *Structure* 4:463–473.

Biochemistry in Focus

Delaney, N. F., Sharma, R., Tadvalkar, L., Clish, C. B., Hallerd, R. G., and Mootha, V. K. 2017. Metabolic profiles of exercise in patients with McArdle disease or mitochondrial myopathy. *Proc. Natl. Acad. Sci. U.S.A.* 114:8402–8407.

CHAPTER 22

Where to Start

Martin, S. A., Brash, A. R., and Robert C. Murphy R. C. 2016. The discovery and early structural studies of arachidonic acid. *J. Lipid Res.* 57:1126–1132

Walther, T. C., and Farese Jr., R. V. 2012. Lipid droplets and cellular lipid metabolism. *Annu. Rev. Biochem.* 81:687–714.

Granneman, J. G., and Moore, H.-P. 2008. Location, location: Protein trafficking and lipolysis in adipocytes. *Trends Endocrinol. Metab.* 19:3–9.

Yang, L., Ding, Y., Chen, Y., Zhang, S., Huo, C., Wang, Y., Yu, J., Zhang, P., Na, H., Zhang, H., et al. 2012. The proteomics of lipid droplets: Structure, dynamics, and functions of the organelle conserved from bacteria to humans. *J. Lipid Res.* 53:1245–1253.

Rinaldo, P., Matern, D., and Bennet, M. J. 2002. Fatty acid oxidation disorders. *Annu. Rev. Physiol.* 64:477–502.

Rasmussen, B. B., and Wolfe, R. R. 1999. Regulation of fatty acid oxidation in skeletal muscle. *Annu. Rev. Nutr.* 19:463–484.

Semenkovich, C. F. 1997. Regulation of fatty acid synthase (FAS). *Prog. Lipid Res.* 36:43–53.

Wolf, G. 1996. Nutritional and hormonal regulation of fatty acid synthase. *Nutr. Rev.* 54:122–123.

Books

Lawrence, G. D. 2010. *The Fats of Life: Essential Fatty Acids in Health and Disease.* Rutgers University Press.

Vance, D. E., and Vance, J. E. (Eds.). 2008. *Biochemistry of Lipids, Lipoproteins, and Membranes.* Elsevier.

Stipanuk, M. H., and Caudill, M. A. (Eds.). 2012. *Biochemical and Physiological Aspects of Human Nutrition* (3rd ed.) Saunders.

Fatty Acid Oxidation

Ross, L. E., Xiao, X., and Lowe, M. E. 2013. Identification of amino acids in human colipase that mediate adsorption to lipid emulsions and mixed micelles. *Biochim. Biophys. Acta* 1831:1052–1059.

Badin, P. M., Loubière, C., Coonen, M., Louche, K., Tavernier, G., Bourlier, V., Mairal, A., Rustan, A. C., Smith, S. R., Langin, D., et al. 2012. Regulation of skeletal muscle lipolysis and oxidative metabolism by the co-lipase CGI-58. *J. Lipid Res.* 53:839–848.

Yang, X., Lu, X., Lombès, M., Rha, G. B., Chi, Y. I., Guerin, T. M., Smart, E. J., Liu, J. 2010. The G_0/G_1 switch gene 2 regulates adipose lipolysis through association with adipose triglyceride lipase. *Cell Metab.* 11:194–205.

Wang, Y., Mohsen, A.-W., Mihalik, S. J., Goetzman, E. S., Vockley, J. 2010. Evidence for physical association of mitochondrial fatty acid oxidation and oxidative phosphorylation complexes. *J. Biol.Chem.* 285:29834–29841.

Ahmadian, M., Duncan, R. E., and Sul, H. S. 2009. The skinny on fat: Lipolysis and fatty acid utilization in adipocytes. *Trends Endocrinol. Metab.* 20:424–428.

Farese, R. V., Jr., and Walther, T. C. 2009. Lipid droplets finally get a little R-E-S-P-E-C-T. *Cell* 139:855–860.

Goodman, J. L. 2008. The gregarious lipid droplet. *J. Biol. Chem.* 283:28005–28009.

Saha, P. K., Kojima, H., Marinez-Botas, J., Sunehag, A. L., and Chan, L. 2004. Metabolic adaptations in absence of perilipin. *J. Biol. Chem.* 279:35150–35158.

Ramsay, R. R. 2000. The carnitine acyltransferases: Modulators of acyl-CoA-dependent reactions. *Biochem. Soc. Trans.* 28:182–186.

Fatty Acid Synthesis

Sun, T., Hayakawa, K., Bateman, K. S., and Fraser, M. E. 2010. Identification of the citrate-binding site of human ATP-citrate lyase using X-ray crystallography. *J. Biol. Chem.* 285:27418–27428.

Fan, F., Williams, H. J., Boyer, J. G., Graham, T. L., Zhao, H., Lehr, R., Qi, H., Schwartz, B., Raushel, F. M., and Meek, T. D. 2012. On the catalytic mechanism of human ATP citrate lyase. *Biochemistry* 51:5198–5211.

Chypre, M., Zaidi, N., and Smans, K. 2012. ATP-citrate lyase: A mini-review. *Biochem. Biophys. Res. Commun.* 422:1–4.

Maier, T., Leibundgut, M., and Ban, N. 2008. The crystal structure of a mammalian fatty acid synthase. *Science* 321:1315–1322.

Ming, D., Kong, Y., Wakil, S. J., Brink, J., and Ma, J. 2002. Domain movements in human fatty acid synthase by quantized elastic deformational model. *Proc. Natl. Acad. Sci. U.S.A.* 99:7895–7899.

Zhang, Y.-M., Rao, M. S., Heath, R. J., Price, A. C., Olson, A. J., Rock, C. O., and White, S. W. 2001. Identification and analysis of the acyl carrier protein (ACP) docking site on β-ketoacyl-ACP synthase III. *J. Biol. Chem.* 276:8231–8238.

Davies, C., Heath, R. J., White, S. W., and Rock, C. O. 2000. The 1.8 Å crystal structure and active-site architecture of β-ketoacyl-acyl carrier protein synthase III (FabH) from *Escherichia coli. Struct. Fold. Design* 8:185–195.

Loftus, T. M., Jaworsky, D. E., Frehywot, G. L., Townsend, C. A., Ronnett, G. V., Lane, M. D., and Kuhajda, F. P. 2000. Reduced food intake and body weight in mice treated with fatty acid synthase inhibitors. *Science* 288:2379–2381.

Acetyl CoA Carboxylase

Kim, C.-W., Moon, Y.-A., Park, S. W., Cheng, D., Kwon, H. J., and Horton, J. D. 2010. Induced polymerization of mammalian acetyl-CoA carboxylase by MIG12 provides a tertiary level of regulation of fatty acid synthesis. *Proc. Natl. Acad. Sci. U.S.A.* 107:9626–9631.

Brownsey, R. W., Boone, A. N., Elliott, J. E., Kulpa, J. E., and Lee, W. M. 2006. Regulation of acetyl-CoA carboxylase. *Biochem. Soc. Trans.* 34:223–227.

Hardie, D. G., Ross, F. A., and Hawley, S. A. 2013. AMP-activated protein kinase: A target for drugs both ancient and modern. *Chem. Biol.* 19:1222–1236.

Munday, M. R. 2002. Regulation of acetyl CoA carboxylase. *Biochem. Soc. Trans.* 30:1059–1064.

Thoden, J. B., Blanchard, C. Z., Holden, H. M., and Waldrop, G. L. 2000. Movement of the biotin carboxylase B-domain as a result of ATP binding. *J. Biol. Chem.* 275:16183–16190.

Eicosanoids

De Caterina, R. 2011. n–3 Fatty acids in cardiovascular disease. *New Engl. J. Med.* 364:2439–2450.

Harizi, H., Corcuff, J.-B., and Gualde, N. 2008. Arachidonic-acid-derived eicosanoids: Roles in biology and immunopathology. *Trends Mol. Med.* 14:461–469.

Nakamura, M. T., and Nara, T. Y. 2004. Structure, function, and dietary regulation of $\Delta 6$, $\Delta 5$, and $\Delta 9$ desaturases. *Annu. Rev. Nutr.* 24:345–376.

Malkowski, M. G., Ginell, S. L., Smith, W. L., and Garavito, R. M. 2000. The productive conformation of arachidonic acid bound to prostaglandin synthase. *Science* 289:1933–1937.

Smith, T., McCracken, J., Shin, Y.-K., and DeWitt, D. 2000. Arachidonic acid and nonsteroidal anti-inflammatory drugs induce conformational changes in the human prostaglandin endoperoxide H2 synthase-2 (cyclooxygenase-2). *J. Biol. Chem.* 275:40407–40415.

Kalgutkar, A. S., Crews, B. C., Rowlinson, S. W., Garner, C., Seibert, K., and Marnett L. J. 1998. Aspirin-like molecules that covalently inactivate cyclooxygenase-2. *Science* 280:1268–1270.

Lands, W. E. 1991. Biosynthesis of prostaglandins. *Annu. Rev. Nutr.* 11:41–60.

Sigal, E. 1991. The molecular biology of mammalian arachidonic acid metabolism. *Am. J. Physiol.* 260:L13–L28.

Weissmann, G. 1991. Aspirin. *Sci. Am.* 264(1):84–90.

Vane, J. R., Flower, R. J., and Botting, R. M. 1990. History of aspirin and its mechanism of action. *Stroke* (12 suppl.):IV12–IV23.

Genetic Diseases and Cancer

Day, E. A., Ford, R. J., and Steinberg G. R. 2017. AMPK as a therapeutic target for treating metabolic diseases. *Trends Endocrinol. Metab.* 28:545–560.

Celestino-Soper, P. B. S., Violante, S., Crawford, E. L., Luo, R., Lionel, A. C., Delaby, E., Cai, G., Sadikovic, B., Lee, K., Lo, C., et al. 2012. A common X-linked inborn error of carnitine biosynthesis may be a risk factor for nondysmorphic autism. *Proc. Natl. Acad. Sci. U.S.A.* 109:7947–7981.

Currie, E., Schulze, A., Zechner, R., Walther, T. C., and Farese, Jr. R. V. 2013. Cellular fatty acid metabolism and cancer. *Cell. Metab.* 18:153–161.

Lutas, A., and Yellen, G. 2013. The ketogenic diet: Metabolic influences on brain excitability and epilepsy. *Trends Neurosci.* 36:32–40.

Beckers, A., Organe, S., Timmermans, L., Scheys, K., Peeters, A., Brusselmans, K., Verhoeven, G., and Swinnen, J. V. 2007. Chemical inhibition of acetyl-CoA carboxylase induces growth arrest and cytotoxicity selectively in cancer cells. *Cancer Res.* 67:8180–8187.

Kuhajda, F. P. 2006. Fatty acid synthase and cancer: New application of an old pathway. *Cancer Res.* 66:5977–5980.

Nyhan, W. L., Barshop, B. A., and Ozand, P. T. 2005. *Atlas of Metabolic Diseases* (2nd ed., pp. 339–300). Hodder Arnold.

Roe, C. R., and Coates, P. M. 2001. Mitochondrial fatty acid oxidation disorders. In *The Metabolic and Molecular Bases of Inherited Diseases* (8th ed., pp. 2297–2326), edited by C. R. Scriver., W. S. Sly, B. Childs, A. L. Beaudet, D. Valle, K. W. Kinzler, and B. Vogelstein. McGraw-Hill.

Brivet, M., Boutron, A., Slama, A., Costa, C., Thuillier, L., Demaugre, F., Rabier, D., Saudubray, J. M., and Bonnefont, J. P. 1999. Defects in activation and transport of fatty acids. *J. Inherit. Metab. Dis.* 22:428–441.

Wanders, R. J., van Grunsven, E. G., and Jansen, G. A. 2000. Lipid metabolism in peroxisomes: Enzymology, functions and dysfunctions of the fatty acid α-and β-oxidation systems in humans. *Biochem. Soc. Trans.* 28:141–149.

Wanders, R. J., Vreken, P., den Boer, M. E., Wijburg, F. A., van Gennip, A. H., and Ijist, L. 1999. Disorders of mitochondrial fatty acyl-CoA β-oxidation. *J. Inherit. Metab. Dis.* 22:442–487.

Kerner, J., and Hoppel, C. 1998. Genetic disorders of carnitine metabolism and their nutritional management. *Annu. Rev. Nutr.* 18:179–206.

Bartlett, K., and Pourfarzam, M. 1998. Recent developments in the detection of inherited disorders of mitochondrial β-oxidation. *Biochem. Soc. Trans.* 26:145–152.

Pollitt, R. J. 1995. Disorders of mitochondrial long-chain fatty acid oxidation. *J. Inherit. Metab. Dis.* 18:473–490.

Industrial Applications

Xu, J., Liu, N., Qiao, K., Vogg, S., and Stephanopoulos, G. 2017. Application of metabolic controls for the maximization of lipid production in semicontinuous fermentation. *Proc. Natl. Acad. Sci. U. S. A.* 114:27–32.

Biochemistry in Focus

Schott, M. B., Rasineni, K., Weller, S. G., Schulze, R. J., Sletten, A. C., Casey, C. A., and McNiven, M. A. 2017. β-Adrenergic induction of lipolysis in hepatocytes is inhibited by ethanol exposure. *J. Biol. Chem.* 292:11815–11828.

CHAPTER 23

Where to Start

Kwon, Y. T., and Ciechanover, A. 2017. The ubiquitin code in the ubiquitin-proteasome system and autophagy. *Trends Biochem. Sci.* 42:873–886.

Varshavsky, A. 2012. The Ubiquitin System, an Immense Realm. *Annu. Rev. Biochem.* 81:167–76.

Ubiquitin-Mediated Protein Regulation. 2009. *Annu. Rev. Biochem.* 78: A series of reviews on the various roles of ubiquitin.

Torchinsky, Y. M. 1989. Transamination: Its discovery, biological and chemical aspects. *Trends Biochem. Sci.* 12:115–117.

Watford, M. 2003. The urea cycle. *Biochem. Mol. Biol. Ed.* 31:289–297.

Books

Magnusson, S. 2010. *Life of Pee: The Story of How Urine Got Everywhere.* Aurum.

Bender, D. A. 2012. *Amino Acid Metabolism* (3rd ed.). Wiley-Blackwell.

Lippard, S. J., and Berg, J. M. 1994. *Principles of Bioinorganic Chemistry.* University Science Books.

Walsh, C. 1979. *Enzymatic Reaction Mechanisms.* W. H. Freeman and Company.

Ubiquitin and the Proteasome

Collins, G. A., and Goldberg, A. L. 2017. The logic of the 26S proteasome. *Cell* 169:792–806

Shemorry, A., Hwang, C.-S., and Varshavsky, A. 2013. Control of protein quality and stoichiometries by N-terminal acetylation and the N-end rule pathway. *Mol. Cell* 50:540–551.

Liu, C.-W., and Jacobson, A. D. 2013. Functions of the 19S complex in proteasomal degradation. *Trends Biochem. Sci.* 38:103–110.

Ehlinger, A., and Walters, K. J. 2013. Structural insights into proteasome activation by the 19S regulatory particle. *Biochemistry* 52:3618–3628.

Peth, A., Nathan, J. A., and Goldberg, A. L. 2013. The ATP costs and time required to degrade ubiquitinated proteins by the 26 S proteasome. *J. Biol. Chem.* 288:29215–29222.

Tomko, Jr., R. J., and Hochstrasser, M. 2013. Molecular architecture and assembly of the eukaryotic proteasome. *Annu. Rev. Biochem.* 82:415–445.

Komander, D., and Rape, M. 2012. The ubiquitin code. *Annu. Rev. Biochem.* 81:203–229.

Greer, P. L., Hanayama, R., Bloodgood, B. L., Mardinly, A. R., Lipton, D. M., Flavell, S. W., Kim, T.-K., Griffith, E. C., Waldon, Z., Maehr, R., et al. 2010. The Angelman syndrome protein Ube3A regulates synapse development by ubiquitinating Arc. *Cell* 140:704–716.

Peth, A., Besche, H. C., and Goldberg A. L. 2009. Ubiquitinated proteins activate the proteasome by binding to Usp14/Ubp6, which causes 20S gate opening. *Mol. Cell* 36:794–804.

Lin, G., Li, D., Carvalho, L. P. S., Deng, H., Tao, H., Vogt, G., Wu, K., Schneider, J., Chidawanyika, T., Warren, J. D., et al. 2009. Inhibitors selective for mycobacterial versus human proteasomes. *Nature* 461:621–626.

Giasson, B. I., and Lee, V. M.-Y. 2003. Are ubiquitination pathways central to Parkinson's disease? *Cell* 114:1–8.

Pagano, M., and Benmaamar, R. 2003. When protein destruction runs amok, malignancy is on the loose. *Cancer Cell* 4:251–256.

Hochstrasser, M. 2000. Evolution and function of ubiquitin-like protein-conjugation systems. *Nat. Cell Biol.* 2: E153–E157.

Pyridoxal Phosphate-Dependent Enzymes

Dajnowicz, S., Parks, J. M., Hu, X., Gesler, K., Kovalevsky, A. Y., and Mueser, T. C. 2017. Direct evidence that an extended hydrogen-bonding network influences activation of pyridoxal 5-phosphate in aspartate aminotransferase. *J. Biol. Chem.* 292:5970–5980.

Eliot, A. C., and Kirsch, J. F. 2004. Pyridoxal phosphate enzymes: Mechanistic, structural, and evolutionary considerations. *Annu. Rev. Biochem.* 73:383–415.

Mehta, P. K., and Christen, P. 2000. The molecular evolution of pyridoxal-5′-phosphate-dependent enzymes. *Adv. Enzymol. Relat. Areas Mol. Biol.* 74:129–184.

Schneider, G., Kack, H., and Lindqvist, Y. 2000. The manifold of vitamin B_6 dependent enzymes. *Structure Fold Des.* 8:R1–R6.

Urea Cycle Enzymes

Haeussinger, D., and Sies, H. 2013. Hepatic encephalopathy: Clinical aspects and pathogenetic concept. *Arch. Biochem. Biophys.* 536:97–100.

Li, M., Li, C., Allen, A., Stanley, C. A., and Smith, T. J. 2012. The structure and allosteric regulation of mammalian glutamate dehydrogenase. *Arch. Biochem. Biophys.* 519:69–80.

Nakagawa, T., Lomb, D. J., Haigis, M. C., and Guarente, L. 2009. SIRT5 deacetylates carbamoyl phosphate synthetase 1 and regulates the urea cycle. *Cell* 137:560–570.

Lawson, F. S., Charlebois, R. L., and Dillon, J. A. 1996. Phylogenetic analysis of carbamoylphosphate synthetase genes: Complex evolutionary history includes an internal duplication within a gene which can root the tree of life. *Mol. Biol. Evol.* 13:970–977.

McCudden, C. R., and Powers-Lee, S. G. 1996. Required allosteric effector site for *N*-acetylglutamate on carbamoyl-phosphate synthetase I. *J. Biol. Chem.* 271:18285–18294.

Amino Acid Degradation

Gerson, A. R., and Guglielmo, C. G. 2011. Flight at Low Ambient Humidity Increases Protein Catabolism in Migratory Birds. *Science* 333:1434–1436.

Li, M., Smith, C. J., Walker, M. T., and Smith, T. J. 2009. Novel inhibitors complexed with glutamate dehydrogenase: Allosteric regulation by control of protein dynamics. *J. Biol. Chem.* 284:22988–23000.

Smith, T. J., and Stanley, C. A. 2008. Untangling the glutamate dehydrogenase allosteric nightmare. *Trends Biochem. Sci.* 33:557–564.

Fusetti, F., Erlandsen, H., Flatmark, T., and Stevens, R. C. 1998. Structure of tetrameric human phenylalanine hydroxylase and its implications for phenylketonuria. *J. Biol. Chem.* 273:16962–16967.

Titus, G. P., Mueller, H. A., Burgner, J., Rodriguez De Cordoba, S., Penalva, M. A., and Timm, D. E. 2000. Crystal structure of human homogentisate dioxygenase. *Nat. Struct. Biol.* 7:542–546.

Erlandsen, H., and Stevens, R. C. 1999. The structural basis of phenylketonuria. *Mol. Genet. Metab.* 68:103–125.

Genetic Diseases

Jayakumar, A. R., Liu, M., Moriyama, M. Ramakrishnan, R., Forbush III, B., Reddy, P. V. V., and Norenberg, M. D. 2008. Na-K-Cl cotransporter-1 in the mechanism of ammonia-induced astrocyte swelling. *J. Biol. Chem.* 283:33874–33882.

Scriver, C. R., and Sly, W. S. (Eds.), Childs, B., Beaudet, A. L., Valle, D., Kinzler, K. W., and Vogelstein, B. 2001. *The Metabolic Basis of Inherited Disease* (8th ed.). McGraw-Hill.

Historical Aspects and the Process of Discovery

Cooper, A. J. L., and Meister, A. 1989. An appreciation of Professor Alexander E. Braunstein: The discovery and scope of enzymatic transamination. *Biochimie* 71:387–404.

Garrod, A. E. 1909. *Inborn Errors in Metabolism*. Oxford University Press (reprinted in 1963 with a supplement by H. Harris).

Childs, B. 1970. Sir Archibald Garrod's conception of chemical individuality: A modern appreciation. *New Engl. J. Med.* 282:71–78.

Holmes, F. L. 1980. Hans Krebs and the discovery of the ornithine cycle. *Fed. Proc.* 39:216–225.

CHAPTER 24

Where to Start

Brewin, N. J. 2013. Legume root nodule symbiosis. *The Biochemist* 35:14–18.

Christen, P., Jaussi, R., Juretic, N., Mehta, P. K., Hale, T. I., and Ziak, M. 1990. Evolutionary and biosynthetic aspects of aspartate aminotransferase isoenzymes and other aminotransferases. *Ann. N. Y. Acad. Sci.* 585:331–338.

Schneider, G., Kack, H., and Lindqvist, Y. 2000. The manifold of vitamin B_6 dependent enzymes. *Structure Fold Des.* 8:R1–R6.

Rhee, S. G., Chock, P. B., and Stadtman, E. R. 1989. Regulation of *Escherichia coli* glutamine synthetase. *Adv. Enzymol. Mol. Biol.* 62:37–92.

Shemin, D. 1989. An illustration of the use of isotopes: The biosynthesis of porphyrins. *Bioessays* 10:30–35.

Books

Wu, G. 2013. *Amino Acids: Biochemistry and Nutrition.* CRC Press.

Bender, D. A. 2012. *Amino Acid Metabolism* (3rd ed.). Wiley-Blackwell.

Scriver, C. R. (Ed.), Sly, W. S. (Ed.), Childs, B., Beaudet, A. L., Valle, D., Kinzler, K. W., and Vogelstein, B. 2001. *The Metabolic Basis of Inherited Disease* (8th ed.). McGraw-Hill.

McMurry, J. E., and Begley, T. P. 2005. *The Organic Chemistry of Biological Pathways.* Roberts and Company.

Blakley, R. L., and Benkovic, S. J. 1989. *Folates and Pterins* (vol. 2). Wiley.

Walsh, C. 1979. *Enzymatic Reaction Mechanisms.* W. H. Freeman and Company.

Nitrogen Fixation

Spatzal, T., Aksoyoglu, M., Zhang, L., Andrade, S. L. A., Schleicher, E., Weber, S., Rees, D. C., Einsle, O. 2011. Evidence for interstitial carbon in nitrogenase FeMo Cofactor. *Science* 334:940.

Lancaster, K. M., Roemelt, M., Ettenhuber, P., Hu, Y., Ribbe, M. W., Neese, F., Bergmann, U., DeBeer, S. 2011. X-ray emission spectroscopy evidences a central carbon in the nitrogenase iron-molybdenum cofactor. *Science* 334:974–977.

Seefeldt, L. C., Hoffman, B. M., and Dean, D. R. 2009. Mechanism of Mo-dependent nitrogenase. *Annu. Rev. Biochem.* 79:701–722.

Halbleib, C. M., and Ludden, P. W. 2000. Regulation of biological nitrogen fixation. *J. Nutr.* 130:1081–1084.

Einsle, O., Tezcan, F. A., Andrade, S. L., Schmid, B., Yoshida, M., Howard, J. B., and Rees, D. C. 2002. Nitrogenase MoFe-protein at 1.16 Å resolution: A central ligand in the FeMo-cofactor. *Science* 297:1696–1700.

Benton, P. M., Laryukhin, M., Mayer, S. M., Hoffman, B. M., Dean, D. R., and Seefeldt, L. C. 2003. Localization of a substrate binding site on the FeMo-cofactor in nitrogenase: Trapping propargyl alcohol with an α-70-substituted MoFe protein. *Biochemistry* 42:9102–9109.

Regulation of Amino Acid Biosynthesis

Li, Y., Zhang, H., Jiang, C., Xu, M., Pang, Y., Feng, J., Xiang, X., Kong, W., Xu, G., Li, Y., et al. 2013. Hyperhomocysteinemia promotes insulin resistance by inducing endoplasmic reticulum stress in adipose tissue. *J. Biol. Chem.* 288:9583–9592.

Eisenberg, D., Gill, H. S., Pfluegl, G. M., and Rotstein, S. H. 2000. Structure-function relationships of glutamine synthetases. *Biochim. Biophys. Acta* 1477:122–145.

Purich, D. L. 1998. Advances in the enzymology of glutamine synthesis. *Adv. Enzymol. Relat. Areas Mol. Biol.* 72:9–42.

Yamashita, M. M., Almassy, R. J., Janson, C. A., Cascio, D., and Eisenberg, D. 1989. Refined atomic model of glutamine synthetase at 3.5 Å resolution. *J. Biol. Chem.* 264:17681–17690.

Schuller, D. J., Grant, G. A., and Banaszak, L. J. 1995. The allosteric ligand site in the V_{max}-type cooperative enzyme phosphoglycerate dehydrogenase. *Nat. Struct. Biol.* 2:69–76.

Rhee, S. G., Park, R., Chock, P. B., and Stadtman, E. R. 1978. Allosteric regulation of monocyclic interconvertible enzyme cascade systems: Use of *Escherichia coli* glutamine synthetase as an experimental model. *Proc. Natl. Acad. Sci. U.S.A.* 75:3138–3142.

Wessel, P. M., Graciet, E., Douce, R., and Dumas, R. 2000. Evidence for two distinct effector-binding sites in threonine deaminase by site-directed mutagenesis, kinetic, and binding experiments. *Biochemistry* 39:15136–15143.

James, C. L., and Viola, R. E. 2002. Production and characterization of bifunctional enzymes: Domain swapping to produce new bifunctional enzymes in the aspartate pathway. *Biochemistry* 41:3720–3725.

Xu, Y., Carr, P. D., Huber, T., Vasudevan, S. G., and Ollis, D. L. 2001. The structure of the PII-ATP complex. *Eur. J. Biochem.* 268:2028–2037.

Krappmann, S., Lipscomb, W. N., and Braus, G. H. 2000. Coevolution of transcriptional and allosteric regulation at the chorismate metabolic branch point of *Saccharomyces cerevisiae*. *Proc. Natl. Acad. Sci. U.S.A.* 97:13585–13590.

Aromatic Amino Acid Biosynthesis

Brown, K. A., Carpenter, E. P., Watson, K. A., Coggins, J. R., Hawkins, A. R., Koch, M. H., and Svergun, D. I. 2003. Twists and turns: A tale of two shikimate-pathway enzymes. *Biochem. Soc. Trans.* 31:543–547.

Pan, P., Woehl, E., and Dunn, M. F. 1997. Protein architecture, dynamics and allostery in tryptophan synthase channeling. *Trends Biochem. Sci.* 22:22–27.

Sachpatzidis, A., Dealwis, C., Lubetsky, J. B., Liang, P. H., Anderson, K. S., and Lolis, E. 1999. Crystallographic studies of phosphonate-based α-reaction transition-state analogues complexed to tryptophan synthase. *Biochemistry* 38:12665–12674.

Weyand, M., and Schlichting, I. 1999. Crystal structure of wild-type tryptophan synthase complexed with the natural substrate indole-3-glycerol phosphate. *Biochemistry* 38:16469–16480.

Crawford, I. P. 1989. Evolution of a biosynthetic pathway: The tryptophan paradigm. *Annu. Rev. Microbiol.* 43:567–600.

Carpenter, E. P., Hawkins, A. R., Frost, J. W., and Brown, K. A. 1998. Structure of dehydroquinate synthase reveals an active site capable of multistep catalysis. *Nature* 394:299–302.

Schlichting, I., Yang, X. J., Miles, E. W., Kim, A. Y., and Anderson, K. S. 1994. Structural and kinetic analysis of a channel-impaired mutant of tryptophan synthase. *J. Biol. Chem.* 269:26591–26593.

Glutathione

Edwards, R., Dixon, D. P., and Walbot, V. 2000. Plant glutathione *S*-transferases: Enzymes with multiple functions in sickness and in health. *Trends Plant Sci.* 5:193–198.

Lu, S. C. 2000. Regulation of glutathione synthesis. *Curr. Top. Cell Regul.* 36:95–116.

Schulz, J. B., Lindenau, J., Seyfried, J., and Dichgans, J. 2000. Glutathione, oxidative stress and neurodegeneration. *Eur. J. Biochem.* 267:4904–4911.

Lu, S. C. 1999. Regulation of hepatic glutathione synthesis: Current concepts and controversies. *FASEB J.* 13:1169–1183.

Ethylene and Nitric Oxide

Hill, B. G., Dranka, B. P., Baily, S. M., Lancaster, Jr., J. R., and Darley-Usmar, V. M. 2010. What part of NO don't you understand? Some answers to the cardinal questions in nitric oxide biology. *J. Biol. Chem.* 285:19699–19704.

Nisoli, E., Falcone, S., Tonello, C., Cozzi, V., Palomba, L., Fiorani, M., Pisconti, A., Brunelli, S., Cardile, A., Francolini, M., et al. 2004. Mitochondrial biogenesis by NO yields functionally active mitochondria in mammals. *Proc. Natl. Acad. Sci U.S.A.* 101:16507–16512.

Bretscher, L. E., Li, H., Poulos, T. L., and Griffith, O. W. 2003. Structural characterization and kinetics of nitric oxide synthase inhibition by novel N_5-(iminoalkyl)- and N_5-(iminoalkenyl)-ornithines. *J. Biol. Chem.* 278:46789–46797.

Haendeler, J., Zeiher, A. M., and Dimmeler, S. 1999. Nitric oxide and apoptosis. *Vitam. Horm.* 57:49–77.

Capitani, G., Hohenester, E., Feng, L., Storici, P., Kirsch, J. F., and Jansonius, J. N. 1999. Structure of 1-aminocyclopropane-1-carboxylate synthase, a key enzyme in the biosynthesis of the plant hormone ethylene. *J. Mol. Biol.* 294:745–756.

Hobbs, A. J., Higgs, A., and Moncada, S. 1999. Inhibition of nitric oxide synthase as a potential therapeutic target. *Annu. Rev. Pharmacol. Toxicol.* 39:191–220.

Stuehr, D. J. 1999. Mammalian nitric oxide synthases. *Biochim. Biophys. Acta* 1411:217–230.

Chang, C., and Shockey, J. A. 1999. The ethylene-response pathway: Signal perception to gene regulation. *Curr. Opin. Plant Biol.* 2:352–358.

Theologis, A. 1992. One rotten apple spoils the whole bushel: The role of ethylene in fruit ripening. *Cell* 70:181–184.

Biosynthesis of Porphyrins

Kaasik, K., and Lee, C. C. 2004. Reciprocal regulation of haem biosynthesis and the circadian clock in mammals. *Nature* 430:467–471.

Leeper, F. J. 1989. The biosynthesis of porphyrins, chlorophylls, and vitamin B_{12}. *Nat. Prod. Rep.* 6:171–199.

Porra, R. J., and Meisch, H.-U. 1984. The biosynthesis of chlorophyll. *Trends Biochem. Sci.* 9:99–104.

CHAPTER 25

Where to Start

Sutherland, J. D. 2010. Ribonucleotides. *Cold Spring Harb. Perspect. Biol.* 2:a005439.

Ipata, P. L. 2011. Origin, utilization, and recycling of nucleosides in the central nervous system. *Adv. Physiol. Educ.* 35:342–346.

Ordi, J., Alonso, P. L., de Zulueta, J., Esteban, J., Velasco, M., Mas, E., Campo, E., and Fernández, P. L. 2006. The severe gout of Holy Roman Emperor Charles V. *New Eng. J. Med.* 355:516–520.

Kappock, T. J., Ealick, S. E., and Stubbe, J. 2000. Modular evolution of the purine biosynthetic pathway. *Curr. Opin. Chem. Biol.* 4:567–572.

Jordan, A., and Reichard, P. 1998. Ribonucleotide reductases. *Annu. Rev. Biochem.* 67:71–98.

Pyrimidine Biosynthesis

Raushel, F. M., Thoden, J. B., Reinhart, G. D., and Holden, H. M. 1998. Carbamoyl phosphate synthetase: A crooked path from substrates to products. *Curr. Opin. Chem. Biol.* 2:624–632.

Huang, X., Holden, H. M., and Raushel, F. M. 2001. Channeling of substrates and intermediates in enzyme-catalyzed reactions. *Annu. Rev. Biochem.* 70:149–180.

Begley, T. P., Appleby, T. C., and Ealick, S. E. 2000. The structural basis for the remarkable proficiency of orotidine 5′-monophosphate decarboxylase. *Curr. Opin. Struct. Biol.* 10:711–718.

Traut, T. W., and Temple, B. R. 2000. The chemistry of the reaction determines the invariant amino acids during the evolution and divergence of orotidine 5′-monophosphate decarboxylase. *J. Biol. Chem.* 275:28675–28681.

Purine Biosynthesis

Pedley, A. M., and Benkovic, S. J. 2017. A new view into the regulation of purine metabolism: The purinosome. *Trends Biochem. Sci.* 42:141–154.

Zhao, H., French, J. B., Fang, Y., and Benkovic, S. J. 2013. The purinosome, a multi-protein complex involved in the de novo biosynthesis of purines in humans. *Chem. Commun.* 49:4444–4452.

Verrier, F., An, S., Ferrie, A. M., Sun, H., Kyoung, M., Deng, H., Fang, Y., and Benkovic, S. J. 2011. GPCRs regulate the assembly of a multienzyme complex for purine biosynthesis. *Nat. Chem. Biol.* 7:909–915.

Mastrangelo, L., Kim, J.-E., Miyanohara, A., Kang, T. H., and Friedmann, T. 2012. Purinergic signaling in human pluripotent stem cells is regulated by the housekeeping gene encoding hypoxanthine guanine phosphoribosyltransferase. *Proc. Natl. Acad. Sci. U.S.A.* 109:3377–3382.

An, S., Kyoung, M., Allen, J. J., Shokat, K. M., and Benkovic, S. J. 2010. Dynamic regulation of a metabolic multi-enzyme complex by protein kinase CK2. *J. Biol. Chem.* 285:11093–11099.

Thoden, J. B., Firestine, S., Nixon, A., Benkovic, S. J., and Holden, H. M. 2000. Molecular structure of *Escherichia coli* PurT-encoded glycinamide ribonucleotide transformylase. *Biochemistry* 39:8791–8802.

McMillan, F. M., Cahoon, M., White, A., Hedstrom, L., Petsko, G. A., and Ringe, D. 2000. Crystal structure at 2.4 Å resolution of *Borrelia burgdorferi* inosine 5′-monophosphate dehydrogenase: Evidence of a sub-

strate-induced hinged-lid motion by loop 6. *Biochemistry* 39:4533–4542.

Levdikov, V. M., Barynin, V. V., Grebenko, A. I., Melik-Adamyan, W. R., Lamzin, V. S., and Wilson, K. S. 1998. The structure of SAICAR synthase: An enzyme in the de novo pathway of purine nucleotide biosynthesis. *Structure* 6:363–376.

Ribonucleotide Reductases

Ahmad, M. F., and Dealwis, C. G. 2013. The structural basis for the allosteric regulation of ribonucleotide reductase. *Prog. Mol. Biol. Transl. Sci.* 117:389–410.

Minnihan, E. C., Nocera, D. G., and Stubbe, J. 2013. Reversible, long-range radical transfer in *E. coli* class Ia ribonucleotide reductase. *Acc. Chem. Res.* 46:2524–2535.

Reichard, P. 2010. Ribonucleotide reductases: Substrate specificity by allostery. *Biochem. Biophys. Res. Commun.* 396:19–23

Avval, F. Z., and Holmgren, A. 2009. Molecular mechanisms of thioredoxin and glutaredoxin as hydrogen donors for mammalian S phase ribonucleotide reductase. *J. Biol. Chem.* 284:8233–8240.

Rofougaran, R., Crona M., Vodnala, M., Sjöberg, B. M., and Hofer, A. 2008. Oligomerization status directs overall activity regulation of the *Escherichia coli* class Ia ribonucleotide reductase. *J. Biol. Chem.* 283:35310–35318.

Nordlund, P., and Reichard, P. 2006. Ribonucleotide reductases. *Annu. Rev. Biochem.* 75:681–706.

Eklund, H., Uhlin, U., Farnegardh, M., Logan, D. T., and Nordlund, P. 2001. Structure and function of the radical enzyme ribonucleotide reductase. *Prog. Biophys. Mol. Biol.* 77:177–268.

Reichard, P. 1997. The evolution of ribonucleotide reduction. *Trends Biochem. Sci.* 22:81–85.

Stubbe, J. 2000. Ribonucleotide reductases: The link between an RNA and a DNA world? *Curr. Opin. Struct. Biol.* 10:731–736.

Logan, D. T., Andersson, J., Sjoberg, B. M., and Nordlund, P. 1999. A glycyl radical site in the crystal structure of a class III ribonucleotide reductase. *Science* 283:1499–1504.

Tauer, A., and Benner, S. A. 1997. The B_{12}-dependent ribonucleotide reductase from the archaebacterium *Thermoplasma acidophila:* An evolutionary solution to the ribonucleotide reductase conundrum. *Proc. Natl. Acad. Sci. U.S.A.* 94:53–58.

Stubbe, J., Nocera, D. G., Yee, C. S., and Chang, M. C. 2003. Radical initiation in the class I ribonucleotide reductase: Long-range proton-coupled electron transfer? *Chem. Rev.* 103:2167–2201.

Stubbe, J., and Riggs-Gelasco, P. 1998. Harnessing free radicals: Formation and function of the tyrosyl radical in ribonucleotide reductase. *Trends Biochem. Sci.* 23:438–443.

Thymidylate Synthase and Dihydrofolate Reductase

Liu, C. T., Hanoian, P., French, J. B., Pringle, T. H., Hammes-Schiffer, S., and Benkovic, S. J. 2013. Functional significance of evolving protein sequence in dihydrofolate reductase from bacteria to humans. *Proc. Natl. Acad. Sci. U.S.A.* 110:10159–10164.

Abali, E. E., Skacel, N. E., Celikkaya, H., and Hsieh, Y.-C. 2008. Regulation of human dihydrofolate reductase activity and expression. *Vitam. Horm.* 79:267–292.

Schnell, J. R., Dyson, H. J., and Wright, P. E. 2004. Structure, dynamics, and catalytic function of dihydrofolate reductase. *Annu. Rev. Biophys. Biomol. Struct.* 33:119–140.

Li, R., Sirawaraporn, R., Chitnumsub, P., Sirawaraporn, W., Wooden, J., Athappilly, F., Turley, S., and Hol, W. G. 2000. Three-dimensional structure of *M. tuberculosis* dihydrofolate reductase reveals opportunities for the design of novel tuberculosis drugs. *J. Mol. Biol.* 295:307–323.

Liang, P. H., and Anderson, K. S. 1998. Substrate channeling and domain-domain interactions in bifunctional thymidylate synthase-dihydrofolate reductase. *Biochemistry* 37:12195–12205.

Miller, G. P., and Benkovic, S. J. 1998. Stretching exercises: Flexibility in dihydrofolate reductase catalysis. *Chem. Biol.* 5:R105–R113.

Carreras, C. W., and Santi, D. V. 1995. The catalytic mechanism and structure of thymidylate synthase. *Annu. Rev. Biochem.* 64:721–762.

Defects in Nucleotide Biosynthesis

Desai, J., Steiger, S., and Anders, H-J. 2017. Molecular pathophysiology of gout. *Trends Mol. Med.* 23:756–768.

Kang, T. H., Park, Y., Bader, J. S., and Friedmann, T. 2013. The housekeeping gene hypoxanthine guanine phosphoribosyltransferase (HPRT) regulates multiple developmental and metabolic pathways of murine embryonic stem cell neuronal differentiation *PLOS ONE* 8:e74967.

Grunebaum, E., Cohen, A., and Roifman, C. M. 2013. Recent advances in understanding and managing adenosine deaminase and purine nucleoside phosphorylase deficiencies. *Curr. Opin. Allergy Clin. Immunol.* 13:630–638.

Fu, R., and Jinnah, H. A. 2012. Genotype-phenotype correlations in Lesch-Nyhan Disease: Moving beyond the gene. *J. Biol. Chem.* 287:2997–3008.

Richette, P., and Bardin, T. 2010. Gout. *Lancet* 375:318–328.

Aiuti, A., Cattaneo, F., Galimberti, S., Benninghoff, U., Cassani, B., Callegaro, L., Scaramuzza, S., Andolfi, G., Mirolo, M., Brigida, I., et al. 2009. Gene therapy for immunodeficiency due to adenosine deaminase deficiency. *New Engl. J. Med.* 360:447–458.

Jurecka, A. 2009. Inborn errors of purine and pyrimidine metabolism. *J. Inherit. Metab. Dis.* 32:247–263.

Nyhan, W. L., Barshop, B. A., and Ozand, P. T. 2005. *Atlas of Metabolic Diseases.* (2nd ed., pp. 429–462). Hodder Arnold.

Scriver, C. R., Sly, W. S., Childs, B., Beaudet, A. L., Valle, D., Kinzler, K. W., and Vogelstein, B. (Eds.). 2001. *The Metabolic and Molecular Bases of Inherited Diseases* (8th ed., pp. 2513–2704). McGraw-Hill.

Nyhan, W. L. 1997. The recognition of Lesch-Nyhan syndrome as an inborn error of purine metabolism. *J. Inherited Metab. Dis.* 20:171–178.

Wong, D. F., Harris, J. C., Naidu, S., Yokoi, F., Marenco, S., Dannals, R. F., Ravert, H. T., Yaster, M., Evans, A., Rousset, O., et al. 1996. Dopamine transporters are markedly reduced in Lesch-Nyhan disease in vivo. *Proc. Natl. Acad. Sci. U.S.A.* 93:5539–5543.

Neychev, V. K., and Mitev, V. I. 2004. The biochemical basis of the neurobehavioral abnormalities in the Lesch-Nyhan syndrome: A hypothesis. *Med. Hypotheses* 63:131–134.

Biochemistry in Focus

Deng, Y., Wang, Z. V., Gordillo, R., An, Y., Zhang, C. et al. 2017. An adipo-biliary-uridine axis that regulates energy homeostasis. *Science* 355: eaaf5375.

CHAPTER 26

Where to Start

Vickers, K. C., and Remaley, A. T. 2014. HDL and cholesterol: Life after the divorce? *J. Lipid Res.* 55:4–12.

Lambert, G., Sjouke, B., Choque, B., Kastelein, J. J. P., and Hovingh, G. K. 2012. The PCSK9 decade. *J. Lipid Res.* 53:2515–2524.

Brown, M. S., and Goldstein, J. L. 2009. Cholesterol feedback: From Schoenheimer's bottle to Scap's MELADL. *J. Lipid Res.* 50:S15–S27.

Gimpl, G., Burger, K., and Fahrenholz, F. 2002. A closer look at the cholesterol sensor. *Trends Biochem. Sci.* 27:595–599.

Oram, J. F. 2002. Molecular basis of cholesterol homeostasis: Lessons from Tangier disease and ABCA1. *Trends Mol. Med.* 8:168–173.

Endo, A. 1992. The discovery and development of HMG-CoA reductase inhibitors. *J. Lipid Res.* 33:1569–1582.

Books

Vance, J. E., and Vance, D. E. (Eds.). 2008. *Biochemistry of Lipids, Lipoproteins and Membranes.* Elsevier.

Nyhan, W. L., Barshop, B. A., and Al-Aqeel, A. I. 2011. *Atlas of Metabolic Diseases.* (3rd ed., pp. 659–780). Hodder Arnold.

Scriver, C. R., Sly, W. S., Childs, B., Beaudet, A. L., Valle, D., Kinzler, K. W., and Vogelstein, B. (Eds.). 2001. *The Metabolic and Molecular Bases of Inherited Diseases* (8th ed., pp. 2707–2960). McGraw-Hill.

Phospholipids and Sphingolipids

Ryan, T., Bamm, V. V., Stykel, M. G., Coackley, C. L., Humphries, K. M., et al. 2018. Cardiolipin exposure on the outer mitochondrial membrane modulates α-synuclein. *Nat. Commun.* 9:817.

Lee, J., Taneva, S. G., Holland, B. W., Tieleman, D. P., and Cornell, R. B. 2014. Structural basis for autoinhibition of CTP: phosphocholine cytidylyltransferase (CCT), the regulatory enzyme in phosphatidylcholine synthesis, by its membrane-binding amphipathic helix. *J. Biol. Chem.* 289:1742–1755.

Tang, W. H. W., Wang, Z., Levison, B. S., Koeth, R. A., Britt, E. B., Fu, F., Wu, Y., and Hazen, S. L. 2013. Intestinal microbial metabolism of phosphatidylcholine and cardiovascular risk. *New Engl. J. Med.* 368:1575–1584.

Pascual, F., and Carman, G. M. 2013. Phosphatidate phosphatase, a key regulator of lipid homeostasis. *Biochim. Biophys. Acta* 1831:514–522.

Bennett, B. J., de Aguiar Vallim, T. Q., Wang, Z., Shih, D. M., Meng, Y., Gregory, J., Allayee, H., Lee, R., Graham, M., Crooke, R., et al. 2013. Trimethylamine-N-oxide, a metabolite associated with atherosclerosis, exhibits complex genetic and dietary regulation. *Cell Metab.* 17:49–60.

Claypool, S. M., and Koehler C. M. 2012. The complexity of cardiolipin in health and disease. *Trends Biochem. Sci.* 37:32–41.

Carman, G. M., and Han, G.-S. 2009. Phosphatidic acid phosphatase, a key enzyme in the regulation of lipid synthesis. *J. Biol. Chem.* 284:2593–2597.

Bartke, N., and Hannun, Y. A. 2009. Bioactive sphingolipids: Metabolism and function. *J. Lipid Res.* 50:S91–S96.

Lee, J., Johnson, J., Ding, Z., Paetzel, M., and Cornell, R. B. 2009. Crystal structure of a mammalian CTP: Phosphocholine cytidylyl-transferase catalytic domain reveals novel active site residues within a highly conserved nucleotidyltransferase fold. *J. Biol. Chem.* 284:33535–33548.

Nye, C. K., Hanson, R. W., and Kalhan, S. C. 2008. Glyceroneogenesis is the dominant pathway for triglyceride glycerol synthesis *in vivo* in the rat. *J. Biol. Chem.* 283:27565–27574.

Biosynthesis of Cholesterol and Steroids

Radhakrishnan, A., Goldstein, J. L., McDonald, J. G., and Brown, M. S. 2008. Switch-like control of SREBP-2 transport triggered by small changes in ER cholesterol: A delicate balance. *Cell Metab.* 8:512–521.

DeBose-Boyd, R. A. 2008. Feedback regulation of cholesterol synthesis: Sterol-accelerated ubiquitination and degradation of HMG CoA reductase. *Cell Res.* 18:609–621.

Hampton, R. Y. 2002. Proteolysis and sterol regulation. *Annu. Rev. Cell Dev. Biol.* 18:345–378.

Kelley, R. I., and Herman, G. E. 2001. Inborn errors of sterol biosynthesis. *Annu. Rev. Genom. Hum. Genet.* 2:299–341.

Istvan, E. S., and Deisenhofer, J. 2001. Structural mechanism for statin inhibition of HMG-CoA reductase. *Science* 292:1160–1164.

Lipoproteins and Their Receptors

Sun, H., Krauss, R. M., Chang, J. T., and Teng, B.-B. 2018. PCSK9 deficiency reduces atherosclerosis, apolipoprotein B secretion, and endothelial dysfunction. *J. Lipid Res.* 59:207–223.

Trinh, M. N., Lu, F., Li, X., Das, A., Liang, Q., et al. 2017. Triazoles inhibit cholesterol export from lysosomes by binding to NPC1. *Proc. Natl. Acad. Sci. U.S.A.* 114:89–94.

Li, X., Lu, F., Trinh, M. N., Schmiege, P., Seemann, J., et al. 2017. 3.3 Å structure of Niemann–Pick C1 protein reveals insights into the function of the C-terminal luminal domain in cholesterol transport. *Proc. Natl. Acad. Sci. U. S. A.* 114:9116–9121.

Gustafsen, C., Kjolby, M., Nyegaard, M., Mattheisen, M., Lundhede, J., Buttenschøn, H., Mors, O., Bentzon, J. F., Madsen, P., Nykjaer, A., et al. 2014. The hypercholesterolemia-risk gene SORT1 facilitates PCSK9 secretion. *Cell Metab.* 19:310–318.

Rye, K-A., Bursill, C. A., Lambert, G., Tabet, F., and Barter, P. J. 2009. The metabolism and anti-atherogenic properties of HDL. *J. Lipid Res.* 50:S195–S200.

Rader, D. J., Alexander, E. T., Weibel, G. L., Billheimer, J., and Rothblat, G. H. 2009. The role of reverse cholesterol transport in animals and humans and relationship to atherosclerosis. *J. Lipid Res.* 50:S189–S194.

Tall, A. R., Yvan-Charvet, L., Terasaka, N., Pagler, T., and Wang, N. 2008. HDL, ABC transporters, and cholesterol efflux: Implications for the treatment of atherosclerosis. *Cell Metab.* 7:365–375.

Jeon, H., and Blacklow, S. C. 2005. Structure and physiologic function of the low-density lipoprotein receptor. *Annu. Rev. Biochem.* 74:535–562.

Beglova, N., and Blacklow, S. C. 2005. The LDL receptor: How acid pulls the trigger. *Trends Biochem. Sci.* 30:309–316.

Oxygen Activation and P450 Catalysis

Stiles, A. R., McDonald, J. G., Bauman, D. R., and Russell, D. W. 2009. CYP7B1: One cytochrome P450, two human genetic diseases, and multiple physiological functions. *J. Biol. Chem.* 284:28485–28489.

Zhou, S.-F., Liu, J.-P., and Chowbay, B. 2009. Polymorphism of human cytochrome P450 enzymes and its clinical impact. *Drug Metab. Rev.* 4:89–295.

Williams, P. A., Cosme, J., Vinkovic, D. M., Ward, A., Angove, H. C., Day, P. J., Vonrhein, C., Tickle, I. J., and Jhoti, H. 2004. Crystal structure of human cytochrome P450 3A4 bound to metyrapone and progesterone. *Science* 305:683–686.

Industrial Applications

Meadows, A., Hawkins, K. M., Tsegaye, Y., Antipov, E., Kim, Y., et al. 2016. Rewriting yeast central carbon metabolism for industrial isoprenoid production. *Nature* 537:694–697.

Biochemistry in Focus

Summers, S. A. 2018. Could ceramides become the new cholesterol? *Cell Metab.* 27:276–280.

Petersen, M. C., and Shulman, G. I. 2017. Roles of diacylglycerols and ceramides in hepatic insulin resistance. *Trends Pharmacol. Sci.* 38:649–665.

CHAPTER 27

Where to Start

Lin, S-C., and Hardie, D. G. 2018. AMPK: Sensing glucose as well as cellular energy status. *Cell Metab.* 27:299–313.

Cahill, Jr. G. F. 2006. Fuel metabolism in starvation. *Annu. Rev. Nutr.* 26:1–22.

Dunn, R. 2013. Everything you know about calories is wrong. *Sci. Am.* (3) 309:57–59.

Kenny, P. J. 2013. The food addiction. *Sci. Am.* (3) 309:44–49.

Taubes, G. 2013. Which one will make you fat? *Sci. Am.* (3) 309:60–65.

Hardie, D. G. 2012. Organismal carbohydrate and lipid homeostasis. *Cold Spring Harb. Perspect. Biol.* 4:a006031.

Books

Kessler, D. A. 2010. *The End of Overeating: Taking Control of the Insatiable American Appetite.* Rodale.

Wrangham, R. 2009. *Catching Fire: How Cooking Made Us Human.* Basic Books.

Stipanuk, M. H., and Caudill, M. A. (Eds.). 2013. *Biochemical, Physiological, & Molecular Aspects of Human Nutrition* (3rd ed.). Saunders-Elsevier.

Fell, D. 1997. *Understanding the Control of Metabolism.* Portland Press.

Frayn, K. N. 1996. *Metabolic Regulation: A Human Perspective.* Portland Press.

Poortmans, J. R. (Ed.). 2004. *Principles of Exercise Biochemistry.* Karger.

Harris, R. A., and Crabb, D. W. 2011. Metabolic interrelationships. In *Textbook of Biochemistry with Clinical Correlations* (pp. 839–882), edited by T. M. Devlin. Wiley-Liss.

Caloric Homeostasis

Perry, R. J., Wang, Y., Cline, G. W., Rabin-Court, A., Song, J. D., et al. 2018. Leptin mediates a glucose-fatty acid cycle to maintain glucose homeostasis in starvation. *Cell* 172:234–248.

Stern, J. H., Rutkowski, J. M., and Scherer, P. E. 2016. Adiponectin, leptin, and fatty acids in the maintenance of metabolic homeostasis through adipose tissue crosstalk. *Cell Metab.* 23:770–784.

Clemmensen, C., Müller, T. D. Woods, S. C., Berthoud, H.-R., Seeley, R. J., and Tschöp, M. H. 2017. Gut-brain cross-talk in metabolic control. *Cell* 168:758–774.

Woods, S. C. 2009. The control of food intake: Behavioral versus molecular perspectives. *Cell Metab.* 9:489–498.

Figlewicz, D. P., and Benoit, S. C. 2009. Insulin, leptin, and food reward: Update 2008. *Am. J. Physiol. Integr. Comp. Physiol.* 296:R9–R19.

Israel, D., and Chua, S. Jr. 2009. Leptin receptor modulation of adiposity and fertility. *Trends Endocrinol. Metab.* 21:10–16.

Meyers, M. G., Cowley, M. A., and Münzberg, H. 2008. Mechanisms of leptin action and leptin resistance. *Annu. Rev. Physiol.* 70:537–556.

Sowers, J. R. 2008. Endocrine functions of adipose tissue: Focus on adiponectin. *Clin. Cornerstone* 9:32–38.

Brehma, B. J., and D'Alessio, D. A. 2008. Benefits of high-protein weight loss diets: Enough evidence for practice? *Curr. Opin. Endocrinol., Diabetes, Obesity* 15:416–421.

Coll, A. P., Farooqi, I. S., and O'Rahillt, S. O. 2007. The hormonal control of food intake. *Cell* 129:251–262.

Muoio, D. M., and Newgard, C. B. 2006. Obesity-related derangements in metabolic regulation. *Annu. Rev. Biochem.* 75:367–401.

Diabetes Mellitus

Schwartz, S. S., Epstein, S., Corkey, B. E., Grant, S. F. A. Gavin III, J. R., Aguilar, R. B., and Herman, M. E. 2017. A unified pathophysiological construct of diabetes and its complications. *Trends Endocrinol. Metab.* 28:645–655.

Franks, P. W., and McCarthy, M. I. 2016. Exposing the exposures responsible for type 2 diabetes and obesity. *Science* 354:69–73.

Lee, J., and Ozcan, U. 2014. Unfolded protein response signaling and metabolic diseases. *J. Biol. Chem.* 289:1203–1211.

Yamauchi, T., and Kadowaki, T. 2013. Adiponectin receptor as a key player in healthy longevity and obesity-related diseases. *Cell Metab.* 17:185–196.

Könner, A. C., and Brüning, J. C. 2012. Selective insulin and leptin resistance in metabolic disorders. *Cell Metab.* 16:144–152.

Zhang, B. B., Zhou, G., and Li, C. 2009. AMPK: An emerging drug target for diabetes and the metabolic syndrome. *Cell Metab.* 9:407–416.

Magkos, F., Yannakoulia, M., Chan, J. L., and Mantzoros, C. S. 2009. Management of the metabolic syndrome and type 2 diabetes through lifestyle modification. *Annu. Rev. Nutr.* 29:8.1–8.34.

Muoio, D. M., and Newgard, C. B. 2008. Molecular and metabolic mechanisms of insulin resistance and β-cell failure in type 2 diabetes. *Nat. Rev. Mol. Cell. Biol.* 9:193–205.

Leibiger, I. B., Leibiger, B., and Berggren, P.-O. 2008. Insulin signaling in the pancreatic β-cell. *Annu. Rev. Nutr.* 28:233–251.

Doria, A., Patti, M. E., and Kahn, C. R. 2008. The emerging architecture of type 2 diabetes. *Cell Metab.* 8:186–200.

Croker, B. A., Kiu, H., and Nicholson, S. E. 2008. SOCS regulation of the JAK/STAT signalling pathway. *Semin. Cell Dev. Biol.* 19:414–422.

Eizirik, D. L., Cardozo, A. K., and Cnop, M. 2008. The role of endoplasmic reticulum stress in diabetes mellitus. *Endocrinol. Rev.* 29:42–61.

Howard, J. K., and Flier, J. S. 2006. Attenuation of leptin and insulin signaling by SOCS proteins. *Trends Endocrinol. Metab.* 9:365–371.

Lowel, B. B., and Shulman, G. 2005. Mitochondrial dysfunction and type 2 diabetes. *Science* 307:384–387.

Taylor, S. I. 2001. Diabetes mellitus. In *The Metabolic Basis of Inherited Diseases* (8th ed., pp. 1433–1469), edited by C. R. Scriver, W. S. Sly, B. Childs, A. L. Beaudet, D. Valle, K. W. Kinzler, and B. Vogelstein. McGraw-Hill.

Exercise Metabolism

Hojman, P., Gehl, J., Christensen, J. F., and Pedersen, B. K. 2018. Molecular mechanisms linking exercise to cancer prevention and treatment. *Cell Metab.* 27:10–21.

Egan, B., and Zierath, J. R. 2013. Exercise metabolism and the molecular regulation of skeletal muscle adaptation. *Cell Metab.* 17:162–184.

Hood, D. A. 2001. Contractile activity-induced mitochondrial biogenesis in skeletal muscle. *J. Appl. Physiol.* 90:1137–1157.

Shulman, R. G., and Rothman, D. L. 2001. The "glycogen shunt" in exercising muscle: A role for glycogen in muscle energetics and fatigue. *Proc. Natl. Acad. Sci. U.S.A.* 98:457–461.

Gleason, T. 1996. Post-exercise lactate metabolism: A comparative review of sites, pathways, and regulation. *Annu. Rev. Physiol.* 58:556–581.

Holloszy, J. O., and Kohrt, W. M. 1996. Regulation of carbohydrate and fat metabolism during and after exercise. *Annu. Rev. Nutr.* 16:121–138.

Hochachka, P. W., and McClelland, G. B. 1997. Cellular metabolic homeostasis during large-scale change in ATP turnover rates in muscles. *J. Exp. Biol.* 200:381–386.

Horowitz, J. F., and Klein, S. 2000. Lipid metabolism during endurance exercise. *Am. J. Clin. Nutr.* 72:558S–563S.

Wagenmakers, A. J. 1999. Muscle amino acid metabolism at rest and during exercise. *Diabetes Nutr. Metab.* 12:316–322.

Metabolic Adaptations in Starvation

Baverel, G., Ferrier, B., and Martin, M. 1995. Fuel selection by the kidney: Adaptation to starvation. *Proc. Nutr. Soc.* 54:197–212.

MacDonald, I. A., and Webber, J. 1995. Feeding, fasting and starvation: Factors affecting fuel utilization. *Proc. Nutr. Soc.* 54:267–274.

Cahill, G. F., Jr. 1976. Starvation in man. *Clin. Endocrinol. Metab.* 5:397–415.

Sugden, M. C., Holness, M. J., and Palmer, T. N. 1989. Fuel selection and carbon flux during the starved-to-fed transition. *Biochem. J.* 263:313–323.

Ethanol Metabolism

Nagy, L. E. 2004. Molecular aspects of alcohol metabolism: Transcription factors involved in early-induced liver injury. *Annu. Rev. Nutr.* 24:55–78.

Molotkov, A., and Duester, G. 2002. Retinol/ethanol drug interaction during acute alcohol intoxication involves inhibition of retinol metabolism to retinoic acid by alcohol dehydrogenase. *J. Biol. Chem.* 277:22553–22557.

Stewart, S., Jones, D., and Day, C. P. 2001. Alcoholic liver disease: New insights into mechanisms and preventive strategies. *Trends Mol. Med.* 7:408–413.

Lieber, C. S. 2000. Alcohol: Its metabolism and interaction with nutrients. *Annu. Rev. Nutr.* 20:395–430.

Niemela, O. 1999. Aldehyde-protein adducts in the liver as a result of ethanol-induced oxidative stress. *Front. Biosci.* 1:D506–D513.

Riveros-Rosas, H., Julian-Sanchez, A., and Pina, E. 1997. Enzymology of ethanol and acetaldehyde metabolism in mammals. *Arch. Med. Res.* 28:453–471.

Biochemistry in Focus

Neufer, P. D., Bamman, M. M., Muoio, D. M., Bouchard, C., Cooper, D. M. et al. 2015. Understanding the cellular and molecular mechanisms of physical activity-induced health benefits. *Cell Metab.* 22:4–11.

Egan, B., and Zierath, J. R. 2013. Exercise metabolism and the molecular regulation of skeletal muscle adaptation. *Cell Metab.* 17:162–184.

CHAPTER 28

Where to Start

Gilman, A. G. 2012. Silver spoons and other personal reflections. *Annu. Rev. Pharmacol. Toxicol.* 52:1–19.

Zhang, H.-Y., Chen, L.-L., Xue-Juan Li, X.-J., and Zhang, J. 2010. Evolutionary inspirations for drug discovery. *Trends Pharmacol. Sci.* 31:443–448.

Books

Kenakin, T. P. 2014. *A Pharmacology Primer: Techniques for More Effective and Strategic Drug Discovery* (4th ed.). Academic Press.

Brunton, L., Knollman, B. C., and Hilal-Dandan, R. 2017. *Goodman and Gilman's The Pharmacological Basis of Therapeutics* (13th ed.). McGraw-Hill Education.

Walsh, C. T., and Schwartz-Bloom, R. D. 2004. *Levine's Pharmacology: Drug Actions and Reactions* (7th ed.). Taylor and Francis Group.

Silverman, R. B., and Holladay, M. W. 2014. *The Organic Chemistry of Drug Design and Drug Action* (3rd ed.). Academic Press.

Walsh, C., and Wencewicz, T. 2016. *Antibiotics: Challenges, Mechanisms, Opportunities*. ASM Press.

Rowland, M., and Tozer, T. N. 2010. *Clinical Pharmacokinetics and Pharmacodynamics: Concepts and Applications* (4th ed.). Lippincott, Williams, & Wilkins.

ADME and Toxicity

Caldwell, J., Gardner, I., and Swales, N. 1995. An introduction to drug disposition: The basic principles of absorption, distribution, metabolism, and excretion. *Toxicol. Pathol.* 23:102–114.

Lee, W., and Kim, R. B. 2004. Transporters and renal drug elimination. *Annu. Rev. Pharmacol. Toxicol.* 44:137–166.

Lin, J., Sahakian, D. C., de Morais, S. M., Xu, J. J., Polzer, R. J., and Winter, S. M. 2003. The role of absorption, distribution, metabolism, excretion and toxicity in drug discovery. *Curr. Top. Med. Chem.* 3:1125–1154.

Poggesi, I. 2004. Predicting human pharmacokinetics from preclinical data. *Curr. Opin. Drug Discov. Devel.* 7:100–111.

Case Histories

Flower, R. J. 2003. The development of COX2 inhibitors. *Nat. Rev. Drug Discov.* 2:179–191.

Tobert, J. A. 2003. Lovastatin and beyond: The history of the HMG-CoA reductase inhibitors. *Nat. Rev. Drug Discov.* 2:517–526.

Vacca, J. P., Dorsey, B. D., Schleif, W. A., Levin, R. B., McDaniel, S. L., Darke, P. L., Zugay, J., Quintero, J. C., Blahy, O. M., Roth, E., et al. 1994. L-735,524: An orally bioavailable human immunodeficiency virus type 1 protease inhibitor. *Proc. Natl. Acad. Sci. U.S.A.* 91:4096–4100.

Wong, S., and Witte, O. N. 2004. The BCR-ABL story: Bench to bedside and back. *Annu. Rev. Immunol.* 22:247–306.

Structure-Based Drug Design

Kuntz, I. D. 1992. Structure-based strategies for drug design and discovery. *Science* 257:1078–1082.

Dorsey, B. D., Levin, R. B., McDaniel, S. L., Vacca, J. P., Guare, J. P., Darke, P. L., Zugay, J. A., Emini, E. A., Schleif, W. A., Quintero, J. C., et al. 1994. L-735,524: The design of a potent and orally bioavailable HIV protease inhibitor. *J. Med. Chem.* 37:3443–3451.

Chen, Z., Li, Y., Chen, E., Hall, D. L., Darke, P. L., Culberson, C., Shafer, J. A., and Kuo, L. C. 1994. Crystal structure at 1.9-Å resolution of human immunodeficiency virus (HIV) II protease complexed with L-735,524, an orally bioavailable inhibitor of the HIV proteases. *J. Biol. Chem.* 269:26344–26348.

Combinatorial Chemistry

Baldwin, J. J. 1996. Design, synthesis and use of binary encoded synthetic chemical libraries. *Mol. Divers.* 2:81–88.

Burke, M. D., Berger, E. M., and Schreiber, S. L. 2003. Generating diverse skeletons of small molecules combinatorially. *Science* 302:613–618.

Edwards, P. J., and Morrell, A. I. 2002. Solid-phase compound library synthesis in drug design and development. *Curr. Opin. Drug Discov. Devel.* 5:594–605.

Genomics

Zambrowicz, B. P., and Sands, A. T. 2003. Knockouts model the 100 best-selling drugs: Will they model the next 100? *Nat. Rev. Drug Discov.* 2:38–51.

Salemme, F. R. 2003. Chemical genomics as an emerging paradigm for postgenomic drug discovery. *Pharmacogenomics* 4:257–267.

Michelson, S., and Joho, K. 2000. Drug discovery, drug development and the emerging world of pharmacogenomics: Prospecting for information in a data-rich landscape. *Curr. Opin. Mol. Ther.* 2:651–654.

Weinshilboum, R., and Wang, L. 2004. Pharmacogenomics: Bench to bedside. *Nat. Rev. Drug Discov.* 3:739–748.

Biochemistry in Focus

Carter, P. J., and Lazar, G. A. 2018. Next generation antibody drugs: pursuit of the 'high-hanging fruit.' *Nature Rev. Drug Discov.* 17:197–223.

CHAPTER 29

Where to Start

O'Donnell, M., Langston, L., and Stillman, B. 2013. Principles and concepts of DNA replication in bacteria, archaea, and eukarya. *Cold Spring Harb. Perspect. Biol.* 5:1–13.

Palermo, G., Cavalli, A., Klein, M. L., Alfonso-Prieto, M., Dal Peraro, K., and De Vivo, M. 2015. Catalytic metal ions and enzymatic processing of DNA and RNA. *Acc. Chem. Res.* 48, 220–228.

Johnson, A., and O'Donnell, M. 2005. Cellular DNA replicases: Components and dynamics at the replication fork. *Annu. Rev. Biochem.* 74:283–315.

Kornberg, A. 1988. DNA replication. *J. Biol. Chem.* 263:1–4.

Wang, J. C. 1982. DNA topoisomerases. *Sci. Am.* 247(1): 94–109.

Lindahl, T. 1993. Instability and decay of the primary structure of DNA. *Nature* 362:709–715.

Greider, C. W., and Blackburn, E. H. 1996. Telomeres, telomerase, and cancer. *Sci. Am.* 274(2):92–97.

Books

Kornberg, A., and Baker, T. A. 1992. *DNA Replication* (2nd ed.). W. H. Freeman and Company.

Bloomfield, V. A., Crothers, D., Tinoco, I., and Hearst, J. 2000. *Nucleic Acids: Structures, Properties and Functions.* University Science Books.

Friedberg, E. C., Walker, G. C., and Siede, W. 1995. *DNA Repair and Mutagenesis.* American Society for Microbiology.

Cozzarelli, N. R., and Wang, J. C. (Eds.). 1990. *DNA Topology and Its Biological Effects.* Cold Spring Harbor Laboratory Press.

DNA Topology and Topoisomerases

Graille, M., Cladiere, L., Durand, D., Lecointe, F., Gadelle, D., Quevillon-Cheruel, S., Vachette, P., Forterre, P., and van Tilbeurgh, H. 2008. Crystal structure of an intact type II DNA topoisomerase: Insights into DNA transfer mechanisms. *Structure* 16:360–370.

Charvin, G., Strick, T. R., Bensimon, D., and Croquette, V. 2005. Tracking topoisomerase activity at the single-molecule level. *Annu. Rev. Biophys. Biomol. Struct.* 34:201–219.

Sikder, D., Unniraman, S., Bhaduri, T., and Nagaraja, V. 2001. Functional cooperation between topoisomerase I and single strand DNA-binding protein. *J. Mol. Biol.* 306:669–679.

Fortune, J. M., and Osheroff, N. 2000. Topoisomerase II as a target for anticancer drugs: When enzymes stop being nice. *Prog. Nucleic Acid Res. Mol. Biol.* 64:221–253.

Isaacs, R. J., Davies, S. L., Sandri, M. I., Redwood, C., Wells, N. J., and Hickson, I. D. 1998. Physiological regulation of eukaryotic topoisomerase II. *Biochim. Biophys. Acta* 1400:121–137.

Wang, J. C. 1998. Moving one DNA double helix through another by a type II DNA topoisomerase: The story of a simple molecular machine. *Q. Rev. Biophys.* 31:107–144.

Baird, C. L., Harkins, T. T., Morris, S. K., and Lindsley, J. E. 1999. Topoisomerase II drives DNA transport by hydrolyzing one ATP. *Proc. Natl. Acad. Sci. U.S.A.* 96:13685–13690.

Vologodskii, A. V., Levene, S. D., Klenin, K. V., Frank, K. M., and Cozzarelli, N. R. 1992. Conformational and thermodynamic properties of supercoiled DNA. *J. Mol. Biol.* 227:1224–1243.

Fisher, L. M., Austin, C. A., Hopewell, R., Margerrison, M., Oram, M., Patel, S., Wigley, D. B., Davies, G. J., Dodson, E. J., Maxwell, A., et al. 1991. Crystal structure of an N-terminal fragment of the DNA gyrase B protein. *Nature* 351:624–629.

Mechanism of Replication

Davey, M. J., and O'Donnell, M. 2000. Mechanisms of DNA replication. *Curr. Opin. Chem. Biol.* 4:581–586.

Keck, J. L., and Berger, J. M. 2000. DNA replication at high resolution. *Chem. Biol.* 7:R63–R71.

Kunkel, T. A., and Bebenek, K. 2000. DNA replication fidelity. *Annu. Rev. Biochem.* 69:497–529.

Waga, S., and Stillman, B. 1998. The DNA replication fork in eukaryotic cells. *Annu. Rev. Biochem.* 67:721–751.

Marians, K. J. 1992. Prokaryotic DNA replication. *Annu. Rev. Biochem.* 61:673–719.

DNA Polymerases and Other Enzymes of Replication

Kurth, I., and O'Donnell, M. 2013. New insights into replisome fluidity during chromosome replication. *Trends Biochem. Sci.* 38:195–203.

Nandakumar, J., and Cech, T. R. 2013. Finding the end: Recruitment of telomerase to telomeres. *Nat. Rev. Mol. Cell. Biol.* 14:69–82.

Singleton, M. R., Sawaya, M. R., Ellenberger, T., and Wigley, D. B. 2000. Crystal structure of T7 gene 4 ring helicase indicates a mechanism for sequential hydrolysis of nucleotides. *Cell* 101:589–600.

Donmez, I., and Patel, S. S. 2006. Mechanisms of a ring shaped helicase. *Nucleic Acids Res.* 34:4216–4224.

Johnson, D. S., Bai, L., Smith, B. Y., Patel, S. S., and Wang, M. D. 2007. Single-molecule studies reveal dynamics of DNA unwinding by the ring-shaped T7 helicase. *Cell* 129:1299–1309.

Lee, S. J., Qimron, U., and Richardson, C. C. 2008. Communication between subunits critical to DNA binding by hexameric helicase of bacteriophage T7. *Proc. Natl. Acad. Sci. U.S.A.* 105:8908–8913.

Toth, E. A., Li, Y., Sawaya, M. R., Cheng, Y., and Ellenberger, T. 2003. The crystal structure of the bifunctional primase-helicase of bacteriophage T7. *Mol. Cell* 12:1113–1123.

Hubscher, U., Maga, G., and Spadari, S. 2002. Eukaryotic DNA polymerases. *Annu. Rev. Biochem.* 71:133–163.

Doublié, S., Tabor, S., Long, A. M., Richardson, C. C., and Ellenberger, T. 1998. Crystal structure of a bacteriophage T7 DNA replication complex at 2.2 Å resolution. *Nature* 391:251–258.

Arezi, B., and Kuchta, R. D. 2000. Eukaryotic DNA primase. *Trends Biochem. Sci.* 25:572–576.

Jager, J., and Pata, J. D. 1999. Getting a grip: Polymerases and their substrate complexes. *Curr. Opin. Struct. Biol.* 9:21–28.

Steitz, T. A. 1999. DNA polymerases: Structural diversity and common mechanisms. *J. Biol. Chem.* 274:17395–17398.

Beese, L. S., Derbyshire, V., and Steitz, T. A. 1993. Structure of DNA polymerase I Klenow fragment bound to duplex DNA. *Science* 260:352–355.

McHenry, C. S. 1991. DNA polymerase III holoenzyme: Components, structure, and mechanism of a true replicative complex. *J. Biol. Chem.* 266:19127–19130.

Kong, X. P., Onrust, R., O'Donnell, M., and Kuriyan, J. 1992. Three-dimensional structure of the β subunit of *E. coli* DNA polymerase III holoenzyme: A sliding DNA clamp. *Cell* 69:425–437.

Polesky, A. H., Steitz, T. A., Grindley, N. D., and Joyce, C. M. 1990. Identification of residues critical for the polymerase activity of the Klenow fragment of DNA polymerase I from *Escherichia coli*. *J. Biol. Chem.* 265:14579–14591.

Lee, J. Y., Chang, C., Song, H. K., Moon, J., Yang, J. K., Kim, H. K., Kwon, S. T., and Suh, S. W. 2000. Crystal structure of NAD$^+$ dependent DNA ligase: Modular architecture and functional implications. *EMBO J.* 19:1119–1129.

Timson, D. J., and Wigley, D. B. 1999. Functional domains of an NAD$^+$-dependent DNA ligase. *J. Mol. Biol.* 285:73–83.

Doherty, A. J., and Wigley, D. B. 1999. Functional domains of an ATP-dependent DNA ligase. *J. Mol. Biol.* 285:63–71.

von Hippel, P. H., and Delagoutte, E. 2001. A general model for nucleic acid helicases and their "coupling" within macromolecular machines. *Cell* 104:177–190.

Tye, B. K., and Sawyer, S. 2000. The hexameric eukaryotic MCM helicase: Building symmetry from nonidentical parts. *J. Biol. Chem.* 275:34833–34836.

Marians, K. J. 2000. Crawling and wiggling on DNA: Structural insights to the mechanism of DNA unwinding by helicases. *Struct. Fold. Des.* 5:R227–R235.

Soultanas, P., and Wigley, D. B. 2000. DNA helicases: "Inching forward." *Curr. Opin. Struct. Biol.* 10:124–128.

de Lange, T. 2009. How telomeres solve the end-protection problem. *Science* 326:948–952.

Bachand, F., and Autexier, C. 2001. Functional regions of human telomerase reverse transcriptase and human telomerase RNA required for telomerase activity and RNA-protein interactions. *Mol. Cell Biol.* 21:1888–1897.

Griffith, J. D., Comeau, L., Rosenfield, S., Stansel, R. M., Bianchi, A., Moss, H., and de Lange, T. 1999. Mammalian telomeres end in a large duplex loop. *Cell* 97:503–514.

McEachern, M. J., Krauskopf, A., and Blackburn, E. H. 2000. Telomeres and their control. *Annu. Rev. Genet.* 34:331–358.

Mutations and DNA Repair

Lindahl, T. 2016. The intrinsic fragility of DNA (Nobel Lecture). *Angew. Chem. Int. Ed. Engl.* 55, 8528–8534.

Modrich, P. 2016. Mechanisms of E. coli and human mismatch repair (Nobel Lecture). *Angew. Chem. Int. Ed. Engl.* 55, 8490–8501.

Sancar, A. 2016. Mechanisms of DNA repair by photolyase and excision nuclease (Nobel Lecture). *Angew. Chem. Int. Ed. Engl.* 55, 8502–8527.

Hu, J., Selby, C. P., Adar, S., Adebali, O., and Sancar, A. 2017. Molecular mechanisms and genomics maps of DNA excision repair in *Escherichia coli* and humans. *J. Biol. Chem.* 292, 15588–15597.

Yang, W. 2003. Damage repair DNA polymerases Y. *Curr. Opin. Struct. Biol.* 13:23–30.

Wood, R. D., Mitchell, M., Sgouros, J., and Lindahl, T. 2001. Human DNA repair genes. *Science* 291:1284–1289.

Shin, D. S., Chahwan, C., Huffman, J. L., and Tainer, J. A. 2004. Structure and function of the double-strand break repair machinery. *DNA Repair (Amst.)* 3:863–873.

Michelson, R. J., and Weinert, T. 2000. Closing the gaps among a web of DNA repair disorders. *BioEssays* 22:966–969.

Aravind, L., Walker, D. R., and Koonin, E. V. 1999. Conserved domains in DNA repair proteins and evolution of repair systems. *Nucleic Acids Res.* 27:1223–1242.

Mol, C. D., Parikh, S. S., Putnam, C. D., Lo, T. P., and Tainer, J. A. 1999. DNA repair mechanisms for the recognition and removal of damaged DNA bases. *Annu. Rev. Biophys. Biomol. Struct.* 28:101–128.

Parikh, S. S., Mol, C. D., and Tainer, J. A. 1997. Base excision repair enzyme family portrait: Integrating the structure and chemistry of an entire DNA repair pathway. *Structure* 5:1543–1550.

Vassylyev, D. G., and Morikawa, K. 1997. DNA-repair enzymes. *Curr. Opin. Struct. Biol.* 7:103–109.

Verdine, G. L., and Bruner, S. D. 1997. How do DNA repair proteins locate damaged bases in the genome? *Chem. Biol.* 4:329–334.

Bowater, R. P., and Wells, R. D. 2000. The intrinsically unstable life of DNA triplet repeats associated with human hereditary disorders. *Prog. Nucleic Acid Res. Mol. Biol.* 66:159–202.

Cummings, C. J., and Zoghbi, H. Y. 2000. Fourteen and counting: Unraveling trinucleotide repeat diseases. *Hum. Mol. Genet.* 9:909–916.

Defective DNA Repair and Cancer

Dever, S. M., White, E. R., Hartman, M. C., and Valerie, K. 2012. BRCA1-directed, enhanced and aberrant homologous recombination: Mechanism and potential treatment strategies. *Cell Cycle* 11:687–94.

Berneburg, M., and Lehmann, A. R. 2001. Xeroderma pigmentosum and related disorders: Defects in DNA repair and transcription. *Adv. Genet.* 43:71–102.

Lambert, M. W., and Lambert, W. C. 1999. DNA repair and chromatin structure in genetic diseases. *Prog. Nucleic Acid Res. Mol. Biol.* 63:257–310.

Buys, C. H. 2000. Telomeres, telomerase, and cancer. *New Engl. J. Med.* 342:1282–1283.

Urquidi, V., Tarin, D., and Goodison, S. 2000. Role of telomerase in cell senescence and oncogenesis. *Annu. Rev. Med.* 51:65–79.

Lynch, H. T., Smyrk, T. C., Watson, P., Lanspa, S. J., Lynch, J. F., Lynch, P. M., Cavalieri, R. J., and Boland, C. R. 1993. Genetics, natural history, tumor spectrum, and pathology of hereditary non-polyposis colorectal cancer: An updated review. *Gastroenterology* 104:1535–1549.

Fishel, R., Lescoe, M. K., Rao, M. R. S., Copeland, N. G., Jenkins, N. A., Garber, J., Kane, M., and Kolodner, R. 1993. The human mutator gene homolog *MSH2* and its association with hereditary nonpolyposis colon cancer. *Cell* 75:1027–1038.

Ames, B. N., and Gold, L. S. 1991. Endogenous mutagens and the causes of aging and cancer. *Mutat. Res.* 250:3–16.

Ames, B. N. 1979. Identifying environmental chemicals causing mutations and cancer. *Science* 204:587–593.

Recombination and Recombinases

Singleton, M. R., Dillingham, M. S., Gaudier, M., Kowalczykowski, S. C., and Wigley, D. B. 2004. Crystal structure of RecBCD enzyme reveals a machine for processing DNA breaks. *Nature* 432:187–193.

Spies, M., Bianco, P. R., Dillingham, M. S., Handa, N., Baskin, R. J., and Kowalczykowski, S. C. 2003. A molecular throttle: The recombination hotspot chi controls DNA translocation by the RecBCD helicase. *Cell* 114:647–654.

Kowalczykowski, S. C. 2000. Initiation of genetic recombination and recombination-dependent replication. *Trends Biochem. Sci.* 25:1562165.

Prevost, C., and Takahashi, M. 2003. Geometry of the DNA strands within the RecA nucleofilament: Role in homologous recombination. *Q. Rev. Biophys.* 36:429–453.

Van Duyne, G. D. 2001. A structural view of Cre-loxP site-specific recombination. *Annu. Rev. Biophys. Biomol. Struct.* 30:87–104.

Chen, Y., Narendra, U., Iype, L. E., Cox, M. M., and Rice, P. A. 2000. Crystal structure of a Flp recombinase-Holliday junction complex: Assembly of an active oligomer by helix swapping. *Mol. Cell* 6:885–897.

Craig, N. L. 1997. Target site selection in transposition. *Annu. Rev. Biochem.* 66:437–474.

Gopaul, D. N., Guo, F., and Van Duyne, G. D. 1998. Structure of the Holliday junction intermediate in Cre-loxP site-specific recombination. *EMBO J.* 17:4175–4187.

Gopaul, D. N., and Van Duyne, G. D. 1999. Structure and mechanism in site-specific recombination. *Curr. Opin. Struct. Biol.* 9:14–20.

Biochemistry in Focus

D. T. Minnick et al. 1999, Side chains that influence the fidelity at the polymerase active site of *Escherichia coli* DNA Polymerase I (Klenow fragment), *J. Biol. Chem.* 274:3067–3075.

CHAPTER 30

Where to Start

Liu, X., Bushnell, D. A., and Kornberg, R. D. 2013. RNA polymerase II transcription: Structure and mechanism. *Biochim. Biophys. Acta* 1829:2–8.

Kornberg, R. D. 2007. The molecular basis of eukaryotic transcription. *Proc. Natl. Acad. Sci. U.S.A.* 104:12955–12961.

Woychik, N. A. 1998. Fractions to functions: RNA polymerase II thirty years later. *Cold Spring Harbor Symp. Quant. Biol.* 63:311–317.

Losick, R. 1998. Summary: Three decades after sigma. *Cold Spring Harbor Symp. Quant. Biol.* 63:653–666.

Sharp, P. A. 1994. Split genes and RNA splicing (Nobel Lecture). *Angew. Chem. Int. Ed. Engl.* 33:1229–1240.

Cech, T. R. 1990. Nobel lecture: Self-splicing and enzymatic activity of an intervening sequence RNA from *Tetrahymena*. *Biosci. Rep.* 10:239–261.

Villa, T., Pleiss, J. A., and Guthrie, C. 2002. Spliceosomal snRNAs: Mg^{2+} dependent chemistry at the catalytic core? *Cell* 109:149–152.

Books

Krebs, J. E., and Goldstein, E. S. 2012. *Lewin's Genes XI* (11th ed.). Jones and Bartlett.

Kornberg, A., and Baker, T. A. 1992. *DNA Replication* (2nd ed.). W. H. Freeman and Company.

Lodish, H., Berk, A., Kaiser, C. A., Krieger, M., Bretscher, A., Ploegh, H., Amon, A., and Scott, M. P. 2012. *Molecular Cell Biology* (7th ed.). W. H. Freeman and Company.

Watson, J. D., Baker, T. A., Bell, S. P., Gann, A., Levine, M., and Losick, R. 2013. *Molecular Biology of the Gene* (7th ed.). Benjamin Cummings.

Gesteland, R. F., Cech, T., and Atkins, J. F. 2006. *The RNA World: The Nature of Modern RNA Suggests a Prebiotic RNA* (3rd ed.). Cold Spring Harbor Laboratory Press.

RNA Polymerases

Liu, X., Bushnell, D. A., Wang, D., Calero, G., and Kornberg, R. D. 2010. Structure of an RNA polymerase II-TFIIB complex and the transcription initiation mechanism. *Science* 327:206–209.

Wang, D., Bushnell, D. A., Huang, X., Westover, K. D., Levitt, M., and Kornberg, R. D. 2009. Structural basis of transcription: Backtracked RNA polymerase II at 3.4 Å resolution. *Science* 324:1203–1206.

Darst, S. A. 2001. Bacterial RNA polymerase. *Curr. Opin. Struct. Biol.* 11:155–162.

Ross, W., Gosink, K. K., Salomon, J., Igarashi, K., Zou, C., Ishihama, A., Severinov, K., and Gourse, R. L. 1993. A third recognition element in bacterial promoters: DNA binding by the alpha subunit of RNA polymerase. *Science* 262:1407–1413.

Cramer, P., Bushnell, D. A., and Kornberg, R. D. 2001. Structural basis of transcription: RNA polymerase II at 2.8 Å resolution. *Science* 292:1863–1875.

Gnatt, A. L., Cramer, P., Fu, J., Bushnell, D. A., and Kornberg, R. D. 2001. Structural basis of transcription: An RNA polymerase II elongation complex at 3.3 Å resolution. *Science* 292:1876–1882.

Zhang, G., Campbell, E. A., Minakhin, L., Richter, C., Severinov, K., and Darst, S. A. 1999. Crystal structure of *Thermus aquaticus* core RNA polymerase at 3.3 Å resolution. *Cell* 98:811–824.

Campbell, E. A., Korzheva, N., Mustaev, A., Murakami, K., Nair, S., Goldfarb, A., and Darst, S. A. 2001. Structural mechanism for rifampicin inhibition of bacterial RNA polymerase. *Cell* 104:901–912.

Darst, S. A. 2004. New inhibitors targeting bacterial RNA polymerase. *Trends Biochem. Sci.* 29:159–160.

Cheetham, G. M., and Steitz, T. A. 1999. Structure of a transcribing T7 RNA polymerase initiation complex. *Science* 286:2305–2309.

Ebright, R. H. 2000. RNA polymerase: Structural similarities between bacterial RNA polymerase and eukaryotic RNA polymerase II. *J. Mol. Biol.* 304:687–698.

Paule, M. R., and White, R. J. 2000. Survey and summary: Transcription by RNA polymerases I and III. *Nucleic Acids Res.* 28:1283–1298.

Initiation and Elongation

Hantsche, M., and Cramer, P. 2017. Conserved RNA polymerase II initiation complex structure. *Curr. Opin. Struct. Biol.* 47, 17–22.

Murakami, K. S., and Darst, S. A. 2003. Bacterial RNA polymerases: The whole story. *Curr. Opin. Struct. Biol.* 13:31–39.

Buratowski, S. 2000. Snapshots of RNA polymerase II transcription initiation. *Curr. Opin. Cell Biol.* 12:320–325.

Conaway, J. W., and Conaway, R. C. 1999. Transcription elongation and human disease. *Annu. Rev. Biochem.* 68:301–319.

Conaway, J. W., Shilatifard, A., Dvir, A., and Conaway, R. C. 2000. Control of elongation by RNA polymerase II. *Trends Biochem. Sci.* 25:375–380.

Korzheva, N., Mustaev, A., Kozlov, M., Malhotra, A., Nikiforov, V., Goldfarb, A., and Darst, S. A. 2000. A structural model of transcription elongation. *Science* 289:619–625.

Reines, D., Conaway, R. C., and Conaway, J. W. 1999. Mechanism and regulation of transcriptional elongation by RNA polymerase II. *Curr. Opin. Cell Biol.* 11:342–346.

Promoters, Enhancers, and Transcription Factors

Vo Ngoc, L., Wang, Y. L., Kassavetis, G. A., and Kadonaga, J. T. 2017. The punctilious RNA polymerase II core promotor. *Genes Dev.* 31, 1289–1301.

Merika, M., and Thanos, D. 2001. Enhanceosomes. *Curr. Opin. Genet. Dev.* 11:205–208.

Park, J. M., Gim, B. S., Kim, J. M., Yoon, J. H., Kim, H. S., Kang, J. G., and Kim, Y. J. 2001. *Drosophila* mediator complex is broadly utilized by diverse gene-specific transcription factors at different types of core promoters. *Mol. Cell. Biol.* 21:2312–2323.

Smale, S. T., and Kadonaga, J. T. 2003. The RNA polymerase II core promoter. *Annu. Rev. Biochem.* 72:449–479.

Gourse, R. L., Ross, W., and Gaal, T. 2000. Ups and downs in bacterial transcription initiation: The role of the alpha subunit of RNA poly-merase in promoter recognition. *Mol. Microbiol.* 37:687–695.

Fiering, S., Whitelaw, E., and Martin, D. I. 2000. To be or not to be active: The stochastic nature of enhancer action. *BioEssays* 22:381–387.

Hampsey, M., and Reinberg, D. 1999. RNA polymerase II as a control panel for multiple coactivator complexes. *Curr. Opin. Genet. Dev.* 9:132–139.

Chen, L. 1999. Combinatorial gene regulation by eukaryotic transcription factors. *Curr. Opin. Struct. Biol.* 9:48–55.

Muller, C. W. 2001. Transcription factors: Global and detailed views. *Curr. Opin. Struct. Biol.* 11:26–32.

Reese, J. C. 2003. Basal transcription factors. *Curr. Opin. Genet. Dev.* 13:114–118.

Kadonaga, J. T. 2004. Regulation of RNA polymerase II transcription by sequence-specific DNA binding factors. *Cell* 116:247–257.

Harrison, S. C. 1991. A structural taxonomy of DNA-binding domains. *Nature* 353:715–719.

Sakurai, H., and Fukasawa, T. 2000. Functional connections between mediator components and general transcription factors of *Saccharomyces cerevisiae. J. Biol. Chem.* 275:37251–37256.

Droge, P., and Muller-Hill, B. 2001. High local protein concentrations at promoters: Strategies in prokaryotic and eukaryotic cells. *Bioessays* 23:179–183.

Smale, S. T., Jain, A., Kaufmann, J., Emami, K. H., Lo, K., and Garraway, I. P. 1998. The initiator element: A paradigm for core promoter heterogeneity within metazoan protein-coding genes. *Cold Spring Harbor Symp. Quant. Biol.* 63:21–31.

Kim, Y., Geiger, J. H., Hahn, S., and Sigler, P. B. 1993. Crystal structure of a yeast TBP/TATA-box complex. *Nature* 365:512–520.

Kim, J. L., Nikolov, D. B., and Burley, S. K. 1993. Co-crystal structure of TBP recognizing the minor groove of a TATA element. *Nature* 365:520–527.

White, R. J., and Jackson, S. P. 1992. The TATA-binding protein: A central role in transcription by RNA polymerases I, II and III. *Trends Genet.* 8:284–288.

Martinez, E. 2002. Multi-protein complexes in eukaryotic gene transcription. *Plant Mol. Biol.* 50:925–947.

Meinhart, A., Kamenski, T., Hoeppner, S., Baumli, S., and Cramer, P. 2005. A structural perspective of CTD function. *Genes Dev.* 19:1401–1415.

Palancade, B., and Bensaude, O. 2003. Investigating RNA polymerase II carboxyl-terminal domain (CTD) phosphorylation. *Eur. J. Biochem.* 270:3859–3870.

Termination

Burgess, B. R., and Richardson, J. P. 2001. RNA passes through the hole of the protein hexamer in the complex with *Escherichia coli* Rho factor. *J. Biol. Chem.* 276:4182–4189.

Yu, X., Horiguchi, T., Shigesada, K., and Egelman, E. H. 2000. Three-dimensional reconstruction of transcription termination factor rho: Orientation of the N-terminal domain and visualization of an RNA-binding site. *J. Mol. Biol.* 299:1279–1287.

Stitt, B. L. 2001. *Escherichia coli* transcription termination factor Rho binds and hydrolyzes ATP using a single class of three sites. *Biochemistry* 40:2276–2281.

Henkin, T. M. 2000. Transcription termination control in bacteria. *Curr. Opin. Microbiol.* 3:149–153.

Gusarov, I., and Nudler, E. 1999. The mechanism of intrinsic transcription termination. *Mol. Cell* 3:495–504.

Riboswitches

Barrick, J. E., and Breaker, R. R. 2007. The distributions, mechanisms, and structures of metabolite-binding riboswitches. *Genome Biol.* 8:R239.

Cheah, M. T., Wachter, A., Sudarsan, N., and Breaker, R. R. 2007. Control of alternative RNA splicing and gene expression by eukaryotic riboswitches. *Nature* 447:497–500.

Serganov, A., Huang, L., and Patel, D. J. 2009. Coenzyme recognition and gene regulation by a flavin mononucleotide riboswitch. *Nature* 458:233–237.

Noncoding RNA

Cech, T. R., and Steitz, J. A. 2014. The noncoding RNA revolution-trashing old rules to forge new ones. *Cell* 157:77–94.

Peculis, B. A. 2002. Ribosome biogenesis: Ribosomal RNA synthesis as a package deal. *Curr. Biol.* 12:R623–R624.

Decatur, W. A., and Fournier, M. J. 2002. rRNA modifications and ribosome function. *Trends Biochem. Sci.* 27:344–351.

Hopper, A. K., and Phizicky, E. M. 2003. tRNA transfers to the limelight. *Genes Dev.* 17:162–180.

Weiner, A. M. 2004. tRNA maturation: RNA polymerization without a nucleic acid template. *Curr. Biol.* 14:R883–R885.

5′-Cap Formation and Polyadenylation

Shatkin, A. J., and Manley, J. L. 2000. The ends of the affair: Capping and polyadenylation. *Nat. Struct. Biol.* 7:838–842.

Bentley, D. L. 2005. Rules of engagement: Co-transcriptional recruitment of pre-mRNA processing factors. *Curr. Opin. Cell Biol.* 17:251–256.

Aguilera, A. 2005. Cotranscriptional mRNP assembly: From the DNA to the nuclear pore. *Curr. Opin. Cell Biol.* 17:242–250.

Ro-Choi, T. S. 1999. Nuclear snRNA and nuclear function (discovery of 5′ cap structures in RNA). *Crit. Rev. Eukaryotic Gene Expr.* 9:107–158.

Bard, J., Zhelkovsky, A. M., Helmling, S., Earnest, T. N., Moore, C. L., and Bohm, A. 2000. Structure of yeast poly(A) polymerase alone and in complex with 3′-dATP. *Science* 289:1346–1349.

Martin, G., Keller, W., and Doublie, S. 2000. Crystal structure of mammalian poly(A) polymerase in complex with an analog of ATP. *EMBO J.* 19:4193–4203.

Zhao, J., Hyman, L., and Moore, C. 1999. Formation of mRNA 3′ ends in eukaryotes: Mechanism, regulation, and interrelationships with other steps in mRNA synthesis. *Microbiol. Mol. Biol. Rev.* 63:405–445.

Minvielle-Sebastia, L., and Keller, W. 1999. mRNA polyadenylation and its coupling to other RNA processing reactions and to transcription. *Curr. Opin. Cell Biol.* 11:352–357.

Small Regulatory RNAs

Winter, J., Jung, S., Keller, S., Gregory, R. I., and Diederichs, S. 2009. Many roads to maturity: MicroRNA biogenesis pathways and their regulation. *Nat. Cell Biol.* 11:228–234.

Ruvkun, G., Wightman, B., and Ha, I. 2004. The 20 years it took to recognize the importance of tiny RNAs. *Cell* 116:S93–S96.

RNA Editing

Gott, J. M., and Emeson, R. B. 2000. Functions and mechanisms of RNA editing. *Annu. Rev. Genet.* 34:499–531.

Simpson, L., Thiemann, O. H., Savill, N. J., Alfonzo, J. D., and Maslov, D. A. 2000. Evolution of RNA editing in trypanosome mitochondria. *Proc. Natl. Acad. Sci. U.S.A.* 97:6986–6993.

Chester, A., Scott, J., Anant, S., and Navaratnam, N. 2000. RNA editing: Cytidine to uridine conversion in apolipoprotein B mRNA. *Biochim. Biophys. Acta* 1494:1–3.

Maas, S., and Rich, A. 2000. Changing genetic information through RNA editing. *BioEssays* 22:790–802.

Splicing of mRNA Precursors

Shi, Y. 2017. Mechanistic insights into precursor messenger RNA splicing by the spliceosome. *Nat. Rev. Mol. Cell Biol.* 18, 655–670.

Shi, Y. 2017. The spliceosome: A protein-directed metalloribozyme. *J. Mol. Biol.* 429, 2640–2653.

Caceres, J. F., and Kornblihtt, A. R. 2002. Alternative splicing: Multiple control mechanisms and involvement in human disease. *Trends Genet.* 18:186–193.

Faustino, N. A., and Cooper, T. A. 2003. Pre-mRNA splicing and human disease. *Genes Dev.* 17:419–437.

Lou, H., and Gagel, R. F. 1998. Alternative RNA processing: Its role in regulating expression of calcitonin/calcitonin gene-related peptide. *J. Endocrinol.* 156:401–405.

Matlin, A. J., Clark, F., and Smith, C. W. 2005. Understanding alternative splicing: Towards a cellular code. *Nat. Rev. Mol. Cell Biol.* 6:386–398.

McKie, A. B., McHale, J. C., Keen, T. J., Tarttelin, E. E., Goliath, R., van Lith-Verhoeven, J. J., Greenberg, J., Ramesar, R. S., Hoyng, C. B., Cremers, F. P., et al. 2001. Mutations in the pre-mRNA splicing factor gene PRPC8 in autosomal dominant retinitis pigmentosa (RP13). *Hum. Mol. Genet.* 10:1555–1562.

Nilsen, T. W. 2003. The spliceosome: The most complex macromolecular machine in the cell? *BioEssays* 25:1147–1149.

Rund, D., and Rachmilewitz, E. 2005. β-Thalassemia. *New Engl. J. Med.* 353:1135–1146.

Patel, A. A., and Steitz, J. A. 2003. Splicing double: Insights from the second spliceosome. *Nat. Rev. Mol. Cell Biol.* 4:960–970.

Sharp, P. A. 2005. The discovery of split genes and RNA splicing. *Trends Biochem. Sci.* 30:279–281.

Valadkhan, S., and Manley, J. L. 2001. Splicing-related catalysis by protein-free snRNAs. *Nature* 413:701–707.

Zhou, Z., Licklider, L. J., Gygi, S. P., and Reed, R. 2002. Comprehensive proteomic analysis of the human spliceosome. *Nature* 419:182–185.

Stark, H., Dube, P., Luhrmann, R., and Kastner, B. 2001. Arrangement of RNA and proteins in the spliceosomal U1 small nuclear ribonu-cleoprotein particle. *Nature* 409:539–542.

Strehler, E. E., and Zacharias, D. A. 2001. Role of alternative splicing in generating isoform diversity among plasma membrane calcium pumps. *Physiol. Rev.* 81:21–50.

Graveley, B. R. 2001. Alternative splicing: Increasing diversity in the proteomic world. *Trends Genet.* 17:100–107.

Newman, A. 1998. RNA splicing. *Curr. Biol.* 8:R903–R905.

Reed, R. 2000. Mechanisms of fidelity in pre-mRNA splicing. *Curr. Opin. Cell Biol.* 12:340–345.

Sleeman, J. E., and Lamond, A. I. 1999. Nuclear organization of pre-mRNA splicing factors. *Curr. Opin. Cell Biol.* 11:372–377.

Black, D. L. 2000. Protein diversity from alternative splicing: A challenge for bioinformatics and post-genome biology. *Cell* 103:367–370.

Collins, C. A., and Guthrie, C. 2000. The question remains: Is the spliceosome a ribozyme? *Nat. Struct. Biol.* 7:850–854.

Self-Splicing and RNA Catalysis

Adams, P. L., Stanley, M. R., Kosek, A. B., Wang, J., and Strobel, S. A. 2004. Crystal structure of a self-splicing group I intron with both exons. *Nature* 430:45–50.

Adams, P. L., Stanley, M. R., Gill, M. L., Kosek, A. B., Wang, J., and Strobel, S. A. 2004. Crystal structure of a group I intron splicing intermediate. *RNA* 10:1867–1887.

Stahley, M. R., and Strobel, S. A. 2005. Structural evidence for a two-metal-ion mechanism of group I intron splicing. *Science* 309:1587–1590.

Carola, C., and Eckstein, F. 1999. Nucleic acid enzymes. *Curr. Opin. Chem. Biol.* 3:274–283.

Doherty, E. A., and Doudna, J. A. 2000. Ribozyme structures and mechanisms. *Annu. Rev. Biochem.* 69:597–615.

Fedor, M. J. 2000. Structure and function of the hairpin ribozyme. *J. Mol. Biol.* 297:269–291.

Hanna, R., and Doudna, J. A. 2000. Metal ions in ribozyme folding and catalysis. *Curr. Opin. Chem. Biol.* 4:166–170.

Scott, W. G. 1998. RNA catalysis. *Curr. Opin. Struct. Biol.* 8:720–726.

Biochemistry in Focus

Thomas R. Cech. 1989. Self-splicing and enzymatic activity from an intervening sequence RNA from *Tetrahymena*

(Nobel lecture). (https://www.nobelprize.org/nobel_prizes/chemistry/laureates/1989/cech-lecture.pdf)

CHAPTER 31

Where to Start

Yusupova, G., and Yusupov, M. 2014. High-resolution structure of the eukaryotic 80S ribosome. *Annu. Rev. Biochem.* 83:467–486.

Anger, A. M., Armache, J.-P., Berninghausen, O., Habeck, M., Subklewe, M., Wilson. D. N., and Beckmann, R. 2013. Structures of the human and Drosophila 80S ribosome. *Nature* 497:80–87.

Novoa, E. M., and Ribas de Pouplana, L. 2012. Speeding with control: Codon usage, tRNAs, and ribosomes. *Trends Genet.* 28:574–581.

Ibba, M., Curnow, A. W., and Söll, D. 1997. Aminoacyl-tRNA synthesis: Divergent routes to a common goal. *Trends Biochem. Sci.* 22:39–42.

Koonin, E. V., and Novozhilov, A. S. 2009. Origin and evolution of the genetic code: The universal enigma. *IUBMB Life* 61:99–111.

Schimmel, P., and Ribas de Pouplana, L. 2000. Footprints of aminoacyl-tRNA synthetases are everywhere. *Trends Biochem. Sci.* 25:207–209.

Books

Rodnina, M. V., Wintermeyer, W., and Green, R. 2011 (Eds.). *Ribosome Structure, Function and Dynamics.* Springer.

Gesteland, R. F., Atkins, J. F., and Cech, T. (Eds.). 2005. *The RNA World* (3rd ed.). Cold Spring Harbor Laboratory Press.

Garrett, R., Douthwaite, S. R., Liljas, A., Matheson, A. T., Moore, P. B., and Noller, H. F. 2000. *The Ribosome: Structure, Function, Antibiotics, and Cellular Interactions.* The American Society for Microbiology.

Aminoacyl-tRNA Synthetases

Kaminska, M., Havrylenko, S., Decottignies, P., Le Maréchal, P., Negrutskii, B., and Mirande, M. 2009. Dynamic organization of aminoacyl-tRNA synthetase complexes in the cytoplasm of human cells. *J. Biol. Chem.* 284:13746–13754.

Park, S. G., Schimmel, P., and Kim, S. 2008. Aminoacyl tRNA synthetases and their connections to disease. *Proc. Natl. Acad. Sci. U.S.A.* 105:11043–11049.

Ibba, M., and Söll, D. 2000. Aminoacyl-tRNA synthesis. *Annu. Rev. Biochem.* 69:617–650.

Sankaranarayanan, R., Dock-Bregeon, A. C., Rees, B., Bovee, M., Caillet, J., Romby, P., Francklyn, C. S., and Moras, D. 2000. Zinc ion mediated amino acid discrimination by threonyl-tRNA synthetase. *Nat. Struct. Biol.* 7:461–465.

Sankaranarayanan, R., Dock-Bregeon, A. C., Romby, P., Caillet, J., Springer, M., Rees, B., Ehresmann, C., Ehresmann, B., and Moras, D. 1999. The structure of threo-

nyl-tRNA synthetase-tRNA^{Thr} complex enlightens its repressor activity and reveals an essential zinc ion in the active site. *Cell* 97:371–381.

Dock-Bregeon, A., Sankaranarayanan, R., Romby, P., Caillet, J., Springer, M., Rees, B., Francklyn, C. S., Ehresmann, C., and Moras, D. 2000. Transfer RNA-mediated editing in threonyl-tRNA synthetase: The class II solution to the double discrimination problem. *Cell* 103:877–884.

de Pouplana, L. R., and Schimmel, P. 2000. A view into the origin of life: Aminoacyl-tRNA synthetases. *Cell. Mol. Life Sci.* 57:865–870.

Transfer RNA

Ibba, M., Becker, H. D., Stathopoulos, C., Tumbula, D. L., and Söll, D. 2000. The adaptor hypothesis revisited. *Trends Biochem. Sci.* 25:311–316.

Weisblum, B. 1999. Back to Camelot: Defining the specific role of tRNA in protein synthesis. *Trends Biochem. Sci.* 24:247–250.

Ribosomes and Ribosomal RNAs

Klinge, S., Voigts-Hoffmann, F., Leibundgut, M., and Ban, N. 2012. Atomic structures of the eukaryotic ribosome. *Trends Biochem. Sci.* 37:189–198.

Jin, H., Kelley, A. C., Loakes, D., and Ramakrishnan, V. 2010. Structure of the 70S ribosome bound to release factor 2 and a substrate analog provides insights into catalysis of peptide release. *Proc. Natl. Acad. Sci. U.S.A.* 107:8593–8598.

Rodnina, M. V., and Wintermeyer, W. 2009. Recent mechanistic insights into eukaryotic ribosomes. *Curr. Opin. Cell Biol.* 21:435–443.

Dinman, J. D. 2008. The eukaryotic ribosome: Current status and challenges. *J. Biol. Chem.* 284:11761–11765.

Wen, J.-D., Lancaster, L., Hodges, C., Zeri, A.-C., Yoshimura, S. H., Noller, H. F., Bustamante, C., and Tinoco, I., Jr. 2008. Following translation by single ribosomes one codon at a time. *Nature* 452:598–603.

Korostelev, A., and Noller, H. F. 2007. The ribosome in focus: New structures bring insights. *Trends Biochem. Sci.* 32:434–441.

Brandt, F., Etchells, S. A., Ortiz, J. O., Elcock, A. H., Hartl, F. U., and Baumeister, W. 2009. The native 3D organization of bacterial polysomes. *Cell* 136:261–271.

Initiation Factors

Søgaard, B., Sørensen, H. P., Mortensen, K. K., and Sperling-Petersen, H. U. 2005. Initiation of protein synthesis in bacteria. *Microbiol. Mol. Biol. Rev.* 69:101–123.

Carter, A. P., Clemons, W. M., Jr., Brodersen, D. E., Morgan-Warren, R. J., Hartsch, T., Wimberly, B. T., and Ramakrishnan, V. 2001. Crystal structure of an initiation factor bound to the 30S ribosomal subunit. *Science* 291:498–501.

Guenneugues, M., Caserta, E., Brandi, L., Spurio, R., Meunier, S., Pon, C. L., Boelens, R., and Gualerzi, C. O.

2000. Mapping the fMet-tRNA$_f^{Met}$ binding site of initiation factor IF2. *EMBO J.* 19:5233–5240.

Meunier, S., Spurio, R., Czisch, M., Wechselberger, R., Guenneugues, M., Gualerzi, C. O., and Boelens, R. 2000. Structure of the fMet-tRNA$_f^{Met}$-binding domain of *B. stearothermophilus* initiation factor IF2. *EMBO J.* 19:1918–1926.

Elongation Factors

Voorhees R. M., and Ramakrishnan, V. 2013. Structural basis of the translational elongation cycle. *Annu. Rev. Biochem.* 82:203–236.

Liu, S., Bachran, C., Gupta, P., Miller-Randolph, S., Wang, H., Crown, D., Zhang, Y., Kavaliauskas, D., Nissen, P., and Knudsen, C. R. 2012. The busiest of all ribosomal assistants: Elongation factor Tu. *Biochemistry* 51:2642–2651.

Schuette, J.-C., Murphy, F. V., Kelley, A. C., Weir, J. R., Giesebrecht, J., Connell, S. R., Loerke, J., Mielke, T., Zhang, W., Penczek, P. A., et al. 2009. GTPase activation of elongation factor EF-Tu by the ribosome during decoding. *EMBO J.* 28:755–765.

Stark, H., Rodnina, M. V., Wieden, H. J., van Heel, M., and Wintermeyer, W. 2000. Large-scale movement of elongation factor G and extensive conformational change of the ribosome during translocation. *Cell* 100:301–309.

Baensch, M., Frank, R., and Kohl, J. 1998. Conservation of the amino-terminal epitope of elongation factor Tu in Eubacteria and Archaea. *Microbiology* 144:2241–2246.

Krasny, L., Mesters, J. R., Tieleman, L. N., Kraal, B., Fucik, V., Hilgenfeld, R., and Jonak, J. 1998. Structure and expression of elongation factor Tu from *Bacillus stearothermophilus*. *J. Mol. Biol.* 283:371–381.

Pape, T., Wintermeyer, W., and Rodnina, M. V. 1998. Complete kinetic mechanism of elongation factor Tu-dependent binding of aminoacyl-tRNA to the A site of the *E. coli* ribosome. *EMBO J.* 17:7490–7497.

Piepenburg, O., Pape, T., Pleiss, J. A., Wintermeyer, W., Uhlenbeck, O. C., and Rodnina, M. V. 2000. Intact aminoacyl-tRNA is required to trigger GTP hydrolysis by elongation factor Tu on the ribosome. *Biochemistry* 39:1734–1738.

Peptide-Bond Formation and Translocation

Kohler, R., Mooney, R. A., Mills, D. J., Landick, R., and Cramer, R. 2017. Architecture of a transcribing-translating expressome. *Science* 356:194–197.

Rodnina, M. V. 2013. The ribosome as a versatile catalyst: Reactions at the peptidyl transferase center. *Curr. Opin. Struct. Biol.* 23:595–602.

Uemura, S., Aitken, C. E., Korlach, J., Flusberg, B. A., Turner, S. W., and Puglisi, J. D. 2010. Real-time tRNA transit on single translating ribosomes at codon resolution. *Nature* 464:1012–1018.

Beringer, M. and Rodnina, M. V. 2007. The ribosomal peptidyl transferase. *Mol. Cell* 26:311–321.

Yarus, M., and Welch, M. 2000. Peptidyl transferase: Ancient and exiguous. *Chem. Biol.* 7:R187–R190.

Vladimirov, S. N., Druzina, Z., Wang, R., and Cooperman, B. S. 2000. Identification of 50S components neighboring 23S rRNA nucleotides A2448 and U2604 within the peptidyl transferase center of *Escherichia coli* ribosomes. *Biochemistry* 39:183–193.

Frank, J., and Agrawal, R. K. 2000. A ratchet-like inter-subunit reorganization of the ribosome during translocation. *Nature* 406:318–322.

Termination

Weixlbaumer, A., Jin, H., Neubauer, C., Voorhees, R. M., Petry, S., Kelley, A. C., and Ramakrishnan, V. 2008. Insights into translational termination from the structure of RF2 bound to the ribosome. *Science* 322:953–956.

Trobro, S., and Åqvist, S. 2007. A model for how ribosomal release factors induce peptidyl-tRNA cleavage in termination of protein synthesis. *Mol. Cell* 27:758–766.

Korostelev, A., Asahara, H., Lancaster, L., Laurberg, M., Hirschi, A., Zhu, J., Trakhanov, S., Scott, W. G., and Noller, H. F. 2008. Crystal structure of a translation termination complex formed with release factor RF2. *Proc. Natl. Acad. Sci. U.S.A.* 105:19684–19689.

Wilson, D. N., Schluenzen, F., Harms, J. M., Yoshida, T., Ohkubo, T., Albrecht, A., Buerger, J., Kobayashi, Y., and Fucini, P. 2005. X-ray crystallography study on ribosome recycling: The mechanism of binding and action of RRF on the 50S ribosomal subunit. *EMBO J.* 24:251–260.

Kisselev, L. L., and Buckingham, R. H. 2000. Translational termination comes of age. *Trends Biochem. Sci.* 25:561–566.

Fidelity and Proofreading

Loveland, A. B., Demo, G. Grigorieff, N., and Korostelev, A. A. 2017. Ensemble cryo-EM elucidates the mechanism of translation fidelity. *Nature* 546:113–119.

Zaher, H. S., and Green, R. 2009. Quality control by the ribosome following peptide bond formation. *Nature* 457:161–166.

Zaher, H. S., and Green, R. 2009. Fidelity at the molecular level: Lessons from protein synthesis. *Cell* 136:746–762.

Ogle, J. M., and Ramakrishnan, V. 2005. Structural insights into translational fidelity. *Annu. Rev. Biochem.* 74:129–177.

Eukaryotic Protein Synthesis

Hinnebusch, A. G. 2014. The scanning mechanism of eukaryotic translation initiation. *Annu. Rev. Biochem.* 83:779–812.

Wein, A. N., Singh, R., Fattah, R., and Leppla, S. H. 2012. Diphthamide modification on eukaryotic elongation factor 2 is needed to assure fidelity of mRNA translation and mouse development. *Proc. Natl. Acad. Sci. U.S.A.* 109:13817–13822.

Rhoads, R. E. 2009. eIF4E: New family members, new binding partners, new roles. *J. Biol. Chem.* 284:16711–16715.

Marintchev, A., Edmonds, K. A., Marintcheva, B., Hendrickson, E., Oberer, M., Suzuki, C., Herdy, B., Sonenberg, N., and Wagner, G. 2009. Topology and regulation of the human eIF4A/4G/4H helicase complex in translation initiation. *Cell* 136:447–460.

Fitzgerald, K. D., and Semler, B. L. 2009. Bridging IRES elements in mRNAs to the eukaryotic translation apparatus. *Biochim. Biophys. Acta* 1789:518–528.

Mitchell, S. F., and Lorsch, J. R. 2008. Should I stay or should I go? Eukaryotic translation initiation factors 1 and 1A control start codon recognition. *J. Biol. Chem.* 283:27345–27349.

Amrani, A., Ghosh, S., Mangus, D. A., and Jacobson, A. 2008. Translation factors promote the formation of two states of the closed-loop mRNP. *Nature* 453:1276–1280.

Sachs, A. B., and Varani, G. 2000. Eukaryotic translation initiation: There are (at least) two sides to every story. *Nat. Struct. Biol.* 7:356–361.

Kozak, M. 1999. Initiation of translation in prokaryotes and eukaryotes. *Gene* 234:187–208.

Bushell, M., Wood, W., Clemens, M. J., and Morley, S. J. 2000. Changes in integrity and association of eukaryotic protein synthesis initiation factors during apoptosis. *Eur. J. Biochem.* 267:1083–1091.

Das, S., Ghosh, R., and Maitra, U. 2001. Eukaryotic translation initiation factor 5 functions as a GTPase-activating protein. *J. Biol. Chem.* 276:6720–6726.

Lee, J. H., Choi, S. K., Roll-Mecak, A., Burley, S. K., and Dever, T. E. 1999. Universal conservation in translation initiation revealed by human and archaeal homologs of bacterial translation initiation factor IF2. *Proc. Natl. Acad. Sci. U.S.A.* 96:4342–4347.

Pestova, T. V., and Hellen, C. U. 2000. The structure and function of initiation factors in eukaryotic protein synthesis. *Cell. Mol. Life Sci.* 57:651–674.

Antibiotics and Toxins

Belova, L., Tenson, T., Xiong, L., McNicholas, P. M., and Mankin, A. S. 2001. A novel site of antibiotic action in the ribosome: Interaction of evernimicin with the large ribosomal subunit. *Proc. Natl. Acad. Sci. U.S.A.* 98:3726–3731.

Brodersen, D. E., Clemons, W. M., Jr., Carter, A. P., Morgan-Warren, R. J., Wimberly, B. T., and Ramakrishnan, V. 2000. The structural basis for the action of the antibiotics tetracycline, pactamycin, and hygromycin B on the 30S ribosomal subunit. *Cell* 103:1143–1154.

Porse, B. T., and Garrett, R. A. 1999. Ribosomal mechanics, antibiotics, and GTP hydrolysis. *Cell* 97:423–426.

Lord, M. J., Jolliffe, N. A., Marsden, C. J., Pateman, C. S., Smith, D. S., Spooner, R. A., Watson, P. D., and Roberts, L. M. 2003. Ricin: Mechanisms of toxicity. *Toxicol. Rev.* 22:53–64.

Protein Transport Across Membranes

Costa, E. A., Subramanian, K., Nunnari, J., and Weissman J. S. 2018. Defining the physiological role of SRP in protein-targeting efficiency and specificity. *Science* 359:689–692.

Akopian, D., Shen, K., Zhang, X., and Shan, S. 2013. Signal recognition particle: An essential protein-targeting machine. *Annu. Rev. Biochem.* 82:693–721.

Nyathi, Y., Wilkinson, B. M., and Pool, M. R. 2013. Co-translational targeting and translocation of proteins to the endoplasmic reticulum. *Biochim. Biophys. Acta* 1833:2392–2402.

Janda, C. Y., Li, J., Oubridge, C., Hernández, H., Robinson, C. V., and Nagai, K. 2010. Recognition of a signal peptide by the signal recognition particle. *Nature* 465:507–510.

Cross, B. C. S., Sinning, I., Luirink, J., and High, S. 2009. Delivering proteins for export from the cytosol. *Nat. Rev. Mol. Cell. Biol.* 10:255–264.

Shan, S., Schmid, S. L., and Zhang, X. 2009. Signal recognition particle (SRP) and SRP receptor: A new paradigm for multistate regulatory GTPases. *Biochemistry* 48:6696–6704.

Johnson, A. E. 2009. The structural and functional coupling of two molecular machines, the ribosome and the translocon. *J. Cell Biol.* 185:765–767.

Pool, R. P. 2009. A trans-membrane segment inside the ribosome exit tunnel triggers RAMP4 recruitment to the Sec61p translocase. *J. Cell Biol.* 185:889–902.

Egea, P. F., Stroud, R. M., and Walter, P. 2005. Targeting proteins to membranes: Structure of the signal recognition particle. *Curr. Opin. Struct. Biol.* 15:213–220.

Halic, M., and Beckmann, R. 2005. The signal recognition particle and its interactions during protein targeting. *Curr. Opin. Struct. Biol.* 15:116–125.

Doudna, J. A., and Batey, R. T. 2004. Structural insights into the signal recognition particle. *Annu. Rev. Biochem.* 73:539–557.

Schnell, D. J., and Hebert, D. N. 2003. Protein translocons: Multifunctional mediators of protein translocation across membranes. *Cell* 112:491–505.

Biochemistry in Focus

Khajuria, R. K., Munschauer, M., Ulirsch, J. C., Fiorini, C., Leif S. Ludwig, L. S. et al. 2018. Ribosome levels selectively regulate translation and lineage commitment in human hematopoiesis. *Cell* 173:90–103.

Mills E. W., and Green, R. 2017. Ribosomopathies: There's strength in numbers. *Science* 358, eaan2755

Simsek, D., Tiu, G. C., Flynn, R. A., Byeon, G. W., Leppek, K., et al. 2017. The mammalian ribo-interactome reveals ribosome functional diversity and heterogeneity. *Cell* 169:1051–1065.

Shi, Z., Fujii, K., Kovary, K. M., Genuth, N. R., Röst, H. L., Teruel, M. N., and Barna, M. 2017. Heterogeneous ribosomes preferentially translate distinct subpools of mRNAs genome-wide. *Mol. Cell* 67:71–83.

CHAPTER 32

Where to Start

Ptashne, M. 2014. The chemistry of regulation of genes and other things. *J. Biol. Chem.* 289:5417–5435.

Pabo, C. O., and Sauer, R. T. 1984. Protein–DNA recognition. *Annu. Rev. Biochem.* 53:293–321.

Ptashne, M., Johnson, A. D., and Pabo, C. O. 1982. A genetic switch in a bacterial virus. *Sci. Am.* 247:128–140.

Ptashne, M., Jeffrey, A., Johnson, A. D., Maurer, R., Meyer, B. J., Pabo, C. O., Roberts, T. M., and Sauer, R. T. 1980. How the lambda repressor and Cro work. *Cell* 19:1–11.

Books

Ptashne, M. 2004. *A Genetic Switch: Phage 3 Revisited* (3rd ed.). Cold Spring Harbor Laboratory Press.

McKnight, S. L., and Yamamoto, K. R. (Eds.). 1992. *Transcriptional Regulation* (vols. 1 and 2). Cold Spring Harbor Laboratory Press.

Lodish, H., Berk, A., Kaiser, C. A., Krieger, M., Bretscher, A., Pleogh, H., Amon, A., and Scott, M. P. 2012. *Molecular Cell Biology* (7th ed.). W. H. Freeman and Company.

DNA-Binding Proteins

Balaeff, A., Mahadevan, L., and Schulten, K. 2004. Structural basis for cooperative DNA binding by CAP and *lac* repressor. *Structure* 12:123–132.

Bell, C. E., and Lewis, M. 2001. The Lac repressor: A second generation of structural and functional studies. *Curr. Opin. Struct. Biol.* 11:19–25.

Lewis, M., Chang, G., Horton, N. C., Kercher, M. A., Pace, H. C., Schumacher, M. A., Brennan, R. G., and Lu, P. 1996. Crystal structure of the lactose operon repressor and its complexes with DNA and inducer. *Science* 271:1247–1254.

Niu, W., Kim, Y., Tau, G., Heyduk, T., and Ebright, R. H. 1996. Transcription activation at class II CAP-dependent promoters: Two interactions between CAP and RNA polymerase. *Cell* 87:1123–1134.

Schultz, S. C., Shields, G. C., and Steitz, T. A. 1991. Crystal structure of a CAP-DNA complex: The DNA is bent by 90 degrees. *Science* 253:1001–1007.

Parkinson, G., Wilson, C., Gunasekera, A., Ebright, Y. W., Ebright, R. E., and Berman, H. M. 1996. Structure of the CAP-DNA complex at 2.5 Å resolution: A complete picture of the protein–DNA interface. *J. Mol. Biol.* 260:395–408.

Busby, S., and Ebright, R. H. 1999. Transcription activation by catabo-lite activator protein (CAP). *J. Mol. Biol.* 293:199–213.

Somers, W. S., and Phillips, S. E. 1992. Crystal structure of the met repressor-operator complex at 2.8 Å resolution reveals DNA recognition by β-strands. *Nature* 359:387–393.

Gene-Regulatory Circuits

Johnson, A. D., Poteete, A. R., Lauer, G., Sauer, R. T., Ackers, G. K., and Ptashne, M. 1981. Lambda repressor and Cro: Components of an efficient molecular switch. *Nature* 294:217–223.

Stayrook, S., Jaru-Ampornpan, P., Ni, J., Hochschild, A., and Lewis, M. 2008. Crystal structure of the lambda repressor and a model for pairwise cooperative operator binding. *Nature* 452:1022–1025.

Arkin, A., Ross, J., and McAdams, H. H. 1998. Stochastic kinetic analysis of developmental pathway bifurcation in phage lambda-infected *Escherichia coli* cells. *Genetics* 149:1633–1648.

Post-transcriptional Regulation

Kolter, R., and Yanofsky, C. 1982. Attenuation in amino acid biosyn-thetic operons. *Annu. Rev. Genet.* 16:113–134.

Yanofsky, C. 1981. Attenuation in the control of expression of bacterial operons. *Nature* 289:751–758.

Miller, M. B., and Bassler, B. L. 2001. Quorum sensing in bacteria. *Annu. Rev. Microbiol.* 55:165–199.

Zhang, R. G., Pappas, T., Brace, J. L., Miller, P. C., Oulmassov, T., Molyneaux, J. M., Anderson, J. C., Bashkin, J. K., Winans, S. C., and Joachimiak, A. 2002. Structure of a bacterial quorum-sensing transcription factor complexed with pheromone and DNA. *Nature* 417:971–974.

Soberon-Chavez, G., Aguirre-Ramirez, M., and Ordonez, L. 2005. Is *Pseudomonas aeruginosa* only "sensing quorum"? *Crit. Rev. Microbiol.* 31:171–182.

Historical Aspects

Lewis, M. 2005. The lac repressor. *C. R. Biol.* 328:521–548.

Jacob, F., and Monod, J. 1961. Genetic regulatory mechanisms in the synthesis of proteins. *J. Mol. Biol.* 3:318–356.

Ptashne, M., and Gilbert, W. 1970. Genetic repressors. *Sci. Am.* 222(6):36–44.

Lwoff, A., and Ullmann, A. (Eds.). 1979. *Origins of Molecular Biology: A Tribute to Jacques Monod*. Academic Press.

Judson, H. 1996. *The Eighth Day of Creation: Makers of the Revolution in Biology*. Cold Spring Harbor Laboratory Press.

Biochemistry in Focus

R. T. Sauer et al. 1982, Cleavage of the λ and P22 repressors by *recA* protein, J. Biol. Chem. 257:4458–4462.

S. Gottesman 1999, Regulation by proteolysis: developmental switches, Curr. Opin. Microbiol. 2:142–147.

CHAPTER 33

Where to Start

Liu, X., Bushnell, D. A., and Kornberg, R. D. 2013. RNA polymerase II transcription: Structure and mechanism. *Biochim. Biophys. Acta* 1829:2–8.

Kornberg, R. D. 2007. The molecular basis of eukaryotic transcription. *Proc. Natl. Acad. Sci. U.S.A.* 104:12955–12961.

Pabo, C. O., and Sauer, R. T. 1984. Protein–DNA recognition. *Annu. Rev. Biochem.* 53:293–321.

Struhl, K. 1989. Helix-turn-helix, zinc-finger, and leucine-zipper motifs for eukaryotic transcriptional regulatory proteins. *Trends Biochem. Sci.* 14:137–140.

Struhl, K. 1999. Fundamentally different logic of gene regulation in eukaryotes and prokaryotes. *Cell* 98:1–4.

Korzus, E., Torchia, J., Rose, D. W., Xu, L., Kurokawa, R., McInerney, E. M., Mullen, T. M., Glass, C. K., and Rosenfeld, M. G. 1998. Transcription factor-specific requirements for coactivators and their acetyltransferase functions. *Science* 279:703–707.

Aalfs, J. D., and Kingston, R. E. 2000. What does "chromatin remodeling" mean? *Trends Biochem. Sci.* 25:548–555.

Books

McKnight, S. L., and Yamamoto, K. R. (Eds.). 1992. *Transcriptional Regulation* (vols. 1 and 2). Cold Spring Harbor Laboratory Press.

Latchman, D. S. 2004. *Eukaryotic Transcription Factors* (4th ed.). Academic Press.

Wolffe, A. 1992. *Chromatin Structure and Function*. Academic Press.

Lodish, H., Berk, A., Kaiser, C. A., Krieger, M., Bretscher, A., Pleogh, H., Amon, A., and Scott, M. P. 2012. *Molecular Cell Biology* (7th ed.). W. H. Freeman and Company.

Chromatin and Chromatin Remodeling

Sadeh, R., and Allis, C. D. 2011. Genome-wide "re"-modeling of nucleosome positions. *Cell* 147:263–266.

Lorch, Y., Maier-Davis, B., and Kornberg, R. D. 2010. Mechanism of chromatin remodeling. *Proc. Natl. Acad. Sci. U.S.A.* 107:3458–3462.

Tang, L., Nogales, E., and Ciferri, C. 2010. Structure and function of SWI/SNF chromatin remodeling complexes and mechanistic implications for transcription. *Prog. Biophys. Mol. Biol.* 102:122–128.

Jenuwein, T., and Allis, C. D. 2001. Translating the histone code. *Science* 293:1074–1080.

Jiang, C., and Pugh, B. F. 2009. Nucleosome positioning and gene regulation: Advances through genomics. *Nat. Rev. Genet.* 10:161–172.

Barski, A., Cuddapah, S., Cui, K., Roh, T. Y., Schones, D. E., Wang, Z., Wei, G., Chepelev, I., and Zhao, K. 2007. High-resolution profiling of histone methylations in the human genome. *Cell* 129:823–837.

Weintraub, H., Larsen, A., and Groudine, M. 1981. β-Globin-gene switching during the development of chicken embryos: Expression and chromosome structure. *Cell* 24:333–344.

Ren, B., Robert, F., Wyrick, J. J., Aparicio, O., Jennings, E. G., Simon, I., Zeitlinger, J., Schreiber, J., Hannett, N., Kanin, E., et al. 2000. Genome-wide location and function of DNA-binding proteins. *Science* 290:2306–2309.

Goodrich, J. A., and Tjian, R. 1994. TBP-TAF complexes: Selectivity factors for eukaryotic transcription. *Curr. Opin. Cell. Biol.* 6:403–409.

Bird, A. P., and Wolffe, A. P. 1999. Methylation-induced repression: Belts, braces, and chromatin. *Cell* 99:451–454.

Cairns, B. R. 1998. Chromatin remodeling machines: Similar motors, ulterior motives. *Trends Biochem. Sci.* 23:20–25.

Albright, S. R., and Tjian, R. 2000. TAFs revisited: More data reveal new twists and confirm old ideas. *Gene* 242:1–13.

Urnov, F. D., and Wolffe, A. P. 2001. Chromatin remodeling and transcriptional activation: The cast (in order of appearance). *Oncogene* 20:2991–3006.

Luger, K., Mader, A. W., Richmond, R. K., Sargent, D. F., and Richmond, T. J. 1997. Crystal structure of the nucleosome core particle at 2.8 Å resolution. *Nature* 389:251–260.

Arents, G., and Moudrianakis, E. N. 1995. The histone fold: A ubiquitous architectural motif utilized in DNA compaction and protein dimerization. *Proc. Natl. Acad. Sci. U.S.A.* 92:11170–11174.

Baxevanis, A. D., Arents, G., Moudrianakis, E. N., and Landsman, D. 1995. A variety of DNA-binding and multimeric proteins contain the histone fold motif. *Nucleic Acids Res.* 23:2685–2691.

Transcription Factors

Green, M. R. 2005. Eukaryotic transcription activation: Right on target. *Mol. Cell* 18:399–402.

Kornberg, R. D. 2005. Mediator and the mechanism of transcriptional activation. *Trends Biochem. Sci.* 30:235–239.

Clements, A., Rojas, J. R., Trievel, R. C., Wang, L., Berger, S. L., and Marmorstein, R. 1999. Crystal structure of the histone acetyltransferase domain of the human PCAF transcriptional regulator bound to coenzyme A. *EMBO J.* 18:3521–3532.

Deckert, J., and Struhl, K. 2001. Histone acetylation at promoters is differentially affected by specific activators and repressors. *Mol. Cell. Biol.* 21:2726–2735.

Dutnall, R. N., Tafrov, S. T., Sternglanz, R., and Ramakrishnan, V. 1998. Structure of the histone acetyltransferase Hat1: A paradigm for the GCN5-related N-acetyltransferase superfamily. *Cell* 94:427–438.

Finnin, M. S., Donigian, J. R., Cohen, A., Richon, V. M., Rifkind, R. A., Marks, P. A., Breslow, R., and Pavletich, N. P. 1999. Structures of a histone deacetylase homologue bound to the TSA and SAHA inhibitors. *Nature* 401:188–193.

Finnin, M. S., Donigian, J. R., and Pavletich, N. P. 2001. Structure of the histone deacetylase SIR2. *Nat. Struct. Biol.* 8:621–625.

Jacobson, R. H., Ladurner, A. G., King, D. S., and Tjian, R. 2000. Structure and function of a human TAFII250 double bromodo-main module. *Science* 288:1422–1425.

Rojas, J. R., Trievel, R. C., Zhou, J., Mo, Y., Li, X., Berger, S. L., Allis, C. D., and Marmorstein, R. 1999. Structure of *Tetrahymena* GCN5 bound to coenzyme A and a histone H3 peptide. *Nature* 401:93–98.

Induced Pluripotent Stem Cells

Takahashi, K., Tanabe, K., Ohnuki, M., Narita, M., Ichisaka, T., Tomoda, K., and Yamanaka, S. 2007. Induction of pluripotent stem cells from adult human fibroblasts by defined factors. *Cell* 131:861–872.

Takahashi, K., and Yamanaka, S. 2006. Induction of pluripotent stem cells from mouse embryonic and adult fibroblast cultures by defined factors. *Cell* 126:663–676.

Park, I. H., Arora, N., Huo, H., Maherali, N., Ahfeldt, T., Shimamura, A., Lensch, M. W., Cowan, C., Hochedlinger, K., and Daley, G. Q. 2008. Disease-specific induced pluripotent stem cells. *Cell* 134:877–886.

Yamanaka, S. 2009. A fresh look at iPS cells. *Cell* 137:13–17.

Yu, J., Hu, K., Smuga-Otto, K., Tian, S., Stewart, R., Slukvin, I. I., and Thomson, J. A. 2009. Human induced pluripotent stem cells free of vector and transgene sequences. *Science* 324:797–801.

Nuclear Hormone Receptors

Downes, M., Verdecia, M. A., Roecker, A. J., Hughes, R., Hogenesch, J. B., Kast-Woelbern, H. R., Bowman, M. E., Ferrer, J. L., Anisfeld, A. M., Edwards, et al. 2003. A chemical, genetic, and structural analysis of the nuclear bile acid receptor FXR. *Mol. Cell* 11:1079–1092.

Evans, R. M. 2005. The nuclear receptor superfamily: A Rosetta stone for physiology. *Mol. Endocrinol.* 19:1429–1438.

Xu, W., Cho, H., Kadam, S., Banayo, E. M., Anderson, S., Yates, J. R., III, Emerson, B. M., and Evans, R. M. 2004. A methylation-mediator complex in hormone signaling. *Genes Dev.* 18:144–156.

Evans, R. M. 1988. The steroid and thyroid hormone receptor superfamily. *Science* 240:889–895.

Yamamoto, K. R. 1985. Steroid receptor regulated transcription of specific genes and gene networks. *Annu. Rev. Genet.* 19:209–252.

Tanenbaum, D. M., Wang, Y., Williams, S. P., and Sigler, P. B. 1998. Crystallographic comparison of the estrogen and progesterone receptor's ligand binding domains. *Proc. Natl. Acad. Sci. U.S.A.* 95:5998–6003.

Schwabe, J. W., Chapman, L., Finch, J. T., and Rhodes, D. 1993. The crystal structure of the estrogen receptor DNA-binding domain bound to DNA: How receptors discriminate between their response elements. *Cell* 75:567–578.

Shiau, A. K., Barstad, D., Loria, P. M., Cheng, L., Kushner, P. J., Agard, D. A., and Greene, G. L. 1998. The structural basis of estrogen receptor/coactivator recognition and the antagonism of this interaction by tamoxifen. *Cell* 95:927–937.

Collingwood, T. N., Urnov, F. D., and Wolffe, A. P. 1999. Nuclear receptors: Coactivators, corepressors and chromatin remodeling in the control of transcription. *J. Mol. Endocrinol.* 23:255–275.

Posttranscriptional Regulation

Rouault, T. A., Stout, C. D., Kaptain, S., Harford, J. B., and Klausner, R. D. 1991. Structural relationship between an iron-regulated RNA-binding protein (IRE-BP) and aconitase: Functional implications. *Cell* 64:881–883.

Klausner, R. D., Rouault, T. A., and Harford, J. B. 1993. Regulating the fate of mRNA: The control of cellular iron metabolism. *Cell* 72:19–28.

Gruer, M. J., Artymiuk, P. J., and Guest, J. R. 1997. The aconitase family: Three structural variations on a common theme. *Trends Biochem. Sci.* 22:3–6.

Theil, E. C. 1994. Iron regulatory elements (IREs): A family of mRNA non-coding sequences. *Biochem. J.* 304:1–11.

MicroRNAs

Ruvkun, G. 2008. The perfect storm of tiny RNAs. *Nat. Med.* 14:1041–1045.

Sethupathy, P., and Collins, F. S. 2008. MicroRNA target site polymorphisms and human disease. *Trends Genet.* 24:489–497.

Adams, B. D., Cowee, D. M., and White, B. A. 2009. The role of miR-206 in the epidermal growth factor (EGF) induced repression of estrogen receptor-α (ERα) signaling and a luminal phenotype in MCF-7 breast cancer cells. *Mol. Endocrinol.* 23:1215–1230.

Jegga, A. G., Chen, J., Gowrisankar, S., Deshmukh, M. A., Gudivada, R., Kong, S., Kaimal, V., and Aronow, B. J. 2007. GenomeTrafac: A whole genome resource for the detection of transcription factor binding site clusters associated with conventional and microRNA encoding genes conserved between mouse and human gene orthologs. *Nucleic Acids Res.* 35:D116–D121.

Biochemistry in Focus

S. Poepsel et al. 2018. Cryo-EM structures of PRC2 simultaneously engaged with two functionally distinct nucleosomes, Nature Structural and Molecular Biology 25:154–162.

CHAPTER 34 [ONLINE ONLY]

Where to Start

Axel, R. 1995. The molecular logic of smell. *Sci. Am.* 273(4):154–159.

Dulac, C. 2000. The physiology of taste, vintage 2000. *Cell* 100:607–610.

Yarmolinsky, D. A., Zuker, C. S., and Ryba, N. J. (2009) Common sense about taste: From mammals to insects. *Cell* 139:234–244.

Stryer, L. 1996. Vision: From photon to perception. *Proc. Natl. Acad. Sci. U.S.A.* 93:557–559.

Hudspeth, A. J. 1989. How the ear's works work. *Nature* 341:397–404.

Olfaction

Buck, L., and Axel, R. 1991. A novel multigene family may encode odorant receptors: A molecular basis for odor recognition. *Cell* 65:175–187.

Saito, H., Chi, Q., Zhuang, H., Matsunami, H., and Mainland, J. D. 2009. Odor coding by a mammalian receptor repertoire. *Sci. Signal.* 2:ra9.

Malnic, B., Hirono, J., Sato, T., and Buck, L. B. 1999. Combinatorial receptor codes for odors. *Cell* 96:713–723.

Zou, D. J., Chesler, A., and Firestein, S. 2009. How the olfactory bulb got its glomeruli: A just so story? *Nat. Rev. Neurosci.* 10:611–618.

De la Cruz, O., Blekhman, R., Zhang, X., Nicolae, D., Firestein, S., and Gilad, Y. 2009. A signature of evolutionary constraint on a subset of ectopically expressed olfactory receptor genes. *Mol. Biol. Evol.* 26:491–494.

Mombaerts, P., Wang, F., Dulac, C., Chao, S. K., Nemes, A., Mendelsohn, M., Edmondson, J., and Axel, R. 1996. Visualizing an olfactory sensory map. *Cell* 87:675–686.

Buck, L. 2005. Unraveling the sense of smell (Nobel lecture). *Angew. Chem. Int. Ed. Engl.* 44:6128–6140.

Belluscio, L., Gold, G. H., Nemes, A., and Axel, R. 1998. Mice deficient in G$_{(olf)}$ are anosmic. *Neuron* 20:69–81.

Vosshall, L. B., Wong, A. M., and Axel, R. 2000. An olfactory sensory map in the fly brain. *Cell* 102:147–159.

Lewcock, J. W., and Reed, R. R. 2003. A feedback mechanism regulates monoallelic odorant receptor expression. *Proc. Natl. Acad. Sci. U.S.A.*101:1069–1074.

Reed, R. R. 2004. After the holy grail: Establishing a molecular mechanism for mammalian olfaction. *Cell* 116:329–336.

Taste

Wang, L., Gillis-Smith, S., Peng, Y., Zhang, J., Chen, X., Salzman, C. D., Ryba, N. J. P., and Zuker, C. S. 2018. The coding of valence and identity in the mammalian taste system. *Nature* 558, 127–131.

Chandrashekar, J., Yarmolinsky, D., von Buchholtz, L., Oka, Y., Sly, W., Ryba, N. J., and Zuker, C. S. 2009. The taste of carbonation. *Science* 326:443–445.

Chandrashekar, J., Hoon, M. A., Ryba, N. J., and Zuker, C. S. 2006. The receptors and cells for mammalian taste. *Nature* 444:288–294.

Huang, A. L., Chen, X., Hoon, M. A., Chandrashekar, J., Guo, W., Tranker, D., Ryba, N. J., and Zuker, C. S. 2006. The cells and logic for mammalian sour taste detection. *Nature* 442:934–938.

Zhao, G. Q., Zhang, Y., Hoon, M. A., Chandrashekar, J., Erlenbach, I., Ryba, N. J. P., and Zuker, C. S. 2003. The receptors for mammalian sweet and umami taste. *Cell* 115:255–266.

Herness, M. S., and Gilbertson, T. A. 1999. Cellular mechanisms of taste transduction. *Annu. Rev. Physiol.* 61:873–900.

Adler, E., Hoon, M. A., Mueller, K. L., Chandrashekar, J., Ryba, N. J., and Zuker, C. S. 2000. A novel family of mammalian taste receptors. *Cell* 100:693–702.

Chandrashekar, J., Mueller, K. L., Hoon, M. A., Adler, E., Feng, L., Guo, W., Zuker, C. S., and Ryba, N. J. 2000. T2Rs function as bitter taste receptors. *Cell* 100:703–711.

Mano, I., and Driscoll, M. 1999. DEG/ENaC channels: A touchy superfamily that watches its salt. *BioEssays* 21:568–578.

Benos, D. J., and Stanton, B. A. 1999. Functional domains within the degenerin/epithelial sodium channel (Deg/ENaC) superfamily of ion channels. *J. Physiol. (Lond.)* 520(part 3):631–644.

McLaughlin, S. K., McKinnon, P. J., and Margolskee, R. F. 1992. Gustducin is a taste-cell-specific G protein closely related to the transducins. *Nature* 357:563–569.

Nelson, G., Hoon, M. A., Chandrashekar, J., Zhang, Y., Ryba, N. J., and Zuker, C. S. 2001. Mammalian sweet taste receptors. *Cell* 106:381–390.

Vision

Stryer, L. 1988. Molecular basis of visual excitation. *Cold Spring Harbor Symp. Quant. Biol.* 53:283–294.

Jastrzebska, B., Tsybovsky, Y., and Palczewski, K. 2010. Complexes between photoactivated rhodopsin and transducin: Progress and questions. *Biochem. J.* 428:1–10.

Wald, G. 1968. The molecular basis of visual excitation. *Nature* 219:800–807.

Ames, J. B., Dizhoor, A. M., Ikura, M., Palczewski, K., and Stryer, L. 1999. Three-dimensional structure of guanylyl cyclase activating protein-2, a calcium-sensitive modulator of photoreceptor guanylyl cyclases. *J. Biol. Chem.* 274:19329–19337.

Nathans, J. 1994. In the eye of the beholder: Visual pigments and inherited variation in human vision. *Cell* 78:357–360.

Nathans, J. 1999. The evolution and physiology of human color vision: Insights from molecular genetic studies of visual pigments. *Neuron* 24:299–312.

Palczewski, K., Kumasaka, T., Hori, T., Behnke, C. A., Motoshima, H., Fox, B. A., LeTrong, I., Teller, D. C., Okada, T., Stenkamp, R. E., et al. 2000. Crystal structure of rhodopsin: A G protein-coupled receptor. *Science* 289:739–745.

Filipek, S, Teller, D. C., Palczewski, K., and Stemkamp, R. 2003. The crystallographic model of rhodopsin and its use in studies of other G protein-coupled receptors. *Annu. Rev. Biophys. Biomol. Struct.* 32:375–397.

Hearing

Furness, D. N., Hackney, C. M., and Evans, M. G. 2010. Localisation of the mechanotransducer channels in mammalian cochlear hair cells provides clues to their gating. *J. Physiol.* 588:765–772.

Lim, K., and Park, S. 2009. A mechanical model of the gating spring mechanism of stereocilia. *J. Biomech.* 42:2158–2164.

Siemens, J., Lillo, C., Dumont, R. A., Reynolds, A., Williams, D. S., Gillespie, P. G., and Muller, U. 2004. Cadherin 23 is a component of the tip link in hair-cell stereocilia. *Nature* 428:950–955.

Spinelli, K. J., and Gillespie, P. G. 2009. Bottoms up: Transduction channels at tip link bases. *Nat. Neurosci.* 12:529–530.

Hudspeth, A. J. 1997. How hearing happens. *Neuron* 19:947–950.

Pickles, J. O., and Corey, D. P. 1992. Mechanoelectrical transduction by hair cells. *Trends Neurosci.* 15:254–259.

Walker, R. G., Willingham, A. T., and Zuker, C. S. 2000. A *Drosophila* mechanosensory transduction channel. *Science* 287:2229–2234.

Hudspeth, A. J., Choe, Y., Mehta, A. D., and Martin, P. 2000. Putting ion channels to work: Mechanoelectrical transduction, adaptation, and amplification by hair cells. *Proc. Natl. Acad. Sci. U.S.A.* 97:11765–11772.

Touch and Pain Reception

Myers, B. R., Bohlen, C. J., and Julius, D. 2008. A yeast genetic screen reveals a critical role for the pore helix domain in TRP channel gating. *Neuron* 58:362–373.

Lishko, P. V., Procko, E., Jin, X., Phelps, C. B., and Gaudet, R. 2007. The ankyrin repeats of TRPV1 bind multiple ligands and modulate channel sensitivity. *Neuron* 54:905–918.

Franco-Obregon, A., and Clapham, D. E. 1998. Touch channels sense blood pressure. *Neuron* 21:1224–1226.

Caterina, M. J., Schumacher, M. A., Tominaga, M., Rosen, T. A., Levine, J. D., and Julius, D. 1997. The capsaicin receptor: A heat-activated ion channel in the pain pathway. *Nature* 389:816–824.

Tominaga, M., Caterina, M. J., Malmberg, A. B., Rosen, T. A., Gilbert, H., Skinner, K., Raumann, B. E., Basbaum, A. I., and Julius, D. 1998. The cloned capsaicin receptor integrates multiple pain-producing stimuli. *Neuron* 21:531–543.

Caterina, M. J., and Julius, D. 1999. Sense and specificity: A molecular identity for nociceptors. *Curr. Opin. Neurobiol.* 9:525–530.

Clapham, D. E. 2003. TRP channels as cellular sensors. *Nature* 426:517–524.

CHAPTER 35 [ONLINE ONLY]

Where to Start

Nossal, G. J. V. 1993. Life, death, and the immune system. *Sci. Am.* 269(3):53–62.

Tonegawa, S. 1985. The molecules of the immune system. *Sci. Am.* 253(4):122–131.

Leder, P. 1982. The genetics of antibody diversity. *Sci. Am.* 246(5):102–115.

Bromley, S. K., Burack, W. R., Johnson, K. G., Somersalo, K., Sims, T. N., Sumen, C., Davis, M. M., Shaw, A. S., Allen, P. M., and Dustin, M. L. 2001. The immunological synapse. *Annu. Rev. Immunol.* 19:375–396.

Books

Punt, J., Stranford, S. A., Jones, P., and Owen, J. 2019. *Kuby Immunology* (8th ed.). W. H. Freeman and Company.

Abbas, A. K., Lichtman, A. H., and Pillai, S. 2017. *Cellular and Molecular Immunology* (9th ed.). Elsevier.

Cold Spring Harbor Symposia on Quantitative Biology. 1989. Volume 54. *Immunological Recognition.* Cold Spring Harbor Laboratory Press.

Weir, D. M. (Ed.). 1996. *Handbook of Experimental Immunology* (5th ed.). Oxford University Press.

Murphy, K., and Weaver, C. 2016. *Janeway's Immunobiology* (9th ed.). W. W. Norton & Company.

Innate Immune System

Janeway, C. A., Jr., and Medzhitov, R. 2002. Innate immune recognition. *Annu. Rev. Immunol.* 20:197–216.

Khalturin, K., Panzer, Z., Cooper, M. D., and Bosch, T. C. 2004. Recognition strategies in the innate immune system of ancestral chordates. *Mol. Immunol.* 41:1077–1087.

Beutler, B., and Rietschel, E. T. 2003. Innate immune sensing and its roots: The story of endotoxin. *Nat. Rev. Immunol.* 3:169–176.

Xu, Y., Tao, X., Shen, B., Horng, T., Medzhitov, R., Manley, J. L., and Tong, L. 2000. Structural basis for signal transduction by the Toll/ interleukin-1 receptor domains. *Nature* 408:111–115.

Jiménez-Dalmaroni, M. J., Gerswhin, M. E., and Adamopoulos, I. E. 2016. The critical role of toll-like receptors —From microbial recognition to autoimmunity: A comprehensive review. *Autoimmunity Rev.* 15:1–8.

Structure of Antibodies and Antibody-Antigen Complexes

Davies, D. R., Padlan, E. A., and Sheriff, S. 1990. Antibody-antigen complexes. *Annu. Rev. Biochem.* 59:439–473.

Poljak, R. J. 1991. Structure of antibodies and their complexes with antigens. *Mol. Immunol.* 28:1341–1345.

Davies, D. R., and Cohen, G. H. 1996. Interactions of protein antigens with antibodies. *Proc. Natl. Acad. Sci. U.S.A.* 93:7–12.

Marquart, M., Deisenhofer, J., Huber, R., and Palm, W. 1980. Crystallographic refinement and atomic models of the intact immunoglobulin molecule Kol and its antigen-binding fragment at 3.0 Å and 1.9 Å resolution. *J. Mol. Biol.* 141:369–391.

Silverton, E. W., Navia, M. A., and Davies, D. R. 1977. Three-dimensional structure of an intact human immunoglobulin. *Proc. Natl. Acad. Sci. U.S.A.* 74:5140–5144.

Padlan, E. A., Silverton, E. W., Sheriff, S., Cohen, G. H., Smith, G. S., and Davies, D. R. 1989. Structure of an antibody-antigen complex: Crystal structure of the HyHEL-10 Fab lysozyme complex. *Proc. Natl. Acad. Sci. U.S.A.* 86:5938–5942.

Rini, J., Schultze-Gahmen, U., and Wilson, I. A. 1992. Structural evidence for induced fit as a mechanism for antibody-antigen recognition. *Science* 255:959–965.

Fischmann, T. O., Bentley, G. A., Bhat, T. N., Boulot, G., Mariuzza, R. A., Phillips, S. E., Tello, D., and Poljak, R. J. 1991. Crystallographic refinement of the three-dimensional structure of the FabD1.3-lysozyme complex at 2.5-Å resolution. *J. Biol. Chem.* 266:12915–12920.

Burton, D. R. 1990. Antibody: The flexible adaptor molecule. *Trends Biochem. Sci.* 15:64–69.

Saphire, E. O., Parren P. W., Pantophlet, R., Zwick, M. B., Morris, G. M., Rudd, P. M., Dwek, R. A., Stanfield, R. L., Burton, D. R., and Wilson, I. A. 2001. Crystal structure of a neutralizing human IgG against HIV-1: A template for vaccine design. *Science* 293:1155–1159.

Calarese, D. A., Scanlan, C. N., Zwick, M. B., Deechongkit, S., Mimura, Y., Kunert R., Zhu, P., Wormald, M. R., Stanfield, R. L., Roux, K. H., et al. 2003. Antibody domain exchange is an immunological solution to carbohydrate cluster recognition. *Science* 300:2065–2071.

Generation of Diversity

Tonegawa, S. 1988. Somatic generation of immune diversity. *Biosci. Rep.* 8:3–26.

Honjo, T., and Habu, S. 1985. Origin of immune diversity: Genetic variation and selection. *Annu. Rev. Biochem.* 54:803–830.

Gellert, M., and McBlane, J. F. 1995. Steps along the pathway of VDJ recombination. *Philos. Trans. R. Soc. Lond. B Biol. Sci.* 347:43–47.

Harris, R. S., Kong, Q., and Maizels, N. 1999. Somatic hypermutation and the three R's: Repair, replication and recombination. *Mutat. Res.* 436:157–178.

Lewis, S. M., and Wu, G. E. 1997. The origins of V(D)J recombination. *Cell* 88:159–162.

Ramsden, D. A., van Gent, D. C., and Gellert, M. 1997. Specificity in V(D)J recombination: New lessons from biochemistry and genetics. *Curr. Opin. Immunol.* 9:114–120.

Roth, D. B., and Craig, N. L. 1998. VDJ recombination: A transposase goes to work. *Cell* 94:411–414.

Sadofsky, M. J. 2001. The RAG proteins in V(D)J recombination: More than just a nuclease. *Nucleic Acids Res.* 29:1399–1409.

MHC Proteins and Antigen Processing

Bjorkman, P. J., and Parham, P. 1990. Structure, function, and diversity of class I major histocompatibility complex molecules. *Annu. Rev. Biochem.* 59:253–288.

Goldberg, A. L., and Rock, K. L. 1992. Proteolysis, proteasomes, and antigen presentation. *Nature* 357:375–379.

Madden, D. R., Gorga, J. C., Strominger, J. L., and Wiley, D. C. 1992. The three-dimensional structure of HLA-B27 at 2.1 Å resolution suggests a general mechanism for tight binding to MHC. *Cell* 70:1035–1048.

Fremont, D. H., Matsumura, M., Stura, E. A., Peterson, P. A., and Wilson, I. A. 1992. Crystal structures of two viral peptides in complex with murine MHC class I H-2Kb. *Science* 257:880–881.

Matsumura, M., Fremont, D. H., Peterson, P. A., and Wilson, I. A. 1992. Emerging principles for the recognition of peptide antigens by MHC class I. *Science* 257:927–934.

Brown, J. H., Jardetzky, T. S., Gorga, J. C., Stern, L. J., Urban, R. G., Strominger, J. L., and Wiley, D. C. 1993. Three-dimensional structure of the human class II histocompatibility antigen HLA-DR1. *Nature* 364:33–39.

Saper, M. A., Bjorkman, P. J., and Wiley, D. C. 1991. Refined structure of the human histocompatibility antigen HLA-A2 at 2.6 Å resolution. *J. Mol. Biol.* 219:277–319.

Madden, D. R., Gorga, J. C., Strominger, J. L., and Wiley, D. C. 1991. The structure of HLA-B27 reveals nonamer self-peptides bound in an extended conformation. *Nature* 353:321–325.

Cresswell, P., Bangia, N., Dick, T., and Diedrich, G. 1999. The nature of the MHC class I peptide loading complex. *Immunol. Rev.* 172:21–28.

Madden, D. R., Garboczi, D. N., and Wiley, D. C. 1993. The antigenic identity of peptide-MHC complexes: A comparison of the conformations of five viral peptides presented by HLA-A2. *Cell* 75:693–708.

T-Cell Receptors and Signaling Complexes

Hennecke, J., and Wiley, D. C. 2001. T-cell receptor-MHC interactions up close. *Cell* 104:1–4.

Ding, Y. H., Smith, K. J., Garboczi, D. N., Utz, U., Biddison, W. E., and Wiley, D. C. 1998. Two human T cell receptors bind in a similar diagonal mode to the HLA-A2/Tax peptide complex using different TCR amino acids. *Immunity* 8:403–411.

Reinherz, E. L., Tan, K., Tang, L., Kern, P., Liu, J., Xiong, Y., Hussey, R. E., Smolyar, A., Hare, B., Zhang, R., et al. 1999. The crystal structure of a T-cell receptor in complex with peptide and MHC class II. *Science* 286:1913–1921.

Davis, M. M., and Bjorkman, P. J. 1988. T-cell antigen receptor genes and T-cell recognition. *Nature* 334:395–402.

Cochran, J. R., Cameron, T. O., and Stern, L. J. 2000. The relationship of MHC-peptide binding and T cell activation probed using chemically defined MHC class II oligomers. *Immunity* 12:241–250.

Garcia, K. C., Teyton, L., and Wilson, I. A. 1999. Structural basis of T cell recognition. *Annu. Rev. Immunol.* 17:369–397.

Garcia, K. C., Degano, M., Stanfield, R. L., Brunmark, A., Jackson, M. R., Peterson, P. A., Teyton, L. A., and Wilson, I. A. 1996. An αβ T-cell receptor structure at 2.5 Å and its orientation in the TCR-MHC complex. *Science* 274:209–219.

Garboczi, D. N., Ghosh, P., Utz, U., Fan, Q. R., Biddison, W. E., Wiley, D. C. 1996. Structure of the complex between human T-cell receptor, viral peptide and HLA-A2. *Nature* 384:134–141.

Gaul, B. S., Harrison, M. L., Geahlen, R. L., Burton, R. A., and Post, C. B. 2000. Substrate recognition by the Lyn protein-tyrosine kinase: NMR structure of the immunoreceptor tyrosine-based activation motif signaling region of the B cell antigen receptor. *J. Biol. Chem.* 275:16174–16182.

Kern, P. S., Teng, M. K., Smolyar, A., Liu, J. H., Liu, J., Hussey, R. E., Spoerl, R., Chang, H. C., Reinherz, E. L., and Wang, J. H. 1998. Structural basis of CD8 coreceptor function revealed by crystallographic analysis of a murine CD8 αβ ectodomain fragment in complex with H-2Kb. *Immunity* 9:519–530.

Konig, R., Fleury, S., and Germain, R. N. 1996. The structural basis of CD4-MHC class II interactions: Coreceptor contributions to T cell receptor antigen recognition and oligomerization-dependent signal transduction. *Curr. Top. Microbiol. Immunol.* 205:19–46.

Davis, M. M., Boniface, J. J., Reich, Z., Lyons, D., Hampl, J., Arden, B., and Chien, Y. 1998. Ligand recognition by αβ T-cell receptors. *Annu. Rev. Immunol.* 16:523–544.

Janeway, C. J. 1992. The T cell receptor as a multicomponent signalling machine: CD4/CD8 coreceptors and CD45 in T cell activation. *Annu. Rev. Immunol.* 10:645–674.

Podack, E. R., and Kupfer, A. 1991. T-cell effector functions: Mechanisms for delivery of cytotoxicity and help. *Annu. Rev. Cell Biol.* 7:479–504.

Davis, M. M. 1990. T cell receptor gene diversity and selection. *Annu. Rev. Biochem.* 59:475–496.

Leahy, D. J., Axel, R., and Hendrickson, W. A. 1992. Crystal structure of a soluble form of the human T cell coreceptor CD8 at 2.6 Å resolution. *Cell* 68:1145–1162.

Bots, M., and Medema, J. P. 2006. Granzymes at a glance. *J. Cell. Sci.* 119:5011–5014.

Lowin, B., Hahne, M., Mattmann, C., and Tschopp, J. 1994. Cytolytic T-cell cytotoxicity is mediated through perforin and Fas lytic pathways. *Nature* 370: 650–652.

Rudolph, M. G., and Wilson, I. A. 2002. The specificity of TCR/ pMHC interaction. *Curr. Opin. Immunol.* 14:52–65.

HIV and AIDS

Fauci, A. S. 1988. The human immunodeficiency virus: Infectivity and mechanisms of pathogenesis. *Science* 239:617–622.

Gallo, R. C., and Montagnier, L. 1988. AIDS in 1988. *Sci. Am.* 259(4): 41–48.

Kwong, P. D., Wyatt, R., Robinson, J., Sweet, R. W., Sodroski, J., and Hendrickson, W. A. 1998. Structure of an HIV gp120 envelope glycoprotein in complex with the CD4 receptor and a neutralizing human antibody. *Nature* 393:648–659.

Vaccines

Johnston, M. I., and Fauci, A. S. 2007. An HIV vaccine—evolving concepts. *New Engl. J. Med.* 356:2073–2081.

Burton, D. R., Desrosiers, R. C., Doms, R. W., Koff, W. C., Kwong, P. D., Moore, J. P., Nabel, G. J., Sodroski, J., Wilson, I. A., and Wyatt, R. T. 2004. HIV vaccine design and the neutralizing antibody problem. *Nature Immunol.* 5:233–236.

Ada, G. 2001. Vaccines and vaccination. *New Engl. J. Med.* 345:1042–1053.

Behbehani, A. M. 1983. The smallpox story: Life and death of an old disease. *Microbiol. Rev.* 47:455–509.

Discovery of Major Concepts

Ada, G. L., and Nossal, G. 1987. The clonal selection theory. *Sci. Am.* 257(2):62–69.

Porter, R. R. 1973. Structural studies of immunoglobulins. *Science* 180:713–716.

Edelman, G. M. 1973. Antibody structure and molecular immunology. *Science* 180:830–840.

Kohler, G. 1986. Derivation and diversification of monoclonal antibodies. *Science* 233:1281–1286.

Milstein, C. 1986. From antibody structure to immunological diversification of immune response. *Science* 231:1261–1268.

Janeway, C. A., Jr. 1989. Approaching the asymptote? Evolution and revolution in immunology. *Cold Spring Harbor Symp. Quant. Biol.* 54:1–13.

Jerne, N. K. 1971. Somatic generation of immune recognition. *Eur. J. Immunol.* 1:1–9.

Biochemistry in Focus

Manier, S., Salem, K. Z., Park, J., Landau, D. A., Getz, G., and Ghobrial, I. M. 2017. Genomic complexity of multiple myeloma and its clinical implications. *Nat. Rev. Clin. Oncol.* 13:100–113.

Kumar, S. K., Rajkumar, V., Kyle, R. A., van Duin, M., Sonneveld, P., Mateos, M.-V., Gay, F., and Anderson, K. C. 2017. Multiple myeloma. *Nat. Rev. Dis. Primers* 3:17046.

CHAPTER 36 [ONLINE ONLY]

Where to Start

Gennerich, A., and Vale, R. D. 2009. Walking the walk: How kinesin and dynein coordinate their steps. *Curr. Opin. Cell Biol.* 21:59–67.

Vale, R. D. 2003. The molecular motor toolbox for intracellular transport. *Cell* 112:467–480.

Vale, R. D., and Milligan, R. A. 2000. The way things move: Looking under the hood of molecular motor proteins. *Science* 288:88–95.

Vale, R. D. 1996. Switches, latches, and amplifiers: Common themes of G proteins and molecular motors. *J. Cell Biol.* 135:291–302.

Mehta, A. D., Rief, M., Spudich, J. A., Smith, D. A., and Simmons, R. M. 1999. Single-molecule biomechanics with optical methods. *Science* 283:1689–1695.

Schuster, S. C., and Khan, S. 1994. The bacterial flagellar motor. *Annu. Rev. Biophys. Biomol. Struct.* 23:509–539.

Books

Howard, J. 2001. *Mechanics of Motor Proteins and the Cytoskeleton.* Sinauer.

Squire, J. M. 1986. *Muscle Design, Diversity, and Disease.* Benjamin Cummings.

Pollack, G. H., and Sugi, H. (Eds.). 1984. *Contractile Mechanisms in Muscle.* Plenum.

Myosin and Actin

Lorenz, M., and Holmes, K. C. 2010. The actin-myosin interface. *Proc. Natl. Acad. Sci. U.S.A.* 107:12529–12534.

Yang, Y., Gourinath, S., Kovacs, M., Nyitray, L., Reutzel, R., Himmel, D. M., O'Neall-Hennessey, E., Reshetnikova, L., Szent-Györgyi, A. G., Brown, J. H., et al. 2007. Rigor-like structures from muscle myosins reveal key mechanical elements in the transduction pathways of this allosteric motor. *Structure* 15:553–564.

Himmel, D. M., Mui, S., O'Neall-Hennessey, E., Szent-Györgyi, A. G., and Cohen, C. 2009. The on-off switch in regulated myosins: Different triggers but related mechanisms. *J. Mol. Biol.* 394:496–505.

Houdusse, A., Gaucher, J. F., Krementsova, E., Mui, S., Trybus, K. M., and Cohen, C. 2006. Crystal structure of apo-calmodulin bound to the first two IQ motifs of myosin V reveals essential recognition features. *Proc. Natl. Acad. Sci. U.S.A.* 103:19326–19331.

Li, X. E., Holmes, K. C., Lehman, W., Jung, H., and Fischer, S. 2010. The shape and flexibility of tropomyosin coiled coils: Implications for actin filament assembly and regulation. *J. Mol. Biol.* 395:327–339.

Fischer, S., Windshugel, B., Horak, D., Holmes, K. C., and Smith, J. C. 2005. Structural mechanism of the recovery stroke in the myosin molecular motor. *Proc. Natl. Acad. Sci. U.S.A.* 102:6873–6878.

Holmes, K. C., Angert, I., Kull, F. J., Jahn, W., and Schroder, R. R. 2003. Electron cryo-microscopy shows how strong binding of myosin to actin releases nucleotide. *Nature* 425:423–427.

Holmes, K. C., Schroder, R. R., Sweeney, H. L., and Houdusse, A. 2004. The structure of the rigor complex and its implications for the power stroke. *Philos. Trans. R. Soc. Lond. B Biol. Sci.* 359:1819–1828.

Purcell, T. J., Morris, C., Spudich, J. A., and Sweeney, H. L. 2002. Role of the lever arm in the processive stepping of myosin V. *Proc. Natl. Acad. Sci. U.S.A.* 99:14159–14164.

Purcell, T. J., Sweeney, H. L., and Spudich, J. A. 2005. A force-dependent state controls the coordination of processive myosin V. *Proc. Natl. Acad. Sci. U.S.A.* 102:13873–13878.

Holmes, K. C. 1997. The swinging lever-arm hypothesis of muscle contraction. *Curr. Biol.* 7:R112–R118.

Berg, J. S., Powell, B. C., and Cheney, R. E. 2001. A millennial myosin census. *Mol. Biol. Cell* 12:780–794.

Houdusse, A., Kalabokis, V. N., Himmel, D., Szent-Györgyi, A. G., and Cohen, C. 1999. Atomic structure of scallop myosin subfrag-ment S1 complexed with MgADP: A novel conformation of the myosin head. *Cell* 97:459–470.

Houdusse, A., Szent-Györgyi, A. G., and Cohen, C. 2000. Three conformational states of scallop myosin S1. *Proc. Natl. Acad. Sci.* 97:11238–11243.

Uyeda, T. Q., Abramson, P. D., and Spudich, J. A. 1996. The neck region of the myosin motor domain acts as a lever arm to generate movement. *Proc. Natl. Acad. Sci. U.S.A.* 93:4459–4464.

Mehta, A. D., Rock, R. S., Rief, M., Spudich, J. A., Mooseker, M. S., and Cheney, R. E. 1999. Myosin-V is a processive actin-based motor. *Nature* 400:590–593.

Otterbein, L. R., Graceffa, P., and Dominguez, R. 2001. The crystal structure of uncomplexed actin in the ADP state. *Science* 293:708–711.

Holmes, K. C., Popp, D., Gebhard, W., and Kabsch, W. 1990. Atomic model of the actin filament. *Nature* 347:44–49.

Schutt, C. E., Myslik, J. C., Rozycki, M. D., Goonesekere, N. C., and Lindberg, U. 1993. The structure of crystalline profilin-β-actin. *Nature* 365:810–816.

van den Ent, F., Amos, L. A., and Lowe, J. 2001. Prokaryotic origin of the actin cytoskeleton. *Nature* 413:39–44.

Schutt, C. E., and Lindberg, U. 1998. Muscle contraction as a Markov process I: Energetics of the process. *Acta Physiol. Scand.* 163:307–323.

Rief, M., Rock, R. S., Mehta, A. D., Mooseker, M. S., Cheney, R. E., and Spudich, J. A. 2000. Myosin-V stepping kinetics: A molecular model for processivity. *Proc. Natl. Acad. Sci. U.S.A.* 97:9482–9486.

Friedman, T. B., Sellers, J. R., and Avraham, K. B. 1999. Unconventional myosins and the genetics of hearing loss. *Am. J. Med. Genet.* 89:147–157.

Kinesin, Dynein, and Microtubules

Reck-Peterson, S. L., Redwine, W. B., Vale, R. D., and Carter, A. P. 2018. The cytoplasmic dynein transport machinery and its many cargoes. *Nature Rev. Mol. Cell Biol.* 19, 382–398.

Bhabha, G., Johnson, G. T., Schroeder, C. M., and Vale, R. D. 2016. How dynein moves along microtubules. *Trends Biochem. Sci.* 41, 94–105.

Cho, C., and Vale, R. D. 2012. The mechanism of dynein motility: Insight from crystal structures of the motor domain. *Biochim. Biophys. Acta* 1823:182–191.

Yildiz, A., Tomishige, M., Gennerich, A., and Vale, R. D. 2008. Intramolecular strain coordinates kinesin stepping behavior along microtubules. *Cell* 134:1030–1041.

Yildiz, A., Tomishige, M., Vale, R. D., and Selvin, P. R. 2004. Kinesin walks hand-over-hand. *Science* 303:676–678.

Rogers, G. C., Rogers, S. L., Schwimmer, T. A., Ems-McClung, S. C., Walczak, C. E., Vale, R. D., Scholey, J. M., and Sharp, D. J. 2004. Two mitotic kinesins cooperate to drive sister chromatid separation during anaphase. *Nature* 427:364–370.

Vale, R. D., and Fletterick, R. J. 1997. The design plan of kinesin motors. *Annu. Rev. Cell. Dev. Biol.* 13:745–777.

Kull, F. J., Sablin, E. P., Lau, R., Fletterick, R. J., and Vale, R. D. 1996. Crystal structure of the kinesin motor domain reveals a structural similarity to myosin. *Nature* 380:550–555.

Kikkawa, M., Sablin, E. P., Okada, Y., Yajima, H., Fletterick, R. J., and Hirokawa, N. 2001. Switch-based mechanism of kinesin motors. *Nature* 411:439–445.

Wade, R. H., and Kozielski, F. 2000. Structural links to kinesin directionality and movement. *Nat. Struct. Biol.* 7:456–460.

Yun, M., Zhang, X., Park, C. G., Park, H. W., and Endow, S. A. 2001. A structural pathway for activation of the kinesin motor ATPase. *EMBO J.* 20:2611–2618.

Kozielski, F., De Bonis, S., Burmeister, W. P., Cohen-Addad, C., and Wade, R. H. 1999. The crystal structure of the minus-end-directed microtubule motor protein ncd reveals variable dimer conformations. *Struct. Fold. Des.* 7:1407–1416.

Lowe, J., Li, H., Downing, K. H., and Nogales, E. 2001. Refined structure of αβ-tubulin at 3.5 Å resolution. *J. Mol. Biol.* 313:1045–1057.

Nogales, E., Downing, K. H., Amos, L. A., and Lowe, J. 1998. Tubulin and FtsZ form a distinct family of GTPases. *Nat. Struct. Biol.* 5:451–458.

Zhao, C., Takita, J., Tanaka, Y., Setou, M., Nakagawa, T., Takeda, S., Yang, H. W., Terada, S., Nakata, T., Takei, Y., et al. 2001. Charcot-Marie-Tooth disease type 2A caused by mutation in a microtubule motor KIF1Bb. *Cell* 105:587–597.

Asai, D. J., and Koonce, M. P. 2001. The dynein heavy chain: Structure, mechanics and evolution. *Trends Cell Biol.* 11:196–202.

Mocz, G., and Gibbons, I. R. 2001. Model for the motor component of dynein heavy chain based on homology to the AAA family of oligo-meric ATPases. *Structure* 9:93–103.

Bacterial Motion and Chemotaxis

Baker, M. D., Wolanin, P. M., and Stock, J. B. 2006. Systems biology of bacterial chemotaxis. *Curr. Opin. Microbiol.* 9:187–192.

Wolanin, P. M., Baker, M. D., Francis, N. R., Thomas, D. R., DeRosier, D. J., and Stock, J. B. 2006. Self-assembly of receptor/signaling complexes in bacterial chemotaxis. *Proc. Natl. Acad. Sci. U.S.A.* 103:14313–14318.

Sowa, Y., Rowe, A. D., Leake, M. C., Yakushi, T., Homma, M., Ishijima, A., and Berry, R. M. 2005. Direct observation of steps in rotation of the bacterial flagellar motor. *Nature* 437:916–919.

Berg, H. C. 2000. Constraints on models for the flagellar rotary motor. *Philos. Trans. R. Soc. Lond. B Biol. Sci.* 355:491–501.

DeRosier, D. J. 1998. The turn of the screw: The bacterial flagellar motor. *Cell* 93:17–20.

Ryu, W. S., Berry, R. M., and Berg, H. C. 2000. Torque-generating units of the flagellar motor of *Escherichia coli* have a high duty ratio. *Nature* 403:444–447.

Lloyd, S. A., Whitby, F. G., Blair, D. F., and Hill, C. P. 1999. Structure of the C-terminal domain of FliG, a component of the rotor in the bacterial flagellar motor. *Nature* 400:472–475.

Purcell, E. M. 1977. Life at low Reynolds number. *Am. J. Physiol.* 45:3–11.

Macnab, R. M., and Parkinson, J. S. 1991. Genetic analysis of the bacterial flagellum. *Trends Genet.* 7:196–200.

Historical Aspects

Huxley, H. E. 1965. The mechanism of muscular contraction. *Sci. Am.* 213(6):18–27.

Summers, K. E., and Gibbons, I. R. 1971. ATP-induced sliding of tubules in trypsin-treated flagella of sea-urchin sperm. *Proc. Natl. Acad. Sci. U.S.A.* 68:3092–3096.

Macnab, R. M., and Koshland, D. E., Jr. 1972. The gradient-sensing mechanism in bacterial chemotaxis. *Proc. Natl. Acad. Sci. U.S.A.* 69:2509–2512.

Taylor, E. W. 2001. 1999. E. B. Wilson lecture: The cell as molecular machine. *Mol. Biol. Cell* 12:251–254.

Biochemistry in Focus

Grati, M., and Kacher, B. 2011. Myosin VIIa and sans localization at stereocilia upper tip-link density implicates these Usher syndrome proteins in mechanotransduction. *Proc. Natl. Acad. Sci. U.S.A.* 108:11476–11481.

Note: Page numbers followed by f, t, and b refer to figures, tables, and boxed material.
Boldface page numbers indicate structural formulas and ribbon diagrams.
Page numbers listed in green refer to terms that appear in the online chapters.

A band, 1160f
A site, ribosomal, 1032, 1034f
A-form DNA, 119–120, 119f, 120f, 120t
Ab initio prediction, 57
AAA ATPases
  in amino acid degradation, 756
  in dynein, 1153, 1154
ABC transporters
  ATP hydrolysis reaction and, 413
  defined, 403, 411
  domain arrangement of, 411, 411f
  in drug resistance, 943
  eukaryotic, 413
  mechanism, 411–412, 412f
  structure of, **412**
ABCA1 (ATP-binding cassette transporter, subfamily A1), 870
ABO blood groups, 359, 359f
Abortive complex, 759
Absorption
  of dietary proteins, 752, 753f
  drug, 924–925, 925f
  light, 623, 623f, 1108–1109
Acarbose, 366, **366**
Acceptor control. *See* Respiratory control
Accessory pigments in photosynthesis, 635–638, 635f, 637f
Accommodation process, 1036
Acetaminophen, 929, 929f
Acetoacetate, 560. *See also* Ketone bodies
  conversion into acetyl CoA, 726
  from phenylalanine, 774
  in respiration, 726
  from tryptophan, 775
  utilization as fuel, 726, 726f
Acetoacetyl CoA, 769–770
Acetyl ACP, 730
Acetyl CoA (acetyl coenzyme A)
  acetoacetate conversion into, 725–726
  acetyl-group-transfer potential, **477**
  in amino acid degradation, 769–770, 772–773
  in citric acid cycle, 523, 544–546, 548–550, 550f, 562
  defined, 477
  in fatty acid metabolism, 715–716
  in fatty acid oxidation, 717–718
  in fatty acid synthesis, 558
  formation from ketone bodies, 724–725, 725f
  formation of, 507, 546
  in glyoxylate cycle, 564, 565f
  pyruvate dehydrogenase complex and, 546
  structure of, 477, **477**
  synthesis of, 544–546
  transfer to cytoplasm, 733–734, 734f
Acetyl CoA acetyltransferase, 560

Acetyl CoA carboxylase, 522
  in citrate concentration, 739, 740f
  in fatty acid metabolism, 729, 730t, 739–741
  filaments of, 739f
  regulation by hormones, 739–740
  regulation of, 739f, 740
Acetyl CoA synthetase, 715
Acetyl groups, 60
Acetyl transacylase, 730, 730t
Acetylation
  defined, 319t
  in gene regulation, 1088–1089, 1090
  in metabolism regulation, 318
  protein, 318–319
Acetylcholine, 423
Acetylcholine receptor, 431
  defined, 423
  as ligand-gated channel, 423
  as nonspecific cation channel, 426
  structure of, 424, 424f
N-acetylglucosamine phosphotransferase, 360
N-acetylglutamate, 765
N-acetylglutamate synthase, 765
Acetyllysine-binding domain, 1088, 1088f
N-acetylneuraminate, 855
N-acetyl-L-phenylalanine *p*–nitrophenyl ester, 277, 277f
Achiral molecules, 32
Acid-base reactions
  in biological processes, 14
  defined, 14
  double helix and, 15–16
  hydrogen ions in, 14
Acids. *See also* Amino acids; Fatty acids; Nucleic acids
  bile, 712, 712f
  protonation/deprotonation of, 16–17
  sialic, 855
Aconitase, 550, 551f, 1091, 1092f
Acquired immune deficiency syndrome (AIDS). *See* Human immunodeficiency virus (HIV) infection
Actin/actin filaments
  barbed, 1157
  decorated, 1158
  defined, 1157
  F, 1157
  formation of, 1157
  G, 1157
  in muscle contraction, 1160–1161
  myosin and, 1158, 1160f
  pointed, 1157
  polymerization of, 1158
  structure of, 195, **195f**, 1156–1157, 1157f
  in thin filaments, 1161

Actinomycin, 994, 995f
Action potentials, 431
  defined, 415
  integration, 424–426
  ion channels and, 415, 415f
  mechanism, 426, 426f
Activated carriers
  as derived from vitamins, 478–480
  of electrons for fuel oxidation, 475–476
  of electrons for reductive biosynthesis, 477–478
  in metabolism, 475–478, 478t
  recurring set of, 478
  of two-carbon fragments, 477–478
Activated glycoaldehyde unit, 664
Activated methyl cycle, 800, 800f
Activation domains, 1080, 1081
Activation energy
  defined, 240
  enzyme decrease of, 240–241, 240f
  symbol, 240
Active membrane transport, 404–406
Active sites
  ATCase, 313–314, **313**, 314f
  chymotrypsin, 278–280, 278f
  concentration of, 249
  as crevice, 242
  defined, 241
  enzyme and substrate interaction, 242
  evolution of, 274
  as microenvironments, 242
  myosin, 302–303, 302f
  proteases, 284f
  residues, 242, 242f
  RNA polymerase, 985
  volume of enzyme, 242
Active transport, 403
Acute intermittent porphyria, 814
Acyclovir monophosphate, 826
Acyl adenylate, 715
Acyl carnitine, 716–717, **716**, 716f
Acyl carrier protein (ACP), 728, **729**, 729f, 731–733
Acyl CoA:cholesterol acyltransferase (ACAT), 866
Acyl intermediates, 259, 259f
Acyl-enzyme intermediates, 277–278, 277f, 756
Adaptive immune system, 1122–1124
Adaptor proteins, 450
ADC. *See* Antibody-drug conjugate (ADC)
Adenine, 5, 5f, 114, **115f**, 118, **118f**, 335
  deamination of, 969, 969f
  methylation of, 297–298, 298f
  resonance structures, 7

Adenine nucleotide translocase (ANT). *See* ATP-ADP translocase
Adenine phosphoribosyltransferase, 830
Adenosine, 116, **116f**
Adenosine 5'-triphosphate. *See* ATP (adenosine triphosphate)
Adenosine deaminase, 178, 839
Adenosine diphosphate. *See* ADP (adenosine diphosphate)
Adenosine triphosphate. *See* ATP (adenosine triphosphate)
S-adenosylhomocysteine, 800, **800**
S-adenosylmethionine (SAM)
  in activated methyl cycle, 800, 800f
  in adenylation of methionine, 771
  in amino acid degradation, 771, **772**
  in amino acid synthesis, 799–801
  as ethylene precursor, 801
  in phospholipid synthesis, 853
Adenylate, 116, 829
Adenylate cyclase, 442–443, 443f
Adenylate kinase, **304f**
Adenylated intermediates, 796
Adenylation
  in amino acid activation, 1027–1028
  in amino acid synthesis, 807–808, 807f
  in translation, 1028–1029
Adenylosuccinate synthetase, 829
Adenylyl transferase, 807, 807f
Adipokines, 893, 914–915
Adiponectin, 882, 894, 914
Adipose cells (fat cells), 711–712
Adipose tissue
  brown, 605, 606f
  fat in, 904
  glucose entry into, 905
  glycerol released from, 906–907
  triacylglycerols in, 907
  white, 605
Adipose triglyceride lipase (ATGL), 714, 714f
ADME properties of drugs, 924–928, 925f
ADP (adenosine diphosphate), **298**
  ATP hydrolyzed to, 466–468
  formation of, 298
  in kinesin movement, 1165–1166, 1165f
  in oxidative phosphorylation, 601, 602f, 604, 605f
  in random sequential reaction, 251
  structure of, 466, **467f**
Adrenaline, 690
Adrenocorticotropic hormone, 61
Advanced glycation end product (AGE), 348
Aequorea victoria, 89
Aerobic glycolysis, 517–518, 557
Affinity chromatography, 74–75, 74f, 79
Affinity labels, 257
Affinity maturation, 1132
Affinity tags, 82, 257, 257f
Aflatoxin activation, 970f
Aggrecan, 355
Agonists, 441, 743, 1086
Agouti-related peptide (AgRP), 894, 894f
Agre, Peter, 1

Agrobacterium tumefaciens, 177, 177f
Alanine
  defined, 32
  formation of, 974
  in gluconeogenesis, 529
  in muscle contraction, 529, 529f
  structure of, 33f
  tryptophan in, 770
Alanine aminotransferase, 759
Alanyl-tRNA, 1023, 1023f
Albinism, 776t
Albumin, 925, 925f
Alcaptonuria, 776
Alcohol dehydrogenase, 910
Alcoholic fermentation, 491, 506, 506f
Alcoholic scurvy, 911
Alcohols
  fermentation of, 505–506, 506f
  formation from pyruvate, 505–507, 506f
  metabolism of, 909–911
  monosaccharides and, 348
  toxicity of, 911–912
Aldehyde dehydrogenase, 743
Aldehydes, oxidation of energy, 499–500
Alder, Julius, 1167
Aldimine, 760
Aldolase, 497
  in Calvin cycle, 651
  in transaldolase reaction, 664–665
Aldoses
  conversion to ketoses, 496
  defined, 342
Aldosterone, 872, 875, **876f**
Allolactose, 1062
Allopurinol, 840
Allosteric constant, 226, 314–315
Allosteric control, 309–310
Allosteric effector, 217
Allosteric enzymes, 252–253, 252f
  allosteric constant, 314–315
  concerted model, 315
  control of, 483–485
  homotropic effects on, 315
  sequential model, 315
  sigmoidal kinetics, 311, 311f
  substrate concentration and, 315
  T-to-R equilibrium in, 316–317, 316f
  T-to-R state transition, 315–316, 316f
  threshold effects, 315f, 316
Allosteric sites, 311
Alloxanthine, 840
α chains, 211, 222–223
α helix, 42–44
  hydrogen-bonding scheme, 43f
  of membrane proteins, 389–390, 389f, 389t
  pitch of, 43
  as right-handed, 43
  schematic view of, 44f
  screw sense, 43
  structure of, **43f**
α₁-antiproteinase, 327
α-amino acids, 31, 37
α-helical coiled coil, 49

α-hemoglobin, 222, **223f**
  alignment with gap insertion, 189f
  amino acid sequences, 187–188, 188f
α-hemoglobin stabilizing protein (AHSP), 222–223, **223f**
α-glucosidase, 492–493
α-glucosidase inhibitors, 365–366
α-keratin, 49
α-sarcin, 1045
α-stabilizing protein (AHSP), 222–223, **223f**
α-synuclein, 852
α-thalassemia, 221–222
αβ dimers, 211
Alternative splicing, 137, 137f
  defects in, 1012t
  defined, 1011
  example of, 1011f
  pre-mRNA, 1011–1012
Altman, Sidney, 984, 995
Alu sequences, 165
Alzheimer's disease, 59–60
Amanita phalloides, 997, 997f
α-amanitin, 998–999
Ames test, 975, 975f
Amide, 24
Amide bonds. *See* Peptide bonds
Amiloride-sensitive sodium channel, 1106, 1106f, 1113–1114
Amines, monosaccharides and, 348
Amino acid degradation, 751–779
  acetoacetyl CoA in, 769–770
  acetyl CoA in, 769–770, 772–773
  amino acid use through, 751
  aminotransferases in, 758–762
  of aromatic amino acids, 773–775
  of branched-chain amino acids, 758–764, 772–773
  coenzymes in, 752
  deamination in, 763
  digestive enzymes in, 752
  fumarate in, 769
  glucose–alanine cycle in, 763–764, 763f
  glutamate dehydrogenase in, 759
  α-ketobutyrate in, 772
  α-ketoglutarate in, 769, 770–771, 770f
  liver in, 758, 763–764
  metabolism errors and, 776–777, 776t
  nitrogen removal in, 778
  overview of, 751–752
  oxaloacetate in, 769, 770
  oxygenases in, 773–775
  propionyl CoA in, 772–773
  proteasomes in, 751, 755–756, 757f
  in protein turnover, 753–758
  pyridoxal phosphate (PLP) in, 752, 762–763
  pyruvate in, 763–764, 769, 769f, 770
  regulation function of, 753–758, 757t
  Schiff-base intermediates in, 759–761
  serine dehydratase in, 763
  succinyl CoA in, 769, 771, 771f
  threonine dehydratase in, 763
  transaminases in, 758–759
  ubiquitin in, 754–755, 754f
  ubiquitination in, 754–755
  urea cycle in, 764–769

Amino acid sequences, 29–30
  alignment of, 187–194, 188f, 189f
  collagen, 49–50, 50f
  comparison of, 95
  determination of, 69–70
  directionality of, 38, 38f
  in DNA probe creation, 96
  DNA sequences and, 95f
  evolution of, 39
  gap insertion, 189, 189f
  of hemoglobin, 187–188, 188f
  homologous, 193–194
  identities, 188
  importance of, 39–40
  information provided by, 95–96
  of insulin, 39f
  of myoglobin, 187–188, 188f
  porin, 385, 385f
  probes generated from, 158, 158f
  in protein identification, 95–96
  in protein structure determination, 52–62
  protein structures and, 39, 39f
  recombinant DNA technology and, 95
  ribonuclease, 52f
  searching for internal repeats, 96, 96f
  shuffled, 189–190, 190f
  as signals, 96
  statistical comparison of alignment, 190, 190f
  substitution in, 190–193, 191f, 192f, 193f
  transmembrane helices from, 388–390, 389f
Amino acid side chains
  aliphatic, 34
  aromatic, 34
  charge of, 35–37, 36f, 37f
  covalent modifications, 61f
  defined, 31
  distribution of, 47
  hydrocarbon, 34
  hydrophilic, 35
  hydrophobic, 34
  hydroxyl-containing, 34, 34f
  ionizable, 35, 36t
  sulfhydryl-containing, 35, 35f
  thiol-containing, 35, 35f
  types of, 32
Amino acid synthesis, 787–788
  S-adenosylmethionine in, 799–801
  adenylation in, 807–808, 807f
  ammonia in, 788–793
  aromatic amino acids, 801–804
  branched pathways in, 805–806
  chorismate in, 801–804, 802f
  enzyme multiplicity in, 806, 806f
  essential amino acids, 793–794, 795f
  feedback inhibition/activation in, 805–807
  glutamine synthetase in, 807–808, 807f
  human, 793–794
  α-ketoglutarate in, 791–793, 794
  nitrogen fixation in, 788–793, 789f
  nonessential amino acids, 793, 794f
  overview of, 787–788
  regulation of, 805–808

  regulatory domains in, 806, 806f
  shikimate in, 801–804
  tetrahydrofolate in, 798–799, 798f, 798t, **799f**
  threonine deaminase in, 805–806, 805f
  by transamination, 794, 795f
Amino acid-attachment site, 131
Amino acids. *See also* Proteins
  abbreviations for, 36–37, 36t
  acetyl groups, 60
  activation by adenylation, 1027–1028
  activation sites, 1028–1029, 1028f
  aromatic, 34, 773–775, 801–804
  basic set of, 794t
  branched-chain, 758–759, 772–773
  carbon skeletons, fates of, 769f
  chirality of, 31, 795, 796f
  code words, 133
  D isomer, 31, 31f
  defined, 21
  dipolar form of, 31–32
  for DNA replication fidelity, 980
  encoding of, 132–135
  essential, 752, 752t, 779, 793, 794f
  excess, 752
  genetic code, 133–134, 133t
  glucogenic, 769–770
  in gluconeogenesis, 519
  hydrophilic, 35, 47
  hydrophobic, 32–34, 33f
  hydroxyl groups of, 34–35, 60
  "inside out" distribution of, 48f
  ionization state of, 32, 32f
  L isomer, 31, 31f
  metabolism, 776–777, 776t
  negatively charged, 35–37, 36f
  nitrogen in, 788
  nonessential, 793, 794f
  peptide bonds, 38, 38f
  polar, 34–35, 34f
  positively charged, 35, 35f
  as precursors, 808–814
  as precursors for neurotransmitters, 810
  pyruvate formation from, 770, 770f
  reactive, 35
  residue, 38, 47
  in secondary structures, 54t
  tRNA attachment, 131, 131f
  undesirable reactivity in, 37f
Amino group, 24
Amino-acid signal sequences, 1046–1048, 1046f
Aminoacrylate, Schiff bases of, 804, **804**
Aminoacyl adenylate, 1027
Aminoacyl-AMP, 1027
Aminoacyl-tRNA, 131, 132f, 1025f
  defined, 1027
  editing of, 1029, 1029f
  elongation factors and, 1036–1037, 1036f
Aminoacyl-tRNA synthetases, 131
  accuracy of, 1036
  activation sites of, 1028–1029, 1028f
  amino acid activation sites, 1028–1029
  classes of, 1030, 1031f
  defined, 1022, 1027

  double sieve in, 1029
  editing sites, 1029, 1029f
  evolution of, 1030
  genetic code and, 1026–1031
  proofreading by, 1029
  subunit structure, 1031, 1031t
Aminoglycoside antibiotics, 1043–1044
δ-aminolevulinate, 811–812
δ-aminolevulinate synthase, 811–812
Aminopeptidase N, 752
Aminopterin, 835–836
Aminotransferases
  in amino acid degradation, 758–761
  blood levels of, 761–762
  defined, 758
  Schiff-base intermediates formation in, 759–761
Ammonia
  in amino acid degradation, 758–759
  in amino acid synthesis, 788
  conversion to urea, 764–769
  formation of, 758–759
  in glutamate synthesis, 791–793
  in glutamine synthesis, 791–793
  in pyrimidine synthesis, 823
  transport of, 763–764
Ammoniotelic organisms, 768–769
Ammonium group, 24
AMP (adenosine monophosphate)
  ATP hydrolyzed to, 466–470
  energy charge and, 525–526
  in purine synthesis, 829–830, 829f
  structure of, 466, **467f**
  synthesis of, 837–838, 837f
AMP-activated protein kinase (AMPK), 739, 740–741, 894, 900
Amphibolic pathways, 465
Amphipathic (amphiphilic) molecules, 379
Amycolatopsis, 994
α-amylase, 350, 492
Amyloid fibers, 59–60, 60f
Amyloid plaques, 60, 60f
Amyloid precursor protein (APP), 60
Amyloidoses, 59
Amylopectin, 350
Amyotrophic lateral sclerosis (ALS), 146, 153, 172
Amytal, 607
Anabolic steroids, 1087
Anabolism, 465, 788b
Analytes, 90
Anaplerotic reaction, 562
Anastrozole, 876–877
Anchor residues, 1136, 1136f
Ancient DNA, 200
Anderson disease, 700t
Androgens
  functions of, 872
  pathways for formation of, 876f
  synthesis of, 875–877, **877**
Androstenedione, 876, **877**
Anemia, 18, 220–221, 222
Anfinsen, Christian, 52, 53
Angelman syndrome, 755
Angiogenin, 185–186, **186f**

Angiotensin II signaling, 439–447
Animal models in drug target testing, 938–939
Animal testing in drug development, 933
Anion exchange, 74
Ankyrin repeats, 1113, 1113f
Annealing, 124
Anomers, 343f, 345
Anosmias, 1099
Antagonists, 1087
Anthranilate, 803–804
Anthrax, 960
Antibiotics
    aminoglycoside, 1043–1044
    protein synthesis inhibition by, 1043–1045, 1044t
    transcription inhibition by, 994–995, 994f
    translocation inhibition by, 1044–1045, 1045f
Antibodies. See also Immunoglobulins
    amino acid sequence data in making, 96
    antibody binding of, 1127–1128, 1128f
    antigen interaction, 84f
    antigen-binding specificity, 1134–1135
    class switching in, 1133–1134, 1134f
    constant region of, 1126, 1127f
    defined, 83, 1122
    diversity of, 1131–1132
    formation of, 1133–1134
    generation of, 83–84
    hypervariable loops of, 1126–1129, 1127f
    monoclonal, 84f, 85–86, 85f, 945
    number of, 1131
    oligomerization of, 1132–1133
    polyclonal, 84, 84f
    primary, 87
    recombination in, 1129–1134, 1130f, 1131f
    secondary, 87–88
    structures of, 83f, 1124–1125
    variable region of, 1126, 1126f, 1127f
Antibody-drug conjugate (ADC), 945
Antibody-protein interactions, 1128–1129, 1129f
Anticancer drugs
    in blocking thymidylate synthesis, 835–836
    targets, 835f
Anticoagulants, 331
Anticodon loop, 1025
Anticodons
    base/base pair, 1026, 1026f
    defined, 131
    in translation, 1024, 1026, 1026f, 1030
Antidiuretic hormone, 98, 98f
Antigenic determinants, 83
Antigens
    ABO blood group, 359, 359f
    antibody interaction, 84, 84f
    cross-linking of, 1125, 1125f
    defined, 83
    synthetic peptides as, 97
Antihemophilic factor, 334
Antiporters, 413, 413f
Antiserum, 84

Antithrombin III, 332–333
AP endonuclease, 971
Apoenzyme, 235
Apolipoprotein B (apo B), 1004, 1005f
Apolipoproteins, 713
Apoptosis
    defined, 324, 609
    mitochondria in, 609
Apoptosome, 609
Apoptotic peptidase-activating factor 1 (APAF-1), 609
Approximation, catalysis by, 275
Aptamers, 202
Apyrase, 166
Aquaporins, 429, 429f
Arabidopsis thaliana, 164–165
Arachidonate, 737, 737f
Archaea, 3, 3f, 4
    membranes of, 379, 379f
    proteasomes of, 758
Arginase, 766
Arginine
    argininosuccinase and, 766
    degradation, 771, 771f
    structure of, 35, 35f
    synthesis of, 796, 796f
Argininosuccinase, 766
Argininosuccinase deficiency, 768, 768f
Argininosuccinate, 765, 767
Argininosuccinate synthetase, 765
Argonaute complex, 1092, 1092f
Aromatase, 876
Aromatase inhibitors, 876–877
Aromatic amino acids
    degradation of, 773–775
    side chains, 34
    synthesis of, 801–804
Arrestin, 1109
Arrow pushing, 24
ARS (autonomously replicating sequence), 158
Arsenite poisoning, 563, 563f
Artificial photosynthetic systems, 639–640
Ascorbic acid (vitamin C)
    deficiency, alcohol related, 911–912
    forms of, 912, 912f
    function of, 479t, 480
ASCT (autologous stem cell transplant), 1148
Asparaginase, 770
Asparagine
    conversion into oxaloacetate, 770
    structure of, 34f, 35
    synthesis of, 796, 796f
Aspartate
    conversion into oxaloacetate, 770
    formation of, 796
    structure of, 35, 36f, 252
    synthesis of, 796, 796f
Aspartate aminotransferase, 759, 761
Aspartate transcarbamoylase. See ATCase
Aspartic acid, 35
Aspartokinases, 806, 806f
Aspartyl proteases, 284, 284f, 285f

Aspirin
    discovery and development of, 932–933
    effect on prostaglandin H_2 synthase-1, 387f
    as prostaglandin inhibitor, 932
Assays
    defined, 71
    enzyme-linked immunosorbent (ELISA), 86–87, 87f
    in protein purification, 71
ATCase (aspartate transcarbamoylase)
    active site of, 313–314, 313, 314f
    allosteric interactions in, 312–316
    catalytic subunit, 312, 313f
    concerted model, 315
    cooperativity, 315
    defined, 310
    feedback inhibition, 311
    inhibition by CTP, 836
    reaction, 310, 311f
    regulatory subunit, 312, 312f
    separable subunits, 312
    sigmoidal curve, 315, 315f
    sigmoidal kinetics, 315f, 316
    structure of, 312–316, 314f
    T-to-R transition, 314f, 315
    ultracentrifugation studies of, 312, 312f
Atheroma, 932
Atherosclerosis, 724, 867–868, 870
Atomic motion, 1108, 1108f
Atorvastatin, 933, 934f
ATP (adenosine triphosphate), 116, 116f, 298, 320t
    as activated carrier, 475
    binding of, 201–202, 201f, 202f
    binding to molecular motors, 1155–1156, 1155f, 1156f
    as biological hydrotrope, 472
    in Calvin cycle, 654
    catabolism generation, 472
    in catabolism stages, 474–475, 475f
    as cellular energy currency, 321
    in citric acid cycle, 542–543, 552–557, 559–560, 561–562
    defined, 464, 466
    effect on ATCase kinetics, 316f, 317
    electrostatic repulsion, 469
    as energy-coupling agent, 468
    as energy-rich molecule, 466
    entropy increase in, 470
    in exercise, 471, 471f, 518–519, 901–902, 903t
    formation of, 464
    generation by NADH, 476
    hydration stabilization in, 470
    hydrolysis of, 274
    as immediate donor of free energy, 472
    in insulin release, 900, 900f
    ion gradients in, 474–475, 474f
    membrane proteins and, 406–413
    in mitochondria, 600–601, 602f
    in muscle, 902
    in muscle contraction, 512–513
    in nitrogen fixation, 788–789

phosphoryl-transfer potential, 469, 470–471, 470f
  in random sequential reaction, 251
  resonance structures, 469–470, 470f
  structure of, 466–468, **467**
  synthesis, 473–474
ATP formation
  1,3-bisphosphoglycerate (1,3-BPG) and, 501–503
  citric acid cycle and, 555–557
  by cyclic photophosphorylation, 634, 634f, 635
  enzyme-bound, 595
  fermentation in, 505
  from glucose to pyruvate conversion, 503
  from glycolysis, 503–504
  from phosphoryl transfer, 501–503
  proton-motive force and, 595–597, 596f
  pyruvate in, 503–504
ATP hydrolysis
  ABC transporters and, 412
  ATP as base to promote, 300, 301f
  in DNA replication, 955, 955f
  in DNA supercoiling, 959–960, 961f
  in driving metabolism, 468–469
  enzyme use of, 299
  as exergonic, 466–468
  free energy from, 466–472
  γ subunit in, 597
  hydrolysis, 299
  in kinesin motion, 1164–1166, 1165f
  magnesium ions in, 299
  myosin and, 299
  myosin conformational changes, 301, 301f
  power stroke in, 1159–1160
  reversible, within myosin active site, 302
  rotational motion, 597
  in transcription, 992–993, 993f
  transition state formation for, 300–301
ATP synthase
  a subunit of, 594–595, **594**, 597–599, 597f, 632
  α subunit of, 594, **594**
  ATP-driven rotation in, 596, 597f
  b subunit of, **594**, 595, 632
  β subunit of, 594, **594**, 596, 632
  bind-change mechanism for, 595–597, 596f
  c ring of, 596, 597–599
  of chloroplasts, 632–633
  defined, 593
  dimer of, 594f
  δ subunit of, 594, **594**
  discovery, 593, 593b
  efficiency of, 599b, 601
  ε subunit of, 634
  F_0 subunit of, 594, **594**, 595
  F_1 subunit of, 594, **594**, 595, 596
  G proteins and, 599
  γ subunit of, 594, **594**, 596, 599, 1167
  γε stalk, 594, 596
  inhibition of, 607
  nucleotide-binding sites, 596, 596f
  proton flow through, 593, 595–597
  proton motion across membrane in, 598, 598f

proton path through membrane in, 598, 598f
  proton-conducting unit, 597–598, 597f
  regulation of, 604–605
  rotational catalysis in, 597, 597f
ATP synthasome, 602
ATP synthesis
  in chloroplasts, 632–633
  efficiency of, 595
  as endergonic process, 592–593
  mechanism, 594–597, 596f
  oxidative phosphorylation uncoupling from, 605–607, 606f
  proton flow around c ring in, 597–599
  proton gradients in, 592–599
  uncoupling electron transport from, 607–608
ATP yield, 603, 604
ATP-ADP cycle, 437, 472, 472f
ATP-ADP translocase, 601–602, 602f, 608
ATP-binding cassettes (ABCs). *See also* ABC transporters
  ATP binding with, 411, 411f
  defined, 411
  distance between, 411
ATP-citrate lyase, 734
ATP-driven pump, 403
ATP-grasp fold, 823
ATPases
  AAA, 756, 1153, 1155–1156
  domains of myosin, 299, 299f, 300, 300f
  mitochondrial. *See* ATP synthase
  Na^+-K^+, 406, 410
  P-type, 403–404, 407–410, 411
  SERCA, 407–410, **408**, 409f
Atractyloside, 608
Atrazine, **638**
Attenuation, 1068–1070, 1069f
Autoimmune diseases, 1143–1144, 1143f
Autoinducers, 1067, 1067f
Autologous stem cell transplant (ASCT), 1148
Autonomously replicating sequence (ARS), 158
Autoradiography, 77
Autotrophs, 620, 646
Axel, Richard, 1099
Axial bonds, 346–347, 347f
Axoneme, 1163
Azothioprine, 940, **940**

B-cell receptor, 1132, 1132f
B cells
  activation of, 1132–1133, 1132f
  defined, 1122
  memory, 1144
B vitamins. *See* Vitamins
B-DNA helix, 119
B-form DNA, 119–120, 119f, 120t
Bacillus amyloliquefaciens, 281
Bacillus anthracis, 960
Bacillus subtilis, 992
Backbone models, 65, 65f
Backbones
  in nucleic acids, 114, 115f
  in polypeptide chains, 38, 38f

Bacteria, 3. *See also Escherichia coli* (*E. coli*)
  carcinogen action on, 975, 975f
  cell membranes of, 393, 393f
  defined, 3
  flagella movement of, 1116–1117
  gene expression in, 1058
  glyoxylate cycle in, 564, 565f
  gram staining of, 394, 394f
  nitrogen fixation in, 788–793
  photosynthetic, 638–639, 639t
  proinsulin synthesis by, 160, 161f
  rapid growth, 136
  restriction enzymes in, 147
  transcription factors in, 1079–1080
  transcription in, 985–995
  translation in, 1035
  upstream promoter elements in, 989
Bacterial artificial chromosomes (BACs), 157–158
Bacterial promoter sites, 130f
Bacterial reaction center
  cyclic electron flow in, 625–626, 626f
  defined, 623
  electron chain in, 623, 624f
Bacteriochlorophylls, 624
Bacteriopheophytins, 624
Bacteriorhodopsin, 385, 385f
Baculovirus, 172
BAL (British anti-lewisite), 563
Ball-and-chain model, 422–423, 423f
Ball-and-stick models, 23f, 24, 65, 65f
Band centrifugation, 81–82, 81f
Barth syndrome, 399
Basal transcription apparatus, 1000
Base-exchange reactions, 853–854
Base-excision repair, 971, 972f
Base-pair substitutions, tests for, 975
Base-paired hairpins, 131
Bases/base pairs, 6–7
  adenine, 5, 5f
  anticodon, 1026, 1026f
  compositions, 118, 119t
  cytosine, 5, 5f
  damage to, 969–970
  double helix, 10
  guanine, 5, 5f
  hydrogen bonds, 5
  minor-groove side of, 953
  mutagenic, 969
  nomenclature of, 822
  non-Watson–Crick, 969, 971, 988
  nucleic acids, 117
  in nucleotides, 116
  purine, 114, **115f**
  pyrimidine, 114, **115f**
  rRNA in, 1031–1032
  rules for, 118
  sequence of, 114, 122, 132
  stacking of, 11, 11f, 119, 119f, 124
  thymine, 5, 5f
  in translation, 1024
  tRNA, 1024
  uracil, 115, **115f**
  Watson–Crick, 118, 118f, 121, 122f, 1026

Basic-leucine zipper (bZip), 1080, 1080f
Bcr-abl gene, 457, 457f
Bcr-Abl kinase, 937–938, 943
Bennett, Claude, 1130
Benson, Andrew, 645
Benzoate, 768
Benzothiazine, **816**
Benzothiazole, **816**
Benzoyl CoA, 768
Berg, Howard, 1168
Beriberi, 562–563, 562b
Beta blockers, 938, 939
β chains, 220, 220f
    of hemoglobin, 207
β oxidation, 902, 904, 908
β pleated sheets, 42, 44–46
    amino acid residues in, 54–55, 54t
    antiparallel, 45f
    defined, 44
    mixed, structure of, 45f
    parallel, 45f
    protein rich, 46f
    twisted, schematic, 46f
β strands
    defined, 44
    DNA recognition through, 1060, 1060f
    of membrane proteins, 385, 385f
    Ramachandran plot for, 44f
    structure of, 44f
β turns, 42, 46, 54–55
β₂-microglobulin, 1135–1136
β-adrenergic receptor, 439–440, 440f
β-adrenergic-receptor kinase, 444
β-carotene, **879f**, 880
β-globin gene
    coding sequences, 136
    structure of, 136, 136f
    transcription and processing of, 136, 136f
β-hemoglobin, 222
β-lactam ring, 259
β-oxidation pathway, 717–718
β-thalassemia, 221–222
Bifunctional enzyme, 526
Bile, 843, 870–871
Bile acids, 712, 712f
Bile salts
    defined, 871
    synthesis of, 871, 872f
Bilirubin, 813
Biliverdin, 813
Biliverdin reductase, 813
Bimolecular reactions, 244
Bimolecular sheets
    defined, 380
    formation of, 381–382
Binding energy, 243–244, 274–275
Binding partners of protein, 88
Binding-change mechanism, 595–597, 596f
Biochemistry
    chemistry concepts in, 6–17
    defined, 2
    environmental factors influencing, 20–21
    evolution timeline, 3f
    as evolving science, 1–24
    functional groups in, 24
    genome sequencing and, 18–20

genomic revolution and, 17–22
    unifying concepts, 2–4
Biofilms, 1067–1068
Biofuels from genetically-engineered algae,
    180–181
Biological diversity
    biochemical unity underlying, 2–4
    similarity and, 3f
Biological macromolecules, 2
Biological membranes. See Membranes
Biologics, 945
Biomolecules, 878–880
Biopterin, 774
Biotin
    defined, 521
    in gluconeogenesis, 519–521, 520f
Biotin carboxyl carrier protein
    (BCCP), 521, 522f
1,3-bisphosphoglycerate (1,3-BPG)
    as acyl phosphate, 499–500
    ATP formation and, 501–503
    in glycolysis, 500–501, 502f
    oxidation of energy, 473–474
    phosphoryl-transfer potential, 470–471
2,3-bisphosphoglycerate (2,3-BPG)
    allosteric effector, 217
    binding of, 216, 216f
    defined, 216
    in oxygen affinity determination,
        216–217, 216f
Bitter taste, 1102, 1103–1104
Bjorkman, Pamela, 1135
Blackburn, Elizabeth, 968
BLAST (Basic Local Alignment Search
    Tool) search, 193, 193f
The Blind Watchmaker (Dawkins), 56
Blood, isozymes in, 318
Blood clotting, 310, 323
    Ca²⁺, 329, 331–332
    clots, 332–333
    enzymatic cascades, 328–329, 328f
    extrinsic pathway of, 328
    hemophilia and, 333–334
    intrinsic pathway of, 328
    regulation of, 332–333
    zymogen activation in, 328–329
Blood glucose homeostasis, 365–366
Blood groups, 359, 359f
Blood–brain barrier, 926
Blosum-62
    alignment of identities only versus, 192f
    defined, 190
    graphic view of, 191f
    myoglobin and leghemoglobin, 193f
Blotting techniques, 146, 149
Boat form, 346–347, 347f
Body mass index (BMI), 890, 891f
Body weight, regulation of, 890–892
Bohr, Christian, 218
Bohr effect, 217–219, 218f
Boltzmann's constant, 240
Bombardment-mediated transformation, 178
Bonds
    axial, 346–347
    covalent, 5, 7
    disulfide, 38–39, 39f

electrostatic interactions, 8
equatorial, 346–347
glycosidic, 348, 348f, 351, 351f, 353
hydrogen. See hydrogen bonds
hydrophobic effect and, 9–10, 10f
hydrophobic interactions and, 9
ionic interactions, 8, 10f
isopeptide, 754
length measurement, 7b
noncovalent, 7–9
peptide. See peptide bonds
units of measure for, 7b
van der Waals interactions, 8–9, 9f,
    11, 381, 1128
Bonitus, Jacob, 562b
Bordetella pertussis, 457
Bortezomib, 758
Bovine ribonuclease, **186f**
Bovine spongiform encephalopathy, 59
Brain
    in caloric homeostasis, 892–896
    glycogen phosphorylase, 690
    leptin effects in, 894–895, 894f
    role in caloric homeostasis, 892–896
    sensory connections to, 1098, 1098f
Branch points, 1005, 1007f
Branched-chain amino acids, 758–761
Branched-chain ketoaciduria, 776
Branching enzyme, 694–695, 695f
Brittle bone disease, 50
Bromodomains, 1088–1089, **1088f**
Brown, Michael, 867
Brown adipose tissue (BAT), 605, 606f
Brown fat mitochondria, 605
Buchner, Edward, 492
Buck, Linda, 1099
Buffers
    action, 16f
    pH and, 16–17
    protonation, 17f

C genes, 1129–1130
C₄ pathway, 656–658, 657f
C₃ plants, 658, 658f
C₄ plants, 658, 658f
¹³C/¹²C ratio, 672
C₄₀ isoprenoids, synthesis of, 879f
C-2'-endo conformation, 120, 120f
C-3'-endo conformation, 120, 120f
C-type lectins, 362–363, 362f
CAAT box, 999, 999f
Caenorhabditis elegans, 70
    genome of, 164
    RNA interference in, 177
Calcineurin, 1133
Calcitriol, 877
Calcium ion channels, 420
Calcium ion pump, 408–410, 409f
Calcium ions
    calmodulin activation, 447
    coordination to multiple ligands,
        446, 446f
    imaging, 446, 446f
    in phosphorylase activation, 689, 689f
    pump structure, **408**
    pumping, 408, 409f

as second messenger, 438–439, **439f**, 445–446
in signal transduction, 445–446
in vision, 1110–1111
Calmodulin (CaM)
defined, 447, 689
in mitochondrial biogenesis, 902
in signal transduction, 447, 447f
Calmodulin-dependent protein kinases (CaM kinases), 447, 447f
Calnexin, 363
Caloric homeostasis, 843, 890–896
in body weight regulation, 890–892
brain role in, 892–896
defined, 890
in diabetes, 896–901
dieting and, 896
evolution and, 890–892
insulin in, 893–894
leptin in, 893–896
obesity and, 889–912
satiation signals in, 892–893, 893f
signaling in, 893–894
suppressors of cytokine signaling (SOCS) in, 895–896, 895f
Calorie (cal), 238
Calreticulin, 363
Calvin, Melvin, 645
Calvin cycle, 645–659
ATP (adenosine triphosphate) in, 654
C_4 pathway and, 656–658, 657f
carbon dioxide in, 646–655, 656–658
defined, 620, 645
five-carbon sugar formation in, 651, 652f
glycolate in, 650–651, 650f
hexose sugar formation in, 651–654, 651f
illustrated, 647f
light regulation of, 655–656, 655f
magnesium ion and, 648–649, 648f, 655
NADPH in, 645–646, 654
overview of, 645–646
oxygenase reaction in, 649–651, 650f
pentose phosphate pathway and, 668
3-phosphoglycerate formation in, 646–648, 649f
photorespiration in, 650f, 651
reactions in, 646–655, 653f
regulation of, 655–656
ribulose 1,5-bisphosphate in, 646–648, 648f
rubisco in, 646–648, 648f, 650f, 655
salvage pathway in, 649–651
stages, 646–648, 647f
thioredoxin in, 655–656, **656f**, 656t
cAMP. *See* Cyclic AMP (cAMP)
cAMP response element binding (CREB) protein, 443
Camptothecin, 961
Cancer
aerobic glycolysis in, 517–518
breast, 876
citric acid cycle defects in, 559–560
colorectal, 974, **974**, 1144
defective DNA repair in, 973–974
drug resistance in, 942–943
fatty acid metabolism in, 735

hypoxia inducible factor 1 (HIF-1) in, 559–560
immune system in, 1144
in Li-Fraumeni syndrome, 974
ovarian, 876–877
pyruvate dehydrogenase kinase (PDK) in, 560
signal transduction abnormalities in, 456, 456f, 457f
skin, in xeroderma pigmentosum, 974
tumor hypoxia in, 518, 518t
tumor suppressor genes and, 973–974
tumor visualization, 517, 517f
Warburg effect in, 517
Cancer therapy, 838
Capillary electrophoresis, 149–150
Capsaicin, 1114–1115
Capsaicin receptor, 1114, 1114f
Captopril, 285
Carbamate
in Calvin cycle, 648
formation of, 219, 219f
Carbamate groups, 219, **219f**
Carbamoyl phosphate
formation of, 763–764
in pyrimidine synthesis, 823
reaction, 764
in urea cycle, 764–769
Carbamoyl phosphate synthetase I, 764–765
Carbamoyl phosphate synthetase II, 823
in pyrimidine synthesis, 823
structure of, 823, **823f**
Carbanion, 664
Carbanion intermediates, 665, 665f
Carbohydrate-asparagine adduct, **61f**
Carbohydrate units, 61
Carbohydrate-asparagine adduct, **61f**
Carbohydrates, 341–365
as building block, 20
disaccharides, 349–350
glucose generation from, 492–493
isomeric forms of, 343f
lectins and, 361–364, 362f
monosaccharide, 342–349
N-linked, 352–353, 353f
O-linked, 352–353
oligosaccharides, 349, 358, 359f
overview of, 341–342
polysaccharides, 350, 351f
proteins linkage, 352–361
Carbon dioxide
binding site, 289, 289f
in Calvin cycle, 646–648, 656–658
defined, 218
hydration acceleration, 287
hydration of, 291, 291f
oxygen release and, 217–219
pH and, 218, 218f, 219
transport from tissue to lungs, 219, 219f
Carbon fuels
free energy of, 472–475, 472f
prominent, 473, 473f
Carbon monoxide
defined, 217
in oxygen transport disruption, 217
poisoning, 227–228, 228f

Carbon-nitrogen bond, 276
Carbonic acid, 218–219
Carbonic anhydrases, 219
bound zinc in, 287–288
in carbon dioxide hydration acceleration, 286
catalytic strategies of, 274
defined, 286
evolution of, 291
mechanism of, 289, 289f
pH effect on, 288, 288f
proton shuttle, 290–291, 291f
structure of, 288, **288**
synthetic analog model system for, 289–290, 289f
Carbonium ion intermediate, 683
Carbonyl group, 24
Carboxamides, 35
γ-carboxyglutamate, 60, 61f, 331–332, 332f
Carboxyhemoglobin, 217, 228
Carboxyl-terminal domain (CTD)
in coupling transcription to pre-mRNA, 1010f
phosphorylation of, 1000
serine residues of, 997
Carboxylation of pyruvate, 521–522, 562
Carcinoembryonic antigen (CEA), 1144
Carcinogens, tests for, 975, 975f
Cardiolipin, 399
Cardiotonic steroids, 410
Cargo receptor, 1048
Carnitine, 716–717, 716f
Carnitine acyltransferase I, 716
Carnitine acyltransferase II, 716
Carotenoids, 635–636, 879–880
Carriers, 403
Cartilage, 355–356, 356f
Caspases, 609
Cassette mutagenesis, 162, 162f
Catabolism
ATP generation by, 472
defined, 465, 788b
energy from, 472–475
free energy of, 471, 472–475, 475f
stages of, 474–475, 475f
Catabolite activator protein (CAP), 1064–1065, 1064f
Catabolite repression, 1064–1065
Catalase, 591
Catalysis
by approximation, 275
by ATCase, 310
Circe effect in, 250, 250b
covalent, 274, 276, 276f, 277f
efficiency of, 249–250
enzyme-substrate binding in, 243–244
enzyme-substrate complex formation in, 241, 241f
general acid-base, 275
maximal rate, 247–248, 249
metal ion, 275
specificity constant, 250
transition state formation and, 240
velocity of, 250
zinc activation of water molecule, 288–290, 289f

Catalytic RNA, 1012–1014
Catalytic strategies, 274–306
    of carbonic anhydrases, 274, 286–291
    in chymotrypsin, 276, 276f, 277f
    of myosins, 274, 298–304
    overview of, 274
    of restriction endonucleases, 274, 291–298
    of serine proteases, 274, 278–280
Catalytic triads, 278
    in chymotrypsin, 277–280, 278f
    in elastase, 281–282, 305f
    in hydrolytic enzymes, 280f, 281–282, 281f
    serine in, 278–280
    site-directed mutagenesis and, 282–283, 282f
    of subtilisin, 281, 282f
    in trypsin, 281–282, 281f
Cataract formation, 511, 511f
Cataracts, 511, 511f
Catecholamines, 810, 810f
Cation exchange, 74
CD3, 1138, 1138f
CD4, 1140, 1140f, 1142–1143
CD8, 1137–1139, 1137f, 1142–1143
CD28, 1138
CD45, 1138
cDNA library, 160
CDP-ethanolamine, 853, **853**
Cech, Thomas, 1017
Cech, Tom, 1017
Celecoxib (Celebrex), 937, **937f**
Cell cycle, 967
Cell-mediated immunity, 1134
Cell-to-cell channels, 427–428, 427f, 428f
Cell-wall biosynthesis, 930–931, 931f
Cells
    defined, 2
    hybridoma, 86
    motion within, 1152, 1152f
Cellular immune response, 1122–1123
Cellular respiration, 605f
    citric acid cycle in, 542–543, 542f
    enzymes in, 235
    in oxidative phosphorylation, 603–609
Cellular RNA, 128–129
Cellulose, 650–651
Central dogma, 114
Centrifugation
    differential, 72, 72f
    sedimentation coefficients in, 80, 80t, 81f
    ultracentrifugation, 80–82, 81f, 312, 312f
    zonal, 81, 81f
Cephalopods, 356
Ceramidase, 857
Ceramide
    insulin insensitivity by excess, 882
    as precursor, 857
    synthesis of, 854–855, 854f
    in tumor growth, 857
Cerebroside, 378, **378**, 855
Cetuximab (Erbitux), 945
cGMP phosphodiesterase, 1108
cGMP-gated ion channels, 1109
Chain, Ernest, 930

Chair form, 346–347, 347f
Chain-elongation reaction, 125, 129f
Chanarin-Dorfman syndrome, 714
Changeux, Jean-Pierre, 214
Channels, 403
Chaperones, 53
Charcot, Jean-Martin, 146
Charcot-Marie-Tooth disease, 1163
Charged tRNA, 1027
Chemical modification reaction, 276
Chemical reactions mechanisms, 24
Chemical shifts, 103
Chemiosmotic hypothesis, 592–593, 593f
Chemoattractants, 1169–1170
Chemokines, 58
Chemorepellant, 1169
Chemotaxis, 1168–1170, 1169f
Chemotroph, 464
Cheng, Yung-Chi, 924
Cheng–Prusoff equation, 924
CheY, 1169
CheZ, 1169
Chirality
    of amino acids, 795, 795f
    establishment of, 791–792, 792f
Chitin, 356, 356f, 357
Chitosan, 357
Chlamydomonas, 1014
Chloramphenicol, 1043, 1044t
Chlorobium thiosulfatophilum, 639
Chlorophyll *a*, 623, 623f, 635f
Chlorophyll *b*, 635–636, 635f
Chlorophylls
    defined, 622
    in photosynthesis, 622–626
    phytol, 622
    structure of, 622–623, 622f
Chlorophylls, light absorption by, 622, 623f
Chloroplasts
    ATP synthase, 632–633
    defined, 621
    electron micrograph of, 621f
    evolution of, 622
    genome of, 622
    in photosynthesis, 621–622
    reaction center, 623, 624f
    redox conditions, 633–634
    structure of, 621, 621f
Chlorpromazine, 931, **931**
Cholecystitis, 871
Cholecystokinin (CCK), 893–894
Cholera, 457
Choleragen, 457
Cholesterol
    "bad," 870
    carbon numbering, 872–873, 873f
    discovery of, 933–935
    excess, effects of, 867, 867f
    fatty acid chains and, 391, 392f
    formation of, 862f
    "good," 870
    levels, clinical management of, 870–871
    lipid rafts, 392–393
    as membrane fluidity regulator, 393
    in membrane lipid, 378–379

    as precursor for vitamin D, 877–878
    structure of, 378–379
Cholesterol synthesis, 858–860
    blocking of, 871
    condensation mechanism of, 860, 860f
    defined, 856, 856b
    isopentenyl pyrophosphate in, 858–859, 859f
    labeling of, 858f
    liver in, 861
    metabolism, 866–867
    mevalonate in, 858–859, 859f
    rate of, 861–862
    regulation of, 861–871
    site of, 864, 865f, 866
    squalene in, 859–860, 861f, 862f
    stages of, 858
    sterol regulatory element binding protein (SREBP) in, 861–862, 863f
Choline, 853
Chorismate in amino acid synthesis, 801–804, 802f
Chromatin, 1077–1079, 1083–1090
    defined, 1077
    in gene expression, 1078, 1083–1090
    higher-order, structure of, 1079, 1079f
    remodeling of, 1083–1090, 1090f
    structure of, 1076, 1077t
Chromatin immunoprecipitation (ChIP), 1083, 1083f
Chromatin-remodeling complexes, 1089
Chromatography
    affinity, 74–75, 74f
    gel-filtration, 73, 73f
    high-performance liquid (HPLC), 75, 75f
    ion-exchange, 73–74, 74f
Chromogenic substrate, 277, 277f
Chromophore, 1107
Chromosomes
    bacterial artificial, 157–158
    defined, 116
    Indian muntjac, 116, 117f
    yeast, 1076, 1076f
    yeast artificial, 157–158, 158f
Chronic myelogenous leukemia (CML), 457
Chronic obstructive pulmonary disease (COPD), 327
Chylomicron remnants, 864
Chylomicrons, 713, 713f, 864–865, 864t, 865
Chymotrypsin, 197, **197f**
    active site, 278, 278f
    catalysis kinetics, 277, 277f
    catalytic mechanism, 276, 276f
    catalytic triad in, 278–280, 278f
    conformations of, 325, 325f
    covalent catalysis in, 276, 276f, 277f
    diisopropylphosphofluoridate (DIPF) and, 276, 276f
    elastase and, 281, 281f
    natural substrate for, **257**
    oxyanion hole, 279, 279f
    peptide hydrolysis by, 278, 279f
    S_1 pockets of, 281f
    serine residue, 276

specificity of, 276f
specificity pocket of, 280, 280f
structure of, 278–280
substrate preferences of, 249–250, 250t
trypsin and, 281, 281f
Chymotrypsinogen, 278
conformations of, 325, 325f
defined, 324
proteolytic activation of, 324f, 325
Cilia, microtubules in, 1163
Ciprofloxacin, 960
Circe effects, 250, 250b
Circular DNA, 120, 121f
in eukaryotic nucleus, 124
Cirrhosis, 911
Cis configuration, peptide bonds, 40, 40t, 41t
Citrate
in fatty acid metabolism, 734, 740f
in glyoxylate cycle, 564, 565f
isomerization of, 550–551, 551f
phosphofructokinase inhibition by, 514
in polymerization, 739
synthesis of, 559
transport to cytoplasm, 559
Citrate synthase
conformational changes on binding, 549, 549f
defined, 549
mechanism of, 549–550, 550f
Citric acid cycle, 541–566, 908
acetyl CoA (acetyl coenzyme A) in, 541, 544–546, 548–550, 550f
aconitase in, 550, 551f
ADP in, 522–523
ATP in, 543, 554–557, 559, 561
biosynthetic roles of, 561–564, 561f
in cancer, 559–560
carbon oxidation in, 548–557
in cellular respiration, 542–543, 543f
citrate isomerization in, 550–551, 551f
citrate synthase in, 548–549, 549f
citryl CoA in, 548–550, 550f
components of, 556t
control sites of, 559, 559f
defects in, 559–560
entry to, 557–561
evolution of, 563–564
in exercise, 562, 562f
glycolysis and, 543–548, 543f
high-energy electron harvesting in, 542–543
illustrated, 555f
intermediates in, 561
isocitrate dehydrogenase in, 551, 559, 559f
isocitrate in, 550–551
α-ketoglutarate dehydrogenase complex in, 552, 559, 559f
α-ketoglutarate in, 551
ketone bodies in, 908, 908f
lipoamide in, 546–548, 548f
in mitochondria, 542, 542f
overview of, 541–542, 542f
oxaloacetate in, 542, 548–549, 554–555
pathway integration, 562, 562f

pattern of, 542, 542f
in photosynthesis. See dark reactions
pyruvate dehydrogenase complex in, 543–548, 544t, 546f, 548f, 557–559, 558f
pyruvate dehydrogenase in, 557–559
pyruvate dehydrogenase phosphatase (PDP) in, 557
rate of the cycle, 559, 559f
reactions of, 554
regulation of, 557–558
replenishment, 561–562
substrate channeling in, 556
succinate dehydrogenase in, 554–555
succinyl CoA in, 552, 553f
succinyl CoA synthetase in, 553
thiamine pyrophosphate (TPP) in, 545, 545f
Citrulline, 765–766
Citrullinemia, 776t
Citryl CoA (citryl coenzyme A), 549, 550, 550f
Cladribine, 838
Clamp loaders, 962, 963
Class I MHC proteins, 1135–1136, 1135f, 1137, 1141, 1141f
Class II MHC proteins, 1139, 1140f
Class switching, 1134, 1147–1148
Clathrin, 395, 866
Clathrin-coated pit, 395
Cleavage
DNA. See DNA, cleavage of
protein, 93–94
Cleland, W. Wallace, 251
Clinical trials, 941–942, 941f
Clofarabine, 838
Cloned genes, 74–75
Cloning
of cDNA, 417
expression, 160
plasmid vectors in, 155–157
in recombinant DNA technology, 158–159
Cloning vectors, 155
mutant λ phage as, 157, 157f
Closed promoter complex, 990, 991f
Clostridium perfringens, 507
Clostridium tetani, 1144
Clotting. See Blood clotting
Cluster of differentiation, 1137
Clustered regularly interspaced palindromic repeats (CRISPRs), 174, 175f
Co-immunoprecipitation, 88, 88f
Coactivators, 1085–1086
Coat proteins (COPs), 1048
Coated pits, 866
Cobalamin, 721, 721f. See also Vitamin B₁₂
Cobalt atom, 721
Codon bias, 134
Codon-anticodon interactions, 1026
Codons
deciphered, 133, 133t
defined, 22, 114, 131
maximizing number of, 1026
of mitochondria, 135, 135t
stop, 1040, 1040f

as synonyms, 133
in translation, 1025–1026, 1026f
Coenzyme A, 477, **477**
Coenzyme B₁₂, 721, 722–723, 722f
Coenzyme Q, 581
Coenzymes
in amino acid degradation, 752
defined, 235–236
prosthetic groups, 236
types of, 236t
vitamin, 478–480, 479t
Cofactors, 235
Cognate DNA, 292, 295–297, **296**, 296f, **297**
Cohesive ends, DNA, 154, 154f
Coiled-coil proteins, 49, **49f**
Collagen
amino acid sequence of, 49–50, 50f
ascorbate and, 912
in cartilage, 355–356
defined, 49
structure of, 50f
triple helix, 50
Color blindness, 1111, 1111f
Color vision
in animals, 1110, 1110f
cone receptors, 1110–1111
defective, 1111
evolution of, 1110, 1110f
Combinational association, antibody diversity and, 1130–1131
Combinatorial chemistry, 200, 933
Combinatorial control, 1081
Committed step, 805
Comparative genomics, 20
Compartmentalization, 485
Compartments, drug target, 926, 926f
Compensatory proliferation, 611
Competitive inhibition. See also Enzyme inhibition
defined, 253
dissociation constant, 254
double-reciprocal plot, 255, 255f
kinetics of, 254–256, 254f
use of, 254
Complement cascade, 1124
Complementarity-determining regions (CDRs), 1127, 1127f
Complementary DNA (cDNA)
cloning and sequencing of, 417
defined, 159
formation of, 159–160, 160f
hybridization with, 146
screening of, 160, 160f
Complementary single-stranded ends, DNA, 154
Complex I mutations, 611, 611f
Concerted model
allosteric enzymes, 314
defined, 214, 314
formulation of, 225–227
modeling oxygen binding with, 227f
T and R states, 214f
tetramers, 215
Condensing enzyme, 730
Cones, 1106, 1110–1111

Conformation selection, 243
Congenital disorders of glycosylation, 359–360
Congenital erythropoietic porphyria, 814
Congenital hemolytic anemia, 532
Congestive heart failure, 410
Conjugation in drug metabolism, 926, 927, 928f
Connexin, 428
Connexons, 428
Consensus sequence, 130b, 136f, 320
Conservative substitutions, 190, 192f
Constitutional isomers, 343, 343f
Controlled termination of replication, 149–150
Conventional kinesin, 1154
Convergent evolution, 196–197, 197f
Cooley anemia, 222
Cooperative binding
    concerted model, 214–215, 214f
    defined, 212
    of oxygen, 212–217
    oxygen delivery and, 212–213, 212f
    sequential model, 215, 215f
    substrates, 252
Cooperative transition, 55
Cooperativity property, 310
Copper ions in cytochrome c oxidase, 586
Coproporphyrinogen III, 812f, 813
Corepressors, 1063
Corey, Robert, 42
Cori, Carl and Gerty, 700, 701
Cori cycle, 529, 529f
Cori disease, 700t
Corrin ring, 721
Corticosteroids, 875
Cortisol, 322
Cosmids, 157
Cotransporters. See Secondary transporters
Coulomb energy, 8
Covalent bonds
    in biological molecules, 7
    defined, 5
    multiple, 7
    as strongest bonds, 7
Covalent catalysis
    chymotrypsin as example, 276, 276f, 277f
    defined, 274
Covalent enzyme-bound intermediate, 500
Covalent modifications
    acetylation, 318, 319t
    common, 319t
    dephosphorylation, 318
    in enzyme regulation, 318–323
    irreversible, 318
    phosphorylation, 319t
COX2 inhibitors, 936–937, 937f
CpG island, 1084
Crassulacea, 359, 359f
Crassulacean acid metabolism (CAM), 659
Creatine, 251
Creatine kinase, 471f
Creatine phosphate, 902, 903t
Creutzfeld–Jakob disease (CJD), 59
Crick, Francis, 5, 6, 17, 114, 118, 950

CRISPR-Cas system, 174–175, 176f
CRISPR-Cas9, 181
CRISPRs (clustered regularly interspaced palindromic repeats), 174, 175f
Critical concentration, 1158
Cro in genetic circuit formation, 1066, 1066f
Cross talk, 439
Cross-links, 38–39, 39f
Cross-peaks, 104
Crown gall, 177
Cryo-electron microscopy (cryo-EM), 70, 105–106, 106f
CTP. See Cytidine triphosphate (CTP)
CTP-phosphocholine cytidylyltransferase (CCT), 853
Cumulative feedback inhibition, 806–807
Cumulative selection, 56
Cushing syndrome, 322–323
Cushing's disease, 322
Cyanidiales, 199
Cyanobacteria, 622–623, 623f
Cyanocobalamin, 721
Cyclic AMP (cAMP)
    binding to regulatory subunit, 322
    defined, 821
    effects in eukaryotic cells, 322
    in eukaryotic cells, 443
    generation by adenylate cyclase, 442–443
    in glycogen metabolism, 690–692
    phosphorylation stimulation, 443
    as second messenger, 438, **439**, 454–455
    in signal transduction, 442–443
    structure of, **322**
Cyclic GMP (cGMP), 459
    defined, 821
    as second messenger, 438, **439**
Cyclic nucleotides, 821
Cyclic photophosphorylation, 634, 634f
Cyclin-dependent protein kinases, 967
Cycloheximide, 1044, 1044t
Cyclooxygenase inhibitors, development of, 937
Cyclosporin, 1133, **1133**
Cys₂His₂ zinc-finger domains, 1080
Cystathionase, 801
Cystathionine, 801
Cystathionine β-synthase, 801
Cysteine, **816**
    linked, 38
    residue modification, 312f
    structure of, 34f, 35
    synthesis of, 797, 801
Cysteine proteases, 283, **284f**
Cysteine thiyl radical, 833
Cystine, 39
Cytidine, 116
Cytidine diphosphodiacylglycerol (CDP-diacylglycerol), 851, **851**
Cytidine monophosphate (CMP), 851, **851**
Cytidine triphosphate (CTP), 316–317, 316f
    ATCase inhibition by, 311, 311f, 836
    defined, 311
    effect on ATCase kinetics, 316–317, 316f
    in pyrimidine synthesis, 825
    T state stabilization, 315, 316f

Cytidine triphosphate synthetase, 825–826
Cytidylate, 116
Cytochrome
    defined, 854
    in photosynthetic reaction center, 623–624, 624f
Cytochrome b₅, 736, 736f
Cytochrome bf complex, 625–626, 629, 629f
Cytochrome c
    conformation of, 592, 592f
    evolution of, 592, 592f
    in oxidative phosphorylation, 580f, 584
Cytochrome c oxidase
    defined, 584
    electron flow in, 607
    heme, 586
    mechanism, 588, 588f
    peroxide bridge, 588, 588f
    proton transport by, 589, 589f
    in respiratory chain, 579–581, 580f, 580t, 584–586
    structure of, 587, 587f
    subunits, 586
Cytochrome P450, 926, 927
    defined, 873
    hydroxylation by, 873
    mechanism, 874
    protective function, 874–875
Cytoglobin, 223
Cytokines, 1140
Cytoplasm, 659, 733, 734f
Cytosine, 5, 5f, 114, **114f**, 118, **118f**, 972–973
Cytoskeleton, 1151–1152, 1162–1163
Cytotoxic (killer) T cells
    CD8 on, 1137–1139
    defined, 1123
    foreign peptide recognition, 1139, 1139f

D amino acids, 31, 31f
D genes, 1131
Dalgarno, Lynn, 134, 1034
Daltons, 38, 38b
Dark reactions, 620
Darst, Seth, 986
Dawkins, Richard, 56
De novo pathways. See also Nucleotide synthesis
    defined, 822
    purine, 826f, 827
    pyrimidine, 822, 822f
Deamination
    in amino acid degradation, 763
    example of, 969, 969f
Death cap, 997f
Debranching enzyme, 701
Decarboxylation
    in citric acid cycle, 544
    in gluconeogenesis, 522
    in pentose phosphate pathway, 659
Decorated actin, 1158
Dedicated protein kinases, 320
Deficiency disorders, 20, 60–61

Degrons, **754**
Dehydratases, 763
7-dehydrocholesterol (pro-vitamin D₃), 877–878, 878f
Deinococcus radiodurans, 950, 950f
Deleterious reactions, 499
Denaturation
    of DNA, 15, 15f
    of proteins, 53
Denisovans, 200, 200f
Density-gradient equilibrium sedimentation, 123, 123f
Dental plaque, 1067
Deoxyadenosine, 116
5′-deoxyadenosyl radical, 722, 722f
Deoxyadenylate, 116
Deoxycytidine, 116
Deoxycytidylate, 116
Deoxyguanosine, 116
Deoxyguanylate, 116
Deoxyhemoglobin
    αβ dimers, 211
    2,3-BPG binding to, 216f
    carbon dioxide and, 219
    defined, 214
    quaternary structure of, **211f**, 213f, 214
    salt bridges in, 218f
Deoxymyoglobin, 208, 209, **209f**
Deoxynucleoside 5′-triphosphates, 125
Deoxyribonucleic acid (DNA). *See* DNA
Deoxyribonucleoside 3′-phosphoramidites, 150, 150f, **151**
Deoxyribonucleoside triphosphate (dNTP), 952
Deoxyribonucleotide synthesis
    control of, 838–839
    deoxyuridylate in, 834, 835
    dihydrofolate reductase in, 834
    overview of, 831
    ribonucleotide reductase in, 831–833, 834f, 837–838, 838f
    thymidylate, 834, 834f, 835–836
Deoxyribose, 5, 114, **114f**, 344, 344f, 346f
Deoxyribose phosphodiesterase, 971
Deoxyuridylate (dUMP), 834, 834f
Dephosphorylation, 318, 320
Deprotonation
    effect of buffer on, 290, 290f
    water, kinetics of, 290, 290f
Designer genes, 162–163
Desmolase, 875
Destroying angel, 997
α-dextrinase, 492
Diabetes mellitus
    defined, 897, 897b
    glucose homeostasis in, 896
    incidence of, 896
    insulin in, 897–900
    insulin resistance and, 897–898, 898f
    ketosis in, 727, 727f
    obesity and, 896–901
    oxaloacetate in, 724
    type 1, 896, 901, 901f
    type 2, 896, 898
Diabetic neuropathy (DN), 566–567

Diacylglycerol (DAG), 438, **439**, 714
    in lipid synthesis, 850
    in signal transduction, 445–446, 445f, 454–455
Diacylglycerol kinase, 851
Dialysis in protein purification, 73, 73f
Diamond-Blackfan anemia, 1051
Diastereoisomers, 343, 343f
Diazotrophic microorganisms, 789
Dicoumarol, 331
Dictyostelium discoideum, 299, 692, 1155
Dicyclohexylcarbodiimide (DCC), 99f, 607
Diels-Alder reaction, 802–803
2,4-dienoyl CoA, 720, **720**
Diet, 20–21, 21f
    fatty acid metabolism and, 740
    ketogenic, 727
    low-carbohydrate, 896
    low-fat, 896
    obesity and, 896
Differential centrifugation, 72, 72f
Diffusion
    facilitated, 403, 405
    lateral, 390
    lipid, 390
    membrane protein, 391
    simple, 405
    transverse, 391, 391f
Diffusion coefficient, 390, 391
Digestion
    of dietary proteins, 752, 753f
    enzymes in, 323–324, 324f
    of lipids, 712–713, 712f
    proteolytic enzymes in, 323–334
    starved-fed cycle and, 905–907
Digestive enzymes, 61
    chymotrypsinogen, 324–325, 324f
    synthesis of, 323
Digitalis, 410
Digitoxigenin, 410, 410f
Diglyceride acyltransferase, 851
Dihydrofolate reductase, 835
Dihydrolipoyl transacetylase, 546, 546f
Dihydropteridine reductase, 774
Dihydrotestosterone (DHT), 876
Dihydroxyacetone, **342**, 343, 343f
Dihydroxyacetone phosphate (DHAP), 348
    in Calvin cycle, 651
    in glycolysis, 494f, 497–498, 498f, 499
    in lipid synthesis, 850, 850f
5,6-Dihydroxyindole, **816**
5,6-Dihydroxyindole-2-carboxylic acid, **816**
Diisopropylphosphofluoridate (DIPF), 256, 256f, **257f**, 276, 276f
Dimer, 51
Dimerization arm, 452, 452f, 453, 453f
Dimethylallyl pyrophosphate, 860
Dimethylbenzimidazole, 721
2,4-dinitrophenol (DNP), 607–608
Dioxygenases, 774–775
Diphosphatidylglycerol (cardiolipin), 399, **399**, 850, **852**
2,3-diphosphoglycerate (2,3-DPG). *See* 2,3-bisphosphoglycerate (2,3-BPG)
Diphthamide, 1045, 1045f

Diphtheria toxin, 1044–1045, 1045f
Dipolar ions, 31–32
Direct DNA repair, 971
Disaccharides
    defined, 349
    structure of, 349–350, 349f
    types of, 349–350, 349f, 350f
Diseases and disorders
    albinism, 776t
    alcaptonuria, 776
    alcohol-related, 910–911
    Alzheimer's disease, 59–60
    amyloidoses, 59
    amyotrophic lateral sclerosis (ALS), 146, 153, 172
    Anderson disease, 700t
    anemia, 353
    Angelman syndrome, 755
    anthrax, 960
    argininosuccinase deficiency, 767, 767f
    arsenite poisoning, 563, 563f
    atherosclerosis, 724, 866–867, 870–871
    autoimmune, 1143, 1143f
    beriberi, 562, 562b
    bovine spongiform encephalopathy, 59
    cancer. *See* cancer
    carbamoyl phosphate synthetase deficiency, 767, 768f
    carnitine deficiency, 716
    cataracts, 511, 511f
    Charcot-Marie-Tooth disease, 1163
    cholera, 457
    citrullinemia, 776
    congenital disorders of glycosylation, 359
    congestive heart failure, 410
    Cooley anemia, 222
    Cori disease, 700t
    coronary artery disease, 867
    Creutzfeldt–Jakob disease (CJD), 59
    deficiency disorders, 20
    diabetes. *See* diabetes mellitus
    drug-resistant, 942–943
    emphysema, 327
    environmental factors in, 20–21
    falciparum malaria, 670
    familial hypercholesterolemia, 849
    galactosemia, 511
    glucose 6-phosphate dehydrogenase deficiency, 668–670
    glycogen-storage, 700–701, 700t, 701f
    gout, 839–840
    Hartnup disease, 752
    heart disease, 871
    hemolytic anemia, 669
    hemophilia, 333
    Hers disease, 700t
    HIV infection. *See* human immunodeficiency virus (HIV) infection
    homocystinuria, 776t
    Huntington disease, 59, 753, 973
    Hurler disease, 355f, 360
    hyperammonemia, 767–768
    hyperlysinemia, 776t
    I-cell disease, 359

Diseases and disorders (*Continued*)
  lactose intolerance, 509–510
  Leber hereditary optic neuropathy
    (LHON), 608
  Lesch–Nyhan syndrome, 840–841
  Li-Fraumeni syndrome, 974
  long QT syndrome (LQTS), 427
  Lynch syndrome, 974, 974f
  mad cow disease, 59
  malaria, 221
  maple syrup urine disease, 776
  McArdle disease, 700t, 701
  mercury poisoning, 562–563
  mitochondrial, 608
  mucopolysaccharidoses, 355
  multidrug resistance in, 411–413
  mutations causing, 153–154
  neurological, protein misfolding in, 59–60
  ornithine transcarbamoylase deficiency,
    767–768, 768f
  osteoarthritis, 355
  osteogenesis imperfecta, 50
  osteomalacia, 877
  Parkinson disease, 59–60, 753, 755
  phenylketonuria, 777
  Pompe disease, 700–701, 700t, 701f
  porphyrias, 614
  predisposition to, 19
  prion, 59–60, 60f
  protein aggregates in, 59–60
  respiratory distress syndrome, 856–857
  retinitis pigmentosa, 1010–1011
  ricin poisoning, 1045
  rickets, 877
  scrapie, 59
  scurvy, 911
  severe combined immunodeficiency
    (SCID), 178, 839
  sickle-cell anemia, 220–221, 220f, 221f
  spina bifida, 841
  Tay–Sachs disease, 856
  thalassemia, 1010, 1010f
  transmissible spongiform
    encephalopathies, 59
  tuberculosis, 758
  tyrosinemia, 776t
  vanishing white matter (VWM) disease,
    1043, 1043f
  vitamin D deficiency, 877
  von Gierke disease, 700, 700t, 701
  Wernicke–Korsakoff syndrome, 911
  whooping cough, 457–458
  xeroderma pigmentosum, 973, 974t
  Zellweger syndrome, 724
Dismutation, 590b
Dispensation sequence, 166
Displacement loop (D-loop), 976
Dissociation constant (K_d), 226, 248, 254,
  1157–1158
Distal histidine, 210, 227
Distant evolutionary relationships, 190–193
Distribution, drug, 925, 925f, 926f
Disulfide bonds
  formation of, 38–39, 39f
  reduction, 94, 95f

Dithiocarbamates, 690
Dithiothreitol, 76
Diuron, 638, **638**
Divergent evolution, 196
DNA
  A-form, 119–120, 119f, 120f, 120t
  ancient, amplification and sequencing of,
    200, 200f
  B-form, 119–120, 119f, 120t
  backbone of, 115f
  bases in, 4, **4**, 10–11, 11f, 117
  building blocks, 4–5
  chemically synthesized linker, 155
  circular, 120, 121f
  coding strand of, 987
  cognate, 292, 295–297, **296**, 296f
  cohesive ends, 154, 154f
  complementary (cDNA), 146, 159–160,
    160f, 423
  complementary single-stranded ends, 154
  covalent structure, 4, 4f
  damage to, 950, 968, 970–972
  defined, 145–146
  denaturation, 15, 15f
  directionality of, 116–117
  distortion of, 297, 297f
  double helix of, 5, 6, 10–11, 15–16, 113,
    117–119
  double-stranded, 129, 163
  electroporation and, 177, 178f
  elongation, 125, 126f
  exchange between species, 199, 199f
  extrachromosomal circular (eccDNA),
    124
  fingerprint, 148
  forensics and, 152–153, 153f
  functions of, 2
  heredity and, 6
  hybrid, 123
  information flow to RNA, 126–127, 127f
  ionic interactions, 10f
  length of, 116–117
  as linear polymer, 4, 113, 114
  linking number of, 957–958, 957f
  melting temperature of, 124
  methylation, 297–298
  microinjection of, 171–172, 171f
  nucleotides, 116
  promoter sites in, 130, 130f, 984
  proofreading of, 971, 971f
  properties of, 4–6
  proviral, 171
  recombinant. *See* recombinant DNA
    technology
  resolution by density-gradient, 123, 123f
  restriction fragment, 147–148
  stability of, 153
  stem-loop structure, 121, 121f
  in storage of genetic information, 4, 6
  structural forms, 119–120, 119f, 120t
  structure of, 4–5, 6, 119f, 121, 121f
  sugars, 4–5
  supercoiled, 120–121, 121f, 956–961
  synthesis of, 150–151, 150f
  topoisomers of, 957–958, 958f

  transferred (T-DNA), 177
  trinucleotide, 117
  unwinding of, 956–961, 984, 990–991,
    991f
  Watson–Crick model, 119, 123
  winding, 1079
  wrapping of, 1078–1079
  X-ray diffraction of, 117f
  Z-form, 119f, 120, 120t
DNA, cleavage of, 154, 154f
  by CRISPR-Cas system, 175, 176f
  EcoRV endonuclease, 292
  metal ion catalysis in, 294–295, 295f
  restriction enzymes and, 291–298
  stereochemistry, 294, 294f
DNA fragments
  covalent insertion of, 154
  insertion and replication of, 155
  insertional inactivation of, 156, 156f
  joined by DNA ligases, 154
  library of, 158–159, 158f
DNA gyrase, 960
DNA ligase, 145, 971
  defined, 954
  DNA fragments joined by, 154
  in DNA replication, 955, 967
  in forming recombinant DNA molecules,
    154–155
  reaction, 954, 954f
DNA microarrays
  defined, 170
  gene-expression analysis with, 170–171,
    171f
  measuring gene expression using, 170,
    170f
DNA photolyase, 971
DNA polymerase α, 966, 967, 967t
DNA polymerase β, 967t
DNA polymerase δ, 967, 967t
DNA polymerase I, 235, 951, **951**, 964, 967t
DNA polymerase II, 967t
DNA polymerase III, 953–954, 962–964,
  963f, 967t
DNA polymerase switching, 966–967
DNA polymerases
  common structural features, 951
  in correcting DNA mistakes, 126
  defined, 125, 951
  DNA fragment of, 952
  error-prone, 969
  eukaryotic, 966–967, 967t
  holoenzyme, 962, 963f, 964–965
  mechanism, 951, 951f
  phosphodiester-bridge formation and,
    125–126
  polymerization reaction catalyzed by, 125,
    125f
  primer, 125, 951
  prokaryotic, 967t
  replication by, 125–127
  specificity of, 952–953
  structure of, 951, **951**
  template, 951
  as template-directed enzymes, 125, 951
  types of, 967t

DNA probes
    amino acid sequences in making, 96
    defined, 148–149
    generated from protein sequences, 158, 158f
    as primer, 151
    synthesis by automated solid-phase
        methods, 150–151
DNA recombination
    in cell division, 978
    defined, 975
    functions of, 976
    Holliday junctions in, 976–978, 977f
    illustrated, 976
    initiation of, 976
    mechanism, 977, 977f
    overview of, 950
    RecA in, 976
    recombinases in, 977
    strand invasion in, 976, 976f
    V(D)J, 1131, 1131f
    VJ, 1130–1131, 1130f
DNA recombination synapse, 977
DNA repair
    base-excision, 971, 972f
    direct, 971
    double-stand, 972
    in E. coli, 970–972
    enzyme complexes in, 971–972, 971f
    example of, 969, 969f
    illustrated, 950f
    ligase in, 972
    mismatch, 971, 971f
    nonhomologous end joining (NHEJ) in, 972
    overview of, 950, 968
    proofreading in, 970, 971f
    single-strand, 962–964
    thymine in, 973–974
    trinucleotide repeats and, 973
    uracil DNA glycosylase in, 973, 973f
DNA replication, 6f, 123f, 124f, 951–968
    ATP hydrolysis in, 955, 955f
    cell cycle and, 967, 967f
    clamp loaders in, 963, 964f
    controlled termination of, 149–150, 149f
    coordinated process in, 962–968
    cross-linking in, 970, 970f
    defective, in cancer, 973–975
    defined, 117
    DNA polymerase III in, 962, 963–965, 963f
    DNA polymerases in, 952, 965–966, 967t
    in E. coli, 122–123, 123f, 962–963,
        965–966, 965f
    error rate, 126
    errors in, 969
    in eukaryotes, 966–967, 966f, 967f
    in eukaryotic cell cycle, 967, 967t
    fidelity, 980
    helicases in, 955–956, 956f
    illustrated, 982f
    lagging strand in, 954, 954f, 962–964, 963f
    leading strand in, 954, 954f, 962–964, 963f
    licensing factors in, 966
    ligase in, 954, 954f, 967
    minor-groove interactions in, 953, 953f
    Okazaki fragments in, 954–955, 954f, 963

origin of, 965–966, 965f, 966f
origin of replication complexes (ORCs),
    966–967
overview of, 950
polymerase switching in, 966–967
by polymerases, 125–127
prepriming complex in, 965, 965f
primer in, 954, 954f
processivity in, 961–962, 961f
replication fork in, 954, 963, 963f
replicon, 966
RNA polymerases in, 954, 954f
semiconservative, 122–123, 123f, 124f
shape complementarity in, 953, 953f
shape selectivity in, 953, 954f
sites of, 965–967
sliding DNA clamp in, 961–962, 961f
specificity of, 952
telomeres in, 967–968, 968f
template in, 950–951
termination of, 965, 965f
trombone model in, 963, 963f
tumor-suppressor genes in, 974
DNA sequencing, 94–95
    amino acid sequences and, 95f
    amplification by PCR, 151–152
    ancient DNA, 200, 200f
    by controlled termination of replication,
        149–150, 149f
    decreasing costs of, 19f
    defined, 145–146
    as gene exploration tool, 146
    human migrations and, 20f
    methods for, 18
    next-generation (NGS), 166–168, 167f
    pyrosequencing, 166
    reversible terminator method, 167
    semiconductor, 167–168
DNA vectors, 154, 155–157
DNA-binding domains
    defined, 1090
    in eukaryotes, 1079–1080
DNA-binding proteins
    basic-leucine zipper (bZip) in, 1080, 1080f
    binding to regulatory sites in operons,
        1060–1065
    in eukaryotes, 1079–1080
    helix-turn-helix motif in, 1059, 1059f
    homeodomains in, 1079, 1079f
    in prokaryotes, 1058–1065
    symmetry matching in, 1059
    transcription inhibition by, 1061
    transcription simulation by, 1063–1064
    zinc-finger domains in, 1080–1081, 1080f
DNA-binding sites, 1063, 1063f
    evolution of, 1063
    GAL4, 1083, 1083f
    hypersensitive, 1083
DnaA, 965, 965f
DnaB, 963
Dolichol phosphate, 357
Domains
    acetyllysine-binding, 1088, 1088f
    defined, 3
    DNA-binding, 1058–1065, 1080–1081

homeodomains, 1080, 1080f
immunoglobulin, 1124, 1126–1127
nuclear-hormone receptor, 1085, 1085f
protein, 48, 48f
Dopamine D2 receptors, 931
Double helix, 5–6
    acid-base reactions and, 15–16
    base pairs, 10
    destabilization, 15–16
    DNA molecule form, 113
    as expression of rules of chemistry, 10–11
    left-handed, 120
    minor groove, 1000
    by nucleic acid strands, 117–122
    reversibly melted, 123–124
    right-handed, 120, 956
    sequence of bases, 122
    stabilization of, 117–119
    structure, 5, **5**
    topoisomerases and, 958–961, 959f, 860f
    in transcription, 999f, 1000
    in transmission of hereditary information,
        122–125
    unwinding of, 958–959
Double helix formation
    from component strands, 6–7, 7f
    entropy and, 13f
    heat release in, 13
    illustrated, 7f
    principles of, 11
Double-blind studies, 941
Double-displacement reactions, 251, 252
Double-reciprocal plot, 248, 248f
    of competitive inhibition, 255f
    of noncompetitive inhibition, 256f
    of uncompetitive inhibition, 256f
Double-stranded DNA, 129, 163
Double-stranded RNA, 176, 176f
Downstream core promoter element (DPE),
    999
Doxycycline, 255
Dreyer, William, 1129
Drosophila melanogaster
    bases and genes, 70
    fluorescence micrograph of, 86f
    in genetic studies, 970
    genome of, 164–165, 1120
    sensory bristles in, 1113
    Toll receptor in, 1120
Drug candidates
    absorption of, 924–925, 925f
    ADME properties of, 924–930, 924f
    administration routes for, 924–925
    characteristics of, 923–930
    distribution of, 925, 925f, 926f
    effective concentrations of, 923, 923f
    excretion of, 929–930
    ligand binding in, 923–924, 923f
    metabolism of, 925–927
    number of, 933–934
    oral bioavailability of, 924, 925f
    potency of, 923–924
    side effects of, 923
    target compartments of, 925, 926f
    therapeutic index of, 929

Drug development, 921–944
  animal testing in, 933
  challenges of, 923–930
  clinical trials in, 940–942, 941f
  combinatorial chemistry in, 933
  criteria, 923–930
  genetic differences and, 939–940
  genomics in, 937–940
  high-throughput screening in, 933
  natural products in, 931–932
  overview of, 921–922
  pharmacogenetics/pharmacogenomics
      in, 939
  phases of, 940–941
  screening libraries in, 933–935, 935f
  serendipitous observations in, 929–931
  7TM receptors in, 938
  split-pool synthesis in, 934, 934f
  structure-based, 935–937, 936f, 937f
Drug targets, 921, 922f
  animal-model testing of, 938
  compartments, 926, 927f
  emerging, 939, 939f
  identified in genomes of pathogens, 939
  identified in human proteome, 937–938
Drugs
  absorption of, 924–925, 925f
  administration routes for, 924–925
  delivery with liposomes, 382
  discovery approaches, 921–922, 922f
  distribution of, 925, 925f, 926f
  excretion of, 928–929, 929f
  half-life of, 929, 929f
  metabolism of, 925–927
  natural products in, 932–933
  resistance to, 942–943
  response to genetic variation in,
      940–941
  side effects of, 923
  targets of, 921, 922f
  therapeutic index of, 930
  toxicity of, 929–930, 929f
Dynamic instability, 1163
Dynein
  AAA ATPases in, 1153, 1154
  ATP binding to, 1154–1156
  structure of, 1154, 1155f

E. coli. See Escherichia coli (E. coli)
E site, ribosomal, 1032, 1034f
Ear, hair cells of, 1112–1113,
      1112f, 1113f
EcoRV endonuclease
  cleavage of DNA, 291–292, 294, 294f
  magnesium ion-binding site in,
      294–295, 295f
  nonspecific and cognate DNA
      with, 297
  recognition site structure, 295f, 296
  twofold rotational symmetry, 295, 295f
Edman degradation, 92, 92f
EF-hand protein family, 447, 447f
Effector functions, 1124
EGF signaling, 451–454
Eicosanoids, 737, 737f, 738, 738f

Elastase
  blocking of, 327
  catalytic triad in, 281, 282f
  S_1 pockets of, 281f
Electrochemical potential. See Membrane
      potential
Electron pushing, 24
Electron transfer
  half-reactions in, 579
  inhibitors, 607–608, 607f
  in oxidative phosphorylation, 576–579,
      591–592, 591f
  in photosynthesis, 622–626
  rate of, 591, 591f
  uncoupling, 606–607
Electron transport chain, 543
  into respirasome, 589, 589f, 590f
Electron-density maps, 101–102, 102f
Electron-proton transfer reactions, 583f
Electron-transferring flavoprotein (ETF),
      717
Electron-transport chain
  components of, 580f, 580t
  in desaturation of fatty acids, 736–737,
      736f
  electron flow through, 573, 574f
  FAD (flavin adenine dinucleotide) and,
      543, 557
  $FADH_2$ (flavin adenine dinucleotide
      reduced) and, 576, 577
  inhibition of, 607
  iron-sulfur clusters in, 581–582, 582f
  NADH (nicotinamide adenine
      dinucleotide reduced) and, 543, 556,
      576, 577, 578
  in oxidative phosphorylation, 573, 579
Electron-transport potential, 576
Electrophilic catalyst, 762
Electrophoresis
  capillary, 149–150
  gel, 75–77, 76f, 77f, 148–149, 148f, 384
  SDS, 76–77, 76f
  two-dimensional, 78, 78f
Electroporation, 177, 178f
Electrospray ionization (ESI), 90
Electrostatic repulsion, 469
Elongation factors
  bacterial versus eukaryotic, 1042
  G (EF-G), 1038–1039, 1039f
  Ts (EF-Ts), 1036–1037
  Tu (EF-Tu), 1036, 1036f, 1037
Embden–Meyerhof pathway, 492
Emphysema, 327
Enantiomers, 343, 346f
Encoding of amino acids, 132–135
Endergonic reaction, 237
Endocytosis, 866
Endoplasmic reticulum (ER), 395
  membrane, 736
  protein glycosylation in, 357–358, 357f
  protein synthesis in, 1044–1049
  protein targeting from, 1045–1048
  ribosome binding to, 1045–1048
  rough, 1046
  smooth, 1046

Endoplasmic reticulum (ER) stress, 900
Endosomes
  defined, 396, 866
  LDL release in, 868, 868f
Endosymbiosis
  mitochondria evolution from, 395
  receptor-mediated, 395–396, 395f
Endosymbiotic event, 575–576
Endotoxin, 1121
Enediol intermediate in glycolysis, 498, 499f
Energy. See also Thermodynamics
  activation, 240–241
  binding, 243, 274
  free. See free energy
  kinetic, 11
  light, 638–639
  potential, 11
  resting human being requirement, 599b
  status, 484
  total, 11
  units of, 238b
Energy charge, 482–483, 485, 485f
  intracellular, 687
  in regulation, 604
Energy coupling agent, 500
Energy homeostasis. See Caloric
      homeostasis
Energy transfer
  from accessory pigments to reaction
      centers, 636, 637f
  resonance, 636–637, 637f
Engrailed, 86f
Enhancers
  defined, 1001, 1082
  experimental demonstration of, 1082f
  in gene expression, 1082
  probe for, 1082f
  in transcription, 1001
Enol intermediates, 550
Enol phosphate, 503
Enolase, 503
Enoyl CoA hydratase, 717
Enoyl reductase, 731
Ensemble studies, 261
Enterohepatic cycling, 928, 928f
Enthalpy, 12
Entropy
  change during chemical reaction, 12
  double helix formation and, 13f
  overall, 13
  total, 11–12
Envelope form, 347, 347f
Environmental factors in biochemistry,
      20–21
Enzymatic cascades, 328–329, 328f
Enzymatic velocity, 249
Enzyme inhibition
  affinity labels, 256–257, 257f
  competitive, 253
  by diisopropylphosphofluoridate (DIPF),
      256, 256f
  illustrated, 256f
  irreversible, 253, 256–258
  mechanism-based (suicide), 258, 259f
  noncompetitive, 254

reversible, 253, 253f
by specific molecules, 253
by transition-state analogs, 260, 260f
transition-state analogs in, 253
uncompetitive, 253
Enzyme kinetics
for allosteric enzyme, 252–253, 252f
chymotrypsin, 277–278, 277f
of competitive inhibitors, 254, 254f
defined, 244
in first-order reactions, 244
Lineweaver–Burk plot, 248, 248f
maximal rate, 247–248, 249
Michaelis constant in, 246, 247–249, 248t
Michaelis–Menten, 248, 252–253, 254, 277
Michaelis–Menten model for, 244–253
of noncompetitive inhibitors, 254, 255f
in pseudo-first-order reactions, 244
in second-order reactions, 244
sigmoidal, 311, 311f
specificity constant, 249–250
steady-state assumption and, 245–247
of uncompetitive inhibitors, 254, 254f
of water deprotonation, 290, 290f
Enzyme multiplicity, 806
Enzyme regulation
by covalent modification, 318–323
of protein kinase A (PKA), 321–322, 322f
strategies, 309–311
Enzyme-linked immunosorbent assay (ELISA)
defined, 86
indirect, 86–87, 87f
sandwich, 87, 87f
Enzyme–pyridoxamine phosphate complex (E-PMP), 761
Enzyme-substrate (ES) binding
in catalysis, 243
conformation selection, 243
induced-fit model of, 243, 243f
lock-and-key model of, 243, 243f
multiple weak attractions, 242
Enzyme-substrate (ES) complex
defined, 241
diffusion-controlled encounter, 250
dissociation constant for, 248
evidence of existence, 241
fates, 245
formation in catalysis, 241, 241f
hydrogen bonds in, 242, 243f
reformed, 245
structure of, **242f**
Enzymes, 30, 233–264
acceleration of reactions, 240–244
activation energy and, 240, 240f, 241
active sites, 234, 241, 242, 242f, 249, 274
activity of, 71
allosteric, 252, 252f, 311
amount control, 310
assay measurement of, 71
ATP hydrolysis and, 299
binding energy between substrate and, 243–244
branching, 694–695, 694f

catalytic activity, control of, 483–485
catalytic efficiency, 249–250, 250t
catalytic power, 234–236
catalytic reaction, 241, 241f
in cellular respiration, 236
classes of, 264, 265t
cofactors for, 235–236, 236t
debranching, 701f
defined, 492b
digestive, 324
energy-transducing, 236
ensemble studies, 262f
Gibbs free energy and, 236–239
of glycolysis, 493
homologous, 767f
inhibition by molecules, 253–261
isozymes, 310, 317–318, 318f
kinetic properties of, 244–253
kinetically perfect, 499
in lipid metabolism, 560–561
metabolism control of, 483
multiple forms of, 310
overview of, 441
P-loop structures, 303–304
peptide-cleaving, 283–285
in photosynthesis, 236
processive, 960b
as proteins, 234
proteolytic, 235, 310, 323–334
of purine nucleotide synthesis, 830
pyridoxal phosphate, 762–763, 762f
rate enhancement by, 234, 234t
reaction rate and, 239, 239f
restriction, 147, 291–298
reversible covalent modification, 310
from RNA, 1017
serpins degradation by, 328
single molecule studies of, 262, 262f
specificity, 234–236, 235f
spectroscopic characteristics of, 241
substituted intermediate, 252
substrate interaction, 242
ternary complex of, 251
turnover number of, 249, 249f
urea cycle, 766–767
waste degradation by, 261
Epidermal growth factor (EGF)
defined, 451
Ras activation and, 453–454, 453f
in signal transduction, 437, 451–454
signaling pathway, 453f, 454
structure of, 451–452, 451f
Epidermal growth factor receptor (EGFR)
defined, 451–452
dimerization of, 452, 452f
modulator structure of, 451–452, 452f
overexpression of, 456
phosphorylation of carboxyl-terminal tail, 452
unactivated, structure of, 452f, 453
Epigenome, 1076
Epimers, 343f, 344
Epinephrine
defined, 439
in fatty acid metabolism, 739–742

in glycogen metabolism, 690–692, 696–698, 696f
in signal transduction, 439–447
synthesis of, 807f, 808
Epitopes, 83
Equatorial bonds, 346–347, 347f
Equilibrium constant, 14
defined, 237
dissociation of water, 14
between enzyme-bound reactants and products, 302
metabolism and, 468
rate constants, 239
of reactions, 238–239
Equilibrium potential, 424–426, 425f
Equilibrium reaction, 239
Erbitux, 945
Error-prone polymerases, 969
Erythropoietin (EPO), 353–354, 354f
Erythrose 4-phosphate, 662, 662f
Escherichia coli (*E. coli*)
amino-terminal protein residues, 1034
DNA polymerase I, 951, **951**
DNA recombination in, 975
DNA repair in, 970–972
DNA replication in, 122–123, 123f, 962, 964–965, 964f, 965f
enterotoxigenic, 855
fatty acid synthesis in, 731
flagella, 1152, 1166–1167
gene expression in, 1057–1060
genome of, 116, 116f
lac operon of, 1061–1062, 1062f, 1063f
membranes, 393
methionine repressor, 1060, 1060f
promoter sequences, 989
pyruvate dehydrogenase complex of, 544t, 546f
RecA protein, 1066
recombinant systems and, 82
restriction enzymes in, 291–292
ribonucleotide reductase of, 831
RNA in, 127, 127t
RNA polymerases in, 131, 131f, 985–988
semiconservative replication of, 123f
two-dimensional electrophoresis and, 78
Essential amino acids, 779
defined, 752, 793
pathway steps, 793–794, 794f
table of, 752t
Estradiol, 876, 877f, 1086
Estrogen response elements (EREs), 1085
Estrogens
defined, 872
DNA-binding receptors and, 1084
pathways for formation of, 877f
synthesis of, 875–877, **878**
Estrone, 876
Ethanol
consumption, 743
fermentation of, 505–506, 506f
in glycolysis, 491
metabolism, 743, 909–912
Ethylene, 800, **801**
Eukarya, 4f

Eukarya domain, 3
Eukaryotes
    defining characteristics, 3
    DNA replication in, 966–967, 966f, 967f
    fatty acid synthesis in, 731
    glycolysis in, 493
    membranes, 393–397
    nuclear envelope, 395, 395f
    protein synthesis inhibition in, 1043
    transcription factors in, 998
    transcription in, 996–1001
    translation in, 996, 996f
    translation initiation in, 1041–1042, 1042f
    tubulins in, 1163
Eukaryotic cell cycle, 967, 967f
Eukaryotic genes
    as discontinuous, 135
    mosaic nature of, 135
    organization of, 1076
    quantitation and manipulation of, 169–178
Eukaryotic promoter sites, 130f
Eukaryotic RNA polymerases, 984, **984**
Eumelanin, 816, 816f
Eversion, 409
Evolution
    of active sites, 274
    of amino acid sequences, 39
    of aminoacyl-tRNA synthetases, 1030
    ancient DNA amplification and
        sequencing in, 200, 200f
    of blood types, 359
    caloric homeostasis and, 890–891
    of carbonic anhydrases, 291
    of chloroplasts, 622
    of citric acid cycle, 563–564
    convergent, 196–197, 197f
    of cytochrome c, 592, 592f
    divergent, 196
    of DNA-binding sites, 1063
    of drug resistance, 942–943
    of globins, 198, 198f
    of glycogen metabolism, 692
    of glycogen phosphorylase, 692
    of immune system, 1119–1120
    of introns and exons, 136
    in laboratory, 201, 201f
    of metabolism, 485–486
    molecular, 200–202
    obesity and, 890–891
    of oxygen, 628b
    of photoreceptors, 1110–1111, 1110f
    of proteasomes, 756, 757f
    of proteins, 136
    of ribosomes, 1031
    of signal transduction pathways, 454
    split genes in, 136–137
    of vitamins, 478–479
Evolutionary relationships, 185, 186
    distant, 190–193
    three-dimensional structures and,
        194–198
Evolutionary trees, 198–200, 198f
Excinuclease, 972
Excretion, drug, 927–928, 928f
Exemestane, 876

Exercise
    ATP in, 471, 471f, 518, 902–905, 903t
    citric acid cycle in, 561, 561f
    creatine phosphate in, 902–903, 903t
    diabetes and, 900–901
    effects on muscle and whole-body
        metabolism, 915
    fatty acid metabolism in, 902, 902f
    fuel sources for, 902–905, 903f, 904f
    glycolysis during, 519
    mitochondrial biogenesis and, 902, 902f
    molecular adaptation to, 915, 915f
    oxidative phosphorylation in, 903–905
    oxygen concentration and, 213f
    in proteins phosphorylation, 323
    super compensation (carbo-loading) and,
        905
Exons
    defined, 114
    in encoding protein domains, 136–137
    shuffling, 137, 137f
Exonuclease, 951
Expression cloning, 160
Expression vectors, 156, 156f
Expressome, 1040
Extracellular signal-regulated kinases
        (ERKs), 453
Extrachromosomal circular DNA
        (eccDNA), 124
Extrinsic pathway, clotting, 328

2-^{18}F-2-D-deoxyglucose (FDG), 517, 517f
F-actin, 1157
F$_{ab}$ fragments, 1124–1126
Facilitated diffusion, 403, 405
Facultative anaerobes, 508
FAD (flavin adenine dinucleotide)
    as catalytic cofactor, 544
    defined, 475
    electron transfer potential of, 546
    structure of oxidized form, 475f, 476f
    structure of reactive components of, 476,
        476f
FADH$_2$ (flavin adenine dinucleotide
        reduced), 717–718
    defined, 476
    electron-transport chain and, 543, 557,
        576, 577
    in oxidation-reduction reactions, 481, 481f
    structure of reactive components of, 476,
        476f
Falciparum malaria, 670
Familial hypercholesterolemia, 849, 867–868
Faraday constant, 406, 577
Farnesene, 880, **880**, 880t
Farnesyl pyrophosphate, 860, **862**
Fast switch, 506
Fasting, 907
Fat
    in adipose tissue, 904
    body, energy storage in, 711
    brown, 605
    as building block, 20
    neutral. See triacylglycerols
Fat cells, 711

Fatty acid chains
    melting temperature, 391–392, 392t
    packing of, 392, 392f
Fatty acid metabolism, 710–785
    acetyl carnitine in, 716–717
    acetyl CoA carboxylase in, 739–741, 739f
    acetyl CoA in, 558, 717–718, 733, 734f,
        735
    acetyl CoA synthase in, 717
    activation in, 715–716
    acyl carrier protein (ACP) in, 728–729,
        729, 729f, 731–732, 733f
    β-oxidation pathway in, 717
    chylomicrons in, 713, 713f
    citrate in, 733–734, 740f
    coenzyme B$_{12}$ in, 720–721, 722f
    degradation in, 710, 711f, 717f, 724f,
        728–736
    diet and, 740
    enoyl CoA hydratase in, 717–718, 719f
    epinephrine in, 740
    in exercise, 901–902, 902f
    FADH$_2$ in, 717–718
    glucagon in, 740
    insulin in, 740
    ketone bodies in, 724–727
    lipases in, 713–714
    lipid mobilization and transport in,
        712–713, 713f
    lipolysis in, 714, 715f
    malonyl CoA in, 729, 739
    methylmalonyl CoA in, 721–723, 722f, **723**
    NADH in, 717–718
    NADPH sources for, 734
    overview of, 709–710
    oxidation in, 715–716, 717–718, 718t,
        723–724
    palmitoyl CoA in, 719, 719f, 739
    pancreatic lipases in, 712, 712f
    pathway integration, 735, 735f
    peroxisomes in, 723–724, 724f
    steps in, 710, 710f
    triacylglycerols in, 710–712, 714f
    in tumor cells, 735
Fatty acid methyl esters (FAME), 180
Fatty acid synthase
    catalytic cycle of, 732, 733f
    defined, 728, 729–731
    structure of, 731, 732f
Fatty acid synthesis
    in animals, 731–732
    in cytoplasm, 733–734, 734f
    defined, 710
    degradation versus, 728–729
    by fatty acid synthase, 728–736
    intermediates in, 742
    malonyl CoA in, 729
    maximal, 739
    reactions in, 729–731
    steps in, 711f, 730–731, 730f
Fatty acid thiokinase, 715
Fatty acids, 61
    animals and, 727–728
    β oxidation of, 908
    chain length in, 375–376, 731

cholesterol and, 392, 392f
covalent attachment of, 710
defined, 374
degree of unsaturation, 375, 392
derivatives, 709
desaturation of, 736–737, 736f
as fuel, 709, 713–719
generation by lipolysis, 715, 715f
in lipids, 374–375, 375f
in muscle, 899, 904
naturally occurring, 376t
in obesity, 898
odd-chain, 720–721, 721f
oxidation, 743
pathological conditions by, 724
physiological roles of, 709
polyunsaturated, 724, 737
saturated, 724
structures of, 375, 375f
unsaturated, 719–728, 719f, 720f,
    736–738
Fatty liver, 743, 910
Favism, 669
F_c fragment, 1124, 1125
Feedforward stimulation, 513
Feedback inhibition, 310, 311, 483
in amino acid synthesis, 805–806
cumulative, 806–807
enzyme multiplicity and, 806, 806f
in purine synthesis control, 837–838, 837f
FeMo cofactor, 790, 790f
Fermentation
    to alcohol, 491, 505–507, 506f
    as anaerobic process, 507
    defined, 505
    in glycolysis, 505–507
    to lactic acid, 491, 506–507, 507f
    start and end points of, 508t
Ferredoxin, 634, 790
Ferredoxin-NADP$^+$, 630–631, 630f
Ferredoxin–thioredoxin reductase, 633, 656
Ferritin, 813, 1090, 1090f, 1091
Ferrochelatase, 813
Fetal hemoglobin, 217, 217f
Fiber types, muscle, 688, 688t
Fibrils, 351
Fibrin, 330, 331f, 333
Fibrin clots, 331f, 332
Fibrin monomer, 330
Fibrinogen
    conversion by thrombin, 329–331
    structure of, 330, 330f
Fibrinopeptides, 329–331
Fibrous proteins, 49–50
50S subunit of ribosomes, 1031, 1032f,
    1037, 1039
Fight-or-flight response, 439
Fingerprint, DNA, 148
First Law of Thermodynamics, 11
First-order reactions, 244
First-pass metabolism, 927
Fischer, Emil, 243, 935
Fischer projections, 23
Five-carbon units, 878–880
5′ cap, 1002–1004, 1005f

Flagella
    microtubules in, 1163
    rotary movement of, 1167–1168
    structure of, 1167, 1167f
Flagellar motor
    chemotaxis and, 1168–1170, 1169f
    components of, 1167, 1168f
    rotation of, 1166–1170
    schematic view of, 1167f
Flagellin, 1166–1167, 1167f
Flavin adenine dinucleotide (FAD), 971
Flavin mononucleotide (FMN), 476, 476f,
    582, 992, 992f
Flavoproteins, 546
Fleming, Alexander, 930
FliG, 1167
FliM, 1167
FliN, 1167
Flip-flop, 391, 391f
Florey, Howard, 930
Fluconazole, 926f
Fluid mosaic model, 391
Fluorescence microscopy, 86f, 89, 89f
Fluorescence recovery after photobleaching
    (FRAP), 390, 390f
Fluorodeoxyuridylate (F-dUMP), 535
Fluorouracil, 835
Folding funnel, 57, 57f
Foodstuffs, energy extraction from,
    474–475, 475f
Formylmethionyl-tRNA, 1035, 1035f
N-formylmethionine (fMet), 134, 1035,
    1036
Formyltetrahydrofolate, 828, 828f
43S preinitiation complex (PIC), 1041
Fourier transform, 101
Foxglove, 410, 410f
Fractional saturation, 212
Fragile X mental retardation protein
    (FMRP), 1051
François, Jacob, 1061
Franklin, Rosalind, 5, 117
Free energy
    of activation, 237
    ATP and, 466–472
    change during chemical reaction, 237–239
    defined, 12
    Gibbs, 12, 236
    in membrane transport, 405f, 406
    negative, 13
    of oxidation of single-carbon compounds,
        472, 472f
    in oxidation-reduction reactions, 577
    of phosphorylated compounds, 471t
    of phosphorylation, 320
    standard change, 237–239
    thermodynamics of metabolism and,
        465–466
Free radicals. See Reactive oxygen species
    (ROS)
Friedreich's ataxia, 582
Fructofuranose, 346, 346f
Fructokinase, 508
Fructopyranose, 346, 346f
Fructose, 344, 344f

excessive consumption of, 508
in glycolysis, 508, 508f
metabolism, 508, 508f
ring structures, 346, 346f
Fructose 1,6-bisphosphate
in gluconeogenesis, 520f
in glycolysis, 494f, 496–497, 496f
Fructose 1-phosphate in glycolysis, 508
Fructose 2,6-bisphosphate (F-2,6-BP)
activation of phosphofructokinase by,
    514, 515f
in gluconeogenesis, 526–527, 526f
in glycolysis, 515, 515f
regulation of phosphofructokinase by,
    513, 514f
Fructose 6-phosphate
in Calvin cycle, 651, 651f, 654, 655f
in gluconeogenesis, 520f
in glycolysis, 494f, 495, 513
in pentose phosphate pathway, 662, 662f
phosphorylation of, 496
Fructose bisphosphatase 2 (FBPase2), 526
FtsZ, 1164
Fumarase, 554
Fumarate
from phenylalanine, 774, 774f
in purine synthesis, 829
in urea cycle, 765, 765f
4-fumarylacetoacetate, 774, 775
Functional groups
defined, 24
of proteins, 30
types, 24
Functional magnetic resonance imaging
    (fMRI), 210, 210f
Funny current (I_f), 431
Furan, 345
Furanose, 345, 345f, 346, 347
Futile cycles, 528, 529

G elongation factor (EF-G), 1038–1039,
    1039f
G proteins
activation of, 440–441, 441f
ATP synthase and, 599
defined, 440
in glycogen metabolism, 690–692
heterotrimeric, 440–441, 441f
resetting of, 443f, 444
role in signaling pathways, 441
small, 453
subunits, 441
G-actin, 1157
G-protein receptor kinase 2 (GRK2), 444
G-protein-coupled receptors (GPCRs), 441
Gain, 692
GAL4 binding sites, 1083, 1083f
Galactitol, 511
Galactokinase, 509
Galactolipids, 621
Galactose, 344, 345f
defined, 509
in glycolysis, 508, 508f
missing transferase and, 511
ring form, 346f

Galactose 1-phosphate uridyl transferase, 509
Galactosemia, 511
β-galactosidase, 156, 360, 1060, 1061f
β-galactosides, 1062
Galdieria sulphuraria, 199, 199f
Gangliosides, 378
    disorders of, 856
    structure of, 855f, 856
    synthesis of, 855, 855f
Gap junctions, 427–428, 427f, 428f
Gas-phase ions, 90
Gasotransmitters, 459
Gastroesophageal reflux disease (GERD), 233–234
Gastrointestinal peptides, 892–893, 893t
GC box, 999, 999f
GDP (guanosine diphosphate)
    in citric acid cycle, 552
    hydrolysis of, 467
    in olfaction, 1099
    in signal transduction, 441, 452, 453, 453f
    in translation, 1036
    in tubulin, 1163
    in vision, 1109
Gel electrophoresis
    defined, 75
    isoelectric focusing and, 77, 77f
    polyacrylamide, 76f
    protein separation by, 75–79
    restriction fragments separation by, 148–149
    SDS, 76–77, 76f
    SDS-PAGE, 77, 78
    two-dimensional electrophoresis and, 78, 78f
Gel-filtration chromatography, 73, 73f, 79
Gellert, Martin, 958
Gene disruption, 172–176
    consequences of, 172–173, 173f
    defined, 172
    by homologous recombination, 172, 172f
Gene expression
    analysis with microarrays, 170–171, 171f
    bacterial, 1040
    constitutive, 1058
    control of, 1051, 1058
    as DNA information transformation, 127–132
    levels of, 169–171
    posttranslational control of, 1070
    regulated, 1058
    RNA in, 127–128
    transcriptional, 1040, 1058
    transcriptional, in eukaryotes, 1090–1093
    transcriptional, in prokaryotes, 1068–1070
    translational, 1040, 1090–1092
Gene expression (eukaryotes), 1076–1079
    activation domains in, 1080, 1081
    basic-leucine zipper (bZip) in, 1080, 1080f
    chromatin in, 1077–1079, 1087–1089
    coactivators in, 1085–1087
    combinatorial control in, 1081
    DNA methylation and, 1084

DNA-binding structures in, 1079–1080
    enhancers in, 1082
    histones in, 1077–1079
    homeodomain in, 1080, 1080f
    hypomethylation in, 1084
    in liver, 1076, 1077t
    mediator in, 1081, 1081f
    nuclear hormone receptors in, 1085–1086, 1085f
    nucleosomes in, 1077–1078, 1078f
    overview of, 1076–1077
    in pancreas, 1076, 1077t
    posttranscriptional, 1090–1093
    versus in prokaryotes, 1076–1078
    steroid-hormone receptors in, 1086–1087, 1087f
    tissue-specific, 1077t
    transcription factors in, 1079–1081
    zinc-finger domains in, 1080–1081, 1080f
Gene expression (prokaryotes), 1058–1060
    attenuation in, 1068–1070, 1069f
    autoinducers in, 1067, 1067f
    biofilms and, 1067–1068
    catabolite activator protein (CAP) in, 1064–1065, 1064f
    catabolite repression in, 1064
    chromatin remodeling in, 1083–1090, 1089f
    corepressors in, 1063
    DNA-binding proteins in, 1058–1060
    DNA-binding sites in, 1063, 1064f
    versus in eukaryotes, 1076–1077
    helix-turn-helix motif, 1059, 1059f
    lac operon in, 1061–1062, 1061f, 1062f
    lac repressor in, 1059, 1061–1062, 1062f, 1063f
    λ repressor in, 1065–1066, 1066f
    ligand binding in, 1062
    operon model, 1061–1062
    overview of, 1058
    posttranscriptional, 1068–1070
    pur repressor in, 1063, 1064f
    quorum sensing in, 1066, 1067f
    regulatory sites in, 1058–1059
    signaling in, 1066–1067
    social interactions in, 1066
    symmetry matching in, 1059–1060
    transcriptional, 1058–1060
"Gene guns," 178
Gene knockdown, 177
Gene knockout. *See* Gene disruption
Gene therapy, 178
Gene transfer, horizontal, 199, 199f
General acid-base catalysis, 275
Genes. *See also specific genes*
    cloned, 74–75
    comparative analysis of, 168–169
    defined, 21
    designer, 162–163
    horizontal transfer, 199, 199f, 298
    number of, in human genome, 165
    protein-coding, 21–22
    regulator, 1061
    reporter, 155
    RNA and, 126–127

split, 136
    structural, 1061
    synthesis by automated solid-phase methods, 150–151
    tumor suppressor, 973
Genetic code
    amino acids, 133, 133t
    aminoacyl-tRNA synthetases and, 1026–1030
    defined, 22, 132
    as degenerate, 133
    directionality of, 132
    features of, 133–134
    as nonoverlapping, 132
    universality of, 135
Genetically modified organisms (GMOs), 178
Genomes
    analysis of, 166–169
    of *Arabidopsis thaliana*, 164–165
    of *Caenorhabditis elegans*, 164
    of chloroplasts, 622
    comparative, 20, 168–169, 168f
    complete, 164f
    defined, 70
    of *Drosophila melanogaster*, 164–165, 1120
    of *E. coli*, 116, 116f
    editing, 173, 174f
    of *Haemophilus influenzae*, 164, 194
    human, 165, 165f, 168, 168f
    of Indian muntjac, 116, 117f
    long interspersed elements, 165
    of mitochondria, 575–576, 575f, 576f
    next-generation sequencing, 166–168, 167f
    noncoding DNA in, 165
    number of genes in, 165
    protein encoding, 21–22
    proteome as functional representation, 70
    of puffer fish, 168–169, 168f
    of *Reclinomonas americana*, 576
    of *Rickettsia prowazekii*, 576
    of *Saccharomyces cerevisiae*, 164, 1076
    sequencing of, 18–20, 163–169
    size of, 116
Genomic library
    creation of, 158, 158f
    defined, 158, 159
    screening of, 159, 159f
Genomic variation, 19
Genomics
    comparison, 20
    in drug discovery, 937–940
    revolution, 17–22
Geranyl pyrophosphate, 860, 860f
Geranyl transferase, 860, **862**
Ghrelin, 893, 893t
Gibbs, Josiah Willard, 12
Gibbs free energy, 12, 236–239, 240
Gibbs free energy of activation, 240
Gigaseal, 416
Gla domain in prothrombin, 329
GlcNAcase, 354
Gleevec (imatinib mesylate), 938
Globin fold, 211

Globins. *See also* Hemoglobin; Myoglobin
　defined, 187
　evolution of, 198–199, 198f
　hexacoordinate, 227
　in human genome, 223
　sequence alignment in, 187–188
Globular proteins, 47
Glomeruli, 927
Glucagon
　in essential enzyme regulation, 527
　excess in type 1 diabetes, 901
　in fatty acid metabolism, 739–740
Glucagon-like peptide 1 (GLP-1), 893, 893t
Glucocorticoids, 872
Glucogenic, amino acids, 769–775
Glucokinase, 515, 905
Gluconeogenesis, 519–524, 533
　defined, 491, 519
　energy charge and, 521, 525–526
　fasting and, 907
　fructose 1,6-bisphosphate in, 520f
　fructose 2,6-bisphosphate in, 526–527,
　　526f
　fructose 6-phosphate in, 520f
　glucose 6-phosphate in, 520f, 523, 523f
　glucose generation and, 523
　glycolysis and, 519, 525–530
　historical perspective on, 669
　in kidney, 521
　in liver, 519, 526–527, 908
　in muscle contraction, 528–530, 529f
　overview of, 491–492, 519
　oxaloacetate in, 520f, 521–522, 908
　pathway integration, 529, 529f
　pathway of, 519, 520f
　phosphoenolpyruvate (PEP) in, 520f
　pyruvate conversion in, 521–522, 522f
　reactions of, 524t
　reciprocal regulation in, 525–530
　regulation of, 525–530, 901, 901f
　stoichiometry of, 524
　substrate cycles in, 528, 528f
　urea cycle in, 766
Glucose, 343–344, 344f
　in adipose tissue, 906
　as cellular fuel, 493
　characteristics of, 493
　complete oxidation of, 603, 604t
　conversion into pyruvate, 504
　in diabetes, 896
　as essential energy source, 343
　formation of, 493
　generation from glucose 6-phosphate,
　　523, 523f
　generation of, 492–493
　in glycogen, 350, 350f, 679, 680–681
　historical perspective on, 639
　homopolymers of, 350–351
　in liver-glycogen metabolism, 698–699
　metabolism, 464f
　as reducing sugar, 347–348
　ring form, 346f
　storage forms of, 350, 350f
　synthesis from noncarbohydrate
　　precursors, 519–524

Glucose 1-phosphate in glycogen
　　metabolism, 681–682, 685
Glucose 6-phosphate (G-6P), 349
　galactose conversion into, 509–510, 509f
　in gluconeogenesis, 520f, 523, 523f, 685
　in glycolysis, 494f
　homeostasis, 905
　isomerization of, 496–497, 496f, 497f
　in liver, 685–686, 700–701
　metabolism of, 665–668, 680, 680f
　in pentose phosphate pathway, 659, 661f,
　　665–668
　transporter, 700
Glucose 6-phosphate dehydrogenase (G6PD)
　deficiency, 669–670
　in pentose phosphate pathway, 668, 673
　in reactive oxygen species protection,
　　668–670
Glucose homeostasis
　blood, 365–366
　in diabetes, 896–901
　in starvation, 907–909
　starved-fed cycle in, 905–907
Glucose transporters (GLUTs), 404
　families, 516–517, 516t
　in glycolysis, 516–517, 516t
Glucose–alanine cycle, 763
Glucose-stimulated insulin secretion (GSIS),
　　899–900
α-1,6-glucosidase, 701
Glutamate, **252**, 770
　chirality of, 792f
　formation of, 758, 762, 794
　in nitrogen fixation, 791–793
　oxidative deamination of, 758–759
　structure of, 35, 36f
　synthesis of, 796, 797f
Glutamate dehydrogenase, 759, 791–793
Glutamate synthase, 792
Glutamic acid, 35
Glutamine, 791–793
　degradation, 772
　in nitrogen transport, 764
　in pyrimidine synthesis, 823
　structure of, 34f, 35
　synthesis of, 796
Glutamine phosphoribosyl
　　amidotransferase, 827
Glutamine synthetase, 791, 792, 807
　regulation of, 807–808, 807f
　structure of, 807–808
Glutathione, 669
　conjugation reaction, 927, **928**
　defined, 809
　structure of, **809**
Glutathione peroxidase, 669, 809, **809f**
Glutathione reductase, 669, 809
Glycan-binding proteins, 361
Glyceraldehyde, 342, **342**, 343, 343f
Glyceraldehyde 3-phosphate (GAP), 348
　in Calvin cycle, 651
　catalytic mechanism of, 501, 502f
　free-energy profiles for, 500, 501f
　in glycolysis, 494f, 497–498, 498f,
　　499–500, 501f, 502f

　oxidation of, 472–473, 500–501
　in pentose phosphate pathway, 660, 661,
　　662f
　structure of, **501f**
Glycerol
　generation by lipolysis, 715, 715f
　in gluconeogenesis, 519–521
　in liver, 714
　in phosphoglycerides, 377
Glycerol 3-phosphate
　in lipid synthesis, 850
　reoxidation of, 600
Glycerol 3-phosphate shuttle, 600–601, 600f
Glycerol phosphate acyltransferase, 850
Glycine
　in Calvin cycle, 649
　conversion into serine, 771
　defined, 32
　in liposomes, 381, 381f
　porphyrins from, 810–813, 812f
　structure of, 33f
　synthesis of, 796, 798f
Glycine synthase, 798
Glycine-glycine-glutamine (GGQ)
　sequence, 1040
Glycoaminoglycans, 352
Glycobiology, 342
Glycocholate, 712f
Glycoforms, 352
Glycogen, 360
　branch point in, 350, 350f
　defined, 350, 679
　glucose storage in, 679, 680–681
　in glycogen metabolism, 359–360, 691f,
　　696
　in liver, 680, 680f, 690
　molecules, 679, 680f, 695f
　in muscle, 680, 680f
　phosphorolytic cleavage of, 682–684
　remodeling, 684, 685f
　storage efficiency, 695
　storage sites of, 680, 680f
　structure of, 680f
Glycogen lakes, 701f
Glycogen metabolism, 679–704
　branching enzyme in, 694–695, 694f
　cAMP in, 691–692
　coordinate control of, 696, 696f
　degradation of, 680
　enzyme breakdown, 681–686
　epinephrine in, 690–692, 691f, 696
　evolution of, 692
　G proteins in, 690–692
　glycogen in, 690–692, 691f, 696
　glycogen phosphorylase in, 681–682
　glycogen synthase in, 693–694, 694f
　glycogen synthase kinase (GSK) in, 695,
　　696, 696f
　glycogenin in, 694
　α-1,6-glucosidase in, 685–686
　hormones in, 690–692, 691f
　insulin in, 698, 698f
　in liver, 698–700, 699f
　overview of, 679–680
　pathway integration, 691f

Glycogen metabolism (*Continued*)
  phosphoglucomutase in, 685, 685f
  phosphoglycerate mutase in, 685
  phosphorolysis in, 681–684
  phosphorylase in, 681–686, 684f, 699, 699f
  protein kinase A (PKA) in, 692, 696, 696f
  protein phosphatase 1 (PP1) in, 692, 696–698, 697f, 699
  pyridoxal phosphate (PLP) in, 683, 683f
  regulation of, 680–681, 695–696
  regulatory cascade for, 690–692, 691f
  7TM receptors in, 690–692, 691f
  termination of, 692
  transferase in, 684, 684f
  UDP-glucose and, 693–694
Glycogen phosphorylase
  a, 686–688, 686f, 689f
  allosteric regulation of, 687, 687f, 688t
  amino-terminal domain of, 682–683, 683f
  b, 687, 687f, 688, 690
  carboxyl-terminal domain of, 682–684, 683f
  defined, 685
  evolution of, 692
  in glycogen metabolism, 681–686
  glycogen-binding site of, 682–683, 683f
  isozymic form, 690
  in liver, 686–687, 697f
  McArdle disease due to skeletal muscle, 702–703
  in muscle, 687, 687f
  regulation of, 686–688, 686f, 688t, 692
  structure of, 682
Glycogen synthase
  α-1,6 linkages, 694–695, 694f
  defined, 693
  in glycogen synthesis, 695
  isozymes of, 693–694
Glycogen synthase kinase (GSK)
  defined, 695
  in glycogen metabolism, 695f, 696
Glycogen-storage diseases, 700–701, 700t, 701f
Glycogenin, 694
Glycolate, 649–650, 650f
Glycolipids, 376, 378, 380–381
N-glycolylneuraminate, 855
Glycolysis
  aerobic, 517–518, 557
  aldehyde oxidation in, 499–500
  anaerobic mode, 557, 903
  ATP formation in, 501–503
  1,3-bisphosphoglycerate (1,3-BPG) in, 473, 499–500, 502f
  in cancer, 517–518
  citric acid cycle and, 543–544, 543f
  covalent enzyme-bound intermediate in, 500
  defined, 491
  dihydroxyacetone phosphate (DHAP) in, 494f, 497–498, 498f, 499
  as Embden–Meyerhof pathway, 492
  enediol intermediate in, 498, 498f
  energy charge and, 525–526
  as energy-conversion pathway, 493–511
  enzymes of, 493

  in eukaryotic cells, 493
  fermentation in, 505–508
  first stage of, 493, 494f, 495f
  fructose 1,6-bisphosphate in, 494f, 496–497, 497f
  fructose 1-phosphate in, 508
  fructose 2,6-bisphosphate in, 514, 515f
  fructose 6-phosphate in, 494f, 496, 497f, 513
  fructose in, 508, 508f
  galactose in, 509–510, 509f
  gluconeogenesis and, 519–521, 525–530
  glucose 6-phosphate (G-6P) in, 494f
  glucose transporters in, 516–517, 516t
  glucose trapping in, 451–430
  glyceraldehyde 3-phosphate (GAP) in, 494f, 497–498, 498f, 499–500, 500f, 502f
  hexokinase in, 495–496, 496f, 513, 515
  historical perspective on, 492
  insulin in, 905
  isomerization of carbon sugars in, 496–497, 498f
  in liver, 513–514, 526–528
  in muscle, 512–513, 514f
  NAD⁺ in, 500, 502f, 505, 508
  NADH in, 501, 505–507
  overview of, 493, 494
  pathway integration, 529, 529f
  pentose phosphate pathway and, 659–662
  phosphofructokinase (PFK) in, 512–513, 512f, 514, 515f
  2-phosphoglycerate in, 502–503, 503f
  3-phosphoglycerate in, 503–504, 503f
  phosphoglycerate kinase in, 470
  phosphoglycerate mutase in, 502
  pyruvate in, 494f, 503–504
  pyruvate kinase in, 513, 515, 516f
  reactions of, 504t–505t
  reciprocal regulation in, 525–530
  regulation of, 901, 901f
  second stage of, 493, 495f, 500f
  stages of, 493–495, 495f, 500f
  thioester intermediate in, 500–501, 501f
  triose phosphate isomerase (TPI) in, 497–498, 498f
Glycolytic intermediates, 668
Glycolytic pathway
  glucokinase in, 515
  hexokinase in, 513, 515
  phosphofructokinase (PFK) in, 470f, 512–513, 514, 515f
  pyruvate kinase in, 513, 516, 516f
  role of, 531
Glycomics, 342
Glycopeptide transpeptidase, 259
Glycophorin, 363, 363f, 364
Glycoproteins, 352–361
Glycosaminoglycans, 354–355, 355f
α-1,6-glucosidase in glycogen metabolism, 684
Glycosidic bonds, 348, 348f, 350, 351f, 353f
Glycosylation, 357, 359, 360f
  congenital disorders of, 359
  errors in, 359
  in nutrient sensing, 354, 354f

Glycosyltransferases, 358–359, 359f
Glyoxylate, 564, 565f, 650
Glyoxylate cycle, 564, 565f
Glyoxysomes, 564
Glyset, 366
GMP (guanosine monophosphate)
  in purine synthesis, 828f, 830
  synthesis of, 836–837, 837f
GMP synthetase, 830
Goldstein, Joseph, 867
Golgi complex, 357, 357f, 360, 1048
Goodall, Jane, 185
Gout, 335–336, 839
Gradient centrifugation, 81–82, 81f
Gram, Hans Christian, 394
Gram negative bacteria, 394
Gram positive bacteria, 394
Gram staining technique, 394, 394f
Granum, 621
Granzymes, 1138
Grb2, 453, 453f
Green fluorescent protein (GFP), 61, 62f, 89, 89f
Greider, Carol, 1, 968
Group-specific reagents, 256
Group-transfer reactions, 482, 482f
GTP (guanosine triphosphate)
  as energy source, 821
  hydrolysis, 443–444
  in olfaction, 1100
  in resetting G proteins, 443–444
  in signal transduction, 441
  in translation, 1036–1037
  in tubulin, 1163
  in vision, 1109
GTP–GDP cycle, 1036–1037, 1037f
GTPase-activating protein (GAP), 454
GTPases
  defined, 453
  Ras family of, 453, 454t
  in signal transduction, 454, 454t
Guanidinium chloride, 52
Guanidinium group of arginine, 35f
Guanine, **5**, 5f, 114, **115f**, 118, **118f**, **335**
Guanine-nucleotide-exchange factor (GEF), 453
Guanosine, 116
Guanylate, 116, 830
Guanylate cyclase, 1109
Guide strand, 177
Gustation. *See* Taste
Gustducin, 1103, 1103f

H zone, 1160, 1160f
Haemophilus influenzae, 164, 194
Hair cells, 1112–1113, 1112f, 1113f
Hairpin turns. *See* Reverse turns
Half-life
  of drugs, 929, 929f
  of proteins, 753, 754, 756f
Half-reactions
  defined, 577
  in electron transport, 579
Haptenic determinants, 1133
Haptens, 1133

Hartnup disease, 752
Hatch, Marshall Davidson, 656
Hatch–Slack pathway, 656
Haworth projections, 345
HCN4 gene, 431–432
Hearing, 1111–1113
    hair cells in, 1112–1113, 1112f, 1113f
    ion channels in, 1113
    mechanism of action, 1112
Heart disease, 871
Heat
    defined, 11
    oxidative phosphorylation uncoupling in
        generation of, 604
    release in double helix formation, 13
    thermodynamics and, 11
Heat shock protein 70 (Hsp70), 195, **195f**
Heat-shock response element, 1000, 1000f
Heat-shock transcription factor (HSTF),
    1000
Heavy (H) chains
    in antibody diversity, 1131–1132
    antigen-binding and, 1131
    defined, 1124–1125
    in immunoglobulin classes, 1125t
Heavy meromyosin (HMM), 1153
Heinz bodies, 670, 670f
Helicase DnaB, 965
Helicases, 124
    defined, 955
    in DNA replication, 955
    mechanism, 955, 956f
    structure of, 955, **955**
    symmetry, 955, 955f
Helix-turn-helix motif, 48, 48f, 1059–1060,
    1059f
Helper T cells
    action, 1140, 1141f
    in cytokine secretion, 1140
    defined, 1123, 1139
    in HIV infection, 1141–1142
    peptide recognition, 1139, 1140f
    T-cell receptors (TCRs) and, 1139–1141
Hemagglutinin, 363–364, 363f
Heme, **209**
    biosynthetic pathway, 812f
    in cytochrome *bf* complex, 629
    in cytochrome *c* oxidase, 586–587
    defined, 46
    intrinsic reactivity of, 210
    labeling of, 809b, 810, 811f
    in oxygen-binding, 208–211, 209f
Heme prosthetic group, 584
Hemiacetal intramolecular, 344
Hemichannels, 428
Hemiketal intramolecular, 345–346
Hemoglobin, 207–228
    $\alpha_2 \beta_2$ tetramer of, 51f
    alignment comparison, 189–190, 190f
    alignment with gap insertion, 189, 189f
    α chains, 211, 222–223
    $\alpha_2 \beta_2$ tetramer of, 51, 51f
    $\alpha\beta$ dimers, 211
    amino acid sequences of, 187–188, 188f
    β chains, 207, 211, 220, 220f

Bohr effect, 217–219
    2,3-BPG binding to, 217
    conformational changes in, 215, 215f
    cooperative oxygen binding by,
        212–217
    cooperativity, 214–215
    defined, 187, 207, 208
    fetal, 217, 217f
    globin curve, 211
    Hill plot, 226f
    mutations and, 220–223
    myoglobin comparison, 208
    oxygen affinity of, 216–217, 218f
    oxygen delivery by, 212, 212f
    oxygen transport by, 217
    oxygen-binding, 208–217
    oxygen-binding curve, 212, 212f
    polypeptide chains, 208, 211
    purified, 216, 216f
    quaternary structure of, 211, **211f**,
        213–214, 213f
    R state, 214, 214f, 215, 225–227
    sequence alignment of, 192, 192f, 193f
    sickle-cell, 220, 220f
    space-filling model, 211f
    T state, 214, 214f, 215, 218, 225–227
    tertiary structure of, 194–195
    thalassemia and, 221–222
Hemoglobin A (HbA), 211
Hemoglobin β-chain gene (HbB), 220
Hemoglobin H (HbH), 221
Hemoglobin S (HbS), 220–221, 220f, 221f
Hemolytic anemia, G6PD deficiency and,
    669–670
Hemophilia, 333–334
Hemorrhagic disposition, 332b
Hemostasis, 328
Henderson–Hasselbalch equation, 16
Heparin, 332
Hepatocytes, 714
Heptad repeat, 49, 49f
Heptoses, 342
HER2 receptor, 452, 456
Herbicides, photosynthesis and, 638
Hereditary nonpolyposis colorectal cancer
    (HNPCC), 974
Hers disease, 700t
Heterodimeric sweet receptor, 1105, 1105f
Heterotrimeric G proteins, 440–441, 440f
Heterotrophs, 620, 646
Heterotropic effects, 317
Hexacoordinate globins, 227
Hexokinase
    defined, 495
    in glycolysis, 495–496, 496f
    inhibition of, 513
    liver and glycolysis and, 515
    metal ion requirement, 495
    muscle and glycolysis and, 512–513
Hexose monophosphate pathway. *See*
    Pentose phosphate pathway
Hexose monophosphate pool, 651
Hexose sugars formation in Calvin cycle,
    651–654, 651f
Hexoses, 343, 495, 652f

High-density lipoproteins (HDLs),
    864, 864t
    defined, 865
    properties of, 870
    protective effects and, 870
    reverse cholesterol transport in, 870
High-performance liquid chromatography
    (HPLC), 75, 75f
High-throughput screening, 933
Hill, Archibald, 225
Hill coefficient, 225, 226f
Hill plot, 225, 226f
His 64 (distal histidine), 197f, 227, **228f**
His tag, 75
Histamine, 808, 808f
Histidine
    degradation, 770, 771f
    distal, 210
    ionization of, 35, 35f
    proton shuttle, 290, 290f
    proximal, 209, 215
    structure of, 35, 35f
Histidine operon, 1069f, 1070
Histone acetyltransferases (HATs), 1087,
    1088
Histone code, 1089
Histone deacetylases, 1089–1090
Histone octamer, 1078–1079
Histones
    acetylation of, 1088
    amino-terminal tails of, 1088–1089
    defined, 1077
    in gene regulation, 1077–1079, 1079f,
        1090t
    homologous, 1078, 1079f
    modifications, 1090, 1090t, 1094
HIV. *See* Human immunodeficiency virus
    (HIV) infection
HMG-CoA reductase, 933
HMG-CoA reductase inhibitor, 933
Hodgkin, Alan, 415
Hoffman, Felix, 932
Hogness box, 130
Holliday, Robin, 977f
Holliday junctions, 976–978, 977f
Holoenzyme, 235
Homeodomain, 1080, 1080f
Homeostasis, 483, 484f. *See also* Caloric
    homeostasis; Glucose homeostasis
Homocysteine
    in activated methyl cycle, 799, 800f
    in amino acid synthesis, 800–801
    in cysteine synthesis, 801
    levels, reducing, 801
    in vascular disease, 801
Homocysteine methyltransferase, 800
Homocystinuria, 776t
Homogenate, 71
Homogenization in protein purification, 71
Homogentisate, 775, **775**, 776
Homogentisate oxidase, 774
Homologies
    databases for, 193–197
    defined, 186–187
    detection of, 187–194

Homologies (*Continued*)
  evolutionary trees and, 197
  repeated motifs and, 196
  sequence alignment in, 189–190,
    190f, 195
  shuffling in, 189–190, 190f
  statistical analysis in, 189–190
  substitution matrices in, 190–193
  three-dimensional structures and, 195
Homologous recombination, 172, 172f
Homologous sequences, 193–194
Homologs, 187, 187f
Homology directed repair (HDR), 173
Homolytic cleavage reaction, 722, 722f
Homopolymers, 351
Homoserine, 37f
Homotropic effects, 315
Hood, Leroy, 1130
Horizontal gene transfer, 199, 199f, 298
Hormone-receptor complex, 441
Hormone receptors, 1085–1086, 1086f
Hormones
  in acetyl CoA carboxylase regulation,
    739–740
  eicosanoid, 737, 737f, **738**, 738f
  in fatty acid metabolism, 738–739
  in gene expression, 1086
  in glucose metabolism, 528
  in glycogen metabolism, 690–692, 691f
  local, 737
  metabolic functions of, 483–484
  in gluconeogenesis, 527
  in signal transduction, 440–441
  steroid, 872–873, 872f
Human genome, 165, 165f, 168, 168f
Human Genome Project, 19f
Human history, genome sequencing and,
  19–20
Human immunodeficiency virus (HIV)
  infection
    classes of, 1142
    drug development effects, 936, 936f
    helper T cells in, 1141–1142
    host cell for, 1142
    immune system and, 1142–1143
    protease, 283, 285f
    protease inhibitors, 285–286, 875, 935,
      935f, 942
    receptor, 1142f
    schematic representation, 1142f
    vaccine for, 1145
Human leukocyte antigen A2 (HLA-A2),
  1135
Human milk oligosaccharides, 352
Human ribonuclease, **186f**
Humoral immune response, 1122
Huntington disease, 59, 753, 973
Hurler disease, 355f, 360
Huxley, Andrew, 415
Hybrid DNA, 123
Hybrid-biological-inorganic (HBI) system,
  639
Hybridoma cells, 86
Hydrogen bonds, 8f
  acceptors, 8, 15–16

in amino group, 24
in base pairs, 5
breaking with membranes, 388
carbonyl group as, 24
covalent bonds, compared to, 8
donors, 8
double helix stabilization by, 117–119
in enzyme-substrate complex, 243, 243f
in hydroxyl group, 24
interaction, 8
as noncovalent bond, 8
strength of, 8
water, 9, 10
Hydrogen ions
  in acid-base reactions, 14
  concentration in solutions, 14
  oxygen release and, 217–219
Hydrogen-bond donors, 8
Hydrolases, 265t
Hydrolysis. *See also* ATP hydrolysis
  GTP (guanosine triphosphate), 443–444
  of lipids, 713–714
  in metabolism, 468–469, 482, 482f
  of peptide bonds, 275, 278, 279f
  of phosphodiester bond, 292–293, 292f
Hydrolytic reaction, 482
Hydronium ions, 14
Hydropathy plots, 389, 390f
Hydrophilic amino acids, 35, 48
Hydrophilic moiety, 379, 379f
Hydrophobic amino acids, 32–34
  structures of, 33f
  types of, 32–34
Hydrophobic effect, 9–10, 10f, 12, 34, 48
Hydrophobic interactions
  defined, 10
  as noncovalent bond, 8
Hydrophobic moiety, 379, 380f
Hydrotrope, 472
3-hydroxyanthranilate, 775, **775**
25-hydroxycholesterol, 775
2-hydroxyglutarate, 560
3-hydroxy-3-methylglutaryl CoA, 773
3-hydroxy-3-methylglutaryl CoA reductase
  (HMG-CoA reductase), 858–859,
  859f, 861, 863f, 871, 871f
Hydroxyl groups, 24, 34–35, 60
Hydroxylation
  cytochrome P450 in, 874–875, 874f
  steroid hormones, 872–873
P-hydroxyphenylpyruvate, 774, **774**, 802
P-hydroxyphenylpyruvate hydroxylase, 774
Hydroxyproline, **61**
Hyperammonemia, 767
Hyperbaric oxygen therapy, 217
Hyperglycemia, 567
Hyperlysinemia, 776t
Hyperpolarization-activated cyclic
  nucleotide-gated (HCN) channels, 431
Hypersensitive sites, 1083
Hypervariable loops, 1127, 1127f
Hypochromism, 124, 124f
Hypoglycemia, 533
Hypomethylation, 1084
Hypoxanthine, 839

Hypoxanthine-guanine
  phosphoribosyltransferase (HGPRT),
  839, 840
Hypoxia
  defined, 518
  tumor, 518, 518t
Hypoxia-inducible factor 1 (HIF-1), 518,
  519f, 559

I band, 1160, 1160f
I-cell disease, 359, 360f
Ibuprofen, 926, 927f
Ice, 9, 9f
IGH locus, 1147
Imatinib mesylate (Gleevec), 938, 943
Imine, 683
Immune response
  basis of, 1124
  cellular, 1122
  development of, critics of, 1123
  humoral, 1122
Immune system, 981–985
  adaptive, 1122–1124
  in autoimmune diseases, 1143–1144, 1143f
  B cells in, 1122, 1132–1133, 1144
  in cancer, 1144
  in disease, 1142–1145
  evolution of, 1120–1122
  innate, 1120–1122
  MHC proteins in, 1135–1136
  overview of, 1119–1120
  T-cell receptors. *See* T-cell receptors
    (TCRs)
  T cells. *See* T cells
Immunity, cell-mediated, 1134
Immunizations, 1144–1145
Immunoelectron microscopy, 89, 89f
Immunoglobulin A (IgA), 1125, 1125t
Immunoglobulin D (IgD), 1125, 1125t
Immunoglobulin domains. *See also*
  Immunoglobulin fold
  constant, 1126, 1126f
  defined, 1124
  variable, 1126, 1127f
Immunoglobulin E (IgE), 1125t, 1126
Immunoglobulin fold
  complementarity-determining regions
    (CDRs), 1127, 1127f
  defined, 1126
  elements of, 1127
  structure of, **1128**
Immunoglobulin G (IgG)
  antigen cross-linking, 1125, 1125f
  cleavage of, 1125, 1125f
  defined, 1124
  properties of, 1125t
  structure of, 1123, 1124f
Immunoglobulin M (IgM)
  defined, 1126
  properties of, 1125t
Immunoglobulins. *See also* Antibodies
  classes of, 1125–1126, 1126f
  defined, 83, 1133
  heavy chains, 1124–1126, 1125t,
    1129–1130

light chains, 1124–1125, 1125t, 1129–1130, 1130f, 1131f
myeloma and, 85
production of, 1123, 1123f
segmental flexibility of, 1125, 1125f
sequence diversity in, 1126f
structure of, 83f
Immunologic techniques
enzyme-linked immunosorbent assay (ELISA), 86–87, 87f
fluorescence microscopy, 89, 89f
hybridoma cells in, 86
monoclonal antibodies in, 85–86, 84f, 85f
polyclonal antibodies in, 84, 84f
in protein studies, 83–89
western blotting, 87–88, 88f
Immunological memory, 1144
Immunoreceptor tyrosine-based activation motif (ITAM), 1132, 1132f
Immunosuppression, 1133
Immunotoxins, 162
In-line displacement, 293
Indian muntjac, 116, 117f
Indinavir (Crixivan), 285, 287f
Indinavir complex, 285, 286f
Indirect ELISA, 86, 87f
Indole groups, 34
Induced pluripotent stem (iPS) cells, 1082, 1082f
Induced-fit model, 243, 243f, 274
Inducer, 1062
Influenza virus, 363–364, 363f
Information science, 18
Ingram, Vernon, 220
Inhibition
competitive, 254, 254f, 255f
noncompetitive, 253–254, 254f, 255f
reversible, 254–255, 255f, 256f
uncompetitive, 254, 254f, 256f
Inhibition constant (K_i), 924
Inhibitory factor 1 (IF1), 604
Innate immune system, 1120–1122
Inosinate, 837
Inosine, 839, 1026
Inositol 1,4,5-trisphosphate (IP_3), 438, **439**, 444, 445f, 485
Insertional inactivation, 156, 156f
In singulo method, 261
Insulin
amino acid sequence of, 39f
binding of, 448–449, 449f
defined, 29
in diabetes, 890–892
in essential enzyme regulation, 527
in fatty acid metabolism, 739–740
in glycogen metabolism, 698, 698f
in glycolysis, 905
MALDI-TOF mass spectrum of, 91f
in obesity, 895–896
pancreatic failure and, 899–900
in protein synthesis, 905
release of, 900, 900f
secretion of, 893
in signal transduction, 447–451, 897–898, 897f

Insulin insensitivity, 882
Insulin receptor
activation of, 448–449, 449f
binding sites for, 897
defined, 448
structure of, 448, 449f
subunits, 448, 448f
Insulin resistance
biochemical basis of, 897–898
by excess ceramides, 882
in muscle, 899–900
in pancreatic failure, 899–900, 900f
Insulin signaling, 447–451
illustrated, 449f
lipid kinase in, 450, 450f
pathway, 450, 451f
termination of, 451
Insulin-receptor substrates (IRS), 449–451, 450f, 698
Integral membrane proteins, 384, 384f
Interatomic distances, 7b
Intercalation, 995
Intermediary metabolism. *See* Metabolism
Internal guide sequence (IGS), 1013
Internal ribosome entry sites (IRES), 1042
Intracellular pathogen, 1134, 1134f
Intramolecular hemiacetal, 344
Intramolecular hemiketal, 345–346
Intramolecular rearrangements, 721, 722f
Intrinsic pathway, clotting, 328
Intrinsically unstructured proteins (IUPs), 58
Introns
defined, 114
detection of, 135–136, 135f
self-splicing, 1013, 1013f
Inverse PCR, 162, 162f
Invertase, 349
Inverted repeats, 295, 295f
Ion channels, 415–416
acetylcholine receptor, 423–424, 424f, 425f
action potentials and, 415, 415f, 424–425
ball-and-chain model for inactivation, 422–423, 423f
calcium, 419
cell-to-cell, 427–428, 427f, 428f
cGMP-gated, 1108, 1109
defined, 404, 415
disruption of, 426–427
energetic basis of selectivity, 419, 419f
equilibrium potential and, 424–426, 425f
gap junction, 427–428, 427f, 428f
in hearing, 1113
inactivation of, 422–423, 423f
ligand-gated, 423–424
membrane permeability and, 422–423
nerve impulses and, 415
patch-clamp technique, 415, 416f
potassium, 417–420, 418f, 419f, 421f, 422f
selectivity filter, 419, 419f
sequence relations of, 417, 417f
shaker, 417
sodium, 415, 419–420, 419f, 421f
in taste, 1105, 1105f
transient receptor potential (TRP), 1144
voltage-gated, 421–422, 422f

Ion gradients, 474, 474f
Ion semiconductor sequencing, 167–168
Ion source, 90
Ion-exchange chromatography
defined, 73–74
illustrated, 74f
Ionic interactions, 8, 10f
IRE-binding protein (IRP), 1091, 1091f
Iron
metabolism of, 1090–1092
oxygen-binding to, 208–211, 209f
structural transition in subunit, 215
Iron center of ribonucleotide reductase, 831, 832f
Iron–oxygen bonding, 210f
Iron-response element (IRE), 1091, 1091f
Iron-sulfur clusters, 581–582, 582f
Iron-sulfur proteins
defined, 581
in nitrogen fixation, 789, 790f
in oxidative phosphorylation, 582
structure of, **789**
Irreversible inhibition. *See also* Enzyme inhibition
defined, 253
in mapping active site, 256–258
mechanism-based (suicide), 258, 259f
Ischemia, 227
Islet-1 gene, 1082
Isocitrate
in citric acid cycle, 551
in glyoxylate cycle, 564, 565f
Isocitrate dehydrogenase
in citric acid cycle, 551, 559, 559f
mutations in, 559
Isocitrate lyase, 564
Isoelectric focusing, 77, 77f
Isoelectric point, 77
Isoform, 317
Isolated congenital asplenia, 1051
Isoleucine, 33f, 34
Isomerases, 480t
Isomerization reactions in metabolism, 482, 482f
Isomers
constitutional, 343, 343f
diastereoisomers, 343, 343f
stereoisomers, 343, 343f
Isopentenyl pyrophosphate, 858–859, **859**, 859f
Isopeptide bonds, 754
Isoprene, 871–880
Isoprenoids, 879, 879f
industrial applications of, 880
Isopropylthiogalactoside (IPTG), 1062
Isotope labeling studies, 556, 810
Isovaleryl CoA, 773
Isovaleryl CoA dehydrogenase, 773
Isozymes
in blood, 318
defined, 317
distinguishing, 317
expression of, 310
as homologous, 310
of lactate dehydrogenase (LDH), 317, 318f
in varying regulation, 310

J genes, 1130–1131
Jagendorf's demonstration, 631, 632f
Jenner, Edward, 1144
Joule (J), 238

K+ ions. *See* Potassium ions
k_{cat} (rate constant), 249
k_{cat}/K_M (specificity constant), 249–250, 250f
K_d (dissociation constant), 248, 254, 1157
Kaplan-Meier curve, 228f
Kendrew, John, 208
Ketimine, **760**
α-ketoacids, 772–773, 794
β-ketoacyl synthase, 730
α-ketobutyrate, 772
Ketogenic amino acids, 769
Ketogenic diets, 727
Ketoglutarate, **252**
α-ketoglutarate
    in amino acid degradation, 769, 770–771,
        770f
    in amino acid synthesis, 791–793, 794
    in citric acid cycle, 551, 551f
    in double-displacement reactions, 252
    in glutamate synthesis, 791, 792
    in nitrogen fixation, 791
    oxidative decarboxylation of, 552
α-ketoglutarate dehydrogenase complex
    in citric acid cycle, 559, 559f
    defined, 552
α-ketoisocaproate, 773
Ketone bodies
    acetyl CoA formation from, 724–725
    in citric acid cycle, 908, 908f
    defined, 724
    in diabetes, 727, 727f
    in fatty acid metabolism, 724
    formation of, 725, 725f
    as fuel source, 724–728
    high levels of, 727
    in liver, 726, 726f, 908, 908f
    as water soluble, 726
Ketoses
    aldoses conversion to, 496
    defined, 342
3-ketosphinganine, 854
β-ketothiolase, 718
Keyhole limpet hemocyanin (KLH), 1133
K_i (inhibition constant), 924
Kidney
    drug excretion in, 927
    gluconeogenesis in, 519
Killed (inactivated) vaccines, 1144
Killer T cells. *See* Cytotoxic (killer) T cells
Kilobase (kb), 128b
Kilocalories (kcal), 238
Kilodaltons (kDa), 38, 38b
Kilojoules (kJ), 238
Kinase fold, 322
Kinases, 495. *See also specific kinases*
Kinesin
    ADP in, 1165f, 1166
    ATP binding to, 1155–1156, 1156f
    ATP hydrolysis in, 1164, 1165f
    defined, 1152

gene mutations, 1163
    monitoring movements mediated by, 1165f
    motion processivity, 1164–1166
    movement along microtubules, 1162,
        1165–1166, 1165f
    neck linker, 1156, 1156f
    P-loop NTPase, 1152
    relay helix, 1156, 1156f
    structure of, 1154, 1154f
    walk, 1166
Kinetic energy, 11
Kinetics
    defined, 244
    enzyme. *See* enzyme kinetics
    as study of reaction rates, 244–245
    of water deprotonation, 290, 290f
Klenow fragment, 951, 980
Klug, Aaron, 1078
K_M. *See* Michaelis constant
Knowledge-based methods, 57–58
Kornberg, Roger, 986
Krebs, Hans, 474, 563b
Krebs cycle. *See* Citric acid cycle
Kringle domain, 329
Kühne, Friedrich Wilhelm, 492b

L amino acids, 31, 31f, 39
L–Deprenyl (Selegiline), **258**
Laborit, Henri, 931
Lac operator, 1062
Lac operon, 1061–1062, 1062f, 1063f
Lac regulatory site, 1058f
Lac repressor
    binding of, 1061
    components of, 1062
    defined, 1059
    effects of IPTG on, 1063f
    inhibition relief, 1064–1065
    structure of, 1061, 1061f, 1063
Lac repressor-DNA complex, 1058–1059,
        1059f
Lactase, 349
    deficiency of, 510
    defined, 493
Lactate, **252**, 567
    in Cori cycle, 529, 529f
    formation from pyruvate, 506–507
    formation of, 506–507
    in gluconeogenesis, 519
    in glycolysis, 491
    in muscle contraction, 528–529, 529f
Lactate dehydrogenase (LDH), 317, 318f,
        529
Lactic acid fermentation, 491, 506–507
Lactic acidosis, 533
Lactobacillus, 510f
β-lactoglobulin, 91f
Lactonases, 660
Lactose, 352
    defined, 349–350
    structure of, 349–350, 350f
Lactose intolerance, 510
Lactose operon. *See Lac* operon; *Lac*
        repressor
Lactose permease, 414, 414f

LacZα gene, 155–156
Lagging strand
    defined, 954
    synthesis of, 953f, 962–964
    template, 963
γ chains, 1125, 1126f
λ phages
    alternative infection modes for, 156–157,
        157f
    for DNA cloning, 155–157
    mutant, as cloning vector, 157, 157f
λ repressor, 1072
    defined, 1065
    in gene expression, 1065–1066
    in genetic circuit formation, 1066, 1066f
    structure of, 1066f
    synthesis control, 1067
Lamprey, 199, 199f
Lanosterol, 860, **862**
Lansoprazole, 938, **938**
Lateral diffusion, 390, 391, 391f
Lavoisier, Antoine, 788
Laws of thermodynamics, 11–13
Leader peptide sequences, 1069, 1069f
Leader sequences, 1068–1069, 1068f
Leading strand
    defined, 954
    synthesis of, 953f, 962–964
Leber, Theodore, 611
Leber hereditary optic neuropathy (LHON),
        608, 611–612, 611f
Lectins
    C-type, 362–363, 362f
    as carbohydrate-binding proteins,
        361–362
    classes, 362–363
    defined, 361
    in interactions between cells, 361–362
Leder, Philip, 1130
Leghemoglobin
    sequence alignment of, 192, 192f, 193f
    tertiary structure of, 194, **194f**
Leptin
    in caloric homeostasis, 893–894
    defined, 893
    effects in brain, 894, 894f
    obesity and, 894f, 895–896
    secretion of, 893, 894
Leptin resistance, 894f, 895–896
Lesch–Nyhan syndrome, 840–841
Leucine, 33f, 34
Leucine-rich repeats (LRRs), 1120
Leukemia, 518
Leukocyte antigen A2 (HLA-A2),
        1135–1136, 1137f
Lever arm, 1155, 1155f
Levinthal, Cyrus, 56
Levinthal's paradox, 56
Li-Fraumeni syndrome, 974
Licensing factors, 966
Licopene, **879f**
Ligand binding
    concerted model, 214–215, 214f
    in gene expression, 930
    sequential model, 215, 215f

Ligand-gated channels, 423
Ligands, 438, 923, 1085
Ligases, 265t
Ligation reactions in metabolism, 483
Light
    absorption, 1108–1109
    electromagnetic spectrum of, 1106, 1106f
    in lowering calcium level, 1109–1110
Light (L) chains, 1124–1125, 1125t,
        1130–1131, 1130f, 1131f
Light absorption, 622–623, 623f
Light energy, 638–640
Light meromyosin (LMM), 1153
Light reactions. *See also* Photosynthesis
    in Calvin cycle, 655, 655f
    defined, 620
    herbicides and, 638
    illustrated, 620f
    stoichiometry for, 634–635
Light-harvesting complexes, 636
Light-independent reactions. *See* Dark
        reactions
Limit dextrin, 492
Limonene, 878, **879f**
Linear polymers, 4, 37
Lineweaver–Burk plot, 248, 248f
α-1,6 linkages, 694–695, 694f
Linking number, 956–958, 957f
Linoleoyl CoA, 720, 720f
Lipases, 712, 712f
Lipid bilayers. *See also* Membrane lipids
    as cooperative structures, 381
    defined, 374, 380
    diagram, 380f
    formation of, 380–381
    as permeability barrier, 382–383, 382f
    self-assembly process, 380
    as self-sealing, 381
    stabilization of, 381
Lipid droplet, 711–712
Lipid metabolism, liver in, 714
Lipid rafts, 392–393
Lipid vesicles, 381–382, 381f
Lipids
    digestion of, 712, 712f
    energy storage in, 710–712
    fatty acids in, 374–376, 375f, 376t
    hydrolysis of, 713–714
    membrane. *See* membrane lipids
    metabolism of. *See* fatty acid metabolism
    transport of, 713, 713f
Lipinski's rules, 925, 925f
Lipoamide in citric acid cycle, 546–548, 548f
Lipoic acid, 544f, 545
Lipolysis, 714, 715f
Lipopolysaccharide (LPS), 1121
Lipoprotein particles, 863–864
Lipoproteins, 863, 865f
    chylomicrons, 864, 864t, 865
    components of, 864
    high-density (HDLs), 864, 864t, 865, 870
    intermediate-density (IDLs), 864,
        864t, 865
    low-density (LDLs), 864, 864t, 865–869,
        865f, 868f

metabolism, 866f
    very-low-density (VLDLs), 864, 864t
Liposomes
    defined, 381
    formation of, 381–382, 381f
    therapeutic applications of, 382
Lipoxygenase, 737, 737f
Live attenuated vaccines, 1144–1145
Liver
    alcoholic injury of, 910–911
    in amino acid degradation, 758, 763–764
    in cholesterol synthesis, 861
    cirrhosis of, 911
    drug metabolism in, 927
    ethanol metabolism in, 909–912
    fasting and, 906
    fatty, 910–911
    fuel metabolism in, 914–915
    gene expression in, 1076, 1077t
    gluconeogenesis in, 519, 521, 526–527,
        908
    glucose 6-phosphate (G-6P) in, 685–686,
        700
    glucose levels and, 685–686
    glucose sensor in, 699
    glycerol in, 714–715
    glycogen in, 680, 680f, 690
    glycogen metabolism in, 698–700, 699f
    glycogen phosphorylase in, 686–687,
        686f, 687f
    glycolysis in, 513–516, 514f, 515f, 516f,
        526–527
    ketone bodies in, 726f
    ketone body synthesis in, 908–909, 908f
    in lipid metabolism, 714
    peroxisomes in, 724f
    in starvation, 907–909
    in starved-fed cycle, 905–907
    triacylglycerol accumulation in, 743
    in triacylglycerol synthesis, 851
L-lectins, 363
Local hormones, 737
Lock-and-key model, 243, 243f
Long interspersed elements (LINES), 165
Long QT syndrome (LQTS), 427
Loops, 46, 46f
Lou Gehrig's Disease, 146
Lovastatin, 871, 871f, 933, **933**
Low-density lipoprotein receptor
    absence of, 867–868
    destruction of, 868–869
    LDL release, 868–869, 868f
    mutations in, 868–869
Low-density lipoproteins (LDLs), 864, 864t
    in cholesterol metabolism, 866–867
    cycling of, 869–870
    defined, 865
    in familial hypercholesterolemia,
        867–868
    metabolism of, 866f
    receptor-mediated endocytosis of, 866,
        866f
    schematic model of, 865f
Lupine leghemoglobin, 193
LuxR, 1067

Lyases
    defined, 482, 734b
    reaction type, 265t
Lymphotactin, 58, 58f
Lynch syndrome, 974, 974f
Lysine
    acetylated, 1088
    structure of, 35, 35f
Lysogenic pathway, 157
Lysosomes
    defined, 359–360
    function of, 360
    in I-cell disease, 360
    with lipids, 856, 856f
Lysozyme
    antibodies against, 1128–1129, 1129f
    backbone model, 65f
    ball-and-stick model, 65f
    ribbon diagram, 66f
    space-filling model, 65f
Lytic pathway, 157

M13 bacteriophage, 980
MacKinnon, Roderick, 417, 422
Macugen, 202
Mad cow disease, 59
Magnesium ions
    in ATP hydrolysis, 299–300
    in Calvin cycle, 648, 648f, 655–656
    in DNA cleavage, 291–293
Main chains. *See* Backbones
Major histocompatible complex (MHC),
        1122–1124, 1135
Major histocompatible complex proteins,
        1134–1142
    anchor residues, 1136, 1136f
    class I, 1135, 1135f, 1136, 1140–1141,
        1141f
    class II, 1139–1142, 1140f
    diversity of, 1140–1141
    peptide presentation by, 1135–1137, 1135f
    in plasma membrane, 1135
    structure of, 1135–1137
    T-cell receptors and, 1136–1142, 1136f
    in transplant rejection, 1141–1142
Malaria, 221, 222f
Malate, 564, 565f
    in Calvin cycle, 657f
    in citric acid cycle, 554
    in gluconeogenesis, 522
Malate dehydrogenase, 555, 734
Malate synthase, 564
Malate–aspartate shuttle, 600, 600f
MALDI-TOF mass spectroscopy, 90–91,
        91f
4-maleylacetoacetate, 775, **775**
Malic enzyme, 734
Malignant hyperthermia, 528
Malonyl ACP, 728, 730
Malonyl CoA, 729, 739
Malonyl transacylase, 730, 730t
Malonyl/acetyl transacylase (MAT), 731,
        732f
Maltase, 350, 365–366, 493
Maltose, 350f, 359

Mammalian fatty acid synthase, 731–732
Mammalian liver homogenate, 975
Manganese, 628, 628f
Manganese center, 628, 628f
Mannose, 344–345, 345f
Mannose 6-phosphate, 360, 360f
Mannose 6-phosphate receptor, 362
Maple-syrup urine disease, 776
Mass spectroscopy
    MALDI-TOF, 90–91, 91f
    peptide sequencing by, 92–93, 93f
    in protein and peptide identification,
        89–97
    proteomic analysis by, 97, 97f
    tandem, 92, 93f
Mass-to-charge ratio, 90
Mast cells, 332, 332f
Matrix, 90
Matrix-assisted laser desorption/ionization
    (MALDI), 90, 360
Maximal rate, 247–248, 249
McArdle disease, 700, 701, 702–703
Mechanism-based (suicide) inhibition, 257,
    257f, 258f
Mechanism-based toxicity, 937
Mediator, 1081, 1081f
Megasynthases, 738
Melanin, 816, 816f
Melanocyte-stimulating hormone (MSH),
    893–894, 894f
Melatonin, 811f
Melting, 124
Melting temperature
    of DNA, 124
    of fatty acid chains, 391, 392t
    of phospholipid membrane, 391, 391f
Membrane anchors, 388, 388f
Membrane channels
    gap junction, 404
    ion, 404, 415–427
    membrane permeability to, 429, 429f
    water, 429, 429f
Membrane diffusion
    facilitated, 403, 405
    lateral, 390, 391, 391f
    lipid, 390
    protein, 391
    simple, 404
Membrane lipids. See also Lipid bilayers
    as amphipathic molecule, 379
    archaeal, 379, 379f
    carbohydrate moieties, 378–379
    cholesterol, 378–379, 858–871
    defined, 374
    diffusion coefficient, 390
    fatty acids in, 374–376, 375f, 376t
    glycolipids, 344, 378–379
    hydrophilicity of, 379, 379f
    hydrophobicity of, 379, 379f, 380
    lateral diffusion, 390, 390f, 391, 391f
    metabolism, 857–858
    movement in membranes, 391, 391f
    permeability barrier, 382
    phospholipids, 376–377
    phospholipids, synthesis of, 850–858

rate of diffusion, 390
    representations of, 379f, 380f
    sphingolipids, synthesis of, 854–857
    synthesis of, 849–883
    triacylglycerols, synthesis of, 850–858
    as two-dimensional solutions, 374
Membrane potential, 405
Membrane proteins
    α helix of, 384–385, 388–390, 389f, 389t
    ATP hydrolysis, 406–413
    β strands of, 386, 386f
    carriers, 403
    channels. See membrane channels
    content variation, 383
    defined, 374
    diffusion, 391
    function of, 383
    gel patterns of, 384, 384f
    hydropathy plots for, 389, 390f
    hydrophobicity of, 388, 388f
    integral, 384, 384f
    linking to membrane surface, 387
    membrane interaction, 384–387
    peripheral, 384, 384f
    pumps. See pumps
    as two-dimensional solutions, 385
Membrane transport
    ABC transporters in, 403, 411–413, 411f,
        412f
    active, 403, 405
    free energy in, 405, 405f
    passive, 405
    potassium ion channels and, 420, 420f
    primary active, 403
    secondary active, 403
    transporters. See transporters
Membrane-bound antibody, 137f
Membrane-spanning α helices, 384–385,
    389, 389f
Membranes, 373–398
    archaeal, 379, 379f
    as asymmetric, 385, 392–393, 393f
    common features of, 385
    defined, 384
    as electrically polarized, 385
    endoplasmic reticulum (ER), 736
    enzymatic activities, 393
    fluidity of, 392–394
    ion gradients across, 474, 474f
    lipid bilayer, 380–382, 385
    mitochondria, 574–575
    as noncovalent assemblies, 385
    overview of, 384–385
    permeability of, 373, 381–382, 382f, 403
    planar bilayer, 381, 381f
    polarity scale for transmembrane helices,
        389, 389t
    processes of, 383–390
    protein interaction, 384–387
    as sheetlike structures, 374
    synthesis of, 393
Memory B cells, 1144
Memory T cells, 1144
Menten, Maud, 245
6-mercaptopurine, 940, **940**

β-mercaptoethanol, 52–53, 52f, 53f, 76
Mercury poisoning, 562–563
Merrifield, R. Bruce, 100
Meselson, Matthew, 122, 123
Mesotrypsin, 328
Metabolic pathways. See also specific
        pathways
    amphibolic, 465
    biosynthetic, 465
    classes of, 465
    defined, 465
    degradative, 465
    formation of, 466
    illustrated, 465f
    motifs, 475–486
    regulation of, 464
    thermodynamically favorable, 465, 466
    thermodynamically unfavorable, 466
Metabolic syndrome, 898–899, 898f
Metabolism, 463–487
    activated carriers in, 475–480
    AMPK in, 740–741
    anabolic reactions in, 468–469
    carbon bonds cleavage, 483, 483f
    catabolic reactions in, 485
    catalytic activity control and, 484–485
    cholesterol synthesis, 866
    common motifs in, 464–465, 475–528
    defined, 464
    drug, 925–927
    enzyme control and, 483
    evolution of, 484–485
    first-pass, 927
    free energy in, 466–467
    fructose, 508, 508f
    glucose, 465f
    of glucose 6-phosphate, 665–668
    group-transfer reactions in, 482, 482f
    hormones and, 483–484
    hydrolytic reactions in, 482–483, 482f
    inborn error of, 779–780
    integration of, 889–912
    interconnected interactions in, 464–466,
        464f
    iron, 1090–1093
    isomerization reactions in, 482, 482f
    ligation reactions in, 481–482, 481f
    lipid, 857–858
    lipoproteins, 865f
    overview of, 464–465
    oxidation-reduction reactions in, 481, 481f
    of pyruvate, 505–507
    regulation of, 483–485
    in starvation, 907–908, 908t
    substrate accessibility control and, 485
    substrate cycles in, 528, 528f
    thermodynamics of, 465–466
    types of reactions in, 481t
Metabolites, 2, 717
Metabolomics, 464
Metal ion catalysis
    in ATP hydrolysis, 299–300
    in carbon dioxide hydrolysis, 288–290
    defined, 274–275
    in DNA cleavage, 294–295, 295f

Metal ions, **952**

Metalloproteases, **284**, 284f, 285

Metals, cofactor, 236, 236t

Metamorphic proteins, 58–59

Metarhodopsin, 1108

Methionine
  degradation, 771
  from homocysteine, 800
  metabolism, 771, 772f
  oxidation of, 328, 328f
  structure of, 33f, 34
  in translation, 1034, 1034f

Methionine sulfoxide, 328, 328f

Methionine synthase, 799

Methotrexate, 835

Methylases, 297

Methylation
  of adenine, 297–298, 298f
  in amino acid synthesis, 798–800
  DNA protection by, 297–298, 297f
  in gene expression, 1084
  of phospholipids, 800
  reaction, 721

Methylcobalamin, 799

β-methylcrotonyl CoA, 772

β-methylglutaconyl CoA, 772

Methylglyoxal, 532, **532**

Methylmalonic acid, 779, **779**

Methylmalonic acidemia/acidosis, 779–780

Methylmalonyl CoA, 721–723, 722f, **723**

Metmyoglobin, 210

Metoprolol, 939

Mevalonate, 858–859, 859f

Micelles, 380, 380f

Michaelis, Leonor, 245

Michaelis constant (K_M)
  in Cheng–Prusoff equation, 924
  defined, 246
  determination of, 247–248
  as enzyme characteristic, 248–249
  physiological consequences of variations, 247
  values of enzymes, 248t

Michaelis–Menten equation, 248, 252

Michaelis–Menten kinetics, 244–253, 253f, 254, 277, 311

Microbiomes, 19, 19f, 727

Microbodies, 649

MicroRNAs, 1004, 1004f, 1092–1093, 1092f

Microsomal ethanol-oxidizing system (MEOS), 910

Microtubules
  arrangement of, 1163, 1163f
  in cilia, 1163
  defined, 1162
  dynamic instability of, 1163
  in flagella, 1163
  function of, 1163
  kinesin movement along, 1162, 1164–1166, 1165f
  structure of, 1162–1163, 1163f
  tubulins in, 1163, 1163f

Miglitol, 366, **366**

Migrations, genome sequencing and, 19–20, 20f

Migratory birds, protein metabolism in, 775

Milk, protein components of, 69

Milstein, César, 85

Mineralocorticoids, 872

Minoxidil, 927

Mismatch DNA repair, 971, 971f

Mitchell, Peter, 593, 593b, 609, 632

Mitochondria
  in apoptosis, 609
  ATP synthase in, 594f, 600–602
  brown fat, 605
  cardiolipin on, 399
  circular DNA from, 121f
  in citric acid cycle, 542, 542f
  codons of, 135, 135t
  electron entry via shuttles, 600–601
  electron micrograph and diagram of, 575f
  endosymbiotic origin of, 575–576
  genome of, 575–576, 575f, 576f
  matrix, 574
  membrane, 574
  overlapping gene complements of, 576, 576f
  oxidative phosphorylation in, 574–576
  properties of, 574–576
  structure of, 574–575, 575f

Mitochondrial ATP-ADP translocase, 601–602, 602f, 608–609

Mitochondrial ATPase. See ATP synthase

Mitochondrial biogenesis, 902, 902f

Mitochondrial diseases, 608

Mitochondrial mutants, 608

Mitochondrial outer membrane permeabilization (MOMP), 609

Mitochondrial porin, 574

Mitochondrial transporters, 602–603, 602f, 603f
  defined, 603f
  illustrated, 603f
  for metabolites, 602–603
  structure of, 602, 602f

Mixed inhibition, 254. See also Enzyme inhibition

Mixed-function oxygenase, 773, 873–874

Mobile genetic elements, 165

Molecular evolution, 200–202

Molecular imaging agents, 446

Molecular models
  backbone, 65, 65f
  ball-and-stick, 23f, 24, 65, 65f
  proteins, 64–66
  ribbon diagrams, 65–66, 66f
  small molecules, 24
  space-filling, 23f, 24, 64–65, 65f

Molecular motors, 1151
  ATP binding to, 1155–1156, 1156f
  dynein, 1152, 1154, 1155f, 1162–1166
  flagella, 1152, 1163, 1166–1170
  kinesin, 1152, 1154, 1162–1166
  myosin, 1152, 1153–1154, 1153f, 1154f, 1158–1162
  operation of, 1151–1152
  overview of, 1151–1152
  P-loop NTPase superfamily, 1151–1155
  structures of, 1152–1156, 1153f, 1154f, 1155f

Molecular oxygen, toxic derivatives of, 590–592

Molecules
  amphipathic nature of, 380
  Fischer projections, 23
  homologous, 186–187
  permeability of, 382, 382f
  stereochemical renderings, 23
  three-dimensional structure of, 23–24

Moloney murine leukemia virus, 171–172

Monoclonal antibodies, 70, 945
  defined, 85
  generation for proteins, 85–86
  illustrated, 84f
  preparation of, 85f
  in signal transduction inhibition, 456

Monod, Jacques, 214, 1061

Monomers, 4

Monooxygenase, 775, 873–874

Monosaccharides. See also Carbohydrates
  alcohols and, 348
  aldoses, 342
  amines and, 348
  boat form of, 346–347, 347f
  chair form of, 346–347, 347f
  defined, 342
  diastereoisomers, 343, 343f
  envelope form of, 347, 347f
  glycosidic bonds, 348, 348f
  isomers of, 343, 343f
  ketoses, 342
  linkage of, 349–351
  modified, 348–349, 348f
  reducing sugars, 347–348
  ring forms of, 345–347, 345f, 346f
  stereoisomers, 343, 343f
  structure of, 342–344

Morphine, 925

MotA–MotB pairs, 1167, 1168f

Mother's curse, 611

Motifs
  defined, 48
  helix-turn-helix, 48, 48f
  metabolism, 463–464
  repeating, 96, 96f

mRNA (messenger RNA)
  cDNA prepared from, 159–160
  complementarity with DNA, 130, 130f
  defined, 113, 128
  5′ cap of, 1002–1003, 1003f
  FMN and, 992
  modification of, 131, 131f
  poly(A) tail of, 1002–1003
  precursors, 1001, 1005–1006, 1007–1008, 1007t
  in protein synthesis, 134
  RNA editing and, 1004–1005, 1004f
  structure of, 1042, 1042f
  transcription and, 1010, 1010f
  transferrin-receptor, 1091, 1091f

Mucins
  defined, 352, 355
  overexpression of, 356
  structure of, 356, 356f

Mucolipidosis, 359

Mucopolysaccharidoses, 355
Mucoprotein, 352
Mucus, 356
Muller, Hermann, 970
Mullis, Kary, 151
Multidrug resistance, 411
Multidrug resistance protein, 411
Multifunctional protein kinases, 320
Multiple myeloma, 85–86, 1147–1148, 1147f
Multiple-substrate reactions
    classes of, 251
    defined, 250–251
    double-displacement, 251, 252
    sequential, 251
Muscle
    ATP in contraction, 512–513
    excess fatty acids in, 899
    fatty acids in, 899, 904
    fiber types, 688, 688t
    glycogen in, 680
    glycogen phosphorylase in, 688, 689f
    glycolysis in, 511–513
    insulin resistance in, 898–900
    myofibrils in, 1160
    PP1 regulation in, 697, 697f
    sarcomere in, 1060, 1060f
Muscle contraction
    actin in, 1159–1161
    alanine in, 529, 530f
    ATP formation in, 902–904
    ATP hydrolysis in, 1159–1162, 1160f
    fuel sources for, 902–904, 903t, 904t
    gluconeogenesis in, 528–530, 529f
    lactate in, 528–529
    myosin in, 1159–1161
    sliding-filament model of, 1161–1162, 1161f
Muscle-relaxation pathway, 932–933, 933f
Mutagenesis
    cassette, 162, 162f
    oligonucleotide-directed, 161, 161f
    by PCR, 162, 162f
    site-directed, 161, 282, 282f
Mutagens, 969–970
Mutant λ phage, 157f
Mutase, 503
Mutations
    deletion, 161, 162, 162f
    disease-causing, functional effects of, 163
    in genes encoding hemoglobin subunits, 220–223
    insertion, 161, 162
    at level of function, 194
    at level of sequence, 194
    point, 161
    in protein kinase A, 322–323
    in recombinant DNA technology, 161–163
    substitution, 161
MWC model, 214–215
Mycobacterium tuberculosis, 153, 567, 758
Myelin, 383, 383f
Myofibrils, 1160
Myogenin, 172, 173f

Myoglobin
    alignment comparison, 190, 190f
    alignment with gap insertion, 189, 189f
    amino acid sequences of, 187–188, 188f
    as compact molecule, 46–47
    defined, 46, 187, 208
    distribution of amino acids in, 47f
    hemoglobin comparison, 208
    Hill plot, 226f
    oxygen-binding curve, 212, 212f
    as single polypeptide, 208
    structure of, 208, 208f, 210
    tertiary structure of, 194
    three-dimensional structure of, 47f
Myopathy, 933
Myosin
    active site, 302, 302f
    altered conformation persistence, 301–302
    ATP binding to, 1155, 1155f, 1156f
    ATP complex structure, 299, 299f
    ATP hydrolysis and, 299–301
    ATPase transition-state analog, 300, 300f
    binding to actin filaments, 1158, 1158f
    catalytic strategies of, 274
    conformational changes, 301, 301f
    defined, 298, 1152
    heavy chains, 1152, 1154, 1154f
    kinetic studies of, 299
    lever arm, 1155, 1155f, 1162, 1162f
    light chains, 1153, 1153f
    motion observation, 1158–1159, 1159f
    movement along actin, 1156–1162, 1160f
    in muscle contraction, 1159–1161
    P-loop NTPase, 1152
    P-loop structures, 303–304
    power stroke release, 1159
    regulatory light chain, 1152
    relay helix, 1156, 1156f
    S1, 1153, 1153f, 1155, 1155f
    S2, 1153–1154, 1153f
    single-molecule motion, 597–598, 598f, 1158–1159, 1159f
    as slow enzymes, 301
    structure of, 1153–1154, 1153f
    switch I and switch II, 1155–1156, 1156f
    in thick filaments, 1060, 1061f
    in thin filaments, 1060
Myosin motors, 1171
Myosin VIIa, 1171
Myrcene, 878, 879f

^{14}N/^{15}N DNA, 122–123, 123f
^{15}N labeling, 809b
N-linkage, 352–353, 353f
N-linked oligosaccharides, 353, 353f, 360
N-terminal rule, 754
Na$^+$ ions. See Sodium ions
Na$^+$-K$^+$ pumps, 406, 410
NAD$^+$ (nicotinamide adenine dinucleotide oxidized)
    in glycolysis, 500, 502f, 505–507
    light absorption and, 71
    from metabolism of pyruvate, 505–507
    reduction potential of, 577
    regeneration of, 507, 507f

structures of oxidized form of, 475, 475f
    synthesis of, 808, 808f
NADH (nicotinamide adenine dinucleotide reduced)
    in ATP generation, 477
    electron-transport chain and, 543, 556, 576, 579
    electrons from, 600–601
    ethanol metabolism in, 910–911
    in fatty acid oxidation, 717–718
    in glycolysis, 501–502, 505–507
    light absorption, 71
    pyruvate reduction by, 506
    in respiratory chain, 582–583, 583f
    structures of oxidized form of, 476, 476f
    transport of, 600–601
NADH-cytochrome b_5 reductase, 736, 736f
NADH-Q oxidoreductase, 579–581, 580f, 580t, 582–583, 583f
NADP$^+$ (nicotinamide adenine dinucleotide phosphate), 660, 665
NADPH (nicotinamide adenine dinucleotide phosphate reduced)
    in Calvin cycle, 645–646, 654
    dihydrofolate reductase and, 834
    in fatty acid synthesis, 734–735
    formation by photosystem I, 632–633
    in glucose 6-phosphate conversion, 659–660
    pathways requiring, 659t
    in pentose phosphate pathway, 666–668
    in reductive biosynthesis, 477
Nalidixic acid, 961–962
Nannochloropsis gaditana, 180
National Center for Biotechnology Information, 193
Neanderthals, 200, 200f
Neck linker, 1156
Negative nitrogen balance, 794
Negatively charged amino acids, 35–37, 36f
Neher, Erwin, 416
Neonatal Fc receptor (FcRn), 945
Nernst equation, 425
Nerve impulses, 415, 423
Neural-tube defects, 841
Neuroglobin, 223, 227, 228f
Neurological diseases. See Diseases and disorders
Neuron, myelination of, 383, 383f
Neuropeptide Y (NPY), 893–894, 894f
Neurotransmitter release, 396, 396f
Neurotransmitters, 423, 810
Neutral fat. See Triacylglycerols
Next-generation sequencing (NGS)
    defined, 166
    detection methods in, 166–167, 167f
    as highly parallel, 166
    ion semiconductor, 167–168
    pyrosequencing, 166
    reversible terminator method, 167
NF-κB, 757
Ngb-H64Q, 227–228, 228f
Niemann-Pick C1 (NPC1) protein, 869
Niemann-Pick C2 (NPC2) protein, 869
Niemann-Pick diseases, 869

9 + 2 array, 1163
Nitric oxide (NO), 459
  synthesis of, 809, 809f
Nitric oxide synthases (NOSs), 459, 810
Nitrogen
  in amino acids, 788
  excess, disposal of, 764–769
  metabolism, 766f, 808
  peripheral tissue transport of, 763–764
Nitrogen fixation
  in amino acid synthesis, 788–793
  ATP in, 789–793
  defined, 788
  FeMo cofactor in, 790, 790f
  glutamate in, 791–793
  glutamine in, 791–793
  iron-sulfur proteins in, 790–791, 790f
  P clusters in, 791, 790
Nitrogenase, 789, 790, **790f**
Nitrogenase complex, 789
Nociceptors, 1114
NompC, 1113
Non-homologous end joining (NHEJ), 173,
    174f
Nonalcoholic fatty liver disease (NAFLD),
    740
Noncompetitive inhibition. See also Enzyme
    inhibition
  defined, 253
  double-reciprocal plot, 255, 255f
  example of, 255
  inhibitor binding, 255
  kinetics of, 254–256, 255f
Nonconservative substitutions, 190, 192, 192f
Noncovalent bonds
  in biological molecules, 7–9
  hydrogen bonds, 5, 8, 8f, 9, 10
  hydrophobic effect and, 9–10, 10f
  types of, 8
  van der Waals interactions, 8–9, 9f, 11
Nonessential amino acids
  defined, 793
  synthesis of, 793, 794
Nonheme iron proteins, 581–582
Nonhomologous end joining (NHEJ), 972
Nonphotochemical quenching (NPQ), 641
Nonpolar molecules, 9
Nonreducing sugar, 347
Nonribosomal peptides, 738
Nonshivering thermogenesis, 606, 606f, 609
Northern blotting, 149
Novobiocin, 960
NTPases, P-loop. See P-loop NTPases
Nuclear envelope, 395, 395f
Nuclear hormone receptors
  defined, 1085
  domain structures, 1085, 1085f
  drug-binding to, 1087
  in gene expression, 1084–1086
  ligand-binding to, 1085, 1086f
Nuclear magnetic resonance (NMR)
    spectroscopy, 70, 102–105, 210
  basis of, 103, 103f
  chemical shifts, 103
  defined, 102–103

NOESY and, 104–105, 104f, 105f
  one-dimensional, 103, 103f
  RNA structure, 202, 202f
  signals, 103f
Nuclear Overhauser effect (NOE), 104
Nuclear Overhauser enhancement
    spectroscopy (NOESY), 104–105,
    104f, 105f
Nuclear pores, 395
Nucleation-condensation model, 56, 57
Nucleic acids. See also DNA; RNA
  backbones in, 114, 115f
  bases, 114–117
  complementary sequences, 117–122
  complex structures, 121, 121f, 122f
  defined, 113
  nucleotides as monomeric units of, 115–116
  polymeric structure, 114f
  single-stranded, 121, 121f, 122f
  solid-phase synthesis of, 147
  structure of, **117f**
Nucleophile, 24
Nucleoside diphosphate, 264
Nucleoside diphosphate kinase, 467, 825
Nucleoside diphosphokinase, 553
Nucleoside monophosphate, 467
Nucleoside monophosphate kinases, 264,
    467, 825
Nucleoside phosphorylases, 838, 839
Nucleoside triphosphates, 116
Nucleosides, 116f
  defined, 116, 822
  nomenclature of, 822
  units, 115–116
Nucleosome core particle, 1077, 1078
Nucleosomes, 1077–1078
Nucleotidases, 838
Nucleotide synthesis, 821–842
  de novo pathways in, 822, 830
  deoxyribonucleotide, 831–836, 837–838,
    838f
  disorders of, 839–841
  inosinate in, 837
  overview of, 821–822
  of purine, 826–830, 826f, 837, 837f
  of pyrimidine, 822–826, 822f, 836–837
  regulation of, 836–838
  salvage pathways in, 822–830, 822f
  substrate channeling in, 824, 824f
Nucleotide triphosphates, 116
Nucleotide-excision repair, 972
Nucleotides, 116f
  cyclic, 821
  defined, 114, 116, 821, 822
  DNA, 116
  in encoding amino acids, 132
  ester linkage, 115
  metabolism disruptions, 838–841
  as monomeric units of nucleic acids,
    115–116
  nomenclature of, 822
  RNA, 116
  sequences in DNA, 146
  in signal transduction pathways, 821
Nutrition, 20–21, 21f

O-GlcNAc transferase, 354
O-linkage, 352
Obesity, 889–901
  causes of, 890–892
  diabetes and, 896–901
  dieting and, 896
  evolution and, 890–892
  health consequences of, 891t
  insulin in, 893–894
  insulin resistance and, 898, 898f
  leptin resistance in, 894f, 895–896
  overview of, 809–810
  prevalence of, 958
Obligate anaerobes, 507, 507t
Octadecanoic acid, 375
Oculocutaneous albinism, type 1, 815
Odd-chain fatty acids, 720, 720f
Odorant receptors (ORs), 204, 204f, 1100
  activation of, 1101f
  conserved and variant regions,
    1099–1100, 1100f
  evolution of, 1099–1100, 1099f
  in olfaction, 1099–1102
Off-diagonal peaks, 104
Off-target toxicity, 929
Okazaki, Reiji, 954
Okazaki fragments, 954–955, 954f, 963
Olfaction, 1098–1102. See also Sensory
    systems
  anosmias, 1099
  combinatorial mechanisms in, 1101–1102
  electronic nose in, 1101–1102, 1102f
  evolution of, 1099–1100, 1099f
  impaired, 1098
  main olfactory epithelium in, 1098, 1098f
  molecule shape and, 1098
  odorant receptors (ORs) in, 1099–1101,
    1099f, 1100f
  7TM receptors in, 1099
  signal transduction in, 1100, 1100f
Olfactory neurons, 1100, 1101f
Oligomeric 7TM receptors, 1105
Oligomerization in B-cell activation,
    1132–1133
Oligomycin, 607
Oligonucleotide-directed mutagenesis, 161,
    161f
Oligopeptides. See Peptides
Oligosaccharide chain, 855
Oligosaccharides, 360f
  antigen structures, 359, 359f
  attached to erythropoietin, 353, 353f
  defined, 349
  enzymes responsible for assembly, 358
  human milk, 352
  N-linked, 352, 352f
  O-linked, 352
  sequencing of, 352–353, 353f
Omega (Ω) loop, 42
Omeprazole (Prilosec), 938, **938**
Oncogenes, 455
One-dimensional NMR, 103, 103f
Open promoter complex, 990, 990f
Open reading frames (ORFs), 194
Operator site, 1061

Operons, 1062f
    defined, 1061
    histidine, 1069–1070, 1069f
    lac, 1061–1062, 1061f, 1062f
    structure of, 1061, 1061f
    threonine, 1069, 1069f
    trp, 1069
Opsin, 1107
Optical traps, 1159f
Oral bioavailability, 925, 926f
Origin of replication (oriC locus), 965
Origin of replication complexes, 966
Ornithine, 765–766
Ornithine transcarbamoylase, 765, 767
Ornithine transcarbamoylase deficiency,
        768–769, 768f
Orotate, 824–825, 824f, 825f
Orotate phosphoribosyltransferase, 824
Orotidylate, 824
Orthologs, 187, 187f
Orthophosphate, 681
Osteoarthritis, 355
Osteogenesis imperfecta, 50
Osteomalacia, 877
Otto, John, 332b
Overlap peptides, 94, 94f
Oxaloacetate, **252**
    in amino acid degradation, 769, 770
    in Calvin cycle, 656, 657f
    in citric acid cycle, 542, 548, 554–555
    conversion into phosphoenolpyruvate,
        522
    in fasting or diabetes, 725
    formation of, 522
    in gluconeogenesis, 519–523, 520f,
        522f, 908
    in glyoxylate cycle, 565f
    synthesis of, 480f, 482f
Oxidation
    activated carriers of electrons for, 475
    of aldehydes, 499
    carbon, 472–473
    in drug metabolism, 926, 927
    energy, 473
    in fatty acid metabolism, 714–715,
        716–719, 718t
    of glyceraldehyde 3-phosphate (GAP),
        459f, 500–501
    of membrane proteins, 719f
    phase I transformations in, 927
    pyruvate dehydrogenase complex and,
        544–545
    of succinate, 553–554
Oxidation-reduction potential.
        See Reduction potential
Oxidation-reduction reactions
    free-energy change in, 577
    in metabolism, 480–481, 481f
Oxidative phosphorylation, 573–614
    ADP in, 601–602, 602f, 603–604, 604f
    ATP yield in, 603–604, 604t
    ATP-ADP translocase in, 601–602,
        602f, 608
    cellular respiration in, 603–609
    chemiosmotic hypothesis in, 592–593,
        593f, 594f

    defined, 474, 573
    electron transfer in, 576–579
    electron-transport chain in, 599f
    in exercise, 902–905
    inhibition of, 607
    iron sulfur clusters in, 581–582, 582f
    in mitochondria, 593–594, 593f, 594f
    mitochondrial transporters in, 602–603,
        602f
    NADH in, 576, 577–579
    overview of, 573–574, 574f, 575f
    photosynthesis comparison, 633f
    proton gradients in, 474, 474f, 574,
        592–599
    proton-motive force in, 574, 593, 593f,
        595–597, 596f
    rate of, 603–607, 607f
    reduction potential in, 577–579, 577f
    respiratory chain in, 579–592
    respiratory control in, 603–604, 605f
    shuttles in, 600–603, 600f, 601f
    transporters in, 600–603, 602f
    uncoupling of, 605–607, 606f
Oxidized lipoamide, 546
Oxidoreductases, 265t
Oxidosqualene cyclase, 860
Oxoguanine-adenine base pair, 969f
Oxyanion hole
    of chymotrypsin, 279, 279f
    defined, 279
    of subtilisin, 281, 281f
Oxygen
    affinity in red blood cells, 216–217, 216f,
        217f
    bound, stabilizing, 211f
    carbon monoxide and, 217
    concentration in tissues, 213, 213f
    cooperative release of, 212
    evolution of, 268b
    in hydroxylation function, 873–874
    iron-oxygen bonding, 210f
    partial pressure, 212
    photons for generation of, 268, 268f
    reactive, 210
    toxic derivatives of, 590–591
Oxygen affinity of hemoglobin, 216–217
Oxygen binding
    with concerted model, 227f
    cooperative, 208, 212–213
    heme electronic structure and, 210
    heme in, 209–210
    by heme iron, 208–211, 209f
    hemoglobin quaternary structure and,
        213–214, 213f
    pure hemoglobin, 216, 216f
    sites, 212
Oxygen therapy, hyperbaric, 217
Oxygen-binding curve
    defined, 212
    fractional saturation, 212
    for hemoglobin, 212, 212f
    for Hill coefficients, 225, 226f
    for myoglobin, 212, 212f
    sigmoid, 212
    T-to-R transition, 213, 215f
Oxygenase reaction, 649–650, 649f

Oxygenases
    in aromatic amino acid degradation, 773–775
    mixed-function, 774
Oxyhemoglobin, **209f**, **213f**, 214
Oxymyoglobin, 208, **209f**

P clusters, 790f, 791
P site, ribosomal, 1032–1033, 1032f
p-aminobenzoic acid (PABA), 253, **253**
P-glycoprotein, 411
P-loop NTPases
    in helicases, 955
    molecular motors, 1152–1156
    proteins containing, 303–304
P-loop structures, 303–304
P-type ATPases
    defined, 403, 407
    as evolutionarily conserved, 410–411
    SERCA, 407–410
p53, 755
p160 family, 1086
P680, 626, 627
P700, 626, 630, 630f
P960, 624–625
PALA, 313–314, **313f**, 314f
Palindromes, 147, 147b
Palmitate, 717f, 718, 733
Palmitoleate, 719–720, 719f
Palmitoyl CoA, 719, 719f, 739, 740
Pamaquine, 669
Pancreas
    α cells, 890
    β cells, 892
    gene expression in, 1076, 1076t
Pancreatic failure, 898–901
Pancreatic lipases, 712, 712f
Pancreatic trypsin inhibitor, 326–328, 326f
Pancreatic zymogens, 325–326, 325f
Papain, 235
Paracoccus denitrificans, 592
Paralogs, 187, 187f
Paraquat, 638
Park, James, 930
Parkinson disease, 59–60, 753, 755
Partial pressure, oxygen, 212
Partition coefficient, 925
Passenger strand, 177
Passive membrane transport, 405
Pasteur, Louis, 492, 1144
Patch-clamp technique, 416, 416f
Pathogen-associated molecular patterns
        (PAMPs), 1121–1122, 1121t, 1122f
Pauling, Linus, 42, 209, 220, 241b, 260
PCSK9 (proprotein convertase subtilisin/
        kexin type), 869, 870
Pectin, 178
Penicillin, 930–931, **930f**
    conformations of, 260, **260f**
    mechanism of action, 259–260, 260f
    reactive site of, 258f, 259
    thiazolidine ring, 258
Penicillium citrinum, 933
Penicilloyl-enzyme derivative, 259, 259f
Pentose phosphate pathway, 658
    Calvin cycle and, 668
    in cell growth, 735

defined, 645, 658
erythrose 4-phosphate in, 662, 662f
fructose 6-phosphate in, 661, 661f
glucose 6-phosphate dehydrogenase in, 668–670
glucose 6-phosphate in, 659–660, 665–668
glyceraldehyde 3-phosphate (GAP) in, 671, 672, 672f
glycolysis and, 660–668
hummingbirds and, 672–673
illustrated, 660f
modes, 666–668
NADP$^+$ in, 665
NADPH in, 666–668
nonoxidative phase of, 659–665, 663t
overview of, 645–646
oxidative phase of, 659, 661f, 663t
rate of, 665
reactions, 662–663, 662f, 663f
ribulose 5-phosphate in, 659–660, 661, 661f, 666–667
sedoheptulose 7-phosphate in, 661, 661f, 662f
tissues with, 667t
transaldolase in, 660–663, 660f
transketolase in, 660–664, 663f
Pentose shunt. *See* Pentose phosphate pathway
Pentoses, 343
Peptide bonds
chemical nature of, 275
cis configuration, 40, 40f, 41f
cleavage of, 61, 280
defined, 37
formation in translation, 1039–1040, 1039f
formation of, 37f
hydrolysis of, 275
lengths, 40, 40f
as planar, 40, 40f
rigidity of, 42
rotation of, 41, 41f
stability of, 37
torsion angles, 41, 41b
trans configuration, 40, 40f, 41f
X-Pro linkages, 41, 41f
Peptide mass fingerprinting, 96
Peptide sequences, 55f
Peptide-ligation methods, 100
Peptides
gastrointestinal, 891t, 892
mass spectroscopy for sequencing, 92–93, 93f
nonribosomal, 738
overlap, 94, 94f
presentation by internalized proteins, 1138
presentation by MHC proteins, 1135–1136, 1135f
synthesis of, 97–100, 99f
Peptidoglycan, 259, 259f
Peptidyl transferase center, 1037
Perforin, 1138
Peripheral membrane proteins, 384, 384f
Periplasm, 394

Periplasmic side, 624
Permeability
lipid bilayers, 381–382, 382f
membrane, 373, 403, 427–428
of potassium ion channels, 419
selective, 373
Permeability coefficients, 382, 382f
Perovskite, 639
Perovskite photovoltaic cells, 639
Peroxide bridge in cytochrome *c* oxidase, 588, 588f
Peroxisome proliferator-activated receptor (PPARα), 911
Peroxisomes
in Calvin cycle, 650
defined, 593
electron micrograph of, 650f, 724f
in fatty acid oxidation, 723–724, 724f
Perutz, Max, 211
PEST sequences, 755
pH
buffers and, 16–17, 16f
carbon dioxide and, 218, 218f, 219
defined, 14
effect on carbonic anhydrase activity, 287–288, 288f
effect on oxygen affinity of hemoglobin, 218f
ionization state as function of, 32f
Phagocyte, 1120
Pharmacogenetics/pharmacogenomics in drug development, 939
Pharmacology, 923b. *See also* Drug development
Phase I transformations in oxidation, 927
Phase II transformations in conjugation, 927
Phenotype-genotype correlation, 939–940, 939f
Phenotypic screening approach, 922f, 923
Phenyl isothiocyanate, 92
Phenylalanine
degradation of, 773, **774**, 775f
structure of, 33f, 34
synthesis of, 801–803, 802f
Phenylalanine hydroxylase, 774
Phenylketonuria, 777
Phenylpyruvate, 802
Pheomelanin, 816, 816f
Phi (φ) angle, 41, 41f, 42f
Phosphatases
deficiency, 559
defined, 320
in phosphorylation, 318–321, 320t
PP2A, 320
Phosphate carrier, 602
Phosphate esters, 474
Phosphates, 17, 474
Phosphatidate, **377**, 851, **851**
Phosphatidic acid phosphatase (PAP), 851, 857–858, 857f
Phosphatidylcholine, synthesis of, 853, 853f
Phosphatidylethanolamine, 853, **853**
Phosphatidylethanolamine methyltransferase, 853
Phosphatidylinositol, 851, **851**

Phosphatidylinositol 3,4,5-trisphosphate (PIP$_3$), 450, 450f, 897
Phosphatidylinositol 4,5-bisphosphate (PIP$_2$), 444, 445f, 450, 450f, 852, 897
Phosphatidylinositol-dependent protein kinase (PDK), 897
Phosphatidylserine, 853
Phosphite triester method, 150, 150f
Phosphocreatine, **251**
Phosphodiester bridge, 125
Phosphoenolpyruvate (PEP)
formation of, 503
in gluconeogenesis, 520f
in glycolysis, 504
NAD$^+$ regeneration in, 505–507
phosphoryl-transfer potential, 470, 503
Phosphoenolpyruvate carboxykinase, 522
Phosphoenolpyruvate carboxylase, 657
Phosphofructokinase (PFK), 497
activation of, 514, 415f
in liver, 513, 514f
in muscle, 511–512, 512f
regulation of, 513, 514f
Phosphofructokinase 2 (PFK2), 526–527, 527f
Phosphoglucomutase, 686f
6-phosphogluconate dehydrogenase, 660
Phosphogluconate pathway. *See* Pentose phosphate pathway
6-phosphoglucono-δ-lactone, 660, 660f
Phosphoglucose isomerase, 496
2-phosphoglycerate in glycolysis, 501–503, 502f
3-phosphoglycerate
in Calvin cycle, 646–647, 647f
in cysteine synthesis, 797–798
in glycine synthesis, 797–798, 798f
in glycolysis, 503–504, 503f
in oxygenase reaction, 649, 650f
in serine synthesis, 797–798, 797f
3-phosphoglycerate dehydrogenase, 805–806
Phosphoglycerate kinase, 501
Phosphoglycerate mutase, 503, 685
Phosphoglycerides, 376, 399, 399f
Phosphoglycolate, 649, 649f, 650f
Phosphoinositide 3-kinases (PI3Ks), 450
Phosphoinositide cascade, 443–445, 444f, 445f. 692
Phospholipase C, 444
Phospholipid synthesis
from activated alcohol, 852–853
activated intermediate, 851–852
phosphatidylcholine, 853, 853f
sources of intermediates in, 850–858, 850f
Phospholipids
base-exchange reactions in generation of, 853–854
in bimolecular sheet formation, 380–383
flip-flop, 391, 391f
lipid vesicles formed from, 381–382
melting temperature, 391, 391f
membrane, 376–377
platform for, 376
Phosphopantetheine, 729f
Phosphopentose epimerase, 653

Phosphopentose isomerase, 193f, 653
Phosphoproteome, 323
Phosphoribomutase, 838
Phosphoribosylpyrophosphate synthetase (PRS), 335–336
5-phosphoribosyl-1-pyrophosphate (PRPP)
  in purine synthesis, 827, 837
  in pyrimidine synthesis, 804, 824–825, 825f
Phosphorolysis in glycogen metabolism, 681–686
Phosphorothioates, 293–294, 294f
Phosphoryl group, 320
Phosphoryl-transfer potential, 469, 470–471, 470f, 473–474
Phosphorylase. *See* Glycogen phosphorylase
Phosphorylase *a*, 686–687, 686f, 689, 689f, 699, 699f
Phosphorylase *b*, 686, 687f, 688, 689, 689f
Phosphorylase kinase
  activation of, 689, 689f
  defined, 689
  subunits, 689, 689f
Phosphorylation
  amplified effects of, 321
  ATP formation by, 634, 634f, 635
  cAMP stimulation of, 443
  of carboxyl-terminal domain (CTD), 1000
  of carboxyl-terminal tail, 452
  in citric acid cycle, 557
  control of, 318–320
  in covalent modification, 318, 319t
  dephosphorylation and, 320
  EGF receptor, 452
  in enzyme regulation, 317, 319t
  free energy of, 321
  of fructose 6-phosphate, 496
  in gluconeogenesis, 522, 527
  in glycogen metabolism, 681–682
  in glycolysis, 493–495, 496
  in insulin signaling, 447–451
  in metabolism, 468, 474
  of nucleoside monophosphates, 467
  oxidative. *See* oxidative phosphorylation
  in phosphorylase activation, 688–689, 689f
  in phosphorylase conversion, 688–689
  of proteins, 323
  in purine nucleotide synthesis, 827–829, 827t
  of pyruvate dehydrogenase, 557
  reactive intermediates, 349
  of serine residue, 526
  substrate-level, 502
  sugars, 348
Phosphorylation potential, 485
Phosphorylethanolamine, 853, **853**
Phosphoserine, **61f**
Phosphothreonine, 61
Phosphotriesters, 150, 150f
Photoinduced charge separation, 623–625, 623f
Photophosphorylation, cyclic, 634, 634f, 635
Photoreceptors, 1106, 1110–1111, 1110f

Photorespiration
  in C₄ pathway, 656–658
  in Calvin cycle, 650, 650f
Photosynthesis, 619–642
  accessory pigments in, 635–640
  artificial photosynthetic systems and, 639–640
  in autotrophs, 646
  bacterial reaction center, 623–626, 624f, 625f
  basic equation of, 620
  chlorophylls in, 622–626
  in chloroplasts, 621–622, 621f
  components of, location of, 637–638, 638f
  cyclic photophoshorylation in, 634, 634f, 635
  dark reactions, 620, 645
  defined, 620
  electron transfer in, 622–626
  energy conversion in, 620–621
  enzymes in, 235
  herbicide inhibition of, 638, **638**
  in heterotrophs, 620
  light absorption in, 623, 623f
  light energy conversion in, 620–621
  light reactions in, 620, 620f, 635, 638, 645
  light-harvesting complexes in, 637, 637f
  overview of, 620–621
  oxidative phosphorylation comparison, 623f
  photoinduced charge separation in, 623–625, 624f
  photosystem I. *See* photosystem I
  photosystem II. *See* photosystem II
  proton-motive force in, 632
  reaction centers in, 623–625, 624f, 626f, 629, 629f, 635–638
  resonance energy transfer in, 636–637, 636f
  in thylakoid membranes, 621–622
  yield in, 620b
  Z scheme of photosynthesis, 631, 631f
Photosynthetic catastrophe, 621f
Photosynthetic systems, 638–639, 639t
Photosystem I. *See also* Photosynthesis
  defined, 621, 626f
  electron flow through, 634
  ferredoxin generated by, 629–630
  illustrated, 625f
  inhibitors of, 638
  link to photosystem II, 629, 629f
  NADPH formation by, 633
  reaction center, 626, 629–630, 630f
  structure of, 629–630, 629f
  Z scheme of photosynthesis, 631, 631f
Photosystem II. *See also* Photosynthesis
  defined, 621
  electron flow through, 627
  electron source, 626–627
  illustrated, 625f
  inhibitors of, 638
  link to photosystem I, 629, 630f
  location of, 627–628
  photochemistry of, 626
  reaction center, 625, 626

  structure of, 626, 626f
  Z scheme of photosynthesis, 631, 631f
Photosystem II subunit S (PsbS), 641
Phototroph, 464
Phytoene, 879, **879f**
Phytol, 622
Pichia pastoris, 82
Pig embryos, UCP-1 into, 607
Ping-pong reactions, 252
Pinopsin, 1110
pKₐ values, 15, 36t
Placebo effect, 941
Planar bilayer membrane, 382, 382f
Planck's constant, 240
Plants
  C₃, 658
  C₄, 658–659
  genetically engineered, 177–178, 177f
  glyoxylate cycle in, 564–565, 565f
  oxaloacetate synthesis, 727
  photosynthesis in. *See* photosynthesis
  starch in, 654
  sucrose synthesis in, 654–655, 655f
  thylakoid membranes of, 637
Plaques in genomic library screening, 159, 159f
Plasma cell, 1141
Plasma membrane, 373
Plasma-membrane proteins, 1046, 1046f
Plasmids
  defined, 154
  engineered, 155
  expression vectors, 156
  genes, 155
  microinjection of, 171f, 172
  polylinker region, 155, 155f
  Ti, 177, 177f
  as vectors for DNA cloning, 155–157
Plasmin, 333
Plasminogen, 333
Plasmodium falciparum, 221, 364, 575
Plastocyanin, 638
Plastoquinone, 626–627, 638
Pleckstrin homology domain, 449
PLP-Schiff-base linkage, 683
Pluripotent stem cells, 1082
Point mutation, 161
Poisoning
  arsenite, 563, 563f
  mercury, 562–563
  ricin, 1045
Polar amino acids, 34–35, 34f
Polarity, 117
Poly(A) tail-binding protein I (PABPI), 1041, 1041f
Polyacrylamide gel electrophoresis (PAGE), 75–76, 76f
Polyacrylamide gels, 148
Polyampholytes, 77
Polycistronic molecules, 1034
Polyclonal antibodies, 84, 84f
Polycomb repressive complex-2 (PRC-2), 1090
Polyenes, 879
Polygalacturonase, 178

Polyketides, 738
Polylinker region, plasmid, 155, 155f
Polymerase chain reaction (PCR)
  cycle steps, 151–152, 151f, 152f
  defined, 151
  DNA sequence amplification by, 151–152, 151f, 152f
  first cycle of, 151f
  as gene exploration tool, 147
  inverse, 162, 162f
  multiple cycles of, 152f
  mutagenesis by, 162, 162f
  quantitative (qPCR), 169–170, 170f
  uses of, 152–153
Polymerase switching, 966–967
Polymerization reactions, 1157
Polymorphisms, 153, 1141, 1141f
Polyoma, 116
Polypeptide chains
  backbone, 38, 38f
  bond rotation in, 41, 41f
  cleavage of, 93, 94t
  components of, 38f
  cross-linked, 38–39, 39f
  directionality of, 38
  disulfide bonds of, 38–39, 39f
  flexibility of, 40–42
  formation of, 37
  loops, 46, 46f
  random-coil conformation, 52
  residue, 38
  reverse turns, 46, 46f
  side chains, 38, 38f, 48
  subunit structures, 51
Polypyrimidine tract, 1005
Polysaccharides
  defined, 350
  glycosidic bonds, 350, 351f
Polysomes, 1040, 1040f
Polyunsaturated fatty acids, 724, 737
Pompe disease, 700–701, 700t, 701f
Porin
  amino acid distribution in, 48, 48f
  amino acid sequence of, 358, 358f
  defined, 358
  hydropathy plot for, 389, 390f
  structure of, **358**
Porphobilinogen, 812f, 813
Porphobilinogen deaminase, 813
Porphyrias, 814
  disorders of, 814
  synthesis of, 810–811, 811f
Porter, Rodney, 1124
Positively charged amino acids, 35, 35f
Posttranscriptional gene expression
  in eukaryotes, 1090–1093
  in prokaryotes, 1068–1070
Posttranslational modifications, 60–62, 95
Potassium ion channels
  as archetypical structure, 417–418
  hERG, 427
  inactivation of, 422–423, 423f
  path through, 418, 418f
  permeability of, 423
  purification of, 417

selectivity filter, 418, 418f, 419, 419f
structure of, 417–420, 418f, 419f
transport model, 420, 420f
voltage-gated, 421–422, 422f
Potassium ions
  action potentials and, 415
  dehydration of, 419, 419f
  hydration of, 420f
Potential energy, 11
Power stroke, 1159–1160
PP2A, 320
Pre-mRNA processing, 1002, 1002t, 1003–1004, 1003f, 1010–1011, 1010f
Precose, 366
Precursor ions, 92
Predisposition, 19
Pregnenolone
  androgen synthesis by, 875–876
  corticosteroids from, 875
  defined, 875
  estrogen synthesis by, 875–877
  progesterone from, 875
  structure of, **876**
  synthesis of, 874–875
Prephenate branch, 802
Pribnow box, 130
Prilosec (omeprazole), 938, **938**
Primaquine, 669
Primary active transport, 404
Primary antibody, 87
Primary messenger, 438
Primary protein structure, 29, 37–42, 51, 83
Primase, 953
Primers, 125
  defined, 950
  DNA probes as, 151
  in DNA replication, 954, 954f
  match stringency, 152
  in polymerase chain reaction (PCR), 151–152, 151f
  RNA, 963
Prion diseases, 59, 60f
Prions, 59, 59f
Pro-N-terminal degrons, 755
Probes. *See* DNA probes
Procaspases, 324
Processivity
  defined, 684, 962, 962b
  in DNA replication, 962–963, 692f
  kinesin motion, 1164–1166
Procollagen, 323
Product ions, 92
Proenzyme, 323
Progesterone, 872, 875, **876**
Progestogen, 872
Programmed cell death. *See* Apoptosis
Proinsulin, synthesis of, 160, 161f
Prokaryotes. *See* Bacteria
Prokaryotic RNA polymerases, 985, **985**
Proliferating cell nuclear antigen (PCNA), 967
Proline
  degradation, 772, 772f
  structure of, 33f, 34
  synthesis of, 796–797

Prolyl hydroxylase, 2, 560, 912
Promoter sites
  base sequences, 130b
  defined, 130
  for transcription, 130–131, 130f
Promoters
  alternative sequences, 990, 990f
  bacterial, 989, 989f
  closed complex, 990–991, 991f
  core, 989
  defined, 989
  open complex, 991, 991f
  strong, 989
  upstream of, 989
  weak, 989
Proofreading
  in DNA repair, 970, 971f
  in transcription, 988
  in translation, 1090–1092
Propionyl CoA, 720, 721f, 771–772
6-*n*-propyl-2-thiouracil, 1103, **1103**
Prostacyclin, 737
Prostacyclin synthase, 737, 737f
Prostaglandin, 737
Prostaglandin H$_2$
  attachment of, **387**
  defined, 387
  formation of, 387, 387f
  synthase-1, 386–387, 386f, 387f
Prosthetic groups, 46, 236
Protease inhibitors
  α$_1$-antiproteinase, 327
  drugs, 285–286, 285f
  in HIV infection treatment, 285, 285f, 875, 935, 936f, 942
  initial design of, 935, 935f
  pancreatic trypsin inhibitor, 326–327, 326f
  serine, 328
Protease-substrate interactions, 280, 280f
Proteases, 360
  active sites, 284f
  applied to glycoproteins, 360
  aspartyl, **283**, 283f, 284, 284f
  catalytic triad in, 281–282
  cysteine, 283–284, **283**, 283f
  metalloproteases, **284f**, 285
  reaction facilitation, 275–286
  serine, 274
Proteasomes. *See also* Amino acid degradation
  19S regulatory unit, 754f, 755
  20S, 754f, 755
  26S, 754f, 755
  in amino acid degradation, 754–755
  defined, 751
  evolution of, 757, 757f
  free amino acid generation, 757f
  prokaryotic, 757, 757f
Protein colipase, 712
Protein Data Bank, 106
Protein domains, 48, 48f

Protein folding, 1033
  essence of, 56
  funnel, 57, 57f
  as highly cooperative process, 55–56
  illustrated, 22f
  Levinthal's paradox in, 56
  misfolding and, 900
  nucleation-condensation model, 56
  pathway of chymotrypsin inhibitor, 57f
  by progressive stabilization, 56–57
  transition from folded to unfolded, 55f
  typing-monkey analogy, 56, 56f
Protein identification
  cleavage in, 93–94, 94f
  co-immunoprecipitation in, 88, 88f
  genomic and proteomic methods as
    complementary, 94–95
  MALDI-TOF mass spectroscopy in,
    90–91, 91f
  mass spectroscopy in, 89–97, 91f, 93f, 97f
  peptide mass fingerprinting, 96
  in protein studies, 70
Protein kinase A (PKA)
  cAMP activation of, 322, 322f
  chains, 443
  consensus sequence, 320
  Cushing syndrome, 322–323
  defined, 310
  gene stimulation, 443
  in glycogen metabolism, 696, 696f
  regulation of, 322, 322f
  in signal transduction, 454
Protein kinase B (PKB), 897
Protein kinase C (PKC), 898
Protein kinase G (PKG), 459
Protein kinase inhibitors, as anticancer
    drugs, 457
Protein kinases
  catalysis, 310
  dedicated, 320
  defined, 318
  multifunctional, 320
  in phosphorylation, 319–321, 320t
  serine, 319–321, 320t
  in signal transduction, 454
  specificity, 320
  threonine, 319–321, 320t
Protein phosphatase, 320
Protein phosphatase 1 (PP1)
  functions of, 692, 696–698
  in glycogen metabolism, 692, 696–698,
    697f, 699
  phosphorylase a and b and, 698–699
  regulation in muscle, 697, 697f
  regulation of glycogen synthesis by,
    696–697, 697f
  substrate for, 699
  subunits, 697
Protein purification, 82
  affinity chromatography in, 74–75, 74f
  assays in, 71
  binding affinity and, 72–75
  cell release and, 71–72
  charge and, 72–75
  dialysis in, 73, 73f

electrophoretic analysis of, 79f
gel electrophoresis in, 75–77, 76f, 77f
gel-filtration chromatography in, 73, 73f
high-performance liquid chromatography
    (HPLC) in, 75, 75f
homogenization in, 71
ion-exchange chromatography in, 73–74,
    74f
isoelectric focusing in, 77, 77f
in protein studies, 70
quantitative evaluation of, 79–80, 79t
recombinant DNA technology in, 82–83
salting out in, 72
SDS-PAGE in, 77, 77f, 78
sedimentation coefficient and, 80, 81f
sedimentation-equilibrium technique in, 82
separation of proteins in, 75–76
size and, 72–75
solubility and, 72–75
two-dimensional electrophoresis in, 78, 78f
ultracentrifugation and, 80–82, 81f
zonal centrifugation in, 81–82, 81f
Protein sorting
  defined, 1045
  pathways, 1048, 1048f
Protein structures, 2, 3f
  amino acid sequence as determinant,
    52–62
  amino acid sequences and, 35, 35f
  complex assembly, 30, 30f
  determination by cryo-EM, 105–106,
    106f
  dictating function, 30f
  elucidation of, 106
  family of, 105, 105f
  models of, 23f, 24, 64–66
  primary, 29, 37–42, 51, 83, 194–195
  quaternary, 29, 51, 51f
  secondary, 29–30, 42–46, 48, 51
  synthetic peptides and, 98
  tertiary, 29, 30, 46–50, 51, 194–195
  three-dimensional prediction from
    sequence, 57–58
Protein studies
  enzyme-linked immunosorbent assay
    (ELISA) in, 86–87, 87f
  fluorescence microscopy in, 89, 89f
  immunology in, 83–89
  mass spectroscopy in, 89–97, 91f, 93f, 97f
  peptide synthesis, 97–100
  purification methods for, 70–83
  steps in, 69–70
  western blotting in, 87–88, 88f
  X-ray crystallography in, 100–102
Protein synthesis, 1021. See also Translation
  accuracy of, 1023, 1036–1037
  adaptors in, 131, 131f
  antibiotic inhibitors of, 1043–1045, 1044f,
    1044t, 1045f
  endoplasmic reticulum (ER) in,
    1045–1048
  eukaryotic versus bacterial, 1041–1043
  insulin in, 905
  mRNA in, 134, 134f
  peptide-ligation methods, 100

polypeptide-chain growth in,
    1022, 1022f
  ribosomes as site of, 1021, 1031–1041
Protein targeting, 1044–1049
Protein turnover. See also Amino acid
    degradation
  defined, 752
  regulation of, 753–758
Proteins, 29–62
  acetylation of, 318–319
  adaptor, 450
  affinity tags and, 82
  aggregated, in neurological diseases,
    59–60, 60f
  aggregation of, 753
  allosteric, 309, 310–317
  alpha helix of, 42–44, 43f
  amino acids. See amino acids
  antibody generation to, 83–84
  β pleated sheets of, 44–46, 45f, 46f
  binding partners of, 88
  as building block, 20
  building blocks of, 2
  carbohydrate units of, 61
  cargo, 1048–1049
  cellular, degradation of, 752–753
  chaperone, 53
  cleavage of, 93–94, 94f
  coat proteins (COPs), 1048
  coiled-coil, 49, 49f
  covalent modification of, 318–323, 319t
  crystallization of, 100
  defined, 2, 38
  degradation of. See amino acid
    degradation
  denatured, 55, 55f
  dietary, digestion and absorption of,
    752, 753f
  disulfide bonds, 38–39
  DNA-binding, 1058–1065
  encoding, 21–22
  enzymes as, 233
  evolution of, 136
  fibrous, 49–50
  flexibility of, 31, 31f
  functional groups of, 30
  functions of, 29–31
  genetically engineered, 162
  glycan-binding, 361
  glycosylation of, 353–360
  half-life of, 754, 755, 755t
  hemoglobin as model of, 207–228
  interaction of, 30
  intrinsically unstructured, 58
  as linear polymers, 29–30, 37
  loops, 46, 46f
  membrane. See membrane proteins
  metabolism, 778
  metamorphic, 58–59
  misfolding of, 59–60
  molecular weights of, 80t
  overview of, 69–70
  phosphorylation of, 323
  plasma-membrane, 1046, 1046f
  posttranslational modifications, 60–62

properties of, 29–31
pure, 71
refolding of, 53
regulatory, 1062
release factors, 134
repressor, 1061
reverse turns, 46, 46f
ribosomal. *See* ribosomes
rigidity of, 31
S values of, 80t
secretory, 1046, 1046f
secretory pathway, 1045–1046
separation of, 69, 75–79
subunits of, 51
tagging for destruction, 753–756
translocation of, 1045–1048
unfolding of, 55, 55f
unstructured, 58–59
Proteoglycans
    in cartilage, 355–356, 356f
    defined, 352–353
    properties of, 354–355
    structural roles, 354–355
Proteolysis, 235, 908, 1072
Proteolytic activation, 310, 323–334
    apoptosis by, 324
    of chymotrypsinogen, 324–325, 324f
    zymogens, 325, 325f
Proteolytic enzymes
    activation of, 323–334
    in blood clotting, 328–329
    chymotrypsinogen as, 324, 324f
    in digestion, 324–328
    function of, 235
    inhibitors of, 326–328
Proteomes, 70
Prothrombin, 328–329, 329f
Proto-oncogenes, 455
Proton abstraction, 290
Proton gradients, 579
    across thylakoid membrane, 631–635
    in ATP synthesis, 592–599
    cytochrome *bf* contribution to, 629, 629f
    directionality of, 629, 629f
    in oxidative phosphorylation, 474, 474f,
        573, 574, 579, 592–599
    power transmission by, 609, 609f
Proton shuttle, 290–291, 291f
Proton transport, 588–589, 589f
Proton-motive force
    ATP forms without, 595–596, 596f
    in oxidative phosphorylation, 573,
        592–593, 593f
    in photosynthesis, 630
Protoplasts, 177
Protoporphyrin, 209
Protoporphyrin IX, 813, 813f
Protospacer-adjacent motif (PAM), 175
Proviral DNA, 171
Proximal histidine, 209, 215
PrP, 59
Prusoff, William, 924
Pseudo-first-order reactions, 244
Pseudogenes, 165
Pseudosubstrate sequence, 322

Psi (ψ) angle, 41, 41f, 42f
Psoralens, 970, 1008
Puffer fish, genome of, 168–169, 168f
Pumps
    action, 407, 407f
    ATP-driven, 403
    calcium ion, 407, 408–411, 408f
    defined, 403
    ion gradients, 405
    MDR, 411
    Na^+-K^+, 406, 410
    purification of, 406
Pur repressor, 1063, 1063f
Purine biosynthetic pathway, 767
Purine catabolism, 839f
Purine nucleotide synthesis, 798
    AMP in, 829–830
    control of, 837–838, 837f
    de novo pathway for, 492f, 826, 826f, 827t
    enzymes of, 830
    GMP in, 829–830
    phosphorylation in, 827–829, 827t, 828f
    ribose phosphate in, 827
    salvage pathway for, 826, 830
Purine nucleotides, 114, **115f**, 826
Purinosomes, 830, 830f
Puromycin, 1044, **1044**, 1044t
Pyran, 344
Pyranose, 344, 345f, 346–347
Pyridoxal phosphate (PLP), 683, 683f
    in amino acid degradation, 752, 761–762
    in Schiff-base intermediates formation,
        759–760
    in Schiff-base linkage, 794
Pyridoxal phosphate enzymes
    bond cleavage in, 762, 762f
    reaction choice, 761–762, 762f
    stereoelectronic effects, 762, 762f
Pyridoxamine phosphate (PMP), 761, 795,
    795f
Pyridoxine (vitamin B_6), 759–760
Pyrimidine biosynthetic pathway, 767
Pyrimidine nucleotide synthesis, 808
    carbamoyl phosphate in, 823, **823**
    control of, 837
    cytidine triphosphate (CTP) in, 825–826
    de novo pathway for, 825, 825f
    glutamine in, 826
    mono-, di-, and triphosphates in, 825
    orotate in, 824–825, 824f
    substrate channeling in, 824–825, 825f
Pyrimidine nucleotides
    defined, 822
    recycling of, 826
Pyrimidines, 114, **115f**
Pyrophosphatase, 125
Pyrosequencing, 166, 167f
Pyrrolidine rings, 34, 50, 50f
Pyruvate, **251**
    in amino acid degradation, 762–763, 769,
        770–771, 770f
    in ATP formation, 502–503
    carboxylation of, 521–522, 562
    conversion into phosphoenolpyruvate,
        521–523, 522f

fates of, 505, 505f
glucose conversion into, 504–505
in glycolysis, 494f
metabolism disruption, 562–563
oxaloacetate synthesis from, 483–484,
    483f
reduction of, 506, 507f
Pyruvate carboxylase
    in citric acid cycle, 561–562
    in gluconeogenesis, 519–522, 522f
Pyruvate carboxylase deficiency (PCD), 533
Pyruvate decarboxylase, 505–506, 560
Pyruvate dehydrogenase, 560–561
    diabetic neuropathy from inhibition of,
        566–567
Pyruvate dehydrogenase complex
    carbon dioxide production, 545–547
    in citric acid cycle, 543–548, 543f, 548f,
        557–558, 557f
    components of, 545–547, 546f
    of *E. coli*, 546, 546f
    mechanism, 544–546, 545f
    reactions of, 546–547, 547f
    regulation of, 557–558, 557f
    response to energy change, 557, 557f
Pyruvate dehydrogenase in citric acid cycle,
    557–556
Pyruvate dehydrogenase kinase (PDK) in
    cancer, 559
Pyruvate dehydrogenase phosphatase, 557
Pyruvate kinase
    liver and glycolysis and, 515, 515f
    muscle and glycolysis and, 513
Pyruvate kinase M, 518

Q cycle
    defined, 584
    in respiratory chain, 585, 585f
Q pool, 581
Q-cytochrome *c* oxidoreductase
    defined, 584–585
    in respiratory chain, 579–581, 580f, 580t,
        584–585
    structure of, 585, 585f
Quantitative PCR (qPCR), 169–170, 170f
Quaternary protein structure, 29, 51f
    complex, 51f
    defined, 51
Quinones
    oxidative states of, 581, 581f
    ubiquitous, 591
Quinonoid intermediate, 761
Quorum sensing, 1066, 1066f

R groups. *See* Amino acid side chains
R state of hemoglobin, 214, 214f, 216, 225–227
Racker, Efraim, 593b
Raf, 453
Raloxifene, **1087**
Ramachandran, Gopalasamudram, 41
Ramachandran plot
    for angles of rotation, 42f
    β stands, 44f
    defined, 41
    for helices, 43, 43f

Random-coil conformation, 52
Ranitidine (Zantac), 938, **938**
Ras
  activation of, 453, 453f
  affixation to cytoplasmic face, 318
  GTPase activity of, 453, 454t
  in protein kinase cascade initiation, 453
Rate constants, 244–245
RBP4, 894–895
Reactants, 237
Reaction centers
  accessory pigments and, 635–638, 637f
  bacterial, 623–625, 624f, 625f
  cyclic electron flow in, 625, 626f
  defined, 623
  photosystem I, 626, 629, 629f
  photosystem II, 626, 627
Reaction rates
  enzymes and, 239–240, 239f
  free-energy difference and, 236
  kinetics as study of, 244–245
  rate constants, 244
Reactions
  acid-base, 14–16
  ATP hydrolysis, 302, 302f
  bimolecular, 244
  in Calvin cycle, 646–654, 653f
  carbonic anhydrases and, 286–291
  of citric acid cycle, 554
  deleterious, 499
  DNA ligase, 954, 954f
  double-displacement, 252, 253
  enzyme acceleration of, 240–244
  equilibrium, 238–239
  equilibrium constant of, 238
  in fatty acid oxidation, 718t
  in fatty acid synthesis, 729–731
  first-order, 244
  free-energy difference of, 236, 237
  of gluconeogenesis, 524t
  group-transfer, 482, 482f
  hydrolytic, 482–483, 483f
  isomerization, 482, 482f
  ligation, 483, 483f
  mechanisms, 24
  multiple-substrate, 251–252
  oxidation-reduction, 481, 481f
  oxygenase, 649–650, 650f
  pentose phosphate pathway, 662–665,
    664f, 665f
  polymerization, 1157–1158
  pseudo-first-order, 244
  reduction potentials of, 578t,
    579–581
  second-order, 244
  sequential, 251
  standard reduction potentials of, 578t
  transaldolase, 664–665, 664f
  transketolase, 662–664, 663f
  velocity versus substrate concentration,
    241, 241f
Reactive oxygen species (ROS), 672–673
  defined, 590
  glucose 6-phosphate dehydrogenase in
    protection against, 669–670

pathological conditions that may entail
    injury, 590t
  release prevention, 210
  source of, 722
Reactive substrate analogs, 256–257, 256f,
    257f
Real-time PCR, 169–170
RecA, 976, 1066, 1072
Receptor tyrosine kinase, 448
Receptor-mediated endocytosis, 395–396,
    395f, 866, 866f
Reclinomonas americana, 576
Recognition helix, 1059
Recognition sequence, 292
Recognition sites (sequences), 292
  in cognate and noncognate DNA,
    295–297, 296f, 297f
  defined, 292
  distortion of, 295–296, 296f
  EcoRV endonuclease, 295, 295f
  structure of, 295, 295t
  tRNA template, 131
Recombinant DNA
  formation of, 154–155
  manipulation of, 154
Recombinant DNA technology, 145–181
  amino acid sequence information, 95
  bacterial artificial chromosomes in,
    157–158
  in biology, 154–163
  blotting techniques in, 146, 148f, 149
  cloning in, 158–159
  cohesive-end method in, 154, 154f
  complementary DNA (cDNA) in,
    159–160, 159f
  complete genome, 163–169, 164f
  defined, 145
  disease-causing mutations, functional
    effects of, 163
  in disease-causing mutations
    identification, 153–154
  DNA probes in, 149, 150–151
  DNA sequencing in, 146, 149–150
  DNA synthesis in, 150–151, 150f
  electroporation in, 177, 178f
  eukaryotic genes, 169–178
  functional effects of disease-causing
    mutations and, 163
  gel electrophoresis in, 148–149, 148f
  gene disruption in, 172–176, 172f, 173f,
    174f, 175f, 176f
  gene expression analysis in, 169–171,
    170f, 171f
  gene insertion into eukaryotic cells, 171–172
  gene therapy in, 178
  genome comparison, 168–169, 168f
  genome sequencing, 164–165
  genomic libraries in, 158–159, 159f
  human genome, 165
  λ phage in, 155–157, 155f, 156f, 157f
  mutation creation, 160–163
  in mutation identification, 153–154
  next-generation sequencing methods,
    166–168
  overview of, 145–146

  in plants, 177–178, 177f
  plasmids in, 155–157
  polymerase chain reaction (PCR) in, 147,
    151–153, 151f, 152f
  in protein purification, 82–83
  requirement, 146–147
  restriction-enzyme analysis in, 146,
    147–148, 154–155
  RNA interference in, 176–177, 176f
  solid-phase approach in, 150–151, 150f
  tools of, 146–154
  transgenic animals in, 171f, 172
  vectors in, 155–157
  yeast artificial chromosomes (YACs) in,
    157–158, 158f
Recombinases, 977
Recombination
  in antibodies, 1129–1134, 1130f, 1131f
  in color blindness, 1111, 1111f
  DNA. *See* DNA recombination
  homologous, 1111b
Recombination signal sequences (RSSs),
    1131
Recombination synapse, 977
Redox balance, 506, 506f
Redox couples, 577
Redox potential. *See* Reduction potential
Reducing sugars, 347–348
5α-reductase, 876
Reductase mRNA in cholesterol regulation,
    862
Reduction potential
  measurement of, 576, 576f
  of NAD⁺, 578
  in oxidative phosphorylation, 576–579,
    577f
  of reactions, 578t, 579
Reed, Randall, 1099
Regulator genes, 1061
Regulatory domains
  in amino acid synthesis, 805–807
  recurring, 806f
  structures of, 806
Regulatory light chain, 1152
Regulatory protein P_{II}, 807–808, 808f
Regulatory proteins, 1062–1063
Regulatory sites, 311
Relaxed molecule, 120
Relay helix, 1155, 1155f
Release factors
  defined, 134
  ribosome (RRF), 1040
  translation termination by, 1040–1041,
    1040f
Repeating motifs, 96, 96f, 196, 196f
Replication factor C (RFC), 967
Replication fork. *See also* DNA replication
  defined, 953
  schematic view of, 963
Replication protein A, 966
Replicon, 966
Reporter genes, 155
Repressors
  corepressors, 1063
  lac, 1061–1062, 1062f, 1063f

λ, 1065–1068, 1068f
  protein, 1061
  pur, 1063, 1064f
Residue, 38, 50
Resistin, 895
Resonance energy transfer, 636–637, 636f
Resonance structures
  ATP (adenosine triphosphate), 469
  defined, 7
  depiction of, 7
  improbable, 470f
Respirasome, 399, 580, 589, 590f
Respiration, 574b
Respiratory chain
  coenzyme Q in, 581–582
  components of, 580f, 580t
  cytochrome *c* oxidase in, 579, 580f, 580t,
    584–588
  defined, 573
  electron transfer in, 591–592, 591f
  NADH-Q oxidoreductase in, 579–581,
    580f, 580t, 582–583, 583f
  in oxidative phosphorylation, 579–592
  Q cycle in, 585, 585f
  Q-cytochrome *c* oxidoreductase in, 580,
    580f, 580t, 584
  succinate-Q reductase in, 580, 580f, 580t,
    584
Respiratory control, 603–604, 604f
Respiratory distress syndrome, 856
Respiratory quotient (RQ), 904
Restriction enzymes (endonucleases), 274
  analysis, 147
  defined, 147, 292
  in DNA cleavage, 147–148, 291–298
  in *E. coli*, 292
  in forming recombinant DNA molecules,
    154–155
  hydrolysis of phosphodiester bond,
    292–293, 292f
  inverted repeats, 295, 295f
  recognition sites, 295, 295f, 296–297, 296f
  restriction-modification systems, 297
  specificity, 147, 147f, 295, 297
  type II, 298
Restriction-fragment-length polymorphisms
    (RFLPs), 153
Restriction fragments
  separation by gel electrophoresis,
    148–149, 148f
  in Southern blotting, 149
Restriction-modification systems, 297
11-*cis*-retinal, 1107, 1108–1109, 1109f
Retinitis pigmentosa, 1010–1011
Retrovir, 285
Retroviruses
  defined, 126
  in drug resistance evolution, 942
  flow of information in, 126–127, 127f
  in gene introduction, 171
Reverse cholesterol transport, 870
Reverse transcriptase, 126–127, 127f
Reverse turns
  amino acid residues in, 54–55
  defined, 46

Reversible covalent modification, 310, 483,
    807–808, 808f
Reversible inhibition, 253, 253f. *See also*
    Enzyme inhibition
Reversible terminator method, 167
Rhabdomylosis, 702
Rhizobium bacteria, 789
Rho, 992, 992f
Rhodopseudomonas viridis, 623, 624f, 638–639
Rhodopsin, 440, 440f, 1107–1108, 1108f
Rhodopsin kinase, 1108
Rhodospirillum rubrum, 592
Ribbon diagrams, 65–66, 66f
Ribonuclease
  amino acid sequences in, 52–53, 52f
  denatured, 53
  reduction and denaturation of, 53f
  sequence comparison of, 185–186, **186f**
  structure of, 52–53, 52f
Ribonuclease III (RNase III), 995
Ribonuclease P (RNase P), 995
Ribonucleic acid (RNA). *See* RNA
Ribonucleotide reductase
  as cancer therapy target, 838–839
  defined, 831
  in deoxyribonucleotide synthesis, 831–833
  elements of, 831, 831f
  mechanism, 832, 832f
  regulation of, 837, 837f
  stable radicals, 833
  structure of, **831**
  tyrosyl radicals, 831, 831f
Ribose, 114, **114f**, 344, 345f, 346f
Ribosomal initiator element (rInr), 998
Ribosome release factor (RRF), 1041
Ribosomes
  bacterial, 1041
  catalysis by proximity and orientation,
    1037
  defined, 134, 1021
  endoplasmic reticulum bound, 1045–
    1048, 1046f
  eukaryotic, 1041
  50S subunit of, 1031–1032, 1032f, 1037,
    1040
  polysomal, 1040, 1040f
  in protein synthesis, 1021, 1031–1041
  schematic representation, 1034f
  70S subunit of, 1031, 1031f
  sites of, 1032, 1032f
  structure of, 1031, 1031f
  subunits of, 1030–1031, 1031f
  30S subunit of, 1030–1031, 1031f, 1036,
    1036f
  in translation, 1026–1041
  tRNA-binding sites, 1032, 1032f
Ribosomopathies, 1051
Riboswitches, 992, 992f
Ribozyme, 1012
Ribulose 1,5-bisphosphate, 646–647, 648f,
  649f, 651–654, 652f, 653f
Ribulose 5-phosphate
  in Calvin cycle, 659, 660f
  in pentose phosphate pathway, 660, 661,
    666–668

Ricin, 1045, 1045b
Ricinus communis, 1045, 1045f
Rickets, 877
Rickettsia prowazekii, 576–577
Rieske center, 584
Rifampicin, 567, 994, 994f
Riociguat, 459
Ritonavir, 875
RNA, 6. *See also* Transcription
  alignment, 204–205
  ATP binding of, 201–202, 201f, 202f
  backbone of, 114, 115f
  base composition of, 129, 129t
  complex structure, 122f
  defined, 115
  double-stranded, 176–177, 176f
  in *E. coli*, 127, 127t
  enzymes from, 1017
  flow of information from DNA to,
    126–127, 127f
  in gene expression, 127–128
  guide strand, 177
  hybridization with, 146
  as linear polymer, 113, 114
  messenger. *See* mRNA (messenger RNA)
  micro, 1004, 1004f, 1092–1093, 1092f
  nucleotides, 116
  passenger strand, 177
  primer, 963
  ribosomal. *See* rRNA (ribosomal RNA)
  single guide (sgRNA), 175, 175f
  small interference (siRNA), 176f, 177
  small nuclear (snRNA), 1007–1008
  small regulatory, 1004, 1004f
  splicing of, 1005–1008, 1005f, 1006f,
    1007t
  stem-loop structure, 121, 121f
  structure of, 121, 121f, 122f, 202, 202f
  synthesis of, 128–129
  transfer. *See* tRNA (transfer RNA)
RNA editing, 1004–1005, 1005f
RNA interference, 176–177, 176f
RNA polymerase holoenzyme complex,
  989, **990f**
RNA polymerase I
  in RNA synthesis, 997, 998, 998f
  rRNA production, 1001–1002
RNA polymerase II
  promoter region, 998
  in RNA synthesis, 997, 998, 998f
RNA polymerase III, 998–999, 998f, 1001,
  1002f
RNA polymerases, 128–129, 998f
  activated precursors, 129
  active site of, 986
  backtracking and, 988–989, 988f
  catalytic action of, 986–995
  components of, 129
  defined, 983, 984
  divalent metal ion, 129
  in DNA replication, 954, 954f
  DNA templates and, 129–130, 129t, 986
  in *E. coli*, 131, 131f, 986–987, 986t
  eukaryotic, 984, **985**, 997–999, 997t, 998f
  eukaryotic promoter elements, 998, 998f

RNA polymerases (*Continued*)
functions of, 984–985
illustrated, 128f
prokaryotic, 984, **984**
promoter sites in, 989–990
in proofreading, 988
RNA synthesis by, 128–129, 129f
RNA-directed, 126
structures of, 984, **984**
subunits of, 986, 986t, 990
in transcription, 986–995
transcription mechanism, 129, 129f
RNA processing
eukaryotic pre-rRNA, 1002, 1002f
of pre-mRNA, 1002f, 1003–1004, 1003f
in transcription of eukaryotes, 996, 1010, 1010f
tRNA, 1002, 1003f
RNA sequences, comparison of, 197–198, 198f
RNA splicing. *See* Splicing
RNA-directed RNA polymerase, 126
RNA-DNA hybrid
backtracking, 988, 988f
during elongation, 994
lengths of, 988
separation, 988, 988f
translocation, 988, 988f
RNA-induced silencing complex (RISC), 176f, 177, 1092
RNA-Seq, 180
Rods, 1106–1107, 1106f
Rofecoxib (Vioxx), 937, **937**, 942
Rotational catalysis in ATP synthase, 597, 597f
Rotenone, 607
rRNA (ribosomal RNA)
in base pairing, 1031
defined, 128
folding of, 1031, 1032
in structural scaffolding, 1031
transcription of, 995–996, 995f, 1002, 1002f
in translation, 1031–1032, 1033f
types of, 128
Rubisco
in Calvin cycle, 646–649, 646f, 647f, 648f, 655–656
defined, 646
magnesium ion in, 647, 647f
oxygenase reaction, 648–649, 648f
structure of, 646, 647f
Rubisco activase, 649

S_1 pockets, 280, 280f
Saccharomyces cerevisiae, 164, 1076
Sakmann, Bert, 416
Salicin, 932
Salicylic acid, 932, **932**
Salmonella test, 975, 975f
Salt bridges, 218, 218f
Salt fractionation, 79
Salting out, 72

Salty taste, 1102, 1105–1106
Salvage pathways. *See also* Nucleotide synthesis
defined, 822
for IMP and GMP synthesis, 841
purine, 826, 830
pyrimidine, 826
Sample half-cell, 577
Sandwich ELISA, 87, 87f
Sanger, Frederick, 39, 149, 163
Sanger dideoxy method, 149, 149f
Sarcin-ricin loop (SRL), 1036
Sarcomere, 1160–1161, 1160f
Sarcoplasmic reticulum Ca^{2+} ATPase. *See* SERCA
Saturated fatty acids, 724
Schiff, Hugo, 665
Schiff bases, 493, 664, 665, 683
in amino acid degradation, 759–761
in amino acid synthesis, 791
in aminoacrylate, 804, **804**
defined, 683b
in retinal, 1107, 1108f
Scoring system, 192, 192b, 192f
Scrapie, 59
Screening libraries, 933–935, 935f
Screw sense, 43, 43b, 43f
Scurvy, 60, 911
SDS (sodium dodecyl sulfate), 76
SDS-PAGE, 77, 77f, 78
SDS-polyacrylamide gel electrophoresis, 383–384
Second Law of Thermodynamics, 11–13
Second messengers, 438–439, **439**, 446, 454–455, 459
Second-order reactions, 244
Secondary active transport, 404
Secondary antibody, 87–88
Secondary protein structure, 29–30, 42–46, 48, 51, 54t
Secondary transporters, 413, 413f
Secretory pathway, 1046
Secretory proteins, 1046, 1046f
Sedimentation coefficients, 80, 81f
Sedimentation equilibrium, 82
Sedoheptulose 7-phosphate, 661, 661f, 662f
Segmental flexibility, 1125, 1125f
Selectins, 362–363
Selective estrogen receptor modulators (SERMs), 1087
Selective permeability, 373
Selectivity filter, 418–419, 419f
Self-splicing
defined, 1013
example of, 1013, 1013f
intron structure, 1013f
mechanism, 1013, 1014f
Self-tolerance, 1143
Semiconservative replication, 122, 123f, 124f
Sensory systems, 1097–1116
brain connection, 1098, 1098f
hearing, 1111–1113
olfaction, 1098–1102
overview of, 1097–1098

taste, 1102–1106
touch, 1113–1115
vision, 1106–1111
Sequence alignment
with conservative substitution, 192f
defined, 187
distant evolutionary relationships, 190–193
with gap insertion, 189, 189f
of hemoglobin, 193f
homologous sequences, 193–194
of identities only versus Blosum-62, 192f
of leghemoglobin, 193f
of repeated motifs, 196, 196f
for residues identifying, 203–204
scoring system, 192, 192b, 192f
shuffling and, 189–190, 189f, 190f
statistical analysis of, 187–194
statistical comparison of, 190, 190f
three-dimensional structures in evaluation of, 195
Sequence comparison methods, 185–186, 194
Sequence identities, 188
Sequence template, 195
Sequential model
allosteric enzymes, 315
of hemoglobin-oxygen binding, 215, 215f
Sequential reactions, 251–252
SERCA
defined, 407
P-type ATPase, 407–410
pumping by, 408–409, 408f
structure of, 408
Serine, 37f
in catalytic triads, 278–280
in chymotrypsin, 276, 276f
in cysteine synthesis, 797, 801
defined, 34–35
from glycine, 770
in glycine synthesis, 797
pyruvate formation from, 770, 770f
in sphingolipid synthesis, 854–855, **854**
sphingosine from, 808
structure of, 34f
synthesis of, 797, 797f, 801
Serine dehydratase, 763
Serine hydroxymethyltransferase, 797
Serine kinases, 319–320, 320t
Serine protease inhibitors, 326
Serine proteases
catalytic strategies of, 274
convergent evolution and, 197
Serotonin, synthesis of, 808, 808f, 810, 811f
Serpins, 326, 328
Seven-transmembrane-helix (7TM) receptor, 204
biological functions mediated by, 440t
defined, 440
in drug development, 938
in glycogen metabolism, 690–692, 691f
ligand binding to, 440–442
in olfaction, 1099
oligomeric, 1105
phosphoinositide cascade activation, 444–445, 445f

in signal termination, 444f
in taste, 1103–1105, 1104f
in vision, 1106–1111
70S initiation complex, 1035–1036, 1036f
70S ribosomes, 1031, 1032f, 1035–1036
Severe acute respiratory syndrome (SARS), 939
Severe combined immunodeficiency (SCID), 178, 839
sGC activators, 459
SgRNA (single guide RNA), 175, 175f
Shaker channel, 417, 422
Shaker gene, 417
Shape complementarity, 952, 952f, 953f
Shemin, David, 809b, 810
Shikimate in amino acid synthesis, 801–804
Shine, John, 134, 1034
Shine–Dalgarno sequences, 134, 1034
Short interspersed elements (SINES), 165
Shuffled sequences, 189–190, 189f, 190f
Sialic acids, 855
Sick sinus syndrome, 432
Sickle-cell anemia, 18, 19, 220–221, 220f, 221f
Sickle-cell trait, 221, 222f
Sickled red blood cells, 220f
Side chains
    amino acid. *See* amino acid side chains
    hydrophilic, 48
    hydrophobic, 48
    polypeptide chains, 38, 38f, 48
Side effect, 924
Sigma (σ) subunit, 989–990
Sigmoid oxygen-binding curve, 212
Sigmoidal curve, 315, 315f
Sigmoidal kinetics, 311, 311f
Signal peptidase, 1046
Signal sequence, 1046
Signal transduction, 437–459
    abnormalities in cancer, 456, 456f, 457f
    abnormalities in cholera, 457
    abnormalities in whooping cough, 457
    adaptor proteins in, 450
    angiotensin II signaling, 439–447
    β-adrenergic receptor in, 439–440, 440f
    calcium ions in, 445–447, 446f, 454–455
    calmodulin in, 447, 447f
    cAMP in, 442–443
    cross talk, 439
    defects, 455–457
    defined, 438
    EGF signaling, 451–454
    element recurrence in, 454–455
    epidermal growth factor (EGF) in, 437–438, 451–454
    epinephrine in, 439–447
    evolution of, 455
    function of, 437, 438f
    G proteins in, 440–443, 441f, 442f
    GTPases and, 454, 454t
    insulin in, 447–451
    molecular circuits, 438–439
    monoclonal antibodies as inhibitors, 456
    nucleotides in, 821
    in olfaction, 1100, 1100f

overview of, 437–439
primary messenger, 438
principles of, 438–439, 438f
protein kinase A in, 443, 454
protein kinases in, 454
Ras in, 453, 453f
second messengers, 438, 446, 454–455
secondary messengers, **439**
7TM receptors, 440–442, 440t, 444f
Src homology domains in, 450, 450f, 453, 455–456, 456f
in vision, 1109, 1109f
Signal-recognition-particle (SRP), 1047, 1047f
Sildenafil, 931–932
Simple diffusion, 405
Single guide RNA (sgRNA), 175, 175f
Single-molecule studies, enzyme, 261–262, 262f
Single-particle analysis, 105
Single-stranded DNA, 148
Single-stranded-binding protein (SSB), 962, 963f
Singulo method, 261
Sinoatrial (SA) node, 431
Site-directed mutagenesis, 161, 161f
Skeletal muscle glycogen phosphorylase, 702–703
Slack, C. Roger, 656
Sliding DNA clamp, 962, 962f
Sliding-filament model, 1161–1162, 1161f
Small G proteins, 453
Small interference RNA (siRNA), 176f, 177
Small nuclear ribonucleoprotein particles (snRNPs), 1007–1008, 1007t
Small nuclear RNA (snRNA), 1007–1008
Small nucleolar ribonucleoprotein (snoRNPs), 1002
Smell. *See* Olfaction
SNARE proteins, 396–397, 397f
Social interactions in gene expression, 1066
SOD1 gene, 163, 171
Sodium dodecyl sulfate (SDS)
    in gel electrophoresis, 76–77, 76f
    PAGE, 77, 77f
Sodium ion channels
    amiloride-sensitive, 1105–1106, 1106f
    inactivation of, 422–423
    paddles, 426
    purification of, 417–418
    in taste, 1105–1106
Sodium ions, 415
Solid-phase DNA synthesis, 150–151, 150f
Solid-phase peptide synthesis, 99f, 100
Soluble antibody molecule, 137f
Soluble guanylate cyclase (sGC), 459
Somatic mutation, antibody diversity and, 1131–1132
Sonicating, 381
Sos, 453, 453f
Sour taste, 1102, 1106
Southern, Edwin, 149
Southern blotting, 148f, 149, 172

Space-filling models, 23f, 24, 64–65, 65f, 211f
Special pair, 624–625
Specific activity, 71
Specificity constant, 249–250
Sperm whale myoglobin, 208
Sphingolipids
    ceramide and, 854–855, 854f
    defined, 854
    diversity and, 856
    gangliosides, 855–856
    synthesis of, 854–856, 855f
Sphingomyelin
    defined, 377–378, 855
    structure of, **378**
Sphingosine, 376
    defined, 377–378
    structure of, **378**
    synthesis of, 808, 808f
Spina bifida, 841
Spleen tyrosine kinase (Syk), 1132
Spliceosomes
    assembly and action, 1007–1010, 1009f
    catalytic center, 1009, 1009f
    defined, 136, 1006
    structure, 1008f
Splicing
    alternative, 137, 137f, 1002, 1011–1012, 1011f, 1012t
    branch points in, 1005, 1007f
    catalytic center, 1009, 1009f
    defined, 136, 1005
    group I, 1014, 1015f
    group II, 1014, 1015f
    mechanism, 1006, 1006f
    mutations affecting, 1005–1006, 1010–1011, 1011f
    pathways comparison, 1015f
    self-splicing and, 1012–1014, 1013f, 1014f
    sites of, 1005–1006, 1005f
    snRNPs in, 1007–1010, 1007t
    spliceosomes in, 1006, 1007–1010, 1008f
    transesterification in, 1006–1007, 1009f
Splicing factors, 1007
Split genes, 136
Split-pool synthesis, 934, 934f
Spudich, James, 1159
Squalene
    cyclization of, 860, 862f
    synthesis of, 859–860, 861f
Squalene synthase, 860
Src
    affixation to cytoplasmic face, 318
    in cancer, 455
Src homology 2 (SH2) domain, 450, 450f, 455–456, 456f
Src homology 3 (SH3) domain, 453, 455, 456f
SREBP (sterol regulatory element binding protein), 861–863, 863f, 867
SREBP cleavage activating protein (SCAP), 862–863, 863f
SRP receptor, 1047, 1047f
Stacking forces, 119

Stahl, Franklin, 122, 123
Standard free-energy change, 237–239
Standard reference half-cell, 577
Staphylococcus aureus, 259, 259f
Starch, 350, 654
Starvation, metabolic adaptations in, 907–909, 908t
Starved-fed cycle, 905–907
Statins, 871
Steady-state assumption
    defined, 246
    in enzyme kinetics, 246
Steady-state system, 246
Stearoyl CoA desaturase, 736, 736f
Steatorrhea, 712
Steitz, Tom, 951
Stem-loop structure, 121, 121f
Stereochemical renderings, 23
Stereochemistry
    of cleaved DNA, 293–294, 294f
    observation of, 293
    of proton addition, 795, 795f
Stereocilia, 1112–1113
Stereoelectronic control, 762, 762f
Stereoisomers
    carbohydrate, 343, 343f
    notation for, 31b
Steric exclusion, 41
Steroid hormones
    anabolic, 1087
    binding and activation, 872
    classes of, 872
    defined, 872
    hydroxylation of, 873–874
    identification of, 872–873
    synthesis of, 872–874, 872f
Sterol regulatory element (SRE), 861
Sterol regulatory element binding protein (SREBP), 861–863, 863f, 867
Sticky ends, DNA, 154, 154f
Stoichiometry
    of gluconeogenesis, 524
    for light reactions, 634
    of palmitate synthesis, 733
Stop codons, 133
Strand invasion, 976, 976f
Strand separation, 151, 151f
Streptococcus pyogenes, 175
Streptomyces achromogenes, 147
Streptomyces coelicolor, 990
Streptomyces lividans, 417–418
Streptomycin, 1043, **1043**, 1044t
Strimvelis, 178
Stroma, 621
Stroma lamellae, 621–622
Strominger, Jack, 930
Structural genes, 1061
Structure–activity relationship (SAR), 935–936
Structure-based drug development, 935–937, 936f, 937f
Substituted enzyme intermediates, 252
Substitution matrix, 190–193, 191f, 192f, 193f

Substitutions
    in amino acid sequences, 190–193, 191f, 192f, 193f
    conservative, 190, 192f
    nonconservative, 190, 192f
Substrate binding
    in catalysis, 243–244
    conformation selection, 243
    cooperative, 252–253
    induced-fit model, 243, 243f, 274
    lock-and-key model, 243, 243f
Substrate channeling
    in amino acid synthesis, 804, 804f
    defined, 556
    in pyrimidine synthesis, 824, 824f
Substrate cycles, 528, 528f
Substrate-induced cleft closing, 496
Substrate-level phosphorylation, 502
Substrates. *See also* Enzyme-substrate (ES) complex
    accessibility of, controlling, 485
    in biochemical reactions, 251–252
    chromogenic, 277, 277f
    concentration, 245, 245f, 247
    defined, 235
    enzyme interaction, 242
    homotropic effect on allosteric enzymes, 315
    multiple, in reactions, 251–252
    reciprocal relation, 837
    spectroscopic characteristics of, 241
Subtilisin
    catalytic triad of, 281–282, 282f
    oxyanion hole of, 281–282, 282f
    site-directed mutagenesis of, 282, 282f
    structure of, 197, **197f**
Subunit, polypeptide chain, 51
Subunit vaccines, 1145
Succinate
    in glyoxylate cycle, 564, 565f
    oxidation of, 554–555
Succinate dehydrogenase in citric acid cycle, 554–555
Succinate-Q reductase, 580–581, 580f, 580t, 584
Succinyl CoA
    in amino acid degradation, 769, 771, 771f
    in citric acid cycle, 552–553, 553f
    formation of, 552, 721–723, 721f, 723f
    porphyrins from, 810–813, 812f
Succinyl CoA synthetase
    in biochemical transformation, 553
    reaction mechanism of, 553, 553f
Sucrase, 349, 493
Sucrose
    defined, 350
    structure of, 349–350, 350f
    synthesis of, 654, 655f
Sucrose 6-phosphate, 654, 655
Sugar pucker, 120f
Sugar–phosphate backbone, 114–117, 118
Sugars
    in cyclic forms, 344–346
    DNA, 5
    five-carbon, 651, 652f, 659–665
    monosaccharides, 342–349

    N-linked, 357
    O-linked, 357
    phosphorylation, 349
    pucker, 120f
    reducing, 347–348
Suicide inhibition, 257, 258f, 835, 840
Suicide inhibitors, 257
Sulfhydryl groups, 35
Sulfanilamide, 253, **253**
Sulfolipids, 621–622
Super compensation (carbo-loading), 905
Supercoiling, 120
    ATP hydrolysis in, 960–961, 960f, 961f
    catalyzation of, 960
    defined, 956
    degree of, 956–958
    DNA condensation from, 958–959
    linking number and, 956–958, 957f
    negative, 958, 958f, 960–961
    positive, 958
    relaxation of, 959–960, 959f
    right-handed, 956–957
    topoisomerases and, 959–961, 959f, **960**, 961f
    twist in, 957–958, 957f
    writhe in, 957–958, 957f
Superhelical cable, 50
Superhelix, 120
Superoxide anion, 210
Superoxide dismutase, 590–591, 590f
Superoxide radicals, 590–591
Supersecondary structures
    defined, 48
    helix-turn-helix, 48f
Suppressors of cytokine signaling (SOCS), 895–896, 895f
Surface complementarity, 11
Surroundings, 11
Svedberg units, 80
Sweet tastants, 1116
Sweet taste, 1102, 1105
Symmetry matching, 1059
Symporters, 413, 413f, 414
Synaptic cleft, 423, 423f
Synchrotron radiation, 101
Synonyms, 133
Synthase, 549b
Synthetic analog system model, 289–290, 289f
Synthetic oligonucleotides, 202
Synthetic peptides
    as antigens, 97
    construction of, 98–100, 99f
    as drugs, 98, 98f
    linking of, 100
    in receptor isolation, 97
    in three-dimensional structure of proteins, 98
Synthetic vaccines, 163
Systems, 11

T-cell receptors (TCRs)
    CD3 and, 1138, 1138f
    CD4 and, 1140, 1140f, 1142–1143
    CD8 and, 1137–1139, 1138f, 1142–1143
    CD28 and, 1138

CD45 and, 1138
defined, 1120, 1136–1137
docking mode, 1124
domains of myosin, 1136–1137, 1136f
genetic diversity of, 1123, 1136–1137
helper T cells and, 1139–1141
killer T cells and, 1137–1139
MHC proteins and, 1137–1141, 1137f
structure of, 1136f
in T-cell activation, 1138, 1138f
T cells
activation of, 1138, 1138f
defined, 1123
helper, 1123, 1139–1141
HIV infection and, 1141–1142
killer, 1123, 1137–1139
memory, 1144
negative selection in, 1142–1143, 1143f
positive selection in, 1142, 1143f
T state of hemoglobin, 214, 214f, 215, 218, 225–227
t-SNARE, 1048
T-to-R equilibrium, 316–317, 316f
T-to-R transition, 215f
T-to-R state transition, ATCase, 314–315, 314f
T1R proteins as sweet receptors, 1105, 1105f
T2R proteins as bitter receptors, 1103–1104, 1104f
Tafazzin, 399
Tamoxifen, **1087**, 1087f
Tandem mass spectroscopy, 92–93, 93f
Tanning process, 816
TAP proteins, 1135
Taq DNA polymerase, 152
Target-based screening approach, 922, 922f
Taste, 1102–1106. *See also* Sensory systems
anatomic structures in, 1103, 1103f
bitter, 1102, 1103–1104
ion channels in, 1105–1106, 1106f
salty, 1102, 1105–1106
7TM receptors in, 1103–1104, 1104f
sour, 1102, 1106
sweet, 1102, 1105
tastants in, 1102–1103, **1102f**
types of, 1102
umami, 1102, 1105
Taste buds, 1103, 1103f
Taste receptor, 1116
TATA box
defined, 130
in transcription of eukaryotes, 998, 998f
TATA-box-binding protein (TBP), 196, 196f, 999–1000, 999f, 1088
TATA-box-binding protein associated factor (TAF), 1088
Taxol, 921, **921**, 1164
Tay–Sachs disease, 856
Telomerase, 968, 968f
Telomeres
defined, 967
in DNA replication, 967–968, 968f
Temperature, 12
Template, DNA
coding strands and, 986, 986f

complementarity with mRNA, 130, 130f
defined, 951b
DNA polymerases, 951
lagging strand, 963
in replication, 951
RNA polymerases and, 129–130, 129t, 985
transcribed regions of, 991–992
transcription bubbles and, 991
transcription in, 131
Template-directed enzyme, DNA polymerase as, 125
Template-recognition site, 131
Termination utilization substance (Tus), 965
Ternary complex, 251
Terpenes, 878
Tertiary protein structure, 29, 30, 46–50, 51, 194–195, 194f
Testosterone, 872, 876, **876**, 877f
Tetrahydrobiopterin, 774, **774**
Tetrahydrofolate, 797f
in amino acid synthesis, 798–799
defined, 798
one-carbon groups carried by, 798t
Tetrahymena, 1012, 1014
Tetrapyrrole, 813
Tetraubiquitin, **754**
Tetroses, 342
Thalassemia, 221–222, 1010, 1011f
Thalassemia major, 222
Therapeutic index, 929–930
Thermodynamics
of coupled reactions, 465–466, 468–469
laws of, 11–13
of metabolism, 465–466
Thermogenin, 605
Thermus aquaticus, 152
Thiamine (vitamin B₁), 479t, 562
Thiamine pyrophosphate (TPP)
in citric acid cycle, 545, 545f
deficiency, 563
defined, 562
Thick filaments in myosin, 1161, 1161f
Thin filaments in actin, 1161
Thioester intermediate, 500–501, 501f
Thioesterase, 731
Thioether groups, 34
6-thioguanine, 940, **940**
Thiol groups, 35
Thiopurine methyltransferase, 940
Thioredoxin, 633
in Calvin cycle, 656, **656**
enzyme activation by, 656f
enzymes regulated by, 656t
Thioredoxin reductase, 833
Thiram, 690, **690**
-35 region, 130
30S initiation complex, 1035, 1036f
30S subunit of ribosomes, 1031, 1032f, 1035
Three-dimensional structure
conservation of, 194–195, 194f
convergent evolution, 196–197
repeated motifs, 196
RNA sequences, 197–198
in sequence alignment evaluation, 195
tertiary structure and, 194–195

in understanding evolutionary relationships, 193
3'-to-5'phosphodiester linkages, 115, 115f
Threonine
defined, 34, 35
degradation of, 772
pyruvate and, 770
structure of, 34f
synthesis of, 806, 806f
Threonine deaminase, 772, 805–806, 805f
Threonine dehydratase, 763, 772
Threonine kinases, 319–320, 320t
Threonine operon, 1069f, 1070
Threonyl-tRNA synthetase, 1028–1029
Threonyl-tRNA synthetase complex, 1030, **1030**
Threshold effects
allosteric enzymes, 315, 315f
defined, 316
Thrombin, 235, 235f
antithrombin and, 332–333
in blood clotting, 329–330, 332
dual function of, 332
inhibitors of, 332
Thromboxane synthase, 737
Thromboxanes, 737
Thylakoid membranes
defined, 621–622
pH gradient across, 632
photosynthesis in, 621–622
photosystem II and, 628
proton gradient across, 631–635
stacked, 637
unstacked, 637
Thylakoid spaces, 621–622, 631–632
Thylakoids, 621
Thymidine, 116, 826
Thymidine kinase, 826
Thymidine monophosphate (TMP), 835
Thymidine phosphorylase, 826
Thymidylate, 116
blocking synthesis of, 835–836
defined, 834
synthesis of, 834, 834f
Thymidylate synthase, 834
Thymine, **5**, 5f, 114, **114f**, 118, **118f**, 826
in DNA repair, 973–974
synthesis of, 798
Thymocytes, 1142
Thyroxine, synthesis of, 808, 808f
Time-of-flight (TOF) mass analyzer, 90, 91f
Tip links, 1112, 1112f, 1171
Tissue factor (TF), 328
Tissue factor pathway inhibitor (TFPI), 332
Tissue plasminogen activator (TPA), 137f, 333, 333f
Titin, 38
Titration, 16
Tobacco mosaic virus, 126
Toll-like receptors (TLRs)
defined, 1120
extracellular domain of, 1121, 1121f
illustrated, 1120f
PAMPs and, 1121–1122, 1121t, 1122f

Tonegawa, Susumu, 1130
Topoisomerases
  bacterial, 960–961
  defined, 958
  type I, 958–960, 959f
  type II, 958, 960–961
Topoisomers, 956–958, 958f
Torpedo marmorata, 423, 424f
Torr, 212b
Torricelli, Evangelista, 212b
Torsion angles, 41, 41b
Tosyl-L-phenylalanine chloromethyl ketone
    (TPCK), 257, **257f**
Touch, 1113–1115
Toxicity, drug, 929–930, 929f
Toxins, 1144
Toxoid vaccines, 1145
Trace elements, 20
Trans configuration, peptide bonds,
    40, 40t, 41t
Trans fat, 724
Trans unsaturated fatty acids, 724
Transaldolase
  in pentose phosphate pathway, 660–662,
    664–665, 664f
  reactions, 664–665, 664f
Transaminases
  in amino acid degradation, 758–761
  defined, 758
  mechanism, 760, 760f
Transamination
  in amino acid degradation, 759–761
  amino acid synthesis by, 794–795, 795f
  mechanism of action, 760, 760f
  Schiff-base intermediates in, 760
Transcription, 128, 983–1017
  ATP hydrolysis in, 993, 993f
  in bacteria, 985–995
  of β-globin gene, 136, 136f
  chemistry of, 985
  defined, 983–984
  in DNA templates, 131
  elongation, 984, 987, 987f, 991, 991f
  in gene expression, 1040
  inhibition by antibiotics, 994–995, 994f
  mRNA processing and, 1010, 1010f
  nuclear hormone receptors in, 1085–1086,
    1086f
  overview of, 983–984
  promoter sites for, 130–131, 130f
  promoters in, 988–989, 989f
  proofreading in, 988
  regulation of, 996–1001
  rho in, 993, 993f
  riboswitches in, 992, 992f
  RNA polymerases in, 129, 129f, 985–995
  of rRNA, 995, 995f, 1001–1002, 1002f
  self-splicing in, 1012–1014, 1013f, 1014f,
    1015f
  splicing in, 984, 1001, 1005–1010, 1005f,
    1006f, 1007f, 1007t
  stages of, 984–985
  start site, 989
  termination of, 984, 991–992, 992f

  of tRNA, 995, 995f
  upstream promoter element (UPE) in,
    989, 998
Transcription activator-like effector
    nucleases (TALENs), 174–175, 175f
Transcription bubbles
  defined, 986
  elongation at, 991, 991f
  schematic representation, 991f
  structure of, **986**
Transcription factors
  activation domain, 1080, 1081
  defined, 172, 1079–1080
  DNA-binding domains of, 1080–1081
  in eukaryotes, 999, 1000, 1079–1083
  in gene expression, 1079–1083
  in induced pluripotent stem (iPS) cells,
    1082, 1082f
  nuclear hormone receptor, 1085–1086,
    1085f
  in prokaryotes, 1079
  regulatory domains of, 1081
Transcription in eukaryotes, 996–1012
  in bacteria versus, 996, 996f
  CAAT box in, 999, 999f
  carboxyl-terminal domain (CTD) in, 997
  downstream core promoter element
    (DPE) in, 999
  enhancers in, 1001
  eukaryotic promoter elements, 998f
  GC box in, 999, 999f
  initiation of, 996, 999–1000, 999f
  initiator element (Inr) in, 999
  microRNAs in, 1004, 1005f
  nuclear membrane in, 996
  pre-mRNA processing and, 1002–1004,
    1003f, 1010–1011, 1010f
  products of, 1001–1012
  RNA editing in, 1004–1005, 1005f
  RNA polymerases in, 997–998, 997t, 998f
  RNA processing and, 996, 1010, 1010f
  splicing in, 1001, 1005–1010, 1005f,
    1006f, 1007t, 1009f
  TATA box in, 998–999, 998f
  transcription factors in, 999, 1000
  translation and, 996, 996f
  upstream promoter element (UPE) in, 998
Transcription initiation, 985
  in bacteria, 986–988
  de novo, 986–988
  in eukaryotes, 996, 999–1000, 999f
Transcription repressors, 1081
Transcriptome, 170
Transducin
  rhodopsin and, 1108
  α subunit of, **304**
Transferases, 265t
Transferred DNA (T-DNA), 177
Transferrin, 396, 1090
Transferrin receptor, 396, 396f, 1090
Transferrin-receptor mRNA, 1091, 1091f
Transgenic mice, 172
Transglutaminase, 331
Transition state

  ATP hydrolysis, 300–301
  in catalysis, 241
  catalytic stabilization of, 234
  defined, 240
  formation facilitation of, 240–244
  symbol, 240
Transition-state analogs, 253, 260, 260f,
    300, 300f
Transketolase
  in Calvin cycle, 652f, 654
  in pentose phosphate pathway, 660–662,
    660f
  reaction, 662–664, 663f
Translation, 43, 128, 1021–1050
  accuracy of, 1022–1023, 1023t
  activation sites, 1028–1029, 1028f
  adenylation in, 1028–1031
  aminoacyl-tRNA synthetase in, 1022,
    1026–1031
  anticodons in, 1024–1025, 1026, 1026f,
    1030
  in bacteria, 1035, 1036f, 1041–1043
  base pairing in, 1024–1025, 1031–1032
  codon-anticodon interactions in, 1026
  codons in, 1025–1026, 1026f
  defined, 1021
  direction of, 1040
  elongation factors in, 1036–1037, 1036f,
    1042
  in eukaryotes, 996, 996f, 1041–1043
  eukaryotic initiation of, 1041–1042, 1042f
  formylmethionyl-tRNA in, 1035, 1035f
  in gene expression, 1040
  initiation factors, 1035–1036, 1036f
  initiation of, 1034–1035, 1036f, 1041
  initiation sites, 1034, 1035f
  mechanism of action, 1039
  mRNA in, 1042
  organization in, 1042
  overview of, 1021–1022
  peptide bond formation in, 1037–1038,
    1038f
  proofreading in, 1029
  reading frame, 1036
  release factors in, 1040–1041, 1040f
  ribosomes in, 1026, 1031–1041
  rRNA in, 1031–1032, 1033f
  Shine–Dalgarno sequences in, 1034
  signaling in, 1045–1048
  termination of, 1040–1041, 1040f, 1042
  translocation in, 1038–1040, 1039f
  tRNA in, 1022, 1023–1026, 1032, 1034f,
    1041–1042
  wobble hypothesis and, 1025, 1026t
Translesion polymerases, 969
Translocase, 1038
Translocation
  inhibition by antibiotics, 1044–1045,
    1045f
  of proteins, 1045–1048
  RNA-DNA hybrid, 987, 988f
  signal sequences in, 1046–1048, 1046f
  in translation, 1038–1040, 1039f
Translocon, 1047

Transmembrane helices, 388–390, 399f, 399t
Transmissible spongiform encephalopathies, 59–60
Transplant rejection, 1141
Transport vesicles, 1048
Transporters
  ABC, 403, 411–413, 411f, 412f, 943
  antiporters, 413, 413f
  ATP-ADP, 602–603, 602f
  dicarboxylate, 602, 602f
  glucose, 404, 516–517, 516t
  glucose 6-phosphate, 701
  mitochondrial, 602–603, 602f
  in oxidative phosphorylation, 600–603
  phosphate, 602, 602f
  in photosynthesis, 622–626, 622f
  pyruvate, 603
  secondary, 413–415
  symporters, 413, 413f
  tricarboxylate, 602, 602f
  uniporters, 413, 413f
Transverse diffusion, 391, 391f
Tree of life, 4f
Triacylglycerol synthesis, 850–858
  liver in, 851
  sources of intermediates in, 850, 850f
Triacylglycerol synthetase complex, 851
Triacylglycerols
  accumulation in liver, 743
  in adipose tissue, 908
  defined, 710–711
  as energy source, 412, 412f, 735–736
  in fatty acid metabolism, 713–714, 714f
Tricarboxylic acid (TCA) cycle. See Citric acid cycle
Triglyceride. See Triacylglycerols
Trimethoprim, 836
Trimethylamine (TMA), 853
Trimethylamine-N-oxide (TMAO), 853
Trinucleotide repeats, 973
Triose phosphate isomerase (TPI), 257, **257f**
  catalytic mechanism of, 498
  in glycolysis, 498–499, 498f
  structure of, 498f
Triose phosphate isomerase deficiency (TPID), 532
Trioses, 342
Triple helix of collagen, 50f
tRNA (transfer RNA)
  acceptor stem, 1025
  as adaptor molecules, 131, 131f
  amino acid attachment to, 131, 131f
  anticodons of, 1023–1024, 1025
  bases, 1024–1025
  CCA terminal region of, 1024f, 1025
  charged, 1027
  codons in, 1025–1026
  common features of, 1023–1025
  defined, 128
  extra arm, 1025
  helix stacking in, 1024, 1025f
  initiator, 1041
  inosine in, 1026
  molecule design, 1023–1025, 1024f

precursors of, 1024
recognition sites on, 1029–1030, 1030f
RNA processing, 1002, 1003f
structure of, 1024, 1024f
synthetase recognition of, 1029–1030, 1030f
template-recognition site, 131
transcription of, 995, 995f
in translation, 1022, 1023–1026, 1032, 1034f
translocation of, 1038–1040, 1039f
wobble hypothesis and, 1025, 1026t
Trombone model, 963–964, 964f
Tropomyosin, 1161
Troponin complex, 1161
TRP channels, 1113
Trp operon, 1069
TRPV1, 106
Trypanosomes, 1005
Trypsin, 235, 235f
  catalytic triad in, 281, 281f
  chymotrypsin and, 281, 281f
  defined, 326
  generation of, 325–326
  S₁ pockets of, 281f
  structure of, 281, 281f
Trypsin inhibitor, 326–327, 327f
Trypsinogen, 325–324
Tryptophan
  in alanine, 770
  degradation of, 775, 775f
  nicotinamide from, 808–809, 808f
  as precursor for neurotransmitters, 810, 810f, 811f
  serotonin from, 808–809, 808f
  structure of, 33f, 34
  synthesis of, 803–804, 803f
Tryptophan synthase, 804
Ts elongation factor (EF-Ts), 1036–1037
Tu elongation factor (EF-Tu), 1036, 1036f
Tuberculosis (TB), 567, 758
Tubulin, 1163, 1164f
Tumor growth, ceramide metabolism as stimulant, 857
Tumor hypoxia, 518
Tumor suppressor genes, 456, 973
Tumor-inducing (Ti) plasmids, 177–178, 177f
Turnover number, 249, 249t
2′, 3′-dideoxy analog, 149, **149**
Twist in supercoiling, 957–958, 957f
Two-dimensional electrophoresis, 78, 78f
Twofold rotational symmetry, 295, 295f
Type 1 diabetes
  defined, 896
  glucagon excess in, 901
  insulin insufficiency in, 901
  metabolic derangements in, 901, 901f
  treatment for, 901
Type 2 diabetes
  defined, 896–897
  metabolic syndrome and, 898, 898f
  treatments for, 900
Type I topoisomerases, 958, 959–960

Type II topoisomerases, 958, 960–961
Type V glycogen storage disease, 702–703
Typing-monkey analogy, 56, 56f
Tyrosine, **774**
  defined, 34, 35
  degradation of, 774, 774f
  as precursor for human pigments, 815–816
  structure of, 34f
  synthesis of, 801–802, 803f
  thyroxine from, 808, 808f
Tyrosine kinases, 319, 448
Tyrosine phosphatase IB, 897
Tyrosinemia, 776t
Tyrosyl radical, 831, 831f
Tyrosyl-tRNA synthetase, 1029

Ubiquinol, 584, 717
Ubiquinone, 581, 717
Ubiquinone reductase, 717
Ubiquitin
  conjugation, 754, 754f
  defined, 751, 753
  pathway, 756–757
  in protein tagging, 754–755
  structure of, 754
Ubiquitination, 754–756
UDP (uridine diphosphate), 695
UDP-galactose 4-epimerase, 509
UDP-glucose, 693–695
  defined, 509, 693, 821
  in galactose conversion, 510
  in glycogen synthesis, 693–694
  synthesis of, 693
UDP-glucose pyrophosphorylase, 693
Ultracentrifugation, 80–82
Umami taste, 1102, 1105
UMP kinase, 825
Uncompetitive inhibition. See also Enzyme inhibition
  defined, 253–254
  double-reciprocal plot, 255, 256f
  example of, 254
  kinetics of, 254, 254f
Uncoupling proteins, 605–607, 606f
Unfolded protein response (UPR), 900
Unicellular red algae, 199f
Uniporters, 413, 413f
Unsaturated fatty acids, 719–720, 719f, 720f, 724, 736–739
Unstructured proteins, 58–59
Upstream promoter element (UPE), 989, 998
Uracil, 115, **115f**
Uracil DNA glycosylase, 973, 973f
Urate, 335, **335**, 840
Urea, 52–53, 764, 766b
Urea cycle
  in amino acid degradation, 764–769
  carbamoyl phosphate in, 764–766
  defined, 752
  disorders of, 767–768, 768f
  enzymes, 767
  in gluconeogenesis, 766, 766f
  illustrated, 764f

Ureotelic, 764
Ureotelic organisms, 764, 769
Uric acid, 335, 769, **769**
Uridine, 116, 843, **843**
Uridine diphosphate glucose (UDP-glucose), 693
Uridine monophosphate synthetase, 825
Uridine triphosphate (UTP), 310, 693
Uridylate, 116, 825
Uroporphyrinogen III, 812f, 813, 814

V genes, 1130–1131
v-SNARE, 1049
Vaccines
    defined, 1144
    in disease prevention/eradication, 1144–1145
    HIV, 1145
    killed (inactivated), 1145
    live attenuated, 1144–1145
    subunit, 1145
    synthetic, 163
    toxoid, 1145
Vaccinia virus, 172
Valine
    degradation of, 773
    in maple syrup urine disease, 776
    structure of, 33f, 34
van der Waals forces, 242–243
van der Waals interactions
    in antigen-antibody binding, 1129
    base stacking, 119, 119f
    contact distance, 9, 11
    defined, 8
    double helix stabilization by, 117–119
    energy of, 9, 9f
    hydrocarbon tails, 381
    maximization of, 11
    as noncovalent bond, 8
    protein stability and, 48
Vanishing white matter (VWM) disease, 1042–1043, 1043f
Variable number of tandem repeats (VNTR) region, 356, 356f
Vascular endothelial growth factor (VEGF), 202, 518
Vasopressin, 98, 98f
VDAC, 575
V(D)J recombination, 1131, 1131f
Vectors
    cloning, 155, 157, 157f
    defined, 154
    expression, 156
    plasmid, 155–157, 155f
    viral, 172
Vertebrates, 208
Very-low-density lipoproteins (VLDLs), 864, 864t
Vibrio fischeri, 1066–1067
Vicine, 669, 669f
Vioxx (rofecoxib), 937, **937**, 942
Viral hepatitis, 761
Viral receptors, 363, 363f
Viral vectors, 172

Virions, 157
Viruses
    coats, 51
    in gene introduction, 171
    human immunodeficiency, 1141–1142, 1142f
    infectious mechanisms of, 363–364
    influenza, 363–364, 363f
    progeny, 157
    retroviruses, 126–127, 127f, 942
    RNA, 126–127
Vision, 1106–1111. *See also* Sensory systems
    in animals, 1110, 1110f
    calcium ion in, 1109–1110
    color, 1106, 1110–1111
    color blindness and, 1111, 1111f
    cones in, 1106, 1110–1111
    evolution of, 1110, 1110f
    light absorption in, 1108–1109
    photoreceptors in, 1106, 1110, 1110f
    retinal-lysine linkage in, 1107, 1108f
    rhodopsin in, 1107, 1107f
    rods in, 1107, 1107f
    7TM receptors in, 1107–1110
    signal transduction in, 1109–1110, 1109f
Visual pigments
    in cones, 1110–1111
    defined, 1107
    in rods, 1107
Vitamin A (retinol), 479t, 480, 480f
Vitamin B_1 (thiamine), 479t, 562
Vitamin B_2 (riboflavin), 479t
Vitamin B_3 (niacin), 479t
Vitamin B_5 (pantothenic acid), 479t
Vitamin B_6 (pyridoxine), 479t
Vitamin B_7 (biotin), 479t
Vitamin B_9 (folic acid), 479t
Vitamin B_{12}, 479t
    as coenzyme, 721, 722f, 723
    corrin ring and cobalt atom, 721
    in fatty acid metabolism, 721
Vitamin C (ascorbic acid)
    deficiency of, alcohol related, 911–912
    forms of, 912, 912f
    function of, 479t, 480
Vitamin D (calcitriol)
    biochemical role, 878
    cholesterol as precursor, 877–878
    deficiency, 877
    function of, 479t, 480
    structure of, 480f
    synthesis of, 877–878, 878f
Vitamin E (α-tocopherol)
    function of, 479t, 480
    structure of, 480f
Vitamin K
    in blood clotting, 329, 331, 479t, 480
    deficiency, 479t
    in γ-carboxyglutamate formation, 331, 332f
    structure of, **331**, 480f
Vitamins
    B, 478, 479t
    coenzyme, 478–480, 479t

    defined, 478
    evolution of, 479
    noncoenzyme, 479, 479t
    in reducing homocysteine levels, 801
    roles of, 20
VJ recombination, 1130–1131, 1130f
V_{max}, 249
Voltage-gated ion channels, 421–422, 421f, 422f
von Gierke disease, 700, 700t, 701
von Gierke, Edgar, 700
VR1 (capsaicin receptor), 1114–1115, 1114f

Wald, George, 1108
Wang, James, 958
Warburg, Otto, 517
Warburg effect, 517, 560, 605
Warfarin, 331
Water
    attack, facilitating, 300, 301f
    concentration of, 14
    dissociation, equilibrium constant of, 14
    as highly cohesive, 9
    hydrogen bonds, 9, 10
    as polar molecule, 9
    properties of, 9
Water-oxidizing complex (WOC), 627, 628f
Watson, James, 5, 6, 17, 118, 950
Watson–Crick base pairs. *See* Bases/base pairs
Watson–Crick DNA model, 118, 118f, 119, 123
Wernicke–Korsakoff syndrome, 911
Wernicke's encephalopathy, 562
Western blotting, 87–88, 88f, 149
White adipose tissue (WAT), 605
Whooping cough, 457
Wiley, Don, 1135
Wilkins, Maurice, 117
Windows, 389
Withe in supercoiling, 957–958, 957f
Withering, William, 410
Wobble hypothesis, 1025, 1026t
Woese, Carl, 3
Wood-Ljungdahl pathway, 736
Wyman, Jeffries, 214

X-ray crystallography, 70
    defined, 100
    diffraction patterns, 101, 101f
    electron-density map, 101–102, 102f
    experiment, 100–101, 100f
    reflections, 101
    resolution, 102, 102f
    synchrotron radiation and, 101
X-ray diffraction of DNA, 117f
Xanthine oxidase, 821, 839–840
Xanthomas, 867
Xanthylate (XMP), 830
Xenobiotic compounds, 926
Xeroderma pigmentosum, 974

Yamanaka, Shinya, 1082
Yanofsky, Charles, 1068
Yeast artificial chromosomes (YACs), 157–158, 158f
Yeast chromosomes, 1076, 1076f

Z line, 1160, 1160f, 1162
Z scheme of photosynthesis, 631, 631f
Z-form DNA, 119f, 120, 120t
Zantac (ranitidine), 938, **938**
ZAP-70, 1138

Zellweger syndrome, 724
Zinc
    activation of water molecule, 288–290, 288f, 289f
    in biological systems, 288
    in carbonic anhydrase, 287–288, 288f
Zinc-finger domains, 1080–1081, 1080f
Zinc-finger nucleases (ZFNs), 174, 175–176, 175f
Zinc-finger protein, 113
Zingiberene, 878, **879f**

ZnCys gene, 181
Zonal centrifugation, 81–82, 81f
Zwitterions, 31, 32f
Zymogens, 325–328, 326f
    cascade of activations, 328–329, 328f
    conversion into proteases, 326
    defined, 310, 323
    developmental process control by, 324
    proteolytic activation, 326, 326f
    secretion of, 324, 324f

pKₐ values of some acids

Acid	pK' (at 25°C)	Acid	pK' (at 25°C)
Acetic acid	4.76	Malic acid, pK_1	3.40
Acetoacetic acid	3.58	pK_2	5.11
Ammonium ion	9.25	Phenol	9.89
Ascorbic acid, pK_1	4.10	Phosphoric acid, pK_1	2.12
pK_2	11.79	pK_2	7.21
Benzoic acid	4.20	pK_3	12.67
n-Butyric acid	4.81	Pyridinium ion	5.25
Cacodylic acid	6.19	Pyrophosphoric acid, pK_1	0.85
Citric acid, pK_1	3.14	pK_2	1.49
pK_2	4.77	pK_3	5.77
pK_3	6.39	pK_4	8.22
Ethylammonium ion	10.81	Succinic acid, pK_1	4.21
Formic acid	3.75	pK_2	5.64
Glycine, pK_1	2.35	Trimethylammonium ion	9.79
pK_2	9.78	Tris (hydroxymethyl) aminomethane	8.08
Imidazolium ion	6.95	Water*	15.74
Lactic acid	3.86		
Fumaric acid, pK_1	3.03		
pK_2	4.44		

*$[H^+] [OH^-] = 10^{-14}$; $[H_2O] = 55.5$ M.

Typical pKₐ values of ionizable groups in proteins

Group	Acid	⇌	Base	Typical pKₐ	Group	Acid	⇌	Base	Typical pKₐ
Terminal α-carboxyl group		⇌		3.1	Cysteine		⇌		8.3
Aspartic acid Glutamic acid		⇌		4.1	Tyrosine		⇌		10.0
Histidine		⇌		6.0	Lysine		⇌		10.4
Terminal α-amino group		⇌		8.0	Arginine		⇌		12.5

Note: pKₐ values depend on temperature, ionic strength, and the microenvironment of the ionizable group.

Bond	Structure	Length (Å)
C—H	R_2CH_2	1.07
	Aromatic	1.08
	RCH_3	1.10
C—C	Hydrocarbon	1.54
	Aromatic	1.40
C=C	Ethylene	1.33
C≡C	Acetylene	1.20
C—N	RNH_2	1.47
	O=C—N	1.34
C—O	Alcohol	1.43
	Ester	1.36
C=O	Aldehyde	1.22
	Amide	1.24
C—S	R_2S	1.82
N—H	Amide	0.99
O—H	Alcohol	0.97
O—O	O_2	1.21
P—O	Ester	1.56
S—H	Thiol	1.33
S—S	Disulfide	2.05